D1543654

Improve Student Success

LEARNSMART®

ADVANTAGE   THE EVOLUTION OF LEARNING

AN INNOVATIVE SUITE OF ADAPTIVE LEARNING PRODUCTS FUELED BY INTELLIGENT AND PROVEN LEARNING TECHNOLOGY

LearnSmart Advantage® is a new series of adaptive learning products fueled by McGraw-Hill LearnSmart®—the most widely used and adaptive learning resource proven to strengthen memory recall, increase retention, and boost grades. Each product in the series helps students study smarter and retain more knowledge.

Products in the LearnSmart Advantage Suite:

> **Leverage Learning Science and Research**
> Technology based on memory research moves students beyond memorizing to truly learning the material.

> **Use Data-driven, Accurate, and Reliable Recommendations**
> Data collected from over 1.5 million student users and more than 1 billion questions answered is leveraged to make the LearnSmart Advantage products intelligent, reliable, and precise.

> **Include Detailed Instructor and Student Reports**
> Valuable reports provide detailed insight into what students are struggling with, while tracking their progress at both the class and individual student level. Students can use powerful reports to pinpoint specific areas to study.

> **Include Current and Accurate Content**
> Years of experience developing content for adaptive learning platforms ensures our Subject Matter Experts are able to leverage the data collected to create the highest quality and most precise content.

WWW.LEARNSMARTADVANTAGE.COM

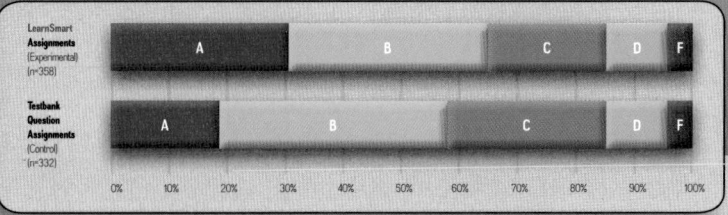

MORE STUDENTS EARN A's AND B's WHEN THEY USE THE LEARNSMART ADVANTAGE TOOLS.

SMARTBOOK®

The first—and only—adaptive reading experience designed to transform the way students read.

> Engages students with a personalized reading experience

> Ensures students retain knowledge

LEARNSMART®

The market-leading **adaptive study tool** proven to strengthen memory recall, increase class retention, and boost grades.

> Moves students beyond memorizing

> Allows instructors to align content with their goals

> Allows instructors to spend more time teaching higher-level concepts

LEARNSMART PREP®

An adaptive course preparation tool that quickly and efficiently helps students prepare for college-level work.

> Levels out student knowledge

> Keeps students on track

LEARNSMART ACHIEVE®

A learning system that continually adapts and provides learning tools to teach students the concepts they don't know.

> Adaptively provides learning resources

> A time management feature ensures students master course material to complete their assignments by the due date

LEARNSMART LABS®

LearnSmart Labs® is a super-adaptive simulated lab experience that brings meaningful scientific exploration to students. Through a series of adaptive questions, LearnSmart Labs identifies a student's knowledge gaps and provides resources to quickly and efficiently close those gaps. Once the student has mastered the necessary basic skills and concepts, they engage in a highly realistic simulated lab experience that allows for mistakes and the execution of the scientific method.

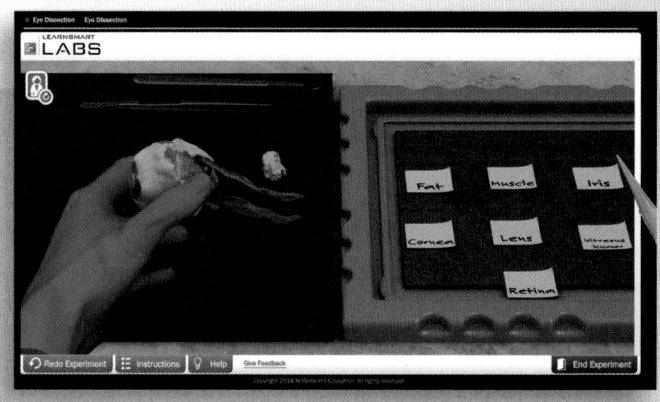

Anatomy & Physiology

AN INTEGRATIVE APPROACH

2e

Michael P. McKinley
Glendale Community College

Valerie Dean O'Loughlin
Indiana University

Theresa Stouter Bidle
Hagerstown Community College

Mc
Graw
Hill
Education

ANATOMY & PHYSIOLOGY: AN INTEGRATIVE APPROACH, SECOND EDITION

Published by McGraw-Hill Education, 2 Penn Plaza, New York, NY 10121. Copyright © 2016 by McGraw-Hill Education. All rights reserved. Printed in the United States of America. Previous edition © 2013. No part of this publication may be reproduced or distributed in any form or by any means, or stored in a database or retrieval system, without the prior written consent of McGraw-Hill Education, including, but not limited to, in any network or other electronic storage or transmission, or broadcast for distance learning.

Some ancillaries, including electronic and print components, may not be available to customers outside the United States.

This book is printed on acid-free paper.

5 6 7 8 9 LWI 21 20 19 18 17

ISBN 978-0-07-802428-3
MHID 0-07-802428-5

Senior Vice President, Products & Markets: *Kurt L. Strand*
Vice President, General Manager, Products & Markets: *Marty Lange*
Vice President, Content Design & Delivery: *Kimberly Meriwether David*
Managing Director: *Michael Hackett*
Director: *James F. Connely*
Director of Digital Content: *Michael G. Koots, PhD*
Brand Manager: *Amy Reed*
Director, Product Development: *Rose Koos*
Product Developer: *Donna Nemmers*
Marketing Manager: *Jessica Cannavo*
Digital Product Developer: *Jake Theobald*
Director, Content Design & Delivery: *Linda Avenarius*
Program Manager: *Angela R. FitzPatrick*
Content Project Managers: *April R. Southwood/Christina Nelson*
Buyer: *Sandy Ludovissy*
Design: *David W. Hash*
Content Licensing Specialists: *Carrie Burger/Leonard Behnke*
Cover Image: *wheelchair athlete © David Moyer*
Art Studio and Compositor: *MPS North America LLC*
Typeface: *10/12 Times LT Std Roman*
Printer: *LSC Communications*

All credits appearing on page or at the end of the book are considered to be an extension of the copyright page.

Library of Congress Cataloging-in-Publication Data

McKinley, Michael P.
  Anatomy & physiology : an integrative approach / Michael P. McKinley, Glendale Community College, Valerie Dean O'Loughlin, Indiana University, Theresa Stouter Bidle, Hagerstown Community College.— [2nd edition].
    pages cm
  Includes bibliographical references and index.
  ISBN 978-0-07-802428-3 (alk. paper)
  1. Human anatomy. 2. Human physiology. I. O'Loughlin, Valerie Dean. II. Bidle, Theresa Stouter. III. Title. IV. Title: Anatomy and physiology.
  QM25.M32 2016
  611—dc23

                                                              2014025205

The Internet addresses listed in the text were accurate at the time of publication. The inclusion of a website does not indicate an endorsement by the authors or McGraw-Hill Education, and McGraw-Hill Education does not guarantee the accuracy of the information presented at these sites.

www.mhhe.com

about the authors

MICHAEL MCKINLEY received his undergraduate degree from the University of California at Berkeley, and both his M.S. and Ph.D. degrees from Arizona State University. He did his postdoctoral fellowship at the University of California Medical School–San Francisco (UCSF) in the laboratory of Dr. Stanley Prusiner, where he worked for 12 years investigating prions and prion diseases. During this time, he was also a member of the UCSF Medical School anatomy faculty, and he taught medical histology for 10 years. In 1991, Michael became a member of the biology faculty at Glendale Community College (GCC) in Glendale, Arizona. He taught undergraduate anatomy and physiology, general biology, and genetics at the GCC Main Campus. In 2009, he moved to the GCC North Campus, where he taught anatomy and physiology courses exclusively until he retired in 2012. Between 1991 and 2000, Michael also participated in Alzheimer disease research and served as director of the Brain Donation Program at the Sun Health Research Institute. During this time he also taught developmental biology and genetics at Arizona State University West Campus. He has been an author and coauthor of more than 80 scientific papers. Mike's vast experience in histology, neuroanatomy, and cell biology greatly shaped the related content in the market-leading McKinley/O'Loughlin/Pennefather-O'Brien/ Harris: *Human Anatomy,* 4th edition textbook. Mike is an active member of the Human Anatomy and Physiology Society (HAPS). He resides in Tempe, Arizona with his wife Jan.

VALERIE DEAN O'LOUGHLIN received her undergraduate degree from the College of William and Mary, and her M.A. and Ph.D. degrees in biological anthropology from Indiana University. She is an associate professor of anatomy at Indiana University, where she teaches human gross anatomy to medical students, basic human anatomy to undergraduates, and human anatomy for medical imaging evaluation to undergraduate and graduate students. She also teaches a pedagogical methods course and mentors M.S. and Ph.D. students pursuing anatomy education research. She is active in the American Association of Anatomists (AAA) and the Society for Ultrasound in Medical Education (SUSME). She recently served as President of the Human Anatomy and Physiology Society (HAPS) and currently is on the Board of Directors. She received the AAA Basmajian Award for excellence in teaching gross anatomy and outstanding accomplishments in scholarship in education. Valerie is co-author of the market-leading McKinley/O'Loughlin/ Pennefather-O'Brien/Harris: *Human Anatomy,* 4th edition textbook.

TERRI STOUTER BIDLE received her undergraduate degree from Rutgers University, her M.S. degree in biomedical science from Hood College in Maryland, and has completed additional graduate coursework in genetics at the National Institutes of Health. She is a professor at Hagerstown Community College, where she teaches anatomy and physiology and genetics to pre–allied health students. Before joining the faculty in 1990, she was the coordinator of the Science Learning Center, where she developed study materials and a tutoring program for students enrolled in science classes. Terri has been a developmental reviewer and has written supplemental materials for both textbooks and lab manuals.

*Author team: Michael McKinley, Valerie Dean O'Loughlin,
and Theresa Bidle*

Dedications

*I am indebted to Jan (my wife); Renee, Ryan, and Shaun
(my children); and Connor, Eric, Patrick,
Keighan, Aydan, and Abbygail (my grandchildren).
They are the love of my life and my inspiration always.*

—Michael P. McKinley

*To my husband Bob and my daughter Erin:
Thank you for always being there for me.*

—Valerie Dean O'Loughlin

*With love and thanks to my husband Jay
and my daughter Stephanie for the many ways
that they have supported me during this project.*

—Terri Stouter Bidle

brief contents

ORGANIZATION OF THE HUMAN BODY

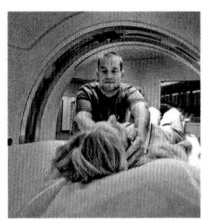

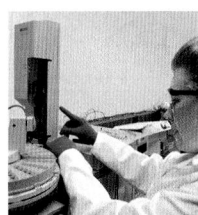

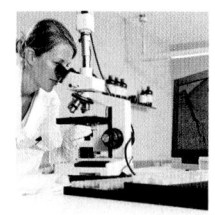
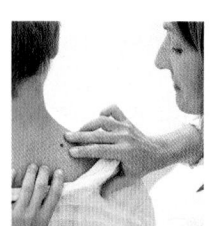
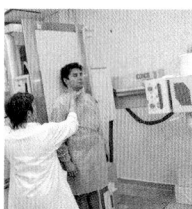

■ COMMUNICATION AND CONTROL

CHAPTER 12

Nervous System: Nervous Tissue 437

CHAPTER 13

Nervous System: Brain and Cranial Nerves 483

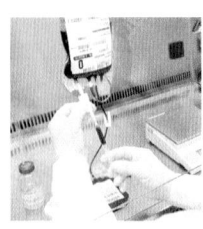
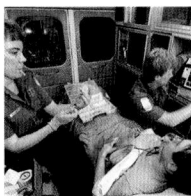
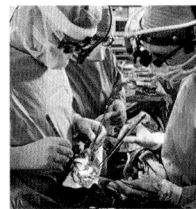

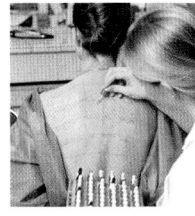

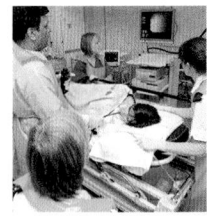

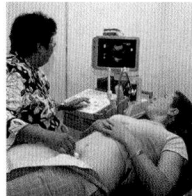

Human anatomy and physiology is a fascinating subject. However, students can be overwhelmed by the complexity, the interrelatedness of concepts from different chapters, and the massive amount of material in the course. Our goal was to create a textbook to guide students on a clearly written and expertly illustrated beginner's path through the human body.

An Integrative Approach

One of the most daunting challenges that students face in mastering concepts in an anatomy and physiology course is integrating related content from numerous chapters. Understanding a topic like blood pressure, for example, requires knowledge from the chapters on the heart, blood vessels, kidneys, and how these structures are regulated by the nervous and endocrine systems. The usefulness of a human anatomy and physiology text is dependent in part on how successfully it helps students integrate these related concepts. Without this, students are only acquiring what seems like unrelated facts without seeing how they fit into the whole.

To adequately explain such complex concepts to beginning students in our own classrooms, we as teachers present multiple topics over the course of many class periods, all the while balancing these detailed explanations with refreshers of content previously covered and intermittent glimpses of the big picture. Doing so ensures that students learn not only the individual pieces, but also how the pieces ultimately fit together. This book represents our best effort to replicate this teach-

ing process. In fact, it is the effective integration of concepts throughout the text that makes this book truly unique from other undergraduate anatomy and physiology texts.

Our goal of emphasizing the interrelatedness of body systems and the connections between form and function necessitates a well-thought-out pedagogical platform to deliver the content. First and foremost, we have written a very user-friendly text with concise, accurate descriptions that are thorough, but don't overwhelm readers with nonessential details. The text narrative is deeply integrated with corresponding illustrations drawn specifically to match the textual explanations. In addition, we have included a set of "Integrate" features that support our theme and work together to give the student a well-rounded introduction to anatomy and physiology. **Integrate: Concept Overview** figures are one- or two-page visual summaries that aggregate related concepts in a big-picture view. These comprehensive figures link multiple sections of a chapter together in a cohesive snapshot ideal for study and review. **Integrate: Concept Connections** boxes provide glimpses of how concepts at hand will play out in upcoming chapters, and also pull vital information from earlier chapters back into the discussion at crucial points when relevant to a new topic. **Integrate: Clinical View** discussions apply concepts from the surrounding narrative to practical or clinical contexts, providing examples of what can go wrong in the human body to help crystallize understanding of the "norm." **Integrate: Learning Strategy** boxes infuse each chapter with practical study tips to understand and remember information. Learning strategies include mnemonics, analogies, and kinesthetic activities that students can perform to relate the anatomy and physiology to their own

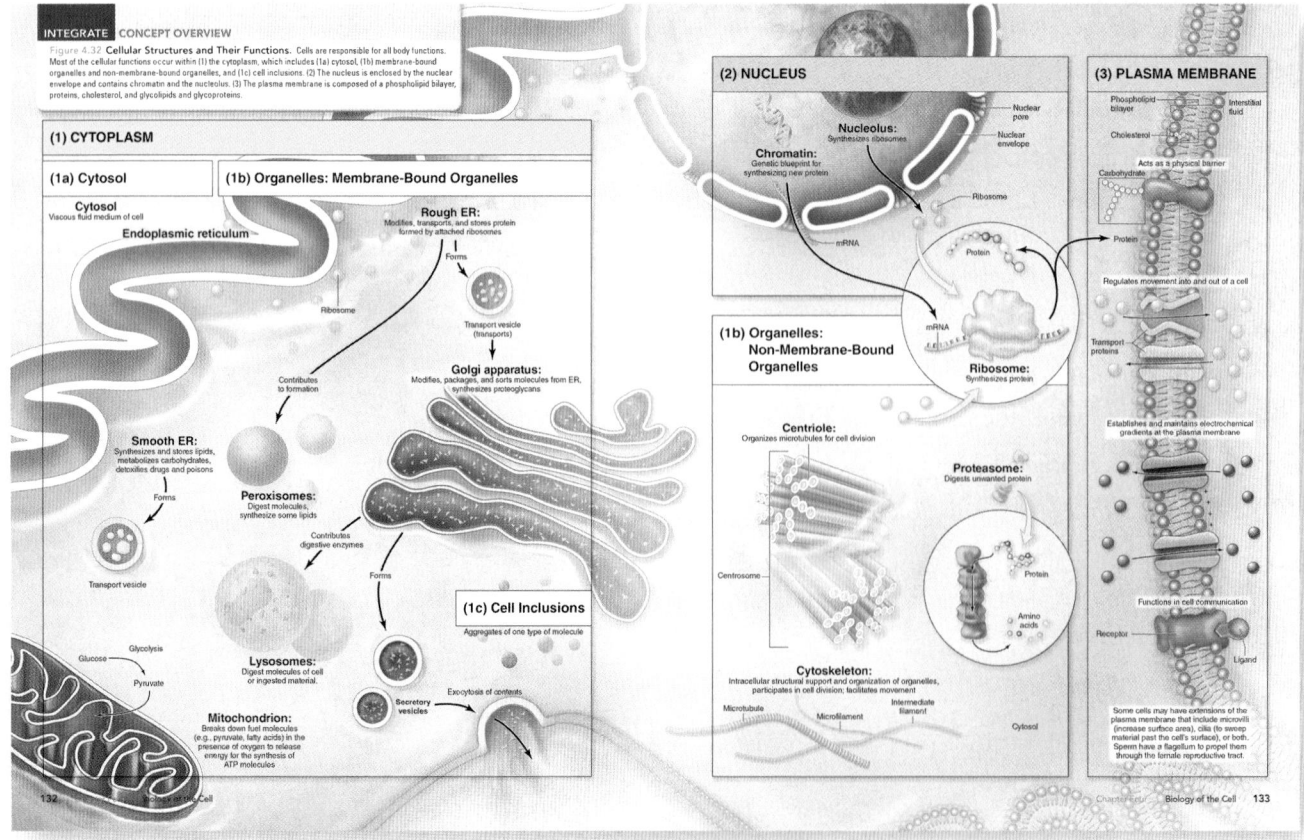

bodies. Finally, the media assets that accompany our book are tied to each section's learning objectives and previewed in the **Integrate: Online Study Tools** boxes at the end of each chapter.

Chapter Organization

In order to successfully execute an integrative approach, foundational topics must be presented at the point when it matters most for understanding. This provides students with a baseline of knowledge about a given concept before it comes time to apply that information in a more complex situation. Topics are thus subdivided and covered in this sequence:

- **Chapter 2: Atoms, Ions, and Molecules** Most students taking an A&P course have limited or no chemistry background, which requires a textbook to provide a detailed, organized treatment of atomic and molecular structure, bonding, water, and biological macromolecules as a basis to understanding physiological processes.

- **Chapter 3: Energy, Chemical Reactions, and Cellular Respiration** ATP is essential to all life processes. A solid understanding of ATP furthers student comprehension of movement of materials across a membrane, muscle contractions, production of needed replacement molecules and structures in cells, action potentials in nerves, pumping of the heart, and removal of waste materials in the kidneys. This textbook elevates the importance of the key concept of ATP by teaching it early. We then utilize this knowledge in later chapters as needed, expanding on what has already been introduced rather than re-teaching it entirely.

- **Chapter 13: Nervous System: Brain and Cranial Nerves and Chapter 14: Nervous System: Spinal Cord and Spinal Nerves** Instead of subdividing the nervous system discussion into separate central nervous system (CNS) and peripheral nervous system (PNS) chapters, nervous system structures are grouped by region. Thus, students can integrate the cranial nerves with their respective nuclei in the brain, and they can integrate the spinal cord regions with the specific spinal nerves that originate from these regions.

- **Chapter 17: Endocrine System** We have organized both the endocrine system chapter and the specific coverage of the many hormones released from endocrine glands to most effectively and efficiently guide students in understanding how this system of control functions in maintaining homeostasis. Within the chapter on the endocrine system, we provide an introduction and general discussion of the endocrine system's central concepts and describe selected representative hormones that maintain body homeostasis. Details of the actions of most other hormones—which require an understanding of specific anatomic structures covered in other chapters—are described in those chapters; for example, sex hormones are discussed in Chapter 28: Reproductive System. Learning the various hormones is facilitated by the inclusion of a "template" figure for each major hormone; each visual template includes the same components (stimulus, receptor, control center, and effectors) organized in a similar layout. In addition, information on each major hormone described in this text can be quickly accessed in the summary tables following chapter 17.

- **Chapter 21: Lymphatic System and Chapter 22: Immune System and the Body's Defense** A single chapter that discusses both the lymphatic system and immune system is overwhelming for most students. Thus, we separated the discussion into two separate chapters. The lymphatic system chapter focuses on the anatomic structures that compose the system, and provides a brief functional overview of each structure. This allows us to provide a thorough discussion and overview of the immune system in a separate chapter, where we frequently reference and integrate material from the earlier chapter.

- **Chapter 29: Development, Pregnancy, and Heredity** Coverage of heredity is included in the chapter on pregnancy and human development as a natural extension of Chapter 28: Reproductive System. This introduction will serve well as a precursor for students who follow their A&P course with a genetics course.

Changes to the Second Edition

The McKinley/O'Loughlin/Bidle textbook remains a resource that guides students on a clearly written and expertly illustrated beginner's path through the human body. Four core principles guided the authors as they made changes for the second edition:

1. Maximize the organization and clarity of the written text to provide instructors an excellent resource in developing their lectures, and provide students with a text that is easy to read and comprehend independently from a classroom environment (an important consideration with an increasing number of students enrolled in online anatomy and physiology courses). To accomplish these goals, 10 sections were added, 7 sections were expanded, and 29 sections were reorganized.

2. Facilitate lecture development and student learning through figures and tables that are well integrated with the text. To this end, 23 new figures and tables were developed, 33 figures and tables were updated, and 12 new photos were added.

3. Further integrate concepts between chapters to best facilitate instructor delivery and student understanding.
   - Some instructors follow the conventional order of teaching muscle physiology prior to neuron physiology, whereas other instructors teach neuron physiology prior to muscle physiology. To provide instructors with the most flexibility in teaching, and provide students with the background concepts that are needed regardless of the order the material is covered, chapter 4 now has a new section on establishing and maintaining resting membrane potential. This section provides the background information for graded potentials and action potentials that are discussed in both chapters 10 and 12.
   - The hormone reference section (10 tables on hormone details) has been relocated to follow immediately after chapter 17. This content applies to the whole book and is now more centrally located for student access.
   - The number and description of both "forward" and "backward" references to other specific chapter sections throughout the book has been increased (thereby increasing the integration of topics in the text).
   - Ten new Concept Connections were added.

4. Aid students in becoming aware of various career paths and clinical applications.

- Chapter opener pages now include a new "Integrate" feature called "Integrate: Career Path," which highlights a career relevant to the chapter material.
- Many students in this course are pursuing a health sciences career. This edition contains 23 added or updated Clinical View boxes to provide further connections to practical applications in the health-care field.

Changes by Chapter

The following changes are not an exhaustive list, but note the most significant changes in this second edition.

Chapter 1

- Updated section 1.1: Anatomy and Physiology Compared
- Updated section 1.2: Anatomy and Physiology Integrated
- New Clinical View: Clinicians' Use of Scientific Method

Chapter 2

- New Concept Connection on electrolytes
- Updated section 2.5c: pH, Neutralization, and the Action of Buffers

Chapter 3

- Updated figure 3.16: Metabolic Pathway of Glycolysis
- Updated figure 3.19: The Electron Transport System
- New figure 3.20: Summary of Stages of Cellular Respiration
- Updated Clinical View: Lactose Intolerance
- Updated section 3.4f: ATP Production

Chapter 4

- New figure 4.7: Membrane Transport—flowchart organized into passive processes and active processes
- New section 4.4: Resting Membrane Potential, which includes new figure 4.20: Resting Membrane Potential (RMP)
- New Concept Overview figure 4.32: Cellular Structures and Their Functions

Chapter 5

- Updated section 5.1: Epithelial Tissue: Surfaces, Linings, and Secretory Functions
- New table 5.1: Overview of Tissues
- New figure 5.3: Organization and Relationship of Epithelia Types
- New photos in expanded table 5.2: Simple Epithelia
- New photo in table 5.3: Stratified Epithelia
- Updated Concept Overview figure 5.4: The Relationship between Epithelial Tissue Type and Function
- New photo in table 5.5: Connective Tissue Proper: Loose Connective Tissue
- Updated Clinical View: Stem Cells
- New photos in table 5.7: Supporting Connective Tissue: Cartilage
- Updated Concept Overview figure 5.10: The Relationship Between Connective Tissue Type and Function

- New photos in table 5.10: Muscle Tissue
- Updated figure 5.12: Body Membranes
- Expanded section 5.6b: Tissue Modification

Chapter 6

- Moved and updated section 6.1d: Functions of the Integument (was previously section 6.3)
- Updated Concept Overview figure 6.8: How Integument Form Influences Its Functions
- Updated section 6.2b: Hair (specifically, Hair Growth and Replacement section)
- New Concept Connection on relationship of wound repair and the immune system
- Updated table 6.2: Skin Cancer

Chapter 7

- New Concept Connection in section 7.2e on stem cells and osteoprogenitor cells
- New photo for bone tissue in figure 7.6: Types of Cells in Bone Connective Tissue
- New photo of spongy bone in figure 7.8: Microscopic Anatomy of Bone
- Updated figure 7.14: Calcitriol Production
- Updated figure 7.16: Classification of Bone Fractures

Chapter 8

- Reversed order of section 8.1a: Axial and Appendicular Skeleton and section 8.1b: Bone Markings
- New Clinical View: Coccyx (Tailbone) Injury

Chapter 9

- Updated figure 9.8: Flexion, Extension, Hyperextension, and Lateral Flexion
- Updated Clinical View: Shoulder Joint Dislocations
- New photo of arthroscopic view of knee joint in Clinical View: Knee Ligament and Cartilage Injuries

Chapter 10

- New section 10.2d: Skeletal Muscle Fibers at Rest, which includes new figure 10.8: Skeletal Muscle Fiber at Rest
- New figure 10.12: Events of an Action Potential at the Sarcolemma
- New figure 10.14: Portion of a Sarcomere (electron micrograph)

Chapter 11

- Expanded section 11.1a: Origin and Insertion
- New Clinical View: Strabismus and Diplopia
- Updated Clinical View: Hernias
- Updated Clinical View: Shin Splints and Compartment Syndrome
- Updated Clinical View: Plantar Fasciitis

Chapter 12

- New section 12.7b: Neurons at Rest, which includes new figure 12.13: Neuron at Rest
- New figure 12.18: Generation of an Action Potential
- New Concept Connection on potential energy and kinetic energy associated with ion gradients
- New Clinical View: Local Anesthetics
- New Clinical View: Neurotoxicity
- Updated Concept Overview figure 12.23: Events of Neuron Physiology
- Revised section 12.9: Characteristics of Action Potentials
- Significantly expanded section 12.10: Neurotransmitters and Neuromodulation, including new figure 12.24: Classification of Neurotransmitters (organizational flowchart), and new figure 12.25: Acetylcholine Release, Removal from Synaptic Cleft, and Action
- Modified table 12.3: Neurotransmitters, now incorporating drugs that influence neurotransmitter release or binding

Chapter 13

- Updated Concept Overview figure 13.12: Anatomic and Functional Areas of the Cerebrum
- Reorganized section 13.3c: Functional Areas of the Cerebrum
- New Concept Connection on action potentials
- New Clinical View: Autism Spectrum Disorder
- New section 13.8b: Electroencephalogram, including new figure 13.28 Electroencephalograms (EEGs)
- New section 13.8c: Sleep, including new figure 13.29: Hypnogram
- Updated Clinical View: Alzheimer Disease: The "Long Goodbye"
- Expanded section 13.8g: Language, including brief discussion of speech disorders
- Updated table 13.5: Cranial Nerves, including specific test to determine nerve damage

Chapter 14

- Reorganized section 14.1: Spinal Cord Gross Anatomy, including modified figure 14.2: Cross Sections of the Spinal Cord
- Reorganized section 14.2: Protection and Support of the Spinal Cord
- New Clinical View: Poliomyelitis
- Reorganized section 14.4a: Overview of Conduction Pathways
- Reorganized section 14.4b: Sensory Pathways
- Updated table 14.1: Functions and Neuron Locations of Principal Sensory Spinal Cord Pathways
- New section 14.6c: Classifying Spinal Reflexes
- Reorganized section 14.6d: Spinal Reflexes
- New Concept Connection on spinal nerves and skeletal muscle innervation

Chapter 15

- Updated section 15.1a: Functional Organization
- Moved CNS Control of the Autonomic Nervous System for earlier chapter position (is now section 15.1c)

- Reorganized table 15.1: Comparison of Somatic and Autonomic Motor Nervous Systems
- New photo for Clinical View: Horner Syndrome
- Reorganized section 15.4b: Sympathetic Pathways
- Expanded section 15.5b: Cholinergic Receptors, including new table 15.4: Cholinergic Receptors
- New table 15.5: Adrenergic Receptors
- New Clinical View: Epinephrine for Treatment of Asthma
- Updated table 15.6: Effects of the Parasympathetic and Sympathetic Divisions
- Updated Concept Overview figure 15.10: Comparison of the Parasympathetic and Sympathetic Divisions of the ANS

Chapter 16

- Reorganized section 16.1: Introduction to Sensory Receptors, including new section 16.1c: Sensory Information Provided by Sensory Receptors
- New Concept Connection on the four cranial nerves associated with the eye
- Reorganized section 16.4: Visual Receptors, discussing the Physiology of Vision: Refraction and Focusing of Light
- Expanded section 16.4d: Physiology of Vision: Phototransduction
- Receptors, on cochlear hair cell stimulation, including new figure 16.28: Inner Hair Cells
- New Clinical View: Deafness
- Updated section 16.5d: Equilibrium and Head Movement

Chapter 17

- Reorganized section 17.2: Endocrine Glands
- Revised table 17.2: Endocrine Glands and Organs Containing Endocrine Cells
- Updated figure 17.5: Eicosanoid Formation
- Expanded section 17.7b: Interactions Between the Hypothalamus and the Anterior Pituitary Gland and revised figure 17.12: Anterior Pituitary Hormones
- Updated figure 17.16: Thyroid Hormone: Synthesis, Storage, and Release
- Added new section 17.10: Other Endocrine Glands (includes pineal gland, parathyroid gland, thymus, heart, kidneys, liver, stomach, small intestine, skin, and adipose connective tissue)
- Tables on major regulatory hormones of the human body now directly follow chapter 17

Chapter 18

- Updated section 18.3a: Hemopoiesis, including new table 18.6: Substances That Influence Hemopoiesis
- Updated Clinical View: Anemia in section 18.3b: Erythrocytes
- Updated Concept Overview figure 18.8: Recycling and Elimination of Erythrocyte Components
- Updated table 18.7: Leukocytes
- New Concept Connection on hemopoiesis and the skeletal system

Chapter 19

- Revised figure 19.14: Anatomic Structures Controlling Heart Activity
- Modified section 19.6a: Nodal Cells at Rest, including modified accompanying figure 19.16: SA Node Cellular Activity
- Modified figure 19.18: Electrical Events of Cardiac Muscle Cells
- Modified section 19.7c: Repolarization and the Refractory Period, including modified figure 19.19: Comparison of Electrical and Mechanical Events in Skeletal Muscle Cells and Cardiac Muscle Cells
- Updated Clinical View: Cardiac Arrhythmias, including added images of abnormal ECGs
- Modified Concept Overview figure 19.22: Changes Associated with a Cardiac Cycle
- New figure 19.23: Sympathetic Innervation of Nodal Cells
- New figure 19.24: Relationship of EDV, ESV, and SV

Chapter 20

- Moved Total Cross-Sectional Area and Velocity of Blood Flow to earlier chapter position (is now section 20.2)
- New section 20.4b: The Myogenic Response
- New image for Clinical View: Tumor Angiogenesis
- Revised figure 20.14: Cardiovascular Center

Chapter 21

- Revised figure 21.1: Lymphatic System
- Revised figure 21.3: Lymphatic Trunks and Ducts
- Modified section 21.4: Secondary Lymphatic Structures, including Lymph Flow Through Lymph Nodes
- Revised Concept Overview figure 21.9: Relationship of the Lymphatic System to Both the Cardiovascular System and Immune System

Chapter 22

- New figure 22.8: Two Branches of Adaptive Immunity (organizational flowchart)
- New figure 22.22: Active and Passive Immunity (organizational flowchart)
- Added information on pattern recognition receptors (e.g., toll-like receptors or TLRs), Tregs, peripheral tolerance, antibody class switching
- New Clinical View: Regulatory T-Lymphocytes and Tumors

Chapter 23

- Reworked section 23.5c: Nervous Control of Breathing
- Modified figure 23.23: Respiratory Center

Chapter 24

- Reorganized section 24.1: Introduction to the Urinary System

Chapter 25

- New figure 25.5: Edema

Chapter 26

- In Clinical View: Reflux Esophagitis and Gastroesophageal Reflux Disease, new photo of normal esophagus to accompany photo of Barrett esophagus
- New Clinical View: Gastric Bypass, including accompanying illustration
- Modified Clinical View: Intestinal Disorders
- New figure 26.14: Regulation of the Digestive Processes in the Stomach

Chapter 27

- Reorganization of section 27.1: Introduction to Nutrition, section 27.2: Macronutrients, section 27.3: Micronutrients, and section 27.4: Guidelines for Adequate Nutrition

Chapter 28

- Updated section 28.3b: Oogenesis and the Ovarian Cycle
- Updated Clinical View: Contraception Methods
- New Clinical View: Paternal Age Risks for Disorders in the Offspring

Chapter 29

- New Concept Connection integrating immunoglobulins (in chapter 22) to specific immunoglobulin classes that are secreted in breast milk

---

We Welcome Your Input!

We hope you enjoy reading this textbook, and that it becomes central to mastering the concepts in your anatomy and physiology course. This text is a product that represents over 75 years of combined teaching experience in anatomy and physiology. We are active classroom instructors, and are well aware of the challenges that current students face in mastering these subjects. We have taken what we have learned in the classroom and have created a textbook truly written for students.

Please let us know what you think about this text. We welcome your thoughts and suggestions for improvement, and look forward to your feedback!

Michael P. McKinley

Glendale Community College, retired

mpmckinley@hotmail.com

Valerie Dean O'Loughlin

Medical Sciences
Indiana University

vdean@indiana.edu

Terri Stouter Bidle

Science Division
Hagerstown Community College

tsbidle@hagerstowncc.edu

ACKNOWLEDGMENTS

Many people have worked with us over the last several years to produce this text. We would like to thank the many individuals at McGraw-Hill who worked with us to create this textbook. We are especially grateful to Donna Nemmers and Kris Queck, our Developmental Editors; Amy Reed, our Brand Manager; April Southwood, Content Project Manager, who expertly guided the project through its production phases; David Hash, Designer, for his beautiful interior and cover designs; Jim Connely, Director of Applied Biology, for his expert guidance on this project; and Jessica Cannavo for her marketing expertise. We would also like to thank our copyeditor, Bea Sussman; our proofreaders, Pat Steele and Carey Lange; and indexer, Maria Coughlin.

Finally, we could not have performed this effort were it not for the love and support of our families. Jan, Renee, Ryan, and Shaun McKinley; Bob and Erin O'Loughlin; and Jay and Stephanie Bidle—thank you and we love you! We are blessed to have you all.

Many instructors and students across the country have positively affected this text through their careful reviews of manuscript drafts, art proofs, and page proofs, as well as through class tests and through their attendance at focus groups and symposia. We gratefully acknowledge their contributions to this text.

Reviewers

Mike Aaron
    Shelton State Community
    College

Pius Aboloye
    North Lake College—
    Dallas County Community
    College District

Mercedes I. Alba
    San Antonio College

Ticiano Alegre
    North Lake College

Karim Alkadhi
    University of Houston–
    College of Pharmacy

Rachel Basco
    Bossier Parish Community
    College

Shawn E. Bearden
    Idaho State University

Rebecca A. Beecroft
    Hagerstown Community
    College

Dena Berg
    Tarrant County College–
    NW

Shelley L. Berg
    Westmoreland County
    Community College

Daniel A. Bergman
    Grand Valley State
    University

Rebecca B. Benard
    Case Western Reserve
    University

Brian T. Berthelsen
    Iowa Western Community
    College

J. Gordon Betts
    Tyler Junior College

Joressia A. Beyer
    John Tyler Community
    College

Evelyn J. Biluk
    CVTC

Jamal I. Bittar
    The University of Toledo

Greta Bolin
    Tarrant County Community
    College

Kristy Boggs
    Spoon River College

Lois Brewer Borek
    Georgia State University

James J. Bottesch
    Eastern Florida State
    College

Molly Brandel
    Northeast Iowa Community
    College

Ty W. Bryan
    Bossier Parish Community
    College

Roger Buchanan
    Arkansas State University

Mickael J. Cariveau
    University of Mount Olive

Carol E. Carr
    John Tyler Community
    College

Weiru Chang
    Estrella Mountain
    Community College

Rebekah Chapman
    Georgia State University

Brendon K. Chastain
    West Kentucky Community
    and Technical College

William M. Clark
    Lone Star College–
    Kingwood

David T. Corey
    Midlands Technical College

Andrew Corless
    Vincennes University

Maurice M. Culver
    Florida State College–
    Jacksonville

George Dalich
    Lake Washington Institute
    of Technology

Tirupapuliyur Damodaran
    North Carolina Central
    University

Edward A. DeGrauw
    Portland Community
    College

Sharon N. DeWitte
    University of South
    Carolina

Cynthia A. Dove
    Hagerstown Community
    College

Joe D'Silva
    Norfolk State University

Arlee Dulak
    University of
    Massachusetts–Lowell

William E. Dunscombe
    Union County College

Kaushik Dutta
    University of New England

Pamela K. Elf
    University of Minnesota–
    Crookston

Aaron M. Elmer
    Murray State College

Cammie Emory
    Bossier Parish Community
    College

Elyce Ervin
    The University of Toledo

Hilary P. Engebretson
    Whatcom Community
    College

Sondra M. Evans
    Florida State College–
    Jacksonville

Rebekah Faber-Starr
    Northwest State
    Community College

Bradley J. Fillmore
    Eastern Washington
    University

Reza Forough
    Bellevue College

Lori Frear
    Wake Technical
    Community College

Dean Furbish
    Wake Technical
    Community College

Sophia E. Garcia
    Tarrant County College

Joseph D. Gar
    West Kentucky Community
    and Technical College

Gary Glaser
    Genesee Community
    College

Peter Germroth
    Hillsborough Community
    College

Tejendra Gill
    University of Houston–
    Central Campus

Wanda L. Goleman
    Northwestern State
    University of Louisiana

Cecilia V. Gonzales
    Palo Alto College

Richard Gonzalez Diaz
    Seminole State College of
    Florida

Richard Griner
    Augusta State University

Anne A. Grippo
    Arkansas State University

Fran Hardin
    Ivy Tech Community
    College

R. Christopher Harvey
    Eastern Florida State
    College

Lesleigh Hastings
    Wake Tech Community
    College

Drew Hataway
    Samford University

Christine N. Haugh
    Iowa Western Community
    College

Barbara R. Heard
 Atlantic Cape Community College

Holly Heckmann
 San Antonio College

Gerald A. Heins
 Northeast Wisconsin Technical College

Carol Hoban
 Kennesaw State University

Regina Neal Hoffman
 Tacoma Community College

Anthony Holt
 University of Arkansas Community College– Morrilton

Diana Homsi
 Tidewater Community College

Mark J. Hubley
 Prince George's Community College

Julie Huggins
 Arkansas State University

Linda E. Ibarra-Gonzales
 Palo Alto College

Alexander Imholtz
 Prince George's Community College

Melissa Iszard
 Eastern Maine Community College

Jean Jackson
 Bluegrass Community and Technical College

Edward Johnson
 Central Oregon Community College

Roishene Johnson
 Bossier Parish Community College

Scott Johnson
 Wake Technical Community College

Wendy E. Johnson
 University of Alaska– Fairbanks

Latoya Jones
 West Kentucky Community and Technical College

Susanne J. Kalup
 Westmoreland County Community College

Marta Klesath
 North Carolina State University

Michelle Klein
 Prince George's Community College

Daniel P. Keogh
 Westmoreland County Community College

Tim Koneval
 Mount Aloysius College

Elizabeth Kozak
 Lewis University

Tyjuanna R.S. LaBennett
 North Carolina Central University

Luis Labiste
 Miami Dade College–North Campus

Stephanie Lee
 Pearl River Community College

Jerri K. Lindsey
 Tarrant County College–NE Campus

Kathryn Link
 Alfred State College

Judy Lonsdale
 Boise State University

Don Loving
 Murray State College

Paul Luyster
 Tarrant County College

Debby Machuca
 Portland Community College

Elizabeth A. Maher
 Ivy Tech Community College

Mara L. Manis
 Hillsborough Community College

Bruce P. Maring
 Daytona State College

Robert E. Martin
 Dyersburg State Community College

Jolie Master
 Atlantic Cape Community College

Craighton S. Mauk
 Gateway Community and Technical College

John McDaniel
 San Jose State University

Nicole L. McDaniels
 Herkimer College

Louis F. McIntyre, Sr.
 Robeson Community College

Karen (Kren) McManus
 North Hennepin Community College

Lauren Aderholt Millstead
 Volunteer State Community College

Evelyn M. Mobley
 Chattanooga State Community College

Dan Moore
 Murray State College

Jane B. N. Moore
 Tarrant County College

Lisa M. Moore
 Iowa Western Community College

Christine Morin
 Prince George's Community College

Serban Andrew Moroianu
 Essex County College

Regi Munro
 Chandler-Gilbert Community College

Chris Nelling
 Hagerstown Community College

Necia Morgan Nicholas
 Calhoun Community College

Jeremy Nicholson
 South Plains College

Jeffrey P. Novack
 Pacific Northwest University, School of Medicine

Maria H. Oehler
 Florida State College– Jacksonville

Jeanine L. Page
 Lock Haven University

Christine Parker
 Finger Lakes Community College

Benjamin Peacock
 Pulaski Technical College

Ryan Peck
 Boise State University

Andrew M. Petzold
 University of Minnesota– Rochester

Jean C. Pfau
 Idaho State University

Marius Pfeiffer
 Tarrant County College

Megan Pickering Stringer
 Jones County Junior College

Gilbert R. Pitts
 Austin Peay State University

John S. Placyk, Jr.
 University of Texas–Tyler

Benjamin Predmore
 University of South Florida

Lynn Preston
 Tarrant County College– NW Campus

Usha K. Raghu
 Westmoreland County Community College

Nancy Rauch
 Merritt College

Amy J. Reber
 Georgia State University

Peter Reuter
 Florida Gulf Coast University

Laura H. Ritt
 Burlington County College

Candice Roberts
 Wake Technical Community College

John C. Robertson
 Westminster College

Robert W. Rothrock
 Camden County College

Grace Rutherford
 El Centro College

Adam B. Safer
 Georgia State University

Joseph F. Shearer
 Gateway Community and Technical College

Melissa S. Shields
 Boise State University; Park University

Girija S. Shinde
 Volunteer State Community College

Nicole L. Shives
 Iowa Western Community College

Michael Silveira
 University of Minnesota– Rochester

Jenna L. Simpson
 Iowa Western Community College

Lisa K. Smith
 Hillsborough Community College

Norman D. Sossong
 Bellevue College

William G. Sproat, Jr.
Walters State Community College

Jeanmarie Stiles
Tarrant County College–NE Campus

Monica R. Storm
Iowa Western Community College

Christina Strandgaard
California State University–Sacramento

Karla Svedarsky
Chippewa Valley Technical College

Yong Tang
Front Range Community College

Rashmi Suni Thakur
American University of Antigua–College of Medicine; Bossier Parish Community College

James F. Thompson
Austin Peay State University

William Ross Tobin, Jr.
Erie Community College–South Campus

Jane Torrie
Tarrant County College–Northwest Campus

Randall L. Tracy
Worcester State University

Padmaja B. Vedartham
Lone Star College–Cy Fair

Corina L. Wack
Chowan University

Kim M. Walsh
Westchester Community College

Chad M. Wayne
University of Houston

John E. Whitlock
Hillsborough Community College

Diane Wilkening Fritz
Gateway Community and Technical College

Denise Williams
Caldwell Community College and Technical Institute

Krista Winters
Spoon River College

Gregory E. J. Wohar
Westmoreland County Community College

Jennifer Wollschlager
University of Minnesota–Rochester

Dean Won
American River College

Timothy John Zehnder
Gardner-Webb University

Bekki Zeigler
Prince George's Community College

Sandy Zetlan
Estrella Mountain Community College

Jay Zimmer
Gardner-Webb University

guided tour

Fully Integrated Content and Pedagogy

Anatomy and Physiology: An Integrative Approach is structured around a tightly integrated learning system that combines illustrations and photos with textual descriptions; focused discussions with big-picture summaries; previously learned material with new content; factual explanations with practical and clinical examples; and bite-sized topical sections with multi-tiered assessment.

Unparalleled Art Program

In a visually oriented subject like A&P, quality illustrations are crucial to understanding and retention. The brilliant illustrations in *Anatomy and Physiology: An Integrative Approach* have been carefully rendered to convey realistic, three-dimensional detail while incorporating pedagogical conventions that help deliver a clear message. Each figure has been meticulously reviewed for accuracy and consistency, and precisely labeled to coordinate with the text discussions.

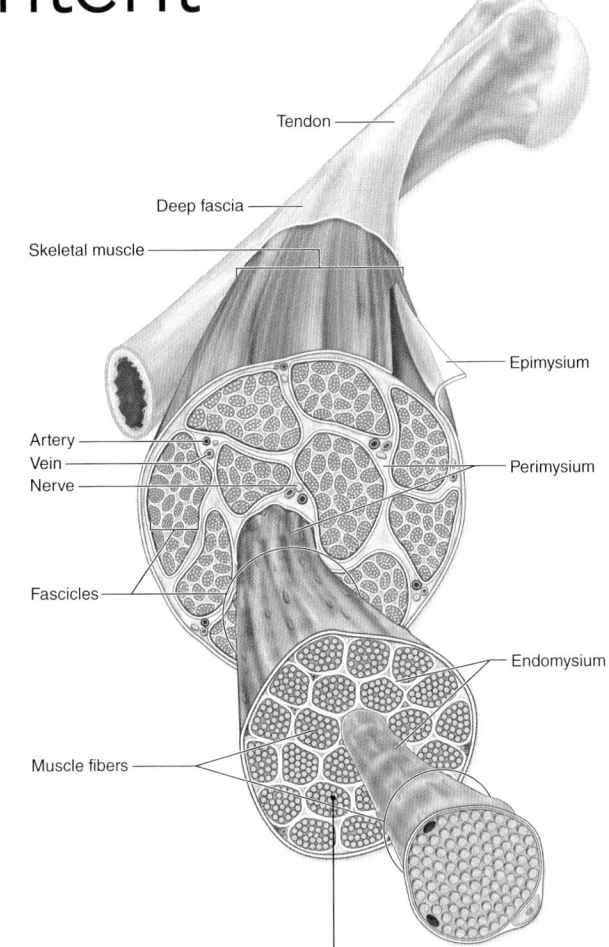

Tendon
Deep fascia
Skeletal muscle
Epimysium
Artery
Vein
Nerve
Perimysium
Fascicles
Endomysium
Muscle fibers

Rich Detail
Vibrant colors and three-dimensional shading make it easy to envision body structures and processes.

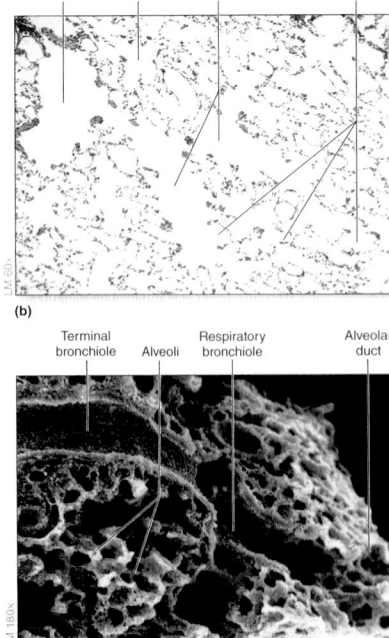

Branch of pulmonary artery
Bronchiole
Terminal bronchiole
Pulmonary arteriole
Branch of pulmonary vein
Pulmonary capillary beds
Respiratory bronchiole
Pulmonary venule
Alveolar duct
Alveoli
Alveolar pores
Interalveolar septum
Alveolar sac
Elastic fibers
Connective tissue
(a)

Respiratory bronchiole   Alveolar sac   Alveolar ducts   Alveoli
(b)

Terminal bronchiole   Alveoli   Respiratory bronchiole   Alveolar duct
(c)

Photographs
Atlas-quality micrographs and cadaver images are frequently paired with illustrations to expose students to the appearance of real anatomic structures.

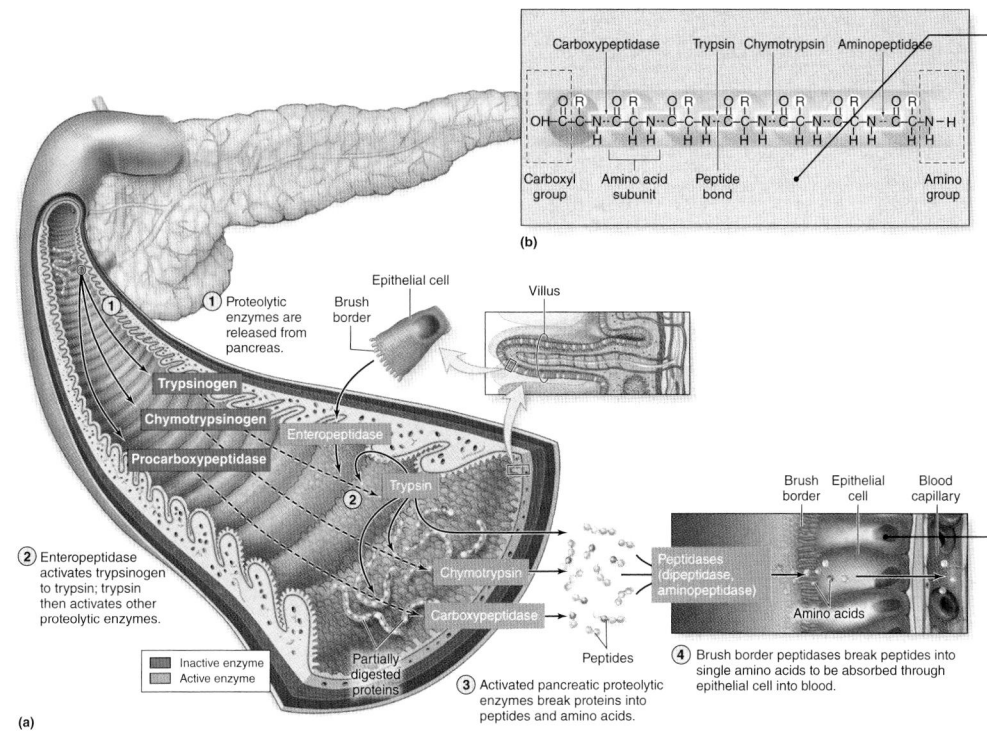

(b)

Multilevel Perspective
Microscopic structures are connected to macroscopic views to show changes in perspective between increasingly detailed drawings.

(a)

Color Coding
Many figures use color coding to organize information and clarify concepts for visual learners.

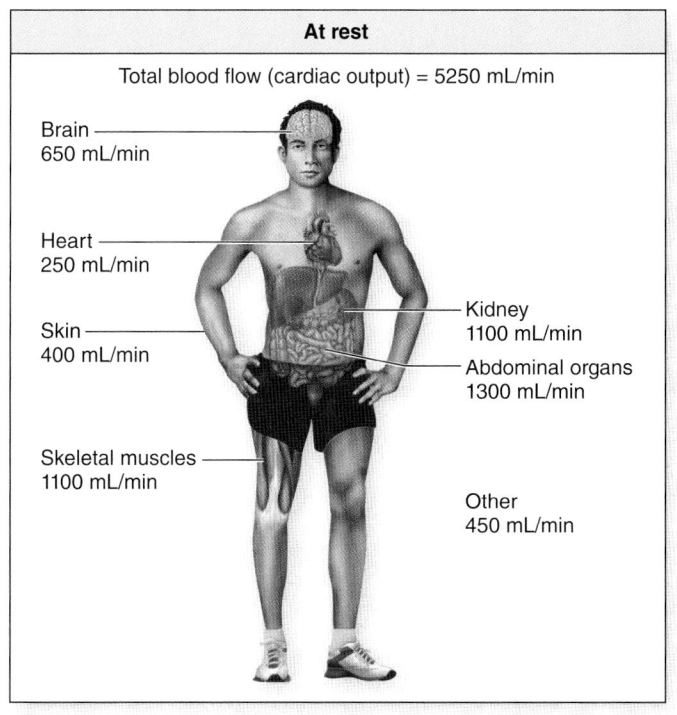

(a)

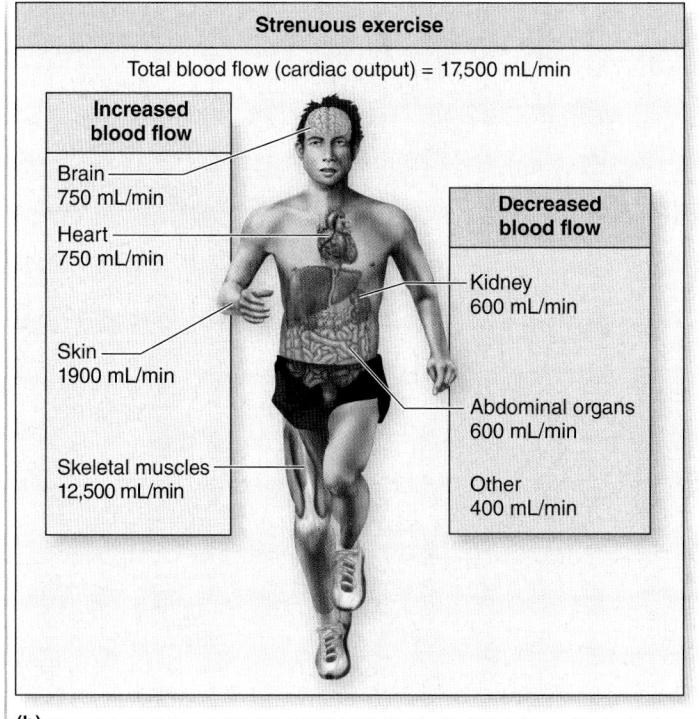

(b)

Real-Life Context
Illustrations include depictions of realistic people and situations to make figures more relevant and memorable.

Integrative Visual Summaries

The groundbreaking **Integrate: Concept Overview** figures combine multiple concepts into one big-picture summary. These striking, visually dynamic presentations offer a review of previously covered material in a creatively designed environment to emphasize how individual parts fit together in the understanding of a larger mechanism or concept.

Integrate: Concept Overview Figures
Multifaceted concepts are brought together in captivating one- or two-page visual presentations.

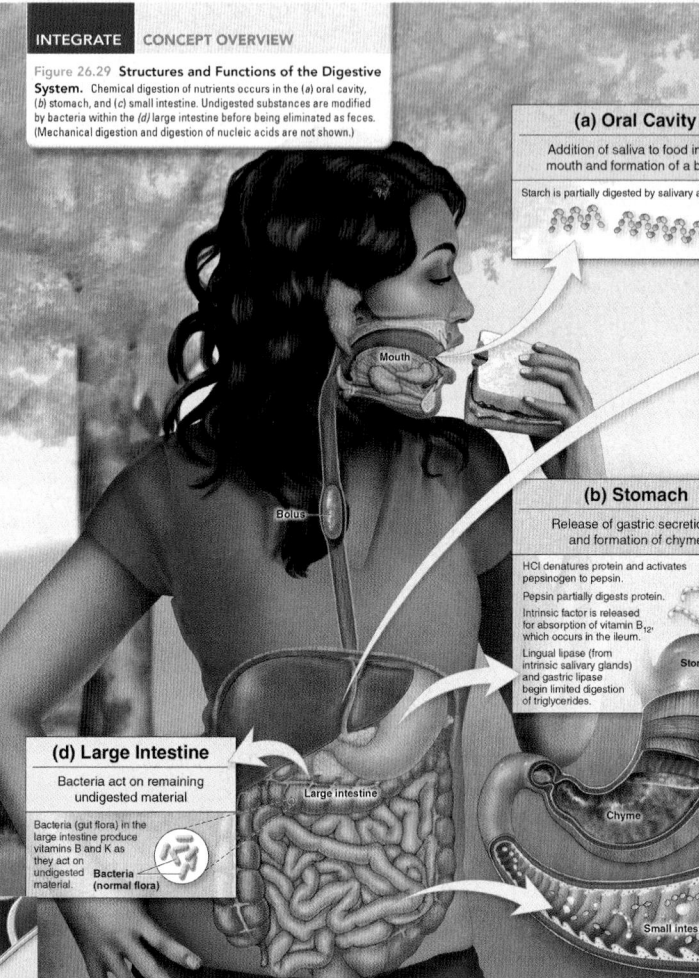

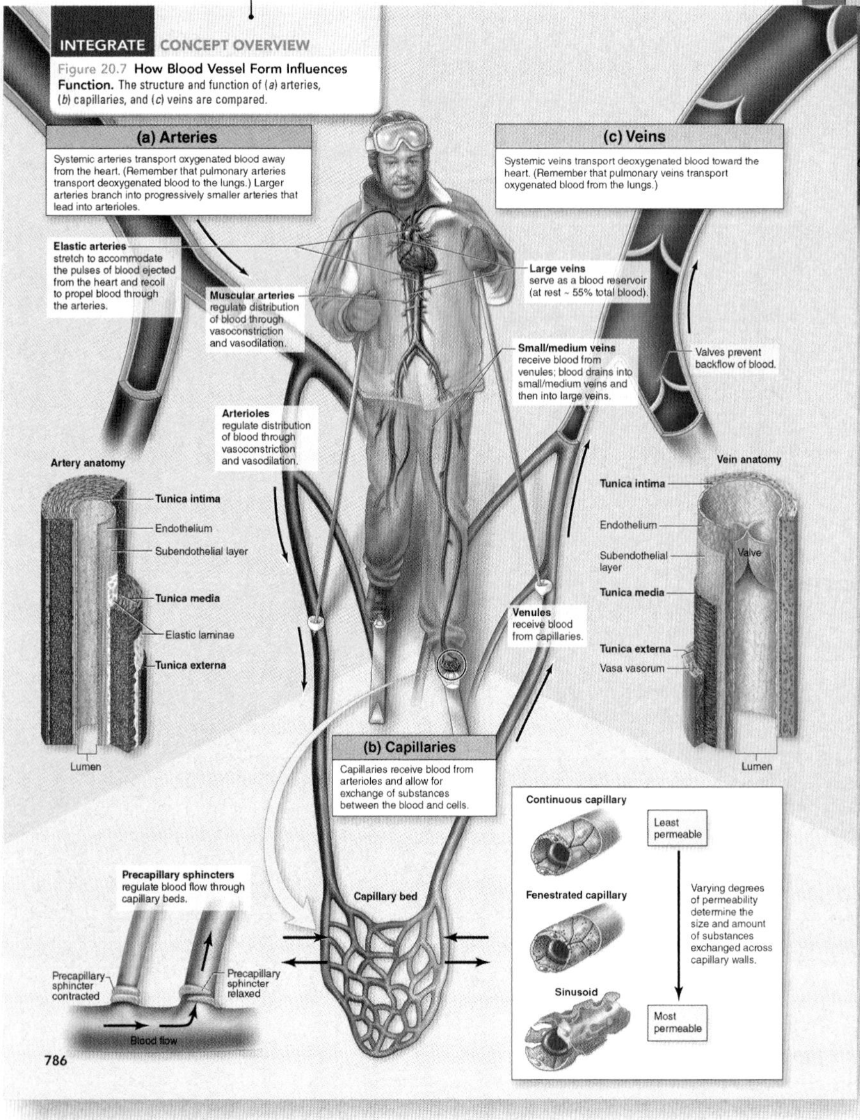

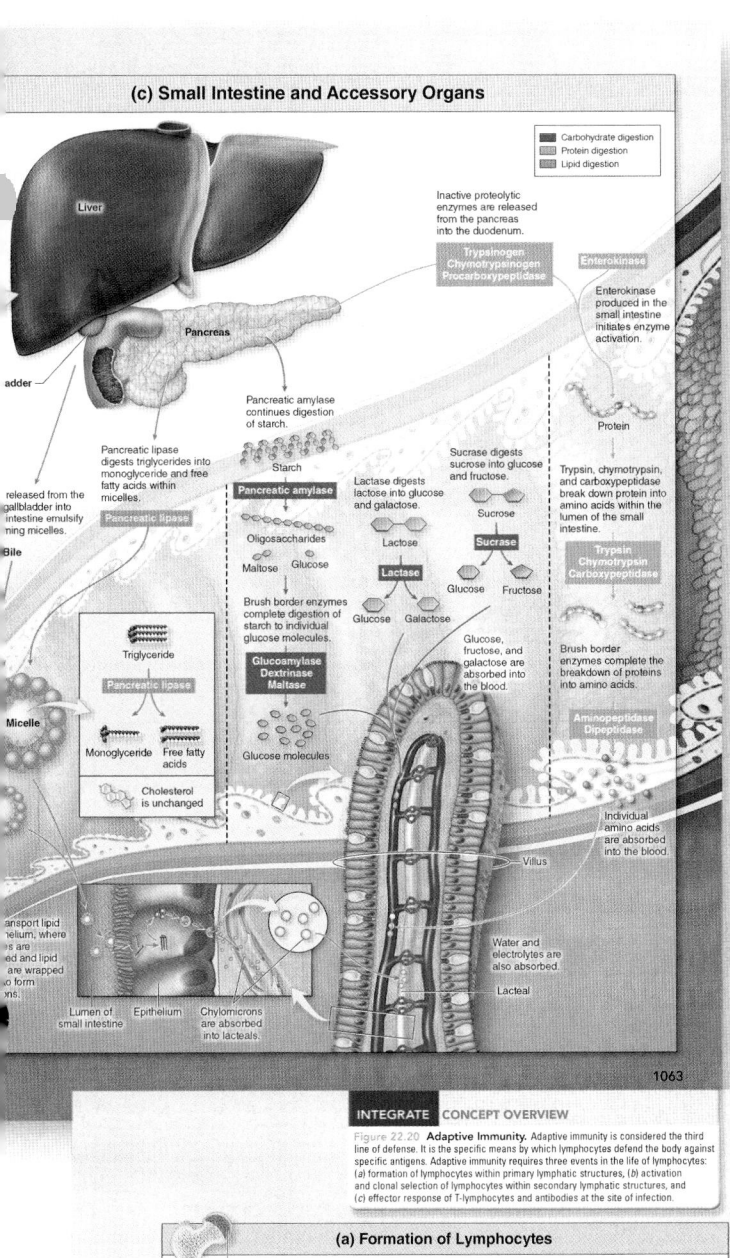

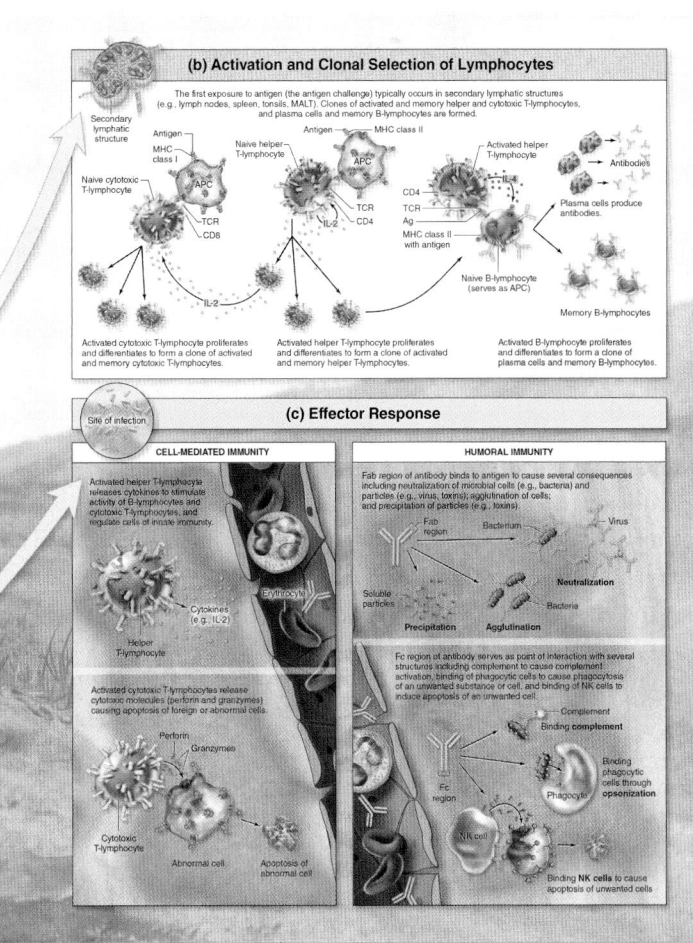

"My students love this artwork, especially the Concept Overview artwork. They use the Concept Overview artwork as a study guide (review) of the text material."

—Jerri K. Lindsey, Tarrant County College–Northeast Campus

Concept Integration

Both backward and forward references are supplied throughout the text to remind the reader of the significance of previously covered material, and to foreshadow how knowledge of a topic at hand will come into play in a later discussion. Simple references appear in the flow of the text, while more detailed refreshers are presented in **Integrate: Concept Connection** boxes.

"I think [Concept Connections] helps the student see connections between the different topics they learned, rather than viewing the content as separate chunks of material."

—Marta Klesath, North Carolina State University

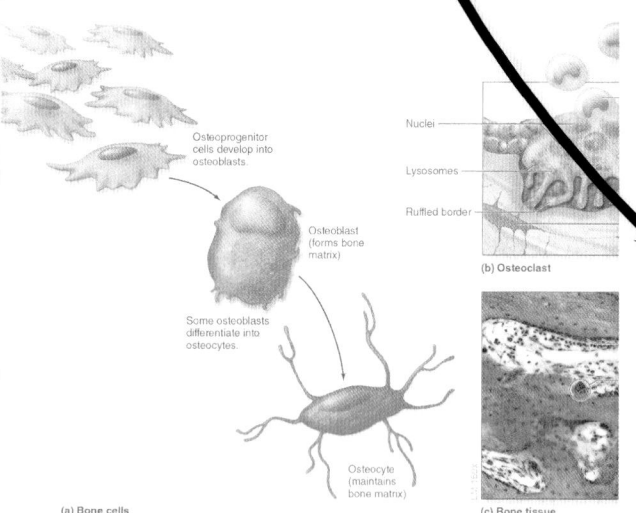

7.2e Microscopic Anatomy: Bone Connective Tissue

LEARNING OBJECTIVES

7. Name the four types of bone cells and their functions.
8. Describe the composition of bone's matrix.
9. Explain bone matrix formation and resorption.
10. Compare the structure of compact bone and spongy bone.

The primary component of bone is **bone connective tissue,** also called *osseous* (os'ē-ŭs; *os* = bone) *connective tissue.* Bone is composed of both cells and extracellular matrix, like all connective tissue. We now describe the cells and matrix that compose bone connective tissue, how the matrix is formed and resorbed, and then the two microscopic arrangements (compact bone and spongy bone).

Cells of Bone

Four types of cells are found in bone connective tissue: osteoprogenitor cells, osteoblasts, osteocytes, and osteoclasts (figure 7.6).

Osteoprogenitor (os'tē-ō-prō-jen'i-ter) **cells** are stem cells derived from mesenchyme (see section 5.2c). When they divide through the process of cellular division, another stem cell is produced along with a "committed cell" that matures to become an osteoblast. As previously described, these stem cells are located in both the periosteum and the endosteum.

Osteoblasts (*blast* = germ) are formed from osteoprogenitor stem cells. Often, osteoblasts are positioned side by side on bone surfaces. Active osteoblasts exhibit a somewhat cuboidal shape and have abundant rough endoplasmic reticulum and Golgi apparatus, reflecting the activity of

Figure 7.6 Types of Cells in Bone Connective Tissue. Four different types of cells are found in b (*a*) Osteoprogenitor cells develop into osteoblasts, many of which differentiate to become osteocytes. (*b*) Some bone osteoclasts. (*c*) A photomicrograph shows osteoblasts, osteocytes, and osteoclasts. **AP|R**

"I think that some of the best parts of this chapter are the Integrate: Clinical View boxes. The author does an excellent job of incorporating relevant clinical examples ...I think the clinical relevance is what truly gets students interested about the material and helps solidify a concept."

—Arlee Dulak, University of Massachusetts, Lowell

INTEGRATE

CONCEPT CONNECTION

In section 5.6b (See Clinical View: "Stem Cells"), we discussed the role of stem cells in potential treatments of disease. Osteoprogenitor cells are one type of adult stem cell that is specific to the skeletal system, and researchers are examining their medical treatment potential.

INTEGRATE

CLINICAL VIEW
Stem Cells

Why all the interest in stem cells?

Stem cells are immature, undifferentiated cells. These cells are able to divide into two cells, the first of which is another stem cell, and the other a cell that could differentiate into a specialized, mature cell with a unique function. Stem cells have generated interest in the scientific and medical communities because of their potential for repair or replacement of damaged or dying tissue.

What are the two basic characteristics of stem cells?

All stem cells exhibit two characteristics: self-renewal and potency. **Self-renewal** refers to their ability to divide repeatedly to produce both new cells for maturation and new stem cells. **Potency** is the potential for differentiation: Different stem cells have varying ability to differentiate into almost any type of cell. Stem cells exhibit the following four levels of potency: totipotency, pluripotency, multipotency, and unipotency:

- **Totipotent** stem cells have a "total potential," meaning that they exhibit the ability to differentiate into any cell type within an organism. A totipotent cell is produced when a secondary oocyte is fertilized by a sperm, giving rise to a zygote. The first few cell divisions of the zygote result in equally totipotent cells. Thus, only embryonic (and not adult) stem cells have the potential to be totipotent.
- **Pluripotent** stem cells are derived from totipotent stem cells. These stem cells are formed from the embryoblast portion (inner cell mass) of the **blastocyst.** The blastocyst is a ball of cells that develops during the first week of development from the zygote. The embryoblast is the portion of the blastocyst that will eventually become an embryo and then a fetus. Pluripotent stem cells can form cells in any of the tissue layers of the embryo, but they cannot form structures such as the placenta. Again, only embryonic stem cells have the potential to be pluripotent.
- **Multipotent** stem cells are derived from pluripotent stem cells. They have the capability to differentiate into a restricted number of some cell types and not others. For example, stem cells in the

bone marrow may be stimulated to mature and differentiate into different types of blood cells, but not into some other types of cells. Some adult stem cells have the potential to be multipotent.

- **Unipotent** stem cells have the ability to differentiate into a single type of cell, yet these cells still retain the ability to renew themselves. Epithelial stem cells (discussed previously) are examples of unipotent stem cells. Many adult stem cells are unipotent.

What are the differences between embryonic and adult stem cells?

Stem cells may be categorized as either **embryonic stem cells** or **adult stem cells.** Embryonic stem cells include those that have begun to divide in the zygote and the cells in the blastocyst. Embryonic stem cells exhibit the greatest degree of potency—and thus, the greatest potential to differentiate into multiple cell types. In contrast, adult stem cells are the immature cells found in postnatal (already born) organisms. Adult stem cells typically are multipotent or unipotent, and thus they exhibit less potency than embryonic stem cells.

How are stem cells harvested?

Most embryonic stem cells must be harvested from a structure no more differentiated than a blastocyst. Most of these blastocysts were donated by families undergoing in vitro fertilization who had stored more blastocysts than needed for a successful pregnancy. If these blastocysts were not used by the family and not donated for research, they typically would be destroyed. Note that opponents of embryonic stem cell research counter that these blastocysts could be implanted and lead to viable infants who could be adopted, and any medical benefit from embryonic stem cells does not justify using them in research. Opponents also maintain that adult stem cell research should be explored instead.

Adult stem cells may be extracted from the bone marrow or tissue of an individual. These adult stem cells have been used to successfully treat certain blood and bone cancers, and research is ongoing about their effectiveness for diseases such as lung inflammation, stroke, and Parkinson disease. The main problem with adult stem cells is their limited potency, which suggests that their use for treatment in diseases is limited. Embryonic stem cells exhibit greater promise for treatment because of their greater potency.

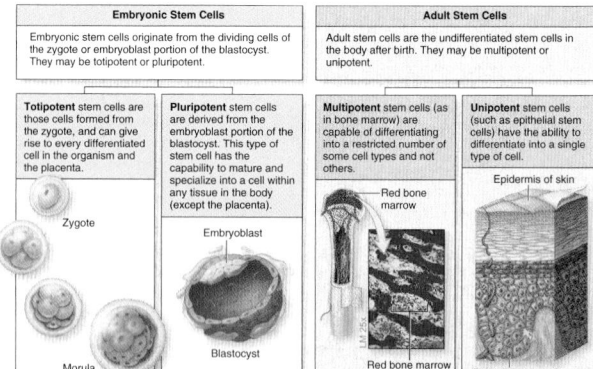

Embryonic Stem Cells		Adult Stem Cells	
Embryonic stem cells originate from the dividing cells of the zygote or embryoblast portion of the blastocyst. They may be totipotent or pluripotent.		Adult stem cells are the undifferentiated stem cells in the body after birth. They may be multipotent or unipotent.	
Totipotent stem cells are those cells formed from the zygote, and can give rise to every differentiated cell in the organism and the placenta.	**Pluripotent** stem cells are derived from the embryoblast portion of the blastocyst. This type of stem cell has the capability to mature and specialize into a cell within any tissue in the body (except the placenta).	**Multipotent** stem cells (as in bone marrow) are capable of differentiating into a restricted number of some cell types and not others.	**Unipotent** stem cells (such as epithelial stem cells) have the ability to differentiate into a single type of cell.

INTEGRATE

LEARNING STRATEGY

The cells associated with innate immunity have a "military-like" function:

- **Neutrophils** are the "foot soldiers" that are the first to arrive at the site of infection.
- **Macrophages** are the "big eaters"—the cleanup crew that arrives at the injured or infected scene late and stays longer.
- **Basophils/mast cells** engage in chemical warfare; causes inflammation.
- **NK cells** serve as "security guards" that "search destroy" unwanted cells.
- **Eosinophils** are the "heavy artillery" to take on th guys" (parasites).

INTEGRATE

LEARNING STRATEGY

Generally, the role of antibodies as weapons is to "tie up the prisoner" until other help arrives. You can remember the six functions of an antibody with the acronyms NAP and CON: Neutralization, Agglutination, and Precipitation (NAP), as well as Complement, Opsonization, and NK cells (CON). Remember—a NAP can help you CONcentrate.

Practical and Clinical Applications

Integrating familiar contexts into the study of A&P makes seemingly abstract concepts more relevant and memorable. **Integrate: Learning Strategy** boxes provide simple, practical advice for learning the material. **Integrate: Clinical View** readings offer insight on how complex physiologic processes or anatomic relationships affect body functioning.

Learning Strategies

Classroom tried-and-tested learning strategies offer everyday analogies, mnemonics, and useful tips to aid understanding and memory.

INTEGRATE

LEARNING STRATEGY

Integrate lab and lecture material: Follow these steps to help you identify the epidermal strata under the microscope:

1. **Determine if the layer is closer to the free surface or is deeper.** Remember the stratum corneum forms the free surface, whereas the stratum basale forms the deepest epidermal layer.

2. **Examine the shape of the keratinocytes.** The stratum basale contains cuboidal to low columnar keratinocytes, the stratum spinosum contains polygonal keratinocytes, and the stratum lucidum and corneum contain squamous keratinocytes.

3. **See if the keratinocytes have a nucleus or are anucleate.** When they are still alive (as in the strata basale, spinosum,

and granulosum), you are able to see nuclei. The stratum lucidum and corneum layers contain dead, anucleate keratinocytes.

4. **Count the layers of keratinocytes in the stratum.** The stratum basale has only one layer of keratinocytes, and the stratum corneum contains 20 to 30 layers of keratinocytes. The other layers contain about two to five layers of keratinocytes.

5. **Determine if the cytoplasm of the keratinocytes contain visible granules.** If the keratinocytes contain visible granules, you likely are looking at the stratum granulosum.

Clinical View

Interesting clinical sidebars reinforce or expand upon the facts discussed within the narrative. The clinical views are adjacent to the facts in the narrative (rather than placed at the end of the chapter) so students may immediately make connections between the narrative and real-life applications.

INTEGRATE

CLINICAL VIEW

Asthma (az′mă) is a chronic condition characterized by episodes of bronchoconstriction coupled with wheezing, coughing, shortness of breath, and excess pulmonary mucus. Typically, the affected person develops sensitivity to an airborne agent, such as pollen, smoke, mold spores, dust mites, or particulate matter. Upon reexposure to this triggering substance, a localized immune reaction occurs in the bronchi and bronchioles, resulting in bronchoconstriction, swollen submucosa, and increased production of mucus. Episodes typically last an hour or two. Continual exposure to the triggering agent increases the severity and frequency of asthma attacks. The walls of the bronchi and bronchioles eventually may become permanently thickened, leading to chronic and unremitting airway narrowing and shortness of breath. If airway narrowing is extreme during a severe asthma attack, death could occur.

The primary treatment for asthma consists of administering inhaled steroids (cortisone-related compounds) to reduce the inflammatory reaction, combined with bronchodilators to alleviate the bronchoconstriction. Allergy shots have proven helpful for some patients. Individuals with severe asthma may need oral doses of steroids to help control the allergic hyper-response and reduce the inflammation. A new treatment called bronchial thermoplasty uses heat to remove some of the outer layers of smooth muscle. This decreases the muscle contractions associated with bronchoconstriction to lessen the severity of asthma.

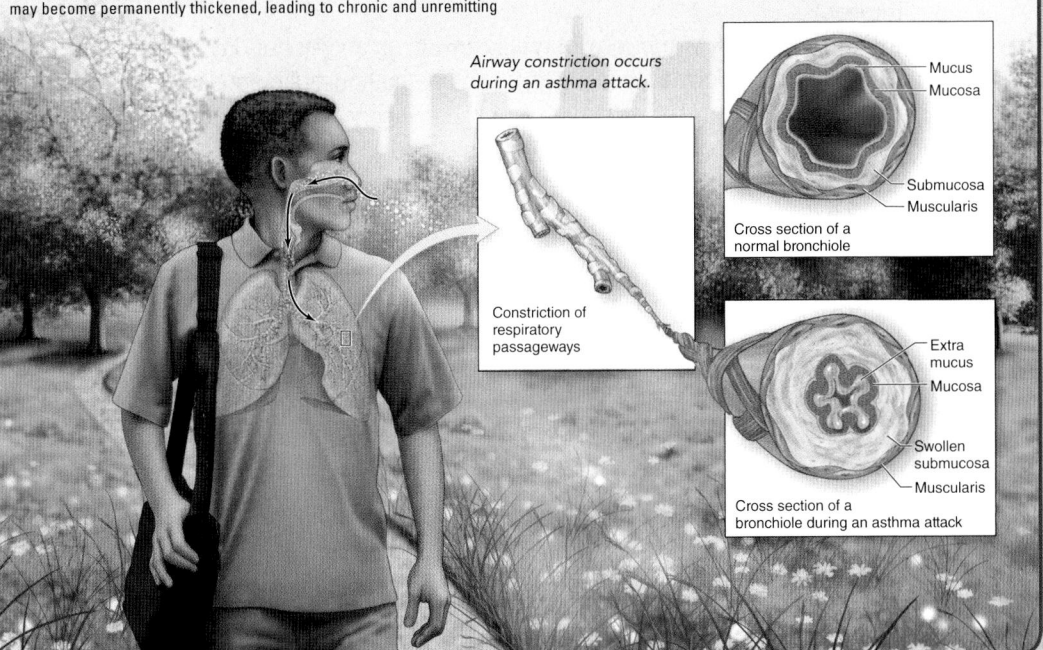

Airway constriction occurs during an asthma attack.

Constriction of respiratory passageways

Cross section of a normal bronchiole
- Mucus
- Mucosa
- Submucosa
- Muscularis

Cross section of a bronchiole during an asthma attack
- Extra mucus
- Mucosa
- Swollen submucosa
- Muscularis

Integrated Assessments

Throughout each chapter, sections begin with learning objectives and end with questions intended to assess whether those objectives have been met. Critical-thinking questions within the narrative prompt students to apply the material as they read. A set of tiered questions at the end of the chapter, as well as additional online problems, further challenge students to master the material.

WHAT DO YOU THINK?

1 Many times during a long-distance race, water stations are positioned along the side of the road so that runners may rehydrate during the race. Sometimes a runner will take a drink, swirl it around in the mouth, and then spit the water back out instead of swallowing it. Do you think this practice should be encouraged? Explain in terms of the effect on the thirst center and the hydrated state of the body.

What Do You Think?
These critical-thinking questions engage students in application or analysis and encourage them to think more globally about the content.

WHAT DID YOU LEARN?

2 Which ions are more prevalent in the intracellular fluid? Which are more prevalent in the extracellular fluid?

3 What is the major distinction in the chemical composition of blood plasma and interstitial fluid?

4 When you are dehydrated, is the net movement of fluid from the blood plasma into the cells or from the cells into the blood plasma?

What Did You Learn?
These mini self-tests at the end of each section help students determine whether they have a sufficient grasp of the information before moving on to the next section.

CHALLENGE YOURSELF

Do You Know the Basics?

Challenge Yourself
Assessments at the end of each chapter progress through knowledge-, application-, and synthesis-level questions. The "Can You Apply ..." and "Can You Synthesize ..." question sets are clinically oriented to encourage concept application, and expose students who may be pursuing health-related careers to problem solving in clinical contexts.

_____ 1. Atoms composed of the same numbers of protons and electrons, but different numbers of neutrons, are called

  a. isome

  b. ions.

Can You Apply What You've Learned?

_____ 2. Substanc
  except

  a. lipids.

  b. glucos

1. Which property of water is significant in children born prematurely because it causes the air sacs to collapse in the lungs, making breathing difficult?

  a. specific heat   c. surface tension

  b. water reactivity   d. capillary action

_____ 3. Tempera
  propertie

  a. cohesi

  b. capill

  c. specif

  d. cohesi

2. A young boy playing outside on a very hot day has become dehydrated. When he enters the house, he appears lethargic. The mother is a nurse and becomes concerned that he may be experiencing a fluid and electr include all of the following *exc*

  a. sodium ion.   c.

  b. glucose.   d.

Can You Synthesize What You've Learned?

_____ 4. All of th
  pH *exce*

  a. acids contain more H⁺ than water.

  b. H⁺ concentration and pH are inversely relate

1. An individual is exposed to high-energy radiation. Which biomolecule that regulates the process of protein synthesis may have been mutated?

2. The lab results from a diabetic patient show a lower than normal pH (a condition referred to as acidosis). Explain the change in H⁺ concentration in the blood, and describe how this change may affect the folding of proteins in the blood plasma (and elsewhere).

3. A patient is given a new drug that decreases blood sugar levels. This drug is regulating which specific molecule?

Integrated Media and Textbook

Print and Digital Study Tools Connected

Each chapter ends with a listing of online tools that may be used to study and master the concepts presented.

INTEGRATE

ONLINE STUDY TOOLS

The following study aids may be accessed through Connect.

Concept Overview Interactive Figure 12.23: Events of Neuron Physiology

Clinical Case Study: A Young Man Who Gets Weak When Overheated

Interactive Questions: This chapter's content is served up in a number of multimedia question formats for student study.

LearnSmart: Topics and terminology include neurons; synapses; glial cells; axon regeneration; ultrastructure of neurons; neuron physiology; physiologic events in the neuron segments; velocity of a nerve signal; neurotransmitters and neuromodulation; neural integration and neuronal pools of the CNS

Anatomy & Physiology Revealed: Topics include multipolar neuron; Schwann cell (neurolemmocyte); unmyelinated axon and myelin sheath; action potential generation and propagation; chemical synapse

Animations: Topics include PSPs (postsynaptic potentials); action potential propagation; action potential generation; action potential propagation in myelinated neurons

Concept Overviews into Digital Learning

Selected **Concept Overview Figures** from the textbook have been transformed into interactive study modules. This digital transformation process was guided by anatomy and physiology professors who reviewed the modules throughout the development process. Interactive Concept Overview Figures also have assessable, autograded learning activities in Connect®, and are also provided separately to instructors as classroom presentation tools.

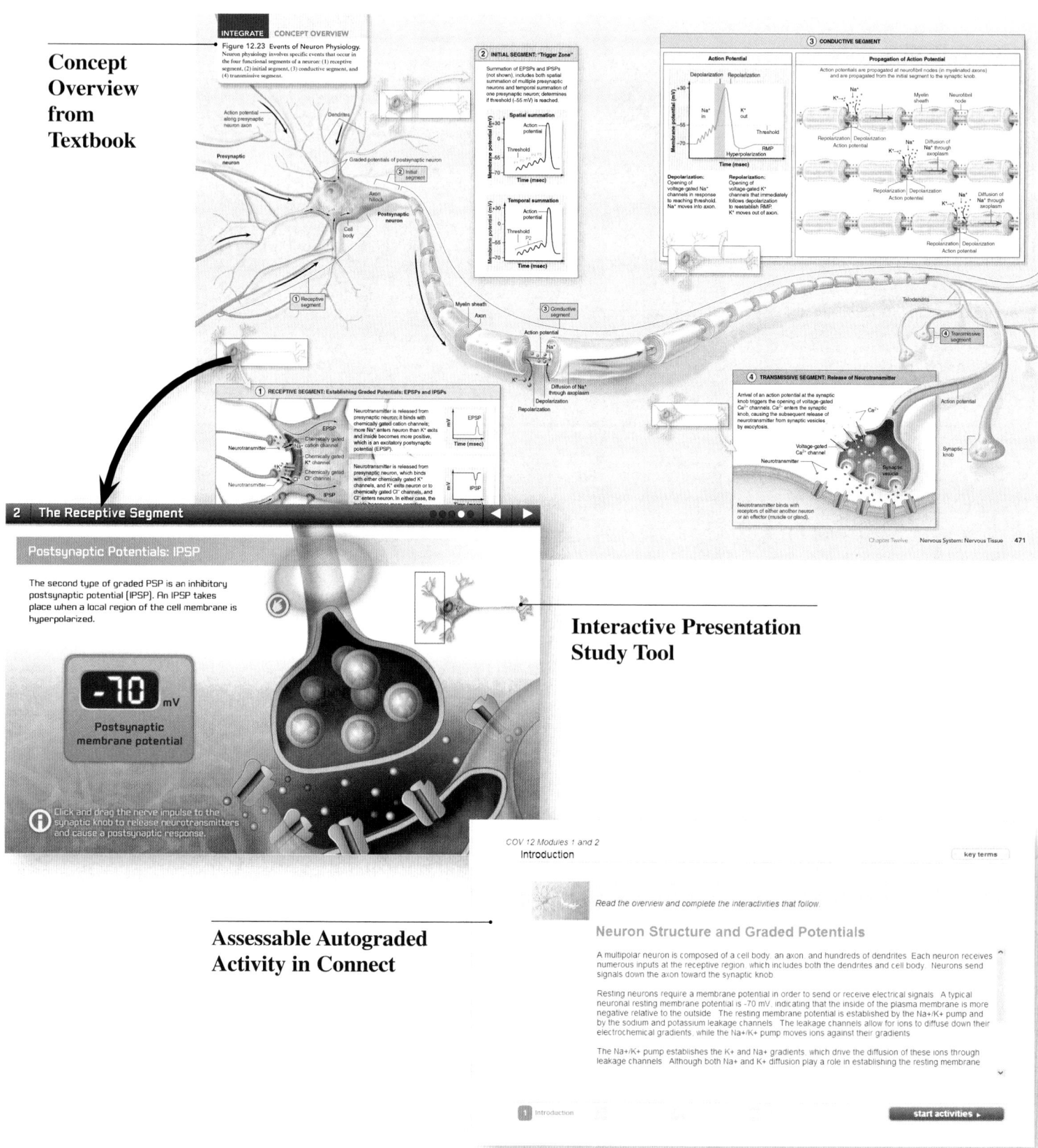

Concept Overview from Textbook

Interactive Presentation Study Tool

Assessable Autograded Activity in Connect

Connect®

McGraw-Hill Connect® interactive learning platform provides:

- Auto-graded assignments
- Adaptive diagnostic tools
- Powerful reporting against learning outcomes and level of difficulty
- An easy-to-use interface
- **McGraw-Hill Tegrity®**, which digitally records and distributes your lectures with a click of a button

Connect includes the full textbook as an integrated, dynamic eBook, which you can also assign. Everything you need . . . in one place!

McGraw-Hill LearnSmart™

Unlike static flashcards or rote memorization, **McGraw-Hill LearnSmart™** ensures your students have mastered course concepts before taking the exam, thereby saving you time and increasing student success.

- The only truly adaptive learning system
- Intelligently identifies course content students have not yet mastered
- Maps out personalized study plans for student success

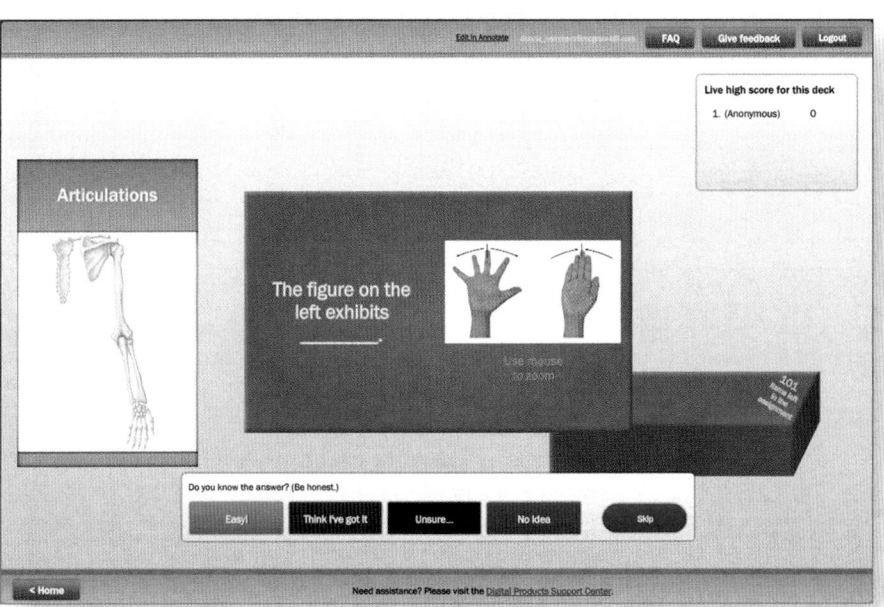

Learning Objectives and Connect®

Each Learning Objective from the textbook is tied to interactive questions in Connect, to assure that all parts of the chapters have adequate coverage within Connect assignments.

6.1 Composition and Functions of the Integument

The integument is the body's largest organ and is composed of all tissue types that function in concert to protect internal body structures. Its surface is an epithelium that protects underlying body layers. The connective tissue that underlies the epithelium provides strength and resilience to the skin. This connective tissue also contains smooth muscle associated with hair follicles (arrector pili) that alters hair position. Finally, nervous tissue detects and monitors sensory stimuli in the skin, which provide information about touch, pressure, temperature, and pain.

The integument accounts for 7% to 8% of the body weight and covers the entire body surface with an area that varies between about 1.5 and 2.0 square meters (m²). Its thickness ranges between 1.5 millimeters (mm) and 4 mm or more, depending on body location. (For comparison, a sheet of copier paper is about 0.1 mm thick, so the thickness of the skin would range between 15 and 40 sheets of paper.) The integument consists of two distinct layers: a layer of stratified squamous epithelium called the epidermis, and a deeper layer of primarily dense irregular connective tissue called the dermis (figure 6.1). Deep to the dermis is a layer of areolar and adipose connective tissue called the subcutaneous layer, or hypodermis. The subcutaneous layer is not part of the integumentary system; however, it is described in this chapter because it is closely involved with both the structure and function of the skin.

6.1a Epidermis

LEARNING OBJECTIVES

1. Describe the five layers (strata) of the epidermis.
2. Differentiate between thick skin and thin skin.
3. Explain what causes differences in skin color.

The epithelium of the integument is called the **epidermis** (ep-i-derm´is; *epi* = on, *derma* = skin). It is a keratinized, stratified squamous epithelium (see section 5.1c).

Careful examination of the epidermis, from the basement membrane to its surface, reveals several specific layers, or strata. From deep to superficial, these layers are the stratum basale, stratum spinosum, stratum granulosum, stratum lucidum (found in thick skin only), and the stratum corneum (figure 6.2). The first three strata listed are composed of living keratinocytes, whereas the most superficial two strata contain dead cells.

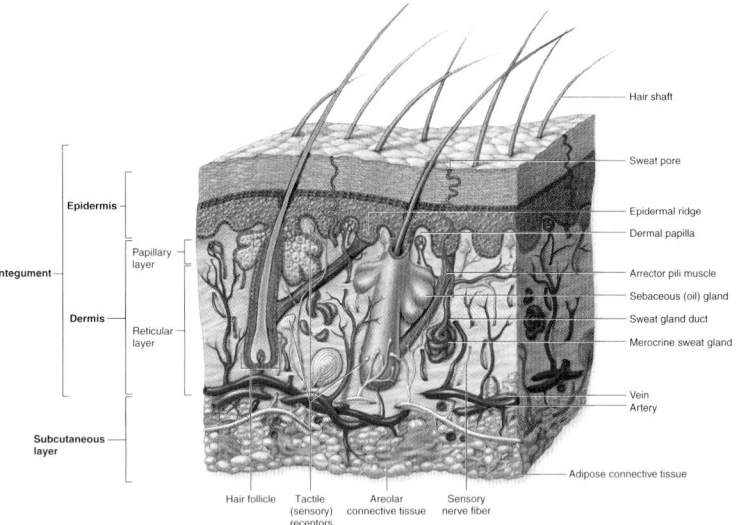

Figure 6.1 **Layers of the Integument.** A diagrammatic sectional view through the integument shows its relationship to the underlying subcutaneous layer. **AP|R**

6.1a Epidermis

LEARNING OBJECTIVES

1. Describe the five layers (strata) of the epidermis.
2. Differentiate between thick skin and thin skin.
3. Explain what causes differences in skin color.

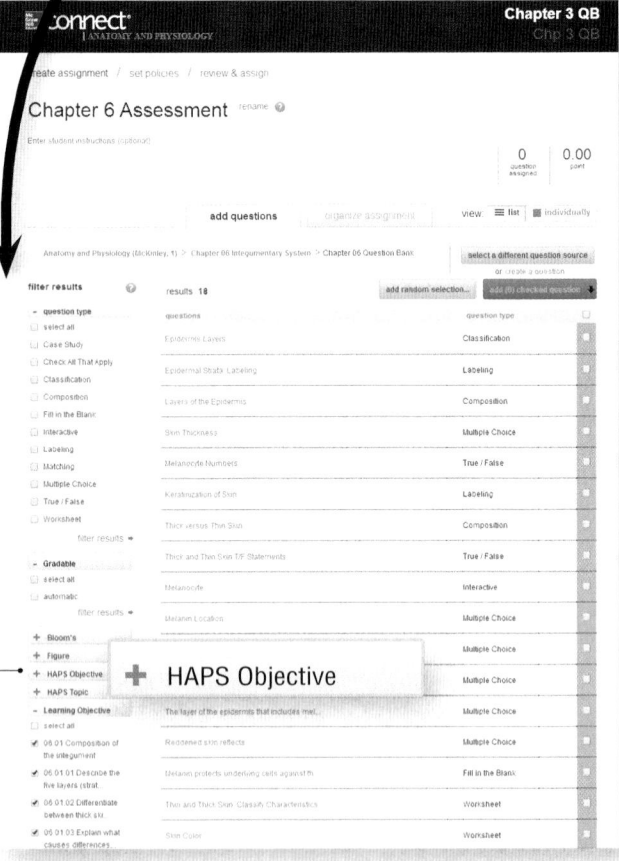

Correlated to HAPS Learning Objectives

Where appropriate, questions in Connect are now tied to the Human Anatomy and Physiology Society (HAPS) Learning Objectives. Instructors may filter assignable questions by HAPS Learning Objectives and see all the corresponding questions.

Anatomy & Physiology REVEALED®

Anatomy & Physiology REVEALED® is an interactive cadaver dissection tool to enhance lecture and lab that students can use anytime, anywhere. Instructors can customize APR 3.0 by indicating the specific content they require in their course through a simple menu selection process. APR 3.0 does the rest. In the course-specific area of the program, students see only the instructor's designated content, and only this content appears in quiz questions.

APR contains all the material covered in an A&P course, including these three modules:

- Body Orientation
- Cells and Chemistry
- Tissues

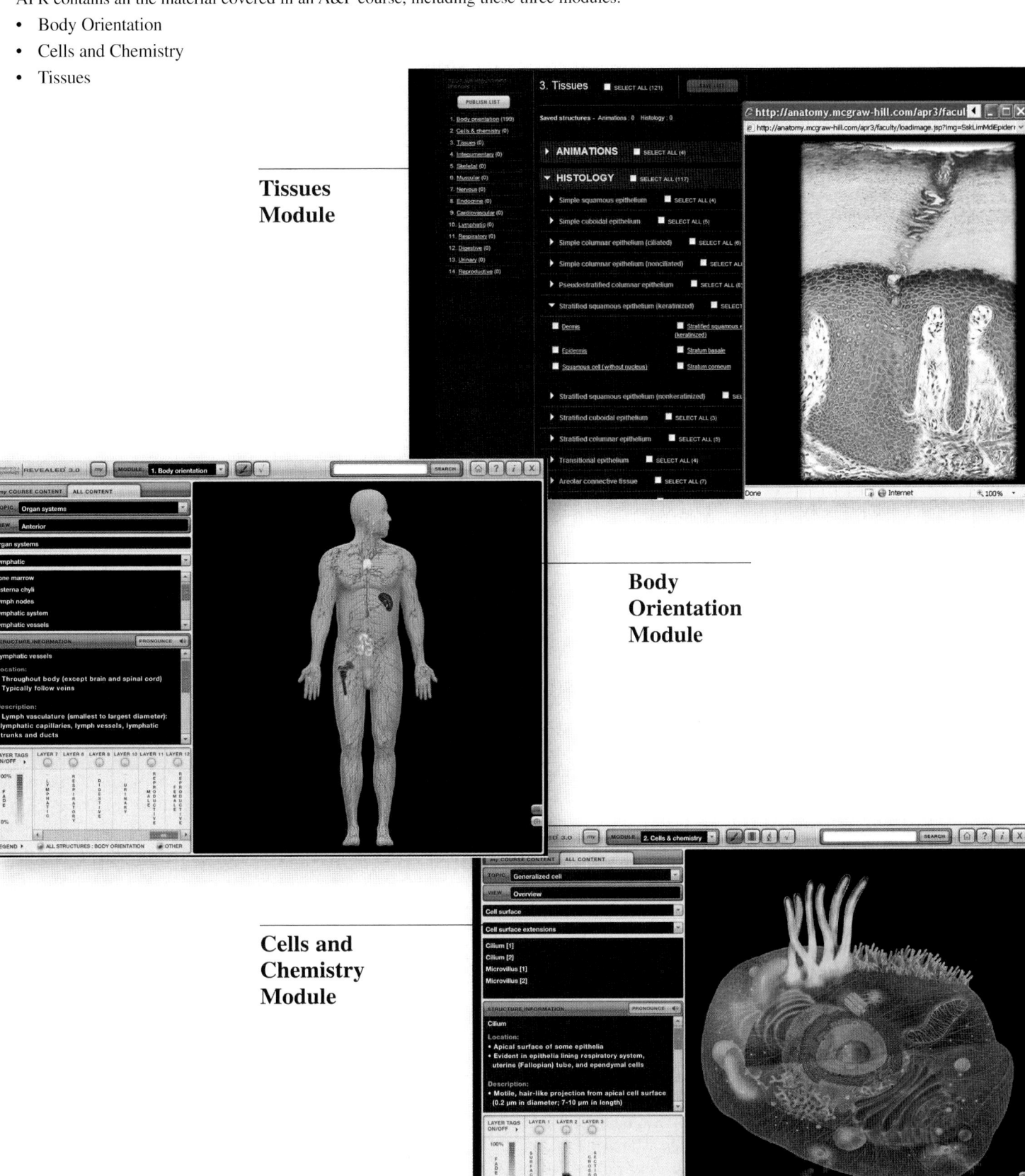

Tissues Module

Body Orientation Module

Cells and Chemistry Module

Presentation Tools

Accessed from the *Anatomy and Physiology: An Integrated Approach* Connect website, **Presentation Tools** is a collection of digital assets including photos, artwork, animations, and other media that can be used to create customized lectures, visually enhanced tests and quizzes, compelling course websites, or attractive printed support materials. All assets are copyrighted by McGraw-Hill Higher Education, but can be used by instructors for classroom purposes. The visual resources in this collection include the following:

- **Art** Full-color digital files of all illustrations in the book can be readily incorporated into lecture presentations, exams, or custom-made classroom materials. In addition, all files are preinserted into PowerPoint slides for ease of lecture preparation.
- **Photos** The photo collection contains digital files of photographs from the text, which can be reproduced for multiple classroom uses.
- **Tables** Every table that appears in the text has been saved in electronic form for use in classroom presentations and/or quizzes.
- **Animations** Numerous full-color animations illustrating important processes are also provided. Harness the visual impact of concepts in motion by importing these files into classroom presentations or online course materials.
- **PowerPoint Lecture Outlines** Ready-made presentations that combine art and lecture notes are provided for each chapter of the text.
- **PowerPoint Slides** For instructors who prefer to create their lectures from scratch, all illustrations, photos, and tables, are preinserted by chapter into blank PowerPoint slides.

Instructors: To access Connect, request registration information from your McGraw-Hill sales representative.

Digital Lecture Capture: Tegrity McGraw-Hill Tegrity® records and distributes your lecture with just a click of a button. Students can view anytime/anywhere via computer, iPod, or mobile device. Tegrity indexes as it records your slideshow presentations and anything shown on your computer, so students can use keywords to find exactly what they want to study.

Computerized Test Bank Updated by Justin York of Glendale Community College, over 3,600 test questions are served up utilizing EZ Test software to accompany *Anatomy and Physiology: An Integrated Approach*. Diploma's software allows you to quickly create a customized test using McGraw-Hill's supplied questions, or by authoring your own questions. Diploma is a downloadable application that allows you to create your tests without an Internet connection—just download the software and question files directly to your computer.

Content Delivery Flexibility *Anatomy and Physiology: An Integrated Approach* by McKinley/O'Loughlin/Bidle is available in many formats in addition to the traditional textbook to give instructors and students more choices when deciding on the format of their A&P text. Choices include the following:

- **Customizable Textbooks: Create** Introducing **McGraw-Hill Create™**—this, self-service website allows you to create custom course materials—print and eBooks—by drawing upon McGraw-Hill's comprehensive, cross-disciplinary content. Add your own content quickly and easily. Tap into other rights-secured third-party sources as well. Then, arrange the content in a way that makes the most sense for your course. Even personalize your book with your course name and information. Choose the best format for your course: color print, black-and-white print, or eBook. The eBook is now viewable on an iPad! And when you are finished customizing, you will receive a free PDF review copy in just minutes! Visit McGraw-Hill Create—*www.mcgrawhillcreate.com*—today and begin building your perfect book.

- **Connect eBook** **McGraw-Hill Connect** eBook takes digital texts beyond a simple PDF. With the same content as the printed book, but optimized for the screen, Connect has embedded media, including animations and videos, which bring concepts to life and provide "just in time" learning for students. Additionally, fully integrated homework allows students to interact with the questions in the text to determine if they're gaining mastery of the content, and can also be assigned by the instructor.

McGraw-Hill and Blackboard®

McGraw-Hill Higher Education and Blackboard have teamed up. What does this partnership mean for you? Blackboard users will find the single sign-on and deep integration of Connect within their Blackboard course an invaluable benefit. Even if your school is not using Blackboard, we have a solution for you. Learn more at www.domorenow.com.

Lab Manual Options to Fit Your Course

Anatomy & Physiology Laboratory Manual by Christine Eckel, Kyla Ross, and Terri Bidle is a laboratory manual specifically developed for the McKinley/O'Loughlin/Bidle *Anatomy and Physiology: An Integrative Approach* text:

- Three versions are available including main, cat, and fetal pig.
- Each chapter opens with a set of learning objectives that are keyed to the post-laboratory worksheet to ensure student understanding of each chapter's objectives.
- The manual includes the highest-quality photographs and illustrations of any laboratory manual in the market.
- Laboratory exercises are "how-to" guides that involve touch, dissection, observation, experimentation, and critical-thinking exercises.
- In-chapter learning activities offer a mixture of labeling exercises, sketching activities, table completion exercises, data recoring, palpation of surface anatomy, and other sources of learning.

- Numerous exercises throughout the manual utilize Physiology Interactive Lab Simulations (Ph.I.L.S.) 4.0 Online to provide additional student understanding of physiology.
- Ph.I.L.S. 4.0 is included with each new laboratory manual.

Laboratory Manual for Human Anatomy & Physiology by Terry Martin is written to coincide with any A&P textbook:

- Three versions available, including main, cat, and fetal pig
- Includes Ph.I.L.S. 4.0 Online
- Outcomes and assessments format
- Clear, concise writing style

Student Supplements

McGraw-Hill offers various tools and technology products to support the textbook. Students can order supplemental study materials by contacting their campus bookstore or online at *www.shopmcgraw-hill.com.*

Instructor Supplements

Instructors can obtain teaching aids by calling the McGraw-Hill Customer Service Department at 1-800-338-3987, vising our online catalog at www.mhhe.com, or by contacting their local McGraw-Hill sales representative.

The Sciences of Anatomy and Physiology

chapter 1

INTEGRATE

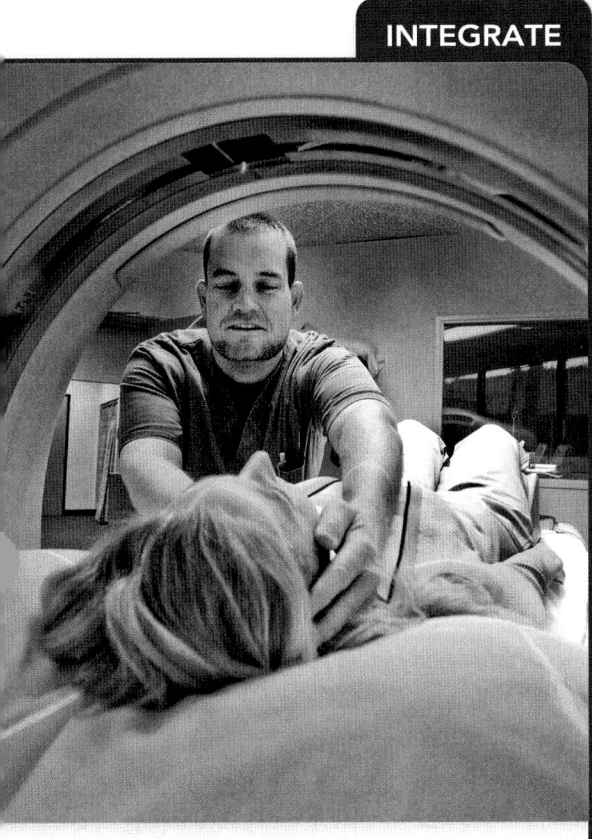

aprevealed.com

Module 1: Body Orientation

CAREER PATH
Medical Imaging Technologist

A medical imaging technologist is trained to utilize a variety of imaging techniques, such as magnetic resonance imaging (MRI), computed tomography (CT), and sonography. The technologist must be able to correctly interpret the physician's instructions, operate the imaging machinery, and communicate with the patient during the procedure. The image above shows a CT technician positioning a patient for a cranial CT scan. This technician must understand relevant brain anatomy and be able to interpret the sectional images produced of the brain.

You are about to embark on an adventure into the amazing world of human anatomy and physiology. Both fields explore the incredible workings of the human body. Anatomy studies the form and structure of the body, whereas physiology examines how the body functions. Together, these applied sciences provide the basis for understanding health and human performance. The basic vocabulary of these sciences is derived from both Greek and Latin.

If you actively practice the vocabulary and descriptive terminology used in this text, your understanding and appreciation of body structure and function will be enhanced significantly. In this book, you will learn that structure and function are inseparable. You will see how the body functions normally, as well as what happens to body function and structure when injury and disease occur.

INTEGRATE

LEARNING STRATEGY

Boxed elements like this provide you with helpful analogies, memory aids, and other study tips to help you better understand and learn the material. Look for these boxes throughout each chapter.

1.1 Anatomy and Physiology Compared

In this section, we compare anatomy and physiology and present the general subdivisions of these sciences.

Anatomy is the study of structure and form. The word *anatomy* is derived from the Greek word *anatome,* which means to cut apart or dissect. Anatomists are scientists who study the structure and form of organisms. Specifically, they examine the relationships among parts of the body as well as the structure of individual organs. **Physiology** is the study of function of the body parts. Physiologists are scientists who examine how organs and body systems function under normal circumstances, as well as how their functioning may be altered via medication or disease. For example, when studying blood capillaries (the smallest of blood vessels), an anatomist may describe the composition of the thin wall. In contrast, a physiologist will explain how the thin wall promotes gas and nutrient exchange between the blood within the capillary and the tissue cells outside of the capillary.

Anatomists and physiologists are professionals who use the scientific method to explain and understand the workings of the body. The **scientific method** refers to a systematic and rigorous process by which scientists:

- Examine natural events (or phenomena) through observation
- Develop a hypothesis (possible explanation) for explaining these phenomena
- Experiment and test the hypothesis through the collection of data
- Determine if the data support the hypothesis, or if the hypothesis needs to be rejected or modified

For example, early anatomists and physiologists used the scientific method to explain how blood circulates through the body. Today, we continue to use the scientific method for a variety of topics, such as to understand how the brain stores memories or explain how cancer may spread throughout the body.

Throughout this text, we have attempted to integrate the study of anatomy and physiology, showing how form and function are interrelated.

1.1a Anatomy: Details of Structure and Form

⟩⟩⟩ LEARNING OBJECTIVES

1. Describe the science of anatomy.
2. List the subdivisions in both microscopic and gross anatomy.

The discipline of anatomy is extremely broad and can be divided into several more specific fields. **Microscopic anatomy** examines structures that cannot be seen by the unaided eye. For most of these studies, scientists prepare individual cells or thin slices of some part of the body and examine these specimens under the microscope. Microscopic anatomy has several subdivisions with two main divisions:

- **Cytology** (sī-tol′ō-jē; *cyto* = cell, *logos* = study) is the study of body cells and their internal structure.
- **Histology** (his-tol′ō-jē; *histos* = web, tissue) is the study of tissues.

Gross anatomy, also called *macroscopic anatomy,* investigates the structure and relationships of body parts that are visible to the unaided eye, such as the intestines, stomach, brain, heart, and kidneys. In these macroscopic investigations, specimens or their parts are often dissected (cut open) for examination. Gross anatomy may be approached in several ways:

- **Systemic anatomy** studies the anatomy of each functional body system. For example, studying the urinary system would involve examining the kidneys (where urine is formed) and the organs of urine transport (ureters and urethra) and storage (urinary bladder). Most undergraduate anatomy and physiology classes use this systemic approach.
- **Regional anatomy** examines all of the structures in a particular region of the body as a complete unit. For example, one may study the axillary (armpit) region of the body, and in so doing examine the blood vessels (axillary artery and vein), nerves (branches of the brachial plexus), lymph nodes (axillary lymph nodes), musculature, connective tissue, and skin. Most medical school gross anatomy courses are taught using a regional anatomy approach.
- **Surface anatomy** focuses on both superficial anatomic markings and the internal body structures that relate to the skin covering them. Health-care providers use surface features to identify and locate important landmarks, such as pulse locations or the proper body region on which to perform cardiopulmonary resuscitation (CPR). Most anatomy and physiology classes also instruct students on important surface anatomy locations.
- **Comparative anatomy** examines similarities and the differences in the anatomy of different species. For example, a comparative anatomy class may examine limb structure in humans, chimps, dogs, and cats.
- **Embryology** (em′brē-ol′o-jē; *embryon* = young one) is the discipline concerned with developmental changes occurring from conception to birth.

Several specialized branches of anatomy focus on the diagnosis of medical conditions or the advancement of basic scientific research. **Pathologic** (path-ō-loj′ik; *pathos* = disease) **anatomy** examines all anatomic changes resulting from disease. Both gross anatomic changes and microscopic structures are examined. **Radiographic anatomy** investigates the relationships among internal structures that may be visualized by specific scanning procedures, such as sonography, magnetic resonance imaging (MRI), or x-ray. (See Clinical View: "Medical Imaging" at the end of this chapter.)

It may seem as though nothing new can be learned about anatomy—after all, the body has been much the same for thousands of years. Yet in fact, new information is being learned from ongoing anatomic studies, some of which displace the traditional thinking about the workings of various organs. Never forget that anatomy is a dynamic, changing science, not a static, unchanging one.

 WHAT DID YOU LEARN?

1 What subdiscipline of anatomy may explore how the lower limb differs between humans and chimpanzees?

1.1b Physiology: Details of Function

 **LEARNING OBJECTIVES**

3. Describe the science of physiology.

4. List the subdivisions in physiology.

Physiologists examine the function of various organ systems, and they typically focus on the molecular or cellular level. Thus, a basic knowledge of both chemistry and cells is essential in understanding physiology, and that's why we've included several early chapters on these topics. Mastery of these early chapters on chemistry and cells is critical to understanding the physiologic concepts that are covered throughout the text.

The discipline of physiology parallels anatomy because it also is very broad and may be subdivided into smaller groups. Many specific physiology subdisciplines focus their studies on a particular body system. For example, **cardiovascular physiology** examines the functioning of the heart, blood vessels, and blood. Cardiovascular physiologists examine how the heart pumps the blood, what are the parameters for healthy blood pressure, and details of the cellular exchange mechanisms by which respiratory gases, nutrients, and wastes move between blood and body structures. Other examples include **neurophysiology** (which examines how nerve impulses travel throughout the nervous system), **respiratory physiology** (which studies how respiratory gases are transferred by gas exchange between the lungs and the blood vessels), and **reproductive physiology** (which explores how the regulation of reproductive hormones can drive the reproductive cycle and influence sex cell production and maturation).

INTEGRATE

CLINICAL VIEW
Etiology and Pathogenesis of Disease

All health-care professionals must understand not only how organ systems function normally, but also how pathology can affect the physiology of this system. Throughout the chapters in this book, Clinical View boxes provide you with selected pathologies and how these pathologies affect the anatomy and physiology of that system.

Pathophysiology investigates the relationship between the functioning of an organ system and disease or injury to that organ system. For example, a pathophysiologist would examine how blood pressure, contractile force of the heart, and both gas and nutrient exchange may be affected in an individual afflicted with heart disease.

 WHAT DID YOU LEARN?

2 What is the relationship between anatomy and physiology?

3 _____ physiology examines how the heart, blood vessels, and blood function.

1.2 Anatomy and Physiology Integrated

 LEARNING OBJECTIVE

1. Explain how the studies of form and function are interrelated.

The sciences of anatomy and physiology are intertwined; one must have some understanding of anatomic form to study physiologic function of a structure. Likewise, one cannot adequately describe and understand the anatomic form of an organ without learning that organ's function. This interdependence of the study of anatomy and physiology reflects the inherent and important interrelationship of how the structure and form of a component of the body determines how it functions. This concept is central to mastering the study of anatomy and physiology.

Integrating the disciplines of anatomy and physiology, rather than trying to separate discussion of form and function, is the easiest way to learn about both fields. Anatomists and physiologists may be describing the organs slightly differently, but both disciplines must use information from the other field for a full understanding of the organ system. You cannot fully understand *how* the small intestine propels food and digests or absorbs nutrients unless you know about the *structure* of the small intestine wall. **Figure 1.1** visually compares how anatomists and physiologists examine the human body, using the small intestine as an example. Note that anatomists (left side of the figure) tend to focus on the form and structure, whereas physiologists (right side of figure) focus on the mechanisms and functions of these structures. However, both anatomists and physiologists understand that the form and function of structures are interrelated. Throughout this text, we integrate these disciplines so you can more easily see that anatomic form and physiologic function are inseparable.

 WHAT DID YOU LEARN?

4 Compare and contrast how anatomists and physiologists describe the small intestine.

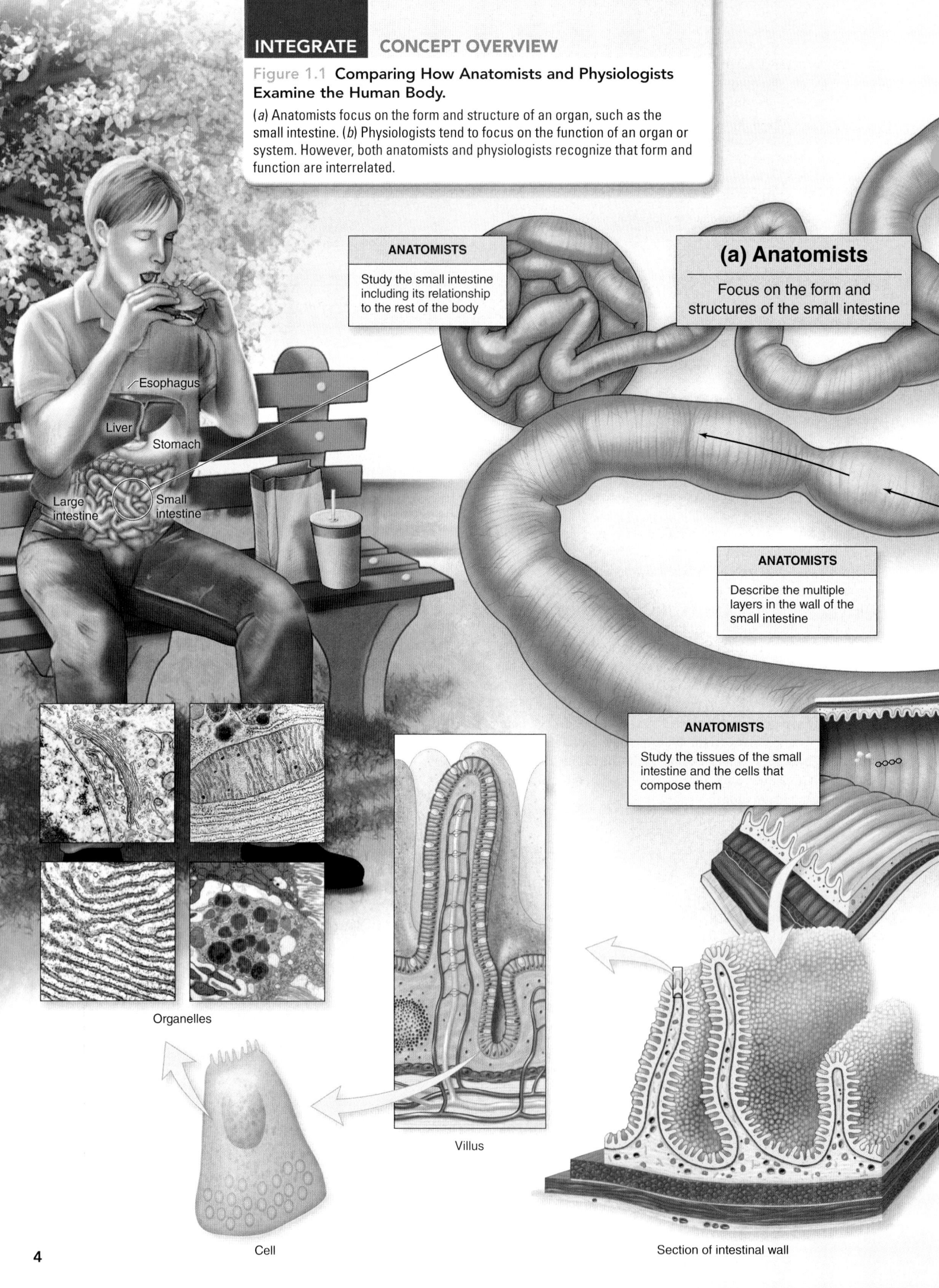

Figure 1.1 Comparing How Anatomists and Physiologists Examine the Human Body.
(*a*) Anatomists focus on the form and structure of an organ, such as the small intestine. (*b*) Physiologists tend to focus on the function of an organ or system. However, both anatomists and physiologists recognize that form and function are interrelated.

ANATOMISTS

Study the small intestine including its relationship to the rest of the body

(a) Anatomists

Focus on the form and structures of the small intestine

ANATOMISTS

Describe the multiple layers in the wall of the small intestine

ANATOMISTS

Study the tissues of the small intestine and the cells that compose them

Esophagus

Liver

Stomach

Large intestine

Small intestine

Organelles

Villus

Cell

Section of intestinal wall

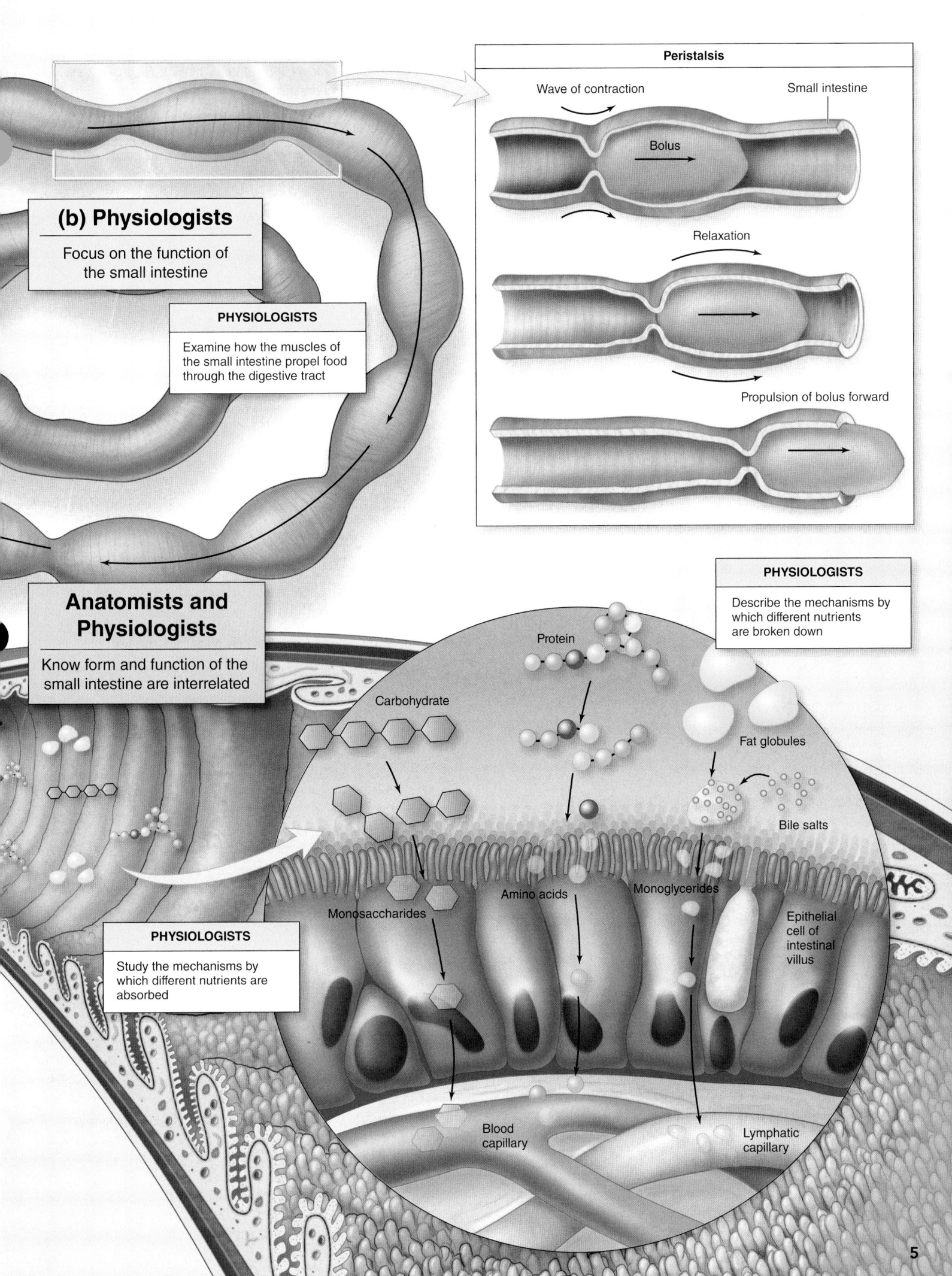

Peristalsis

Wave of contraction

Small intestine

Bolus

Relaxation

Propulsion of bolus forward

(b) Physiologists

Focus on the function of the small intestine

PHYSIOLOGISTS

Examine how the muscles of the small intestine propel food through the digestive tract

Anatomists and Physiologists

Know form and function of the small intestine are interrelated

PHYSIOLOGISTS

Describe the mechanisms by which different nutrients are broken down

PHYSIOLOGISTS

Study the mechanisms by which different nutrients are absorbed

Protein

Carbohydrate

Fat globules

Bile salts

Monosaccharides

Amino acids

Monoglycerides

Epithelial cell of intestinal villus

Blood capillary

Lymphatic capillary

1.3 The Body's Levels of Organization

Scientists group the body's components into an organizational hierarchy of form and function. In thinking about these levels, it is helpful to know the characteristics common to living things and how each level supports these characteristics. For example, the organ system concept allows functions to be considered as an interaction between many organs.

1.3a Characteristics That Describe Living Things

LEARNING OBJECTIVE

1. List the characteristics common to all living things.

Several properties are common to all organisms, including humans:

- **Organization.** All organisms exhibit a complex structure and order. In the next section, we note that the human body has several increasingly complex levels of organization.
- **Metabolism.** All organisms engage in **metabolism** (mĕ-tab′ō-lizm; *metabole* = change), which is defined as the sum of all of the chemical reactions that occur within the body. Metabolism consists of both **anabolism** (ă-nab′ō-lizm, *anabole* = a raising up), in which small molecules are joined to form larger molecules, and **catabolism** (kă-tab′ō-lizm; *katabole* = a casting down), in which large molecules are broken down into smaller molecules. An example of a metabolic reaction is the use of cellular energy (called ATP, see section 2.7) for muscle contraction (see section 10.3).

WHAT DO YOU THINK?

1. When you digest a meal, what type of metabolic reactions do you think you are utilizing primarily: *anabolic* or *catabolic* chemical reactions? Why?

- **Growth and Development.** During their lifetime, organisms assimilate materials from their environment and often exhibit increased size (growth) and increased specialization as related to form and function (development). As the human body grows and develops, structures such as the brain become more complex and sophisticated.
- **Responsiveness.** All organisms exhibit **responsiveness,** which is the ability to sense and react to **stimuli** (changes in the external or internal environment). A stimulus to the skin of the hands, such as an extremely hot temperature, causes the human to withdraw the hand from the stimulus so as to prevent injury or damage. Responsiveness occurs at almost all levels of organization.
- **Regulation.** An organism must be able to adjust or direct internal bodily function in the face of environmental changes. When body temperature rises, the body regulates this change by circulating more blood near its surface to facilitate heat loss, and thus return the body to within normal range. (The process of maintaining body structures and function is called homeostasis, which is discussed in greater depth in section 1.5)
- **Reproduction.** All organisms produce new cells for growth, maintenance, and repair. The somatic (body) cells divide by a process called mitosis, whereas sex cells (called gametes) are produced by another type of cell division called meiosis. The sex cells, under the right conditions, have the ability to develop into a new living organism.

WHAT DID YOU LEARN?

5. What does it mean if an organism is "responsive," and how does this characteristic relate to the survival of this organism?

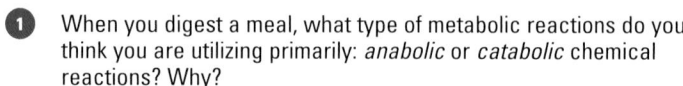

Atom Molecule

Macromolecule (e.g., DNA)   Organelle (e.g., Golgi apparatus)

Chemical level

1.3b The View from Simplest to Most Complex

LEARNING OBJECTIVE

2. Describe the levels of organization in the human body.

Anatomists and physiologists recognize several levels of increasingly complex organization in humans, as illustrated in **figure 1.2.** These levels, from simplest to most complex, are the chemical level, cellular level, tissue level, organ level, organ system level, and organismal level.

The **chemical level** is the simplest level, and it involves atoms and molecules. **Atoms** are the smallest units of matter that exhibit the characteristics of an element, such as carbon and hydrogen. When two or more atoms combine they form a **molecule.** Examples of molecules include a sugar, a water molecule, or a vitamin. More complex molecules are called **macromolecules** and include some proteins and the deoxyribonucleic acid (DNA) molecules. Macromolecules form specialized microscopic subunits in cells called **organelles,** which are microscopic structures found within cells.

The **cellular level** consists of **cells,** which are the smallest living structures and serve as the basic units of structure and function in organisms. Cells and their components are formed from the atoms and molecules from the chemical level. The structures of cells vary widely, reflecting the specializations needed for their different functions. For example, a skeletal muscle cell may be very long and contain numerous organized protein filaments that aid in muscle contraction, whereas a red blood cell is small and has a flattened disc shape that facilitates the quick and effective exchange of respiratory gases.

The **tissue level** consists of **tissues,** which are groups of similar cells that perform common functions. There are four major types of tissues.

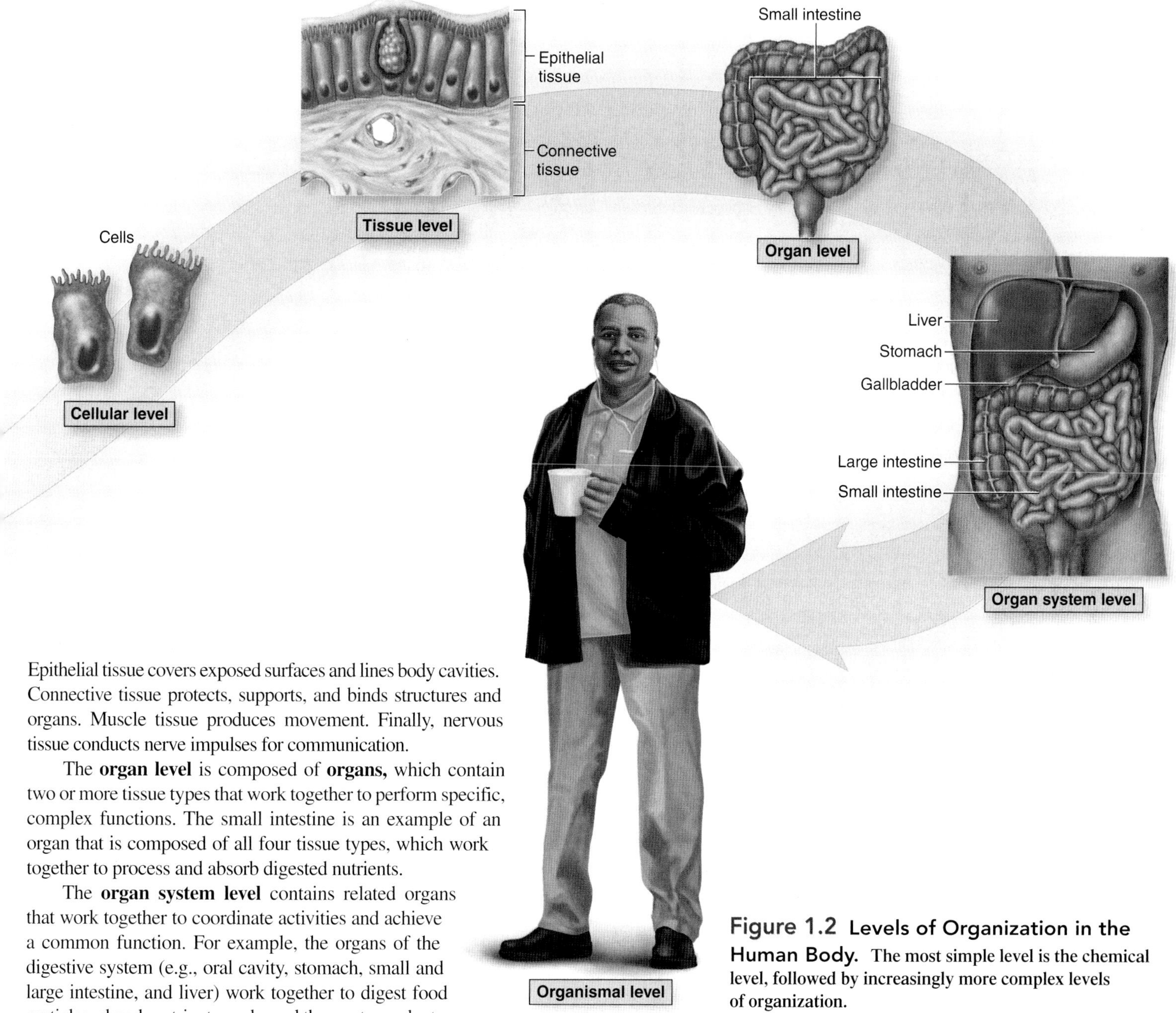

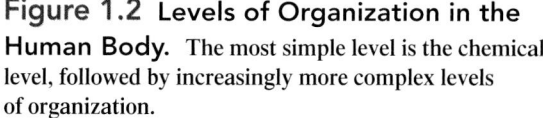

Epithelial tissue

Connective tissue

Tissue level

Cells

Cellular level

Small intestine

Organ level

Liver

Stomach

Gallbladder

Large intestine

Small intestine

Organ system level

Organismal level

Epithelial tissue covers exposed surfaces and lines body cavities. Connective tissue protects, supports, and binds structures and organs. Muscle tissue produces movement. Finally, nervous tissue conducts nerve impulses for communication.

The **organ level** is composed of **organs,** which contain two or more tissue types that work together to perform specific, complex functions. The small intestine is an example of an organ that is composed of all four tissue types, which work together to process and absorb digested nutrients.

The **organ system level** contains related organs that work together to coordinate activities and achieve a common function. For example, the organs of the digestive system (e.g., oral cavity, stomach, small and large intestine, and liver) work together to digest food particles, absorb nutrients, and expel the waste products.

The highest level of structural organization in the body is the **organismal level.** All body systems function interdependently in an **organism,** which is the living being.

Figure 1.2 Levels of Organization in the Human Body. The most simple level is the chemical level, followed by increasingly more complex levels of organization.

 WHAT DID YOU LEARN?

6 Does a higher level of organization contain all the levels beneath it? Explain.

INTEGRATE

CONCEPT CONNECTION

Throughout future chapters, boxes like this one will highlight how various organ systems do not work in isolation, but rather are interconnected to carry out overlapping functions. For example, the cardiovascular system and respiratory system work together in the transport of respiratory gases (oxygen and carbon dioxide) by the blood throughout the body.

1.3c Introduction to Organ Systems

 LEARNING OBJECTIVE

3. Compare the organ systems of the human body.

All organisms must exchange nutrients, wastes, and gases with their environment to carry on their metabolism. Simple organisms (e.g., bacteria) may exchange these substances directly across their surface membranes. In contrast, complex, multicellular organisms require sophisticated organ systems with specialized structures and functions to perform the myriad of activities required for the routine events of life. In humans, 11 **organ systems** are commonly denoted, each composed of interrelated organs that work in concert to perform specific functions (**figure 1.3**). A person maintains a healthy body through the intricate interworkings of all of its organ systems. Subsequent chapters examine each of these organ systems in detail.

 WHAT DID YOU LEARN?

7 Which organ system is responsible for filtering the blood and removing the waste products of the blood in the form of urine?

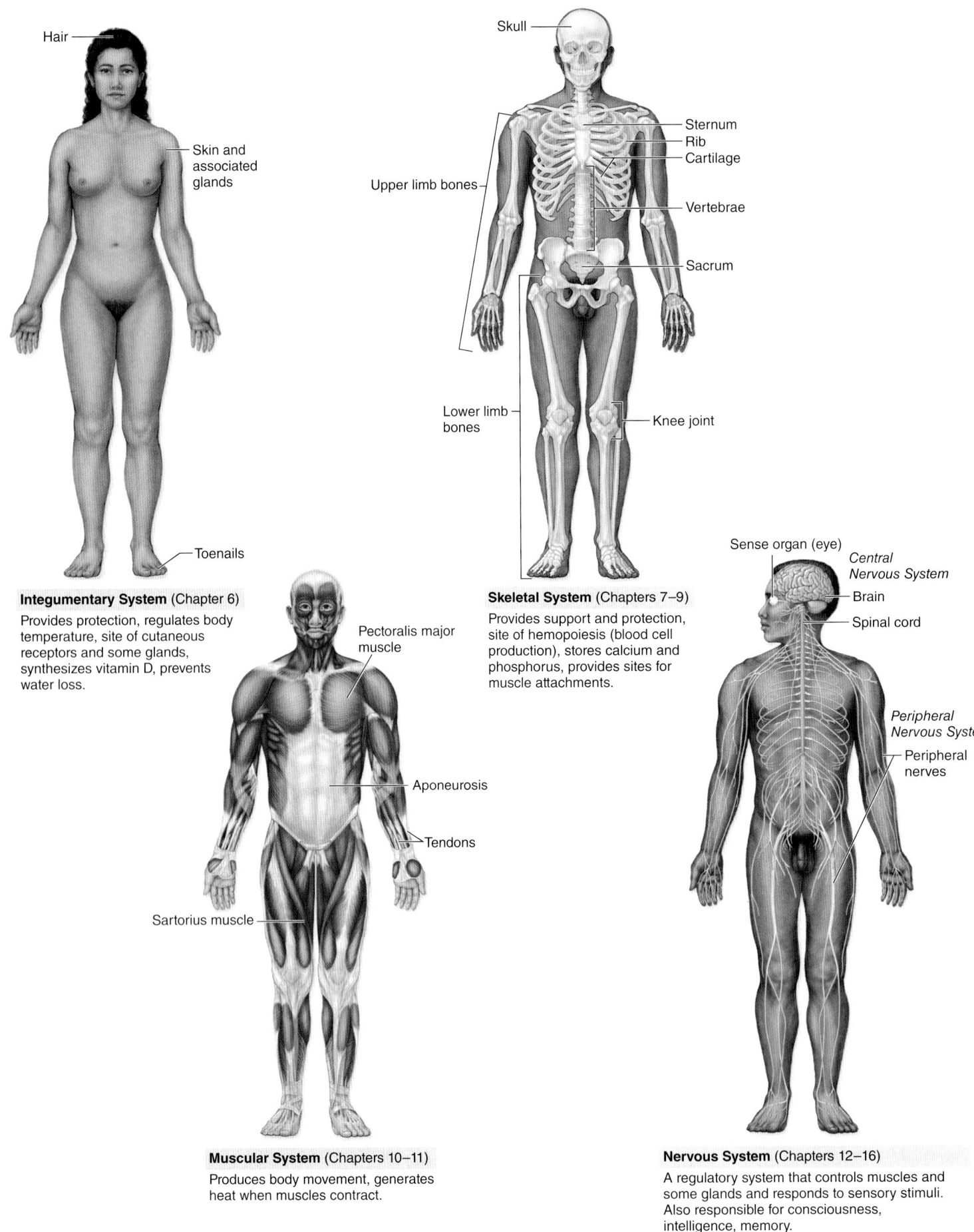

Integumentary System (Chapter 6)
Provides protection, regulates body temperature, site of cutaneous receptors and some glands, synthesizes vitamin D, prevents water loss.

Hair

Skin and associated glands

Toenails

Skull

Sternum
Rib
Cartilage

Upper limb bones

Vertebrae

Sacrum

Lower limb bones

Knee joint

Skeletal System (Chapters 7–9)
Provides support and protection, site of hemopoiesis (blood cell production), stores calcium and phosphorus, provides sites for muscle attachments.

Pectoralis major muscle

Aponeurosis

Tendons

Sartorius muscle

Muscular System (Chapters 10–11)
Produces body movement, generates heat when muscles contract.

Sense organ (eye)

Central Nervous System

Brain

Spinal cord

Peripheral Nervous System

Peripheral nerves

Nervous System (Chapters 12–16)
A regulatory system that controls muscles and some glands and responds to sensory stimuli. Also responsible for consciousness, intelligence, memory.

Figure 1.3 Organ Systems. Major components and characteristics of the 11 organ systems of the human body are presented. **AP|R**

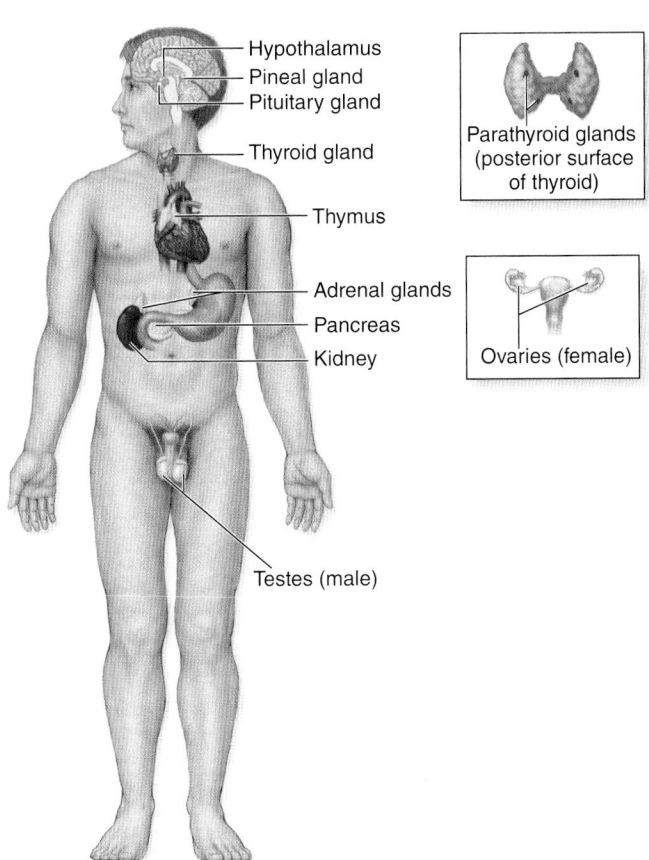

Hypothalamus
Pineal gland
Pituitary gland
Thyroid gland
Thymus
Adrenal glands
Pancreas
Kidney
Testes (male)

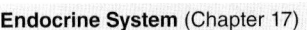

Parathyroid glands
(posterior surface
of thyroid)

Ovaries (female)

Endocrine System (Chapter 17)

Consists of glands and cell
clusters that secrete hormones,
which regulate development,
growth and metabolism; maintain
homeostasis of blood composition
and volume, control digestive
processes, and control
reproduction.

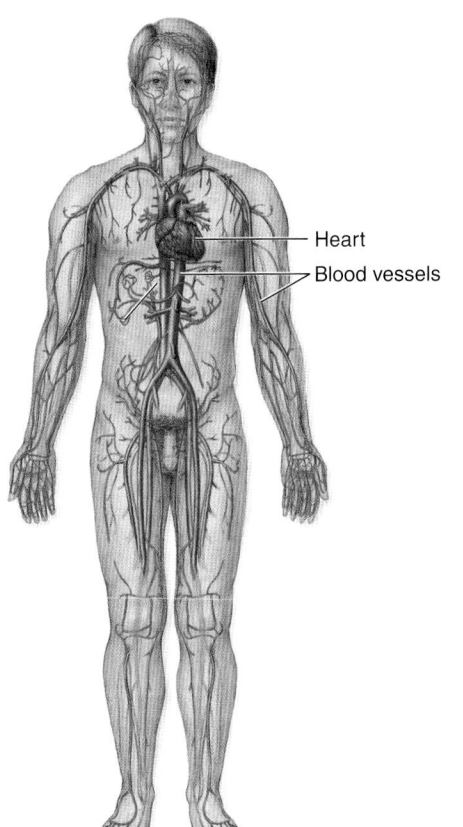

Heart
Blood vessels

Cardiovascular System (Chapters 18–20)

Consists of the heart and blood vessels; the
heart moves blood through blood vessels in
order to distribute hormones, nutrients,
gases, and pick up waste products.

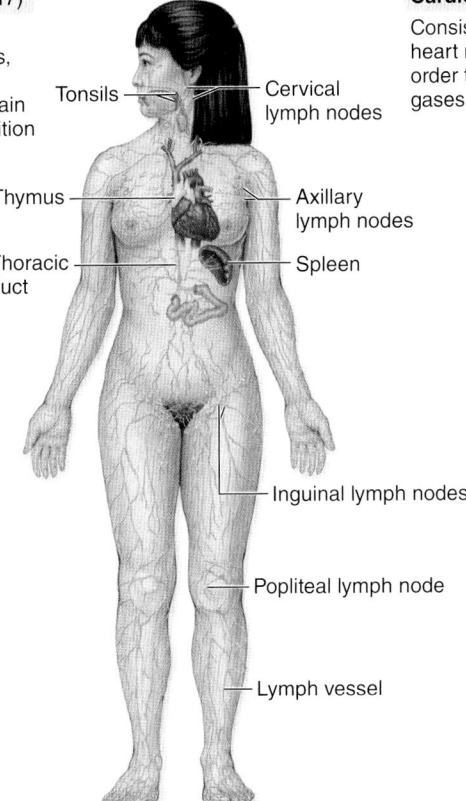

Tonsils
Cervical
lymph nodes
Thymus
Axillary
lymph nodes
Thoracic
duct
Spleen
Inguinal lymph nodes
Popliteal lymph node
Lymph vessel

Lymphatic System (Chapters 21–22)

Transports and filters lymph (interstitial
fluid transported through lymph vessels)
and participates in an immune response
when necessary.

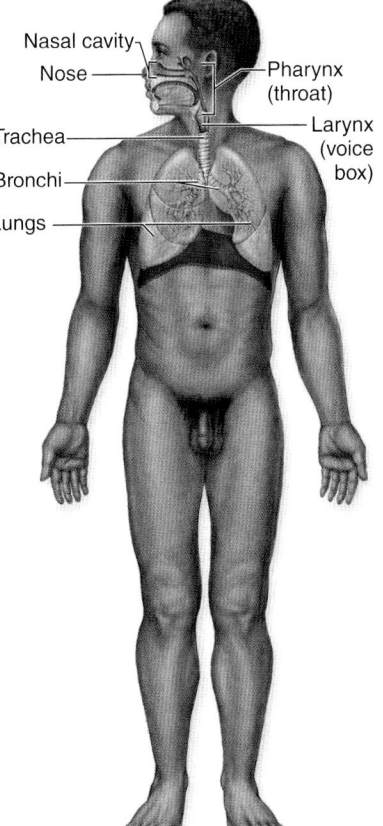

Nasal cavity
Nose
Pharynx
(throat)
Larynx
(voice
box)
Trachea
Bronchi
Lungs

Respiratory System (Chapter 23)

Responsible for exchange of gases
(oxygen and carbon dioxide)
between blood and the air in the
lungs.

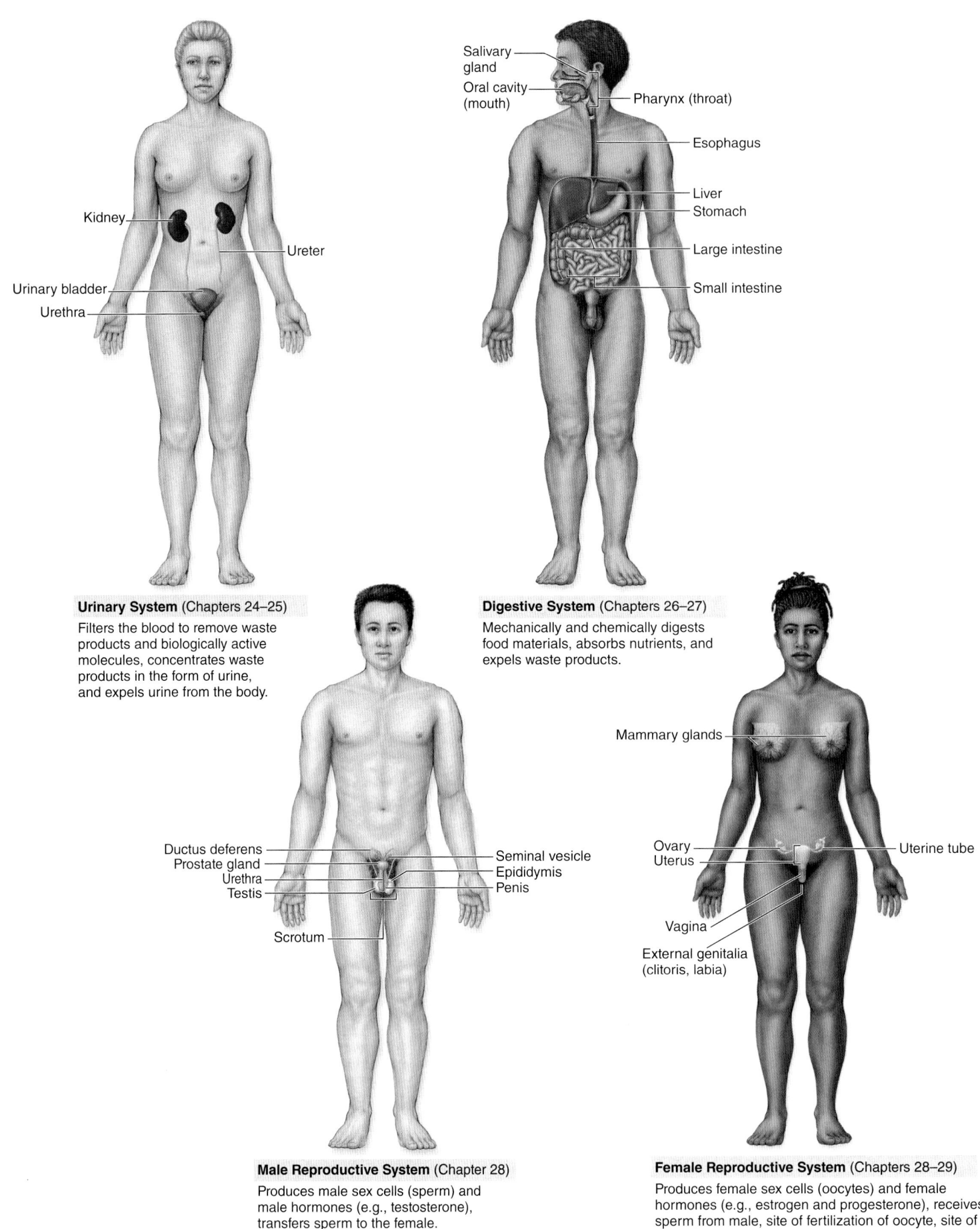

Urinary System (Chapters 24–25)

Filters the blood to remove waste products and biologically active molecules, concentrates waste products in the form of urine, and expels urine from the body.

Digestive System (Chapters 26–27)

Mechanically and chemically digests food materials, absorbs nutrients, and expels waste products.

Male Reproductive System (Chapter 28)

Produces male sex cells (sperm) and male hormones (e.g., testosterone), transfers sperm to the female.

Female Reproductive System (Chapters 28–29)

Produces female sex cells (oocytes) and female hormones (e.g., estrogen and progesterone), receives sperm from male, site of fertilization of oocyte, site of growth and development of embryo and fetus, produces and secretes breast milk for nourishment of newborn.

Figure 1.3 Organ Systems. *(continued)* **AP|R**

1.4 The Language of Anatomy and Physiology

Clinicians and researchers in anatomy and physiology require a precise language to ensure that they are all discussing the same features and functions. A technical terminology has been developed that describes body position, direction, regions, and body cavities. These technical terms are different from those used in everyday conversation, because the more conversational terms often do not accurately describe location and position or identify structures. For example, the term *arm* in everyday conversation refers to the entire upper limb, but in anatomy the specific portions of the upper limb are named, and the term *arm* or *brachium* refers only to that part of the upper limb between the shoulder and the elbow.

Most anatomic and physiologic terms are derived from Greek or Latin, and we frequently provide word origins, pronunciations, and definitions of terms where appropriate throughout this text. We've used *Stedman's Medical Dictionary* (which defines all medical terms) and *Terminologia Anatomica* (which lists and categorizes the modern, proper anatomic terms) as references.

1.4a Anatomic Position

LEARNING OBJECTIVE

1. Describe the anatomic position and its importance in the study of anatomy.

Descriptions of any body region or part require a common initial point of reference. Note that terms such as "superior" and "inferior" can be relative terms. For example, when a person is standing it would be accurate to say "the heart is superior to the stomach," yet if that person is lying down in a supine position, this statement would seem not to be true. For accuracy and clarity, anatomists and physiologists describe these parts based on the premise that the body is in what is termed the anatomic position, which is then the point of common reference. An individual in the **anatomic position** stands upright with the feet parallel and flat on the floor, the upper limbs are at the sides of the body, and the palms face anteriorly (toward the front); the head is level, and the eyes look forward toward the observer **(figure 1.4)**. All of the anatomic and directional terms used in this book refer to the body in anatomic position.

INTEGRATE

LEARNING STRATEGY

Breaking a word into smaller parts can help you understand and remember its meaning. In this book, we provide word derivations for new terms following their pronunciations. For example, in the case of histology, the study of tissues, we give (*histos* = web, tissue, *logos* = study).

Many biological terms share some of the same prefixes, suffixes, and word roots, so learning the meanings of these common terms can help you figure out the meanings of unfamiliar terms. A review of prefixes, suffixes, and word roots also appears on the inside of the back cover of this book.

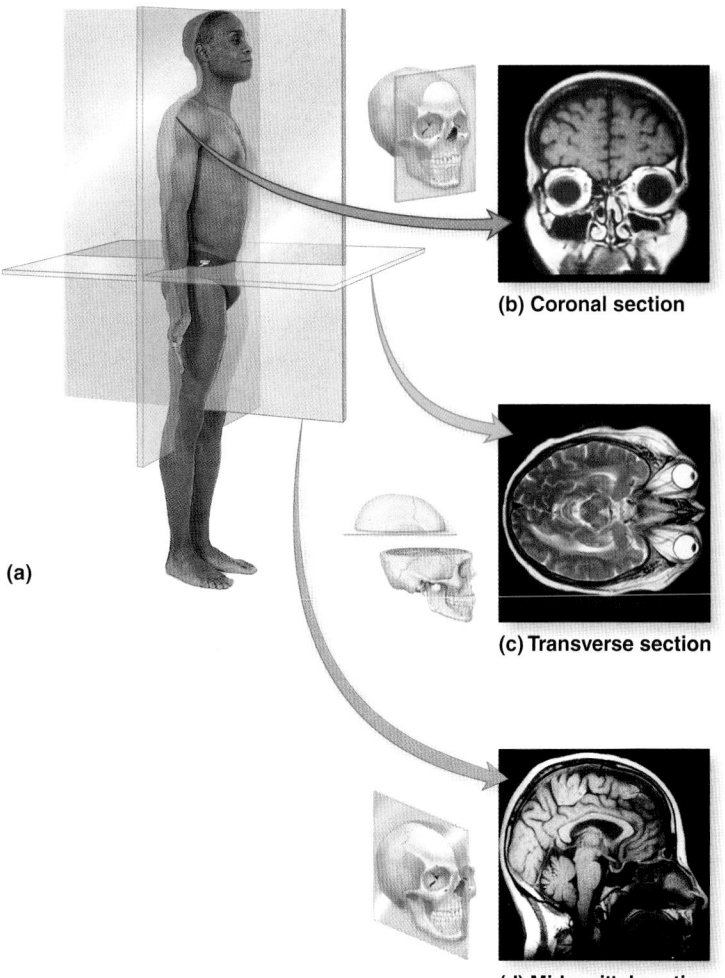

(a)

(b) Coronal section

(c) Transverse section

(d) Midsagittal section

Figure 1.4 Anatomic Position and Body Planes. (*a*) In the anatomic position, the body is upright, and the forearms are positioned so the palms are facing anteriorly. A plane is an imaginary surface that slices the body into specific sections. Sections are shown from each of the three major anatomic planes of reference: (*b*) coronal, (*c*) transverse, and (*d*) midsagittal. AP|R

1.4b Sections and Planes

LEARNING OBJECTIVE

2. Describe the anatomic sections and planes through the body.

Anatomists and physiologists refer to real or imaginary "slices" of the body, called sections or planes, to examine the internal anatomy and describe the position of one body part relative to another. The term **section** implies an actual cut or slice to expose the internal anatomy, whereas the word **plane** implies an imaginary flat surface passing through the body. The three major anatomic planes are the coronal, transverse, and midsagittal planes (figure 1.4).

A **coronal** (kōr′ō-năl; *korone* = crown) **plane,** also called a *frontal plane,* is a vertical plane that divides the body or organ into *anterior* (front) and *posterior* (back) parts. When a coronal plane is taken through the trunk, the anterior portion contains the chest and the posterior portion contains the back and buttocks.

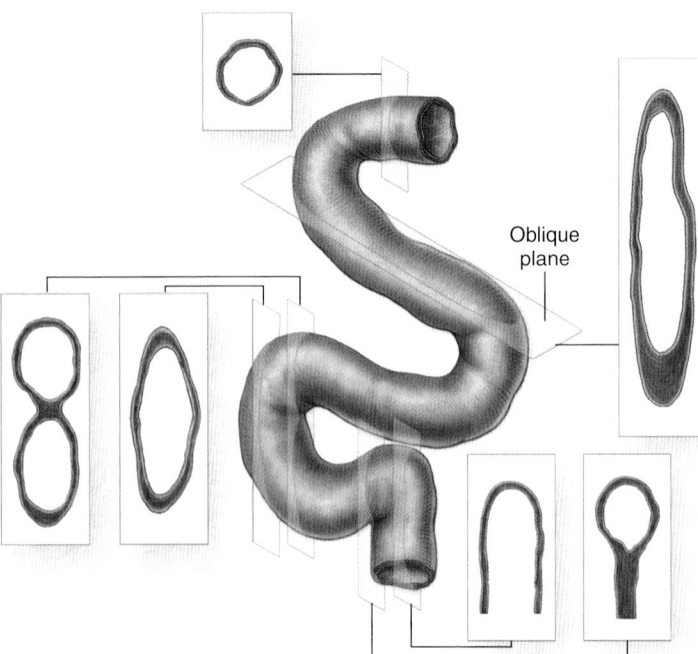

Oblique plane

Figure 1.5 Sections from a Three-Dimensional Structure. Serial sections through an object are used to reconstruct its three-dimensional structure, as in these sections of the small intestine. Often a single section, such as the plane at the lower part of this figure, misrepresents the complete structure of the object. An oblique plane is labeled for reference.

A **transverse plane,** also called a *horizontal plane* or *cross-sectional plane,* divides the body or organ into *superior* (top) and *inferior* (bottom) parts. If a transverse plane is taken through the middle of the trunk, the superior portion contains the chest and the inferior portion contains the abdomen.

A **midsagittal** (mid-saj′ĭ-tăl; *sagitta* = arrow) **plane,** or *median plane,* is a vertical plane and divides the body or organ into *left* and *right halves.* A midsagittal plane through the head will split it into a left half and a right half (each containing one eye, one ear, and half of the nose and mouth). A plane that is parallel to the midsagittal plane, but either to the left or right of the midsagittal plane, is termed a **sagittal plane.** A sagittal plane divides a structure into left and right portions that are not equal. Although there is only one midsagittal plane, an infinite number of sagittal planes are possible.

In addition to these major planes, there are numerous minor planes called **oblique** (ob-lēk′) **planes** that pass through a structure at an angle (**figure 1.5**).

Interpreting body sections has become increasingly important for health-care professionals. Technical advances in medical imaging have produced sectional images of internal body structures (figure 1.4b). To determine the shape of any object within a section, we must be able to reconstruct its three-dimensional shape by observing many continuous sections.

Sectioning the body or an organ along different planes often results in very different views of that organ or region. For example, different sections through the abdominal cavity exhibit multiple profiles of the long, twisted tube that is the small intestine. These sections may appear as circles, ovals, a figure eight, or maybe a long tube with parallel sides, depending on where the section was taken (figure 1.5). Being able to convert and interpret two-dimensional images into three-dimensional structures is especially important when comparing and understanding histologic and gross anatomic views of the same organ.

WHAT DID YOU LEARN?

8 What type of plane would separate the nose and mouth into superior and inferior structures?

1.4c Anatomic Directions

LEARNING OBJECTIVE

3. Define the different anatomic directional terms.

Once the body is in the anatomic position, we can precisely describe the relative positions of structures by using specific directional terms. These directional terms are precise and brief, and most of them are presented in opposing pairs. Examples include **anterior** and **posterior**, **dorsal** (toward the back) and **ventral** (toward the belly), and **proximal** (nearer to the trunk) and **distal** (farther from the trunk). **Table 1.1** and **figure 1.6** describe some commonly used directional terms. Studying the table and figure together, and referring back to them as needed, will maximize your understanding of anatomic directions and aid your study of anatomy throughout the rest of this book.

WHAT DID YOU LEARN?

9 Which directional term would be most appropriate in the sentence "The elbow is _____ to the wrist"?

Table 1.1	Anatomic Directional Terms		
Direction	**Term**	**Meaning**	**Example**
Relative to front (belly side) or back of the body	Anterior	In front of; toward the front surface	The stomach is *anterior* to the spinal cord.
	Posterior	In back of; toward the back surface	The heart is *posterior* to the sternum.
	Dorsal	At the back side of the human body	The spinal cord is on the *dorsal* side of the body.
	Ventral	At the belly side of the human body	The umbilicus (navel, belly button) is on the *ventral* side of the body.
Relative to the head or bottom of the body	Superior	Closer to the head	The chest is *superior* to the pelvis.
	Inferior	Closer to the feet	The stomach is *inferior* to the heart.
	Cranial (cephalic)	At the head end	The shoulders are *cranial* to the feet.
	Caudal	At the rear or tail end	The buttocks are *caudal* to the head.
	Rostral	Toward the nose or mouth	The frontal lobe of the brain is *rostral* to the back of the head.
Relative to the midline or center of the body	Medial	Toward the midline of the body	The lungs are *medial* to the shoulders.
	Lateral	Away from the midline of the body	The arms are *lateral* to the heart.
	Deep	On the inside, internal to another structure	The heart is *deep* to the rib cage.
	Superficial	On the outside	The skin is *superficial* to the biceps brachii muscle.
Relative to point of attachment of appendage	Proximal	Closer to point of attachment to trunk	The elbow is *proximal* to the hand.
	Distal	Farther away from point of attachment to trunk	The wrist is *distal* to the elbow.

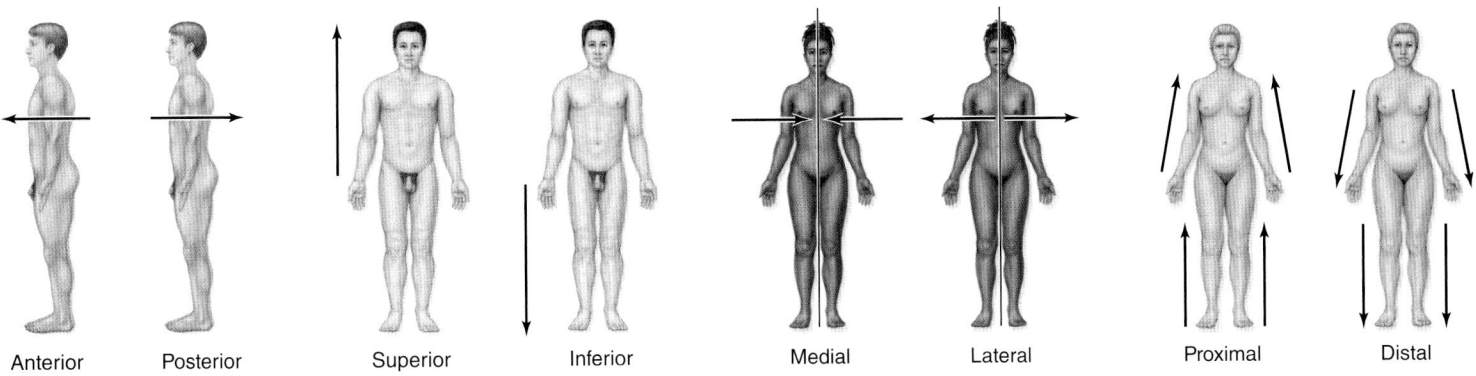

| Anterior | Posterior | Superior | Inferior | Medial | Lateral | Proximal | Distal |

Figure 1.6 Directional Terms in Anatomy. Directional terms precisely describe the location and relative relationships of body parts. (See also table 1.1.) AP|R

1.4d Regional Anatomy

LEARNING OBJECTIVE

4. Identify the major regions of the body, using proper anatomic terminology.

The human body is partitioned into two main regions, the axial and appendicular regions. The **axial** (ak′sē-ăl) **region** includes the head, neck, and trunk; it forms the main vertical axis of the body. The **appendicular** (ap′en-dik′ū-lăr) **region** is composed of the upper and lower limbs, which attach to the axial region. Several more specific regions are located within these two main ones, and they are identified by proper anatomic terminology. **Figure 1.7** and **table 1.2** identify the major regional terms and some additional minor ones. Not all regions are shown in figure 1.7.

WHAT DID YOU LEARN?

⑩ The term *antebrachial* refers to which body region?

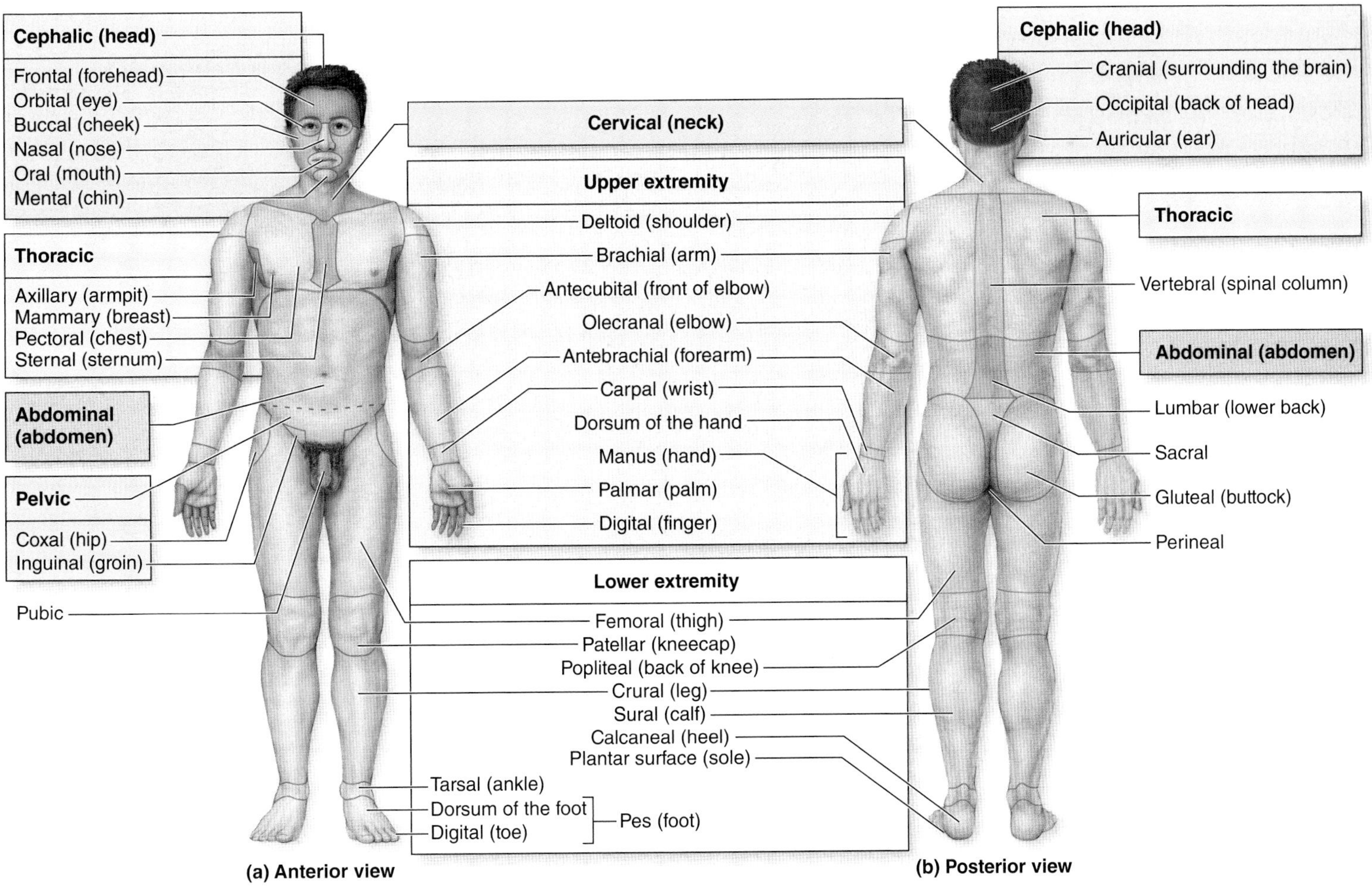

Cephalic (head)
Frontal (forehead)
Orbital (eye)
Buccal (cheek)
Nasal (nose)
Oral (mouth)
Mental (chin)

Thoracic
Axillary (armpit)
Mammary (breast)
Pectoral (chest)
Sternal (sternum)

Abdominal (abdomen)

Pelvic
Coxal (hip)
Inguinal (groin)

Pubic

Cervical (neck)

Upper extremity
Deltoid (shoulder)
Brachial (arm)
Antecubital (front of elbow)
Olecranal (elbow)
Antebrachial (forearm)
Carpal (wrist)
Dorsum of the hand
Manus (hand)
Palmar (palm)
Digital (finger)

Lower extremity
Femoral (thigh)
Patellar (kneecap)
Popliteal (back of knee)
Crural (leg)
Sural (calf)
Calcaneal (heel)
Plantar surface (sole)
Tarsal (ankle)
Dorsum of the foot
Digital (toe)
Pes (foot)

(a) Anterior view

Cephalic (head)
Cranial (surrounding the brain)
Occipital (back of head)
Auricular (ear)

Thoracic
Vertebral (spinal column)

Abdominal (abdomen)
Lumbar (lower back)
Sacral
Gluteal (buttock)
Perineal

(b) Posterior view

Figure 1.7 Regional Terms. (*a*) Anterior and (*b*) posterior views show key regions of the body. Their common names appear in parentheses. AP|R

Table 1.2 — Human Body Regions

Regional Name	Description	Regional Name	Description
Abdominal	Region inferior to the thorax (chest) and superior to the hip bones	Manus	Hand
Antebrachial	Forearm (the portion of the upper limb between the elbow and the wrist)	Mental	Chin
		Nasal	Nose
Antecubital	Region anterior to the elbow; also known as the cubital region	Occipital	Posterior aspect of the head
		Olecranal	Posterior aspect of the elbow
Auricular	Visible surface structures of the ear	Oral	Mouth
Axillary	Armpit	Orbital	Eye
Brachial	Arm (the portion of the upper limb between the shoulder and the elbow)	Palmar	Palm (anterior surface) of the hand
		Patellar	Kneecap
Buccal	Cheek	Pectoral	Chest
Calcaneal	Heel of the foot	Pelvic	Pelvis
Carpal	Wrist	Perineal	Diamond-shaped region between the legs that contains the anus and external reproductive organs
Cephalic	Head		
Cervical	Neck	Pes	Foot
Coxal	Hip	Plantar	Sole of the foot
Cranial	Skull	Pollex	Thumb
Crural	Leg (the portion of the lower limb between the knee and the ankle)	Popliteal	Area posterior to the knee
		Pubic	Anterior region of the pelvis
Deltoid	Shoulder	Radial	Lateral (thumb side) aspect of forearm
Digital	Fingers or toes (also called phalangeal)	Sacral	Posterior region between the hip bones
Dorsal/ Dorsum	Back	Scapular	Shoulder blade
Facial	Face	Sternal	Anterior middle region of the thorax
Femoral	Thigh	Sural	Calf (posterior part of the leg)
Fibular	Lateral aspect of the leg	Tarsal	Ankle, root of the foot
Frontal	Forehead	Thoracic	Chest or thorax
Gluteal	Buttock	Tibial	Medial aspect of leg
Hallux	Great toe	Ulnar	Medial aspect of the forearm
Inguinal	Groin (sometimes used to indicate the crease or junction of the thigh with the trunk)	Umbilical	Navel
		Vertebral	Spinal column
Lumbar	Relating to the loins, or the inferior part of the back, between the ribs and pelvis		
Mammary	Breast		

1.4e Body Cavities and Membranes

 **LEARNING OBJECTIVES**

5. Describe the body cavities and their subdivisions.

6. Explain the role of serous membranes in the ventral cavities.

Internal organs and organ systems are housed within enclosed spaces, or cavities. These body cavities are named either according to the bones that surround them or the organs they contain. For purposes of discussion, these body cavities are grouped into a posterior aspect and a ventral cavity.

Posterior Aspect

The **posterior aspect** of the body is different from the ventral cavity, in that the posterior aspect contains cavities that are completely encased in bone and are physically and developmentally different from the ventral cavity. The term *dorsal body cavity* has been used by others to describe this posterior aspect, but is not used here because of these differences between the ventral cavity and posterior aspect.

The posterior aspect is subdivided into two enclosed cavities **(figure 1.8a)**. A **cranial cavity** is formed by the bones of the cranium, and so it also goes by the name *endocranium*. The cranial cavity houses the brain. The second cavity is the **vertebral** (ver'te-brăl) **canal,** which is formed by the bones of the vertebral column. The vertebral canal houses the spinal cord.

Ventral Cavity

The **ventral cavity** is the larger, anteriorly placed cavity in the body (figure 1.8). Unlike the posterior aspect, the ventral cavity and its subdivisions do not completely encase their organs in bone. The ventral cavity is partitioned by the diaphragm into a superior **thoracic** (thō-ras'ik) **cavity** and an inferior **abdominopelvic** (ab-dom'i-nō-pel'vik) **cavity.**

Another significant difference between the posterior aspect and the ventral cavity is that the subdivisions of the ventral cavity are lined with thin **serous membranes.** (Posterior aspect cavities have no serous membranes.) In this usage a *membrane* is a continuous layer of cells, as compared to the plasma membrane that surrounds a single cell. Serous membranes form two layers: (1) a **parietal** (pă-rī'ĕ-tăl) **layer** that typically lines the internal surface of the body wall and (2) a **visceral** (vis'er-ăl) **layer** that covers the external surface of the organs (collectively called the **viscera**) within that cavity. Between the parietal and visceral serous membrane layers is a potential space called the **serous cavity.** Serous membranes secrete a liquid called **serous fluid** within a serous cavity. Serous fluid has the consistency of oil and serves as a lubricant. In a living human, organs (e.g., heart, lungs, intestines) are moving and rubbing against each other and the body wall. Friction caused by this movement is reduced

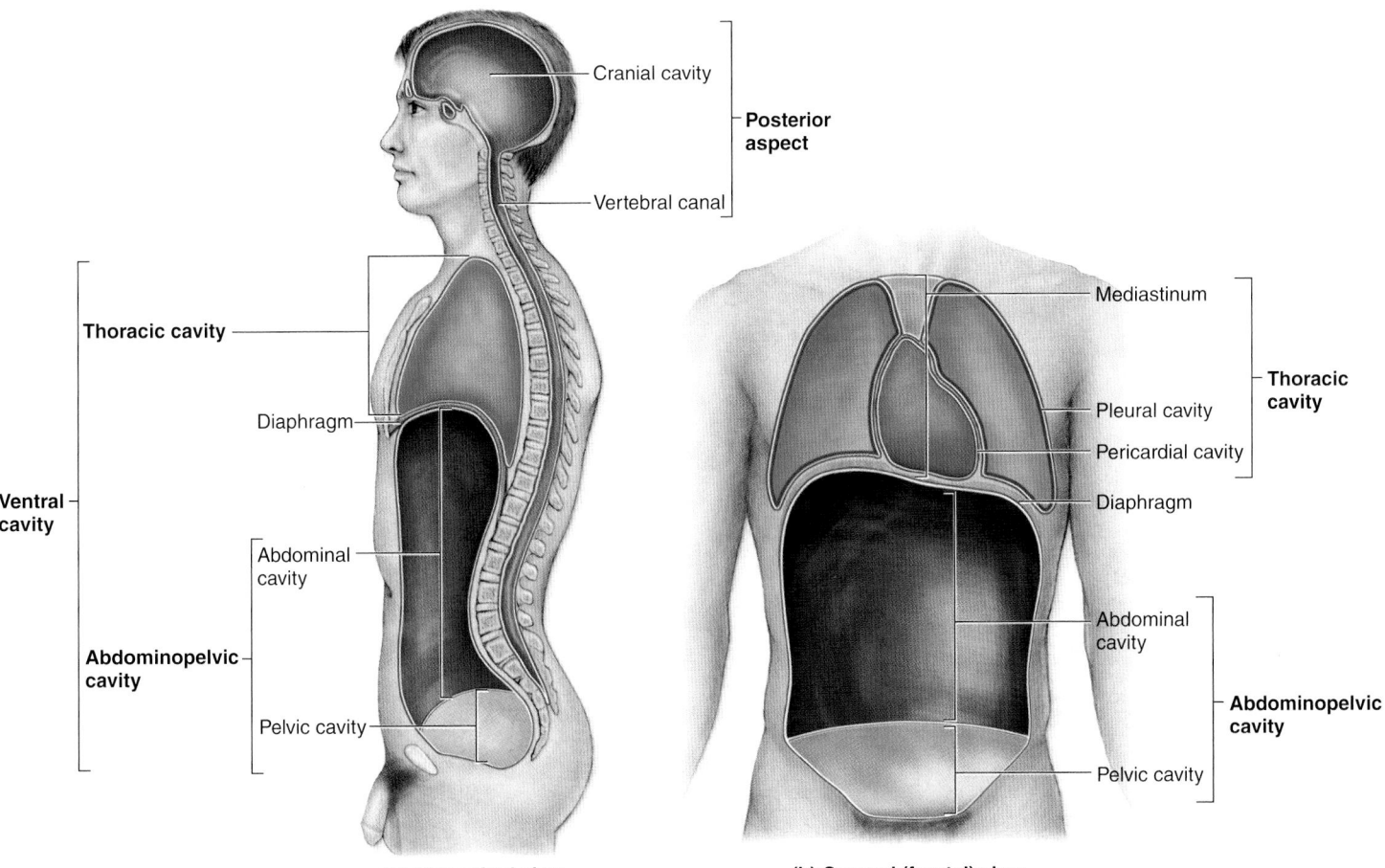

(a) Midsagittal view   (b) Coronal (frontal) view

Figure 1.8 Body Cavities. The body is composed of two main spaces: the posterior aspect and the ventral cavity. (*a*) A midsagittal view shows both the posterior aspect and the ventral cavity. (*b*) A coronal view shows the relationship between the thoracic and abdominopelvic cavities within the ventral cavity. **AP|R**

by the serous fluid so the organs move more smoothly against one another and the body walls. Serous membranes will be discussed again in section 5.5b.

 WHAT DO YOU THINK?

2 What do you think would happen to your body organs if there were no serous fluid between the parietal and visceral layers?

Figure 1.9a provides a helpful analogy for visualizing the serous membrane layers. The closed fist is comparable to an organ, and the balloon is comparable to a serous membrane. When a fist is pushed against the wall of the balloon, the inner balloon wall that surrounds the fist is comparable to the visceral layer of the serous membrane. The outer balloon wall is comparable to the parietal layer of the serous membrane. The thin, air-filled space within the balloon, between the two "walls," is comparable to the serous cavity. Note that the organ is not *inside* the serous cavity; it is actually *outside* the cavity and merely covered by the serous membrane.

Thoracic Cavity   The median space in the thoracic cavity is called the **mediastinum** (mē-dē-as-tī′nŭm; *medius* = middle) (figure 1.8b). It contains the heart, thymus, esophagus, trachea, and major blood vessels that connect to the heart.

Within the mediastinum, the heart is enclosed by a two-layered serous membrane called the serous **pericardium** (per-ĭ-kar′dē-ŭm; *peri* = around, *kardia* = heart). The **parietal pericardium** is the outermost layer of the serous membrane and forms the sac around the heart, whereas the **visceral pericardium** forms the heart's external surface (figure 1.9b). The **pericardial cavity** is the potential space between the parietal and visceral layers of the pericardium, and it contains serous fluid.

The right and left sides of the thoracic cavity house the lungs, which are associated with a two-layered serous membrane called the **pleura** (plūr′ă; = a rib) (figure 1.9c). The **parietal pleura** is the outer layer of the serous membrane and lines the internal surface of the thoracic wall. The inner layer is the **visceral pleura,** which covers the external surface of each lung. The **pleural cavity** is the potential space between these parietal and visceral layers, and it contains serous fluid.

Abdominopelvic Cavity   The abdominopelvic cavity may be subdivided into two smaller cavities by a horizontal plane at the level of the superior aspects of the hip bones. The area superior to this plane is the **abdominal cavity;** the **pelvic cavity** lies inferior to

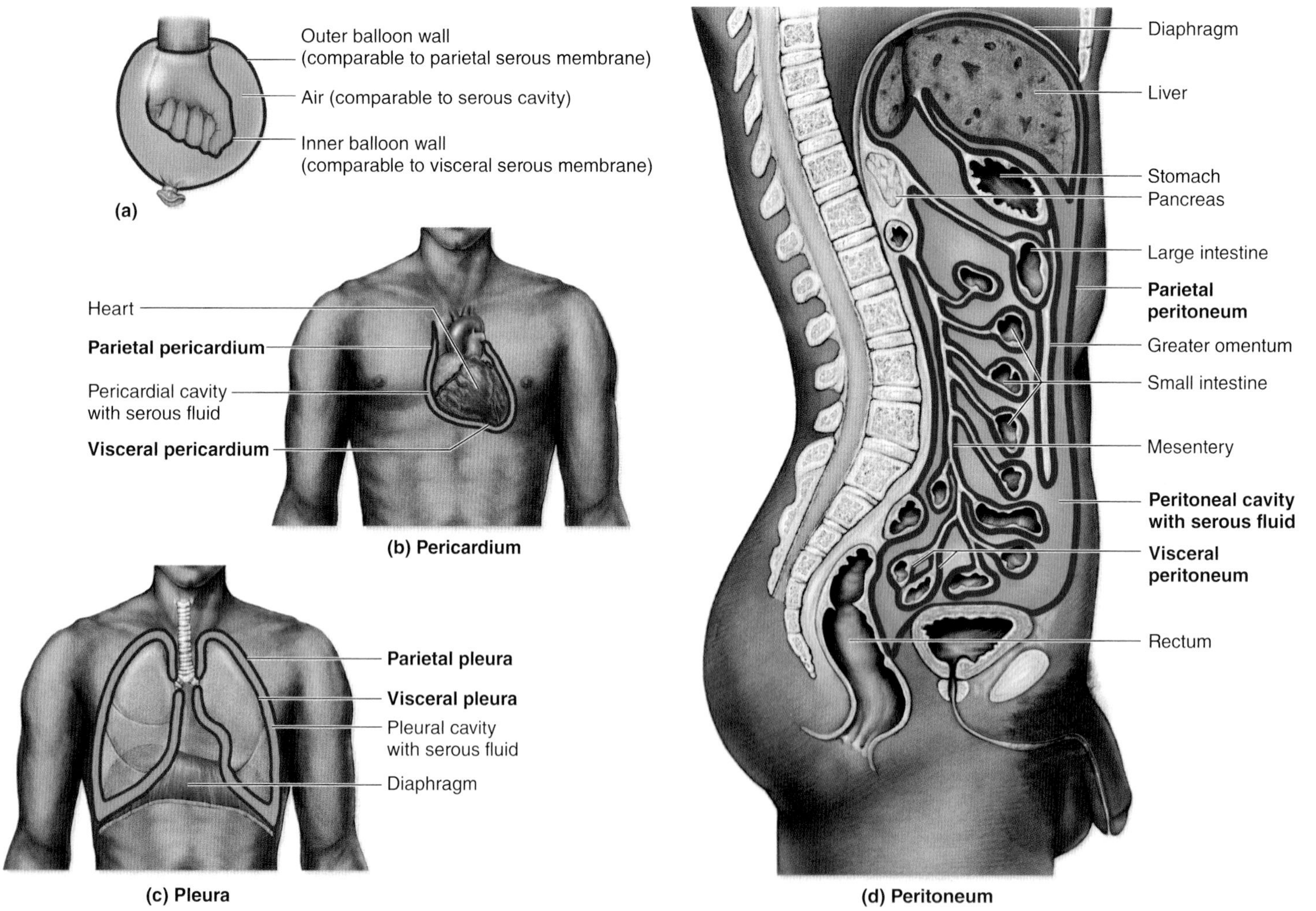

Outer balloon wall
(comparable to parietal serous membrane)

Air (comparable to serous cavity)

Inner balloon wall
(comparable to visceral serous membrane)

(a)

Heart

Parietal pericardium

Pericardial cavity
with serous fluid

Visceral pericardium

(b) Pericardium

Parietal pleura

Visceral pleura

Pleural cavity
with serous fluid

Diaphragm

(c) Pleura

Diaphragm

Liver

Stomach
Pancreas

Large intestine

Parietal peritoneum

Greater omentum

Small intestine

Mesentery

Peritoneal cavity with serous fluid

Visceral peritoneum

Rectum

(d) Peritoneum

Figure 1.9 Serous Membranes in the Thoracic and Abdominopelvic Body Cavities. Serous membranes line the inside of the cavity (parietal layer) and cover the outside of an organ (visceral layer) within the cavity. (*a*) The parietal and visceral serous membranes are similar to the inner and outer balloon walls that wrap around a fist, where the fist represents the body organ. (*b*) Parietal and visceral layers of the pericardium line the pericardial cavity around the heart. (*c*) Parietal and visceral layers of the pleura line the pleural cavity between the lungs and the chest wall. (*d*) Parietal and visceral layers of the peritoneum line the peritoneal cavity that lies between the abdominopelvic organs and the body wall. AP|R

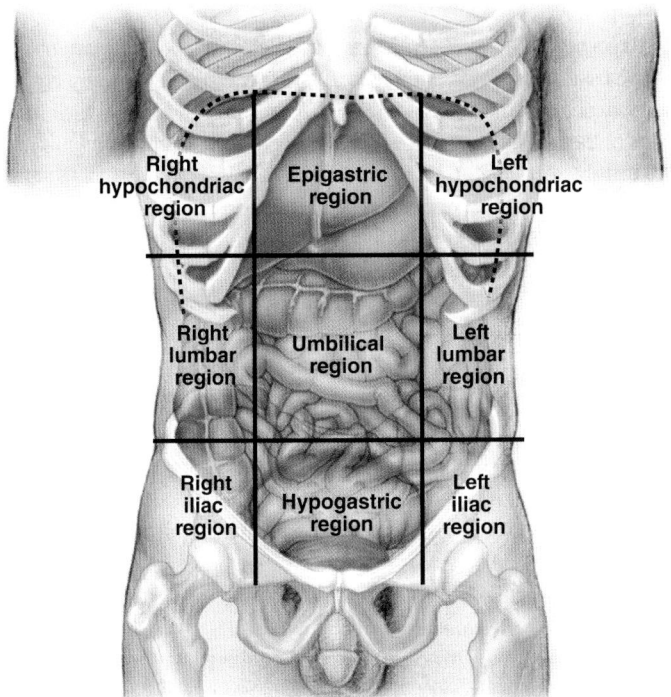

(a) Abdominopelvic regions

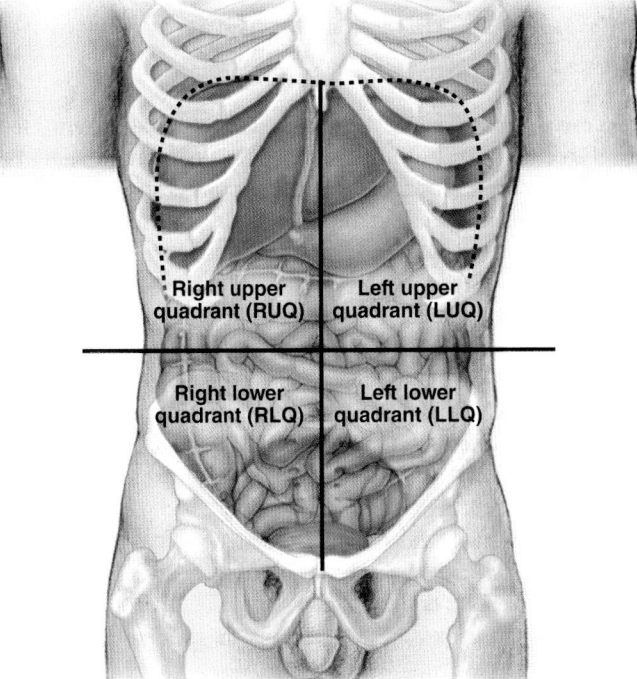

(b) Abdominopelvic quadrants

Figure 1.10 Abdominopelvic Regions and Quadrants. The abdominopelvic cavity can be subdivided into (*a*) nine regions or (*b*) four quadrants for purposes of description or identification. AP|R

this plane where it is wedged between the two hip bones. You can locate the division between these two cavities by palpating (feeling for) the superior ridges of your hip bones. The abdominal cavity contains most of digestive system organs, as well as the kidneys and most of the ureters. The pelvic cavity contains the distal part of the large intestine, the remainder of the ureters and the urinary bladder, and the internal reproductive organs.

The **peritoneum** (per′i-tō-nē′um; *periteino* = to stretch over) is the two-layered serous membrane that lines the abdominopelvic cavity (figure 1.9*d*). The **parietal peritoneum,** the outer layer of this serous membrane, lines the internal walls of the abdominopelvic cavity. The **visceral peritoneum** is the inner layer of this serous membrane, and it covers the external surfaces of most abdominal and pelvic organs. The potential space between these serous membrane layers is the **peritoneal cavity,** which contains and is lubricated by serous fluid.

 WHAT DID YOU LEARN?

 Which body cavity is associated with the lungs, and what are the names of its serous membranes?

1.4f Abdominopelvic Regions and Quadrants

LEARNING OBJECTIVE

7. Compare the terms used to subdivide the abdominopelvic region into nine regions or four quadrants.

To more accurately describe organ location, anatomists and health-care professionals commonly partition the large abdominopelvic cavity into smaller compartments. Nine compartments, called **abdominopelvic regions,** are delineated by using two transverse planes and two sagittal planes.

These nine regions are shown in **figure 1.10*a*** and summarized here:

- The **umbilical** (ŭm-bil′i-kăl; = navel) **region** is the middle region and is named for the umbilicus, or navel (belly button) that lies in its center.
- The **epigastric** (ep-ĭ-gas′trik; *epi* = above, *gaster* = belly) **region** is the superior region above the umbilical region.
- The **hypogastric** (hī-pō-gas′trik; *hypo* = under) **region** lies inferior to the umbilical region.
- The **right** and **left hypochondriac** (hī-pō-kon′drē-ak; *chondr* = cartilage) **regions** are inferior to the costal cartilages and lateral to the epigastric region.
- The **right** and **left lumbar regions** are lateral to the umbilical region.
- The **right** and **left iliac** (il′ē-ak; *eileo* = to twist) **regions** are lateral to the hypogastric region.

Some health-care professionals prefer to partition the abdomen more simply into four quadrants, using the umbilicus as the central point and having imaginary transverse and midsagittal planes pass through the umbilicus (figure 1.10*b*). The quadrants are named **right upper quadrant (RUQ), left upper quadrant (LUQ), right lower quadrant (RLQ),** and **left lower quadrant (LLQ).** These quadrants, like the abdominopelvic regions, are used to accurately locate and describe various aches, pains, injuries, or other abnormalities.

 WHAT DID YOU LEARN?

⓬ If a physician makes an incision into the abdomen along the midsagittal plane, superior to the umbilicus and just inferior to the diaphragm, then the skin of the _____ abdominopelvic region was incised.

1.5 Homeostasis: Keeping Internal Conditions Stable

Have you ever noticed that your body maintains an average internal temperature of about 37°C (98.6°F), regardless of the outside temperature? Perhaps you also have noticed that the size of your pupil is altered in response to light intensity entering your eye, or that your breathing returns to normal shortly after exercise. Likewise, your heart rate, blood pressure, and blood levels of sugar (glucose) and oxygen (O_2) are also regulated and maintained within certain parameters. In fact, there are hundreds of anatomic structures and physiologic processes that are continuously monitored and adjusted within your body so that they are kept within normal limits.

The term **homeostasis** (*homoios* = similar, *stasis* = standing) refers to the ability of an organism to maintain consistent internal environment, or "steady state," in response to changing internal or external conditions. Homeostasis is a central theme throughout this text, and you will be learning the appropriate details about homeostasis in each chapter. In this section, we introduce you to the general concept of homeostasis. We describe the general components of homeostatic systems, provide specific examples of these regulatory processes, and then describe the relationship of homeostasis, health, and disease.

1.5a Components of Homeostatic Systems

LEARNING OBJECTIVES

1. Define the components of a homeostatic system.
2. Be able to recognize each of the components in representative systems.

The body maintains homeostasis by utilizing homeostatic control systems. Three components are associated with each homeostatic system: receptor, control center, and effector **(figure 1.11)**.

Receptor

The **receptor** is the body structure that detects changes in a variable, which is a substance or process that is regulated. A receptor typically consists of sensory neurons (nerve cells). These neurons may be in the skin, internal organs of the body, or specialized organs such as the eye, ear, tongue, or nose. A **stimulus** is a change in the variable (a physical or chemical factor), such as a change in light, temperature, chemicals (e.g., glucose or oxygen levels), or stretch in muscle. For example, the retina of the eye (receptor) detects a change in light (stimulus) entering the eye.

Control Center

The **control center** is the structure that interprets input from the receptor and initiates changes through the effector. You can think of it as the "go between" for the other two components of a homeostatic system. The control center is generally a portion of the nervous system (brain or spinal cord) or an endocrine organ (such as the thyroid gland). A homeostatic system involving the nervous system provides a relatively quick means of responding to change. An example is regulating blood pressure when you rise from bed in the morning. In contrast, the endocrine system usually provides a means of a more sustained response over several hours or days through the release of hormones. An example is when the parathyroid hormone continuously regulates blood calcium levels, a process that is essential for the normal function of both muscles and nerves. Note that the control center is sometimes the same structure as the receptor because it both detects the stimulus and causes a response to regulate it. For example, the pancreas acts as a receptor because it detects an increase in blood glucose and also acts as a control center because it releases the hormone insulin in response.

Effector

The **effector** is the structure that brings about the change to alter the stimulus. Most body structures can serve as effectors. The most common effectors are muscles and glands. For example, smooth muscle in the

Figure 1.11 Components of a Homeostatic Control Mechanism.
A homeostatic control mechanism consists of a receptor (detects a stimulus), a control center (integrates input and initiates change through the effector), and an effector (brings about a change in response to the stimulus).

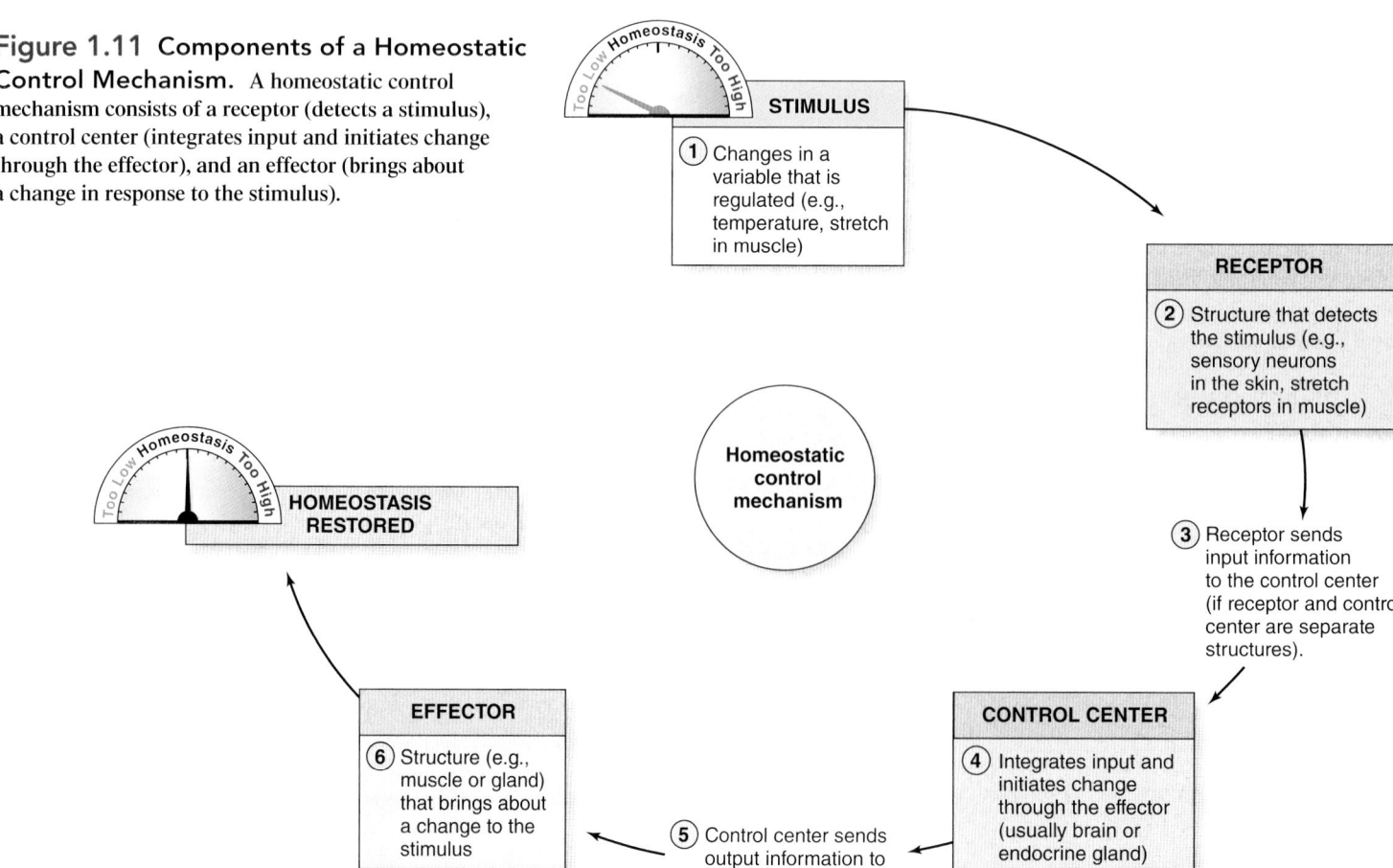

STIMULUS
1. Changes in a variable that is regulated (e.g., temperature, stretch in muscle)

RECEPTOR
2. Structure that detects the stimulus (e.g., sensory neurons in the skin, stretch receptors in muscle)

Homeostatic control mechanism

3. Receptor sends input information to the control center (if receptor and control center are separate structures).

HOMEOSTASIS RESTORED

CONTROL CENTER
4. Integrates input and initiates change through the effector (usually brain or endocrine gland)

EFFECTOR
6. Structure (e.g., muscle or gland) that brings about a change to the stimulus

5. Control center sends output information to an effector.

walls of air passageways (bronchioles) regulate airflow into and out of the lungs. Glands, such as the pancreas, release hormones (e.g., insulin).

As you view figure 1.11, notice that the response of a homeostatic system occurs through a feedback loop that includes the following:

- A stimulus
- The detection of the stimulus by a receptor
- Input information relayed to the control center (if a separate structure)
- Integration of the input by the control center and initiation of a change through effectors
- Return of homeostasis by the actions of effectors

Homeostatic control systems are separated into two broad categories based on whether the system maintains the variable within a normal range by moving the stimulus in the opposite direction, or amplifies the stimulus in the same direction. These two types of feedback control are called negative feedback and positive feedback, respectively.

 WHAT DID YOU LEARN?

13 List and describe the three components of a homeostatic system, and give examples of each in the human body.

1.5b Homeostatic Systems Regulated by Negative Feedback

 LEARNING OBJECTIVES

3. Define negative feedback.

4. Explain how homeostatic mechanisms regulated by negative feedback detect and respond to environmental changes.

Most processes in the body are controlled by negative feedback. If a homeostatic system is controlled by **negative feedback,** the resulting action will always be in the *opposite* direction of the stimulus. In this way, the variable is maintained within a normal level, or what is called its **set point.**

How a variable that is regulated by negative feedback fluctuates over time can be viewed in **figure 1.12.** Notice that the variable does not remain constant over time but rather it fluctuates, and its fluctuation occurs around the set point. If the stimulus increases, the homeostatic system is activated to cause a decrease in the stimulus until it returns to the set point. In contrast, if the stimulus decreases, the homeostatic system causes an increase in the stimulus until it returns to normal. This idea is generally better understood by describing a specific example, such as temperature regulation.

Temperature Regulation

We begin by first explaining how a negative feedback mechanism works to maintain the temperature of your home at a set point of 70°F. On a very cold day, the indoor temperature drops. This drop in temperature is detected by the thermostat. The drop in temperature is relayed through the electrical wiring of your home to the heat pump, which is then activated. The heat pump continues to heat your home until the thermostat reaches 70°F. An electrical signal is then sent from the thermostat to shut off the heat pump.

Body temperature is regulated in an analogous way to how the temperature of your home is regulated (**figure 1.13a**). If you venture outside on a cold day, body temperature may begin to drop. This

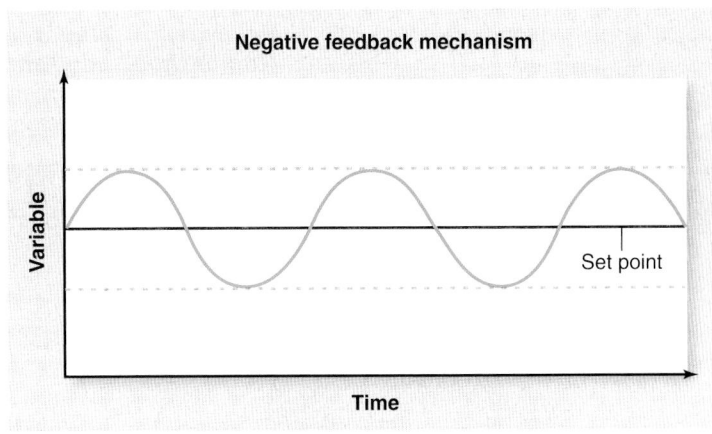

Figure 1.12 Negative Feedback. Note that when a variable is regulated by negative feedback, the variable fluctuates around a set point (rather than being a constant).

decrease in body temperature is detected by the sensory receptors of the skin, which send nerve impulses to the hypothalamus (a component of the brain). (The hypothalamus can also directly detect changes in body temperature by monitoring blood temperature as it passes through this region of the brain.) The hypothalamus compares sensory input to body temperature set point, and initiates motor output responses to blood vessels in the skin to decrease the diameter of the inside opening (lumen) of the vessels, thus decreasing the amount of blood circulating to the surface of the body. As a result, less heat is released through the skin. Nerve impulses are also sent to skeletal muscles, which cause shivering, and perhaps to smooth muscle associated with hair follicles of the skin, causing "goose bumps."

In contrast, on a very hot day (figure 1.13b), or when you are engaging in strenuous exercise, an increase in body temperature is detected by the sensory receptors of the skin or hypothalamus. The hypothalamus detects the difference between the increased body temperature and the original temperature set point, and transmits motor output to the blood vessels of the skin. This change increases the lumen diameters of blood vessels so that additional blood is brought near the surface of the body for the release of heat through the skin. Nerve impulses are also sent from the hypothalamus to the sweat glands to initiate sweating. Both responses help cool the body by the loss of heat from its surface. In these examples, regulation occurs through the nervous system.

Other examples of homeostatic regulation through the nervous system include the withdrawal reflex in response to injury from stepping on glass or burning your hand (see section 14.6), regulating heart rate and blood pressure when you exercise (see section 20.6), or changing breathing rate in response to an increase in carbon dioxide levels (see section 23.5).

Recall that the control center may also be the endocrine system. Examples of homeostatic systems that regulate through the endocrine system include the parathyroid gland release of parathyroid hormone in response to a decrease in blood calcium (see section 7.6) or pancreas release of insulin in response to an increase in blood glucose (see section 17.9).

 WHAT DID YOU LEARN?

14 On a cold day, what are some of the strategies the body uses to conserve heat?

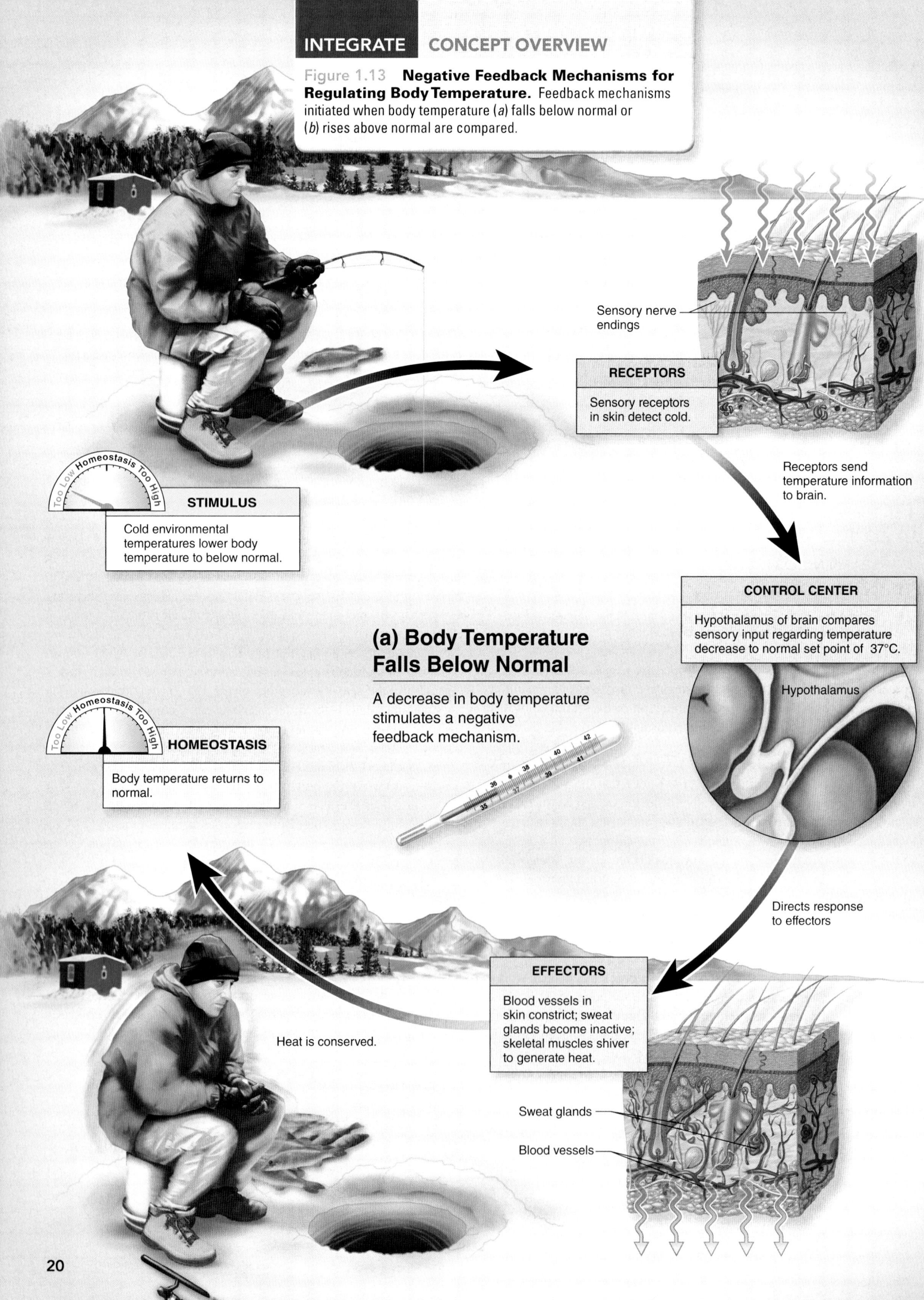

Figure 1.13   **Negative Feedback Mechanisms for Regulating Body Temperature.** Feedback mechanisms initiated when body temperature (*a*) falls below normal or (*b*) rises above normal are compared.

STIMULUS

Cold environmental temperatures lower body temperature to below normal.

RECEPTORS

Sensory receptors in skin detect cold.

Sensory nerve endings

Receptors send temperature information to brain.

(a) Body Temperature Falls Below Normal

A decrease in body temperature stimulates a negative feedback mechanism.

CONTROL CENTER

Hypothalamus of brain compares sensory input regarding temperature decrease to normal set point of 37°C.

Hypothalamus

HOMEOSTASIS

Body temperature returns to normal.

Heat is conserved.

Directs response to effectors

EFFECTORS

Blood vessels in skin constrict; sweat glands become inactive; skeletal muscles shiver to generate heat.

Sweat glands

Blood vessels

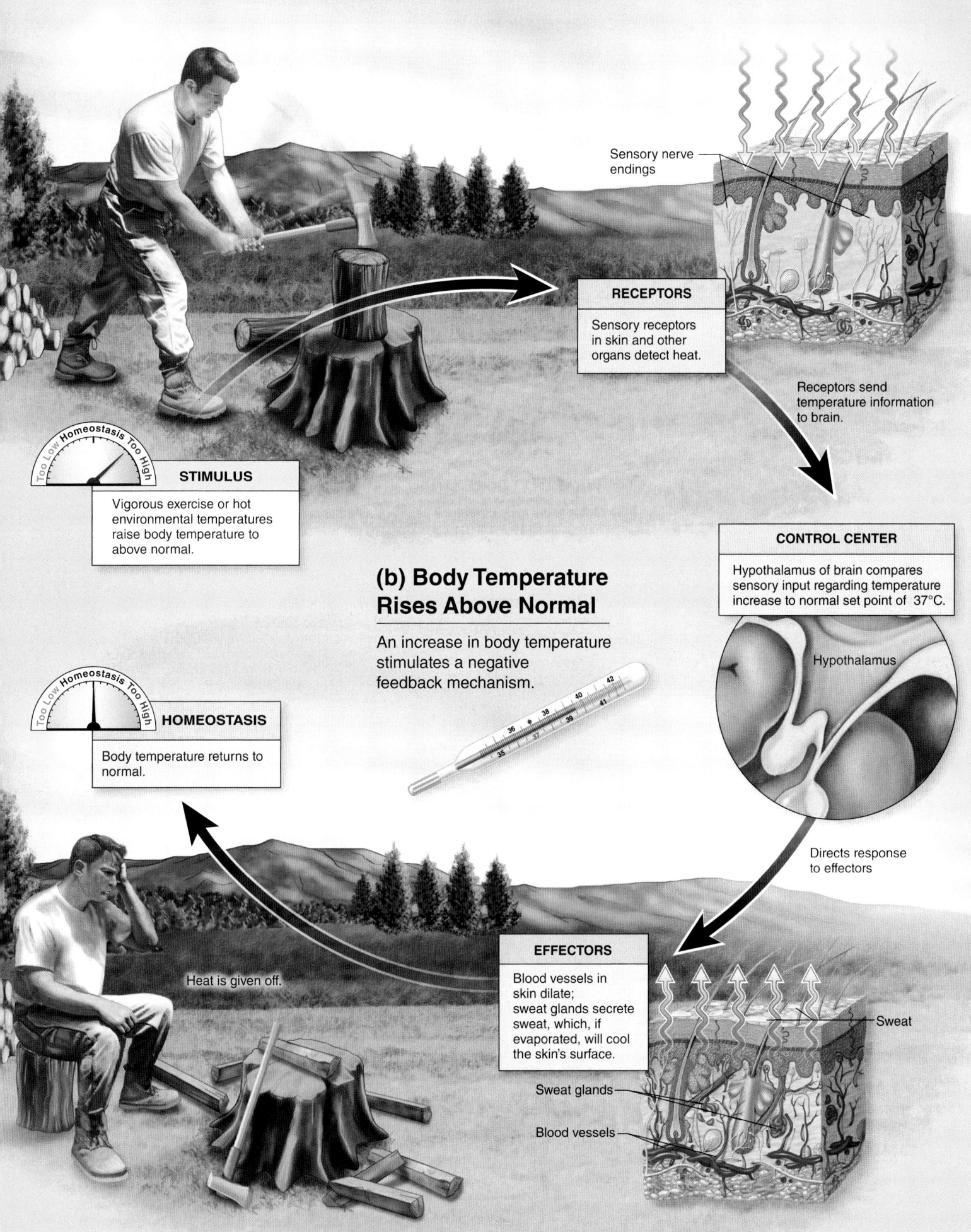

Sensory nerve endings

RECEPTORS

Sensory receptors in skin and other organs detect heat.

Receptors send temperature information to brain.

STIMULUS

Vigorous exercise or hot environmental temperatures raise body temperature to above normal.

Too Low Homeostasis Too High

(b) Body Temperature Rises Above Normal

An increase in body temperature stimulates a negative feedback mechanism.

CONTROL CENTER

Hypothalamus of brain compares sensory input regarding temperature increase to normal set point of 37°C.

Hypothalamus

HOMEOSTASIS

Body temperature returns to normal.

Too Low Homeostasis Too High

Directs response to effectors

Heat is given off.

EFFECTORS

Blood vessels in skin dilate; sweat glands secrete sweat, which, if evaporated, will cool the skin's surface.

Sweat

Sweat glands

Blood vessels

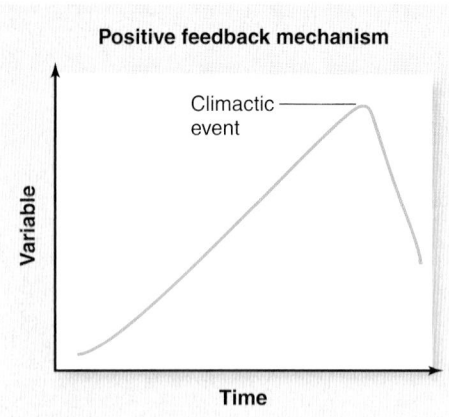

Positive feedback mechanism

Variable

Climactic event

Time

Figure 1.14 Positive feedback.
Positive feedback results in the stimulus being reinforced until a climactic event occurs, and then the body returns to homeostasis.

1.5c Homeostatic Systems Regulated by Positive Feedback

LEARNING OBJECTIVES

5. Define positive feedback.

6. Describe the actions of a positive feedback loop.

A homeostatic system may also be controlled by **positive feedback.** The stimulus here is reinforced to continue in the *same* direction until a climactic event occurs **(figure 1.14).** Following the climactic event, the body again returns to homeostasis. Because their end result is to increase the activity (instead of initially returning the body to homeostasis), positive feedback mechanisms occur much less frequently than negative feedback mechanisms.

Figure 1.15 illustrates one example of a positive feedback mechanism in the human body, when a mother breastfeeds her baby. The baby suckling at the breast is the initial stimulus detected by sensory receptors in the skin of the nipple region. The receptors transmit this input to the control center, which is the hypothalamus of the brain. The hypothalamus signals the posterior pituitary to release the hormone oxytocin into the blood. Oxytocin is the "output" that is sent to the effector, which is the glandular tissue of the breast. Oxytocin stimulates the mammary gland to eject the breast milk. The baby feeds and the cycle repeats as long as the baby suckles. Once the baby stops suckling (and thus the initial stimulus is removed), then the cycle will stop.

Other examples of positive feedback mechanisms include the blood clotting cascade (see section 18.4) and uterine contractions involved in labor and childbirth (see section 29.6).

WHAT DID YOU LEARN?

15 What is the main difference between a homeostatic system regulated by negative feedback versus positive feedback?

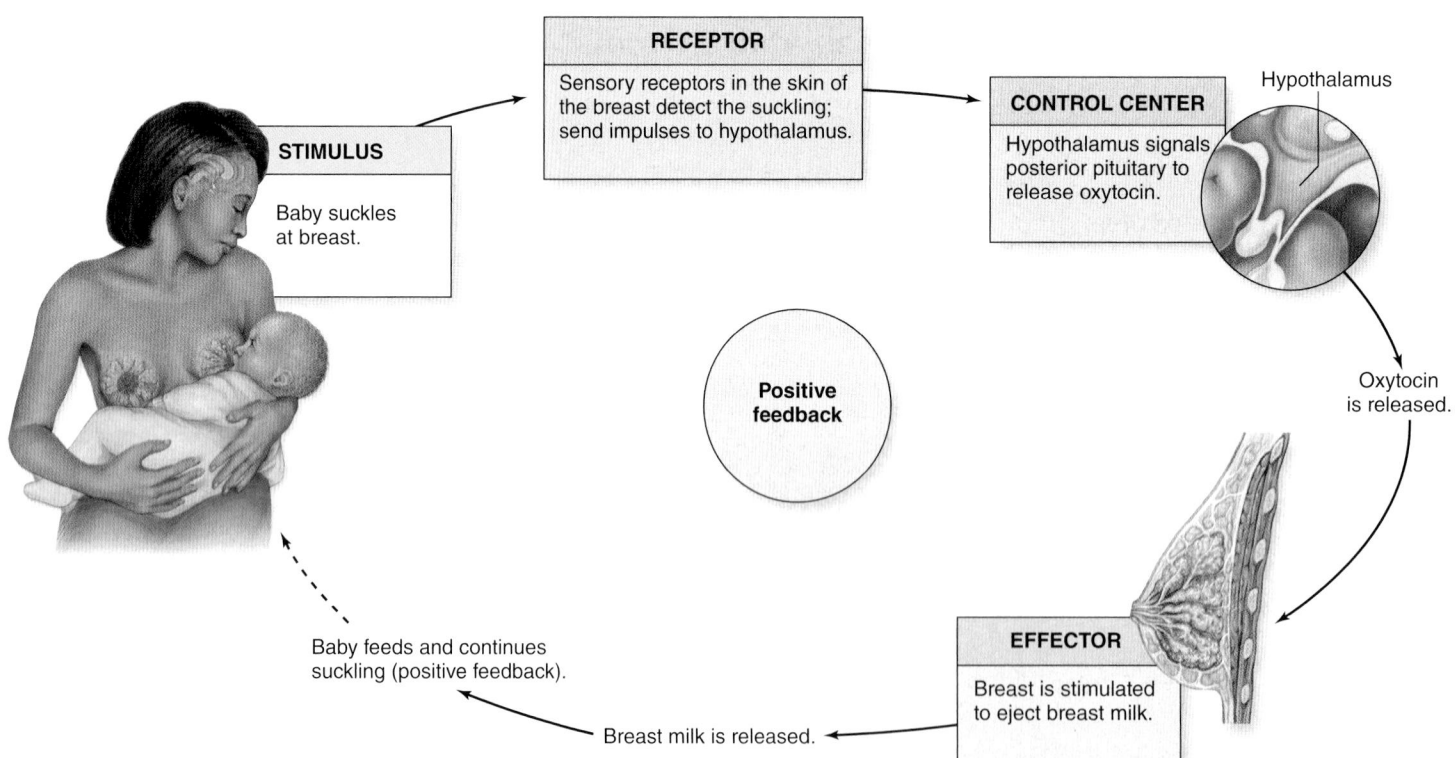

Figure 1.15 Positive Feedback. Positive feedback mechanisms often work in loops, where the initial step in the pathway is the stimulus, and the end product of the pathway is to stimulate (not turn off) the pathway activity. In this example of a mother breastfeeding her child, the stimulus of the baby suckling encourages release of hormones that stimulate the breast to secrete more breast milk.

1.6 Homeostasis, Health, and Disease

LEARNING OBJECTIVE

1. Explain the general relationship of maintaining homeostasis to health and disease.

In summary, homeostasis is a term that describes the many physiologic processes to maintain the health of the body. These characteristics are noted about homeostatic systems:

- They are dynamic.
- The control center is generally the nervous system or the endocrine system.
- There are three components: receptor, control center, and effector.
- They are typically regulated through negative feedback to maintain a normal value or set point.
- It is when these systems fail that a homeostatic imbalance or disease results, ultimately threatening an individual's survival.

Diabetes is an example of a homeostatic imbalance. Diabetes occurs when the homeostatic mechanisms for regulating blood glucose are not functioning normally, and blood glucose fluctuates out of the normal range, sometimes resulting in extremely high blood glucose readings. High blood glucose results in damage to anatomic structures throughout the body. Patients with diabetes must rely on other methods, such as diet restriction, exercise, and perhaps a medication to lower blood glucose.

Sometimes a homeostatic imbalance results when critical changes from aging or disease cause a variable that is normally controlled by negative feedback, to be abnormally controlled by positive feedback. An example is when there is extensive damage to the heart, perhaps from a heart attack. This heart is less able to pump blood to the structures of the body including the heart itself. Consequently, the heart receives reduced amounts of nutrients and oxygen. The heart becomes progressively weaker, and even less able to pump blood to the body's structure. Ultimately, the heart becomes so weak that the heart stops beating.

Treating patients generally involves determining a **diagnosis,** or the specific cause of the homeostatic imbalance. Once diagnosed, the patient is treated through the administration of medications or through other therapeutic avenues to facilitate the body in maintaining homeostasis.

Health-care practitioners also need to understand how the drugs patients are taking may affect the normal homeostatic control mechanisms. For example, one type of medication for the treatment of depression is an SSRI, which stands for selective serotonin reuptake inhibitor. Paroxetine (Paxil), fluoxetine (Prozac), and sertraline (Zoloft) are examples of SSRIs. Serotonin is a type of neurotransmitter. Normally, a neurotransmitter is released from one nerve cell in response to a nerve impulse. The neurotransmitter accomplishes its communication task, and then is taken up again by the nerve cell for future use. Some depressed individuals may have lower levels of serotonin, so an SSRI blocks the reuptake of serotonin into the nerve cell. Therefore, serotonin stays outside the nerve cell for a longer period of time and its effects are prolonged, which may elevate the mood of the patient taking the SSRI.

However, like all drugs, SSRIs come with some drawbacks. Some SSRI side effects include digestive system distress, such as nausea, upset stomach, diarrhea, or combinations of all three. As it turns out, serotonin is also used in the nerve cells of the digestive system. By tinkering with the serotonin reuptake in the brain, the drug also affects serotonin reuptake in the digestive system. Essentially, the digestive system becomes a bit more excitable due to the intake of the SSRI drug, with the symptoms just described.

Virtually all medications have some benefits and some side effects, many of which can be explained by examining the homeostatic control mechanisms with which they interact. Thus, an understanding of these mechanisms is a must for anatomists, physiologists, and health-care practitioners.

WHAT DID YOU LEARN?

16 What is an example of a disease process by which homeostasis is disrupted?

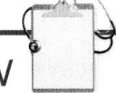

INTEGRATE

CLINICAL VIEW

Establishing Normal Ranges for Clinical Practice

It is interesting to know that what is clinically accepted as the "normal range" for a variable, such as body temperature of 98.6°F, blood glucose of 80–110 milligrams/deciliter (mg/dL), or blood pressure of 90–120/60–80 mm Hg is determined by sampling healthy individuals in a population. A normal range for a variable is determined by the value for 95% of the individuals sampled. Health-care practitioners should be aware that this means that 5% of the population, although healthy, will have values for a given variable considered outside of the normal range.

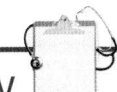

INTEGRATE

CLINICAL VIEW

Clinicians' Use of Scientific Method

Clinicians regularly apply the principles of the scientific method when interacting with patients. Consider what typically occurs when a patient with a health problem or complaint goes in for a doctor's appointment. First, information is gathered. The nurse obtains the patient's weight, blood pressure, and other vital signs. The physician solicits the patient's medical history, asks about his or her specific complaint(s), and completes a physical examination. Based on the information gathered, the clinician forms a hypothesis or a tentative explanation of any specific symptoms the patient may be experiencing. As a follow-up to the initial hypothesis, the clinician orders tests and evaluates the test results. After all information is gathered, the clinician draws a conclusion to make a diagnosis. (Sometimes additional tests may be ordered if the test results are inconclusive.) Following a definitive diagnosis, the clinicians treat the patient and additional information is gathered as the patient's response to the treatment is monitored.

CLINICAL VIEW
Medical Imaging

Health-care professionals have taken advantage of sophisticated medical imaging techniques to extend their ability to visualize internal body structures noninvasively (e.g., without inserting an instrument into the body). Some of the most common techniques are radiography, sonography, computed tomography, digital subtraction angiography, dynamic spatial reconstruction, magnetic resonance imaging, and positron emission tomography.

Radiography

Radiography (rā-dē-og′ra-fē; *radius* = ray, *grapho* = to write) is the primary method of obtaining an image of a body part for diagnostic purposes. A beam of **x-rays,** which are a form of high-energy radiation, penetrates solid structures within the body. X-rays can pass through soft tissues but they are absorbed by dense tissues, including bone, teeth, and tumors. Film images produced by x-rays passing through soft tissues leave the film lighter in the areas where x-rays are absorbed. Hollow organs can be visualized if they are filled with a radiopaque (rā-dē-ō-pāk; *opacus* = shady) substance that absorbs x-rays.

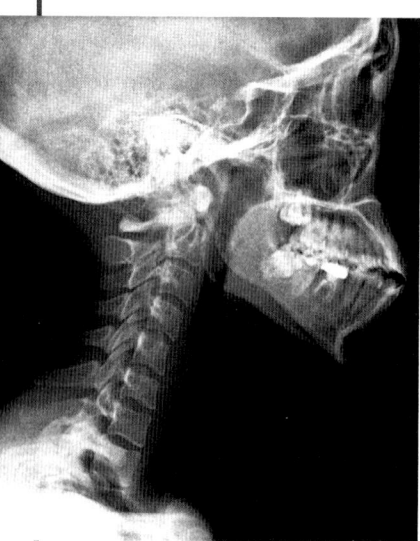

Radiograph (x-ray) of the head and neck.

The term x-ray also applies to the photograph (radiograph) made by this technique. Originally, x-rays got their name because they were an unknown type of radiation, but they are also called roentgen rays in honor of Wilhelm Roentgen, the German physicist who accidentally discovered them. Radiography is commonly used in dentistry, mammography, diagnosis of fractures, and chest examination. Disadvantages of x-rays are that they are difficult to interpret when organs overlap in the images, and they are unable to reveal slight differences in tissue density. In addition, the radiation of an x-ray is not without risk.

Sonography

The second most widely used imaging method is **sonography** (sŏ-nog′ră-fē; *sonus* = sound, *grapho* = to write), also known as *ultrasound*. A technician slowly moves a small, handheld device across the body surface. This device

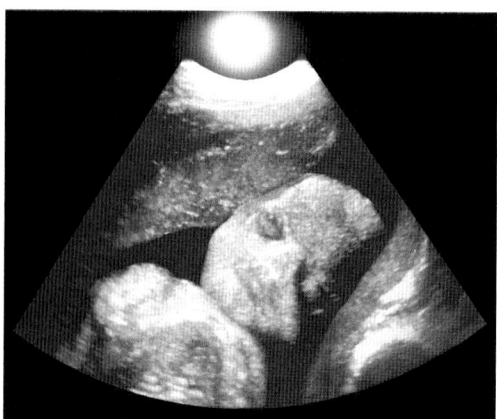

Sonogram of a fetus.

produces high-frequency ultrasound waves and then receives signals that are reflected from internal organs. The image produced is called a *sonogram.* Sonography is the method of choice in obstetrics, where a sonogram can visualize the placenta, examine the fetus and evaluate fetal age, position, and development. Sonography avoids the harmful effects of x-rays, and the equipment is inexpensive and portable.

Improvements include three-dimensional and four-dimensional ultrasound. In three-dimensional ultrasound, sound waves are emitted in various angles and processed in a computer. This creates a three-dimensional view. A two-dimensional ultrasound is a flat image and a three-dimensional ultrasound shows depth, contour, and detail. Four-dimensional ultrasound shows movement using a compilation of three-dimensional images. Movements like heart motion and yawning can be seen in real time. When radiography or sonography fail to produce the desired images, other more detailed but much more expensive imaging techniques are available.

Computed Tomography (CT)

A **computed tomography (CT)** (tō′mō-graf-ē; *tomos* = a section) scan, previously termed a computerized axial tomography (CAT) scan, is a more sophisticated application of x-rays. A patient is slowly moved through a cylindrical, doughnut-shaped machine while low-intensity x-rays are emitted on one side the cylinder, passed through the body, collected by detectors, and then processed and analyzed by a computer. These signals produce an image of the body that is about the thickness of a dime. Continuous thin "slices" can be used to reconstruct a three-dimensional image of the body. Little overlap of organs occurs in these thin sections, and the image is much sharper than one obtained by a conventional x-ray. CT scanning is useful for identifying tumors, aneurysms, kidney stones, cerebral hemorrhages, and other abnormalities. A drawback to CTs is that they expose the patient to higher doses of radiation than a traditional x-ray.

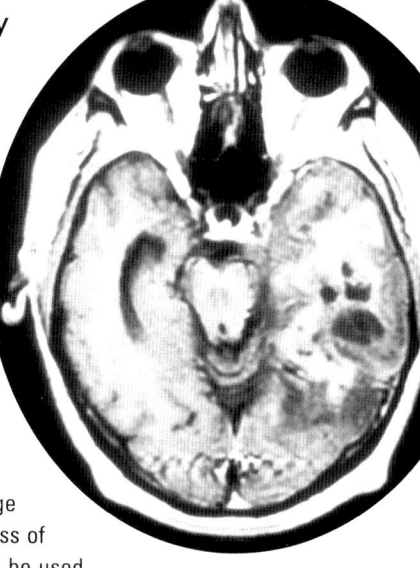

Computed tomography (CT) scan of the head at the level of the eyes.

Digital Subtraction Angiography (DSA)

Digital subtraction angiography (DSA) is a modified three-dimensional x-ray technique used primarily to view blood vessels. It involves taking radiographs both prior to and after injecting an opaque medium into a blood vessel. The computer compares the before and after images, and removes or subtracts the data from the before image from the data generated by the after image, thus leaving an image that may indicate evidence of vessel blockages. DSA is useful in the procedure in which a physician directs a catheter through a blood vessel and puts a stent in the area where a blood vessel is blocked. The image produced by the DSA allows the physician to accurately guide the catheter to the blockage.

Dynamic Spatial Reconstruction (DSR)

Using modified CT scanners, a special technique called **dynamic spatial reconstruction (DSR)** provides two important pieces of medical information: (1) three-dimensional images of body organs, and (2) information about the normal organ movement as well as changes in its internal volume. Unlike traditional static CT scans, DSR allows the physician to see the movement

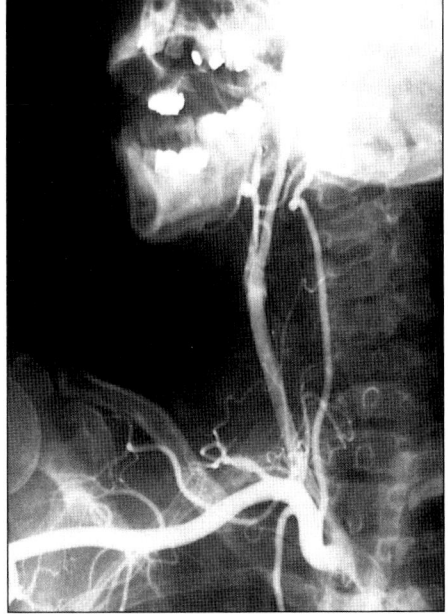

Digital subtraction angiography (DSA) shows three-dimensional images of blood vessels and normal changes in these vessels.

A specific type of MRI, called **functional MRI (fMRI)**, provide the means to map brain function based on local oxygen concentration differences in blood flow. Increased blood flow relates to increased brain activity and is detected by a decrease in deoxyhemoglobin (the form of hemoglobin lacking oxygen) in the blood.

Positron Emission Tomography (PET)

The **positron emission tomography (PET)** scan is used both to analyze the metabolic state of a tissue at a given moment in time and to determine which tissues are most active. The procedure begins with an injection of radioactively labeled glucose (sugar), which emits particles called positrons (like electrons, but with a positive charge). Collisions between a positron and electron cause the release of gamma rays that can be detected by sensors and analyzed by computer. The result is a brilliant color image that shows which tissues were using the most glucose at that moment. In cardiology, the image can reveal the extent of damaged heart tissue—because damaged heart tissue consumes little or no glucose, the damaged tissue will appear dark. PET scans have also been used to illustrate activity levels in the brain. PET scans also may detect whether certain cancers have metastasized throughout the body, because cancerous cells will take up more glucose and show up as a "hot spot" on the scan. The PET scan is an example of nuclear medicine, which uses radioactive isotopes to form anatomic images of the body.

of an organ. This type of observation, at slow speed or halted in time completely, has been invaluable in observations of the heart and the flow of blood through blood vessels.

Magnetic Resonance Imaging (MRI)

Magnetic resonance imaging (MRI), previously called *nuclear magnetic resonance (NMR) imaging,* was developed as a noninvasive technique to visualize soft tissues. The patient is placed in a supine position within a cylindrical chamber that is surrounded by a large electromagnet. The magnet generates a strong magnetic field that causes protons in the nuclei of hydrogen atoms in the tissues to align. Thereafter, upon exposure to radio waves, the protons absorb additional energy and align in a different direction. The hydrogen atoms then abruptly realign themselves to the magnetic field immediately after the radio waves are turned off. This results in the release of the atoms' excess energy at different rates, depending on the type of tissue. A computer analyzes the emitted energy to produce an image of the body. MRI is better than CT for distinguishing between soft tissues, such as the white and gray matter of the nervous system. However, dense structures (e.g., bone) do not show up well in MRI. Formerly, another disadvantage of MRI was that patients felt claustrophobic while isolated in the closed cylinder. However, newer MRI technology has improved the hardware and lessened this effect.

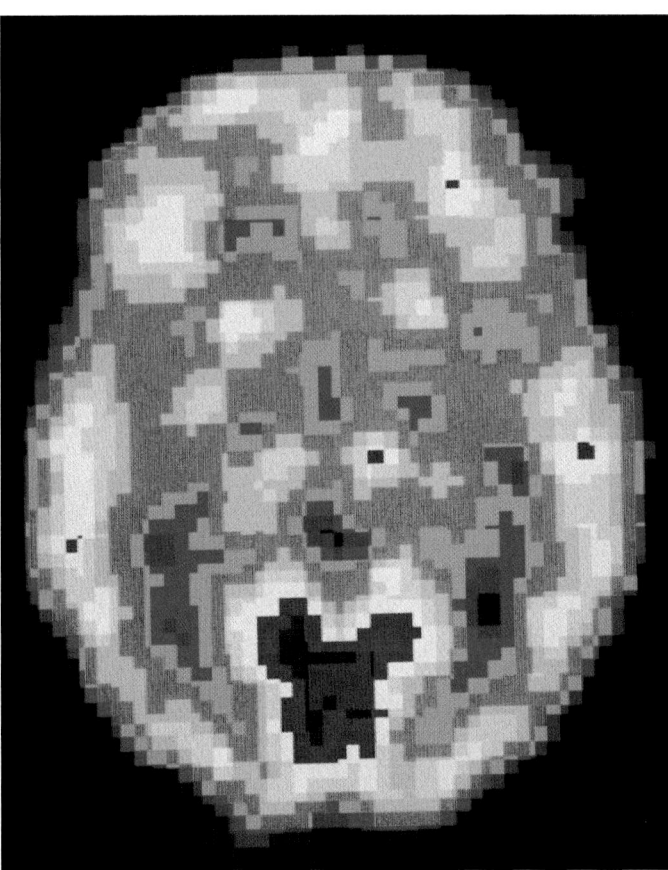

Magnetic resonance imaging (MRI) scan of the head at the level of the eyes.

Positron emission tomography (PET) scan of the brain of an unmedicated schizophrenic patient. Red areas indicate high glucose use (metabolic activity). The visual center at the posterior region of the brain was especially active when the scan was made.

CHAPTER SUMMARY

1.1 Anatomy and Physiology Compared 2

- Anatomy is the study of structure and form of the human body, whereas physiology is the study of function of these parts.

1.1a Anatomy: Details of Structure and Form 2

- Anatomy may be subdivided into microscopic anatomy (anatomic study of materials using the microscope) and gross anatomy (the study of structures visible to the unaided eye).

1.1b Physiology: Details of Function 3

- Physiologists may examine the function of specific body systems (e.g., cardiovascular physiology) and may focus on problems or pathologies of such systems (pathophysiology).

1.2 Anatomy and Physiology Integrated 3

- Form and function are interrelated. Anatomists cannot gain a full appreciation of anatomic form without first understanding the structure's function. Likewise, physiologists cannot fully appreciate body functions without learning about the structure's form.
- It is easiest to learn anatomy and physiology by integrating the two disciplines, rather than trying to separate the study of form from the study of function.

1.3 The Body's Levels of Organization 6

- Scientists group the body's components into an organizational hierarchy of form and function.

1.3a Characteristics That Describe Living Things 6

- All living organisms exhibit several common properties: organization, metabolism, growth and development, responsiveness, adaptation, and reproduction.

1.3b The View from Simplest to Most Complex 6

- Anatomic structure is organized in an increasingly complex series of levels: the chemical level, cellular level, tissue level, organ level, organ system level, and the organismal level.

1.3c Introduction to Organ Systems 7

- The human body contains 11 organ systems: integumentary, skeletal, muscular, nervous, endocrine, cardiovascular, lymphatic, respiratory, urinary, digestive, and reproductive.

1.4 The Language of Anatomy and Physiology 11

- Clear, exact terminology accurately describes body structures and helps us identify and locate them.

1.4a Anatomic Position 11

- The anatomic position is used as a standard reference point for the human body.

1.4b Sections and Planes 11

- Three planes section the body and help describe relationships among the parts of the three-dimensional human body: the coronal plane, the transverse plane, and the midsagittal plane.

1.4c Anatomic Directions 12

- Specific directional terms indicate body structure locations.

1.4d Regional Anatomy 13

- Specific anatomic terms identify body regions.

1.4e Body Cavities and Membranes 14

- Body cavities are spaces that enclose organs and organ systems.
- The posterior aspect of the body contains the cranial cavity and the vertebral canal.
- The ventral cavity is subdivided into a thoracic cavity (which contains the pleural cavities, mediastinum, and pericardial cavity) and an abdominopelvic cavity (which is partitioned into an abdominal cavity and a pelvic cavity).
- The ventral cavity is lined by thin serous membranes. A parietal layer lines the internal body wall surface, and a visceral layer covers the organs.

1.4f Abdominopelvic Regions and Quadrants 17

- Regions and quadrants are two aids for describing locations of the abdominopelvic viscera.
- There are nine abdominopelvic regions and four abdominopelvic quadrants.

1.5 Homeostasis: Keeping Internal Conditions Stable 18

- Homeostasis refers to the body's ability to maintain a relatively stable internal environment, even in the face of changing internal or external environmental factors.

1.5a Components of Homeostatic Systems 18

- The three components are the receptor (detects a stimulus), control center (interprets input from the receptor and initiates changes through the effector), and effector (the structure that brings about a change to the stimulus).

1.5b Homeostatic Systems Regulated by Negative Feedback 19

- Negative feedback mechanisms or loops are initiated by either an increase or a decrease in the stimulus, and the end result is to return the stimulus to within its normal range or set point. Most feedback mechanisms in the human body work by negative feedback.

1.5c Homeostatic Systems Regulated by Positive Feedback 22

- Positive feedback mechanisms are initiated by a stimulus, and they maintain or increase the activity of the original stimulus.

1.6 Homeostasis, Health, and Disease 23

- An understanding of the concept of homeostasis is essential when understanding the structure and function of a normal, healthy body, the mechanisms of disease, and how the body reacts to pharmaceutical agents.

CHALLENGE YOURSELF

Do You Know the Basics?

_____ 1. Examining the superficial anatomic markings and internal body structures as they relate to the covering skin is called
- a. regional anatomy.
- b. surface anatomy.
- c. pathologic anatomy.
- d. comparative anatomy.

_____ 2. The _____ level of organization is composed of two or more tissue types that work together to perform a common function.
- a. cellular
- b. molecular
- c. organ
- d. organismal

_____ 3. The term _____ refers to the sum of all chemical reactions in the body.
- a. metabolism
- b. responsiveness
- c. stimulus
- d. reproduction

_____ 4. A midsagittal plane separates the body into
- a. anterior and posterior portions.
- b. superior and inferior portions.
- c. right and left halves.
- d. unequal right and left portions.

_____ 5. The term used to describe an appendage structure that is closest to its point of attachment to the trunk is
- a. distal.
- b. lateral.
- c. superior.
- d. proximal.

_____ 6. The _____ region is the anterior part of the knee.
- a. patellar
- b. popliteal
- c. pes
- d. inguinal

_____ 7. Which body cavity is located inferior to the diaphragm and superior to a horizontal line drawn between the superior edges of the hip bones?
- a. abdominal cavity
- b. pelvic cavity
- c. pleural cavity
- d. pericardial cavity

_____ 8. The _____ is the serous membrane layer that covers the surface of the lungs.
- a. parietal pleura
- b. visceral pericardium
- c. visceral peritoneum
- d. visceral pleura

_____ 9. The state of maintaining a constant internal environment within an organism is called
- a. reproduction.
- b. homeostasis.
- c. imbalance.
- d. life.

_____ 10. In a negative feedback mechanism, which of the following events does _not_ occur?
- a. A stimulus is a change in some variable (e.g., rise in blood glucose levels).
- b. A receptor perceives a stimulus.
- c. The control center sends output to an effector.
- d. The effector stimulates or increases the stimulus, so the cycle continues.

11. What are the similarities and differences between anatomy and physiology?

12. List the levels of organization in a human, starting at the simplest level and proceeding to the most complex. Give an example of a structure in each level.

13. What properties are common to all living things?

14. Name the organ systems in the human body.

15. Describe the body in the anatomic position. Why is the anatomic position used?

16. List the anatomic term that describes each of the following regions: forearm, wrist, chest, armpit, thigh, and foot.

17. What are the two body cavities within the posterior aspect, and what does each cavity contain?

18. Describe the structure and function of serous membranes in the body.

19. What are the main components in a homeostatic control system?

20. Compare and contrast negative and positive feedback mechanisms.

Can You Apply What You've Learned?

Use the following paragraph to answer questions 1–3.

Your friend Eric complains of some pain in his "belly area." You ask him to point to the precise location of the pain. He points to a region that is below his umbilicus, on the lower right side of his abdomen, and just medial to his hip bones.

1. Which abdominal quadrant contains Eric's pain?
   a. Right upper quadrant
   b. Right lower quadrant
   c. Left upper quadrant
   d. Left lower quadrant

2. You also could describe the pain as being in the _____ abdominopelvic region.
   a. right lumbar
   b. right hypochondriac
   c. right umbilical
   d. right iliac region

3. Eric goes to the doctor to determine the cause and source of the pain. The physician orders a CT scan, which shows that Eric has an enlarged and inflamed appendix (an organ associated with the digestive system). Eric asks the physician why she didn't just take an x-ray of his belly region. She explains an x-ray would not be the best diagnostic imaging tool in this case because
   a. x-ray images are more expensive to produce than CT scans.
   b. soft-tissue structures don't show up well on basic x-rays.
   c. the x-rays could inflame the appendix further and cause it to burst.
   d. x-rays now are used primarily for bone injuries only.

4. When you are outside on a hot, humid day, what body changes occur to help your body temperature return to normal?
   a. The blood vessels in your skin constrict.
   b. The sweat glands release sweat.
   c. Nerve impulses are sent to muscles to cause shivering.
   d. The smooth muscle associated with hair follicles contracts, causing goose bumps.

5. A friend just started taking Zoloft (an SSRI) and is experiencing an upset stomach and diarrhea. Your friend asks if the drug is causing her symptoms and you respond:
   a. Yes, because the drug is irritating her stomach lining and that explains her symptoms.
   b. Yes, because serotonin is located both in the brain and the digestive tract, so the drug is altering digestive system functioning.
   c. No, because the drug is supposed to elevate mood and affect brain function, and it shouldn't have any effect on the digestive system.
   d. No, because the drug is quickly absorbed from the digestive tract and does not remain in the digestive system long enough to have any effect there.

Can You Synthesize What You've Learned?

1. Lynn was knocked off her bicycle during a race. She broke some bones in her right antebrachial region, suffered an abrasion on her mental region, and had severe bruising on her right gluteal and femoral regions. Explain where each of these injuries is located.

2. Carly was stung by a bee and was taken to the emergency room because she was undergoing anaphylactic shock (e.g., her breathing became more rapid and more difficult, her heartbeat increased). She was given a shot of epinephrine, which reduced her allergic reactions and brought her breathing and heartbeat back to normal. Did the dose of epinephrine result in a negative feedback mechanism occurring, or a positive feedback mechanism occurring? Explain your answer.

3. Your grandmother is being seen by a radiologist to diagnose a possible tumor in her small intestine. Explain to your grandmother what imaging techniques would best determine whether a tumor exists, and which techniques would be inadequate for determining the placement of the tumor.

INTEGRATE

ONLINE STUDY TOOLS

The following study aids may be accessed through Connect.

Clinical Case Study: A Clinical Case of Hyperthermia in an Athlete

Interactive Questions: This chapter's content is served up in a number of multimedia question formats for student study.

LearnSmart: Topics and terminology include anatomy and physiology and details of form and function; the body's levels of organization; terminology for anatomic position, sections, and planes; homeostasis, health, and disease

Anatomy & Physiology Revealed: Topics include organ systems, planes of section, directional terms, body regions, body cavities, pleura and pericardium; abdominal quadrants and regions

Animations: Topics include positive and negative feedback

Atoms, Ions, and Molecules

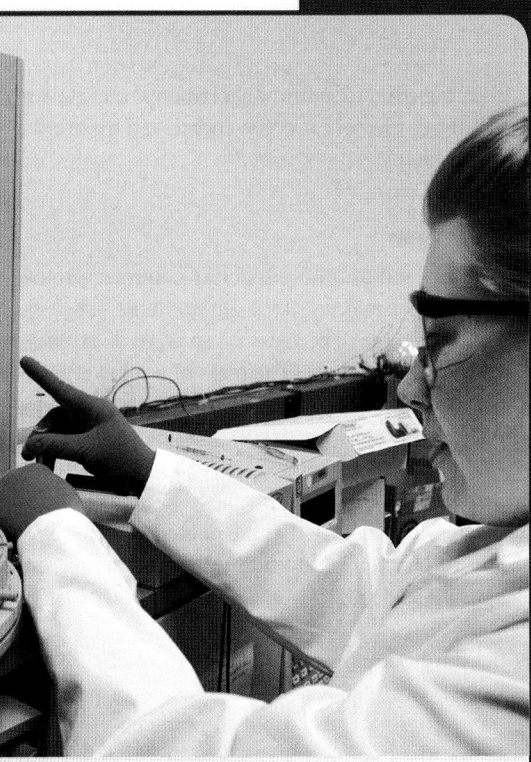

INTEGRATE

CAREER PATH
Clinical Chemist

A clinical chemist utilizes laboratory tests to evaluate blood samples, study DNA, and examine both tissues and cells. Samples obtained from patients are sent to the laboratory and evaluated by a clinical chemist. The results of these tests provide more precise information to the physician for both determining a diagnosis and implementing an appropriate treatment.

Anatomy & Physiology REVEALED®
aprevealed.com

Module 2:
Cells and Chemistry

We describe many fascinating physiologic processes of the human body in later chapters, including the transmission of an impulse along a nerve, the transport of oxygen in the blood, and nutrient digestion in the gastrointestinal tract. By reading and comprehending the material in this chapter on atoms, ions, and molecules—and information on energy, enzymes, and metabolism in chapter 3—you will develop a working knowledge of the basic chemical concepts needed to ➡

understand these processes. We will refer back to these concepts throughout the rest of the text, so that you'll more easily see the connections between chemistry and life processes.

Our coverage of chemistry is not comprehensive. Rather, our discussion is tailored to focus primarily on chemical structures relevant to the study of anatomy and physiology, and we have included examples of their application in the human body. The general characteristics of atoms, ions, and molecules are presented in the earlier sections of this chapter. Then we describe specific molecules, including water and water mixtures, and the four major organic biological macromolecules: lipids, carbohydrates, nucleic acids, and proteins. The chapter concludes with a section devoted to proteins, the most versatile macromolecules of living systems. Now, let's embark upon an adventure to explore the body's simplest constituents: atoms and molecules.

2.1 Atomic Structure

The human body at its simplest level of organization is composed of chemical structures that include atoms, ions, and molecules. We begin our discussion on the human body's chemical composition by describing matter, atoms, elements, and the position of each element in the periodic table. This information allows us to make certain predictions regarding the chemical properties of each element, including whether and how the element forms ions and molecules.

2.1a Matter, Atoms, Elements, and the Periodic Table

LEARNING OBJECTIVES

1. Define matter, and list its three forms.
2. Describe and differentiate among the subatomic particles that compose atoms.
3. Explain the arrangement of elements in the periodic table based on atomic number.
4. Diagram the structure of an atom.

The human body is composed of **matter,** generally defined as a substance that has mass and occupies space. Matter is present in the body in three forms: solid, liquid, and gas. For example, bone is a solid, blood is a liquid, and oxygen (O_2) and carbon dioxide (CO_2) are gases.

All matter is composed of atoms. An **atom** is the smallest particle that exhibits the chemical properties of an **element.** There are 92 naturally occurring elements. Hydrogen is the smallest and lightest element; uranium is the largest and heaviest element. Technical advances both in chemistry and physics have resulted in the ability to produce "ultraheavy" elements that are larger than uranium. All elements, including the scientifically manufactured elements, are organized into chart form in the **periodic table of elements (figure 2.1a)**.

Elements are grouped into major, minor, and trace elements based on the percentage each composes by weight in the human body. Major elements make up over 98% and minor elements less than 1% (figure 2.1b). In comparison, trace elements appear in the body in only limited amounts (less than 0.01%). Only 12 elements occur in living organisms in greater than trace amounts: oxygen, carbon, hydrogen, nitrogen, calcium, phosphorus, sulfur, potassium, sodium, chlorine, magnesium, and iron. All of these elements except iron have atomic numbers (described shortly) between 1 and 20 in the periodic table. Consequently, our emphasis in this chapter will be on only the first 20 elements.

Note in figure 2.1a that the 12 common elements are elevated above the other elements. Each element is color-coded, and these color-codes are used throughout the text.

The Components of an Atom

Atoms are composed of three **subatomic particles:** neutrons, protons, and electrons **(figure 2.2)**. Two major criteria differentiate subatomic particles—namely, mass and charge. The mass of an atom is expressed as the **atomic mass unit (amu)**, or *dalton*. (The mass of a subatomic particle is very small: consider that 1 amu is equal to 1.66×10^{-27} kilograms.)

Neutrons and protons each have a mass of 1 amu. A **neutron** is uncharged (meaning it is neutral), whereas a **proton** has a positive charge of one (+1). Neutrons and protons compose almost the entire mass or weight of an atom and are located at the center, or core, of the atom, called the **atomic nucleus.** The atomic nucleus contains mass because it has both neutrons and protons, and it is positively charged because of the protons.

The **electron** is the third component of an atom. An electron has a very small mass—only about 1/1800th of the mass of a proton or neutron—and makes a negligible (very small) contribution to total mass. Each electron has a negative charge of one (−1). Electrons are located at varying distances from the nucleus in regions called *orbitals*, often depicted as either an electron cloud or as discrete energy shells. Both the cloud model and shell model indicate where the electrons are most likely found as depicted in figure 2.2.

The Periodic Table

Elements differ in the number of subatomic particles. The periodic table may be used to obtain the number of subatomic particles in an atom of a specific element. Several important features for each element appear in the periodic table, including the element's symbol, atomic number, and average atomic mass.

A unique **chemical symbol** has been assigned to each element. An element is usually identified either by its first letter, or by its first letter plus an additional letter of its English name. For example, H is for hydrogen, C is for carbon, and O is for oxygen. If more than one element begins with the same letter, the symbol will also include a second letter written in lowercase (e.g., He for helium and Cl for chlorine). The symbol of some elements, in rare cases, is derived from their Latin name. For instance, the symbol for sodium is Na (natrium) and potassium is K (kalium).

The **atomic number** of an element indicates the number of protons in an atom of that element and is located above its symbol in the

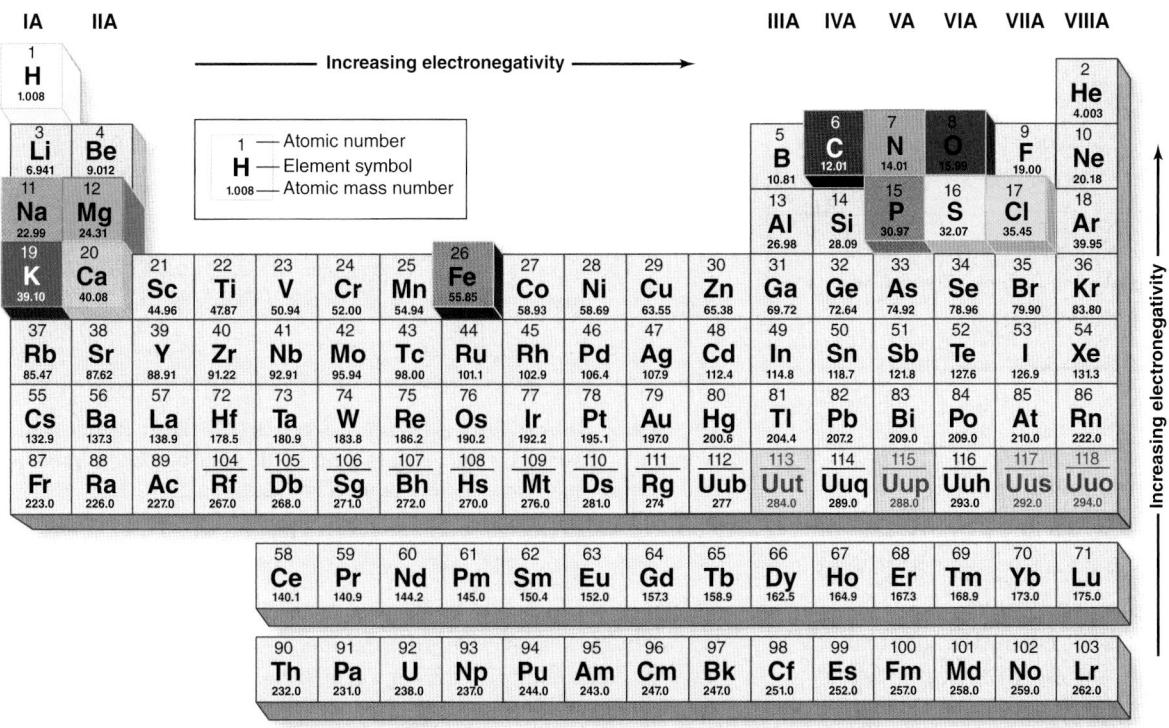

IA	IIA
1 **H** 1.008	
3 **Li** 6.941	4 **Be** 9.012
11 **Na** 22.99	12 **Mg** 24.31
19 **K** 39.10	20 **Ca** 40.08

——— Increasing electronegativity ———

1 —— Atomic number
H —— Element symbol
1.008 —— Atomic mass number

IIIA	IVA	VA	VIA	VIIA	VIIIA

| | | | | | 2 **He** 4.003 |

| 5 **B** 10.81 | 6 **C** 12.01 | 7 **N** 14.01 | 8 **O** 15.99 | 9 **F** 19.00 | 10 **Ne** 20.18 |
| 13 **Al** 26.98 | 14 **Si** 28.09 | 15 **P** 30.97 | 16 **S** 32.07 | 17 **Cl** 35.45 | 18 **Ar** 39.95 |

21 **Sc** 44.96	22 **Ti** 47.87	23 **V** 50.94	24 **Cr** 52.00	25 **Mn** 54.94	26 **Fe** 55.85	27 **Co** 58.93	28 **Ni** 58.69	29 **Cu** 63.55	30 **Zn** 65.38	31 **Ga** 69.72	32 **Ge** 72.64	33 **As** 74.92	34 **Se** 78.96	35 **Br** 79.90	36 **Kr** 83.80
37 **Rb** 85.47	38 **Sr** 87.62	39 **Y** 88.91	40 **Zr** 91.22	41 **Nb** 92.91	42 **Mo** 95.94	43 **Tc** 98.00	44 **Ru** 101.1	45 **Rh** 102.9	46 **Pd** 106.4	47 **Ag** 107.9	48 **Cd** 112.4	49 **In** 114.8	50 **Sn** 118.7	51 **Sb** 121.8	52 **Te** 127.6	53 **I** 126.9	54 **Xe** 131.3
55 **Cs** 132.9	56 **Ba** 137.3	57 **La** 138.9	72 **Hf** 178.5	73 **Ta** 180.9	74 **W** 183.8	75 **Re** 186.2	76 **Os** 190.2	77 **Ir** 192.2	78 **Pt** 195.1	79 **Au** 197.0	80 **Hg** 200.6	81 **Tl** 204.4	82 **Pb** 207.2	83 **Bi** 209.0	84 **Po** 209.0	85 **At** 210.0	86 **Rn** 222.0
87 **Fr** 223.0	88 **Ra** 226.0	89 **Ac** 227.0	104 **Rf** 267.0	105 **Db** 268.0	106 **Sg** 271.0	107 **Bh** 272.0	108 **Hs** 270.0	109 **Mt** 276.0	110 **Ds** 281.0	111 **Rg** 274	112 **Uub** 277	113 **Uut** 284.0	114 **Uuq** 289.0	115 **Uup** 288.0	116 **Uuh** 293.0	117 **Uus** 292.0	118 **Uuo** 294.0

| 58 **Ce** 140.1 | 59 **Pr** 140.9 | 60 **Nd** 144.2 | 61 **Pm** 145.0 | 62 **Sm** 150.4 | 63 **Eu** 152.0 | 64 **Gd** 157.3 | 65 **Tb** 158.9 | 66 **Dy** 162.5 | 67 **Ho** 164.9 | 68 **Er** 167.3 | 69 **Tm** 168.9 | 70 **Yb** 173.0 | 71 **Lu** 175.0 |
| 90 **Th** 232.0 | 91 **Pa** 231.0 | 92 **U** 238.0 | 93 **Np** 237.0 | 94 **Pu** 244.0 | 95 **Am** 243.0 | 96 **Cm** 247.0 | 97 **Bk** 247.0 | 98 **Cf** 251.0 | 99 **Es** 252.0 | 100 **Fm** 257.0 | 101 **Md** 258.0 | 102 **No** 259.0 | 103 **Lr** 262.0 |

Increasing electronegativity

(a)

Most Common Elements of the Human Body				
Major elements (collectively compose more than 98% of body weight)			**Minor elements** (collectively compose less than 1% of body weight)	
Symbol		% Body weight	Symbol	% Body weight
O	Oxygen	65.0	S Sulfur	0.25
C	Carbon	18.0	K Potassium	0.20
H	Hydrogen	10.0	Na Sodium	0.15
N	Nitrogen	3.0	Cl Chlorine	0.15
Ca	Calcium	1.5	Mg Magnesium	0.05
P	Phosphorus	1.0	Fe Iron	0.006

(b)

Figure 2.1 Periodic Table of Elements. (*a*) The periodic table includes all of the elements arranged in chart form. The grayed out boxes at the bottom of the table are for proposed elements that as of yet do not have sufficient evidence to support their official inclusion as elements. (*b*) Twelve of the elements that compose the human body are present in greater than trace amounts. These twelve elements are color coded in both (*a*) and (*b*), and this color scheme is used throughout the book to represent these elements.

periodic table. Note that elements are arranged by the atomic number in consecutive order within rows (figure 2.1*a*). The atomic number is designated as a subscript at the left of the chemical symbol when it is written. For example, $_1$H shows that the nucleus of a hydrogen atom has one proton, and $_6$C indicates that the carbon nucleus has six protons.

The **atomic mass** (or *atomic weight*) indicates the mass of both protons and neutrons in the atomic nucleus, and it reflects the "heaviness" of an element's atoms relative to atoms of other elements. (The electrons are not included because of their relatively small mass.) It is shown below the element's symbol on the periodic table. The average atomic mass is rounded to the nearest whole number and it is designated by a superscript to the left of the chemical symbol when it is written. For example, a sodium atom, with an atomic number of 11 and an average atomic mass of 22.99, is designated as $_{11}^{23}$Na.

WHAT DO YOU THINK?

1 How would the chemical shorthand for oxygen be written?

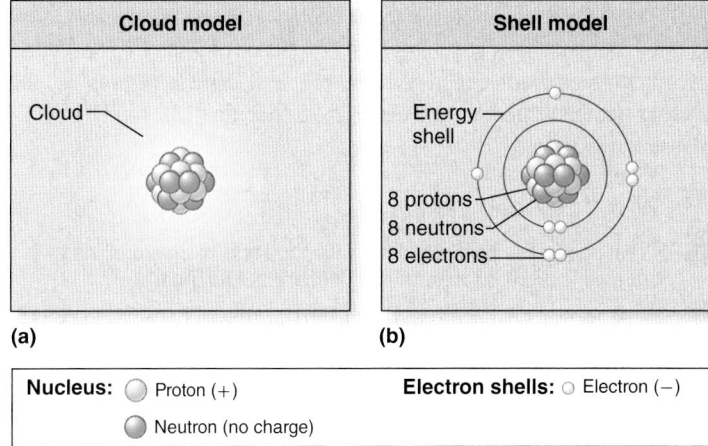

Cloud model	Shell model
Cloud	Energy shell, 8 protons, 8 neutrons, 8 electrons

(a) **(b)**

Nucleus: ○ Proton (+) ● Neutron (no charge)   **Electron shells:** ○ Electron (−)

Figure 2.2 General Atomic Structure. Oxygen atoms are depicted with protons and neutrons in the nucleus. Electrons are shown in both a cloud model (*a*) and an energy shell model (*b*). **AP|R**

INTEGRATE

LEARNING STRATEGY

To diagram an atom, the number of protons, neutrons, and electrons must be known. How these are determined can be summarized as follows:

- **Proton number (p)** = Atomic number
- **Neutron number (n)** = Atomic mass (p + n) – atomic number (p)
- **Electron number (e)** = Proton number (p)

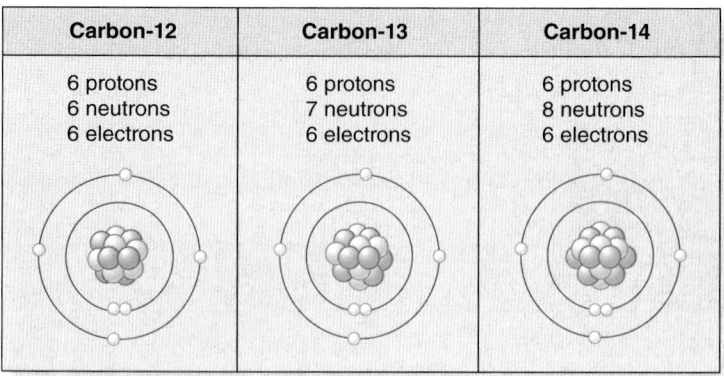

Carbon-12	Carbon-13	Carbon-14
6 protons 6 neutrons 6 electrons	6 protons 7 neutrons 6 electrons	6 protons 8 neutrons 6 electrons

Figure 2.3 The Three Most Common Carbon Isotopes. Notice that all three carbon isotopes contain the same number of protons and electrons but have different numbers of neutrons.

Determining the Number of Subatomic Particles

The number of each type of subatomic particle is determined as follows:

- The **number of protons** is the atomic number—thus, carbon has six protons and oxygen has eight protons.
- The **number of neutrons** can be determined by subtracting the atomic number (number of protons) from the atomic mass (protons and neutrons). For example, to calculate the number of neutrons in sodium ($_{11}^{23}Na$), you would take the total of 23 protons and neutrons, and subtract the number of protons = 11; thus, sodium has 12 neutrons (23 − 11 = 12).
- The **number of electrons** in an atom is determined indirectly by the atomic number. This is possible because all atoms are *neutral*. The number of negatively charged electrons must equal the number of positively charged protons for an atom to be neutral. Each electron has a charge of −1 and counters the +1 charge of a proton in the nucleus—therefore, the atom has zero charge, meaning it has no net charge. Atoms that have lost or gained electron(s) do not have the same number of protons and electrons and are referred to as ions (described later).

Diagramming Atomic Structures

An atom is represented as having shells of electrons that surround the atomic nucleus. The electrons within a shell have a given energy level. Each shell can hold only a limited number of electrons. The innermost shell may hold up to two electrons and the second shell up to eight electrons. All subsequent shells have a capacity of eight electrons but may also house more than eight electrons (the conditions for when the third and higher shells hold more than eight electrons are beyond our purposes here). The electron shells closest to the nucleus must be filled prior to filling any potential shells at some greater distance from the nucleus when diagramming an atom. Figure 2.2 shows the placement of protons and neutrons within the nucleus and electrons in specific shells for the atomic structure of oxygen.

 WHAT DID YOU LEARN?

1 What subatomic particles determine the mass of an atom? What subatomic particles determine the charge of an atom?

2 Diagram the atomic structure of chlorine—atomic number is 17 and the mass number is 35.

2.1b Isotopes

 LEARNING OBJECTIVES

5. Describe an isotope.
6. Explain how radioisotopes differ from other types of isotopes.

Elements in nature usually occur as a mixture of isotopes. **Isotopes** (ī′sō-tōp; *iso* = same) are different atoms of the same element that have the same number of protons and electrons but differ in the number of neutrons. Isotopes exhibit essentially identical chemical characteristics but have different atomic masses.

For example, carbon exists in three isotopes: carbon-12, carbon-13, and carbon-14 (**figure 2.3**). All isotopes of carbon have six protons in their nuclei—however, carbon-12 has six neutrons, carbon-13 has seven neutrons, and carbon-14 has eight neutrons. Generally, one isotope is usually more common than the others. Carbon-12 is the most common isotope for carbon.

The weighted average of the atomic mass for all isotopes of an element is the **average atomic mass.** This value is included on the periodic table below each element. The average atomic mass for carbon, for example, is 12.01 amu.

Some isotopes are referred to as **radioisotopes.** They are unstable because they contain an excess number of neutrons. Radioisotopes usually lose nuclear components in the form of high-energy radiation that includes alpha particles, beta particles, or gamma rays as they decay or break down into a more stable isotope. The time it takes for 50% of the radioisotope to become stable is its **physical half-life.** This time may vary from a few hours to thousands of years. Radioisotopes produced in a nuclear power plant, for example, have a half-life of at least 10,000 years. In comparison, the time required for half of the radioactive material (e.g., from a medical test using radioactive contrast material) to be eliminated from the body is the **biological half-life.** Biological half-life is also applied to substances that are nonradioactive, such as hormones or drugs (see section 17.4b).

 WHAT DID YOU LEARN?

3 Do isotopes represent the same element? Do they have the same number of protons, neutrons, or electrons? Describe a radioisotope.

2.1c Chemical Stability and the Octet Rule

LEARNING OBJECTIVES

7. Describe how elements are organized in the periodic table based on the valence electron number.
8. State the octet rule.

Recall that the periodic table is organized into rows based on atomic number. It is also organized into columns based on the number of electrons in the outer shell, or what is referred to as the **valence shell.**

This organization is presented in **figure 2.4**, showing the atomic structure of elements 1 to 20 arranged as they appear in a periodic table. Column IA shows hydrogen, lithium, sodium, and potassium: All the atomic structures of these elements contain one electron in their outer shell. Each consecutive column thereafter (columns IIA–VIIIA) has one additional electron in its outer shell for each element in that column. This organization allows us to make predictions about the chemical characteristics of a given element simply by its location in the periodic table.

Notice that the elements in column VIIIA each have a valence shell that is full or complete. These elements (only helium, neon, and argon are shown in figure 2.4) have an outermost shell housing eight electrons, except for helium, which has only two electrons. A complete outer shell results in chemical stability. These stable atoms are chemically inert and do not combine with other elements. The atoms in column VIIIA are referred to as *noble gases* because they do not react with the "common" elements in the rest of the periodic table.

Examine the atomic structure of the other elements in figure 2.4 and note that they lack a full outer shell with eight electrons. As we will see, these elements tend to lose, gain, or share electrons to obtain a complete outer shell. Chemists call this tendency the **octet rule:** Atoms obtain an outer shell with eight electrons and gain chemical stability through the loss, gain, or sharing of electrons. Not all elements follow the octet rule, but it is accurate for our purposes here. Ions are formed from the loss or gain of electrons, and covalently bonded molecules are formed by the sharing of electrons. We describe the formation of ions and covalently bonded molecules in the next two sections.

WHAT DID YOU LEARN?

4 What is the relationship of the octet rule and chemical stability?

INTEGRATE

CLINICAL VIEW
Medical Imaging of the Thyroid Gland Using Iodine Radioisotopes

Radioisotopes including ^{14}C, ^{32}P, and ^{123}I have been extensively used in basic biology research efforts, medical investigations, and diagnoses. Radioisotopes introduced into the body during medical procedures are used by cells in a manner similar to a substance that is a nonradioisotope (e.g., ^{12}C, ^{31}P, and ^{127}I). Thus, products of metabolic reactions that include these elements may be traced and quantified by the high-energy radiation the radioisotopes emit. Individuals who work with radioactive materials must protect themselves from the potentially damaging high-energy radiation that may harm living cells and produce genetic damage. The image below is of a thyroid gland that has taken up radioactive ^{123}I. The entire thyroid appears whitish. The darker colored area is a benign nodule that is less metabolically active than the surrounding thyroid tissue.

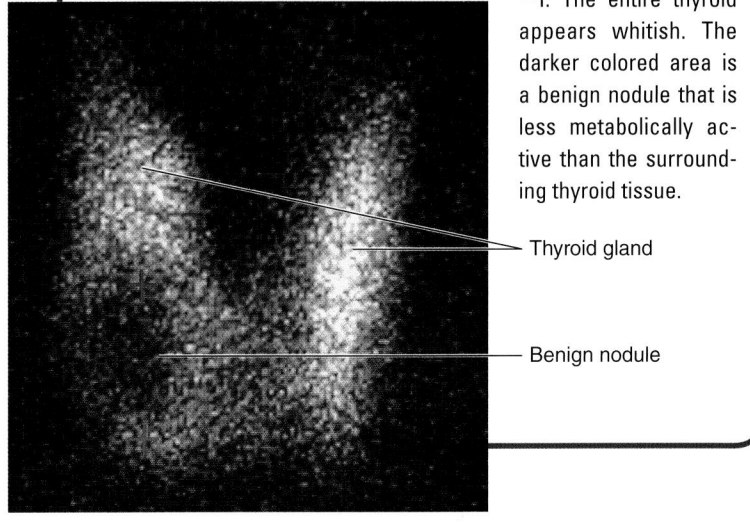

Thyroid gland

Benign nodule

Number of valence electrons

1	2	3	4	5	6	7	8
IA	IIA	IIIA	IVA	VA	VIA	VIIA	VIIIA

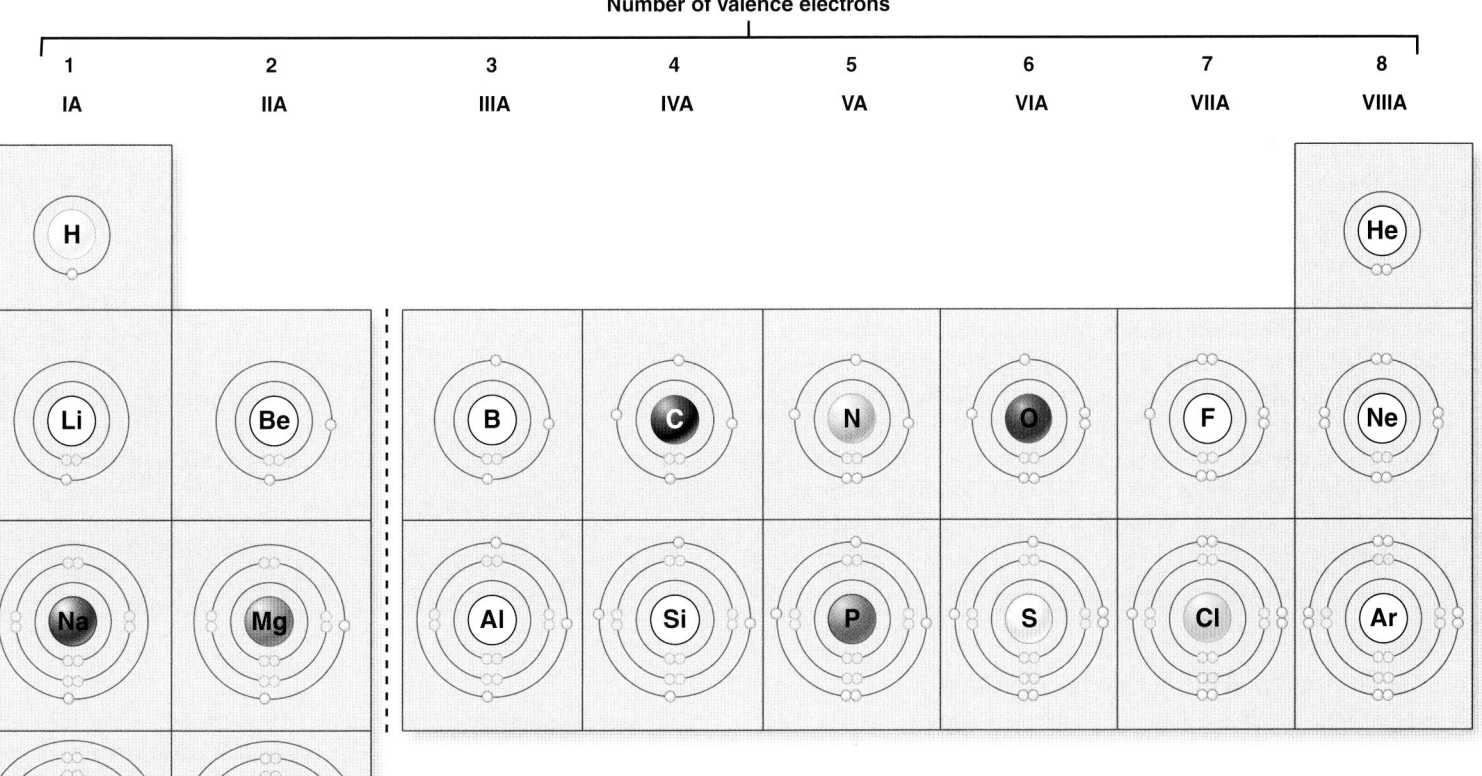

Figure 2.4 Organization of the Periodic Table Based on Valence Shell Electrons. The atomic structures of elements 1 to 20 of the periodic table are depicted, showing the nucleus and electrons in energy shells. The valence shell electrons are shown in yellow. Valence shell number increases by one in each column as you move from left to right from columns IA through VIIIA.

2.2 Ions and Ionic Compounds

The body consists mostly of chemical compounds. **Chemical compounds** are stable associations between two or more elements combined in a fixed ratio. These associations are classified as either ionic compounds or molecular compounds. Here we describe ionic compounds, which are structures composed of ions that are held together in a lattice by ionic bonds. We first define ions and list some common ions of the human body, then explain how ions are formed and their charge determined, and finally we describe ionic bonds and the electrostatic interaction between ions within an ionic compound. Molecular compounds are described in section 2.3.

2.2a Ions

LEARNING OBJECTIVES

1. Define an ion.
2. List some common ions in the body.
3. Differentiate between cations and anions.
4. Describe how charges are assigned to ions.

Ions are atoms or groups of atoms with either a positive charge or a negative charge. They are produced from the loss or gain of one or more electrons, respectively. **Table 2.1** lists the most common ions in the human body along with their significant physiologic functions—notice the diversity of body structures (e.g., nerves, muscles, liver, stomach) that require specific ions to function normally. Maintaining homeostatic blood concentration of each of these ions is critical to health because it preserves or sustains what is called electrolyte balance. For example, consider the importance of ion levels with respect to the following: A relatively small quantity of K^+ is used in sports drinks (e.g., Gatorade) to replace the K^+ lost in sweat, whereas a relatively large dose is used by some states for lethal injection.

INTEGRATE

CONCEPT CONNECTION

Ions, including the common ions in the human body, function as electrolytes (substances that when dissolved in water form cations and anions and can conduct an electric current; see section 25.3a). Optimal health requires **electrolyte balance,** which is the state of maintaining homeostatic blood levels of the different ions (see section 25.3b). An **electrolyte imbalance** occurs when the blood concentration of an electrolyte becomes either too high or too low. These changes, in some cases, may be fatal. Good nursing care includes knowing the causes, effects, and symptoms associated with electrolyte imbalances of the common ions (see table 25.3), and understanding how to treat them.

Table 2.1	Common Ions in the Human Body and Their Physiologic Significance

COMMON CATIONS (POSITIVELY CHARGED IONS)

Cation	Structure	Physiologic Significance
Sodium ion	Na^+	• Most common extracellular cation • Participant in conducting electrical signals in nerves and muscle • Most important in osmotic movement of water • Sodium gradient involved in cotransport of other substances across a plasma membrane
Potassium ion	K^+	• Most common intracellular cation • Participant in conducting electrical signals in nerves and muscle • Role in glycogen storage in liver and muscle • Function in pH balance
Calcium ion	Ca^{2+}	• Hardness of bone and teeth • Muscle contraction • Exocytosis (including release of neurotransmitter) • Blood clotting • Second messenger in hormonal stimulation of cells
Magnesium ion	Mg^{2+}	• Required for ATP production
Hydrogen ion	H^+	• Concentration determines pH of blood and other fluids of the body

COMMON ANIONS (NEGATIVELY CHARGED IONS)

Anion	Structure	Physiologic Significance
Chloride ion	Cl^-	• Alters nerve cell responsiveness to stimulation • Component of stomach acid (HCl) • Chloride shift in erythrocytes
Bicarbonate ion	HCO_3^- 	• Conversion of CO_2 gas to HCO_3^-, which is transported in the blood • Buffering of pH in blood
Phosphate ion	PO_4^{3-}	• As $Ca_3(PO_4)_2$, it hardens bone and teeth • Component of phospholipids (membranes) • Component of nucleotides, including ATP and nucleic acids (DNA and RNA) • Most common intracellular anion • Intracellular buffer

Losing Electrons and the Formation of Cations

A sodium atom, found in column IA of the periodic table, is a good example of an element that can reach stability by losing an electron. The atomic structure of sodium (**figure 2.5a**) has one electron in its outer shell. By giving up or donating that electron, sodium now satisfies the octet rule and becomes stable. But is the structure still neutral? Recall that "neutral" means that the number of positively charged protons is equal to the number of negatively charged electrons. Because an electron has been donated by the sodium atom, it now has 11 protons but only 10 electrons, and the charge is calculated as follows: 11(+) and 10 (–) = +1. Ions with a positive charge are called **cations.** Thus, sodium ion is a cation with a +1 charge and is designated as Na^+. Other examples of common cations, as listed in table 2.1, include K^+, Ca^{2+}, Mg^{2+}, and H^+.

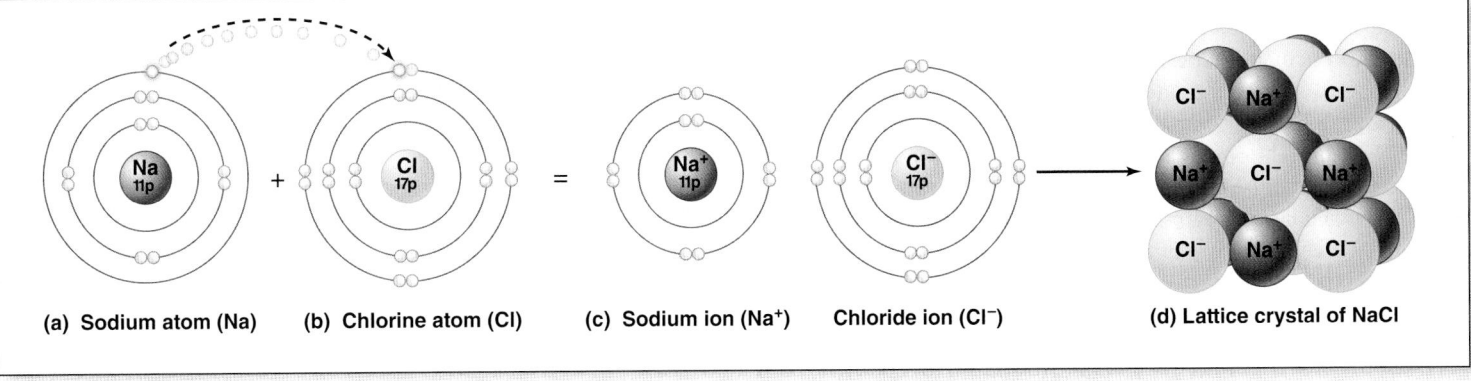

(a) Sodium atom (Na) (b) Chlorine atom (Cl) (c) Sodium ion (Na⁺) Chloride ion (Cl⁻) (d) Lattice crystal of NaCl

Figure 2.5 Formation of an Ionic Bond Involving Sodium and Chloride. (*a*) A sodium atom donates an electron to a chlorine atom (*b*). (*c*) The loss of an electron from a sodium atom results in the formation of a positively charged sodium ion (Na⁺), and the gain of an electron by chlorine results in the formation of a negatively charged chloride ion (Cl⁻). (*d*) The NaCl ionic compound (a salt) is composed of a lattice crystal formed by the electrostatic attractions between Na⁺ and Cl⁻. AP|R

Gaining Electrons and the Formation of Anions

A chlorine atom is a good example of an element that can reach stability by gaining an electron (figure 2.5*b*). The atomic structure of chlorine contains seven electrons in its outer shell, and by gaining one electron, stability is reached. The structure formed is referred to as a chloride ion. Because an electron has been gained, the chloride ion has 17 protons and now has 18 electrons, and the charge is calculated as follows: 17(+) and 18(−) = −1, written as Cl⁻. Negatively charged ions are called **anions.** Some simple and complex anions are included in table 2.1.

Two of the common anions listed in table 2.1, bicarbonate ion (HCO_3^-) and phosphate ion (PO_4^{3-}) are composed of more than one atom and are referred to as **polyatomic ions.** These form when one or more atoms in a structure composed of many atoms has lost or gained electrons.

General Rules for Assigning Charges

A general rule helps you determine which atoms gain or lose electrons and the charge that they develop (figure 2.4). Atoms with one, two, or three electrons in the outer shell generally donate electrons and become positively charged cations. The amount of charge depends upon the number of electrons donated—namely, one, two, or three. For example, because calcium has two electrons in its outer shell, it reaches stability by losing two negatively charged electrons, develops a charge of +2, and in ionic form is written as Ca^{2+}.

In comparison, atoms with five, six, or seven electrons in the outer shell tend to gain electrons and become negatively charged anions. The amount of charge depends upon the number of electrons gained to meet the octet rule criterion—namely, three, two, or one. An atom with seven electrons in the outer shell, such as chlorine, reaches stability and becomes chloride (Cl⁻) when it gains one electron and develops a −1 charge.

Note that carbon (atomic number 6), with four electrons in its outer shell, neither gains nor loses electrons to become an ion. It only forms covalent bonds in molecules, as described in section 2.3.

WHAT DO YOU THINK?

2 Can you predict the specific charge for magnesium ion (atomic number 12) based on its position in the periodic table?

Using the Periodic Table to Assign Charges

The periodic table (see figure 2.1*a*) may be used to quickly assess whether an atom will become a cation or anion and the amount of its specific charge. Usually, elements on the left side (termed the

"metallic" side) of the periodic table, and in column IIIA, tend to lose electrons. The specific positive charge is dependent upon the position of the element in the periodic table: Group IA = +1, group IIA = +2, and group IIIA = +3. In contrast, elements on the right side (called the "nonmetallic" side) of the periodic table (columns VA–VIIA) tend to gain electrons. The specific amount of negative charge is as follows: Group VA = −3, group VIA = −2, and group VIIA = −1. Sometimes there are anomalies, such as iron (Fe), which is the metal in the hemoglobin molecule within red blood cells (erythrocytes). It can form more than one type of ion, either a ferrous (Fe^{2+}) or ferric (Fe^{3+}) ion.

WHAT DID YOU LEARN?

5 List the common cations and anions of the human body, including their name and symbol.

6 On the periodic table (see figure 2.1), highlight the elements that form the common ions of the human body (do not include the polyatomic ions).

7 Explain how and why ions form based on the octet rule.

2.2b Ionic Bonds

LEARNING OBJECTIVES

5. Define an ionic bond.
6. Describe an ionic compound of NaCl.
7. List other examples of ionic compounds.

Positively charged cations and negatively charged anions may bind together by electrostatic interactions called **ionic bonds.** The structures

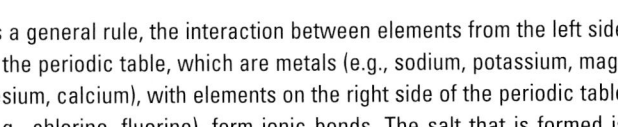

LEARNING STRATEGY

As a general rule, the interaction between elements from the left side of the periodic table, which are metals (e.g., sodium, potassium, magnesium, calcium), with elements on the right side of the periodic table (e.g., chlorine, fluorine), form ionic bonds. The salt that is formed is composed of ions held together in a lattice arrangement by ionic bonds.

formed are **salts.** A classic example involves the formation of common table salt from metallic atoms of sodium and nonmetallic atoms of chlorine. Each sodium atom loses one outer shell electron to a chlorine atom. The sodium atom then becomes a sodium ion (Na^+), and the chlorine atom becomes a chloride ion (Cl^-). The oppositely charged Na^+ and Cl^- ions are held together by ionic bonds in a precise, lattice crystal structure composing an **ionic compound** (figure 2.5d). NaCl is the smallest repeating structure of this ionic compound and represents its chemical formula (the chemical constituents in a compound and their ratios).

Consider also the ionic compound magnesium chloride ($MgCl_2$). Notice that the chemical formula contains one magnesium and two chloride ions. This is because magnesium, which is in column IIA on the periodic table, has two electrons in its outer shell. It becomes stable by losing one electron to each of the two chlorine atoms.

Other examples of ionic compounds include those involving polyatomic anions, sodium bicarbonate ($NaHCO_3$), and the most common ionic compound in the body, calcium phosphate, $Ca_3(PO_4)_2$, which helps harden bones and teeth. Notice that each of these examples is a combination of a common cation and common anion (table 2.1).

 WHAT DID YOU LEARN?

8 Could an ionic bond form between two cations or between two anions? Explain.

2.3 Covalent Bonding, Molecules, and Molecular Compounds

Instead of losing or gaining electrons to form ionic compounds that are arranged in a lattice, atoms also have the possibility of reaching chemical stability by sharing electrons. The sharing of electrons between atoms results in a **covalently bonded molecule.** Most molecules are composed of two or more different elements and are called **molecular compounds.** Thus, molecules of carbon dioxide (CO_2) and water (H_2O) are molecular compounds. However, molecules composed of only one element are not, such as molecular oxygen (O_2) and molecular hydrogen (H_2).

Here we describe covalent bonds, the different types of covalently bonded molecules, and interactions between covalently bonded molecules, after first discussing how molecules are represented with molecular and structural formulas.

2.3a Chemical Formulas: Molecular and Structural

 **LEARNING OBJECTIVES**

1. Define a molecular formula.
2. Describe a structural formula, and explain its use in differentiating isomers.

Molecular Formula

The **molecular formula** is the number and types of atoms composing a molecule. An example of a molecular formula would be carbonic acid (H_2CO_3), where the molecule contains two hydrogen atoms, one carbon atom, and three oxygen atoms.

Structural Formula

The **structural formula** of a molecule is complementary to its molecular formula and exhibits not only the numbers and types of atoms but also their arrangements within the molecule. Atoms are always arranged in a defined manner within each molecule. The molecular formula for carbon dioxide (CO_2) thus is complemented by its structural formula (O=C=O).

Structural formulas provide a means for differentiating **isomers,** which are molecules composed of the same number and types of elements but arranged differently in space. Two important isomers in humans are glucose and galactose **(figure 2.6)**. These sugar molecules both have a molecular formula of $C_6H_{12}O_6$, indicating that each molecule has six carbon, twelve hydrogen, and six oxygen atoms. However, the atoms are arranged differently in space as shown in their structural formulas. Notice the different arrangement of atoms on carbon four in the six-sided ring structure of glucose and galactose. (Sugars in a ring-structure have carbon atoms that are numbered beginning with the carbon to the right of the oxygen as carbon one, and then are numbered consecutively clockwise around the ring.) Although fructose is a five-sided ring, it is also an isomer of glucose and galactose because it contains the same number and types of elements, but it has a different structural arrangement.

Isomers may have very different properties from one another— therefore, structural formulas are an important piece of chemical information.

 WHAT DID YOU LEARN?

9 What information about a molecule is gained by a structural formula? How does this differ from a molecular formula?

10 What is an isomer?

	Glucose	Galactose	Fructose
Molecular formula	$(C_6H_{12}O_6)$	$(C_6H_{12}O_6)$	$(C_6H_{12}O_6)$
Structural formula	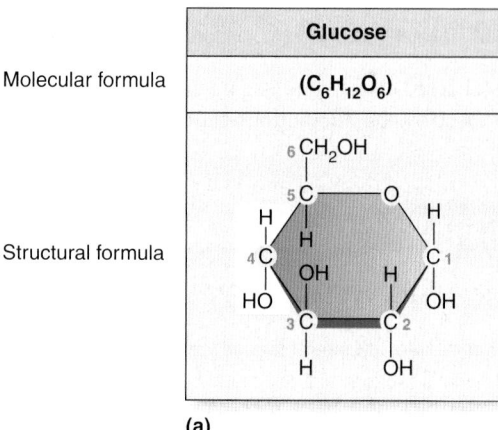	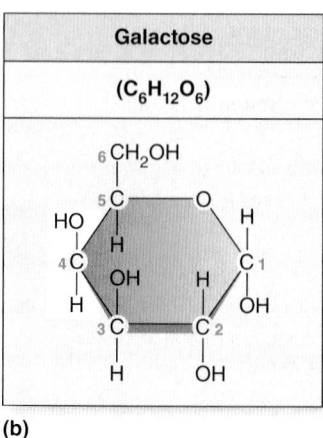	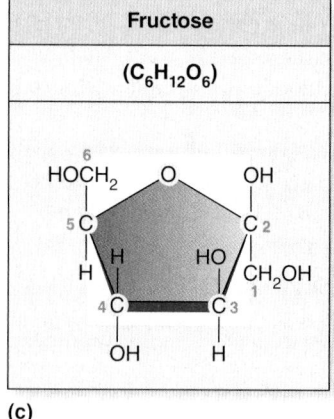
	(a)	(b)	(c)

Figure 2.6 Isomers. Isomers contain the same number and types of elements that are arranged differently in space. (*a*) Glucose, (*b*) galactose, and (*c*) fructose are isomers that share the same chemical formulas ($C_6H_{12}O_6$) but have different structural formulas (highlighted in yellow).

2.3b Covalent Bonds

LEARNING OBJECTIVES

3. Describe a covalent bond and explain its formation based on the octet rule.

4. List the four most common elements in the human body.

5. Distinguish between single, double, and triple covalent bonds.

6. Explain polar and nonpolar covalent bonds.

The bond that is formed when atoms *share* electrons is a **covalent bond.** A covalent bond forms when both atoms require electrons to become stable. This takes place when the participating atoms that form the chemical bond have four, five, six, or seven electrons in the outer shell. View figure 2.4 and note that this applies to the nonmetal elements on the right side of the periodic table. (The exception is hydrogen because only two electrons are needed to complete its outermost electron shell.)

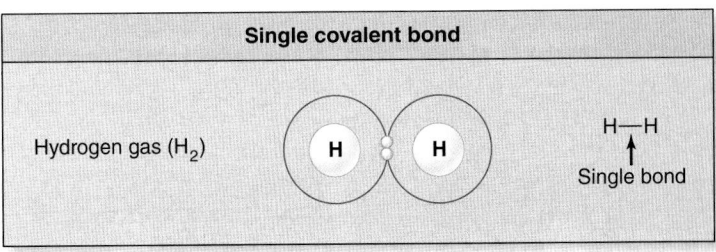
(a)

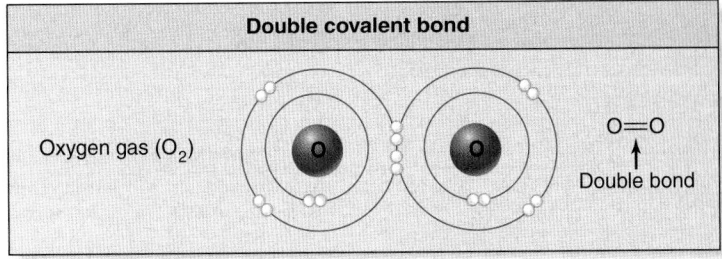
(b)

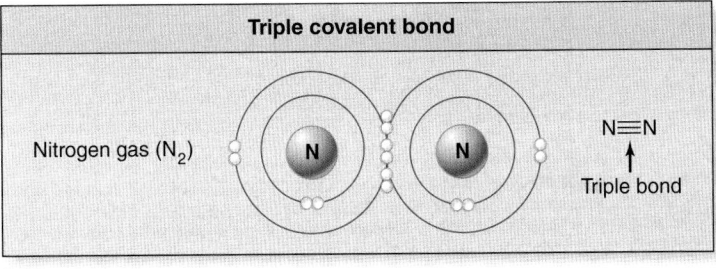
(c)

Figure 2.7 Single, Double, and Triple Covalent Bonds.
Covalent bonds involve the sharing of electrons. (*a*) One pair of electrons is shared in a single covalent bond, (*b*) two pairs of electrons are shared in a double covalent bond, and (*c*) three pairs of electrons are shared in a triple covalent bond.

INTEGRATE

LEARNING STRATEGY

The interaction between elements from the right side of the periodic table (e.g., carbon, oxygen, nitrogen), or a hydrogen atom typically form covalent bonds. The resulting structure is a molecule.

The four most common elements of the human body that form covalent bonds are hydrogen (H), oxygen (O), nitrogen (N), and carbon (C). These four elements account for over 96% of the body's weight. The simplest example of covalent bond formation occurs when hydrogen gas is formed from two hydrogen atoms. The covalent bond formation between two hydrogen atoms fills the outer shell of each atom because each hydrogen atom shares its single electron (and only two electrons are required for the first shell to be complete).

The Number of Bonds an Atom Can Form

Although hydrogen can share only one pair of electrons and thus produce one covalent bond to become stable, some elements are able to share more than one pair. The number of covalent bonds formed by an atom may be determined by examining the number of electrons needed to complete the outer shell (see figure 2.4). The atomic structure of the four most common elements that compose molecules show that hydrogen needs one electron, oxygen needs two electrons, nitrogen needs three electrons, and carbon needs four electrons. Thus, hydrogen can form one covalent bond, oxygen two, nitrogen three, and carbon can form four covalent bonds.

INTEGRATE

LEARNING STRATEGY

The number of bonds formed by the four most common elements can be remembered with the acronym HONC: hydrogen = 1, oxygen = 2, nitrogen = 3, and carbon = 4.

Single, Double, and Triple Covalent Bonds

Atoms of elements that can form more than one covalent bond may do so through combinations of single, double, or triple covalent bonds (**figure 2.7**). A **single covalent bond** is one pair of electrons shared between two atoms. The bond just described between two hydrogen

atoms is a single covalent bond. A **double covalent bond** involves the sharing of two pairs of electrons between two atoms. An example is the double covalent bond between two oxygen atoms. The sharing of two pairs of electrons between oxygen atoms is necessary to achieve stability because each oxygen atom has only six electrons in its outer shell and needs eight electrons to satisfy the octet rule. Three pairs of electrons are shared between atoms in some molecules, forming a **triple covalent bond.** A prime example is the triple covalent bond between two nitrogen atoms. The order of stability (i.e., the more energy required to break the bonds) is from least to greatest: single covalent bond, double covalent bond, triple covalent bond.

An atom of a specific element might share electrons in a variety of ways to satisfy the octet rule. For example, a carbon atom contains four electrons in the outer electron shell and thus needs four electrons to satisfy the octet rule. These four electrons can be obtained in a number of ways to form different types of molecules (**figure 2.8**).

Carbon Skeleton Formation

Numerous carbon atoms are sometimes bonded together to form a "carbon skeleton." Three possible arrangements may occur: a straight chain, branched chain, or a ring (**figure 2.9**). Notice that in a structural formula of a carbon chain or ring skeleton, the letter "C" is often not included in the structural formula with the understanding that a carbon atom lies where lines meet at an angle. In addition, if a carbon atom does not have its required four bonds shown in the structural formula, the additional atoms are assumed to be hydrogen atoms.

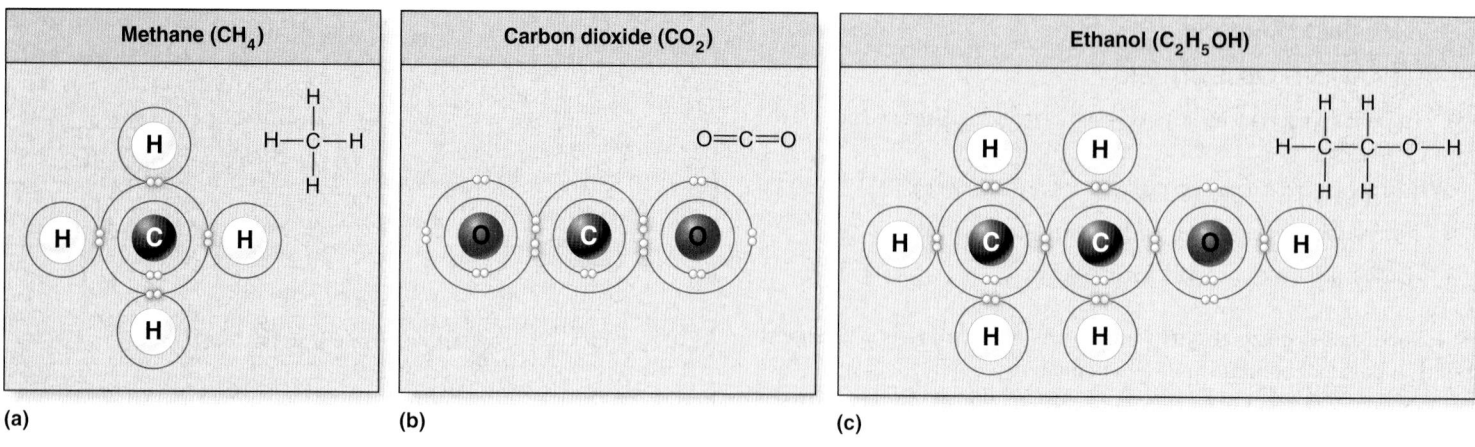

Figure 2.8 Carbon-Containing Molecules. Carbon can form four covalent bonds and may do so in many ways. For example, carbon can form (*a*) four single covalent bonds with four hydrogen atoms in methane; (*b*) two double bonds with oxygen atoms in carbon dioxide; and (*c*) four single covalent bonds in a more complex structure like ethanol.

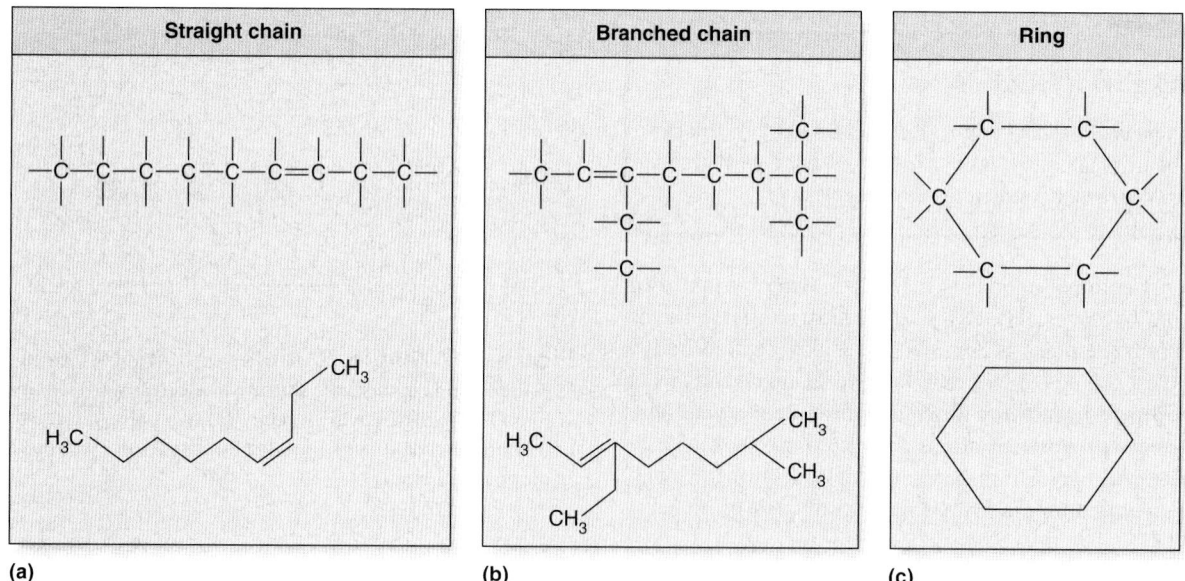

Figure 2.9 Carbon Skeleton. A carbon skeleton has three major types of arrangements including (*a*) a straight chain, (*b*) a branched chain, and (*c*) a ring. Remember the convention that a carbon is present where the lines meet at an angle.

Nonpolar and Polar Covalent Bonds

Atoms share electrons in a covalent bond either equally or unequally between the atoms. How they share is determined by the relative attraction each atom has for electrons, a concept referred to as **electronegativity.** Because two atoms of the same element, such as two hydrogen atoms, two oxygen atoms, or two carbon atoms, have equal attraction for electrons, they share the electrons equally. The resultant bond is a **nonpolar covalent bond.**

Different types of atoms have varying degrees of electronegativity, or attraction for electrons, and thus may share the electrons unequally. The resultant bond is a **polar covalent bond** (an important exception will be described shortly). As a general rule, electronegativity increases both from left to right across a row of the periodic table and from bottom to top in a column (see figure 2.1*a*). These trends reflect that, for elements across a specific row, a greater number of protons in the nucleus of the atom are "pulling" on the electrons. For elements in a column, the elements closer to the top of the periodic table have electrons in shells closer to the nucleus (see figure 2.4). Thus, electronegativity is determined both by the number

of protons in the nucleus and the proximity of the valence electron shells to the nucleus.

Consequently, for the four most common elements composing living organisms, the order of electronegativity from least to greatest is as follows: hydrogen < carbon < nitrogen < oxygen. Therefore, of these four elements, oxygen has the greatest attraction for electrons, and hydrogen has the least. The electrons, in each case, spend a greater amount of time orbiting the nucleus of the more electronegative atom.

Because electrons have a negative charge, the more electronegative atom develops a partial negative charge, and the less electronegative atom develops a partial positive charge. Partial charges are written using the Greek letter delta (δ) followed by a superscript + or – to designate the relative charge. For example, a polar covalent bond formed between oxygen and hydrogen would be assigned and written as $^{\delta-}O\!-\!H^{\delta+}$. Note that the term **polar** (or *dipole*) refers to the "poles" of partial electrical charges, which are analogous to poles of a magnet.

Polar bonds have varying degrees of unequal sharing of electrons. Thus, polar bonds are along a continuum with ionic bonds (which involve a complete donation of electrons) on one end and nonpolar bonds (with equal sharing of electrons) on the other.

One important exception exists to the rule that a polar bond generally forms between two different atoms. The exception is carbon bonding with hydrogen. Because the electronegativity difference between carbon and hydrogen is relatively small, a covalent bond produced between carbon and hydrogen (C—H) has approximately equal sharing of electrons and forms an essentially *nonpolar* covalent bond between them.

WHAT DID YOU LEARN?

11 Explain covalent bond formation in terms of chemical stability.

12 Assign the partial charges between nitrogen and hydrogen (N—H) in a polar covalent bond.

13 Why are some covalent bonds nonpolar and others polar? Identify the exception to the rule that polar covalent bonds are formed between two different types of atoms.

2.3c Nonpolar, Polar, and Amphipathic Molecules

LEARNING OBJECTIVES

7. Describe the difference between a nonpolar molecule and a polar molecule.

8. Define an amphipathic molecule.

Covalent bonds may be nonpolar or polar. To determine whether an entire molecule is nonpolar or polar, the molecule as a whole is evaluated to determine the prevalence (and relative strength) of its nonpolar and polar bonds. The most important concept to remember is that **nonpolar molecules** contain nonpolar covalent bonds, which are bonds formed between the same elements (e.g., C—C, O—O), by C—H bonds, or both. Oxygen (O_2) and triglyceride (fat) molecules are examples of nonpolar molecules **(figure 2.10)**.

In comparison, **polar molecules** contain polar covalent bonds. These molecules contain different elements that are bonded together, such O—H, C—O, N—H, and N—O. Water (H_2O) and glucose ($C_6H_{12}O_6$) are both polar molecules. Notice that oxygen is bonded to

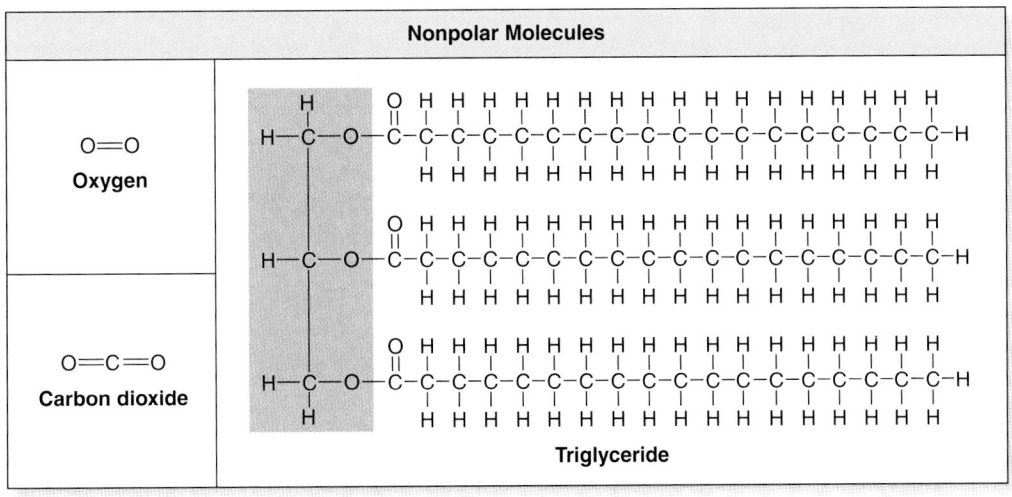

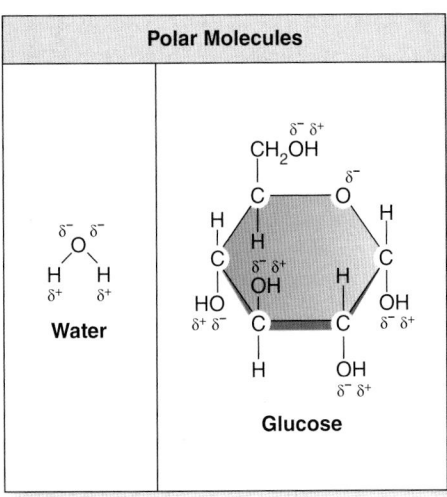

(a) (b)

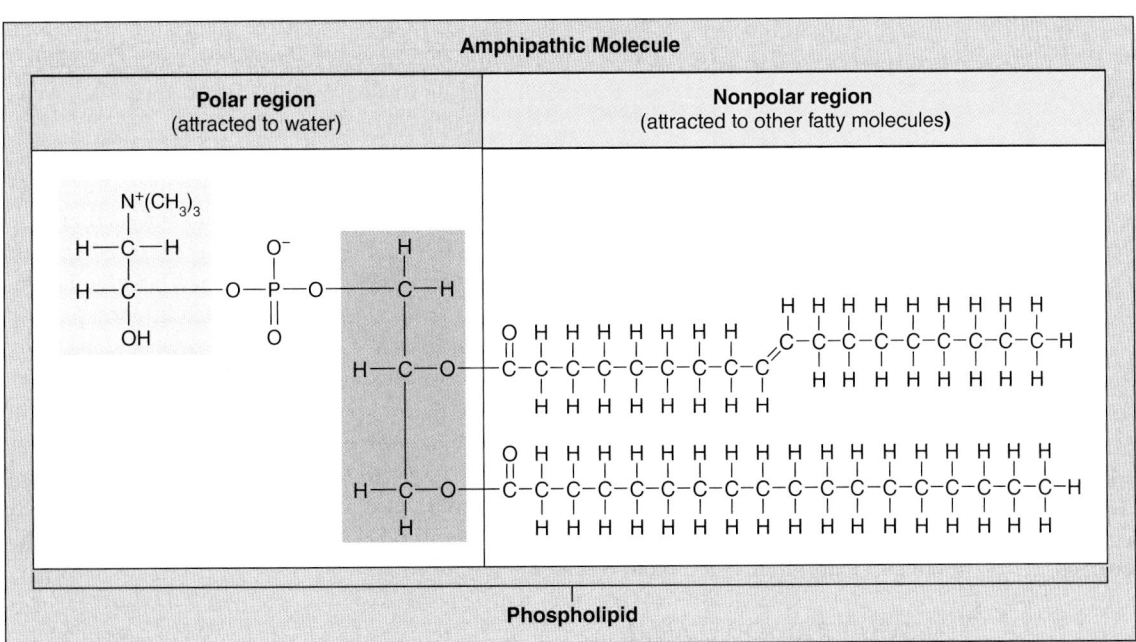

(c)

Figure 2.10 Nonpolar, Polar, and Amphipathic Molecules. (*a*) Oxygen, carbon dioxide, and a triglyceride are examples of nonpolar molecules. (*b*) Water and glucose are examples of polar molecules. (*c*) A phospholipid is an example of an amphipathic molecule.

INTEGRATE

CONCEPT CONNECTION

The two respiratory gases, O_2 and CO_2, are both nonpolar molecules. This chemical characteristic—and the fact that they are small in size—permits both to easily cross cell membranes including the plasma membrane, which forms the outer barrier of a cell. The transport of respiratory gases is discussed in other chapters of the text, including section 4.3a on membrane transport and the exchange of O_2 and CO_2 between the blood and cells in section 23.6.

two hydrogen atoms in a water molecule, and several C—O and O—H bonds are within a glucose molecule. One exception to this general pattern exists: A molecule containing polar covalent bonds that extend in opposite directions can be nonpolar because the partial charges cancel each other. Carbon dioxide ($^{\delta-}O{=}C{=}O^{\delta-}$) is an example.

Sometimes a molecule is large enough that it can have one major part that is nonpolar and another part that is polar. Molecules that contain both nonpolar and polar components are called **amphipathic** (*amphi* = both, *patheia* = feeling) **molecules.** A phospholipid molecule is an example of an amphipathic molecule.

WHAT DO YOU THINK?

3 Is a fatty acid (see figure 2.18) a nonpolar or polar molecule? Explain your answer. Would you predict that it will or will not dissolve in water?

WHAT DID YOU LEARN?

14 Are O_2 and CO_2 nonpolar or polar molecules?

2.3d Intermolecular Attractions

LEARNING OBJECTIVES

9. Describe hydrogen bonding between polar molecules.
10. List and define the intermolecular attractions between nonpolar molecules.

Molecules sometimes have weak chemical attractions to other molecules called **intermolecular** (*inter* = between) **attractions.** One important intermolecular attraction is called a hydrogen bond. A **hydrogen bond** forms between polar molecules. It is a weak attraction between a partially positive (δ^+) hydrogen atom within a polar molecule and a partially negative (δ^-) atom within a polar molecule. The partially negative atom is usually oxygen, but sometimes nitrogen. A hydrogen bond is designated in this text with a dotted or broken line. A hydrogen bond is shown in **figure 2.11** between the partially positive charge of a hydrogen atom within a glucose molecule and the partially negative charge of an oxygen atom within a water molecule. Although an individual hydrogen bond is a weak bond or attraction (approximately 5–10% the strength of a covalent bond), hydrogen bonds collectively are strong.

Other types of intermolecular attractions involve nonpolar molecules. These interactions are collectively called **van der Waals forces.** They occur when electrons orbiting the nucleus of an atom of a nonpolar molecule are, for a brief instant, distributed unequally. One portion of the atom is slightly negative and another portion is slightly positive. This momentary unequal distribution of charge in the atom induces an unequal distribution of electrons in an adjacent atom of another nonpolar molecule. Individual van der Waals forces are weak (about 1% of the strength of a covalent bond).

A third type of interaction is **hydrophobic interactions,** which results when nonpolar molecules are placed in water or another polar substance. (Hydrophobic interactions are described in section 2.4c.) Note that all of these attractions may also occur between different portions of a large molecule. In this case, they are more appropriately called **intramolecular** (*intra* = within) **attractions.**

Hydrogen bonds form between water molecules and have significant influence in how water molecules behave as described in section 2.4. Hydrogen bonding, van der Waals forces, and hydrophobic interactions are all important in establishing and maintaining the three-dimensional shape of complex molecules such as DNA and proteins (described later), as well as the temporary binding of molecular structures to one another, such as the binding of a hormone to a protein receptor.

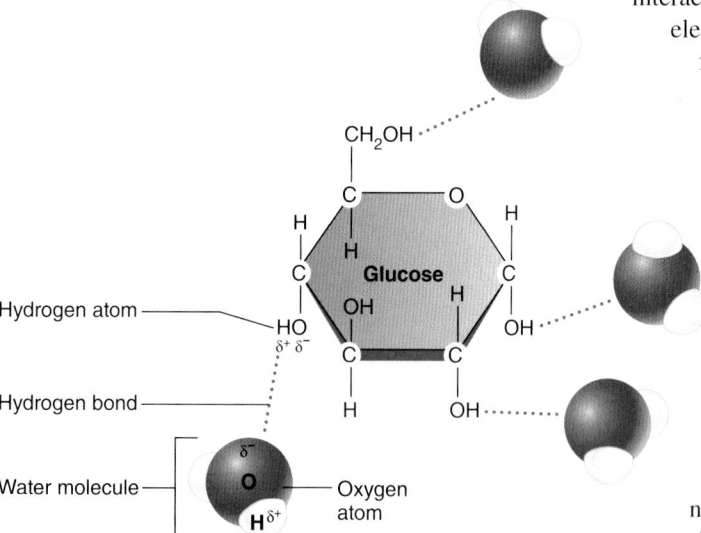

Hydrogen atom

Hydrogen bond

Water molecule — Oxygen atom

Figure 2.11 Hydrogen Bonding. A hydrogen bond formed between a partially positive charged hydrogen atom of a glucose molecule and the partially negative charged oxygen atom of a water molecule.

WHAT DID YOU LEARN?

15 What is the name of the intermolecular attraction between a partially charged hydrogen of one polar molecule with a partially negative atom of another polar molecule?

2.4 Molecular Structure of Water and the Properties of Water

Water is the first molecule that we examine in detail. It is appropriate to begin with water because it is the substance that composes approximately two-thirds of the human body by weight. We first look closely at the molecular structure of water molecules and then describe several important water properties. The relevance to normal body activities is included for each property.

2.4a Molecular Structure of Water

 LEARNING OBJECTIVE

1. Describe the molecular structure of water and how water molecules form four hydrogen bonds.

Water is a polar molecule composed of one oxygen atom bonded to two hydrogen atoms. It exhibits polarity because there is an unequal sharing of electrons between the oxygen atom and each of the two hydrogen atoms **(figure 2.12a)**. The oxygen atom is more electronegative, and it has two partial negative charges. In contrast, each hydrogen atom exhibits a single partial positive charge.

Every water molecule has the ability to form four hydrogen bonds with adjacent water molecules. This is because each of the two hydrogen atoms forms one hydrogen bond, and each oxygen atom forms two hydrogen bonds (figure 2.12b). Recall from our earlier discussion of hydrogen bonds that these intermolecular attractions are individually weak but collectively very strong. Thus, hydrogen bonding between water molecules is central to the properties exhibited by water.

WHAT DID YOU LEARN?

16   What is the intermolecular bond that is significant in determining the properties of water?

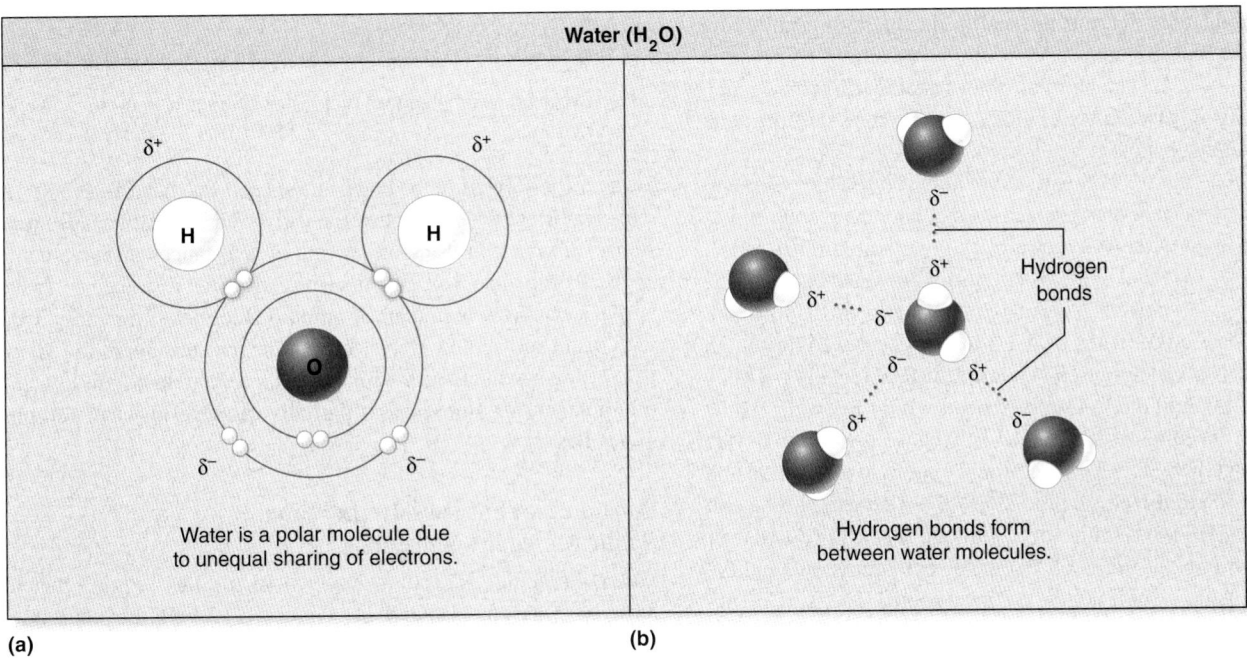

Figure 2.12 Water Molecule. (*a*) A water molecule is a polar molecule due to the unequal sharing of electrons between the oxygen atom and each of the two hydrogen atoms. Partial charges are shown. (*b*) Hydrogen bonds form between the partially positive (δ^+) hydrogen atom of one water molecule with the partially negative (δ^-) oxygen atom of a different water molecule.

2.4b Properties of Water

2.   List the different properties of water and provide an example of the importance of each property within the body.

Phases of Water

Water is present in three phases, depending upon the temperature: a gas (water vapor), a liquid (water), and a solid (ice). Substances that have a low molecular mass (described in section 2.6b) such as water generally are present in the gaseous phase at room temperature. However, water is liquid at room temperature because hydrogen bonds hold the water molecules in the liquid phase and limit their escape into the gaseous phase. Almost all water within the human body occurs within the liquid phase, although small amounts of water are present as water vapor within the air passageways. As a liquid, water serves the following functions:

- **Transports.** Substances are dissolved in water and moved throughout the body in water-based fluids (e.g., blood and lymph).
- **Lubricates.** Water-based fluids located between body structures decrease friction (e.g., serous fluid between the heart and its sac, synovial fluid within joints).
- **Cushions.** The force of sudden body movements is absorbed by water-based fluids (e.g., cerebrospinal fluid surrounding the brain and spinal cord).
- **Excretes Wastes.** Unwanted substances are eliminated in the body dissolved in water (e.g., urine).

Cohesion, Surface Tension, and Adhesion

Cohesion is the attraction between water molecules. They are inclined to "stick together" because hydrogen bonds form between these molecules. **Surface tension** is the inward pulling of cohesive forces at the surface of water. This inward attraction occurs because water molecules at the surface are pulled by hydrogen bonds in only three directions, whereas water molecules that are internal in the liquid are pulled by hydrogen bonds in four directions. **Adhesion** is the attraction between water molecules and a substance other than water. This occurs when hydrogen bonds form between water molecules and the molecules that compose those other substances.

Surface tension can be demonstrated readily. Try this experiment: First, stack two flat plates of glass, such as two clean microscope slides, and then lift them apart. Note that they are easily separated. Then repeat this experiment, but first place one or two drops of water between the slides before they are stacked. Note it is much more difficult, if not impossible, to separate the plates of glass without first wedging them apart. This is because water causes increased surface tension between the surfaces. A similar situation is experienced within the walls of the moist air sacs of the lungs (called alveoli) because opposing air sac walls will stick together. However, we produce a mixture of lipids and proteins called *surfactant* (see section 23.3e) that prevents the alveoli from collapsing when we breathe out. Without surfactant (a risk of some premature infants), alveoli collapse with each breath out, and the two moist sides of the alveoli adhere to one another. The next breath in requires breaking of the surface tension within alveoli and their reinflation—a situation that requires much greater effort.

High Specific Heat and High Heat of Vaporization

Temperature is a measure of the kinetic energy, or random movement, of atoms or molecules within a substance. The relationship between temperature and kinetic energy is direct—the temperature is higher when there is a greater amount of kinetic energy in an object. Two properties of water influence water temperature: its specific heat and the heat of vaporization.

Specific heat is the amount of energy (measured in calories) required to increase the temperature of 1 gram of a substance by 1 degree Celsius (C). The specific heat of water has one of the highest values of any substance (1 calorie/gram/°C). This is because most of the energy imparted into water during heating is first used to break hydrogen bonds. Thereafter, the energy from heating increases the kinetic energy of water molecules. As we change between cool and warm environments, or generate large amounts of heat during physical exertion, much of this heat (energy) is used to break hydrogen bonds and not increase the random movement of molecules (kinetic energy) within the body. Thus, body temperature remains relatively constant.

Heat of vaporization is the energy required for the release of molecules from a liquid phase into the gaseous phase for 1 gram of a substance. Water has a high heat of vaporization because the hydrogen bonds between individual water molecules must first be broken before these molecules can be released from the liquid phase into the gaseous phase. This is the reason why sweating is an effective measure in helping to cool the body. As water molecules evaporate from the surface of the skin, excess heat is dissipated from the body as liquid water is changed to a gas.

≋❔ **WHAT DID YOU LEARN?**

17 Which property of water contributes to the need to produce surfactant and prevent collapse of the alveoli? Which property contributes to body temperature regulation through sweating? Why is sweating less effective in cooling the body on a humid day?

2.4c Water as the Universal Solvent

≋✺ **LEARNING OBJECTIVES**

3.   Compare substances that dissolve in water with those that both dissolve and dissociate in water. Distinguish between electrolytes and nonelectrolytes.

4.   Describe the chemical interactions of nonpolar substances and water.

5.   Explain how amphipathic molecules interact in water to form chemical barriers.

Water is the **solvent** of the body, and substances that dissolve in water are called **solutes.** Water is called the **universal solvent** because most substances dissolve in it. However, not all substances dissolve completely or at all in water. The chemical properties of a substance (whether it is polar, charged, nonpolar, or amphipathic) determine how it interacts with water molecules. Here we examine the substances that dissolve in water (polar molecules and ions) and those that do not dissolve in water (nonpolar molecules) or that partially dissolve in water (amphipathic molecules).

Substances That Dissolve in Water (Polar Molecules and Ions)

Water molecules are polar. Each molecule has regions that are differently charged: $H^{\delta+}$ and $O^{\delta-}$ (figure 2.12). Some polar molecules (e.g., glucose) and other charged substances or ions (e.g., Na^+, HCO_3^-) interact with and **dissolve** within water. In other words, they disperse in the water. The substances that dissolve in water are appropriately called **hydrophilic** (meaning "water-loving"). Many water molecules surround this substance and form a **hydration shell** around it (**figure 2.13***a*).

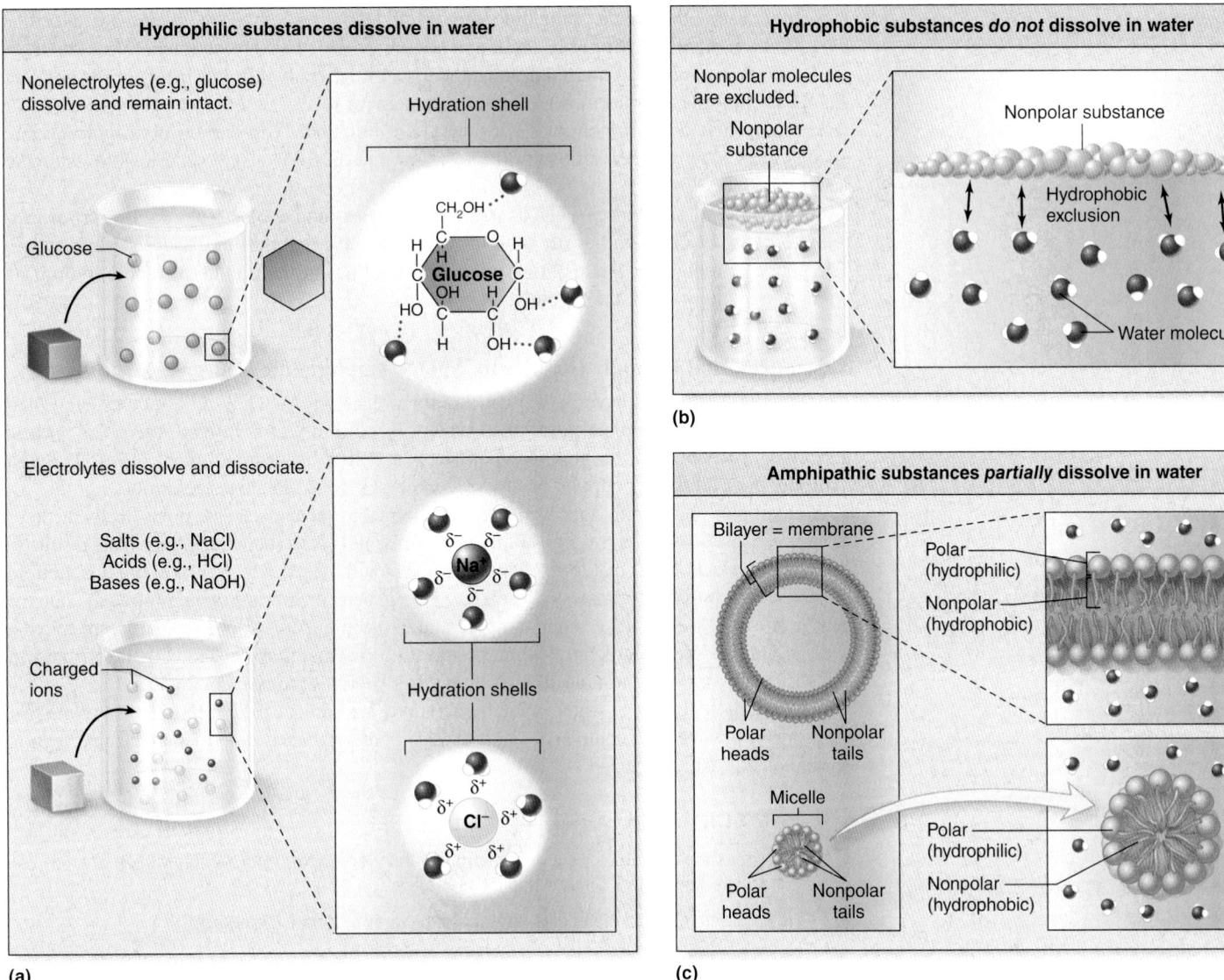

Figure 2.13 Substance Interaction with Water. Chemical properties of the substance combined with water will determine their interaction. (*a*) Hydrophilic substances dissolve in water. Nonelectrolytes dissolve and remain intact within water as water molecules form a hydration shell around each molecule. Electrolytes, which include salts, acids, and bases, dissolve and dissociate in water to a certain extent as water molecules form hydration shells around each ion. (*b*) Hydrophobic molecules are nonpolar substances that do not dissolve in water; rather, the nonpolar molecules are "pushed" out of the water by hydrophobic exclusion. (*c*) Amphipathic molecules are unique in that the polar portion dissolves in water and the nonpolar portion does not. Membranes (e.g., plasma membrane, nuclear envelope) and micelles are two common arrangements formed within the body by amphipathic molecules.

Polar molecules such as glucose and alcohol dissolve in water as a result of the hydrogen bonds that form between those molecules and water. However, here the solute remains intact as individual molecules are surrounded by water molecules.

Some substances dissolve in water but do not remain intact. These substances both dissolve and **dissociate,** meaning that they pull apart or separate. Salts are ionic compounds that both dissolve and dissociate when placed into water. When added to water, the cations and anions of the salt are pulled apart. For example, sodium chloride (NaCl) dissociates in water to form both Na^+ and Cl^- ions. Hydration shells form around each ion, causing their separation. The water molecule's partial positive charges ($H^{\delta+}$) interact with the released anions (Cl^-) of the salt, whereas the water molecule's partial negative charges ($O^{\delta-}$) interact with the released cations (Na^+).

Acids and bases also dissociate in water. Hydrochloric acid (HCl) dissociates to form both H^+ and Cl^-, whereas sodium bicarbonate dissociates to form both Na^+ and HCO_3^- ions. Acids and bases are described in detail in the next section.

Substances that both dissolve and dissociate in water, such as salts, acids, and bases, can readily conduct an electric current. For this reason, they are called **electrolytes.** In contrast, substances that remain intact when introduced into water, such as glucose, do not conduct an electric current and are called **nonelectrolytes.** Maintaining normal levels of electrolytes including salts, acids, and bases is discussed in chapter 25.

Substances That Do Not Dissolve in Water (Nonpolar Molecules)

Nonpolar molecules do not dissolve in water, and so they are called **hydrophobic** (meaning "water-fearing"). The hydrogen bonds between water molecules cause the water molecules to be cohesive and attract each other; at the same time, they exclude, or "force out," the nonpolar molecules by a process called **hydrophobic exclusion** (figure 2.13b). The interaction between the molecules of the "excluded" nonpolar substance is termed **hydrophobic interaction** because it appears that these molecules are avoiding water. In this case, contact between the polar water molecules and the nonpolar substance is minimized. You can observe hydrophobic exclusion by placing a few drops of oil into water; the oil forms small spherical drops on the water's surface.

Hydrophobic substances, such as triglycerides (fats) and cholesterol, require carrier proteins to be transported within the blood because of their inability to dissolve within water. The nonpolar molecules are enclosed within the confines of the protein molecule to limit its contact with the water in the blood (section 18.2a).

Substances That Partially Dissolve in Water (Amphipathic Molecules)

Amphipathic molecules have both polar and nonpolar regions. They do not completely dissolve, nor are they completely excluded when placed into water. Instead, the polar portion dissolves in water, and the nonpolar portion is repelled by water (figure 2.13c). Through hydrophobic exclusion, nonpolar portions become positioned in close proximity.

Recall that phospholipid molecules are amphipathic molecules. The polar heads of these molecules have contact with water, but their nonpolar tails are grouped together. This results in bilayers of phospholipids that form chemical barriers within the body. A bilayer of phospholipid molecules composes membranes of a cell (e.g., the plasma membrane, which forms the outside barrier of a cell; section 4.2a). Other amphipathic molecules form a spherical structure called a *micelle.* This is a special structure within the digestive tract that is associated with the breakdown and absorption of nonpolar molecules including triglycerides (see section 26.4c).

The many functions of water in the human body are summarized in **figure 2.14**. Note that the property of water as a neutral solvent (included in this figure) is described in the next section on acidic and basic solutions.

WHAT DID YOU LEARN?

18 How does the interaction of a nonelectrolyte and water differ from the interaction of an electrolyte and water? Give examples of each.

19 How do phospholipid molecules interact with water to form a membrane?

2.5 Acidic and Basic Solutions, pH, and Buffers

Acidic and basic solutions occur specifically when an acid or base is added to water. Here we describe why water is neutral; define an acid and a base; and explain pH, neutralization, and the action of buffers.

2.5a Water: A Neutral Solvent

LEARNING OBJECTIVE

1. Describe what is formed when water dissociates.

Another property of water is that it spontaneously dissociates to form ions. The covalent chemical bond between oxygen and either of the two hydrogen atoms in a water molecule spontaneously breaks apart at a low rate (about two dissociations occur per billion water molecules). This number is about 10^{-7} ions (or 1/10,000,000 ions) per liter.

A hydrogen ion (H^+) transfers to a second water molecule during dissociation. The water molecule that picked up the extra hydrogen ion is called a *hydronium ion,* and it is represented as H_3O^+. The electron of this "transferred" hydrogen remains associated with the original water molecule, which now is deficient in one hydrogen ion but still has the original electron. This molecule is called a *hydroxide ion* (OH^-). This dissociation reaction is written as

$$H_2O + H_2O \longrightarrow H_3O^+ + OH^-$$

simplified to

$$H_2O \longrightarrow H^+ + OH^-$$

Figure 2.14 **Water's Roles in the Body.**
Water serves several critical functions within the body. It helps regulate body temperature, serves as the universal solvent, cushions, transports, lubricates, and creates high surface tension that allows body structures to cling to each other. The neutral pH of water is changed by the addition of an acid or base.

CUSHIONS

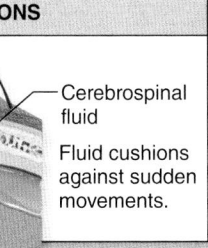

Cerebrospinal fluid

Skull

Brain

Fluid cushions against sudden movements.

REGULATES BODY TEMPERATURE

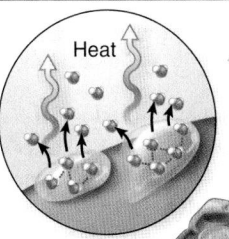

Water helps regulate body temperature due to its high specific heat and high heat of vaporization.

Heat

TRANSPORTS

Water is the fluid medium to transport substances in blood and other body fluids (e.g., lymph, urine).

UNIVERSAL SOLVENT

Hydrophilic substance

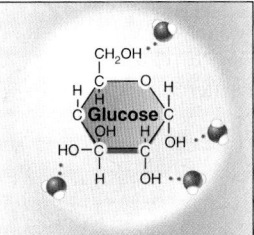

Glucose

Nonelectrolytes dissolve and remain intact.

Na⁺

Cl⁻

Electrolytes dissolve and dissociate.

Hydrophobic molecules

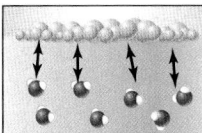

Water molecules exclude (or force out) nonpolar molecules, thus proteins are required for their transport within the body.

Amphipathic molecules

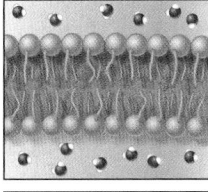

Polar portion dissolves, nonpolar portion excluded.

Amphipathic molecules form chemical barriers (e.g., cell membranes, micelles).

LUBRICATES

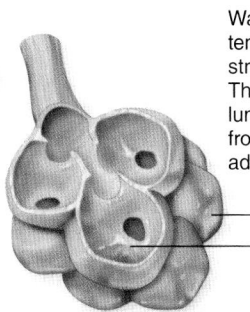

Fluid serves as a lubricant to decrease friction.

Heart

Pericardial sac

Pericardial fluid

HIGH SURFACE TENSION

Water's high surface tension causes structures to adhere. The moist alveoli in the lungs are prevented from collapsing and adhering by surfactant.

Alveolus

Surfactant

NEUTRAL pH

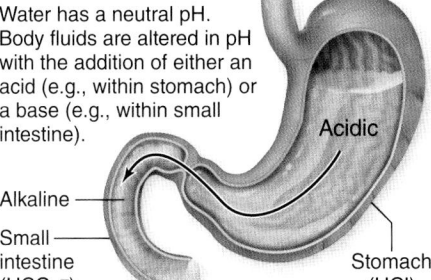

Water has a neutral pH. Body fluids are altered in pH with the addition of either an acid (e.g., within stomach) or a base (e.g., within small intestine).

Acidic

Alkaline

Small intestine (HCO_3^-)

Stomach (HCl)

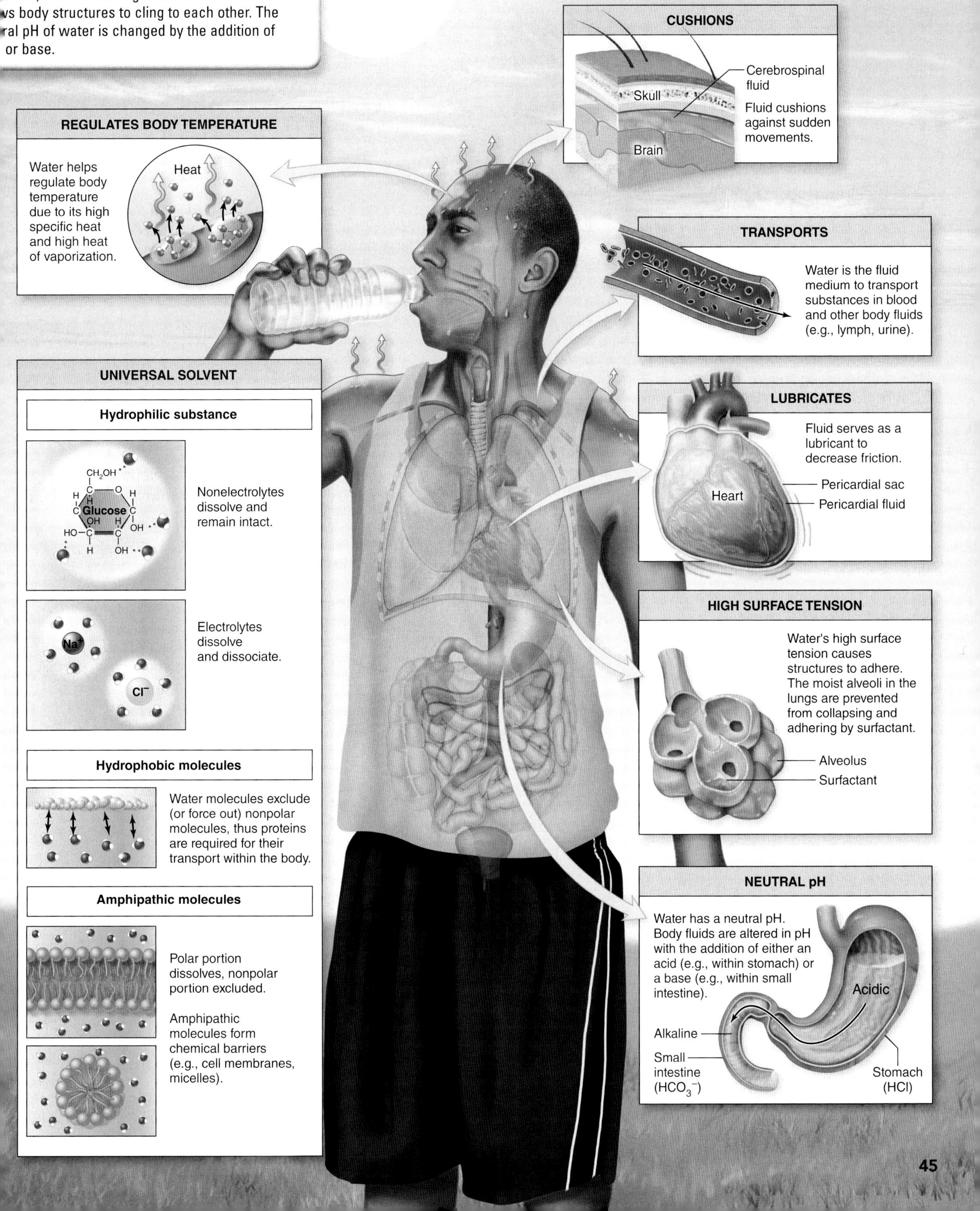

Equal numbers of positively charged hydrogen ions (H^+) and negatively charged hydroxide ions (OH^-) are produced from the dissociation of water. Consequently, water has no net charge and is neutral.

 WHAT DID YOU LEARN?

20 Explain why water is neutral.

2.5b Acids and Bases

 LEARNING OBJECTIVE

2. Explain the difference between an acid and a base.

An **acid** is a substance that dissociates in water to produce both an H^+ and an anion. An acid increases the concentration of H^+ (written as [H^+]) that are free in solution **(figure 2.15)**. Because H^+ is a proton, an acid is also called a **proton donor.** The equation is as follows:

Substance A (an acid in water) $\longrightarrow$ H^+ + Anion

Strong acids dissociate to a greater extent and produce more H^+. Hydrochloric acid (HCl), secreted by cells lining the stomach, is a good example of a strong acid. Weak acids, such as carbonic acid (H_2CO_3) in the blood, dissociate to a lesser extent, producing fewer H^+ ions.

In contrast, a **base** accepts H^+ when added to a solution. Therefore, a base is also called a **proton acceptor.** A base decreases the concentration of H^+ free in solution. Thus,

Substance B (a base in water) + H^+ $\longrightarrow$ B—H

Stronger bases dissociate to a greater extent and bind more H^+ than do weak bases, leaving less H^+ in solution. Sodium hydroxide (NaOH) is an example of a strong base. Ammonia and bleach are effective cleaning agents because they are strong bases. Weak bases bind less H^+, leaving more H^+ in solution. Bicarbonate (HCO_3^-) is one of the most important weak bases in the body; it is both transported in the blood and released into secretions from the liver and pancreas that enter the small intestine.

 WHAT DID YOU LEARN?

21 Which type of substance releases H^+ when added to water?

2.5c pH, Neutralization, and the Action of Buffers

 LEARNING OBJECTIVES

3. Define pH and explain the relative pH values of both acids and bases.

4. Explain the term neutralization, and describe how the neutralization of both an acid and a base occur.

5. Describe the action of a buffer.

The pH of a solution is a measure of the relative amounts of H^+ it contains; it is expressed as a number between 0 and 14. (The unit of measurement for the pH scale is moles/liter.) pH is the inverse relationship of the logarithmic values for a given *hydrogen ion concentration* [H^+]:

$$\text{Negative log of } [H^+] = \frac{1}{\log \text{ of } [H^+]}$$

Recall that water readily dissociates to produce 10^{-7} H^+ and OH^- ions (or 1/10,000,000) per liter. If this concentration of H^+ is placed into the formula just given for pH, then the value for the pH of water is calculated to be equal to 7. In other words:

$$[H^+] = 1 \times 10^{-7}$$

$$[H^+] = 0.0000001$$

$$pH = 7$$

Notice that as [H^+] changes, pH changes. For example, if the [H^+] increases so that [H^+] = 1×10^{-4}, [H^+] = 0.0001, and the pH decreases to 4. Conversely, if the [$H+$] decreases so that [H^+] = 1×10^{-9}, [H^+] = 0. 000000001, and the pH increases to 9. Thus, [H^+] and pH value are *inversely related.* The inverse relationship between [H^+] and pH is an extremely important concept. Remember, as [H^+] increases, pH decreases, and as [$H+$] decreases, pH increases.

Interpreting the pH Scale

Pure water and other solutions that have equal concentrations of H^+ and OH^- are neutral and have a pH of 7. Solutions with a pH below 7 are acidic, and solutions with a pH above 7 are basic, or alkaline.

Moving from one increment to another (e.g., 7 to 6) represents a 10-fold change in [H^+]. Therefore, a pH 6 solution has a [H^+] that is

Figure 2.15 pH. pH scale is a measure of the relative concentration of H^+ and OH^-. A neutral solution has equal amounts of H^+ and OH^-, acidic solutions contain greater amounts of H^+ than OH^-, and basic solutions contain lesser amounts of H^+ than OH^-. Examples of common solutions that exhibit a specific pH are shown.

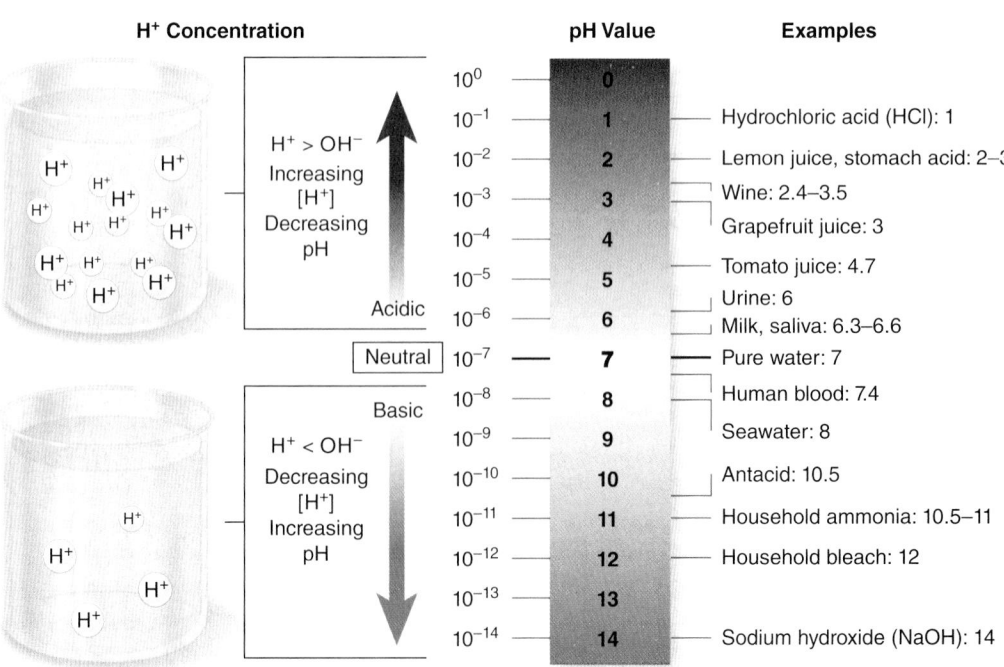

10 times greater than pure water. Changing by two units (e.g., 7 to 5) is a 100-fold increase in [H⁺]. This relationship is accurate for changes above 7 as well, except that each increment is a 10-fold decrease in [H⁺].

 WHAT DO YOU THINK?

4 If stomach acid has a pH of 2, how much more acidic is stomach acid than water (pH 7)? Predict what would happen to the stomach if it were not protected from the effects of hydrochloric acid.

Neutralization

Neutralization occurs when a solution that is either acidic or basic is returned to neutral (pH 7). The neutralization of an acidic solution is accomplished by adding a base, whereas a basic solution is neutralized by adding an acid. Thus, medications to neutralize stomach acid, such as Rolaids or Tums, must contain a base.

Buffers

A **buffer** is either a single substance or an associated group of substances that helps prevent pH changes if either excess acid or base is added. A buffer acts either to accept H⁺ from excess acid or donate H⁺ to neutralize excess base. Both carbonic acid (H_2CO_3), a weak acid, and bicarbonate (HCO_3^-), a weak base, are present within the blood and serve as buffers to maintain the pH of the blood within the normal range of 7.35 to 7.45. It is critical to maintain acid-base balance because even small changes in pH (called an acid-base disturbance or imbalance) can be fatal. Acid-base balance and acid-base imbalance are described in detail in sections 25.5 and 25.6, respectively.

   WHAT DID YOU LEARN?

22 What is the general relationship of [H⁺] and pH?

23 Why are buffers important, and how do they function to help maintain pH?

INTEGRATE

LEARNING STRATEGY

You may find it helpful to think of buffers as H⁺ sponges. If acid is added, the buffer "absorbs" the H⁺, and if a base is added, the buffer releases H⁺. In either case, this helps maintain the H⁺ in solution, thus helping prevent changes in pH.

2.6 Water Mixtures

Mixtures are formed from the combining or mixing of two or more substances. Several types of mixtures have already been described, including sugar water, salt water, and acidic solutions. Two defining features of mixtures are (1) the substances that are mixed are not chemically changed; and (2) the substances in the mixture can be separated by physical means, such as by evaporation or by filtering. Our emphasis here is on water mixtures. We describe how water mixtures are categorized into three types and explain how solution concentration is expressed.

2.6a Categories of Water Mixtures

 LEARNING OBJECTIVES

1. Compare and contrast the three different types of water mixtures.
2. Explain how an emulsion differs from other types of mixtures.

Water mixtures are placed into three categories based on the relative size of the substance mixed with water and include suspensions, colloids, and solutions (**figure 2.16**).

- **Suspension.** A suspension occurs when material that is larger in size than 100 nanometers is mixed with water. Unlike a colloid or a solution, a suspension does not remain mixed together

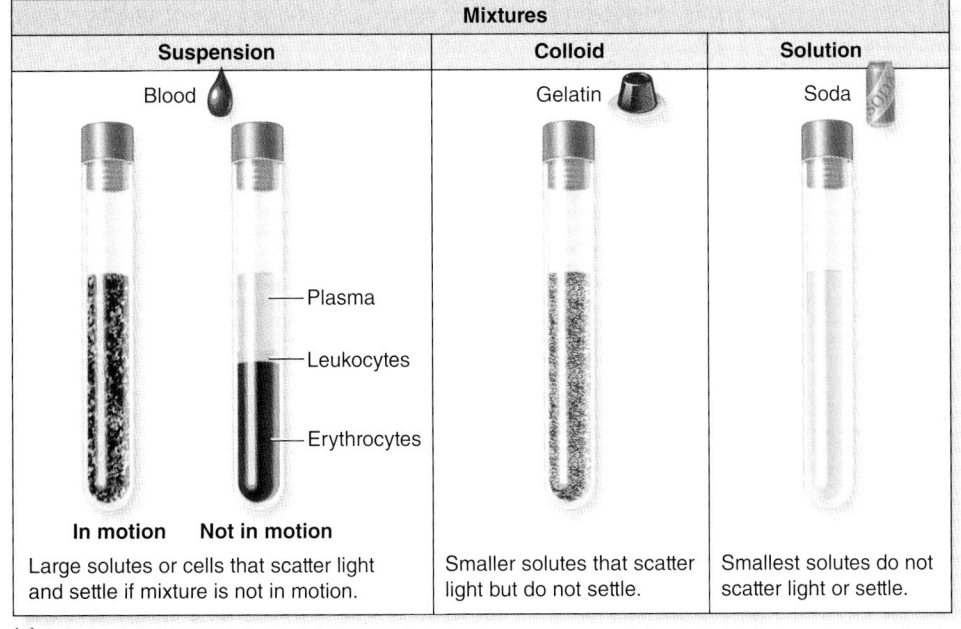

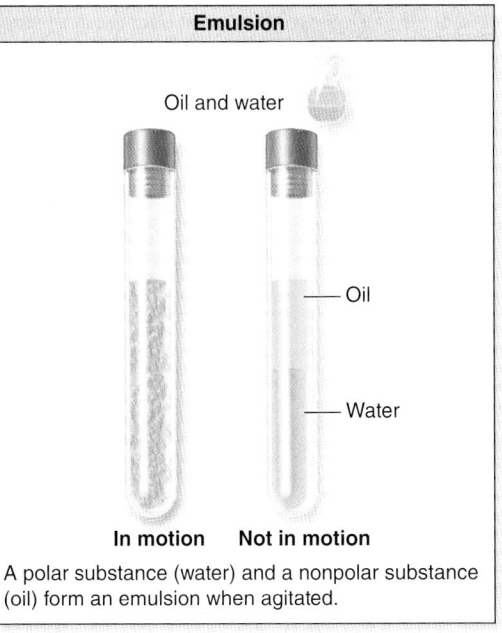

(a)   **(b)**

Figure 2.16 Mixtures and Emulsions. (*a*) Mixtures include suspensions, colloids, and solutions. (*b*) An emulsion is a suspension specifically composed of water (or other polar substance) and a nonpolar liquid, such as vegetable oil, that mix only when agitated.

unless it is in motion. A suspension appears cloudy or opaque until the particles drop out of the liquid. Sand in water is an example of a suspension. Blood cells within the plasma (the liquid portion) of blood form a suspension.

- **Colloid.** A colloid is a mixture composed of protein within water, where the protein ranges in size from 1 to 100 nanometers. A colloid may appear either opaque or milky and scatters light like a suspension—but, unlike a suspension, remains mixed when not in motion. Some colloids, including gelatin and agar media (used in the microbiology laboratory), display an interesting feature—namely, they become liquid when heated but change to a gel-like state when left standing and cooled. Colloids within the body include fluid within a cell (cytosol) and fluid within the blood (plasma).

- **Solution.** A solution is a homogeneous mixture in which the substance is smaller than 1 nanometer, and it dissolves in water. The water is the solvent, and the dissolved substance is the solute. Both salt water and sugar water are examples of solutions. The small size of the solutes in a solution results in a mixture with these characteristics: The solutes are not visible, do not scatter light, and do not settle if the solution is not in motion. Blood plasma is an example of a body solution, with salts, glucose, HCO_3^-, and other dissolved nonprotein substances.

A suspension composed specifically of water and a nonpolar (hydrophobic) liquid substance such as vegetable oil does not "mix" unless

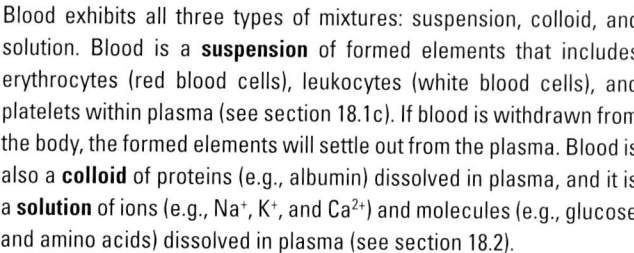

INTEGRATE

CONCEPT CONNECTION

Blood exhibits all three types of mixtures: suspension, colloid, and solution. Blood is a **suspension** of formed elements that includes erythrocytes (red blood cells), leukocytes (white blood cells), and platelets within plasma (see section 18.1c). If blood is withdrawn from the body, the formed elements will settle out from the plasma. Blood is also a **colloid** of proteins (e.g., albumin) dissolved in plasma, and it is a **solution** of ions (e.g., Na^+, K^+, and Ca^{2+}) and molecules (e.g., glucose and amino acids) dissolved in plasma (see section 18.2).

shaken or agitated (e.g., oil and vinegar salad dressing). The combination of water and a nonpolar liquid when forcibly mixed is classified instead as an **emulsion.** Breast milk is an example of an emulsion in the body.

 WHAT DID YOU LEARN?

24 When left standing, erythrocytes (red blood cells) settle to the bottom of a tube of blood. Based on this characteristic alone, blood would be characterized as a (a) suspension, (b) colloid, or (c) solution?

25 Why is blood also considered the other two types of water mixtures?

2.6b Expressions of Solution Concentration

LEARNING OBJECTIVE

3. Explain the different ways to express the concentration of solute in a solution.

The amount of solute dissolved in a solution determines the concentration of a solution. Concentration of a solution may be expressed in several ways, including mass/volume, mass/volume percent, molarity, and molality. The units for expressing concentration are summarized in **table 2.2**, which includes a description, units of measurement, and an example for each.

Mass/volume is mass of solute per volume of solution. Results from a blood test are often expressed in mass/volume. **Mass/volume percent** is grams of solute per 100 milliliters of solution. For example, mass/volume percent is the unit of measurement for intravenous (IV) solutions.

Molarity is a measure of number of moles per *liter of solution.* A molar solution of glucose is made by placing 180.10 grams (its molecular mass) of glucose into a container and adding enough water until it measures 1 liter. (Both moles and molecular mass are described shortly.) **Molality** is the moles *per kilogram of solvent.* A solution of one molality is made by placing 180.10 grams of glucose into a container and adding 1 kilogram of water. Molarity and molality may be used interchangeably, subject to this caveat: The two values are the closest when the measurements are taken at 4°C. At this temperature, 1 liter of water is at its most dense, and its mass is exactly equal to 1 kilogram of water.

Table 2.2	Expressing Solution Concentrations		
Solution Concentration	**Expressed As**	**Unit of Measurement**	**Examples**
Mass/volume	Mass of solute per volume of solution	μg solute/dL solution	Normal blood concentration of iron is within the range of 40 to 150 μg/dL.
		mg solute/dL solution	Normal blood concentration of glucose is between 70 and 110 mg/dL.
Mass/volume percent	Grams of solute per 100 milliliters (mL) of solution	grams/100 mL	5% dextrose intravenous (IV) solution (D5W) has a concentration of 5 grams of dextrose (glucose) per 100 mL of solution.
			Physiologic saline (0.9% NaCl) has 0.9 grams of NaCl per 100 mL of solution.
Molarity	Moles of solute per liter of solution	moles solute/L solution	0.164 mol/L solution
Molality	Moles of solute per kilogram of solvent	moles solute/kg solvent	0.164 mol/kg solvent

mg = milligrams; dL = deciliters; kg = kilograms; μg = micrograms

Molarity alters with changes in temperature (and to a limited degree with changes in pressure). When water temperature increases above 4°C, the solution expands, and molarity (moles per liter of solution) decreases slightly because it is based on volume (liter) of solution. In contrast, molality (moles per kilogram of solvent) does not alter with changes in temperature because it is based on mass (kilograms) of solvent. Consequently, molality is the more accurate of the two values. However, molality is more difficult to measure in the human body; thus, molarity—the slightly less accurate unit of concentration—is more often used.

Osmoles, Osmolarity, and Osmolality

Another means of expressing concentration is with **osmoles (osm),** which reflects whether a substance either dissolves, or dissolves and dissociates, when placed into a solution (i.e., whether it is a nonelectrolyte or an electrolyte). It is the unit of measurement for the number of particles in solution. The term osmole generally is used to reflect the extent a solution is able to alter water movement through osmosis (a concept described in section 4.3b). If a solute dissolves but does not dissociate, such as occurs when glucose, amino acids, or proteins are placed in water, then the osmolarity = 1 osmole (osm). However, if the solute both dissolves and dissociates, such as occurs with both NaCl and $CaCl_2$, then there is a change in the number of particles after the solute is in solution. A one molar (1 M) solution of NaCl dissociates in solution to form both 1 M Na^+ and 1 M Cl^-. The osmoles of 1 M of a NaCl solution = 2 osm. What would you predict for the osmoles of $CaCl_2$? Each molecule of $CaCl_2$ dissolves into three particles: 1 Ca^{2+} and 2 Cl^-; thus, 1 M $CaCl_2$ = 3 osm.

Osmoles can be expressed as either osmolarity or osmolality. **Osmolarity** is the number of particles in a 1 liter solution, whereas **osmolality** is the number of particles in 1 kilogram of water. Osmolality more accurately reflects the osmotic movement of water (section 4.3b) in the body but it is difficult to measure. Osmolarity is a close approximation to osmolality. Although osmolarity is less accurate, it is often the preferred unit of measurement for the same reason that molarity is used over molality.

The range of values associated with the body is generally much smaller than osmoles and given as milliosmoles (mOsm). Note that 1 osm = 1000 mOsm. Thus, normal blood serum may be expressed as either 275 to 295 mOsm per liter or 275 to 295 mOsm per kilogram.

Moles and Molecular Mass

Notice in table 2.2 that molarity and molality are based on the number of particles in units called a **mole.** The specific value of 1 mole is 6.022×10^{23} atoms, ions, or molecules. The number of particles may seem huge, but the particles themselves are minuscule.

A mole is the mass in grams that is equal either to the atomic mass of an element or the molecular mass of a compound. For example, a mole of carbon is equal to 12.01 grams. **Molecular mass** is determined using the molecular formula and the average atomic mass for each element. To find a compound's molecular mass, multiply the number of units of each element by its average atomic mass and add together the totals. The molecular mass of glucose ($C_6H_{12}O_6$), for example, is determined as follows:

$$6 \text{ carbon atoms} \times 12.01 \text{ amu} = 72.06 \text{ amu}$$
$$12 \text{ hydrogen atoms} \times 1.008 \text{ amu} = 12.10 \text{ amu}$$
$$6 \text{ oxygen atoms} \times 15.99 \text{ amu} = 95.94 \text{ amu}$$
$$\text{Molecular mass} = 180.10 \text{ amu}$$

Thus, 1 mole of glucose ($C_6H_{12}O_6$) would be equal to 180.10 grams (with some variation due to isotopes).

WHAT DID YOU LEARN?

26 What are four ways solution concentration may be expressed?

2.7 Biological Macromolecules

Four classes of organic biological macromolecules (biomolecules) can be distinguished in living systems: lipids, carbohydrates, nucleic acids, and proteins. We begin this section by discussing the similarities among the four classes, then describe each class in detail.

INTEGRATE

LEARNING STRATEGY

You can use the acronym **CLaPiN** to remember the four major classes of macromolecules—carbohydrates, lipids, proteins, and nucleic acids.

2.7a General Characteristics

LEARNING OBJECTIVES

1. Differentiate between an organic molecule and an inorganic molecule.
2. Describe the general chemical composition of biomolecules.
3. Define a monomer and polymer.
4. Describe the role of water in both dehydration and hydrolysis reactions in altering biomolecules.

Biomolecules are a subset of organic molecules. **Organic molecules** are defined as molecules that contain carbon (e.g., methane, glucose). Most organic molecules are a component of living organisms or have been produced from them. All other types of molecules are **inorganic molecules** including water, salts (e.g., sodium chloride), acids (e.g., carbonic acid), and bases (e.g., sodium hydroxide).

Chemical Composition

Biomolecules always contain carbon, hydrogen, and generally oxygen. Some biomolecules may also have one or more of the following: nitrogen (N), phosphorus (P), or sulfur (S). Notice that all of these elements (except hydrogen) are clustered on the right side of the periodic table (see figure 2.1a).

The carbon component of biomolecules may simply be an individual carbon atom or numerous carbon atoms arranged in a "carbon skeleton." The structure of a carbon skeleton may be a chain, branch, or ring (see figure 2.9).

The single carbon atom or the carbon skeleton within some biomolecules only have hydrogen atoms attached. These molecules are more specifically called **hydrocarbons.** They are nonpolar molecules because they contain only C—C and C—H bonds. Consequently, hydrocarbons are hydrophobic and not soluble in water. Methane gas (CH_4) is an example of a hydrocarbon.

In comparison, the single carbon atom or the carbon skeleton of other biomolecules may contain functional groups. A functional group is two or more atoms that display specific chemical characteristics. Some common examples of these groups include hydroxyl (—OH), amine (—NH_2), and carboxyl (—COOH). Almost all functional groups are polar and able to form hydrogen bonds, increasing the biomolecule's

INTEGRATE

LEARNING STRATEGY

A polymer is like a necklace. This necklace might be a pearl necklace, where every subunit is identical, or it might be a charm necklace, where every unit is slightly different. However, like the individual pieces in a necklace, the chemical subunits combine to produce the polymer that is the finished product.

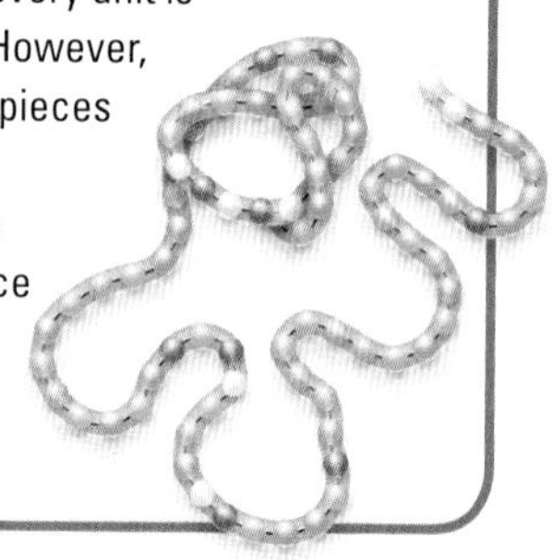

solubility in water. In addition, some functional groups may act like an acid and release H^+, such as a carboxyl group, whereas others may act like a base by binding H^+, such as an amine group. Biomolecules that contain functional groups typically have more than one within it. Some of the important functional groups are presented in **table 2.3**.

Polymers

Many important biomolecules are called polymers. **Polymers** are molecules that are made up of repeating subunits called **monomers,** and each monomer is either identical or similar in its chemical structure. Some important carbohydrates (e.g., glycogen, starch), nucleic acids, and proteins are polymers, whereas lipids are not. Carbohydrate polymers contain sugar monomers, nucleic acids have nucleotide monomers, and proteins are composed of amino acid monomers.

Process of Dehydration Synthesis and Hydrolysis

Two processes are associated with both the synthesis and breakdown of complex biomolecules: dehydration synthesis and hydrolysis. During the synthesis of complex molecules from simpler subunits, one specific subunit loses an —H, and the other subunit loses an —OH, to form a water molecule as a new covalent bond is produced. This type of reaction is called **dehydration**

Table 2.3	Functional Groups and Their Chemical Properties			
Functional Group	**Structural Formula**	**Properties**	**Representative Molecules**	**Structural Diagram of Example Molecule**
Hydroxyl	—OH	• Polar • Forms hydrogen bonds • Increases molecules' solubility in water	• Carbohydrates • Proteins • Nucleic acids • Lipids	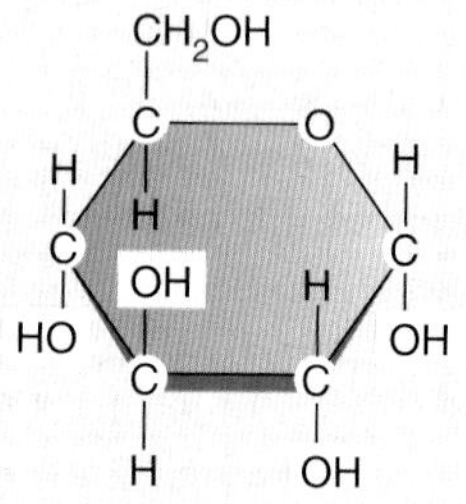 Glucose
Carboxyl (carboxylic acid)	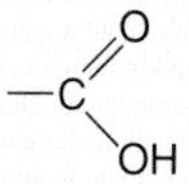	• Polar • Forms hydrogen bonds • Increases molecules' solubility in water • Acts as an acid	• Proteins • Lipids	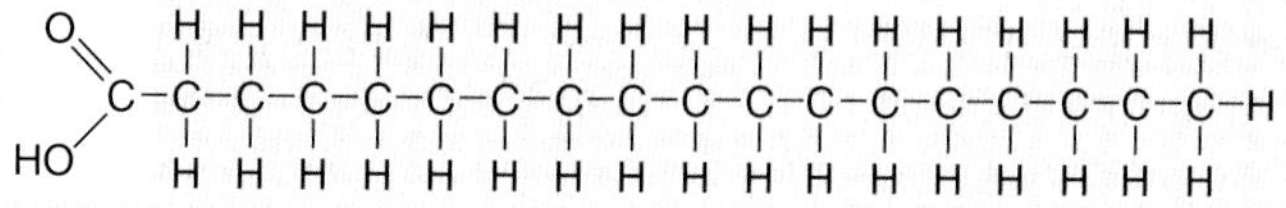 Fatty acid
Amine	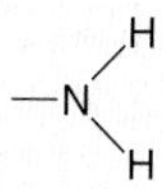	• Polar • Forms hydrogen bonds • Increases molecules' solubility in water • Acts as a base	• Proteins • Nucleic acids	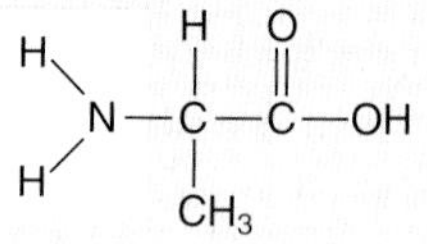 Alanine (an amino acid)
Phosphate	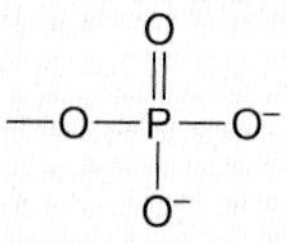	• Polar • Forms hydrogen bonds • Increases molecules' solubility in water • Forms phosphodiester bonds • Acts as an acid (shown here with hydrogens released)	• Nucleic acids • Phospholipids • ATP	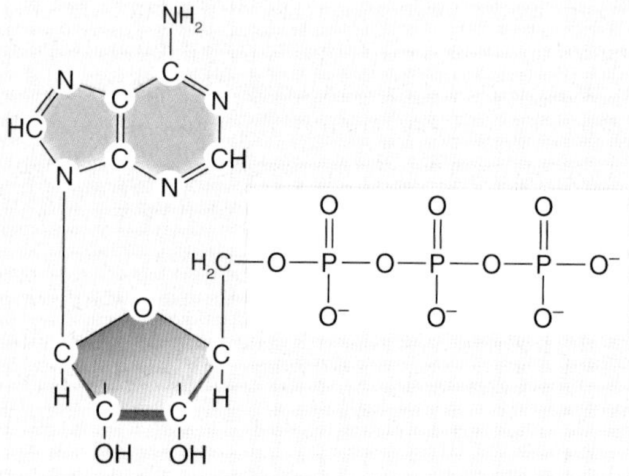 Adenosine triphosphate (ATP)

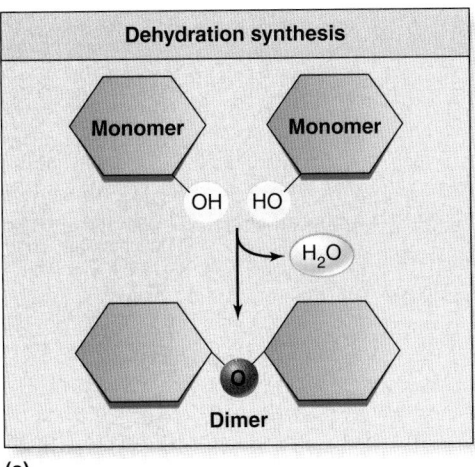

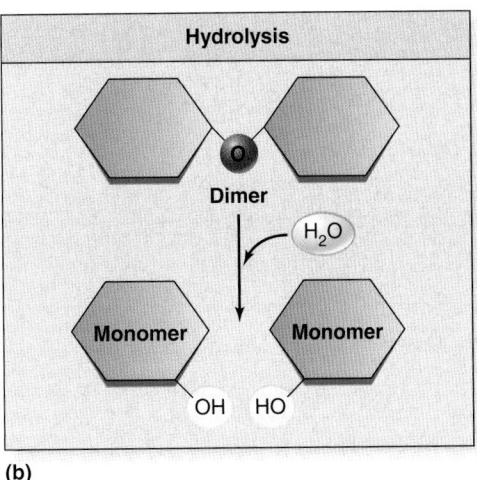

Figure 2.17 Dehydration Synthesis and Hydrolysis. (*a*) Dehydration synthesis involves the loss of a water molecule from simpler components as they are formed into a complex molecule. (*b*) Hydrolysis occurs with the addition of a water molecule to a complex molecule as it is digested into simpler components.

(*de* = away, *hydro* = water) **synthesis** or *condensation* because the equivalent of a water molecule is "lost" from the original structures **(figure 2.17)**.

During the breakdown of complex molecules, H_2O molecules are split. An —H is added to one subunit, and an —OH is added to another subunit in the complex molecule, and the chemical bond is broken between them. Because the equivalent of water is added to digest the molecule, this process is referred to as **hydrolysis** (hī-drol′i-sis; *lysis* = destruction) or a *hydrolysis reaction*. We will see examples of both of these types of reactions in the following sections on biomolecules.

WHAT DID YOU LEARN?

27 Using a different color than you did for highlighting the common ions on the periodic table, highlight the six common elements that form organic molecules. Which element both (a) forms a common ion, and (b) is a common element in biomolecules?

28 What functional groups may act as an acid?

29 What defines a polymer? List the three biomolecules that are polymers and the monomers that compose them.

2.7b Lipids

LEARNING OBJECTIVES

5. Describe the general characteristics of a lipid.

6. Identify the four types of lipids and their physiologic roles.

Lipids are the only category of biomolecules that are not polymers, because they are not formed from repeating monomers. Rather, they are a very diverse group of fatty, water-insoluble (hydrophobic) molecules that function as stored energy, components of cellular membranes, and hormones. Triglycerides (neutral fats), phospholipids, steroids, and eicosanoids are the four primary classes of lipids. They are summarized in **table 2.4**.

Triglycerides: Energy Storage

Triglycerides (trī-glis′ĕr-idz; *tri* = three), or *triacylglycerols*, are the most common form of lipids in living things. They are used for long-term energy storage in adipose connective tissue (see section 5.2) and for structural support, cushioning, and insulation of the body. Adipose connective tissue deep to the skin in the abdomen, for example, serves as long-term energy storage and insulates the abdomen against heat loss. Adipose connective tissue posterior to the eye cushions the eye within the eye's bony orbit.

Triglycerides are formed from a glycerol molecule and three fatty acids. Glycerol is a three-carbon molecule with a hydroxyl functional group attached to each carbon. A fatty acid is composed of a long chain of hydrocarbons with a carboxylic acid functional group on one end.

Table 2.4 Major Classes of Lipids[1]

Class	Structure	Description	Function
Triglycerides	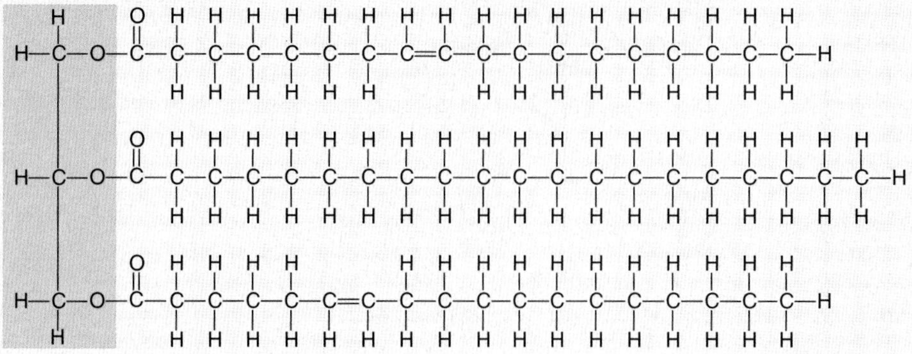	Composed of glycerol and three fatty acids Fatty acids may be saturated or unsaturated	Long-term energy storage in adipose connective tissue Structural support, cushioning, and insulation of the body
Phospholipids	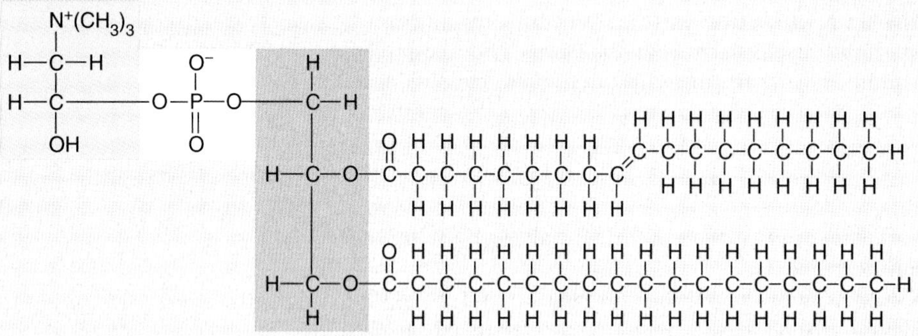	Composed of glycerol, two fatty acids, a phosphate, and various organic groups Glycerol, phosphate, and organic groups form a polar head, and fatty acids form two nonpolar tails	Major component of membranes including the plasma membrane, which forms the chemical barrier between the inside and outside of a cell
Steroids (includes cholesterol, steroid hormones, and bile salts)	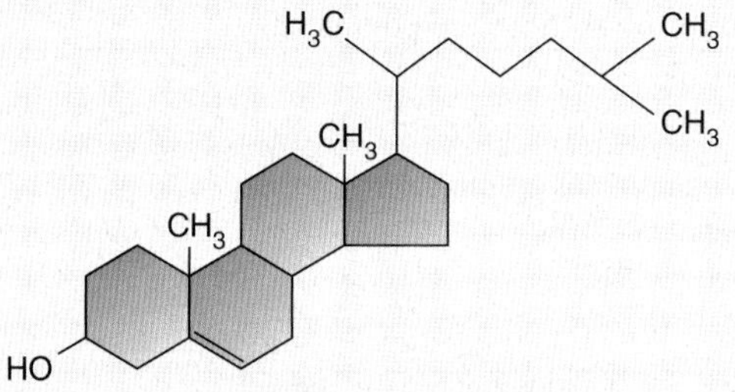	Four rings composed predominantly of hydrocarbons that differ in the side chains extending from the rings	Cholesterol is a component of plasma membranes and is the precursor molecule for synthesis of other steroids Steroid hormones are regulatory molecules released by certain endocrine glands Bile salts facilitate micelle formation in the digestive tract
Eicosanoids (includes prostaglandins, prostacyclins, thromboxanes, and leukotrienes)	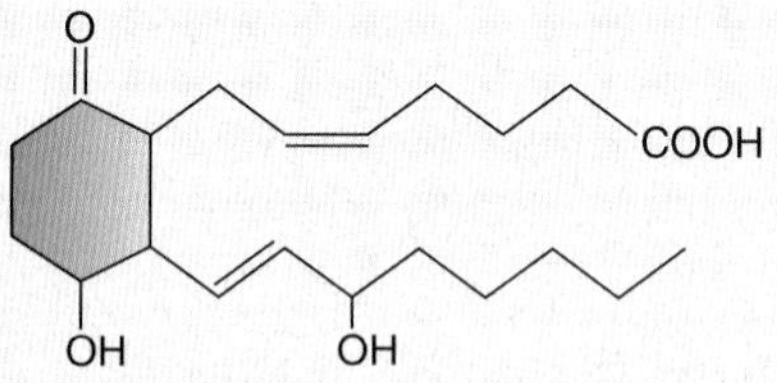	Modified 20-carbon fatty acids synthesized from arachidonic acid	Local acting signaling molecules associated with all body systems, with primary functions in the inflammatory response of the immune system and communication within the nervous system

1. Glycolipids and fat-soluble vitamins are also types of lipids.

Triglyceride molecules form by the process of dehydration synthesis, during which the equivalent of a water molecule is lost for each fatty acid added to the glycerol: an —H from the glycerol and an —OH from the fatty acid (figure 2.18).

Fatty acids may vary in length—commonly ranging in even numbers from 14 to 20 carbons—and may differ in the number and position of double bonds between the carbons in the chain. The fatty acid is **saturated** if it lacks double bonds—that is, every carbon has the maximum number of hydrogen atoms bound to it. An **unsaturated** fatty acid has one double bond, and a **polyunsaturated** fatty acid has two or more double bonds.

Adipose connective tissue is used to store triglycerides. When conditions of excess nutrients exist, adipose connective tissue binds fatty acids to glycerol to form triglycerides in a dehydration synthesis process called **lipogenesis** (līp′o-jĕn′e-sis; lipos = fat, genesis = production). Adipose connective tissue breaks down triglycerides and releases the products into the blood when nutrients are needed. This process is a hydrolysis reaction called **lipolysis** (li-pol′i-sis).

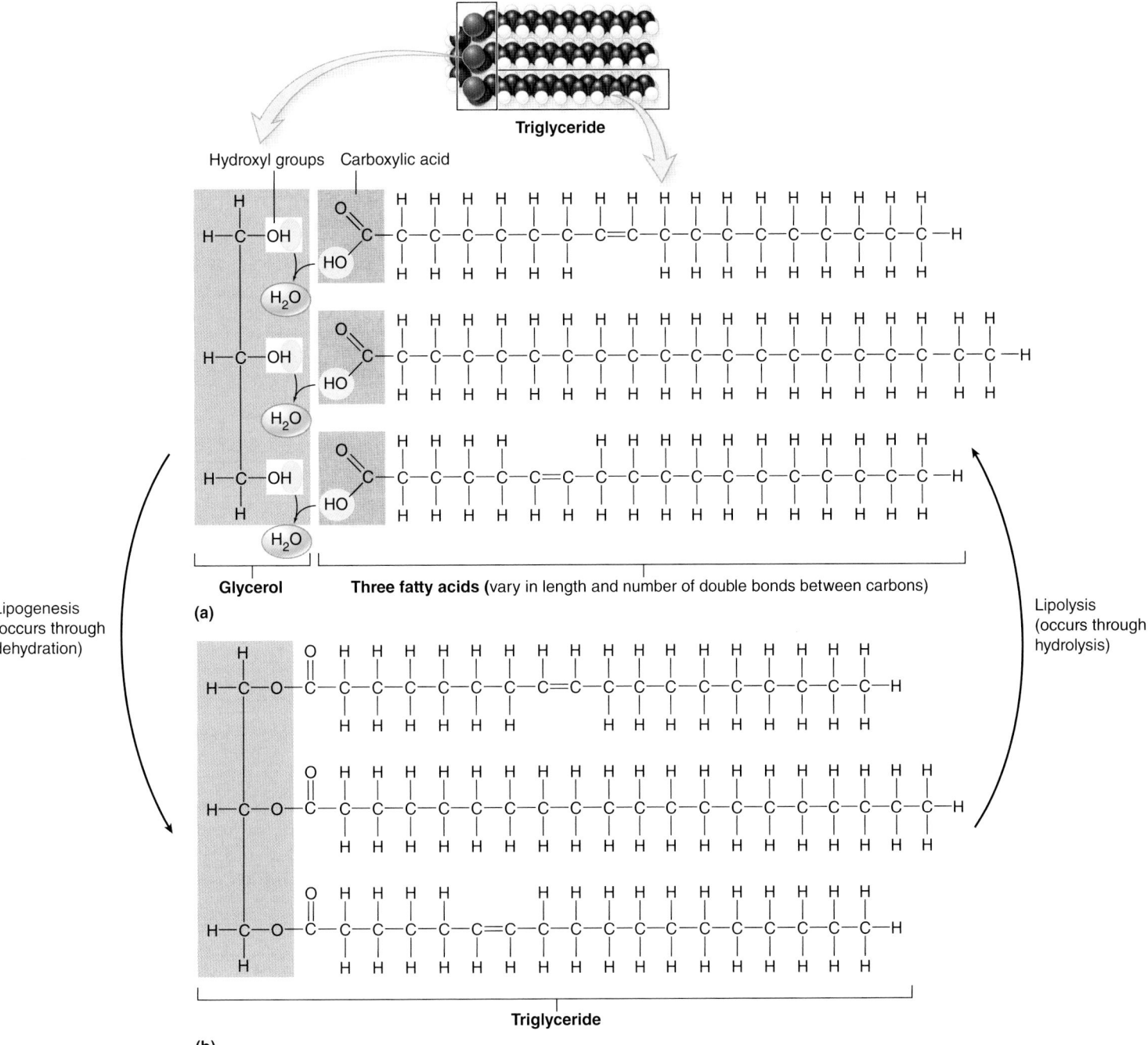

Figure 2.18 Triglycerides. (*a*) A glycerol and three fatty acid molecules. (*b*) A triglyceride molecule. Lipogenesis occurs through a dehydration synthesis reaction that involves the removal of a water molecule between each fatty acid and a glycerol. Lipolysis is a hydrolysis reaction that splits the glycerol and three fatty acids by the addition of water at each of the fatty acids.

Phospholipids: Membranes

Phospholipids were previously described as amphipathic molecules that form chemical barriers of cell membranes including plasma membranes that form the outer barrier of a cell (see figure 2.13*c*). The chemical structure of a phospholipid is similar to a triglyceride, except that one end of the glycerol has a polar phosphate group with various organic groups (choline, ethanolamine, or the amino acid serine) attached to it instead of a fatty acid (see table 2.4). The glycerol, phosphate, and organic groups are polar and form the water-soluble hydrophilic part of the molecule referred to as the **hydrophilic (polar) head.** The two fatty acid molecules attached to the glycerol form a water-insoluble hydrophobic end called the **hydrophobic (nonpolar) tails.**

Steroids: Ringed Structures Including Some Hormones

Steroids are composed predominantly of hydrocarbons arranged in a distinct multiringed structure. A steroid has four attached carbon rings; three rings have six carbon atoms and one ring has five carbon atoms. Steroids differ in the side chains extending from their rings. Steroids include cholesterol, steroid hormones (e.g., testosterone, estrogen), and bile salts. Cholesterol is a component of animal plasma membranes and is also the precursor used to synthesize other steroids. Cholesterol is synthesized in the liver from fatty acids and also may be obtained from eating animal products such as meat, eggs, and milk.

CLINICAL VIEW

Fatty Acids: Saturated, Unsaturated, and Trans Fats

Most naturally occurring animal fats are **saturated,** whereas most naturally occurring vegetable fats (except palm and coconut tropical oils) are **unsaturated.** You can usually distinguish between a saturated and unsaturated fat by whether it is solid (most saturated fats) or liquid (most unsaturated fats) at room temperature. Unsaturated fats are generally healthier than saturated fats—which is why we are encouraged to use olive oil instead of butter for our cooking.

Unsaturated fats are converted to saturated fats through the process of **hydrogenation.** For example, vegetable oil may be hydrogenated to form margarine. Studies have shown that partial hydrogenation of fatty acids may lead to the production of **trans fats.** These fats are solid at a lower temperature than their counterparts, *cis fats.* Studies have shown that trans fats in the diet increase the risk of developing lipid buildup on the inside of blood vessels (atherosclerosis), a condition that increases the risk of heart attack or stroke.

Eicosanoids: Locally Acting Hormones

Eicosanoids (ī′kō-să-noydz; *eicosa* = twenty; *eido* = form) are modified 20-carbon fatty acids that are synthesized as needed from arachidonic acid, a common component of plasma membranes. Four classes of eicosanoids are produced and include prostaglandins, prostacyclins, thromboxanes, and leukotrienes. These molecules act locally and are signaling molecules associated with all body systems: they have primary functions in the inflammatory response of the immune system and communication within the nervous system. How eicosanoids are synthesized from arachidonic acid is described in section 17.3b.

Other Lipids

Glycolipids—lipid molecules with carbohydrate attached—are also lipids. These molecules are associated with plasma membranes and serve several roles, including cellular binding to form tissues (see section 4.2a). Fat-soluble vitamins such as vitamins A, E, and K (see section 27.3a) are also lipids.

WHAT DID YOU LEARN?

30 Do lipid molecules typically dissolve in water? Explain.

31 Which class of lipids forms cell membranes? What characteristic allows it to perform this function?

2.7c Carbohydrates

LEARNING OBJECTIVES

7. Describe the distinguishing characteristics of carbohydrates.

8. Explain the relationship between glucose and glycogen.

9. Name some other carbohydrates found in living systems.

The term **carbohydrate** means "hydrated carbon." Nearly every carbon is hydrated with the equivalent of a water molecule—that is, both an —H and an —OH are usually attached to every carbon. The general chemical formula for carbohydrates is $(CH_2O)_n$, where n indicates the number of carbon atoms that are in a molecule. The number of carbon atoms typically ranges from 3 to 7. The least complex carbohydrates are simple sugar monomers called **monosaccharides.** All monosaccharides have between three and seven carbon atoms. Carbohydrates formed from two monosaccharides are **disaccharides,** and those with many monosaccharides are **polysaccharides.**

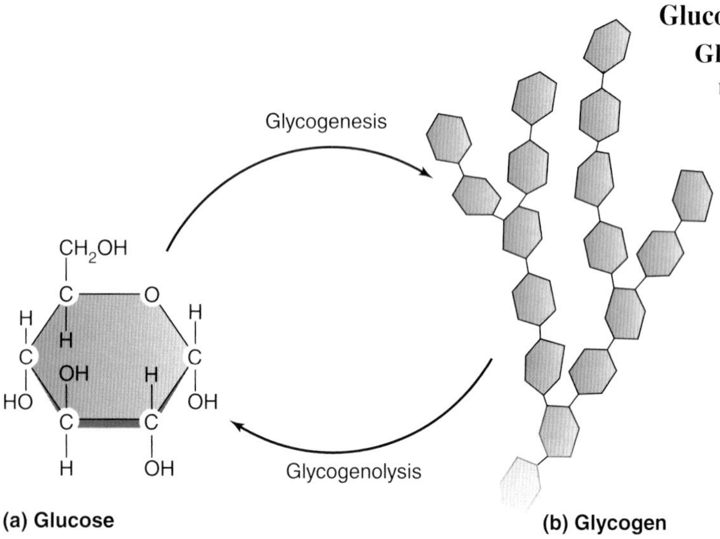

Glucose and Glycogen

Glucose is a six-carbon (hexose) carbohydrate that is the most common monosaccharide. It is shown here (and throughout the text) in the ring form (**figure 2.19**). Glucose is crucial to life processes because it is the primary nutrient supplying energy to cells. In fact, the brain and other nervous tissue use glucose almost exclusively as their nutrient fuel molecule. The concentration of blood glucose must be carefully maintained by homeostasis (see section 1.5) to ensure a continual, adequate energy supply for cellular activities. One way the body ensures its supply is to store excess glucose immediately following a meal. Liver and skeletal muscle tissue absorb the excess glucose—and then bind the glucose monomers together to form a polysaccharide called **glycogen** by a process called **glycogenesis** (glī′kō-jen′ĕ-sis). The glycogen polymer is shown in figure 2.19*b*—although only several glucose molecules are shown, glycogen may contain thousands of glucose monomers.

When blood glucose levels drop between meals, the liver hydrolyzes some of the glycogen into glucose and releases it into the blood. This process is called **glycogenolysis** (glī′kō-jĕ-nol′i-sis). Thus, the liver serves the role of a "glucose bank," storing glycogen, then breaking down glycogen as needed to release glucose. Nutrient storage and release are important in the endocrine regulation of nutrient blood levels of both triglycerides and glucose. Note: The liver can also form glucose from noncarbohydrate sources (e.g., fats, proteins) through a process called **gluconeogenesis.**

Figure 2.19 Glucose and Glycogen.
(*a*) The monomer glucose is typically shown in ring form. (*b*) Glycogen is a polysaccharide composed of multiple glucose monomers. Glycogen is formed from glucose molecules through glycogenesis, whereas glycogen is digested into glucose through glycogenolysis.

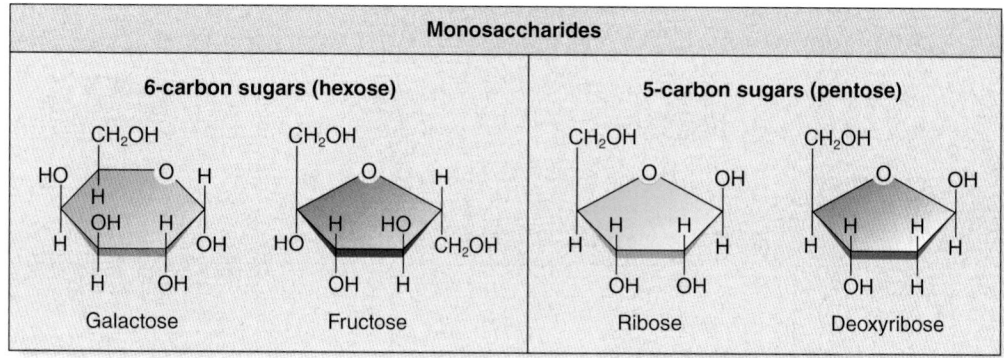

Monosaccharides

6-carbon sugars (hexose)

Galactose

Fructose

5-carbon sugars (pentose)

Ribose

Deoxyribose

(a)

Figure 2.20 Other Simple Carbohydrates. (*a*) Hexose monosaccharides include galactose and fructose, and pentose monosaccharides include ribose and deoxyribose. (*b*) Disaccharides include sucrose, lactose, and maltose.

Disaccharides

Sucrose

Lactose

Maltose

(b)

Other Types of Carbohydrates

Other hexose monosaccharides, including galactose and fructose, are glucose isomers (**figure 2.20***a*). Some monosaccharides, such as ribose and deoxyribose, are composed of five carbons. These five-carbon monosaccharides are called *pentose sugars.* The pentose sugars ribose and deoxyribose are structural components of nucleic acids, which are discussed in the next section. The only structural difference between these pentose sugars is the lack of an oxygen atom on carbon two of the deoxyribose sugar.

Disaccharides are composed of two simple sugars bonded together (figure 2.20*b*). The most common disaccharides are sucrose (table sugar), lactose (milk sugar), and maltose (malt sugar, which is found in sprouting grains). All three disaccharides contain a glucose monosaccharide bonded to a second hexose monosaccharide.

Polysaccharides are composed of three or more sugars. The common polysaccharide in animals is glycogen. Polysaccharides in plants include starch and cellulose, which are also composed of repeating glucose monomers. Plant starch is a major nutritional source of glucose for humans. It is found in potatoes, grains, and many other plant foods. Glucose released from the breakdown of starch within the digestive tract is absorbed into the blood. Cellulose, a structural polysaccharide of plant cell walls, however, is a source of fiber (nondigestible substances). Humans cannot digest cellulose because of the unique chemical bonds between the glucose molecules. The breakdown of both disaccharides and polysaccharides is described in section 26.4a.

 WHAT DID YOU LEARN?

32 What is the repeating monomer of glycogen? Where is glycogen stored in the body?

33 For each of the following, indicate if it is a monosaccharide, disaccharide, or polysaccharide: fructose, galactose, glucose, glycogen, lactose, maltose, starch, and sucrose.

2.7d Nucleic Acids

 **LEARNING OBJECTIVES**

10. Describe the general structure of a nucleic acid.

11. Describe the structure of a nucleotide monomer.

12. Distinguish between DNA and RNA.

13. Name other important nucleotides.

Nucleic acids (**figure 2.21***c, d*) are macromolecules within cells that store and transfer genetic or hereditary information. Originally discovered within the cell nucleus, nucleic acids ultimately

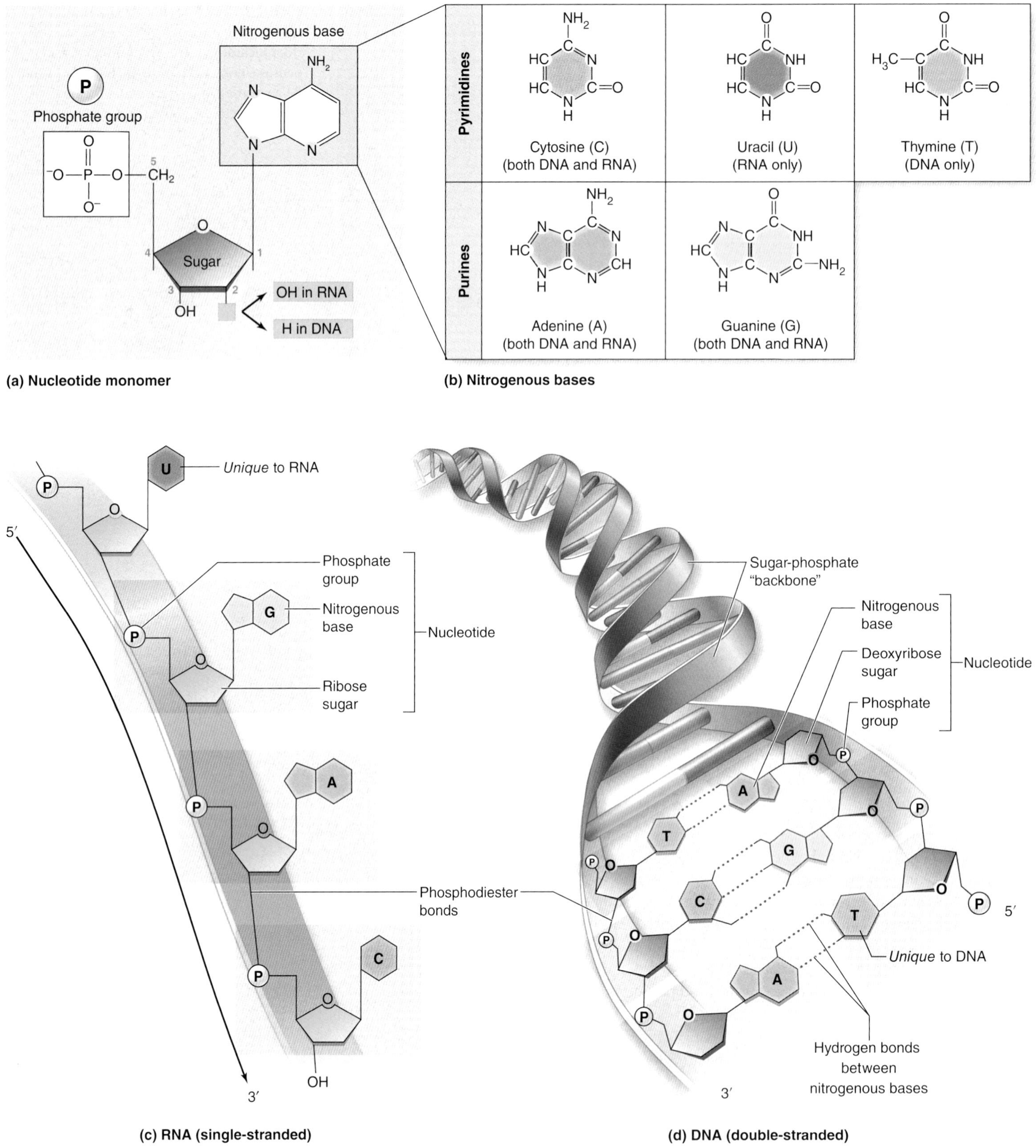

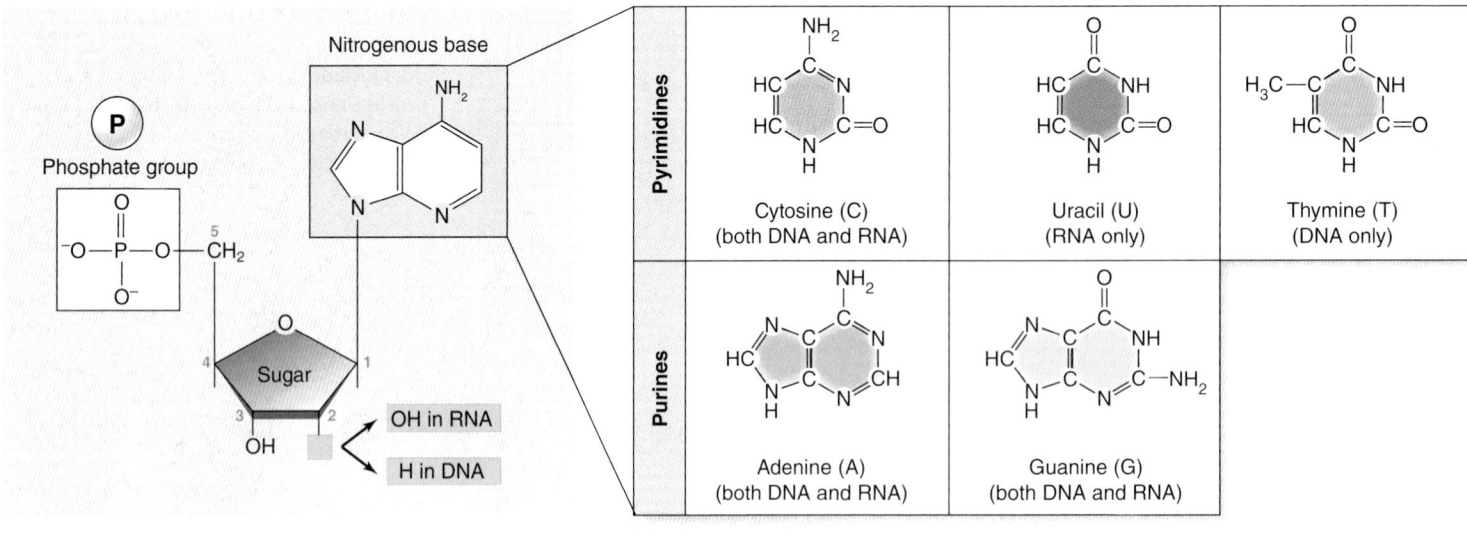

(a) Nucleotide monomer

(b) Nitrogenous bases

Pyrimidines

Cytosine (C)
(both DNA and RNA)

Uracil (U)
(RNA only)

Thymine (T)
(DNA only)

Purines

Adenine (A)
(both DNA and RNA)

Guanine (G)
(both DNA and RNA)

P Phosphate group

Nitrogenous base

Sugar

OH in RNA

H in DNA

Unique to RNA

5′

Phosphate group

Nitrogenous base

Nucleotide

Ribose sugar

Phosphodiester bonds

3′

OH

(c) RNA (single-stranded)

Sugar-phosphate "backbone"

Nitrogenous base

Deoxyribose sugar

Nucleotide

Phosphate group

5′

Unique to DNA

Hydrogen bonds between nitrogenous bases

3′

(d) DNA (double-stranded)

Figure 2.21 Nucleic Acids. (*a*) A general representation of a nucleotide monomer, which is composed of a pentose sugar (ribose or deoxyribose), a phosphate functional group, and a nitrogenous base. Nucleotides that contain ribose are ribonucleotides, and nucleotides that contain deoxyribose are deoxyribonucleotides. (*b*) The five nitrogenous bases. (*c*) RNA is a single-stranded nucleic acid formed from repeating units of ribonucleotides linked together by phosphodiester bonds. Each ribonucleotide contains one of the nitrogenous bases uracil, guanine, adenine, or cytosine. (*d*) DNA is a double-stranded nucleic acid with each strand composed of repeating units of deoxyribonucleotides linked together by phosphodiester bonds. Each deoxyribonucleotide contains one of the nitrogenous bases thymine, guanine, adenine, or cytosine. Hydrogen bonds between complementary bases (T:A and C:G) hold the two strands together. Both RNA and DNA participate in protein formation.

determine the types of proteins synthesized within cells—a process described in detail in section 4.8.

The two classes of nucleic acid are deoxyribonucleic acid (DNA) and ribonucleic acid (RNA). Both DNA and RNA are polymers composed of **nucleotide** monomers. These monomers are linked together through covalent bonds between one nucleotide monomer and an adjacent nucleotide monomer. This covalent linkage is called a **phosphodiester bond.**

The Nucleotide Monomer

A nucleotide monomer has three components: a sugar, a phosphate functional group, and a nitrogenous base (figure 2.21a). The sugar is a five-carbon pentose sugar (deoxyribose for DNA and ribose for RNA). A phosphate functional group is attached at carbon number five; a nitrogenous base is attached to the same sugar but at carbon number one. A **nitrogenous base** has either a single-ring or a double-ring structure that contains both carbon and nitrogen within the ring.

Five different nitrogenous bases commonly occur in nucleic acids (figure 2.21b). Single-ring nitrogenous bases are called **pyrimidines,** and they include cytosine (C), uracil (U), and thymine (T). (You can remember these bases with the acronym CUT.) Double-ring nitrogenous bases are called **purines,** which include adenine (A) and guanine (G). Nitrogenous bases within either group—pyrimidines or purines—differ in the functional groups attached to the ring.

Deoxyribonucleic Acid (DNA)

Deoxyribonucleic acid (DNA) is a double-stranded nucleic acid (figure 2.21d); it can be found as a component of chromosomes within the nucleus. A small circular strand of DNA is also within mitochondria. (These cellular components are described in chapters 3 and 4.) The nucleotides that form DNA have a deoxyribose sugar, a phosphate, and one of four nitrogenous bases: adenine, guanine, cytosine, or thymine. Notice that DNA does not contain uracil. The double strands of nucleic acid are held together by hydrogen bonds formed between complementary nitrogenous bases: thymine with adenine and guanine with cytosine.

Ribonucleic Acid (RNA)

Ribonucleic acid (RNA) is a single-stranded nucleic acid located both within the cell nucleus and within the cytoplasm of the cell (figure 2.21c). The nucleotides that are part of RNA molecules are composed of the sugar ribose, a phosphate, and one of four nitrogenous bases: adenine, guanine, cytosine, or uracil. RNA does not contain thymine. **Table 2.5** compares the chemical structural differences between RNA and DNA.

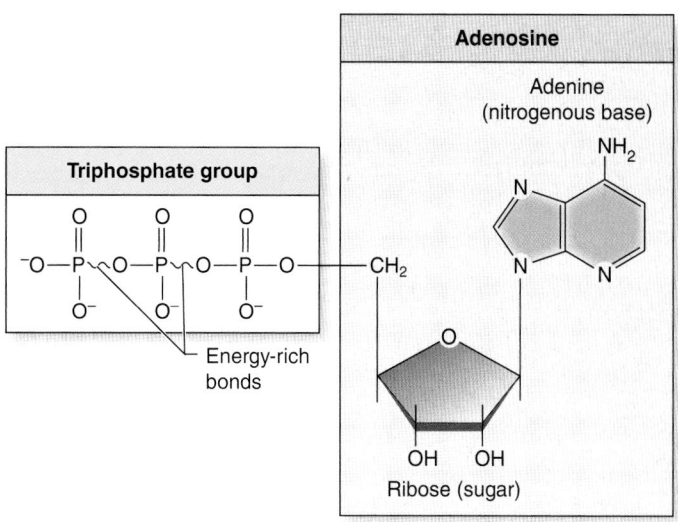

Figure 2.22 ATP. Adenosine triphosphate (ATP) is composed of the sugar ribose and the nitrogenous base adenine, which together are referred to as adenosine. Three phosphates attached to adenosine form adenosine triphosphate.

Other Important Nucleotides and Nucleotide-containing Molecules

An important nucleotide is **adenosine triphosphate** (a-den′ō-sēn trī-fos′fāt), or **ATP.** It is composed of the nitrogenous base adenine, a ribose sugar, and three phosphate groups covalently linked **(figure 2.22).** ATP is the central molecule in the transfer of chemical energy within cells. Biologists often refer to this molecule as the "energy currency" of a cell. The covalent phosphate bond linkages between the last two phosphate groups are unique, energy-rich bonds. ATP is stored in all cells in limited amounts and is produced continuously. It is generally used immediately. When ATP molecules are split into adenosine diphosphate (ADP) and phosphate, energy is released. This energy is then used by the cell (e.g., muscle contraction).

Two important nucleotide-containing molecules are *nicotinamide adenine dinucleotide (NAD⁺)* and *flavin adenine dinucleotide (FAD).* Both molecules participate in the production of ATP that occurs in cellular mitochondria. (ATP, NAD⁺, and FAD are discussed in more detail in chapter 3.)

WHAT DID YOU LEARN?

34 What is the general function of nucleic acids?

35 What are the structural differences between RNA and DNA?

2.7e Proteins

LEARNING OBJECTIVES

14. List the general functions of proteins.

15. Describe the general structure of amino acids and proteins.

Scientists estimate that between 50,000 and 100,000 different proteins are synthesized by cells. Following their synthesis they function within a cell, within a plasma membrane, blood plasma, and other body fluids. Proteins serve a vast array of functions. For example, proteins

- Serve as *catalysts* (enzymes) in most metabolic reactions of the body (section 3.3)

Table 2.5	Differences Between RNA and DNA	
Characteristic	**RNA**	**DNA**
Number of strands	1	2
Sugar	Ribose	Deoxyribose
Nitrogenous base	Uracil (unique to RNA)	Thymine (unique to DNA)

- Act in *defense,* which occurs, for example, when immunoglobulins (antibodies) attach to foreign substances for their elimination (section 22.8)
- Aid in *transport,* such as hemoglobin molecules transporting respiratory gases within the blood (section 18.3b)
- Contribute to structural *support,* such as collagen, a major component of ligaments and tendons (section 5.2d)
- Cause *movement,* when myosin and actin proteins interact during contraction of muscle tissue (section 10.2b)
- Perform *regulation,* such as occurs when insulin helps control blood glucose levels (section 17.9)
- Provide *storage,* such as ferritin, which stores iron in liver cells (section 18.3b)

Table 2.6 organizes the functions of proteins into several categories providing the general function, class of protein, and examples for each.

General Protein Structure

Proteins are polymers composed of one or more linear strands of **amino acid** monomers that may number in the thousands **(figure 2.23)**. Twenty different amino acids are normally found in the proteins of living organisms. Each amino acid has both an amine ($-NH_2$) functional group

Table 2.6	Protein Functions	
Function	**Class of Protein**	**Examples of Proteins and Their Functions**
Catalysts	Enzymes	Hydrolytic enzymes: Cleave polysaccharides
		Isomerases: Convert a molecule to its isomer
		DNA polymerase: Synthesizes DNA
		Kinases: Transfer phosphate functional groups
Defense	Immunoglobulins	Antibodies: "Tag" foreign proteins for elimination
	Cell surface antigens	Major histocompatibility complex (MHC) proteins: Serve as "self" recognition molecules
Transport	Circulating transporters	Hemoglobin: Carries O_2 and CO_2 in blood
		Transferrin: Carries iron in the blood
	Membrane transporters	Cytochromes: Participate in electron transport
		Sodium–potassium pump: Participates in establishing a resting membrane potential
		Glucose transporter: Transports glucose across plasma membranes
Support	Supporting proteins	Collagen: Forms ligaments, tendons
		Keratin: Forms hair, nails
		Fibrin: Forms blood clots
Movement	Contractile proteins	Actin: Participates in contraction of muscle fibers
		Myosin: Participates in contraction of muscle fibers
Regulation	Osmotic proteins	Albumin: Maintains osmotic concentration of blood
	Hormones	Insulin: Controls blood glucose levels
		Antidiuretic hormone (ADH): Increases water retention by kidneys
		Oxytocin: Stimulates uterine contractions and milk ejection
	Molecular chaperones	Protein disulfide isomerase: Participates in folding of proteins that involve intramolecular interactions
Storage	Metal binding proteins	Ferritin: Stores iron in liver cells
		Casein: Binds iron in breast milk
	Ion binding proteins	Calmodulin: Binds calcium ions in sarcoplasmic reticulum (SR) of muscle cells

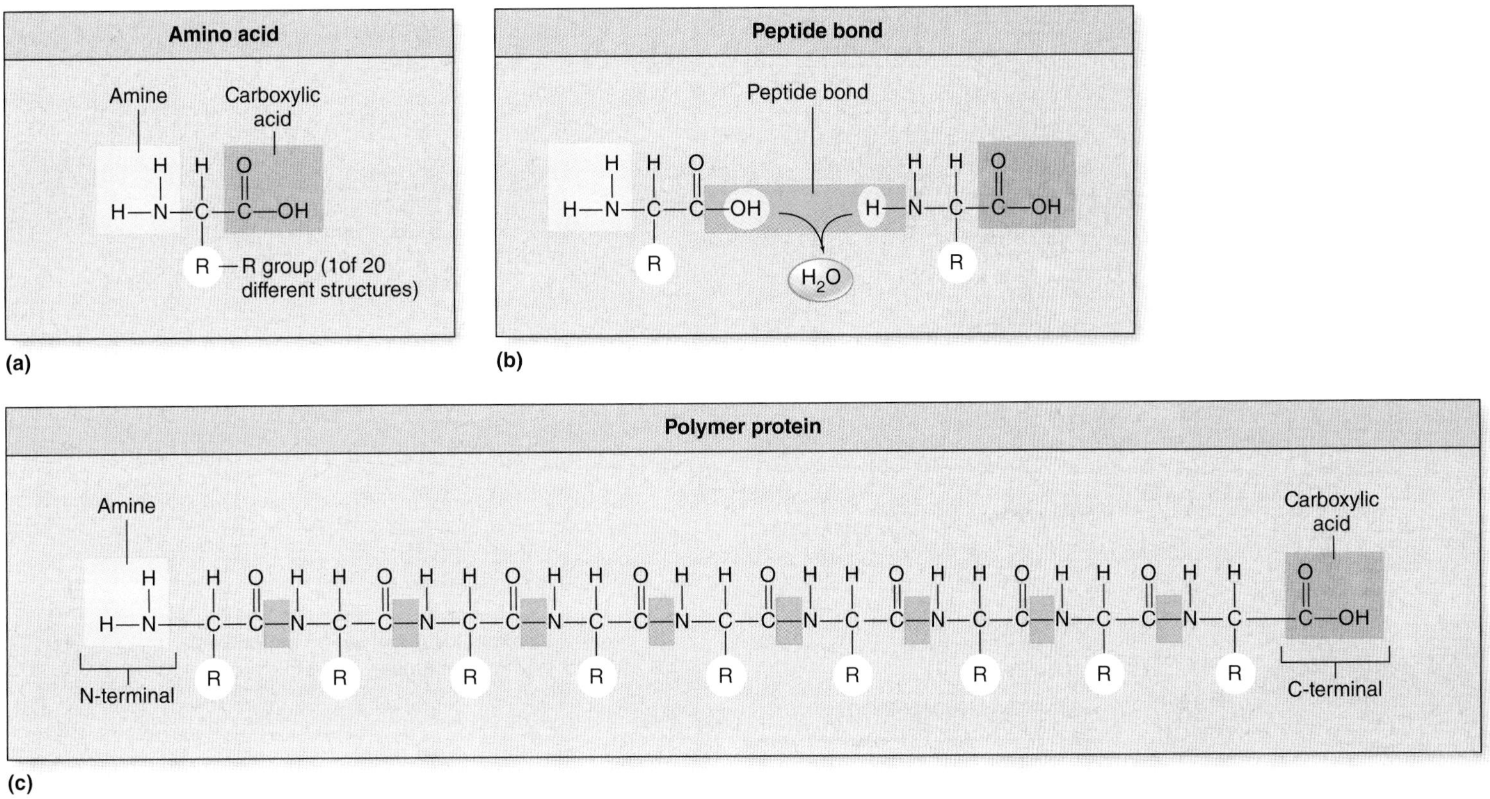

Amino acid

Amine   Carboxylic acid

$$H-N-C-C-OH$$

R — R group (1of 20 different structures)

(a)

Peptide bond

Peptide bond

H_2O

(b)

Polymer protein

Amine   Carboxylic acid

N-terminal   C-terminal

(c)

Figure 2.23 Proteins. (*a*) Amino acids are the monomers of proteins. (*b*) Dehydration synthesis reaction occurs as a hydrogen atom is removed from the amine of one amino acid and a hydroxyl functional group is removed from the carboxylic acid of another amino acid to form a peptide bond between two amino acids. A molecule of water is formed during the process. (*c*) The polymer protein is composed of repeating amino acid monomers that are held together by peptide bonds.

and a carboxyl (—COOH) functional group, which is also called a carboxylic acid functional group (see table 2.3). Both functional groups are covalently linked to the same carbon atom, which accounts for the general name "amino acid" for these monomers. The other two covalent bonds to this carbon are a hydrogen (—H) and different side-chain structures that are simply referred to as the **R (Remainder) group.** The R groups distinguish different amino acids from one another. Properties of the R groups form the basis for classifying amino acids (described in section 2.8).

Amino acids are linked covalently by **peptide bonds** that form during dehydration synthesis reactions between the amine functional group of one amino acid and the carboxylic acid functional group of a second amino acid. An —H is lost from the amine group and an —OH from the carboxylic acid of another amino acid. The ends of a protein are distinguished as the **N-terminal** end, which has a free amine group, and the **C-terminal** end, which has a free carboxyl group.

A strand of amino acids that includes between 3 and 20 amino acids is termed an **oligopeptide,** whereas a **polypeptide** has a strand composed of between 21 and 199 amino acids. If more than 200 amino acids are linked, the structure is called a **protein.** ((Note that the specific numbers of amino acids for each category of protein are debated by experts.) We use the general term *protein,* in this text, to apply to all three types of structures.)

Proteins with carbohydrate attached are called **glycoproteins.** ABO blood groups are based on structural differences in glycoproteins within the plasma membrane of erythrocytes (see section 18.3b).

Lipids, carbohydrates, nucleic acids, and proteins are four major classes of organic biomolecules that compose the human body. They are summarized in **figure 2.24.**

INTEGRATE

CONCEPT CONNECTION

The amine group of an amino acid can be removed from the amino acid by a process known as **deamination.** Nitrogen in the amine group is converted into **urea,** one form of nitrogenous waste, through a metabolic process called the urea cycle. Other types of nitrogenous wastes include uric acid, a waste product of the breakdown of nucleic acid, and creatinine, a waste product of muscle tissue breakdown. Nitrogenous waste is eliminated from the body by the urinary system (see chapter 24).

WHAT DID YOU LEARN?

36 What are the monomers of proteins and the name of the bond between them?

37 What are the names of structures that contain: 2 amino acids, 3 to 20 amino acids, 21 to 199 amino acids, and 200 or more amino acids? What general term is used to refer to any of these structures, except a structure composed of 2 amino acids?

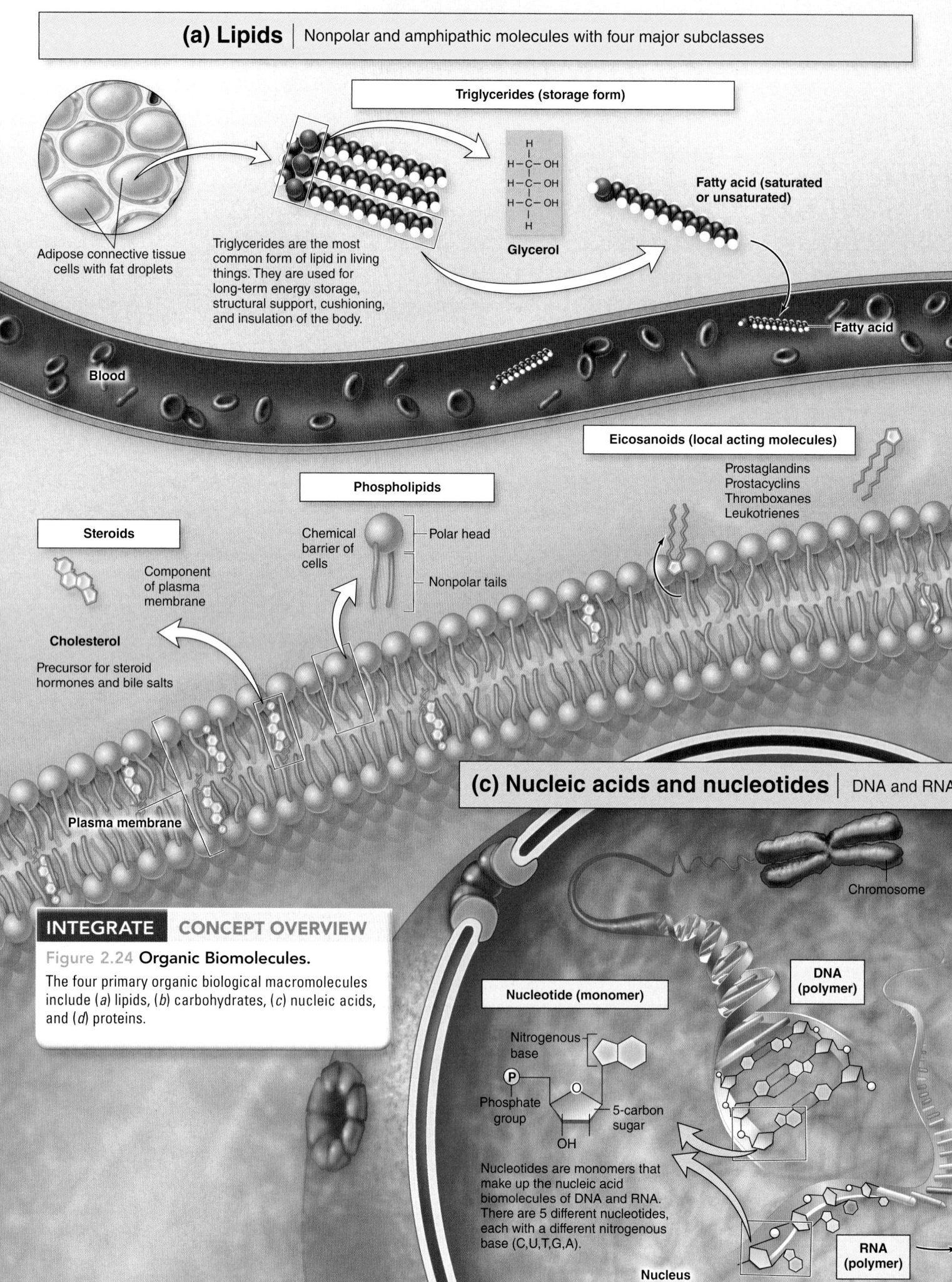

(a) Lipids | Nonpolar and amphipathic molecules with four major subclasses

Triglycerides (storage form)

Adipose connective tissue cells with fat droplets

Triglycerides are the most common form of lipid in living things. They are used for long-term energy storage, structural support, cushioning, and insulation of the body.

Glycerol

Fatty acid (saturated or unsaturated)

Blood

Fatty acid

Eicosanoids (local acting molecules)

Prostaglandins
Prostacyclins
Thromboxanes
Leukotrienes

Phospholipids

Chemical barrier of cells

Polar head

Nonpolar tails

Steroids

Component of plasma membrane

Cholesterol

Precursor for steroid hormones and bile salts

Plasma membrane

(c) Nucleic acids and nucleotides | DNA and RNA

Chromosome

DNA (polymer)

INTEGRATE CONCEPT OVERVIEW

Figure 2.24 **Organic Biomolecules.**

The four primary organic biological macromolecules include (*a*) lipids, (*b*) carbohydrates, (*c*) nucleic acids, and (*d*) proteins.

Nucleotide (monomer)

Nitrogenous base

Phosphate group

5-carbon sugar

OH

Nucleotides are monomers that make up the nucleic acid biomolecules of DNA and RNA. There are 5 different nucleotides, each with a different nitrogenous base (C,U,T,G,A).

Nucleus

RNA (polymer)

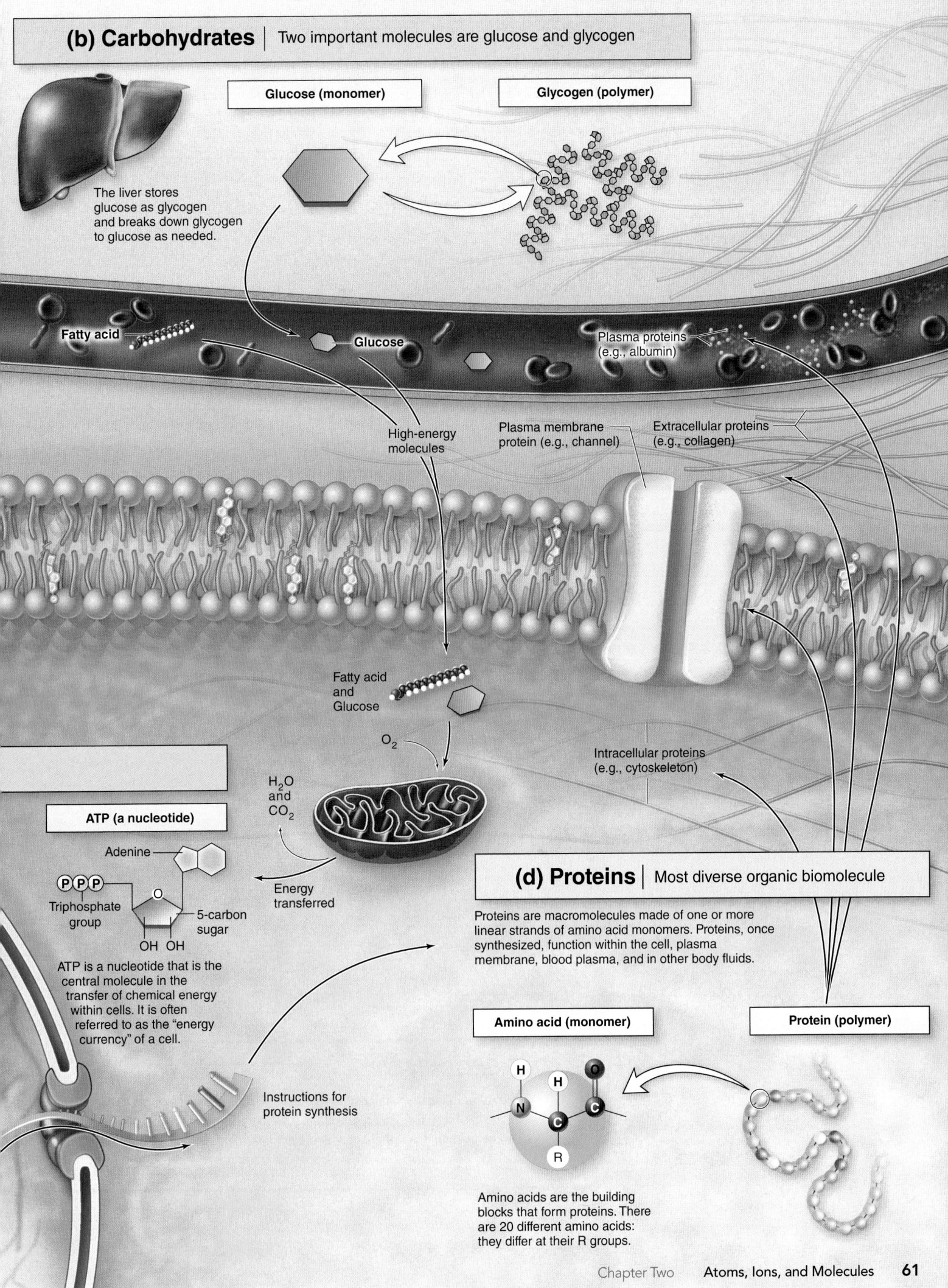

(b) Carbohydrates | Two important molecules are glucose and glycogen

Glucose (monomer)

Glycogen (polymer)

The liver stores glucose as glycogen and breaks down glycogen to glucose as needed.

Fatty acid

Glucose

Plasma proteins (e.g., albumin)

High-energy molecules

Plasma membrane protein (e.g., channel)

Extracellular proteins (e.g., collagen)

Fatty acid and Glucose

O_2

H_2O and CO_2

ATP (a nucleotide)

Adenine

(P)(P)(P)

Triphosphate group

5-carbon sugar

OH OH

ATP is a nucleotide that is the central molecule in the transfer of chemical energy within cells. It is often referred to as the "energy currency" of a cell.

Energy transferred

Instructions for protein synthesis

Intracellular proteins (e.g., cytoskeleton)

(d) Proteins | Most diverse organic biomolecule

Proteins are macromolecules made of one or more linear strands of amino acid monomers. Proteins, once synthesized, function within the cell, plasma membrane, blood plasma, and in other body fluids.

Amino acid (monomer)

Protein (polymer)

H

H

O

N

C

C

R

Amino acids are the building blocks that form proteins. There are 20 different amino acids; they differ at their R groups.

2.8 Protein Structure

Here we describe protein structure in more depth, including categories of amino acids, amino acid sequence, and three-dimensional protein shape. We also discuss how a protein can lose its three-dimensional shape and the consequences of this occurrence.

2.8a Categories of Amino Acids

LEARNING OBJECTIVES

1. Name the categories of amino acids.
2. Distinguish between nonpolar, polar, and charged amino acids.
3. Give examples of amino acids with special characteristics.

Amino acids are organized into groups based on the chemical characteristics of their R group. They are nonpolar, polar, charged, and those having special functions (**figure 2.25**).

- **Nonpolar amino acids** contain R groups with either hydrogen (glycine) or hydrocarbons (alanine, valine, isoleucine, leucine, phenylalanine, and tryptophan). They tend to group with other nonpolar amino acids by hydrophobic interactions within the body's aqueous environment.

- **Polar amino acids** contain R groups with elements in addition to carbon and hydrogen (e.g., O, N, or S) (serine, threonine, asparagine, glutamine, and tyrosine). They form interactions with other polar amino acids and with water molecules.

- **Charged amino acids** can have either a negative charge or a positive charge. Those with a negatively charged R group include glutamate and aspartate and those with a positively charged R group include histidine, lysine, and arginine. An ionic bond (electrostatic interaction) can form between an R group with a negative charge and an R group with a positive charge. Groups of amino acids that are either polar or charged are hydrophilic, and their presence increases the solubility of the protein in water.

- **Amino acids with special functions**—three amino acids have unique characteristics. The R group in proline attaches to the amine group, forming a ring. Proline amino acids cause a bend in the protein chain. The sulfhydryl (—S—H) functional groups of two cysteine amino acids form **disulfide bonds.** These covalent bonds are significant in stabilizing the folding of a protein (described next). Methionine is always the first amino acid positioned when a protein is synthesized (see section 4.8b), and it may or may not be removed later.

WHAT DID YOU LEARN?

 Why is the amino acid leucine (figure 2.25) classified as a nonpolar amino acid?

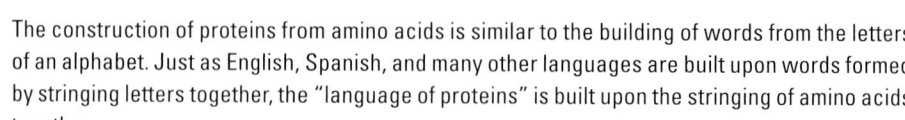

INTEGRATE

LEARNING STRATEGY

The construction of proteins from amino acids is similar to the building of words from the letters of an alphabet. Just as English, Spanish, and many other languages are built upon words formed by stringing letters together, the "language of proteins" is built upon the stringing of amino acids together.

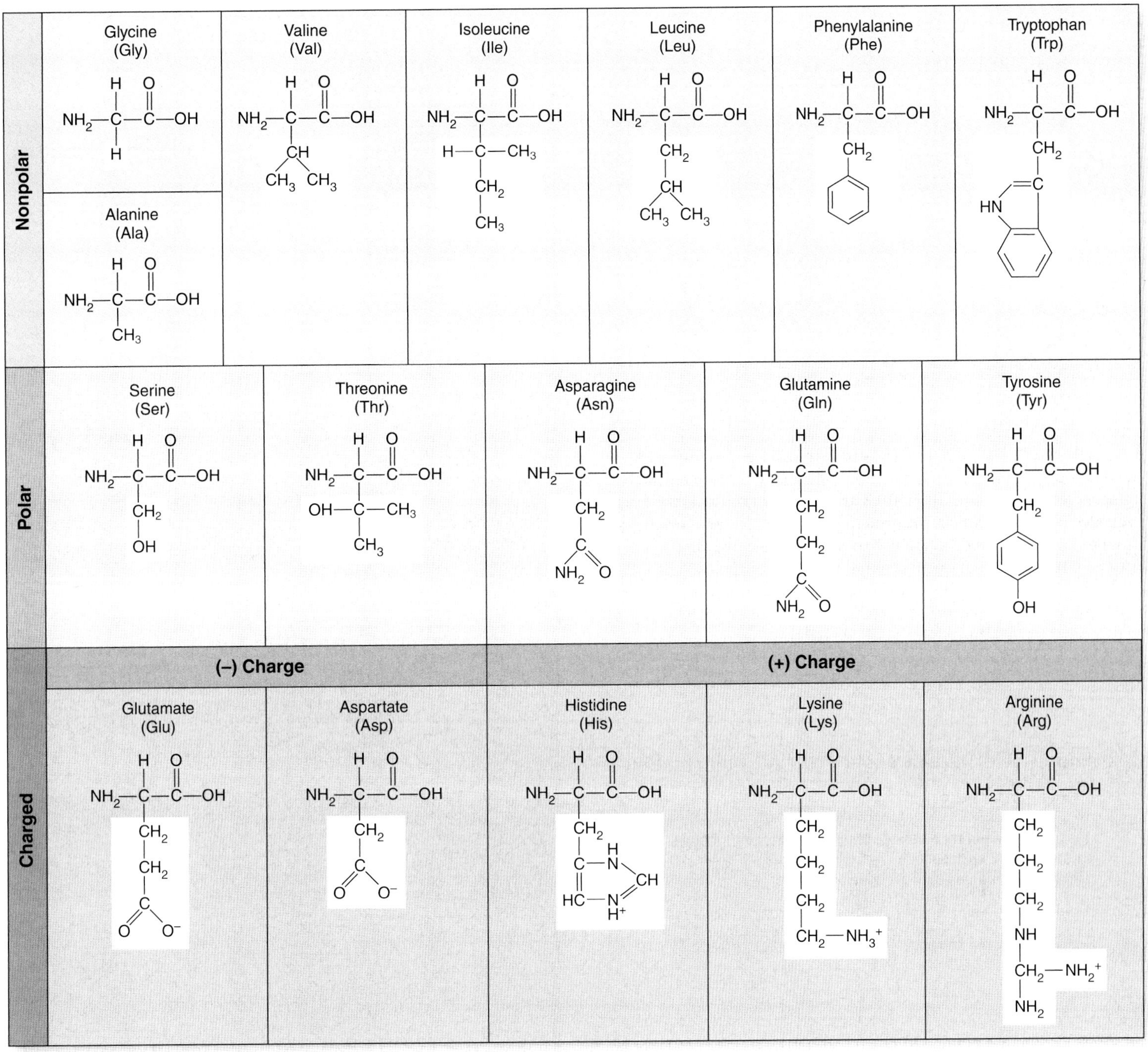

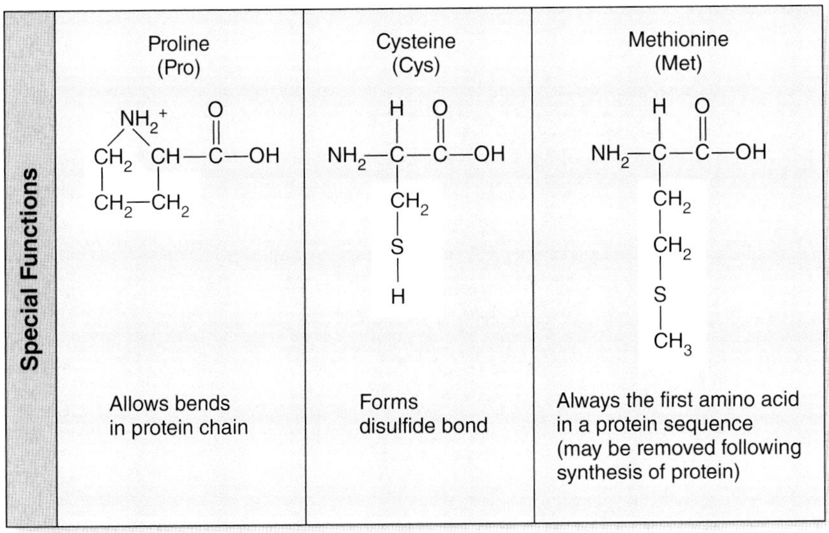

Figure 2.25 Amino Acids. Amino acids are categorized based on the chemical properties of the R groups. The categories include nonpolar, polar, charged, and those with special functions.

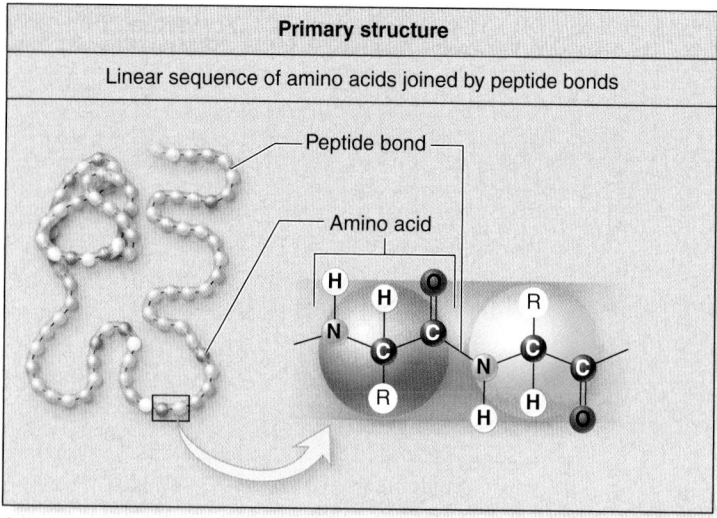

Primary structure

Linear sequence of amino acids joined by peptide bonds

Peptide bond

Amino acid

(a)

Figure 2.26 Levels of Protein Structure. Amino acids bonded together form protein polymers. There are four increasingly complex levels of organization. (*a*) The primary structure is the linear sequence of amino acids in the protein. (*b*) The secondary structures of a protein may include alpha helixes and beta-pleated sheets. (*c*) The tertiary structure is the completed 3-dimensional shape or conformation of the protein, which may be a globular or fibrous protein. (*d*) A quaternary structure is formed in some complex proteins when two or more protein molecules associate to form the final protein.

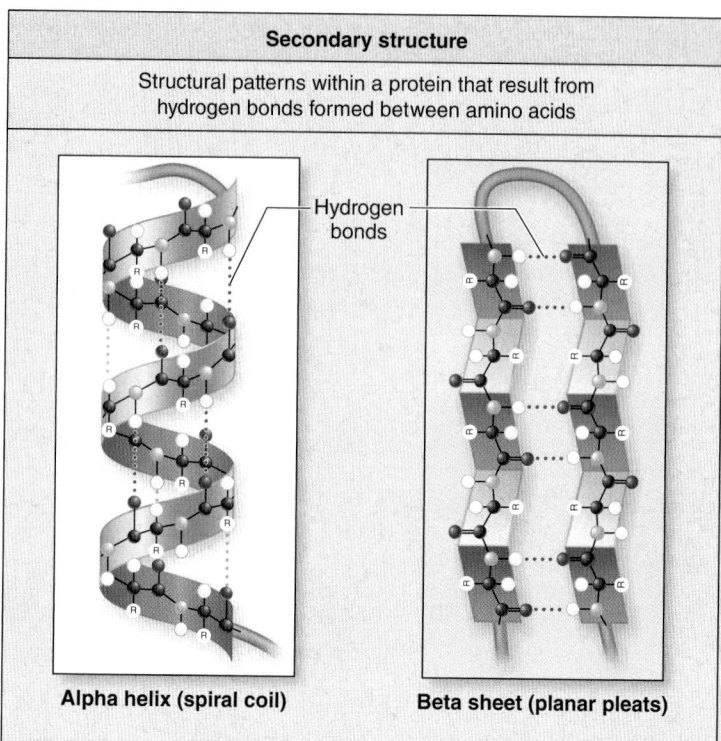

Secondary structure

Structural patterns within a protein that result from hydrogen bonds formed between amino acids

Hydrogen bonds

Alpha helix (spiral coil)   **Beta sheet (planar pleats)**

(b)

2.8b Amino Acid Sequence and Protein Conformation

LEARNING OBJECTIVES

4. Describe the different types of intramolecular (or intermolecular) attractions that participate in both the folding of a protein and in maintaining its three-dimensional shape.

5. Distinguish between the four structural hierarchy levels of proteins.

6. Explain what is meant by denaturation and list factors that can cause it.

We describe a protein as being composed of a linear sequence of amino acids that are bonded together through covalent peptide bonds. This sequence is called its **primary structure (figure 2.26*a*).** The protein then folds to form into its three-dimensional shape, or **conformation.** Protein conformation is crucial for its proper function.

The conformational structure of proteins has increasingly complex hierarchies, or levels, of organization beyond a protein's primary structure, including secondary, tertiary, and perhaps quaternary structural levels. These more complex structural organizations are dependent upon intramolecular attractions between the amino acids in the linear sequence for the proper folding and maintaining of a protein's conformation. The process of protein folding is assisted by specialized proteins called *chaperones* that "direct" the folding process.

The intramolecular interactions that contribute to the final conformation of a protein are as follows:

- The primary protein structure is forced into its initial shape as hydrophobic exclusion "tucks" amino acids with nonpolar R groups into a more central location, limiting their contact with water.

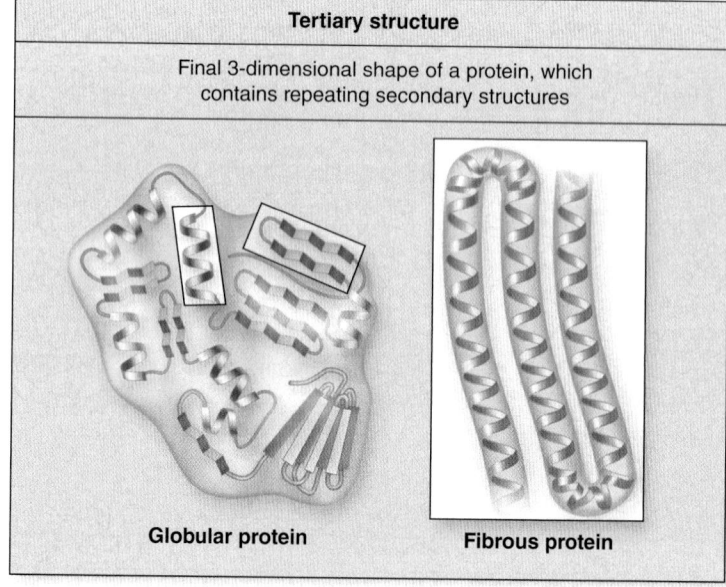

Tertiary structure

Final 3-dimensional shape of a protein, which contains repeating secondary structures

Globular protein   **Fibrous protein**

(c)

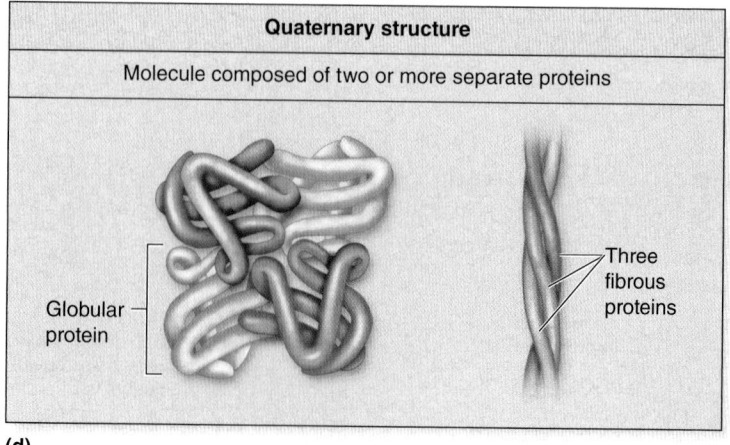

Quaternary structure

Molecule composed of two or more separate proteins

Globular protein

Three fibrous proteins

(d)

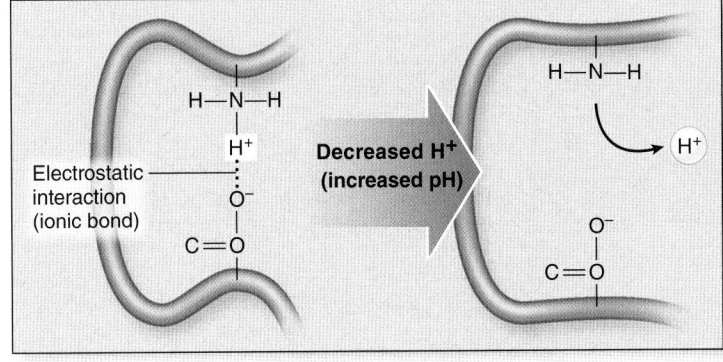

Figure 2.27 Denaturation. Proteins denature in response to (*a*) increased H⁺ (decrease in pH) as H⁺ is added to the negatively charged R group or (*b*) decreased H⁺ (increase in pH) as H⁺ is removed from the positively charged R group.

Protein
(tertiary
structure)

Electrostatic interaction (ionic bond)   H—N—H   H⁺   O⁻   C=O

Increased H⁺ (decreased pH)

H—N—H   H⁺   OH   C=O   H⁺

(a)

Electrostatic interaction (ionic bond)   H—N—H   H⁺   O⁻   C=O

Decreased H⁺ (increased pH)

H—N—H   O⁻   C=O   H⁺

(b)

- Hydrogen bonds form between polar R groups, and between the amine and carboxylic acid functional groups of closely positioned amino acids.
- Electrostatic interactions (ionic bonds) form between negatively charged and positively charged R groups.
- Disulfide bonds form between the sulfhydryl (—S—H) groups of two cysteine amino acids.

Higher levels of protein organization result from folding. **Secondary structures** are patterns within a protein that may repeat several times. The two distinctive secondary levels of organization are a spiral coil, called an **alpha helix** and a planar pleat arrangement, called a **beta sheet.** These secondary structures confer unique characteristics on the regions of the protein where they occur. Alpha helix gives some elasticity to fibrous proteins that are located, for example, in skin or hair. In contrast, beta sheets give some degree of flexibility to many globular proteins (e.g., enzymes).

Tertiary structure is the final three-dimensional shape exhibited by one completed polypeptide chain. Two categories of proteins—either fibrous or globular—are distinguished by their molecular shape. For example, **globular proteins** fold into a compact, often nearly spherical shape such as enzymes, antibodies, and some hormones. In comparison, **fibrous proteins** are extended linear molecules that are constituents of ligaments, tendons, and contractile proteins within muscle cells.

The **quaternary structure** of a protein is present only in those proteins with two or more polypeptide chains. The protein hemoglobin is an example because it is composed of four polypeptide chains. Each of these chains has its own primary, secondary, and tertiary structures. Only when the constituent polypeptides associate through intermolecular attractions to form the quaternary structure does the biological molecule become active. Therefore, hemoglobin is functional only when all four polypeptide chains are present in the correct association.

The normal function of a protein may also require a **prosthetic group.** These are nonprotein structures covalently bonded to the protein. The lipid heme group in the hemoglobin protein is an example of a prosthetic group (see figure 18.6).

WHAT DO YOU THINK?

5 Why might proteins be rendered nonfunctional by exposure to high temperatures?

The biological activity of a protein is usually disturbed or terminated when its conformation is changed. This is called **denaturation** (dē-na'tyū-rā"shŭn). The tertiary structure of the protein is disrupted if the protein is heated or chemically altered. Usually, the loss of biological activity cannot be reversed. Denaturation may occur as a consequence of increased temperature because this increase weakens the intramolecular interactions that hold the protein in its three-dimensional shape. Alternatively, denaturation may take place in response to changes in pH.

Changes in pH can denature proteins because the alteration of hydrogen ion concentration [H⁺] interferes with electrostatic interactions (and other intramolecular bonds) that hold the protein in its three-dimensional shape. If pH decreases as a consequence of an increase in [H⁺], the excess H⁺ binds to negatively charged R groups, as shown in **figure 2.27a.** The electrostatic interaction is disrupted because an H⁺ attaches to the negatively charged R group, with an accompanying loss of the negatively charged R group that participated in the electrostatic interaction. If pH increases, reflecting a decrease in [H⁺], the electrostatic interaction is disrupted because the H⁺ that previously participated in the electrostatic interaction has been removed, as shown in figure 2.27b. For this reason, changes in blood pH out of the normal homeostatic range can be lethal (acid-base balance is described in section 25.5.) Proteins are essential in the normal processes of the body and the structure and function of many different proteins will be discussed throughout this text.

WHAT DID YOU LEARN?

39 What distinguishes the tertiary and quaternary level of organization of a protein?

40 What happens to a protein when it denatures? How does a protein denature when it is exposed to higher than normal H⁺ concentration?

- Atoms, ions, and molecules form the human body at its simplest level.

2.1 Atomic Structure 30

- Diagramming atomic structure from the periodic table provides the foundation to understand isotopes, ions, and molecules.

2.1a Matter, Atoms, Elements, and the Periodic Table 30

- An atom is the smallest particle that exhibits the chemical properties of an element.
- Atomic structure—composed of protons, neutrons, and electrons—can be drawn from the information obtained from the periodic table.

2.1b Isotopes 32

- Atoms having the same number of protons and electrons—but differing in the number of neutrons and thus atomic mass—are called isotopes.
- Unstable isotopes resulting from an excess number of neutrons are called radioisotopes.

2.1c Chemical Stability and the Octet Rule 32

- Atoms with a complete outer shell of eight electrons are chemically stable. Atoms bond to reach chemical stability.

2.2 Ions and Ionic Compounds 34

- Chemical compounds, which include ionic compounds and molecular compounds, are stable associations between two or more elements combined in a fixed ratio.

2.2a Ions 34

- An ion is a charged atom that is either positively charged or negatively charged as a result of having lost or gained electrons, respectively.
- Common ions of the human body include sodium (Na^+), potassium (K^+), calcium (Ca^{2+}), magnesium (Mg^{2+}), hydrogen (H^+), chloride (Cl^-), bicarbonate (HCO_3^-), and phosphate (PO_4^{3-}).
- Cations are positively charged ions that are formed by the loss of electron(s) from atoms with one, two, or three electrons in their outer shell, or elements in columns IA, IIA, or IIIA in the periodic table, respectively.
- Anions are negatively charged ions that are formed by the gain of electron(s) by atoms typically with five, six, or seven electrons in the outer shell, or elements in columns VA, VIA, or VIIA in the periodic table, respectively.

2.2b Ionic Bonds 35

- Ionic bonds are electrostatic attractions between positively charged cations and negatively charged anions that hold the ions in a lattice crystal (or salt).

2.3 Covalent Bonding, Molecules, and Molecular Compounds 36

- Molecules formed with two or more atoms of the same or different elements are covalently bonded together.
- Molecular compounds are formed by two different elements covalently bonded together.

2.3a Chemical Formulas: Molecular and Structural 36

- The molecular formula is the number and kinds of atoms within a molecule.
- The structural formula demonstrates both the number and kind of atoms within a molecule and their arrangement within the molecule—and can be used to distinguish isomers.
- Isomers have the same number and kind of atoms, or molecular formula, but differ in their arrangement in space.

2.3b Covalent Bonds 37

- Covalent bonds are formed between two atoms in which both atoms have four, five, six, or seven electrons in their outer shell (except hydrogen).
- A single covalent bond is the typical covalent bond and involves the sharing of a pair of electrons. A double or triple covalent bond can be formed if both atoms, respectively, need at least two or three electrons to become stable.
- Nonpolar covalent bonds occur when the two atoms share electrons equally, as occurs between atoms of the same element, or almost equally, as occurs when hydrogen bonds to carbon. Polar covalent bonds occur between atoms of different elements that share electrons unequally.

2.3c Nonpolar, Polar, and Amphipathic Molecules 39

- Nonpolar molecules generally include those molecules composed of nonpolar bonds.
- Polar molecules generally include those molecules composed predominantly of polar bonds.
- Amphipathic molecules are large molecules that have both nonpolar and polar regions.

2.3d Intermolecular Attractions 40

- Intermolecular attractions occur between molecules, and intramolecular attractions occur between regions within large molecules.
- Hydrogen bonds occur between a partially positively charged hydrogen atom and a partially negatively charged atom of polar molecules.
- Intermolecular attractions between nonpolar molecules include van der Waals forces and hydrophobic interactions.

(continued on next page)

2.7 Biological Macromolecules (continued)	**2.7c Carbohydrates 54**
	• Carbohydrates are molecules with the following chemical formula: $(CH_2O)_n$. Carbohydrates exist in increasing levels of complexity that include monosaccharides, disaccharides, and polysaccharides.
	• Glucose is the most common monosaccharide in the human body and is used for energy. When in excess, glucose is stored as the polysaccharide called glycogen in liver and skeletal muscle tissue.
	• Other types of carbohydrates include the monosaccharides galactose, fructose, ribose, and deoxyribose; the disaccharides sucrose, lactose, and maltose; and the polysaccharides glycogen, starch, and cellulose.
	2.7d Nucleic Acids 55
	• Nucleic acids include deoxyribonucleic acid (DNA) and ribonucleic acid (RNA), polymers formed from nucleotide monomers. These molecules ultimately determine the type of proteins synthesized by cells.
	• Adenosine triphosphate (ATP) is the energy currency molecule of a cell.
	2.7e Proteins 57
	• Proteins serve many different functions in the body.
	• Proteins are polymers that differ in the number and sequence of 20 different amino acid monomers arranged in a unique linear sequence.
2.8 Protein Structure 62	• The three-dimensional structure of proteins is dependent upon the linear sequence of its amino acids.
	2.8a Categories of Amino Acids 62
	• The 20 amino acids can be categorized as nonpolar, polar, charged, and those with special functions.
	2.8b Amino Acid Sequence and Protein Conformation 64
	• Protein organization includes the primary, secondary, tertiary structures—and a quaternary structure if there are two or more protein chains. These levels of organization ultimately determine the structure and function of a protein.
	• Denaturation usually results in the loss of biological activity of a protein in response to the change in its three-dimensional shape that may have occurred through an increase in temperature or a change in pH.

CHALLENGE YOURSELF

Do You Know the Basics?

_____ 1. Atoms composed of the same numbers of protons and electrons, but different numbers of neutrons, are called

 a. isomers. c. isotopes.

 b. ions. d. organic atoms.

_____ 2. Substances that dissolve in water include all of the following *except*

 a. lipids. c. proteins.

 b. glucose. d. salts.

_____ 3. Temperature stabilization is dependent upon which properties of water?

 a. cohesion and adhesion

 b. capillary action and adhesion

 c. specific heat and heat of vaporization

 d. cohesion and specific heat

_____ 4. All of the following are accurate about H^+ concentration and pH *except*

 a. acids contain more H^+ than water.

 b. H^+ concentration and pH are inversely related.

 c. neutralizing an acidic solution requires that base is added.

 d. a pH of 6 is basic or alkaline.

_____ 5. Blood is a mixture that is more specifically described as a

 a. suspension. c. solution.

 b. colloid. d. All of these are correct.

_____ 6. Which organic biomolecule is *not* a polymer?

 a. triglyceride c. glycogen

 b. protein d. DNA

_____ 7. Glucose is stored as which molecule within the liver and skeletal muscle tissue?

 a. starch c. glycogen

 b. phospholipid d. glucagon

_____ 8. All of the following are common ions of the human body *except*

 a. Na^+. c. Ca^{2+}.

 b. P^+. d. Cl^-.

_____ 9. Intermolecular attractions between polar molecules are

 a. hydrophobic interactions.

 b. hydrogen bonds.

 c. van der Waals forces.

 d. ionic bonds.

_____ 10. When a protein permanently unfolds, it has been

 a. polymerized. c. converted to nucleic acids.

 b. denatured. d. made more efficient.

11. List the common ions of the human body by name, symbol, and charge.

12. Describe a polar bond and a polar molecule.

13. Diagram two water molecules and label the polar covalent bonds and a hydrogen bond.

14. Compare and contrast what occurs when a substance dissolves in water with a substance that dissolves and dissociates in water. Include examples of specific substances.

15. Define the terms acid, base, pH, and buffers.

16. Explain the units for expressing a concentration that are included in this chapter.

17. List the four organic biomolecules and the building blocks that compose them.

18. Which two organic biomolecules contain nitrogen atoms and contribute to the formation of nitrogenous waste that must be eliminated by the urinary system?

19. Describe how phospholipid molecules form the plasma membrane of a cell.

20. Explain protein denaturation, including how it occurs and the consequences in response to an increase in temperature and a change in pH.

Can You Apply What You've Learned?

1. Which property of water is significant in children born prematurely because it causes the air sacs to collapse in the lungs, making breathing difficult?
   a. specific heat
   c. surface tension
   b. water reactivity
   d. capillary action

2. A young boy playing outside on a very hot day has become dehydrated. When he enters the house, he appears lethargic. The mother is a nurse and becomes concerned that he may be experiencing a fluid and electrolyte imbalance. Electrolytes include all of the following *except*
   a. sodium ion.
   c. potassium ion.
   b. glucose.
   d. chloride ion.

3. A young woman has noticed that her thyroid appears enlarged. One of the diagnostic procedures used to produce an image of her thyroid requires this type of substance that emits high-energy radiation.
   a. ions
   c. radioisomers
   b. radioisotopes
   d. isomers

4. The condition of rickets involves bones that have insufficient amounts of this common ion, resulting in the bones bending under a child's weight.
   a. Na^+
   c. Cl^-
   b. K^+
   d. Ca^{2+}

5. The hormone insulin is a _____ composed of repeating units of amino acids and cannot be administered orally because the enzymes of the gastrointestinal tract will break the peptide bond through the process of hydrolysis, releasing individual amino acids.
   a. nucleic acid
   c. protein
   b. glycogen
   d. steroid

Can You Synthesize What You've Learned?

1. An individual is exposed to high-energy radiation. Which biomolecule that regulates the process of protein synthesis may have been mutated?

2. The lab results from a diabetic patient show a lower than normal pH (a condition referred to as acidosis). Explain the change in H^+ concentration in the blood, and describe how this change may affect the folding of proteins in the blood plasma (and elsewhere).

3. A patient is given a new drug that decreases blood sugar levels. This drug is regulating which specific molecule?

INTEGRATE

ONLINE STUDY TOOLS  connect |ANATOMY & PHYSIOLOGY LEARNSMART® AP|R

The following study aids may be accessed through Connect.

Clinical Case Study: A True Story of Killing Patients with Potassium Chloride

Interactive Questions: This chapter's content is served up in a number of multimedia question formats for student study.

LearnSmart: Topics and terminology include atomic structure; ions and ionic compounds; covalent bonding, molecules, and molecular compounds; molecular structure of water and the properties of

water; acidic and basic solutions, pH, and buffers; water mixtures; biological macromolecules; and protein structure

Anatomy & Physiology Revealed: Topics include atomic structure, bonds

Animations: Topics include ionic solutions; ionic bond; solutions; making solutions; buffers

Energy, Chemical Reactions, and Cellular Respiration

Anatomy & Physiology **REVEALED**
aprevealed.com

Module 2:
Cells and Chemistry

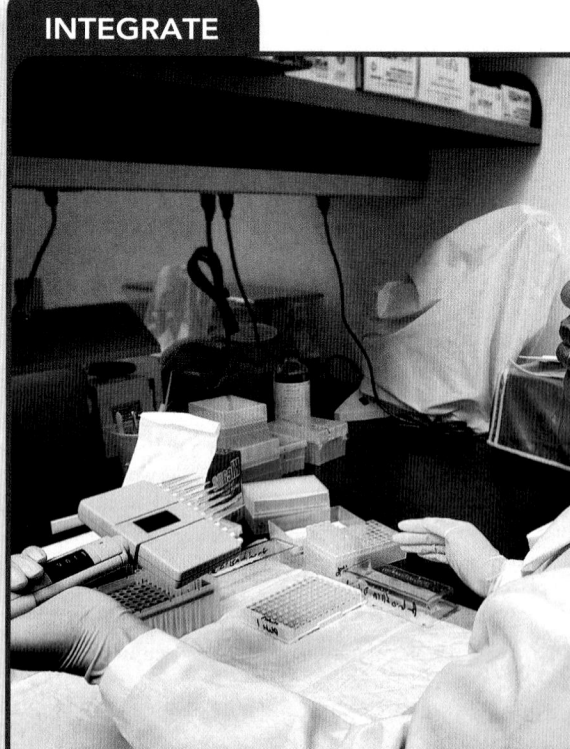

INTEGRATE

All living organisms require energy. In humans, energy is needed to power muscle, pump blood through the body, absorb nutrients from the gastrointestinal tract, and exchange respiratory gases. Energy is also required both to synthesize new molecules for maintenance, growth and repair, and to establish cellular ion concentrations. Here we discuss the chemical principles of energy, chemical reactions, enzymes, and metabolic pathways. Many of these concepts are then integrated to describe the metabolic pathway that breaks down glucose molecules. It is during the breakdown of glucose (and other fuel molecules) that energy is released to synthesize adenosine triphosphate (ATP). ATP is the molecule that serves as the "energy currency" of a cell for movement, synthesis, and for all other energy-requiring cellular processes.

CAREER PATH
Biochemists

Biochemists study the chemical composition and physical principles of living cells and organisms, including those associated with cell metabolism, development, growth, reproduction, and heredity. Their work may involve determining the effects of substances (e.g., foods, vitamins, enzymes, hormones, allergens) on living organisms or studying the effects of mutations that lead to cancer.

3.1 Energy

Energy is defined as the capacity to do work. Energy differs from matter in that it has no mass and does not take up space. It is invisible except for the effects it has on matter. This section describes the two major classes of energy, the various forms of energy, and physical laws that govern energy.

3.1a Classes of Energy

LEARNING OBJECTIVE

1. Describe the two classes of energy.

Two classes of energy exist: potential energy and kinetic energy. **Potential energy** is the energy of position or stored energy. **Kinetic energy** is the energy of motion. Potential energy can be converted or changed to kinetic energy and vice versa. For example, water at the top of a dam has potential energy because of its position; when the water falls over the dam it now has kinetic energy because of its motion. The kinetic energy of falling water can be harnessed to do work if it drives a water wheel at the bottom of the dam.

A bow and arrow also provides an example of an energy conversion from potential energy to kinetic energy. When the arrow is pulled back in the bow, it has potential energy because of the tension of the bowstring. This potential energy is converted to kinetic energy as the string is released and the arrow flies. The kinetic energy of the flying arrow can do work when it knocks an apple from a tree.

Potential energy is exhibited in cells of living organisms when a concentration gradient exists across the plasma membrane, which is the boundary between the inside and outside of a cell **(figure 3.1a)**. Sodium ion (Na^+) concentration typically is greater outside the cell than inside. This difference in Na^+ concentration across the membrane is analogous to the water at the top of a dam because it represents potential energy. The movement of Na^+ from a high concentration outside the cell to a low concentration inside the cell is an example of kinetic energy. Like water falling over a dam, the kinetic energy of Na^+ movement may be harnessed to do work. You will see applications of this principle both in this chapter and in later chapters (e.g., see section 24.6).

Potential energy is also exhibited by the position of electrons in electron shells relative to an atom's nucleus (figure 3.1b). Electrons can move from a higher-energy shell to a lower-energy shell. Note that when electrons move during a chemical reaction, they may do so either within the same chemical structure or from one chemical structure to another (e.g., in the electron transport chain of a mitochondrion, a cell organelle that synthesizes ATP). The kinetic energy of electron movement can be harnessed to do work—movement of electrons is critical to the formation of ATP molecules.

Note that potential energy has the ability, or *potential*, to do work because of its position. Potential energy must be converted to kinetic energy to be actively engaged in doing work.

WHAT DID YOU LEARN?

1 Both the movement of Na^+ down its concentration gradient and the movement of an electron from a higher energy to a lower energy state are examples of (a) potential energy or (b) kinetic energy?

3.1b Forms of Energy

LEARNING OBJECTIVES

2. Describe chemical energy (one form of potential energy) and the various forms of kinetic energy.

3. List the three important molecules within the body that function primarily in chemical energy.

Potential and kinetic energy exist in several forms. We first describe chemical energy, which is one form of potential energy, and then describe several forms of kinetic energy.

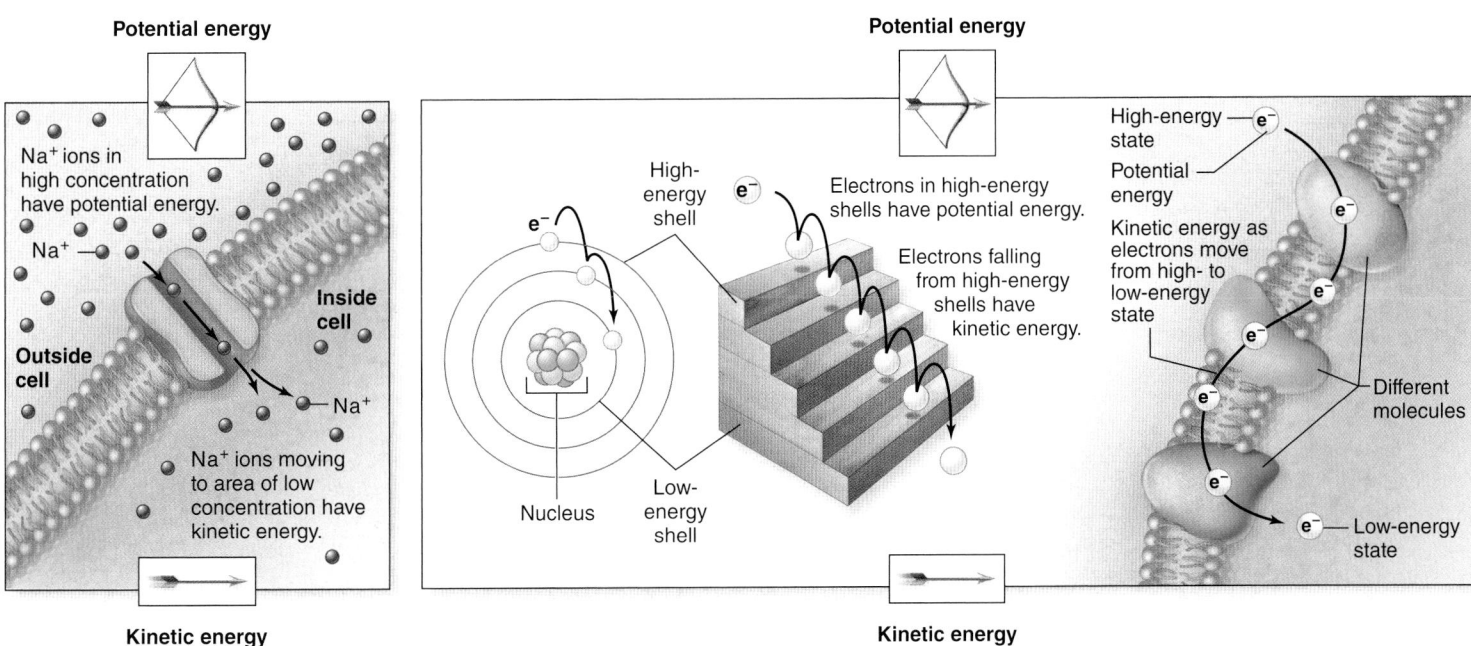

(a) Concentration gradients **(b) Movement of electrons**

Figure 3.1 Conversion of Potential Energy to Kinetic Energy. Potential energy that is converted to kinetic energy may be harnessed to do work. (*a*) The Na^+ gradient across the plasma membrane has potential energy that changes to kinetic energy when Na^+ moves from where it is in high concentration outside of the cell to where it is in low concentration inside the cell. (*b*) A high-energy electron has potential energy and may be converted to kinetic energy either in the same atom or be transferred to different molecules as the electron "falls" from high-energy shells to low-energy shells.

Chemical Energy (A Form of Potential Energy)

Chemical energy is one form of potential energy. **Chemical energy** is the energy stored in a molecule's chemical bonds, and is the most important form of energy in the human body. It specifically is used for the energy-requiring processes of movement, synthesis of molecules, and establishment of concentration gradients. The chemical bonds of all molecules have chemical energy. This energy is released when bonds are broken during chemical reactions.

Three important molecules in the human body function primarily in chemical energy storage: triglycerides, glucose, and adenosine triphosphate (ATP). These molecules differ in their chemical structure, where they are stored, and the length of time each generally stores energy. Recall the following from chapter 2:

- Triglycerides are involved in long-term energy storage in adipose connective tissue.
- Glucose is stored in the liver and muscle tissue in the form of the polymer glycogen.
- ATP is stored in all cells in limited amounts and is produced continuously and used immediately for cells' energy-requiring processes.

Note that protein also stores chemical energy and can be used as a fuel molecule. However, as described in section 2.7e, proteins are primarily important as the structural and functional components of the body.

Kinetic Energy Forms

Other forms of energy—electrical, mechanical, sound, radiant, and heat—exist as kinetic energy.

Electrical energy is the movement of charged particles. Examples include electricity, which is the movement of electrons along a wire, and the propagation of an impulse due to the movement of ions across the plasma membrane of a neuron (nerve cell).

Mechanical energy is exhibited by an object in motion due to an applied force. Examples of mechanical energy include muscle contraction for walking and the pumping action of the heart to circulate blood.

Sound energy occurs when the compression of molecules that move in a solid, liquid, or gas is caused by a vibrating object, such as the head of a drum or the vibration of the vocal cords. The sense of hearing is initiated when sound waves cause vibration of the tympanic membrane (eardrum) in the ear.

Radiant energy is the energy of electromagnetic waves traveling in the universe. Radiant energy consists of a spectrum of different energy forms that vary in wavelength and frequency (**figure 3.2**). The higher the frequency in the spectrum, the greater the amount of radiant energy

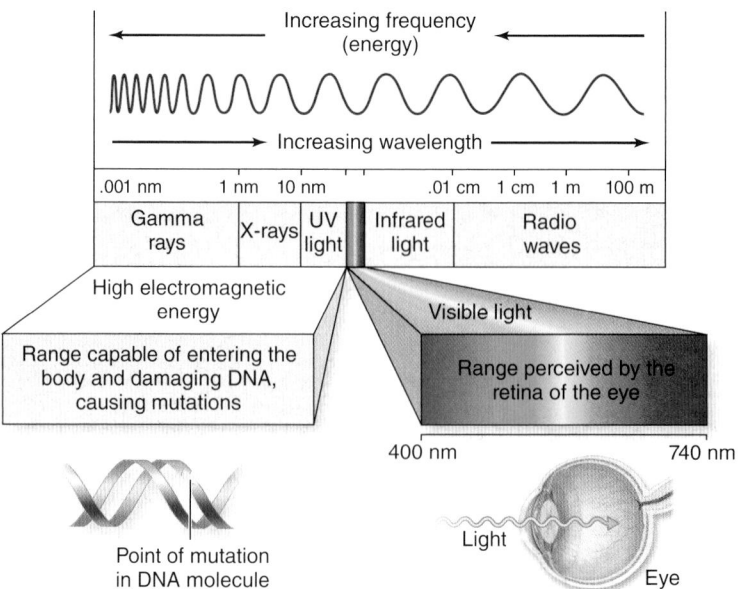

Figure 3.2 The Electromagnetic Spectrum. Different forms of radiant energy are represented on an electromagnetic spectrum. The forms of energy are arranged from highest energy (shortest waves) to lowest energy (longest waves).

associated with it. Gamma rays have the highest amount of radiant energy, whereas radio waves have the lowest. All forms of radiant energy with a frequency higher than visible light (gamma rays, x-rays, and ultraviolet [UV] light) have sufficient energy to penetrate the body and mutate (change) the DNA of living organisms. Cells of the skin normally protect themselves from everyday UV light exposure by producing the pigment melanin (see section 6.1a). This process commonly darkens skin color and is referred to as tanning. Visible light is a lower frequency radiant energy that is detected by retinal cells of the eye. This visual input is then relayed along the optic nerve to the brain for interpretation.

Heat is the kinetic energy associated with random motion of atoms, ions, or molecules. It is usually considered an unusable form of energy, or a "waste product" that accompanies all changes in energy form because heat is the only type of energy that is not available to do work. (An exception is the energy available from a heat gradient, such as occurs in a steam engine.) Heat is measured as the **temperature** of a substance.

 WHAT DID YOU LEARN?

 Muscle contraction is an example of what form of energy?

3.1c Laws of Thermodynamics

 LEARNING OBJECTIVES

4. State the first law and second law of thermodynamics.

5. Explain why energy conversion is always less than 100%.

Energy can be converted from one form to another. For example:

- When a candle is burned, chemical energy in the burning wax is converted to both light and heat.

- Light energy from the sun is converted by the retinal cells into electrical energy associated with a nerve impulse involved in sight.

- Chemical energy in the food we eat is first converted into another chemical form (ATP), which is then converted in our cells into mechanical energy used to power muscle contraction.

In these examples, energy is simply changing from one form to another. The study of energy transformations is called **thermodynamics** (ther′mō-dī-nam′iks; *thermo* = heat, *dynamis* = force).

Two laws describe energy transformations: the first and second laws of thermodynamics. The **first law of thermodynamics** states that energy can neither be created nor destroyed—it can only be transformed or converted from one form to another.

The **second law of thermodynamics** states that every time energy is transformed from one form to another, some of that energy is converted to heat. That means there is never 100% conversion of one form of usable energy to another. Energy conversions have a price, and it always appears as heat. Because heat is not available to do work, the usable amount of energy is decreased each time an energy conversion occurs. For example, the conversion of the chemical energy in gasoline to the mechanical energy (movement) of a car is approximately 25%. This means that approximately 75% of gasoline's chemical energy is converted to sound and heat.

Heat is produced when the chemical energy stored in the foods we eat is used to power our muscle contractions. One of the functions of muscle tissue is to produce heat that keeps the body warm (see section 10.1a). When the environmental temperature drops, and we begin to move around in the hope of generating enough heat to keep ourselves warm, we are applying the second law of thermodynamics.

Figure 3.3 summarizes the two major classes of energy, the various forms of energy as they relate to the body, and the laws of energy.

 WHAT DID YOU LEARN?

 Energy can neither be created nor destroyed. However, according to the first and second laws of thermodynamics, what can happen to it, and what is always generated?

Figure 3.3 **Energy as It Relates to Human Body Function.**
(*a*) Potential and kinetic energy are the two classes of energy. (*b*) Forms of usable energy include chemical energy and various forms of kinetic energy. (*c*) Laws of energy describe how energy can be converted from one form to another and that heat is always produced during the process.

(a) Potential and Kinetic Energy | The Two Classes of Energy

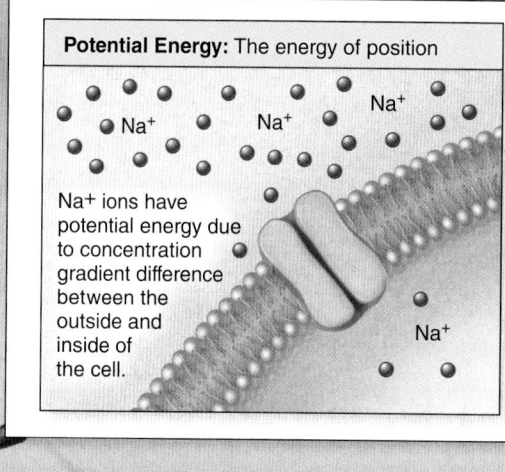

Potential Energy: The energy of position

Na^+ ions have potential energy due to concentration gradient difference between the outside and inside of the cell.

Kinetic Energy: The energy of motion

Na^+ ions exhibit kinetic energy as they move down the concentration gradient.

A perched eagle has potential energy.

When the eagle flies, it converts its potential energy to kinetic energy.

(b) Forms of Usable Energy Available to Do Work

Potential Energy

Chemical Energy: Energy stored in chemical bonds of molecules

Glycogen

Potential energy in chemical bonds

CH_2OH

Example: Glucose, a high-energy molecule, can be stored as glycogen within liver and muscle cells for later use.

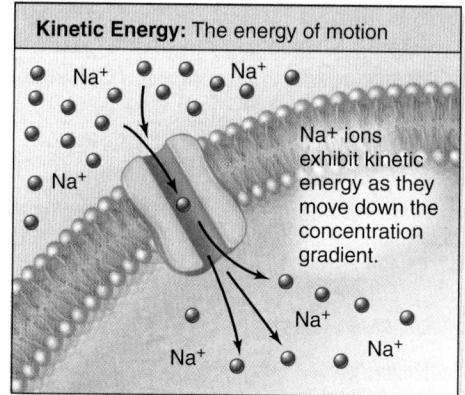

Electrical Energy: Movement of charged particles

Example: The propagation of an impulse in a neuron is due to the movement of charged ions across the plasma membrane.

(c) Laws of Energy

Conversion of energy from one form to another in the human body produces heat and helps to maintain homeostasis.

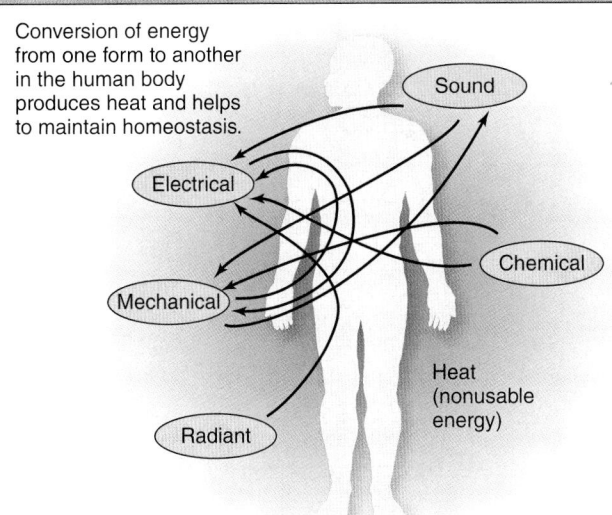

Sound
Electrical
Chemical
Mechanical
Radiant
Heat (nonusable energy)

First Law of Thermodynamics

Energy cannot be created or destroyed, it can only be converted from one form to another.

Second Law of Thermodynamics

Every time energy is transformed from one form to another, some of that energy is converted to heat.

Energy is constantly being converted from one form to another in order to perform the tasks that keep the human body alive and active.

(Heat radiating from forearm)

Kinetic Energy

Mechanical Energy: Movement of a structure or a substance due to an applied force

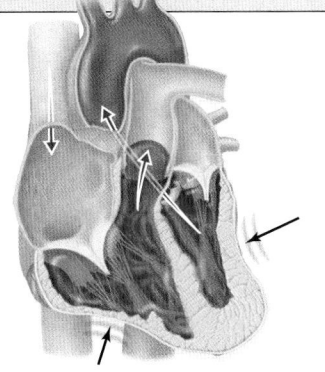

Example: The pumping action of the heart to circulate blood is a form of mechanical energy.

Sound Energy: Movement of compressed molecules through a medium initiated by a vibrating object

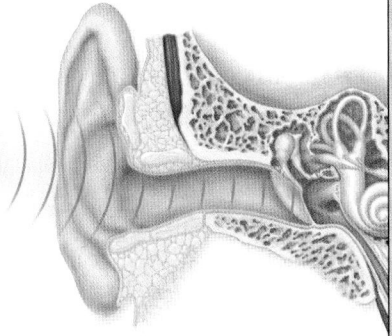

Example: Sound waves vibrating the tympanic membrane of the ear stimulate sensory receptors for hearing.

Radiant Energy: Movement of electromagnetic waves that travel in the universe and vary in wavelength and frequency

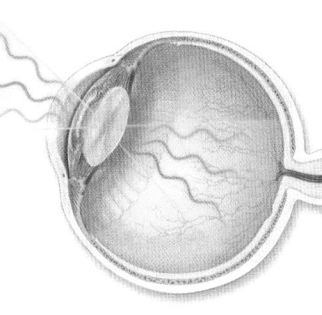

Example: Visible light, a form of radiant energy, is focused on the retina of the eye for vision.

3.2 Chemical Reactions

Chemical energy is used in the energy-requiring processes of cells. To understand the central role of chemical energy in living cells, we need to know about chemical reactions and how they occur.

3.2a Chemical Equations

LEARNING OBJECTIVES

1. Explain what occurs in a chemical reaction.
2. Distinguish between reactants and products.

Millions of chemical reactions are occurring in living organisms at any given time. **Metabolism** is the collective term for all biochemical reactions in living organisms. A **chemical reaction** occurs when chemical bonds in an existing molecular structure are broken and new ones formed to produce a different structure. When chemical structures are changed, a summary of their changes is written as a **chemical equation.** The components of a chemical equation are called reactants and products.

Reactants are the substrates, or substances, that are present prior to the start of the chemical reaction; they are usually written on the left side of the equation. **Products** are substances that are formed by the subsequent chemical reaction and they are generally written on the right side of the equation. For example, a generic chemical reaction is written

$$A + B \longrightarrow C$$

A and B are reactants in this reaction and C is the product. An arrow is used to indicate the reaction direction. A chemical reaction typically has the arrow drawn to the right, indicating a net change of reactants to products.

The number of elements on one side of the reaction is equal to the number on the other side in a balanced chemical equation. For example:

$$Ca^{2+} + 2\ Cl^- \longrightarrow CaCl_2$$

which indicates that one calcium ion combines with two chloride ions to form calcium chloride.

WHAT DID YOU LEARN?

4 What are the differences between reactants and products in a chemical equation?

3.2b Classification of Chemical Reactions

LEARNING OBJECTIVES

3. Describe the three classifications of chemical reactions.
4. Distinguish between catabolism and anabolism.
5. Discuss the exchange that takes place in an oxidation-reduction reaction.
6. Explain ATP cycling.

Chemical reactions are classified based upon three different criteria that include (1) changes in chemical structure, (2) changes in chemical energy, and (3) whether the reaction is irreversible or reversible.

Classifying Changes in Chemical Structure

The general categories of chemical reactions based on changes in chemical structure include decomposition, synthesis, and exchange reactions **(figure 3.4)**. The first category is referred to as a **decomposition** (dē′kom-pō-zish′ŭn) **reaction** because the initial large molecule is digested or broken down into smaller structures. A simplified equation for a decomposition reaction is

$$AB \longrightarrow A + B$$

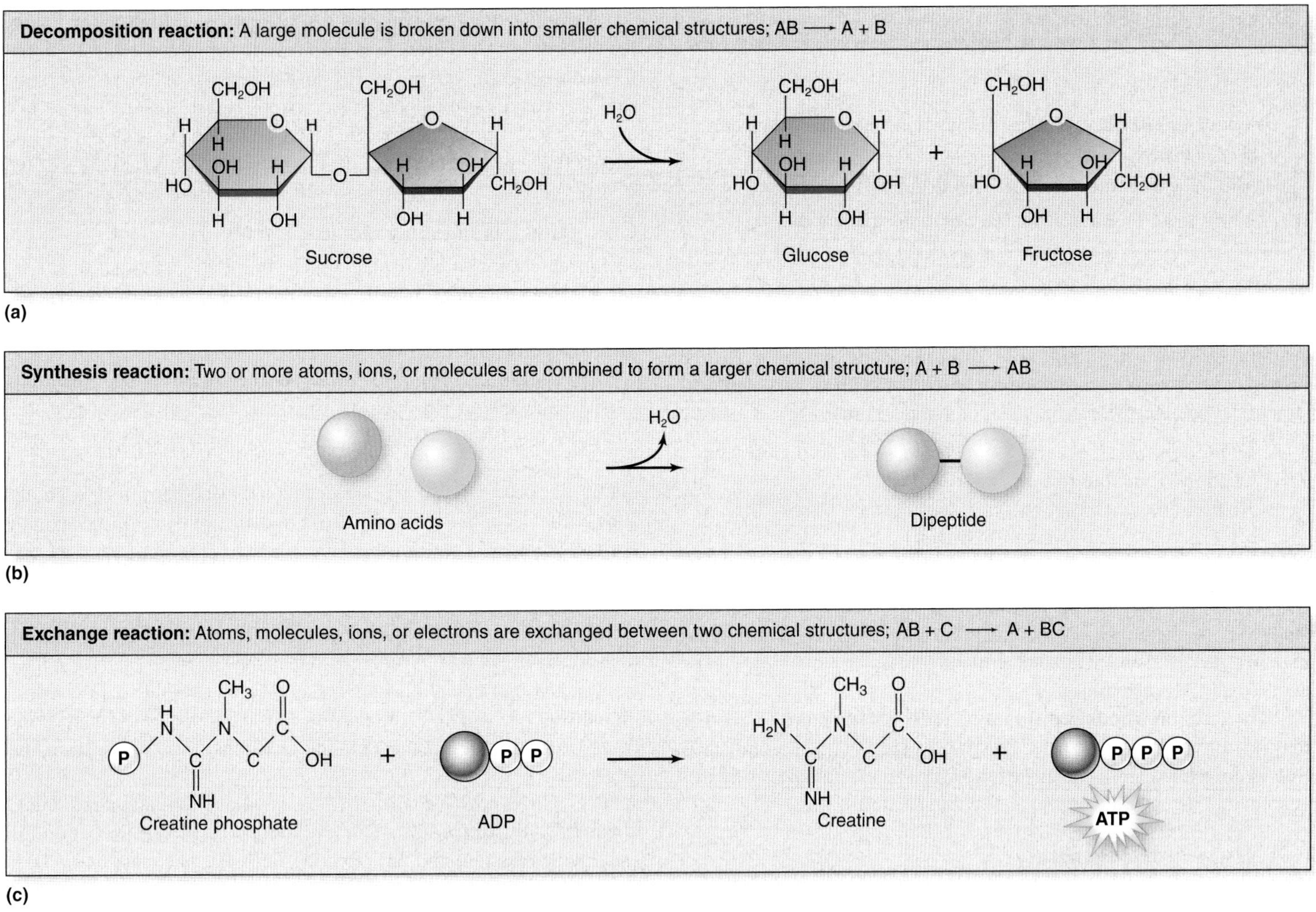

Decomposition reaction: A large molecule is broken down into smaller chemical structures; AB ⟶ A + B

Sucrose   Glucose   Fructose

(a)

Synthesis reaction: Two or more atoms, ions, or molecules are combined to form a larger chemical structure; A + B ⟶ AB

Amino acids   Dipeptide

(b)

Exchange reaction: Atoms, molecules, ions, or electrons are exchanged between two chemical structures; AB + C ⟶ A + BC

Creatine phosphate   ADP   Creatine   ATP

(c)

Figure 3.4 Classification of Chemical Reactions. Chemical reactions can be classified according to the chemical changes that occur. These reactions include (*a*) decomposition, (*b*) synthesis, and (*c*) exchange reactions.

Decomposition reactions occur, for example, during the *hydrolysis* reaction (see section 2.7a) of sucrose into glucose and fructose molecules in the digestive tract (figure 3.4*a*). All of the decomposition reactions in the body are collectively referred to as either **catabolism** (kă-tab′ō-lizm; *katabole* = a casting down) or *catabolic reactions*.

A **synthesis** (sin′thĕ-sis; *syn* = together, *thesis* = arranging) **reaction** occurs when two or more atoms, ions, or molecules are combined to form a larger chemical structure as existing bonds are broken and new bonds are formed. A simplified equation for a synthesis reaction is

$$A + B \longrightarrow AB$$

An example of a synthesis reaction is the *dehydration synthesis* reaction (see section 2.7a) that occurs during the formation of a dipeptide from two amino acids (figure 3.4*b*). **Anabolism** (ă-nab′ō-lizm; *ana* = up) is the collective term for all synthesis reactions in the body. These are also called *anabolic reactions*.

The third category of reactions based upon changes in chemical structure is the **exchange reaction,** in which atoms, molecules, ions, or electrons are exchanged between two chemical structures; such a reaction has both decomposition and synthesis components. This type of reaction is the most prevalent type in the human body. A simplified equation for an exchange reaction is

$$AB + C \longrightarrow A + BC$$

The production of adenosine triphosphate (ATP) in muscle tissue is an example of an exchange reaction:

$$\text{Creatine phosphate} + \text{Adenosine diphosphate (ADP)} \longrightarrow \text{Creatine} + \text{ATP}$$

The bond between phosphate and creatine is broken in this reaction. The creatine becomes a free molecule, while the phosphate is transferred and bonded to ADP to form ATP (figure 3.4*c*).

INTEGRATE

LEARNING STRATEGY

- **Catabolism** involves complex molecules being broken down, or digested, into simpler molecules. We can remember catabolism by thinking of a "cat" eating from a "bowl," which leads to food molecules being digested.

- **Anabolism** is the reverse reaction, namely synthesis or the building of complex molecules from simple molecules. For example, some athletes illegally use anabolic steroids to stimulate synthesis of contractile proteins within muscle tissue.

- **Metabolism** is the collective term for all biochemical reactions that occur within the human body; these include the processes of both catabolism and anabolism.

INTEGRATE

LEARNING STRATEGY

Use the phrase "**LEO** says **GER**" to remember the movement of electrons in redox reactions.

LEO = chemical structure that **L**oses **E**lectrons is **O**xidized.

GER = chemical structure that **G**ains **E**lectrons is **R**educed.

The movement of electrons can be harnessed to do work. Thus, the electrons in oxidation-reduction reactions represent energy transfer. Consequently, when we say that glucose is oxidized, we indicate that the glucose is losing its electrons and releasing its chemical energy. Other molecules are reduced and gain both electrons and energy (e.g., such as NAD^+ becoming NADH).

Classifying Changes in Chemical Energy

Chemical reactions are also classified by the relative amounts of chemical energy associated with the reactants and products. These two categories are based upon energy change and are termed exergonic reactions and endergonic reactions.

Exergonic (ek′sĕr-gon′ik; *exo* = outside, *ergon* = work) **reactions** involve reactants at the start of a reaction that have *more* potential energy within their chemical bonds than do the products that are formed **(figure 3.6a)**. Exergonic means that energy "goes out," or is released during the course of breakdown reactions. Decomposition reactions, such as the decomposition of glucose to carbon dioxide and water, are normally exergonic reactions.

Endergonic (en′dĕr-gon′ik; *endo* = within) **reactions** involve reactants that have *less* energy within their chemical bonds than do the products. Endergonic means that energy must be "put in," or supplied, to proceed (figure 3.6b). Endergonic reactions yield products that have a net increase in potential energy as compared to what was present in the substrates. Synthesis reactions, such as the formation of a dipeptide from two amino acids, are endergonic reactions.

Oxidation-Reduction Reactions An **oxidation-reduction reaction** (also termed a *redox reaction*) is a specific type of exchange reaction that involves the movement of electrons from one chemical structure to another. The term oxidation-reduction refers to

- **Oxidation,** which occurs as a molecule, atom, or ion loses an electron(s) and thus becomes **oxidized**
- **Reduction**, which occurs as a molecule, atom, or ion gains an electron(s) and thus becomes **reduced**

The term "reduced" is used because the chemical structure gains an electron, a *negatively* charged particle, so it can be thought of as "reduced" in charge. Oxidation and reduction reactions always occur together because one chemical structure loses the electron(s) and another gains the electron(s).

The electron transferred from one chemical structure to another may be moved alone, designated chemically as e⁻—or be accompanied by a hydrogen ion (H⁺), designated as H (i.e., an electron plus a hydrogen ion that together are shown as a hydrogen atom, H).

WHAT DO YOU THINK?

1 Electron transfer occurs when NAD^+ becomes NADH. Is the NAD^+ oxidized or reduced? Explain.

An example of an oxidation-reduction reaction in the cells involves the **nicotinamide adenine dinucleotide (NAD^+)** molecule, which is a modified dinucleotide that is linked at the phosphates and contains nicotinamide. NAD^+ is important in ATP synthesis. **Figure 3.5** shows an example in which an energy-rich molecule, such as glucose, is oxidized because it has given up two hydrogen atoms, while NAD^+ has gained both a hydrogen ion (H⁺) and two electrons (e⁻) and is reduced. (Note that one H⁺ is released into the surrounding local environment during this reaction.)

ATP Cycling **ATP cycling** is the continuous formation and breakdown of ATP **(figure 3.7)**. This cycling involves ATP formation (an endergonic reaction) and ATP splitting (an exergonic reaction). ATP is formed when energy is released in exergonic reactions using glucose or other fuel molecules from the foods we eat. These molecules undergo oxidation, and energy stored within their chemical bonds is transferred to ADP and P_i (free phosphate) to form ATP. In turn, ATP is then split into ADP and P_i, and the energy released is used for endergonic reactions, as well as other energy-requiring cellular processes. Thus, energy released in exergonic reactions in the body is coupled to endergonic reactions that require energy input so that endergonic reactions can proceed. A cell cannot stockpile ATP, so typically only a few seconds worth of ATP is present. Instead, the formation of ATP must occur continuously through the processes of breakdown of glucose (and other fuel molecules) to provide energy for endergonic reactions.

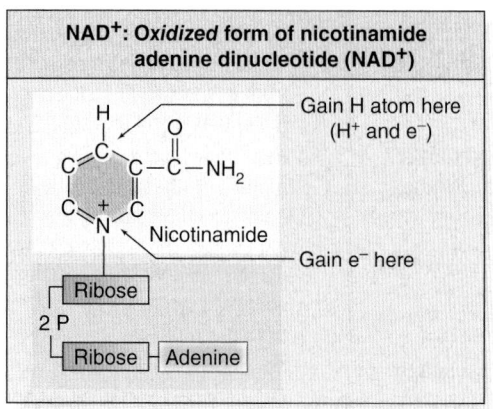

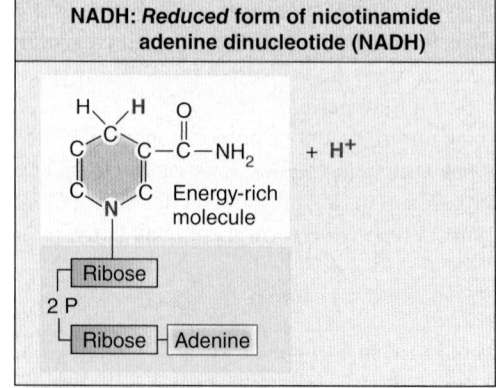

Figure 3.5 Different Forms of Nicotinamide Adenine Dinucleotide. Two hydrogen (H) atoms are donated from an energy-rich molecule (e.g., glucose) to the oxidized form of nicotinamide adenine dinucleotide (NAD^+). NAD^+ accepts one hydrogen atom (H) and one electron (e⁻). One H⁺ ion is released into the surrounding local environment. The energy-rich molecule (e.g., glucose) has released (lost) its electrons and is oxidized, and NAD^+ has gained electrons and is reduced to become NADH. NADH is later oxidized when it releases the H⁺ and 2 electrons. AP|R

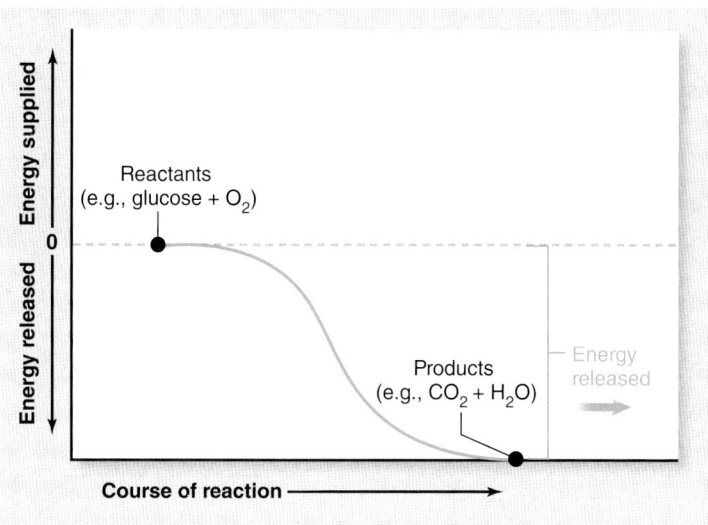

(a) Exergonic reaction

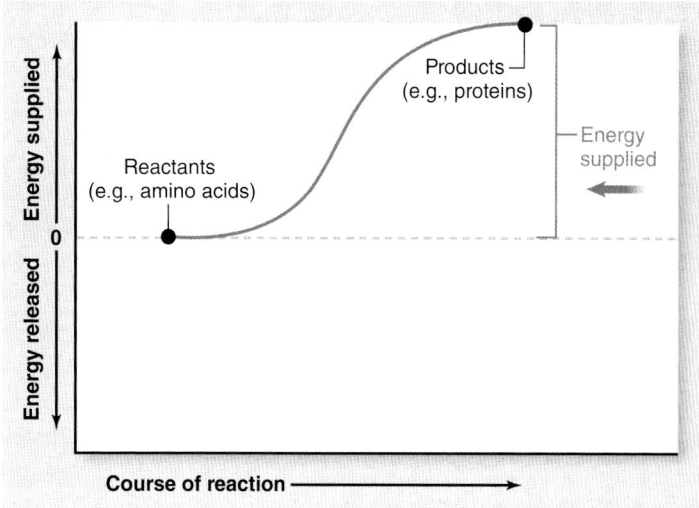

(b) Endergonic reaction

Figure 3.6 Exergonic and Endergonic Reactions. Chemical reactions are classified based on the change in chemical energy of the reactants and products. (*a*) The reactants have more energy than the products, and energy is released, during an exergonic reaction. (*b*) The reactants have less energy than the products, and energy must be supplied, in an endergonic reaction.

Classifying Reactions as Irreversible or Reversible

A third way of classifying chemical reactions is based upon whether they are irreversible or reversible. An **irreversible reaction** involves reactants converted to product at a rate that yields a net loss of reactants and a net gain in product. Many reactions are irreversible, and are written with the arrow to the right:

$$A + B \longrightarrow AB \quad \text{or} \quad AB \longrightarrow A + B$$

A **reversible reaction** differs from an irreversible reaction because it does not proceed only to the right with reactants becoming products over time, but instead reactants become products at a rate equal to products becoming reactants (once equilibrium is reached). Consequently, there is no net change in concentration in either reactants or products, and the reaction is in a state of equilibrium. The relationship of the reactants and products in a reversible reaction is shown with arrows in both directions:

$$A + B \rightleftarrows AB$$

A reversible reaction remains in equilibrium if left undisturbed. However, the equilibrium can be disturbed in a reversible reaction if a change in either the amount of reactant or the amount of product occurs. For example, either an increase in reactants or a decrease in products drives the equation to the right, which contributes to the formation of additional product until a new equilibrium is reached. In contrast, a decrease in reactants or an increase in product drives the equation to the left, which contributes to the formation of additional reactants until a new equilibrium is reached.

An important example of a reversible reaction in the human body occurs when carbon dioxide (CO_2) and water (H_2O) combine to form carbonic acid (H_2CO_3). The reaction is

$$CO_2 + H_2O \rightleftarrows H_2CO_3$$

The newly formed carbonic acid is unstable, and it dissociates to yield both hydrogen ion (H^+) and bicarbonate ion (HCO_3^-). The equation for the complete chemical reaction is

$$CO_2 + H_2O \rightleftarrows H_2CO_3 \rightleftarrows H^+ + HCO_3^-$$

This reversible reaction occurs in different locations within the human body and is important in several critical physiologic processes,

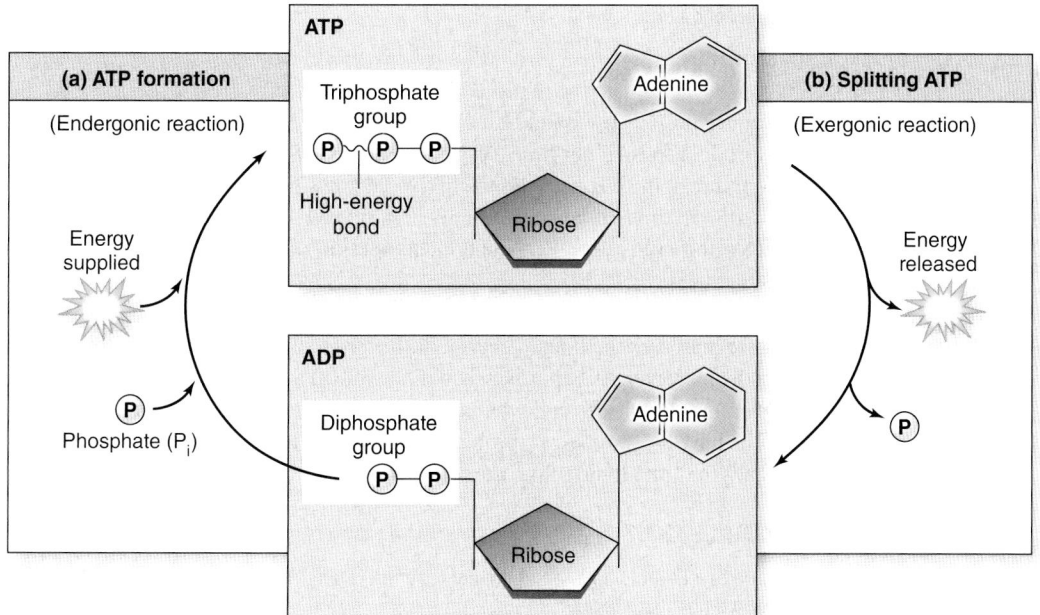

Figure 3.7 ATP Cycling. (*a*) The high-energy chemical bond of ATP is formed by a dehydration reaction between ADP and P_i. Energy is required and is supplied by the energy-releasing oxidation of fuel molecules. (*b*) The high-energy chemical bond of ATP is split by hydrolysis to form ADP and P_i.

Table 3.1 Classification of Chemical Reactions

Type of Chemical Reaction	Definition	Example
CHANGE IN CHEMICAL STRUCTURE		
Decomposition	Complex chemical structure broken down into simpler structures	Sucrose → glucose and fructose 
Synthesis	Simple chemical structures bonded together into a more complex structure	Amino acids → dipeptide 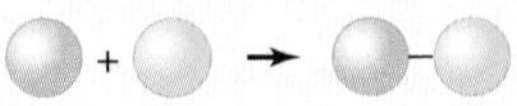
Exchange	Atoms, molecules, ions, or electrons exchanged between two chemical structures	Creatine phosphate + ADP → Creatine + ATP 
CHANGES IN CHEMICAL ENERGY		
Exergonic	Energy released	Glucose and oxygen → carbon dioxide and water 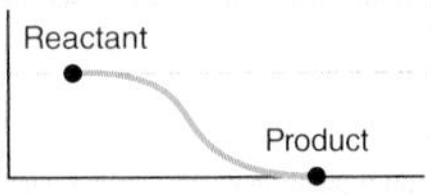
Endergonic	Energy required	Amino acids → dipeptide 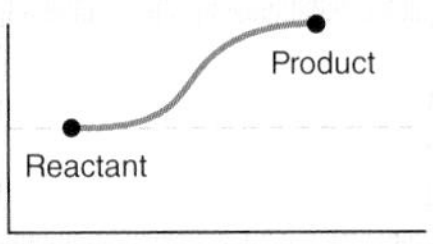
NET DIRECTION OF REACTION		
Irreversible	Net change of reactants to products	Most chemical reactions A + B → AB, or AB → A + B
Reversible	Formation of products = formation of reactants (once equilibrium is reached)	$CO_2 + H_2O \leftrightarrows H_2CO_3 \leftrightarrows H^+ + HCO_3^-$

including blood transport of carbon dioxide (see section 23.7b) and maintaining acid-base balance (see section 25.5).

Table 3.1 summarizes the three major ways chemical reactions are classified.

WHAT DID YOU LEARN?

5 For a biochemical reaction that involves simple chemical structures bonded together into a more complex molecule, choose the more accurate term in each pair that best describes this type of chemical reaction: (a) synthesis or decomposition reaction; (b) exergonic or endergonic reaction; (c) collective term for this type of reaction (catabolism or anabolism).

6 What molecule is formed from exergonic reactions and used as the energy currency for endergonic reactions and other energy-requiring processes within the cell?

7 Explain what occurs when the equilibrium is disturbed in reversible reactions by changes in reactants and products.

3.2c Reaction Rates and Activation Energy

LEARNING OBJECTIVES

7. Define chemical reaction rate.

8. Explain activation energy.

Reaction rate is the measure of how quickly a chemical reaction takes place; this rate determines the amount of product formed per time. A primary factor that influences the reaction rate is the energy required to break the chemical bonds in a molecule so that new bonds can form the product. The energy required to break existing chemical bonds for the chemical reaction to proceed is called the **activation energy,** or E_a. A chemical reaction occurs when sufficient energy is supplied to overcome the E_a.

In the laboratory setting, merely heating a mixture of reactants often overcomes the E_a. An elevation in temperature increases the kinetic energy of the molecules, providing enough energy to break chemical bonds. However, this approach is not feasible in a living cell. The consequence of an increase in temperature within the cell would denature all of its proteins and cause its death (see section 2.8b). Cells have solved their E_a problem through the use of biologically active protein catalysts called enzymes, the topic of the next section.

WHAT DID YOU LEARN?

8 Explain the effect a fever would have on chemical reaction rates within the body. What is the risk to protein structure with a high fever?

3.3 Enzymes

Chemical reactions must proceed at a rate that is sufficient to sustain life. Enzymes are the chemical structures that facilitate the millions of chemical changes that occur within the human body every second, and their importance cannot be overstated.

3.3a Function of Enzymes

LEARNING OBJECTIVE

1. Describe the general function of enzymes.

Enzymes are biologically active **catalysts** that function to accelerate normal physiologic activities by decreasing the activation energy (E_a) of chemical reactions. **Figure 3.8** shows the difference in E_a of an **uncatalyzed reaction** (chemical reaction without an enzyme) and **catalyzed reaction** (chemical reaction with an enzyme) of sucrose decomposition to yield glucose and fructose. Notice the following:

- The reaction is exergonic because the reactant sucrose has higher potential energy than the combined potential energy of the products glucose and fructose.
- Activation energy is required to initiate the reaction to occur even though it is exergonic.
- The presence of an enzyme lowers the required E_a.

More glucose and fructose are formed in a given period of time in the presence of an enzyme than would be formed without the enzyme.

Note that enzymes only facilitate chemical reactions to proceed that would already occur; their presence increases the rate of product formation by lowering the E_a. For example, let's compare the uncatalyzed and catalyzed rate of production for the reversible **carbonic acid reaction:**

$$CO_2 + H_2O \rightleftharpoons H_2CO_3 \rightleftharpoons H^+ + HCO_3^-$$

This reaction is catalyzed by the enzyme carbonic anhydrase. The chemical reaction still proceeds when carbonic anhydrase is absent, however only about 100 H_2CO_3 molecules are formed per hour. In contrast, when the enzyme carbonic anhydrase is present, up to 2.16 billion H_2CO_3 molecules may be formed per hour—a tremendously greater rate.

WHAT DID YOU LEARN?

9 What is the relationship of enzymes and activation energy?

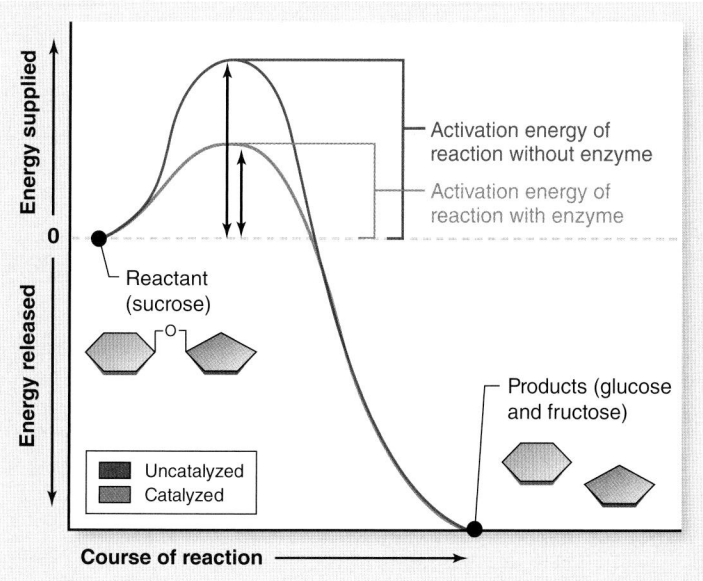

Figure 3.8 Activation Energy (E_a). The energy barrier that must be overcome for a reaction to proceed is the activation energy (E_a). Comparison of the E_a of an uncatalyzed reaction and a catalyzed reaction.

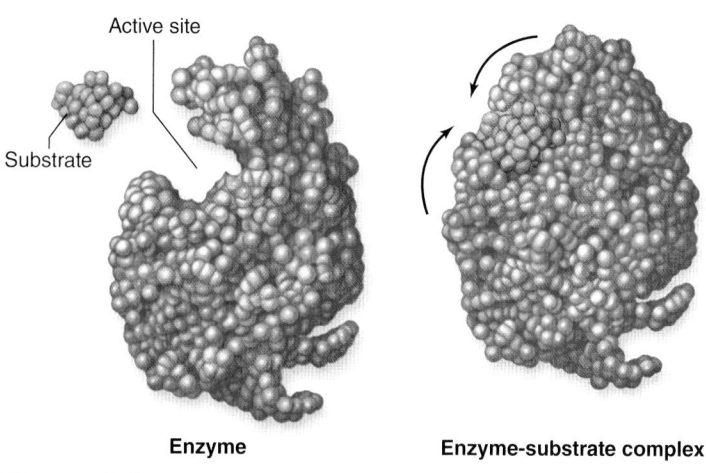

Active site

Substrate

Enzyme   Enzyme-substrate complex

Figure 3.9 Enzyme Structure. An enzyme is a globular protein with a depressed or grooved region that serves as the active site. The substrate binds to a specific active site forming the enzyme-substrate complex. **AP|R**

3.3b Enzyme Structure and Location

LEARNING OBJECTIVES

2. Describe the key structural components of enzymes.

3. Identify places in the body where enzymes may be found.

Most enzymes are globular proteins (see section 2.8b) that range in size from relatively small proteins composed of about 60 amino acids to very large proteins of more than 2500 amino acids. The amino acids in the protein chain form a unique three-dimensional molecular structure with a region called the **active site.** The active site accommodates the substrate(s) of a reaction to temporarily form an **enzyme-substrate complex (figure 3.9).** The **specificity** in the shape of the active site permits only a single substrate, or type of substrate, to bind to the active site, and thus the enzyme is capable of catalyzing only one specific reaction.

Enzymes are produced by normal protein synthesis processes within cells (see section 4.8). Once formed, the location of the enzyme varies. Enzymes may

- Remain within the cell; an example is DNA polymerase, which helps form new DNA.

- Become embedded within the plasma membrane (the outer boundary of a cell); an example is lactase, which digests the milk sugar lactose and is found in plasma membranes of cells that line the small intestine.

- Be secreted from the cell; an example is pancreatic amylase released from the pancreas into the small intestine to participate in the digestion of starch.

WHAT DID YOU LEARN?

10 What is the active site of an enzyme and how does it relate to a substrate?

3.3c Mechanism of Enzyme Action

LEARNING OBJECTIVES

4. Explain the steps by which an enzyme catalyzes a reaction.

5. Describe cofactors and their role in reactions.

Examples of enzymatic activity are diagrammed in **figure 3.10.** We show both an example of a decomposition reaction involving the

enzyme lactase and a synthesis reaction involving the enzyme glycogen synthetase. In both cases, the enzyme facilitates the reaction as follows:

(1) The substrate enters the active site of the enzyme, and the enzyme temporarily binds with the substrate to form an enzyme-substrate complex.

(2) Entry of the substrate into the active site induces the conformation (structure) of the enzyme to change slightly, resulting in an even closer fit between substrate and enzyme. This response is referred to as the *induced-fit model* of enzyme function. The interaction is analogous to giving someone a hug.

(3) Stress on chemical bonds in the substrate molecule is caused by the change in enzyme shape. Consequently, this stress lowers E_a, and the bonds in the substrates are more easily broken, permitting new chemical bonds to be formed.

(4) The newly formed molecule, now called the product, is released from the enzyme. The enzyme is then free to repeat the process again and again with other substrates.

INTEGRATE

CONCEPT CONNECTION

Until recently, all enzymes were thought to be proteins. Research data now suggests that some RNA molecules, which are termed *ribozymes* (or *RNA enzymes*), can function as enzymes. Many investigations have shown that the RNA of ribosomes, the cellular organelle described in section 4.8b, acts as an enzyme that catalyzes the synthesis of amino acids into a protein.

Cofactors

Enzymes often require **cofactors** that are "helper" ions or molecules to ensure that a reaction occurs. A cofactor is a *nonprotein* structure that may be either an inorganic or organic substance (see section 2.7a) associated with a particular enzyme or enzymatic reaction. Inorganic cofactors are attached to the enzyme and are required for their normal function. For example, zinc ion is bound to carbonic anhydrase enzyme; without zinc, carbonic anhydrase is unable to function. In addition, many vitamins (e.g., B_6 and B_{12}), derivatives of vitamins, or modified nucleotides such as NAD^+ may serve as organic cofactors. Organic cofactors are not attached to enzymes and have specific functions in assisting enzymes. For example, the NAD^+ coenzyme accepts hydrogen during chemical reactions to become NADH. Organic cofactors are more specifically referred to as **coenzymes** in some sources, and this term is also used in this text.

WHAT DID YOU LEARN?

11 What is the mechanism of enzyme action, including the role of cofactors?

3.3d Classification and Naming of Enzymes

LEARNING OBJECTIVES

6. Identify the six major classes of enzymes and the general functions of enzymes in each class.

7. Describe the naming conventions for enzymes.

There are thousands of different enzymes. To help make sense of the many different types of enzymes, biochemists organize them into six major functional classes, with subclasses within each category. **Table 3.2** briefly describes the six major classes of enzymes, including oxidoreductase, transferase, hydrolase, isomerase, ligase, and lyase.

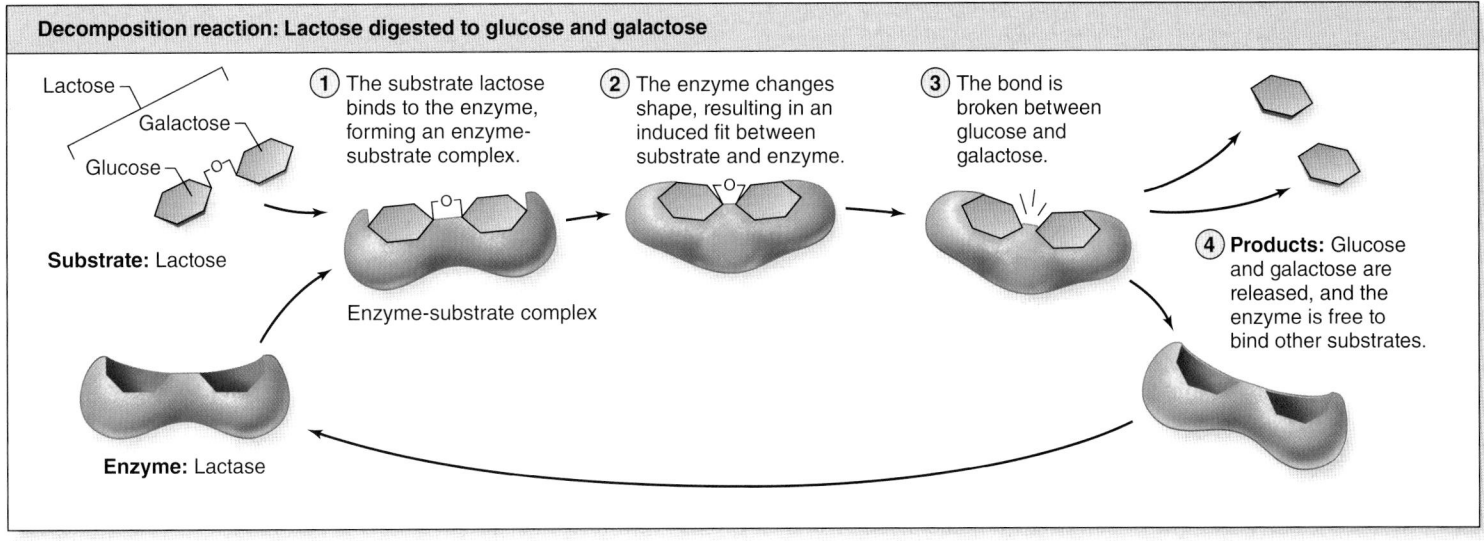

(a)

Decomposition reaction: Lactose digested to glucose and galactose

Lactose — Galactose — Glucose

Substrate: Lactose

Enzyme: Lactase

Enzyme-substrate complex

(1) The substrate lactose binds to the enzyme, forming an enzyme-substrate complex.

(2) The enzyme changes shape, resulting in an induced fit between substrate and enzyme.

(3) The bond is broken between glucose and galactose.

(4) **Products:** Glucose and galactose are released, and the enzyme is free to bind other substrates.

(b)

Synthesis reaction: Glucose molecules synthesized into a glycogen molecule

Glucose

Substrate: Glucose monomers

Enzyme: Glycogen synthetase

(1) The glucose substrate binds to the enzyme, forming an enzyme-substrate complex.

(2) The enzyme changes shape, resulting in an induced fit between substrate and enzyme.

(3) Bonds are broken and a new bond is formed between the new glucose molecule and the growing glycogen molecule.

(4) **Product:** Glycogen is released, and the enzyme is free to bind other substrates.

Figure 3.10 Mechanism of Action for Enzymes in Decomposition and Synthesis Reactions. (*a*) Enzymes may decompose larger complex molecules into simpler chemical structures, such as when lactose is broken down into glucose and galactose. (*b*) Enzymes may synthesize simple chemical structures into larger complex structures, such as when glucose molecules are bonded together to form glycogen.

Table 3.2	Major Classes of Enzymes	
Enzyme Class	**Description**	**Examples**
Oxidoreductase	Transfers electrons from one substance to another	Dehydrogenase uses NAD^+ or a molecule other than oxygen as electron acceptor. Peroxidase uses hydrogen peroxide (H_2O_2) as electron acceptor.
Transferase	Transfers a functional group	Phosphorylase transfers a phosphate (PO_4^{3-}) to a different substance. Kinase transfers a phosphate (PO_4^{3-}), usually from ATP to a different substance.
Hydrolase	Splits a chemical bond using water	Phosphatase removes phosphate. Protease digests proteins. Lipase splits lipids (e.g., triglyceride). Sucrase splits sucrose.
Isomerase	Converts one isomer to another	Mutase transfers atoms within a molecule.
Ligase	Bonds two molecules together	Synthetase bonds two molecules using ATP.
Lyase	Splits a chemical bond in the absence of water	Decarboxylase cleaves a molecule to release carbon dioxide.

Enzymes in the oxidoreductase class, for example, participate in oxidation-reduction reactions. **Dehydrogenase** (dē-hī′drō-jen-ās) enzymes are a subcategory of enzymes within the oxidoreductase class. These enzymes participate in oxidation-reduction reactions by moving hydrogen between molecules.

Another example of a class of enzymes is the transferase class. All enzymes in this class transfer atoms or molecules between chemical structures. **Kinase** (kī′nās) enzymes belong to this class because they specifically transfer a phosphate functional group, usually from ATP to another molecule. You will learn about specific examples of both dehydrogenase and kinase enzymes in section 3.4.

WHAT DO YOU THINK?

2 Given what you already know about isomers (see section 2.3a), what can you predict that an enzyme in the isomerase class would do?

The name of a given enzyme is generally based upon the name of the substrate or product involved in the chemical reaction, sometimes the name of the subclass, and the suffix *-ase*, added to the final word of the name. Here are three examples:

- **Pyruvate dehydrogenase** is an enzyme that transfers a hydrogen, specifically from a pyruvate molecule (see section 3.4c).
- **DNA polymerase** is central to the formation of the polymer DNA from deoxyribonucleotides (see section 4.9).
- **Lactase** digests the disaccharide lactose. (See Clinical View: "Lactose Intolerance" included in figure 3.14.)

Although the name of an enzyme generally contains an *-ase* suffix and the rest of the name reflects its function, there are exceptions. For example, pepsin, trypsin, and chymotrypsin (see section 26.4b) are all protein-digesting enzymes with names that do not provide clear clues to these molecules' enzymatic nature or specific activity.

WHAT DID YOU LEARN?

12 Explain how enzymes are generally named.

3.3e Enzymes and Reaction Rates

LEARNING OBJECTIVES

8. Define how enzyme and substrate concentration affect reaction rates.

9. Explain the effect of temperature on enzymes.

10. Describe how pH changes affect enzymes.

Several conditions influence the reaction rates catalyzed by enzymes. The most significant factors are enzyme and substrate concentration, temperature, and pH.

Effect of Enzyme and Substrate Concentration

The rate of a chemical reaction may be accelerated by either an increase in enzyme concentration or an increase in substrate concentration. An increase in substrate concentration, however, increases the rate of reaction only up to the point of saturation of the enzyme (**figure 3.11a**). **Saturation** occurs when so much substrate is present that all enzyme molecules are actively engaged in the chemical reaction, resulting in no further (notable) increase in reaction rate.

Effect of Temperature

Enzymes are proteins, and their three-dimensional shape is dependent upon environmental variables, including temperature and pH as

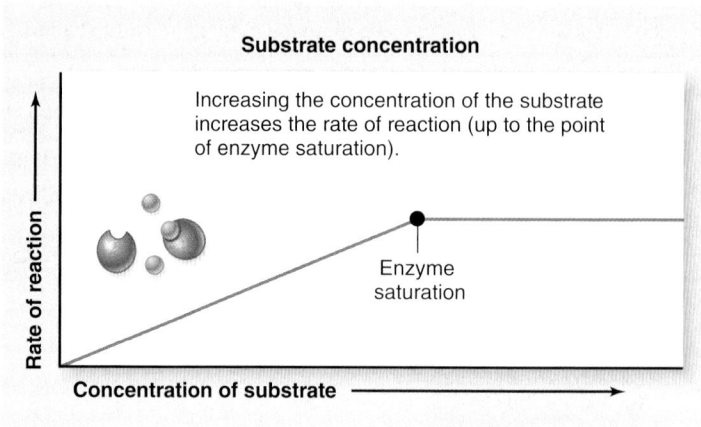

Substrate concentration

Increasing the concentration of the substrate increases the rate of reaction (up to the point of enzyme saturation).

Rate of reaction

Enzyme saturation

Concentration of substrate

(a)

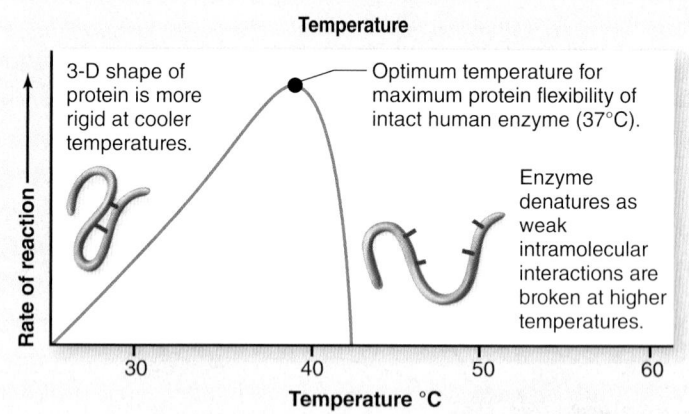

Temperature

3-D shape of protein is more rigid at cooler temperatures.

Optimum temperature for maximum protein flexibility of intact human enzyme (37°C).

Enzyme denatures as weak intramolecular interactions are broken at higher temperatures.

Rate of reaction

30   40   50   60
Temperature °C

(b)

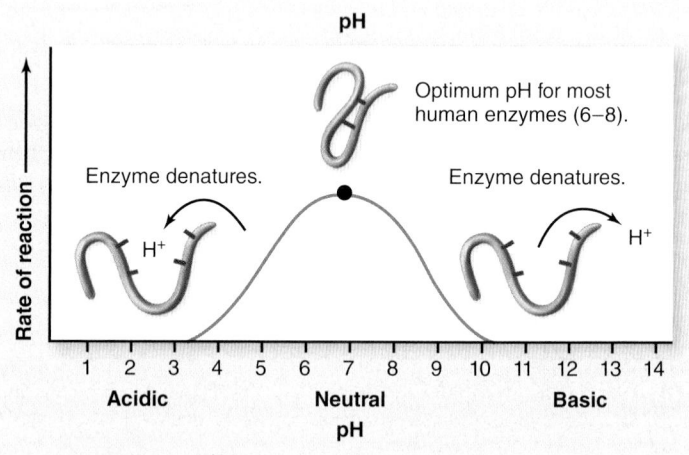

pH

Optimum pH for most human enzymes (6–8).

Enzyme denatures.

Enzyme denatures.

Rate of reaction

H^+

H^+

1   2   3   4   5   6   7   8   9   10 11 12 13 14
Acidic   **Neutral**   **Basic**
pH

(c)

Figure 3.11 Environmental Conditions That Influence Reaction Rates of Enzymes. Reaction rates are influenced by changes in (a) the concentration of substrates, (b) temperature, and (c) pH. AP|R

described in section 2.8b. Each enzyme has a specific environment in which it can most effectively participate in a chemical reaction. Human enzymes function efficiently at their **optimal temperature,** usually 37°C (98.6°F), which is the normal body temperature (figure 3.11b). Temperature increases in the body increase enzymatic activity; one advantage is that this enhances the body's ability to fight off infectious agents (see section 22.3e).

More severe increases in temperature—meaning temperatures greater than 40°C (104°F) in humans—weaken the intramolecular bonds that hold an enzyme's protein structure in its three-dimensional shape. The protein subsequently denatures, permanently losing function. The greater the increase in temperature, the more likely this is to occur.

Effect of pH

Enzymes function most efficiently at their **optimal pH.** Optimal pH for most human enzymes is between pH 6 and 8, and changes in pH can affect the enzyme (figure 3.11c). An increase in H^+ (which causes a decrease in pH) results in additional H^+ binding to the enzyme. In contrast, a decrease in H^+ (which causes an increase in pH) results in the release of H^+ from an enzyme. In either case, the change in amount of H^+ attached to the enzyme disrupts the electrostatic interactions that hold the enzyme protein in its shape. A significant disruption results in denaturation of the enzyme (see section 2.8b).

Not all enzymes have an optimal pH between 6 and 8. The pH in the stomach, for example, is between 2 and 4; thus, the optimal pH for the stomach enzyme pepsin corresponds to this pH range. In comparison, the pH in the small intestine is between 6 and 9. Thus, when stomach contents are moved into the small intestine, the stomach enzyme pepsin is inactivated (see section 26.4b).

 WHAT DID YOU LEARN?

13 How do changes in substrate concentration, temperature, and pH affect the reaction rate of enzyme-catalyzed chemical reactions?

3.3f Controlling Enzymes

LEARNING OBJECTIVE

11. Describe how competitive and noncompetitive inhibitors control enzyme action.

An enzyme continues to facilitate the conversion of its substrate(s) to product as long as ample substrate is present and environmental conditions are close to normal. However, uncontrolled enzymes would result in depleted substrate levels and concentration of products that exceeds what is needed. Thus, enzymes must be temporarily "turned off" to prevent overproduction. Control of enzymes occurs through **inhibitors** that are substances that bind to an enzyme and turn it off, thus preventing it from catalyzing the reaction (**figure 3.12**). Later, the release of the inhibitor from the enzyme allows it to function and continue catalyzing the reaction. This switching occurs in different ways, depending upon whether the inhibitor is competitive or noncompetitive.

A **competitive inhibitor** resembles the substrate and binds to the active site of the enzyme. Consequently, the substrate and the regulatory compound compete with each other for occupation of the enzyme's active site (figure 3.12b). The amount of substrate relative to the amount of competitive inhibitor determines the degree of inhibition. The greater the concentration of the substrate, the less likely the competitive inhibitor will occupy the enzyme's active site. In contrast, if substrate concentration decreases, the competitive inhibitor is more likely to occupy the enzyme's active site, and lower amounts of product are formed.

Noncompetitive inhibitors do not resemble the substrate. They inhibit an enzyme by binding to a site on the enzyme other than the active site, a site termed the **allosteric** (al'ō-ster'ik; *allo* = other, *stereo* = three-dimensionality) **site.** Binding of a noncompetitive inhibitor to the allosteric site induces a conformational change in the enzyme with an accompanying change in the shape of the enzyme's active site (figure 3.12c). Noncompetitive inhibitors are also called **allosteric inhibitors** because they bind to the allosteric site. This type of inhibition is not influenced by the concentration of substrate.

 WHAT DID YOU LEARN?

14 How are enzymes regulated through competitive and noncompetitive inhibitors?

3.3g Metabolic Pathways and Multienzyme Complexes

LEARNING OBJECTIVES

12. Distinguish between a metabolic pathway and a multienzyme complex.

13. Explain the role of negative feedback in enzyme regulation.

14. Identify and explain the processes involving phosphate that commonly are used to regulate enzymes.

Usually multiple enzymes are required to convert an initial substrate to a final product. Depending upon both the substrate and sequence of conversion, these multiple enzymes are arranged either in a metabolic pathway or as a multienzyme complex.

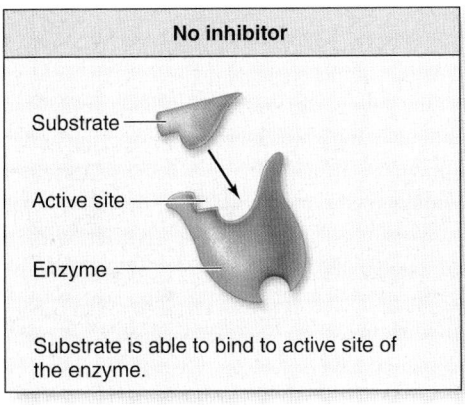

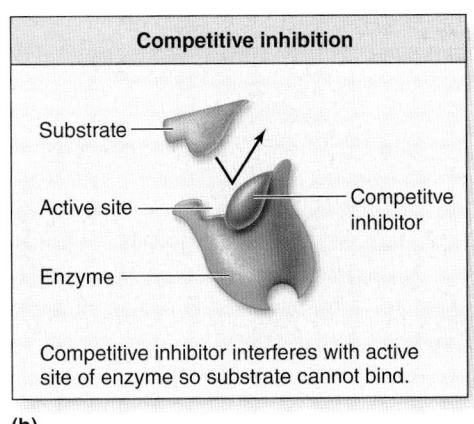

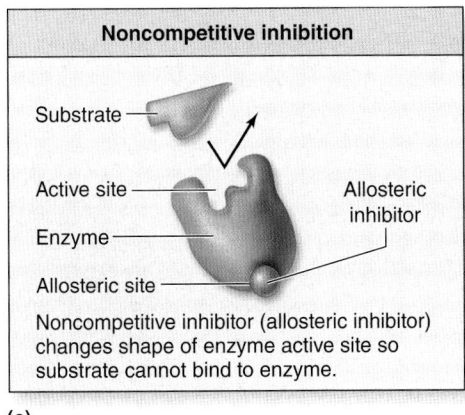

Figure 3.12 Enzyme Inhibition. (*a*) The substrate binds to the active site of the enzyme with no inhibitor present. A substrate can be prevented from binding to an active site by (*b*) a competitive inhibitor that enters the active site, or (*c*) a noncompetitive inhibitor (allosteric inhibitor) that binds to a site other than the active site to induce a conformational change in the enzyme with an accompanying change in its active site.

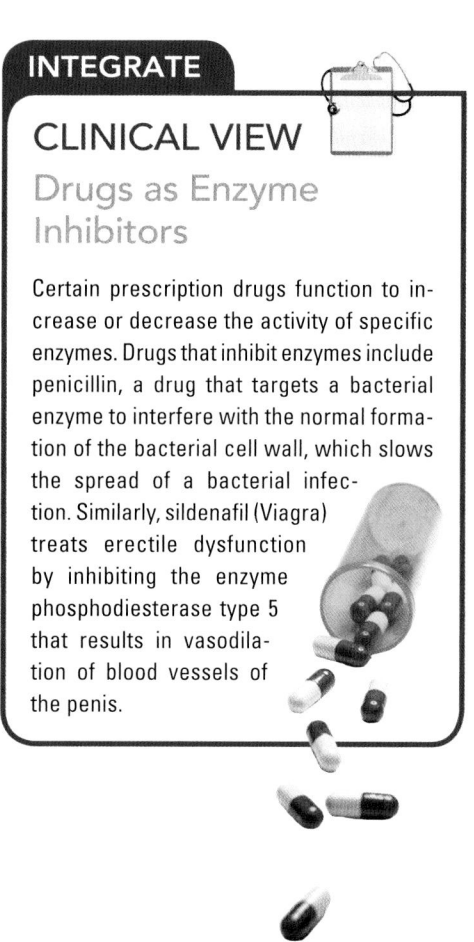

INTEGRATE

CLINICAL VIEW

Drugs as Enzyme
Inhibitors

Certain prescription drugs function to increase or decrease the activity of specific
enzymes. Drugs that inhibit enzymes include
penicillin, a drug that targets a bacterial
enzyme to interfere with the normal formation of the bacterial cell wall, which slows
the spread of a bacterial infec-
tion. Similarly, sildenafil (Viagra)
treats erectile dysfunction
by inhibiting the enzyme
phosphodiesterase type 5
that results in vasodila-
tion of blood vessels of
the penis.

A **metabolic pathway** is formed by numerous enzymes **(figure 3.13)**. Each enzyme catalyzes one progressive change to its specific substrate molecule and then releases the product.
In turn, the product of one enzyme becomes the substrate of the next enzyme. For example,
there are the numerous enzymes involved in the chemical breakdown of glucose to produce
carbon dioxide and water during the production of ATP (as described in section 3.4).

A **multienzyme complex** is a group of enzymes that are physically attached to each other
through noncovalent bonds to form the complex. These attached enzymes work in a sequence
of reactions. Pyruvate dehydrogenase, the multienzyme complex involved in breakdown of
glucose, is an example as described in section 3.4c.

A multienzyme complex has two major advantages. First, the product from one chemical
reaction is immediately bound to the next enzyme in the multienzyme complex. This makes
it more likely that the needed product is formed and less likely that the substance will diffuse
away and come into contact with an enzyme from a different biochemical pathway. Second,
the enzymatic pathway can be regulated by controlling the single complex rather than multiple
individual enzymes.

Metabolic pathways and multienzyme complexes must be regulated to prevent overproduction of an unneeded product and exhaustion of substrates that could be used elsewhere.
This regulation occurs through the process of negative feedback. The product from a metabolic pathway acts as an allosteric inhibitor to turn off an enzyme early in the metabolic
pathway, for example. As the product accumulates, it is more likely to become bound to the
enzyme and inhibit the metabolic pathway, with progressively less and less product being
formed. Over time, as the amount of product decreases, the amount of the allosteric inhibitor bound to the enzyme decreases, and activity of that enzymatic pathway increases once
again. In this way, a steady state of product is produced.

One specific mechanism for regulating enzymes is by either phosphorylation or dephosphorylation of the enzyme. **Phosphorylation** (fos′fŏr-i-lā′shŭn) is the addition of a phosphate
group, whereas **dephosphorylation** is the removal of a phosphate group. Note that phosphorylation may turn on some enzymes but turn off other enzymes. Equally, dephosphorylation
may cause opposite effects in activity by different types of enzymes. The enzymes that add
phosphate are generally called **protein kinases,** whereas enzymes that remove the phosphates
are called **phosphatases** (see figure 4.21). This concept is described in more detail in section 17.5b in discussion of the endocrine system.

Figure 3.13 Metabolic Pathway. A metabolic pathway is composed of numerous enzymes to convert a specific substrate to the final product.
The product of one enzyme is the substrate for the next enzyme in the pathway. Metabolic pathways can be regulated by negative feedback that
involves a product that serves as an allosteric inhibitor binding to an enzyme early in the pathway.

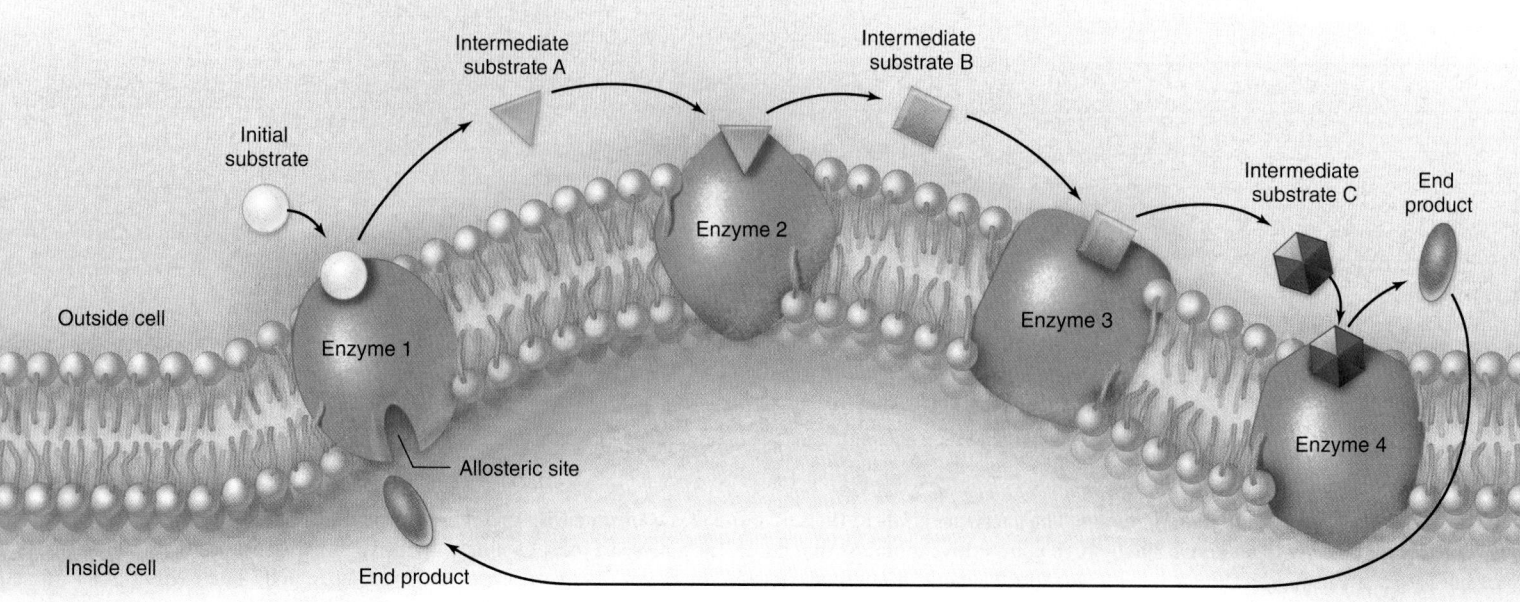

Figure 3.14 summarizes some important concepts of enzymes, including their function, structure and location, mechanism of action, and other features.

 WHAT DID YOU LEARN?

15 What is a metabolic pathway? Explain the role of negative feedback in enzyme regulation.

16 What two processes involve phosphate and are commonly used to regulate enzymes in a metabolic pathway or a multienzyme complex?

3.4 Cellular Respiration

Cellular respiration is a multistep metabolic pathway whereby organic molecules (e.g., glucose, fatty acids, amino acids) are disassembled (broken down) in a controlled manner by a series of enzymes. During this disassembly, potential energy stored in the molecule's chemical bonds is released; the energy is then used to make new bonds between ADP and P_i (phosphate) to form ATP (see figure 3.7). It is important to note the following about the processes of cellular respiration:

- These processes are exergonic or energy-releasing.
- The organic molecule that has given up its energy has done so by releasing high-energy electrons; thus the molecule is said to be oxidized.
- The energy released is used to synthesize ATP, which is an endergonic or energy-requiring process.
- Oxygen is required for maximum ATP production.

Although different types of organic molecules may be chemically digested during the collective processes of cellular respiration, our discussion here focuses on the oxidation of glucose.

3.4a Overview of Glucose Oxidation

 LEARNING OBJECTIVES

1. Write the overall formula for glucose oxidation.

2. Name the two pathways that generate ATP.

3. List the four stages of glucose oxidation and where each stage occurs within a cell.

Glucose oxidation occurs within cells and is a step-by-step enzymatic breakdown of glucose with the accompanying release of energy to synthesize ATP. If oxygen is available, glucose is completely broken down and carbon dioxide and water are formed. Here we describe several significant features of glucose oxidation.

Overall Chemical Reaction   Glucose has the chemical formula $C_6H_{12}O_6$. It is an energy-rich molecule because of its many C—C, C—H, and C—O chemical bonds. When enzymes completely disassemble glucose, the net chemical reaction for the process is

$$C_6H_{12}O_6 + 6\,O_2 \longrightarrow 6\,CO_2 + 6\,H_2O$$

Pathways for ATP Production   Glucose oxidation is an exergonic reaction. During the many enzymatic reaction steps that accomplish the breakdown of glucose, some of the energy of the broken bonds is captured to attach P_i to ADP to synthesize ATP: The energy transfer from bonds in the glucose molecule can be used either directly (least common way) or indirectly (most common way) to form ATP. The direct method of synthesizing ATP is called **substrate-level phosphorylation.** The indirect method—in which the energy is first released to coenzymes (i.e., NAD^+, FAD) that then transfer the energy to form ATP—is called **oxidative phosphorylation.**

INTEGRATE

CONCEPT CONNECTION

Cellular respiration typically requires an uninterrupted supply of O_2 and the continuous removal of CO_2. In chapter 23, we describe the systemic processes of **respiration,** which involve both the respiratory system and cardiovascular system. These systems function in both the delivery of O_2 from the atmosphere to the body's cells and the movement of CO_2 from the body's cells to the atmosphere. Individuals with either impaired respiratory function (e.g., emphysema) or cardiovascular disease (e.g., congestive heart failure) may experience difficulty in the delivery of O_2 to the body's cells. These individuals often experience "energy problems" and feel lethargic and tired. This decrease in energy results from the dependence of cellular respiration on O_2 delivery for maximum production of ATP as described in this chapter.

Figure 3.14 **How Enzymes Work.** Characteristics of enzymes include (*a*) enzyme function; (*b*) structure and location; (*c*) naming of enzymes; (*d*) mechanism of action; (*e*) reaction rate; (*f*) metabolic pathway, multienzyme complex and their regulation; and (*g*) controlling enzymes.

(a) Enzyme function

Enzymes decrease the activation energy (E_a) so reaction rate increases and more product is formed in a given time period.

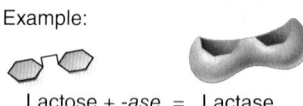

- E_a without enzyme
- E_a with enzyme

Energy supplied / Energy released
Time →

(b) Structure and location

Most enzymes are globular proteins with a unique active site.

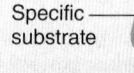
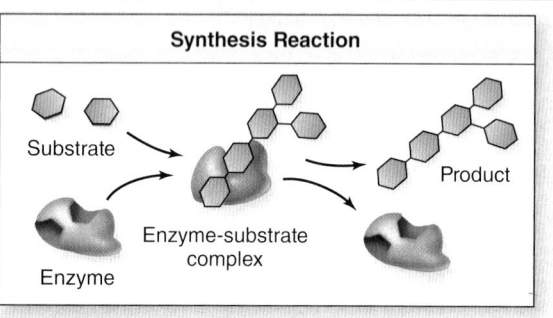

Specific substrate
Active site
Enzyme
Outside of cell
Plasma membrane
Inside of cell

Enzymes can be located inside a cell, outside a cell, or embedded in a cell's plasma membrane.

(c) Naming of enzymes

The name of an enzyme typically includes the name of the substrate or product, or sometimes the name of the class or subclass involved in the chemical reaction with an *-ase* ending.

Example:

Lactose + *-ase* = Lactase

(d) Mechanism of action

Enzymes participate in either decomposition or synthesis reactions.

Decomposition Reaction

Substrate
Products
Enzyme-substrate complex
Enzyme

Synthesis Reaction

Substrate
Product
Enzyme-substrate complex
Enzyme

(e) Reaction rate (speed of a chemical reaction)

Reaction rate is influenced by substrate or enzyme concentration, temperature, and pH.

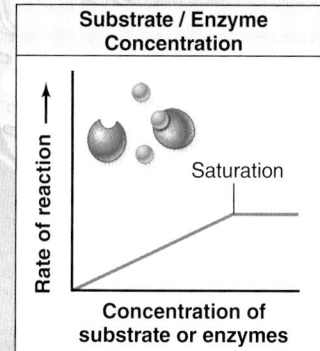

Substrate / Enzyme Concentration

Rate of reaction
Saturation
Concentration of substrate or enzymes

Increasing the substrate or enzyme concentration (to the point of enzyme saturation) increases the rate of reaction.

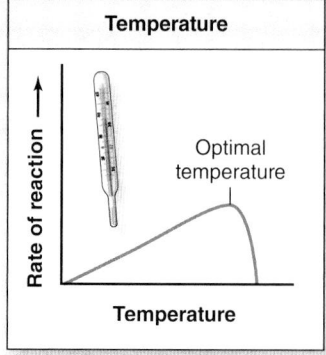

Temperature

Rate of reaction
Optimal temperature
Temperature

Increasing the temperature increases the rate of reaction up to the point of denaturation of the enzyme.

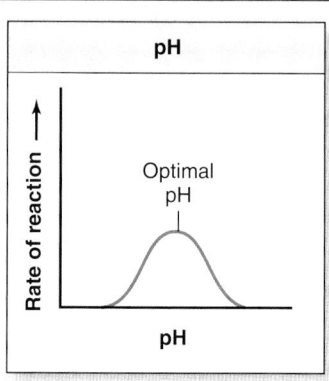

pH

Rate of reaction
Optimal pH
pH

Enzymes are most effective at their optimal pH. Either an increase or decrease from the optimal pH decreases the rate of the reaction.

(f) Metabolic pathway and multienzyme complex

Metabolic pathway: Series of enzymes

Many metabolic pathways are regulated by negative feedback by a product.

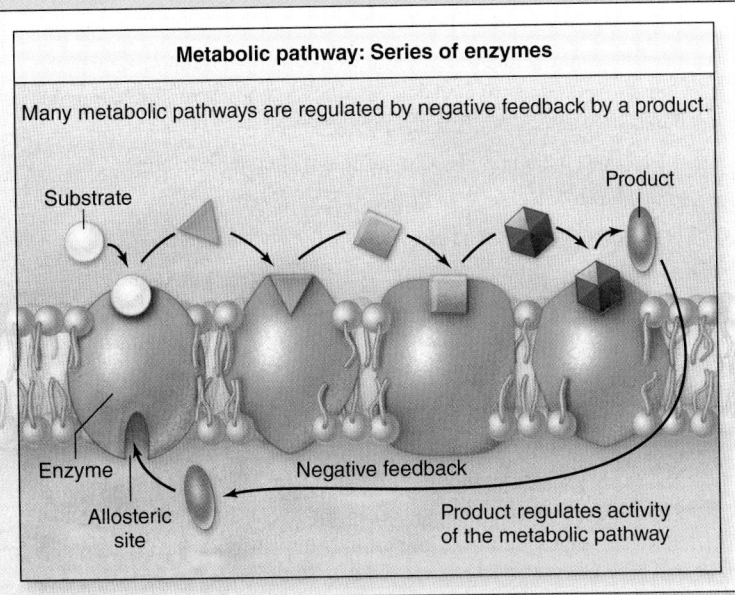

Substrate

Product

Enzyme

Allosteric site

Negative feedback

Product regulates activity of the metabolic pathway

Multienzyme complex: Physically linked enzymes

The product of one enzyme becomes the substrate of a different enzyme in the complex.

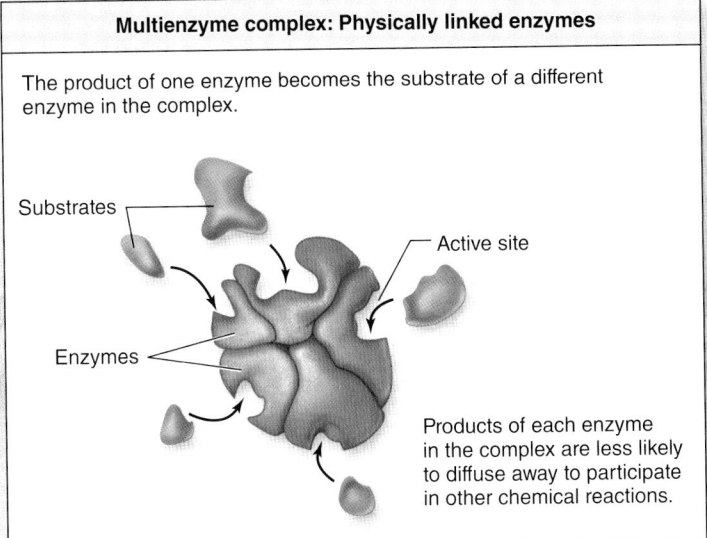

Substrates

Active site

Enzymes

Products of each enzyme in the complex are less likely to diffuse away to participate in other chemical reactions.

(g) Controlling enzymes

Competitive inhibitor

Competitive inhibitors interfere with active site directly.

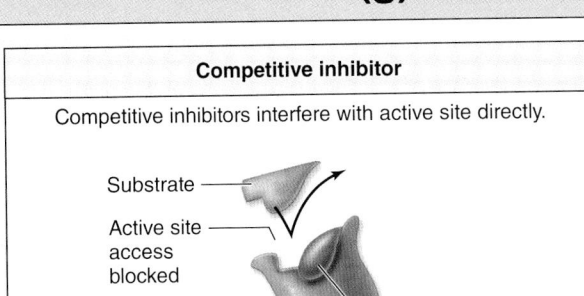

Substrate

Active site access blocked

Enzyme

Competitive inhibitor

Noncompetitive inhibitor

Allosteric inhibitors change shape of enzyme so the substrate cannot bind to the active site.

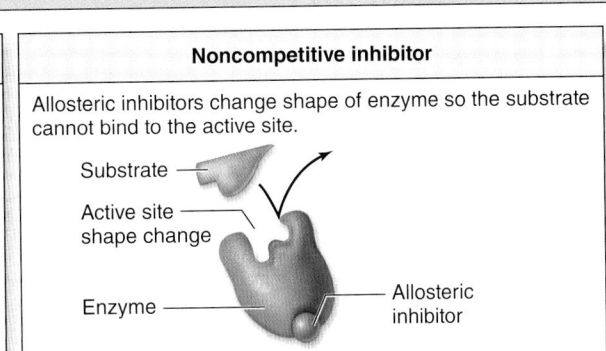

Substrate

Active site shape change

Enzyme

Allosteric inhibitor

INTEGRATE

CLINICAL VIEW
Lactose Intolerance

Lactase is an enzyme required to break the bond in the disaccharide lactose (milk sugar) into glucose and galactose for its absorption from the digestive tract into the blood. Lactose intolerance is caused by either a deficiency in the enzyme lactase or an abnormal (low-functioning) lactase enzyme. Differences in lactase enzyme function are due to genetic variation within the population. For example, individuals of northern European descent have a low incidence of lactose intolerance. Lactose intolerance is also more common in older adults because they produce less lactase over time. Abdominal upset including nausea, diarrhea, bloating, and gas are the most common symptoms. Avoidance of foods containing milk, drinking milk with lactose removed (lactose-free milk), or the oral administration of products containing lactase enzymes are recommended to avoid symptoms of lactose intolerance.

Lactose

Glucose

Galactose

Lactase

Lactase embedded in plasma membrane of cells lining the small intestine

REDUCED FAT MILK

REDUCED FAT MILK

MILK

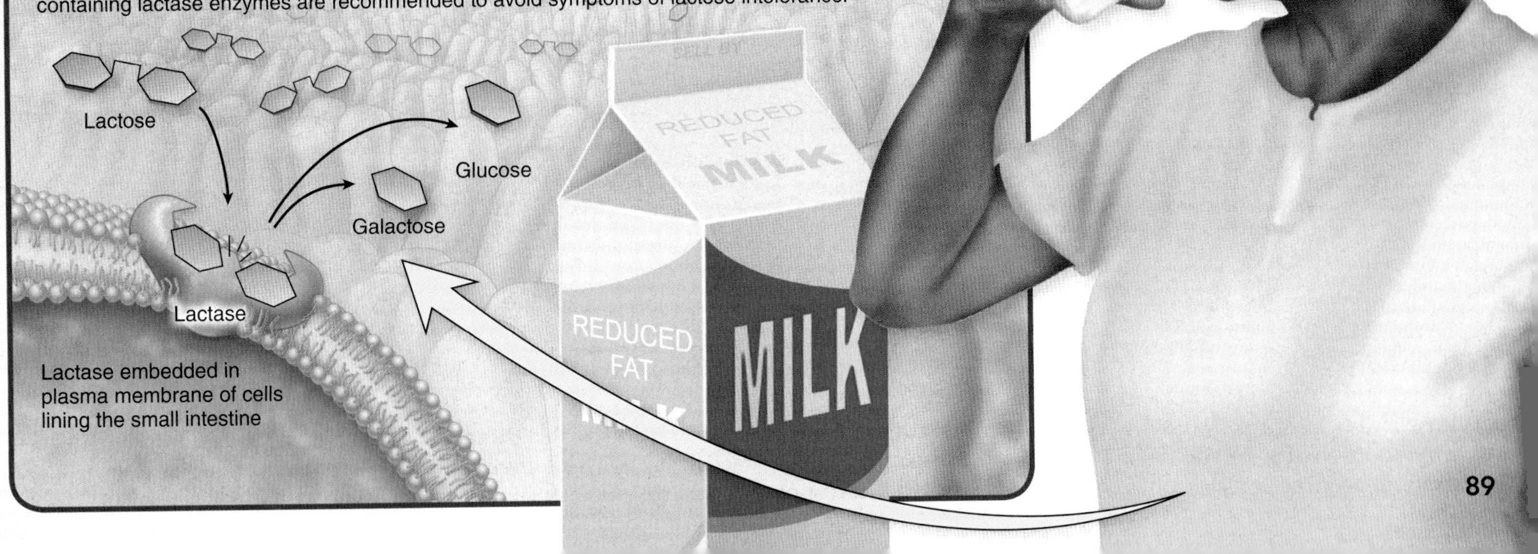

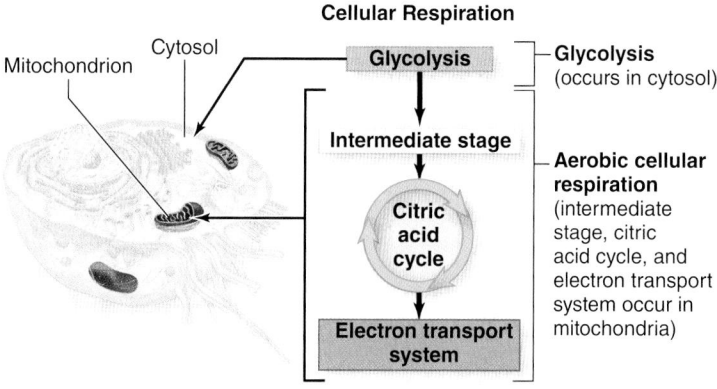

Cellular Respiration

Glycolysis — Glycolysis (occurs in cytosol)

Intermediate stage
Citric acid cycle
Electron transport system

— Aerobic cellular respiration (intermediate stage, citric acid cycle, and electron transport system occur in mitochondria)

Figure 3.15 Cellular Structures Required for Cellular Respiration. Components of the cell associated with glucose oxidation include the cytosol, where the enzymes for glycolysis are located, and mitochondria where the enzymes for aerobic cellular respiration (intermediate stage, citric acid cycle, and electron transport system) are housed. **AP|R**

INTEGRATE

LEARNING STRATEGY

Keep in mind the following questions as you read through each of the first three stages of cellular respiration (the stages required for completely oxidizing [disassembling] of glucose):

1. Does it occur in the cytosol or mitochondria of a cell?
2. Does it require oxygen (i.e., is it aerobic)?
3. What is the initial substrate and the final product?
4. Is energy released to produce ATP directly (substrate-level phosphorylation), or transferred to a coenzyme that serves as a "temporary holder" that will participate in oxidative phosphorylation, or both?

Cellular Location The complete oxidation of glucose requires at least 20 different enzymes that are located in both the cell's cytosol and its mitochondria **(figure 3.15)**. **Cytosol** is the semifluid contents of the cell. **Mitochondria** (sing., mitochondrion) are small organelles within the cell that are described in section 3.4c.

Four Stages of Cellular Respiration We separate the processes of glucose oxidation into four stages: glycolysis, intermediate stage, citric acid cycle, and the electron transport system. Glycolysis occurs in the cytosol and does not require oxygen; thus, glycolysis can occur either in the presence of oxygen or in the absence of oxygen. The other three stages occur in the mitochondria and require oxygen to proceed.

WHAT DID YOU LEARN?

17 Write the overall chemical reaction for glucose oxidation, and explain the general process of what is occurring.

18 What are the four stages of cellular respiration for glucose oxidation, and identify the cellular location in which each occurs?

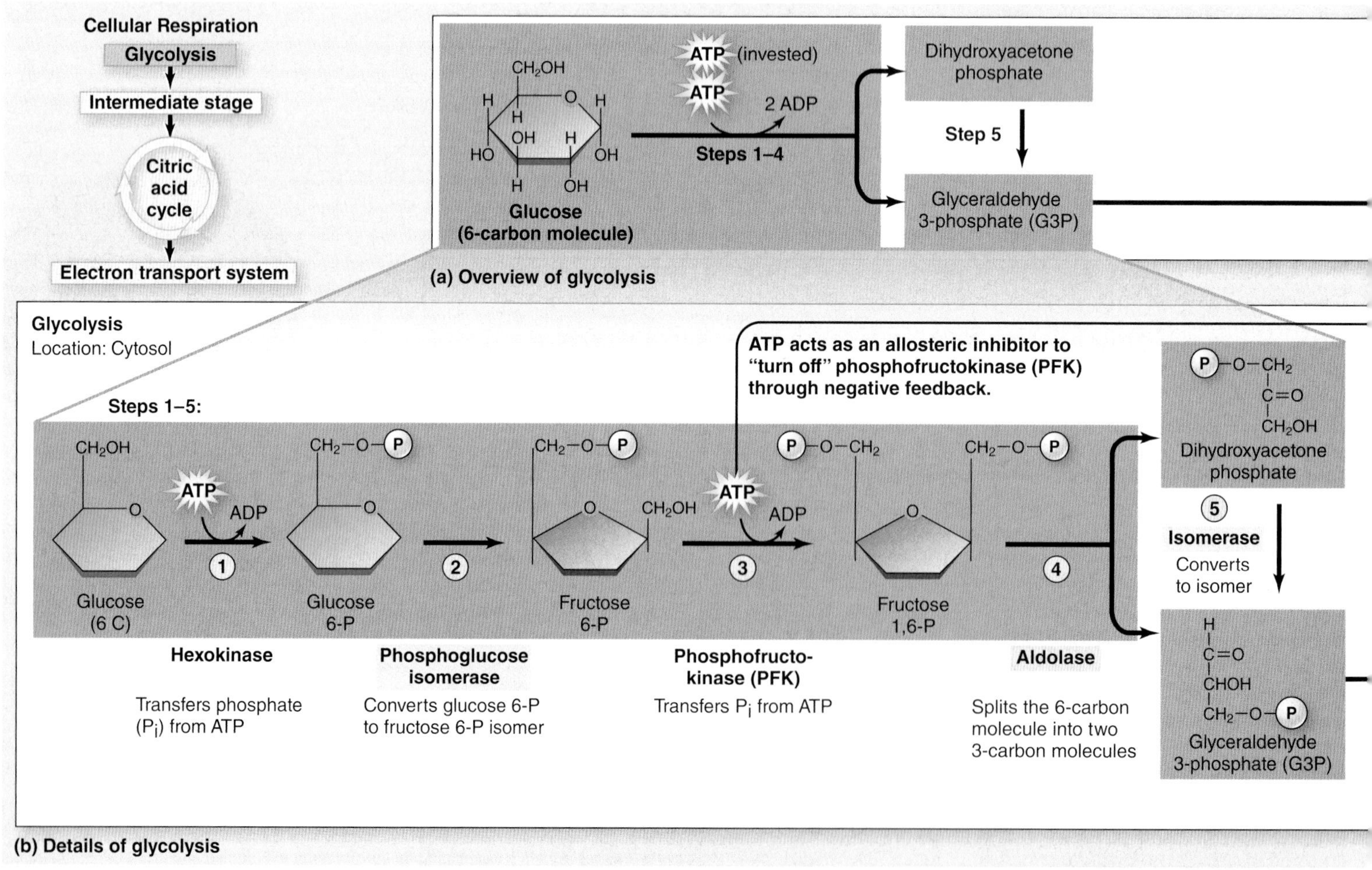

Cellular Respiration

Glycolysis
Intermediate stage
Citric acid cycle
Electron transport system

ATP (invested)
ATP — 2 ADP
Steps 1–4

Glucose (6-carbon molecule)

Dihydroxyacetone phosphate
Step 5
Glyceraldehyde 3-phosphate (G3P)

(a) Overview of glycolysis

Glycolysis
Location: Cytosol

ATP acts as an allosteric inhibitor to "turn off" phosphofructokinase (PFK) through negative feedback.

Dihydroxyacetone phosphate

Steps 1–5:

Glucose (6 C) — ATP/ADP — (1) — Glucose 6-P — (2) — Fructose 6-P — ATP/ADP — (3) — Fructose 1,6-P — (4)

(5) Isomerase — Converts to isomer

Glyceraldehyde 3-phosphate (G3P)

Hexokinase
Transfers phosphate (Pᵢ) from ATP

Phosphoglucose isomerase
Converts glucose 6-P to fructose 6-P isomer

Phosphofructo-kinase (PFK)
Transfers Pᵢ from ATP

Aldolase
Splits the 6-carbon molecule into two 3-carbon molecules

(b) Details of glycolysis

Figure 3.16 Metabolic Pathway of Glycolysis. Glycolysis is a metabolic pathway that occurs within the cytosol. The pathway requires ten enzymes involved in the conversion of glucose to pyruvate with a net production of 2 ATP molecules and 2 NADH molecules. The fate of pyruvate is dependent upon the availability of oxygen to the cell. (*a*) An overview of glycolysis. (*b*) The detailed pathway of glycolysis. **AP|R**

3.4b Glycolysis

⟩⟩⟨ LEARNING OBJECTIVE

4. Summarize the metabolic pathway of glycolysis, including (a) where it occurs in a cell, (b) if it requires oxygen, (c) the initial substrate and final product, and (d) the molecules formed during energy transfer.

Glycolysis (glī-kol′i-sis; *glykys* = sweet, *lysis* = loosening) does not require oxygen. Ten enzymes within the cytosol of a cell participate in the metabolic pathway of glycolysis. Glucose is broken down in this pathway into two pyruvate molecules with an accompanying energy transfer to form a net production of 2 ATP molecules and 2 NADH molecules.

Steps of Glycolysis

The ten enzymatically regulated chemical reactions of glycolysis are shown in **(figure 3.16)**; both an overview figure (figure a) and the detailed steps (figure b) are included. Note that steps 1 through 5 occur once per glucose, and steps 6 through 10 occur *twice* per glucose because glucose is split into *two* three-carbon molecules.

1–5 Steps 1 through 5 of glycolysis involve splitting glucose into two molecules of glyceraldehyde 3-phosphate (G3P) through the action of the first five enzymes. ATP is "invested" when kinase enzymes transfer P_i from ATP to glucose and the breakdown products of glucose (steps 1 and 3). Thus, an investment of 2 ATP molecules occurs at these two steps.

6–7 Steps 6 and 7 of glycolysis occur twice in oxidation of a glucose molecule. Step 6 involves transferring an unattached P_i to the substrate (so this molecule now has two phosphates),

and two hydrogen atoms are released to NAD^+ to form an NADH (and H^+). This transfer of hydrogen is catalyzed by a hydrogenase enzyme. In step 7, the original P_i is transferred to ADP to form ATP through substrate-level phosphorylation by a kinase enzyme.

8–10 Steps 8 through 10 of glycolysis also occur twice in oxidation of a glucose molecule. These steps involve converting the molecule produced in step 7 to an isomer (step 8) and then the loss of a water molecule (step 9). The remaining P_i is transferred to ADP to form ATP through substrate-level phosphorylation by a kinase enzyme (step 10), forming the final product of pyruvate.

⟩⟩⟨ WHAT DO YOU THINK?

3 What is the net energy transfer during glycolysis?

Summary of Glycolysis

Glycolysis is a metabolic process that occurs in the cytosol without the requirement of oxygen. Glucose is the initial substrate and pyruvate is the final product. The net transfer of energy is used in the formation of 2 ATP and 2 NADH molecules.

- **Formation of ATP.** Two ATP molecules are invested early in glycolysis (steps 1 and 3). Four ATP molecules are formed during glycolysis (steps 7 and 10, which occur twice per the original glucose molecule). Thus, there is a net of 2 ATP molecules formed during glycolysis (2 ATP invested, 4 ATP formed).

- **Formation of NADH.** Two NADH molecules are formed from glucose breakdown during glycolysis (step 6, which occurs twice per the original glucose molecule).

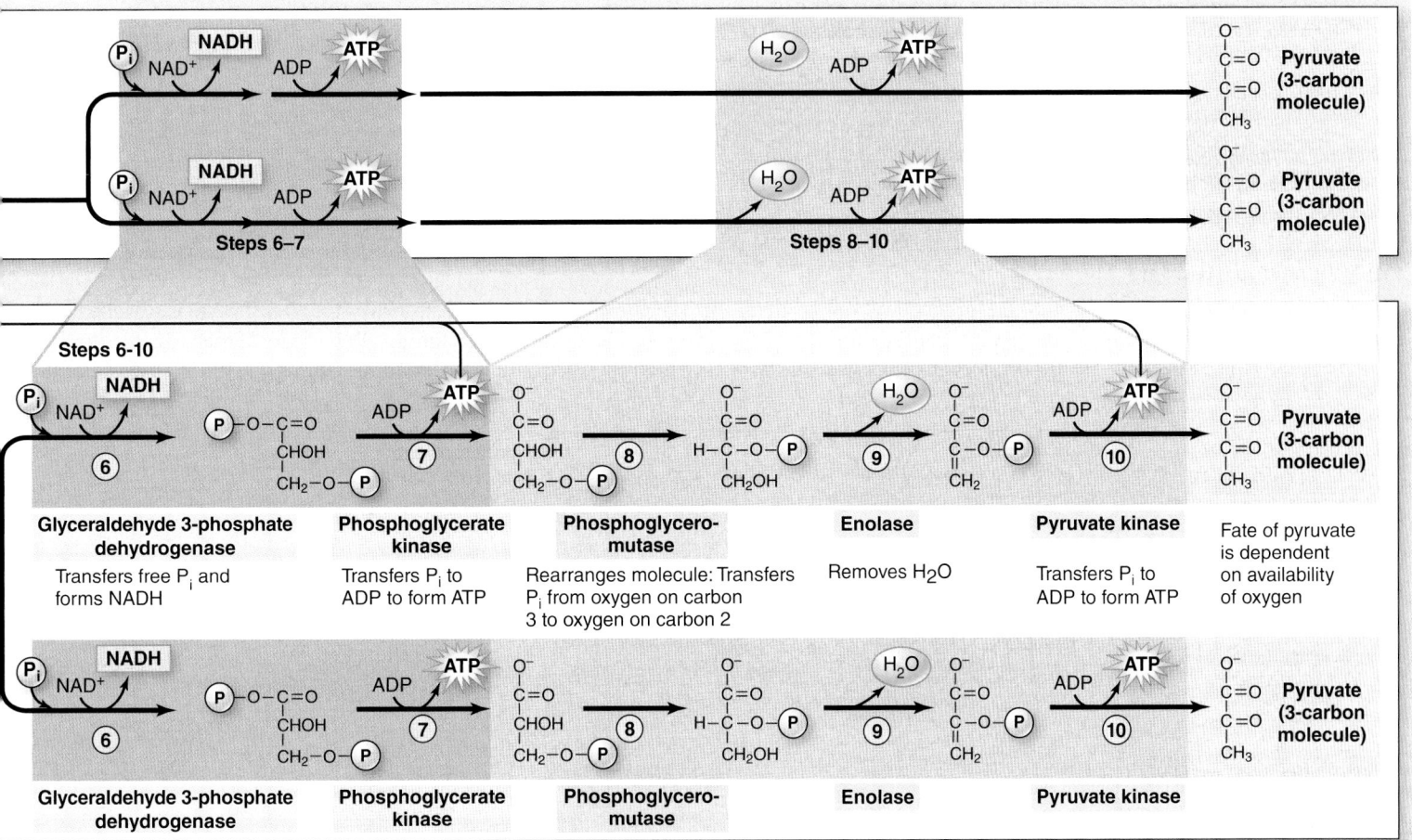

Regulation of Glycolysis

Glycolysis is regulated through negative feedback like many metabolic pathways. ATP acts as an allosteric inhibitor to "turn off" phosphofructokinase (PFK) (step 3).

As ATP levels increase in the cell cytosol, ATP binding inhibits PFK, and the glycolytic pathway is progressively shut down. In contrast, as ATP decreases, glycolysis increases. Phosphofructokinase is also regulated in a similar way by other substances that indicate the "energy status" of the cell, including NADH, citrate (an intermediate in the citric acid cycle), fatty acids, and other fuel molecules. Increased levels of these substances results in a decrease in the processes of glycolysis.

The Fate of Pyruvate

Pyruvate is the final product of glycolysis. What chemical changes are then made to pyruvate depend upon the availability of oxygen. If sufficient oxygen is available, pyruvate enters a mitochondrion to complete its aerobic breakdown yielding carbon dioxide and water (described next). In contrast, if sufficient oxygen is not available, pyruvate is converted to lactate (described in section 3.4g).

WHAT DID YOU LEARN?

19 Describe glycolysis—where it occurs, if the process requires oxygen (i.e., is aerobic), the net chemical reaction, and the net energy transfer.

20 What are the two general fates of pyruvate, and what is the criterion that determines its fate?

3.4c Intermediate Stage

LEARNING OBJECTIVES

5. Explain the enzymatic reaction of the intermediate stage, including (a) where it occurs in a cell, (b) if it requires oxygen, (c) the initial substrate and final product, and (d) the molecules formed during energy transfer.

6. Define decarboxylation.

The remaining stages of cellular respiration, including the intermediate stage, citric acid cycle, and the electron transport system are aerobic processes that occur within mitochondria.

Mitochondrion Structure

A mitochondrion is a double-membrane organelle, composed of an outer membrane and an inner membrane that has inward folds called **cristae** (**figure 3.17a**). The fluid-filled space between the two membranes is the **outer compartment.** The innermost space in a mitochondrion is called the **matrix.** Both the multienzyme complex of the intermediate stage and the enzymes of the citric acid cycle metabolic pathway reside in the matrix. The significant molecules that participate in the electron transport system are embedded in the cristae.

Intermediate Stage and Pyruvate Dehydrogenase

The **intermediate stage** (figure 3.17b) is the "link" between the multistep metabolic processes of glycolysis (first stage) with the multistep metabolic processes that occur in the citric acid cycle (third stage). The intermediate stage is catalyzed by a multienzyme complex called **pyruvate dehydrogenase.**

During the intermediate stage, pyruvate dehydrogenase brings together pyruvate and a molecule of coenzyme A (CoA) that is already present within the matrix to form acetyl CoA (a two-carbon molecule with CoA attached). Concurrently, a carboxyl group, consisting of one carbon atom and two oxygen atoms, is released from the pyruvate as CO_2. This process is termed **decarboxylation** (dē'kar-bok'si-lā'shŭn; _de_ = away). Energy is released during decarboxylation as two hydrogen atoms (two electrons plus two hydrogen ions) are transferred to the coenzyme NAD$^+$ to form NADH (and H$^+$) during this process. The acetyl CoA then enters the third stage of glucose oxidation, termed the citric acid cycle.

Note that the intermediate stage must occur twice for the complete digestion of the original glucose molecule, because two pyruvate molecules were produced from each glucose that went through glycolysis. Thus, 2 NADH are produced from the original glucose molecule.

WHAT DID YOU LEARN?

21 Explain the enzymatic reaction involving pyruvate dehydrogenase in the intermediate stage—where it occurs, if the process requires oxygen, the net chemical reaction, and the net energy transfer.

3.4d Citric Acid Cycle

LEARNING OBJECTIVE

7. Summarize the metabolic pathway of the citric acid cycle including (a) where it occurs in a cell, (b) if it requires oxygen, (c) the initial substrate and final product, and (d) the molecules formed during energy transfer.

The **citric acid cycle** (also known as the _Krebs cycle_) is a cyclic metabolic pathway that occurs through the activity of nine enzymes located within the matrix of mitochondria. During the citric acid cycle, the

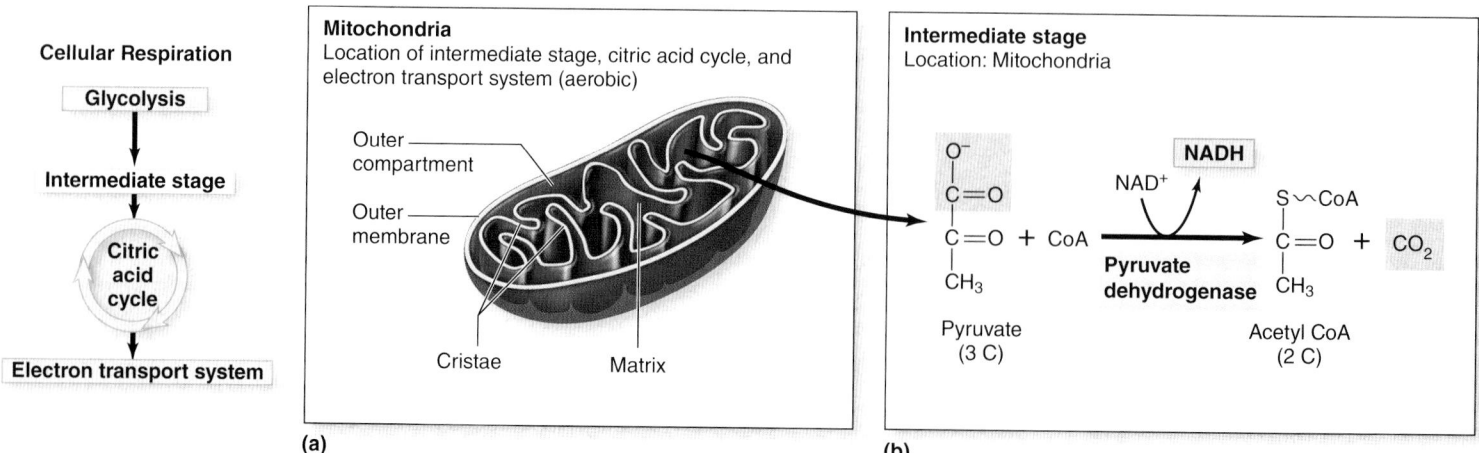

(a)

(b)

Figure 3.17 Intermediate Stage. (_a_) Mitochondria are the cellular organelles where aerobic cellular respiration occurs. (_b_) The intermediate stage involves a multienzyme complex called pyruvate dehydrogenase, which converts pyruvate to acetyl CoA and carbon dioxide with formation of NAD$^+$ to NADH.

acetyl CoA produced in the intermediate stage is converted to two CO_2 molecules and a CoA molecule is released. Energy is transferred to form one ATP molecule, three NADH molecules, and 1 $FADH_2$ molecule during one "turn" of the citric acid cycle.

Steps of the Citric Acid Cycle

The nine enzymatic steps of the citric acid cycle are shown in **(figure 3.18)**; both an overview figure and the detailed steps are included.

1. Step 1 of the citric acid cycle uses the first enzyme to combine an acetyl CoA molecule produced in the intermediate stage with a molecule of **oxaloacetate** (ok′să-lō-ă-sē′tāt) to form **citrate**. (Note that the addition of a hydrogen ion to citrate forms citric acid.) The first step of this cycle gives this enzymatic pathway its name.

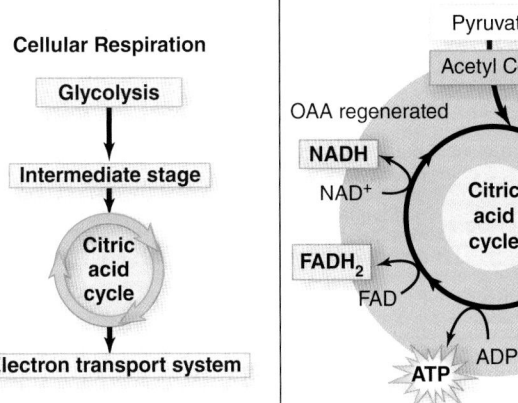

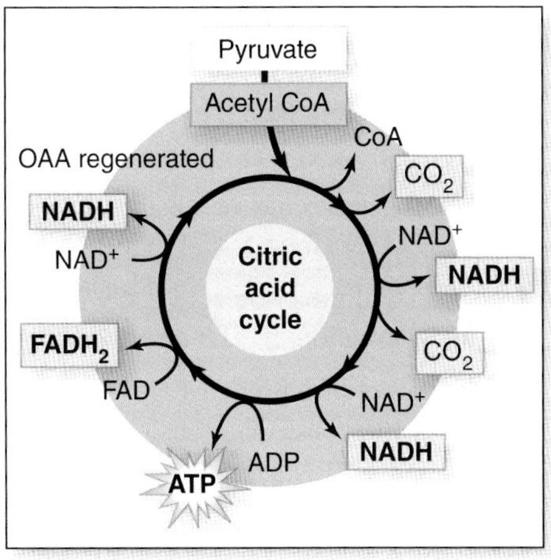

(a) Net chemical reaction of citric acid cycle

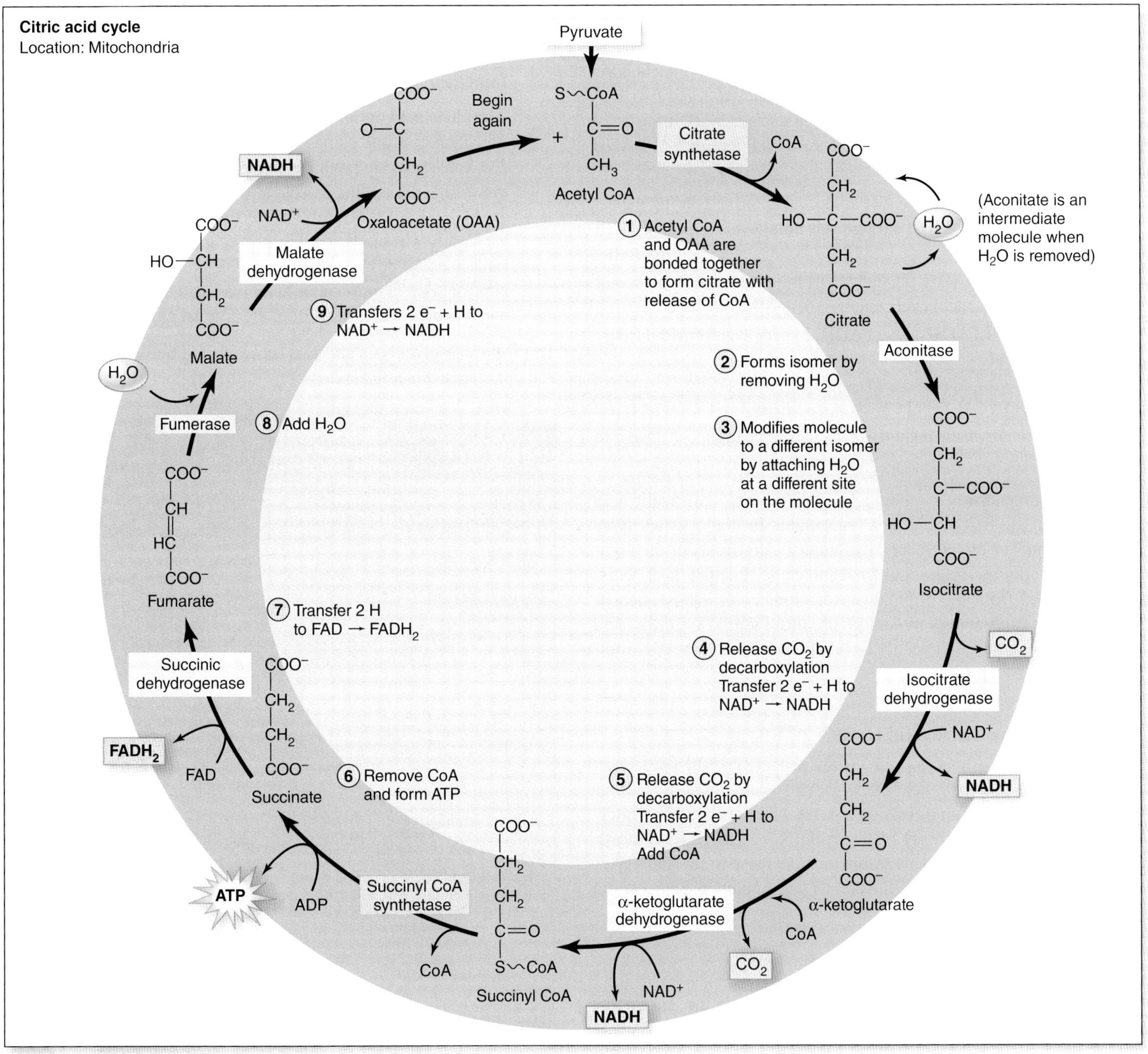

Citric acid cycle
Location: Mitochondria

(b) Details of citric acid cycle

Figure 3.18 Citric Acid Cycle. The citric acid cycle is the metabolic pathway that occurs within the matrix of mitochondria for the chemical breakdown of acetyl CoA with the net transfer of energy to form 1 ATP, 3 NADH, and 1 $FADH_2$ molecules. (*a*) An overview of the citric acid cycle. (*b*) The detailed pathway of the citric acid cycle. AP|R

(2–3) Steps 2 and 3 of the citric acid cycle form an isomer by removing a water molecule from citrate and then reattaching it to a different location on the molecule.

(4–5) Steps 4 and 5 of the citric acid cycle occur through two different hydrogenase enzymes that participate in the transfer of hydrogen to NAD^+ to form NADH. CoA is also attached during step 5.

(6) Step 6 of the citric acid cycle involves the removal of CoA and the formation of ATP through substrate-level phosphorylation.

(7) Step 7 of the citric acid cycle occurs through the action of a dehydrogenase that transfers hydrogens to FAD to form $FADH_2$.

(8) Step 8 of the citric acid cycle is the removal of water.

(9) Step 9 of the citric acid cycle is catalyzed by a dehydrogenase that transfers hydrogen to NAD^+ to form NADH. Oxaloacetic acid (OAA) is regenerated in this final step.

Summary of the Citric Acid Cycle

The citric acid cycle is a metabolic process that occurs in mitochondria and requires oxygen. Acetyl CoA is the initial substrate and two CO_2 molecules and one CoA molecule are the products. The net transfer of energy produced in this cycle is used to form 1 ATP molecule, 3 NADH molecules, and 1 $FADH_2$ molecule.

- **Formation of ATP.** One ATP molecule is formed during the citric acid cycle (step 6) through substrate-level phosphorylation.
- **Formation of NADH.** Three NADH molecules are formed during the citric acid cycle (steps 4, 5, and 9).
- **Formation of $FADH_2$.** One $FADH_2$ molecule is formed during the citric acid cycle (step 7).

WHAT DO YOU THINK?

(4) Why is the enzymatic pathway of the citric acid cycle considered to be a "cycle"?

This enzymatic pathway is called a "cycle" because oxaloacetic acid is involved in the first step and is regenerated in the last step. Note that two "turns" of the citric acid cycle must also occur for the complete digestion of the original glucose molecule (one per each acetyl CoA generated from the original glucose molecule). Consequently, the number of high-energy molecules produced within the citric acid cycle from one glucose molecule is 2 ATP molecules, 6 NADH molecules, and 2 $FADH_2$ molecules.

Regulation of the Citric Acid Cycle

Regulation of the citric acid cycle occurs primarily at the enzyme in the first step of the citric acid cycle (citrate synthetase). If cellular energy demands are high, levels of NADH, ATP, and pathway intermediates are low, the activity of the cycle is increased. In contrast, an increase in these substances results in a decrease in the activity of the cycle. These physiologic adjustments help maintain homeostatic levels of ATP molecules.

Completion of Glucose Digestion

Following glycolysis and two "turns" through both the intermediate stage (which generates 2 CO_2 molecules) and the citric acid cycle (which produces 4 CO_2 molecules), glucose has been completely digested and the six carbon atoms of glucose ($C_6H_{12}O_6$) have been released as 6 CO_2 molecules. Notice that the carbon atoms in the glucose (and other fuel molecules) that you eat are converted to carbon dioxide within the mitochondria of your cells!

Summary of the Chemical Breakdown of Glucose

Table 3.3 provides a summary of each of the first three stages of glucose oxidation that results in the chemical breakdown of glucose into carbon dioxide. The following critical aspects of this process include

- **Glycolysis.** Glycolysis occurs in the cytosol and does not require oxygen. Energy is transferred to form 2 ATP molecules (net) and 2 NADH molecules. If sufficient oxygen is available, the pyruvate formed enters a mitochondrion and is further metabolized in the intermediate stage and the citric acid cycle.
- **The intermediate stage.** The intermediate stage occurs in a mitochondrion. It involves a multienzyme complex that converts pyruvate to acetyl CoA and 1 CO_2 molecule. Energy is transferred to form 1 NADH molecule. Remember, one NADH molecule is formed per pyruvate entering the intermediate stage. Recall that two pyruvates are produced from one glucose molecule. Thus, the intermediate stage must occur twice—so a total of 2 NADH molecules are formed from the original glucose molecule.
- **The citric acid cycle.** The citric acid cycle also occurs in a mitochondrion and completes the breakdown of glucose. Two CO_2 molecules are produced per turn of the cycle. Energy is transferred during this process to form 1 ATP, 3 NADH, and 1 $FADH_2$. Remember, this reflects the energy transferred per each acetyl CoA entering the citric acid cycle. Two acetyl CoA molecules are produced from one glucose molecule. Thus, the citric acid cycle must occur twice—so a total of 2 ATP, 6 NADH, and 2 $FADH_2$ are formed from the original glucose molecule.

Thus, through glycolysis and two turns of both the intermediate stage and citric acid cycle, the six carbons in the original glucose have been

Table 3.3	Comparison of the First Three Stages of Glucose Breakdown		
Characteristics	**Glycolysis**	**Intermediate Stage**	**Citric Acid Cycle**
Where it occurs	Cytosol	Mitochondria	Mitochondria
Requires oxygen (is aerobic)?	No	Yes (aerobic)	Yes (aerobic)
Substrate	Glucose	Pyruvate (2 pyruvates from each glucose)	Acetyl CoA (2 acetyl CoA from each glucose)
Product	2 pyruvate molecules	Acetyl CoA and 1 CO_2 per pyruvate	2 CO_2 per acetyl CoA
Pathway or complex	Metabolic pathway	Multienzyme complex	Metabolic pathway
Net energy molecules produced	2 ATP (net) and 2 NADH	1 NADH per pyruvate	1 ATP per acetyl CoA 3 NADH per acetyl CoA 1 $FADH_2$ per acetyl CoA
How lack of oxygen affects the stage	Lactate produced (to regenerate NAD^+ so glycolysis can continue)	Pathway inhibited by lack of oxygen	Pathway inhibited by lack of oxygen

released as 6 CO_2 molecules and the energy has been transferred to form:

Glycolysis   → 2 ATP   2 NADH
Intermediate stage →   2 NADH
Citric acid cycle   → 2 ATP   6 NADH   2 $FADH_2$

WHAT DID YOU LEARN?

22   Summarize the metabolic pathway of the citric acid cycle—where it occurs, if the process requires oxygen, the net chemical reaction, and the net energy transfer.

23   What energy molecules are produced from the chemical breakdown of glucose during each of the three steps of cellular respiration?

3.4e The Electron Transport System

LEARNING OBJECTIVES

8.   Describe the importance of NADH and $FADH_2$ in energy transfer.

9.   Explain the actions that take place in the electron transport system.

Given that the breakdown of glucose is completed by the end of the citric acid cycle, what is the function of the electron transport system, the final stage of cellular respiration? The **electron transport system** involves the transfer of electrons (energy) from the coenzymes NADH and $FADH_2$ that are produced during the first three stages of cellular respiration. The energy released from these coenzymes is used to form ATP. This is a critical stage in cellular respiration because most of the energy captured in glucose oxidation is initially transferred to form NADH from NAD^+, as well as the smaller amounts of energy to form $FADH_2$ from FAD. These processes involve structures located in the inner folded membrane (or cristae) of mitochondria.

Structures of the Electron Transport System

Several significant types of molecules are embedded within the cristae of the mitochondria: H^+ pumps, electron carriers, and ATP synthetase enzymes **(figure 3.19)**. H^+ pumps are proteins that transport H^+ from the matrix to the outer membrane compartment. This maintains a H^+ gradient between the outer compartment and the matrix within a mitochondrion, with more H^+ in the outer compartment than in the matrix. H^+ pumps also bind and release electrons (e^-).

Electron carriers ubiquinone (Q) and cytochrome c (C) are located between proteins serving as H^+ proton pumps. Electron carriers transport electrons (e^-) between the H^+ pumps. You may be familiar with these electron carriers because they are sold as antioxidant supplements to cyclists, bodybuilders, and other athletes (e.g., CoQ10 as a ubiquinone supplement).

This series of H^+ pumps and electron carriers is collectively called the **electron transport chain**. ATP synthetase allows for the passage of H^+ from the outer compartment back into the matrix. During this process, the flow of H^+ down its concentration gradient is harnessed to bond P_i to ADP to form ATP.

Steps of the Electron Transport System

The processes of the electron transport system are organized into three major steps (figure 3.19):

①   **Electrons are transferred from coenzymes to O_2.** The coenzyme, either NADH or $FADH_2$, releases hydrogen (e^- and H^+) and it is oxidized. The released electrons are passed through the electron transport chain to O_2, which serves as the final electron acceptor. O_2 combines with 4 e^- + 4 H^+ to produce

---

Cellular Respiration

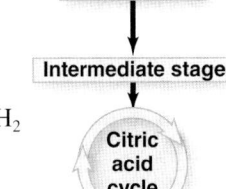

Glycolysis → Intermediate stage → Citric acid cycle → Electron transport system

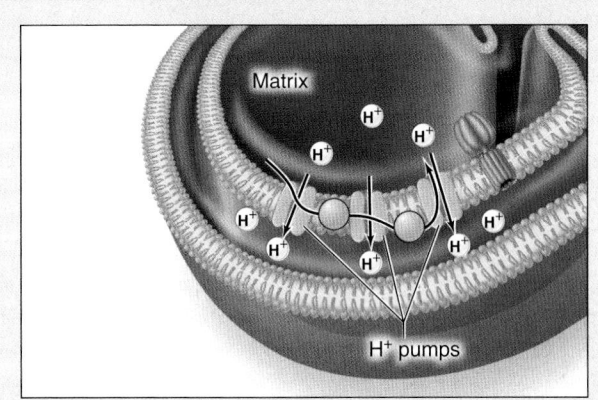

① Electrons are transferred from NADH and $FADH_2$ through a series of electron carriers within the cristae. O_2 is the final electron acceptor. (O_2, e^- and H^+ join to form H_2O.)

② Energy of electrons "falling" (in step 1) is used by H^+ pumps to move H^+ up its concentration gradient from the matrix to the outer compartment.

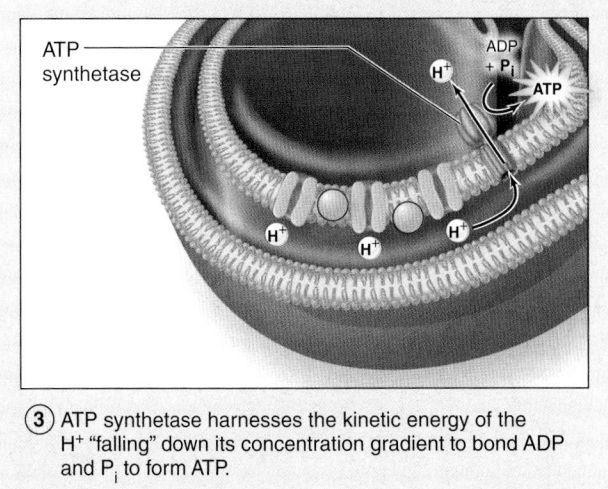

③ ATP synthetase harnesses the kinetic energy of the H^+ "falling" down its concentration gradient to bond ADP and P_i to form ATP.

Figure 3.19 The Electron Transport System. The processes of the electron transport system can be organized into three steps in which energy temporarily held by the coenzymes NADH and $FADH_2$ is transferred to form a bond between ADP and P_i, yielding ATP. **AP|R**

two molecules of H_2O. Thus, oxygen is a reactant in cellular respiration, and it becomes part of the water that is produced. Notice that the oxygen you breathe is converted to water within the mitochondria of your cells!

② **Proton gradient is established.** As electrons are "falling" and passed through the electron transport chain, their kinetic energy is harnessed by H^+ pumps to move H^+ from the mitochondrial matrix into the outer compartment, maintaining a proton gradient.

③ **Proton gradient is harnessed to form ATP.** H^+ moves down its concentration gradient as it is transported across the inner membrane by ATP synthetase. It moves from the outer compartment into the matrix. (Note that H^+ moves back into the area of the mitochondrion from which it was just pumped.) This process is analogous to water falling over a dam and turning a water wheel. The kinetic energy of the falling H^+ is harnessed by ATP synthetase to form a new bond between ADP and P_i, producing an ATP.

This process of forming ATP is referred to as **oxidative phosphorylation** because it involves oxygen as the final electron acceptor, and ATP is formed from the phosphorylation of ADP. This process is distinguished from substrate-level phosphorylation, which forms ATP from energy directly released from a substrate, as occurs in specific steps of glycolysis (see figure 3.16, steps 7 and 10) and the citric acid cycle (see figure 3.18, step 6). The steps of cellular respiration are summarized in **figure 3.20.**

WHAT DID YOU LEARN?

㉔ What is the importance of NADH and $FADH_2$ in energy transfer?

㉕ What are the three primary steps that take place in the electron transport system?

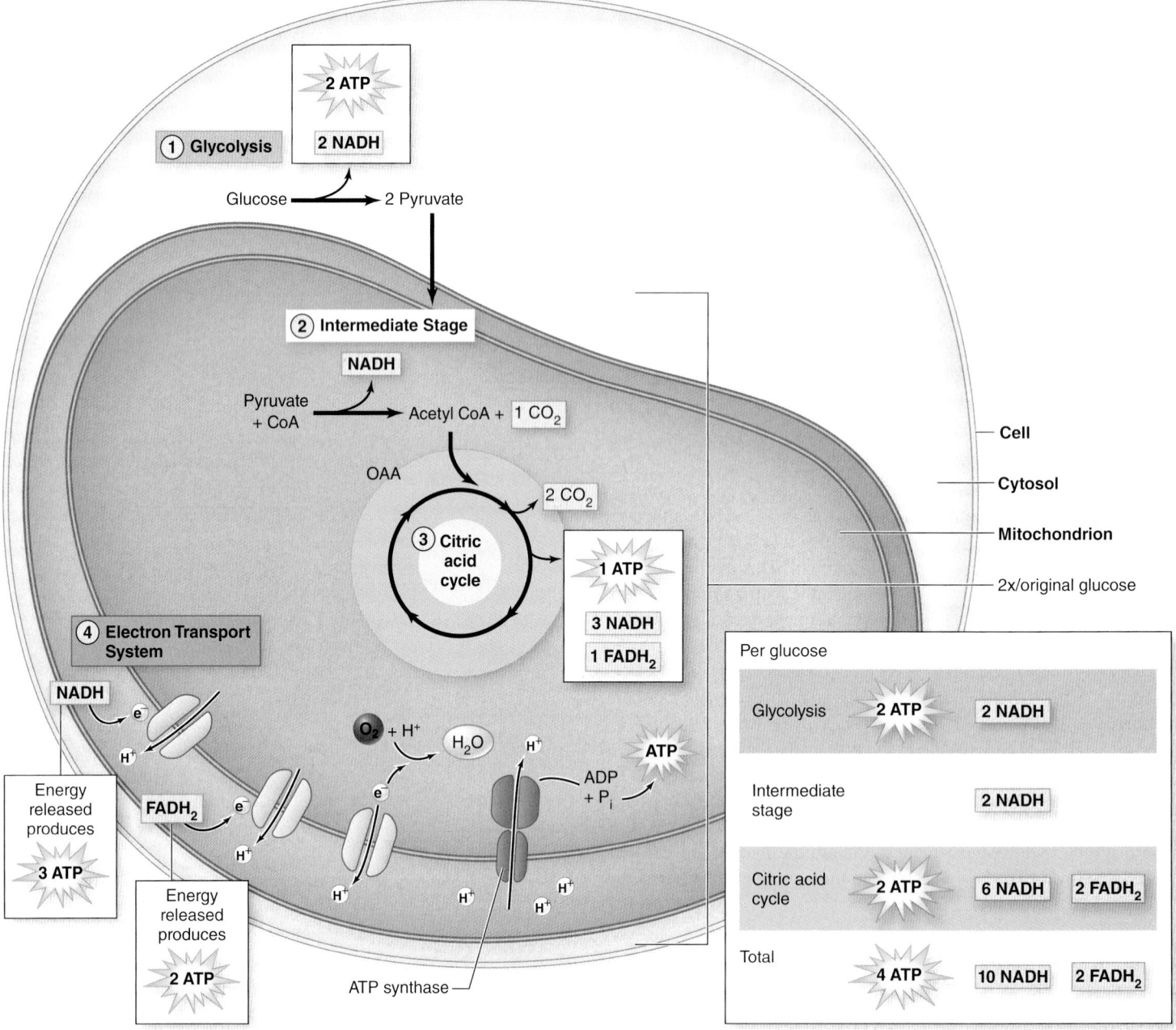

Figure 3.20 Summary of Stages of Cellular Respiration. (*1*) Glycolysis (an enzymatic pathway within the cytosol that initiates the breakdown of glucose to two pyruvate molecules). (*2*) Intermediate stage (a multienzyme complex within mitochondria that converts pyruvate to acetyl CoA). (*3*) Citric acid cycle (an enzymatic pathway within mitochondria that completes the breakdown of glucose to CO_2). (*4*) Electron transport system (specialized structures in cristae of mitochondria that provide the means to transfer energy from NADH and $FADH_2$ coenzymes to bond P_i to ADP to form ATP molecules—a process in which oxygen accepts electrons and combines with H^+ to form H_2O.

INTEGRATE

CLINICAL VIEW
Cyanide Poisoning

Cyanide binds with a specific cristae electron carrier of the electron transport chain (called cytochrome oxidase). This binding inhibits the processes of the electron transport chain and the subsequent production of ATP. Although oxygen is available for "catching electrons" within the electron transport chain, the inhibition of one of the electron carriers within the electron transport chain prevents the electrons from reaching oxygen. Treatment for nonlethal doses (doses as low as 1.5 mg/kg body weight can be fatal) involves the administration of substances that bind cyanide (e.g., nitrites), which are then eliminated in the urine.

3.4f ATP Production

 LEARNING OBJECTIVE

10. Calculate the number of ATP molecules produced in cellular respiration if oxygen is not available and if oxygen is available.

The number of ATP molecules generated when electrons are released from coenzymes is dependent upon the entry point of the electrons into the electron transport chain (figure 3.19). The electrons from NADH enter at the top of the electron transport chain and are passed through three H^+ pumps, which results in enough energy being released to generate 3 ATP molecules. In contrast, the electrons from $FADH_2$ enter at the second H^+ pump; this results in the generation of 2 ATP molecules. Consequently, each NADH generates 3 ATP and each $FADH_2$ generates 2 ATP.

 WHAT DO YOU THINK?

5 Given that energy from each NADH produces 3 ATP molecules and each $FADH_2$ produces 2 ATP molecules, calculate the number of ATP molecules generated from glucose during cellular respiration.

We are able to calculate the specific number of ATP molecules produced in breakdown of a glucose molecule by knowing the following: (1) the specific number of energy molecules (i.e., ATP, NADH, and $FADH_2$) that are generated from glucose breakdown in each of the first three stages of cellular respiration, and (2) the specific number of ATP generated by the oxidation of each type of coenzyme in the electron transport system (NADH = 3 ATP molecules and FADH = 2 ATP molecules).

Below is a summary of the number of ATP molecules produced by substrate-level phosphorylation and the number of ATP produced by oxidative phosphorylation from glucose oxidation.

Stage/Total	Substrate-Level Phosphorylation	Oxidative Phosphorylation
Glycolysis	2 ATP	2 NADH → 6 ATP
Intermediate Stage	—	2 NADH → 6 ATP
Citric Acid Cycle	2 ATP	6 NADH → 18 ATP
		2 $FADH_2$ → 4 ATP
Total Number of ATP/ Based on Method of Formation	**4 ATP**	**34 ATP**

In theory, the total ATP produced from one glucose molecule is 38 ATP. However—and this is very important for energy totals—the actual yield is lower. There are several energy-requiring steps for transporting molecules during cellular respiration that decrease the actual net ATP produced. These energy-requiring processes include moving (1) pyruvate from the cytosol into mitochondria, (2) phosphate and ADP into mitochondria for its use in ATP synthesis, and (3) NADH produced during glycolysis into mitochondria for oxidative phosphorylation in the electron transport system. Note: the net ATP produced from a glucose molecule is 30 ATP.

 WHAT DID YOU LEARN?

26 How many net ATP molecules are generated through glycolysis (i.e., without participation of mitochondria)? How many net ATP molecules are formed through the combined processes that occur in the cytosol and those that occur within mitochondria?

3.4g The Fate of Pyruvate with Insufficient Oxygen

 LEARNING OBJECTIVES

11. Explain the fate of pyruvate when oxygen is in short supply.

12. Describe the impact on ATP production if there is insufficient oxygen.

We followed pyruvate in the previous discussion assuming that sufficient oxygen was present for oxidative phosphorylation to occur. If sufficient oxygen is not available we must consider the following:

1. Cellular respiration processes requiring oxygen (i.e., aerobic cellular respiration) decrease, including the activity of the electron transport chain. Electrons remain with the NADH and $FADH_2$ molecules and NADH and $FADH_2$ accumulate. This is accompanied by decreased levels of NAD^+ and FAD.
2. The cell becomes increasingly dependent upon glycolysis, a metabolic pathway that requires NAD^+ to continue.
3. Extended low-oxygen conditions would ultimately result in the complete shutdown of glycolysis within the cell because of the lack of NAD^+.
4. NAD^+ must be regenerated if glycolysis is to continue.

Regenerating NAD^+ involves the transfer of hydrogen from NADH. Two electrons and hydrogen are transferred from NADH to pyruvate, which is converted to lactate. (Note that the addition of a hydrogen ion to lactate forms lactic acid.) This enzymatic reaction is catalyzed by lactate dehydrogenase (**figure 3.21**).

Although an effective means to permit glycolysis to continue (because it makes NAD^+ available), we must keep in mind that without the availability and use of mitochondria, only 2 ATP are produced per

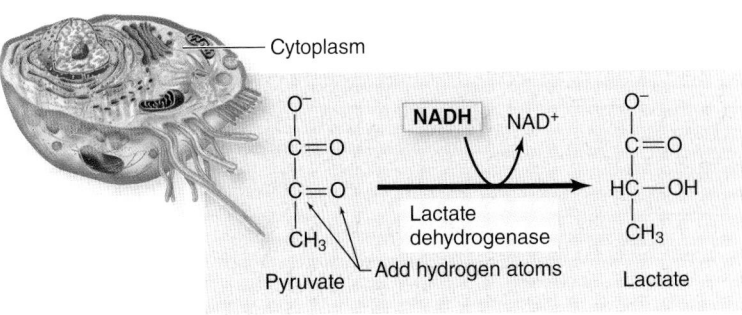

Figure 3.21 Conversion of Pyruvate to Lactate. Lactate dehydrogenase converts pyruvate to lactate to regenerate NAD^+ molecules.

glucose. Compared to the net 30 ATP produced with sufficient oxygen, there is significantly less ATP generated (2 ATP versus 30 ATP). This is a 15-fold difference! Remember this important association: low O_2 equals low energy. So, in clinical practice when you work with individuals with decreased ability to deliver oxygen to cells (e.g., those with impaired respiratory or cardiovascular function), keep in mind that they will have less available ATP to meet the body's energy needs.

 WHAT DID YOU LEARN?

27 Pyruvate is converted to what molecule if there is insufficient oxygen? Explain why this occurs.

3.4h Other Fuel Molecules That Are Oxidized in Cellular Respiration

 LEARNING OBJECTIVE

13. Describe the entry point in the metabolic pathway of cellular respiration for both fatty acids and amino acids.

There are other fuel molecules, such as fatty acids and amino acids that may be oxidized to generate ATP (see figure 27.8).

Fatty acids are enzymatically changed two carbon units at a time to form acetyl CoA. This process is called **beta-oxidation.** Acetyl CoA then enters the cell respiration metabolic pathway at the citric acid cycle. Because fatty acids enter this metabolic pathway in the mitochondria, they can only be oxidized aerobically. (Note that a by-product of fatty acid metabolism is the production of ketoacids, with significant amounts produced in individuals with uncontrolled diabetes mellitus; see Clinical View: "Lactic Acidosis and Ketoacidosis" in section 25.6c.)

A different pathway is employed if protein is used for fuel. The point of entry of deaminated amino acids (amino acids with the amine group [$–NH_2$] removed) is dependent upon the specific type of amino acids. Different amino acids enter the metabolic pathway at glycolysis, the intermediate stage, or the citric acid cycle. The amine group is a waste product that is converted to urea and excreted by the kidneys.

 WHAT DID YOU LEARN?

28 Why is oxygen required to burn fatty acids?

INTEGRATE

CONCEPT CONNECTION

What happens to the increased levels of lactate produced by skeletal muscle tissue? Lactate is either absorbed by surrounding muscle tissue or transported by the blood to the liver. Muscle cells may either use lactate immediately for ATP synthesis by converting it back to pyruvate or converting it to glucose and storing it as glycogen. Liver cells convert lactate into glucose. Glucose formed in the liver is either stored as glycogen in the liver or released back into the blood (for muscles or other cells to take up). This cycling of lactate from muscles to the liver, the conversion of lactate to glucose, and the subsequent transport of glucose from the liver to muscle is called the Cori cycle.

CHAPTER SUMMARY

	• The concepts discussed are energy, chemical reactions, enzymes, metabolic pathways, and the production of ATP through cellular respiration.
3.1 Energy 71	• Energy is the capacity to do work.
	3.1a Classes of Energy 71 • Energy exists in two classes: energy based upon position, called potential energy, and energy of motion, called kinetic energy. • Energy can be converted from potential energy to kinetic energy (or vice versa). Examples are the movement of a substance down its concentration gradient, or the movement of electrons from a higher-energy state to a lower-energy state.
	3.1b Forms of Energy 71 • Energy exists in different forms, including chemical, electrical, mechanical, sound, radiant, and heat.
	3.1c Laws of Thermodynamics 73 • The first law of thermodynamics states that energy cannot be created or destroyed, only converted from one form to another. • The second law of thermodynamics states that some energy is lost as heat with every energy conversion.
3.2 Chemical Reactions 76	• Chemical reactions are expressed in chemical equations and are classified using various criteria.
	3.2a Chemical Equations 76 • Metabolism is the collective term for all biochemical reactions that occur within the body. • Reactants become products in a chemical reaction, and an arrow indicates the direction of change.
	3.2b Classification of Chemical Reactions 76 • Chemical reactions can be classified using different criteria: change in chemical structure, change in chemical energy, and if the reaction is irreversible or reversible.
	3.2c Reaction Rates and Activation Energy 80 • Reaction rate is the measure of how quickly a chemical reaction takes place; this determines the amount of product formed per time. • Activation energy is the energy required to break existing chemical bonds for the chemical reaction to proceed.

CHALLENGE YOURSELF

Do You Know the Basics?

_____ 1. Energy in ATP is used to power skeletal muscle contraction. This is an example of what type of energy conversion?

a. chemical energy to mechanical energy

b. light energy to mechanical energy

c. chemical energy to light energy

d. electrical energy to chemical energy

_____ 2. Oxidation-reduction can be best classified as a(n) _____ reaction.

a. exchange

b. endergonic

c. synthesis

d. reversible

_____ 3. All of the following increase enzymatic activity *except*

a. an increase in temperature.

b. an increase in pH.

c. an increase in concentration of the substrate.

d. an increase in concentration of the enzyme that catalyzes the reaction.

_____ 4. ATP inhibits phosphofructokinase by binding to an allosteric site in glycolysis. ATP is functioning as a

a. competitive inhibitor.

b. competitive activator.

c. noncompetitive inhibitor.

d. noncompetitive activator.

_____ 5. All of the following are accurate about enzymes *except*

a. Enzymes are typically globular proteins with an active site.

b. Enzymes decrease activation energy.

c. Enzymes can be used over and over to catalyze a substrate to a product.

d. Enzymes are versatile and can catalyze different types of chemical reactions.

_____ 6. Glucose is converted to pyruvate in which stage of cellular respiration?

a. glycolysis

b. intermediate stage

c. citric acid cycle

d. electron transport system

_____ 7. NAD$^+$ and FAD are examples of

a. enzymes.

b. high-energy organic molecules that are digested in cellular respiration.

c. allosteric inhibitors.

d. coenzymes.

_____ 8. All stages of cellular respiration are decreased in conditions of insufficient oxygen *except*

a. glycolysis.

b. the intermediate stage.

c. the citric acid cycle.

d. the electron transport system.

_____ 9. In glycolysis, _____ ATP are formed, and if sufficient oxygen is present, _____ ATP are formed.

a. 2, 2

b. 36, 38

c. 2, 30

d. 10, 30

_____ 10. Oxidative phosphorylation involves

a. electrons transported in the electron transport chain and accepted by O_2.

b. ATP synthetase harnessing the energy in a proton gradient.

c. coenzymes NADH and FADH$_2$ giving up their electrons.

d. all of the above.

11. List and define the different forms of energy, and give one example each of their application in the human body.

12. Describe the different ways of classifying chemical reactions, and explain the category to which oxidation-reduction belongs.

13. Explain ATP cycling.

14. Describe the structure and mechanism of enzymes.

15. Describe a metabolic pathway, and explain how it is controlled by negative feedback.

16. Summarize glycolysis, including where it occurs in a cell, if it requires oxygen, the substrate and final product, and the formation of energy containing molecules (i.e., ATP, NADH, FADH$_2$).

17. In general terms, explain the fate of pyruvate if there is (a) sufficient oxygen, and (b) insufficient oxygen.

18. Describe how oxygen becomes part of water during cellular respiration.

19. Identify the source of carbon in carbon dioxide.

20. Based on what you know about glycolysis and aerobic cellular respiration, explain the advantage in terms of ATP production of a healthy respiratory and cardiovascular system.

Can You Apply What You've Learned?

1. Albinism (achromia) is a genetic condition in which an individual cannot synthesize melanin from tyrosine (an amino acid), a brown pigment of the hair, skin, and eyes. These individuals lack

a. specific fatty acids.

b. a protein that contains tyrosine.

c. an enzyme that converts tyrosine to melanin.

d. cofactors that convert tyrosine to melanin.

2. If an individual has impaired respiratory function, such as occurs with emphysema, you would expect all of the following *except*

   a. production of additional lactate.

   b. an impaired ability to make ATP.

   c. low energy levels and complaints of being tired.

   d. increased aerobic cellular respiration.

3. Another challenge to a patient with impaired respiratory function is the buildup of CO_2 in the blood. What would you predict given the following enzymatic reaction that occurs in the blood?

   $$H_2O + CO_2 \rightleftharpoons H_2CO_3 \rightleftharpoons H^+ + HCO_3^-$$

   a. increased production of H_2O

   b. increased production of H^+ (with an accompanying decrease in pH)

   c. decreased production of H^+ (with an accompanying increase in pH)

   d. All of the above.

4. You would expect decreased production of ATP in all of the following individuals *except*

   a. an individual with impaired ability to transport oxygen in the blood, such as a person with anemia.

   b. an individual with severe asthma.

   c. an individual in congestive heart failure.

   d. an athlete.

5. Brown adipose tissue contains cells that allow H^+ to fall down the concentration gradient in the electron transport chain without producing ATP. Instead, all of the kinetic energy is converted to heat. If scientists could increase the amount of brown adipose tissue in our bodies, then

   a. our body temperature would be cooler.

   b. these cells would be more efficient at producing ATP.

   c. we could eat more and not gain weight.

   d. we would be able to run faster.

Can You Synthesize What You've Learned?

1. Tiffany had returned to her college dorm and was having difficulty breathing. She knew she was having an asthma attack. What changes in her energy level are predicted?

2. Provide a general explanation to a patient on the advantages of aerobic fitness in terms of ATP production.

3. What occurs to the amount of product formed in a metabolic pathway if inhibition does not occur?

INTEGRATE

ONLINE STUDY TOOLS

The following study aids may be accessed through Connect.

Clinical Case Study: Mitochondrial Dysfunction: A Special Form of Inheritance

Interactive Questions: This chapter's content is served up in a number of multimedia question formats for student study.

LearnSmart: Topics and terminology include energy; chemical reactions; enzymes; cellular respiration

Anatomy & Physiology Revealed: Topics include NADH oxidative-reduction reactions; enzymes; glycolysis; citric acid cycle (Kreb's cycle); electron transport and ATP synthesis

Animations: Topics include enzyme structure and action; coenzymes; metabolic pathways; glycolysis

chapter 4

Biology of the Cell

Module 2:
Cells and Chemistry

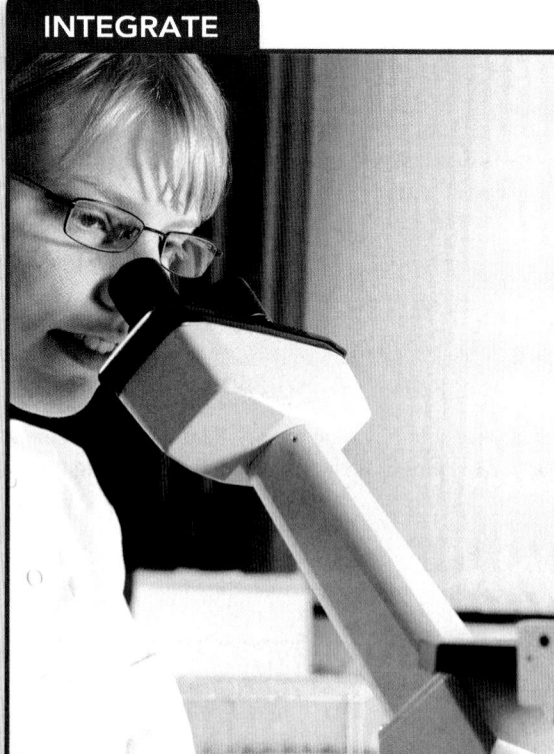

INTEGRATE

CAREER PATH
Cytologist

A cytologist examines cells under a microscope to detect abnormalities that may indicate cancer or other diseases. Cytologists prepare cell samples using specialized equipment and apply stains or other techniques to enhance specimens for detailed observation. Drawing upon an extensive knowledge of cell structure and function, they then analyze cell samples and provide their findings to a pathologist, who ultimately makes a diagnosis.

Heart cells contract to pump blood out of the heart chambers; retinal cells of the eye detect light; phagocytic white blood cells engulf foreign substances (e.g., bacteria, viruses); and pancreatic cells synthesize and secrete the hormone insulin. All human body processes are ultimately dependent upon cells and their activities. For this reason, the cell is often referred to as the "functional unit of the body." Knowledge of cellular structure and function is critical for understanding the concepts of all later chapters.

Throughout this chapter, we present a broad discussion of a cell by describing how cells are studied and the general structural components and functions of cells. Subsequent chapters examine specialized cells and provide details of their unique functions.

4.1 Introduction to Cells

Our examination of cells begins with a description of how we study them. We then describe how the cells that compose the human body vary in both size and shape, and that these differences reflect their specific function. This section concludes with a discussion of the common structural features all cells possess and the general functions that all cells must perform.

4.1a How Cells Are Studied

 LEARNING OBJECTIVE

1. Distinguish among light microscopy, transmission electron microscopy, and scanning electron microscopy.

The study of cells is called **cytology** (sī-tol'ō-jē; *kytos* = cell). The small size of cells is the greatest obstacle to determining their nature. Cells were discovered after microscopes were invented because high-magnification microscopes are required to see the smallest human body cells. The dimensional unit often used to measure cell size is the micrometer (μm). One micrometer is equal to 1/10,000 of a centimeter (about 1/125,000 of an inch).

Microscopy is the use of a microscope to view small-scale structures, and it is an invaluable asset in anatomic investigations. The most commonly used instruments are the light microscope, the transmission electron microscope, and the scanning electron microscope.

Microscopy samples have no inherent contrast (difference between specimen and background) so structures cannot be seen clearly. To provide contrast, colored-dye stains are used with light microscopes, and heavy-metal stains are used with both transmission electron and scanning electron microscopes. **Figure 4.1** compares the images produced when each of these types of microscopes is used to examine the same specimen—in this case, cilia on the surface of epithelial tissue lining the respiratory tract.

The **light microscope (LM)** produces a two-dimensional image by passing visible light through the specimen. Glass lenses focus and magnify the image as it is projected toward the eye (figure 4.1*a*).

The **electron microscope (EM)** uses a beam of electrons to "illuminate" the specimen. Electron microscopes easily exceed the magnification obtained by light microscopy—but more importantly, they improve the resolution (ability to see details) by more than a thousand-fold over the light microscope. A **transmission electron microscope (TEM)** directs an electron beam through a thin-cut section of the specimen. The resultant two-dimensional image is focused onto a screen for viewing or onto photographic film to record the image. The TEM image in figure 4.1*b* shows a close-up section of the cilia on the surface of the epithelial cells.

For detailed three-dimensional study of a specimen's surface, **scanning electron microscope (SEM)** analysis is the method of choice (figure 4.1*c*). Here, the electron beam is moved across the surface of the specimen, and reflected electrons generate a surface-topography image captured on a television screen.

 WHAT DID YOU LEARN?

1 What is the advantage of using a TEM instead of an LM to study intracellular structure?

4.1b Cell Size and Shape

LEARNING OBJECTIVES

2. Describe the range in size of human cells.

3. Name some of the shapes cells may exhibit.

Cells are typically depicted as being of one size and either spherical or cubelike in shape, when in reality the structure of the approximately 75 trillion cells of the adult human shows great variety. Most cells are microscopic in size, but some are large enough to be seen with the naked eye **(figure 4.2)**. For example, red blood cells (erythrocytes) are relatively small with a diameter of about 7–8 μm, whereas

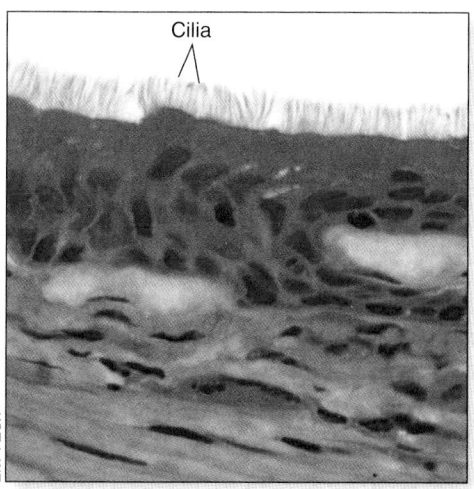

(a) Light microscopy

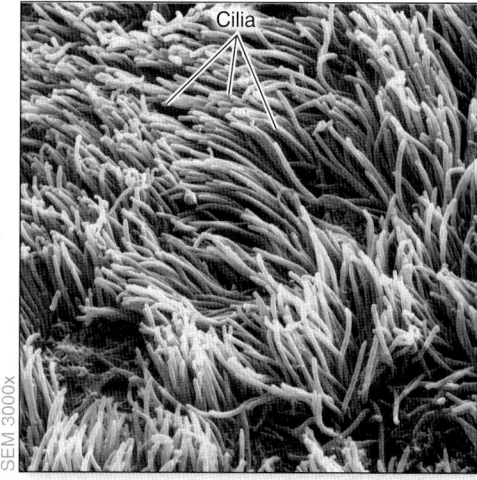

(b) Transmission electron microscopy

(c) Scanning electron microscopy

Figure 4.1 Microscopic Techniques for Cellular Studies. Different techniques are used to investigate cellular anatomy. (*a*) A light microscope (LM) shows hairlike structures, termed cilia, that project from cells lining the respiratory tract. (*b*) A transmission electron microscope (TEM) reveals the ultrastructure, and (*c*) a scanning electron microscope (SEM) shows the three-dimensional image of the cilia of the same type of cells.

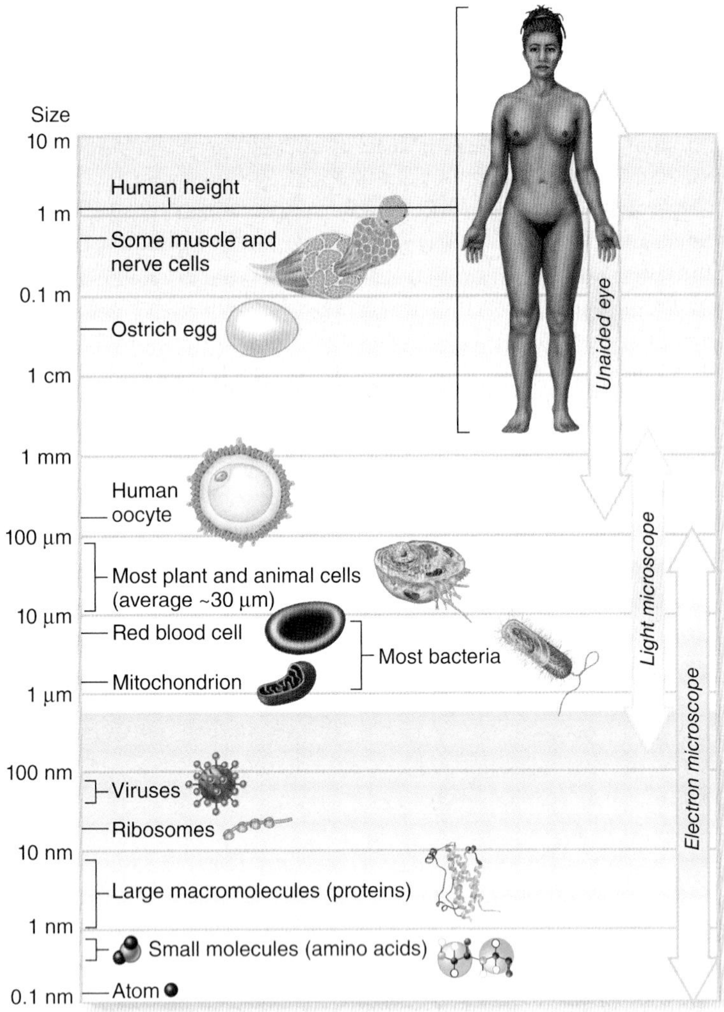

Figure 4.2

Size

10 m	
1 m	Human height
0.1 m	Some muscle and nerve cells
	Ostrich egg
1 cm	
1 mm	
100 μm	Human oocyte
10 μm	Most plant and animal cells (average ~30 μm)
1 μm	Red blood cell
	Mitochondrion
100 nm	Viruses
	Ribosomes
10 nm	
1 nm	Large macromolecules (proteins)
	Small molecules (amino acids)
0.1 nm	Atom

Unaided eye

Light microscope

Electron microscope

Most bacteria

Figure 4.2 The Range of Cell Sizes. Most cells in the human body are between 1 micrometer (μm) and 100 μm in diameter.

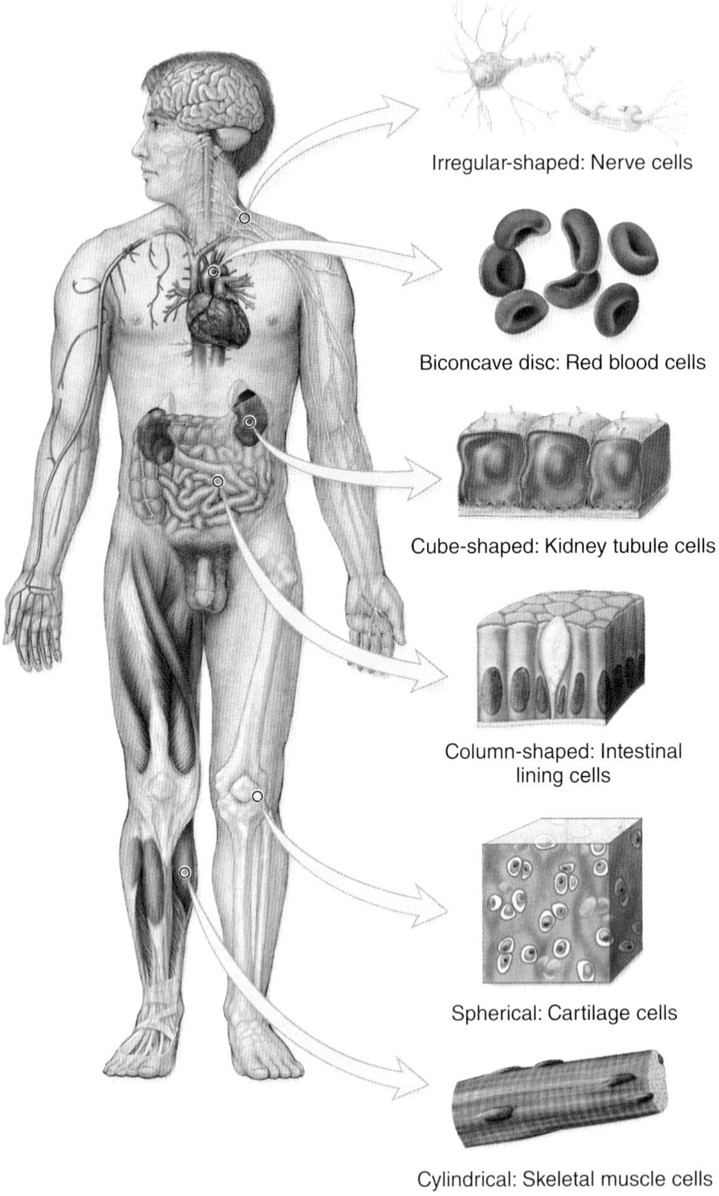

Irregular-shaped: Nerve cells

Biconcave disc: Red blood cells

Cube-shaped: Kidney tubule cells

Column-shaped: Intestinal lining cells

Spherical: Cartilage cells

Cylindrical: Skeletal muscle cells

Figure 4.3 The Variety of Cell Shapes. Cells throughout the body exhibit different shapes that support various functions.

a human oocyte has a diameter of about 120 μm. To help you to relate to how small some cells are, consider that about 5 million erythrocytes would fit on the head of a pin. Cells also vary greatly in shape **(figure 4.3)**. Although some cells are spherical or cubelike, others may be columnlike, cylindrical, disc-shaped, or have an irregular form. Note that a relationship exists between the size and shape of a cell and its function in the body.

WHAT DID YOU LEARN?

2 Which cell is larger, an erythrocyte or a human oocyte? What are their respective sizes?

4.1c Common Features and General Functions

LEARNING OBJECTIVES

4. Describe the three main structural features of a cell.
5. Identify the membrane-bound and non-membrane-bound organelles.
6. Distinguish between organelles and cell inclusions.
7. Explain the general functions that cells must perform.

Most cells are composed of characteristic parts that work together to allow each cell type in the body to perform certain common functions.

Overview of Cellular Components

The generalized cell shown in **figure 4.4** isn't an actual body cell, but rather a representation of a cell that combines features of different types of body cells. The common features include the following:

- **Plasma membrane.** The plasma (plaz′mă; *plasso* = to form) membrane is the cell membrane that forms the outer, limiting barrier separating the internal contents of the cell from the external environment. Modified extensions of the plasma membrane include cilia, a flagellum, and microvilli.

- **Nucleus.** The nucleus (nū′klē-ūs; *nux* = the kernel) is the largest structure within the cell and is enclosed by a nuclear envelope. Much of the internal content of the nucleus is the genetic material, deoxyribonucleic acid (DNA). The fluid within the nucleus is called the nucleoplasm. Within the nucleus is a dark-staining body called the nucleolus.

- **Cytoplasm.** Cytoplasm (sī′tō-plazm; *plasma* = a thing formed) is a general term for all cellular contents located between the plasma membrane and the nucleus. The three primary components of the cytoplasm are the cytosol, organelles, and inclusions.

Cytoplasmic Components

The **cytosol** (sī′tō-sol; *sol* = soluble), also called the *intracellular fluid* (ICF) or *cytoplasmic matrix,* is the viscous, syruplike fluid of the cytoplasm. It has a high water content and contains many dissolved macromolecules that include carbohydrates, lipids, proteins, and small molecules such as glucose and amino acids. Cytosol also contains various types of ions that are used in cellular functions.

Organelles (or′gă-nel; *organon* = organ, *elle* = the diminutive suffix), meaning "little organs," are complex, organized structures within cells that have unique characteristic shapes and functions. Two categories of organelles are recognized: membrane-bound organelles and

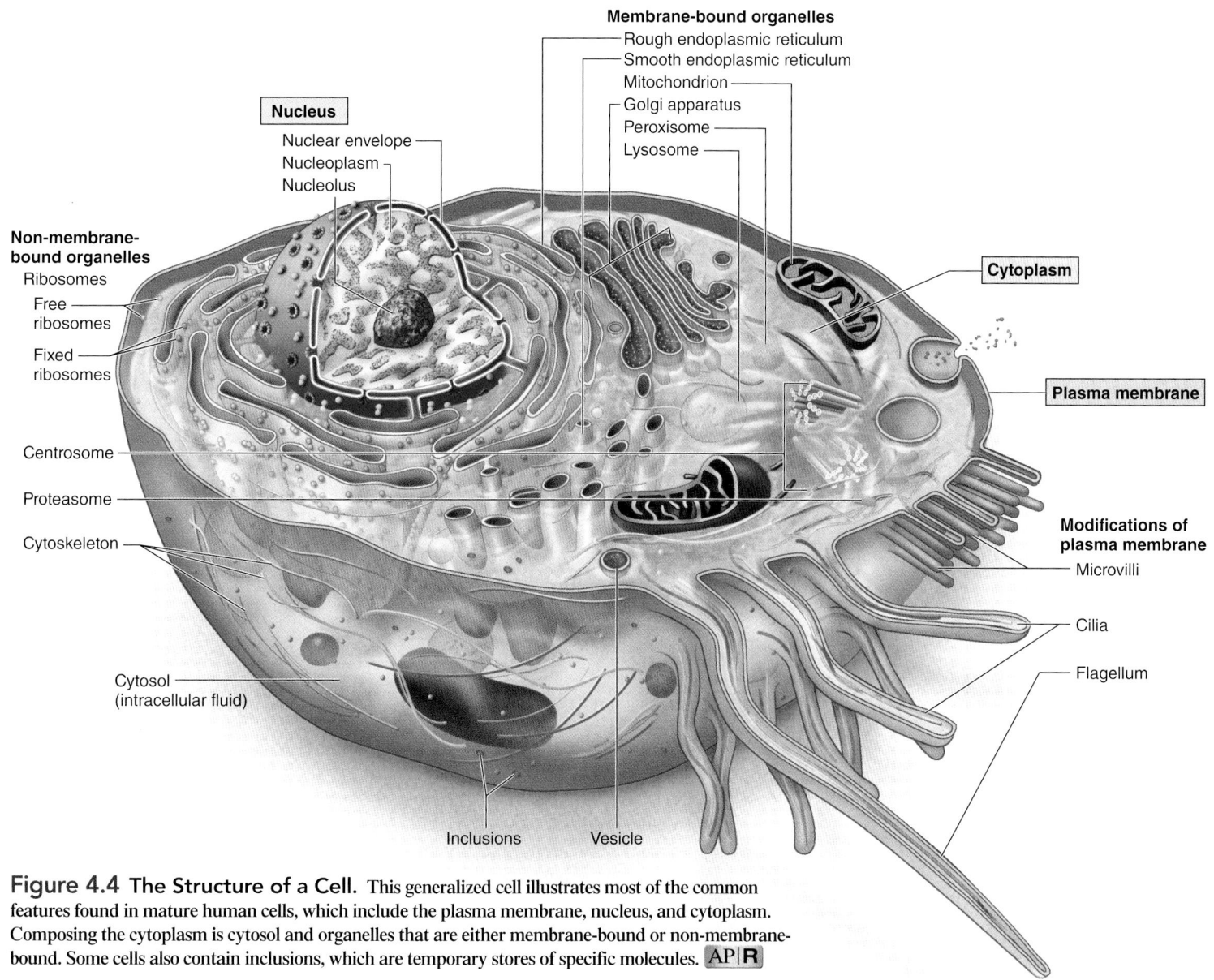

Figure 4.4 The Structure of a Cell. This generalized cell illustrates most of the common features found in mature human cells, which include the plasma membrane, nucleus, and cytoplasm. Composing the cytoplasm is cytosol and organelles that are either membrane-bound or non-membrane-bound. Some cells also contain inclusions, which are temporary stores of specific molecules. AP|R

non-membrane-bound organelles. **Membrane-bound organelles,** or *membranous organelles,* are enclosed by a membrane similar to the plasma membrane. The membrane separates the organelle's contents from the cytosol so that the specific activities of the organelle can proceed without disruption from other cellular activities. Membrane-bound organelles include the endoplasmic reticulum (rough and smooth), Golgi apparatus, lysosomes, peroxisomes, and mitochondria (figure 4.4). Vesicles are temporary membrane-bound structures formed from the endoplasmic reticulum, Golgi apparatus, and plasma membrane. The **non-membrane-bound organelles,** or *nonmembranous organelles,* are not enclosed within a membrane. These structures are generally composed of protein and include ribosomes (either fixed to a membrane or free within the cytosol), the cytoskeleton, the centrosome, and proteasomes. Each of these organelles is discussed in detail in section 4.6.

The cytosol of some cells temporarily store **inclusions.** Cell inclusions are not considered organelles, but are rather aggregates (clusters) of a single type of molecule. Molecules are continuously being added to and removed from inclusions. Pigments (e.g., melanin, a stored pigment in some skin, hair, and eye cells) and nutrient stores (e.g., glycogen and triglycerides) are examples of cell inclusions.

General Cell Functions

Cells must perform general functions that include:

- **Maintain integrity and shape of a cell.** The integrity and shape of a cell is dependent upon both the plasma membrane, which forms the external boundary of the cell and the internal contents, which function to support the cell.
- **Obtain nutrients and form chemical building blocks.** Each cell must get nutrients and other needed substances from its surrounding fluid. Cells form new chemical structures and harvest the energy necessary for survival through diverse metabolic processes.
- **Dispose of wastes.** Cells must dispose of the waste products they produce, so they do not accumulate and disrupt normal cellular activities.

In addition, some cells are capable of undergoing cell division to make more cells of the same type as described in section 4.9. These new cells help to maintain the tissue or organ to which they belong by providing cells for new growth and replacing cells that die. However, during development, many cells lose their ability to divide.

 WHAT DID YOU LEARN?

3 What are the three main structural features of a cell?

4 What cellular structure is responsible for forming the boundary of a cell and maintaining its integrity?

4.2 Chemical Structure of the Plasma Membrane

The plasma membrane is not a rigid boundary, but rather is a fluid matrix composed of approximately an equal mixture, by weight, of lipids and proteins. It regulates the movement of most substances both into and out of a cell.

4.2a Lipid Components

 LEARNING OBJECTIVE

1. List the lipid components of the plasma membrane, and explain the actions of each component.

The plasma membrane contains several different types of lipids including phospholipids, cholesterol, and glycolipids **(figure 4.5)**.

Most of the plasma membrane lipids are **phospholipids** (see section 2.3c). Often these molecules are artistically portrayed in the membrane as a "balloon with two tails." The balloonlike "head" is polar and hydrophilic. In contrast, the two "tails" are nonpolar and hydrophobic. Phospholipid molecules readily associate to form two parallel sheets of molecules lying tail-to-tail, with the hydrophobic tails forming the internal environment of the membrane and their hydrophilic polar heads directed outward. This basic structure of the plasma membrane framework is called the **phospholipid bilayer.** The phospholipid bilayer ensures that cytosol remains inside the cell, and **interstitial fluid** (the fluid that surrounds cells) remains outside.

Cholesterol is scattered within the hydrophobic regions of the phospholipid bilayer. It strengthens the membrane and stabilizes it at temperature extremes.

Glycolipids (glī′kō-lip′id; *glykys* = sweet) are lipids with attached carbohydrate groups. They are located only on the outer phospholipid layer of the membrane, where they are exposed to the interstitial fluid. Together, the carbohydrate portion of the glycolipid molecules and the glycoprotein molecules (described in the next section) help to form the **glycocalyx** (glī′kō-kā′liks; *kalyx* = husk), a "coating of sugar" on the cell's surface. It is interesting to note that the pattern of sugars of the glycocalyx is unique to each individual except identical twins.

The lipid portion of the plasma membrane is insoluble in water, which ensures that the plasma membrane will not simply dissolve when it comes into contact with water. Rather, this boundary is an effective nonpolar physical barrier to most substances. Only small and nonpolar substances can readily penetrate (move through) this barrier without assistance.

INTEGRATE

CONCEPT CONNECTION

Cholesterol is a component of plasma membranes found only in animal cells, not in plant cells. Consequently, any animal-derived food, such as eggs, milk, and meat, contains cholesterol. Foods obtained from plants, including carrots, corn, and potato chips cooked in vegetable oil, do not contain cholesterol.

 WHAT DID YOU LEARN?

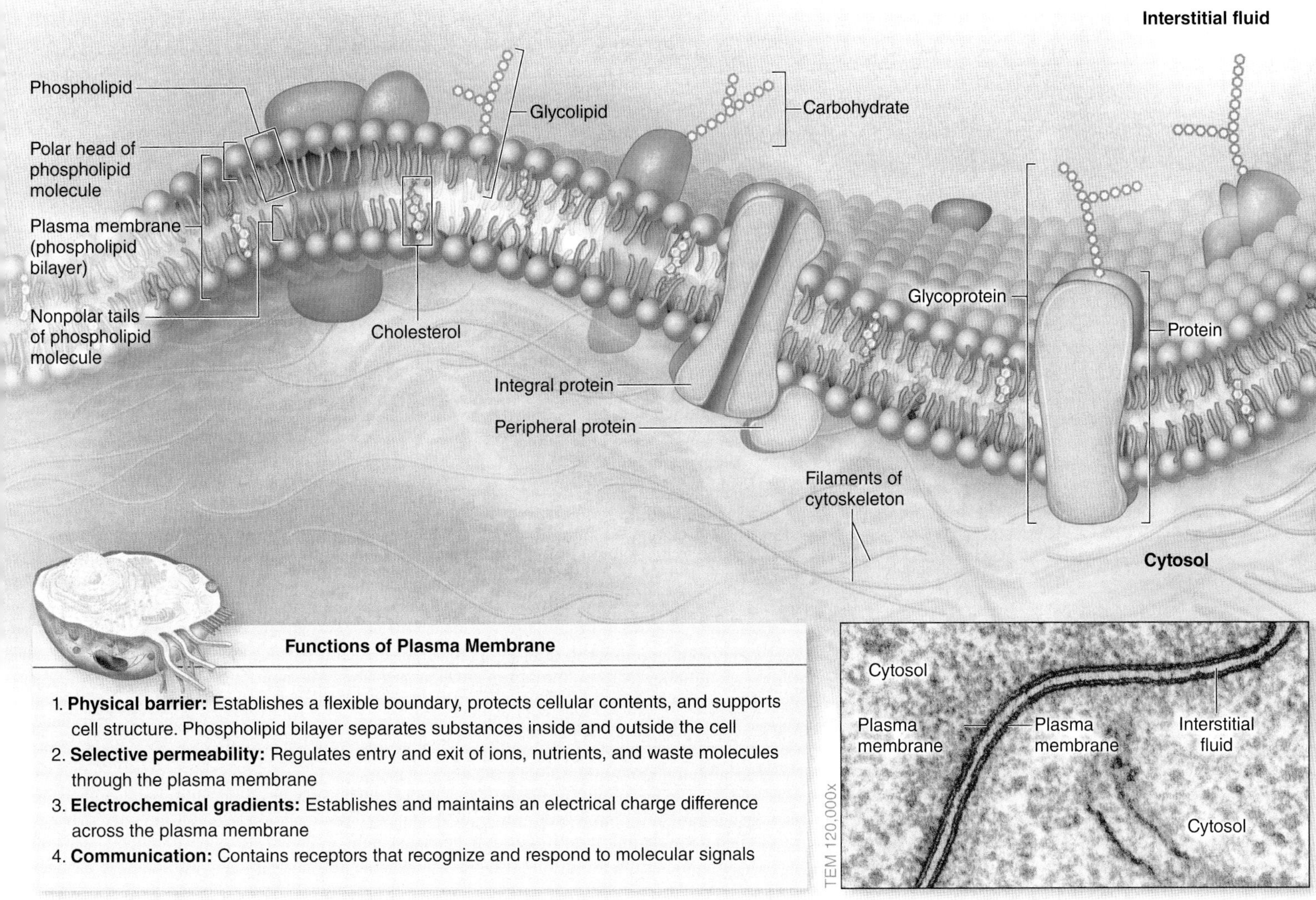

5 How do lipids maintain the basic physical barrier of the plasma membrane?

Phospholipid

Polar head of phospholipid molecule

Plasma membrane (phospholipid bilayer)

Nonpolar tails of phospholipid molecule

Glycolipid

Carbohydrate

Interstitial fluid

Cholesterol

Integral protein

Peripheral protein

Glycoprotein

Protein

Filaments of cytoskeleton

Cytosol

Functions of Plasma Membrane

1. **Physical barrier:** Establishes a flexible boundary, protects cellular contents, and supports cell structure. Phospholipid bilayer separates substances inside and outside the cell
2. **Selective permeability:** Regulates entry and exit of ions, nutrients, and waste molecules through the plasma membrane
3. **Electrochemical gradients:** Establishes and maintains an electrical charge difference across the plasma membrane
4. **Communication:** Contains receptors that recognize and respond to molecular signals

TEM 120,000×

Cytosol

Plasma membrane

Plasma membrane

Interstitial fluid

Cytosol

(a) Plasma membrane

(b) Adjacent plasma membranes

Figure 4.5 Structure and Functions of the Plasma Membrane. (*a*) The plasma membrane is a phospholipid bilayer, with cholesterol and proteins scattered throughout. (*b*) The plasma membranes (phospholipid bilayers) of two adjacent cells is visible in a TEM. AP|R

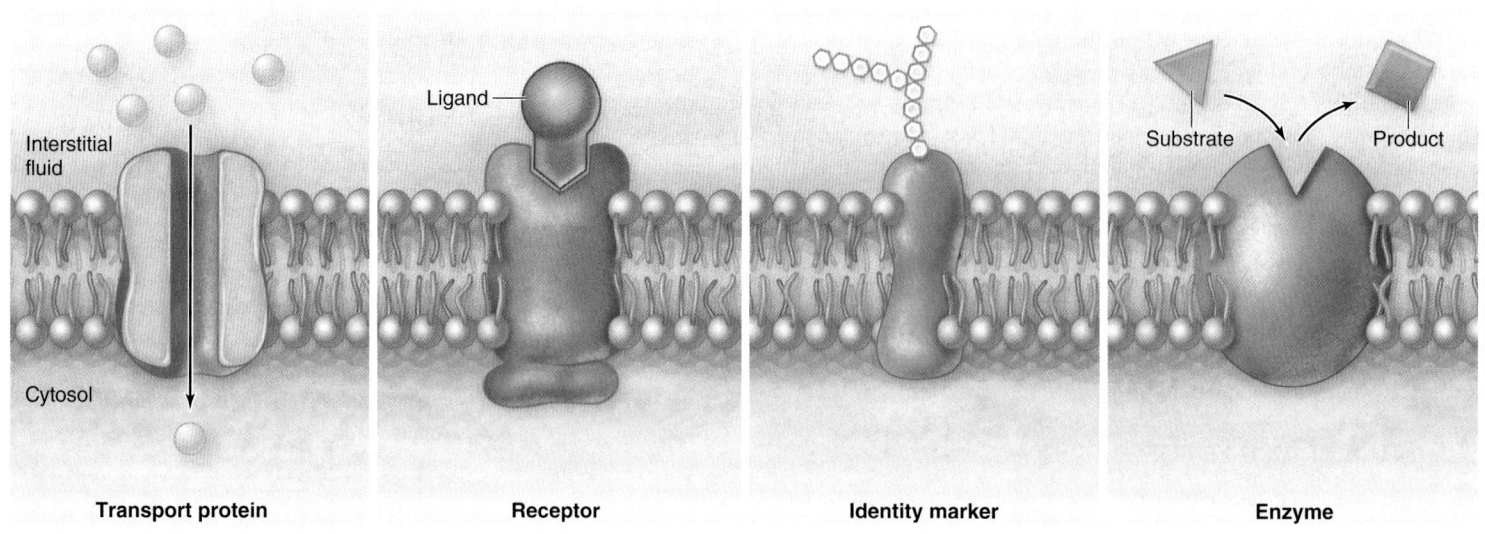

| Transport protein | Receptor | Identity marker | Enzyme |

Figure 4.6 Plasma Membrane Proteins. The major functional categories of plasma membrane proteins include the several types of transport proteins (e.g., channels, carriers, pumps), cell surface receptors, identity markers, enzymes, anchoring sites for the cytoskeleton, and cell-adhesion proteins.

4.2b Membrane Proteins

✎ LEARNING OBJECTIVES

2. Differentiate between the two types of membrane proteins based upon their relative position in the plasma membrane.

3. Name the six major roles played by membrane proteins.

Although lipids form the main component of the plasma membrane, the proteins dispersed within the lipids make up about half of the plasma membrane by weight. Proteins can "float" and move about the fluid

bilayer, much like a beach ball floating on the water surface in a swimming pool. Most of the membrane's specific functions are determined by its resident proteins.

Membrane proteins are classified as one of two structural types: integral or peripheral. **Integral proteins** are embedded within, and extend across, the phospholipid bilayer (figure 4.5). Hydrophobic regions within the integral proteins interact with the hydrophobic interior of the membrane. In contrast, the hydrophilic regions of the integral proteins are exposed to the aqueous environments on either side of the membrane. Many integral membrane proteins are **glycoproteins** that have

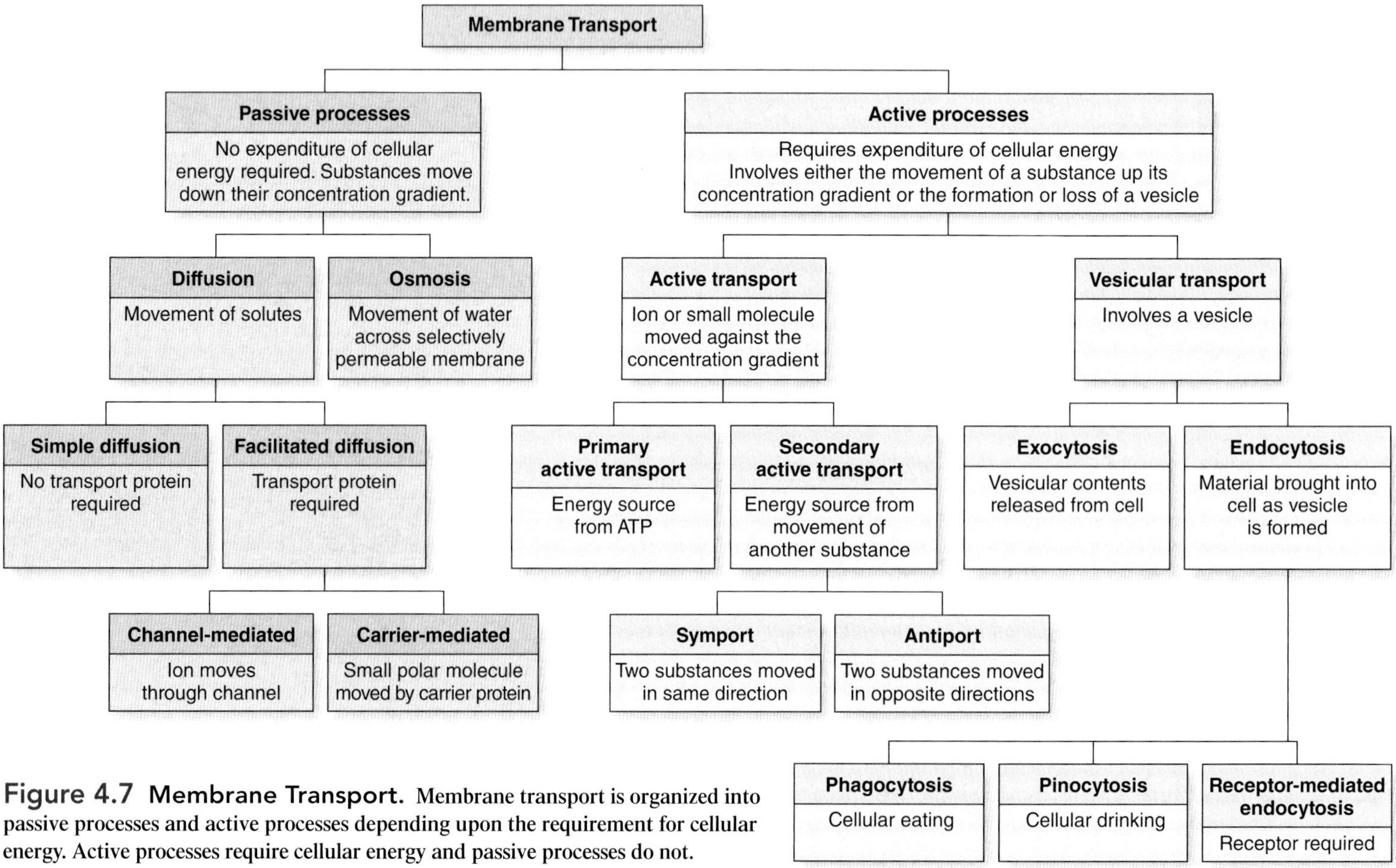

Membrane Transport

Passive processes
No expenditure of cellular energy required. Substances move down their concentration gradient.

Active processes
Requires expenditure of cellular energy
Involves either the movement of a substance up its concentration gradient or the formation or loss of a vesicle

Diffusion
Movement of solutes

Osmosis
Movement of water across selectively permeable membrane

Active transport
Ion or small molecule moved against the concentration gradient

Vesicular transport
Involves a vesicle

Simple diffusion
No transport protein required

Facilitated diffusion
Transport protein required

Primary active transport
Energy source from ATP

Secondary active transport
Energy source from movement of another substance

Exocytosis
Vesicular contents released from cell

Endocytosis
Material brought into cell as vesicle is formed

Channel-mediated
Ion moves through channel

Carrier-mediated
Small polar molecule moved by carrier protein

Symport
Two substances moved in same direction

Antiport
Two substances moved in opposite directions

Phagocytosis
Cellular eating

Pinocytosis
Cellular drinking

Receptor-mediated endocytosis
Receptor required

Figure 4.7 Membrane Transport. Membrane transport is organized into passive processes and active processes depending upon the requirement for cellular energy. Active processes require cellular energy and passive processes do not.

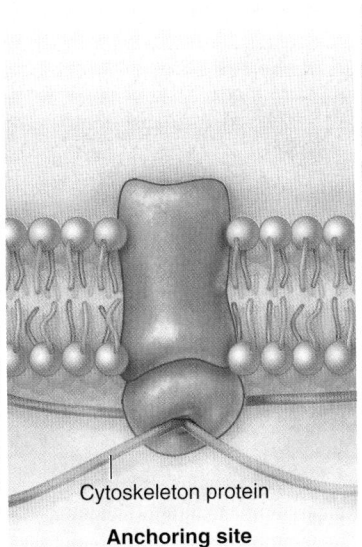

Cytoskeleton protein

Anchoring site

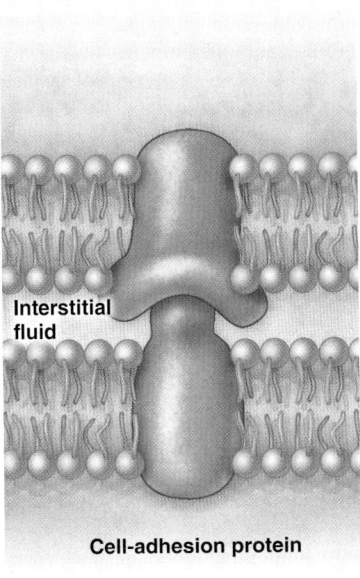

Interstitial fluid

Cell-adhesion protein

carbohydrates exposed to the interstitial fluid. In contrast, **peripheral proteins** are not embedded within the lipid bilayer. They are attached loosely either to the external or internal surfaces of the membrane and are often "anchored" to the exposed parts of an integral protein.

Membrane proteins are also categorized functionally based upon the specific role they serve **(figure 4.6)**.

- **Transport proteins** provide a means of regulating the movement of substances across the plasma membrane. Different types of transport proteins include **channels, carriers, pumps, symporters,** and **antiporters** (see section 4.3).
- **Cell surface receptors** bind specific molecules that are called ligands. **Ligands** are molecules that bind to macromolecules (e.g., binding to a receptor). An example of a ligand is a neurotransmitter released from a nerve cell that binds to the cell surface receptor of a muscle cell to initiate contraction (see section 10.3a).
- **Identity markers** communicate to other cells that they belong to the body. Cells of the immune system use identity markers to distinguish normal, healthy cells from foreign, damaged, or infected cells that are to be destroyed (see section 22.4c).
- **Enzymes** may be attached either to the internal or external surface of a cell for catalyzing chemical reactions (see section 3.3b).
- **Anchoring sites** secure the cytoskeleton (the internal, protein support of a cell) to the plasma membrane (see section 4.6b).
- **Cell-adhesion proteins** are for cell-to-cell attachments. Proteins that form membrane junctions perform a number of functions including binding cells to one another (see section 22.3d).

WHAT DID YOU LEARN?

6 What type of plasma membrane protein provides the means for moving materials across the plasma membrane? What are three subtypes?

4.3 Membrane Transport

The plasma membrane serves as the physical barrier between a cell and the fluid that surrounds it called the interstitial fluid, as described in section 4.2. In addition, the plasma membrane regulates movement of materials into and out of a cell through membrane transport, establishes and maintains electrochemical gradients across the plasma membrane, and functions in cell communication (see figure 4.32). In this section we discuss its role in membrane transport. Its functions in establishing electrochemical gradients and in cell communication are described in sections 4.4 and 4.5, respectively.

Obtaining and eliminating substances across the plasma membrane occur through several different processes that are collectively called **membrane transport**. These processes are organized into two major categories based upon the requirement for expending cellular energy. **Passive processes** do not require cellular energy expenditure. Instead they simply depend upon the kinetic energy inherent within a substance as it moves down its concentration gradient (i.e., from where there is more of a substance to where there is less). Diffusion and osmosis are the two major types of passive processes. **Active processes** differ because they require cells to expend energy. This involves either a substance being pumped up its concentration gradient (i.e., from where there is less of a substance to where there is more) or the release (or formation) of a membrane-bound vesicle. The major energy-requiring processes are active transport and vesicular transport, respectively. **Figure 4.7** is an organizational flow chart of membrane transport processes to refer to as you read through this section.

4.3a Passive Processes: Diffusion

LEARNING OBJECTIVES

1. Summarize the general concept of diffusion.
2. Distinguish between the cellular processes of simple diffusion and facilitated diffusion.

Molecules and ions are in constant motion due to their kinetic energy. They randomly move about and when they strike obstacles such as other molecules and ions they bounce off, moving in a different direction. If a concentration gradient exists, the substance becomes more evenly distributed over time. This net movement of a substance from where it is more concentrated to where it is less concentrated is called **diffusion** (di-fū′zhun; *diffundo* = to pour in different directions). Diffusion, if unopposed, occurs until the substance reaches **equilibrium** (i.e., the molecules become evenly distributed throughout a given area(s)) **(figure 4.8)**.

Figure 4.8 Diffusion. When a drop of dye is placed into a beaker of water, the dye molecules diffuse within the water down their concentration gradient, spreading out until equilibrium has been reached. AP|R

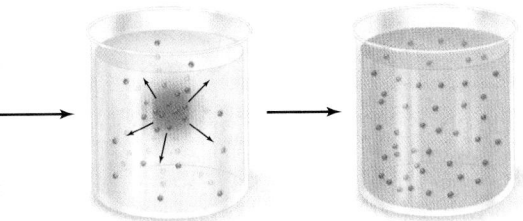

The rate that substances diffuse is not constant, but is dependent upon environmental conditions including

- **The "steepness" of its concentration gradient.** Steepness of a concentration gradient is a measure of the difference in concentration of a substance between two areas. A steeper concentration gradient causes a faster rate of diffusion.

- **Temperature.** Temperature reflects the kinetic energy (or random movement) of a substance. When the temperature is higher, there is a greater random movement of the molecules and ions composing a substance, resulting in a faster rate of diffusion.

Cellular Diffusion

Diffusion across the plasma membrane occurs as a solute moves from an area of high concentration to an area of low concentration (i.e., the solute moves down its concentration gradient). These processes are dependent upon concentration gradients that generally exist between the cytosol and interstitial fluid for different solutes (e.g., O_2, CO_2, glucose, ions). Whether a solute diffuses unassisted through the plasma membrane or is facilitated by a plasma membrane protein determines if the substance moves by simple diffusion or facilitated diffusion.

Simple Diffusion Solutes that are small and nonpolar move into or out of a cell down their concentration gradient by **simple diffusion.** These molecules do not require a transport protein. They simply pass between the phospholipid molecules that form the plasma membrane (**figure 4.9**). Materials that move via simple diffusion include respiratory gases (O_2 and CO_2), small nonpolar fatty acids, ethanol, and urea (a nitrogenous waste produced from amino acids).

The plasma membrane cannot regulate simple diffusion—rather, the movement of these substances is dependent only upon the concentration gradient. The substance continues to move across the plasma membrane as long as a concentration gradient exists. Impaired respiratory and cardiovascular function can alter the concentration gradients of oxygen and carbon dioxide, resulting in decreased diffusion of these gases.

Facilitated Diffusion Small solutes that are charged or polar are effectively blocked from passing through the plasma membrane

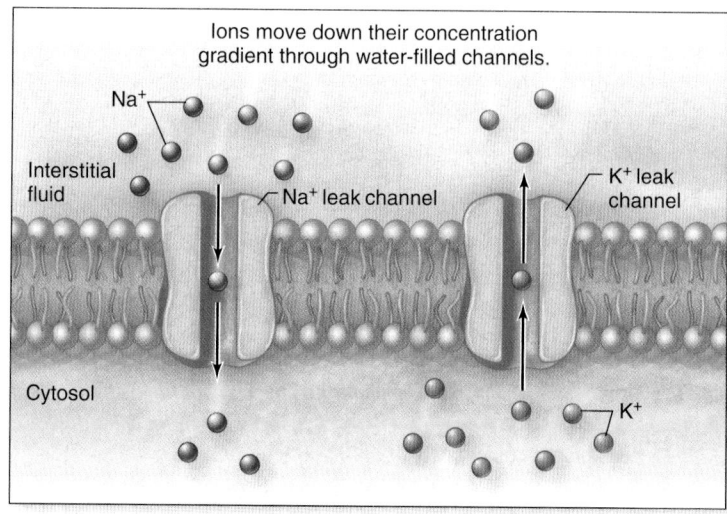

(a) Channel-mediated diffusion

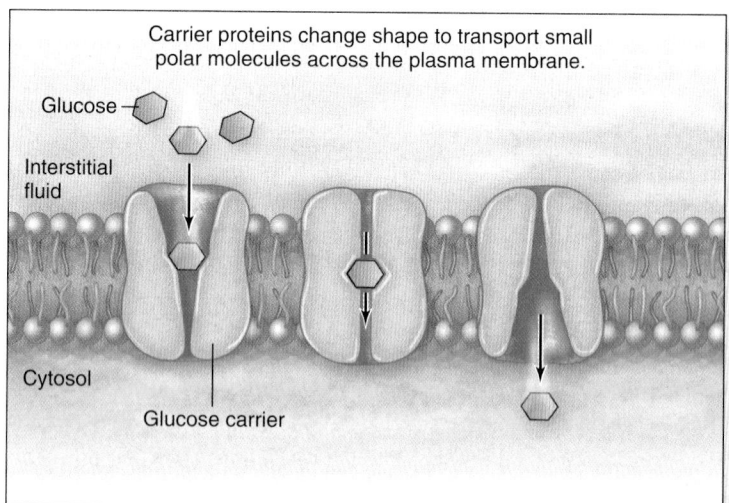

(b) Carrier-mediated diffusion

Figure 4.10 Facilitated Diffusion of Solutes. Facilitated diffusion occurs when ions or small polar molecules are transported down their concentration gradient either through or by plasma membrane proteins. (*a*) Channel-mediated diffusion: Ions (e.g., Na⁺ and K⁺) move through specific water-filled protein channels. (*b*) Carrier-mediated diffusion: Small polar molecules (e.g., glucose) are transported by protein carriers. AP|R

by the nonpolar phospholipid bilayer. Their transport either into or out of the cell must be assisted by plasma membrane proteins in a process called **facilitated** (fa-sil′i-tā-ted) **diffusion.** Two types of facilitated diffusion—channel-mediated diffusion and carrier-mediated diffusion—are distinguished by the type of transport protein used to move the substance across the membrane.

Channel-mediated diffusion is the movement of small ions across the plasma membrane through water-filled protein channels (**figure 4.10a**). Each channel is typically specific for one type of ion. The channel is either a **leak channel,** which (as a general rule) is continuously open, or a **gated channel.** A gated channel is usually closed, opens only in response to a stimulus (e.g., chemical, light, voltage change), and then stays open for just a fraction of second before it closes. For example, Na⁺ leak channels allow Na⁺ to pass through continuously. In contrast, chemically gated Na⁺ channels open only to allow Na⁺ to move through the channel in response to the presence of a particular chemical (e.g., neurotransmitter). Channel-mediated diffusion is important in the normal function of both muscle cells and nerve cells.

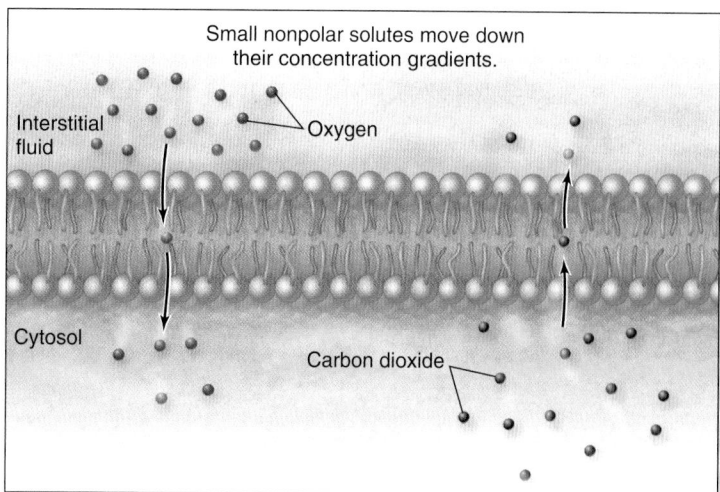

Figure 4.9 Simple Diffusion of Solutes. Simple diffusion occurs when small, nonpolar molecules pass between plasma membrane phospholipid molecules. Each type of molecule moves down its concentration gradient. Here, oxygen diffuses into a cell and carbon dioxide diffuses out of a cell.

Carrier-mediated diffusion is the movement of small polar molecules, such as simple sugars or amino acids. They are assisted across the plasma membrane by a **carrier protein.** Carrier proteins transport substances such as glucose. The binding of the substance induces the carrier protein to change shape and move it to the other side of the membrane. Like channels, a carrier moves a substance *down* its gradient; but, note that carrier-mediated diffusion involves a conformational change in the carrier protein for the transport of the substance across the plasma membrane. Figure 4.10*b* depicts how a carrier protein binds the substance, changes shape, and then releases the substance on the other side of the membrane. A carrier that transports only one substance is called a **uniporter.** Glucose carriers are uniporters that normally prevent the loss of glucose in the urine (see section 24.6c).

The number of channels and carriers in a plasma membrane determines the maximum rate at which a substance can be transported. Thus, a cell can alter the transport rate of a given substance down its concentration gradient by changing the number of channel or carrier proteins in the plasma membrane. A greater rate occurs with increased numbers of these transport proteins, and a lesser rate with decreased numbers.

WHAT DID YOU LEARN?

7 How does O_2 diffuse into a cell and CO_2 diffuse out of a cell?

8 Compare and contrast how an ion is transported across the plasma membrane versus how a small, polar molecule is transported.

4.3b Passive Processes: Osmosis

LEARNING OBJECTIVES

3. Define osmosis.

4. Define osmotic pressure.

5. Describe the relationship of osmosis and tonicity.

Osmosis is unlike the other types of passive membrane transport, because it involves water movement and does not involve the movement of solutes (see figure 4.7). **Osmosis** (os-mō'sis; *osmos* = a thrusting) is the passive movement of water through a **selectively permeable** (or *semipermeable*) membrane. This movement occurs in response to a difference in relative concentration of water on either side of a membrane. Please refer to **figure 4.11** as you read through this section.

Plasma Membrane: A Selectively Permeable Membrane

The plasma membrane is a selectively permeable membrane that allows the passage of water, but its phospholipid bilayer prevents the movement of most solutes.

Water molecules cross the plasma membrane in one of two ways: They either "slip between" the molecules of the phospholipid bilayer (limited amounts) or they move through integral protein water channels called **aquaporins** (ak-kwă-pōr'in; *aqua* = water; *porus* = channel). Thus, cells can alter the amount of water that crosses the plasma membrane by changing the number of aquaporins.

The phospholipid bilayer of the plasma membrane is nonpermeable to most solutes. In the context of osmosis, solutes are classified into two categories based upon whether their passage across the plasma membrane is prevented by the phospholipid bilayer. **Permeable** solutes (e.g., small and nonpolar solutes such as oxygen, carbon dioxide, and urea) pass through the bilayer, and **non-permeable** solutes (e.g., charged, polar, or large solutes such as ions, glucose, and proteins) are prevented from crossing the bilayer. (The term "solutes" in this discussion on osmosis will refer to "non-permeable solutes.")

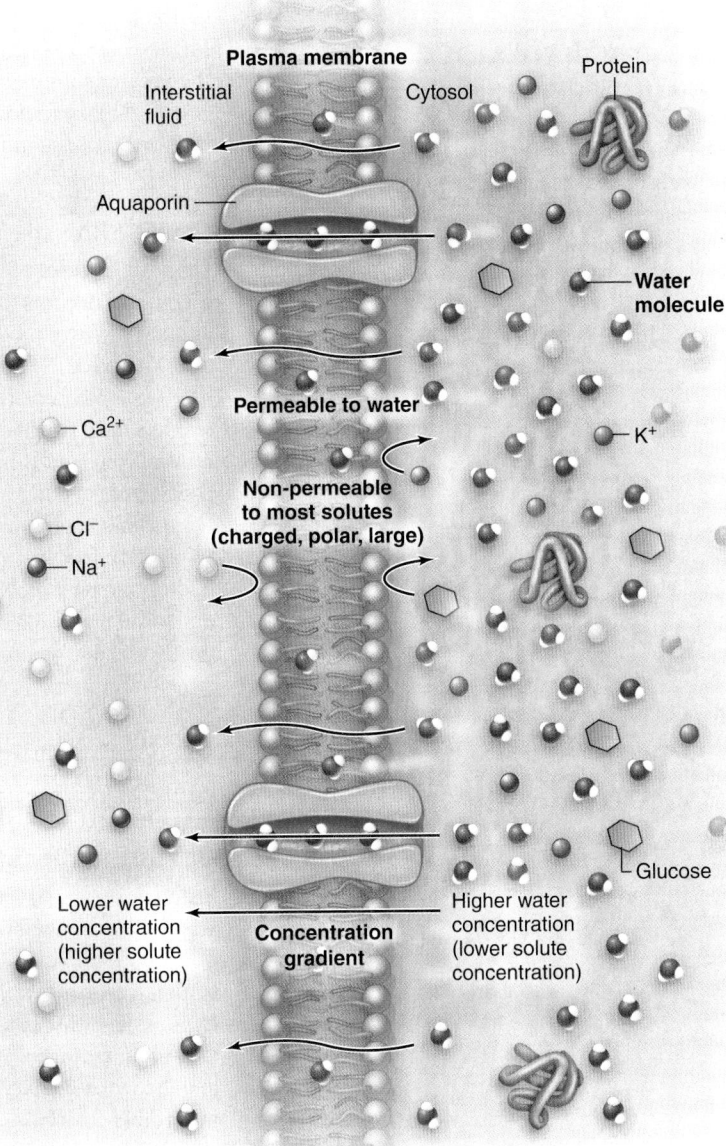

Figure 4.11 Osmosis in Cells. Osmosis occurs in cells across the plasma membrane, which is permeable to water and non-permeable to most solutes. Water always moves across the plasma membrane from an area of high water concentration to an area of low water concentration until equilibrium is reached. AP|R

Concentration Gradients Across the Plasma Membrane

A difference in solute concentration can exist between the cytosol and the interstitial fluid because solutes are prevented from moving across the bilayer of the plasma membrane. Note that when a solute concentration exists, a water concentration also exists. A solution with a greater concentration of solutes contains a lower concentration of water. For example, a solution containing 3% solutes has a lower water concentration (97% water) than a solution with 1% solutes (99% water).

Movement of Water Into or Out of a Cell by Osmosis

The net movement of water by osmosis is dependent upon the concentration gradient between the cytosol and the solution in which the cell is immersed. For example, water moves down its concentration gradient from the solution containing 1% solutes (and 99% water) into the solution containing 3% solutes (and 97% water). Water continues to move until equilibrium is reached (the concentration of water in the

INTEGRATE

LEARNING STRATEGY

You may find it helpful to think of osmotic pressure as the pressure that is exerted by solutes to draw or "pull" water into an area where there is a higher concentration of solutes. The higher the relative solute concentration, the greater the osmotic pressure—thus, a greater amount of water is drawn into an area by osmosis.

cell and surrounding fluid are equal). Note that water moves toward the solution with the lower water concentration (or stated another way, water moves toward the solution with the greater solute concentration).

Figure 4.11 shows water moving across the plasma membrane by osmosis from an area of high water concentration to an area of low water concentration.

Osmotic Pressure

Osmotic pressure is the pressure exerted by the movement of water across a semipermeable membrane due to a difference in solution concentration. The steeper the gradient, the greater the amount of water moved by osmosis and the higher the osmotic pressure.

Figure 4.12 helps us to visualize the movement of water by osmosis. Each U-shaped tube has two areas separated by a semipermeable membrane that allows the passage of water, but restricts the passage of solutes. Initially side A has more solutes and less water than side B. Water moves from side B into side A by osmosis (against the force of gravity) until the two fluids are equal in concentration.

Osmotic pressure can be measured indirectly. Imagine placing a stopper on side A in figure 4.12 and exerting force to return the fluid to its original level. The force exerted increases hydrostatic pressure within the U-shaped tube. (**Hydrostatic pressure** is the pressure exerted by a fluid on the inside wall of its container.) The osmotic pressure exerted in this setup is equal to the hydrostatic pressure produced to return the fluid to its original level.

WHAT DO YOU THINK?

 Which setup would exhibit the greater osmotic pressure: a cell with a cytosol concentration of 0.9% immersed in (a) pure water, or (b) a 0.2% NaCl solution? Explain.

Osmosis and Tonicity

When water crosses the plasma membrane of a cell by osmosis, the cell either gains or loses water with an accompanying change in the cell's volume and osmotic pressure. The ability of a solution to change the volume or pressure (or the "tone") of the cell by osmosis is called **tonicity.** Three terms are used to describe the relative concentration of solutions: isotonic, hypotonic, or hypertonic **(figure 4.13)**.

In an **isotonic** (ī′sō-ton′ik; *iso* = equal; *tonus* = stretching) solution, both the solution and the cytosol have the same relative concentration of solutes. An example of a solution that is isotonic to erythrocytes is normal saline with a concentration of 0.9% NaCl. Under these conditions, the relative amounts of water inside and outside the cell are equal, and no net movement of water occurs (figure 4.13a). (Normal saline is used commonly in intravenous [IV] solutions to maintain a patient's fluid balance.)

In a **hypotonic** (hī′pō-ton′ik; *hypo* = under) solution, the solution has a lower concentration of solutes, and there is a higher concentration of water than in the cytosol. Pure water contains no solutes, so it is the most extreme example of solution that is hypotonic to

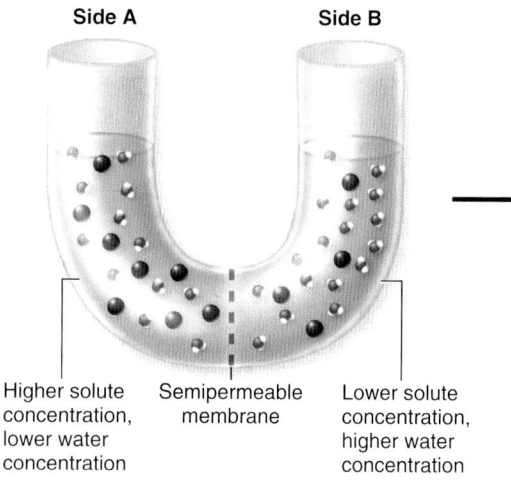

Side A   Side B   Side A   Side B

Higher solute   Semipermeable   Lower solute
concentration,   membrane   concentration,
lower water   higher water
concentration   concentration

Semipermeable
membrane

Initial setup: Side A contains proportionately more solute and less water.

Final setup: Water moved by osmosis from side B down the water gradient to side A until the concentrations of side A and side B are equal.

Figure 4.12 Osmotic Pressure. The semipermeable membrane allows the passage of water but restricts the passage of solutes. If a water gradient exists, water moves by osmosis from where it is more concentrated (side B) to where it is less concentrated (side A) until equilibrium is reached. Osmotic pressure is the pressure exerted by this movement of water.

● Non-permeable solutes (e.g., glucose, Na⁺, protein)
● Water molecules

Isotonic solution	Hypotonic solution	Hypertonic solution
Interstitial fluid is the same concentration as cytosol.	Interstitial fluid is less concentrated than cytosol.	Interstitial fluid is more concentrated than cytosol.

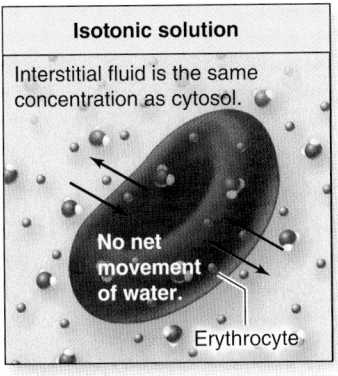

No net movement of water.

Erythrocyte

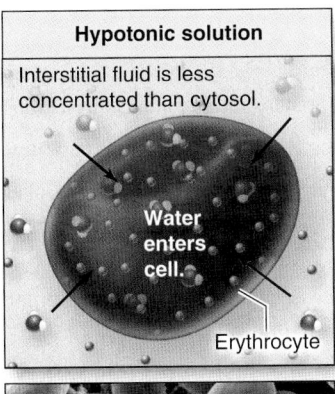

Water enters cell.

Erythrocyte

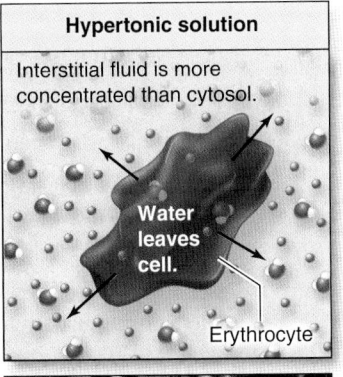

Water leaves cell.

Erythrocyte

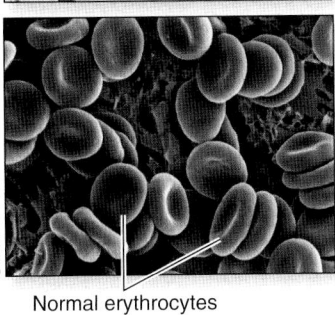

SEM 6900x

Normal erythrocytes

(a)

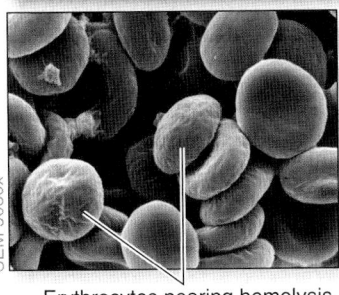

SEM 9030x

Erythrocytes nearing hemolysis

(b)

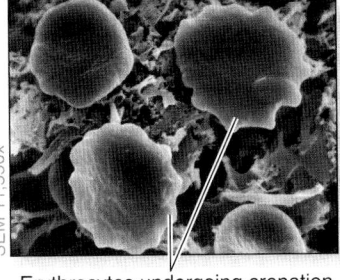

SEM 11,550x

Erythrocytes undergoing crenation

(c)

Figure 4.13 Tonicity.
Tonicity has an effect on water movement; it refers to the relative solute concentration strength of a solution. (*a*) In an isotonic solution (e.g., 0.9% NaCl), there is no net movement of water. The cell shape remains unchanged. (*b*) In a hypotonic solution (e.g., pure water), water moves into the cell. (*c*) In a hypertonic solution (e.g., 3% NaCl), water moves out of the cell. AP|R

erythrocytes. Under these conditions, water moves down its concentration gradient from where there is more water (outside the cell) to where there is less water (inside the cell). The entry of water increases both the volume and pressure within the cell (figure 4.13*b*). **Lysis** (or rupture) of the cell can occur if the difference in concentration is large enough. **Hemolysis** (hē-mol'i-sis; *hem* = blood, *lysis* = destruction) is the specific term for rupturing erythrocytes.

A **hypertonic** (hī'pĕr-ton'ik; *hyper* = above) solution has a higher concentration of solutes, and thus a lower concentration of water than in the cytosol. For example, a solution that contains 3% NaCl is hypertonic to red blood cells. In this case, water moves out of the cell into the surrounding fluid where the water concentration is lower. Thus, a decrease in cell volume and pressure occurs (figure 4.13*c*). If the difference in concentration is large, the cell shrinks, a process called **crenation** (krē-nā'shŭn; *crena* = notch). This prevents us from being able to replace fluid by drinking seawater, which has an average salt concentration of 3.5%.

WHAT DID YOU LEARN?

9 Define osmosis.

10 What occurs to the tonicity of a cell when it is placed into an isotonic, hypotonic, or hypertonic solution?

11 What general conclusion can you make concerning the movement of water? There is always a net movement of water by osmosis toward (a) an isotonic solution, (b) a hypotonic solution, or (c) a hypertonic solution.

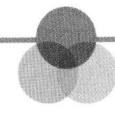

INTEGRATE

CONCEPT CONNECTION

Osmosis is important in several significant physiologic processes that are discussed later in the text, including capillary exchange between the blood and body cells (see section 20.3) formation of urine (see section 24.6), and maintaining fluid balance (see section 25.2).

INTEGRATE

LEARNING STRATEGY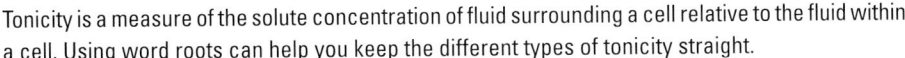

Tonicity is a measure of the solute concentration of fluid surrounding a cell relative to the fluid within a cell. Using word roots can help you keep the different types of tonicity straight.

Iso means "same as." The solute concentration of an isotonic solution is the same as that of the cytosol, and so there is no net movement of water.

Hypo means "under." The solute concentration of a hypotonic solution is lower than that of the cytosol, and so there is relatively more water in the solution and water moves into the cell.

Hyper means "more than." The solute concentration of a hypertonic solution is higher than that of the cytosol, and so there is relatively less water in the solution and water moves out the cell.

4.3c Active Processes

LEARNING OBJECTIVES

6. Compare and contrast primary and secondary active transport.

7. Explain the difference between exocytosis and endocytosis.

8. Describe the endocytotic processes of phagocytosis, pinocytosis, and receptor-mediated endocytosis.

Active processes of membrane transport are those that require the expenditure of cellular energy (see figure 4.7). These processes occur only in living organisms. Active processes are organized into active transport and vesicular transport.

Active Transport

Active transport is the movement of a solute against its concentration gradient (i.e., from a low concentration to a high concentration) across a cellular membrane. These processes are responsible for maintaining concentration gradients between the cytosol of a cell and interstitial fluid. The direct source of energy for active transport determines whether the movement is called primary active transport or secondary active transport.

Primary Active Transport   **Primary active transport** uses energy derived directly from the breakdown of ATP (see figure 3.7). This breakdown also provides the phosphate group that is added to the membrane transport pump, resulting in a change in the protein's shape and the subsequent movement of a solute across the membrane. The addition of the phosphate to a protein is called *phosphorylation* (see section 3.3g).

Cellular protein pumps that move ions across the membrane are more specifically called **ion pumps.** Ion pumps are a major factor in a cell's ability to maintain its internal concentrations of ions. As an example, Ca^{2+} pumps embedded in the plasma membranes of erythrocytes move calcium out of the cell to prevent the cell from becoming rigid due to the accumulation of calcium **(figure 4.14)**. Therefore, the erythrocyte remains flexible enough to move through capillaries (the smallest blood vessels; see section 18.3b). H^+ pumps are another type of transport protein that function in maintaining cellular pH.

The **sodium-potassium (Na^+/K^+) pump** is a special type of ion pump. It is specifically called an *exchange pump* because it moves one type of ion into a cell against its concentration gradient, while moving another type of ion out of the cell against its concentration gradient. (You may find it helpful to think of the Na^+/K^+ pump as a "dual pump" because it moves two different ions against their respective concentration gradients.) The plasma membrane preserves steep concentration gradient differences for these ions by continuously exporting Na^+ out of the cell and moving K^+ into the cell.

Figure 4.15 shows the steps in the process in which three Na^+ ions are pumped *out of* a cell for every two K^+ ions that are pumped *into* a cell. The cell must expend ATP to maintain the levels of these ions on each side of the membrane. The Na^+/K^+ pump is also called a *sodium-potassium ATPase* because the protein pump is an enzyme that splits ATP to power the pump. There is a 1:2:3 ratio for this pump: 1 ATP is required to pump 2 K^+ ions into the cell and 3 Na^+ ions out of the cell.

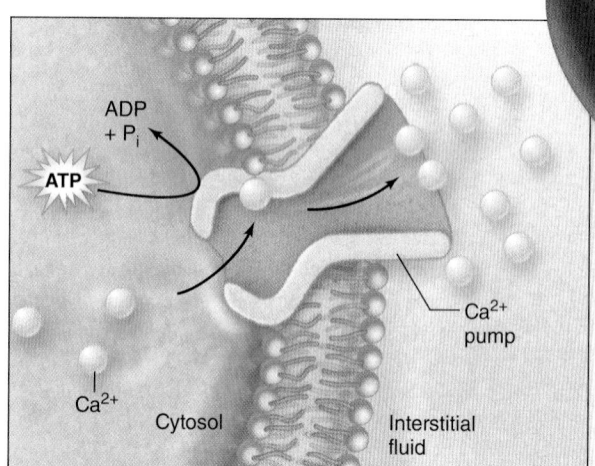

Figure 4.14 Ca^{2+} Pump. A Ca^{2+} pump uses ATP to move calcium ions against the calcium gradient from the inside to the outside of the cell.

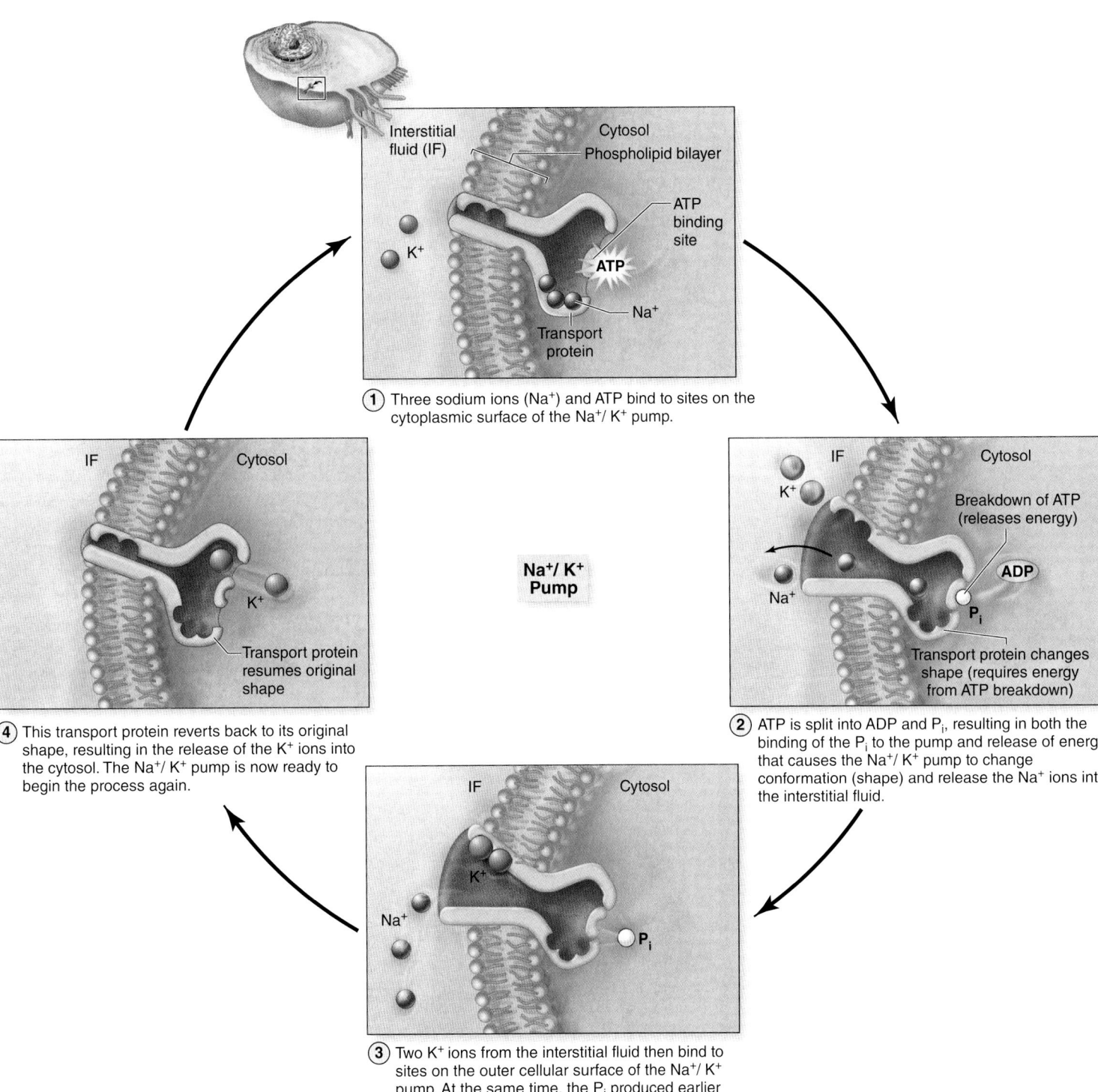

① Three sodium ions (Na^+) and ATP bind to sites on the cytoplasmic surface of the Na^+/K^+ pump.

Na^+/K^+ Pump

② ATP is split into ADP and P_i, resulting in both the binding of the P_i to the pump and release of energy that causes the Na^+/K^+ pump to change conformation (shape) and release the Na^+ ions into the interstitial fluid.

④ This transport protein reverts back to its original shape, resulting in the release of the K^+ ions into the cytosol. The Na^+/K^+ pump is now ready to begin the process again.

③ Two K^+ ions from the interstitial fluid then bind to sites on the outer cellular surface of the Na^+/K^+ pump. At the same time, the P_i produced earlier by ATP hydrolysis is released into the cytosol.

Figure 4.15 Na^+/K^+ Pump. The Na^+/K^+ pump is a plasma membrane transport protein that uses ATP to move both Na^+ and K^+ ions through the membrane in opposite directions from their region of low concentration to their region of high concentration (1 ATP is split for moving 3 Na^+ out of the cell and 2 K^+ into the cell). AP|R

Secondary Active Transport **Secondary active transport** is also called *cotransport*, or *coupled transport*. It moves a substance against its concentration gradient using energy provided by the movement of a second substance down its specific concentration gradient. Put another way, the kinetic energy of one substance moving down its concentration gradient across the membrane provides the "power" to pump the other substance against its concentration gradient across the membrane (much like water moving over a dam and turning a water wheel can generate electricity; see section 3.1a). The substance that moves down its concentration gradient often is Na^+. The two types of secondary active transport include

- **Symport.** If the two substances are moved in the same direction, these transport proteins are called **symporters** (or *cotransporters*), and the process is **symport secondary active transport.**
- **Antiport.** If the two substances are moved in opposite directions, then the transport proteins are called **antiporters** (or *countertransporters*) and the process is **antiport secondary active transport.**

Figure 4.16 compares the processes of transporting a substance by a symporter or an antiporter. In the symport example, glucose binds to the transport protein in the membrane (figure 4.16*a*). This binding helps alter the shape of the transport protein, and then both glucose and Na^+ are transported into the cell. The Na^+ moves down its concentration gradient into the cell and provides the energy to move glucose up its concentration gradient into the cell. Notice that both Na^+ and glucose are moved in the *same* direction. In contrast, an antiporter moves the two substances in the *opposite* direction (figure 4.16*b*). The H^+ is moved out of the cell as Na^+ is transported into the cell. The movement of Na^+ down its concentration gradient has provided the energy to move H^+ up its concentration gradient, but in the opposite direction of the Na^+ movement.

Secondary active transport mechanisms are ultimately dependent upon the primary active transport mechanisms of Na^+/K^+ pumps (described earlier). The activity of these pumps produce and sustain a distinct concentration gradient difference between Na^+ on opposite sides of the plasma membrane, with more Na^+ in the interstitial fluid and less Na^+ in the cytosol.

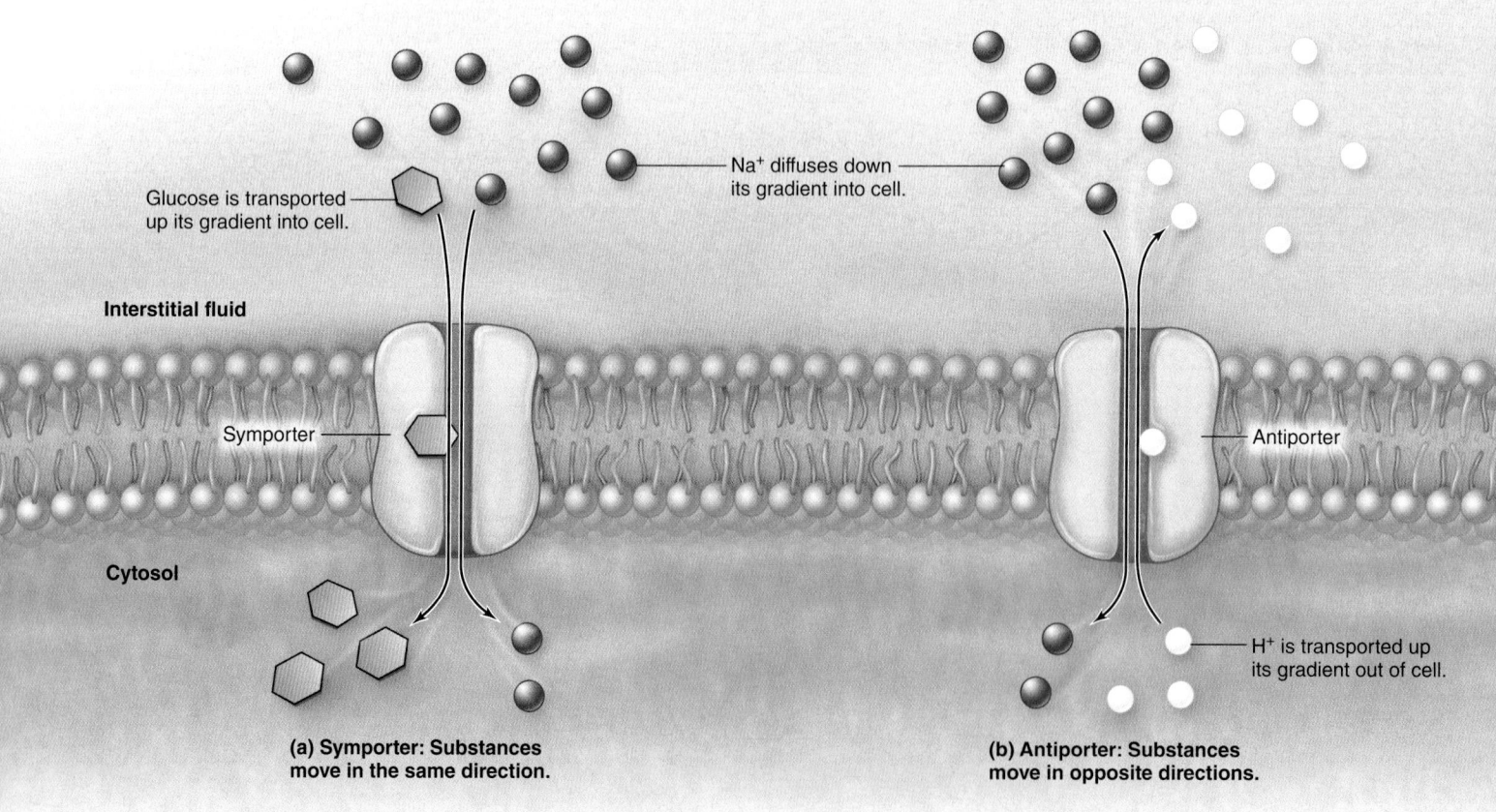

Glucose is transported up its gradient into cell.

Na^+ diffuses down its gradient into cell.

Interstitial fluid

Symporter

Antiporter

Cytosol

H^+ is transported up its gradient out of cell.

(a) Symporter: Substances move in the same direction.

(b) Antiporter: Substances move in opposite directions.

Figure 4.16 Secondary Active Transport. Secondary active transport moves one substance against its concentration gradient, and is powered by the movement of a second substance (usually Na^+) down its concentration gradient. (*a*) A symporter transports both substances in the same direction. (*b*) An antiporter transports the two substances in opposite directions. Remember, in both symport and antiport, the kinetic energy of Na^+ moving down its concentration gradient is harnessed to move a different substance (e.g., glucose, H^+) up its concentration gradient.

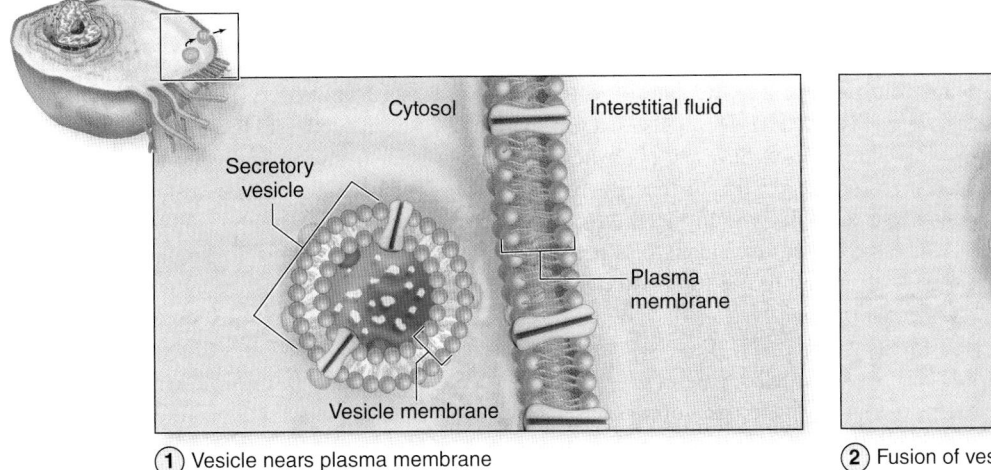

Cytosol   Interstitial fluid

Secretory
vesicle

Plasma
membrane

Vesicle membrane

1 Vesicle nears plasma membrane

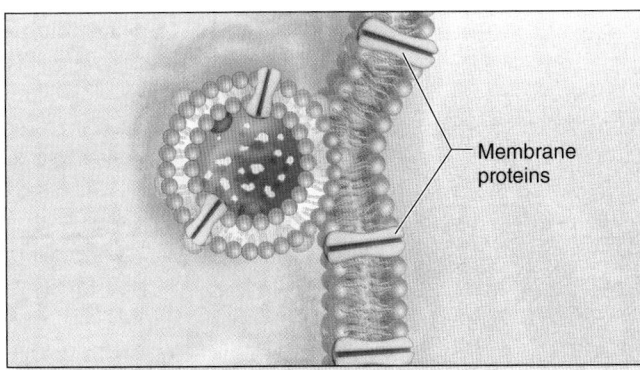

Membrane
proteins

2 Fusion of vesicle membrane with plasma membrane

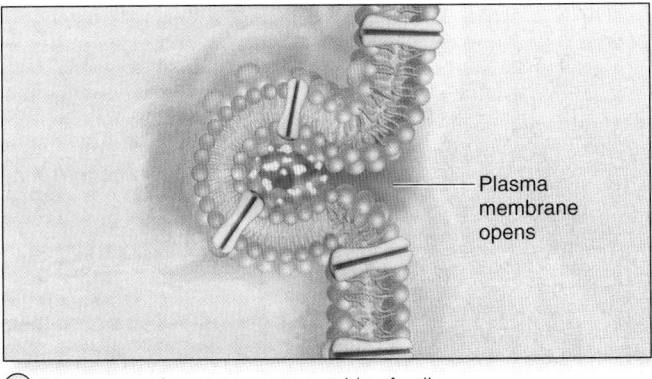

Plasma
membrane
opens

3 Plasma membrane opens to outside of cell

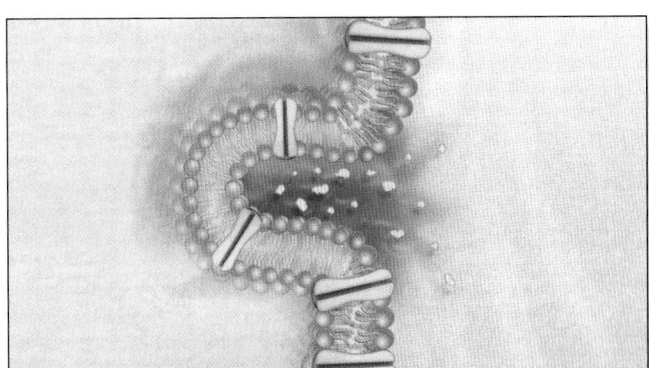

4 Release of vesicle components into the interstitial fluid and integration of vesicle membrane components into the plasma membrane

Figure 4.17 Exocytosis. In exocytosis, the cell secretes bulk volumes of materials from the cell into the interstitial fluid as a vesicle fuses with the plasma membrane. Fusion of the vesicle to the plasma membrane is the energy-requiring step.

An enormous Na^+ gradient results (about 99% of Na^+ is in the interstitial fluid with only 1% in the cytosol). The Na^+ concentration gradient has potential energy that is harnessed as Na^+ moves down its concentration gradient in secondary active transport.

Vesicular Transport

Vesicular transport is also called *bulk transport.* It involves energy input to transport large substances (or large amounts of a substance) across the plasma membrane by a **vesicle** (ves'i-kl; *vesica* = bladder), which is a membrane-bounded sac filled with materials. Vesicular transport is organized into processes of exocytosis and endocytosis (see figure 4.7).

Exocytosis   The means by which either large substances or large amounts of substances are secreted *from* the cell is called **exocytosis** (ek'sō-sī-tō'sis; *exo* = outside, *kytos* = cell, *osis* = condition of) **(figure 4.17)**. Macromolecules, such as large proteins and polysaccharides, are too large to be moved across the membrane, even with the assistance of transport proteins. The material for secretion typically is packaged within intracellular transport vesicles. When the vesicle and plasma membrane come into contact, the phospholipid molecules of the vesicle and plasma membrane bilayers rearrange themselves so that the two membranes fuse. The fusion of these lipid bilayers requires the cell to expend energy in the form of ATP. Following fusion, the vesicle contents are released to the outside of the cell. An example of exocytosis is the release of neurotransmitter molecules from nerve cells (see section 12.8d).

Endocytosis   The cellular uptake of large substances or large amounts of substances from the external environment into the cell is called **endocytosis** (en'dō-sī-tō'sis; *endon* = within). Endocytosis is used for the uptake of nutrients and extracellular debris for digestion, retrieval of membrane regions added to the plasma membrane during exocytosis, and regulation of composition of membrane proteins to alter cellular processes (e.g., membrane transport and communication).

The steps of endocytosis are similar to the exocytosis steps, only in reverse. Endocytosis occurs when substances within the interstitial fluid are packaged into a vesicle that forms at the cell surface for internalization into the cell (figure 4.18). A small area of plasma membrane folds inward into the cytosol to form a pocket, or **invagination** (in-vaj′i-nā-zhun; *in* = in, *vagina* = a sheath). The pocket deepens as endocytosis proceeds and then it pinches off when the lipid bilayer fuses. Severing of the newly forming vesicle from the plasma membrane requires specialized proteins and is the energy-expending step. The new intracellular vesicle now present contains material that was formerly outside the cell.

The three types of endocytosis include phagocytosis, pinocytosis, and receptor-mediated endocytosis. They are differentiated based upon the specific material being transported and the mechanism involved.

Phagocytosis (fag′ō-sī-tō′sis; *phago* = to eat) means *cellular eating*. It is a nonspecific process that occurs when a cell engulfs or captures a large particle external to the cell by forming membrane extensions that are called **pseudopodia** (sū-dō-pō′dē-ă; *pseudes* = false, *pous* = foot) or *false feet*, to surround the particle (figure 4.18*a*). Once the particle is engulfed by the pseudopodia, it is enclosed within a membrane sac. When the sac is internalized, its contents are broken down chemically (digested) after it fuses with a lysosome (a cellular organelle containing digestive

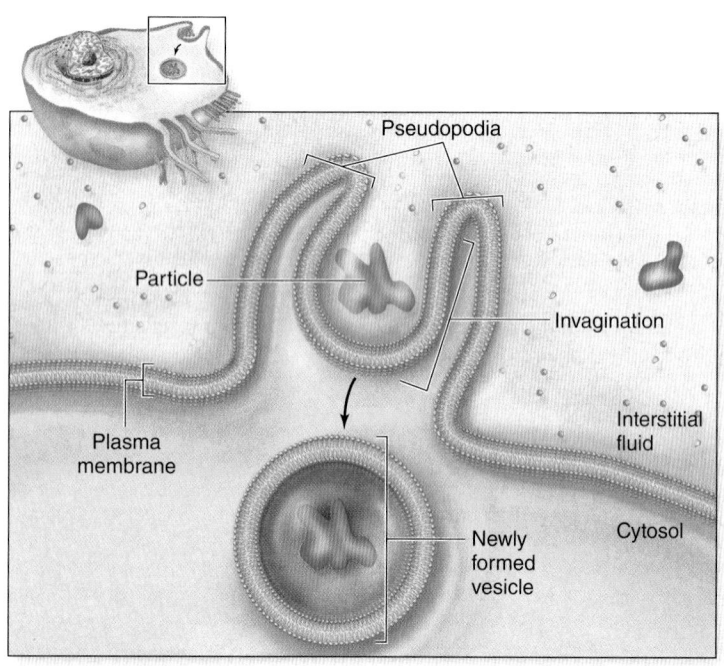

(a) Phagocytosis (cellular eating)

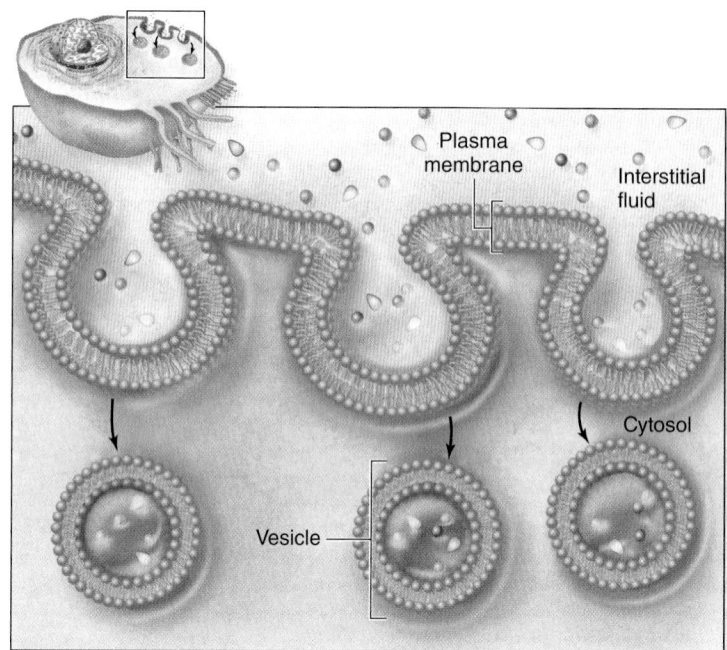

(b) Pinocytosis (cellular drinking)

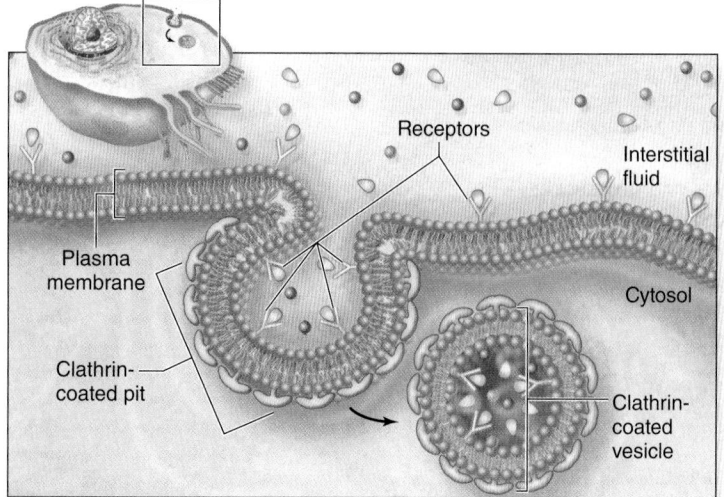

(c) Receptor-mediated endocytosis

Figure 4.18 Three Forms of Endocytosis. Endocytosis is the process whereby a vesicle is formed as the cell acquires materials from the interstitial fluid. (*a*) Phagocytosis occurs when membrane extensions called pseudopodia engulf a relatively large particle and internalize it into a vesicle. (*b*) Pinocytosis is the incorporation of numerous droplets of interstitial fluid into the cell in small vesicles. (*c*) Receptor-mediated endocytosis occurs when specific molecules bind to receptors in the plasma membrane. These receptors with bound molecules then aggregate within the membrane, and are internalized when the membrane invaginates, forming a vesicle. The formation of the vesicle is the energy-requiring step.

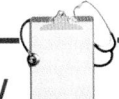
CLINICAL VIEW

Familial Hypercholesteremia

Familial hypercholesteremia is an inherited genetic disorder that involves either defective or absent cellular receptor proteins that bind low-density lipoproteins (LDLs), defects in the proteins of the LDLs, or other possible mutations. Defects in either the LDL receptor or the proteins of the LDLs interfere with the normal process of receptor-mediated endocytosis of cholesterol into cells. LDLs that contain cholesterol remain in the blood, resulting in greatly elevated levels of blood cholesterol. Consequently, cholesterol accumulates in the blood vessels, causing plaque buildup and narrowing of the blood vessels (i.e., atherosclerosis), especially those providing blood to the heart (coronary vessels). Individuals with this genetic defect are likely to experience blockage of the coronary arteries, resulting in a heart attack. The age of occurrence of a heart attack depends upon the severity of the protein defect. In severe cases, an individual may experience a heart attack during the teen years.

enzymes, described in section 4.6a). Only a few types of cells are able to perform phagocytosis. For example, it occurs regularly when a white blood cell engulfs and digests a microbe (e.g., bacterium; section 22.3b).

Pinocytosis (pin′ō-sī-tō′sis or pī′nō-; *pineo* = to drink) is known as *cellular drinking*. This process occurs when the cell internalizes droplets of interstitial fluid that contain dissolved solutes. Multiple, small vesicles are formed (figure 4.18*b*). This process is considered nonspecific because all solutes dissolved within the droplet are taken into the cell. Most cells perform this type of membrane transport.

Receptor-mediated endocytosis uses receptors on the plasma membrane to bind molecules within the interstitial fluid and bring the molecules into the cell. This enables the cell to obtain bulk quantities of certain substances, even though those substances may not be very concentrated in the interstitial fluid.

Receptor-mediated endocytosis begins when specific molecules in the interstitial fluid attach to their distinct integral membrane protein receptors in the plasma membrane to form a ligand-receptor complex (figure 4.18*c*). Following the binding of the ligand, the ligand-receptor complexes move laterally in the plane of the plasma membrane and accumulate at special membrane regions that contain **clathrin** protein on the internal surface of the membrane. The clathrin-coated regions of the plasma membrane housing the ligand-receptor complex folds inward to form an invagination called a *clathrin-coated pit*. This invagination deepens, pinches off, and the lipid bilayer of the plasma membrane fuses to form a *clathrin-coated vesicle* that then moves into the cytosol. Following the formation of the clathrin-coated vesicles, the clathrin coat must be enzymatically removed before the vesicle may proceed to its intracellular destination. Again, it is the fusion of these lipid bilayers that requires the cell to expend energy in the form of ATP. Following entry, receptors and ligands are uncoupled. Ligands may be stored, modified, or destroyed, and receptors (unless damaged) are returned to the plasma membrane.

The transport of cholesterol from the blood to a cell is an example of receptor-mediated endocytosis. When cholesterol travels in the blood it is bound to protein molecules in structures called low-density lipoproteins (or LDLs). LDLs move from the blood into the interstitial fluid and then bind to LDL receptors in the cell's plasma membrane. LDLs are then internalized by the process of receptor-mediated endocytosis just described (see section 27.6c).

The various types of membrane transport mechanisms are integrated in **figure 4.19**. Passive processes are depicted on the left and active processes on the right in this two-page summary figure.

WHAT DID YOU LEARN?

12 What transport process involved in the movement of Na⁺ down its gradient is used to power another substance up its gradient?

13 Engulfing of a bacterium by a white blood cell occurs by what type of cellular transport?

Figure 4.19 **Passive and Active Processes of Membrane Transport.**

Transport processes are separated into two major categories. (*a*) Passive processes, which do not require expenditure of cellular energy, include simple diffusion, facilitated diffusion (channel-mediated and carrier-mediated), and osmosis. (*b*) Active processes, which require cellular energy, include active transport (primary and secondary) and vesicular transport (exocytosis and various forms of endocytosis).

(a) Passive Processes | Do not require expenditure of cellular energy; substance moves into or out of a cell down its concentration gradient.

DIFFUSION: Movement of a solute from an area of higher concentration to an area of lower concentration across a plasma membrane.

Simple Diffusion: Small and nonpolar solutes move unassisted between phospholipid molecules of the plasma membrane; no transport protein required.

Oxygen

Carbon dioxide

Interstitial fluid   Cytosol

Facilitated Diffusion: Ions and small polar molecules are assisted across the plasma membrane by a transport protein (channel or carrier).

Na⁺

Channel-Mediated: Ion (e.g., Na^+) movement is facilitated by channels across the plasma membrane.

Channel

Carrier

Glucose

Carrier-Mediated: Small polar molecule movement (e.g., glucose) is facilitated by protein carriers across the plasma membrane.

Interstitial fluid   Cytosol

OSMOSIS: Movement of water across a selectively permeable membrane from an area of higher water concentration to an area of lower water concentration (through either the phospholipid bilayer or aquaporins).

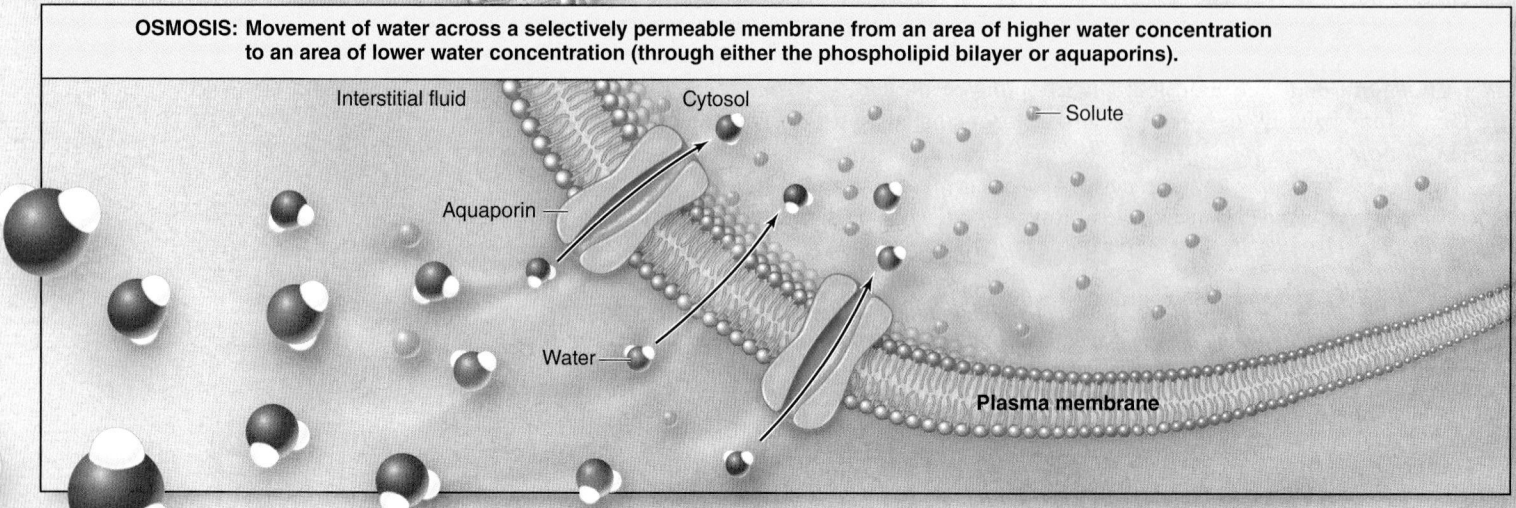

Interstitial fluid   Cytosol   Solute

Aquaporin

Water

Plasma membrane

(b) Active Processes | Require expenditure of cellular energy; substance moves up its concentration gradient or involves a vesicle.

ACTIVE TRANSPORT: Movement of a substance against its concentration gradient via a protein pump, symporter, or antiporter.

Primary Active Transport: Pumps are powered by splitting an ATP molecule.

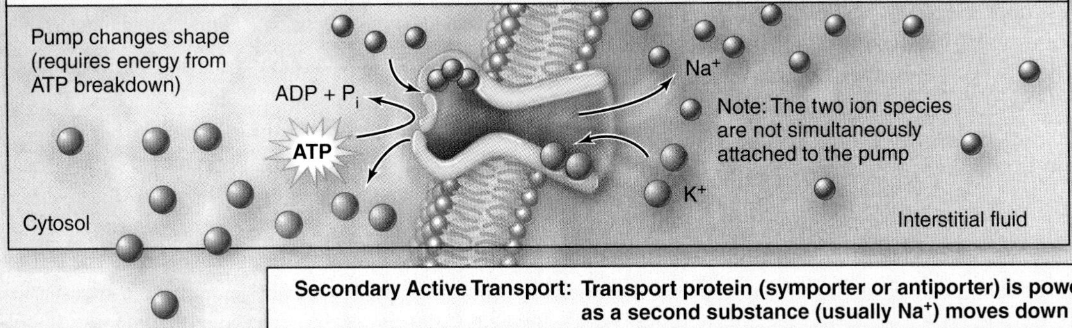

Pump changes shape (requires energy from ATP breakdown)

ADP + P_i

ATP

Na^+

Note: The two ion species are not simultaneously attached to the pump

K^+

Cytosol

Interstitial fluid

Secondary Active Transport: Transport protein (symporter or antiporter) is powered by energy harnessed as a second substance (usually Na^+) moves down a concentration gradient.

Cytosol

Interstitial fluid

Glucose

Na^+

Symport: Two substances are moved in the same direction by a symporter protein.

Antiport: Two substances are moved in opposite directions by an antiporter protein.

H^+

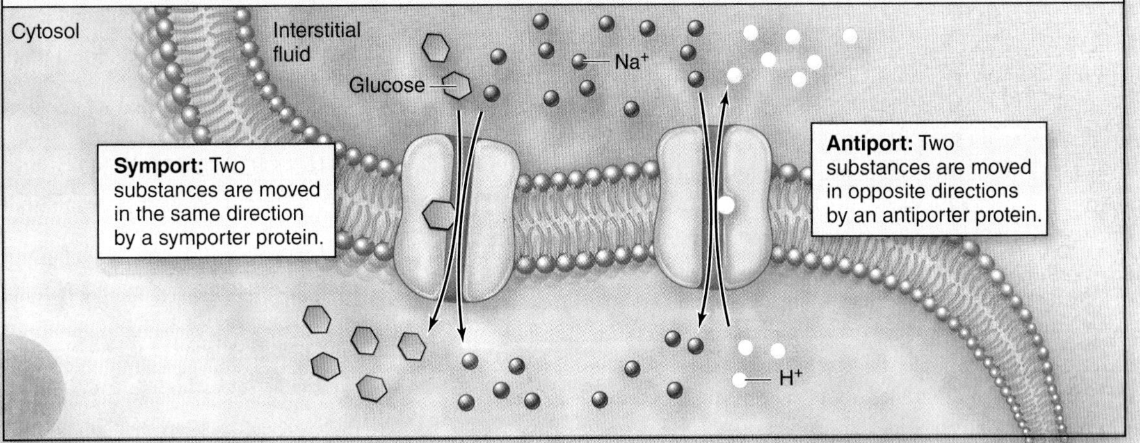

VESICULAR TRANSPORT: Movement of a substance across the plasma membrane via a vesicle.

Exocytosis: Vesicular content is released from a cell.

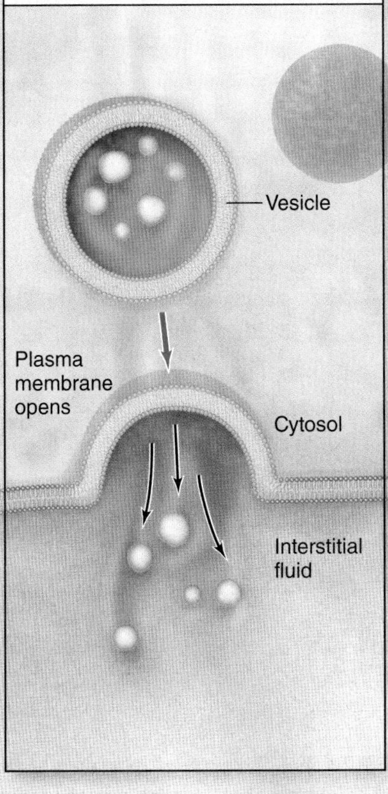

Vesicle

Plasma membrane opens

Cytosol

Interstitial fluid

Endocytosis: Material is brought into a cell as vesicle is formed. The three types of endocytosis include phagocytosis, pinocytosis, and receptor-mediated endocytosis.

Vesicles

Receptors

Receptor-Mediated Endocytosis: Movement of a specific substance (e.g., cholesterol) into a cell following the binding of the substance to a receptor.

Particle

Pinocytosis: Interstitial fluid is taken into a cell.

Phagocytosis: Particulate matter external to the cell is engulfed by pseudopodia (e.g., white blood cells engulfing a bacterium).

Pseudopodia

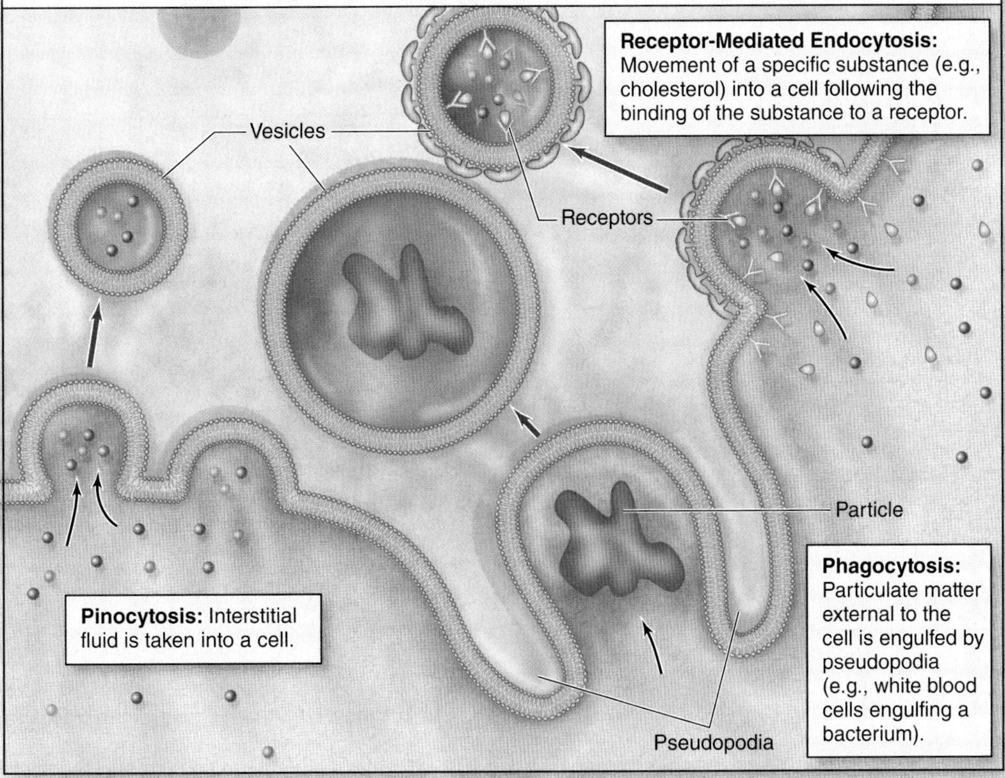

4.4 Resting Membrane Potential

The plasma membrane also functions in establishing and maintaining an electrochemical gradient at the plasma membrane called the *resting membrane potential* (RMP), which is essential in the normal function of both muscle cells (see chapter 10) and nerve cells (see chapter 12). We first define an RMP and then discuss how resting membrane potentials are established and maintained. Refer to **figure 4.20** as you read through this section.

4.4a Introduction

LEARNING OBJECTIVES

1. Define a resting membrane potential.
2. Describe the cellular conditions that are significant for establishing and maintaining a resting membrane potential.

Cells have an electrical charge difference at the plasma membrane. This electrical charge difference represents potential energy (see section 3.1a), and thus is appropriately called the **membrane potential.** The membrane potential when a cell is at rest is more specifically called the **resting membrane potential (RMP).** Two cellular conditions are significant in establishing and maintaining an RMP.

First, a cell has an unequal distribution of ions and molecules across the plasma membrane. The cytosol close to the plasma membrane contains relatively more K^+ than does the surrounding interstitial fluid that is close to the plasma membrane. In comparison, the interstitial fluid close to the plasma membrane contains relatively more Na^+ than the cytosol close to the plasma membrane. These relative distributions of K^+ (more inside) and Na^+ (more outside) are the result of the activity of Na^+/K^+ pumps (described in section 4.3c). In addition, the cytosol has negatively charged protein molecules, which are formed by protein synthesis (a cellular process described in section 4.8). Note that these negatively charged proteins are too large to pass through the plasma membrane.

INTEGRATE

LEARNING STRATEGY

To understand the concept of "relatively negative," consider the following simple arrangement. There are 100 positive charges (+) outside of a membrane and 30 positive charges (+) inside of a membrane—thus, the inside is less positive, or it is *relatively* negative to the outside.

Second, the relative amounts of positive and negative charges are not equally distributed at the plasma membrane. There is relatively more positive charge on the outside of a cell than on the inside of a cell. Thus, the inside of the cell is *relatively negative* compared to the outside of the cell at the plasma membrane. This can be measured by electrodes, which are positioned with one just inside the cell and the other outside the cell. Cell types vary in the specific value of their resting membrane potential, which typically range between –50 millivolts (mV) and –100 mV. Neurons, for example, have a resting membrane potential of –70 mV (see section 12.7b).

WHAT DID YOU LEARN?

14 Define a resting membrane potential.

4.4b Establishing and Maintaining an RMP

LEARNING OBJECTIVES

3. Explain the role of both K^+ and Na^+ in establishing an RMP.
4. Discuss how Na^+/K^+ pumps are necessary in maintaining an RMP.

A resting membrane potential is primarily a consequence of the relative movement of ions across the plasma membrane. The two most significant ions are K^+ and Na^+. The net movement of each of these is dependent upon

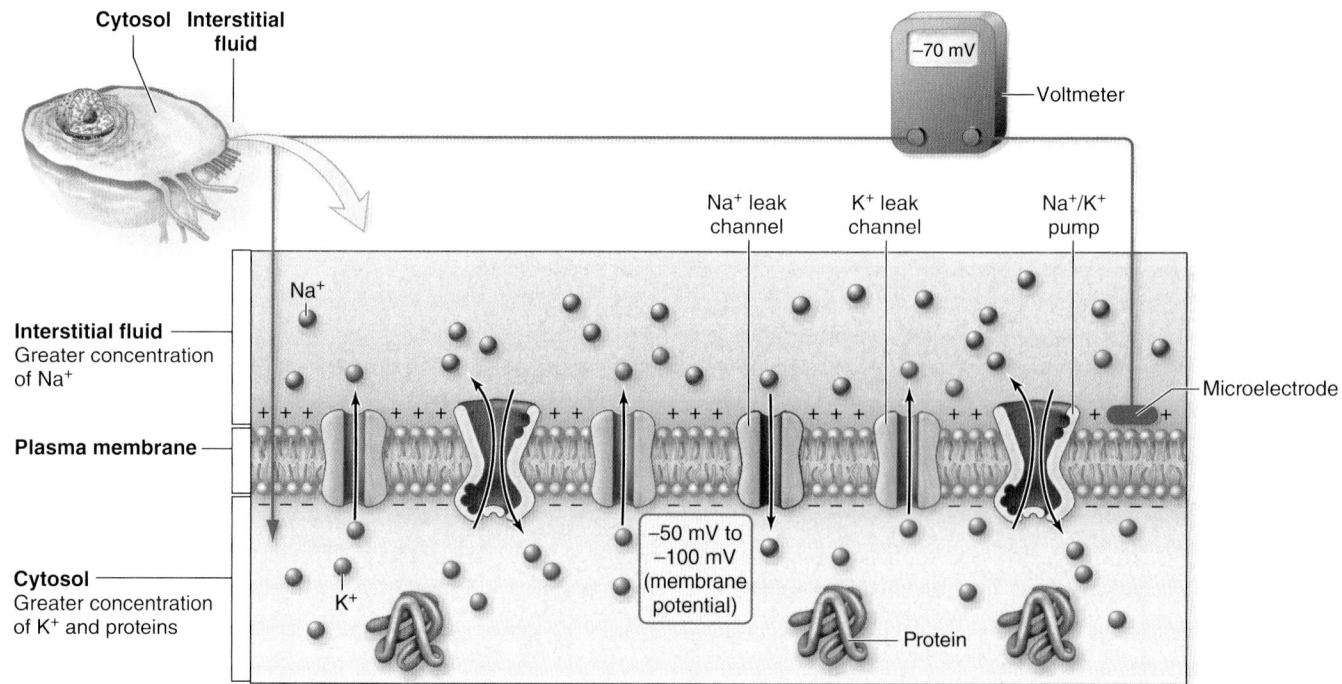

Figure 4.20 Resting Membrane Potential (RMP). The resting membrane potential is the electrical difference across the plasma membrane when the cell is at rest. This difference is due to the relative concentration of charged substances including Na^+, K^+, and negatively charged proteins.

both the number of its leak channels and the **electrochemical gradient,** which is the combination of the electrical gradient at the plasma membrane and the chemical concentration gradient of the specific ion. Here we first discuss the role of K^+ and the role of Na^+ in establishing a resting membrane potential. We then describe the role of Na^+/K^+ pumps.

Establishing an RMP

The Role of K^+   Potassium diffusion is the most important factor in establishing the specific value of the resting membrane potential. Movement of K^+ is dependent upon its electrochemical gradient. Potassium ions exit the cell through K^+ leak channels moving down their relatively steep chemical concentration gradient from the cytosol into the interstitial fluid. This loss of K^+ leaves relatively more negatively charged structures (e.g., proteins) inside the cell. They remain within the cell because they are too large to cross the plasma membrane. The movement of K^+ to the outside of a cell is, however, opposed by the electrical gradient. The positive charge on the outside of the cell repels the movement of K^+, and the negative charge on the inside of the cell attracts K^+. Thus, K^+ movement out of the cell is facilitated by its chemical concentration gradient but opposed by the electrical gradient. As additional K^+ diffuses out of the cell into the interstitial fluid, the inside becomes more negative. Consequently, the pull to keep K^+ within the cell is greater. At some point, the force of the chemical concentration gradient allowing K^+ out of a cell becomes equal to the electrical gradient that opposes this movement. Thus, K^+ movement has reached equilibrium. In neurons, for example, if only K^+ leak channels were present, the loss of the K^+ from the neuron would result in a resting membrane potential with a specific value of –90 mV.

The Role of Na^+   Sodium diffusion into cells occurs simultaneously to the loss of K^+, and is also dependent upon its electrochemical gradient. Sodium ions enter the cell through Na^+ leak channels moving down their chemical concentration gradient from the interstitial fluid into the cytosol. Sodium ions are also "pulled" into the cell by the electrical gradient. Both of these forces (the chemical gradient and the electrical gradient) facilitate the movement of Na^+ into a cell. However, limited numbers of Na^+ leak channels prevent as much Na^+ movement into the neuron as K^+ out. This movement of Na^+ results in the inside becoming more positive. Thus, in a neuron, this would account for the resting membrane potential being –70 mV (instead of the –90 mV if only K^+ leak channels were present).

Maintaining an RMP

The Na^+/K^+ pumps are significant in maintaining the gradients of both K^+ and Na^+ following their diffusion. Each type of ion is pumped back up its concentration gradient by Na^+/K^+ pumps. Na^+ and K^+ are moved in opposite directions; Na^+ are pumped out of the cell and K^+ are transported into the cell. We will see in later chapters that changes in the resting membrane potentials caused by the regulated opening and closing of *gated* channels can alter the passage of specific ions, and these movements are central to both contraction of muscle cells (chapter 10) and relaying an impulse in nerve cells (chapter 12).

INTEGRATE

CONCEPT CONNECTION

Resting membrane potentials are central in several important physiologic processes, including stimulating both skeletal muscle cells (see chapter 10) and cardiac muscle cells (see chapter 19) to contract, nerve impulse propagation (see chapter 12), and sensory receptors to respond to stimuli (chapter 16).

 WHAT DID YOU LEARN?

15 Explain how the resting membrane potential is established and maintained.

4.5 Cell Communication

The plasma membrane plays a significant role in communication between cells in addition to serving as a boundary, functioning in membrane transport, and establishing and maintaining a resting membrane potential. Numerous structures in the membrane, including glycoproteins and glycolipids, facilitate both direct interaction with other cells and recognition and response to certain ligand signals external to the cell.

4.5a Direct Contact Between Cells

 LEARNING OBJECTIVE

1. Explain how cells communicate through direct contact.

Physical or direct contact between two cells is important in the normal functioning of some cells, especially those of the immune system. One of the primary functions of the immune system is to make contact with and destroy both unhealthy cells (e.g., infected cells, cancer cells) and foreign cells (e.g., bacterial cells, transplanted cells). Body cells communicate to our immune cells that they both belong to the body and are healthy through direct contact that involves the glycocalyx. Recall that the glycocalyx is the coating of carbohydrates on the external surface of a cell. These carbohydrates extend from the proteins and lipids of the plasma membrane. The pattern of sugars is unique to each individual except identical twins. The immune system is able to distinguish normal, healthy cells from unwanted cells by making direct contact with a cell to determine if it exhibits the same pattern of sugars of the glycocalyx as the body's cells. It is because unhealthy cells and foreign cells express a different pattern that they are subsequently destroyed.

Another example of direct contact between cells is the contact that occurs between sperm and an oocyte during the process of fertilization. The sperm recognizes and binds to the egg by its unique glycocalyx (see section 29.2a).

Direct contact is also critical in the process of development and in cellular regrowth following injury. If you cut the skin of your finger, the cells in the upper layer of the skin (the epidermis) begin to divide (see section 6.3). Cell division continues to fill in the gap created by the injury. When the damaged tissue has been replaced, overgrowth of skin tissue is prevented by inhibition caused by cellular contact.

 WHAT DID YOU LEARN?

16 What are some examples of how cells communicate through direct contact?

4.5b Ligand-Receptor Signaling

 LEARNING OBJECTIVE

2. Describe the three general mechanisms of response to the binding of a ligand with a receptor.

Most communication between cells occurs through ligands. Recall ligands are molecules that bind with macromolecules (e.g., receptors). Ligands involved in communication include neurotransmitters from nerve cells and hormones from endocrine cells. The cell that receives the information has a receptor that can bind the ligand. Binding initiates mechanisms for controlling the growth, reproduction, and other cellular processes of individual cells.

There are three general types of receptors that bind ligands. They differ in their response to ligand binding as follows (**figure 4.21**):

- **Channel-linked receptors** (or **chemically gated channels**) permit ion passage either into or out of a cell in response to ligand (e.g., neurotransmitter) binding (figure 4.21*a*). Channel-linked receptors are required to initiate electrical changes to the resting membrane potential in both muscle cells (see section 10.2d) and nerve cells (see section 12.8).

- **Enzymatic receptors** function as protein kinase enzymes and are activated to directly phosphorylate (add a phosphate to) other enzymes within the cell (figure 4.21*b*). Recall that enzymes can be either turned on or turned off through phosphorylation (see section 3.3g). This provides a mechanism for altering enzymatic activity within a cell in response to external signals.

- **G protein–coupled receptors** also involve protein kinase activation; note that these protein kinase enzymes are activated indirectly through the G protein that serves as an intermediate molecule. The general steps of activation are described in figure 4.21*c*, and a detailed discussion of G proteins is provided in section 17.5b.

WHAT DID YOU LEARN?

 How do action of enzymatic receptors and G protein–coupled receptors differ?

4.6 Cellular Structures

The cellular structures described in this section include membrane-bound organelles, non-membrane-bound organelles, vesicles for transport, and structures that extend from the cell's surface.

4.6a Membrane-Bound Organelles

LEARNING OBJECTIVES

1. List the membrane-bound organelles of a typical human cell.
2. Describe the structure and main function(s) of each.

Membrane-bound organelles within the cytoplasm are surrounded by a membrane (similar in composition to the plasma membrane) that separates the organelle's contents from the cytosol. This allows the activities

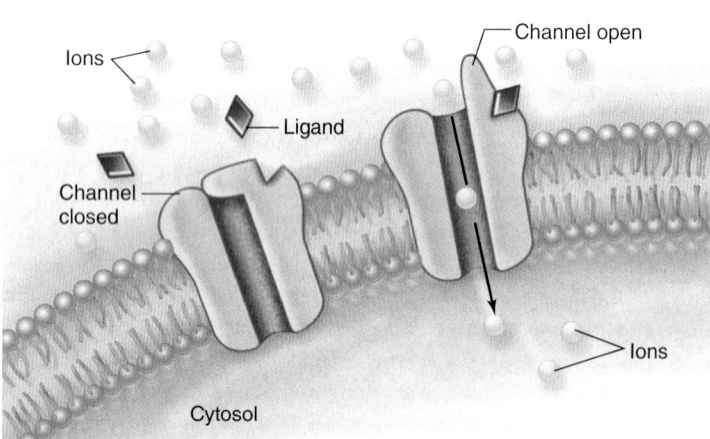

(a) Channel-linked receptors

(b) Enzymatic receptors

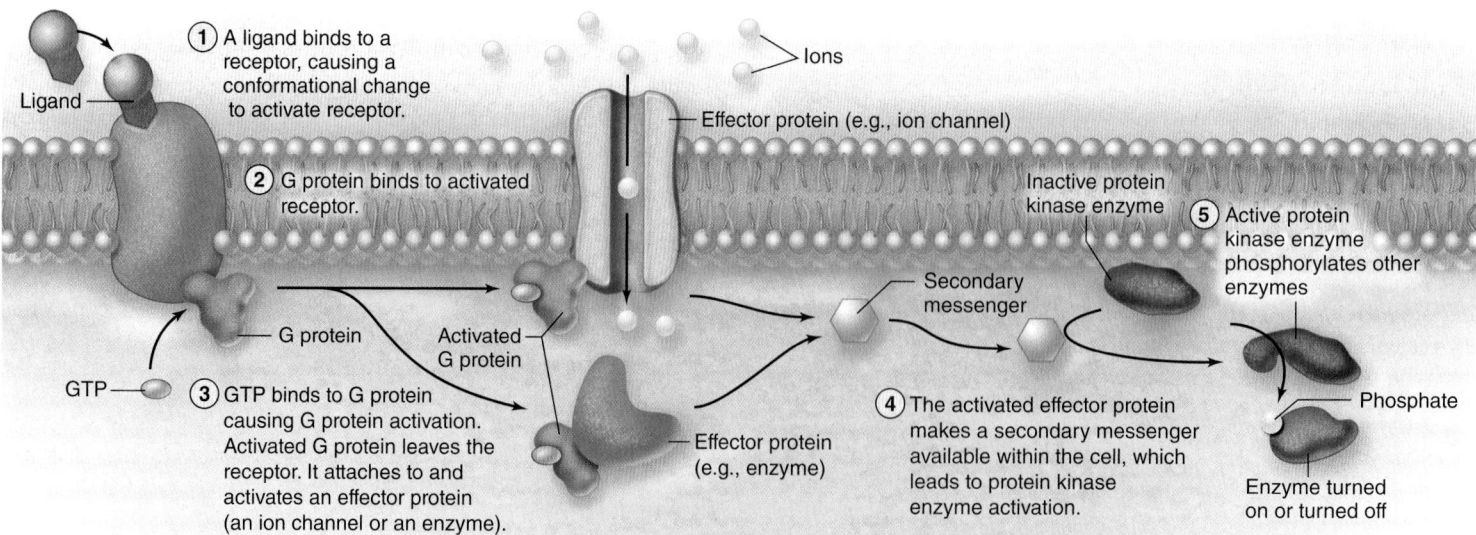

(c) G protein–coupled receptors

Figure 4.21 Membrane Receptors. Receptors bind ligands that will initiate a cellular change. (*a*) Channel-linked receptors bind a ligand and open to allow a specific ion to move down its concentration gradient. (*b*) Enzymatic receptors (generally protein kinase enzymes) bind a ligand and are activated to phosphorylate other enzymes. (*c*) G protein–coupled receptors bind ligand and activate protein kinase enzymes indirectly through a G protein, as described in steps 1–5.

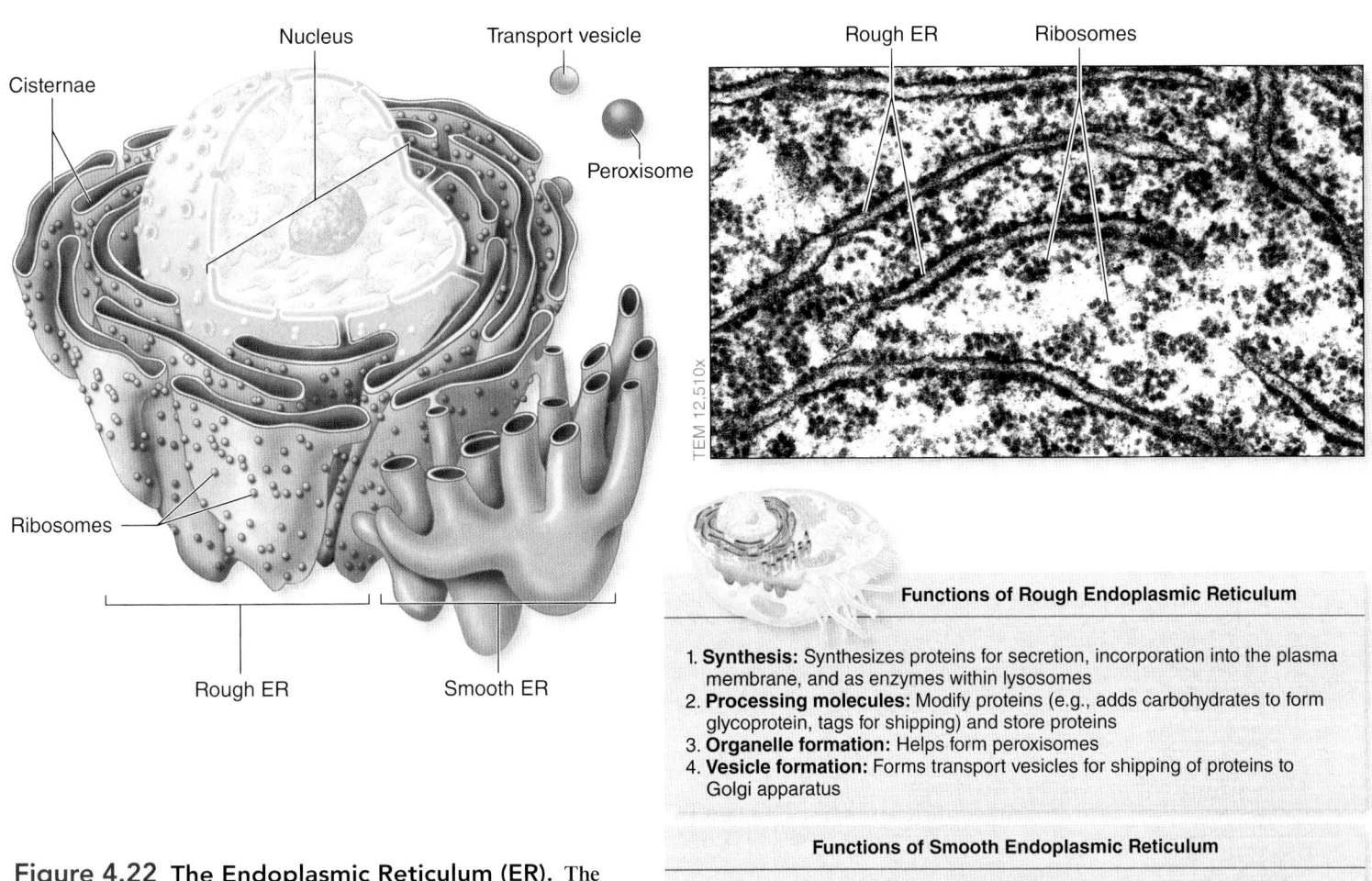

Functions of Rough Endoplasmic Reticulum

1. **Synthesis:** Synthesizes proteins for secretion, incorporation into the plasma membrane, and as enzymes within lysosomes
2. **Processing molecules:** Modify proteins (e.g., adds carbohydrates to form glycoprotein, tags for shipping) and store proteins
3. **Organelle formation:** Helps form peroxisomes
4. **Vesicle formation:** Forms transport vesicles for shipping of proteins to Golgi apparatus

Functions of Smooth Endoplasmic Reticulum

1. **Synthesis:** Site of lipid (e.g., steroid) synthesis
2. **Processing molecules:** Carbohydrate metabolism (e.g., glycogen synthesis)
3. **Detoxification:** Detoxifies drugs and poisons
4. **Vesicle formation:** Forms transport vesicles for shipping to Golgi apparatus

Figure 4.22 The Endoplasmic Reticulum (ER). The rough ER is composed of membranes with ribosomes attached to their cytoplasmic surface. It is readily distinguishable from the even-surfaced, interconnected tubules of the smooth ER, which lacks associated ribosomes. However, the two are continuous. AP|R

of each organelle to proceed in a relatively isolated and controlled environment. Each organelle differs in its shape, membrane composition, and associated enzymes. These differences account for the unique functions of each. Membrane-bound intracellular organelles include the endoplasmic reticulum, Golgi apparatus, lysosomes, peroxisomes, and mitochondria (see figure 4.4). We first describe the structure and then the function of each.

Endoplasmic Reticulum

The **endoplasmic reticulum** (en'dō-plas'mik re-tik'ū-lum; *rete* = net) **(ER)** is an extensive interconnected membrane network that varies in shape (e.g., tubules, sacs), but with one continuous lumen **(figure 4.22)**. The ER typically extends from the nuclear envelope to the plasma membrane and composes about one-half of the membrane within a cell. The extensive ER membrane surface serves as a point of attachment for ribosomes (an organelle described in section 4.6b) and various types of enzymes. Endoplasmic reticulum with ribosomes attached is referred to as **rough ER,** whereas ER portions without ribosomes attached is referred to as **smooth ER.**

Rough ER The ribosomes of rough ER produce proteins that will be released from the cell (e.g., insulin from the pancreas), incorporated into the plasma membrane (see figure 4.6), and serve as digestive enzymes of lysosomes. These proteins are inserted either into the ER

membrane or through the ER membrane into the lumen of the ER as they are synthesized. The protein may be changed by either the addition of other molecules (e.g., carbohydrates to form glycoproteins) or removal of part of what was originally synthesized. A molecular tag (called a *signal sequence*) can be added to the protein that determines its destination. (This is much like adding an address to a letter.) The modified proteins are packaged and then stored until their release. Transport from the ER occurs when small, enclosed membrane sacs pinch off from the ER. These sacs are termed **transport vesicles** (figure 4.22). They shuttle proteins from the rough ER lumen to another organelle called the Golgi apparatus for further modification. The amount of rough ER is greater in cells producing large amounts of protein for secretion, such as a cell in the pancreas that releases insulin to control blood glucose (sugar). The ER also helps form peroxisomes (described later in this section).

Smooth ER The smooth ER carries out diverse metabolic processes that vary by cell type. Smooth ER functions include synthesis, transport, and storage of different types of lipids (e.g., phospholipids, steroids); carbohydrate metabolism (e.g., glycogen synthesis and breakdown in liver cells); and detoxification of drugs and poisons. Abundant amounts of smooth ER occur, for example, within the cells of the testes to produce the steroid hormone testosterone, and in the cells of the liver to detoxify alcohol.

Golgi Apparatus

The **Golgi apparatus,** also called the *Golgi complex* or *Golgi body,* is typically composed of several (e.g., about four or five) elongated, flattened saclike membranous structures called *cisternae* (**figure 4.23**). The Golgi apparatus exhibits a distinct polarity. The two poles are called the *cis*-face and *trans*-face. The ***cis*-face** is closer in proximity to the ER and the diameter of its flattened sac is larger compared to the ***trans*-face.**

One of the primary functions of the Golgi apparatus is to modify, package, and sort proteins (and glycoproteins) that are made by the rough ER (figure 4.22b). Transport vesicles arrive from the ER and fuse with *cis*-face of the Golgi apparatus. Materials move between cisternae from the *cis*-face to the *trans*-face. Within the lumen of the Golgi apparatus, molecules are modified; this may involve removal of portions of a molecule, or additions

to a molecule (e.g., addition of a carbohydrate or a phosphate group). As in the ER, a signal sequence may be added. The Golgi apparatus also synthesizes its own molecules including long, unbranched polysaccharides (see section 2.7c), which are modified with the addition of small amounts of protein to form proteoglycans (e.g., glycosaminoglycans; see section 5.2a). At the *trans*-face, **secretory vesicles** form and carry both the modified and newly formed molecules away from the Golgi apparatus for different fates. Some secretory vesicles become components of the plasma membrane and others release their contents from the cell into the interstitial fluid by exocytosis. Thus, the Golgi apparatus is especially extensive and active in cells specialized for protein, glycoprotein, and proteoglycan secretion.

Vesicles released from the *trans*-face of the Golgi apparatus may also contain hydrolytic digestive enzymes required for lysosomes (described next). You may find it helpful to think of the Golgi

Functions of Golgi Apparatus

1. **Synthesis:** Forms proteoglycans
2. **Processing molecules:** Modify and store protein (that was formed by RER)
3. **Organelle formation:** Synthesizes digestive enzymes for lysosomes
4. **Vesicle formation:** Forms secretory vesicles for delivering components of the plasma membrane and releasing contents from the cell by exocytosis

Trans-face — Secretory vesicles

Cis-face

Lysosome

Transport vesicle

Lumen of cisterna filled with secretory product

TEM 96,000×

Transport vesicle

(a) Cisternae

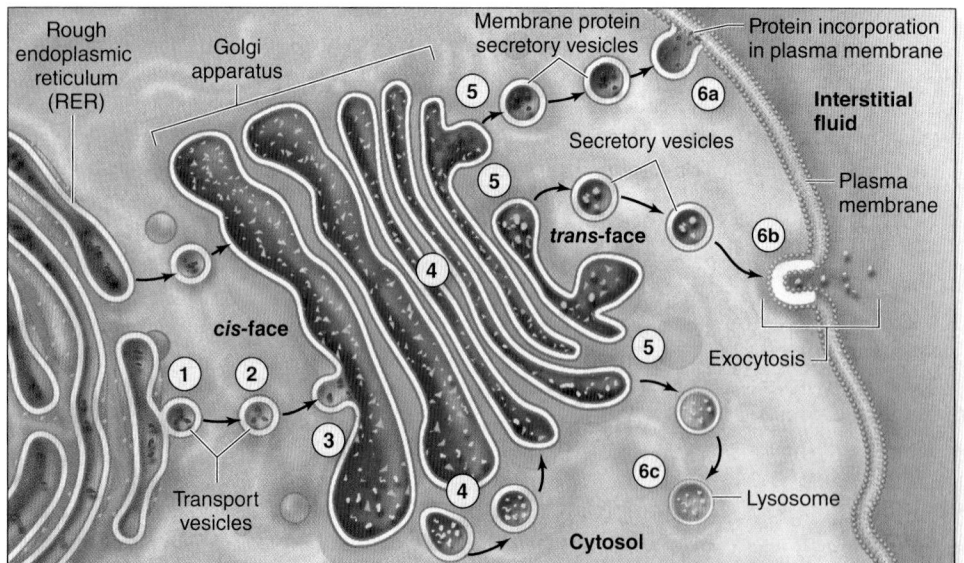

(b) Endomembrane System: Movement of materials through the Golgi apparatus

Rough endoplasmic reticulum (RER)

Golgi apparatus

Membrane protein secretory vesicles

Protein incorporation in plasma membrane

Interstitial fluid

Secretory vesicles

Plasma membrane

trans-face

Exocytosis

cis-face

Transport vesicles

Lysosome

Cytosol

Synthesis, Modification, and Shipping of Proteins

1. RER synthesizes protein that is released in a transport vesicle.

2. Vesicle from the RER moves to the Golgi apparatus.

3. Vesicle fuses with Golgi apparatus at the *cis*-face.

4. Proteins are modified as they move through Golgi apparatus (e.g., addition of carbohydrate).

5. Modified proteins are packaged and released within secretory vesicle from the *trans*-face.

6. Secretory vesicles merge with the plasma membrane to either insert molecules into the plasma membrane (6a) or release contents by exocytosis (6b). Vesicles also provide digestive enzymes to lysosomes (6c).

Figure 4.23 The Golgi Apparatus and Endomembrane System. Each Golgi apparatus is composed of several flattened membrane sacs (cisternae). The arrangement of these sacs exhibits both structural and functional polarity. (*a*) A TEM and a drawing provide different views of the Golgi apparatus. (*b*) The Golgi apparatus is part of the endomembrane system, which is a system of membranous structures of a cell that provide a means of transporting substances into, out of, and within a cell. AP|R

apparatus as the "warehouse" center. Molecules arrive at the *receiving region* (*cis*-face), are modified and packaged within the lumen (and along with some new structures that are produced), are then shipped out at the *shipping region* (*trans*-face).

Lysosomes

Lysosomes (lī'sō-sōm; *soma* = body) are small, membranous sacs that contain digestive enzymes, which are immersed in acidic fluid (pH 5) **(figure 4.24)**. These enzymes are contributed to lysosomes as portions of the Golgi apparatus containing digestive enzymes pinch off to form vesicles, and these vesicles then fuse with the lysosome.

Lysosomes participate in digestion of unneeded or unwanted substances. Within a healthy cell, lysosomes digest the contents of endocytosed vesicles. For example, following phagocytosis of microorganisms by certain white blood cells, the vesicle containing the microorganism fuses with a lysosome. Its digestive enzymes break down the large biomolecules composing the microorganism (e.g., proteins, polysaccharides) into smaller molecules (see figure 22.3a).

Lysosomes also digest molecular structures of damaged organelles in a similar fashion; this process is specifically called **autophagy** (aw-tōf'ă-jē; *autos* = self, *phago* = to eat). When a cell is damaged or dies, enzymes from its lysosomes are eventually released into the cytosol, resulting in the rapid digestion of the molecular components of the cell itself. This process is called **autolysis** (aw-tol'i-sis; *lysis* = dissolution). Two nicknames have been given to lysosomes: (1) "garbagemen" because of their "cleanup" activities of eliminating unwanted structures and (2) "suicide packets" because of their function in autolysis.

WHAT DO YOU THINK?

2 What would happen to a cell if it did not contain any lysosomes (or if its lysosomes were not functioning)? Would the cell be able to survive?

INTEGRATE

CLINICAL VIEW
Lysosomal Storage Diseases

Lysosomal storage diseases are an extensive group of heritable disorders that are characterized by accumulation of incompletely digested biomolecules within lysosomes. Lysosomal storage diseases occur because of mutations in the genes that code for one of the more than 40 different lysosomal enzymes. **Tay-Sachs disease** is one example of a lysosomal storage disease. Lysosomes in affected individuals lack an enzyme needed to break down complex membrane lipids (gangliosides). As a result, these complex lipids accumulate within nerve cells.

The cellular signs of Tay-Sachs disease are swollen lysosomes due to accumulation of the lipid. Affected infants appear normal at birth, but begin to show signs of the disease by the age of 6 months. The nervous system bears the brunt of the damage. Paralysis, blindness, and deafness typically develop over a period of 1 or 2 years, followed by death, usually by the age of 4. Unfortunately, there is no treatment or cure for this fatal disease.

Peroxisomes

Peroxisomes (per-ok'si-sōm) are membrane-enclosed sacs that contain over 50 different enzymes that vary by cell type. They are usually smaller in diameter than lysosomes **(figure 4.25)**. Peroxisomes are initially formed by vesicles first pinching off from the rough ER. Proteins (formed from ribosomes free in the cytosol) are then incorporated into the peroxisomes to serve as their enzymes. As additional

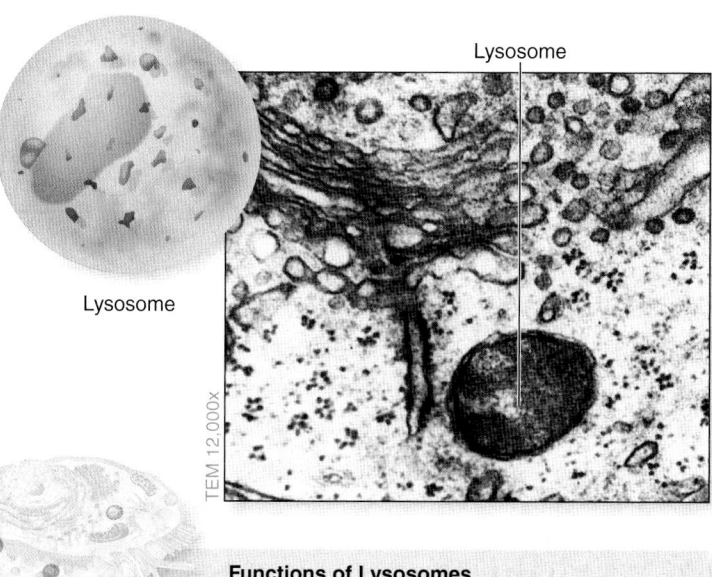

Functions of Lysosomes

Digestion: Break down molecules within vesicles that enter cell by endocytosis, remove damaged organelles and cellular components (autophagy), and break down cellular components following cellular death (autolysis)

Figure 4.24 Lysosomes. These organelles are small, spherical membrane-bound organelles that house enzymes for intracellular digestion. Both a drawing and TEM are shown. AP|R

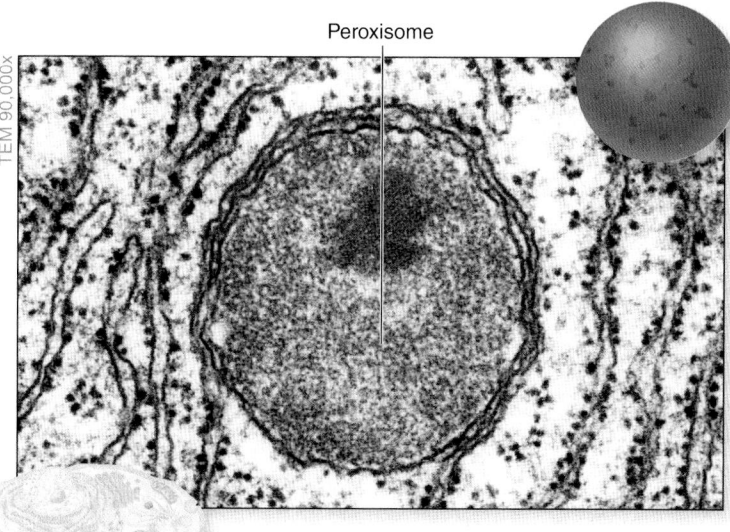

Functions of Peroxisomes

1. **Digestion:** Break down molecules (e.g., fatty acids, amino acids, uric acid) with hydrogen peroxide produced during the process
2. **Synthesis:** Form specific types of lipids (e.g., plasmalogens, bile salts)

Figure 4.25 Peroxisomes. A TEM and drawing show a peroxisome within a cell. Peroxisomes are small, spherical membrane-bound organelles that contain enzymes that function in both digestion and synthesis of molecules.

membrane (formed by the ER) and proteins are added, a peroxisome increases in size. When a critical size is reached, it splits forming two peroxisomes.

Although peroxisomes appear simple in structure, they engage in numerous metabolic functions. They were first named based on their role in chemical digestion, which involves removal of hydrogen from a molecule with the accompanying production of hydrogen peroxide. The hydrogen peroxide is subsequently broken down into water and oxygen (by catalase enzyme). Molecules broken down within peroxisomes by this process include fatty acids, amino acids, and uric acid (a waste product of nucleic acid breakdown). The breakdown of fatty acids is a process that removes two hydrogen-carbon units at a time from the fatty acid chain (a process called *beta oxidation*). These units are then converted to acetyl CoA and may be taken up by the mitochondria within the cell, where they are oxidized to transfer energy to form ATP (see section 3.4h).

Peroxisomes also engage in lipid synthesis (a role it shares with the ER). Lipids formed by peroxisomes include specialized phospholipids (e.g., plasmalogens) within the cells of the heart and brain and bile salts within the cells of the liver. Notice that peroxisomes function in both digestion and synthesis of molecules.

Endomembrane System

The **endomembrane system** is an extensive array of membrane-bound structures that includes the endoplasmic reticulum, Golgi apparatus, vesicles, lysosomes, and peroxisomes. The plasma membrane and nuclear envelope are also considered part of this membrane system. All of these structures are either directly attached to one another or are connected through vesicles that move between them. They are involved in various forms of metabolic processes that occur within a cell, and provide a means of transporting substances within the cell, as shown in figure 4.23b. Note that mitochondria, described next, are the only membrane-bound organelles that are not components of the endomembrane system.

Mitochondria

Mitochondria (mī′tō-kon′drē-ă; sing., *mitochondrion*, mī-tō-kon′ drē-on; *mitos* = thread, *chondros* = granule) were first described in chapter 3. They are oblong-shaped organelles with a double membrane with the folds of the inner membrane called cristae (**figure 4.26**). The matrix, which is the inner region of a mitochondrion, contains a small, unique circular fragment of DNA that has genes for producing mitochondrial proteins (not shown in figure 4.26). Mitochondria engage in aerobic cellular respiration to complete the digestion of glucose and other fuel molecules, such as fatty acids, for the transfer of energy to synthesize ATP molecules, the cell's energy currency. For this reason, mitochondria are called the "powerhouses" of the cell. Mitochondria numbers within cells increase (by fission) with greater demands for ATP production. (See section 3.4 for a detailed discussion of cellular respiration.)

WHAT DID YOU LEARN?

18 Describe the general structure of both the endoplasmic reticulum and Golgi apparatus, and discuss their functional relationship in the synthesis, modification, storage, and release of molecules.

19 Lysosomes and peroxisomes are both small membranous sacs. Which of these functions in (a) both digestion and synthesis of molecules, and (b) digesting unwanted organelles.

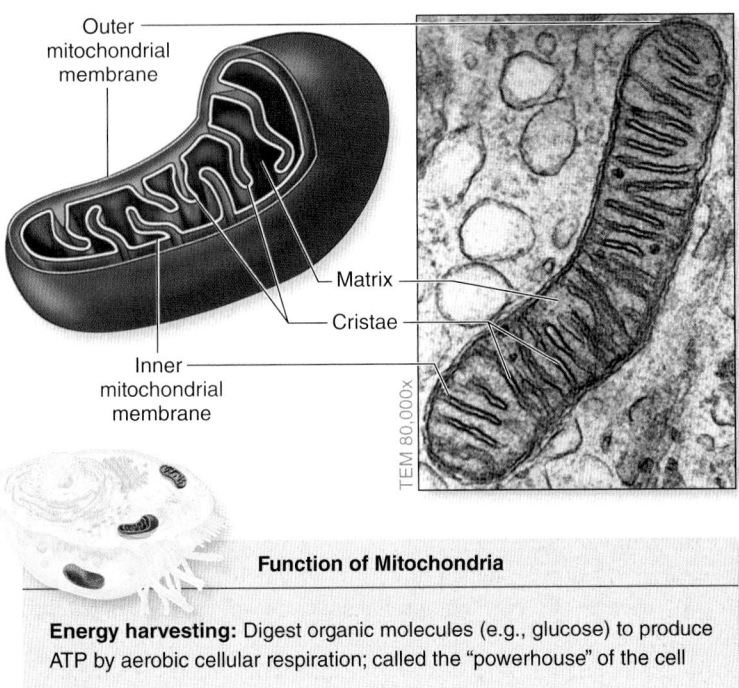

Function of Mitochondria

Energy harvesting: Digest organic molecules (e.g., glucose) to produce ATP by aerobic cellular respiration; called the "powerhouse" of the cell

Figure 4.26 Mitochondria. A drawing and TEM show the parts of a mitochondrion. Mitochondria are the double-membrane-bound organelles within the cytoplasm that engage in aerobic cellular respiration to produce ATP for energy-requiring cellular processes.

4.6b Non-Membrane-Bound Organelles

LEARNING OBJECTIVES

3. List the non-membrane-bound organelles of a typical human cell.

4. Describe the structure and main function(s) of each.

Non-membrane-bound organelles are composed of either protein alone or protein and RNA. They include ribosomes, the cytoskeleton, centrosome, and proteasomes.

Ribosomes

Ribosomes are non-membrane-bound organelles containing protein and ribonucleic acid (RNA) that are arranged into both a large and small subunit. The large subunit has three hollow areas designated as the *A, P, and E sites* (**figure 4.27**). You can think of the ribosomal subunits as puzzle pieces that are made within the nucleolus and then moved into the cytosol where the pieces are put together into one complete puzzle (a ribosome).

Ribosomes are either bound or free. **Bound ribosomes** are attached to the external surface of the ER membrane to form rough ER. Recall that bound ribosomes are used to synthesize proteins destined for export out of the cell, to become an integral part of the plasma membrane, or to serve as enzymes within lysosomes. **Free ribosomes** are suspended within the cytosol. In general, all other proteins that function within the cell are synthesized by free ribosomes. The details of how ribosomes function in synthesizing protein are described in section 4.8.

Cytoskeleton

The **cytoskeleton** has a central role in numerous cellular activities that include intracellular structural support and organization of organelles, cell division, and movement of materials. Formed by a framework of diverse fibrous proteins, the cytoskeleton extends throughout the interior

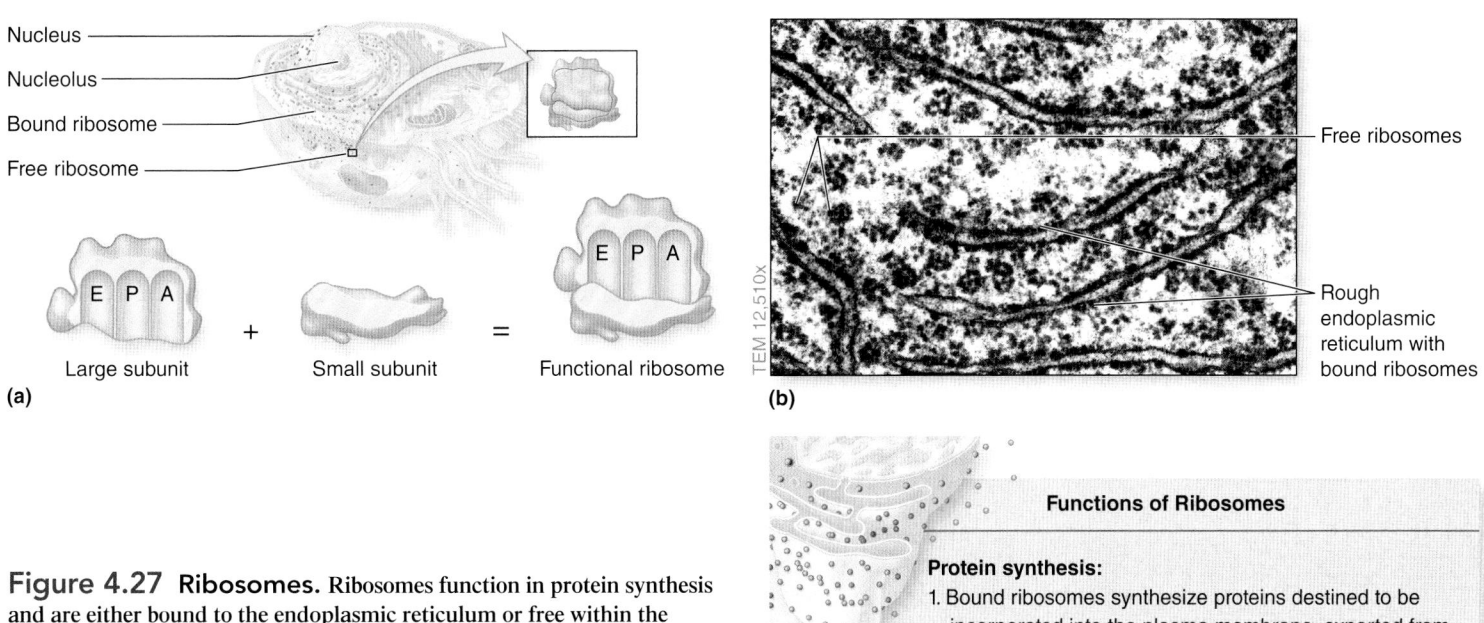

(a)

Nucleus
Nucleolus
Bound ribosome
Free ribosome

E P A
Large subunit

+

Small subunit

=

E P A
Functional ribosome

(b)

TEM 12,510x

Free ribosomes

Rough endoplasmic reticulum with bound ribosomes

Functions of Ribosomes

Protein synthesis:
1. Bound ribosomes synthesize proteins destined to be incorporated into the plasma membrane, exported from the cell, or housed within lysosomes
2. Free ribosomes synthesize proteins for use within the cell

Figure 4.27 Ribosomes. Ribosomes function in protein synthesis and are either bound to the endoplasmic reticulum or free within the cytosol. (*a*) Ribosomes consist of both small and large subunits with each subunit composed of protein and RNA. (*b*) A TEM shows both free and bound ribosomes in the cell cytosol. AP|R

of a cell and anchors to proteins in the plasma membrane. Three separate types of protein molecules form the cytoskeleton—microfilaments, intermediate filaments, and microtubules **(figure 4.28)**.

Microfilaments (mī-krō-fil′ă-ment; *micros* = small) are the smallest components of the cytoskeleton with a diameter of about 7 nanometers. They are composed of *actin* protein monomers that are organized into two thin, intertwined protein filaments (actin filaments) similar to two twisted pearl strands. They form an interlacing network on the

cytoplasmic side of the plasma membrane. Microfilaments help maintain cell shape, form internal support of microvilli (described shortly), separate the two cells formed during cytokinesis (a process of cell division described in section 4.9), facilitate cytoplasmic streaming (the movement of the cytoplasm associated with changing cell shape), and participate in muscle contraction. Individual globular actin proteins are added to one end of the microfilament for growth in a particular direction and removed from the other end for shortening.

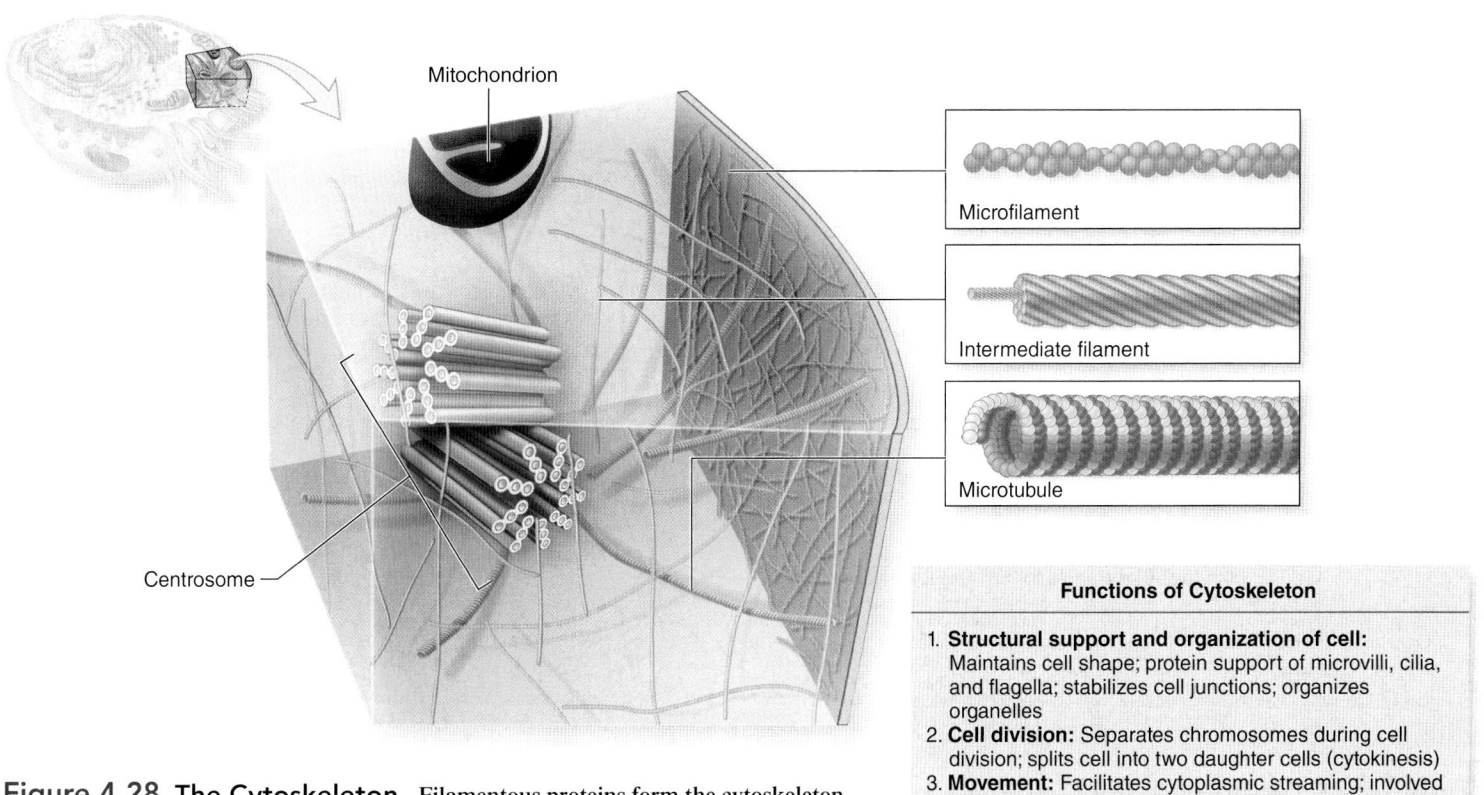

Mitochondrion

Microfilament

Intermediate filament

Microtubule

Centrosome

Functions of Cytoskeleton

1. **Structural support and organization of cell:** Maintains cell shape; protein support of microvilli, cilia, and flagella; stabilizes cell junctions; organizes organelles
2. **Cell division:** Separates chromosomes during cell division; splits cell into two daughter cells (cytokinesis)
3. **Movement:** Facilitates cytoplasmic streaming; involved in movement of vesicles within a cell; participates in muscle contraction

Figure 4.28 The Cytoskeleton. Filamentous proteins form the cytoskeleton. They help give the cell its shape and coordinate cellular movements. The three cytoskeletal elements are microfilaments, intermediate filaments, and microtubules. AP|R

INTEGRATE

CONCEPT
CONNECTION

A specific example of proteasome action occurs when a virus or other infectious agent enters a cell. Proteins of the infectious agent are cleaved by a proteasome. These degraded peptide fragments of the infectious agent are considered "nonself" and are presented to specialized white blood cells, alerting the immune system that the body has been "invaded" (see section 22.4c).

Intermediate filaments are between 8 and 12 nanometers in diameter. They are more rigid than microfilaments, and they support cells structurally and stabilize junctions between them. Their protein composition differs, depending upon the cells in which they occur. Keratin, a protein of the skin, hair, and nails, is an example of one type of intermediate filament; another type forms neurofilaments of nerve cells.

Microtubules (mī-krō-tū′būl; *tubus* = tube) are hollow cylinders that are approximately 25 nanometers in diameter. They are composed of long chains of a globular protein called *tubulin*. Microtubules are not permanent structures. They may be elongated or shortened as needed by the addition or removal of tubulin monomers. Microtubules help maintain cell shape, organize and move organelles within a cell, form protein components of cilia and flagella, participate in cellular transport of vesicles, and separate chromosomes during cell division.

Centrosome

The **centrosome** is a structure typically in close proximity to the nucleus. It contains a pair of perpendicularly oriented cylindrical **centrioles** (sen′trē-ōl; *kentron* = a point, center) surrounded by protein that is amorphous (without a distinctive shape) (**figure 4.29**). The paired centrioles are positioned perpendicular to each other with each composed of triplets of microtubules arranged in a circle. The primary function of a centrosome is organizing microtubules within the cytoskeleton. The centrosome is best known for its function in cellular division, during which microtubules form spindle fibers to facilitate chromosome movement, as described in section 4.9.

Proteasomes

Large, barrel-shaped protein complexes called **proteasomes** (prō′tē-ă-sōm) are major protein-digesting organelles located within both the cytosol and nucleus of cells (**figure 4.30**). Proteasomes degrade cell proteins through an ATP-dependent pathway; these proteins include damaged proteins, incorrectly folded proteins, and normal proteins that are no longer

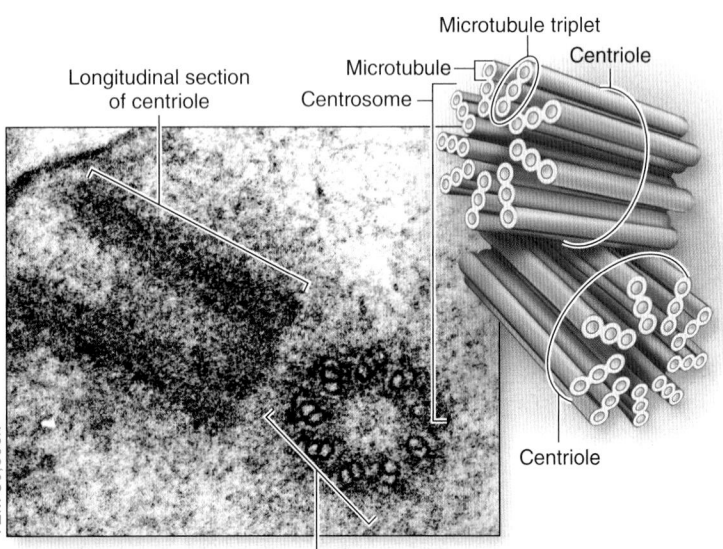

Functions of Centrosomes and Centrioles

1. **Synthesis:** Organizes microtubules (proteins of cytoskeleton) and supports their growth in nondividing cells
2. **Cell division:** Directs formation of mitotic spindle in dividing cells

Figure 4.29 Centrosome and Centrioles. A TEM and a drawing show that a region of the cytoplasm called the centrosome contains a centriole pair immediately adjacent to the nucleus.

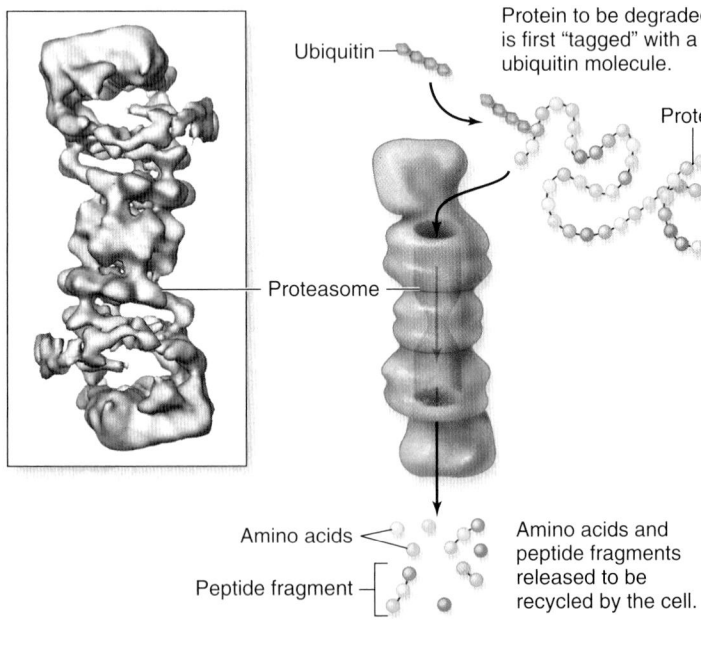

Functions of Proteasomes

1. **Protein digestion:** Degrades proteins that are damaged, incorrectly folded, or those that are no longer needed
2. **Quality assurance:** Controls the quality of exported cell proteins

Figure 4.30 Proteasomes. These organelles maintain order within the cell by digesting abnormal and unwanted cellular proteins. A sketch of a TEM and drawing are both shown.

needed by the cell. The action of proteasomes also provides a means to control the quality of exported cell proteins. This latter function is especially critical during the regulation of cellular metabolism, cell division, and activities associated with cell signaling. When a protein is targeted for removal by proteasomes, it usually is marked for destruction by having a protein called ubiquitin bound to it. This is the first step in the ultimate degradation of the protein by proteasomes. You may find it helpful to think of proteasomes as "garbage disposals" for unwanted protein.

Recent research has shown that significant age-related alterations in proteasome structure and function may prevent or inhibit the normal removal of proteins from the cell.

 WHAT DID YOU LEARN?

20 Which non-membrane-bound organelle functions (a) to digest unwanted proteins, (b) form the structural support of cell, (c) synthesize proteins, and (d) participate in cell division?

4.6c Structures of the Cell's External Surface

 LEARNING OBJECTIVES

5. Distinguish between cilia and flagella in terms of both structure and function.

6. Describe the structure and function of microvilli.

The structures that extend from the surface of some cells include cilia, flagella, and microvilli. Cilia and flagella are extensions of the plasma membrane involved in movement, whereas microvilli are structures that increase the surface area of the plasma membrane.

Cilia and Flagella

Cilia (sil′ē-ă; sing., sil′ē-ŭm; *cilium* = an eyelash) are small (5 μm to 10 μm in length and 0.2 μm in width) hairlike projections extending from the exposed surfaces of some cells. They contain both cytoplasm and supportive microtubule proteins, and are enclosed by the plasma membrane. Cilia are usually found in large numbers on the exposed surfaces of specific cells such as those that line portions of the respiratory passageways (see figure 4.1). The beating of these cilia moves mucus and any adherent substances along the cell surface toward the throat, where it may then be expelled from the respiratory system (see section 23.1c).

Flagella (flă-jel′ă; sing., *flagellum,* flă-jel′ŭm; a whip) are similar to cilia in basic structure—however, they are longer (50 μm in length and about 0.5 μm in width), and when present, there is usually only one. The function of a flagellum is to help propel an entire cell. In humans, the only example of a cell with a flagellum is sperm (see figure 28.18), which moves through the female reproductive tract to reach the oocyte. The movement of both cilia and flagella occurs through the microtubules within their core; a process that requires energy provided by the splitting of ATP molecules.

Microvilli

Microvilli are thin, microscopic membrane extensions from the surface of the plasma membrane. Microvilli are shorter and more narrow than cilia (average about 1 μm high and 0.08 μm wide), more densely packed together, and lack powered movement **(figure 4.31)**. Each microvillus is supported by microfilaments (actin proteins are cross-linked into a dense bundle that serves as its structural core). Microvilli provide a more extensive plasma membrane surface area for more efficient membrane transport. Just as not all cells have cilia, not all cells have microvilli. Cells with microvilli occur, for example, throughout the small intestine, where increased surface area is needed to absorb digested nutrients. Cellular structures and their associated functions are integrated in **figure 4.32**.

 WHAT DID YOU LEARN?

21 Which cellular surface structure functions in (a) increasing the cell's surface area, and (b) movement of material past the cell?

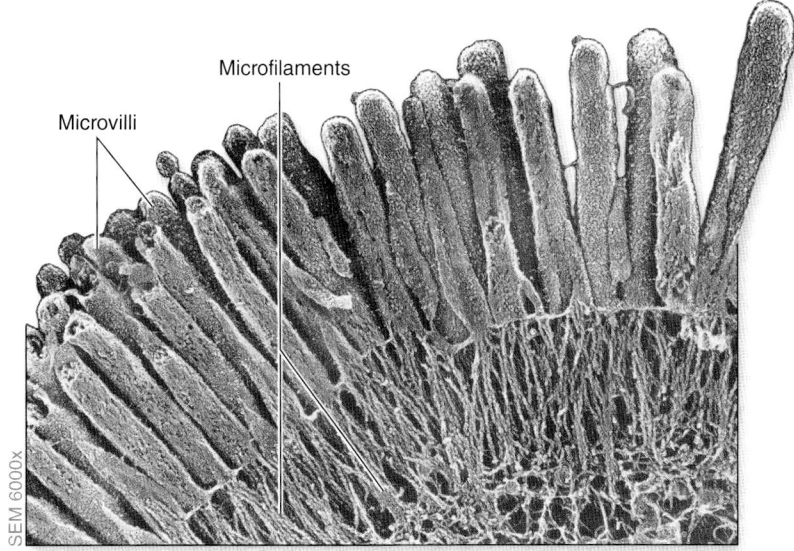

Figure 4.31 Microvilli. Microvilli are thin, microscopic projections. They extend from the surface of the plasma membrane and are supported by microfilaments. Microvilli function to increase the surface area of the plasma membrane for more efficient membrane transport. **AP|R**

Figure 4.32 **Cellular Structures and Their Functions.** Cells are responsible for all body functions. Most of the cellular functions occur within (1) the cytoplasm, which includes (1a) cytosol, (1b) membrane-bound organelles and non-membrane-bound organelles, and (1c) cell inclusions. (2) The nucleus is enclosed by the nuclear envelope and contains chromatin and the nucleolus. (3) The plasma membrane is composed of a phospholipid bilayer, proteins, cholesterol, and glycolipids and glycoproteins.

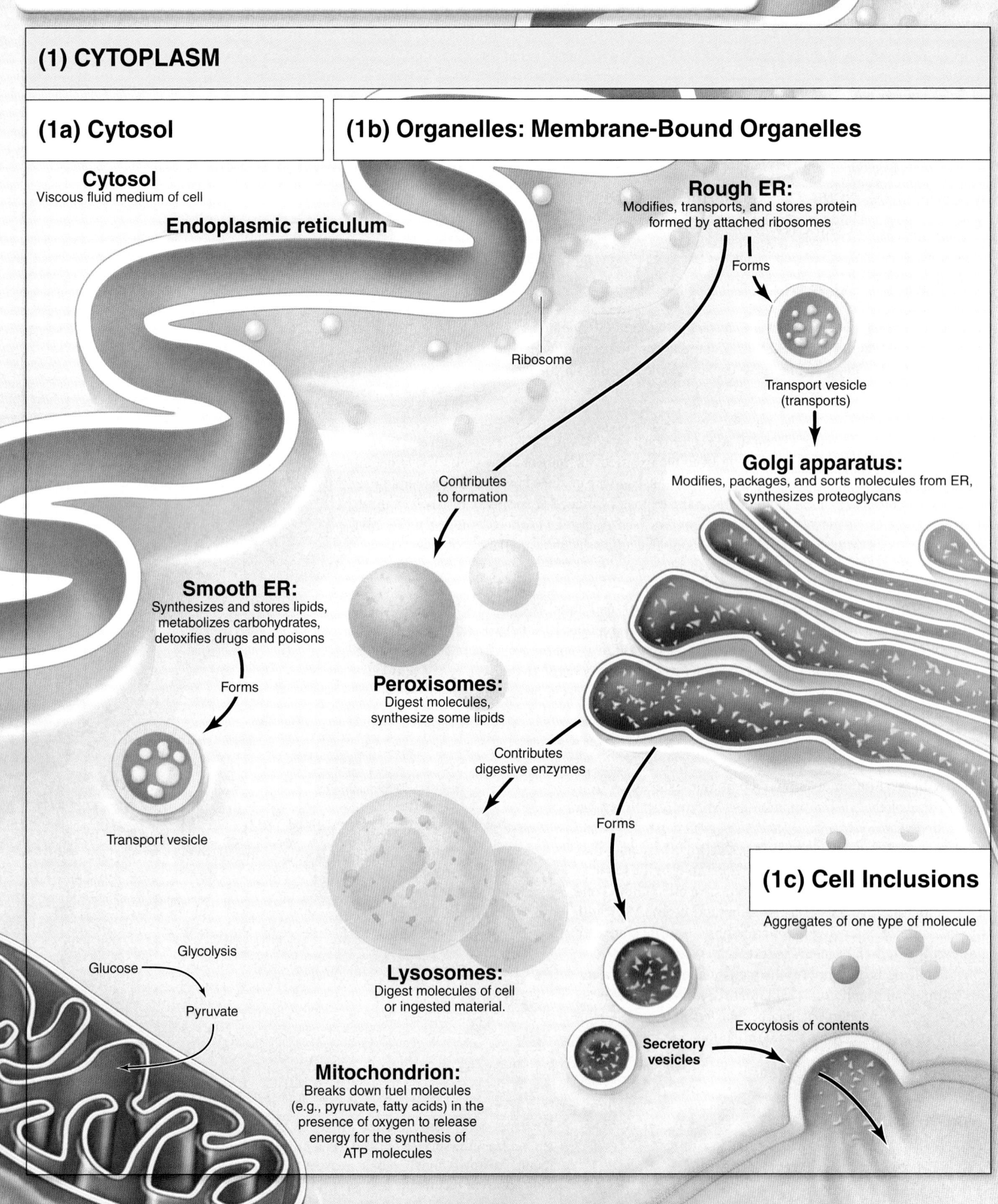

(1) CYTOPLASM

(1a) Cytosol

Cytosol
Viscous fluid medium of cell

Endoplasmic reticulum

Ribosome

Glycolysis

Glucose — Pyruvate

Smooth ER:
Synthesizes and stores lipids, metabolizes carbohydrates, detoxifies drugs and poisons

Forms

Transport vesicle

(1b) Organelles: Membrane-Bound Organelles

Rough ER:
Modifies, transports, and stores protein formed by attached ribosomes

Forms

Transport vesicle (transports)

Golgi apparatus:
Modifies, packages, and sorts molecules from ER, synthesizes proteoglycans

Contributes to formation

Peroxisomes:
Digest molecules, synthesize some lipids

Contributes digestive enzymes

Forms

Lysosomes:
Digest molecules of cell or ingested material.

Mitochondrion:
Breaks down fuel molecules (e.g., pyruvate, fatty acids) in the presence of oxygen to release energy for the synthesis of ATP molecules

(1c) Cell Inclusions

Aggregates of one type of molecule

Exocytosis of contents

Secretory vesicles

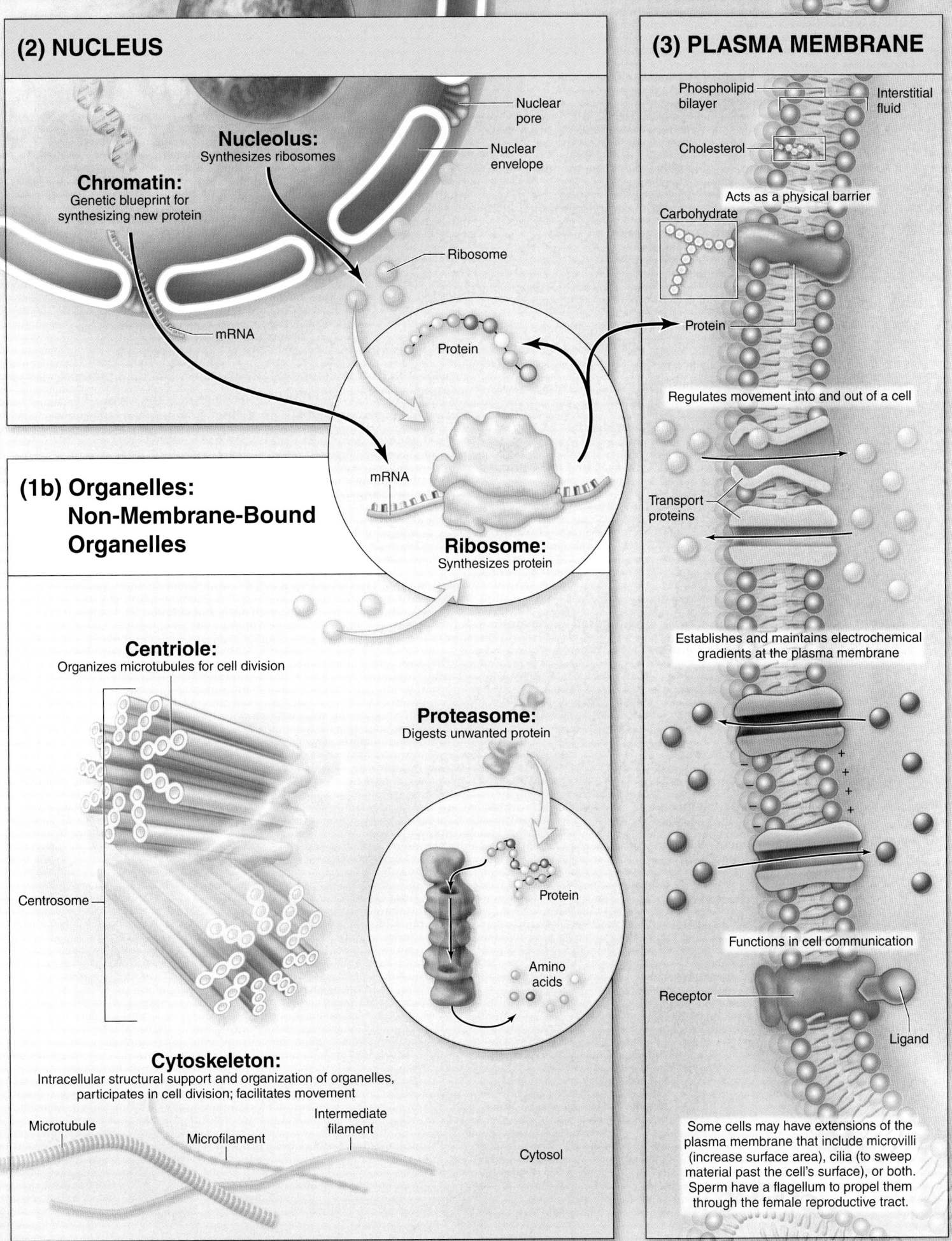

(2) NUCLEUS

Nucleolus:
Synthesizes ribosomes

Nuclear pore

Nuclear envelope

Chromatin:
Genetic blueprint for synthesizing new protein

mRNA

Ribosome

Protein

mRNA

Ribosome:
Synthesizes protein

(1b) Organelles: Non-Membrane-Bound Organelles

Centriole:
Organizes microtubules for cell division

Centrosome

Proteasome:
Digests unwanted protein

Protein

Amino acids

Cytoskeleton:
Intracellular structural support and organization of organelles, participates in cell division; facilitates movement

Microtubule

Microfilament

Intermediate filament

Cytosol

(3) PLASMA MEMBRANE

Phospholipid bilayer

Interstitial fluid

Cholesterol

Acts as a physical barrier

Carbohydrate

Protein

Regulates movement into and out of a cell

Transport proteins

Establishes and maintains electrochemical gradients at the plasma membrane

Functions in cell communication

Receptor

Ligand

Some cells may have extensions of the plasma membrane that include microvilli (increase surface area), cilia (to sweep material past the cell's surface), or both. Sperm have a flagellum to propel them through the female reproductive tract.

4.6d Membrane Junctions

LEARNING OBJECTIVE

7. Compare and contrast the structure and function of the three major types of membrane junctions.

Membrane junctions connect and support cells. Most of our cells are arranged into structural units called *tissues* (see chapter 5) that act together in a common function. To provide an orderly arrangement between some cells and coordinate their interactions, membrane junctions are located between adjacent cells. There are three major types of membrane junctions: tight junctions, desmosomes, and gap junctions **(figure 4.33)**.

Tight Junctions

A **tight junction** (or a *zonula occludens*) is composed of plasma membrane proteins that form strands or rows of proteins (e.g., claudins and occludins). These cell membrane junctions are positioned at the apical surfaces around the circumference of each of the adjacent cells. Tight junctions function like spot welds to seal off the intercellular space and prevent substances from passing between the cells; this requires all materials to move through, rather than between, the cells. For example, tight junctions associated with epithelial cells that form the lining of the small intestine prevent corrosive digestive enzymes that are within the lumen of the intestine from moving between cells and damaging other body structures (see section 26.3b). These junctions also prevent leakage of urine through the urinary bladder wall (see section 24.8b).

Desmosomes

A **desmosome** (dez'mō-sōm; *desmos* = a band, *soma* = body), also called *macula adherens* ("adhering spot"), is composed of several different proteins that bind neighboring cells. A thickened protein called a protein plaque is located on the internal surface of the plasma membrane of adjoining cells. Numerous protein filaments (e.g., desmoglein) extend from the plaque through the plasma membrane of both neighboring cells. These filamentous proteins extend across the small space between the two cells and are anchored to one another. Also anchored to each protein plaque are intermediate filaments of the cytoskeleton that extend throughout the cell to provide support and strength. Observe the pattern in figure 4.33 of these proteins forming a desmosome: intermediate filament, protein plaque, adjoining protein filaments, protein plaque, and intermediate filaments. This anatomic arrangement provides structural integrity to cells that are exposed to stress, such as the external layer of the skin (see section 5.1a) and cardiac muscle (see section 19.3e). Hemidesmosomes ("half of a desmosome") anchor the basal layer (attached surface) of cells of the epidermis to the underlying basement membrane (see section 5.1a).

Gap Junctions

A **gap junction** is composed of six transmembrane proteins, called **connexons** (kon-neks'onz), that form a tiny, fluid-filled tunnel or pore that extends across a small gap (about 2 nanometers) between adjacent cells. Gap junctions provide a direct passageway for substances to travel between neighboring cells. Ions, glucose, amino acids, and other small solutes can pass directly from the cytoplasm of one cell into the neighboring cell through these channels. The flow of ions between cells allows spread of electrical activity in cardiac muscle of the heart (see section 19.3e).

WHAT DID YOU LEARN?

22 Which cellular junction (a) provides resistance to mechanical stress, (b) allows the passage of ions between cells, and (c) prevents leakage between cells?

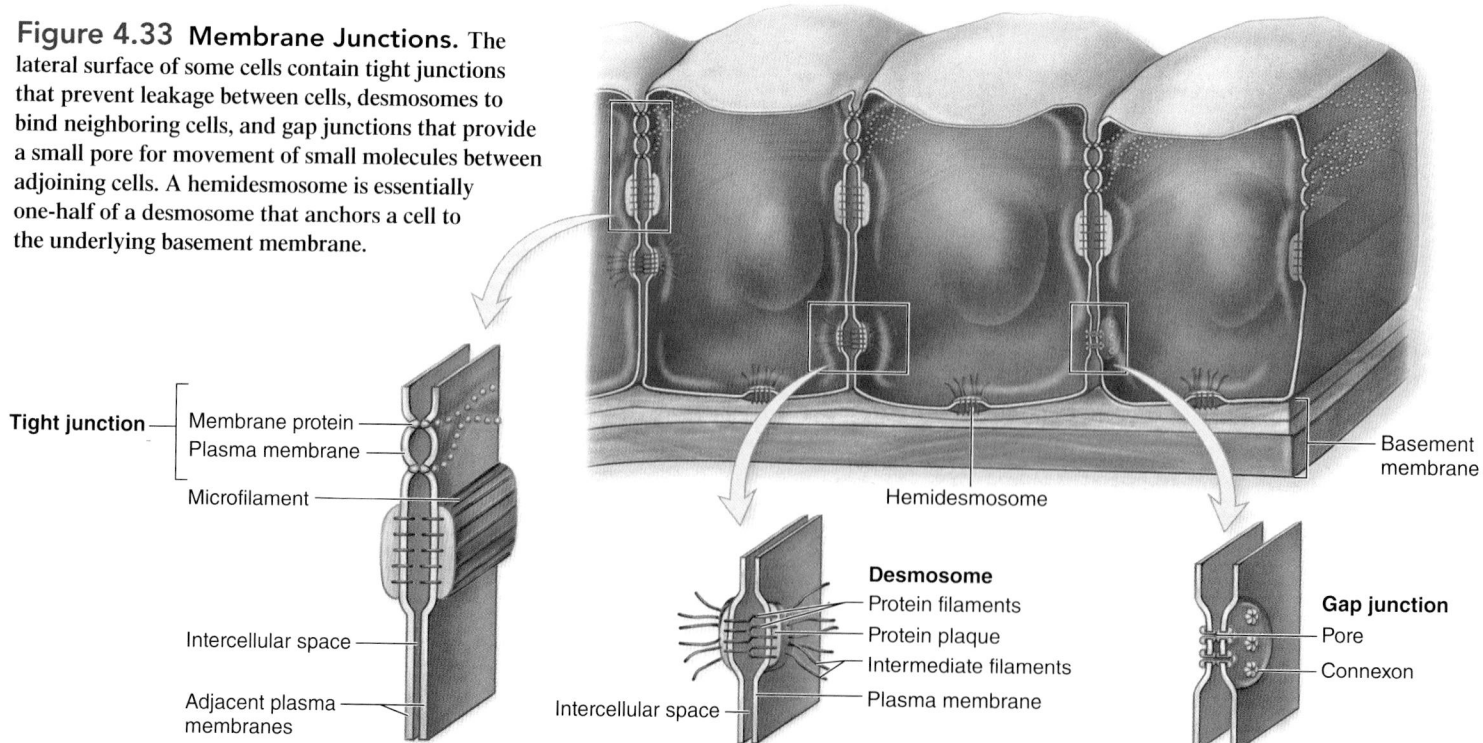

Figure 4.33 Membrane Junctions. The lateral surface of some cells contain tight junctions that prevent leakage between cells, desmosomes to bind neighboring cells, and gap junctions that provide a small pore for movement of small molecules between adjoining cells. A hemidesmosome is essentially one-half of a desmosome that anchors a cell to the underlying basement membrane.

Tight junction
— Membrane protein
— Plasma membrane
— Microfilament
— Intercellular space
— Adjacent plasma membranes

Hemidesmosome

Basement membrane

Desmosome
— Protein filaments
— Protein plaque
— Intermediate filaments
— Plasma membrane

Intercellular space

Gap junction
— Pore
— Connexon

4.7 Structure of the Nucleus

The nucleus is the largest structure in the cell, typically averaging about 5 μm to 7 μm in diameter. It is often called the cell's control center (**figure 4.34**). A cell typically has one nucleus. However, erythrocytes contain no nucleus, and skeletal muscle cells have many nuclei. The shape of the nucleus generally mirrors the shape of a cell. For example, a cuboidal cell has a spherical nucleus in the center of the cell, whereas a thin, flattened cell has a nucleus elongated in the same direction as the cell. Some cells contain a uniquely shaped nucleus. For example, some white blood cells have a multilobed nucleus that may have two or more segments (see figures in table 18.7).

4.7a Nuclear Envelope and Nucleolus

LEARNING OBJECTIVES

1. Describe the nuclear envelope.
2. Explain the structure and function of a nucleolus.

The nucleus is enclosed within a double membrane that is called the **nuclear envelope** (or *nuclear membrane*). It separates the cytoplasm from the **nucleoplasm** (the fluid within the nucleus). This boundary controls the movement of materials between the nucleus and the surrounding cytoplasm. Each of the two layers of the nuclear envelope is a phospholipid bilayer, similar in structure to the plasma membrane. Externally it is continuous with the rough ER in the cytoplasm. **Nuclear pores** are open passageways formed by proteins that extend through fused regions of the nuclear envelope. They are larger than ion channels and allow for the passage of larger molecules both into the nucleus (e.g., proteins) and out of the nucleus (e.g., messenger RNA). Ions and water-soluble molecules also pass through nuclear pores.

The cell nucleus typically contains one dark-staining, usually spherical, body called a **nucleolus** (nū-klē-ō-lŭs) (figure 4.34*a*). (The plural term is **nucleoli** [nū-klē-ō-lī].) A nucleolus is not membrane-bound. It is composed of protein and RNA, and is responsible for producing the large and small subunits of ribosomes.

Not all cells contain a nucleolus. Its presence and relative number indicate the protein synthesis activity of a cell. For example, nerve cells contain more than one nucleolus because numerous ribosomes are needed for the production of its many proteins. In contrast, sperm have no nucleoli because they produce no proteins.

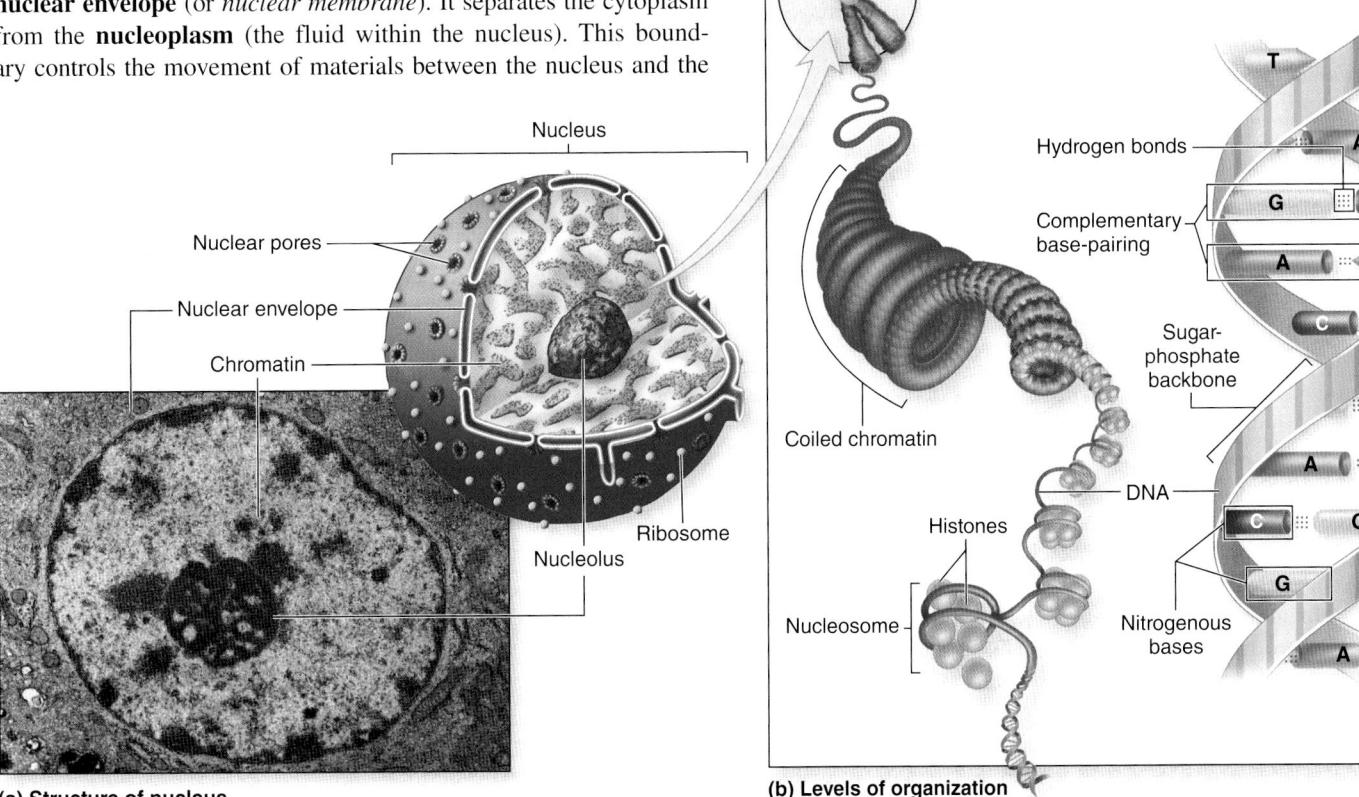

(a) Structure of nucleus

(b) Levels of organization

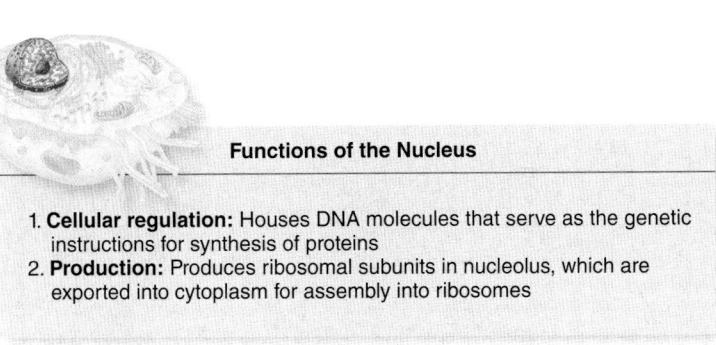

Functions of the Nucleus

1. **Cellular regulation:** Houses DNA molecules that serve as the genetic instructions for synthesis of proteins
2. **Production:** Produces ribosomal subunits in nucleolus, which are exported into cytoplasm for assembly into ribosomes

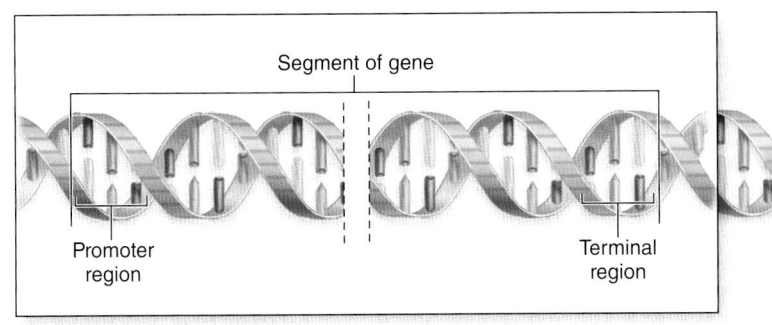

(c) Functional unit: A gene

Figure 4.34 Nucleus, DNA and Chromatin Structure, and Genes. (*a*) A TEM and drawing show the structural characteristics of the nucleus within a cell. DNA is the genetic material housed within the cell nucleus. (*b*) DNA is a polymer of nucleotides in the shape of a double helix. Strands of DNA and histone proteins associate to form chromatin, which forms chromosomes when cell division occurs. (*c*) The functional unit of DNA is the gene, which is a sequence of DNA that directs the synthesis of a specific protein. AP|R

23 What is the function of nuclear pores within the nuclear envelope?

24 What is the function of the nucleolus?

4.7b DNA, Chromatin, and Chromosomes

LEARNING OBJECTIVE

3. Describe the relationship of DNA, chromatin, and genes.

The nucleus houses the nuclear DNA (a nucleic acid biomolecule) along with the nucleolus and nucleoplasm. Recall that a nucleic acid biomolecule is formed by repeating monomers called **nucleotides** (see figure 2.21). DNA is specifically composed of **deoxyribonucleotides**, which include the five-carbon sugar deoxyribose, a phosphate, and one of four nitrogenous bases: adenine (A), cytosine (C), guanine (G), and thymine (T). Nucleotide monomers are linked through the phosphate groups by **phosphodiester bonds** to form a polymer strand. Each DNA molecule contains two complementary strands of nucleotides. They are bonded together by weak hydrogen bonds between the nucleotide bases to form a double helix structure. Note that the base adenine pairs only with the base thymine, whereas the base guanine pairs only with the base cytosine. This specific interaction between bases is called **complementary base pairing**: A to T, and C to G.

Think of the DNA as a spiral ladder, where the sugar and phosphate components of the nucleotide monomers form the vertical "struts" of the ladder. The horizontal "rungs" of the ladder are formed by pairs of nucleotide bases interconnected by weak hydrogen bonds to the complementary strand. The order of the bases in the nucleotides ultimately codes for specific proteins that the body needs.

DNA is an enormous macromolecule that comprises most of the genetic material of the cell (a small amount is found within mitochondria). The amount of DNA in a human cell contains over 3 billion pairs of nucleotides. A human somatic (body) cell nucleus has 46 separate double-stranded DNA molecules. To help package the DNA within the nucleus, each long DNA double helix winds around a cluster of special nuclear proteins, called **histones,** forming a complex called a **nucleosome** (figure 4.34b). When a cell is not dividing, the DNA and its associated proteins are in the form of a finely filamented mass called **chromatin** (krō′ma-tin; *chroma* = color), which resembles an unrolled spool of thread. Chromatin becomes tightly coiled masses called **chromosome** (krō′mō-sōm; *chromo* = color, *soma* = body) only when a cell is dividing.

DNA is organized functionally into discrete units called **genes** (figure 4.34c). Genes are segments of nucleotides within DNA that provide the instructions for the synthesis of specific proteins. About 1–2% of the total amount of DNA composes genes. The average length of a gene is about 3000 nucleotide base pairs. Associated with each gene is a *promoter* region that is analogous to a "start" signal, and a *terminal* region that is analogous to a "stop" signal for the transcription, or copying, of a gene into an RNA molecule to direct the synthesis of a protein (as described in section 4.8). Although you are probably more familiar with the term chromosome (than chromatin), our genetic DNA is typically present in our cells as chromatin. This more loose arrangement of DNA allows cellular structures access to genes. As we will see, the tightly coiled mass of chromosomes is needed only to prevent the DNA from becoming tangled during cell division (as described in section 4.9).

WHAT DID YOU LEARN?

25 Describe the structural relationship of DNA and chromatin, and the functional relationship of DNA and genes.

4.8 Function of the Nucleus and Ribosomes

We continue our discussion of cellular functions by describing protein synthesis, the central process upon which all other cellular activities ultimately depend. The DNA directs protein synthesis, which occurs at ribosomes in the cytoplasm. Consequently, two major events are involved: (1) **transcription,** which is the formation of a ribonucleic acid (RNA) copy of a gene from DNA in the nucleus; and (2) **translation,** which uses the information coded in RNA for the synthesis of the protein by ribosomes in the cytosol **(figure 4.35)**. For both processes we first describe the required structures and then the mechanism involved.

4.8a Transcription: Synthesizing RNA

LEARNING OBJECTIVES

1. List the required structures for transcription.

2. Explain the three steps of transcription.

Transcription occurs within the nucleus of the cell. This process occurs when a segment of DNA is "read" and copied to produce a newly formed strand of RNA.

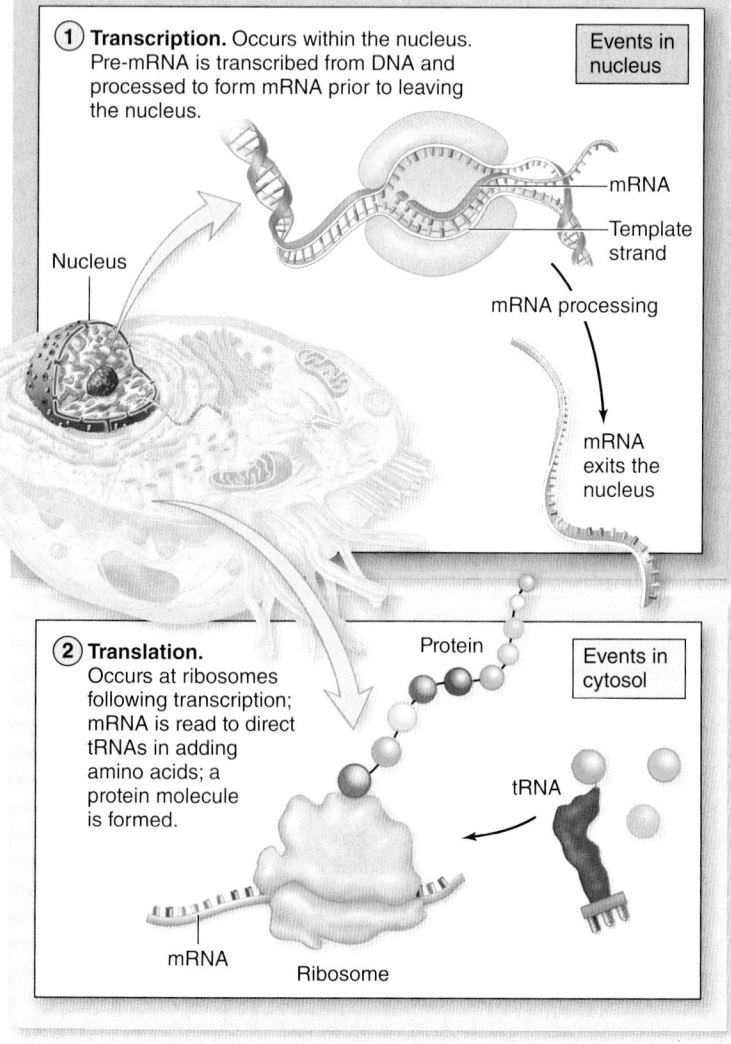

① Transcription. Occurs within the nucleus. Pre-mRNA is transcribed from DNA and processed to form mRNA prior to leaving the nucleus.

Events in nucleus

Nucleus

mRNA

Template strand

mRNA processing

mRNA exits the nucleus

② Translation. Occurs at ribosomes following transcription; mRNA is read to direct tRNAs in adding amino acids; a protein molecule is formed.

Protein

Events in cytosol

tRNA

mRNA

Ribosome

Figure 4.35 Steps Required for Protein Synthesis.

Required Structures

DNA is the major structure required in transcription. DNA serves as the template to form an RNA molecule that is complementary to the sequence of nucleotides in the gene. RNA (as described in section 2.7d) is a nucleic acid composed of repeating ribonucleotide monomers. Each **ribonucleotide** has a five-carbon sugar ribose, a phosphate, and one of four nitrogenous bases that include adenine (A), cytosine (C), guanine (G), and uracil (U). Unlike DNA, RNA is only a single strand of nucleotides (see figure 2.21).

Forming RNA during transcription requires both large numbers of ribonucleotides and the enzyme RNA polymerase. These structures are located within the nucleoplasm in the nucleus. **RNA polymerase** assembles the ribonucleotides by complementary base pairing ribonucleotides with DNA as follows:

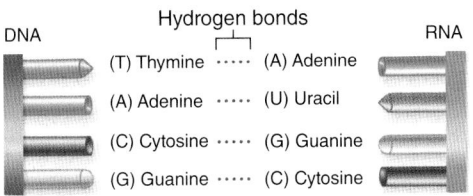

Although other enzymes and many regulatory factors are involved in the process, we limit our discussion to the basic process of transcription that involves DNA, ribonucleotides, and the enzyme RNA polymerase.

Process of Transcription

Three functional types of RNA are produced during transcription: messenger RNA (mRNA), transfer RNA (tRNA), and ribosomal (rRNA). The general process of transcription includes three major events: *initiation, elongation,* and *termination* (**figure 4.36**).

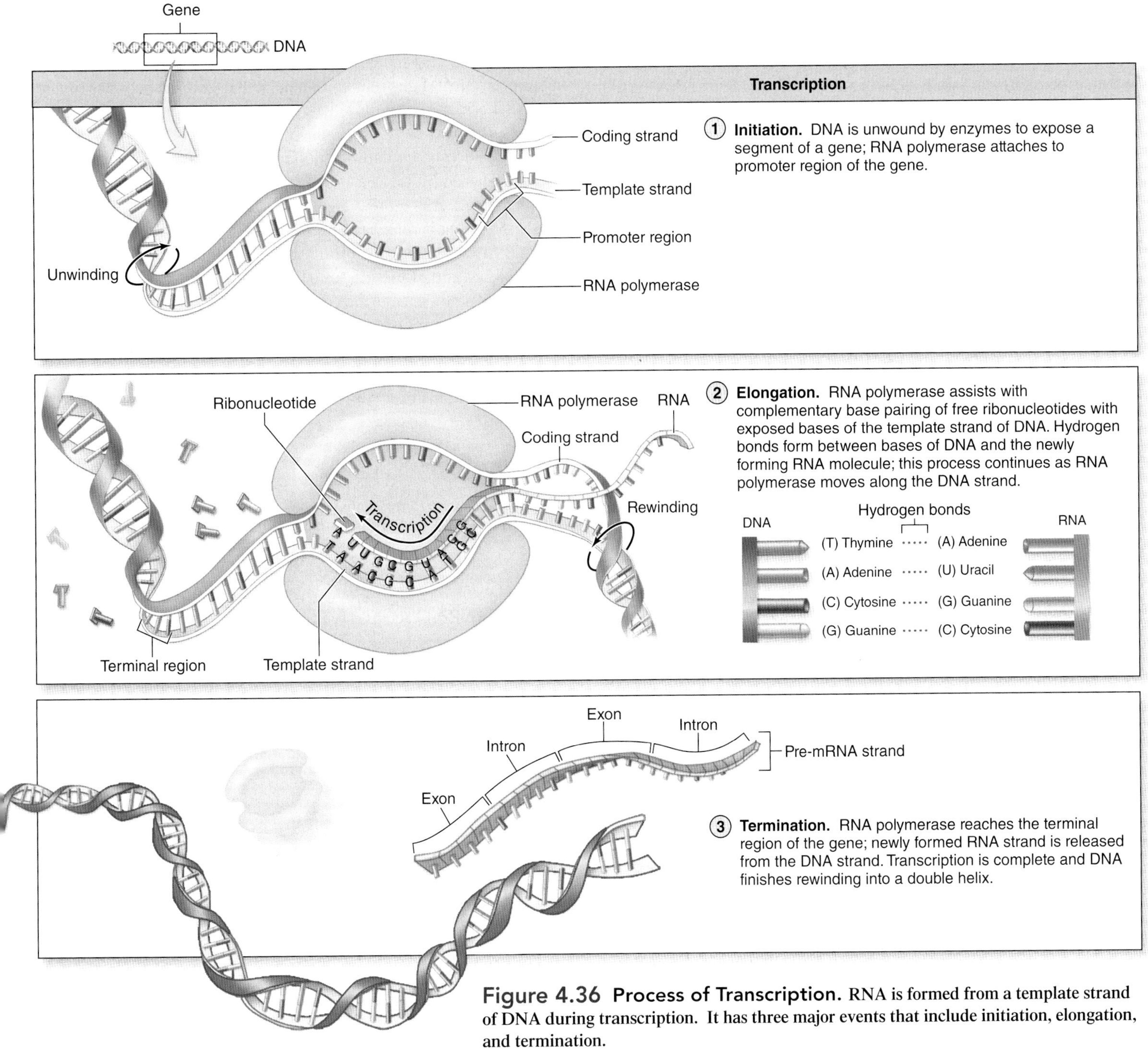

Gene

DNA

Transcription

Coding strand

Template strand

Promoter region

RNA polymerase

Unwinding

(1) **Initiation.** DNA is unwound by enzymes to expose a segment of a gene; RNA polymerase attaches to promoter region of the gene.

Ribonucleotide

RNA polymerase   RNA

Coding strand

Transcription

AUUGGGUAGG
TAAGGC ATC

Rewinding

Terminal region

Template strand

(2) **Elongation.** RNA polymerase assists with complementary base pairing of free ribonucleotides with exposed bases of the template strand of DNA. Hydrogen bonds form between bases of DNA and the newly forming RNA molecule; this process continues as RNA polymerase moves along the DNA strand.

Hydrogen bonds

DNA   RNA

(T) Thymine ····· (A) Adenine
(A) Adenine ····· (U) Uracil
(C) Cytosine ····· (G) Guanine
(G) Guanine ····· (C) Cytosine

Exon

Intron

Intron

Exon

Pre-mRNA strand

(3) **Termination.** RNA polymerase reaches the terminal region of the gene; newly formed RNA strand is released from the DNA strand. Transcription is complete and DNA finishes rewinding into a double helix.

Figure 4.36 Process of Transcription. RNA is formed from a template strand of DNA during transcription. It has three major events that include initiation, elongation, and termination.

INTEGRATE

LEARNING STRATEGY

The process of transcription may be compared to writing down a recipe. The DNA is the recipe book, and a gene is a specific recipe. The recipe book is opened (initiation), the specific recipe is written down (elongation), and the recipe book is closed (termination).

Initiation DNA is usually coiled into a double helix, so it first must be unwound in the region of the gene so its information is available for "reading" (copying). Specific enzymes help to partially unwind the DNA and make it accessible to RNA polymerase, the enzyme that catalyzes the synthesis of an **mRNA (messenger RNA)** molecule. After the partial unwinding of DNA, RNA polymerase attaches to the DNA strand and moves along its length until it comes in contact with the promoter region (the "start" region) associated with a gene.

The gene is marked for transcription by a number of regulatory factors that reflect a requirement for the specific protein coded for by that gene. The promoter serves as the start point for gene transcription. When identification of the gene and binding of the appropriate factors occurs, the hydrogen bonds between the two strands of DNA break, thus allowing expansion or "bubbling out" in that region. That permits the nitrogenous bases in the region to be exposed. Because RNA is a single-stranded molecule, only one strand of DNA is copied. This DNA strand is referred to as the **template strand.** The other DNA strand is the **coding strand,** and is not copied.

Elongation Free ribonucleotides are base-paired in a complementary way with the exposed bases in the DNA template strand during the elongation process. The enzyme RNA polymerase assists in this base pairing. Base pairing involves the formation of hydrogen bonds between a base of a ribonucleotide and its complementary base of the DNA strand. For example, if the base sequence of a DNA template strand is TTAGC-TAGC, then the base sequence of the newly formed RNA strand will be AAUCGAUCG. (Recall that RNA contains uracil instead of thymine.) A phosphodiester bond is formed between each of the ribonucleotides to form the RNA polymer. RNA polymerase continues to move along the length of DNA until the entire gene has been transcribed. As a result, a new mRNA is formed from the "information" in the gene.

Termination When the terminal region is reached at the end of the gene, RNA polymerase is released from the DNA as hydrogen bonds are broken between the DNA strand and newly formed mRNA strand. The DNA rewinds into a double helix. The newly formed mRNA is a "recipe" copied from DNA for synthesizing a specific protein (e.g., insulin).

Modifications to mRNA

Several significant changes are made to the newly formed mRNA before it leaves the nucleus. The initially synthesized strand of mRNA is more specifically called the **pre-mRNA** (or primary transcript). The changes result in the formation of a mature **mRNA,** which is then used as the recipe to make the protein.

Splicing Pre-mRNA includes **introns,** which are noncoding regions. These introns are removed (most are degraded, but some have specialized functions in regulating gene expression). **Exons** are the nucleotide sequences in the mRNA that were transcribed from DNA and are subsequently **spliced** together. This process is catalyzed by a ribonucleoprotein molecule complex (composed of RNA and protein) called

INTEGRATE

LEARNING STRATEGY

You may find it helpful to think of the splicing of pre-mRNA like the splicing of film that occurs with producing a movie. Unwanted segments of the film are cut out, and the remaining segments are spliced together to produce the final version of the movie. Interestingly, the same pre-mRNA can be spliced in various ways to produce different mature mRNA (much like a film can be spliced in different ways to create a different story).

a **spliceosome.** The pattern of splicing varies depending on several factors, including the organism's stage of development and the cell type. Thus, splicing provides a means of producing a larger number of proteins from the available DNA.

Other Changes Further modifications to form the mature mRNA include capping and addition of a polyA tail. **Capping** involves the unique bonding of a ribonucleotide containing guanine to the lead end of the mRNA. This increases the stability of an mRNA strand helping to prevent its digestion by nucleic acid digesting enzymes (nucleases) that are present within the cytoplasm. The **polyA tail** addition involves removal of terminal segments of the mRNA and placing numerous adenine-containing ribonucleotides at the tail end of the mRNA. The addition of a polyA tail, like the splicing, provides a means of producing more than one mature mRNA transcript, because the segment that is removed and the addition of the polyA tail can be done at different sites. One function of the polyA tail is to serve as a measure of age of the mRNA. These nucleotides are subsequently removed over time, and the tail is shortened. When only a certain amount of the tail remains, nuclease enzymes will destroy the mRNA.

The newly formed mature mRNA exits the nucleus following its modification (figure 4.35). It leaves through nuclear pores to enter the cytoplasm as it makes it way to a ribosome for translation (the second event of protein synthesis).

WHAT DID YOU LEARN?

26 What are the three major structures required for transcription? Explain where and how transcription occurs.

4.8b Translation: Synthesizing Protein

LEARNING OBJECTIVES

3. List the required structures for translation.

4. Name the three functional forms of RNA, explain what is meant by codon, and identify three types of codon sequences.

5. Describe the three steps of translation.

INTEGRATE

LEARNING STRATEGY

The required components of translation may be compared to a chef cooking up a masterpiece. The ribosome is the kitchen, the mRNA is the recipe, the tRNAs are the chef's assistants, the amino acids are the ingredients, and the protein is the completed dish. The tRNA assistants bring the amino acid ingredients to the kitchen, as directed by the mRNA recipe, to produce a protein masterpiece.

Translation is the synthesis of a new protein and takes place at ribosomes within the cytosol. The mRNA is threaded through a ribosome, and the information it contains is "read." The code in the nucleotide sequence of mRNA is translated, meaning that it is converted into the language of amino acids to produce newly formed strands of protein.

Required Structures

Translation requires ribosomes, mRNA, tRNA, and large numbers of free amino acids. Protein is the product formed.

Ribosomes were introduced earlier in the section on non-membrane-bound organelles (see figure 4.27). These globular organelles are composed of a large subunit and a small subunit synthesized by the nucleolus. Each subunit is considered to be a ribonucleoprotein because each is assembled from one to two RNA molecules and various proteins. The large subunit has three associated grooves or spaces: (1) the A (aminoacyl) site, which is the place where new amino acids are added; (2) the P (peptidyl) site that holds the newly forming polypeptide (protein); and (3) the E (exit) site for the tRNA that is exiting the ribosome.

Three functional types of RNA are required for protein synthesis: ribosomal RNA, messenger RNA, and transfer RNA (**figure 4.37**). **Ribosomal RNA (rRNA)** is so named because it is the specific type of RNA forming ribosomes. Research has shown that rRNAs within the ribosomal structure serve as the catalysts during assembly of amino acids into a protein molecule.

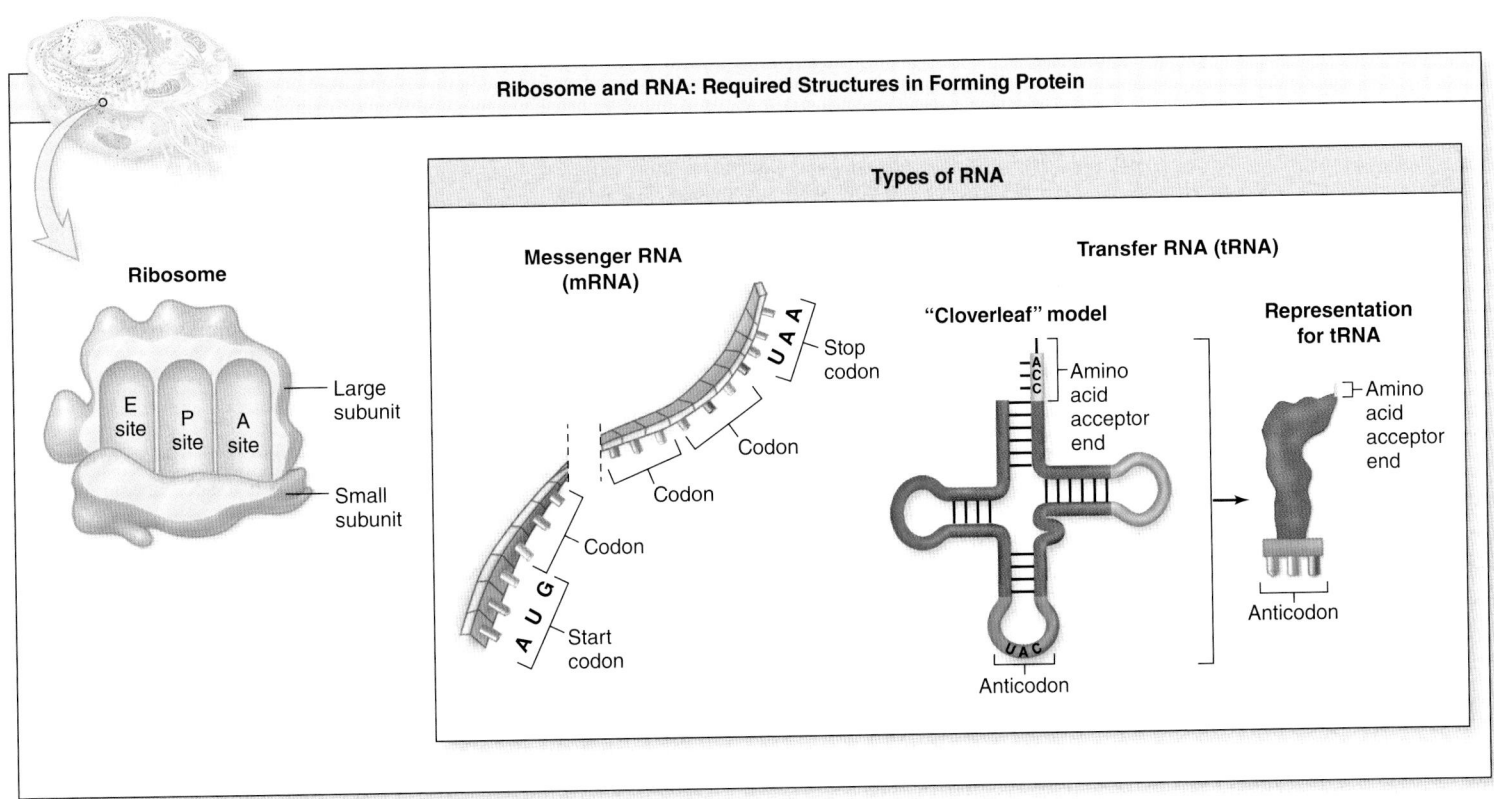

(a)

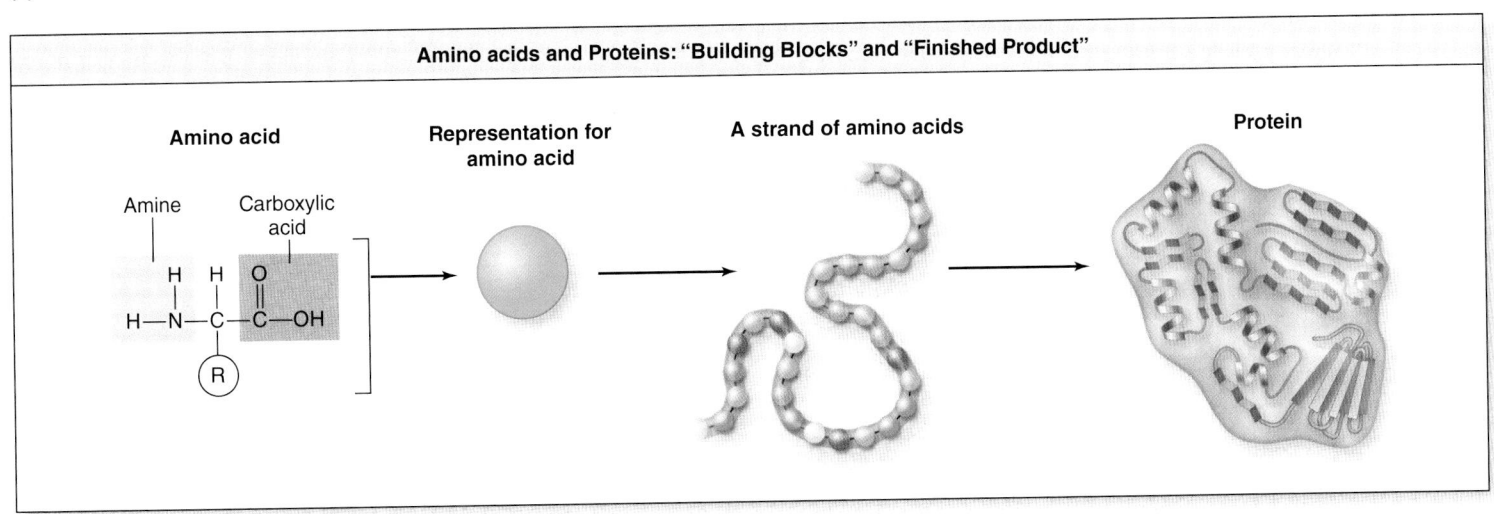

(b)

Figure 4.37 Required Structures for Translation. The process of translation uses the information in the mRNA to direct protein synthesis. (*a*) Translation occurs at ribosomes (composed of protein and rRNA) and requires both messenger RNA and transfer RNA. (*b*) Amino acids are the building blocks used to produce the newly formed protein molecule.

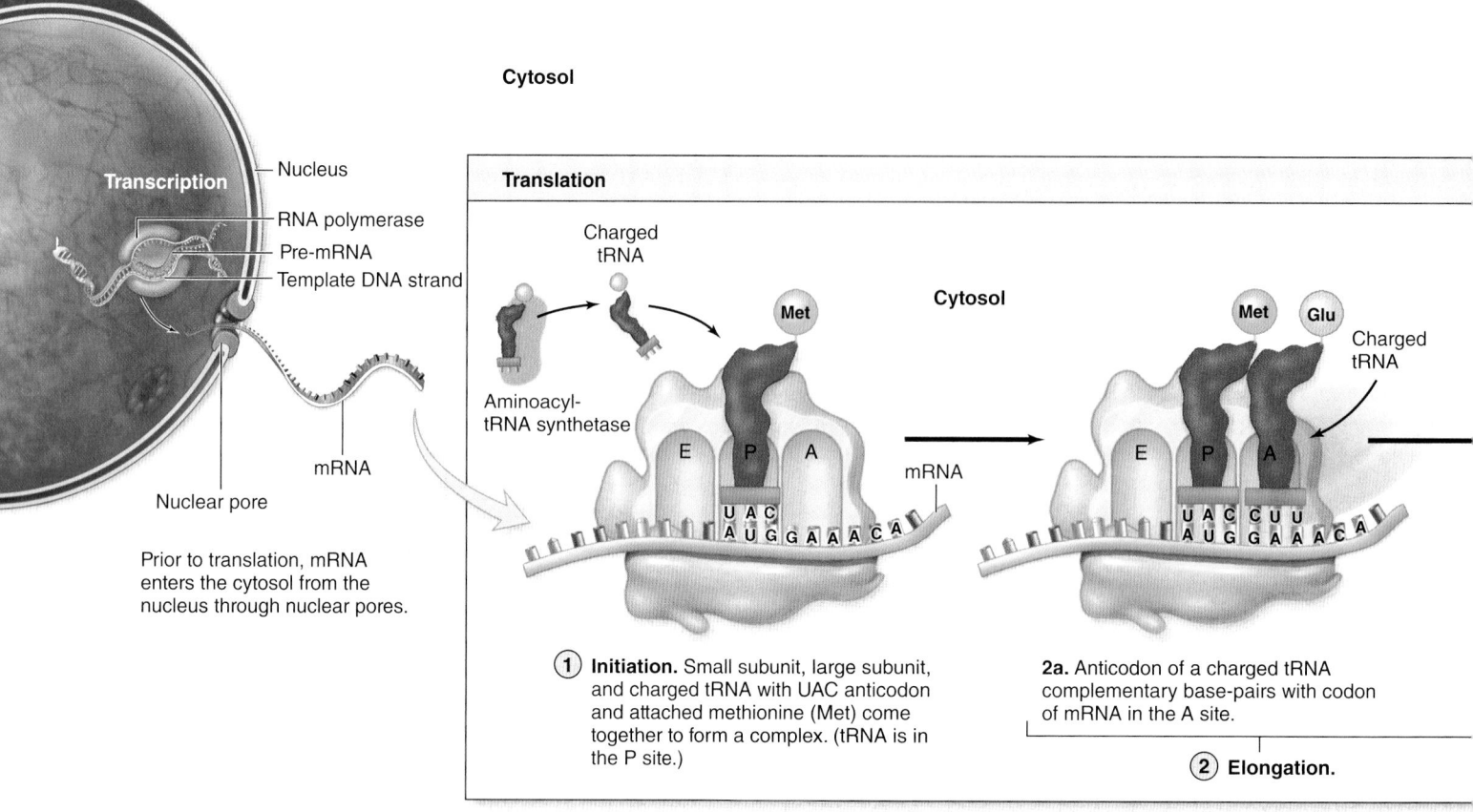

Figure 4.38 Process of Translation. Protein is synthesized at ribosomes by translation of the mRNA. It is directed by mRNA and facilitated by tRNA. The three major events include initiation, elongation, and termination. **AP|R**

Messenger RNA is the molecule that was transcribed from a gene. It carries the "instructions" for synthesizing a protein. Messenger RNA is a linear sequence of nucleotides varying in length, depending upon the size of the protein to be made. The mRNA is read three nucleotide bases at a time. Each three-base unit is called a **codon.** A molecule of mRNA contains the following three categories of codons:

- A **start codon** that always contains the three bases AUG; this is the signal that indicates where protein synthesis begins.
- The consecutive codons following the start codon and before the stop codon serve to direct the assembly of amino acids into a newly synthesized protein.
- The **stop codon** follows the codons used to assemble the new protein, and it is always one of these three sequences of bases: UAA, UAG, or UGA. Collectively they serve as the point where the reading of mRNA ends.

Transfer RNA (tRNA) is the third functional type of RNA. It serves as an "adapter" to bring a specific amino acid to a specific mRNA codon. Transfer RNAs are typically composed of between 70 and 100 nucleotides: When shown in simplified form, it is shaped like a cloverleaf. A tRNA molecule has two significant regions: the amino acid acceptor and the anticodon. The amino acid acceptor provides an attachment site for a specific amino acid. The **anticodon** is a specific sequence of three adjacent nucleotides for each type of tRNA. The anticodon serves two primary functions. Its sequence (e.g., UAC) determines the specific amino acid (e.g., methionine) to which a given tRNA attaches. The process of the tRNA binding its designated amino acid is catalyzed by a specific enzyme called **aminoacyl-tRNA synthetase.** There are 20 aminoacyl-tRNA synthetase enzymes (because there are 20 different amino acids; see figure 2.25). Prior to translation, an amino acid is attached to the correct tRNA by its own specific aminoacyl-tRNA synthetase. The tRNA is now called

a *charged* tRNA after the binding of the amino acid has occurred. The second function of the anticodon of each tRNA is to serve as the "adapter site" for binding a tRNA to its complementary codon of a mRNA.

As noted, there are 20 different amino acids normally found in the proteins of living organisms. We discussed in section 2.8 that the properties of the R groups form the basis for organizing and grouping amino acids. Thus, to synthesize a new protein, which may contain hundreds to thousands of individual amino acids, the necessary amounts of the different amino acids must be available in the cytosol in proximity to ribosomes.

Process of Translation

Translation of the mRNA sequence into a functional protein also involves three major events, as those identified for transcription: *initiation, elongation,* and *termination* (**figure 4.38**).

Initiation   Initiation (figure 4.38, step 1) requires the formation of a complex composed of a small subunit of a ribosome, a large subunit of a ribosome, the newly formed mRNA, and a tRNA. The small subunit

---

INTEGRATE

CONCEPT CONNECTION

Chemical reactions are generally catalyzed by enzymes that are globular proteins (see section 3.3). Ribosomes are composed of both protein and rRNA. It is the rRNA of the ribosome—and not the protein—that catalyzes the synthesis of protein. For this reason, the rRNA of the ribosome is called a **ribozyme,** a catalytic RNA molecule. The protein of the ribosome in this case primarily serves a structural role to hold the rRNA molecules in the correct orientation.

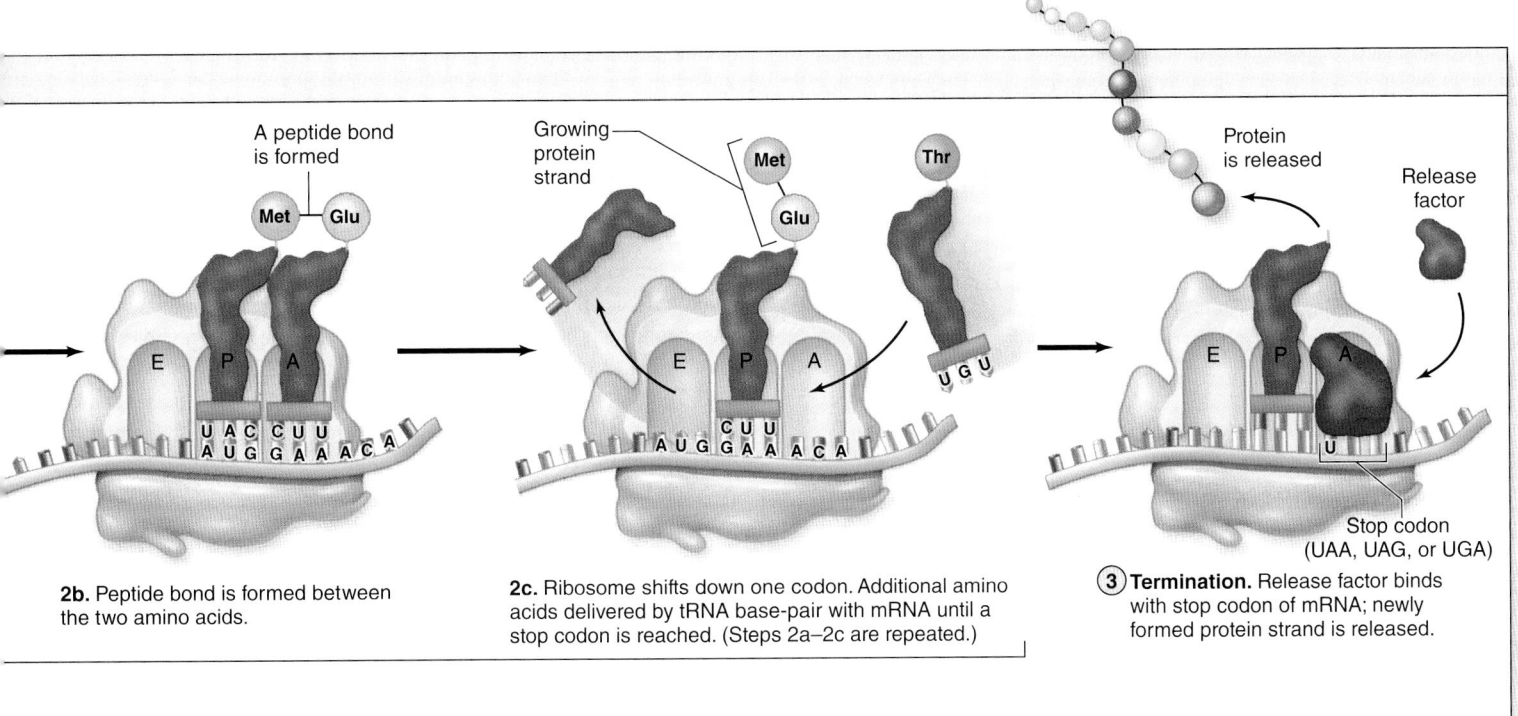

2b. Peptide bond is formed between the two amino acids.

2c. Ribosome shifts down one codon. Additional amino acids delivered by tRNA base-pair with mRNA until a stop codon is reached. (Steps 2a–2c are repeated.)

③ Termination. Release factor binds with stop codon of mRNA; newly formed protein strand is released.

of a ribosome moves along the mRNA until it reaches the start codon (AUG). A charged tRNA that possesses the anticodon UAC then base-pairs to the start codon AUG in the mRNA. The amino acid methionine is bound to this tRNA. Methionine is always the first amino acid to be used in the synthesis of a protein, but later it may be removed as protein synthesis continues and the protein matures. The large subunit then joins with the small subunit of the ribosome. The start codon now occupies the P site of the ribosome.

Elongation   Elongation (figure 4.38, step 2) involves the orderly delivery of all subsequent amino acids by specific tRNAs to form the protein. Three primary steps occur, as follows:

2a. A charged tRNA with a complementary anticodon base-pairs with the codon of the mRNA in the A site.

2b. A peptide bond is formed between the amino acid in the P site and the amino acid in the A site (as the bond of amino acid to tRNA in P site breaks).

2c. The ribosome then moves three nucleotide positions (the equivalent of a codon) "downstream" (beyond) the start codon on the mRNA. There is, of course, an accompanying change in position of the tRNAs. The tRNA that was in the A site is now in the P site, and the A site is again open. (The E site is the exit site from which the now uncharged tRNA is released.)

These processes (steps 2a–2c) repeat until the entire mRNA sequence has been translated. The product is a protein composed of a linear strand of amino acids.

Termination   Termination (figure 4.38, step 3) is when a stop codon (UAA, UAG, or UGA) enters the A site to end translation. A release factor enters the A site at this point instead of a charged tRNA. When the ribosome hits the factor bound to the mRNA stop codon, the two subunits of the ribosome are separated from the mRNA and the newly synthesized protein is released.

Note that a single mRNA can be read by more than one ribosome simultaneously, and thus many copies of that protein can be made

quickly. An mRNA with many ribosomes attached along its length is called a **polyribosome.**

WHAT DO YOU THINK?

3 What might be the potential consequences to protein structure and ability to function if DNA is mutated in a specific gene?

WHAT DID YOU LEARN?

27 What is a codon and an anticodon?

28 How is mRNA attached to ribosomes and translated into the language of protein?

4.8c DNA as the Control Center of a Cell

LEARNING OBJECTIVE

6. Explain why DNA is considered the cell's control center.

Recall from section 2.7e that the human body is estimated to contain between 50,000 to 100,000 different proteins. These proteins serve a vast array of functions that include catalyzing reactions, defense, transport, support, movement, regulation, and storage. DNA is responsible for directing the synthesis of the proteins that carry out body functions.

Additionally, DNA is indirectly responsible for other metabolic changes that occur within a cell—including the synthesis of steroids and other lipids, and the enzymatic pathway of glucose oxidation—because DNA controls synthesis of the enzymes that are responsible for catalyzing both the decomposition and synthesis of chemical structures. All of these roles explain why DNA is considered the control center of the cell and sometimes is referred to as the "boss" of the cell.

WHAT DID YOU LEARN?

29 The genetic code of DNA is the specific instructions to make what biomolecule?

4.9 Cell Division

Cells divide their nucleus in two ways, depending upon the cell type. **Mitosis** (mī-tō′sis; *mitos* = thread) occurs in somatic cells, and **meiosis** occurs in sex cells. **Somatic cells** are all of the cells in the body other than the sex cells, and **sex cells** form either sperm or oocytes. Meiosis is discussed in detail in section 28.2. Here we define and describe somatic cell division that involves mitosis.

Cell division occurs when one cell divides to produce two cells. It is essential to produce, replace, and maintain the trillions of cells that form a normally functioning human body. It is a necessary process for development, tissue growth, replacement of old or dying cells, and tissue repair required by loss of tissue from trauma or disease.

4.9a Cellular Structures

 LEARNING OBJECTIVES

1. Explain the structure and function of centrioles in cell division.
2. Describe the structural difference between chromatin and chromosomes, and note when each is present in a cell.

We saw earlier that a centrosome is a structure containing a pair of perpendicularly oriented cylindrical centrioles located in close proximity to the nucleus (see figure 4.29). The centrosome organizes the microtubules that facilitate movement of chromosomes during cell division. In addition, the nucleus of human somatic cells typically has 46 separate DNA molecules that are organized as either loosely coiled chromatin or tightly coiled chromosomes (see figure 4.34).

WHAT DID YOU LEARN?

30 How is chromatin distinguished from a chromosome?

4.9b The Cell Cycle

LEARNING OBJECTIVES

3. Summarize the phases of the cell cycle and the activities that occur in each phase.
4. Name and explain the four main stages of mitosis.
5. Explain the function of cytokinesis.

The **cell cycle** depicts the steps in the replication of a somatic cell. It consists of all changes the cell undergoes as it divides into two identical cells called **daughter cells.** There are two major phases in the cell cycle: **interphase** and the **mitotic (M) phase.** The phases and events of the somatic cell cycle are described here and summarized in **figure 4.39.** On average, the cell is in interphase for approximately 23 hours, and going through the mitotic phase for about 1 hour.

INTEGRATE

CONCEPT CONNECTION

During the initial stages of development, all cells have the ability to reproduce. However, during development as cells change into their specific types (a process called *differentiation*), some cells retain the ability to replicate, whereas others decrease or lose the ability. Damaged or diseased cells are not replaced, and results in a lack of function. For example, epithelial skin cells are replaced frequently. If you cut your finger, the cells go through cell division replacing the damaged cells of the skin (see section 6.4). However, other cells, such as cardiac muscle cells, undergo cell division infrequently or not at all. Thus, if an individual has a heart attack, and cardiac muscle cells die from lack of oxygen, these cells are not replaced or replaced to a limited extent (see section 19.4).

Interphase

Interphase is the time the cell prepares for division. This phase is distinctive in a light microscope because the DNA within the nucleus remains in the form of loosely coiled chromatin.

Interphase has three distinct phases: G_1, S, and G_2. During the **G_1 phase** (called the *first gap stage*) of the cell cycle cells grow and produce new organelles, and other structures needed for DNA replication. Replication of the centrioles to produce two centriole pairs is also initiated in this phase.

During the **S phase** (called *synthesis*) the 46 double helix strands of DNA are replicated. Forming DNA requires large numbers of deoxyribonucleotides and the enzyme **DNA polymerase.** All of these components are located in the nucleoplasm within the nucleus.

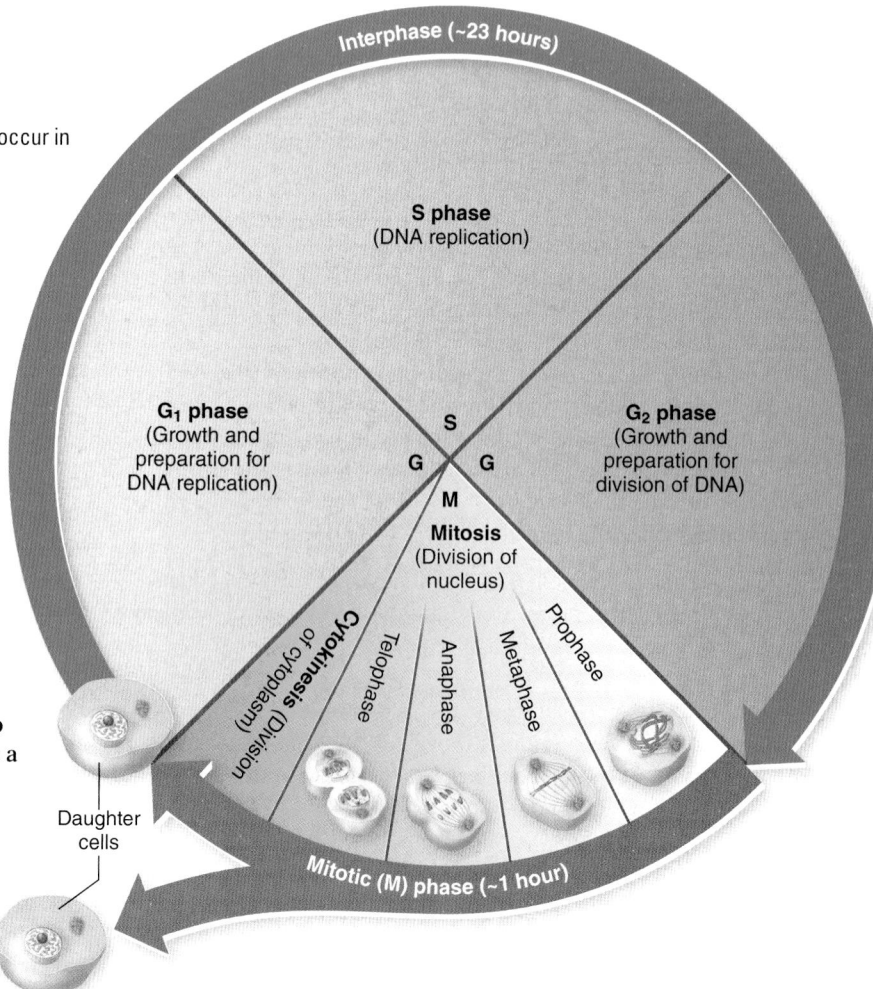

Figure 4.39 The Cell Cycle. The cell cycle includes two basic phases: interphase and the mitotic (M) phase. Interphase is a growth period that is subdivided into G_1, S, and G_2. Synthesis of DNA occurs during the S phase of interphase. The mitotic phase includes mitosis, the process of nuclear division, and cytokinesis, the division of the cytoplasm. Note that this figure does not accurately reflect the relative time a cell would spend in each phase. **AP|R**

The following steps in DNA replication **(figure 4.40)** involve unwinding, breaking, assembly, and restoration:

① **Unwinding of DNA molecule.** The spiral, complementary DNA strands are unwound from each other by specific enzymes.

② **Breaking the parent strands apart.** The hydrogen bonds holding the complementary bases together in the DNA strands are broken. Once the portions of strands are separated, binding proteins (not shown) ensure that the strands remain separated.

③ **Assembly of new DNA strands.** Both strands of DNA are read as templates by DNA polymerase enzymes that move along both parental strands (one called the *leading strand* and the other the *lagging strand* because of how transcription takes place on each). DNA polymerase assembles new strands of DNA as complementary deoxyribonucleotides are paired. For example, if the base sequence of a small portion of a DNA strand is TTAGCTAGC, then the base sequence of the newly formed complementary DNA strand assembled by DNA polymerase would be AATCGATCG. Complementary base pairs are held together by hydrogen bonds. The bond between nucleotides in the DNA polymer is a phosphodiester bond.

④ **Restoration of DNA double helix.** The DNA double strands are returned to their coiled, helix structure.

The process continues until the entire length of both strands of DNA are replicated. The replicated DNA strands, now called **sister chromatids,** remain attached at a region called a **centromere** (sen′trō-mēr; *kentron* = center, *meros* = part). The sister chromatids are separated at the centromeres during mitosis; following their separation, each is called a **chromosome.**

WHAT DO YOU THINK?

④ Describe the difference in DNA replication and transcription in terms of (a) the type of nucleic acid formed, and (b) the amount of DNA copied.

The last part of interphase, called the **G₂ phase** (or the *second gap phase*), is brief (see figure 4.39). During this phase, centriole replication is completed (having produced two centriole pairs present within the cell) and enzymes and other structures needed for cell division are synthesized.

INTEGRATE

LEARNING STRATEGY

To prevent confusion between the process of DNA replication and transcription (formation of RNA from DNA), remember that transcription of mRNA is analogous to copying *one* recipe from a cookbook. The recipe is written as RNA. In contrast, DNA replication is analogous to printing a replica of an *entire* cookbook. The cookbook is printed as DNA.

Figure 4.40 DNA Replication. The two helical strands of the double-stranded parent DNA molecule are separated and unwound to make both parental strands available as templates for construction of new DNA strands. All 46 DNA strands are copied during DNA replication. AP|R

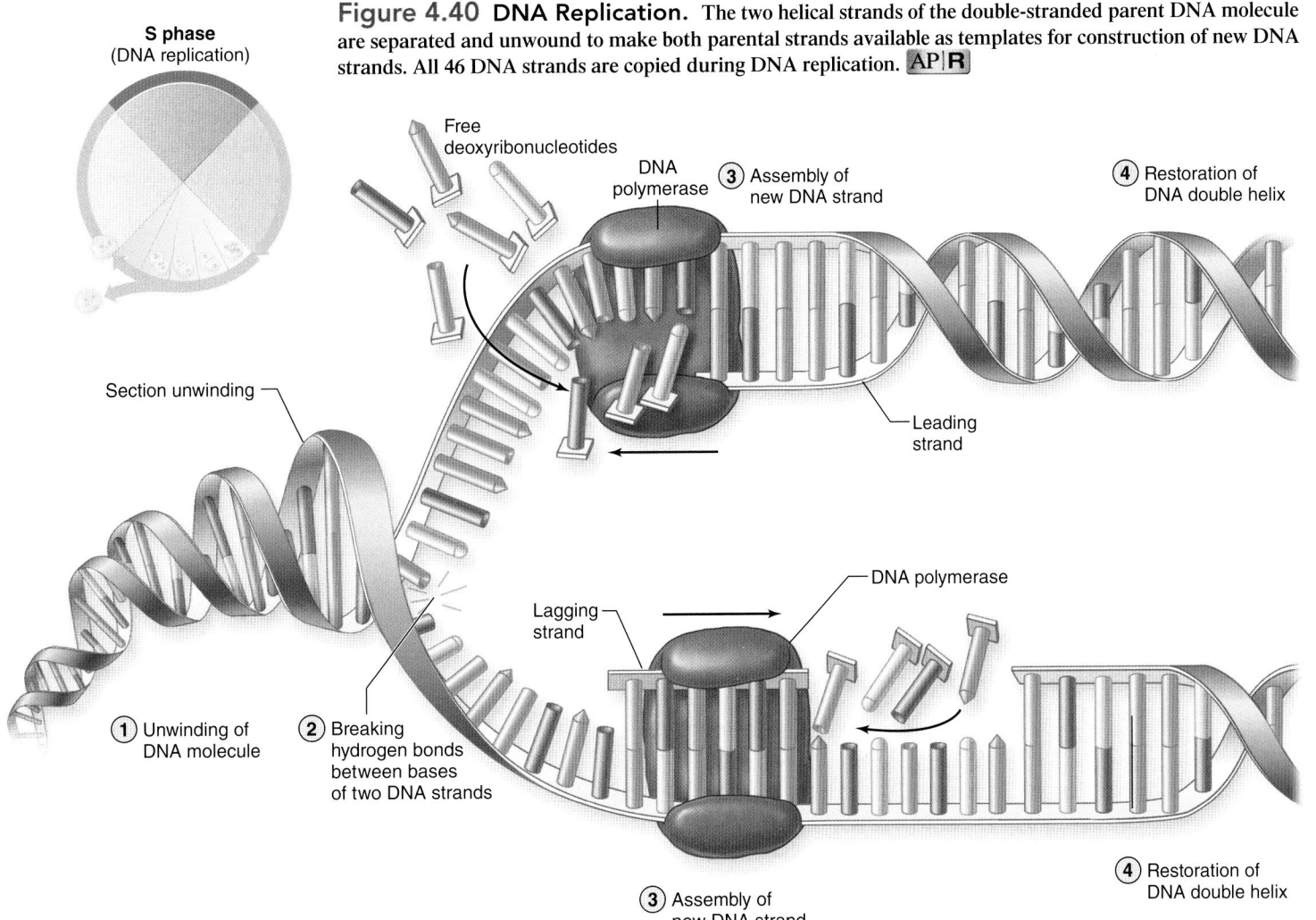

S phase
(DNA replication)

Free deoxyribonucleotides

DNA polymerase

③ Assembly of new DNA strand

④ Restoration of DNA double helix

Section unwinding

Leading strand

① Unwinding of DNA molecule

② Breaking hydrogen bonds between bases of two DNA strands

Lagging strand

DNA polymerase

③ Assembly of new DNA strand

④ Restoration of DNA double helix

Figure 4.41 Interphase, Mitosis, and Cytokinesis. Drawings and micrographs depict what happens inside a cell during the stages of (*a*) interphase and (*b–e*) the mitotic phase of cell division. Cytokinesis overlaps with mitosis and generally begins during anaphase.

Interphase (G₁, S, G₂)	Mitotic Phase (Mitosis and Cytokinesis)

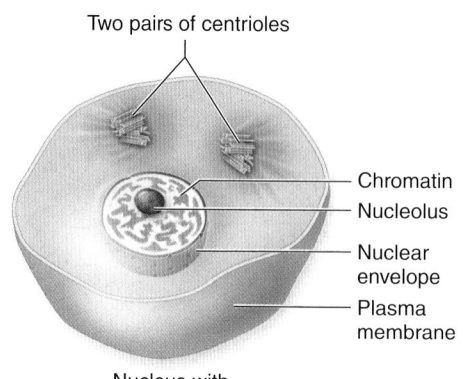

Two pairs of centrioles

Chromatin
Nucleolus
Nuclear envelope
Plasma membrane

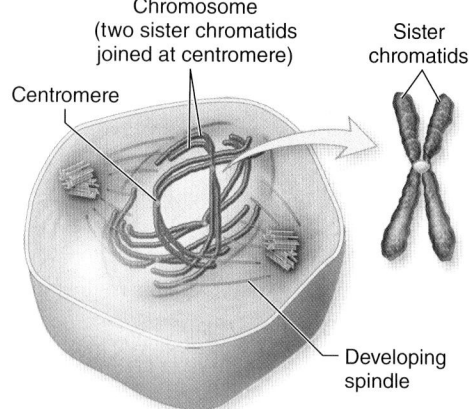

Chromosome (two sister chromatids joined at centromere)
Sister chromatids
Centromere
Developing spindle

Nucleus with chromatin

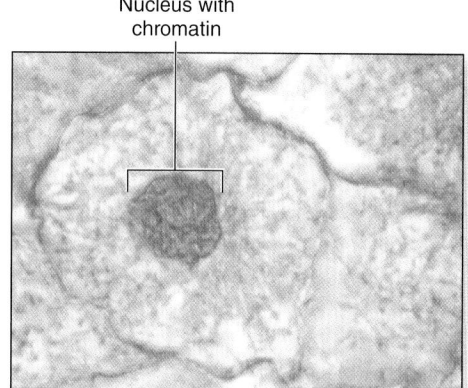

Nucleus with dispersed chromosomes

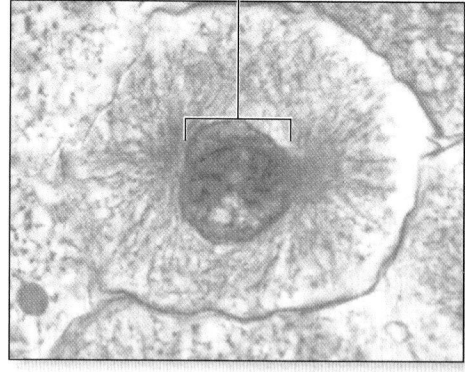

(a) Interphase

Synthesis of cellular components needed for cell division, including synthesis of DNA which occurs during the S phase and duplication of centrioles, which occurs throughout interphase.

(b) Prophase

Chromosomes appear due to coiling of chromatin.

Nucleolus breaks down.

Spindle fibers begin to form from centrioles.

Centrioles move toward opposing cell poles.

Nuclear envelope breaks down at the end of this stage.

M Phase (Mitotic Phase)

Following interphase, cells enter the M (mitotic) phase. Two distinct events occur in this phase to produce two new cells: Mitosis, which is division of the nucleus, and cytokinesis, which is the division of the cytoplasm. **Mitosis** begins first with cytokinesis starting and overlapping with later stages of mitosis.

Four consecutive phases take place during mitosis: prophase, metaphase, anaphase, and telophase, which can be remembered with the acronym P-MAT. Each phase merges smoothly into the next in a nonstop process. The events of each stage are summarized in **figure 4.41**.

Prophase is the first stage of mitosis (step b). Chromatin becomes supercoiled into the chromosomes that are more maneuverable and are less likely to become tangled during cell division. The DNA and protein within the chromatin coil, wrap, and twist, forming the chromosomes. Chromosomes are composed of the two sister chromatids that resemble relatively short, thick rods, and become noticeable with a light microscope during the prophase stage as dark-staining structures within the nucleus.

In addition, the nucleolus breaks down and disappears. Elongated microtubules called *spindle fibers* begin to grow from the centrioles. The two centriole pairs are pushed apart by the elongating microtubules composing the spindle fibers; eventually, the centriole pairs come to lie at opposite poles (ends) of the cell. The end of prophase is marked by the dissolution (disassembly) of the nuclear envelope. This permits the chromosomes to be moved by spindle fibers through the cytoplasm during the next stages of mitosis.

Metaphase is the second stage of mitosis (step c), during which the chromosomes are aligned along an imaginary line in the middle of the cell (region called the equatorial plate). This alignment occurs through the growth of spindle fibers from each centriole toward the

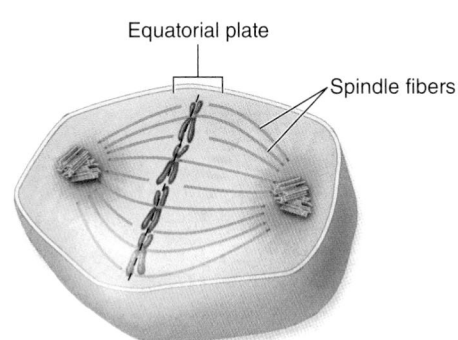

Equatorial plate — Spindle fibers

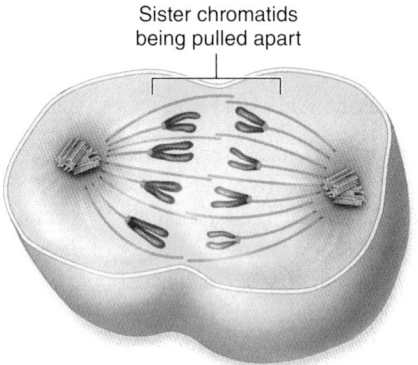

Sister chromatids being pulled apart

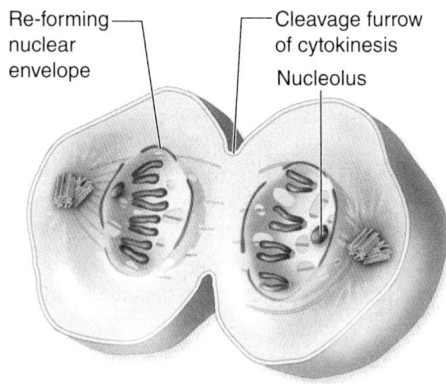

Re-forming nuclear envelope — Cleavage furrow of cytokinesis — Nucleolus

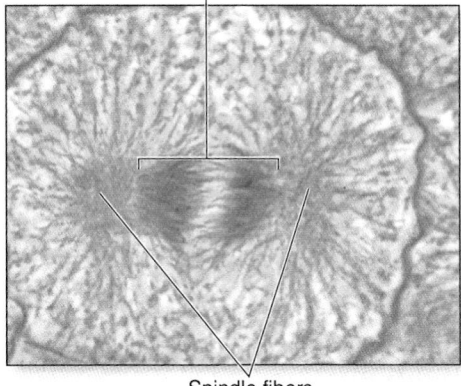

Chromosomes aligned on equatorial plate — Spindle fibers

Sister chromatids being pulled apart

Spindle fibers

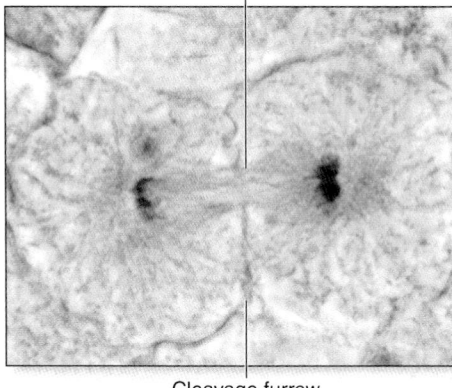

Cytokinesis occurring

Cleavage furrow

(c) Metaphase

Spindle fibers attach to the centromeres of the chromosomes extending from the centrioles.

Chromosomes are aligned at the equatorial plate of the cell by spindle fibers.

(d) Anaphase

Sister chromatids are separated by spindle fibers and moved toward opposite ends of the cell.

During this process centromeres that held sister chromatids together separate; each sister chromatid is now a chromosome with its own centromere.

Cytokinesis begins.

(e) Telophase

Chromosomes uncoil to form chromatin.

A nucleolus reforms within each nucleus.

Spindle fibers break up and disappear.

New nuclear envelope forms around each set of chromosomes.

Cytokinesis continues as cleavage furrow deepens.

chromosomes. Some fibers attach to the centromere of each chromosome, directing their movement to the equatorial plate.

Anaphase, which is the third stage of mitosis, initiates as the spindle fibers cause the sister chromatids to be moved apart toward the cell's poles; the centromere leads the way, and its "arms" trail behind (step d). Each chromatid is now a chromosome composed of one DNA double helix with its own centromere.

Telophase begins with the arrival of a group of chromosomes at each cell pole. Essentially, the processes of prophase are reversed in telophase. The chromosomes begin to uncoil and return to the form of dispersed threads of chromatin, each new nucleus forms a nucleolus, the mitotic spindle breaks up and disappears, and a new nuclear envelope forms around each set of chromosomes. Telophase signals the end of nuclear division.

Cytokinesis   Cytokinesis (sī′tō-ki-nē′sis; *kinesis* = movement) is the other major event in the mitotic phase, and it is the division of the cytoplasm between the two newly forming cells. This phase usually begins early and overlaps with anaphase and telophase of mitosis. A ring of microfilament proteins on the inner surface of the cell's plasma membrane contracts at the cell's equator. It pinches the mother cell into two separate cells in a manner analogous to the tightening of a belt. The resulting **cleavage furrow** that appears indicates where the cytoplasm is dividing. Two new daughter cells are formed and cell division is complete.

WHAT DID YOU LEARN?

 Describe the process of DNA replication that occurs during the S phase of interphase.

 What are the events that occur during the mitotic phase (mitosis and cytokinesis)? Explain each.

INTEGRATE

LEARNING STRATEGY

These tips should help you remember the major events of each phase of mitosis:

- The **p** in **p**rophase stands for the **puffy** ball of chromosomes that forms within the nucleus.
- The **m** in **m**etaphase stands for **middle:** During this phase, the chromosomes align along the **middle** of the cell.
- The **a** in **a**naphase stands for **apart:** During this phase, the sister chromatids are pulled **apart**.
- The **t** in **t**elophase stands for **two:** During this phase, **two** new cells become noticeable as a cleavage furrow deepens to divide the cytoplasm.

INTEGRATE

CLINICAL VIEW

Tumors

Normally, many regulatory mechanisms signal a cell when it should divide and when to stop dividing. Tumors arise when cells either proceed through the cell cycle without a start signal or fail to respond to signals that normally stop cell division. The tumor may, due to its size, interfere with the function of the normal surrounding cells. A cancerous tumor is invasive, and cells may enter the blood or lymph and metastasize to other areas of the body and establish secondary tumors. Various types of cancer are discussed in later chapters of this text (e.g., skin cancer in chapter 6, leukemia in chapter 7, brain tumors in chapter 13, lymphomas in chapter 21, lung cancer in chapter 23, prostate cancer, cervical cancer, and breast cancer in chapter 28).

4.10 Cell Aging and Death

LEARNING OBJECTIVES

1. Define apoptosis.
2. List the actions that occur in a cell during apoptosis.

Aging is a normal, continuous process that often exhibits obvious body signs. In contrast, changes within cells at the molecular level due to aging are neither obvious nor well understood. The reduced metabolic functions of normal cells often have wide-ranging effects throughout the body, including a reduced ability of cells to maintain homeostasis. These signs of aging reflect a lower number of normally functioning body cells, and may even suggest abnormal functions of some remaining cells.

Cells affected by aging may exhibit alteration in either the structure or number of specific organelles. For example, if mitochondrial function begins to fail, the cell's ability to synthesize ATP diminishes. Additionally, changes in the distribution and structure of the chromatin and chromosomes within the nucleus may occur. Often, both chromatin and chromosomes clump, shrink, or fragment as a result of repeated divisions.

Essentially, cells die by one of two mechanisms: (1) they are killed by harmful agents or mechanical damage; or (2) they are induced to commit suicide, a process of programmed cell death called **apoptosis** (ap′op-tō′sis; *apo* = off, *ptosis* = a falling).

Apoptosis occurs in orderly, well-defined continuous degradative steps to destroy and remove cellular components and eventually cell remnants. This biochemical mechanism is initiated by ligand-receptor signaling. Upon binding of ligand to a receptor, inactive, self-destructive enzymes within the cytoplasm are turned on and initiate the following actions:

- Destruction of DNA polymerase to prevent the synthesis of new DNA
- Digestion of the DNA into small fragments
- Digestion of the cytoskeleton, thus destroying structural support for organelles and the nucleus—the cell appears to shrink and become rounded and the nuclear shape changes
- Condensation of the cytosol and destruction of organelles—specifically, the mitochondria to deprive the cell of ATP needed for energy-requiring processes
- Formation of small irregular blebs (bubbles) on the plasma membrane surface.
- Initiation of other plasma membrane signals to stimulate destruction of the cell externally by phagocytic cells

Programmed cell death occurs both to promote proper development and to remove harmful cells. For example, the proper development of fingers and toes begins with the formation of a paddlelike structure at the distal end of the developing limb. Programmed cell death removes the cells and tissues between the fingers and toes developing within this paddle structure.

Programmed cell death sometimes destroys harmful cells, reducing potential health threats. Cells of our immune system promote programmed cell death in some virus-infected cells to reduce the further spread of infection (see section 22.3c). Cells that have damage to their DNA often appear to promote events leading to apoptosis, presumably to prevent these cells from causing developmental defects or becoming cancerous. Some cancer therapy treatments lead to apoptosis in certain types of cancer cells.

WHAT DID YOU LEARN?

 33 What are the specific changes that occur to DNA during apoptosis?

CHAPTER SUMMARY

- Cells are the structural and functional units of the body.

- Cells vary in size and shape, but have certain common features and functions.

4.1a How Cells Are Studied   103

- Cells are microscopic and can be studied using a light microscope (LM), transmission electron microscope (TEM), and scanning electron microscope (SEM).

4.1b Cell Size and Shape   103

- Although some cells are round or cubelike, other cells are flat, cylindrical, oval, or quite irregular in shape.

4.1c Common Features and General Functions   104

- The three major structural components of a cell include the nucleus, plasma membrane, and cytoplasm (composed of cytosol, organelles, and perhaps cell inclusions).
- All cells must maintain their integrity and shape, obtain nutrients and form chemical building blocks, dispose of wastes, and if possible, replace cells.

- The plasma membrane is a fluid matrix that has approximately an equal mixture of lipids and proteins by weight.

4.2a Lipid Components   106

- The plasma membrane is composed of a bilayer of phospholipids with embedded cholesterol molecules. Glycolipids are lipids with carbohydrates extending from the outer surface of the cell.

4.2b Membrane Proteins   108

- Plasma membrane proteins are integral proteins that extend through the plasma membrane, whereas peripheral proteins reside on either the internal or external surface of the plasma membrane.
- Functionally, the plasma membrane proteins include several types of transport proteins, receptors, identity markers, enzymes, attachment sites for the cytoskeleton, and cell-adhesion proteins.

- Substances are moved into and out of cells by processes of membrane transport, and are organized into passive processes and active processes. Active processes require cellular energy and passive processes do not.

4.3a Passive Processes: Diffusion   109

- Diffusion is the movement of a solute from an area where it is more concentrated to an area where it is less concentrated.
- Simple diffusion is the unassisted movement of small nonpolar molecules through the phospholipid bilayer.
- Channel-mediated facilitated diffusion is the transport of ions through channels that are either always open (leak channels), or they open and close as a result of a stimulus (gated channels).
- Carrier-mediated facilitated diffusion is the transport of small polar molecules through a carrier that is induced to change shape to move the molecules across the plasma membrane.

4.3b Passive Processes: Osmosis   111

- Osmosis is the passive movement of water across a semipermeable membrane down a water concentration gradient.
- Osmotic pressure is the pressure exerted by the movement of water across a selectively permeable membrane due to a difference in solution concentration; the greater the difference, the greater the osmotic pressure.
- Isotonic, hypotonic, or hypertonic describe the relative concentration of solutions.

4.3c Active Processes   114

- Active processes require the expenditure of cellular energy and include both active transport and vesicular transport.
- The two types of active transport are primary active transport, which obtains its energy directly from ATP; and secondary active transport, which is "powered" by the movement of a second substance (usually sodium ion) down its gradient.
- Vesicular transport occurs through energy-requiring processes that involve a vesicle for transporting large materials or relatively large amounts of a substance across the plasma membrane.
- Exocytosis moves material out of a cell, and endocytosis moves substances into a cell.
- The three types of endocytosis are phagocytosis, pinocytosis, and receptor-mediated endocytosis.

- The plasma membrane functions in establishing and maintaining the resting membrane potential (RMP).

4.4a Introduction   122

- The RMP is the electrical charge difference at the plasma membrane when a cell is at rest; it typically ranges between −50 millivolts (mV) and −100 mV.

4.4b Establishing and maintaining an RMP   122

- K^+ leaks out of a cell through K^+ leak channels and Na^+ leaks into a cell through Na^+ leak channels. This movement of ions is primarily responsible for establishing an RMP. The Na^+/K^+ pumps maintain ion gradients following movement of these ions.

(continued on next page)

4.5 Cell Communication 123	• Cell communication occurs either through direct contact or through binding of ligands released from other cells.
	4.5a Direct Contact Between Cells 123
	• Direct contact is used as a means of cell communication by the cells of the immune system in protecting the body from potentially harmful substances. It is also used during fertilization, development, and cellular repair.
	4.5b Ligand-Receptor Signaling 123
	• Three general types of receptors are distinguished by their different response to the binding of a ligand: they include channel-linked receptors, enzymatic receptors, and G protein–coupled receptors.
4.6 Cellular Structures 124	• Cellular structures include membrane-bound and non-membrane-bound organelles, vesicles, and structures extending from the cell surface.
	4.6a Membrane-Bound Organelles 124
	• Membrane-bound organelles are surrounded by a membrane that separates the organelle's contents from the cytosol so that the specific activities of the organelle can proceed without being disrupted by other cellular activities.
	• The membrane-bound organelles include the endoplasmic reticulum, the Golgi apparatus, lysosomes, peroxisomes, and mitochondria. They are involved in various forms of metabolic processes, including synthesis and degradation processes that occur within a cell.
	4.6b Non-Membrane-Bound Organelles 128
	• Non-membrane-bound organelles are composed of either protein alone or protein and RNA; they include ribosomes, the cytoskeleton, centrosomes with centrioles, and proteasomes.
	4.6c Structures of the Cell's External Surface 131
	• Cilia and flagella are extensions of the plasma membrane supported by microtubules; cilia sweep materials along the cell's outer surface; a flagellum, located only on sperm, moves the sperm through the female reproductive tract.
	• Microvilli are extensions of the plasma membrane supported by microfilaments, which serve to increase cell surface area for more efficient membrane transport.
	4.6d Membrane Junctions 134
	• Membrane junctions include tight junctions, desmosomes, and gap junctions.
4.7 Structure of the Nucleus 135	• The nucleus is the largest structure within a cell; it has a shape that typically mirrors the cell's shape.
	4.7a Nuclear Envelope and Nucleolus 135
	• The nuclear envelope is a double phospholipid bilayer that serves as the boundary between nucleoplasm and the cytoplasm.
	• A cell typically contains one nucleolus within its nucleus. It is a structure responsible for synthesizing the large and small subunits of ribosomes.
	4.7b DNA, Chromatin, and Chromosomes 136
	• DNA is wrapped around histone proteins and packaged as chromatin.
	• Chromatin is supercoiled into chromosomes only when a cell is proceeding through cell division.
	• DNA contains functional units called genes; a gene is a segment of DNA that carries instructions for making a specific protein.
4.8 Function of the Nucleus and Ribosomes 136	• The nucleus and ribosomes are required to synthesize proteins, a process that involves transcription and translation.
	4.8a Transcription: Synthesizing RNA 136
	• RNA is formed from DNA through transcription, a process that occurs in the nucleus and requires DNA, free ribonucleotides, and the enzyme RNA polymerase.
	4.8b Translation: Synthesizing Protein 138
	• Translation occurs in the cytosol and requires ribosomes (composed of protein and rRNA), messenger (mRNA), transfer RNA (tRNA), and large numbers of free amino acids and results in the synthesis of a new protein.
	4.8c DNA as the Control Center of a Cell 141
	• DNA is responsible for directing the synthesis of proteins.
4.9 Cell Division 142	• Mitosis is one of two types of dividing a nucleus during cell division that occurs in cells.
	• Cell division involving mitosis produces two identical cells when one divides, and is a necessary process for development, tissue growth, replacement of old or dying cells, and tissue repair.
	4.9a Cellular Structures 142
	• The major structures required in cell replication are chromatin (chromosomes), centrioles, free deoxyribonucleotides, and the enzyme DNA polymerase.
	4.9b The Cell Cycle 142
	• The cell cycle consists of a series of changes the cell undergoes between its formation and the time it divides into two identical cells, called daughter cells. It is divided into two major phases: interphase and mitotic phase.
4.10 Cell Aging and Death 146	• Cellular changes associated with aging are neither obvious nor well understood.
	• Cell death occurs by harmful agents or mechanical damage, or through induction that leads to suicide, a process called apoptosis.

Do You Know the Basics?

_____ 1. All of the following general functions are carried out by all cells *except*

    a. obtaining nutrients and chemical building blocks.

    b. maintaining integrity of the plasma membrane.

    c. replacing cells through cell division.

    d. disposing of wastes.

_____ 2. The molecule that is responsible for most functions of the plasma membrane is

    a. the phospholipid bilayer.

    b. cholesterol.

    c. glycolipid.

    d. protein.

_____ 3. All of the following are active processes, requiring cellular energy, *except*

    a. primary active transport.

    b. carrier-mediated facilitated diffusion.

    c. endocytosis.

    d. exocytosis.

_____ 4. One substance flows down its concentration gradient, and the energy is harnessed to move another substance up its gradient in the same direction. This best describes

    a. primary active transport.

    b. endocytosis.

    c. symport secondary active transport.

    d. antiport secondary active transport.

_____ 5. All of the following are membrane-bound organelles *except*

    a. ribosomes.

    b. lysosomes.

    c. the Golgi apparatus.

    d. the endoplasmic reticulum.

_____ 6. This organelle is composed of extensive amounts of membrane. It synthesizes lipids and detoxifies harmful substances such as alcohol. This organelle is

    a. the smooth endoplasmic reticulum.

    b. a mitochondrion.

    c. a proteasome.

    d. a lysosome.

_____ 7. Which of the following organelles destroys malformed proteins, proteins that do not fold normally, and proteins that are no longer needed by a cell?

    a. centrioles

    b. peroxisomes

    c. proteasomes

    d. nucleolus

_____ 8. The process of forming RNA from DNA is called

    a. mitosis.

    b. DNA replication.

    c. translation.

    d. transcription.

_____ 9. During this stage of mitosis, the chromatin coils to form chromosomes, the nuclear envelope disappears, the nucleolus dissolves, spindle fibers are formed, and the centrioles migrate to the poles.

    a. prophase

    b. metaphase

    c. anaphase

    d. telophase

_____ 10. Erythrocytes do not have a nucleus. In what two cellular processes can they *not* engage?

    a. apoptosis and synthesis of lipid

    b. protein synthesis and cell division

    c. digestion of unwanted protein and cell division

    d. formation of vesicles and protein synthesis

11. Describe the general structure and function of the three major structures of a cell.

12. List and describe the functions of plasma membrane proteins.

13. Describe the passive processes of membrane transport including simple diffusion, facilitated diffusion, and osmosis.

14. Describe the active processes of membrane transport including primary active transport, secondary active transport, and the various forms of vesicular transport.

15. List the membrane-bound structures, and describe the structure and function of each.

16. Describe the three types of proteins that form the cytoskeleton, and explain the general function of each.

17. Compare and contrast the structure and function of cilia and microvilli.

18. Describe the processes of transcription and translation.

19. Explain how DNA either directly or indirectly controls the processes of a cell.

20. Explain the processes that occur in the different stages of the cell cycle, including DNA replication, mitosis, and cytokinesis.

Can You Apply What You've Learned?

1. Michael was born with Tay-Sachs disease. Which of the following organelles in Michael's cells lacks a specific enzyme that digests organic molecules?

    a. mitochondria

    b. lysosomes

    c. the Golgi apparatus

    d. centrioles

2. A young man in his 20s has a heart attack and is rushed to the hospital. Blood is drawn, and his cholesterol level is tested and found to be very high. The doctor tells him that he has a genetic condition in which he is unable to effectively remove LDL particles containing cholesterol from his blood and into his cells. Which cellular process is not functioning normally?

   a. channel-mediated facilitated diffusion

   b. receptor-mediated endocytosis

   c. exocytosis

   d. simple diffusion

3. Tumors involve a malfunction in this cellular process.

   a. transcription

   b. translation

   c. phagocytosis

   d. mitosis

4. A rare genetic disease involves the inability of the cells to respond to testosterone. The cells are lacking

   a. enzymes for testosterone.

   b. protein carriers for testosterone.

   c. receptors for testosterone.

   d. all plasma membrane proteins.

5. The hormone insulin is a protein composed of repeating units of amino acids. It is produced through the process(es) of

   a. transcription and translation.

   b. DNA replication.

   c. mitosis.

   d. differentiation.

Can You Synthesize What You've Learned?

1. The liver produces a protein called albumin. The major function of albumin is to exert osmotic pressure to pull fluid back into the blood. Predict what happens to osmotic pressure in a patient who has cirrhosis of the liver and is not producing adequate levels of albumin.

2. In a patient with pneumonia (a respiratory condition that results in lower levels of oxygen in the blood), will diffusion of oxygen increase, decrease, or stay the same in comparison to normal? Explain.

3. Explain to a young man, with reduced numbers of receptors for LDL, why his cholesterol level is elevated.

INTEGRATE

ONLINE STUDY TOOLS

The following study aids may be accessed through Connect.

Concept Overview Interactive: Figure 4.19 Passive and Active Processes of Membrane Transport

Clinical Case Study: A Young Man with Scaly Skin

Interactive Questions: This chapter's content is served up in a number of multimedia question formats for student study.

LearnSmart: Topics and terminology include chemical structure of plasma membrane; membrane transport; cell communication; cellular structures; structure and function of the nucleus and ribosomes; cell division; cell aging and death

Anatomy & Physiology Revealed: Topics include structure of plasma membrane; diffusion; osmosis and tonicity; sodium-potassium pump; smooth and rough endoplasmic reticulum; Golgi apparatus; lysosome; intermediate filament; microvilli; nucleus; protein synthesis; cell cycle and mitosis; DNA replication

Animations: Topics include osmosis; primary active transport; sodium-potassium exchange pump; endocytosis and exocytosis; lysosomes; transcription; how translation works; how the cell cycle works

Tissue Organization

chapter

5

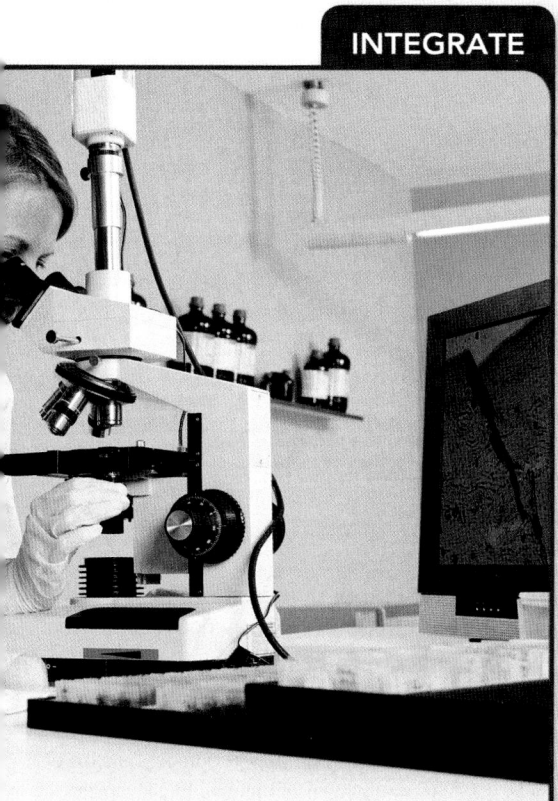

INTEGRATE

Anatomy & Physiology **REVEALED**
aprevealed.com

Module 3: Tissues

CAREER PATH
Histologist

The microscopic anatomy of cells and tissues is studied by a histologist. This professional uses various microscopy techniques such as light microscopy (LM) and electron microscopy (EM). In the hospital setting, a histologist may prepare frozen tissue sections taken from patient biopsies for rapid analysis by a pathologist. A thorough understanding of the characteristics of the four basic tissue types is essential for this health-care professional. Having an in-depth knowledge base of normal tissue structure ensures the histologist also can recognize possible tissue abnormalities, and thus possible evidence of disease or infection.

The trillions of cells in the human body are organized into more complex units called tissues. **Tissues** typically are groups of similar cells and extracellular material that perform a common function, such as providing protection or facilitating body movement. The study of tissues is called **histology** (his-tol′ō-jē; *histos* = web). Histological images are viewed using various types of microscopes, which provide different levels of detail (see section 4.1).

Tissues in the body are classified into four major types: epithelial tissue, connective tissue, muscle tissue, and nervous tissue (**table 5.1**). You can remember the four major tissue types with the acronym Con-MEN. These four tissue types vary in the structure of their cells, the functions of these cells, and the composition of an **extracellular matrix** (mā′triks; *matrix* = womb). The extracellular matrix is composed of varying amounts of protein fibers, water, and dissolved molecules (e.g., glucose, oxygen). Its consistency ranges from fluid to semisolid to solid.

Table 5.1 Overview of Tissues

	Epithelial Tissue	Connective Tissue	Muscle Tissue	Nervous Tissue
Tissue Type	Epithelial Tissue	Connective Tissue	Muscle Tissue	Nervous Tissue
Composition	Tightly packed cells with minimal extracellular matrix	Contains cells, protein fibers, and ground substance	Cells that may be spindle-shaped, branching, or cylindrical; contain myofilaments	Contains neurons and glial cells
Functions	Covers body and organ surfaces, lines body cavities and organ cavities, forms some glands	Binds, supports, and protects other tissues and organs	Moves the skeleton or organ walls	Transmits nerve impulses and processes information

Subtypes	Simple Epithelium
	   Simple squamous
	   Simple cuboidal
	   Simple columnar
	   Pseudostratified columnar
	Stratified Epithelium
	   Stratified squamous
	   Stratified cuboidal
	   Stratified columnar
	   Transitional
		Connective Tissue Proper
		   Loose (areolar, adipose, reticular)
		   Dense (regular, irregular, elastic)
		Supporting Connective Tissue
		   Cartilage (hyaline, elastic, fibrocartilage)
		   Bone
		Fluid Connective Tissue
		   Blood
		   Lymph
			Skeletal
			Cardiac
			Smooth
				(None)

5.1 Epithelial Tissue: Surfaces, Linings, and Secretory Functions

An **epithelium** (ep-i-thē′lē-um; *epi* = upon, *thele* = nipple; pl., epithelia) also referred to as **epithelial tissue**, is composed of one or more layers of closely packed cells, and it contains little to no extracellular matrix between these cells. Epithelial tissue covers the body surfaces, lines the body cavities and organ cavities, and forms most glands.

5.1a Characteristics of Epithelial Tissue

LEARNING OBJECTIVE

1. Describe the common features of epithelial tissue.

All epithelia exhibit the following common characteristics, some of which are shown in **figure 5.1**:

- **Cellularity.** Epithelial tissue is composed almost entirely of tightly packed cells. There is a minimal amount of extracellular matrix between the cells.
- **Polarity.** An epithelium has an **apical** (āp′i-kăl) **surface** (*free,* or *superficial*), which is exposed either to the external environment or to some internal body space. The apical surface may have either microvilli or cilia. Recall from section 4.6c that microvilli are small membranous projections on the apical surface of the cell that increase its surface area for secretion and absorption, whereas cilia are numerous, slightly longer, membranous projections that move fluid, mucus, and materials past the cell surface. The lateral surfaces may contain membrane (intercellular) junctions (see section 4.6d). Additionally, each epithelium

has a **basal** (bā′săl) **surface** (a fixed or deep surface), where the epithelium is attached to the underlying connective tissue.

- **Attachment to a basement membrane.** The epithelial layer is bound at its basal surface to a thin **basement membrane.** It may be seen as a single noncellular (or molecular) layer using the light microscope—however, in reality it consists of three molecular layers that can be viewed using an electron microscope: the *lamina lucida,* the *lamina densa,* and the *reticular*

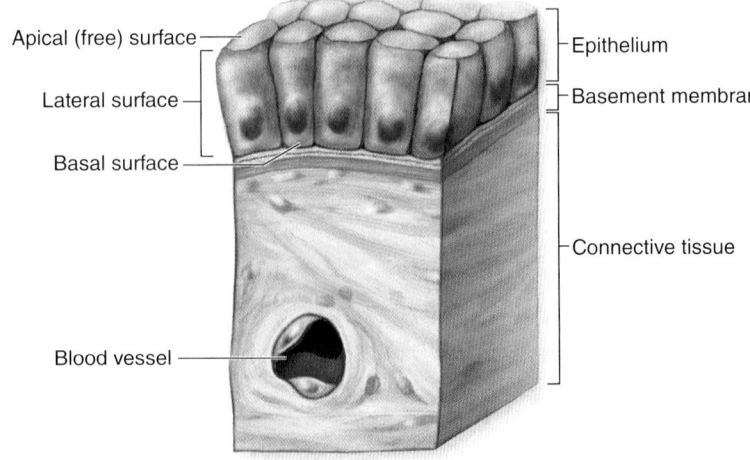

Figure 5.1 Characteristics of Epithelia. An epithelium is composed mostly of cells, exhibits its polarity, and the lateral surfaces of cells are connected by membrane junctions (see figure 4.33). An epithelium attaches to underlying tissue via a basement membrane. **AP|R**

Apical (free) surface — Epithelium
Lateral surface — Basement membrane
Basal surface
Connective tissue
Blood vessel

lamina. These molecular layers are formed by secretions of both the epithelium and the underlying connective tissue, and are composed of collagen, glycoproteins (e.g., laminin, fibronectin), and proteoglycans. The two laminae closest to the epithelium (lamina lucida and lamina densa) contain collagen fibers as well as specific proteins and carbohydrates, some of which are secreted by the epithelial cells. Cells in the underlying connective tissue secrete the reticular lamina, which contains protein fibers and carbohydrates. Together, these basement membrane components strengthen the attachment and form a selective molecular barrier between the epithelium and the underlying connective tissue.

- **Avascularity.** All epithelial tissues lack blood vessels. Nutrients for epithelial cells are obtained either directly across the apical surface or by diffusion across the basal surface from blood vessels within the underlying connective tissue.
- **Extensive innervation.** Epithelia are richly innervated (supplied with nerves) to detect changes in the environment at that body or organ region.
- **High regeneration capacity.** Epithelial cells undergo cell division frequently. This characteristic allows this tissue to regenerate itself at a high rate; a necessary condition for a tissue that is often exposed to the environment and lost by abrasion and damage. The continual replacement occurs through cell division of the deepest epithelial cells (called stem cells), which are adjacent to the basement membrane.

WHAT DID YOU LEARN?

1 Why does an epithelium need to be highly regenerative?

5.1b Functions of Epithelial Tissue

LEARNING OBJECTIVE

2. Explain the four functions of epithelial tissues.

Epithelia have several functions, although no one epithelium performs all of them. These functions include the following:

- **Physical protection.** Epithelial tissues protect both external and internal surfaces from dehydration, abrasion, and destruction by physical, chemical, or biological agents.
- **Selective permeability.** All substances that enter or leave the body must pass through an epithelium, and thus epithelial cells act as "gatekeepers." An epithelium typically exhibits a range of permeability; it may be relatively non-permeable to some substances, while promoting and assisting the passage of other ions and molecules.
- **Secretions.** Some epithelial cells are specialized to produce and release secretions. Individual gland cells may be scattered among other cell types in an epithelium or arranged in small, organized clusters within a gland (see section 5.1d).
- **Sensations.** Epithelial tissues are innervated by sensory nerve endings to detect changes in the external environment at the epithelial surface. These nerve endings—and those in the underlying connective tissue—continuously relay sensory input to the nervous system concerning touch, pressure, temperature, and pain. Additionally, several organs contain a specialized epithelium, called a *neuroepithelium,* that houses specific cells responsible for the senses of sight, taste, smell, hearing, and equilibrium (as described in chapter 16).

WHAT DO YOU THINK?

1 Why do you think epithelial tissue does not contain any blood vessels? Can you think of any epithelial function that could be compromised if blood vessels were running through the tissue?

WHAT DID YOU LEARN?

2 Why is an epithelium considered selectively permeable?

5.1c Classification of Epithelial Tissue

LEARNING OBJECTIVES

3. Name the classes of epithelia based on cell layers and cell shapes.
4. Give examples of each type of epithelium.

The body contains many different types of epithelia, and the classification of each type is indicated by a two-part name. The first part of the name refers to the *number* of epithelial cell layers, and the second part describes the *shape* of cells at the apical (superficial) surface of the epithelium.

Classification by Number of Cell Layers

Epithelia may be classified as either simple or stratified **(figure 5.2a)**. A **simple epithelium** is one cell layer thick, and all of the epithelial cells are in direct contact with the basement membrane. A simple epithelium is found in areas where stress is minimal and filtration, absorption, or secretion is the primary function. Examples of locations include the lining of the air sacs in the lung, the intestines, and blood vessels.

A **stratified epithelium** contains two or more layers of epithelial cells. Only the cells in the deepest (basal) layer are in direct contact with the basement membrane. A stratified epithelium resembles a brick wall, where the bricks in contact with the ground represent the basal layer and the bricks at the top of the wall represent the apical (superficial) layer. This tissue provides either more structural support or better protection for underlying tissue. A stratified epithelium is found in areas likely to be subjected to abrasive activities or mechanical stresses, as multiple layers of cells are better able to resist the wear and tear (e.g., the skin, internal lining of the esophagus, and the internal lining of the urinary bladder). Cells in the basal layer continuously regenerate as the cells in the apical layer are lost due to abrasion or stress.

A **pseudostratified** (sū′dō-strat′i-fīd; *pseudo* = false, *stratum* = layer) **epithelium** appears layered (stratified) because the cells' nuclei are distributed at different levels between the apical and basal surfaces. Although all of these epithelial cells are attached to the basement membrane, some of them do not reach its apical surface. For our purposes, we have classified pseudostratified epithelium as a type of simple epithelium, because all of the cells are attached to the basement membrane.

Classification by Cell Shape

Epithelia are also classified by the shape of the cell at the apical surface. In a simple epithelium, all of the cells display the same shape, whereas in a stratified epithelium, a difference in shape can be seen between cells within the basal layer and those within the apical layer. Figure 5.2b shows the three common cell shapes seen in epithelia: squamous, cuboidal, and columnar. (Note that the cells in this figure all appear hexagonal when viewed from their apical surface. Thus, these terms describe the cells' shapes when viewed laterally, or from the side.)

Figure 5.2 Classification of Epithelia. Two criteria are used to classify epithelia: the number of cell layers and the shape of the cell at the apical surface. (*a*) An epithelium is simple if it is one cell layer thick, and stratified if it has two or more layers of cells. (*b*) Epithelial cell shapes include squamous (thin, flattened cells), cuboidal (cells about as tall as they are wide), and columnar (cells taller than they are wide).

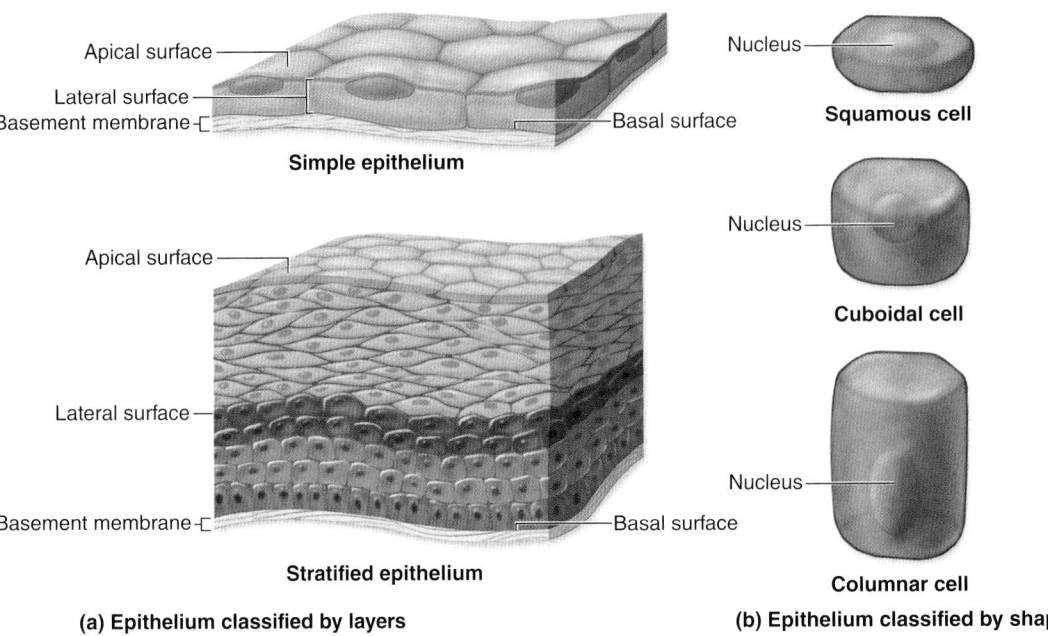

Simple epithelium

Stratified epithelium

(a) Epithelium classified by layers

Squamous cell

Cuboidal cell

Columnar cell

(b) Epithelium classified by shapes

Squamous (skwā′mŭs; *squamosus* = scaly) **cells** are flat, wide, and somewhat irregular in shape. The cells are arranged like floor tiles, and the nucleus is somewhat flattened. **Cuboidal** (kū-boy′dăl) **cells** are about as tall as they are wide. The cells do not resemble perfect cubes because their edges are somewhat rounded. The cell nucleus is spherical and located within the center of the cell. **Columnar** (kol-ŭm′năr) **cells** are slender and taller than they are wide. The cell nucleus is oval and usually oriented lengthwise and in the basal region of the cell. Another shape classification that occurs in epithelial cells is called **transitional.** These cells can readily change their shape from polyhedral to more flattened, depending upon the degree to which the epithelium is stretched. The shape change occurs when the epithelium cycles between distended and relaxed states, such as in the lining of the bladder, which fills with urine and is later emptied.

Using the classification system just described, epithelium can be broken down into the primary types shown in **figure 5.3.**

Simple Squamous Epithelium

A **simple squamous epithelium** consists of a single layer of flattened cells (**table 5.2a**). When viewed "en face" (looking onto the surface), the irregularly shaped cells display a spherical to oval nucleus, and the cells are tightly bound together. Each squamous cell resembles a fried egg, with the slightly bulging nucleus of the cell representing the yolk. This epithelium is extremely delicate and represents the thinnest possible barrier to allow rapid movement of molecules and ions by membrane transport processes (see section 4.3). Simple squamous epithelium forms the lining of the air sacs (alveoli) of the lung, where this thin epithelium is well suited for the exchange of oxygen and carbon dioxide between the blood and the inhaled air. Simple squamous epithelium also is found lining the lumen (inside space) of blood vessel walls, where it allows for rapid exchange of nutrients and waste between the blood and the interstitial fluid surrounding the blood vessels. Serous membranes, which cover body organs and secrete serous fluid, are also formed by a simple squamous epithelium.

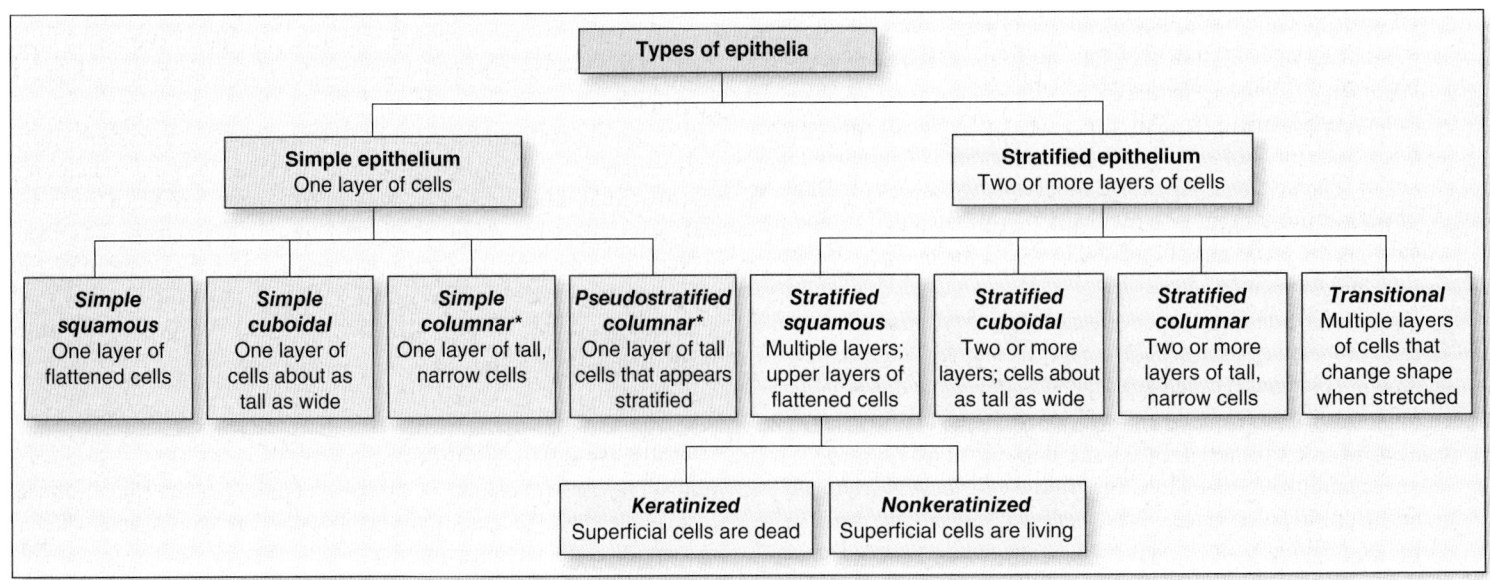

* Ciliated and nonciliated subtypes

Figure 5.3 Organization and Relationship of Epithelia Types. Epithelia are classified by (1) the number of cell layers and (2) the cell shape at the surface.

Table 5.2   Simple Epithelia

(a) SIMPLE SQUAMOUS EPITHELIUM AP|R

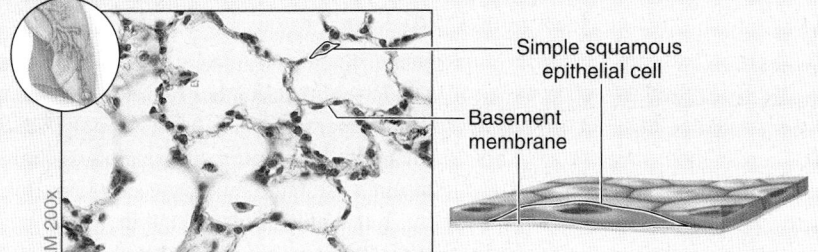

Simple squamous epithelial cell

Basement membrane

LM 200×

Structure
Single layer of thin, flat cells resembling irregular floor tiles; the single nucleus of each cell bulges at its center

Function
Rapid diffusion, filtration, and some secretion in serous membranes

Location
Air sacs in lungs (alveoli); lining of lumen of lymph vessels and blood vessels (endothelium); serous membranes of body cavities (mesothelium)

(b) SIMPLE CUBOIDAL EPITHELIUM AP|R

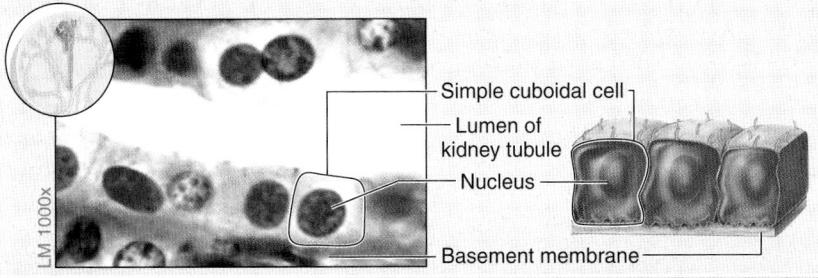

Simple cuboidal cell

Lumen of kidney tubule

Nucleus

Basement membrane

LM 1000×

Structure
Single layer of cells about as tall as they are wide; spherical and centrally located nucleus

Function
Absorption and secretion, forms glands and small ducts

Location
Thyroid gland follicles; surface of ovary; kidney tubules; secretory regions and ducts of most glands

(c) NONCILIATED SIMPLE COLUMNAR EPITHELIUM AP|R

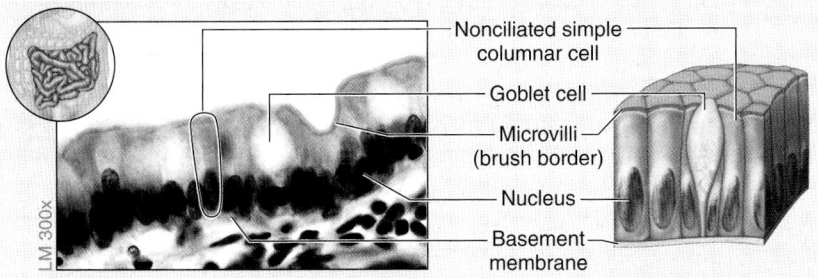

Nonciliated simple columnar cell

Goblet cell

Microvilli (brush border)

Nucleus

Basement membrane

LM 300×

Structure
Single layer of cells taller than they are wide; oval-shaped nucleus oriented lengthwise in basal region of cell; apical regions of cell may have microvilli; may contain goblet cells that secrete mucin

Function
Absorption and secretion; secretion of mucin

Location
Lining of most of digestive tract

(d) CILIATED SIMPLE COLUMNAR EPITHELIUM AP|R

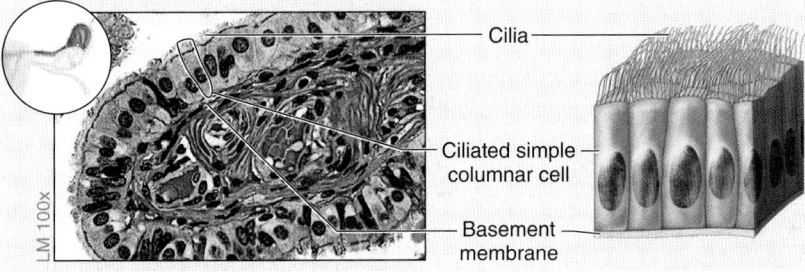

Cilia

Ciliated simple columnar cell

Basement membrane

LM 100×

Structure
Single layer of ciliated cells taller than they are wide; oval-shaped nucleus oriented lengthwise in basal region of cell; may contain goblet cells

Function
Secretion of mucin and movement of mucus along apical surface of epithelium by cilia; oocyte movement through uterine tube

Location
Lining of the larger bronchioles of respiratory tract and the uterine tubes

(e) CILIATED PSEUDOSTRATIFIED COLUMNAR EPITHELIUM AP|R

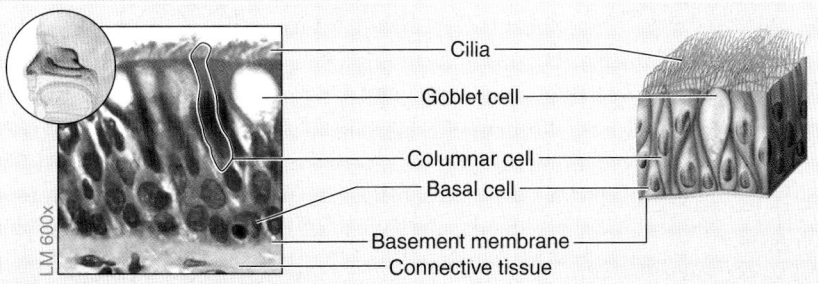

Cilia

Goblet cell

Columnar cell

Basal cell

Basement membrane

Connective tissue

LM 600×

Structure
Single layer of cells with varying heights; all cells connect to the basement membrane, but not all cells reach the apical surface; has goblet cells and cilia

Function
Protection; also involved in movement of mucus across surface by ciliary action

Location
Lining of the larger airways of respiratory tract, including nasal cavity, part of pharynx, larynx, trachea, and bronchi

(f) NONCILIATED PSEUDOSTRATIFIED COLUMNAR EPITHELIUM AP|R

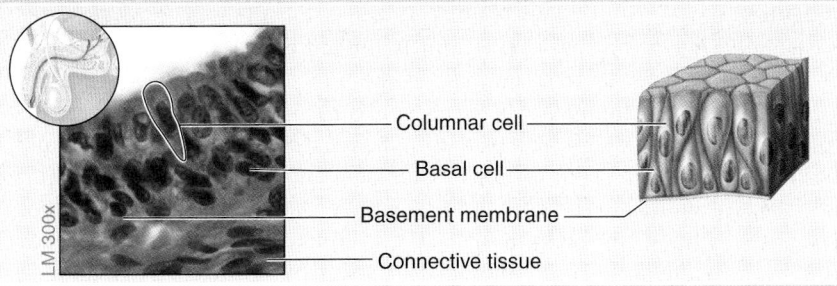

Columnar cell

Basal cell

Basement membrane

Connective tissue

LM 300×

Structure
Single layer of cells with varying heights; all cells connect to the basement membrane, but not all cells reach the apical surface; lacks goblet cells and cilia

Function
Protection

Location
Rare—lining of part of the male urethra and epididymis

Specific names are used to refer to the simple squamous epithelia in certain locations within the body. **Endothelium** (en-dō-thē′lē-ŭm; *endon* = within) is the name of the simple squamous epithelium that lines both blood vessels and lymph vessels (see sections 20.1a and 21.1), and **mesothelium** (mez-ō-thē′lē-ŭm; *mesos* = middle) is the name given to the simple squamous epithelium that forms the serous membranes of body cavities (see section 1.4e). Mesothelium gets its name from the embryonic primary germ layer called mesoderm, from which it is derived (see section 5.6a).

Simple Cuboidal Epithelium

A **simple cuboidal epithelium** contains one layer of uniformly shaped cells that are about as tall as they are wide with a centrally located spherical nucleus. This epithelium is designed for absorption and secretion. Its cells' uniformity in shape makes them ideal to form the structural components of glands. For example, a simple cuboidal epithelium forms the follicles (spherical structures) of the thyroid gland and covers each ovary. Simple cuboidal epithelium also composes the walls of small ducts (or tubules), including those of kidney tubules.

Simple Columnar Epithelium

A **simple columnar epithelium** is composed of a single layer of cells that are taller than they are wide. The nucleus is oval, oriented lengthwise, and located in the basal region of the cell. This type of epithelium is ideal for both secretory and absorptive functions. Simple columnar epithelium has two forms: One type has no cilia, whereas the apical surface of the other type is covered with cilia.

Nonciliated simple columnar epithelium often contains **microvilli** (see section 4.6c) and a scattering of unicellular glands called **goblet cells** (table 5.2c). Individual microvilli cannot be distinguished under the microscope; rather, the microvilli collectively appear as a bright, fuzzy structure known as a **brush border.** Goblet cells secrete **mucin** (mū′sin), which is a glycoprotein that when hydrated (mixed with water) forms **mucus.** Nonciliated simple columnar epithelium lines most of the digestive tract, from the stomach to the anal canal.

Ciliated simple columnar epithelium has cilia that project from the apical surfaces of the cells (table 5.2d). Mucus covers these apical surfaces and is moved along by the beating of the cilia. Goblet cells typically are interspersed throughout this epithelium. Ciliated columnar epithelium lines the larger bronchioles (air passageways) in the lung. It also lines the luminal (internal) surface of the uterine tubes, where it helps move an oocyte from the ovary to the uterus.

Pseudostratified Columnar Epithelium

A **pseudostratified columnar epithelium** (sū′dō-strat′i-fīd; *pseudes* = false, *stratum* = layer) is so named because upon first glance, it *appears* to consist of multiple layers of cells. However, this epithelium is not really stratified because all of its cells are in direct contact with the basement membrane. Although it may look stratified because the nuclei are scattered at different distances from the basal surface, not all of the cells reach the apical surface in this epithelium. Its columnar cells always reach the apical surface, and the shorter cells are stem cells that give rise to the columnar cells. Pseudostratified columnar epithelium consists of two forms: **pseudostratified ciliated columnar epithelium,** which contains cilia on its apical surface (table 5.2e), and **pseudostratified nonciliated columnar epithelium,** which lacks cilia (table 5.2f). Both types perform protective functions. The ciliated form houses goblet cells that secrete mucin, which hydrates to become the mucus that traps foreign particles and is moved by the beating cilia. This type is found in the larger air passageways of the respiratory system (e.g., the nasal cavity, part of the pharynx [throat], larynx [voice box], trachea, and bronchi).

The nonciliated form is rare, lacks goblet cells and cilia, and occurs primarily in part of the male urethra and epididymis.

Stratified Squamous Epithelium

A **stratified squamous epithelium** has multiple cell layers, and only the deepest layer of cells is in direct contact with the basement membrane. The cells in the basal layers have a cuboidal or polyhedral shape, whereas the apical cells display a flattened, squamous shape. A stratified squamous epithelium is so named because of its multiple cell layers and the shape of its apical cells. This epithelium is adapted to protect underlying tissues from damage caused by abrasion and friction. Stem cells in the basal layer continuously divide, to produce a new stem cell and a committed cell that is gradually displaced toward the surface to replace those cells that have been lost. This type of epithelium exists in two forms: keratinized and nonkeratinized.

In **keratinized stratified squamous epithelium**, the superficial layers are composed of cells that are dead. These cells lack nuclei and all organelles, and instead are filled with the protein **keratin** (ker′ă-tin; *keras* = horn), which is a tough, protective protein that strengthens the tissue **(table 5.3a)**. New cells produced in the basal region of the epithelium migrate toward the apical surface of the tissue. During their migration, the cells fill with keratin they produce, which makes them very strong, but as a consequence the cells lose their organelles and nuclei, and die. Thus, the strength of keratin has a trade-off. The epidermis (outer layer) of the skin consists of keratinized stratified squamous epithelium.

The cells in **nonkeratinized stratified squamous epithelium** remain alive including those at the tissue's apical surface, and they are kept moist with secretions such as saliva or mucus. These cells lack keratin. Because all of the cells are alive, the flattened nuclei characteristic of squamous cells are visible throughout the tissue (table 5.3b). Nonkeratinized stratified squamous epithelium lines the oral cavity (mouth), part of the pharynx (throat), the esophagus, the vagina, and the anus.

Stratified Cuboidal Epithelium

A **stratified cuboidal epithelium** contains two or more layers of cells, and the superficial cells tend to be cuboidal in shape (table 5.3c). Stratified cuboidal epithelium, like simple cuboidal epithelium, forms tubes and coverings. However, stratified cuboidal epithelium is thicker and functions in protection and secretion. This tissue forms the walls of the ducts of most exocrine glands (see section 5.1d), such as the ducts of the sweat glands in the skin, the lining of some parts of the male urethra, and the periphery of ovarian follicles.

Stratified Columnar Epithelium

A **stratified columnar epithelium** is relatively rare in the body. It consists of two or more layers of cells, but only the cells at the apical surface are columnar in shape (table 5.3d). This type of epithelium protects and secretes. It is found in the large ducts of salivary glands and in some segments of the male urethra.

Transitional Epithelium

A **transitional epithelium** is limited to the urinary tract (urinary bladder, ureters, and part of the urethra). It varies in appearance, depending upon whether it is in a relaxed state or a stretched state (table 5.3e). In a relaxed state, the basal cells appear cuboidal or polyhedral, and the apical cells are large and rounded. When transitional epithelium stretches, it thins and the apical cells flatten and become almost squamous in shape. One distinguishing feature of transitional epithelium is the presence of some **binucleated** (containing two

Table 5.3 Stratified Epithelia

(a) KERATINIZED STRATIFIED SQUAMOUS EPITHELIUM AP|R

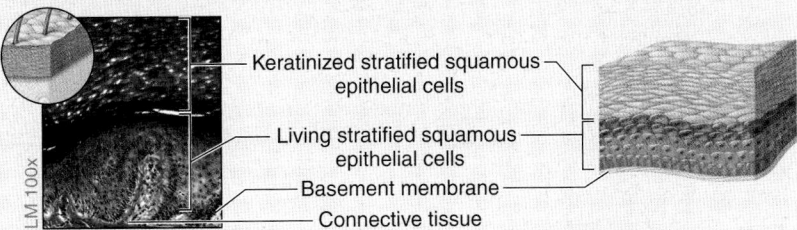

Keratinized stratified squamous epithelial cells
Living stratified squamous epithelial cells
Basement membrane
Connective tissue

LM 100x

Structure
Multiple cell layers; basal cells are cuboidal or polyhedral, whereas apical cells are squamous; apical cells are dead and filled with the protein keratin

Function
Protection of underlying tissue from abrasion

Location
Epidermis of skin

(b) NONKERATINIZED STRATIFIED SQUAMOUS EPITHELIUM AP|R

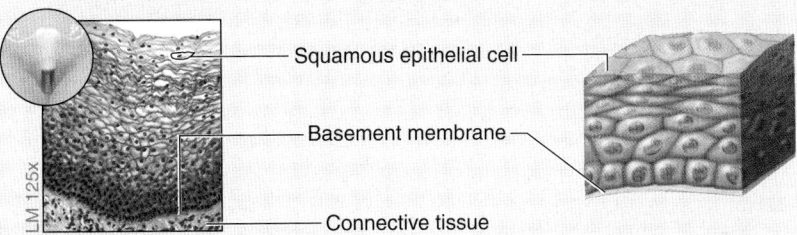

Squamous epithelial cell
Basement membrane
Connective tissue

LM 125x

Structure
Multiple cell layers; basal cells are cuboidal or polyhedral, whereas apical (superficial) cells are squamous; superficial cells are alive and kept moist

Function
Protection of underlying tissue from abrasion

Location
Lining of oral cavity, part of pharynx, esophagus, lining of vagina, and anus

(c) STRATIFIED CUBOIDAL EPITHELIUM AP|R

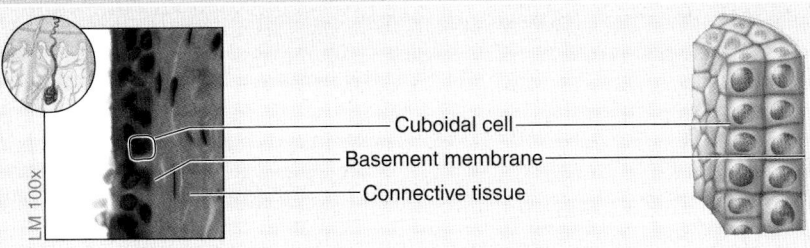

Cuboidal cell
Basement membrane
Connective tissue

LM 100x

Structure
Two or more layers of cells; cells at the apical surface are about as tall as they are wide

Function
Protection and secretion

Location
Ducts of most exocrine glands, some regions of the male urethra, and ovarian follicles

(d) STRATIFIED COLUMNAR EPITHELIUM AP|R

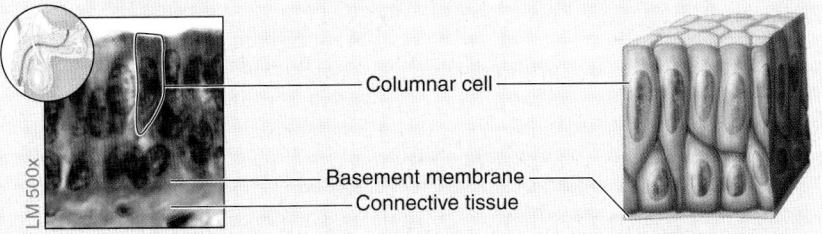

Columnar cell
Basement membrane
Connective tissue

LM 500x

Structure
Two or more layers of cells; cells at the apical surface are taller than they are wide

Function
Protection and secretion

Location
Large ducts of salivary glands and in membranous part of male urethra

(e) TRANSITIONAL EPITHELIUM AP|R

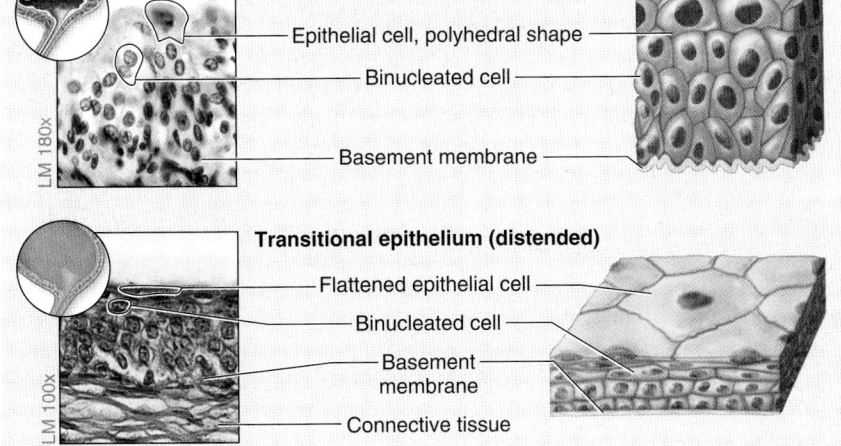

Transitional epithelium (relaxed)
Epithelial cell, polyhedral shape
Binucleated cell
Basement membrane

LM 180x

Transitional epithelium (distended)
Flattened epithelial cell
Binucleated cell
Basement membrane
Connective tissue

LM 100x

Structure
Epithelial appearance varies, depending upon whether tissue is stretched or relaxed; relaxed epithelium (*top*) has polyhedral, rounded cells at the apical surface, whereas distended epithelium (*bottom*) has flattened cells at the apical surface; some cells are binucleated

Function
Distension (stretching) and relaxation to accommodate urine volume changes in the urinary bladder, ureters, and part of urethra

Location
Lining of urinary bladder, ureters, and part of urethra

---

nuclei) cells. By being able to stretch as the bladder fills, this tissue ensures that urine does not seep into the underlying tissues of these organs.

 WHAT DO YOU THINK?

 Which types of epithelia are well suited for protection and why?

Figure 5.4 **The Relationship Between Epithelial Tissue Type and Function.** (*a*) Simple epithelium is designed for the functions of absorption, secretion, and diffusion. In contrast, (*b*) stratified epithelium's many layers make it best suited for protection.

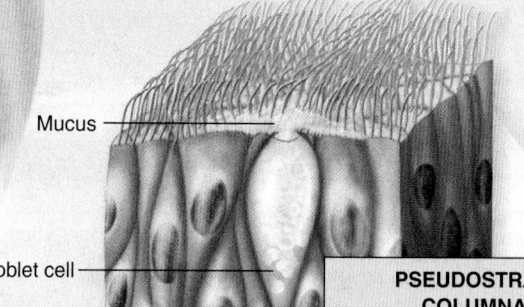

Mucus

Goblet cell

Epithelial cells

PSEUDOSTRATIFIED CILIATED COLUMNAR EPITHELIUM

Location: Ciliated form in most of upper respiratory tract, including the trachea

Function: Protection, secretion of mucus; cilia propels mucus along epithelial surface

(a) Simple Epithelia | Best suited for absorption, secretion, or diffusion

Blood capillary

Erythrocyte

Simple squamous epithelium of capillary wall

Simple squamous epithelium of alveolus wall

O_2

CO_2

Alveolus

SIMPLE SQUAMOUS EPITHELIUM

Location: Alveolus walls in the lung and in capillary walls

Function: A single thin layer of cells allows for rapid diffusion of gases between an alveolus of the lung and a capillary

Microvilli

Simple columnar cell

Nutrients

Mucus

Goblet cell

Blood capillary

Lymphatic capillary

SIMPLE COLUMNAR EPITHELIUM

Location: Small intestine

Function: The microvilli and single layer of cells facilitates absorption of nutrients and the goblet cells secrete mucus

Convoluted tubule in the kidney

Simple cuboidal epithelial cells

Blood capillary

Exchange of filtrate

SIMPLE CUBOIDAL EPITHELIUM

Location: Convoluted tubules of the kidney

Function: A single layer of cuboidal cells functions in absorption and secretion of materials between filtrate and the blood

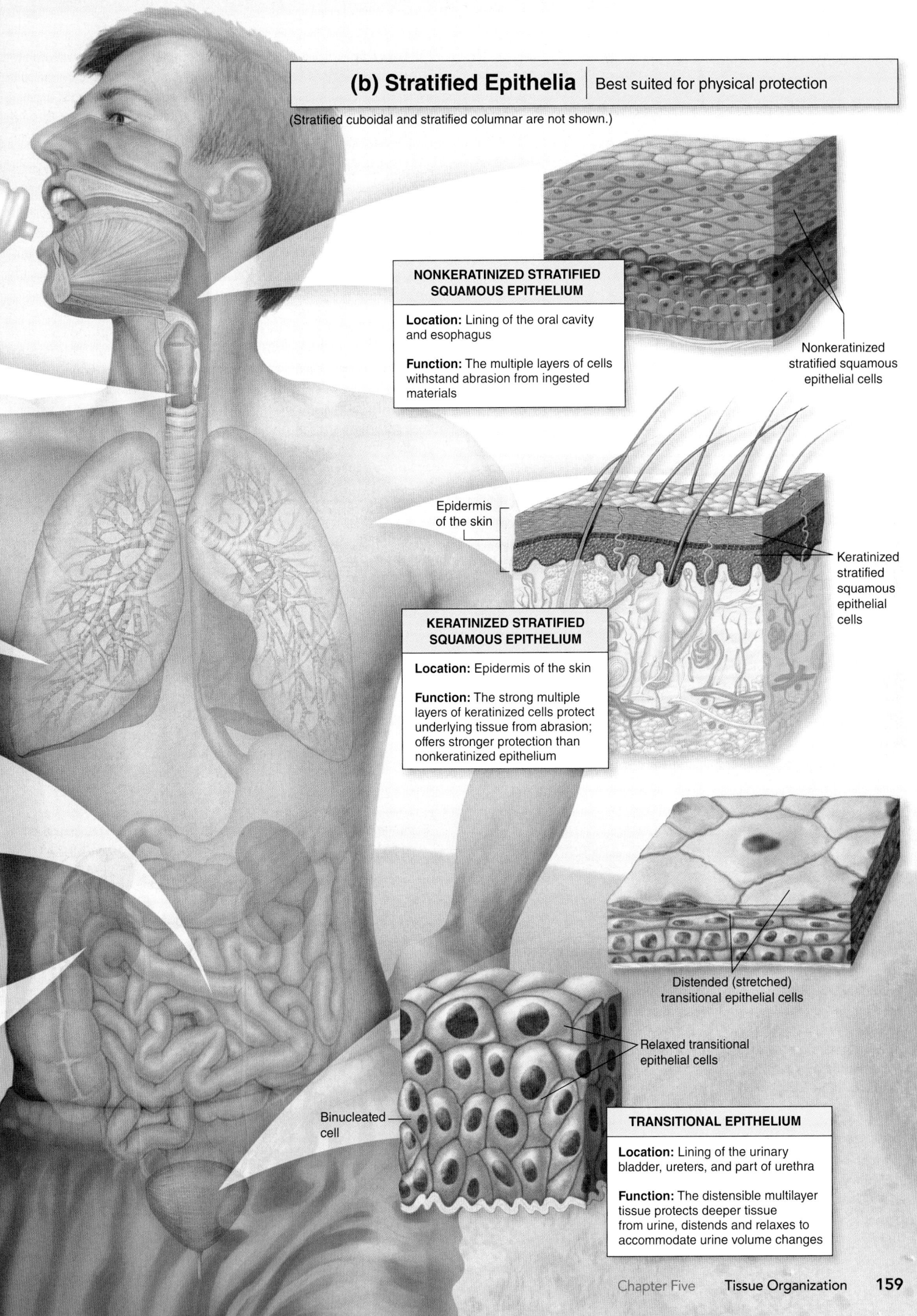

(b) Stratified Epithelia | Best suited for physical protection

(Stratified cuboidal and stratified columnar are not shown.)

Nonkeratinized stratified squamous epithelial cells

NONKERATINIZED STRATIFIED SQUAMOUS EPITHELIUM

Location: Lining of the oral cavity and esophagus

Function: The multiple layers of cells withstand abrasion from ingested materials

Epidermis of the skin

Keratinized stratified squamous epithelial cells

KERATINIZED STRATIFIED SQUAMOUS EPITHELIUM

Location: Epidermis of the skin

Function: The strong multiple layers of keratinized cells protect underlying tissue from abrasion; offers stronger protection than nonkeratinized epithelium

Distended (stretched) transitional epithelial cells

Relaxed transitional epithelial cells

Binucleated cell

TRANSITIONAL EPITHELIUM

Location: Lining of the urinary bladder, ureters, and part of urethra

Function: The distensible multilayer tissue protects deeper tissue from urine, distends and relaxes to accommodate urine volume changes

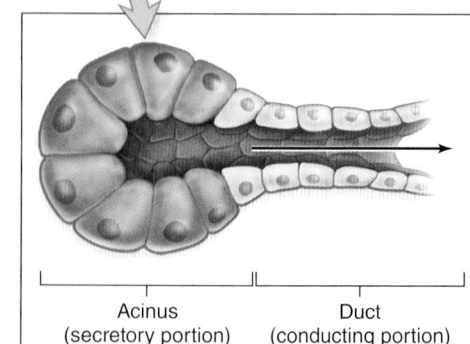

INTEGRATE

LEARNING STRATEGY

Integrate lab and lecture material: Ask yourself the following questions to help distinguish the type of epithelium under the microscope:

1. *Is the epithelium one layer or many layers thick?* If it is one layer thick, then you are looking at some type of simple epithelium. If it appears many layers thick, you are looking at either some type of stratified epithelium or pseudostratified epithelium.

2. *What is the shape of the cells at the apical surface of the epithelium?* If the cells are flattened, it is a squamous epithelium; if they are about as tall as they are wide, the cells are cuboidal, and if the cells are tall and narrow they are columnar.

Your answer for question 1 gives you the first part of the epithelium name (e.g., simple). Your answer for question 2 gives you the second part of the epithelium name (e.g., squamous). Put these answers together, and you get the name of the tissue (e.g., simple squamous).

Now that you have examined all the different types of epithelia, refer to **figure 5.4** to reexamine the relationship between epithelial type and function. Note in this figure that the simple epithelia are better designed for diffusion, absorption, and secretion functions, because these epithelia are thinner than stratified epithelium. Stratified epithelia are best suited for protective functions. Thus, when you examine organs with different epithelia, you will have a clue about the organ's function based on its epithelium.

WHAT DID YOU LEARN?

3 How does simple epithelium differ from stratified epithelium?

4 What epithelial tissue lines the air sacs of the lungs?

5 What epithelial tissue contains multiple layers of cells, and the most superficial cells are squamous, dead, and filled with the protein keratin?

5.1d Glands

LEARNING OBJECTIVES

5. Define glands.

6. Distinguish between endocrine and exocrine glands.

7. List exocrine gland types based on both anatomic form and physiologic method of secretion.

Glands are either individual cells or multicellular organs composed predominantly of epithelial tissue. They secrete substances either for use elsewhere in the body or for elimination from the body. Glandular secretions may include mucin, electrolytes, hormones, enzymes, or urea (a nitrogenous waste produced by the body).

Endocrine and Exocrine Glands

Endocrine (en'dō-krin; *endon* = within, *krino* = to separate) **glands** lack ducts and secrete their products, called *hormones*, directly into the blood. Hormones act as chemical messengers to influence cell activities elsewhere in the body. Endocrine glands, such as the thyroid and adrenal glands, are discussed in depth in chapter 17.

Exocrine (ek'sō-krin) **glands** typically originate from an invagination of epithelium that burrows into the deeper connective tissues. These glands usually maintain their connection with the epithelial surface by means of a **duct**, an epithelium-lined tube through which the gland secretions are discharged onto the epithelial surface. Examples of exocrine glands include sweat glands, mammary glands, and salivary glands.

Exocrine glands may be unicellular (one-celled) or multicellular. **Unicellular exocrine glands** typically do not contain a duct, and they are located close to the surface of the epithelium in which they reside. The most common type of unicellular exocrine gland is the goblet cell, which is commonly found in both simple columnar epithelium and pseudostratified ciliated columnar epithelium (refer to tables 5.2c, d for examples). In contrast, **multicellular exocrine glands** contain numerous cells that work together to produce a secretion (**figure 5.5**). The gland often consists of **acini** (as'i-nī; *acinus* = grape), which are the clusters of cells that produce the secretion, and one or more smaller **ducts**, which merge to form a larger duct that transports the secretion to the epithelial surface. Multicellular exocrine glands typically are surrounded by a fibrous capsule, and extensions of the capsule called *septa* partition the gland into **lobes.**

Classification of Exocrine Glands

Multicellular exocrine glands may be classified either by anatomic form or by method of secretion, which may be thought of as a physiologic classification.

Classification by Anatomic Form Exocrine glands may be classified anatomically based on the structure and complexity of their ducts. **Simple glands** have a single, unbranched duct; **compound glands** have branched ducts. In addition, glands may be classified according to the shape of their secretory portions. The gland is called **tubular** if the secretory portion and the duct have the same diameter. If the secretory portion forms an expanded sac, the gland is called **acinar.** Finally, a gland with both tubules and acini is called

Figure 5.5 General Structure of Multicellular Exocrine Glands. Exocrine glands may contain secretory portions called acini, and conducting portions composed of many ducts that merge to form a larger duct that transports the secretion to the epithelial surface.

Exocrine gland

Duct

Lobe

Acinus (secretory portion)

Duct (conducting portion)

a **tubuloacinar gland.** **Figure 5.6** shows several types of exocrine glands based on their anatomic form.

Classification by Method of Secretion   Glands may be classified physiologically by their method of secretion. The three basic types of glands in this classification are merocrine glands, apocrine glands, and holocrine glands **(figure 5.7).**

Merocrine (mer′ō-krin; *meros* = share) **glands** package their secretions into secretory vesicles and release the secretions by exocytosis (see section 4.3c). The glandular cells remain intact and are not damaged in any way by producing the secretion. Examples of merocrine glands include lacrimal (tear) glands; salivary glands; some sweat glands, also known as *eccrine glands*; the exocrine glands of the pancreas (see section 26.3c); and the gastric glands of the stomach (see section 26.2d).

Apocrine (ap′ō-krin; *apo* = away from, off) **glands** produce their secretion in the following way: The apical membrane around a portion of the glandular cell cytoplasm with the secretory product pinches off and becomes the secretion. The glandular cells repair the damage and then continue to produce new secretions in the same manner. Examples include the mammary glands (see section 28.3f) and ceruminous glands of the ear (see section 16.5a).

Holocrine (hōl′ō-krin; *holos* = whole) **glands** are formed from cells that accumulate a product; the entire cell then disintegrates. Thus, a holocrine secretion is a viscous mixture of both cell fragments and the

INTEGRATE

LEARNING STRATEGY

The *apo* part of "apocrine" sounds like "a part." Apocrine gland secretions are produced when *a part* of the cell is pinched off and becomes the secretion. The *hol* part of "holocrine" sounds like the word "whole." Holocrine gland secretions are produced when the *whole* cell ruptures, dies, and becomes the secretion. And merocrine glands "merrily" form vesicles to release secretions.

product the cell produced prior to its disintegration. The ruptured, dead cells are continuously replaced by other epithelial cells undergoing cellular division. The oil-producing glands (sebaceous glands) in the skin are examples of holocrine glands (see section 6.2c).

WHAT DID YOU LEARN?

6 What are the two basic parts of a multicellular exocrine gland?

7 What are the differences between holocrine and merocrine glands?

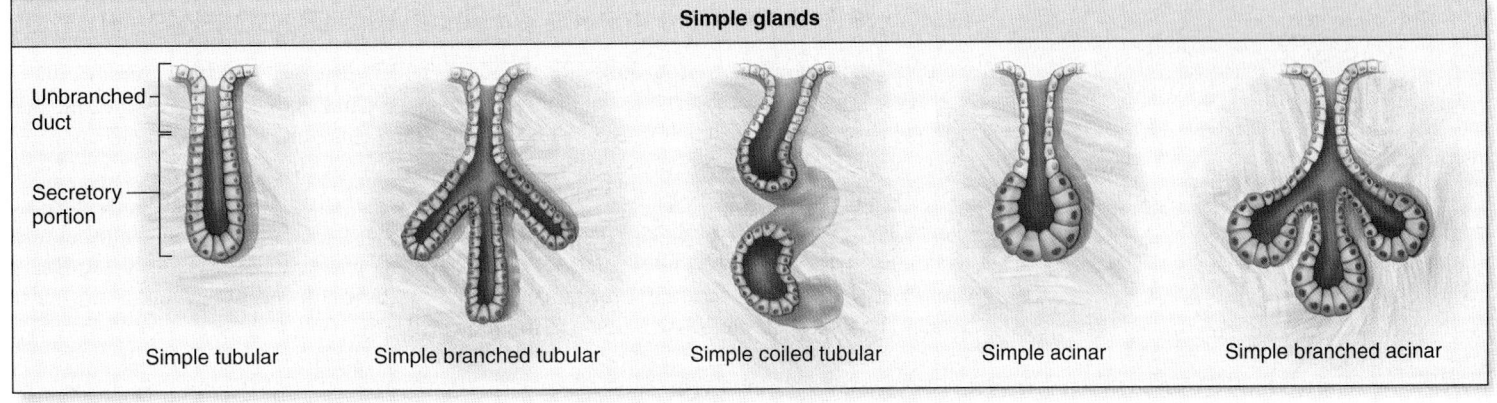

Simple glands

Unbranched duct

Secretory portion

Simple tubular   Simple branched tubular   Simple coiled tubular   Simple acinar   Simple branched acinar

(a)

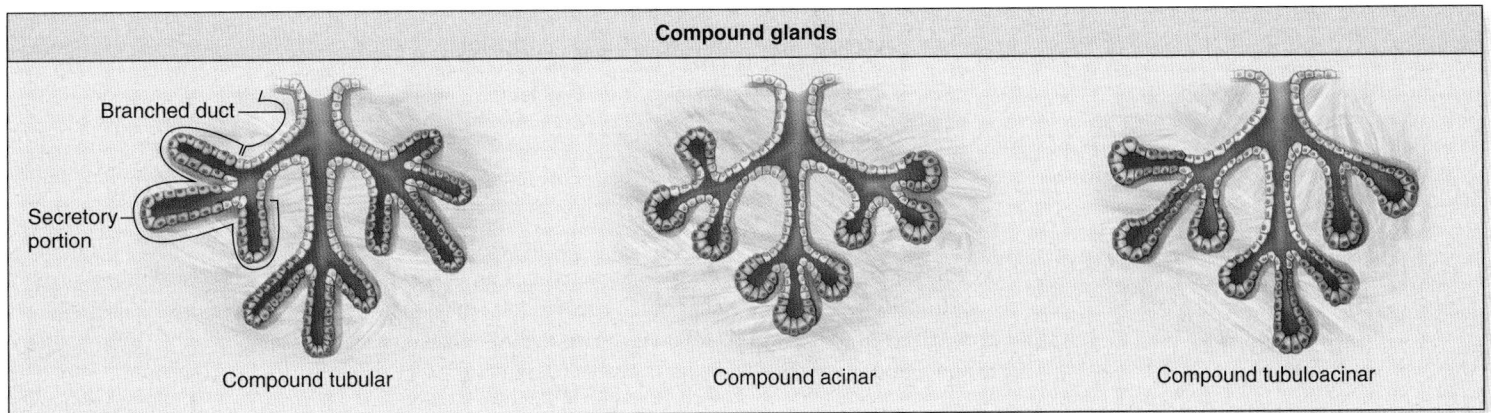

Compound glands

Branched duct

Secretory portion

Compound tubular   Compound acinar   Compound tubuloacinar

(b)

Figure 5.6 Structural Classification of Multicellular Exocrine Glands. Simple glands (*a*) have unbranched ducts, whereas compound glands (*b*) have ducts that branch. These glands also exhibit different forms: Tubular glands have secretory cells in a space with a uniform diameter, acinar glands have secretory cells arranged in saclike acini, and tubuloacinar glands have secretory cells in both the tubular and acinar regions.

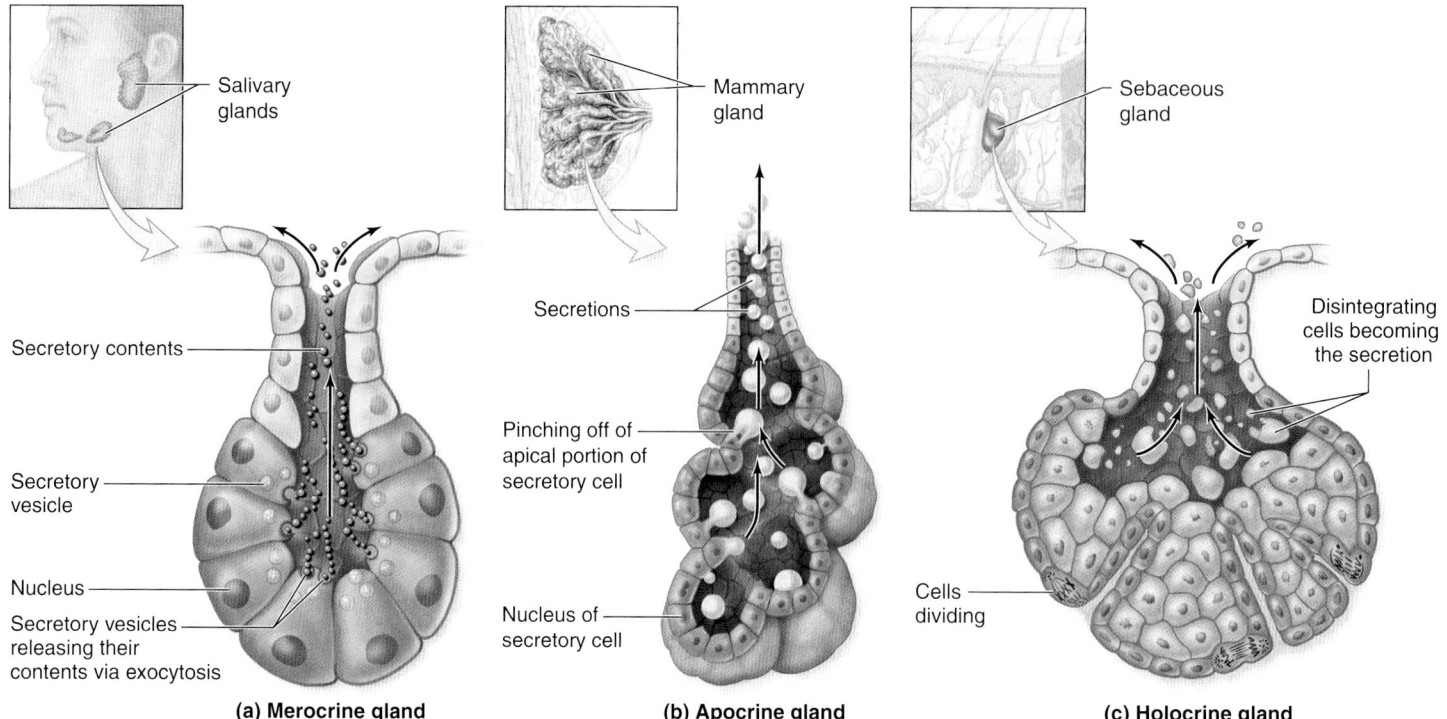

Figure 5.7 Methods of Exocrine Gland Secretion. Exocrine glands use different processes to release their secretory product. (*a*) Merocrine glands secrete products by means of exocytosis at the apical surface of the secretory cells. (*b*) Apocrine gland secretion is produced by a pinching off of the apical surface of the cell. (*c*) Holocrine gland secretion is produced through the destruction of the entire secretory cell. Lost cells are replaced by cell division at the base of the gland.

5.2 Connective Tissue: Cells in a Supportive Matrix

Connective tissue is the most diverse, abundant, and widely distributed of the tissues. Connective tissue functions to support, protect, and bind organs. Examples of connective tissue include tendons (structures that attach muscle to bone) and ligaments (structures that attach bone to bone), adipose tissue (fat), cartilage, bone, and blood. (See table 5.1.)

All connective tissues share a common origin; they all originated from an embryonic connective tissue called mesenchyme (discussed in section 5.2c). In addition, while almost all connective tissue is vascular, the different types of connective tissue exhibit a range of vascularity, from very vascular (in areolar connective tissue) to poorly vascular (in dense regular connective tissue) to avascular (in mature cartilage).

5.2a Characteristics of Connective Tissue

LEARNING OBJECTIVES

1. Describe the three components of connective tissue.
2. Give examples of resident cells and wandering cells in connective tissue proper.
3. Name three types of protein fibers found in connective tissue.
4. Identify three types of molecules that may be found in ground substance.

All connective tissues share three basic components: cells, protein fibers, and ground substance (**figure 5.8**). Together, the ground substance and the protein fibers it houses form an extracellular matrix. The specific types of cells may vary between the various classes of connective tissue. However, diversity in connective tissue is due primarily to the different types and amounts of protein fibers, as well as the varying proportions of the ground substance.

Cells

Each class of connective tissue contains specific types of cells. For example, dense regular connective tissue, which forms ligaments and tendons, contains fibroblasts, adipose connective tissue (fat) contains adipocytes, and cartilage is composed of chondrocytes. Unlike cells of epithelial tissue, most connective tissue cells are not in direct contact with each other and usually are widely scattered throughout the tissue.

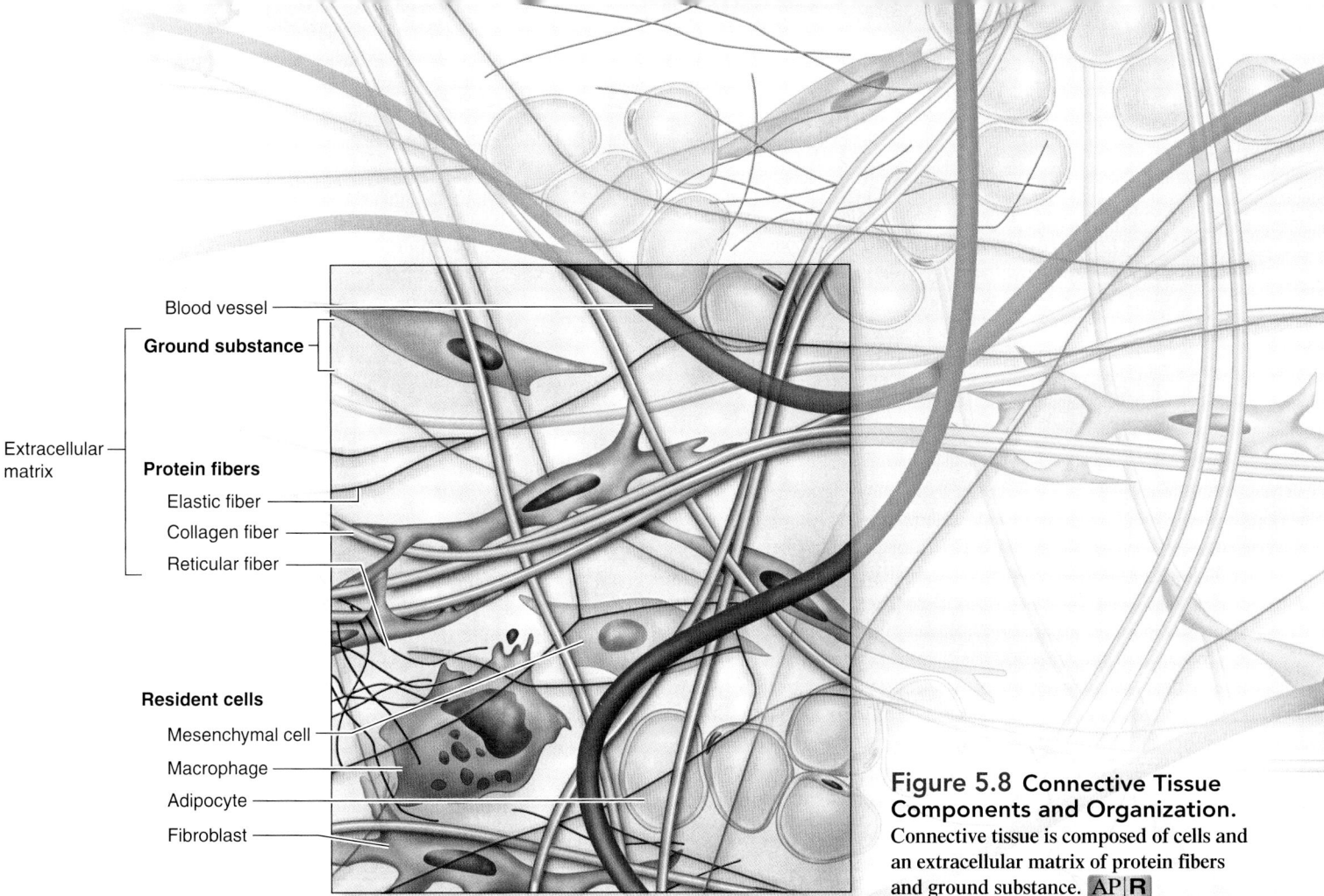

Blood vessel

Ground substance

Extracellular matrix

Protein fibers
Elastic fiber
Collagen fiber
Reticular fiber

Resident cells
Mesenchymal cell
Macrophage
Adipocyte
Fibroblast

Figure 5.8 Connective Tissue Components and Organization. Connective tissue is composed of cells and an extracellular matrix of protein fibers and ground substance. AP|R

Connective tissue proper contains two classes of cells: resident cells and wandering cells. **Resident cells** are stationary cells that are permanently housed within the connective tissue. They help support, maintain, and repair the extracellular matrix. Examples of resident cells include the following:

- **Fibroblasts** (fī′brō-blast; *fibra* = fiber, *blastos* = germ) are relatively flat cells with tapered ends and are the most abundant resident cells in connective tissue proper. They produce the fibers and ground substance components of the extracellular matrix.

- **Adipocytes** (ad′i-pō-sīt; *adip* = fat), also called *fat cells,* appear in small clusters within some types of connective tissue proper. If large clusters of these cells dominate an area, the connective tissue is called adipose connective tissue.

- **Mesenchymal cells** (me-seng′ki-mal) are a type of embryonic stem cell within connective tissue. If the tissue becomes damaged, these cells will divide. One cell that is produced replaces the mesenchymal stem cell, while the other cell becomes a committed cell that moves into the damaged area and differentiates into the type of connective tissue cell that is needed. (More discussion about stem cells is in the Clinical View: "Stem Cells" in section 5.6b.)

- **Fixed macrophages** are relatively large, irregular-shaped cells that are derived from a type of white blood cell called a monocyte (see section 18.3c). They are dispersed throughout the matrix, where they phagocytize (engulf) damaged cells or pathogens. When they encounter foreign materials, the cells also release chemicals that stimulate the immune system and attract numerous wandering cells to the tissue.

Wandering cells continuously move throughout the connective tissue proper and are components of the immune system (see chapter 22). They also may help repair damaged extracellular matrix. These cells are primarily types of **leukocytes** (lū′kō-sīt; *leukos* = white), also known as *white blood cells,* and protect the body against harmful agents. Examples of wandering cells and their specific functions include the following:

- **Mast cells** are small, mobile cells that usually are found close to blood vessels; they secrete heparin to inhibit blood clotting and histamine to dilate blood vessels and increase blood flow, which is significant in the inflammatory response.

- **Plasma cells** are formed when B-lymphocytes are activated by exposure to foreign materials. Plasma cells produce **antibodies**, which are proteins that immobilize a foreign material and prevent it from causing further damage.
- **Free macrophages** are mobile, phagocytic cells that wander through the connective tissue. They function like fixed macrophages, yet they are able to move throughout the tissue.
- **Other leukocytes** also migrate through the blood vessel walls into the connective tissue. These include neutrophils, a type of leukocyte that phagocytizes bacteria, and T-lymphocytes, a type of leukocyte which attacks foreign materials.

Protein Fibers

The protein fibers in connective tissue usually strengthen and support the tissue. Three basic types of protein fibers may be found in connective tissue: collagen fibers, reticular fibers, and elastic fibers.

Collagen fibers are unbranched, "cablelike" long fibers that are strong, flexible, and resistant to stretching. These fibers are stronger than steel of the same diameter. Collagen forms about 25% of the body's protein, and the fibers appear white in fresh tissue, so they often are called *white fibers*. In tissue sections stained with hematoxylin and eosin, they appear pink. Collagen fibers are numerous in structures such as tendons and ligaments.

Reticular fibers are similar to collagen fibers but much thinner. They contain the same protein subunits found in collagen, but their subunits are combined in a different way. These fibers form a branching, interwoven framework that is tough but flexible. Reticular fibers are especially abundant in the **stroma** (connective tissue framework) of organs such as the lymph nodes, spleen, and liver.

Finally, **elastic fibers** contain the protein elastin. The fibers branch and rejoin, and appear wavy. Elastic fibers stretch and recoil easily. Fresh elastic fibers have a yellowish color and often are called *yellow fibers*. These fibers are visible only in tissue sections that have been stained with special stains, which make the elastic fibers appear black. Elastic fibers are abundant in the skin, arteries, and lungs, to allow them to return to their normal shape after being stretched.

INTEGRATE

CLINICAL VIEW

Scurvy

Collagen is an important protein that strengthens and supports almost all body tissues, especially connective tissue. Vitamin C (ascorbic acid) is essential for the production and maintenance of healthy collagen fibers. **Scurvy**, a disease caused by vitamin C deficiency, is marked by weakness, ulceration of gums resulting in tooth loss, hemorrhages, abnormal bone growth, and easily ruptured capillaries. Scurvy was prevalent among nineteenth-century sailors who, on long sea voyages, lacked vitamin C in their food. Sailors eventually learned that eating citrus fruits, such as limes and lemons, on their voyages prevented scurvy (this also explains why sailors received the nickname *limeys*). Today, collagen production disorders that are caused by nutritional deficiencies are treated by consuming foods high in vitamin C, such as citrus fruits, broccoli, cauliflower, peppers, spinach, and tomatoes, or by taking vitamin C supplements.

Ground Substance

Ground substance is a noncellular material produced by the connective tissue cells, and it is within this ground substance that the connective tissue cells and protein fibers reside. The ground substance may be viscous (as in blood), semisolid (as in cartilage), or solid (as in bone). Together, the ground substance and the protein fibers it houses form an extracellular matrix.

Ground substance contains different large molecules as well as varying amounts of water. **Glycosaminoglycans** (glī'kōs-am-i-nō-glī'kan; *glykys* = sweet, *glycan* = saccharide), or **GAGs**, are one type of large molecule in the ground substance. A GAG is a polysaccharide that is composed completely of carbohydrate building blocks, some of which have an attached amine group. GAGs are negatively charged and hydrophilic. The negative charges attract cations, such as sodium (Na^+), and as a result water follows the movement of the positive ion. Thus, GAGs are able to attract and absorb water. Different GAGs attract varying amounts of water, depending on their number of negative charges, so the fluidity of the ground substance varies as a result. Different types of GAGs include chondroitin sulfate, heparan sulfate, and hyaluronic acid.

When a GAG is linked to a protein, it forms an even larger molecule within the ground substance called a **proteoglycan.** Proteoglycans have over 90% of their structure composed of carbohydrates, in the form of GAGs. The large structure of a proteoglycan is due primarily to the large number of negative charges in its GAGs, which then repel each other and cause the molecule to spread out and occupy more space. As we will see in this and future chapters, GAGs and proteoglycans perform numerous important functions in the body.

The ground substance includes other molecules like **adherent glycoproteins** (proteins with carbohydrates attached; see section 4.2b), which act like glue to bond connective tissue cells and fibers to the ground substance. Examples of adherent glycoproteins include fibronectin, fibrillin, and laminin.

WHAT DID YOU LEARN?

8 What are the basic functional differences between resident cells and wandering cells in connective tissue?

9 What are the function of GAGs in ground substance?

5.2b Functions of Connective Tissue

LEARNING OBJECTIVE

5. Describe the functions of connective tissue.

The many types of connective tissue collectively perform a wide variety of functions, including

- **Physical protection.** Bones of the skull and the thoracic cage protect delicate organs such as the brain, heart, and lungs; adipose connective tissue packed both around the kidneys and posterior to the eyes help protect these organs.
- **Support and structural framework.** Bones serve as the framework for the adult body and provide a place for muscle attachment; cartilage keeps air tubes like the trachea and bronchi patent (open); and connective tissue proper forms supportive capsules around organs such as the kidney and spleen.
- **Binding of structures.** Ligaments bind bone to bone, tendons bind muscle to bone, and dense irregular connective tissue anchors the skin to the underlying muscle and bone.

- **Storage.** Adipose connective tissue is the major energy reserve in the body; bone is the primary reservoir for calcium and phosphorus.
- **Transport.** Blood carries nutrients, gases, and wastes between different regions of the body.
- **Immune protection.** Many connective tissues contain leukocytes that protect the body against disease and mount an immune response when necessary. Additionally, the viscous nature of the extracellular matrix restricts the movement and spread of disease-causing organisms.

 WHAT DID YOU LEARN?

10 In what ways does connective tissue provide a structural framework?

5.2c Embryonic Connective Tissue

 LEARNING OBJECTIVE

6. Compare and contrast mesenchyme and mucous connective tissue.

Two types of embryonic connective tissue have been identified: mesenchyme and mucous connective tissue. They have different names because they occupy different locations, but both are embryonic connective tissues. **Mesenchyme** (mez′en-kīm; *enkyma* = infusion) is the first type of connective tissue to emerge in the developing embryo. It has star-shaped (stellate) or spindle-shaped mesenchymal cells dispersed within a gel-like ground substance that contains fine, immature protein fibers (**table 5.4a**). In fact, ground substance makes up a larger proportion than mesenchymal cells in this type of tissue. Mesenchyme is the source of all other connective tissues. Adult connective tissues often house numerous mesenchymal (stem) cells that provide support in the repair of the tissue following damage or injury.

INTEGRATE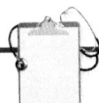

CLINICAL VIEW

What Are You Planning to Do with Your Baby's Umbilical Cord?

A fetus's blood contains stem cells that are the same as those found in a child's bone marrow, and these cells can be used to treat a variety of life-threatening diseases. Cord blood can be harvested immediately following the birth of a baby, and the blood specimen can be shipped to a cord blood bank for testing, processing, and storage. Some conditions successfully treated to date with cord blood stem cells include lymphoma (cancer of the lymph nodes); leukemia (cancer of the blood); anemia resulting from bone marrow damage, which may happen as a complication of cancer chemotherapy; and even sickle-cell disease.

Although this technology is hopeful, it is expensive and the cost is not covered by most insurance plans. Further, each cord blood sample contains relatively few stem cells. Although these stem cells can be used in unrelated recipients, it takes longer for the cells to complete the engraftment process, which leaves the patient vulnerable to infections for a longer amount of time than when using bone marrow–derived stem cells.

A second type of embryonic connective tissue is **mucous connective tissue**, also known as *Wharton's jelly* (table 5.4b). The immature protein fibers in this tissue are more numerous than those within mesenchyme. Mucous connective tissue is located within the umbilical cord only.

 WHAT DID YOU LEARN?

11 What is the composition of mesenchyme and what is its function?

Table 5.4	Embryonic Connective Tissue

(a) MESENCHYME

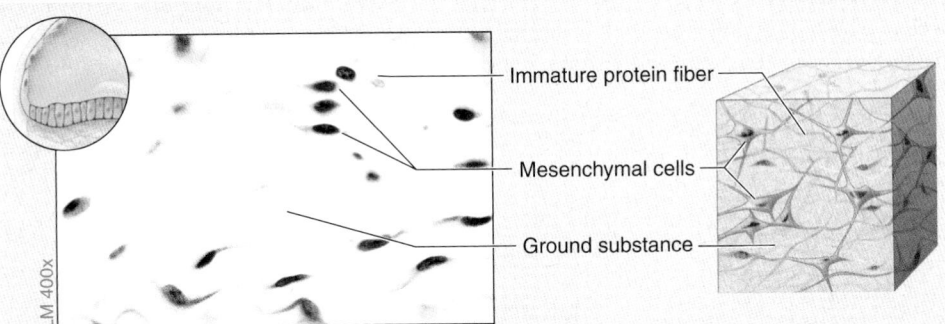

Immature protein fiber
Mesenchymal cells
Ground substance

Structure
Ground substance is a viscous fluid with some immature protein fibers; mesenchymal cells are stellate or spindle-shaped

Function
Common origin for all other connective tissue types

Location
Throughout the body of the embryo and fetus

LM 400x

(b) MUCOUS CONNECTIVE TISSUE

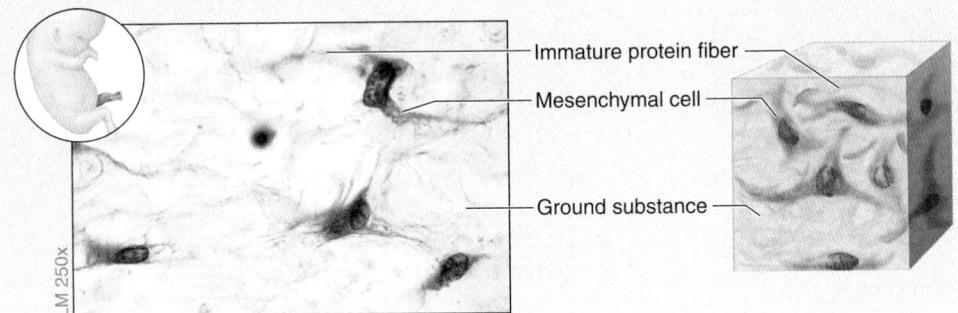

Immature protein fiber
Mesenchymal cell
Ground substance

Structure
Mesenchymal cells scattered within a viscous ground substance; immature protein fibers are more abundant here than in mesenchyme

Function
Support of structures in umbilical cord

Location
Umbilical cord of fetus

LM 250x

5.2d Classification of Connective Tissue

LEARNING OBJECTIVE

7. Distinguish the types of connective tissue and the locations where each type is found.

All connective tissue is ultimately derived from mesenchyme. Mesenchyme begins to differentiate in the developing fetus as it forms the connective tissues that ultimately are found in the adult body. The connective tissue types present after birth are classified into three broad categories: connective tissue proper, supporting connective tissue, and fluid connective tissue **(figure 5.9)**.

Connective Tissue Proper

Connective tissue proper is divided into two broad groups: loose connective tissue and dense connective tissue. This classification is based upon the relative proportions of cells, fibers, and ground substance.

Loose Connective Tissue **Loose connective tissue** contains relatively fewer cells and protein fibers than dense connective tissue. The protein fibers are sparse and irregularly arranged (hence, the name "loose connective tissue"), and there is abundant, viscous ground substance. Loose connective tissues act as the body's "packing material" by supporting and surrounding structures and organs. There are three types of loose connective tissue: areolar connective tissue, adipose connective tissue, and reticular connective tissue.

Areolar (ă-rē′ō-lăr) **connective tissue** has a loosely unconfined organization of collagen and some elastic fibers and is highly vascularized **(table 5.5a)**. This connective tissue type contains all of the fixed and wandering cells of connective tissue proper, although the predominant cell is the fibroblast. The ground substance is abundant and viscous. Areolar connective tissue is found nearly everywhere in the body. It is found in the skin (papillary layer of the dermis) and is a major component of the subcutaneous layer that is deep to the skin. It binds skin and some epithelia to deeper tissues. It also surrounds organs, individual nerve and muscle cells, and blood vessels.

Adipose connective tissue (commonly known as *fat*) is a highly vascularized loose connective tissue composed primarily of adipocytes (table 5.5*b*). Adipocytes are filled with lipid droplets, with the nucleus pushed to the inside edge of the plasma membrane. On a histology slide, the lipid is extracted during tissue processing so all that is left is the plasma membrane and nucleus of the adipocyte. There are two types of adipose connective tissue: white and brown. Brown adipose tissue is found in newborns and is designed to generate heat. As we age, we lose most of our brown adipose tissue and instead predominantly have white adipose tissue. White adipose tissue stores energy, acts as an insulator, and serves both as packing around structures as well as a cushion against shocks. It is located throughout the body in places such as the subcutaneous layer deep to the skin and surrounding various organs. Typically, the number of adipocytes

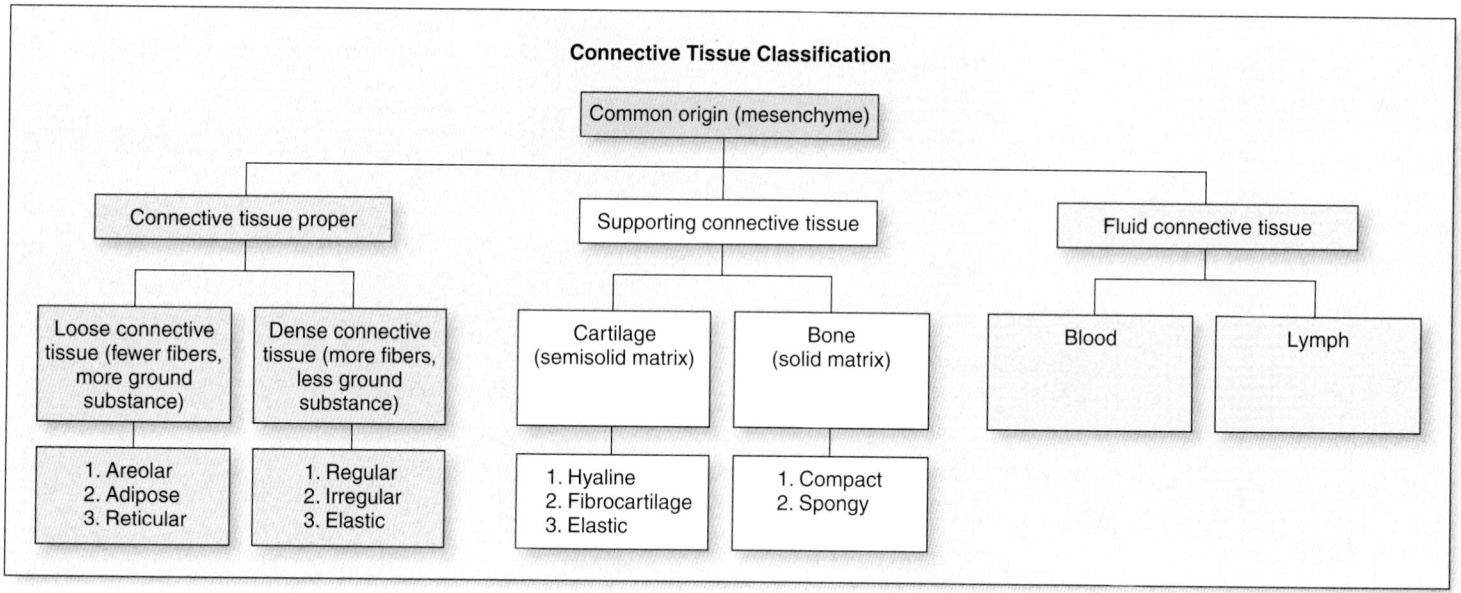

Figure 5.9 Connective Tissue Classification. Mesenchymal cells are the origin of all connective tissue cell types. The three classes of connective tissue are connective tissue proper, supporting connective tissue, and fluid connective tissue.

Table 5.5 Connective Tissue Proper: Loose Connective Tissue

(a) AREOLAR CONNECTIVE TISSUE   AP|R

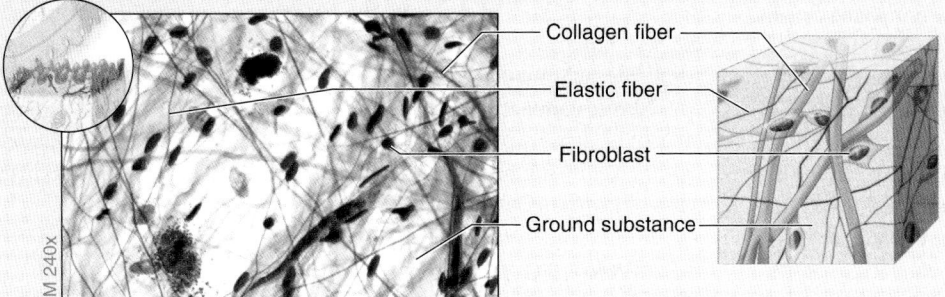

- Collagen fiber
- Elastic fiber
- Fibroblast
- Ground substance

LM 240x

Structure
Abundant, viscous ground substance; scattered fibroblasts; many blood vessels

Function
Protects tissues and organs; binds skin and some epithelia to deeper tissue

Location
Papillary layer of the dermis (skin); subcutaneous layer (deep to skin); surrounds organs, nerve cells, some muscle cells, and blood vessels

(b) ADIPOSE CONNECTIVE TISSUE   AP|R

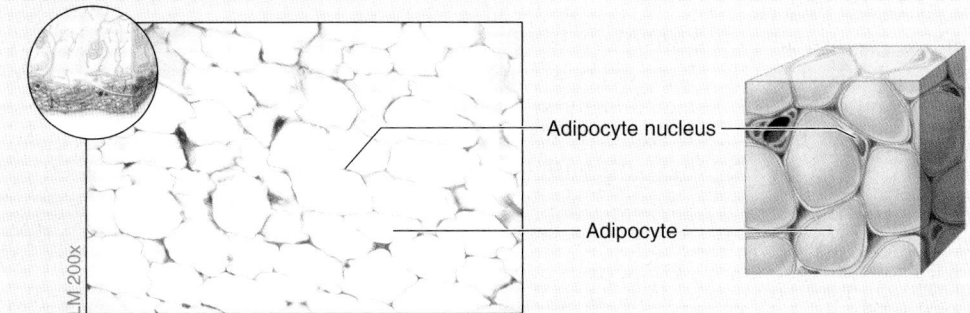

- Adipocyte nucleus
- Adipocyte

LM 200x

Structure
Closely packed adipocytes; nucleus pushed to edge of cell by large fat droplet; contains many blood vessels

Function
Stores energy; insulates, cushions, and protects

Location
Subcutaneous layer; surrounds and covers some organs

(c) RETICULAR CONNECTIVE TISSUE   AP|R

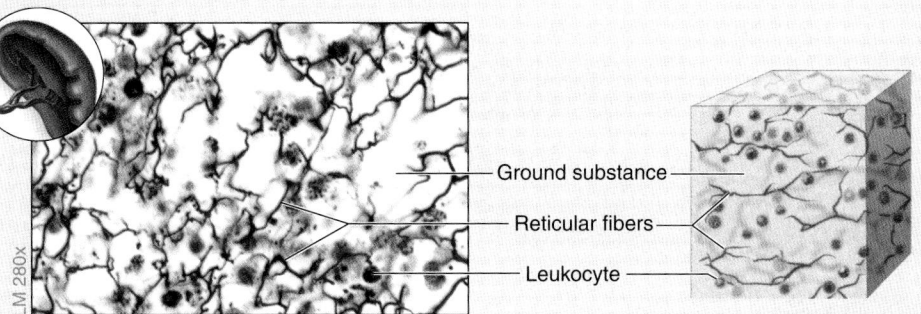

- Ground substance
- Reticular fibers
- Leukocyte

LM 280x

Structure
Viscous ground substance; scattered arrangement of reticular fibers, leukocytes, and some fibroblasts

Function
Provides stroma (supportive framework) to lymphatic organs

Location
Spleen, lymph nodes, and red bone marrow

INTEGRATE

CLINICAL VIEW
Marfan Syndrome

Marfan syndrome is a rare genetic disease of connective tissue that causes skeletal, cardiovascular, and visual system abnormalities. It results from an abnormal gene on chromosome 15. Patients with Marfan syndrome tend to have (1) abnormally long fingers, toes, and upper and lower limbs; (2) malformation of the thoracic cage, vertebral column, or both as a result of excessive growth of ribs; and (3) easily dislocated joints, resulting from weak ligaments, tendons, and joint capsules. Cardiovascular system problems involve a weakness in the aorta and abnormal heart valves. Visual system abnormalities develop because the thin fibers that hold the lens of the eye in place are weak, allowing the lens to slip out of place.

Patients usually exhibit symptoms of Marfan syndrome by age 10. Marfan syndrome symptoms may range from mild to severe. In severe

Individuals with Marfan syndrome may have abnormally long fingers and highly flexible joints.

cases, individuals may die of cardiovascular-related problems before age 50. However, those with milder symptoms and who receive early diagnosis and medical management may have long life spans.

remains relatively stable in an individual, and weight gain or loss is due to the adipocytes enlarging or shrinking in size, respectively.

Reticular connective tissue houses abundant leukocytes and some fibroblasts within a meshwork of reticular fibers (table 5.5*c*). This tissue forms the stroma (structural framework) of many lymphatic organs, such as the spleen, lymph nodes, and red bone marrow.

Dense Connective Tissue   **Dense connective tissue** is composed primarily of protein fibers and has proportionately less ground substance than loose connective tissue. It also is known as *collagenous tissue* because collagen fibers usually are the dominant fiber type. There are three categories of dense connective tissue: dense regular connective tissue, dense irregular connective tissue, and elastic connective tissue.

Dense regular connective tissue contains limited ground substance yet abundant collagen fibers that are packed tightly and align parallel to one another. The fibers resemble lasagna noodles stacked one on top of another (**table 5.6*a***). This tissue type is found in tendons and ligaments, where stress typically is applied in a single direction. Dense regular connective tissue has few blood vessels, and thus it takes a long time to heal following injury, because a rich blood supply is necessary for quick healing.

Dense irregular connective tissue contains bundles and clumps of collagen fibers that extend in all directions (table 5.6*b*). This tissue provides support and resistance to stress in multiple directions, and has an extensive blood supply. Dense irregular connective tissue is found in most of the dermis of the skin, the periosteum surrounding bone, and the perichondrium surrounding cartilage. It also forms capsules around some internal organs, such as the liver, kidneys, and spleen.

Elastic connective tissue is composed of numerous fibroblasts among branching, densely packed elastic fibers (table 5.6*c*). The elastic fibers provide the ability for the tissue to stretch and recoil. This tissue is found in the walls of large arteries, the trachea and vocal cords.

WHAT DO YOU THINK?

3 What type of connective tissue have you damaged when you sprain your ankle?

Supporting Connective Tissue

There are two types of supporting connective tissue: cartilage and bone. Both form a strong, durable framework that protects and supports the soft body tissue. The extracellular matrix contains many protein

---

Table 5.6	Connective Tissue Proper: Dense Connective Tissue

(a) DENSE REGULAR CONNECTIVE TISSUE AP|R

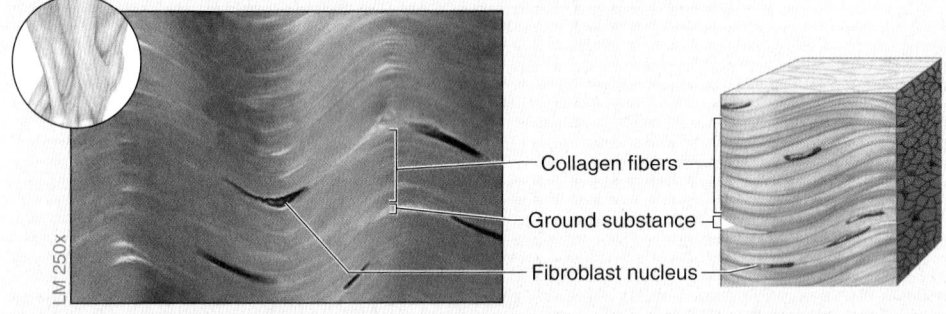
Collagen fibers — Ground substance — Fibroblast nucleus

Structure
Densely packed, parallel arrays of collagen fibers; fibroblasts squeezed between layers of fibers; scarce ground substance; greatly reduced blood supply

Function
Attaches bone to bone (ligament) as well as muscle to bone (tendon); resists stress applied in one direction

Location
Tendons; ligaments (e.g., interosseous membrane between radius and ulna)

(b) DENSE IRREGULAR CONNECTIVE TISSUE AP|R

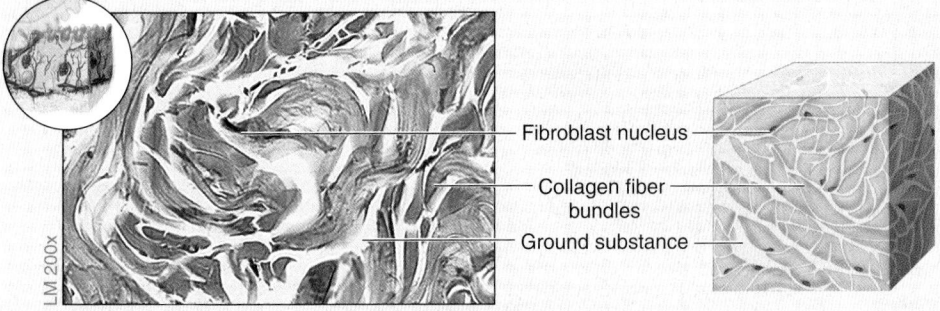
Fibroblast nucleus — Collagen fiber bundles — Ground substance

Structure
Collagen fibers randomly arranged and clumped together; fibroblasts in spaces among fibers; more ground substance than in dense regular connective tissue; extensive blood supply

Function
Withstands stresses applied in all directions; durable

Location
Most of dermis of skin; periosteum covering bone; perichondrium covering cartilage, some organ capsules

(c) ELASTIC CONNECTIVE TISSUE AP|R

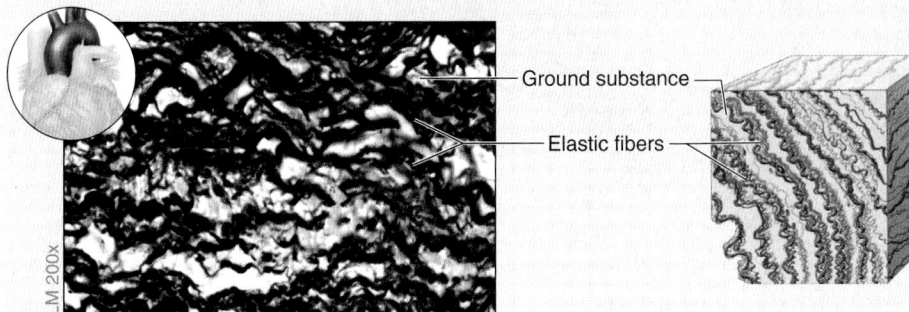
Ground substance — Elastic fibers

Structure
Predominantly composed of elastic fibers; fibroblasts occupy some spaces between fibers

Function
Allows for stretching and recoil

Location
Walls of elastic arteries (such as the aorta), trachea, vocal cords

fibers and a ground substance that ranges from semisolid (cartilage) to solid (bone).

Cartilage **Cartilage** has a firm, semisolid extracellular matrix that contains variable amounts of collagen and elastic protein fibers. Mature cartilage cells are called **chondrocytes** (kon′drō-sīt; *chondros* = gristle, cartilage). These cells occupy small spaces called **lacunae** (lă-kū′nē; *lacus* = a hollow, a lake) within the extracellular matrix. Most cartilage is surrounded by a dense irregular connective tissue covering called the **perichondrium** (per-i-kon′drē-ŭm; *peri* = around). The perichondrium has two distinct layers: an outer fibrous layer and an inner cellular layer. Cartilage is stronger and more resilient than previously discussed connective tissue types, and it provides more flexibility than bone. It occurs in areas of the body that need support and must withstand deformation, such as the tip of the nose or the auricle (external part) of the ear.

Chondrocytes produce and secrete a chemical that prevents blood vessel growth and formation within the extracellular matrix. Thus, mature cartilage is avascular, and as a result the chondrocytes must exchange nutrients and waste products by diffusion with blood vessels outside of the cartilage.

Three major types of cartilage are found in the body: hyaline cartilage, fibrocartilage, and elastic cartilage. They exhibit both differences in density and dispersal of chondrocytes within the extracellular matrix.

Hyaline (hī′ă-lin; *hyalos* = glass) **cartilage** is the most common type of cartilage. It is named for its clear, glassy appearance when viewed under the microscope (**table 5.7a**). Its chondrocytes are irregularly scattered: The collagen within the extracellular matrix is not readily observed by light microscopy. Hyaline cartilage is surrounded by a perichondrium. If this tissue type is stained with hematoxylin and eosin and examined under the microscope, the tissue resembles carbonated grape soda, where the lacunae represent the bubbles in the soda. Hyaline cartilage is found in many areas of the body, including structures of the respiratory tract (nose, trachea, most of the larynx), costal cartilage (cartilage attached to ribs), and the articular ends of long bones. It also forms most of the fetal skeleton.

Fibrocartilage (fī′brō-kār′ti-lăj; *fibro* = fiber) is a weight-bearing cartilage. It has numerous coarse, readily visible protein fibers that are arranged as irregular bundles between large chondrocytes (table 5.7b). There is only a sparse amount of ground substance. The densely interwoven collagen fibers contribute to the durability of this cartilage.

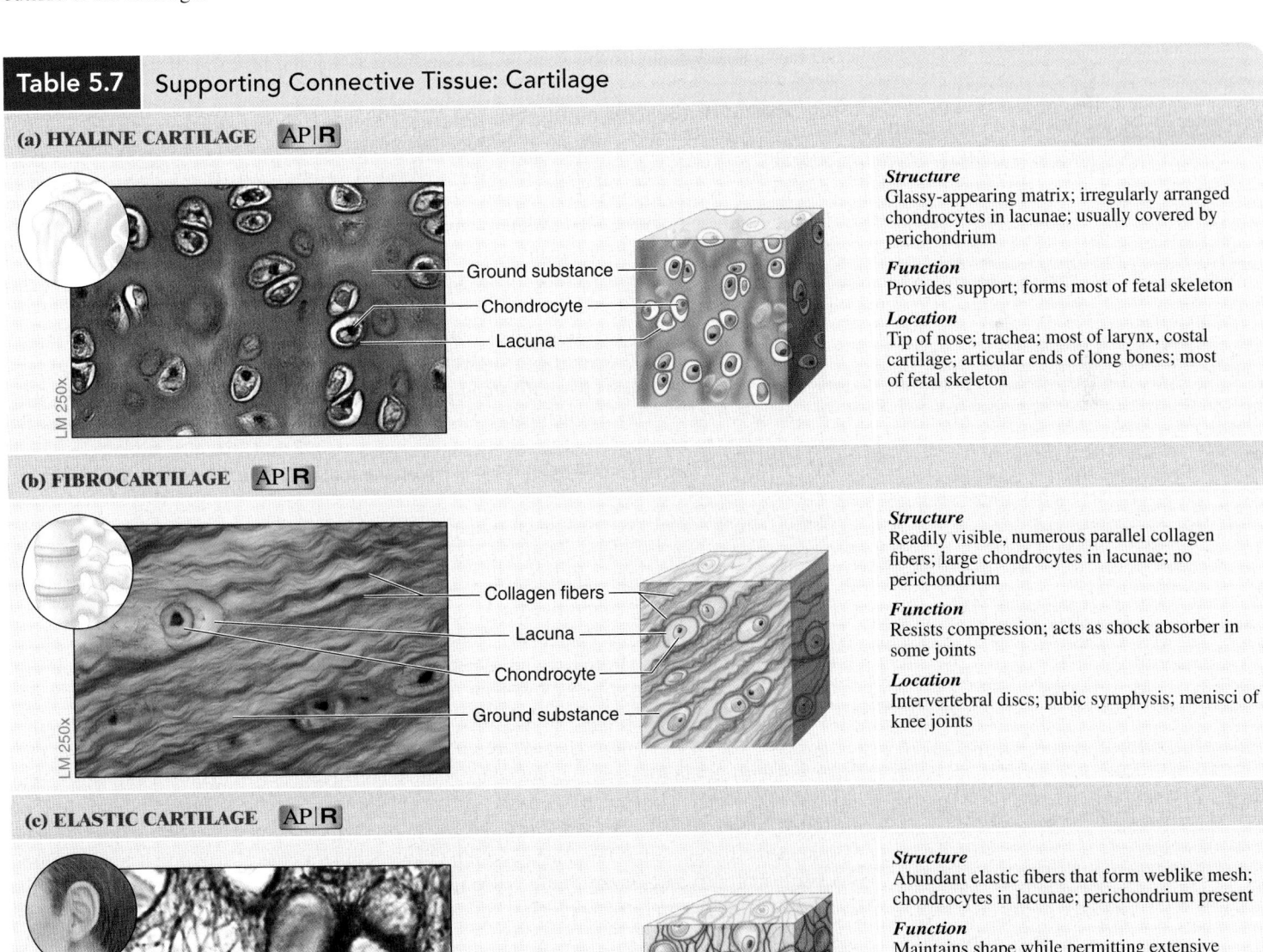

Table 5.7	Supporting Connective Tissue: Cartilage

(a) HYALINE CARTILAGE AP|R

Ground substance
Chondrocyte
Lacuna

LM 250x

Structure
Glassy-appearing matrix; irregularly arranged chondrocytes in lacunae; usually covered by perichondrium

Function
Provides support; forms most of fetal skeleton

Location
Tip of nose; trachea; most of larynx, costal cartilage; articular ends of long bones; most of fetal skeleton

(b) FIBROCARTILAGE AP|R

Collagen fibers
Lacuna
Chondrocyte
Ground substance

LM 250x

Structure
Readily visible, numerous parallel collagen fibers; large chondrocytes in lacunae; no perichondrium

Function
Resists compression; acts as shock absorber in some joints

Location
Intervertebral discs; pubic symphysis; menisci of knee joints

(c) ELASTIC CARTILAGE AP|R

Chondrocyte
Elastic fibers
Ground substance

LM 200x

Structure
Abundant elastic fibers that form weblike mesh; chondrocytes in lacunae; perichondrium present

Function
Maintains shape while permitting extensive flexibility

Location
External ear; epiglottis of larynx

INTEGRATE

LEARNING STRATEGY

Integrate your learning of lecture and lab material: Ask yourself the following questions to help distinguish the types of connective tissue proper under the microscope:

1. *Is the connective tissue loose or dense?* Loose connective tissue has fewer protein fibers and both more space and ground substance, whereas dense connective tissue has densely packed protein fibers.

2. *If the tissue is loose, what types of cells are there?* Areolar connective tissue primarily contains fibroblasts, whereas adipose connective tissue contains adipocytes. The presence of numerous leukocytes may indicate reticular connective tissue.

3. *If the tissue is dense, are the protein fibers oriented in parallel (dense regular connective tissue) or different-sized clumps (which would indicate dense irregular connective tissue)?* Elastic connective tissue may resemble dense regular connective tissue, but its fibers are not as neatly arranged.

There is no perichondrium. Fibrocartilage acts as a good shock absorber and resists compression. It is located in the intervertebral discs (circular supportive structures between adjacent vertebrae), pubic symphysis (between the anterior parts of the hip bones), and the menisci of the knee joint.

Elastic cartilage is the flexible, springy cartilage. It is so named because it contains numerous elastic fibers within its extracellular matrix (table 5.7c). The chondrocytes are closely packed and surrounded by a small amount of extracellular matrix. The elastic fibers are densely packed together and ensure that this tissue is both resilient and very flexible. Elastic cartilage is surrounded by a perichondrium. Note that both elastic cartilage and elastic connective tissue contain abundant amounts of elastic fibers. However, elastic cartilage has a semisolid ground substance and contains chondrocytes, whereas elastic connective tissue has a fluid ground substance formed by fibroblasts.

Elastic cartilage is found in the external ear and the epiglottis (a structure of the larynx that prevents swallowed materials from entering the trachea). You can see for yourself how flexible elastic cartilage is by performing this experiment: Fold your external ear over your finger, hold for 10 seconds, and release. Your ear springs back to its original shape because the elastic cartilage resists the deformational pressure you applied. (This also explains why our ears aren't permanently misshapen if we sleep on them in an unusual way!)

Bone Bone connective tissue is also known as *osseous connective tissue* and makes up the mass of most of the structures referred to as "bones." Bone is more solid than cartilage and provides greater support, although it is not as flexible. Section 7.2e provides a detailed description of the histology of bone connective tissue.

The extracellular matrix of bone connective tissue consists of organic components (collagen fibers and glycoproteins) and inorganic components composed of a mixture of calcium salts, primarily calcium phosphate. The bone cells are called **osteocytes** (os′tē-ō-sīt) and are housed within spaces in the extracellular matrix called **lacunae.**

The two forms of bone tissue are compact bone and spongy (cancellous, trabecular) bone. **Compact bone** appears completely solid, but is in fact perforated by a number of neurovascular canals **(table 5.8).** It has a uniform histologic pattern. Compact bone is formed from cylindrical structures called **osteons** (os′tē-on), which display concentric rings of bone connective tissue called lamellae. The lamellae encircle a central canal that houses blood vessels and nerves. **Spongy bone** is located within the interior of a bone, and it contains a latticework structure of bone connective tissue that is very strong, yet lightweight (see figure 7.8).

Bone serves a variety of functions. As an organ, bones provide levers for movement, and they support soft tissues as well as protect vital body organs. The hard extracellular matrix of bone connective tissue stores important minerals, such as calcium and phosphorus. Finally, some spongy bone houses **hemopoietic** (hē′mō-poy-et′ik; *hemat* = blood) cells, which form a type of reticular connective tissue that makes blood cells (a process called hemopoiesis).

Table 5.8 | Supporting Connective Tissue: Bone

BONE AP|R

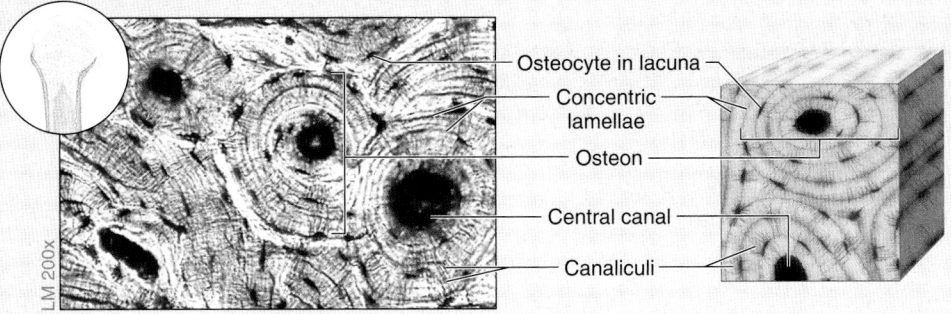

Osteocyte in lacuna
Concentric lamellae
Osteon
Central canal
Canaliculi

LM 200x

Structure
Calcified extracellular matrix containing osteocytes trapped in lacunae; compact bone arranged in osteons (concentric lamellae arranged around a central canal); spongy bone (not shown) is a meshwork that has a different organization from compact bone

Function
Provides levers for body movement, supports soft structures, protects organs, stores calcium and phosphorus; spongy bone contains hemopoietic tissue and is the site for hemopoiesis

Location
Bones of the body

Table 5.9 | Fluid Connective Tissue: Blood

BLOOD AP|R

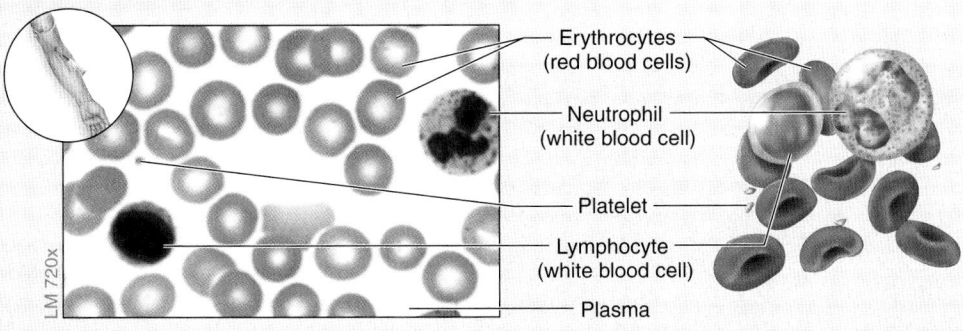

Erythrocytes (red blood cells)
Neutrophil (white blood cell)
Platelet
Lymphocyte (white blood cell)
Plasma

LM 720x

Structure
Contains formed elements (erythrocytes, leukocytes, and platelets); dissolved protein within a liquid ground substance called plasma

Function
Erythrocytes transport respiratory gases (oxygen and carbon dioxide); leukocytes help protect the body from infectious agents and platelets help with blood clotting. Dissolved protein fibers will coalesce and help with blood clotting when needed. Plasma transports nutrients, wastes, and hormones throughout the body.

Location
Primarily within blood vessels and in the heart; leukocytes also located in lymphatic structures and organs, and can migrate to infected or inflamed tissues of the body

Fluid Connective Tissue

There are two types of fluid connective tissue: blood and lymph. **Blood** is a fluid connective tissue composed of **formed elements.** Formed elements include cells, both erythrocytes (red blood cells) and leukocytes (white blood cells), and cellular fragments called platelets (table 5.9). The liquid ground substance is called **plasma**, and within it are dissolved proteins.

Blood has numerous functions. The erythrocytes transport respiratory gases (oxygen and carbon dioxide), and the leukocytes protect the body from infectious agents. Platelets and the protein fibers help clot the blood. Plasma transports nutrients, wastes, and hormones throughout the body. Blood is discussed in greater detail in chapter 18. **Lymph** is derived from blood plasma, but it contains no cellular components or fragments (which is why we don't examine it histologically here). Ultimately, lymph is returned to the bloodstream. We discuss it in greater detail in section 21.1.

Figure 5.10 summarizes the relationships between connective tissue types and their functions.

 WHAT DID YOU LEARN?

12 Compare loose connective tissue to dense connective tissue with respect to fiber density and distribution, and the amount of ground substance.

13 Describe the composition and location of fibrocartilage.

14 Why is blood considered a connective tissue?

Figure 5.10 **The Relationship Between Connective Tissue Type and Function.** (*a*) Connective tissue proper binds structures together, whereas (*b*) supporting connective tissue either provides a framework for or protects underlying soft tissues. (*c*) Fluid connective tissue is responsible for fluid, nutrient, gas, and waste transport.

(a) Connective Tissue Proper | Binds structures together

DENSE CONNECTIVE TISSUE

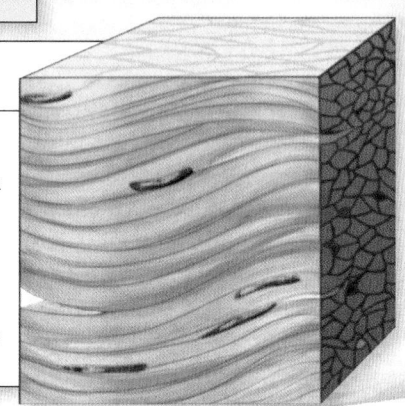

Dense Regular Connective Tissue

Location: Interosseous membrane (ligament between radius and ulna bones in forearm)

Main function: Forms tendons that bind muscle to bone and ligaments that bind bone to bone

Dense Irregular Connective Tissue

Location: Dermis of the skin

Main function: Binds the epidermis to the underlying subcutaneous layer inferior to skin

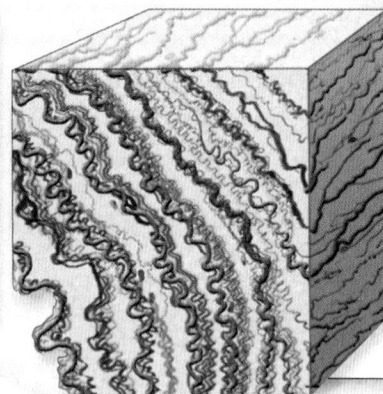

Elastic Connective Tissue

Location: Aorta

Main function: Allows for stretching and recoil in some organs (e.g., aorta)

LOOSE CONNECTIVE TISSUE

Adipose Connective Tissue

Location: Subcutaneous layer inferior to skin

Main function: Supports and surrounds structures and organs; forms the subcutaneous layer, which binds the skin and muscle

Other functions: stores lipids, insulates and cushions

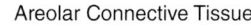

Areolar Connective Tissue

Location: Subcutaneous layer inferior to skin

Main function: Protects tissues and organs; binds skin and some epithelia to deeper tissue

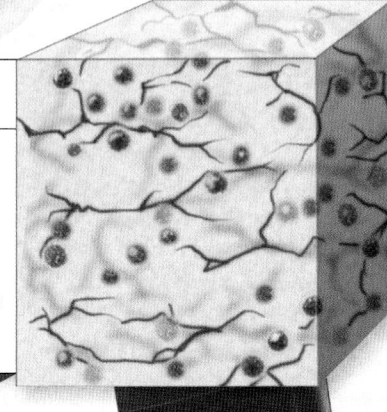

Reticular Connective Tissue

Location: Spleen

Main function: Supports and surrounds structures and organs

Other functions: Houses leukocytes, which offer protection

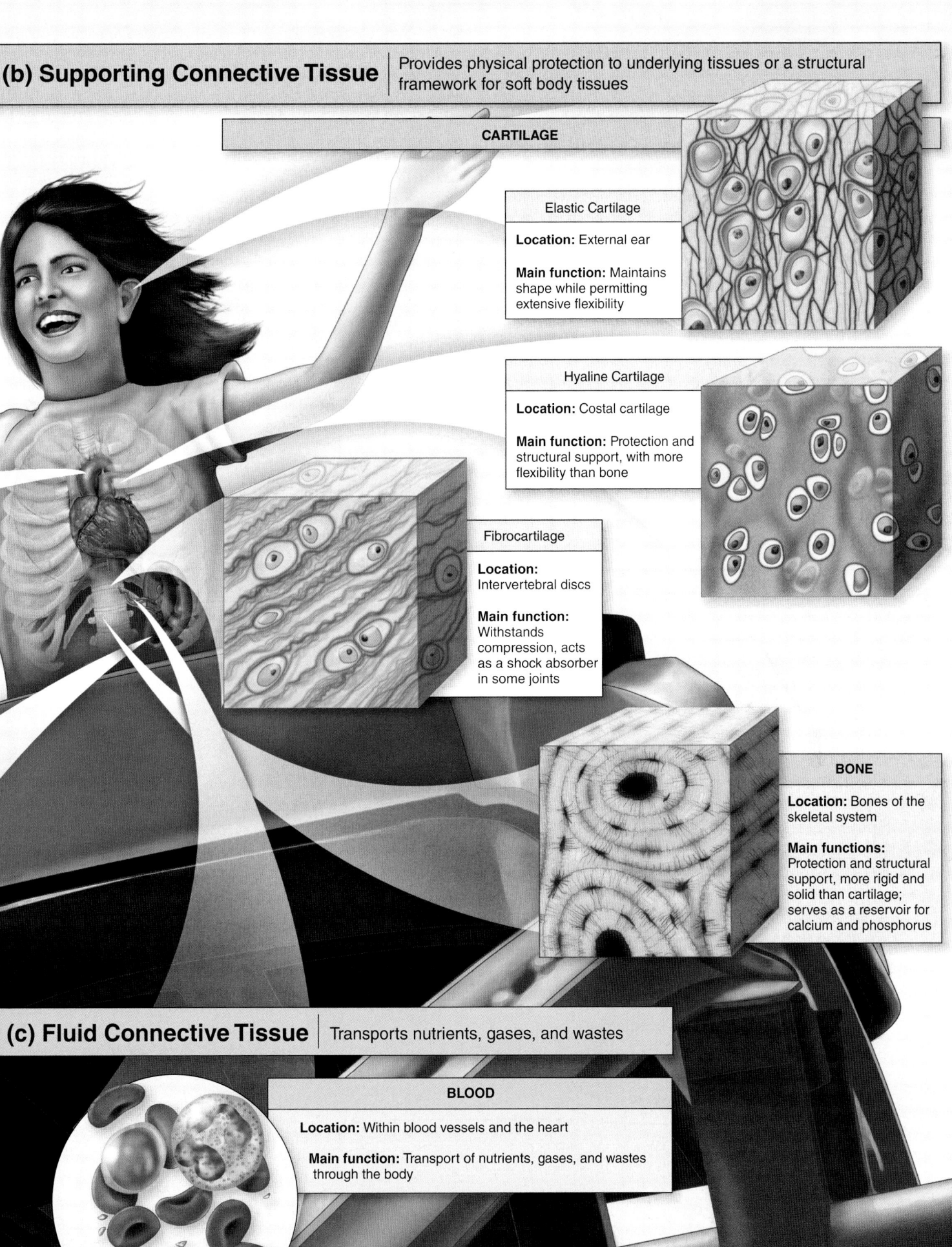

(b) Supporting Connective Tissue
Provides physical protection to underlying tissues or a structural framework for soft body tissues

CARTILAGE

Elastic Cartilage

Location: External ear

Main function: Maintains shape while permitting extensive flexibility

Hyaline Cartilage

Location: Costal cartilage

Main function: Protection and structural support, with more flexibility than bone

Fibrocartilage

Location: Intervertebral discs

Main function: Withstands compression, acts as a shock absorber in some joints

BONE

Location: Bones of the skeletal system

Main functions: Protection and structural support, more rigid and solid than cartilage; serves as a reservoir for calcium and phosphorus

(c) Fluid Connective Tissue
Transports nutrients, gases, and wastes

BLOOD

Location: Within blood vessels and the heart

Main function: Transport of nutrients, gases, and wastes through the body

173

5.3 Muscle Tissue: Movement

LEARNING OBJECTIVES

1. Describe the structure of skeletal, cardiac, and smooth muscle.

2. Compare the functions of each type of muscle and where each type is found.

Muscle tissue is composed of specialized cells that can contract when stimulated. When this tissue contracts, it produces movement, such as the voluntary motion of body parts, contraction of the heart, and propulsion of materials through the digestive and urinary tracts. The three types of muscle tissue are skeletal muscle, cardiac muscle, and smooth muscle. These tissues are described briefly here and discussed in greater detail in sections 10.2, 10.10, and 19.3e.

Skeletal muscle tissue, also known as *striated* or *voluntary muscle tissue*, is primarily responsible for movement of the skeleton (although this tissue also moves some nonskeletal structures, such as the skin of the face). It is composed of long cylindrical cells called skeletal muscle fibers. These fibers are arranged in parallel bundles that typically run the length of the entire muscle. Such long fibers need more than one nucleus to control and carry out all cellular functions; thus, each skeletal muscle fiber is multinucleated (see figure 10.2), with nuclei located at the periphery of the fiber (**table 5.10a**). Under the light microscope, skeletal muscle fibers exhibit alternating light and dark bands, called **striations**, that reflect the overlapping pattern of parallel thick and thin contractile protein filaments. Additionally, skeletal muscle is considered **voluntary** because it usually does not contract unless stimulated by the somatic (voluntary) nervous system (see section 12.1).

Cardiac muscle tissue is confined to the thick middle layer of the heart wall, called the myocardium; it is responsible for the contraction of the heart to pump blood. Cardiac muscle tissue contains visible striations, but unlike skeletal muscle, the cardiac muscle cells are short and often bifurcating (branching) (table 5.10b). Cardiac muscle cells contain one or two centrally located nuclei. In addition, the cells are connected by **intercalated** (in-ter′kă-lā-ted; *intercalates* = inserted between) **discs**, which are intercellular junctions between the cells composed of desmosomes and gap junctions.

Table 5.10 — Muscle Tissue

(a) SKELETAL MUSCLE TISSUE AP|R

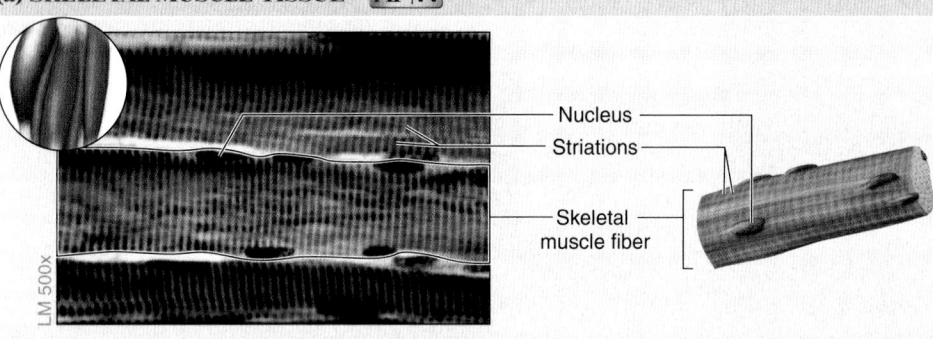

Nucleus — Striations — Skeletal muscle fiber

Structure and characteristics
Long, cylindrical, striated fibers (cells) arranged parallel and unbranched; fibers are multinucleated; fiber is under voluntary control

Function
Primarily responsible for moving skeleton and selected other components of the body

Location
Attaches to bones or sometimes to skin (e.g., facial muscles)

(b) CARDIAC MUSCLE TISSUE AP|R

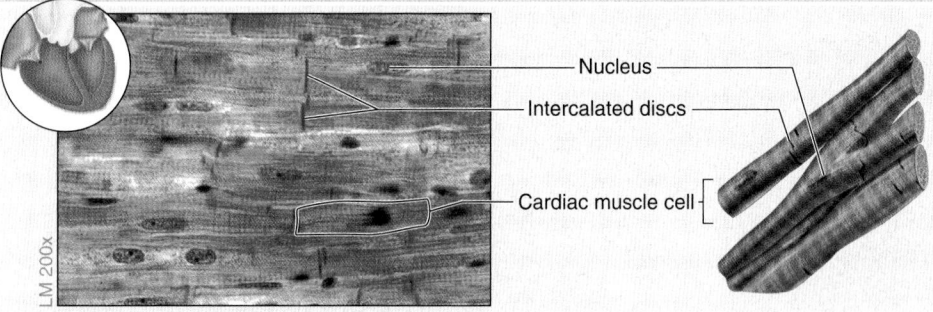

Nucleus — Intercalated discs — Cardiac muscle cell

Structure and characteristics
Short, striated cells typically branching; cells contain one or two centrally located nuclei; intercalated discs between cells; under involuntary control

Function
Pumps blood through heart

Location
Heart wall (myocardium)

(c) SMOOTH MUSCLE TISSUE AP|R

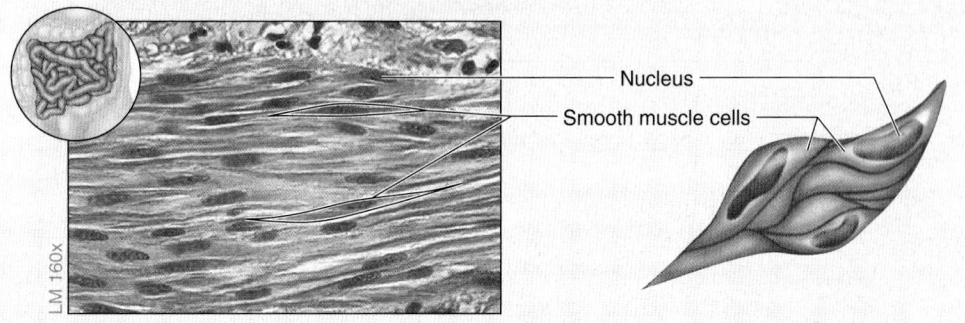

Nucleus — Smooth muscle cells

Structure and characteristics
Nonstriated cells that are short and fusiform in shape; contain one centrally located nucleus; under involuntary control

Function
Moves and propels materials through internal organs; controls the size of the lumen

Location
Walls of hollow internal organs, such as intestines, stomach, airways, urinary bladder, uterus, and blood vessels

Intercalated discs appear as dark, thick lines when viewed in the microscope. Intercalated discs strengthen the connection between cells and promote the rapid conduction of electrical activity through many cells at once, allowing the cells of a heart chamber to contract as a unit. Cardiac muscle cells are considered **involuntary** because they cannot be controlled by the somatic (voluntary) nervous system activity to initiate a contraction; instead, specialized cardiac muscle cells (pacemaker cells) in the heart wall initiate the contraction (see section 19.7).

Smooth muscle tissue, also called *visceral* or *involuntary muscle tissue,* is so named because it lacks the striations seen in other muscle tissue, and so this tissue appears smooth (table 5.10*c*). Smooth muscle cells are fusiform (spindle-shaped), which means they are thick in the middle and tapered at their ends. These cells are relatively short and contain one centrally located oval nucleus. Smooth muscle tissue is also called *visceral muscle tissue* because it is found in the walls of most viscera, such as the intestines, stomach, airways, urinary bladder, uterus, and blood vessels. The contraction of smooth muscle helps propel material movement through these organs or controls the size of the lumen. This tissue is considered involuntary because we do not have voluntary control over the muscle.

 WHAT DO YOU THINK?

4 Why do you think smooth muscle lacks striations?

 WHAT DID YOU LEARN?

15 Compare and contrast the structure of skeletal and cardiac muscle tissue.

5.4 Nervous Tissue: Information Transfer and Integration

LEARNING OBJECTIVES

1. Describe the structure of nervous tissue.
2. List the functions of nervous tissue.

Nervous tissue is located within the brain, spinal cord, and the nerves that traverse through the body. It consists of cells called **neurons** (nūr′onz) that receive, transmit, and process nerve impulses. It also contains a larger number of cells called **glial cells** (or *supporting*

INTEGRATE

LEARNING STRATEGY

Integrate your learning of lecture and lab material: Ask the following questions to help distinguish the types of muscle tissue viewed microscopically:

1. *What is the shape of the cell?* Skeletal muscle cells are long and cylindrical, cardiac muscle cells are short and bifurcated, and smooth muscle cells are spindle-shaped.

2. *How many nuclei are present, and are the nuclei centrally located or at the periphery of the cell?* Skeletal muscle has multiple nuclei that are located at the periphery of the cell. Cardiac muscle has one or two centrally located nuclei. Smooth muscle has a single centrally located nucleus.

3. *Do the cells have striations?* Skeletal and cardiac muscle have striations, but smooth muscle does not. Further, only cardiac muscle has intercalated discs between the cells.

cells), which do not transmit nerve impulses but instead are responsible for the protection, nourishment, and support of the neurons **(table 5.11)**.

Each neuron has a prominent **cell body** that houses both the nucleus and other organelles. Extending from the cell body are branches called **nerve cell processes.** The shorter and more numerous processes are **dendrites** (den′drītes; *dendrites* = relating to a tree), that receive incoming signals and transmit the information to the cell body. The single long process extending from the cell body is the **axon** (ak′son; *axon* = axis), which carries outgoing signals to other cells. Because of the extensive lengths of some axons, neurons are usually the longest cells in the body; some are longer than 1 meter. We discuss nervous tissue in detail in chapter 12.

 WHAT DID YOU LEARN?

16 What is the difference between a neuron and a glial cell?

Table 5.11	Nervous Tissue

NERVOUS TISSUE AP|R

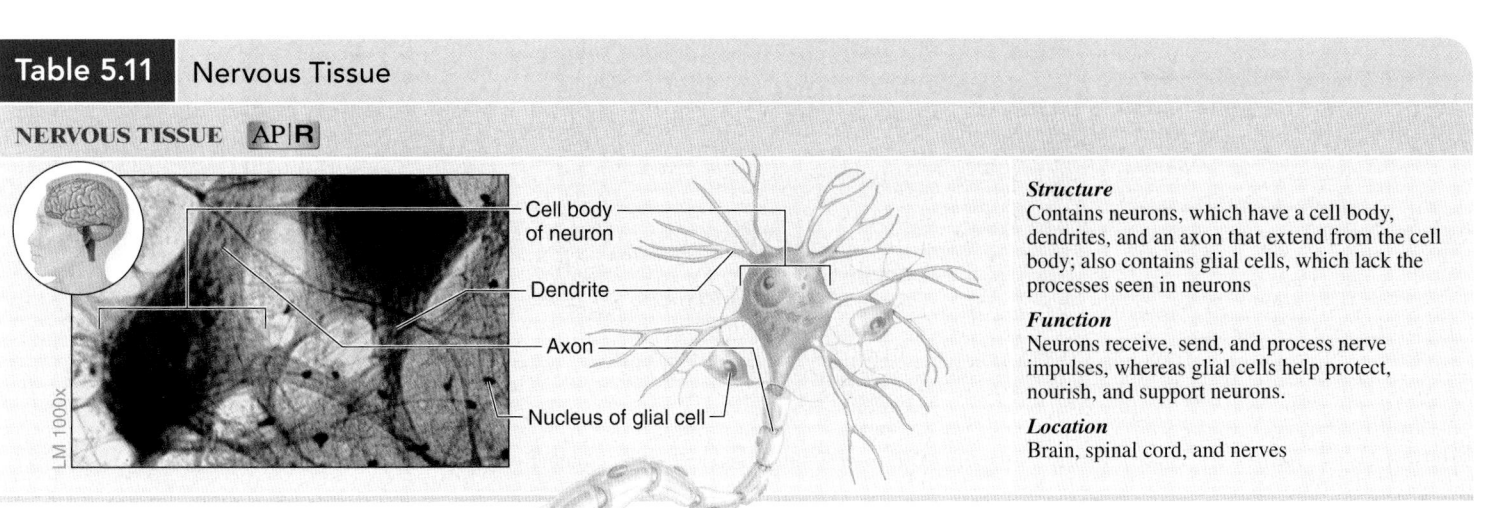

Labels: Cell body of neuron; Dendrite; Axon; Nucleus of glial cell. LM 1000×

Structure
Contains neurons, which have a cell body, dendrites, and an axon that extend from the cell body; also contains glial cells, which lack the processes seen in neurons

Function
Neurons receive, send, and process nerve impulses, whereas glial cells help protect, nourish, and support neurons.

Location
Brain, spinal cord, and nerves

5.5 Integration of Tissues in Organs and Body Membranes

We have learned that a tissue is a group of similar cells that perform a common function. Organs and body membranes also perform certain specialized functions, and they consist of two or more tissues that support their function.

5.5a Organs

LEARNING OBJECTIVES

1. Define an organ.
2. Explain the roles of different tissues in an organ.

An **organ** is a structure that is composed of two or more tissue types that work together to perform specific, complex functions. The key to organ structure is that the different tissue types must work in concert. For example, the stomach contains all four types of tissue (**figure 5.11**). It is lined by an epithelium, has both areolar and dense connective tissue in its walls, contains three layers of smooth muscle in those walls, and possesses abundant nervous tissue.

All these tissues work together to perform the functions of the stomach. Glands associated with epithelial tissue secrete substances for chemical digestion of ingested nutrients. Connective tissue houses the blood vessels and nerves that supply the stomach as well as provides shape and support. Smooth muscle contracts and relaxes so that contents within the stomach may be mechanically mixed and broken down. Nervous tissue is responsible for both regulating contraction of muscle and the stimulation of secretion by glands.

WHAT DID YOU LEARN?

17. Describe why the stomach is considered an organ.

5.5b Body Membranes

LEARNING OBJECTIVES

3. Explain the structure and functions of mucous, serous, cutaneous, and synovial membranes.
4. Identify the locations of these membranes.

Epithelial and connective tissue together form structures called body membranes, which should not be confused with the plasma membranes of cells. **Body membranes** are formed from an epithelial layer that is bound to an underlying connective tissue. These membranes line body cavities, cover the viscera, or cover the body's external surface. There are four types of body membranes: mucous, serous, cutaneous, and synovial, shown in **figure 5.12**.

A **mucous membrane**, also called a **mucosa** (mū-kō′să), lines passageways and compartments that eventually open to the external environment; these include the digestive, respiratory, urinary, and reproductive tracts. Mucous membranes perform absorptive, protective, and secretory functions, or a combination of these functions. A mucous membrane is formed by an epithelium and an underlying connective tissue called the lamina propria. Often, this membrane is covered with a layer of mucus derived from goblet cells, multicellular glands, or both.

A **serous membrane** lines body cavities that typically do not open to the external environment, and their locations were first introduced in section 1.4e. The membrane is composed of a simple squamous epithelium called mesothelium. Serous membranes produce a thin, watery

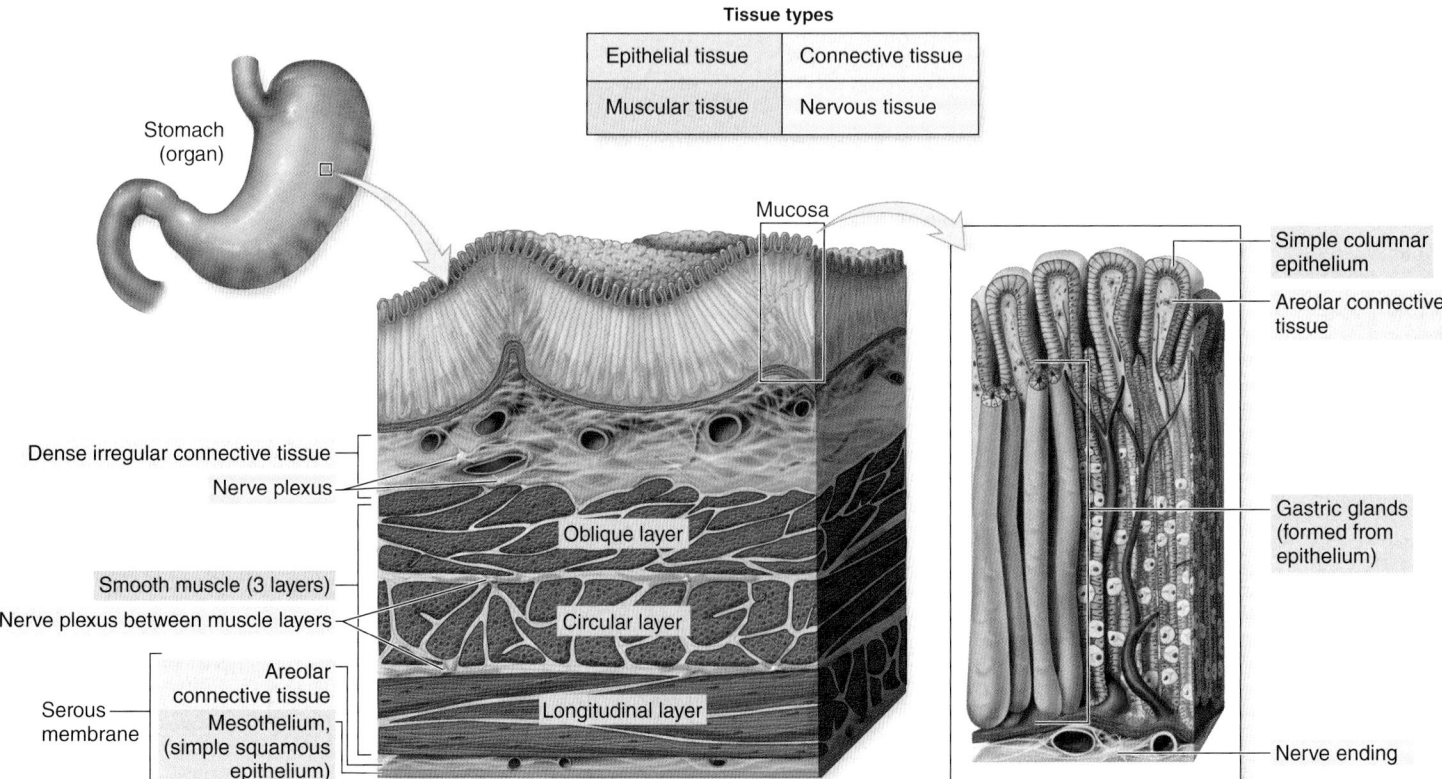

Figure 5.11 Roles of Tissues in an Organ. Different tissue types (epithelial, connective, muscle, and nervous) work together to perform the functions of the stomach.

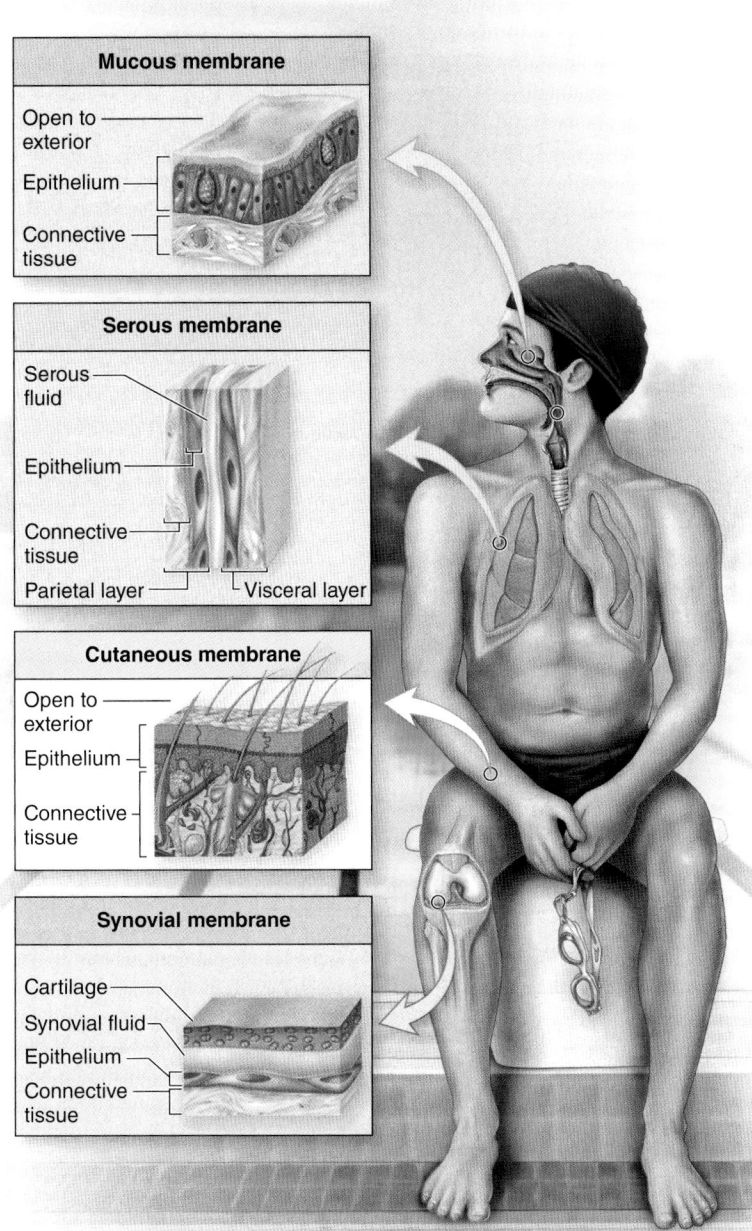

Mucous membrane

Open to exterior
Epithelium
Connective tissue

Serous membrane

Serous fluid
Epithelium
Connective tissue
Parietal layer — Visceral layer

Cutaneous membrane

Open to exterior
Epithelium
Connective tissue

Synovial membrane

Cartilage
Synovial fluid
Epithelium
Connective tissue

Figure 5.12 Body Membranes. Body membranes are formed from epithelial and connective tissue and are designed to line body cavities, cover the viscera, or cover the external surface of the body. The four types of body membranes are mucous, serous, cutaneous, and synovial.

INTEGRATE

CONCEPT CONNECTION

Serous membranes reduce friction in a similar manner for many body systems. In the cardiovascular system, the serous fluid produced by the pericardium prevents friction from the movement of the beating heart (see section 19.2b). The serous fluid secreted by the pleura of the respiratory system prevents painful abrasion when the lungs are expanding with air during inspiration (see section 23.4c). Serous fluid secreted by the peritoneum prevents friction between the moving abdominal organs and the abdominal wall during the digestion of ingested food (see section 26.1d).

serous fluid, or *transudate* (tran′sū-dāt; *trans* = across, *sudo* = to sweat), which is derived from blood plasma. Serous membranes form two associated layers: a **parietal layer** that lines the inside of the body cavity and a **visceral layer** that covers the surface of the internal organs. Between these two layers is a **serous cavity**, which is a potential space into which the serous fluid is secreted. The serous fluid reduces the friction between their opposing surfaces. Examples of serous membranes include part of the pericardium (which is associated with the heart), the pleura (associated with the lungs) and the peritoneum (associated with abdominal organs).

The largest body membrane is the **cutaneous** (kū-tā′nē-ŭs; *cutis* = skin) **membrane**, also known as the *skin,* which covers the external surface of the body. The cutaneous membrane is composed of a keratinized stratified squamous epithelium (called the epidermis) and an underlying layer of connective tissue (called the dermis). Its many functions include protecting internal organs and preventing water loss. This membrane is discussed in greater detail in chapter 6.

Some joints in the body are lined by a **synovial** (si-nō′vē-ăl) **membrane** that is composed of a superficial layer of squamous epithelial cells that lack a basement membrane and rests on an areolar connective tissue layer. The epithelial cells secrete a **synovial fluid** that reduces friction among the moving bone parts and distributes nutrients to the cartilage on the articular surfaces of bone (see section 9.4a).

WHAT DID YOU LEARN?

18 What are the differences between the parietal and visceral layers of the serous membrane?

5.6 Tissue Development and Aging

The process by which the many different tissues of the body are generated from a single cell, the fertilized egg, could take up an entire book. Here we describe the milestones of this complex process; you will learn more in chapter 29. In addition, we consider how tissues change over time and with different conditions. We conclude with a brief summary of how the aging process affects the body's tissues.

5.6a Tissue Development

LEARNING OBJECTIVES

1. Explain the stages of tissue development in the embryo.
2. Describe the three primary germ layers and the tissues to which they give rise.

To understand how all tissues form, some basic information about the human embryo is required. When a secondary oocyte (egg) is fertilized by a sperm, it forms a diploid cell called a **zygote.** The zygote undergoes multiple cell divisions; eventually, a multicellular structure called a **blastocyst** is formed. The cells that form the embryo are collectively known as the **embryoblast.**

Embryoblast cells differentiate in the second and third weeks of development. By the third week of development, three **primary germ layers** are formed. All body tissues develop from these layers (**figure 5.13**). The three primary germ layers are called ectoderm, mesoderm, and endoderm. When these three primary germ layers have formed, the growing structure may now be referred to as an **embryo.**

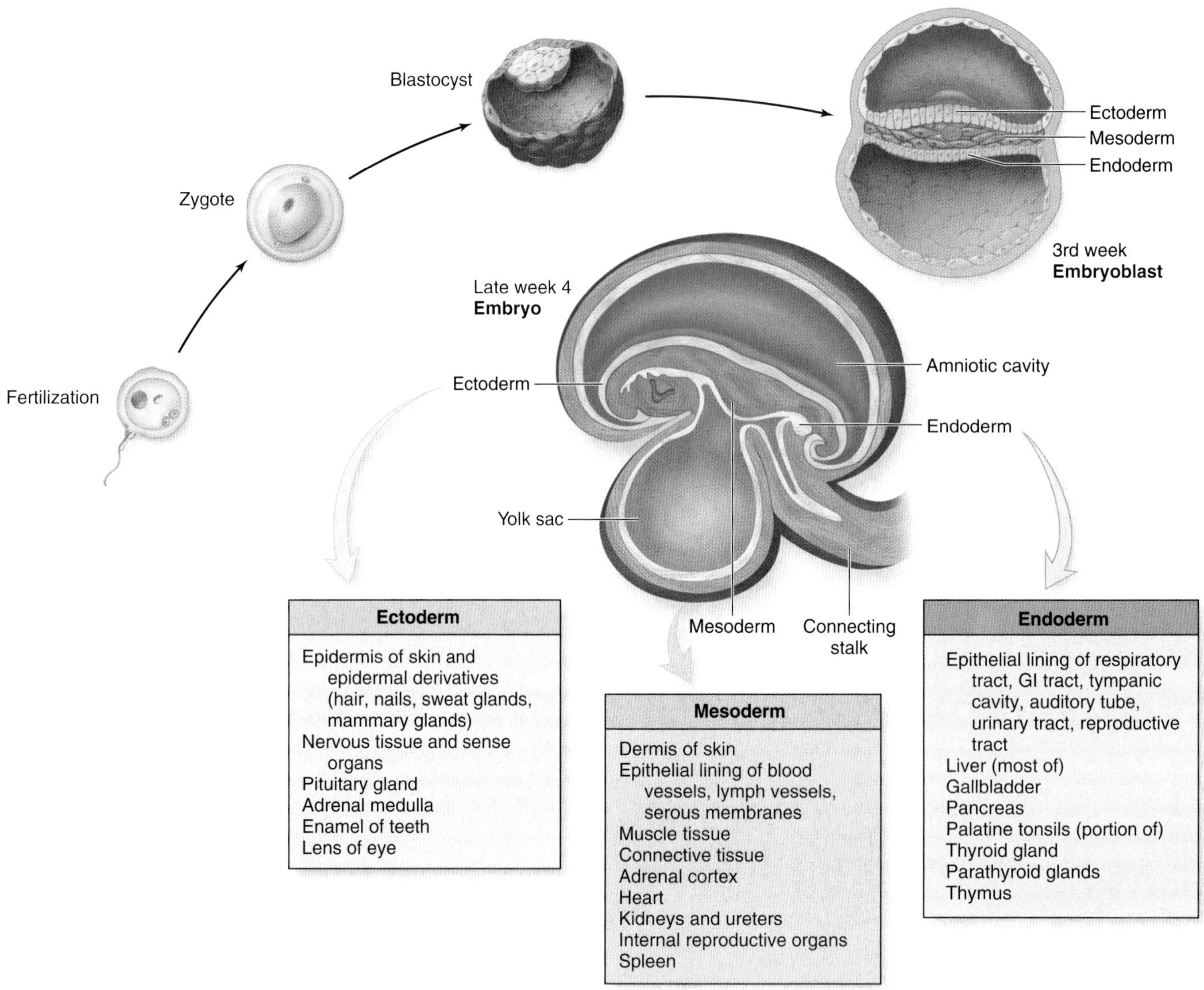

Figure 5.13 **Primary Germ Layers and Their Derivatives.** The primary germ layers (ectoderm, mesoderm, and endoderm) are formed during the third week of development and will give rise to all of the tissues in the body.

Ectoderm is initially located on the dorsal and external surfaces of the embryo. It is responsible for forming many externally placed structures, such as the epidermis of the skin, hair, nails, and the exocrine glands of the skin. Thus, some but not all epithelial tissues are derived from ectoderm. Tooth enamel, the lens of the eye, and the adrenal medulla are derived from ectoderm, as is all nervous tissue such as the brain, spinal cord, and nerves.

Mesoderm is the middle primary germ layer. It forms all muscle tissue and both the epithelial lining of vessels and serous membranes that line the body cavities. Mesoderm becomes mesenchyme, which then goes on to form all connective tissues in the body. The dermis of the skin, adrenal cortex, heart, spleen, kidneys and ureters, and internal reproductive structures are mesoderm-derived.

Endoderm becomes the innermost germ layer when the embryo undergoes shape changes. It forms the epithelial linings of the tympanic cavity (middle ear) and auditory tube, as well as the digestive, respiratory, reproductive, and urinary tracts. Endoderm also forms organs such as the thyroid gland, parathyroid glands, the thymus and portions of the palatine tonsils, as well as most of the liver, gallbladder, and pancreas.

WHAT DID YOU LEARN?

19 What are the three primary germ layers, and when do they form?

5.6b Tissue Modification

LEARNING OBJECTIVE

3. Describe how tissues may change in form, size, or number of cells.

Tissues may change in size, form, or number of cells in response to a stimulus. **Hypertrophy** refers to an increase in the size of the existing cells in a tissue, although the *number* of cells remains constant. For example, skeletal muscle cells may hypertrophy when a person undergoes a long-term rigorous exercise regimen. **Hyperplasia** (*plasso* = to form) is an increase in the number of cells in a tissue. Developing a "callus" on the palm of your hand is an example of these skin cells undergoing hyperplasia.

Shrinkage of tissue by a decrease in either cell size or cell number is called **atrophy** (at'rō-fē). Atrophy may result from normal aging (senile atrophy) or from failure to use an organ or tissue (disuse atrophy). If an individual becomes bedridden or must wear a cast for a broken bone, the affected muscles exhibit disuse atrophy as the skeletal muscle fibers become smaller. If the atrophy is not due to long-term problems, then typically physical therapy and a reuse of the tissues can minimize or reverse the atrophic changes.

As we place different stresses on our bodies, our tissues may actually transform into another type of tissue. Sometimes a mature

CLINICAL VIEW
Stem Cells

Why all the interest in stem cells?

Stem cells are immature, undifferentiated cells. These cells are able to divide into two cells, the first of which is another stem cell, and the other a cell that could differentiate into a specialized, mature cell with a unique function. Stem cells have generated interest in the scientific and medical communities because of their potential for repair or replacement of damaged or dying tissue.

What are the two basic characteristics of stem cells?

All stem cells exhibit two characteristics: self-renewal and potency. **Self-renewal** refers to their ability to divide repeatedly to produce both new cells for maturation and new stem cells. **Potency** is the potential for differentiation: Different stem cells have varying ability to differentiate into almost any type of cell. Stem cells exhibit the following four levels of potency: totipotency, pluripotency, multipotency, and unipotency:

- **Totipotent** stem cells have a "total potential," meaning that they exhibit the ability to differentiate into any cell type within an organism. A totipotent cell is produced when a secondary oocyte is fertilized by a sperm, giving rise to a zygote. The first few cell divisions of the zygote result in equally totipotent cells. Thus, only embryonic (and not adult) stem cells have the potential to be totipotent.

- **Pluripotent** stem cells are derived from totipotent stem cells. These stem cells are formed from the embryoblast portion (inner cell mass) of the **blastocyst**. The blastocyst is a ball of cells that develops during the first week of development from the zygote. The embryoblast is the portion of the blastocyst that will eventually become an embryo and then a fetus. Pluripotent stem cells can form cells in any of the tissue layers of the embryo, but they cannot form structures such as the placenta. Again, only embryonic stem cells have the potential to be pluripotent.

- **Multipotent** stem cells are derived from pluripotent stem cells. They have the capability to differentiate into a restricted number of some cell types and not others. For example, stem cells in the bone marrow may be stimulated to mature and differentiate into different types of blood cells, but not into some other types of cells. Some adult stem cells have the potential to be multipotent.

- **Unipotent** stem cells have the ability to differentiate into a single type of cell, yet these cells still retain the ability to renew themselves. Epithelial stem cells (discussed previously) are examples of unipotent stem cells. Many adult stem cells are unipotent.

What are the differences between embryonic and adult stem cells?

Stem cells may be categorized as either **embryonic stem cells** or **adult stem cells.** Embryonic stem cells include those that have begun to divide in the zygote and the cells in the blastocyst. Embryonic stem cells exhibit the greatest degree of potency—and thus, the greatest potential to differentiate into multiple cell types. In contrast, adult stem cells are the immature cells found in postnatal (already born) organisms. Adult stem cells typically are multipotent or unipotent, and thus they exhibit less potency than embryonic stem cells.

How are stem cells harvested?

Most embryonic stem cells must be harvested from a structure no more differentiated than a blastocyst. Most of these blastocysts were donated by families undergoing in vitro fertilization who had stored more blastocysts than needed for a successful pregnancy. If these blastocysts were not used by the family and not donated for research, they typically would be destroyed. Note that opponents of embryonic stem cell research counter that these blastocysts could be implanted and lead to viable infants who could be adopted, and any medical benefit from embryonic stem cells does not justify using them in research. Opponents also maintain that adult stem cell research should be explored instead.

Adult stem cells may be extracted from the bone marrow or tissue of an individual. These adult stem cells have been used to successfully treat certain blood and bone cancers, and research is ongoing about their effectiveness for diseases such as lung inflammation, stroke, and Parkinson disease. The main problem with adult stem cells is their limited potency, which suggests that their use for treatment in diseases is limited. Embryonic stem cells exhibit greater promise for treatment because of their greater potency.

Embryonic Stem Cells	Adult Stem Cells
Embryonic stem cells originate from the dividing cells of the zygote or embryoblast portion of the blastocyst. They may be totipotent or pluripotent.	Adult stem cells are the undifferentiated stem cells in the body after birth. They may be multipotent or unipotent.

Totipotent	Pluripotent	Multipotent	Unipotent
Totipotent stem cells are those cells formed from the zygote, and can give rise to every differentiated cell in the organism and the placenta.	**Pluripotent** stem cells are derived from the embryoblast portion of the blastocyst. This type of stem cell has the capability to mature and specialize into a cell within any tissue in the body (except the placenta).	**Multipotent** stem cells (as in bone marrow) are capable of differentiating into a restricted number of some cell types and not others.	**Unipotent** stem cells (such as epithelial stem cells) have the ability to differentiate into a single type of cell.

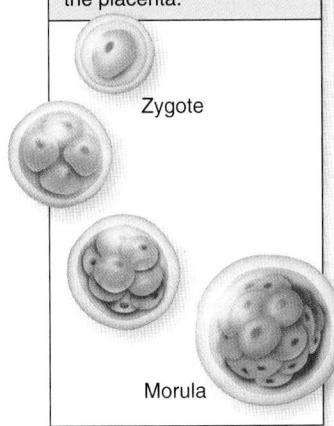

Zygote

Morula

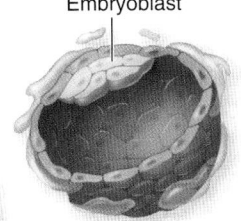

Embryoblast

Blastocyst

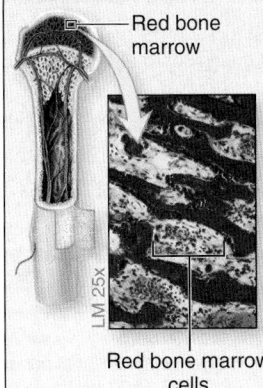

Red bone marrow

LM 25×

Red bone marrow cells

Epidermis of skin

Epithelial stem cell

epithelium changes to a different form of mature epithelium, a phenomenon called **metaplasia** (met-ă-plā′zē-ă; *metaplasis* = transformation). Metaplasia may occur as an epithelium adapts to environmental conditions. For example, smokers typically experience metaplastic changes in the epithelium of the trachea (windpipe). The smoke and its by-products are the environmental stressors that change the normal pseudostratified ciliated columnar epithelium lining the trachea to a nonkeratinized stratified squamous epithelium. (If a person quits smoking, their metaplastic epithelium will fairly quickly revert back to its pseudostratified ciliated columnar epithelium.) Another example occurs in some individuals with chronic acid reflux, also known as heartburn. Here, the nonkeratinized stratified squamous epithelium of the inferior esophagus may transform to a simple columnar epithelium, like that seen in the stomach.

Dysplasia (dis-plā′zē-ă; *dys* = bad) refers to abnormal tissue development. For example, cervical dysplasia may develop when a woman is exposed to the human papillomavirus. Dysplasias have the potential to turn into cancer (and thus are sometimes referred to as precancerous), but they also have the potential to revert back to normal tissue. Thus, cervical dysplasia has the potential to either turn into cancer or revert back to normal tissue, which is why dysplastic cells are closely monitored by health-care professionals.

When tissue growth proceeds out of control, a tumor composed of abnormal tissue develops, and the condition is called **neoplasia** (nē-ō-plā′zē-ă; *neo* = new, *plasis* = molding). Neoplasms (tumors) may be benign or malignant. A **benign** (bě-nīn′; *benignus* = kind) neoplasm typically is localized in its growth and does not spread, whereas a **malignant** neoplasm is characterized by invading local tissues and potentially **metastasizing**, or spreading, to other tissues of the body. A malignant neoplasm is commonly known as **cancer.** It is believed that most cancers are a result of DNA damage, either through environmental factors or genetics or a

INTEGRATE

CLINICAL VIEW
Gangrene

Gangrene is the necrosis (death) of the soft tissues of a body part due to a diminished or obstructed arterial blood supply to that region. The body parts most commonly affected are the limbs, fingers, or toes. Gangrene is a major complication of diabetics, who often suffer from diminished blood flow to their extremities as a consequence of their disease. Gangrene occurs in several different forms.

Intestinal gangrene usually follows an obstruction of the blood supply to the intestines. Without a sufficient blood supply, the tissue will undergo necrosis and gangrene. Untreated intestinal gangrene leads to death.

Dry gangrene is where the involved body part is desiccated, sharply demarcated, and shriveled, usually due to constricted blood vessels as a result of exposure to extreme cold. Dry gangrene can be a complication of frostbite or of a variety of cardiovascular diseases that restrict blood flow, primarily to the hands and feet.

Wet gangrene is caused by a bacterial infection of tissues that have lost their blood and oxygen supply. The cells in the dying tissue rupture and release fluid. The wet environment allows bacteria to flourish, and they often produce a foul-smelling pus. The most common bacteria associated with wet gangrene include *Streptococcus, Staphylococcus, Enterobacter,* and *Klebsiella.* Wet gangrene must be treated quickly with antibiotics and removal of the necrotic tissue.

Gas gangrene most often affects muscle tissue, and the bacteria associated with gas gangrene typically are *Clostridium.* As the bacteria invade the necrotic tissue, a release of gases from the tissue produces gas bubbles. These bubbles make a crackling sound in the tissue, especially if the patient is moved. Symptoms of fever, pain, and edema (localized swelling) occur within 72 hours of the initial trauma to the region. The treatment for gas gangrene is similar to that of wet gangrene.

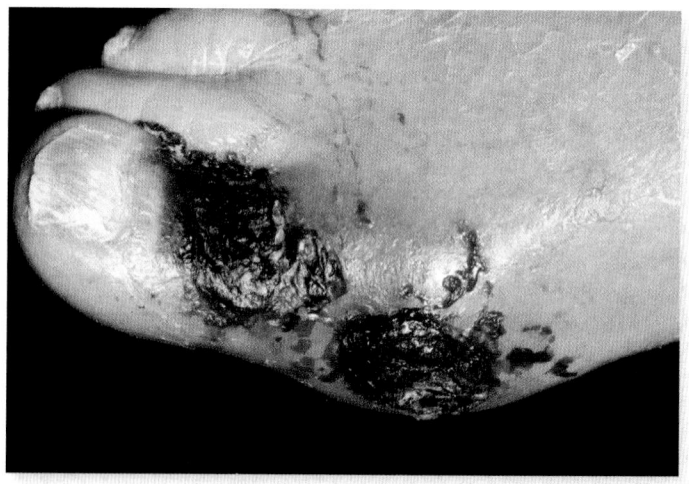

Dry gangrene of the foot.

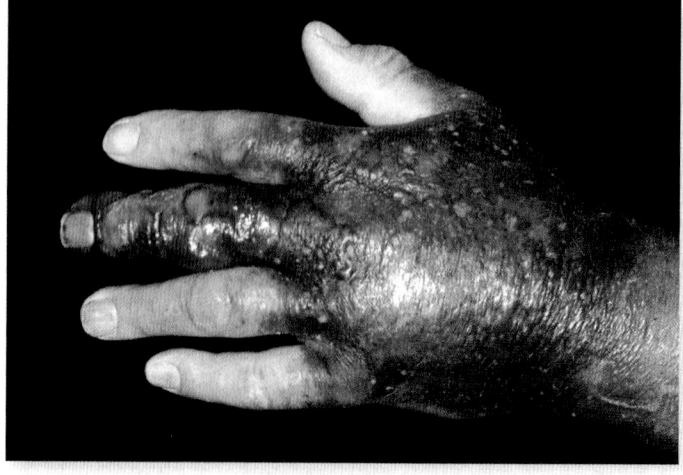

Gas gangrene in a recently amputated limb.

combination thereof. The growth and proliferation of malignant cells can interfere with the normal functioning of other tissues and organs, leading to the morbidity and mortality of the individual.

Necrosis is the term for tissue death. Necrosis typically occurs due to tissue damage that is not reversible, and an inflammatory response (see section 22.3d) usually occurs in the tissue in response to the damage. Gangrene is an example of tissue necrosis, and is described in detail (see Clinical View: "Gangrene").

 WHAT DID YOU LEARN?

20 What is the difference between metaplasia, dysplasia, and a malignant neoplasia?

5.6c Aging of Tissues

 LEARNING OBJECTIVE

4. List some changes that occur in tissues with age.

All tissues change as a result of aging. Proper nutrition, good health, normal circulation, and relatively infrequent wounds promote continued normal tissue functioning past middle age. Thereafter, the support, maintenance, and replacement of cells and extracellular matrix become less efficient. Physical damage and physiologic changes can alter the structure and chemical composition of many tissues. For example, as individuals age, epithelia become thinner and connective tissues lose their pliability and resiliency. The amount of collagen in the body declines with age, so tissue repair takes longer. Bones become brittle, and muscle and nervous tissue begin to atrophy. Poor diet and circulation problems accelerate these tissue declines. Eventually, cumulative losses from relatively minor damage or injury may contribute to major health problems.

 WHAT DID YOU LEARN?

21 How do epithelia and connective tissue change when we age?

INTEGRATE

CLINICAL VIEW
Tissue Transplant

Grafting is the process of surgically transplanting healthy tissue to replace diseased, damaged, or defective tissue. Tissue transplant may be characterized in four ways: an autograft, a syngenetic graft, an allograft, or a heterograft.

An **autograft** (aw'tō-graft; *autos* = self) is a tissue transplant from one site to a different site on the same individual. Autografts are often performed with skin, as healthy skin from one part of the body is grafted to another part of the body where the skin has been damaged by burns or chemicals. Because an autograft is a person's own body tissue, the body will not reject the tissue as foreign. However, autografts may not be feasible in certain situations, such as when the amount of skin damaged is so great that it would not be possible to transplant such a large portion of tissue.

A **syngenetic** (sin-jĕ-net'ik; *syn* = together) **graft**, also called an *isograft,* is a tissue transplant from one person to a genetically identical person (i.e., identical twins). Very few of us, however, have an identical twin, so this type of graft is not possible for most people.

An **allograft** (al'ō-graft; *allos* = other) is a tissue transplant from one person to another person who is genetically different. Many tissue types have been used as allografts, including skin, muscle, bone, and cartilage. The term allograft also is used for the transplantation of organs or parts of organs, such as heart valves, kidneys, and the liver. The patient and the organ donor must be as genetically similar as possible as the closer the match, the less likely the allograft will be rejected. The recipient of the transplanted organ(s) must take powerful immunosuppressant drugs, which help prevent the body from rejecting the organ. Unfortunately, these same drugs work by suppressing the entire immune system, making the transplant patient more susceptible to illness.

A **xenograft** (zē'nō-graft; *xeno* = foreign), also called a *heterograft* (hĕ'ter-ō-graft; *heteros* = other), is a tissue transplant from an animal into a human being. For example, porcine (pig) and bovine (cattle) tissues have been used as replacements for heart valves, blood vessels, and bone. As with an allograft, xenografts may be prone to rejection and thus may not last as long as a synergetic graft would.

- Tissues are classified into four general types: epithelial tissue, connective tissue, muscle tissue, and nervous tissue.

5.1 Epithelial Tissue: Surfaces, Linings, and Secretory Functions 152

- Epithelial tissue covers the surface of the body, lines body cavities, and forms secretory structures called glands.

5.1a Characteristics of Epithelial Tissue 152

- The characteristics of epithelia include cellularity, polarity, attachment to a basement membrane, avascularity, rich innervation, and high regeneration ability.

5.1b Functions of Epithelial Tissue 153

- Epithelia provide physical protection, are selectively permeable, produce secretions, and contain nerve endings to detect sensations.

5.1c Classification of Epithelial Tissue 153

- Epithelia are classified by the number of cell layers and the shape of the surface (apical) cells.
- A simple epithelium has only one layer of cells that is in direct contact with the basement membrane; a stratified epithelium has two or more layers of cells and only the deepest (basal) layer is in direct contact with the basement membrane.
- Examples of cell shape include squamous (cells are flattened), cuboidal (cells are about as tall as they are wide), and columnar (cells are taller than they are wide).
- Pseudostratified columnar epithelium appears stratified but is not; all cells are in contact with the basement membrane.
- Transitional epithelium contains several layers of rounded cells, and the epithelium appearance changes between a relaxed and distended state.

5.1d Glands 160

- Endocrine glands secrete hormones into the blood, whereas exocrine glands secrete their products onto the epithelial surface.
- Multicellular glands may be classified either by anatomic form (simple vs. compound and tubular vs. acinar) or physiologically by their method of secretion (merocrine, apocrine, or holocrine).

5.2 Connective Tissue: Cells in a Supportive Matrix 162

- Connective tissue supports, protects, and binds the body organs.

5.2a Characteristics of Connective Tissue 162

- Connective tissue contains cells, protein fibers, and a ground substance. The protein fibers and ground substance together form the extracellular matrix.

5.2b Functions of Connective Tissue 164

- Connective tissue provides physical protection, support, structural framework, binding of structures, storage, transport, and immune protection.

5.2c Embryonic Connective Tissue 165

- All connective tissues are derived from an embryonic connective tissue called mesenchyme.
- Mucous connective tissue is found only in the umbilical cord.

5.2d Classification of Connective Tissue 166

- Loose connective tissue has a high volume of ground substance; it is easily distorted and serves to cushion shocks.
- Dense connective tissue consists primarily of large amounts of protein fibers and relatively little ground substance.
- Supporting connective tissue (cartilage and bone) provides support and protection to the soft tissues and organs of the body.
- Fluid connective tissue (blood) contains formed elements, dissolved protein fibers, and a watery ground substance.

5.3 Muscle Tissue: Movement 174

- Skeletal muscle tissue is composed of long, cylindrical, multinucleated fibers that are striated. The nuclei are at the periphery of the fiber, and the tissue is under voluntary control.
- Cardiac muscle tissue is located in the heart wall. Its cells are branched, short, striated, and contain one or two centrally located nuclei. The tissue is under involuntary control.
- Smooth muscle tissue is found in the walls of internal organs; cells are fusiform (spindle-shaped), contain one centrally located nucleus, have no striations, and are under involuntary control.

5.4 Nervous Tissue: Information Transfer and Integration 175

- Nervous tissue contains neurons and glial cells, and forms the brain, spinal cord and nerves.
- Neurons receive stimuli and transmit nerve impulses.
- Glial cells support, protect, and nourish the neurons.

5.5 Integration of Tissues in Organs and Body Membranes 176

5.5a Organs 176

- An organ contains two or more tissue types that work together to perform specific, complex functions.

5.5b Body Membranes 176

- Body membranes line body cavities, cover the viscera, or cover the external surface of the body.
- Mucous membranes secrete mucus and line body cavities that communicate with the exterior.
- Serous membranes secrete serous fluid and line internal cavities that do not open to the exterior.
- The cutaneous membrane is the skin, and it protects internal body structures.
- Synovial membranes secrete synovial fluid and line the inner surfaces of synovial joint cavities.

5.6 Tissue Development and Aging   177

5.6a Tissue Development   177

- In the embryo, the primary germ layers (ectoderm, mesoderm, and endoderm) give rise to all of the tissues in the body.
- Ectoderm forms the epidermis, nervous tissue, adrenal medulla, lens of the eye, and tooth enamel.
- Mesoderm forms the dermis, all connective and muscle tissue, serous membranes, the heart, adrenal cortex, kidneys and ureters, and internal reproductive organs.
- Endoderm forms the epithelial linings of the digestive, respiratory, reproductive, and urinary tracts as well as the thyroid gland, parathyroid glands, the thymus and portions of the palatine tonsils, and most of the liver, gallbladder, and pancreas.

5.6b Tissue Modification   178

- Metaplasia is a change from one mature epithelial type to another epithelium in response to injury or stress.
- Hypertrophy is an increase in cell size; hyperplasia is an increase in cell number.
- Atrophy is a decrease in tissue size, caused either by a decrease in the number of cells, a decrease in the size of the remaining cells, or both.

5.6c Aging of Tissues   181

- When tissues age, repair and maintenance become less efficient, and the structure of many tissues is altered.

CHALLENGE YOURSELF

Do You Know the Basics?

_____ 1. Which tissue contains a calcified ground substance and is specialized for structural support?

    a. muscle tissue

    b. dense regular connective tissue

    c. areolar connective tissue

    d. bone connective tissue

_____ 2. Which of the following is *not* a characteristic of areolar connective tissue?

    a. predominant cell type is the fibroblast

    b. abundant ground substance

    c. densely packed protein fibers

    d. occurs in the subcutaneous layer of the skin

_____ 3. _____ membranes line body cavities that typically open to the exterior, such as the nasal cavity.

    a. Mucous   b. Serous

    c. Cutaneous   d. Synovial

_____ 4. Which of the following is a correct statement about a simple epithelium?

    a. All of the cells are in direct contact with the basement membrane.

    b. It is designed for protection against mechanical abrasion.

    c. It is formed from multiple layers of cells.

    d. It may contain the protein keratin.

_____ 5. All of the following are characteristics of an epithelium *except*

    a. It is selectively permeable.

    b. It may form exocrine glands.

    c. Its cells are highly regenerative.

    d. It contains abundant blood vessels.

_____ 6. Which connective tissue type is composed of cells called chondrocytes and may be surrounded by a covering called perichondrium?

    a. bone   b. dense irregular

    c. cartilage   d. areolar

_____ 7. Which tissue type is formed from mesoderm?

    a. epidermis (outer layer) of skin

    b. nervous tissue

    c. smooth muscle tissue

    d. epithelial lining of the urinary bladder

_____ 8. Which muscle type consists of long, cylindrical, striated fibers with multiple nuclei located at the periphery of the fiber?

    a. smooth muscle   b. skeletal muscle

    c. cardiac muscle   d. All of these are correct.

_____ 9. Which epithelial tissue type lines the trachea (air tube)?

    a. simple columnar epithelium

    b. pseudostratified ciliated columnar epithelium

    c. simple squamous epithelium

    d. transitional epithelium

_____ 10. A gland that releases its secretion by exocytosis out of secretory vesicles is called a(n) _____ gland.

    a. merocrine   b. apocrine

    c. holocrine   d. All of these are correct.

11. What are some characteristics of all types of epithelium?

12. Describe the two main criteria by which epithelia are classified.

13. List the epithelia type that lines (a) the lumen of the stomach, (b) the oral cavity, (c) the urinary bladder, and (d) the air sacs (alveoli) of the lungs.

14. What are the types of exocrine glands, classified by method of secretion, and how does each method of secretion work?

15. Name the four types of body membranes, and cite a location of each type.

16. What characteristics are common to all connective tissues?

17. What are the main structural differences between dense regular and dense irregular connective tissue?

18. In what regions of the body would you expect to find hyaline cartilage, fibrocartilage, and elastic cartilage, and why would these types be located in these regions?

19. What are the similarities and differences between skeletal muscle, cardiac muscle, and smooth muscle?

20. What is the difference between neurons and glial cells in nervous tissue?

Can You Apply What You've Learned?

1. John is a 53-year-old construction worker who has come into your office complaining of a sore knee joint. You see a buildup of fluid close to the patella (kneecap) but deep to the skin and suspect the soreness is due to bursitis, an inflammation of membranes that surround some joints. Which type of body membrane is inflamed?

    a. cutaneous membrane   b. serous membrane

    c. synovial membrane   d. mucous membrane

2. Your optometrist shines a light in your eye and notices your pupil constricts (becomes smaller) in response to the light. She tells you the iris (the colored part of the eye) is a muscle that adjusts the size of the pupil automatically in response to the amount of light entering the eye. Based on this information, which type of muscle do you think forms the iris?

    a. skeletal   b. cardiac   c. smooth   d. visual

Use the following paragraph to answer questions 3–5.

During a biology lab, Erin used a cotton swab to remove some tissue from the inner side of her cheek. She then placed the tissue on a slide to examine it under the microscope.

3. Why was the tissue able to be removed so easily without causing injury to the rest of Erin's cheek?

    a. The tissue contained abundant amounts of ground substance to keep the tissue puffy and relatively intact.

    b. The tissue contained multiple layers of cells, so removing a few cells wouldn't harm the rest of the tissue.

    c. The tissue contained lots of blood vessels, so blood filled any gaps left when the original cells were removed.

    d. The remaining cells were interconnected by intercalated discs, which formed a strong bond between the cells.

4. When Erin examined the cells under the microscope, what shape were the cells?

    a. squamous   b. cuboidal

    c. columnar   d. circular

5. If Erin removed a large chunk of this tissue from the same site, the shape and characteristics of the deepest cells would be

    a. the same shape as the original cells under the microscope.

    b. cuboidal.

    c. binucleated and circular.

    d. squamous.

Can You Synthesize What You've Learned?

1. During a microscopy exercise in the anatomy laboratory, a student makes the following observations about a tissue section: (a) The section contains some different types of scattered protein fibers—that is, they exhibit different widths, some are branched, some are long and unbranched. (b) The observed section has some "open spaces"—that is, places between both cells and the fibers that appear clear with no recognizable features. (c) Several connective tissue cell types are scattered throughout the section, but these cells are not grouped tightly together. What type of tissue is the student observing? Where might this tissue be found in the body?

2. Your father is suffering from a painful knee joint. He has been told that he either has the early stages of arthritis or some inherent joint problems. His friend recommends that he take a chemical supplement with his meals, called chondroitin sulfate, which has been shown to help some people with joint aches and pains. Specifically, this supplement may alleviate symptoms caused by degenerated cartilage on the surfaces on bones in joints. Based on your knowledge of connective tissues, do you think the chondroitin sulfate supplements could help your father's knee problems?

INTEGRATE

ONLINE STUDY TOOLS  ■ connect |ANATOMY & PHYSIOLOGY  ■LEARNSMART® AP|R

The following study aids may be accessed through Connect.

Clinical Case Study: Did Abraham Lincoln Suffer from Marfan Syndrome?

Interactive Questions: This chapter's content is served up in a number of multimedia question formats for student study.

LearnSmart: Topics and terminology include characteristics and functions of epithelial tissue; characteristics and functions of connective tissue; embryonic connective tissue; movement of muscle tissue; information transfer and integration of nervous

tissue; tissue integration in organs and body membranes; tissue development and aging

Anatomy & Physiology Revealed: Topics include simple epithelia; pseudostratified columnar epithelia; stratified columnar epithelia; loose connective tissue types; dense connective tissue types; supporting connective tissue types; compact bone; fluid connective tissue: blood; muscle tissue types; nervous tissue

Animation: Pleural membranes

Integumentary System

Module 4: Integumentary System

INTEGRATE

CAREER PATH
Dermatologist

A dermatologist is a physician that cares for the integumentary system (e.g., skin, hair, nails, and exocrine glands) and treats its diseases. The skin is our interface with the world around us and provides clues about our health. The dermatologist must have a thorough understanding of the tissues that compose the skin and how the integumentary system repairs itself when exposed to trauma. Here, a dermatologist is inspecting a patient's skin for evidence of melanoma, which is a malignant cancer of melanocytes (one cell type found in the epidermis of the skin).

The **integument** (in-teg'ū-ment; *integumentium* = a covering) is the skin that covers your body. Skin is also known as the **cutaneous** (kū-tā'nē-ŭs; *cutis* = skin) **membrane.** The **integumentary** (in-teg-ū-men'tă-rē) **system** consists of the skin and its derivatives: nails, hair, sweat glands, and sebaceous (oil) glands. On average, each square inch of skin contains up to 20 feet of blood vessels, 650 sweat glands, 100 sebaceous glands, and over 1000 nerve endings. Our skin is a barrier to the outside world and is subjected to trauma, harmful chemicals, pollutants, microbes, and damaging sunlight. Changes in the color of the skin may reflect body disorders or anomalies; skin changes or lesions sometimes reflect systemic infections or disease. The scientific study and treatment of the integumentary system is called **dermatology** (der-mă-tol'ō-jē; *derma* = skin, *logos* = study). In this chapter, we examine the specific layers of the integument and the tissues that compose each layer. We also will see how the composition of the integumentary system is related to its functions. We conclude by discussing integument repair and how the integumentary system ages.

6.1 Composition and Functions of the Integument

The integument is the body's largest organ and is composed of all tissue types that function in concert to protect internal body structures. Its surface is an epithelium that protects underlying body layers. The connective tissue that underlies the epithelium provides strength and resilience to the skin. This connective tissue also contains smooth muscle associated with hair follicles (arrector pili) that alters hair position. Finally, nervous tissue detects and monitors sensory stimuli in the skin, which provide information about touch, pressure, temperature, and pain.

The integument accounts for 7% to 8% of the body weight and covers the entire body surface with an area that varies between about 1.5 and 2.0 square meters (m^2). Its thickness ranges between 1.5 millimeters (mm) and 4 mm or more, depending on body location. (For comparison, a sheet of copier paper is about 0.1 mm thick, so the thickness of the skin would range between 15 and 40 sheets of paper.) The integument consists of two distinct layers: a layer of stratified squamous epithelium called the epidermis, and a deeper layer of primarily dense irregular connective tissue called the dermis (**figure 6.1**). Deep to the dermis is a layer of areolar and adipose connective tissue called the subcutaneous layer, or hypodermis. The subcutaneous layer is not part of the integumentary system; however, it is described in this chapter because it is closely involved with both the structure and function of the skin.

6.1a Epidermis

LEARNING OBJECTIVES

1. Describe the five layers (strata) of the epidermis.
2. Differentiate between thick skin and thin skin.
3. Explain what causes differences in skin color.

The epithelium of the integument is called the **epidermis** (ep-i-derm'is; *epi* = on, *derma* = skin). It is a keratinized, stratified squamous epithelium (see section 5.1c).

Careful examination of the epidermis, from the basement membrane to its surface, reveals several specific layers, or strata. From deep to superficial, these layers are the stratum basale, stratum spinosum, stratum granulosum, stratum lucidum (found in thick skin only), and the stratum corneum (**figure 6.2**). The first three strata listed are composed of living keratinocytes, whereas the most superficial two strata contain dead cells.

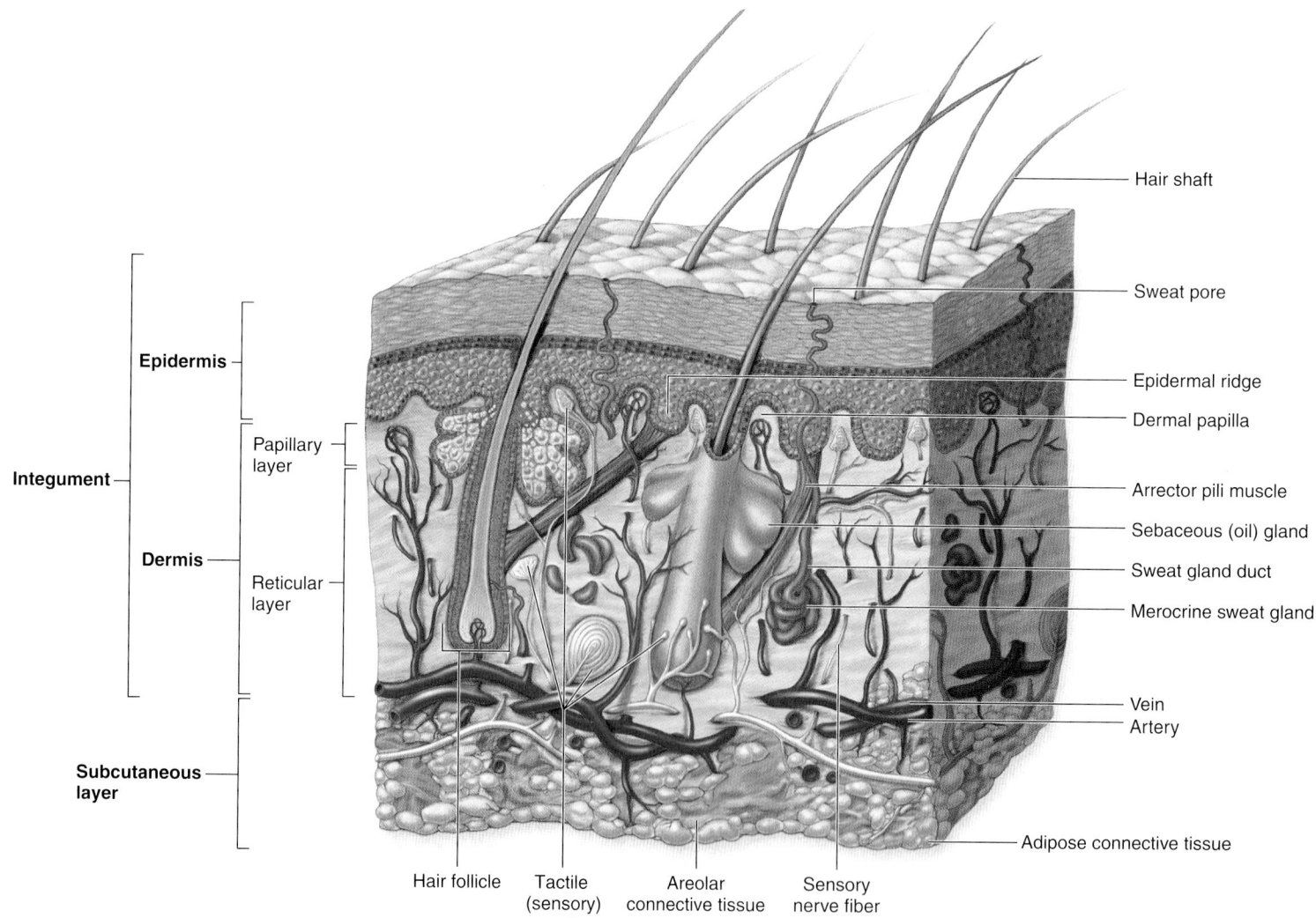

Figure 6.1 Layers of the Integument. A diagrammatic sectional view through the integument shows its relationship to the underlying subcutaneous layer. AP|R

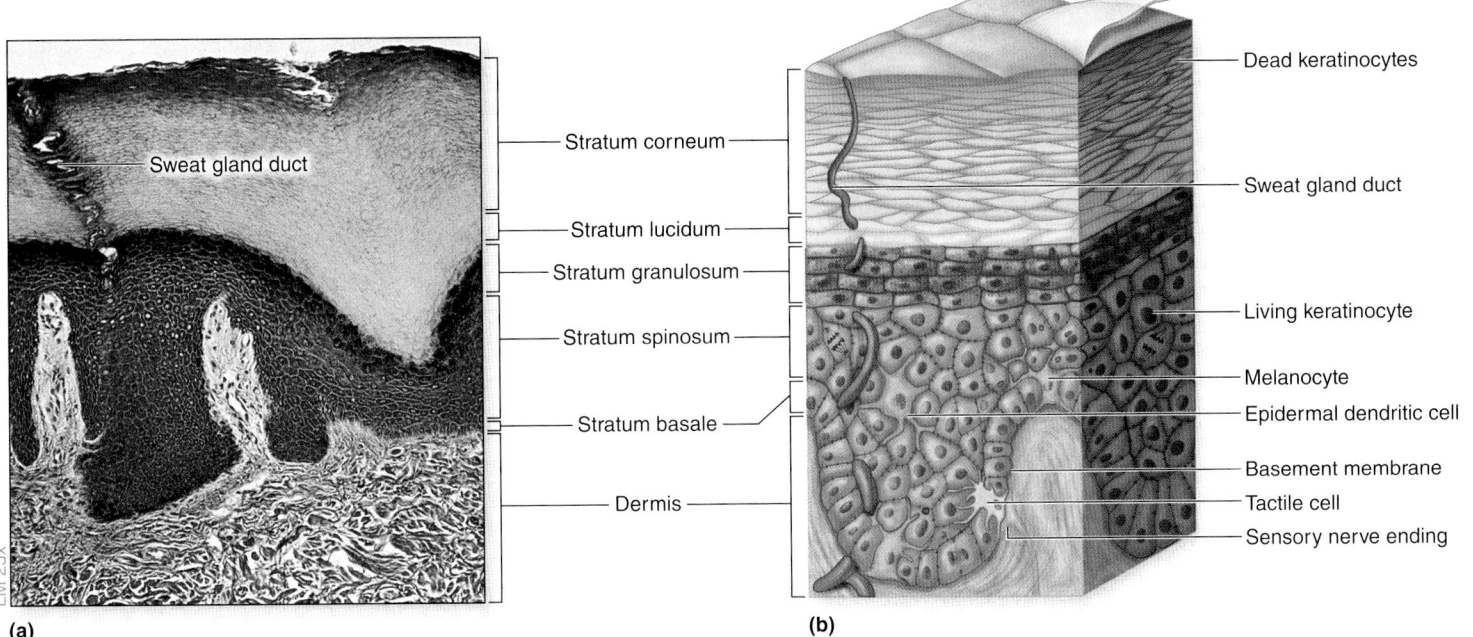

(a)

(b)

Figure 6.2 Epidermal Strata. (*a*) Photomicrograph and (*b*) diagram compare the order and relationship of the epidermal strata in thick skin. AP|R

Stratum Basale

The deepest epidermal layer is the **stratum basale** (strat′ŭm bah-sā′lē), also known as the *stratum germinativum*, or *basal layer*. This single layer of cuboidal to low columnar cells is tightly attached to an underlying basement membrane that separates the epidermis from the connective tissue of the adjacent dermis. Three types of cells occupy the stratum basale (figure 6.2*b*):

1. **Keratinocytes** (ke-ra′ti-nō-sīt; *keras* = horn) are the most abundant cell type in the epidermis and are found throughout all epidermal strata. The stratum basale is dominated by large keratinocyte stem cells, which divide to generate new cells that replace dead keratinocytes shed from the surface. Their name is derived from their synthesis of **keratin,** a protein that strengthens the epidermis considerably. Keratin is a family of fibrous structural proteins that are both tough and insoluble. Fibrous keratin molecules can twist and intertwine around each other to form helical intermediate filaments of the cytoskeleton. The keratins found in epidermal cells of the skin are called *cytokeratins*. Their structure in these cells gives skin its strength and makes the epidermis almost waterproof.

2. **Melanocytes** (mel′ă-nō-sīt; *melano* = black) have long, branching processes and are scattered among the keratinocytes of the stratum basale. They produce and store the pigment **melanin** (mel′ă-nin) in response to ultraviolet light exposure. Their cytoplasmic processes transfer melanin pigment within **melanosomes** (mel′ă-nō-sō mes) to the keratinocytes within the basal layer and sometimes in more superficial layers. This pigment (which includes the colors black, brown, tan, or yellow-brown) accumulates around the nucleus of the keratinocyte and shields the nuclear DNA from ultraviolet radiation. The darker tones of the skin result from melanin produced by the melanocytes. Thus, "tanning" is the result of the melanocytes attempting to block UV light from causing mutations in the DNA of your skin cells!

3. **Tactile cells,** also called *Merkel cells,* are few in number and found scattered among the cells within the stratum basale. Tactile cells are sensitive to touch and, when compressed, they release chemicals that stimulate sensory nerve endings, providing information about objects touching the skin (see section 16.2a).

Stratum Spinosum

Several layers of polygonal keratinocytes form the **stratum spinosum** (spī-nō′sŭm), or *spiny layer.* Each time a keratinocyte stem cell in the stratum basale divides, a daughter cell is pushed toward the external surface from the stratum basale, while the other cell remains as a stem cell in the stratum basale. Once this new cell enters the stratum spinosum, it begins to differentiate into a non-dividing, highly specialized keratinocyte. Sometimes the deepest cells in this layer still undergo mitosis to help replace epidermal cells that exfoliate from the surface. The nondividing keratinocytes in the stratum spinosum attach to their neighbors by many membrane junctions called desmosomes (described in section 4.6d).

The process of preparing epidermal tissue for observation on a microscope slide shrinks the cytoplasm of the cells in the stratum spinosum. Because the cytoskeletal elements and desmosomes remain intact, the shrunken stratum spinosum cells resemble miniature porcupines that are attached to their neighbors. This spiny appearance accounts for the name of this layer.

In addition to the keratinocytes, the stratum spinosum also contains the fourth epidermal cell type called **epidermal dendritic** (*Langerhans*) **cells** (figure 6.2*b*). Epidermal dendritic cells are immune cells that help fight infection in the epidermis. These immune cells are often present in the stratum spinosum and stratum granulosum, but they are not identifiable in standard histologic preparations. Their phagocytic activity initiates an immune response to protect the body against pathogens that have penetrated the superficial epidermal layers as well as epidermal cancer cells. (See chapter 22 for a detailed discussion of the immune system.)

Stratum Granulosum

The **stratum granulosum** (gran-ū-lō′sum), or *granular layer*, consists of three to five layers of keratinocytes superficial to the stratum spinosum. Within this stratum begins a process called **keratinization** (ker′ă-tin-i-zā′shŭn), where the keratinocytes fill up with the protein keratin, and in so doing, cause both the cell's nucleus and organelles to disintegrate and the cell dies. Keratinization is not complete until the cells reach the more superficial epidermal layers. A fully keratinized cell is dead (because it has neither a nucleus nor organelles), but it is structurally strong because of the keratin it contains.

Stratum Lucidum

The **stratum lucidum** (lū′sĭ-dum), or *clear layer*, is a thin, translucent region of about two to three cell layers that is superficial to the stratum granulosum. This stratum is found only in the thick skin within the palms of the hands and the soles of the feet. Cells occupying this layer appear pale and featureless and have indistinct boundaries. The keratinocytes within this layer are flattened and filled with the translucent protein called **eleidin** (ē-lē′i-din), which is an intermediate product in the process of keratin maturation. This layer helps protect the skin from ultraviolet light.

Stratum Corneum

The **stratum corneum** (kōr′nē-ŭm; *corneus* = horny), or *hornlike layer*, is the most superficial layer of the epidermis. It is the stratum you see when you look at your skin. The stratum corneum consists of about 20 to 30 layers of dead, scaly, interlocking keratinized cells. The dead cells are **anucleate** (lacking a nucleus) and are tightly packed together.

A keratinized, or *cornified*, epithelium contains large amounts of keratin. After keratinocytes are formed from stem cells within the stratum basale, they change in structure and in their relationship to their neighbors as they progress through the different strata until they eventually reach the stratum corneum and are sloughed off its external surface. Major changes during keratinocyte migration include synthesis of keratin and the loss of the nucleus and organelles as described. What remains of these cells in the stratum corneum is essentially keratin protein enclosed in a thickened plasma membrane. Migration of the keratinocyte to the stratum corneum occurs during the first 2 weeks of the

keratinocyte's life. The dead, keratinized cells usually remain for an additional 2 weeks in the exposed stratum corneum layer, providing a barrier before they are shed. Overall, individual keratinocytes are present in the integument for about 1 month following their formation.

The normally dry stratum corneum presents a thickened surface unsuitable for the growth of many microorganisms. Additionally, some secretions onto the surface of the epidermis from exocrine glands help prevent the growth of microorganisms on the epidermis, thus supporting its barrier function.

Variations in the Epidermis

The epidermis exhibits variations both between different body regions within one individual and differences between individuals. The epidermis varies in its thickness, coloration, and skin markings.

Thick Skin Versus Thin Skin Over most of the body, the skin ranges in thickness between 1 mm and 2 mm. Skin is classified as either thick or thin based on the number of epidermal strata and the relative thickness of the epidermis, rather than the thickness of the entire integument **(figure 6.3)**.

Thick skin is found on the palms of the hands, and the soles of the feet. All five epidermal strata occur in the thick skin. The epidermis of thick skin ranges between 0.4 mm and 0.6 mm thick. It houses sweat glands but has no hair follicles or sebaceous (oil) glands.

Thin skin covers most of the body. It lacks a stratum lucidum, so it has only four specific layers in the epidermis. Thin skin contains the following structures: hair follicles, sebaceous glands, and sweat glands. The epidermis of thin skin ranges from 0.075 mm to 0.150 mm thick.

WHAT DO YOU THINK?

1 Why do you think thick skin lacks hair follicles and sebaceous glands? Think about the locations of where thick skin is found, and how that may interfere with the function of skin in that area.

Skin Color Normal skin color results from a combination of the colors of hemoglobin, melanin, and carotene. **Hemoglobin** (hē-mō-glō′bin; *haima* = blood) is an oxygen-binding protein present in red blood cells (see section 18.3). It exhibits a bright red color upon binding

---

INTEGRATE

LEARNING STRATEGY

Integrate lab and lecture material: Follow these steps to help you identify the epidermal strata under the microscope:

1. **Determine if the layer is closer to the free surface or is deeper.** Remember the stratum corneum forms the free surface, whereas the stratum basale forms the deepest epidermal layer.

2. **Examine the shape of the keratinocytes.** The stratum basale contains cuboidal to low columnar keratinocytes, the stratum spinosum contains polygonal keratinocytes, and the stratum lucidum and corneum contain squamous keratinocytes.

3. **See if the keratinocytes have a nucleus or are anucleate.** When they are still alive (as in the strata basale, spinosum,

and granulosum), you are able to see nuclei. The stratum lucidum and corneum layers contain dead, anucleate keratinocytes.

4. **Count the layers of keratinocytes in the stratum.** The stratum basale has only one layer of keratinocytes, and the stratum corneum contains 20 to 30 layers of keratinocytes. The other layers contain about two to five layers of keratinocytes.

5. **Determine if the cytoplasm of the keratinocytes contain visible granules.** If the keratinocytes contain visible granules, you likely are looking at the stratum granulosum.

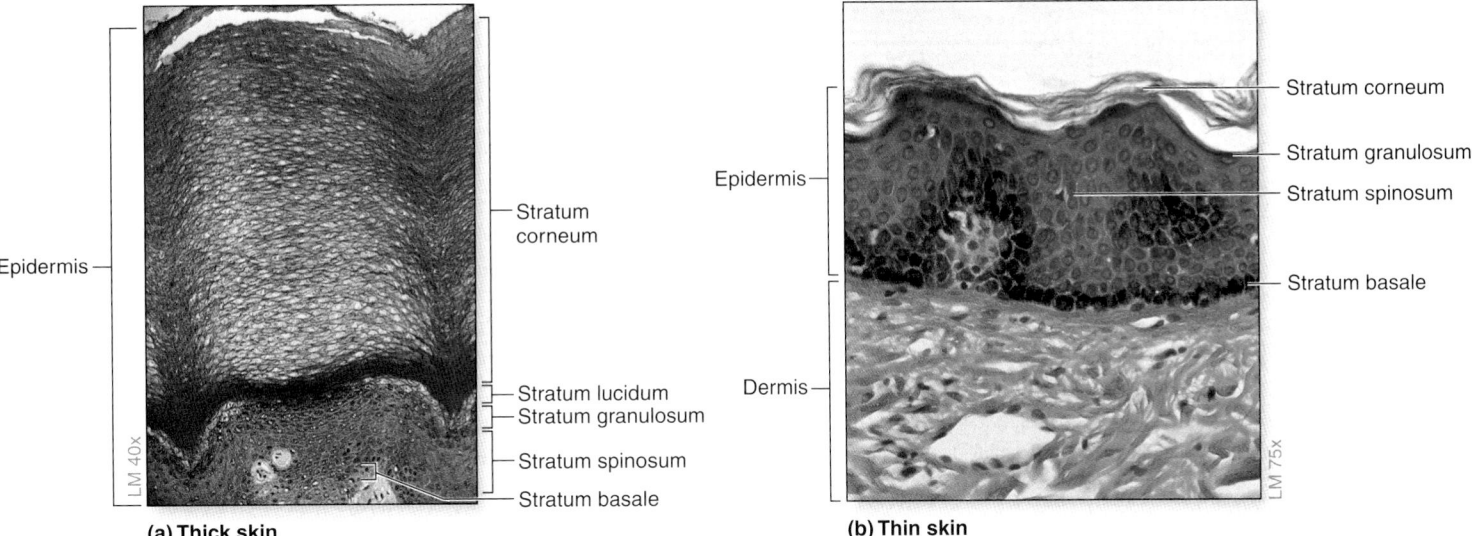

(a) Thick skin

Epidermis

Stratum corneum

Stratum lucidum
Stratum granulosum
Stratum spinosum
Stratum basale

LM 40x

(b) Thin skin

Epidermis

Stratum corneum
Stratum granulosum
Stratum spinosum

Stratum basale

Dermis

LM 75x

Figure 6.3 Thick Skin and Thin Skin. The stratified squamous epithelium of the epidermis varies in thickness, depending upon the region of the body in which it is located. (*a*) Thick skin contains all five epidermal strata and covers the soles of the feet and the palms of the hands. (*b*) Thin skin covers most body surfaces; it contains only four epidermal strata and lacks a stratum lucidum. AP|R

oxygen, thus giving blood vessels in the dermis a reddish tint that is seen most easily in lightly pigmented individuals. If the blood vessels in the superficial layers vasodilate (i.e., the blood vessel diameter increases), such as during physical exertion, then the red tones are much more visible.

Melanin is a pigment produced and stored in melanocytes, and occurs in a variety of black, brown, tan, and yellow-brown shades as mentioned earlier. Recall that melanin is transferred in melanosomes from melanocytes to keratinocytes in the stratum basale. Because keratinocytes are displaced toward the stratum corneum, melanocyte activity affects the color of the entire epidermis (**figure 6.4**).

The amount of melanin in the skin is determined by both heredity and light exposure. All people have about the same *number* of melanocytes. However, melanocyte *activity* and the *color* of the melanin produced by these cells varies among individuals and races, resulting in different skin color tones. Darker-skinned individuals have melanocytes that produce relatively more and darker melanin than lighter-skinned individuals. Further, these more active melanocytes tend to

package melanin into cells in the more superficial epidermal layers, such as the stratum granulosum. Recall that exposure to ultraviolet light stimulates melanocytes to make more melanin.

Carotene (kar′ō-tēn) is a yellow-orange pigment that is acquired from various yellow-orange vegetables, such as carrots, corn, and squashes. Normally, it accumulates inside keratinocytes of the stratum corneum and in the subcutaneous fat. Within the body, carotene is converted into vitamin A, which plays an important role in normal vision. Vitamin A has also been thought to reduce potentially dangerous molecules called *free radicals* that form during normal metabolic activity in the body. Additionally, carotene may improve immune cell number and activity.

Albinism (al′bi-nizm) is an inherited recessive condition where the enzyme needed to produce melanin is nonfunctional. As a result, melanocytes are unable to produce melanin. Individuals who have albinism typically have white hair, pale skin, and pink irises.

Skin Markings   A **nevus** (nē-vŭs; pl., *nevi*), commonly called a *mole,* is a harmless localized overgrowth of melanocytes. On rare

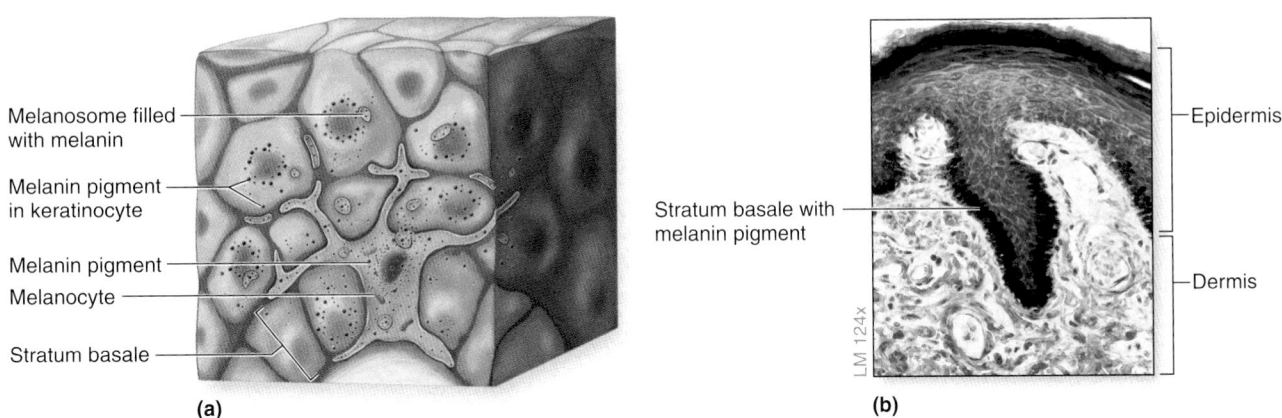

Melanosome filled with melanin

Melanin pigment in keratinocyte

Melanin pigment
Melanocyte

Stratum basale

Stratum basale with melanin pigment

Epidermis

Dermis

LM 124x

(a)

(b)

Figure 6.4 Production of Melanin by Melanocytes. Melanin gives a yellow-brown to tan to brown or black color to the skin. (*a*) Melanosomes in melanocytes transport the melanin pigment to the keratinocytes, where the pigment surrounds the nucleus. (*b*) Melanin is incorporated into the keratinocytes primarily of the stratum basale.

INTEGRATE

CLINICAL VIEW
UV Radiation, Sunscreens, and Sunless Tanners

The sun generates three forms of ultraviolet radiation: UVA (ultraviolet A), UVB (ultraviolet B), and UVC (ultraviolet C). The wavelength of UVA ranges between 320 and 400 nanometers (nm), that of UVB ranges between 290 and 320 nm, and UVC ranges between 100 and 280 nm. In contrast, visible light ranges begin at about 400 nm (the deepest violet). UVC rays are absorbed by the upper atmosphere and do not reach the earth's surface; UVA and UVB rays do reach the surface and can affect individuals' skin color.

UVA light commonly is termed *tanning rays,* whereas UVB often is called *burning rays.* Many tanning salons claim to provide a safe tan because they use only UVA rays. However, UVA rays can cause burning as well as tanning, and they also inhibit the immune system. Both UVA and UVB rays are believed to cause skin cancer. Thus, there is no such thing as a healthy suntan.

Sunscreens are lotions that contain materials that absorb or block UVA and UVB rays. Sunscreens can help protect the skin, not only for light-skinned people but also for those with darker skin colors—but *only* if they are used correctly. First, sunscreen must be applied liberally over all exposed body surfaces, and reapplied after being in the water or perspiring. Second, a sunscreen must have a high enough **SPF (sun protection factor)**. SPF is a number determined experimentally by exposing subjects to a light spectrum. The amount of light that induces redness in sunscreen-protected skin, divided by the amount of light that induces redness in unprotected skin, equals the SPF. For example, a sunscreen with an SPF of 15 will delay the onset of sunburn in a person who would otherwise burn in 10 minutes to 150 minutes. However, it is never safe to assume that a sunscreen will protect you completely from the sun's harmful rays.

Sunless tanners create a tanned, bronzed skin without UV light exposure. The most effective ones contain **dihydroxyacetone (DHA)** as their active ingredient. Sunless tanners do not affect melanin production. Instead, when applied to the epidermis, DHA interacts with the amino acids in the cells to produce a darkened, brownish color. Because only the most superficial epidermal cells are affected, the color change is temporary and lasts about 5 to 7 days. There are some sunless tanners on the market that contain other chemicals, but they do not appear to be as effective.

Most sunless tanners contain no sunscreen and offer no protection against UV rays because protective melanin has not been produced. Thus, individuals who use sunless tanners should also apply sunscreen to protect their skin.

---

occasions, a nevus may become malignant, typically as a consequence of excessive UV light exposure. Thus, nevi should be monitored for changes that may suggest malignancy. **Freckles** are yellowish or brown spots that represent localized areas of increased melanocyte activity, not an increase in melanocyte numbers. A freckle's degree of pigmentation varies and is dependent upon both sun exposure and heredity.

A **hemangioma** (he-man′jē-ō′mă) is an anomaly that results in skin discoloration due to blood vessels that proliferate to form a benign tumor. **Capillary hemangiomas,** or *strawberry-colored birthmarks,* appear in the skin as bright red to deep purple nodules that are usually present at birth and disappear in childhood. However, their development may occur in adults. **Cavernous hemangiomas,** also known as *port-wine stains,* involve larger dermal blood vessels and may last a lifetime.

Friction ridges are another type of skin marking. These ridge patterns are along the contours of the skin, varying from small conical pegs (in thin skin) to the complex arches and whorls. Friction ridges are found on the fingers (fingerprints), palms, soles, and toes **(figure 6.5)**. These ridges are formed from large folds and valleys of both dermis and epidermis. They help increase friction on contact, so that our hands can firmly grasp items and our feet do not slip when we walk barefoot. Some researchers have suggested friction ridges also provide flexibility to the skin and allow it to deform without being damaged. When sweat glands and oil glands release their secretions, noticeable fingerprints may be left on touched surfaces. Each individual has a unique pattern of friction ridges, allowing matching of prints and identification of individuals.

WHAT DID YOU LEARN?

1 As you trim your roses, a thorn penetrates your palm through all epidermal strata. What are the layers of the epidermis penetrated, starting from the surface of the skin?

2 Briefly describe the process of keratinization. Where does it occur? Why is it important?

3 How does hemoglobin contribute to skin color?

4 What is the function of friction ridges?

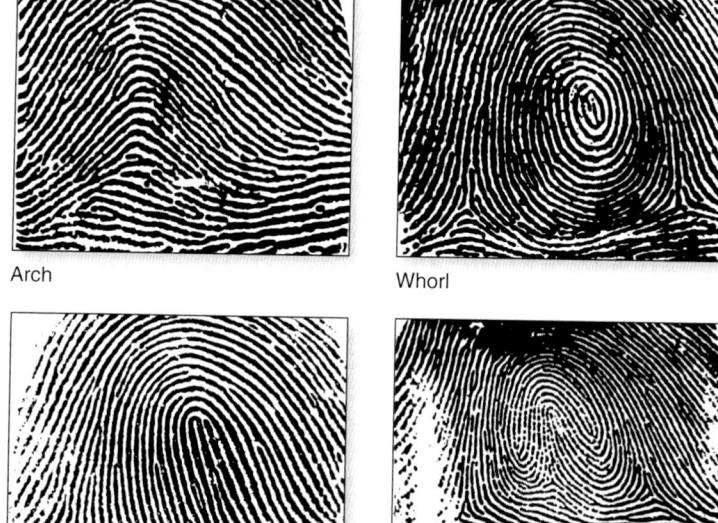

Arch

Whorl

Loop

Combination

Figure 6.5 Friction Ridges of Thick Skin. Friction ridges form fingerprints, palm prints, and toe prints. Shown here are four basic fingerprint patterns.

6.1b Dermis

LEARNING OBJECTIVES

4. Characterize the two layers of the dermis.

5. Explain the significance of cleavage lines.

6. Describe how dermal blood vessels function in temperature regulation.

The **dermis** (der′mis) is deep to the epidermis and ranges in thickness from 0.5 mm to 3.0 mm. This layer of the integument is composed of connective tissue proper (see section 5.2d), and contains primarily collagen fibers, although both elastic and reticular fibers also are found within the dermis. Additionally, researchers recently have discovered motile cells in the dermis called *dendritic cells*. These cells are similar to the epidermal dendritic cells in that they serve an immune function, except they are located in the dermis. Other structures within the dermis are blood vessels, sweat glands, sebaceous glands, hair follicles, nail roots, sensory nerve endings, and arrector pili. Two major regions of the dermis can be distinguished: a superficial papillary layer and a deeper reticular layer **(figure 6.6)**.

Papillary Layer of the Dermis

The **papillary** (pap′i-lār-ē) **layer** is the superficial region of the dermis that is deep to the epidermis. It is composed of areolar connective tissue, and it derives its name from the projections of the dermis called **dermal papillae** (der′măl pă-pil′ē; *papilla* = a nipple). The dermal papillae interdigitate with deep projections of the epidermis called **epidermal ridges,** much like two sets of egg crate foam stacked on top of one another. Together, the epidermal ridges and dermal papillae increase the area of contact between the two layers and interlock them. Each dermal papilla contains the capillaries that supply nutrients to the cells of the epidermis. Additionally,

dermal papillae contain sensory nerve endings that serve as tactile receptors (see figure 6.1); these receptors continuously monitor touch on the surface of the epidermis. Tactile receptors are discussed in detail in section 16.2a.

Reticular Layer of the Dermis

The **reticular layer** forms the deeper, major portion of the dermis that extends from the papillary layer to the underlying subcutaneous layer. The reticular layer consists primarily of dense irregular connective tissue through which large bundles of collagen fibers project in all directions. These fibers are interwoven into a meshwork that surrounds structures in the dermis, such as the hair follicles, sebaceous glands and sweat glands, nerves, and blood vessels. The word *reticular* means "network" and refers to this meshwork of collagen fibers. Note that the reticular layer is different from reticular connective tissue, described in section 5.2d.

Lines of Cleavage and Stretch Marks

The majority of the collagen and elastic fibers in the skin are oriented in parallel bundles at specific body locations. The alignment of fiber bundles within the dermis is a result of the direction of applied stress during routine movement; therefore, the function of the bundles is to resist stress. **Lines of cleavage** (*tension lines*) in the skin identify the predominant orientation of collagen fiber bundles **(figure 6.7)**. These are clinically and surgically significant because any procedure resulting in a cut perpendicular to a cleavage line usually is pulled open as

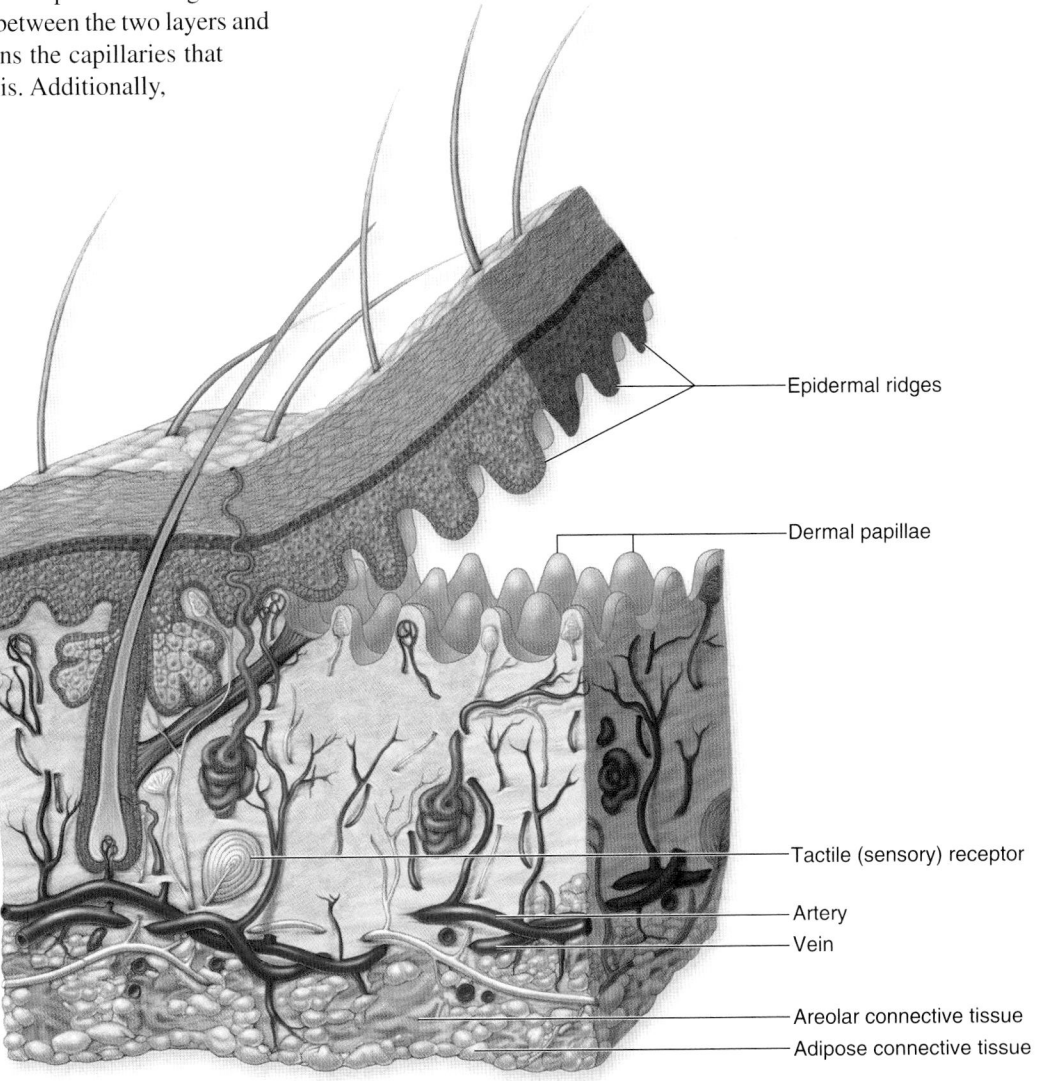

Figure 6.6 Layers of the Dermis.
The dermis is composed of a papillary layer and a reticular layer. AP|R

- Epidermis
- Dermis
  - Papillary layer
  - Reticular layer
- Subcutaneous layer

- Epidermal ridges
- Dermal papillae
- Tactile (sensory) receptor
- Artery
- Vein
- Areolar connective tissue
- Adipose connective tissue

a result of the recoil resulting from cut elastic fibers. This often results in slow healing and increased scarring. In contrast, an incision made parallel to a cleavage line usually will remain closed. Therefore, surgical procedures should be planned to consider lines of cleavage, thus ensuring rapid healing and prevention of scarring.

Collagen fibers and elastic fibers together contribute to the visible physical characteristics of the skin. Whereas the collagen fibers impart tensile strength, elastic fibers allow some stretch and recoil in the dermis during normal movement activities. Stretching of the skin that may occur as a result of excessive weight gain or pregnancy often exceeds the elastic capabilities of the skin. When the skin is stretched beyond its capacity, some collagen fibers are torn and result in stretch marks, called **striae** (strī′ē; *stria* = furrow). Both the flexibility and thickness of the dermis are diminished by effects of exposure to ultraviolet light and aging, causing either sagging or wrinkled skin.

WHAT DID YOU LEARN?

5 Compare and contrast the papillary versus reticular layer of the dermis, with respect to their tissue type and the structures they contain.

6 What is indicated by the lines of cleavage in the skin, and why is this medically important?

INTEGRATE

CLINICAL VIEW
Tattoos

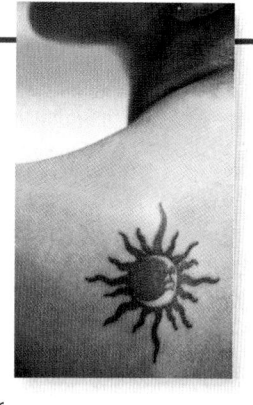

Tattoos are permanent images produced on the integument through the process of injecting a dye into the dermis. The dermis doesn't have a rapid cell turnover, so the injected dye remains in this area for a long time. Scar tissue surrounds the dye granules, which are too large for dendritic cells to ingest, and they become a permanent part of the dermis layer.

It is usually impossible to completely remove a tattoo, and some scarring may occur. Older methods include excision (cutting out the tattoo), dermabrasion (sanding down the tattooed skin), and cryosurgery (freezing the area of tattooed skin prior to its removal). Lasers are now used in some cases to break down the tattoo pigments. Newer tattoo inks have been introduced that allow for easier tattoo removal.

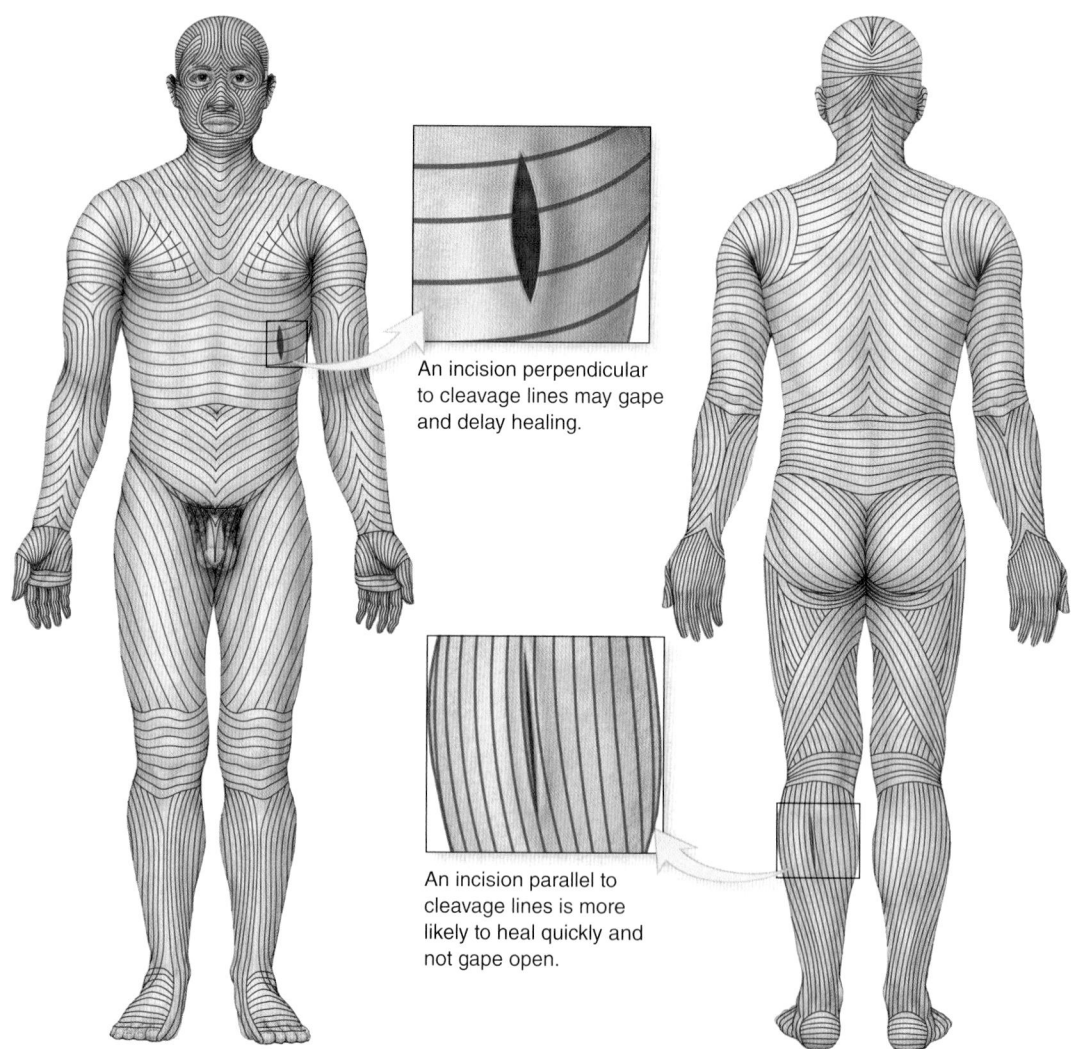

An incision perpendicular to cleavage lines may gape and delay healing.

An incision parallel to cleavage lines is more likely to heal quickly and not gape open.

Figure 6.7 Lines of Cleavage. Lines of cleavage partition the skin and indicate the predominant direction of underlying collagen fibers in the reticular layer of the dermis.

6.1c Subcutaneous Layer

LEARNING OBJECTIVE

7. List the functions of the subcutaneous layer.

Deep to the integument is the **subcutaneous** (sŭb-kū-tā′nē-ŭs; *sub* = beneath, *cutis* = skin) **layer,** also called the *hypodermis,* or *superficial fascia.* It is not considered a part of the integument. This layer consists of both areolar connective tissue and adipose connective tissue (see figure 6.1). In some locations of the body, adipose connective tissue predominates, thus the subcutaneous layer is called **subcutaneous fat.** The connective tissue fibers of the reticular layer of the dermis are extensively interwoven with those of the subcutaneous layer to stabilize the position of the skin and bind it to the underlying structures. The subcutaneous layer pads and protects the body, acts as an energy reservoir, and provides thermal insulation.

Drugs often are injected into the subcutaneous layer because its extensive vascular network promotes rapid absorption. Normally, the subcutaneous layer is thicker in women than in men, and its regional distribution also differs between the sexes. Adult males tend to accumulate subcutaneous fat primarily at the neck, upper arms, abdomen, lower back, and buttocks, whereas adult females accumulate adipose connective tissue primarily in the breasts, buttocks, hips, and thighs.

Table 6.1 reviews the layers of the integument and the subcutaneous layer.

WHAT DID YOU LEARN?

7 What types of tissue form the subcutaneous layer?

Table 6.1 Integument Layers and the Subcutaneous Layer AP|R

Layer	Specific Layer	Description
INTEGUMENT: EPIDERMIS		
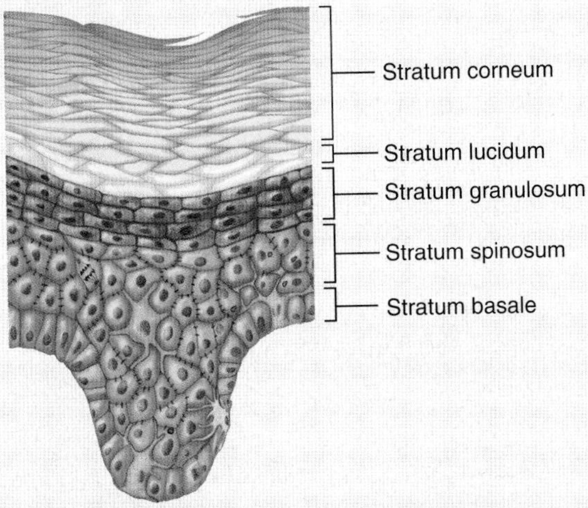 Stratum corneum / Stratum lucidum / Stratum granulosum / Stratum spinosum / Stratum basale	Stratum corneum	Most superficial layer of epidermis; 20–30 layers of dead, flattened, anucleate, keratin-filled keratinocytes
	Stratum lucidum	2–3 layers of anucleate, dead keratinocytes; seen only in thick skin (e.g., palms of hands, soles of feet)
	Stratum granulosum	3–5 layers of keratinocytes with distinct granules in cytoplasm; keratinization begins in this layer
	Stratum spinosum	Several layers of keratinocytes attached to neighbors by desmosomes; epidermal dendritic cells present
	Stratum basale	Deepest, single layer of cuboidal to low columnar keratinocytes in contact with basement membrane; cell division occurs here; also contains melanocytes and tactile cells
INTEGUMENT: DERMIS		
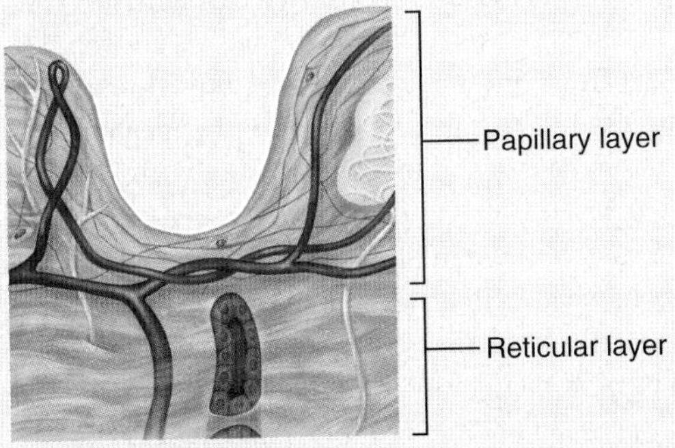 Papillary layer / Reticular layer	Papillary layer	Superficial layer of dermis; composed of areolar connective tissue; forms dermal papillae
	Reticular layer	Deeper layer of dermis; composed of dense irregular connective tissue surrounding hair follicles, sebaceous glands and sweat glands, nerves, and blood vessels
SUBCUTANEOUS LAYER		
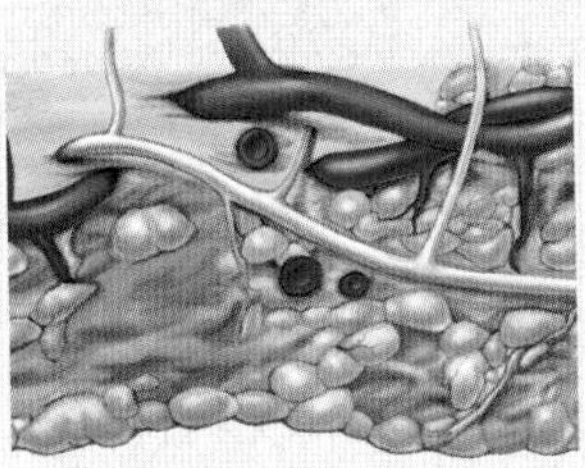	No specific layers	Not considered part of the integument; deep to dermis; composed of areolar and adipose connective tissue

CONCEPT CONNECTION

The integumentary system protects virtually all body systems:

- The integument provides the first line of defense against pathogens and toxins trying to enter the body (see section 22.3a).

- The integument prevents fluid loss, and thus helps the cardiovascular system maintain blood volume (see section 25.2a).

- When muscular exertion generates heat (see section 10.1a), the integumentary system helps release excess heat via sweating and vasodilation.

- The integument provides abundant sensory information to the nervous system (see section 16.2a).

- The integument can synthesize the vitamin D that is required for calcium homeostasis (see section 7.6a).

- Dendritic cells work with the rest of the immune system in initiating an immune response (see section 22.2a).

- Hairs in the nasal cavity help the respiratory system filter inspired air (see section 23.2a).

- The integumentary system and the urinary system both excrete nitrogenous waste products (see section 24.1).

6.1d Functions of the Integument

LEARNING OBJECTIVES

8. Name ways in which the integument protects the body and prevents water loss.
9. Describe the integument's involvement in calcium and phosphorus utilization.
10. Describe the integument's role in secretion and absorption.
11. Identify the immune cells that reside in the integument, and describe their actions.
12. Explain how the skin helps cool the body or retain warmth.
13. List the sensations detected by the skin's sensory receptors.

The epidermis serves a protective function, helps prevent water loss, and is involved with metabolic regulation. The epidermis and the dermis together secrete and absorb materials, and both play a role in immunity. We explore these functions in detail.

Protection

The epidermis acts as a physical barrier that protects the entire body from injury and trauma. It offers protection against harmful chemicals, toxins, microbes, and excessive heat or cold. The skin also protects deeper tissues from solar radiation, especially UV rays. When exposed to the sun, the melanocytes become more active and produce more melanin, thus giving the skin a tanned look. Even when you get a sunburn, the deeper tissues (muscles and internal organs) remain unaffected.

Prevention of Water Loss and Water Gain

The epidermis is water resistant, but not entirely waterproof. Some water is always lost through the skin when you sweat. More water is typically lost through transpiration, a process in which fluids slowly penetrate through the epidermis and then evaporate into the surrounding air (see section 25.2a). We realize how important the skin is in preventing water loss when treating individuals with severe burns. One of the main dangers is dehydration, because without the protective skin barrier, much larger amounts of water can escape from body tissues.

The skin also helps prevent water gain. If the skin were *not* water resistant, then each time you took a bath you would swell up like a sponge as the skin absorbed water!

Metabolic Regulation

Vitamin D₃, also called **cholecalciferol** (kō'l ē-kal-sif'er-ol), is synthesized from a steroid precursor by the keratinocytes when they are exposed to ultraviolet radiation. Vitamin D_3 is then released into the blood and transported to the liver, where it is converted to another intermediate molecule (calcidiol), and then transported to the kidney, where it is converted to **calcitriol** (kal-si-trī'ol). Calcitriol is the active form of vitamin D and is considered a hormone (see section 7.6a). It increases absorption of calcium and phosphate from the small intestine into the blood. Thus, the synthesis of vitamin D_3 is important in regulating the levels of calcium and phosphate in the blood. As little as 10 to 15 minutes of direct sunlight a day provides your body with its daily vitamin D through this process.

The skin also is involved in other forms of metabolic regulation. It is able to convert some compounds to slightly different forms that may be used by the skin. For example, when topical corticosteroids (e.g., hydrocortisone) are applied to the skin, the corticosteroid medication enters the keratinocytes, where the cells convert and use the medication to stop the inflammation and itching.

WHAT DO YOU THINK?

 During the Industrial Revolution, many children in cities spent little time outdoors and most of their time working in factories, leading to an increase in a disorder called rickets. Rickets is a bone disorder caused by inadequate vitamin D. Based on your knowledge of skin function, why do you think these children developed rickets?

Secretion and Absorption

Skin exhibits a secretory function when it discharges substances from the body during sweating. Sweating occurs when the body needs to cool itself off. Notice that sweat sometimes feels "gritty" because of the waste products being secreted onto the skin surface. The substances secreted in sweat include water, salts, and *urea*, a nitrogen waste product of body cells. The amount of urea, salts, and water secreted can be adjusted by the skin, and in so doing, the skin alters electrolyte levels in the body (see section 25.3). Thus, the skin also plays a role in electrolyte homeostasis. Sebum, excreted by sebaceous glands, lubricates the epidermis and hair, and helps make the integument water resistant.

The skin can absorb certain chemicals and drugs, such as estrogen from a birth control patch or nicotine from a nicotine patch. We refer to the skin as being **selectively permeable** because some materials are able to pass through it, whereas others are effectively blocked. In a process called **transdermal administration**, drugs that are soluble either in oils or lipid-soluble carriers may be administered transdermally by an adhesive patch that keeps the drug in contact with the skin surface. Drugs administered in this way slowly penetrate the epidermis and can be absorbed into the blood vessels of the dermis. Transdermal patches are especially useful because they release a continual, slow absorption of the drug into the blood over a relatively long period of time. The barrier to drug diffusion through the epidermis requires that the concentration of the drug in the patch be relatively high.

Immune Function

Earlier we described a small population of immune cells within the stratum spinosum called epidermal dendritic cells. These cells play an important role in initiating an immune response against pathogens that have penetrated the skin (see Chapter 22). These cells, along with dendritic cells of the dermis, also mount an attack against epidermal cancer cells.

Temperature Regulation

We previously discussed temperature regulation in section 1.5. Body temperature can be influenced by the vast capillary networks and sweat glands in the dermis. Dermal blood vessels have an important role in body temperature and blood pressure regulation. **Vasoconstriction** (vā′sō; *vas* = a vessel) means that the diameters of the vessels narrow, so relatively less blood can travel through them. Because less blood can flow through these dermal blood vessels, relatively more blood must travel through blood vessels deeper under the skin. The net effect is a shunting of blood *away* from the periphery of the body and toward deeper structures. Vasoconstriction of dermal blood vessels occurs if the body is trying to conserve heat. This is why we look paler when we are exposed to cold temperatures.

Conversely, **vasodilation** of the dermal blood vessels means that the diameter of the vessels increases, so relatively more blood can travel through them. The dermal blood vessels vasodilate so that more blood will travel close to the body surface and excess heat may be lost if the body is too warm. This additional blood flow through the dermis results in a more reddish/pinkish hue to the skin. Your face may become flushed when you exercise, for example, because your dermal blood vessels dilate in an attempt to release the excess heat. Conversely, note how on a cold day your face may be paler than normal. When you come in from the cold, your face may become flushed as your body adjusts to the warmer temperature.

Sensory Reception

The dermis of the skin has an extensive **innervation**, which refers to its distribution of nerve fibers. Sensory nerve fibers in the skin monitor stimuli in both the dermis and epidermis. Touch receptors (e.g., tactile corpuscles) detect stimuli such as cold, pressure, and vibration and then initiate sensory input to the brain (see section 16.2a). This rich innervation allows us to be very aware of our surroundings and to differentiate among the different kinds of sensory signals from receptors in the skin. In addition, motor nerve fibers extend to the skin to control both blood flow and gland secretions.

Figure 6.8 illustrates how the anatomy of the integument is designed to help with these various functions.

WHAT DID YOU LEARN?

8 How does the skin produce vitamin D?

9 Is the skin entirely waterproof? Explain.

10 What are some ways the skin can dissipate excess heat?

(a) Epidermis Functions

PROTECTION

Epidermal strata provide layers of protection against harmful chemicals, toxins, microbes, and excessive heat or cold. Skin also protects deeper tissues from UV radiation as melanocytes are stimulated to produce more melanin.

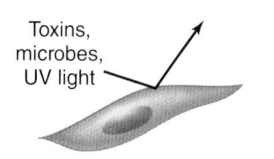

Toxins, microbes, UV light

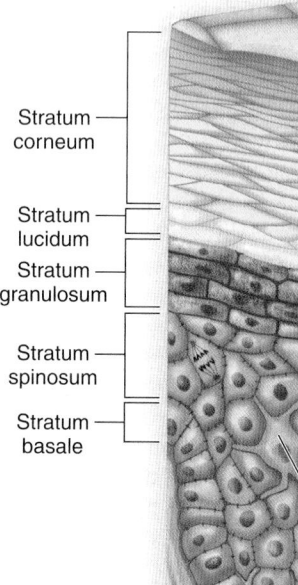

Stratum corneum

Stratum lucidum

Stratum granulosum

Stratum spinosum

Stratum basale

PREVENTION OF WATER LOSS AND WATER GAIN

The epidermis is water resistant and keeps water from either exiting or entering the skin easily.

METABOLIC REGULATION

Sunlight

Upon exposure to UV rays, keratinocytes produce vitamin D_3 and melanocytes are stimulated to produce more melanin, giving the skin a more tanned look.

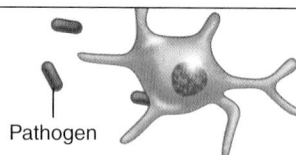

Melanocyte

Epidermal dendritic cell

SECRETION AND ABSORPTION

Materials (e.g., sebum, sodium, water, urea) secreted by dermal structures are released onto the epidermal surface. The skin is selectively permeable because some materials (e.g., certain drugs, like nicotine and estrogen patches) may be absorbed while others are blocked.

Transdermal nicotine patch

IMMUNE FUNCTION

Pathogen

Epidermal dendritic cells engulf and destroy pathogens, alert the immune system to the presence of pathogens, and initiate an immune response. (Note: The dermis also contains its own dendritic cells.)

Figure 6.8 **How Integument Form Influences Its Functions.**
(*a*) The epidermis is composed of multiple layers of keratinized epithelial cells that make it well suited for protection and prevention of water loss. It also participates in secretion and absorption, metabolic regulation, and immunity. (*b*) The dermis is composed of well-vascularized connective tissue that participates in temperature regulation, secretion, absorption, and sensory reception.

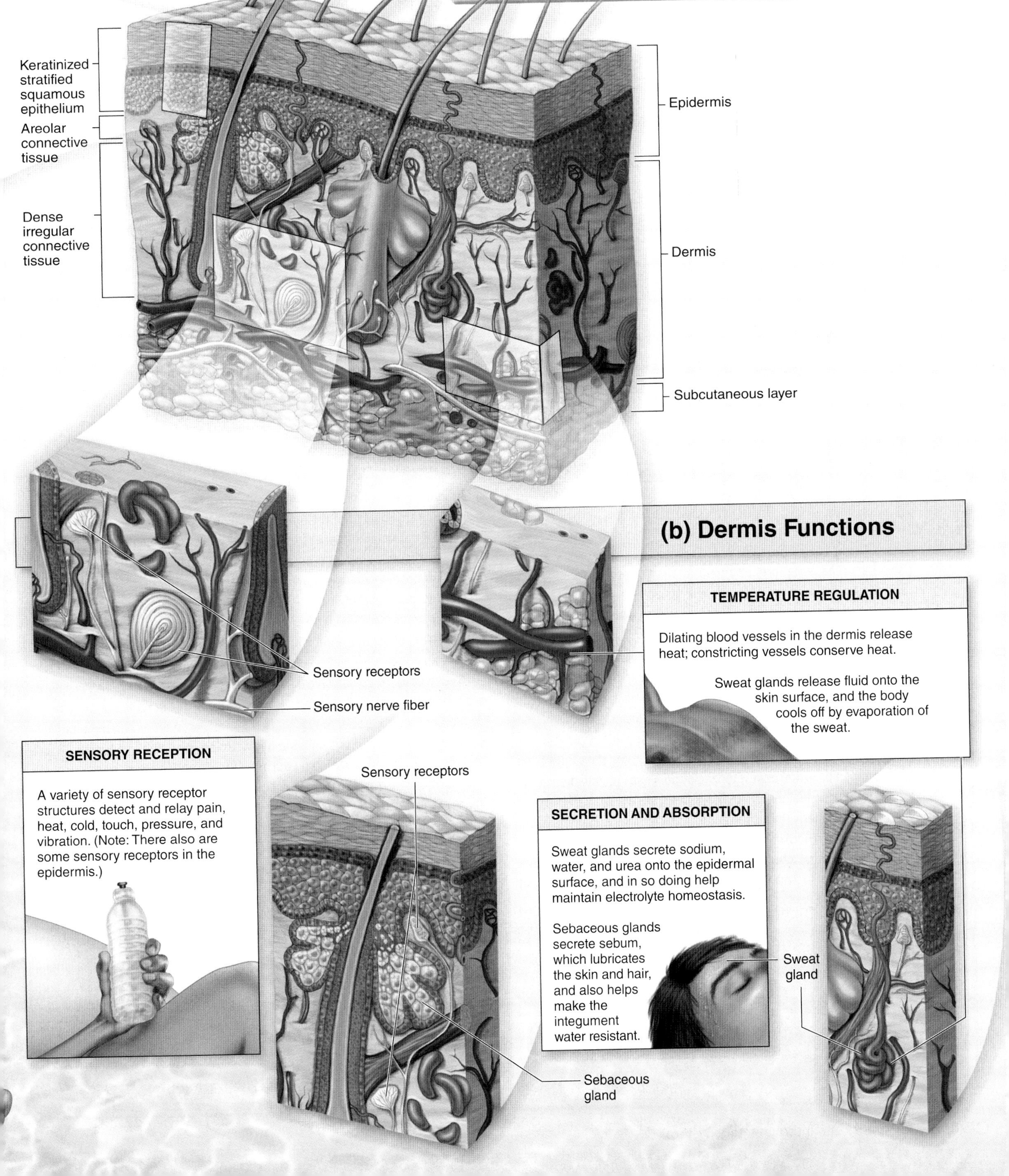

Keratinized stratified squamous epithelium

Areolar connective tissue

Dense irregular connective tissue

Epidermis

Dermis

Subcutaneous layer

(b) Dermis Functions

Sensory receptors

Sensory nerve fiber

TEMPERATURE REGULATION

Dilating blood vessels in the dermis release heat; constricting vessels conserve heat.

Sweat glands release fluid onto the skin surface, and the body cools off by evaporation of the sweat.

SENSORY RECEPTION

A variety of sensory receptor structures detect and relay pain, heat, cold, touch, pressure, and vibration. (Note: There also are some sensory receptors in the epidermis.)

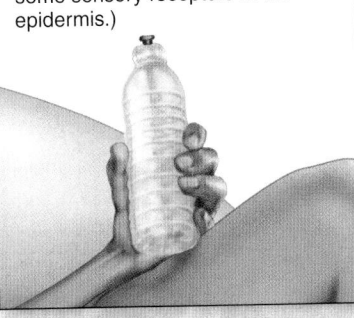

Sensory receptors

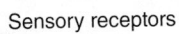

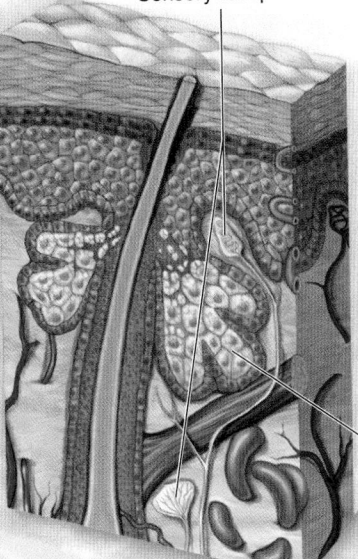

SECRETION AND ABSORPTION

Sweat glands secrete sodium, water, and urea onto the epidermal surface, and in so doing help maintain electrolyte homeostasis.

Sebaceous glands secrete sebum, which lubricates the skin and hair, and also helps make the integument water resistant.

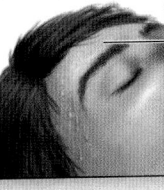

Sweat gland

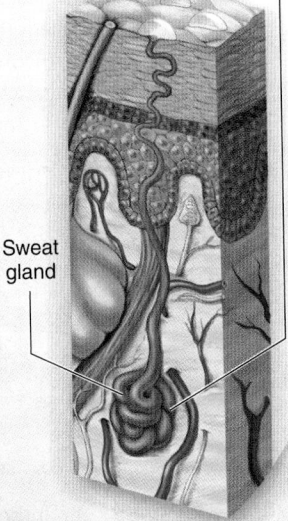

Sebaceous gland

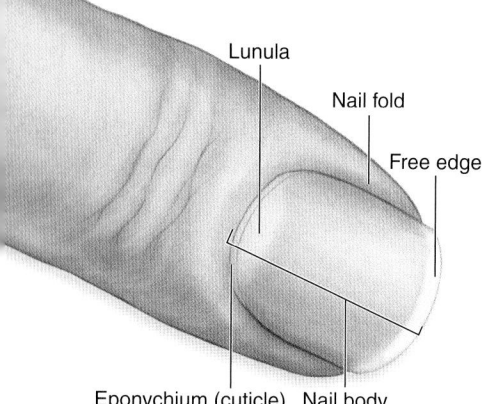

Lunula
Nail fold
Free edge
Eponychium (cuticle) Nail body

(a)

Figure 6.9 Structure of a Fingernail. Nails are a hard derivative of the stratum corneum that protect sensitive fingertips. (*a*) Surface view of a fingernail. (*b*) Sagittal section showing the internal details of a fingernail. **AP|R**

Nail matrix
Nail root
Nail bed

Hyponychium
Nail plate
Dermis
Epidermis
Phalanx (finger bone)

(b)

6.2 Integumentary Structures Derived from Epidermis

The nails, hair, and exocrine glands of the skin all are derived from the epidermal epithelium. These structures also are known as **epidermal derivatives**, or *appendages*, of the integument. They formed during embryologic development as portions of the epidermis invaginated into the dermis. Both nails and hairs are composed primarily of dead, keratinized epithelial cells, whereas the exocrine glands are composed of living epithelial cells.

6.2a Nails

LEARNING OBJECTIVES

1. Describe the function of nails.
2. List the main components of the nail.

Nails are scalelike modifications of the stratum corneum layer of the epidermis that form on the dorsal edges of the fingers and toes. They protect the distal tips of the digits and prevent damage or extensive distortion during jumping, kicking, or catching. Fingernails also assist us in grasping objects.

Each nail has a distal whitish **free edge,** a pinkish **nail body,** and a **nail root,** which is the proximal part embedded in the skin **(figure 6.9).** Together, these parts form the **nail plate.** The nail body covers a layer of epidermis that is called the **nail bed,** which contains only the deeper, living cell layers of the epidermis.

INTEGRATE

CLINICAL VIEW

Nail Disorders

Changes in the shape, structure, or appearance of the nails are clinically significant. A change may indicate the existence of a disease process affecting metabolism throughout the body. In fact, the state of a person's fingernails and toenails can be indicative of his or her overall health. Nails are subject to various disorders.

Brittle nails are prone to vertical splitting and separation of the nail plate layers at the free edge. Overexposure to water or to certain household chemicals can cause brittle nails. Keeping the nails moisturized and limiting exposure to water and chemicals can alleviate brittle nails.

An **ingrown nail** occurs when the edge of a nail digs into the skin around it. This painful condition is first characterized by pain and inflammation at the site. If left untreated, some ingrown toenails can cause infection. Ingrown nails may result from overly tight shoes and improper trimming of the nails.

Onychomycosis (on'i-kō-mī-kō'sis; *onych* = nail, *mykes* = fungus, *osis* = condition) is a fungal infection that occurs in nails constantly exposed to warmth and moisture. The fungus starts to grow under the nail and

eventually causes a yellowish discoloration, a thickened nail, and brittle, cracked edges. Fungal infections can result in permanent damage to the nail and a potential for spread of infection. Treatment involves taking oral fungal medications for long periods of time (a minimum of 6 to 12 weeks, and in some cases up to a year) to eradicate the fungal infection.

Yellow nail syndrome occurs when growth and thickening of the nail slows or stops completely. As the nail growth slows, the nails become yellowish or sometimes greenish. Yellow nail syndrome often, but not always, may be an outward sign of respiratory disease, such as chronic bronchitis.

Spoon nails, or *koilonychia* (koy-lō-nik'ē-ă; *koilos* = hollow, *onych* = nail), are a nail malformation where the outer surface of the nails are concave instead of convex. Spoon nails frequently are a sign of an iron deficiency. Treating the iron deficiency should alleviate the spoon nail condition.

Beau's lines run horizontally across the nail and indicate a temporary interference with nail growth at the time this portion of the nail was formed. Injury to the nail or severe illness can cause Beau's lines. They also may be seen in individuals suffering from chronic malnutrition.

Vertical ridging of the nails is common and usually does not indicate any serious medical problem. The condition occurs more frequently as we get older.

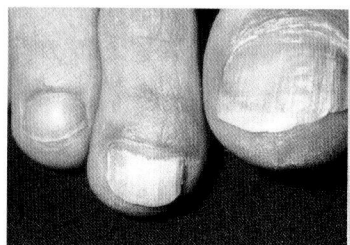
Onychomycosis

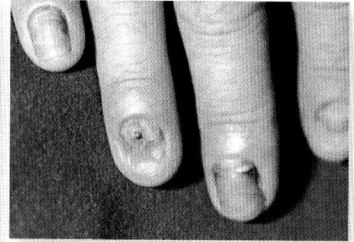
Yellow nail syndrome

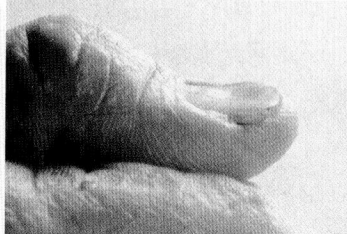
Spoon nails

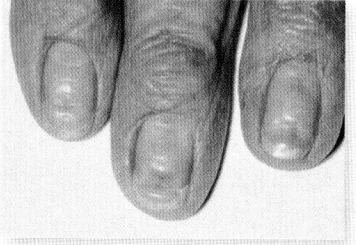
Beau's lines

Most of the nail body appears pink because of the blood flowing in the underlying capillaries; the free edge of the nail appears white because there are no underlying capillaries. At the nail root and the proximal end of the nail body, the nail bed thickens to form the **nail matrix,** which is the actively growing part of the nail. The **lunula** (lū′nū-lă; *luna* = moon) is the whitish semilunar area of the proximal end of the nail body. It has a whitish appearance because a thickened stratum basale obscures the underlying blood vessels.

Along the lateral and proximal borders of the nail, folds of skin called **nail folds** overlap the nail. The **eponychium** (ep-ō-nik′ē-ŭm; *epi* = upon, *onyx* = nail), also known as the **cuticle** (kū′ti-kl), is a narrow band of epidermis extending from the margin of the nail wall onto the nail body. The **hyponychium** (hi-pō-nik′ē-ŭm; *hypo* = below) is a region of thickened stratum corneum over which the free nail edge projects.

WHAT DID YOU LEARN?

11 What is the difference between the eponychium and the hyponychium of a fingernail?

6.2b Hair

LEARNING OBJECTIVES

3. Describe the structure of a hair and a follicle.
4. List the functions of hair.

Hair is found almost everywhere on the body except the palms of the hands and palmar surface of the fingers, the sides and soles of the feet and toes, the lips, and portions of the external genitalia. The general structure of hair and its relationship to the integument is shown in **figure 6.10.**

Hair Type and Distribution

A single hair, or *pilus*, has the shape of a slender filament; it is composed of keratinized cells growing from hair follicles that extend deep into the dermis, often projecting into the underlying subcutaneous layer. Differences in hair density are due primarily to differences in the texture and pigmentation of the hair.

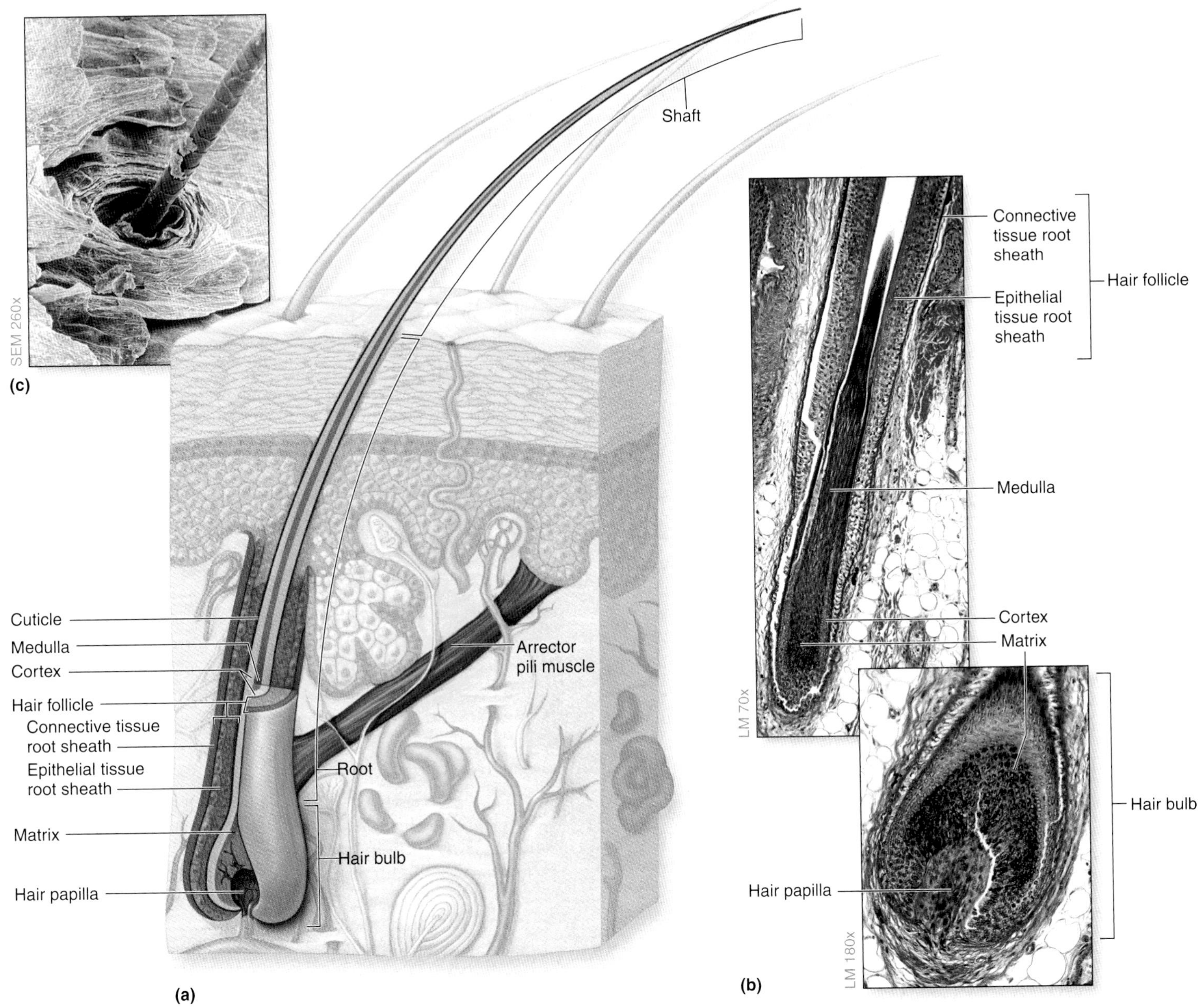

Figure 6.10 Hair. Hair is a derivative of the epithelium. (*a*) A hair grows from a follicle extending from the epidermis into the dermis. (*b*) Photomicrographs of a hair follicle. (*c*) SEM of a hair emerging from its follicle. AP|R

We produce three kinds of hair during our lives: lanugo, vellus, and terminal hair. **Lanugo** (lă-nū′gō) is a fine, unpigmented, downy hair that first appears on the fetus in the last trimester of development. At birth, most of the lanugo has been replaced by similarly fine unpigmented or lightly pigmented hair called **vellus** (vel′ŭs; *vellus* = fleece). Vellus is the primary human hair and is found on the upper and lower limbs. **Terminal hair** is usually coarser, pigmented, and longer than vellus. It grows on the scalp, and it is also the hair of eyebrows and eyelashes. At puberty, terminal hair replaces vellus hair in the axillary and pubic regions, and it forms the beard on the faces of males.

Hair Structure and Follicles

Three zones can be recognized along the length of a hair: the hair bulb, root, and shaft. The **hair bulb** consists of epithelial cells and is a swelling at the base where the hair originates in the dermis. The epithelium at the base of the bulb surrounds a small **hair papilla,** which is composed of a small amount of connective tissue containing tiny blood vessels and nerves. The **root** is the zone of the hair extending from the bulb to the skin surface. The **shaft** is the third portion of the hair that extends beyond the skin surface. The hair bulb contains living epithelial cells, whereas the root and shaft consist of dead epithelial cells.

Hair production involves a specialized type of keratinization that occurs within the **hair matrix,** a structure immediately adjacent to the hair papilla in the hair bulb. Epithelial cells near the center of the hair matrix divide, producing new cells that are gradually pushed toward the surface. The **medulla,** not found in all hair types, is a remnant of the soft core of the matrix. It is composed of loosely arranged cells and air spaces, and contains flexible, soft keratin. Several layers of flattened cells closer to the outer surface of the developing hair form the relatively hard **cortex.** A single cell layer around the cortex forms the **cuticle,** which coats the hair.

The **hair follicle** (fol′i-kl) is an oblique tube that surrounds the hair root. It always extends into the dermis and sometimes into the subcutaneous layer. The cells of the follicle walls are organized into two principal concentric layers: an outer **connective tissue root sheath,** which originates from the dermis; and an inner **epithelial tissue root sheath,** which originates from the epidermis (figure 6.9b). Extending from the hair follicle to the dermal papillae are thin ribbons of smooth muscle that collectively are called the **arrector pili** (ă-rek′tōr pī′lī). Stimulation of the arrector pili is usually a result of an emotional state, such as fear or rage, or as a response to exposure to cold temperatures. Upon stimulation, the arrector pili contracts, pulling on the hair follicle and elevating the hair, producing "goose bumps."

Functions of Hair

The millions of hairs distributed on the surface of the human body have important functions that include

- **Protection.** The hair on the head protects the scalp from sunburn and injury. Hair within the nostrils entraps particles before they travel deeper into the respiratory system, whereas hairs within the external ear canal protect the ear from insects and foreign particles. Eyelashes protect the eyes.
- **Facial expression.** The hairs of the eyebrows function primarily to enhance facial expression.
- **Heat retention.** Hair on the head prevents the loss of conducted heat from the scalp to the surrounding air. Individuals who have lost their scalp hair release much more heat through the scalp than those who have a full head of hair.

- **Sensory reception.** Hair has associated tactile receptors (hair root plexuses) that detect light touch (see section 16.2a).
- **Visual identification.** Hair characteristics are important in determining age and sex, and in identifying individuals. (Hair analysis also assists in determining animal species.)
- **Chemical signal dispersal.** Hairs help disperse *pheromones,* which are chemical signals involved in attracting members of the opposite sex and in sex recognition. Pheromones are secreted by specific sweat glands onto hairs in the axillary and pubic regions, as described shortly.

Hair Color

Hair color is a result of the synthesis of melanin in the matrix adjacent to the hair papillae. Variations in hair color reflect genetically determined differences in the structure of the melanin. Additionally, environmental and hormonal factors may influence the color of the hair. As people age, the production of pigment decreases, and thus hair becomes lighter in color. Gray hair results from the gradual reduction of melanin production within the hair follicle; white hair occurs due to a complete stoppage of melanin production.

Hair Growth and Replacement

There are three stages of the hair growth cycle: anagen, catagen, and telogen:

1. The **anagen phase** is the active phase of growth where living cells of the hair bulb are rapidly growing, dividing, and transforming into hair. It is the longest part of the growth cycle and lasts from about 18 months to as much as 7 years, depending on the genetics of the person. During the anagen phase, each hair strand grows about one-third of a millimeter per day, which equals 0.5 to 1.0 cm per month. On a normal scalp, 80–95% of follicles are in anagen phase.
2. The **catagen phase** is a brief regression period where cell division ceases and the follicle undergoes involution. This very short phase lasts for about 3 to 4 weeks.
3. The **telogen phase** is the resting phase and is usually the phase where the hair is shed (these hairs are the ones we find in our comb or brush). After 3 to 4 months in the telogen phase, the cells of the hair bulb start regrowing, and the follicle reenters the anagen phase.

The hair growth rate and the duration of the hair growth cycle vary; however, the scalp normally loses between 10 and 100 hairs per day. Continuous losses that exceed 100 hairs per day often indicate a health problem. Sometimes hair loss may be temporary as a result of one or more of the following factors: exposure to drugs, dietary factors, radiation, high fever, or stress. Thinning of the hair, called **alopecia** (al-ō-pē′shē-ă; *alopekia* = a disease like fox mange), can occur in both sexes, usually as a result of aging. In **diffuse hair loss**, a condition that is both dramatic and distressing, hair is shed from all parts of the scalp. Women primarily suffer from this condition, which may be due to hormones, drugs, or iron deficiency.

In males, the condition called **male pattern baldness** causes loss of hair first from only the crown region of the scalp rather than uniformly. It is caused by a combination of genetic and hormonal influences. The relevant gene for male pattern baldness has two alleles, one for uniform hair growth and one for baldness. The baldness allele is dominant in males and is expressed only in the presence of a high level of testosterone, which causes the terminal hair of the scalp to be replaced by thinner vellus, beginning on the top of the head and later at the sides. In females, the baldness allele is recessive.

Excessive male pattern hairiness in areas of the body that normally do not have terminal hair is called **hirsutism** (her′sū-tizm; *hirsutus* = shaggy). This hair growth typically occurs on the face, chest, and back and may affect both sexes, but is especially distressing for women. Hirsutism most commonly is caused by an excess of male sex hormones called androgens, either through a medical condition (such as polycystic ovarian syndrome) or by certain medications that cause a rise in androgens.

💡 WHAT DID YOU LEARN?

12 What are the three zones of a hair?

13 How does hair function in protection and heat retention?

6.2c Exocrine Glands of the Skin

✍ LEARNING OBJECTIVES

5. Differentiate between the two types of sweat glands.

6. Describe the function of sebaceous glands.

7. Name two other modified integumentary glands.

The skin houses many types of exocrine glands. The two most common types of exocrine glands are **sweat** (*sudoriferous*) **glands** and **sebaceous glands;** these are shown in **figure 6.11**.

Sweat Glands

The two different groups of sweat glands in the skin are merocrine (*eccrine*) sweat glands and apocrine sweat glands (see section 5.1d). Both have a coiled, tubular secretory portion that is located in the reticular layer of the dermis, and a **sweat gland duct** that transports the secretion to the surface of the epidermis (in a merocrine sweat gland) or into a hair follicle (in an apocrine sweat gland). The opening of the sweat gland duct on the epidermal surface is an indented region called a **sweat pore.**

Both types of sweat glands contain myoepithelial cells. These specialized epithelial cells are sandwiched between the secretory gland cells and the underlying basement membrane. In response to nervous system stimulation, myoepithelial cells contract to squeeze the gland, causing it to discharge its accumulated secretions.

✍ WHAT DO YOU THINK?

3 The autonomic nervous system is a part of the nervous system that can be activated when we are frightened or nervous. What would you expect to happen to sweat gland production and secretion when we are frightened or nervous?

Sweat pore

Sweat gland duct

Hair follicle

Sebaceous gland

Merocrine sweat gland

Arrector pili muscle

Apocrine sweat gland

(a)

LM 100×

Merocrine sweat gland duct

(b) **Merocrine sweat gland**

LM 80×

Apocrine sweat gland duct

(c) **Apocrine sweat gland**

LM 40×

Sebaceous glands

Hair follicle

(d) **Sebaceous glands**

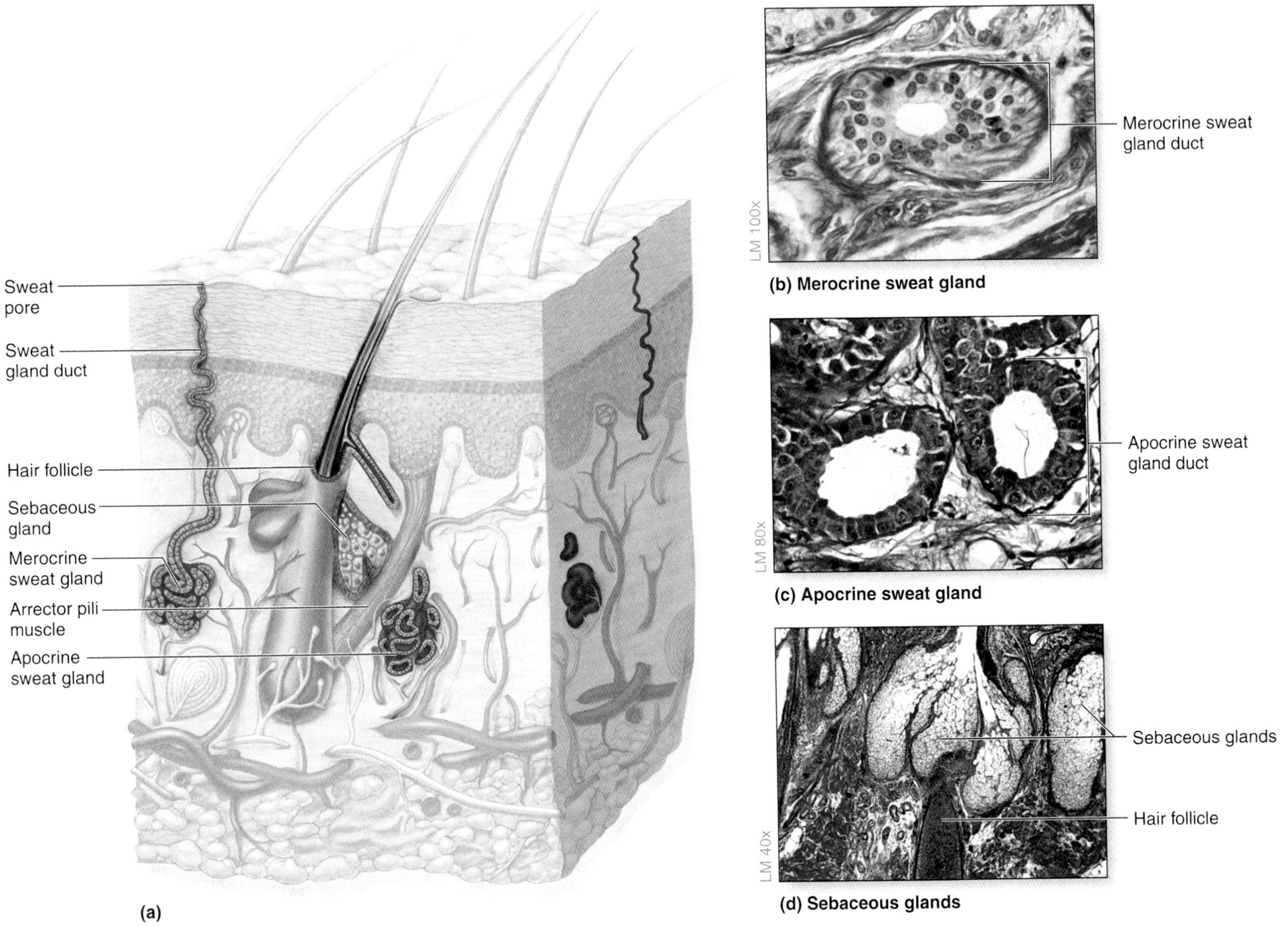

Figure 6.11 Exocrine Glands of the Skin. (*a*) The integument contains sweat glands and sebaceous glands. (*b*) Merocrine sweat glands have a duct with a narrow lumen that opens onto the skin surface through a pore. (*c*) Apocrine sweat glands exhibit a duct with a larger lumen to convey secretion products into a hair follicle. (*d*) The cells of sebaceous glands are destroyed during the release of their oily secretion into the hair follicle. AP|R

Merocrine Sweat Glands **Merocrine** (*eccrine*) **sweat glands** (figure 6.11*b*) are the most numerous and widely distributed sweat glands. The adult integument contains between 3 and 4 million merocrine sweat glands. They are simple, coiled tubular glands that discharge their secretions directly onto the surface of the skin. The clear secretion they release by exocytosis (see section 5.1d) is called **sweat;** it consists of approximately 99% water and 1% other chemicals that include electrolytes (primarily sodium and chloride), metabolites (e.g., lactate), and waste products (urea and ammonia).

The major function of merocrine sweat glands is **thermoregulation,** which is the regulation of body temperature by evaporation of fluid from the skin (see sections 1.5b, 25.2a). Merocrine sweat gland secretions provide a means for the loss of both water and electrolytes. The secretions also may help eliminate a number of ingested drugs. Finally, merocrine sweat gland secretions provide some protection from environmental hazards both by diluting harmful chemicals and by preventing the growth of microorganisms (antibacterial and antifungal activity).

Apocrine Sweat Glands **Apocrine sweat glands** (figure 6.11*c*) are coiled, tubular glands that release their secretion into hair follicles in the axillae, around the nipples, in the pubic region, and in the anal region. Originally, these glands were called *apocrine* because their cells were thought to secrete their product by an apocrine mechanism (meaning that the apical portion of the cell's cytoplasm pinches off and, along with cellular components of the apical region, becomes the secretory product—see section 5.1d). Now, researchers have shown that both apocrine and merocrine sweat glands produce their secretion

by exocytosis. However, the secretory portion of an apocrine gland has a much larger lumen than that of a merocrine gland, and the secretions they produce are different, so these glands continue to be called apocrine glands. The secretion they produce is viscous and cloudy, and it contains both proteins and lipids that are acted upon by bacteria to produce a distinct, noticeable odor. These odorous secretions are the pheromones discussed in section 6.2b. (Underarm deodorant is designed to mask the odor, whereas a deodorant with antiperspirant also helps prevent the formation of sweat.) These sweat glands become active and produce secretions beginning around puberty.

Sebaceous Glands

Sebaceous glands are holocrine glands that produce an oily, waxy secretion called **sebum** (sē'bŭm; *sebum* = tallow) that is usually discharged into a hair follicle and onto the hair itself. Sebum acts as a lubricant to keep the skin and hair from becoming dry, brittle, and cracked. Sebum also has some bactericidal (bacteria-killing) properties. Several sebaceous glands may open onto a single follicle.

The secretion of sebum in both sexes is stimulated by hormones, especially androgens (male sex hormones). Sebaceous glands are relatively inactive during childhood; however, they are activated during puberty in both sexes, when the production of sex hormones increases (see section 28.1b).

Other Integumentary Glands

Some specialized glands of the integument are restricted to specific locations. Two important examples are the ceruminous glands and the mammary glands.

INTEGRATE

CLINICAL VIEW

Acne and Acne Treatments

Acne (ak'nē) is the term used to describe plugged sebaceous ducts. Acne typically becomes abundant beginning at puberty, because increases in hormone levels stimulate sebaceous gland secretion, making the pores more prone to blockage. Acne is prevalent during the teenage years, although any age group may experience acne.

The types of acne lesions include

- **Comedo** (kom'ē-dō). A sebaceous gland plugged with sebum. An open comedo is called a **blackhead,** because the plugged material has a dark, blackish appearance. A closed comedo is called a **whitehead,** because the top surface is whitish in color.

- **Papule** (pap'ūl) and **pustule** (pŭs'tyūl). Both are dome-shaped lesions. Papules typically are fluid-filled, form red elevations on the skin and do not contain pus. Papules may become pustules,

which are filled with a mixture of white blood cells, dead skin cells, and bacteria (called pus).

- **Nodule** (nod'ūl). Similar to a pustule, but extending into the deeper skin layers and usually rupturing the hair follicle wall. Nodules can be prone to scarring.

- **Cyst.** A large, fluid-filled nodule that can become severely inflamed and painful and can lead to scarring of the skin.

Many medicinal treatments are available for acne, depending upon the type and severity. The effectiveness of the following medications varies from individual to individual. These treatments include benzoyl peroxide, salicylic (sal-i-sil'ik) acid, topical and oral antibiotics, topical vitamin A–like compounds (e.g., retinoids such as tretinoin [Retin-A]), and systemic retinoids such as isotretinoin (e.g., Accutane).

Other treatments include light chemical skin peels and comedo extraction by a dermatologist. Untreated acne, if severe, can lead to scarring, as can picking at the acne lesions.

Blackhead (open comedo)

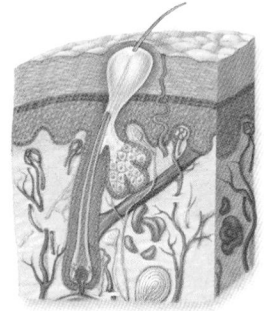
Whitehead (closed comedo)

Pustule

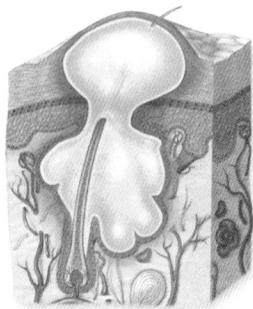
Nodule

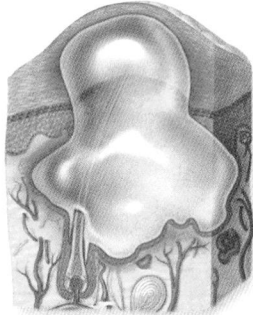
Cyst

Ceruminous (sĕ-rū′mi-nŭs) **glands** are modified apocrine sweat glands located only in the external acoustic meatus (ear canal), where their secretion forms a waterproof earwax called **cerumen** (sĕ-rū′men). Both cerumen and the tiny hairs in the meatus help trap foreign particles or small insects and keeps them from reaching the eardrum. Cerumen also helps lubricate the external acoustic meatus and eardrum (see section 16.5a).

The **mammary glands** of the breasts are modified apocrine sweat glands. Both males and females have mammary glands, but these glands become functional only in pregnant and lactating females, when they produce milk, a secretion that nourishes offspring. The development of the glands and its secretions are controlled by a complex interaction between gonadal and pituitary hormones, discussed in section 28.3.

WHAT DID YOU LEARN?

14 How do apocrine sweat glands differ from merocrine sweat glands in terms of their location, secretions, and function?

15 What do sebaceous glands secrete, and where is this material secreted?

6.3 Repair and Regeneration of the Integumentary System

LEARNING OBJECTIVES

1. Distinguish between regeneration and fibrosis.
2. Describe the process of wound healing.

The components of the integumentary system exhibit a tremendous ability to respond to stressors, trauma, and damage. Repetitive mechanical stresses applied to the integument stimulate cell division in the stem cells of the stratum basale, resulting in a thickening of the epidermis and an improved ability to withstand stress. For example, walking about without shoes causes the soles of the feet to thicken, thus providing more protection for the underlying tissues.

Damaged tissues are normally repaired in one of two ways. The replacement of damaged or dead cells with the same cell type is called **regeneration.** This restores organ function. When regeneration is not possible because part of the organ is too severely damaged or its cells lack the capacity to divide, the body fills in the gap with scar (fibrous) tissue. This process of scar tissue deposition during healing is referred to as **fibrosis,** and serves to bind the damaged parts together. The replacement scar tissue is produced by fibroblasts and is composed primarily of collagen fibers. Some structural restoration occurs; however, functional activities are not restored. Fibrosis is the repair response in tissues subjected to severe injuries or burns.

Both regeneration and fibrosis may occur in the healing of damage to the skin. **Figure 6.12** illustrates stages in wound healing of the skin:

(1) Cut blood vessels initiate bleeding into the wound. The blood brings clotting proteins, numerous leukocytes (white blood cells), and antibodies.

INTEGRATE

CONCEPT CONNECTION

The process of wound repair requires stimulation and activation of the immune system, described in detail in chapter 22.

INTEGRATE

CLINICAL VIEW
Psoriasis

Psoriasis (sō-rī′-ă-sis) is a chronic autoimmune skin disease that has periods of flare-ups and remissions throughout a person's lifetime. A type of leukocyte called a T-lymphocyte mistakenly attacks the keratinocytes, causing rapid overgrowth and overproduction of new skin cells. The normal sloughing-off cycle of keratinocytes is thrown out of balance, and the proliferation of cells develops into patches of typically whitish, scaly skin (called *plaques*) that form on the epidermal surface. The plaques are not contagious, but for the affected individual they may cause severe itching, pain, and skin cracking and bleeding. Any part of the skin may be affected, but common areas of involvement include the scalp, the limbs, and the buttocks. Treatments for psoriasis include topical corticosteroids, ultraviolet light therapy, and certain oral medications (e.g., methotrexate) that may interfere with the production of skin cells.

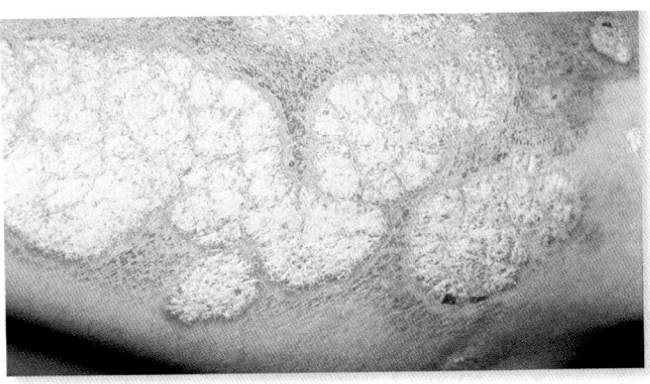

Multiple whitish plaques on the forearm of a person with psoriasis.

(2) A blood clot forms, temporarily patching the edges of the wound together and acting as a barrier to prevent the entry of pathogens into the body. Internal to the clot, macrophages and neutrophils (two types of leukocytes, see section 18.3c) clean the wound of cellular debris. (For more information about blood clotting, see section 18.4.)

(3) The cut blood vessels regenerate and grow in the wound. A soft mass deep in the wound becomes **granulation** (gran′ū-lā-shŭn) **tissue,** which is a vascular connective tissue that initially forms in a healing wound. Macrophages within the wound begin to remove the clotted blood. Fibroblasts produce new collagen fibers in the region.

(4) Epithelial regeneration of the epidermis occurs due to division of epithelial cells at the edge of the wound. These new epithelial cells migrate over the wound, moving internally to the now superficial remains of the clot (the scab). The connective tissue is replaced by fibrosis.

The skin repair and regeneration process is dependent on the extent of the injury. The wider and deeper the surface affected, the longer it takes for skin to be repaired. Additionally, the area under repair usually is more susceptible to complications due to fluid loss and infection. As the severity of the damage increases, the repair and regeneration ability of the integument is strained and its return to its original condition becomes much less likely. Some integumentary system components

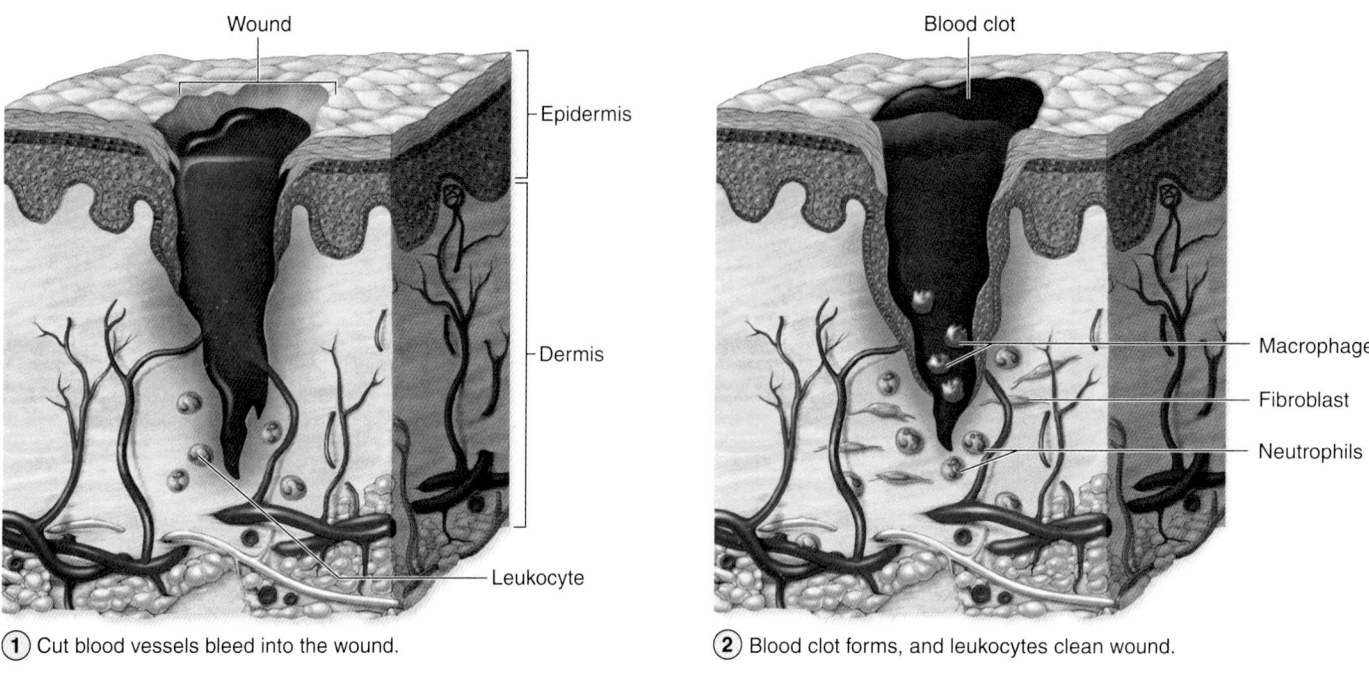

① Cut blood vessels bleed into the wound.

② Blood clot forms, and leukocytes clean wound.

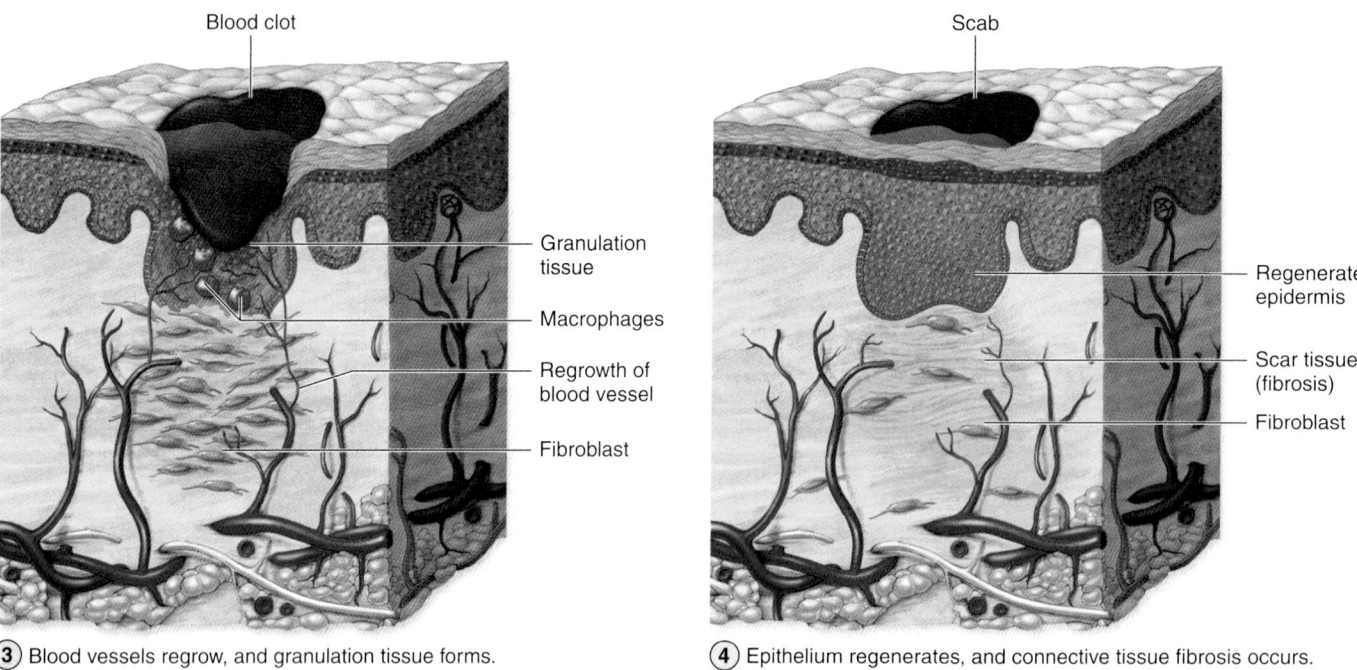

③ Blood vessels regrow, and granulation tissue forms.

④ Epithelium regenerates, and connective tissue fibrosis occurs.

Figure 6.12 Stages in Wound Healing. Cut blood vessels in tissue initiate the process of wound healing.

that are not repaired following severe damage to the integument include hair follicles, exocrine glands, nerves, and the arrector pili muscle cells.

 **WHAT DID YOU LEARN?**

16 What is granulation tissue, and when does it appear during wound healing of the skin?

6.4 Development and Aging of the Integumentary System

The integumentary system structures are derived from both ectoderm and mesoderm germ layers (see section 5.6a). The ectoderm is the origin of the epidermis, whereas the mesoderm is the origin of the dermis.

CLINICAL VIEW

Burns

Burns are a major cause of accidental death and are usually caused by heat, radiation, harmful chemicals, sunlight, or electrical shock. The immediate threat to life results primarily from fluid loss, infection, and the effects of burned, dead tissue.

Burns are classified by depth of tissue involvement. First- and second-degree burns are called *partial-thickness burns;* third-degree burns are called *full-thickness burns.*

How are first-, second-, and third-degree burns distinguished?

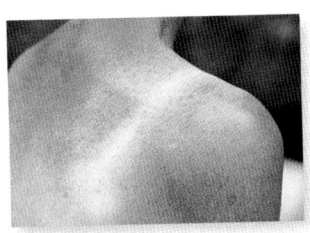

First-degree burn

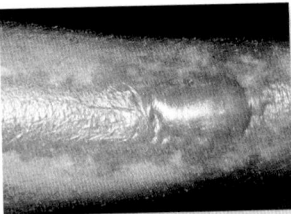

Second-degree burn

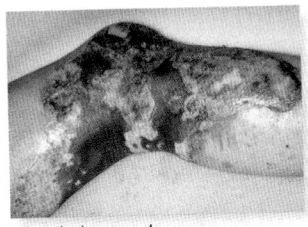

Third-degree burn

First-degree burns, often referred to as *superficial burns,* involve only the epidermis and are characterized by redness, pain, and slight edema. An example is a mild sunburn. Treatment involves immersing the burned area in cool water or applying cool, wet compresses, possibly followed by covering the burn with sterile, nonadhesive bandages. The healing time averages about 3 to 5 days, and typically there is no scarring.

Second-degree burns involve the epidermis and part of the dermis. The skin appears red, tan, or white. The skin also is blistered and painful. Examples include very severe sunburns (where the skin also blisters) or scalding from hot liquids or chemicals. The treatment is similar to that for first-degree burns, and care must be taken not to break the blisters, which would increase risk of infection. Applying ointments to the blisters is not recommended, because the ointments can retain heat in the burned area. In addition, burned limbs should be elevated to prevent swelling. Healing times are approximately 2 to 4 weeks, and slight scarring may occur.

Third-degree burns involve the epidermis, dermis, and subcutaneous layer, which often are destroyed. Third-degree burns typically are caused by contact with corrosive chemicals or fire, or prolonged contact with extremely hot water. Dehydration is a major concern with a third-degree burn, because the entire portion of skin has been lost, and water cannot be retained in the area. Third-degree burn victims must be aggressively treated for dehydration, or they may die. In addition, patients typically are given antibiotics because risk of infection is very great. The treatment may vary slightly, depending upon what caused the burn. Most third-degree burns require hospitalization. Skin grafting typically is needed for patients with third-degree burns, because the entire dermis and its vasculature are destroyed and regeneration is limited. A **skin graft** is a piece of skin transplanted from one part of the body to another to cover a destroyed area. Skin grafts help prevent infection and dehydration in the affected area, and they also help minimize abnormal connective tissue fibrosis and disfigurement.

How is the overall severity of a burn injury determined?

The severity of a burn injury is measured not only by the degree of the burn, but also by the age of the patient, the general size of the burn, and

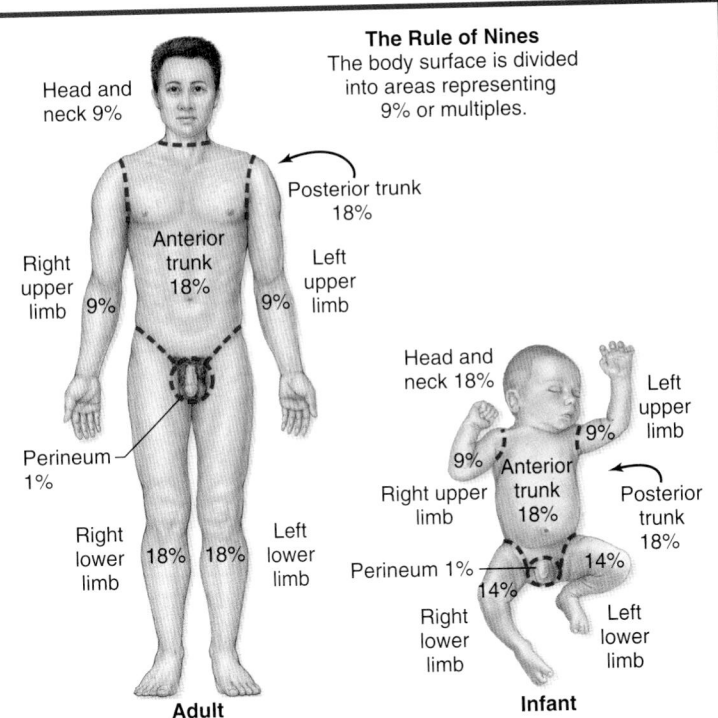

The Rule of Nines
The body surface is divided into areas representing 9% or multiples.

Rule of nines method to determine the extent of burn in adults and infants.

the location of the burn. For example, a burn on the face may require more extensive treatment than a similar burn on an extremity. The **rule of nines** is used to estimate surface area of a burn. Simply put, most (but not all) major body areas approximately account for some factor of 9% of the total body surface area. In adults, the anterior and posterior parts of the head and neck count 9% of the total body surface area, each upper limb counts 9%, each lower limb and gluteal region counts 18%, the anterior trunk counts 18%, the posterior trunk counts 18%, and the perineum is 1%. Estimating surface area of a burn is critical for determining appropriate fluid replacement. The greater the surface area of the burn, the greater the volume of fluids that are lost, and these fluids must be replaced, either orally or intravenously.

Burns are considered very severe or critical if one of the following criteria is met:

1. Over 25% of the body has second-degree burns.
2. Over 10% of the body has third-degree burns.
3. Third-degree burns are present on the hands, feet, face, or perineum.

What treatments are used for burn injuries?

In general, acute treatment involves managing fluid loss, relieving swelling, pain management, removing dead tissue and foreign material from the wound (debridement), controlling infection, and increasing caloric intake.

Swelling also may occur as the blood capillaries become more permeable ("leaky"), and fluid may collect in localized tissues, which exacerbates the overall fluid loss in the circulation. In severe cases, a procedure called an **escharotomy** (es-kă-rot′ō-mē) is performed, where an incision is made in the dermis to lessen the constriction caused by the swelling.

Pain medications may be given to alleviate any discomfort from the burn and resulting swelling. Antibiotics and other medications may be given to help limit and prevent infection.

Finally, individuals with severe burns become hypermetabolic as the body attempts to heal, and so their demand for nutrition greatly increases. The burn patient must be given additional caloric intake, sometimes as much as two to three times their normal caloric intake, to meet the demands. Typically, this supplementary nutrition is given through feeding tubes, IVs, or both.

INTEGRATE

CLINICAL VIEW
Botox and Wrinkles

Many individuals have explored ways to diminish the appearance of age-related wrinkles. One popular treatment for wrinkles caused by repeated facial muscle expression is botulinum toxin type A (Botox). This medicine is derived from the toxin produced by the bacterium *Clostridium botulinum*. Although large doses of this toxin may be deadly, small therapeutic doses temporarily block nerve impulses to the facial expression muscles, thereby decreasing or eliminating the wrinkles they produce.

The procedure is done in a doctor's office, where the doctor injects Botox in the specific facial muscles responsible for the wrinkles, such as frown lines and crows' feet. The effect is temporary, and an individual must repeat the procedure after about four months, as the muscles regain their function.

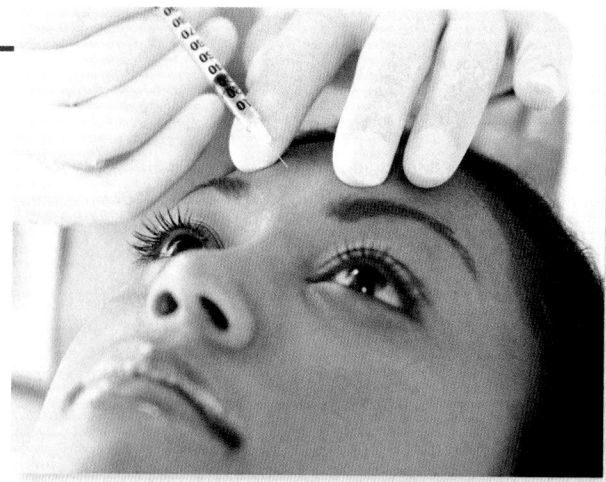

Botox is relatively safe, but some individuals may experience adverse effects, and overuse of Botox can produce a face that appears frozen and devoid of much facial expression.

6.4a Integument and Integumentary Derivatives Development

LEARNING OBJECTIVES

1. Describe how integument develops from two germ layers.
2. Explain the developmental origins of nails, hair, and glands.

By the end of the seventh week of development, the ectoderm forms a layer of squamous epithelium that flattens and then becomes both a covering layer called the **periderm,** and an underlying basal layer. The basal layer will form the stratum basale and all other epidermal layers. By the twenty-first week, the stratum corneum and friction ridges form. During the fetal period, the periderm is eventually sloughed off. The sloughed-off cells mix with sebum secreted by the sebaceous glands, producing a waterproof protective coating called the **vernix caseosa** that coats the skin of the fetus.

The dermis is derived from mesoderm. During the embryonic period, this mesoderm becomes **mesenchyme.** The mesenchymal cells begin to form the components of the dermis at about eleven weeks.

Fingernails and toenails start to form in the tenth week of development. The fingernails reach the tips of the fingers by 32 weeks, whereas the toenails become fully formed by about 36 weeks.

Hair follicles begin to appear between 9 and 12 weeks of development as pockets of cells, called **hair buds,** invade the dermis from the overlying stratum basale of the epidermis. These hairs do not become easily recognizable in the fetus until about the twentieth week. Finally, sweat and sebaceous glands develop from the stratum basale of the epidermis and first appear at about 20 weeks on the palms and soles and later in other regions.

WHAT DID YOU LEARN?

17 What two primary germ layers form the integument?

6.4b Aging of the Integument

LEARNING OBJECTIVES

3. Explain changes to the skin with age.
4. List factors that contribute to skin aging.

Although some adolescents develop acne when they enter puberty, most skin changes do not become obvious until an individual reaches middle age. Then, the skin repair processes take longer to complete because of a reduced number and activity of stem cells. Skin repair and regeneration activities that took 3 weeks in a healthy young person often take twice that time for a person in the seventh decade of life. Additionally, the reduced stem cell activity in the epidermis results in a thinner skin that is less likely to protect against abrasive, mechanical trauma.

Collagen fibers in the dermis decrease in number and organization, and elastic fibers lose elasticity. Years of particular facial expressions (squinting, smiling) produce crease lines in the integument. As a result, the skin forms wrinkles and becomes less resilient. In addition, the skin's immune responsiveness is diminished by a decrease in the number and efficiency of epidermal dendritic cells. Also, hair follicles either produce thinner hairs or stop production entirely.

Chronic overexposure to UV rays can damage the DNA in epidermal cells and accelerate aging, and it is the predominant factor in the development of nearly all skin cancers. Skin cancer is the most common type of cancer. It occurs most frequently on the head and neck regions, followed by other regions commonly exposed to the sun. Fair-skinned individuals, especially those who had severe sunburns as children, are most at risk for skin cancer.

Skin cancer can arise in anyone of any age. Individuals should use sunscreen regularly and avoid prolonged exposure to the sun. An individual should regularly and thoroughly inspect his or her skin for any changes, such as an increase in the number or size of moles, or appearance of new skin lesions. In addition, a person should be examined routinely by a dermatologist. Table 6.2 compares and describes the three main types of skin cancer.

  WHAT DID YOU LEARN?

18 How do UV rays contribute to skin aging?

Table 6.2	Skin Cancer

BASAL CELL CARCINOMA

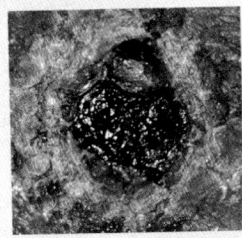

- Most common type of skin cancer
- Least dangerous type as it seldom metastasizes
- Originates in stratum basale
- First appears as small, shiny elevation that enlarges and develops central depression with pearly edge
- Usually occurs on face
- Treated by surgical removal of lesion

SQUAMOUS CELL CARCINOMA

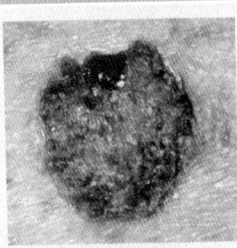

- Arises from keratinocytes of stratum spinosum
- Lesions usually appear on scalp, ears, lower lip, or dorsum of hand
- Early lesions are raised, reddened, scaly; Later lesions form concave ulcers with elevated edges
- Treated by early detection and surgical removal of lesion
- May metastasize to other parts of the body

MALIGNANT MELANOMA

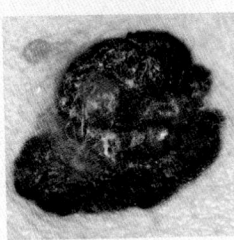

- Most deadly type of skin cancer due to aggressive growth and metastasis
- Arises from melanocytes, usually in a preexisting mole
- Individuals at increased risk include those who have had severe sunburns, especially as children
- Characterized by change in mole diameter, color, shape of border, and symmetry
- Survival rate improved by early detection and surgical removal of lesion
- Advanced cases (metastasis of disease) difficult to cure and are treated with chemotherapy, interferon therapy, and radiation therapy

The usual signs of melanoma may be easily remembered using the ABCDE rule. Report any of the following changes in a birthmark or mole to your physician:

A = Asymmetry: One-half of a mole or birthmark doses not match the other.

B = Border: Edges are notched, irregular, blurred, or ragged.

C = Color: Color is not uniform; differing shades (usually brown or black and sometimes patches of white, blue, or red) may be seen.

D = Diameter: Affected area is larger than 6 mm (about 1/4 inch) or is growing larger.

E = Evolving: Change in the size, shape, or color of a mole. Or a change in symptoms, such as how a mole feels (how itchy or tender it feels) or what happens on the surface of a mole (especially bleeding).

CHAPTER SUMMARY

- The integumentary system includes both the skin (integument) and its derivatives (nails, hair, sweat glands, and sebaceous glands).

6.1 Composition and Functions of the Integument 186

- The integument has a superficial epidermis (composed of keratinized stratified squamous epithelium) and a deeper dermis (that is primarily dense irregular connective tissue).
- Deep to the integument is the subcutaneous layer, which helps adhere the integument to underlying structures.

6.1a Epidermis 186

- Cell types in the epidermis include keratinocytes (the most abundant cell type), melanocytes (produce melanin), epidermal dendritic cells (initiate an immune response), and tactile cells (sensitive to touch).
- The epidermis is organized into specific layers called strata. From deepest to most superficial, they are the stratum basale (deepest layer, with actively dividing keratinocytes), stratum spinosum, stratum granulosum, stratum lucidum, and stratum corneum (many layers of dead keratinocytes).
- Keratinization is the process by which keratinocytes fill up with the protein keratin, and as a result the cell dies. Keratinization begins in the stratum granulosum.
- Thick skin (palms of hands, soles of feet) has five epidermal strata, whereas thin skin (on the rest of the body) has four.
- Skin color is a result of hemoglobin in the blood vessels of the dermis, melanin pigment, and carotene pigment.

6.1b Dermis 191

- The dermis has a superficial papillary layer and a deep reticular layer.
- The papillary layer is primarily areolar connective tissue. Its dermal papillae interlock with the epidermal ridges of the epidermis.
- The reticular layer is composed of dense irregular connective tissue, and contains the hair follicles and the secretory portions of glands.

6.1c Subcutaneous Layer 193

- The subcutaneous layer protects body parts, acts as an energy reservoir, and provides thermal insulation.

6.1d Functions of the Integument 194

- The integument has numerous functions including protection, prevention of water loss, and water gain, metabolic regulation, secretion and absorption, immune function, temperature regulation, and sensory reception.

6.2 Integumentary Structures Derived from Epidermis 198

- Nails, hair and the exocrine glands of the skin are all derived from the epidermis, and are known as epidermal derivatives.

6.2a Nails 198

- Nails are formed from the stratum corneum layer of the epidermis; they protect the digits and aid in grasping objects.

6.2b Hair 199

- Hair consists of a bulb (a swelling where the hair originates in the dermis), a root (the portion of the hair deep to the skin surface), and the shaft (the portion of the hair that extends beyond the skin surface).
- Arrector pili muscles can contract to elevate the hair.
- Among the functions of hair are protection, facial expression, heat retention, sensory reception, visual identification, and chemical signal (pheromone) dispersal.

6.2c Exocrine Glands of the Skin 201

- Merocrine sweat glands produce a watery secretion called sweat.
- Apocrine sweat glands produce a viscous secretion that, when acted upon by bacteria, produces a noticeable odor.
- Sebaceous glands discharge sebum onto hair follicles by holocrine secretion.
- Ceruminous glands produce cerumen (earwax) to lubricate the external acoustic meatus and eardrum.
- Mammary glands are modified apocrine sweat glands to produce breast milk.

6.3 Repair and Regeneration of the Integumentary System 203

- Regeneration refers to the replacement of damaged or dead cells. Fibrosis is the replacement of damaged tissues with scar tissue.
- The wider and deeper the surface affected, the longer it takes for the skin to be repaired.

6.4 Development and Aging of the Integumentary System 204

- The epidermis and epidermal derivates are derived from ectoderm, and the dermis is derived from mesoderm.

6.4a Integument and Integumentary Derivatives Development 206

- The integument begins to form during the embryonic period.
- Nails start to form in the tenth week of development, hair follicles first appear between 9 and 12 weeks of development, and exocrine glands first appear during the fetal period.

6.4b Aging of the Integument 206

- As an individual ages, the skin repair processes take longer, wrinkles appear, the epidermis becomes thinner (due to reduced stem cell activity), and skin cancers may become more likely.

Do You Know the Basics?

_____ 1. Which statement is false about sebaceous glands?

    a. They release their secretion onto a hair follicle.

    b. They release their product by apocrine secretion.

    c. They are located in the dermis.

    d. The product they secrete acts as a lubricant and waterproofer.

_____ 2. The layer of the epidermis in which cells begin the process of keratinization is the

    a. stratum corneum.

    b. stratum basale.

    c. stratum lucidum.

    d. stratum granulosum.

_____ 3. The sweat glands that communicate with skin surfaces throughout the body, producing a secretion that is primarily water, are

    a. apocrine glands.

    b. merocrine glands.

    c. sebaceous glands.

    d. ceruminous glands.

_____ 4. Which of the following is *not* a function of the integument?

    a. acts as a physical barrier

    b. stores calcium in the dermis

    c. regulates temperature through vasoconstriction and vasodilation of dermal blood vessels

    d. participates in immune defense

_____ 5. What layer is correctly matched with the tissue that forms it?

    a. papillary layer of dermis; areolar connective tissue

    b. subcutaneous layer; dense irregular connective tissue

    c. reticular layer of dermis; stratified squamous epithelium

    d. epidermis; dense irregular connective tissue.

_____ 6. Which statement is true about melanin and melanocytes?

    a. It is produced by cells that are located in the stratum spinosum.

    b. It is a pigment that accumulates inside keratinocytes.

    c. Darker-skinned individuals have more melanocytes than lighter-skinned individuals.

    d. Albinism is caused by a lack of melanocytes in the body.

_____ 7. A _____ degree burn typically involves the epidermis and part of the dermis. The subcutaneous layer is not affected.

    a. first-

    b. second-

    c. third-

    d. fourth-

_____ 8. The cells in a hair follicle that are responsible for forming hair are the

    a. hair papilla cells.

    b. matrix cells.

    c. medullary cells.

    d. cortex cells.

_____ 9. Which epidermal cell type is responsible for detecting touch sensations?

    a. keratinocyte

    b. melanocyte

    c. tactile cell

    d. epidermal dendritic cell

_____ 10. At what stage of wound healing does granulation tissue first form?

    a. after scar tissue forms along the wound

    b. before the blood completely clots

    c. before leukocytes enter the site and clean the wound

    d. after a blood clot forms and prior to scar tissue forming.

11. Describe the composition of the layers of the epidermis.

12. List the four main cell types in the epidermis, their function, and the layer(s) of the integument in which they reside.

13. Describe the tissue type and structure of the two specific layers of the dermis.

14. Describe how the skin is involved in vitamin D production.

15. Compare the structure and composition of the following nail parts: nail body, nail bed, eponychium, and lunula.

16. What are the three types of hair?

17. Where are ceruminous glands located, and what do they secrete?

18. Discuss the steps involved in wound repair of the integument.

19. What embryonic tissues form the integument?

20. What are some effects of aging on the integument?

Can You Apply What You've Learned?

1. Alexander is a 15-year-old boy with extensive acne on his nose, forehead, and cheeks. He reached puberty at age 14, at which point his acne became more abundant in these areas. What is the anatomic basis for his acne?

    a. His merocrine sweat glands have begun producing abundant amounts of sweat.

b. His sebaceous gland ducts have become blocked.

c. His sebaceous glands are not producing enough sebum to lubricate his skin.

d. His apocrine sweat glands are producing a secretion that has a marked odor.

2. During anatomy lab, Susan scratched her arm and noticed that some skin cells had sloughed off. She decided to prepare a slide of these cells and examine them under a microscope. What characteristics do you expect the cells to have?

a. polygonal cells with prominent nuclei

b. cuboidal cells, some of which were undergoing mitosis

c. flattened anucleate cells

d. oval cells surrounded by abundant collagen

3. While running to class, Jennifer slipped and skinned her knee. The wound appeared superficial, yet there was extensive bleeding. Based on this information, and her knowledge of the composition of the integument, she determined that the wound penetrated

a. the stratum corneum layer of the epidermis only.

b. all layers of the epidermis but not the dermis.

c. all layers of the epidermis and part of the dermis.

d. all layers of the epidermis, dermis, and the subcutaneous layer.

Can You Synthesize What You've Learned?

1. When you are outside on a cold day, your skin is much paler than normal. Later, when you enter a warm room, your face becomes flushed. What are the reasons for the change in color of your face?

2. Teri was involved in a chemical accident where she suffered third-degree burns on 30% of her body. What potential complications could Teri develop as a result of these burns? As Teri's physician, how would you help minimize these complications?

3. At the age of 50, John noticed that one of the moles on his face appeared different than normal. The mole appeared larger than normal, darkened, and asymmetrical. John's dermatologists suspected skin cancer. Based on the above description, what type of skin cancer is most likely, and should John be concerned?

INTEGRATE

ONLINE STUDY TOOLS

The following study aids may be accessed through Connect.

Clinical Case Study: A Case of Malignant Melanoma in a Young Woman

Interactive Questions: This chapter's content is served up in a number of multimedia question formats for student study.

LearnSmart: Topics and terminology include composition of the integument; integumentary structures derived from the epidermis; functions of the epidermis; repair and regeneration of the

integumentary system; development and aging of the integumentary system

Anatomy & Physiology Revealed: Topics include thick skin and subcutaneous tissue; thick and thin skin magnifications of dermis and epidermis; fingernail structure; hair follicle structure; sebaceous glands

Skeletal System: Bone Structure and Function

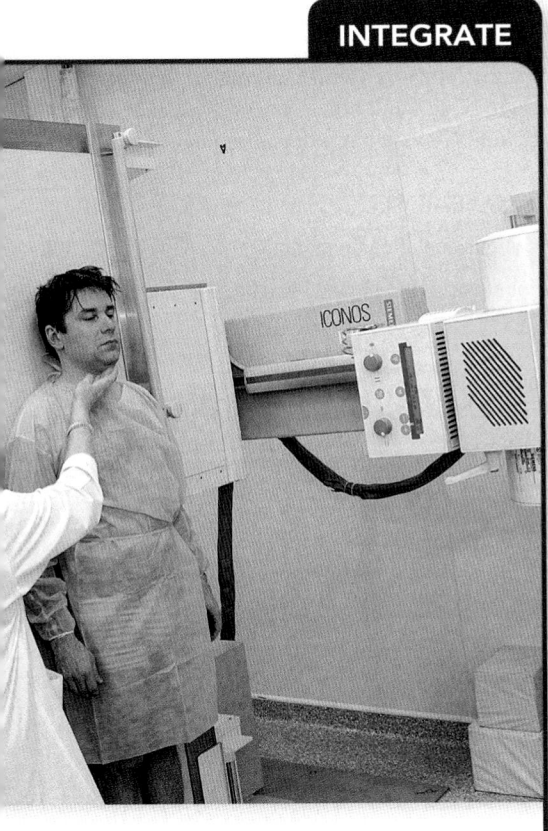

INTEGRATE

aprevealed.com

Module 5: Skeletal System

CAREER PATH
Radiologist

Radiologists and radiology technicians use an array of medical imaging techniques (such as radiography, CT and MRI) to produce images that view details of internal body structures. These imaging techniques tend to be noninvasive and help in the diagnosis of internal pathology and trauma. All radiologists must be able to accurately identify skeletal structures and distinguish them from associated muscles and soft tissues. In addition, the radiologist must be able to tell the difference between a broken bone and a growing bone, because both may appear similar in radiographs to the layperson.

Mention of the skeletal system conjures up images of dry, supposedly lifeless bones in various sizes and shapes. But the **skeleton** (skel′ĕ-ton; *skeletos* = dried) is much more than a supporting framework for the soft tissues of the body. The skeletal system is composed of dynamic living tissues; it interacts with all of the other organ systems and continually rebuilds and remodels itself.

We begin this chapter with a brief description of the skeletal system and a detailed discussion of bone anatomy, the primary organ of this system. Then we examine several important concepts of bone physiology including cartilage growth; bone formation; bone growth and bone remodeling; the regulation of blood calcium; and the effects of aging on the skeletal system. Our chapter concludes with a discussion of bone fracture and repair.

7.1 Introduction to the Skeletal System

LEARNING OBJECTIVES

1. List the structures of the skeletal system.
2. Compare and contrast compact and spongy bone.
3. Identify the types and locations of cartilage within the skeletal system.

Our skeletal system includes the bones of the skeleton as well as cartilage, ligaments, and other connective tissues that stabilize or connect the bones.

Bones of the skeleton are the primary organs of the skeletal system. They form the rigid framework of the body and perform other functions described shortly. Two types of bone connective tissue are present in most of the bones of the body: compact bone and spongy bone (see section 5.2d). **Compact bone** (also called *dense* or *cortical* bone) is a relatively dense connective bone tissue that appears white, smooth, and solid. It makes up approximately 80% of the total bone mass. **Spongy bone** (also called *cancellous* or *trabecular* bone) is located internal to compact bone, appears porous, and makes up approximately 20% of the total bone mass.

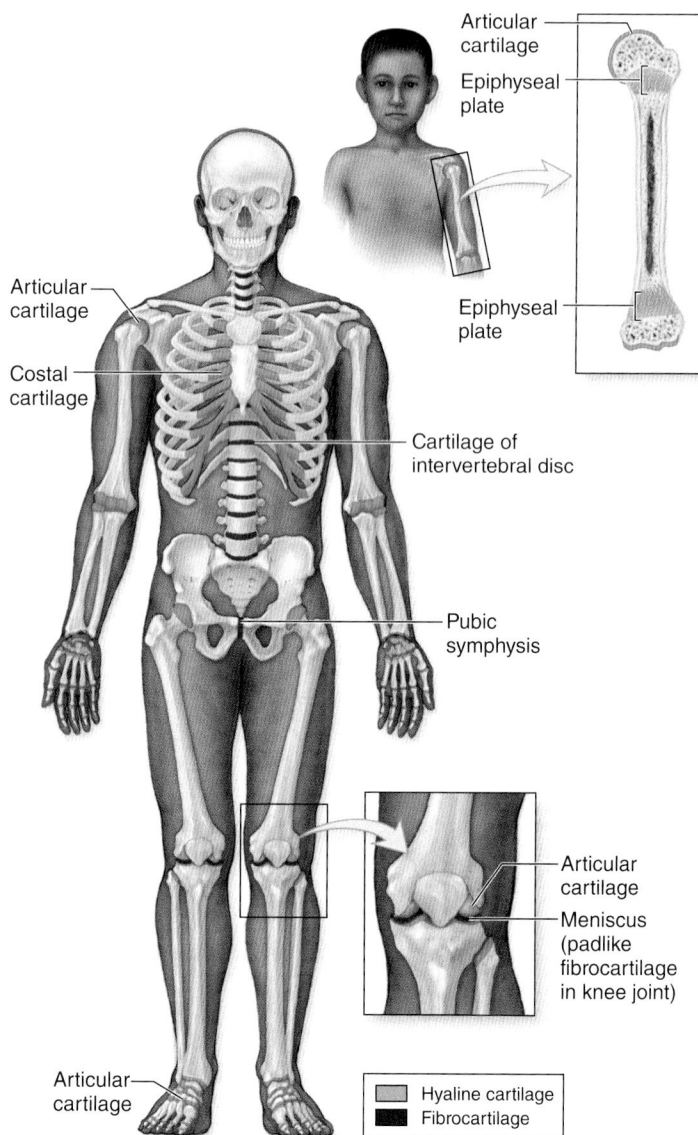

Figure 7.1 Distribution of Cartilage in the Adult and Juvenile Skeletons. Three types of cartilage are found in the adult and juvenile skeletons; the two more common subtypes (hyaline cartilage and fibrocartilage) are shown here.

Cartilage is a semirigid connective tissue that is more flexible than bone. Recall from section 5.2d that there are three subtypes of cartilage; the two most common subtypes are described below (**figure 7.1**).

- **Hyaline cartilage** attaches ribs to the sternum (costal cartilage), covers the ends of some bones (articular cartilage), and is the cartilage within growth plates (epiphyseal plates). Hyaline cartilage also provides a model for the formation of most of the bones in the body.
- **Fibrocartilage** is a weight-bearing cartilage that withstands compression. It forms the intervertebral discs, the pubic symphysis (cartilage between bones of the pelvis), and cartilage pads of the knee joints (menisci).

The role of **ligaments** (dense regular connective tissue that anchors bone to bone), **tendons** (dense regular connective tissue that connects muscle to bone), and other connective tissue structures associated with the skeletal system are described in chapter 9.

WHAT DID YOU LEARN?

❶ Describe the composition of compact bone and spongy bone.

❷ In what three locations of the body do you find fibrocartilage?

7.2 Bone: The Major Organ of the Skeletal System

Our bones—such as the bone of the thigh (femur) or bone of the upper arm (humerus)—are organs. Here we describe the general functions, classification based upon shape, gross anatomy, and histology of bone.

7.2a General Functions

LEARNING OBJECTIVE

1. Describe the general functions of bone.

Bones perform several basic functions: support and protection, movement, hemopoiesis, and storage of mineral and energy reserves.

Support and Protection

Bones provide structural support and serve as a framework for the entire body. Bones also protect many delicate tissues and organs from injury and trauma. The rib cage protects the heart and lungs; the cranial bones enclose and protect the brain; the vertebrae enclose the spinal cord; and the pelvis cradles urinary and reproductive organs, as well as the terminal end of the gastrointestinal tract.

INTEGRATE

CONCEPT CONNECTION

Both the muscular system and the nervous system need calcium to properly function. Luckily, the skeletal system typically houses a sufficient supply of calcium that may be "tapped" by both of these systems when blood calcium levels are low.

Movement

Bones serve as attachment sites for skeletal muscles, other soft tissues, and some organs. Muscles attached to the bones of the skeleton contract and exert a pull on the skeleton that then functions as a system of levers. The bones of the skeleton can alter the direction and magnitude of the forces generated by the skeletal muscles. Potential movements range from powerful contractions needed for running and jumping to delicate and precise movements required to remove a splinter from the finger.

Hemopoiesis

Hemopoiesis (hē′mō-poy-ē′sis; *haima* = blood, *poiesi* = making) is the process of blood cell production. It occurs in red bone marrow connective tissue that contains stem cells that form blood cells and platelets. (The process of hemopoiesis is described in greater detail in section 18.3a.)

Storage of Mineral and Energy Reserves

Most of the body's reserves of the minerals calcium and phosphate are stored within and then released from bone. Calcium is an essential mineral for such body functions as muscle contraction (see section 10.3), blood clotting (see section 18.4), and release of neurotransmitter from nerve cells (section 12.8d). Phosphate is a structural component of ATP, nucleotides, and phospholipids (see chapter 2), and is an important component of the plasma membrane (see section 4.2).

When calcium or phosphate is needed by the body, some bone connective tissue is broken down, and the minerals are released into the blood. In addition, potential energy in the form of lipids is stored in yellow bone marrow in the shafts of some adult bones.

WHAT DID YOU LEARN?

❸ What two minerals are stored in bone, and what are their functions in the body?

7.2b Classification of Bones

LEARNING OBJECTIVE

2. Describe the four major classes of bones as determined by shape.

Bones appear in various shapes and sizes, depending upon their function. The four classes of bone as determined by shape are long bones, short bones, flat bones, and irregular bones **(figure 7.2)**.

Long bones are greater in length than width. These bones have an elongated, cylindrical shaft (diaphysis). This is the most common bone shape. Long bones are found in the upper limbs (namely, the arm, forearm, palm, and fingers) and lower limbs (thigh, leg, sole of the foot, and toes). Long bones vary in size. The small bones in the fingers and toes are long bones, as are the larger tibia and fibula of the lower limb.

Short bones have a length nearly equal to their width. Examples of short bones include the carpals (wrist bones) and tarsals (bones in the foot). *Sesamoid bones,* which are small, sesame seed-shaped bones along the tendons of some muscles, are also classified as short bones. The patella (kneecap) is the largest sesamoid bone.

Flat bones are so named because they have flat, thin surfaces that may be slightly curved. They provide extensive surfaces for muscle attachment and protect underlying soft tissues. Flat bones form the roof of the skull, the scapulae (shoulder blades), the sternum (breastbone), and the ribs.

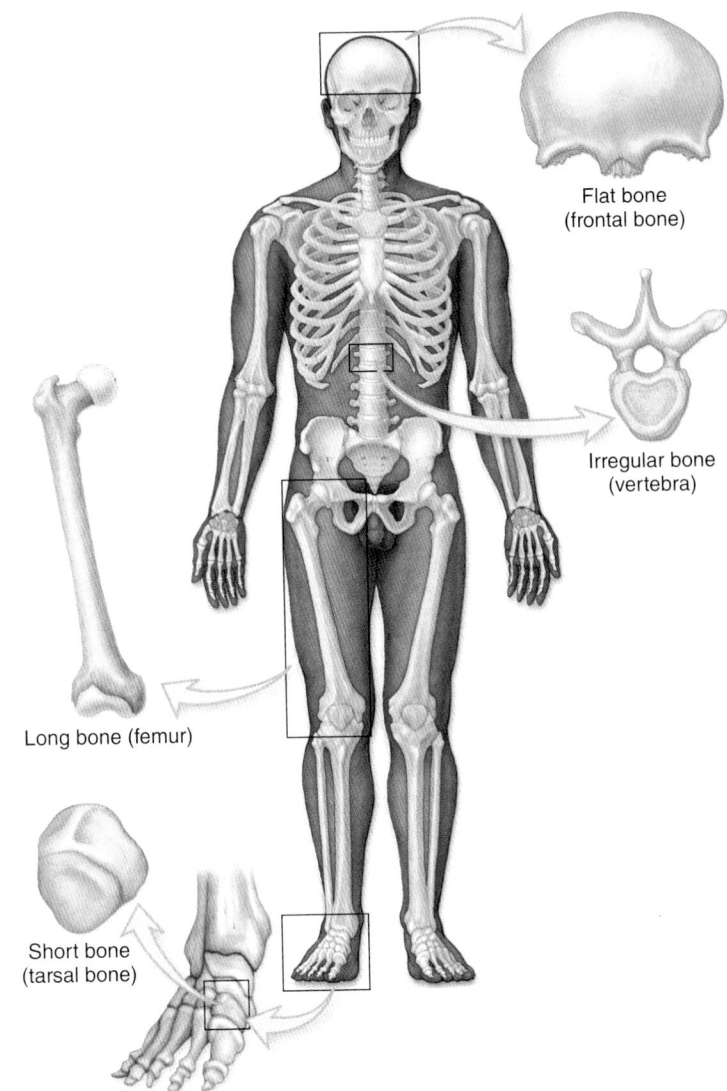

Figure 7.2 Classification of Bone by Shape. Four different classes of bone are recognized according to shape: long, short, flat, and irregular.

- Flat bone (frontal bone)
- Irregular bone (vertebra)
- Long bone (femur)
- Short bone (tarsal bone)

Irregular bones have elaborate, sometimes complex shapes and do not fit into any of the preceding categories. The vertebrae; ossa coxae (hip bones); and several bones in the skull, such as the ethmoid, sphenoid, and sutural bones, are examples of irregular bones.

WHAT DID YOU LEARN?

❹ What are several examples of flat bones in the body?

7.2c Gross Anatomy of Bones

LEARNING OBJECTIVES

3. Explain the structural components of a long bone.

4. Compare the gross anatomy of other bones to that of a long bone.

5. Explain the general function of blood vessels and nerves that serve a bone.

Our discussion continues with details of the gross anatomy of a long bone. We compare it to other classes of bones, and discuss the vascularization and innervation of bone.

Gross Anatomy of a Long Bone

Long bones are the most common bone shape in the body and thus serve as a useful model of bone structure (**figure 7.3a**).

Regions of a Long Bone One of the principal gross features of a long bone is its shaft, which is called the **diaphysis** (dī-af′i-sis; pl., *diaphyses,* dī-af′i-sēz; growing between). The elongated, usually cylindrical, diaphysis provides for the leverage and major weight support of a long bone. Extending inward from the compact bone along the length of the diaphysis are spicules (think spike-like structures) of spongy bone. The hollow, cylindrical space within the diaphysis is called the **medullary** (*marrow*) **cavity.** In children, this cavity contains red bone marrow, which later is replaced by yellow bone marrow in adults.

An expanded knobby region called the **epiphysis** (e-pif′i-sis; pl., *epiphyses,* e-pif′i-sēz; *epi* = upon, *physis* = growth) is at each end of a long bone. A **proximal epiphysis** is the end of the bone closest to the body trunk, and a **distal epiphysis** is the end farthest from the trunk. An epiphysis is composed of an outer thin layer of compact bone and an inner, more extensive

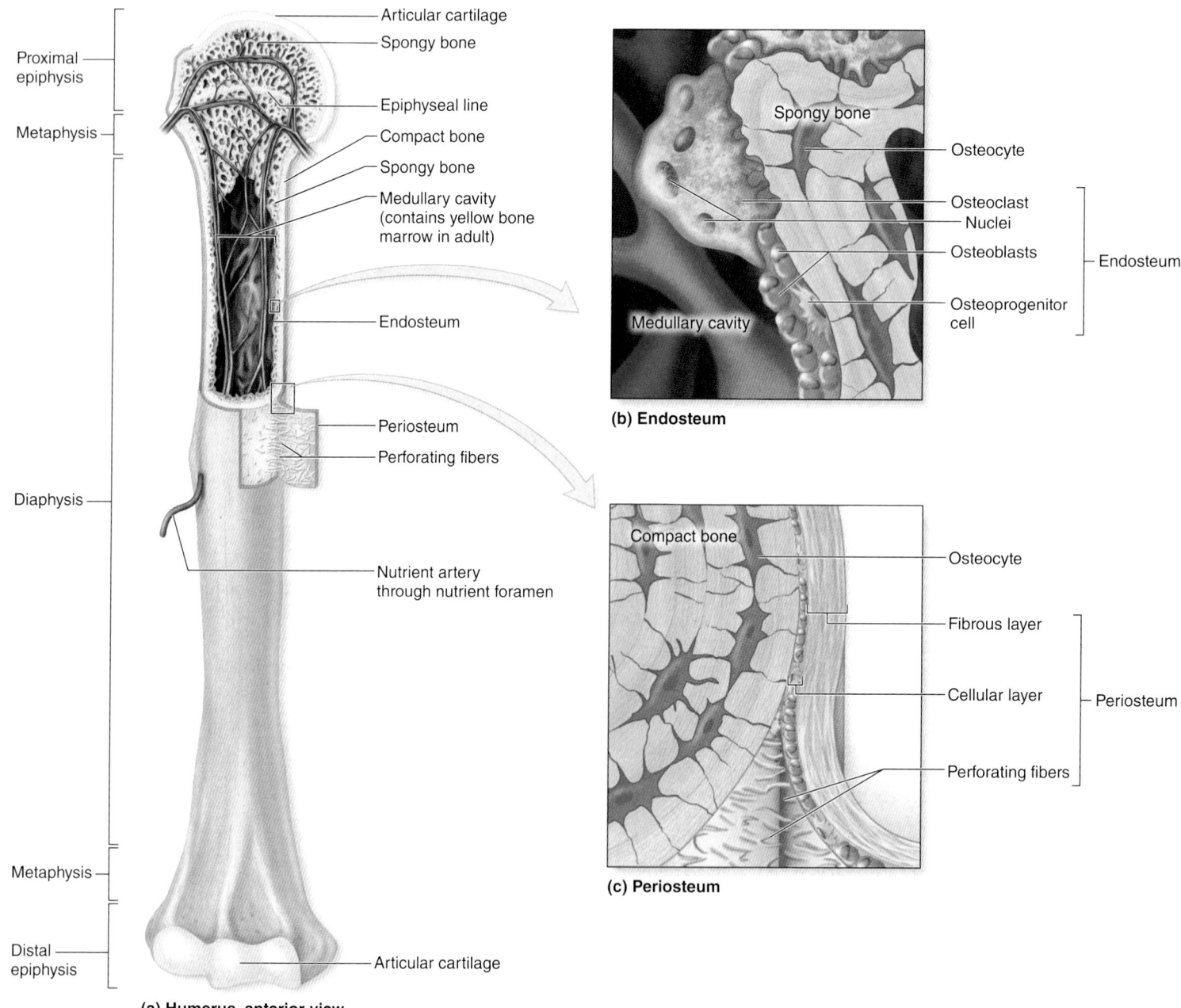

Figure 7.3 Gross Anatomy of a Long Bone. Long bones support soft tissues in the limbs. (*a*) A typical long bone, such as the humerus, contains both compact and spongy bone. (*b*) The endosteum lines the internal surface of the bone along the edge of the medullary cavity. (*c*) The periosteum lines the external surface of the bone shaft. AP|R

region of spongy bone. Spongy bone within the epiphysis resists stress that is applied from many directions. Covering the joint surface of an epiphysis is a thin layer of hyaline cartilage called the **articular cartilage.** This cartilage helps reduce friction and absorb shock in moveable joints.

The **metaphysis** (mě-taf′i-sis) is the region in a mature bone sandwiched between the diaphysis and the epiphysis. This region contains the **epiphyseal** (ep-i-fiz′ē-ăl), or *growth,* **plate** in a growing bone. It is a thin layer of hyaline cartilage that provides for the continued lengthwise growth of the bone. The remnant of the epiphyseal plate in adults is a thin, defined area of compact bone called the **epiphyseal line.**

Coverings and Linings of Bone   A tough sheath called **periosteum** (per-ē-os′tē-ŭm; *peri* = around, *osteon* = bone) covers the outer surface of the bone except for the areas covered by articular cartilage (figure *7.3a, c*). The periosteum consists of two layers. The outer fibrous layer of dense irregular connective tissue protects the bone from surrounding structures, anchors blood vessels and nerves to the surface of the bone, and serves as an attachment site for ligaments and tendons. The inner cellular layer includes osteoprogenitor cells, osteoblasts, and osteoclasts. The function of these cells is described shortly. The periosteum is anchored to the bone by numerous collagen fibers called **perforating fibers,** or *Sharpey's fibers,* which run perpendicular to the diaphysis.

The **endosteum** (en-dos′tē-ŭm; *endo* = within) is an incomplete layer of cells that covers all internal surfaces of the bone within the medullary cavity (figure *7.3a, b*). The endosteum, like the periosteum, contains osteoprogenitor cells, osteoblasts, and osteoclasts.

Gross Anatomy of Other Bone Classes

Short, flat, and irregular bones differ in their gross anatomic structure from long bones. The external surface generally is composed of compact bone, the interior is composed entirely of spongy bone, and there is no medullary cavity. **Figure 7.4** shows the compact and spongy bone arrangement in the skull bone. Observe the roughly parallel compact bone with a layer of internally placed spongy bone. In a flat bone of the skull, the spongy bone is also called **diploë** (dip′lō-ē; *diplous* = double).

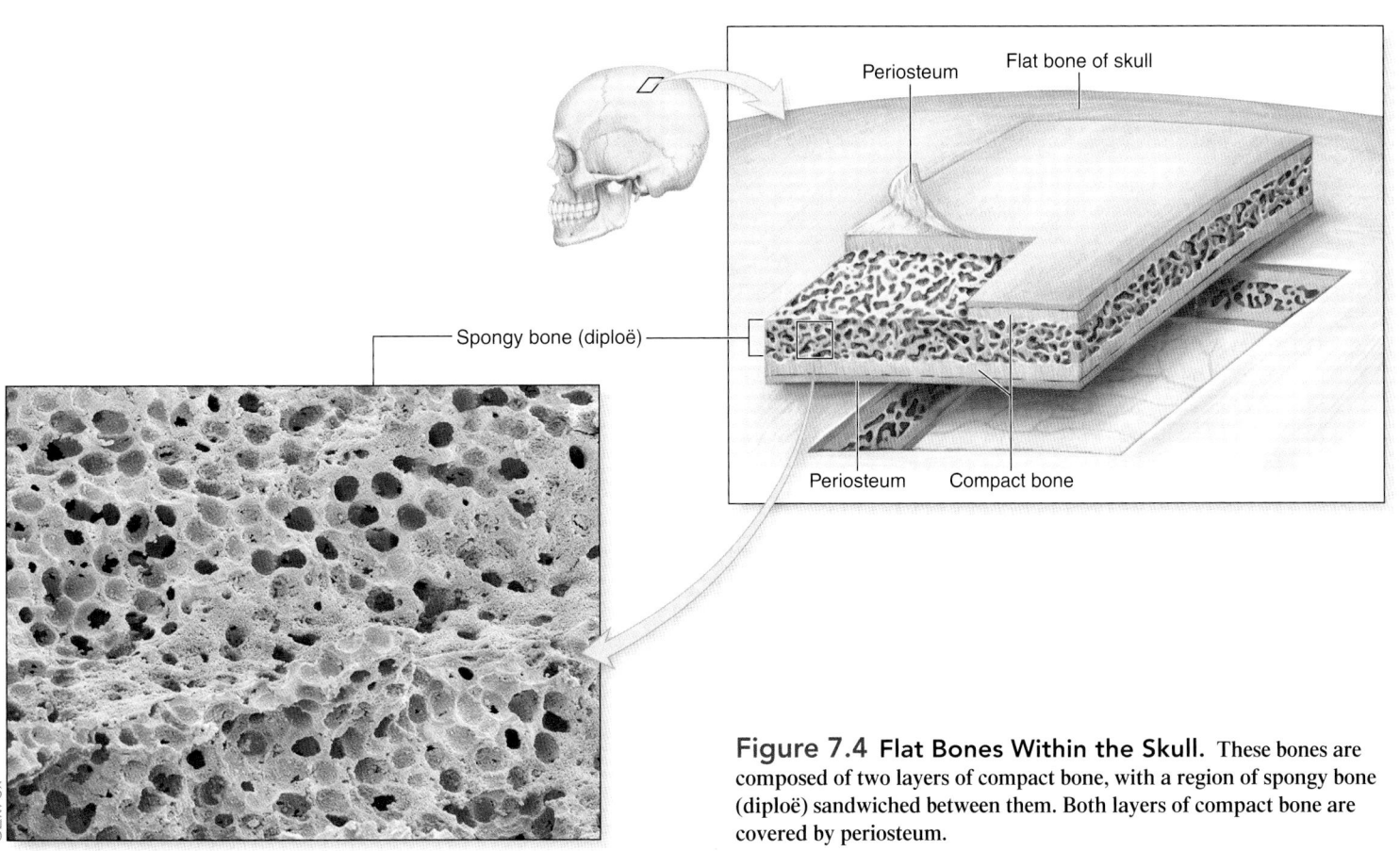

Spongy bone (diploë)

Periosteum   Flat bone of skull

Periosteum   Compact bone

Figure 7.4 Flat Bones Within the Skull. These bones are composed of two layers of compact bone, with a region of spongy bone (diploë) sandwiched between them. Both layers of compact bone are covered by periosteum.

Blood Supply and Innervation of Bone

Bone is highly **vascularized** (meaning it is supplied by many blood vessels), especially in regions containing spongy bone (see figure 7.3*a*). Blood vessels enter bones from the periosteum. Typically, only one nutrient artery enters and one nutrient vein exits the bone via a small opening or hole in the bone called a **nutrient foramen.** Blood vessels supply nutrients and oxygen required by cells and remove waste products from bone cells.

Nerves that supply bones accompany blood vessels through the nutrient foramen and innervate the bone as well as its periosteum, endosteum, and marrow cavity. These are mainly sensory nerves that signal injuries of the skeleton.

WHAT DID YOU LEARN?

5 How do the diaphysis and epiphysis of a bone differ in structure?

6 What is the function of a nutrient foramen in bone?

7.2d Bone Marrow

LEARNING OBJECTIVE

6. Compare and contrast the structure and location of the two types of bone marrow.

Bone marrow is the soft connective tissue of bone that includes red bone marrow and yellow bone marrow **(figure 7.5)**. **Red bone marrow** (myeloid tissue) is hemopoietic (blood cell forming) and contains reticular connective tissue, immature blood cells, and fat.

The locations of red bone marrow differ between children and adults. In children, red bone marrow is located in the spongy bone of most of the bones of the body as well as the medullary cavity of long bones. Much of the red bone marrow degenerates as children mature into adults, and the marrow primarily in the medullary cavities of long bones turns into a fatty substance called **yellow bone marrow.** As a result, adults have red bone marrow only in selected portions of the axial skeleton, such as the flat bones of the skull, the vertebrae, the ribs, the sternum, and the ossa coxae (hip bone). Adults also have red bone marrow in the proximal epiphyses of each humerus and femur.

Note that severe anemia—a condition in which erythrocyte (red blood cell) numbers are lower than normal, resulting in insufficient oxygen reaching the cells of the body—may trigger conversion of yellow bone marrow back to red bone marrow, a change that facilitates the production of additional erythrocytes. (See Clinical View: "Anemia", on page 714.)

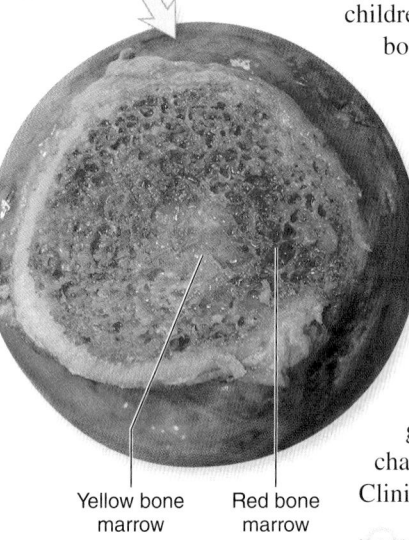

(a) Red bone marrow distribution in the adult

Yellow bone marrow   Red bone marrow

(b) Head of femur, sectioned

Figure 7.5 Red Bone Marrow.
(*a*) Distribution of red bone marrow in the adult skeleton. (*b*) Sectional view of a femoral head, comparing red and yellow bone marrow.

WHAT DID YOU LEARN?

7 Where is red bone marrow found in the adult skeleton?

INTEGRATE

CLINICAL VIEW
Bone Marrow Transplant

Red bone marrow may be transplanted in an individual whose bone marrow was destroyed by radiation or chemotherapy, or who has abnormally functioning bone marrow (as is the case in an individual with leukemia, where the marrow produces abnormal blood cells)—see Clinical View: "Leukemia," on page 722. Donor bone marrow is most commonly harvested from the hip bone, or less commonly, from the sternum. The harvested cells are injected into the blood of the recipient, where they will be transported to the normal locations for red bone marrow. Bone marrow must be "matched" between donor and recipient, just like blood types must be matched, so the immune system will not attack the tissue as something foreign (see chapter 22). Thus, an individual can receive bone marrow only from a donor that is a close match.

7.2e Microscopic Anatomy: Bone Connective Tissue

LEARNING OBJECTIVES

7. Name the four types of bone cells and their functions.
8. Describe the composition of bone's matrix.
9. Explain bone matrix formation and resorption.
10. Compare the structure of compact bone and spongy bone.

The primary component of bone is **bone connective tissue,** also called *osseous* (os′ē-ŭs; *os* = bone) *connective tissue*. Bone is composed of both cells and extracellular matrix, like all connective tissue. We now describe the cells and matrix that compose bone connective tissue, how the matrix is formed and resorbed, and then the two microscopic arrangements (compact bone and spongy bone).

Cells of Bone

Four types of cells are found in bone connective tissue: osteoprogenitor cells, osteoblasts, osteo-cytes, and osteoclasts **(figure 7.6)**.

Osteoprogenitor (os′tē-ō-prō-jen′i-ter) **cells** are stem cells derived from mesenchyme (see section 5.2c). When they divide through the process of cellular division, another stem cell is produced along with a "committed cell" that matures to become an osteoblast. As previously described, these stem cells are located in both the periosteum and the endosteum.

Osteoblasts (*blast* = germ) are formed from osteoprogenitor stem cells. Often, osteoblasts are positioned side by side on bone surfaces. Active osteoblasts exhibit a somewhat cuboidal shape and have abundant rough endoplasmic reticulum and Golgi apparatus, reflecting the activity of

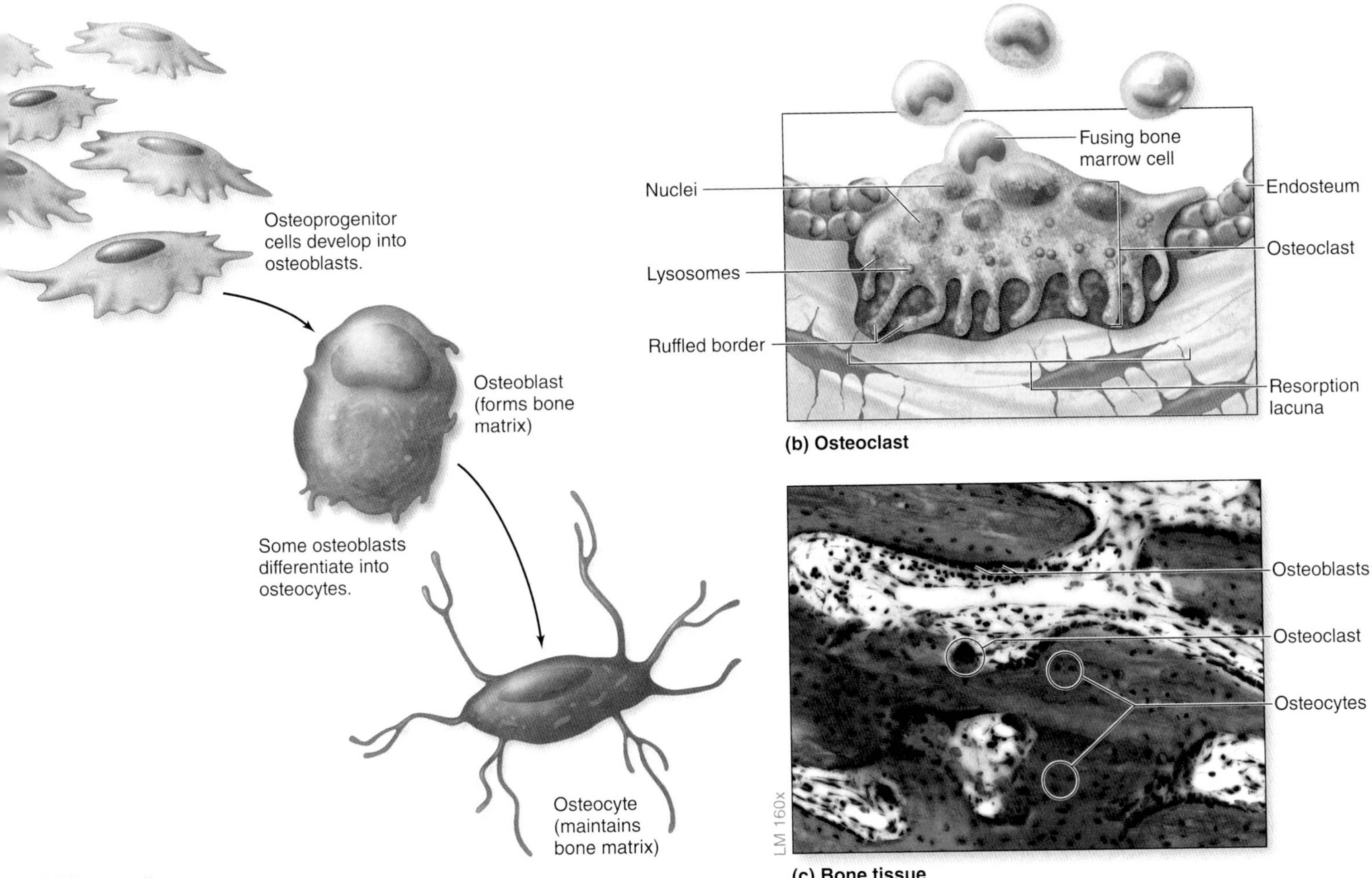

INTEGRATE

CONCEPT CONNECTION

In section 5.6b (See Clinical View: "Stem Cells"), we discussed the role of stem cells in potential treatments of disease. Osteo-progenitor cells are one type of adult stem cell that is specific to the skeletal system, and researchers are examining their medical treatment potential.

Osteoprogenitor cells develop into osteoblasts.

Osteoblast (forms bone matrix)

Some osteoblasts differentiate into osteocytes.

Osteocyte (maintains bone matrix)

(a) Bone cells

Fusing bone marrow cell

Nuclei

Lysosomes

Ruffled border

Endosteum

Osteoclast

Resorption lacuna

(b) Osteoclast

LM 160x

Osteoblasts

Osteoclast

Osteocytes

(c) Bone tissue

Figure 7.6 Types of Cells in Bone Connective Tissue. Four different types of cells are found in bone connective tissue. (*a*) Osteoprogenitor cells develop into osteoblasts, many of which differentiate to become osteocytes. (*b*) Some bone marrow cells fuse to form osteoclasts. (*c*) A photomicrograph shows osteoblasts, osteocytes, and osteoclasts. **AP|R**

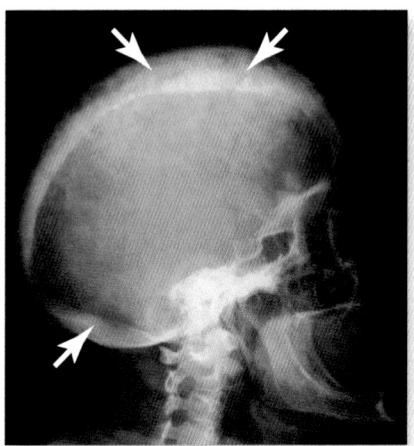

INTEGRATE

CLINICAL VIEW
Osteitis Deformans

Lateral x-ray of a skull with osteitis deformans. White arrows indicate areas of excessive bone deposition.

Osteitis deformans (*Paget disease of bone*) results from a disruption in the balance between osteoclast and osteoblast function. It is characterized by excessive bone resorption (excessive osteoclast activity) followed by excessive bone deposition (excessive osteoblast activity). In osteitis deformans, the osteoclasts are five times larger than normal and may contain 20 or more nuclei (compared to about 3 to 5 nuclei in normal osteoclasts). As a result, these larger osteoclasts resorb bone at a higher rate. In response to this excessive bone resorption, the osteoblasts deposit additional bone, but this new bone is poorly formed and unstable, making it more susceptible to deformation and fractures. The bones most commonly affected include the pelvis, skull, vertebrae, femur (thigh bone), and tibia (leg bone). Initial symptoms include bone deformity and pain. Eventually, the lower limb bones may bow, and the skull often becomes thicker and enlarged.

Composition of the Bone Matrix

The matrix of bone connective tissue has both organic and inorganic components. The organic component is osteoid, which is produced by osteoblasts. Osteoid is composed of both collagen protein plus a semisolid ground substance of proteoglycans (including chondroitin sulfate) and glycoproteins that suspends and supports the collagen fibers. These organic components give bone tensile strength by resisting stretching and twisting, and contribute to its overall flexibility.

The inorganic portion of the bone matrix is made up of salt crystals that are primarily calcium phosphate, $Ca_3(PO_4)_2$. Calcium phosphate and calcium hydroxide interact to form crystals of **hydroxyapatite** (hī-drok'sē-ap-a-tīt), which is $Ca_{10}(PO_4)_6(OH)_2$. The crystals also incorporate other salts (e.g., calcium carbonate) and ions (e.g., sodium, magnesium, sulfate, and fluoride) during the process of calcification. These crystals deposit around the long axis of collagen fibers in the extracellular matrix. The crystals harden the matrix and account for the rigidity or relative inflexibility of bone that provide its compressional strength.

The correct proportion of organic and inorganic substances in the matrix of bone allows it to function optimally. A loss of protein, or the presence of abnormal protein, results in brittle bones; insufficient calcium results in soft bones.

Bone Matrix: Its Formation and Resorption

Bone formation begins when osteoblasts secrete the initial semisolid organic form of bone matrix called osteoid. **Calcification** (kal'si-fi-kā'shŭn), or *mineralization*, subsequently occurs to the osteoid when hydroxyapatite crystals deposit in the bone matrix. Calcification is initiated when the concentration of calcium ions and phosphate ions reach critical levels and precipitate out of solution, thus forming the hydroxyapatite crystals that deposit in and around the collagen fibers. The entire process of bone formation requires a number of substances, including vitamin D (which enhances calcium absorption from the

these cells. Osteoblasts perform the important function of synthesizing and secreting the initial semisolid organic form of bone matrix called **osteoid** (os'tē-oyd; *eidos* = resemblance). Osteoid later calcifies as a result of salt crystal deposition. As a consequence of this mineral deposition on osteoid, osteoblasts become entrapped within the matrix they produce and secrete, and thereafter they differentiate into osteocytes.

Osteocytes (*cyt* = cell) are mature bone cells derived from osteoblasts that have lost their bone-forming ability when enveloped by calcified osteoid. Connections between the original neighboring osteoblasts are maintained as they become osteocytes. Osteocytes maintain the bone matrix and detect mechanical stress on a bone. If stress is detected, osteoblasts are signaled, and it may result in the deposition of new bone matrix at the surface.

Osteoclasts (os'tē-ō-klast; *klastos* = broken) are large, multinuclear, phagocytic cells. They are derived from fused bone marrow cells similar to those that produce monocytes (described in section 18.3c). These cells exhibit a ruffled border where they contact the bone, which increases their surface area exposure to the bone. An osteoclast is often located within or adjacent to a depression or pit on the bone surface called a **resorption lacuna** (*Howship's lacuna*). Osteoclasts are involved in breaking down bone in an important process called bone resorption (described shortly).

INTEGRATE

LEARNING STRATEGY

Think of bone matrix like concrete poured among rebar (interlinking grid of steel bars) to form a basement floor. The collagen fibers are the rebar, and the ground substance with hydroxyapatite is the concrete. Without the concrete, the rebar bends; and without the rebar, the concrete crumbles.

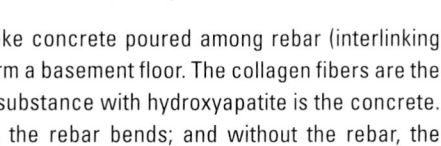

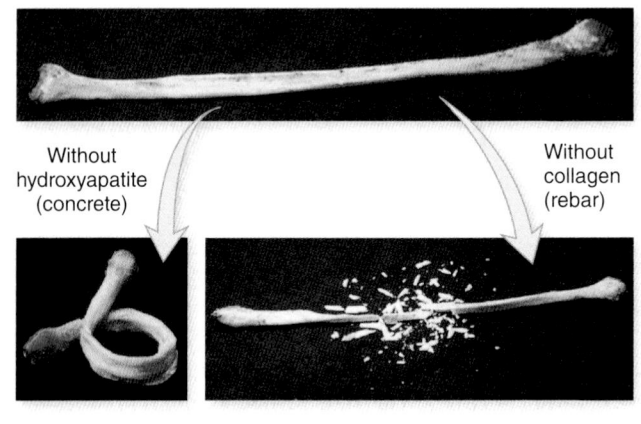

Without hydroxyapatite (concrete)

Without collagen (rebar)

gastrointestinal tract) and vitamin C (which is required for collagen formation), as well as calcium and phosphate for calcification.

Bone resorption is a process whereby bone matrix is destroyed by substances released from osteoclasts into the extracellular space adjacent to the bone. Proteolytic enzymes released from lysosomes within the osteoclasts chemically digest the organic components (collagen fibers and proteoglycans) of the matrix, while hydrochloric acid (HCl)

dissolves the mineral parts (calcium and phosphate) of the bone matrix. The liberated calcium and phosphate ions enter the blood. Bone resorption may occur when blood calcium levels are low (described in detail in section 7.6b).

Comparison of Compact and Spongy Bone Microscopic Anatomy

Compact bone and spongy bone have unique and differing microscopic architecture (**figure 7.7**).

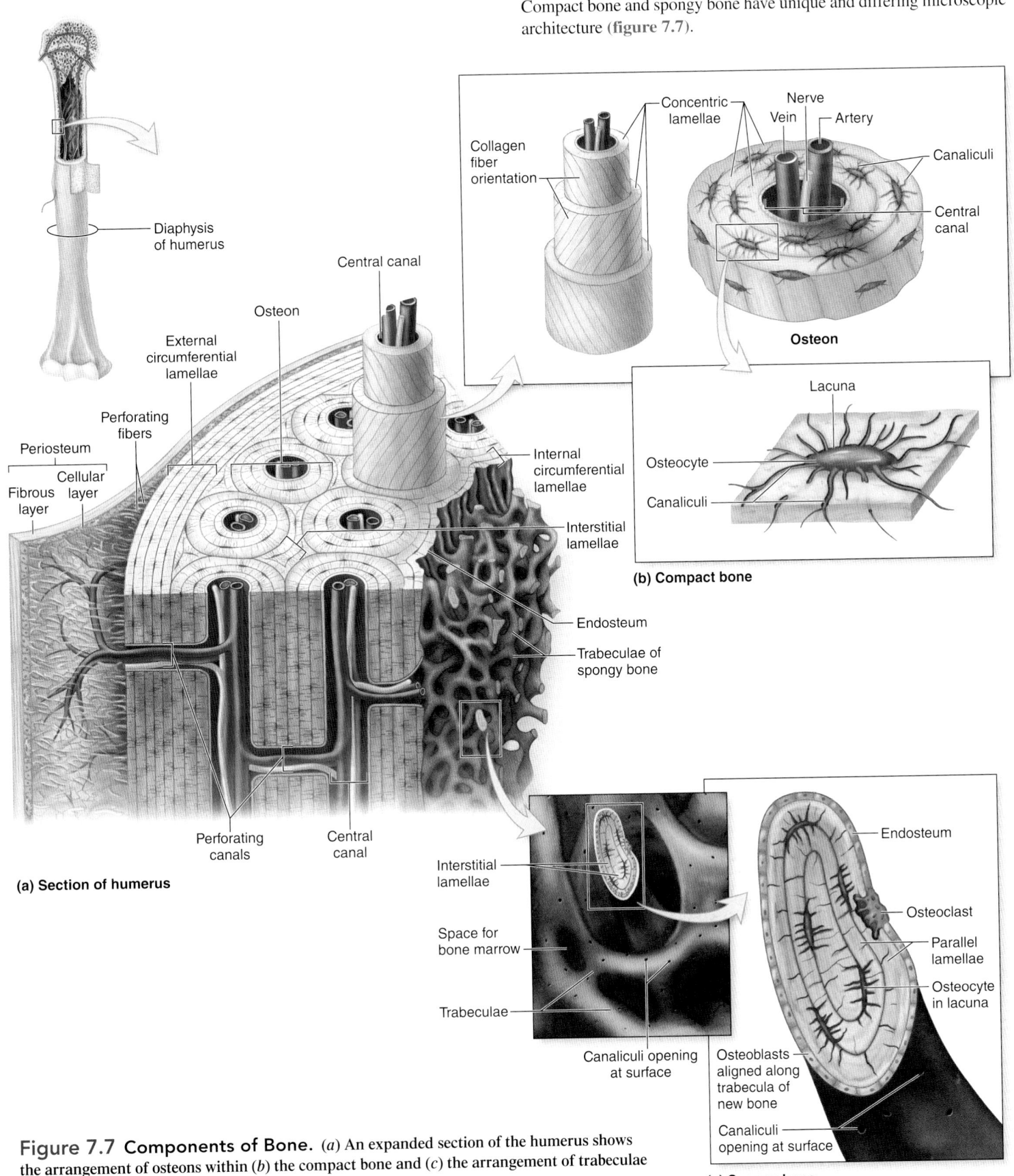

Figure 7.7 Components of Bone. (*a*) An expanded section of the humerus shows the arrangement of osteons within (*b*) the compact bone and (*c*) the arrangement of trabeculae within spongy bone. AP|R

(a) Section of humerus
(b) Compact bone
(c) Spongy bone

INTEGRATE

LEARNING STRATEGY

The analogy of an archery target can help you remember the following components of an osteon:

- The entire *target* represents the *osteon.*
- The *bull's-eye* of the target is the *central canal.*
- The *rings* of the target are the *concentric lamellae.*

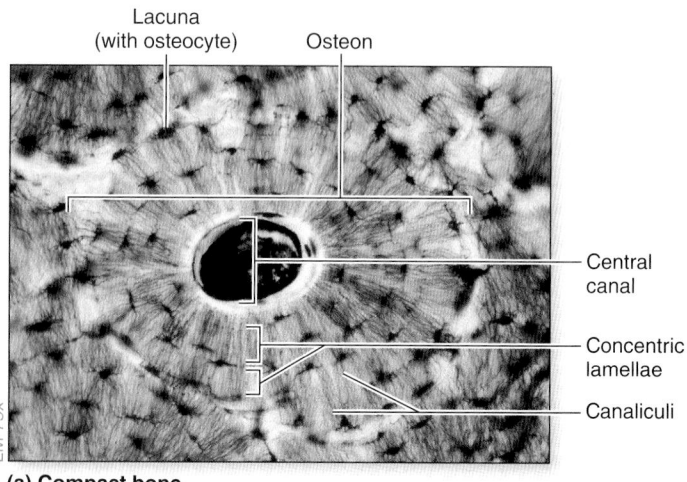

(a) Compact bone

Compact Bone Microscopic Anatomy Compact bone is composed of small cylindrical structures called **osteons,** or *Haversian systems.* An osteon is the basic functional and structural unit of mature compact bone (figure 7.7*a, b*). Osteons are oriented parallel to the diaphysis of the long bone. When an osteon is viewed in cross section, it has the appearance of a bull's-eye target. An osteon has several components as follows:

- **The central** (*Haversian*) **canal** is a cylindrical channel that lies in the center of the osteon and runs parallel to it. Extending through the central canal are the blood vessels and nerves that supply the bone.
- **Concentric lamellae** (lă-mel′-ē; sing., *lamella,* lă-mel′ă; *lamina* = plate, leaf) are rings of bone connective tissue that surround the central canal and form the bulk of the osteon. The numbers of concentric lamellae vary among osteons. Each lamella contains collagen fibers oriented at an angle in one direction; adjacent lamellae contain collagen fibers oriented at an angle that is 90 degrees different from both the previous and next lamellae. This alternating pattern of collagen fiber direction gives bone part of its strength and resilience.
- **Osteocytes** are mature bone cells found in small spaces (see next) between adjacent concentric lamellae. These cells maintain the bone matrix.
- **Lacunae** are the small spaces that each house an osteocyte.
- **Canaliculi** (kan-ă-lik′ū-lī; sing., *canaliculus,* kan-ă-lik′ū-lŭs; *canalis* = canal) are tiny, interconnecting channels within the bone connective tissue that extend from each lacuna, travel through the lamellae, and connect to other lacunae and the central canal. Canaliculi house osteocyte cytoplasmic projections that permit intercellular contact and communication. Nutrients, minerals, gases, and wastes are transported through the cytoplasmic extensions within these passageways, allowing their exchange between the blood vessels of the central canal and the osteocytes.

Figure 7.8 shows cross sections of osteons as viewed through a light microscope and a scanning electron microscope. Several other structures occur in compact bone, but are not part of the osteon proper, including the following (see figure 7.7*a*):

- **Perforating** (*Volkmann*) **canals** resemble central canals in that they also contain blood vessels and nerves. However, perforating canals run perpendicular to the central canals and help connect multiple central canals within different osteons, thus forming a vascular and innervation connection among the multiple osteons.

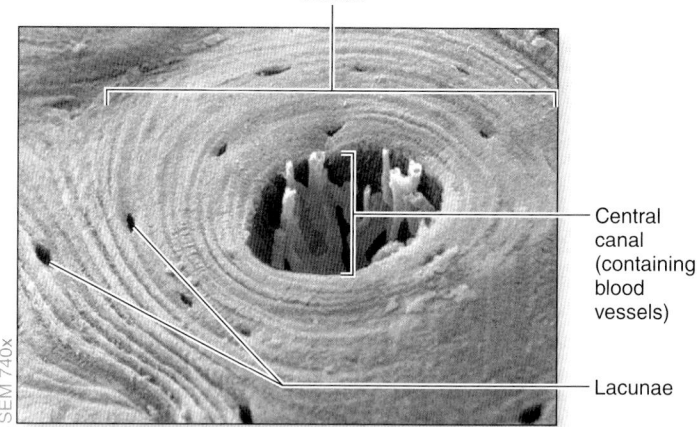

(b) Compact bone

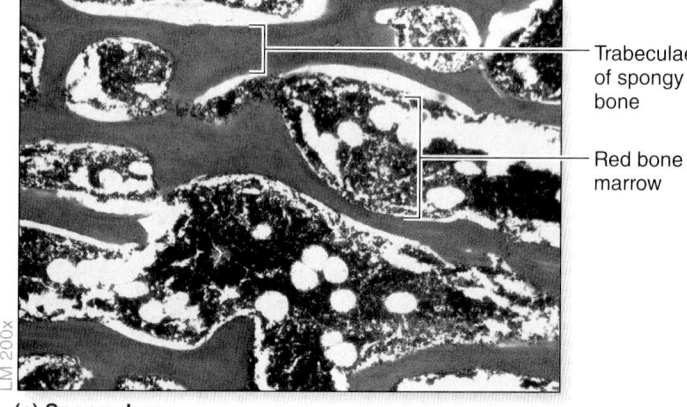

(c) Spongy bone

Figure 7.8 Microscopic Anatomy of Bone. (*a*) Light micrograph and (*b*) SEM of osteons in a cross section of compact bone. (*c*) Light micrograph of spongy bone. **AP|R**

- **Circumferential lamellae** are rings of bone immediately internal to the periosteum of the bone **(external circumferential lamellae)** or internal to the endosteum **(internal circumferential lamellae).** Both external and internal circumferential lamellae run the entire circumference of the bone itself (hence, their name).
- **Interstitial lamellae** (*interstitial systems*) are either the components of compact bone that are between osteons or are the leftover parts of osteons that have been partially resorbed—thus they often look like a "bite" has been taken out of them. The interstitial lamellae are incomplete and typically have no central canal.

Spongy Bone Microscopic Anatomy Unlike compact bone, spongy bone contains no osteons (figures 7.7*c*, 7.8*c*) Instead, its structure is an open lattice of narrow rods and plates of bone, called **trabeculae** (tră-bek′ū-lē; sing., *trabecula*, tră-bek′ū-lă; *trabs* = a beam). Bone marrow (when present) fills in between the trabeculae. When a segment of spongy bone is examined microscopically, you can see **parallel lamellae** composed of bone matrix. Between adjacent lamellae are osteocytes resting in lacunae, with numerous canaliculi radiating from the lacunae. Nutrients reach the osteocytes by diffusion through cytoplasmic processes within the canaliculi that open onto the surfaces of the trabeculae.

Note that the trabeculae often form a meshwork of crisscrossing bars and plates of small bone pieces. This structure provides great resistance to stresses applied in many directions by distributing the stress throughout the entire framework. As an analogy, visualize the jungle gym climbing apparatus on a children's playground. It is capable of supporting the weight of numerous children whether they are distributed throughout its structure or all localized in one area. This is accomplished because stresses and forces are distributed throughout the structure.

WHAT DID YOU LEARN?

8 What are the functions of the osteoprogenitor cell, osteoblast, osteocyte, and osteoclast?

9 What organic and inorganic substances compose bone matrix?

10 What are the major components of an osteon?

7.2f Microscopic Anatomy: Hyaline Cartilage Connective Tissue

LEARNING OBJECTIVE

11. Analyze the structure of hyaline cartilage and the cells in its matrix.

Hyaline cartilage contains a population of cells scattered throughout a glassy-appearing matrix of protein fibers (primarily collagen) embedded within a gel-like ground substance (see section 5.2d). This ground substance is similar to that of bone in that it includes proteoglycans, such as chondroitin sulfate, but it differs from bone because its inorganic salts do not include calcium. This makes hyaline cartilage both resilient and flexible. Additionally, cartilage also contains a high percentage of water (60% to 70% by weight). The high water content makes it highly compressible, allowing hyaline cartilage to function as a good shock absorber.

Chondroblasts (kon′drō-blast; *chondros* = grit or gristle) are derived from mesenchymal cells and they produce the cartilage matrix. Once chondroblasts become encased within the matrix they have produced and secreted, the cells are called **chondrocytes** (kon′drō-sīt; *cyte* = cell) and occupy small spaces called **lacunae.** These mature cartilage cells maintain the matrix. Hyaline cartilage—except the articular cartilage—is covered by a dense irregular connective tissue sheet called the **perichondrium** that helps maintain its shape. Mature cartilage is avascular (not penetrated by blood vessels) and contains no nerves. Nutrients and oxygen are supplied to the cartilage by diffusion from blood vessels in the perichondrium. Several important differences between bone connective tissue and hyaline cartilage connective tissue are summarized in **table 7.1**.

WHAT DID YOU LEARN?

11 What are the primary ways that hyaline cartilage tissue differs from bone tissue?

Table 7.1	Comparison of Bone Connective Tissue and Hyaline Cartilage Connective Tissue	
Characteristics	**Bone Connective Tissue**	**Hyaline Cartilage Connective Tissue**
Cells that form matrix	Osteoblasts	Chondroblasts
Mature cells	Osteocytes	Chondrocytes
Mature cells in lacunae	Yes	Yes
Calcium present in matrix	Yes	No
Blood supply in mature tissue	Extensive	Avascular

7.3 Cartilage Growth

LEARNING OBJECTIVE

1. Compare interstitial and appositional growth of cartilage.

We focus on the process of cartilage growth before discussing bone growth because certain types of bone formation and bone growth are dependent upon the growth of hyaline cartilage.

Cartilage development and growth begins during embryologic development. Cartilage can grow both in length through the process of interstitial growth, or can grow in width by appositional growth (**figure 7.9**).

Interstitial (in-ter-stish′ăl) **growth** occurs within the internal regions of cartilage through the following series of four steps (figure 7.9*a*):

① Chondrocytes housed within lacunae are stimulated to undergo mitotic cell division.

② Following cell division, two cells occupy a single lacuna; they are now called chondroblasts.

③ As chondroblasts begin to synthesize and secrete new cartilage matrix, they are pushed apart. These cells now reside in their own lacuna and are called chondrocytes.

④ The cartilage continues to grow in the internal regions as chondrocytes continue to produce more matrix.

Appositional (ap-ō-zish′ŭn-ăl) **growth** is an increase in width along the cartilage's outside edge, or periphery. The following three steps occur in this process (figure 7.9*b*):

① Undifferentiated stem cells at the internal edge of the perichondrium begin to divide. (Note the perichondrium contains mesenchymal cells as well as these stem cells.)

② New undifferentiated stem cells and committed cells that differentiate into chondroblasts are formed. These chondroblasts are located at the periphery of the old cartilage, where they begin to produce and secrete new cartilage matrix.

③ The chondroblasts, as a result of matrix formation, push apart and become chondrocytes, with each occupying its own lacuna. The cartilage continues to grow at the periphery as chondrocytes continue to produce more matrix.

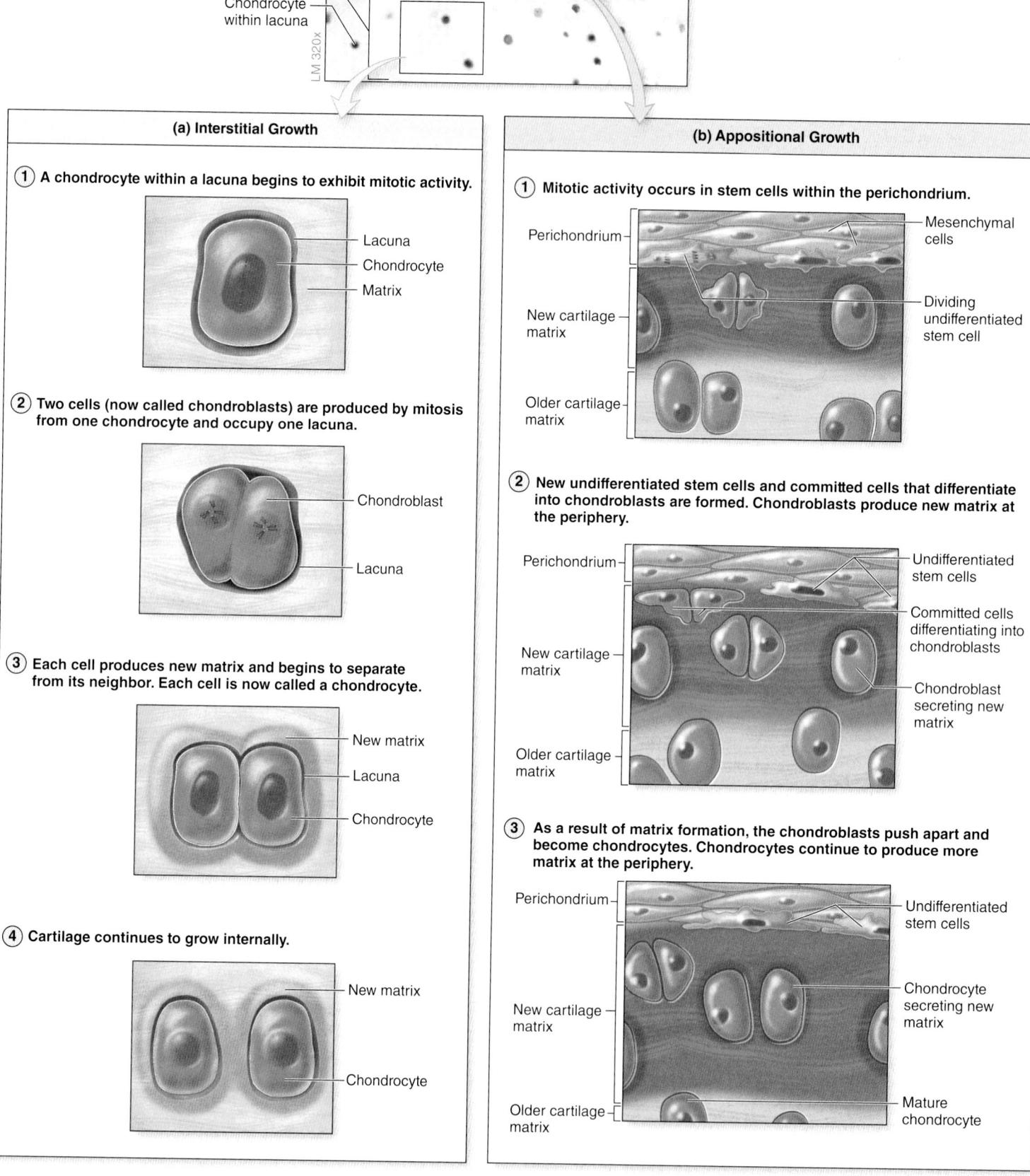

Matrix

Chondrocyte within lacuna

Hyaline cartilage

Perichondrium

LM 320x

(a) Interstitial Growth

(1) A chondrocyte within a lacuna begins to exhibit mitotic activity.

- Lacuna
- Chondrocyte
- Matrix

(2) Two cells (now called chondroblasts) are produced by mitosis from one chondrocyte and occupy one lacuna.

- Chondroblast
- Lacuna

(3) Each cell produces new matrix and begins to separate from its neighbor. Each cell is now called a chondrocyte.

- New matrix
- Lacuna
- Chondrocyte

(4) Cartilage continues to grow internally.

- New matrix
- Chondrocyte

(b) Appositional Growth

(1) Mitotic activity occurs in stem cells within the perichondrium.

- Perichondrium
- New cartilage matrix
- Older cartilage matrix
- Mesenchymal cells
- Dividing undifferentiated stem cell

(2) New undifferentiated stem cells and committed cells that differentiate into chondroblasts are formed. Chondroblasts produce new matrix at the periphery.

- Perichondrium
- New cartilage matrix
- Older cartilage matrix
- Undifferentiated stem cells
- Committed cells differentiating into chondroblasts
- Chondroblast secreting new matrix

(3) As a result of matrix formation, the chondroblasts push apart and become chondrocytes. Chondrocytes continue to produce more matrix at the periphery.

- Perichondrium
- New cartilage matrix
- Older cartilage matrix
- Undifferentiated stem cells
- Chondrocyte secreting new matrix
- Mature chondrocyte

Figure 7.9 Formation and Growth of Cartilage. Cartilage grows either (a) from within by interstitial growth or (b) at its periphery or edge by appositional growth. AP|R

During early embryonic development, both interstitial and appositional cartilage growth occur simultaneously. Note that interstitial growth declines rapidly as the cartilage matures because the cartilage becomes semirigid, and it is no longer able to expand. Further growth can occur only at the periphery of the tissue, so later growth is primarily appositional. Once the cartilage is fully mature, new cartilage growth typically stops. Thereafter, cartilage growth usually occurs only after injury to the cartilage, yet this growth is limited due to the lack of blood vessels in the tissue.

 WHAT DID YOU LEARN?

12 Where do interstitial and appositional growth of cartilage occur?

7.4 Bone Formation

Ossification (os′i-fi-kā′shŭn; *facio* = to make), or **osteogenesis** (os′tē-ō-jen′ĕ-sis; *genesis* = beginning), refers to the formation and development of bone connective tissue. Ossification begins in the embryo and continues as the skeleton grows during childhood and adolescence. By the eighth through twelfth weeks of embryonic development, the skeleton begins forming from either thickened condensations of mesenchyme (intramembranous ossification) or a hyaline cartilage model of bone (endochondral ossification).

7.4a Intramembranous Ossification

 LEARNING OBJECTIVES

1. Identify bones that are produced by intramembranous ossification.
2. Explain the four main steps in intramembranous ossification.

Intramembranous (in′tră-mem′brā-nŭs) **ossification** literally means "bone growth within a membrane." It is so named because the thin layer of mesenchyme in these areas is sometimes referred to as a membrane. Intramembranous ossification also is called *dermal ossification* because the mesenchyme that is the source of these bones is in the area of the future dermis. Recall from section 5.2 that mesenchyme is an embryonic connective tissue that has mesenchymal cells and abundant ground substance.

Intramembranous ossification produces the flat bones of the skull, some of the facial bones (zygomatic bone, maxilla), the mandible (lower jaw), and the central part of the clavicle (collarbone). It begins when mesenchyme becomes thickened and condensed with a dense supply of blood capillaries, and it continues in the following steps **(figure 7.10)**:

① **Ossification centers form within thickened regions of mesenchyme beginning at the eighth week of development.** Some cells in the thickened, condensed mesenchyme divide, and the committed cells that are formed then differentiate into osteoprogenitor cells. Some osteoprogenitor cells become osteoblasts and begin to secrete osteoid. Multiple ossification centers develop within the thickened mesenchyme as the number of osteoblasts increases.

② **Osteoid undergoes calcification.** Osteoid formation is quickly followed by calcification, as calcium salts are deposited onto the osteoid and then they crystallize (solidify). When calcification entraps osteoblasts within lacunae in the matrix, the entrapped cells become osteocytes.

③ **Woven bone and its surrounding periosteum form.** Initially, the newly formed bone connective tissue is immature and not well organized, a type called **woven bone,** or *primary bone*. Eventually woven bone is replaced by **lamellar bone,** or

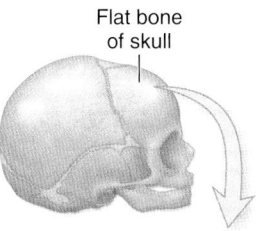

Flat bone of skull

Figure 7.10 Intramembranous Ossification. A flat bone in the skull forms from mesenchymal cells in a series of continuous steps.

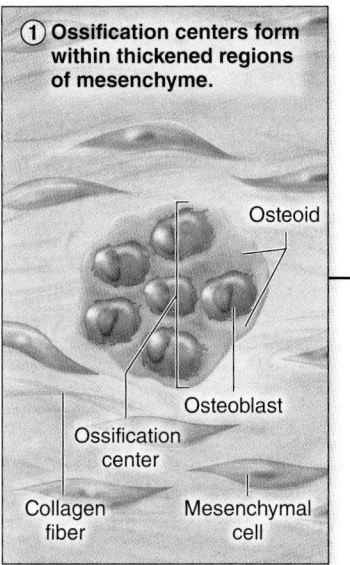

① Ossification centers form within thickened regions of mesenchyme.

Osteoid

Osteoblast

Ossification center

Collagen fiber

Mesenchymal cell

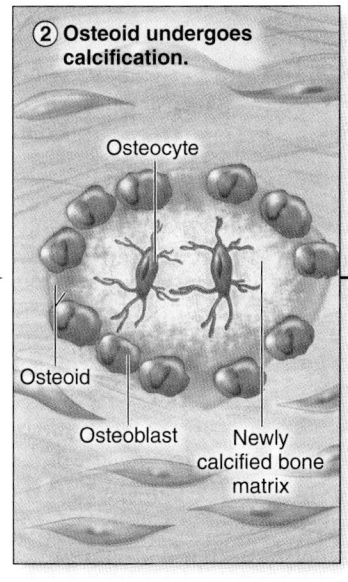

② Osteoid undergoes calcification.

Osteocyte

Osteoid

Osteoblast

Newly calcified bone matrix

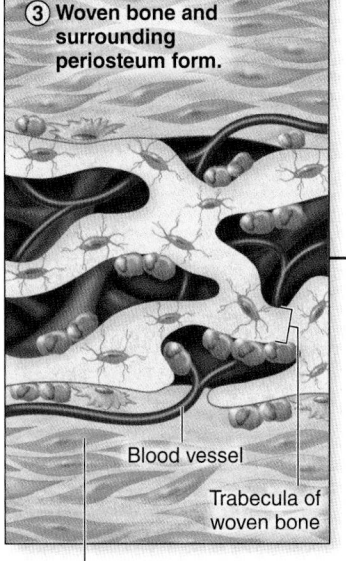

③ Woven bone and surrounding periosteum form.

Blood vessel

Trabecula of woven bone

Mesenchyme condensing to form the periosteum

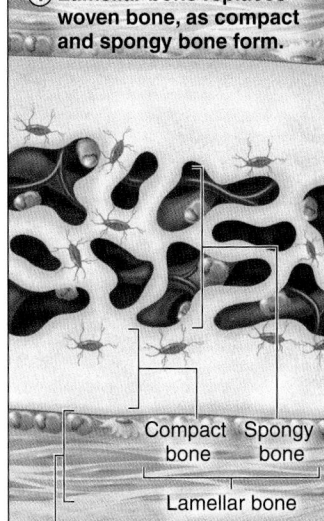

④ Lamellar bone replaces woven bone, as compact and spongy bone form.

Compact bone

Spongy bone

Lamellar bone

Periosteum

LEARNING STRATEGY

Endochondral bone growth is a complex process. Before trying to remember every detail, first learn the following basics:

1. A hyaline cartilage model of bone forms.
2. Bone first replaces hyaline cartilage in the diaphysis.
3. Next, bone replaces hyaline cartilage in the epiphyses.
4. Eventually, bone replaces hyaline cartilage everywhere, except the epiphyseal plates and articular cartilage.
5. By a person's late 20s, all epiphyseal plates typically have ossified, and lengthwise bone growth is complete.

secondary bone (see next). The mesenchyme that still surrounds the woven bone begins to thicken and eventually organizes to form the periosteum. Mesenchymal cells grow and develop to produce additional osteoblasts. Newly formed blood vessels also branch throughout this region. The calcified trabeculae and intertrabecular spaces are composed of spongy bone.

④ **Lamellar bone replaces woven bone, as compact bone and spongy bone form.** Lamellar bone replaces the trabeculae of woven bone. On the internal and external surfaces, spaces between the trabeculae are filled and the bone becomes compact bone. Internally, the trabeculae are modified slightly and produce spongy bone. The typical structure of a flat cranial bone is composed of two external layers of compact bone with a layer of spongy bone in between (see figure 7.4).

WHAT DID YOU LEARN?

13 When does intramembranous ossification begin? What bones are formed from this method?

7.4b Endochondral Ossification

LEARNING OBJECTIVES

3. Explain the steps in endochondral ossification of a long bone.
4. Differentiate between intramembranous ossification and endochondral ossification.

Endochondral (en-dō-kon′drāl; *endo* = within, *chondral* = cartilage) **ossification** begins with a hyaline cartilage model and produces most bones of the skeleton, including those of the upper and lower limbs, the pelvis, the vertebrae, and the ends of the clavicle.

Long bone development in the limb is a good example of this process, which takes place in the following six steps **(figure 7.11)**:

① **The fetal hyaline cartilage model develops.** During the eighth to twelfth week of development, chondroblasts secrete cartilage matrix and a hyaline cartilage model forms. Chondrocytes are trapped within lacunae, and a perichondrium surrounds the cartilage.

② **Cartilage calcifies, and a periosteal bone collar forms.** Within the center of the cartilage model (future diaphysis), chondrocytes start to hypertrophy (enlarge) and resorb (eat away) some of the surrounding cartilage matrix, producing larger holes in the matrix. As these chondrocytes enlarge, the cartilage matrix begins to calcify. Chondrocytes in this region die and disintegrate because nutrients cannot diffuse to them through this calcified matrix. The result is a calcified cartilage shaft with large holes where living chondrocytes had been.

As the cartilage in the shaft is calcifying, blood vessels grow toward the cartilage and start to penetrate the perichondrium around the shaft. Stem cells within the perichondrium divide to form osteoblasts. The osteoblasts develop as this supporting connective tissue becomes highly vascularized, and the perichondrium becomes a periosteum. The osteoblasts within the internal layer of the periosteum start secreting a layer of osteoid around the calcified cartilage shaft. The osteoid hardens and forms a periosteal bone collar around this shaft.

(3) The primary ossification center forms in the diaphysis. A growth of capillaries and osteoblasts, called a **periosteal bud,** extends from the periosteum into the core of the cartilage shaft, invading the spaces where the living chondrocytes had been. The remains of the calcified cartilage serve as a template on which osteoblasts begin to produce osteoid. This region is called the **primary ossification center** because it is the first major center of bone formation. Bone development extends in both directions toward the epiphyses from the primary ossification center. Healthy bone connective tissue quickly displaces the calcified, degenerating cartilage in the shaft. Most, but not all, primary ossification centers have formed by the twelfth week of development.

(4) Secondary ossification centers form in the epiphyses. The same basic process that formed the primary ossification center occurs later in the epiphyses. Beginning around the time of birth, the hyaline cartilage in the center of each epiphysis calcifies and begins to degenerate. Epiphyseal blood vessels and osteoprogenitor cells enter each epiphysis. **Secondary ossification centers** form as bone displaces calcified cartilage. Note that not all secondary ossification centers form at birth; some form later in childhood. As the secondary ossification centers form, osteoclasts resorb some bone matrix within the diaphysis, creating a hollow medullary cavity.

INTEGRATE

CLINICAL VIEW

Forensic Anthropology: Determining Age at Death

During endochondral ossification (discussion begins on page 224), the epiphyseal plates ossify and fuse to the rest of the bone in an orderly manner. The timing of such fusion is well known. If an epiphyseal plate has not yet ossified, the diaphysis and epiphysis are still two separate pieces of bone. Thus, a skeleton that displays separate epiphyses and diaphyses (as opposed to whole fused bones) is that of a juvenile rather than an adult. Forensic anthropologists use this anatomic information to help determine the age of skeletal remains.

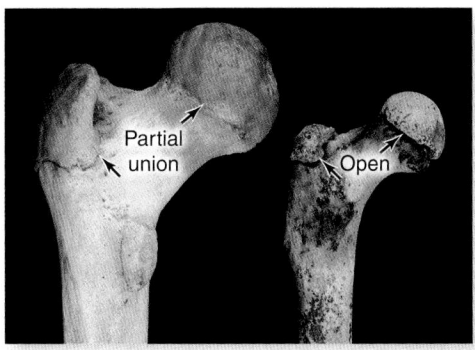

(Left) *Partial union—a femur with partially fused epiphyses.*
(Right) *Open—no fusion between the epiphyses and the diaphysis.*

Fusion of an epiphyseal plate is progressive, and is usually scored as follows:
- **Open** (no bony fusion or union between the epiphysis and the other bone end)
- **Partial union** (some fusion between the epiphysis and the rest of the bone, but a distinct line of separation may be seen)
- **Complete union** (all visible aspects of the epiphysis are united to the rest of the bone)

When determining the age at death from skeletal remains, the skeleton will be older than the oldest complete union and younger than the youngest open center. For example, if one epiphyseal plate that typically fuses at age 17 is completely united, but another plate that typically fuses at age 19 is open, the skeleton is that of a person between the ages of 17 and 19.

Current standards for estimating age based upon epiphyseal plate fusion have primarily used male skeletal remains. Female epiphyseal plates tend to fuse approximately 1 to 2 years earlier than those of males, so this fact needs to be considered when estimating the age of a female skeleton. Further, population differences may exist with some epiphyseal plate unions. With these caveats in mind, the following table lists standards for selected epiphyseal plate unions.

Bone	Male Age at Complete Epiphyseal Union
Humerus, lateral epicondyle	11–16 (female: 9–13)
Humerus, medial epicondyle	11–16 (female: 10–15)
Humerus, head	14.5–23.5
Proximal radius	14–19
Distal radius	17–22
Distal fibula and tibia	14.5–19.5
Proximal tibia	15–22
Femur, head	14.5–21.5
Distal femur	14.5–21.5
Clavicle	19–30

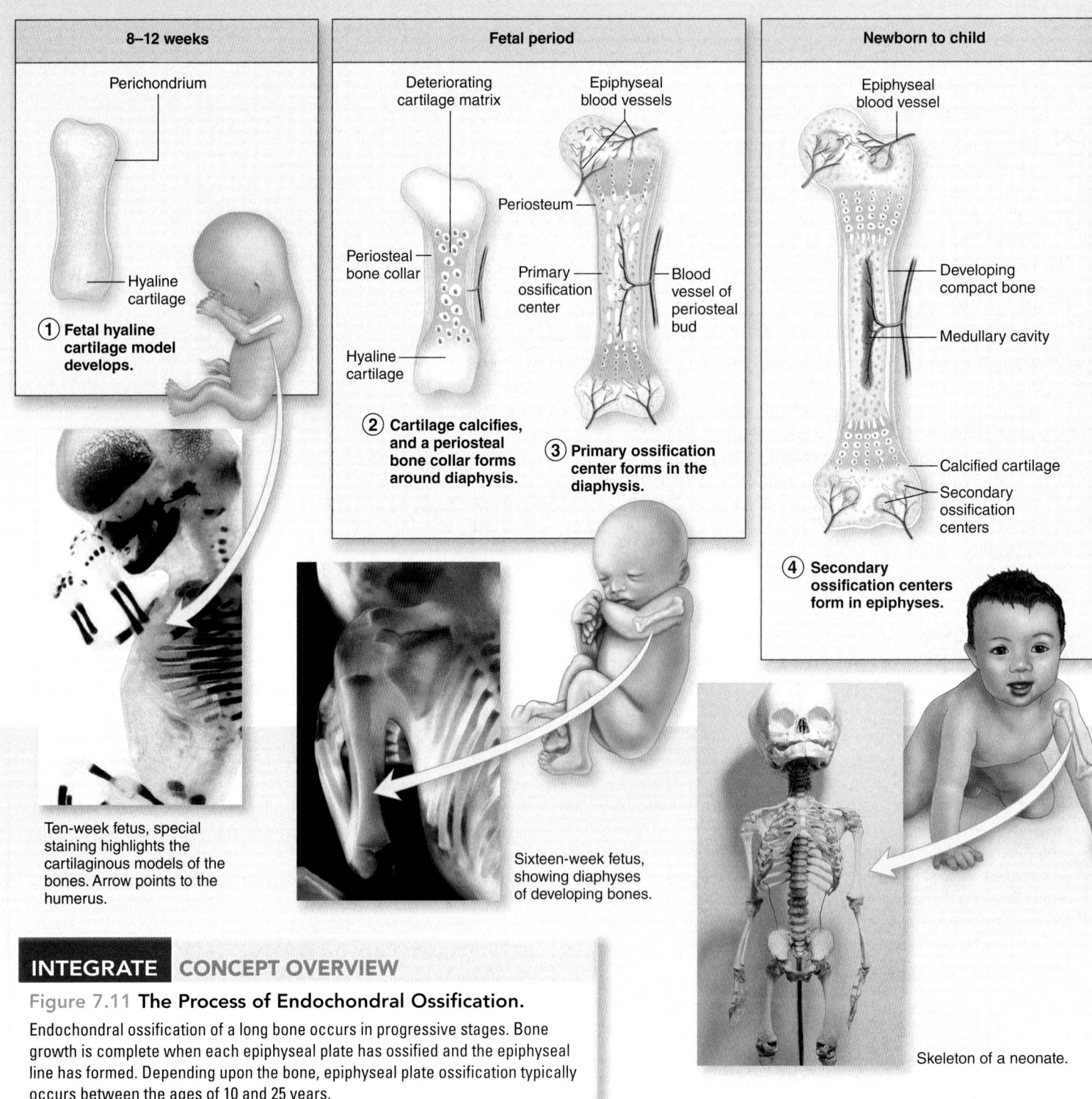

8–12 weeks

Perichondrium

Hyaline cartilage

① **Fetal hyaline cartilage model develops.**

Ten-week fetus, special staining highlights the cartilaginous models of the bones. Arrow points to the humerus.

Fetal period

Deteriorating cartilage matrix

Epiphyseal blood vessels

Periosteum

Periosteal bone collar

Primary ossification center

Blood vessel of periosteal bud

Hyaline cartilage

② **Cartilage calcifies, and a periosteal bone collar forms around diaphysis.**

③ **Primary ossification center forms in the diaphysis.**

Sixteen-week fetus, showing diaphyses of developing bones.

Newborn to child

Epiphyseal blood vessel

Developing compact bone

Medullary cavity

Calcified cartilage

Secondary ossification centers

④ **Secondary ossification centers form in epiphyses.**

Skeleton of a neonate.

INTEGRATE CONCEPT OVERVIEW

Figure 7.11 **The Process of Endochondral Ossification.**
Endochondral ossification of a long bone occurs in progressive stages. Bone growth is complete when each epiphyseal plate has ossified and the epiphyseal line has formed. Depending upon the bone, epiphyseal plate ossification typically occurs between the ages of 10 and 25 years.

⑤ **Bone replaces almost all cartilage, except the articular cartilage and epiphyseal cartilage.** By late bone development, almost all of the hyaline cartilage is displaced by bone. Hyaline cartilage remains as articular cartilage only on the articular surface of each epiphysis and at the epiphyseal plates.

⑥ **Lengthwise growth continues until the epiphyseal plates ossify and form epiphyseal lines.** Lengthwise bone growth continues into puberty until the epiphyseal plate is converted to the epiphyseal line, indicating that the bone has reached its adult length. Depending upon the bone, most epiphyseal plates ossify to become epiphyseal lines

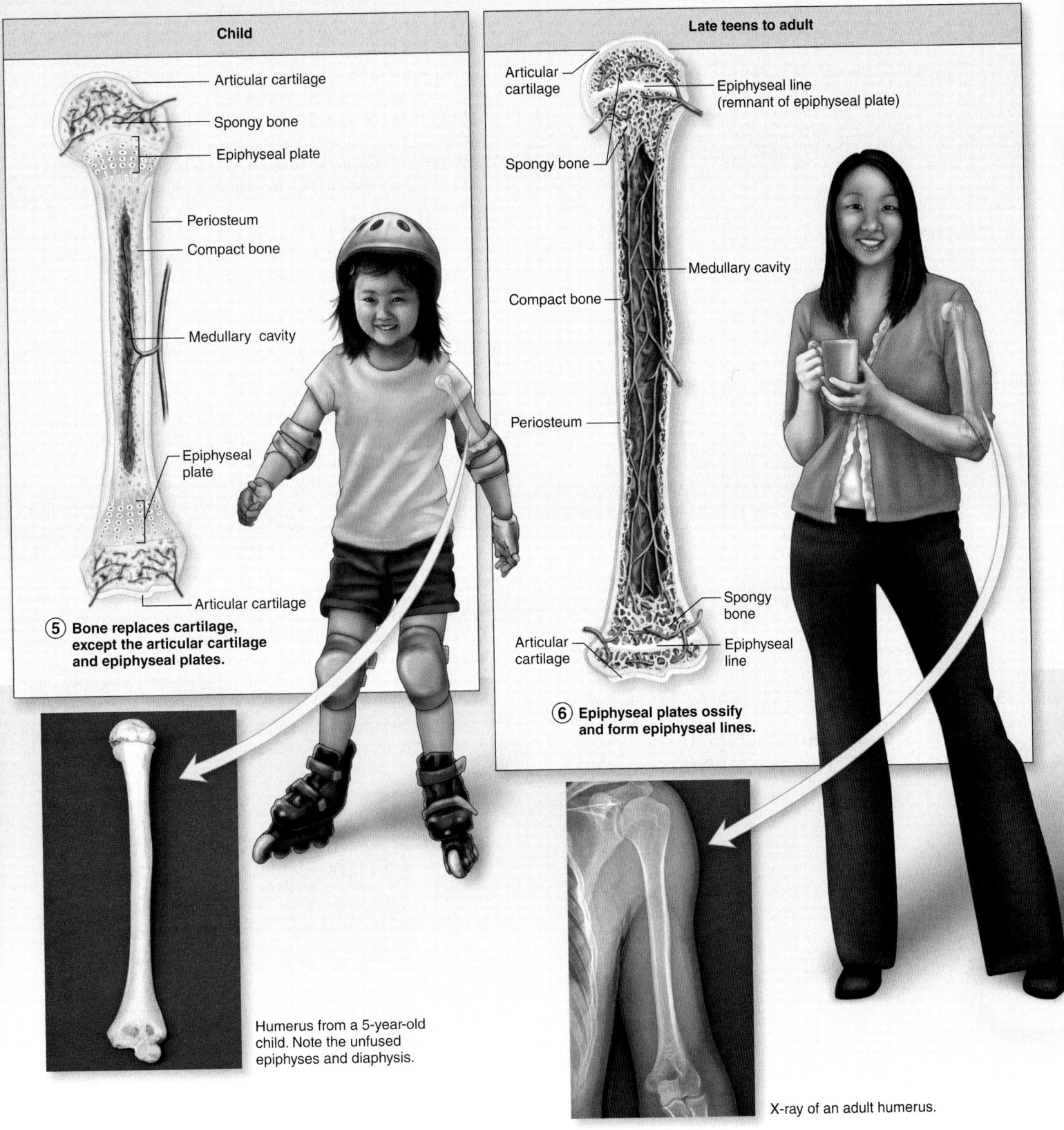

Child

Articular cartilage

Spongy bone

Epiphyseal plate

Periosteum

Compact bone

Medullary cavity

Epiphyseal plate

Articular cartilage

⑤ Bone replaces cartilage, except the articular cartilage and epiphyseal plates.

Humerus from a 5-year-old child. Note the unfused epiphyses and diaphysis.

Late teens to adult

Articular cartilage

Epiphyseal line (remnant of epiphyseal plate)

Spongy bone

Medullary cavity

Compact bone

Periosteum

Spongy bone

Articular cartilage

Epiphyseal line

⑥ Epiphyseal plates ossify and form epiphyseal lines.

X-ray of an adult humerus.

between the ages of 10 and 25. (The last epiphyseal plates to ossify are those of the clavicle in the late 20s.)

WHAT DO YOU THINK?

❶ Why does endochondral bone formation involve so many complex steps? Instead of having the hyaline cartilage model followed by the separate formation of the diaphysis and epiphyses, why can't bone simply be completely formed in the fetus?

WHAT DID YOU LEARN?

⓮ Briefly describe the process by which a long bone forms by endochondral ossification.

7.5 Bone Growth and Bone Remodeling

Bone growth and bone remodeling both begin during embryologic development. We examine both processes here and the major hormones that regulate them.

7.5a Bone Growth

LEARNING OBJECTIVES

1. Compare and contrast the five zones of the epiphyseal plate, and describe how growth in length occurs there.

2. Describe the steps of appositional growth.

As with cartilage growth, a long bone's growth in length is called interstitial growth, and its growth in diameter or thickness is termed appositional growth.

Interstitial Growth

Interstitial growth is dependent upon growth of cartilage within the epiphyseal plate. The epiphyseal plate exhibits five distinct microscopic zones that are continuous from the first zone nearest the epiphysis to the last zone nearest the diaphysis (figure 7.12):

1. **Zone of resting cartilage.** This zone is farthest from the medullary cavity of the diaphysis and nearest the epiphysis. It is composed of small chondrocytes distributed throughout the cartilage matrix. It resembles mature and healthy hyaline cartilage. This region secures the epiphysis to the epiphyseal plate.

2. **Zone of proliferating cartilage.** Chondrocytes in this zone undergo rapid mitotic cell division, enlarge slightly, and become aligned like a stack of coins into longitudinal columns of flattened lacunae. These columns are parallel to the diaphysis.

3. **Zone of hypertrophic cartilage.** Chondrocytes cease dividing and begin to hypertrophy (enlarge in size) in this zone. The walls of the lacunae become thin because the chondrocytes resorb matrix as they hypertrophy.

4. **Zone of calcified cartilage.** This zone usually is composed of 2 to 3 layers of chondrocytes. Minerals are deposited in the matrix between the columns of lacunae; this calcification destroys the chondrocytes and makes the matrix appear opaque.

5. **Zone of ossification.** The walls break down between lacunae in the columns, forming longitudinal channels. These spaces are invaded by capillaries and osteoprogenitor cells from the medullary cavity. New matrix of bone is deposited on the remaining calcified cartilage matrix.

Growth in bone length occurs specifically within both zone 2 as chondrocytes undergo mitotic cell division, and in zone 3 as chondrocytes hypertrophy. These activities combine to push the zone of resting cartilage toward the epiphysis. Note that it is the flexible matrix of hyaline cartilage, and not the hard, calcified matrix of bone, that permits this growth. Once growth in length has occurred, new bone connective tissue is then produced at the same rate in zone 5. Thus, growth in length is due to growth in hyaline cartilage connective tissue that is later replaced with bone. This process is similar to the endochondral ossification process that occurs during bone development.

The epiphyseal plate maintains its thickness during childhood as it is pushed away from the center of the shaft. At maturity, the rate of epiphyseal cartilage production slows, and the rate of osteoblast activity accelerates. As a result, the epiphyseal plate continues to narrow until it ultimately disappears, and interstitial growth completely stops. Eventually, the only remnant of each epiphyseal plate is an internal thin line of compact bone called an **epiphyseal line.** The loss of the hyaline cartilage and the appearance of the remnant epiphyseal line signals the end of interstitial growth.

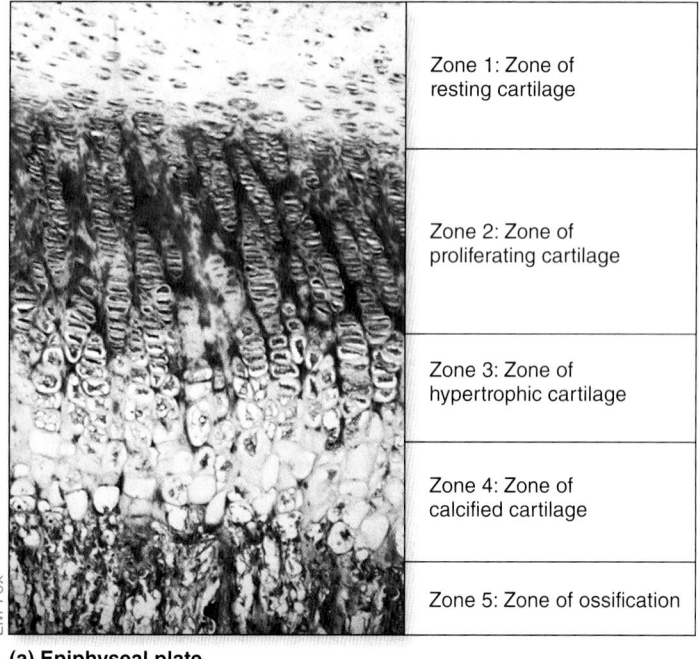

LM 70x

(a) Epiphyseal plate

Zone 1: Zone of resting cartilage

Zone 2: Zone of proliferating cartilage

Zone 3: Zone of hypertrophic cartilage

Zone 4: Zone of calcified cartilage

Zone 5: Zone of ossification

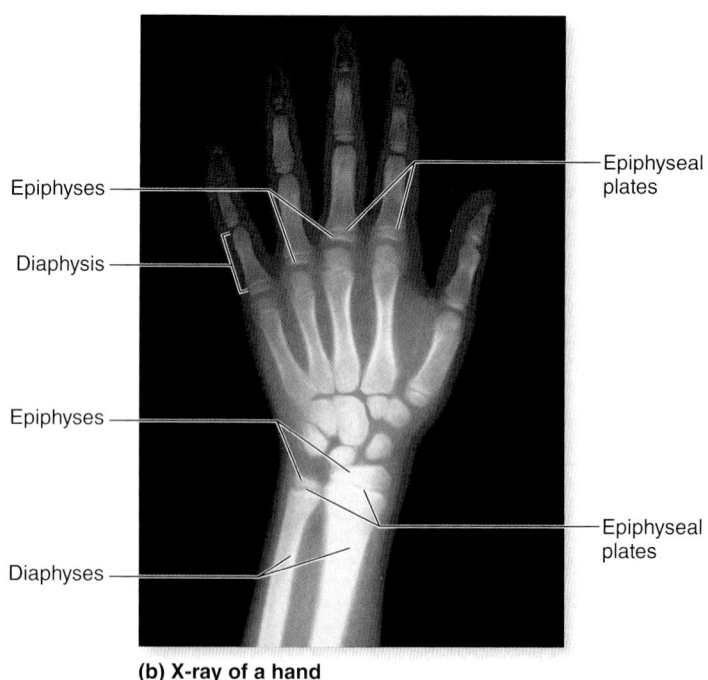

Epiphyses

Diaphysis

Epiphyses

Diaphyses

Epiphyseal plates

Epiphyseal plates

(b) X-ray of a hand

Figure 7.12 Epiphyseal Plate. (*a*) In a growing long bone, the epiphyseal plate, located at the boundary between the diaphysis and the epiphysis, exhibits five distinct but continuous zones. Zones 1–4 are cartilage, and zone 5 is bone. (*b*) An x-ray of a child's hand shows the cartilaginous epiphyseal plate as a dark line between the epiphysis and the diaphysis of a long bone. AP|R

INTEGRATE

CLINICAL VIEW
Achondroplastic Dwarfism

Achondroplasia (ā-kon-drō-plā′zē-ǎ) is characterized by abnormal conversion of hyaline cartilage to bone. The most common form is **achondroplastic dwarfism,** in which the long bones of the limbs stop growing in childhood, whereas the other bones usually continue to grow normally. Thus, an individual with achondroplastic dwarfism is short in stature but generally has a large head. Those affected may have bowed lower limbs and lordosis (exaggerated curvature of the lumbar spine). Achondroplastic dwarfism results from a failure of chondrocytes in the second and third zones of the epiphyseal plate (figure 7.12*a*) to multiply and enlarge. As a result, there is inadequate endochondral ossification. Most cases result from a spontaneous mutation during DNA replication, whereas other cases are due to inheriting the disorder from an affected parent.

WHAT DO YOU THINK?

2 How could a physician determine whether a person had reached full height by examining x-rays of the patient's bones?

Appositional Growth

Appositional growth occurs within the periosteum **(figure 7.13)**. In this process, osteoblasts in the inner cellular layer of the periosteum produce and deposit bone matrix within layers parallel to the surface, called external circumferential lamellae. These lamellae are analogous to tree rings: As they increase in number, the structure increases in diameter. Thus, the bone becomes wider as new bone is laid down at its periphery. As this new bone is being laid down, osteoclasts along the medullary cavity resorb bone matrix, creating an expanding medullary cavity. The combined effects of bone growth at the periphery and bone resorption within the medullary cavity transform an infant bone into a larger version called an adult bone.

WHAT DID YOU LEARN?

15 How does a bone grow in width?

7.5b Bone Remodeling

LEARNING OBJECTIVES

3. Define bone remodeling, and give examples of how it varies in different bones and different portions of the same bone.

4. Explain the effect of mechanical stress on bone remodeling.

Even when adult bone size has been reached, the bone continues to renew and reshape itself throughout a person's lifetime. This constant, dynamic process of continual addition of new bone tissue (bone deposition) and removal of old bone tissue (bone resorption) is a process called **bone remodeling.** This ongoing process occurs at both the periosteal and endosteal surfaces of a bone.

It is estimated that about 20% of the adult human skeleton is replaced yearly. However, bone remodeling does not occur at the same rate everywhere in the skeleton. For example, the compact bone in our skeleton is replaced at a slower rate than the spongy bone. The distal part of the femur (thigh bone) is replaced every 4 to 6 months, whereas the diaphysis of this bone may not be completely replaced during an individual's lifetime.

Clearly, bone remodeling is dependent upon the coordinated activities of osteoblasts, osteocytes, and osteoclasts. The relative activities of these cells is influenced by two primary factors: hormones (described in the next section) and mechanical stress to the bone.

Mechanical stress occurs in the form of weight-bearing movement and exercise, and it is required for normal bone remodeling. Stress is detected by osteocytes and communicated to osteoblasts. Osteoblasts increase synthesis of osteoid, and this is followed by deposition of mineral salts. Bone strength increases over a period of time in response to mechanical stress.

Mechanical stresses that significantly affect bone result from skeletal muscle contraction and gravitational forces. Typically, the bones of athletes become noticeably thicker as a result of repetitive and stressful exercise. Weight-bearing activities, such as weight lifting, walking, or running help build and retain bone mass. Research has shown that regular weight-bearing exercise can increase total bone mass in adolescents and young adults prior to its inevitable reduction later in life. In fact, research suggests that even 70- and 80-year-olds who perform moderate weight training can increase their bone mass.

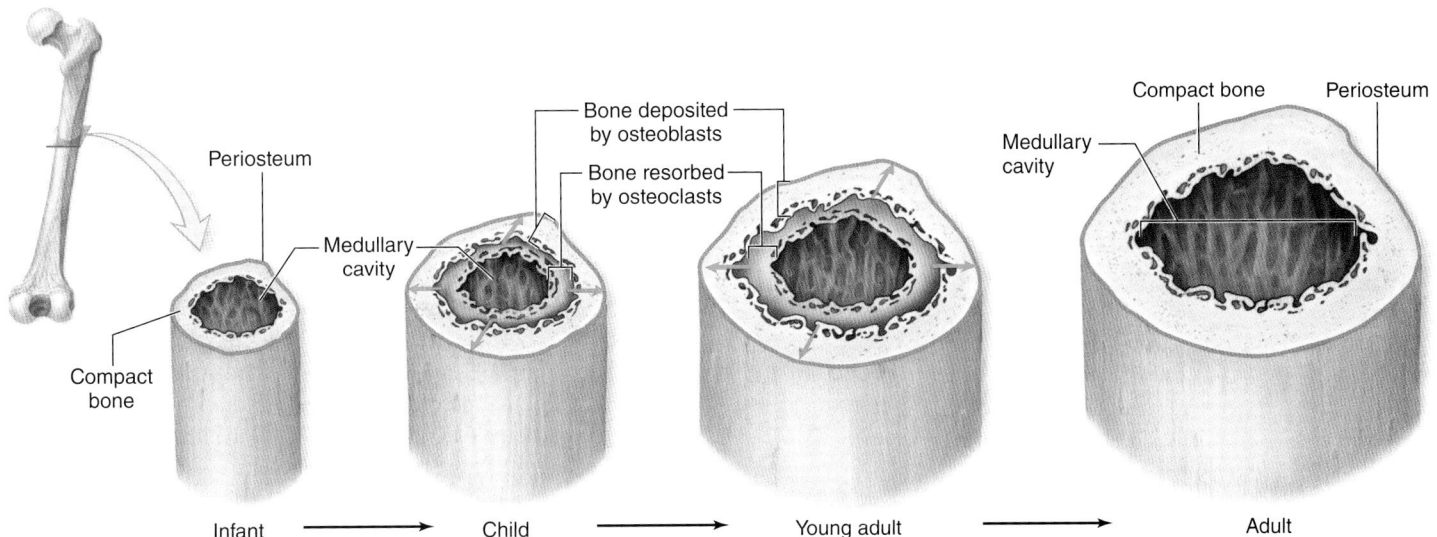

Figure 7.13 Appositional Bone Growth. A bone increases in diameter as new bone is added to the surface. At the same time, some bone may be removed from the inner surface to enlarge the medullary cavity. AP|R

INTEGRATE

CLINICAL VIEW
Why Are Males Typically Taller Than Females?

Two processes determine why males typically are taller than females. First, the growth spurt in females is triggered primarily by estrogen, and in males it is caused primarily by testosterone. Ironically, estrogen is a more powerful growth stimulant than testosterone. However, the growth plate closes more quickly in response to estrogen, which consequently provides a smaller window of time for accelerated growth. Second, females enter puberty about 2 to 3 years earlier than males, so females have fewer years of childhood growth prior to puberty than do males.

In contrast, removal or significant decrease of mechanical stress weakens bone through both reduction of collagen formation and demineralization. When a person has a fractured bone and wears a cast or is bedridden, the strength of the unstressed bone decreases in the immobilized limb. While in space, astronauts must exercise so that the lack of gravity won't cause their bones to lose mass.

 WHAT DID YOU LEARN?

16 What is bone remodeling, where does it occur, and when does it occur?

7.5c Hormones That Influence Bone Growth and Bone Remodeling

 LEARNING OBJECTIVE

5. Identify the hormones that influence bone growth and bone remodeling, and describe their effects.

Hormones are defined as molecules released from one cell into the blood, and travel throughout the body to affect other cells (see chapter 17). Certain hormones influence bone composition and growth patterns by altering the rates of chondrocyte, osteoblast, and osteoclast activity **(table 7.2)**.

Table 7.2	Effects of Hormones on Bone Maintenance and Growth
Hormone	**Effect on Bone**
Growth hormone	Stimulates liver to produce the hormone IGF, which causes cartilage proliferation at epiphyseal plate and resulting bone elongation
Thyroid hormone	Stimulates bone growth by stimulating metabolic rate of osteoblasts
Calcitonin	Promotes calcium deposition in bone and inhibits osteoclast activity
Parathyroid hormone	Increases blood calcium levels by encouraging bone resorption by osteoclasts
Sex hormones (estrogen and testosterone)	Stimulate osteoblasts; promote epiphyseal plate growth and closure
Glucocorticoids	Increase bone loss and, in children, impair bone growth when there are chronically high levels of glucocorticoids
Serotonin	Inhibits osteoprogenitor cells from differentiating into osteoblasts when there are chronically high levels of serotonin

Growth hormone is also called *somatotropin* (sō'mă-tō-trō-pin), and is produced by the anterior pituitary gland (see section 17.8a). It affects bone growth by stimulating the liver to form another hormone called **insulin-like growth factor (IGF)** (also called *somatomedin*; sō'mă-tō-mē'din). Both growth hormone and IGF directly stimulate growth of cartilage in the epiphyseal plate.

Thyroid hormone is secreted by the thyroid gland and stimulates bone growth by influencing the basal metabolic rate of bone cells (see section 17.8b). If maintained in proper balance, growth hormone and thyroid hormone regulate and maintain normal activity at the epiphyseal plates until puberty.

Sex hormones (**estrogen** and **testosterone**), which begin to be secreted in relatively large amounts at puberty (see section 28.1b), dramatically accelerate bone growth. Sex hormones increase the rate of both cartilage growth and bone formation within the epiphyseal plate. Ironically, the appearance of high levels of sex hormones at puberty also signals the beginning of the end for growth at the epiphyseal plate. This happens because the rate of bone formation occurs at a faster rate than the rate of cartilage growth. Bone growth eventually overcomes the region of cartilage, replacing all cartilage with bone at the epiphyseal plates.

INTEGRATE

CONCEPT CONNECTION

Many hormones secreted by the endocrine system (see chapter 17) are responsible for normal growth and homeostasis of bone tissue. Growth hormone, thyroid hormone, calcitonin, and sex hormones help promote bone growth, whereas parathyroid hormone, calcitriol, glucocorticoids, and serotonin may either inhibit bone growth or increase bone resorption. Disorders of the endocrine system, as a result, often are manifested in part by skeletal system disorders.

 WHAT DO YOU THINK?

3 Given what you know about the effects of testosterone, explain why there is a risk of stunted growth to a young boy (pre-puberty) taking anabolic steroids (substances that have similar effects to testosterone).

Glucocorticoids are a group of steroid hormones that are released from the adrenal cortex and regulate blood glucose levels. High amounts increase bone loss and, in children, impairs growth at the epiphyseal plate. It is because of this relationship that a child's growth is monitored if receiving high doses of glucocorticoids as an anti-inflammatory, such as a treatment for children with severe asthma.

Serotonin (ser-ō-tō'nin) was previously discussed in section 1.6. Researchers have discovered that most bone cells have serotonin receptors, and specifically that when levels of circulating serotonin are too high, osteoprogenitor cells are prevented from differentiating into osteoblasts. Thus, serotonin appears to play a role in the rate and regulation of normal bone remodeling because it affects osteoblast differentiation. Further research is ongoing to see if abnormally high levels of serotonin are linked to low bone density disorders.

Three additional hormones—parathyroid hormone, calcitriol, and calcitonin—participate in both regulating bone remodeling and regulating blood calcium levels. These hormones are discussed in detail in the next section.

 WHAT DID YOU LEARN?

17 What are the effects of growth hormone and thyroid hormone on bone growth and bone mass?

7.6 Regulating Blood Calcium Levels

Regulating calcium concentration in blood (between 8.9 and 10.1 milligrams per deciliter [mg/dL]) is essential because calcium is required for numerous physiologic processes such as initiation of muscle contraction; exocytosis of molecules from cells, including nerve cells (neurons); stimulation of the heart by pacemaker cells; and blood clotting. The two primary hormones that regulate blood calcium are calcitriol (an active form of vitamin D) and parathyroid hormone. We describe blood calcium regulation here because of the role of the skeleton in storage of calcium. (Also see Table R.2 for information about the hormones involved in regulating blood calcium levels.)

7.6a Activation of Vitamin D to Calcitriol

LEARNING OBJECTIVE

1. Explain the activation of vitamin D to calcitriol.

To effectively describe the actions of calcitriol and parathyroid hormone we first describe the enzymatic pathway of activating vitamin D to calcitriol. The three steps are as follows (**figure 7.14**):

1. Ultraviolet light converts the precursor molecule in keratinocytes of the skin (7-dehydrocholesterol, a modified cholesterol molecule) to **vitamin D$_3$ (cholecalciferol),** which is released into the blood. Vitamin D$_3$ also is absorbed from the small intestine into the blood from the diet.

2. Vitamin D$_3$ circulates throughout the blood. As it passes through the blood vessels of the liver, it is converted by liver enzymes to **calcidiol** by the addition of a hydroxyl group (—OH). Both steps 1 and 2 occur continuously with limited regulation.

3. Calcidiol circulates in the blood: As it passes through blood vessels of the kidney, it is converted to **calcitriol** by kidney enzymes (when another —OH group is added). Calcitriol is the active form of Vitamin D$_3$. The presence of parathyroid hormone increases the rate of this final enzymatic step in the kidney. Thus, greater amounts of calcitriol are formed when PTH is present.

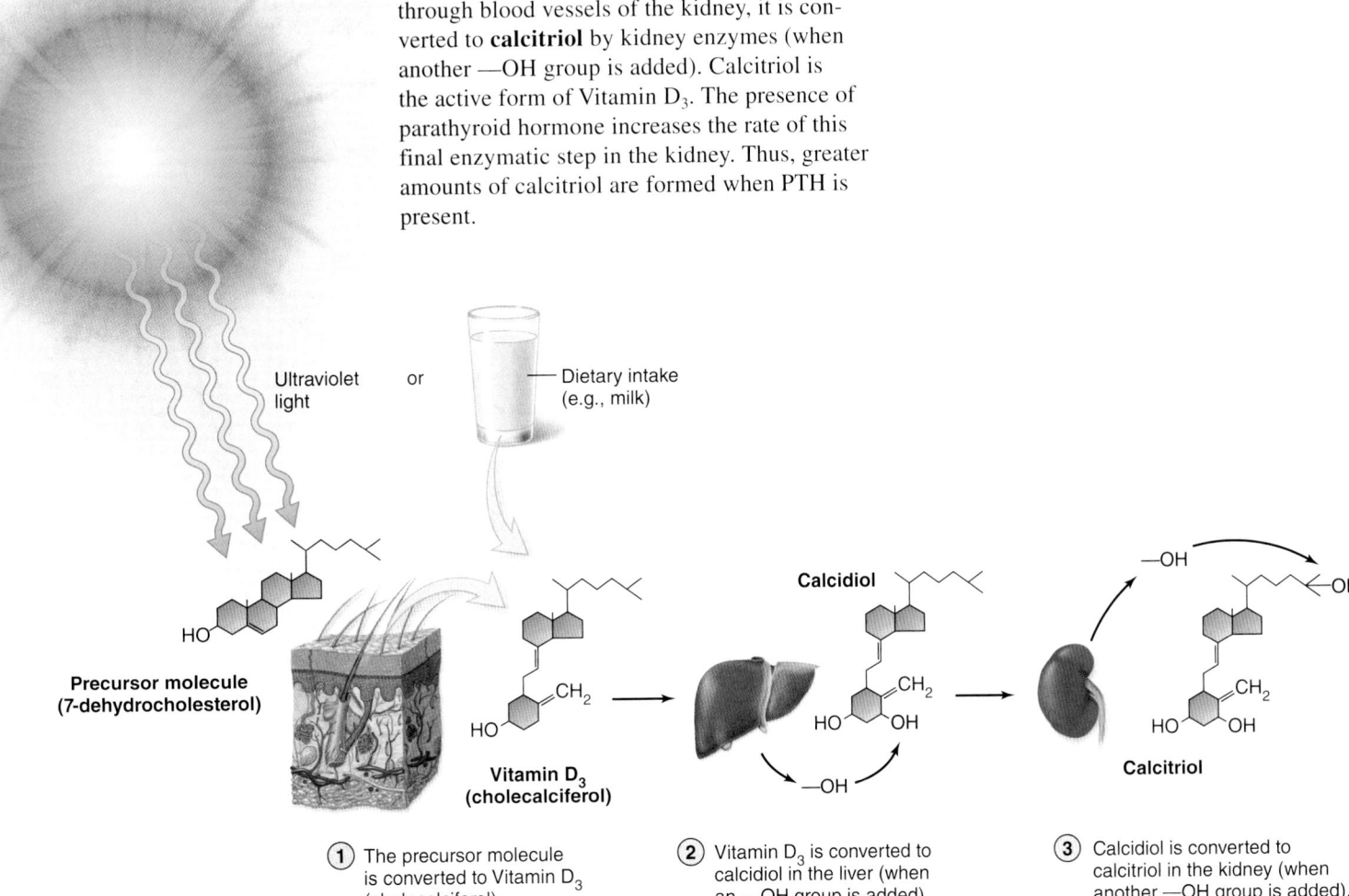

Figure 7.14 Calcitriol Production. Calcitriol is produced as follows: When keratinocytes are exposed to UV rays, a precursor molecule (7-dehydrocholesterol) in keratinocytes is transformed to vitamin D$_3$ (cholecalciferol). Humans also may obtain vitamin D$_3$ from dietary sources, such as milk. The liver then synthesizes calcidiol from the vitamin D$_3$. Finally, the kidneys will convert calcidiol to calcitriol.

Vitamin D in its active form of calcitriol hormone has the unique function of stimulating absorption of calcium ions (Ca^{2+}) from the small intestine into the blood.

☀ WHAT DID YOU LEARN?

❶⑧ What organs are involved in activating vitamin D_3 to calcitriol?

7.6b Parathyroid Hormone and Calcitriol

☀ LEARNING OBJECTIVES

2. Discuss the release of parathyroid hormone.

3. Explain how parathyroid hormone and calcitriol function together to regulate blood calcium levels.

Parathyroid hormone (PTH) is secreted and released by the parathyroid glands (see section 17.10b) in response to reduced blood calcium levels **(figure 7.15)**. The final enzymatic step converting calcidiol to calcitriol in the kidney occurs more readily in the presence of PTH.

PTH and calcitriol interact with selected major organs as follows:

- **Bone connective tissue of the skeleton.** PTH and calcitriol act synergistically (their combined effect is greater than the sum of their individual effects) to increase the release of calcium from the bone into the blood, by increasing osteoclast activity.
- **Kidneys.** PTH and calcitriol act synergistically to stimulate the kidney to excrete less calcium in the urine (and thus retain more calcium in the blood). This occurs by increasing calcium reabsorption in the tubules in the kidney (see section 24.6).

INTEGRATE

CLINICAL VIEW
Rickets

Rickets is a disease caused by a vitamin D deficiency in childhood and characterized by overproduction and deficient calcification of osteoid tissue. Patients with rickets acquire a bowlegged appearance as their weight increases and the bones in their lower limbs bend. The disease also is characterized by disturbances in growth, hypocalcemia, and sometimes tetany (cramps and muscle twitches), usually caused by low blood calcium.

During the Industrial Revolution, the incidence of rickets increased as children in cities were forced to work indoors in factories. Rickets continues to occur in some developing nations. The incidence recently has increased among urban U.S. children, who spend much of their time indoors and typically do not drink enough milk, opting instead for soft drinks.

Bowing lower limb long bones

Colorized radiograph of a child with rickets.

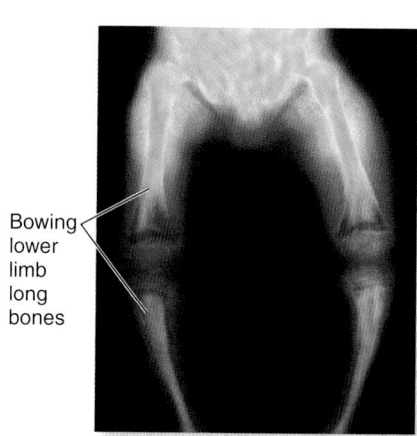

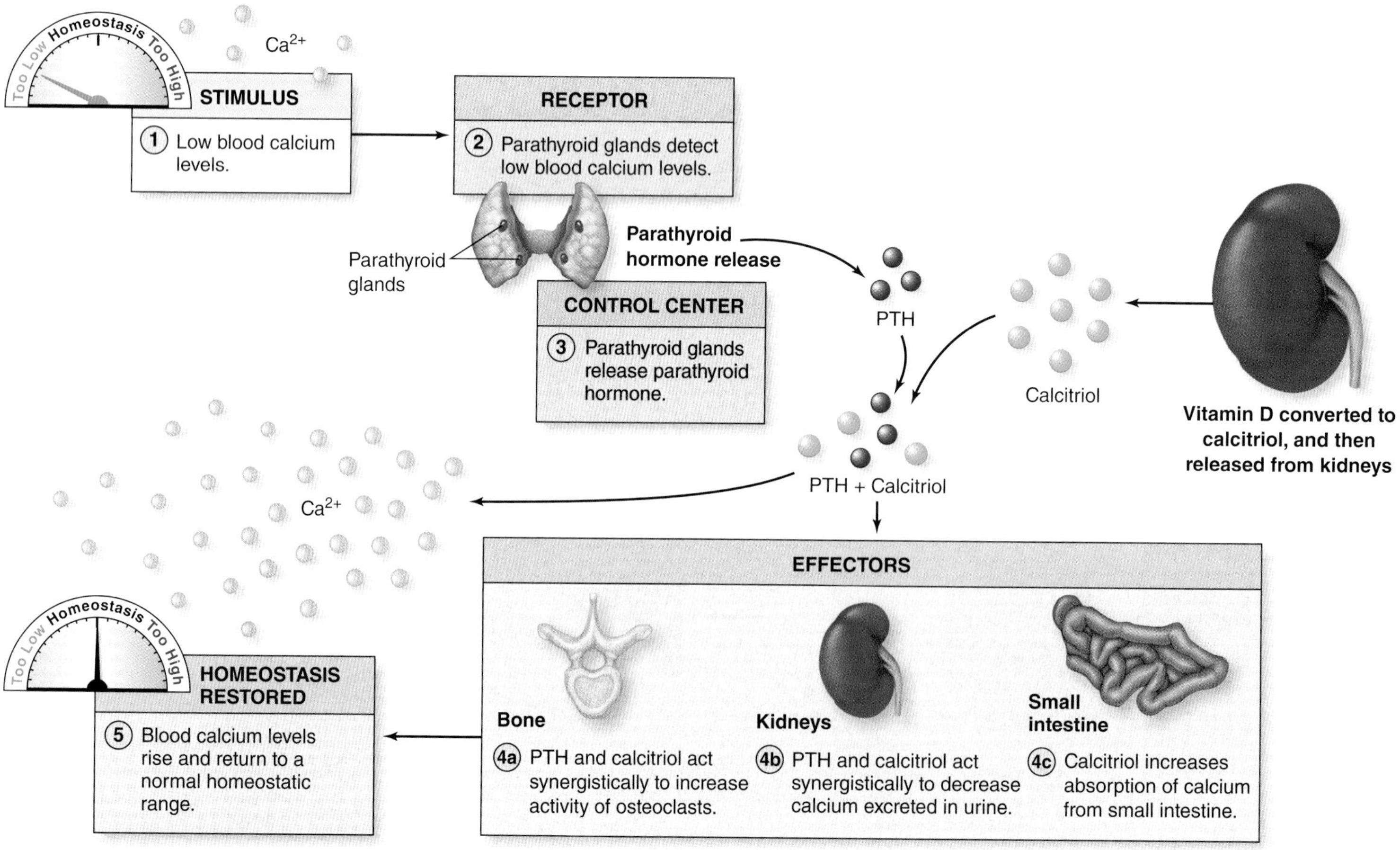

Figure 7.15 Effects of Parathyroid Hormone and Calcitriol on Blood Calcium Levels. Blood calcium levels are closely regulated by a negative feedback mechanism that involves the parathyroid gland, calcitriol, and various effectors (bone, kidneys, and small intestine). A low blood calcium level is the initial stimulus for the parathyroid glands to release parathyroid hormone. Together, PTH and calcitriol target various effectors to ultimately instigate a rise in blood calcium levels and return to homeostasis.

STIMULUS
① Low blood calcium levels.

RECEPTOR
② Parathyroid glands detect low blood calcium levels.

Parathyroid glands

Parathyroid hormone release

CONTROL CENTER
③ Parathyroid glands release parathyroid hormone.

PTH

Calcitriol

Vitamin D converted to calcitriol, and then released from kidneys

PTH + Calcitriol

Ca^{2+}

HOMEOSTASIS RESTORED
⑤ Blood calcium levels rise and return to a normal homeostatic range.

EFFECTORS

Bone
④a PTH and calcitriol act synergistically to increase activity of osteoclasts.

Kidneys
④b PTH and calcitriol act synergistically to decrease calcium excreted in urine.

Small intestine
④c Calcitriol increases absorption of calcium from small intestine.

- **Small intestine.** A function unique to calcitriol is to increase absorption of calcium from the small intestine into the blood.

The removal of calcium from bone, the decrease in loss of calcium from the kidney, and increase in calcium absorption from the gastrointestinal tract result in elevating blood calcium and returning it to within the normal homeostatic range. Subsequently, the release of additional PTH is inhibited by negative feedback.

WHAT DID YOU LEARN?

19 When are parathyroid hormone and calcitriol secreted, and what organs respond to them?

7.6c Calcitonin

LEARNING OBJECTIVE

4. Discuss the homeostatic system involving the hormone calcitonin and its effect on blood calcium levels.

Calcitonin (kal-si-tō′nin; *calx* = lime, *tonos* = stretching) is another hormone that aids in regulating blood calcium levels—however, it has a less significant role than either PTH or calcitriol. Calcitonin is released from the thyroid gland—specifically, from its parafollicular cells (see section 17.8b) in response to high blood calcium levels; it is also secreted in response to stress from exercise.

- Although the entire function of calcitonin is unclear, it is known that calcitonin primarily inhibits osteoclast activity. In addition, calcitonin stimulates the kidneys to increase the loss of calcium in the urine. The result is a reduction in blood calcium levels.

The following limitations are observed with calcitonin:

- Calcitonin seems to have the greatest effect under conditions where there is the greatest turnover of bone, such as in growing children.

INTEGRATE
CONCEPT CONNECTION

Blood calcium levels are regulated via the interactions of the endocrine system (see Chapter 17), the skeletal system, and the kidneys (urinary system—see Chapter 24). When blood calcium levels are low, parathyroid hormone and calcitriol stimulate bone to release calcium into the blood, the kidneys excrete less calcium in the urine, and the small intestine absorbs more calcium into the blood. When blood calcium levels are high, the thyroid gland secretes calcitonin to stimulate bone deposition and increase calcium excretion from the kidneys.

- If high doses of calcitonin are administered, blood calcium levels decrease only temporarily. Thus, therapeutic injections of calcitonin cannot provide long-term decrease in blood calcium.

WHAT DID YOU LEARN?

20 How does calcitonin regulate blood calcium levels?

7.7 Effects of Aging

LEARNING OBJECTIVE

1. Describe how age influences bone structure.

Aging affects bone connective tissue in two ways. First, the tensile strength of bone decreases due to a reduced rate of protein synthesis by osteoblasts. Consequently, the relative amount of inorganic minerals in the bone matrix increases (due to decreased matrix protein), and the bones of the skeleton become brittle and susceptible to fracture.

INTEGRATE
CLINICAL VIEW
Osteoporosis

Osteoporosis, meaning *porous bones,* is a disease that results in decreased bone mass and leads to weakened bones that are prone to fracture. The occurrence of osteoporosis is greatest among the elderly, especially Caucasian women, and the severity is closely linked to both age and onset of menopause. Smoking is also a risk factor for osteoporosis. Postmenopausal women are at risk because (1) women have less bone mass than men, (2) women begin losing bone mass earlier and faster in life (sometimes around 35 years of age), and (3) postmenopausal women no longer produce significant amounts of estrogen, which appears to help protect against osteoporosis by stimulating bone growth. As a result of osteoporosis, the incidence of fractures increases, most frequently in the wrist, hip, and vertebral column.

The best treatment for osteoporosis seems to be prevention. Young adults should maintain good nutrition and physical activity to ensure adequate bone density, thus allowing for the normal, age-related loss later in life. Calcium supplements with vitamin D help maintain bone health, but by themselves will not stimulate new bone growth. Medical treatments involve two strategies: (1) slowing the rate of bone loss and (2) attempting to stimulate new bone growth.

A class of medications called bisphosphonates (e.g., alendronate [Fosamax], risedronate [Actonel], ibandronate [Boniva]) currently are prescribed to slow the progression of osteoporosis. These drugs work by interfering with osteoclast function and thus retarding the removal of bone during remodeling. Unfortunately, these drugs have also been implicated in increased risk of *osteonecrosis* (bone death) of the jaw and some other bone growth abnormalities, so patients are recommended not to take the drugs for longer than 5 years at a time.

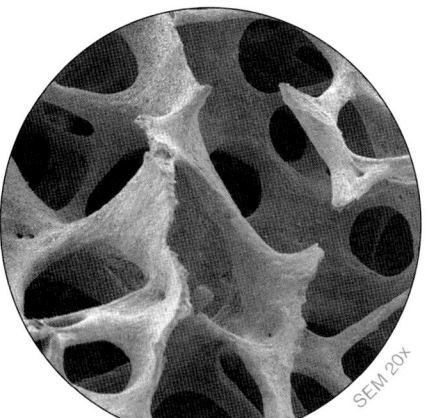

(a) Normal bone

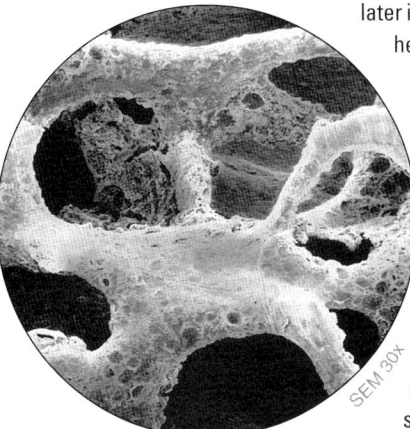

(b) Osteoporotic bone

Second, bone loses calcium and other minerals (demineralization). The bones of the skeleton become thinner and weaker, resulting in insufficient ossification, a condition called **osteopenia** (os'tē-ō-pen'ē-ǎ; *penia* = poverty). Aging causes all people to become slightly osteopenic. This reduction in bone mass may begin as early as 35–40 years of age, when osteoblast activity declines, while osteoclast activity continues at previous levels. Different parts of the skeleton are affected unequally. Vertebrae, jaw bones, and epiphyses lose large amounts of mass, resulting in reduced height, loss of teeth, and fragile limbs.

Every decade, women lose roughly more of their skeletal mass than do men. A significant percentage of older women and a smaller proportion of older men suffer from **osteoporosis** (os'tē-ō-pō-rō'sis; *poros* = pore, *osis* = condition), a condition characterized by reduction in bone mass sufficient to compromise normal function (see Clinical View: "Osteoporosis").

In addition, vitamin D and numerous hormones including growth hormone, estrogen, and testosterone decrease with age. This decrease in hormone levels contributes to reduction in bone mass.

WHAT DID YOU LEARN?

21 Explain why women are more likely to develop osteoporosis than men.

7.8 Bone Fracture and Repair

LEARNING OBJECTIVE

1. Explain the four steps by which fractures heal.

Bone has great mineral strength, but it may break as a result of unusual stress or a sudden impact. Breaks in bones are called **fractures** and are classified in several ways. A **stress fracture** is a thin break caused by increased physical activity in which the bone experiences repetitive loads (e.g., as seen in some runners). A **pathologic fracture** usually occurs in bone that has been weakened by disease. In a **simple fracture,** the broken bone does not penetrate the skin, whereas in a **compound fracture,** one or both ends of the broken bone pierce the overlying skin. **Figure 7.16** shows the different classifications of fractures.

The healing of a simple fracture takes about 2 to 3 months, whereas a compound fracture takes longer to heal. Fractures heal much more quickly in young children (average healing time, 3 weeks) and become slower to heal as we age. In the elderly, the normal thinning and weakening of bone increases the incidence of fractures, and

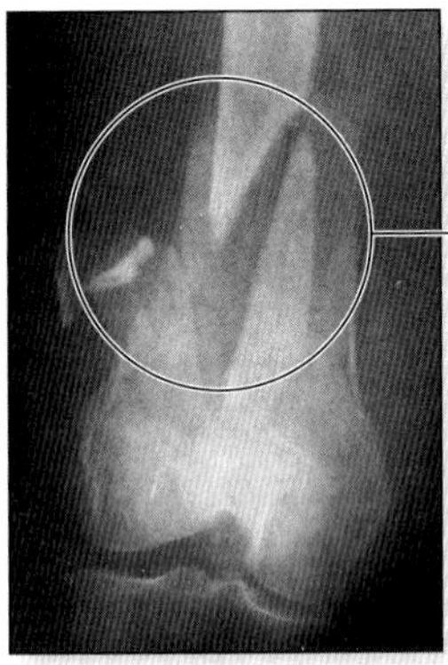

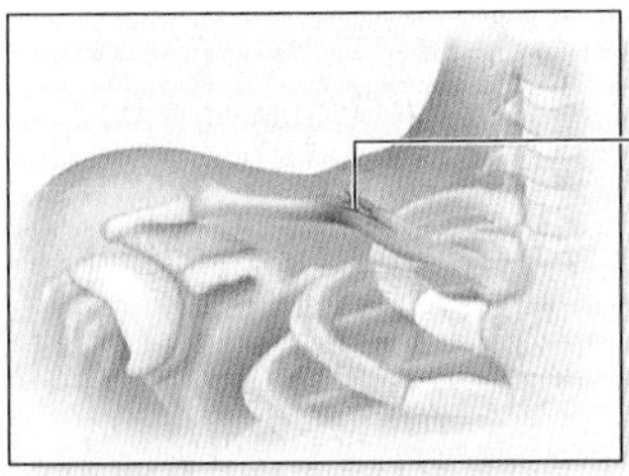

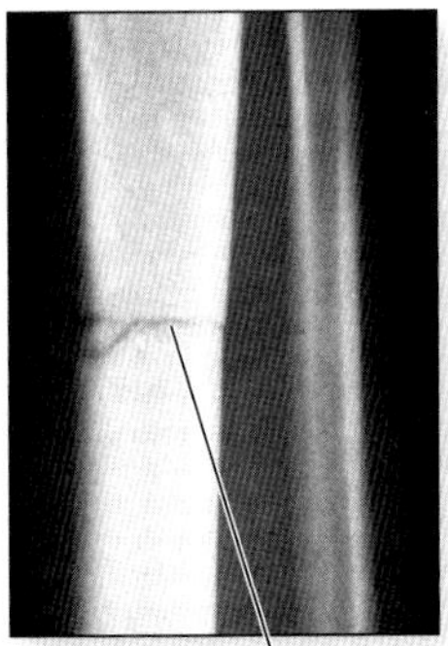

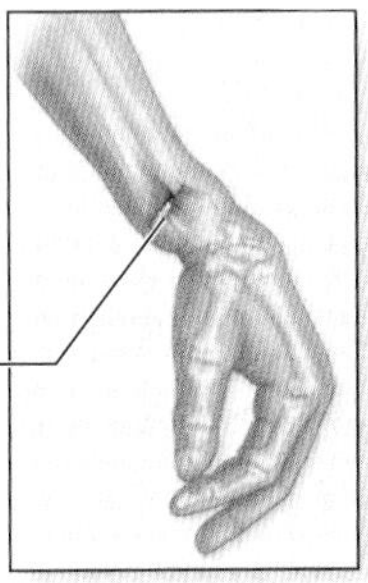

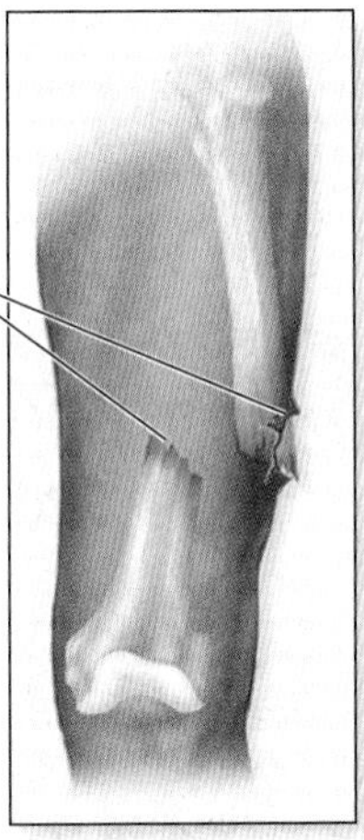

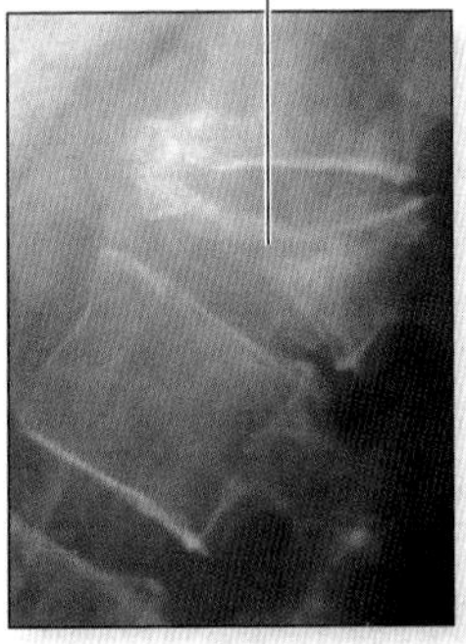

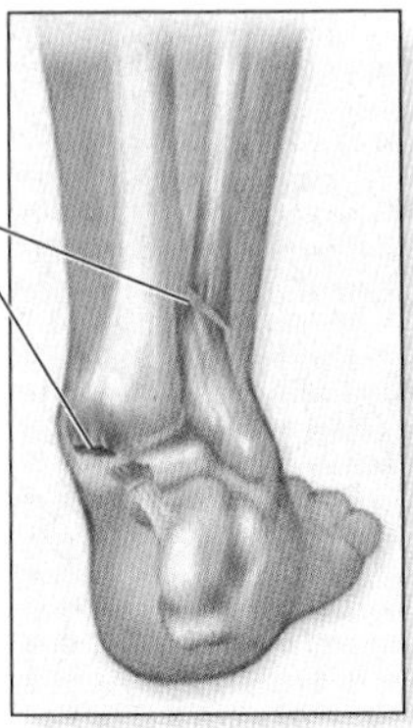

Classification of Bone Fractures	
Fracture	**Description**
Avulsion	Complete severing of a body part (typically a toe or finger)
Colles	Fracture of the distal end of the lateral forearm bone (radius); produces a "dinner fork" deformity
Comminuted	Bone is splintered into several small pieces between the main parts
Complete	Bone is broken into two or more pieces
Compound (open)	Broken ends of the bone protrude through the skin
Compression	Bone is squashed (may occur in a vertebra during a fall)
Depressed	Broken part of the bone forms a concavity (as in skull fracture)
Displaced	Fractured bone parts are out of anatomic alignment
Epiphyseal	Epiphysis is separated from the diaphysis at the epiphyseal plate
Greenstick	Partial fracture; one side of bone breaks—the other side is bent
Hairline	Fine crack in which sections of bone remain aligned (common in skull)
Impacted	One fragment of bone is firmly driven into the other
Incomplete	Partial fracture extends only partway across the bone
Linear	Fracture is parallel to the long axis of the bone
Oblique	Diagonal fracture is at an angle
Pathologic	Weakening of a bone caused by disease process (e.g., cancer)
Pott	Fracture is at the distal ends of the tibia and fibula
Simple (closed)	Bone does not break through the skin
Spiral	Fracture spirals around axis of long bone; results from twisting stress
Stress	Thin fractures due to repeated, stressful impact such as running (These fractures often are difficult to see on x-rays, and a bone scan may be necessary to accurately identify their presence.)
Transverse	Fracture is at right angles to the long axis of the bone

Figure 7.16 Classification of Bone Fractures.

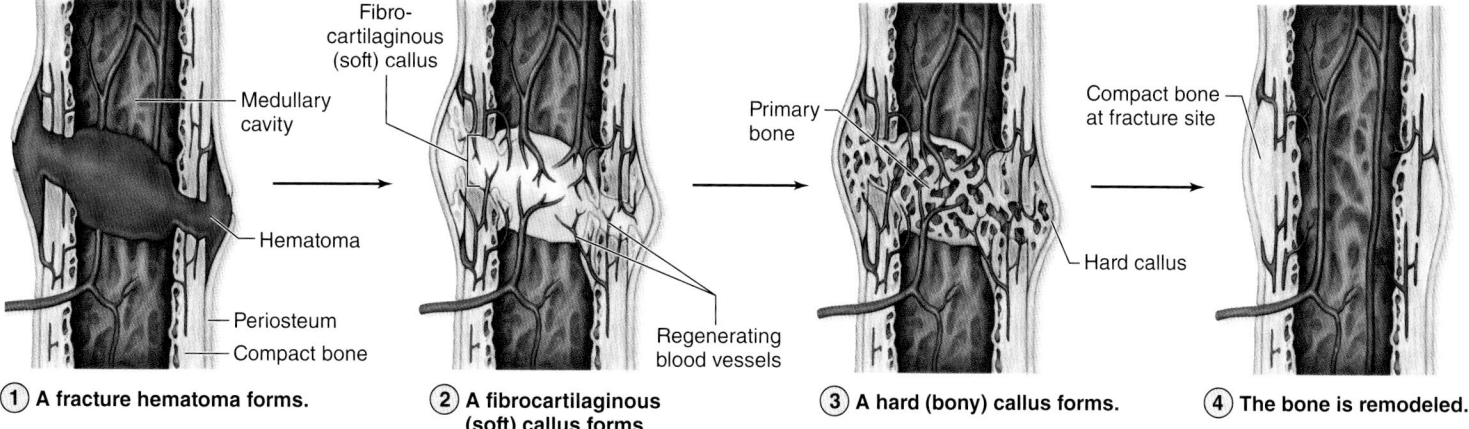

Fibro-cartilaginous (soft) callus			

Medullary cavity

Hematoma

Periosteum

Compact bone

Primary bone

Regenerating blood vessels

Compact bone at fracture site

Hard callus

① A fracture hematoma forms.

② A fibrocartilaginous (soft) callus forms.

③ A hard (bony) callus forms.

④ The bone is remodeled.

Figure 7.17 Fracture Repair. The repair of a bone fracture occurs in a series of steps.

some complicated fractures require surgical intervention to heal properly. Bone fracture repair can be described as a series of four steps (**figure 7.17**):

① **A fracture hematoma forms.** A bone fracture tears blood vessels inside the bone and within the periosteum, causing bleeding. This bleeding results in a **fracture hematoma** that forms from the clotted blood.

② **A fibrocartilaginous (soft) callus forms.** Regenerated blood capillaries infiltrate the fracture hematoma. First, the fracture hematoma is reorganized into an actively growing connective tissue called a *procallus.* Fibroblasts within the procallus produce collagen fibers that help connect the broken ends of the bones. Chondroblasts in the newly growing connective tissue form a dense regular connective tissue associated with the cartilage. Eventually, the procallus becomes a **fibrocartilaginous (soft) callus** (kal′ŭs; hard skin). The fibrocartilaginous callus stage lasts at least 3 weeks.

③ **A hard (bony) callus forms.** Within a week after the injury, osteoprogenitor cells in areas adjacent to the fibrocartilaginous callus become osteoblasts and produce trabeculae of primary bone. The fibrocartilaginous callus is then replaced by this bone, which forms a **hard (bony) callus.** The trabeculae of the hard callus continue to grow and thicken for several months.

④ **The bone is remodeled.** Remodeling is the final phase of fracture repair. The hard callus persists for at least 3 to 4 months, as osteoclasts remove excess bony material from both exterior and interior surfaces. Compact bone replaces primary bone. The fracture usually leaves a slight thickening of the bone (as detected by x-ray); however, in some instances healing occurs with no persistent obvious thickening.

WHAT DID YOU LEARN?

22 What are the four basic steps in fracture repair?

23 Explain the risk of bearing weight on a bone when the fibrocartilage callus is forming.

INTEGRATE

CLINICAL VIEW

Bone Scans

Bone scans are tests that can detect bone pathologies. The patient often is injected intravenously with a small amount of a radioactive tracer compound that is absorbed by bone. A scanning camera then detects and measures the radiation emitted from the bone. The images produced show normal bone tissue with an even distribution of gray coloring (although the axial skeleton typically appears uniformly darker than the appendicular skeleton). An abnormal bone scan may have focal darker areas called *hot spots,* or lighter areas called *cold spots.* Hot spots typically indicate increased metabolism or greater turnover of the bone tissue, as may be experienced by a stress fracture or cancer metastasis to bone. Cold spots indicate less metabolic activity of bone tissue, as may occur with avascular necrosis (death due to lack of blood supply) of a bone. **AP|R**

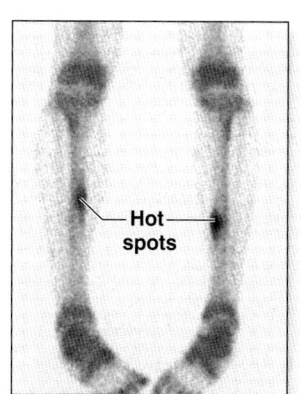

Hot spots

"Hot spots" in both tibiae indicative of stress fractures.

CHAPTER SUMMARY

- The skeletal system is composed of dynamic living tissue.

7.1 Introduction to the Skeletal System 212

- Bones, cartilage, ligaments, and other connective tissue that stabilize or connect bones compose the skeletal system.

7.2 Bone: The Major Organ of the Skeletal System 212

- Bones are organs that contain all tissue types, the most abundant being bone (osseous) connective tissue.

7.2a General Functions 212
- The functions of bone include support and protection, movement, hemopoiesis, and storage of minerals and energy reserves.

7.2b Classification of Bones 213
- Bones are classified by shape as long, short, flat, or irregular.

7.2c Gross Anatomy of Bones 213
- A long bone contains the following regions: diaphysis, epiphysis, metaphysis, articular cartilage, medullary cavity.
- A long bone is covered externally by the periosteum and lined internally by the endosteum.
- All bones contain a rich blood supply and innervation.

7.2d Bone Marrow 216
- Bone marrow fills the internal spaces of bone and includes red bone marrow (hemopoietic tissue) and yellow bone marrow (fat).

7.2e Microscopic Anatomy: Bone Connective Tissue 217
- Osteoprogenitor cells are bone stem cells; osteoblasts produce osteoid; osteocytes maintain the bone matrix; and osteoclasts resorb bone.
- The bone matrix is made up of collagen protein fibers and ground substance (composed of glycoproteins, proteoglycans, and hydroxyapatite crystals).
- Compact bone forms the dense outer, solid region of bone, whereas spongy bone is located internally.

7.2f Microscopic Anatomy: Hyaline Cartilage Connective Tissue 221
- Hyaline cartilage is composed of chondrocytes in lacunae within a semirigid matrix.

7.3 Cartilage Growth 221

- Cartilage growth includes both interstitial growth (growth from within preexisting cartilage) and appositional growth (growth around the periphery of cartilage).

7.4 Bone Formation 223

- Ossification, or osteogenesis, is the process of bone connective tissue formation.

7.4a Intramembranous Ossification 223
- In intramembranous ossification, bone forms from a thin layer of mesenchyme (sometimes called a membrane).

7.4b Endochondral Ossification 224
- Endochondral ossification uses a hyaline cartilage model that is gradually replaced by newly formed bone tissue.

7.5 Bone Growth and Bone Remodeling 228

- Bone growth and bone remodeling begin during development.

7.5a Bone Growth 228
- Bone growth occurs in length through interstitial growth and in width through appositional growth at the periosteum.
- The epiphyseal plate contains five zones where cartilage grows and eventually is replaced by bone.

7.5b Bone Remodeling 229
- The continual deposition of new bone tissue by osteoblasts and resorption of bone by osteoclasts is called bone remodeling.

7.5c Hormones That Influence Bone Growth and Bone Remodeling 230
- Growth hormone, thyroid hormone, and sex hormones stimulate bone growth by increasing osteoblast activity.
- High doses of glucocorticoids interfere with normal bone growth and high doses of serotonin interfere with normal bone remodeling.
- Calcitonin inhibits osteoclast activity and stimulates osteoblast activity, whereas parathyroid hormone and calcitriol stimulate osteoclast activity.

7.6 Regulating Blood Calcium Levels 231

- Calcium homeostasis requires precise controls over calcium uptake, calcium loss, and calcium storage.

7.6a Activation of Vitamin D to Calcitriol 231
- Vitamin D is a pre-hormone that is activated to calcitriol through several enzymatic steps.

7.6b Parathyroid Hormone and Calcitriol 232
- Parathyroid hormone is released from the parathyroid gland in response to decreased blood calcium levels, and its release increases the final step in the synthesis of calcitriol.
- The combined actions of parathyroid hormone and calcitriol increase blood calcium levels to within the normal range.

7.6c Calcitonin 233
- Calcitonin is a hormone released from the thyroid gland in response to increased blood calcium levels. In regulating blood calcium is thought to be less significant than parathyroid hormone and calcitriol, at least in adults.

7.7 Effects of Aging 233

- Due to aging, the tensile strength of bone decreases, and bone loses calcium and other minerals (demineralization).

7.8 Bone Fracture and Repair 234

- A fracture is a break in a bone that can usually be healed if portions of the blood supply, endosteum, and periosteum remain intact.

CHALLENGE YOURSELF

Do You Know the Basics?

_____ 1. Which bone is formed from intramembranous bone growth?

    a. femur

    b. rib

    c. os coxae (hip bone)

    d. frontal bone

_____ 2. All of the following are functions of cartilage _except_

    a. cartilage serves as a site for hemopoiesis.

    b. cartilage provides support for soft tissue.

    c. cartilage forms the initial model in endochondral ossification.

    d. cartilage provides a smooth gliding surface at the end of bones in freely movable joints.

_____ 3. Which is _not_ a function of bone?

    a. It protects some internal organs, such as the brain, heart, and lungs.

    b. It helps move the body by serving as levers.

    c. It stores phosphorus within the bone connective tissue.

    d. Its yellow bone marrow forms blood cells.

_____ 4. The femur is an example of a(n)

    a. short bone.

    b. flat bone.

    c. long bone.

    d. irregular bone.

_____ 5. Which cell type is most likely to have created the medullary cavity in a long bone?

    a. osteocytes

    b. osteoblasts

    c. osteoclasts

    d. osteoprogenitor cells

_____ 6. Which long bone structure is correctly matched with its description or function?

    a. epiphysis; the end of a bone that is composed of compact bone only

    b. articular cartilage; fibrocartilage located at the ends of a bone

    c. periosteum; responsible for growth in bone width

    d. perforating fibers; blood vessels that penetrate the bone.

_____ 7. Which statement is correct about an osteon?

    a. The circumferential lamellae surround the blood vessels and nerves within an osteon.

    b. Canaliculi allow for nutrient and waste exchange among the osteocytes.

    c. The middle region of an osteon is called the perforating canal.

    d. They are oriented perpendicular to the diaphysis of a long bone.

_____ 8. All of the following accurately describe the hyaline cartilage _except_

    a. the matrix of hyaline cartilage contains calcium.

    b. hyaline cartilage is avascular.

    c. hyaline cartilage lacks nerves.

    d. hyaline cartilage is a flexible, semirigid connective tissue.

_____ 9. To elevate blood calcium levels, all of these must occur _except_

    a. calcitonin is secreted by the thyroid gland.

    b. less calcium is excreted in the urine.

    c. parathyroid hormone is secreted by the parathyroid glands.

    d. increased calcium is absorbed from the small intestine into the blood.

_____ 10. An epiphyseal line appears when

    a. epiphyseal plate growth has ended.

    b. epiphyseal plate growth is just beginning.

    c. growth in bone diameter is just beginning.

    d. the bone is fractured at that location.

11. Describe the structure of a typical long bone.

12. Describe the general function of both osteoblast and osteoclast activity.

13. Describe the microscopic anatomy of compact bone.

14. Compare and contrast interstitial growth versus appositional growth of cartilage.

15. List the steps involved in endochondral ossification.

16. Which of the five zones in the epiphyseal plate cartilage are specifically responsible for bone growth in length? Explain.

17. Discuss the effect of exercise on bone mass.

18. Compare and contrast the effects of growth hormone and glucocorticoids on bone growth.

19. Describe how parathyroid hormone regulates blood calcium concentration.

20. What are the steps in fracture repair?

Can You Apply What You've Learned?

1. Jorge is donating bone marrow to a friend of his who has leukemia (a type of blood cell cancer). Jorge is 30 years old, so the doctor knows she must insert the needle in

    a. the diaphysis of the femur.

    b. the hip bone.

    c. the distal epiphysis of the tibia.

    d. the diaphysis of the humerus.

2. You are given the following assignment. Obtain two small chicken or turkey bones. Bake one bone in the oven at a high temperature for approximately 30 minutes. Place the other in vinegar (acidic pH) for several days. Which of the following would most accurately reflect what occurred?

   a. Proteins in the bone matrix have denatured in the bone that is baked, and the vinegar has denatured the proteins in the soaked bone, so both bones are flexible.

   b. Proteins in the bone matrix are lost from the baked bone, and the bone becomes flexible. The soaked bone loses calcium and results in a brittle bone.

   c. Proteins in matrix have denatured from high temperature, and the bone is brittle; calcium has been removed from the bone soaked in vinegar and it has become flexible.

   d. Proteins are lost in the baked bone matrix, and this bone becomes flexible. Calcium loss from the bone soaked in vinegar results in a flexible bone.

3. Your dog is chewing on an adult cow long bone and breaks it open. He pulls out some fleshy material within the diaphysis of the bone. You know that this fleshy material

   a. produced red blood cells for the adult cow.

   b. contains fat.

   c. was formed from osteoblasts.

   d. is abnormal and is not normally found in bone.

4. To identify the approximate age of skeletal remains of a body found in the forest, the forensic anthropologists are most interested in which of the following?

   a. whether the epiphyses have fused to the diaphyses

   b. number of bones in the skeleton

   c. length of the long bones in the legs

   d. presence or absence of articular cartilage

5. In your anatomy and physiology laboratory, you look at prepared slides of developing bone. In the epiphyseal plate region, you note the chondrocytes are slightly enlarged and stacked in a longitudinal array. What epiphyseal plate zone is in your field of view?

   a. zone of rest

   b. zone of proliferation

   c. zone of hypertrophy

   d. zone of calcification

Can You Synthesize What You've Learned?

1. The traditional surgical procedure to treat advanced thyroid gland tumors is to remove the affected organ. Some concerns with the results of this surgery were raised regarding the parathyroid glands, which are small attachments on the posterior side of the thyroid gland. Why should the surgeon be concerned about the removal of these glands? A new procedure has been developed to sequester some parathyroid gland tissue into a plastic mesh holder and implant this small holder back into the body. Why is this implant an advantage to the patient?

2. A fireman fell from a ladder while fighting a fire and severely fractured the thigh and leg bones in his right leg. He was hospitalized for several weeks, then he was wheelchair-bound for several months while his leg bones healed. Upon removal of the cast, it was apparent that his right thigh and leg bones were thin and weak. What factors contributed to this thinning and weakening of the bones, and what should the fireman do to improve the strength of these bones?

3. Elise is 14 and lives in an apartment in the city. She does not like outdoor activities, so she spends most of her spare time watching TV, playing video games, drinking soft drinks, and talking to friends on the phone. One afternoon, Elise tries to run down the stairs while talking on the phone, and falls, breaking her leg. Although she appears healthy, her leg takes longer to heal than expected. What might cause the longer healing time?

INTEGRATE

ONLINE STUDY TOOLS  connect |ANATOMY & PHYSIOLOGY LEARNSMART AP|R

The following study aids may be accessed through Connect.

Clinical Case Study: A Clinical Case of Postmenopausal Osteoporosis

Interactive Questions: This chapter's content is served up in a number of multimedia question formats for student study.

LearnSmart: Topics and terminology include bone formation; bone structure; cartilage growth; bone growth and remodeling; regulating blood calcium levels; effects of aging; bone fracture and repair

Anatomy & Physiology Revealed: Topics include humerus anterior and posterior views; compact bone; hyaline cartilage; epiphyseal plate in hand; appositional bone growth; bone scan

Animations: Topics include bone growth in width; osteoporosis

Skeletal System: Axial and Appendicular Skeleton

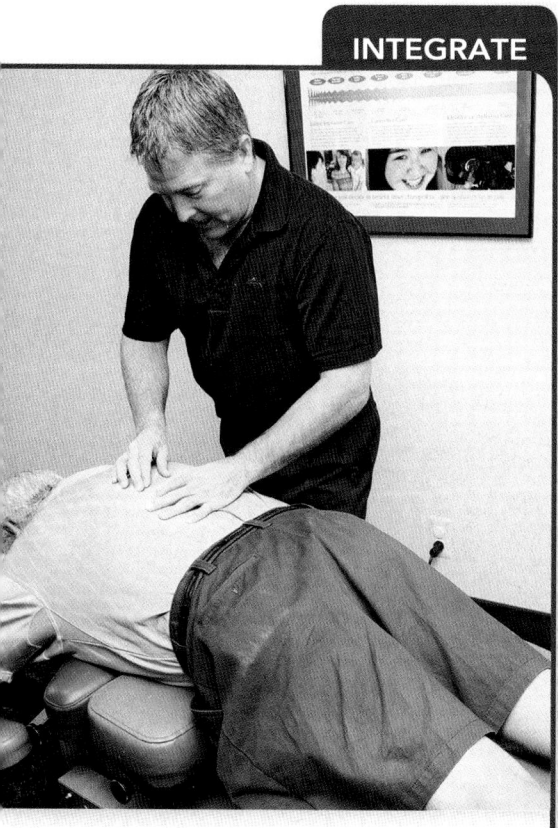

CAREER PATH
Chiropractor

Chiropractors are health-care professionals who focus on the musculoskeletal system. These individuals palpate bony surface landmarks to pinpoint both the bones and the overlying muscle and connective tissue by performing manual therapy. By manipulating the spine and other bony regions, chiropractors may adjust various muscle imbalances and may improve the functioning of the nerves that may be impinged by the spine (vertebral column). Here, a chiropractor is palpating the spinous and transverse processes of the patients back to perform a spinal adjustment.

Anatomy & Physiology | REVEALED®
aprevealed.com

Module 5: Skeletal System

The bones of the skeleton form an internal framework to support soft tissues, protect vital organs, bear the body's weight, and function as levers to help us move. An adult skeleton typically has 206 named bones, although this number varies in some individuals. Bones differ in size, shape, and weight, and this diversity is related directly to the skeleton's many functions.

For criminologists, pathologists, and anthropologists, bones can tell an intricate anatomic story. The skeleton can provide information about an individual such as their sex, age at death, and possible pathologies. In this chapter, not only will you learn the names of and features on the individual bones, but we also provide an overview of how you can determine age at death from the skeletal remains, as well as how you can tell the difference in sex between certain bones of the skeleton. You can practice determining age at death and determining sex using the bones in your anatomy lab. This will permit you to learn some of the variations seen in bone morphology so you can start to unlock the mysteries of the bones in front of you in the lab.

8.1 Components of the Skeleton

We first examine the two subdivisions of the skeleton: the axial skeleton and the appendicular skeleton. Then we discuss the names for characteristic markings on bones.

8.1a Axial and Appendicular Skeleton

LEARNING OBJECTIVE

1. Compare and contrast the functions and composition of the axial and appendicular skeletons.

The skeletal system is organized into two divisions: the axial skeleton and the appendicular skeleton (**figure 8.1**).

The **axial skeleton** is so named because it is composed of the bones along the central axis of the body, which are commonly divided into three regions—the skull, the vertebral column, and the thoracic cage. The main function of the axial skeleton is to form a framework that supports and protects the organs. Additionally, the spongy bone of most of the axial skeleton contains hemopoietic tissue that is responsible for blood cell formation (see section 18.3a).

The **appendicular skeleton** includes the bones of the upper and lower limbs, and the girdles of bones that attach the upper and lower limbs to the axial skeleton. The pectoral girdle consists of bones that

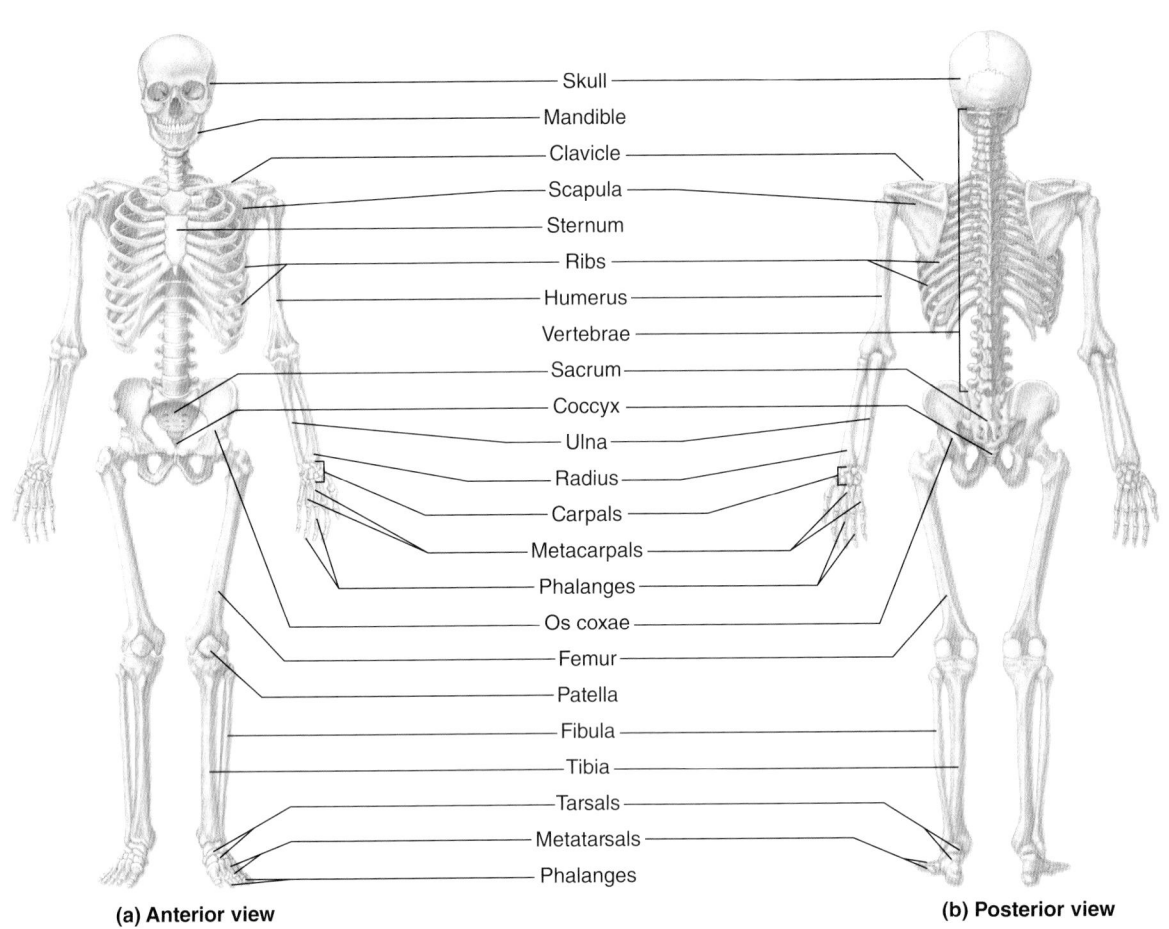

(a) Anterior view **(b) Posterior view**

Bones of the Axial Skeleton (80)				
Skull (22)	Cranial bones (8) Frontal bone (1), parietal bones (2), temporal bones (2), occipital bone (1), sphenoid bone (1), ethmoid bone (1)	**Vertebral column (26)**	Cervical vertebrae (7)	
			Thoracic vertebrae (12)	
	Facial bones (14) Zygomatic bones (2), lacrimal bones (2), nasal bones (2), vomer (1), inferior nasal conchae (2), palatine bones (2), maxillae (2), mandible (1)		Lumbar vertebrae (5)	
			Sacrum (1)	
			Coccyx (1)	
Associated bones of the skull (7)	Auditory ossicles (6) Malleus (2), incus (2), stapes (2)	**Thoracic cage (25)**	Sternum (1)	
	Hyoid bone (1)		Ribs (24)	
Bones of the Appendicular Skeleton (63 bones per each side of the body, 126 bones total)				
Pectoral girdle (4 bones total)	Clavicle (2)	Scapula (2)	**Pelvic girdle (2 bones total)**	Os coxae (2)	
Upper limbs (30 bones per each upper limb, 60 bones total)	Humerus (2)	Carpals (16)	**Lower limbs (30 bones per each lower limb, 60 bones total)**	Femur (2)	Tarsals (14)
	Radius (2)	Metacarpals (10)		Patella (2)	Metatarsals (10)
	Ulna (2)	Phalanges (28)		Tibia (2)	Phalanges (28)
				Fibula (2)	

Figure 8.1 Axial and Appendicular Skeleton. (*a*) Anterior and (*b*) posterior views compare the axial and appendicular components of the skeleton. The axial skeleton is colored blue and the appendicular skeleton is colored tan.

hold the upper limbs in place, whereas the pelvic girdle consists of bones that hold the lower limbs in place.

WHAT DID YOU LEARN?

❶ What is the general function of the axial skeleton, and which bones are considered part of the axial skeleton?

8.1b Bone Markings

LEARNING OBJECTIVE

2. Become familiar with terminology for common bone markings.

Distinctive **bone markings** are the surface features that characterize each bone in the body. Projections from the bone surface mark the point where muscles, tendons, and ligaments attach. Sites of articulation between adjacent bones tend to be smooth areas. Depressions, grooves, and openings through bones indicate sites where blood vessels and nerves travel. Anatomists use specific terms to describe these characteristics (figure 8.2).

Knowing the names of bone markings will help you learn about specific bones described in this chapter. For example, when trying to locate the foramen magnum of the skull, you have an advantage if you know that foramen means "hole" or "passageway."

WHAT DID YOU LEARN?

❷ What is the difference between a foramen and a fissure?

General Structure	Anatomic Term	Description
Articulating surfaces	Condyle	Large, smooth, rounded oval structure
	Facet	Small, flat, shallow surface
	Head	Prominent, rounded epiphysis
	Trochlea	Smooth, grooved, pulley-like process
Depressions	Alveolus (pl., *alveoli*)	Deep pit or socket in the maxillae or mandible
	Fossa (pl., *fossae*)	Flattened or shallow depression
	Sulcus	Narrow groove
Projections for tendon and ligament attachment	Crest	Narrow, prominent, ridgelike projection
	Epicondyle	Projection adjacent to a condyle
	Line	Low ridge
	Process	Any marked bony prominence
	Ramus (pl., *rami*)	Angular extension of a bone relative to the rest of the structure
	Spine	Pointed, slender process
	Trochanter	Massive, rough projection found only on the femur
	Tubercle	Small, round projection
	Tuberosity	Large, rough projection
Openings and spaces	Canal	Passageway through a bone
	Fissure	Narrow, slitlike opening through a bone
	Foramen (pl., *foramina*)	Rounded passageway through a bone
	Meatus	Passageway through a bone
	Sinus	Cavity or hollow space in a bone

Figure 8.2 Bone Markings. Specific anatomic terms describe the characteristic features on bones.

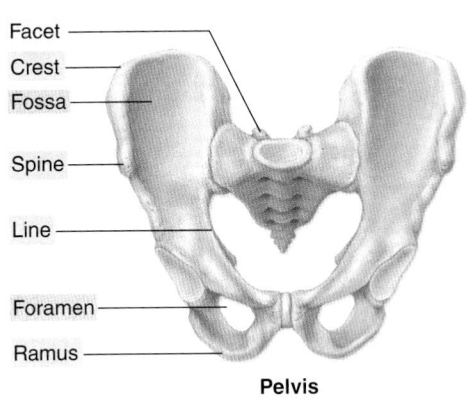

Pelvis

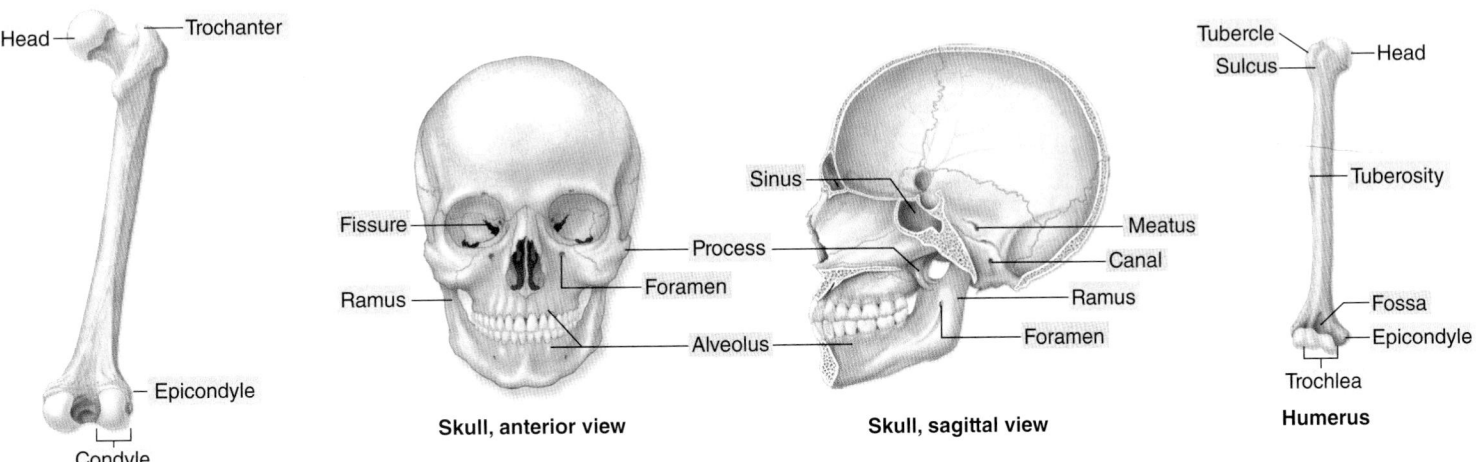

Femur

Skull, anterior view

Skull, sagittal view

Humerus

INTEGRATE

LEARNING STRATEGY

Many bones of the body have the same names as the body regions where they are found. Before you begin learning the bones of the skeleton, it may help you to review table 1.1 (Anatomic Directional Terms) and table 1.2 (Human Body Regions). In addition, remember that this book uses the anatomic terms published in *Terminologia Anatomica* (see section 1.4), which have been reviewed and approved by the International Federation of Associations of Anatomists.

INTEGRATE

LEARNING STRATEGY

As you examine a bone in lab, try to palpate (feel) the same bone on your own body. In this way, you will start to understand how the bone is placed, how it associates with other bones, how it moves in a living body, and how we use it. In effect, you can use your body as a "bone study guide."

8.2 Bones and Features of the Skull

We begin our examination of the skeleton by discussing its most complex structure, the skull. The skull is made up of 22 bones. Here we describe the anatomy and landmarks of the skull, the sutures (fibrous joints) that connect the bones of the cranium, and the specialized features of the orbital and nasal complexes and paranasal sinuses.

8.2a General Anatomy of the Skull

 **LEARNING OBJECTIVE**

1. Distinguish between the cranial and the facial bones.

The **skull** is composed of both cranial and facial bones. **Cranial bones** form the rounded **cranium** (krā′nē-um; *kranion* = skull), which completely surrounds and encloses the brain.[1] The cranium consists of eight bones that form a roof and a base. The roof of the cranium, called the **calvaria** (kal-vā′rē-ă), or is composed of part of the frontal bone, the parietal bones, and part of the occipital bone. The **base** of the cranium is composed of portions of the ethmoid, sphenoid, occipital, and temporal bones. Some skulls in the anatomy lab have had their calvariae cut away, making the distinction between the calvaria and base easier to distinguish.

 Facial bones form the face. They also protect the entrances to the digestive and respiratory systems. Touch your cheeks, your jaws, and the bridge of your nose; these bones are facial bones. The facial bones give shape and individuality to the face, form part of the orbit and nasal cavities, support the teeth, and provide for the attachment of muscles involved in facial expression and mastication (chewing). There are 14 facial bones, including the paired zygomatic bones, lacrimal bones, nasal bones, inferior nasal conchae, palatine bones, maxillae, and the unpaired vomer and mandible.

 The skull contains several prominent cavities (**figure 8.3**). The largest is the **cranial cavity** (or *endocranium*) that encloses, protects, and supports the brain. (Volume of an adult cranial

---

1. Osteologists (scientists who study bones) define the cranium as the entire skull minus the mandible. In this text, we use the term *cranium* to denote the bones that directly surround the brain only.

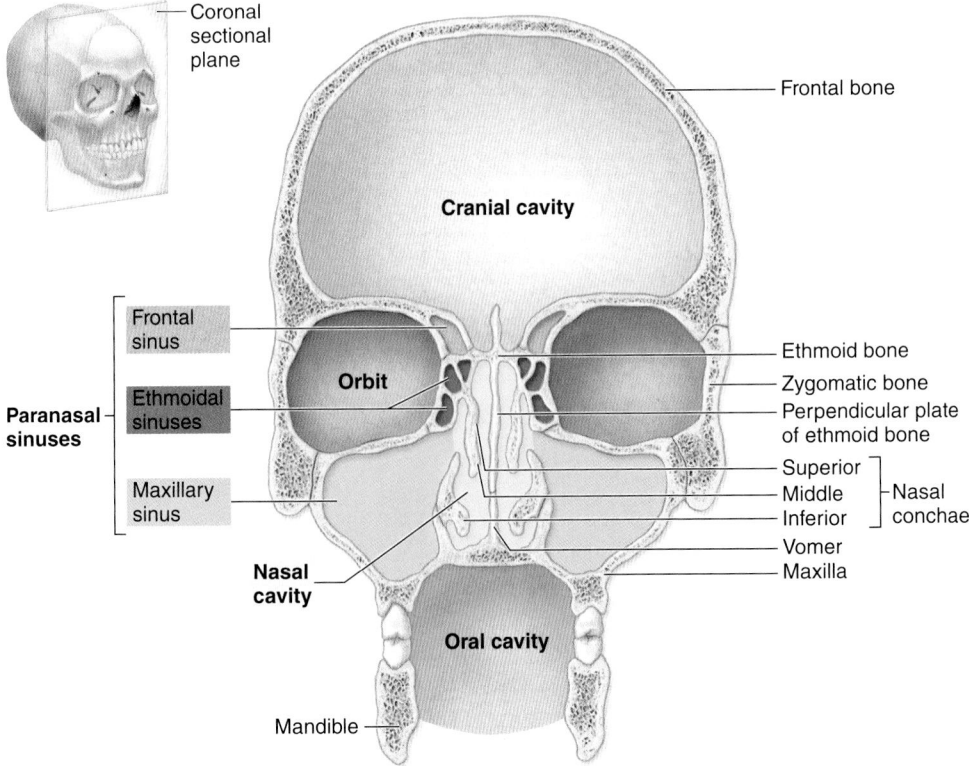

Figure 8.3 Major Cavities of the Skull. A coronal section diagram highlights the cranial cavity, orbits, three of the four sets of paranasal sinuses, nasal cavity, and oral cavity.

cavity ranges from approximately 1300 to 1500 cubic centimeters, which is about 50 fluid ounces.) The skull also forms and has several smaller cavities, including the **orbits** (eye sockets), the oral cavity, the nasal cavity, and the paranasal sinuses.

 WHAT DID YOU LEARN?

3 What bones form the skull? Which of these bones are cranial bones? Which of these bones are facial bones?

8.2b Views of the Skull and Landmark Features

LEARNING OBJECTIVES

2. Identify the locations of cranial and facial bones in various views of the skull.
3. Learn key bone markings and features of each of the bones of the cranium.
4. Compare and contrast the locations and contents of three cranial fossae.

To best understand the complex nature of the skull, we first examine the skull as a whole and learn which bones are best seen from a particular view. Note that only some major features will be mentioned in this section. Later in the chapter, we examine the individual skull bones in detail.

WHAT DO YOU THINK?

1 What is the benefit of the skull's being made of multiple smaller bones, rather than one big bone?

A cursory glance at the skull reveals numerous bone markings, such as canals, fissures, and foramina that serve as passageways for blood vessels and nerves. The major foramina of the cranial and facial bones are summarized in **table 8.1**. Refer to this table as we examine the skull from various directions. (This table also will be important when we study cranial nerves in section 13.9 and blood vessels in chapter 20.)

Table 8.1	Passageways Within the Skull	
Passageway	**Location**	**Primary Structures That Pass Through**
CRANIAL BONES		
Carotid canal	Petrous part of temporal bone	Internal carotid artery
Cribriform foramina	Cribriform plate of ethmoid bone	Olfactory nerves (CN I)
Foramen lacerum	Between petrous part of temporal bone, sphenoid bone, and occipital bone	None
Foramen magnum	Occipital bone	Vertebral arteries; spinal cord; accessory nerves (CN XI)
Foramen ovale	Greater wing of sphenoid bone	Mandibular branch of trigeminal nerve (CN V$_3$)
Foramen rotundum	Greater wing of sphenoid bone	Maxillary branch of trigeminal nerve (CN V$_2$)
Foramen spinosum	Greater wing of sphenoid bone	Middle meningeal vessels
Hypoglossal canal	Anteromedial to occipital condyle of occipital bone	Hypoglossal nerve (CN XII)
Inferior orbital fissure	Junction of maxilla, sphenoid, and zygomatic bones	Infraorbital nerve (branch of CN V$_2$)
Jugular foramen	Between temporal bone and occipital bone (posterior to carotid canal)	Internal jugular vein; glossopharyngeal nerve (CN IX); vagus nerve (CN X); accessory nerve (CN XI)
Mastoid foramen	Posterior to mastoid process of temporal bone	Mastoid emissary veins
Optic canal	Posteromedial part of orbit in lesser wing of sphenoid bone	Optic nerve (CN II)
Stylomastoid foramen	Between mastoid and styloid processes of temporal bone	Facial nerve (CN VII)
Superior orbital fissure	Posterior part of orbit between greater and lesser wings of sphenoid bone	Ophthalmic veins; oculomotor nerve (CN III); trochlear nerve (CN IV); ophthalmic branch of trigeminal nerve (CN V$_1$); abducens nerve (CN VI)
Supraorbital foramen	Supraorbital margin of orbit in frontal bone	Supraorbital artery; supraorbital nerve (branch of CN V$_1$)
FACIAL BONES		
Greater and lesser palatine foramina	Palatine bone	Palatine vessels; greater and lesser palatine nerves (branches of CN V$_2$)
Incisive foramen	Posterior to incisor teeth in hard palate of the maxilla	Nasopalatine nerve (branch of CN V$_2$)
Infraorbital foramen	Inferior to orbit in maxilla	Infraorbital artery; infraorbital nerve (branch of CN V$_2$)
Lacrimal groove	Lacrimal bone	Nasolacrimal duct
Mandibular foramen	Medial surface of ramus of mandible	Inferior alveolar blood vessels; inferior alveolar nerve (branch of CN V$_3$)
Mental foramen	Inferior to second premolar on anterolateral surface of mandible	Mental blood vessels; mental nerve (branch of CN V$_3$)

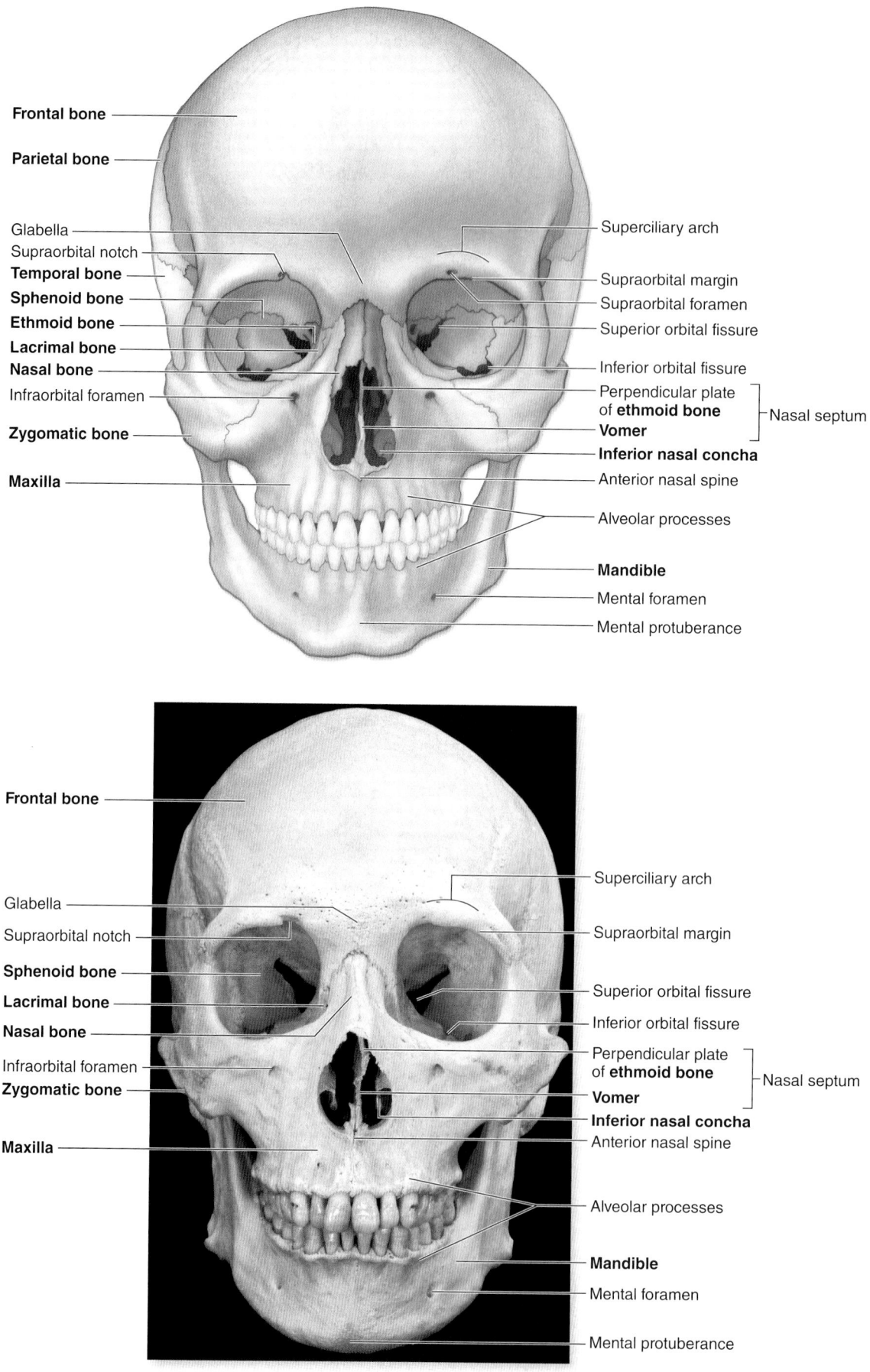

Frontal bone

Parietal bone

Glabella

Supraorbital notch

Temporal bone

Sphenoid bone

Ethmoid bone

Lacrimal bone

Nasal bone

Infraorbital foramen

Zygomatic bone

Maxilla

Superciliary arch

Supraorbital margin

Supraorbital foramen

Superior orbital fissure

Inferior orbital fissure

Perpendicular plate
of **ethmoid bone**

Vomer

] Nasal septum

Inferior nasal concha

Anterior nasal spine

Alveolar processes

Mandible

Mental foramen

Mental protuberance

Frontal bone

Glabella

Supraorbital notch

Sphenoid bone

Lacrimal bone

Nasal bone

Infraorbital foramen

Zygomatic bone

Maxilla

Superciliary arch

Supraorbital margin

Superior orbital fissure

Inferior orbital fissure

Perpendicular plate
of **ethmoid bone**

Vomer

] Nasal septum

Inferior nasal concha

Anterior nasal spine

Alveolar processes

Mandible

Mental foramen

Mental protuberance

Anterior view

Figure 8.4 Anterior View of the Skull. The frontal bone, nasal bones, maxillae, and mandible are prominent in this view. **AP|R**

Anterior View

An anterior view (figure 8.4) shows several major bones of the skull. The **frontal bone** forms the forehead. The left and right orbits (see figure 8.3) are formed from a complex articulation of multiple skull bones. There are two large openings within each orbit called the **superior orbital fissure** and the **inferior orbital fissure** (figure 8.4). Superior to the orbits on the anterior surface of the frontal bone are the **superciliary** (sū per-sil′ē-ār-ē; *super* = above, *cilium* = eyelid) **arches,** otherwise known as the brow ridges. Male skulls tend to have larger and more pronounced superciliary arches than do female skulls. The left and right **nasal bones** form the bony bridge of the nose. Superior to the nasal bones and between the orbits is a landmark area called the **glabella** (glă-bel′ă; *glabellus* = smooth).

The left and right **maxillae** (mak-sil′ē; sing., *maxilla,* mak-sil-ă; jawbone), also called *maxillary bones,* fuse in the midline to form most of the upper jaw and the lateral boundaries of the nasal cavity. The maxillae also help form a portion of both the floor of each orbit and the roof of the oral cavity. Inferior to each orbit in the maxilla is an **infraorbital foramen,** which is a passageway for blood vessels and nerves to the face.

The lower jaw is formed by the **mandible.** The prominent *chin* of the mandible is called the **mental protuberance.** The oral margins of the maxillae and mandible each have **alveolar** (al-vē′ō-lăr) **processes** that contain the teeth.

The nasal cavity is also seen in an anterior view. Its inferior border is marked by a prominent **anterior nasal spine.** The thin ridge of bone that divides the nasal cavity into left and right halves is called the **nasal septum.** Along the inferolateral walls of the nasal cavity are two scroll-shaped bones called the **inferior nasal conchae** (kon′kē; sing., *concha,* kon′kă; shell).

Superior View

The superior view of the skull in **figure 8.5a** primarily shows four of the cranial bones: the frontal bone, both **parietal** (pă-rī′ē-tăl; *paries* = wall) **bones,** and the **occipital** (ok-sip′i-tăl; *occiput* = back of head) **bone.** The articulation between the frontal and parietal bones is the coronal suture, so named because it runs along a coronal plane. The sagittal suture connects the left and right parietal bones along the midline of the skull.

Along the posterior one-third of the sagittal suture is either a single **parietal foramen** or paired **parietal foramina,** which serve as the passage of small veins between the brain and the scalp. The lateral surface of each parietal bone exhibits a rounded, smooth area called the **parietal eminence.** The superior part of the lambdoid suture represents the articulation of the occipital bone with both parietal bones.

Posterior View

The posterior view of the skull in figure 8.5b shows a portion of the occipital, parietal, and temporal bones, as well as the lambdoid suture between the occipital and parietal bones. Within the lambdoid suture there may be one or more sutural bones. The **external occipital protuberance** (prō-tū′ber-ans) is a prominence on the posterior aspect of the skull. Palpate the back of your head; males tend to have a prominent, pointed external occipital protuberance, whereas females have a more subtle, rounded protuberance. Intersecting the external occipital protuberance are two horizontal ridges, the **superior** and **inferior nuchal** (nū′kăl) **lines** (see figure 8.7).

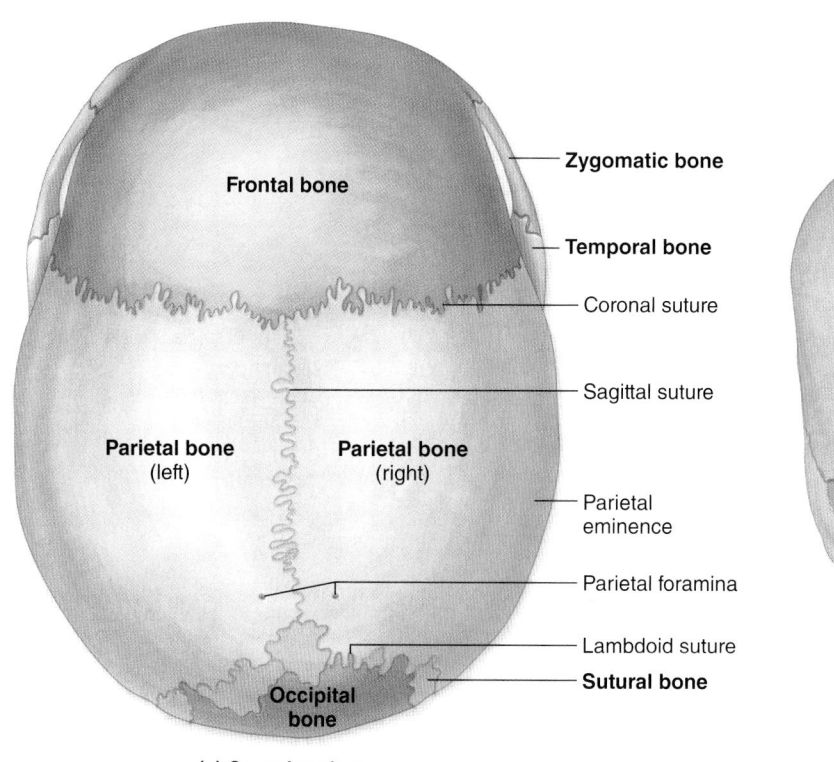

(a) Superior view

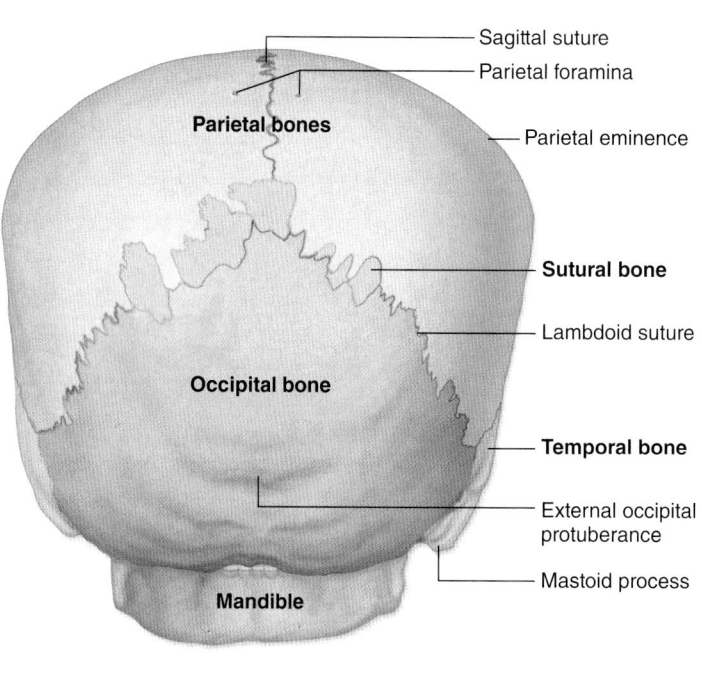

(b) Posterior view

Figure 8.5 Superior and Posterior Views of the Skull. (*a*) The superior aspect of the skull shows the major sutures and the flat bones of the skull. (*b*) The posterior view is dominated by the occipital and parietal bones.

Lateral View

A lateral view of the skull (**figure 8.6**) shows one parietal bone, **temporal bone,** and **zygomatic** (zī′gō-mat′ik; *zygoma* = a joining, a yoke) **bone.** This view also shows part of the maxilla, mandible, frontal bone, and occipital bone. The **superior** and **inferior temporal lines** arc across the surface of the parietal and frontal bones and mark the attachment site

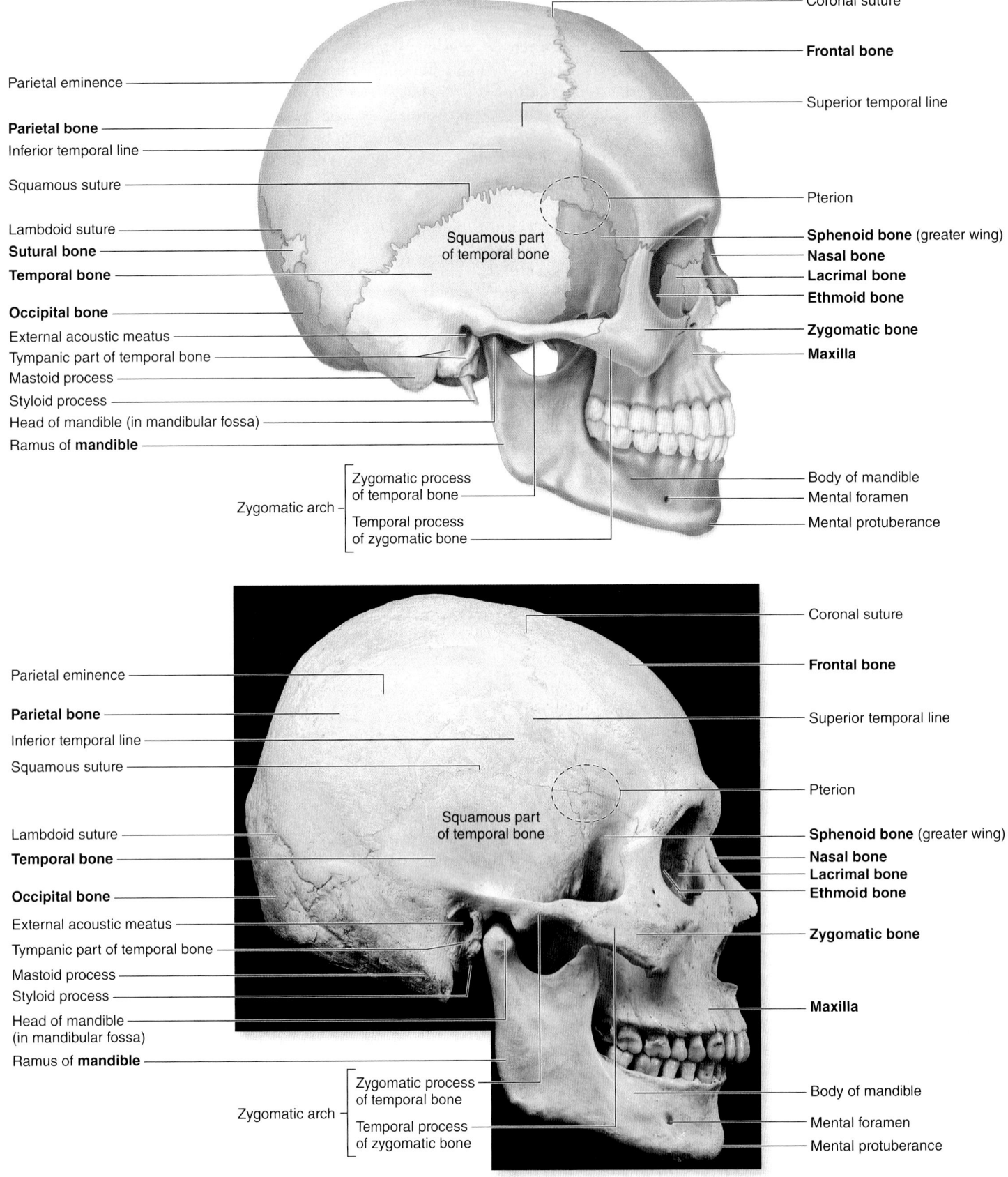

Lateral view

Figure 8.6 Lateral View of the Skull. The parietal, temporal, zygomatic, frontal, and occipital bones, as well as the maxilla and mandible, are prominent in this view. AP|R

of the temporalis muscle (see section 11.3c). The small **lacrimal** (lak′ri-măl; *lacrima* = a tear) **bone** articulates with the maxilla anteriorly and with the ethmoid bone posteriorly. A portion of the **sphenoid** (sfē′noyd; wedge-shaped) **bone** articulates with the frontal, parietal, and temporal bones. This region is called the **pterion** (tĕ′rē-on; *ptéron* = wing) and is circled on figure 8.6. Pterion includes the H-shaped set of sutures of these four articulating bones.

The **temporal process** of the **zygomatic bone** and the **zygomatic process** of the **temporal bone** fuse to form the **zygomatic arch.** Put your fingers along the bony prominences ("apples") of your cheeks and move your fingers posteriorly toward your ears; you are feeling the zygomatic arch. The zygomatic arch terminates superior to the point where the mandible articulates with the **mandibular** (man-dib′ū-lăr) **fossa** of the temporal bone. This articulation is called the **temporomandibular joint (TMJ)** and is described further in section 9.7a. By putting your finger anterior to your external ear opening and then opening and closing your jaw, you can feel that joint moving.

The **squamous part** of the temporal bone lies directly inferior to the squamous suture. Immediately posterolateral to the mandibular fossa is the **tympanic** (tim-pan′ik; *tympanon* = drum) **part** of the temporal bone. This is a small, bony ring surrounding the external ear opening called the **external acoustic meatus** (mē-ā′tŭs; a passage), or *external auditory canal.* Inferior and posterior to this meatus is the **mastoid** (mas′toyd; *masto* = breast; *eidos* = resemblance) **process,** the bump you feel behind your external ear opening. The **styloid** (stī′loyd; *stylos* = pillar, post) **process** is a thin, pointed projection of bone located anteromedial to the mastoid process. It serves as an attachment site for several hyoid and tongue muscles.

Sagittal Sectional View

Cutting the skull along a sagittal sectional plane reveals bones that form the endocranium and the nasal cavity **(figure 8.7a)**. The cranial cavity is formed from a complex articulation of the frontal, parietal, temporal, occipital, **ethmoid** (eth′moyd; *ethmos* = sieve), and sphenoid bones.

Vessel impressions may be visible on the internal surface of the skull. The **frontal sinus** (a space in the frontal bone) and the **sphenoidal sinus** (open space in the sphenoid bone) are visible in a sagittal view.

A sagittal sectional view also shows the bones that form the nasal septum more clearly. The **perpendicular plate** of the ethmoid forms the posterosuperior portion of the nasal septum, whereas the **vomer** (vō′mer; plowshare) forms the posteroinferior portion. (The anterior part of the nasal septum is cartilaginous.) The ethmoid bone serves as the division between the anterior floor of the cranial cavity and the roof of the nasal cavity. The **palatine process** of the maxillae and the **palatine** (pal′-a-tīn) **bones** form the hard palate (figure 8.7b), which acts as both the floor of the nasal cavity and a portion of the roof of the mouth. Move your tongue along the roof of your mouth; you are palpating the maxillae anteriorly and palatine bones posteriorly.

Inferior (Basal) View

In an inferior (basal) view, the most anterior structure is the hard palate (figure 8.7b). On the posterior aspect of either side of the palate are the **medial** and **lateral pterygoid** (ter′i-goyd; *pteryx* = winglike) **plates** of the sphenoid bone. Together, both plates form a **pterygoid process.** Medially adjacent to these structures are the internal openings of the nasal cavity, called the **choanae,** (kō′an-ē; sing., *choana*, kō′an-ă; funnel).

Between the mandibular fossa and the pterygoid processes are several paired foramina and canals. Typically, these openings provide passage for specific blood vessels and nerves. For example, the **jugular** (jŭg′ū-lar; *jugulum* = throat) **foramen** is an opening between the temporal and occipital bones that provides passageway for the internal jugular vein and several nerves. The **foramen lacerum** (anteromedial to the carotid canal) extends between the occipital and temporal bones. This opening is covered by cartilage in a living individual. The entrance to the **carotid** (ka-rot′id; *karoo* = to put to sleep) **canal** is anteromedial to the jugular foramen; the internal carotid artery passes through this canal.

The **stylomastoid foramen** lies between the mastoid process and the styloid process. The facial nerve (CN VII) extends through the stylomastoid foramen to innervate the facial muscles.

The largest foramen of all is the **foramen magnum,** literally meaning "big hole." Through this opening, the spinal cord enters the cranial cavity and is continuous superiorly with the brainstem. On either side of the foramen magnum are the rounded **occipital condyles,** which articulate with the first cervical vertebra of the vertebral column. At the anteromedial edge of each condyle is a **hypoglossal canal** through which the hypoglossal nerve (CN XII) extends to innervate tongue muscles.

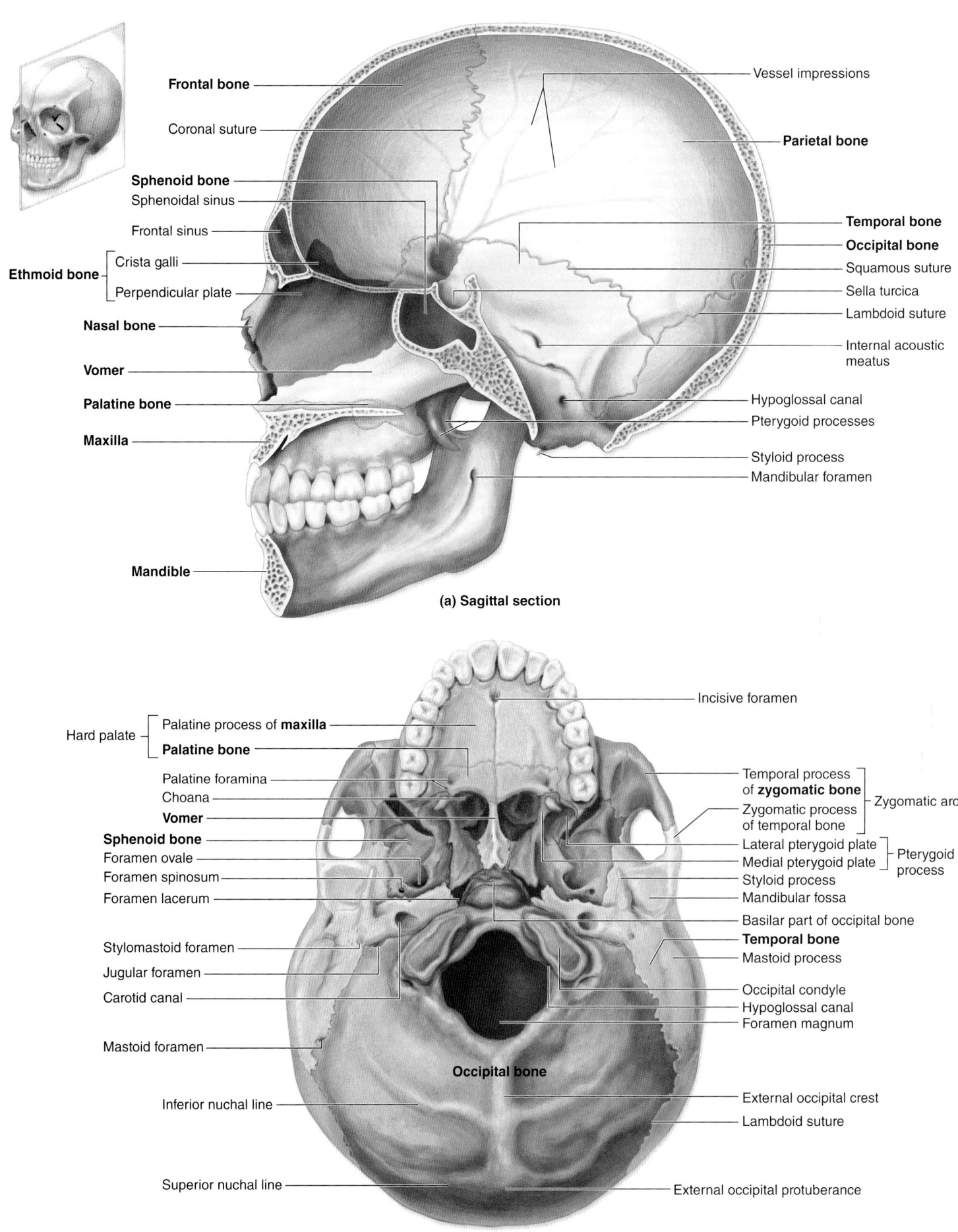

Frontal bone
Coronal suture
Sphenoid bone
Sphenoidal sinus
Frontal sinus
Ethmoid bone
  Crista galli
  Perpendicular plate
Nasal bone
Vomer
Palatine bone
Maxilla
Mandible

Vessel impressions
Parietal bone
Temporal bone
Occipital bone
Squamous suture
Sella turcica
Lambdoid suture
Internal acoustic meatus
Hypoglossal canal
Pterygoid processes
Styloid process
Mandibular foramen

(a) Sagittal section

Hard palate
  Palatine process of **maxilla**
  Palatine bone
Palatine foramina
Choana
Vomer
Sphenoid bone
Foramen ovale
Foramen spinosum
Foramen lacerum
Stylomastoid foramen
Jugular foramen
Carotid canal
Mastoid foramen
Inferior nuchal line
Superior nuchal line

Incisive foramen
Temporal process of **zygomatic bone**   Zygomatic arch
Zygomatic process of temporal bone
Lateral pterygoid plate   Pterygoid process
Medial pterygoid plate
Styloid process
Mandibular fossa
Basilar part of occipital bone
Temporal bone
Mastoid process
Occipital condyle
Hypoglossal canal
Foramen magnum
Occipital bone
External occipital crest
Lambdoid suture
External occipital protuberance

(b) Inferior view

Figure 8.7 Sagittal Section and Inferior View of the Skull. (*a*) Features such as the perpendicular plate of the ethmoid bone, the vomer, and the frontal and sphenoidal sinuses, as well as the internal relationships of the skull bones, are best seen in sagittal section. (*b*) The hard palate, sphenoid bone, parts of the temporal bone, and the occipital bone with its foramen magnum may be seen in the inferior view. AP|R

Internal View of Cranial Base

When the top of the skull is cut and removed, the internal view of the cranial base (figure 8.8) is revealed. Here we see the frontal bone surrounding the delicate **cribriform** (krib′ri-fōrm; *cribrum* = sieve, *forma* = form) **plate** of the ethmoid bone. The plate has numerous perforations called the **cribriform foramina,** which provide passageways for the olfactory nerves (CN I) into the superior portion of the nasal cavity. The anteromedial part of the cribriform plate exhibits a midsagittal elevation called the **crista galli** (kris′tă = crest; gal′lē = of a rooster), to which the cranial dural septa of the brain attach (see section 13.2a).

The relatively large sphenoid is located posterior to the frontal bone. It is often referred to as a "bridging bone" because it unites the cranial and facial bones. The lateral expansions of the sphenoid bone are called the **greater wings** and the **lesser wings** of the **sphenoid.** The bony depression on the sphenoid called the **sella turcica** (sel′ă = saddle; tur′si-ka = Turkish) houses the pituitary gland, which is suspended inferiorly from the base of the brain (see figure 13.1c). Anterior to the sella turcica are the **optic canals** through which the optic nerves (CN II) extend from the eyes in the orbits to the brain (figure 8.8).

The lateral regions of the cranial base are formed by the **petrous** (pet′rŭs; *petra* = a rock) **part** of each temporal bone, whereas the posterior region is formed by the occipital bone. The **internal acoustic meatus** (also called the *internal auditory canal*) opens in the more medial portion of the temporal bone and contains the proximal part of the facial nerve (CN VII) and the vestibulocochlear nerve (CN VIII).

An internal landmark of the occipital bone is the **internal occipital protuberance.** The **internal occipital crest** extends from the protuberance to the posterior border of the foramen magnum. Large grooves within the cranium are formed from impressions from the venous sinuses of the brain that lie within them (see section 13.2a).

Each bone of the cranium has specific surface features, and each specific bone is shown and summarized in **table 8.2.** The facial bones are summarized in **table 8.3.**

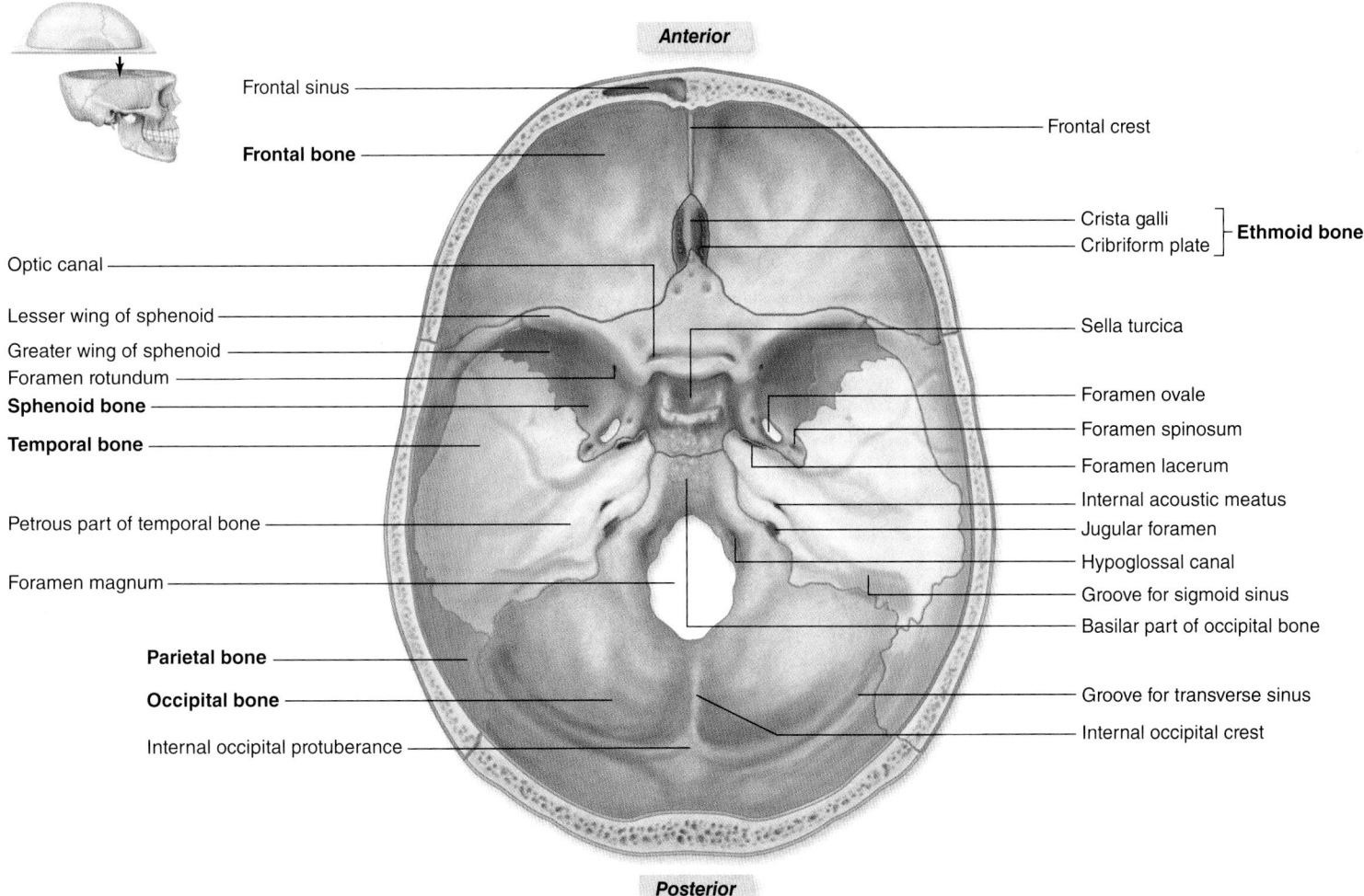

Figure 8.8 Internal View of the Cranial Base. In this transverse section, the internal portions of the frontal, ethmoid, sphenoid, temporal, and occipital bones are prominent. AP|R

Table 8.2 Cranial Bones and Selected Features[1]

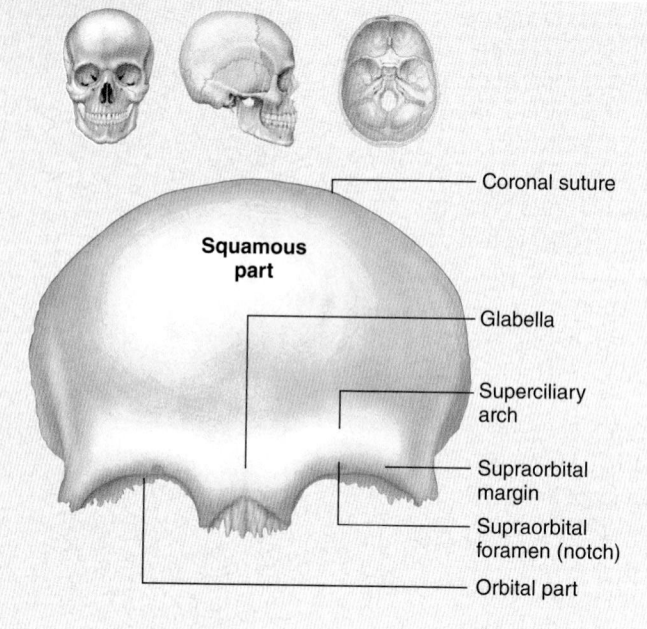

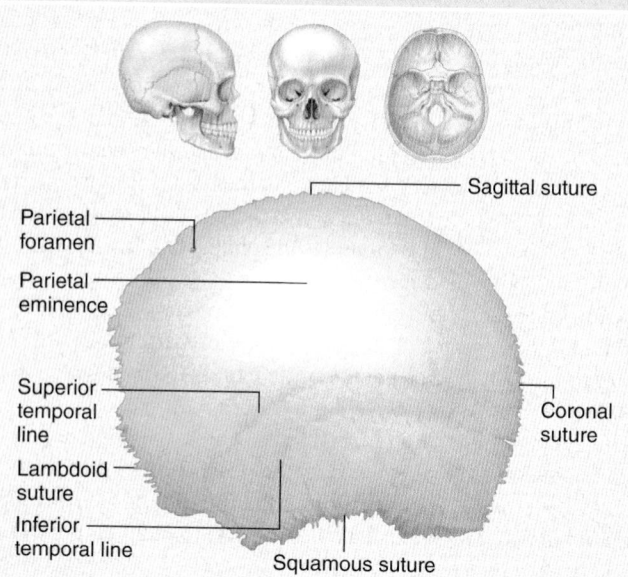

(a) Frontal Bone AP|R		**(b) Parietal Bone** AP|R	
Associated Passageways	Supraorbital foramen or notch	*Associated Passageways*	Parietal foramen
Description and Boundaries of Bone	Forms the superior and anterior parts of the skull; part of anterior cranial fossa and orbit	*Description and Boundaries of Bone*	Each forms most of lateral and superior walls of the skull
Selected Features and Their Functions	**Frontal crest:** Attachment site for meninges to help stabilize brain within the skull **Frontal sinuses:** Lighten bone, moisten inhaled air, and give resonance to voice **Orbital part:** Forms roof of orbit **Squamous part:** Attachment of scalp muscles **Supraorbital margin:** Forms protective superior border of orbit	*Selected Features and Their Functions*	**Superior** and **inferior temporal lines:** Attachment sites for temporalis muscle **Parietal eminence:** Forms rounded prominence on each side of the skull

1. Note: Not all features may be shown in this table; please refer to figures 8.4–8.8 in this chapter.

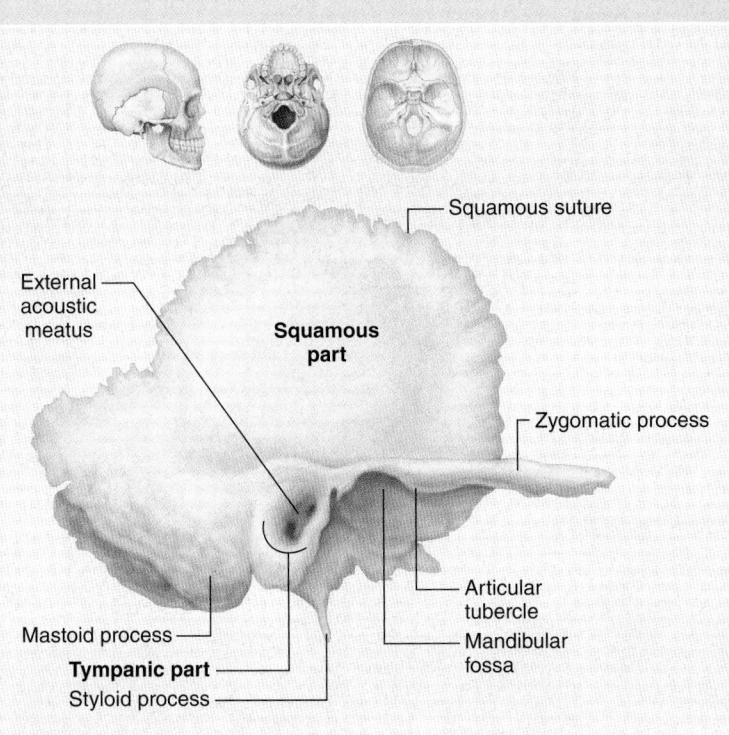

External acoustic meatus

Squamous suture

Squamous part

Zygomatic process

Articular tubercle

Mandibular fossa

Mastoid process

Tympanic part

Styloid process

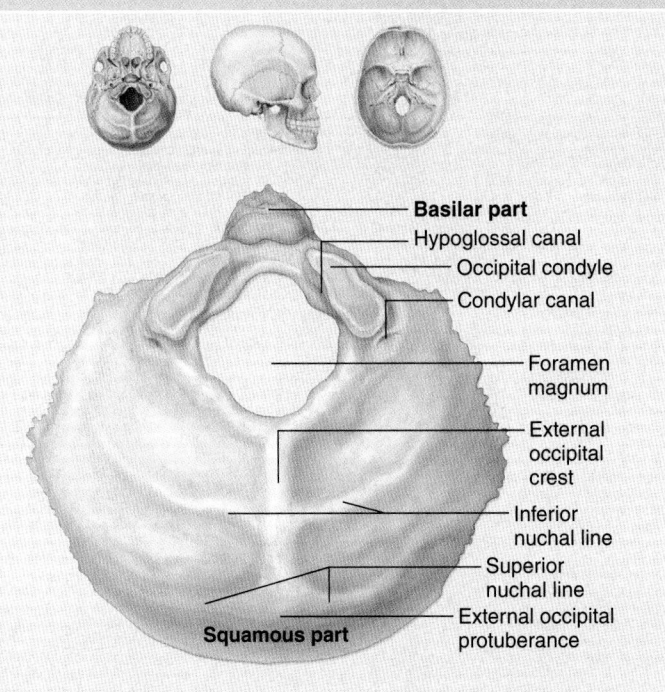

Basilar part

Hypoglossal canal

Occipital condyle

Condylar canal

Foramen magnum

External occipital crest

Inferior nuchal line

Superior nuchal line

External occipital protuberance

Squamous part

(c) Temporal Bone AP|R		(d) Occipital Bone AP|R	
Associated Passageways	Stylomastoid foramen Carotid canal External acoustic meatus Internal acoustic meatus Mastoid foramen Jugular foramen (with occipital bone)	*Associated Passageways*	Foramen magnum Hypoglossal canal Jugular foramen (with temporal bone) Condylar canal
Description and Boundaries of Bone	Each forms inferolateral wall of the skull; forms part of middle cranial fossa. Three parts are included below.	*Description and Boundaries of Bone*	Forms posteroinferior part of the skull, including most of posterior cranial fossa; forms part of base of the skull
Selected Features and Their Functions	**Petrous part:** Protects sensory structures in inner ear **Squamous part:** Attachment site of some jaw muscles **Tympanic part:** Houses external acoustic meatus **Mastoid process:** Attachment site of some neck muscles to extend or rotate head **Styloid process:** Attachment site for hyoid bone ligaments and muscles **Zygomatic process:** Articulates with zygomatic bone to form zygomatic arch **Mandibular fossa:** Articulates with mandible **Articular tubercle:** Limits displacement of head of mandible within mandibular fossa	*Selected Features and Their Functions*	**External occipital crest:** Attachment site for ligaments **External occipital protuberance:** Attachment site for neck ligaments and muscles **Inferior** and **superior nuchal lines:** Attachment sites for neck ligaments and muscles **Occipital condyles:** Articulate with first cervical vertebra

(continued on next page)

Table 8.2 Cranial Bones and Selected Features *(continued)*

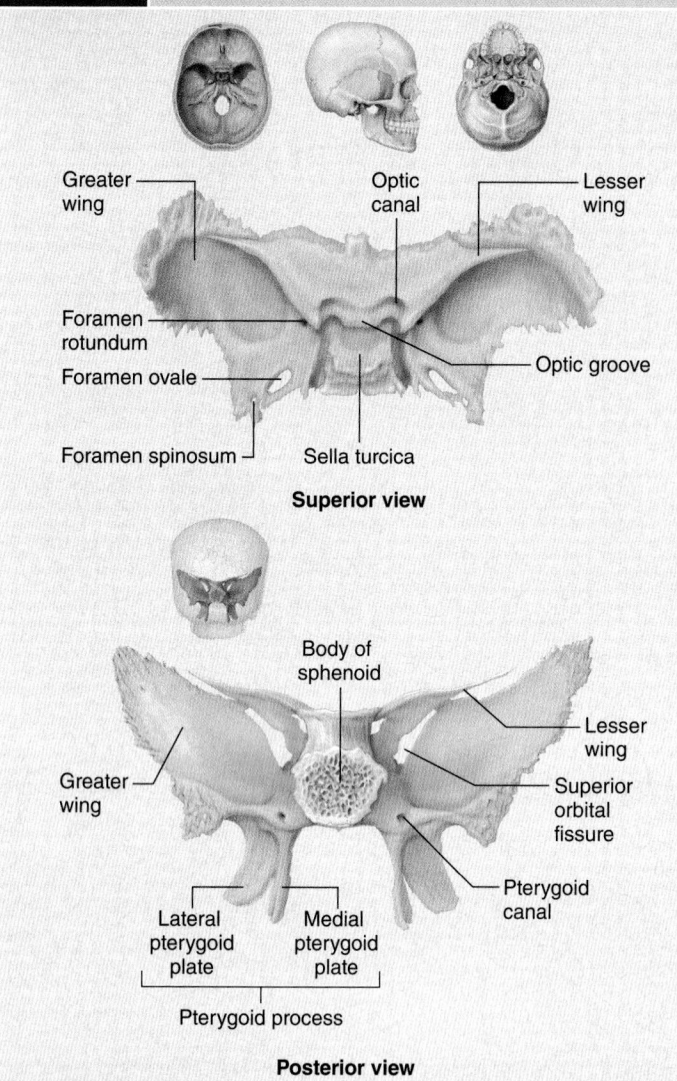

Superior view

Greater wing
Optic canal
Lesser wing
Foramen rotundum
Foramen ovale
Optic groove
Foramen spinosum
Sella turcica

Posterior view

Body of sphenoid
Lesser wing
Greater wing
Superior orbital fissure
Lateral pterygoid plate
Medial pterygoid plate
Pterygoid canal
Pterygoid process

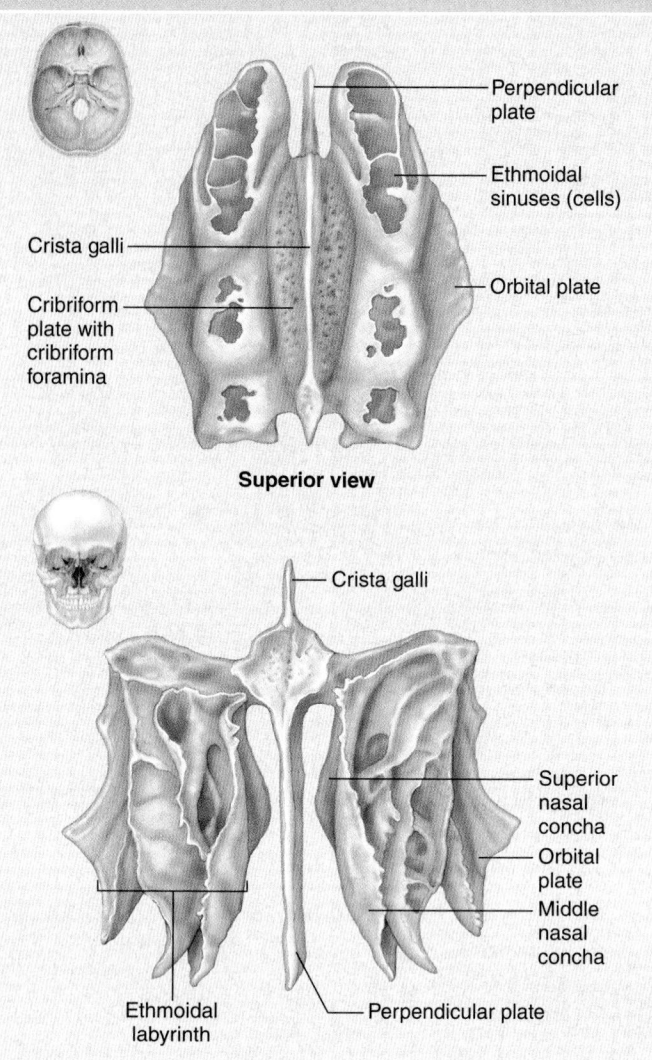

Superior view

Perpendicular plate
Ethmoidal sinuses (cells)
Crista galli
Cribriform plate with cribriform foramina
Orbital plate

Anterior view

Crista galli
Superior nasal concha
Orbital plate
Middle nasal concha
Ethmoidal labyrinth
Perpendicular plate

(e) Sphenoid Bone AP|R		(f) Ethmoid Bone AP|R	
Associated Passageways	Foramen lacerum (with temporal and occipital bones) Foramen ovale Foramen rotundum Foramen spinosum Optic canal Pterygoid canal Superior orbital fissure	*Associated Passageways*	Cribriform foramina
Description and Boundaries of Bone	Forms part of the base of the skull; posterior part of orbit; part of anterior and middle cranial fossae	*Description and Boundaries of Bone*	Forms part of anterior cranial fossa; part of nasal septum, roof and lateral walls of nasal cavity; part of medial wall of orbit
Selected Features and Their Functions	**Body:** Houses sphenoidal sinuses **Sella turcica:** Houses pituitary gland **Optic groove:** Depression on body between the optic canals **Medial** and **lateral pterygoid plates:** Attachment sites for chewing muscles **Lesser wings:** Form part of anterior cranial fossa; contain optic canals **Greater wings:** Form part of middle cranial fossa, lateral surface of skull, and orbits **Sphenoidal sinuses:** Lighten bone, moisten inhaled air, and give resonance to voice	*Selected Features and Their Functions*	**Cribriform plate:** Contains cribriform foramina for passageway of olfactory nerves (CN I) **Crista galli:** Attachment site for cranial dural septa to help stabilize brain within the skull **Ethmoidal labyrinths:** Contain ethmoidal sinuses and nasal conchae **Ethmoidal sinuses (cells):** Lighten bone, moisten inhaled air, and give resonance to voice **Nasal conchae (superior** and **middle):** Increase airflow turbulence through nasal cavity so air can be adequately moistened and cleaned by nasal mucosa **Orbital plate:** Forms part of medial wall of orbit **Perpendicular plate:** Forms superior part of nasal septum

Table 8.3 Facial Bones and Selected Features[1]

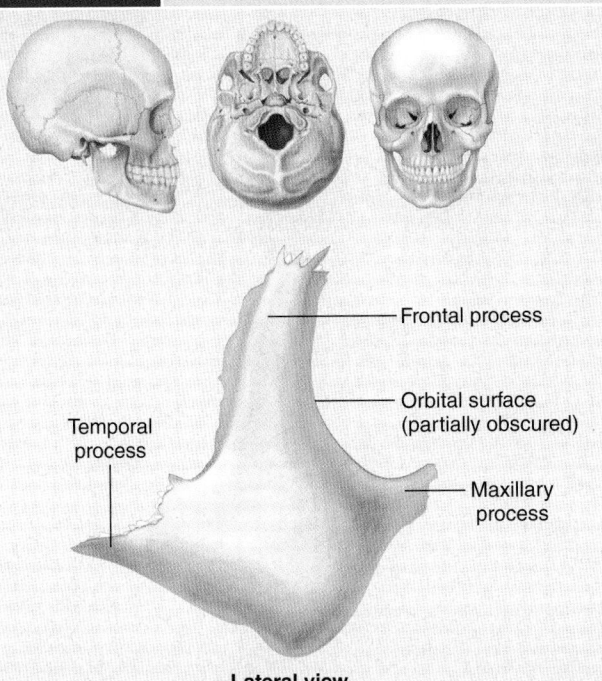

Temporal process

Frontal process

Orbital surface (partially obscured)

Maxillary process

Lateral view

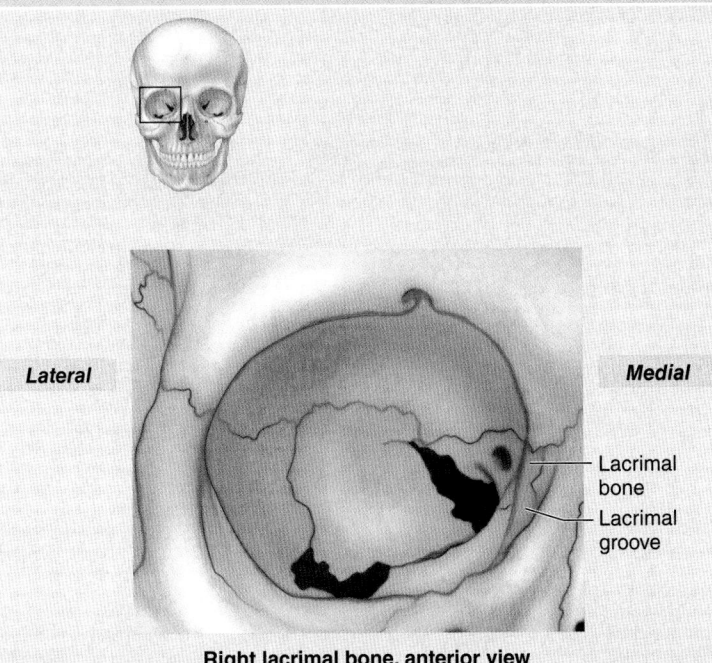

Lateral

Medial

Lacrimal bone

Lacrimal groove

Right lacrimal bone, anterior view

(a) Zygomatic Bone

Associated Passageways	none
Description and Boundaries of Bone	Each forms a cheek and lateral part of the orbit
Selected Features and Their Functions	**Frontal process:** Articulates with frontal bone **Temporal process:** Articulates with temporal bone to form zygomatic arch **Maxillary process:** Articulates with maxilla

(b) Lacrimal Bone

Associated Passageways	Lacrimal groove (see also figure 8.6)
Description and Boundaries of Bone	Each forms part of the medial wall of the orbit
Selected Features and Their Functions	**Lacrimal groove:** Contains nasolacrimal duct

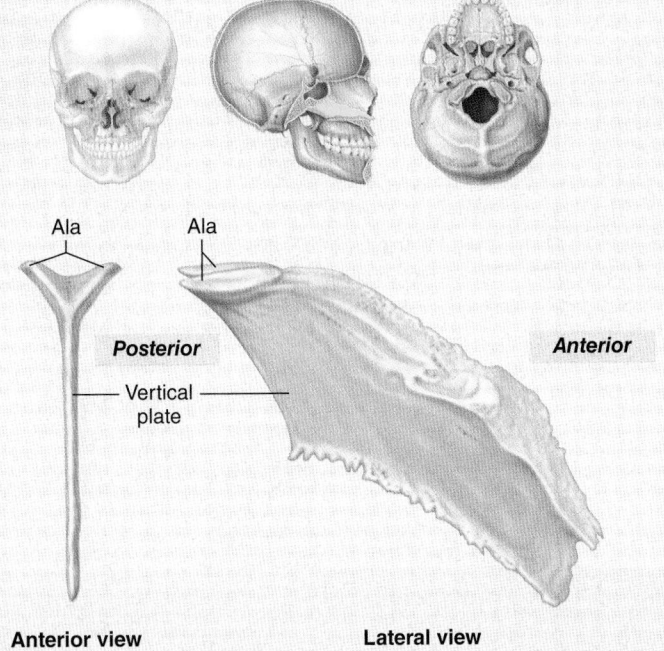

Ala

Ala

Posterior

Anterior

Vertical plate

Anterior view **Lateral view**

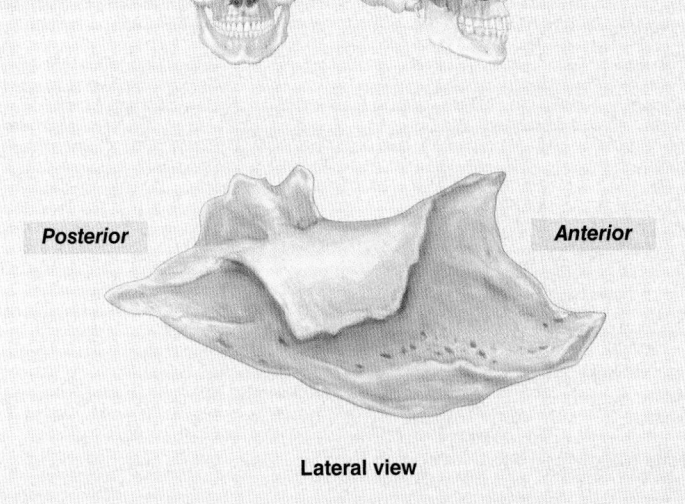

Posterior

Anterior

Lateral view

(c) Vomer

Associated Passageways	none
Description and Boundaries of Bone	Forms inferior and posterior part of nasal septum
Selected Features and Their Functions	**Ala:** Articulates with sphenoid bone **Vertical plate:** Articulates with perpendicular plate of ethmoid bone

(d) Inferior Nasal Concha

Associated Passageways	none
Description and Boundaries of Bone	Curved bone that projects medially from lateral walls of the nasal cavity
Selected Features and Their Functions	Increase airflow turbulence in nasal cavity

1. Nasal bones are not in the table and are shown on figures 8.4 and 8.6.

(continued on next page)

Table 8.3   Facial Bones and Selected Features *(continued)*

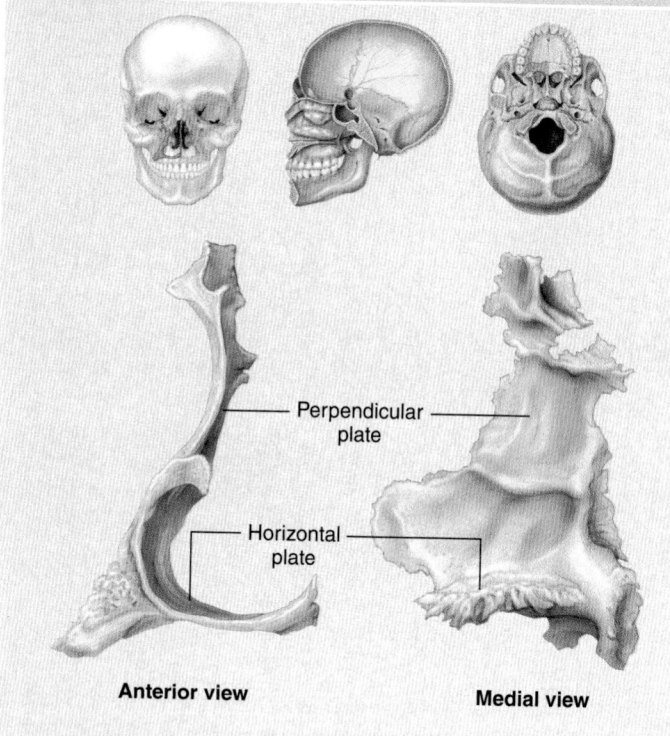

Anterior view · Medial view

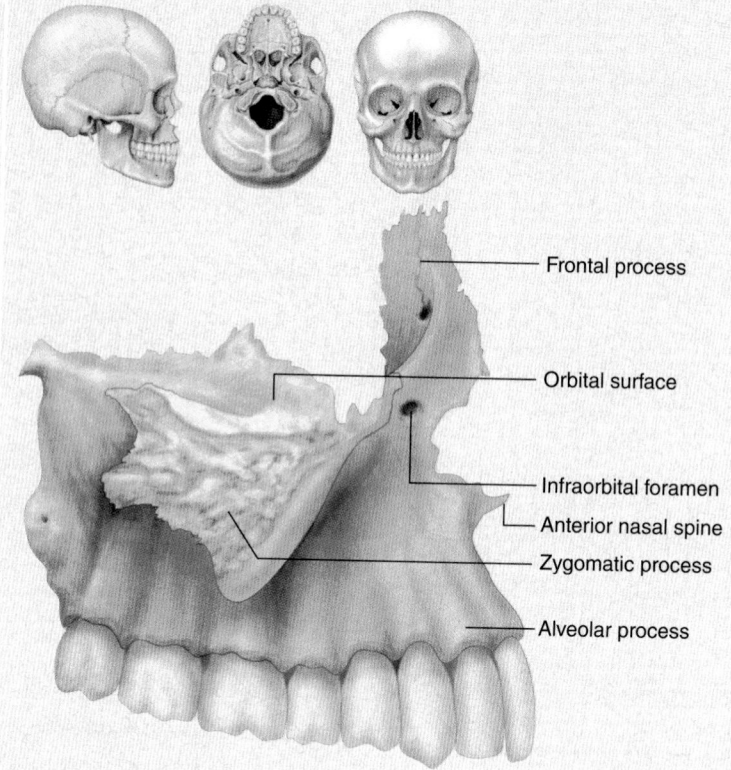

Frontal process

Orbital surface

Infraorbital foramen

Anterior nasal spine

Zygomatic process

Alveolar process

Perpendicular plate

Horizontal plate

Right maxilla, lateral view

(e) Palatine Bone		**(f) Maxilla** AP|R	
Associated Passageways	Greater and lesser palatine foramina	*Associated Passageways*	Incisive foramen
			Infraorbital foramen
Description and Boundaries of Bone	Each forms posterior part of hard palate; also forms small part of nasal cavity and orbit wall	*Description and Boundaries of Bone*	Each forms anterior portion of face; upper jaw and parts of hard palate, inferior parts of orbits, and part of the walls of nasal cavity
Selected Features and Their Functions[2]	**Horizontal plate:** Forms posterior part of hard palate **Perpendicular plate:** Forms part of nasal cavity and orbit	*Selected Features and Their Functions[2]*	**Anterior nasal spine:** Anterior projection formed by union of left and right maxillae **Alveolar process:** Houses the teeth **Frontal process:** Forms part of lateral aspect of the nasal bridge **Infraorbital margin:** Forms inferolateral border of orbit **Maxillary sinus:** Lightens bone **Orbital surface:** Forms part of orbit **Palatine process:** Forms most of bony palate **Zygomatic process:** Articulates with zygomatic bone

2. Note: Not all features listed in the table may be shown in above art; please refer to the other images in this chapter.

Mandibular fossa of temporal bone

Temporomandibular joint

Head of mandible

Coronoid process

Mandibular notch

Mandibular foramen

Condylar process

Ramus

Mylohyoid line

Alveolar process

Mental foramen

Body

Mental protuberance

Angle of mandible

Lateral view

(g) Mandible	AP R	
Associated Passageways	Mandibular foramen Mental foramen
Description and Boundaries of Bone	Forms the lower jaw
Selected Features and Their Functions[2]	**Alveolar process:** Houses the teeth **Angle of the mandible:** Junction between the body and ramus **Body:** Horizontal portion of mandible **Condylar** (kon′di-lăr) **process:** Posterior projection off ramus; contains head of mandible **Coronoid** (kōr′ŏ-noyd; *korone* = a crow) **process:** Anterior projection off ramus **Head of mandible:** Articulates with temporal bone **Mandibular notch:** U-shaped depression between coronoid and condylar processes **Mental protuberance:** Forms the chin **Mylohyoid** (mī′lō-hī-oyd) **line:** Attachment site for mylohyoid muscle **Ramus** (rā′mŭs): Vertical portion of mandible

2. Note: Not all features listed in the table may be shown in above art; please refer to the other images in this chapter.

Articulations of selected cranial and facial bones that are ob-scured by more superficial bones are shown in **figure 8.9.**

INTEGRATE

CLINICAL VIEW
Cleft Lip and Palate

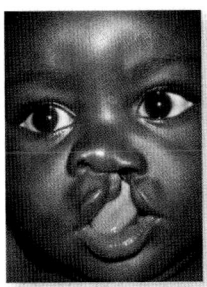

Cleft lip

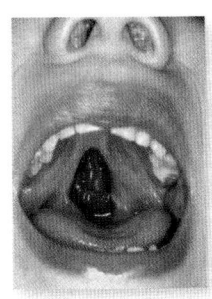

Cleft palate

A **cleft lip** is the incomplete fusion of upper jaw components of the developing embryo, resulting in a split upper lip extending from the mouth to the side of one nostril. Cleft lip appears in 1 per 1000 births and tends to be more common in males. The etiology of cleft lip is multifactorial, in that both genetic and environmental factors (such as cigarette smoking or alcohol ingestion during pregnancy) appear to contribute to the condition.

Another anomaly that can develop is **cleft palate**, a congenital fissure in the midline of the palate. A cleft palate results when the left and right maxillary and palatine bones fuse incompletely or do not fuse at all. In the more severe cases, children have swallowing and feeding problems because food can easily travel from the oral cavity into the nasal cavity. Cleft palate occurs in about 1 per 2500 births and tends to be more common in females. Like cleft lip, the etiology of cleft palate is multifactorial. Cleft palate sometimes occurs in conjunction with cleft lip.

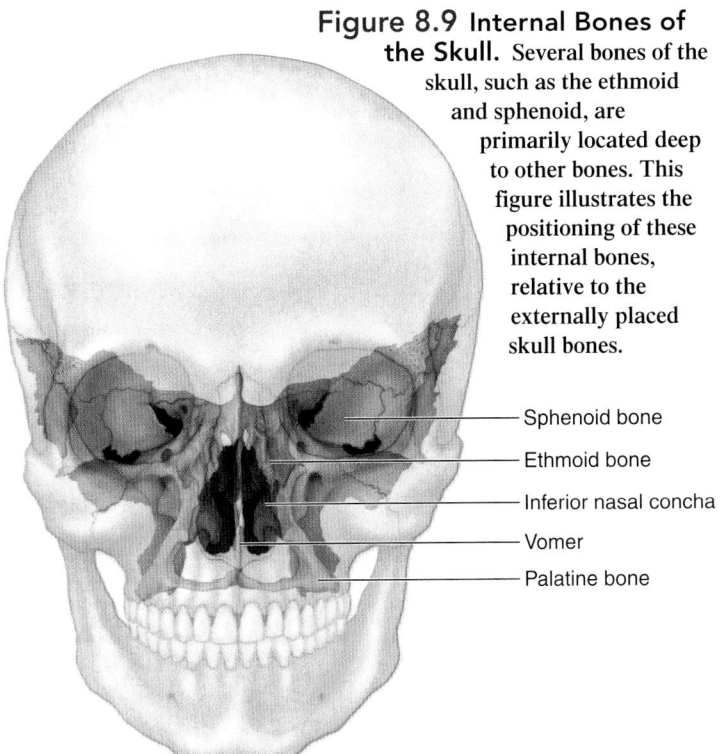

Figure 8.9 Internal Bones of the Skull. Several bones of the skull, such as the ethmoid and sphenoid, are primarily located deep to other bones. This figure illustrates the positioning of these internal bones, relative to the externally placed skull bones.

Sphenoid bone

Ethmoid bone

Inferior nasal concha

Vomer

Palatine bone

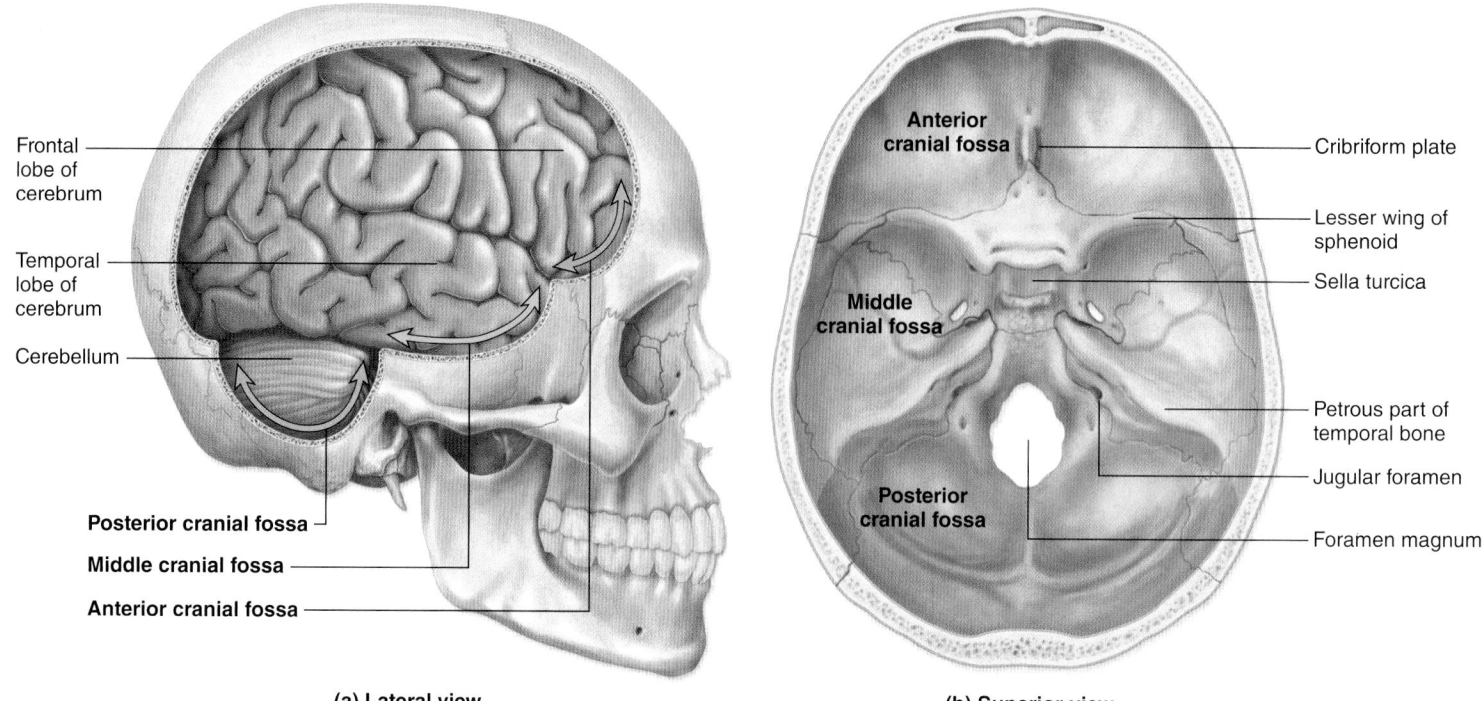

Frontal lobe of cerebrum

Temporal lobe of cerebrum

Cerebellum

Posterior cranial fossa

Middle cranial fossa

Anterior cranial fossa

(a) Lateral view

Anterior cranial fossa

Middle cranial fossa

Posterior cranial fossa

Cribriform plate

Lesser wing of sphenoid

Sella turcica

Petrous part of temporal bone

Jugular foramen

Foramen magnum

(b) Superior view

Figure 8.10 Cranial Fossae. (*a*) Lateral and (*b*) superior views show the three levels of depression in the cranium (anterior, middle, and posterior cranial fossae) that parallel the contours of the ventral surface of the brain.

Cranial Fossae   The contoured floor of the cranial cavity exhibits three curved depressions called the **cranial fossae** (figure 8.10).

The **anterior cranial fossa** is the shallowest of the three depressions. It is formed by the frontal bone, the ethmoid bone, and the lesser wings of the sphenoid bone. The anterior cranial fossa houses the frontal lobes of the brain (see section 13.3b).

The **middle cranial fossa** is inferior and posterior to the anterior cranial fossa. It ranges from the posterior edge of the lesser wings of the sphenoid bone (anteriorly) to the anterior region of the petrous part of the temporal bone (posteriorly). It houses the temporal lobes of the brain and the pituitary gland.

The **posterior cranial fossa** is the most inferior and posterior cranial fossa and extends from the posterior region of the petrous part of the temporal bones to the occipital bone. This fossa supports part of the brainstem and the cerebellum (see sections 13.5 and 13.6).

WHAT DID YOU LEARN?

4 What bones may be prominently seen in an anterior view of the skull?

5 What bones form the middle cranial fossa, and which part of the brain resides in this fossa?

8.2c Sutures

LEARNING OBJECTIVE

5. Describe the locations of the sutures between the cranial bones.

Sutures (sū′chūr; *sutura* = a seam) are immovable joints that form the boundaries between the cranial bones (see figures 8.5–8.7). Dense regular connective tissue connects cranial bones firmly together at a suture. The sutures often have intricate interlocking forms, like puzzle pieces.

Numerous sutures are present in the skull, each with a specific name. Many of the smaller sutures are named for the bones or features they interconnect. For example, the occipitomastoid suture connects the occipital bone with the portion of the temporal bone that houses the mastoid process. Here we discuss only the largest sutures: the coronal, lambdoid, sagittal, and squamous sutures:

- The **coronal** (kō-rō′nal; *coron* = crown) **suture** extends laterally across the superior surface of the skull along a coronal plane. It represents the articulation between the anterior frontal bone and the more posterior parietal bones.
- The **lambdoid** (lam′doyd) **suture** extends like an arc across the posterior surface of the skull, articulating with the parietal bones and the occipital bone. It is named for the Greek letter "lambda," which its shape resembles.
- The **sagittal** (saj′i-tăl; *sagitta* = arrow) **suture** extends between the coronal and lambdoid sutures along the midsagittal plane. It articulates the right and left parietal bones.
- A **squamous** (skwā′mus) **suture** (or *squamosal suture*) on each side of the skull articulates the temporal bone and the parietal bone of that side. The squamous part of the temporal bone typically overlaps the parietal bone.

One common variation in sutures is the presence of **sutural bones** (*Wormian bones*) (see figure 8.5*b*). Sutural bones may range in size from a tiny pebble to a quarter, but they can be much larger. Sutural bones represent independent bone ossification centers and are most common and numerous in the lambdoid suture.

In adulthood, the sutures typically are obliterated (closed) as the adjoining bones fuse. This fusion starts internally and is followed by fusion on the skull's external surface. Although the timing of suture closure can be highly variable, the coronal suture typically is the first to fuse, usually in the late 20s to early 30s, followed by the sagittal suture (usually in the 30s or later) and then the lambdoid suture (usually in the 40s). The squamous suture usually does not fuse until late adulthood (60-plus years), or it may not fuse at all. Osteologists can estimate the approximate age at death of an individual by examining the extent of suture closure in the skull.

WHAT DID YOU LEARN?

6 What bones articulate at the lambdoid suture? When does this suture typically fuse?

INTEGRATE

CLINICAL VIEW

Craniosynostosis and Plagiocephaly

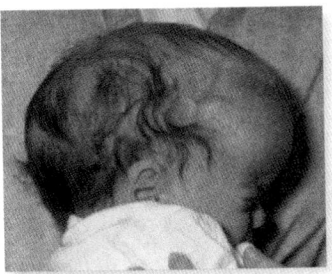

Sagittal synostosis

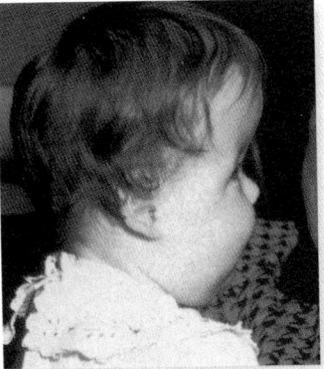

Coronal synostosis

When cranial growth has stopped in adulthood, the sutures gradually fuse and are obliterated. **Craniosynostosis** (krā′nē-ō-sin′os-tō′sis) refers to the premature fusion or closing of one or more of these cranial sutures, and may be due to multiple causes. If this premature fusion occurs early in life or *in utero*, skull shape is dramatically affected. If not surgically treated, a craniosynostotic individual typically grows up with an unusual craniofacial shape. For example, if the sagittal suture fuses prematurely (a condition called **sagittal synostosis**), the skull cannot grow and expand laterally as the brain grows, and compensatory skull growth occurs in an anterior-posterior fashion. A child with sagittal synostosis develops a very elongated, narrow skull shape, a condition called **scaphocephaly**, or *dolichocephaly*. **Coronal synostosis** refers to premature fusion of the coronal suture, which causes the skull to be abnormally short and wide.

Plagiocephaly is the term used to describe an asymmetric head shape, where one part of the skull (usually the frontal or occipital region) has an oblique flattening. Plagiocephaly may be caused by unilateral coronal cranio-synostosis or asymmetric lambdoid synostosis. It also is commonly caused by normal deformational factors, such as consistently sleeping on the same side of the head. Incidence of plagio-cephaly has risen in the United States since the 1990s, primarily due to the National Institute of Child Health and

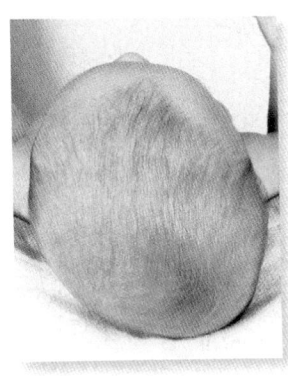

Plagiocephaly

Human Development Safe to Sleep Campaign (formerly called the Back to Sleep Campaign), which encourages parents to place children on their backs to sleep (instead of on their stomachs) so as to reduce the incidence of SIDS. Mild forms of plagiocephaly may be corrected by wearing a corrective helmet; more severe forms may necessitate surgery.

8.2d Orbital and Nasal Complexes, Paranasal Sinuses

LEARNING OBJECTIVES

6. List the bones that form the orbital and nasal complexes.
7. Describe the location and function of the paranasal sinuses.

The bony cavities called orbits enclose and protect the eyes and the muscles that move them. The **orbital complex** consists of multiple bones that form each orbit. The borders of the orbital complex are shown and listed in **figure 8.11**.

The **nasal complex** is composed of bones and cartilage that enclose the nasal cavity and the paranasal sinuses. Most of these bones are best seen in sagittal section, as shown in **figure 8.12**.

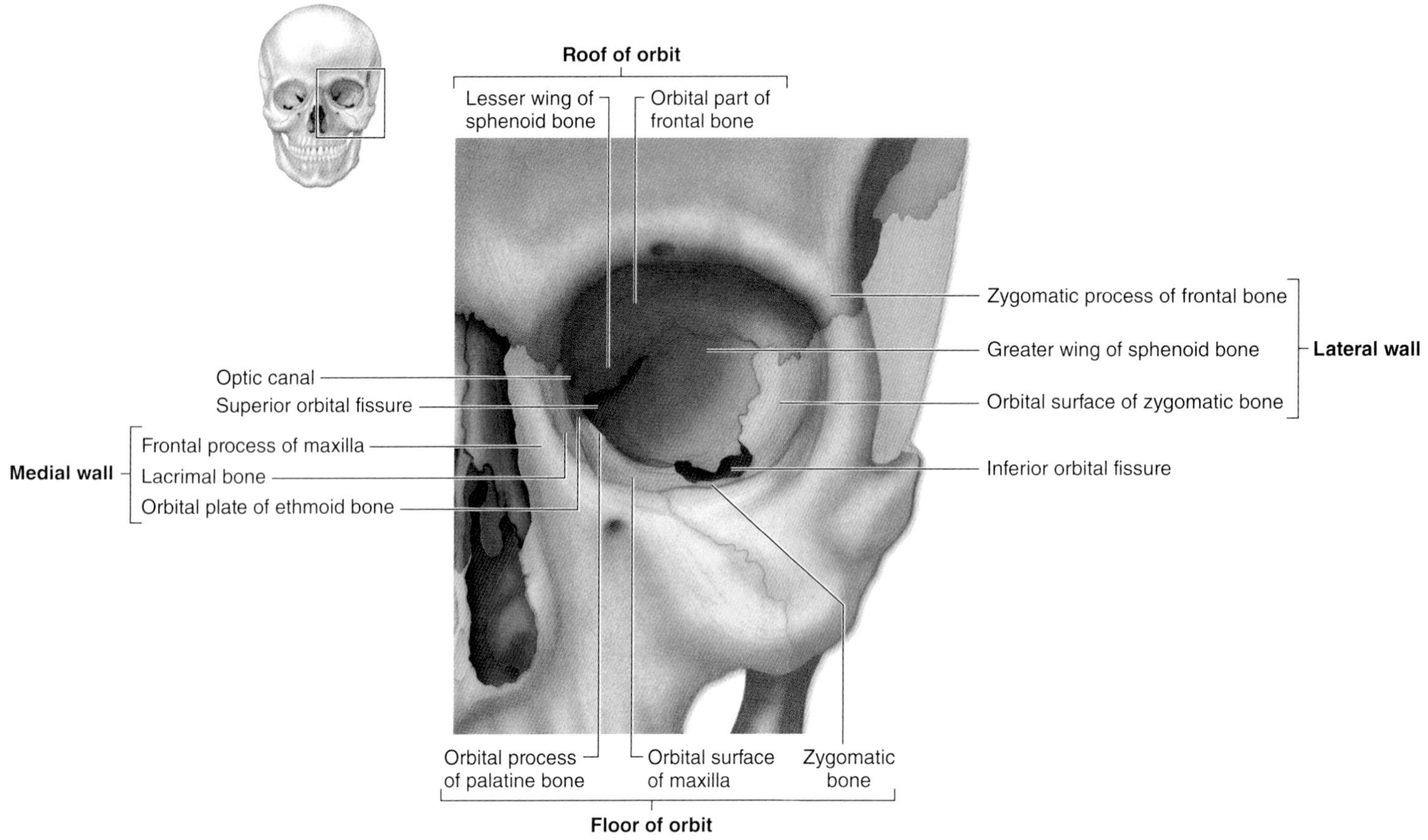

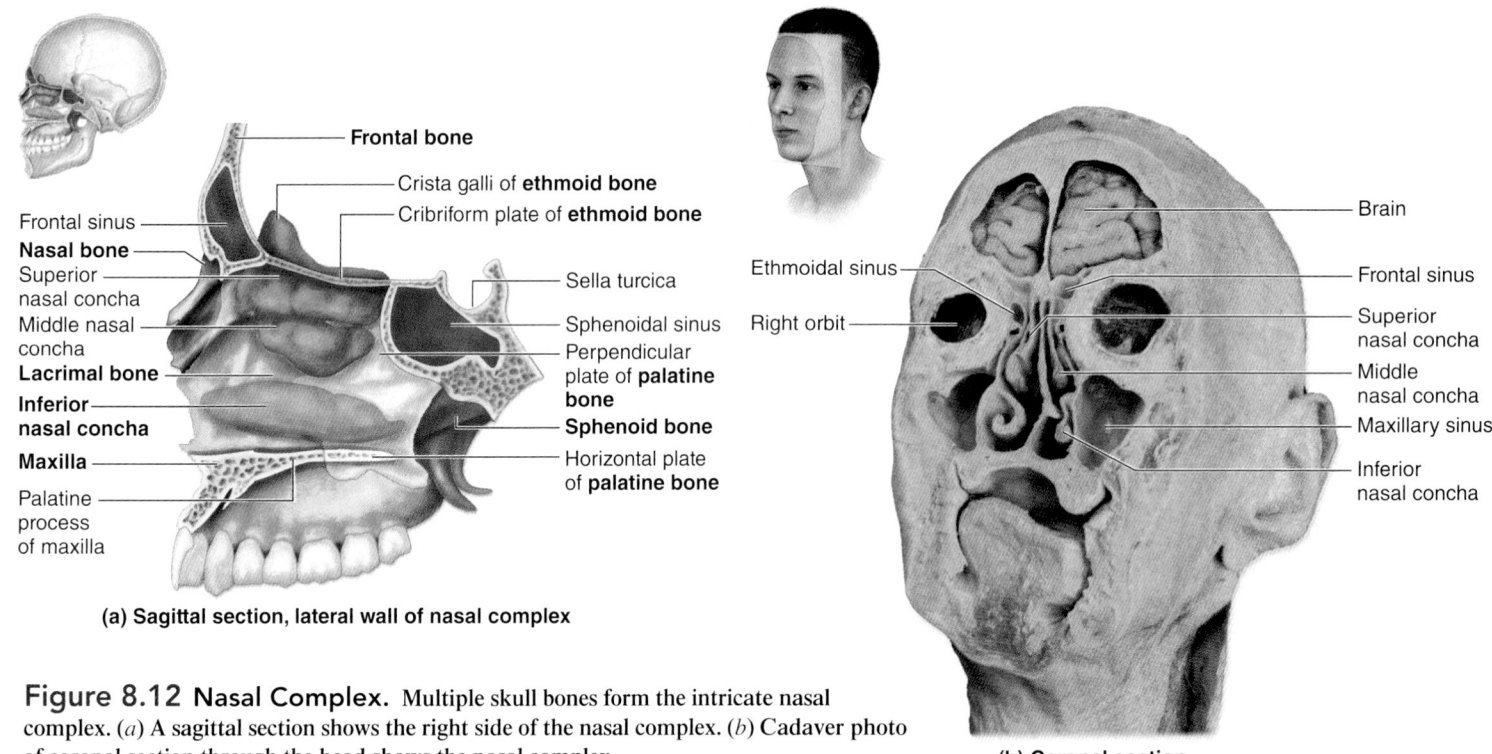

Roof of orbit

Lesser wing of sphenoid bone — Orbital part of frontal bone

Zygomatic process of frontal bone
Greater wing of sphenoid bone — **Lateral wall**
Orbital surface of zygomatic bone

Optic canal
Superior orbital fissure

Inferior orbital fissure

Medial wall — Frontal process of maxilla
Lacrimal bone
Orbital plate of ethmoid bone

Orbital process of palatine bone — Orbital surface of maxilla — Zygomatic bone

Floor of orbit

Figure 8.11 Left Orbit. Several bones compose the orbit of the eye and collectively form the orbital complex. **AP|R**

Frontal bone

Crista galli of **ethmoid bone**
Cribriform plate of **ethmoid bone**

Frontal sinus
Nasal bone
Superior nasal concha
Middle nasal concha
Lacrimal bone
Inferior nasal concha
Maxilla
Palatine process of maxilla

Sella turcica
Sphenoidal sinus
Perpendicular plate of **palatine bone**
Sphenoid bone
Horizontal plate of **palatine bone**

(a) Sagittal section, lateral wall of nasal complex

Ethmoidal sinus
Right orbit

Brain
Frontal sinus
Superior nasal concha
Middle nasal concha
Maxillary sinus
Inferior nasal concha

(b) Coronal section

Figure 8.12 Nasal Complex. Multiple skull bones form the intricate nasal complex. (*a*) A sagittal section shows the right side of the nasal complex. (*b*) Cadaver photo of coronal section through the head shows the nasal complex.

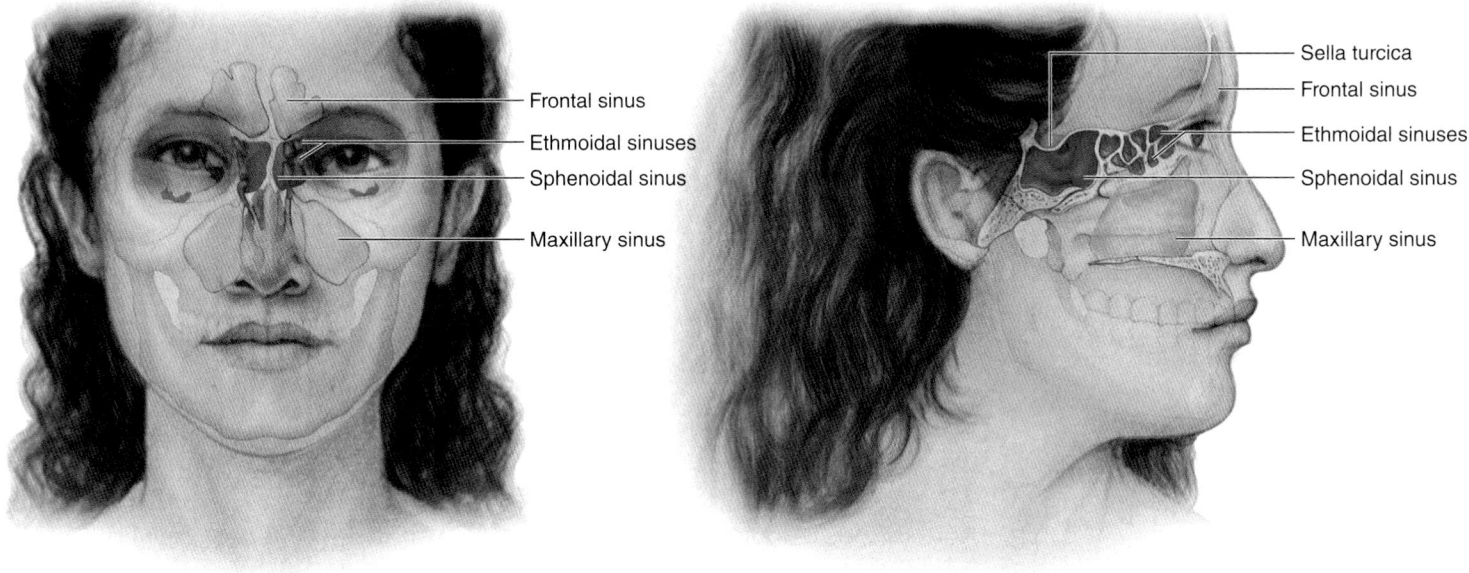

(a) Anterior view

Frontal sinus
Ethmoidal sinuses
Sphenoidal sinus
Maxillary sinus

(b) Lateral view

Sella turcica
Frontal sinus
Ethmoidal sinuses
Sphenoidal sinus
Maxillary sinus

Figure 8.13 Paranasal Sinuses. The paranasal sinuses are air-filled chambers within the frontal, ethmoid, and sphenoid bones and the maxillae. They are shown in (*a*) anterior and (*b*) lateral views. They are lined with a mucous membrane and act as extensions of the nasal cavity.

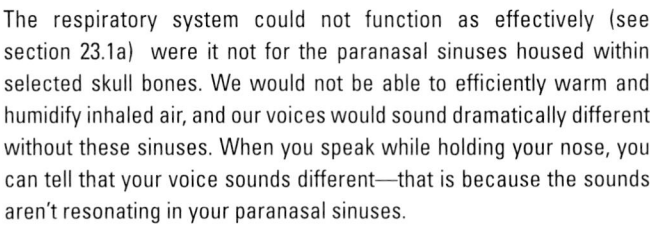

INTEGRATE

CONCEPT CONNECTION

The respiratory system could not function as effectively (see section 23.1a) were it not for the paranasal sinuses housed within selected skull bones. We would not be able to efficiently warm and humidify inhaled air, and our voices would sound dramatically different without these sinuses. When you speak while holding your nose, you can tell that your voice sounds different—that is because the sounds aren't resonating in your paranasal sinuses.

We have already described the ethmoidal, frontal, maxillary, and sphenoidal sinuses in connection with the bones where they are found. As a group, these air-filled chambers that open into the nasal cavities are called the **paranasal sinuses** (sī′nŭs; cavity, hollow) **(figure 8.13)**. The sinuses have a mucous membrane lining that helps to humidify and warm inhaled air. Additionally, the sinus spaces cause the skull bones in which they are located to be lighter, and also provide resonance to the voice.

 WHAT DID YOU LEARN?

7 What bones form the floor of the orbit?

8 In which four bones are the paranasal sinuses located?

8.3 Bones Associated with the Skull

 LEARNING OBJECTIVES

1. Locate and identify the auditory ossicles.
2. Describe the structure and function of the hyoid bone.

The auditory ossicles and the hyoid bone are bones of the axial skeleton associated with the skull. **Auditory ossicles** (os′i-kl) are three tiny

ear bones housed within the petrous part of each temporal bone. These bones—the **malleus** (mal′ē-us), the **incus** (ing′kŭs), and the **stapes** (stā′pēz)—are discussed in depth in section 16.5a.

The **hyoid bone** is a slender, curved bone located inferior to the skull between the mandible and the larynx (voice box) **(figure 8.14)**. It does not articulate with any other bone in the skeleton. The hyoid has a medial **body** and two paired hornlike processes, the **greater cornua** (kor′nū-ă = horn; sing. *cornu*, kōr′nū) and the **lesser cornua.** The cornua and body serve as attachment sites for tongue and larynx (voice box) muscles and ligaments.

 WHAT DID YOU LEARN?

9 What are the names of auditory ossicles and in which specific bone are they found?

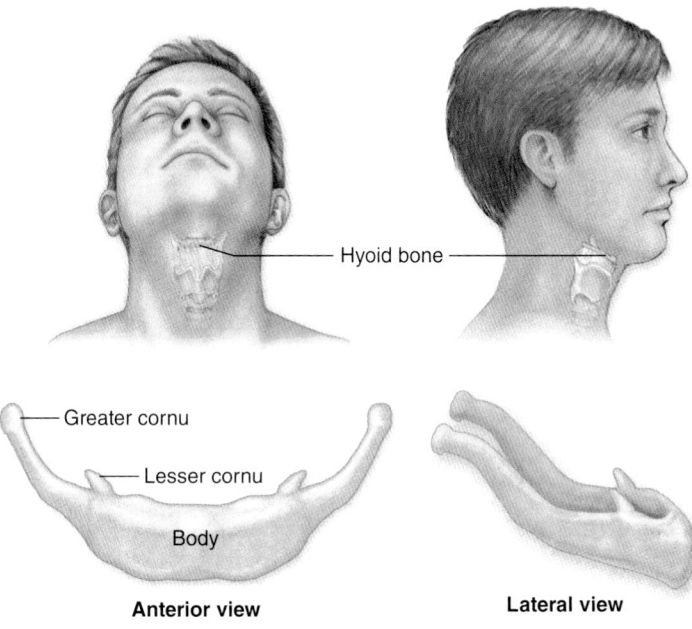

Hyoid bone

Greater cornu
Lesser cornu
Body

Anterior view   **Lateral view**

Figure 8.14 Hyoid Bone. The hyoid bone is inferior to the mandible and is not in direct contact with any other bone.

8.4 Sex and Age Determination from Analysis of the Skull

The skull can provide insight about the sex and age of an individual. We first describe some diagnostic features of the skull used to determine the sex of an individual. Then we compare how the skull changes through the fetal period, childhood, early adulthood, and old age.

8.4a Sex Differences in the Skull

LEARNING OBJECTIVE

1. Identify the similarities and differences between male and female skulls.

Human female and male skulls display some obvious differences in general shape and size, a phenomenon known as **sexual dimorphism.** Typical "female" features tend to be gracile (delicate, small), whereas "male" features tend to be more robust (larger, sturdier, bulkier). **Table 8.4** summarizes the general sex differences seen in the skull.

However, caution is required when a skull and other skeletal remains are used to determine an individual's sex. Both skeletons and skeletal features vary in their general size and robusticity among populations. For example, some male Asian skeletal remains may be less robust than those of a female Native American. Further, it often is difficult or impossible to determine the sex of infant and juvenile remains, because skull characteristics appear female-like until well after puberty.

The most accurate method of determining sex is to look at multiple skeletal features and make a judgment based on the majority of features present. For example, if a skull displays two femalelike characteristics and four malelike characteristics, the skull will likely be classified as male. If your anatomy lab uses real skulls, use table 8.4 to try to determine the sex of the skull you are studying.

WHAT DO YOU THINK?

2 It is difficult to determine the sex of a young child's skull because both male and female skulls at this stage of development appear femalelike. What factors do you think cause those features to change in males by adulthood?

WHAT DID YOU LEARN?

10 What are some features that differ between female and male skulls?

8.4b Aging of the Skull

LEARNING OBJECTIVES

2. Compare the structure of fetal, child, and adult skulls.

3. List the fontanelles and the ages at which they close.

The shape and structure of cranial elements differ between infants and adults, causing variations in their proportions and size. The most significant growth in the skull occurs before age 5, when the brain is still growing and exerting pressure against the developing skull bones' internal surface. Brain growth is 90–95% complete by age 5, at which time cranial bone growth is close to completion, and the cranial sutures are almost fully developed. Note that early in life the skull grows at a much faster rate than does the rest of the body. Thus, a young child's cranium is relatively larger compared to the rest of its body than that of an adult.

Table 8.4	Sex Differences in the Skull	

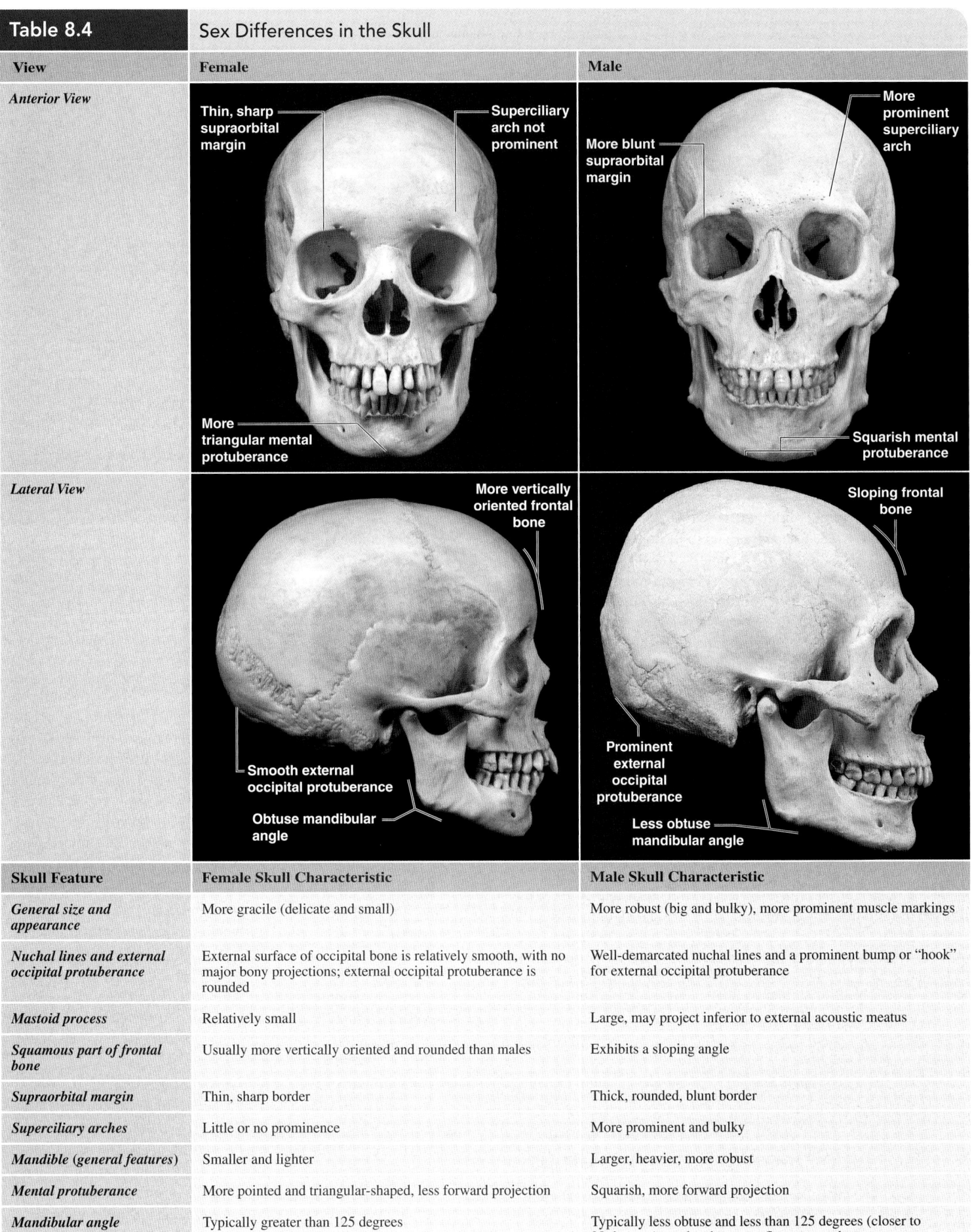

View	Female	Male
Anterior View	Thin, sharp supraorbital margin Superciliary arch not prominent More triangular mental protuberance	More prominent superciliary arch More blunt supraorbital margin Squarish mental protuberance
Lateral View	More vertically oriented frontal bone Smooth external occipital protuberance Obtuse mandibular angle	Sloping frontal bone Prominent external occipital protuberance Less obtuse mandibular angle

Skull Feature	Female Skull Characteristic	Male Skull Characteristic
General size and appearance	More gracile (delicate and small)	More robust (big and bulky), more prominent muscle markings
Nuchal lines and external occipital protuberance	External surface of occipital bone is relatively smooth, with no major bony projections; external occipital protuberance is rounded	Well-demarcated nuchal lines and a prominent bump or "hook" for external occipital protuberance
Mastoid process	Relatively small	Large, may project inferior to external acoustic meatus
Squamous part of frontal bone	Usually more vertically oriented and rounded than males	Exhibits a sloping angle
Supraorbital margin	Thin, sharp border	Thick, rounded, blunt border
Superciliary arches	Little or no prominence	More prominent and bulky
Mandible (general features)	Smaller and lighter	Larger, heavier, more robust
Mental protuberance	More pointed and triangular-shaped, less forward projection	Squarish, more forward projection
Mandibular angle	Typically greater than 125 degrees	Typically less obtuse and less than 125 degrees (closer to 90 degrees); angle edges may flare outward

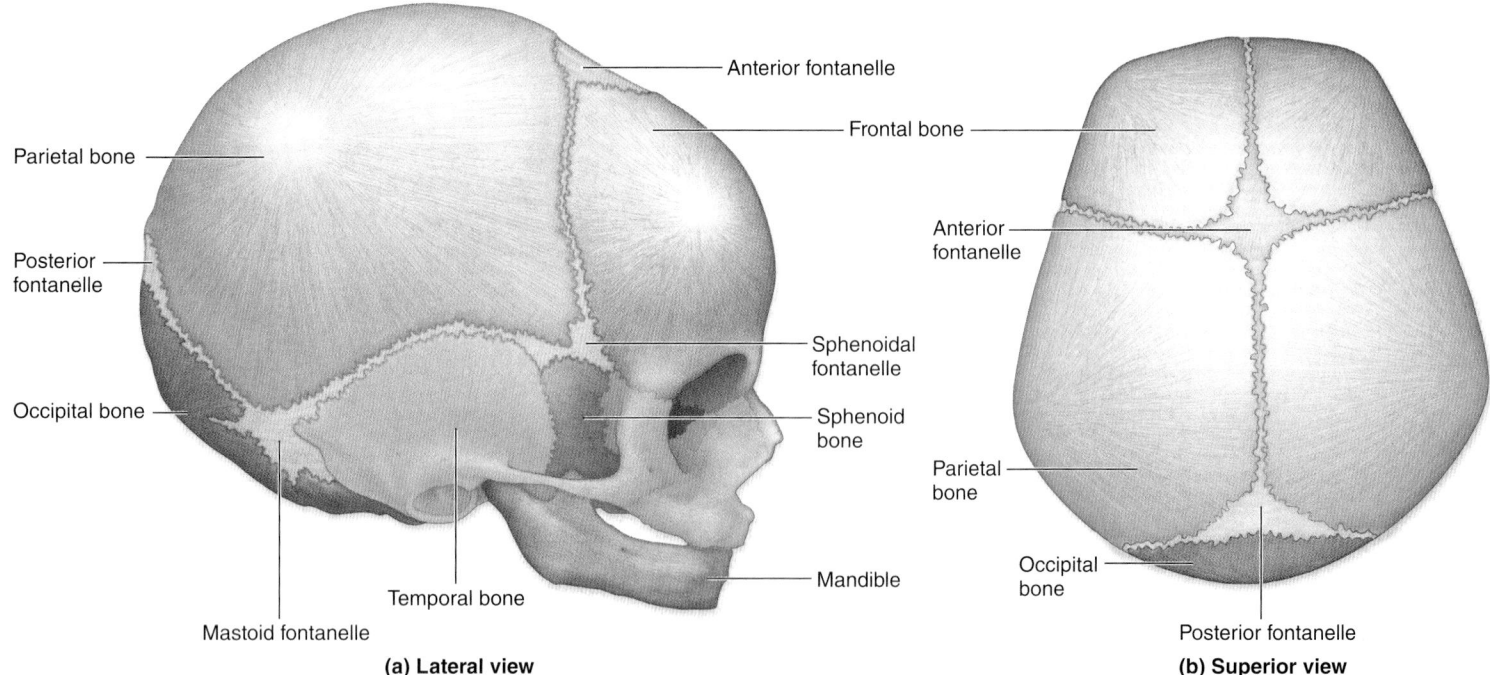

Parietal bone

Posterior fontanelle

Occipital bone

Mastoid fontanelle

Temporal bone

(a) Lateral view

Anterior fontanelle

Frontal bone

Sphenoidal fontanelle

Sphenoid bone

Mandible

Anterior fontanelle

Parietal bone

Occipital bone

Posterior fontanelle

(b) Superior view

Figure 8.15 Fetal Skull. (*a*) Lateral and (*b*) superior views show the flat bones in an infant skull, which are separated by fontanelles. These allow for the distortion of the skull during birth and the growth of the brain after birth.

A neonatal (infant) cranium is shown in lateral and superior views in **figure 8.15**. The infant's cranial bones are not yet large enough to surround the brain completely, so some cranial bones are interconnected by flexible areas of dense regular connective tissue membrane in regions called **fontanelles** (fon'tă-nel = little spring; sometimes spelled *fontanels*). Fontanelles are sometimes referred to as the "soft spots" on a baby's head. The fontanelles enable some flexion in the bony plates within the skull during birth, thus allowing the child's head to pass through the birth canal to ease the baby's passage (see section 29.6d). Newborns frequently have a "cone-shaped" head due to this temporary deformation, but the cranial bones usually return to their normal position by a few days after birth. Some fontanelles, such as the small **mastoid** and **sphenoidal fontanelles,** close relatively quickly after birth. However others are present until many months after birth, when skull bone growth finally starts to keep pace with brain growth. The **posterior fontanelle** normally closes around 9 months of age; the larger **anterior fontanelle** doesn't close until about 15 months of age.

The skull undergoes many more changes as we age. The maxillary sinus becomes more prominent after age 5, and by age 10 the frontal sinus is becoming well formed. Later, the cranial sutures start to fuse and ossify. As a person ages, the teeth start to wear down from use, a process called *dental attrition.* Finally, if an individual loses some or all of his or her teeth, the alveolar processes of the maxillae and mandible regress and eventually disappear.

 WHAT DID YOU LEARN?

 11 What are the two largest fontanelles, and when do they disappear?

8.5 Bones of the Vertebral Column

The adult **vertebral column** is composed of 26 bones, including 24 individual **vertebrae** (ver'tĕ-brā; sing., *vertebra,* ver'tĕ-bră) and the fused vertebrae that form both the sacrum and the coccyx. Each vertebra (except the first and the last) articulates with one superior vertebra and one inferior vertebra. Here we consider the vertebral column's general function and regions, the curves of the spine, anatomy of a generalized

vertebra, and anatomic details of the components of the five regions of the vertebral column.

8.5a Types of Vertebrae

 LEARNING OBJECTIVES

1. Describe the functions of the vertebral column.
2. List the five types of vertebrae.

The vertebral column provides vertical support for the body and supports the weight of the head. It helps maintain an upright body position. Most important, it houses and protects the delicate spinal cord.

The vertebral column is partitioned into five divisions or regions (**figure 8.16**). Vertebrae are identified by using a capital letter to denote their region, followed by a numerical subscript that indicates their sequence going from a superior to an inferior location:

- Seven **cervical** (ser'vĭ-kal; *cervix* = neck) **vertebrae** (designated C_1–C_7) form the bones of the neck (cervical region). The first cervical vertebra (C_1) articulates superiorly with the occipital condyles of the skull. The seventh cervical vertebra articulates inferiorly with the first thoracic vertebra.

- Twelve **thoracic vertebrae** (designated T_1–T_{12}) form the superior region of the back (thoracic region). Each thoracic vertebra articulates laterally with one or two pairs of ribs. The twelfth thoracic vertebra articulates inferiorly with the first lumbar vertebra.

- Five **lumbar vertebrae** (designated L_1–L_5) form the inferior concave region ("small") of the back (lumbar region). The fifth lumbar vertebra articulates inferiorly with the first sacral vertebra.

- The **sacrum** (sā'krŭm) is formed from five sacral vertebrae (designated S_1–S_5) that fuse into a single bony structure by the mid to late 20s. The sacrum articulates with L_5 superiorly, the first coccygeal vertebra inferiorly, and laterally with the two ossa coxae (hip bones).

- The **coccyx** (kok'siks) is commonly called the *tailbone* and is formed from four **coccygeal vertebrae** (designated Co_1–Co_4) that start to unite during puberty. The first coccygeal vertebra

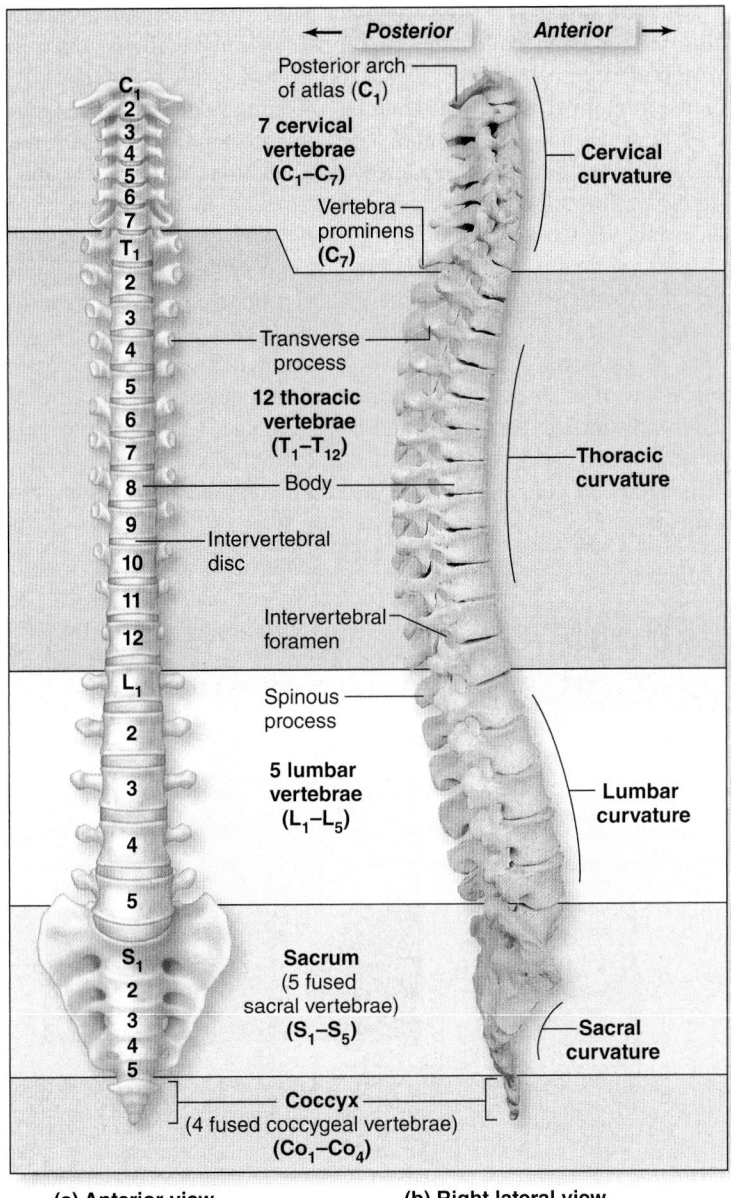

7 cervical
vertebrae
(C₁–C₇)

Posterior arch
of atlas (C₁)

Cervical
curvature

Vertebra
prominens
(C₇)

Transverse
process

12 thoracic
vertebrae
(T₁–T₁₂)

Body

Thoracic
curvature

Intervertebral
disc

Intervertebral
foramen

Spinous
process

5 lumbar
vertebrae
(L₁–L₅)

Lumbar
curvature

Sacrum
(5 fused
sacral vertebrae)
(S₁–S₅)

Sacral
curvature

Coccyx
(4 fused coccygeal vertebrae)
(Co₁–Co₄)

(a) Anterior view   (b) Right lateral view

Figure 8.16 Vertebral Column. (*a*) Anterior and (*b*) right lateral views show the types of vertebrae and curvatures in the vertebral column.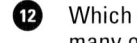

(Co₁) articulates with the inferior end of the sacrum. In much later years, the coccyx also may fuse to the sacrum.

WHAT DID YOU LEARN?

12 Which vertebrae are located in the "small" of the back, and how many of these vertebrae are there?

8.5b Spinal Curvatures

LEARNING OBJECTIVES

3. Name the four spinal curvatures of an adult vertebral column.

4. Explain the sequence of curvature development.

The vertebral column has some flexibility because it is not straight and rigid. When viewed from a lateral perspective, the adult vertebral column has four **spinal curvatures:** the **cervical, thoracic, lumbar,** and **sacral curvatures.** This arrangement better supports the weight of the body when standing than could a straight spine.

INTEGRATE

CLINICAL VIEW

Spinal Curvature Abnormalities

There are three main spinal curvature deformities: kyphosis, lordosis, and scoliosis.

Kyphosis (kī-fō'sis) is an exaggerated thoracic curvature that is directed posteriorly, producing a *hunchback* look. Kyphosis often results from osteoporosis, but also may occur due to a vertebral compression fracture, osteomalacia (a disease in which adult bones become demineralized), abnormal vertebral growth, or chronic contractions in muscles that insert on the vertebrae.

Lordosis (lōr-dō'sis) is an exaggerated lumbar curvature, often called *swayback,* that is seen as a protrusion of the abdomen and buttocks. Lordosis may have the same causes as kyphosis, or it may result from the added abdominal weight associated with pregnancy or obesity.

Scoliosis (skō-lē-ō'sis) is the most common spinal curvature deformity. It is an abnormal lateral curvature that sometimes results during development when both the vertebral arch and body fail to form, or form incompletely, on one side of a vertebra. It can also be caused by unilateral muscular paralysis, or spasm, in the back. Mild cases of scoliosis may be treated in adolescence by wearing a back brace, whereas more severe cases may require surgical intervention.

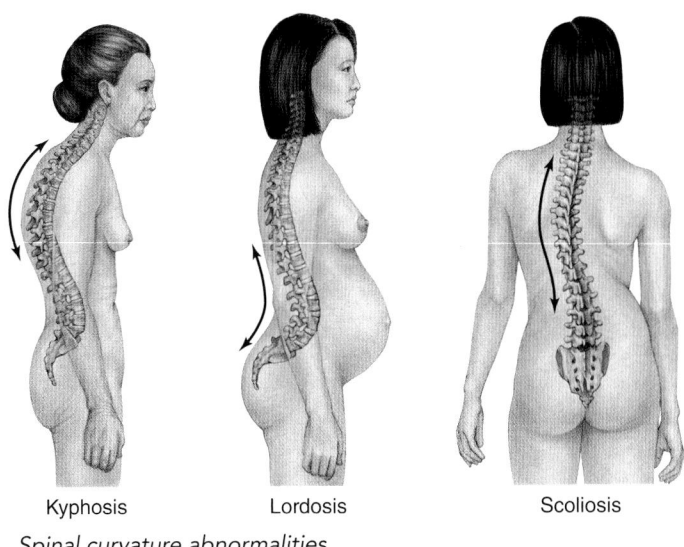

Kyphosis   Lordosis   Scoliosis

Spinal curvature abnormalities.

The spinal curvatures appear sequentially during fetal, newborn, and child developmental stages. The **primary curves** are the thoracic and sacral curvatures, and they are present at birth. These curvatures arch posteriorly and result in the vertebral column being C-shaped.

The **secondary curves** are called the cervical and lumbar curvatures, and they appear after birth. These curvatures arch anteriorly and are also known as *compensation curves* because they help shift the trunk weight over the legs. The cervical curvature appears when the child is first able to hold up its head without support (usually around 3–4 months of age). The lumbar curvature appears when the child is learning to stand and walk (typically by the first year of life). These curvatures become accentuated as the child becomes more adept at walking. The sacral curvature is less pronounced in females than in males, to allow for a greater pelvic outlet to accommodate the passage of an infant through the birth canal.

WHAT DID YOU LEARN?

13 What are the secondary curves, and when do they appear? What is their general function?

8.5c Vertebral Anatomy

LEARNING OBJECTIVES

5. Identify the parts of a typical vertebra.

6. Compare and contrast the different types of vertebrae.

All vertebrae share some common structural features (**figure 8.17**). The anterior region of each vertebra is a thick, cylindrical **body,** or *centrum,* which is the weight-bearing structure of each vertebra. Posterior to the vertebral body is the **vertebral arch,** also called

the *neural arch.* The body together with the vertebral arch enclose an opening called the **vertebral foramen.** All the stacked vertebral foramina collectively form a superior-to-inferior directed **vertebral canal** that contains the spinal cord. Lateral openings between adjacent vertebrae are the **intervertebral foramina.** The intervertebral foramina provide a horizontally directed passageway through which spinal nerves extend to various parts of the body (see section 14.5).

The vertebral arch is composed of two pedicles and two laminae. The **pedicles** (ped′ĭ-kĕl; *pes* = foot) originate from the posterolateral margins of the body, whereas the **laminae** (lam′i-nē; sing., *lamina,* lam′i-nă = layer) extend posteromedially from the posterior edge of each pedicle. A **spinous process** projects posteriorly from the junction of the left and right laminae. Most of these spinous processes can be palpated along the skin of the back. Lateral projections on both sides of the vertebral arch are called **transverse processes.**

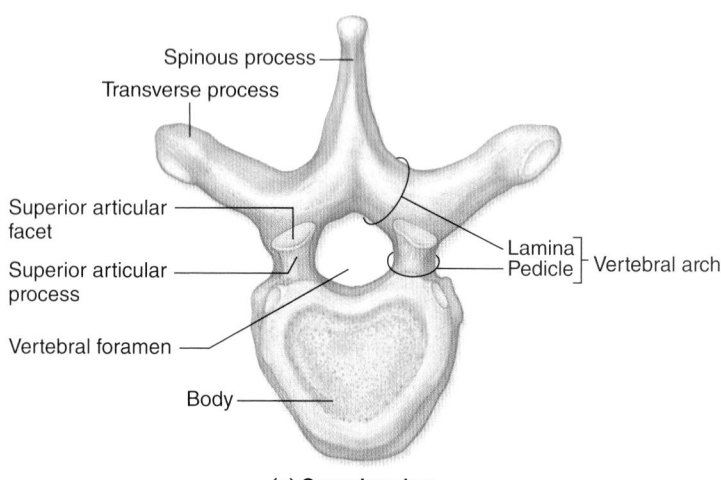

(a) Superior view

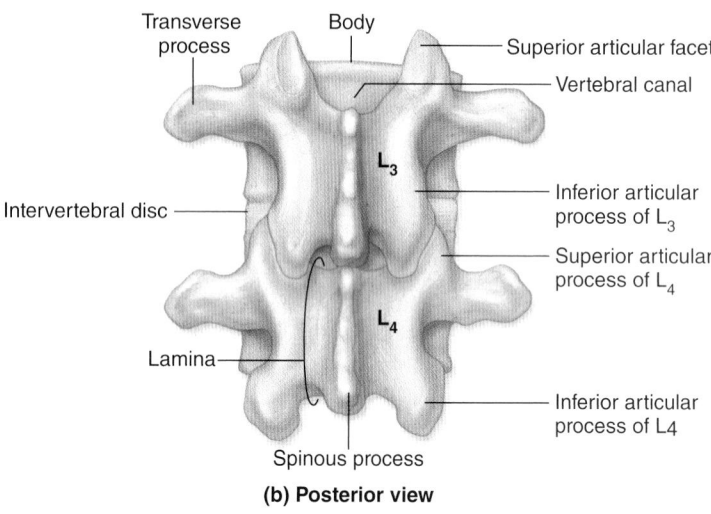

(b) Posterior view

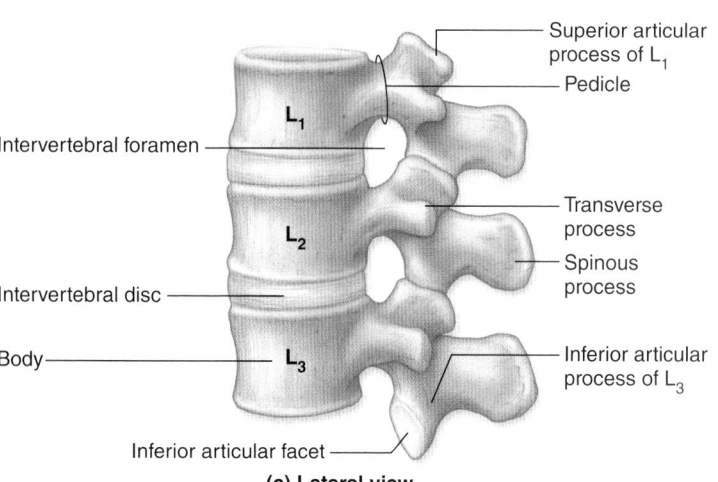

(c) Lateral view

Figure 8.17 Vertebral Anatomy. (*a*) Superior view of a thoracic vertebra. (*b*) Articulation between lumbar vertebrae, posterior view. (*c*) Articulation between lumbar vertebrae, lateral view. AP|R

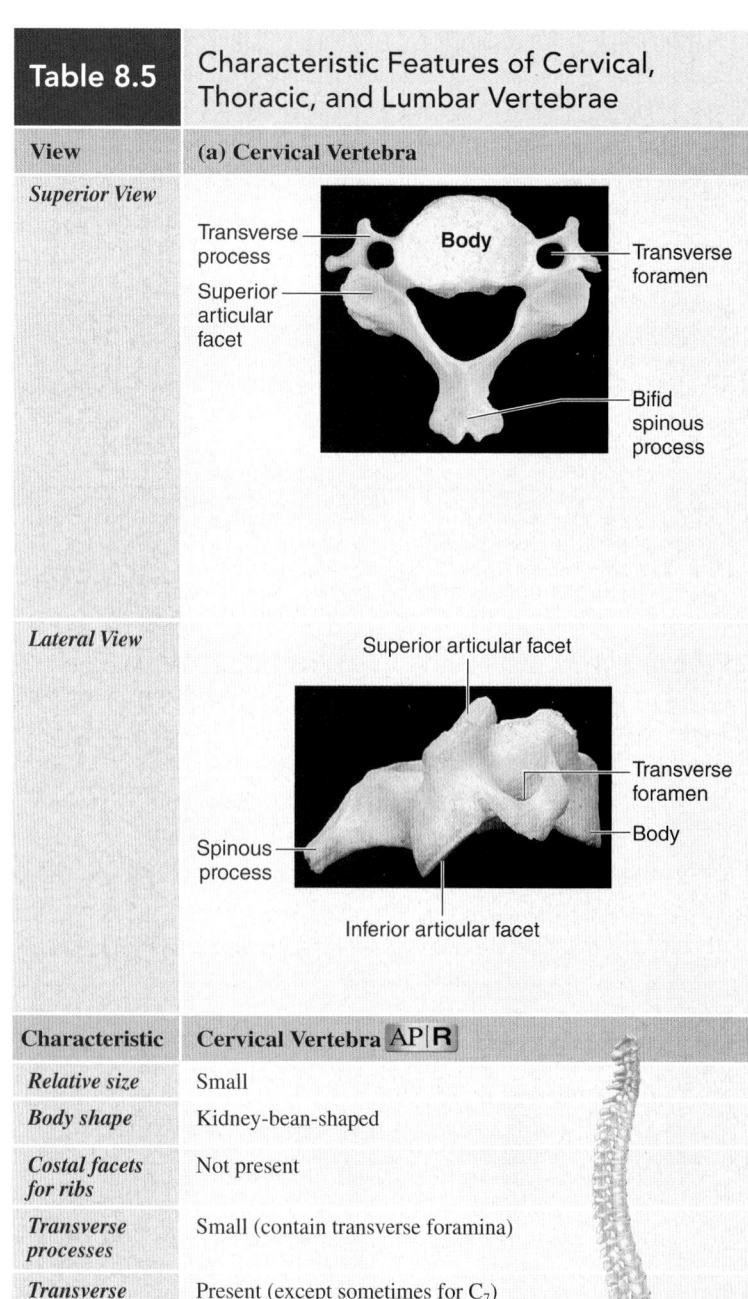

Table 8.5	Characteristic Features of Cervical, Thoracic, and Lumbar Vertebrae	
View	**(a) Cervical Vertebra**	
Superior View		
Lateral View		

Characteristic	Cervical Vertebra AP|R
Relative size	Small
Body shape	Kidney-bean-shaped
Costal facets for ribs	Not present
Transverse processes	Small (contain transverse foramina)
Transverse foramina	Present (except sometimes for C₇)
Spinous process	Slender: C₂–C₆ are often bifid (Note: C1 has no spinous process)

Each vertebra has **superior** and **inferior articular processes** that originate at the junction between the pedicles and laminae. Each articular process has a smooth surface called an **articular facet** (fas′et, fă-set′). The facets on the inferior articular processes of each vertebra articulate with the facets on the superior articular processes of the vertebra immediately inferior to it.

The stack of vertebral bodies is stabilized and interconnected by ligaments. Adjacent vertebral bodies are separated by pads of fibrocartilage, called the **intervertebral** (in-ter-ver′te-brăl) **discs.** Intervertebral discs are composed of an outer ring of fibrocartilage, called the **anulus fibrosus** (an′ū-lŭs fī-brō′sus), and an inner gelatinous circular region, called the **nucleus pulposus** (shown in Clinical View: "Herniated Discs"). Intervertebral discs make up approximately one-quarter of the entire vertebral column length. They act as shock absorbers between the vertebral bodies and also permit the vertebral column to bend. For example, when you bend your torso anteriorly, the intervertebral discs are compressed at the bending (anterior) surface and pushed out at the opposite (posterior) surface.

Over the course of a day, as body weight and gravity act on the vertebral column, the intervertebral discs become compressed and flattened. But while a person is lying horizontally during sleep, the intervertebral discs are able to expand and spring back to their original shape.

In general, the vertebrae are smallest near the skull. They become gradually larger moving inferiorly through the body trunk as weight-bearing increases. Although vertebrae are divided into regions, there are no anatomically discrete "cutoffs" between the regions. For example, the most inferior cervical vertebra has some structural similarities to the most superior thoracic vertebra, as the two vertebrae are adjacent to one another. Likewise, the most inferior thoracic vertebra may look similar to the first lumbar vertebra. **Table 8.5** compares the characteristics of the cervical, thoracic, and lumbar vertebrae and lists unique features of each regional group of vertebrae.

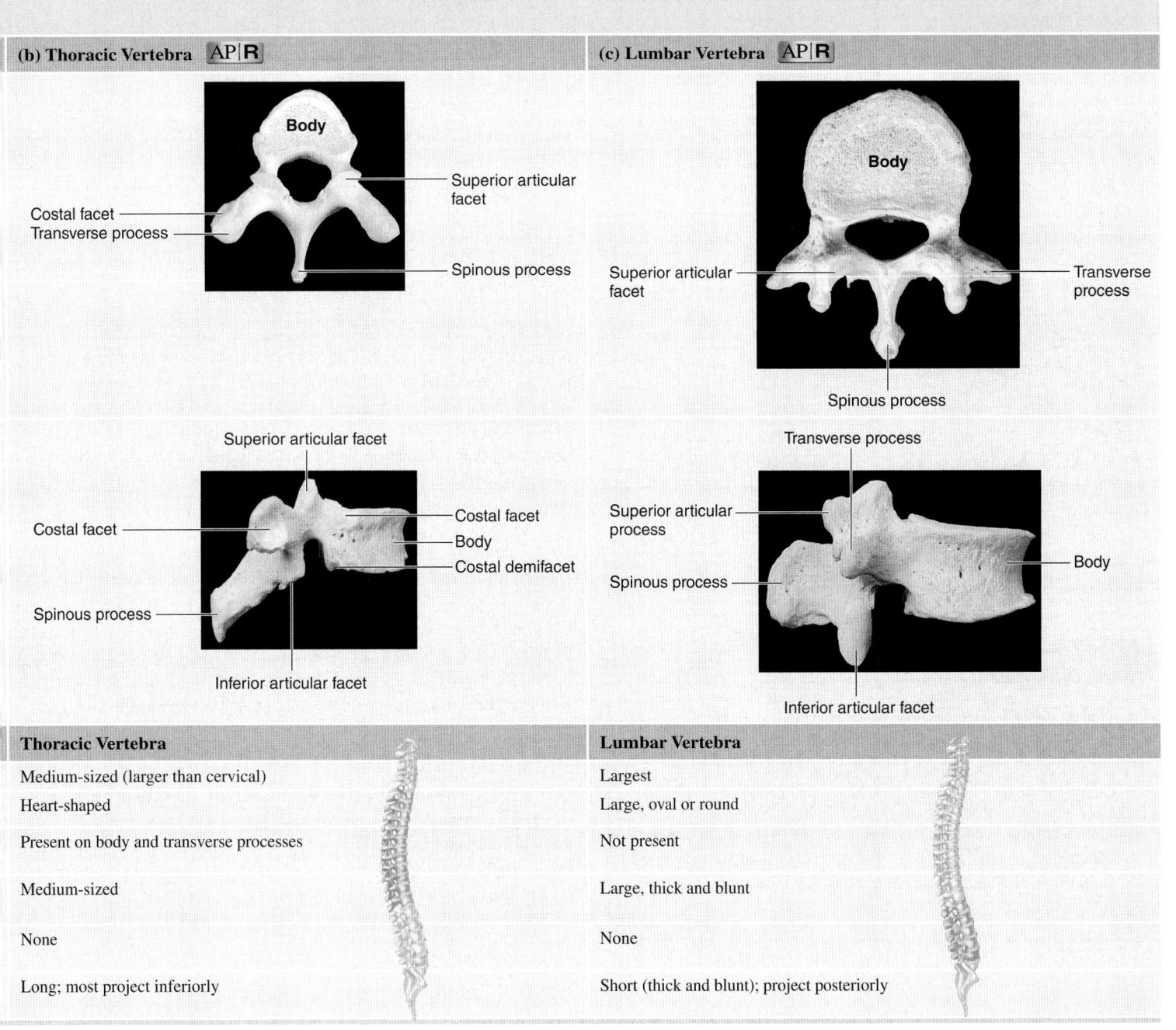

(b) Thoracic Vertebra AP|R

Body
Costal facet
Transverse process
Superior articular facet
Spinous process

Superior articular facet
Costal facet
Costal facet
Body
Costal demifacet
Spinous process
Inferior articular facet

(c) Lumbar Vertebra AP|R

Body
Superior articular facet
Transverse process
Spinous process

Transverse process
Superior articular process
Superior articular process
Body
Spinous process
Inferior articular facet

Thoracic Vertebra	Lumbar Vertebra
Medium-sized (larger than cervical)	Largest
Heart-shaped	Large, oval or round
Present on body and transverse processes	Not present
Medium-sized	Large, thick and blunt
None	None
Long; most project inferiorly	Short (thick and blunt); project posteriorly

INTEGRATE

CLINICAL VIEW
Herniated Discs

A **herniated** (her'nē-ā-ted) **disc** occurs when the gelatinous nucleus pulposus protrudes into or through the anulus fibrosus. This herniation produces a bulging of the disc contents posterolaterally into the vertebral canal and pinches the spinal cord, nerves of the spinal cord, or both. The cervical and lumbar intervertebral discs are the most common discs to be injured, because the vertebral column has more mobility in these regions, and the lumbar region bears increased weight. Cervical herniated discs can cause neck pain and pain down the upper limb, whereas lumbar herniated discs frequently cause low back pain. If a herniated lumbar disc starts to pinch nerve roots, the patient may feel pain down the entire lower limb, a condition known as **sciatica.**

Treatment options include "wait-and-see" if the disc heals on its own, nonsteroidal anti-inflammatory drugs (NSAIDs) such as ibuprofen, steroidal drugs, and physical therapy. Surgical treatments include **microdiscectomy,** a microsurgical technique whereby the portion of the herniated disc is removed, or **discectomy,** a more invasive technique in which the laminae of the nearby vertebrae and the back muscles are incised before removing the herniated portions of the disc.

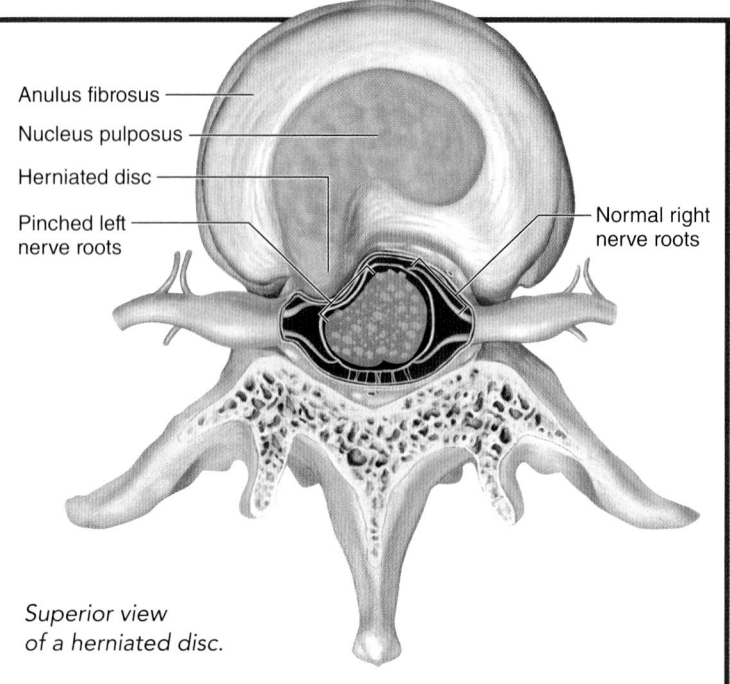

Anulus fibrosus

Nucleus pulposus

Herniated disc

Pinched left nerve roots

Normal right nerve roots

Superior view of a herniated disc.

Cervical Vertebrae

The cervical vertebrae are the most superiorly located vertebrae. They typically have kidney-bean-shaped bodies and extend inferiorly from the occipital bone of the skull through the neck to the thorax. Because cervical vertebrae support only the weight of the head, their vertebral bodies are relatively small and light. Most cervical vertebrae are distinguished from other vertebrae by the presence of **transverse foramina** in their transverse processes that house the vertebral artery and vein (sometimes C_7 does not have these foramina). Table 8.5 summarizes the key features of the typical cervical vertebra (C_3–C_6); the other cervical vertebrae are described here.

Atlas (C_1) The first cervical vertebra, called the **atlas** (at'las), supports the head through its articulation with the occipital condyles of the occipital bone **(figure 8.18a)**. This vertebra is named for the Greek mythological figure Atlas, who carried the world on his shoulders. The articulation between the occipital condyles and the atlas, called the atlanto-occipital joint, permits us to nod our heads "yes." The atlas is readily distinguished from the other vertebrae because it lacks both

INTEGRATE

CONCEPT CONNECTION

Healthy, normal vertebrae are vital for the proper functioning of our nervous system. If intervertebral discs herniate, they may pinch on the spinal cord (see section 14.1) or portions of spinal nerves (see section 14.5) and cause pain or numbness. Severe abnormal spinal curvatures may similarly impinge on the spinal cord. Finally, malformed intervertebral foramina may pinch on exiting spinal nerves, resulting in pain from those nerves.

a body and a spinous process. Instead, the atlas has lateral masses that are connected by semicircular **anterior** and **posterior arches,** each containing slight protuberances, the **anterior** and **posterior tubercles** (tū'běr-kĕl). The atlas has depressed, oval **superior** and **inferior articular facets** that articulate with the occipital condyles and the axis, respectively. Finally, the atlas has an **articular facet for dens** on its anterior arch (see next).

Axis (C_2) The body of the atlas separates from the atlas and fuses during development to the body of the second cervical vertebra, called the **axis** (ak'sis) (figure 8.18b). This fusion produces the most distinctive feature of the axis, the prominent **dens,** or *odontoid* (ō-don'toyd; *odont* = tooth) . The dens acts as a pivot for the lateral rotation of both the atlas and the skull. This articulation between the atlas and axis, called the atlantoaxial joint, permits us to shake our heads "no" (figure 8.18c). This joint is stabilized by a transverse ligament.

Vertebra Prominens (C_7) The seventh cervical vertebra represents a transition from cervical to the thoracic vertebral region (see figure 8.16). The spinous process of both C_7 and all the thoracic vertebrae are nonbifid (not forked)—however, this process in C_7 is much longer than it is within the other cervical vertebrae. It is easily palpated through the skin between the shoulder blades and inferior to the neck. Thus, C_7 also is called the **vertebra prominens** (prom'ĭ-nens; prominent).

Thoracic Vertebrae

There are 12 thoracic vertebrae, and each articulates with the ribs (table 8.5). Thoracic vertebrae typically have heart-shaped bodies and are distinguished from all other types of vertebrae by the presence of **costal facets** or **costal demifacets** (semicircular facets) on the lateral side of the body and on the sides of the transverse processes. The head of the rib articulates with the costal facet or demifacet on the body of

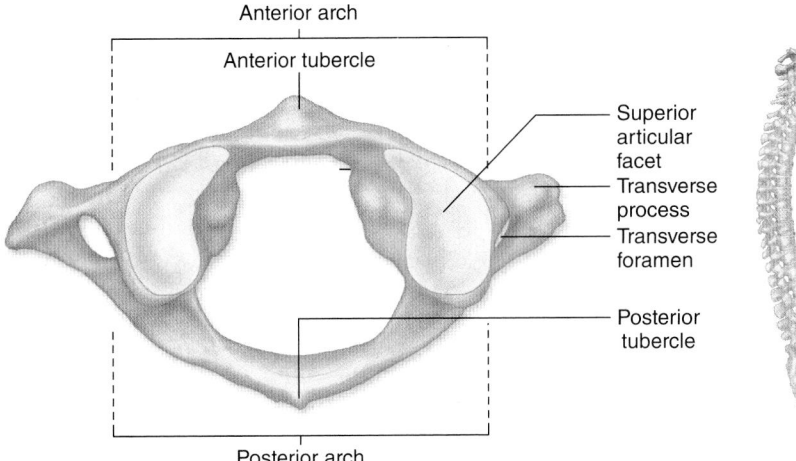

Anterior arch

Anterior tubercle

Superior articular facet

Transverse process

Transverse foramen

Posterior tubercle

Posterior arch

(a) Atlas (C₁), superior view

Figure 8.18 Cervical Vertebrae C₁ and C₂. The atlas (C₁) and the axis (C₂) differ in structure from a typical cervical vertebra. (*a*) A superior view of the atlas shows the atlas lacks a body and a spinous process. (*b*) Posterosuperior view of the axis clearly illustrates the dens. (*c*) The articulation of the atlas and axis, called the atlantoaxial joint, allows partial rotation of the atlas. **AP|R**

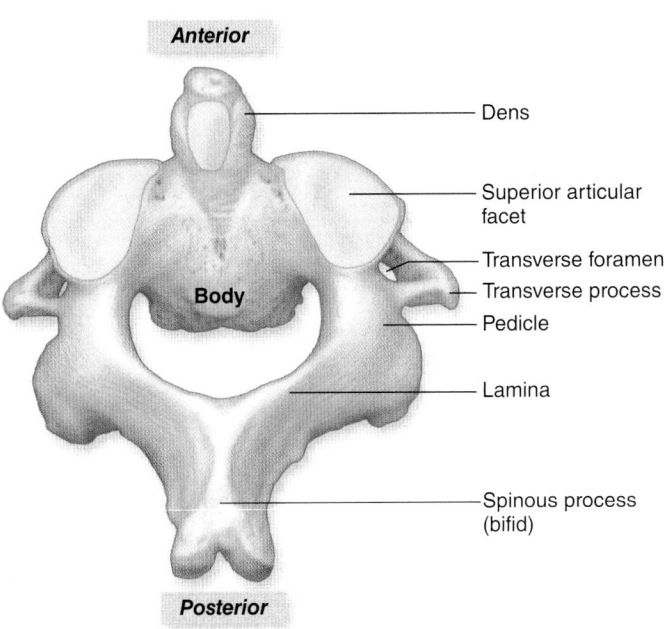

Anterior

Dens

Superior articular facet

Transverse foramen

Transverse process

Pedicle

Body

Lamina

Spinous process (bifid)

Posterior

(b) Axis (C₂), posterosuperior view

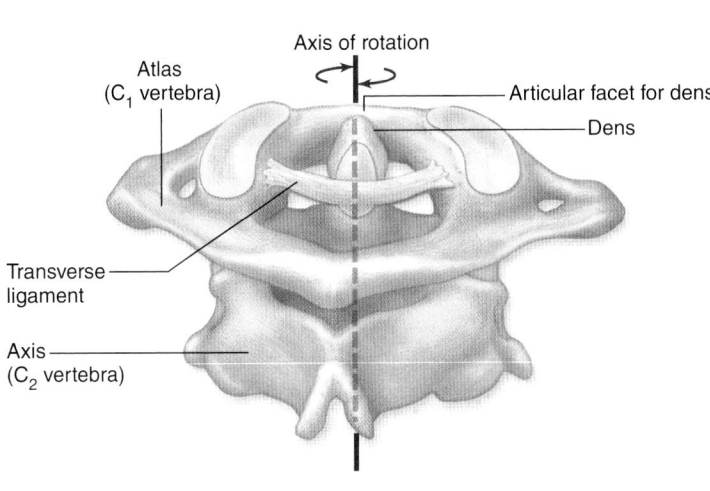

Axis of rotation

Atlas (C₁ vertebra)

Articular facet for dens

Dens

Transverse ligament

Axis (C₂ vertebra)

(c) Atlas and axis, posterosuperior view

the thoracic vertebra. The tubercle of the rib articulates with the costal facets on the transverse processes of the vertebra.

The thoracic vertebrae vary slightly with respect to their transverse costal facets. Vertebrae T_1–T_{10} have transverse costal facets on their transverse processes; T_{11} and T_{12} lack these transverse costal facets because the eleventh and twelfth ribs do not have tubercles (and thus do not articulate with the transverse processes). The costal facets on the bodies of the thoracic vertebrae also display variations. Some vertebrae may have a single whole facet; others may have two demifacets.

Lumbar Vertebrae

The largest vertebrae are the lumbar vertebrae. A typical lumbar vertebra body is thicker than that of all the other vertebrae, and its body is oval or round (table 8.5). The lumbar vertebrae are distinguished by the features they lack—that is, lumbar vertebrae have neither transverse foramina nor costal facets.

The lumbar vertebrae bear most of the weight of the body. The thick spinous processes provide extensive surface area for the attachment of inferior back muscles that reinforce or adjust the lumbar curvature.

 WHAT DO YOU THINK?

3 You are given a vertebra to identify. It has transverse foramina and a bifid spinous process. Is this a cervical, thoracic, or lumbar vertebra?

Sacrum

The sacrum is an anteriorly curved, somewhat triangular bone that forms the posterior wall of the pelvic cavity (**figure 8.19**). The **apex** of the sacrum is a narrow, pointed portion of the bone that projects inferiorly, whereas the bone's broad superior surface forms the **base.**

The sacrum is composed of five fused sacral vertebrae. These vertebrae start to fuse shortly after puberty and are usually completely fused between ages 20 and 30. The horizontal lines of fusion that remain are called **transverse ridges.** Superiorly, the sacrum articulates with L_5 via a pair of **superior articular processes.** The vertebral canal becomes much narrower and continues through the sacrum on its posterior side as the **sacral canal.** The sacral canal terminates in an inferior opening called the **sacral hiatus** (hī-ā′tŭs; *hio* = to yawn). On either side of the sacral hiatus are bony projections called the **sacral cornua.**

The anterosuperior edge of the first sacral vertebra bulges anteriorly into the pelvic cavity and is called the **promontory.** The paired **anterior** and **posterior sacral foramina** permit the passage of nerves to the pelvic organs and the gluteal region, respectively. A dorsal ridge, termed the **median sacral crest,** is formed by the fusion of the spinous processes of individual sacral vertebrae. On each lateral surface of the sacrum is the **ala** (meaning

Figure 8.19 Sacrum and Coccyx. The sacrum is formed by the fusion of five sacral vertebrae, and the coccyx is formed by the fusion of four coccygeal vertebrae. (*a*) The promontory of the sacrum is seen in anterior view, and (*b*) a posterior view shows the median sacral crest and sacral hiatus. AP|R

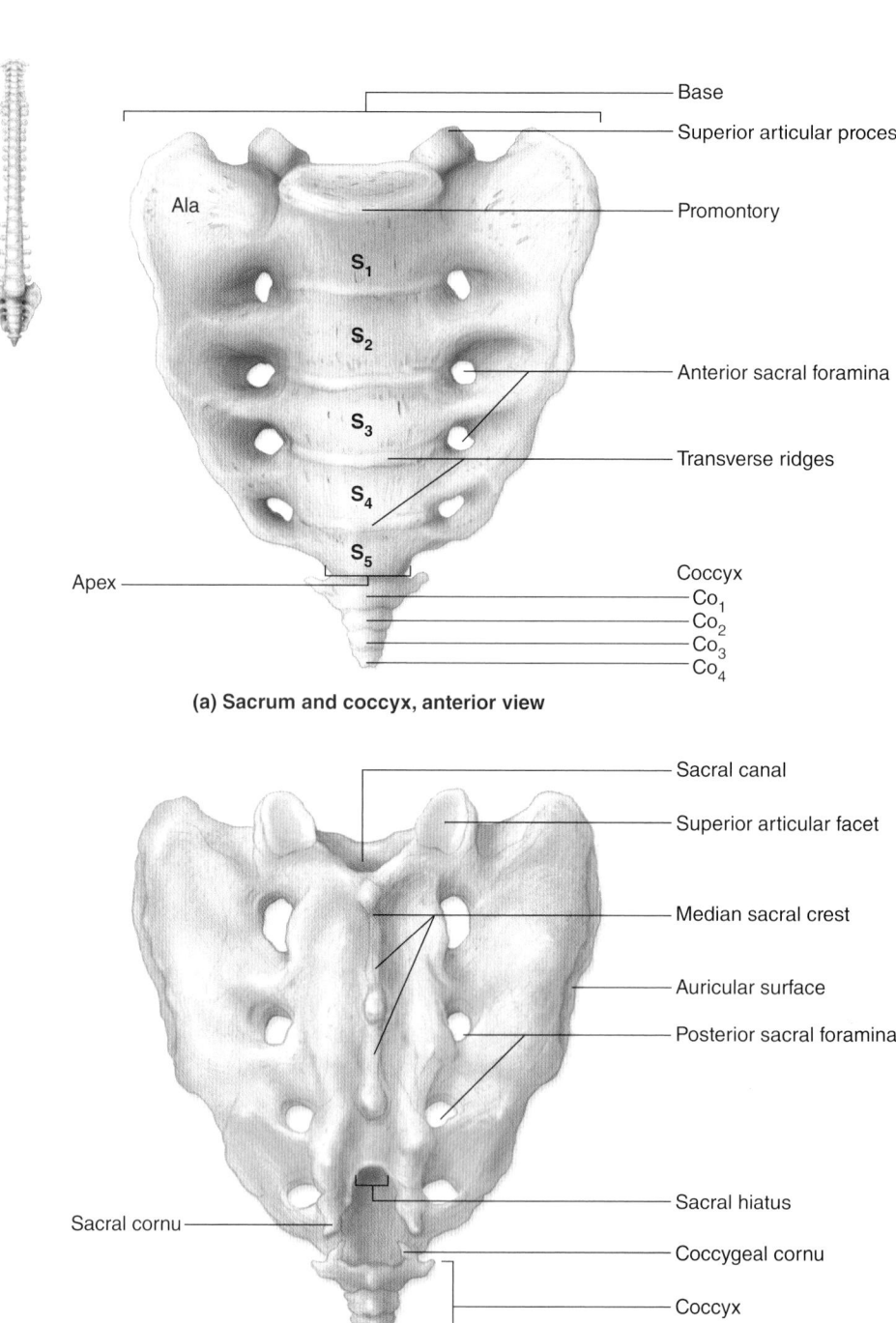

(a) Sacrum and coccyx, anterior view

(b) Sacrum and coccyx, posterior view

"wing"). On the lateral surface of the ala is the **auricular surface,** which marks the site of articulation with the os coxae of the pelvic girdle, forming the strong, stable **sacroiliac** (sā-krō-il′ē-ak) **joint.**

Coccyx

Four small coccygeal vertebrae fuse to form the coccyx. These individual vertebrae begin to fuse by about age 25. The coccyx is an attachment site for several ligaments and some muscles. The first and second coccygeal vertebrae have unfused vertebral arches and transverse processes. The prominent laminae of the first coccygeal vertebrae are known as the **coccygeal cornua,** which curve to meet the sacral cornua. In males, the coccyx tends to project anteriorly, but in females it tends to project more inferiorly. In very elderly individuals, the coccyx may fuse with the sacrum.

WHAT DID YOU LEARN?

14 Compare the locations and functions of the transverse foramina, intervertebral foramina, and the vertebral foramen.

15 How do the atlas and axis differ from other cervical vertebrae?

INTEGRATE

CLINICAL VIEW
Coccyx (Tailbone) Injury

Although the coccyx is small, it may be prone to bruising or even fracture. Activities that can injure the coccyx include a fall on the buttocks, sitting down abruptly on a chair, a direct hit from contact sports, or frequent cycling. In addition, the coccyx may be injured during childbirth. The severe pain from a coccyx injury is referred to as *coccydynia.* Both bruising and fractures of the coccyx take many weeks to heal, but often are treated conservatively with rest, ice, and non-steroidal anti-inflammatory drugs (NSAIDs).

8.6 Bones of the Thoracic Cage

The bony framework of the chest is called the **thoracic cage** and consists of the thoracic vertebrae posteriorly, the ribs laterally, and the sternum anteriorly **(figure 8.20).** The thoracic cage acts as a protective enclosure around the thoracic organs and also provides attachment points for many muscles.

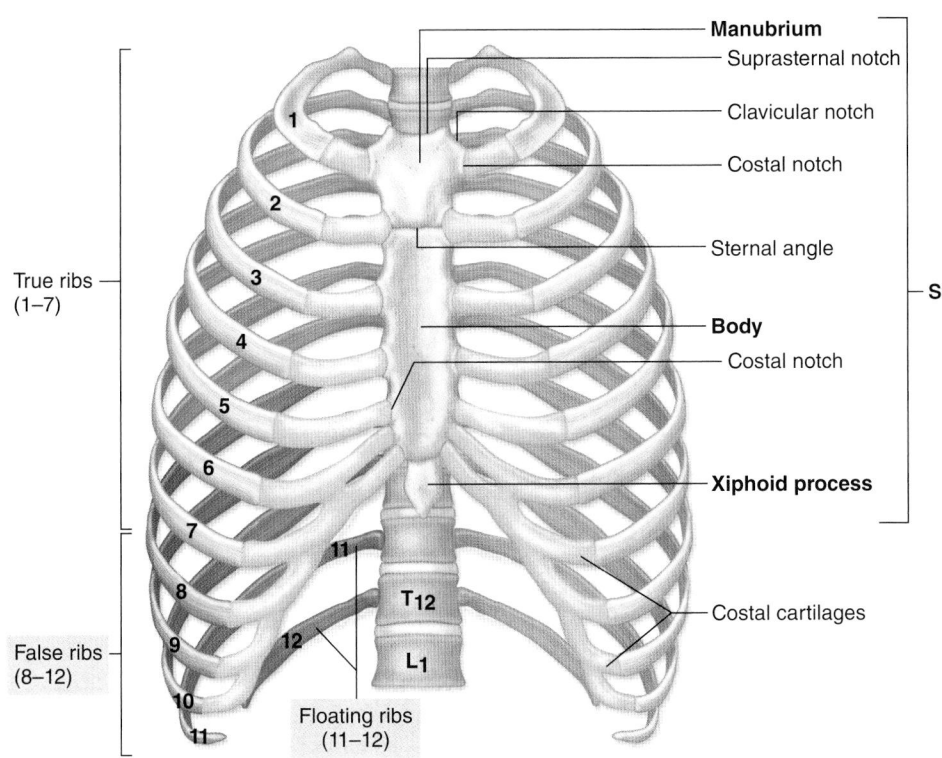

Manubrium
Suprasternal notch
Clavicular notch
Costal notch
Sternal angle
Body
Costal notch
Xiphoid process
Costal cartilages
Sternum

True ribs (1–7)
False ribs (8–12)
Floating ribs (11–12)

T12
L1

Figure 8.20 Thoracic Cage. The thoracic cage is composed of the thoracic vertebrae, ribs, and sternum. It protects and encloses the organs in the thoracic cavity.
AP|R

INTEGRATE

CLINICAL VIEW
Sternal Foramen

In up to 4–10% of all adults, a midline **sternal foramen** is present in the body of the sternum. The sternal foramen represents failure of the left and right ossification centers of the sternal body to fuse completely. Sometimes, this opening may be misidentified as a bullet wound. Thus, a crime scene investigator must be aware of this congenital anomaly when examining skeletal remains. In rare instances, individuals with previously undetected sternal foramina have died after an acupuncture session, when the acupuncture needle was unknowingly inserted through the sternal foramen and into the heart.

Sternal foramen

8.6a Sternum

LEARNING OBJECTIVE

1. Identify the three main components of the sternum and their features.

The **sternum** (ster'nŭm; *sternon* = the chest), also referred to as the *breastbone,* is a flat bone that forms in the anterior midline of the thoracic wall. Its shape has been likened to that of a sword. The sternum is composed of three parts: the manubrium, the body, and the xiphoid process.

The **manubrium** (mă-nū'brē-ŭm) is the widest and most superior portion of the sternum (the "handle" of the bony sword). Two **clavicular notches** articulate the sternum with the left and right clavicles. The shallow superior indentation between the clavicular notches is called the **suprasternal** (or *jugular*) **notch.** A single pair of **costal notches** represent articulations for the first ribs' costal cartilages.

The **body** is the longest part of the sternum and forms its bulk (the "blade" of the bony sword). Individual costal cartilages from ribs 2–7 are attached to the body at indented articular costal notches. The body and the manubrium articulate at the **sternal angle,** a horizontal ridge that may be palpated under the skin. The sternal angle is an important landmark in that the costal cartilages of the second ribs attach there; thus, it may be used to count the ribs.

The **xiphoid** (zi'foyd; *xiphos* = sword) **process** represents the very tip of the "sword." This small, inferiorly pointed cartilaginous projection often doesn't ossify until after age 40.

WHAT DID YOU LEARN?

16 What sternal structures form the sternal angle, and what is its clinical significance?

8.6b Ribs

LEARNING OBJECTIVES

2. Describe the features found on all ribs.

3. Differentiate between true ribs and false ribs.

The ribs are elongated, curved, flattened bones that originate on or between the thoracic vertebrae and end in the anterior wall of the thorax **(figure 8.21a)**. Both males and females have 12 pairs of ribs. Ribs 1–7 are called **true ribs.** True ribs connect individually to the sternum by separate cartilaginous extensions called **costal** (kos'tal; *costa* = rib) **cartilages** (figure 8.20). The smallest true rib is the first.

Ribs 8–12 are called **false ribs** because their costal cartilages do not attach directly to the sternum. The costal cartilages of ribs 8–10 fuse to the costal cartilage of rib 7 and thus indirectly articulate with the sternum. The last two pairs of false ribs (ribs 11 and 12) are called **floating ribs** because they have no connection with the sternum.

The vertebral end of a typical rib articulates with the vertebral column at the **head** (figure 8.21). The articular surface of the head is divided into **superior** and **inferior articular facets** by an interarticular **crest.** The surfaces of these facets articulate with the costal facets or demifacets on the bodies of the thoracic vertebrae. The **neck** of the rib lies between the head and the tubercle. The **tubercle** of the rib has an articular facet for the costal facet on the transverse process of the thoracic vertebra. Figure 8.21b, c illustrates how most of the ribs articulate with the thoracic vertebrae.

INTEGRATE

CLINICAL VIEW
Variations in Rib Development

In 1 out of every 200 people, the costal element of the seventh cervical vertebra elongates and forms a rudimentary **cervical rib.** Cervical ribs may compress the artery and nerves extending toward the upper limb, producing tingling or pain. Less commonly, an extra pair of ribs may form from the costal elements of the first lumbar vertebra. Some individuals lack a pair of twelfth ribs, because their costal elements from the twelfth thoracic vertebra failed to elongate. Another rib development anomaly is fused (bicipital) ribs. Finally, **bifid ribs** occur in 1.2% of the world's population (and up to 8.4% of Samoans). A bifid rib splits into two separate portions when it reaches the sternum.

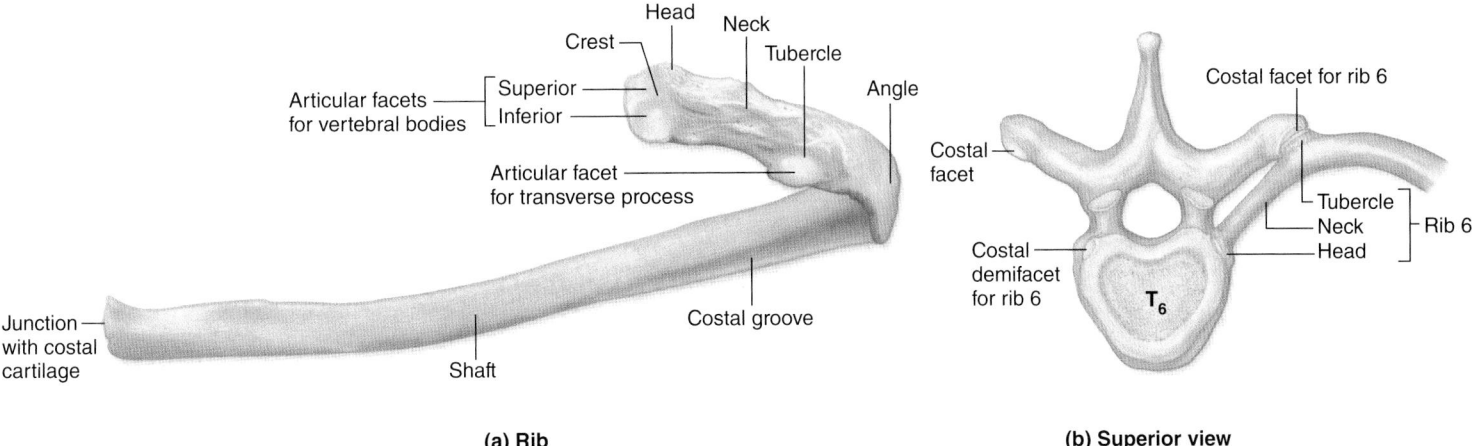

(a) Rib

(b) Superior view

Figure 8.21 Rib Anatomy and Articulation with Thoracic Vertebrae. Paired ribs attach to thoracic vertebrae posteriorly and extend anteroinferiorly to the anterior chest wall. (*a*) Features of ribs 2–10. (*b*) Superior and (*c*) lateral views show the articulation of a rib with two vertebrae.

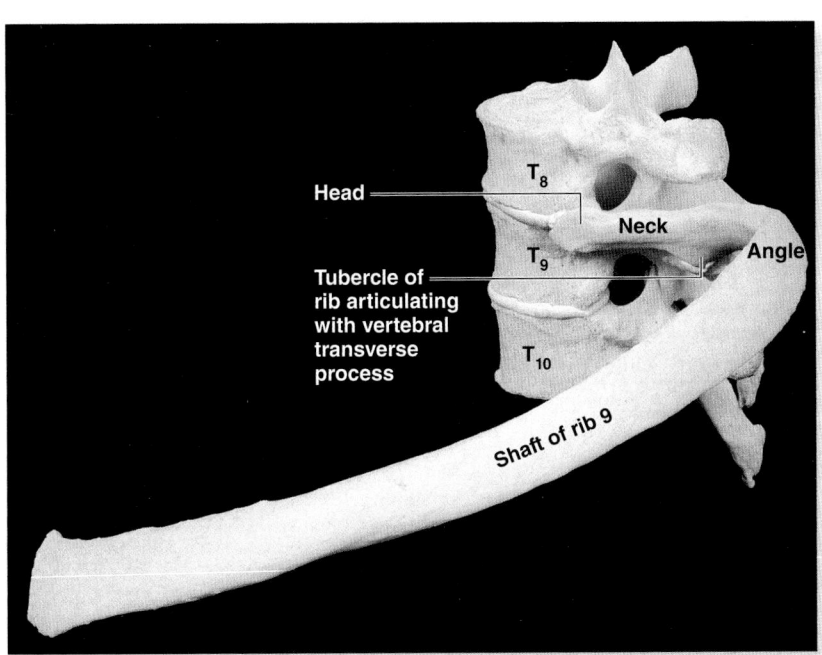

(c) Lateral view

The **angle** (*border*) of the rib indicates the site where the tubular **shaft** begins to curve anteriorly toward the sternum. A prominent **costal groove** along its inferior internal border marks the path of nerves (see section 14.5c) and blood vessels to the thoracic wall.

WHAT DID YOU LEARN?

17 Where specifically do the head and tubercle of a rib each articulate?

8.7 The Upper and Lower Limbs: A Comparison

LEARNING OBJECTIVES

1. Identify skeletal features common to the upper and lower limbs.
2. Describe the functional reasons for differences between the upper and lower limb skeletons.

Humans evolved from quadrupeds, which are animals that move on four feet. Quadruped limbs are very similar because all of the limbs are structured to support the body weight and move the animal. However, as our ancestors evolved into modern human beings, we became bipedal. Only our lower limbs normally support our body weight and are responsible for moving our bodies when we walk or run. In contrast, our upper limbs have been freed from these functions and are able to do other things, such as grasping objects and utilizing tools with our hands.

Our upper and lower limb skeletons share some common features based on this evolutionary history, and they exhibit some differences based on the primary functions of each limb. **Figure 8.22** summarizes the similarities. The proximal part of both upper and lower limbs are supported by a "girdle" of bones; the pectoral girdle (clavicles and scapulae) holds the upper limbs in place, while

Figure 8.22 **Similarities Between the Upper Limb and Lower Limb Skeletons.** (*a*) The proximal part of each limb has a girdle that holds the limb in place. (*b*) The distal part of each limb contains two long bones, followed by multiple short bones and then numerous long bones in the hand and foot.

(a) Proximal part of the limb

A "girdle" supports each limb.

Each girdle has a rounded, cuplike depression (socket) in which the head of the proximal part of each limb bone fits.

Pectoral girdle = left and right clavicles and scapulae

Upper Limb

Pelvic girdle = left and right ossa coxae

Lower Limb

The proximal part of the limb contains a single bone with a rounded head.

The rounded heads of the humerus and femur fit within their respective girdles, and allow for a wide range of movement at the shoulder and hip joints.

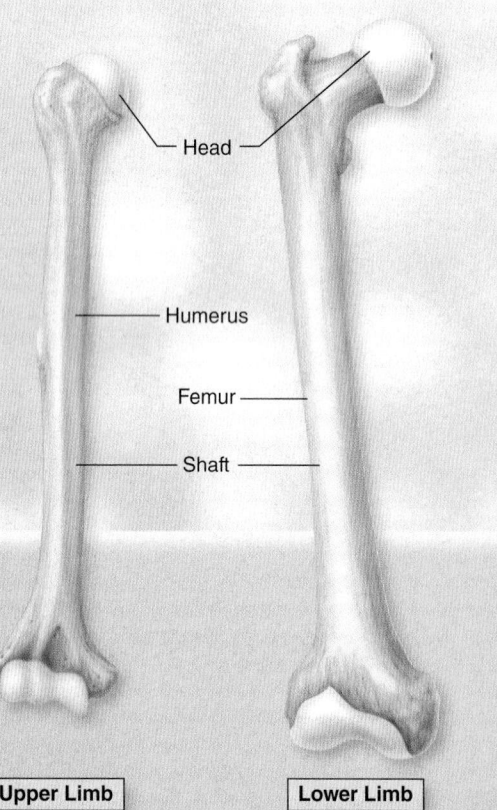

Head

Humerus

Femur

Shaft

Upper Limb

Lower Limb

The distal part of each limb contains two bones connected by an interosseous membrane.

The interosseous membrane keeps the bones a fixed distance apart and allows these bones to pivot about one another. (Note: Pivoting is much more limited in the lower limb.)

The styloid processes of the radius and ulna are structurally similar to the malleoli of the tibia and fibula.

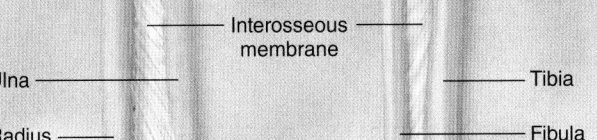

Interosseous membrane

Ulna

Radius

Tibia

Fibula

Styloid process

Lateral malleolus

Medial malleolus

Upper Limb

Lower Limb

(b) Distal part of the limb

The hands and the feet have similar arrangements of bones.

Both the hand and the foot have 5 metacarpals or metatarsals respectively, and 14 phalanges. Note the thumb and the big toe are the most robust of the digits, yet they each have only 2 phalanges.

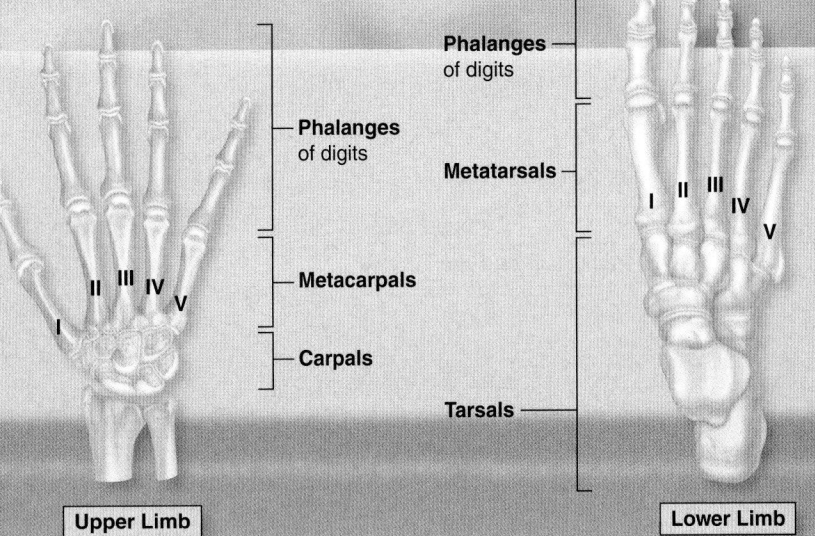

Phalanges of digits

Phalanges of digits

Metatarsals

Metacarpals

Carpals

Tarsals

Upper Limb

Lower Limb

The multiple carpal and tarsal bones allow for a wide range of movement at either the wrist or the ankle joints.

the pelvic girdle (both ossa coxae) articulates with the lower limb. The proximal part of each limb has one large bone; the humerus in the upper limb and the femur in the lower limb. The distal part of each limb contains two bones; these bones are able to pivot slightly about one another. Both the wrist and the proximal foot contain multiple bones (carpal and tarsal bones, respectively) that allow for a range of movement. Finally, the feet and hands are very similar in that both contain either 5 metacarpals (palm of hand) or 5 metatarsals (arch of foot), and each contains a total of 14 phalanges (bones of the fingers/toes).

The structural differences between the upper and lower limb skeletons arise from the functional differences. Understanding these general differences between upper and lower limbs will make the study of their individual bones easier. Because the lower limb is weight bearing and is used for locomotion, some mobility at specific joints has been lost for greater stability. The upper limb is not weight bearing, so both arm and forearm bones are relatively smaller and lighter than the similar respective lower limb bones. Additionally, the upper limb joints are relatively more mobile than the respective lower limb joints, so we may utilize the upper limbs for a wide range of activities. Unfortunately, more mobile joints are less stable, and that is why some of the upper limb joints (such as the shoulder joints) are the most frequently injured.

WHAT DID YOU LEARN?

18 What are some of the functional differences between the upper and lower limbs?

8.8 The Pectoral Girdle and Its Functions

The **pectoral** (pek'tŏ-răl; *pectus* = breastbone) **girdle** articulates with the trunk and supports the upper limbs. A pectoral girdle consists of the clavicles and the scapulae.

8.8a Clavicle

LEARNING OBJECTIVE

1. Identify and locate the clavicle and its landmarks.

The **clavicle** (klav'i-kĕl; *clavis* = key), commonly known as the collarbone, is an elongated S-shaped bone that extends between the manubrium of the sternum and the acromion of the scapula (**figure 8.23**). Its **sternal end** (medial end) is roughly pyramidal in shape and articulates with the manubrium of the sternum, forming the sternoclavicular joint (see section 9.7b). The **acromial end** (lateral end) of the clavicle is broad and flattened. The acromial end articulates with the acromion of the scapula, forming the acromioclavicular joint. You can palpate your own clavicle by first locating the superior aspect of your sternum and then moving your hand laterally. The curved bone you feel under your skin, and close to the neck opening of your shirt is your clavicle.

The superior surface of the clavicle is relatively smooth and the inferior surface is roughened (figure 8.23). On the inferior surface, near the acromial end, is a rough tuberosity called the **conoid** (kˉo'noyd; *konoeides* = cone-shaped) **tubercle** for the conoid ligament (part of the coracoclavicular ligament of the shoulder joint; see section 9.7b). The inferiorly located prominence at the sternal end of the clavicle is the **costal tuberosity,** for the attachment of the shoulder's costoclavicular ligament.

WHAT DID YOU LEARN?

19 How do the sternal end and acromial end of the clavicle differ?

8.8b Scapula

LEARNING OBJECTIVE

2. Describe the landmarks and features of the scapula.

The **scapula** (skap'yū-lă) is a broad, flat, triangular bone that forms the *shoulder blade* (**figure 8.24**). You can palpate your scapula by putting your hand on your superolateral back region and moving your

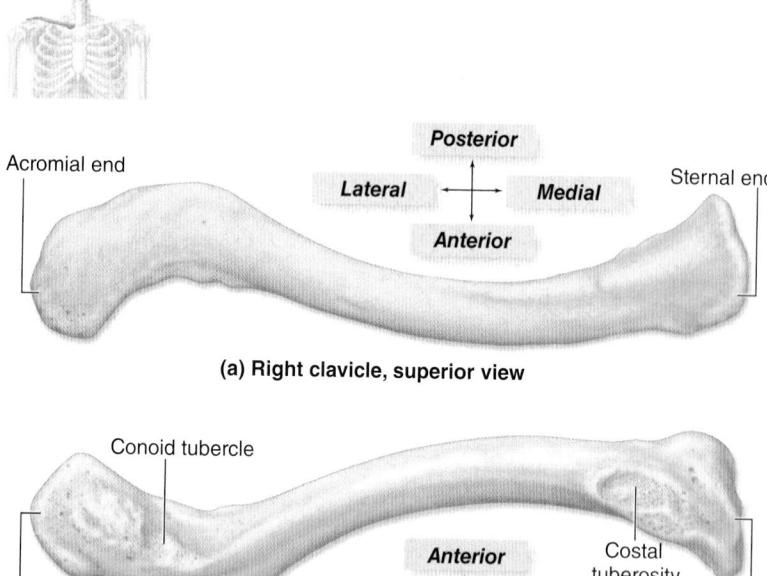

(a) **Right clavicle, superior view**

(b) **Right clavicle, inferior view**

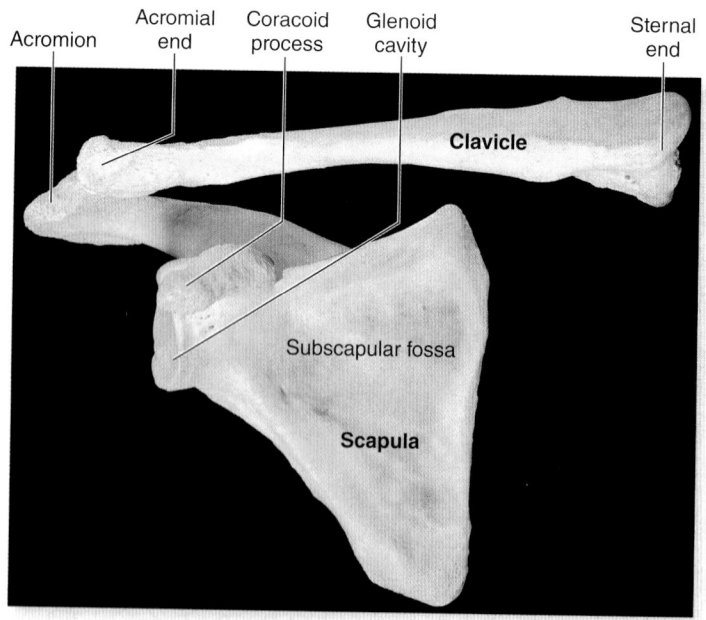

(c) **Right scapula and clavicle articulation, anterior view**

Figure 8.23 Clavicle. The S-shaped clavicle represents the only direct connection between the pectoral girdle and the axial skeleton. (*a*) Superior and (*b*) inferior views of the right clavicle. (*c*) Anterior view of an articulated right clavicle and scapula. AP|R

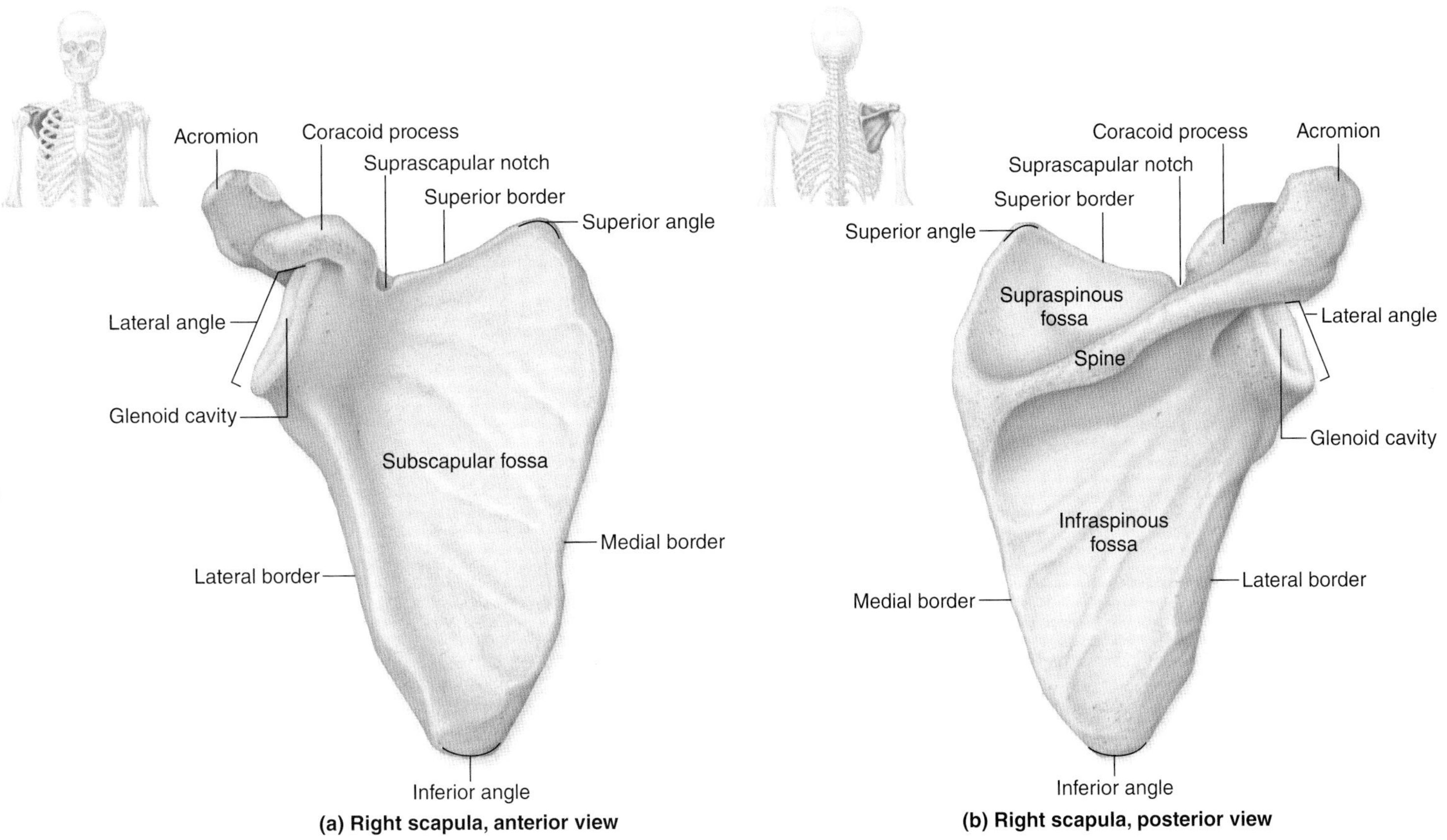

Acromion Coracoid process
Suprascapular notch
Superior border
Superior angle

Lateral angle

Glenoid cavity

Subscapular fossa

Lateral border

Medial border

Inferior angle

(a) Right scapula, anterior view

Coracoid process Acromion
Suprascapular notch
Superior border
Superior angle

Supraspinous
fossa

Lateral angle

Spine

Glenoid cavity

Infraspinous
fossa

Lateral border

Medial border

Inferior angle

(b) Right scapula, posterior view

Figure 8.24 Scapula. The right scapula is known as the "shoulder blade" and is shown in (a) anterior and (b) posterior views.

upper limb; the bone you feel moving is the scapula. The **spine** of the scapula is a ridge of bone on the posterior aspect of the scapula. It is easily palpated under the skin. The spine is continuous with a larger, posterior process called the **acromion** (a-krō′mē-on; *akron* = tip, *omos* = shoulder), which forms the bony tip of the shoulder. Palpate your upper shoulder; the prominent bump you feel is the acromion. The **coracoid** (kōr′ă-koyd) **process** is the smaller, more anterior, hook-shaped projection that is a site for muscle attachment.

The triangular shape of the scapula forms three sides, or borders. The **superior border** is the horizontal edge of the scapula superior to the spine of the scapula; the **medial border** (also called the *vertebral border*) is the edge of the scapula closest to the vertebrae; and the **lateral border** (also called the *axillary border*) is closest to the axilla. A **suprascapular notch** (which in some individuals is a **suprascapular foramen**) in the superior border provides passage for the suprascapular nerve and blood vessels.

Between these borders are the superior, inferior, and lateral angles. The **superior angle** is located between the superior and medial while the **inferior angle** is positioned between the medial and lateral borders. The **lateral angle** is primarily made up of the cup-shaped, shallow **glenoid** (glē′noyd; glen′oyd; resembling a socket) **cavity,** or *glenoid fossa*, which articulates with the humerus, the bone of the arm.

The broad, relatively smooth, anterior surface of the scapula is called the **subscapular** (sŭb-skap′yū-lăr; *sub* = under) **fossa**. A large muscle called the subscapularis overlies this fossa. The spine subdivides the posterior surface of the scapula into two shallow fossae. The depression superior to the spine is the **supraspinous** (sū-pră-spī′nŭs; *supra* = above) **fossa;** inferior to the spine is a broad, extensive surface called the **infraspinous fossa.** The supraspinatus

and infraspinatus muscles, respectively, occupy these fossae (see section 11.8b).

WHAT DID YOU LEARN?

20 What fossae are located on the scapula, and what is located in each fossa?

8.9 Bones of the Upper Limb

The upper limb consists of the brachium (arm), antebrachium (forearm), and hand. The complex structure of the hand in particular gives humans capabilities beyond those of most other vertebrates.

Each upper limb contains a total of 30 bones:

- 1 humerus, located in the brachium region
- 1 radius and 1 ulna, located in the antebrachium region
- 8 carpal bones, which form the wrist
- 5 metacarpal bones, which form the palm of the hand
- 14 phalanges, which form the fingers

8.9a Humerus

LEARNING OBJECTIVES

1. Describe the articulations of the humerus.
2. List landmarks and features of the humerus.

The **humerus** (hyū′mĕr-ŭs) is the longest and largest upper limb bone (**figure 8.25**). Its proximal end has a hemispherical **head** that

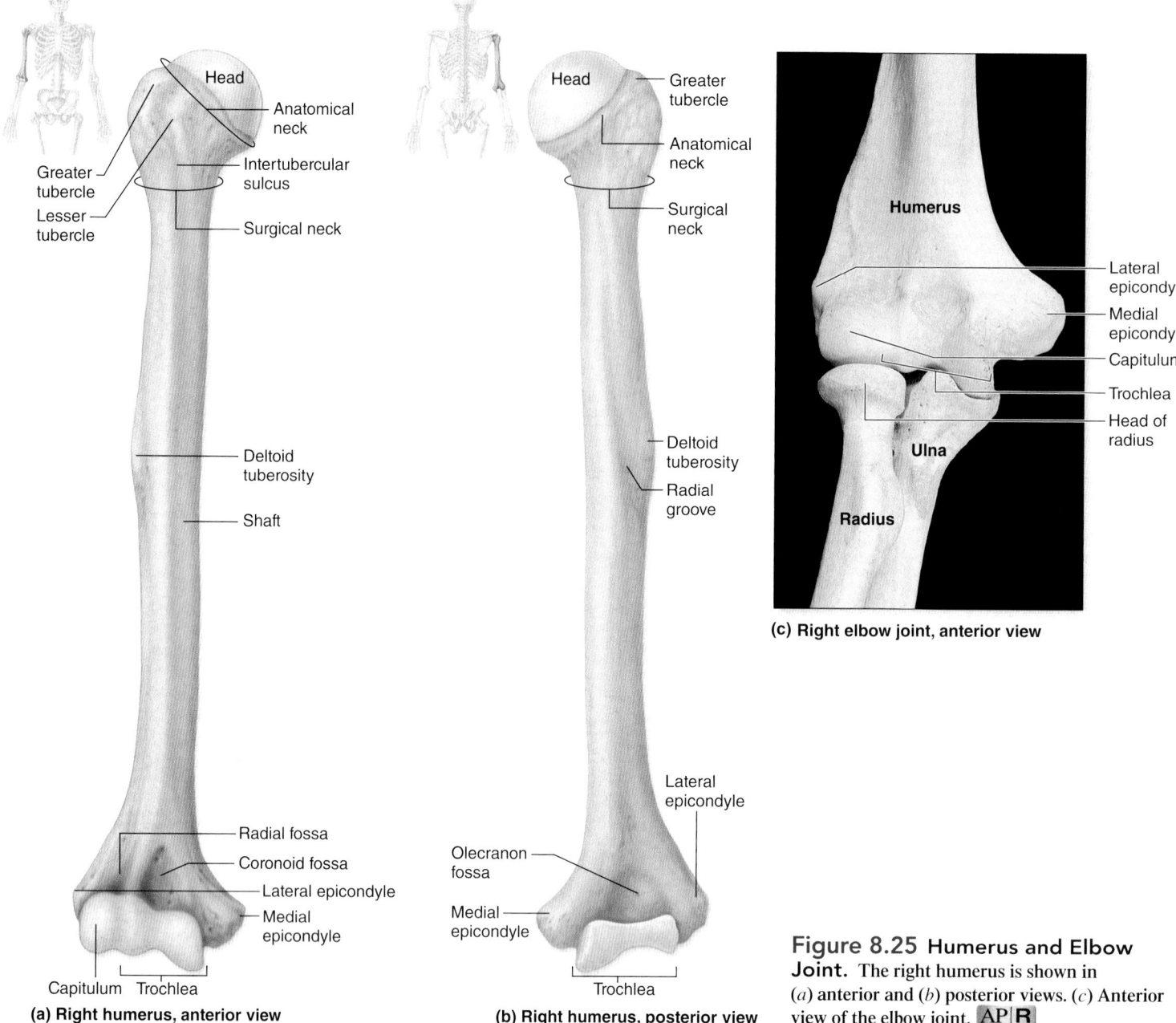

Head
Anatomical neck
Greater tubercle
Intertubercular sulcus
Lesser tubercle
Surgical neck

Deltoid tuberosity

Shaft

Radial fossa
Coronoid fossa
Lateral epicondyle
Medial epicondyle

Capitulum   Trochlea

(a) Right humerus, anterior view

Head
Greater tubercle
Anatomical neck
Surgical neck

Deltoid tuberosity
Radial groove

Lateral epicondyle

Olecranon fossa

Medial epicondyle

Trochlea

(b) Right humerus, posterior view

Humerus

Lateral epicondyle
Medial epicondyle
Capitulum
Trochlea
Head of radius

Ulna

Radius

(c) Right elbow joint, anterior view

Figure 8.25 Humerus and Elbow Joint. The right humerus is shown in (*a*) anterior and (*b*) posterior views. (*c*) Anterior view of the elbow joint. AP|R

articulates with the glenoid cavity of the scapula. The prominent **greater tubercle** is positioned lateral to the head and helps form the rounded contour of the shoulder. The **lesser tubercle** is smaller and located more medial to the head. Between the two tubercles is the **intertubercular sulcus** (or *bicipital sulcus,* or *bicipital groove*), a depression that contains the tendon of the long head of the biceps brachii muscle (see section 11.8c).

Between the tubercles and the head of the humerus is the **anatomical neck,** an almost indistinct groove that marks the location of the former epiphyseal plate. The **surgical neck** is a narrowing of the bone immediately distal to the tubercles, at the transition from the head to the shaft. This feature is called the "surgical" neck because it is a common fracture site.

The **shaft** of the humerus has a roughened area, termed the **deltoid** (del′toyd; *deltoides* = like the Greek letter Δ) **tuberosity** (tū′bĕr-os′i-tē), which extends along its lateral surface for about half the length of the humerus. The deltoid muscle of the shoulder attaches to this roughened surface (see section 11.8b). The **radial groove** (or *spiral groove*) is located adjacent to the deltoid tuberosity and is the location of the radial nerve (see section 14.5e) and some blood vessels.

Together, the bones of the humerus, radius, and ulna form the elbow joint (figure 8.25c). The **medial** and **lateral epicondyles** (ep-i-kon′d¯ıl; *epi* = upon, *kondylos* = a knuckle) are bony side projections on the distal humerus that provide surfaces for muscle attachment. Palpate the sides of your elbow; the bumps you feel are the medial and lateral epicondyles. Placed posterior to the medial epicondyle is the ulnar nerve (see section 14.5e). (You actually are hitting this nerve when you hit your "funny bone.")

The distal end of the humerus has two smooth, curved surfaces for articulation with the bones of the forearm. The rounded **capitulum** (kă-pit′yū-lŭm; *caput* = head) is located laterally and articulates with the head of the radius. The pulley-shaped **trochlea** (trok′lē-ă; *trochileia* = a pulley) is located medially and articulates with the trochlear notch of the ulna. Additionally, the distal end of the humerus exhibits three depressions, two on its anterior surface and one on its posterior surface. The anterolaterally placed **radial fossa** accommodates the head of the radius; the anteromedially placed **coronoid** (kōr′ŏ-noyd; *korone* = a crown; *eidos* = resembling) **fossa** accommodates the coronoid process of the ulna. The posterior depression, called the **olecranon** (ō-lek′ră-non; *olene* = ulna, *kranion* = head) **fossa,** accommodates the olecranon of the ulna when the elbow is extended (straight).

WHAT DID YOU LEARN?

21 What is the difference between the anatomical and surgical neck of the humerus?

22 What features on the humerus articulate with the radius and ulna?

8.9b Radius and Ulna

LEARNING OBJECTIVES

3. Compare and contrast the features of the radius and the ulna.

4. Explain how the radius, ulna, and humerus articulate.

5. Differentiate between supination and pronation of the forearm.

The radius and ulna form the forearm **(figure 8.26)**. In anatomic position, these bones are parallel, and the **radius** (rā′dē-ŭs; spoke of

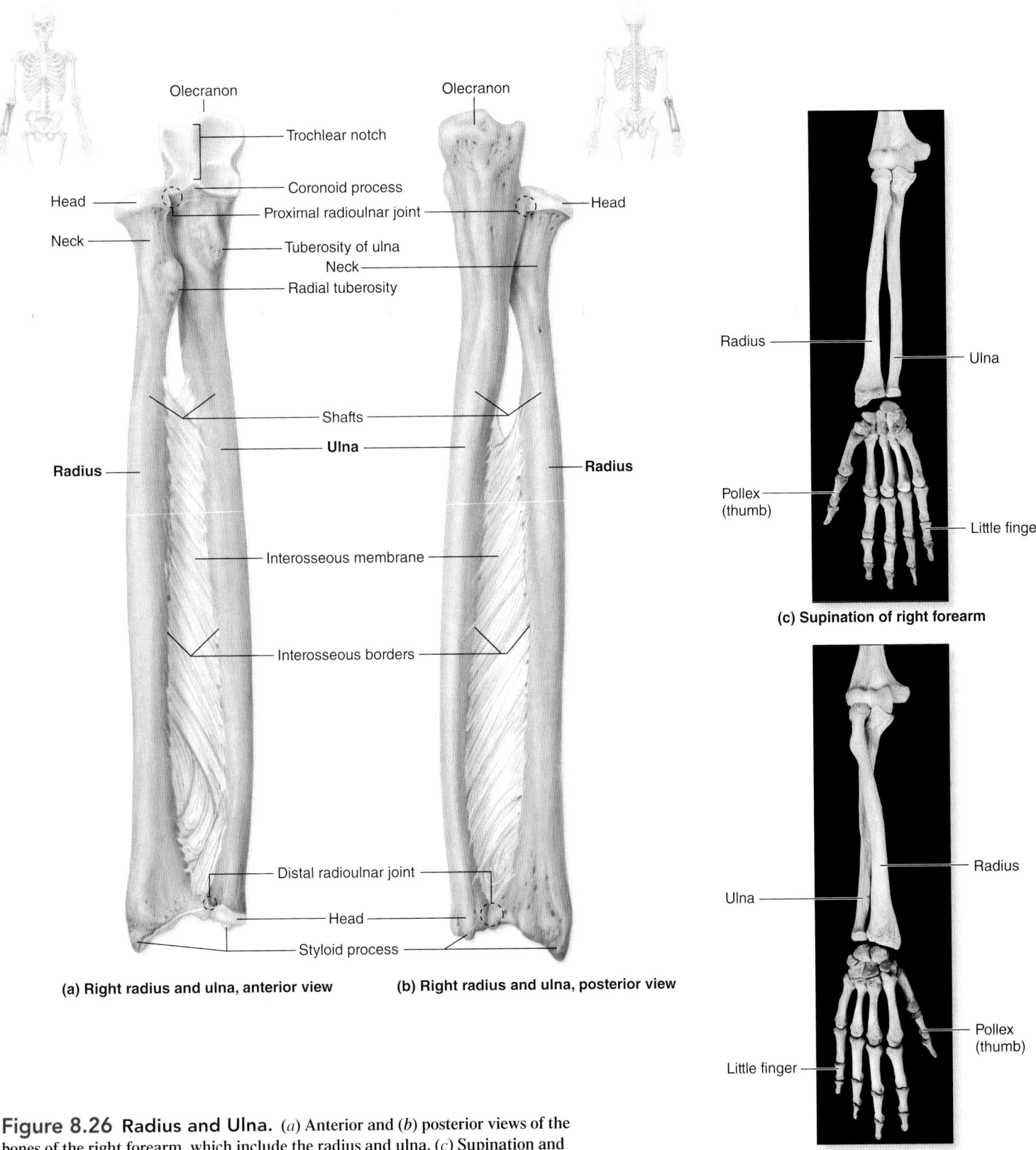

(a) Right radius and ulna, anterior view

(b) Right radius and ulna, posterior view

(c) Supination of right forearm

(d) Pronation of right forearm

Figure 8.26 Radius and Ulna. (*a*) Anterior and (*b*) posterior views of the bones of the right forearm, which include the radius and ulna. (*c*) Supination and (*d*) pronation of the right forearm. AP|R

a wheel, ray) is located more laterally. The proximal end of the radius has a distinctive disc-shaped **head** that articulates with the capitulum of the humerus. A narrow **neck** extends from the radial head to the **radial tuberosity** (or *bicipital tuberosity*). The radial tuberosity is an attachment site for the biceps brachii muscle.

The **shaft** of the radius curves slightly and leads to a wider distal end where there is a laterally placed **styloid** (stī′loyd) **process.** This bony projection can be palpated on the lateral side of the wrist, just proximal to the thumb. On the distal medial surface of the radius is an **ulnar notch,** which articulates with the medial surface of the distal end of the ulna.

The **ulna** (ŭl′nǎ; *olene* = elbow) is the longer, medially placed bone of the forearm. At the proximal end of the ulna, a C-shaped **trochlear notch** interlocks with the trochlea of the humerus. The posterosuperior aspect of the trochlear notch has a prominent projection called the **olecranon.** The olecranon articulates with the olecranon fossa of the humerus and forms the posterior "bump" of the elbow. The inferior lip of the trochlear notch, called the **coronoid process,** articulates with the humerus at the coronoid fossa. Lateral to the coronoid process, a smooth, curved radial notch accommodates the head of the radius and helps form the proximal radioulnar joint. Also at the proximal end of this bone is the **tuberosity of ulna.** At the distal end of the ulna, the shaft narrows and terminates in a knoblike **head** that has a posteromedial **styloid process.** The styloid process of the ulna may be palpated on the medial ("little finger") side of the wrist.

Both the radius and the ulna exhibit **interosseous borders,** which face each other; the ulna's interosseous border faces laterally, while the radius' interosseous border faces medially. These interosseous borders are connected by an **interosseous membrane** (*interosseous ligament*) composed of dense regular connective tissue. This membrane helps keep the radius and ulna a fixed distance apart from one another and provides a pivot of rotation for the forearm. The bony joints that move during this rotation are the proximal and distal radioulnar joints.

In anatomic position, the palm of the hand is facing anteriorly, and the bones of the forearm are said to be in **supination** (sū′pi-nā′shǔn) (figure 8.26c). Note that the radius and the ulna are parallel with one another. If you view your own supinated forearm, the radius is on the lateral (thumb) side of the forearm, and the ulna is on the medial (little finger) side.

Pronation (prō-nā′shǔn) of the forearm requires that the radius cross over the ulna and that both bones pivot along the interosseous membrane (figure 8.26d). When the forearm is pronated, the palm of the hand is facing posteriorly and the head of the radius is still along the lateral side of the elbow, but the distal end of the radius has crossed over and become a more medial structure.

When an individual has the upper limbs extended and forearms supinated, note that the bones of the forearm may angle laterally from the elbow joint. This positioning is referred to as the **carrying angle** of the elbow, and this angle positions the bones of the forearms such that the forearms will clear the hips during walking as the forearms swing. Females have wider carrying angles than males, presumably because they have wider hips than males.

INTEGRATE

LEARNING STRATEGY ✐

No matter what the position of the forearm (pronated or supinated), the distal end of the radius is always near the thumb, and the distal end of the ulna is near the little finger.

WHAT DID YOU LEARN?

23 What are some bony features that the radius and ulna share?

24 Describe how the radius and ulna are positioned when the forearm is pronated.

8.9c Carpals, Metacarpals, and Phalanges

LEARNING OBJECTIVES

6. Locate and identify the carpals and metacarpals.
7. Describe the phalanges and their relative locations.

The bones that form the wrist and hand are the carpals, metacarpals, and phalanges **(figure 8.27)**. The **carpals** (kar′pǎl; *karpus* = wrist) are small, short bones that form the wrist. They are arranged in two rows (a proximal row and a distal row) of four bones each and allow for the multiple movements possible at the wrist.

The *proximal row* of carpal bones, listed from lateral to medial, are the **scaphoid** (skaf′oyd; *skaphe* = boat), **lunate** (lū′nāt; *luna* = moon), **triquetrum** (trī-kwē′trŭm; *triquetrus* = three-cornered), and **pisiform** (pis′i-fōrm; *pisum* = pea, *forma* = appearance).

The *distal row* of the carpal bones, listed from lateral to medial, are the **trapezium** (tra-pē′zē-ŭm; *trapeza* = table), **trapezoid** (trap′ě-zoyd), **capitate** (kap′i-tāt), and **hamate** (ha′māt; *hamus* = hook).

Bones in the palm of the hand are called **metacarpals** (met′ă-kar′pǎl; *meta* = beyond). Five metacarpal bones articulate with the distal carpal bones and support the palm. Roman numerals I–V denote the metacarpal bones, with metacarpal I located at the base of the thumb, and metacarpal V at the base of the little finger.

A total of 14 bones are present in the digits; these are called **phalanges** (fǎ-lan′jēz; sing., *phalanx,* fā′langks; line of soldiers). Three phalanges are found in each of the second through fifth fingers,

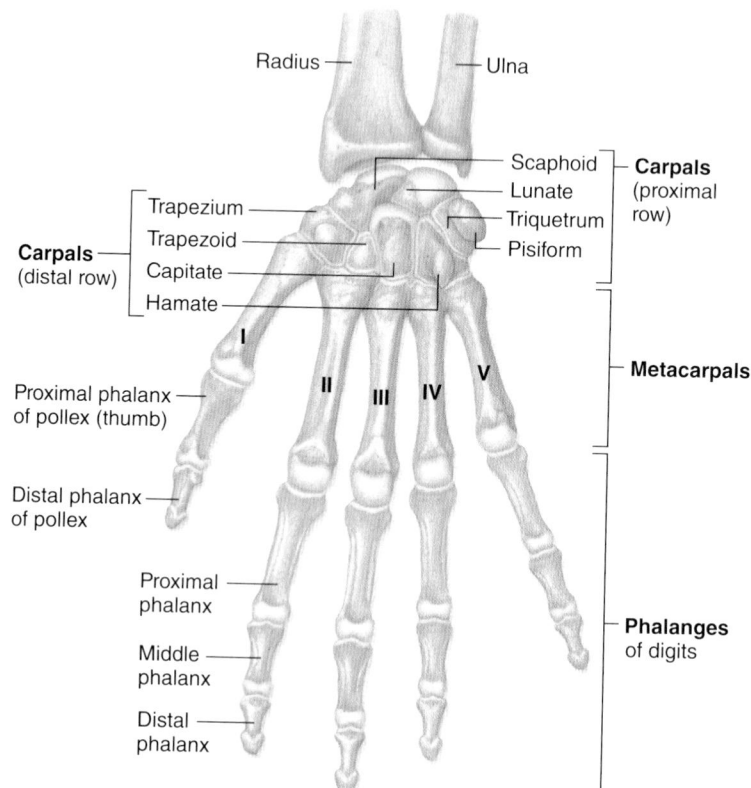

Radius — Ulna

Trapezium
Trapezoid
Carpals (distal row)
Capitate
Hamate

Scaphoid
Lunate — **Carpals** (proximal row)
Triquetrum
Pisiform

I
II III IV V

Proximal phalanx of pollex (thumb)

Distal phalanx of pollex

Proximal phalanx

Middle phalanx

Distal phalanx

Metacarpals

Phalanges of digits

Right wrist and hand, anterior view

Figure 8.27 Bones of the Carpals, Metacarpals, and Phalanges. The carpal bones form the wrist, and the metacarpals and phalanges form the hand. Anterior (palmar) view of the right wrist and hand. AP|R

INTEGRATE

CLINICAL VIEW
Scaphoid Fractures

The scaphoid bone is one of the more commonly fractured carpal bones. A fall on the outstretched hand may cause the scaphoid to fracture into two separate pieces. Usually, blood vessels are torn on the proximal part of the scaphoid, resulting in **avascular necrosis**, which is death of the bone tissue due to inadequate blood supply. Scaphoid fractures take a very long time to heal properly due to this complication.

but only two phalanges are present within the thumb, also known as the **pollex** (pol′eks; thumb). The **proximal phalanx** articulates with the head of a metacarpal, whereas the **distal phalanx** is the bone in the very tip of the finger. The **middle phalanx** of each finger lies between the proximal and distal phalanges; however, a middle phalanx is not present in the pollex.

WHAT DID YOU LEARN?

25 List the eight carpal bones. Which of these bones may be prone to developing avascular necrosis if fractured?

8.10 The Pelvic Girdle and Its Functions

The adult **pelvis** (pel′vis; pl., *pelves,* pel′vēz; basin) is composed of four bones: the sacrum, the coccyx, and the right and left **ossa coxae** (os′ă kok′să; sing., *os coxae;* hip bone) **(figure 8.28)**. The pelvis protects and supports the viscera in the inferior part of the ventral body cavity.

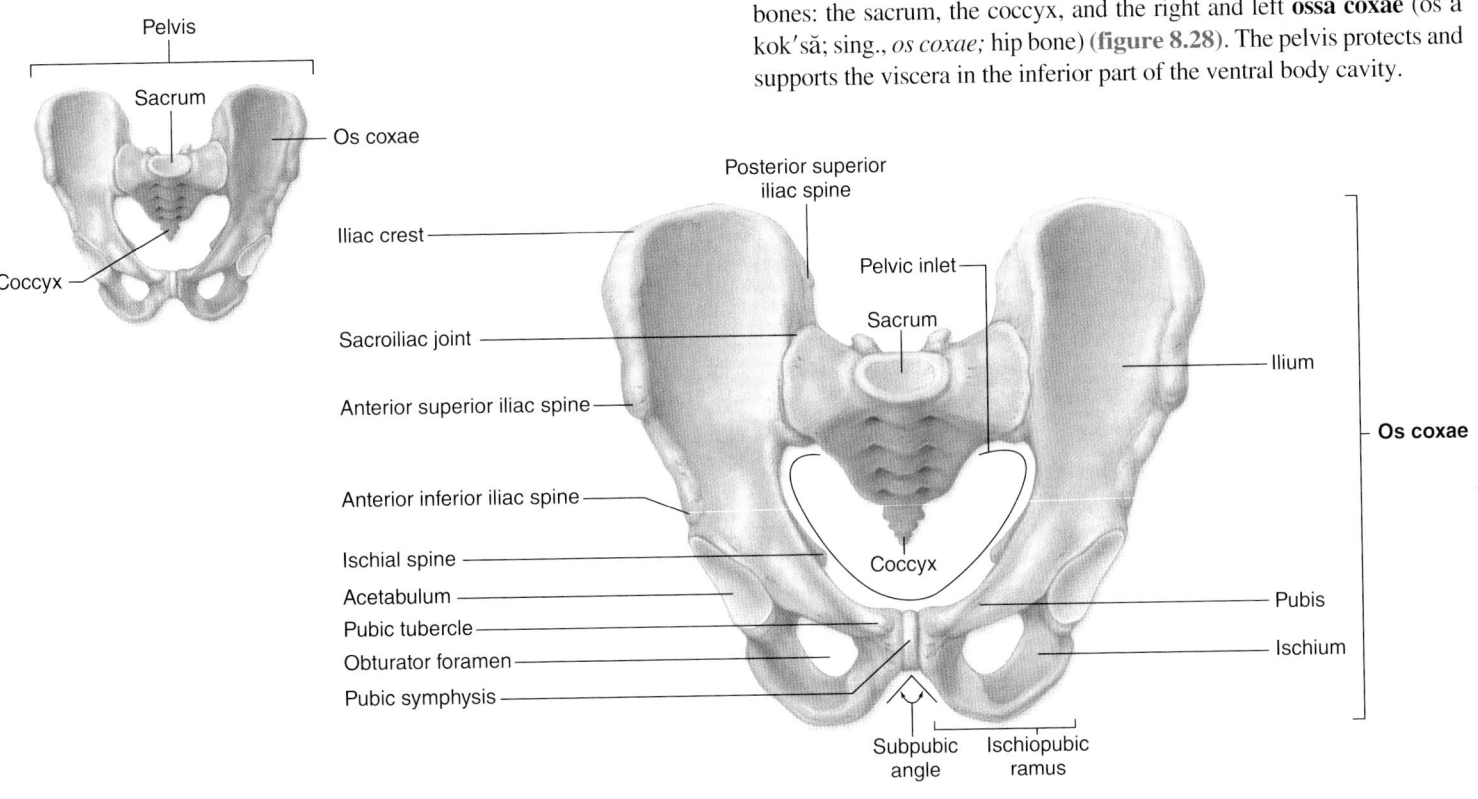

(a) Anterior view

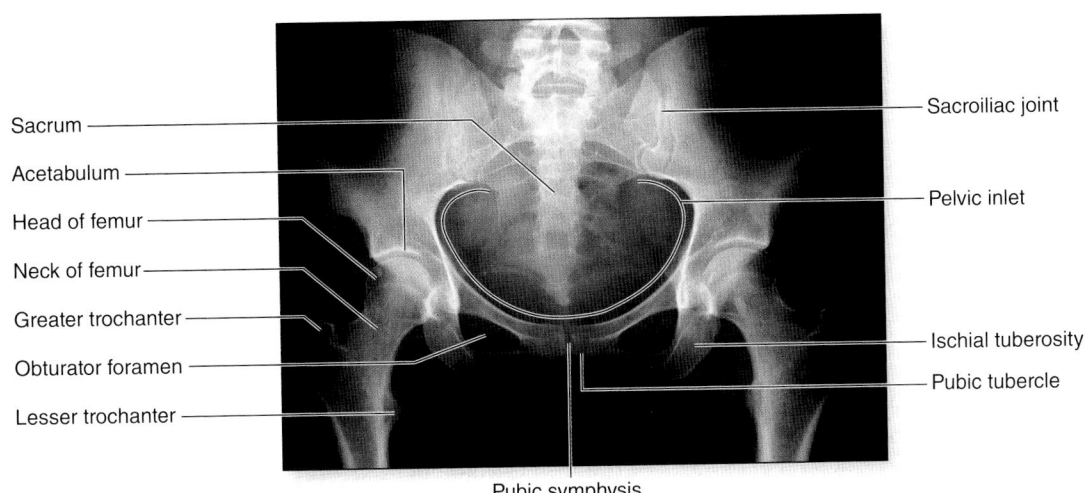

(b) Pelvis radiograph, anterior view

Figure 8.28 Pelvis. The complete pelvis consists of (*a*) the two ossa coxae, the sacrum, and the coccyx. (*b*) A radiograph shows an anterior view of the articulation between the pelvis and the femur. AP|R

The term **pelvic girdle** refers to both the left and right ossa coxae only. The pelvic girdle articulates with the trunk and provides an attachment point for each lower limb. When a person is standing upright, the pelvis is angled (or 'tipped') slightly anteriorly.

8.10a Os Coxae

 LEARNING OBJECTIVES

1. Name the three bones that make up each os coxae.
2. Describe how the ossa coxae articulate with each femur and sacrum.
3. Describe landmarks and features of an os coxae.

The os coxae is commonly referred to as the *hip bone* (and sometimes as the *coxal bone* or the *innominate bone*). Each os coxae is formed from three separate bones: the ilium, ischium, and pubis **(figure 8.29)**. These three bones fuse between the ages of 13 and 15 years to form the single os coxae.

Each os coxae articulates posteriorly with the sacrum at the sacroiliac joint. The femur articulates with a deep, curved depression on the lateral surface of the os coxae called the **acetabulum** (as-ĕ-tab′yū-lŭm; shallow cup). The acetabulum contains a smooth curved surface, called the **lunate surface,** which is C-shaped and articulates with the femoral head. The ilium, ischium, and pubis all contribute a portion to the acetabulum—thus, it represents a region where these bones have fused.

WHAT DO YOU THINK?

4 Compare and contrast the glenoid cavity of the scapula with the acetabulum of the os coxae. Which girdle maintains stronger, more tightly fitting bony connections with its respective limb—the pectoral girdle or the pelvic girdle? Explain.

The largest of the three coxal bones is the **ilium** (il′ē-ŭm; groin, flank), which forms the superior region of the os coxae and part of the acetabular surface. The wide, fan-shaped portion of the ilium is called the **ala** (ā′lă; wing). The ala terminates inferiorly at a ridge called the **arcuate** (ar′kyū-āt; *arcuatus* = bowed) **line** on the medial surface of the ilium. On the medial side of the ala is a depression termed the **iliac fossa.** On the lateral surface of the ilium, the **anterior, posterior,** and **inferior gluteal** (glū′tē-ăl; *gloutos* = buttock) **lines** are attachment sites for the gluteal muscles of the buttock (see section 11.9a). The posteromedial side of the ilium exhibits a large, roughened area called the **auricular** (aw-rik′yū-lăr; *auris* = ear) **surface,** where the ilium articulates with the sacrum.

The superiormost ridge of the ilium is the **iliac crest.** Palpate the posterosuperior edges of your hips; the ridge of bone you feel on each side is the iliac crest. The iliac crest arises anteriorly from a projection called the **anterior superior iliac spine** and extends posteriorly to the **posterior superior iliac spine.** Located inferiorly to the ala of the ilium are the **anterior inferior iliac spine** and the **posterior inferior iliac spine.** The posterior inferior iliac spine is adjacent to a prominent **greater sciatic notch** (sī-at′ik; *sciaticus* = hip joint), through which the sciatic nerve extends to the lower limb (see section 14.5g).

The ilium fuses with the **ischium** (is′kē-ŭm; *ischion* = hip) near the superior and posterior margins of the acetabulum. Posterior to the acetabulum, the prominent triangular **ischial** (is′kē-ăl) **spine** projects medially. The bulky bone superior to the ischial spine is called the **body** of the ischium. The **lesser sciatic notch** is a semicircular depression inferior to the ischial spine. The posterolateral border of the ischium is a roughened projection called the **ischial tuberosity.** The ischial tuberosities also are called the *sitz bones* by some health professionals and fitness instructors because they support the weight of the body when seated. If you palpate your buttocks while in a sitting position, you can feel the large ischial tuberosities. An elongated **ramus** (rā′mŭs; pl., *rami*, rā′mē) of the ischium extends from the ischial tuberosity toward its anterior fusion with the pubis.

The **pubis** (pyū′bis) fuses with the ilium and ischium at the acetabulum. The **ramus of ischium** fuses anteriorly with the **inferior pubic ramus** to form the **ischiopubic ramus** (figure 8.28). The **superior pubic ramus** originates at the anterior margin of the acetabulum. Between the superior and inferior pubic rami is an anteriorly placed mass of bone called the **body** of the pubis. The **obturator** (ob′tū-rā-tŏr; *obturo* = to occlude) **foramen** is a space in the os coxae that is encircled by both pubic and ischial rami. A roughened ridge, called the **pubic crest,** is located on the anterosuperior surface of the superior pubic ramus, and it ends at the **pubic tubercle.** A roughened area on the body of the pubis, called the **symphysial** (sim-fiz′ē-ăl; growing together) **surface** or *pubic symphysis*, denotes the site of articulation between the pubic bones. On the medial surface of the pubis, the **pectineal** (pek-tin′ē-ăl) **line** originates and extends diagonally across the pubis to merge with the arcuate line.

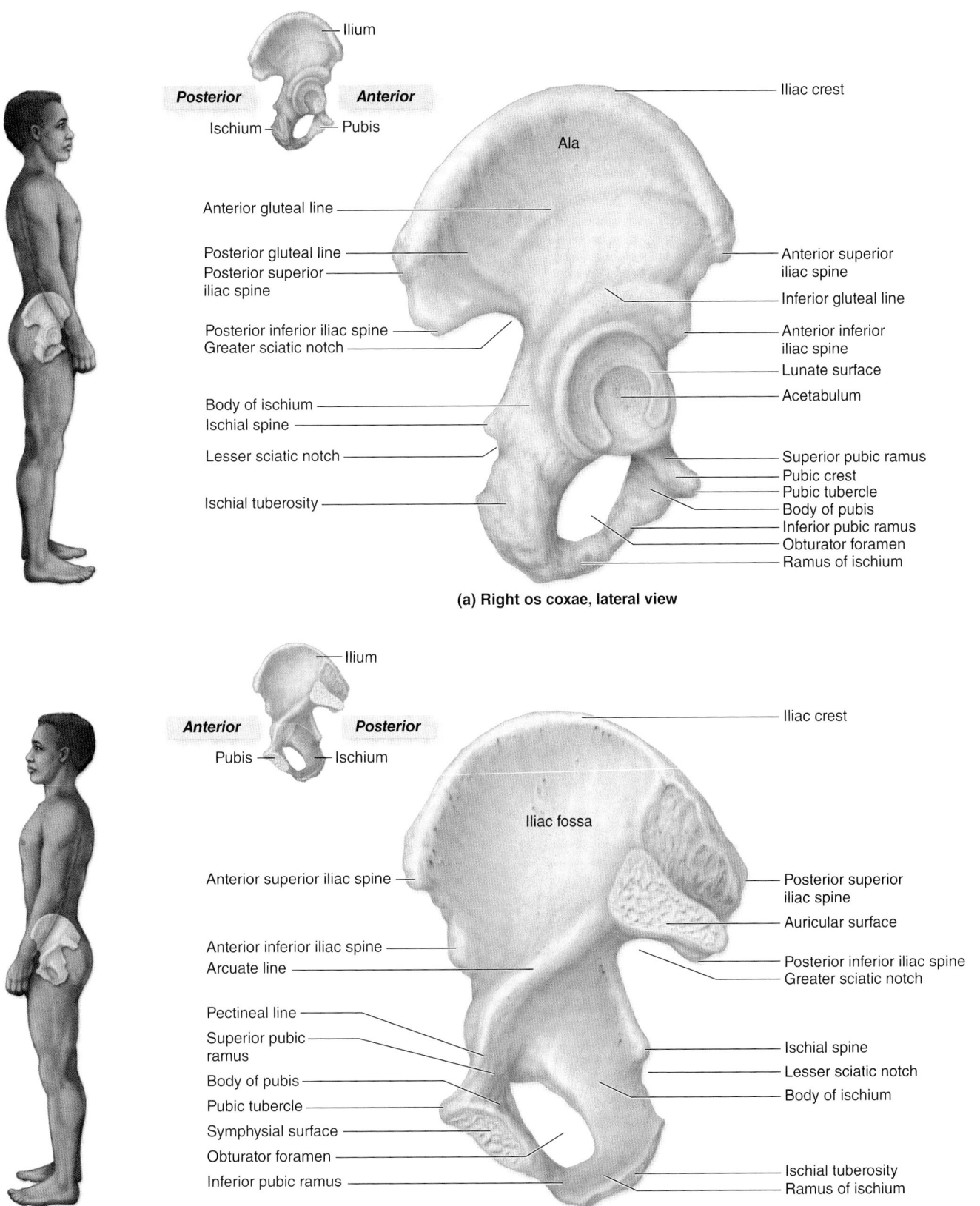

Ilium

Posterior   Anterior

Ischium   Pubis

Ilium

Iliac crest

Ala

Anterior gluteal line

Posterior gluteal line
Posterior superior
iliac spine

Posterior inferior iliac spine
Greater sciatic notch

Body of ischium
Ischial spine

Lesser sciatic notch

Ischial tuberosity

Anterior superior
iliac spine

Inferior gluteal line

Anterior inferior
iliac spine

Lunate surface

Acetabulum

Superior pubic ramus
Pubic crest
Pubic tubercle
Body of pubis
Inferior pubic ramus
Obturator foramen
Ramus of ischium

(a) Right os coxae, lateral view

Ilium

Anterior   Posterior

Pubis   Ischium

Iliac crest

Iliac fossa

Anterior superior iliac spine

Anterior inferior iliac spine
Arcuate line

Pectineal line
Superior pubic
ramus
Body of pubis
Pubic tubercle
Symphysial surface
Obturator foramen
Inferior pubic ramus

Posterior superior
iliac spine

Auricular surface

Posterior inferior iliac spine
Greater sciatic notch

Ischial spine

Lesser sciatic notch

Body of ischium

Ischial tuberosity
Ramus of ischium

(b) Right os coxae, medial view

Figure 8.29 Os Coxae. Each os coxae is formed by the fusion of three bones: an ilium, an ischium, and a pubis. Diagrams show the features of these bones in (*a*) lateral and (*b*) medial views. AP|R

WHAT DID YOU LEARN?

26 What three bones fuse to form the os coxae?

27 Where are the ischial tuberosities located, what is an alternative name for them, and what is their function?

8.10b True and False Pelves

LEARNING OBJECTIVES

4. Differentiate between the true and false pelves.
5. Compare and contrast the pelvic inlet and pelvic outlet.

The **pelvic brim** is a continuous oval ridge that extends from the pubic crest, pectineal line, and arcuate line to the rounded inferior edges of the sacral ala and promontory. This pelvic brim helps subdivide the entire pelvis into a true pelvis and a false pelvis **(figure 8.30)**. The **true pelvis**, also known as the *lesser pelvis*, lies *inferior* to the pelvic brim. It encloses the pelvic cavity and forms a deep bowl that contains the pelvic organs. The **false pelvis**, also known as the *greater pelvis*, lies *superior* to the pelvic brim. It is enclosed by the alae of the ilia. It forms the inferior region of the abdominal cavity and houses the inferior abdominal organs.

The pelvis also has a superior and an inferior opening, and each has clinical significance. The **pelvic inlet,** also known as the *superior pelvic aperture,* is the superiorly positioned space enclosed by the pelvic brim. In other words, the pelvic brim is the bony oval *ridge* of bone, whereas the pelvic inlet is the *space* surrounded by the pelvic brim. The pelvic inlet is the opening at the boundary between the true pelvis and the false pelvis.

The **pelvic outlet,** also known as the *inferior pelvic aperture,* is the inferiorly placed opening bounded by the coccyx, the ischial tuberosities, and the inferior border of the symphysial surface. In males, the ischial spines commonly project into the pelvic outlet, thereby narrowing the diameter of this outlet. In contrast, female ischial spines less frequently project into the pelvic outlet (so the birth canal will not be obstructed by these bony prominences). The pelvic outlet is covered with muscles and skin, and it forms the body region called the **perineum** (per′i-nē′ŭm). The width and size of the pelvic outlet is especially important in females, because the opening must be wide enough to accommodate the fetal head during childbirth (see section 29.6).

WHAT DID YOU LEARN?

28. How is the pelvic inlet distinguished from the pelvic outlet?

8.10c Sex Differences in the Pelvis

LEARNING OBJECTIVE

6. Compare and contrast the anatomy of male and female pelves.

Although it is possible to determine the sex of a skeleton by examining the skull, the most reliable indicator of sex is the pelvis, primarily the ossa coxae. The ossa coxae are the most sexually dimorphic bones of the body due to the demands of pregnancy and childbirth in females. For example, the female pelvis is shallower and wider than the pelvis

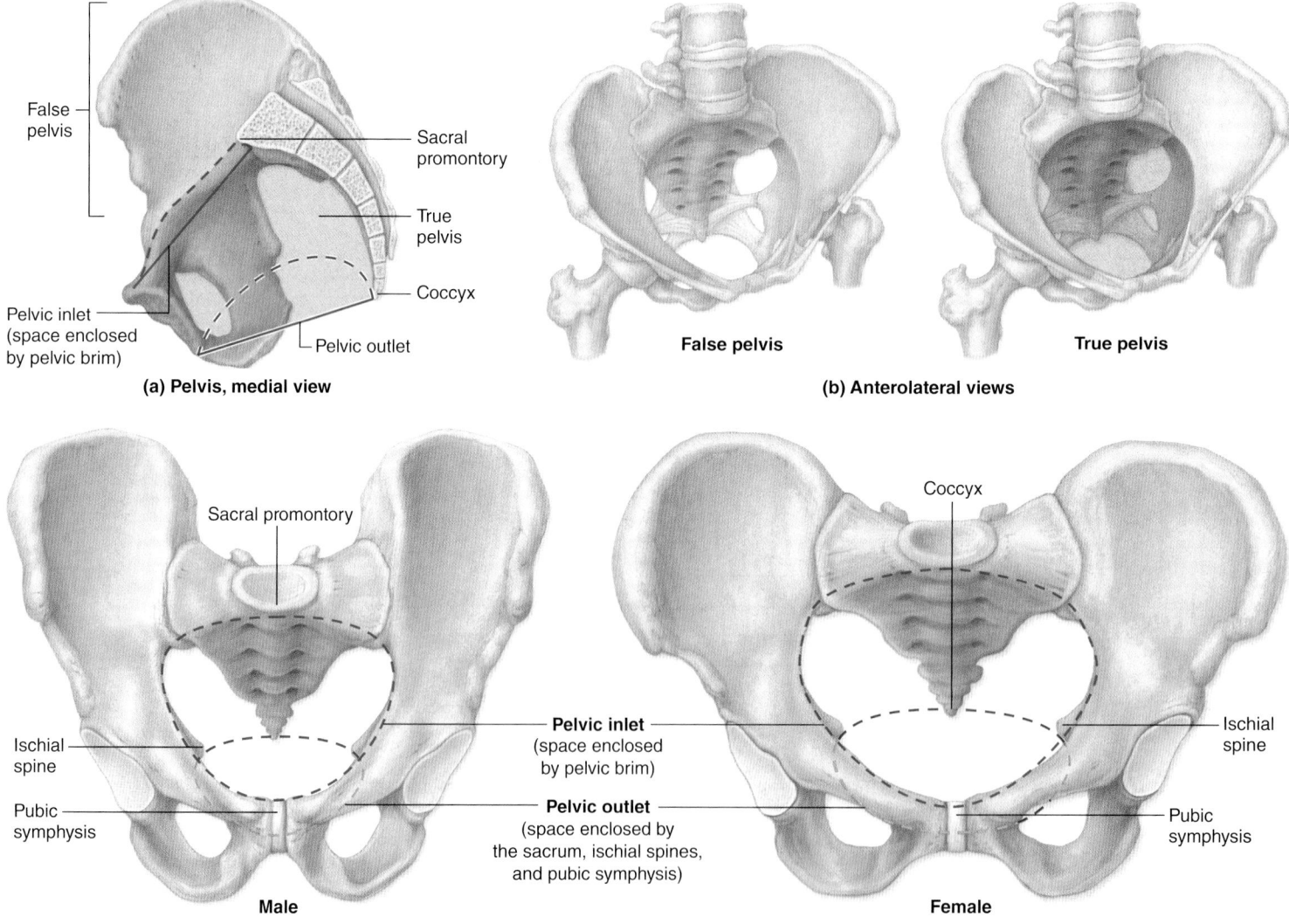

(a) Pelvis, medial view

(b) Anterolateral views

(c) Pelvis, Anterosuperior views

Figure 8.30 Features of the Pelvis. The pelvic brim is the oval bony ridge that subdivides the pelvis into a true pelvis and a false pelvis. The pelvic inlet is the space enclosed by the pelvic brim, whereas the pelvic outlet is the inferior opening in the true pelvis. (*a*) Medial and (*b*) anterolateral views of the true and false pelves. (*c*) Anterosuperior view of male and female pelves, demonstrating the sex differences between the pelvic inlet and outlet.

of a male to accommodate the infant's head as it passes through the birth canal.

Some of these differences are obvious, such as that males have narrower hips than females do. But we can find many other differences by examining the shapes and orientations of the pelvic bones. For example, the female ilium flares more laterally, whereas the male ilium projects more superiorly, which is why males typically have narrower hips. Because the female pelvis is wider, the acetabulum projects more laterally, and the greater sciatic notch is much wider as well. In contrast, the male acetabulum projects more anteriorly, and the male greater sciatic notch is much narrower, deeper and U-shaped. Females tend to have a **preauricular sulcus,** which is a depression or groove between the greater sciatic notch and the sacroiliac articulation. Males tend not to have this sulcus. The sacrum tends to be shorter and wider in females.

The body of the pubis in females is much longer and almost rectangular in shape, compared to the shorter, triangular-shaped male pubic body. The **subpubic angle** (or *pubic arch*) is the angle formed when the left and right pubic bones are aligned at their symphysial surfaces. Because females have much longer pubic bones, the corresponding subpubic angle is much wider and more convex, usually much greater than 100 degrees. The male subpubic angle is much narrower and typically does not extend past 90 degrees.

Several significant differences between the female and male pelves are shown and listed in **table 8.6.**

INTEGRATE

CONCEPT CONNECTION

The skeletal system and female reproductive system (see section 28.3) are linked by the fact that the shape of the bony pelvis has a direct relation on whether a female will have a difficult labor and delivery.

Table 8.6	Sex Differences Between the Female and Male Pelvis	
View	**Female**	**Male**
Medial View	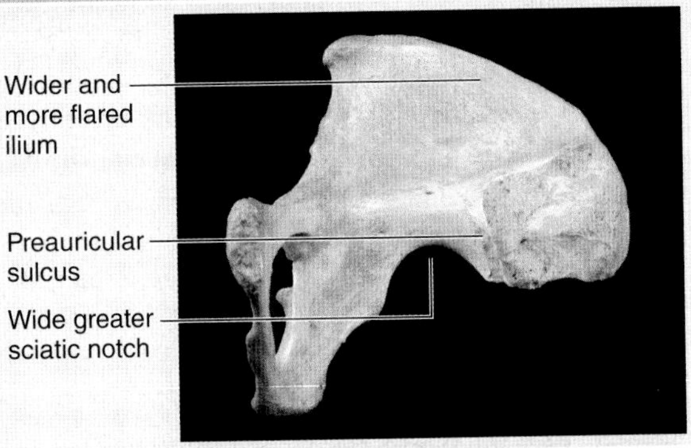 Wider and more flared ilium / Preauricular sulcus / Wide greater sciatic notch	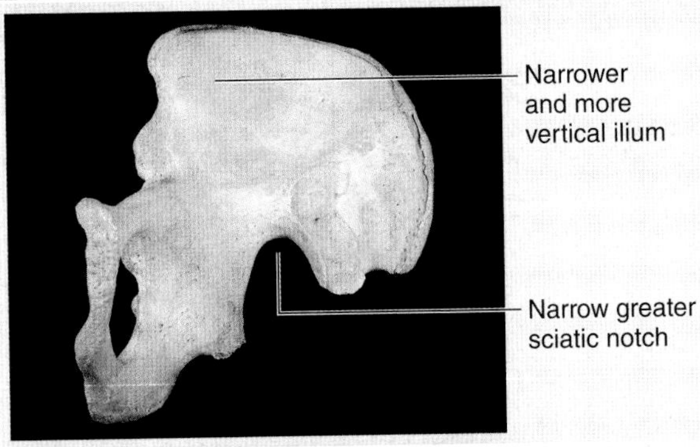 Narrower and more vertical ilium / Narrow greater sciatic notch
Anterior View	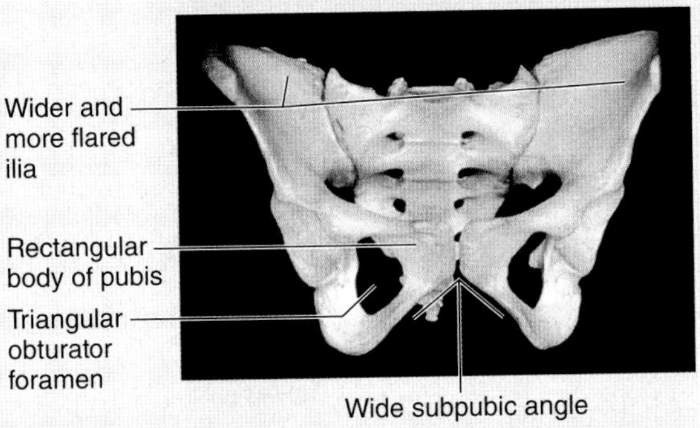 Wider and more flared ilia / Rectangular body of pubis / Triangular obturator foramen / Wide subpubic angle	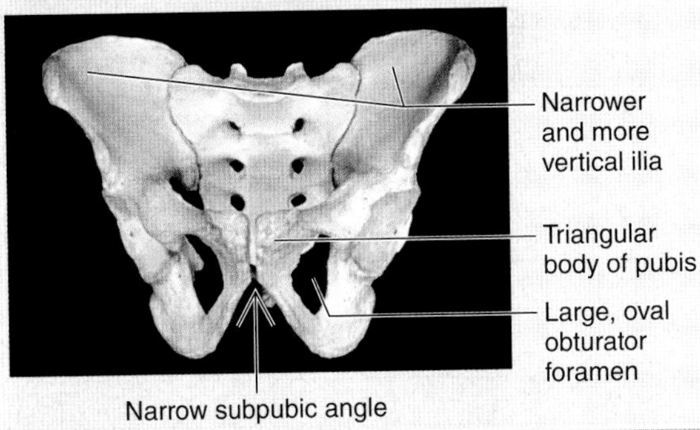 Narrower and more vertical ilia / Triangular body of pubis / Large, oval obturator foramen / Narrow subpubic angle

Feature	Female Characteristic	Male Characteristic
General appearance	Less massive; gracile processes, less prominent muscle markings	More massive; more robust processes, more muscle markings
General width	Ilia wider, more flared	Ilia narrower and more vertically oriented, less flared
Pelvic inlet	Spacious, wide and oval	Heart-shaped
Greater sciatic notch	Wide and shallow	Narrow and U-shaped, deep
Obturator foramen	Smaller and triangular	Larger and oval
Subpubic angle	Broader, more convex, usually greater than 100 degrees	Narrow, V-shaped, usually less than 90 degrees
Body of pubis	Longer, more rectangular	Short, triangular
Preauricular sulcus	Usually present	Usually absent
Sacrum	Shorter and wider; flatter sacral curvature	Narrower and longer; more curved (greater sacral curvature)
Ischial spine	Rarely projects into pelvic outlet	Frequently rotated inward, projects into pelvic outlet

WHAT DID YOU LEARN?

29 How do male and female pelves differ with respect to the shape of the pubis, subpubic angle, greater sciatic notch, and overall shape of the pelvis?

8.10d Age Differences in the Ossa Coxae

 LEARNING OBJECTIVE

7. Describe changes to the ossa coxae as a person ages.

The ossa coxae are an excellent indicator of both sex and age, and they also can provide a reliable estimate of age at death. These estimates are given in age ranges (as opposed to precise numbers) because some variation may occur in how an os coxae exhibits the age-related changes.

Osteologists have noted age-related changes to the auricular surface of the ilium. The auricular surface of a young adult typically has some billowing texture to it (e.g., appears to have "hills and valleys"), and the surface is fine-grained. As the auricular surface ages, the billowing flattens out and the surface becomes more coarse and granular. In much older individuals, the surface may develop some bony lipping (evidence of osteoarthritic changes) and the surface becomes even more rough and irregular.

Osteologists also have documented that the symphysial surface of the pubis undergoes uniform, age-related changes as well. In fact, the symphysial surface has become one of the most reliable indicators for estimating age at death. In a young adult (age range from 15–24), the symphysial surface is billowed, and no well-formed rim is found around the surface. As the person ages, this billowing becomes more flattened, and a bony rim begins to form around the circumference of the symphysial surface. This rim is completed about ages 35–50 for most individuals. Once the rim is complete, the symphysial surface becomes depressed and concave and may become pitted in much older individuals. The rim or border may start to break down, and bony lipping (arthritis) develops along the edges of the symphysial surface. These last stages typically occur after age 50.

WHAT DID YOU LEARN?

30 What are some differences in the symphysial surface between a young adult and an older (50-plus) adult?

8.11 Bones of the Lower Limb

The lower limb makes up the thigh, leg, and foot. The structure of the foot enables it to support the body during bipedal walking and running.

The arrangement and numbers of bones in the lower limb are similar to those of the upper limb. Each lower limb contains a total of 30 bones:

- 1 femur, located in the femoral region
- 1 patella (kneecap), located in the patellar region
- 1 tibia and 1 fibula, located in the crural region
- 7 tarsal bones, which form the bones of the ankle and proximal foot
- 5 metatarsal bones, which form the arched part of the foot
- 14 phalanges, which form the toes

8.11a Femur and Patella

LEARNING OBJECTIVES

1. Describe the articulations of the femur.

2. Identify key landmarks and features of the femur.

3. Describe the location and function of the patella.

The **femur** (fē′mŭr; thigh) is the longest bone in the body as well as the strongest and heaviest **(figure 8.31)**. The nearly spherical **head** of the femur articulates with the os coxae at

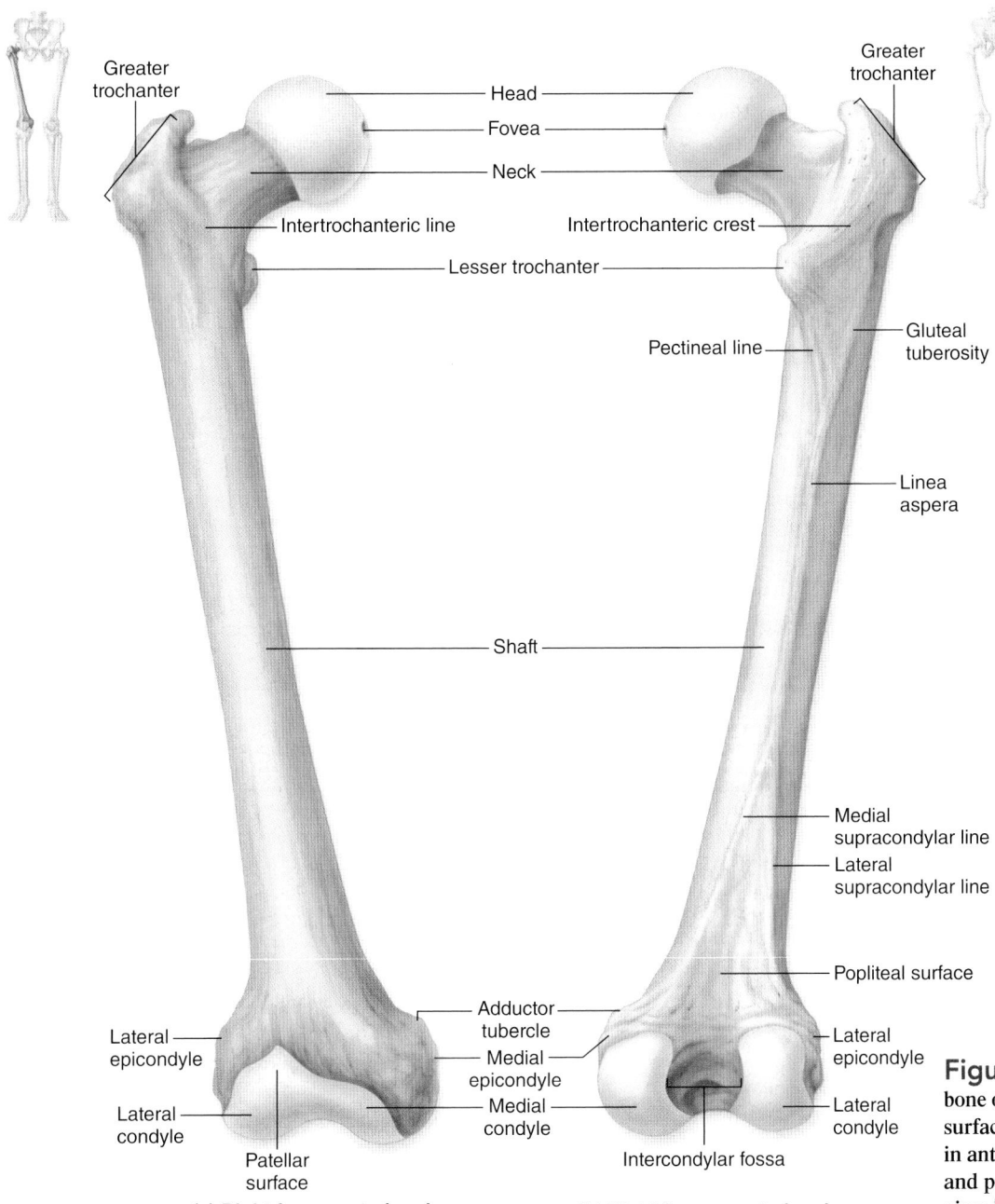

Greater trochanter

Head

Fovea

Neck

Intertrochanteric line

Lesser trochanter

Greater trochanter

Intertrochanteric crest

Pectineal line

Gluteal tuberosity

Linea aspera

Shaft

Medial supracondylar line

Lateral supracondylar line

Popliteal surface

Lateral epicondyle

Lateral condyle

Patellar surface

Adductor tubercle

Medial epicondyle

Medial condyle

Lateral epicondyle

Lateral condyle

Intercondylar fossa

(a) Right femur, anterior view

(b) Right femur, posterior view

Figure 8.31 Femur. The femur is the bone of the femoral region. (*a*) The patellar surface and intertrochanteric line are best seen in anterior view, and (*b*) the lesser trochanter and popliteal surface are best seen in posterior view. **AP|R**

the acetabulum. There is a small depression within the head of the femur, called the **fovea** (fō′vē-ă; a pit), or *fovea capitis*. Here a small ligament connects the head of the femur to the acetabulum. Distal to the head, an elongated, constricted **neck** joins the **shaft** of the femur at an angle. This results in a medial angling of the femur, which brings the knees closer to the midline.

The **greater trochanter** (trō-kan′ter; a runner) projects laterally from the junction of the neck and shaft. A **lesser trochanter** is located on the femur's posteromedial surface. These are rough processes that serve as insertion sites for powerful gluteal and thigh muscles. The greater and lesser trochanters are connected on the posterior surface of the femur by a thick oblique ridge of bone called the **intertrochanteric** (in′ter-trō-kan-tār′ik) **crest.** Anteriorly, a raised **intertrochanteric line** extends between the two trochanters and marks the distal edge of the hip joint capsule. Inferior to the intertrochanteric crest, the **pectineal line** marks the attachment of the pectineus muscle; the **gluteal** (glŭ′tē-ăl; *gloutos* = buttock) **tuberosity** marks the attachment of the gluteus maximus muscle (see section 11.9a).

The gluteal tuberosity and pectineal line merge into an elevated, midline ridge called the **linea aspera** (lin′ē-ă as′pĕr-ă;), where many thigh muscles attach. Distally, the linea aspera branches into **medial** and **lateral supracondylar lines.** A flattened triangular area, called the

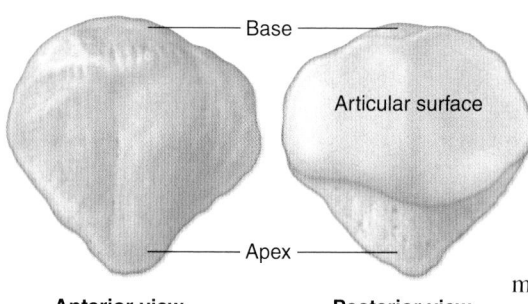

Base

Articular surface

Apex

Anterior view   Posterior view

Figure 8.32 Patella. The patella is a sesamoid bone located within the tendon of the quadriceps femoris muscle. These views show the right patella. **AP|R**

popliteal (pop-lit′ē-ăl; *poples* = ham of knee) **surface,** is bordered by these ridges. The medial supracondylar ridge terminates in the **adductor tubercle.** This is a rough, raised projection that is the site of attachment for the adductor magnus muscle (see section 11.9).

On the distal, inferior surface of the femur there are two smooth, oval articulating surfaces called the **medial** and **lateral condyles.** Superior to each condyle are projections called the **medial epicondyle** and **lateral epicondyle,** respectively. When you flex your knee, you can palpate these epicondyles in the thigh on the sides of your knee joint. The medial and lateral supracondylar lines terminate at these epicondyles. On the distal posterior surface of the femur, a deep **intercondylar fossa** separates the two condyles. A smooth medial depression on the anterior surface, called the **patellar surface,** is the place where the patella articulates with the femur.

The **patella** (pa-tel′ă; *patina* = shallow disk), or *kneecap* is a large, roughly triangular sesamoid bone housed within the tendon of the quadriceps femoris muscle (**figure 8.32**). The patella allows the tendon to glide more smoothly, and it protects the knee joint. The superior **base** of the patella is broad, whereas its inferior **apex** is pointed. The posterior aspect of the patella has an **articular surface** that articulates with the patellar surface of the femur.

WHAT DID YOU LEARN?

31 What are the locations and functions of the greater and lesser trochanter?

32 Where does the patella articulate with the femur?

8.11b Tibia and Fibula

LEARNING OBJECTIVES

4. Describe the features of the tibia and fibula.

5. Explain how the function of the tibia differs from that of the fibula.

6. Describe how the tibia and fibula articulate.

The skeleton of the leg (crural region) has two parallel bones, the thick, strong tibia and a slender fibula (**figure 8.33**). Like the radius and ulna, these two bones are connected by an interosseous membrane that extends between their **interosseous borders.** The interosseous membrane stabilizes the relative positions of the tibia and fibula, and additionally provides a pivot of minimal rotation for the two bones.

The **tibia** (tib′ē-ă; large shinbone) is the medially placed bone and the only weight-bearing bone of the crural region. Its broad, superior head has two relatively flat surfaces, the **medial** and **lateral condyles,** which articulate with the medial and lateral condyles of the femur, respectively. Separating the condyles of the tibia is a prominent ridge called the **intercondylar eminence** (em′i-nens). On the proximal posterolateral side of the tibia is a **fibular articular surface,** where the head of the fibula articulates to form the **superior** (or *proximal*) **tibiofibular joint.**

The rough anterior surface of the tibia near the proximal condyles is the **tibial tuberosity,** which can be palpated just inferior to the patella and marks the attachment site for the patellar ligament. The **anterior border** (or *margin*), often referred to as the *shin,* is a prominent ridge that extends distally along the anterior tibial surface from the tibial tuberosity.

The tibia narrows distally, but at its medial border it forms a large, prominent process called the **medial malleolus** (ma-lē′ō-lŭs; *malleus* = hammer). Palpate the medial side of your ankle; the bump you feel is your medial malleolus. There is a **fibular notch** on the distal posterolateral side of the tibia where the fibula articulates and forms the **inferior** (or *distal*) **tibiofibular joint.** On the inferior distal surface of the tibia is the smooth **inferior articular surface** for the talus, one of the tarsal bones.

The **fibula** (fib′yū-lă; buckle, clasp) is the long, thin, laterally placed bone of the leg. The fibula does not bear any weight, but several muscles originate from it. The rounded, knoblike **head** of the fibula is slightly inferior and posterior to the lateral condyle of the tibia. Distal to the fibular head is the **neck** of the fibula, followed by its **shaft.** The fibula's distal tip, called the **lateral malleolus,** extends laterally to the ankle joint, where it provides lateral stability. Palpate the lateral side of your ankle; the bump you feel is your lateral malleolus.

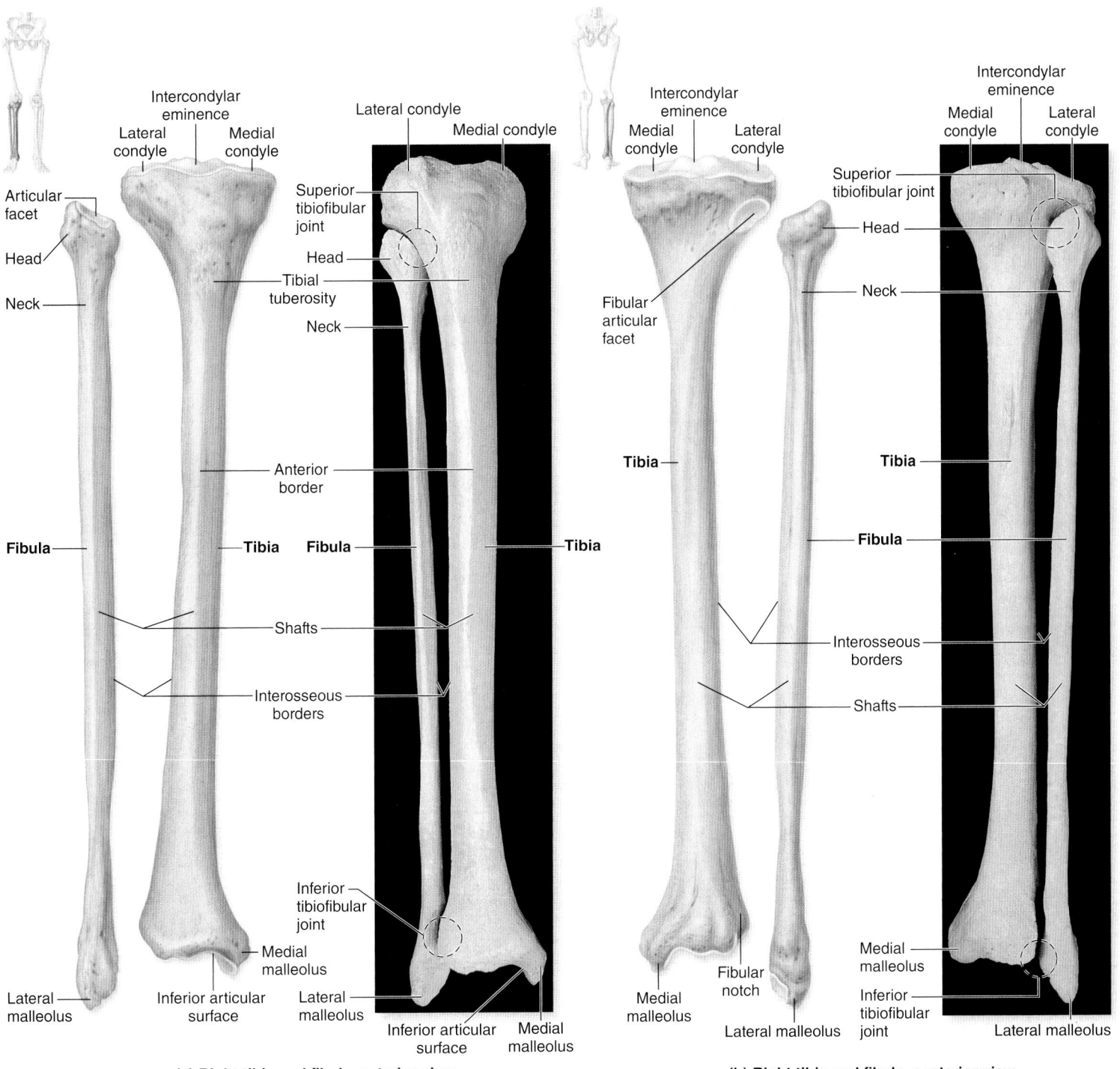

Figure 8.33 Tibia and Fibula. The tibia and fibula are the bones of the crural (leg) region. Diagram and photo show the right tibia and fibula in (a) an anterior view and (b) a posterior view. AP|R

(a) Right tibia and fibula, anterior view

(b) Right tibia and fibula, posterior view

WHAT DO YOU THINK?

5 The medial and lateral malleoli of the leg are similar to what bony features of the forearm?

WHAT DID YOU LEARN?

33 What are some bony features that are similar or the same between the tibia and fibula?

34 What is the primary function of the tibia?

CLINICAL VIEW
Pathologies of the Foot

A **bunion** (bŭn'yŭn; *buigne* = bump) is a localized swelling at the first metatarsophalangeal joint. This bump causes that toe to point toward the second toe instead of in a purely anterior direction. Bunions are usually caused by wearing shoes that fit too tightly, and are among the most common foot problems.

Pes cavus (pes că'vus), or *clawfoot*, is characterized by excessively high longitudinal arches. The joints between the metatarsals and proximal phalanges are often overly extended, and the joints between the different phalanges are bent so that they appear clawed.

Talipes equinovarus (tal'i-pēz ē-kwē-nō-vā'rŭs) is usually referred to as *congenital clubfoot*. This foot deformity typically occurs when there isn't enough room in the womb. In this condition, the feet are permanently inverted (the soles of the feet are twisted medially), and the ankles are plantar flexed (the soles of the feet are twisted more posteriorly), as if the patient were trying to stand on tiptoe.

Pes planus (pla'nus), commonly known as *flat feet*, is where the medial longitudinal arch is flattened so that the entire sole touches the ground. Pes planus is often caused by excessive weight, postural abnormalities, or weakened supporting tissue. Individuals who spend most of their day standing may have slightly fallen arches by the end of the day, but with proper rest their arches can return to normal shape.

A **metatarsal stress fracture** usually results when repetitive pressure or stress on the foot causes a small crack to develop in the outer surface of a metatarsal. Runners are especially prone to this injury because they put repetitive stress on their feet.

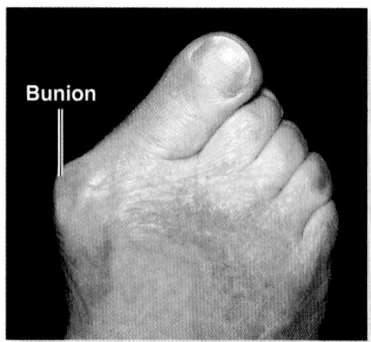

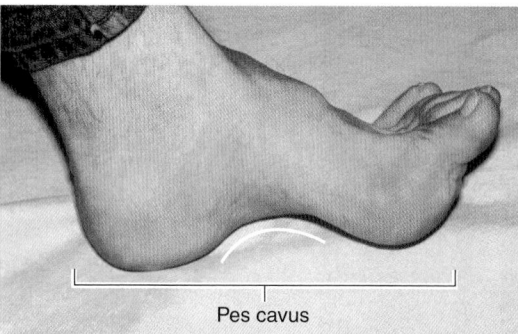

Pes cavus

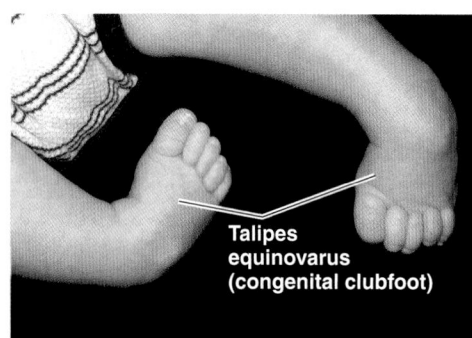

Talipes equinovarus (congenital clubfoot)

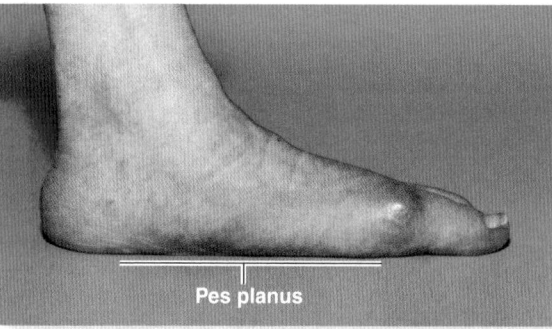

Pes planus

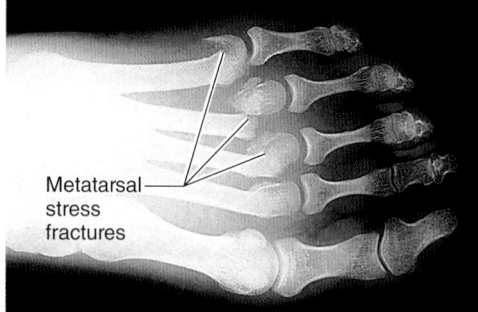

Metatarsal stress fractures

8.11c Tarsals, Metatarsals, and Phalanges

LEARNING OBJECTIVES

7. Locate and identify the tarsals and metatarsals.

8. Describe the phalanges and their relative locations.

The bones that form the ankle and foot are the tarsals, metatarsals, and phalanges (**figure 8.34**). The seven **tarsals** (tar'săl; *tarsus* = flat surface) of the ankle and proximal foot are similar to the eight carpal bones of the wrist in some respects, although their shapes and arrangement are different from those of their carpal bone counterparts.

The talus, calcaneus, and navicular bone are considered the *proximal row* of tarsal bones. The superiormost and second largest tarsal bone is the **talus** (tā'lŭs; ankle bone), which articulates with the tibia. The **calcaneus** (kal-kā'nē-ŭs) is the largest tarsal bone and forms the heel. Its posterior end is a rough, knob-shaped projection that is the point of attachment for the calcaneal (Achilles) tendon extending from the strong posterior leg muscles (see section 11.9c). The **navicular** (nă-vik'yū-lăr; *navis* = ship) **bone** is on the medial side of the ankle.

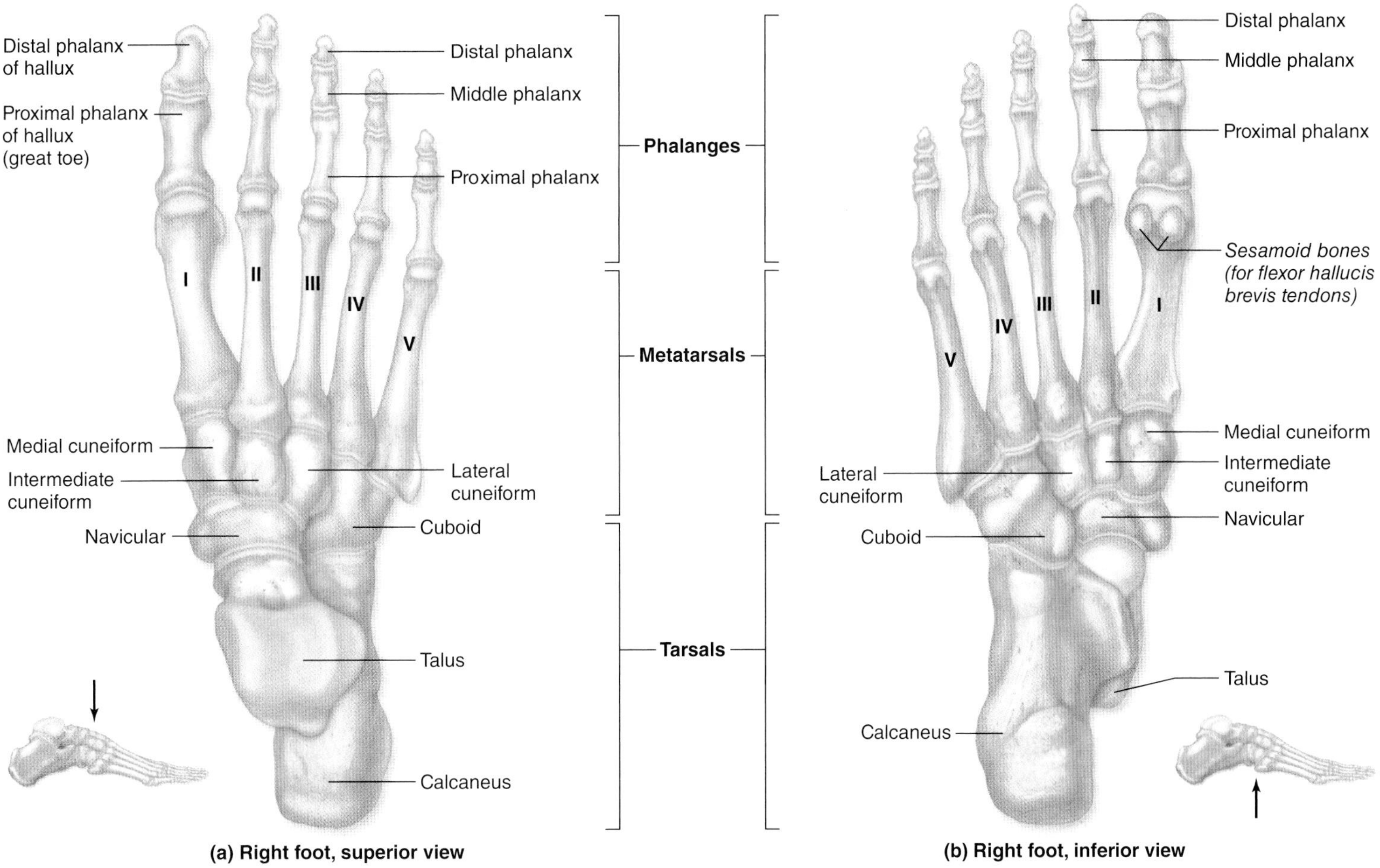

Distal phalanx — of hallux

Proximal phalanx — of hallux (great toe)

Distal phalanx
Middle phalanx
Proximal phalanx

I II III IV V

Phalanges

Medial cuneiform —
Intermediate — cuneiform
Navicular —

Lateral — cuneiform
Cuboid —

Metatarsals

Talus —

Calcaneus —

Tarsals

Distal phalanx
Middle phalanx
Proximal phalanx

Sesamoid bones (for flexor hallucis brevis tendons)

III II I
IV
V

Medial cuneiform
Intermediate cuneiform
Navicular

Lateral — cuneiform
Cuboid —

Talus

Calcaneus —

(a) Right foot, superior view

(b) Right foot, inferior view

Figure 8.34 Bones of the Tarsals, Metatarsals, and Phalanges. Tarsal bones form the ankle and proximal foot, metatarsals form the arched sole of the foot, and phalanges form the toes. Diagrams show (*a*) superior and (*b*) inferior views of the right foot. **AP|R**

The *distal row* of four tarsal bones includes the cuneiforms and the cuboid bone. The **medial cuneiform** (kū′nē-i-fōrm; *cuneus* = wedge), **intermediate cuneiform,** and **lateral cuneiform bones** are wedge-shaped bones that articulate with and are positioned anterior to the navicular bone. The laterally placed **cuboid** (kyū′boyd; *kybos* = cube) **bone** articulates at its medial surface with the lateral cuneiform and at its posterior surface with the calcaneus.

The **metatarsals** (met′ă-tar′săl) of the foot are five long bones similar in arrangement and name to the metacarpal bones of the hand. They form the arched sole of the foot and are identified with Roman numerals I–V, proceeding medially to laterally. The metatarsals articulate proximally either with the cuneiform bones or the cuboid bone. Distally, each metatarsal bone articulates with a proximal phalanx. At the head of the first metatarsal are two tiny sesamoid bones, which insert on the tendons of the flexor hallucis brevis muscle and help these tendons move more freely (see section 11.9d).

The bones of the toes (like the bones of the fingers and pollex) are called **phalanges.** The toes contain a total of 14 phalanges. The great toe is the **hallux** (hal′ŭks; *hallex* = great toe), and it has only two phalanges (proximal and distal); each of the other four toes has three phalanges (proximal, middle, and distal).

WHAT DID YOU LEARN?

35 What are the seven tarsal bones?

8.11d Arches of the Foot

LEARNING OBJECTIVE

9. Describe the three arches of the foot and their functions.

Normally, the sole of the foot is arched, which helps it support the weight of the body and ensures that the blood vessels and nerves on the sole of the foot are not pinched when we are standing. The three arches of the foot are the medial longitudinal, lateral longitudinal, and transverse arches (**figure 8.35**).

The **medial longitudinal arch** (*arcus* = bow) is the highest of the three arches and extends from the heel to the great toe. It is formed from the calcaneus, talus, navicular, and cuneiform bones, and metatarsals I–III. The medial longitudinal arch prevents the medial side of the foot from touching the ground and gives our footprint its characteristic shape (figure 8.35*d*).

The **lateral longitudinal arch** is not as high as the medial arch, so the lateral part of the foot *does* contribute to a footprint. This arch extends between the little toe and the heel, and it is formed from the calcaneus and cuboid bones, and metatarsals IV and V.

The **transverse arch** runs perpendicular to the longitudinal arches. It is formed from the distal row of tarsals and the bases of all five metatarsals.

The shape of the foot arches is maintained primarily by the foot bones themselves. These bones are shaped so that they can interlock and support their weight in an arch, much like the wedge-shaped blocks of an arched bridge can support the bridge without other mechanical supports. Secondarily, strong ligaments that attach to the bones and contracting muscles pull on the tendons, thereby helping to maintain the arches' shapes.

WHAT DID YOU LEARN?

36 Why is it preferable to have an arched (versus a flat) foot?

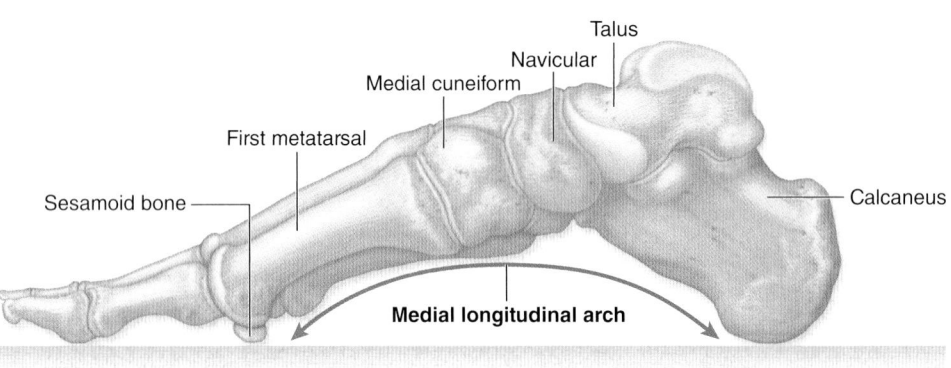

(a) Right foot, medial view

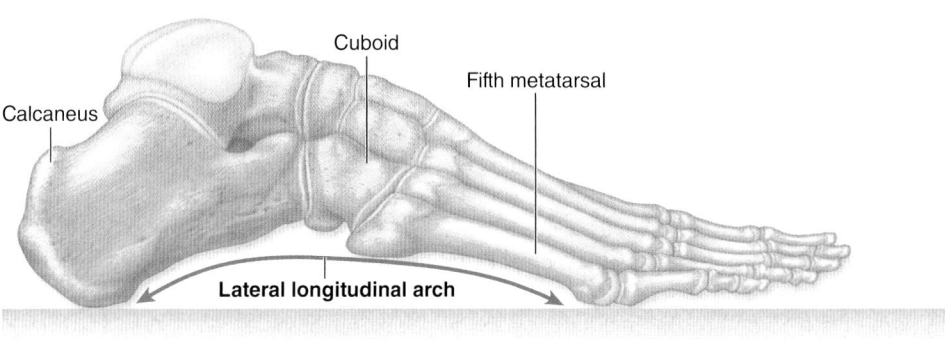

(b) Right foot, lateral view

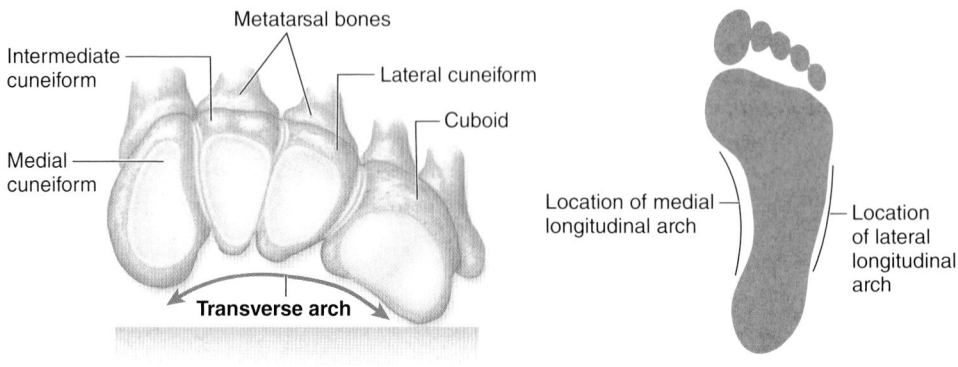

(c) Right foot, distal row of tarsals and metatarsals

(d) Footprint of right foot

Figure 8.35 Arches of the Foot. The foot's two longitudinal arches and one transverse arch allow for better weight support. (*a*) Medial longitudinal arch. (*b*) Lateral longitudinal arch. (*c*) Transverse arch as seen in cross-sectional view. (*d*) A footprint illustrates the placement of the longitudinal arches.

8.12 Development of the Skeleton

1. Describe how the limb buds form.
2. Compare and contrast upper and lower limb bud development.

As mentioned in section 7.4, bone forms by either intramembranous ossification or endochondral ossification. Many of the bones of the skull are formed from intramembranous ossification, whereas almost all of the remaining bones of the skeleton form through endochondral ossification. (The exception is the clavicle, which is formed from a central membranous ossification center; its ends are formed from endochondral ossification centers.)

The appendicular skeleton begins to develop during the fourth week, when **limb buds** appear as small ridges along the lateral sides of the embryo. The upper limb buds appear early in the fourth week (approximately day 26), and the lower limb buds appear a few days later (day 28) **(figure 8.36)**. Lower limb development lags behind upper limb development by about 2 to 4 days. The upper and lower limbs form proximodistally, meaning that the more proximal parts of the limbs form first (in weeks 4–5), whereas the more distal parts differentiate later.

Early limb buds are composed of lateral plate mesoderm and covered by a layer of ectoderm. The musculature of the limbs forms from somitic mesoderm that migrates to the developing limbs during the fifth week of development.

At the apex of each limb bud, part of the ectoderm forms an elevated thickening called the **apical ectodermal** (ek-tō-der′măl) **ridge.** By mechanisms not completely understood, this ridge "signals" the underlying tissue to form the various components of the limb.

Initially, the limb buds are cylindrical. The distal portion of the upper limb bud forms a rounded, paddle-shaped **hand plate** by the early fifth week. It later becomes both the palm and fingers. In the lower limb bud, a corresponding **foot plate** forms during the sixth week. These plates develop longitudinal thickenings called **digital rays,** which eventually form the digits. The digital rays in the hand plate appear in the late sixth week, and the foot digital rays appear during the early seventh week. The digital rays initially are connected by intermediately placed tissue, which later undergoes programmed cell death (apoptosis; see section 4.10). Thus, as this intermediate tissue dies, notching occurs between the digital rays, and separate digits are formed. This process occurs in the seventh week and is complete by the eighth week for both the fingers and the toes.

WHAT DID YOU LEARN?

 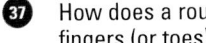

37 How does a rounded hand plate (or foot plate) evolve into separate fingers (or toes)?

Figure 8.36 Development of the Appendicular Skeleton. The upper and lower limbs develop between weeks 4 and 8. Upper limb development precedes corresponding lower limb development by 2 to 4 days.

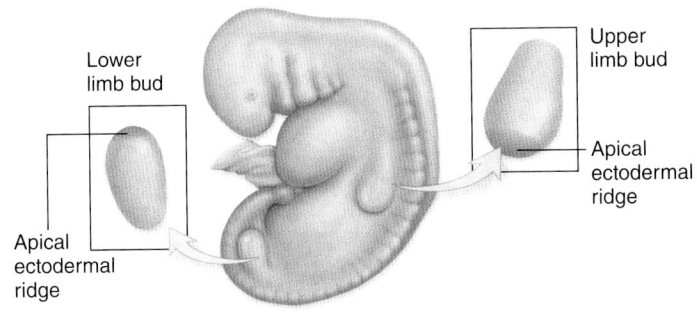

Week 4: Upper and lower limb buds form.

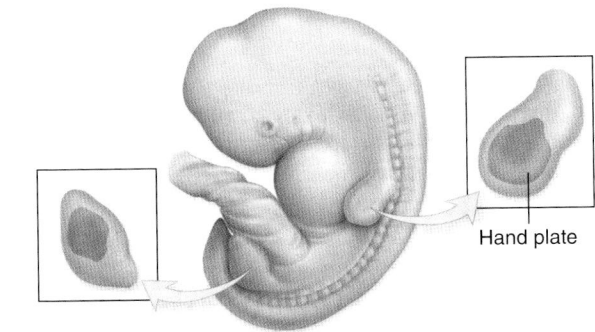

Week 5: Hand plate forms.

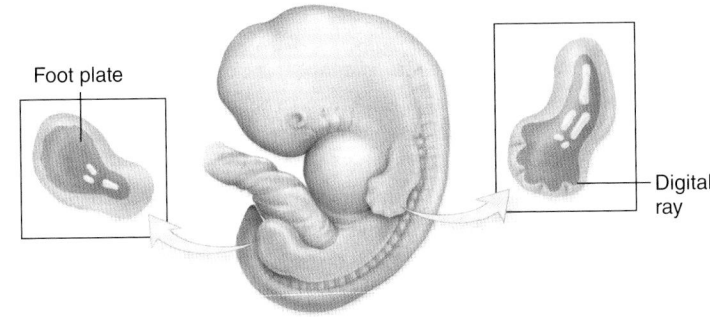

Week 6: Digital rays appear in hand plate. Foot plate forms.

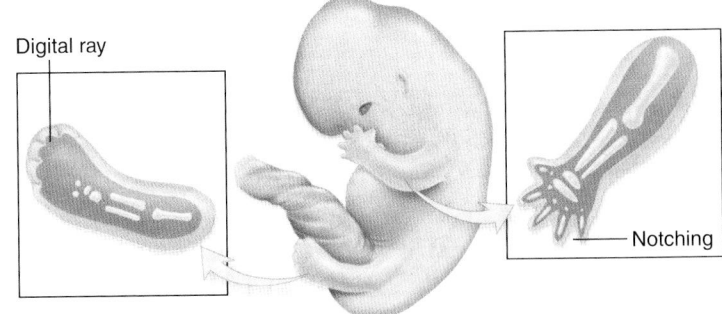

Week 7: Notching develops between digital rays of hand plate. Digital rays appear in foot plate.

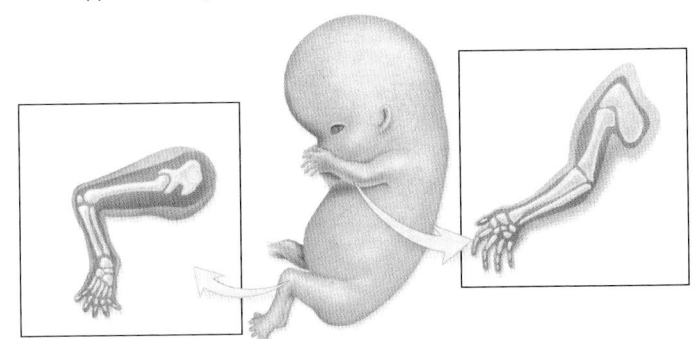

Week 8: Separate fingers and toes formed.

INTEGRATE

CLINICAL VIEW
Limb Malformations

Limb and finger malformations may occur due to genetic or environmental influences. Some limb and finger malformations include the following:

- **Polydactyly** (pol′ē-dak′ti-lē; *poly* = many, *daktylos* = finger) is the condition of having extra digits. Polydactyly tends to run in families and appears to have a genetic component.

- **Ectrodactyly** (ek-trō-dak′ti-lē; *ectro* = congenital absence of a part) is the absence of a digit. Like polydactyly, ectrodactyly runs in families.

- **Syndactyly** (sin-dak′ti-lē; *syn* = together) refers to "webbing" or abnormal fusion of the digits. It occurs when the intermediate tissue between the digital rays fails to undergo normal programmed cell death (apoptosis). In mild cases, some extra tissue is found between the digits, whereas more severe cases may have two or more digits that are completely fused.

- **Amelia** (ă-mē′lē-ă; *a* = without, *melos* = a limb) refers to the complete absence of a limb, whereas **meromelia** (mer-ō-mē′lē-ă; *mero* = part) refers to the partial absence of a limb.

- **Phocomelia** (fō-kō-mē′lē-ă; *phoke* = a seal) refers to a short, poorly formed limb that resembles the flipper of a seal.

All of these conditions may be caused by genetic factors, but most often they result from environmental influences, such as medications or drugs ingested by the pregnant mother. A notable instance of limb malformation involved the drug **thalidomide**, which was first marketed in Europe in 1954 as a nonbarbiturate sleep aid. Physicians later discovered that the drug also helped quell nausea, and so some prescribed it for their pregnant patients who were experiencing morning sickness.

In the late 1950s and early 1960s, the incidence of limb malformations in Europe and Canada skyrocketed (thalidomide was not approved for use in the United States). Researchers later found that thalidomide binds to particular regions of chromosomal DNA, effectively "locking up" specific genes and preventing their expression. Among the genes most affected in the fetus were those responsible for blood vessel growth. In the absence of an adequate vascular network, limb bud formation is disrupted. Researchers and clinicians discovered that if a pregnant female took thalidomide during weeks 4–8 of embryonic development (the time when the limbs are at their critical stage of development), limb formation was severely disrupted.

Thalidomide was taken off the market in the 1960s, but recently it has made a comeback as a treatment of multiple myeloma (a type of cancer of bone marrow cells), treating the symptoms of AIDS, and treating lupus (an autoimmune disease). It also is a very effective treatment for the more devastating effects of leprosy (which is prevalent in South America and other parts of the world). Unfortunately, although the drug now comes with extensive warnings and precautions, some patients ignore the warnings and others take the medication prescribed for someone else. As a result, there has been a resurgence of thalidomide babies born in South America in recent years. Thalidomide is the classic example of how teratogens can affect the delicate cycle of embryonic development, and why females of childbearing age should be sure they aren't pregnant before taking any medication.

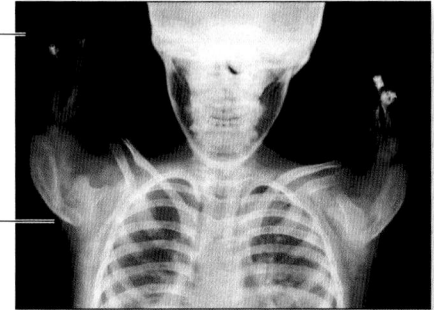

Shortened flipperlike upper limb

Radiograph of a child with phocomelia.

CHAPTER SUMMARY

8.1 Components of the Skeleton 240	• The adult skeleton is typically composed of 206 bones; skeletal features can be used to determine height, age at death, sex, and general health.
	8.1a Axial and Appendicular Skeleton 240
	• The axial skeleton includes the skull, vertebral column, and thoracic cage.
	• The appendicular skeleton includes the pectoral and pelvic girdles, the bones of the upper limb, and the bones of the lower limb.
	8.1b Bone Markings 241
	• Specific anatomic terms are used to describe various features on bones.
8.2 Bones and Features of the Skull 242	**8.2a General Anatomy of the Skull 242**
	• Cranial bones enclose the cranial cavity; facial bones support the entrances to the digestive and respiratory systems.
	8.2b Views of the Skull and Landmark Features 243
	• Specific bones, foramina, processes, and bone landmarks of the skull are seen in multiple views of the skull.
	• The cranial bones include the frontal bone, paired parietal bones, occipital bone, paired temporal bones, sphenoid bone, and ethmoid bone.
	• The facial bones include the paired zygomatic bones, lacrimal bones, nasal bones, inferior nasal conchae, palatine bones, maxillae, a single vomer bone, and a single mandible.
	• The anterior, middle, and posterior cranial fossae are located within the cranial cavity and house specific regions of the brain.

8.2c Sutures 256
- Sutures are immobile joints between skull bones. They allow for growth of the skull bones during childhood.

8.2d Orbital and Nasal Complexes, Paranasal Sinuses 258
- Seven bones form the orbital complex: the maxilla, frontal, lacrimal, ethmoid, sphenoid, palatine, and zygomatic bones.
- The nasal complex is composed of bones and cartilage that enclose the nasal cavity and paranasal sinuses.
- Paranasal sinuses function to lighten the skull and give resonance to the voice.

- Auditory ossicles (malleus, incus, stapes) are tiny ear bones housed in each temporal bone.
- The hyoid bone does not articulate with any other bone but serves as an attachment site for several muscles and ligaments.

- Diagnostic features of the skull may be used to determine the sex and age at death.

8.4a Sex Differences in the Skull 260
- Female skulls tend to be more gracile, have more pointed (versus squared-off) chins, and have sharper supraorbital margins.
- Male skulls tend to be more robust, have more prominent bone markings—such as nuchal lines and external occipital protuberance—and have squared-off chins and angles of the mandible.

8.4b Aging of the Skull 260
- Fontanelles permit the skulls of infants to distort during birth as well as expand as the brain grows.
- In adulthood, the sutures begin to fuse and ossify at regular rates, permitting determination of age at death.

- The vertebral column is composed of 26 vertebrae.

8.5a Types of Vertebrae 262
- There are 7 cervical vertebrae, 12 thoracic vertebrae, 5 lumbar vertebrae, the sacrum, and the coccyx.

8.5b Spinal Curvatures 263
- Spinal curvatures help support the weight of the body better than a straight spine.

8.5c Vertebral Anatomy 264
- A typical vertebra has a body and a vertebral arch. Together they enclose a vertebral foramen, which houses the spinal cord.
- Cervical vertebrae typically have both transverse foramina and bifid spinous processes.
- Thoracic vertebrae have costal facets on their bodies and on most of their transverse processes.
- Lumbar vertebrae are more massive than cervical and thoracic vertebrae. They lack costal facets and transverse foramina.
- The sacrum is a triangular-shaped bone with five fused vertebrae.
- Four small coccygeal vertebrae fuse to form the coccyx.

- The thoracic cage is composed of the thoracic vertebrae, the ribs, and the sternum.

8.6a Sternum 270
- The sternum is composed of the manubrium, body, and xiphoid process.

8.6b Ribs 270
- Ribs 1–7 are called true ribs, and ribs 8–12 are called false ribs. The last two false ribs (11–12) are floating ribs.

- Each limb is held in place by a girdle (pectoral girdle for the upper limbs, pelvic girdle for the lower limbs).
- The arm and thigh each contain one bone, the forearm and leg contain two bones that pivot against one another, there are multiple short bones in the wrist and proximal foot, and both the hand and foot contain 14 phalanges each.

- The pectoral girdle is composed of the clavicle and scapula; it articulates with the axial skeleton and supports the upper limb.

8.8a Clavicle 274
- The clavicle is the S-shaped bone that articulates with the sternum.

8.8b Scapula 274
- The scapula forms the "shoulder blade." Its glenoid cavity articulates with the head of the humerus.

- Each upper limb contains a humerus, radius, ulna, 8 carpals, 5 metacarpals, and 14 phalanges.

8.9a Humerus 275
- The humerus is the bone of the arm. It articulates distally with the radius and ulna at the elbow.

8.9b Radius and Ulna 277
- The radius and ulna are the bones of the forearm.

8.9c Carpals, Metacarpals, and Phalanges 278
- The 8 carpal bones form the wrist, the 5 metacarpals form the bones of the palm, and the 14 phalanges form the bones of the fingers.

(continued on next page)

8.10 The Pelvic Girdle and Its Functions 279	• The pelvic girdle consists of two ossa coxae, while the pelvis is composed of the ossa coxae, sacrum, and the coccyx.

8.10a Os Coxae   280
- Each os coxae forms through the fusion of an ilium, an ischium, and a pubis. The acetabulum of the os coxae articulates with the head of the femur.

8.10b True and False Pelves   282
- The pelvic brim is an oval ridge of bone that divides the pelvis into a true pelvis and a false pelvis.

8.10c Sex Differences in the Pelvis   282
- The ossa coxae are the most sexually dimorphic bones of the body.
- The female pelvis is wider, has a broader greater sciatic notch, and a more rectangular-shaped pubis than the male pelvis.

8.10d Age Differences in the Ossa Coxae   284
- As a person ages, the symphysial surface transforms from a billowed surface to more oval and flattened.

8.11 Bones of the Lower Limb 284	• Each lower limb is composed of the femur, patella, tibia, fibula, 7 tarsals, 5 metatarsals, and 14 phalanges.

8.11a Femur and Patella   284
- The femur has a rounded head and an elongated neck.
- The medial and lateral condyles of the femur articulate with the condyles on the tibia.
- The patella is the kneecap and inserts within the quadriceps femoris tendon.

8.11b Tibia and Fibula   286
- The tibia is the medially located thick and strong bone of the leg, and its medial malleolus forms the medial bump of the ankle.
- The fibula is the laterally located slender bone, and its lateral malleolus forms the lateral bump of the ankle.

8.11c Tarsals, Metatarsals, and Phalanges   288
- The tarsal bones form the proximal foot, the 5 metatarsals form the bones of the arch of the foot, and the 14 phalanges form the bones of the toes.

8.11d Arches of the Foot   289
- The three arches of the foot support the body's weight and ensure plantar structures do not get compressed when we are standing.

8.12 Development of the Skeleton 291	• The appendicular skeleton forms from limb buds beginning in the fourth week. Development of the limbs is mostly complete by week 8.

CHALLENGE YOURSELF

Do You Know the Basics?

_____ 1. The bony portion of the nasal septum is formed by the
   a. perpendicular plate of the ethmoid bone and vomer.
   b. perpendicular plate of the ethmoid bone only.
   c. nasal bones and perpendicular plate of the ethmoid bone.
   d. vomer and sphenoid bones.

_____ 2. Which bone marking is matched with its correct description?
   a. foramen; bony prominence
   b. facet; opening in the bone
   c. tubercle; small, bony projection
   d. alveolus; narrow groove

_____ 3. The frontal and parietal bones articulate at the _____ suture.
   a. coronal
   b. sagittal
   c. lambdoid
   d. squamosal

_____ 4. The compression of an infant's skull bones at birth is facilitated by spaces between unfused cranial bones called
   a. ossification centers.
   b. fontanelles.
   c. foramina.
   d. fossae.

_____ 5. Most _____ vertebrae have a heart-shaped body and a long spinous process that is angled inferiorly.
   a. cervical
   b. thoracic
   c. lumbar
   d. sacral

_____ 6. The female pelvis typically has which of the following characteristics?
   a. narrow, U-shaped greater sciatic notch
   b. wide subpubic angle, greater than 100 degrees
   c. short, triangular pubic body
   d. smaller, heart-shaped pelvic inlet

_____ 7. When the forearm is supinated,

    a. the pollex is laterally placed.

    b. the radius and ulna are crossed.

    c. the little finger is laterally placed.

    d. the pisiform is facing posteriorly.

_____ 8. The spine of the scapula separates which two fossae?

    a. supraspinous, subscapular

    b. subscapular, infraspinous

    c. infraspinous, supraspinous

    d. supraspinous, glenoid

_____ 9. The femur articulates with the tibia at the

    a. linea aspera.

    b. medial and lateral condyles.

    c. head of the femur.

    d. greater trochanter of the femur.

_____ 10. When sitting upright, you are resting on your

    a. pubic bones.

    b. ischial tuberosities.

    c. sacroiliac joints.

    d. iliac crest.

11. What are sutures, and how do they affect skull shape and growth?

12. What cranial and facial bones may be seen easily on the inferior surface of the skull?

13. What are the functions of the paranasal sinuses?

14. What are the spinal curvatures, when do they form, and what are their functions?

15. Describe similarities and differences among true, false, and floating ribs.

16. Compare and contrast the anatomic and functional features of the pectoral and pelvic girdles.

17. What are the primary similarities and differences between the upper and lower limbs?

18. Distinguish between the true and false pelves. What bony landmark separates the two?

19. What are the functions of the arches of the foot?

20. Discuss the development of the limbs. What primary germ layers form the limb bud? List the major events during each week of limb development.

Can You Apply What You've Learned?

Use the following paragraph to answer questions 1–5.

You are called to a crime scene reported in the woods. A hiker found a skeleton underneath some leaves—and as the osteologist of the team, it is up to you to identify the bones, as well as determine the age and sex of the skeleton. You begin examining the skeleton.

1. The first bone you identify is long and rather large. It has a rounded head, an elongated neck, and smooth condyles on its distal surface. There also are large bony projections near the neck of this bone. Based on these features, you determine that the bone in question is a

    a. humerus.

    b. radius.

    c. femur.

    d. tibia.

2. As you examine the rest of the skeleton, you notice an S-shaped bone that appears to have been fractured prior to death. This bone likely was fractured due to a fall on the outstretched hand. What bone is it?

    a. clavicle

    b. metacarpal

    c. rib

    d. phalanx

3. You pick up the skull and begin examining it. The mastoid process is rather small, and the external occipital protuberance is not well defined. The supraorbital margins are rather sharp, and the mental protuberance is pointed (instead of squared-off). All of these features lead you to the following conclusion:

    a. The skeleton has been buried for a long time.

    b. The skull is from a female.

    c. The skull is from a male.

    d. The skull is from a young child.

4. Based on the answer you gave to question 3, what feature(s) would you expect to see in the pelvis?

    a. narrow pelvic inlet

    b. elongated, rectangular pubic bones

    c. narrow, U-shaped sciatic notch

    d. 90-degree subpubic arch

5. The police officers also want to know if you can determine the age at death of the skeleton. You determine that all long bone epiphyses have fused to their diaphyses, and all permanent teeth are erupted. The cranial sutures are still open, and the symphysial surface is flattened, but there is no complete rim around the symphysis. Based on these features, a likely age range for the skeleton would be

    a. younger than 10 years.

    b. 10–20 years.

    c. 20–35 years.

    d. 35–50 years.

Can You Synthesize What You've Learned?

1. Paul viewed his newborn daughter through the nursery window at the hospital and was distressed because the infant's skull was badly misshapen. A nurse told him not to worry—the shape of the infant's head would return to normal in a few days. What caused the misshapen skull, and what anatomic feature of the neonatal skull allows it to return to a more rounded shape?

2. A female in her first trimester of pregnancy sees her physician. She suffers from lupus and has read that the drug thalidomide has shown remarkable promise in treating the symptoms. Should the physician prescribe this drug for her at this time? Why or why not?

3. Forensic anthropologists are investigating portions of a human pelvis found in a cave. How can they tell the sex, relative age, and some physical characteristics of the individual based on the pelvis only?

INTEGRATE

ONLINE STUDY TOOLS

The following study aids may be accessed through Connect.

Clinical Case Study: A Historical Case of Spinal Involvement with Tuberculosis

Interactive Questions: This chapter's content is served up in a number of multimedia question formats for student study.

LearnSmart: Topics and terminology include skull bones and features; sex and age determination from skull analysis; bones of the vertebral column; bones of the thoracic cage; upper and lower limbs; pectoral girdle; pelvic girdle; development of the skeleton

Anatomy & Physiology Revealed: Topics include skull and skull bones; vertebral column; cervical, thoracic, and lumbar vertebrae; sacrum; thoracic cage; bones of the arm and hand; bones of the pelvis; bones of the leg, ankle, and foot

Animation: The skull

Skeletal System: Articulations

INTEGRATE

aprevealed.com

Module 5: Skeletal System

CAREER PATH
Sports Medicine Physician

A sports medicine physician can diagnose musculoskeletal injuries, especially those involving the articulations (joints) we discuss in this chapter. This physician must be aware of the normal range of movement at each joint, the muscular and ligamentous structures that support the joint, and how injury may impact mobility and ultimate healing of the joint. Here, a sports medicine physician is examining and taping a soccer player's ankle after an injury on the field.

Our skeleton protects vital organs and supports soft tissues. Its marrow cavity is the source of new blood cells. When it interacts with the muscular system, the skeleton helps the body move. Bones are too rigid to bend—but they meet at joints that anatomists call articulations. In this chapter, we examine how bones articulate and permit varying degrees freedom of movement, depending upon both the shapes and supporting structures of the different joints.

9.1 Classification of Joints

LEARNING OBJECTIVES

1. Define a joint.
2. Compare the structural and the functional classification of joints.
3. Explain the inverse relationship between mobility and stability within a joint.

A **joint,** or **articulation** (ar-tik´-yū-lā´shŭn), is the place of contact between bones, between bone and cartilage, or between bones and teeth. Bones are said to **articulate** with each other at a joint. The scientific study of joints is called **arthrology** (ar-throl´ō-jē; *arthron* = joint, *logos* = study).

Joints are classified by both their structural characteristics and the movements they allow **(table 9.1).** Joints are categorized structurally on the basis of whether a space occurs between the articulating bones and the type of connective tissue that binds the articulating surfaces of the bones:

- A **fibrous** (fī´brŭs) **joint** has no joint cavity and occurs where bones are held together by dense regular (fibrous) connective tissue.
- A **cartilaginous** (kar-ti-laj´i-nŭs) **joint** has no joint cavity and occurs where bones are joined by cartilage.
- A **synovial** (si-nō´vē-ăl) **joint** has a fluid-filled joint cavity that separates the articulating surfaces of the bones. The articulating surfaces are enclosed within a connective tissue capsule, and the bones are attached to each other by various ligaments.

Joints are classified functionally based on the extent of movement they permit.

- A **synarthrosis** (sin´ar-thrō´sis; pl., *synarthroses,* -sēz; *syn* = joined, together; *osis* = condition) is an immobile joint. Two types of fibrous joints and one type of cartilaginous joint are synarthroses.
- An **amphiarthrosis** (am´fē-ar-thrō´sis; pl., *amphiarthroses,* -sēz; *amphi* = around) is a slightly mobile joint. One type of fibrous joint and one type of cartilaginous joint are amphiarthroses.
- A **diarthrosis** (dī-ar-thrō´sis; pl., *diarthroses,* -sēz; *di* = two) is a freely mobile joint. All synovial joints are diarthroses.

The motion permitted at a joint ranges from no movement, such as where some skull bones interlock at a suture, to extensive movement like that seen at the shoulder, where the humerus articulates with the scapula. The structure of each joint determines both its mobility and its stability. There is an inverse relationship between mobility and stability in articulations. When the mobility of a joint increases, its stability decreases. In contrast, if a joint is immobile, it has maximum stability. **Figure 9.1** illustrates the "tradeoff" between mobility and stability for various joints. It allows you to view and compare the structural versus functional classification of some common joints.

Table 9.1	**Joint Classifications**			
Structural Classification	**Structural Characteristics**	**Structural Categories**	**Example(s)**	**Functional Classification**
Fibrous (see figure 9.2)	Dense regular connective tissue holds together the ends of bones and bone parts; no joint cavity	**Gomphosis:** Periodontal membranes hold tooth to bony jaw	Tooth to jaw	Synarthrosis (immobile)
		Suture: Dense regular connective tissue connects skull bones	Lambdoid suture (connects occipital and parietal bones)	Synarthrosis (immobile)
		Syndesmosis: Dense regular connective tissue fibers (interosseous membrane) between bones	Articulation between radius and ulna, and between tibia and fibula	Amphiarthrosis (slightly mobile)
Cartilaginous (see figure 9.3)	Pad of cartilage is wedged between the ends of bones; no joint cavity	**Synchondrosis:** Hyaline cartilage between bones	Epiphyseal plates in growing bones; costochondral joints	Synarthrosis (immobile)
		Symphysis: Fibrocartilage pad between bones	Pubic symphysis; intervertebral disc articulations	Amphiarthrosis (slightly mobile)
Synovial (see figure 9.6)	Ends of bones covered with articular cartilage; joint cavity separates the articulating bones; joint enclosed by an articular capsule, lined by a synovial membrane; contains synovial fluid	*Uniaxial* **Plane joint:** Flattened or slightly curved faces slide across one another	**Plane joint:** Intercarpal joints, intertarsal joints	Diarthrosis (freely mobile)
		Hinge joint: Convex feature of one bone fits into concave depression of another bone	**Hinge joint:** Elbow joint, knee joint, IP (interphalangeal) joints	
		Pivot joint: Bone with a rounded surface fits into a ring formed by a ligament and another bone	**Pivot joint:** Atlantoaxial joint	
		Biaxial **Condylar joint:** Oval articular surface on one bone closely interfaces with a depressed oval surface on another bone	**Condylar joint:** MP (metacarpophalangeal or metatarsophalangeal) joints	Diarthrosis (freely mobile)
		Saddle joint: Saddle-shaped articular surface on one bone closely interfaces with a saddle-shaped surface on another bone	**Saddle joint:** Articulation between carpal and first metacarpal bone	
		Multiaxial (*Triaxial*) **Ball-and-socket joint:** Round head of one bone rests within cup-shaped depression in another bone	**Ball-and-socket joint:** Glenohumeral (shoulder) joint, hip joint	Diarthrosis (freely mobile)

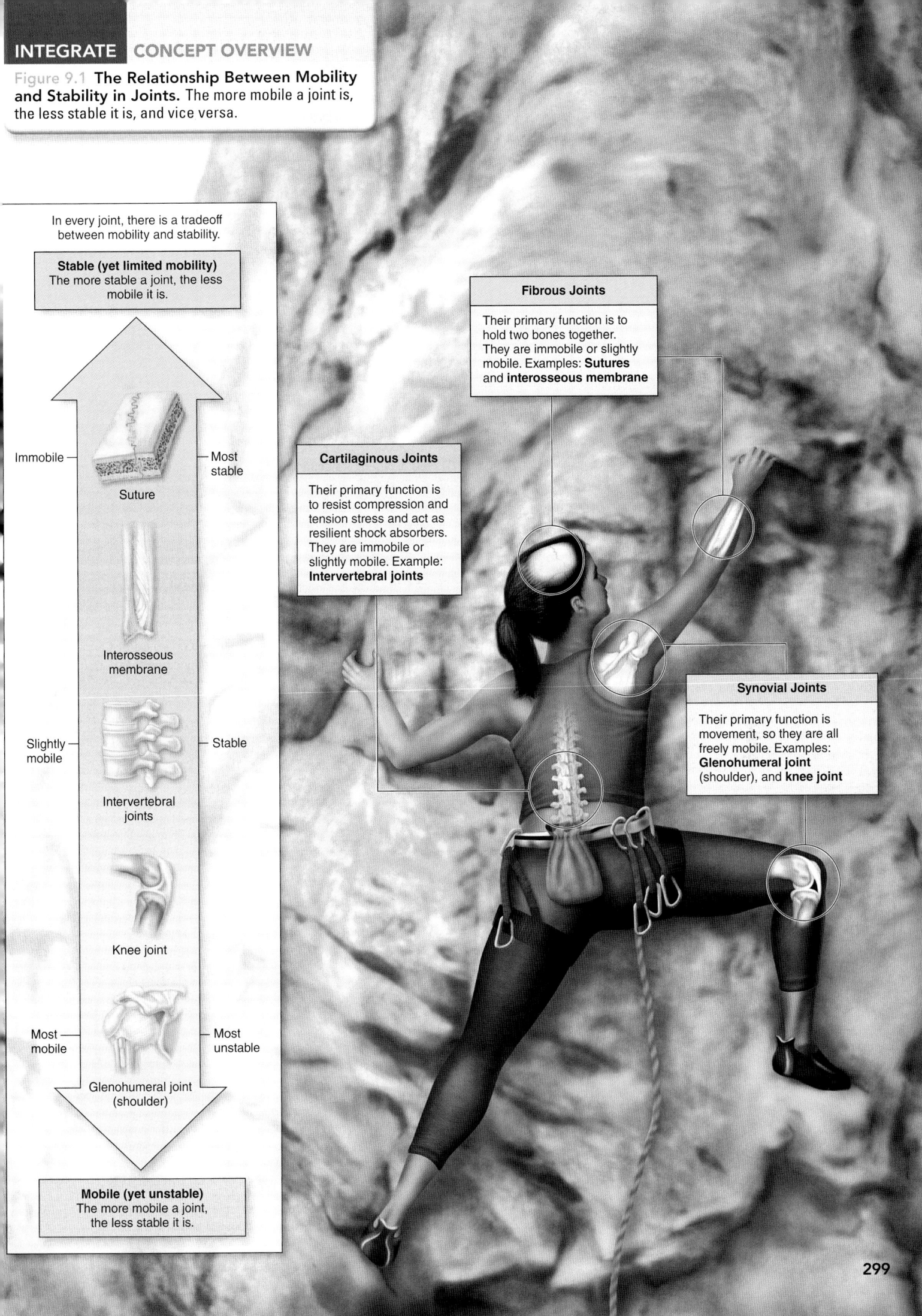

Figure 9.1 **The Relationship Between Mobility and Stability in Joints.** The more mobile a joint is, the less stable it is, and vice versa.

In every joint, there is a tradeoff between mobility and stability.

Stable (yet limited mobility)
The more stable a joint, the less mobile it is.

Immobile — Suture — Most stable

Interosseous membrane

Slightly mobile — Intervertebral joints — Stable

Knee joint

Most mobile — Glenohumeral joint (shoulder) — Most unstable

Mobile (yet unstable)
The more mobile a joint, the less stable it is.

Fibrous Joints

Their primary function is to hold two bones together. They are immobile or slightly mobile. Examples: **Sutures** and **interosseous membrane**

Cartilaginous Joints

Their primary function is to resist compression and tension stress and act as resilient shock absorbers. They are immobile or slightly mobile. Example: **Intervertebral joints**

Synovial Joints

Their primary function is movement, so they are all freely mobile. Examples: **Glenohumeral joint** (shoulder), and **knee joint**

INTEGRATE

LEARNING STRATEGY

You can logically figure out the names of most joints by putting together the names of the bones that form them. For example, the *glenohumeral* joint is where the glenoid cavity of the scapula meets the head of the humerus, and the *sternoclavicular* joint is where the manubrium of the sternum articulates with the sternal end of the clavicle.

The following detailed discussion of articulations is based upon their structural classification, with functional categories included as appropriate.

WHAT DID YOU LEARN?

❶ What is the relationship between mobility and stability in a joint?

❷ Are all fibrous joints also synarthroses? Explain why or why not.

9.2 Fibrous Joints

Articulating bones in fibrous joints are connected by dense regular connective tissue. Fibrous joints have no joint cavity; thus, they lack a space between the articulating bones. Most fibrous joints are immobile or at most only slightly mobile; their primary function is to hold together two bones. Examples include the articulations of the teeth in their sockets, sutures between skull bones, and the articulations between either the radius and ulna or the tibia and fibula. In this section, we look at the three most common types of fibrous joints: gomphoses, sutures, and syndesmoses **(figure 9.2)**.

9.2a Gomphoses

LEARNING OBJECTIVE

1. Explain the location and characteristics of gomphoses.

A **gomphosis** (gom-fō'sis; pl., *gomphoses*, -sēz; *gomphos* = bolt) resembles a "peg in a socket." The only gomphoses in the human body are the articulations of the roots of individual teeth with the alveolar processes (sockets) of the mandible and the maxillae. A tooth is held firmly in place by fibrous **periodontal** (per'ē-ō-don'tăl; *peri* = around, *odous* = tooth) **membranes.** This joint is functionally classified as a synarthrosis.

The reasons orthodontic braces can be painful and take so long to correctly position the teeth are directly related to the architecture of the gomphosis. The orthodontist's job is to reposition these normally immobile joints through the use of clamps, bands, rings, and braces. In response to these mechanical stressors, osteoblasts and osteoclasts work together to modify the alveolar process, resulting in the remodeling of the joints and the slow repositioning of the teeth.

WHAT DID YOU LEARN?

❸ Where are gomphoses located, and what type of movement do they allow?

9.2b Sutures

LEARNING OBJECTIVE

2. Describe the location and functions of sutures.

Sutures (sū'chūr; *sutura* = a seam) are immobile fibrous joints (synarthroses) that are found only between certain bones of the skull. Sutures have distinct, interlocking, usually irregular edges that both increase their

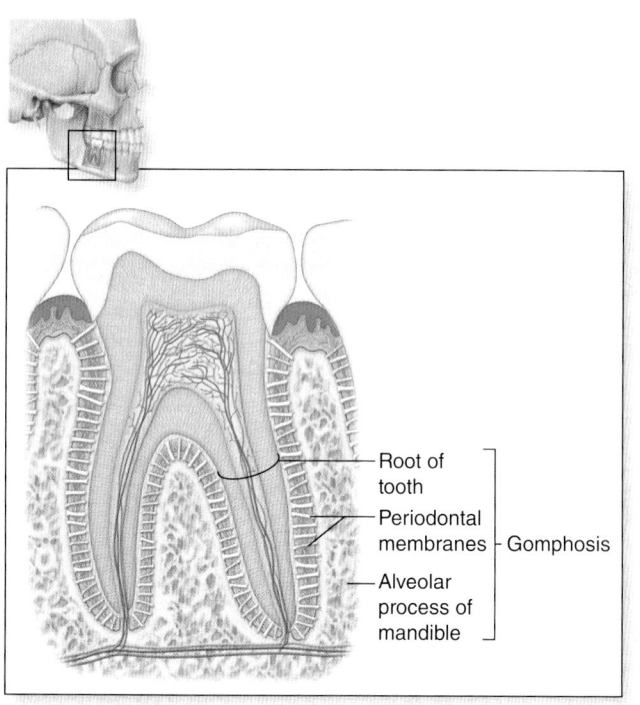

(a) Gomphosis

— Root of tooth
— Periodontal membranes — Gomphosis
— Alveolar process of mandible

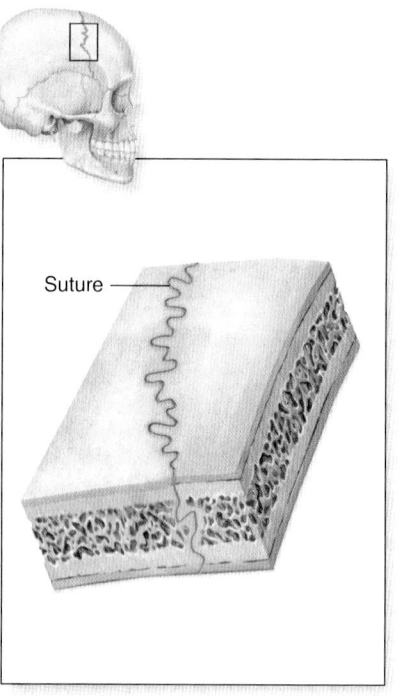

Suture

(b) Suture

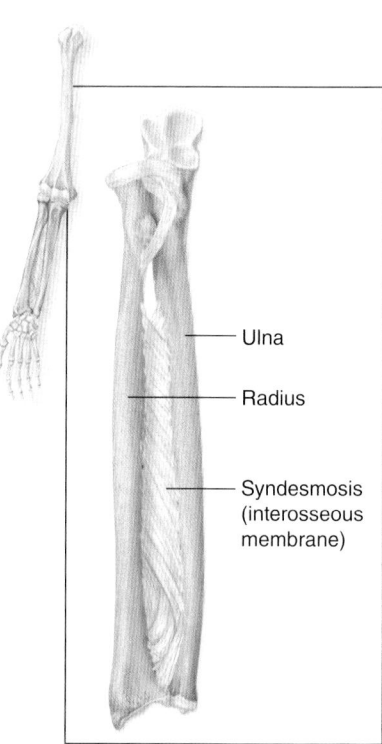

Ulna

Radius

Syndesmosis (interosseous membrane)

(c) Syndesmosis

Figure 9.2 Fibrous Joints. Dense regular connective tissue binds the articulating bones in fibrous joints to prevent or restrict movement. (*a*) A gomphosis is the immobile joint between a tooth and the jaw. (*b*) A suture is an immobile joint between bones of the skull. (*c*) A syndesmosis permits slight mobility between the radius and the ulna. AP|R

strength and decrease the number of fractures at these articulations. In addition to joining bones, sutures permit the skull to grow as the brain increases in size during childhood. In an older adult, the dense regular connective tissue in the suture becomes ossified, fusing the skull bones together. When the bones have completely fused across the suture line, these obliterated sutures become **synostoses** (sin-os-tō'sēz; sing., -sis; *osteon* = bone).

 WHAT DID YOU LEARN?

4 What is the composition of a suture, and where in the body is it found?

9.2c Syndesmoses

 **LEARNING OBJECTIVE**

3. List the locations of syndesmoses and describe their function.

Syndesmoses (sin'dez-mō'sēz; sing., *syndesmosis*; *syndesmos* = a fastening) are fibrous joints in which articulating bones are joined by long strands of dense regular connective tissue only. Because syndesmoses allow for slight mobility, they are classified functionally as amphiarthroses. Syndesmoses are found both between the radius and ulna and between the tibia and fibula. The shafts of the two articulating bones are bound by a broad ligamentous sheet called an **interosseous membrane** (or *interosseous ligament*). The interosseous membrane provides a pivot where the radius and ulna (or the tibia and fibula) can move against one another.

 WHAT DID YOU LEARN?

5 What type of movement is allowed at a syndesmosis?

9.3 Cartilaginous Joints

Cartilaginous joints have cartilage between the articulating bones. Like fibrous joints, cartilaginous joints also lack a joint cavity. They may be either immobile or slightly mobile. The cartilage found between the articulating bones is either hyaline cartilage or fibrocartilage. The two types of cartilaginous joints are synchondroses and symphyses **(figure 9.3)**.

CLINICAL VIEW
Costochondritis

Costochondritis refers to inflammation and irritation of the costochondral joints, resulting in localized chest pain. The cause of costochondritis is usually unknown, but some documented cases include repeated minor trauma to the chest wall (e.g., from forceful repeated coughing or from overexertion during exercise) or either bacterial or viral infection of the joints themselves. The localized chest pain may be mistaken for pain resulting from a myocardial infarction (heart attack). Costochondritis may be treated with NSAIDs (nonsteroidal anti-inflammatory drugs, such as aspirin). With proper rest and treatment, symptoms typically disappear after several weeks.

9.3a Synchondroses

LEARNING OBJECTIVE

1. Describe the locations and functions of synchondroses.

An articulation in which bones are joined by hyaline cartilage is called a **synchondrosis** (sin'kon-drō'sis; pl., *synchondroses*, -sēz; *chondros* = cartilage). Functionally, all synchondroses are immobile and thus are classified functionally as synarthroses.

WHAT DO YOU THINK?

1 Why is a synchondrosis also a synarthrosis? Why would you want a synchondrosis to be immobile?

The hyaline cartilage of epiphyseal plates in children forms synchondroses that bind the epiphyses and diaphysis of long bones (figure 9.3*a*). When the hyaline cartilage stops growing, bone replaces the cartilage and a synchondrosis no longer exists (see section 7.4b).

The spheno-occipital synchondrosis is found between the body of the sphenoid and the basilar part of the occipital bone. This synchondrosis typically fuses between 18 and 25 years of age, making it a useful tool for assessing the age of the skull (see section 8.4b).

Other examples of synchondroses are formed from costal cartilage. The **costochondral** (kos-tō-kon'drăl; *costa* = rib) **joint,** the joint

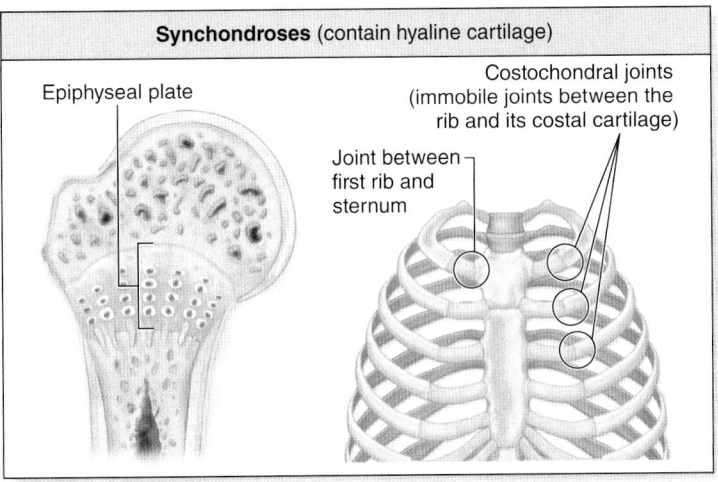

Synchondroses (contain hyaline cartilage)

Epiphyseal plate

Costochondral joints (immobile joints between the rib and its costal cartilage)

Joint between first rib and sternum

(a)

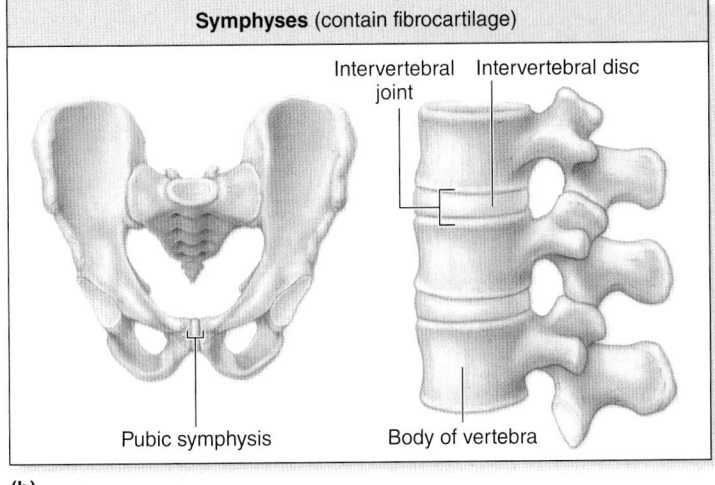

Symphyses (contain fibrocartilage)

Intervertebral joint   Intervertebral disc

Pubic symphysis   Body of vertebra

(b)

Figure 9.3 Cartilaginous Joints. Articulating bones are joined by cartilage in cartilaginous joints. (*a*) Synchondroses are immobile joints that occur in an epiphyseal plate in a long bone, between each rib and its respective costal cartilage, and in the joint between the first rib and the sternum. (*b*) Symphyses are slightly mobile and occur in the pubic symphysis and the intervertebral joints. **AP|R**

between each bony rib and its respective costal cartilage, is a synchondrosis. Finally, the attachment of the first rib to the sternum by costal cartilage (called the *first sternocostal joint*) is another synchondrosis. Here, the first rib and its costal cartilage are united firmly to the manubrium of the sternum to provide stability to the rib cage. (Note that the sternocostal joints between the sternum and the costal cartilage of ribs 2–7 are synovial joints and not synchondroses.)

WHAT DID YOU LEARN?

6 What is the composition of a synchondrosis, and where in the body is it found?

9.3b Symphyses

LEARNING OBJECTIVE

2. Name the locations of symphyses and their functions in these locations.

A **symphysis** (sim′fi-sis; pl., *symphyses*, -sēz; growing together) has a pad of fibrocartilage between the articulating bones (figure 9.3b). The fibrocartilage resists both compression and tension stresses and acts as a resilient shock absorber. All symphyses are amphiarthroses—thus they allow slight mobility.

One example of a symphysis is the pubic symphysis, which is located between the right and left pubic bones. In pregnant females, the pubic symphysis becomes more mobile to allow the pelvis to change shape slightly as the fetus passes through the birth canal.

Other examples of symphyses are the intervertebral joints, where the bodies of adjacent vertebrae are both separated and united by intervertebral discs. Individual intervertebral discs allow only slight movements between the adjacent vertebrae; however, the collective movements of all the intervertebral discs afford the spine considerable flexibility.

WHAT DID YOU LEARN?

7 Into what functional category is a symphysis placed? Why is it in this category?

9.4 Synovial Joints

Synovial joints are freely mobile articulations. Most of the commonly known joints in the body are synovial joints, including the glenohumeral (shoulder) joint, the temporomandibular joint, the elbow joint, and the knee joint.

9.4a Distinguishing Features and Anatomy of Synovial Joints

LEARNING OBJECTIVES

1. Describe the characteristics common to all synovial joints.

2. List the basic features of a synovial joint.

3. Explain the composition and function of synovial fluid in a typical synovial joint.

Unlike the joints previously discussed, the bones in a synovial joint are separated by a space called a joint cavity. Functionally, all synovial joints are classified as diarthroses, because all are freely mobile. Often, the terms *synovial joint* and *diarthrosis* are equated. All synovial joints include several basic features: an articular capsule, a joint cavity, synovial fluid, articular cartilage, ligaments, nerves, and blood vessels **(figure 9.4).**

Each synovial joint is composed of a double-layered capsule called the **articular** (ar-tik′yū-lăr) **capsule,** or *joint capsule.* Its outer layer is called the **fibrous layer,** and the inner layer is a **synovial membrane** (or *synovium*). The fibrous layer is formed from dense connective tissue. It strengthens the joint to prevent the bones from being pulled apart. The synovial membrane is composed of squamous epithelial cells (that lack a basement membrane) resting on an areolar connective tissue layer. This membrane covers all the internal joint surfaces not covered by cartilage and lines the articular capsule, and helps produce synovial fluid (described shortly).

All articulating bone surfaces in a synovial joint are covered by a thin layer of hyaline cartilage called **articular cartilage.** This cartilage has numerous functions: It reduces friction in the joint during movement, acts as a spongy cushion to absorb compression placed on the joint, and prevents damage to the articulating ends of the bones. This special hyaline cartilage lacks a perichondrium. Mature cartilage is avascular, so it does not have blood vessels to bring nutrients to and remove waste products from the cartilage. The repetitious compression and expansion that occurs during exercise is vital to maintaining healthy articular cartilage because this action enhances its obtaining nutrition and its waste removal.

Only synovial joints house a **joint cavity** (or *articular cavity*), a space that permits separation of the articulating bones. The articular cartilage and synovial fluid (described next) within the joint cavity together reduce friction as bones move at a synovial joint.

Synovial fluid is a viscous, oily substance located within a synovial joint. It is produced both from the synovial membrane cells and filtrate formed from blood plasma. Synovial fluid has three functions:

1. Synovial fluid lubricates the articular cartilage on the surface of articulating bones (in the same way that oil in a car engine lubricates the moving engine parts).

2. Synovial fluid nourishes the articular cartilage's chondrocytes. The relatively small volume of synovial fluid must be circulated

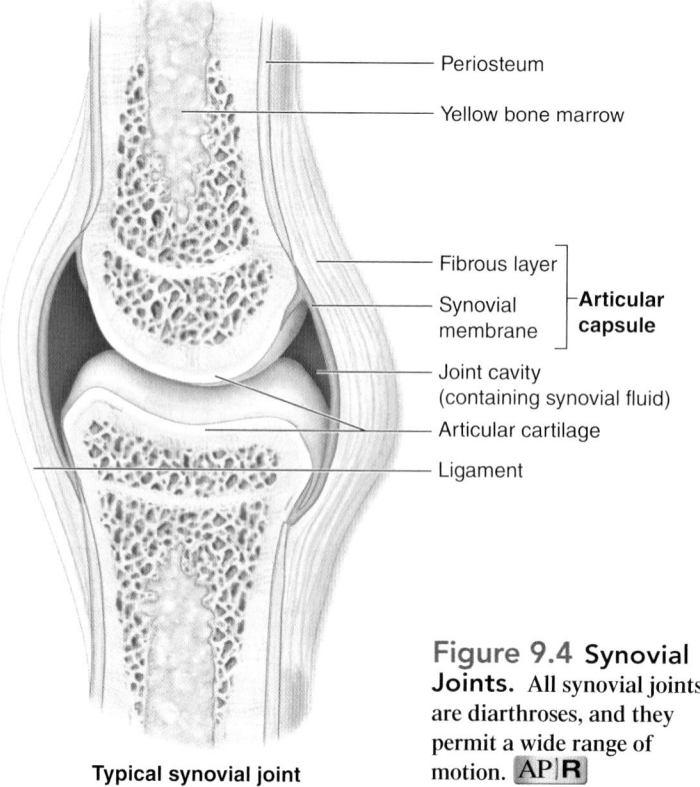

Periosteum

Yellow bone marrow

Fibrous layer

Synovial membrane — Articular capsule

Joint cavity (containing synovial fluid)

Articular cartilage

Ligament

Figure 9.4 Synovial Joints. All synovial joints are diarthroses, and they permit a wide range of motion. AP|R

Typical synovial joint

continually to provide nutrients to and remove wastes from these cells. Whenever movement occurs at a synovial joint, the combined compression and re-expansion of the articular cartilage circulates the synovial fluid into and out of the cartilage matrix.

3. Synovial fluid acts as a shock absorber, distributing stresses and force evenly across the articular surfaces when the pressure in the joint suddenly increases.

Ligaments (lig′ă-ment; *ligamentum* = a band) are composed of dense regular connective tissue, and they connect one bone to another bone. Ligaments function to stabilize, strengthen, and reinforce most synovial joints. **Extrinsic ligaments** are outside of, and physically separate from, the joint capsule, whereas **intrinsic ligaments** represent thickenings of the articular capsule itself. Intrinsic ligaments include *extracapsular ligaments* outside the joint capsule and *intracapsular ligaments* within the joint capsule.

All synovial joints have numerous **sensory nerves** and **blood vessels** that innervate and supply the articular capsule and associated ligaments. The sensory nerves detect painful stimuli in the joint and report on the amount of movement and stretch within the joint. By monitoring stretching at a joint, the nervous system can detect changes in our posture and adjust body movements.

Tendons (ten′dŏn; *tendo* = extend) are like ligaments and are composed of dense regular connective tissue but they are not part of the synovial joint itself. Whereas a ligament binds bone to bone, a tendon attaches a muscle to a bone. When a muscle contracts, the tendon from that muscle moves the bone to which it is attached, thus causing movement at the joint. Tendons help stabilize joints because they pass across or around a joint to provide mechanical support, and sometimes they limit the range or amount of movement permitted at a joint.

Synovial joints usually have bursae and fat pads as accessory structures in addition to the main components just described. A **bursa** (ber′să; pl., *bursae*, ber′sē; a purse) is a fibrous, saclike structure that

INTEGRATE

CLINICAL VIEW
"Cracking Knuckles"

Stretching or pulling on a synovial joint causes the joint volume to immediately expand and the pressure on the fluid within the joint to decrease, so that a partial vacuum exists within the joint. As a result, the gases dissolved in the fluid become less soluble, and they form bubbles, a process called *cavitation*. When the joint is stretched to a certain point, the pressure in the joint drops even lower, so the bubbles in the fluid burst, resulting in a popping or cracking sound. It typically takes about 20 to 30 minutes for the gases to dissolve back into the synovial fluid. You cannot crack your knuckles again until these gases dissolve. Contrary to popular belief, cracking your knuckles does *not* cause arthritis.

contains synovial fluid and is lined internally by a synovial membrane **(figure 9.5a)**. Bursae are associated with most synovial joints and are where bones, ligaments, muscles, skin, or tendons overlie each other and rub together. Bursae may be either connected to the joint cavity or completely separate from it. They serve to alleviate the friction resulting from the various body movements, such as where a tendon or ligament rubs against bone. An elongated bursa called a **tendon sheath** wraps around a tendon where there may be excessive friction. Tendon sheaths are especially common in the confined spaces of the wrist and ankle (figure 9.5b).

Fat pads are often distributed along the periphery of a synovial joint. They act as packing material and provide some protection for the joint. Often fat pads fill the spaces that form when bones move and the joint cavity changes shape (figure 9.5a).

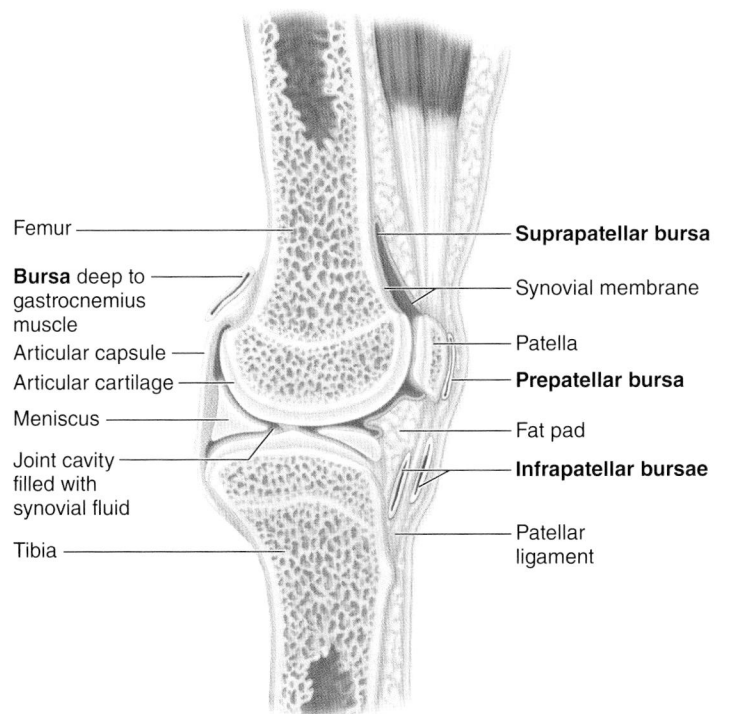

(a) **Bursae of the knee joint, sagittal section**

Femur
Bursa deep to gastrocnemius muscle
Articular capsule
Articular cartilage
Meniscus
Joint cavity filled with synovial fluid
Tibia
Suprapatellar bursa
Synovial membrane
Patella
Prepatellar bursa
Fat pad
Infrapatellar bursae
Patellar ligament

(b) **Tendon sheaths of wrist and hand, anterior view**

Tendon sheath (opened)
Tendon of flexor digitorum profundus
Tendon of flexor digitorum superficialis
Digital tendon sheaths
Tendon sheath around flexor pollicis longus tendon
Common flexor tendon sheath
Tendon of flexor carpi radialis
Tendon of flexor pollicis longus
Tendons of flexor digitorum superficialis and flexor digitorum profundus

Figure 9.5 Bursae and Tendon Sheaths. Synovial fluid–filled structures called bursae and tendon sheaths reduce friction where ligaments, muscles, tendons, and bones rub together. (*a*) The knee joint contains a number of bursae (blue and purple). (*b*) The wrist and hand contain numerous tendon sheaths (blue).

8 What are the basic characteristics of all types of synovial joints?

9 What is the purpose of synovial fluid in the joint?

9.4b Classification of Synovial Joints

LEARNING OBJECTIVES

4. Explain the movement of a joint with respect to the three perpendicular axes of space.

5. Compare and contrast the six types of synovial joints.

Synovial joints are classified by the shapes of their articulating surfaces and the types of movement they allow. Movement of a bone at a synovial joint is best described with respect to three intersecting perpendicular planes or axes:

- A joint is said to be **uniaxial** (yū-nē-ak′sē-ăl; *unus* = one) if the bone moves in just one plane or axis.
- A joint is **biaxial** (bī-ak′sē-ăl; *bi* = double) if the bone moves in two planes or axes.
- A joint is **multiaxial** (or **triaxial** (trī-ak′sē-ăl; *tri* = three) if the bone moves in multiple planes or axes.

All synovial joints are diarthroses, as mentioned, but some are more mobile than others. From least mobile to most freely mobile, the six specific types of synovial joints are plane joints, hinge joints, pivot joints, condylar joints, saddle joints, and ball-and-socket joints. These joints and examples of where they are found in the body are shown in **figure 9.6**.

WHAT DO YOU THINK?

2 If a ball-and-socket joint is more mobile than a gliding joint, which of these two joints is the more stable?

A **plane** (*planus* = flat) **joint,** also called a *planar* or *gliding joint,* is the simplest synovial articulation and the least mobile type of diarthrosis. Anatomists have debated how to describe the movement of this joint with respect to perpendicular planes. We describe this type of synovial joint as a uniaxial joint because it usually allows only limited side-to-side movements in a single plane, and because there is no rotational or angular movement with this joint. The articular surfaces of the bones are flat, or planar. Examples of plane joints include the intercarpal and intertarsal joints (the joints between the carpal bones and tarsal bones, respectively).

A **hinge joint** is formed by the convex surface of one articulating bone fitting into a concave depression on the other bone in the joint. Movement is confined to a single axis, like the movement seen at the hinge of a door, so a hinge joint is considered a uniaxial joint. An example is the elbow joint. The trochlear notch of the ulna fits directly into the trochlea of the humerus, so the forearm can be moved only anteriorly toward the arm or posteriorly away from the arm. Other hinge joints occur in the knee and the finger (interphalangeal [IP]) joints.

A **pivot joint** is a uniaxial joint in which one articulating bone with a rounded surface fits into a ring formed by a ligament and another bone. The first bone rotates on its longitudinal axis relative to the second bone. An example is the proximal radio-ulnar joint, where the rounded head of the radius pivots along the ulna and permits the radius to rotate. Another example is the atlanto-axial joint between the first two cervical vertebrae. The rounded dens of the axis fits snugly against an articular facet on the anterior arch of the atlas. This joint pivots when you shake your head "no."

Condylar (kon′di-lar) **joints,** also called *condyloid* or *ellipsoid joints,* are biaxial joints with an oval, convex surface on one bone that articulates with a concave articular surface on the second bone of the joint. Biaxial joints can move in two axes, such as back-and-forth and side-to-side. Examples of condylar joints are the metacarpophalangeal (met′ă-kar′pō-fă-lan′jē-ăl) (MP) joints of fingers 2 through 5. The MP joints are commonly referred to as "knuckles." Examine your hand and look at the movements along the MP joints; you can flex and extend the fingers at this joint, which is one axis of movement. You also can move your fingers apart from one another and move them closer together, which is the second axis of movement.

A **saddle joint** is so named because the articular surfaces of the bones have convex and concave regions that resemble the shape of a saddle. This biaxial joint allows a greater range of movement than either a condylar or hinge joint. The carpometacarpal joint of the thumb (between the trapezium and the first metacarpal) is an example of a saddle joint. This joint permits the thumb to move toward the other fingers so that we can grasp objects.

Ball-and-socket joints are multiaxial joints in which the spherical articulating head of one bone fits into the rounded, cuplike socket of a second bone. Examples of these joints are the coxal (hip) and glenohumeral (shoulder) joints. The multiaxial nature of these joints permits movement in three planes. Move your arm at your shoulder, and note the wide range of movements that can be produced. The ball-and-socket joint is considered the most freely mobile type of synovial joint.

WHAT DID YOU LEARN?

10 What types of movements do each of the six kinds of joints allow?

9.5 Synovial Joints and Levers

When analyzing synovial joint movement and muscle contraction, anatomists often compare the movement to the mechanics of a lever; this practice of applying mechanical principles to biology is known as **biomechanics.**

9.5a Terminology of Levers

LEARNING OBJECTIVES

1. Define a lever.

2. Discriminate between the effort arm and the resistance arm in a lever.

A **lever** (lev′er, le′ver; to lift) is an elongated, rigid object that rotates around a fixed point called the **fulcrum** (ful′krum). A seesaw is a familiar example of a lever. Levers have the ability to alter or change the speed and distance of movement produced by a force, the direction of an applied force, and the force strength.

Movement occurs when an **effort** applied to one point on the lever exceeds a **resistance** located at some other point. The part of a lever from the fulcrum to the point of effort is called the **effort arm,** and the lever part from the fulcrum to the point of resistance is the **resistance arm.** In the body, a long bone acts as a lever, a joint serves as the fulcrum, and the effort is generated by a muscle attached to the bone.

WHAT DID YOU LEARN?

11 What is the difference between the effort arm and the resistance arm in a lever?

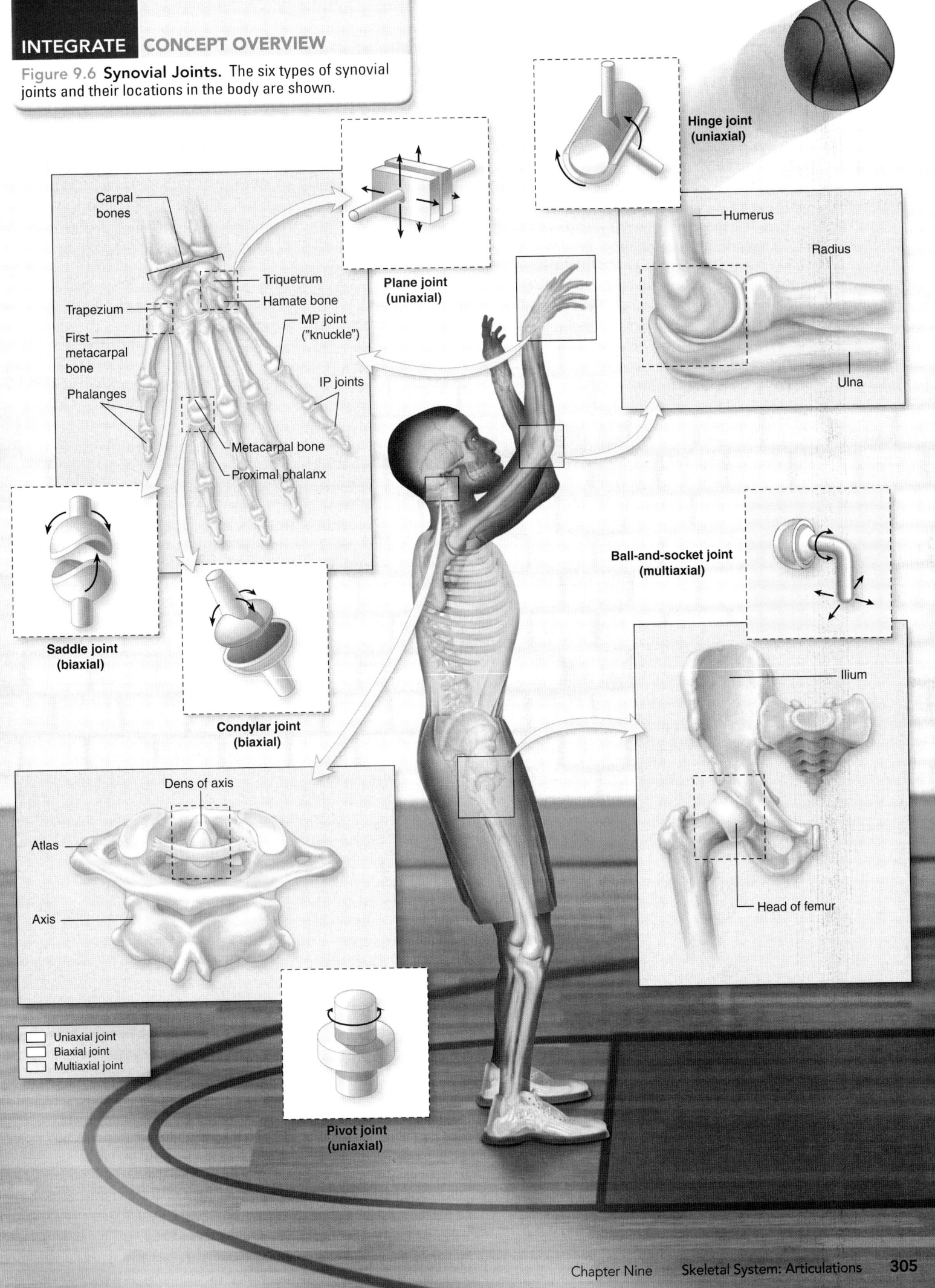

Figure 9.6 **Synovial Joints.** The six types of synovial joints and their locations in the body are shown.

Hinge joint (uniaxial)

Humerus

Radius

Ulna

Carpal bones

Triquetrum

Hamate bone

Trapezium

MP joint ("knuckle")

First metacarpal bone

IP joints

Phalanges

Metacarpal bone

Proximal phalanx

Plane joint (uniaxial)

Saddle joint (biaxial)

Condylar joint (biaxial)

Ball-and-socket joint (multiaxial)

Ilium

Head of femur

Dens of axis

Atlas

Axis

Uniaxial joint
Biaxial joint
Multiaxial joint

Pivot joint (uniaxial)

9.5b Type of Levers

☀ LEARNING OBJECTIVE

3. Compare and contrast the three types of levers in the human body.

Three classes of levers are found in the human body: first-class, second-class, and third-class (**figure 9.7**).

First-Class Levers

A **first-class lever** has a fulcrum in the middle, between the effort (force) and the resistance. An example of a first-class lever is a pair of scissors. The effort is applied to the handle of the scissors while the resistance is at the cutting end of the scissors. The fulcrum (pivot for movement) is along the middle of the scissors, between the handle and the cutting ends. In the body, an example of a first-class lever is the atlanto-occipital joint of the neck, where the muscles on the posterior side of the neck (effort) pull inferiorly on the nuchal lines of the skull and oppose the tendency of the head (resistance) to tip anteriorly.

Second-Class Levers

The resistance in a **second-class lever** is between the fulcrum and the applied effort. A common example of this type of lever is lifting the handles of a wheelbarrow, allowing it to pivot on its wheel at the opposite end and lift a load in the middle. The load weight is the resistance, and the upward lift on the handle is the effort. A small force can balance a larger weight in this type of lever, because the effort is always farther from the fulcrum than the resistance. Second-class levers are rare in the body, but one example occurs when the foot is depressed (plantar

INTEGRATE

LEARNING STRATEGY ✎

Use the acronym **FRE** to help you remember which part of the lever system is in between the other two parts:

- In a first-class lever, the **f**ulcrum (and the first letter of this acronym) is in between the resistance and the effort.
- In a second-class lever, the **r**esistance (and the second letter of the acronym) is in between the fulcrum and the effort.
- In a third-class lever, the **e**ffort (and the third letter of this acronym) is in between the fulcrum and the resistance.

flexed) so that a person can stand on tiptoe. The contraction of the calf muscle causes a pull superiorly by the calcaneal tendon attached to the heel (calcaneus).

Third-Class Levers

A **third-class lever** is noted when the effort is applied between the resistance and the fulcrum, as when picking up a small object with a pair of forceps. Third-class levers are the most common levers in the body. A third-class lever is found at the elbow where the fulcrum is the joint between the humerus and ulna, the effort is applied by the biceps brachii muscle, and the resistance is provided by any weight

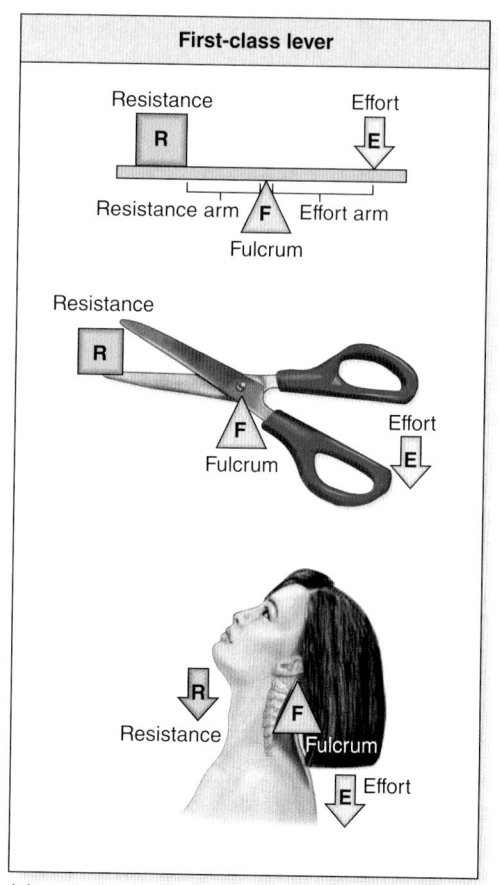

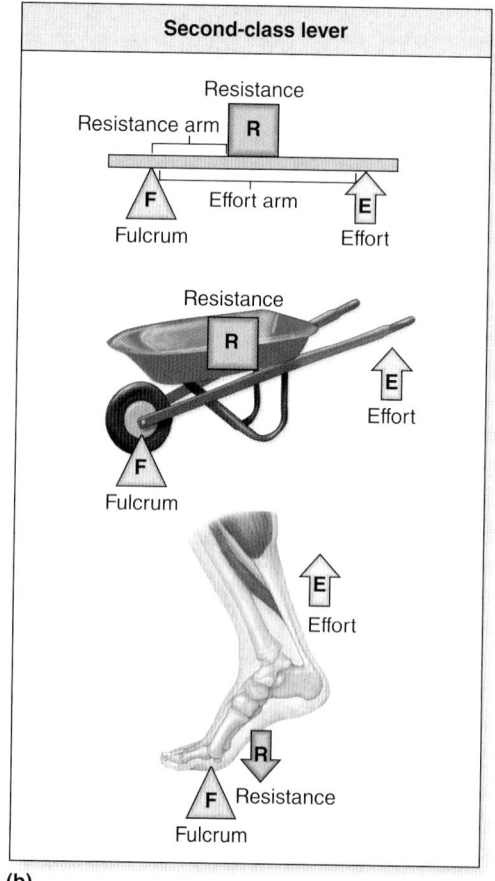

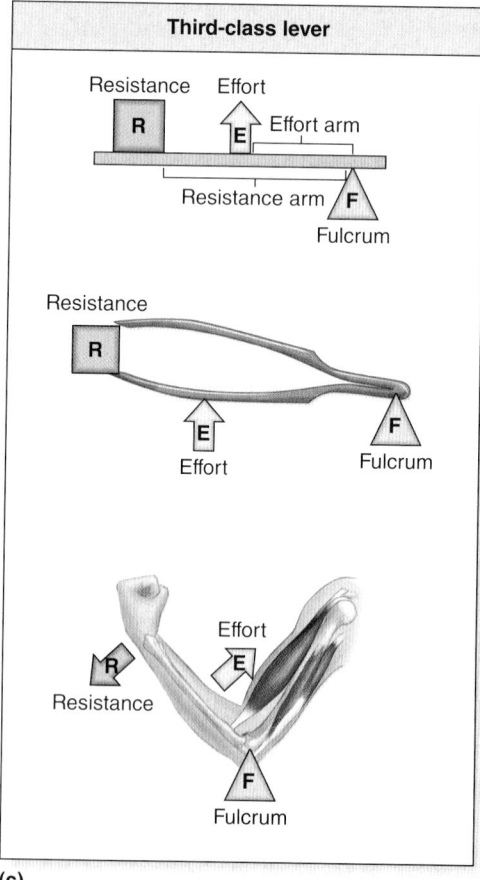

(a) **(b)** **(c)**

Figure 9.7 Classes of Levers. (*a*) In a first-class lever, the fulcrum is located between the resistance and effort, such as with a pair of scissors or the trapezius muscle (in the neck). (*b*) In a second-class lever, the resistance is between the fulcrum and effort, such as with a wheelbarrow or the calf muscles. (*c*) The most common type of lever is the third-class lever, where effort is applied between the resistance and the fulcrum, such as with forceps (tweezers) or the arm muscles.

in the hand or by the weight of the forearm itself. The mandible acts as a third-class lever when you bite with your incisors on a piece of food. The temporomandibular joint is the fulcrum, and the temporalis muscle exerts the effort, whereas the resistance is the item of food being bitten.

 WHAT DID YOU LEARN?

12 How does the position of the fulcrum, resistance, and effort vary in first-class, second-class, and third-class levers?

9.6 The Movements of Synovial Joints

Four types of motion occur at synovial joints: gliding, angular, rotational, and special movements (motions that occur only at specific joints) (table 9.2).

9.6a Gliding Motion

 **LEARNING OBJECTIVE**

1. Describe gliding motion, and name joints in which it occurs.

Gliding is a simple movement in which two opposing surfaces slide slightly back-and-forth or side-to-side with respect to one another. In a gliding motion, the angle between the bones does not change, and only limited movement is possible in any direction. Gliding motion typically occurs along plane joints, such as between the carpals or the tarsals.

 WHAT DID YOU LEARN?

13 What joints typically use gliding motion?

9.6b Angular Motion

 LEARNING OBJECTIVES

2. Describe angular motion.

3. Name the specific types of angular motion.

4. Give examples of joints that exhibit angular motion.

Angular motion either decreases or increases the angle between two bones. These movements may occur at many of the synovial joints. They include the following specific types: flexion and extension, hyperextension, lateral flexion, abduction and adduction, and circumduction.

Table 9.2	Movements at Synovial Joints AP|R		
Movement	**Description**		**Opposing Movement[1]**
GLIDING MOTION	Two opposing articular surfaces slide past each other in almost any direction; the amount of movement is slight		
ANGULAR MOTION	The angle between articulating bones increases or decreases		
Flexion	The angle between articulating bones decreases in an anterior-posterior (AP) plane		Extension
Extension	The angle between articulating bones increases in an anterior-posterior (AP) plane		Flexion
Hyperextension	Extension movement continues past the anatomic position		Flexion
Lateral flexion	The vertebral column moves (bends) in a lateral direction along a coronal plane		None
Abduction	Lateral movement of a body part away from the midline		Adduction
Adduction	Lateral movement of a body part toward the midline		Abduction
Circumduction	A continuous movement that combines flexion, abduction, extension, and adduction in succession; the distal end of the limb or digit moves in a circle		None
ROTATIONAL MOTION	A bone pivots around its own longitudinal axis		
Pronation	Rotation of the forearm where the palm is turned posteriorly		Supination
Supination	Rotation of the forearm in which the palm is turned anteriorly		Pronation
SPECIAL MOVEMENTS	Types of movement that do not fit in the previous categories		
Depression	Movement of a body part inferiorly		Elevation
Elevation	Movement of a body part superiorly		Depression
Dorsiflexion	Ankle joint movement where the dorsum (superior surface) of the foot is brought toward the anterior surface of the leg		Plantar flexion
Plantar flexion	Ankle joint movement where the sole of the foot is brought toward the posterior surface of the leg		Dorsiflexion
Eversion	Twisting motion of the foot that turns the sole laterally or outward		Inversion
Inversion	Twisting motion of the foot that turns the sole medially or inward		Eversion
Protraction	Anterior movement of a body part from anatomic position		Retraction
Retraction	Posterior movement of a body part from anatomic position		Protraction
Opposition	Special movement of the thumb across the palm toward the fingers to permit grasping and holding of an object		Reposition

1. Some movements (e.g., circumduction) do not have an opposing movement.

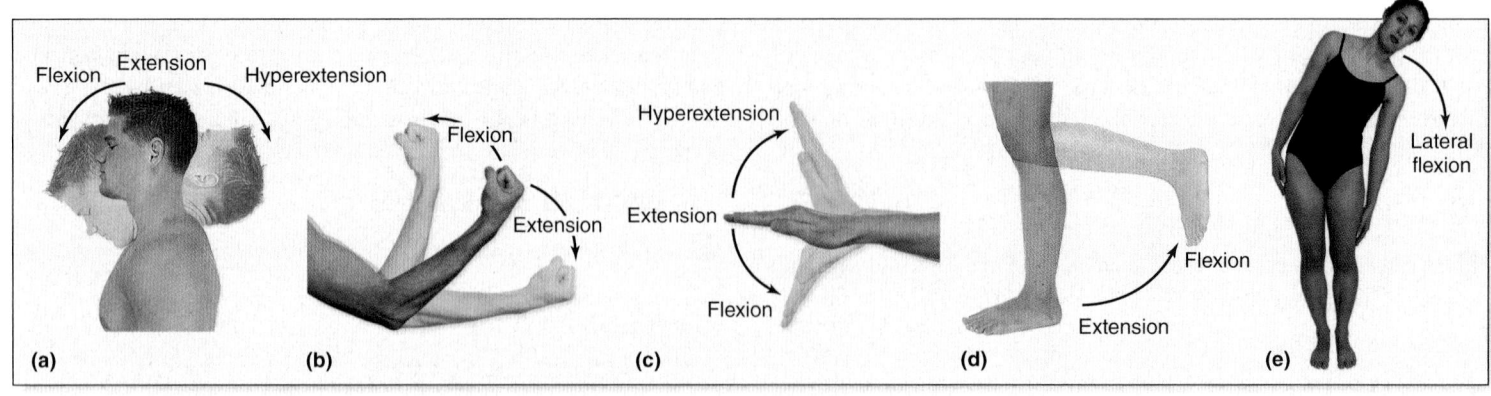

Figure 9.8 Flexion, Extension, Hyperextension, and Lateral Flexion. Flexion decreases the joint angle in an anterior-posterior (AP) plane, whereas extension increases the joint angle in the AP plane. Hyperextension is extension of a joint beyond 180 degrees. Lateral flexion decreases a joint angle, but in a coronal plane. Examples of joints that allow some of these movements are (*a*) the atlanto-occipital joint, (*b*) the elbow joint, (*c*) the radiocarpal joint, (*d*) the knee joint, and (*e*) the intervertebral joints. AP|R

Flexion (flek'shŭn; *flecto* = to bend) is movement in an anterior-posterior (AP) plane of the body that *decreases* the angle between the bones. Bones are brought closer together as the angle between them decreases. Examples are the bending of the fingers toward the palm to make a fist, the bending of the forearm toward the arm at the elbow, flexion at the shoulder when the arm is raised anteriorly, and flexion of the neck when the head is bent anteriorly and you look down at your feet.

The opposite of flexion is **extension** (eks-ten'shŭn; *extensio* = a stretching out), which is movement in an anterior-posterior (AP) plane that *increases* the angle between the articulating bones. Extension is a straightening action that occurs in an AP plane. Straightening the arm and forearm until the upper limb projects directly away from the anterior side of your body, or straightening the fingers after making a clenched fist, are examples of extension.

When a joint is extended more than 180 degrees, the movement is termed **hyperextension** (hī'per-eks-ten'shŭn; *hyper* = above normal). For example, if you extend your arm and hand with the palm facing inferiorly, and then raise the back of your hand as if admiring a new ring on your finger, the wrist is hyperextended. If you glance up at the ceiling while standing, your neck is hyperextended. Flexion, extension, and hyperextension of various body parts are illustrated in figure 9.8*a–d*.

Lateral flexion occurs when the trunk of the body moves in a coronal plane laterally away from the body. This type of movement occurs primarily between the vertebrae in the cervical and lumbar regions of the vertebral column (figure 9.8*e*).

WHAT DO YOU THINK?

3 When sitting upright in a chair, are your hip and knee joints flexed or extended?

Abduction (ab-dŭk'shŭn; *duco* = to draw) means to "move away," and it is a lateral movement of a body part *away from* the body midline. Abduction occurs when either the arm or the thigh is moved laterally away from the body midline. Abduction of either the fingers or the toes

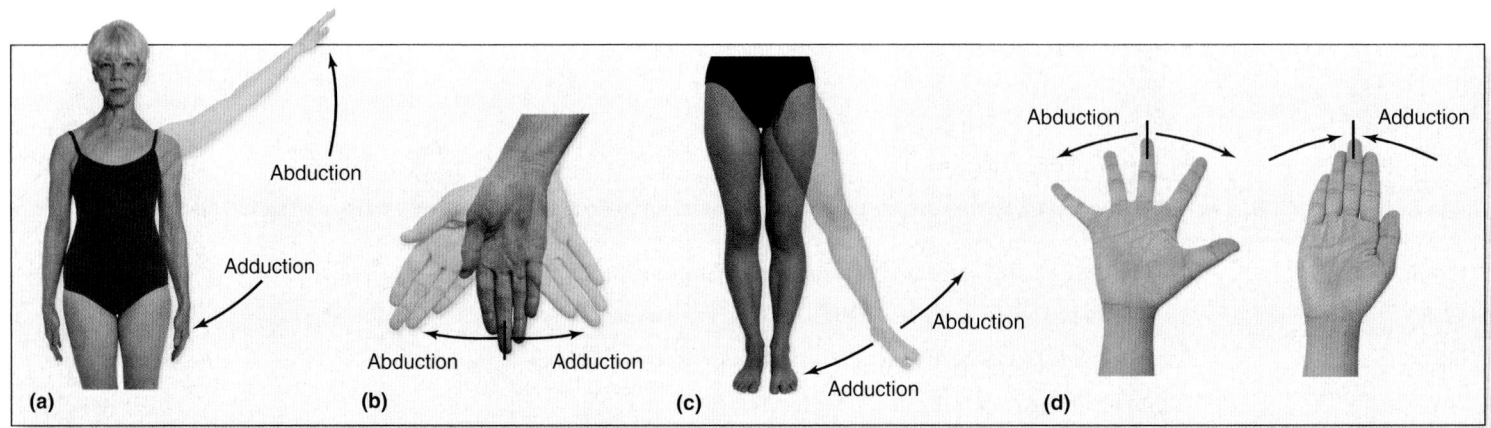

Figure 9.9 Abduction and Adduction. Abduction moves a body part away from the trunk in a lateral direction, whereas adduction moves the body part toward the trunk in a medial direction. Some examples occur at (*a*) the glenohumeral joint, (*b*) the radiocarpal joint, (*c*) the hip joint, and (*d*) the metacarpophalangeal (MP) joints. AP|R

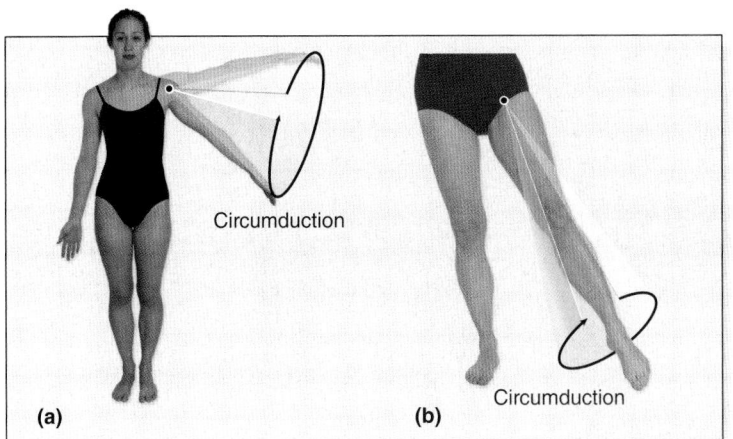

Figure 9.10 Circumduction. Circumduction is a complex movement that involves flexion, abduction, extension, and adduction in succession. Examples of joints that allow this movement are (a) the glenohumeral joint and (b) the hip joint.

means that you spread them apart, away from the longest digit that acts as the midline. Abducting the wrist (also known as *radial* deviation) involves pointing the hand and fingers laterally, away from the body. The opposite of abduction is **adduction** (ad-dŭk′shŭn), meaning to "move toward." This is the medial movement of a body part *toward* the body midline. Adduction occurs when the raised arm or thigh is brought back toward the body midline, or in the case of the digits, toward the midline of the hand. Adducting the wrist (also known as *ulnar* deviation) involves pointing the hand and fingers medially, toward the body. Abduction and adduction of various body parts are shown in **figure 9.9**.

Circumduction (ser-kŭm-dŭk′shŭn; *circum* = around) is a sequence of movements in which the proximal end of an appendage remains relatively stationary while the distal end makes a circular motion **(figure 9.10)**. The resulting movement makes an imaginary cone shape. This is demonstrated when you draw a circle on the blackboard. The shoulder remains stationary while your hand moves. The tip of the imaginary cone is the stationary shoulder, while the rounded "base" of the cone is the circle made by the hand. Circumduction is a complex movement that occurs as a result of a continuous sequence of flexion, abduction, extension, and adduction.

WHAT DID YOU LEARN?

14 How do flexion and extension differ? What movements are involved in circumduction?

9.6c Rotational Motion

LEARNING OBJECTIVE

5. Explain rotational motion, and name joints in which it occurs.

Rotation is a pivoting motion in which a bone turns on its own longitudinal axis **(figure 9.11)**. Rotational movement occurs at the atlantoaxial joint, which pivots when you rotate your head

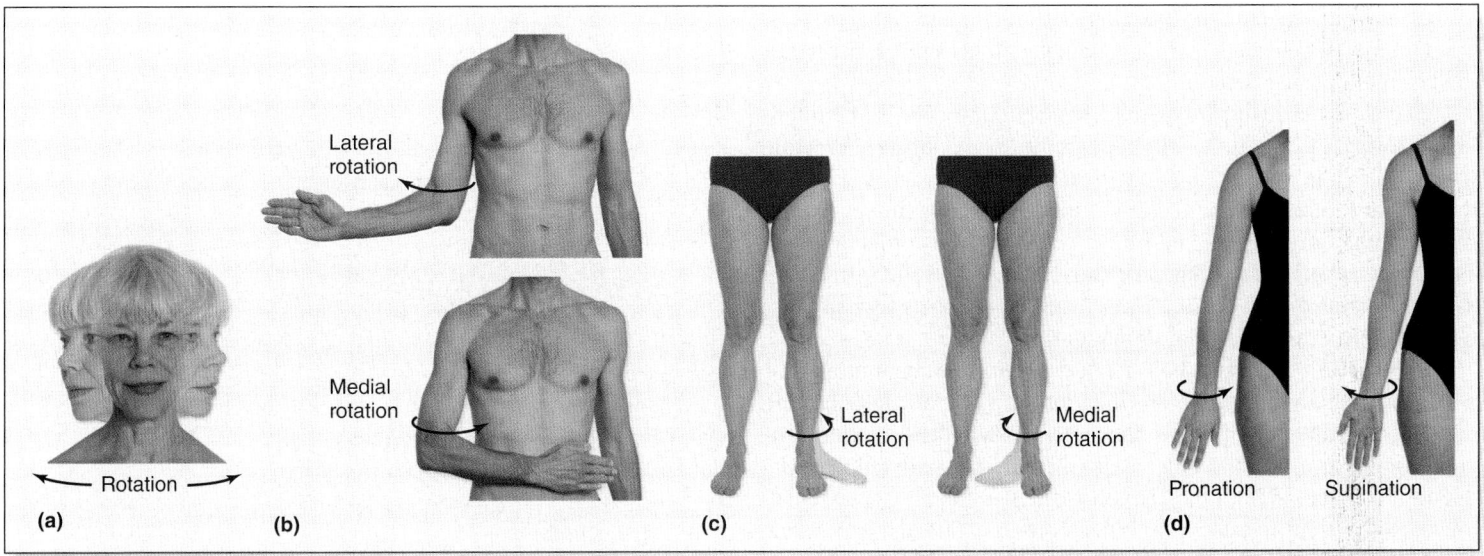

Figure 9.11 Rotational Movements. Rotation allows a bone to pivot on its longitudinal axis. Examples of joints that allow this movement are (a) the atlantoaxial joint, (b) the glenohumeral joint, and (c) the hip joint. (d) Pronation and supination occur at the forearm. AP|R

to gesture "no." Some limb rotations are described as either away from the median plane or toward it. For example, **lateral rotation** (or *external* rotation) turns the anterior surface of the femur or humerus laterally, whereas **medial rotation** (or *internal* rotation) turns the anterior surface of the femur or humerus medially.

Pronation (prō-nā′shŭn) is the medial rotation of the forearm so that the palm of the hand is directed posteriorly or inferiorly. The radius and ulna are crossed to form an X (see section 8.9b). **Supination** (sū′pi-nā′shŭn) occurs when the forearm rotates laterally so that the palm faces anteriorly or superiorly. In the anatomic position, the forearm is supinated. Figure 9.11d illustrates pronation and supination.

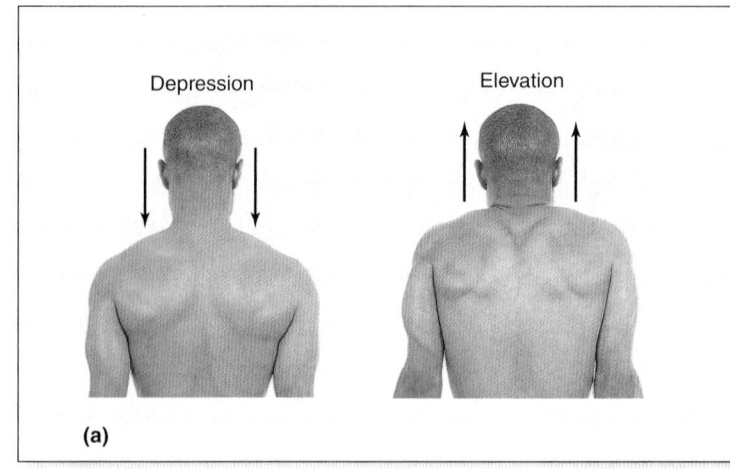

(a)

 WHAT DID YOU LEARN?

 15 What is pronation, and where in the body may this type of movement be performed?

9.6d Special Movements

 LEARNING OBJECTIVE

6. Explain what is meant by special movements, and give examples of joints at which they occur.

Some movements occur only at specific joints and do not readily fit into any of the functional categories previously discussed. These special movements include depression and elevation, dorsiflexion and plantar flexion, eversion and inversion, protraction and retraction, and opposition.

Depression (dē-presh′ŭn, *de* = away, down, *presso* = to press) is the inferior movement of a part of the body. Examples of depression include the movement of the mandible while opening the mouth to chew food and the movement of your shoulders in an inferior direction. **Elevation** (el-ĕ-vā′shŭn) is the superior movement of a body part. Examples of elevation include the superior movement of the mandible while closing the mouth and the movement of the shoulders in a superior direction (shrugging your shoulders). **Figure 9.12a** illustrates depression and elevation at the glenohumeral joints.

Dorsiflexion and plantar flexion are limited to the ankle joint (figure 9.12b). **Dorsiflexion** (dōr-si-flek′shŭn) occurs when the talocrural (ankle) joint is bent such that the dorsum (superior surface) of the foot and the toes moves toward the leg. This movement occurs when you dig in your heels, and it prevents your toes from scraping the ground when you take a step. **Plantar flexion** (plan′tăr; *planta* = sole of foot) is a movement of the foot at the talocrural joint so that the toes point inferiorly. When a ballerina is standing on her tiptoes, her ankle joint is in full plantar flexion.

Eversion and inversion are movements that occur at the intertarsal joints of the foot only (figure 9.12c). During **eversion** (ē-ver′zhŭn) the sole of the foot turns to face laterally, whereas the sole of the foot turns medially during **inversion** (in-ver′zhŭn). (*Note:* Some orthopedists and runners use the terms pronation and supination when describing foot movements as well, instead of using inversion and eversion. Simply put, inversion is foot supination, whereas eversion is foot pronation.)

Protraction (prō-trak′shŭn) is the anterior movement of a body part from anatomic position, as when jutting your jaw anteriorly at the temporomandibular joint or hunching the shoulders anteriorly by crossing the arms. In the latter case, the clavicles move anteriorly due to movement at both the acromioclavicular and sternoclavicular joints. **Retraction** (rē-trak′shŭn) is the posteriorly directed movement of a body part from the anatomic position. Figure 9.12d illustrates protraction and retraction at the temporomandibular joint.

At the carpometacarpal joint, the thumb moves toward the palmar tips of the fingers as it crosses the palm of the hand. This movement is called **opposition** (op′ō-si′shŭn) (figure 9.12e). It enables the hand to grasp objects and is the most distinctive digital movement in humans. The opposite movement is called **reposition.**

 WHAT DID YOU LEARN?

 16 What is the difference between inversion and eversion, and which joints allow these movements?

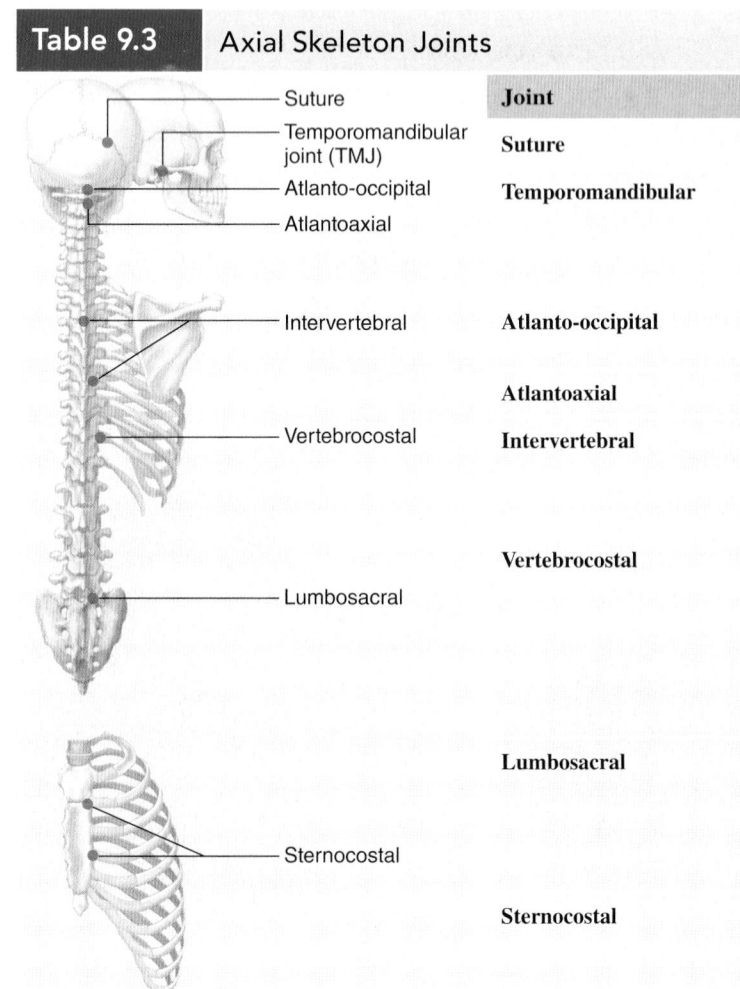

Table 9.3	Axial Skeleton Joints
	Joint
	Suture
	Temporomandibular
	Atlanto-occipital
	Atlantoaxial
	Intervertebral
	Vertebrocostal
	Lumbosacral
	Sternocostal

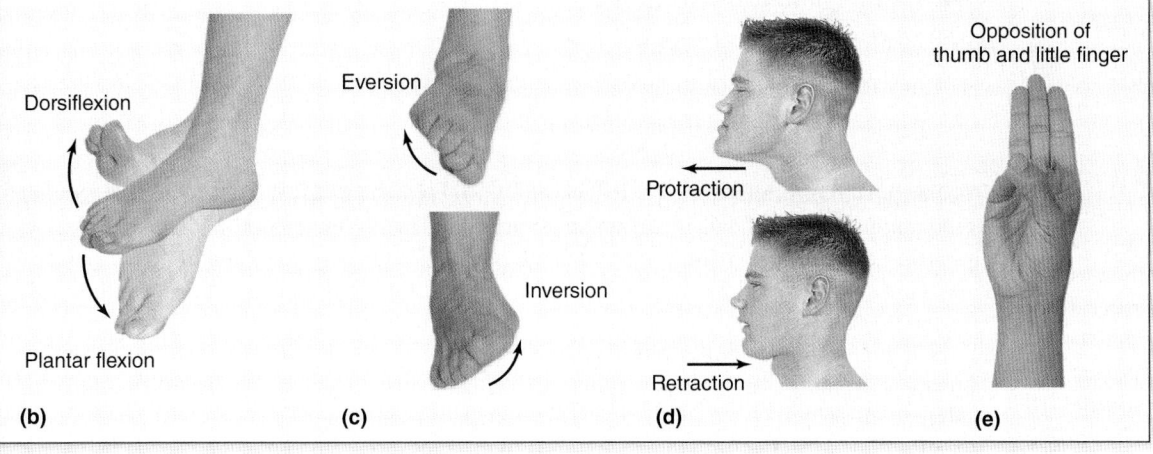

Figure 9.12 Special Movements Allowed at Synovial Joints. (*a*) Depression and elevation at the glenohumeral joint. (*b*) Dorsiflexion and plantar flexion at the talocrural joint. (*c*) Eversion and inversion at the intertarsal joints. (*d*) Protraction and retraction at the temporomandibular joint. (*e*) Opposition at the carpometacarpal joints.

AP|R

9.7 Features and Anatomy of Selected Joints

Both the axial skeleton and appendicular skeleton exhibit many more joints than are individually discussed here. Table 9.3 summarizes the main features of major joints of the axial skeleton.

The structure and function of the more commonly known articulations of the axial and appendicular skeletons are examined in this section. These are the temporomandibular joint of the skull; the shoulder joint and elbow joint; and the hip joint, knee joint, and talocrural (ankle) joint.

Articulation Components	Structural Classification	Functional Classification	Description of Movement
Adjacent skull bones	Fibrous joint	Synarthrosis	None allowed
Head of mandible and mandibular fossa of temporal bone Head of mandible and articular tubercle of temporal bone	Synovial (hinge, plane) joints	Diarthrosis	Depression, elevation, lateral displacement, protraction, retraction, slight rotation
Superior articular facets of atlas and occipital condyles of occipital bone	Synovial (condylar) joint	Diarthrosis	Extension and flexion of the head; slight lateral flexion of head to sides
Anterior arch of atlas and dens of axis	Synovial (pivot) joint	Diarthrosis	Head rotation
Vertebral bodies of adjacent vertebrae Superior and inferior articular processes of adjacent vertebrae	Cartilaginous joint (symphysis) between vertebral bodies; synovial (plane) joint between articular processes	Amphiarthrosis between vertebral bodies; diarthrosis between articular processes	Extension, flexion, lateral flexion of vertebral column
Facets of heads of ribs and bodies of adjacent thoracic vertebrae and intervertebral discs between adjacent vertebrae Articular part of tubercles of ribs and facets of transverse processes of thoracic vertebra	Synovial (plane) joint	Diarthrosis	Some slight gliding
Body of the fifth lumbar vertebra and the base of the sacrum Inferior articular facets of fifth lumbar vertebra and superior articular facets of first sacral vertebra	Cartilaginous joint (symphysis) between lumbar body and base of sacrum; synovial (plane) joint between articular processes	Amphiarthrosis between lumbar body and base of sacrum; diarthrosis between articular processes	Extension, flexion, lateral flexion of vertebral column
Sternum and first seven pairs of ribs	Cartilaginous joint (synchondrosis) between sternum and first ribs; synovial (plane) joint between sternum and ribs 2–7	Synarthrosis between sternum and first ribs; diarthrosis between sternum and ribs 2–7	No movement between sternum and first ribs; some gliding movement permitted between sternum and ribs 2–7

INTEGRATE

CLINICAL VIEW

TMJ Disorders

The temporomandibular joint (TMJ) is subject to various disorders. TMJ disorders are often seen in people who habitually chew gum or grind or clench their teeth. The most common TMJ disorder occurs as a result of alterations in the ligaments that secure the joint, causing progressive internal displacement of the articular disc. As the articular disc is forced out of its normal position, a clicking or popping noise may be heard as the person opens or closes the mouth. Pain from the TMJ disorder may be felt not only within the joint but also in such areas as the paranasal sinuses, tympanic membrane (eardrum), oral cavity, eyes, and teeth. The widespread distribution of pain occurs because all of these structures, including the muscle and jaws, are innervated by numerous sensory branches of the trigeminal nerve (see section 13.9).

9.7a Temporomandibular Joint

LEARNING OBJECTIVES

1. Describe the features of the temporomandibular joint (TMJ).
2. List the movements of the TMJ.

The **temporomandibular** (tem′pŏ-rō-man-dib′yū-lăr) **joint (TMJ)** is the articulation formed at the point where the head of the mandible articulates with the temporal bone; specifically, the articular tubercle of the temporal bone anteriorly and the mandibular fossa posteriorly. This small, complex articulation is the only mobile joint between bones in the skull (**figure 9.13**; table 9.3).

The temporomandibular joint has several unique anatomic features. A loose **articular capsule** surrounds the joint and promotes an extensive range of motion. It contains an **articular disc** that is a thick pad of fibrocartilage separating the articulating bones and extending horizontally to divide the synovial cavity into two separate chambers. As a result, the TMJ is really two synovial joints—one between the temporal bone and the articular disc, and a second between the articular disc and the mandible.

Several ligaments support the TMJ. The **sphenomandibular ligament** (an extracapsular ligament) is a thin band that extends anteriorly and inferiorly from the sphenoid to the medial surface of the mandibular ramus. The **temporomandibular ligament** (or *lateral ligament*) is composed of two short bands that extend inferiorly and posteriorly from the articular tubercle of the temporal bone to the mandible.

The temporomandibular joint exhibits hinge, gliding, and some pivot joint movements. It functions like a hinge during jaw depression and elevation while chewing. It glides slightly forward during protraction of the jaw for biting, and glides slightly from side to side to grind food between the teeth during chewing.

WHAT DID YOU LEARN?

17 What movements are allowed at the temporomandibular joint?

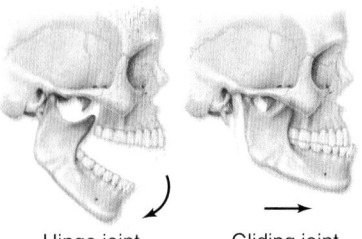

Hinge joint   Gliding joint

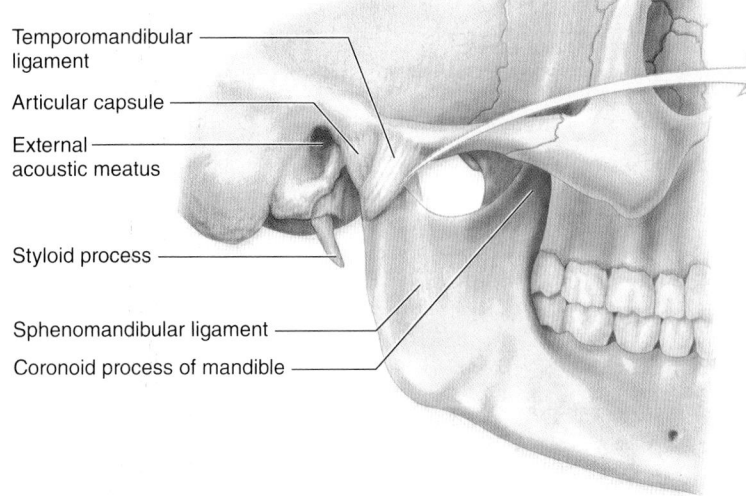

Temporomandibular ligament

Articular capsule

External acoustic meatus

Styloid process

Sphenomandibular ligament

Coronoid process of mandible

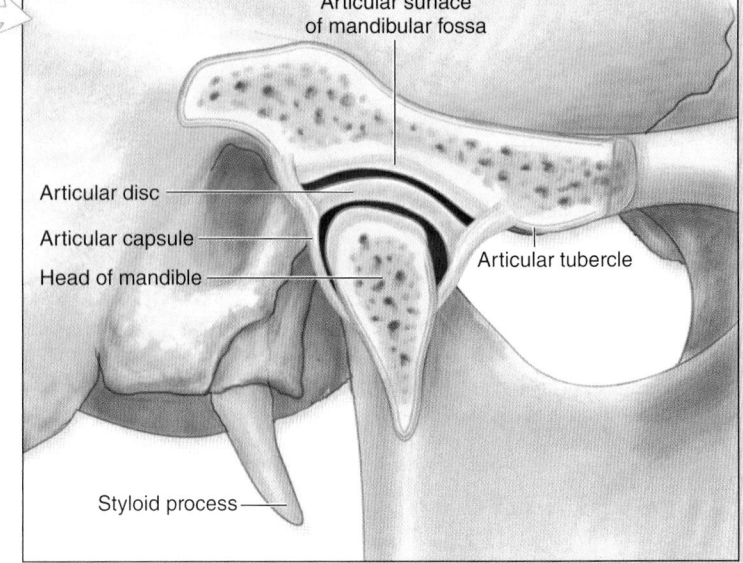

Articular surface of mandibular fossa

Articular disc

Articular capsule

Head of mandible

Articular tubercle

Styloid process

Figure 9.13 Temporomandibular Joint. The articulation between the head of the mandible and the mandibular fossa of the temporal bone exhibits a wide range of movements. **AP|R**

9.7b Shoulder Joint

✓ LEARNING OBJECTIVES

3. Describe the three individual joints that make up the shoulder articulation.

4. Explain why the glenohumeral joint is relatively unstable.

The joints associated with movement at the shoulder include the sternoclavicular joint, the acromioclavicular joint, and the glenohumeral joint. (Table 9.4 lists the features of the major joints of the pectoral girdle and upper limbs.)

Table 9.4 Pectoral Girdle and Upper Limb Joints

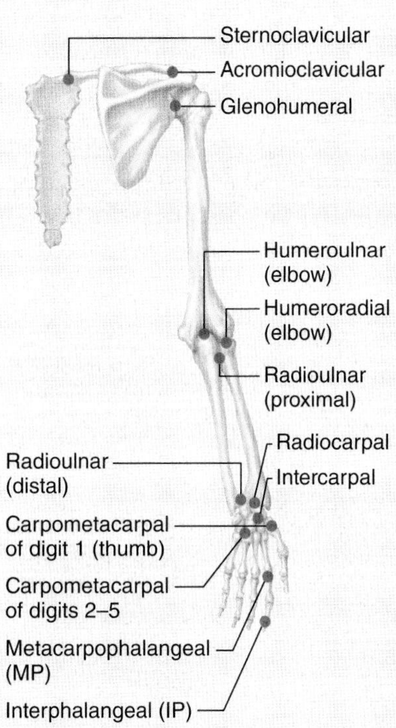

Sternoclavicular
Acromioclavicular
Glenohumeral

Humeroulnar (elbow)
Humeroradial (elbow)
Radioulnar (proximal)
Radiocarpal
Intercarpal
Radioulnar (distal)
Carpometacarpal of digit 1 (thumb)
Carpometacarpal of digits 2–5
Metacarpophalangeal (MP)
Interphalangeal (IP)

Joint	Articulation Components	Structural Classification	Functional Classification	Description of Movement
Sternoclavicular	Manubrium of sternum and sternal end of clavicle	Synovial (saddle)	Diarthrosis	Depression, elevation, and circumduction
Acromioclavicular	Acromion of scapula and acromial end of clavicle	Synovial (plane)	Diarthrosis	Gliding of scapula on clavicle
Glenohumeral	Glenoid cavity of scapula and head of humerus	Synovial (ball-and-socket)	Diarthrosis	Abduction, adduction, circumduction, flexion, extension, lateral rotation, and medial rotation of arm
Elbow	**Humeroulnar joint:** Trochlea of humerus and trochlear notch of ulna **Humeroradial joint:** Capitulum of humerus and head of radius	Synovial (hinge)	Diarthrosis	Flexion and extension of forearm
Radioulnar	**Proximal joint:** Head of radius and radial notch of ulna **Distal joint:** Distal end of ulna and ulnar notch of radius	Synovial (pivot)	Diarthrosis	Rotation of radius with respect to ulna
Radiocarpal	Distal end of radius; lunate, scaphoid, and triquetrum	Synovial (condylar)	Diarthrosis	Abduction, adduction, circumduction, flexion and extension of wrist
Intercarpal	Adjacent bones in proximal row of carpal bones Adjacent bones in distal row of carpal bones Adjacent bones between proximal and distal rows (midcarpal joints)	Synovial (plane)	Diarthrosis	Gliding
Carpometacarpal	**Thumb:** Trapezium and first metacarpal **Other digits:** Carpals and metacarpals II–V	Synovial (saddle) at thumb; synovial (plane) at other digits	Diarthrosis	Flexion, extension, abduction, adduction, circumduction and opposition at thumb; gliding at other digits
Metacarpophalangeal (MP joints, "knuckles")	Heads of metacarpals and bases of proximal phalanges	Synovial (condylar)	Diarthrosis	Flexion, extension, abduction, adduction, and circumduction of phalanges
Interphalangeal (IP joints)	Heads of proximal and middle phalanges with bases of middle and distal phalanges, respectively	Synovial (hinge)	Diarthrosis	Flexion and extension of phalanges

Sternoclavicular Joint

The **sternoclavicular** (ster'nō-kla-vik'yū-lǎr) **joint** is a saddle joint formed by the articulation between the manubrium of the sternum and the sternal end of the clavicle **(figure 9.14)**. An **articular disc** partitions the sternoclavicular joint into two parts and forms two separate synovial cavities. As a result, a wide range of movement is possible, including depression, elevation, and circumduction.

Support and stability are provided to this articulation by the fibers of the articular capsule and by multiple extracapsular ligaments, such as the sternoclavicular and costoclavicular ligaments. This anatomical arrangement makes the sternoclavicular joint very stable. If you fall on an outstretched hand so that force is applied to the joint, the clavicle will fracture before this joint dislocates.

Acromioclavicular Joint

The **acromioclavicular** (ă-krō'mē-ō-kla-vik'yū-lǎr) **joint** is a plane joint between the acromion of the scapula and the lateral end of the clavicle **(figure 9.15)**. A fibrocartilaginous articular disc lies within the joint cavity between these two bones. This joint works with both the sternoclavicular joint and the glenohumeral joint to give the upper limb a full range of movement.

Several ligaments provide great stability to this joint. The fibrous joint capsule is strengthened superiorly by an **acromioclavicular ligament.** In addition, a very strong **coracoclavicular** (kōr'ă-kō-kla-vik'yū-lǎr) **ligament** binds the clavicle to the coracoid process of the scapula. If this ligament is torn (as occurs in severe shoulder separations; see Clinical View: "Shoulder Joint Dislocations" on p. 316), the acromion and clavicle no longer align properly.

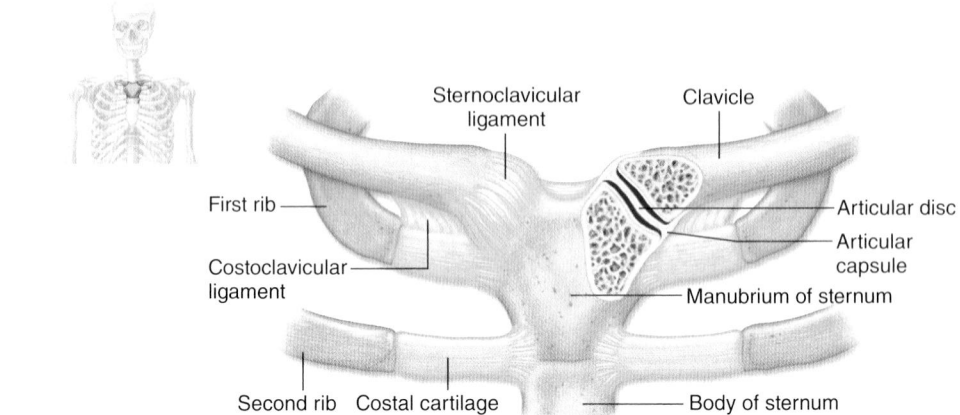

(a) Anterior view with upper limbs in anatomic position

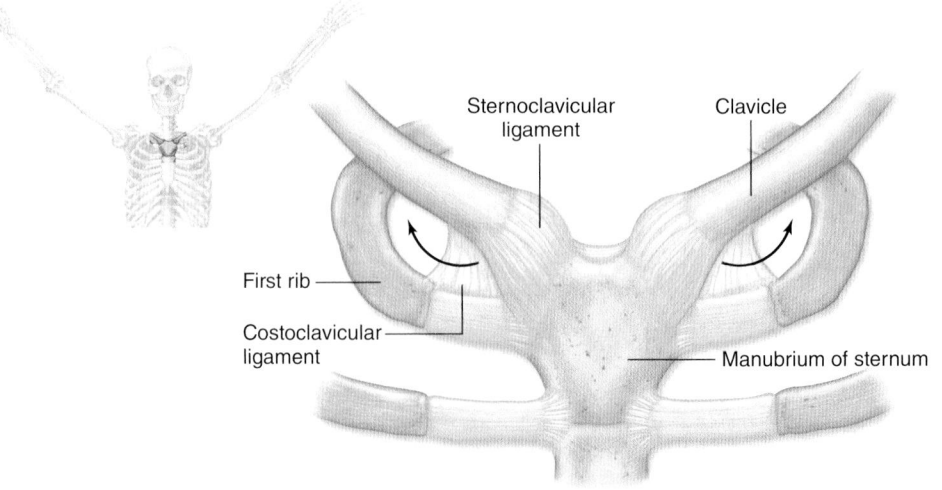

(b) Anterior view with upper limbs raised

Figure 9.14 Sternoclavicular Joint. The sternoclavicular joint helps stabilize movements of the entire shoulder. The positioning of the joint is compared (*a*) with the upper limbs by the sides of the body and (*b*) with the upper limbs raised.

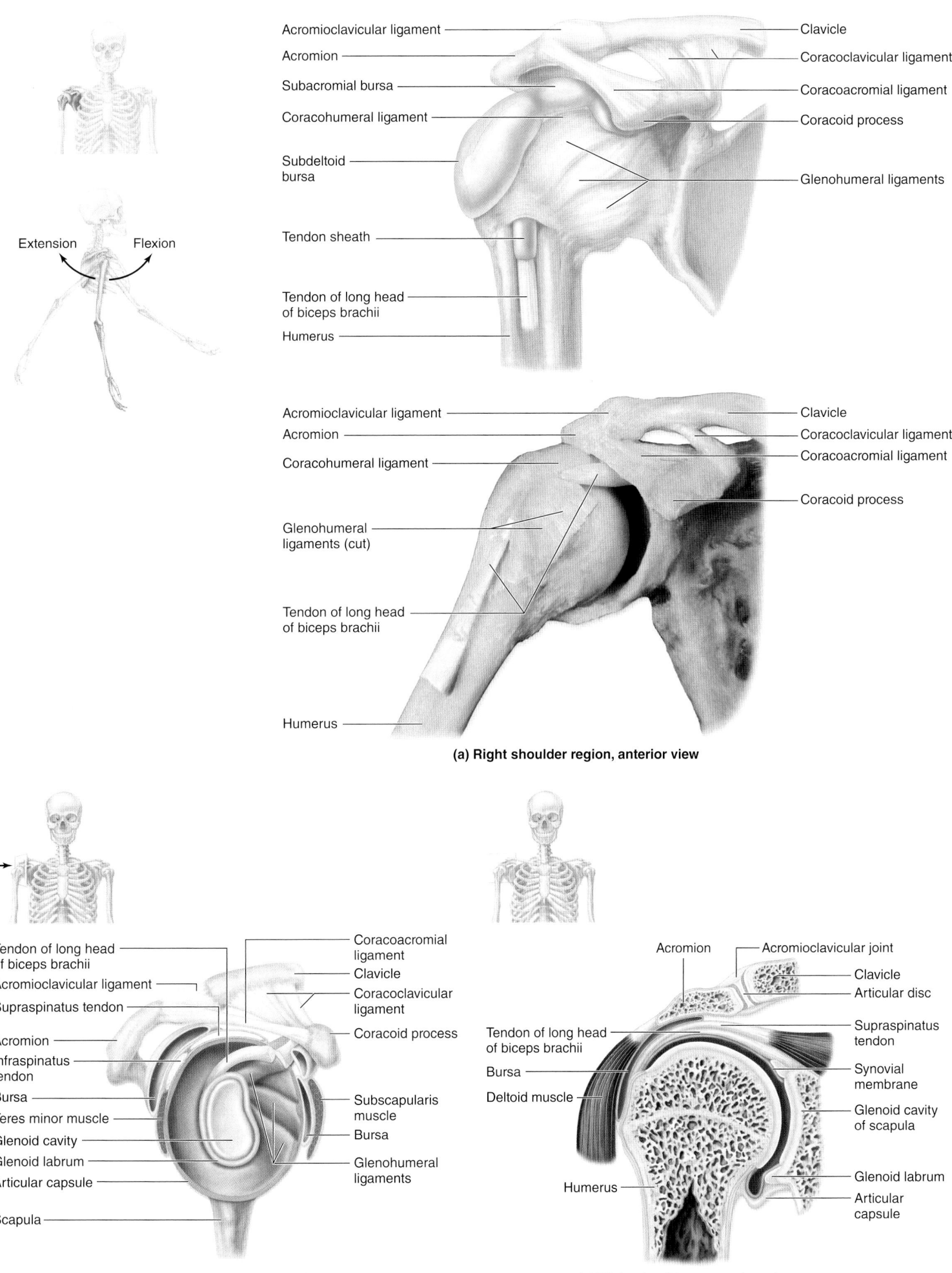

Extension Flexion

Acromioclavicular ligament
Acromion
Subacromial bursa
Coracohumeral ligament
Subdeltoid bursa
Tendon sheath
Tendon of long head of biceps brachii
Humerus

Clavicle
Coracoclavicular ligament
Coracoacromial ligament
Coracoid process
Glenohumeral ligaments

Acromioclavicular ligament
Acromion
Coracohumeral ligament
Glenohumeral ligaments (cut)
Tendon of long head of biceps brachii
Humerus

Clavicle
Coracoclavicular ligament
Coracoacromial ligament
Coracoid process

(a) Right shoulder region, anterior view

Tendon of long head of biceps brachii
Acromioclavicular ligament
Supraspinatus tendon
Acromion
Infraspinatus tendon
Bursa
Teres minor muscle
Glenoid cavity
Glenoid labrum
Articular capsule
Scapula

Coracoacromial ligament
Clavicle
Coracoclavicular ligament
Coracoid process
Subscapularis muscle
Bursa
Glenohumeral ligaments

(b) Right shoulder, lateral view

Acromion
Tendon of long head of biceps brachii
Bursa
Deltoid muscle
Humerus

Acromioclavicular joint
Clavicle
Articular disc
Supraspinatus tendon
Synovial membrane
Glenoid cavity of scapula
Glenoid labrum
Articular capsule

(c) Right shoulder, coronal section

Figure 9.15 Acromioclavicular and Glenohumeral Joints. (*a*) Anterior diagrammatic view and cadaver photo of both joints on the right side of the body. (*b*) Right lateral view and (*c*) right coronal section show the articulating bones and supporting structures at the shoulder. AP|R

INTEGRATE

CLINICAL VIEW
Shoulder Joint Dislocations

Dislocation (dis′lō-kā′shun; *dis* = apart, *locatio* = placing), a joint injury in which the articulating bones have separated, are common in the shoulder. Although these injuries can occur at any of the three shoulder joints described in the chapter, they are more common at the acromioclavicular joint or the glenohumeral joint.

The term **shoulder separation** refers to a dislocation of the acromioclavicular joint. This injury often results from a hard blow to the joint, as when a hockey player is "slammed into the boards" or one falls onto the shoulder. Symptoms include tenderness and edema (swelling) in the area of the joint and pain when the arm is abducted more than 90 degrees, because in this position significant movement occurs between the separated bone surfaces. Additionally, the acromion will appear very prominent and pointed. Treatment can range from rest to surgery, depending upon the severity of the dislocation. Glenohumeral joint dislocations are very common because this joint is very mobile and yet unstable. These dislocations

usually occur when a fully abducted humerus is struck hard—for example, when a quarterback is hit as he is about to release a football, or when a person falls on an outstretched hand. The initial blow pushes the humerus into the inferior part of the articular capsule and tears the capsule as the humerus dislocates. (Recall the inferior part of the capsule is relatively weak and not protected by muscle tendons as are the other surfaces of the capsule.) Once the humeral head is no longer held in place by the capsule, the anterior thorax (chest) muscles pull superiorly and medially on the humeral head, causing it to lie just inferior to the coracoid process. The result is that the shoulder appears flattened and "squared-off," because the humeral head is dislocated anteriorly and inferiorly to the glenohumeral joint capsule. Some glenohumeral dislocations can be repaired by "popping" the humerus back into the glenoid cavity. More severe dislocations may need surgical repair.

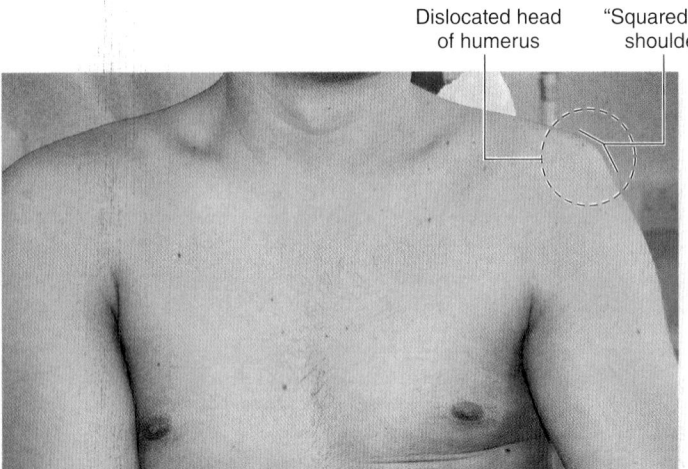

Glenohumeral joint dislocation.

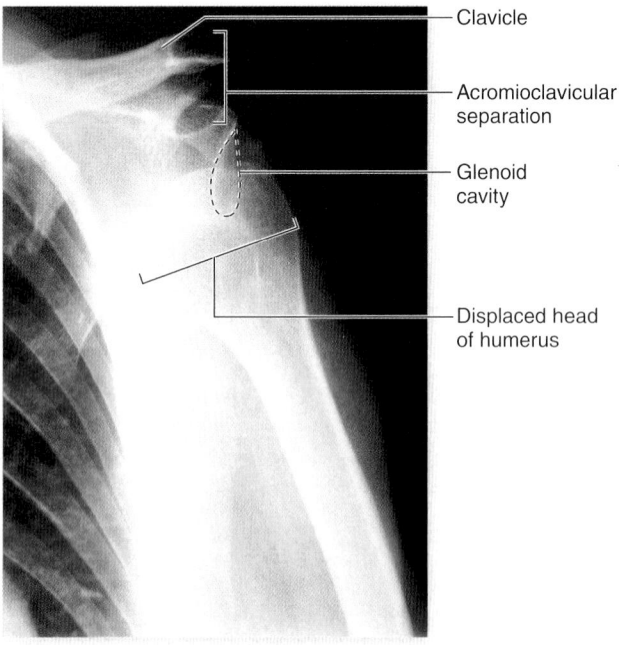

Radiograph of acromioclavicular and glenohumeral dislocations.

Glenohumeral (Shoulder) Joint

The **glenohumeral** (glē′nō-hyū′mer-ăl) **joint** is commonly referred to as the shoulder joint. It is a ball-and-socket joint formed by the articulation of the head of the humerus and the glenoid cavity of the scapula (figure 9.15). It permits the greatest range of motion of any joint in the body, and so it is also both the most unstable joint in the body and the one most frequently dislocated.

The fibrocartilaginous **glenoid labrum** encircles and covers the surface of the glenoid cavity. A relatively loose articular capsule attaches to the surgical neck of the humerus. The glenohumeral joint has several major ligaments. The **coracoacromial** (kōr′ă-kō-ă-krō′mē-ăl) **ligament** extends across the space between the coracoid process and the acromion. The large **coracohumeral** (kōr′ă-kō-hyū′mer-ăl) **ligament** is a thickening of the superior part of the joint capsule. It extends from the coracoid process to the humeral head. The **glenohumeral ligaments** are three thickenings of the anterior portion of the articular capsule. These ligaments are often indistinct or absent and provide only minimal support. In addition, the tendon of the long head of biceps brachii travels within the articular capsule and helps stabilize the humeral head in the joint.

Ligaments of the glenohumeral joint strengthen the joint only minimally. Most of the joint's strength is due to the **rotator cuff** muscles surrounding it (see section 11.8b). The rotator cuff muscles (subscapularis, supraspinatus, infraspinatus, and teres minor) work as a group to hold the head of the humerus in the glenoid cavity. The tendons of these muscles encircle the joint (except for its inferior portion) and fuse with the articular capsule. Because the inferior portion of the joint lacks support from rotator cuff muscles, this area is weak and is the most likely site of injury.

Bursae help decrease friction at the specific places on the shoulder where both tendons and large muscles extend across the joint capsule. The shoulder has a relatively large number of bursae.

 WHAT DID YOU LEARN?

18 Why is the shoulder joint considered the most mobile and at the same time the most unstable joint in the human body?

9.7c Elbow Joint

 LEARNING OBJECTIVES

5. Describe the elbow joint and its motion.
6. Explain why the elbow joint is relatively stable.

The **elbow joint** is a hinge joint composed of two articulations: (1) the humeroulnar joint, where the trochlea of the humerus articulates with the trochlear notch of the ulna; and (2) the humeroradial joint, where the capitulum of the humerus articulates with the head of the radius. Both joints are enclosed within a single articular capsule (**figure 9.16**; table 9.4).

The elbow is an extremely stable joint for several reasons. First, the articular capsule is fairly thick, and thus effectively protects the articulations. Second, the bony surfaces of the humerus and ulna interlock very well, and thus provide a solid bony support. Finally, multiple strong supporting ligaments help reinforce the articular capsule. Because of the tradeoff between stability and mobility, the elbow joint is very stable but is not as mobile as some other joints, such as the glenohumeral joint.

The elbow joint has two main supporting ligaments. The **radial collateral ligament** (or *lateral collateral ligament*) is responsible for stabilizing the joint at its lateral surface; it extends around the head of the radius between the anular ligament and the lateral epicondyle of the humerus. The **ulnar collateral ligament** (or *medial collateral ligament*) stabilizes the medial side of the joint and extends from the medial epicondyle of the humerus both to the coronoid process and the olecranon of the ulna. In addition, an **anular** (an′ū-lăr; *anulus* = ring) **ligament** surrounds the neck of the radius and binds the proximal head of the radius to the ulna. The anular ligament helps hold the head of the radius in place.

Despite the support from the capsule and ligaments, the elbow joint is subject to damage from severe impacts or unusual stresses. For example, if you fall on an outstretched hand and the elbow joint is partially flexed, the posterior stress on the ulna combined with contractions of muscles that extend the elbow may break the ulna at the center of the trochlear notch. Sometimes dislocations result from stresses to the elbow. This is particularly true when growth is still occurring at the epiphyseal plate, so children and teenagers may be prone to humeral epicondyle dislocations or fractures.

 WHAT DID YOU LEARN?

19 What is the function of the anular ligament in the elbow joint, and what injury may occur to this ligament and joint in young children?

9.7d Hip Joint

 LEARNING OBJECTIVES

7. Describe the hip joint and its motions.
8. Explain why the hip joint is more stable than the glenohumeral joint.

The **hip joint,** also known as the *coxal joint,* is the articulation between the head of the femur and the relatively deep, concave acetabulum of the os coxae (**figure 9.17**; **table 9.5**). A fibrocartilaginous **acetabular labrum** further deepens this socket. The hip joint's more extensive bony architecture is therefore much stronger and more stable than that of the glenohumeral joint. Conversely, the hip joint's increased stability means that

INTEGRATE

CONCEPT CONNECTION

The shoulder joint illustrates the interrelatedness of the skeletal system (see chapter 8) and muscular system (see chapter 11). The stability of the glenohumeral joint primarily comes from the musculature, as opposed to the skeletal elements. Injury to this musculature will affect the workings of this joint.

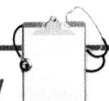

INTEGRATE

CLINICAL VIEW
Subluxation of the Head of the Radius

The term **subluxation** refers to an incomplete dislocation, in which the contact between the bony joint surfaces is altered, but they are still in partial contact. In subluxation of the head of the radius, the head is pulled out of the anular ligament. The layman's terms for this injury include *pulled elbow, nursemaid's elbow,* or *slipped elbow*. This injury occurs commonly and almost exclusively in children (typically those younger than age 5), because a child's anular ligament is thin and the head of the radius is not fully formed. After age 5, both the ligament and the radial head are more fully formed so the risk of this type of injury lessens dramatically. However, it may occur if an individual pulls suddenly on a child's pronated forearm.

Luckily, treatment is simple: The pediatrician applies posteriorly placed pressure to the head of the radius while slowly supinating and extending the child's forearm. This movement literally screws the radial head back into the anular ligament. In most cases, this manual treatment brings immediate relief.

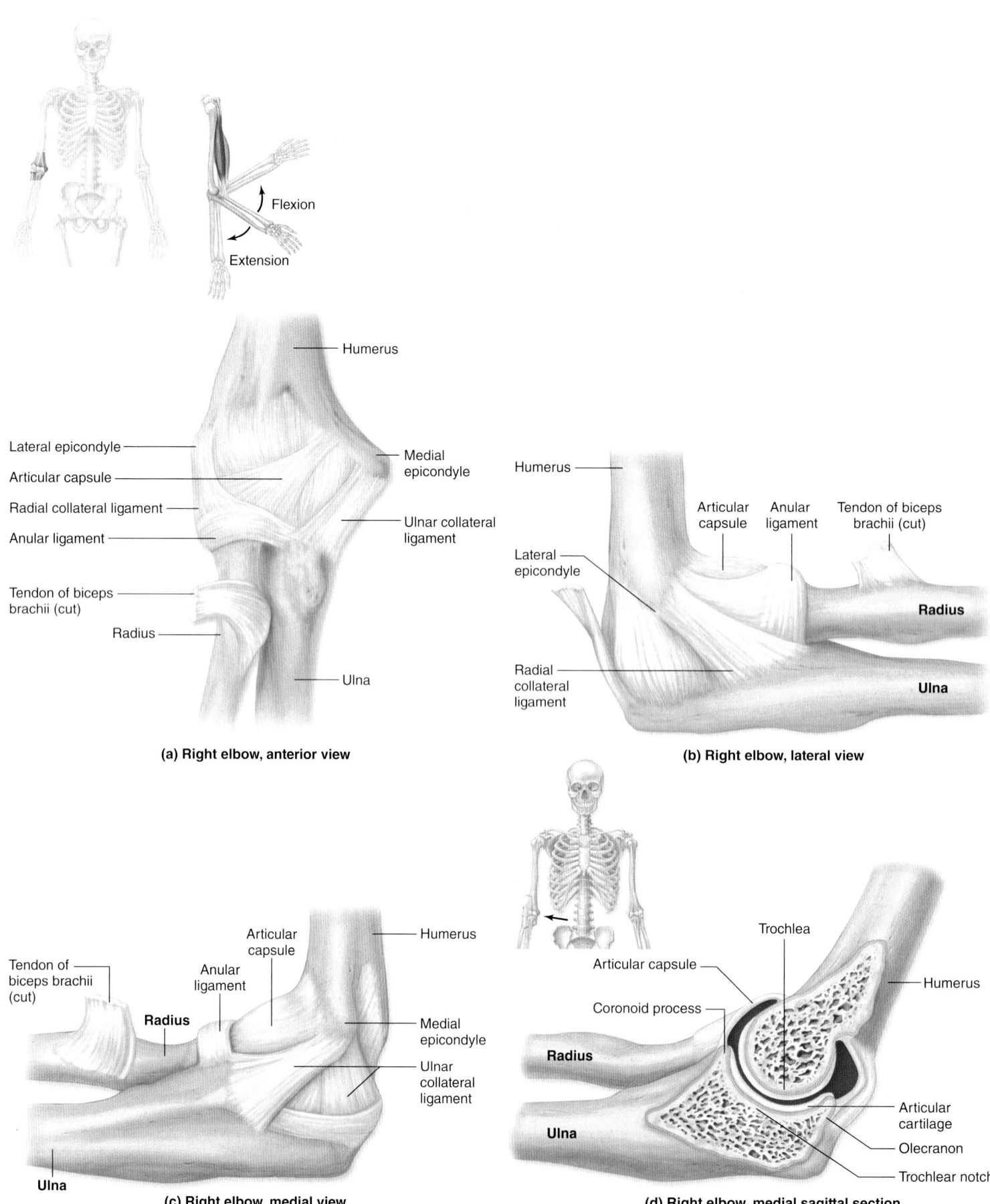

Flexion

Extension

Humerus

Lateral epicondyle

Articular capsule

Radial collateral ligament

Anular ligament

Tendon of biceps brachii (cut)

Radius

Medial epicondyle

Ulnar collateral ligament

Ulna

(a) Right elbow, anterior view

Humerus

Lateral epicondyle

Radial collateral ligament

Articular capsule

Anular ligament

Tendon of biceps brachii (cut)

Radius

Ulna

(b) Right elbow, lateral view

Tendon of biceps brachii (cut)

Radius

Anular ligament

Articular capsule

Ulna

Humerus

Medial epicondyle

Ulnar collateral ligament

(c) Right elbow, medial view

Trochlea

Articular capsule

Coronoid process

Radius

Ulna

Humerus

Articular cartilage

Olecranon

Trochlear notch

(d) Right elbow, medial sagittal section

Figure 9.16 Elbow Joint. The elbow joint is a hinge joint. The right elbow is shown here in (*a*) anterior view, (*b*) lateral view, (*c*) medial view, and (*d*) sagittal section. AP|R

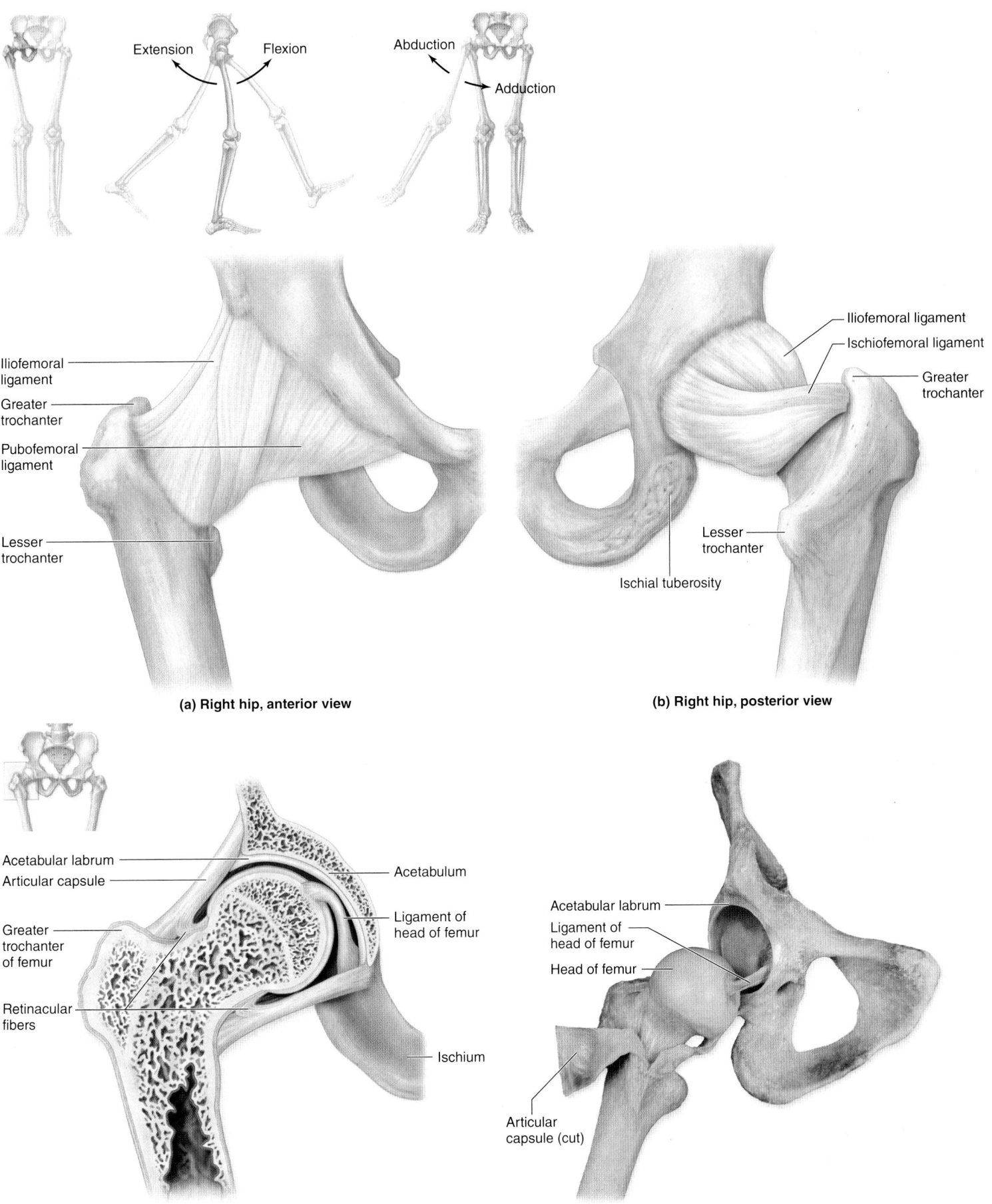

(a) Right hip, anterior view

Extension Flexion

Abduction

Adduction

Iliofemoral ligament

Greater trochanter

Pubofemoral ligament

Lesser trochanter

(b) Right hip, posterior view

Iliofemoral ligament

Ischiofemoral ligament

Greater trochanter

Lesser trochanter

Ischial tuberosity

(c) Right hip, coronal section

Acetabular labrum

Articular capsule

Greater trochanter of femur

Retinacular fibers

Acetabulum

Ligament of head of femur

Ischium

(d) Right hip, anterior view, internal aspect of joint

Acetabular labrum

Ligament of head of femur

Head of femur

Articular capsule (cut)

Figure 9.17 Hip Joint. The hip joint is formed by the head of the femur and the acetabulum of the os coxae. The right hip joint is shown in (*a*) anterior view, (*b*) posterior view, and (*c*) coronal section. (*d*) Cadaver photo of the coxal joint, with the articular capsule cut to show internal structures. **AP|R**

Table 9.5   Pelvic Girdle and Lower Limb Joints

Joint	Articulation Components	Structural Classification	Functional Classification	Description of Movement
Sacroiliac	Auricular surfaces of sacrum and ilium	Synovial (plane)	Diarthrosis	Slight gliding; more movement during pregnancy and childbirth
Hip (coxal)	Head of the femur and acetabulum of os coxae	Synovial (ball-and-socket)	Diarthrosis	Flexion, extension, abduction, adduction, circumduction, medial and lateral rotation of thigh
Pubic symphysis	Two pubic bones	Cartilaginous (symphysis)	Amphiarthrosis	Very slight gliding; more movement during childbirth
Knee	**Tibiofemoral joint:** Medial condyle of femur, medial meniscus, and medial condyle of tibia **Patellofemoral joint:** Patella and patellar surface of femur	Synovial (hinge) at tibiofemoral joint,[1] both synovial (hinge) and synovial (plane) at patellofemoral joint	Diarthrosis	Flexion, extension, lateral rotation of leg in flexed position, slight medial rotation
Tibiofibular	**Superior joint:** Head of fibula and lateral condyle of tibia **Inferior joint:** Distal end of fibula and fibular notch of tibia	**Superior joint:** Synovial (plane) **Inferior joint:** Fibrous (syndesmosis)	Amphiarthrosis	Slight rotation of fibula during dorsiflexion of foot
Talocrural (ankle)	Distal end of tibia and medial malleolus of tibia with talus Lateral malleolus of fibula and talus	Synovial (hinge)	Diarthrosis	Dorsiflexion and plantar flexion
Intertarsal	Between the tarsal bones	Synovial (plane)	Diarthrosis	Eversion and inversion of foot
Tarsometatarsal	Three cuneiforms and cuboid (tarsals) and bases of five metatarsals	Synovial (plane)	Diarthrosis	Slight gliding
Metatarso-phalangeal (MP joints)	Heads of metatarsals and bases of proximal phalanges	Synovial (condylar)	Diarthrosis	Flexion, extension, abduction, adduction, and circumduction of phalanges
Interphalangeal (IP joints)	Heads of proximal and middle phalanges with bases of middle and distal phalanges, respectively	Synovial (hinge)	Diarthrosis	Flexion and extension of phalanges

1. Although anatomists classify the tibiofemoral joint as a hinge joint, some kinesiologists and exercise scientists prefer to classify the tibiofemoral joint as a modified condylar joint.

INTEGRATE

CLINICAL VIEW
Fracture of the Femoral Neck

Fracture of the femoral neck is often incorrectly referred to as a "fractured hip," because the break is in the femoral neck, not the os coxae. With this injury, the pull of the lower limb muscles causes the leg to rotate laterally and appear shorter by several inches. Fractures of the femoral neck are of two types: intertrochanteric and subcapital.

Intertrochanteric fractures of the femoral neck occur distally to or outside the hip joint capsule—in other words, these fractures are *extracapsular.* The fracture line runs between the greater and lesser trochanters. This type of injury typically occurs in younger and middle-aged individuals. It usually occurs in response to trauma.

Subcapital (or *intracapsular*) **fractures** of the femoral neck occur within the hip articular capsule, very close to the head of the femur itself. This type of fracture usually occurs in elderly people whose bones have been weakened by osteoporosis and thus are more susceptible to fracture.

Subcapital fractures result in a tearing of the retinacular arteries that supply the head and neck of the femur. The ligament of the head of the femur may be torn as well. As a result, the head and neck of the femur lose their blood supply and may develop **avascular necrosis,** which is death of the bone due to lack of blood. Frequently, hip replacement surgery is needed, whereby a metal femoral head and neck replace the dying bone. This surgery is not without risk, however, and many elderly individuals do not survive because of surgical complications.

---

it is less mobile than the glenohumeral joint. The hip joint must be more stable (and thus less mobile) because it supports the body weight.

The hip joint is secured by a strong articular capsule, several ligaments, and a number of powerful muscles. The articular capsule extends from the acetabulum to the trochanters of the femur, enclosing both the femoral head and neck. This arrangement prevents the head from moving away from the acetabulum. The ligamentous fibers of the articular capsule reflect around the neck of the femur. These reflected fibers, called **retinacular** (ret-i-nak′yū-lăr) **fibers,** provide additional stability to the capsule. Located within the retinacular fibers are retinacular arteries (branches of the deep femoral artery), which supply almost all of the blood to the head and neck of the femur.

The articular capsule is reinforced by three spiraling intracapsular ligaments: The **iliofemoral** (il′ē-ō-fem′ŏ-răl) **ligament** is a Y-shaped ligament that provides strong reinforcement for the anterior region of the articular capsule. The **ischiofemoral** (is-kē-ō-fem′ŏ-răl) **ligament** is a spiral-shaped, posteriorly located ligament. The **pubofemoral** (pyū′bō-fem′ŏ-răl) **ligament** is a triangular thickening of the capsule's inferior region. All of these spiraling ligaments become taut when the hip joint is extended, so the hip joint is most stable in the extended position. Try this experiment: Flex your hip joint, and try to move the femur; you may notice a great deal of mobility. Now extend your hip joint (stand up), and try to move the femur. Because those ligaments are taut, you don't have as much mobility in the joint as you did when the hip joint was flexed.

Another tiny ligament, the **ligament of head of femur,** also called the *ligamentum teres,* originates along the acetabulum. Its attachment point is the fovea of the head of the femur. This ligament does not provide strength to the joint; rather, it typically contains a small artery that supplies the head of the femur.

The combination of a deep bony socket, a strong articular capsule, supporting ligaments, and muscular padding gives the hip joint its stability. Movements possible at the hip joint include flexion, extension, abduction, adduction, circumduction, and medial and lateral rotation.

WHAT DID YOU LEARN?

20 How do the glenohumeral and hip joints compare with respect to their mobility and stability?

9.7e Knee Joint

LEARNING OBJECTIVES

9. Describe the knee joint and its motion.
10. Name the ligaments that support the knee joint.

The **knee joint** is the largest and most complex diarthrosis of the body (**figure 9.18**; table 9.5; see also figure 9.5). It is primarily a hinge joint, but when the knee is flexed, it is also capable of slight rotation and lateral gliding. Structurally, the knee is composed of two separate

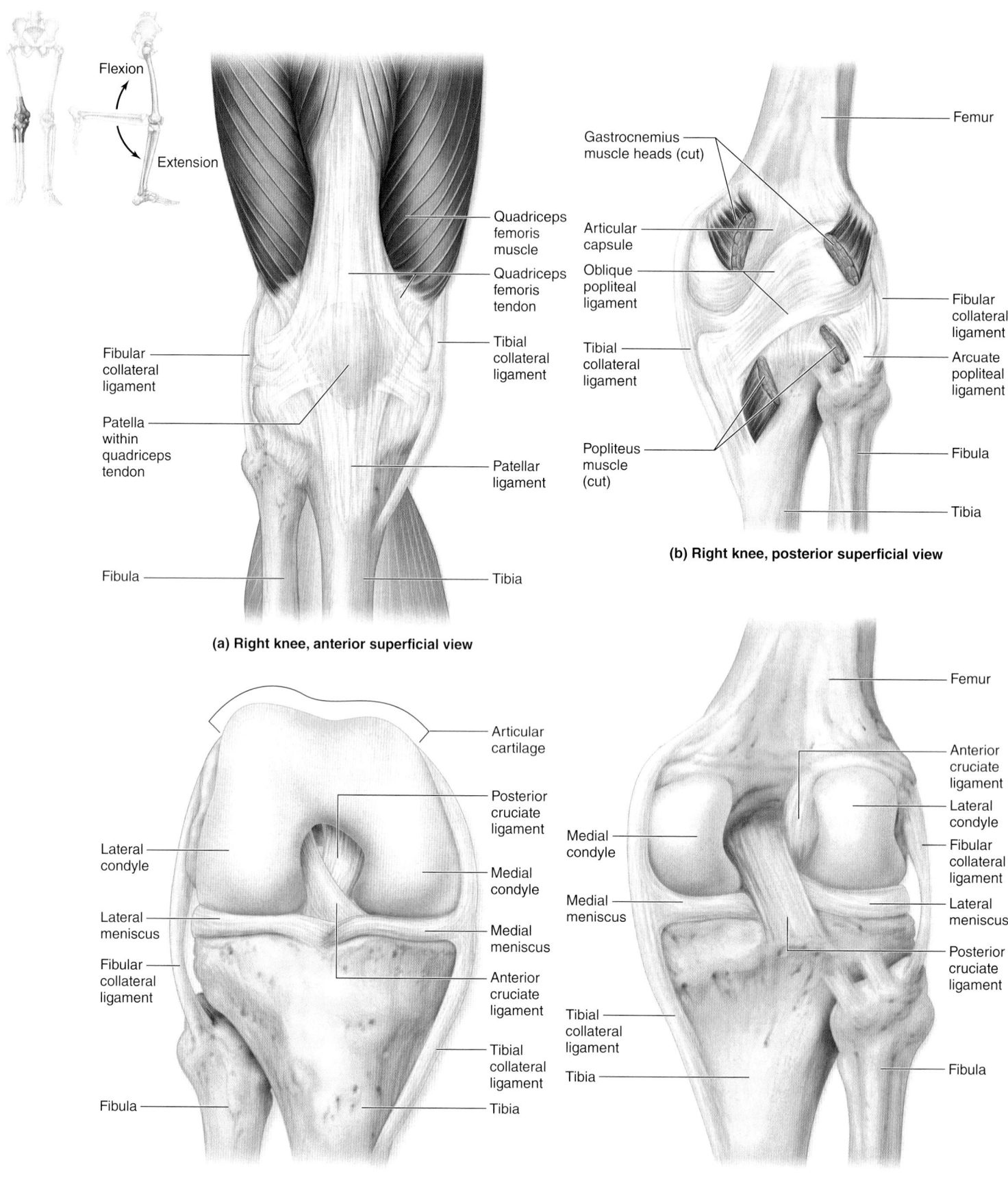

Flexion

Extension

Quadriceps femoris muscle

Quadriceps femoris tendon

Tibial collateral ligament

Fibular collateral ligament

Patella within quadriceps tendon

Patellar ligament

Fibula

Tibia

(a) Right knee, anterior superficial view

Gastrocnemius muscle heads (cut)

Femur

Articular capsule

Oblique popliteal ligament

Fibular collateral ligament

Tibial collateral ligament

Arcuate popliteal ligament

Popliteus muscle (cut)

Fibula

Tibia

(b) Right knee, posterior superficial view

Articular cartilage

Lateral condyle

Posterior cruciate ligament

Lateral meniscus

Medial condyle

Fibular collateral ligament

Medial meniscus

Anterior cruciate ligament

Fibula

Tibial collateral ligament

Tibia

(c) Right knee, anterior deep view

Femur

Medial condyle

Anterior cruciate ligament

Lateral condyle

Medial meniscus

Fibular collateral ligament

Lateral meniscus

Tibial collateral ligament

Posterior cruciate ligament

Tibia

Fibula

(d) Right knee, posterior deep view

Figure 9.18 Knee Joint. This joint is the most complex diarthrosis of the body. (*a*) Anterior superficial, (*b*) posterior superficial, (*c*) anterior deep, and (*d*) posterior deep views reveal the complex interrelationships of the parts of the right knee. AP|R

articulations: (1) The **tibiofemoral** (tib-ē-ō-fem′ŏ-răl) **joint** is between the condyles of the femur and the condyles of the tibia, and (2) the **patellofemoral joint** is between the patella and the patellar surface of the femur.

The knee joint has an articular capsule that encloses only the medial, lateral, and posterior regions of the knee joint. The articular capsule does not cover the anterior surface of the knee joint; rather, the quadriceps femoris muscle tendon passes over the knee joint's anterior surface. The patella is embedded within this tendon, and the **patellar ligament** extends beyond the patella and continues to where it attaches on the tibial tuberosity of the tibia. Thus, there is no single unified capsule in the knee, nor is there a common joint cavity. Posteriorly the capsule is strengthened by several popliteal ligaments.

INTEGRATE

CLINICAL VIEW
Knee Ligament and Cartilage Injuries

Although the knee is capable of bearing much weight and has numerous strong supporting ligaments, it is highly vulnerable to injury, especially among athletes. Because the knee is reinforced by tendons and ligaments only, ligamentous injuries to the knee are very common.

The tibial collateral ligament is frequently injured when the leg is forcibly abducted at the knee, such as when a person's knee is hit on the lateral side. Because the tibial collateral ligament is attached to the medial meniscus, the medial meniscus may be injured as well.

Injury to the fibular collateral ligament can occur if the medial side of the knee is struck, resulting in hyperadduction of the leg at the knee. This type of injury is fairly rare, in part because the fibular collateral ligament is very strong and also because medial blows to the knee are not common.

The anterior cruciate ligament (ACL) can be injured when the leg is hyperextended—for example, if a runner's foot hits a hole. Because the ACL is rather weak compared to the other knee ligaments, it is especially prone to injury. To test for ACL injury, a physician gently tugs anteriorly on the tibia. In this so-called **anterior drawer test**, too much forward movement indicates an ACL tear.

Posterior cruciate ligament (PCL) injury may occur if the leg is hyperflexed or if the tibia is driven posteriorly on the femur. PCL injury occurs rarely, because this ligament is rather strong. To test for PCL injury, a physician gently pushes posteriorly on the tibia. In this **posterior drawer test**, too much posterior movement indicates a PCL tear.

The menisci also may be prone to injury. Tears in the meniscus may occur due to blows to the knee or due to general overuse of the joint. Because the menisci are composed of fibrocartilage, they cannot regenerate and often must be surgically treated.

The **unhappy triad** of injuries refers to a triple injury of the tibial collateral ligament, medial meniscus, and anterior cruciate ligament. This is the most common type of football injury. It occurs when a player is illegally "clipped" by a lateral blow to the knee, and the leg is forcibly abducted and laterally rotated. If the blow is severe enough, the tibial collateral ligament tears, followed by tearing of the medial meniscus, as these two structures are connected. The force that tears the tibial collateral ligament and the medial meniscus is thus transferred to the ACL. Because the ACL is relatively weak, it tears as well.

The treatment of ligamentous knee injuries depends upon the severity and type of injury. Conservative treatment involves immobilizing the knee for a period of time to rest the joint. Surgical treatment can include repairing the torn ligaments or replacing the ligaments with a graft from another tendon or ligament (such as the quadriceps tendon). Many knee surgeries now may be performed with arthroscopy. **Arthroscopy** is a type of conservative surgical treatment where a small incision is made in the knee and then an **arthroscope** (an instrument with a camera and light source) is inserted in the knee, allowing the surgeon to clearly see the surgical area without having to make large incisions.

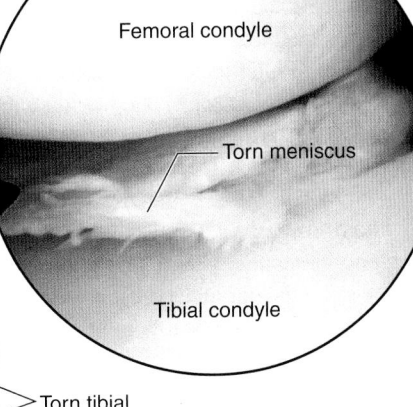

Femoral condyle

Torn meniscus

Tibial condyle

Arthroscopic view of knee joint, showing torn meniscus

Lateral blow to knee

Torn anterior cruciate ligament

Torn tibial collateral ligament

Torn medial meniscus

"Unhappy triad" of injuries to the right knee.

CLINICAL VIEW

Ankle Sprains and Pott Fractures

A **sprain** is a stretching or tearing of ligaments, without fracture or dislocation of the joint. An ankle sprain results from twisting of the foot, almost always due to *overinversion.* Fibers of the lateral ligament are either stretched (in mild sprains) or torn (in more severe sprains), producing localized swelling and tenderness anteroinferior to the lateral malleolus. *Overeversion* sprains rarely occur due to the strength of the deltoid (medial) ligament. Recall that ligaments are composed of dense regular connective tissue, which is poorly vascularized (see section 5.2d). Poorly vascularized

tissue takes a long time to heal, and that is the case with ankle sprains. These structures are also prone to reinjury.

If overeversion *does* occur, the injury that usually results is called a **Pott fracture** (see section 7.8). If the foot is overeverted, it pulls on the deltoid ligament, which is very strong and doesn't tear. Instead, the pull can avulse (pull off) the medial malleolus of the tibia. The force from the injury then continues to move the talus laterally, as the medial malleolus can no longer restrict side-to-side movements of the ankle. As the talus moves laterally and puts force on the fibula, the fibula fractures as well (usually at its distal end or by the lateral malleolus). Thus, both the tibia and the fibula fracture in this injury, and yet the deltoid ligament remains intact.

On either side of the knee joint are two collateral ligaments that become taut on extension and provide additional stability to the joint. The **fibular collateral ligament** (*lateral collateral ligament*) reinforces the lateral surface of the joint. This ligament extends from the femur to the fibula and prevents hyperadduction of the leg at the knee. (In other words, it prevents the leg from moving too far medially relative to the thigh). The **tibial collateral ligament** (*medial collateral ligament*) reinforces the medial surface of the knee joint. This ligament runs from the femur to the tibia and prevents hyperabduction of the leg at the knee. (In other words, it prevents the leg from moving too far laterally relative to the thigh.) This ligament is attached to the medial meniscus of the knee joint as well, so an injury to the tibial collateral ligament usually affects the medial meniscus.

Deep to the articular capsule and within the knee joint itself are a pair of C-shaped fibrocartilage pads positioned on the condyles of the tibia. These pads are called the **medial meniscus** and the **lateral meniscus.** They partially stabilize the joint medially and laterally, act as cushions between articular surfaces, and continuously change shape to conform to the articulating surfaces as the femur moves.

Two **cruciate** (krū′shē-āt) **ligaments** are deep to the articular capsule of the knee joint. They limit the anterior and posterior movement of the femur on the tibia. These ligaments cross each other in the form of an X, hence the name "cruciate" (which means "cross"). The **anterior cruciate ligament (ACL)** extends from the posterior femur to the anterior side of the tibia. When the knee is extended, the ACL is pulled tight and prevents hyperextension. The ACL prevents the tibia from moving too far anteriorly relative to the femur. The **posterior cruciate ligament (PCL)** attaches from the anteroinferior femur to the posterior side of the tibia. The PCL becomes taut on flexion, and so it prevents hyperflexion of the knee joint. The PCL also prevents posterior displacement of the tibia relative to the femur.

Humans are bipedal animals, meaning that we walk on two feet. An important aspect of bipedal locomotion is the ability to "lock" the knees in the extended position and stand erect without tiring the leg muscles. At full extension, the tibia rotates laterally so as to tighten the anterior cruciate ligament and squeeze the menisci between the tibia and femur. Muscular contraction by the popliteus muscle unlocks and flexes the knee joint. (This ability should be distinguished from being unable to bend or straighten the knee because of injury or disease, which is sometimes also called locking.)

WHAT DID YOU LEARN?

21 What are the functions of each of the intracapsular ligaments of the knee joint?

9.7f Talocrural (Ankle) Joint

LEARNING OBJECTIVE

11. Describe the talocrural joint and its motion.

The **talocrural (ankle) joint** is a highly modified hinge joint that permits both dorsiflexion and plantar flexion. It includes two articulations within one joint capsule. One articulation is

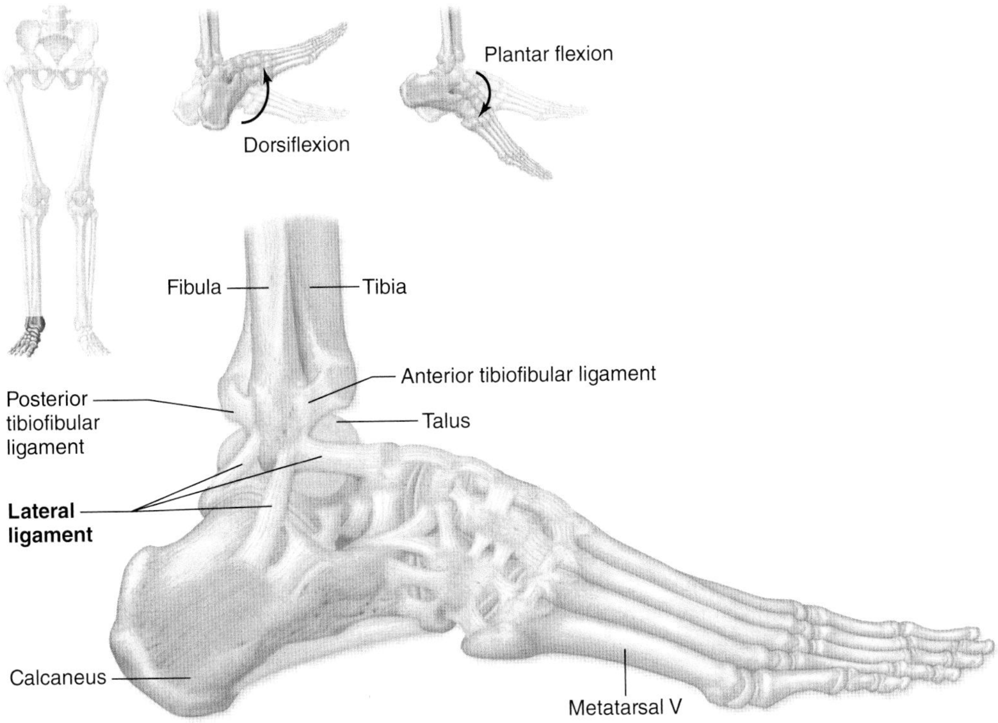

Plantar flexion

Dorsiflexion

Fibula —— —— Tibia

—— Anterior tibiofibular ligament

Posterior
tibiofibular
ligament

—— Talus

**Lateral
ligament**

Calcaneus

Metatarsal V

(a) Right foot, lateral view

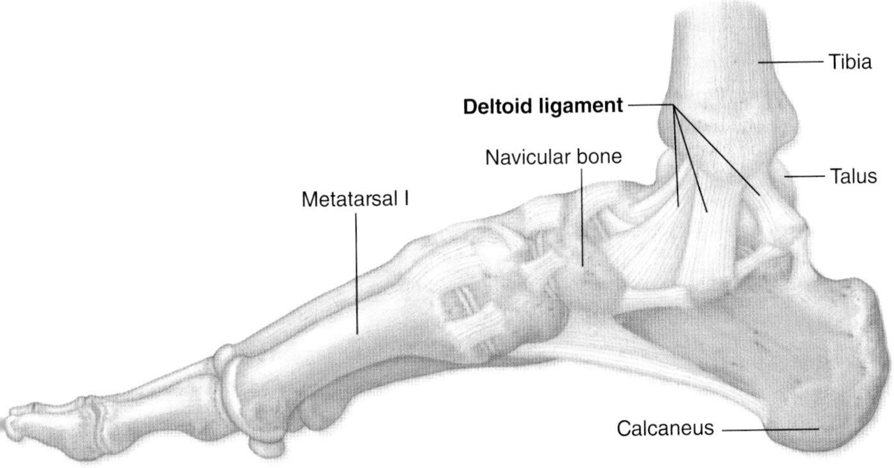

Tibia

Deltoid ligament

Navicular bone

Talus

Metatarsal I

Calcaneus

(b) Right foot, medial view

Figure 9.19 Talocrural Joint. (*a*) Lateral and (*b*) medial views of the right foot show that the talocrural joint contains articulations among the tibia, fibula, and talus. This joint permits dorsiflexion and plantar flexion only. AP|R

between the distal end of the tibia and the talus; the other is between the distal end of the fibula and the lateral aspect of the talus (**figure 9.19**; table 9.5). The medial and lateral malleoli of the tibia and fibula, respectively, form extensive medial and lateral margins and prevent the talus from sliding side-to-side.

The talocrural joint includes several distinctive anatomic features. Its articular capsule covers the distal surfaces of the tibia, the medial malleolus, the lateral malleolus, and the talus. A multipart **deltoid ligament** (or *medial ligament*) binds the tibia to the foot on the medial side. This ligament prevents overeversion of the foot. It is incredibly strong and rarely tears; in fact, it typically will pull off the medial malleolus before it ever tears! A much thinner multipart **lateral ligament** binds the fibula to the foot on the lateral side. This ligament prevents overinversion of the foot. It is not as strong as the deltoid ligament, and is prone to sprains and tears. Two **tibiofibular** (tib′ē-ō-fib′yū-lăr) **ligaments** (**anterior** and **posterior**) bind the tibia to the fibula.

WHAT DID YOU LEARN?

22 What bones articulate at the talocrural joint, and what movements are permitted at this joint?

9.8 Development and Aging of the Joints

LEARNING OBJECTIVES

1. Explain how the three major types of joints form in the embryo and fetus.

2. Describe some of the common age-related changes in joints.

Joints start to form by the sixth week of development and progressively differentiate during the fetal period. In the area of future fibrous joints, the mesenchyme around the developing bone differentiates into dense regular connective tissue, whereas in cartilaginous joints, it differentiates either into fibrocartilage or hyaline cartilage.

The development of the synovial joints is more complex. The most *laterally* placed mesenchyme forms the articular capsule and supporting ligaments of the joint. Just medial to this region, the mesenchyme forms the synovial membrane, which then starts secreting synovial fluid into the joint cavity. The *centrally* located mesenchyme may be reabsorbed or can form menisci or articular discs, depending upon the type of synovial joint.

Prior to the closure of the epiphyseal plates, some injuries to a young person may result in subluxation or fracture of an epiphysis, with potential adverse effects on the future development and health of the joint. Some adverse effects include the bone not reaching its potential full length, or the individual may develop arthritic-like changes in the joint.

INTEGRATE

CLINICAL VIEW
Arthritis

Arthritis (ar-thrī'tis) is a group of inflammatory or degenerative diseases of joints that occur in various forms. Each form presents the same symptoms: swelling of the joint, pain, and stiffness. It is the most prevalent crippling disease in the United States. Some common forms of arthritis are gouty arthritis, osteoarthritis, and rheumatoid arthritis.

Gouty arthritis is typically seen in middle-aged and older individuals, and is more common in males. Often called *gout*, this disease occurs as a result of an increased level of *uric acid* (a normal cellular waste product) in the blood. This abnormal level causes urate crystals to accumulate in the blood, synovial fluid, and synovial membranes. The body's inflammatory response to the urate crystals results in joint pain. Gout usually begins with an attack on a single joint (often in the great toe), and later progresses to other joints. Eventually, gouty arthritis may immobilize joints by causing fusion between the articular surfaces of the bones.

Osteoarthritis (OA) is the most common type of arthritis. This chronic degenerative joint condition also is termed *wear-and-tear arthritis* because repeated use of a joint gradually wears down the articular cartilage, much like the repeated use of a pencil eraser wears down the eraser. If the cartilage is worn down enough, osteoarthritis results. Eventually, bone rubs against bone, causing abrasions on the bony surfaces. Without the protective articular cartilage, movements at the joints become stiff and painful. The joints most affected by osteoarthritis are those of the fingers, knuckles, hips, knees, and shoulders. Osteoarthritis is typically seen in older individuals, although more and more athletes are experiencing arthritis at an earlier age due to the repetitive stresses placed on their joints.

Rheumatoid (rū'mă-toyd) **arthritis (RA)** is typically seen in younger and middle-aged adults, and it is much more prevalent in women. The common age at onset is approximately 40–50 years of age (although individuals as young as their teens have been diagnosed with the disorder). Symptoms include pain and joint swelling, muscle weakness, osteoporosis, and assorted problems with both the heart and the blood vessels. Rheumatoid arthritis is an *autoimmune disorder* in which the body's immune system targets its own tissues for attack. Rheumatoid arthritis starts with synovial membrane inflammation. Fluid and white blood cells leak from small blood vessels into the joint cavity, causing an increase in synovial fluid volume. As a consequence, the joint swells, and the inflamed synovial membrane thickens; eventually, the articular cartilage and, often, the underlying bone become eroded. Scar tissue later forms and ossifies, and bone ends fuse together—a process called **ankylosis** (ang'ki-lō'sis)—immobilizing the joint. Two types of medications often are prescribed to treat RA. The faster-acting first-line medications include NSAIDs and corticosteroids and are used to relieve joint pain. The slower, but longer-acting second-line medications, such as methotrexate and hydroxychloroquine, help put the disease into remission and slow down the destruction of the joints.

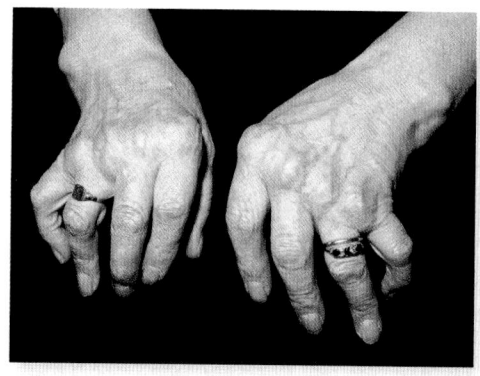 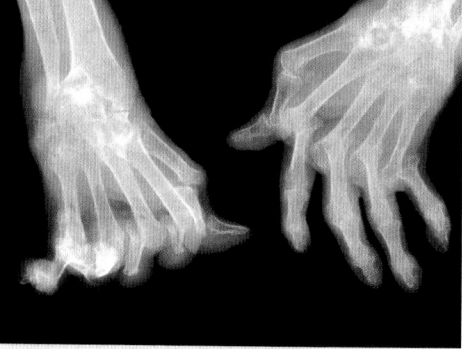

A photograph and colorized radiograph of hands with rheumatoid arthritis.

Arthritis is a rheumatic (i.e., referring to the joints or muscles) disease that involves damage to articular cartilage (see Clinical View: "Arthritis"). The primary problem that develops in an aging joint is osteoarthritis, also known as degenerative arthritis. The cause of the damage may vary, but it usually results from cumulative wear and tear at the joint surface.

Just as the strength of a bone is maintained by continual application of stress, the health of joints is directly related to moderate exercise. Exercise compresses the articular cartilages, causing synovial fluid to be squeezed out of the cartilage and then pulled back inside the cartilage matrix. This flow of fluid gives the chondrocytes within the cartilage the nourishment required to maintain their health. Exercise also strengthens the muscles that support and stabilize the joint. However, extreme exercise should be avoided, because it aggravates potential joint problems and may worsen osteoarthritis.

 WHAT DID YOU LEARN?

 What are some ways that joints change as a person ages?

CHAPTER SUMMARY

	• Articulations are the joints where bones are in contact. Joints differ in structure, function, and the amount of movement they allow.
9.1 Classification of Joints 298	• The three structural categories of joints are fibrous, cartilaginous, and synovial. • The three functional categories of joints are synarthroses (immobile joints), amphiarthroses (slightly mobile joints), and diarthroses (freely mobile joints).
9.2 Fibrous Joints 300	• Fibrous joints lack a joint cavity and they interconnect articulating bones by dense regular connective tissue.

9.2a Gomphoses 300
- A gomphosis is a synarthrosis between the tooth and either the mandible or the maxillae.

9.2b Sutures 300
- A suture is a synarthrosis that tightly binds the bones of the skull. Fused sutures are called synostoses.

9.2c Syndesmoses 301
- A syndesmosis is an amphiarthrosis, and the bones are connected by interosseous membranes.

9.3 Cartilaginous Joints 301	• Cartilaginous joints lack a joint space; the cartilage may be either hyaline cartilage or fibrocartilage.

9.3a Synchondroses 301
- A synchondrosis is a synarthrosis where hyaline cartilage is wedged between the articulating bones.

9.3b Symphyses 302
- A symphysis is an amphiarthrosis, where a disc of fibrocartilage is wedged between the articulating bones.

9.4 Synovial Joints 302	• All synovial joints are diarthroses.

9.4a Distinguishing Features and Anatomy of Synovial Joints 302
- Synovial joints contain an articular capsule, a joint cavity, synovial fluid, articular cartilage, ligaments, nerves, and blood vessels.

9.4b Classification of Synovial Joints 304
- The six types of synovial joints are plane, hinge, pivot, condylar, saddle, and ball-and-socket.

9.5 Synovial Joints and Levers 304	• Biomechanics is the practice of applying mechanical principles to biology.

9.5a Terminology of Levers 304
- Synovial joints may be compared to levers, which have a fixed point (fulcrum) around which movement occurs when effort applied to one point exceeds resistance at another point.

9.5b Types of Levers 306
- A first-class lever has a fulcrum between the effort and resistance.
- A second-class lever has the resistance placed between the fulcrum and the effort.
- A third-class lever, the most common type of lever in the human body, has the effort applied between the resistance and the fulcrum.

(continued on next page)

9.6 The Movements of Synovial Joints 307	• Motions that occur at synovial joints include gliding, angular, rotational, and special.

9.6a Gliding Motion   307
• Gliding is a simple movement where two opposing surfaces slide back-and-forth or side-to-side against one another.

9.6b Angular Motion   307
• Angular movements include flexion, extension, hyperextension, lateral flexion, abduction, adduction, and circumduction.

9.6c Rotational Motion   309
• Rotational movements involve a pivoting motion. Examples of rotational movements are lateral rotation, medial rotation, pronation, and supination.

9.6d Special Movements   310
• Special movements include depression and elevation, dorsiflexion and plantar flexion, eversion and inversion, protraction and retraction, and opposition.

9.7 Features and Anatomy of Selected Joints 311	• In each articulation, unique features of the articulating bones support the intended movements.

9.7a Temporomandibular Joint   312
• The temporomandibular joint is an articulation between the head of the mandible and the mandibular fossa of the temporal bone.

9.7b Shoulder Joint   313
• The sternoclavicular joint and acromioclavicular joint support movement of the shoulder.
• The glenohumeral joint is a ball-and-socket joint between the glenoid cavity of the scapula and the head of the humerus.

9.7c Elbow Joint   317
• The elbow is a hinge joint among the humerus, radius, and ulna.

9.7d Hip Joint   317
• The hip joint is a ball-and-socket joint between the head of the femur and the acetabulum of the os coxae.

9.7e Knee Joint   321
• The knee joint is primarily a hinge joint but is capable of slight rotation and gliding.

9.7f Talocrural (Ankle) Joint   324
• The talocrural joint is a hinge joint that permits dorsiflexion and plantar flexion.

9.8 Development and Aging of the Joints 326	• Joints begin to form during week 6 of development. • Osteoarthritis is a common joint problem that occurs with aging.

CHALLENGE YOURSELF

Do You Know the Basics?

_____ 1. The greatest range of mobility of any joint in the body is found in the

a. knee joint.
b. hip joint.
c. glenohumeral joint.
d. elbow joint.

_____ 2. A movement of the foot that turns the sole outward or laterally is called

a. dorsiflexion.
b. inversion.
c. eversion.
d. plantar flexion.

_____ 3. A _____ is formed when two bones previously connected by a suture fuse.

a. gomphosis
b. synostosis
c. symphysis
d. syndesmosis

_____ 4. The ligament that helps to maintain the alignment of the condyles between the femur and tibia and to limit the anterior movement of the tibia on the femur is the

a. tibial collateral ligament.
b. posterior cruciate ligament.
c. anterior cruciate ligament.
d. fibular collateral ligament.

_____ 5. Which joint is a diarthrosis?

a. symphysis

b. synchondrosis

c. syndesmosis

d. saddle

_____ 6. In this type of lever, the effort is located between the resistance and the fulcrum. An example would be your knee joint.

a. first-class

b. second-class

c. third-class

d. Two of the above are correct.

_____ 7. A metacarpophalangeal (MP) joint, which has oval articulating surfaces and permits movement in two planes, is what type of synovial joint?

a. condylar

b. plane

c. hinge

d. saddle

_____ 8. All of the following ligaments provide stability to the hip joint *except* the

a. ischiofemoral ligament.

b. pubofemoral ligament.

c. iliofemoral ligament.

d. ligament of the head of the femur.

_____ 9. Which of the following is a function of synovial fluid?

a. lubricates the joint

b. provides nutrients for the articular cartilage

c. absorbs shock within the joint

d. All of these are correct.

_____ 10. Plantar flexion and dorsiflexion are movements permitted at the _____ joint.

a. hip

b. knee

c. sternoclavicular

d. talocrural

11. Discuss the factors that influence both the stability and mobility of a joint. What is the relationship between a joint's mobility and its stability?

12. Both fibrous joints and synovial joints have dense regular connective tissue holding the bones together. So how are these two joints different both structurally and functionally?

13. List and describe all joints that are functionally classified as synarthroses.

14. How do a hinge joint and a pivot joint compare with respect to structure, function, and location within the body?

15. Compare and contrast first-, second-, and third-class levers.

16. Describe and compare the movements of abduction, adduction, pronation, and supination.

17. Most ankle sprains are overinversion injuries. What are the anatomic reasons why overeversion ankle sprains are relatively uncommon? Are there any overeversion injuries that occur to the ankle?

18. What are the main supporting ligaments of the elbow joint?

19. Compare the functions of the tibial and the fibular collateral ligaments in the knee joint. Which of the two is injured more frequently and why?

20. What are the similarities and differences between osteoarthritis and rheumatoid arthritis?

Can You Apply What You've Learned?

Use the following paragraph to answer questions 1–3.

A mother and her 4-year-old son were visiting a toy store, and the child did not want to leave. As the child threw a temper tantrum and resisted his mother, the mother pulled on the boy's arm to lead him out of the store. Immediately after the pull, the boy cried out in pain and displayed a prominent bump on the lateral side of the elbow. The mother drove the boy to the doctor in a panic. The doctor examined the boy's elbow and determined the boy had a subluxated head of the radius.

1. Which ligament failed to keep the head of the radius in place when the mother pulled on the boy's elbow?

a. anular ligament

b. ulnar collateral ligament

c. radial collateral ligament

d. coronoid ligament

2. The doctor mentions this type of injury is common in children younger than 5 years of age. What is one reason for this?

a. The olecranon of the ulna does not fit in properly with the olecranon fossa of the radius.

b. The head of the radius is not fully formed.

c. The medial and lateral epicondyle epiphyseal plates have not yet fused to the rest of the humerus.

d. The articular capsule of the elbow joint is weak in its anterior surface.

3. What bony feature caused the prominent "bump" on the lateral side of the elbow?

a. lateral epicondyle of the humerus

b. coronoid process of the ulna

c. head of the radius

d. radial collateral ligament

4. While Robert was running, he stepped into a pothole and he twisted his right ankle. A swelling appeared along the lateral side of this ankle. Which ligament was injured, and what movement resulted in the injury?

a. deltoid ligament, caused by overeversion of the foot

b. lateral ligament, caused by overeversion of the foot

c. deltoid ligament, caused by overinversion of the foot

d. lateral ligament, caused by overinversion of the foot

5. Most knee ligaments become taut upon extension of the joint, except for one. Which knee ligament becomes taut upon *flexion* of the joint, and prevents hyperflexion of the joint?

    a. anterior cruciate ligament
    b. posterior cruciate ligament
    c. patellar ligament
    d. tibial collateral ligament

Can You Synthesize What You've Learned?

1. During soccer practice, Erin tripped over the outstretched leg of a teammate and fell directly onto her shoulder. She was taken to the hospital in excruciating pain. Examination revealed that the head of the humerus had moved inferiorly and anteriorly into the axilla. What happened to Erin in this injury?

2. While Lucas and Omar were watching a football game, a player was penalized for "clipping," meaning that he had hit an opposing player on the lateral knee, causing hyperabduction at the knee joint. Lucas asked Omar what the big deal was about clipping. What joint is most at risk, and what kind of injuries can occur if a player gets clipped?

3. Jackie visits her physician because she is experiencing pain by her right ear. The doctor checks her ears and sees no sign of infection. She asks Jackie to open and close her mouth while she palpates the portions of her face adjacent to her ears. Why is the doctor having Jackie move her mouth, when she is experiencing ear pain? How may the two be related? What do you think the doctor will discover when Jackie opens and closes her mouth?

INTEGRATE

ONLINE STUDY TOOLS

The following study aids may be accessed through Connect.

Clinical Case Study: A Historical Case of Spinal Involvement with Tuberculosis

Interactive Questions: This chapter's content is served up in a number of multimedia question formats for student study.

LearnSmart: Topics and terminology include classification of joints; fibrous joints; cartilaginous joints; synovial joints; synovial joints and levers; movements of synovial joints; features and anatomy of selected joints; development and aging of joints

Anatomy & Physiology Revealed: Topics include skull, synovial joint; temporomandibular joint; glenohumeral joint; elbow joint; hand; hip joint; knee joint; tibiofibular joint

Animation: Synovial joints

INTEGRATE

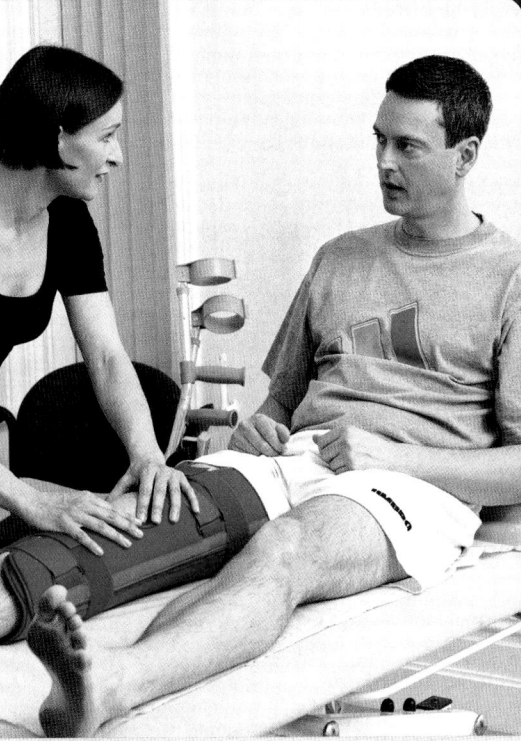

CAREER PATH
Athletic Trainers

Athletic trainers are allied health-care professionals that assess and diagnose injuries, develop safe and effective strength and physical fitness enhancement programs, and are interested in educating others in the prevention and treatment of athletic injuries. They work with athletes, coaches, and other health-care professionals under the direction of a physician.

The majority have earned a master's degree or higher and are board certified. Their educational background includes human anatomy and physiology, biomechanics, methods in physical training, rehabilitation of musculoskeletal injuries, first aid, and nutrition.

Anatomy &
Physiology | **REVEALED**®
aprevealed.com

Module 6: Muscular System

When we hear the word "muscle," most of us think of the muscles that move the skeleton. Over 700 skeletal muscles have been named, and together they form the **muscular system.** Skeletal muscles, however, are not the only places where muscle tissue is found. Muscle tissue is distributed almost everywhere in the body and is responsible for the movement of materials within and throughout the body. This vital tissue propels the food we eat along the gastrointestinal tract, expels the waste products we produce, changes the amount of air that moves into and out of the lungs, and pumps blood to body tissues.

➡

The three types of muscle tissue—skeletal muscle, cardiac muscle, and smooth muscle—were first introduced and compared in section 5.3. Here we describe the details of skeletal muscle anatomy and physiology. The chapter finishes with a brief description of cardiac muscle (which is described in detail in chapter 19) and a general discussion about smooth muscle. The naming of individual skeletal muscles of the muscular system is covered in chapter 11.

10.1 Introduction to Skeletal Muscle

Skeletal muscle typically composes 40–50% of the weight of an adult. It is primarily attached to the skeleton but is also found, for example, at the openings of the gastrointestinal and urinary tracts. We begin our discussion on skeletal muscle by describing both the general functions of skeletal muscle organs and characteristics of skeletal muscle tissue.

10.1a Functions of Skeletal Muscle

LEARNING OBJECTIVE

1. Explain the five general functions of skeletal muscle.

The hundreds of skeletal muscles within the body perform the following functions:

- **Body movement.** Contraction of muscles attached to bones of the skeleton produce both the large movements involved in running and localized movements such as underlining a word in a book. Other types of body movements include those associated with producing facial expressions, speaking, breathing, and swallowing.
- **Maintenance of posture.** Contraction of specific skeletal muscles stabilizes joints and helps maintain the body's posture, such as holding the head and trunk erect. These postural muscles contract continuously when you are awake to keep you from collapsing.
- **Protection and support.** Some skeletal muscle is arranged in layers along the walls of the abdominal cavity and the floor of the pelvic cavity. These layers of muscle protect the internal organs and support their normal position within the abdominopelvic cavity.
- **Regulating elimination of materials.** Circular muscle bands, called **sphincters** (sfingk′ter; *sphincter* = a band) contract and relax to regulate passage of material. These skeletal muscle sphincters at the **orifices** (or′i-fis; *orificium* = opening) of the gastrointestinal and urinary tracts allow you to voluntarily control the expulsion of feces and urine, respectively.
- **Heat production.** Energy is required for muscle tissue contraction, and heat is always produced by this energy use (the second law of thermodynamics; see section 3.1c). Thus, muscles are like small furnaces that continuously generate heat and function to help maintain your normal body temperature.

You shiver when you are cold because involuntary muscle contraction gives off heat. Likewise, you sweat during exercise to release the additional heat produced by your working muscles (see sections 1.5b and 6.1d).

WHAT DID YOU LEARN?

1. What are the five major functions of skeletal muscle?

10.1b Characteristics of Skeletal Muscle Tissue

LEARNING OBJECTIVE

2. Describe the five characteristics of skeletal muscle tissue.

Skeletal muscle tissue is composed of muscle cells that exhibit these characteristics—excitability, conductivity, contractility, elasticity, and extensibility:

- **Excitability** is the ability of a cell to respond to a stimulus (e.g., chemical, stretch). The stimulus causes a local change in the resting membrane potential in the excitable cell (see section 4.4). Skeletal muscle cells are specifically stimulated by neurotransmitter released from neurons (see section 10.3a).
- **Conductivity** involves an electrical change that travels along the plasma membrane as voltage-gated channels open sequentially during an action potential. Both muscle cells and neurons exhibit conductivity.
- **Contractility** is exhibited when contractile proteins within skeletal muscle cells slide past one another. Contractility is what enables muscle cells to cause body movement and to perform the other functions of muscles.
- **Elasticity** is the ability of a muscle to return to its original length following either shortening or lengthening of the muscle.
- **Extensibility** is the lengthening of a muscle cell. For example, when you flex your elbow joint, you are contracting the biceps brachii on the anterior side of your arm, while the triceps brachii on the posterior side is extended with the motion. The reverse is true when you straighten your elbow joint.

WHAT DID YOU LEARN?

2. Explain the skeletal muscle characteristics of contractility, elasticity, and extensibility. How do these differ?

10.2 Anatomy of Skeletal Muscle

A single muscle, such as the gracilis muscle on the medial thigh, may be composed of thousands of muscle cells that are typically as long as the entire muscle. Because of their potentially extraordinary length, skeletal muscle cells are often referred to as **muscle fibers** (or *myofibers*) (see section 5.3). Here we describe the gross anatomy of skeletal muscle, the microscopic anatomy of individual skeletal muscle fibers, and innervation of skeletal muscle fibers.

10.2a Gross Anatomy of Skeletal Muscle

 LEARNING OBJECTIVES

1. Identify and describe the three connective tissue layers associated with a muscle.
2. Describe the structure and function of a tendon and an aponeurosis.
3. Explain the function of blood vessels and nerves serving a muscle.

A skeletal muscle is an organ composed of skeletal muscle fibers, connective tissue layers, blood vessels, and nerves. The organization of a muscle is shown in **figure 10.1**. The specific anatomic arrangement of muscle fibers within a muscle (organ) is important to understand as you begin your study of muscle. As you view figure 10.1, notice that within the muscle many muscle fibers are organized into bundles called **fascicles** (fas´i-kl; *fascis* = bundle); and fascicles are bundled within the whole skeletal muscle organ.

WHAT DO YOU THINK?

1 List the structures of skeletal muscle from largest to smallest: muscle fiber, muscle, and fascicle.

Connective Tissue Components

Three layers of connective tissue occur within muscles: the epimysium, the perimysium, and the endomysium. These layers provide protection, sites for distribution of blood vessels and nerves, and a means of attachment to the skeleton or other structures within the body:

- The **epimysium** (ep-i-mis´ē-ŭm; *epi* = upon; *mys* = muscle) is a layer of dense irregular connective tissue that surrounds the whole skeletal muscle.
- The **perimysium** (per-i-mis´ē-ŭm; *peri* = around) surrounds the fascicles. The dense irregular connective tissue sheath of the perimysium contains extensive arrays of blood vessels and nerves that branch to supply muscle fibers within each individual fascicle.
- The **endomysium** (en´dō-mis´ē-ŭm; *endon* = within) is the innermost connective tissue layer. It is a delicate, areolar connective tissue layer that surrounds and electrically insulates each muscle fiber. It contains reticular protein fibers to help bind together neighboring muscle fibers and to support capillaries near these fibers.

A **tendon** is a thick, cordlike structure composed of dense regular connective tissue (see section 5.2d). It is formed by the three connective tissue layers as they merge and extend past the muscle fibers. Tendons attach the muscle to bone. Sometimes, the connective tissue

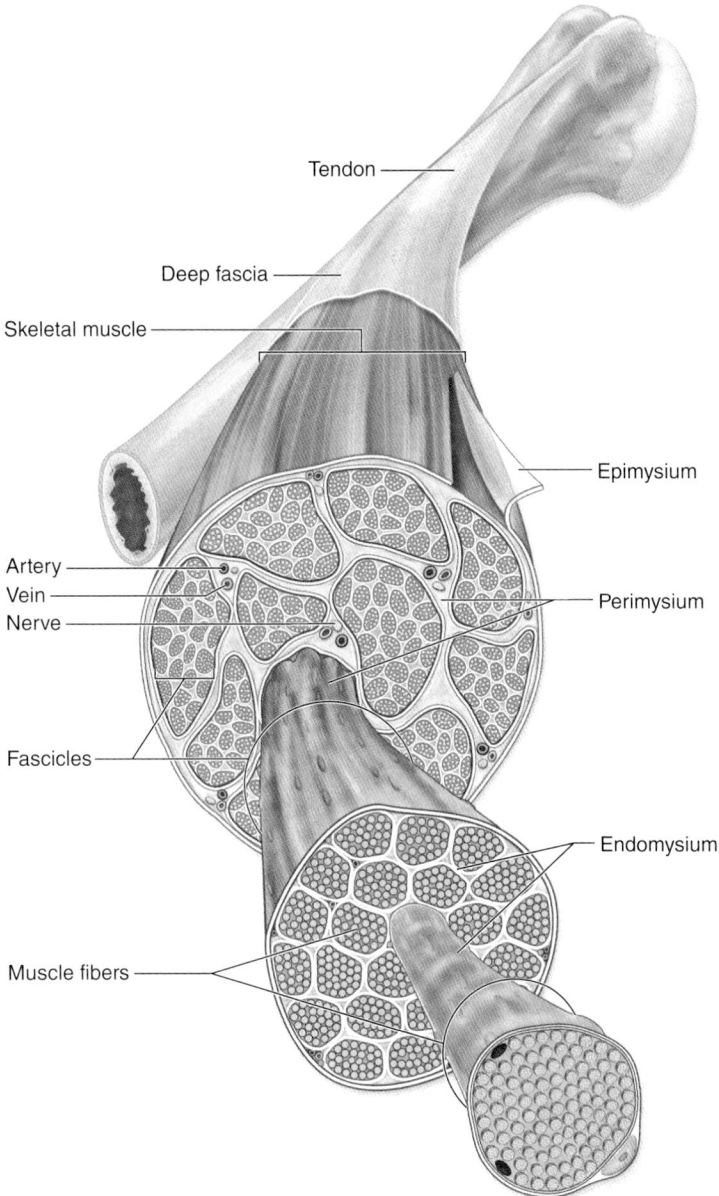

Figure 10.1 Structural Organization of Skeletal Muscle.
A skeletal muscle consists of multiple fascicles (bundles of muscle fibers) ensheathed in a tough outer connective tissue membrane called the epimysium. Each fascicle is wrapped in a connective tissue layer called the perimysium. Within fascicles, each muscle fiber is surrounded by a delicate connective tissue layer termed endomysium. **AP|R**

layers instead form a thin, flattened sheet of dense irregular tissue, termed an **aponeurosis** (ap´ō-nū-rō´sis; *apo* = from, *neuron* = sinew) (see figures 11.5 and 11.16).

Deep fascia (fash´ē-ă; band or filler), also called *visceral* or *muscular fascia,* is an additional expansive sheet of dense irregular connective tissue (see section 5.2d) that is external to the epimysium. Deep fascia separates individual muscles; binds together muscles with similar functions; contains nerves, blood vessels, and lymph vessels; and serves to fill spaces between muscles. The deep fascia is internal or deep to a layer called the **superficial fascia** (or *subcutaneous layer;* see section 6.1c). The superficial fascia is composed of areolar connective tissue and adipose connective tissue that separates muscle from skin.

Blood Vessels and Nerves

Skeletal muscle is **vascularized** (vas′kyū-lăr-īzd; *vas* = vessel) by an extensive network of blood vessels. The blood vessels deliver both oxygen and nutrients to the muscle fibers. They also remove waste products produced by the muscle fibers.

Skeletal muscle is **innervated** (in′ĕr-vāt′ed; *nervus* = nerve) by or functionally connected to and controlled by motor neurons (of the somatic nervous system; see section 12.1b). **Somatic motor neurons** extend from the brain and spinal cord to skeletal muscle fibers. Each motor neuron has a long extension called an **axon** (nerve fiber) that branches extensively at its terminal end (see section 12.2b). The axon extends through all three connective tissue layers to almost make contact with an individual muscle fiber (i.e., there is a very small gap of about 30 nanometers between the motor neuron and muscle fiber). The junction between the axon and the muscle fiber itself is called a neuromuscular junction, which is discussed later in this chapter. Skeletal muscle is classified as **voluntary muscle** because the muscle fibers can be consciously controlled by the nervous system.

WHAT DID YOU LEARN?

3 Identify the location and function of these connective tissue structures associated with muscle: endomysium, perimysium, epimysium, deep fascia, and superficial fascia.

10.2b Microscopic Anatomy of Skeletal Muscle

LEARNING OBJECTIVES

4. Explain how a skeletal muscle fiber becomes multinucleated.
5. Describe the sarcolemma, T-tubules, and sarcoplasmic reticulum of a skeletal muscle fiber.
6. Distinguish between thick and thin filaments.
7. Explain the organization of myofibrils, myofilaments, and sarcomeres.
8. List and describe the structures associated with energy production within skeletal muscle fibers.

Skeletal muscle fibers (as shown in figure 10.1) are the cells forming a muscle. Skeletal muscle fibers, like other cells, contain cytoplasm with the typical cellular structures such as the Golgi apparatus, ribosomes, and vesicles (see section 4.6). Note that the cytoplasm in skeletal muscle fibers is more specifically called **sarcoplasm** (sar′kō-plazm; *sark* = flesh). In addition, skeletal muscle fibers have several specialized features that we describe here including the details of its contractile proteins.

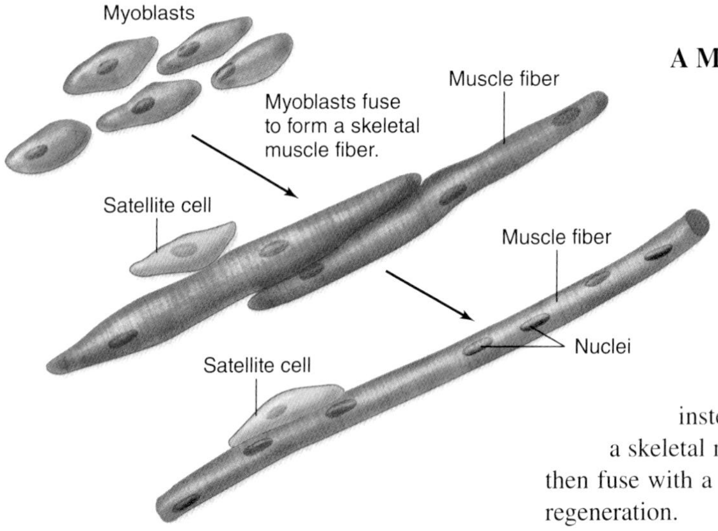

A Multinucleated Cell

A skeletal muscle fiber is typically between 10 and 500 micrometers (μm) in diameter, and as previously noted may extend the length of the entire muscle, ranging from about 100 μm to 30 centimeters. To reach this length, groups of embryonic cells termed **myoblasts** (mī′ō-blast; *blastos* = germ) fuse to form single skeletal muscle fibers during development **(figure 10.2)**. During this fusion process, each myoblast nucleus contributes to the eventual total number of nuclei in the fiber. Consequently, skeletal muscle fibers are **multinucleated** (i.e., they have numerous nuclei) **(figure 10.3a)**.

Some myoblasts do not fuse with muscle fibers during development and instead remain in adult skeletal muscle tissue as **satellite cells** (figure 10.2). If a skeletal muscle is injured, some satellite cells may be stimulated to differentiate and then fuse with a damaged skeletal muscle fiber to assist to a limited extent in its repair and regeneration.

Figure 10.2 Development of Skeletal Muscle. Embryonic muscle cells called myoblasts fuse to form a single skeletal muscle fiber. After development, both muscle fibers and satellite cells are present. Satellite cells are myoblasts that do not go on to form the skeletal muscle fiber. Instead, satellite cells remain with postnatal skeletal muscle tissue and assist in repair of muscles.

Sarcolemma and T-tubules

The plasma membrane of a skeletal muscle fiber is called the **sarcolemma** (sar′kō-lem′ă; *lemma* = husk) (figure 10.3a). Deep invaginations of the sarcolemma, called **T-tubules** or **transverse** (trans-vers′; *trans* = across, *versus* = to turn) **tubules,** extend into the skeletal muscle fiber as a network of narrow membranous tubules to the sarcoplasmic reticulum, which is the endoplasmic reticulum (ER) of the muscle (described shortly). Located within the membrane of both the sarcolemma (along its length) and the T-tubules are voltage-gated channels (figure 10.3c).

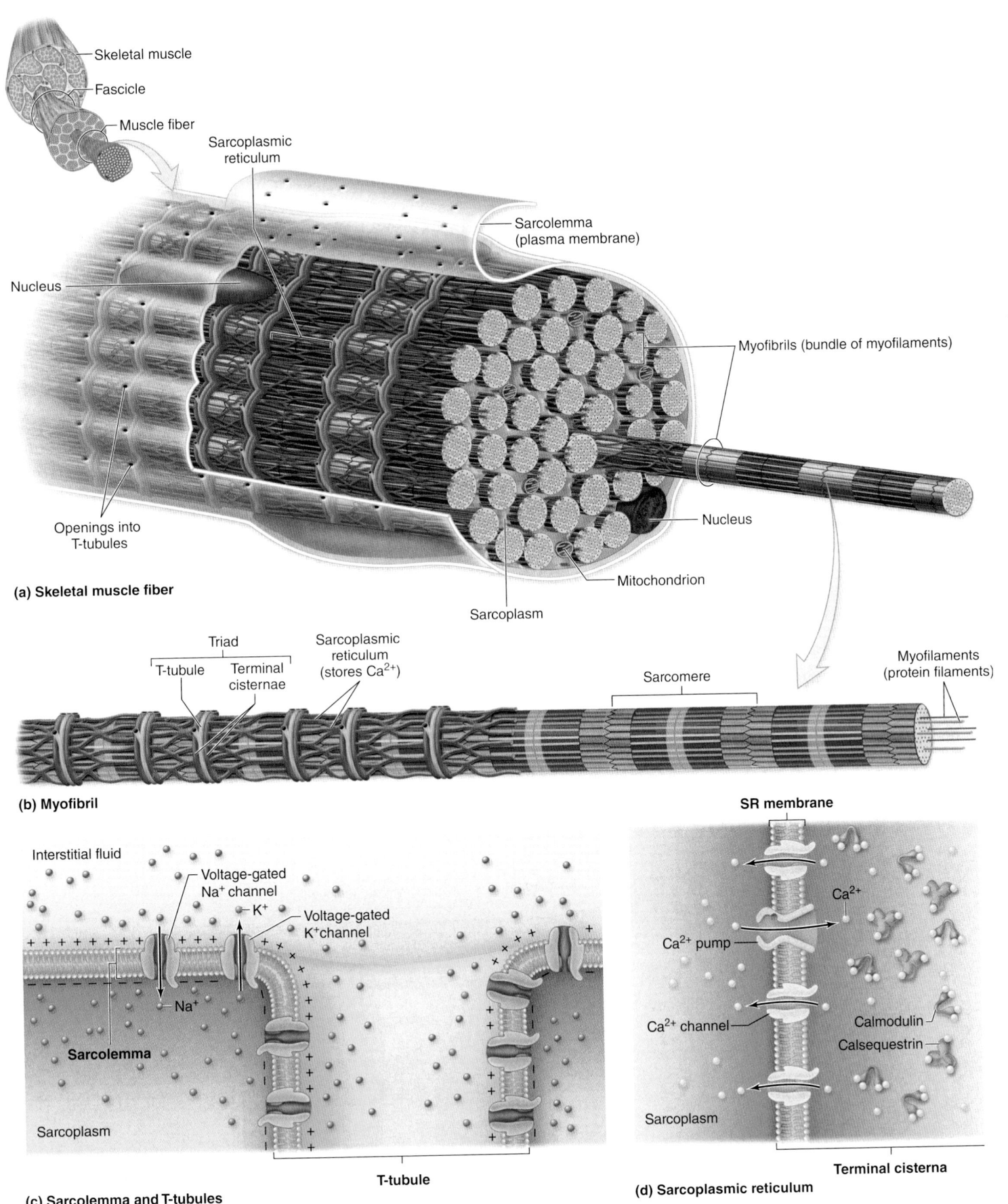

(a) Skeletal muscle fiber

- Skeletal muscle
- Fascicle
- Muscle fiber
- Sarcoplasmic reticulum
- Sarcolemma (plasma membrane)
- Nucleus
- Myofibrils (bundle of myofilaments)
- Openings into T-tubules
- Nucleus
- Mitochondrion
- Sarcoplasm

(b) Myofibril

- Triad
- T-tubule
- Terminal cisternae
- Sarcoplasmic reticulum (stores Ca^{2+})
- Sarcomere
- Myofilaments (protein filaments)

(c) Sarcolemma and T-tubules

- Interstitial fluid
- Voltage-gated Na$^+$ channel
- K$^+$
- Voltage-gated K$^+$ channel
- Na$^+$
- Sarcolemma
- Sarcoplasm
- T-tubule

(d) Sarcoplasmic reticulum

- SR membrane
- Ca^{2+}
- Ca^{2+} pump
- Ca^{2+} channel
- Calmodulin
- Calsequestrin
- Sarcoplasm
- Terminal cisterna

Figure 10.3 Structure and Organization of a Skeletal Muscle Fiber. (*a*) A muscle fiber is composed predominantly of myofibrils, which extend the length of the muscle fiber. (*b*) A myofibril is composed of bundles of myofilaments (protein filaments), and is enclosed within segments of the sarcoplasmic reticulum. The sarcoplasmic reticulum is a reservoir for calcium ions (Ca^{2+}). (*c*) The sarcolemma is physically connected to the sarcoplasmic reticulum by invaginations of the sarcolemma called T-tubules; both contain voltage-gated Na$^+$ channels, and voltage-gated K$^+$ channels. (*d*) The sarcoplasmic reticulum membrane contains both Ca^{2+} pumps and voltage-gated Ca^{2+} channels. AP|R

These channels include both voltage-gated Na⁺ channels and voltage-gated K⁺ channels, which participate in conducting an electrical signal (an action potential) as described in section 10.3b.

Myofibrils

Approximately 80% of the volume of a skeletal muscle fiber is composed of long, cylindrical structures termed **myofibrils** (mī′ō-fī′bril) (figure 10.3a). A skeletal muscle fiber contains hundreds to thousands of myofibrils. Each myofibril extends the length of the entire skeletal muscle fiber (and is about 1 to 2 micrometers in diameter). Note that each myofibril is composed of bundles of contractile proteins called myofilaments and is enclosed in portions of the sarcoplasmic reticulum (figure 10.3b).

Sarcoplasmic Reticulum

The **sarcoplasmic reticulum** (sar-kō′-plaz′mik re-tik′ū-lŭm; *rete* = a net) is an internal membrane complex that is similar to the smooth endoplasmic reticulum of other cells. Segments of the sarcoplasmic reticulum fit around the myofibril like a sleeve of membrane netting. At either end of individual sections of the sarcoplasmic reticulum are blind sacs called **terminal cisternae** (sis-ter′nē; sing., sis-ter′nă; *cista* = a box), which are much like the hem of a sleeve; they serve as the reservoirs for calcium ions (Ca²⁺). Terminal cisternae are immediately adjacent to each T-tubule. Together, two terminal cisternae and a centrally located T-tubule form a structure called a **triad** that functions during muscle contraction.

Two types of transport proteins are embedded within the membrane of the sarcoplasmic reticulum (figure 10.3d): Ca²⁺ pumps and Ca²⁺ channels. Calcium pumps move Ca²⁺ into the sarcoplasmic reticulum where it is stored bound to specialized proteins called **calmodulin** (kal-mod′ū-lin) and **calsequestrin** (kal′sē-kwes′trin). Calcium channels open to release Ca²⁺ from the sarcoplasmic reticulum into the sarcoplasm. Release of Ca²⁺ from the sarcoplasmic reticulum causes muscle contraction, as described in section 10.3c.

Myofilaments

Myofilaments (mī′ō-fil′ă-ment; *filum* = thread) are contractile proteins that are bundled within myofibrils (figure 10.3b). A myofilament is not as long as a myofibril; rather it takes *many* successive units of myofilaments to extend the entire length of the myofibril. Myofibril bundles contain two types of myofilaments: thick filaments and thin filaments **(figure 10.4)**.

Thick Filaments   **Thick filaments** (or *thick myofilaments*) are assembled from bundles of 200 to 500 **myosin** protein molecules (figure 10.4a). Each myosin protein consists of two strands; each strand has a globular head and an elongated tail. The head contains a binding site for actin of the thin filaments. The head also has a catalytic site where adenosine triphosphate (ATP) attaches and is split into adenosine diphosphate (ADP) and phosphate (Pᵢ). (It is because the head of myosin functions as an ATPase enzyme that myosin is often referred to more specifically as **myosin ATPase**.) The tails of two strands of a myosin molecule are intertwined. The myosin molecules are oriented so that the long tails point toward the center of the thick filaments and the heads point toward the ends of the thick filaments.

Thin Filaments   **Thin filaments** (or *thin myofilaments*) are approximately half of the diameter of thick filaments (about 5 to 6 nanometers). Thin filaments are primarily composed of two strands of **actin** protein that are twisted around each other to form a helical shape (figure 10.4b). In each strand of actin, many (about 300 to 400) small, spherical molecules (G, or globular, actin) are connected to form a fibrous strand (F, or filamentous, actin). F-actin resembles two beaded necklaces that are twisted and intertwined together, with G-actin as the individual beads. Each globular G-actin molecule has a significant feature called a **myosin binding site.** The myosin head attaches to the myosin binding site of actin during muscle contraction.

Tropomyosin and troponin are regulatory proteins associated with thin filaments. Together they are called the troponin-tropomyosin complex. **Tropomyosin** (trō-pō-mī′ō-sin) is a short, thin, twisted filament that is a "stringlike" protein. Consecutive tropomyosin molecules cover small

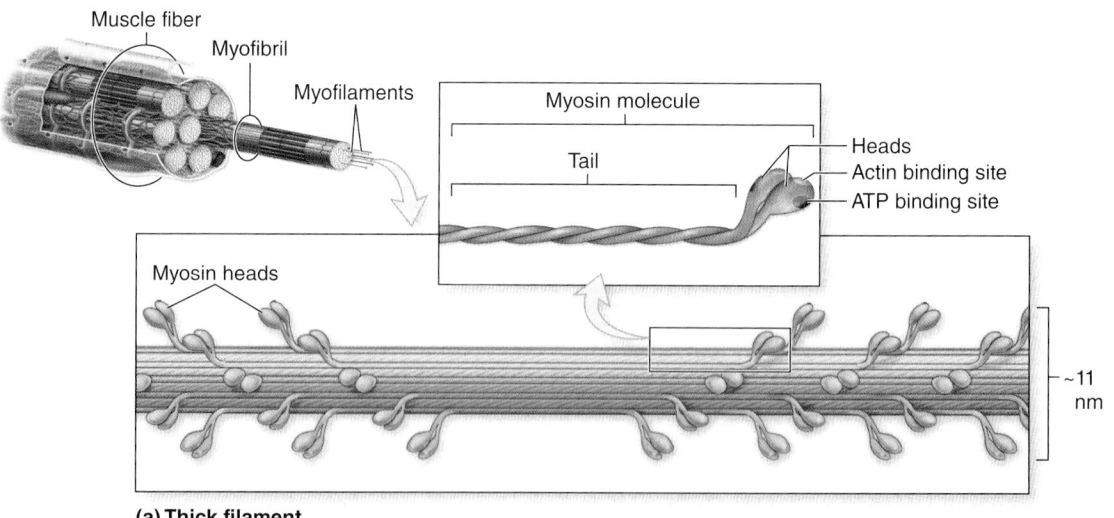

Figure 10.4 Molecular Structure of Thick and Thin Filaments.
Myofilaments, which include thick filaments and thin filaments, are the contractile proteins bundled within myofibrils. (*a*) A thick filament consists of 200 to 500 myosin protein molecules. (*b*) A thin filament is composed of actin, tropomyosin, and troponin proteins.

AP|R

(a) Thick filament

(b) Thin filament

regions of the actin strands, including the myosin binding sites in a non-contracting muscle. **Troponin** (trō′pō-nin) is a globular or "ball-like" protein attached to tropomyosin. Troponin contains the binding site for Ca^{2+}.

Organization of a Sarcomere

Myofilaments within myofibrils are arranged in repeating microscopic cylindrical units (2 micrometers in length) called **sarcomeres** (sar′kō-mēr; *meros* = part). **Figure 10.5a** shows several repeating sarcomeres in a section of a myofibril within a muscle fiber. The number of sarcomeres will vary with the length of the myofibril within the muscle fiber. Each sarcomere is composed of overlapping thick filaments and thin filaments.

A two-dimensional longitudinal view of a cylindrical sarcomere is shown in figure 10.5b. Here we see that each sarcomere is delineated at both ends by Z discs. **Z discs** (also called *Z lines*) are composed of specialized proteins that are positioned perpendicular to the myofilaments and serve as anchors for the thin filaments. Although the Z disc appears as a flat disc when the myofibril is viewed from its "end," only the edge of the disc is visible in a side view, and it sometimes looks like a zigzagged line.

The thick filaments and thin filaments overlap within a sarcomere, forming the following regions:

- **I bands** extend from both directions of a Z disc and are bisected by the Z disc. These end regions contain only thin filaments; this region appears light when viewed with a microscope. At maximal muscle shortening, the thin filaments are pulled parallel along the thick filaments, causing the I band to disappear.

- The **A band** is the central region of a sarcomere that contains the entire thick filament. Thin filaments partially overlap the thick filament on each end of an A band. The A band appears dark when viewed with a microscope. The A band does not change in length during muscle contraction.

- The **H zone** (also called the *H band*) is the most central portion of the A band in a resting sarcomere. This region does not have thin filament overlap; only thick filaments are present. During maximal muscle shortening, this zone disappears when the thin filaments are pulled past thick filaments.

- The **M line** is a thin transverse protein meshwork structure in the center of the H zone. It serves as an attachment site for the thick filaments and keeps the thick filaments aligned during contraction and relaxation events.

INTEGRATE

LEARNING STRATEGY

You can remember A bands and I bands in the following manner: The word *dark* contains the letter "A," just as A bands are *dark* colored. In contrast, the word *light* contains the letter "I," reminding you that I bands are *light* colored.

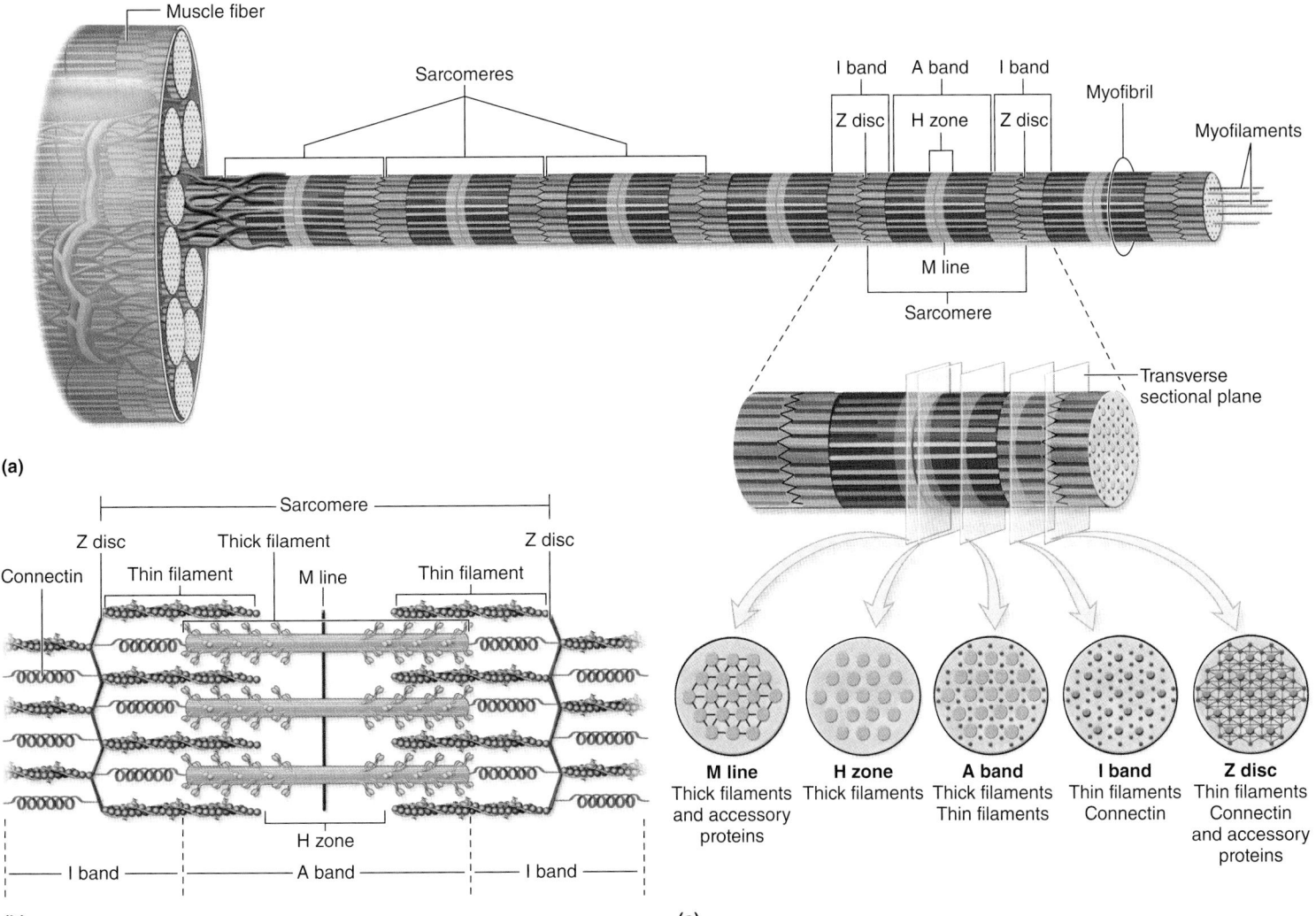

Figure 10.5 Structure of a Sarcomere. (*a*) Numerous sarcomeres extend the length of a myofibril. (*b*) Longitudinal section of a sarcomere. (*c*) Cross sections of different regions of a sarcomere. AP|R

INTEGRATE

LEARNING STRATEGY

To help remember what shortens (and what stays the same length) during skeletal muscle shortening, try this demonstration. Place your hands with your palms facing you and your thumbs pointed upward as shown in the figure. Now imagine there are cylinders placed between and partially overlapping your fingers. Your fingers represent the thin filaments, the cylinders the thick filaments, your thumbs the Z discs, and the distance between your thumbs is the sarcomere. Now slide your fingers toward each other. Both your fingers (the thin filaments) and the cylinders (the thick filaments) remain the same length. However, the distance between your thumbs (the sarcomere) shortens. Notice that both the space between your fingers (the H zone) and where your fingers did not overlap with the cylinders (the I band) disappear during muscle shortening. Thus, note that the length of each individual structure stays the same, but the relative relationship between structures changes.

(a) Relaxed sarcomere   (b) Contracted sarcomere

The overlapping myofilaments form unique patterns of alternating light and dark regions within a skeletal muscle fiber. When a longitudinal section of skeletal muscle tissue is viewed in a microscope, it appears striated (striped) (see figure 10.15). This striated appearance is due to both the size and density differences between thin filaments and thick filaments.

Figure 10.5c shows cross sections through various regions in a sarcomere. It presents the relative sizes, arrangements, and organization of thick and thin filaments at different locations within the sarcomere. Notice that in a cross section of an A band the arrangement of thick filaments relative to thin filaments is the following: Each thin filament has three thick filaments around it that form a triangle at its periphery, and each thick filament is sandwiched by six thin filaments.

WHAT DO YOU THINK?

2 If a muscle is contracted and shortening, what happens to the following: (a) width of the A band, (b) width of the H zone, (c) relationship of the Z discs, and (d) width of the I band?

Other Structural and Functional Proteins   There are other proteins that have structural and functional roles within muscle fibers. These include connectin and dystrophin (only connectin is shown in figure 10.5).

Connectin (kon-nek′tin), also called *titin*, is a "cablelike" protein that extends from the Z discs to the M line through the core of each thick filament (figure 10.5b). It stabilizes the position of the thick filament and maintains thick filament alignment within a sarcomere. Additionally, portions of the connectin molecules are coiled and "springlike" so that during sarcomere shortening they are compressed to produce passive tension. This passive tension is then released to return the sarcomere to its normal resting length. Thus, connectin contributes to muscle fiber elasticity.

Dystrophin (dis-trō′fin) is part of a protein complex that anchors myofibrils that are adjacent to the sarcolemma to proteins within the sarcolemma. These proteins of the sarcolemma also extend to the connective tissue of the endomysium that encloses the muscle fiber. Thus, dystrophin links internal myofilament proteins of a muscle fiber to external proteins. The genetic disorder of *muscular dystrophy* is caused by abnormal structure, or amounts, of dystrophin protein.

Mitochondria and Other Structures Associated with Energy Production

Skeletal muscle fibers have a great demand for energy and contain several components that facilitate the production of ATP. Skeletal muscle fibers have abundant mitochondria for aerobic cellular respiration (see section 3.4); a typical skeletal muscle fiber contains approximately 300 mitochondria. The fibers also contain glycogen stores (granules called glycosomes) for use as an immediate fuel molecule. **Myoglobin** (mī-ō-glō′bin) is a molecule unique to muscle tissue. Myoglobin is a reddish, globular protein that is somewhat similar to hemoglobin. It binds oxygen when the muscle is at rest and releases it for use during muscular contraction. This additional source of oxygen provides the means to enhance aerobic cellular respiration and the production of ATP. Skeletal muscle fibers also contain another type of molecule called creatine phosphate. Creatine phosphate provides muscle fibers with a very rapid means of supplying ATP (creatine phosphate is described in section 10.4a).

WHAT DID YOU LEARN?

4 Draw and label a diagram of a sarcomere.

5 Place the following gross anatomic and microscopic anatomic structures in order from largest to smallest: fascicle, myofibril, myofilament, muscle, muscle fiber, and sarcomere and describe their anatomic relationship.

10.2c Innervation of Skeletal Muscle Fibers

LEARNING OBJECTIVES

9. Define a motor unit, and describe its distribution in a muscle and why it varies in size.

10. Describe the three components of a neuromuscular junction.

The anatomic relationship of skeletal muscle fibers and somatic motor neurons that control them is described in this section.

Motor Unit

Recall that somatic motor neurons are nerve cells that transmit nerve signals from the brain or spinal cord to control skeletal muscle activity (see section 12.1b). The axon of each motor neuron splits to form many individual branches to innervate numerous skeletal muscle fibers. A single motor neuron and the muscle fibers it controls is called a **motor unit** (**figure 10.6**).

The number of skeletal muscle fibers a single motor neuron innervates—and thus the size of the motor unit—varies and can range from small motor units that have less than five muscle fibers to large motor units that have several thousand muscle fibers. The size of the motor unit determines the degree of control. There is an inverse relationship between the size of a motor unit and the degree of control. For example, motor neurons innervating eye muscles are small because greater control is essential in the muscles that move the eye. In contrast, a single motor neuron controls several thousand individual skeletal muscle fibers in the power-generating muscles in our lower limbs, where less precise control is required.

The skeletal muscle fibers of a motor unit are not clustered within one area of a muscle, but rather are dispersed throughout most of a muscle. Normally, the stimulation of a motor unit does not produce a strong contraction in a localized area within the muscle, but a weak contraction over a wide area.

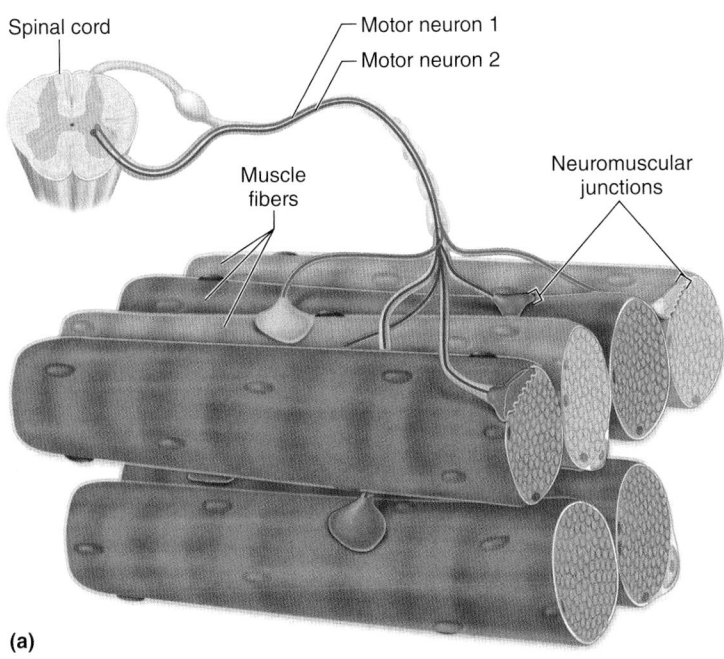

(a)

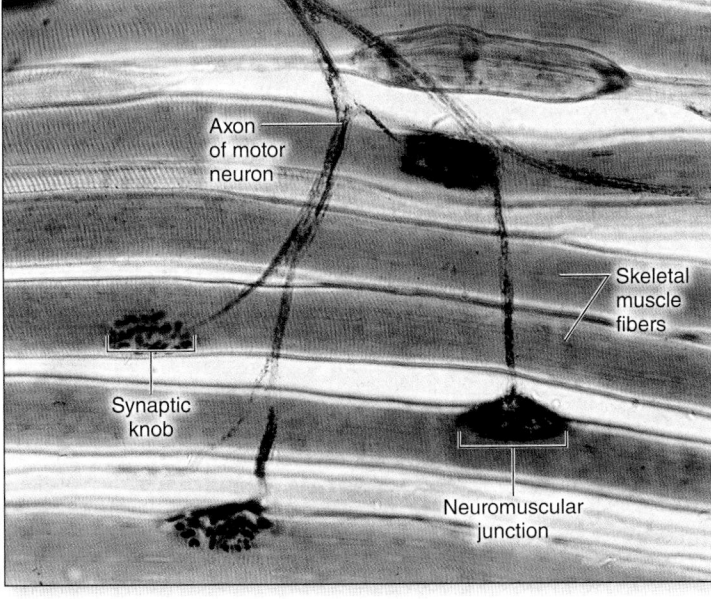

(b)

Figure 10.6 A Motor Unit. (*a*) A motor unit is a motor neuron and all of the skeletal muscle fibers it innervates. Different colors distinguish the two motor units in this figure. (*b*) Light micrograph of the terminal ends of a motor neuron axon and skeletal muscle fibers in a motor unit. AP|R

Neuromuscular Junctions

Each skeletal muscle fiber is typically described as having one neuromuscular junction. A **neuromuscular** (nūr-ō-mŭs′kū-lăr) **junction** is the specific location, usually in the mid-region of the skeletal muscle fiber where it is innervated by a motor neuron (**figure 10.7a**). The neuromuscular junction has the following parts: synaptic knob, motor end plate, and synaptic cleft.

Synaptic Knob   The **synaptic** (si-nap′tik) **knob** of a motor neuron is an expanded tip of an axon. Where the axon nears the sarcolemma of a muscle fiber, the synaptic knob enlarges and flattens to cover a relatively large surface area of the sarcolemma. The synaptic knob cytosol houses numerous **synaptic vesicles** (small membrane sacs) filled with molecules of the neurotransmitter **acetylcholine** (a-sē′til-kō′lēn) **(ACh).**

Several points can be made about synaptic knobs (figure 10.7b). First, Ca^{2+} pumps are embedded within the plasma membrane of the synaptic knob. Prior to the arrival of the nerve signal at the synaptic knob, Ca^{2+} pumps within the axonal membrane have established a Ca^{2+} concentration gradient, with more Ca^{2+} outside the synaptic knob than inside it. Second, voltage-gated Ca^{2+} channels are also embedded in the membrane of the synaptic knob. Opening of these channels allows Ca^{2+} to flow down its concentration gradient from the interstitial fluid into the synaptic knob, which will trigger exocytosis of acetylcholine from the vesicles. Third, vesicles are normally repelled from the synaptic knob plasma membrane.

Motor End Plate   The **motor end plate** is a specialized region of the sarcolemma of a muscle fiber. (Its name "motor end" reflects that it is located at the end of a motor neuron and "plate" because of its large, saucerlike appearance.) It has numerous folds and indentations (junction folds) to increase the membrane surface area covered by the synaptic knob. The motor end plate has vast numbers of **ACh receptors.** These plasma membrane protein channels are chemically gated ion channels (see section 4.3a). Binding of ACh opens these channels, allowing Na^+ entry into the muscle fiber and K^+ to exit. ACh receptors are like doors; ACh is the only "key" to open these receptor doors.

Synaptic Cleft   The **synaptic cleft** is an extremely narrow (30 nanometers), fluid-filled space separating the synaptic knob and the motor end plate. The enzyme **acetylcholinesterase** (a-sē′til-kō′lēn-es′ter-ās) **(AChE)** resides within the synaptic cleft (not shown in figure 10.7) and quickly breaks down ACh molecules following their release into the synaptic cleft. See electron micrograph of neuromuscular junction (see figure 12.22). A detailed view of the breakdown of acetylcholine by acetylcholinesterase is shown in figure 12.25.

WHAT DID YOU LEARN?

6   What is a motor unit, and why does it vary in size?

7   Diagram and label the anatomic structures of a neuromuscular junction.

10.2d Skeletal Muscle Fibers at Rest

LEARNING OBJECTIVE

11.   Describe a skeletal muscle fiber at rest.

Skeletal muscle fibers exhibit several significant features when the muscle is at rest. See **figure 10.8** as your read through this section.

One essential feature of skeletal muscle fibers is the electrical charge difference across the sarcolemma; the intracellular fluid right inside the plasma membrane is relatively negative in comparison to the fluid outside of the cell. This electrical charge difference when the cell is at rest is called the **resting membrane potential (RMP)** (see

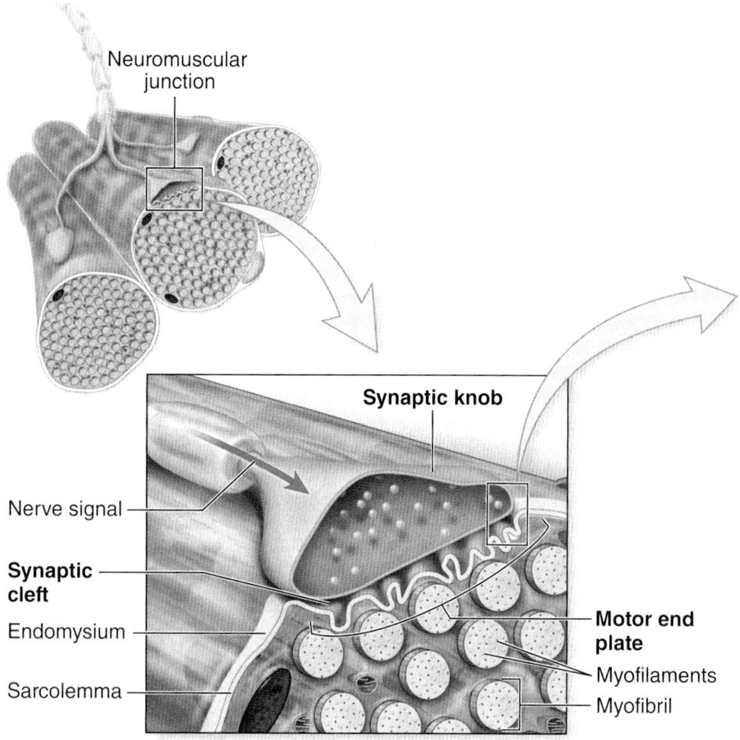

(a) Neuromuscular junction

Neuromuscular junction

Synaptic knob

Nerve signal

Synaptic cleft

Endomysium

Sarcolemma

Motor end plate

Myofilaments

Myofibril

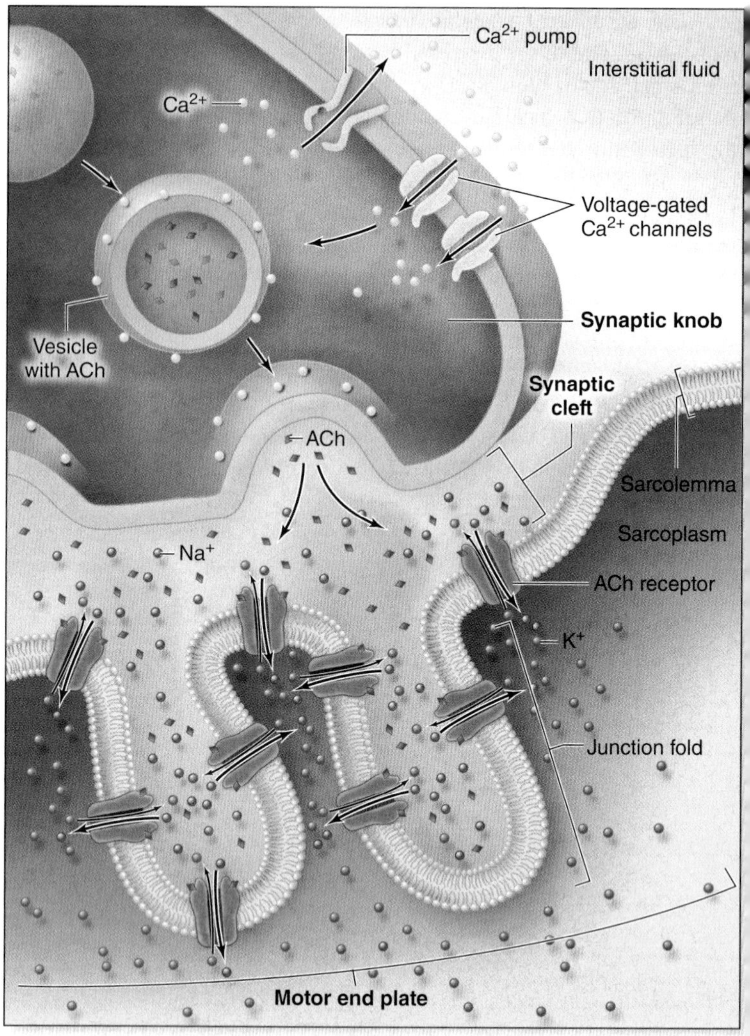

(b) Close-up of neuromuscular junction

Ca²⁺ pump

Interstitial fluid

Ca²⁺

Voltage-gated Ca²⁺ channels

Synaptic knob

Vesicle with ACh

Synaptic cleft

ACh

Sarcolemma

Sarcoplasm

Na⁺

ACh receptor

K⁺

Junction fold

Motor end plate

Figure 10.7 Structure and Organization of a Neuromuscular Junction. The synaptic knob of an axon meets a skeletal muscle fiber to form a neuromuscular junction. (*a*) Three primary components are within a neuromuscular junction and include a synaptic knob, motor end plate, and synaptic cleft. (*b*) Synaptic knobs house synaptic vesicles containing the neurotransmitter acetylcholine (ACh). Embedded in the plasma membrane of the synaptic knob are Ca²⁺ pumps and voltage-gated Ca²⁺ channels. The motor end plate contains ACh receptors, which are chemically gated ion channels. **AP|R**

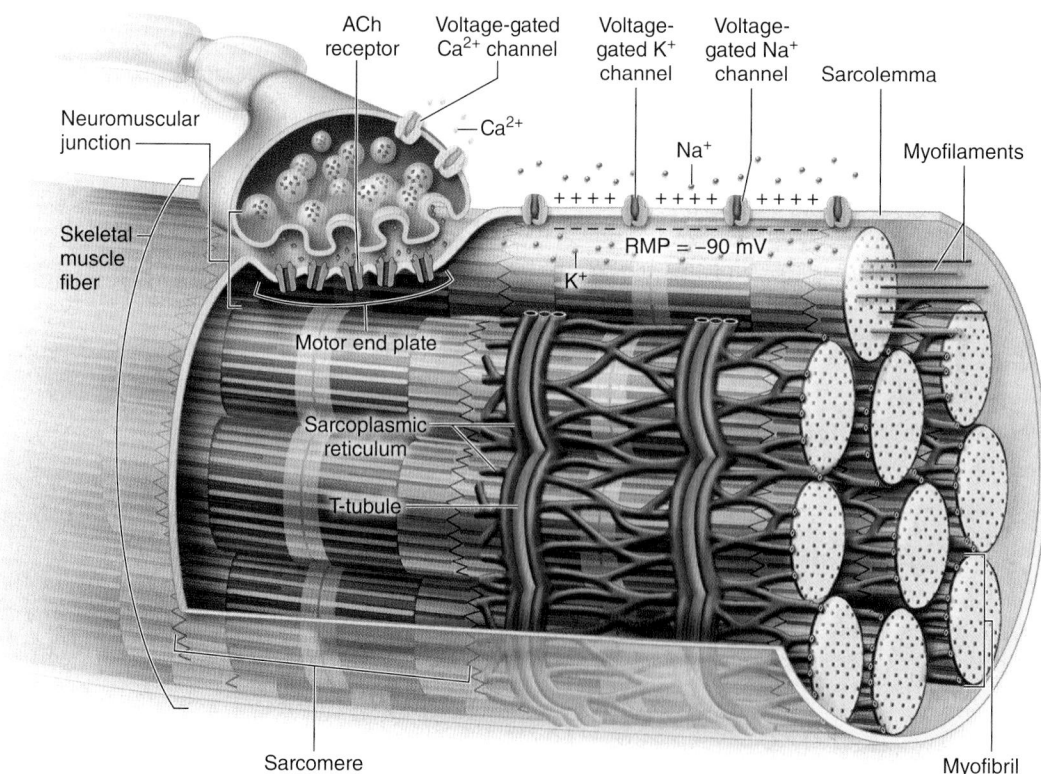

ACh receptor
Voltage-gated Ca²⁺ channel
Voltage-gated K⁺ channel
Voltage-gated Na⁺ channel
Sarcolemma

Neuromuscular junction

Ca²⁺

Na⁺

Myofilaments

Skeletal muscle fiber

++++ ++++ ++++

RMP = −90 mV

K⁺

Motor end plate

Sarcoplasmic reticulum

T-tubule

Sarcomere

Myofibril

Figure 10.8 Skeletal Muscle Fiber at Rest. At rest, a skeletal muscle fiber has a negative resting membrane potential (−90 mV) with more Na⁺ outside the cell and more K⁺ inside the cell. All gated channels are closed, Ca²⁺ is stored within the sarcoplasmic reticulum, and the contractile proteins (myofilaments) are in their relaxed position.

section 4.4). Skeletal muscle fibers have an RMP of about –90 millivolts (mV). An RMP is established and maintained by both leak channels and Na⁺/K⁺ pumps (not shown in figure 10.8). The primary function of the Na⁺/K⁺ pumps is to maintain the concentration gradients for Na⁺ (with more Na⁺ outside the cell) and K⁺ (with more K⁺ inside the cell).

The acetylcholine receptors (chemically gated ion channels) within the motor end plate and the voltage-gated Na^+ channels and voltage-gated K^+ channels in the sarcolemma and T-tubules are closed, calcium ion is stored within the terminal cisternae of the sarcoplasmic reticulum, and the contractile proteins (myofilaments) within the sarcomeres are in their relaxed position.

 **WHAT DID YOU LEARN?**

8   Describe the distribution or Na⁺ and K⁺ at the sarcolemma.

10.3 Physiology of Skeletal Muscle Contraction

A motor neuron stimulates skeletal muscle fibers. This stimulation ultimately results in the interaction between myofilaments within the skeletal muscle fibers to produce tension. The resulting tension is exerted on the portions of the skeleton (or other body structures) where the muscle is attached to cause movement in the body.

CLINICAL VIEW
Myasthenia Gravis (MG)

Myasthenia (mī′as-thē′nē-ă) **gravis (MG)** is an autoimmune disease that occurs in about 1 in 10,000 people, primarily women between 20 and 40 years of age. Antibodies attack the neuromuscular junctions, binding ACh receptors together into clusters. The abnormally clustered ACh receptors are removed from the muscle fiber sarcolemma by endocytosis, thus significantly diminishing the number of receptors within the sarcolemma. The resulting decreased muscle stimulation causes rapid fatigue and muscle weakness. Eye and facial muscles are often attacked first, producing double vision and drooping eyelids. These symptoms are usually followed by swallowing problems, limb weakness, and overall low physical stamina. Some patients with MG have a normal life span, whereas others die quickly from paralysis of the respiratory muscles.

The anatomic structures and associated physiologic processes of skeletal muscle contraction include the events that occur at the (1) neuromuscular junction; (2) sarcolemma, T-tubules, and sarcoplasmic reticulum; and (3) sarcomeres. An overview of these processes is included in **figure 10.9**.

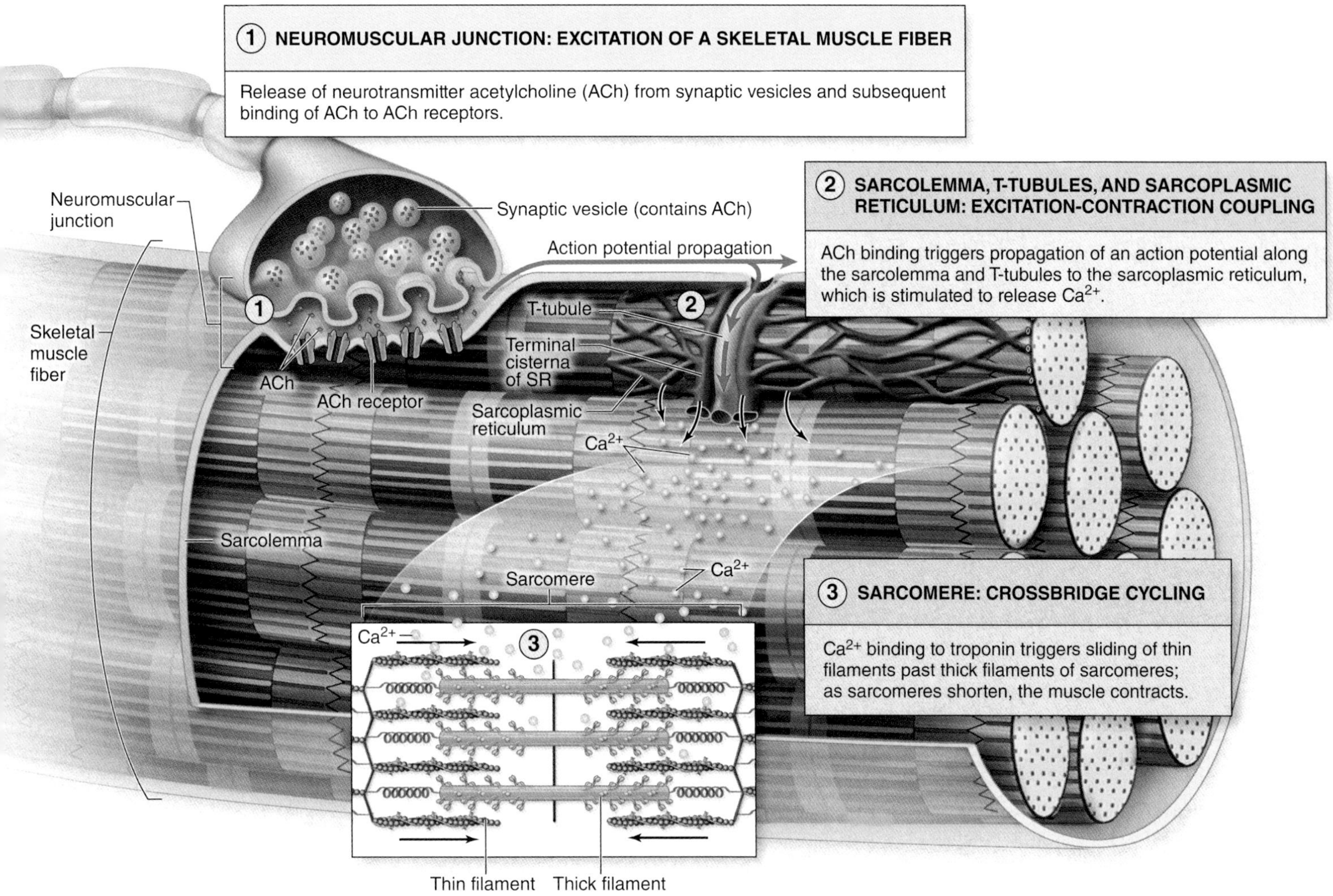

① NEUROMUSCULAR JUNCTION: EXCITATION OF A SKELETAL MUSCLE FIBER

Release of neurotransmitter acetylcholine (ACh) from synaptic vesicles and subsequent binding of ACh to ACh receptors.

② SARCOLEMMA, T-TUBULES, AND SARCOPLASMIC RETICULUM: EXCITATION-CONTRACTION COUPLING

ACh binding triggers propagation of an action potential along the sarcolemma and T-tubules to the sarcoplasmic reticulum, which is stimulated to release Ca^{2+}.

③ SARCOMERE: CROSSBRIDGE CYCLING

Ca^{2+} binding to troponin triggers sliding of thin filaments past thick filaments of sarcomeres; as sarcomeres shorten, the muscle contracts.

Neuromuscular junction · Synaptic vesicle (contains ACh) · Action potential propagation · Skeletal muscle fiber · ACh · ACh receptor · T-tubule · Terminal cisterna of SR · Sarcoplasmic reticulum · Ca²⁺ · Sarcolemma · Sarcomere · Ca²⁺ · Ca²⁺ · Thin filament · Thick filament

Figure 10.9 Overview of Events in Skeletal Muscle Contraction. Skeletal muscle contraction involves the physiologic events that occur (1) at the neuromuscular junction, (2) along the sarcolemma and T-tubules to the sarcoplasmic reticulum of a skeletal muscle fiber, and (3) within a sarcomere. AP|R

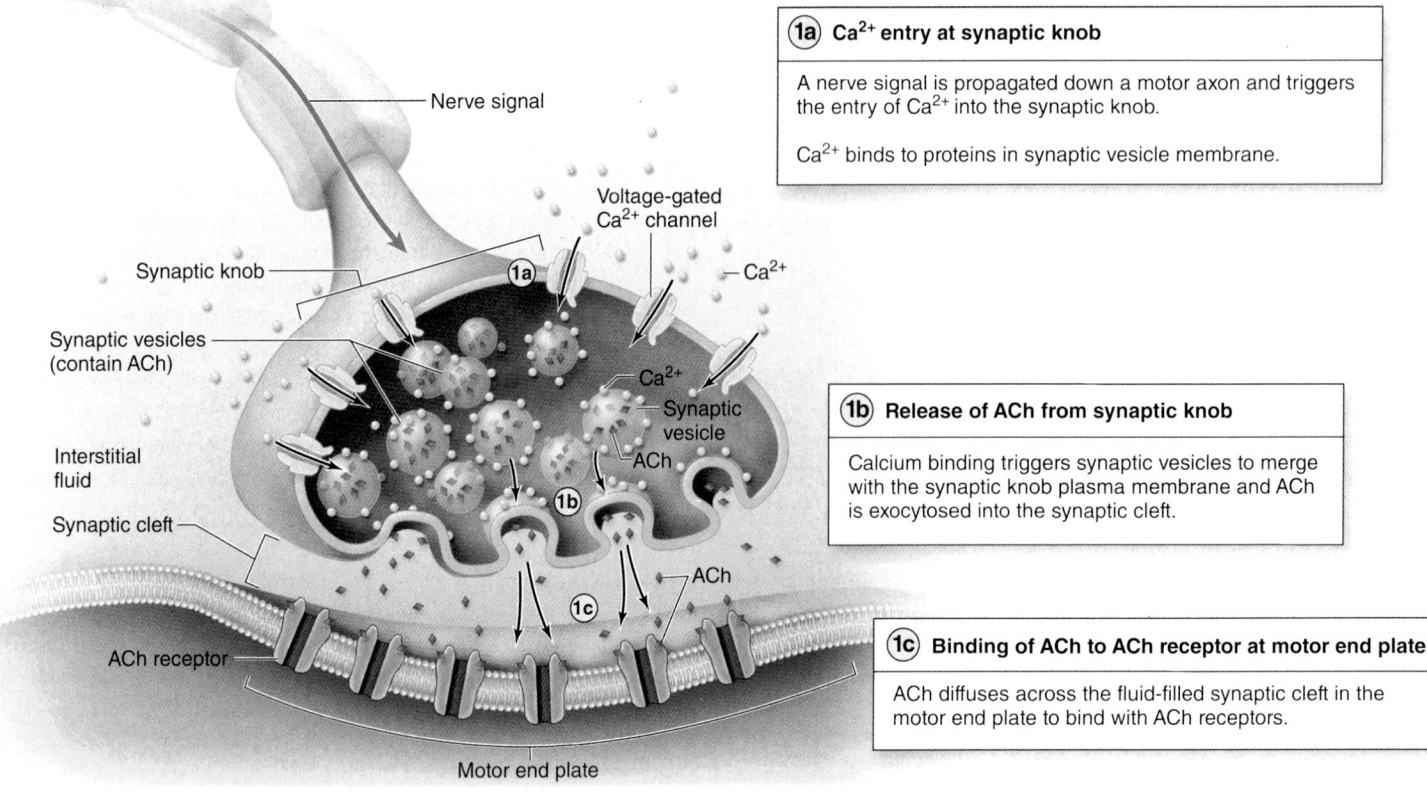

① NEUROMUSCULAR JUNCTION: EXCITATION OF A SKELETAL MUSCLE FIBER

Nerve signal

Synaptic knob

Synaptic vesicles (contain ACh)

Interstitial fluid

Synaptic cleft

ACh receptor

Voltage-gated Ca²⁺ channel

Ca²⁺

Ca²⁺

Synaptic vesicle

ACh

ACh

Motor end plate

①a Ca²⁺ entry at synaptic knob

A nerve signal is propagated down a motor axon and triggers the entry of Ca²⁺ into the synaptic knob.

Ca²⁺ binds to proteins in synaptic vesicle membrane.

①b Release of ACh from synaptic knob

Calcium binding triggers synaptic vesicles to merge with the synaptic knob plasma membrane and ACh is exocytosed into the synaptic cleft.

①c Binding of ACh to ACh receptor at motor end plate

ACh diffuses across the fluid-filled synaptic cleft in the motor end plate to bind with ACh receptors.

Figure 10.10 Neuromuscular Junction: Excitation of a Skeletal Muscle Fiber. A skeletal muscle fiber is excited by the release of the neurotransmitter acetylcholine (ACh) from a motor neuron synaptic knob. AP|R

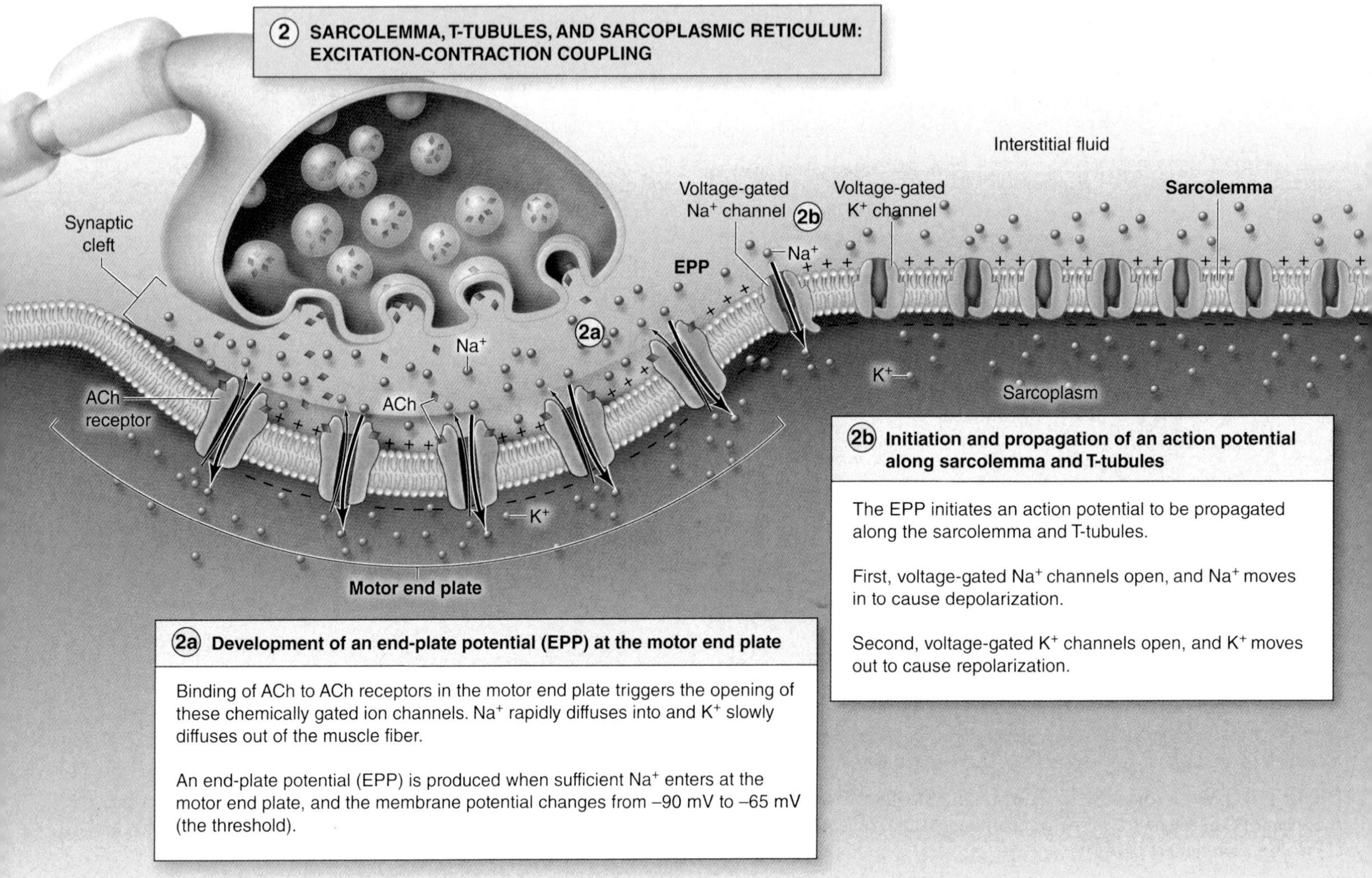

② SARCOLEMMA, T-TUBULES, AND SARCOPLASMIC RETICULUM: EXCITATION-CONTRACTION COUPLING

Synaptic cleft

ACh receptor

ACh

Motor end plate

Na⁺

K⁺

EPP

Voltage-gated Na⁺ channel

Voltage-gated K⁺ channel

Interstitial fluid

Sarcolemma

Na⁺

K⁺

Sarcoplasm

②a Development of an end-plate potential (EPP) at the motor end plate

Binding of ACh to ACh receptors in the motor end plate triggers the opening of these chemically gated ion channels. Na⁺ rapidly diffuses into and K⁺ slowly diffuses out of the muscle fiber.

An end-plate potential (EPP) is produced when sufficient Na⁺ enters at the motor end plate, and the membrane potential changes from –90 mV to –65 mV (the threshold).

②b Initiation and propagation of an action potential along sarcolemma and T-tubules

The EPP initiates an action potential to be propagated along the sarcolemma and T-tubules.

First, voltage-gated Na⁺ channels open, and Na⁺ moves in to cause depolarization.

Second, voltage-gated K⁺ channels open, and K⁺ moves out to cause repolarization.

10.3a Neuromuscular Junction: Excitation of a Skeletal Muscle Fiber

LEARNING OBJECTIVE

1. Explain the events that lead to release of the neurotransmitter ACh from a motor neuron.

The first physiologic event of skeletal muscle contraction is muscle fiber *excitation* by a somatic motor neuron—an event that occurs at the neuromuscular junction and results in release of ACh and its subsequent binding to ACh receptors. These events are summarized in **figure 10.10**.

Calcium Entry at Synaptic Knob

A nerve signal is propagated down a motor axon. (You will learn more about nerve signals in section 12.8c.) The nerve signal triggers the opening of voltage-gated Ca^{2+} channels within the synaptic knob, and calcium moves down its concentration gradient from the interstitial fluid through the open channels into the synaptic knob. Calcium binds with membrane proteins (synaptotagmin) exposed on the external surface of synaptic vesicles (step 1a).

Release of ACh from Synaptic Knob

The binding of Ca^{2+} to synaptic vesicles triggers the merging of synaptic vesicles with the synaptic knob plasma membrane, resulting in exocytosis of ACh into the synaptic cleft. Acetylcholine is released from approximately 300 vesicles per nerve signal with each vesicle releasing thousands of molecules of ACh (step 1b).

Binding of ACh at Motor End Plate

ACh diffuses across the fluid-filled synaptic cleft to bind with ACh receptors within the motor end plate. This causes excitation of a skeletal muscle fiber (step 1c).

Note that nerve signals are repeatedly propagated along the motor axon (at about 10 to 40 times per second). Thus, these events (steps 1a–1c) will continue until stimulation of the skeletal muscle fiber by the neuron ceases (stops).

WHAT DID YOU LEARN?

9. What triggers the binding of synaptic vesicles to the synaptic knob membrane to cause exocytosis of ACh?

10.3b Sarcolemma, T-tubules, and Sarcoplasmic Reticulum: Excitation-Contraction Coupling

LEARNING OBJECTIVE

2. Describe the steps in excitation-contraction coupling.

The second physiologic event of muscle contraction is excitation-contraction coupling—an event that involves the sarcolemma, T-tubules, and sarcoplasmic reticulum. This event "couples" or links the events of skeletal muscle stimulation at the neuromuscular junction (first step) to the events of contraction caused by sliding myofilaments within the sarcomeres of skeletal muscle fiber (third step). Three events occur during excitation-contraction coupling: development of an end-plate potential at the motor end plate, initiation and propagation of an action potential along the sarcolemma and T-tubules, and release of Ca^{2+} from the sarcoplasmic reticulum. These events are summarized in **figure 10.11.**

Figure 10.11 Skeletal Muscle Fiber: Excitation-Contraction Coupling. Skeletal muscle fiber excitation by a motor neuron is coupled to the contraction of myofilaments within the muscle fiber. (Note: Normally, there is no space between the T-tubule and terminal cisternae. Space is for illustration purposes only.)

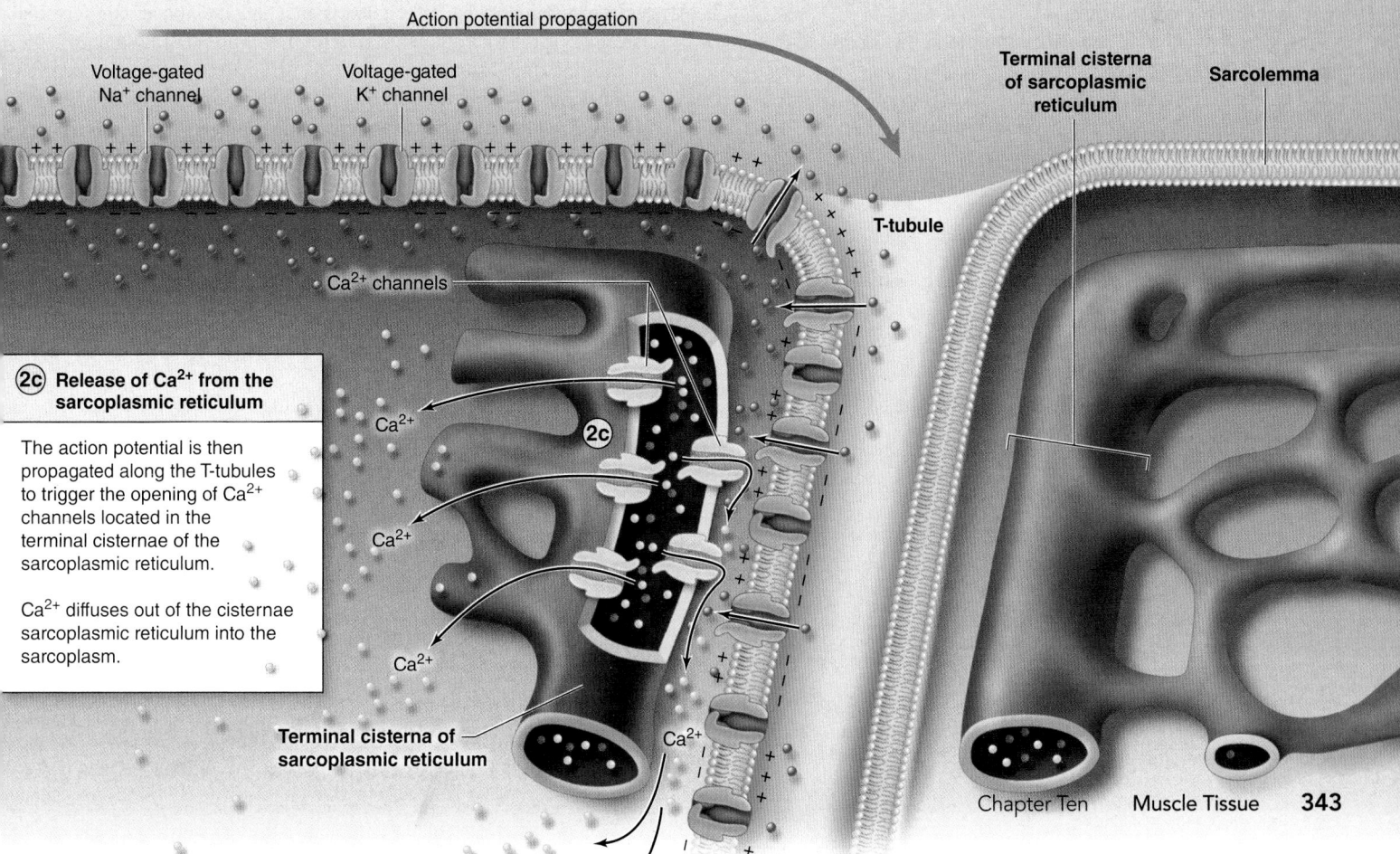

Action potential propagation

Voltage-gated Na⁺ channel Voltage-gated K⁺ channel

Terminal cisterna of sarcoplasmic reticulum   Sarcolemma

T-tubule

Ca²⁺ channels

2c Release of Ca²⁺ from the sarcoplasmic reticulum

The action potential is then propagated along the T-tubules to trigger the opening of Ca²⁺ channels located in the terminal cisternae of the sarcoplasmic reticulum.

Ca²⁺ diffuses out of the cisternae sarcoplasmic reticulum into the sarcoplasm.

Ca²⁺

Ca²⁺

2c

Ca²⁺

Terminal cisterna of sarcoplasmic reticulum

Ca²⁺

Development of an End-Plate Potential at the Motor End Plate

The ACh receptors, which are *chemically gated* ion channels, are stimulated to open temporarily when ACh binds to them (step 2a). The opening of these channels allows relatively small amounts of both Na^+ to rapidly diffuse into the skeletal muscle fiber and K^+ to slowly diffuse out of the skeletal muscle fiber. More Na^+ diffuses in then K^+ diffuses out, and there is a net gain of positive charge on the inside of the skeletal muscle fiber. The flow of both Na^+ and K^+ ions quickly slows and then ceases as the ions meet with resistance. Thus, these changes in membrane potential in the motor end plate are both transient (short-lived) and local. However, if there is sufficient gain of positive charge to change the RMP of about –90 mV to –65 mV, an end-plate potential is produced. An **end-plate potential (EPP)** is the minimum voltage change (or threshold) in the motor end plate that can trigger opening of voltage-gated channels in the sarcolemma to initiate an action potential.

Initiation and Propagation of Action Potential Along the Sarcolemma and T-tubules

The EPP triggers an action potential that is propagated along the sarcolemma and T-tubules of the skeletal muscle fiber (step 2b). An **action potential** involves two events: depolarization that causes the inside of the sarcolemma of the skeletal muscle fiber to become positive due to the influx of Na^+; and repolarization, which is the returning of the inside of the sarcolemma to its relatively negative resting membrane potential due to the outward flow of K^+.

The electrical change of the EPP in the motor end plate stimulates the opening of voltage-gated Na^+ channels in the adjacent area of the sarcolemma. The opening of voltage-gated Na^+ channels allows Na^+ to rapidly move across the sarcolemma down its concentration gradient into the skeletal muscle fiber. Sufficient Na^+ enters to cause a reversal of the membrane potential of the sarcolemma. The inside, which was relatively negative, becomes relatively positive with a change in the membrane potential from the threshold value of –65 mV to +30 mV. This reversal in polarity at the sarcolemma is referred to as **depolarization.**

The propagation of depolarization along the length of the sarcolemma and T-tubules involves the sequential opening of voltage-gated Na^+ channels. The inflow of Na^+ at the initial portion of the sarcolemma causes adjacent regions of the sarcolemma to experience electrical changes that initiate voltage-gated Na^+ channels in these areas to open. Sodium flows in to cause depolarization in this region of the sarcolemma. Adjacent depolarization is repeated rapidly down the sarcolemma and T-tubules. The propagation of an action potential along the sarcolemma and T-tubule is similar to the falling of a series of stacked dominoes—once started, it does not stop until it reaches the end.

Voltage-gated K^+ channels located along the sarcolemma and T-tubules open immediately following the opening of the voltage-gated Na^+ channels. The opening of voltage-gated K^+ channels allows K^+ to move across the sarcolemma down its concentration gradient and out of the skeletal muscle fiber. Sufficient K^+ exits so that the membrane potential at the sarcolemma and T-tubules reverses and the negative resting membrane potential (–90 mV) is reestablished. This process, which changes the membrane potential from +30 mV to reestablish the RMP of –90 mV, is referred to as **repolarization.** The opening of voltage-gated K^+ channels also occurs sequentially, and repolarization is propagated along the sarcolemma and T-tubules. Repolarization allows the muscle fiber to propagate a new action potential when stimulated again by a motor neuron.

Note that an action potential is a self-sustaining electrical change in the membrane potential that is propagated along the sarcolemma and is caused by the sequential opening of *voltage*-gated channels. Action potential propagation at the sarcolemma is similar to action potential propagation that occurs in neurons (see section 12.8c).

Figure 10.12 is a graph of the electrical changes at the sarcolemma. These electrical changes include reaching the threshold, depolarization, and repolarization. The period of time that includes depolarization and repolarization is called the **refractory period.** The refractory period is significant because during this brief period of time the muscle cannot be restimulated. A new action potential can occur only when the resting membrane potential at the sarcolemma has been reestablished.

Release of Calcium from the Sarcoplasmic Reticulum

When the action potential reaches the sarcoplasmic reticulum it triggers the opening of Ca^{2+} channels that are located in the terminal cisternae of the sarcoplasmic reticulum (figure 10.11, step 2c). The opening of Ca^{2+} channels allows Ca^{2+} to diffuse out of the cisternae of the sarcoplasmic reticulum into the sarcoplasm. Calcium now "mingles" with the thick filaments and thin filaments of myofibrils.

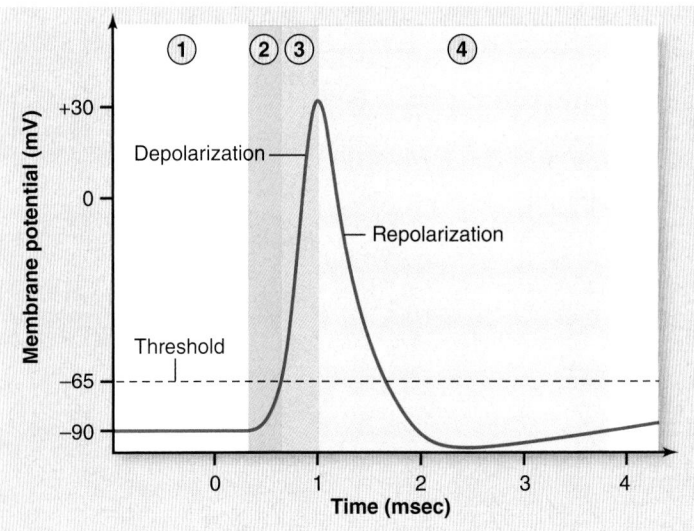

① The sarcolemma of an unstimulated skeletal muscle fiber has a resting membrane potential (RMP) of –90 mV.

② The threshold is reached when an end-plate potential (EPP) is produced as sufficient Na^+ enters the motor end plate to change the RMP from –90 mV to –65 mV (the threshold value).

③ **Depolarization** occurs as voltage-gated Na^+ channels on the sarcolemma open and Na^+ enters rapidly, reversing the polarity from negative to positive (–65 mV to +30 mV).

④ **Repolarization** occurs due to closure of voltage-gated Na^+ channels and opening of voltage-gated K^+ channels on the sarcolemma. K^+ moves out of the cell and the polarity is reversed from positive to negative (+30 mV to –90 mV).

Figure 10.12 Events of an Action Potential at the Sarcolemma. A tracing of the membrane voltage (mV) changes associated with an action potential initiated at the neuromuscular junction. Changes occur in just a few milliseconds, and result from the opening and closing of voltage-gated Na^+ channels and voltage-gated K^+ channels in the membrane of the sarcolemma.

WHAT DID YOU LEARN?

10 What two events are linked in the physiologic process called excitation-contraction coupling?

11 Provide a description of the events of excitation-contraction coupling.

10.3c Sarcomere: Crossbridge Cycling

LEARNING OBJECTIVE

3. Summarize the changes that occur within a sarcomere during contraction.

The third physiologic event in skeletal muscle contraction involves binding of Ca^{2+} and crossbridge cycling. These events are summarized in **figure 10.13**.

Calcium Binding

Calcium released from the sarcoplasmic reticulum binds to a subunit of globular troponin, a component of thin filaments. This induces a conformational change in troponin. Recall that troponin is attached to tropomyosin, forming the troponin-tropomyosin complex. When troponin changes shape, the entire troponin-tropomyosin complex is moved and the myosin binding sites of actin are exposed. Crossbridge cycling is initiated (step 3a).

Crossbridge Cycling

Four steps are then repeated in crossbridge cycling: (1) crossbridge formation (attaching of myosin head to actin), (2) power stroke (pulling thin filament by movement of myosin head), (3) release of myosin head from actin, and (4) resetting of myosin head (step 3b–e):

Crossbridge formation Myosin heads, which are in the "cocked," or ready, position attach to exposed myosin binding sites of actin. Binding of each myosin head results in formation of a **crossbridge** between the thick and thin filament (step 3b).

Power stroke After forming a crossbridge, the myosin head swivels (or ratchets) in what is called a **power stroke.** The swiveling of a myosin head pulls the thin filament a small

distance past the thick filament toward the center of the sarcomere. ADP and P_i are released during this process, and the ATP binding site becomes available again (step 3c).

Release of Myosin Head ATP then binds to the ATP binding site of a myosin head, which causes the release of the myosin head from the binding site on actin (step 3d).

Reset Myosin Head Myosin ATPase splits ATP into ADP and P_i, providing the energy to reset the myosin head in the cocked position (step 3e).

If Ca^{2+} is still present, and the myosin binding sites are still exposed, then these four steps involving the myosin heads continue: attach, pull, release, and reset. It is the repetitive action of these steps that results in sarcomere shortening, and a sarcomere moves from its relaxed state into a contracted state. Note that calcium levels remain elevated because the skeletal muscle fiber is repeatedly stimulated by the motor neuron at a very rapid rate. **Figure 10.14** is an electron micrograph of crossbridges between myosin heads and myosin binding sites in actin.

A sarcomere in both a relaxed and contracted skeletal muscle fiber that has shortened is shown in **figure 10.15**. The following changes to the sarcomere occur in the contracted muscle: The H zone disappears, the I band narrows in width and may disappear, and the Z discs in one sarcomere move closer together. However, the thin and thick filaments do not shorten. A description of the repetitive movement of thin filaments sliding past thick filaments is called the **sliding filament theory**. The three major events of skeletal muscle contraction are integrated in **figure 10.16**.

WHAT DO YOU THINK?

3 Calcium is released from the sarcoplasmic reticulum following death. Without available ATP, rigor mortis (stiffening of the body) occurs. Explain why, if ATP is unavailable, muscles remain in a contracted state.

WHAT DID YOU LEARN?

12 What is the function of Ca^{2+} in skeletal muscle contraction?

13 Describe the four processes that repeat in crossbridge cycling to cause sarcomere shortening.

14 What causes the release of the myosin head from actin? What resets the myosin head?

INTEGRATE

CLINICAL VIEW

Muscular Paralysis and Neurotoxins

Muscular paralysis may occur if either nervous system function at the neuromuscular junction or excitation-contraction coupling is impaired. Two such conditions are tetanus and botulism.

Tetanus is a form of spastic paralysis caused by a toxin produced by the bacterium *Clostridium tetani*. The toxin blocks the release of glycine (an inhibitory neurotransmitter in the spinal cord), resulting in overstimulation of the muscles and excessive muscle contractions. Penetrating wounds contaminated with soil and vegetable matter are especially prone to developing *C. tetani* infection. This condition is potentially life-threatening, and so we routinely are vaccinated against it.

Botulism (bot′ū-lizm), a potentially fatal muscular paralysis, is caused by a toxin produced by a related bacterium, *Clostridium botulinum*. The toxin prevents the release of acetylcholine (ACh) at synaptic knobs and

leads to muscular paralysis. Like *C. tetani*, *C. botulinum* is common in the environment and also produces its toxin only under anaerobic conditions. Most cases of botulism poisoning result from ingesting the toxin in canned foods that were not processed at temperatures high enough to kill the botulism spores. Similarly, ingestion of unpasteurized honey by infants in the first year of life can introduce *C. botulinum* spores into their immature gastrointestinal tracts.

The Food and Drug Administration (FDA) approved the use of botulinum toxin type A (Botox) for temporary diminishing of wrinkles. Botox is also used clinically to help reduce overcontraction of muscle (or spasticity) associated with certain disorders or conditions (e.g., cerebral palsy, multiple sclerosis, torticollis, changes following a stroke or spinal cord injury). Botox injections have become one of the most important treatments for spasticity, and are most effective 1 to 2 weeks after the injections, with spasticity reduced for up to 3 to 6 months. Treatments may be repeated as often as every 3 months.

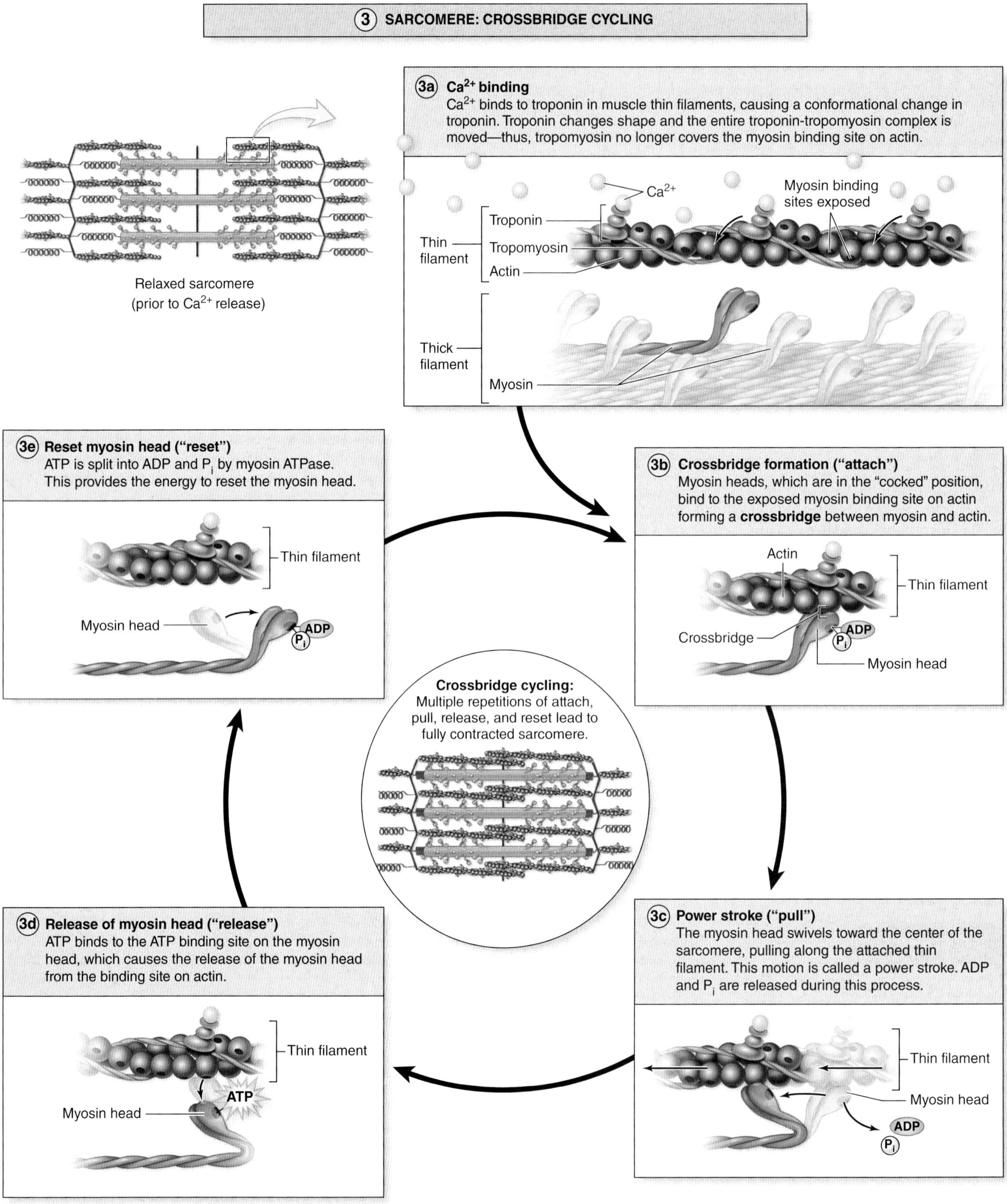

③a Ca²⁺ binding
Ca^{2+} binds to troponin in muscle thin filaments, causing a conformational change in troponin. Troponin changes shape and the entire troponin-tropomyosin complex is moved—thus, tropomyosin no longer covers the myosin binding site on actin.

Relaxed sarcomere
(prior to Ca^{2+} release)

Ca^{2+}
Myosin binding sites exposed
Troponin
Thin filament
Tropomyosin
Actin
Thick filament
Myosin

③e Reset myosin head ("reset")
ATP is split into ADP and P_i by myosin ATPase. This provides the energy to reset the myosin head.

Thin filament
Myosin head
ADP
P_i

③b Crossbridge formation ("attach")
Myosin heads, which are in the "cocked" position, bind to the exposed myosin binding site on actin forming a **crossbridge** between myosin and actin.

Actin
Thin filament
Crossbridge
ADP
P_i
Myosin head

Crossbridge cycling:
Multiple repetitions of attach, pull, release, and reset lead to fully contracted sarcomere.

③d Release of myosin head ("release")
ATP binds to the ATP binding site on the myosin head, which causes the release of the myosin head from the binding site on actin.

Thin filament
ATP
Myosin head

③c Power stroke ("pull")
The myosin head swivels toward the center of the sarcomere, pulling along the attached thin filament. This motion is called a power stroke. ADP and P_i are released during this process.

Thin filament
Myosin head
ADP
P_i

Figure 10.13 Sarcomere: Skeletal Muscle Contraction. Contractile proteins slide past each other toward the center of the sarcomere, and the sarcomere shortens. AP|R

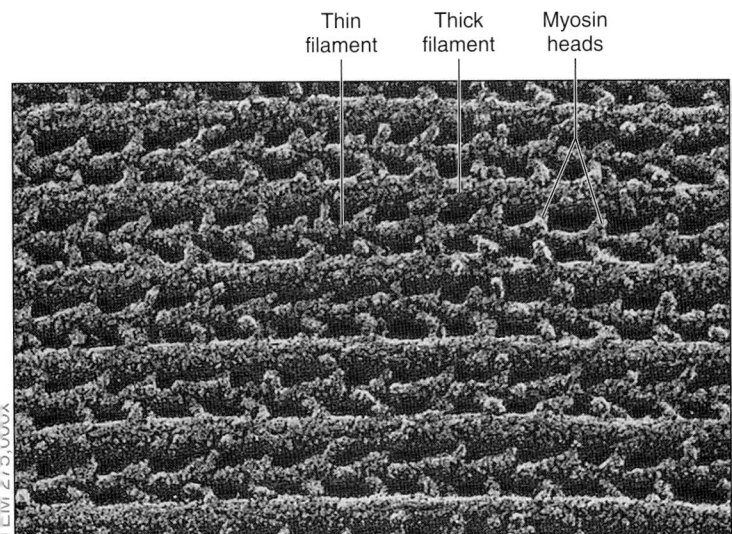

Thin filament | Thick filament | Myosin heads

Figure 10.14 Portion of a Sarcomere. Electron micrograph of a sarcomere shows interaction between the myosin heads of the thick filament with the thin filament (TEM 275,000×).

10.3d Skeletal Muscle Relaxation

LEARNING OBJECTIVES

4. Discuss what happens to each of the following to allow for skeletal muscle relaxation: ACh, action potential, Ca^{2+} concentration in sarcoplasm, and troponin-tropomyosin complex.

5. Explain the relationship of skeletal muscle elasticity and muscle relaxation.

The first step in skeletal muscle relaxation is the termination of the rapid nerve signals propagated along the motor neuron. When the nerve signals stop, there is no additional release of acetylcholine, and the acetylcholine remaining in the synaptic cleft is hydrolyzed by acetylcholinesterase (see figure 12.25). The ACh receptors close, and both the end-plate potentials at the motor end plate and the action potentials along the sarcolemma and T-tubules cease.

Calcium channels in the sarcoplasmic reticulum close. The Ca^{2+} already released from the sarcoplasmic reticulum is continuously returned into the terminal cisternae by Ca^{2+} pumps. After cessation of skeletal muscle fiber stimulation, the remaining Ca^{2+} in the sarcoplasm is transported back into storage within the sarcoplasmic reticulum, where it is bound by both calmodulin and calsequestrin proteins.

Troponin returns to its original shape when Ca^{2+} is removed, and simultaneously the tropomyosin moves over the myosin binding sites on actin. This prevents myosin-actin crossbridge formation. Through

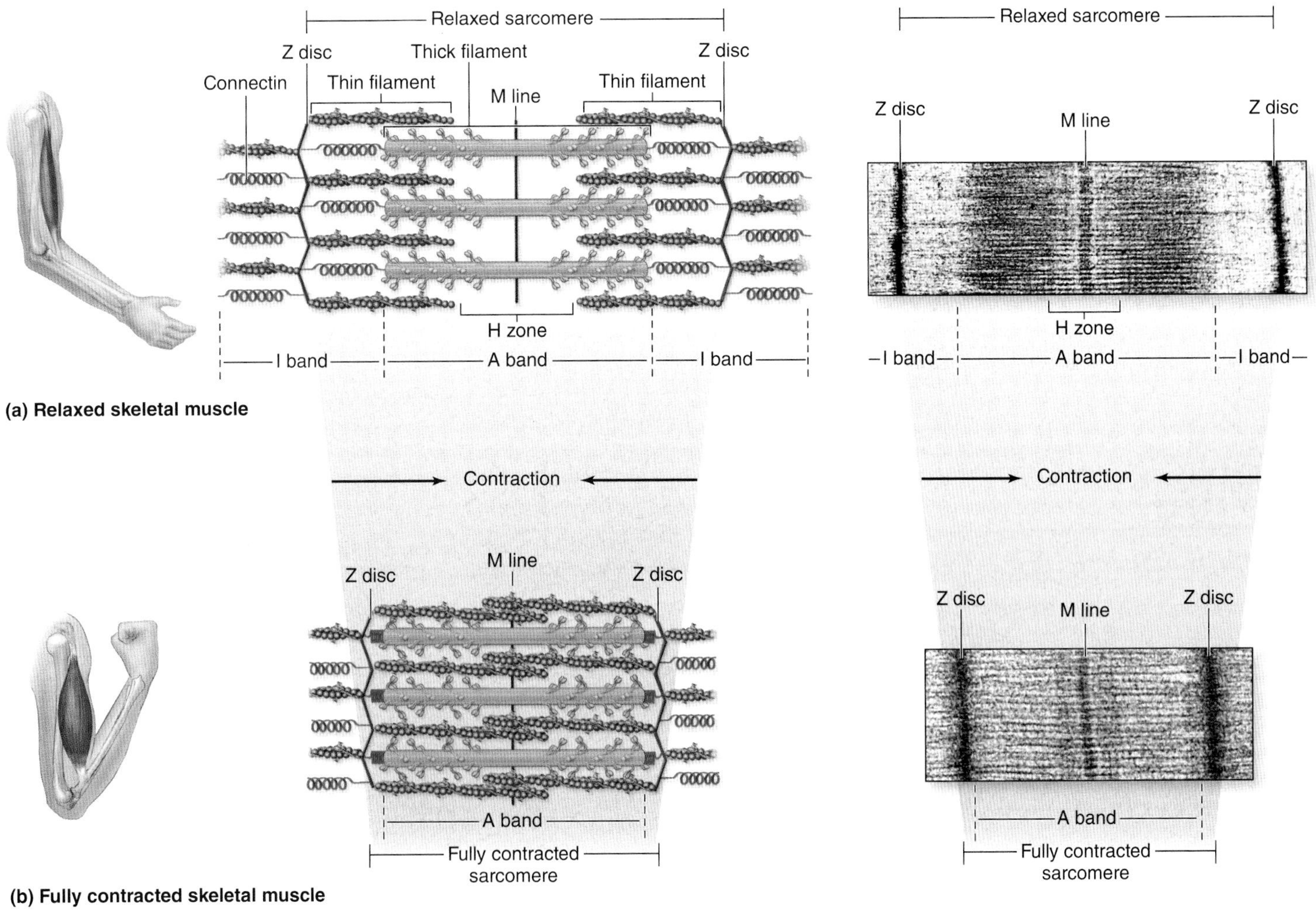

(a) Relaxed skeletal muscle

Relaxed sarcomere

Connectin | Z disc | Thin filament | Thick filament | M line | Z disc | Thin filament

I band — A band — I band

H zone

Relaxed sarcomere

Z disc | M line | Z disc

I band — A band — I band

H zone

(b) Fully contracted skeletal muscle

Contraction

Z disc | M line | Z disc

A band

Fully contracted sarcomere

Contraction

Z disc | M line | Z disc

A band

Fully contracted sarcomere

Figure 10.15 Sarcomere Shortening. Illustrations and corresponding electron micrographs demonstrate sarcomere shortening during skeletal muscle contraction. (*a*) In a relaxed skeletal muscle, the A band, I band, and H zone are all visible. (*b*) In a fully contracted muscle, the sarcomere shortens, the Z discs move closer together, the I band narrows and may disappear, and the H zone disappears.

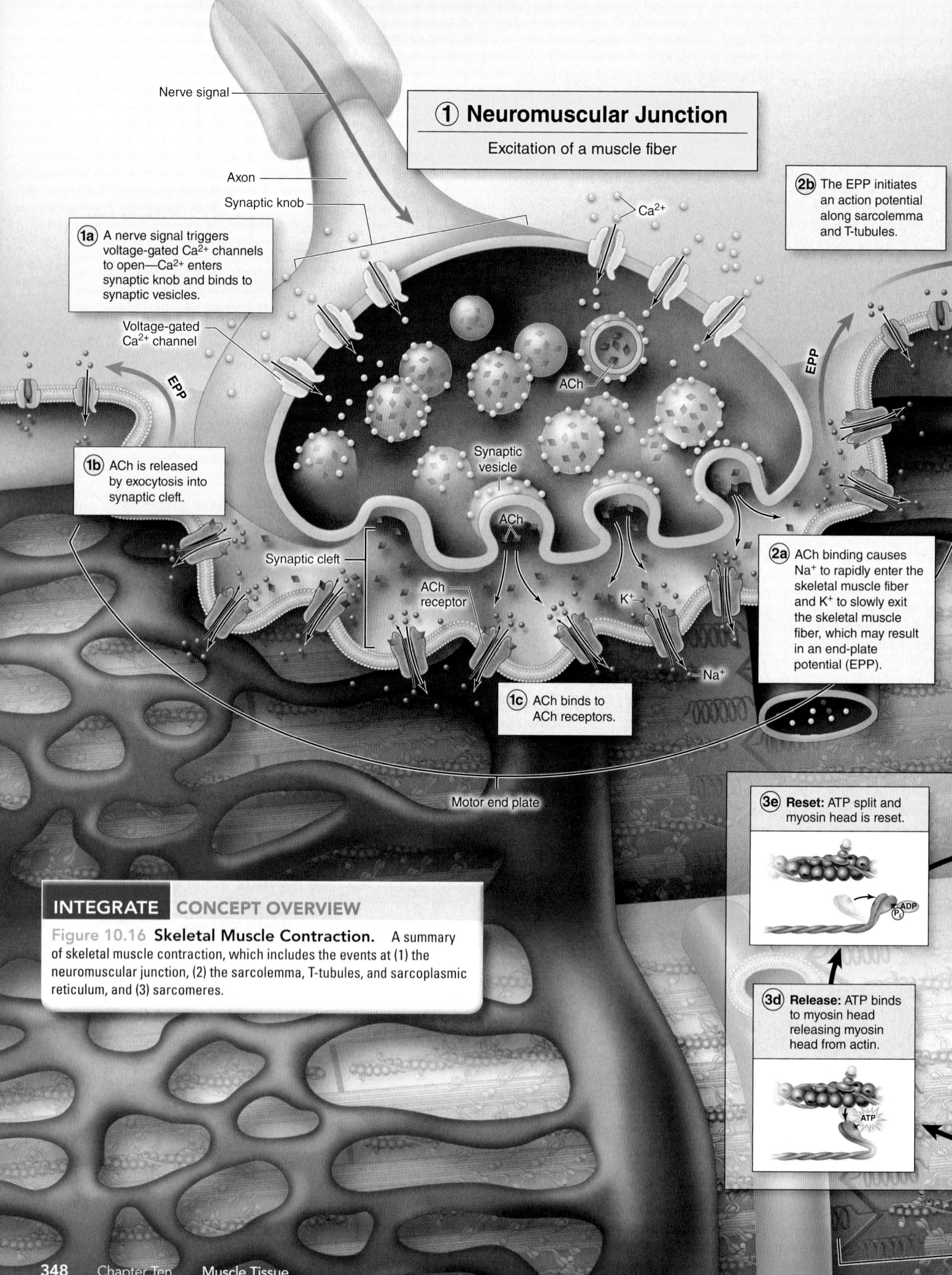

Nerve signal

Axon

Synaptic knob

① Neuromuscular Junction

Excitation of a muscle fiber

1a A nerve signal triggers voltage-gated Ca^{2+} channels to open—Ca^{2+} enters synaptic knob and binds to synaptic vesicles.

Voltage-gated Ca^{2+} channel

EPP

1b ACh is released by exocytosis into synaptic cleft.

Synaptic cleft

Ca^{2+}

ACh

Synaptic vesicle

ACh

ACh receptor

K^+

1c ACh binds to ACh receptors.

Na^+

EPP

2b The EPP initiates an action potential along sarcolemma and T-tubules.

2a ACh binding causes Na^+ to rapidly enter the skeletal muscle fiber and K^+ to slowly exit the skeletal muscle fiber, which may result in an end-plate potential (EPP).

Motor end plate

INTEGRATE **CONCEPT OVERVIEW**

Figure 10.16 **Skeletal Muscle Contraction.** A summary of skeletal muscle contraction, which includes the events at (1) the neuromuscular junction, (2) the sarcolemma, T-tubules, and sarcoplasmic reticulum, and (3) sarcomeres.

3e **Reset:** ATP split and myosin head is reset.

ADP
P_i

3d **Release:** ATP binds to myosin head releasing myosin head from actin.

ATP

② Sarcolemma, T-tubules, and Sarcoplasmic Reticulum
Excitation-contraction coupling

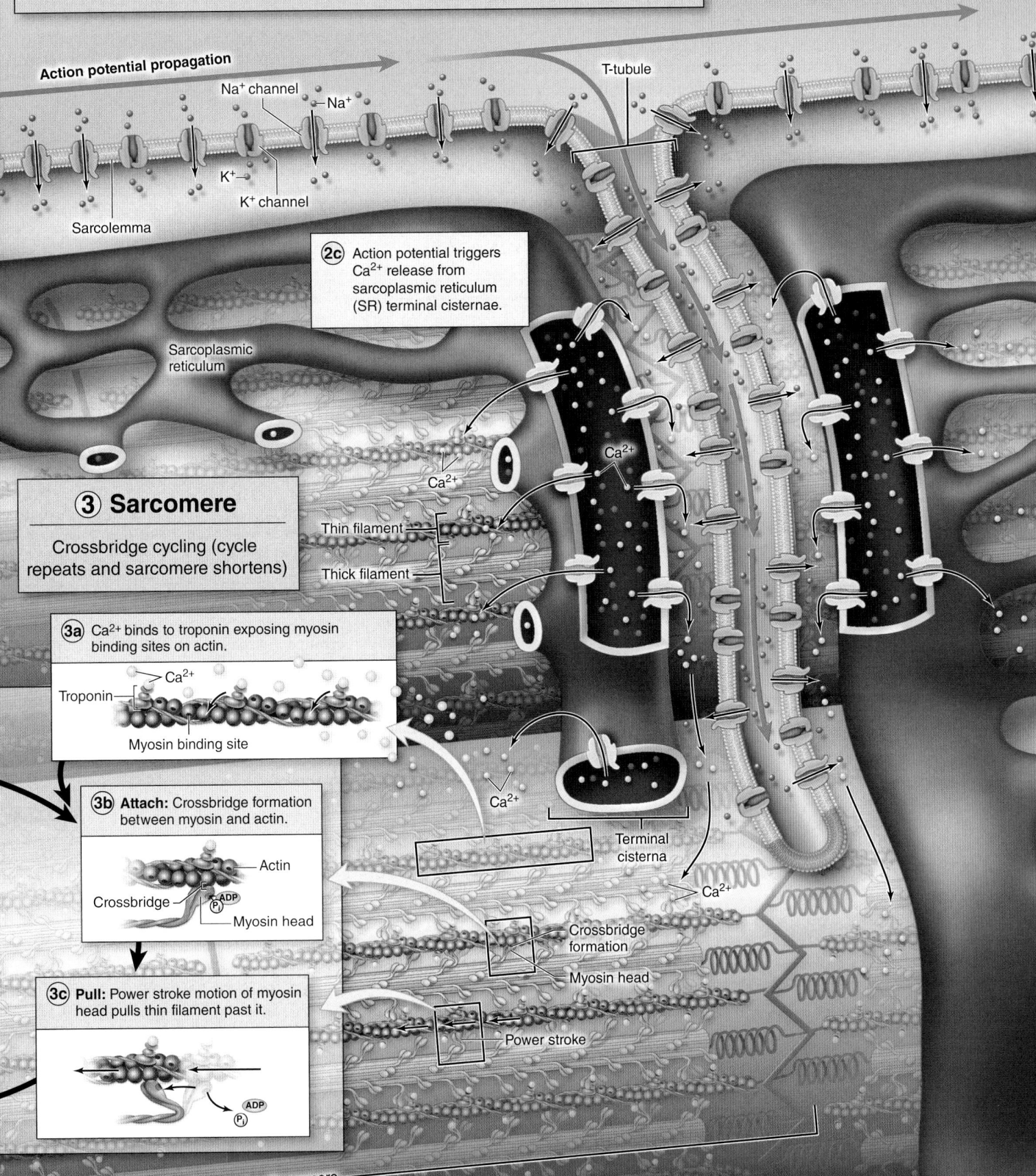

Action potential propagation

Na⁺ channel

Na⁺

K⁺

K⁺ channel

Sarcolemma

T-tubule

2c Action potential triggers Ca^{2+} release from sarcoplasmic reticulum (SR) terminal cisternae.

Sarcoplasmic reticulum

Ca^{2+}

③ Sarcomere
Crossbridge cycling (cycle repeats and sarcomere shortens)

Thin filament

Thick filament

Ca^{2+}

Ca^{2+}

3a Ca^{2+} binds to troponin exposing myosin binding sites on actin.

Troponin

Ca^{2+}

Myosin binding site

3b **Attach:** Crossbridge formation between myosin and actin.

Actin

Crossbridge

ADP

P_i

Myosin head

3c **Pull:** Power stroke motion of myosin head pulls thin filament past it.

ADP

P_i

Terminal cisterna

Ca^{2+}

Crossbridge formation

Myosin head

Power stroke

Contracting Sarcomere

INTEGRATE

CLINICAL VIEW
Rigor Mortis

Within a few hours after the heart stops beating, ATP levels in skeletal muscle fibers have been completely exhausted. The sarcoplasmic reticulum loses its ability to return Ca^{2+} from the sarcoplasm and move it back into the sarcoplasmic reticulum. As a result, the Ca^{2+} already present in the sarcoplasm, as well as the Ca^{2+} that continues to leak out of the sarcoplasmic reticulum, trigger a sustained contraction in the fibers. Remember that ATP is needed to detach the myosin head of the thick filament from the myosin binding site of actin on the thin filaments. Because ATP is no longer available, the crossbridges between thick and thin filaments cannot detach. All skeletal muscles lock into a contracted position and the body of the deceased individual becomes rigid. This physiologic state, termed **rigor mortis** (rig′er mōr′tis), continues for about 15 to 24 hours. Rigor mortis then disappears because lysosomal enzymes are released within the muscle fibers, causing autolysis (self-destruction and breakdown) of the myofibrils.

Forensic pathologists often use the development and resolution of rigor mortis to establish an approximate time of death. Because a number of factors affect the rate of development and resolution of rigor mortis, environmental conditions need to be taken into consideration. For example, a warmer body will develop and resolve rigor mortis much more quickly than a body of normal temperature. The following chart provides rough guidelines for estimating the death interval, assuming that body temperature and ambient (surrounding environment) temperature are within normal range.

Death Interval	Body Temperature	Stiffness
Dead less than 3 hours	Warm	No stiffness
Dead 3–8 hours	Warm, but cooling	Developing stiffness
Dead 8–24 hours	Ambient temperature	Stiff, but resolving
Dead 24–36 hours	Ambient temperature	No stiffness

the natural elasticity of the muscle fiber, the muscle may return to its original relaxed position, a process facilitated by the release of passive tension that developed in connectin proteins that were compressed during shortening.

It is interesting to note that a significant amount of ATP required by skeletal muscle fibers is used by the Ca^{2+} pumps of the sarcoplasmic reticulum. Thus, ATP is required for both contraction (by myosin ATPase) and relaxation. In fact, if sufficient ATP is not available (such as occurs following death), muscle relaxation cannot occur and the muscle remains in a contracted state (see Clinical View: "Rigor Mortis").

WHAT DID YOU LEARN?

15 How do acetylcholinesterase and Ca^{2+} pumps function in the relaxation of a muscle?

10.4 Skeletal Muscle Metabolism

We discussed how cells form ATP through the process of cellular respiration in section 3.4. Here those concepts are integrated to describe specifically how a muscle fiber meets its very-high-energy needs—and also how its various means of supplying ATP are used to classify skeletal muscle fibers into three primary types.

10.4a Supplying Energy for Skeletal Muscle Contraction

LEARNING OBJECTIVES

1. Describe immediate, short-term, and long-term means that muscle fibers use to supply ATP for muscle contraction.
2. Explain how the means of supplying ATP is related to intensity and duration of exercise.

ATP that is already present within the skeletal muscle fiber is hydrolyzed by **ATPase** into ADP and P_i. This usually provides only enough energy for about 5 to 6 seconds of maximal exertion. A skeletal muscle fiber has several means of supplying ATP to meet its high demand for energy. These are differentiated based on the relative amount of time required to generate ATP and are appropriately called the immediate, short-term, and long-term means for supplying ATP.

Immediate Means of Supplying ATP: Phosphate Transfer

An additional few seconds of energy is generated in muscle by the transfer of a high-energy phosphate. These processes are not dependent upon the presence of oxygen. One means occurs through the enzyme **myokinase,** which transfers a phosphate (P_i) from one ADP to another ADP, yielding ATP and adenosine monophosphate (AMP) (**figure 10.17a**).

Another means of supplying ATP involves the enzyme **creatine kinase,** which instead transfers P_i from **creatine phosphate** to ADP, yielding ATP and creatine. This provides an additional 10 to 15 seconds of energy during maximum exertion. During times of rest, as small amounts of ATP accumulate, the pathway described in figure 10.17b is reversed. Creatine kinase transfers P_i from ATP to creatine, yielding creatine phosphate and ADP for the purpose of increasing the amount of creatine phosphate for the next contraction.

WHAT DO YOU THINK?

4 When skeletal muscle tissue is damaged, creatine kinase is released. What general conclusions can be drawn with increasing blood levels of creatine kinase?

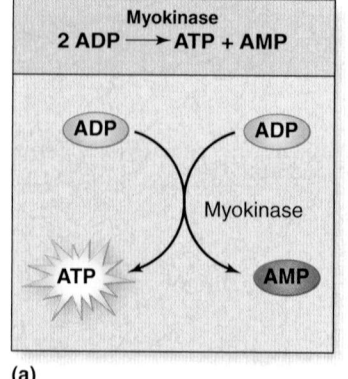
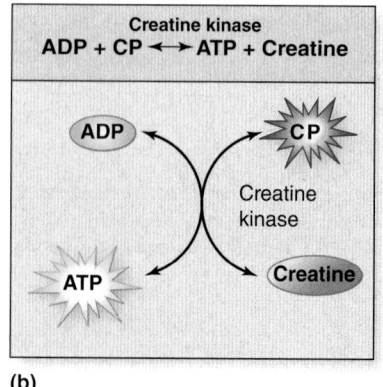

(a) (b)

Figure 10.17 Phosphate Transfer. Immediate sources of additional ATP are available for skeletal muscle fibers from (*a*) the transfer of phosphate (P_i) from one ADP to another ADP that is catalyzed by myokinase enzyme, and (*b*) the transfer of P_i from creatine phosphate (CP) to ADP that is a reversible reaction catalyzed by creatine kinase (CK) enzyme.

INTEGRATE

CLINICAL VIEW
Creatine Kinase

Creatine kinase, also called *creatine phosphokinase,* is the enzyme that helps transfer a phosphate between creatine and ATP. Two different forms of creatine kinase occur in cardiac muscle and skeletal muscle. The heart muscle form of creatine kinase is found in elevated blood levels in patients suffering from a myocardial infarction (heart attack). This provides a specific diagnostic tool for identifying damage to the heart. In contrast, elevated levels of the skeletal muscle form of creatine kinase are used to diagnose degenerative skeletal muscle disease, such as muscular dystrophy. However, note that elevated levels of the skeletal muscle form may also occur after intense exercise, and therefore is not always a sign of disease.

Short-Term Means of Supplying ATP: Glycolysis

Recall from section 3.4 that the metabolic processes for generating ATP involve both glycolysis, which occurs within the cytosol and does not require oxygen, and aerobic metabolic pathways, which occur within mitochondria and require oxygen. The short-term means of supplying ATP is through glycolysis **(figure 10.18a)**. Glucose is made available

either directly from glycogen stores within the muscle fiber or delivered by the blood; through many enzymatic steps glucose is broken down to two pyruvate molecules. The net energy released during glycolysis is 2 ATP molecules per glucose molecule (see section 3.4b).

Long-Term Means of Supplying ATP: Aerobic Cellular Respiration

The long-term means of supplying ATP is through **aerobic cellular respiration** that occurs within mitochondria (figure 10.18b). The nutrient source for aerobic production of ATP is the pyruvate made available following glycolysis. Aerobic cellular respiration requires the presence of oxygen that is both delivered by the blood and provided by myoglobin (described in section 10.2b).

Pyruvate enters a mitochondrion, where it is oxidized completely to carbon dioxide (CO_2) through the metabolic pathways of the intermediate stage and citric acid cycle (see sections 3.4c and d). The oxidation of pyruvate results in the transfer of chemical bond energy to coenzymes, NADH and $FADH_2$. The energy transferred to NADH and $FADH_2$ is then used to generate ATP in the electron transport system by a process called *oxidative phosphorylation* (see section 3.4e). The net amount of energy produced through oxidative phosphorylation is 30 ATP molecules.

ATP is also generated in aerobic cellular respiration from triglycerides, structures composed of glycerol and three fatty acids (see section 2.7b). The amount of ATP generated is dependent upon the length of the fatty acid chain. The longer the chain, the more ATP

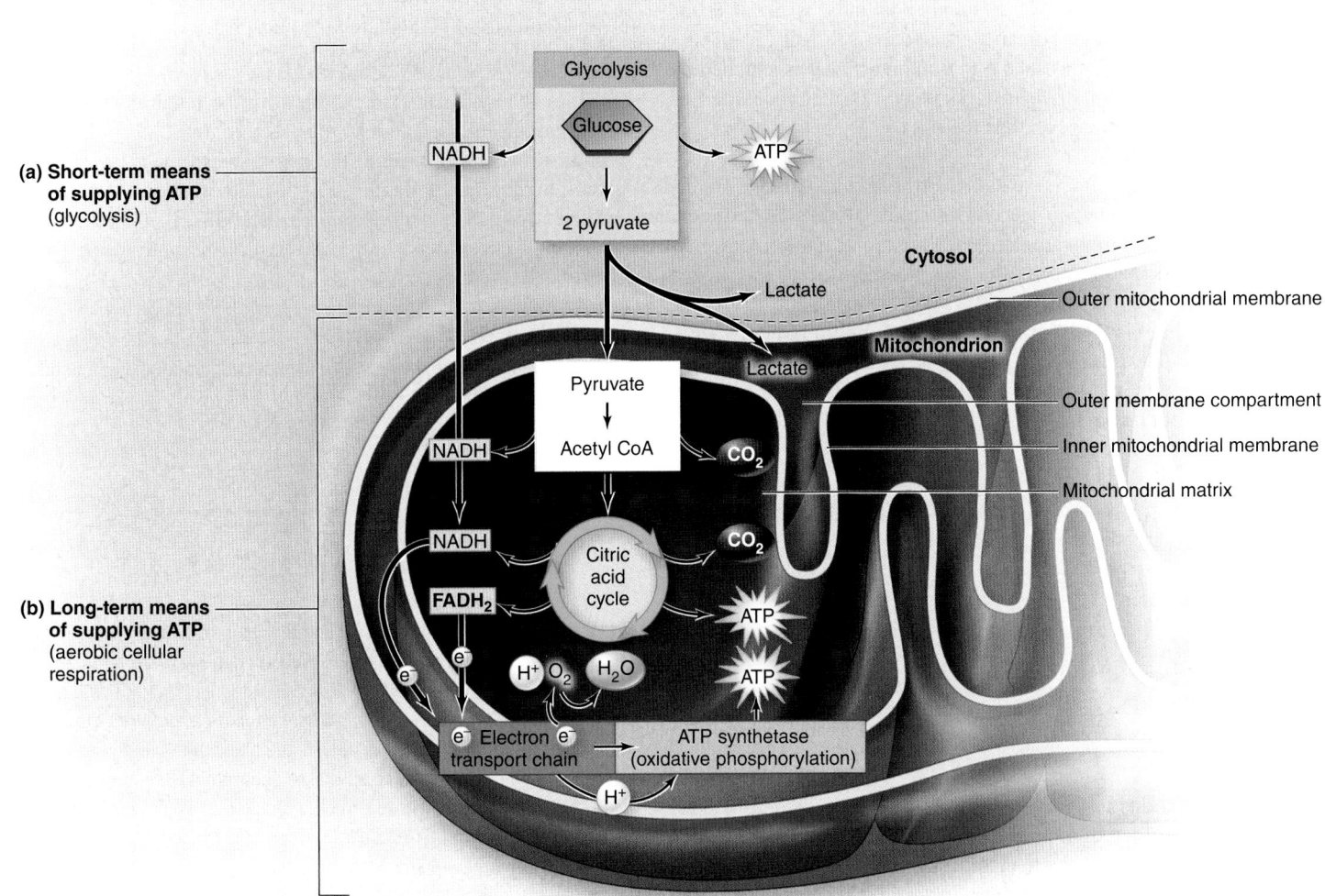

Figure 10.18 Metabolic Processes for Generating ATP. (*a*) Glycolysis, which occurs within the cytosol, is the short-term means for ATP production. (*b*) Aerobic cellular respiration that occurs within mitochondria is the long-terms means for ATP production.

Figure 10.19 **Utilization of Immediate, Short-Term, and Long-Term Energy Sources.** The intensity and duration of an activity are important factors in energy utilization. For short sprints, the available ATP and the ATP made available through phosphate transfer are primarily used, whereas for longer runs, glycolysis is used initially but will be replaced by aerobic cellular respiration.

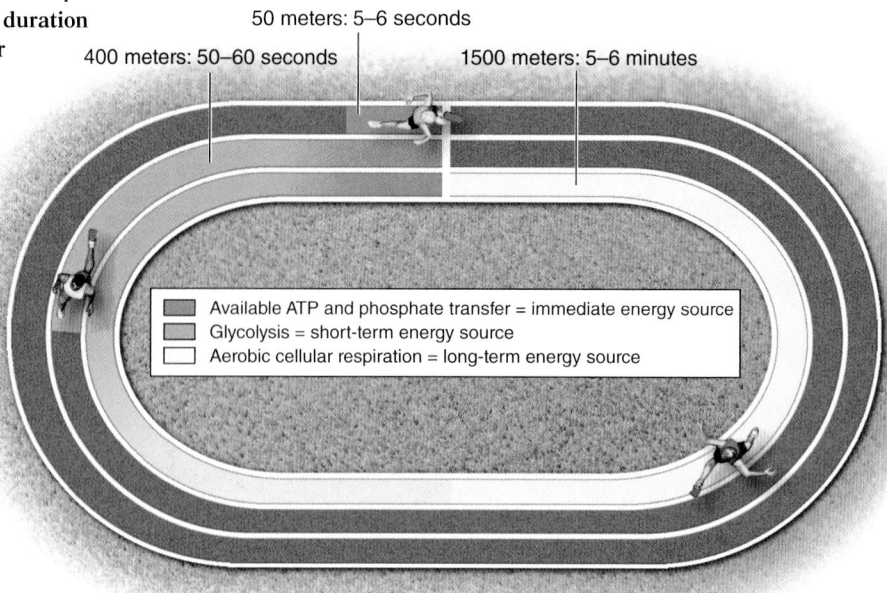

50 meters: 5–6 seconds
400 meters: 50–60 seconds
1500 meters: 5–6 minutes

■ Available ATP and phosphate transfer = immediate energy source
■ Glycolysis = short-term energy source
□ Aerobic cellular respiration = long-term energy source

1500-meter track

produced. Fatty acids are the preferential fuel molecule for generating ATP in most skeletal muscle tissue. However, one of the major drawbacks to the use of fatty acids is that oxygen must be present. Thus, sufficient oxygen must be delivered continuously during sustained exercise for skeletal muscle tissue to use fatty acids to generate ATP.

Energy Supply and Varying Intensity of Exercise

The use of immediate, short-term, and long-term sources for supplying ATP is dependent upon both the intensity and duration of an activity. To illustrate the use of energy, we describe the primary means of supplying ATP for runners at a track meet in which individuals run different distances **(figure 10.19)**.

When an individual participates in a 50-meter sprint, an event that may take 5 to 6 seconds, ATP is supplied primarily by available ATP and phosphate transfer. In a longer sprint of 400 meters, an event that may take 50 to 60 seconds, ATP is supplied initially by the ATP and phosphate transfer and then primarily by glycolysis. Finally, in a 1500-meter run, an event that may take 5 to 6 minutes, ATP is supplied by all three means, but primarily by aerobic processes after about the first minute. Keep in mind, however, that there is overlap between the three different energy sources.

Intense exercise that is sustained longer than approximately 1 minute is dependent upon the body's ability to deliver sufficient oxygen through the cardiovascular and respiratory systems. One consequence of participating in aerobic exercise (defined as a sustained exercise of moderate intensity that involves raising the heart rate above the baseline) is that it produces changes within both the respiratory and cardiovascular systems that enhance oxygen delivery, thus allowing an individual to exercise both at greater levels of intensity and for longer periods of time.

WHAT DID YOU LEARN?

16 Additional ATP is made immediately available in skeletal muscle through which phosphate-containing molecules?

17 What are the various means for making ATP available in a 1500-meter race?

10.4b Oxygen Debt

LEARNING OBJECTIVE

3. Define oxygen debt, and explain why it occurs.

There are limitations to how much oxygen can be supplied to skeletal muscle in a given time period. When an individual participates in exercise during which the demand for oxygen exceeds the availability of oxygen, an oxygen debt is incurred. **Oxygen debt** is the amount of additional oxygen that is consumed following exercise to

restore pre-exercise conditions. This additional oxygen is required primarily to

- Replace oxygen on hemoglobin molecules in the blood and myoglobin molecules in skeletal muscle
- Replenish glycogen stored in skeletal muscle fibers
- Replenish ATP and creatine phosphate
- Convert lactate back to glucose

Additional oxygen is also required by the respiratory muscles that are engaging in forced breathing (see section 23.5b), the heart as it pumps more blood through the body, and to meet the needs of the overall higher metabolic rate. The next time you see someone breathing hard following exercise you will realize that they are "paying off" their oxygen debt, helping to return the body to its state prior to exercise.

WHAT DID YOU LEARN?

18 What is oxygen debt, and how is the additional oxygen used following intense exercise?

INTEGRATE

CONCEPT CONNECTION

What happens to lactate following its formation? Increasing amounts of pyruvate are converted to lactate when insufficient oxygen is delivered to skeletal muscle tissue (see section 3.4g). Lactate can either enter a mitochondrion, where it is converted back to pyruvate and oxidized to carbon dioxide within a mitochondrion, or leave the cell and enter the blood. Lactate that enters the blood can then either be taken up by the heart to be used as fuel to generate ATP (see section 19.3e) or taken up by the liver to produce glucose through gluconeogenesis (see section 27.6d). This cycling of lactate to the liver for conversion to glucose and the subsequent transport of glucose from the liver to the muscle is called the **Cori cycle**.

10.5 Skeletal Muscle Fiber Types

Skeletal muscle fiber types are organized into three primary categories. Here we discuss the criteria that may be used to classify them, and then describe the three primary muscle fiber types.

10.5a Criteria for Classification of Muscle Fiber Types

LEARNING OBJECTIVE

1. Explain the two primary criteria used to classify skeletal muscle fiber types.

Skeletal muscle fibers that compose a muscle are differentiated into three categories based on two criteria: (1) the type of contraction generated, and (2) the primary means used for supplying ATP.

Type of Contraction Generated

Skeletal muscle fibers differ in the power, speed, and duration of the muscle contraction generated. **Power** is related to the diameter of a muscle fiber; large muscle fibers have a larger number of myofibrils in parallel, allowing them to produce a more powerful contraction.

Speed has traditionally been described based on whether the skeletal muscle fiber expresses the relatively slow or fast genetic variant of myosin ATPase, the enzyme that splits ATP. Those with a fast variant are called **fast-twitch fibers,** and those with the slow variant are called **slow-twitch fibers.** However, recent evidence shows that fast-twitch fibers also have both a fast rate of action potential propagation along the sarcolemma and are quick in their Ca^{2+} release and reuptake by the sarcoplasmic reticulum in comparison to slow-twitch fibers. Thus, these fibers initiate a contraction more quickly following stimulation than a slow-twitch fiber (0.01 milliseconds [msec] versus at least 0.02 msec), and produce a contraction of shorter **duration** (7.5 msec versus 100 msec).

Fast-twitch fibers typically have all three characteristics: They produce a strong contraction, initiate a contraction more quickly following stimulation, and produce a contraction of shorter duration. These characteristics account for why fast-twitch fibers exhibit both power and speed in comparison to slow-twitch fibers.

Means for Supplying ATP

The second criterion to differentiate skeletal muscle fibers is whether the primary means the muscle uses to supply ATP is either aerobic cellular respiration or glycolysis. **Oxidative fibers** specialize in providing ATP through aerobic cellular respiration and have several features that support these processes, including an extensive capillary network, large numbers of mitochondria, and a large supply of the red pigment myoglobin. (The presence of both myoglobin and mitochondria gives these fibers a red appearance, and are sometimes called **red fibers.**) The higher levels of ATP generated provide energy for oxidative fibers to continue contracting for extended periods of time without tiring or fatiguing—thus, these fibers are also called **fatigue-resistant.**

In contrast, **glycolytic fibers** specialize in providing ATP through glycolysis. Generally, they have fewer structures needed for aerobic cellular respiration—thus, they have less extensive capillary networks, fewer mitochondria, and smaller amounts of myoglobin. (The relatively small amount of myoglobin and mitochondria is why these fibers have a white appearance and are sometimes called **white fibers.**) However, they do have large glycogen reserves for supplying glucose for glycolysis, which is useful when oxygen stores are low. Glycolytic fibers generally tire easily after a short time of sustained muscular activity—thus, these fibers are also called **fatigable.**

WHAT DID YOU LEARN?

19 Explain how a fast-twitch fiber differs from a slow-twitch fiber, and how an oxidative fiber differs from a glycolytic fiber.

10.5b Classification of Muscle Fiber Types

LEARNING OBJECTIVE

2. Compare and contrast the three muscle fiber types.

Physiologists use both the type of contraction generated and type of energy used to supply ATP to differentiate skeletal muscle fibers into three subtypes **(table 10.1):**

- **Slow oxidative (SO) fibers,** also called *type I fibers,* typically have half the diameter of other skeletal muscle

Table 10.1 Structural and Functional Characteristics of Different Types of Skeletal Muscle Fibers

Fiber Characteristics	Slow Oxidative (SO) Fibers (Type I Fibers)	Fast Oxidative (FO) Fibers (Type IIa Fibers)	Fast Glycolytic (FG) Fibers (Type IIb Fibers)
ATP Use	Slow	Fast	Fast
Capacity to Make ATP	High	Moderate	Limited
Concentration of Capillaries	Extensive	Moderately extensive	Sparse
Color of Fibers	Red	Lighter red	White (pale)
Contraction Velocity	Slow	Fast	Fast
Resistance to Fatigue	Highest	High	Low
Fiber Diameter	Smallest	Intermediate	Largest
Number of Mitochondria	Many	Many	Few
Amount of Myoglobin	Large	Medium	Small
Primary Fiber Function	Endurance (e.g., maintaining posture, marathon running)	Medium duration, moderate movement (e.g., walking, biking)	Short duration, intense movement (e.g., sprinting, lifting weights)
Muscles with a Large Abundance of Fiber Type	Trunk and lower limb muscles	Lower limb muscles	Upper limb muscles

fibers and contain slow myosin ATPase. These cells produce contractions that are slower and less powerful. However, they can contract over long periods of time without fatigue because ATP is supplied primarily through aerobic cellular respiration. These fibers appear dark red because of the presence of large amounts of both myoglobin molecules and mitochondria.

- **Fast oxidative (FO) fibers,** also called *intermediate fibers,* or *type IIa,* are the least numerous of the skeletal muscle fiber types. They are intermediate in size and contain fast myosin ATPase. They produce a fast, powerful contraction with ATP provided primarily through aerobic respiration. However, the vascular supply to fast oxidative fibers is less extensive than the network of capillaries serving SO fibers—thus, the delivery rate of nutrients and oxygen is lower. These fibers also contain myoglobin, but less than the amount found in SO fibers. Consequently, these fibers can be distinguished from SO fibers on a microscopic image because they appear a lighter red than SO fibers.

- **Fast glycolytic (FG) fibers,** also called *fast anaerobic fibers,* or *type IIb,* are the most prevalent skeletal muscle fiber type. They are largest in diameter, contain fast myosin ATPase, and provide both power and speed. However, they can contract for only short bursts because ATP is provided primarily through glycolysis. These fibers appear white because of the relative lack of myoglobin and mitochondria.

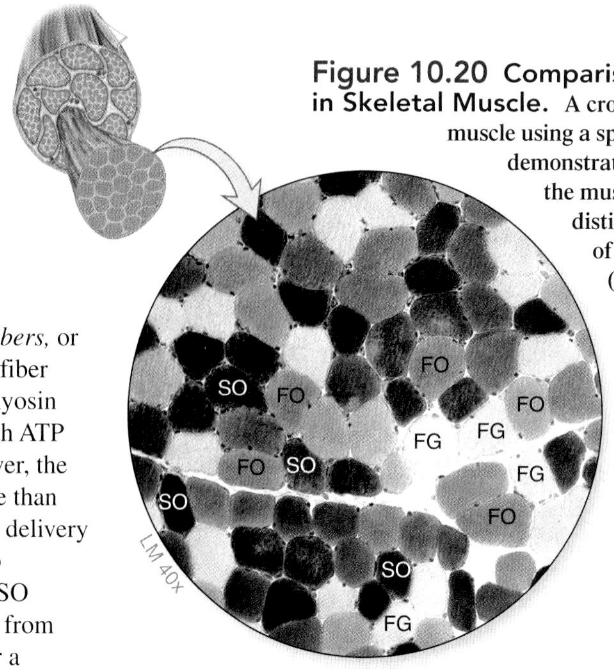

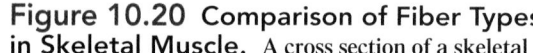

Figure 10.20 Comparison of Fiber Types in Skeletal Muscle. A cross section of a skeletal muscle using a specific staining technique demonstrates the types of fibers in the muscle. The fibers are distinguished by their shade of color. Fast glycolytic (FG) fibers are the lightest; slow oxidative (SO) fibers are the darkest; and fast oxidative (FO) fibers are less dark than the SO fibers.

WHAT DID YOU LEARN?

21 Which muscle fiber type primarily composes muscles that maintain posture?

10.6 Measurement of Skeletal Muscle Tension

Muscle tension is the force generated when a skeletal muscle is stimulated to contract. The term "tension" is used to describe the force that a muscle exerts because a muscle can only pull on a structure. Consider how you place tension on a rubber band when you pull it.

Muscle tension produced in a contracting muscle is measured in several classic laboratory experiments. One variation of these experiments uses the specimen of a gastrocnemius (calf) muscle (see section 11.9c) with an attached sciatic nerve that is excised from a frog. The gastrocnemius muscle is then anchored to an apparatus that produces a *myogram,* a graphic recording of changes in muscle tension when it is stimulated. Here we describe the generation and graphic recording of (1) a muscle twitch; (2) motor unit recruitment; and (3) treppe, wave summation, incomplete tetany, and tetany.

10.6a Muscle Twitch

LEARNING OBJECTIVE

1. Describe what occurs in a muscle during a single twitch, and relate each event to a graph of a twitch.

Electrodes that either are in direct contact with the muscle or the sciatic nerve are used to apply single, brief episodes of stimulation to the muscle and the resulting muscle contraction is recorded using a myograph **(figure 10.21)**. The voltage is increased in increments until the muscle responds, producing a twitch. A **twitch** is defined as a single, brief contraction period and then relaxation period of a skeletal muscle in response to a single stimulation. The minimum voltage needed to stimulate the skeletal muscle to generate a twitch is the **threshold.** The voltage below the threshold is called a *subthreshold stimulus.*

There is a delay called a **latent period** (*lag period*) that occurs after the stimulus is applied and before the contraction of the muscle fiber begins. There is no change in fiber length during the latent

WHAT DID YOU LEARN?

20 Which muscle fiber type is slow and fatigue-resistant? What is the advantage of this muscle fiber type?

10.5c Distribution of Muscle Fiber Types

LEARNING OBJECTIVE

3. Describe the distribution of muscle fiber types in a muscle and how this relates to function.

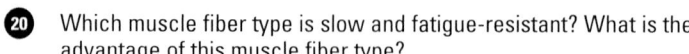

A mixture of muscle fiber types in a typical skeletal muscle is shown in **figure 10.20**. Although most muscles contain a mixture of all three fiber types, the relative percentage of the muscle fiber types varies among different skeletal muscles of the body and reflects the function of the muscle. For example, the muscles of the eye and hand require swift but brief contractions and so they contain a high percentage of FG fibers. In contrast, SO fibers dominate many postural back and calf muscles, which contract almost continually to help us maintain an upright posture.

Variations are also present between individuals, and this is seen most dramatically in high-caliber athletes. Elite distance runners have higher proportions of SO fibers in their lower limb muscles, and top athletes that participate in brief periods of intense activity, such as sprinting or weight lifting, have a higher percentage of FG fibers. These variations in the proportion of the muscle fiber types are determined primarily by a person's genes and less so by the type of training. A proportion of FG fibers may develop the appearance and functional capabilities of FO fibers with physical conditioning if the muscle is used repeatedly for endurance events. Whether this shift actually represents a change of muscle fiber type—or is simply a temporary alteration to the muscle that reverts back when the training ceases—remains controversial.

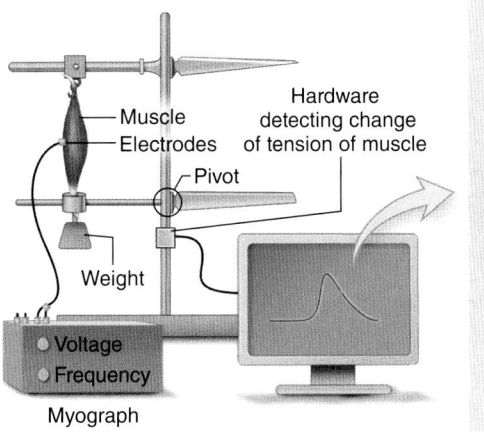
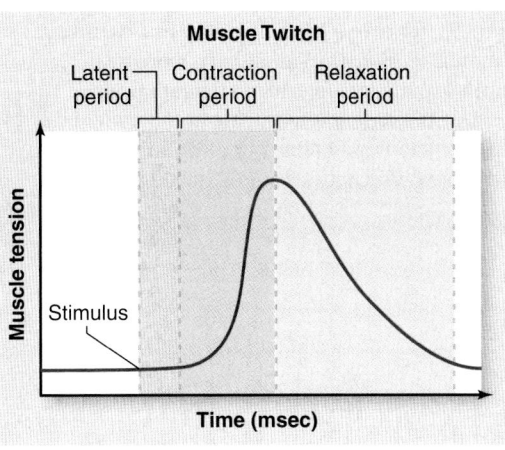

Muscle Twitch

Latent period — Contraction period — Relaxation period

Muscle tension

Stimulus

Time (msec)

Figure 10.21 Muscle Twitch. A myogram of a single brief stimulation of a skeletal muscle results in a single contraction event called a twitch, which is recorded using a myograph. The latent period is the elapsed time between stimulation of the muscle fiber and the generation of contractile force. The contraction period is the time during which there is an increase in muscle tension. The relaxation period is the time when there is a decrease in muscle tension.

period. This delay can be accounted for by the time necessary for the occurrence of all of the events in excitation-contraction coupling, Ca^{2+} release from the sarcoplasmic reticulum into the cytosol, and the beginning of tension generation within the muscle fiber. The **contraction period** begins as repetitive power strokes pull the thin filaments past the thick filaments, shortening the sarcomeres; muscle tension increases during muscle contraction. The **relaxation period** begins with release of crossbridges resulting from return of Ca^{2+} back into the sarcoplasmic reticulum; muscle tension decreases during muscle relaxation. Relaxation depends upon the elasticity of connectin within muscle tissue to return to its original length following shortening of the muscle.

 WHAT DO YOU THINK?

5 Based on skeletal muscle fiber type distribution described earlier, would you predict that the duration of a muscle twitch in an extrinsic eye muscle is shorter or longer than it is in the gastrocnemius muscle? Explain.

WHAT DID YOU LEARN?

22 What events are occurring in a muscle that produce the different components of a muscle twitch (latent period, contraction, and relaxation)?

10.6b Changes in Stimulus Intensity: Motor Unit Recruitment

LEARNING OBJECTIVE

2. Explain the events that occur in motor unit recruitment as the intensity of stimulation is increased.

The gastrocnemius muscle is stimulated repeatedly in a set of experiments to demonstrate motor unit recruitment, and each stimulation event is at a greater voltage. The frequency of stimulation remains the same, and the time between stimulation events is sufficient for the muscle to contract and relax before it is stimulated again. Because motor units vary in their sensitivity to stimulation, each increase in voltage causes a greater number of motor units to contract **(figure 10.22)**. Consequently, the tension generated with each muscle contraction increases until the point of maximum contraction is reached when all motor units have been stimulated. This increase in muscle tension that occurs with an increase in stimulus intensity is called **recruitment,** or **multiple motor unit summation.**

Recruitment helps to account for how our muscles can exhibit both the all-or-none law and exert varying degrees of force. The **all-or-none law** states that if a muscle fiber contracts in response to stimulation,

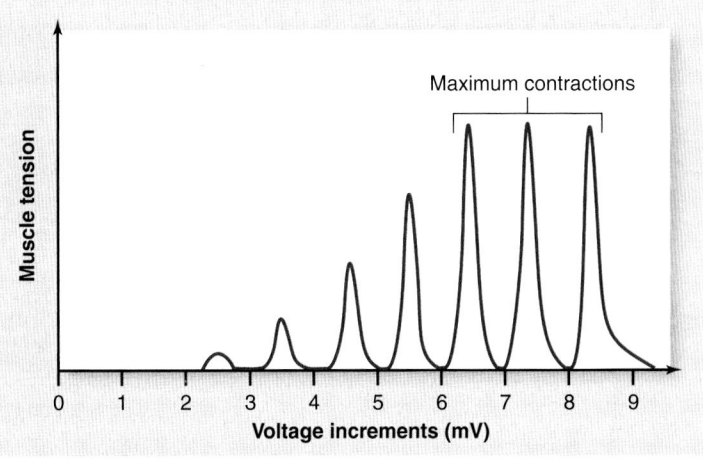

Maximum contractions

Muscle tension

Voltage increments (mV)

Figure 10.22 Skeletal Muscle Response to Change in Stimulus Intensity. Increasing the intensity of stimulation increases the number of motor units involved, resulting in recruitment, or multiple motor unit summation.

it will contract completely (all)—and if the stimulus is not sufficient, it will not contract (none). This means that the muscle fiber contracts maximally or not at all.

The difference in the force and precision of skeletal muscle movement is varied primarily by changing the number of motor units that are activated. If a reduced number of motor units are activated, then fewer muscle fibers contract and less force is exerted. In contrast, if a greater number of motor units are activated, more muscle fibers contract and a greater force is exerted. This permits us to control the contraction of the muscles of our arm to exert the force needed to lift a pencil or to lift a suitcase.

 WHAT DID YOU LEARN?

23 What is recruitment? Explain its importance in the body.

10.6c Changes in Stimulus Frequency: Treppe, Wave Summation, Incomplete Tetany, and Tetany

LEARNING OBJECTIVE

3. Distinguish between treppe, wave summation, incomplete tetany, and tetany that occur with an increase in frequency of stimulation.

A different set of experiments subjects the skeletal muscle to increasing frequency of stimulation, while the voltage remains the same.

Figure 10.23 Skeletal Muscle Response to Change in Stimulus Frequency.

(*a*) A twitch always produces the same amount of muscle tension. (*b*) Treppe is an increase in muscle tension that occurs when not all Ca^{2+} is returned to the sarcoplasmic reticulum prior to the next stimulus and muscle temperature increases. Stimulation is occurring at the same frequency and voltage. (*c*) Wave summation, incomplete tetany, and tetany are seen when the muscle is stimulated at varying frequencies that allow different degrees of relaxation.

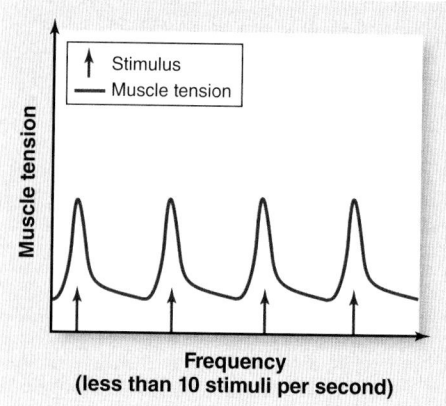

(a) Twitch

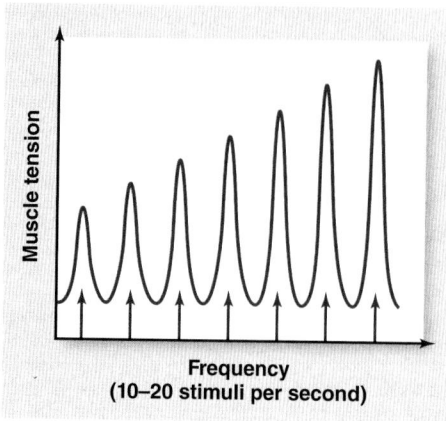

(b) Treppe

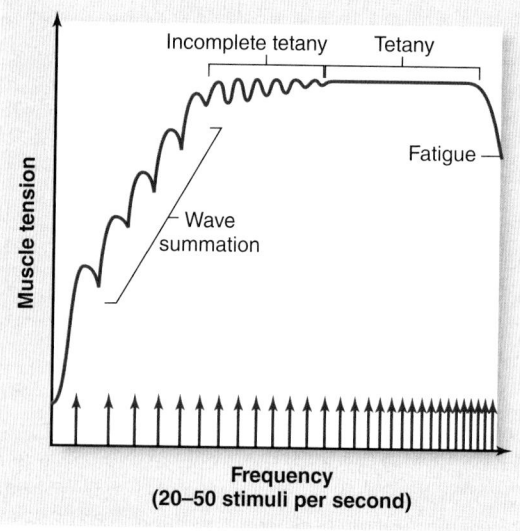

(c) Wave summation, incomplete tetany, and tetany

Note that each of the three graphs in **figure 10.23** has a different frequency of stimulation; each subsequent graph shows a greater frequency than the previous graph.

The first image in the figure illustrates the stimulation frequency of a skeletal muscle that occurs at a relatively slow rate (less than 10 stimuli per second) (figure 10.23*a*). At this rate, each muscle twitch shows that the muscle is contracting and completely relaxing before the next stimulation event. The muscle tension produced in each muscle twitch is the same.

The frequency of stimulation is increased (between 10 and 20 stimuli per second), as shown in figure 10.23*b*. Again, the rate of frequency allows time for muscle contraction and complete relaxation. However, the generated muscle tension increases. This increase is not due to a higher voltage because voltage remains the same. Rather, the increase in muscle tension may be produced because there is insufficient time to remove all of the Ca^{2+} from the sarcoplasm and return it back into the sarcoplasmic reticulum. Consequently, more crossbridges may form upon subsequent stimulation, causing a stronger contraction. The heat generated by muscle contraction also results in more efficient molecular interactions or functions, such as the activity of the enzyme ATPase. The result is a stepwise increase in strength of contraction termed **treppe** (trĕp′e; from German *treppe* = staircase), sometimes referred to as the "warming-up" effect.

Stimulation can occur so rapidly (20 to 50 stimuli per second) that complete relaxation of the muscle does not occur before the next stimulation event (figure 10.23*c*). The restimulated muscle displays a summation of contractile forces as the effect of each new wave is added to the previous wave. This effect is often called either **wave summation** because contraction waves are added together or "summed," or **temporal summation** because it depends upon increasing frequency of stimulation.

Further increases in stimulation frequency allow less time for relaxation between contraction cycles, and now **incomplete tetany** (the tension tracing continues to increase and the distance between waves decreases) is noted. Stimulation frequency is further increased (40 to 50 stimuli per second) until ultimately the contractions of the muscle fiber "fuse" and form a continuous contraction that lacks any relaxation. This continuous contraction is called **tetany** (the tension tracing is a smooth line). If stimulation continues, the muscle reaches **fatigue**—a decrease in muscle tension occurs from repetitive stimulation (muscle fatigue is described in section 10.7d). Changes in muscle stimulation frequency by the nervous system primarily allow skeletal muscle contraction to exert a coordinated action that gradually increases in force.

Nervous stimulation of muscles in the human body usually does not exceed 25 stimuli per second. Therefore, muscle tetany is seen only in laboratory experiments. Sustained contractions in the body occur when you are holding something you do not want to drop because the nervous system stimulates different motor units within the same muscle in an overlapping pattern, so the muscle tension can be maintained for a longer period.

 WHAT DID YOU LEARN?

 What happens during wave summation? Explain its importance in the body.

10.7 Factors Affecting Skeletal Muscle Tension Within the Body

The discussion of muscle tension continues with a description of several factors that influence the action of muscles within the human body, including resting tension in the muscle, the relationship between muscle tension and resistance, how contraction force generated is dependent upon myofilament overlap, and the factors that influence muscle fatigue.

10.7a Muscle Tone

LEARNING OBJECTIVE
1. Describe muscle tone, and explain its significance.

Skeletal muscles do not completely relax even when at rest. **Muscle tone** is the resting tension in a muscle generated by involuntary nervous stimulation of the muscle. Limited numbers of motor units within a muscle are usually stimulated randomly at any given time to maintain a constant tension; the specific motor units being stimulated during rest change continuously so motor units do not become fatigued.

This random contraction of small numbers of motor units causes the muscle to develop tension, called **resting muscle tone.** These random contractions do not generate enough tension to cause movement. The resting muscle tone establishes constant tension on the muscle's tendon, thus stabilizing the position of the bones and joints. Note that muscle tone decreases during deep sleep (sleep associated with rapid eye movement, or REM, sleep).

WHAT DID YOU LEARN?
25 What is the function of skeletal muscle tone?

10.7b Isometric Contractions and Isotonic Contractions

LEARNING OBJECTIVE
2. Distinguish between isometric and isotonic contractions, and give examples of both.

Two primary factors must be considered when describing the consequences of consciously initiating muscle contraction: (1) force generated by the muscle, and (2) resistance (load) that must be overcome. When muscle tension is insufficient to overcome the resistance (i.e., force generated is less than the load), there is no movement of the muscle. This type of muscle contraction is called an **isometric** (ī-sō-met′rik; *iso* = same, *metron* = measure) **contraction.** Thus, the muscle contracts and muscle tension increases, but muscle length stays the same. Some examples of isometric contractions include pushing on a wall (a posture for stretching one's leg muscles), holding a very heavy weight in the gym while your arm does not move, attempting to move a shovel load of snow that is too heavy, or holding a baby in one position (**figure 10.24a**).

When muscle tension results in movement of the muscle, this type of muscle contraction is called an **isotonic** (ī-sō-ton′ik; *tonos* = tension) **contraction.** The tone in the muscle remains the same as the length of muscle changes. Examples of an isotonic contraction include walking, lifting a baby, or swinging a tennis racket. Isotonic contractions are differentiated into two subclasses based on whether the muscle is shortening or lengthening as it contracts (figure 10.24b). The shortening of muscle

INTEGRATE

CLINICAL VIEW
Isometric Contraction and Increase in Blood Pressure

An increase in blood pressure is generally associated with sustained isometric contractions of skeletal muscles. Thus, individuals who already have high blood pressure may experience an additional spike in pressure during intensive exercise, placing them at an increased risk of having a heart attack. This is something for those at risk of having a heart attack to consider before shoveling snow, because both the sustained isometric contractions and the general peripheral vasoconstriction (that results from being in the cold) can raise blood pressure to unhealthy levels.

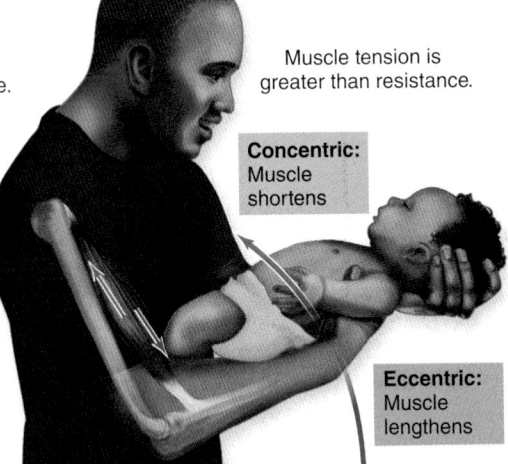

Muscle tension is less than resistance.

Muscle tension is greater than resistance.

Concentric: Muscle shortens

Eccentric: Muscle lengthens

Isometric contraction
Muscle tension is less than the resistance. Although tension is generated, the muscle does not shorten, and no movement occurs.

(a)

Isotonic contraction
Muscle tension is greater than the resistance. The muscle shortens (concentric) or lengthens (eccentric), and movement occurs.

(b)

Figure 10.24 Isometric Versus Isotonic Contraction.

length is a **concentric contraction.** This occurs because the muscle tension is greater than the resistance. It may occur in the biceps brachii (muscle of the anterior arm) when lifting a baby**.** In contrast, lengthening of muscle is an **eccentric contraction.** During an eccentric contraction, the muscle exerts *less* force than that needed to move the load, and the muscle lengthens. If you are holding a 10-pound weight in your hand, and the muscles of your anterior upper arm (e.g., biceps brachii) exert 5 pounds of force, then the biceps muscle lengthens in an eccentric contraction (as your arm extends). This also occurs, for example, in the biceps brachii when placing a baby into a crib.

 WHAT DID YOU LEARN?

26 When you flex your biceps brachii while doing "biceps curls," what is the type of movement?

10.7c Length-Tension Relationship

 LEARNING OBJECTIVE

3. Explain the length-tension relationship in skeletal muscle contraction.

One factor that influences the amount of tension a muscle can generate when stimulated is the amount of overlap of thick and thin filaments when the muscle begins its contraction. This principle is termed the **length-tension relationship.** A muscle generates different amounts of tension dependent upon its length at the time of stimulation. The graphical presentation of the length-tension relationship is called the **length-tension curve (figure 10.25)**.

A skeletal muscle fiber stimulated when it is at a normal resting length generates a maximum contractile force because there is optimal overlap of thick and thin filaments. In contrast, a muscle that is either already contracted, or one that is overly stretched, produces a weaker contraction when stimulated. Weaker contractions in muscles that are already contracted occur because the thick filaments are close to the Z discs, and sliding filaments are limited in their movement. Weaker contractions occur in muscles that are overly stretched because there is minimal thick and thin filament overlap for crossbridge formation. So, for example, you may be able to lift a heavier dumbbell when your elbow is partially flexed than when your elbow is fully extended, because there is minimal overlap of thick and thin filaments during the full extension.

 WHAT DID YOU LEARN?

27 Describe the relative force of contraction that can be developed in your back muscles when you bend at the knees to lift an object and when you bend at the waist to lift an object, based on the length-tension relationship. Explain the significance.

10.7d Muscle Fatigue

 LEARNING OBJECTIVE

4. Define muscle fatigue, and explain some of its causes.

Muscle fatigue is the reduced ability or inability of the muscle to produce muscle tension. The primary cause of muscle fatigue during excessive or sustained exercise (e.g., running a marathon) is caused by a decrease in glycogen stores. However, there are many other causes of muscle fatigue, with the different causes still being debated. Here they are organized by the specific physiologic event of muscle contraction that is affected.

- **Excitation at the neuromuscular junction.** Muscle fatigue may be caused either by insufficient free Ca^{2+} at the neuromuscular junction to enter the synaptic knob or by a decreased number of synaptic vesicles to release neurotransmitter. Both limit the ability of somatic motor neurons to stimulate a muscle.

- **Excitation-contraction coupling.** Muscle fatigue may be due to a change in ion concentration (e.g., Na^+, K^+) that interferes with the ability of the muscle fiber to conduct an action potential along the sarcolemma. This interferes with stimulating release of Ca^{2+} from the sarcoplasmic reticulum.

- **Crossbridge cycling.** Muscle fatigue may result from increased phosphate ion (P_i) concentration. Elevated P_i concentration in the muscle sarcoplasm interferes with P_i release from the myosin head during crossbridge cycling, and this slows the rate

Figure 10.25 Length-Tension Curve.
The tension generated by a muscle is graphically related to its precontraction resting length. (*a*) If the muscle is already contracted at the time of stimulation, it does not have the ability to shorten much more, and it exhibits a weak contraction. (*b*) Muscle at its normal resting length is generally capable of exhibiting the strongest contraction because of optimal overlap of myofilaments. (*c*) If the muscle is very stretched when stimulated, relatively little contraction may occur because myofilaments have minimal overlap.

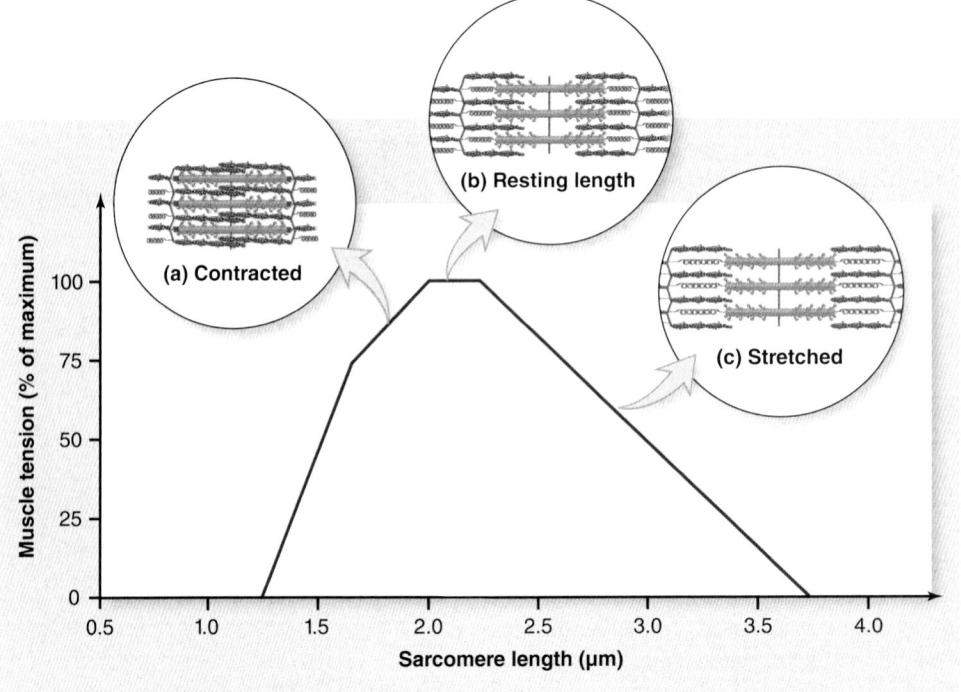

INTEGRATE

CLINICAL VIEW
Muscle Pain Associated with Exercise

Accumulation of lactate produced from glycolysis is commonly (but, incorrectly) thought to cause muscle pain as a result of exercise. Studies have shown, however, that muscle pain is due in part to some minor tearing of the skeletal muscle fibers, which results in a buildup of fluid and inflammation.

of cycling. Muscle fatigue also may occur when lower amounts of Ca^{2+} are available for release from the sarcoplasmic reticulum (part of which is due to its binding with the excess P_i). Lower Ca^{2+} levels results in less Ca^{2+} binding to troponin, producing a weaker muscle contraction. Thus, both an increase in P_i concentration and lower Ca^{2+} levels result in a weaker force generated during muscle contraction.

Lack of ATP is not currently thought to be a primary cause of muscle fatigue. This is because ATP levels are generally maintained through aerobic cellular respiration in mitochondria during sustained exercise. It remains to be determined if ATP may still be a factor because of its location in the cell—that is, within the mitochondria and not in proximity to myofilaments.

 WHAT DID YOU LEARN?

 28 How can muscle fatigue result from changes in each of the three primary events of skeletal muscle contraction?

10.8 Effects of Exercise and Aging on Skeletal Muscle

Skeletal muscle is affected by the process of exercise and aging. Here we describe the effects on our skeletal muscle of both a sustained exercise program and a lack of exercise, along with the changes that occur as we age.

10.8a Effects of Exercise
 LEARNING OBJECTIVE

1. Compare and contrast the changes in skeletal muscle that occur as a result of an exercise program or from the lack of exercise.

Changes in Muscle from a Sustained Exercise Program

The outcome of exercise that involves repetitive stimulation of skeletal muscle fibers depends upon the type of exercise. Endurance (or aerobic) exercise (e.g., running 10 miles) results in skeletal muscle fibers synthesizing cellular structures needed for energy production (e.g., glycolytic enzymes, mitochondria). These changes primarily enhance

ATP production and the ability of the individual to perform for a longer period of time. In comparison, resistance exercise (e.g., lifting heavy weights in the gym) stimulates skeletal muscle fibers to increase contractile proteins (myosin, actin). These changes primarily result in an increase in muscle size, which is called **hypertrophy** (hī-pĕr′trō-fē; *hyper* = above or over, *trophe* = nourishment) (see section 5.6b). Hypertrophy also results from an increase in the number of mitochondria and larger glycogen reserves, which results in an increased ability to produce ATP. An athlete who competes as a bodybuilder or weight lifter exhibits obvious hypertrophied muscular development; however, healthy muscle can be maintained without going to such an extreme. Recent evidence suggests that some (limited) increase in the *number* of muscle fibers also may occur, a process called **hyperplasia.**

Changes in Muscle from Lack of Exercise

Lack of exercise or muscle use results primarily in decreasing the muscle fiber size, a process called **atrophy** (at′rō-fē; *a* = without). This causes a decrease in muscle fiber size, tone, and power, and the muscle becomes flaccid. Even a temporary reduction in muscle use can lead to muscular atrophy. Comparing limb muscles before and after a cast that has been worn for a fracture reveals the loss of muscle tone and size for the casted limb. Individuals who suffer damage to the nervous system or are paralyzed by spinal injuries gradually lose skeletal muscle tone and size in the areas affected. Although muscle atrophy is initially reversible, dead or dying muscle fibers are not replaced. When extreme atrophy occurs, the loss of gross skeletal muscle function is permanent because muscle is replaced with connective tissue including adipose connective tissue. For these reasons, physical therapy is required for patients who suffer temporary loss of mobility.

 WHAT DID YOU LEARN?

 29 What anatomic changes occur in a skeletal muscle fiber when it undergoes hypertrophy?

10.8b Effects of Aging
 LEARNING OBJECTIVE

2. Summarize the effects of aging on skeletal muscle.

A slow, progressive loss of skeletal muscle mass typically begins in a person's mid-30s as a direct result of decreased activity. The size and power of skeletal muscles also decrease. Often, the lost muscle mass is replaced by either adipose or fibrous connective tissue. There is a loss both in fiber number and fiber diameter. A decrease in fiber diameter is due to a reduction in both the size and number of myofibrils and the number of myofilaments. Oxygen storage capacity decreases because there is less myoglobin, glycogen reserves are reduced, and a decreased ability to produce ATP also occurs with aging. Overall, muscle strength and endurance are impaired, and the individual has a tendency to fatigue more quickly. Decreased cardiovascular performance often accompanies aging; thus, the blood supply to active skeletal muscles is much less in elderly people when they exercise.

Skeletal tissue has a reduced capacity to recover from disease or injury as a person grows older. The number of satellite cells in skeletal muscle steadily decreases, and scar tissue often forms due to diminished repair capabilities. Finally, the elasticity of skeletal muscle also decreases. Muscle mass is often replaced by dense regular (fibrous) connective tissue, a process called **fibrosis** (fī-brō′sis). The increasing amounts of this connective tissue decrease the flexibility of muscle; an increase in collagen fibers can restrict movement and circulation.

INTEGRATE

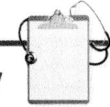

CLINICAL VIEW
Anabolic Steroids as Performance-Enhancing Compounds

Anabolic (an-ă-bol'ik) **steroids** are synthetic substances that mimic the actions of natural testosterone. Recall from section 3.2b that the term "anabolism" refers to the synthesis of complex molecules (e.g., protein) from simple molecules (e.g., amino acids). To date, over 100 compounds have been developed with anabolic properties, but they all require a prescription for legal use in the United States. Anabolic steroids have only a few accepted medical uses—among them, the treatment of delayed puberty, certain types of impotence, and the wasting condition associated with HIV infection and other diseases. Because anabolic steroids stimulate the manufacture of muscle proteins, these compounds have become popular with some athletes as performance enhancers.

Relatively large doses of anabolic steroids are needed to stimulate excessive muscle development that results in extra strength and speed. But this enhanced muscle power and speed comes at a price. Medical professionals have reported many devastating side effects associated with extended anabolic steroid use, including increased risk of heart disease and stroke, kidney damage, liver tumors, testicular atrophy and a reduced sperm count, gynecomastia (development of fatty breast tissue), acne, high blood pressure, aggressive behavior, and personality aberrations. Because anabolic steroids mimic the effects of testosterone, female athletes who use them often experience menstrual irregularities, growth of facial hair, and in extreme circumstances atrophy of the uterus and mammary glands; sterility has even been reported. Adding to these problems is the route of administration. Because many of these steroid preparations must be injected, the improper use or sharing of needles raises the possibility of transferring pathogens that cause disease (e.g., AIDS or hepatitis). For all of these reasons, the use of anabolic steroids as performance enhancers has been widely banned.

All of us will eventually experience a decline in muscular performance, regardless of our lifestyle or exercise patterns. But we can improve our chances of being in good shape later in life by striving for physical fitness throughout life.

WHAT DID YOU LEARN?

30 What changes in skeletal muscle accompany aging?

heartbeat. The autonomic nervous system (a division of the nervous system that controls cardiac muscle, smooth muscle, and glands involuntarily, discussed in chapter 15) controls the rate and force of contraction of cardiac muscle.

WHAT DID YOU LEARN?

31 What are three anatomic or physiologic differences between skeletal muscle and cardiac muscle?

10.9 Cardiac Muscle Tissue

 LEARNING OBJECTIVE

1. List and describe the similarities and differences between skeletal muscle and cardiac muscle.

In addition to skeletal muscle, two other types of muscle occur in the body: cardiac muscle and smooth muscle (see section 5.3). Here we briefly review the characteristics of cardiac muscle (for a detailed description of cardiac muscle, see section 19.3e).

Cardiac muscle cells are individual muscle cells arranged in thick bundles within the heart wall (**figure 10.26**). Cardiac muscle cells branch and are both shorter and thicker than skeletal muscle fibers. (They typically have a diameter of about 15 μm and range in length from 50 to 100 μm.) Individual cells are joined to adjacent muscle cells at junctions termed intercalated discs. **Intercalated** (in-ter'kă-lā'ted) **discs** are unique to cardiac muscle; they are composed of desmosomes and gap junctions (see section 4.6d). These cells have only one or two nuclei. Cardiac muscle cells are striated because, like skeletal muscle fibers, cardiac muscle cells also contain sarcomeres. Cardiac muscle cells contain a large number of mitochondria, and use aerobic respiration almost exclusively to generate the ATP required for their unceasing work.

Cardiac muscle cells are stimulated by a specialized **autorhythmic** pacemaker. This feature is responsible for the repetitious, rhythmic

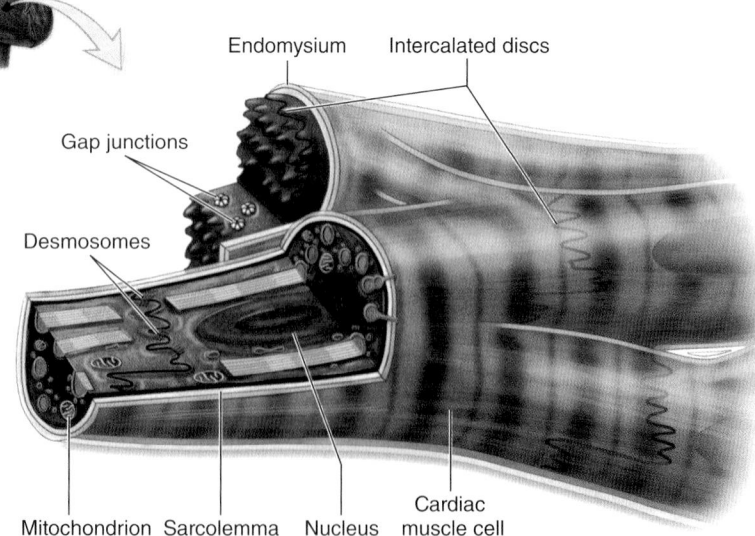

Figure 10.26 Cardiac Muscle. Cardiac muscle is found only in the heart walls. Cardiac muscle cells branch and are connected by intercalated discs. **AP|R**

10.10 Smooth Muscle Tissue

Smooth muscle tissue is located throughout the body typically composing approximately 2% of the weight of an adult. Here we describe the general features of smooth muscle, including locations, microscopic anatomy, mechanism of contraction, how it is controlled, and its functional categories.

10.10a Location of Smooth Muscle

☑ LEARNING OBJECTIVE

1. Identify organs of various body systems where smooth muscle is located.

Smooth muscle tissue is found in the walls of organs of many different body systems. Smooth muscle function is determined by its location. For example (figure 10.27):

- **Cardiovascular system:** Blood vessels regulate blood pressure and the distribution of blood flow.
- **Respiratory system:** Bronchioles (air passageways) control the resistance to airflow that enters and exits the air sacs (alveoli) of the lungs.
- **Digestive system:** The stomach, small intestine, and large intestine mix and propel ingested material as it is moved along the gastrointestinal tract.
- **Urinary system:** Ureters propel urine from the kidney to the urinary bladder, which eliminates urine from the body.
- **Female reproductive system:** The uterus helps expel the baby during delivery.

These are but a few examples of smooth muscle distribution in our bodies. Smooth muscle also composes other specialized structures, including the iris of the eye to control the amount of light entering the eye, the ciliary body of the eye for focusing on an object (see section 16.4b), and the arrector pili muscles that produce "goose bumps" (see section 6.2b).

Smooth muscle resembles both skeletal muscle and cardiac muscle because it has cells that can increase in size (hypertrophy). Both skeletal muscle and cardiac muscle, however, have limited ability to increase in number through mitosis (hyperplasia), whereas smooth muscle retains this ability. The smooth muscle in the wall of the uterus exemplifies this characteristic as it increases in thickness during pregnancy from a combination of both hypertrophy and hyperplasia (see section 29.5c). There is a significant advantage for any tissue that retains mitotic ability because following an injury the damaged tissue is replaced with the original tissue (not scar tissue) (see section 4.9). Thus, as you consider the various locations of smooth muscle in the body, keep in mind that if smooth muscle is damaged, it is typically replaced with new smooth muscle tissue and has the potential to continue functioning as it did before.

💡 WHAT DID YOU LEARN?

 Where is smooth muscle located in the human body?

10.10b Microscopic Anatomy of Smooth Muscle

☑ LEARNING OBJECTIVE

2. Compare the microscopic anatomy of smooth muscle to skeletal muscle.

Smooth muscle cells are small and fusiform shaped (widest in the middle and tapered on the ends) with a centrally located nucleus (figure 10.28). They typically have a diameter of 5 to 10 micrometers and a length of 50 to 200 micrometers. Thus, their diameter is

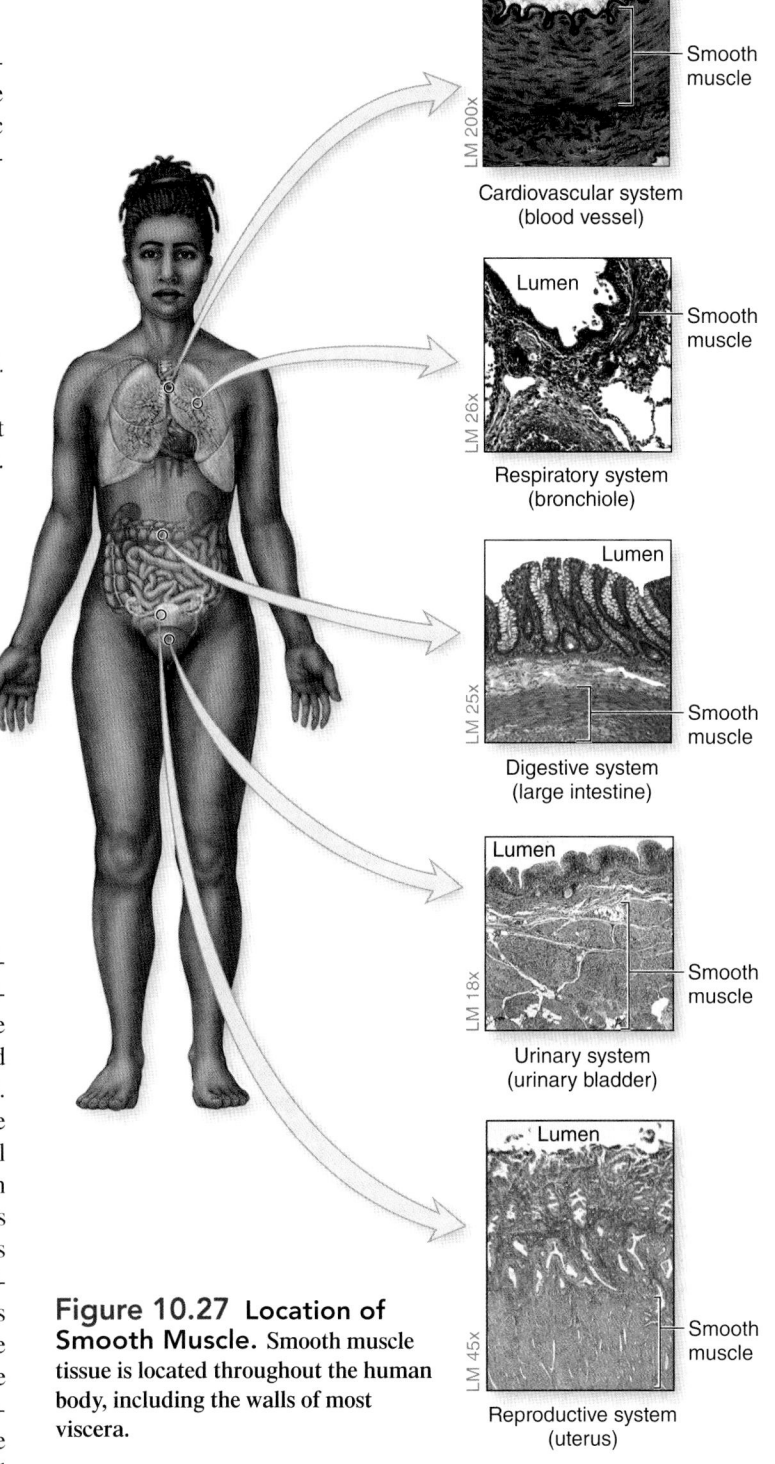

Figure 10.27 Location of Smooth Muscle. Smooth muscle tissue is located throughout the human body, including the walls of most viscera.

up to ten times smaller and their length thousands of times shorter than a skeletal muscle fiber. An endomysium wraps around each smooth muscle cell. The small tapered ends of the cells overlap the larger middle area of adjacent cells to provide for close packing of cells.

The sarcolemma contains various types of Ca^{2+} channels (e.g., voltage-gated, chemically gated, modality gated) that allow these cells to respond to different types of stimuli. Transverse tubules are absent in smooth muscle cells. Instead, the sarcolemmal surface area is increased by "flasklike" invaginations called *caveolae* (kav'ē-ŏ-lē; pl. *caveola;* sing., small pocket). The sarcoplasmic reticulum is sparse and located close to the sarcolemma with some caveolae in contact with it. The source of Ca^{2+} comes both from the interstitial fluid outside the cell and the sarcoplasmic reticulum.

Figure 10.28 Microscopic Anatomy of Smooth Muscle. (*a*) Smooth muscle cells are fusiform cells that overlap to provide close packing of cells. (*b*) Relaxed smooth muscle cells are elongated in the fusiform shape. (*c*) Contracted smooth muscle cells are compact due to the tension exerted by the contractile proteins on the intermediate filaments that pull on both dense bodies within the sarcoplasm and dense plaques within the plasma membrane. AP|R

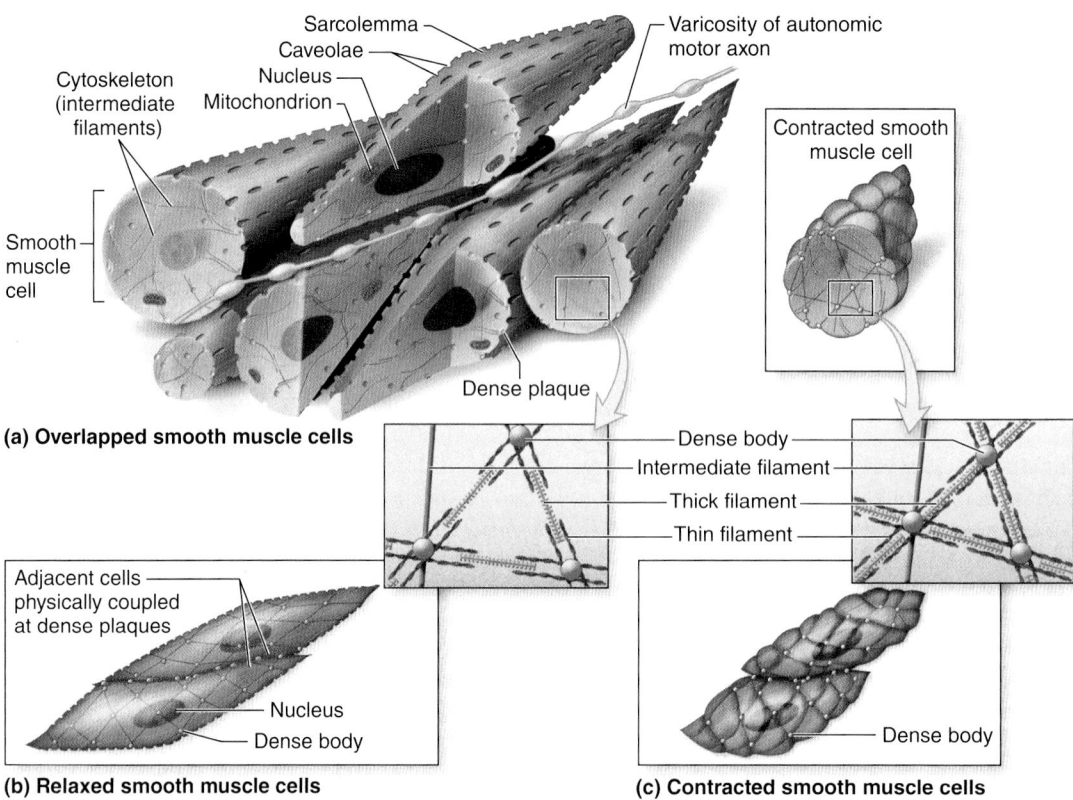

(a) Overlapped smooth muscle cells

(b) Relaxed smooth muscle cells

(c) Contracted smooth muscle cells

Arrangement of Anchoring Proteins and Contractile Proteins of Smooth Muscle

Smooth muscle contains a unique arrangement of anchoring protein structures that includes the cytoskeleton, dense bodies, and dense plaques. The cytoskeletal network is composed of an extensive array of **intermediate filaments** (see section 4.6b). The intermediate filaments are linked with **dense bodies** at points where they interact within the sarcoplasm of the smooth muscle cell, whereas intermediate filaments are linked by **dense plaques** at points where they attach on the inner surface of the sarcolemma. Thus, intermediate filaments extend across the cell with dense bodies as "spot welds" that anchor intermediate filaments to each other and dense plaques that anchor the intermediate filaments to the plasma membrane.

The contractile proteins in smooth muscle are arranged between dense bodies and dense plaques rather than in sarcomeres as they are in both skeletal and cardiac muscle. The Z discs that anchor the sarcomere on either end in skeletal muscle fibers are also absent. Lack of sarcomeres and Z discs contributes to the absence of striations, giving these muscle cells their "smooth" appearance.

The contractile proteins are oriented at oblique angles to the longitudinal axis of the smooth muscle cell and appear to spiral. Consequently, contraction (as described in the next section) results in a twisting of the smooth muscle, which is similar to a corkscrew (figure 10.28c).

Comparison of Myofilaments of Smooth Muscle and Skeletal Muscle

Thick filaments in smooth muscle have myosin heads along their entire length, rather than only at the ends as in skeletal muscle. The more numerous heads can form additional crossbridges with actin to produce a powerful muscle contraction. Additionally, these myosin heads have modifications that allow them to "latch on" to the actin of thin filaments and remain attached without using additional ATP. This mechanism is called the **latchbridge mechanism.**

Smooth muscle thin filaments are composed of actin and tropomyosin, but they do not contain troponin molecules as in skeletal

muscle fibers. Instead, two other proteins are required for initiation of smooth muscle contraction: (1) calmodulin, a protein that binds Ca^{2+} to form a **Ca^{2+}-calmodulin complex;** and (2) **myosin light-chain kinase (MLCK),** an enzyme that is activated by the Ca^{2+}-calmodulin complex to phosphorylate the smooth muscle myosin head. The phosphorylation of the smooth muscle myosin head causes activation of its ATPase activity.

A third protein called **myosin light-chain phosphatase** is an enzyme that dephosphorylates the myosin head, resulting in the inactivation of the ATPase activity. This inactivation is required for relaxation of smooth muscle.

WHAT DID YOU LEARN?

33 How are anchoring proteins and contractile proteins in smooth muscle cells arranged?

34 What is the specific role of the following structures within smooth muscle cells: calmodulin, myosin light-chain kinase, and myosin light-chain phosphatase?

10.10c Smooth Muscle Contraction

LEARNING OBJECTIVE

3. Explain the sequence of steps in smooth muscle contraction.

Smooth muscle contraction resembles skeletal muscle contraction in that it (1) is initiated by Ca^{2+}, (2) involves the sliding of thin filaments past thick filaments, and (3) requires ATP. There are significant differences, however, as shown and described in **figure 10.29**.

In response to stimulation, Ca^{2+} enters the sarcoplasm from both the interstitial fluid and sarcoplasmic reticulum. It binds to calmodulin to form a Ca^{2+}-calmodulin complex that then binds to MLCK, resulting in its activation. The activated kinase (MLCK) phosphorylates the myosin head to both activate the myosin ATPase activity of the myosin head and to allow the myosin head to bind to actin forming a crossbridge. Myosin ATPase hydrolyzes ATP to produce the power stroke. The myosin head releases and reattaches to the actin

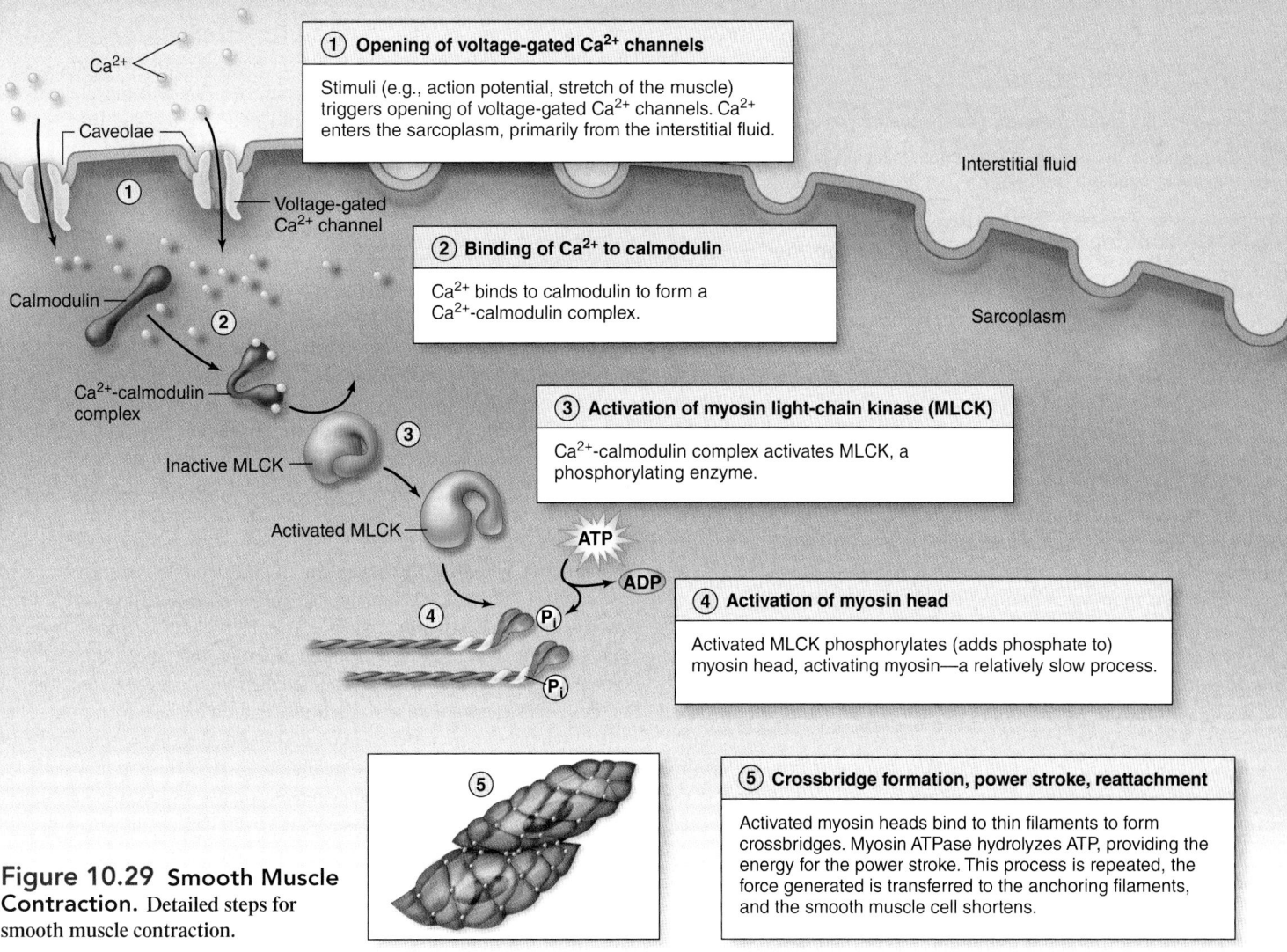

1 Opening of voltage-gated Ca²⁺ channels

Stimuli (e.g., action potential, stretch of the muscle) triggers opening of voltage-gated Ca^{2+} channels. Ca^{2+} enters the sarcoplasm, primarily from the interstitial fluid.

2 Binding of Ca²⁺ to calmodulin

Ca^{2+} binds to calmodulin to form a Ca^{2+}-calmodulin complex.

3 Activation of myosin light-chain kinase (MLCK)

Ca^{2+}-calmodulin complex activates MLCK, a phosphorylating enzyme.

4 Activation of myosin head

Activated MLCK phosphorylates (adds phosphate to) myosin head, activating myosin—a relatively slow process.

5 Crossbridge formation, power stroke, reattachment

Activated myosin heads bind to thin filaments to form crossbridges. Myosin ATPase hydrolyzes ATP, providing the energy for the power stroke. This process is repeated, the force generated is transferred to the anchoring filaments, and the smooth muscle cell shortens.

Figure 10.29 Smooth Muscle Contraction. Detailed steps for smooth muscle contraction.

repetitively, causing the thin filament to slide past the thick filament. This sliding results in a pull on the attached dense bodies anchored to the intermediate filaments of the cytoskeleton and the dense plaques attached to the sarcolemma. The anchoring filaments move inward and the entire smooth muscle cell shortens.

Relaxation of smooth muscle is more complex than relaxation of skeletal muscle. It requires, in addition to both cessation of stimulation and the removal of Ca^{2+} from the sarcoplasm, the dephosphorylation of myosin by myosin light-chain phosphatase. Note that smooth muscle may remain in the contracted state following the removal of Ca^{2+} and dephosphorylation of the myosin head due to the special latchbridge mechanism that keeps the myosin attached to the thin filament.

Characteristics of Smooth Muscle Contraction

Smooth muscle contraction exhibits three characteristics that allow it to effectively fulfill its functions.

Initiation and Duration of Contraction Smooth muscle contraction is usually slow to develop with maximum tension at about 500 milliseconds after stimulation. The relatively long latent period is due primarily to both the requirement for phosphorylating the myosin head by MLCK enzymes and variations in the speed of myosin ATPase activity. The duration of contraction typically extends over a 1- to 2-second period due to the slowness of Ca^{2+} pumps in removing Ca^{2+} from the sarcoplasm, the requirement for dephosphorylation of the myosin head by phosphatase, and the possibility of myosin locking to actin (latchbridge mechanism). Contraction of smooth muscle does not generally require a rapid onset, but it does require the ability to remain in the contracted

state for extended periods of time. This characteristic is important because smooth muscle must maintain continuous tone (tonic contraction) in visceral walls, such as the gastrointestinal tract and blood vessels.

Fatigue-Resistant The energy requirements for smooth muscle contraction are relatively low in comparison to skeletal muscle, and ATP is generally supplied through aerobic cellular respiration. The latchbridge mechanism provides the means of maintaining muscle contraction without use of additional ATP. Consequently, smooth muscle may contract for extended periods of time without becoming fatigued—obviously a necessary requirement for maintaining the tonic contractions just described.

Length-Tension Curve Smooth muscle exhibits a broader length-tension curve than skeletal muscle. Recall that the force of muscle contraction generated by skeletal muscle is dependent upon its muscle length at the time of stimulation. It shows maximum force at its optimal resting length, but a decreased force of contraction if it is either shortened or lengthened (see figure 10.25). These limitations are due to the Z discs that prevent additional shortening and lack of myosin heads in the center of thick filaments, respectively. Smooth muscle has neither of these limitations—thus, it can contract forcefully when compressed to approximately half its resting length or stretched to twice its resting length. Consider, for example, that the storage of urine in the urinary bladder results in stretching of the urinary bladder wall (see section 24.8c). The greater the amount of urine, the greater the amount of stretch of the smooth muscle in the urinary bladder wall. The ability of smooth muscle to contract forcefully at varying degrees of stretch allows us to easily empty our bladder regardless of the amount of urine it is holding.

35 What are the steps of smooth muscle contraction?

36 What unique characteristics of smooth muscle allow it to fulfill its functions? Explain.

10.10d Controlling Smooth Muscle

LEARNING OBJECTIVE

4. Briefly explain the different means of controlling smooth muscle.

We cannot voluntarily control the contraction of the smooth muscle in the wall of the digestive tract, as we find when our stomach "growls" at an inappropriate time. Smooth muscle, like cardiac muscle, is controlled by the autonomic nervous system. The response of smooth muscle to stimulation by the nervous system—that is, whether it contracts or relaxes—is dependent upon the specific neurotransmitter that is released and the receptors to which the neurotransmitter binds. Smooth muscle within the walls of bronchioles, for example, contracts in response to the release of ACh, and relaxes in response to norepinephrine.

Smooth muscle also contracts in response to being stretched. This physiologic response is called the **myogenic** (mī′o-jen′ik; *genesis* = origin) **response.** The myogenic response occurs, for example, in smooth muscle in the walls of blood vessel, stomach, and urinary bladder. Its response, however, is not continuous if the stretch is prolonged. Instead, the smooth muscle exhibits what is called the **stress-relaxation response.** This occurs when smooth muscle is "stressed" by being stretched. It responds by contracting, but after a given period of time, it relaxes. For example, swallowed materials entering the stomach cause its wall to stretch, and the smooth muscle in the wall initially contracts. After a period of time it relaxes, allowing additional food to more easily enter the stomach.

Smooth muscle is also stimulated to contract by various hormones, a decrease in pH, low oxygen concentration, increased carbon dioxide levels, certain drugs, and pacemaker cells. For example, the hormone oxytocin causes contraction of smooth muscle in the uterus to expel the baby at childbirth (see section 29.6c). A pacemaker (similar to the pacemaker in the heart) stimulates smooth muscle in the walls of the stomach and small intestine to contract rhythmically to mix and propel the contents through the lumen of these organs (see section 26.2d).

WHAT DID YOU LEARN?

37 What are the various forms of stimulation for controlling smooth muscle?

38 Explain the stress-relaxation response of smooth muscle.

10.10e Functional Categories

LEARNING OBJECTIVES

5. Explain the primary functional difference between multiunit and single-unit smooth muscle.

6. Compare the location and regulation of both multiunit and single-unit smooth muscle.

Smooth muscle is classified into two broad groups based upon whether the smooth muscle fibers are either stimulated to contract independently or as one unit. Multiunit smooth muscle cells receive stimulation to contract individually, whereas single-unit smooth muscle cells are stimulated to contract in unison (syncytium) **(figure 10.30).**

Multiunit smooth muscle is found within the eye in both the iris and ciliary muscles, composing the arrector pili

muscles in the skin, the wall of larger air passageways within the respiratory system, and the walls of larger arteries. Smooth muscle cells in these body structures are arranged into motor units, and they have a neuromuscular junction. These two features are similar to skeletal muscle, except that the motor neuron here is a component of the autonomic nervous system. The degree of contraction of this smooth muscle is dependent upon the number of motor units activated, thus facilitating increasing degrees of tension as more motor units are stimulated.

Most smooth muscle belongs to the category of **single-unit smooth muscle.** The smooth muscle cells of single unit muscle typically form two or three sheets. These sheets of smooth muscle are within the walls of the digestive, urinary, and reproductive tracts, as well as smaller portions of the respiratory tract, and most blood vessels. These large sheets of smooth muscle are functionally linked by gap junctions between cells. The smooth muscle cells of single-unit smooth muscle are also called **visceral smooth muscle** because they occur in the walls of most viscera (internal organs).

Nerve stimulation of single-unit smooth muscle occurs through numerous swellings of the autonomic motor neurons that pass in close proximity to several smooth muscle cells; these swellings are called **varicosities** (figure 10.30b). Synaptic vesicles within the varicosities contain one type of neurotransmitter (e.g., ACh or norepinephrine). Receptors in smooth muscle cells are scattered diffusely across the sarcolemma of these cells. This contrasts to their distribution in skeletal muscle cells, where receptors are clustered in a motor end plate. This scattered and loose arrangement of receptors in single-unit smooth muscle is called a *diffuse junction.* Neurotransmitter released from varicosities stimulates numerous smooth muscle cells simultaneously. This occurs much like the extensions of a sprayer releasing water onto a lawn or garden. The stimulation may subsequently be spread from cell to cell via gap junctions, and smooth muscle cells contract synchronously as one unit.

Significant features of skeletal muscle, cardiac muscle, and smooth muscle are compare in **table 10.2.**

Figure 10.30 Multiunit and Single-Unit Smooth Muscle.
(a) Smooth muscle fibers of multiunit smooth muscle are stimulated to contract independently. Anatomic arrangement is in motor units similar to skeletal muscle fibers, except the motor neuron is part of the autonomic nervous system. (b) Single-unit smooth muscle cells are stimulated to contract as a group. This is facilitated by both numerous varicosities of autonomic motor neurons that stimulate smooth muscle and gap junctions located between cells that spread the stimulation.

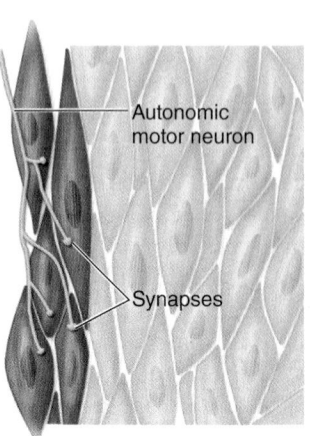

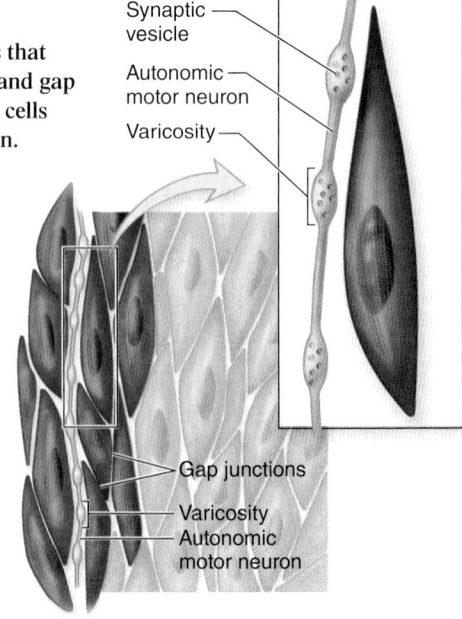

(a) Multiunit smooth muscle   (b) Single-unit smooth muscle

| Table 10.2 | Muscle Tissue Types: *General Comparisons* | | |

Muscle Tissue	Skeletal Muscle	Cardiac Muscle	Smooth Muscle
Location	Generally attached to the skeleton (or to subcutaneous tissue); voluntary movement of body components	Heart only; pumps blood through blood vessels	Walls of hollow organs (e.g., intestines, blood vessels); also located in iris and ciliary body of eye, and arrector pili of the integument; involuntary movement of body components
Associated Connective Tissue	Epimysium, perimysium, endomysium — Epimysium — Perimysium — Endomysium	Endomysium only — Endomysium — Cardiac muscle cell	Endomysium only — Endomysium — Smooth muscle cells
Cell Appearance and Shape	Long cylindrical fibers; with multiple peripheral nuclei; striated; T-tubules *Diameter:* Large (10–500 μm) *Length:* Long (100 μm–30 cm) LM 500x	Medium-sized branching cells with one or two centrally located nuclei; striated; T-tubules; intercalated discs *Diameter:* Medium (about 15 μm) *Length:* Short (50–100 μm) LM 200x	Small, overlapping fusiform cells with a single centrally located nucleus; nonstriated; caveolae *Diameter:* Small (5–10 μm) *Length:* Short (50–200 μm) LM 160x
Calcium Source	Well-developed sarcoplasmic reticulum	Sarcoplasmic reticulum not as well developed as in skeletal muscle; most Ca^{2+} from interstitial fluid	Sarcoplasmic reticulum not well developed; most Ca^{2+} from interstitial fluid
Contractile Unit and Ca^{2+} Binding	Sarcomere; Ca^{2+} binds to troponin	Sarcomere; Ca^{2+} binds to troponin	No sarcomeres; calcium binds calmodulin, not troponin
Stimulation	Nervous control is voluntary (by somatic nervous system); excitatory	Autorhythmic due to pacemaker within heart; spread by gap junctions; nervous control is involuntary (by the autonomic nervous system); excitatory or inhibitory	Multiunit smooth muscle: Regulated by autonomic nervous system; excitatory or inhibitory; no gap junctions Single-unit smooth muscle: Autonomic nervous stimulation through varicosities spread by gap junctions; other stimuli (e.g., stretch, pH)
Response and Primary Energy Source	Slow oxidative (SO): Slow; aerobic production of ATP Fast oxidative (FO): Rapid and powerful; aerobic production of ATP Fast glycolytic (FG): Rapid and powerful; glycolytic production of ATP	Slow; aerobic production of ATP	Very slow and long duration; aerobic production of ATP

 WHAT DID YOU LEARN?

39 Explain why smooth muscle of the eye is multiunit smooth muscle and the smooth muscle in the wall of digestive organs is single-unit smooth muscle.

- Muscle tissue moves the skeleton and materials within and throughout the body.

10.1 Introduction to Skeletal Muscle 332

- Skeletal muscles have a number of functions and exhibit certain properties.

10.1a Functions of Skeletal Muscle 332
- Skeletal muscles produce body movement, maintain posture, protect and support body structures, move and eliminate materials, and produce heat to help maintain body temperature.

10.1b Characteristics of Skeletal Muscle Tissue 332
- Skeletal muscle tissue exhibits excitability, conductivity, contractility, elasticity, and extensibility.

10.2 Anatomy of Skeletal Muscle 333

- Individual skeletal muscles may extend the entire length of a muscle and are called muscle fibers.

10.2a Gross Anatomy of Skeletal Muscle 333
- Skeletal muscle is ensheathed by three connective tissue layers: an outer epimysium, a middle perimysium, and an inner endomysium.
- Tendons or aponeuroses are extensions of these three layers of connective tissue that attach muscle ends to other structures.
- Skeletal muscle is highly vascularized, and it is innervated by motor neurons that exert voluntary control of the muscle.

10.2b Microscopic Anatomy of Skeletal Muscle 334
- A skeletal muscle fiber is a multinucleated cell.
- The sarcolemma, T-tubules, and sarcoplasmic reticulum have specialized membrane pumps and channels that participate in muscle excitability, conductivity, and initiation of muscle contraction.
- Skeletal muscle fibers are filled with myofibrils that house thick and thin protein myofilaments, which are composed of myosin and actin protein, respectively.
- Myofilaments are arranged in repeating, functional units called sarcomeres.
- Other specialized structural and functional proteins of muscle tissue fibers include connectins (titin) and dystrophin.
- Numerous mitochondria, glycogen stores, myoglobin, and creatine phosphate all function to help meet the high-energy demands of skeletal muscle tissue.

10.2c Innervation of Skeletal Muscle Fibers 338
- A motor unit consists of a motor neuron and all the muscle fibers it innervates and controls.
- The neuromuscular junction is the location where a motor neuron innervates a muscle fiber.

10.2d Skeletal Muscle Fibers at Rest 339
- At rest, skeletal muscle fibers have an RMP of –90 mV with more Na^+ outside the cell and more K^+ inside the cell.

10.3 Physiology of Skeletal Muscle Contraction 341

- Skeletal muscle physiology involves three major events: excitation, excitation-contraction coupling, and crossbridge cycling.

10.3a Neuromuscular Junction: Excitation of a Skeletal Muscle Fiber 343
- Excitation involves the arrival of a nerve signal to stimulate release of the neurotransmitter acetylcholine (ACh) contained within synaptic vesicles.

10.3b Sarcolemma, T-tubules, and Sarcoplasmic Reticulum: Excitation-Contraction Coupling 343
- Excitation-contraction coupling links excitation of the muscle by the motor neuron to muscle contraction through the sarcolemma, T-tubules, and sarcoplasmic reticulum.

10.3c Sarcomere: Crossbridge Cycling 345
- Crossbridge cycling is initiated by the release of Ca^{2+} from the sarcoplasmic reticulum. This allows myosin heads to bind to actin to pull thin filaments past thick filaments. This process is the sliding filament theory.

10.3d Skeletal Muscle Relaxation 347
- Muscle returns to the resting state through the natural elasticity of the muscle fiber.

10.4 Skeletal Muscle Metabolism 350

- Skeletal muscle tissue exhibits a high metabolic demand for energy.

10.4a Supplying Energy for Skeletal Muscle Contraction 350
- The three major means to supply energy to skeletal muscle are immediate (phosphate transfer), short term (glycolysis), and long term (aerobic cellular respiration).
- The duration and intensity of activity determines the primary means of supplying adenosine triphosphate (ATP).

10.4b Oxygen Debt 352
- Oxygen debt is the additional oxygen that must be taken in after exercise to restore pre-exercise conditions.

10.5 Skeletal Muscle Fiber Types 353

- Two criteria are used to classify skeletal muscle fiber types into three categories.

10.5a Criteria for Classification of Muscle Fiber Types 353
- The criteria for classifying muscle fiber types include the type of contraction generated (power, speed, duration) and the primary means of supplying ATP (glycolytic and aerobic cellular respiration).

10.5b Classification of Muscle Fiber Types 353
- The three subtypes of skeletal muscle fibers are slow oxidative (SO) fibers, fast oxidative (FO) fibers, and fast glycolytic (FG) fibers.

10.5c Distribution of Muscle Fiber Types 354
- Skeletal muscles typically contain all muscle fiber types; however, the relative percentage will vary between different skeletal muscles of the body and between individuals for certain muscles, such as the muscles of the leg.

CHALLENGE YOURSELF

Do You Know the Basics?

_____ 1. The unit of skeletal muscle structure that is composed of bundles of myofibrils, enclosed within a sarcolemma, and surrounded by a connective tissue covering called endomysium is a

a. myofibril.

b. fascicle.

c. myofilament.

d. myofilament.

_____ 2. The physiologic event that takes place at the plasma membrane of a skeletal muscle fiber is

a. release of calcium.

b. propagation of an action potential.

c. binding of calcium by troponin.

d. crossbridge cycling.

_____ 3. In a skeletal muscle fiber, Ca^{2+} is released from

    a. ACh receptors.

    b. the motor end plate.

    c. the sarcoplasmic reticulum.

    d. the sarcolemma and T-tubules.

_____ 4. The bundle of dense regular connective tissue that attaches a skeletal muscle to bone is called a(n)

    a. tendon.

    b. ligament.

    c. endomysium.

    d. fascicle

_____ 5. In excitation-contraction coupling, the transverse tubules function to

    a. conduct an action potential into the sarcoplasmic reticulum to cause release of calcium.

    b. uptake and store excess Na^+ and K^+ from the sarcoplasm.

    c. keep the thin and thick myofilaments separated.

    d. provide structural support for sarcomeres.

_____ 6. During muscle contraction, the I band

    a. hides the H zone.

    b. shortens or narrows.

    c. overlaps the Z line.

    d. always remains the same length.

_____ 7. During a concentric contraction of a muscle fiber, myofibrils

    a. lengthen.

    b. remain the same length.

    c. increase in diameter.

    d. shorten.

_____ 8. What event causes a troponin-tropomyosin complex to regain its original shape in muscle relaxation?

    a. stimulation of ACh receptors

    b. diffusion of Na^+ back into transverse tubules

    c. return of Ca^{2+} into the sarcoplasmic reticulum

    d. breaking the bond with tropomyosin

_____ 9. In sustained, moderate exercise, skeletal muscle is predominantly using _____ as its energy source.

    a. amino acids

    b. glucagon

    c. fatty acids

    d. creatine phosphate

_____ 10. Skeletal muscle and cardiac muscle are similar in that both types of muscle

    a. have cells that branch.

    b. contain intercalated discs.

    c. are under involuntary control.

    d. are striated.

11. Explain the structural relationship between a sarcomere, myofibril, myofilament, and a muscle fiber.

12. Diagram and label a sarcomere including a thick filament, thin filament, A band, H zone, I band, Z disc, and M line.

13. Explain why the ratio of motor neurons to skeletal muscle fibers is greater in muscles that control eye movement than in postural muscles of the leg.

14. Put the following skeletal muscle contraction events in the order that they occur:

    a. The myosin head swivels toward the center of the sarcomere.

    b. Calcium ions are released from the sarcoplasmic reticulum and bind to troponin.

    c. An action potential is propagated along the sarcolemma and transverse tubules.

    d. Myosin binds to actin, forming crossbridges.

    e. Myosin heads bind ATP molecules and releases from actin.

    f. Tropomyosin molecules are moved off active sites on actin.

    g. ATPase splits ATP, providing the energy to reset the myosin head.

15. Explain the various means of providing ATP for skeletal muscle contraction.

16. Explain why athletes who excel at short sprints probably have fewer slow-twitch fibers in their lower limb muscles.

17. Explain why skeletal muscle generates the most force when it is at its resting length at the time of stimulation based on the length-tension relationship.

18. Describe the characteristics of smooth muscle that allow it to contract for extended periods of time and not fatigue.

19. Describe the response of smooth muscle to sustained stretch.

20. Identify the location of both multiunit smooth muscle and single-unit smooth muscle, and explain the difference in how each is regulated.

Can You Apply What You've Learned?

1. A bacterial toxin is known to block the release of ACh at the motor end plate of skeletal muscle. Consequently,

    a. the skeletal muscle contracts with increasing force.

    b. the skeletal muscle contracts with increasing frequency.

    c. the ability to stimulate the muscle is impaired.

    d. other neurotransmitters would stimulate the muscle.

2. One of the primary reasons that one individual is faster in a 50-meter sprint than another is due to

    a. a greater number of muscle fibers of smaller diameter.

    b. more oxidative fibers in the lower limb muscles.

    c. an enhanced ability to deliver oxygen to the muscles.

    d. a greater percentage of fast-twitch fibers in the lower limb muscles.

3. Which electrolyte imbalance is least likely to impair muscle contraction because it is not required in muscle contraction?

 a. F^-

 b. Na^+

 c. K^+

 d. Ca^{2+}

4. Rigor mortis occurs following death because

 a. tropomyosin remains over the myosin binding sites of actin.

 b. myosin heads attach to actin and are not released due to lack of ATP.

 c. the myosin becomes misshapen.

 d. all of the Ca^{2+} remains within the sarcoplasmic reticulum.

5. An athlete participates in aerobic exercise three times a week. One of the changes is an increased ability to deliver oxygen to her skeletal muscles. Over time she notices that she can continue the exercise with greater intensity and duration. The reason for this change is that there is a(n)

 a. greater response from phosphate transfer.

 b. greater production of ATP from glycolysis and less from aerobic cellular respiration.

 c. greater production of ATP from aerobic cellular respiration.

 d. increased production of lactate.

Can You Synthesize What You've Learned?

1. Your anatomy and physiology class is required for a career in forensics, and one of the short essays is an explanation for why the body becomes stiff after death. Provide an answer for an individual that has some understanding of skeletal muscle physiology.

2. Describe the effect of the botulinum toxin, which inhibits the release of acetylcholine at the neuromuscular junction. Would the poison curare, which competes for acetylcholine receptors have a similar effect? Explain

3. Smooth muscle is within the urinary bladder wall. Explain why, if you initially have the sensation of having to urinate, the sensation sometimes passes. Base your answer on the stress-relaxation response.

INTEGRATE

ONLINE STUDY TOOLS

The following study aids may be accessed through Connect.

Concept Overview Interactive: Figure 10.16: Skeletal Muscle Contraction

Clinical Case Study: Progressive Weakness in a Young Woman

Interactive Questions: This chapter's content is served up in a number of multimedia question formats for student study.

LearnSmart: Topics and terminology include introduction to skeletal muscle; anatomy and physiology of skeletal muscle; skeletal muscle metabolism; skeletal muscle fiber types; measurement of skeletal muscle tension; factors affecting skeletal muscle tension; effects of exercise and aging on skeletal muscle; cardiac muscle tissue; smooth muscle tissue

Anatomy & Physiology Revealed: Topics include skeletal muscle; skeletal muscle striations; sarcomere; sliding filament; neuromuscular junction; excitation-contraction coupling; cross bridge cycle; cardiac muscle; smooth muscle

Animations: Topics include skeletal muscle; function of the neuromuscular junction; action potentials and muscle contraction; sarcomere shortening; breakdown of ATP and crossbridge movement during muscle contraction; mechanics of single fiber contraction; activation of contraction in smooth vs. skeletal muscle

chapter
11

Muscular System: Axial and Appendicular Muscles

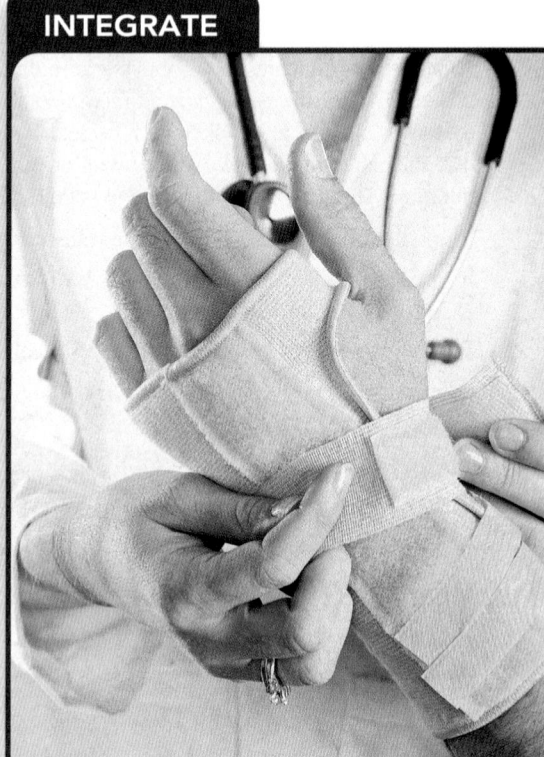

INTEGRATE

CAREER PATH
Physical Therapist

Physical therapists help injured individuals achieve greater mobility and improvement of their quality of life. Orthopedic physical therapists (a specialty of physical therapy) treat musculoskeletal disorders and help patients regain mobility after orthopedic surgery. An understanding of skeletal muscle function and which muscles work synergistically and antagonistically is essential for orthopedic physical therapists. These health-care professionals utilize this knowledge to develop treatment plans for the patient, such as the patient shown here (who is recovering from a wrist injury).

Anatomy & Physiology | **REVEALED**®
aprevealed.com

Module 6: Muscular System

The partitioning of the skeletal system into axial and appendicular divisions provides a useful guideline for subdividing the muscular system. **Axial muscles** have both their origins and insertions on parts of the axial skeleton. Axial muscles support and move the head and vertebral column, function in nonverbal communication by affecting facial features, move the mandible during chewing, assist in food processing and swallowing, aid in breathing, and both support and protect the abdominal and pelvic organs. The **appendicular muscles** control the movements of the upper and lower limbs, and stabilize and control the movements of the pectoral and pelvic girdles. These muscles are organized into groups based upon their location in the body or the part of the skeleton they move. Some muscles of both divisions are shown in **figure 11.1**.

The muscles in this chapter have been organized into groups according to their location in the body. For each group, tables provide descriptions of the muscles as well as information about their action, origin, insertion, and innervation. (Note: The word *innervation* refers to the nerve(s) that controls a muscle and stimulates it to contract. For further information about the nerves listed in the tables, see sections 13.9 and 14.5.)

Figure 11.1 Body Musculature. (*a*) Anterior view shows superficial muscles on the right side of the body and some deeper muscles on the left side. (*b*) Posterior view shows superficial muscles on the left side of the body and some deeper muscles on the right side. Labels for the axial muscles are in bold; not all muscles shown in the figure are identified.

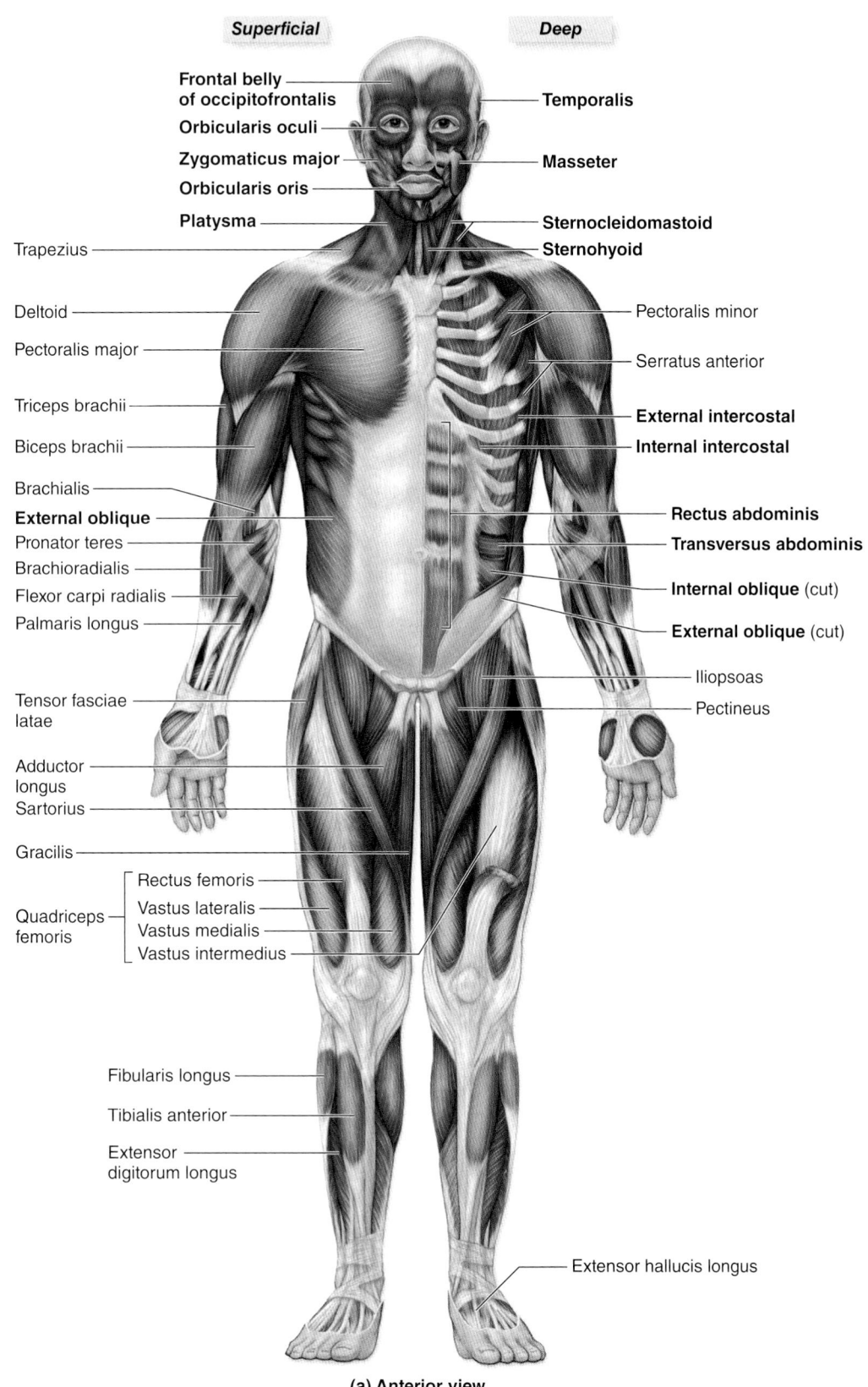

Superficial

Deep

Frontal belly of occipitofrontalis

Temporalis

Orbicularis oculi

Zygomaticus major

Masseter

Orbicularis oris

Platysma

Sternocleidomastoid

Trapezius

Sternohyoid

Deltoid

Pectoralis minor

Pectoralis major

Serratus anterior

Triceps brachii

External intercostal

Biceps brachii

Internal intercostal

Brachialis

External oblique

Rectus abdominis

Pronator teres

Transversus abdominis

Brachioradialis

Flexor carpi radialis

Internal oblique (cut)

Palmaris longus

External oblique (cut)

Iliopsoas

Tensor fasciae latae

Pectineus

Adductor longus

Sartorius

Gracilis

Rectus femoris

Vastus lateralis

Quadriceps femoris

Vastus medialis

Vastus intermedius

Fibularis longus

Tibialis anterior

Extensor digitorum longus

Extensor hallucis longus

(a) Anterior view

(continued on next page)

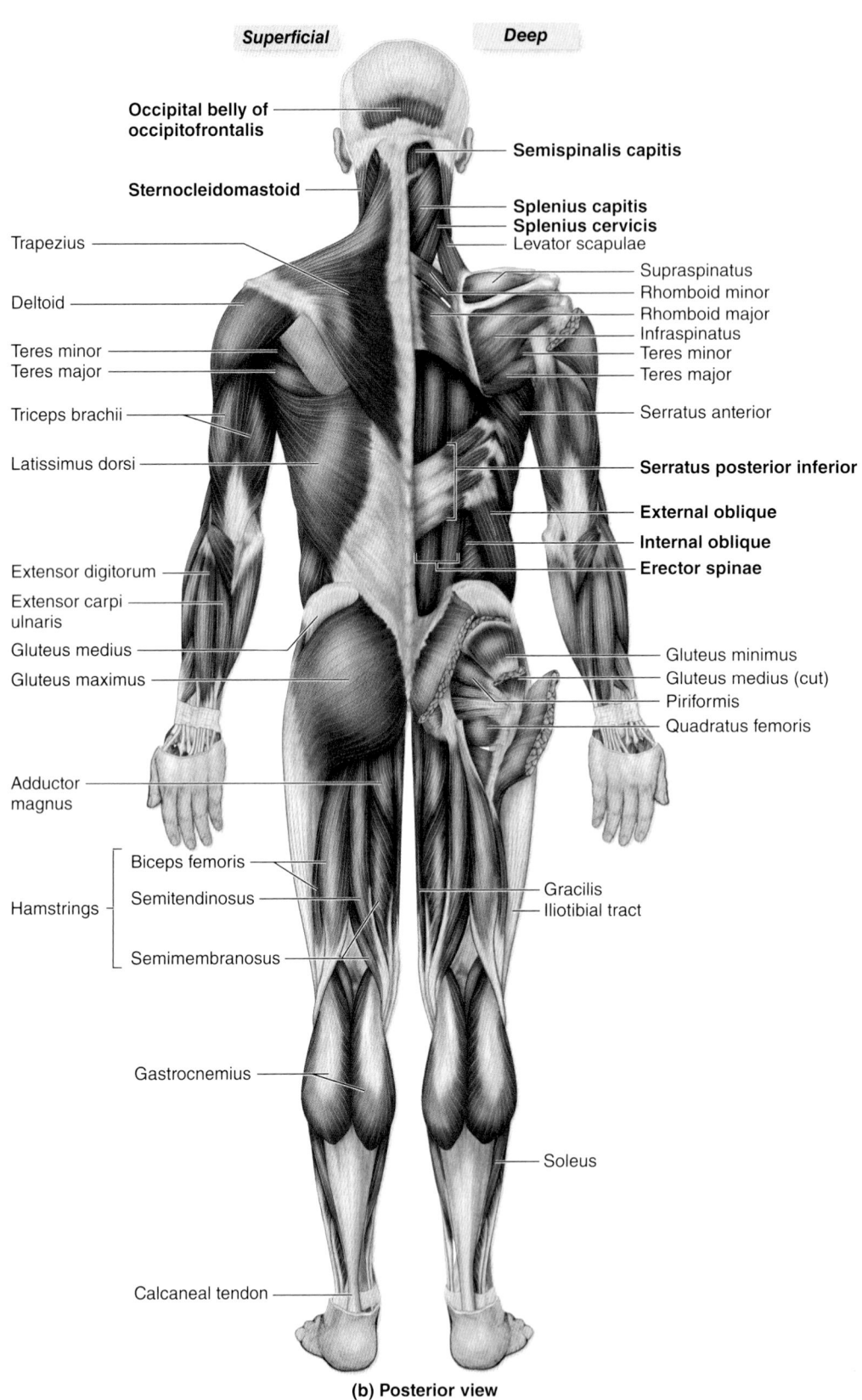

Superficial   **Deep**

Occipital belly of occipitofrontalis

Semispinalis capitis

Sternocleidomastoid

Splenius capitis
Splenius cervicis
Levator scapulae

Trapezius

Supraspinatus
Rhomboid minor
Rhomboid major
Infraspinatus
Teres minor
Teres major

Deltoid

Teres minor
Teres major

Serratus anterior

Triceps brachii

Serratus posterior inferior

Latissimus dorsi

External oblique

Internal oblique
Erector spinae

Extensor digitorum

Extensor carpi ulnaris

Gluteus minimus
Gluteus medius (cut)
Piriformis
Quadratus femoris

Gluteus medius

Gluteus maximus

Adductor magnus

Biceps femoris

Gracilis
Iliotibial tract

Hamstrings

Semitendinosus

Semimembranosus

Gastrocnemius

Soleus

Calcaneal tendon

(b) Posterior view

11.1 Skeletal Muscle Composition and Actions

In section 10.2, we examined the macroscopic and microscopic anatomy of skeletal muscle. Here we compare the origin versus the insertion of a skeletal muscle, the organizational patterns exhibited by skeletal muscle fibers, and the general actions of skeletal muscles.

11.1a Origin and Insertion

✓ LEARNING OBJECTIVE

1. Compare and contrast the origin and the insertion of a skeletal muscle.

Many skeletal muscles extend between bones and cross at least one moveable joint. Upon contraction, one bone moves while the other bone usually remains fixed. The less moveable attachment of a muscle is called its **origin** (ōr′i-jin; *origo* = source). The more moveable attachment of the muscle is its **insertion** (in-ser′shŭn; *inserto* = to plant in) (**figure 11.2**). Usually, the insertion is pulled toward the origin. In the limbs, the origin typically lies proximal to the insertion. For example, the biceps brachii muscle originates on the scapula and inserts on the radius. The contraction of this muscle pulls the forearm toward the shoulder.

Sometimes, neither the origin nor the insertion can be determined easily either by movement or position. In this case, other criteria are used. For example, if a muscle extends between a broad aponeurosis and a narrow tendon, the aponeurosis is considered the origin, and the tendon is attached to the insertion. If there are several tendons at one end of the muscle and just one tendon at the other end, each of the multiple tendons is considered an origin, and the single tendon at the other end of the bone is considered the insertion.

Some anatomists and clinicians have started replacing the terms "origin" and "insertion" with the terms *proximal attachment* and *distal attachment* (or *superior attachment* and *inferior attachment* for axial muscles). The reasons for this change are because the latter terms are easier to identify on the body, whereas origin and insertion may vary depending upon the movement of the muscle. These new terms are not yet widely used, which is why in this text we continue to use origin and insertion.

≋✓ WHAT DID YOU LEARN?

❶ What is the difference between the origin and insertion of a skeletal muscle?

11.1b Organizational Patterns of Skeletal Muscle Fibers

≋✓ LEARNING OBJECTIVE

2. Describe and differentiate between the organizational patterns in muscle fascicles.

As mentioned in section 10.2a, bundles of muscle fibers, termed fascicles, lie parallel to each other within each muscle. However, the organization of fascicles in different muscles often varies. There are four different patterns of fascicle arrangement: circular, parallel, convergent, and pennate (**figure 11.3**).

Circular Muscles

A **circular muscle** has concentrically arranged fibers around an opening or recess. A circular muscle is also called a *sphincter*, and its contraction decreases the passageway diameter. An example is the orbicularis oris muscle that encircles the opening of the mouth.

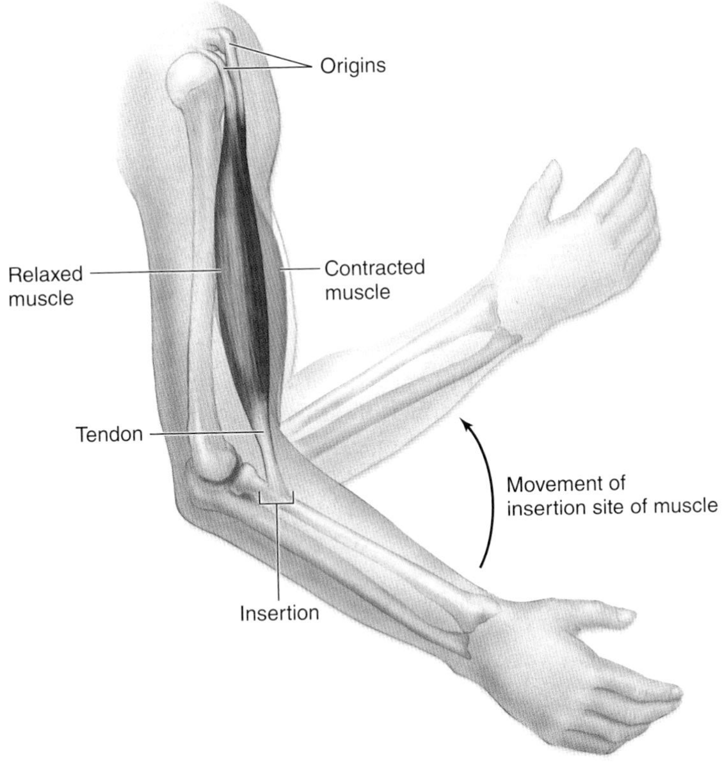

Figure 11.2 Muscle Origin and Insertion. The origin is the less mobile point of attachment of the muscle, whereas the insertion is more mobile, as shown in this view of the biceps brachii muscle.

Origins · Relaxed muscle · Contracted muscle · Tendon · Movement of insertion site of muscle · Insertion

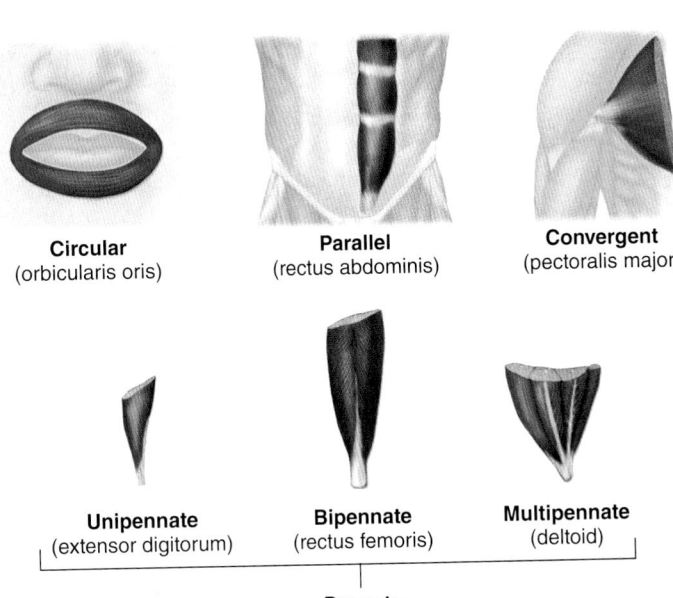

Circular (orbicularis oris) · **Parallel** (rectus abdominis) · **Convergent** (pectoralis major)

Unipennate (extensor digitorum) · **Bipennate** (rectus femoris) · **Multipennate** (deltoid)

Pennate

Figure 11.3 Organization of Muscle Fibers. Muscle fascicles may be organized into one of four basic patterns: circular, parallel, convergent, or pennate.

INTEGRATE

CLINICAL VIEW
Intramuscular Injections

An **intramuscular injection** is one way that a medication may be administered. Since skeletal muscle has a rich blood supply, the medication may be inserted into the muscle with a syringe. The medication enters the cardiovascular system through the muscle's blood vessels and from there is transported throughout the body. This method of medication delivery allows a large amount of medication to be delivered at once without significant discomfort, and it ensures the delivery is slower and more uniform than if the medication is taken orally or intravenously. Further, some medications may be better absorbed via the muscle than orally, and some medications may not be tolerated when taken orally.

Examples of medications that are taken intramuscularly include most vaccines, insulin, some fertility and contraceptive medications, and large doses of selected antibiotics (such as penicillin). Common intramuscular injection sites include the deltoid, gluteal, and quadriceps muscles.

Parallel Muscles

The fascicles in a **parallel muscle** run parallel to its long axis. Sometimes these muscles are cylindrical with an expanded central region. In this case, parallel muscles have a central body, called the **belly**, or *gaster*. When this muscle contracts and shortens, the muscle increases in diameter. Parallel muscles have high endurance but are not strong. Examples of parallel muscle include the rectus abdominis (an anterior abdominal muscle) or the biceps brachii of the arm.

Convergent Muscles

A **convergent** (kon-ver′jent) **muscle** has widespread muscle fibers over a broad area that converge on a common attachment site. This attachment site may be a single tendon, a tendinous sheet, or a slender band of collagen fibers known as a **raphe** (rā′fē; *rhaphe* = seam). These muscle fibers are often triangular in shape, resembling a broad fan with a tendon at the tip. A convergent muscle is versatile—that is, the direction of its pull can be modified merely by activating a specific, single group of muscle fibers at any one time. However, when the fibers in a convergent muscle all contract at once, they do not pull as hard on the tendon as a parallel muscle of the same size because the muscle fibers on opposite sides of the tendon are not working together; rather, they are pulling in different directions. An example of a convergent muscle is the pectoralis major of the chest.

Pennate Muscles

Pennate (pen′āt; *penna* = feather) **muscles** are so named because the fascicles exhibit the same angle with respect to their tendon—that is, they resemble a large feather. Pennate muscles have one or more tendons extending through their body, and the fascicles are arranged at an oblique angle to the tendon. Because pennate muscle fibers pull at an angle to the tendon, this type of muscle does not move its tendons as far as parallel muscles move their tendons. However, the contraction of a pennate muscle generates more tension than does a parallel muscle of the same size, and thus it is stronger.

There are three types of pennate muscles:

- In a **unipennate muscle,** all of the muscle fibers are on the same side of the tendon. The extensor digitorum, a long muscle that extends the fingers, is a unipennate muscle.
- A **bipennate muscle,** the most common type, has muscle fibers on both sides of the tendon. The interosseous muscles on both the palmar and dorsal sides of the metacarpals are composed of bipennate muscles that help adduct and abduct the digits.
- A **multipennate muscle** has branches of the tendon within the muscle. The triangular deltoid that covers the superior surface of the shoulder joint is a multipennate muscle.

WHAT DID YOU LEARN?

2 Which muscle is stronger—a pennate muscle or a parallel muscle?

11.1c Actions of Skeletal Muscles

LEARNING OBJECTIVE

3. Differentiate between agonists, antagonists, and synergists.

Skeletal muscles generally do not work in isolation; rather, they work together to produce movements. Muscles are grouped according to their primary actions into three types: agonists, antagonists, and synergists.

An **agonist** (ag′on-ist; *agon* = a contest), also called a *prime mover,* is a muscle that contracts to produce a particular movement, such as extending the forearm. The triceps brachii of the forearm is an agonist that causes extension of the forearm.

An **antagonist** (an-tag′ō-nist; *anti* = against) is a muscle whose actions oppose those of the agonist. For example, if the agonist produces extension, the antagonist produces flexion. The contraction of the agonist lengthens the antagonist, and vice versa. As this movement occurs, the lengthened muscle usually does not relax completely. Instead, the tension within the muscle being lengthened is adjusted to control the speed of the movement and ensure that it is smooth. When the triceps brachii acts as an agonist to extend the forearm, the biceps brachii muscle on the anterior side of the humerus acts as an antagonist to stabilize the movement and to produce the opposing action, which is flexion of the forearm.

A **synergist** (sin′er-jist; *syn* = with, *ergo* = work) is a muscle that assists the agonist in performing its action. The contraction of a synergist usually either contributes to tension exerted close to the insertion of the muscle or stabilizes the point of origin. Usually, synergists are most useful at the start of a movement when the agonist is lengthened and cannot exert much power. Examples of synergistic muscles are the biceps brachii and the brachialis muscles of the arm. Both muscles work synergistically (together) to flex the elbow joint. Synergists may also assist an agonist by preventing movement at a joint and thereby stabilizing the origin of the agonist. In this case, these synergistic muscles are called *fixators.*

WHAT DID YOU LEARN?

3 What is the difference between an agonist and a synergist?

11.2 Skeletal Muscle Naming

LEARNING OBJECTIVES

1. List the seven aspects of muscles that may contribute to their names.

2. Give examples of muscles whose names contain an indication of action, specific body region, attachments, orientation of muscle fibers, shape, size, and muscle heads.

In section 1.4, you learned some of the anatomic terminology used to describe the body; in chapter 8, you saw how anatomic terms are applied to the bones of the skeleton. Naming of skeletal muscles follows similar conventions, and usually muscle names provide clues to their identification. As **figure 11.4** shows, skeletal muscles are named according to the following criteria:

- **Muscle action.** Names that indicate the primary function or movement of the muscle include *flexor, extensor,* and *pronator.* These are such common actions that the names almost always contain other clues to the appearance or location of the muscle. For example, the flexor digitorum longus muscle is a long muscle responsible for flexing the digits.

- **Specific body regions.** The rectus femoris is on the thigh (*femur*), and the tibialis anterior is on the *anterior* surface of the *tibia.* Muscles that are close to the body surface often are termed *superficialis* (sū′per-fish-ē-ă′lis) or *externus* (eks-ter′nŭs). In contrast, deeper placed muscles may have names such as *profundus* (prō-fŭn′dŭs; deep) or *internus.*

- **Muscle attachments.** Many muscle names identify their origins, insertions, or other prominent attachments. In these cases, the first part of the name indicates the origin and the second part the insertion. For example, the sternocleidomastoid has origins on the *sternum* and the *clavicle* (*cleido*) and an insertion on the *mastoid* process of the temporal bone.

- **Orientation of muscle fibers.** The rectus abdominis muscle is named for its lengthwise-running muscle fibers; *rectus* means "straight." Similarly, names such as *oblique* or *obliquus* (ob-lī′kwŭs) indicate muscles with fibers extending at an angle to the longitudinal axis of the body. The internal and external oblique muscles are abdominal muscles that have angled muscle fibers.

- **Muscle shape.** Examples of shape in muscle names include *deltoid* (shaped like a triangle), *orbicularis* (circular muscle fibers), *rhomboid* (shaped like a rhombus), and *trapezius* (shaped like a trapezoid). Short muscles may be called *brevis* (brev′is); long muscles are called *longus* (lon′gŭs) or *longissimus* (lon-jīs′ĭ-mŭs; longest). *Teres* (ter′ēz) muscles are both long and round.

- **Muscle size.** Large muscles may be called *magnus* (big), *major* (bigger), or *maximus*

(biggest). Small muscles may be called *minor* (smaller) or *minimus* (smallest). Examples of muscles named by size include the buttocks muscles: gluteus maximus, gluteus medius, and gluteus minimus.

- **Muscle heads/tendons of origin.** Some muscles are named after specific features; for example, how many tendons of origin or how many muscle bellies or heads each contains. A *biceps* muscle has two tendons of origin, a *triceps* muscle has three heads or tendons, and a *quadriceps* muscle has four heads or tendons.

 WHAT DID YOU LEARN?

4 What are some words used in muscle names that refer to muscle shape?

5 The gluteus maximus muscle gets its name from which categories for naming muscles?

Muscle action	Example
Adductor (adducts body part)	Adductor magnus
Abductor (abducts body part)	Abductor pollicis longus
Flexor (flexes a joint)	Flexor carpi radialis
Extensor (extends a joint)	Extensor hallucis longus
Specific body regions	**Example**
Oris (mouth)	Orbicularis oris
Cervicis (neck)	Semispinalis cervicis
Brachial (arm)	Biceps brachii
Carpi (wrist)	Flexor carpi ulnaris
Pollicis (thumb)	Opponens pollicis
Gluteal (buttocks)	Gluteus medius
Femoris (thigh)	Quadratus femoris
Hallucis (great toe)	Extensor hallucis longus
Anterior (toward the front of the body)	Tibialis anterior
Posterior or dorsal/dorsi (toward the back of the body)	Tibialis posterior Latissimus dorsi
Superior (closer to the head)	Serratus posterior superior
Inferior (closer to the feet)	Serratus posterior inferior
Superficialis (superficial)	Flexor digitorum superficialis
Profundus (deep)	Flexor digitorum profundus
Muscle attachments	**Example**
Sternum and clavicle (cleido)	Sternocleidomastoid
Between the ribs	Intercostal
Subscapular fossa	Subscapularis
Fibula	Fibularis longus
Zygomatic bone	Zygomaticus major
Orientation of muscle fibers	**Example**
Rectus (straight)	Rectus abdominis
Oblique (angled)	External oblique
Orbicularis (circular)	Orbicularis oculi
Muscle shape	**Example**
Deltoid (triangular)	Deltoid
Quadratus (rectangular)	Pronator quadratus
Trapezius (trapezoidal)	Trapezius
Longus (long)	Abductor pollicis longus
Brevis (short)	Abductor pollicis brevis
Muscle size	**Example**
Major (larger of two muscles)	Pectoralis major
Minor (smaller of two muscles)	Pectoralis minor
Maximus (largest)	Gluteus maximus
Medius (medium sized)	Gluteus medius
Minimus (smallest)	Gluteus minimus
Muscle heads / tendons of origin	**Example**
Biceps (two heads)	Biceps femoris
Triceps (three heads)	Triceps brachii
Quadriceps (four heads)	Quadriceps femoris

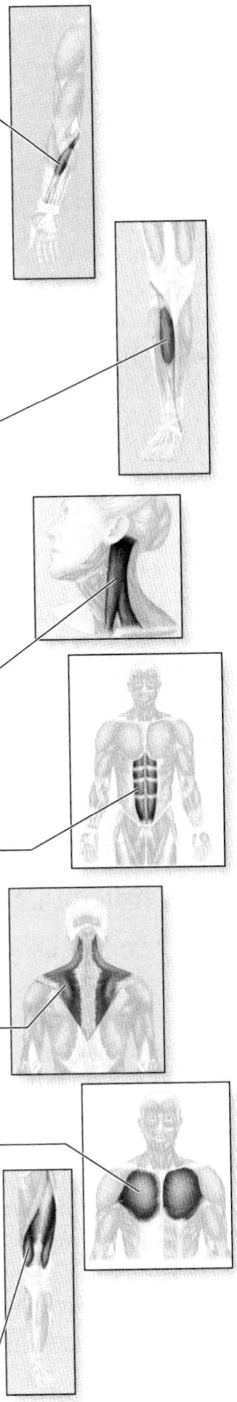

Figure 11.4 Muscle Naming. Muscles are named according to a variety of features.

INTEGRATE

LEARNING STRATEGY

The following suggestions may help you learn the muscles:

- Muscles may be organized into groups. Learning the muscles in groups is easiest.
- When studying a particular muscle, try to palpate it on yourself, and look in a mirror so you can better visualize the muscle's location. Contract the muscle to sense its action.
- Repeat the name of a muscle aloud to become familiar with it.
- Associate visual images from models, cadavers, a photographic atlas, or dissected animals with muscle names.
- Locate the origins and insertions of muscles on an articulated skeleton to understand how they produce their actions.
- Learn the derivation of each muscle name.

INTEGRATE

CONCEPT CONNECTION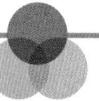

Vision problems may be due to musculoskeletal issues, nervous system issues, or both. For example, if a person cannot abduct his right eye, that indicates that CN VI (abducens nerve) may be injured (see section 13.9). In addition, some vision problems may be due to a single weak extraocular muscle, and correcting this muscle imbalance (with either exercises or wearing a patch over the stronger eye) helps alleviate the vision problem. Thus, a physician must integrate both the muscle and innervation information to properly diagnose a patient's visual disturbance.

11.3 Muscles of the Head and Neck

The muscles of the head and neck are separated into several specific groups. Almost all of these muscles (except for a few muscles of the anterior neck) originate on either the bones of the skull or the hyoid bone.

11.3a Muscles of Facial Expression

LEARNING OBJECTIVES

1. Name the muscles that move the forehead, the skin around the eyes and the nose, and describe their actions.
2. List the muscles that move the mouth and cheeks and their actions.

The muscles of facial expression originate in the superficial fascia or on the skull bones **(figure 11.5)**. These muscles insert into the superficial fascia of the skin, so when they contract, they pull on the skin, causing it to move. Most of these muscles are innervated by the seventh cranial nerve (CN VII), the facial nerve (see section 13.9).

The **epicranius** is composed of the **occipitofrontalis muscle** and a broad **epicranial aponeurosis,** also called the *galea aponeurotica.* The **frontal belly** of the occipitofrontalis is superficial to the frontal bone on the forehead. When this muscle contracts, it raises the eyebrows and wrinkles the skin of the forehead. The **occipital belly** of the occipitofrontalis covers the posterior aspect of the skull. When this muscle contracts, it retracts the scalp slightly.

Deep to the frontal belly of the occipitofrontalis is the **corrugator supercilii.** This muscle draws the eyebrows together and creates vertical wrinkle lines around the nose. The **orbicularis oculi** consists of circular muscle fibers that surround the eye's orbit. When this muscle contracts, the eyelid closes, as when you wink, blink, or squint. The **levator palpebrae superioris** elevates the upper eyelid when you open your eyes.

Several muscles of facial expression are associated with the nose. The **nasalis** elevates the corners of the nostrils. When you "flare" your nostrils, you are using the nasalis muscles. If you wrinkle your nose in distaste after smelling a foul odor, you have used your **procerus** muscle. This muscle is continuous with the frontal belly of the occipitofrontalis muscle and runs over the bridge of the nose, where it produces transverse wrinkles when it contracts.

INTEGRATE

CLINICAL VIEW

Idiopathic Facial Nerve Paralysis (Bell Palsy)

Unilateral paralysis of the muscles of facial expression is termed **facial nerve (CN VII) paralysis.** If the cause of the condition is unknown, doctors refer to it as **idiopathic** (id'ē-ō-path'ik; *idios* = one's own, *pathos* = suffering) **facial nerve paralysis,** or *Bell palsy.* CN VII paralysis has sometimes been associated with herpes simplex 1 viral infection. Facial nerve paralysis is also associated with exposure to cold temperatures, and is commonly seen in individuals who sleep with one side of their head facing an open window. Whatever the underlying cause, the nerve becomes inflamed and compressed within the narrow stylomastoid foramen, through which the facial nerve travels. The muscles on the same side of the face are paralyzed as a result.

Treatment of facial nerve paralysis usually means alleviating the symptoms. Doctors often use prednisone (a type of steroid) to reduce the inflammation

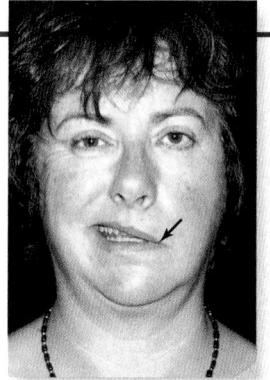

Facial nerve (CN VII) paralysis on the left side of the face. Note the drooping left side of the mouth (arrow) and the lack of contraction by the left orbicularis oris when the woman tries to smile.

and swelling of the nerve. If herpes simplex infection is suspected, an antiviral medication known as acyclovir (Zovirax) is also given. Like its underlying cause, recovery from idiopathic facial nerve paralysis is equally mysterious. Over 50% of all patients experience a complete, spontaneous recovery within 30 days of their first symptoms. Recovery may take longer for some, whereas other patients may never recover.

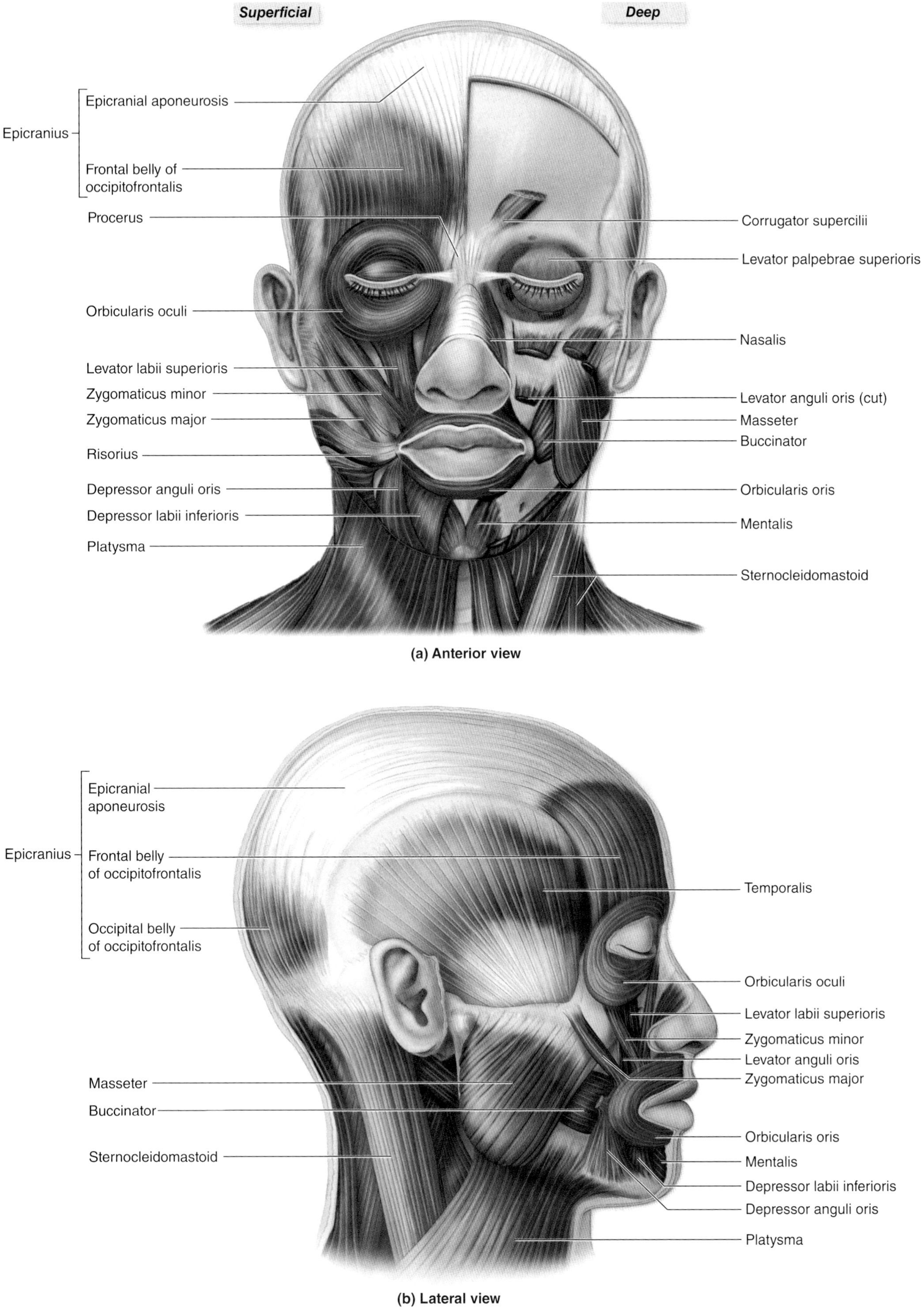

Superficial **Deep**

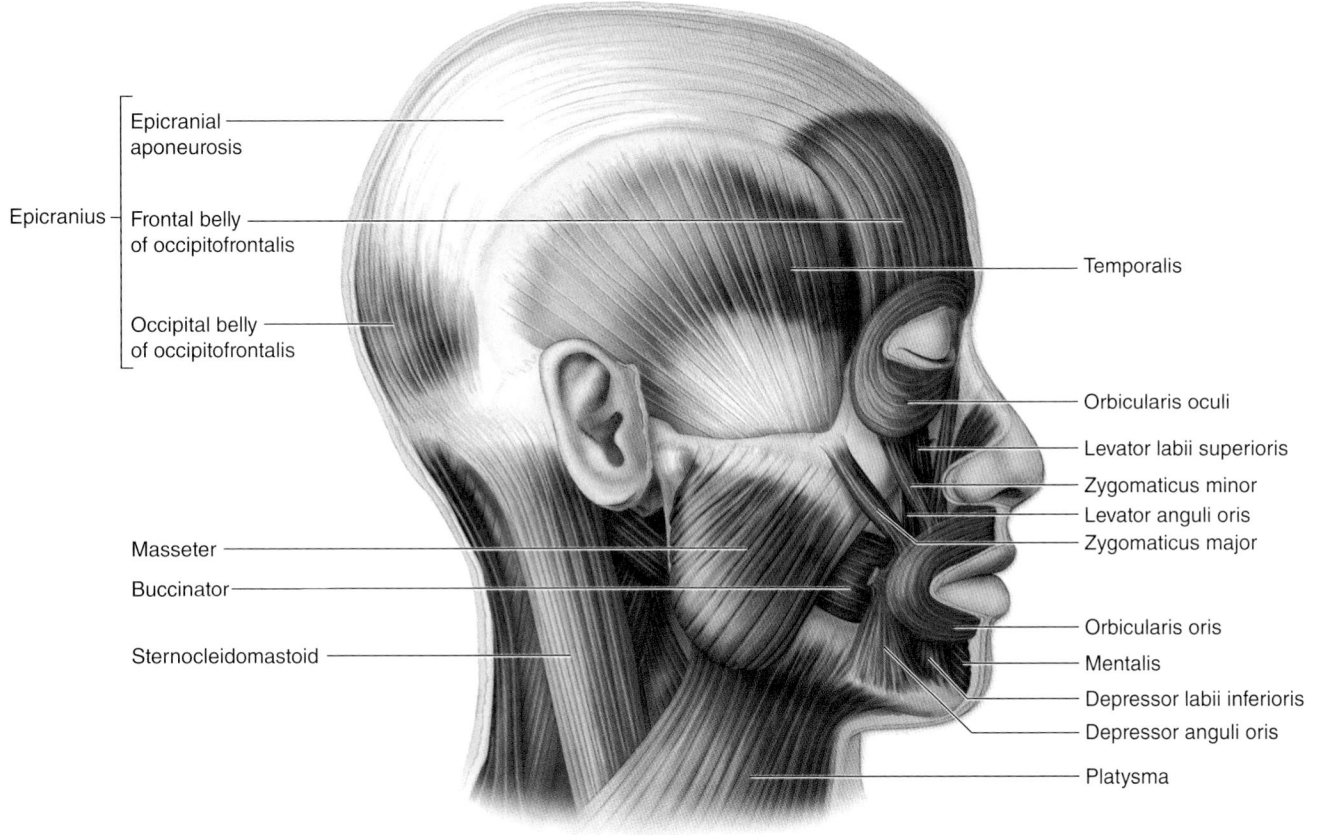

Epicranius
- Epicranial aponeurosis
- Frontal belly of occipitofrontalis

Procerus

Orbicularis oculi

Levator labii superioris
Zygomaticus minor
Zygomaticus major
Risorius
Depressor anguli oris
Depressor labii inferioris
Platysma

Corrugator supercilii
Levator palpebrae superioris
Nasalis
Levator anguli oris (cut)
Masseter
Buccinator
Orbicularis oris
Mentalis
Sternocleidomastoid

(a) Anterior view

Epicranius
- Epicranial aponeurosis
- Frontal belly of occipitofrontalis
- Occipital belly of occipitofrontalis

Masseter
Buccinator
Sternocleidomastoid

Temporalis

Orbicularis oculi
Levator labii superioris
Zygomaticus minor
Levator anguli oris
Zygomaticus major

Orbicularis oris
Mentalis
Depressor labii inferioris
Depressor anguli oris
Platysma

(b) Lateral view

Figure 11.5 Muscles of Facial Expression. AP|R

The mouth is the most expressive part of the face. The **orbicularis oris** consists of muscle fibers that encircle the opening of the mouth. When this muscle contracts, you close your mouth. In addition, when you "pucker up" for a kiss, you are using this muscle. The **depressor labii inferioris** does what its name suggests—it pulls the lower lip inferiorly. The **depressor anguli oris** is considered the "frown" muscle, because it pulls the corners of the mouth inferiorly. (Note, however, that it takes more muscles than just this one to produce a frown.)

In contrast, some muscles of the mouth elevate part or all of the upper lips. The **levator labii superioris** pulls the upper lip superiorly, as if a person is sneering or snarling. The **levator anguli oris** pulls the corners of the mouth superiorly and laterally. The **zygomaticus major** and **zygomaticus minor** work with the levator anguli oris muscles. You use all of these muscles when you smile. The **risorius** pulls the corner of the lips laterally; you use this muscle if you make a closed-mouth smile.

The **mentalis** attaches to the lower lip, and when it contracts, it protrudes the lower lip (as when a person "pouts"). The **platysma** tenses the skin of the neck and pulls the lower lip inferiorly. If you stand in front of a mirror and tense the skin of your neck, you can see these thin muscles bulging out.

The **buccinator** compresses the cheek against the teeth when we chew (and is the reason our cheeks don't "bulge out" like a squirrel's when we eat). Infants use the buccinator when they suckle at the breast. Some trumpet players (such as Dizzy Gillespie) have stretched out their buccinator muscles, allowing their cheeks to be "puffy" with air when they play the trumpet.

Table 11.1 summarizes the attachments and movements of the muscles of facial expression. **Figure 11.6** illustrates how these muscles produce some of the more characteristic expressions.

 WHAT DID YOU LEARN?

6 What muscles of facial expression must contract for you to smile?

7 The corners of the mouth are pulled inferiorly into a frown position by the contraction of what muscle?

Table 11.1	Muscles of Facial Expression		
Region/Muscle	**Action**	**Origin/Insertion**	**Innervation (see section 13.9)**
SCALP			
Epicranius (ep′ĭ-kră′nē-us; *epi* = over, *cran* = skull): Composed of an epicranial aponeurosis and the two bellies of the occipitofrontalis muscle			
Frontal belly of occipitofrontalis (ok-sip′i-tō-fron-tā′lis) *front* = forehead	Moves scalp, eyebrows; wrinkles skin of forehead	O: Epicranial aponeurosis I: Skin and subcutaneous layer of forehead and eyebrows	CN VII (facial nerve)
Occipital belly of occipitofrontalis *occipito* = base of skull	Retracts scalp	O: Superior nuchal line I: Epicranial aponeurosis	CN VII (facial nerve)
NOSE			
Nasalis (nā′ză-lis)	Compresses bridge and depresses tip of nose; elevates corners of nostrils	O: Maxillae and alar cartilage of nose I: Dorsum of nose	CN VII (facial nerve)
Procerus (prō-sē′rŭs) *procerus* = long	Moves and wrinkles nose	O: Nasal bone and lateral nasal cartilage I: Aponeurosis at bridge of nose and skin of forehead	CN VII (facial nerve)
MOUTH			
Buccinator (buk′sĭ-nā′tōr) *bucco* = cheek	Compresses cheek, holds food between teeth during chewing	O: Alveolar processes of mandible and maxillae I: Orbicularis oris	CN VII (facial nerve)
Depressor anguli oris (dĕ-pres′ŏr ang′gyū-lī ōr′ŭs) *depressor* = depresses *angul* = angle *oris* = mouth	Draws corners of mouth inferiorly and laterally ("frown" muscle)	O: Body of mandible I: Skin at inferior corner (angle) of mouth	CN VII (facial nerve)
Depressor labii inferioris (lā′bē-ī in-fĕr′ē-ōr-is) *labi* = lip *infer* = below	Draws lower lip inferiorly	O: Body of mandible lateral to midline I: Skin at inferior lip	CN VII (facial nerve)
Levator anguli oris (lē-vā′tor, le-vā′ter) *leva* = raise	Draws corners of mouth superiorly and laterally ("smile" muscle)	O: Lateral maxilla I: Skin at superior corner (angle) of mouth	CN VII (facial nerve)
Levator labii superioris (sū-pĕr′ē-ōr-is)	Opens lips; raises and furrows the upper lip ("Elvis" lip snarl)	O: Zygomatic bone; maxilla I: Skin and muscle of superior lip	CN VII (facial nerve)

| Table 11.1 | Muscles of Facial Expression (continued) |

Region/Muscle	Action	Origin/Insertion	Innervation (see section 13.9)
Mentalis (men-tă′lis) *ment* = chin	Protrudes lower lip ("pout"); wrinkles chin	O: Central mandible I: Skin of chin	CN VII (facial nerve)
Orbicularis oris (ŏr-bik′yū-lā′ris) *orb* = circular	Compresses and purses lips ("kiss" muscle)	O: Maxilla and mandible; blend with fibers from other facial muscles I: Encircling mouth; skin and muscles at angles to mouth	CN VII (facial nerve)
Risorius (ri-sōr′ē-ūs) *risor* = laughter	Draws corner of lip laterally; tenses lips	O: Deep fascia associated with masseter muscle I: Skin at angle of mouth	CN VII (facial nerve)
Zygomaticus major (zī′gō-mat′i-kūs) *zygomatic* = cheekbone *major* = greater	Elevates corner of the mouth ("smile" muscle)	O: Zygomatic bone I: Skin at superolateral edge of mouth	CN VII (facial nerve)
Zygomaticus minor *minor* = lesser	Elevates corner of the mouth ("smile" muscle)	O: Zygomatic bone I: Skin of superior lip	CN VII (facial nerve)
EYE			
Corrugator supercilii (kōr′ū-gā-ter sū′per′sil′ē-ī) *corrugo* = to wrinkle *cilium* = eyelid	Pulls eyebrows inferiorly and medially; creates vertical wrinkles above nose	O: Medial end of superciliary arch I: Skin superior to supraorbital margin and superciliary arch	CN VII (facial nerve)
Levator palpebrae superioris (pal-pē′brā) *palpebra* = eyelid	Elevates superior eyelid	O: Lesser wing of sphenoid bone I: Superior tarsal plate and skin of superior eyelid	CN III (oculomotor nerve)
Orbicularis oculi (ok′yū-lī) *ocul* = eye	Closes eye; produces winking, blinking, squinting ("blink" muscle)	O: Medial wall or margin of the orbit I: Skin surrounding eyelids	CN VII (facial nerve)
NECK			
Platysma (plă-tiz′mă) *platy* = flat	Pulls lower lip inferiorly; tenses skin of neck	O: Fascia of deltoid and pectoralis major muscles and acromion of scapula I: Skin of cheek and mandible	CN VII (facial nerve)

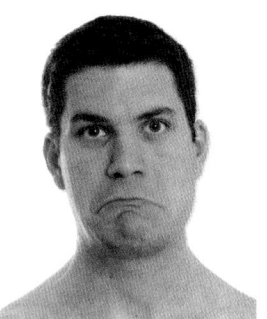

Depressor anguli oris
(frown)

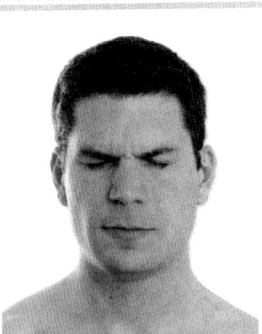

Orbicularis oculi
(blink/close eyes)

Zygomaticus major
(smile)

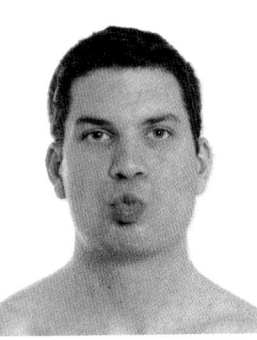

Orbicularis oris
(close mouth/kiss)

Frontal belly of occipitofrontalis
(wrinkle forehead, raise eyebrows)

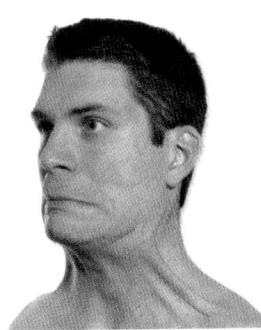

Platysma
(tense skin of neck)

Figure 11.6 Surface Anatomy of Some Muscles of Facial Expression. These muscles permit complex expressions that often are used as a means of communication.

11.3b Extrinsic Eye Muscles

LEARNING OBJECTIVES

3. Become familiar with the six extrinsic muscles of the eye, and describe how each affects eye movement.

4. Name the three cranial nerves that innervate the extrinsic eye muscles, and identify which muscles they act upon.

The **extrinsic eye muscles,** often called *extraocular muscles,* move the eyes. They are termed extrinsic because they insert onto the outer white surface of the eye, called the sclera. There are six extrinsic eye muscles: the four rectus muscles (medial, lateral, inferior, and superior) and the two oblique muscles (inferior and superior) **(figure 11.7)**.

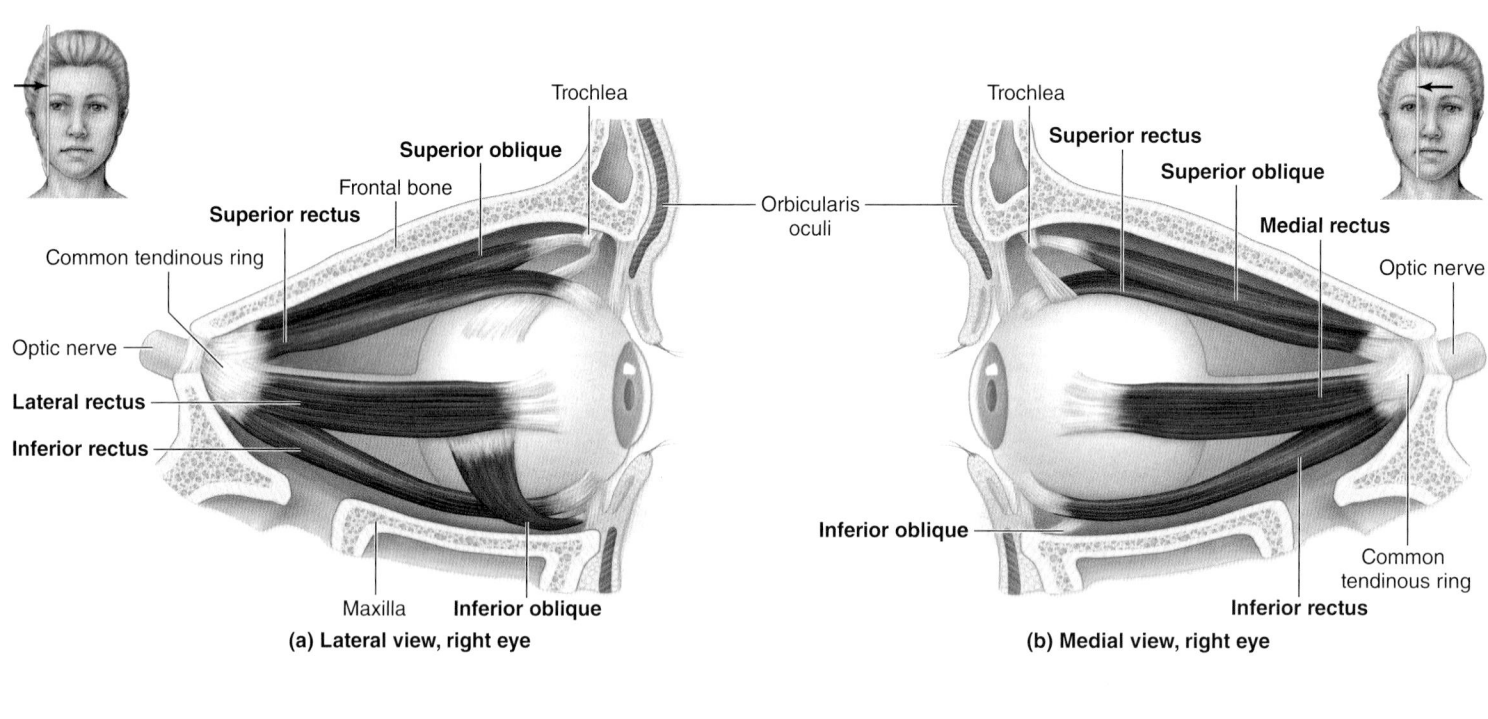

(a) Lateral view, right eye

(b) Medial view, right eye

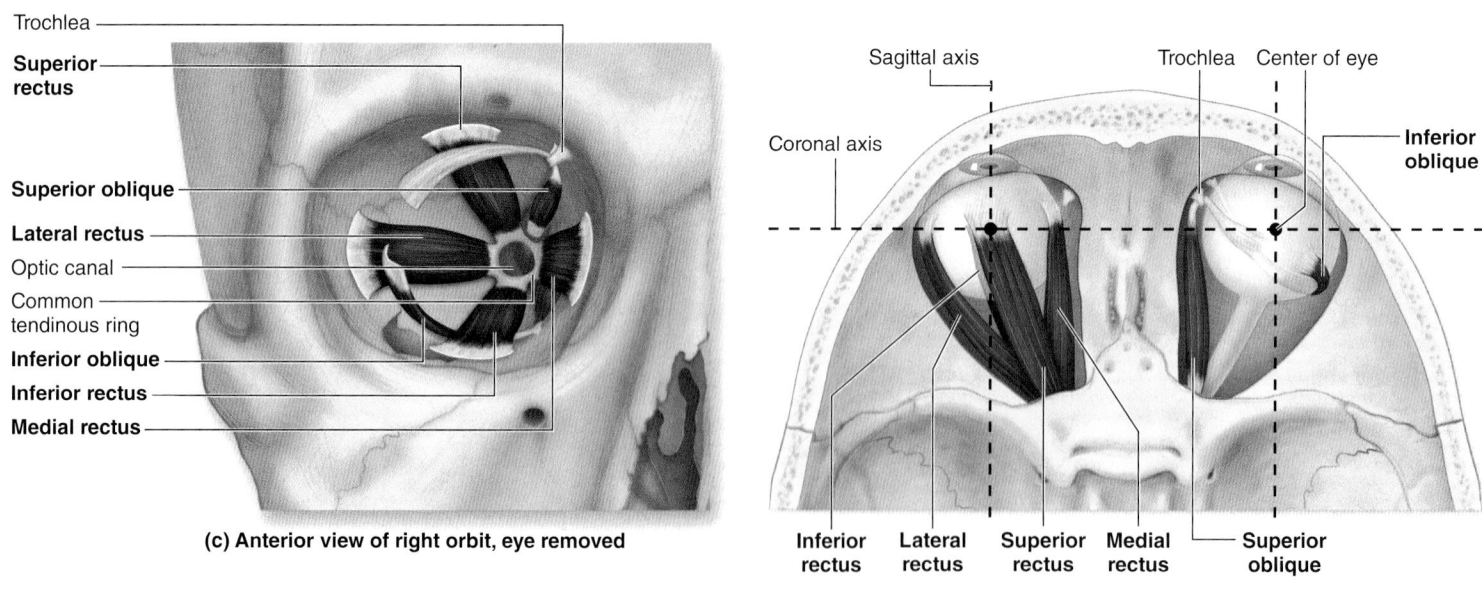

(c) Anterior view of right orbit, eye removed

(d) Superior view, left and right orbits

Figure 11.7 Extrinsic Muscles of the Eye. The extrinsic eye muscles control movements of the eye. The names of these muscles are in bold in the figure. (*a*) The insertions for most of the extrinsic eye muscles are noticeable in a lateral view of the right eye. (*b*) The medial rectus muscle appears prominently in a medial view of the right eye. (*c*) Most eye muscles originate from a common tendinous ring, shown here in an anterior view of the right orbit. (*d*) A superior view of the left and right orbits illustrates the insertion differences between the rectus and oblique muscles and how these insertion differences affect their movement of the eye. **AP|R**

The rectus eye muscles originate from a **common tendinous ring** in the orbit. These muscles attach to the *anterior* part of the eye, and are named according to which side of the eye they are located (medial, lateral, inferior, or superior).

The **medial rectus** attaches to the anteromedial surface of the eye and pulls the eye medially (adducts the eye). It is innervated by CN III (oculomotor nerve). The **lateral rectus** attaches to the anterolateral surface of the eye and pulls the eye laterally (abducts the eye). This muscle is innervated by CN VI (abducens). (Notice that this nerve's name tells you what muscle it innervates—the muscle that abducts the eye.) The **inferior rectus** attaches to the anteroinferior part of the eye. The inferior rectus pulls the eye inferiorly (as when you look down) and medially (as when you look at your nose). The **superior rectus** is located superiorly and attaches to the anterosuperior part of the sclera. The superior rectus pulls the eye superiorly (as when you look up) and medially (as when you look at your nose). The inferior and superior rectus muscles are innervated by CN III. Figure 11.7*d* illustrates that the superior and inferior rectus muscles do not pull directly parallel to the long axis of the eye; that is why both muscles also move the eye slightly in the medial direction.

The oblique eye muscles originate from within the orbit and insert on the *posterolateral* part of the sclera of the eye. The **inferior oblique** elevates the eye and turns the eye laterally. Since this muscle attaches to the inferior *posterior* part of the eye, contracting this muscle pulls the posterior part of the eye inferiorly (but *elevates* the *anterior* part of the eye). This muscle is innervated by CN III. The **superior oblique** depresses the eye and turns the eye laterally. This muscle passes through a pulleylike loop, called the **trochlea,** in the anteromedial orbit. This muscle attaches to the superior *posterior* part of the eye, so contracting this muscle pulls the posterior part of the eye superiorly (but *depresses* the *anterior* surface of the eye). This muscle is innervated by CN IV (trochlear). (Notice that this nerve's name is derived from the trochlea that holds the superior oblique in place.)

Table 11.2 compares the extrinsic muscles of the eye. Refer to section 13.9 to review the cranial nerves mentioned in this section.

 WHAT DID YOU LEARN?

8 Which extrinsic eye muscles abduct the eye (move the eye laterally)?

INTEGRATE

LEARNING STRATEGY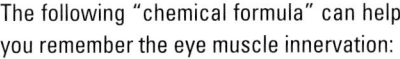

The following "chemical formula" can help you remember the eye muscle innervation:

$$[(SO_4)(LR_6)]_3$$

In words, the superior oblique (**SO**) is innervated by cranial nerve IV (**4**), the lateral rectus (**LR**) is innervated by cranial nerve VI (**6**), and the rest of the eye muscles are innervated by cranial nerve III (**3**).

INTEGRATE

CLINICAL VIEW
Strabismus and Diplopia

The extrinsic eye muscles are designed to move the left and right eyes in unison, so both eyes focus on the same image. **Strabismus** (stră-biz'mŭs; *strabismos* = a squinting) refers to a condition where both eyes can't focus on the same image, and the gaze of one eye is displaced. Strabismus may be caused by cranial nerve injury (see section 13.9), weaker eye muscles in one eye, or may occur when the brain favors vision in the stronger eye (which is the cause of *lazy eye*). Individuals with strabismus may experience **diplopia** (dip'lō'pē'ă; *diploos* = double, *ops* = eye), or double vision.

Table 11.2	Extrinsic Eye Muscles		
Group/Muscle	**Action**	**Origin/Insertion**	**Innervation (see section 13.9)**
RECTUS MUSCLES			
Medial rectus (mē'dē-ăl rek'tus) *rectus = straight*	Moves eye medially (adducts eye)	O: Common tendinous ring I: Anteromedial surface of eye	CN III (oculomotor nerve)
Lateral rectus (lat'er-ăl)	Moves eye laterally (abducts eye)	O: Common tendinous ring I: Anterolateral surface of eye	CN VI (abducens nerve)
Inferior rectus	Moves eye inferiorly (depresses eye) and medially (adducts eye)	O: Common tendinous ring I: Anteroinferior surface of eye	CN III (oculomotor nerve)
Superior rectus	Moves eye superiorly (elevates eye) and medially (adducts eye)	O: Common tendinous ring I: Anterosuperior surface of eye	CN III (oculomotor nerve)
OBLIQUE MUSCLES			
Inferior oblique *obliquus = slanting*	Moves eye superiorly (elevates eye) and laterally (abducts eye)	O: Anterior orbital surface of maxilla I: Posteroinferior, lateral surface of eye	CN III (oculomotor nerve)
Superior oblique	Moves eye inferiorly (depresses eye) and laterally (abducts eye)	O: Sphenoid bone I: Posterosuperior, lateral surface of eye	CN IV (trochlear nerve)

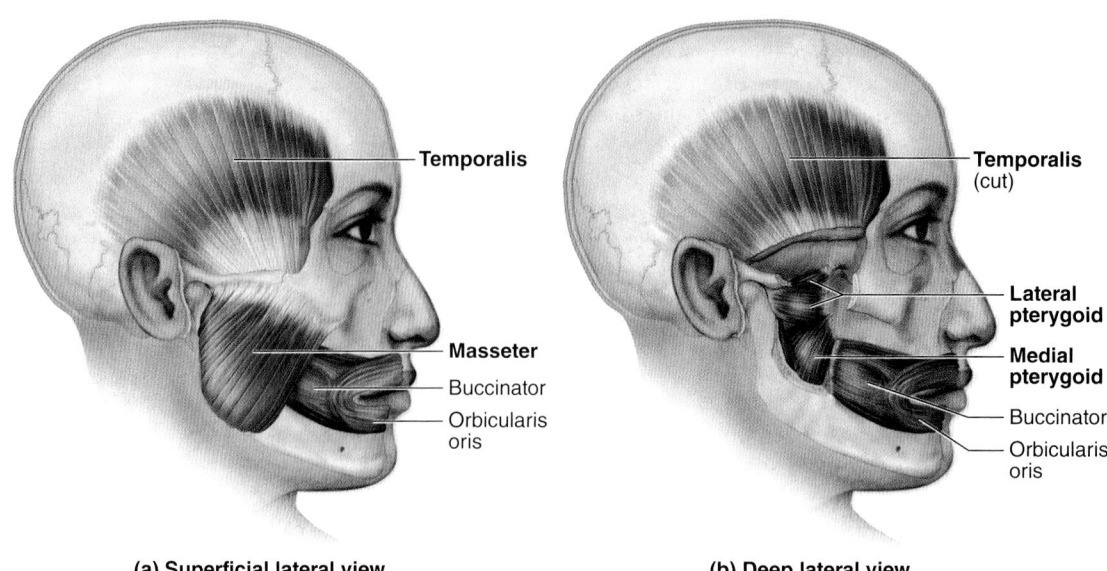

Figure 11.8 Muscles of Mastication. (*a*) Superficial and (*b*) deep lateral views of the muscles of mastication (shown in bold), which move the mandible.

AP|R

(a) Superficial lateral view

Temporalis
Masseter
Buccinator
Orbicularis oris

(b) Deep lateral view

Temporalis (cut)
Lateral pterygoid
Medial pterygoid
Buccinator
Orbicularis oris

11.3c Muscles of the Oral Cavity and Pharynx

LEARNING OBJECTIVES

5. Describe how each of the four muscles of mastication affects mandibular movement.
6. Describe the actions of the intrinsic muscles and the four paired extrinsic muscles of the tongue.
7. Explain what function the three primary muscles of the pharynx accomplish.

The muscles of the oral cavity and pharynx include muscles that help us chew, move the tongue, and swallow.

Muscles of Mastication

The term **mastication** (mas-ti-kā′shŭn) refers to the process of chewing. These muscles move the mandible at the temporomandibular joint (TMJ). There are four paired muscles of mastication: the temporalis, the masseter, and the lateral and medial pterygoids **(figure 11.8)**. The muscles of mastication are innervated by the mandibular division of CN V (trigeminal).

The **temporalis** (or *temporal* muscle) is a broad, fan-shaped muscle that extends from the temporal lines of the skull and inserts on the coronoid process of the mandible. It elevates and retracts (pulls posteriorly) the mandible. You can palpate the temporalis by placing your fingers along your temple (lateral skull at same level of orbits) as you open and close your jaw. The muscle you feel contracting is the temporalis.

The **masseter** elevates and protracts (pulls anteriorly) the mandible. It is the most powerful and important of the masticatory muscles. This short, thick muscle is superficial to the temporalis. You can feel the movement of the masseter by palpating near the angle of the mandible as you open and close your mouth.

The **lateral** and **medial pterygoid** muscles arise from the pterygoid processes of the sphenoid bone and insert on the mandible. Both pterygoids protract the mandible and move it from side to side during chewing. These movements maximize the effectiveness of the teeth while chewing or grinding foods of various consistencies. The medial pterygoid also elevates the mandible.

Table 11.3 summarizes the characteristics of the muscles of mastication.

Table 11.3	Muscles of Mastication		
Muscle	**Action**	**Origin/Insertion**	**Innervation (see section 13.9)**
Temporalis (tem-pŏ-rā′lis) *tempora* = pertaining to temporal bone	Elevates and retracts mandible	O: Superior and inferior temporal lines I: Coronoid process of mandible	CN V₃ (trigeminal nerve, mandibular division)
Masseter (mas′ĕ-tĕr) *maseter* = chewer	Elevates and protracts mandible; prime mover of mandible elevation	O: Zygomatic arch I: Coronoid process, lateral surface and angle of mandible	CN V₃ (trigeminal nerve, mandibular division)
Medial pterygoid (ter′i-goyd) *pterygoid* = winglike	Elevates and protracts mandible; produces side-to-side movement of mandible	O: Maxilla, palatine, and medial surface of lateral pterygoid plate I: Medial surface of mandibular ramus	CN V₃ (trigeminal nerve, mandibular division)
Lateral pterygoid	Protracts mandible; produces side-to-side movement of mandible	O: Greater wing of sphenoid and lateral surface of lateral pterygoid plate I: Condylar process of mandible	CN V₃ (trigeminal nerve, mandibular division)

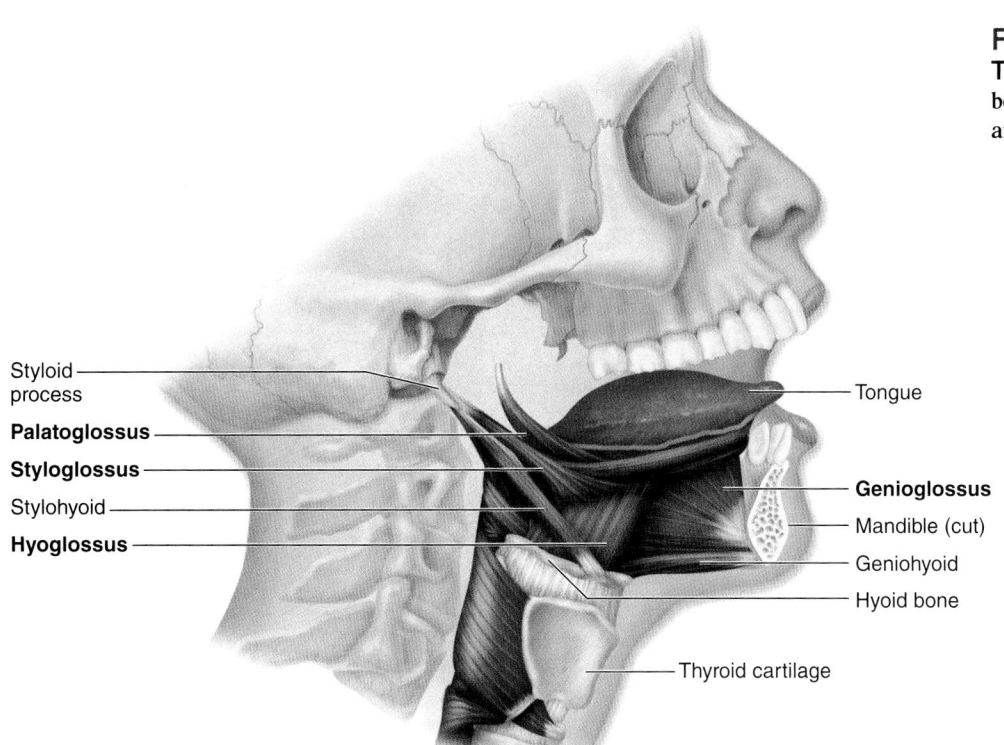

Figure 11.9 Muscles That Move the Tongue. Extrinsic tongue muscles (shown in bold) originate on structures other than the tongue and insert onto it to allow gross tongue movement.

Styloid process
Palatoglossus
Styloglossus
Stylohyoid
Hyoglossus

Tongue
Genioglossus
Mandible (cut)
Geniohyoid
Hyoid bone
Thyroid cartilage

Right lateral view

Muscles That Move the Tongue

The tongue is an agile, highly mobile organ. It consists of **intrinsic muscles** that curl, squeeze, and fold the tongue during chewing and speaking.

The **extrinsic muscles** of the tongue have their origin on other head and neck structures and insert on the tongue. The extrinsic muscles end in the suffix –*glossus,* meaning "tongue" (**figure 11.9**). These extrinsic tongue muscles are used in various combinations to accomplish the precise, complex, and delicate tongue movements required for proper speech and manipulating food within the mouth. Most of these muscles are innervated by CN XII, the hypoglossal nerve.

The left and right **genioglossus** muscles originate on the mandible and protract the tongue. You use these muscles when you stick out your tongue. The left and right **styloglossus** muscles originate on the styloid processes of the temporal bone. These muscles elevate and retract the tongue (pull the tongue posteriorly, back into the mouth). The left and right **hyoglossus** muscles originate on the hyoid bone and insert on the sides of the tongue. These muscles depress and retract the tongue. The left and right **palatoglossus** muscles originate on the soft palate and elevate the posterior portion of the tongue.

Table 11.4 summarizes the characteristics of the muscles that move the tongue.

Table 11.4	Muscles That Move the Tongue		
Muscle	**Action**	**Origin/Insertion**	**Innervation (see section 13.9)**
Genioglossus (jē′nē-ō-glos′ŭs) *geni* = chin *glossus* = tongue	Protracts tongue	O: Mental spines of mandible I: Inferior region of tongue; hyoid bone	CN XII (hypoglossal nerve)
Styloglossus (stī′lō-glos′ŭs) *stylo* = pertaining to styloid process of temporal bone	Elevates and retracts tongue	O: Styloid process of temporal bone I: Side and inferior aspect of tongue	CN XII (hypoglossal nerve)
Hyoglossus (hī′ō-glos′ŭs) *hyo* = pertaining to hyoid bone	Depresses and retracts tongue	O: Hyoid bone I: Inferolateral side of tongue	CN XII (hypoglossal nerve)
Palatoglossus (pal-ă-tō-glos′ŭs) *palato* = palate	Elevates posterior part of tongue	O: Anterior surface of soft palate I: Side and posterior aspect of tongue	CN X (vagus nerve) via pharyngeal plexus of nerves

Figure 11.10 **Pharyngeal Constrictors, Palate Muscles, and Laryngeal Elevators.** A right lateral view reveals some of the muscles that constrict the pharynx when swallowing, move the palate, and elevate the larynx (palatopharyngeus and salpingopharyngeus not shown). AP|R

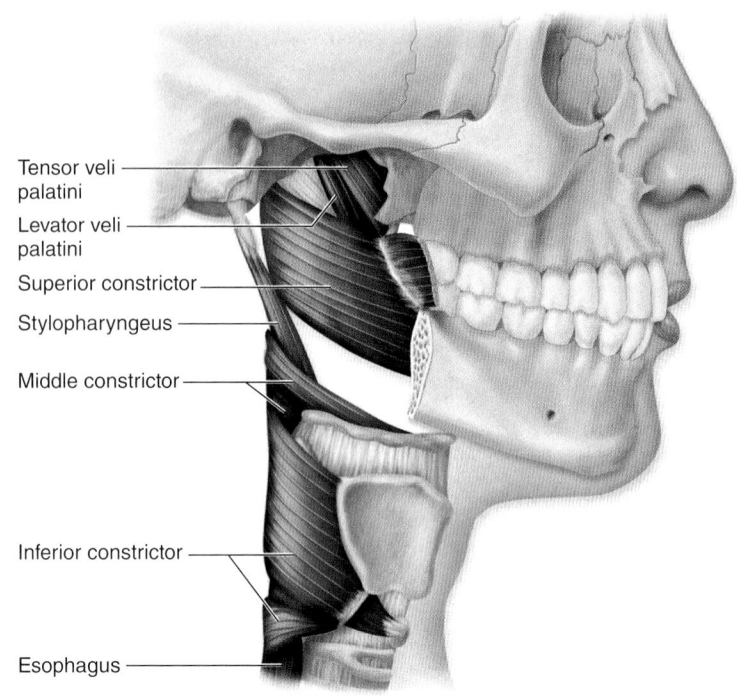

Tensor veli palatini
Levator veli palatini
Superior constrictor
Stylopharyngeus
Middle constrictor
Inferior constrictor
Esophagus

Right lateral view

Table 11.5	Muscles of the Pharynx		
Region/Muscle	**Action**	**Origin/Insertion**	**Innervation (see section 13.9)**
PALATE MUSCLES			
Levator veli palatini (vel′ī pal′ă-tē′nī) *velum* = veil	Elevates soft palate when swallowing	O: Petrous part of temporal bone I: Soft palate	CN X (vagus nerve)
Tensor veli palatini (ten′sōr) *tensus* = to stretch	Tenses soft palate and opens auditory tube when swallowing or yawning	O: Sphenoid bone; region around auditory tube I: Soft palate	CN V₃ (trigeminal nerve, mandibular division)
PHARYNGEAL CONSTRICTORS			
Superior constrictor (kon-strik′ter, -tōr) *constringo* = to draw together	Constricts pharynx in sequence to force bolus into esophagus; superior is innermost	O: Pterygoid process of sphenoid bone; medial surface of mandible I: Posterior median raphe (muscle fiber union from both sides)	CN X (vagus nerve) via branches of pharyngeal plexus
Middle constrictor	Constricts pharynx in sequence	O: Hyoid bone I: Posterior median raphe	CN X (vagus nerve) via branches of pharyngeal plexus
Inferior constrictor	Constricts pharynx in sequence; inferior is outermost	O: Thyroid and cricoid cartilage I: Posterior median raphe	CN X (vagus nerve) via branches of pharyngeal plexus
LARYNGEAL (VOICE BOX) ELEVATORS			
Palatopharyngeus (pal′ă-tō-fă-rin′jē′ŭs) *pharynx* = pharynx	Elevates pharynx and larynx	O: Soft palate I: Side of pharynx and thyroid cartilage of larynx	CN X (vagus nerve) via branches of pharyngeal plexus
Salpingopharyngeus (sal-ping′gō-fă-rin′jē-ŭs) *salpinx* = trumpet	Elevates pharynx and larynx	O: Auditory tube I: Blends with palatopharyngeus on lateral wall of pharynx	CN X (vagus nerve) via branches of pharyngeal plexus
Stylopharyngeus (stī′lō-fă-rin′jē-ŭs)	Elevates pharynx and larynx	O: Styloid process of temporal bone I: Side of pharynx and thyroid cartilage of larynx	CN IX (glossopharyngeal nerve) via branches of pharyngeal plexus

Pharynx Muscles

The **pharynx,** commonly known as the *throat*, is a funnel-shaped tube that lies posterior to both the oral and nasal cavities (see figure 26.4). Several muscles help form or attach to this tube and aid in swallowing **(figure 11.10).** Most pharyngeal muscles are innervated by CN X (vagus nerve).

The primary pharynx muscles are the **pharyngeal constrictors (superior, middle,** and **inferior).** When a bolus of food enters the pharynx, these muscles contract sequentially to initiate swallowing and force the bolus inferiorly into the esophagus. Other pharyngeal muscles help elevate or tense the palate when swallowing. These muscles are summarized in **table 11.5.**

WHAT DID YOU LEARN?

9 What movements do the medial and lateral pterygoid muscles perform?

10 What is the general function of the extrinsic muscles of the tongue?

11.3d Muscles of the Anterior Neck: The Hyoid Muscles

LEARNING OBJECTIVE

8. Contrast the actions of the four suprahyoid muscles and the four infrahyoid muscles.

The muscles of the anterior neck are divided into the suprahyoid muscles, which are superior to the hyoid bone, and the infrahyoid muscles, which are inferior to the hyoid bone **(figure 11.11).**

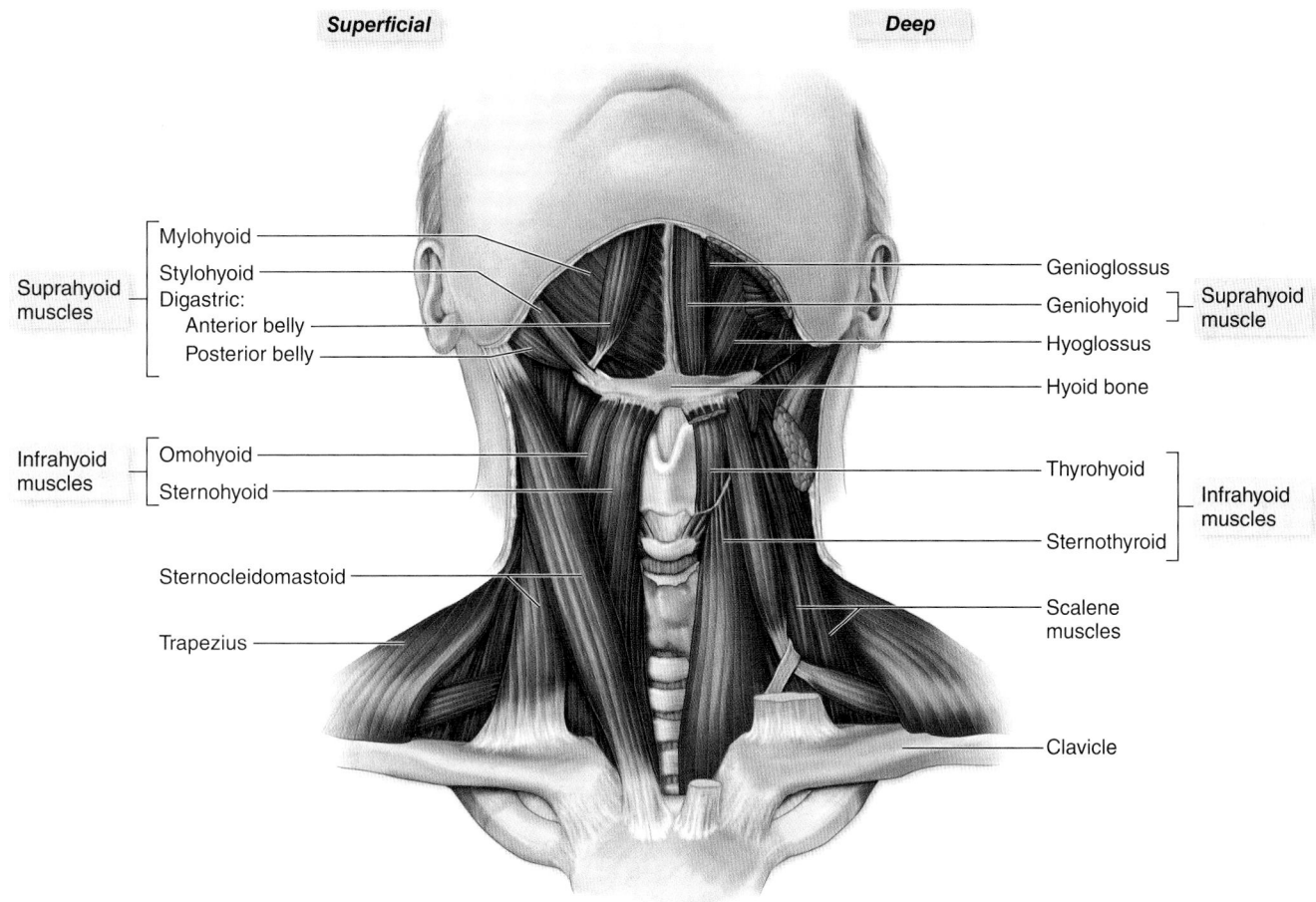

Anterior view

Figure 11.11 Muscles of the Anterior Neck. The anterior neck muscles move the hyoid bone and thyroid cartilage. Superficial muscles are shown on the right side of the body, and deeper muscles are shown on the left side of the body. **AP|R**

The **suprahyoid muscles** are associated with the floor of the mouth. In general, these muscles act as a group to elevate the hyoid bone during swallowing or speaking. Some of these muscles perform additional functions: The **digastric** has two bellies, anterior and posterior. The anterior belly extends from the mental protuberance to the hyoid, and the posterior belly continues from the hyoid to the mastoid part of the temporal bone. The two bellies are united by an intermediate tendon that is held in position by a fibrous loop. In addition to elevating the hyoid bone, the digastric muscle can also depress the mandible. The **geniohyoid** attaches to the mental spines of the mandible and the hyoid bone. This muscle elevates the hyoid bone. The broad, flat **mylohyoid** provides a muscular floor to the mouth. When this muscle contracts, it both elevates the hyoid bone and raises the floor of the mouth. The muscle fibers of the left and right mylohyoid are aligned in a V shape. The **stylohyoid** connects the styloid process of the skull and the hyoid bone. Upon contraction, it elevates the hyoid bone, causing the floor of the oral cavity to elongate during swallowing.

As swallowing is completed, the **infrahyoid muscles** contract to influence the position of the hyoid bone and the larynx. In general, these muscles either depress the hyoid bone or depress the thyroid cartilage of the larynx. The **omohyoid** contains two thin muscle bellies anchored in place by a fascia "sling." This muscle is lateral to the sternohyoid and extends from the superior border of the scapula to insert on the hyoid, where it depresses the hyoid bone. The **sternohyoid** extends from the sternum to the hyoid bone, where it depresses the hyoid bone. The **sternothyroid** is deep to the sternohyoid. It extends from the sternum to the thyroid cartilage of the larynx. It depresses the thyroid cartilage to return it to its original position after swallowing. The **thyrohyoid** extends from the thyroid cartilage of the larynx to the hyoid bone. It depresses the hyoid bone and elevates the thyroid cartilage to close off the larynx during swallowing. In addition, the omohyoid, sternohyoid, and thyrohyoid help anchor the hyoid so the digastric can depress the mandible.

Table 11.6 summarizes the characteristics of these muscles.

Table 11.6	Muscles of the Anterior Neck		
Region/Muscle	**Action**	**Origin/Insertion**	**Innervation (see sections 13.9 and 14.5d)**
SUPRAHYOID MUSCLES			
Digastric (dī-gas′trik) *di* = two *gaster* = belly	Depresses mandible; elevates hyoid bone	O: Anterior belly: Mandible near mental protuberance; posterior belly, mastoid process I: Hyoid bone via fascia sling	Anterior belly: CN V₃ (trigeminal nerve, mandibular division) Posterior belly: CN VII (facial nerve)
Geniohyoid (jĕ-nē-ō-hī′oyd)	Elevates hyoid bone	O: Mental spines of mandible I: Hyoid bone	First cervical spinal nerve (C1) via CN XII (hypoglossal nerve)
Mylohyoid (mī′lō-hī′oyd) *myle* = molar	Elevates hyoid bone; elevates floor of mouth	O: Mylohyoid line of mandible I: Hyoid bone	CN V₃ (trigeminal nerve, mandibular division)
Stylohyoid (stī-lō-hī′oyd)	Elevates hyoid bone	O: Styloid process of temporal bone I: Hyoid bone	CN VII (facial nerve)
INFRAHYOID MUSCLES			
Omohyoid (ō-mō-hī′oyd) *omo* = shoulder	Depresses hyoid bone; fixes hyoid during opening of mouth	O: Superior border of scapula I: Hyoid bone	Cervical spinal nerves C1–C3 through ansa cervicalis (from cervical plexus)
Sternohyoid (ster′nō-hī′oyd) *sterno* = sternum	Depresses hyoid bone	O: Manubrium of sternum and medial end of clavicle I: Hyoid bone	Cervical spinal nerves C1–C3 through ansa cervicalis (from cervical plexus)
Sternothyroid (ster′nō-thī′royd) *thyro* = thyroid cartilage	Depresses thyroid cartilage of larynx	O: Posterior surface of manubrium of sternum I: Thyroid cartilage of larynx	Cervical spinal nerves C1–C3 through ansa cervicalis (from cervical plexus)
Thyrohyoid (thī-rō-hī′oyd)	Depresses hyoid bone and elevates thyroid cartilage of larynx	O: Thyroid cartilage of larynx I: Hyoid bone	First cervical spinal nerve C1 via CN XII (hypoglossal nerve)

WHAT DO YOU THINK?

① Since muscles frequently are named for their attachment sites, what do you think the prefix *-omo* in "omohyoid" means?

WHAT DID YOU LEARN?

⑪ List the four suprahyoid muscles, and describe a common function for them.

11.3e Muscles That Move the Head and Neck

✓ **LEARNING OBJECTIVE**

9. Compare and contrast the actions of the anterolateral neck muscles and the posterior neck muscles.

Muscles that move the head and neck originate on the vertebral column, the thoracic cage, and the pectoral girdle, and insert on bones of the cranium (**figure 11.12**; see also figure 11.11).

Anterolateral Neck Muscles

The anterolateral neck muscles all flex the head and/or neck. The main muscles in this group are the sternocleidomastoid and the three scalenes.

The **sternocleidomastoid** is a thick, cordlike muscle that extends from the sternum and clavicle to the mastoid process posterior to the ear. Contraction of both sternocleidomastoid muscles (called **bilateral contraction**) flexes the neck. Contraction of just one sternocleidomastoid muscle (termed **unilateral contraction**) results in lateral flexion of the head to its own side, accompanied by rotation of the head to the opposite side. Thus, if the left sternocleidomastoid muscle contracts, it rotates the head to the right side of the body. The three **scalene muscles** (**anterior, middle,** and **posterior**)

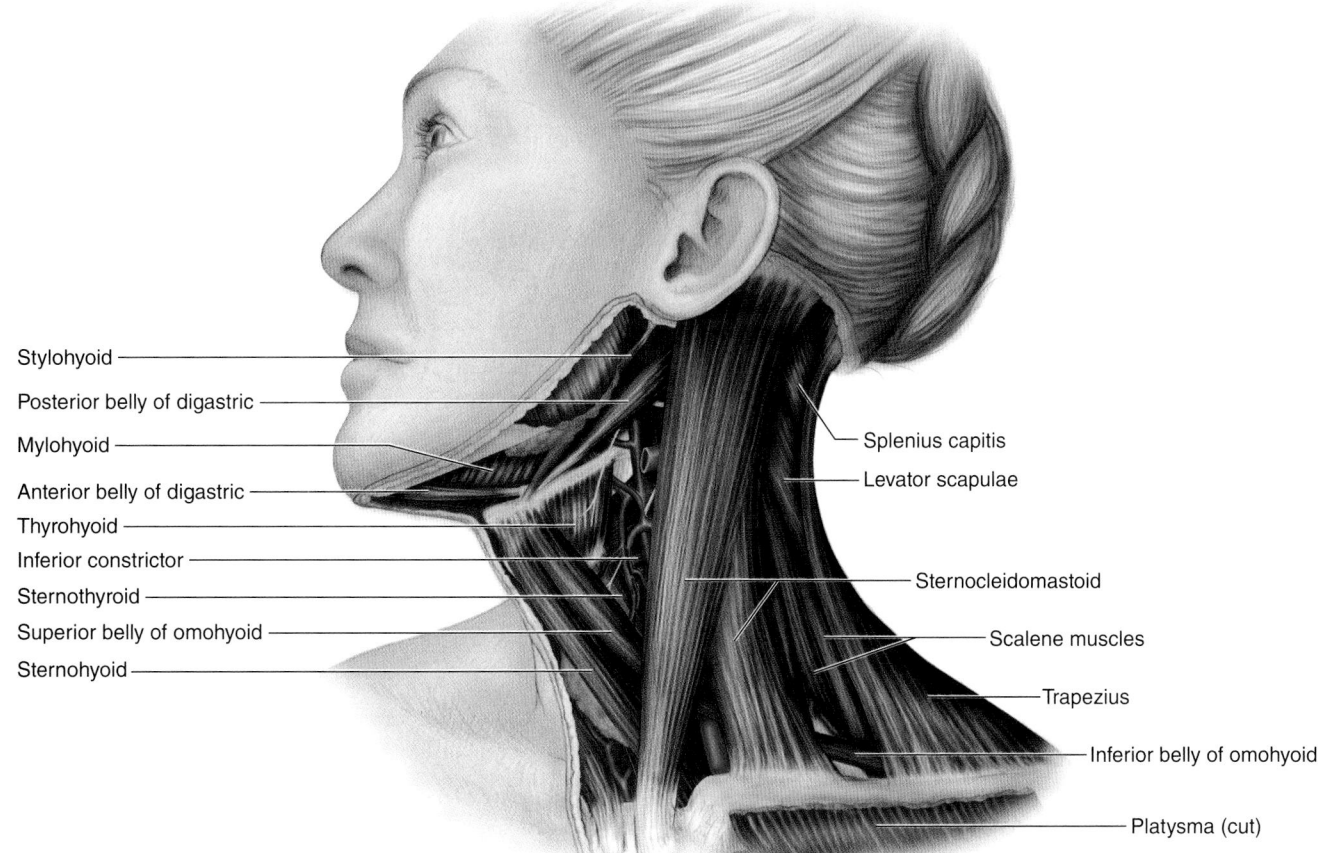

Stylohyoid
Posterior belly of digastric
Mylohyoid
Anterior belly of digastric
Thyrohyoid
Inferior constrictor
Sternothyroid
Superior belly of omohyoid
Sternohyoid

Splenius capitis
Levator scapulae
Sternocleidomastoid
Scalene muscles
Trapezius
Inferior belly of omohyoid
Platysma (cut)

Anterolateral view

Figure 11.12 Muscles That Move the Head and Neck. Anterolateral muscles collectively flex the neck, whereas posterior neck muscles extend the head and/or neck. AP|R

work with the sternocleidomastoid to flex the neck. In addition, the scalene muscles elevate the first and second ribs during forced inspiration (see section 23.5b).

Posterior Neck Muscles

Many of the posterior neck muscles work together to extend the head and/or neck **(figure 11.13)**. The trapezius attaches to the skull and helps extend the head, neck, or both, but its primary function is to help move the pectoral girdle.

When the left and right **splenius capitis, splenius cervicis, semispinalis capitis,** and **longissimus capitis** muscles bilaterally contract, they extend the neck. Unilateral contraction turns the head and neck to the same side.

A group of muscles called the suboccipital muscles includes the obliquus capitis superior, obliquus capitis inferior, rectus capitis posterior major, and rectus capitis posterior minor. The oblique muscles turn the head to the same side, whereas the rectus muscles extend the head and neck.

Table 11.7 summarizes the characteristics of the muscles of the head and neck.

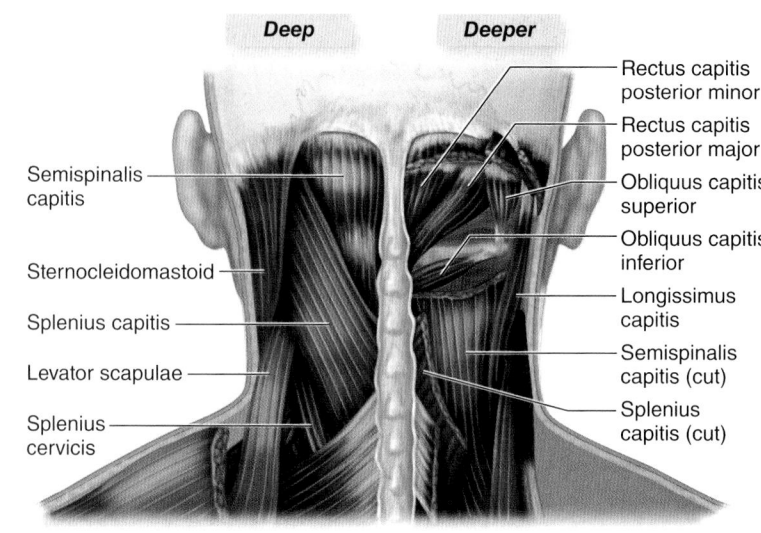

Deep   **Deeper**

Semispinalis capitis
Sternocleidomastoid
Splenius capitis
Levator scapulae
Splenius cervicis

Rectus capitis posterior minor
Rectus capitis posterior major
Obliquus capitis superior
Obliquus capitis inferior
Longissimus capitis
Semispinalis capitis (cut)
Splenius capitis (cut)

Posterior view

Figure 11.13 Posterior Neck Muscles. An illustration shows the deep and deeper muscles that extend and rotate the head and neck. AP|R

 WHAT DID YOU LEARN?

12 Which neck muscles extend the neck? Which neck muscles flex the neck?

Table 11.7	Muscles That Move the Head and Neck

Muscle	Action	Origin/Insertion	Innervation (see sections 13.9 and 14.5c, d)
Sternocleidomastoid (ster′nō-klī′dō-măs-toyd) *sterno* = sternum *cleido* = clavicle *masto* = mastoid process	Unilateral action[1]: Lateral flexion, rotation of head to opposite side Bilateral action[2]: Flexes neck	O: Manubrium and sternal end of clavicle I: Mastoid process	CN XI (accessory nerve)
Scalene muscles (anterior, middle, posterior) (see also table 11.9) (skā′lēnz) *scalene* = uneven	Flex neck (when 1st rib is fixed); elevate 1st and 2nd ribs during forced inspiration when neck is fixed	O: Transverse processes of cervical vertebrae I: Superior surface of 1st and 2nd ribs	Cervical spinal nerves
Splenius capitis and **cervicis** (splē′nē-ŭs ka-pī-tis) (ser′vi′sis) *splenion* = bandage	Unilateral action: Turns head to same side Bilateral action: Extends head/neck	O: Ligamentum nuchae I: Occipital bone and mastoid process of temporal bone	Cervical spinal nerves
Longissimus capitis (lon-gis′i-mŭs) *longissimus* = longest *caput* = head	Unilateral action: Turns head to same side Bilateral action: Extends head/neck	O: Transverse process of T_1–T_4 and articular processes of C_4–C_7 vertebrae I: Mastoid process	Cervical and thoracic spinal nerves
Obliquus capitis superior (ob-lī′kwŭs) *obliquus* = slanting	Turns head to same side	O: Transverse process of atlas I: Inferior nuchal line of occipital bone	Suboccipital nerve (posterior ramus of C1 spinal nerve)
Obliquus capitis inferior	Turns head to same side	O: Spinous process of axis I: Transverse process of atlas	Suboccipital nerve (posterior ramus of C1 spinal nerve)
Rectus capitis posterior major	Extends head/neck	O: Spinous process of axis I: Inferior nuchal line of occipital bone	Suboccipital nerve (posterior ramus of C1 spinal nerve)
Rectus capitis posterior minor	Extends head/neck	O: Posterior tubercle of atlas I: Inferior nuchal line of occipital bone	Suboccipital nerve (posterior ramus of C1 spinal nerve)

1. *Unilateral action* means only one muscle (either the left or right muscle) is contracting.
2. *Bilateral action* means both the left and right muscles are contracting together.

INTEGRATE

CLINICAL VIEW
Congenital Muscular Torticollis

Photo of 7-year-old boy with CMT.

Congenital muscular torticollis (CMT), often known as *wryneck*, is a condition where a newborn presents with a shortened and tightened sternocleidomastoid muscle that may persist into childhood. It is thought to be a result of trauma resulting from either a difficult birth or prenatal position of the fetus. The trauma causes a hematoma and fibrosing of the muscle tissue. Pediatricians also have seen an increase in acquired muscular torticollis among newborns who are kept in infant seats for extended periods of time outside of the car. Children with CMT often tilt their heads to the affected side and their chins to the unaffected side. Because the child typically favors a particular head position when sleeping, plagiocephaly (flattening of the head) often accompanies CMT.

CMT treatment typically involves stretching the affected muscle several times a day, changing sleeping positions, and making the child use the affected side more frequently. A newer approach to treatment of CMT is the use of botulinum toxin type A (Botox), which impairs contraction of the affected muscle), combined with stretching. In cases that do not respond to the treatments mentioned, sternocleidomastoid release surgery is recommended. The sternocleidomastoid is cut from at least one attachment point, repositioned, reattached, and Botox may be injected into the muscle. Recovery includes wearing a neck brace and physical therapy.

11.4 Muscles of the Vertebral Column

LEARNING OBJECTIVES

1. Name and describe the three groups of erector spinae muscles.
2. Describe the actions of the transversospinalis and quadratus lumborum muscles.

The muscles of the vertebral column are very complex; they have multiple origins and insertions, and they exhibit extensive overlap **(figure 11.14)**.

All of these muscles are covered by the most superficial back muscles, which actually move the upper limb.

Note that the neck is actually the cervical portion of the vertebral column. Thus, the posterior muscles discussed previously in connection with neck extension (splenius cervicis, splenius capitis, longissimus capitis, semispinalis capitis) extend the *cervical* region of the vertebral column.

The **erector spinae** are used to maintain posture and to help an individual stand erect. When the left and right erector spinae muscles contract together, they extend the vertebral column. If the erector spinae muscles on only one side contract, the vertebral column flexes laterally toward that same side.

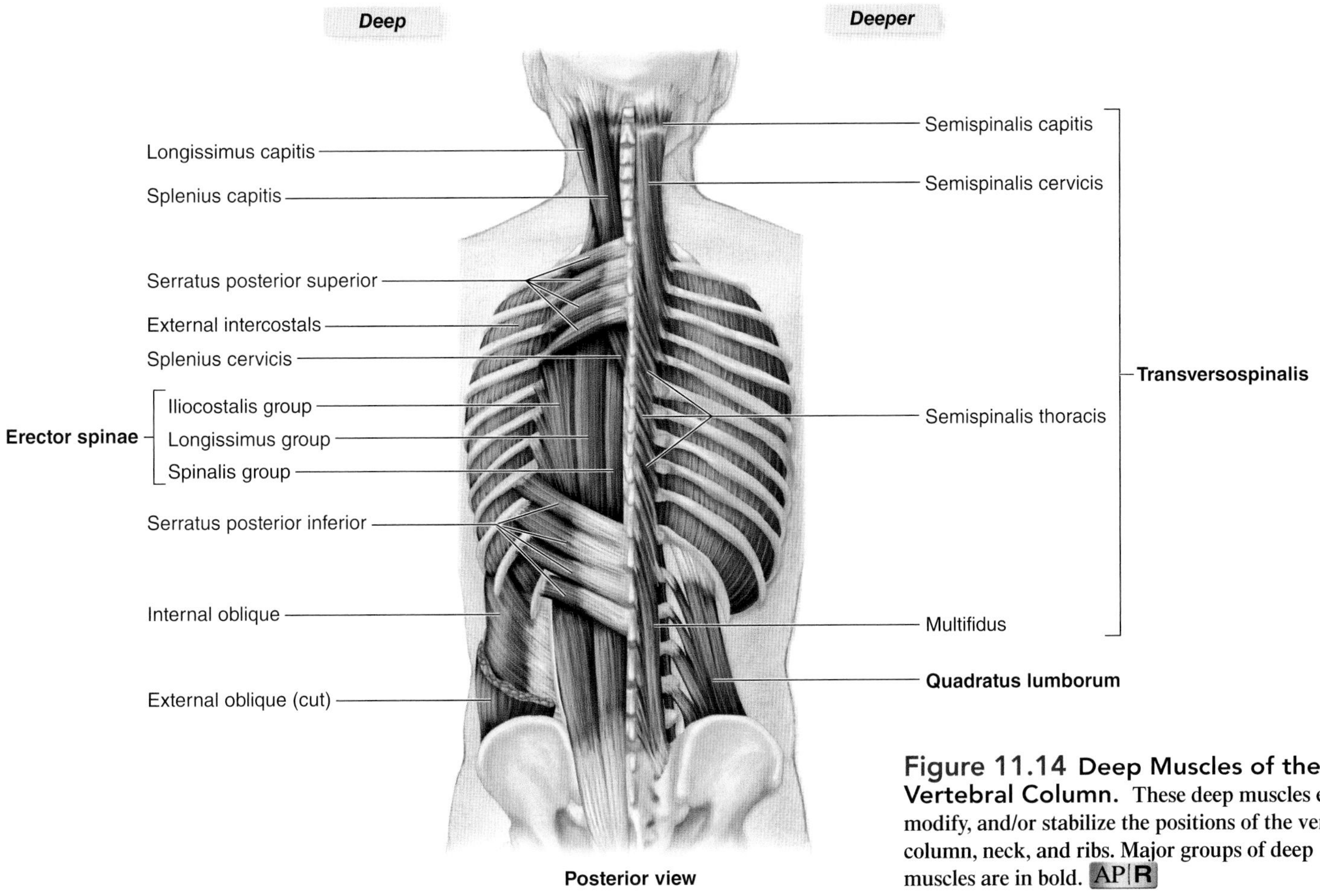

Deep   **Deeper**

Longissimus capitis

Splenius capitis

Serratus posterior superior

External intercostals

Splenius cervicis

Erector spinae — Iliocostalis group / Longissimus group / Spinalis group

Serratus posterior inferior

Internal oblique

External oblique (cut)

Semispinalis capitis

Semispinalis cervicis

Transversospinalis

Semispinalis thoracis

Multifidus

Quadratus lumborum

Posterior view

Figure 11.14 Deep Muscles of the Vertebral Column. These deep muscles extend, modify, and/or stabilize the positions of the vertebral column, neck, and ribs. Major groups of deep muscles are in bold. AP|R

The erector spinae muscles are organized into three groups. The muscles are named based upon the body region with which they are associated.

- The **iliocostalis group** is the most laterally placed of the three erector spinae components. It is composed of three parts: cervical, thoracic, and lumbar.

- The **longissimus group** is medial to the iliocostalis group. The fibers of the longissimus muscle group insert on the transverse processes of the vertebrae. The longissimus group is composed of three parts: capitis, cervical, and thoracic.

- The **spinalis group** is the most medially placed of the erector spinae muscles. The spinalis muscle fibers insert on the spinous processes of the vertebrae. The spinalis group is composed of cervical and thoracic parts.

Deep to the erector spinae, a group of muscles collectively called the **transversospinalis muscles** connect and stabilize the vertebrae. This group includes several specific muscles, which are listed in the accompanying table.

A final pair of muscles helps move the vertebral column. The **quadratus lumborum muscles** are located primarily in the lumbar region. When the left and right quadratus lumborum muscles bilaterally contract, they extend the vertebral column. When either the left or right quadratus lumborum muscle unilaterally contracts, it laterally flexes the vertebral column.

Table 11.8 summarizes the characteristics of the muscles of the vertebral column.

 WHAT DID YOU LEARN?

13 Which muscles form the erector spinae, and what are the general actions of the erector spinae?

Table 11.8	Muscles of the Vertebral Column		
Group/Muscle	**Action**	**Origin/Insertion**	**Innervation (see sections 14.5c, d, and f)**
ERECTOR SPINAE			
Iliocostalis group (il-ē-ō-kos-tă′lis) *ilio* = ilium *cost* = rib	Bilateral action: Extends neck and vertebral column; maintains posture Unilateral action: Laterally flexes vertebral column	O: Tendon from posterior part of iliac crest, posterior sacrum, and lumbar spinous processes I: Angles of ribs; transverse processes of cervical vertebrae	Cervical, thoracic, and lumbar spinal nerves
Longissimus group (lon-gis′i-mŭs)	Bilateral action: Extends neck and vertebral column; maintains posture Unilateral action: Rotates head and laterally flexes vertebral column	O: Tendon from posterior part of iliac crest, posterior sacrum, and lumbar spinous processes I: Mastoid process of temporal bone and transverse processes of cervical and thoracic vertebrae	Cervical and thoracic spinal nerves
Spinalis group (spī-nā′lis) *spin* = spine	Bilateral action: Extends neck and vertebral column; maintains posture Unilateral action: Laterally flexes vertebral column	O: Lumbar spinous processes (thoracic part) and C_7 spinous process (cervical part) I: Spinous process of axis and thoracic vertebrae	Cervical and thoracic spinal nerves
TRANSVERSOSPINALIS GROUP			
Multifidus (mul-tif′i-dŭs) *multus* = much *findo* = to cleave	Bilateral action: Extends vertebral column Unilateral action: Rotates vertebral column toward opposite side	O: Sacrum and transverse processes of each vertebra I: Spinous processes of vertebrae located 2–4 segments superior to origin	Cervical, thoracic, and lumbar spinal nerves
Rotatores (rō-tā′tōrz)	Bilateral action: Extends vertebral column Unilateral action: Rotates vertebral column toward opposite side	O: Transverse processes of each vertebra I: Spinous process of immediately superior vertebra	Cervical, thoracic, and lumbar spinal nerves
Semispinalis group (sem′ē-spī-nā′lis)	Bilateral action: Extends vertebral column and neck Unilateral action: Laterally flexes vertebral column and neck	O: Transverse processes of C_4–T_{12} vertebrae I: Occipital bone and spinous processes of cervical and thoracic vertebrae	Cervical and thoracic spinal nerves
SPINAL EXTENSORS AND LATERAL FLEXORS			
Quadratus lumborum (kwah-drā′tūs lŭm-bōr′ŭm) *quad* = four-sided	Bilateral action: Extends vertebral column Unilateral action: Laterally flexes vertebral column	O: Iliac crest and iliolumbar ligament I: 12th rib; transverse processes of lumbar vertebrae	Thoracic and lumbar spinal nerves

11.5 Muscles of Respiration

LEARNING OBJECTIVES

1. List the posterior and anterior thoracic muscle groups involved in respiration, and describe their actions.

2. Describe the role of the diaphragm in breathing and in raising intra-abdominal pressure.

The process of respiration involves inspiration and expiration. During **inspiration,** several muscles contract to increase the dimensions of the thoracic cavity to allow the lungs to fill with air. During **expiration,** some respiratory muscles contract and others relax, collectively decreasing the dimensions of the thoracic cavity and forcing air out of the lungs (see section 23.5b).

The muscles of respiration are on the posterior and anterior surfaces of the thorax. These muscles are covered by more superficial muscles (such as the pectoral muscles, trapezius, and latissimus dorsi) that move the upper limb.

Two posterior thorax muscles assist with respiration. The **serratus posterior superior** attaches to ribs 2–5 (see figure 11.14) and elevates these ribs during forced inspiration, thereby increasing the lateral dimensions of the thoracic cavity. The **serratus posterior inferior** attaches to ribs 8–12 and depresses those ribs during forced expiration. (Normal, quiet expiration takes no active muscular effort.)

Several groups of anterior thorax muscles change the dimensions of the thorax during respiration (**figure 11.15**). The scalene muscles (discussed previously with other neck muscles) help elevate the first and second ribs during forced inspiration, thereby increasing the dimensions of the thoracic cavity.

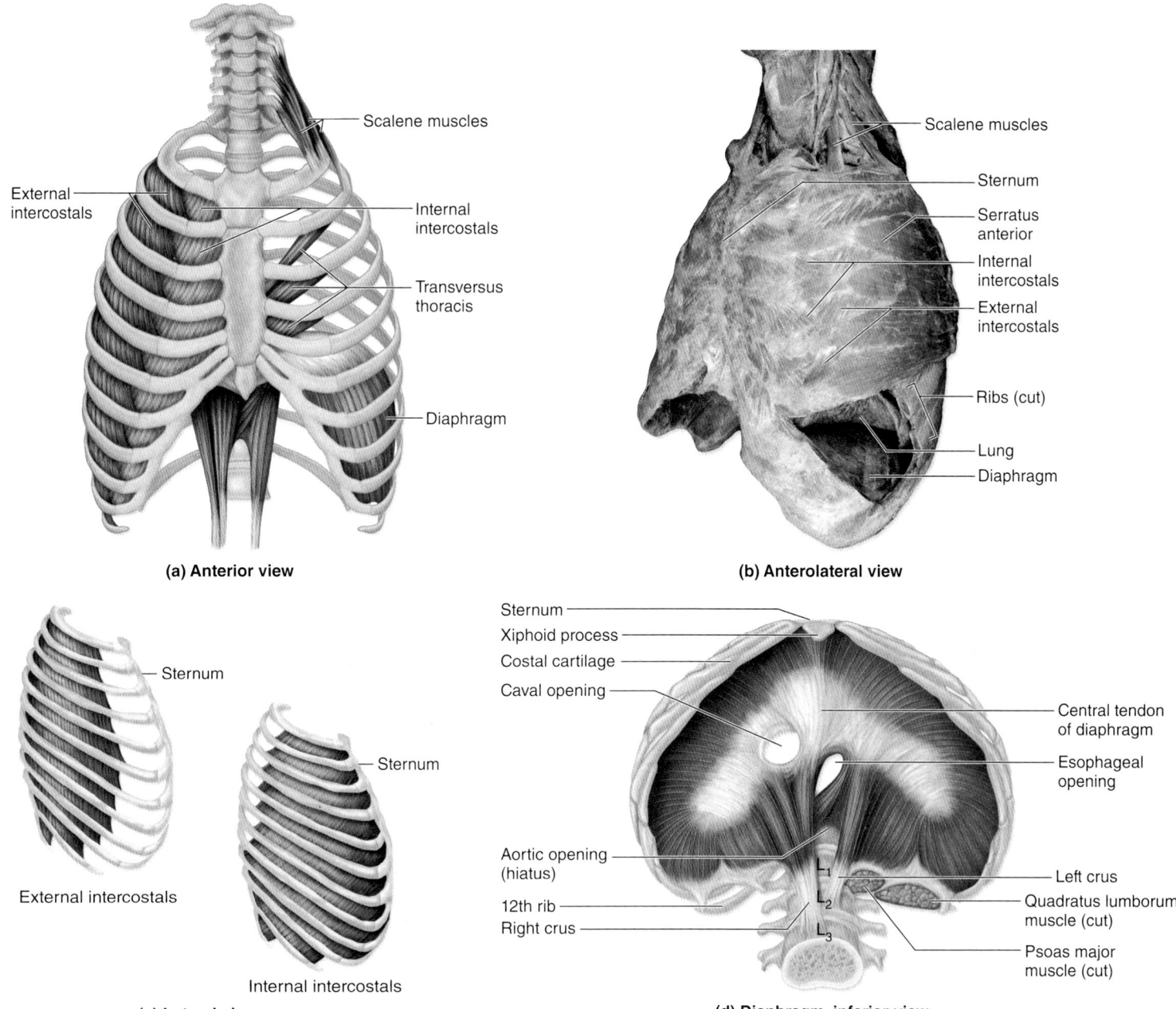

(a) Anterior view

(b) Anterolateral view

(c) Lateral view

(d) Diaphragm, inferior view

Figure 11.15 Muscles of Respiration. These skeletal muscles contract rhythmically to alter the size of the thoracic cavity and facilitate respiration. (*a*) Anterior view. (*b*) A cadaver photo provides an anterolateral view, with the inferior ribs cut to expose the thoracic cavity and the superior surface of the diaphragm. (*c*) Lateral views demonstrate fiber directions of the external and internal intercostals. (*d*) Inferior view of the diaphragm.

AP|R

| Table 11.9 | Muscles of Respiration |

Muscle	Action	Origin/Insertion	Innervation (see sections 14.5c, d)
Serratus posterior superior (sĕr-ā′tŭs) *serratus* = a saw	Elevates ribs during forced inspiration	O: Spinous processes of C_7–T_3 vertebrae I: Lateral borders of ribs 2–5	Thoracic spinal nerves
Serratus posterior inferior	Depresses ribs during forced expiration	O: Spinous processes of T_{11}–L_3 vertebrae I: Inferior borders of ribs 8–12	Thoracic spinal nerves
Scalene muscles (see also table 11.7)	Elevate 1st and 2nd ribs during forced inspiration when neck is fixed	O: Transverse processes of cervical vertebrae I: Superior surface of 1st and 2nd ribs	Cervical spinal nerves
External intercostals (in′ter-kos′talz) *inter* = between *cost* = rib	Elevate ribs during quiet and forced inspiration	O: Inferior border of superior rib I: Superior border of inferior rib	Thoracic spinal nerves
Internal intercostals	Depresses ribs during forced expiration	O: Superior border of inferior rib I: Inferior border to superior rib	Thoracic spinal nerves
Transversus thoracis (trans-ver′sŭs thō-ra′sis)	Depresses ribs during forced expiration	O: Posterior surface of xiphoid process and inferior region of sternum I: Costal cartilages 2–6	Thoracic spinal nerves
Diaphragm (dī′ă-fram) *dia* = across *phragm* = partition	Contraction causes flattening of diaphragm (moves inferiorly) during inspiration and thus expands thoracic cavity; increases pressure in abdominopelvic cavity	O: Inferior internal surface of ribs 7–12; xiphoid process of sternum and costal cartilages of inferior 6 ribs; lumbar vertebrae I: Central tendon	Phrenic nerves (C_3–C_5)

The **external intercostals** extend inferomedially from the superior rib to the adjacent inferior rib. The external intercostals assist in expanding the thoracic cavity by elevating the ribs during inspiration. This movement is like lifting a bucket handle—that is, as the bucket handle (rib) is elevated, its distance from the center of the bucket (thorax) increases. Thus, contraction of the external intercostals increases the transverse dimensions of the thoracic cavity. The **internal intercostals** lie deep to the external intercostals, and their muscle fibers are at right angles to the external intercostals. The internal intercostals depress the ribs only during forced expiration; normal, quiet expiration takes no active muscular effort. A small **transversus thoracis** extends across the inner surface of the thoracic cage and attaches to ribs 2–6. It helps depress the ribs during forced expiration.

Finally, the **diaphragm** is an internally placed, dome-shaped muscle that forms a partition between the thoracic and abdominal cavities. (The term *diaphragm* refers to a muscle or group of muscles that covers or partitions an opening.) The diaphragm is the most important muscle associated with breathing. The muscle fibers of the diaphragm converge from its margins toward a fibrous

central tendon. During inspiration, the diaphragm contracts and the central tendon is pulled inferiorly toward the abdominal cavity, thereby increasing the vertical dimensions of the thoracic cavity.

Table 11.9 summarizes the characteristics of the muscles of respiration. Further details about muscles of respiration are found in section 23.5b.

WHAT DO YOU THINK?

2 After you've eaten a very large meal, it is sometimes difficult to take a big, deep breath. Why is it more difficult to breathe deeply with a full GI tract?

WHAT DID YOU LEARN?

14 Compare the actions of the external intercostals and the internal intercostals.

15 How is the diaphragm involved in respiration?

11.6 Muscles of the Abdominal Wall

LEARNING OBJECTIVES

1. List the four pairs of abdominal muscles.
2. Compare the actions of the rectus abdominis muscle with the oblique muscles and transversus abdominis.

The anterolateral wall of the abdomen is reinforced by four pairs of muscles that collectively compress and hold the abdominal organs in place: the external oblique, internal oblique, transversus abdominis, and rectus abdominis **(figure 11.16)**. These muscles also work together to flex and stabilize the vertebral column.

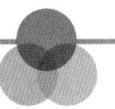

INTEGRATE

CONCEPT CONNECTION

When the diaphragm contracts, it compresses the abdominal cavity, thus increasing intra-abdominal pressure. Increased intra-abdominal pressure is necessary for urination (see section 24.8c), defecation (see section 26.3d), and childbirth (see section 29.6). Beyond respiration, diaphragm movements are also important in helping return venous blood to the heart from the inferior half of the body (see section 20.5a).

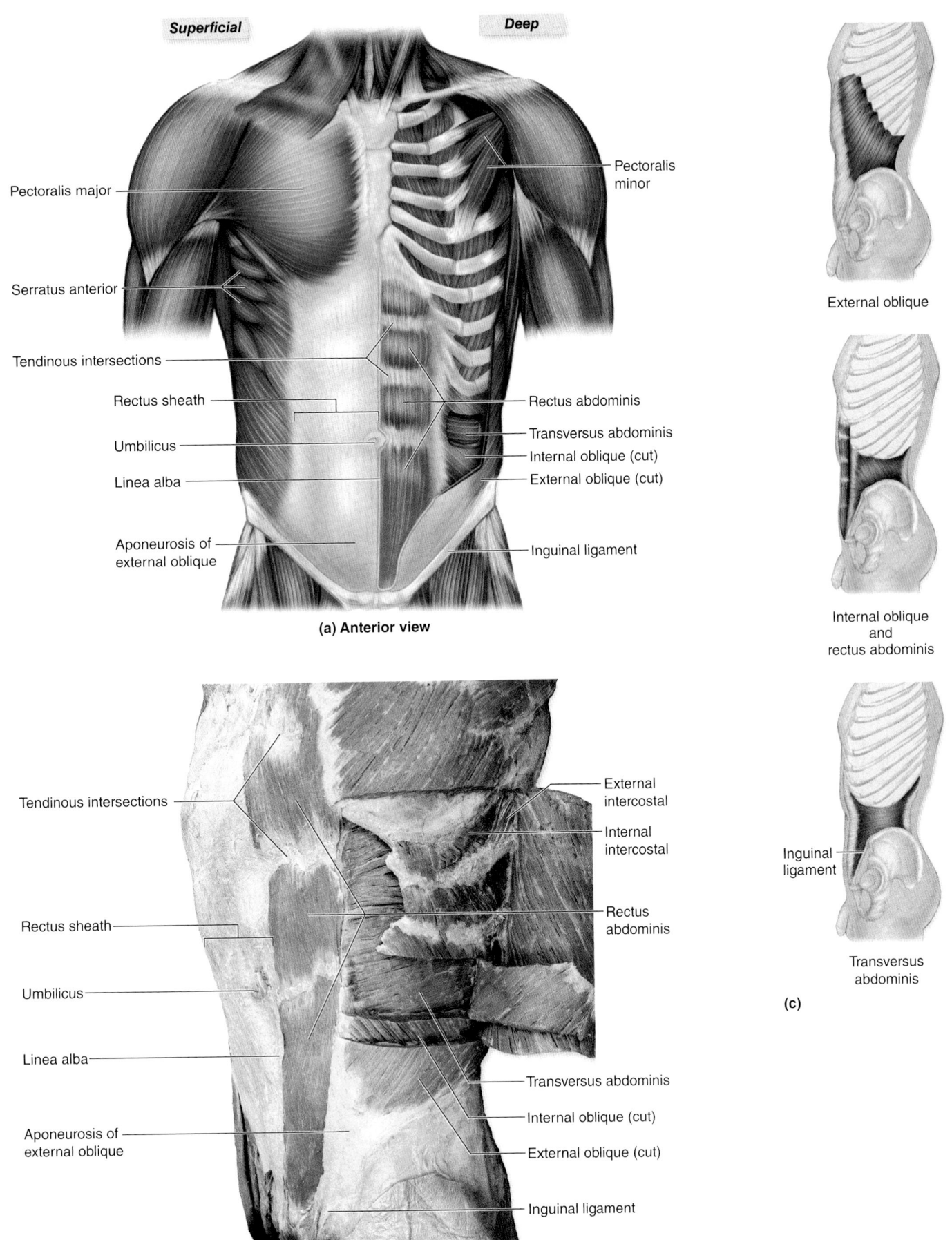

Superficial

Deep

Pectoralis major

Serratus anterior

Tendinous intersections

Rectus sheath

Umbilicus

Linea alba

Aponeurosis of external oblique

Pectoralis minor

Rectus abdominis

Transversus abdominis

Internal oblique (cut)

External oblique (cut)

Inguinal ligament

(a) Anterior view

External oblique

Internal oblique and rectus abdominis

Tendinous intersections

Rectus sheath

Umbilicus

Linea alba

Aponeurosis of external oblique

External intercostal

Internal intercostal

Rectus abdominis

Transversus abdominis

Internal oblique (cut)

External oblique (cut)

Inguinal ligament

Inguinal ligament

Transversus abdominis

(c)

(b) Anterolateral view

Figure 11.16 Muscles of the Abdominal Wall. The abdominal muscles compress abdominal contents and flex the vertebral column. (*a*) An illustration depicts the anterior view of some superficial and deep muscles. (*b*) A cadaver photo provides an anterolateral view of the muscles of the abdominal wall. (*c*) Diagrams show some individual abdominal muscles, ranging from superficial to deep. APǀR

LEARNING STRATEGY

The fibers of the external intercostals and external oblique muscles run in the same direction—inferomedially. This is the same direction that you put your hands in your pockets.

The fibers of the internal intercostals and internal oblique muscles run perpendicular (in the opposite direction) to the external muscles—superomedially.

The muscle fibers of the superficial **external oblique** are directed inferomedially. The external oblique is composed of muscle along the lateral abdominal wall and forms an aponeurosis as it projects anteriorly. Inferiorly, the aponeurosis of the external oblique forms a strong, cordlike **inguinal ligament** that extends from the anterior superior iliac spine to the pubic tubercle. Immediately deep to the external oblique is the **internal oblique.** Its muscle fibers project superomedially, which is at right angles to the external oblique. Like the external oblique, this muscle forms an aponeurosis as it projects anteriorly. The deepest muscle is the **transversus abdominis,** whose fibers project transversely across the abdomen and has an aponeurosis as it projects anteriorly. When these three sets of muscles unilaterally contract, they laterally flex the vertebral

column. These muscles also rotate the vertebral colum to the opposite side of the contracting muscle.

The **rectus abdominis** is a long, straplike muscle that extends vertically the entire length of the anteromedial abdominal wall between the sternum and the pubic symphysis. It is partitioned into four segments by three fibrous **tendinous intersections,** which form the traditional "six-pack" of a muscular, toned abdominal wall. The rectus abdominis is enclosed within a fibrous sleeve called the **rectus sheath,** which is formed from the aponeuroses of the external oblique, internal oblique, and transversus abdominis muscles. The left and right rectus sheaths are connected by a vertical fibrous strip termed the **linea alba.**

Table 11.10 summarizes the characteristics of the muscles of the abdominal wall.

You have probably noticed that multiple muscles may work together to perform a common function. For example, several neck muscles and back muscles work together to extend the vertebral column. Learning muscles in groups according to common function helps most students assimilate the anatomy information. **Table 11.11** summarizes the actions of various axial muscles and groups them according to common function. Note that a muscle that has multiple functions is listed in more than one group.

WHAT DID YOU LEARN?

16 What are the main actions of the abdominal muscles?

CLINICAL VIEW

Hernias

The condition in which a portion of the viscera, particularly the intestine, protrudes through a weakened point of the muscular wall of the abdominopelvic cavity is called a **hernia** (her′nē-ă; rupture). A significant medical problem may develop if the herniated portion of the intestine swells, becoming trapped. Blood flow to the trapped segment may diminish, causing that portion of the intestine to die.

How do inguinal hernias form, and why are males more prone to get them?

An **inguinal hernia** is the most common type of hernia. The inguinal region is one of the weakest areas of the abdominal wall. Within this region is a canal (inguinal canal) that allows the passage of the spermatic cord in males, and a smaller structure called the *round ligament*

in females. The inguinal canal, or the **superficial inguinal ring** associated with it, is often the site of a rupture or separation of the abdominal wall. Males are more likely to develop inguinal hernias than females because their inguinal canals and superficial inguinal rings are larger to allow room for the spermatic cord. Rising pressure in the abdominal cavity, as might develop while straining to lift a heavy object, provides the force to push a segment of the small intestine into the canal.

How do physicians test for an inguinal hernia?
The physician inserts a finger in the depression formed by the superficial inguinal ring and asks the patient to turn his head and cough, because the act of coughing increases intra-abdominal pressure and would potentially induce a portion of intestine to poke through the ring if there was a problem. While the patient coughs, the physician palpates the superficial inguinal ring to make sure no intestine is protruding through the ring.

Testing for a potential inguinal hernia on the right side. Note the herniated intestine protruding through the left superficial inguinal ring.

Anterior superior iliac spine

Inguinal ligament

Superficial inguinal ring

Pubic tubercle

Herniated intestine

Table 11.10 — Muscles of the Abdominal Wall

Muscle	Action	Origin/Insertion	Innervation (see sections 14.5c, f)
External oblique	Unilateral action[1]: Laterally flexes vertebral column; rotation of vertebral column Bilateral action[2]: Flexes vertebral column and compresses abdominal wall	O: External and inferior borders of the inferior 8 ribs I: Linea alba by a broad aponeurosis; some to iliac crest	Spinal nerves T8–T12, L1
Internal oblique	Unilateral action: Laterally flexes vertebral column; rotation of vertebral column to opposite side Bilateral action: Flexes vertebral column and compresses abdominal wall	O: Lumbar fascia, inguinal ligament, and iliac crest I: Linea alba, pubic crest, inferior rib surfaces (last 4 ribs); costal cartilages of ribs 8–10	Spinal nerves T8–T12, L1
Transversus abdominis (ab-dom′i-nis) *transverse* = across	Unilateral action: Laterally flexes vertebral column Bilateral action: Flexes vertebral column and compresses abdominal wall	O: Iliac crest, cartilages of inferior 6 ribs; lumbar fascia; inguinal ligament I: Linea alba and pubic crest	Spinal nerves T8–T12, L1
Rectus abdominis	Flexes vertebral column and compresses abdominal wall	O: Superior surface of pubis near pubic symphysis I: Xiphoid process of sternum; inferior surfaces of ribs 5–7	Spinal nerves T7–T12

1. *Unilateral action* means only one muscle (either the left or right muscle) is contracting.
2. *Bilateral action* means both the left and right muscles are contracting together.

Table 11.11 — Muscle Actions on the Axial Skeleton

Extend the Head, Neck, and/or Vertebral Column	Flex the Head, Neck, and/or Vertebral Column	Laterally Flex the Vertebral Column	Rotate the Head and/or Neck to One Side	Elevate the Ribs	Depress the Ribs
Splenius muscles[2]	Sternocleidomastoid[2]	Quadratus lumborum[1]	Sternocleidomastoid[1]	Serratus posterior superior	Serratus posterior inferior
Erector spinae[2] (iliocostalis, longissimus, spinalis)	Scalene muscles[2]	External oblique[1]	Splenius muscles[1]	External intercostals	Internal intercostals
Quadratus lumborum[2]	External oblique[2]	Internal oblique[1]	Longissimus capitis[1]	Scalene muscles (1st and 2nd ribs only)	Transversus thoracis
Transversospinalis group[2]	Internal oblique[2]	Transversus abdominis[1]	Obliquus capitis inferior[1]		
Rectus capitis posterior major and minor[2]	Transversus abdominis[2]		Obliquus capitis superior[1]		
	Rectus abdominis[2]				

1. Unilateral action of muscles
2. Bilateral action

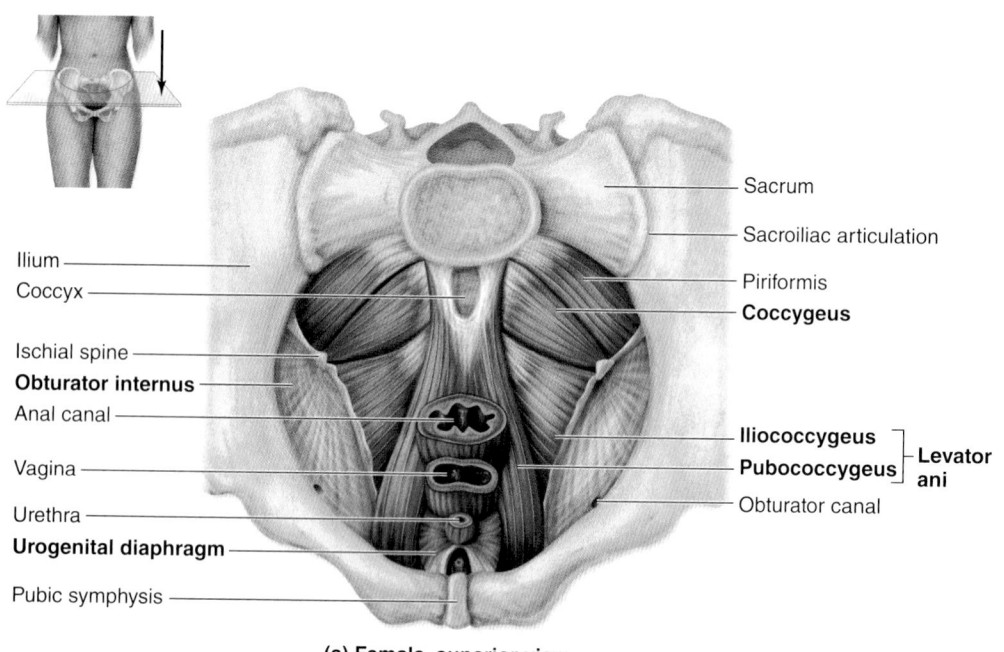

Figure 11.17 Muscles of the Pelvic Floor. The pelvic cavity floor is composed of muscle layers that form the urogenital and anal triangles, extend across the pelvic outlet, and support the organs in the pelvic cavity (puborectalis not shown). (*a*) Superior view of the female pelvic cavity. Inferior views show (*b*) male and (*c*) female perineal regions. Muscles of the pelvic floor are in bold.

(a) Female, superior view

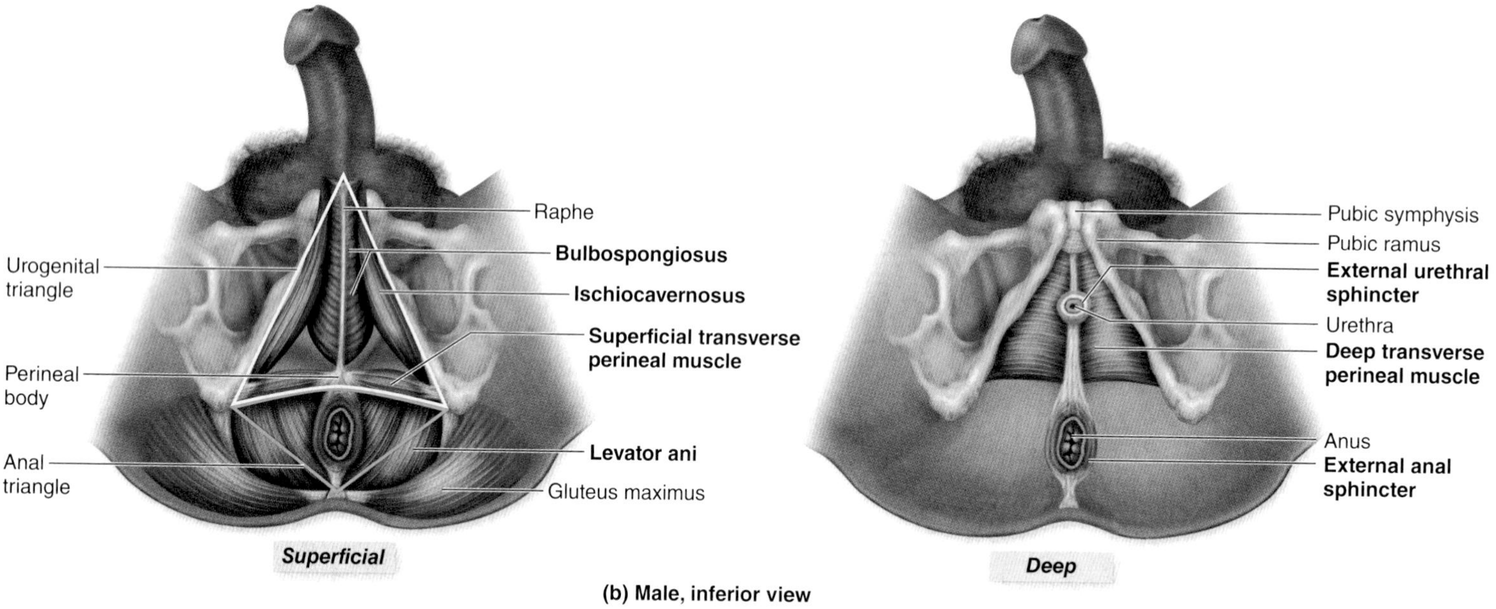

Sacrum
Sacroiliac articulation
Ilium
Coccyx
Piriformis
Coccygeus
Ischial spine
Obturator internus
Anal canal
Iliococcygeus
Pubococcygeus } **Levator ani**
Vagina
Urethra
Obturator canal
Urogenital diaphragm
Pubic symphysis

(b) Male, inferior view

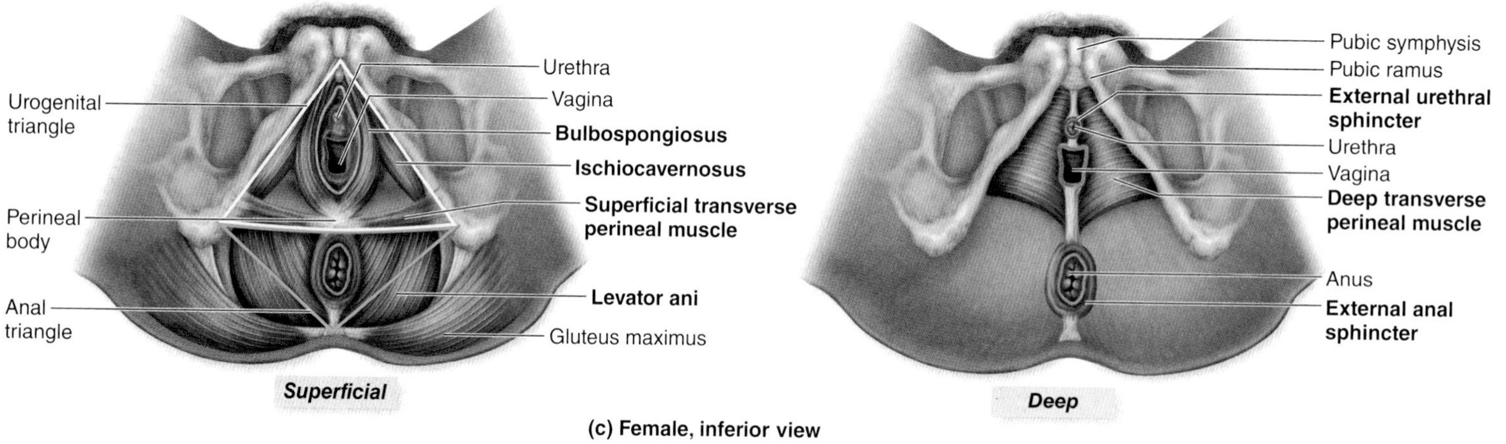

Raphe
Bulbospongiosus
Ischiocavernosus
Superficial transverse perineal muscle
Levator ani
Gluteus maximus
Urogenital triangle
Perineal body
Anal triangle
Superficial

Pubic symphysis
Pubic ramus
External urethral sphincter
Urethra
Deep transverse perineal muscle
Anus
External anal sphincter
Deep

(c) Female, inferior view

Urethra
Vagina
Bulbospongiosus
Ischiocavernosus
Superficial transverse perineal muscle
Levator ani
Gluteus maximus
Urogenital triangle
Perineal body
Anal triangle
Superficial

Pubic symphysis
Pubic ramus
External urethral sphincter
Urethra
Vagina
Deep transverse perineal muscle
Anus
External anal sphincter
Deep

11.7 Muscles of the Pelvic Floor

LEARNING OBJECTIVES

1. Describe the functions of the pelvic floor muscles.
2. Identify the boundaries of the perineum.

The floor of the pelvic cavity is formed by three layers of muscles and associated fasciae, collectively known as the **pelvic diaphragm.** The pelvic diaphragm extends from the ischium and pubis of the ossa coxae across the pelvic outlet to the sacrum and coccyx. These muscles collectively form the pelvic floor and support the pelvic viscera (**figure 11.17**).

The diamond-shaped region between the lower appendages is called the **perineum** (per'i-nē'ŭm). The perineum has four significant bony landmarks: the pubic symphysis anteriorly, the coccyx posteriorly, and both ischial tuberosities laterally. A transverse line drawn between the ischial tuberosities partitions the perineum into an anterior **urogenital triangle** that contains the external genitalia and urethra, and a posterior **anal triangle** that contains the anus (figure 11.17*b*, *c*).

INTEGRATE

CONCEPT CONNECTION

The muscles of the pelvic floor may become stretched or torn after childbirth (see section 29.6), so women may not have adequate support for the pelvic organs. These issues may become more problematic as a woman ages. As a result, incontinence (leakage of urine) is a common complaint among women who have given birth. Strengthening the pelvic floor muscles (with exercises such as Pilates or Kegel exercises) often alleviates these problems.

Table 11.12 describes the specific muscles of the pelvic floor and perineum, and summarizes their characteristics.

WHAT DID YOU LEARN?

17 What are the functions of the pelvic floor muscles?

Table 11.12	Muscles of the Pelvic Floor		
Group/Muscle	**Action**	**Origin/Insertion**	**Innervation (see section 14.5g)**
ANAL TRIANGLE			
Coccygeus (kok-sij'ē-ŭs) *coccy* = coccyx	Forms pelvic floor and supports pelvic viscera	O: Ischial spine I: Lateral and inferior borders of sacrum	Spinal nerves S4–S5
External anal sphincter (ā'năl sfingk'tĕr) *anal* = referring to anus *sphin* = squeeze	Closes anal opening; must relax to defecate	O: Perineal body I: Encircles anal opening	Pudendal nerve (S2–S4)
Levator ani (lē-vā'tor, lē-vā-ter ā'nī; *levator* = raises; *ani* = anus): Group of muscles that form the anterior and lateral parts of the pelvic diaphragm			
Iliococcygeus (il'ē-ō-kok-si'jē-ŭs)	Forms pelvic floor and supports pelvic viscera	O: Pubis and ischial spine I: Coccyx and median raphe	Pudendal nerve (S2–S4)
Pubococcygeus (pyū'bō-kok-si'jē-ŭs)	Forms pelvic floor and supports pelvic viscera	O: Pubis and ischial spine I: Coccyx and median raphe	Pudendal nerve (S2–S4)
Puborectalis (pyū'bō-rek'tăl-is) *rectal* = rectum	Supports anorectal junction; must relax to defecate	O: Pubis and ischial spine I: Coccyx and median raphe	Pudendal nerve (S2–S4)
UROGENITAL TRIANGLE			
Superficial layer			
Bulbospongiosus (female) (bul'bō-spŭn'jē-ō'sŭs) *bulbon* = bulb *spongio* = sponge	Narrows vaginal opening; compresses and stiffens clitoris	O: Sheath of collagen fibers at base of clitoris I: Perineal body	Pudendal nerve (S2–S4)
Bulbospongiosus (male)	Ejects urine or semen; compresses base of penis; stiffens penis	O: Sheath of collagen fibers at base of penis I: Median raphe and perineal body	Pudendal nerve (S2–S4)
Ischiocavernosus (ish'ē-ō-kav'er-nō-sŭs) *caverna* = hollow chamber	Assists with erection of penis or clitoris	O: Ischial tuberosities and ischial ramus I: Pubic symphysis	Pudendal nerve (S2–S4)
Superficial transverse perineal muscle (per'i-nē'ăl)	Supports pelvic organs	O: Ramus of ischium I: Perineal body	Pudendal nerve (S2–S4)
Deep layer (urogenital diaphragm)			
Deep transverse perineal muscle	Supports pelvic organs	O: Ischial ramus I: Median raphe of urogenital diaphragm	Pudendal nerve (S2–S4)
External urethral sphincter (yū-rē'thrăl)	Constricts urethra so as to voluntarily inhibit urination	O: Rami of ischium and pubis I: Median raphe of urogenital diaphragm	Pudendal nerve (S2–S4)

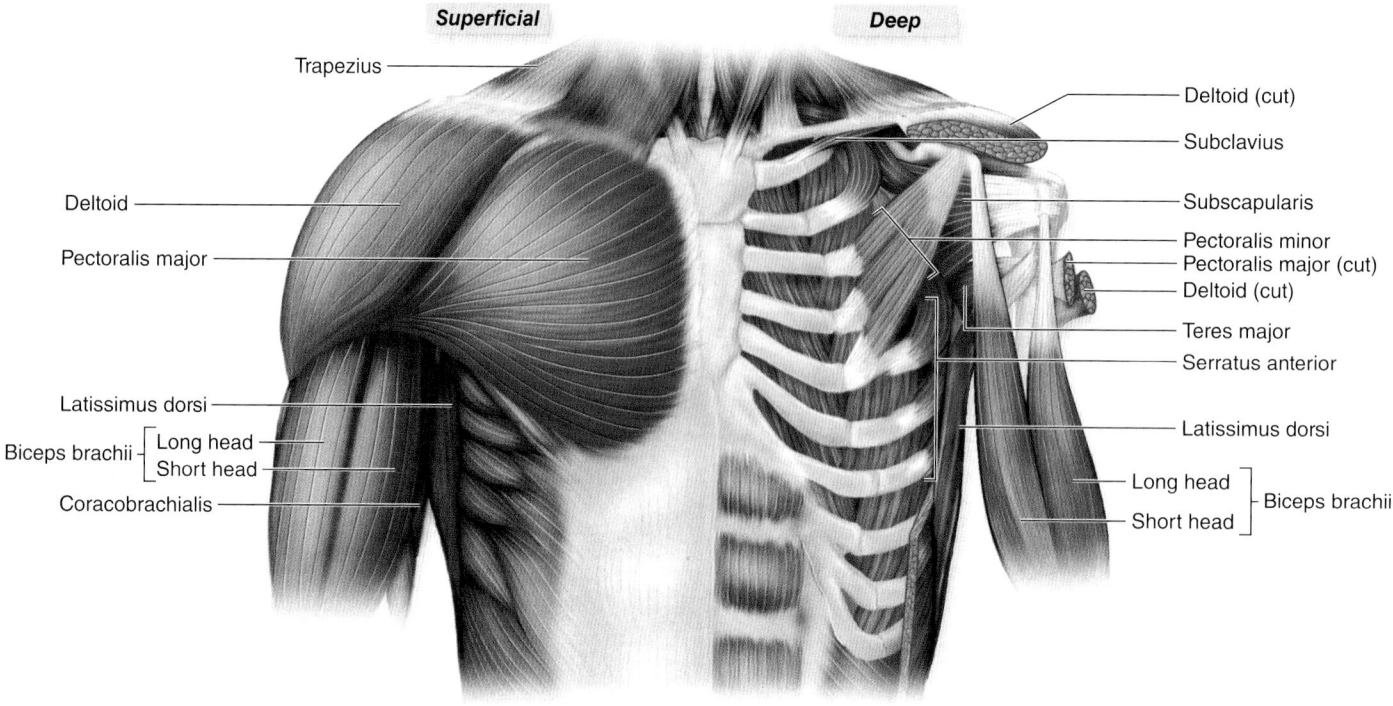

Superficial   **Deep**

Trapezius

Deltoid (cut)

Subclavius

Subscapularis

Deltoid

Pectoralis minor

Pectoralis major

Pectoralis major (cut)

Deltoid (cut)

Teres major

Serratus anterior

Latissimus dorsi

Latissimus dorsi

Biceps brachii — [Long head
                    Short head]

Long head

Short head —] Biceps brachii

Coracobrachialis

Anterior view

Figure 11.18 Anterior Muscles Associated with the Proximal Upper Limb. This anterior view compares some components of both the axial and appendicular musculature. Only those muscles that move the upper limb are labeled. Superficial muscles are shown on the right side of the body, and deep muscles are shown on the left side. AP|R

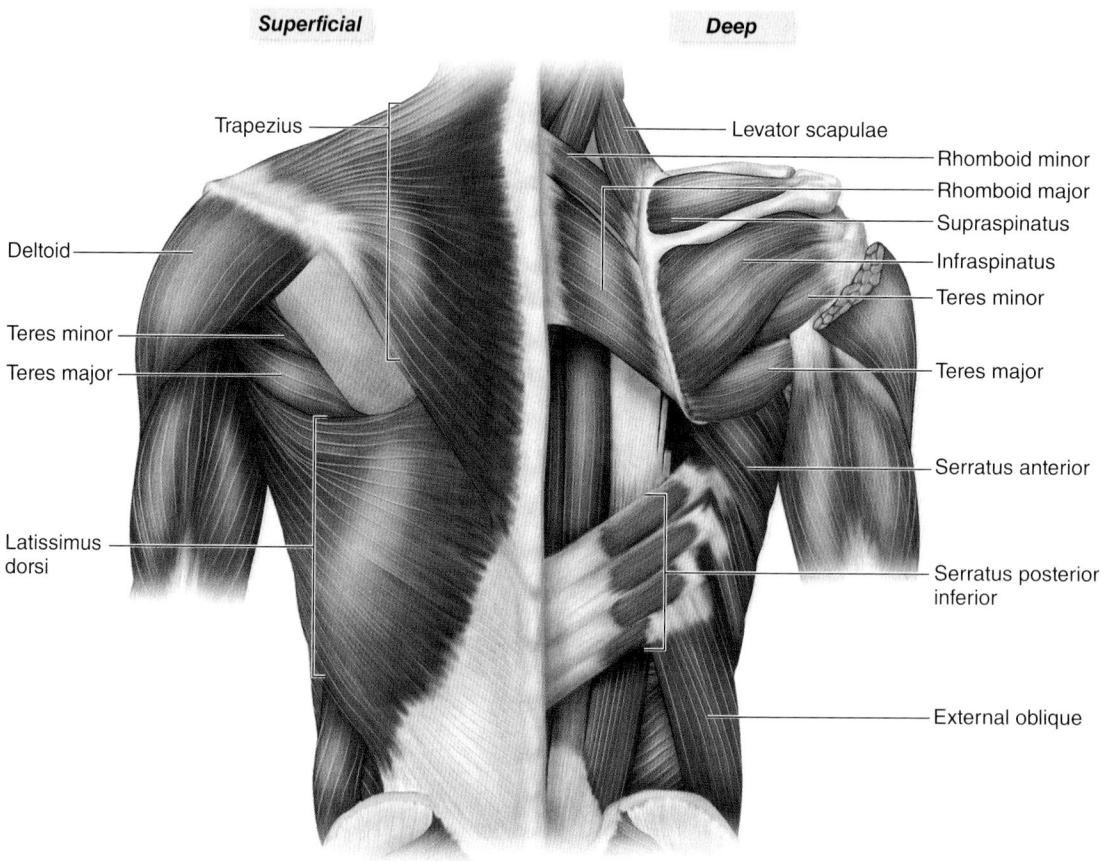

Superficial   **Deep**

Trapezius

Levator scapulae

Rhomboid minor

Rhomboid major

Supraspinatus

Deltoid

Infraspinatus

Teres minor

Teres minor

Teres major

Teres major

Serratus anterior

Latissimus dorsi

Serratus posterior inferior

External oblique

Posterior view

Figure 11.19 Posterior Muscles Associated with the Proximal Upper Limb. This posterior view compares some components of both the axial and appendicular musculature. Only those muscles that move the upper limb are labeled. Superficial muscles are shown on the left side of the body, and deep muscles are shown on the right. AP|R

11.8 Muscles of the Pectoral Girdle and Upper Limb

Muscles that move the pectoral girdle and upper limbs are organized into specific groups:

- Muscles that move the pectoral girdle
- Muscles that move the glenohumeral joint/arm
- Arm and forearm muscles that move the elbow joint/forearm
- Forearm muscles that move the wrist joint, hand, and fingers
- Intrinsic muscles of the hand

Some of these muscles are superficial, and others are deep (figures 11.18 and 11.19).

11.8a Muscles That Move the Pectoral Girdle

LEARNING OBJECTIVE

1. Compare and contrast how the anterior and posterior thoracic muscles move the pectoral girdle.

The muscles of the pectoral girdle originate on the axial skeleton and insert on the scapula and clavicle. These muscles both stabilize the scapula and move it to increase the arm's angle of movements. Some of the superficial muscles of the thorax are grouped together according to the scapular movement they direct: elevation, depression, protraction, or retraction (figure 11.20).

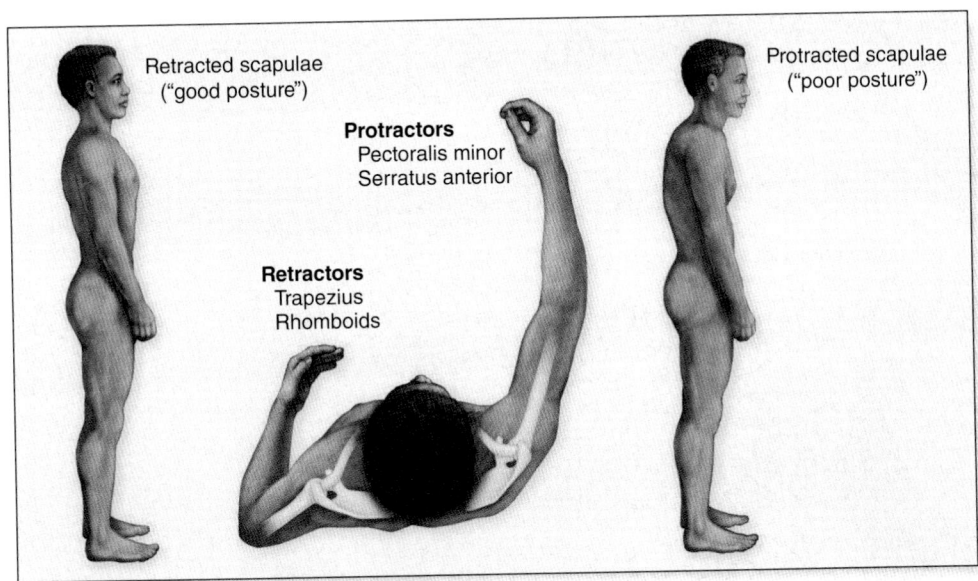

(a) Retraction and protraction of scapula

Figure 11.20 Actions of Some Thoracic Muscles on the Scapula. Individual muscles may contribute to different, multiple actions. (*a*) The scapula can be retracted or protracted. When you are standing upright and have good posture, your scapulae are retracted. Conversely, poor posture demonstrates scapular protraction. (*b*) Muscles that elevate and depress the scapula. (*c*) Muscles that rotate the scapula.

Retracted scapulae ("good posture")

Protracted scapulae ("poor posture")

Protractors
Pectoralis minor
Serratus anterior

Retractors
Trapezius
Rhomboids

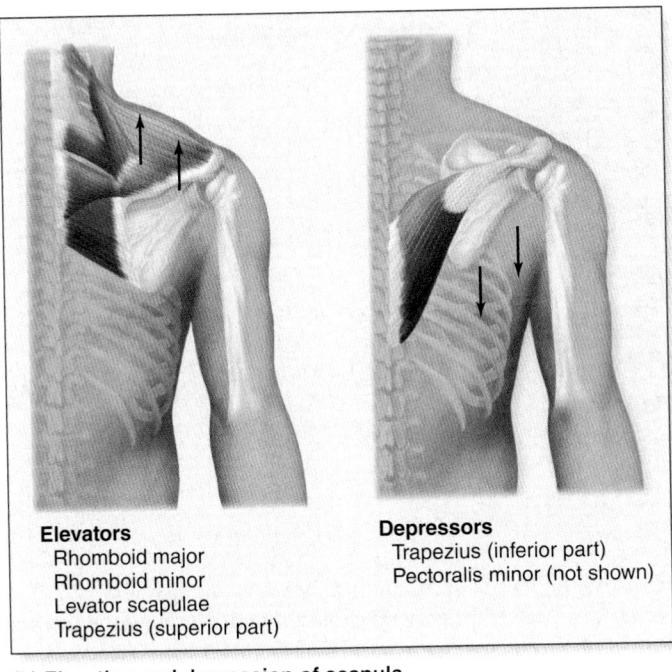

Elevators
Rhomboid major
Rhomboid minor
Levator scapulae
Trapezius (superior part)

Depressors
Trapezius (inferior part)
Pectoralis minor (not shown)

(b) Elevation and depression of scapula

Superior rotators
Serratus anterior
Trapezius (superior part)

Inferior rotators
Rhomboid major
Rhomboid minor
Levator scapulae

(c) Superior and inferior rotation of scapula

The muscles that move the pectoral girdle are classified according to their location in the thorax as either anterior or posterior thoracic muscles. The anterior thoracic muscles are the pectoralis minor, serratus anterior, and subclavius (shown in **figure 11.21a**).

The **pectoralis minor** is deep to the pectoralis major. This muscle helps depress and protract (pull anteriorly) the scapula. When your shoulders are hunched forward, the pectoralis minor muscle is contracting. The **serratus anterior** is a large, flat, fan-shaped muscle

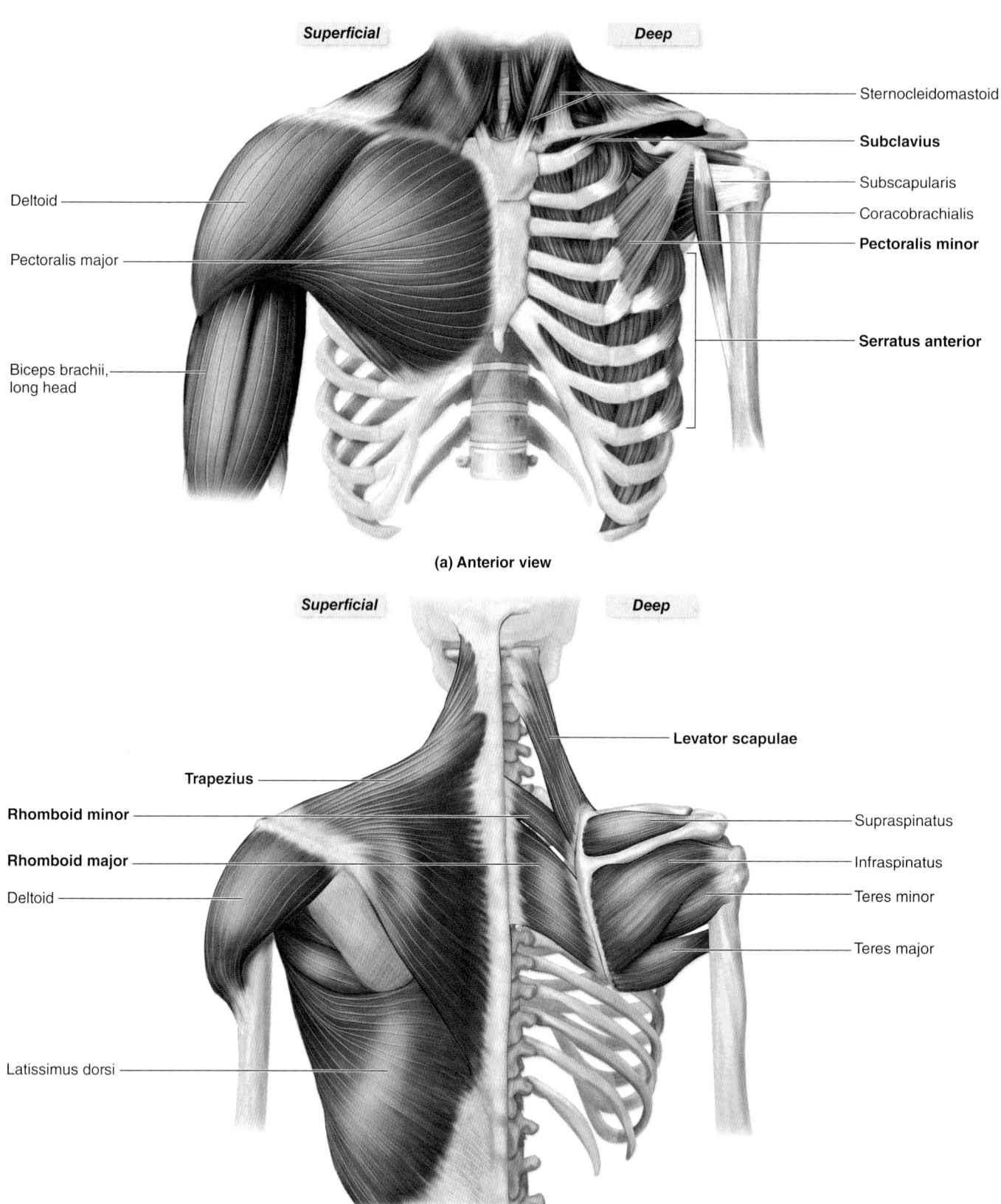

(a) Anterior view

(b) Posterior view

Figure 11.21 Muscles That Move the Pectoral Girdle and the Glenohumeral Joint/Arm. Illustrations show the muscles whose primary function is to move the pectoral girdle (scapula or clavicle), labeled in bold. Muscles that attach to the pectoral girdle but primarily move the arm are labeled but not in bold. (*a*) Anterior view. Superficial muscles are shown on the right side of the body, and deep muscles on the left. (*b*) Posterior view. Superficial muscles are shown on the left side of the body and deep muscles on the right. AP|R

INTEGRATE

LEARNING STRATEGY

When studying appendicular muscle function, remember these two basic rules:

1. If a muscle (or its tendons) crosses over or spans a joint, then the muscle must move that joint. For example, the biceps brachii crosses the elbow joint and therefore must move it.

2. Conversely, if a muscle (or its tendons) does not cross over or span a joint, it cannot move that joint. For example, the deltoid is found in the shoulder, and no part of it crosses the elbow joint. Therefore, the deltoid does not move the elbow joint.

positioned between the ribs and the scapula. Its name is derived from the saw-toothed (serrated) appearance of its origins on the ribs. This muscle is the agonist (prime mover) in scapula protraction, and helps stabilize the scapula against the posterior side of the rib cage. It also is a powerful superior rotator of the scapula by moving the glenoid cavity superiorly, as occurs when you abduct the upper limb. The

subclavius extends from the clavicle to the first rib, and its main action is to stabilize and depress the clavicle.

The posterior thoracic muscles are the levator scapulae, rhomboid major, rhomboid minor, and trapezius (shown in figure 11.21*b*). The **levator scapulae** originates from multiple heads on the transverse processes of the cervical vertebrae and inserts on the superior angle of the scapula. As its name implies, its primary action is to elevate the scapula. It can also inferiorly rotate the scapula so that the glenoid cavity points inferiorly.

Both the **rhomboid major** and **rhomboid minor** are located deep to the trapezius. These rhomboid muscles are parallel bands that run inferolaterally from the vertebrae to the scapula. They help elevate and retract (adduct) the scapula, as when you are standing up straight. The rhomboid muscles also inferiorly rotate the scapula.

The **trapezius** is a large, diamond-shaped muscle that extends from the skull and vertebral column to the pectoral girdle laterally. The trapezius can elevate, depress, retract, or rotate the scapula, depending upon which fibers of the muscle are contracting.

Table 11.13 summarizes the characteristics of the thoracic muscles that move the pectoral girdle.

 WHAT DID YOU LEARN?

18 List the posterior thoracic muscles that move the pectoral girdle, and describe their common action(s).

Table 11.13	Thoracic Muscles That Move the Pectoral Girdle		
Group	**Actions**	**Origin/Insertion**	**Innervation (see sections 13.9 and 14.5d, e)**
ANTERIOR MUSCLES			
Pectoralis minor (pek'tō-ra'lis mī'nŏr) *pectus* = chest	Protracts and depresses scapula	O: Ribs 3–5 I: Coracoid process of scapula	Medial pectoral nerve (C8–T1)
Serratus anterior (ser-ā'tŭs an-tēr'ē-ōr) *serratus* = saw	Agonist in scapula protraction; superiorly rotates scapula (so glenoid cavity moves superiorly); stabilizes scapula	O: Ribs 1–8, anterior and superior margins I: Medial border of scapula; anterior surface	Long thoracic nerve (C5–C7)
Subclavius (sŭb-klā'vē-ŭs) *sub* = under *clav* = clavicle	Depresses and stabilizes clavicle	O: Rib 1 I: Inferior surface of clavicle	Nerve to subclavius (C5–C6)
POSTERIOR MUSCLES			
Levator scapulae (lē-vā'tor, lē-vā-ter skap'yū-lē) *levator* = raises	Elevates scapula; inferiorly rotates scapula (pulls glenoid cavity inferiorly)	O: Transverse processes of C_1–C_4 I: Superior part of medial border of scapula	Cervical nerves (C3–C4) and dorsal scapular nerve (C5)
Rhomboid major (rom-bōid' mā'jŏr) *rhomboid* = diamond shaped	Elevates and retracts (adducts) scapula; inferiorly rotates scapula	O: Spinous processes of T_2–T_5 I: Medial border of scapula from spine to inferior angle	Dorsal scapular nerve (C5)
Rhomboid minor	Elevates and retracts (adducts) scapula; inferiorly rotates scapula	O: Spinous processes of C_7–T_1 I: Medial border of scapula superior to spine	Dorsal scapular nerve (C5)
Trapezius (tra-pē'zē-ŭs) *trapezion* = irregular four-sided figure	Superior fibers: Elevate and superiorly rotate scapula Middle fibers: Retract scapula Inferior fibers: Depress scapula	O: Superior nuchal line of occipital bone; ligamentum nuchae; spinous processes of C_7–T_{12} I: Clavicle; acromion process and spine of scapula	Accessory nerve (CN XI)

11.8b Muscles That Move the Glenohumeral Joint/Arm

✎ LEARNING OBJECTIVES

2. List the muscles that extend, flex, adduct, and abduct the glenohumeral joint.

3. Compare the actions of the four scapular muscles of the rotator cuff.

The phrases "moving the glenohumeral joint" and "moving the arm or humerus" mean the same thing. A movement such as flexion of the arm requires movement at the glenohumeral joint. Throughout this text, we refer to both the joint where the movement is occurring and the body region that is being moved to minimize any confusion.

The glenohumeral joint is crossed by 11 muscles that insert on the arm (humerus) or forearm (radius and/or ulna) (figure 11.21). The **latissimus dorsi** is a broad, triangular muscle located on the inferior part of the back. Often, it is referred to as the "swimmer's muscle," because its actions are required for many swimming strokes. It is the prime arm extensor and also adducts and medially rotates the arm. The **pectoralis major** is a large, thick, fan-shaped muscle that covers the superior part of the thorax. It is the principal flexor of the arm and also adducts and medially rotates the arm.

The latissimus dorsi and pectoralis major muscles are the primary attachments of the arm to the trunk, and they are the prime movers of the glenohumeral joint. These muscles are antagonists with respect to arm flexion and arm extension. However, these same two muscles work together (synergistically) when performing other movements, such as adducting and medially rotating the humerus.

The triceps brachii and biceps brachii, discussed in detail with the muscles that move the elbow joint, also participate in the glenohumeral joint. Specifically, the long head of the triceps brachii originates on the infraglenoid tubercle, spans the glenohumeral joint, and

INTEGRATE

LEARNING STRATEGY ✎

Generally speaking, muscles that originate anterior to the glenohumeral joint flex the arm (move it anteriorly), and those that originate posterior to the joint extend the arm (move it posteriorly).

helps extend and adduct the arm. The long head of the biceps brachii originates on the supraglenoid tubercle of the scapula and assists in flexing the arm.

The seven remaining muscles that move the humerus at the glenohumeral joint are termed the scapular muscles, because they originate on the scapula. They include the deltoid, coracobrachialis, teres major, and the four muscles of the rotator cuff.

The **deltoid** is a thick, powerful muscle that functions as a prime abductor of the arm and forms the rounded contour of the shoulder. Note that the fibers of the deltoid originate from three different points, and these different fiber groups all perform different functions: (1) The anterior fibers flex and medially rotate the arm. (2) The lateral fibers abduct the arm; in fact, the deltoid is the prime abductor of the arm. (3) The posterior fibers extend and laterally rotate the arm. The **coracobrachialis** is a synergist to the pectoralis major in flexing and adducting the arm. The **teres major** works synergistically with the latissimus dorsi by extending, adducting, and medially rotating the arm.

Among the scapular muscles, four **rotator cuff muscles** (subscapularis, supraspinatus, infraspinatus, and teres minor) provide the strength and stability of the glenohumeral joint **(figure 11.22)**. These muscles attach the scapula to the humerus (see also figure 9.15).

Figure 11.22 Rotator Cuff Muscles. The rotator cuff muscles reinforce the glenohumeral joint and secure the head of the humerus in the glenoid cavity. (*a*) The subscapularis is best seen in anterior view and medially rotates the humerus (as in winding up for a pitch). (*b*) The supraspinatus abducts the humerus (as in executing the pitch), while the infraspinatus and teres minor laterally rotate the humerus (as in completing the pitch and slowing down the pitching arm). These three muscles are located along the posterior aspect of the scapula. **AP|R**

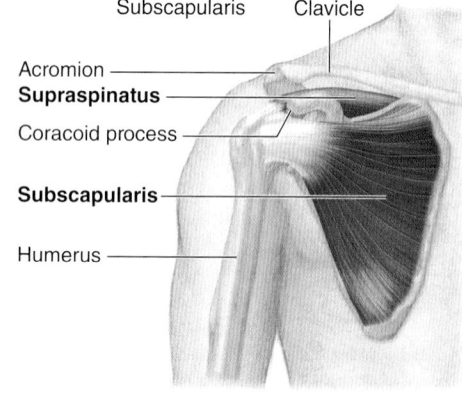

(a) Anterior view

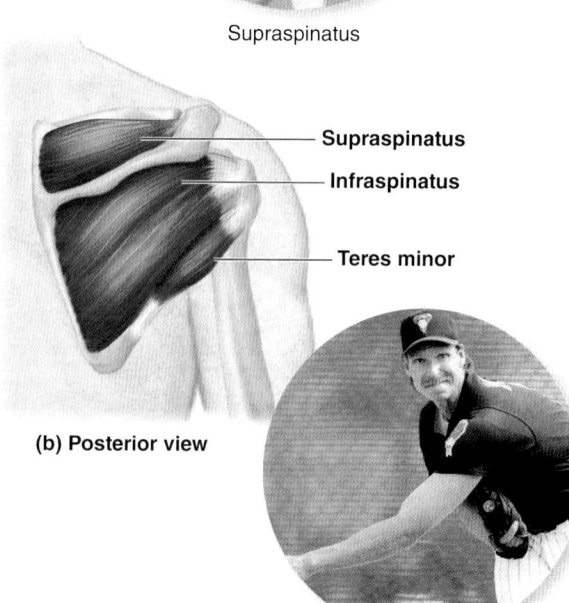

(b) Posterior view

Infraspinatus and teres minor

The specific movements of each muscle are best learned when relating their actions to pitching a ball:

- The **subscapularis** is used when you wind up for a pitch. It medially rotates the arm.
- The **supraspinatus** is used when you start to execute the pitch, by fully abducting the arm.

- The **infraspinatus** and **teres minor** help slow down the pitching arm upon completion of the pitch. These two muscles adduct and laterally rotate the arm.

Table 11.14 summarizes the characteristics of the muscles that move the glenohumeral joint and arm.

Table 11.14	Muscles That Move the Glenohumeral Joint/Arm		
Group/Muscle	**Action**	**Origin/Insertion**	**Innervation (see section 14.5e)**
MUSCLES ORIGINATING ON AXIAL SKELETON			
Latissimus dorsi (lă-tis′i-mŭs dōr′sī) *latissimus* = widest *dorsi* = back	Agonist of arm extension; also adducts and medially rotates arm ("swimmer's muscle")	O: Spinous processes of T_7–T_{12}; ribs 8–12; iliac crest; thoracolumbar fascia I: Intertubercular groove of humerus	Thoracodorsal nerve (C6–C8)
Pectoralis major	Agonist of arm flexion; also adducts and medially rotates arm	O: Medial clavicle; costal cartilages of ribs 2–6; body of sternum I: Lateral part of intertubercular groove of humerus	Lateral pectoral (C5–C7) and medial pectoral (C8–T1) nerves
MUSCLES ORIGINATING ON SCAPULA			
Deltoid (del′toyd) *delta* = triangular	Anterior fibers: Flex and medially rotate arm Middle fibers: Prime mover of arm abduction Posterior fibers: Extend and laterally rotate arm	O: Acromial end of clavicle; acromion and spine of scapula I: Deltoid tuberosity of humerus	Axillary nerve (C5–C6)
Coracobrachialis (kōr′ă-kō-brā-kē-a′lis) *coraco* = coracoid *brachi* = arm	Adducts and flexes arm	O: Coracoid process of scapula I: Middle medial shaft of humerus	Musculocutaneous nerve (C5–C6 nerve fibers)
Teres major (ter′ēz) *teres* = round	Extends, adducts, and medially rotates arm	O: Inferior lateral border and inferior angle of scapula I: Lesser tubercle and intertubercular groove of humerus	Lower subscapular nerve (C5–C6)
Triceps brachii (long head) (trī′seps brā′kē-ī) *triceps* = three heads *brachi* = arm	Extends and adducts arm	O: Infraglenoid tubercle of scapula I: Olecranon of ulna	Radial nerve (C5–C7 nerve fibers)
Biceps brachii (long head) (bī′seps) *biceps* = two heads	Flexes arm	O: Supraglenoid tubercle of scapula I: Radial tuberosity and bicipital aponeurosis	Musculocutaneous nerve (C5–C6 nerve fibers)
Rotator cuff muscles (rō-tā′ter, tōr kŭf; *rotatio* = to revolve): Collectively, these four muscles stabilize the glenohumeral joint			
Subscapularis (sŭb-skap′yū-lār′ris)	Medially rotates arm	O: Subscapular fossa of scapula I: Lesser tubercle of humerus	Upper and lower subscapular nerves (C5–C6)
Supraspinatus (sū-pră-spī-nā′tŭs) *supra* = above, over *spin* = spine	Abducts arm	O: Supraspinous fossa of scapula I: Greater tubercle of humerus	Suprascapular nerve (C5–C6)
Infraspinatus (in-fră-spī-nā′tŭs) *infra* = below	Adducts and laterally rotates arm	O: Infraspinous fossa or scapula I: Greater tubercle of humerus	Suprascapular nerve (C5–C6)
Teres minor	Adducts and laterally rotates arm	O: Upper dorsal lateral border of scapula (superior to teres major origin) I: Greater tubercle of humerus	Axillary nerve (C5–C6)

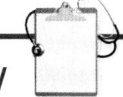

INTEGRATE

CLINICAL VIEW
Rotator Cuff Injuries

A **rotator cuff injury** is the result of trauma or disease that affects any portion of the rotator cuff musculature or tendons. Extensive and repetitive use of the rotator cuff muscles can cause tearing of muscle fibers or rupture of tendon attachments. The rotator cuff muscles also may be injured upon a fall on the shoulder or by trying to lift an object that is too heavy. The supraspinatus muscle is most commonly involved, likely because the tendon may become impinged (pinched) inferior to the acromion during use of the muscle. The risk of rotator cuff injuries increases with age due to years of use, reduced blood flow to the muscles and tendons as we age, and the increased likelihood of developing bone spurs in the shoulder that impinge on the tendons.

Common symptoms of a rotator cuff injury include swelling and tenderness in the area of the shoulder, as well as pain with specific shoulder movements, especially abduction. The pain may range from moderate to severe. This syndrome is especially common in baseball players because the repetitive shoulder movements while pitching and throwing the ball can impinge the supraspinatus tendon against the acromion. Painters also may experience rotator cuff injuries due to the repetitive overhead upper limb movements involved with their work. Diagnosis may be confirmed with both a physical (that asks about a patient history and performs range of movement activities to the affected upper limb) and imaging studies, such as MRI or ultrasound.

Treatment is dependent upon on the severity of the injury. Initially, the pain may be controlled with icing, NSAIDs (e.g., aspirin), or with corticosteroid shots to the affected region. Physical therapy is used to restore the shoulder's range of motion and strengthen affected muscles. Severe rotator cuff injuries that have not been helped with nonsurgical therapies typically require surgical repair, which may include removal of bone spurs as well as repair of the torn tendon. This surgery may be done **arthroscopically** (where a small camera called an **arthroscope** is inserted through a small incision into the joint to guide the physician's use of small surgical instruments), or a more invasive open repair surgery may be required. Physical therapy is required after surgery to restore range of motion.

The muscles that move the arm at the glenohumeral joint are grouped in **table 11.15** according to different types of actions. Note that a muscle that has multiple functions is listed in more than one group. For example, the deltoid may abduct, extend, or flex the humerus, depending upon whether its middle, posterior, or anterior fibers are contracting. Pectoralis major and coracobrachialis both adduct and flex the humerus, so they are listed in both the adduction and flexion columns. We recommend that you copy the columns multiple times and then test your knowledge by trying to write out all of the muscles in a group without looking at your notes.

WHAT DID YOU LEARN?

19 Which muscles extend the arm at the glenohumeral (shoulder) joint?

20 Identify the rotator cuff muscles, and describe their actions.

Table 11.15	Summary of Muscle Actions at the Glenohumeral Joint/Arm				
Abduction	**Adduction**	**Extension**	**Flexion**	**Lateral Rotation**	**Medial Rotation**
Deltoid (middle fibers)*	**Latissimus dorsi**	**Latissimus dorsi**	**Pectoralis major**	**Infraspinatus**	**Subscapularis**
Supraspinatus	**Pectoralis major**	**Deltoid (posterior fibers)**	**Deltoid (anterior fibers)**	**Teres minor**	Deltoid (anterior fibers)
	Coracobrachialis	Teres major	Coracobrachialis	Deltoid (posterior fibers)	Latissimus dorsi
	Teres major	Long head of triceps brachii	(Long head of biceps brachii)		Pectoralis major
	Teres minor				Teres major
	Infraspinatus				

*Boldface indicates an agonist; others are synergists. Parentheses around an entire muscle name, such as the long head of the biceps brachii under Flexion, indicate only a slight effect.

11.8c Arm and Forearm Muscles That Move the Elbow Joint/Forearm

LEARNING OBJECTIVES

4. Name the muscles in the arm's anterior and posterior compartments, and contrast their common functions.

5. Describe the muscles that pronate and supinate the forearm.

When you move the elbow joint, you move the bones of the forearm. Thus, the phrase "flexing the elbow joint" is synonymous with "flexing the forearm."

The muscles in the limbs are organized into **compartments,** which are surrounded by deep fascia. Each compartment houses functionally related skeletal muscles, as well as their associated nerves and blood vessels. In general, muscles in the same compartment tend to perform similar functions. **Figure 11.23** provides a visual overview of how the muscles are organized into compartments. Note how muscles in opposite compartments tend to be antagonists. For example, the anterior forearm muscles are primarily flexors and pronators, whereas the posterior forearm muscles are primarily extensors and supinators. Likewise, in the lower limb, knee extensors are in the anterior compartment of the thigh, whereas knee flexors are in the posterior compartment of the thigh. Hip adductors are in the medial compartment of the thigh, whereas a hip abductor is in the lateral compartment of the thigh. These compartments may help you to learn the muscles in common functional groups. If you know what compartment a muscle is in, you likely can figure out what action that muscle performs, and vice versa.

The muscles of the arm may be subdivided into an anterior compartment and a posterior compartment. The **anterior compartment** primarily contains elbow flexors, so it is also known as the *flexor compartment.* Muscles in this compartment are supplied by the deep brachial artery and are innervated by the musculocutaneous nerve. Muscles in this compartment include the coracobrachialis (note that this muscle is an arm flexor and not an elbow flexor), biceps brachii, brachialis, and brachioradialis muscles. Put your hand on your anterior arm, and then flex your elbow. Observe how these muscles bulge as they contract, affirming that these anterior compartment arm muscles flex the elbow.

The **posterior compartment** contains elbow extensors, so this compartment is also called the *extensor compartment.* These muscles are innervated by the radial nerve and receive their blood supply from the deep brachial artery. The primary muscle in this compartment is the triceps brachii. Put your hand on your posterior arm, and then extend your elbow. You can feel your muscles contracting as you perform this action, affirming that posterior arm muscles extend the elbow. Perform these activities each time we discuss muscles in a particular limb compartment, so you can see and feel for yourself how these muscles move.

Muscles of the Arm's Anterior Compartment

On the anterior side of the humerus are the principal flexors of the forearm: the biceps brachii and brachialis **(figure 11.24).** The **biceps brachii** is a large, two-headed muscle on the anterior surface of the humerus. This muscle flexes the elbow joint, and it is a powerful supinator of the forearm when the elbow is flexed. (An example of this supination movement occurs when you tighten a screw with your right hand.) The tendon of the long head of the biceps brachii crosses the shoulder joint, and so this muscle helps flex the humerus as well (albeit weakly).

The **brachialis** is deep to the biceps brachii on the anterior surface of the humerus. It is the most powerful flexor of the forearm at the elbow. The **brachioradialis** is another prominent muscle on the anterolateral surface of the forearm. It is a synergist in elbow flexion, effective primarily when the prime movers of forearm flexion have already partially flexed the elbow.

Muscles of the Arm's Posterior Compartment

The posterior compartment of the arm contains two muscles that extend the forearm at the elbow: the triceps brachii and the anconeus **(figure 11.25).** The **triceps brachii** is the large, three-headed muscle on the posterior surface of the arm. The long head of the triceps brachii also crosses the glenohumeral joint, where it helps extend the humerus. All three parts of this muscle merge to form a common insertion on the olecranon of the ulna. A weak elbow extensor is the small **anconeus** that crosses the posterolateral region of the elbow.

Figure 11.23 Muscle Compartmentalization.

In the upper limb, the (*a*) arm and (*b*) forearm both may be divided into anterior "flexor" compartments and posterior "extensor" compartments. (*c*) The thigh may be split into four compartments, whereas the (*d*) leg is split into three compartments. Each compartment contains muscles that tend to perform similar movements.

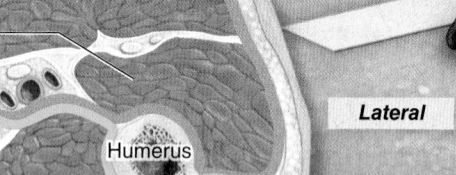

(a) Left arm

Anterior

Medial

Lateral

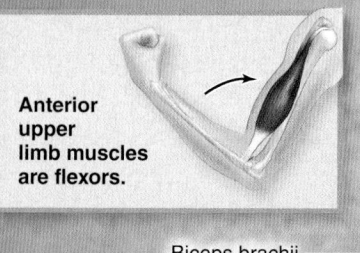

Anterior upper limb muscles are flexors.

Biceps brachii

Brachialis

Humerus

Triceps brachii
- Medial head
- Lateral head
- Long head

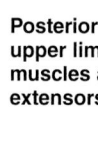

Posterior upper limb muscles are extensors.

Posterior

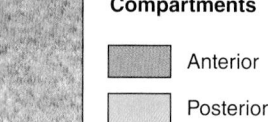

Muscle Compartments
- Anterior
- Posterior
- Medial
- Lateral

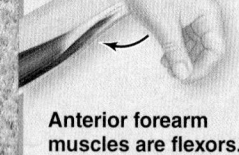

Anterior forearm muscles are flexors.

(b) Left forearm

Anterior

Palmaris longus
Flexor carpi radialis
Flexor digitorum superficialis
Flexor carpi ulnaris
Flexor digitorum profundus
Flexor pollicis longus

Brachioradialis

Extensor carpi radialis longus

Extensor carpi radialis brevis

Radius

Medial

Ulna

Lateral

Abductor pollicis longus
Extensor pollicis longus
Extensor digitorum
Extensor digiti minimi
Extensor carpi ulnaris

Posterior

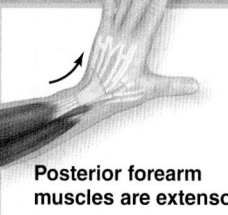

Posterior forearm muscles are extensors.

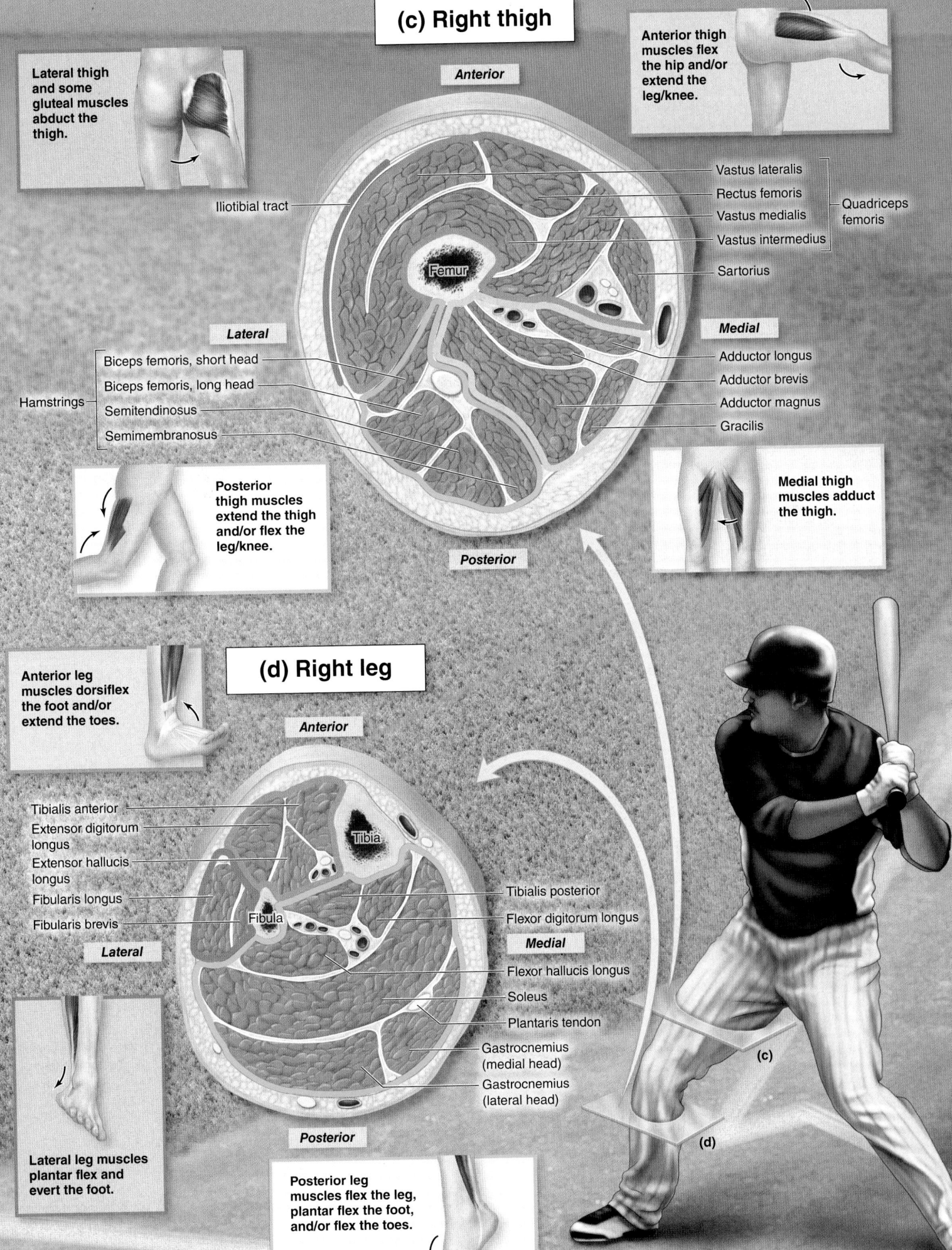

(c) Right thigh

Lateral thigh and some gluteal muscles abduct the thigh.

Anterior thigh muscles flex the hip and/or extend the leg/knee.

Anterior

Iliotibial tract

Vastus lateralis
Rectus femoris
Vastus medialis
Vastus intermedius
Sartorius

Quadriceps femoris

Femur

Lateral

Hamstrings
- Biceps femoris, short head
- Biceps femoris, long head
- Semitendinosus
- Semimembranosus

Medial

Adductor longus
Adductor brevis
Adductor magnus
Gracilis

Posterior thigh muscles extend the thigh and/or flex the leg/knee.

Medial thigh muscles adduct the thigh.

Posterior

(d) Right leg

Anterior leg muscles dorsiflex the foot and/or extend the toes.

Anterior

Tibialis anterior
Extensor digitorum longus
Extensor hallucis longus
Fibularis longus
Fibularis brevis

Tibia

Fibula

Tibialis posterior
Flexor digitorum longus

Medial

Flexor hallucis longus
Soleus
Plantaris tendon
Gastrocnemius (medial head)
Gastrocnemius (lateral head)

Lateral

Lateral leg muscles plantar flex and evert the foot.

Posterior

Posterior leg muscles flex the leg, plantar flex the foot, and/or flex the toes.

(c)

(d)

407

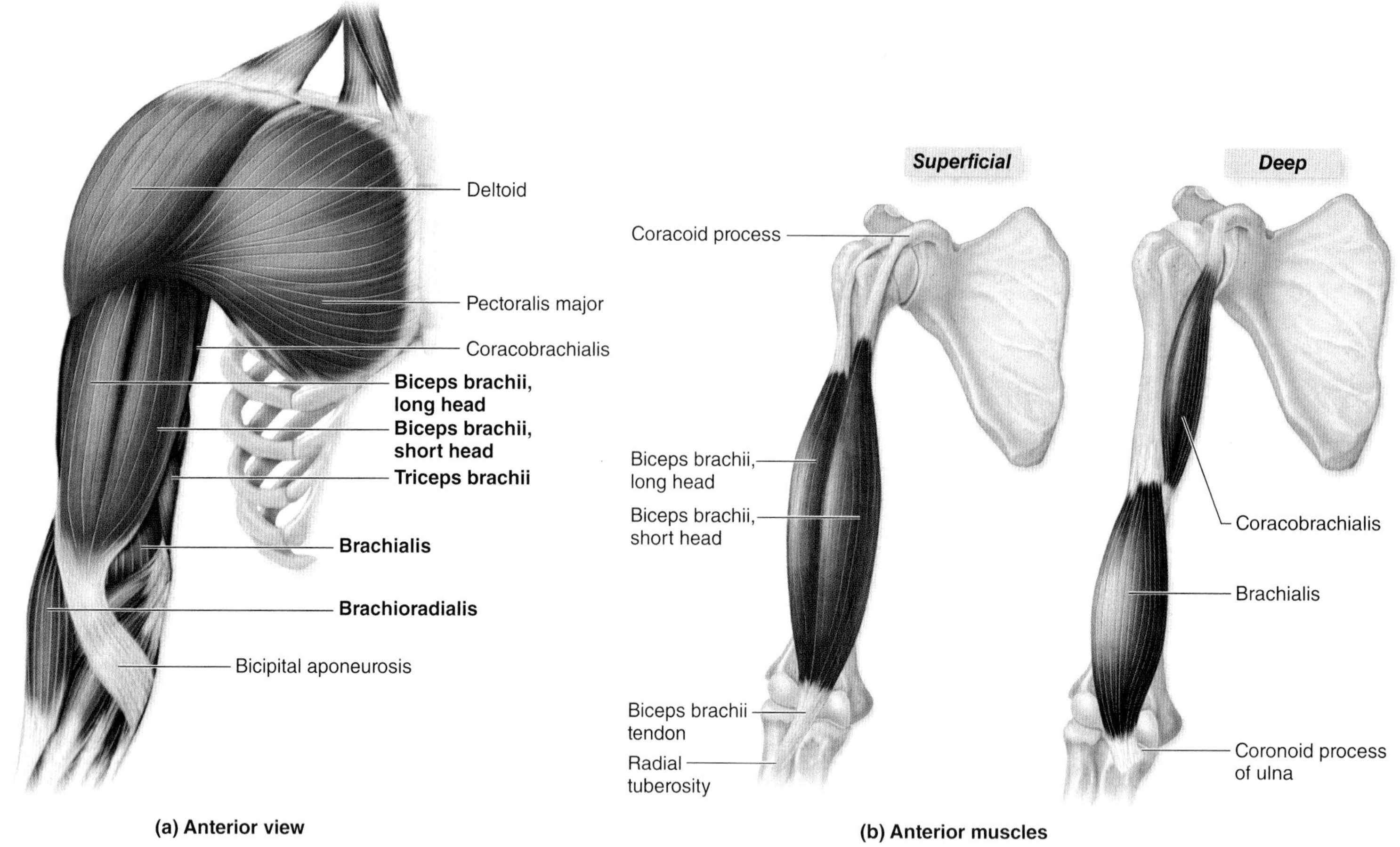

(a) Anterior view

- Deltoid
- Pectoralis major
- Coracobrachialis
- **Biceps brachii, long head**
- **Biceps brachii, short head**
- **Triceps brachii**
- **Brachialis**
- **Brachioradialis**
- Bicipital aponeurosis

(b) Anterior muscles

Superficial

- Coracoid process
- Biceps brachii, long head
- Biceps brachii, short head
- Biceps brachii tendon
- Radial tuberosity

Deep

- Coracobrachialis
- Brachialis
- Coronoid process of ulna

Figure 11.24 Anterior Muscles That Move the Elbow Joint/Forearm. (*a*) The right arm and shoulder show the anterior muscles that produce movements at the elbow joint, labeled in bold. (*b*) Superficial and deep anterior arm muscles. AP|R

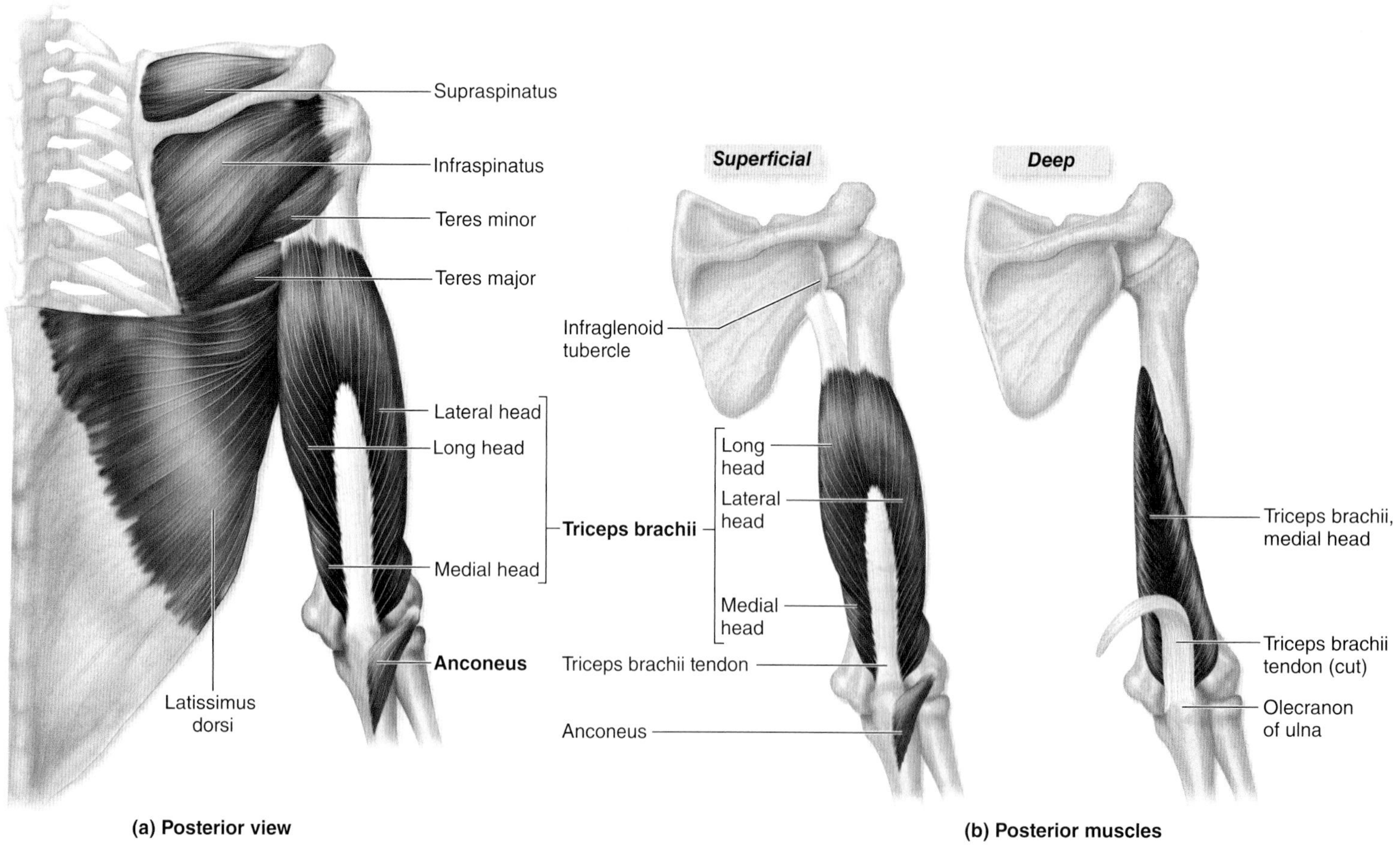

(a) Posterior view

- Supraspinatus
- Infraspinatus
- Teres minor
- Teres major
- Lateral head ⎫
- Long head ⎬ **Triceps brachii**
- Medial head ⎭
- **Anconeus**
- Latissimus dorsi

(b) Posterior muscles

Superficial

- Infraglenoid tubercle
- Long head
- Lateral head
- Medial head
- Triceps brachii tendon
- Anconeus

Deep

- Triceps brachii, medial head
- Triceps brachii tendon (cut)
- Olecranon of ulna

Figure 11.25 Posterior Muscles That Move the Elbow Joint/Forearm. (*a*) The right arm and shoulder show the posterior muscles that produce movements at the elbow joint, labeled in bold. (*b*) Superficial and deep posterior arm muscles.

WHAT DO YOU THINK?

3 The brachialis is on the anterior surface of the arm. Without looking at the muscle tables, determine whether this muscle flexes or extends the elbow joint. How did you reach your conclusion?

Muscles of the Forearm That Act on the Elbow Joint

Some forearm muscles pronate or supinate the forearm (figure 11.26). As their names imply, both the **pronator teres** and **pronator quadratus** rotate the radius across the surface of the ulna to pronate the forearm. These muscles are located in the anterior compartment of the forearm. They are antagonistic to the **supinator** in the posterior compartment of the forearm. The supinator works synergistically with the biceps brachii to supinate the forearm.

Table 11.16 summarizes the characteristics of the muscles that move the forearm, and table 11.17 groups them according to common function. By learning these muscles as groups, you can gain a better understanding of how they work together to perform specific functions.

WHAT DID YOU LEARN?

21 What are the muscles in the anterior compartment of the arm, and what common action(s) do they perform?

22 What muscles pronate the forearm and which supinate the forearm?

INTEGRATE

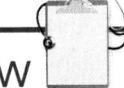

CLINICAL VIEW
Lateral Epicondylitis ("Tennis Elbow")

Lateral epicondylitis (ep'i-kon-di-lī'tis), or *tennis elbow,* is a painful condition resulting from trauma or overuse of the common extensor tendon of the posterior forearm muscles. The pain arises from the lateral epicondyle of the humerus, the attachment site of the common extensor tendon. Lateral epicondylitis most often results from the repeated forceful contraction of the forearm extensors, as when pulling a heavy object from an overhead shelf, shoveling snow, or hitting a backhand shot in tennis.

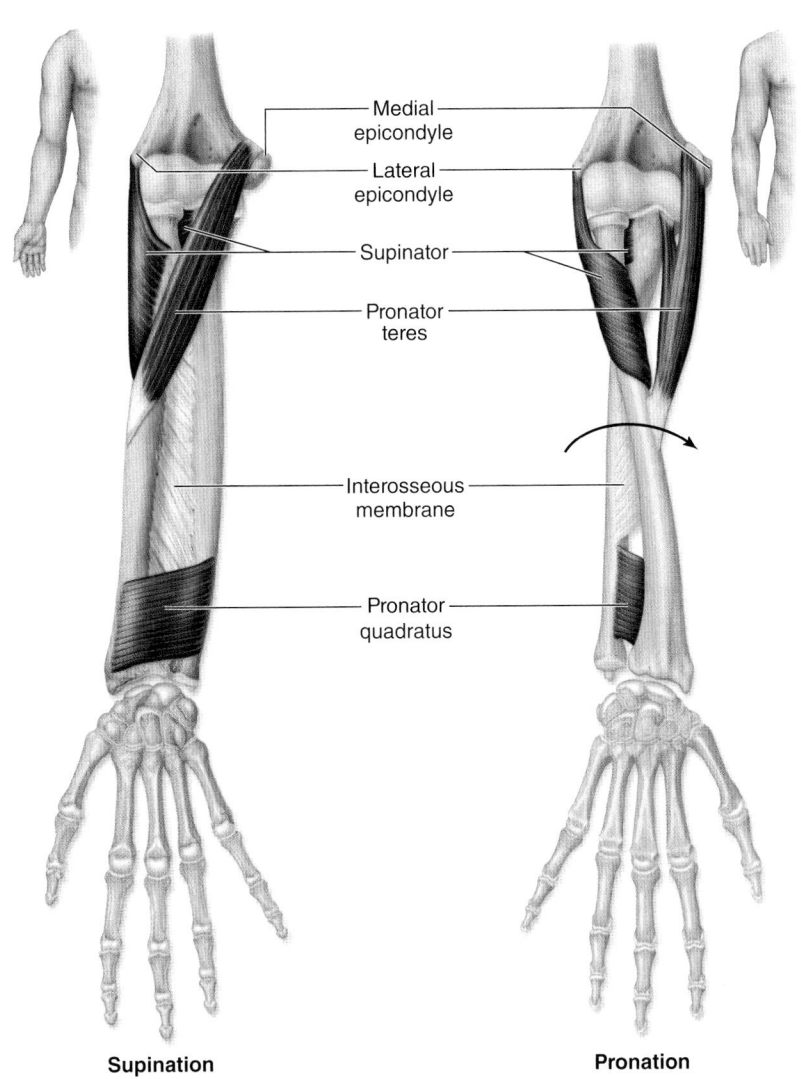

Supination   Pronation

Figure 11.26 Forearm Muscles That Supinate and Pronate. A view of the right upper limb shows the supinator muscle supinates the forearm, whereas pronator teres and pronator quadratus pronate the forearm. (The biceps brachii, an arm muscle not shown here, also supinates the forearm.) **AP|R**

Table 11.16

Muscles That Move the Forearm

Muscle	Action	Origin/Insertion	Innervation (see section 14.5e)
FLEXORS (ANTERIOR ARM)			

Biceps brachii
  Long head
  Short head
	Flexes forearm, powerful supinator of forearm
	
	Long head also flexes arm
		O: Long head: Supraglenoid tubercle of scapula
		  Short head: Coracoid process of scapula
		I: Radial tuberosity and bicipital aponeurosis
			Musculocutaneous nerve (C5–C6 nerve fibers)

Brachialis (brā′kē-al′is)	Primary flexor of forearm	O: Distal anterior surface of humerus I: Tuberosity and coronoid process of ulna	Musculocutaneous nerve (C5–C6 nerve fibers)
Brachioradialis (brā′kē-ō-rā′dē-al′is)	Flexes forearm	O: Lateral supracondylar ridge of humerus I: Styloid process of radius	Radial nerve (C6–C7 nerve fibers)
EXTENSORS (POSTERIOR ARM)			

Triceps brachii
  Long head
  Lateral head
  Medial head
	Primary extensor of forearm
	
	Long head of triceps also extends and adducts arm
		O: Long head: Infraglenoid tubercle of scapula
		  Lateral head: Posterior humerus above radial groove
		  Short head: Posterior humerus below radial groove
		I: Olecranon of ulna
			Radial nerve (C5–C7 nerve fibers)

Anconeus (ang-kō′nē-ŭs) *ankon* = elbow	Extends forearm	O: Lateral epicondyle of humerus I: Olecranon of ulna	Radial nerve (C6–C8 nerve fibers)
PRONATORS (ANTERIOR FOREARM MUSCLES)			
Pronator quadratus (prō-nā-tōr kwah-drā′tŭs)	Pronates forearm	O: Distal ¼ of ulna I: Distal ¼ of radius	Median nerve (C8–T1 nerve fibers)
Pronator teres	Pronates forearm	O: Medial epicondyle of humerus and coronoid process of ulna I: Lateral surface of radius	Median nerve (C6–C7 nerve fibers)
SUPINATOR (POSTERIOR FOREARM MUSCLE)			
Supinator (sū′pi-nā-tōr)	Supinates forearm	O: Lateral epicondyle of humerus and ulna distal to radial notch I: Anterolateral surface of radius distal to radial tuberosity	Radial nerve (C6–C8 nerve fibers)

11.8d Forearm Muscles That Move the Wrist Joint, Hand, and Fingers

LEARNING OBJECTIVES

6. Describe the muscles of the anterior compartment and their actions, and identify the layer in which each resides.

7. Explain the actions of the muscles of the posterior compartment, and identify the layer in which each resides.

Most muscles in the forearm move the hand at the wrist, the fingers, or both. These muscles are called extrinsic muscles of the wrist and hand, because the muscles originate on the forearm, not the wrist or hand. Palpate your own forearm; it is bigger near the elbow because the bellies of these forearm muscles form the bulk of this region, whereas distally at the wrist, only long tendons of these muscles are present.

Deep fascia partitions the forearm muscles into an anterior (flexor) compartment and a posterior (extensor) compartment (figure 11.27). Most of the **anterior compartment** muscles originate on the medial epicondyle of the humerus via a common flexor tendon. Muscles in the anterior compartment of the forearm generally tend to flex the wrist, the metacarpophalangeal (MP) joints. Some of these also flex the interphalangeal (IP) joints of the fingers. Most of the **posterior compartment** muscles originate on the lateral epicondyle of the

Table 11.17 — Summary of Muscle Actions at the Elbow Joint/Forearm

Extension	Flexion	Pronation	Supination
Triceps brachii*	**Brachialis**	**Pronator teres**	**Biceps brachii**
(Anconeus)	Biceps brachii	**Pronator quadratus**	Supinator
	Brachioradialis		

*Boldface indicates an agonist; others are synergists. Parentheses around an entire muscle name indicate only a slight effect.

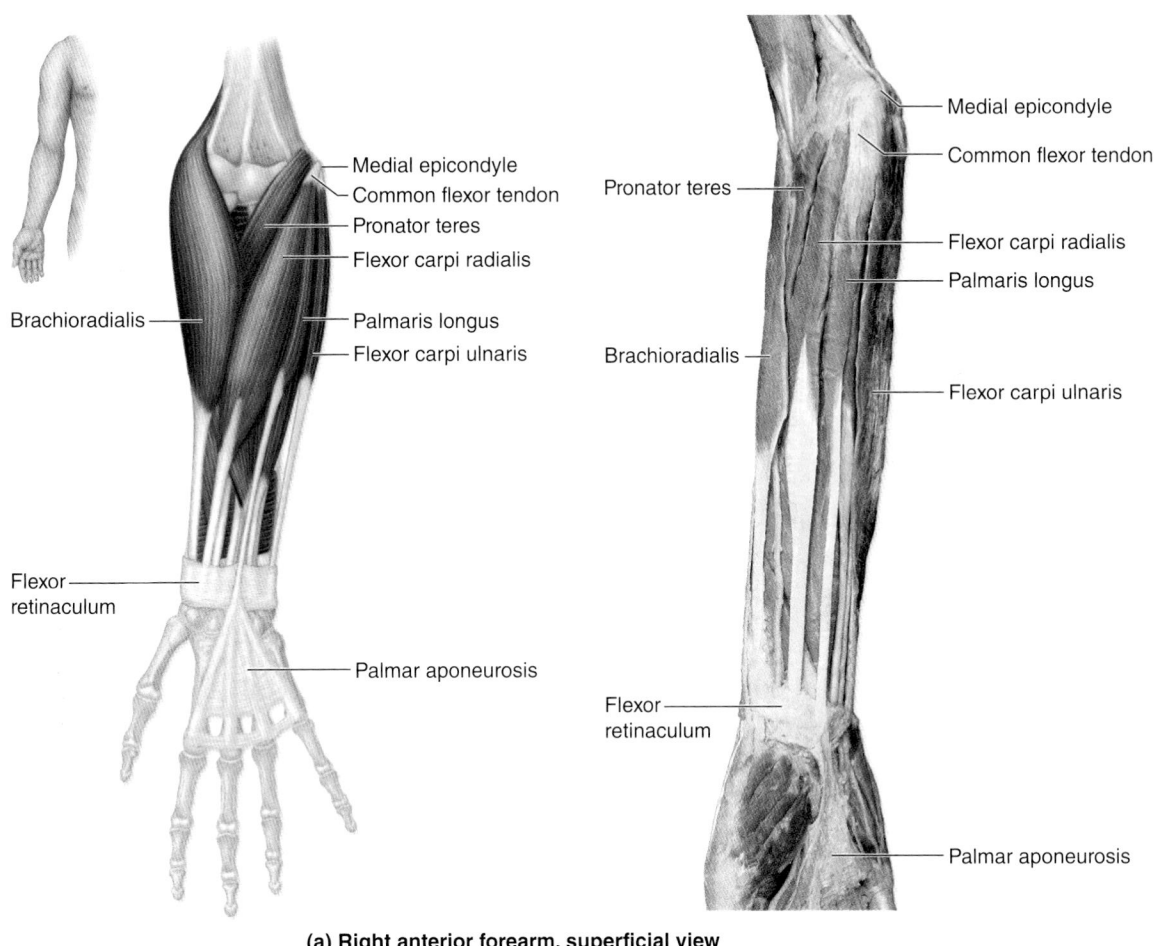

(a) Right anterior forearm, superficial view

Medial epicondyle
Common flexor tendon
Pronator teres
Flexor carpi radialis
Brachioradialis
Palmaris longus
Flexor carpi ulnaris
Flexor retinaculum
Palmar aponeurosis

Pronator teres
Brachioradialis
Medial epicondyle
Common flexor tendon
Flexor carpi radialis
Palmaris longus
Flexor carpi ulnaris
Flexor retinaculum
Palmar aponeurosis

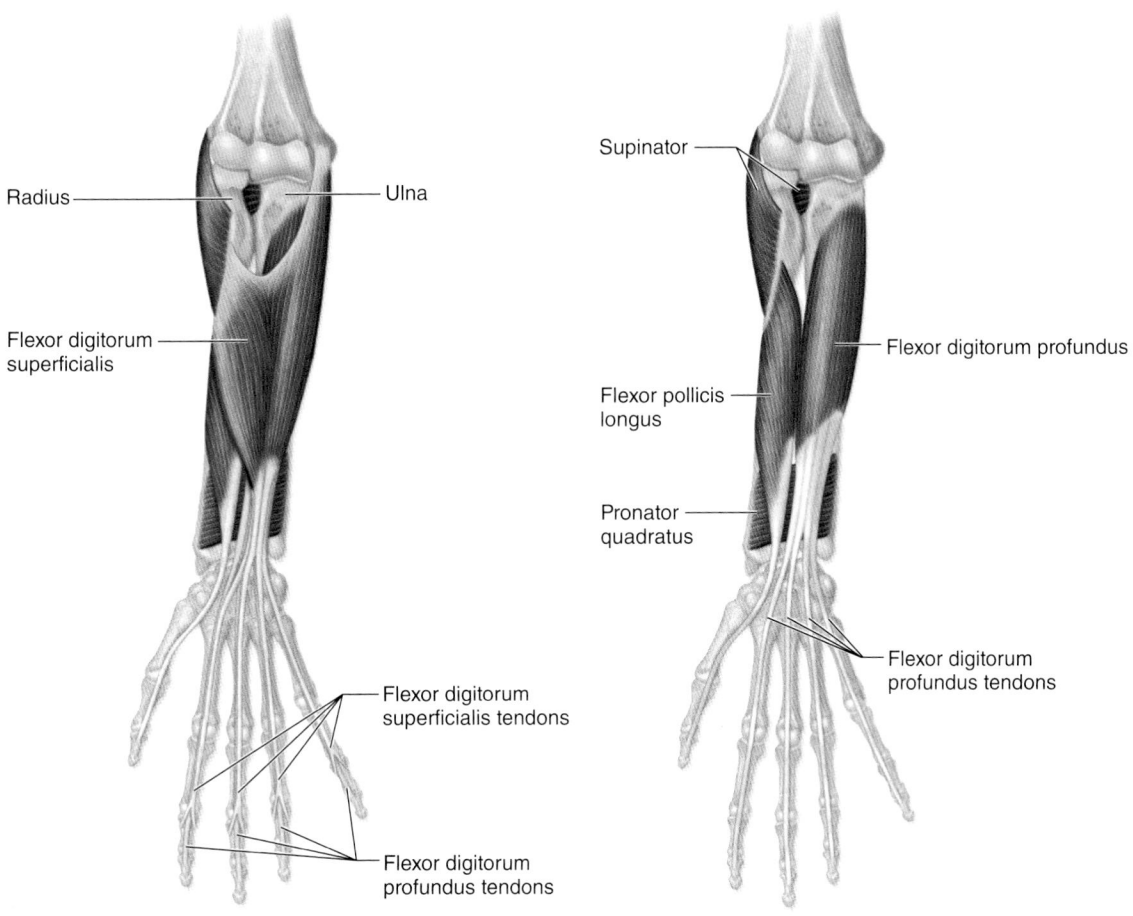

(b) Right anterior forearm, intermediate view

Radius
Ulna
Flexor digitorum superficialis
Flexor digitorum superficialis tendons
Flexor digitorum profundus tendons

(c) Right anterior forearm, deep view

Supinator
Flexor digitorum profundus
Flexor pollicis longus
Pronator quadratus
Flexor digitorum profundus tendons

Figure 11.27 Anterior Forearm Muscles. The anterior forearm muscles pronate the forearm or flex the wrist and fingers. They may be subdivided into superficial, intermediate, and deep layers. (*a*) Illustration and cadaver photo show the superficial muscles of the right anterior forearm. Illustrations of the (*b*) intermediate and (*c*) deep muscles of the right anterior forearm. **AP|R**

humerus via a common extensor tendon. Muscles in the posterior compartment of the forearm tend to extend the wrist. Some also extend the MP joints and the IP joints.

Retinacula of the Forearm

At the wrist, the deep fascia of the forearm forms thickened, fibrous bands termed retinacula (ret-i-nak′ū-la; *retineo* = to hold back). The retinacula help hold the tendons close to the bone and prevent the tendons from "bowstringing" outward. The palmar (anterior) surface of the carpal bones is covered by the **flexor retinaculum** (figure 11.27a). Flexor tendons of the digits and the median nerve pass through the tight space between the bones and the flexor retinaculum, which is called the **carpal tunnel**. The **extensor retinaculum** is superficial to the dorsal surface of the carpal bones. Extensor tendons of the wrist and digits pass between the bones and the extensor retinaculum.

Muscles of the Forearm's Anterior Compartment

The muscles of the **anterior compartment** of the forearm may be subdivided into a superficial layer, an intermediate layer, and a deep layer. The muscles of the superficial and intermediate layers originate from the common flexor tendon that attaches to the medial epicondyle of the humerus. The deep layer of muscles originates directly on the bones of the forearm.

Note that not all anterior forearm muscles cause flexion. Both the pronator teres and the pronator quadratus, discussed previously, are located in the anterior compartment of the forearm but their primary function is pronation. Likewise, the supinator muscle is in the posterior compartment of the forearm, yet its primary function is supination.

The **superficial layer** of anterior forearm muscles is arranged from the lateral to the medial surface of the forearm in the following order: pronator teres (described previously), flexor carpi radialis, palmaris longus, and flexor carpi ulnaris. The **flexor carpi radialis** tendon is prominent on the lateral side of the forearm. This muscle flexes the wrist and abducts the hand at the wrist. The **palmaris longus** is absent in some individuals. This narrow, superficial muscle on the anterior surface of the forearm weakly assists in wrist flexion. The **flexor carpi ulnaris** flexes the wrist and adducts the hand at the wrist.

You can determine the positioning of the three superficial muscles of the anterior forearm and the pronator teres muscle on your own body by performing the exercise shown in **figure 11.28**. Wrap your thumb around the medial epicondyle of the other arm, so your thumb is positioned behind the elbow. Align your little finger along the medial border of your forearm. The natural placement of your four fingers overlies the placement of superficial layer muscles.

The **intermediate layer** in the anterior compartment of the forearm contains a single muscle (see figure 11.27b). The **flexor digitorum superficialis** splits into four tendons that each insert on the middle

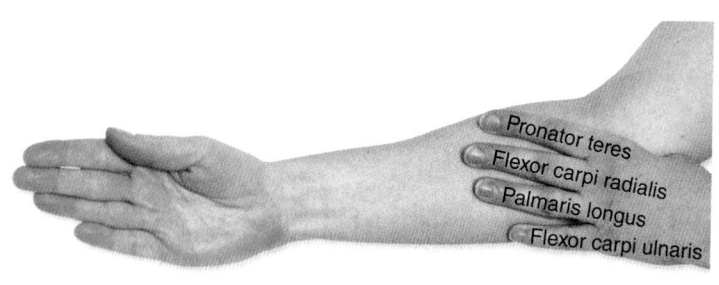

(Left hand covers medial epicondyle)

Figure 11.28 Positioning of the Superficial Anterior Forearm Muscles. By positioning the left hand at the medial epicondyle of the right humerus, fingers 2–5 lay in the approximate position of the superficial muscles of the anterior forearm.

phalanges of fingers 2–5. This muscle crosses over the wrist, MP joints, and PIP (proximal interphalangeal) joints of fingers 2–5; thus, it flexes all of these joints. Since the flexor digitorum superficialis does not cross over the DIP (distal interphalangeal) joints of these fingers, it cannot move the DIP joints.

The **deep layer** of the forearm anterior compartment muscles includes the flexor pollicis longus (lateral side), the flexor digitorum profundus (medial side), and the pronator quadratus (deep) (see figure 11.27c). The **flexor pollicis longus** attaches to the distal phalanx of the thumb and flexes the MP and IP joints of the thumb. In addition, because this muscle crosses the wrist joint, it can weakly flex the wrist. The **flexor digitorum profundus** lies deep to the flexor digitorum superficialis. This muscle splits into four tendons that insert on the distal phalanges of fingers 2–5. The flexor digitorum profundus flexes the wrist, MP joints, PIP joints, and DIP joints of fingers 2–5.

Muscles of the Forearm's Posterior Compartment

Muscles of the **posterior compartment** of the forearm are primarily wrist and finger extensors. An exception is the supinator, which helps supinate the forearm. The posterior compartment muscles may be subdivided into a superficial layer and a deep layer.

The **superficial layer** of posterior forearm muscles originates from a common extensor tendon on the lateral epicondyle of the humerus (**figure 11.29a**). These muscles are positioned laterally to medially as follows: The **extensor carpi radialis longus** is medial to the brachioradialis. It extends the wrist and abducts the hand at the wrist. The **extensor carpi radialis brevis** works synergistically with the extensor carpi radialis longus. The **extensor digitorum** splits into four tendons that insert on the distal phalanges of fingers 2–5. It extends the wrist, MP joints, PIP joints, and DIP joints of fingers 2–5. The **extensor digiti minimi** attaches to the distal phalanx of the pinky (finger 5). It works with the extensor digitorum to extend the little finger. On the medial surface of the posterior forearm, the **extensor carpi ulnaris** inserts on the fifth metacarpal bone, where it acts to extend the wrist and adduct the hand.

The **deep layer** muscles originate directly on the bones of the posterior forearm and insert on the wrist or hand (figure 11.29b). These muscles weakly extend the wrist and do the following other functions: (1) The **abductor pollicis longus** abducts the thumb. (2) The **extensor pollicis brevis** attaches to the proximal phalanx of the thumb and helps extend the MP joint of the thumb. (3) The **extensor pollicis longus** inserts on the distal phalanx of the thumb, so it extends the MP and IP joints of the thumb. (4) The **extensor indicis** extends the MP, PIP, and DIP joints of the index finger (finger 2).

INTEGRATE

CONCEPT CONNECTION

Because the median nerve (see section 14.5e) passes deep to the flexor retinaculum, this nerve may be pinched within the carpal tunnel. Thus, the musculoskeletal anatomy is related to the proper functioning of selected components of the nervous system.

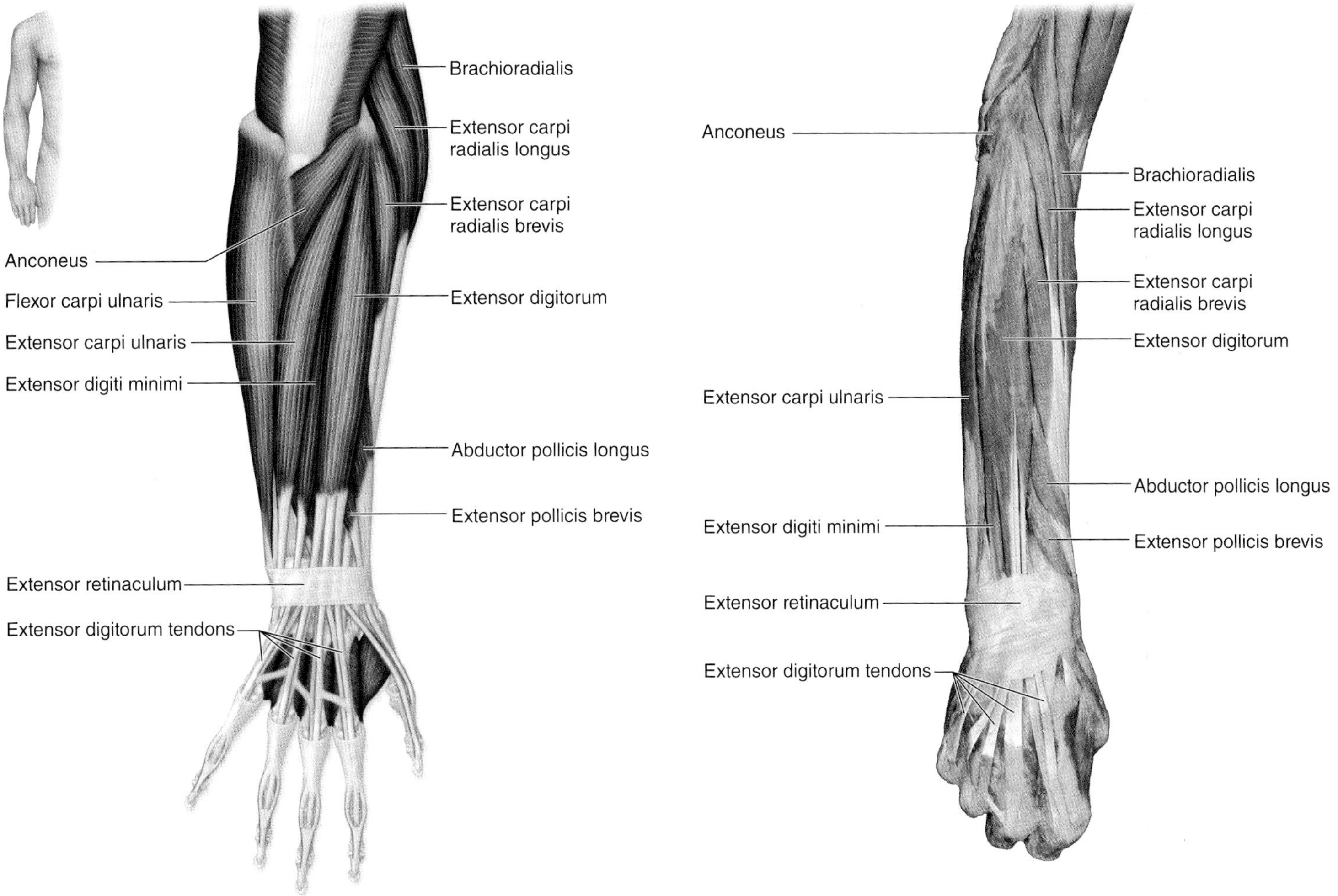

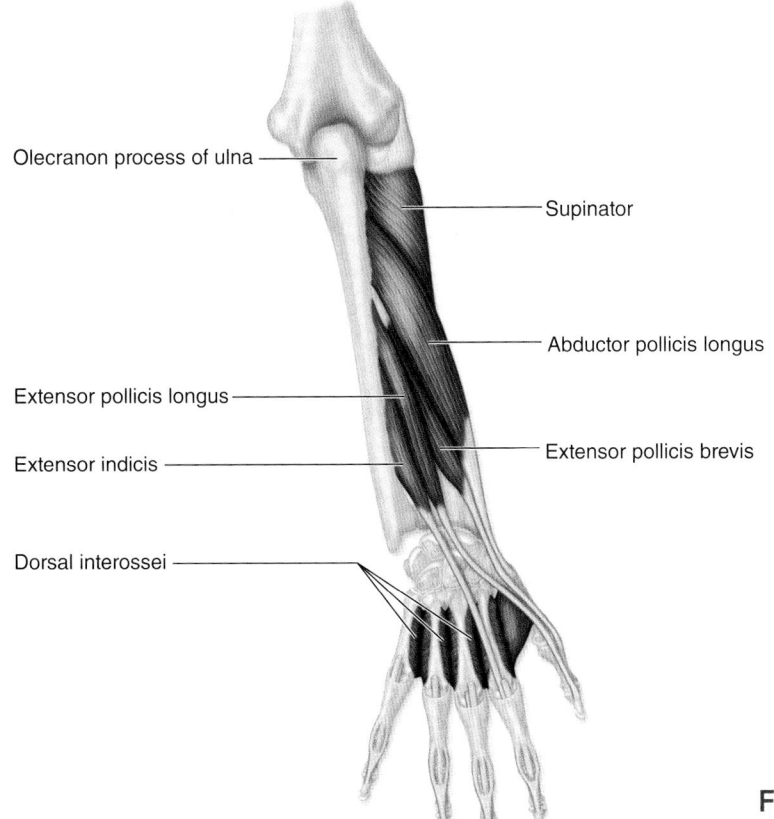

Brachioradialis

Extensor carpi radialis longus

Extensor carpi radialis brevis

Anconeus

Flexor carpi ulnaris

Extensor carpi ulnaris

Extensor digiti minimi

Extensor digitorum

Abductor pollicis longus

Extensor pollicis brevis

Extensor retinaculum

Extensor digitorum tendons

Anconeus

Brachioradialis

Extensor carpi radialis longus

Extensor carpi radialis brevis

Extensor digitorum

Extensor carpi ulnaris

Abductor pollicis longus

Extensor digiti minimi

Extensor pollicis brevis

Extensor retinaculum

Extensor digitorum tendons

(a) Right posterior forearm, superficial views

Olecranon process of ulna

Supinator

Abductor pollicis longus

Extensor pollicis longus

Extensor indicis

Extensor pollicis brevis

Dorsal interossei

(b) Right posterior forearm, deep view

Figure 11.29 Posterior Forearm Muscles. The posterior forearm muscles supinate the forearm or extend the wrist or fingers. They may be subdivided into (*a*) superficial and (*b*) deep layers, as shown in these views of the right forearm.

CLINICAL VIEW
Carpal Tunnel Syndrome

The space between the carpal bones and the flexor retinaculum is the **carpal tunnel.** Numerous finger flexor tendons extend through this tunnel as well as the median nerve, which innervates the skin on the lateral palmar region of the hand and the muscles that move the thumb. Any compression of either the median nerve or the tendons in the tunnel results in carpal tunnel syndrome. The syndrome is characterized by pain and **paresthesia** (par-es-thē′zē-ǎ; *aisthesis* = sensation), which is the feeling of "pins and needles." Sometimes, more extensive sensory loss as well as motor loss occurs in the muscles of the hand supplied by the median nerve.

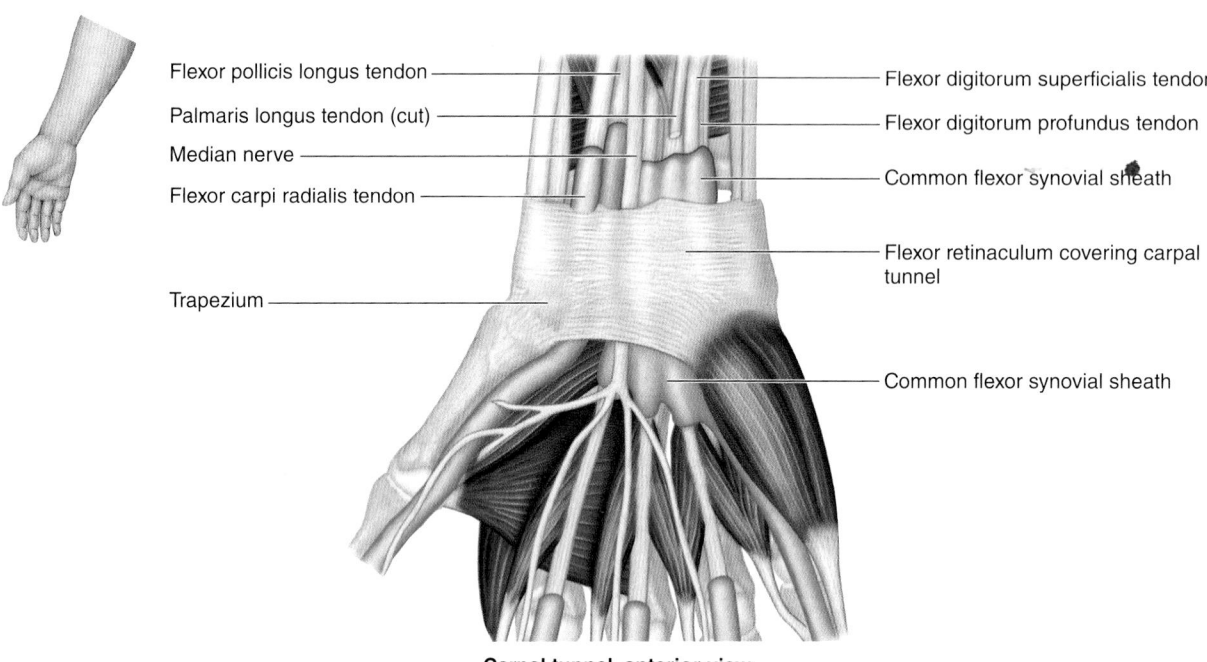

Flexor pollicis longus tendon
Palmaris longus tendon (cut)
Median nerve
Flexor carpi radialis tendon
Trapezium
Flexor digitorum superficialis tendon
Flexor digitorum profundus tendon
Common flexor synovial sheath
Flexor retinaculum covering carpal tunnel
Common flexor synovial sheath

Carpal tunnel, anterior view

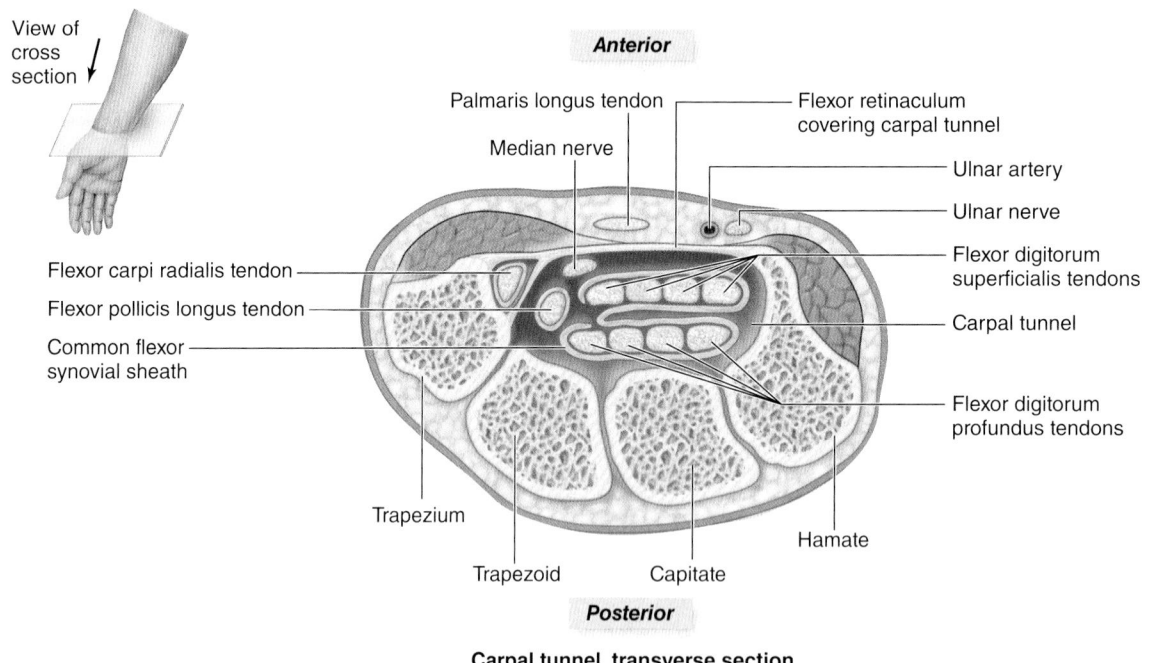

View of cross section
Anterior
Palmaris longus tendon
Median nerve
Flexor retinaculum covering carpal tunnel
Ulnar artery
Ulnar nerve
Flexor digitorum superficialis tendons
Carpal tunnel
Flexor digitorum profundus tendons
Flexor carpi radialis tendon
Flexor pollicis longus tendon
Common flexor synovial sheath
Trapezium
Trapezoid
Capitate
Hamate
Posterior

Carpal tunnel, transverse section

Table 11.18 summarizes the characteristics of the muscles that move the wrist joint, hand, and fingers.

WHAT DID YOU LEARN?

23 What are the common actions of muscles in the anterior compartment of the forearm?

24 What muscles in the posterior compartment move the thumb?

Table 11.18	Forearm Muscles That Move the Wrist Joint, Hand, and Digits		
Group/Muscle	**Action**	**Origin/Insertion**	**Innervation (see section 14.5e)**
ANTERIOR MUSCLES: SUPERFICIAL			
Pronator teres (previously described in table 11.16)			
Flexor carpi radialis (flek′sŏr kar′pī rā-dē-ā′lis) *carpi* = wrist	Flexes wrist and abducts hand	O: Medial epicondyle of humerus I: Base of metacarpals II and III	Median nerve (C6–C7 nerve fibers)
Palmaris longus (pahl-mār′is) *palma* = palm	Weak wrist flexor	O: Medial epicondyle of humerus I: Flexor retinaculum and palmar aponeurosis	Median nerve (C6–C7 nerve fibers)
Flexor carpi ulnaris (ŭl-nā′ris) *ulnar* = ulna	Flexes wrist and adducts hand	O: Medial epicondyle of humerus; olecranon and posterior surface of ulna I: Pisiform and hamate bones; base of metacarpal V	Ulnar nerve (C8–T1)
ANTERIOR MUSCLES: INTERMEDIATE			
Flexor digitorum superficialis (dij′i-tōr′ŭm sū′per-fish-ē-ā′lis) *digit* = finger, toe *superficial* = close to surface	Flexes wrist; 2nd–5th MP joints and PIP joints	O: Medial epicondyle of humerus, coronoid process of ulna I: Middle phalanges of fingers 2–5	Median nerve (C6–C7 nerve fibers)
ANTERIOR MUSCLES: DEEP			
Flexor pollicis longus (pol′i-sis) *pollex* = thumb	Flexes MP joint of thumb, IP joint of thumb; weakly flexes wrist	O: Anterior shaft of radius; interosseous membrane I: Distal phalanx of thumb	Median nerve (C6–C7 nerve fibers)
Flexor digitorum profundus (prō-fŭn′dŭs) *profound* = deep	Flexes wrist; 2nd–5th MP joints, PIP joints, and DIP joints	O: Anteromedial surface of ulna; interosseous membrane I: Distal phalanges of fingers 2–5	Lateral ½ of muscle innervated by median nerve (C6–C8 nerve fibers), medial ½ of muscle innervated by ulnar nerve (C8 nerve fibers)
Pronator quadratus (previously described in table 11.16)			
POSTERIOR MUSCLES: SUPERFICIAL			
Extensor carpi radialis longus (eks-ten′sŏr,–sōr)	Extends wrist, abducts hand	O: Lateral supracondylar ridge of humerus I: Base of metacarpal II	Radial nerve (C6–C7 nerve fibers)
Extensor carpi radialis brevis (brev′is) *brevis* = short	Extends wrist, abducts hand	O: Lateral epicondyle of humerus I: Base of metacarpal III	Radial nerve (C6–C7 nerve fibers)
Extensor digitorum	Extends wrist; extends 2nd–5th MP joints, PIP joints and DIP joints	O: Lateral epicondyle of humerus I: Distal and middle phalanges of fingers 2–5	Radial nerve (C6–C8 nerve fibers)
Extensor digiti minimi *digitus minimus* = little finger	Extends MP and PIP joints of finger 5; extends wrist weakly	O: Lateral epicondyle of humerus I: Proximal phalanx of finger 5	Radial nerve (C6–C8 nerve fibers)
Extensor carpi ulnaris	Extends wrist, adducts hand	O: Lateral epicondyle of humerus; posterior border of ulna I: Base of metacarpal V	Radial nerve (C6–C8 nerve fibers)

(continued on next page)

Table 11.18	Forearm Muscles That Move the Wrist Joint, Hand, and Digits (continued)		
Group/Muscle	Action	Origin/Insertion	Innervation (see section 14.5e)
POSTERIOR MUSCLES: DEEP			
Abductor pollicis longus (ab-dŭk'ter, tōr)	Abducts thumb; extends wrist (weakly)	O: Proximal dorsal surfaces of radius and ulna; interosseous membrane I: Lateral edge of metacarpal I	Radial nerve (C6–C8 nerve fibers)
Extensor pollicis brevis	Extends MP joint of thumb; extends wrist (weakly)	O: Posterior surface of radius; interosseous membrane I: Proximal phalanx of thumb	Radial nerve (C6–C8 nerve fibers)
Extensor pollicis longus	Extends MP and IP joints of thumb, extends wrist (weakly)	O: Posterior surface of ulna; interosseous membrane I: Distal phalanx of thumb	Radial nerve (C6–C8 nerve fibers)
Extensor indicis (in'di-sis) *index* = forefinger	Extends MP PIP and DIP joints of finger 2; extends wrist (weakly)	O: Posterior surface of ulna; interosseous membrane I: Tendon of extensor digitorum	Radial nerve (C6–C8 nerve fibers)
Supinator (previously described in table 11.16)			

INTEGRATE

LEARNING STRATEGY

To remember the functions of the palmar and dorsal interosseous muscles, use this mnemonic phrase:

PAD-DAB

(**P**almar interossei **AD**duct the fingers, while **D**orsal interossei **AB**duct the fingers.)

11.8e Intrinsic Muscles of the Hand

LEARNING OBJECTIVE

8. Compare the actions of the three groups of intrinsic muscles of the hand.

The intrinsic muscles of the hand are small muscles that both originate and insert on the hand; they are housed entirely within the palm (figure 11.30). These muscles are divided into three groups: The **thenar** (thē'nar) **group** forms the thick, fleshy mass (thenar eminence) at the base of the thumb; the **hypothenar group** forms a smaller fleshy mass (hypothenar eminence) at the base of the little finger; and the **midpalmar group** occupies the space between the first two groups.

The thenar and hypothenar groups contain smaller muscles:

- Small flexors (**flexor pollicis brevis** in the thenar group and **flexor digiti minimi brevis** in the hypothenar group) flex the thumb and the little finger, respectively.

- Abductors (**abductor pollicis brevis** in the thenar group and **abductor digiti minimi** in the hypothenar group) abduct the thumb and little finger, respectively.

- Opponens muscles (**opponens pollicis** in the thenar group and **opponens digiti minimi** in the hypothenar group) assist in the opposition of the thumb and little finger, respectively.

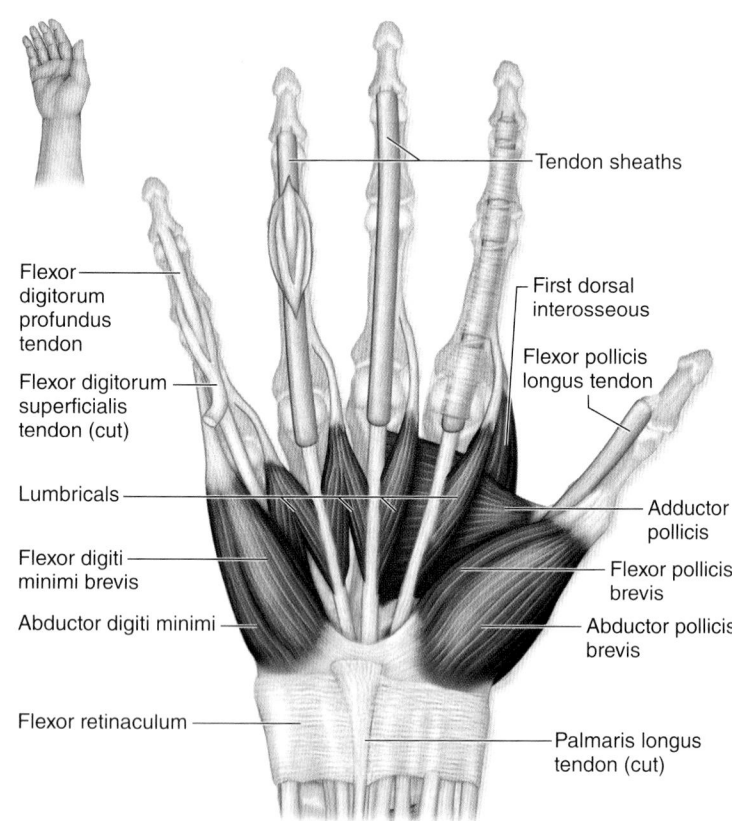

Flexor digitorum profundus tendon
Flexor digitorum superficialis tendon (cut)
Lumbricals
Flexor digiti minimi brevis
Abductor digiti minimi
Flexor retinaculum
Tendon sheaths
First dorsal interosseous
Flexor pollicis longus tendon
Adductor pollicis
Flexor pollicis brevis
Abductor pollicis brevis
Palmaris longus tendon (cut)

(a) Right hand, superficial palmar view

Figure 11.30 Intrinsic Muscles of the Hand. These muscles allow the fine, controlled movements necessary for such activities as writing, typing, and playing a guitar. (*a*) Palmar (anterior) views of the superficial muscles of the right hand. (*b*) Palmar view of the deep muscles. (*c*) Posterior (dorsal) view of the superficial muscles. **AP|R**

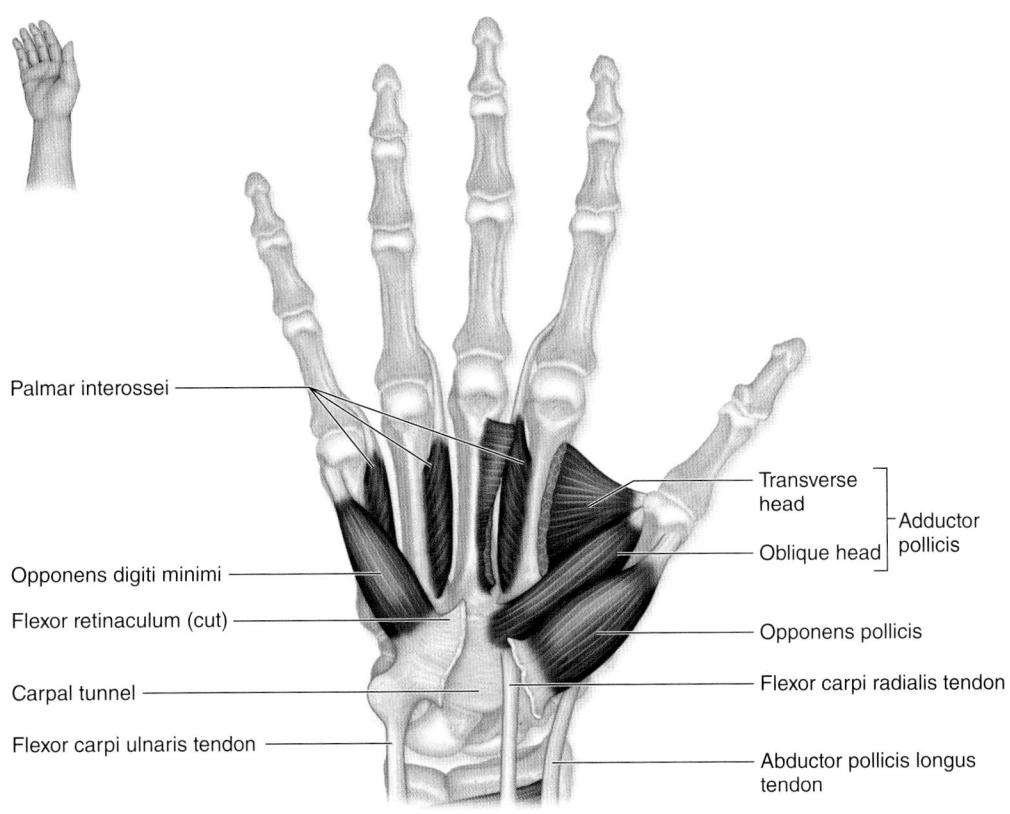

Palmar interossei

Opponens digiti minimi

Flexor retinaculum (cut)

Carpal tunnel

Flexor carpi ulnaris tendon

Transverse head

Oblique head — Adductor pollicis

Opponens pollicis

Flexor carpi radialis tendon

Abductor pollicis longus tendon

(b) Right hand, deep palmar view

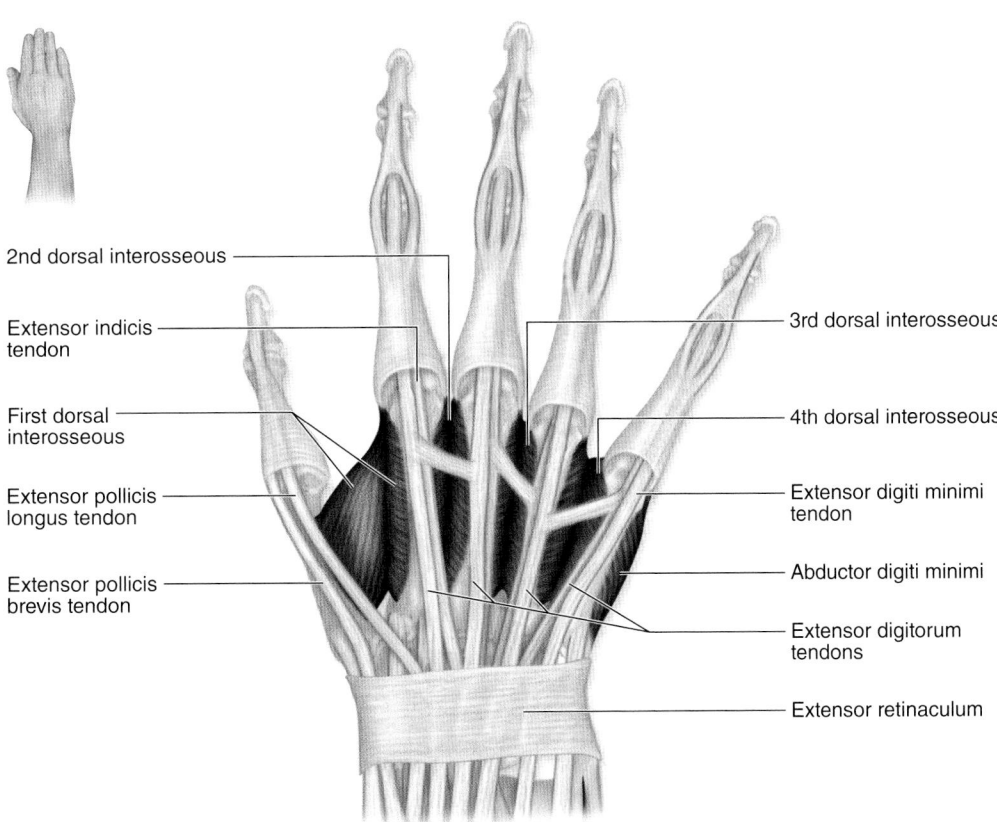

2nd dorsal interosseous

Extensor indicis tendon

First dorsal interosseous

Extensor pollicis longus tendon

Extensor pollicis brevis tendon

3rd dorsal interosseous

4th dorsal interosseous

Extensor digiti minimi tendon

Abductor digiti minimi

Extensor digitorum tendons

Extensor retinaculum

(c) Right hand, posterior view

The midpalmar group contains the following muscles: lumbricals, dorsal interossei, palmar interossei, and adductor pollicis. The **lumbrical muscles** are four worm-shaped muscles. These muscles flex the MP joints and at the same time extend the PIP and DIP joints of fingers 2–5. The **dorsal interossei** are four deep bipennate muscles located between the metacarpals. They flex the MP joints and at the same time extend the PIP and DIP joints of fingers 2–5. In addition, the dorsal interossei abduct fingers 2–5. The **palmar interossei** are three small muscles that adduct the fingers. In addition, these muscles work with the lumbricals and dorsal interossei to flex the MP joints and at the same time extend the PIP and DIP joints of fingers 2–5. The **adductor pollicis** is sometimes incorrectly classified as a palmar interosseous muscle. As its name suggests, this muscle adducts the thumb.

The muscles that control the specific movements of the fingers, hand, and wrist, are summarized in table 11.19 and grouped according to common muscle actions in table 11.20.

 WHAT DID YOU LEARN?

 Identify the intrinsic muscles of the hand that abduct the fingers.

Table 11.19	Intrinsic Muscles of the Hand		
Group/Muscle	**Action**	**Origin/Insertion**	**Innervation (see section 14.5e)**
THENAR GROUP			
Flexor pollicis brevis	Flexes thumb	O: Flexor retinaculum; trapezium I: Proximal phalanx of thumb	Median nerve (C8–T1 nerve fibers)
Abductor pollicis brevis	Abducts thumb	O: Flexor retinaculum, scaphoid, trapezium I: Lateral side of proximal phalanx of thumb	Median nerve (C8–T1 nerve fibers)
Opponens pollicis (ō-pō′nens) *opponens* = to place against	Opposition of thumb	O: Flexor retinaculum and trapezium I: Lateral side of metacarpal I	Median nerve (C8–T1 nerve fibers)
HYPOTHENAR GROUP			
Flexor digiti minimi brevis	Flexes finger 5	O: Hamate bone, flexor retinaculum I: Proximal phalanx of finger 5	Ulnar nerve (C8–T1)
Abductor digiti minimi	Abducts finger 5	O: Pisiform bone, tendon of flexor carpi ulnaris I: Proximal phalanx of finger 5	Ulnar nerve (C8–T1)
Opponens digiti minimi	Opposition of finger 5	O: Hamate bone, flexor retinaculum I: Metacarpal bone V	Ulnar nerve (C8–T1)
MIDPALMAR GROUP			
Lumbricals (lŭm′bri-kălz) *lumbricus* = earthworm	Flexes 2nd–5th MP joints and extends the 2nd–5th PIP and DIP joints	O: Tendons of flexor digitorum profundus I: Dorsal tendons on fingers 2–5	Median nerve (lateral two lumbricals 1, 2) and ulnar nerve (medial two lumbricals 3, 4)
Dorsal interossei (dōr′săl in′ter-os′ē-ī) *interossei* = between bones	Abducts fingers 2–5; flexes MP joints 2–5 and extends the PIP and DIP joints	O: Adjacent, opposing faces of metacarpals I: Dorsal tendons on fingers 2–5	Ulnar nerve (C8–T1)
Palmar interossei	Adducts fingers 2–5; flexes MP joints 2–5 and extends the PIP and DIP joints	O: Metacarpal bones II, IV, V I: Sides of proximal phalanges bases for fingers 2, 4, and 5	Ulnar nerve (C8–T1)
Adductor pollicis (ă-dŭk′tŏr)	Adducts thumb	O: Oblique head: Capitate bone, bases of metacarpals II, III Transverse head: Metacarpal III I: Medial side of proximal phalanx of thumb	Ulnar nerve (C8–T1)

Table 11.20		Summary of Muscle Actions at the Wrist and Hand				
Hand Abduction	**Hand Adduction**	**Wrist Extension**	**Wrist Flexion**
Flexor carpi radialis	Extensor carpi ulnaris	Extensor digitorum	Flexor carpi radialis
Extensor carpi radialis brevis	Flexor carpi ulnaris	Extensor carpi radialis brevis	Flexor carpi ulnaris
Extensor carpi radialis longus		Extensor carpi radialis longus	Flexor digitorum superficialis
		Extensor carpi ulnaris	Flexor digitorum profundus
		(Extensor indicis)*	(Palmaris longus)
		(Extensor pollicis longus)	(Flexor pollicis longus)
		(Extensor pollicis brevis)	
		(Abductor pollicis longus)	
Finger Abduction	**Finger Adduction**	**IP Joint Extension**	**IP Joint Flexion**
Dorsal interossei	Palmar interossei	Extensor digitorum	Flexor digitorum profundus
Abductor pollicis longus	Adductor pollicis	Extensor indicis	Flexor digitorum superficialis
Abductor pollicis brevis		Extensor pollicis brevis	Flexor pollicis longus
Abductor digiti minimi		Extensor pollicis longus	Flexor pollicis brevis
		Extensor digiti minimi	Flexor digiti minimi
		Lumbricals	
		Dorsal interossei	
		Palmar interossei	

*Parentheses around a muscle name indicate only a slight effect.

11.9 Muscles of the Pelvic Girdle and Lower Limb

The most powerful and largest muscles in the body are those of the lower limb. These muscles are designed to support the weight of the body and move the lower limbs during locomotion. Like the upper limb, the lower limb muscles also are organized in compartments (see figure 11.23).

As with the muscles of the pectoral girdle and upper limb, the pelvic girdle and lower limb muscles can be organized into groups:

- Muscles that move the hip joint/thigh
- Thigh muscles that move the knee joint/leg
- Muscles of the leg that move the ankle, foot, and toes
- Intrinsic muscles of the foot

INTEGRATE

CLINICAL VIEW

Shin Splints and Compartment Syndrome

Shin splints, also called *shin splint syndrome,* refers specifically to soreness or pain somewhere along the length of the tibia. Shin splints often occur in runners or joggers who are either new to the sport or poorly conditioned. Some healthcare professionals consider shin splints a type of compartment syndrome of the anterior compartment of the leg.

Generally, **compartment syndrome** is a condition in which the blood vessels within a limb compartment become compressed as a result of inflammation and swelling secondary to muscle strain, contusion, or overuse. For example, compartment syndrome can occur in an individual who suddenly embarks on an intensive exercise regimen. More severe compartment syndrome can occur due to trauma to the limb compartment, such as a bone fracture or rupture of a blood vessel.

Since the deep fascia is tight and cannot stretch, swelling from muscles or any accumulating fluid or blood increases pressure in the compartment. Both the blood vessels and the nerves of the compartment become compressed and compromised. Mild cases of compartment syndrome may be treated by immobilizing and resting the affected limb. In more severe cases, the fascia may have to be incised (cut) to relieve the pressure and decompress the affected compartment.

11.9a Muscles That Move the Hip Joint/Thigh

LEARNING OBJECTIVES

1. Compare and contrast the functions of the muscles in the anterior, medial, lateral, and posterior compartments of the thigh.

2. Describe the actions of the three gluteal muscles.

Note that in the subsequent discussion the phrases "moving the hip joint" and "moving the thigh" mean the same thing. The **fascia lata,** the deep fascia of the thigh, encircles the thigh muscles like a supportive stocking and tightly binds them. The fascia lata partitions the thigh muscles into compartments, each with its own blood and nerve supply. The *anterior compartment* muscles either extend the knee or flex the thigh; these are discussed in the next section. The muscles of the *medial compartment* adduct the thigh. The single muscle in the *lateral compartment* abducts the thigh. Most muscles of the *posterior compartment* act as both flexors of the knee and extensors of the thigh. Some of these muscles also abduct the thigh. We discuss the muscles that move the thigh first.

Multiple muscles insert on the anterior thigh and flex the hip joint **(figure 11.31a):** The **psoas major** and the **iliacus** have different origins (they originate on the lumbar vertebrae and ilium, respectively), but they share the common insertion at the lesser trochanter of the femur. Collectively, the two muscles merge and insert on the femur as the **iliopsoas.** Together, these muscles work synergistically to flex the thigh. The rectus femoris and a long, thin muscle called the sartorius flex the thigh. Both are examined later in this chapter in connection with the thigh muscles that move the leg.

Six muscles are located in the **medial compartment** of the thigh. Most of these muscles adduct the thigh, and some of them perform additional functions. The **adductor longus, adductor brevis, gracilis,** and **pectineus** also flex the thigh. A fifth muscle, the **adductor magnus,** also extends and laterally rotates the thigh. The **obturator externus** does not adduct the thigh, but it laterally rotates the thigh.

On the lateral thigh is a single muscle called the **tensor fasciae latae** (figure 11.31b). It attaches to a lateral thickening of the fascia lata, called the **iliotibial tract** (or *iliotibial band*), which extends from the iliac crest to the lateral condyle of the tibia. The tensor fasciae latae abducts and medially rotates the thigh.

The posterior muscles that move the thigh include three gluteal muscles and the "hamstring" muscle group (which are described in a later section). The **gluteus maximus** is the largest and heaviest of the three gluteal muscles; it is the chief extensor of the thigh and it laterally rotates the thigh. Deep to the gluteus maximus are the **gluteus medius** and **gluteus minimus,** which abduct and medially rotate the thigh (figure 11.31c).

Deep to the gluteal muscles are a group of muscles that collectively laterally rotate the thigh/hip joint, as when the legs are crossed with one ankle resting on the knee. These muscles are organized from superior to inferior within the posterior thigh as the **piriformis, superior gemellus, obturator internus, inferior gemellus,** and **quadratus femoris.**

Finally, the posterior thigh contains a group of muscles that are collectively referred to as the **hamstrings** because a ham is strung up by these muscles while being smoked. The hamstring muscles are the biceps femoris, semimembranosus, and semitendinosus. These muscles share a common origin on the ischial tuberosity of the os coxae, and insert on the leg. Thus, these muscles move both the thigh and the knee. Their primary thigh movement is extension. These muscles are discussed again in the next section on movement at the leg.

Table 11.21 summarizes the characteristics of the muscles that move the hip joint and thigh, and **table 11.22** groups these muscles according to their common actions on the thigh.

 WHAT DID YOU LEARN?

26 List the compartments of the thigh, and describe the common action of the muscles in each.

Figure 11.31 Muscles That Act on the Hip and Thigh. (*a*) Anterior, (*b*) lateral, and (*c*) deep posterior views of the right thigh. Most muscles that act on the thigh (femur) originate from the os coxae. (Obturator externus not shown.) AP|R

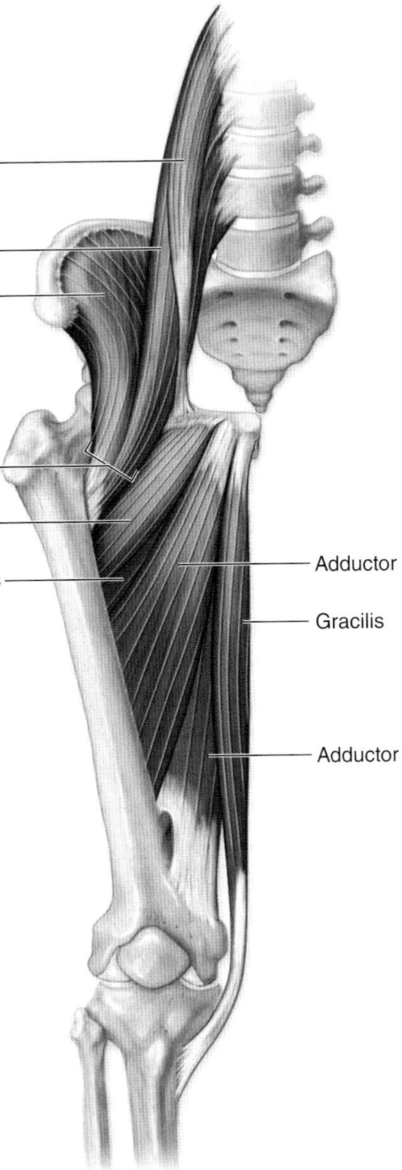

Psoas minor

Psoas major

Iliacus

Iliopsoas

Pectineus

Adductor brevis

Adductor longus

Gracilis

Adductor magnus

(a) Right thigh, anterior view

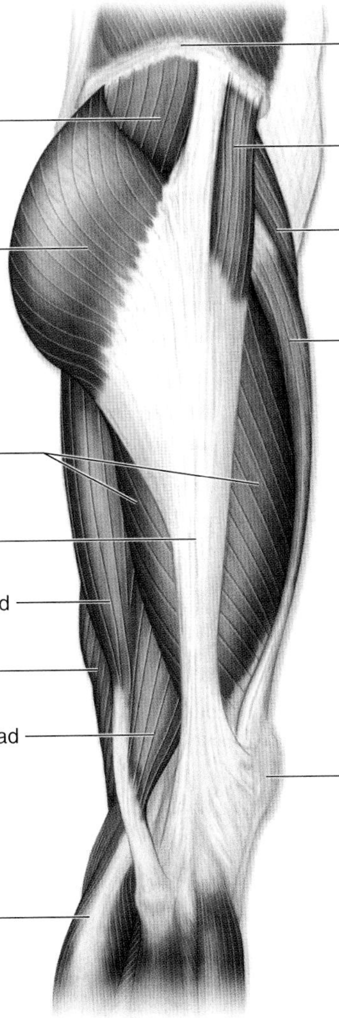

Gluteus medius

Gluteus maximus

Vastus lateralis

Iliotibial tract

Biceps femoris, long head

Semimembranosus

Biceps femoris, short head

Gastrocnemius

Iliac crest

Tensor fasciae latae

Sartorius

Rectus femoris

Patella

(b) Right thigh, lateral view

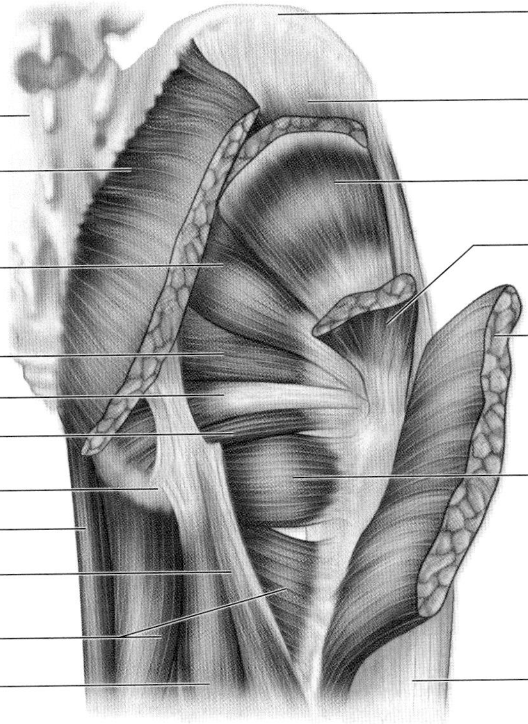

Sacrum

Gluteus maximus (cut)

Piriformis

Superior gemellus

Obturator internus

Inferior gemellus

Ischial tuberosity

Gracilis

Biceps femoris, long head

Adductor magnus

Semitendinosus

Iliac crest

Gluteus medius (cut)

Gluteus minimus

Gluteus medius (cut)

Gluteus maximus (cut)

Quadratus femoris

Iliotibial tract

(c) Right thigh, deep posterior view

Table 11.21	Muscles That Move the Hip Joint/Thigh		
Group/Muscle	**Action**	**Origin/Insertion**	**Innervation (see sections 14.5f, g)**
ANTERIOR THIGH COMPARTMENT (THIGH FLEXORS)			
Psoas major (sō′as) *psoa* = loin muscle	Flexes thigh	O: Transverse processes and bodies of T_{12}–L_5 vertebrae I: Lesser trochanter of femur with iliacus	Branches of lumbar plexus (L2–L3)
Iliacus (il-ī′ă′kŭs) *iliac* = ilium	Flexes thigh	O: Iliac fossa I: Lesser trochanter of femur with psoas major	Femoral nerve (L2–L3 nerve fibers)
Sartorius (sar-tōr′ē-ŭs) *sartor* = tailor	Flexes thigh and laterally rotates thigh; flexes leg and medially rotates leg	O: Anterior superior iliac spine I: Tibial tuberosity, medial side	Femoral nerve (L2–L3 nerve fibers)
Rectus femoris (fem′ō-ris) *femoris* = femur	Flexes thigh; extends leg	O: Anterior inferior iliac spine I: Quadriceps tendon to patella and then patellar ligament to tibial tuberosity	Femoral nerve (L2–L4)
MEDIAL THIGH COMPARTMENT (THIGH ADDUCTORS)			
Adductor longus	Adducts thigh; flexes thigh	O: Pubis near pubic symphysis I: Linea aspera of femur	Obturator nerve (L2–L4)
Adductor brevis	Adducts thigh; flexes thigh	O: Inferior ramus and body of pubis I: Upper third of linea aspera of femur	Obturator nerve (L2–L3 nerve fibers)
Gracilis (gras′i-lis) *gracilis* = slender	Adducts and flexes thigh; flexes leg	O: Inferior ramus and body of pubis I: Upper medial surface of tibia	Obturator nerve (L2–L4)
Pectineus (pek′ti-nē-ŭs) *pectin* = comb	Flexes thigh; adducts thigh	O: Pectineal line of pubis I: Pectineal line of femur	Femoral nerve (L2–L4) or obturator nerve (L2–L4)
Adductor magnus (mag′nŭs) *magnus* = large	Adducts thigh; adductor part of muscle flexes thigh; hamstring part of muscle extends and laterally rotates thigh	O: Inferior ramus of pubis and ischial tuberosity I: Hamstring part: Linea aspera of femur Adductor part: Adductor tubercle of femur	Adductor part: Obturator nerve (L2–L4) Hamstring part: Tibial division of sciatic nerve (L2–L4 nerve fibers)
Obturator externus (ob′tū-rā-tŏr eks-ter′nŭs) *obturator* = any structure that occludes an opening *externus* = outside	Laterally rotates thigh	O Margins of obturator foramen and obturator membrane I: Trochanteric fossa of posterior femur	Obturator nerve (L3–L4 nerve fibers)
LATERAL THIGH COMPARTMENT (THIGH ABDUCTOR)			
Tensor fasciae latae (ten′sŏr fash′ă lā′tē) *tensor* = to make tense *fascia* = band *lata* = wide	Abducts thigh; medially rotates thigh	O: Iliac crest and lateral surface of anterior superior iliac spine I: Iliotibial band	Superior gluteal nerve (L4–S1)
GLUTEAL GROUP			
Gluteus maximus (glū-tē′ŭs mak′si-mŭs) *glutos* = buttock *maximus* = largest	Extends thigh; laterally rotates thigh	O: Iliac crest, sacrum, coccyx I: Iliotibial tract of fascia lata; linea aspera and gluteal tuberosity of femur	Inferior gluteal nerve (L5–S2)

Table 11.21		Muscles That Move the Hip Joint/Thigh (continued)		
Group/Muscle	**Action**	**Origin/Insertion**		**Innervation (see section 14.5g)**
Gluteus medius (mē′dē-ŭs) *medius* = middle	Abducts thigh; medially rotates thigh	O: Posterior iliac crest; lateral surface between posterior and anterior gluteal lines I: Greater trochanter of femur		Superior gluteal nerve (L4–S1)
Gluteus minimus (min-i-mŭs) *minimus* = smallest	Abducts thigh; medially rotates thigh	O: Lateral surface of ilium between inferior and anterior gluteal lines I: Greater trochanter of femur		Superior gluteal nerve (L4–S1)
DEEP MUSCLES OF THE GLUTEAL REGION (LATERAL THIGH ROTATORS)				
Piriformis (pir′i-fōr′mis) *pirum* = pear *forma* = form	Laterally rotates thigh	O: Anterolateral surface of sacrum I: Greater trochanter		Nerve to piriformis (S1–S2)
Superior gemellus (jē-mel′ŭs) *gemin* = twin, double	Laterally rotates thigh	O: Ischial spine and tuberosity I: Greater trochanter		Nerve to obturator internus (L5–S1)
Obturator internus (in-ter′nŭs) *internus* = inside	Laterally rotates thigh	O: Posterior surface of obturator membrane; margins of obturator foramen I: Greater trochanter		Nerve to obturator internus (L5–S1)
Inferior gemellus	Laterally rotates thigh	O: Ischial tuberosity I: Obturator internus tendon		Nerve to quadratus femoris (L5–S1)
Quadratus femoris	Laterally rotates thigh	O: Lateral border of ischial tuberosity I: Intertrochanteric crest of femur		Nerve to quadratus femoris (L5–S1)
HAMSTRING GROUP				
Biceps femoris Long head Short head	Extends thigh (long head only); flexes leg (both long head and short head); laterally rotates leg	O: Long head: Ischial tuberosity Short head: Linea aspera of femur I: Head of fibula		Long head: Tibial division of sciatic nerve (L4-S1 nerve fibers) Short head: Common fibular division of sciatic nerve (L5–S1 nerve fibers)
Semimembranosus (sem′ē-mem-bră-nō-sŭs) *semi* = half *membranosus* = membrane	Extends thigh and flexes leg; medially rotates leg	O: Ischial tuberosity I: Posterior surface of medial condyle of tibia		Tibial division of sciatic nerve (L4–S1 nerve fibers)
Semitendinosus (sem′ē-ten-di-nō′sŭs) *tendinosus* = tendon	Extends thigh and flexes leg; medially rotates leg	O: Ischial tuberosity I: Proximal medial surface of tibia		Tibial division of sciatic nerve (L4–S1 nerve fibers)

Table 11.22		Summary of Muscle Actions at the Hip Joint/Thigh			
Abduction	**Adduction**	**Extension**	**Flexion**	**Lateral Rotation**	**Medial Rotation**
Gluteus medius	Adductor brevis, longus, magnus	**Gluteus maximus***	**Iliopsoas**	Adductor magnus (hamstring part)	Gluteus medius
Gluteus minimus	Gracilis	Adductor magnus (hamstring part)	Adductor brevis, longus, magnus (adductor part)	Gluteus maximus	Gluteus minimus
Tensor fasciae latae	Pectineus	Biceps femoris (long head)	Pectineus	Sartorius	Tensor fasciae latae
		Semimembranosus	Sartorius	Obturator externus	
		Semitendinosus	Rectus femoris	Obturator internus	
			Gracilis	Piriformis	
				Superior gemellus	
				Inferior gemellus	
				Quadratus femoris	

*Boldface indicates an agonist; others are synergists.

11.9b Thigh Muscles That Move the Knee Joint/Leg

LEARNING OBJECTIVES

3. List muscles of the thigh's anterior compartment that move the knee joint.

4. Describe the muscles of the thigh that flex the knee joint.

The muscles that act on the knee form most of the mass of the thigh. They include muscles of the anterior compartment and posterior compartment of the thigh as well as certain muscles already described in the preceding section.

Muscles of the Thigh's Anterior Compartment

The **anterior (extensor) compartment** of the thigh is composed of the large **quadriceps femoris,** the prime mover of knee extension **(figure 11.32)**. The quadriceps femoris is a composite muscle with four heads: the **rectus femoris, vastus lateralis, vastus medialis,** and **vastus intermedius.** All four muscles converge on a single **quadriceps tendon,** which extends to the patella and then continues inferiorly as the **patellar ligament** and inserts on the tibial tuberosity.

Also within the anterior compartment is the **sartorius,** which acts on both the hip and knee joints, flexing and laterally rotating the thigh while flexing and medially rotating the leg. This muscle is the longest in the body and is termed the "tailor's muscle" because it helps us sit cross-legged, as tailors used to do.

Muscles of the Thigh's Medial Compartment

The **gracilis** muscle (in the medial compartment of the thigh) not only adducts the thigh, as described earlier, but also flexes the leg, since it spans the knee joint.

Muscles of the Thigh's Posterior Compartment

The **posterior (flexor) compartment** of the thigh contains the three hamstring muscles discussed previously **(figure 11.33)**. These muscles also flex the leg. The **biceps femoris** is a two-headed muscle that inserts on the lateral side of the leg. This muscle also can laterally rotate the leg when the leg is flexed. The **semimembranosus** and **semitendinosus** insert on the medial side of the leg. The semimembranosus and semitendinosus also medially rotate the leg when the leg is flexed.

Finally, several leg muscles span the knee joint and work to flex the leg. These muscles (gastrocnemius, plantaris, and popliteus) are discussed in the next section, as we examine muscles of the leg.

Table 11.23 summarizes the characteristics of the thigh muscles that move the knee joint and leg.

WHAT DID YOU LEARN?

27 List the thigh muscles that flex the knee joint.

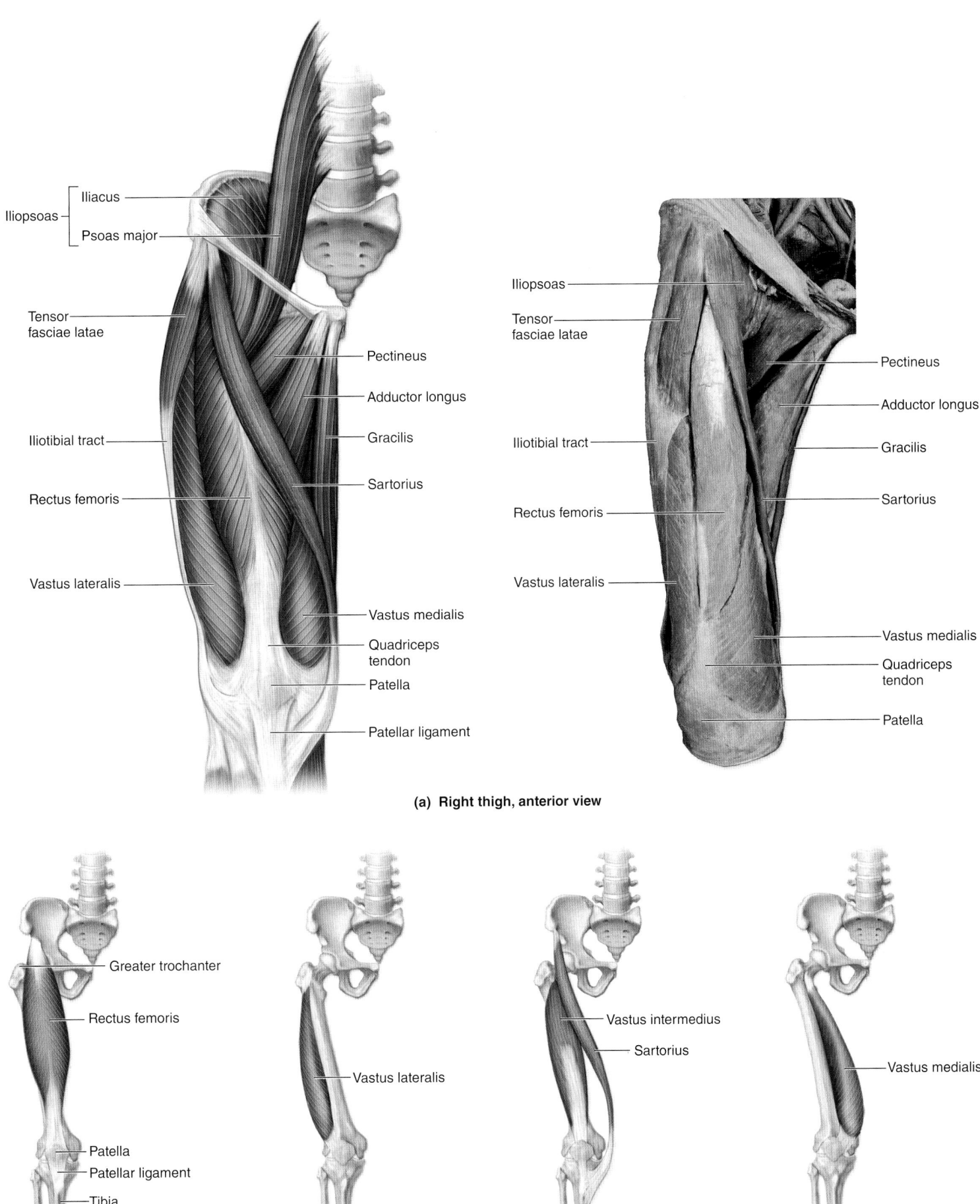

(a) Right thigh, anterior view

Iliopsoas {
- Iliacus
- Psoas major
}

Tensor fasciae latae

Iliotibial tract

Rectus femoris

Vastus lateralis

Pectineus

Adductor longus

Gracilis

Sartorius

Vastus medialis

Quadriceps tendon

Patella

Patellar ligament

Iliopsoas

Tensor fasciae latae

Iliotibial tract

Rectus femoris

Vastus lateralis

Pectineus

Adductor longus

Gracilis

Sartorius

Vastus medialis

Quadriceps tendon

Patella

Greater trochanter

Rectus femoris

Patella

Patellar ligament

Tibia

Vastus lateralis

Vastus intermedius

Sartorius

Vastus medialis

(b) Anterior thigh muscles

Figure 11.32 Muscles of the Anterior Thigh. Muscles of the anterior thigh flex the thigh and extend the leg. (*a*) Illustration and cadaver photo show an anterior view of the right thigh. (*b*) Individual muscles of the right anterior thigh. AP|R

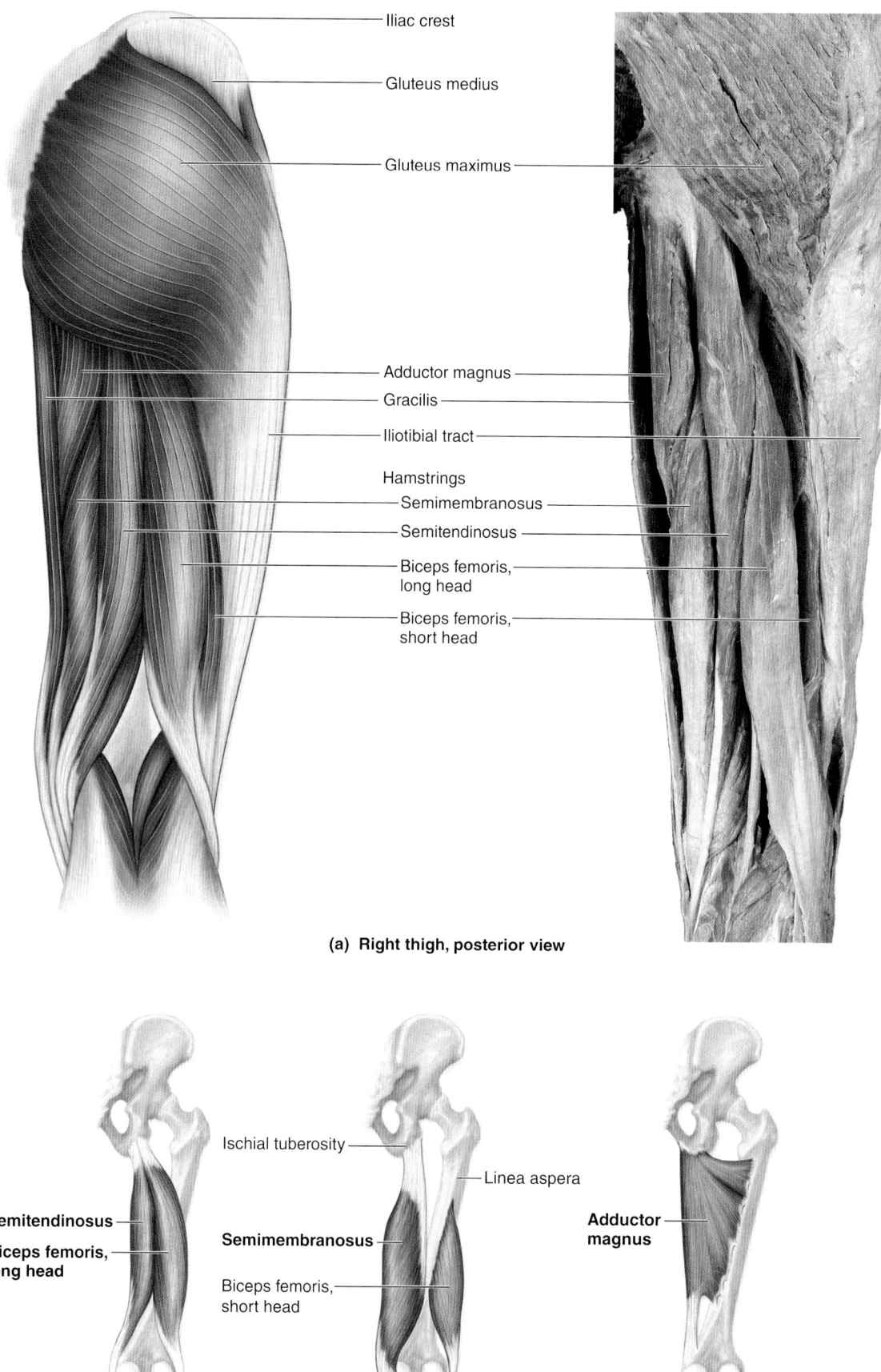

(a) Right thigh, posterior view

- Iliac crest
- Gluteus medius
- Gluteus maximus
- Adductor magnus
- Gracilis
- Iliotibial tract
- Hamstrings
  - Semimembranosus
  - Semitendinosus
  - Biceps femoris, long head
  - Biceps femoris, short head

Semitendinosus
Biceps femoris, long head

- Ischial tuberosity
- Linea aspera

Semimembranosus
Biceps femoris, short head

Adductor magnus

- Head of fibula

(b) Thigh extensors

Figure 11.33 Muscles of the Gluteal Region and Posterior Thigh. Muscles of the posterior thigh extend the thigh and flex the leg. (*a*) Illustration and cadaver photo show the gluteal and posterior muscles of the right thigh. (*b*) Individual muscles that extend the thigh are shown in bold (note the short head of biceps femoris does not participate in thigh extension). AP|R

Table 11.23	Thigh Muscles That Move the Knee Joint/Leg		
Group/Muscle	Action	Origin/Insertion	Innervation (see section 14.5f)
LEG EXTENSORS (ANTERIOR THIGH MUSCLES)			
Quadriceps femoris			
Rectus femoris	Extends leg; flexes thigh	O: Anterior inferior iliac spine I: Quadriceps tendon to patella and then patellar ligament to tibial tuberosity	Femoral nerve (L2–L4)
Vastus intermedius (vas'tŭs in-ter-mē'dē-ŭs) *vastus* = great *intermedius* = intermediate	Extends leg	O: Anterolateral surface of femur I: Quadriceps tendon to patella and then patellar ligament to tibial tuberosity	Femoral nerve (L2–L4)
Vastus lateralis (lat-er-ă'lis)	Extends leg	O: Greater trochanter and linea aspera I: Quadriceps tendon to patella and then patellar ligament to tibial tuberosity	Femoral nerve (L2–L4)
Vastus medialis (mē-dē-ă'lis)	Extends leg	O: Intertrochanteric line and linea aspera of femur I: Quadriceps tendon to patella and then patellar ligament to tibial tuberosity	Femoral nerve (L2–L4)
LEG FLEXORS			
Sartorius	Flexes thigh and rotates thigh laterally; flexes leg and rotates leg medially	See table 11.21	See table 11.21
Gracilis	Flexes and adducts thigh; flexes leg	See table 11.21	See table 11.21
Hamstrings (biceps femoris, semimembranosus, semitendinosus)	Extend thigh and flex leg; rotate leg laterally	See table 11.21	See table 11.21

11.9c Leg Muscles That Move the Ankle, Foot, and Toes

LEARNING OBJECTIVES

5. Compare and contrast the muscles of the three compartments of the leg and their actions.
6. Distinguish between the muscles of the superficial layer and deep layer of the leg's posterior compartment.

The muscles that move the ankle, foot, and toes are housed within the leg and are called the **crural muscles.** Some of these muscles also help flex the leg. The deep fascia partitions the leg musculature into three compartments (anterior, lateral, and posterior), each with its own nerve and blood supply, and muscles in the same compartment tend to share common functions (see figure 11.23).

INTEGRATE

CONCEPT CONNECTION

You will learn in section 20.5a that venous circulation of the lower limbs is reliant upon the muscular system. Specifically, the regular contraction and relaxation of the leg muscles works as a skeletal muscle "pump" to propel venous blood from the lower limb back to the torso. When the lower limbs are immobile for long periods of time (e.g., during long plane rides or when a person is bedridden), the skeletal muscle pump is inactive, and the risk of developing a blood clot in the lower limb veins increases.

Muscles of the Leg's Anterior Compartment

Anterior compartment leg muscles dorsiflex the foot, extend the toes, or both **(figure 11.34)**. The **extensor digitorum longus** dorsiflexes the foot and extends toes 2–5. The **extensor hallucis longus** sends a tendon to the dorsum of the great toe (hallux), and so it dorsiflexes the foot and extends the great toe. The **fibularis tertius** *(peroneus tertius)* dorsiflexes and weakly everts the foot. The **tibialis anterior** is the primary dorsiflexor of the foot. This muscle attaches to the medial plantar side of the foot, so it also inverts the foot. Analogous to tendons of the wrist, tendons of the muscles within the anterior compartment are held tightly against the ankle by multiple deep fascia thickenings, collectively referred to as **extensor retinaculum.**

Muscles of the Leg's Lateral Compartment

The **lateral compartment** leg muscles contain two synergistic muscles that are very powerful evertors of the foot and weak plantar flexors **(figure 11.35)**. The long, flat **fibularis longus** *(peroneus brevis)* inserts on the plantar side of the foot. The **fibularis brevis** *(peroneus brevis)* lies deep to the fibularis longus, and its tendon inserts onto the base of the fifth metatarsal.

Muscles of the Leg's Posterior Compartment

The **posterior compartment** of the leg is composed of seven muscles that are separated into superficial and deep groups **(figure 11.36)**. The superficial muscles and most of the deep muscles plantar flex the foot at the ankle.

The **superficial layer** of the posterior compartment contains the gastrocnemius, soleus, and plantaris. The **gastrocnemius** has two thick muscle bellies that form the prominence on the posterior part of the leg often referred to as the "calf." This muscle spans both the knee and the ankle joints; it flexes the leg and plantar flexes the foot. The **soleus** is a broad, flat muscle deep to the gastrocnemius *(solea* = sandal). This muscle plantar flexes the foot. The gastrocnemius and soleus are collectively known as the **triceps surae,** and together they are the most powerful plantar flexors of all of the leg muscles. These two muscles share a common tendon of insertion, the **calcaneal tendon** *(Achilles tendon).* The **plantaris** is a small muscle that is absent in some individuals. It is a weak leg flexor and plantar flexor of the foot.

The **deep layer** of the posterior compartment contains four muscles. The **flexor digitorum longus** attaches to the distal phalanges of toes 2–5, plantar flexes the foot, and flexes the MP, PIP, and DIP joints of these toes. The **flexor hallucis longus** plantar flexes the foot and flexes the great toe. The **tibialis posterior** is the deepest of the posterior compartment muscles. It plantar flexes and inverts the foot. The **popliteus** flexes the leg and medially rotates the tibia slightly to "unlock" the fully extended knee joint. This muscle originates and inserts in the popliteal region, so it only moves the knee, not the foot.

Table 11.24 summarizes the characteristics of the muscles that move the leg. **Table 11.25** groups muscles according to their action on the leg. Note that many thigh and leg muscles are involved with leg flexion.

WHAT DID YOU LEARN?

28 What are the common actions of each of the compartments of the leg?

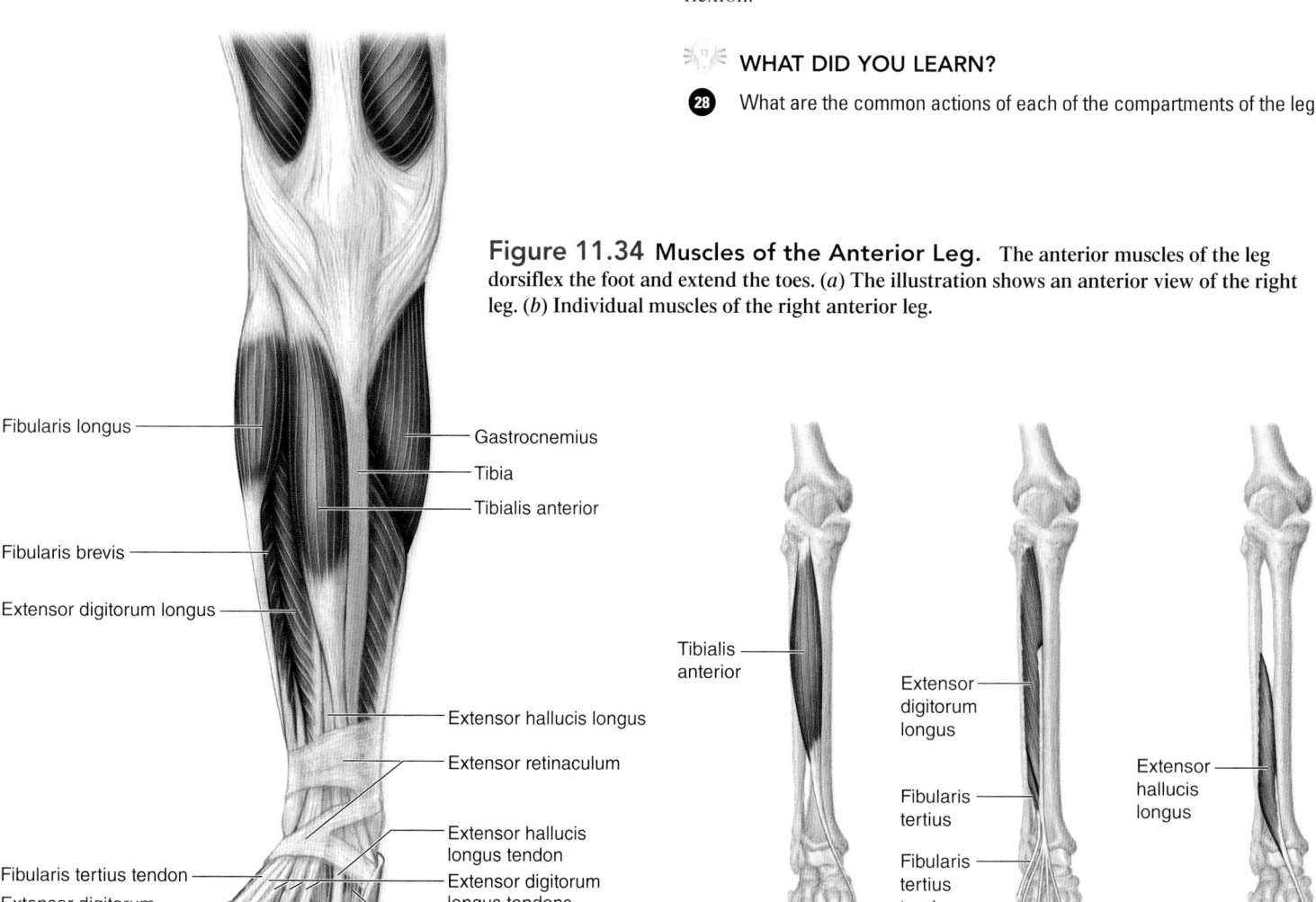

Figure 11.34 Muscles of the Anterior Leg. The anterior muscles of the leg dorsiflex the foot and extend the toes. (*a*) The illustration shows an anterior view of the right leg. (*b*) Individual muscles of the right anterior leg.

Fibularis longus

Gastrocnemius

Tibia

Tibialis anterior

Fibularis brevis

Extensor digitorum longus

Extensor hallucis longus

Extensor retinaculum

Extensor hallucis longus tendon

Extensor digitorum longus tendons

Fibularis tertius tendon

Extensor digitorum brevis

Extensor hallucis brevis

(a) Right leg, anterior view

Tibialis anterior

Extensor digitorum longus

Fibularis tertius

Fibularis tertius tendon

Extensor hallucis longus

(b) Anterior leg muscles

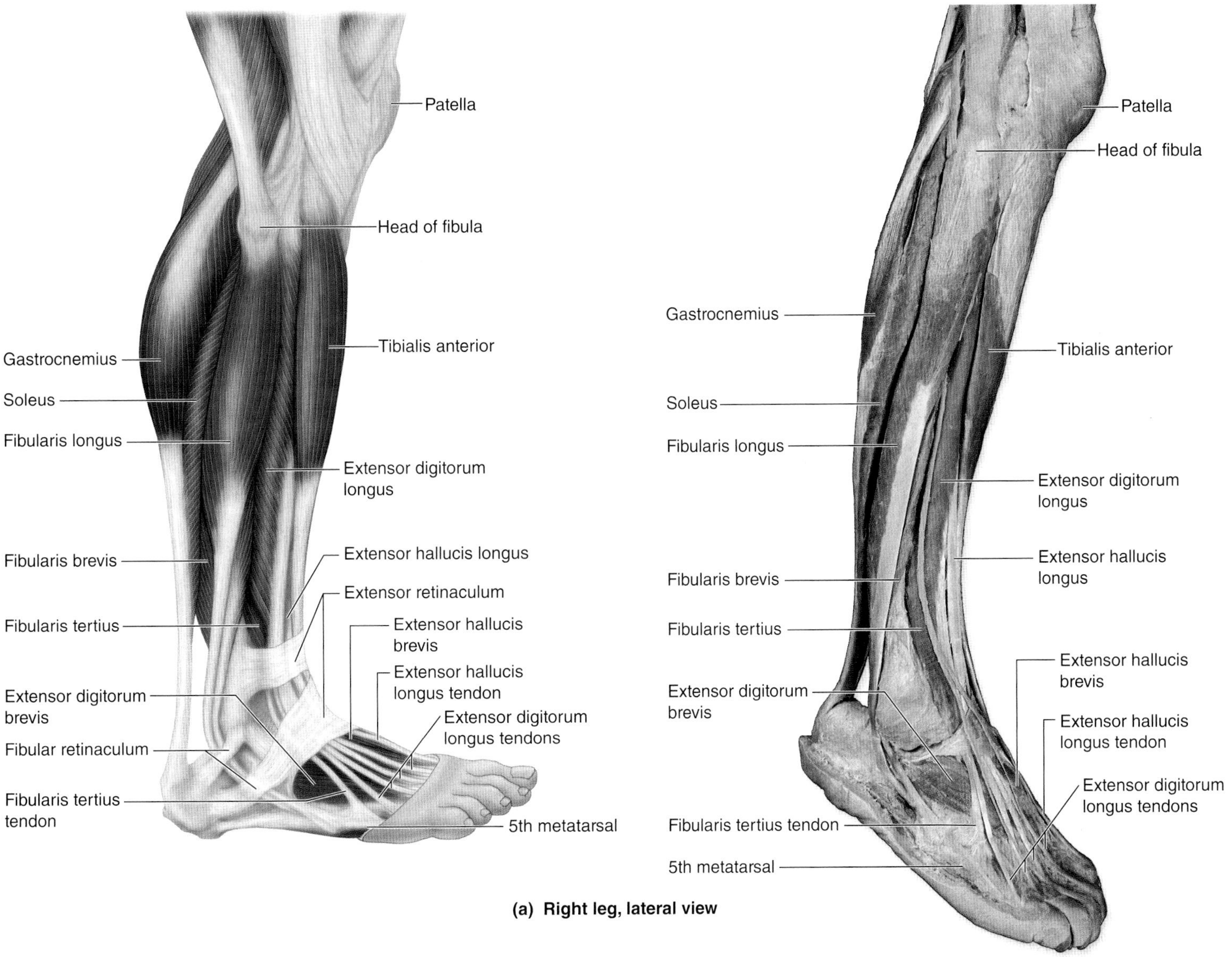

Patella

Head of fibula

Gastrocnemius

Tibialis anterior

Soleus

Fibularis longus

Extensor digitorum longus

Fibularis brevis

Extensor hallucis longus

Fibularis tertius

Extensor retinaculum

Extensor hallucis brevis

Extensor hallucis longus tendon

Extensor digitorum brevis

Extensor digitorum longus tendons

Fibular retinaculum

Fibularis tertius tendon

5th metatarsal

Patella

Head of fibula

Gastrocnemius

Tibialis anterior

Soleus

Fibularis longus

Extensor digitorum longus

Fibularis brevis

Extensor hallucis longus

Fibularis tertius

Extensor digitorum brevis

Extensor hallucis brevis

Extensor hallucis longus tendon

Extensor digitorum longus tendons

Fibularis tertius tendon

5th metatarsal

(a) Right leg, lateral view

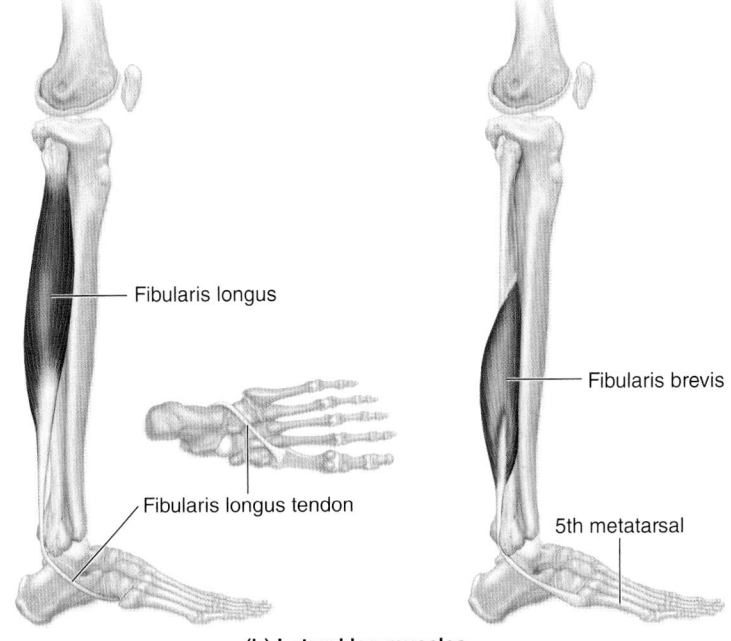

Fibularis longus

Fibularis brevis

Fibularis longus tendon

5th metatarsal

(b) Lateral leg muscles

Figure 11.35 Muscles of the Lateral Leg.
(*a*) Illustration and cadaver photo show a lateral view of the right leg. (*b*) The fibularis longus and the fibularis brevis evert and plantar flex the foot. AP|R

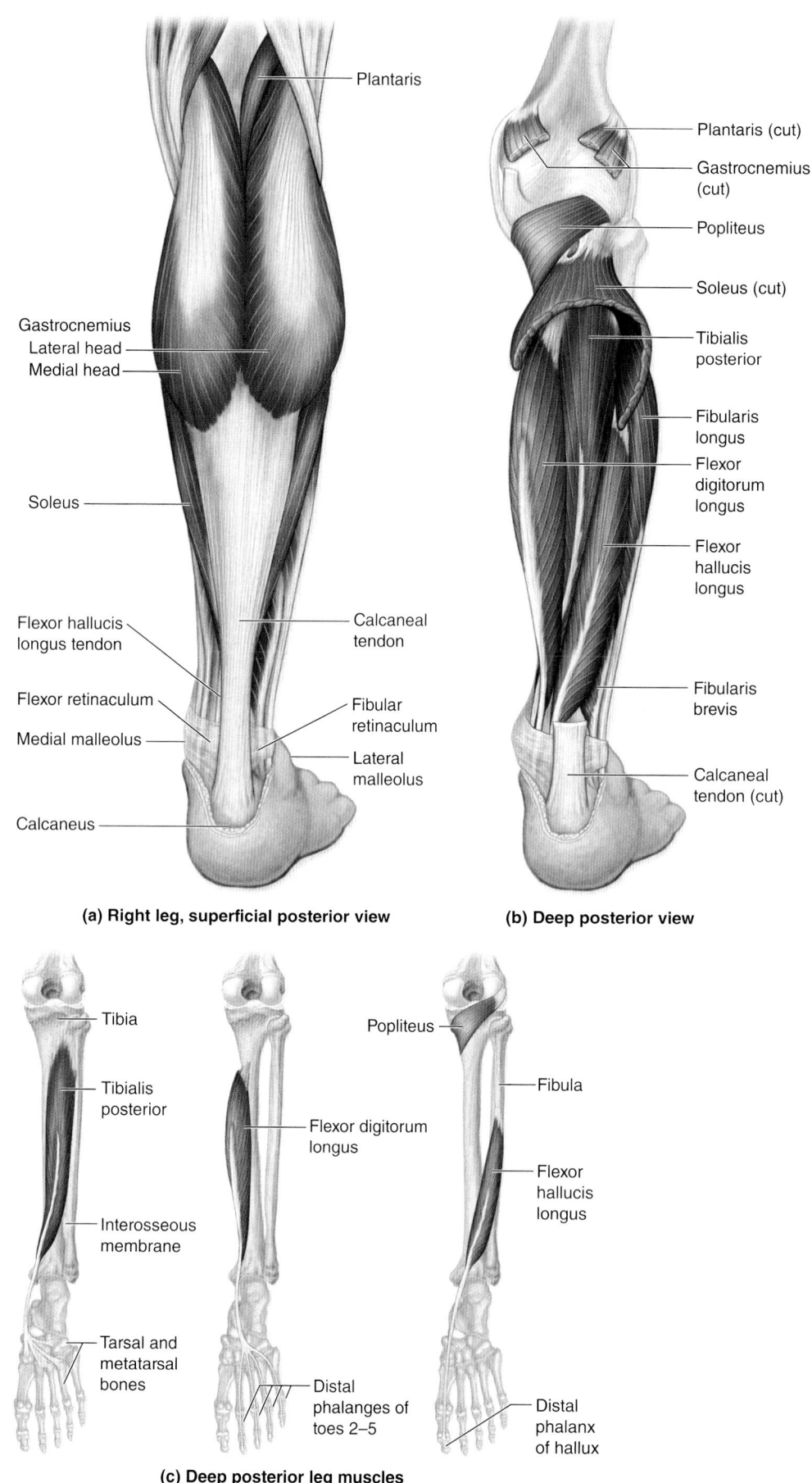

(a) Right leg, superficial posterior view

(b) Deep posterior view

(c) Deep posterior leg muscles

Figure 11.36 Muscles of the Posterior Leg. The posterior muscles of the leg plantar flex the foot and flex the toes. (*a*) Superficial and (*b*) deep views of the posterior right leg. (*c*) Selected individual muscles of the deep posterior compartment.

Table 11.24 — Leg Muscles

Group/Muscle	Action	Origin/Insertion	Innervation (see section 14.5g)
ANTERIOR COMPARTMENT (DORSIFLEXORS AND TOE EXTENSORS)			
Extensor digitorum longus	Extends toes 2–5; dorsiflexes foot	O: Lateral condyle of tibia; anterior surface of fibula; interosseous membrane I: Distal phalanges of toes 2–5	Deep fibular nerve (L4–S1)
Extensor hallucis longus (hal'ĭ-sis) *hallux* = great toe	Extends great toe (1); dorsiflexes foot	O: Anterior surface of fibula; interosseous membrane I: Distal phalanx of great toe (1)	Deep fibular nerve (L4–S1)
Fibularis tertius (fib'yū-lā'ris ter'she-ŭs) *fibularis* = fibula *tertius* = third	Dorsiflexes and weak evertor of foot	O: Anterior distal surface of fibula; interosseous membrane I: Base of metatarsal V	Deep fibular nerve (L5–S1 fibers)
Tibialis anterior (tib-ē-a'lis) *tibial* = tibia	Dorsiflexes foot; inverts foot	O: Lateral condyle and proximal shaft of tibia; interosseous membrane I: Metatarsal I and first (medial) cuneiform	Deep fibular nerve (L4–S1)
LATERAL COMPARTMENT (EVERTORS AND PLANTAR FLEXORS)			
Fibularis longus	Everts foot; weak plantar flexor	O: Head and upper ⅔ of shaft fibula; lateral condyle of tibia I: Base of metatarsal I; medial cuneiform	Superficial fibular nerve (L5–S2)
Fibularis brevis	Everts foot; weak plantar flexor	O: Midlateral shaft of fibula I: Base of metatarsal V	Superficial fibular nerve (L5–S2)
POSTERIOR COMPARTMENT (PLANTAR FLEXORS, FLEXORS OF THE LEG AND TOES)			
Superficial layer			
Triceps surae (sŭr'ē)			
Gastrocnemius (gas-trok-nē'mē-ŭs) *gaster* = belly *kneme* = leg	Flexes leg; plantar flexes foot	O: Superior and posterior surfaces of lateral and medial condyles of femur I: Calcaneus (via calcaneal tendon)	Tibial nerve (L4–S1 nerve fibers)
Soleus (sō-lē'ŭs) *solea* = sandal	Plantar flexes foot	O: Head and proximal shaft of fibula; medial border of tibia I: Calcaneus (via calcaneal tendon)	Tibial nerve (L4–S1 nerve fibers)
Plantaris (plan-tar'is) *planta* = sole of foot	Weak leg flexor and plantar flexor	O: Lateral supracondylar ridge of femur I: Posterior region of calcaneus	Tibial nerve (L4–S1 nerve fibers)
Deep layer			
Flexor digitorum longus	Plantar flexes foot; flexes MP, PIP and DIP joints of toes 2–5	O: Posteromedial surface of tibia I: Distal phalanges of toes 2–5	Tibial nerve (L5–S1 nerve fibers)
Flexor hallucis longus	Plantar flexes foot; flexes MP and IP joints of great toe (1)	O: Posterior lower ⅔ of fibula I: Distal phalanx of great toe (1)	Tibial nerve (L5–S1 nerve fibers)
Tibialis posterior	Plantar flexes foot; inverts foot	O: Fibula, tibia, and interosseous membrane I: Metatarsals II–IV; navicular bone; cuboid bone; all cuneiforms	Tibial nerve (L5–S1 nerve fibers)
Popliteus (pop-li-tē'ŭs) *poplit* = back of knee	Flexes leg; medially rotates tibia to unlock the knee	O: Lateral condyle of femur I: Posterior, proximal surface of tibia	Tibial nerve (L4–L5 nerve fibers)

Table 11.25 — Summary of Muscle Actions at the Knee Joint/Leg

Extension	Flexion
Quadriceps femoris	Sartorius
Rectus femoris	Gracilis
Vastus lateralis	Adductor longus, brevis, magnus
Vastus intermedius	Biceps femoris
Vastus medialis	Semimembranosus
	Gastrocnemius
	Popliteus
	(Plantaris)*

*Parentheses around an entire muscle name indicate only a slight effect.

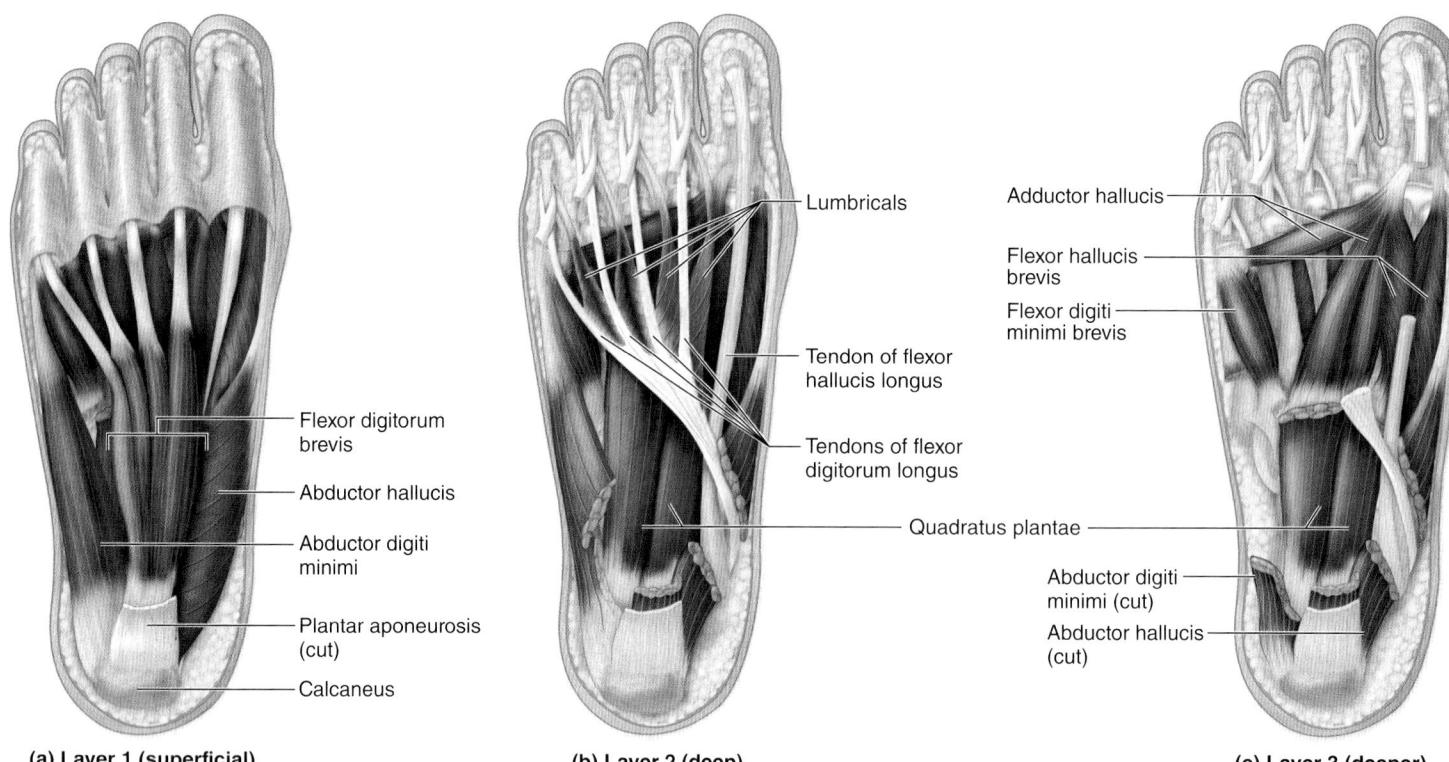

- Lumbricals
- Tendon of flexor hallucis longus
- Tendons of flexor digitorum longus
- Quadratus plantae

- Adductor hallucis
- Flexor hallucis brevis
- Flexor digiti minimi brevis
- Abductor digiti minimi (cut)
- Abductor hallucis (cut)

- Flexor digitorum brevis
- Abductor hallucis
- Abductor digiti minimi
- Plantar aponeurosis (cut)
- Calcaneus

(a) Layer 1 (superficial)   **(b) Layer 2 (deep)**   **(c) Layer 3 (deeper)**

Figure 11.37 Plantar Intrinsic Muscles of the Foot.
These muscles move the toes. (*a*) Superficial, (*b*) deep, and (*c*) deeper layers of the intrinsic muscles of the right foot. (*d*) Plantar and (*e*) dorsal views of the deepest layers. **AP|R**

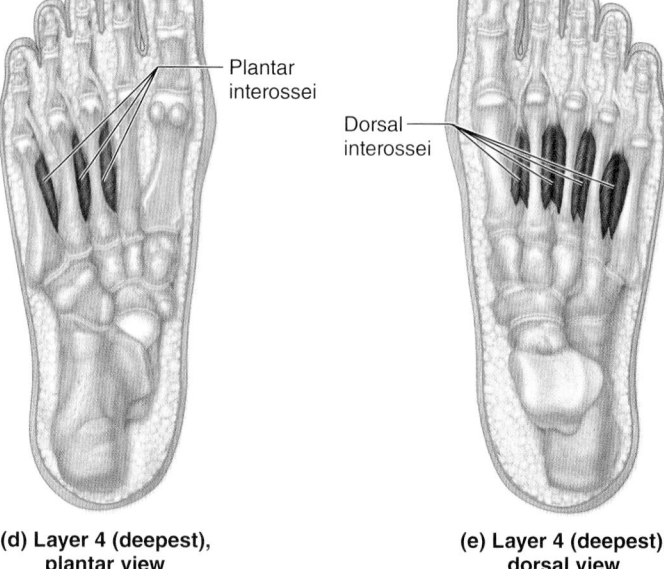

- Plantar interossei
- Dorsal interossei

(d) Layer 4 (deepest), plantar view   **(e) Layer 4 (deepest), dorsal view**

11.9d Intrinsic Muscles of the Foot

 LEARNING OBJECTIVE

7. Identify the muscles of each group and their actions.

The intrinsic muscles of the foot both originate and insert within the foot. They support the arches and move the toes to aid locomotion. Most of these muscles are comparable to the intrinsic muscles of the hand, meaning that they have similar names and locations. However, the intrinsic muscles of the foot rarely perform all the precise movements their names suggest.

The intrinsic foot muscles form a dorsal group and a plantar group. The dorsal group contains only two muscles: the extensor hallucis brevis and the extensor digitorum brevis (see figure 11.34). The **extensor hallucis brevis** extends the MP joint of the great toe, while the **extensor digitorum brevis** extends the MP and PIP joints of toes 2–4.

 WHAT DO YOU THINK?

4 The extensor digitorum brevis only goes to toes 2–4, so how is it possible to extend your little toe (toe 5)?

The plantar surface of the foot is supported by the **plantar aponeurosis** formed from the deep fascia of the foot. This aponeurosis extends between the phalanges of the toes and the calcaneus. It also encloses the plantar muscles of the foot. The plantar muscles are grouped into four layers (**figure 11.37**) and are described in detail in **table 11.26**.

Table 11.27 groups the leg and intrinsic foot muscles according to their common actions on the foot.

 WHAT DID YOU LEARN?

29 Identify the intrinsic muscles of the foot that extend the toes.

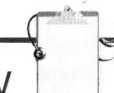

INTEGRATE

CLINICAL VIEW
Plantar Fasciitis

Plantar fasciitis (fas′ē-ī′tis, fash′ē-ī′tis) is an inflammation of the plantar aponeurosis. Factors associated with plantar fasciitis include overexertion that stresses the fascia, weight-bearing activities (lifting heavy objects, running, or walking), excessive body weight, improperly fitting shoes, and poor biomechanics (e.g., wearing high-heeled shoes or having flat feet).

Table 11.26 — Intrinsic Muscles of the Foot

Group/Muscle	Action	Origin/Insertion	Inversion (see section 14.5g)
DORSAL SURFACE (TOE EXTENSORS)			
Extensor hallucis brevis	Extends MP joint of great toe (1)	O: Calcaneus and inferior extensor retinaculum I: Proximal phalanx of great toe (1)	Deep fibular nerve (S1–S2 nerve fibers)
Extensor digitorum brevis	Extends MP and PIP joints of toes 2–4	O: Calcaneus and inferior extensor retinaculum I: Middle phalanges of toes 2–4	Deep fibular nerve (S1–S2 nerve fibers)
PLANTAR SURFACE (TOE FLEXORS, ABDUCTORS, ADDUCTORS)			
Layer 1 (superficial)			
Flexor digitorum brevis	Flexes MP and PIP joints of toes 2–5	O: Calcaneus I: Middle phalanges of toes 2–5	Medial plantar nerve (S2–S3)
Abductor hallucis	Abducts great toe (1)	O: Calcaneus I: Medial side of proximal phalanx of great toe (1)	Medial plantar nerve (S2–S3)
Abductor digiti minimi	Abducts toe 5	O: Calcaneus (inferior surface tuberosity) I: Lateral side of proximal phalanx of toe 5	Lateral plantar nerve (S2–S3)
Layer 2 (deep)			
Quadratus plantae (plan′tē) *planta* = sole of foot	Pulls on flexor digitorum longus tendons to flex toes 2–5	O: Calcaneus, long plantar ligament I: Tendons of flexor digitorum longus	Lateral plantar nerve (S2–S3)
Lumbricals	Flexes MP joints and extends PIP and DIP joints of toes 2–5	O: Tendons of flexor digitorum longus I: Tendons of extensor digitorum longus	Medial plantar nerve (1st lumbrical); lateral plantar nerve (2nd through 4th lumbricals)
Layer 3 (deeper)			
Adductor hallucis	Adducts great toe (1)	O: Transverse head: Capsules of MP joints III–V ; Oblique head: Bases of metatarsals II–IV I: Lateral side of proximal phalanx of great toe (1)	Lateral plantar nerve (S2–S3)
Flexor hallucis brevis	Flexes MP joint of great toe (1)	O: Cuboid and lateral cuneiform bones I: Proximal phalanx of great toe (1)	Medial plantar nerve (S2–S3)
Flexor digiti minimi brevis	Flexes MP joint of toe 5	O: Metatarsal V I: Proximal phalanx of toe 5	Lateral plantar nerve (S2–S3)
Layer 4 (deepest)			
Dorsal interossei	Abducts toes	O: Adjacent sides of metatarsals I: Sides of proximal phalanges of toes 2–4	Lateral plantar nerve (S2–S3)
Plantar interossei	Adducts toes	O: Sides of metatarsals III–V I: Medial side of proximal phalanges of toes 3–5	Lateral plantar nerve (S2–S3)

Table 11.27 — Summary of Leg and Foot Muscle Actions at the Foot and Toes

FOOT				TOES			
Dorsiflexion	Plantar Flexion	Eversion	Innervation	Extension	Flexion	Abduction	Adduction
Tibialis anterior*	**Gastrocnemius**	**Fibularis longus**	**Tibialis posterior**	Extensor digitorum longus	Flexor digitorum longus	Abductor hallucis	Adductor hallucis
Extensor digitorum longus	**Soleus**	**Fibularis brevis**	**Tibialis anterior**	Extensor hallucis longus	Flexor hallucis longus	Abductor digiti minimi	Plantar interossei
(Extensor hallucis longus)	Flexor digitorum longus	(Fibularis tertius)		Extensor digitorum brevis	Flexor digitorum brevis	Dorsal interossei	
(Fibularis tertius)	Flexor hallucis longus			Extensor hallucis brevis	Flexor hallucis brevis		
	Tibialis posterior				Flexor digiti minimi brevis		
	(Fibularis longus)						
	(Fibularis brevis)						

*Boldface indicates an agonist; others are synergists. Parentheses around an entire muscle name indicate only a slight effect.

- Axial muscles attach to components of the axial skeleton, whereas appendicular muscles stabilize or move components of the appendicular skeleton.

11.1 Skeletal Muscle Composition and Actions 373

- Skeletal muscle typically has origins and insertions, and the fascicles are organized in one of four basic patterns.

11.1a Origin and Insertion 373

- The origin is the less movable attachment of a muscle, whereas the insertion is the more movable attachment.

11.1b Organizational Patterns of Skeletal Muscle Fibers 373

- Muscle fibers may be arranged in circular, parallel, convergent, or pennate patterns.

11.1c Actions of Skeletal Muscles 374

- An agonist is a prime mover, whereas an antagonist opposes the agonist.
- A synergist assists an agonist.

11.2 Skeletal Muscle Naming 375

- Muscles receive their name according to muscle action, body region, muscle attachment, direction of the muscle fibers, shape, size, and muscle heads/origins.

11.3 Muscles of the Head and Neck 376

- Muscles of the head and neck are separated into groups based upon their specific activities.

11.3a Muscles of Facial Expression 376

- The muscles of facial expression arise from the skull and often attach to the skin.

11.3b Extrinsic Eye Muscles 380

- The six extrinsic eye muscles attach to the external surface of the eye and control the eye's movement.

11.3c Muscles of the Oral Cavity and Pharynx 382

- The muscles of mastication elevate and move the mandible during chewing.
- Intrinsic tongue muscles form the tongue itself, whereas extrinsic tongue muscles move the tongue during food manipulation, swallowing, and speech.
- Pharynx muscles function during swallowing.

11.3d Muscles of the Anterior Neck: The Hyoid Muscles 385

- The suprahyoid muscles elevate the hyoid bone, whereas the infrahyoid muscles depress the hyoid bone, move the thyroid cartilage during swallowing or speaking, or both.

11.3e Muscles That Move the Head and Neck 387

- Anterolateral neck muscles flex the head and neck, whereas posterior neck muscles extend the head and neck.

11.4 Muscles of the Vertebral Column 389

- The erector spinae muscles and other deep back muscles extend the vertebral column.

11.5 Muscles of Respiration 391

- The intercostal muscles, transversus thoracis, and diaphragm change the shape of the thoracic cavity when we breathe.

11.6 Muscles of the Abdominal Wall 392

- The abdominal wall muscles compress the abdomen, hold the abdominal organs in place, and flex the vertebral column.

11.7 Muscles of the Pelvic Floor 397

- The muscles of the pelvic floor support the pelvic organs and form a muscular wall that covers the inferior pelvic opening.

11.8 Muscles of the Pectoral Girdle and Upper Limb 399

- Five groups of muscles are associated with pectoral girdle and upper limb movement: muscles that move (1) the pectoral girdle; (2) the glenohumeral joint/arm; (3) the elbow joint/forearm; (4) the wrist joint, hand, and fingers; and (5) the intrinsic muscles of the hand.

11.8a Muscles That Move the Pectoral Girdle 399

- Anterior thoracic muscles tend to depress the scapula or clavicle, protract the scapula or clavicle, or both. The posterior thoracic muscles, in comparison, elevate the scapula, retract the scapula, or both.

11.8b Muscles That Move the Glenohumeral Joint/Arm 402

- The pectoralis major flexes the arm, and the latissimus dorsi and teres major extend it, whereas all adduct and medially rotate the arm.
- The deltoid flexes, extends, and abducts the arm.
- The rotator cuff muscles provide strength and stability to the glenohumeral joint.

11.8c Arm and Forearm Muscles That Move the Elbow Joint/Forearm 405

- The principal flexors are on the anterior side of the arm, and the principal extensors are on the posterior side of the arm.
- The pronator teres and pronator quadratus pronate the forearm, whereas the supinator and biceps brachii supinate the forearm.

11.8d Forearm Muscles That Move the Wrist Joint, Hand, and Fingers 410

- Anterior forearm muscles flex the wrist and finger joints, whereas posterior forearm muscles extend the wrist and joints of the fingers.
- The tendons of the anterior and posterior forearm muscles are held in place by bands of dense regular connective tissue called retinacula.

11.8e Intrinsic Muscles of the Hand 416
- The intrinsic muscles may be divided into three groups: (1) the thenar group (moves the thumb), (2) hypothenar group (moves the little finger), and (3) the midpalmar group (moves fingers 2–5).

- Four groups of muscles are associated with the pelvis and lower limb: (1) muscles that move the hip joint/thigh, (2) thigh muscles that move the knee joint/leg, (3) leg muscles, and (4) intrinsic muscles of the foot.

11.9a Muscles That Move the Hip Joint/Thigh 420
- Anterior thigh muscles flex the thigh.
- Gluteus maximus and the posterior thigh muscles (hamstrings) extend the thigh.
- Gluteus medius, gluteus minimus, and tensor fasciae latae abduct the thigh.
- Medial thigh muscles flex and adduct the thigh.

11.9b Thigh Muscles That Move the Knee Joint/Leg 424
- The quadriceps femoris extends the leg.
- Several medial thigh muscles, sartorius, and posterior thigh muscles (hamstrings) flex the leg.

11.9c Leg Muscles That Move the Ankle, Foot, and Toes 427
- Anterior leg muscles dorsiflex the foot, extend the toes, or both. One muscle also inverts the foot.
- Lateral leg muscles evert the foot and plantar muscles flex the foot.
- Posterior leg muscles plantar flex the foot, flex the toes, or both. One muscle inverts the foot.

11.9d Intrinsic Muscles of the Foot 432
- Dorsal muscles extend the toes.
- The four layers of plantar muscles can potentially flex, abduct, or adduct the toes.

CHALLENGE YOURSELF

Do You Know the Basics?

____ 1. Which statement is true about an agonist?
   a. It opposes the function of the prime mover.
   b. It is the primary muscle that produces a movement.
   c. It functions primarily to stabilize a joint.
   d. Its muscle fibers are always circular in structure.

____ 2. When the left and right _____ contract, they flex the head and neck.
   a. sternocleidomastoid
   b. longissimus group
   c. splenius
   d. rectus abdominis

____ 3. When this large muscle contracts, the vertical dimensions of the thoracic cavity increase.
   a. external intercostal
   b. internal intercostal
   c. diaphragm
   d. transversus thoracis

____ 4. This muscle depresses and adducts the eye.
   a. inferior rectus
   b. inferior oblique
   c. lateral rectus
   d. superior oblique

____ 5. Each of these muscles can flex the vertebral column *except* the
   a. external oblique.   c. spinalis.
   b. transversus abdominis.   d. internal oblique.

____ 6. The dorsal interossei muscles in the hand
   a. adduct fingers 2–5.
   b. abduct fingers 2–5.
   c. flex the PIP and DIP joints.
   d. extend the MP joints.

____ 7. Muscles in the anterior compartment of the leg
   a. evert the foot.
   b. dorsiflex the foot and extend the toes.
   c. plantar flex the foot.
   d. flex the toes.

____ 8. All of the following muscles flex the forearm *except* the
   a. brachialis.
   b. biceps brachii.
   c. brachioradialis.
   d. anconeus.

____ 9. Which muscles originate on the ischial tuberosity and extend the thigh plus flex the leg?
   a. adductor
   b. fibularis
   c. hamstring
   d. quadriceps

____ 10. The _____ plantar flexes the foot.
   a. iliopsoas
   b. gastrocnemius
   c. fibularis tertius
   d. vastus intermedius

11. What are some of the ways that muscles are named?

12. Which muscles of facial expression do you use to (a) smile, (b) close your eyes, and (c) close your mouth?

13. Distinguish between suprahyoid and infrahyoid muscles, and describe the functions of each group.

14. What is the effect of contracting the abdominal oblique muscles?

15. What movements are possible at the glenohumeral joint, and which muscles are the prime movers for each of these movements?

16. Identify the compartments of the arm (brachium), the muscles in each compartment, and their function.

17. Compare and contrast the flexor digitorum superficialis and the flexor digitorum profundus; where does each insert, how are their tendons interrelated, and what muscle actions do they perform?

18. What muscles are responsible for thigh extension? Which of these is the prime mover of thigh extension?

19. What leg muscles allow a ballet dancer to rise up and balance on her toes?

20. Which muscles are responsible for foot inversion?

Can You Apply What You've Learned?

1. A 50-year-old woman was concerned about the appearance of "crow's-feet" by her eyes. She was told by her physician that these wrinkles were caused by years of squinting and blinking using which muscle?

   a. frontal belly of occipitofrontalis

   b. orbicularis oculi

   c. risorius

   d. orbicularis oris

2. Eliza complained of double vision and went to see her optometrist. The optometrist tested the function of various eye muscles and discovered that Eliza could not move her right eye medially. Which muscle may be injured?

   a. superior oblique   c. medial rectus

   b. lateral rectus   d. inferior oblique

3. After an intensive workout, George felt especially sore in his posterior arm regions. What repetitive workout activity most likely caused the soreness?

   a. flexing the humerus   c. pronating the forearm

   b. flexing the forearm   d. extending the forearm

4. While Carly was playing soccer, she was kicked in the anterior thigh by an opposing teammate. Due to this injury, what muscle function may have been difficult to perform?

   a. extending the knee   c. extending the thigh

   b. flexing the knee   d. dorsiflexing the foot

5. Joshua broke his fibula and had to wear a cast for 6 weeks. The muscles attaching to this bone atrophied during this time. What muscle function was Joshua unable to perform as a result?

   a. plantar flexing the foot   c. everting the foot

   b. flexing the knee   d. dorsiflexing the foot

Can You Synthesize What You've Learned?

1. Albon is a 45-year-old male who characterizes himself as a "couch potato." He exercises infrequently and has a rounded abdomen ("beer belly"). While helping a friend move some heavy furniture, he felt a sharp pain deep within his abdomino-pelvic cavity. An emergency room resident told Albon that he had suffered an inguinal hernia. What is this injury, how did it occur, and how might Albon's poorly developed abdominal musculature have contributed to it?

2. While training on the balance beam, Pat slipped during her landing from a back flip and fell, straddling the beam. Although only slightly sore from the fall, she became concerned when she suddenly lost the ability to completely control her urination. What might have happened to Pat's pelvic floor structures during the fall?

3. After falling while skateboarding, Karen had surgery on her elbow. During her recovery, she must visit the physical therapist to improve muscle function around the elbow. Develop a series of exercises that may improve all of Karen's elbow movements, and determine which muscles are being helped by each exercise.

4. Why is it more difficult for Eric to lift a heavy weight when his forearm is pronated than when it is in the supine position?

INTEGRATE

ONLINE STUDY TOOLS

 connect |ANATOMY & PHYSIOLOGY   LEARNSMART® AP|R

The following study aids may be accessed through Connect.

Clinical Case Study: A Young Boy Who Is Losing His Ability to Walk

Interactive Questions: This chapter's content is served up in a number of multimedia question formats for student study.

LearnSmart: Topics and terminology include skeletal muscle composition and actions; skeletal muscle naming; muscles of the head and neck; muscles of the vertebral column; muscles of

respiration; muscles of the abdominal wall; muscles of the pelvic floor; muscles of the pectoral girdle and upper and lower limbs

Anatomy & Physiology Revealed: Topics include dissection of the head and neck, orbit, back, thorax, abdomen, shoulder, forearm, arm, wrist, hand, hip, thigh, and foot; muscle actions of pectoralis major, latissimus dorsi, flexor digitorum superficialis, and gastrocnemius

Nervous System: Nervous Tissue

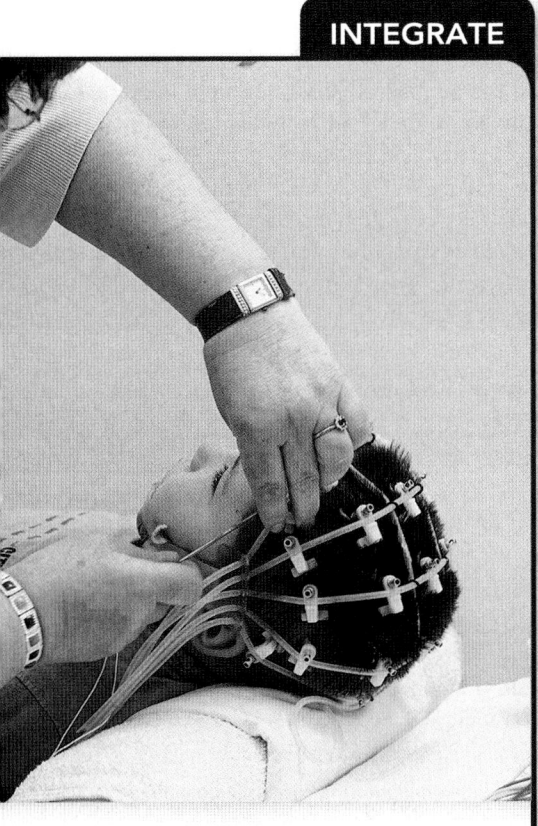

CAREER PATH
EEG Technician

An electroencephalogram (EEG) technician monitors a patient's brain waves by attaching electrodes to the scalp. Results from an EEG help aid in the diagnoses of impaired brain function resulting from sleep disturbance, epilepsy, and infectious disease. They also help measure progress in the treatment of stroke and brain trauma. An essential function performed by EEG technicians is the definitive confirmation that brain functional activities have ceased in a patient. These professional duties are performed in neurologic clinics and research facilities as well as hospitals and physicians' offices.

Module 7: Nervous System

Throughout the day, your body perceives and responds to multiple sensations. You smell spring flowers, feel the touch of a hand on your shoulder, and are visually aware of your environment. You control multiple muscle movements to walk, talk to the person sitting next to you, and hold this textbook. Other muscle movements occur without your voluntary input: Your heart beats, your stomach churns to digest your breakfast, and you jump at the sound of a honking horn. All of these sensations and muscle movements are interpreted and controlled by your **nervous system.**

Here we introduce the study of the nervous system by first describing its general functions and overall organization and then discussing the components of nervous tissue, which is the primary tissue of the nervous system. The next several chapters investigate different aspects of the nervous system, including the structure and function of the brain and cranial nerves (see chapter 13), spinal cord and spinal nerves (see chapter 14), organization and function of the autonomic nervous system (see chapter 15), and the senses (see chapter 16).

12.1 Introduction to the Nervous System

The nervous system is composed of the brain, spinal cord, nerves, and ganglia. Here we examine the general functions of the nervous system and how it is organized both structurally and functionally.

12.1a General Functions of the Nervous System

LEARNING OBJECTIVE

1. Describe the three general functions of the nervous system.

The nervous system serves as the body's primary communication and control system. It provides a rapid means of integrating and regulating body functions through electrical activity transmitted along specialized nervous system cells called neurons to accomplish the following:

- **Collect information. Receptors** are specialized nervous system structures that monitor changes in both the internal and external environment called **stimuli** (sing., *stimulus*). For example, receptors in the skin detect stimuli associated with touch—this sensory input then is relayed along nerves to the spinal cord and brain.
- **Process and evaluate information.** After processing sensory input, the brain and spinal cord determine what response, if any, is required.
- **Initiate response to information.** The brain and spinal cord initiate a response as motor output via nerves to **effectors.** Effectors include all three types of muscle tissue and glands. The result, or effect, may be either muscle contraction (or relaxation) or a change in gland secretion.

WHAT DID YOU LEARN?

❶ What is the function of receptors? What are the different types of effectors controlled by the nervous system?

12.1b Organization of the Nervous System

LEARNING OBJECTIVES

2. Identify the structural components included in the CNS and those in the PNS.
3. Explain the functional organization of the nervous system.

Anatomists and physiologists have devised various ways to organize the structural and functional components of the nervous system. However, always keep in mind that such artificial divisions are merely intended to simplify discussion—there is only one nervous system.

Structural Organization: Central Versus Peripheral Nervous System

The nervous system consists of two anatomic divisions: the central nervous system and the peripheral nervous system (**figure 12.1a**). The **central nervous system (CNS)** includes the brain and spinal cord. The brain is protected and enclosed within the skull, whereas the spinal cord is housed and protected within the vertebral canal.

The **peripheral** (pĕ-rif′ĕr-ăl) **nervous system (PNS)** includes **nerves,** which are bundles of neuron processes (axons) and **ganglia** (gang′glē-ă; sing., *ganglion;* swelling) that are clusters of neuron cell bodies located along nerves.

Functional Organization: Sensory Versus Motor Nervous System

The nervous system has two functional divisions: the sensory nervous system and the motor nervous system (figure 12.1b). Both the sensory nervous system and the motor nervous system have CNS and PNS components.

Sensory Nervous System The **sensory nervous system** or *afferent* (af′er-ent; *afferens* = to bring to) *nervous system* is responsible for receiving sensory information *from* receptors that detect stimuli and transmitting this information *to* the CNS.

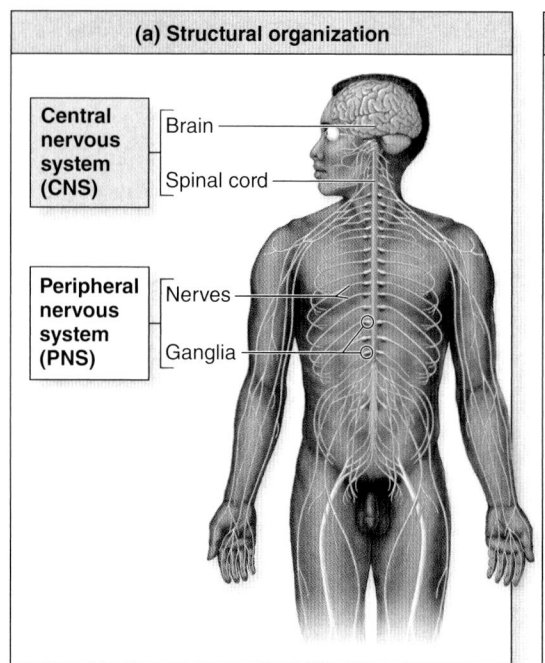

(a) Structural organization

| Central nervous system (CNS) | Brain |
| | Spinal cord |

| Peripheral nervous system (PNS) | Nerves |
| | Ganglia |

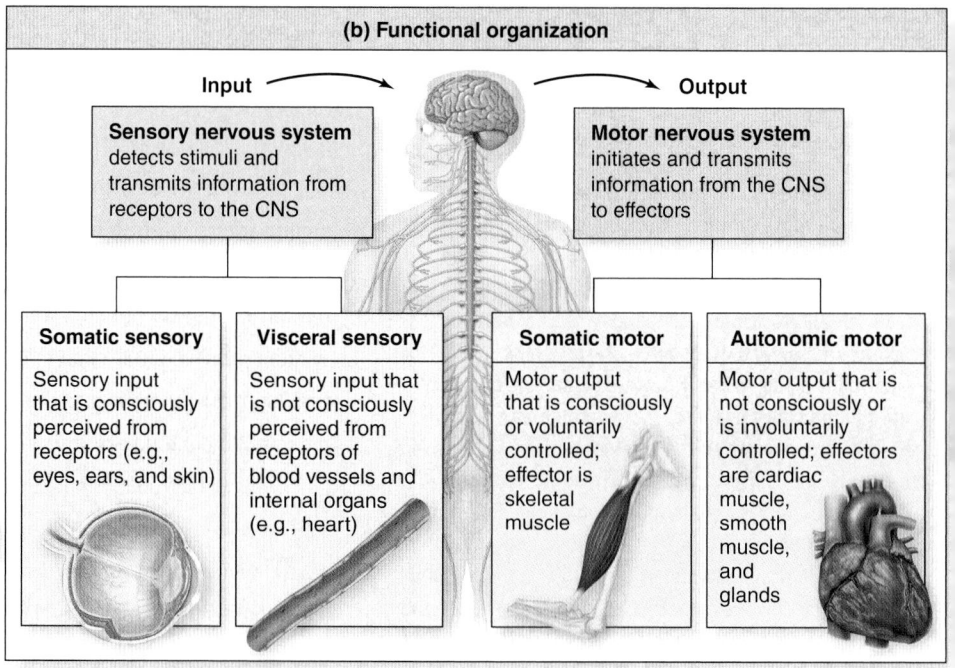

(b) Functional organization

Input → | Output →

Sensory nervous system detects stimuli and transmits information from receptors to the CNS

Motor nervous system initiates and transmits information from the CNS to effectors

Somatic sensory
Sensory input that is consciously perceived from receptors (e.g., eyes, ears, and skin)

Visceral sensory
Sensory input that is not consciously perceived from receptors of blood vessels and internal organs (e.g., heart)

Somatic motor
Motor output that is consciously or voluntarily controlled; effector is skeletal muscle

Autonomic motor
Motor output that is not consciously or is involuntarily controlled; effectors are cardiac muscle, smooth muscle, and glands

Figure 12.1 Organization of the Nervous System. The nervous system is organized into both structural and functional categories. (*a*) Structural divisions include the central nervous system (CNS), which is composed of the brain and spinal cord, and the peripheral nervous system (PNS), which is composed of nerves and ganglia. (*b*) Functionally, the nervous system consists of sensory (input) and motor (output) divisions, both of which are further divided into somatic and visceral (autonomic) components. AP|R

The sensory nervous system may be further subdivided based upon whether we are consciously aware of the stimulus that is detected. The two components are the somatic sensory and the visceral sensory. The **somatic** (sō-mat′ik; soma = body) **sensory** components detect stimuli that we can consciously perceive. Receptors of the somatic sensory nervous system include the receptors associated with the five senses (i.e., eyes, nose, tongue, ears, skin) and proprioceptors (receptors in joints and muscles that detect body position).

In comparison, the **visceral** (vis′ĕr-ăl; viscus = internal organ) **sensory** components detect stimuli that we typically do not consciously perceive. Receptors of the visceral sensory nervous system include structures located within blood vessels and internal organs (e.g., heart, stomach, kidneys). Visceral receptors detect chemical composition of the blood or stretch of an organ wall, for example. The various types of receptors are described in detail in chapter 16.

Motor Nervous System The **motor nervous system** or *efferent* (ef′er-ent; *efferens* = to bring out) *nervous system* is responsible for initiating and transmitting motor output *from* the CNS *to* effectors. This system controls muscle tissue and glands.

The motor nervous system, like the sensory nervous system, is subdivided into somatic and visceral parts. Here the distinction is based upon whether the effector stimulated can be controlled consciously (voluntarily). The **somatic motor** component initiates and transmits motor output from the CNS to skeletal muscles. For example, you exert voluntary control over your leg muscles as you press on the accelerator of your car.

The **autonomic** (aw-tō-nom′ik; *auto* = self, *nomos* = law) **motor** (or *visceral motor*) component innervates and regulates cardiac muscle, smooth muscle, and glands without our conscious control. We can neither voluntarily make our heart stop beating nor prevent our stomachs from growling. The autonomic motor component has two further subdivisions—sympathetic and parasympathetic—which are described in chapter 15.

Certain diseases are associated with specific components of the nervous system. For example, shingles (a painful skin rash caused by varicella zoster virus) infects somatic sensory neurons extending from receptors in the skin, whereas polio (caused by the poliovirus) preferentially infects somatic motor neurons to skeletal muscle, which in some cases may result in muscle weakness and paralysis.

 WHAT DID YOU LEARN?

 ❷ What are the two primary functional divisions of the nervous system? How do they differ?

12.2 Nervous Tissue: Neurons

Nervous tissue is the primary tissue of the nervous system and is composed of two distinct cell types: neurons and glial cells (first introduced in section 5.4). Neurons are excitable cells that initiate and transmit electrical signals and glial cells are nonexcitable cells that primarily support and protect the neurons. We describe neurons here and glial cells in detail in section 12.4.

12.2a General Characteristics of Neurons

 LEARNING OBJECTIVE

  1. Describe five distinguishing features common to all neurons.

The basic structural unit of the nervous system is the **neuron** (nū′ron). These cells have several special characteristics including:

- **Excitability.** This is responsiveness to a stimulus (e.g., chemical, stretch, pressure change). The stimulus causes a local change in the resting membrane potential in the excitable cell. Local electrical changes are called graded potentials (described in section 12.8a).

- **Conductivity.** This involves an electrical change that is quickly propagated along the plasma membrane as voltage-gated channels open sequentially during an action potential (described in section 12.8c).

- **Secretion.** Neurons release neurotransmitters in response to conductive activity. Neurotransmitters are stored in vesicles and when released may have either an excitatory or an inhibitory effect on their target structures (other neurons or effectors).

- **Extreme longevity.** Most neurons formed during fetal development are still functional in very elderly individuals.
- **Amitotic.** During fetal development of neurons, mitotic activity (see section 4.9) is lost in most neurons, except those in the olfactory epithelium of the nose and in certain areas of the brain.

Prevailing medical wisdom for years has maintained that the number of neurons in your brain shortly after birth is set for your lifetime. Recent studies, however, have shown that this is not always the case. Researchers investigating the brain's hippocampus (a region involved in memory processing; see section 13.7a) have found that this region of the brain contains a population of neural stem cells. These stem cells were once thought to give rise only to new glial cells in adults, but it is now clear that under special circumstances they can mature into neurons. The new neurons appear able to incorporate themselves, at least to some degree, into the brain circuitry. Researchers have learned that the surrounding glial cells provide the chemical signals that direct a stem cell down the path of neuron maturation.

WHAT DID YOU LEARN?

3 Explain the neuron characteristics of excitability, conductivity, and secretion.

12.2b Neuron Structure

LEARNING OBJECTIVES

2. Describe the three basic anatomic features common to most neurons.
3. Identify and describe the structures unique to neurons.

Neurons come in many shapes and sizes, but they typically share certain basic structural features that include a cell body, dendrites, and an axon (**figure 12.2**).

The **cell body** is also called the *soma,* and it is enclosed by a plasma membrane and contains cytoplasm surrounding a nucleus. Cell bodies serve as the neuron's control center. They also transmit graded potentials to the axon. The graded potential is either received from the dendrites or initiated within the cell body.

The cytoplasm within the cell body is called the **perikaryon** (per′i-kar′ē-on; *peri* = around, *karyon* = kernel), although some anatomists use that term to describe the whole cell body. The nucleus accommodates a prominent nucleolus, which form ribosomes (see section 4.6b). Free and bound ribosomes together are referred to as either: (1) **chromatophilic** (krō-mă-tō-fil′ik; *chromo* = color, *phileo* = to love) **substance,** because they stain darkly with basic dyes; or (2) *Nissl bodies,* because they were first described by the German microscopist Franz Nissl. Cytologists consider that the gray color of gray matter seen in gross dissections of the brain and spinal cord is due to the chromatophilic substance, along with the absence of myelin, a glistening coat of insulating material (described in section 12.4c).

Dendrites (den′drīt; *dendrites* = relating to a tree) tend to be relatively short, small, tapering, unmyelinated processes that branch off the cell body. Some neurons have only one dendrite; others have many. Dendrites transmit graded potentials toward the cell body; in essence, they receive input and then transfer it to the cell body for processing. The greater the number of dendrites, the more input a neuron may receive.

The **axon** (ak′son) (or *nerve fiber*) is typically a longer process emanating from the cell body to make contact with other neurons, muscle cells, or gland cells. The axon extends from a triangular region of the cell body called the **axon hillock** (hil′lok). The cytoplasm within an axon is called **axoplasm,** and the plasma membrane of an axon is called an **axolemma.** Unlike the cell body, the axon is devoid of chromatophilic substance. This distinctive difference allows the cell body to be distinguished from the axon when nervous tissue is viewed with a microscope.

Axons give rise to a few side branches called **axon collaterals.** Most axons and their collaterals branch extensively at their distal end into an array of fine terminal extensions called **telodendria** (tel-ō-den′drē-a; sing., *telodendrion; telos* = end), or *axon terminals.* The extreme tips of these fine extensions are slightly expanded regions called **synaptic** (si-nap′tik) **knobs,** also called *synaptic bulbs, end bulbs,* or *terminal boutons.* Within the synaptic knobs are numerous **synaptic vesicles** containing neurotransmitter. A synaptic knob ends at a functional junction called a synapse (described shortly). Axons function in the initiation and propagation of action potentials, which trigger synaptic vesicles to release neurotransmitter from the synaptic knobs.

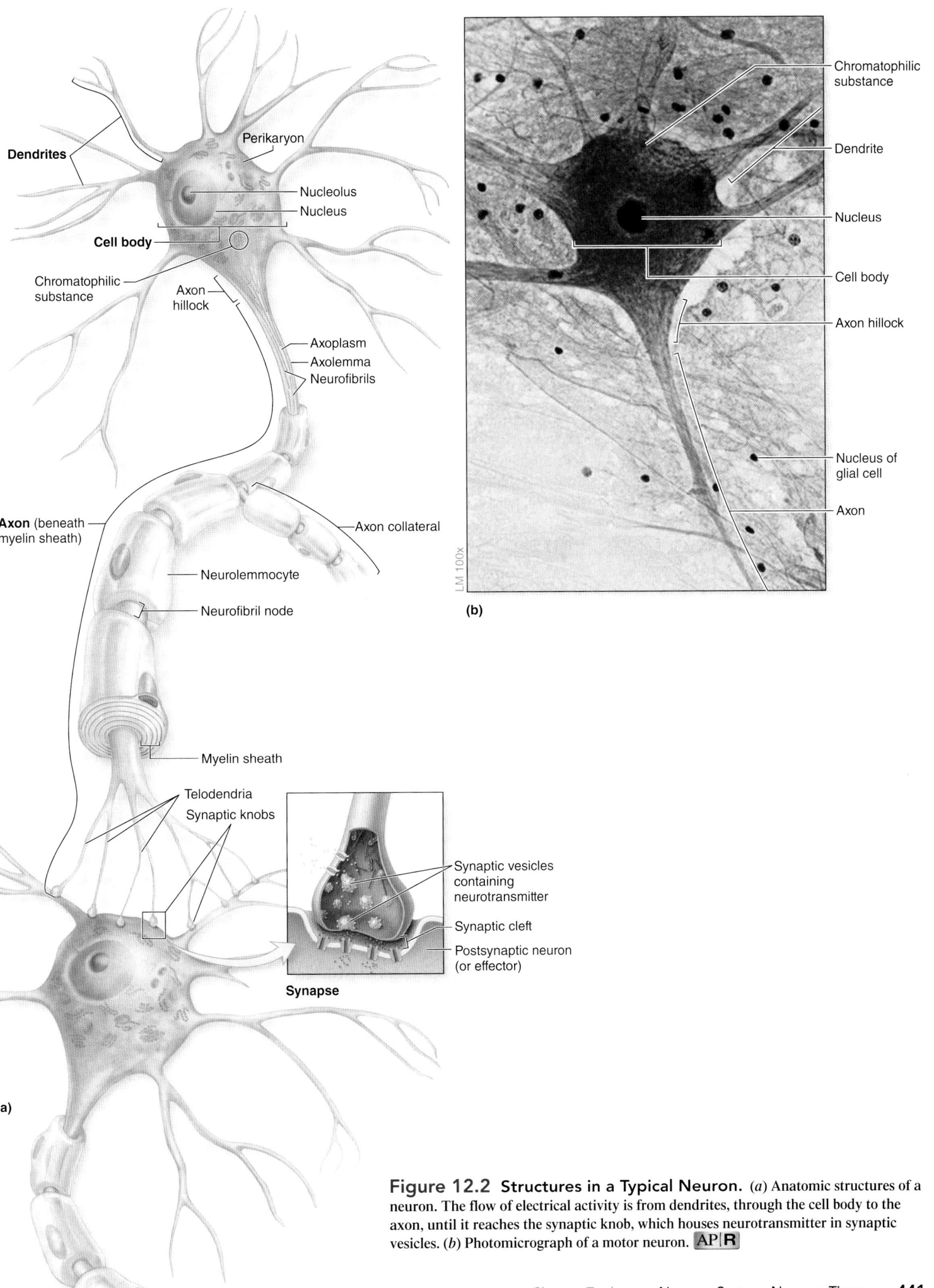

(a)

Dendrites

Perikaryon

Nucleolus

Nucleus

Cell body

Chromatophilic substance

Axon hillock

Axoplasm

Axolemma

Neurofibrils

Axon (beneath myelin sheath)

Axon collateral

Neurolemmocyte

Neurofibril node

Myelin sheath

Telodendria

Synaptic knobs

Synapse

Synaptic vesicles containing neurotransmitter

Synaptic cleft

Postsynaptic neuron (or effector)

(b)

LM 100x

Chromatophilic substance

Dendrite

Nucleus

Cell body

Axon hillock

Nucleus of glial cell

Axon

Figure 12.2 Structures in a Typical Neuron. (*a*) Anatomic structures of a neuron. The flow of electrical activity is from dendrites, through the cell body to the axon, until it reaches the synaptic knob, which houses neurotransmitter in synaptic vesicles. (*b*) Photomicrograph of a motor neuron. AP|R

INTEGRATE

CLINICAL VIEW

Pathogenic Agents and Fast Axonal Transport

Several pathogenic agents—including herpesvirus, rabies virus, poliovirus, and the tetanus toxin—enter a neuron at the synaptic knob. Once inside, they "hitch a ride" to the cell body by fast axonal transport. The pathogenic agents ultimately cause destruction of these neurons, producing the signs and symptoms associated with each disease.

Cytoskeleton

The **cytoskeleton** in a neuron is composed of microfilaments, intermediate filaments, and microtubules (see section 4.6b). The intermediate filaments, called **neurofilaments** (nūr-ō-fil′ă-ment; *filamentum* = thread), aggregate to form bundles called **neurofibrils** (nūr-ō-fī′bril; *fibrilla* = fiber). Neurofibrils extend as a complex network through the neuron, where their tensile strength provides support. A protein (called tau) that stabilizes microtubules of the neuron cytoskeleton is associated with Alzheimer disease. (See Clinical View: "Alzheimer Disease: The 'Long Goodbye'" in section 13.8d.)

 WHAT DID YOU LEARN?

4 What are the functions of these neuron structures: dendrites, axon, synaptic vesicles, and neurofibrils?

12.2c Neuron Transport

 LEARNING OBJECTIVE

4. Distinguish between fast axonal transport and slow axonal transport, and give examples of the different substances moved by each.

Axons are generally dependent upon the cell body to provide them with newly synthesized materials and to break down or recycle their used materials. To accomplish this, substances are moved in both directions through an axon. **Anterograde** (an′ter-ō-grād; *ante* = before, *gradior* = to step) **transport** is the movement of materials from the cell body to synaptic knobs, and **retrograde** (ret′rō-grād; *retro* = backward) **transport** is the movement of materials from synaptic knobs to the cell body. The transport processes are classified as either fast axonal transport or slow axonal transport, depending upon the relative speed of movement.

Fast Axonal Transport

Fast axonal transport occurs at approximately 400 millimeters per day. The mechanism involves movement along microtubules. You may find it helpful to think of this process as similar to a substance being pulled along a train track. The power for this movement comes from specialized motor proteins (e.g., kinesin, dynein) that split ATP to supply the energy needed.

Substances can be moved in either direction (anterograde or retrograde). Cellular structures formed in the cell body are moved by anterograde transport toward the synaptic knobs and include vesicles, organelles, and glycoproteins required at the synapse. Used vesicles to be broken down and recycled, and potentially harmful agents, are moved via retrograde transport from the synaptic knob to the cell body. Interestingly, new evidence supports the idea that some of these vesicles are transporting hormone-like molecules for the purposes of communication between neurons. This represents a means of neurons communicating information in a retrograde fashion from synaptic knobs to the cell body. Research continues in this area.

Slow Axonal Transport

Slow axonal transport occurs at approximately 0.1 to 3 millimeters per day. This type of movement results from the flow of the axoplasm, and is also called *axoplasmic flow*. The substances are only moved from the cell body towards the synaptic knob. These substances include enzymes, cytoskeletal components, and new axoplasm for regenerating axons.

 WHAT DID YOU LEARN?

5 Which type of axonal transport is both anterograde and retrograde? Give examples of substances transported by this method.

12.2d Classification of Neurons

 LEARNING OBJECTIVES

5. Name and describe the four structural categories of neurons.
6. Identify the three functional categories of neurons and where each is primarily located.

Neurons vary in morphology and location. Similar to the components of the entire nervous system, they are classified according to both their structure and their function.

Table 12.1 Structural Classification of Neurons

Neuron Type	Structure	Description	Examples of Functional Types
Multipolar Neuron	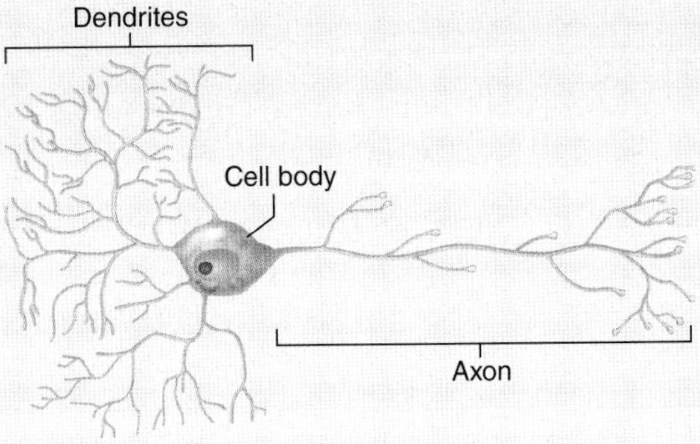	Multiple processes extend directly from the cell body; typically many dendrites and one axon; most common type of neuron	All motor neurons; most interneurons
Bipolar Neuron	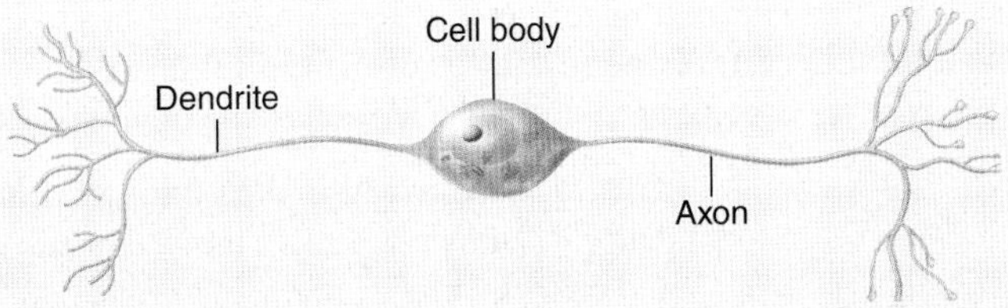	Two processes extend directly from the cell body; one dendrite and one axon; relatively uncommon	Some special sense neurons (e.g., retina of eye, olfactory epithelium in nose)
Unipolar Neuron	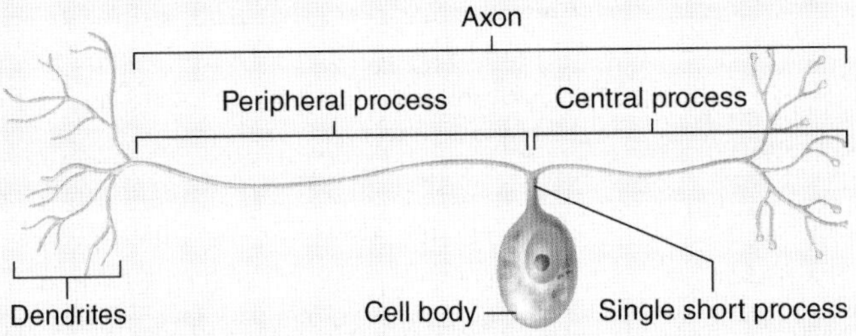	Single short process extends directly from the cell and looks like a T as a result of the fusion of two processes into one long axon	Most sensory neurons
Anaxonic Neuron	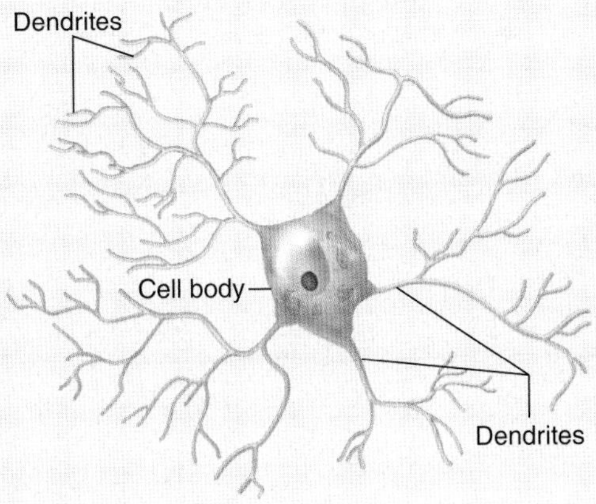	Processes are only dendrites; no axon present	Interneurons of the central nervous system (CNS)

Structural Classification

Neurons are classified structurally based upon the number of neuron processes emanating directly from the cell body. Thus, they may be classified as multipolar, bipolar, unipolar, or anaxonic neurons (table 12.1).

Multipolar neurons are the most common type of neuron. These neurons have many dendrites and a single axon that extends from the cell body.

Bipolar neurons have two processes that extend from the cell body—one dendrite and one axon. The location of these neurons is relatively limited in humans (e.g., the retina of the eye and the olfactory epithelium in the nose).

Unipolar neurons have a single, short neuron process that emerges from the cell body and branches like a T. These neurons are also called **pseudounipolar** (sū′dō-yū-ni-pō-lăr; *pseudo* = false, *uni* = one) because they start out as bipolar neurons during development, but their two processes fuse into a single process. The naming of the branched processes in unipolar neurons has been a source of confusion with regard to the common definitions of dendrites and axons. It seems most appropriate to call the short, multiple-branched receptive endings dendrites. The other portion is called the axon because these processes, like other axons, generate and conduct action potentials. These axons are composed of the combined **peripheral process** (from dendrites to the cell body) and **central process** (from the cell body into the CNS).

Anaxonic (an-aks'on-ic; *an* = without) **neurons** have only dendrites and no axons. They are different from other types of neurons because they produce graded potentials, but they do not produce action potentials.

Functional Classification

Neurons are classified functionally according to the direction the action potential travels relative to the CNS. The three categories are sensory neurons, motor neurons, and interneurons **(figure 12.3)**.

Sensory neurons (*afferent neurons*) are the neurons of the sensory nervous system. They are responsible for conducting sensory input from both somatic sensory and visceral sensory receptors to the CNS. Most sensory neurons are unipolar. However, a few somatic sensory neurons are bipolar, such as those in the retina of the eye (see figure 16.12) and olfactory epithelium of the nose (see figure 16.5).

Motor neurons (*efferent neurons*) are the neurons of the motor nervous system, conducting motor output from the CNS to both somatic effectors and visceral effectors. All motor neurons are multipolar.

Interneurons (*association neurons*) lie entirely within the CNS. They receive stimulation from many other neurons and carry out the integrative function of the nervous system—that is, they receive, process, and store information and "decide" how the body responds to stimuli. Interneurons facilitate communication between sensory and motor neurons. Interneurons outnumber all other neurons; it is estimated that 99% of our neurons are interneurons. Neuron classification based upon structure and function is integrated in Table 12.1.

WHAT DID YOU LEARN?

6 How are the different processes that extend from a cell body used to structurally classify neurons?

7 Where are interneurons located, and what is their function?

12.2e Relationship of Neurons and Nerves

LEARNING OBJECTIVES

7. Describe the structure of a nerve, including the three layers of connective tissue wrappings.

8. Explain how nerves are classified structurally and functionally.

A **nerve** is a cablelike bundle of parallel axons that are components of the peripheral nervous system. A single axon typically must be viewed using a microscope, whereas a nerve tends to be a macroscopic structure. **Figure 12.4** shows a typical nerve. Like a muscle, a nerve has three successive connective tissue wrappings:

- The **epineurium** (ep-i-nū'rē-ŭm; *epi* = upon) is a thick layer of dense irregular connective tissue that encloses the entire nerve and provides both support and protection.

- The **perineurium** (per-i-nū'rē-ŭm; *peri* = around) is a layer of dense irregular connective tissue that wraps **fascicles** (fas'i-kl), which are bundles of axons. This layer supports blood vessels.

- An individual axon in a myelinated neuron is surrounded by neurolemmocytes (a type of glial cell described in section 12.4) and wrapped in the **endoneurium** (en-dō-nū'rē-ŭm; *endon* = within), a delicate layer of areolar connective tissue that separates and electrically insulates each axon. Also within this connective tissue layer are capillaries that supply each axon.

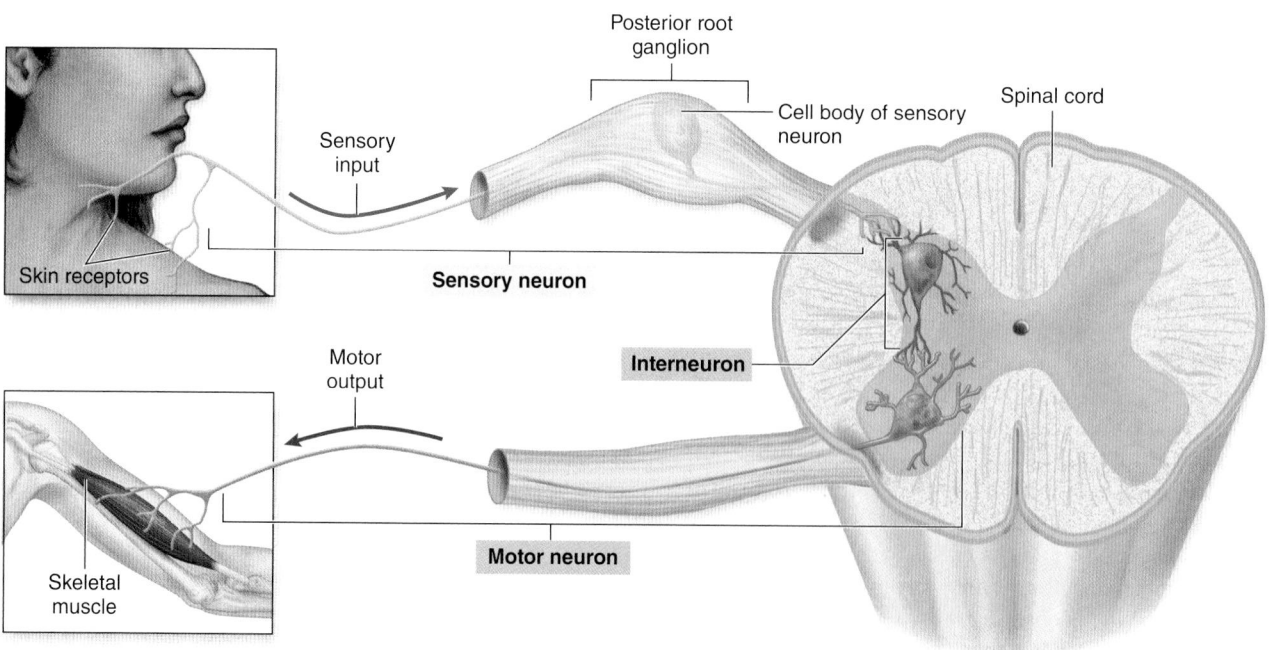

Figure 12.3 Functional Classification of Neurons. Sensory neurons transmit sensory input from receptors to the central nervous system. Interneurons are completely within the CNS and process information within the CNS. Motor neurons transmit motor output from the CNS to effectors (muscle or gland). Notice that the most common structural type is drawn for each functional class: a unipolar sensory neuron, a multipolar interneuron, and a multipolar motor neuron. Note that the relative sizes of the cell bodies are depicted larger than normal. APǀR

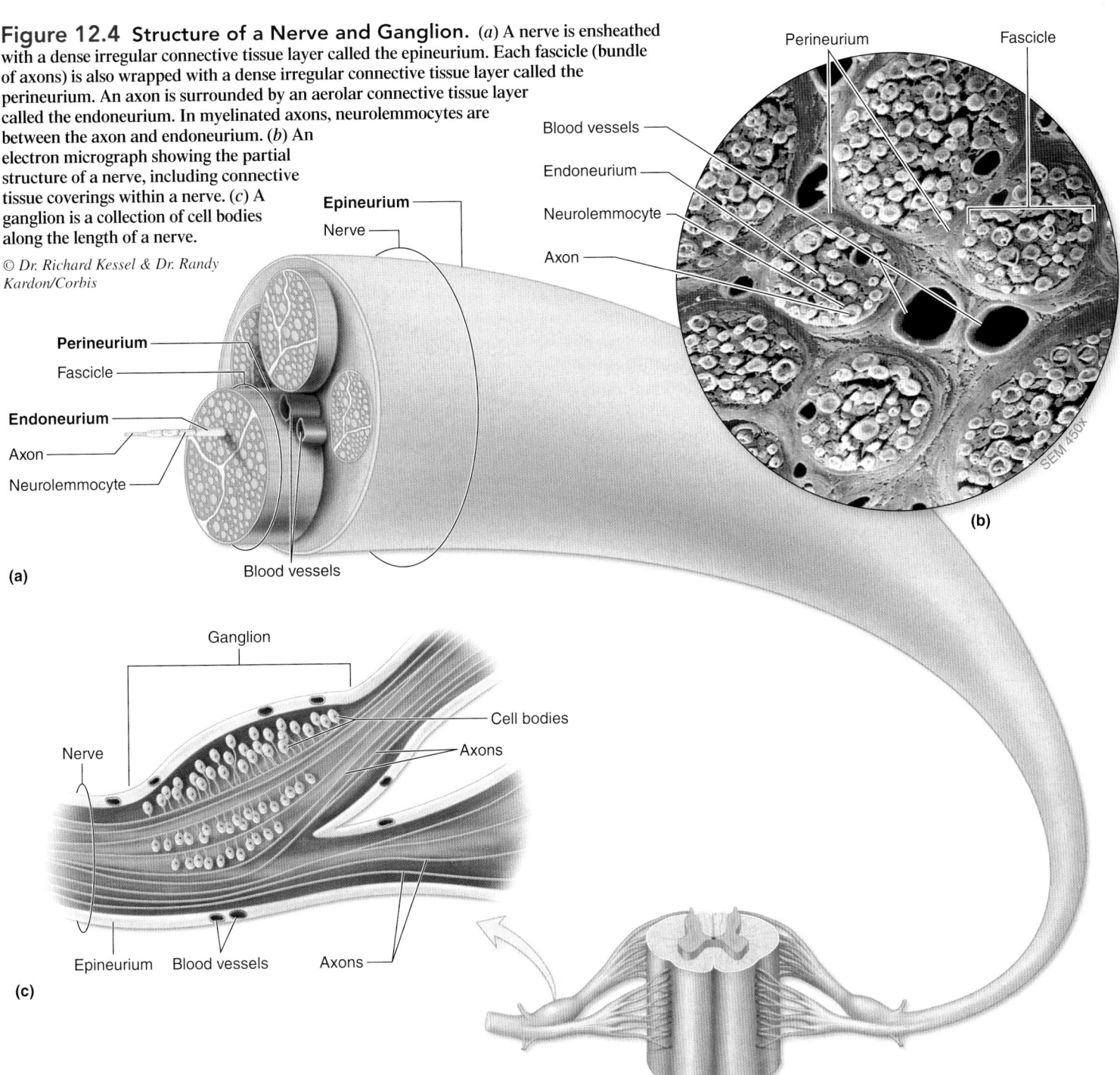

Figure 12.4 Structure of a Nerve and Ganglion. (*a*) A nerve is ensheathed with a dense irregular connective tissue layer called the epineurium. Each fascicle (bundle of axons) is also wrapped with a dense irregular connective tissue layer called the perineurium. An axon is surrounded by an aerolar connective tissue layer called the endoneurium. In myelinated axons, neurolemmocytes are between the axon and endoneurium. (*b*) An electron micrograph showing the partial structure of a nerve, including connective tissue coverings within a nerve. (*c*) A ganglion is a collection of cell bodies along the length of a nerve.

© *Dr. Richard Kessel & Dr. Randy Kardon/Corbis*

Classification of Nerves

Structural classification of nerves is based upon the CNS component from which the nerve extends: **Cranial nerves** extend from the brain (discussed in detail in section 13.9), and **spinal nerves** extend from the spinal cord (discussed in detail in section 14.5).

Functional classification of nerves is based upon the functional type of neuron (sensory neuron or motor neuron) a nerve contains. **Sensory nerves** contain sensory neurons that relay information to the CNS, and **motor nerves** contain motor neurons that relay information from the CNS. In contrast, **mixed nerves** contain both sensory and motor neurons. Most named nerves are mixed nerves. However, note that in mixed nerves individual sensory or motor neurons still transmit only one type of information.

WHAT DID YOU LEARN?

8 What are the three connective tissue wrappings in a nerve, and what specific structure does each ensheathe?

12.3 Synapses

LEARNING OBJECTIVES

1. Define a synapse.
2. Describe the essential structural and functional differences between a chemical synapse and an electrical synapse.

A **synapse** (sin′aps; *syn* = together, *hapto* = to clasp) is the specific location where a neuron is functionally connected either to another neuron or an effector (muscle or gland). There are two types of synapses in the human body: chemical synapses and electrical synapses. Most synapses within the nervous system are chemical synapses.

A **chemical synapse** between two neurons is composed of a **presynaptic** (prē-si-nap′tik; *pre* = before) **neuron,** which is the signal producer, and a **postsynaptic** (pōst-si-nap′tik; *post* = after) **neuron,** which is the signal receiver or target. The synapse may be between the axon of the presynaptic neuron and any portion of the surface of a postsynaptic neuron (dendrite, cell body, or axon), except those regions that are covered by a myelin sheath (described in section 12.4c). Most commonly a synapse is with a dendrite of the postsynaptic neuron. The synaptic knob of the presynaptic neuron does not quite make contact with the postsynaptic neuron (see inset, figure 12.2). The two neurons are separated by an extremely narrow fluid-filled gap (of about 30 nanometers) called the **synaptic cleft.**

Transmission between a presynaptic and postsynaptic neuron occurs when **neurotransmitter** (nū′rō-trans′mit-ĕr) molecules stored in synaptic vesicles are released from the synaptic knob of a presynaptic neuron into the synaptic cleft. Some of the neurotransmitter diffuses across the synaptic cleft to bind to receptors within the plasma membrane of the postsynaptic neuron to initiate a graded potential. There is a **synaptic delay** associated with neurotransmitter release at chemical synapses. The delay is the time between the neurotransmitter release from the presynaptic cell, its diffusion across the synaptic cleft, and neurotransmitter binding to receptors in the postsynaptic plasma membrane. This delay is usually between 0.3 and 0.5 milliseconds. Note that one postsynaptic neuron may, and often does, receive signals from more than one presynaptic neuron simultaneously.

A second, and much less common, type of synapse is an electrical synapse. An **electrical synapse** is composed of a presynaptic neuron and a postsynaptic neuron physically bound together. Gap junctions (see section 4.6d) are present in the plasma membranes of both neurons and facilitate the flow of ions between the cells. The cells act as though they share a common plasma membrane. Thus, the electrical signal passes between the cells with essentially no synaptic delay. Electrical synapses are located within limited regions of the brain and the eyes.

 WHAT DID YOU LEARN?

9 What is a chemical synapse within the nervous system, and how does it function?

12.4 Nervous Tissue: Glial Cells

Glial cells are the other distinct cell type within nervous tissue. These nonexcitable cells serve primarily to support and protect the neurons.

12.4a General Characteristics of Glial Cells

LEARNING OBJECTIVE

1. List the distinguishing features of glial cells.

Glial (glī′ăl; *glia* = glue) **cells** are sometimes referred to as *neuroglia* (nū-rog′lē-a). They are found within both the CNS and the PNS. Glial cells are both smaller than neurons and capable of mitosis. Glial cells do not transmit electrical signals, but they do assist neurons with their functions. The glial cells cooperate to physically protect and help nourish neurons as well as provide an organized, supporting scaffolding for all the nervous tissue. During development, glial cells form the framework

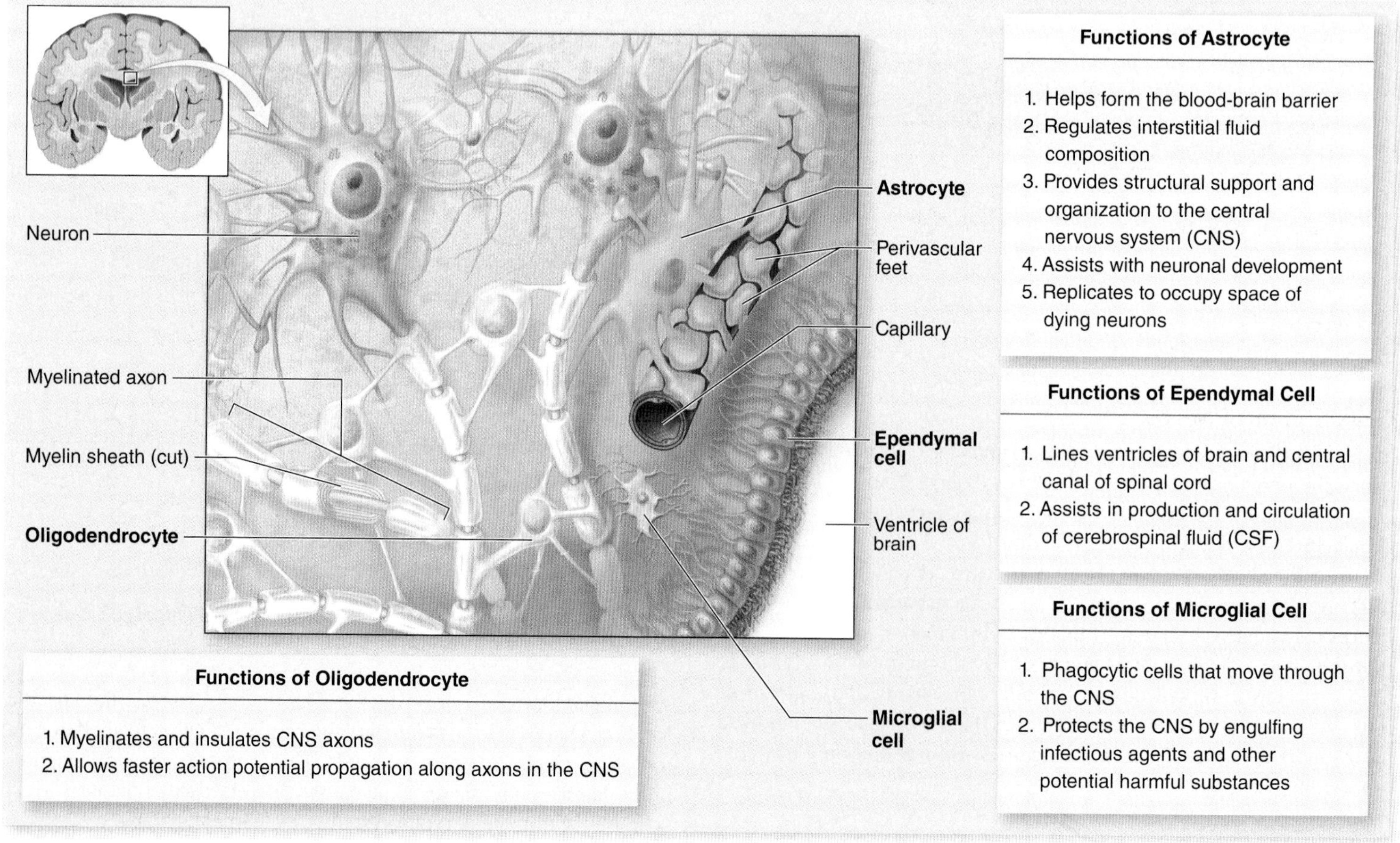

Functions of Astrocyte

1. Helps form the blood-brain barrier
2. Regulates interstitial fluid composition
3. Provides structural support and organization to the central nervous system (CNS)
4. Assists with neuronal development
5. Replicates to occupy space of dying neurons

Functions of Ependymal Cell

1. Lines ventricles of brain and central canal of spinal cord
2. Assists in production and circulation of cerebrospinal fluid (CSF)

Functions of Microglial Cell

1. Phagocytic cells that move through the CNS
2. Protects the CNS by engulfing infectious agents and other potential harmful substances

Functions of Oligodendrocyte

1. Myelinates and insulates CNS axons
2. Allows faster action potential propagation along axons in the CNS

Labels: Neuron, Myelinated axon, Myelin sheath (cut), Oligodendrocyte, Astrocyte, Perivascular feet, Capillary, Ependymal cell, Ventricle of brain, Microglial cell

Figure 12.5 Glial Cells of the Central Nervous System (CNS). Four types of glial cells are located within the CNS, including astrocytes, ependymal cells, microglia, and oligodendrocytes. These cells differ in both their structure and function.

that guides young, migrating neurons to their final destinations. Recent evidence has shown that glial cells are critical for the normal function at neural synapses, both maintaining the anatomic structure of synapses and modifying transmission that occurs there.

Glial cells far outnumber neurons. The nervous tissue of a young adult may contain 35 to 100 billion neurons and 100 billion to 1 trillion glial cells. Collectively, glial cells account for roughly half the volume of the nervous system.

 WHAT DID YOU LEARN?

⑩ If a person has a brain tumor, is it more likely to have developed from neurons or from glial cells? Why?

12.4b Types of Glial Cells

 LEARNING OBJECTIVE

2. Describe structure and function of the four types of glial cells within the CNS, and the two types of glial cells of the PNS.

Glial Cells of the CNS

Four types of glial cells are found in the central nervous system (CNS). These different cells include astrocytes, ependymal cells, microglia, and oligodendrocytes **(figure 12.5)**. They can be distinguished based upon size, intracellular organization, and the presence of specific cytoplasmic processes.

Astrocytes (as′trō-sīt; *astron* = star) exhibit a starlike shape due to projections from their surface. These numerous cell processes have contact with both capillaries (smallest blood vessels) and neurons. Astrocytes are the most abundant glial cell in the CNS and constitute over 90% of the nervous tissue in some areas of the brain. Astrocytes nurture, protect, support, and guide neurons, as follows:

- **Help form the blood-brain barrier.** The ends of astrocyte processes are called **perivascular feet:** They both cover and wrap around capillaries in the brain. The perivascular feet and the brain capillaries together contribute to a **blood-brain barrier (BBB).** The BBB strictly controls movement of substances from exiting the blood and entering the nervous tissue in the brain. The BBB protects the delicate neurons of the brain from toxins, but at the same time allows needed nutrients to pass through (see section 13.2d).

- **Regulate interstitial fluid composition.** Astrocytes help maintain an optimal chemical composition of the interstitial fluid (fluid around cells) within the brain. For example, astrocytes regulate potassium ion concentration by absorbing these ions to sustain a constant potassium ion concentration that is critical to electrical activity of neurons.

- **Form structural support.** The cytoskeleton in astrocytes strengthens these cells to provide a structural framework to support and organize neurons within the CNS.

- **Assist neuronal development.** Astrocytes help direct the development of neurons in the fetal brain by secreting chemicals that regulate the formation of connections between neurons.

- **Occupy the space of dying neurons.** When neurons are damaged and die, the space they formerly occupied is often filled by astrocytes that replicate through cell division.

Ependymal (e-pen′di-măl) **cells** are ciliated simple cuboidal or simple columnar epithelial cells that line the internal cavities (ventricles) of the brain (see figure 13.7) and the central canal of the spinal

INTEGRATE

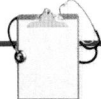

CLINICAL VIEW
Tumors of the Central Nervous System

Neoplasms resulting from unregulated cell growth, commonly known as **tumors,** sometimes occur within the central nervous system (CNS). A tumor that originates within the organ where it is found is called a primary tumor. Because most mature neurons cannot undergo mitosis, primary CNS tumors typically originate in supporting tissues within the brain or spinal cord that have retained the capacity to undergo mitosis: the meninges (protective membranes of the CNS) or the glial cells. Glial cell tumors, termed **gliomas,** may be either relatively benign and slow-growing or malignant (capable of metastasizing or spreading to distant sites).

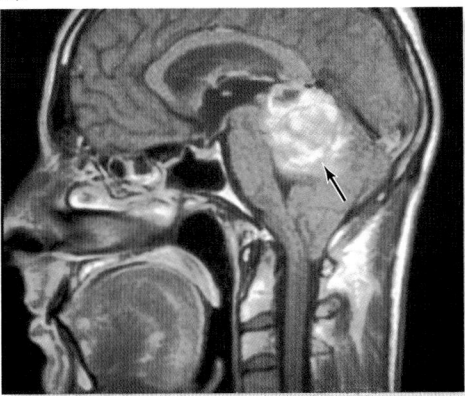

An MRI shows a glioma (arrow).

cord (see figure 14.3). These cells have slender processes that branch extensively to make contact with other glial cells in the surrounding nervous tissue.

Ependymal cells and nearby blood capillaries together form a network called the *choroid* (ko′royd) *plexus* (see figure 13.8). The choroid plexus helps produce cerebrospinal fluid (CSF), a clear liquid that bathes the external surfaces of the CNS and fills its internal cavities. The cilia of ependymal cells help circulate the CSF (see section 13.2c).

Microglia (mī-krog′lē-ă; *micros* = small) are typically small cells that have slender branches extending from the main portion of the cell. They represent the smallest percentage of CNS glial cells with some estimates of their prevalence as low as 5%. They are classified as phagocytic cells of the immune system (see section 22.2a). Microglial cells wander through the CNS and replicate in response to an infection. They protect the CNS against microorganisms and other potentially harmful substances by engulfing infectious agents and removing debris from dead or damaged nervous tissue that results from infections, inflammation, trauma, and brain tumors.

Oligodendrocytes (ol′i-gō-den′drō-sīt; *oligos* = few) are large cells with a bulbous body and slender cytoplasmic extensions or processes. The extensions of oligodendrocytes wrap around and insulate axons within the CNS to form a myelin sheath. This insulation allows for faster propagation of action potentials.

Glial Cells of the PNS

Two types of glial cells are found in the peripheral nervous system (PNS). These specialized glial cells include satellite cells and neurolemmocytes **(figure 12.6)**.

Functions of Satellite Cell

1. Electrically insulates PNS cell bodies.
2. Regulates nutrient and waste exchange for cell bodies in ganglia

Functions of Neurolemmocyte

1. Myelinates and insulates PNS axons
2. Allows for faster action potential propagation along an axon in the PNS

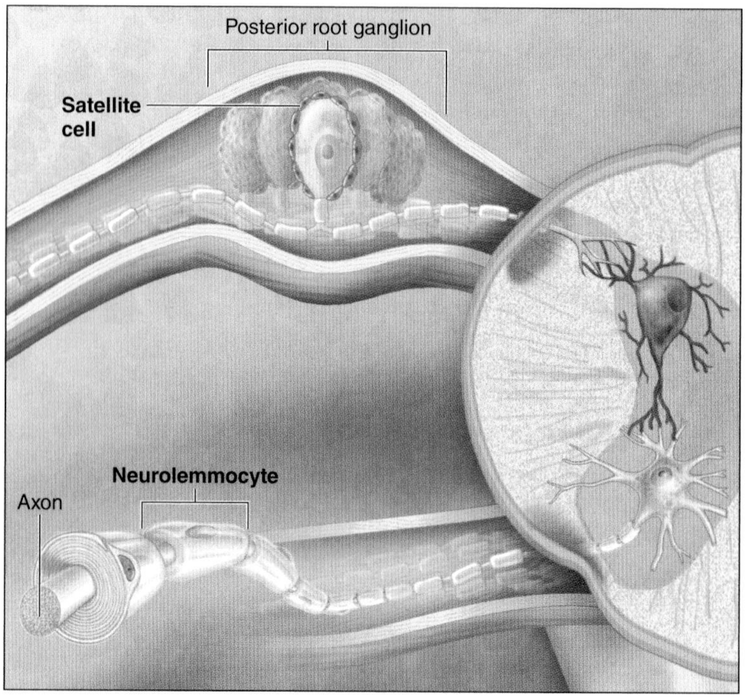

Figure 12.6 Glial Cells of the Peripheral Nervous System (PNS). Two primary types of glial cells are located within the PNS, including satellite cells and neurolemmocytes.

Satellite cells are flattened cells arranged around neuronal cell bodies in a ganglion that physically separate cell bodies from their surrounding interstitial fluid. They both electrically insulate the cell body and regulate the continuous exchange of nutrients and waste products between neuron cell bodies and their environment.

Neurolemmocytes (nūr-ō-lem′ō-sīt) are also called *Schwann cells.* These elongated and flattened cells wrap around and insulate axons within the PNS to form a myelin sheath. As with myelin sheaths formed by oligodendrocytes in the CNS, this allows for faster propagation of action potentials. The process of forming the myelin sheath by both neurolemmocytes and oligodendrocytes is described next.

WHAT DID YOU LEARN?

11 If a person suffers from meningitis (an inflammation of the meningeal coverings around the brain), which type of glial cell usually replicates in response to the infection?

12 Which specific type of glial cells ensheaths axons in the PNS?

12.4c Myelination

LEARNING OBJECTIVES

3. Define myelination, and describe the composition and function of a myelin sheath.

4. Distinguish between the myelination process carried out by neurolemmocytes in the PNS and by oligodendrocytes in the CNS.

Myelination (mī′ĕ-li-nā′shŭn) is the process by which part of an axon is wrapped with **myelin** (mī′e-lin). Myelin is the insulating covering around the axon that consists of repeating concentric layers of plasma membrane of glial cells. Myelination is completed by neurolemmocytes in the PNS and by oligodendrocytes in the CNS. Myelin mainly consists of the plasma membrane of these glial cells and contains a large proportion of lipids and a lesser amount of proteins. The high lipid content of the myelin gives an axon a distinct, glossy-white appearance and serves to effectively insulate an axon.

Figure 12.7 illustrates the process of myelinating a PNS axon. The neurolemmocyte starts to encircle a 1-millimeter portion of an axon. As the neurolemmocyte continues to wrap around the axon, the cytoplasm and nucleus of the neurolemmocyte are squeezed to the periphery of the neurolemmocyte (the outside edge). The overlapping inner layers of the plasma membrane form the **myelin sheath.** The periphery of the neurolemmocyte contains the cytoplasm and nucleus and is called the **neurilemma** (nūr-i-lem′ă; *lemma* = husk). This process is similar to what would happen if you were to take a balloon with a small amount of water in it and wrap it numerous times around a pencil. The balloon is wrapped over and over around a section of your pencil and the part of the balloon containing water is pushed to the outside. The wrapped layers of balloon represent the myelin sheath, and the external portion of the balloon with the water represents the neurilemma.

A neurolemmocyte in the PNS can myelinate only a 1-millimeter portion of a single axon. Thus, if an axon is longer than 1 millimeter (and most PNS axons are), it takes many neurolemmocytes to myelinate the entire axon. **Figure 12.8a** shows an axon that has seven neurolemmocytes wrapped around it. The axons in many of the nerves in the body have hundreds or thousands of neurolemmocytes along their entire length. The gaps between the neurolemmocytes are called **neurofibril** (nū′rō-fī′bril) **nodes,** or *nodes of Ranvier.*

An oligodendrocyte in the CNS, in comparison, can myelinate a 1-millimeter portion of *many* axons at the same time and not just one. Figure 12.8b shows oligodendrocytes myelinating portions of three different axons. The cytoplasmic extensions of the oligodendrocyte wrap repeatedly around a portion of each axon where plasma membrane

INTEGRATE

CLINICAL VIEW
Nervous System Disorders
Affecting Myelin

Multiple sclerosis (MS) is progressive demyelination of neurons in the central nervous system accompanied by the destruction of oligodendrocytes. MS is an autoimmune disorder, because the body's immune cells mistake the oligodendrocytes as foreign and attack them. As a result, the conduction of action potentials is disrupted, leading to impaired sensory perception and motor coordination. Repeated inflammatory events at myelinated sites cause scarring (sclerosis), and in time some function is permanently lost. The disease usually affects young adults between the ages of 18 and 40. It is five times more prevalent in individuals of European descent than it is in African-Americans. Among the typical symptoms are vision problems, muscle weakness and spasms, urinary infections and bladder incontinence, and drastic mood changes.

Guillain-Barré syndrome (GBS) is a disorder in which inflammation causes loss of myelin from the peripheral nerves and spinal nerve roots. It is characterized by muscle weakness that begins in the distal limbs, but rapidly advances to involve proximal muscles as well (ascending paralysis). Most cases of GBS are preceded by an acute, flulike illness, although no specific infectious agent has ever been identified. The condition in rare instances may follow an immunization. Even though GBS appears to be an immune-mediated condition, the use of steroids provides little if any measurable improvement. In fact, most people recover almost all neurologic function on their own with little medical intervention.

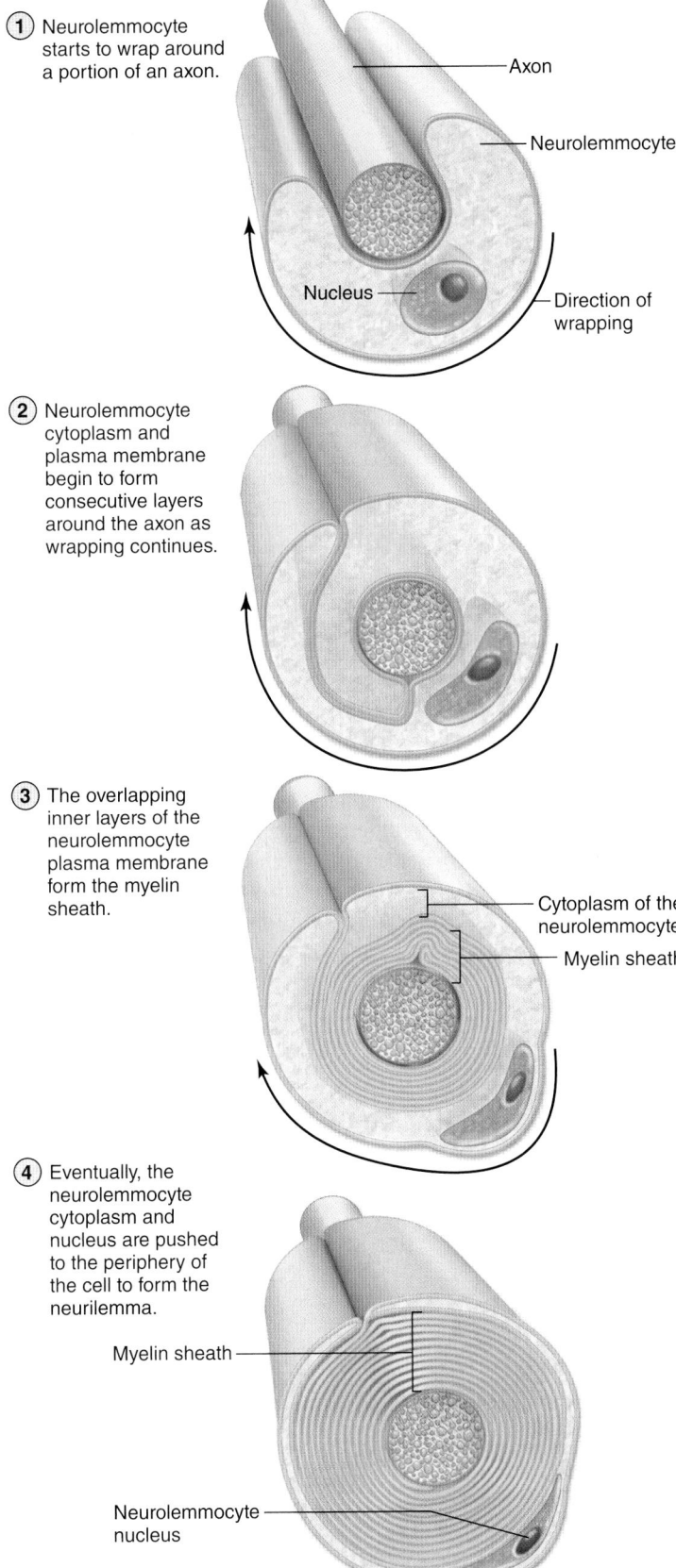

1. Neurolemmocyte starts to wrap around a portion of an axon.

Axon

Neurolemmocyte

Nucleus

Direction of wrapping

2. Neurolemmocyte cytoplasm and plasma membrane begin to form consecutive layers around the axon as wrapping continues.

3. The overlapping inner layers of the neurolemmocyte plasma membrane form the myelin sheath.

Cytoplasm of the neurolemmocyte

Myelin sheath

4. Eventually, the neurolemmocyte cytoplasm and nucleus are pushed to the periphery of the cell to form the neurilemma.

Myelin sheath

Neurolemmocyte nucleus

Neurilemma

Figure 12.7 Myelination of PNS Axons. A myelin sheath surrounds most axons. In the PNS, neurolemmocytes form both a myelin sheath and neurilemma in a series of sequential stages. **AP|R**

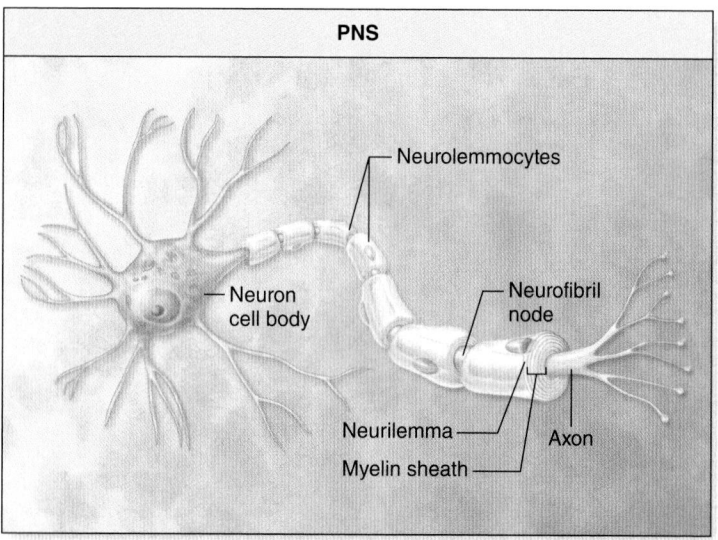

PNS

Neurolemmocytes

Neuron
cell body

Neurofibril
node

Neurilemma

Axon

Myelin sheath

(a) Myelination by neurolemmocytes

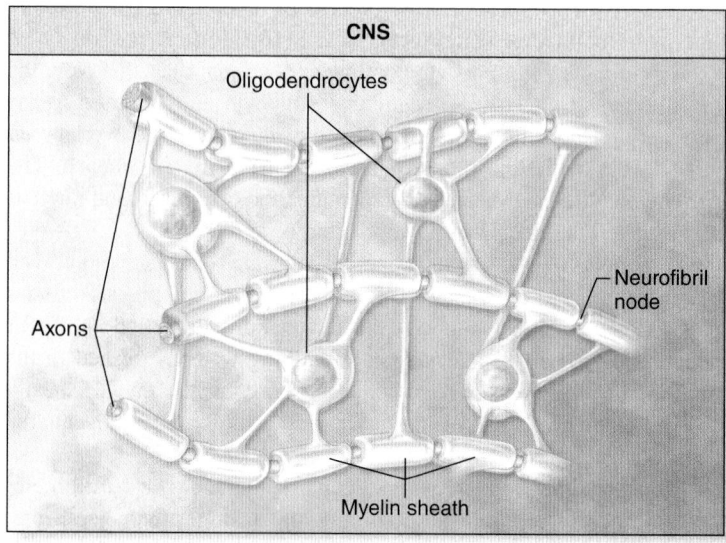

CNS

Oligodendrocytes

Neurofibril
node

Axons

Myelin sheath

(b) Myelination by oligodendrocytes

Figure 12.8 Myelin Sheaths in the PNS and CNS. (*a*) In the PNS, the neurolemmocyte ensheathes only one small part of a single axon to form both a myelin sheath and a neurilemma. (*b*) In the CNS, an oligodendrocyte wraps around a small part of multiple axons, forming a myelin sheath (but no neurilemma is formed).

layers of the oligodendrocyte form the myelin sheath. Note that no neurilemma is formed as the CNS neurons are myelinated. Neurofibril nodes are also located between adjacent oligodendrocyte "wraps." You may find it helpful to imagine that an oligodendrocyte is like a latex glove with fluid added to it. Each finger of the glove is wrapped numerous times around the axon of different neurons, and the fluid is pushed into the hand of the glove. Remember, myelination in both the PNS and CNS allows for faster propagation of action potentials.

Not all axons are myelinated. **Unmyelinated axons** in the PNS (figure 12.9) are also associated with neurolemmocytes, which help

to protect and support the axon. However, no myelin sheath covers them. Thus, the axon merely rests in a depressed portion of the neurolemmocyte, but its plasma membrane does not form repeated layers around the axon. In the CNS, unmyelinated axons are not associated with oligodendrocytes.

WHAT DID YOU LEARN?

13 What is the function of the myelin sheath? How does myelination of axons occur in the PNS?

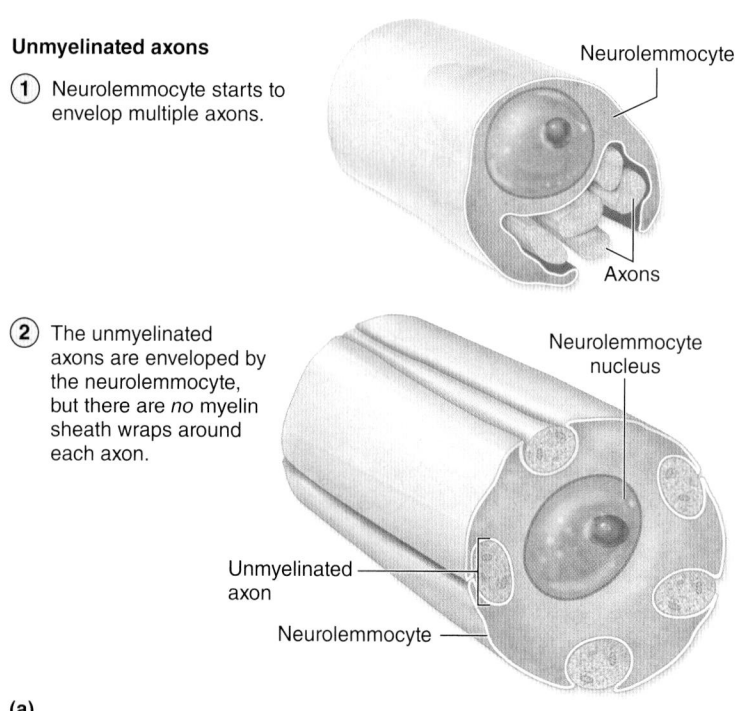

Unmyelinated axons

1 Neurolemmocyte starts to envelop multiple axons.

Neurolemmocyte

Axons

2 The unmyelinated axons are enveloped by the neurolemmocyte, but there are *no* myelin sheath wraps around each axon.

Neurolemmocyte
nucleus

Unmyelinated
axon

Neurolemmocyte

(a)

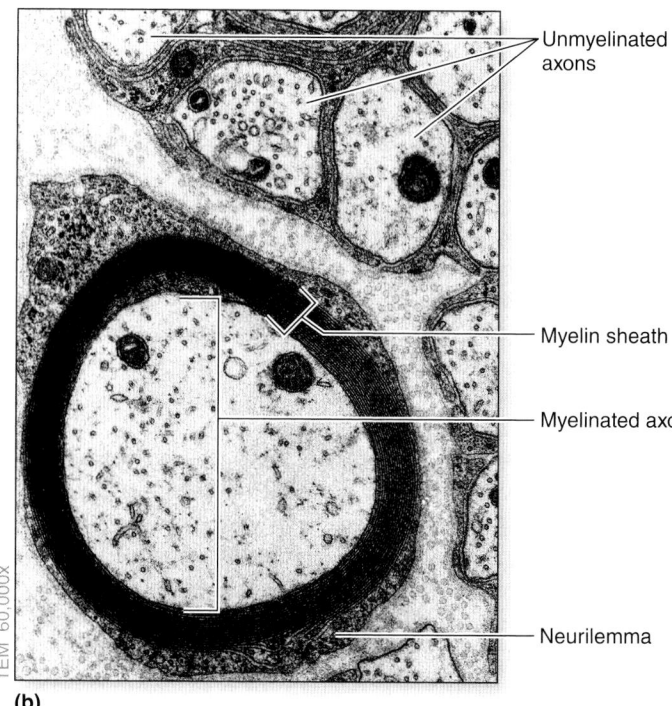

Unmyelinated
axons

Myelin sheath

Myelinated axon

Neurilemma

TEM 60,000x

(b)

Figure 12.9 Unmyelinated Axons. (*a*) Unmyelinated axons are surrounded by a neurolemmocyte but are not wrapped in a myelin sheath. (*b*) An electron micrograph shows a myelinated axon and several unmyelinated axons.

12.5 Axon Regeneration

LEARNING OBJECTIVES

1. Identify factors that influence regeneration of PNS axons, and explain why axon regeneration in the CNS is limited.

2. Describe the events of Wallerian degeneration and axon regrowth.

PNS axons are vulnerable to cuts, crushing injuries, and other types of trauma. However, a damaged axon can regenerate if the cell body remains intact and a critical amount of neurilemma remains. The degree of success of PNS axon regeneration depends upon two primary factors: (1) the amount of damage, and (2) the distance between the site of the damaged axon and the structure it innervates. The possibility of repair is decreased with an increase in either of these two factors.

Neurolemmocytes play an active role in regeneration. This process is illustrated in **figure 12.10** and follows these stages:

1. The axon is severed by some type of trauma.

2. The portion proximal to the trauma seals off by membrane fusion and swells. The swelling is a result of axoplasmic flow (slow transport) from the neuron cell body through the axon. At the same time, the axon severed from the cell body and the myelin sheath surrounding the axon breaks down— a process called **Wallerian** (waw-lē′rē-ăn) **degeneration.** Macrophages (phagocytic cells) remove the debris. However, the neurilemma in the distal region survives.

3. The neurilemma in conjunction with the remaining endoneurium forms a regeneration tube.

4. The axon regenerates and remyelination occurs. The regeneration tube guides the axon sprout as it begins to grow rapidly through the regeneration tube at a rate of about 2 to 5 millimeters per day. This occurs under the influence of nerve growth factors released by the neurolemmocytes.

5. Innervation is restored as the axon reestablishes contact with its original structure. The structure is either a receptor for sensory neurons to regain sensory perception or an effector for motor neurons to regain function of a muscle or gland.

Potential regeneration of damaged neurons within the CNS is very limited for several reasons. First, oligodendrocytes do not release a nerve growth factor, and in fact they actively inhibit axon growth by producing and secreting several growth-inhibitory molecules. Second, the large number of axons crowded within the CNS tends to complicate regrowth activities. Finally, both astrocytes and connective tissue coverings may form some scar tissue that obstructs axon regrowth. Medical researchers are attempting to overcome these limitations to treat patients with spinal cord injuries. (See Clinical View: "Spinal Cord Injuries" in section 14.3b.)

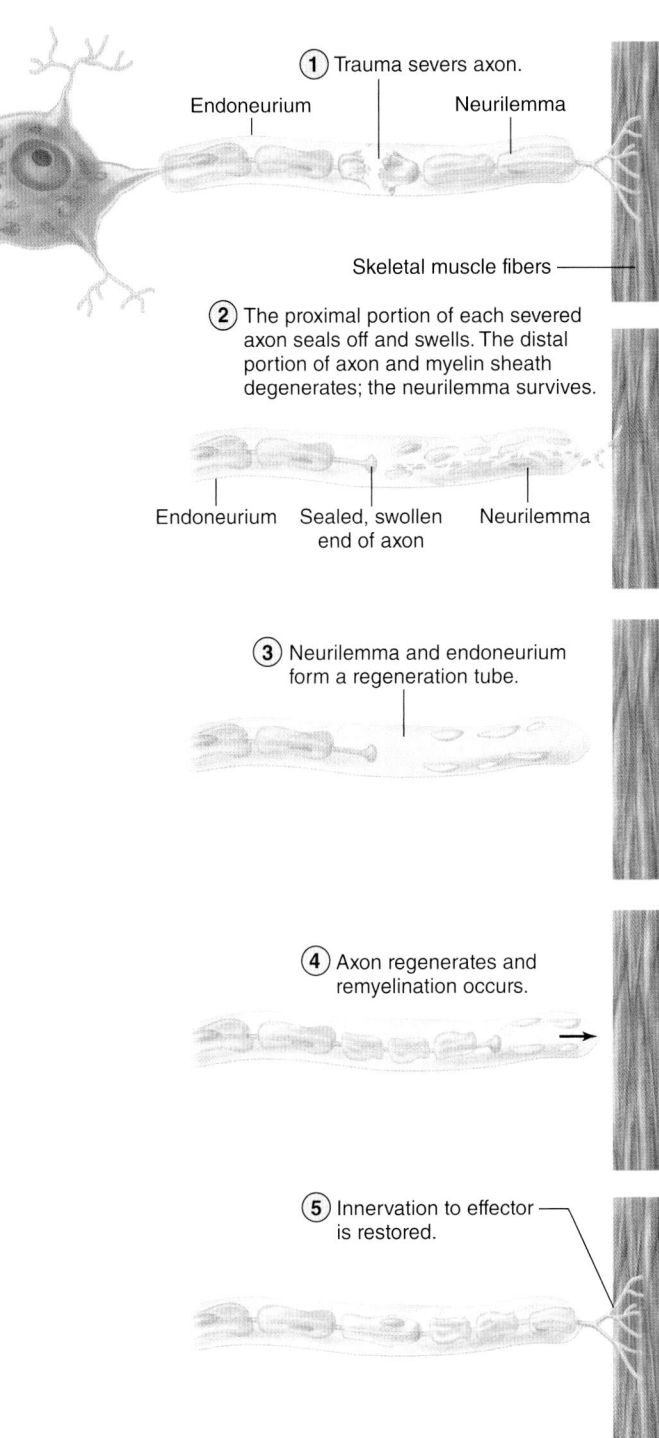

1. Trauma severs axon.
   - Endoneurium
   - Neurilemma
   - Skeletal muscle fibers

2. The proximal portion of each severed axon seals off and swells. The distal portion of axon and myelin sheath degenerates; the neurilemma survives.
   - Endoneurium
   - Sealed, swollen end of axon
   - Neurilemma

3. Neurilemma and endoneurium form a regeneration tube.

4. Axon regenerates and remyelination occurs.

5. Innervation to effector is restored.

Figure 12.10 Regeneration of PNS Axons. Following injury to a peripheral nerve, the severed axon may be repaired and grow out to reinnervate the receptor or effector (in this case, a skeletal muscle fiber).

WHAT DID YOU LEARN?

14. What two primary factors determine the effectiveness of PNS axon regeneration?

15. How does the process of nerve regeneration occur in the PNS?

12.6 Plasma Membrane of Neurons

The ability of the nervous system to integrate and regulate body functions is ultimately dependent upon the transmission of electrical activity (i.e., graded potentials and action potentials) by neurons. Here we discuss the details of the plasma membrane of neurons that makes transmitting electrical activity possible. Included are descriptions of the various types of pumps and channels in a neuron's plasma membrane and their typical distribution within a neuron.

12.6a Types of Pumps and Channels

LEARNING OBJECTIVE

1. Distinguish between a pump and a channel, and describe the three specific states of a voltage-gated Na^+ channel.

Neurons contain transport proteins for moving substances across the plasma membrane. These include both pumps and various types of channels. Transport proteins were first discussed in section 4.3 and are reviewed here.

Pumps maintain specific concentration gradients by moving substances up (against) a concentration gradient, a process that requires cellular energy. The plasma membrane of neurons contains both sodium-potassium (Na^+/K^+) pumps (see figure 4.15) and calcium (Ca^{2+}) pumps (see figure 4.14). A great deal of energy is required to power the vast number of Na^+/K^+ pumps in a neuron's plasma membrane—approximately two-thirds of a neuron's energy expenditure!

Channels provide the means for a substance to move down (with) its concentration gradient. Neurons contain the following major types of channels:

- **Leak** (*passive*) **channels.** These channels are always open, allowing continuous diffusion of a specific type of ion from a region of high concentration to a region of low concentration. Examples of leak channels are sodium ion (Na^+) leak channels and potassium ion (K^+) leak channels.
- **Chemically gated channels.** These channels are normally closed. They open in response to binding of a neurotransmitter. When open, they allow a specific type of ion (or ions) to diffuse across the plasma membrane. Examples of chemically gated channels include chemically gated K^+ channels and chemically gated chloride ion (Cl^-) channels.
- **Voltage-gated channels.** These channels are also normally closed, but they open in response to changes in electrical charge (potential) across the plasma membrane. When open, they allow a specific type of ion to diffuse across the membrane. Examples of voltage-gated channels include voltage-gated Na^+ channels, voltage-gated K^+ channels, and voltage-gated Ca^{2+} channels. Most gated channels have one gate that is in either one of two states: closed or open. Voltage-gated Na^+ channels are unique in that they have two gates (an *activation gate* and an *inactivation gate*) and thus exhibit three states **(figure 12.11)**.

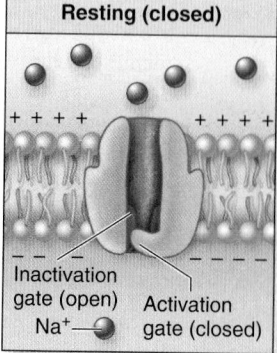 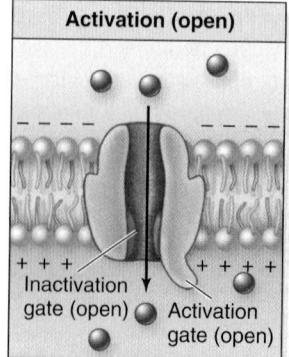 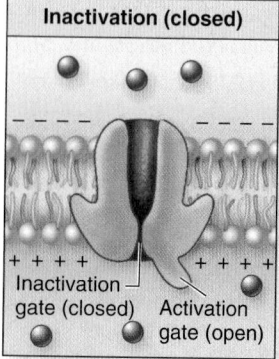

Figure 12.11 Voltage-Gated Na^+ Channels. A voltage-gated channel opens to permit ion movement when a specific transmembrane voltage is reached. The voltage-gated Na^+ channels exhibit three different states: resting, activation, and inactivation.

Three States of Voltage-Gated Na⁺ channels

1. **Resting state.** Although the inactivation gate is open, the activation gate is closed, and entry of Na⁺ is prevented.
2. **Activation state.** Both the inactivation gate (which remains open) and the activation gate are open (activation gate opens in response to a voltage change); Na⁺ moves into the cell through the open channel.
3. **Inactivation state.** Although the activation gate is open, the inactivation gate is *temporarily* closed (for several milliseconds) following activation of the Na⁺ channel—during this time, it cannot be stimulated to reopen, and entry of Na⁺ is prevented. (The resting state of voltage-gated Na⁺ channels is reestablished as the inactivation gate opens and the activation closes.)

Modality gated channels are an additional type of channel that is normally closed. These channels open (or close) in response to a stimulus other than a chemical or a voltage change (see section 16.1d). Modality gated channels are components of sensory neurons, which detect changes in the external or internal environment. For example, receptor cells of the skin contain modality gated channels that are stimulated by mechanical pressure to open (see section 16.2a) and receptor cells of the eye (photoreceptors) contain modality gated channels that are stimulated by light to close (see section 16.4d). These types of channels are described in more detail in chapter 16.

 WHAT DID YOU LEARN?

 16 Describe the three states of voltage-gated Na⁺ channels in neurons.

12.6b Distribution of Pumps and Channels

 LEARNING OBJECTIVES

2. List the channels and pumps that are located along the entire neuron, and identify the general function of each.
3. Identify and describe the four functional neuron segments, including the distribution of channels and pumps in each.

Some pumps and channels are located throughout the entire neuron plasma membrane, whereas others are primarily located only in specific segments of a neuron's plasma membrane. Their distribution is related to function. Here we discuss the distribution of the types of pumps and channels within the neuron plasma membrane **(figure 12.12)**. Note that the distribution of channels and pumps shown has been simplified for the purposes of this discussion and reflects their primary location in a typical multipolar neuron.

Entire Plasma Membrane of a Neuron

Na⁺ leak channels, K⁺ leak channels, and Na⁺/K⁺ pumps are located throughout the entire neuron plasma membrane. These are important in establishing and maintaining the resting membrane potential (RMP) of neurons (described in section 12.7b).

Plasma Membrane of Functional Segments in a Neuron

A typical neuron is functionally organized into four segments: receptive segment, initial segment, conductive segment, and transmissive segment. Each region differs in the primary types of channels and pumps located within its plasma membrane:

- The **receptive segment** includes both dendrites and the cell body, which are the regions of the neuron that receive stimuli to excite the neuron. Chemically gated channels (cation channels, K⁺ channels, and Cl⁻ channels) are located in this segment; no significant numbers of voltage-gated channels are present. (Note that cation channels allow the passage of both Na⁺ into the neuron and K⁺ out of the neuron. However, more Na⁺ moves into the neuron than K⁺ moves out.)
- The **initial segment** is conventionally considered to be composed of the *axon hillock*. This segment contains both voltage-gated Na⁺ channels and voltage-gated K⁺ channels. (Note: Recent studies have shown that specific location of the initial segment may vary and in some cases may be located within the axon. We will use the convention of equating the initial segment with the axon hillock.)

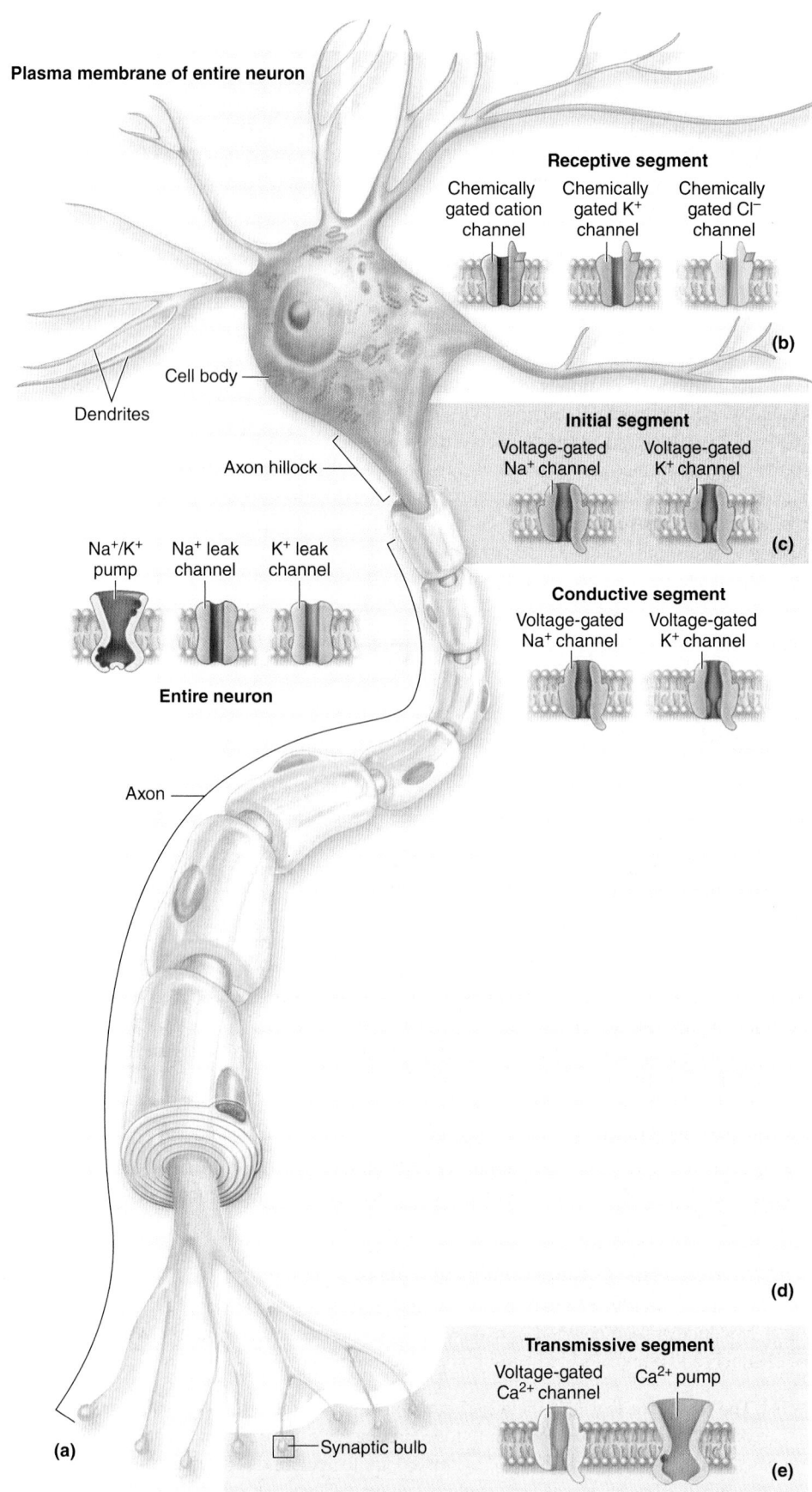

Plasma membrane of entire neuron

Receptive segment

Chemically gated cation channel

Chemically gated K^+ channel

Chemically gated Cl^- channel

(b)

Dendrites

Cell body

Axon hillock

Na^+/K^+ pump

Na^+ leak channel

K^+ leak channel

Entire neuron

Initial segment

Voltage-gated Na^+ channel

Voltage-gated K^+ channel

(c)

Conductive segment

Voltage-gated Na^+ channel

Voltage-gated K^+ channel

Axon

(d)

Transmissive segment

Voltage-gated Ca^{2+} channel

Ca^{2+} pump

(a)

Synaptic bulb

(e)

Figure 12.12 Distribution of Pumps and Channels in the Plasma Membrane of a Neuron. (*a*) Na^+/K^+ pumps, Na^+ leak channels, and K^+ leak channels are found throughout the entire plasma membrane of a neuron. Additional types of channels and pumps are present only in specific functional segments of the neuron as shown (*b–e*). Note that this represents a simplified depiction of the distribution of channels and pumps in a multipolar neuron plasma membrane. AP|R

- The **conductive segment** is equivalent to the length of the axon and its branches called telodendria. Like the initial segment, it also contains both voltage-gated Na^+ channels and voltage-gated K^+ channels.
- The **transmissive segment** includes the synaptic knobs and contains voltage-gated Ca^{2+} channels and Ca^{2+} pumps.

 WHAT DID YOU LEARN?

17 Which functional segment of a neuron contains chemically gated channels? Which functional segments contains voltage-gated channels?

12.7 Introduction to Neuron Physiology

Initiating and transmitting electrical currents is central to the function of neuron physiology. Necessary to this process is establishing and maintaining a resting membrane potential that can be changed from its resting value. Here we first describe both the basic principles of electrical currents (and Ohm's law) and the physiologic conditions of a neuron at rest.

12.7a Neurons and Ohm's Law

 LEARNING OBJECTIVE

1. Integrate the concepts of voltage, current, and resistance with neuron structure and function.

Recall from section 3.1 that electrical energy is the movement of charged particles, and that all usable forms of energy are available to do work. Neuron activity is dependent upon electrical energy, specifically electrical currents. Three important and relevant characteristics are associated with electrical currents: voltage, current, and resistance.

- **Voltage** is the measure of the *amount of difference* in electrical charge between two areas and represents potential energy. The unit of measurement is volts (V), or millivolts (mV); 1 V = 1000 mV. The larger the difference in charge, the higher the voltage. You may have noticed that batteries are available in different sizes (e.g., a small [1.5 V] or a large battery [12 V]). The voltage is an indication of the relative potential energy stored in the battery.
- **Current** is the *movement of charged particles* across the barrier that separates this charge difference. The greater the movement of charged particles, the greater the current. Movement of charged particles (or current) can be harnessed to do work (e.g., a current of electrons occurs between the positive and negative terminals in a battery when placed in a flashlight and the flashlight is turned on).
- **Resistance** is the *opposition to the movement of charged particles*. This is the barrier between the charged areas. The larger the resistance, the lower the current.

The relationship of voltage, current, and resistance is expressed by **Ohm's law:**

$$\text{Current} = \frac{\text{Voltage}}{\text{Resistance}}$$

This expression shows that current is directly related to voltage and inversely related to resistance. Thus, a greater current is possible with a larger voltage difference and a lower resistance.

Let us relate the general concepts of electrical currents to neurons. In neurons:

- Charged particles are ions (not like in a battery or with electricity, which involves the flow of electrons).
- There is a difference in charge on either side of the plasma membrane (voltage) due to an unequal distribution of ions.
- The plasma membrane phospholipid bilayer offers resistance because it generally does not allow the passage of ions.
- Resistance is altered across the plasma membrane through ion channels that open and close. Resistance is decreased when ion channels are open and increased when ion channels are closed.
- Current is generated when either positively charged ions or negatively charged ions diffuse across the plasma membrane through open channels.

INTEGRATE

CONCEPT CONNECTION

Potential energy is the energy due to position (see section 3.1a). Recall that ion gradients represent potential energy because of their position relative to the plasma membrane. **Kinetic energy** is the energy of motion. Ions can flow across the plasma membrane through open channels, establishing ion currents. It is this movement of ions (which is a type of kinetic energy) that is harnessed when electrical signals (i.e., graded potentials and action potentials) are transmitted in a neuron.

WHAT DID YOU LEARN?

18 What is the role of ions, the phospholipid bilayer, and plasma membrane channels in neurons relative to the concepts of current, voltage, and resistance?

12.7b Neurons at Rest

LEARNING OBJECTIVES

2. Describe the conditions of a neuron at rest.

3. Define resting membrane potential, and state its typical value for neurons.

4. Explain how the resting membrane potential is established and maintained in neurons.

Neurons at rest have several important characteristics **(figure 12.13a)**:

- Ion concentration gradients exists for K⁺, Na⁺, and Cl⁻ across the plasma membrane along the entire neuron. At the plasma membrane there is relatively more K⁺ in the cytosol than in the interstitial fluid (IF) surrounding the neuron, whereas there is more

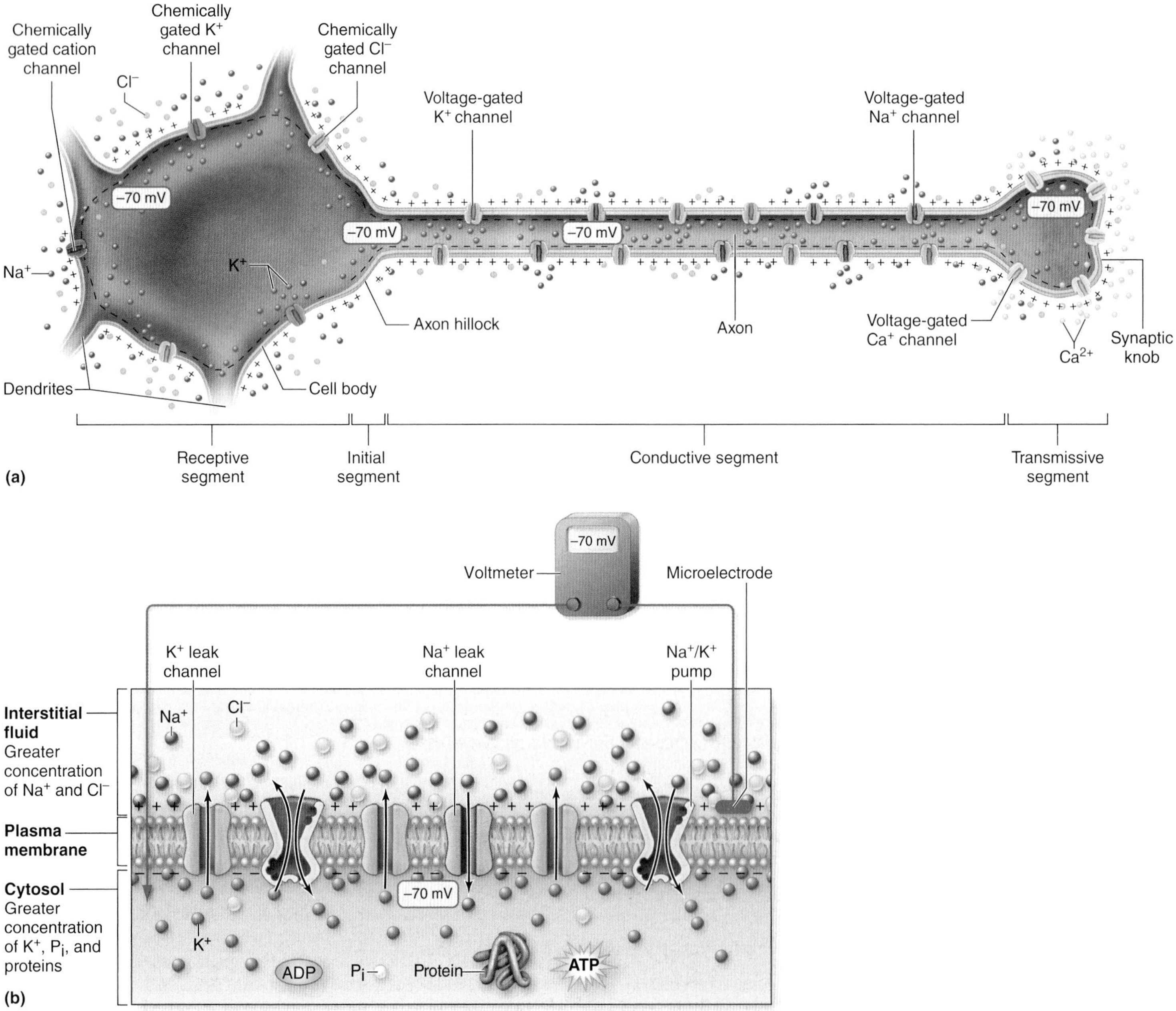

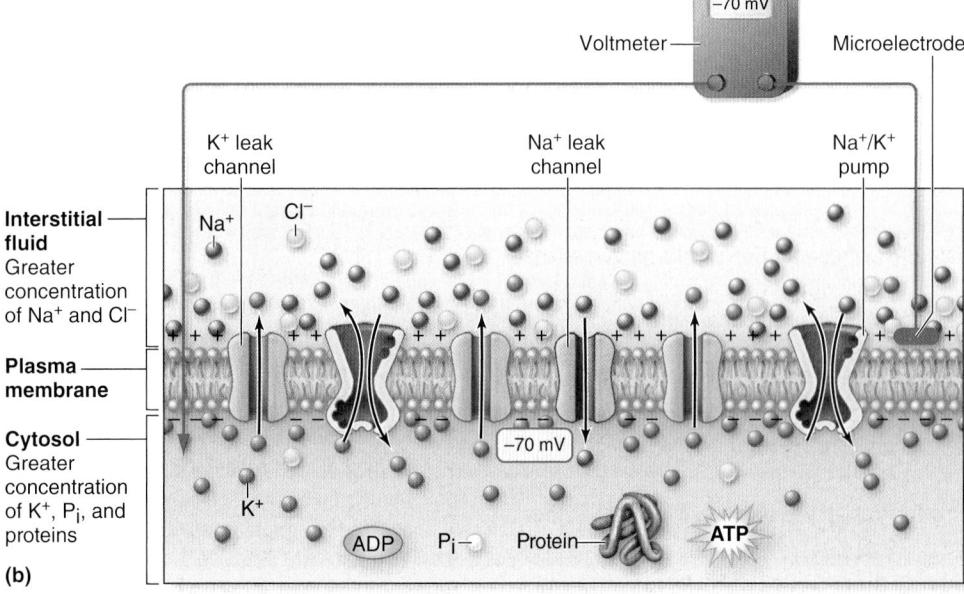

Figure 12.13 Neuron at Rest. (*a*) At rest, a neuron has a negative resting membrane potential (RMP), which on average is −70 mV. All gated channels are closed, and concentration gradients exist for Na⁺, K⁺, and Cl⁻ along the length of the neuron, and for Ca²⁺ at the synaptic knob. (*b*) The RMP is established and maintained by K⁺ leak channels, Na⁺ leak channels, and Na⁺/K⁺ pumps.

Na$^+$ and Cl$^-$ in the IF than in the cytosol. These gradients are established by Na$^+$/K$^+$ pumps that move three Na$^+$ out of the neuron for every two K$^+$ moved in. (Chloride ion follows the movement of Na$^+$.)

- A Ca^{2+} concentration gradient exists at the synaptic knob. Calcium pumps within this segment continuously pump Ca^{2+} from within the synaptic knob to the surrounding IF. Thus, there is more Ca^{2+} in the IF outside the synaptic knob than within the cytosol in the synaptic knob.

- Gated channels are *closed* including the chemically gated channels in the receptive segment, the voltage-gated Na$^+$ channels and voltage-gated K$^+$ channels in both the initial segment and conductive segment, and voltage-gated Ca^{2+} channels in the transmissive segment.

- There is an electrical charge difference (an electrical gradient) across the plasma membrane; the cytosol adjacent to the plasma membrane is relatively negative in comparison to the IF outside of the cell. This electrical charge difference is called a **membrane potential.** When the neuron is a rest, the membrane potential is more specifically called the **resting membrane potential (RMP).** The RMP of a neuron is typically –70 millivolts (mV), but can range between –40 mV and –90 mV.

Note that a voltmeter is used to measure the voltage difference across the plasma membrane. This is done by placing one microelectrode into the neuron and the other microelectrode outside the neuron in the interstitial fluid (figure 12.13*b*). The value of the voltage difference is negative (typically –70 mV in neurons) because the voltage of the cytosol at the plasma membrane is *relatively* negative compared to the voltage measured in the IF outside the plasma membrane— that is, more positive ions reside outside a neuron than just inside a neuron when it is at rest. (Consider the simplified example of having 100+ on the outside of a cell and 30+ on the inside of the cell, the inside has a relatively negative value of –70.)

Membrane potentials are first described in detail in section 4.4. Here we review how a resting membrane potential is both established and maintained and apply these concepts to neurons. See figure 12.13*b* as you read through this section.

Establishing and Maintaining the Resting Membrane Potential

Establishing and maintaining the RMP is dependent upon the distribution of ions (as just described) as well as additional substances. These include negatively charged phosphate ions (Pi) as components of organic molecules (e.g., ATP) and negatively charged protein molecules. Both are more prevalent within a neuron's cytosol than in the surrounding IF.

An RMP is chiefly a consequence of the movement of ions across the plasma membrane through leak channels (both K$^+$ leak channels and Na$^+$ leak channels).

The Role of K$^+$ Potassium diffusion is the most important factor in establishing the specific value of the resting membrane potential. K$^+$ movement is dependent upon its electrochemical gradient, which is the combination of the electrical gradient at the plasma membrane and the K$^+$ chemical concentration gradient. Potassium ions exit the neuron moving down their relatively steep chemical concentration gradient moving into the IF. The loss of K$^+$ leaves relatively more negatively charged structures (e.g., phosphate and proteins) inside the cell. These structures remain within the cell because they are too large to cross the plasma membrane. The movement of K$^+$ to the outside of a cell is, however, opposed by the electrical gradient. The positive charge on the outside of the cell repels the movement of K$^+$, and the negative charge on the inside of the cell attracts K$^+$. Thus, K$^+$ movement is facilitated by its chemical concentration gradient but opposed by the electrical gradient. As additional K$^+$ diffuses out of the neuron, the inside becomes more negative. Consequently, the pull to keep K$^+$ in the cell is greater. At some point, the force of the chemical gradient allowing K$^+$ out of a cell becomes equal to the electrical gradient that opposes this movement. Thus, K$^+$ movement has reached equilibrium. If only K$^+$ leak channels were present in neurons, the loss of the K$^+$ would result in an RMP with a value of –90 mV.

The Role of Na$^+$ The typical neuron RMP is –70 mV, as noted earlier. The difference between –90 mV established by K$^+$ movement only and the RMP of –70 mV is primarily the result of Na$^+$ movement into a neuron. The Na$^+$ enters the cell through Na$^+$ leak channels. It moves down its chemical concentration gradient, and it is also "pulled" into the cell by the electrical gradient. Both forces facilitate the movement of Na$^+$ into a neuron. However, limited numbers of Na$^+$ leak channels prevent as much Na$^+$ moving into the neuron as K$^+$ moves out.

The Role of Na⁺/K⁺ Pumps $\quad$ The Na^+/K^+ pumps play a relatively small role in establishing a resting membrane potential. These pumps contribute approximately -3 mV of the total -70 mV difference by moving more Na^+ out of the neuron than K^+ is pumped into the neuron. The Na^+/K^+ pumps have a more significant role in maintaining the concentration gradients of both K^+ and Na^+. These concentration gradients allow for the diffusion of Na^+ and K^+ as part of the neuron's generation of an electrical current, as described later.

WHAT DID YOU LEARN?

19 Describe the conditions of a neuron at rest in terms of the RMP, concentration gradients for Na^+, K^+, Cl^- along the entire neuron and Ca^{2+} at the synaptic knob, and the state of the gated channels.

20 Explain how an RMP is established and maintained in neurons.

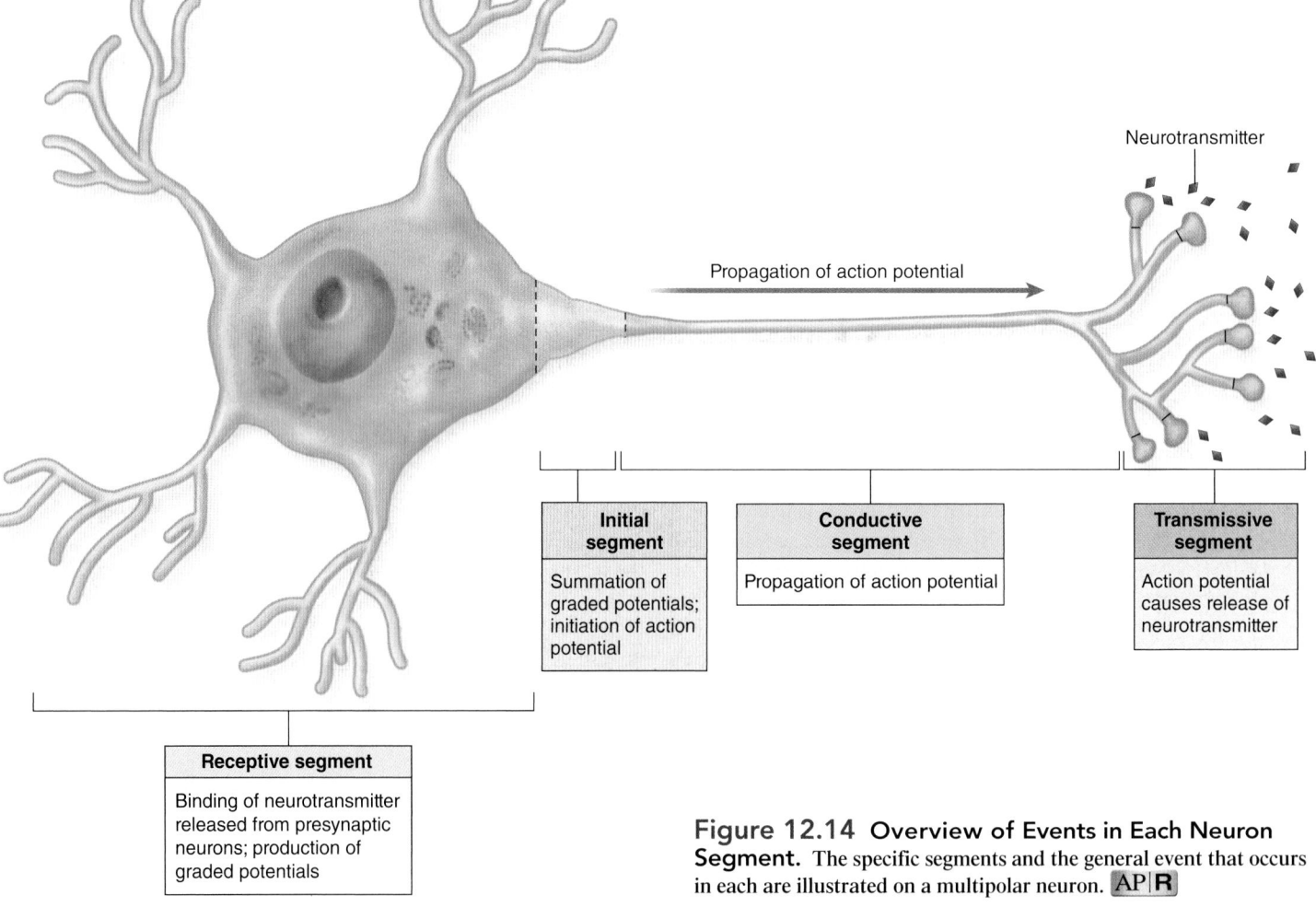

Neurotransmitter

Propagation of action potential

Initial segment	**Conductive segment**	**Transmissive segment**
Summation of graded potentials; initiation of action potential	Propagation of action potential	Action potential causes release of neurotransmitter

Receptive segment

Binding of neurotransmitter released from presynaptic neurons; production of graded potentials

Figure 12.14 Overview of Events in Each Neuron Segment. The specific segments and the general event that occurs in each are illustrated on a multipolar neuron. AP|R

12.8 Physiologic Events in the Neuron Segments

Here we discuss the physiologic events that occur in the functional neuron segments from the time of initial stimulation at the dendrites and cell body until release of a neurotransmitter from a synaptic knob. An overview of the events that occur in the receptive, initial, conductive, and transmissive segments is provided in **figure 12.14**.

12.8a Receptive Segment

LEARNING OBJECTIVES

1. Describe a postsynaptic potential.
2. Compare and contrast the action of neurotransmitters in developing both excitatory and inhibitory postsynaptic potentials (graded potentials) in the receptive segment.
3. Graph and explain an excitatory postsynaptic potential (EPSP) and an inhibitory postsynaptic potential (IPSP).

We begin our discussion with **figure 12.15**, which shows the arrangement of several presynaptic neurons in close proximity to one postsynaptic neuron. Between each presynaptic and postsynaptic neuron is a synaptic cleft that is the fluid-filled space between them. Each presynaptic neuron releases neurotransmitter that binds with receptors (chemically gated channels) in the receptive segment—dendrites and cell body—of a postsynaptic neuron.

The establishment of graded potentials is the significant event that occurs in the receptive segment of a neuron. **Graded potentials** are relatively small, short-lived changes in the resting membrane potential that are caused by the movement of small amounts of ion across the plasma membrane.

Graded potentials are established in the receptive segment by the opening of *chemically gated channels*. Recall that there are three different types of chemically gated channels in the receptive segment of a neuron: chemically gated cation channels, chemically gated K^+ channels, and chemically gated Cl^- channels. Neurotransmitter released from a presynaptic neuron binds with a specific type of chemically gated channel triggering it to open, which temporarily allows passage of a small amount of a specific type of ion (or ions) across the plasma membrane. The ions then move along the plasma membrane in a local current.

The local currents associated with graded potentials are *short-lived* (1 millisecond to a few milliseconds) because the flow or current of ions along the plasma membrane experiences resistance. Consider that when cation channels open, and Na^+ enters a cell and moves along the inside of the plasma membrane, Na^+ experiences resistance by the contents of the cytosol. Consequently, the local current of Na^+ becomes weaker and eventually ceases. Thus, a graded potential lasts only as long as the channels are open and until the local ion current ceases.

All graded potentials cause relatively small changes (less than 1 mV) in the resting membrane potential. However, graded potentials vary in both the degree of change and the direction of change. The *degree of change* is dependent upon the magnitude of the stimulus. A larger stimulus opens more chemically gated channels, and more ions flow across the plasma membrane than occurs during a weaker stimulus. Thus, the larger the stimulus, the more channels that open, the greater the flow of ions and the larger the current. (The term "graded" in graded potentials reflects this difference in magnitude.) The *direction of change* (i.e., whether the membrane potential becomes more positive or more negative) is dependent upon the type of chemically gated channel that opens. For example, the opening of chemically gated cation channels allows more Na^+ (positively charged ion) to enter the neuron (than K^+ to exit), causing the inside of the neuron to become more positive (e.g., −70 mV to −69 mV). This change in the membrane potential in the positive direction is called **depolarization** (dē-pō′lăr-i-zā′shŭn; *de* = away). In contrast, the change in the membrane potential in the negative direction (e.g., −70 mV to −71 mV), which is caused by the opening of either chemically gated K^+ channels to allow K^+ to exit the neuron or chemically gated Cl^- channels for Cl^- to enter the neuron, is called **hyperpolarization** (hī′pĕr-pō′lăr-i-zā′shŭn; *hyper* = above).

Graded potentials that occur in postsynaptic neurons are specifically called **postsynaptic potentials.** Postsynaptic potentials that result in the neuron becoming more positive are more specifically called **excitatory postsynaptic potentials (EPSPs),** whereas those that result in the neuron becoming more negative are called **inhibitory postsynaptic potentials (IPSPs).** Note that numerous postsynaptic potentials—both excitatory and inhibitory—are typically generated because a postsynaptic neuron can bind many neurotransmitter molecules simultaneously.

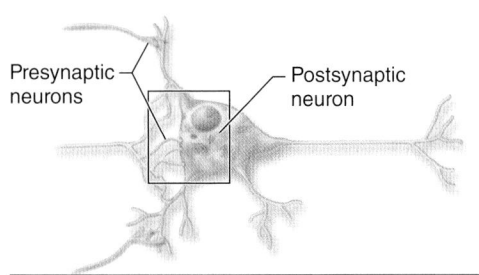

Figure 12.15 Postsynaptic Potentials in the Receptive Segment.

Neurotransmitter released from presynaptic neurons crosses the synaptic cleft and initiates a graded (postsynaptic) potential. Binding of a neurotransmitter causes either (*a*) an excitatory postsynaptic potential (EPSP) or (*b*) an inhibitory postsynaptic potential (IPSP), depending upon the neurotransmitter released and the specific type of receptor to which it binds.

Generation of EPSP

1 Neurotransmitter released from presynaptic neurons binds to postsynaptic neuron receptors, which are chemically gated cation channels, causing them to open.

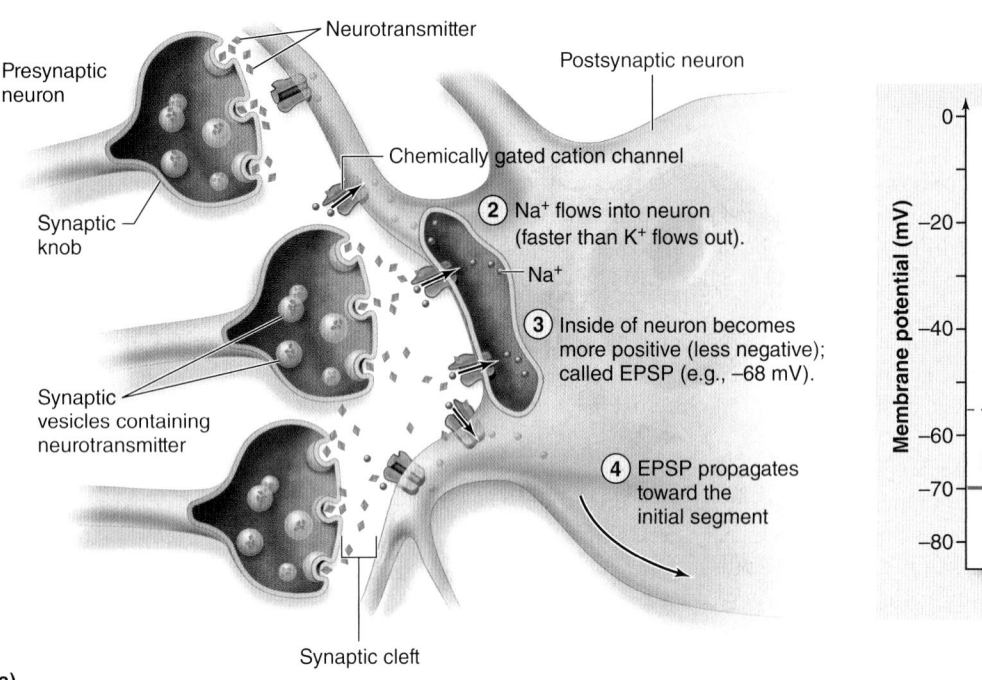
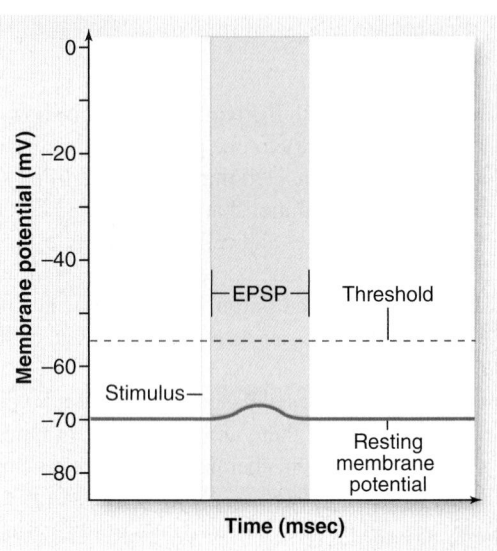

2 Na⁺ flows into neuron (faster than K⁺ flows out).

3 Inside of neuron becomes more positive (less negative); called EPSP (e.g., −68 mV).

4 EPSP propagates toward the initial segment

(a)

Generation of IPSP

1 Neurotransmitter released from presynaptic neurons binds to postsynaptic neuron receptors, which are either chemically gated K⁺ channels or chemically gated Cl⁻ channels, causing them to open.

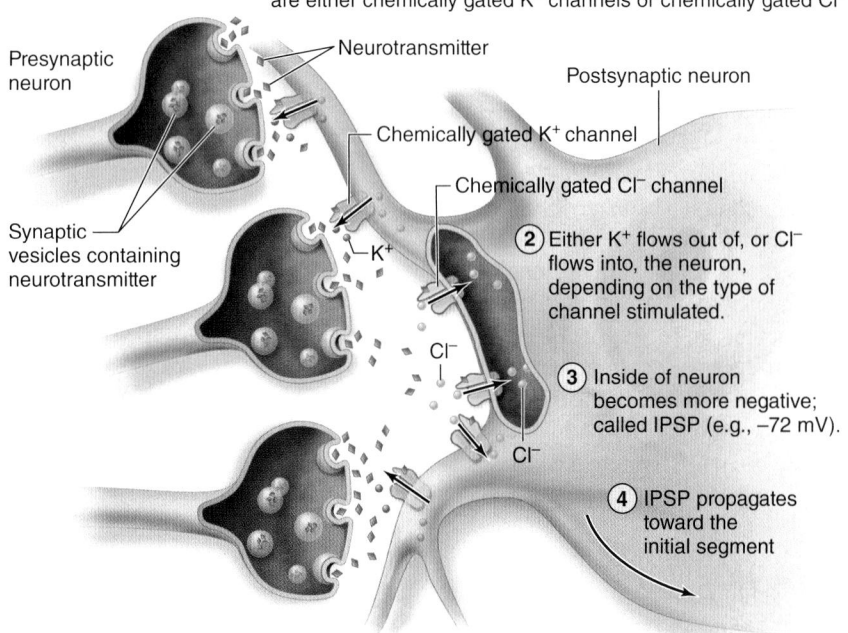
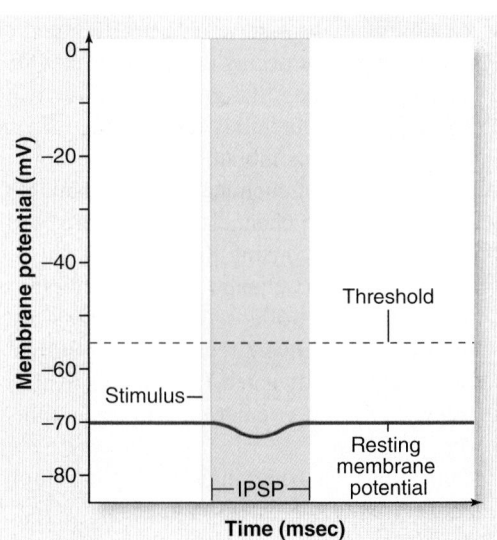

2 Either K⁺ flows out of, or Cl⁻ flows into, the neuron, depending on the type of channel stimulated.

3 Inside of neuron becomes more negative; called IPSP (e.g., −72 mV).

4 IPSP propagates toward the initial segment

(b)

Generation of an EPSP

The release of neurotransmitter from a presynaptic neuron may result in an EPSP (figure 12.15a):

1. The neurotransmitter crosses the synaptic cleft and binds specifically to a receptor that is a chemically gated cation channel, causing it to open.

2. These channels allow both Na^+ and K^+ to move down their respective concentration gradients. However, more Na^+ moves down its concentration gradient into the neuron than K^+ moves out. The amount of neurotransmitter determines the number of chemically gated cation channels that open.

3. Consequently, the inside of the neuron becomes slightly more positive, or less negative, by gaining these positively charged ions. This temporary, less negative state is called an excitatory postsynaptic potential (EPSP). (See graph in figure 12.15a.)

4. The local current of Na^+ becomes weaker as it moves along the neuron plasma membrane toward the initial segment and decreases in intensity with the distance traveled.

Generation of an IPSP

The release of neurotransmitter from a presynaptic neuron may instead result in an IPSP (figure 12.15b):

1. The neurotransmitter crosses the synaptic cleft and binds either to a chemically gated K^+ channel or a chemically gated Cl^- channel, depending upon the neurotransmitter and channels present.

2. If the neurotransmitter binds to a receptor that is a chemically gated K^+ channel, this channel opens and K^+ moves out of the neuron down its concentration gradient, causing a loss of positively charged ions. In contrast, if the neurotransmitter binds to a receptor that is a chemically gated Cl^- channel, opening of this channel allows Cl^- to flow down its concentration gradient to move into the neuron, causing the gain of negatively charged ions. The amount of neurotransmitter determines the number of either chemically gated K^+ or chemically gated Cl^- channels that open.

3. Consequently, the inside of the cell becomes slightly *more negative* if either chemically gated K^+ channels open and K^+ exits, or chemically gated Cl^- channels open and Cl^- enters. This temporary, more negative state is called an inhibitory postsynaptic potential (IPSP). (See graph in figure 12.15b.)

4. The local current of ions becomes weaker as it moves along the neuron plasma membrane toward the initial segment and decreases in intensity with the distance traveled.

The degree of change in the RMP is dependent upon the amount of neurotransmitter bound per unit of time. As more neurotransmitter is released by presynaptic neurons, then more channels open in the receptive segment of the postsynaptic neuron, and there is a greater change in the membrane potential.

Figure 12.16 shows both an illustration and a photo of numerous presynaptic neurons with a postsynaptic neuron. Neurotransmitter is typically being released both from multiple presynaptic neurons and quickly released from the same presynaptic neuron over a very short period of time. The result is many EPSPs, many IPSPs, or both, being generated simultaneously (or within a narrow period of time) within the receptive segment. The outcome of these EPSPs and IPSPs is determined in the initial segment.

WHAT DO YOU THINK?

1. Does the generation of IPSPs make it more likely or less likely that a nerve signal will be sent?

WHAT DID YOU LEARN?

21. How are EPSP and IPSP graded potentials established in the receptive segment of a neuron?

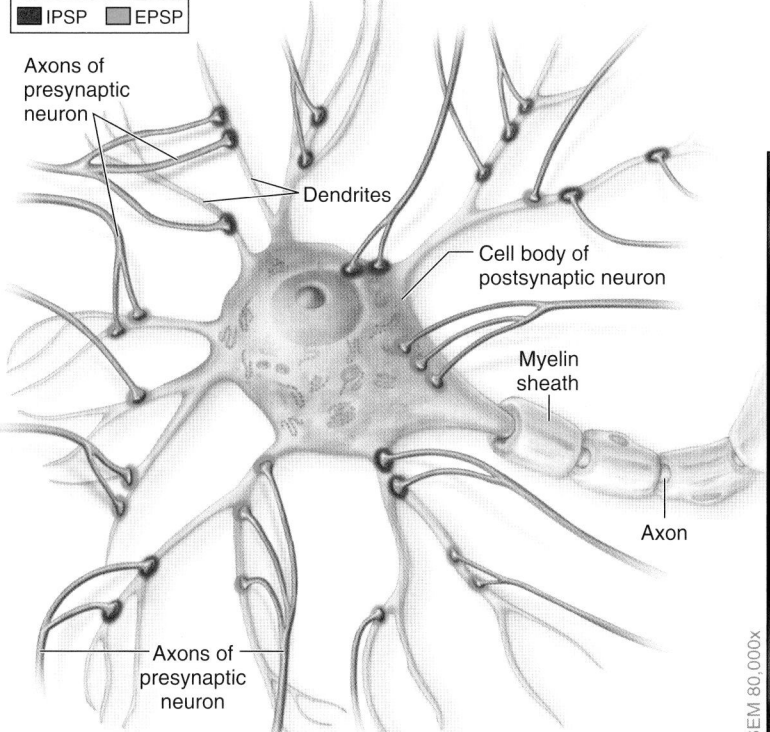

■ IPSP ☐ EPSP

Axons of presynaptic neuron

Dendrites

Cell body of postsynaptic neuron

Myelin sheath

Axon

Axons of presynaptic neuron

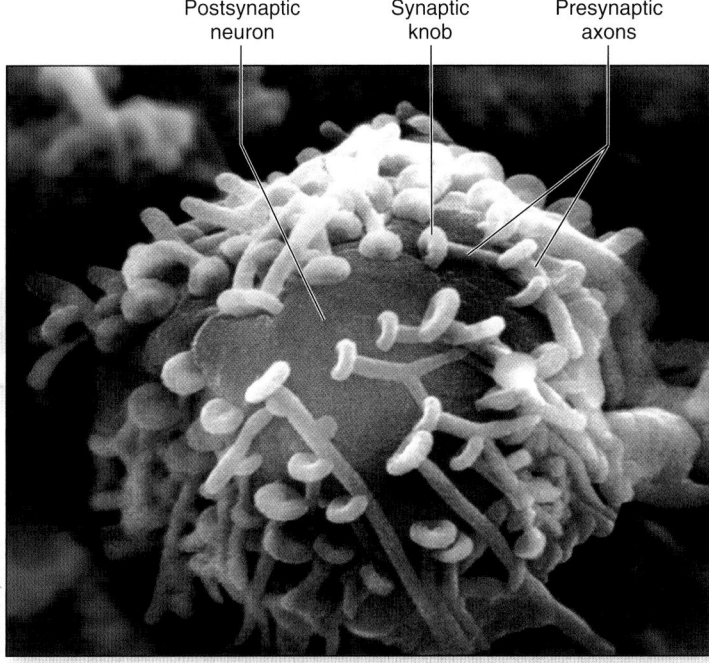

Postsynaptic neuron

Synaptic knob

Presynaptic axons

SEM 80,000x

Figure 12.16 Several Presynaptic Neurons with a Postsynaptic Neuron. A diagram and photograph of numerous axons of presynaptic neurons with the receptive segment of a postsynaptic neuron. This anatomic arrangement permits release of neurotransmitters from many presynaptic neurons and the subsequent, and simultaneous generation of both EPSPs and IPSPs in the postsynaptic neuron. AP|R

12.8b Initial Segment

LEARNING OBJECTIVE

4. Define summation, and describe the two types of summation that can occur in the initial segment.

EPSPs and IPSPs are local currents of ions that move along the plasma membrane toward the initial segment. The changes in the membrane potential associated with these graded postsynaptic potentials are "added" in the initial segment; this process is called **summation.** The sensitivity of voltage-gated channels to open in response to a minimum voltage change in membrane potential is the determining factor if an action potential is initiated. The minimum voltage change is called the **threshold membrane potential.** On average the value for the threshold membrane potential is about –55 mV (although the specific value does vary somewhat between different types of neurons). This is a change of +15 mV from the RMP. When this threshold is reached, the voltage-gated channels are stimulated to open, which will initiate the generation of an action potential that will be propagated along the axon.

A single EPSP is incapable of causing the postsynaptic neuron to reach threshold. Additionally, IPSPs negate the effect of EPSPs. The occurrence of both EPSPs and IPSPs together results in a "tug-of-war" as to whether the threshold is reached. Thus, numerous EPSPs must be generated in the receptive segment and arrive at the initial segment simultaneously, or nearly at the same time, if the threshold is to be reached.

Threshold can be reached through two types of summation called spatial summation and temporal summation, which may act in concert to produce an effect:

- **Spatial summation** occurs when *multiple* presynaptic neurons release neurotransmitter at various locations onto the receptive segment, thus generating EPSPs, IPSPs, or both in the postsynaptic neuron (**figure 12.17a**).

- **Temporal summation** occurs when a *single* presynaptic neuron repeatedly releases neurotransmitter to produce multiple EPSPs in the postsynaptic neuron at the same location repeatedly within a very short period of time (figure 12.17b).

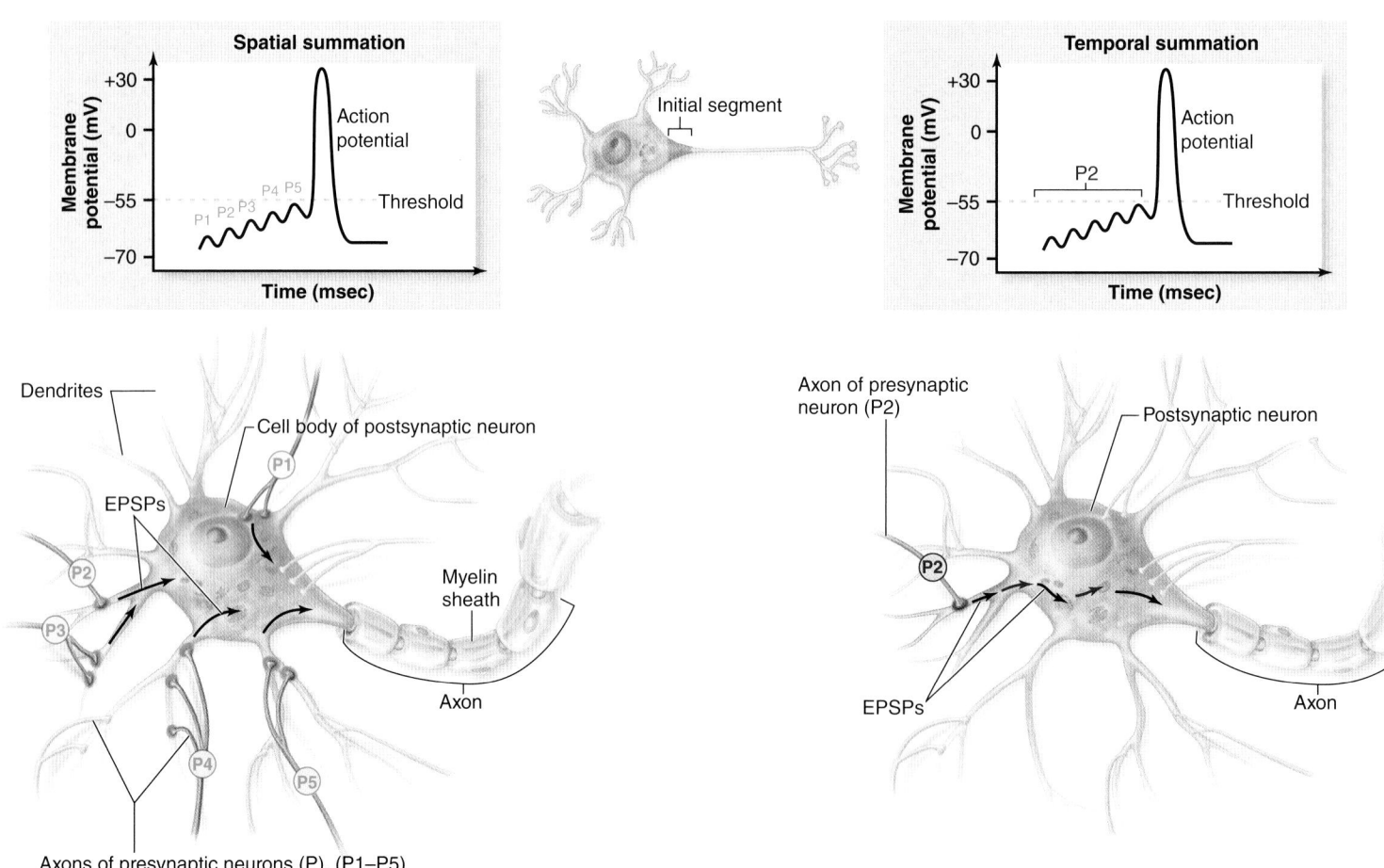

(a) Spatial summation

(b) Temporal summation

Figure 12.17 Initial Segment: Summation. Summation is the integration of the effects of postsynaptic potentials arriving in the initial segment. (*a*) Spatial summation involves numerous presynaptic neurons (P1–P5) collectively stimulating the postsynaptic neuron within a given time period; whereas (*b*) temporal summation involves a single presynaptic neuron (P2) rapidly and intensely stimulating the postsynaptic neuron in a given time period. Note that negating effects of IPSPs are not shown.

Typically, both spatial summation and temporal summation are occurring simultaneously. When the graded potentials arrive at the initial segment within a small period of time, they can either contribute (if EPSPs) or interfere with (if IPSPs) to the threshold value being reached. If the threshold value is reached, an action potential is initiated. (Graded potentials typically last for approximately 15 milliseconds.) Any change in voltage below the threshold value is not sufficient to open voltage-gated channels and is called a **subthreshold value;** these channels remain closed and no action potential is initiated. (See the Learning Strategy on this page to help you with this concept.)

The **all or none law** (see section 10.6b) applies to action potentials propagated along plasma membrane of neurons. If the threshold is reached, an action potential is initiated and propagated along the axon without decreasing in intensity (all). If only a subthreshold value is reached, it is not initiated (none). Additionally, voltage changes with a value greater than the threshold (e.g., voltage change of +20 mV) result in the same intensity in the action potential that is initiated as occurs when the threshold is reached. The initial segment is sometimes called the *trigger zone* because what occurs there is similar to what happens with a gun. Sufficient pressure is applied to the trigger of a gun and the bullet is fired from the chamber, or insufficient pressure is placed on the trigger and the gun is not fired. Likewise, the bullet travels at the same velocity regardless of whether the pressure placed on the trigger is greater than needed.

 WHAT DID YOU LEARN?

22 What is the significance of the threshold membrane potential in the initial segment of a neuron?

12.8c Conductive Segment

 LEARNING OBJECTIVES

5. Describe and graph an action potential.

6. Explain propagation of an action potential in both unmyelinated and myelinated axons.

7. Define refractory period, and explain the difference between the absolute refractory period and relative refractory period associated with transmitting an action potential.

The conductive segment is equivalent to the total length of the axon. The main activity of the conductive segment is propagation of an action potential along the axolemma (the axon plasma membrane). An **action potential** involves two processes: depolarization and repolarization. We first introduced the term *depolarization* in section 12.8a as the gain of positive ions that results in the membrane potential becoming more positive (e.g., -70 mV to -69 mV). Here, the term **depolarization** more specifically refers to the gain of positive charge within a neuron that occurs to such an extent to change the plasma membrane potential from negative to positive. This reversal of polarity is due to the opening of *voltage-gated* Na^+ channels and the subsequent movement of Na^+ into the cell. **Repolarization** (rē-pō′lăr-i-zā-shŭn) is the return of polarity from positive back to negative (the RMP). Repolarization is due to the opening of *voltage-gated* K^+ channels and the subsequent movement of K^+ out of the cell.

Once initiated, action potentials are propagated along an axon to the synaptic knob as both voltage-gated Na^+ channels and voltage-gated K^+ channels open sequentially along the length of the axolemma. The propagation of an action potential is called a **nerve signal** or *nerve impulse*. The details of both depolarization and its propagation and repolarization and its propagation are described here in detail. Refer to **figure 12.18** as you read through this section.

 **INTEGRATE**

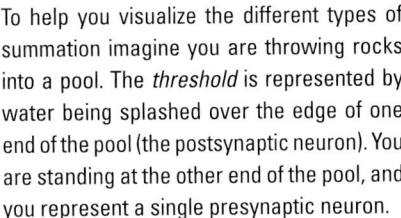 **LEARNING STRATEGY**

To help you visualize the different types of summation imagine you are throwing rocks into a pool. The *threshold* is represented by water being splashed over the edge of one end of the pool (the postsynaptic neuron). You are standing at the other end of the pool, and you represent a single presynaptic neuron.

You throw one rock into the pool creating a small ripple, representing an EPSP. Water does not splash out of the other end of the pool (the threshold is not reached). *Spatial summation* is demonstrated by several individuals throwing rocks into different areas of the pool. The collective "ripples" added together cause water to splash over the end of the pool (threshold is reached).

Likewise, *temporal summation* is illustrated if you throw rocks into the same area of the pool repeatedly and quickly, and the collective ripples are added together to cause water to splash over the end of the pool (threshold is reached).

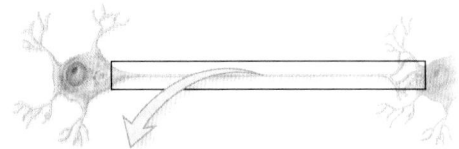

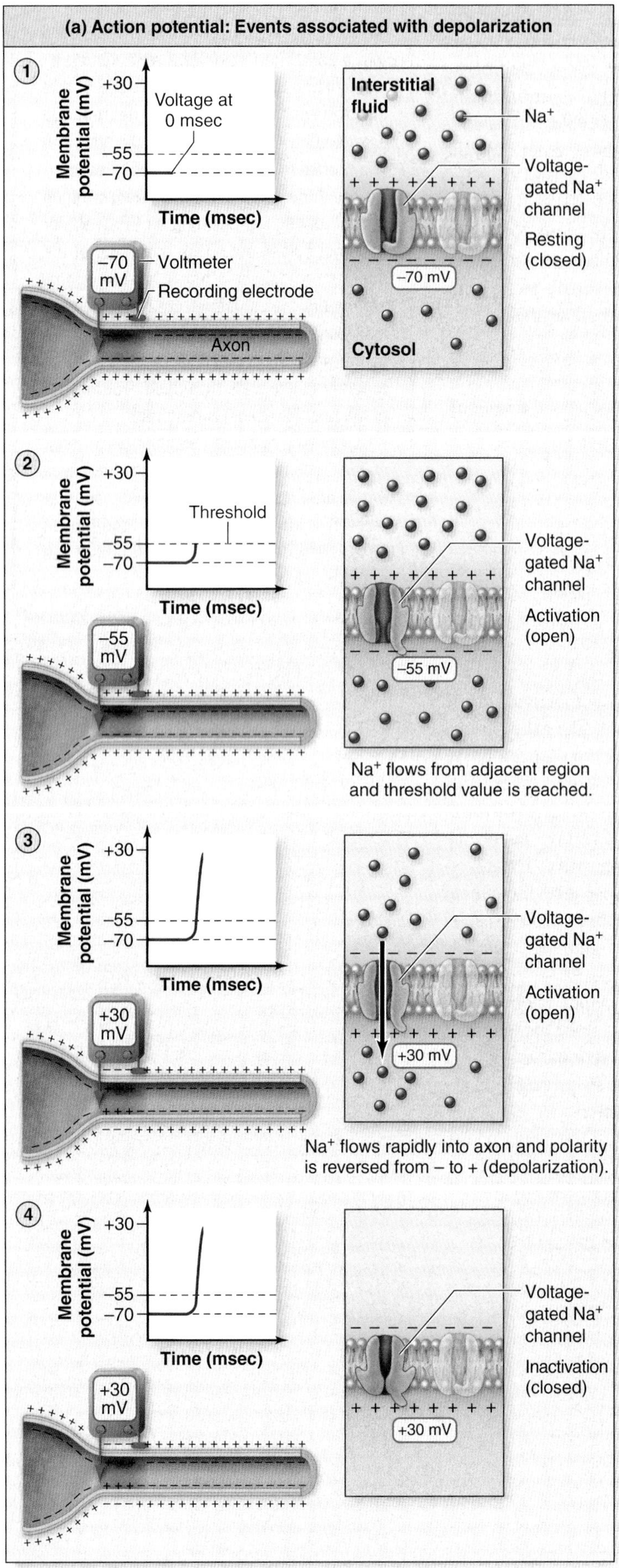

(a) Action potential: Events associated with depolarization

①

Membrane potential (mV) / Time (msec)
+30, −55, −70

Voltage at 0 msec

−70 mV Voltmeter
Recording electrode
Axon

Interstitial fluid
Na⁺
Voltage-gated Na⁺ channel
Resting (closed)
−70 mV
Cytosol

②

Membrane potential (mV) / Time (msec)
+30, −55, −70

Threshold

−55 mV

Voltage-gated Na⁺ channel
Activation (open)
−55 mV

Na⁺ flows from adjacent region and threshold value is reached.

③

Membrane potential (mV) / Time (msec)
+30, −55, −70

+30 mV

Voltage-gated Na⁺ channel
Activation (open)
+30 mV

Na⁺ flows rapidly into axon and polarity is reversed from − to + (depolarization).

④

Membrane potential (mV) / Time (msec)
+30, −55, −70

+30 mV

Voltage-gated Na⁺ channel
Inactivation (closed)
+30 mV

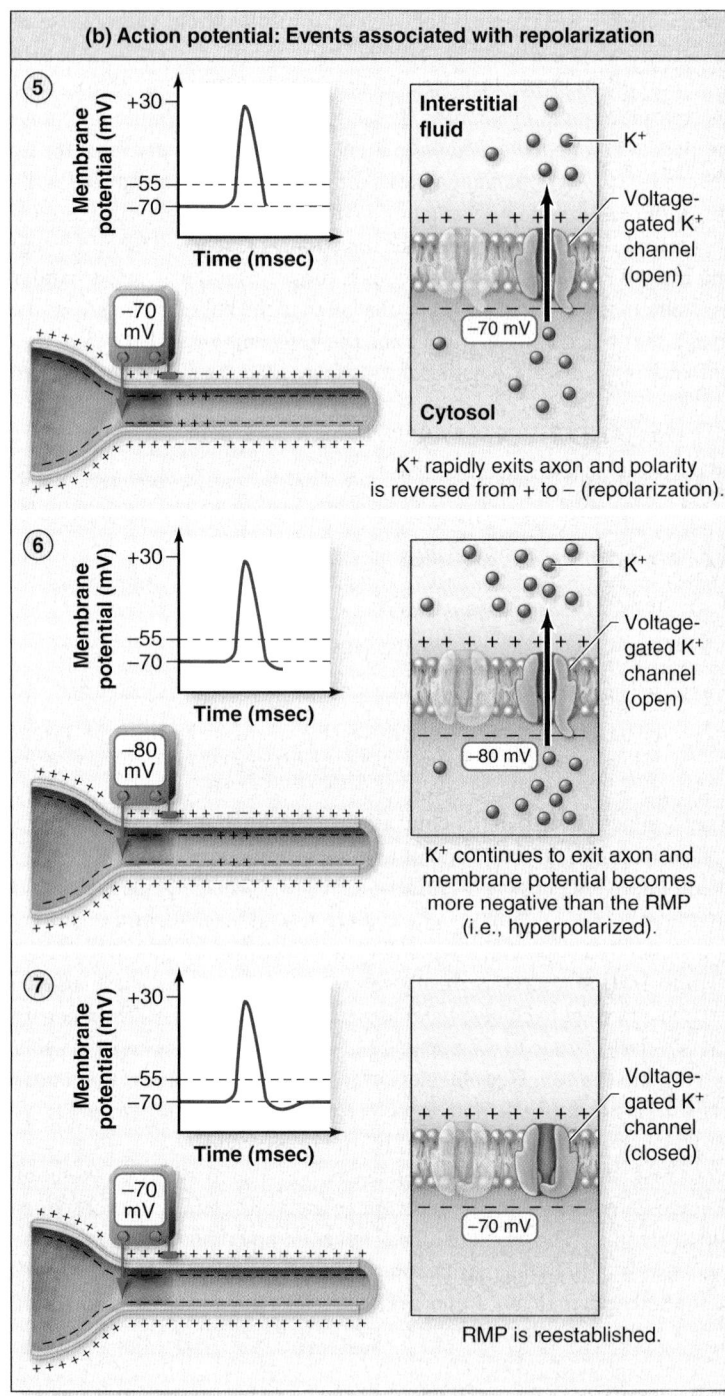

(b) Action potential: Events associated with repolarization

⑤

Membrane potential (mV) / Time (msec)
+30, −55, −70

−70 mV

Interstitial fluid
K⁺
Voltage-gated K⁺ channel (open)
−70 mV
Cytosol

K⁺ rapidly exits axon and polarity is reversed from + to − (repolarization).

⑥

Membrane potential (mV) / Time (msec)
+30, −55, −70

−80 mV

K⁺
Voltage-gated K⁺ channel (open)
−80 mV

K⁺ continues to exit axon and membrane potential becomes more negative than the RMP (i.e., hyperpolarized).

⑦

Membrane potential (mV) / Time (msec)
+30, −55, −70

−70 mV

Voltage-gated K⁺ channel (closed)
−70 mV

RMP is reestablished.

Figure 12.18 Generation of an Action Potential. An action potential involves (*a*) depolarization, which occurs as Na⁺ moves into the axon through open voltage-gated Na⁺ channels to change the membrane potential from −70 mV to +30 mV, and (*b*) repolarization, which occurs as K⁺ moves out of the axon through open voltage-gated K⁺ channels to reverse the polarity from +30 mV to −70 mV. (Refer to figure 12.11 to review states of voltage-gated Na⁺ channels.)

Depolarization and Its Propagation

The following events are associated with depolarization and its propagation (figure 12.18a):

1. Initially, the voltage-gated Na⁺ channels are closed and the membrane potential is −70 mV (the resting membrane potential).

2. Sodium ions flow into the region from adjacent areas. The membrane potential becomes more positive moving away from −70 mV. Voltage-gated Na⁺ channels are triggered to open when sufficient Na⁺ flows into the region to change the membrane potential from −70 mV to −55 mV (the threshold value).

3. Voltage-gated Na⁺ channels remain open to allow rapid Na⁺ entry into the axon to cause depolarization. Sufficient Na⁺ enters the axon to reverse the membrane potential from negative (−55 mV) to positive (+30 mV). (Note that the exact value can vary from 0 mV to +50 mV.) This movement of Na⁺ is extremely small, representing a change of approximately 0.01% in the Na⁺ concentration, but is sufficient to cause depolarization at the plasma membrane.

4. The voltage-gated Na⁺ channels are opened for only a very short duration and then close, changing from the activation state to the temporary inactivation state. This temporarily prevents their reopening.

Steps 1–4 are repeated in adjacent regions downstream from the cell body (farther away from the cell body) as Na⁺ flows to adjacent regions of the axon's plasma membrane, causing these regions downstream to also become more positive and to reach threshold value (−55 mV). (Note that only the voltage-gated Na⁺ channels downstream are triggered to open. The voltage-gated Na⁺ channels upstream are temporarily in the inactivation state and are prevented from reopening; thus, the action potential is propagated in only one direction—away from the cell body.)

Adjacent depolarization is repeated rapidly down the plasma membrane of the axon toward the synaptic knob, and **propagation of depolarization** occurs. The propagation of depolarization is analogous to the tipping of a row of standing dominoes. Once the first domino is initiated to fall, each domino falls in sequence until the last domino is reached.

Repolarization and Its Propagation

The following events are associated with repolarization and its propagation (figure 12.18b):

5. The reaching of the threshold value (−55 mV) also triggers voltage-gated K⁺ channels to open. These channels are relatively slow to open and are not completely open until about the point that depolarization has ended. Voltage-gated K⁺ channels remain open to allow rapid K⁺ exit from the axon to cause repolarization. Sufficient K⁺ exits the axon to change the membrane potential from positive (+30 mV) to its negative RMP (−70 mV). Note that repolarization triggers the voltage-gated Na⁺ channels to change from the inactivation state to the resting state. This allows voltage-gated Na⁺ channels to be stimulated to open again to send a new nerve signal.

6. The voltage-gated K⁺ channels typically remain open longer than the time needed to reestablish the resting membrane potential (−70 mV). The inside of the neuron during this brief time is more negative than the RMP, or hyperpolarized (decreasing to approximately −80 mV).

7. Voltage-gated K⁺ channels close and the RMP is then reestablished by leak channels and pumps back to −70 mV.

Steps 5–7 are repeated in adjacent regions downstream from the cell body (as adjacent voltage-gated K⁺ channels open and then close). Consider that propagation of repolarization is similar to the resetting up of a row of dominoes so that the process can happen again. The electrical changes associated with an action potential, sometimes called a *spike potential*, are shown and described in **figure 12.19**.

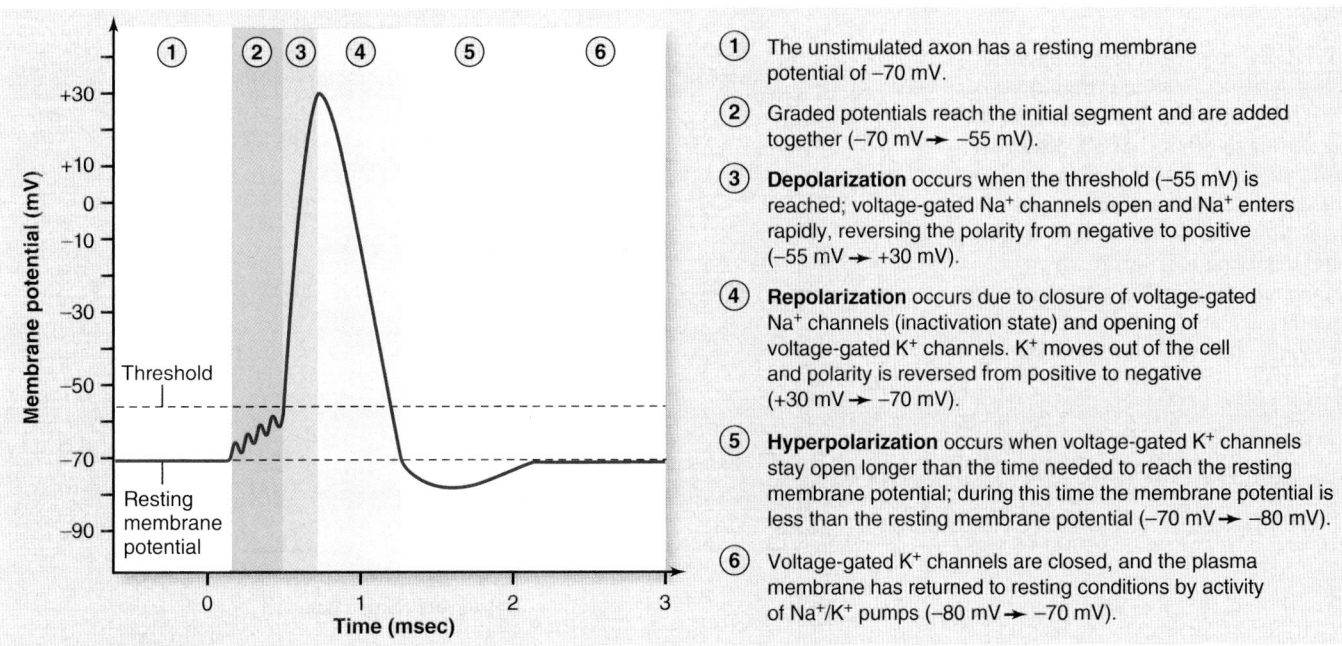

1. The unstimulated axon has a resting membrane potential of −70 mV.

2. Graded potentials reach the initial segment and are added together (−70 mV → −55 mV).

3. **Depolarization** occurs when the threshold (−55 mV) is reached; voltage-gated Na⁺ channels open and Na⁺ enters rapidly, reversing the polarity from negative to positive (−55 mV → +30 mV).

4. **Repolarization** occurs due to closure of voltage-gated Na⁺ channels (inactivation state) and opening of voltage-gated K⁺ channels. K⁺ moves out of the cell and polarity is reversed from positive to negative (+30 mV → −70 mV).

5. **Hyperpolarization** occurs when voltage-gated K⁺ channels stay open longer than the time needed to reach the resting membrane potential; during this time the membrane potential is less than the resting membrane potential (−70 mV → −80 mV).

6. Voltage-gated K⁺ channels are closed, and the plasma membrane has returned to resting conditions by activity of Na⁺/K⁺ pumps (−80 mV → −70 mV).

Figure 12.19 Events of an Action Potential. A tracing of the membrane voltage changes (measured in millivolts [mV]) associated with an action potential initiated in the initial segment. Changes occur in just a few milliseconds, and result from the opening and closing of voltage-gated Na⁺ channels and voltage-gated K⁺ channels in the plasma membrane of the axon, and are sometimes referred to as a *spike potential*, reflecting the shape of the recorded potential.

INTEGRATE

CLINICAL VIEW
Local Anesthetics

Local anesthetics, such as lidocaine, inhibit the action of voltage-gated Na^+ channels, effectively blocking the nerve signal. Pain associated with local medical procedures—for example, administering stitches or filling a tooth cavity—that is normally detected and conducted along an axon of sensory neurons from the body's pain receptors is effectively blocked from reaching the CNS. Even applying ice to a painful area can help reduce pain sensation by slowing the transmission of sensory action potentials initiated by painful stimuli.

Refractory Period

A **refractory period** is the brief time period after an action potential has been initiated during which an axon is either incapable of generating another action potential or a greater than normal amount of stimulation is required to generate another action potential. The excitable plasma membrane recovers at this time and becomes ready to respond to another stimulus. The refractory period has two phases: the absolute refractory period and the relative refractory period (**figure 12.20**).

The **absolute refractory period** is the time (about 1 millisecond) after an action potential onset when no amount of stimulus, no matter how strong, can initiate a second action potential. During this time, the voltage-gated Na^+ channels are first opened, and then the voltage-gated Na^+ channels are closed in the inactivation state. They generally remain in the inactivation state until the membrane potential has almost returned to the resting membrane potential through repolarization. Consequently, no voltage difference across the plasma membrane of an axon during this time can open the voltage-gated Na^+ channels for the next action potential. The absolute refractory period ensures that the action potential moves along the axon in only one direction toward the synaptic knobs. (See steps 2 to 4 in figure 12.20.)

The **relative refractory period** occurs immediately after the absolute refractory period. Another action potential may now be initiated in an axon only if the stimulation of the plasma membrane is greater than the stimulus normally needed to generate an action potential. At this time, voltage-gated Na^+ channels have returned to their resting state, but the neuron is hyperpolarized due to the slightly extended time that voltage-gated K^+ channels remain open during repolarization. (See step 5 in figure 12.20.)

Continuous Conduction and Saltatory Conduction

Specifically how an action potential is propagated along the axon is dependent upon whether the axon is unmyelinated or myelinated. **Continuous conduction** occurs in unmyelinated axons and involves the sequential opening of voltage-gated Na^+ channels and voltage-gated K^+ channels located within the axon plasma membrane along the entire length of the axon. When previously discussing the process of action potential conduction, we described it as it would occur in an unmyelinated axon (**figure 12.21a**).

Saltatory (sal'tă-tōrē; *saltare* = to jump) **conduction** occurs in myelinated axons (figure 12.21b). Here, action potentials do not occur in regions of the axon that are myelinated—rather they are propagated only at neurofibril nodes. This is due to anatomic differences in the two types of regions of a myelinated axon. Myelinated regions of an axon contain limited numbers of voltage-gated Na^+ channels and voltage-gated K^+ channels, and myelin is a great insulator that prevents ion movement even if additional channels are present. In contrast, neurofibril

Figure 12.20 Refractory Periods.
The absolute refractory period is from the point of depolarization until repolarization is almost complete. During this time, it is not possible to initiate another action potential, no matter how strong the stimulus. The relative refractory period is the time immediately following the absolute refractory period, during which a stronger than normal stimulus is required to initiate another action potential because the neuron is hyperpolarized (below the RMP of −70 mV).

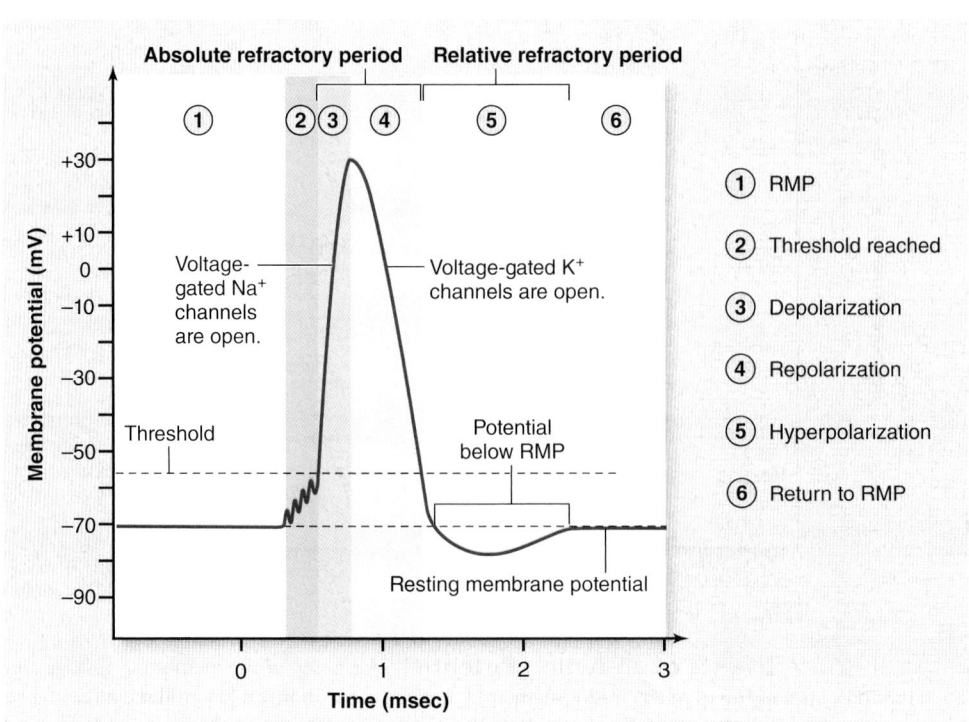

(a) Continuous conduction	(b) Saltatory conduction

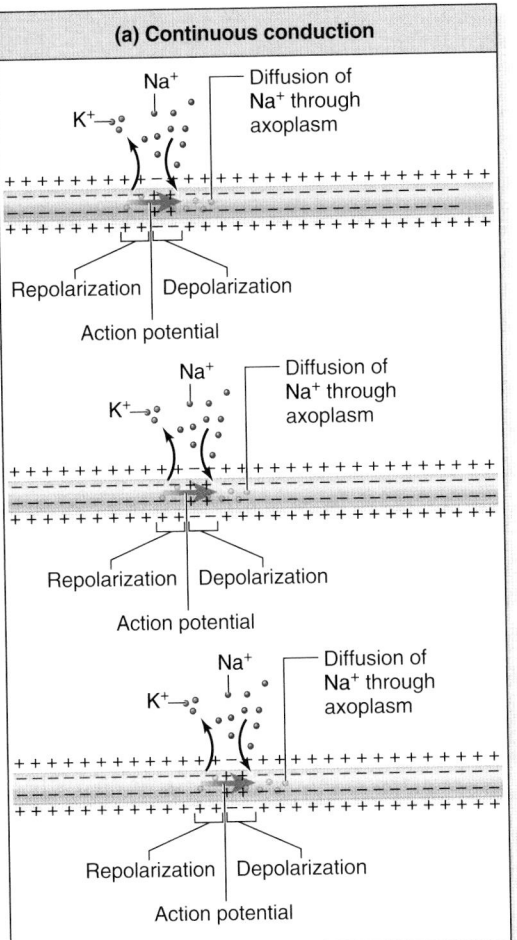

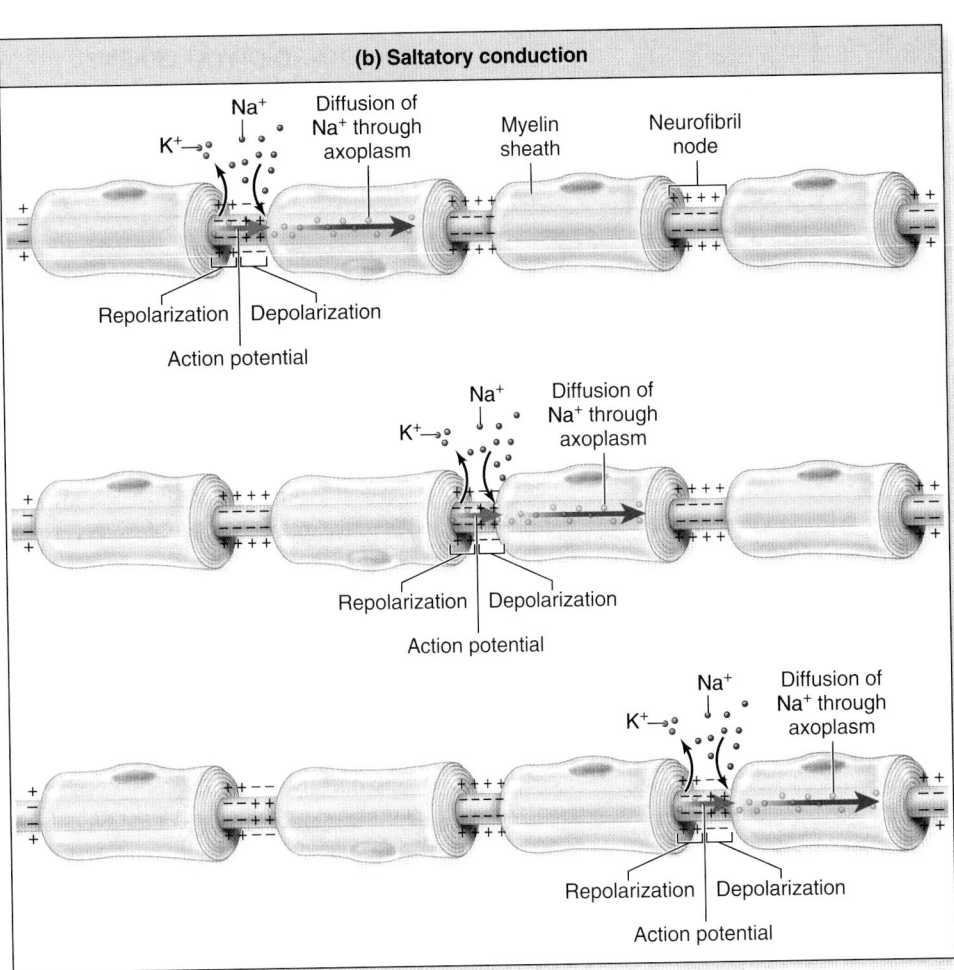

Figure 12.21 Propagation of an Action Potential. Propagation of an action potential occurs from the initial segment to the synaptic knob. (*a*) In unmyelinated axons, voltage-gated channels open sequentially along the entire length of an axon, and the process is called continuous conduction. (*b*) In myelinated axons, voltage-gated channels open only at the neurofibril nodes, and the process is referred to as saltatory conduction. **AP|R**

nodes have a relatively large number of both voltage-gated Na⁺ channels and voltage-gated K⁺ channels and lack myelin insulation. Ions are relatively free to flow into and out of the axon in these regions when the channels are open. A nerve signal is transmitted along a myelinated axon, as follows:

- **Neurofibril node.** An action potential occurs only at neurofibril nodes. Depolarization is due to opening of voltage-gated Na⁺ channels; Na⁺ diffuses into the axon. (This is followed by repolarization as voltage-gated K⁺ channels open, and K⁺ diffuses out.)

- **Myelinated regions.** The Na⁺ diffuses through axoplasm of the axon internal to the axolemma (which is insulated by myelin). Two critical aspects should be noted regarding this Na⁺ diffusion: (1) It is relatively fast, faster than the events at the neurofibril nodes, and (2) as Na⁺ diffuses through the axoplasm it experiences resistance and the local current becomes weaker with distance.

- **Next neurofibril node.** The arrival of the relatively weak Na⁺ current at the next neurofibril node is sufficient to cause the opening of voltage-gated Na⁺ channels located there. This results in the establishing of a new action potential as Na⁺ enters the axon and a new local current is established. This process repeats as the nerve signal continues down the length of an axon until it reaches the synaptic knobs. The transmission of a nerve signal along an axon is called *saltatory conduction* because the action potential occurs only at neurofibril nodes; thus, it seems to "jump" from node to node (a distance of approximately 1 mm). Transmission of a nerve signal in a myelinated axon is much faster (120 meters per second) than in an unmyelinated axon (2 meters per second), because an action potential is generated only at the neurofibril nodes of myelinated axon rather than along the entire length of unmyelinated axon. Saltatory conduction also is more efficient because less energy is required by Na⁺/K⁺ pumps to maintain the RMP.

 **WHAT DID YOU LEARN?**

23 How does depolarization and repolarization occur in the conductive segment of a neuron?

24 Explain propagation of an action potential—in both an unmyelinated axon and in a myelinated axon.

12.8d Transmissive Segment

 **LEARNING OBJECTIVES**

8. Describe events that occur when the propagated action potential reaches the transmissive segment.

9. Explain the general role of Ca^{2+} in neurotransmitter release.

Recall that the transmissive segment is the synaptic knob. The main activity that occurs at the transmissive segment is the release of neurotransmitter from synaptic vesicles (**figure 12.22**). Prior to the arrival of the action potential, Ca^{2+} pumps embedded in the plasma membrane of a synaptic knob (not shown in figure) establish a calcium concentration gradient by pumping Ca^{2+} out to the IF. Consequently, there is more calcium outside of the synaptic knob than inside it.

When the nerve signal reaches the synaptic knob at the end of the axon, the voltage change associated with depolarization (+30 mV) triggers the opening of voltage-gated Ca^{2+} channels. Calcium ions move down their concentration gradient from the IF into the synaptic knob. Calcium ions bind to proteins associated with synaptic vesicles, which triggers a series of events, resulting in fusion of synaptic vesicles with the plasma membrane of synaptic knob. Neurotransmitter is subsequently released into the synaptic cleft by exocytosis (see section 4.3c). Approximately 300 vesicles are released per action potential. The neurotransmitter then diffuses across the cleft between the synaptic knob and the cell to be stimulated. There it binds to specific cellular protein receptors of another neuron or an effector organ (muscle or gland). Note that exocytosis of neurotransmitter is facilitated by numerous proteins (e.g., synaptotagmin, SNARE proteins), and research continues in this area to determine the exact functional role of each of these proteins.

INTEGRATE

CLINICAL VIEW
Neurotoxicity

Neurotoxicity is the damage caused to nervous tissue (neurons or glial cells) by neurotoxins. **Neurotoxins** can be substances produced within the body (e.g., beta amyloid, oxygen free radicals), substances from microbes (e.g., botulinum toxin, tetanus toxin) or synthetic substances (e.g., pesticides, industrial solvents), ethanol, chemicals used in medical treatment for chemotherapy, radiation treatment, and organ transplants.

Neurotoxins cause harm by several mechanisms, including the following:

- Interfering with transmission of action potentials (e.g., tetrodotoxin, found in some fish and amphibians, blocks Na^+ channels; agitoxin from scorpions interferes with K^+ channels; lead causes loss of myelination primarily in somatic motor neurons that control skeletal muscle)

- Altering events that occur at a synapse (e.g., botulinum toxin, which causes botulism, blocks the release of acetylcholine from synaptic vesicles at the neuromuscular junction of skeletal muscle fibers; toxin produced by *Clostridium tetani*, which causes tetanus, blocks the release of inhibitory transmitters from motor neurons that control skeletal muscle)

- Inducing detrimental structural changes to a neuron that can potentially result in the death of the neuron cell (e.g., mercury decreases protein synthesis; Alzheimer disease involves formation of neurofibrillary tangles)

Symptoms of neurotoxicity may include muscle weakness or spasticity, numbness (decreased ability to feel), loss of memory, impaired vision, decreased mental ability, headache, and behavioral problems.

Figure 12.22 Transmissive Segment: Release of Neurotransmitter.
(*a*) Diagram depicting sequential steps involved in release of neurotransmitter from synaptic vesicles by exocytosis at the synaptic knob (transmissive segment). (*b*) An electron micrograph of synaptic vesicles in a synaptic knob.
AP|R

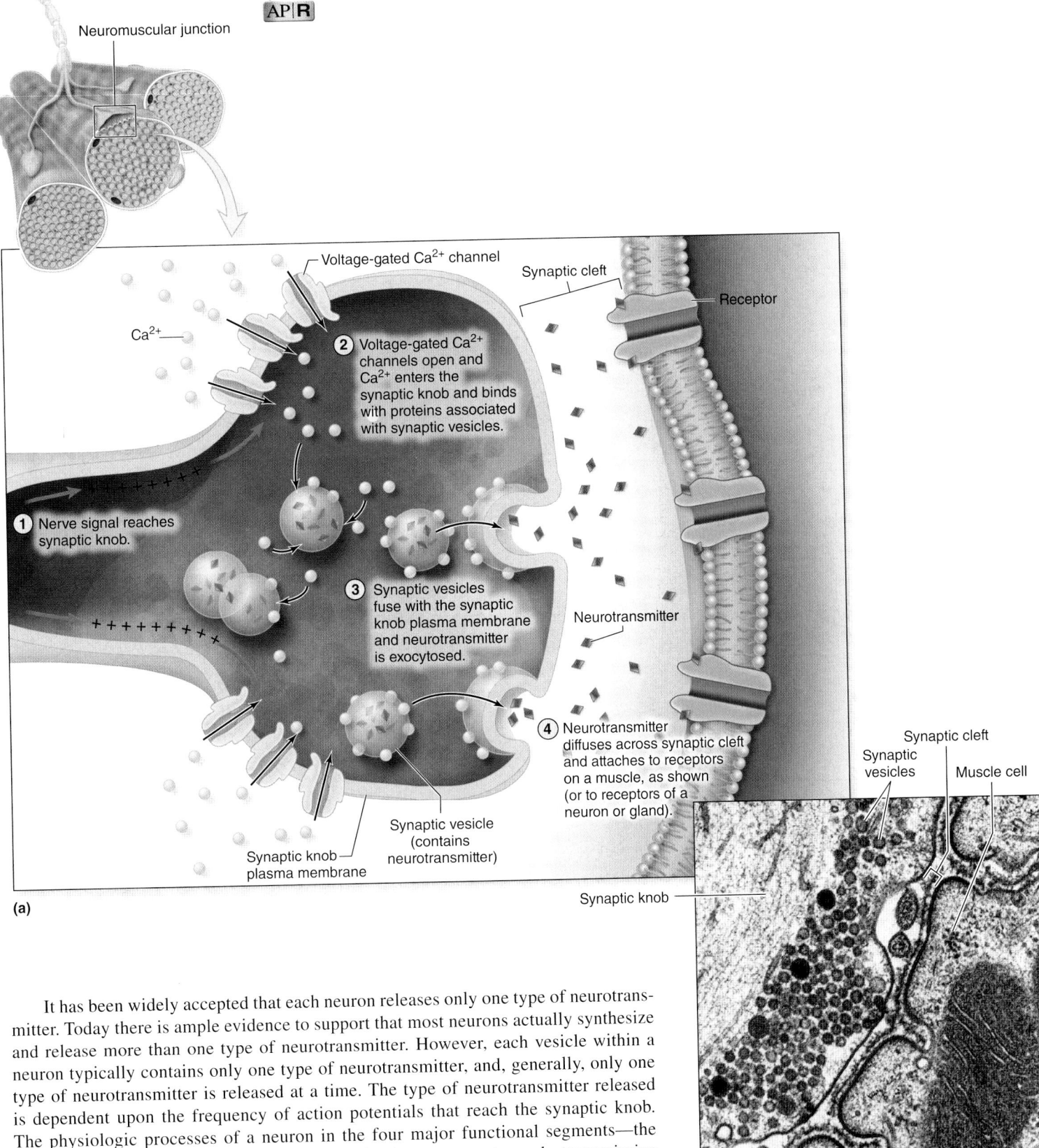

Neuromuscular junction

Voltage-gated Ca^{2+} channel

Synaptic cleft

Receptor

Ca^{2+}

2 Voltage-gated Ca^{2+} channels open and Ca^{2+} enters the synaptic knob and binds with proteins associated with synaptic vesicles.

1 Nerve signal reaches synaptic knob.

3 Synaptic vesicles fuse with the synaptic knob plasma membrane and neurotransmitter is exocytosed.

Neurotransmitter

4 Neurotransmitter diffuses across synaptic cleft and attaches to receptors on a muscle, as shown (or to receptors of a neuron or gland).

Synaptic vesicle (contains neurotransmitter)

Synaptic knob plasma membrane

(a)

Synaptic cleft
Synaptic vesicles
Muscle cell
Synaptic knob

(b)

It has been widely accepted that each neuron releases only one type of neurotransmitter. Today there is ample evidence to support that most neurons actually synthesize and release more than one type of neurotransmitter. However, each vesicle within a neuron typically contains only one type of neurotransmitter, and, generally, only one type of neurotransmitter is released at a time. The type of neurotransmitter released is dependent upon the frequency of action potentials that reach the synaptic knob. The physiologic processes of a neuron in the four major functional segments—the receptive segment, initial segment, conductive segment, and transmissive segment—are summarized in **figure 12.23**.

WHAT DID YOU LEARN?

25 What is the sequence of events from the arrival of an action potential at the synaptic knob until the release of neurotransmitter into the synaptic cleft?

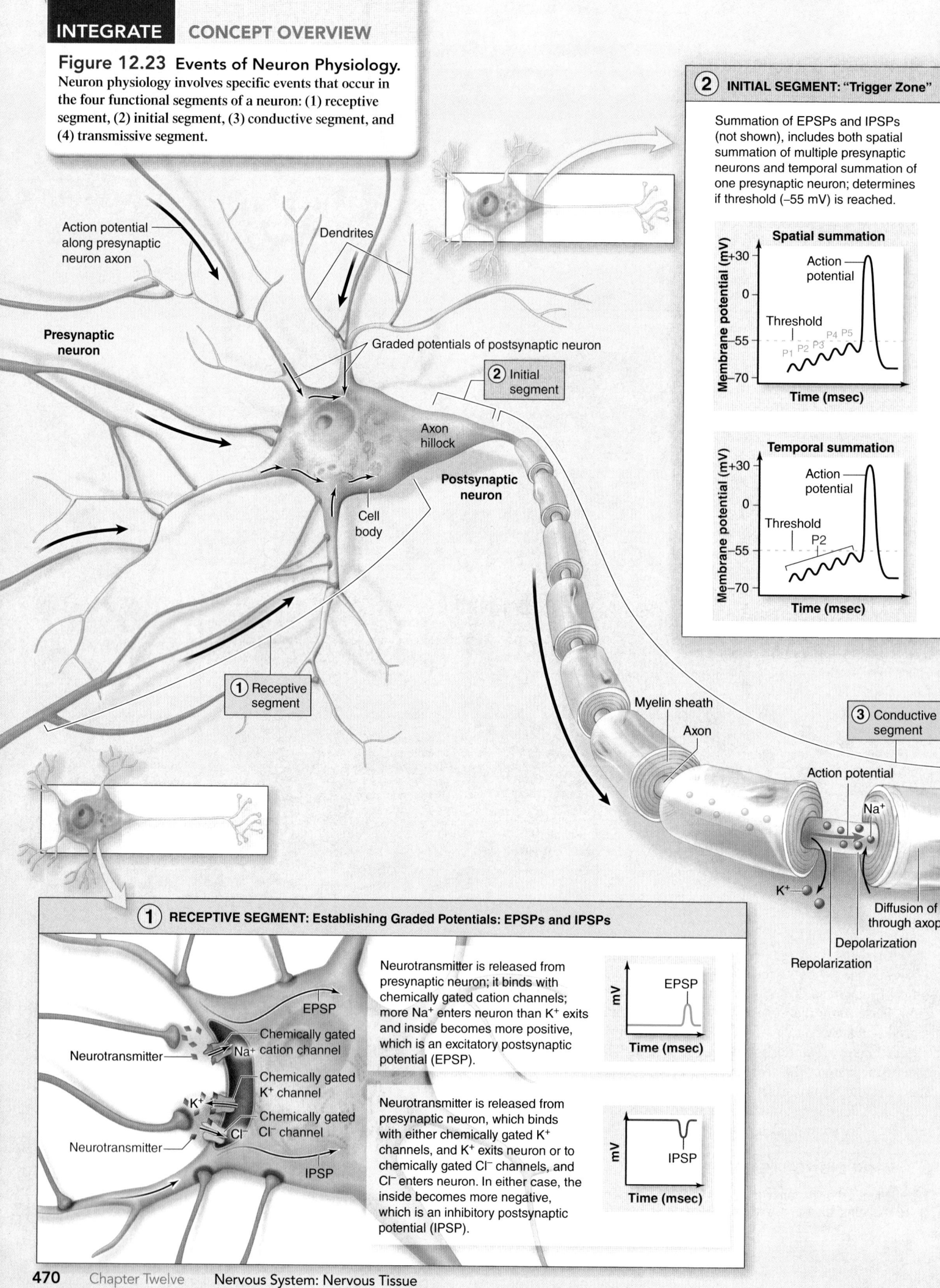

Figure 12.23 Events of Neuron Physiology.
Neuron physiology involves specific events that occur in the four functional segments of a neuron: (1) receptive segment, (2) initial segment, (3) conductive segment, and (4) transmissive segment.

Action potential along presynaptic neuron axon

Dendrites

Presynaptic neuron

Graded potentials of postsynaptic neuron

② Initial segment

Axon hillock

Postsynaptic neuron

Cell body

① Receptive segment

Myelin sheath

Axon

③ Conductive segment

Action potential

Na⁺

K⁺

Diffusion of N through axopl

Depolarization

Repolarization

② INITIAL SEGMENT: "Trigger Zone"

Summation of EPSPs and IPSPs (not shown), includes both spatial summation of multiple presynaptic neurons and temporal summation of one presynaptic neuron; determines if threshold (–55 mV) is reached.

Spatial summation

Membrane potential (mV): +30, 0, –55, –70

Action potential

Threshold

P1 P2 P3 P4 P5

Time (msec)

Temporal summation

Membrane potential (mV): +30, 0, –55, –70

Action potential

Threshold

P2

Time (msec)

① RECEPTIVE SEGMENT: Establishing Graded Potentials: EPSPs and IPSPs

EPSP

Neurotransmitter

Chemically gated Na⁺ cation channel

Chemically gated K⁺ channel

Chemically gated Cl⁻ channel

K⁺

Neurotransmitter

IPSP

Neurotransmitter is released from presynaptic neuron; it binds with chemically gated cation channels; more Na⁺ enters neuron than K⁺ exits and inside becomes more positive, which is an excitatory postsynaptic potential (EPSP).

mV / EPSP / Time (msec)

Neurotransmitter is released from presynaptic neuron, which binds with either chemically gated K⁺ channels, and K⁺ exits neuron or to chemically gated Cl⁻ channels, and Cl⁻ enters neuron. In either case, the inside becomes more negative, which is an inhibitory postsynaptic potential (IPSP).

mV / IPSP / Time (msec)

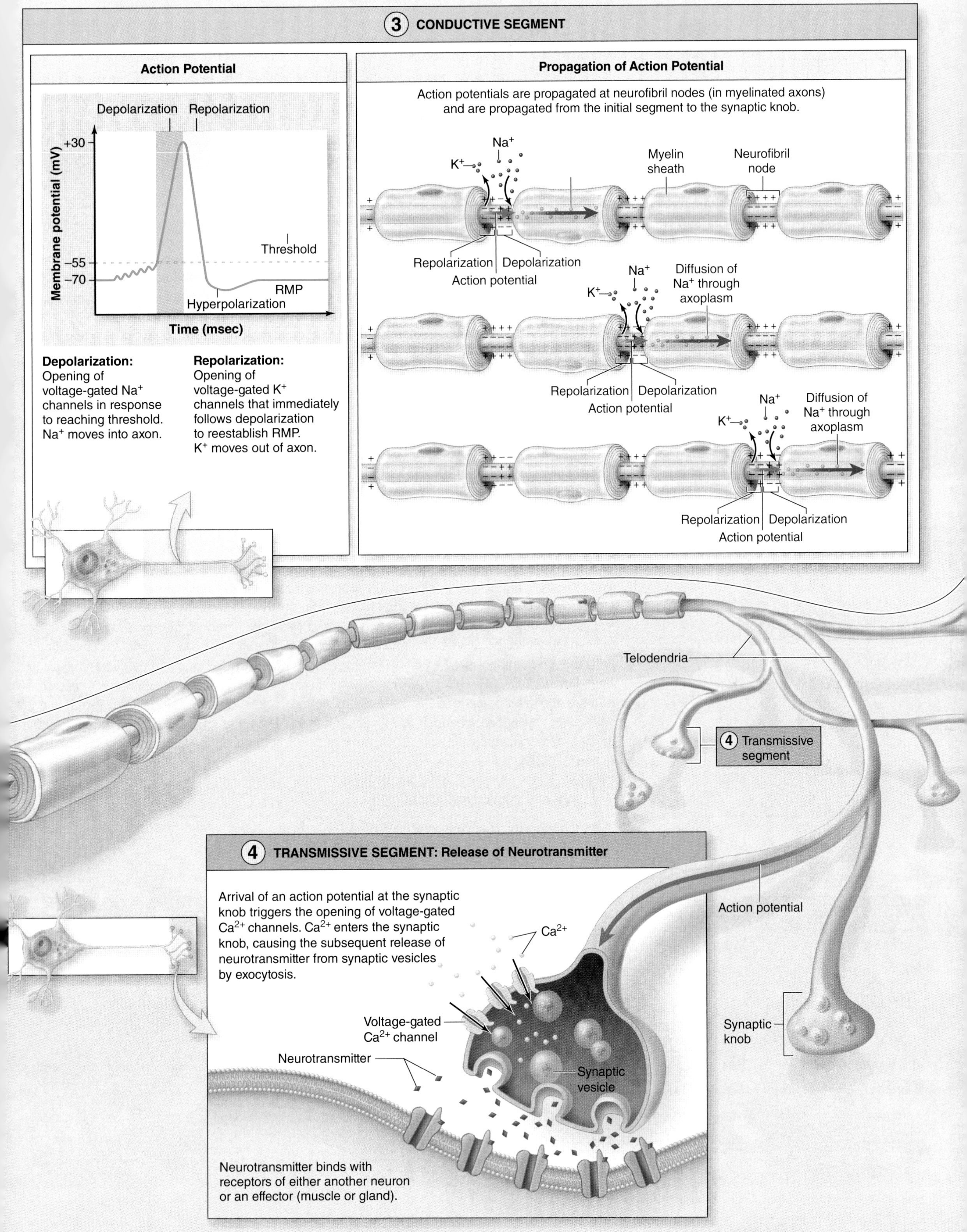

Action Potential

Depolarization:
Opening of voltage-gated Na⁺ channels in response to reaching threshold. Na⁺ moves into axon.

Repolarization:
Opening of voltage-gated K⁺ channels that immediately follows depolarization to reestablish RMP. K⁺ moves out of axon.

Propagation of Action Potential

Action potentials are propagated at neurofibril nodes (in myelinated axons) and are propagated from the initial segment to the synaptic knob.

④ Transmissive segment

Telodendria

Action potential

Synaptic knob

④ TRANSMISSIVE SEGMENT: Release of Neurotransmitter

Arrival of an action potential at the synaptic knob triggers the opening of voltage-gated Ca^{2+} channels. Ca^{2+} enters the synaptic knob, causing the subsequent release of neurotransmitter from synaptic vesicles by exocytosis.

Voltage-gated Ca^{2+} channel

Neurotransmitter

Synaptic vesicle

Neurotransmitter binds with receptors of either another neuron or an effector (muscle or gland).

CONCEPT CONNECTION

Both graded potentials and action potentials also occur in skeletal muscle fibers. Graded potentials occur in the motor end plate of skeletal muscle fibers, and when the threshold is reached are called an end-plate potential. An end-plate potential initiates an action potential along the sarcolemma of a skeletal muscle fiber to trigger the release of calcium from the sarcoplasmic reticulum (see section 10.3b). Calcium ion binds to troponin to initiate the sliding of contractile proteins within the skeletal muscle fiber (see section 10.3c).

12.9 Characteristics of Action Potentials

Here we compare graded potentials and actions potentials (the two types of electrical signals in neurons) and describe several aspects of action potential propagation, including velocity of action potentials and frequency of action potentials.

12.9a Graded Potentials Versus Action Potentials

LEARNING OBJECTIVE

1. Compare graded potentials and action potentials.

Recall from earlier sections that two types of electrical signals are associated with neurons—graded potentials and action potentials. **Graded potentials** (described in section 12.8a) occur in the receptive segment of a neuron (dendrites and cell bodies) and are due to the opening of *chemically gated* channels. The chemically gated channels open temporarily to allow passage of a relatively small amount of a specific type of ion across the plasma membrane. This results in the membrane potential either becoming more positive (depolarization) or more negative (hyperpolarization) than the resting membrane potential. The degree of change is dependent upon the magnitude of the stimulus; thus, it is graded. A larger stimulus opens more chemically gated channels, and more ions flow across the plasma membrane than occurs during a weaker stimulus. The established **local current** of ions loses its intensity (loses charge) as it moves along the plasma membrane; thus, graded potentials are short-lived (1 millisecond to a few milliseconds) and travel relatively short distances.

An **action potential,** in comparison, is generated within the initial segment (see section 12.8b) and propagated along the conductive segment of the neuron (see section 12.8c). An action potential is initiated when *voltage-gated* channels open in response to a minimum voltage change (threshold value). First, voltage-gated Na^+ channels open, allowing Na^+ into a neuron to cause depolarization (reversal of membrane potential from negative to positive). These voltage-gated channels then close and voltage-gated K^+ channels open, allowing K^+ out to cause repolarization (return of membrane potential from positive to negative). An action potential is self-propagating and maintains its intensity (charge difference) as it moves along the axon to the synaptic knob because of the successive opening of voltage-gated channels. Propagation of an action potential is called a nerve signal (or nerve impulse). Action potentials obey the all or none law because any voltage sufficient to open the voltage-gated channels (threshold value) initiates an action potential (all), whereas any voltage below the threshold (subthreshold) is not sufficient to open these channels and an action potential is not sent (none). The characteristics of these two very distinctive electrical events that occur at the plasma membrane are summarized in **table 12.2**.

 WHAT DID YOU LEARN?

 26 Explain how action potentials differ from graded potentials.

Table 12.2	Graded Potential Versus Action Potential	
Characteristics	**Graded Potential**	**Action Potential**
*Neuron segment**	Dendrites and cell body	Axon
Channels	Chemically gated channels	Voltage-gated channels
Direction of voltage change	Positive or negative	Positive then negative
Amount of voltage change	Relatively small change	Relatively large change that causes temporary reversal of polarity
Degree of voltage change	Dependent upon magnitude of the stimulus	Generally does not vary
Duration	1 msec to a few msec	Self-propagating along axon
Distance traveled	Relatively short distance	Length of axon
Change in intensity	Decreases with distance	Same intensity (because voltage-gated channels continue to open in sequence)

*The regions listed reflect the most common location for that type of potential.

12.9b Velocity of Action Potential Propagation

 LEARNING OBJECTIVES

2. Describe the two primary factors that influence the velocity of action potential propagation.
3. Identify the criteria used to distinguish the groups of nerve fibers.

Propagation of an action potential along an axon plasma membrane varies in its velocity and is influenced primarily by two factors—the diameter of an axon and the myelination of an axon:

- **Diameter of the axon.** Nerve signal velocity is generally faster in axons with a larger diameter. This is because there is less resistance to the movement of ions within the larger axon, allowing these axons to reach threshold more rapidly than smaller axons.
- **Myelination of the axon.** Myelination of the axon was described earlier (see section 12.8c) and is the more important factor influencing nerve signal velocity. Propagation of an action potential occurs more rapidly in myelinated axons than in unmyelinated axons.

A nerve fiber is an axon and its myelin sheath. Nerve fibers are classified into three major groups called A, B, and C, based upon their conduction velocity. Group A has a conduction velocity that may be as fast as 150 meters per second; these fibers have both a large diameter and are myelinated. Most somatic sensory neurons that extend from sensory receptors to the CNS, and all somatic motor neurons that extend from the CNS to skeletal muscles are included in this group. Group B conducts at approximately 15 meters per second, and group C conducts at 1 meter per second; group B and group C nerve fibers are generally small in diameter, unmyelinated, or both. Sensory and motor visceral (autonomic) neurons, as well as small somatic sensory neurons that extend from the receptors of the skin to the CNS, are included in groups B and C.

 **WHAT DID YOU LEARN?**

27 What are the general characteristics of group A nerve fibers, and what functions do they normally serve?

12.9c Frequency of Action Potentials

 LEARNING OBJECTIVE

4. Describe how action potentials vary in frequency.

Action potentials are always propagated along an axon at the same amplitude (change in voltage). However, their frequency can vary, and is dependent upon the stimulus strength. As the stimulus strength increases, the frequency of action potentials increases (up to the point of maximum frequency). Consider the following: a brighter light initiates more nerve signals to be relayed from the eye along the optic nerve to the brain, and a loud sound initiates more nerve signals to be relayed along the vestibulocochlear nerve from the inner ear to the brain. The brain then interprets the increased frequency of nerve signals as a more intense stimulus (see section 16.1c). In addition, frequency of action potentials relayed along somatic motor neurons to skeletal muscle increases muscle tension as described in section 10.6c and shown in figure 10.23.

Recall from section 12.8d that varying frequency of action potentials can also influence the type of neurotransmitter released from the synaptic knobs for those neurons that store and release more than one type of neurotransmitter.

 WHAT DID YOU LEARN?

28 Explain how frequency of action potentials differs from the velocity of an action potential propagation.

12.10 Neurotransmitters and Neuromodulation

Neurotransmitters are released at the synaptic cleft and their action is modified by neuromodulation. Here we describe the different means of classifying neurotransmitters based upon chemical structure and function, prior to reviewing critical features of acetylcholine and other specific types of neurotransmitters. We then discuss how the action of neurotransmitters can be altered through the process of neuromodulation.

12.10a Classification of Neurotransmitters

LEARNING OBJECTIVES

1. Identify the four classes of neurotransmitters based upon chemical structure.
2. Describe how neurotransmitters are classified based upon function.

Conventionally, neurotransmitters have been defined as small organic molecules that (1) are synthesized by neurons and stored within vesicles in synaptic knobs; (2) are released from the vesicles when an action potential triggers calcium entry into the synaptic knob; (3) bind to a specific receptor in a target cell (neuron, muscle, or gland); and (4) trigger a physiologic response in the target cell. Today there are estimated to be approximately 100 known neurotransmitters. However, it should be noted that some molecules that are called neurotransmitters (e.g., nitric oxide) do not meet all of these criteria.

Neurotransmitters are classified based upon their chemical structure and function (**figure 12.24**). There are four categories based upon the chemical structure and include

- **Acetylcholine (ACh).** The structure of ACh is significantly different from the other neurotransmitters and for this reason is placed in its own category.
- **Biogenic amines** (also called *monoamines*). They are derived from certain amino acids (see figure 2.25) by the removal of a

carboxyl group (—COOH) and the addition of another functional group (e.g., an hydroxyl group) by enzymatic pathways within the cytoplasm. The functional group added determines whether the molecule belongs to either **catecholamines** (dopamine, norepinephrine, and epinephrine) that are synthesized from the amino acid tyrosine or **indolamines,** which include histamine and serotonin that are synthesized from histidine and tryptophan, respectively.

- **Amino acids.** These include glutamate, aspartate, serine, glycine, and gamma aminobutyric acid (GABA; a modified amino acid). Some controversy remains about how chemical structures that are so plentiful in the cell for protein synthesis can serve the neurotransmitter communication function.
- **Neuropeptides** (*or peptides*). These are chains of amino acids that range in length from 2 to 40 amino acids. Examples of neuropeptides include the natural opiates (e.g., enkephalins, beta-endorphins), and substance P.

Classification based upon function reflects the specific effect that a neurotransmitter has on the membrane potential of a target cell. Neurotransmitters are considered **excitatory** if they induce an EPSP, whereas they are **inhibitory** if they induce an IPSP (see section 12.8a). (Note that some neurotransmitters may be either excitatory or inhibitory depending upon the specific response they cause in their target organs.)

Another classification based upon function reflects whether the target cell response is either **direct** (i.e., the neurotransmitter directly binds to the receptor of the target cell to cause opening of an ion channel) or **indirect** (i.e., the neurotransmitter binds to a receptor that activates the second messenger pathway involving G protein (see G proteins in section 4.5b). The second messenger ultimately can trigger much more diverse effects including the opening of ion channels, the activation of an existing enzymatic pathway, or transcription of genes for the synthesis of new proteins.

WHAT DID YOU LEARN?

29. Describe how neurotransmitters are classified based upon structure and function.

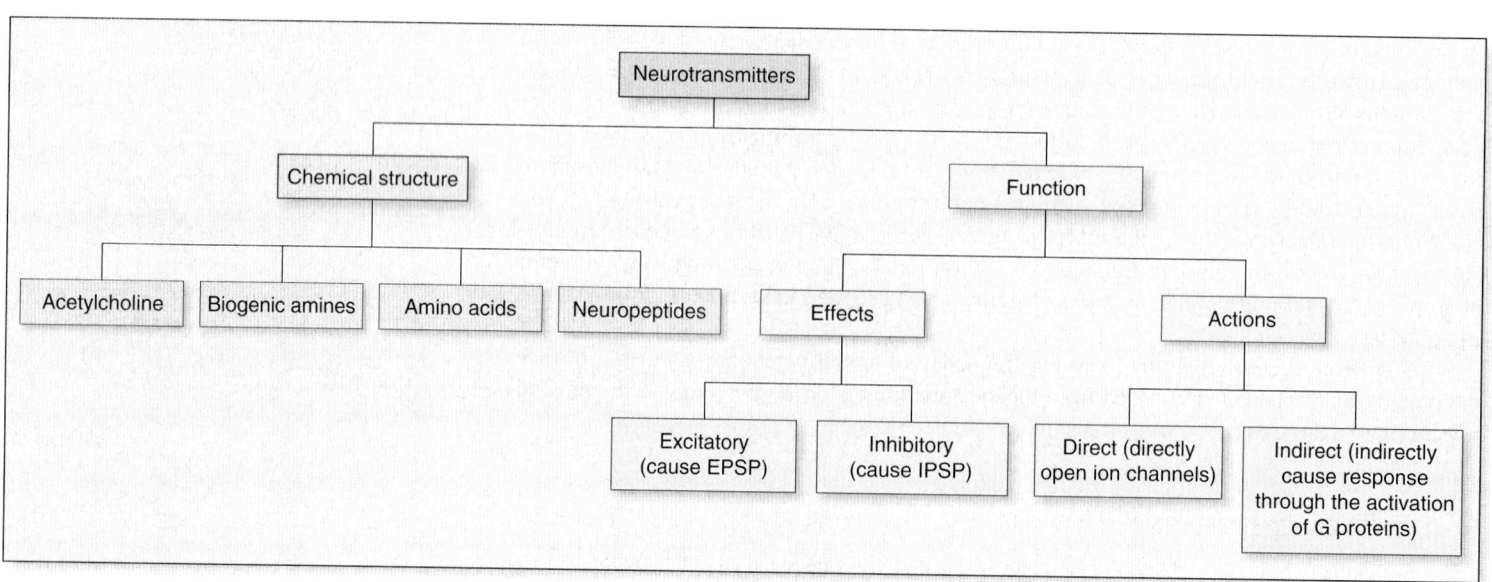

Figure 12.24 Classification of Neurotransmitters. Neurotransmitters are classified based upon both their chemical structure and their function. Three of the chemical structures (except acetylcholine) are amino acids or synthesized from amino acids (biogenic amines and peptides). Neurotransmitters are also organized by function—either by their effects (produce an EPSP or an IPSP) or their action on the target organ (directly opens ion channels or indirectly by acting through G proteins to cause more diverse effects).

12.10b Features of Neurotransmitters

LEARNING OBJECTIVE

3. Describe how acetylcholine functions as a neurotransmitter.
4. Discuss the different mechanisms for removing neurotransmitter from the synaptic cleft.

Here we discuss the synthesis and function of acetylcholine (ACh) in detail because it exhibits all of the classical features of neurotransmitters (as described in the previous section) and is the most understood. Several attributes about acetylcholine are considered, including its (1) synthesis, (2) removal from the synaptic cleft, and (3) interaction with target cells. Please view **figure 12.25** as you read through this section. Specific diseases, drugs, and poisons that alter the normal function of ACh are also included in this figure.

The neurotransmitter acetylcholine is released from neurons located throughout the body. These include the somatic motor neurons at the neuromuscular junction (as described in section 10.3a) and many neurons of the autonomic nervous system (see chapter 15). Acetylcholine also acts as a neuromodulator (described later) within the central nervous system, where it acts to increase attention and arousal.

Synthesis and release (step a). Acetylcholine is synthesized from acetate and choline and then stored within synaptic vesicles within the synaptic knob of a neuron. Thousands of molecules of ACh are released by exocytosis into the synaptic cleft in response to the arrival of an action potential in the presynaptic neuron. The more frequent the action potentials the greater the amount of acetylcholine released.

Removal from synaptic cleft (step b). Some ACh molecules will be immediately digested by **acetylcholinesterase,** an enzyme that resides in the synaptic cleft. Acetylcholine is digested into choline and acetate, and the choline is taken up into the neuron that released it. Some of the ACh molecules cross the synaptic cleft and bind to target cell receptors (and will quickly dissociate usually within 1 millisecond to then be digested by acetylcholinesterase).

WHAT DO YOU THINK?

2. Predict the general effect of a drug that crosses the blood brain barrier and inhibits the action of acetylcholinesterase.

We do want to mention that there are other means of removing different types of neurotransmitter from the synaptic cleft. These include (1) the reuptake of the neurotransmitter into the synaptic knob and subsequent enzymatic digestion (e.g., amino acids and biogenic amines are digested by monoamine oxidase [MAO]); or (2) diffusion from the synaptic cleft and its uptake by surrounding glial cells.

Certain prescription medications were developed based upon their ability to influence the amount of neurotransmitter in a synaptic cleft (e.g., selective serotonin reuptake inhibitors [SSRIs] block the reuptake of serotonin and are used in the treatment of depression). Other antidepressants function by inhibiting MAO enzymes. The result is that neurotransmitter reuptake or breakdown is slowed and the neurotransmitter remains active in the synaptic cleft for a longer period of time. The role of regulating acetylcholine levels in the treatment of Alzheimer disease is discussed in the Clinical View: "Acetylcholinesterase Inhibitor."

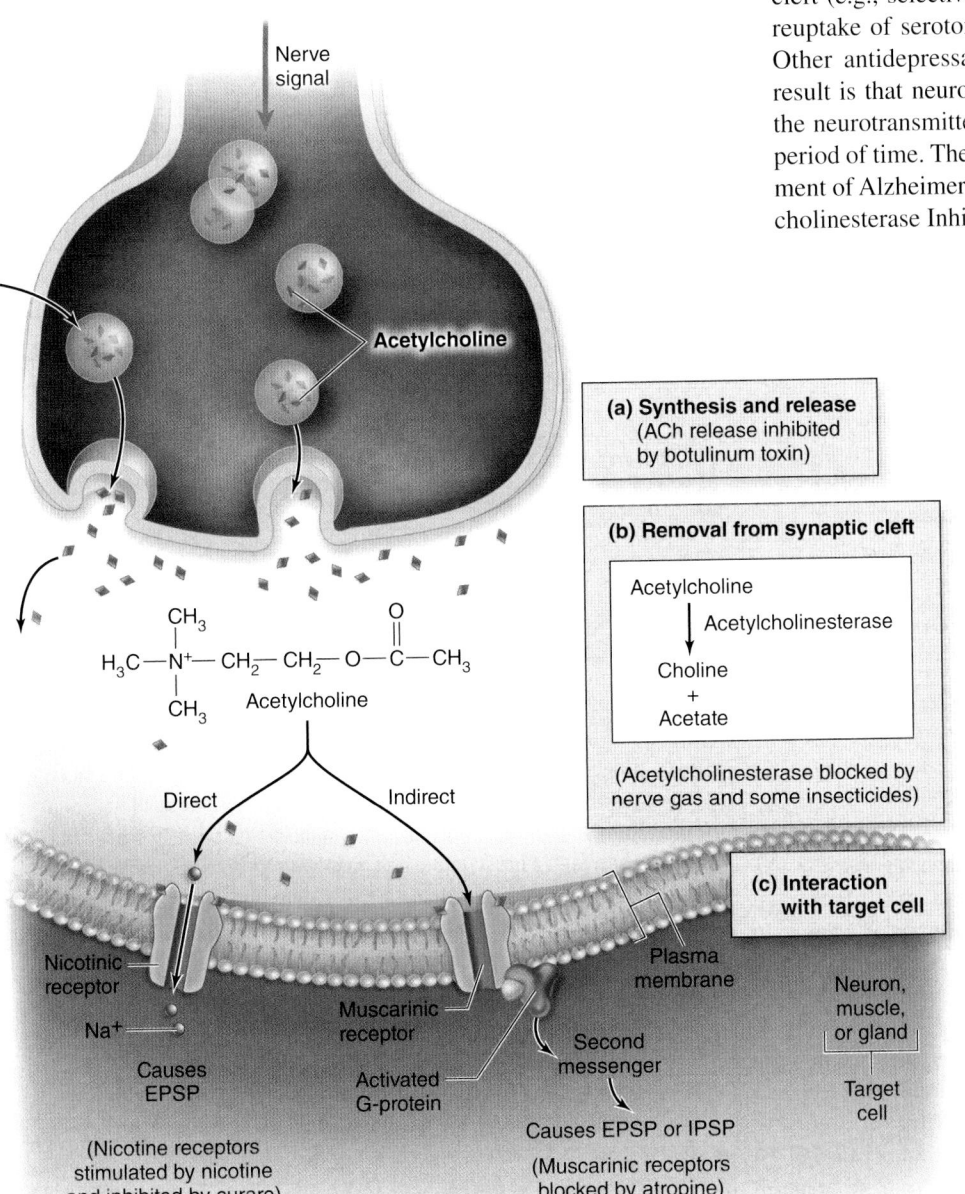

(a) **Synthesis and release**
(ACh release inhibited by botulinum toxin)

(b) **Removal from synaptic cleft**

Acetylcholine
→ Acetylcholinesterase
Choline
+
Acetate

(Acetylcholinesterase blocked by nerve gas and some insecticides)

(c) **Interaction with target cell**

Figure 12.25 Acetylcholine Release, Removal from Synaptic Cleft, and Action. Acetylcholine (*a*) is synthesized and released from synaptic knobs into a synaptic cleft, (*b*) is removed from the synaptic cleft through enzymatic breakdown by acetylcholinesterase, and (*c*) interacts with target cells either directly or indirectly. Each of these processes is altered by specific types of toxins, chemical poisons, or drugs.

INTEGRATE

CLINICAL VIEW
Acetylcholinesterase Inhibitor

One neurologic change associated with Alzheimer disease is the decrease in acetylcholine receptors. A specific type of medication (galantamine hydrobromide) acts as a reversible competitive inhibitor (see section 3.3f) to temporarily bind and inhibit acetylcholinesterase. Acetylcholine molecules remain longer in the synaptic cleft, making it more likely that they will interact with the acetylcholine receptors that are still present on the target cells.

Interaction with Target Cells (step c). The effect ACh has on a target cell depends upon the specific type of receptor embedded within the plasma membrane of the target cell. The receptors are either nicotinic receptors or one of several subtypes of muscarinic receptors. ACh interacts directly with **nicotinic receptors**, causing the opening of ion channels and the production of an EPSP. In comparison, the interaction of ACh with **muscarinic receptors** causes ion channels to open indirectly through the second messenger pathway that involves G protein (see section 4.5b). Interestingly, the result may be the formation of an EPSP or an IPSP depending upon the specific subtype of muscarinic receptor to which ACh binds.

 WHAT DID YOU LEARN?

 How is it possible for acetylcholine to generate either an EPSP or an IPSP?

INTEGRATE

CONCEPT CONNECTION

Other neurotransmitters include **adenosine triphosphate (ATP;** a molecule first described in section 2.7d) and adenosine (which is merely adenosine without the three phosphates). **Adenosine,** for example, when bound to adenosine receptors in the brain has an inhibitory effect. Caffeine acts as a stimulant by blocking adenosine receptors and acting as a competitive inhibitor (see section 3.3f).

12.10c Neuromodulation

 LEARNING OBJECTIVES

5. Define neuromodulation including its function in facilitation and inhibition.

6. Describe how nitric oxide and endocannabinoids function as neuromodulators.

Neuromodulation (nūr'ō-mod-yū-lā'shŭn) is the release of chemicals (other than classical neurotransmitters) from cells that locally regulate or alter the response of neurons to neurotransmitters. The substances released are called **neuromodulators,** and they become participants in the "decision making" that occurs during the transmission of information by the nervous system. Neuromodulation generally results in either facilitation or inhibition. **Facilitation** occurs when there is greater response from a postsynaptic neuron because of the release of neuromodulators. Facilitation may result from either an increased amount of neurotransmitter in the synaptic cleft (greater release, slower breakdown, or slower reuptake) or an increased number of receptors on postsynaptic neurons. **Inhibition** occurs when there is less response from a postsynaptic neuron because of the release of neuromodulators. This results from either a decreased amount of neurotransmitter in the synaptic cleft (decreased release, faster breakdown, or faster reuptake) or decreased number of receptors on postsynaptic neurons. Enkephalins, for example, act as neuromodulators to decrease the perception of pain by blocking transmission of the sensory input at the level of the spinal cord.

Nitric oxide (named "molecule of the year" in 1992 by the *Science* journal) is a unusual neurotransmitter, and is classified by some experts as a neuromodulator. Nitric oxide is different from the classical neurotransmitter for several reasons. It is a short-lived gas that is not stored in a vesicle, but synthesized from the amino acid arginine on an "as needed" basis. This small, nonpolar molecule is produced and released from *postganglionic* neurons. It diffuses (to more than 100 μm) and enters cells in all directions. Its entry into presynaptic neurons provides a retrograde means of communication (i.e., a means of communication in the reverse direction that typically occurs between a presynaptic and postsynaptic neuron). Nitric oxide functions within the brain in developing memories by stimulating presynaptic neurons to increase the release of neurotransmitter. Nitric oxide is also released from motor neurons that innervate blood vessels, causing relaxation of smooth muscle within the blood vessel wall. This results in vasodilation (increasing blood vessel diameter) and increased blood flow. (Endothelial cells that line blood vessels also release nitric oxide to cause vasodilation; see section 20.3b.)

Endocannabinoids (en'dō-ka'nab'i-noydz) are molecules that bind with the same receptors as the active ingredient tetrahydrocannabinol (THC) within cannabis (i.e., marijuana). Endocannabinoids are similar to nitric oxide in that they are small, nonpolar molecules, which are produced and released on demand from postsynaptic neurons. Binding of endocannabinoid decreases neurotransmitter release from presynaptic neurons, altering learning and memory, affecting appetite, and suppressing nausea. Neurotransmitters and neuromodulators are summarized in **table 12.3**.

 WHAT DID YOU LEARN?

 How does nitric oxide act as a neuromodulator?

INTEGRATE

CONCEPT CONNECTION
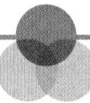

A region of the brain called the hippocampus and its role in memory is described in section 13.8e. Numerous cellular changes to neurons are associated with memory. These occur in response to increased activity between a presynaptic and postsynaptic neuron. Two of these changes include (1) an increase in Ca^{2+} concentration in the synaptic knobs of presynaptic neurons, which causes release of additional neurotransmitter; and (2) an increase in both the number and sensitivity of receptors that bind neurotransmitter in the postsynaptic neurons. These modifications enhance the magnitude of graded potentials established in the postsynaptic neurons—making it either more likely (if EPSPs are generated) or less likely (if IPSPs are generated) that the postsynaptic neuron will reach the threshold and initiate an action potential along the axon.

Table 12.3 Neurotransmitters

Neurotransmitter	Description/Action
ACETYLCHOLINE (ACh)	
$H_3C-\overset{\overset{CH_3}{\mid}}{\underset{\underset{CH_3}{\mid}}{N^+}}-CH_2-CH_2-O-\overset{\overset{O}{\parallel}}{C}-CH_3$	It has a chemical structure significantly different from other neurotransmitters and for this reason is placed in its own structural category; it is released into the neuromuscular junction to stimulate skeletal muscle contraction, inhibits cardiac muscle, either inhibits or excites smooth muscle and glands; released from preganglionic and postganglionic neurons of the parasympathetic nervous system and from all preganglionic neurons and some of the postganglionic neurons of the sympathetic nervous system; acts as a neuromodulator in the central nervous system (CNS); acetylcholine activity is altered by certain drugs and poisons (see figure 12.25).
BIOGENIC AMINES	
[structure: NH$_2$—CH$_2$—CH— with OH and Aromatic ring bearing OH, OH groups]	Molecules synthesized from an amino acid by removal of the carboxyl group and retaining the single amine group; also called monoamines
Catecholamines	A very distinct group of biogenic amines so named for a specific chemical structural similarity; originally described as hormones (chemicals produced by a gland in one part of the body that affect cells in other parts of the body)
Dopamine	Produces inhibitory activity in the brain; important roles in cognition (learning, memory), motivation, behavior, and mood; decreased levels in Parkinson disease; amphetamines increase release; cocaine decreases removal from synaptic cleft; ecstasy increases release
Norepinephrine (noradrenaline)	Neurotransmitter of peripheral autonomic nervous system (sympathetic division) and various regions of the CNS; amphetamines increase release; cocaine decreases removal from synaptic cleft; ecstasy increases release
Epinephrine (adrenaline)	Has various effects in the thalamus, hypothalamus, and spinal cord
Indolamines	
Histamine	Neurotransmitter of the CNS; plays a role in sleep and memory
Serotonin	Has various functions in the brain related to sleep, appetite, cognition (learning, memory), and mood; fluoxetine (Prozac) decreases reuptake; ecstasy increases the release; LSD binds to most serotonin receptors
AMINO ACIDS	
$NH_2-\overset{\overset{}{\mid}}{\underset{\underset{R}{\mid}}{CH_2}}-\overset{\overset{O}{\parallel}}{C}-OH$	Molecules with both carboxyl (—COOH) and amine (—NH$_2$) groups and various R groups; building blocks of proteins; act as signaling molecules in the nervous system
Glutamate	Excites activity in nervous system to promote cognitive function in the brain (learning and memory); most common neurotransmitter in the brain; stroke causes excessive release resulting in neuron death
Aspartate	Excites activity primarily in descending motor pathways through the spinal cord to skeletal muscle
Serine	Activates diverse areas in the brain
Gamma-aminobutyric acid (GABA)	Modified amino acid that is synthesized from glutamate; primary inhibitory neurotransmitter in the brain; also influences muscle tone; alcohol, diazepam (Valium), and barbiturates increase inhibitory effects of GABA
Glycine	Inhibits activity between neurons in the brain, spinal cord, and eye; strychnine blocks receptors that bind glycine
NEUROPEPTIDES	
Tyr—Gly—Gly—Phe—Met	Small molecules made of chains of amino acids; generally act through G proteins to cause more diverse effects; opioids include enkephalins, endorphins, endomorphins, dynorphins, and nociceptin; methadone binds at opiate receptors
Enkephalin	Helps regulate response to something that is perceived to be noxious or potentially painful
Beta-endorphin	Prevents release of pain signals from neurons and fosters a feeling of well-being; morphine mimics endorphins; heroin is converted to morphine in the body
Neuropeptide Y	Involved in memory regulation and energy balance (increased food intake and decreased physical activity)
Somatostatin	Inhibits activities of neurons in specific brain areas
Substance P	Assists with pain information transmission into the brain
Cholecystokinin	Stimulates neurons in the brain to help mediate satiation (fullness) and repress hunger
Neurotensin	Helps control and moderate the effects of dopamine
OTHERS	
Adenosine	Part of a nucleotide (a building block of nucleic acid); has an inhibitory effect on neurons in the brain and spinal cord
Nitric oxide	Involved in learning and memory; relaxation of muscle in the digestive tract; relaxation of smooth muscle in blood vessels
Endocannabinoids	Most prevalent receptors in the brain

12.11 Neural Integration and Neuronal Pools of the CNS

LEARNING OBJECTIVE

1. Identify the four different types of neuronal pools, and explain how they function.

The nervous system coordinates and integrates neuronal activity in part because billions of interneurons within the nervous system are grouped in complex patterns called **neuronal pools,** or *neuronal circuits* or *pathways*. Neuronal pools are identified based upon function into four types of circuits: converging, diverging, reverberating, and parallel-after-discharge **(figure 12.26).** A pool may be localized, with its neurons confined to one specific location, or its neurons may be distributed in several different regions of the CNS. However, all neuronal pools are restricted in their number of input sources and output destinations.

The **converging circuit** involves inputs that come together (converge) at a single postsynaptic neuron (figure 12.26*a*). This neuron receives input from several presynaptic neurons. For example, multiple sensory neurons synapse on the neurons in the salivary nucleus in the brainstem, causing the salivary nucleus to alter activity of salivary glands to produce saliva at mealtime. The various inputs originate from more than one stimulus: smelling food, seeing dinnertime on the clock, hearing food preparation activities, or seeing pictures of food in a magazine. These multiple inputs lead to a single output: the production of saliva.

A **diverging circuit** spreads information from one presynaptic neuron to several postsynaptic neurons, or from one pool to multiple pools (figure 12.26*b*). The neurons in the brain control the movements of skeletal muscles in the legs during walking and also stimulate the muscles in the back to maintain posture and balance while walking. In this case, a single or a few inputs lead to multiple outputs.

Reverberating circuits utilize feedback to produce a repeated, cyclical stimulation of the circuit: This is termed *reverberation* (figure 12.26*b*). Once activated, a reverberating circuit may continue to function until the cycle is broken by either inhibitory stimuli or synaptic fatigue. The repetitious nature of a reverberating circuit ensures that we continue breathing while we are asleep.

In a **parallel-after-discharge circuit,** input is transmitted simultaneously along several neuron pathways to a common postsynaptic cell (figure 12.26*d*). Note that neuron pathways in a parallel-after-discharge circuit vary in the number of neurons within the pathway and thus the number of synapses within the pathway. Recall that neuron to neuron communication at a synapse involves a synaptic delay (the time delay for the events at a synapse). Consequently, the greater the number of neurons in the pathway, the greater the number of synapses and the greater the amount of time required to transmit the information. This results in the information arriving from the point of stimulus input to the common postsynaptic cell at varying times. You might find it helpful to think of the arrival of information from each group of neurons to the common postsynaptic cell as an "echo" of the original stimulus input. This type of circuit is believed to be involved in higher-order thinking; for example, it reinforces the repetitive neural activity needed for performing precise mathematical calculations.

WHAT DID YOU LEARN?

32 How are neurons arranged in a converging circuit?

33 What are the differences between a reverberating circuit and a parallel-after-discharge circuit?

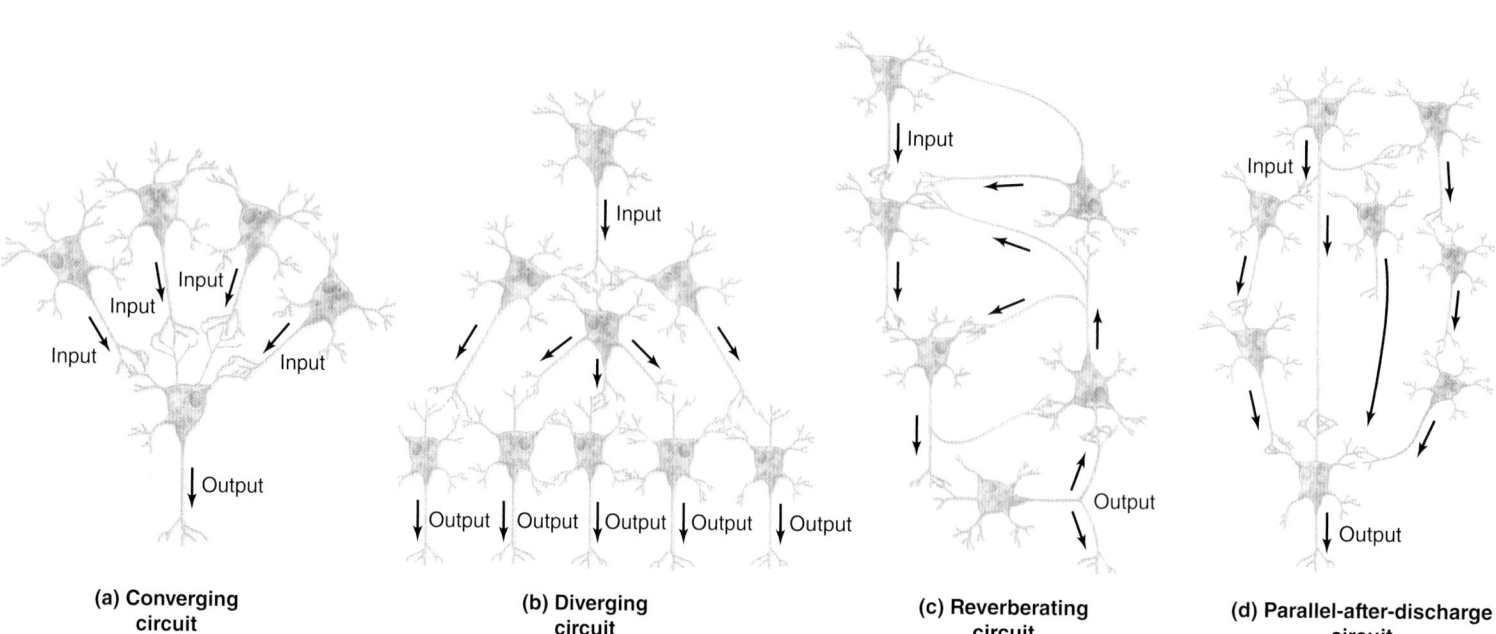

(a) Converging circuit **(b) Diverging circuit** **(c) Reverberating circuit** **(d) Parallel-after-discharge circuit**

Figure 12.26 Neuronal Pools. Neuronal pools are groups of neurons arranged in specific patterns (circuits) through which input is conducted and distributed. Four types of neuronal pools are recognized.

- The nervous system interprets and controls all sensory input from receptors and motor output to effectors.

12.1 Introduction to the Nervous System 438

- The nervous system is composed of the brain, spinal cord, nerves, and ganglia.

12.1a General Functions of the Nervous System 438

- The nervous system collects information through receptors and sensory input, processes and evaluates information, and responds through motor output to effectors (muscles or glands).

12.1b Organization of the Nervous System 438

- The nervous system is organized structurally into the central nervous system (CNS) and the peripheral nervous system (PNS).
- The nervous system is organized functionally into a sensory component and a motor component.

12.2 Nervous Tissue: Neurons 439

- Nervous tissue is composed of excitable neurons that initiate and transmit action potentials, and glial cells that support and protect them.

12.2a General Characteristics of Neurons 439

- General characteristics of neurons include excitability, conductivity, secretion, and longevity; in addition, they are typically amitotic.

12.2b Neuron Structure 440

- A generalized neuron has a cell body. Processes typically extending from the cell body are numerous short, tapering dendrites and a single and often long axon.

12.2c Neuron Transport 442

- Neurons transport substances between the cell body and synaptic knobs by fast axonal transport and slow axonal transport.

12.2d Classification of Neurons 442

- Neurons are classified structurally as multipolar, bipolar, unipolar, and anaxonic by the number of processes attached to the cell body.
- The three functional categories of neurons are sensory neurons, motor neurons, and interneurons.

12.2e Relationship of Neurons and Nerves 444

- A nerve is a collection of axons that are wrapped in connective tissue.
- Each axon (and its surrounding neurolemmocyte) is wrapped with an endoneurium, fascicles of axons are ensheathed with a perineurium, and the entire nerve is enclosed with an epineurium.

12.3 Synapses 445

- A synapse is the functional junction of a neuron with either another neuron or an effector.
- Synapses are either chemical synapses or electrical synapses.

12.4 Nervous Tissue: Glial Cells 446

- Glial cells are the other distinct cell type of nervous tissue.

12.4a General Characteristics of Glial Cells 446

- Glial cells are nonexcitable cells that primarily support and protect the neurons.

12.4b Types of Glial Cells 447

- Four types of glial cells within the central nervous system include astrocytes, ependymal cells, microglia, and oligodendrocytes.
- Two types of glial cells within the peripheral nervous system include satellite cells and neurolemmocytes.

12.4c Myelination 448

- Myelination is the process by which part of an axon is wrapped and insulated with myelin.
- Neurolemmocytes myelinate axons in the PNS, and oligodendrocytes myelinate axons in the CNS.

12.5 Axon Regeneration 451

- Regeneration of damaged neurons is limited to PNS axons.
- A PNS axon can regrow to reestablish innervation if the cell body is intact and a critical amount of neurilemma remains.

12.6 Plasma Membrane of Neurons 452

- Establishing and changing a resting membrane potential is dependent upon various types of pumps and channels within a neuron's plasma membrane.

12.6a Types of Pumps and Channels 452

- Pumps and channels are membrane proteins that facilitate movement of ions across the neuron plasma membrane.

12.6b Distribution of Pumps and Channels 453

- Some membrane transport proteins are located along the entire neuron, and some are primarily in specific functional neuron segments.

12.7 Introduction to Neuron Physiology 455

- Neuron physiology involves the initiation and transmission of electrical currents.

12.7a Neurons and Ohm's Law 455

- Ohm's law (current = voltage/resistance) has application in the principles of neuron physiology.

12.7b Neurons at Rest 456

- Neurons at rest have a negative resting membrane potential (RMP), which on average is –70 mV.
- Other characteristics of neurons at rest include closed gated channels, Na^+, K^+, and Cl^- concentration gradients along the length of the axon, and a Ca^{2+} concentration gradient at the synaptic knob.

(continued on next page)

12.8 Physiologic Events in the Neuron Segments 459	• Physiologic events that occur in a neuron's functional segments begin with stimulation of the receptive segment until the release of neurotransmitter from the transmissive segment.

12.8a Receptive Segment 459

• The receptive segment includes the dendrites and cell body. This segment involves the formation and propagation of graded potentials: both excitatory postsynaptic potentials (EPSPs) and inhibitory postsynaptic potentials (IPSPs).

12.8b Initial Segment 462

• The summation (or adding together) of EPSPs and IPSPs that occurs in the initial segment determines whether the threshold value (–55 mV) is reached and an action potential is initiated.

12.8c Conductive Segment 463

• The conductive segment is involved in the propagation of an action potential (nerve signal), a process that involves depolarization and repolarization.

• The brief period of time that an axon is either incapable of generating an action potential or a greater than normal amount of stimulation is required to generate another action potential is called the refractory period.

• Saltatory conduction occurs if the axon is myelinated.

12.8d Transmissive Segment 468

• The transmissive segment involves exocytosis of neurotransmitter from synaptic vesicles, which are located within synaptic knobs.

12.9 Characteristics of Action Potentials 472	• Action potentials differ from graded potentials and can vary in both velocity and frequency.

12.9a Graded Potentials Versus Action Potentials 472

• Graded potentials are short-lived electrical signals that occur in the dendrites and cell body, whereas action potentials are self-propagating electrical signals that are initiated in the initial segment and propagated along an axon.

12.9b Velocity of Action Potential Propagation 473

• Nerve fibers, which are axons and their myelin sheath, are classified into three groups based upon the velocity of the nerve signal conduction.

• The velocity of action potentials is greater in larger and myelinated axons.

12.9c Frequency of Action Potentials 473

• Frequency of action potentials occurs with increased stimulation of the neuron.

12.10 Neurotransmitters and Neuro-modulation 474	• Neurotransmitters are released at synaptic clefts and their action is modified by neuromodulation.

12.10a Classification of Neurotransmitters 474

• Neurotransmitters are conventionally described as molecules synthesized by neurons, which are then stored within vesicles in synaptic knobs and when released bind to specific receptors in a target cell to trigger a physiologic response.

• Major classes of neurotransmitters include acetylcholine, biogenic amines, amino acids, and neuropeptides.

12.10b Features of Neurotransmitters 475

• Acetylcholine (ACh) is discussed in detail because it exhibits all of the classical features of neurotransmitters and is the most understood.

12.10c Neuromodulation 476

• Neuromodulation is the release of chemicals other than neurotransmitters that either increase the responsiveness to a neurotransmitter (facilitation) or decrease the responsiveness to a neurotransmitter (inhibition).

12.11 Neural Integration and Neuronal Pools of the CNS 478	• Interneurons are organized into neuronal pools, which are groups of interconnected neurons with specific functions and are classified as converging, diverging, reverberating, and parallel-after-discharge.

CHALLENGE YOURSELF

Do You Know the Basics?

_____ 1. The cell body of a neuron does all of the following *except*

a. release neurotransmitter into the synaptic cleft.

b. produce synaptic vesicles containing neurotransmitter that are subsequently transported to the synaptic knob.

c. conduct graded potentials to the initial segment.

d. receive graded potentials from dendrites.

_____ 2. Neurons that have only two processes attached to the cell body are called

a. unipolar.

b. bipolar.

c. multipolar.

d. efferent.

_____ 3. Which neurons are located only within the CNS?

    a. afferent neurons

    b. glial cells

    c. sensory neurons

    d. interneurons

_____ 4. EPSPs are caused by the movement of

    a. Na^+ out of the cell.

    b. Na^+ into the cell.

    c. K^+ into the cell.

    d. both Na^+ and K^+ into the cell.

_____ 5. The glial cells that help produce and circulate cerebrospinal fluid in the CNS are

    a. satellite cells.

    b. microglia.

    c. ependymal cells.

    d. astrocytes.

_____ 6. Which of the following is a part of the PNS?

    a. microglia

    b. spinal cord

    c. brain

    d. neurolemmocyte

_____ 7. An action potential is generated when threshold is reached, at which time

    a. voltage-gated K^+ channels close.

    b. voltage-gated Na^+ channels open.

    c. chemically gated Na^+ channels open.

    d. Ca^{2+} enters the cell.

_____ 8. Which type of neuronal pool utilizes nerve impulse feedback to repeatedly stimulate the circuit?

    a. converging circuit

    b. diverging circuit

    c. reverberating circuit

    d. parallel-after-discharge circuit

_____ 9. At an electrical synapse, presynaptic and postsynaptic membranes interface through

    a. neurofibril nodes.

    b. gap junctions.

    c. telodendria.

    d. neurotransmitters.

_____ 10. The two primary factors that influence the speed of an action potential propagation are axon diameter and

    a. myelination.

    b. the type of associated glial cell(s).

    c. concentration of K^+ in the cell.

    d. the length of the axon.

11. What are the four structural types of neurons? How do they compare to the three functional types of neurons?

12. Identify the principal glial cell types, and briefly discuss the function of each type.

13. How does myelination differ between the CNS and the PNS?

14. Describe the procedure by which a PNS axon may repair itself (axon regeneration).

15. Describe how the resting membrane potential is established and maintained in a neuron.

16. Compare and contrast graded potentials and action potentials.

17. Explain summation of EPSPs and IPSPs and the relationship to the initiation of an action potential.

18. Graph and explain the events associated with an action potential.

19. Explain the mechanism for the release of neurotransmitter from the synaptic knob.

20. List and briefly describe the major types of neurotransmitters.

Can You Apply What You've Learned?

1. Andrew was brought to the doctor's office after he was bitten by a stray dog. The concern was that the dog might be infected with the rabies virus. The rabies virus infects neurons by using which method that normally transports materials from the synaptic knob to the cell body?

    a. anterograde transport

    b. fast axonal transport

    c. slow axonal transport

    d. All of these are correct.

2. An elderly neighbor was diagnosed with an astrocytoma tumor in the brain. This cancer affects what types of cell?

    a. ependymal cells

    b. microglia

    c. astrocytes

    d. satellite cells

3. Cynthia has received her lab results and is told that her blood calcium levels are abnormal. The event in neuron transmission that would most likely be affected would be

    a. summation of graded potentials in the initial segment.

    b. production of graded potentials in the dendrites and cell body.

    c. release of neurotransmitter from the synaptic knob.

    d. propagation of an action potential in the axon.

4. Heidi's physician prescribed a medication that is known to block the reuptake of serotonin neurotransmitter from the synaptic cleft. This medication affects what segment in neuron transmission that is responsible for releasing the neurotransmitter?

    a. initial segment

    b. conductive segment

    c. transmissive segment

    d. receptive segment

5. Sarah wants to call her new friend Julie and needs to write down her phone number but cannot find a pen. She continues to repeat the number over and over. This is most likely occurring in what type of neuronal pool?

   a. reverberating circuit

   b. divergent circuit

   c. convergent circuit

   d. parallel-after-discharge circuit

Can You Synthesize What You've Learned?

1. Over a period of 6 to 9 months, Marianne began to experience vision problems as well as weakness and loss of fine control of the skeletal muscles in her leg. Blood tests revealed the presence of antibodies (immune system proteins) that attack myelin. Beyond the presence of the antibodies, what was the cause of Marianne's vision and muscular difficulties?

2. Surgeons were able to reattach Irving's amputated limb, sewing both the nerves and the blood vessels back together. After the surgery, which proceeded very well, the limb regained its blood supply almost immediately, but the limb remained motionless and Irving had no feeling in it for several months. Why did it take longer to reestablish innervation than circulation?

3. Certain types of neurotoxins prevent depolarization of the axon. What specific type of channel is impaired?

INTEGRATE

ONLINE STUDY TOOLS

The following study aids may be accessed through Connect.

Concept Overview Interactive Figure 12.23: Events of Neuron Physiology

Clinical Case Study: A Young Man Who Gets Weak When Overheated

Interactive Questions: This chapter's content is served up in a number of multimedia question formats for student study.

LearnSmart: Topics and terminology include neurons; synapses; glial cells; axon regeneration; ultrastructure of neurons; neuron physiology; physiologic events in the neuron segments; velocity of a nerve signal; neurotransmitters and neuromodulation; neural integration and neuronal pools of the CNS

Anatomy & Physiology Revealed: Topics include multipolar neuron; Schwann cell (neurolemmocyte); unmyelinated axon and myelin sheath; action potential generation and propagation; chemical synapse

Animations: Topics include PSPs (postsynaptic potentials); action potential propagation; action potential generation; action potential propagation in myelinated neurons

chapter 13

Nervous System: Brain and Cranial Nerves

INTEGRATE

CAREER PATH
Speech Language Pathologist

A speech language pathologist (SLP) assesses and diagnoses a range of communication and swallowing disorders. This health professional evaluates and develops a plan to improve a patient's speech and language skills. Patients range from the very young to elderly, and the communication difficulties vary from stuttering, difficulties in understanding and producing language, and abnormal intonations in speech. SLPs work in a variety of health-care settings, such as hospitals, nursing homes, day care centers, and schools. An SLP must understand relevant brain anatomy related to speech and communication, as well as be aware of the anatomic structures and neurologic pathways involved in the production of speech and language.

Module 7: Nervous System

The human brain, while weighing on average about 3 pounds, is able to simultaneously process billions of bits of information. It is continuously receiving sensory input from sensory receptors and initiating motor output to control effectors. This organ allows us to comprehend complex information, compute mathematical problems, and write poetry. The brain, which is part of the central nervous system (CNS), is associated with the 12 pairs of cranial nerves, which are considered part of the peripheral nervous system (PNS). The brain and cranial nerves are described in this chapter.

13.1 Brain Organization and Development

We begin our study of the brain by introducing its four major regions and surface landmark structures. Next we provide the essentials of embryonic brain development that help clarify how the structures of the adult brain are named and related. Then we examine the distribution of gray and white matter in the brain as a whole.

13.1a Overview of Brain Anatomy

 **LEARNING OBJECTIVE**

1. Describe the general anatomic features of the brain.

The brain is composed of four major regions: the cerebrum, diencephalon, brainstem, and cerebellum. **Figure 13.1** shows the major parts of the adult brain from several views. The cerebrum is divided into two halves, called the left and right cerebral hemispheres. Each hemisphere may be further subdivided into five functional areas called lobes.

Our skull volume limits the size of the brain, so the outer brain tissue is folded on itself so that more neurons could fit within the endocranium. These folds of brain tissue are called **gyri** (jī′rī; sing., *gyrus; gyros* = circle). The shallow depressions between those folds are called **sulci** (sŭl′sī; sing., sŭl′kŭs; furrow, ditch), and the deeper grooves are named **fissures** (fish′ŭr).

Two directional terms are often used to describe relative positions of brain anatomy. Anterior is synonymous with **rostral** (meaning "toward the nose"), and posterior is synonymous with **caudal** (meaning "toward the tail").

 WHAT DID YOU LEARN?

 1 What are the four major regions of the brain?

13.1b Development of Brain Divisions

LEARNING OBJECTIVES

2. Describe the process of neurulation.
3. Provide the scientific names for the embryonic forebrain, midbrain, and hindbrain.
4. Name the five secondary brain vesicles, and describe their embryonic origins.

Before we can understand how the divisions of the brain form, we first must explore how the nervous system is derived from ectoderm, one of the three primary germ layers (see figure 5.13). This process is called neurulation, and is described next.

Neurulation

Nervous tissue development begins in the embryo during the third week of development with a thickening of a portion of the ectoderm that overlies the notochord (see section 5.6a) **(figure 13.2)**. This thickened ectoderm is called the **neural plate.** The neural plate is induced by the notochord (see section 29.3b) to form a neural tube,

---

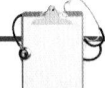

INTEGRATE

CLINICAL VIEW

Traumatic Brain Injuries: Concussion and Contusion

Traumatic brain injury (TBI) refers to the acute brain damage that occurs as a result of an accident or trauma. A **concussion** is the most common type of TBI. It is characterized by temporary, abrupt loss of consciousness after a blow to the head or the sudden stop of a moving head. Headache, drowsiness, lack of concentration, confusion, and amnesia (memory loss) may occur. Multiple concussions have a cumulative effect, causing the affected person to lose a small amount of mental ability with each episode. In fact, a history of multiple concussions has been related to long-term personality changes, depression, and intellectual decline. Athletes who are prone to concussions (such as football and soccer players) are at greater risk for these detrimental changes, so coaches and athletic trainers are being educated to be more cautious about letting an athlete play if a concussion is suspected.

A **contusion** is a TBI where there is bruising of the brain due to trauma that causes blood to leak from small vessels into the subarachnoid space

(a fluid-filled space surrounding the brain). The bruising may appear on a computed tomography (CT) scan of the head. Usually, the person immediately loses consciousness (normally for no longer than 5 minutes). Respiration abnormalities and decreased blood pressure sometimes occur as well.

Of particular concern is a rare but serious condition called **second impact syndrome (SIS)**, where an individual experiences a second brain injury prior to the resolution of the first injury, and develops severe brain swelling and possible death as a result. For this reason, it is essential that the original TBI completely heals before an individual is allowed to resume a behavior that may put the individual at risk for another TBI. Both severe traumatic brain injury and repetitive TBIs may cause long-term cognitive deficits and motor impairment. Individuals may need physical, occupational, and speech therapy to regain a portion of these functions.

Interestingly, preliminary research has shown that TBI patients who received therapeutic progesterone made a greater and faster recovery than individuals with similar TBIs who did not receive the therapy. Thus, a reproductive hormone (progesterone) also appears to help the nervous system with its healing.

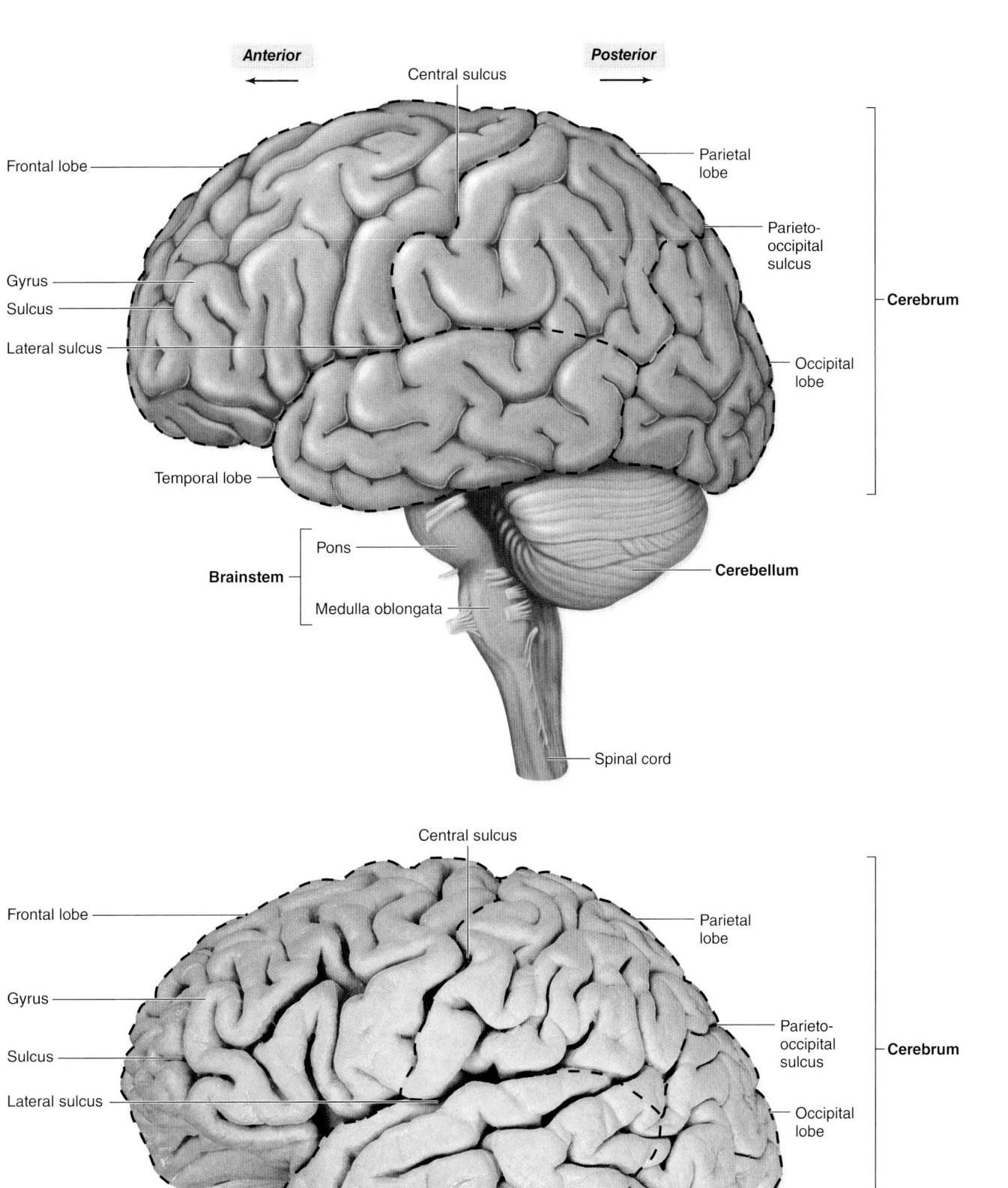

(a) Left lateral view

(continued on next page)

Figure 13.1 The Human Brain. The brain is a complex organ that is formed from several subdivisions. (*a*) An illustration and cadaver photo of a left lateral view show the cerebrum, cerebellum, and portions of the brainstem (in bold); the diencephalon is not visible. AP|R

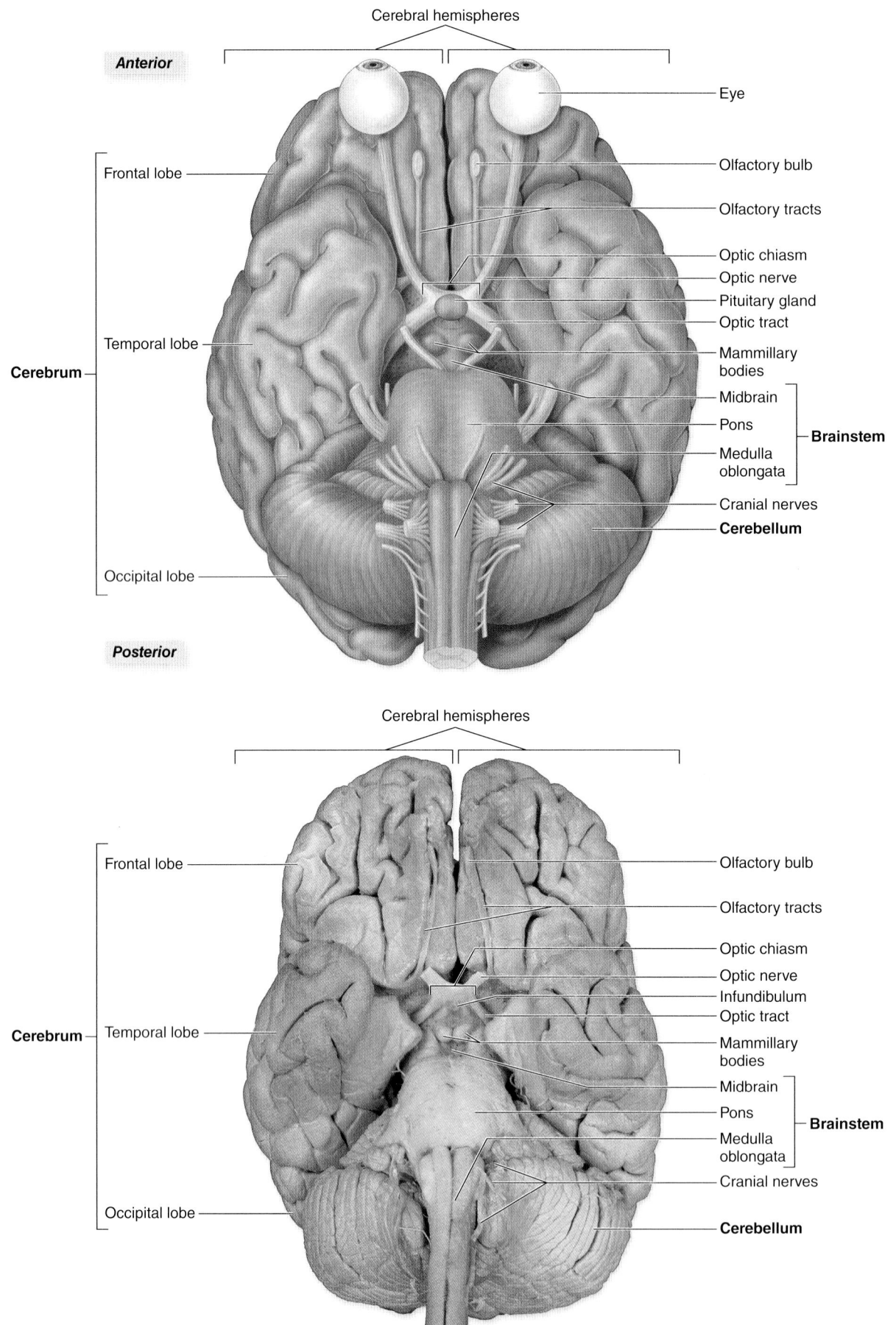

Cerebral hemispheres

Anterior

Eye

Frontal lobe

Olfactory bulb

Olfactory tracts

Optic chiasm
Optic nerve
Pituitary gland
Optic tract

Temporal lobe

Mammillary bodies

Cerebrum

Midbrain
Pons
Brainstem
Medulla oblongata

Cranial nerves
Cerebellum

Occipital lobe

Posterior

Cerebral hemispheres

Frontal lobe

Olfactory bulb

Olfactory tracts

Optic chiasm

Optic nerve
Infundibulum
Optic tract

Cerebrum

Temporal lobe

Mammillary bodies

Midbrain
Pons
Brainstem
Medulla oblongata

Cranial nerves

Occipital lobe

Cerebellum

(b) Inferior view

Figure 13.1 The Human Brain (continued). (*b*) An illustration and cadaver photo of the brain in inferior view best demonstrate the cranial nerves arising from the base of the brain. The major regions of the brain are in bold. AP|R

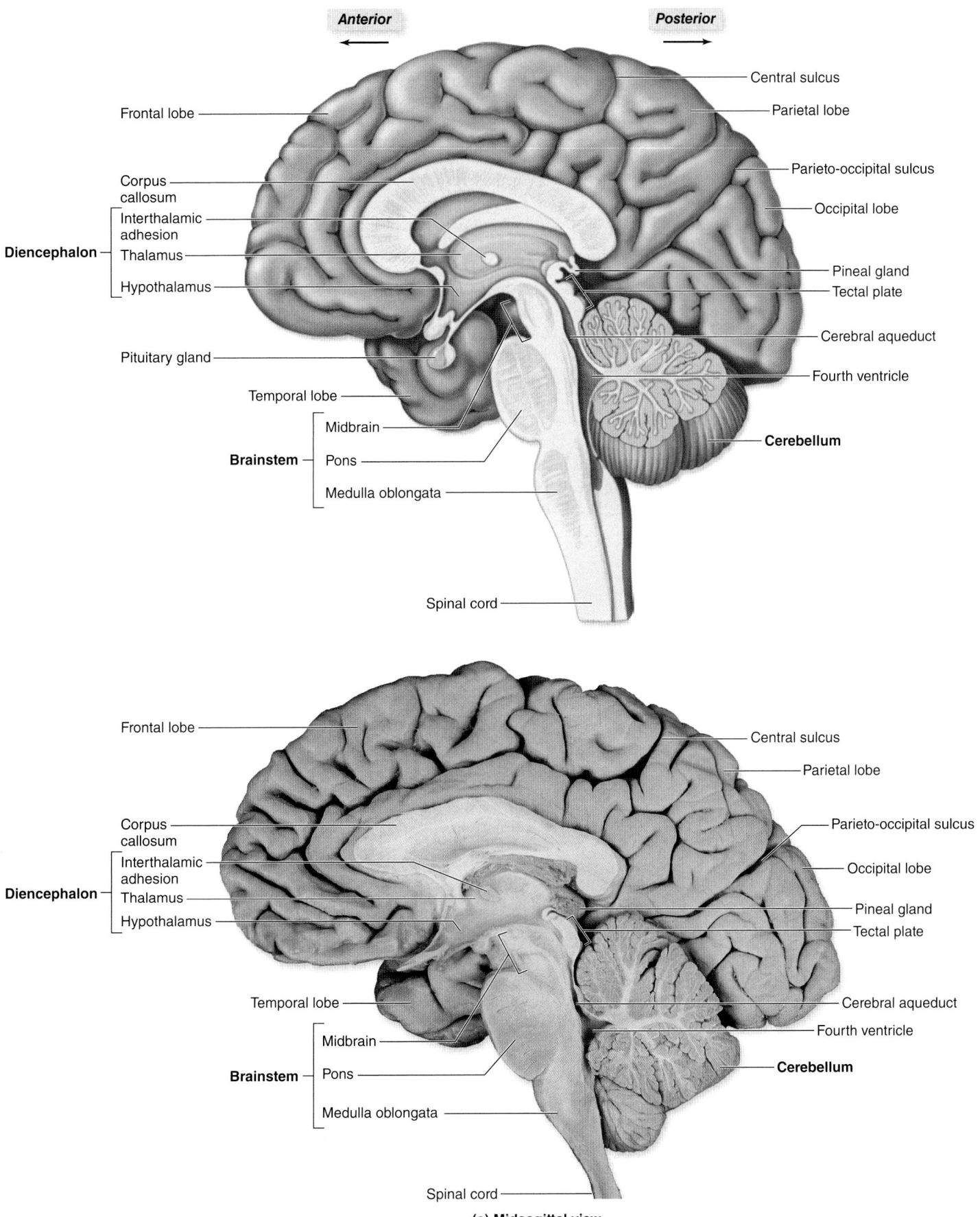

Anterior ←

Posterior →

Frontal lobe

Corpus callosum

Diencephalon {
Interthalamic adhesion
Thalamus
Hypothalamus
}

Pituitary gland

Temporal lobe

Brainstem {
Midbrain
Pons
Medulla oblongata
}

Spinal cord

Central sulcus

Parietal lobe

Parieto-occipital sulcus

Occipital lobe

Pineal gland

Tectal plate

Cerebral aqueduct

Fourth ventricle

Cerebellum

(c) Midsagittal view

Figure 13.1 The Human Brain (continued). (*c*) Internal structures such as the thalamus and hypothalamus are best seen in midsagittal view in an illustration and cadaver photo. The major regions of the brain are in bold. AP|R

in a process called **neurulation** (nū′rū-lā′shŭn). Neurulation ultimately forms all nervous tissue structures. The process of neurulation is shown in **figure 13.2** and explained here:

① The neural plate develops a central longitudinal indentation called the **neural groove.** As this is occurring, cells along the lateral margins of the neural plate proliferate, becoming the thickened **neural folds.** The tips of the neural folds form **neural crest cells** (or simply, the *neural crest*).

② The neural folds elevate and approach one another as the neural groove continues to deepen. The neural crest cells are now at the very highest point of the neural groove. When viewed from a superior angle, the neural folds resemble the sides of a hot dog roll, with the neural groove represented by the opening in the roll.

③ The neural crest cells begin to pinch off from the neural folds and form other structures.

④ By the end of the third week, the neural folds have met and fused at the midline as the neural groove starts to form a **neural tube,** which has an internal space called the **neural canal.** The neural tube initially fuses at its midline, and later the portions of the neural folds slightly superior and inferior to this midline fuse as well. Thus, the neural tube forms as the neural folds "zip" together both superiorly and inferiorly.

For a short time, the neural tube is open at both its ends. These openings, called **neuropores** (nū′rō-pōr), close during the end of the fourth week. The opening closest to the future head is the **cranial neuropore,** while the opening closest to the future buttocks region is the **caudal neuropore.** If these openings do not close, the developing human will have a neural tube defect (see Clinical View: "Neural Tube Defects"). The developing neural tube forms the central nervous system. In particular, the cranial part of the neural tube expands to form the brain, while the caudal part of the neural tube expands to form the spinal cord (see section 14.7).

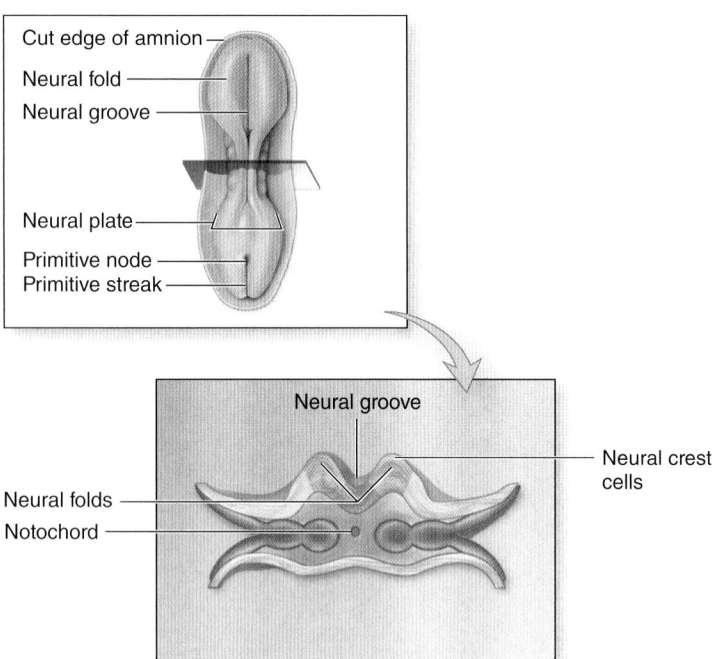

Figure 13.2 Nervous System Development. The process of neurulation begins in the third week, and the neural tube finishes closing by the end of week 4.

① Neural folds and neural groove form from the neural plate.

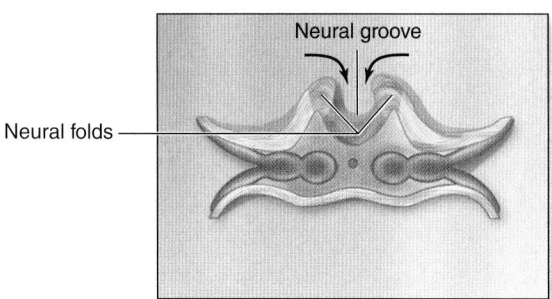

② Neural folds elevate and approach one another.

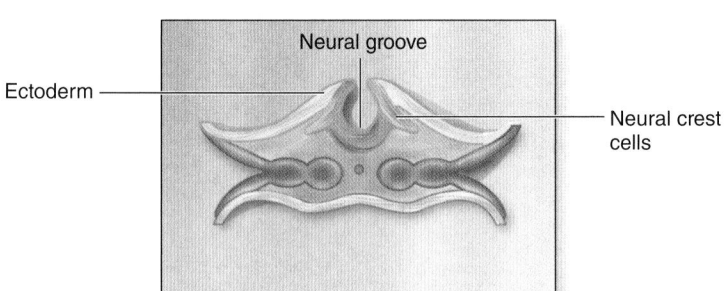

③ Neural crest cells begin to "pinch off" from the neural folds and form other structures.

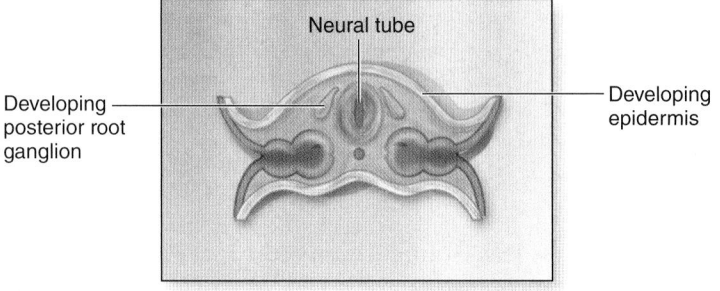

④ Neural folds fuse to form the neural tube.

CLINICAL VIEW
Neural Tube Defects

Neural tube defects (NTDs) are serious developmental deformities of the brain, spinal cord, and meninges (protective membranes around the brain and spinal cord). The two basic categories of NTDs are anencephaly and spina bifida. Both conditions result from localized failure of the developing neural tube to close.

Anencephaly (an'en-sef-ă-lē; *an* = without, *enkephalos* = brain) is substantial or complete absence of a brain as well as the bones making up the cranium. Infants with anencephaly rarely live longer than a few hours following birth. Fortunately, neural tube defects of this magnitude are rare, and are easily detected with prenatal ultrasound, thus alerting the parents to the condition.

Spina bifida (spī'nă bĭf'ĭ-dă; *spina* = spine, *bifidus* = cleft in two parts) occurs more frequently than anencephaly. This defect results when the caudal portion of the neural tube, often in the lumbar or sacral region, fails to close. Two forms of spina bifida occur: the more severe spina bifida cystica and the less severe spina bifida occulta. In *spina bifida cystica*, almost no vertebral arch forms, so the posterior aspect of the spinal cord in this region is left unprotected (*figures a and b*). Typically, there is a large cystic structure in the back, filled with cerebrospinal fluid (CSF) and covered by a thin layer of skin or in some cases only by meninges. Paralysis of the lower limbs is often part of the spina bifida syndrome.

Spina bifida occulta is less serious, but much more common than the cystica form. This condition is characterized by a partial defect of the vertebral arch, typically involving the vertebral lamina and spinous process (*figure c*). Often there is a tuft of hair in the region of the bony defect, which may alert

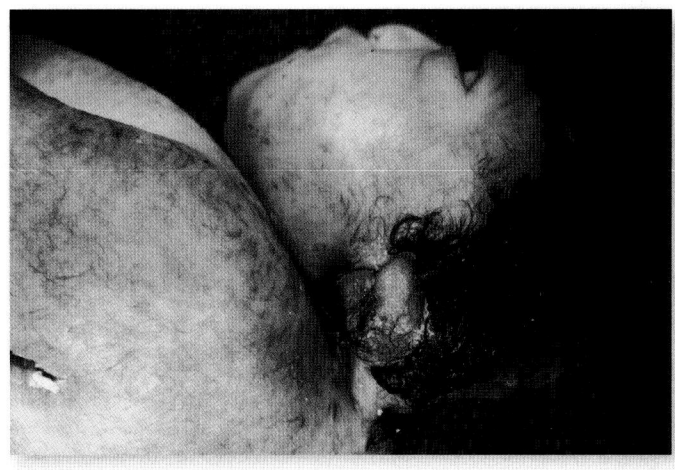

A newborn with anencephaly.

the physician of the location of the possible defect. Most people with this condition are otherwise asymptomatic, and it is generally detected incidentally during an x-ray for an unrelated reason.

Researchers have discovered that increased intake of vitamin B_{12} and vitamin B_9 (generally referred to as either **folic acid** or **folate**) by pregnant women is correlated with a decreased incidence of neural tube defects. Both vitamin B_{12} and folic acid are critical to DNA formation and are necessary for cellular division and tissue differentiation. Thus, women of childbearing age are encouraged to have sufficient levels of folic acid in their diet, because these women may not realize they are pregnant until after week 4, when the neural tube formation is complete.

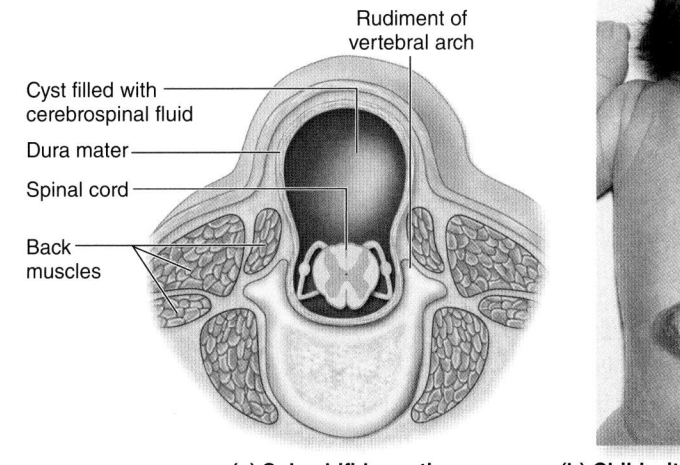

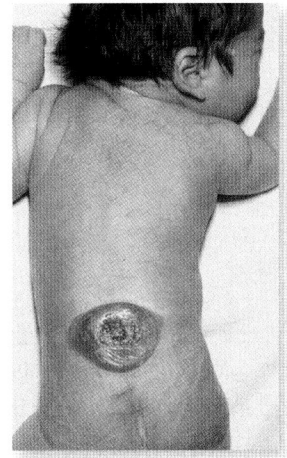

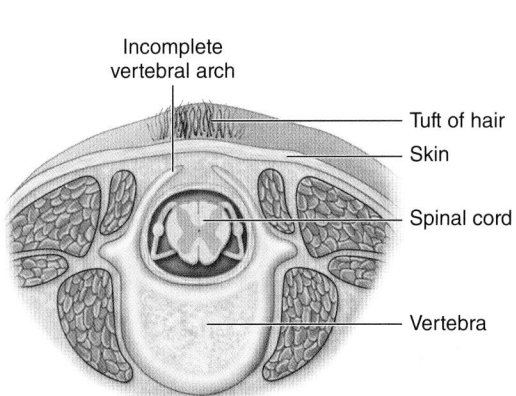

(a) Spina bifida cystica   (b) Child with spina bifida cystica   (c) Spina bifida occulta

Spina bifida is a neural tube disorder that occurs in two forms: (a, b) spina bifida cystica; (c) spina bifida occulta.

Development of the Brain

The brain develops from the cranial (rostral) part of the neural tube in the human embryo. The neural tube undergoes disproportionate growth rates in different regions. This growth has formed three **primary brain vesicles** by the late fourth week of development, which eventually give rise to all the different regions of the adult brain (**table 13.1**). The names of these vesicles describe their relative positions in the developing head: The forebrain is called the **prosencephalon** (pros-en-sef'ă-lon; *proso* = forward, *enkephalos* = brain); the midbrain is called the **mesencephalon** (mez-en-sef'ă-lon; *mes* = middle); and the hindbrain is called

Table 13.1 Major Brain Structures: Embryonic Through Adult

EMBRYONIC DEVELOPMENT ⟶ **ADULT STRUCTURE**

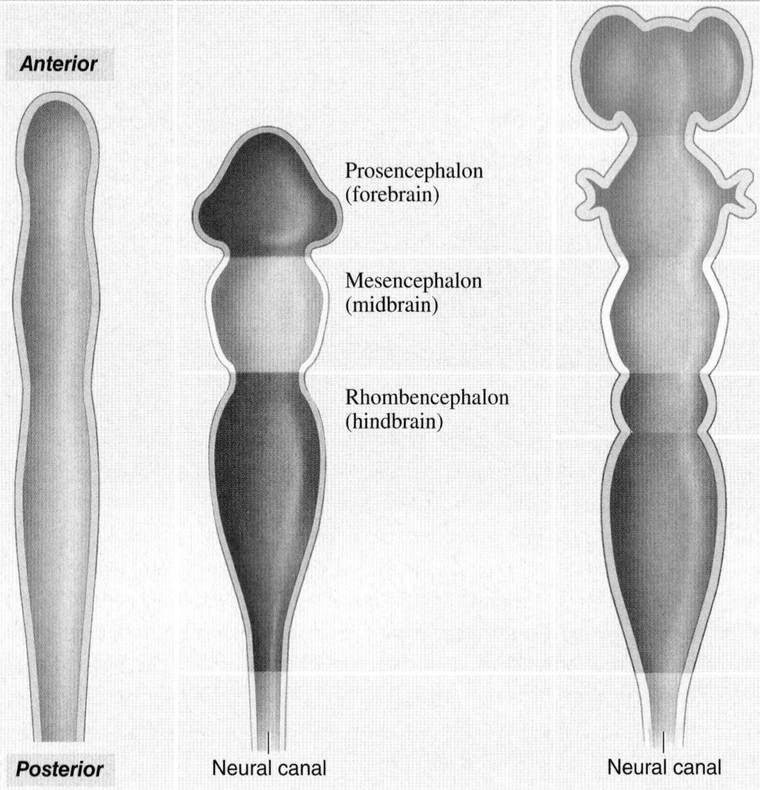

Neural Tube	Primary Brain Vesicles	Secondary Brain Vesicles (Future Adult Brain Regions)[1]	Neural Canal Derivative[2]	Structures Within Brain Regions
Anterior	Prosencephalon (forebrain)	Telencephalon	Lateral ventricles	Cerebrum
		Diencephalon	Third ventricle	Epithalamus, Thalamus, Hypothalamus
	Mesencephalon (midbrain)	Mesencephalon (midbrain)	Cerebral aqueduct	Midbrain
	Rhombencephalon (hindbrain)	Metencephalon	Fourth ventricle (superior part)	Pons, cerebellum
Posterior		Myelencephalon	Fourth ventricle (inferior part); part of central canal	Medulla oblongata
Neural canal		Neural canal		

1. The embryonic secondary vesicles form the adult brain regions—thus, they share the same names.
2. The neural canal in each specific brain region will form its own named space.

the **rhombencephalon** (rom-ben-sef'ă-lon; *rhombo* = rhomboid) for its rhomboidal shape (**figure 13.3a**).

By the fifth week of development, the three primary vesicles further develop into a total of five secondary brain vesicles (figure 13.3b):

- The **telencephalon** (tel-en-sef'ă-lon; *tel* = head end) arises from the prosencephalon and eventually forms the cerebrum.
- The **diencephalon** (dī-en-sef'ă-lon; *dia* = through) also derives from the prosencephalon, and it eventually forms the thalamus, hypothalamus, and epithalamus.
- The **mesencephalon** is the only primary vesicle that does not form a new secondary vesicle. It becomes the midbrain.
- The **metencephalon** (met'-en-sef'ă-lon; *meta* = after) arises from the rhombencephalon and eventually forms the pons and cerebellum.
- The **myelencephalon** (mī'el-en-sef'ă-lon; *myelos* = medulla) also derives from the rhombencephalon, and it eventually forms the medulla oblongata.

The telencephalon grows rapidly and envelops the diencephalon during the embryonic and fetal periods. As the future brain develops, its surface folds especially in the telencephalon, leading to the formation of the adult sulci and gyri. Most of the sulci and gyri develop late in the fetal period, so that by the time the fetus is born, its brain closely resembles that of an adult even though its functional development takes much longer (figure 13.3c–e).

WHAT DID YOU LEARN?

2 How does the neural plate form a neural tube?

3 Identify the five secondary vesicles, and list the adult brain structures they form.

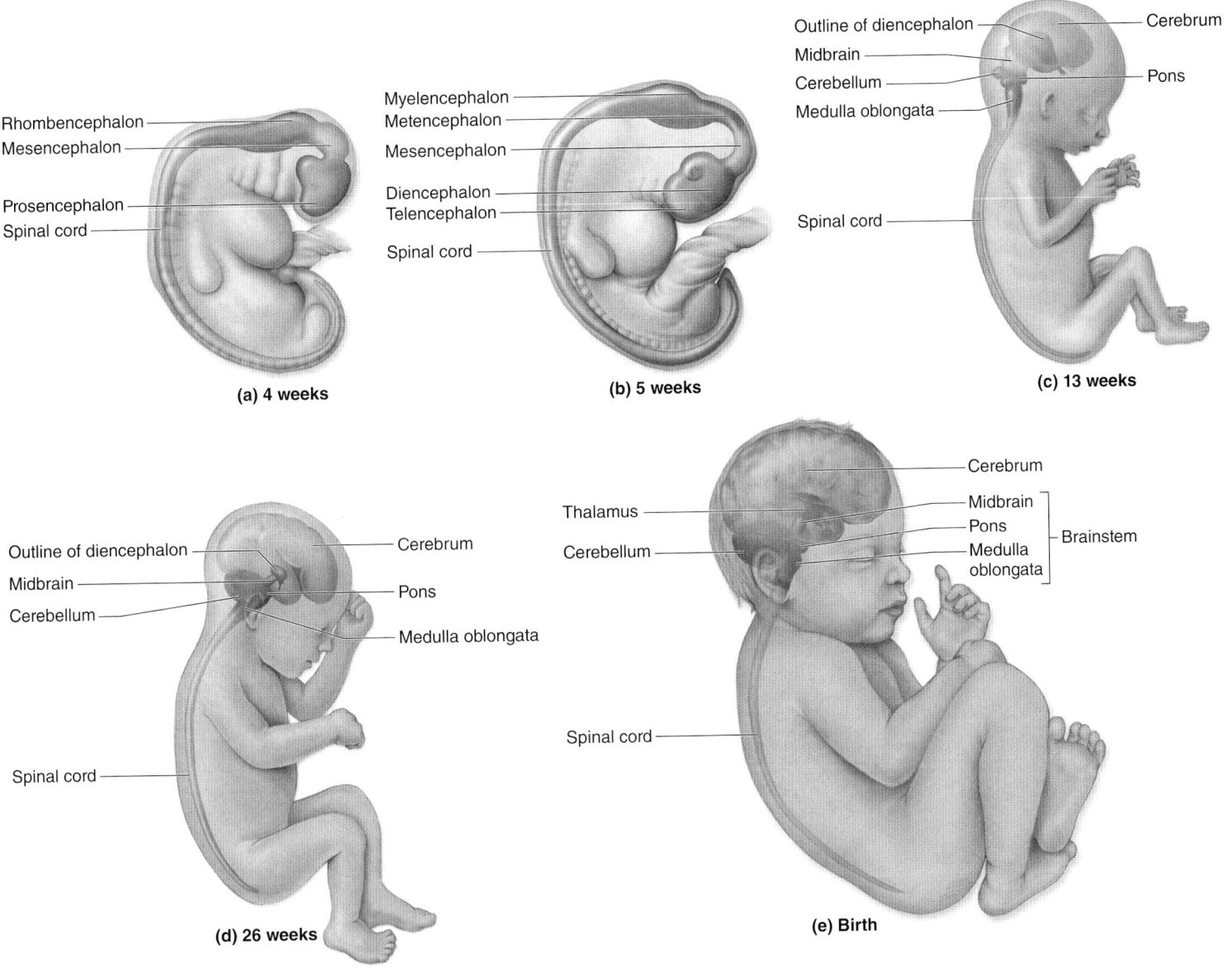

Figure 13.3 Structural Changes in the Developing Brain. (*a*) As early as 4 weeks, the growing brain is curled because of space restrictions in the developing head. (*b*) At 5 weeks, the secondary brain vesicles appear. (*c*) By 13 weeks, the telencephalon grows rapidly and envelops the diencephalon. (*d*) Some major sulci and gyri are present by 26 weeks. (*e*) The features of an adult brain are present at birth. AP|R

13.1c Gray Matter and White Matter Distribution

LEARNING OBJECTIVE

5. Compare and contrast the distribution of gray and white matter throughout the brain divisions.

Two distinct tissue areas are recognized within the brain and spinal cord: gray matter and white matter. The **gray matter** primarily derives its color from the motor neuron and interneuron cell bodies and their associated capillary beds, as well as the dendrites and some unmyelinated axons. The **white matter** derives its color from the myelin on the abundant myelinated axons.

An outer, superficial region of gray matter in the cerebrum forms during brain development from migrating peripheral neurons. These external sheets of gray matter, called the **cerebral cortex** (se-rē′bral kor′teks; *cerebro* = brain, *cortex* = bark), cover the surface of the cerebrum. The white matter lies deep to the gray matter of the cortex. However, within the masses of white matter, the brain also contains discrete internal clusters of gray matter called **cerebral nuclei.** These are oval, spherical, or sometimes irregularly shaped clusters of neuron cell bodies. **Figure 13.4** shows the distribution of gray matter and white matter in various regions of the CNS. **Table 13.2** is a glossary of nervous system structures.

WHAT DID YOU LEARN?

❹ Where is gray matter located within the cerebrum and spinal cord?

Figure 13.4 Gray and White Matter in the CNS. The gray matter represents regions containing neuron cell bodies, dendrites, and unmyelinated axons, whereas the white matter derives its color from myelinated axons. The distribution of gray and white matter is compared in (*a*) the cerebrum and diencephalon, (*b*) the cerebellum and brainstem, (*c*) the medulla oblongata (which is the inferior portion of the brainstem), and (*d*) the spinal cord. AP|R

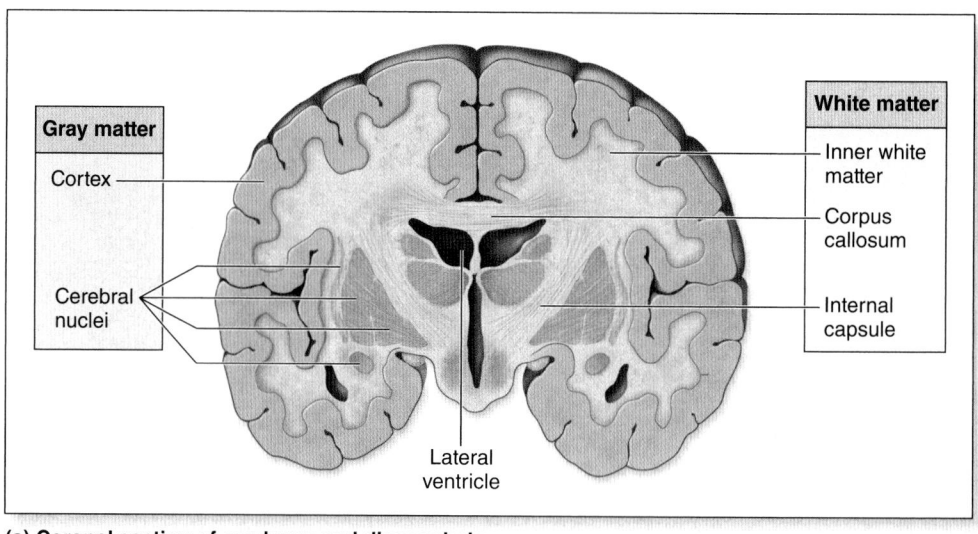

(a) Coronal section of cerebrum and diencephalon

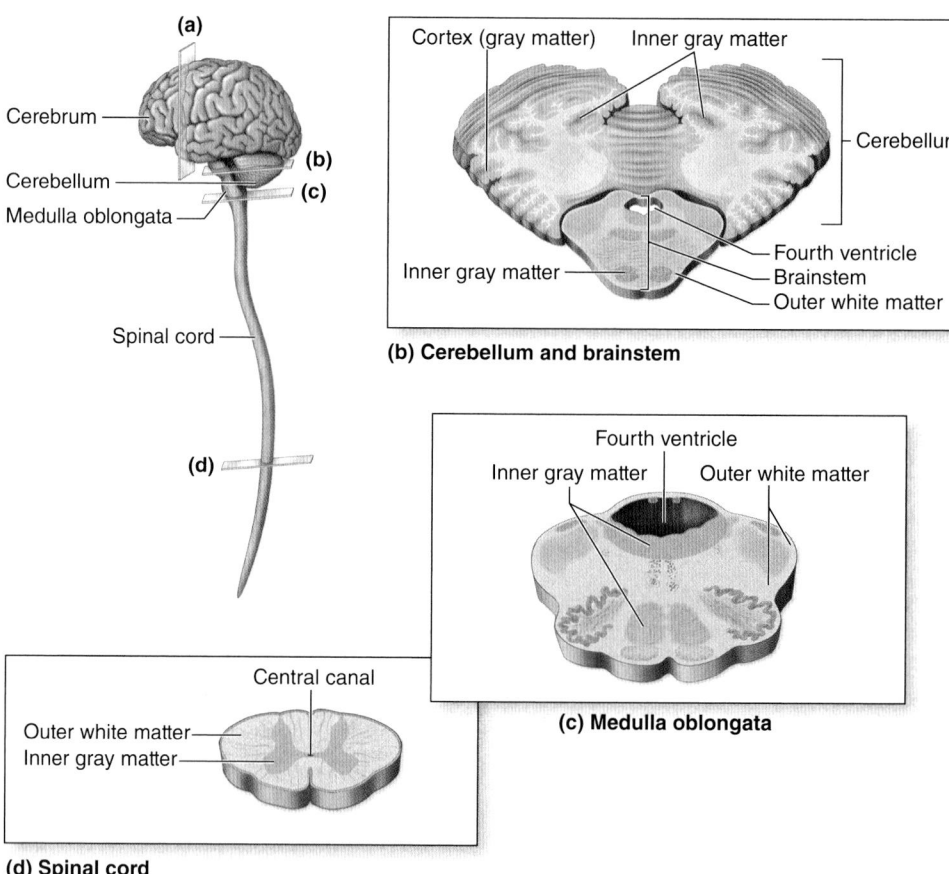

(b) Cerebellum and brainstem

(c) Medulla oblongata

(d) Spinal cord

Table 13.2	Glossary of Nervous System Structures
Structure	**Description**
Ganglion	Cluster of neuron cell bodies in the peripheral nervous system (PNS)
Center	Cluster of neuron cell bodies with a common function in the central nervous system (CNS)
Nucleus	Center that displays discrete anatomic boundaries
Nerve	Bundle of axons in the PNS
Nerve plexus	Network of nerves
Tract	Bundles of axons within the CNS in which the axons have a similar function and share a common origin and destination
Funiculus	Group of tracts in a specific area of the spinal cord
Pathway	Centers and tracts that connect the CNS with body organs and systems

13.2 Protection and Support of the Brain

The brain is both protected and isolated by multiple structures. The bony cranium provides rigid support, while protective connective tissue membranes called meninges surround and partition portions of the brain. Cerebrospinal fluid (CSF) acts as a cushioning fluid between specific layers of meninges. Finally, the brain has a unique blood-brain barrier to prevent entry of harmful materials into the brain from the blood.

13.2a Cranial Meninges

LEARNING OBJECTIVES

1. Compare and contrast the structure and locations of the three meninges, and list the spaces found between the meninges.

2. Describe the four cranial dural septa, and give their locations.

The **cranial meninges** (mĕ-nin′jes, mē′nin-jēz; sing., *meninx*, men′ingks; membrane) are three connective tissue layers that separate and support the soft tissue of the brain from the bones of the cranium, enclose and protect some of the blood vessels that supply the brain, and help contain and circulate cerebrospinal fluid. From deep (closest to the brain) to superficial (farthest from the brain), the cranial meninges are the pia mater, the arachnoid mater, and the dura mater (**figure 13.5**).

Pia Mater

The **pia mater** (pī′ă mā′ter, pē′ă mah′ter; *pia* = tender, *mater* = mother) is the innermost of the cranial meninges. It is a thin layer of delicate areolar connective tissue that tightly adheres to the brain and follows every contour of the brain surface.

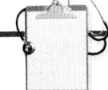

INTEGRATE

CLINICAL VIEW

Epidural and Subdural Hematomas

A pooling of blood outside of a vessel is referred to as a **hematoma** (hē -mă -tō′mă; *hemato* = blood, *oma* = tumor). An **epidural hematoma** is a pool of blood forming in the epidural space of the brain, usually due to a severe blow to the head. The adjacent brain tissue becomes distorted and compressed as a result of the hematoma continuing to increase in size. Severe neurologic injury and death may occur if the bleeding is not stopped and the accumulated blood removed by surgically drilling a hole in the skull, suctioning out the blood, and ligating (tying off) the bleeding vessel.

A **subdural hematoma** is a hemorrhage that occurs in the subdural space between the dura mater and arachnoid mater. These hematomas typically result from ruptured veins caused by either fast or violent rotational motion of the head. Blood pools in this space and compresses the brain, although usually these events occur more slowly than with an epidural hematoma. Subdural hematomas are treated similarly to epidural hematomas.

Arachnoid Mater

The **arachnoid** (ă-rak′noyd) **mater**, also called the *arachnoid membrane*, lies external to the pia mater. The term *arachnoid* means "resembling a spider web," and this meninx is so named because it is partially composed of a delicate web of collagen and elastic fibers, termed the **arachnoid trabeculae**. Immediately deep to the arachnoid mater is the **subarachnoid space**, which contains cerebrospinal fluid (to be discussed in section 13.2c). The arachnoid trabeculae extend through this space from the arachnoid to the underlying pia mater. Both the arachnoid trabeculae and cerebrospinal fluid support cerebral arteries and veins within the subarachnoid space. Between the arachnoid mater and the overlying dura mater is a potential space, the **subdural space**. The subdural space becomes an actual space if blood or fluid accumulates there, a condition called a subdural hematoma (see Clinical View: "Epidural and Subdural Hematomas").

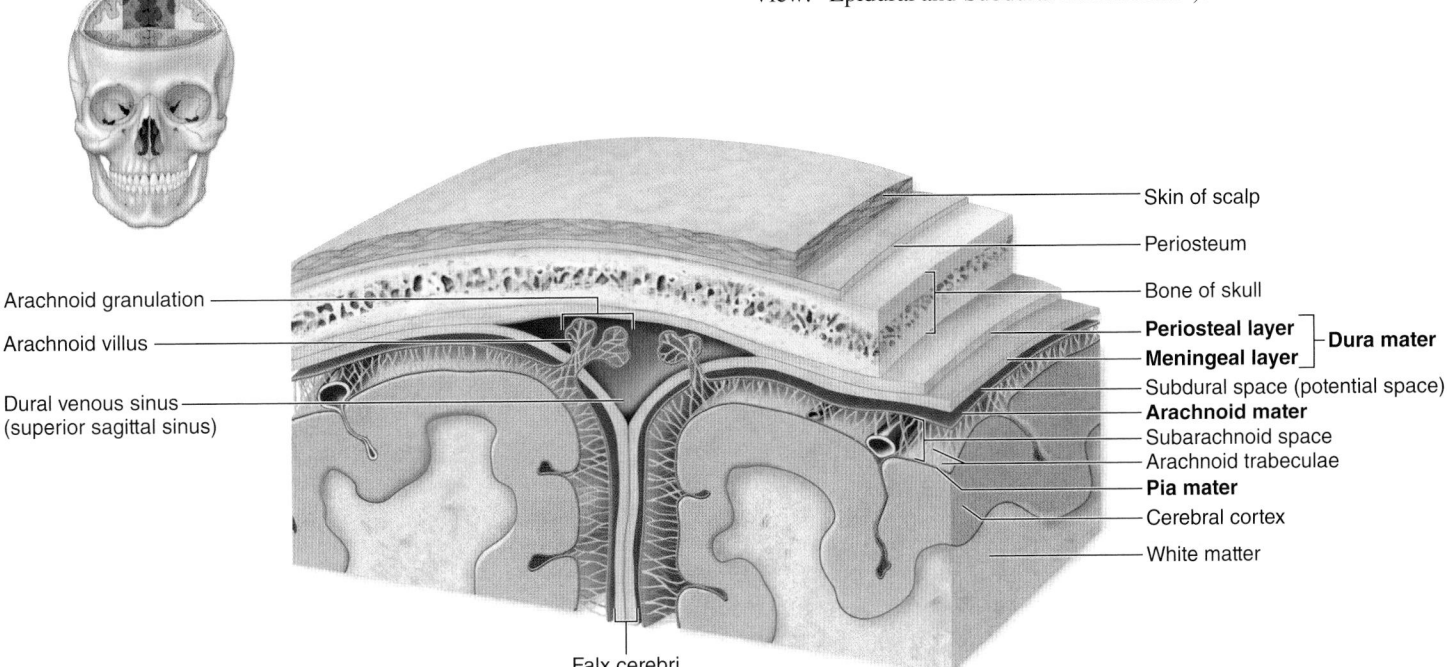

Arachnoid granulation

Arachnoid villus

Dural venous sinus (superior sagittal sinus)

Falx cerebri

Skin of scalp

Periosteum

Bone of skull

Periosteal layer / Meningeal layer — **Dura mater**

Subdural space (potential space)

Arachnoid mater

Subarachnoid space

Arachnoid trabeculae

Pia mater

Cerebral cortex

White matter

Figure 13.5 Cranial Meninges. A coronal section of the superior portion of the head depicts the organization of the three meningeal layers: the dura mater, the arachnoid mater, and the pia mater. In the midline, folds of the inner meningeal layer of the dura mater form the falx cerebri, which partitions the two cerebral hemispheres. The inner meningeal layer and the outer periosteal layer separate at various locations to form the dural venous sinuses, such as the superior sagittal sinus (shown here), which drain blood away from the brain.

Dura Mater

The **dura mater** (dū′ră; *dura* = tough) is the strongest of the meninges, as its Latin name indicates. This outer, dense irregular connective tissue covering is composed of two layers. The **meningeal** (mě-nin′jē-ăl, men′in-jē′ăl) **layer** is immediately superficial to the arachnoid. The **periosteal** (per-ē-os′tē-ăl) **layer,** the more superficial layer, forms the periosteum on the internal surface of the cranial bones. The meningeal layer is usually fused to the periosteal layer, except in specific areas where these two layers separate to form large, blood-filled spaces called **dural venous sinuses.** (A sinus is a modified vein.) Dural venous sinuses are typically triangular in cross section, and unlike most other veins, they do not have valves to regulate venous blood flow. The dural venous sinuses drain blood from the brain, and most of the specific sinuses are shown in **figure 13.6.**

The dura mater and the bones of the skull may be separated by a potential **epidural** (ep-i-dū′răl) **space,** which contains the arteries and veins that nourish the meninges and bones of the cranium. Like the subdural space, the epidural space may become a real space if blood or fluid accumulates within it.

Cranial Dural Septa

The meningeal layer of the dura mater extends as flat partitions into the cranial cavity at four locations. Collectively, these double layers of dura mater are called **cranial dural septa** (sing., *septum* = wall). These membranous partitions separate specific parts of the brain and provide additional stabilization and support to the brain. There are four cranial dural septa: the falx cerebri, tentorium cerebelli, falx cerebelli, and diaphragma sellae (figure 13.6). The **falx cerebri** (falks sē-rē′brī; *falx* = sickle) is the largest of the four dural septa. This large, sickle-shaped vertical fold of dura mater is located in the midsagittal plane and projects into the longitudinal fissure between the left and right cerebral hemispheres. Anteriorly, its inferior portion attaches to the crista galli of the ethmoid bone; posteriorly, its inferior portion attaches to the internal occipital crest (see section 8.2). Running within the margins of either end of this dural septum are two dural venous sinuses: the **superior sagittal sinus** and the **inferior sagittal sinus.**

The **tentorium cerebelli** (ten-tō′rē-ŭm ser-e-bel′ī) is a horizontally oriented fold of dura mater that separates both the occipital and temporal lobes of the cerebrum from the cerebellum. It is named for

INTEGRATE

CLINICAL VIEW
Meningitis

Meningitis is the inflammation of the meninges, and typically it is caused by viral or bacterial infection. Early symptoms may include fever, severe headache, vomiting, and a stiff neck (because pain from the meninges may be referred [see section 16.2b] to the posterior neck). Bacterial meningitis typically produces more severe symptoms and may result in brain damage and death if left untreated. Both viral and bacterial meningitis are contagious and may be spread through respiratory droplets or oral secretions, so it is a disease that may spread rapidly through college dormitories or military barracks (where individuals live in close quarters). Thus, most teenagers are recommended to get the bacterial meningitis vaccine (which protects them against the most common bacterial strains that cause meningitis) prior to attending college.

the fact that it forms a dural "tent" over the cerebellum. The **transverse sinuses** run within its posterior border, whereas the **straight sinus** is found along its midsagittal plane. The anterior surface of the tentorium cerebelli has a gap or opening, called the **tentorial notch** (or *tentorial incisure*), to allow for the passage of the brainstem.

Extending into the midsagittal line inferior to the tentorium cerebelli is the **falx cerebelli,** a sickle-shaped vertical partition that divides the left and right cerebellar hemispheres. A tiny **occipital** (ok-sip′i-tăl; *occiput* = back of head) **sinus** (another dural venous sinus) runs in its posterior vertical border.

The **diaphragma sellae** (dī-ă-frag′mă sel′ē; *sella* = saddle) is the smallest of the dural septa. It forms a "roof" over the sella turcica of the sphenoid bone. A small opening within it allows for the passage of a thin stalk, called the infundibulum, that attaches the pituitary gland to the base of the hypothalamus (described in section 13.4c).

WHAT DO YOU THINK?

1 Which meningeal layer provides the most support and physical protection to the brain? Why?

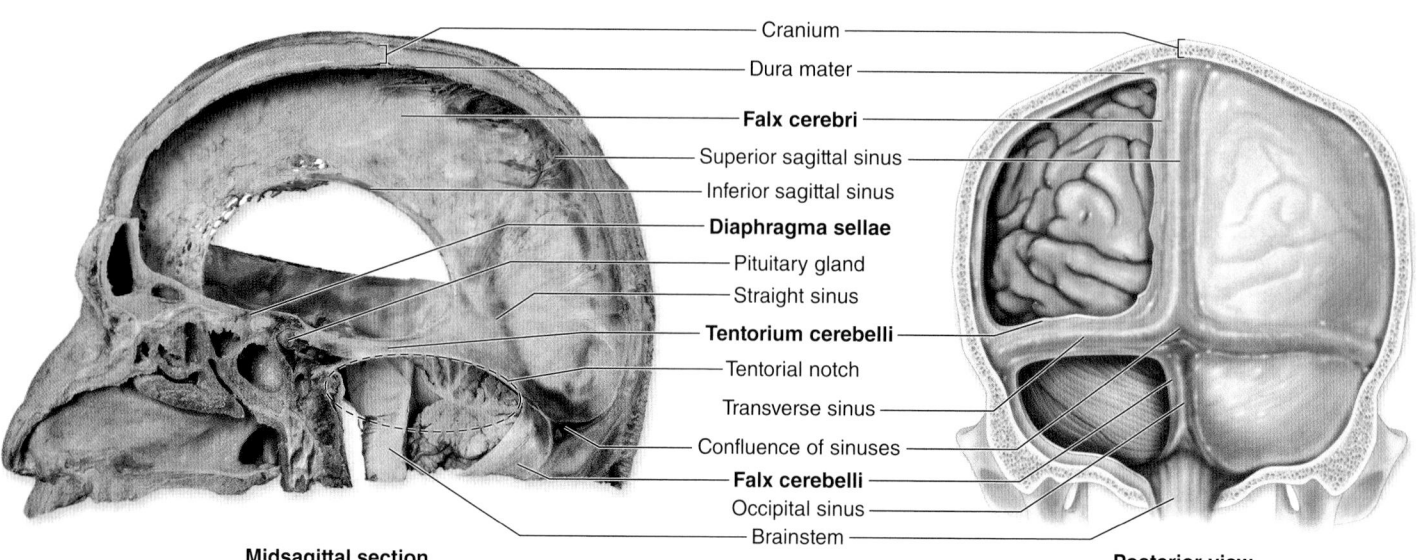

Midsagittal section

Posterior view

Figure 13.6 Cranial Dural Septa. A midsagittal section and posterior view of the head show the orientation of the falx cerebri, falx cerebelli, tentorium cerebelli, and diaphragma sellae.

💡 WHAT DID YOU LEARN?

5 From deepest (closest to the brain) to superficial (farthest away from the brain), name the meninges and the spaces between the meninges.

6 Where is the falx cerebri located, and what is its role?

💡 WHAT DID YOU LEARN?

7 Where is the fourth ventricle located, and how does it connect with the subarachnoid space?

13.2b Brain Ventricles

✓ **LEARNING OBJECTIVE**

3. Describe the anatomy and location of the ventricles.

Ventricles (ven′tri-kl; little cavity) are cavities or expansions within the brain that are derived from the neural canal (the lumen of the embryonic neural tube). All of the ventricles are lined with ependymal cells (see section 12.4b) and contain cerebrospinal fluid. The ventricles are connected with one another as well as with the central canal of the spinal cord (**figure 13.7**).

There are four ventricles within the brain: Two **lateral ventricles** are in the cerebrum, separated by a thin medial partition called the **septum pellucidum** (pe-lū′si-dum; *pellucid* = transparent) (see figure 13.4). Within the diencephalon is a smaller, thinner ventricle called the **third ventricle** (figure 13.7). Each lateral ventricle is connected with the third ventricle through an opening called the **interventricular foramen** (formerly called the *foramen of Munro*). A narrow canal called the **cerebral aqueduct** (ak′we-dŭkt) (also called the *mesencephalic aqueduct* and formerly called the *aqueduct of Sylvius*) passes through the midbrain and connects the third ventricle with the sickle-shaped **fourth ventricle.** The fourth ventricle is located between the pons, medulla oblongata, and cerebellum. It opens to the subarachnoid space via a single *median aperture* and paired *lateral apertures.* The fourth ventricle narrows at its inferior end before it merges with the slender **central canal** in the spinal cord.

13.2c Cerebrospinal Fluid

✓ **LEARNING OBJECTIVES**

4. Explain the three functions of cerebrospinal fluid.

5. Trace the circulation of cerebrospinal fluid, beginning with its origin and ending with its removal.

Cerebrospinal (sĕ-rē′brō-spī-năl) **fluid (CSF)** is a clear, colorless liquid that circulates within the ventricles and subarachnoid space. CSF bathes the exposed surfaces of the central nervous system and completely surrounds it. CSF performs several important functions:

- **Buoyancy.** The brain floats within the CSF, thereby reducing its apparent weight by more than 95%; this prevents the brain from being crushed under its own weight. Without CSF to support it, portions of the brain would sink through the foramen magnum.

- **Protection.** CSF provides a liquid cushion to protect delicate neural structures from sudden movements. When you try to walk quickly in a swimming pool, your movements are slowed as the water acts as a "movement buffer." CSF likewise helps slow movements of the brain if the skull or body moves suddenly and forcefully.

- **Environmental stability.** CSF transports nutrients and chemical messengers to the brain and removes waste products from the brain. Additionally, CSF protects nervous tissue from chemical fluctuations that would disrupt neuron function. The waste products and excess CSF are eventually transported into the venous circulation.

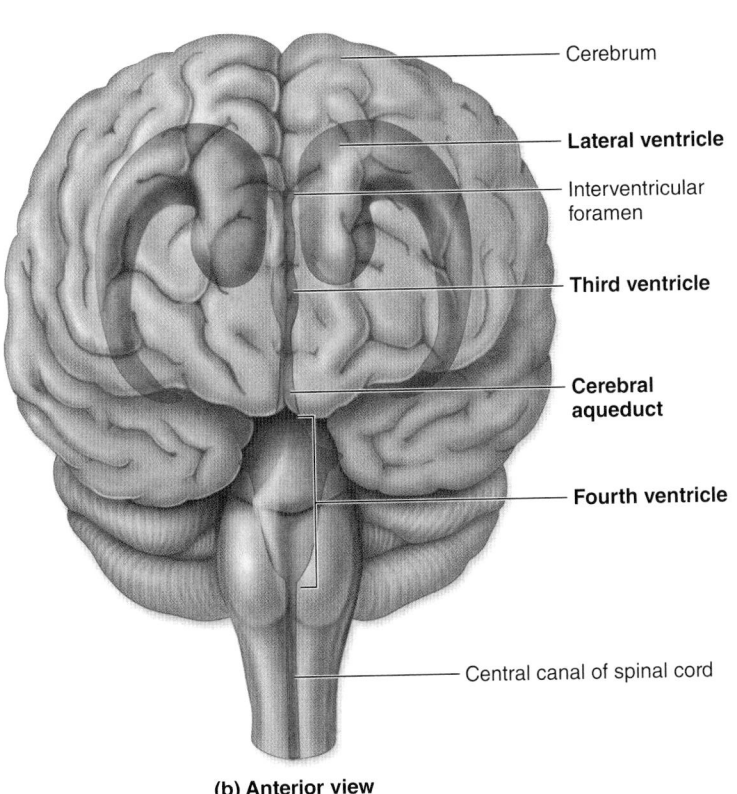

(a) Lateral view

Posterior ← / Anterior →

Third ventricle — Interventricular foramen — Lateral ventricles — Cerebral aqueduct — Fourth ventricle — Lateral aperture — Median aperture — Central canal of spinal cord

(b) Anterior view

Cerebrum — Lateral ventricle — Interventricular foramen — Third ventricle — Cerebral aqueduct — Fourth ventricle — Central canal of spinal cord

Figure 13.7 Ventricles of the Brain. The ventricles are formed from the embryonic neural canal. They contain cerebrospinal fluid (CSF), which transports chemical messengers, nutrients, and waste products. (*a*) Lateral and (*b*) anterior views show the positioning and relationships of the ventricles. AP|R

CSF Formation, Circulation, and Removal

The production of CSF by the brain occurs at a rate of about 500 milliliters (mL) (or 1/2 liter) per day, with the volume of CSF in the subarachnoid space at any given time ranging between 100 mL and 160 mL. Cerebrospinal fluid is initially formed by the **choroid plexus** (ko′royd plek′sŭs; *chorioeides* = membrane, *plexus* = a braid), a region of specialized tissue in each ventricle. The choroid plexus is composed of a layer of glial cells called **ependymal** (ep-en′di-măl; *ependyma* = an upper garment) **cells** (see section 12.4b) and the blood capillaries that lie within the pia mater (**figure 13.8**).

The formation of CSF by the choroid plexus occurs as follows: blood plasma (a watery fluid containing glucose and ions such as K^+, Na^+, Cl^-, and Ca^{2+}) is filtered from the blood capillaries. Its composition is modified by the ependymal cells to produce CSF, which is released into a ventricle. The chemical composition of CSF that enters the ventricles is slightly different from the filtered blood plasma with relatively more Na^+ and Cl^- and relatively less glucose, K^+, and Ca^{2+}. Following its formation, the CSF circulates through the ventricles, where additional fluid is added by ependymal cells (see figure 12.5). The CSF then circulates from the ventricles into the subarachnoid space. Once CSF enters the subarachnoid space, its volume increases with the addition of excess interstitial fluid from the brain (which is formed during capillary exchange; see section 20.3). Thus, CSF is collectively formed by the choroid plexus (about 30%), ependymal cells lining the ventricles (about 30%), and fluid added into the subarachnoid space (about 40%). (As a result, the brain does not need any lymph vessels [see section 21.1], because it rids itself of excess fluid through the CSF.)

The CSF must be continuously reabsorbed from the subarachnoid space at the same rate as its formation so that the fluid does not accumulate and compress or damage the nervous tissue. Reabsorption of CSF occurs at **arachnoid villi** (vil′ī; *villi* = shaggy hair), which are fingerlike extensions of the arachnoid mater that project through the dura mater into the

INTEGRATE

CLINICAL VIEW
Hydrocephalus

Hydrocephalus (hī-drō-sef′a-lŭs; *hydro* = water, *kephale* = head) refers to the pathologic condition of excessive CSF, which often leads to brain distortion. Most cases of hydrocephalus result from either an obstruction in CSF flow that restricts its reabsorption into the venous bloodstream or some intrinsic problem with the arachnoid villi themselves.

If hydrocephalus develops in a young child, the head becomes enlarged and neurologic damage may result. If hydrocephalus develops after the cranial sutures have closed, the brain may be compressed within the fixed cranium as the ventricles expand, resulting in permanent brain damage. Hydrocephalus may be treated surgically by implanting shunts (tubes) that drain excess CSF to other body regions. The fluid is then absorbed into the blood.

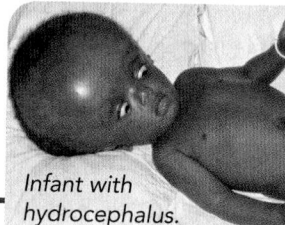

Infant with hydrocephalus.

dural venous sinuses. A collection of arachnoid villi forms an **arachnoid granulation.** As additional CSF is formed, fluid pressure rises within the subarachnoid space, which forces CSF from the subarachnoid space across the arachnoid villi to return to the blood within the dural venous sinuses. As a result, the arachnoid villi provide a conduit for a one-way flow of excess CSF to be returned to the blood within the dural venous sinuses. **Figure 13.9** describes in detail and shows the process of CSF production, circulation, and removal.

WHAT DID YOU LEARN?

8 What are the three main functions of cerebrospinal fluid?

9 Where is CSF first produced, where does it circulate, and how does CSF get removed?

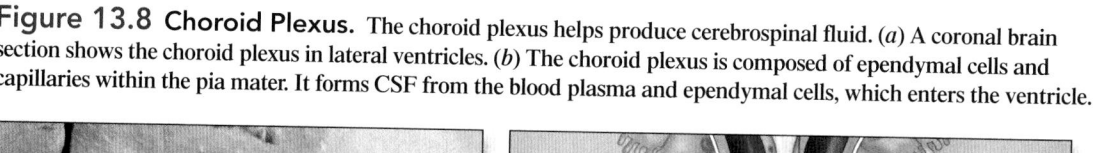

Figure 13.8 Choroid Plexus. The choroid plexus helps produce cerebrospinal fluid. (*a*) A coronal brain section shows the choroid plexus in lateral ventricles. (*b*) The choroid plexus is composed of ependymal cells and capillaries within the pia mater. It forms CSF from the blood plasma and ependymal cells, which enters the ventricle.

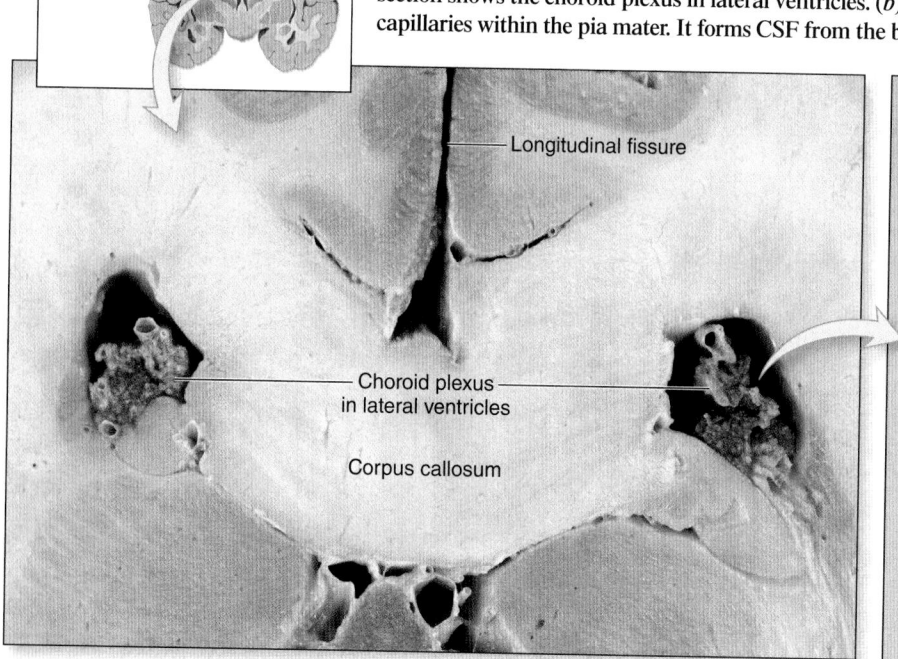

Longitudinal fissure

Choroid plexus in lateral ventricles

Corpus callosum

(a) Coronal section of the brain, close-up

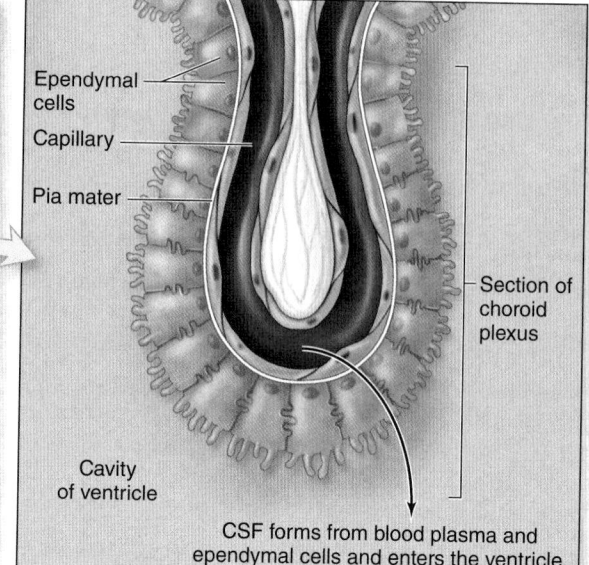

Ependymal cells

Capillary

Pia mater

Section of choroid plexus

Cavity of ventricle

CSF forms from blood plasma and ependymal cells and enters the ventricle.

(b) Choroid plexus

① CSF is produced by the **choroid plexus** in the ventricles.

② CSF flows from the **lateral ventricles,** through the **interventricular foramen** into the **third ventricle,** and then through the **cerebral aqueduct** into the **fourth ventricle.**

③ CSF in the fourth ventricle passes through the paired lateral apertures or the single median aperture, and into the **subarachnoid space** as well as the central canal of the spinal cord.

④ As the CSF flows through the subarachnoid space, it provides buoyancy to support the brain.

⑤ Excess CSF flows into the **arachnoid villi,** then drains into the **dural venous sinuses.** The greater pressure on the CSF in the subarachnoid space ensures that CSF moves into the venous sinuses without permitting venous blood to enter the subarachnoid space.

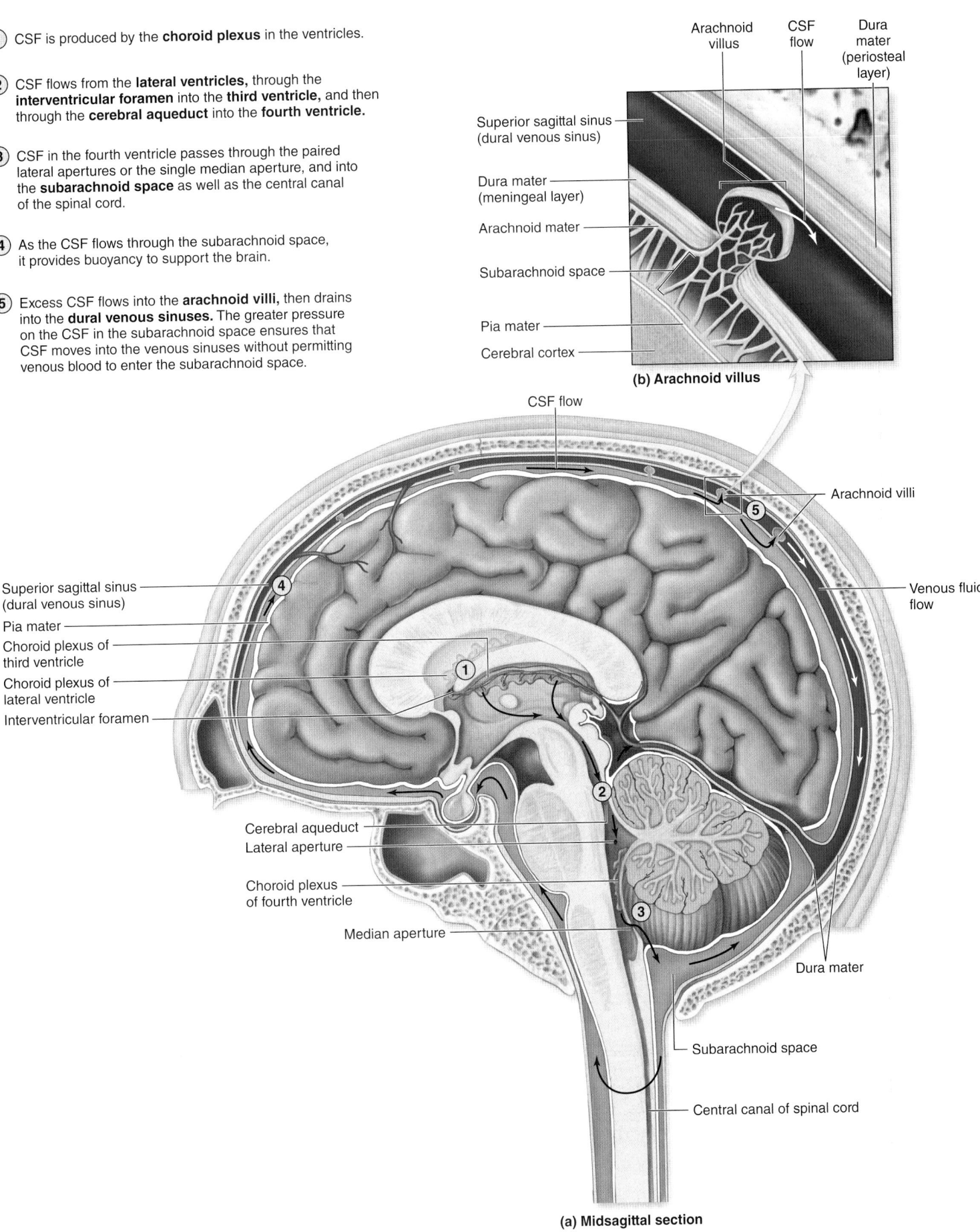

Arachnoid villus   CSF flow   Dura mater (periosteal layer)

Superior sagittal sinus (dural venous sinus)

Dura mater (meningeal layer)

Arachnoid mater

Subarachnoid space

Pia mater

Cerebral cortex

(b) Arachnoid villus

CSF flow

Superior sagittal sinus (dural venous sinus)

Pia mater

Choroid plexus of third ventricle

Choroid plexus of lateral ventricle

Interventricular foramen

Cerebral aqueduct

Lateral aperture

Choroid plexus of fourth ventricle

Median aperture

Arachnoid villi

Venous fluid flow

Dura mater

Subarachnoid space

Central canal of spinal cord

(a) Midsagittal section

Figure 13.9 Production and Circulation of Cerebrospinal Fluid. (*a*) A midsagittal section identifies the sites where cerebrospinal fluid (CSF) is formed and the pathway of its circulation toward the arachnoid villi. (*b*) CSF flows from the arachnoid villi into the dural venous sinuses. AP|R

INTEGRATE

CONCEPT CONNECTION

As just mentioned, waste products and excess CSF are transported into the blood of the cardiovascular system (see chapters 19 and 20). These waste products (along with other waste products picked up from other tissues) will be filtered from the blood specifically by the kidneys of the urinary system (see chapter 24). The kidneys remove these waste products and secrete them in the form of urine, which is expelled from the body.

13.2d Blood-Brain Barrier

LEARNING OBJECTIVES

6. Describe the components that form the blood-brain barrier.
7. Explain how the blood-brain barrier protects the brain.

Nervous tissue is protected from the general circulation by the **blood-brain barrier (BBB)**. This barrier strictly regulates which substances can and cannot enter the interstitial fluid of the brain to help prevent exposure of neurons in the brain to drugs, waste products in the blood, and variations in levels of normal substances (e.g., ions, hormones) that could adversely affect brain function.

The BBB is formed by specialized capillaries. Capillaries are typically composed of an endothelial lining resting on a basement membrane (see table 20.2 that shows types of capillaries). Capillaries forming the BBB exhibit three significant structural differences from other capillaries (**figure 13.10**). (1) The endothelial cells contain tight junctions, which prevent passage of materials between cells (see section 4.6d). Thus, most substances are forced through the endothelial cells and their movement is controlled by membrane transport processes (see section 4.3). (2) The capillary wall is made more substantial by a thickened basement membrane that further restricts the passage of substances from the blood into the brain. (3) The capillaries forming the BBB are wrapped in the **perivascular feet** of **astrocytes** (previously discussed in section 12.4b), which form the outermost portion of the BBB. The BBB acts as a "gatekeeper" to control which materials pass from the blood into the brain.

Even so, the barrier is not absolute. Lipid-soluble compounds, such as nicotine, alcohol, and some anesthetics, can diffuse across the endothelial plasma membranes (and thus past this barrier) and into the interstitial fluid of the CNS to reach the brain neurons. In addition, drugs such as cocaine and methamphetamine can damage this barrier.

There are several important exceptions to the presence of the BBB in the brain. It is markedly reduced or missing in three distinct locations in the CNS: the choroid plexus, hypothalamus, and pineal gland. The capillaries of the choroid plexus must be permeable to produce CSF, and the hypothalamus and pineal gland produce certain hormones that must have ready access to the blood.

WHAT DID YOU LEARN?

10. How does the blood-brain barrier protect nervous tissue?

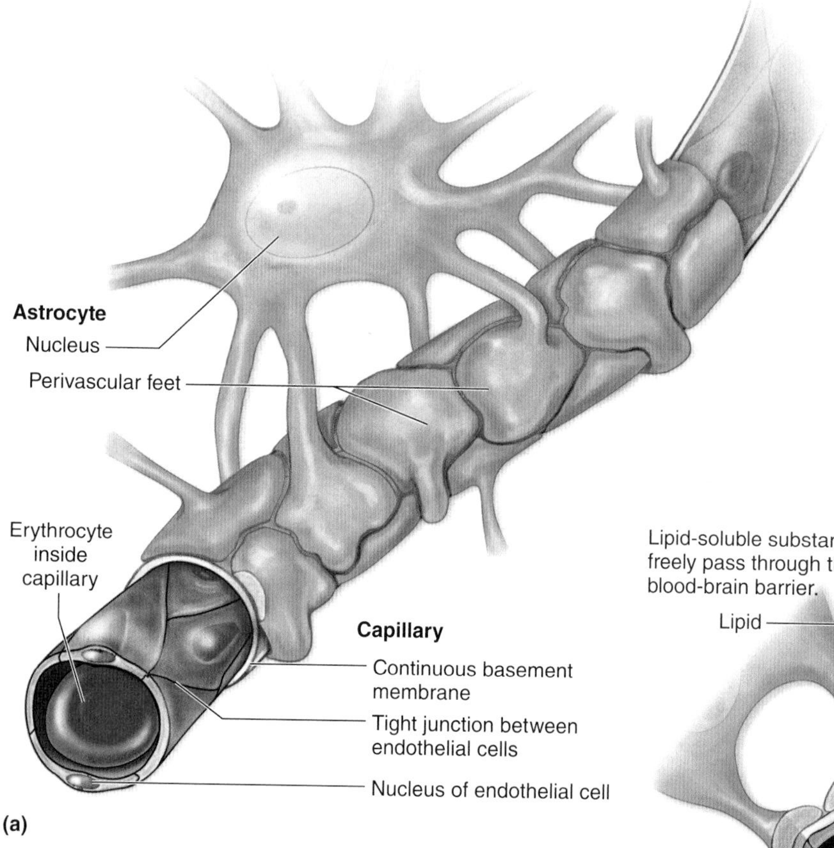

Astrocyte
Nucleus
Perivascular feet

Erythrocyte inside capillary

Capillary
Continuous basement membrane
Tight junction between endothelial cells
Nucleus of endothelial cell

(a)

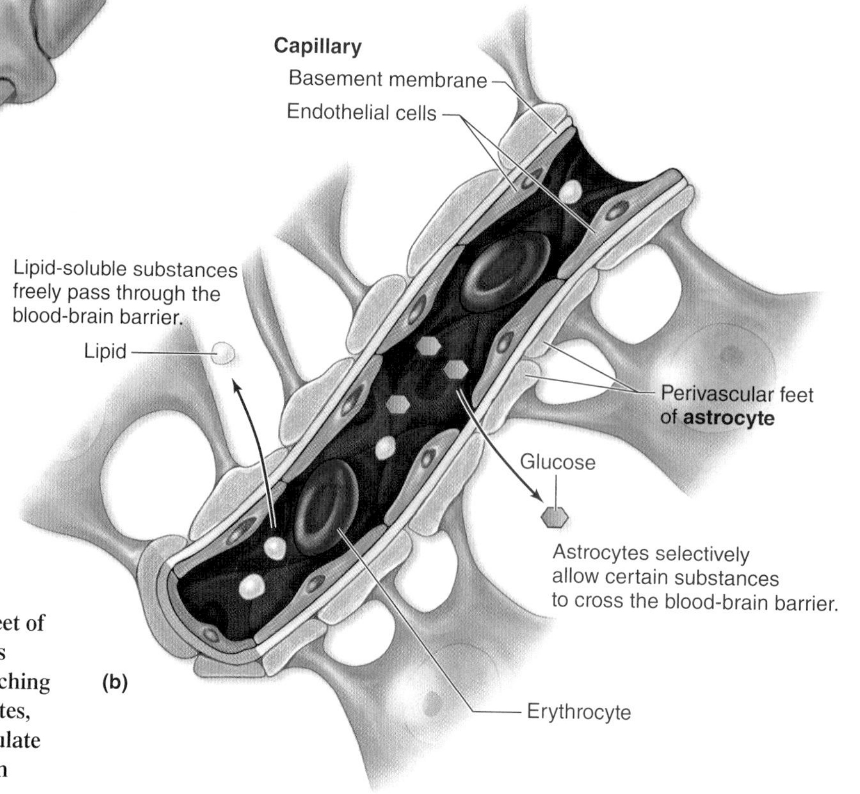

Capillary
Basement membrane
Endothelial cells

Lipid-soluble substances freely pass through the blood-brain barrier.
Lipid

Perivascular feet of **astrocyte**

Glucose

Astrocytes selectively allow certain substances to cross the blood-brain barrier.

(b)

Erythrocyte

Figure 13.10 Blood-Brain Barrier. The perivascular feet of the astrocytes and the tight endothelial junctions of the capillaries work together to prevent harmful materials in the blood from reaching the brain. (*a*) Here we show just a few perivascular feet of astrocytes, so that their structure may be appreciated. (*b*) The astrocytes regulate the movement of most substances, but lipid-soluble substances can pass through the barrier freely.

13.3 Cerebrum

The **cerebrum** is the location of conscious thought processes and the origin of all complex intellectual functions. It is readily identified as the two large hemispheres on the superior aspect of the brain (see figure 13.1). The functional activities in your cerebrum enable you to read and comprehend the words in this textbook, turn its pages, form and remember ideas, and talk about what you've learned with your peers. It is the center of your intelligence, reasoning, sensory perception, thought, memory, and judgment, as well as your voluntary control of skeletal muscle movement, and conscious perception of your special senses (e.g., vision, hearing, balance, smell, and taste).

13.3a Cerebral Hemispheres

LEARNING OBJECTIVES

1. Discriminate between the left and right cerebral hemispheres regarding their general functions.
2. Identify the role of the corpus callosum.

The cerebrum is composed of two halves, called the left and right **cerebral hemispheres** (hem'i-sfēr; *hemi* = half, *sphaira* = ball) **(figure 13.11)**. These paired cerebral hemispheres are separated by a narrow, deep cleft called the **longitudinal fissure** that extends along the midsagittal plane. The cerebral hemispheres are separate from one another, except at a few locations where bundles of axons called **tracts** form white matter regions that allow for communication between them.

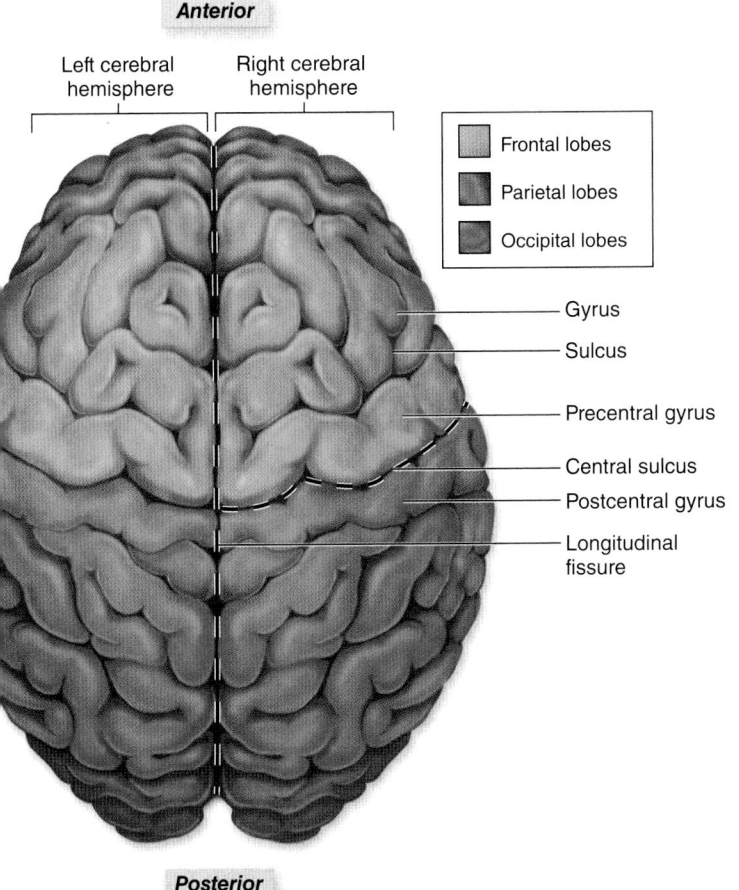

Anterior

Left cerebral hemisphere | Right cerebral hemisphere

- Frontal lobes
- Parietal lobes
- Occipital lobes

- Gyrus
- Sulcus
- Precentral gyrus
- Central sulcus
- Postcentral gyrus
- Longitudinal fissure

Posterior

Superior view

Figure 13.11 Cerebral Hemispheres. A superior view shows the cerebral hemispheres, where our conscious activities, memories, behaviors, plans, and ideas are initiated and controlled. **AP|R**

The largest of these white matter tracts, the **corpus callosum** (kōr'pŭs kal-lō'sŭm; *corpus* = body, *callosum* = hard) connects the hemispheres (see a midsagittal section of the corpus callosum in figure 13.1*c*). The corpus callosum provides the main method of communication between these hemispheres.

Three points should be kept in mind regarding the cerebral hemispheres:

- It is difficult to assign a precise function to a specific region of the cerebral cortex in most cases. Considerable overlap and indistinct boundaries permit a single region of the cortex to exhibit several different functions. Additionally, some aspects of cortical function, such as memory or consciousness, cannot easily be assigned to any single region.

- As a general rule, both cerebral hemispheres receive their sensory information from, and project motor commands to, the opposite side of the body. The right cerebral hemisphere controls the left side of the body, and vice versa.

- The two hemispheres appear as anatomic mirror images, but they display some functional differences, termed *cerebral lateralization*. For example, the regions of the brain that are responsible for controlling speech and understanding verbalization are frequently located in the left cerebral hemisphere. These types of differences primarily affect higher-order functions, which are addressed later in this chapter.

WHAT DO YOU THINK?

2 One past treatment for severe epilepsy was to cut the corpus callosum, thus confining epileptic seizures to just one cerebral hemisphere. How would cutting the corpus callosum affect communication between the left and right hemispheres?

WHAT DID YOU LEARN?

11 What is the general function of each cerebral hemisphere?

12 What is the function of the corpus callosum?

13.3b Lobes of the Cerebrum

LEARNING OBJECTIVE

3. Explain the physical boundaries, important features, and functions of each cerebral lobe.

Each cerebral hemisphere is divided into five anatomically distinct lobes. Four of these lobes are visible on the external surface and are named for the overlying cranial bones: the frontal, parietal, temporal, and occipital lobes **(figure 13.12*a*)**. The fifth lobe, called the insula, is not visible at the surface of the hemispheres. Each lobe exhibits specific cortical regions and association areas.

The **frontal lobe** lies deep to the frontal bone and forms the anterior part of the cerebral hemisphere. The frontal lobe ends posteriorly at a deep groove called the **central sulcus** that marks the boundary with the parietal lobe. The inferior border of the frontal lobe is marked by the **lateral sulcus,** a deep groove that separates the frontal and parietal lobes from the temporal lobe. An important anatomic feature of the frontal lobe is the **precentral gyrus,** which is the mass of nervous tissue immediately anterior to the central sulcus. The frontal lobe is primarily concerned with voluntary motor functions, concentration, verbal communication, decision making, planning, and personality.

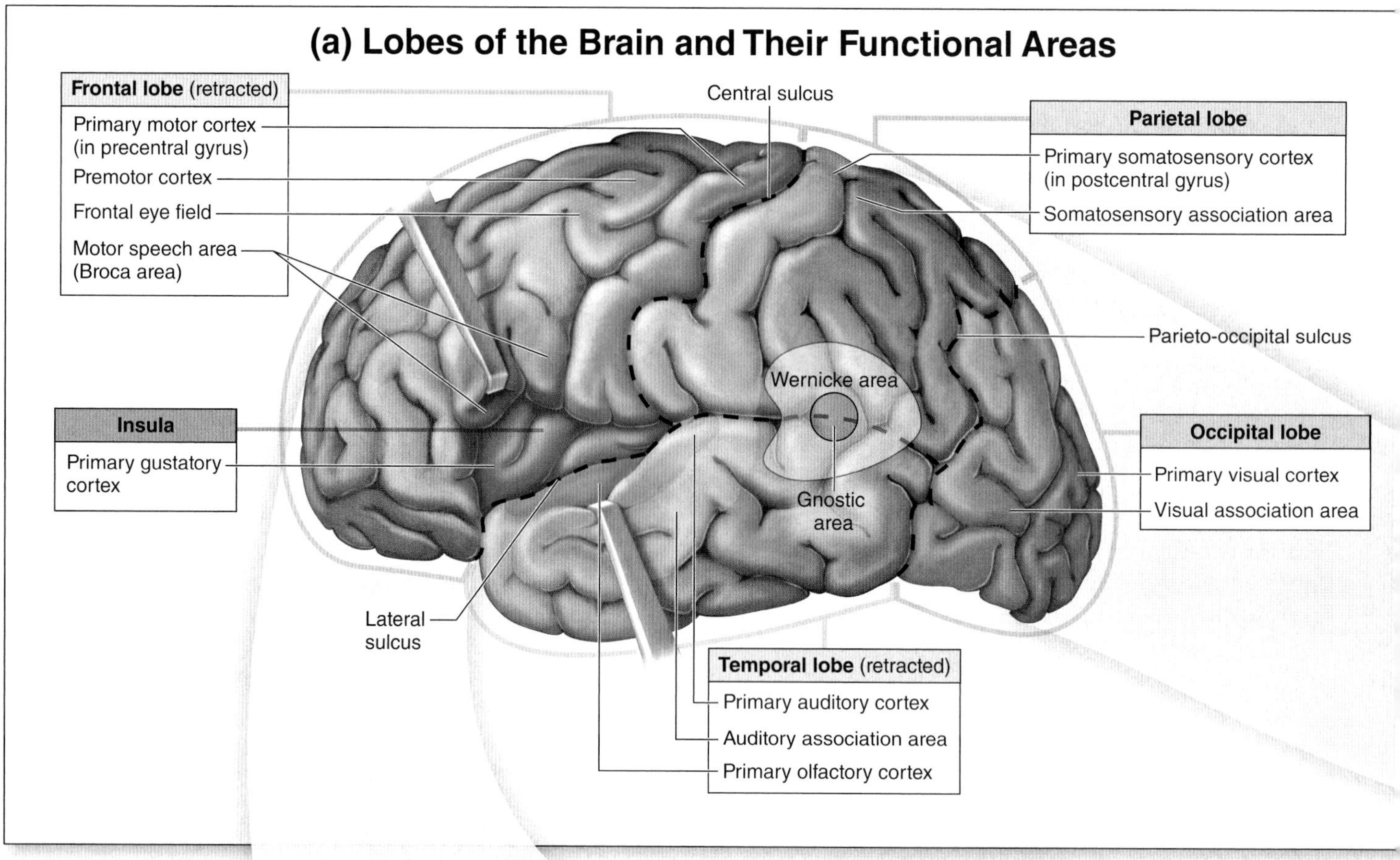

(a) Lobes of the Brain and Their Functional Areas

Frontal lobe (retracted)

Primary motor cortex (in precentral gyrus)

Premotor cortex

Frontal eye field

Motor speech area (Broca area)

Central sulcus

Parietal lobe

Primary somatosensory cortex (in postcentral gyrus)

Somatosensory association area

Wernicke area

Parieto-occipital sulcus

Insula

Primary gustatory cortex

Occipital lobe

Primary visual cortex

Visual association area

Gnostic area

Lateral sulcus

Temporal lobe (retracted)

Primary auditory cortex

Auditory association area

Primary olfactory cortex

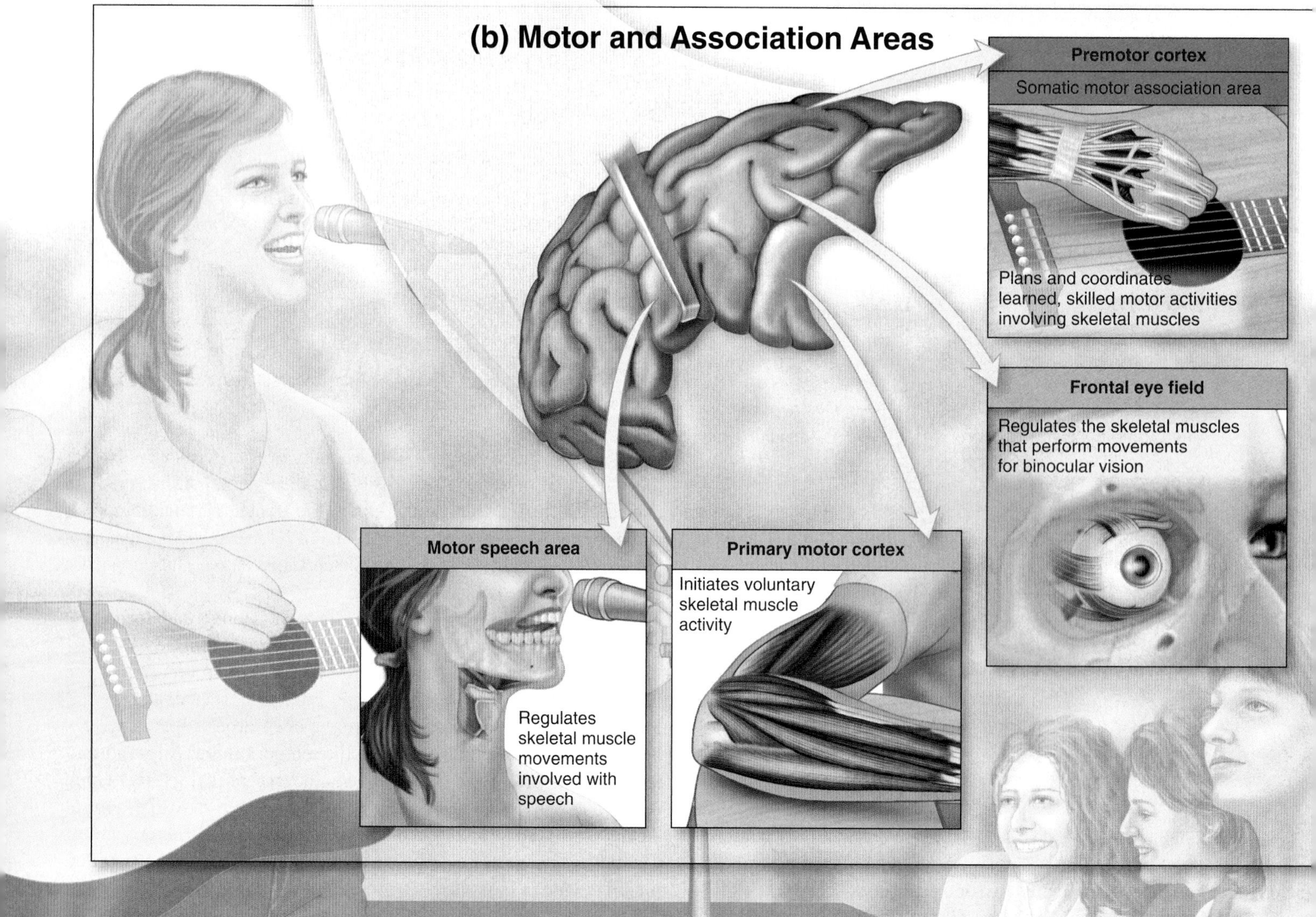

(b) Motor and Association Areas

Premotor cortex

Somatic motor association area

Plans and coordinates learned, skilled motor activities involving skeletal muscles

Frontal eye field

Regulates the skeletal muscles that perform movements for binocular vision

Motor speech area

Regulates skeletal muscle movements involved with speech

Primary motor cortex

Initiates voluntary skeletal muscle activity

(d) Functional Brain Regions

Prefrontal cortex
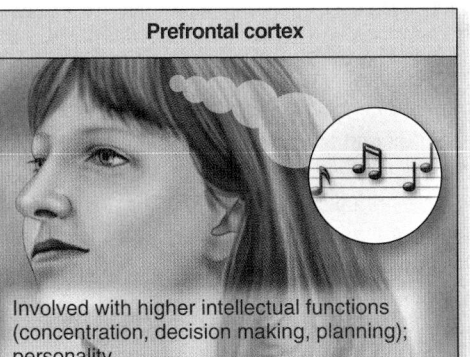
Involved with higher intellectual functions (concentration, decision making, planning); personality

Wernicke area

This multi-association area helps understand spoken and written language

Gnostic area
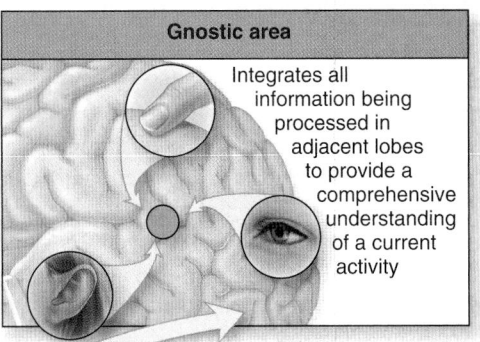
Integrates all information being processed in adjacent lobes to provide a comprehensive understanding of a current activity

Primary gustatory cortex
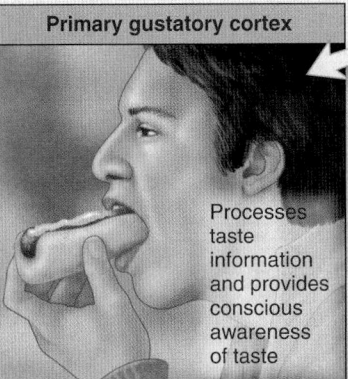
Processes taste information and provides conscious awareness of taste

(c) Sensory Areas

Primary somatosensory cortex
Somatosensory association area
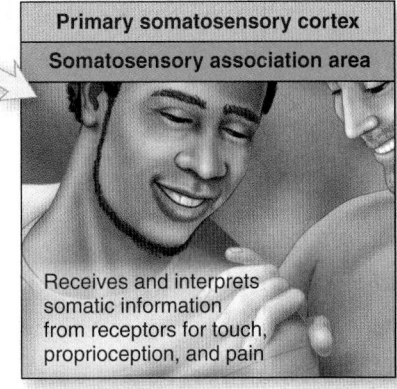
Receives and interprets somatic information from receptors for touch, proprioception, and pain

Primary olfactory cortex
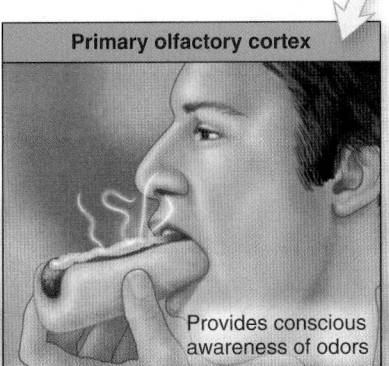
Provides conscious awareness of odors

Primary auditory cortex
Auditory association area
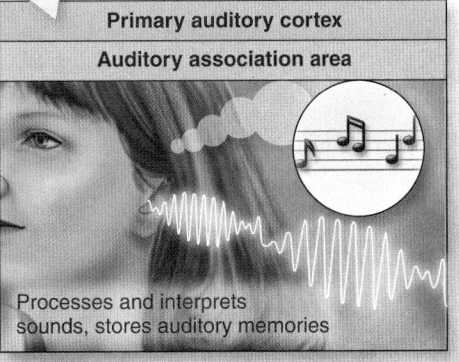
Processes and interprets sounds, stores auditory memories

Primary visual cortex
Visual association area
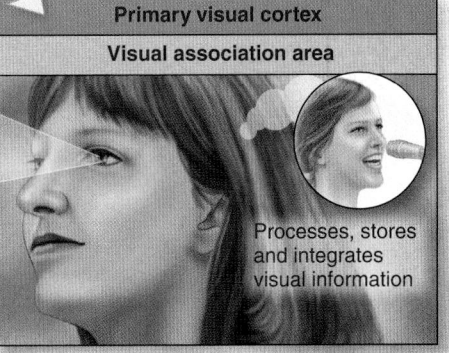
Processes, stores and integrates visual information

INTEGRATE CONCEPT OVERVIEW

Figure 13.12 Anatomic and Functional Areas of the Cerebrum. (*a*) Each cerebral hemisphere is partitioned into five lobes that each contain specific cortical regions and association areas. (*b*) The major motor areas and their association cortices are highlighted. (*c*) The major sensory areas and their association cortices are shown. (*d*) A functional brain region acts like a multi-association area to integrate information from several association areas.

The **parietal** (pă-rī′ĕ-tăl) **lobe** lies deep to the parietal bone and forms the superoposterior part of each cerebral hemisphere. It terminates anteriorly at the central sulcus, posteriorly at a relatively indistinct **parieto-occipital sulcus,** and laterally at the lateral sulcus. An important anatomic feature of this lobe is the **postcentral gyrus,** which is the mass of nervous tissue immediately posterior to the central sulcus. The parietal lobe is involved with general sensory functions, such as evaluating the shape and texture of objects being touched.

The **temporal lobe** lies internal to the temporal bone and inferior to the lateral sulcus. This lobe is involved with hearing and smell.

The **occipital lobe** lies internal to the occipital bone and forms the posterior region of each hemisphere. This lobe is responsible for processing incoming visual information and storing visual memories.

The **insula** (in′sū-lă; *insula* = island) is a small lobe deep to the lateral sulcus. It can be observed by laterally reflecting (pulling aside) the temporal lobe. The insula's lack of accessibility has prevented aggressive studies of its function, but it is apparently involved in memory and the interpretation of taste.

WHAT DID YOU LEARN?

13 List the five cerebral lobes and the main functions of each.

13.3c Functional Areas of the Cerebrum

LEARNING OBJECTIVES

4. Locate and list the functions of the motor cortices and association areas.

5. Differentiate among the sensory cortical regions and their association areas.

6. Explain the functions of the prefrontal cortex, and hypothesize why this brain region may function differently in adults and teenagers.

7. Compare and contrast the main actions of the Wernicke area and the gnostic area.

Research has shown that specific structural areas of the cerebral cortex have distinct motor and sensory functions. In contrast, some higher mental functions, such as language and memory, are dispersed over large areas. We have organized the functions of the cerebral cortex into motor functions and their association areas and sensory functions and their association areas. Please note that the central sulcus serves as an anatomic landmark that separates motor functions that are controlled by the frontal lobe, and sensory input that is relayed to the parietal lobe, occipital lobe, and temporal lobe, as well as the internal insula.

Motor Areas

The cortical areas that control motor functions are housed within the frontal lobes. The **primary motor cortex,** also called the *somatic motor area*, is located within the precentral gyrus of the frontal lobe (figure 13.12). Neurons in this area control voluntary skeletal muscle activity. The axons of these neurons project contralaterally (to the opposite side) either within the brainstem or spinal cord. Thus, the left primary motor cortex controls the right-side voluntary muscles, and vice versa.

The distribution of the primary motor cortex innervation to various body parts can be diagrammed as a **motor homunculus** (hō-mŭngk′ū-lŭs; *little man*) on the precentral gyrus **(figure 13.13a)**. The bizarre, distorted proportions of the homunculus body reflect the

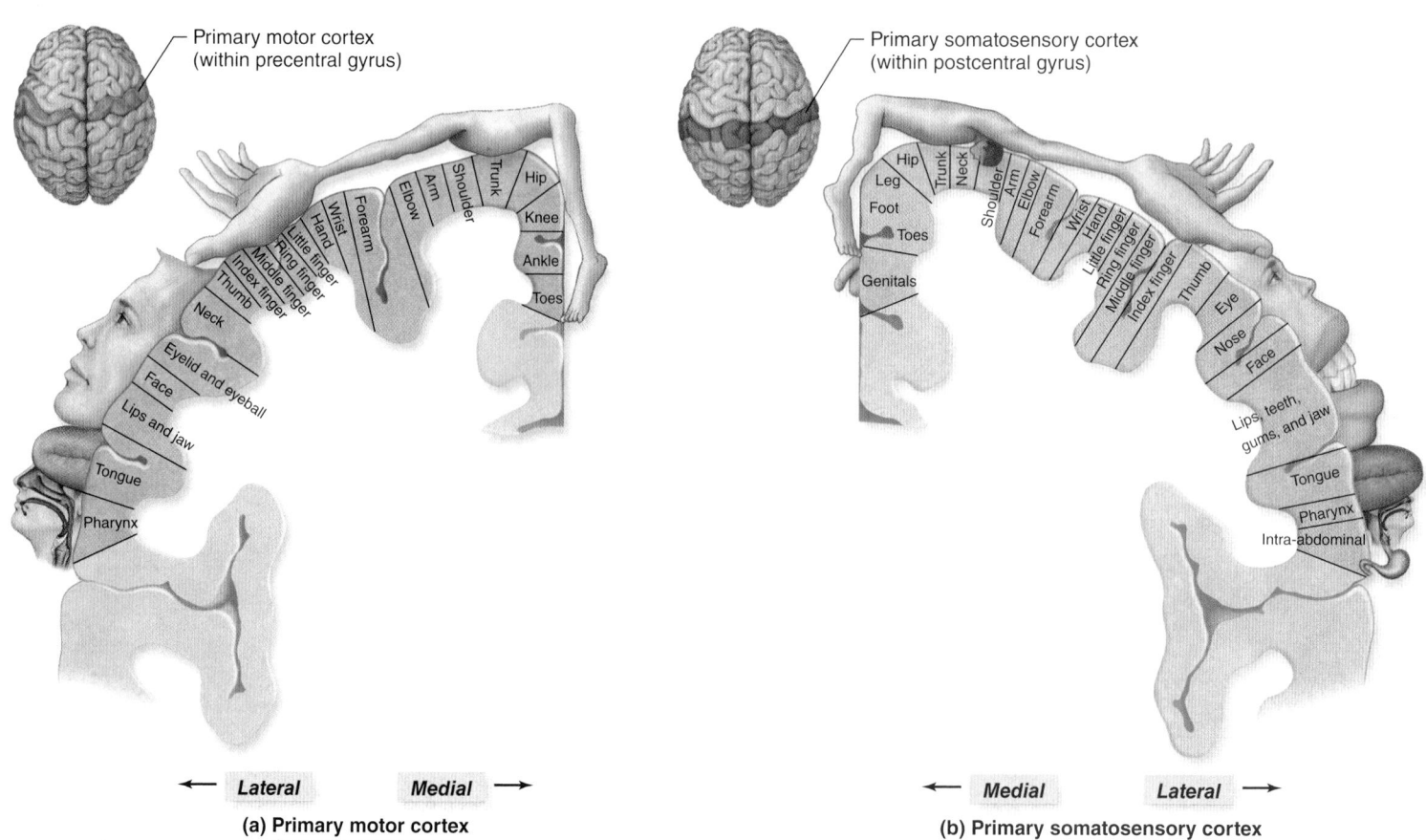

(a) Primary motor cortex
(somatic motor area)

(b) Primary somatosensory cortex

Figure 13.13 Primary Motor and Somatosensory Cortices. Body maps called the motor homunculus and the sensory homunculus illustrate the topography of (*a*) the primary motor cortex and (*b*) the primary somatosensory cortex in coronal section. The figure of the body (homunculus) depicts the nerve distributions; the size and location of each body region indicates relative innervation. AP|R

INTEGRATE

CLINICAL VIEW
Brodmann Areas

Korbinian Brodmann studied the comparative anatomy of the mammalian brain cortex in the early 1900s and was among the first to correlate physiologic activities with previously determined anatomic locations of the brain. Brodmann produced a map that shows the specific areas of the cerebral cortex where certain functions occur. Brodmann developed a numbering system that correlates with his map and shows that similar cognitive functions are usually sequential. For example, areas 3, 1, and 2 represent the primary somatosensory cortex; area 17 overlaps the primary visual cortex; and areas 44 and 45 form the motor speech area. Recent technological improvements allow neuroscientists to more precisely pinpoint the location of physiologic activities in the cortex, but the Brodmann Areas map is still useful for historical perspective and early views of the brain.

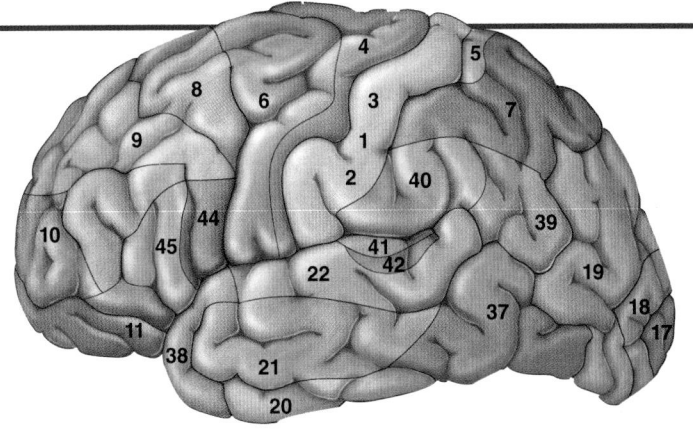

Modern rendition of Korbinian Brodmann's map of the brain, showing selected Brodmann areas.

amount of cortex dedicated to the motor activity of each body part. For example, the hands are represented by a much larger area of cortex than the trunk, because the hand muscles perform much more detailed, precise movements than do the trunk muscles. From a functional perspective, more motor activity is devoted to the human hand than in other animals because our hands are adapted for the precise, fine motor movements needed to manipulate the environment, and many motor units are devoted to muscles that move the hand and fingers.

Certain motor functions have been mapped to specific areas of the frontal lobe, including the motor speech area and the frontal eye field. The **motor speech area** (*Broca area*) is located in most individuals within the inferolateral portion of the left frontal lobe (figure 13.12). This region is responsible for regulating the patterns of breathing and controlling the muscular movements necessary for vocalization. The **frontal eye field** is on the superior surface of the middle frontal gyrus. It is immediately anterior to the premotor cortex in the frontal lobe. This cortical area controls and regulates the eye movements needed for reading and coordinating binocular vision. Some investigators include the frontal eye fields within the premotor area, which is described next.

The primary motor cortical regions are connected to adjacent **association areas** that coordinate a motor response (figure 13.12). The **premotor cortex** also is called the *somatic motor association area,* and it is located within the frontal lobe immediately anterior to the precentral gyrus. It is primarily responsible for coordinating learned, skilled motor activities, such as moving the eyes in a coordinated fashion when reading a book or playing the piano. An individual who has sustained trauma to this area would still be able to understand written letters and words, but would have difficulty reading because his or her eyes couldn't follow the lines on a printed page.

Sensory Areas

The cortical areas within the parietal, temporal, and occipital lobes and the insula are involved with conscious awareness of sensation, as described previously. Each of the major senses has a distinct cortical area. Keep in mind as you read through this section that each primary sensory cortical region is the specific area of the cortex that receives sensory input from a specific type of receptor. In addition, each primary cortical region typically has an association area. The general function of association areas is to receive input from the primary region and integrate the current sensory input with previous experiences.

The **primary somatosensory cortex** is housed within the postcentral gyrus of the parietal lobes. Neurons within this cortex receive general somatic sensory information from the skin regarding touch, pressure, pain, and temperature receptors, as well as sensory input from proprioceptors from the joints and muscles regarding the conscious interpretation of body position. We typically are conscious of the sensations received by this cortex. A **sensory homunculus** may be traced on the postcentral gyrus surface, similar to the a motor homunculus just described (figure 13.13*b*). The surface area of somatosensory cortex devoted to a body region indicates the amount of sensory information collected within that region. Thus, the lips, fingers, and genital region occupy larger portions of the homunculus, whereas the trunk of the body has proportionately fewer receptors, so its associated homunculus region is smaller.

The **somatosensory association area** is located within the parietal lobe and lies immediately posterior to the primary somatosensory cortex. It integrates sensory information and interprets sensations to determine the texture, temperature, pressure, and shape of objects. The somatosensory association area allows us to identify known objects without seeing them or when our eyes are closed. For example, even when our eyes are closed, we can tell the difference between the coarse feel of a handful of dirt; the smooth and round shape of a marble; and the thin, flat, rounded surface of a coin because those interpretations of the textures and shapes have already been stored in the somatosensory association area.

WHAT DO YOU THINK?

 Predict the difference in loss of function that might accompany brain damage (perhaps from a cerebrovascular accident [CVA] or stroke) to the primary somatosensory cortex versus the somatosensory association area.

Sensory information for sight, sound, taste, and smell arrives at cortical regions other than the parietal lobe (figure 13.12). The **primary visual cortex** is located within the occipital lobe, where it receives and processes incoming visual information. The **visual association area** is located within the occipital lobe and it surrounds the primary visual area. It enables us to process visual information by analyzing color, movement, and form, and to use this information to identify the things we see. For example, when we look at a face, the primary visual cortex receives bits of visual information, but the visual association area is responsible for integrating all of this information into a recognizable picture of a face.

The **primary auditory cortex** is located within the temporal lobe, where it receives and processes auditory information. The **auditory association area** is located within the temporal lobe, posteroinferior to the primary auditory cortex. Within this association area, the cortical neurons interpret the characteristics of sound and store memories of sounds heard in the past. The next time a song is playing over and over in your head, you will know that this auditory association area is responsible.

The **primary olfactory** (ol-fak′to-rē; *olfactus* = smell) **cortex** is also located within the temporal lobe and provides conscious awareness of smells. Finally, the **primary gustatory** (gŭs′ tă-tōr-ē; *gustatio* = taste) **cortex** is within the insula and is involved in processing taste information.

Functional Brain Regions

A **functional brain region** acts like a multi-association area between lobes for integrating information from individual association areas. One functional brain region is the **prefrontal cortex,** located in the rostral portions of the frontal lobes and anterior to both the primary motor and premotor cortices. The prefrontal cortex is associated with many higher intellectual functions such as complex thought, judgment, expression of personality, planning future behaviors, and decision making. By retrieving and coordinating information from multiple areas of the brain, the prefrontal cortex also will evaluate potential consequences of one's actions, and in so doing will modulate one's behavior based on societal norms. Interestingly, the prefrontal cortex continues to develop into our teens and 20s, as the axons continue to myelinate in this region and unnecessary synapses are removed. As a result, neuroscientists hypothesize that the reason many teenagers may have difficulty in planning, are impulsive, emotional, and risk takers is because the prefrontal cortex has not fully matured.

Another functional brain region is the **Wernicke area**, which is typically located only within the left hemisphere. The Wernicke area is involved in recognizing, understanding, and comprehending spoken or written language. As you may expect, the Wernicke area and the motor speech area must work together for fluent communication to occur.

The **gnostic** (nos′tik) **area** (or *common integrative area*) is composed of regions of the parietal, occipital, and temporal lobes. This region integrates all somatosensory, visual, and auditory information being processed by the association areas within these lobes. Thus, it provides comprehensive understanding of a current activity. For example, suppose you awaken from a daytime nap: The hands on the clock indicate that it is 12:30, you smell food cooking, and you hear your friends talking about being hungry. The gnostic area then interprets this information to mean it is lunchtime.

Figure 13.12 summarizes how the anatomic and functional areas of the cerebrum are distributed.

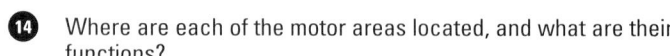

WHAT DID YOU LEARN?

14 Where are each of the motor areas located, and what are their functions?

15 In general, what is the purpose of the association areas?

16 Why does the prefrontal cortex perform differently in a teenager versus an adult?

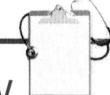

INTEGRATE

CLINICAL VIEW
Autism Spectrum Disorder

Autism spectrum disorder (ASD), also known simply as *autism*, is a widely variable disorder of neural development that affects 1 in 88 children in the United States alone. It typically is recognized in early childhood, but diagnosis may be difficult until a child is older. Since 2013, the phrase *autism spectrum disorder* is used to group and describe a variety of similar disorders including autistic disorder, childhood disintegrative disorder, and Asperger syndrome. ASD varies in severity among those affected (hence the term *spectrum* in its name), but all are characterized by some form of social and communication difficulties. Some children may experience delays in language acquisition or may be completely nonverbal. Social interaction is difficult, ranging from inability to reciprocate interest during a conversation to being withdrawn into the child's "own world." Intelligence also varies widely, from severe cognitive delay to possessing savantlike skills in focused areas like math or music.

Individuals with ASD often are highly sensitive to stimuli such as loud noises or unfamiliar people, and may struggle in adjusting to changes in routine. Discomfort due to overstimulation or frustration in the inability to communicate can lead to tantrums or "meltdowns." Other behaviors and traits commonly associated with ASD include repetitive motions like hand flapping or rocking, resistance to changes in routine (e.g., insisting on wearing the same shirt or eating the same meal each day), inability to engage in pretend play, inability to gauge the feelings of others, and intense interest in a particular activity or subject.

ASD is believed to stem from an inability of the brain to process information between neurons. However, the specific mechanisms and causes of the condition are not well understood or agreed upon. Genetic factors are thought to be involved, in part because autism affects males four times more often than females, and it often manifests in siblings. Biochemical and environmental factors have also been explored as potential causes, but few definitive answers exist. The disturbing aspect of this condition is that the number of cases has steadily increased since the late 1980s. The ability to detect the condition has improved, which may have increased the incidence of diagnosis.

A fraudulent paper published in 1988 claimed that the measles, mumps, and rubella (MMR) vaccine was linked to an increased risk of developing autism. In the years that followed, the paper was shown to have manipulated data and the study was inherently flawed, resulting in a retraction of the paper and the author (who was an MD) losing his medical license for serious professional misconduct. Numerous studies since then have shown no link between vaccines and developing autism. Unfortunately, the misconception that vaccines cause ASD still persists among some and has led to both a decline in vaccination rates and an increase in disease outbreaks as a result.

Treatment for ASD includes proven methods of speech and behavioral therapy, as well as holistic approaches that involve various diets, supplements, and experimental procedures. Some children with autism will go on to develop skills and live independent lives, whereas others will not. The biggest predictors for independence in adulthood are level of intelligence and ability to communicate.

13.3d Central White Matter

8. Identify the three main tracts of the central white matter.

The **central white matter** lies deep to the gray matter of the cerebral cortex. It is composed primarily of myelinated axons. Most of these axons are grouped into bundles called **tracts,** which are classified as association tracts, commissural tracts, or projection tracts **(figure 13.14).**

Association tracts connect different regions of the cerebral cortex within the same hemisphere. Short association tracts are composed of **arcuate** (ar'kū-āt; *arcuatus* = bowed) **fibers;** they connect neighboring gyri within the same lobe. (Fibers are bundles of axons.) An example of an association tract that is composed of arcuate fibers would be the tract that connects the premotor or motor association area with the primary motor cortex, both of which are in the frontal lobe. The longer association tracts, called the **longitudinal fasciculi** (fă-sik'-ū-lī; *fascis* = bundle), connect gyri in different lobes of the same hemisphere. An example of a longitudinal fasciculus would be the tract that connects the Wernicke area to the motor speech area.

Commissural (kom-i-sūr'ăl) **tracts** extend between the cerebral hemispheres through axonal bridges called commissures. The prominent commissural tracts that link the left and right cerebral hemispheres include the large, C-shaped **corpus callosum** and the smaller **anterior** and **posterior commissure** (see figure 13.17).

Projection tracts link the cerebral cortex to both the inferior brain regions and the spinal cord (figure 13.14). Examples of projection tracts are the corticospinal tracts that carry motor signals from the cerebrum to the brainstem and spinal cord (see section 14.4c). The packed group of axons in these tracts passing in between the cerebral nuclei and the thalamus is called the **internal capsule** (see figure 13.4).

WHAT DID YOU LEARN?

17. What portions of the brain are linked by each type of tract: (a) association, (b) commissural, and (c) projection tracts?

INTEGRATE

CONCEPT CONNECTION

Recall from section 12.8c that action potentials are propagated along axons. The tracts in the brain contain thousands of axons, which transmit action potentials to allow for communication among the different regions of the brain.

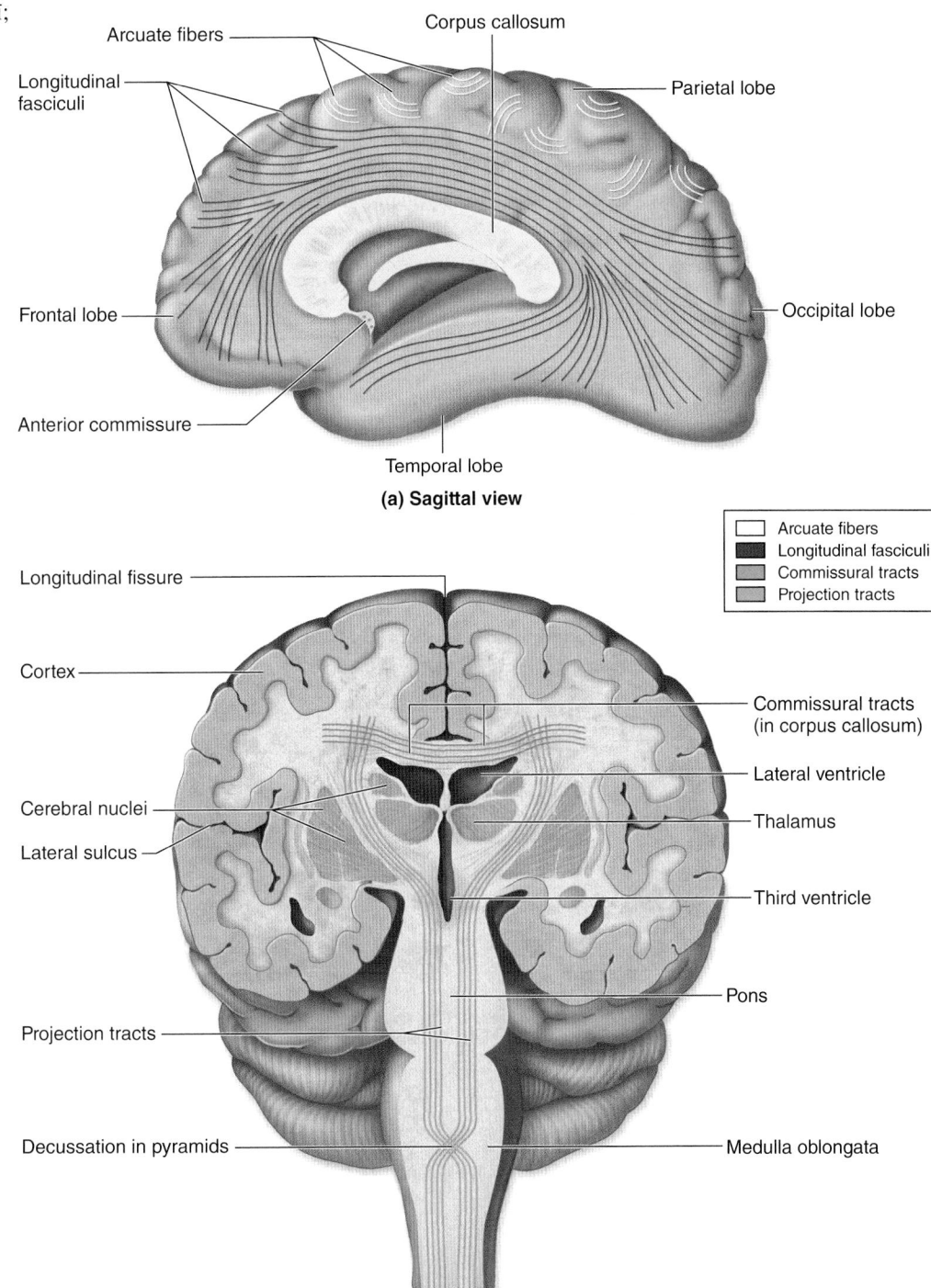

(a) Sagittal view

Arcuate fibers — Corpus callosum — Parietal lobe — Longitudinal fasciculi — Frontal lobe — Occipital lobe — Anterior commissure — Temporal lobe

- Arcuate fibers
- Longitudinal fasciculi
- Commissural tracts
- Projection tracts

Longitudinal fissure — Cortex — Cerebral nuclei — Lateral sulcus — Projection tracts — Decussation in pyramids — Commissural tracts (in corpus callosum) — Lateral ventricle — Thalamus — Third ventricle — Pons — Medulla oblongata

Figure 13.14 Cerebral White Matter Tracts. White matter tracts are composed primarily of myelinated axons. Three major groups of axons are recognized based on their distribution. (*a*) A sagittal view shows arcuate fibers and longitudinal fasciculi association tracts, which extend between gyri within one cerebral hemisphere. (*b*) A coronal view shows how commissural tracts extend between cerebral hemispheres, whereas projection tracts extend between the hemispheres and the brainstem. AP|R

(b) Coronal section

13.3e Cerebral Lateralization

LEARNING OBJECTIVES

9. Explain the phenomenon of cerebral lateralization.

10. Identify the functions of left and right hemispheres in most individuals.

Anatomically, the left and right cerebral hemispheres appear identical, but careful examination reveals a number of differences. Humans tend to have shape asymmetries of the frontal and occipital lobes of the brain, called **petalias** (pe′tal-ē-ă). Right-handed individuals typically have right frontal petalias, meaning that the right frontal lobe projects farther than the left frontal lobe, and left occipital petalias, meaning that the left occipital lobe projects farther than the right occipital lobe. Conversely, left-handed individuals are inclined to have the reverse pattern (left frontal and right occipital petalias). The hemispheres also differ with respect to some of their functions. Each hemisphere tends to be specialized for certain tasks, a phenomenon called **cerebral lateralization** (lat′er-al-ī-zā′shŭn). Higher-order centers in both hemispheres tend to have different but complementary functions (**figure 13.15**).

In most people, the left hemisphere is the **categorical hemisphere.** It usually contains the Wernicke area and the motor speech area. It is specialized for language abilities, and is also important in performing sequential and analytical reasoning tasks, such as those required in science and mathematics. This hemisphere appears to direct or partition information into smaller fragments for analysis. The term *categorical hemisphere* reflects this hemisphere's function in categorization and identification.

The other hemisphere (the right in most people) is called the **representational hemisphere,** because it is concerned with visuospatial relationships and analyses. It is the seat of imagination and insight, musical and artistic skill, perception of patterns and spatial relationships, and comparison of sights, sounds, smells, and tastes.

Both cerebral hemispheres remain in constant communication through commissures, especially through the corpus callosum, which contains hundreds of millions of axons that project between the hemispheres.

Lateralization of the cerebral hemispheres develops early in life (prior to 5–6 years of age). In a young child, the functions of a damaged or removed hemisphere are often taken over by the other hemisphere before lateralization is complete. Some aspects of lateralization differ between the sexes. Women have a thicker posterior part of the corpus callosum due to additional commissural axons in this region. Adult males typically exhibit more lateralization than females and suffer more functional loss when one hemisphere is damaged.

Cerebral lateralization is highly correlated with handedness. Right-handed individuals tend to have a slightly different lateralization pattern than those who are left-handed. In about 95% of the population, the left hemisphere is the categorical hemisphere, thus correlating with the 90% incidence of right-handed individuals in the population. However, the correlation is not nearly as strict among left-handed people, who may have either hemisphere as their categorical hemisphere. Interestingly, a thicker corpus callosum in left-handers suggests that more signals may be relayed between their hemispheres. Finally, the left hemisphere is

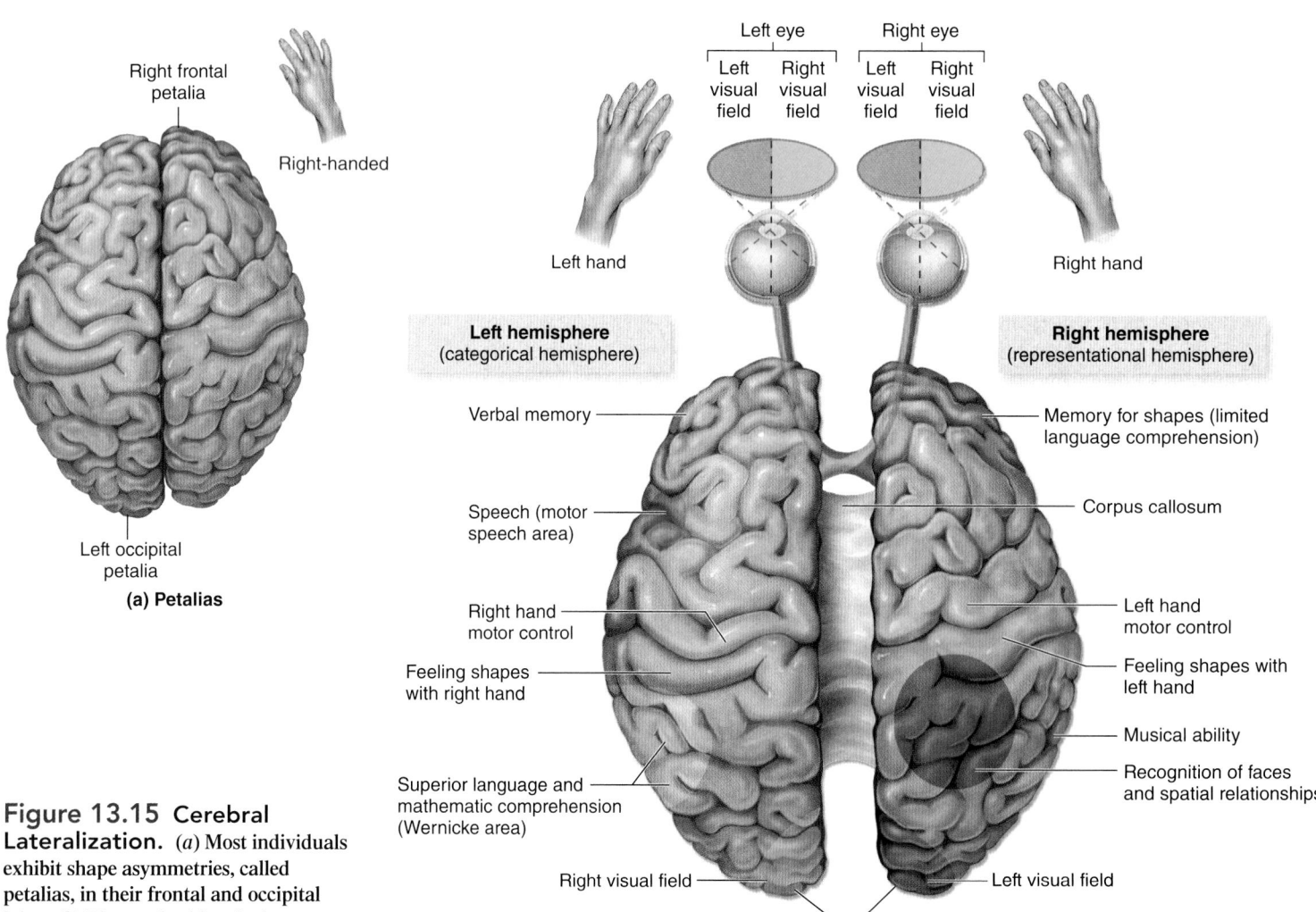

Figure 13.15 Cerebral Lateralization. (*a*) Most individuals exhibit shape asymmetries, called petalias, in their frontal and occipital lobes. (*b*) The cerebral hemispheres exhibit functional differences as a result of some specialization.

INTEGRATE

CLINICAL VIEW
Hemispherectomies and Cerebral Lateralization

Epilepsy is a disorder in which neurons transmit action potentials too frequently and rapidly, causing seizures that detrimentally affect motor and sensory function. Most seizures may be controlled by medications, but if medications are ineffective then surgery may be deemed necessary. Surgical removal of the brain part that is the source of the seizures often eliminates the seizures. In the most severe cases, a drastic form of therapy for the epilepsy is a cerebral **hemispherectomy** (hem′ē-sfēr-ek′tō-mē) in which the side of the brain responsible for the seizure activity is surgically removed.

Although brain function does not return to complete normalcy following a hemispherectomy, the remaining hemisphere amazingly takes over some of the functions of the missing hemisphere. The younger the individual, the better the chances of functional recovery.

INTEGRATE

CLINICAL VIEW
Cerebrovascular Accident

A **cerebrovascular accident** (**CVA,** or *stroke*) is caused by reduced blood supply to a part of the brain due to either a blocked arterial blood vessel or hemorrhage from a blood vessel. If impaired blood flow lasts longer than about 10 minutes, brain tissue may die. Symptoms of a CVA include loss or blurring of vision, weakness or slight numbness, headache, dizziness, and walking difficulties. Each side of the brain controls the opposite side of the body, so a CVA on the left side of the brain presents with symptoms on the right side of the body, and vice versa.

A brief episode of lost sensation or motor ability or "tingling" in the limbs is called a **transient ischemic attack** (**TIA**) or *mini-stroke*. It results from a temporary clot in a blood vessel that dissolves in a matter of minutes. However, TIAs can indicate increased risk for a more serious vessel blockage in the future.

the speech-dominant hemisphere; it controls speech in almost all right-handed people as well as in many left-handed ones.

WHAT DID YOU LEARN?

18 What is cerebral lateralization?

19 What are the primary functions of the left cerebral hemisphere? Of the right cerebral hemisphere?

13.3f Cerebral Nuclei

LEARNING OBJECTIVE

11. Identify the four cerebral nuclei, and explain their functions.

The **cerebral nuclei** (also called the *basal nuclei*) are paired, irregular masses of gray matter buried deep within the central white matter in the basal (deepest) region of the cerebral hemispheres inferior to the floor of the lateral ventricle (**figure 13.16**; see also figure 13.4). (These masses of gray matter also are sometimes incorrectly called the basal ganglia. However, the term *ganglion* [sing.] is best restricted to clusters of neuron cell bodies *outside* the CNS, whereas a *nucleus* is a collection of cell bodies *within* the CNS.) In general, the cerebral nuclei help regulate motor output initiated by the cerebral cortex, to help inhibit unwanted movements. Diseases that affect the cerebral nuclei (such as Huntington disease—see Clinical View: "Brain Disorders") often are associated with jerky, involuntary movements.

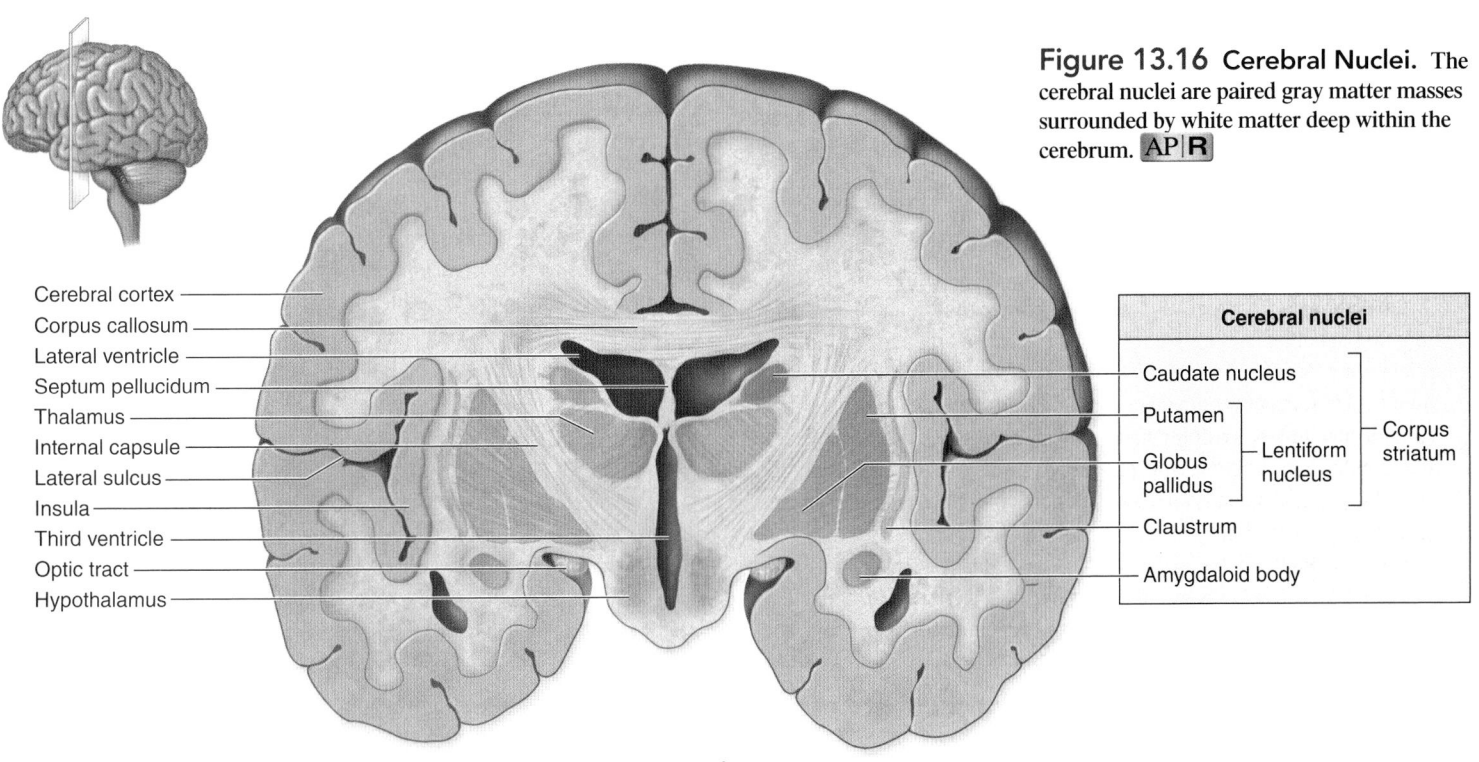

Figure 13.16 Cerebral Nuclei. The cerebral nuclei are paired gray matter masses surrounded by white matter deep within the cerebrum. AP|R

Cerebral cortex
Corpus callosum
Lateral ventricle
Septum pellucidum
Thalamus
Internal capsule
Lateral sulcus
Insula
Third ventricle
Optic tract
Hypothalamus

Cerebral nuclei
Caudate nucleus
Putamen
Globus pallidus — Lentiform nucleus
Claustrum
Amygdaloid body
Corpus striatum

Coronal section

Cerebral nuclei have multiple components, and each component has its own specific functions related to the overall function of the cerebral nuclei. These components include

- The C-shaped **caudate** (kaw′dāt; *cauda* = tail) **nucleus** has an enlarged head and a slender, arching tail that parallels the swinging curve of the lateral ventricle. When a person begins to walk, the neurons in this nucleus stimulate the appropriate skeletal muscles to produce the pattern and rhythm of arm and leg movements associated with walking.
- The **lentiform** (len′ti-fōrm; *lenticula* = lentil, *forma* = shape) **nucleus** is a compact, triangular mass, made up of both the **putamen** (pū-tā′men; shell) and **globus pallidus** (glō′bŭs pal′i-dŭs; *globus* = ball, *pallidus* = pale), two masses of gray matter positioned between the insula and the lateral wall of the diencephalon. The putamen functions in controlling skeletal muscular movement at the subconscious level, while the globus pallidus both excites and inhibits the activities of the thalamus to control and adjust skeletal muscle tone.
- The **amygdaloid** (ā-mig′dă-loyd; *amygdala* = almond) **body** (often just called the *amygdala*) is an expanded region at the tail of the caudate nucleus. It participates in the expression of emotions, control of behavioral activities, and development of moods (see the limbic system in section 13.7a).
- The **claustrum** (klaws′trŭm; barrier) is a thin sliver of gray matter formed by a layer of neurons located immediately internal to the cortex of the insula and derived from that cortex. It processes visual information at a subconscious level.

The term **corpus striatum** (strī-ā′tŭm; striped) describes the striated or striped appearance of the internal capsule as it passes between the caudate nucleus and the lentiform nucleus.

 WHAT DID YOU LEARN?

 20 What is the general function of the cerebral nuclei, and what are the anatomic components that make up the nuclei?

 **INTEGRATE**

CLINICAL VIEW
Brain Disorders

Brain disorders may be expressed either by a malfunction in processing of sensory input, transmitting motor output, or by some combination of both activities. Some disturbances of the brain include headache, cerebral palsy, encephalitis, Huntington disease, and Parkinson disease.

Headache typically is due either to dilated blood vessels in the skull or muscle contraction (such as may occur when an individual develops eyestrain from staring at a computer screen too long). **Migraine headaches** are severe, recurring headaches that often affect only one side of the head. Headaches are not a brain disorder, but they sometimes accompany other diseases or brain disorders.

Cerebral palsy (pawl′zē) is actually a group of neuromuscular disorders that usually result from damage to an infant's brain before, during, or immediately after birth. Three forms of cerebral palsy involve impairment of skeletal motor activity to some degree: *athetoid,* characterized by slow, involuntary, writhing hand movements; *ataxic,* marked by lack of muscular coordination; and *spastic,* exhibiting increased muscular tone. Intellectual impairment and speech difficulties sometimes accompany this disorder.

Encephalitis (en-sef-ă-lī′tis; *enkephalos* = brain, *itis* = inflammation) is an acute inflammatory disease of the brain, most often due to viral infection. Symptoms include drowsiness, fever, headache, neck pain, coma, and paralysis. Death may occur.

Huntington disease is an autosomal dominant hereditary disease that affects the cerebral nuclei. It causes rapid, jerky, involuntary movements that usually start unilaterally in the face, but over months and years progress to the arms and legs. Progressive intellectual deterioration also occurs, including personality changes, memory loss, and irritability. The disease has an onset age of mid 30s–40s, and is fatal within 10 to 20 years.

Parkinson disease is a slow-progressing neurologic condition that affects muscle movement and balance. Parkinson patients exhibit stiff posture, an expressionless face, slow voluntary movements, a resting tremor (especially in the hands), and a shuffling gait. The disease is caused by a deficiency of the neurotransmitter dopamine, which results from decreased dopamine production of degenerating neurons in the substantia nigra (nuclei in the midbrain). Dopamine deficiency prevents brain cells from performing their usual inhibitory functions within the cerebral nuclei. By the time symptoms develop, the person has lost 80–90% of the cells responsible for producing dopamine. Current treatments include medications that enhance the amount of dopamine in the remaining cells of the substantia nigra, and medications (e.g., rasagiline) to treat the symptoms.

Boxer Muhammad Ali and actor Michael J. Fox are two famous individuals with Parkinson disease.

13.4 Diencephalon

The **diencephalon** is sandwiched between the inferior regions of the cerebral hemispheres and for this reason is often referred to as the "in-between brain." The diencephalon components include the epithalamus, the thalamus, and the hypothalamus **(figure 13.17)**. The diencephalon provides the relay centers for some sensory and motor pathways, and for control of visceral activities.

13.4a Epithalamus

LEARNING OBJECTIVES

1. List components located in the epithalamus, and describe their functions.
2. Explain how circadian rhythm is regulated.

The **epithalamus** (ep′i-thal′ă-mŭs) partially forms the posterior roof of the diencephalon and covers the third ventricle. The posterior portion of the epithalamus houses the pineal gland and the habenular nuclei.

The **pineal** (pin′ē-ăl; *pineus* = pineconelike) **gland,** or *pineal body*, is an endocrine gland (see section 17.10a). It secretes the hormone **melatonin,** which appears to help regulate day-night cycles known as the body's **circadian rhythm.** (Some companies are marketing the sale of melatonin in pill form as a cure for jet lag and insomnia, although this cure has yet to be proven.)

The **habenular** (hă-ben′ū-lăr; *habena* = strap) **nuclei** help relay signals from the limbic system (see section 13.7a) to the midbrain and are involved in visceral and emotional responses to odors.

WHAT DID YOU LEARN?

21 What is the location and function of the pineal gland?

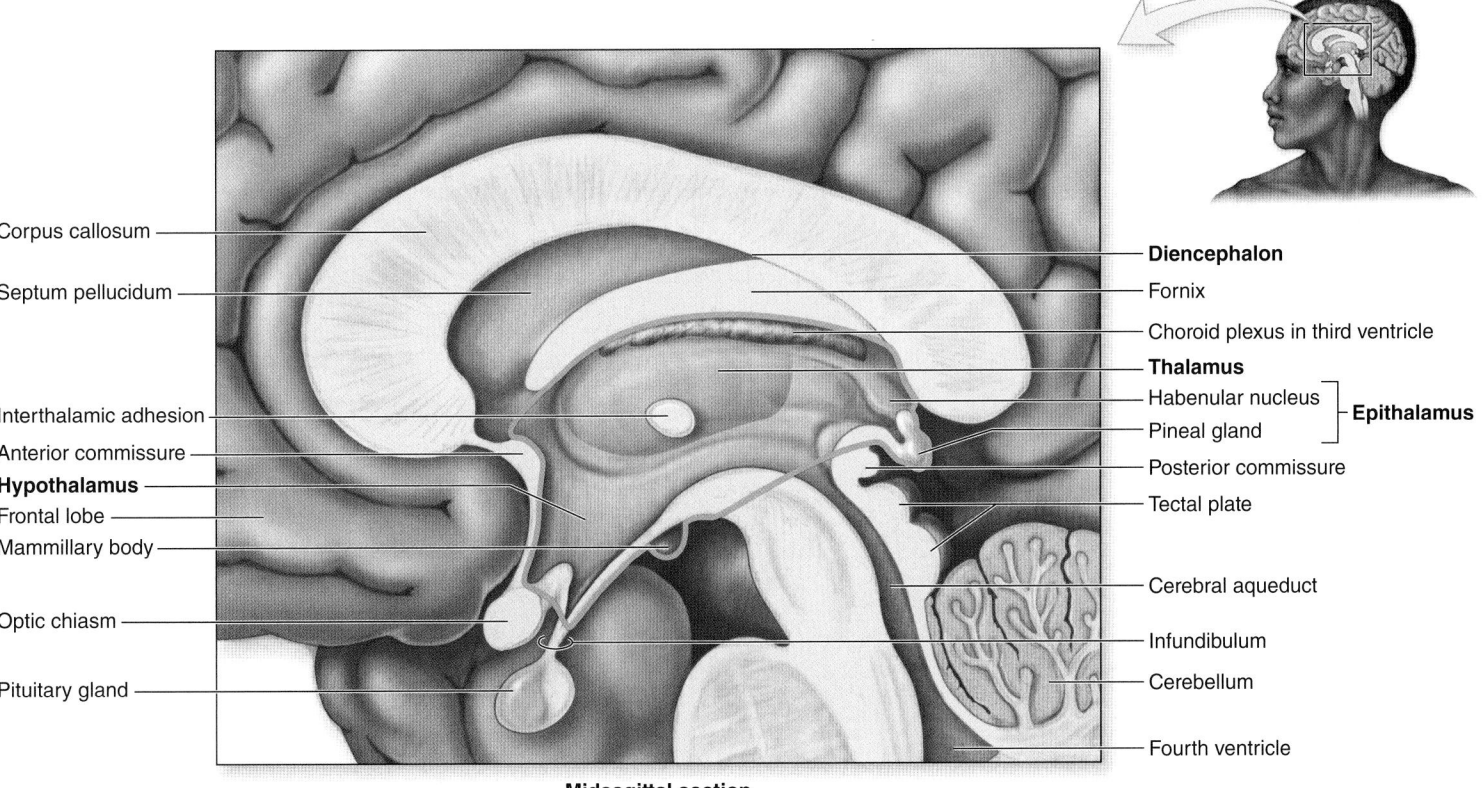

Corpus callosum
Septum pellucidum
Interthalamic adhesion
Anterior commissure
Hypothalamus
Frontal lobe
Mammillary body
Optic chiasm
Pituitary gland

Diencephalon
Fornix
Choroid plexus in third ventricle
Thalamus
Habenular nucleus ⎤ **Epithalamus**
Pineal gland ⎦
Posterior commissure
Tectal plate
Cerebral aqueduct
Infundibulum
Cerebellum
Fourth ventricle

Midsagittal section

Figure 13.17 Diencephalon. The diencephalon (outlined in purple) encloses the third ventricle and connects the cerebral hemispheres to the brainstem. The right portion of the diencephalon is outlined and shown here in midsagittal section. The diencephalon and its major subdivisions are listed in bold. AP|R

13.4b Thalamus

LEARNING OBJECTIVE

3. Discuss the action of the thalamus on sensory information.

The **thalamus** (thal´ă-mŭs; *thalamus* = bed) is identified as the paired oval masses of gray matter that lie on either side of the third ventricle (**figure 13.18**), where they form the superolateral walls of the third ventricle. When viewed in midsagittal section, the thalamus is located between the anterior commissure and the pineal gland. The **interthalamic adhesion** (or *intermediate mass*) is a small, midline mass of gray matter that connects the right and left thalamic bodies.

Each part of the thalamus is a gray matter mass composed of about a dozen major **thalamic nuclei** that are organized into groups; axons from these nuclei project to particular regions of the cerebral cortex. Sensory nerve signals from all the conscious senses except olfaction converge on the thalamus and synapse in at least one of its nuclei. For example, the ventral posterior nuclei relay sensory information to the primary somatosensory cortex of the parietal lobe, while auditory information is relayed through the medial geniculate nuclei.

The thalamus is the principal and final relay point for incoming sensory information that is processed and then projected to the appropriate lobe of the cerebral cortex. Only a relatively small portion of the sensory information that arrives at the thalamus is forwarded to the cerebrum because the thalamus acts as an information filter. For example, the thalamus is responsible for filtering out the sounds and sights in a busy dorm cafeteria when you are trying to study. The thalamus also "clues in" the cerebrum about where this sensory information came from. For example, the thalamus lets the cerebrum know that sensory information it receives came from the eye, indicating that the information is visual.

WHAT DO YOU THINK?

4 If there were no thalamus, how would this affect the cerebrum's interpretation of sensory stimuli?

WHAT DID YOU LEARN?

22 What is the general function of the thalamus?

13.4c Hypothalamus

LEARNING OBJECTIVE

4. Describe seven functions of the hypothalamus.

The **hypothalamus** (hī´pō-thal´ă-mŭs; *hypo* = under) is the anteroinferior region of the diencephalon. A thin, stalklike **infundibulum** (in-fŭn-dib´ū-lŭm; funnel) extends inferiorly from the hypothalamus to attach to the pituitary gland (**figure 13.19**).

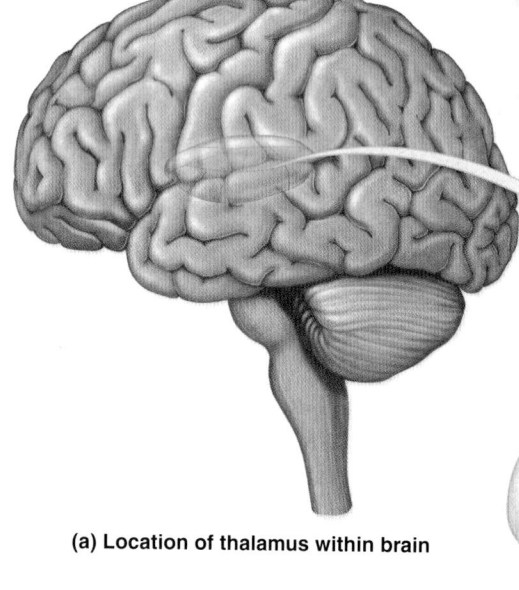

(a) Location of thalamus within brain

Figure 13.18 Thalamus. (*a*) Lateral view of the brain identifies the approximate location of the thalamus. (*b*) The thalamus is composed of clusters of nuclei organized into groups, as shown in this enlarged view. Not all nuclei may be seen from this angle. AP|R

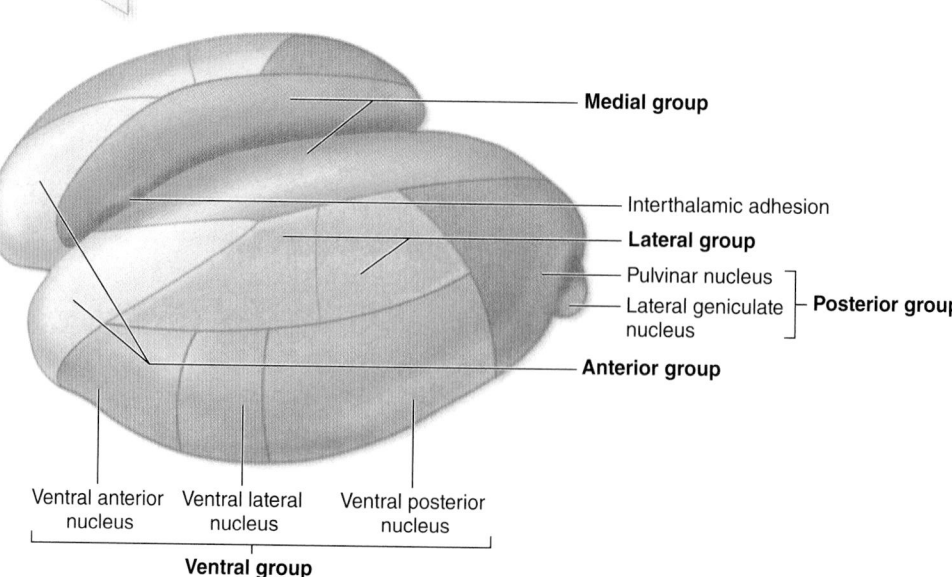

Medial group

Interthalamic adhesion

Lateral group

Pulvinar nucleus ⎤
Lateral geniculate nucleus ⎦ — **Posterior group**

Anterior group

Ventral anterior nucleus   Ventral lateral nucleus   Ventral posterior nucleus

Ventral group

(b) Thalamus, superolateral view

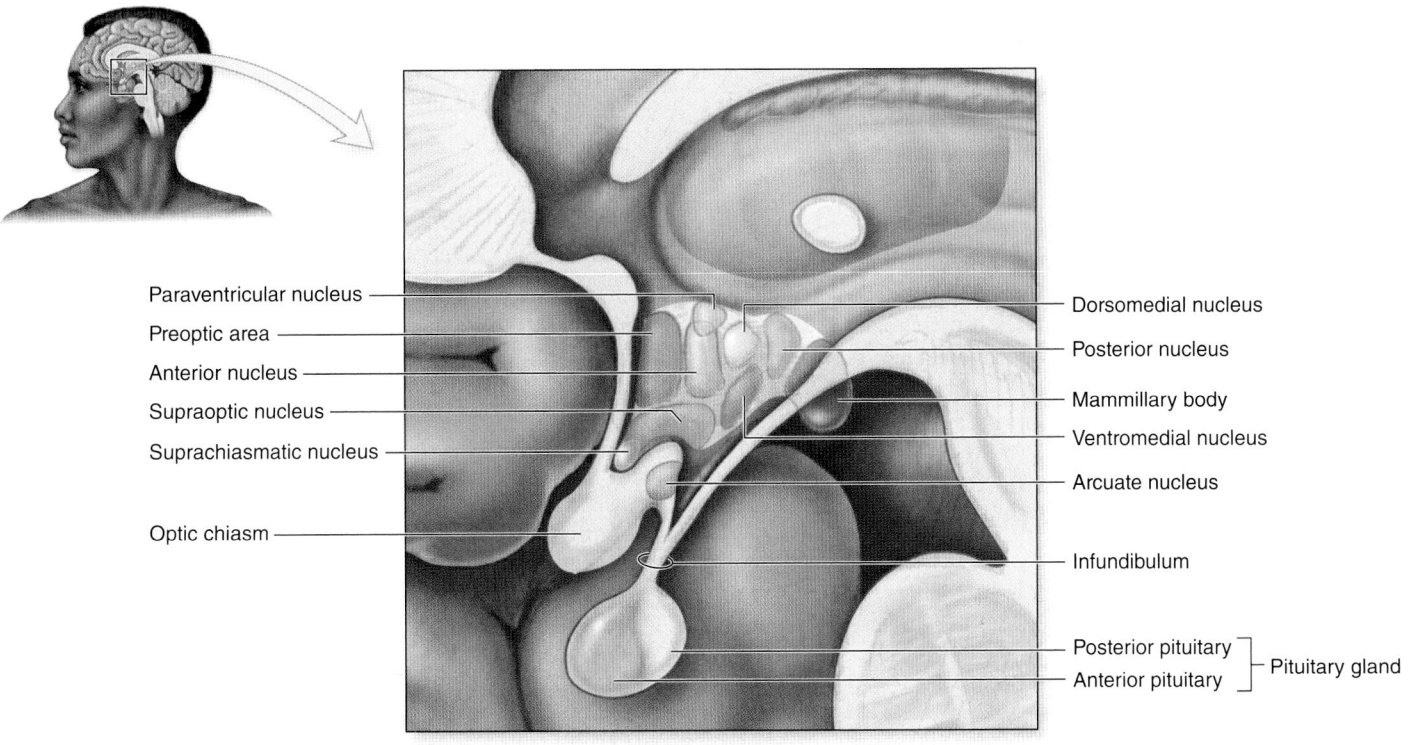

Paraventricular nucleus
Preoptic area
Anterior nucleus
Supraoptic nucleus
Suprachiasmatic nucleus

Optic chiasm

Dorsomedial nucleus
Posterior nucleus
Mammillary body
Ventromedial nucleus
Arcuate nucleus

Infundibulum

Posterior pituitary
Anterior pituitary
} Pituitary gland

Sagittal section of hypothalamus

Figure 13.19 Hypothalamus. The hypothalamus is located anteroinferior to the thalamus and is organized into multiple nuclei. AP|R

The hypothalamus has numerous functions that are controlled by specific nuclei as listed in table 13.3. Functions of the hypothalamus include

- **Master control of the autonomic nervous system.** The hypothalamus is a major autonomic integration center. In essence, it is the "president" of the corporation known as the autonomic nervous system (described in detail in chapter 15). It projects descending axons to autonomic nuclei in the inferior brainstem that influence heart rate, blood pressure, digestive activities, and respiration.

- **Master control of the endocrine system.** The hypothalamus is also "president" of another corporation—the endocrine system—overseeing most but not all of that system's functions. The hypothalamus secretes hormones that control secretory activities in the anterior pituitary gland, and it produces both antidiuretic hormone and oxytocin. Its function in the endocrine system is described in detail in section 17.7b.

- **Regulation of body temperature.** The body's thermostat is located within the hypothalamus. Neurons in the preoptic area detect altered blood temperatures and signal other hypothalamic nuclei, which control the mechanisms that heat or cool the body (see sections 1.5b, 6.1d, and 27.8b).

- **Control of emotional behavior.** The hypothalamus is located at the center of the limbic system, the part of the brain that controls emotional responses, such as pleasure, aggression, fear, rage, contentment, and the sex drive.

- **Control of food intake.** Neurons within the ventromedial nucleus monitor levels of nutrients such as glucose and amino acids in the blood and produce sensations of hunger.

- **Control of water intake.** Specific neurons within the anterior nucleus of the hypothalamus continuously monitor the concentration of dissolved substances in the blood to regulate the sensation of thirst (see section 25.2). When the hypothalamus detects dehydration from the blood, our feelings of thirst are stimulated.

- **Regulation of sleep-wake (circadian) rhythms.** The suprachiasmatic nucleus directs the pineal gland to secrete melatonin at certain times of the day. Thus, both work to regulate circadian rhythms.

Table 13.3	Functions Controlled by Selected Hypo-thalamic Nuclei
Nucleus or Hypothalamic Region	**Function**
Anterior nucleus	"Thirst center" (stimulates fluid intake); autonomic control center
Arcuate nucleus	Regulates appetite, release of gonadotropin-releasing hormone, release of growth hormone–releasing hormone, and release of prolactin-inhibiting hormone
Mammillary body	Directs sensations related to olfaction; controls swallowing
Paraventricular nucleus	Produces oxytocin primarily
Preoptic area	"Thermostat" (regulates body temperature)
Suprachiasmatic nucleus	Regulates sleep-wake (circadian) rhythm
Supraoptic nucleus	Produces antidiuretic hormone (ADH) primarily
Ventromedial nucleus	"Satiety center" (produces hunger and satiety sensations)

WHAT DID YOU LEARN?

23 How does the hypothalamus control our feelings of hunger and thirst?

13.5 Brainstem

The **brainstem** connects the cerebrum, diencephalon, and cerebellum to the spinal cord. Three regions form the brainstem. From superior to inferior, these include the midbrain, the pons, and the medulla oblongata (figure 13.20).

The brainstem is a bidirectional passageway for all tracts extending between the major regions of the brain and the spinal cord. It also contains many autonomic centers and reflex centers required for our survival, and it houses nuclei of many of the cranial nerves.

13.5a Midbrain

LEARNING OBJECTIVES

1. List the major features of the midbrain.
2. Identify the locations and functions of structures that are visible in a cross-sectional view of the midbrain.
3. Explain the involuntary actions produced by the superior and inferior colliculi of the tectal plate.

The **midbrain,** or *mesencephalon,* is the superior portion of the brainstem. The midbrain has several major components which are

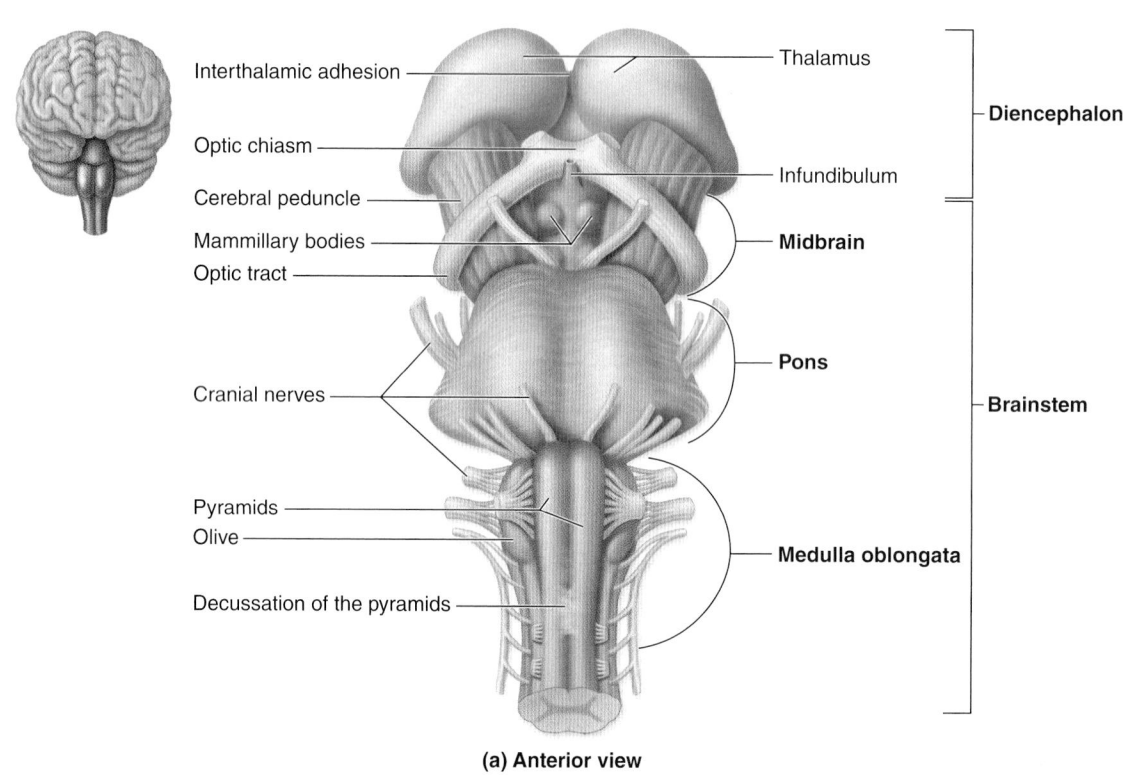

(a) Anterior view

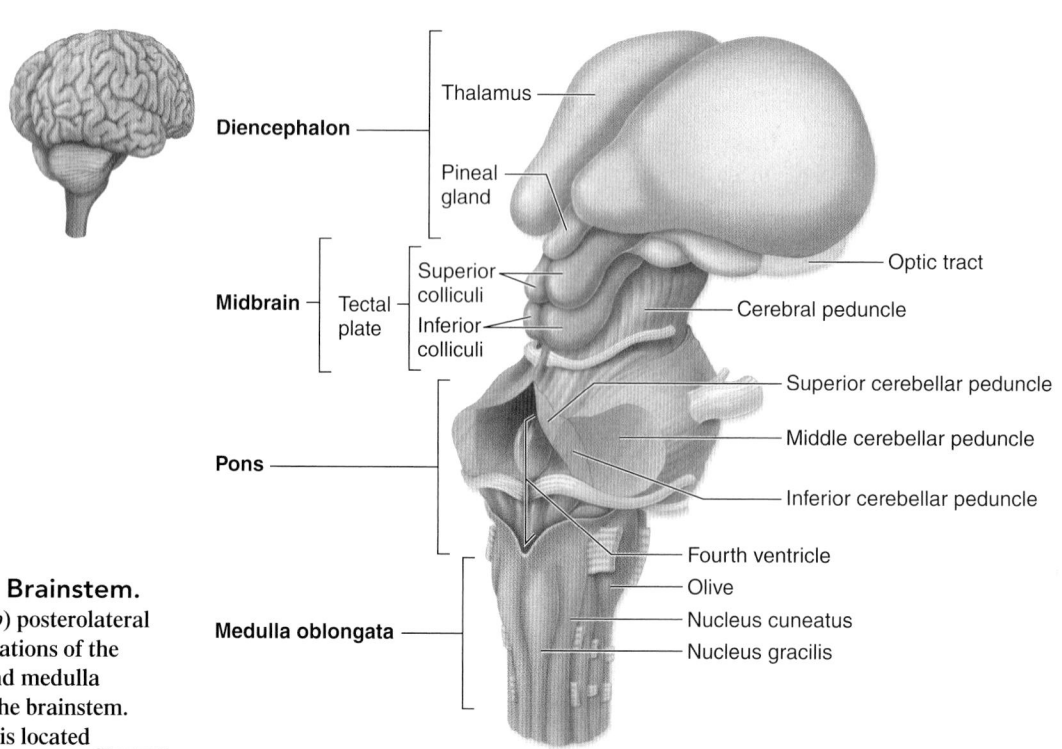

Figure 13.20 Brainstem.
(*a*) Anterior and (*b*) posterolateral views show the locations of the midbrain, pons, and medulla oblongata within the brainstem. The diencephalon is located superior to the brainstem. **AP|R**

(b) Posterolateral view

visible externally in figure 13.20. **Figure 13.21** shows cross sections of the midbrain and allows us to view its internal structures. These structures are discussed based on their position from anterior to posterior.

Cerebral peduncles (pĕ´dŭng-kl; *pedunculus* = little foot) are motor tracts located on the anterolateral surfaces of the midbrain. Descending axon bundles of the pyramidal system (corticospinal tracts) project through the cerebral peduncles and carry voluntary motor commands from the primary motor cortex of each cerebral hemisphere. Additionally, the midbrain is the final destination of the **superior cerebellar peduncles** connecting the cerebellum to the midbrain (figure 13.20). Bands of myelinated axons composing a **medial lemniscus** exit the medulla oblongata, extend through the pons and midbrain, and end in the thalamus.

The **substantia nigra** (sŭb-stan´shē-ă nī´gră; *substantia* = substance, *niger* = black) consists of bilaterally symmetrical nuclei within the midbrain (figure 13.21). Its name derives from its almost black appearance that is due to melanin pigmentation. The substantia nigra

houses clusters of neurons that produce the neurotransmitter dopamine, which affects brain processes to control movement, emotional response, and ability to experience pleasure and pain. Degeneration of these cells in the substantia nigra is a pathology that underlies Parkinson disease (see Clinical View: "Brain Disorders").

The **tegmentum** (teg-men´tŭm; covering structure) is sandwiched between the bilaterally symmetrical nuclei of the substantia nigra and the periaqueductal gray matter (described in the next paragraph). The tegmentum contains the pigmented **red nuclei** and the reticular formation. The reddish color of the nuclei is due to both blood vessel density and iron pigmentation in the neuronal cell bodies. The tegmentum integrates information from the cerebrum and cerebellum and issues involuntary motor commands to the erector spinae muscles of the back (see section 11.4) to help maintain posture while standing, bending at the waist, or walking.

Extending through the midbrain is the **cerebral aqueduct** connecting the third and fourth ventricles; it is surrounded by a region called the **periaqueductal gray matter.** The nuclei of the oculomotor nerve (CN III)

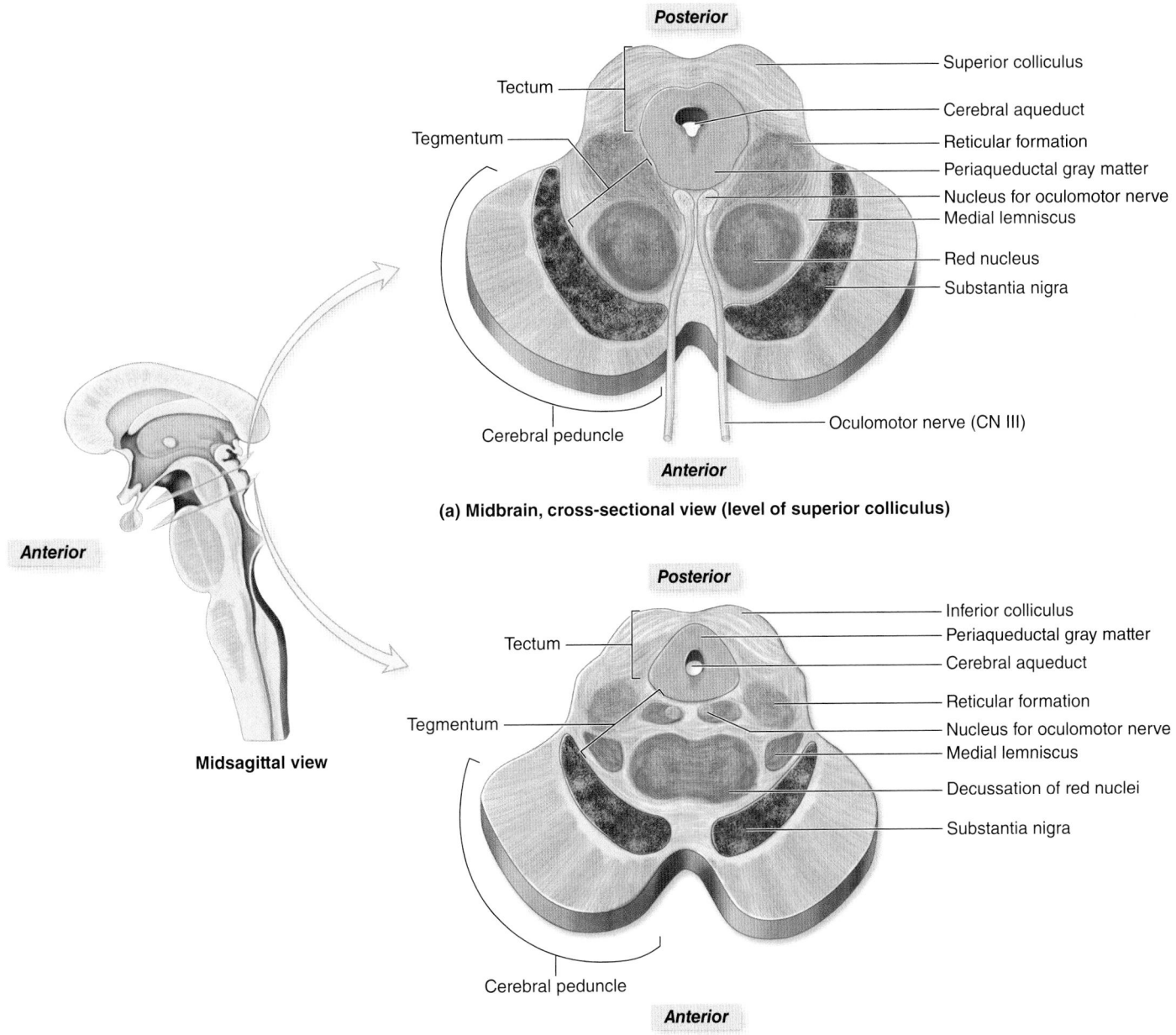

Midsagittal view

(a) Midbrain, cross-sectional view (level of superior colliculus)

Posterior
Tectum
Tegmentum
Cerebral peduncle
Anterior

Superior colliculus
Cerebral aqueduct
Reticular formation
Periaqueductal gray matter
Nucleus for oculomotor nerve
Medial lemniscus
Red nucleus
Substantia nigra
Oculomotor nerve (CN III)

(b) Midbrain, cross-sectional view (level of inferior colliculus)

Posterior
Tectum
Tegmentum
Cerebral peduncle
Anterior

Inferior colliculus
Periaqueductal gray matter
Cerebral aqueduct
Reticular formation
Nucleus for oculomotor nerve
Medial lemniscus
Decussation of red nuclei
Substantia nigra

Figure 13.21 Midbrain. Components of the midbrain at (*a*) the level of the superior colliculus and (*b*) the level of the inferior colliculus are shown in cross-sectional view. The nuclei for CN IV are not shown. AP|R

and the trochlear nerve (CN IV) are housed in the midbrain (these nerves are described in detail later in this chapter).

The **tectum** (tek'tŭm; roof) is the posterior region of the midbrain dorsal to the cerebral aqueduct. It contains two pairs of sensory nuclei, the superior and inferior colliculi, which are collectively called the **tectal plate** (*quadrigeminal* [kwah'dri-jem'i-năl] *plate*, or *corpora quadrigemina*). These nuclei are relay stations in the processing pathway of visual and auditory sensations. The **superior colliculi** (ko-lik'yū-lī; sing., *colliculus* = mound) are the superior nuclei. They are called *visual reflex centers* because they help visually track moving objects and control reflexes such as turning the eyes and head in response to a visual stimulus. For example, the superior colliculi are at work when you think you see a large animal running at you and turn suddenly toward the image. The paired **inferior colliculi** are the *auditory reflex centers*, meaning that they control reflexive turning of the head and eyes in the direction of a sound, such as a sudden loud bang.

WHAT DID YOU LEARN?

24 What is the function of the substantia nigra, and what disease may affect its proper working?

25 What parts of the midbrain contain paired visual and auditory sensory nuclei?

13.5b Pons

LEARNING OBJECTIVES

4. Identify the respiratory center located in the pons.

5. Identify the actions of the superior olivary complex.

The **pons** (ponz; bridge) is a bulging region on the anterior part of the brainstem that forms from part of the metencephalon (**figure 13.22**; see also figure 13.20). Sensory and motor tracts are located within and extend through it to connect to the brain and spinal cord. Additionally, the **middle cerebellar peduncles** are transverse axons that connect the pons to the cerebellum.

The pons houses autonomic nuclei in the **pontine respiratory center** (previously called the *pneumotaxic* [nū-mō-tăk'sik] *center*). This vital center, along with the medullary respiratory center within the medulla oblongata, regulates the skeletal muscles of breathing.

The **superior olivary nuclei** are located in the inferior pons. Each nucleus receives auditory input and is involved in the pathway for sound localization.

The pons also houses sensory and motor **cranial nerve nuclei** for the trigeminal (CN V), abducens (CN VI), and facial (CN VII) cranial nerves. Some of the nuclei for the vestibulocochlear cranial nerve (CN VIII) also are located there.

WHAT DID YOU LEARN?

26 What is the name and the general function of the autonomic respiratory center in the pons?

13.5c Medulla Oblongata

LEARNING OBJECTIVES

6. Describe the main features of the medulla oblongata.

7. List the autonomic centers of the medulla and the actions they control.

The **medulla oblongata** (me-dūl'ā ob-long-gah'tă; *medulla* = marrow or middle; *oblongus* = rather long) is often simply called the **medulla.** It is the most caudal (inferior) part of the brainstem and is continuous with the spinal cord inferiorly. The caudal portion of the medulla has a flattened, rounded shape and narrow central canal. As this tubelike opening

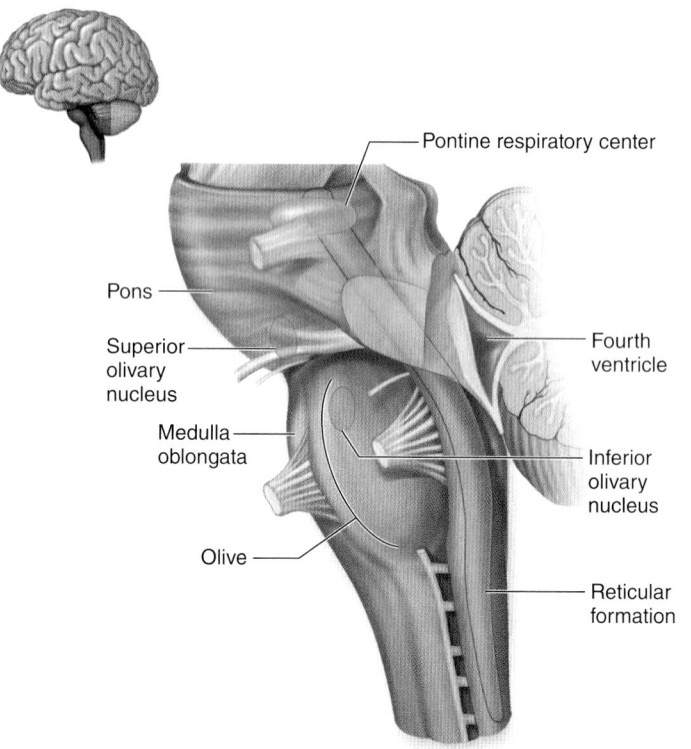

(a) Longitudinal section (cut-away)

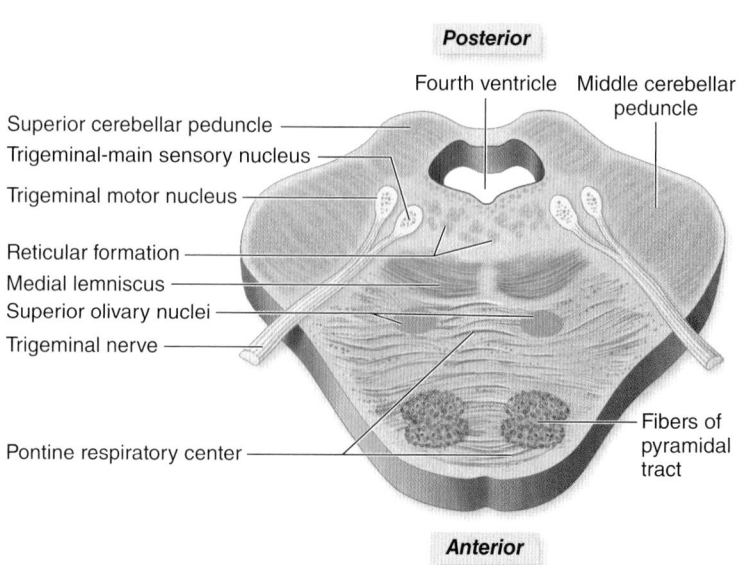

(b) Pons, cross-sectional view

Figure 13.22 Pons. The pons is a bulge on the ventral side of the brainstem that contains nerve tracts, nuclei, and part of the reticular formation. (*a*) A partially cut-away longitudinal section identifies the pontine respiratory center and the superior olivary nucleus. (*b*) A cross section through the pons shows the pontine respiratory center, fiber tracts, and some cranial nerve nuclei.

extends rostrally toward the pons, the central canal enlarges and becomes the inferior portion of the fourth ventricle. All communication between the brain and spinal cord involves tracts that ascend or descend through the medulla oblongata (figure 13.23; see also figures 13.20 and 13.22).

Several external landmarks are readily visible on the medulla oblongata. The anterior surface exhibits two longitudinal ridges called the **pyramids** (pir′ă-mid), which house the motor projection tracts called the corticospinal (pyramidal) tracts that extend through the medulla oblongata (see section 14.4c). In the caudal region of the medulla, most of these axons cross to the opposite side of the brain at a point called the **decussation** (dē-kŭ-sā′shŭn; *decussate* = to cross in the form of an X) **of the pyramids.** As a result of the crossover, each cerebral hemisphere controls the voluntary movements of the opposite side of the body. Immediately lateral to each pyramid is a distinct bulge, called the **olive,** which contains a large fold of gray matter called the **inferior olivary nucleus.** The inferior olivary nuclei relay ascending sensory nerve signals, especially proprioceptive information, to the cerebellum. Additionally, paired **inferior cerebellar peduncles** are tracts that connect the medulla oblongata to the cerebellum. The reticular formation (shown in figure 13.23 and discussed in detail in section 13.7b) originates in the medulla oblongata, and extends through the brainstem to the diencephalon.

The medulla oblongata contains several autonomic nuclei that group together to form centers that regulate functions vital for life. The most important autonomic centers in the medulla oblongata and their functions include

- The **cardiac center,** which regulates both the heart's rate and its strength of contraction (see section 19.5b).

- The **vasomotor center,** which controls blood pressure by regulating the contraction and relaxation of smooth muscle in the walls of the smallest arteries (the *arterioles*) to alter these vessels' diameter. Blood pressure increases when vessel walls constrict and lowers when vessel walls dilate (see section 20.6).

- The **medullary respiratory center,** which regulates the respiratory rate. It is composed of a ventral respiratory group and a dorsal respiratory group. These groups are influenced by the pontine respiratory center (see section 23.5c).

- Other nuclei in the medulla, which are involved in coughing, sneezing, salivation, swallowing, gagging, and vomiting.

WHAT DO YOU THINK?

5 Based on your understanding of the medulla oblongata's functions, would you expect severe injury to the medulla oblongata to cause death or merely be disabling? Why?

Finally, the medulla oblongata contains cranial nerve nuclei that are associated with the vestibulocochlear (CN VIII), glossopharyngeal (CN IX), vagus (CN X), accessory (CN XI), and hypoglossal (CN XII) cranial nerves. The medulla oblongata also contains the **nucleus cuneatus** (kū-nē-ā′tŭs; wedge) and the **nucleus gracilis** (gras-i′lis; slender), which relay somatic sensory information to the thalamus. The medial lemniscus exits from these nuclei and projects through the brainstem (where it is found posterior to the substantia nigra) to the ventral posterior nucleus of the thalamus.

WHAT DID YOU LEARN?

27 Where are the pyramids located, and what is their function?

28 What are the three main autonomic centers located in the medulla?

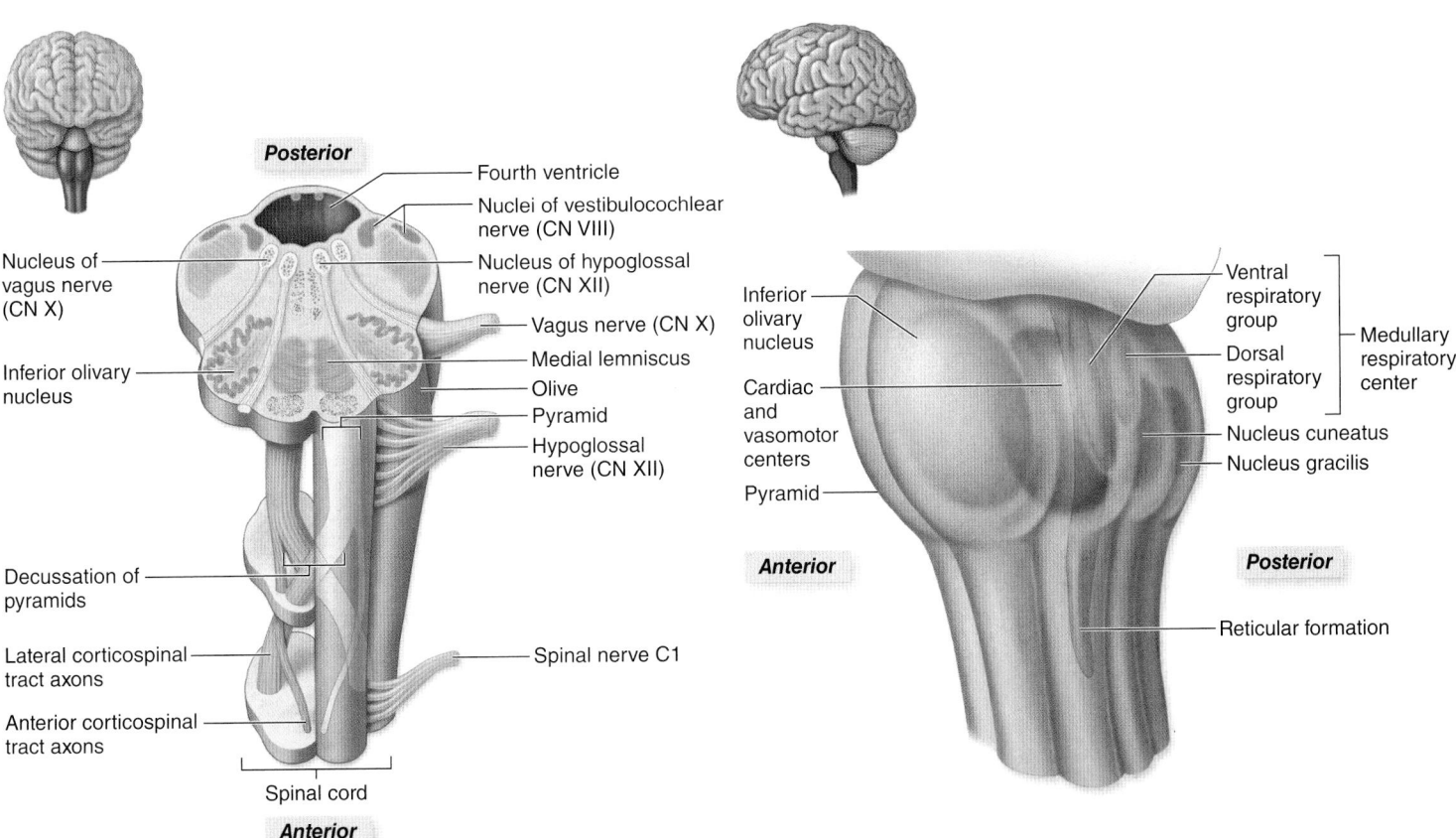

(a) Medulla oblongata, cross-sectional view

(b) Medulla oblongata, lateral view

Figure 13.23 Medulla Oblongata. The medulla oblongata connects the brain to the spinal cord. (*a*) A cross section illustrates important internal structures and decussations of the pyramids. (*b*) The medulla contains several nuclei that are involved in the regulation of heart rate, respiratory rate, and receiving and sending sensory information about limb movement. AP|R

13.6 Cerebellum

The **cerebellum** (ser-e-bel′ŭm; = little brain) is the second largest part of the brain. It produces fine control over muscular actions and stores memories of movement patterns, such as the playing of scales on a piano.

13.6a Structural Components of the Cerebellum

LEARNING OBJECTIVES

1. Name the parts and landmarks of the cerebellum.
2. Identify the three tracts through which the brainstem is linked to the cerebellum.

The cerebellum has a complex, highly convoluted surface covered by a layer of cerebellar cortex. The folds of the cerebellar cortex are called **folia** (fō′lē-ă; *folium* = leaf) **(figure 13.24)**. The cerebellum is composed of left and right **cerebellar hemispheres.** Each hemisphere consists of two lobes, the **anterior lobe** and the **posterior lobe,** which are separated by the **primary fissure.**

A narrow band of cortex known as the **vermis** (ver′mis; worm) lies along the midline between the left and right cerebellar lobes. The vermis receives sensory input reporting torso position and balance.

The cerebellum is partitioned internally into three regions: an outer gray matter layer of cortex called the **cerebellar cortex,** an internal region of white matter, and the deepest gray matter layer that is composed of cerebellar nuclei. The internal region of white matter is called the **arbor vitae** (ar′bōr vī′tē; *arbor* = tree, *vita* = life) because its distribution pattern resembles the branches of a tree.

Three thick nerve tracts, called peduncles, connect the cerebellum with the brainstem (see figure 13.20*b*). The **superior cerebellar peduncles** connect the cerebellum to the midbrain. The **middle**

cerebellar peduncles connect the pons to the cerebellum. The **inferior cerebellar peduncles** connect the cerebellum to the medulla oblongata. These extensive communication lines enable the cerebellum to "fine-tune" skeletal muscle movements and interpret all body proprioceptive movement.

WHAT DID YOU LEARN?

29 What are the primary anatomic components of the cerebellum?

30 The middle cerebellar peduncles connect the cerebellum to which part of the brainstem?

13.6b Functions of the Cerebellum

LEARNING OBJECTIVE

3. Explain the functions of the cerebellum.

The cerebellum does not *initiate* skeletal muscle movement. Rather, it coordinates and "fine-tunes" skeletal muscle movements that were initiated by the cerebrum, and ensures that skeletal muscle contraction follows the correct pattern, leading to smooth, coordinated movements. The cerebellum stores memories of previously learned movement patterns. This function is performed indirectly, by regulating activity along both the voluntary and involuntary motor pathways at the cerebral cortex, cerebral nuclei, and motor centers in the brainstem. The cerebrum initiates a movement and sends a "rough draft" of the movement to the cerebellum, which then coordinates and adjusts it. For example, the controlled, precise movements a classical guitarist makes when playing a concerto result from fine-tuning by the cerebellum. Without the cerebellum, the guitarist's movements would be choppy and sloppy, without precise coordination between the two hands.

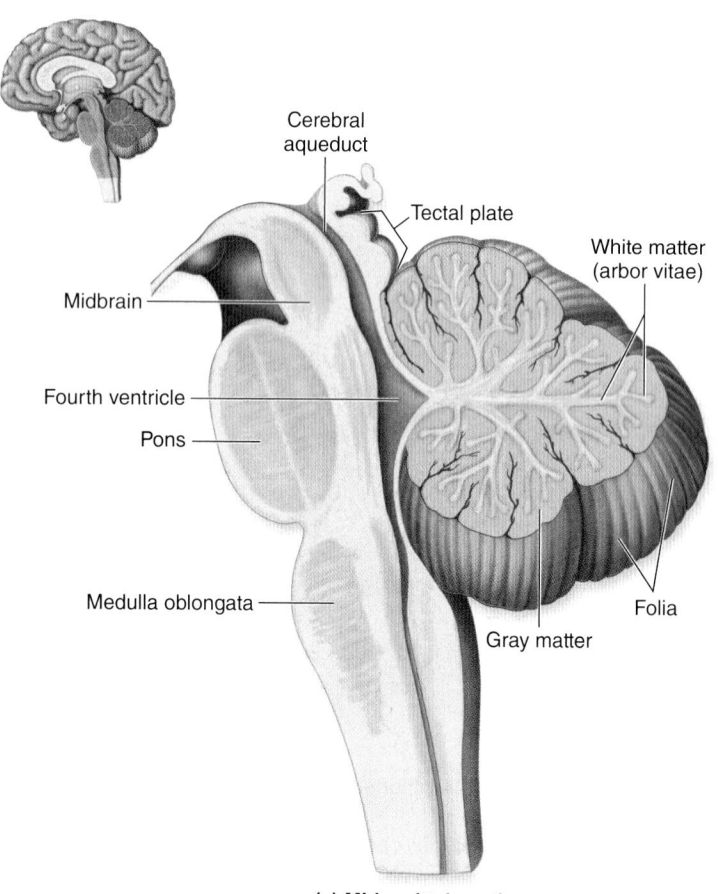

(a) Midsagittal section

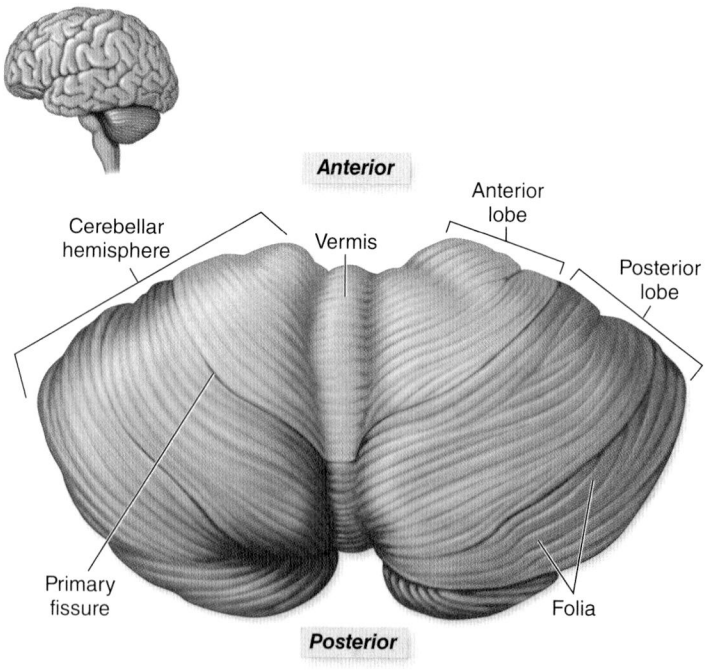

(b) Cerebellum, superior view

Figure 13.24 Cerebellum. The cerebellum lies posterior to the pons and medulla oblongata of the brainstem. (*a*) A midsagittal section shows the relationship of the cerebellum to the brainstem. (*b*) A superior view compares the anterior and posterior lobes of the cerebellum. (Note: The cerebrum and diencephalon have been removed.) **AP|R**

INTEGRATE

CONCEPT CONNECTION

The precise movements of our fingers rely on multiple systems. The skeletal system provides structural support for the muscles and tissues within the fingers, the muscular system is responsible for the movement of the fingers, and the nervous system sends the nerve signals to the muscular system to control the patterned movements.

The cerebellum has several additional functions. Skeletal muscle activity is adjusted to maintain equilibrium and posture. It also receives proprioceptive (sensory) information from the muscles and joints and uses this information to regulate the body's position. For example, you are able to balance on one foot because the cerebellum takes the proprioceptive information from the body joints and "maps out" a muscle tone plan to keep the body upright. Finally, because proprioceptive information from both the body's muscles and joints is sent to the cerebellum, the cerebrum knows the position of each body joint and its muscle tone, even if the person is not looking at the joint. For example, if you close your eyes, you are still aware of which body joints are flexed and which are extended.

The cerebellum continuously receives convergent input from both the various sensory pathways and from the motor pathways in the brain (figure 13.25). In this way, the cerebellum unconsciously perceives the state of the body, receives the plan for movement, and then follows the activity to see if it was carried out correctly. When the cerebellum detects a disparity between the intended and actual movement, it may generate error-correcting nerve signals. These nerve signals are transmitted to both the premotor and primary motor cortices via the

INTEGRATE

CLINICAL VIEW
Effects of Alcohol and Drugs on the Cerebellum

A variety of drugs, and alcohol in particular, can temporarily or permanently impair cerebellar function. Alcohol intoxication leads to the following symptoms of impaired cerebellar function that are used in the classic sobriety tests performed by police officers:

- **Disturbance of gait.** A person under the influence of alcohol rarely walks in a straight line, but appears to sway and stagger. In addition, falling and bumping into objects are likely, due to the temporary cerebellar disturbance.
- **Loss of balance and posture.** When attempting to stand on one foot, a person who is intoxicated usually tips and falls over.
- **Inability to detect proprioceptive information.** When asked to close the eyes and touch the nose, an intoxicated person frequently misses the mark. This reaction is due to reduced ability to sense proprioceptive information, compounded by uncoordination of skeletal muscles.

brainstem and the thalamus. Descending pathways then transmit these error-correcting signals to the motor neurons. Thus, the cerebellum influences and controls movement by indirectly affecting the excitability of motor neurons.

WHAT DID YOU LEARN?

31 What are the functions of the cerebellum?

Voluntary movements
The primary motor cortex and the cerebral nuclei within the cerebrum send nerve signals through the nuclei of the pons to the cerebellum.

Assessment of voluntary movements
Proprioceptors in skeletal muscles and joints send sensory input regarding the degree of movement to the cerebellum.

Integration and analysis
The cerebellum compares the planned movements (motor nerve signals) against the results of the actual movements (sensory nerve signals).

Corrective feedback
The cerebellum sends nerve signals through the thalamus to the primary motor cortex and to motor nuclei in the brainstem.

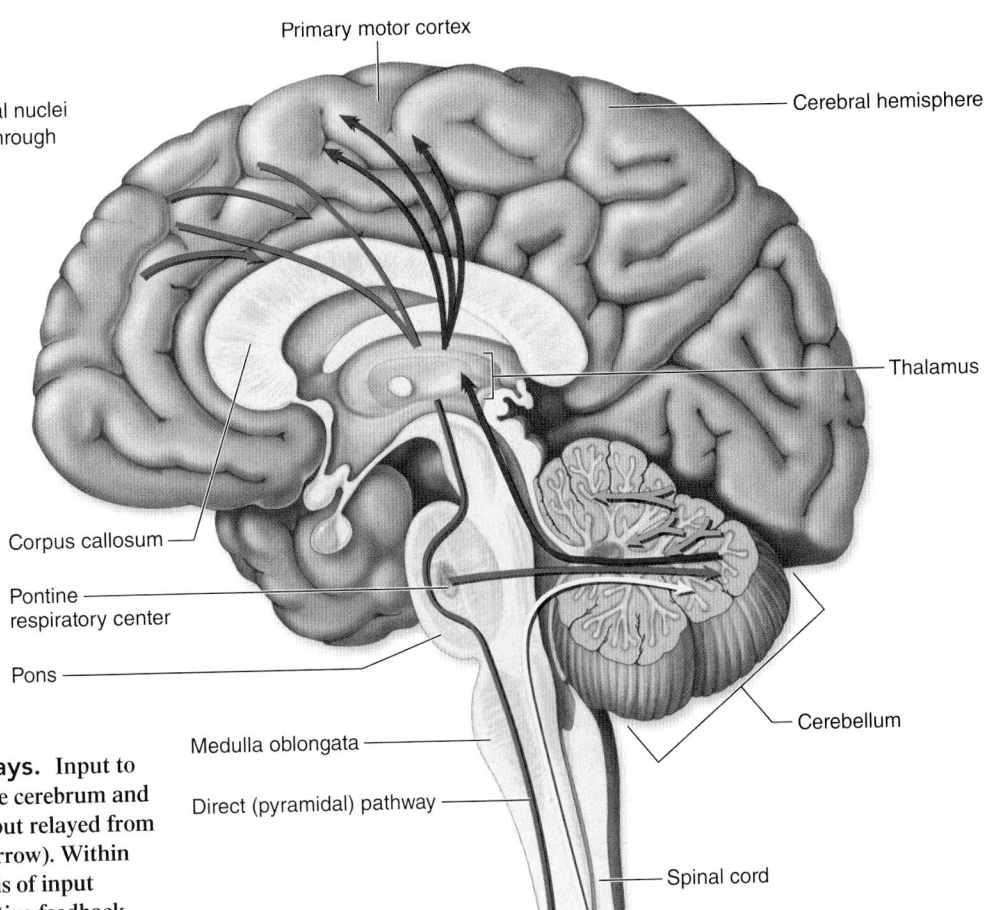

Figure 13.25 Cerebellar Pathways. Input to the cerebellum from the motor cortex of the cerebrum and the pons (blue arrows), and the sensory input relayed from proprioceptors to the cerebellum (yellow arrow). Within the cerebellum, the integration and analysis of input information occurs (green arrows). Corrective feedback output from the cerebellum (red arrows) extends to the cerebrum.

Labels: Primary motor cortex, Cerebral hemisphere, Thalamus, Corpus callosum, Pontine respiratory center, Pons, Medulla oblongata, Direct (pyramidal) pathway, Cerebellum, Spinal cord

Sagittal section

13.7 Functional Brain Systems

The brain has two important functional systems that work together for a common purpose, even though their structures may be scattered throughout the brain. These systems are the limbic system and the reticular formation.

13.7a Limbic System

LEARNING OBJECTIVES

1. Describe the main functions of the limbic system.
2. List the seven structures that compose the limbic system, and summarize their actions.

The **limbic** (lim′bik) **system** is composed of multiple cerebral and diencephalic structures that collectively process and experience emotions. Thus, the limbic system is sometimes referred to as the *emotional brain*. The structures of the limbic system form a ring or border around the diencephalon (*limbus* = border). Although neuro-anatomists continue to debate the components of the limbic system, the brain structures commonly recognized are shown in **figure 13.26** and listed here:

1. The **cingulate** (sin′gū-lāt; *cingulum* = girdle, to surround) **gyrus** is an internal mass of cerebral cortex located within the longitudinal fissure and superior to the corpus callosum. This cortical mass may be seen only in sagittal section, and it surrounds the diencephalon. It receives input from the other components of the limbic system.
2. The **parahippocampal gyrus** is a mass of cerebral cortical tissue in the temporal lobe. Its function is associated with the hippocampus.
3. The **hippocampus** (hip-ō-kam′pŭs; seahorse) is a nucleus located superior to the parahippocampal gyrus that connects to the diencephalon via a structure called the fornix. As its

name implies, this nucleus is shaped like a seahorse. Both the hippocampus and the parahippocampal gyrus are essential in storing memories and forming long-term memory.
4. The **amygdaloid body** connects to the hippocampus. The amygdaloid body is involved in several aspects of emotion, especially fear. It can also help store and code memories based on how a person emotionally perceives them—for example, as related to fear, extreme happiness, or sadness.
5. The **olfactory bulbs, olfactory tracts,** and **olfactory cortex** are part of the limbic system as well. You have probably experienced how particular odors can provoke certain emotions or be associated with certain memories.
6. The **fornix** (fōr′niks; arch) is a thin tract of white matter that connects the hippocampus with other diencephalon limbic system structures.
7. Various nuclei in the diencephalon, such as the **anterior thalamic nuclei,** the **habenular nuclei** of the epithalamus, the **septal nuclei,** and the **mammillary** (mam′i-lār-ē; *mammilla* = nipple) **bodies** of the hypothalamus, interconnect other parts of the limbic system and contribute to its overall function.

WHAT DID YOU LEARN?

32 What are the components of the limbic system?

33 What are the main functions of the limbic system?

13.7b Reticular Formation

LEARNING OBJECTIVES

3. Describe the components and function of the reticular formation.
4. Explain anatomy and function of the reticular activating system (RAS).

Projecting vertically through the core of the midbrain, pons, and medulla is a loosely organized mass of gray matter called the **reticular formation** (**figure 13.27**). The reticular formation extends

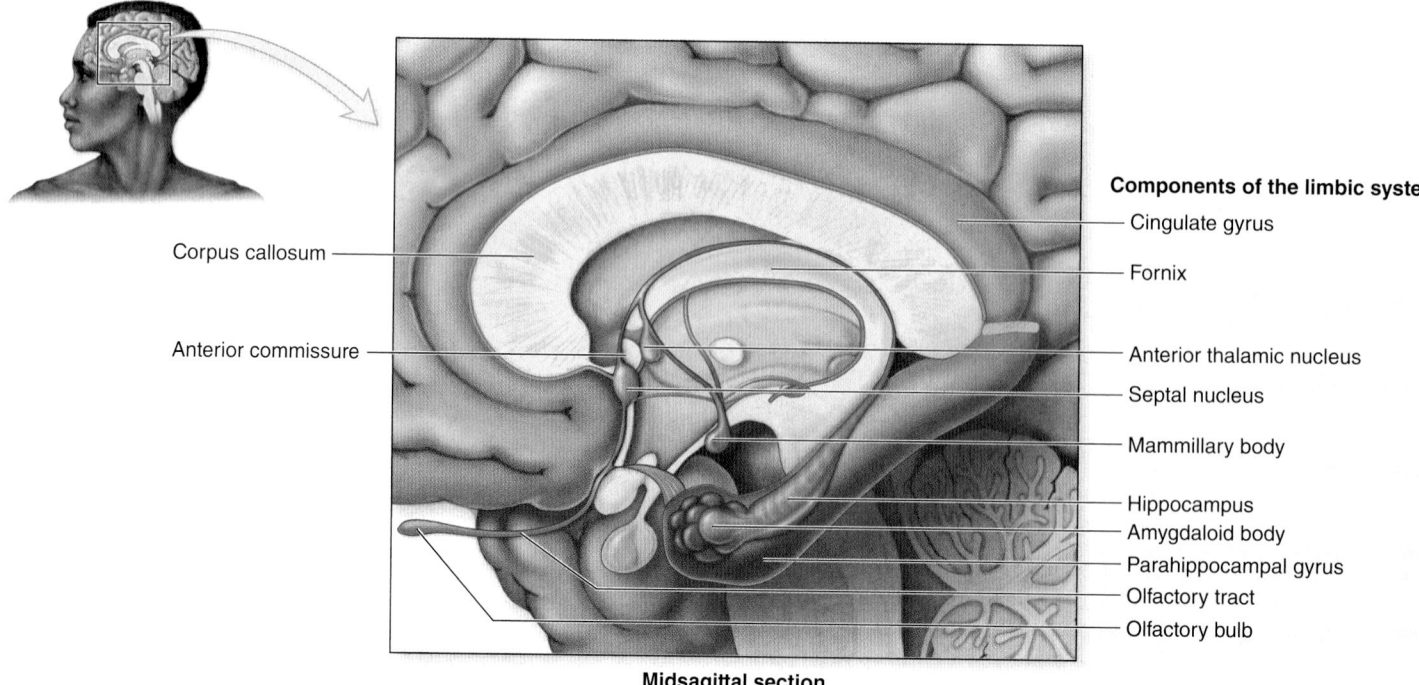

Corpus callosum

Anterior commissure

Components of the limbic system
Cingulate gyrus
Fornix
Anterior thalamic nucleus
Septal nucleus
Mammillary body
Hippocampus
Amygdaloid body
Parahippocampal gyrus
Olfactory tract
Olfactory bulb

Midsagittal section

Figure 13.26 Limbic System. The components of the limbic system affect behavior and emotions. The olfactory cortex of the temporal lobe is not shown. AP|R

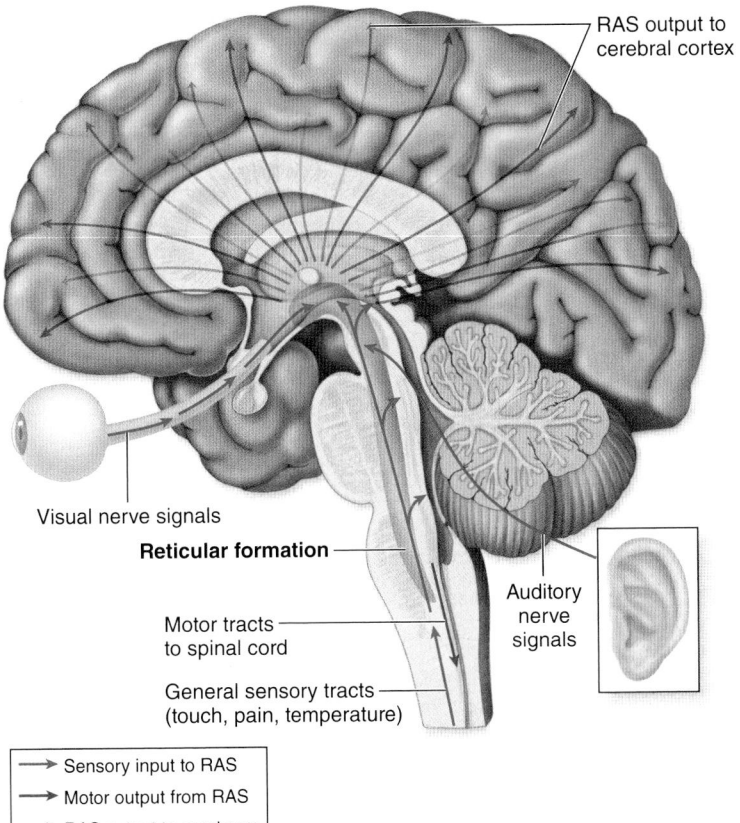

Figure 13.27 **The Reticular Formation.** The reticular formation receives and processes various types of input from sensory receptors (blue arrows). It participates in cyclic activities such as arousing the cortex to consciousness (purple arrows) and controlling the sleep-wake cycle. Some motor output from the reticular formation influence muscle activity (red arrow).

slightly into the diencephalon and the spinal cord as well. This functional brain system has both motor and sensory components.

The motor component of the reticular formation communicates with the spinal cord and is responsible for regulating muscle tone (especially when the muscles are at rest). This motor component also assists in autonomic motor functions, such as respiration, blood pressure, and heart rate, by working with the autonomic centers in the medulla and pons.

The sensory component of the reticular formation is responsible for alerting the cerebrum to incoming sensory information. This sensory component is called the **reticular activating system (RAS),** and it contains sensory axons that project to the cerebral cortex. The RAS processes visual, auditory, and touch stimuli and uses this information to keep us in a state of mental alertness. Additionally, the RAS arouses us from sleep. The sound of an alarm clock can awaken us because the RAS receives this sensory stimulus and sends it to the cerebrum. Conversely, under conditions of little or no stimuli, such as when you are in bed with the lights out, and no sounds are disturbing you, the RAS is not stimulated and you find it easier to sleep.

Consciousness includes an awareness of sensation, voluntary control of motor activities, and the activities necessary for higher mental processing. It involves the simultaneous activity of large areas of the cerebral cortex. Levels of consciousness exist on a continuum. The highest state of consciousness and cortical activity is **alertness,** in which the individual is responsive, aware of self, and well-oriented to person, place, and time.

WHAT DID YOU LEARN?

34 How is the reticular activating system related to the reticular formation?

13.8 Integrative Functions and Higher-Order Brain Functions

Higher-order mental functions include learning, memory, and reasoning. These functions occur within the cortex of the cerebrum and involve multiple brain regions connected by complicated networks and arrays of axons. Both conscious and unconscious processing of information are involved in higher-order mental functions, and this processing may be continually adjusted or modified.

13.8a Development of Higher-Order Brain Functions

LEARNING OBJECTIVE

1. Describe the relationship between age and higher-order brain functioning.

From infancy on, our motor control and processing capabilities become increasingly complex as we grow and mature. During the first year of life, the number of cortical neurons continues to increase. The myelination of most CNS axons continues throughout the first 2 years. (As a result, pediatricians recommend that infants and toddlers drink whole milk instead of skim milk, so their bodies will have adequate fat intake to support the development of myelin and the brain in general.) The brain grows rapidly in size and complexity so that by the age of 5, brain growth is 95% complete. (The rest of the body doesn't reach its adult size until puberty.)

As the CNS continues to develop, some neurons expand their number of connections, providing the increased number of synaptic junctions required for increasingly complex reflex activities and processing. During this same period, the brain will "prune" various synaptic connections, so only the most commonly used connections will remain. Some CNS axons remain unmyelinated until the teenage years (e.g., some of the axons in the prefrontal cortex). In general, the axons of PNS neurons continue to myelinate past puberty. A person's ability to carry out higher-order mental functions is a direct result of the level of nervous system maturation.

WHAT DID YOU LEARN?

35 What are some implications of the brain's development not being complete until the mid-teens?

13.8b Electroencephalogram

LEARNING OBJECTIVE

2. Describe how an electroencephalogram examines brain activity.

An **electroencephalogram (EEG)** is a diagnostic test where electrodes are attached to the head to record the electrical activity of the brain (**figure 13.28**). This procedure is performed to investigate sleep disorders, brain lesions, and to determine if an individual is in a coma or a persistent vegetative state (see Clinical View: "Pathologic States of Consciousness"). EEGs also may evaluate a **seizure,** which is an event of abnormal electrical activity in the brain. There are different types of seizures, some of which may result in a brief blackout, and others that may result in shaking and muscle spasms. **Epilepsy** is the condition where a person experiences repeated seizures over time.

An EEG measures and plots four types of brain waves (i.e., alpha, beta, theta, and delta). The distribution and frequency of these waves varies, depending upon whether the person is a child or an adult and

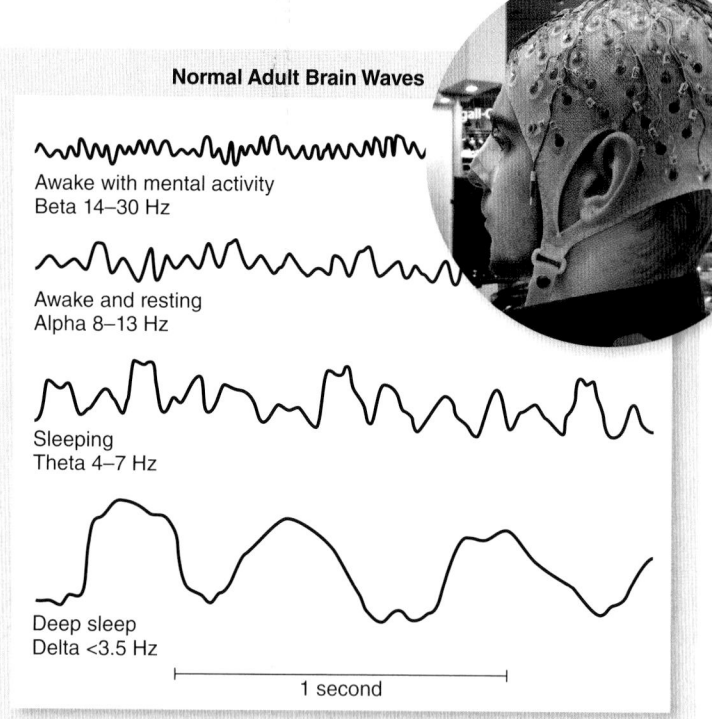

Normal Adult Brain Waves

Awake with mental activity
Beta 14–30 Hz

Awake and resting
Alpha 8–13 Hz

Sleeping
Theta 4–7 Hz

Deep sleep
Delta <3.5 Hz

1 second

Figure 13.28 Electroencephalograms (EEGs). An individual wears an EEG cap that contains multiple electrodes to record brain wave activity, and a sample EEG recording is shown.

INTEGRATE

CLINICAL VIEW
Pathologic States of Unconsciousness

When a person is asleep, he or she is technically unconscious, but not pathologically so. However, other unconscious conditions are pathologic.

A brief loss of consciousness, termed **fainting,** or **syncope** (sin'kŏ-pē; *cutting short*), often signals inadequate cerebral blood flow due to low blood pressure, as might follow hemorrhage or sudden emotional stress.

Stupor (stū'per; *stupeo* = to be stunned) is a moderately deep level of unconsciousness from which the person can be aroused only by extreme repeated or painful stimuli. A stupor may be associated with metabolic disorders such as low blood sugar, diseases of the liver or kidney, CVA or other brain trauma, or drug use.

A **coma** is a deep and profound state of unconsciousness from which the person cannot be aroused, even by repeated or painful stimuli. A person in a coma is alive but unable to respond to the environment. The causes of a coma include a severe head injury or CVA, marked metabolic failure (as occurs in advanced liver and kidney disease), very low blood sugar, or drug use.

A **persistent vegetative state** is a condition in which the person has lost his or her thinking ability and awareness of the environment, but noncognitive brain functions continue, such as the brainstem's monitoring of heart rate, breathing, and the sleep-wake cycle. Some people in this state exhibit spontaneous movements, such as moving their eyes, grimacing, crying, and even laughing.

---

if the individual is in a deep sleep, having a seizure, or is experiencing a pathologic state of consciousness. For example, alpha and beta waves are typically seen in an awake or alert state, whereas theta and delta waves are more common during sleep. The presence of theta and delta waves in an awake adult is suggestive of a brain abnormality. Each electrode attached to a person's head will register a brain wave over that region of the head, so a patient's EEG printout will show multiple brain waves over a period of time.

WHAT DID YOU LEARN?

36 An EEG may be used to evaluate what clinical conditions?

13.8c Sleep

LEARNING OBJECTIVES

3. Describe the main characteristics of sleep.

4. Compare and contrast non-REM and REM sleep.

People normally alternate between periods of alertness and **sleep,** which is the natural, temporary absence of consciousness from which a person can be aroused by normal stimulation. Cortical activity is depressed during sleep, but functions continue in the vital centers within the brainstem. Sleep is a natural, repeated condition of rest for both the body and mind. During periods of sleep, the decreased body activities are accompanied by conservation of energy, increasing growth, and renewal of strength in an individual.

Sleep may be subdivided into two main types, **non-REM** (non-rapid eye movement) and **REM** (rapid eye movement) **sleep.** Both types are distinguished by their EEG patterns and the absence or presence of rapid eye movements. In addition, it is during REM sleep that we have our most memorable dreams (although we may not remember all of our dreams when we awake). We spend about 75% of our total sleep time in non-REM sleep, and the remaining 25% in REM sleep. Some sleep scientists believe that non-REM sleep is meant for body repair, whereas REM sleep is consolidating and organizing memories, because the brain is very active during this period and it uses as much oxygen as when an individual is awake.

Non-REM sleep may be further subdivided into four stages. The EEG has helped scientists detect these four stages. We cycle through these non-REM stages and REM sleep multiple times throughout a normal-length sleep cycle, as shown in **figure 13.29.** The different

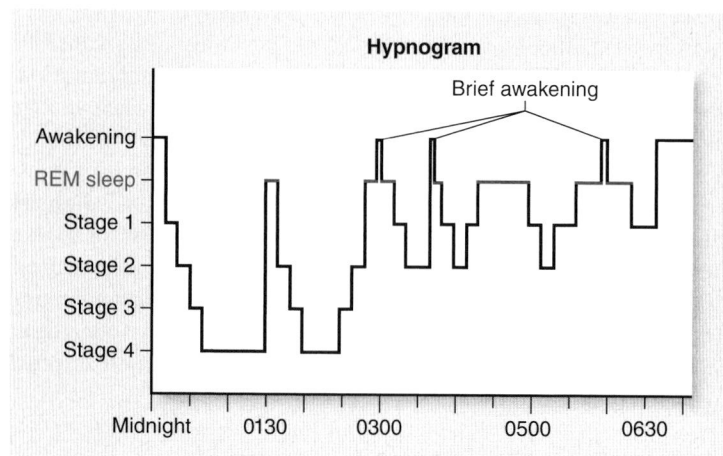

Hypnogram

Brief awakening

Awakening
REM sleep
Stage 1
Stage 2
Stage 3
Stage 4

Midnight   0130   0300   0500   0630

Figure 13.29 Hypnogram. A hypnogram documents the length of time a person is in each sleep stage.

stages of non-REM sleep differ in the types of brain waves present (e.g., alpha, beta, theta, and delta) and the ease at which one may be awakened. After about 90 minutes of non-REM sleep, the first incidence of REM sleep occurs and typically lasts about 10 minutes. The body then cycles back into non-REM sleep and then a longer period of REM sleep.

The amount of sleep a person needs varies based on our age and our health. Infants typically need up to 17 to 18 hours of sleep a day, and this number drops as we get older. Teens typically need between 8.5 to 9.5 hours of sleep a night, whereas the average adult needs about 7 to 8 hours of sleep. Lack of sleep has been associated with depression, impaired memory, and decreased immune function.

The term **insomnia** refers to the difficulty in falling asleep and staying asleep. Insomnia becomes more prevalent as we age, and certain medications may interfere with sleep (including how frequently we experience REM sleep) as well. **Sleep apnea** is where a person experiences repeated breathing interruptions during sleep. These breathing interruptions cause the individual to wake repeatedly throughout the night, so the individual is at risk to a variety of ailments associated with lack of sleep. Sleep apnea may be treated with a **CPAP (continuous positive airway pressure) machine,** where air is pumped through a mask that the patient wears during sleep so as to keep the airways open and allow the individual to sleep uninterrupted.

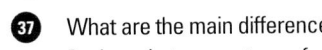 **WHAT DID YOU LEARN?**

37 What are the main differences between non-REM and REM sleep? During what percentage of our sleep cycle are we in each type of sleep?

13.8d Cognition

LEARNING OBJECTIVES

5. Identify the brain areas in which cognition occurs.

6. Explain how lesions to different regions of the cortex affect cognition.

Mental processes such as awareness, knowledge, memory, perception, and thinking are collectively called **cognition** (kog-ni′shŭn). The association areas of the cerebrum, which form about 70% of the nervous tissue in the brain, are responsible for both cognition and the processing and integration of information between sensory input and motor output areas.

Various studies of individuals suffering from brain lesions (caused by cancer, infection, stroke, and trauma) have provided insight into the functions of these areas of the brain. For example, the frontal association area (prefrontal cortex) integrates information from the sensory, motor, and association areas to enable the individual to think, plan, and execute appropriate behavior. Thus, an individual with a frontal lobe lesion exhibits personality abnormalities.

If an individual loses the ability to detect and identify stimuli (termed loss of awareness) on one side of the body, or on the limbs on that side, the primary somatosensory area in the hemisphere opposite the affected side of the body has been damaged.

An individual who has **agnosia** (ag-nō′zē-ă; *a* = without, *gnosis* = knowledge) displays an inability either to recognize or to understand the meaning of various stimuli. For example, a lesion in the temporal lobe may result in an inability to recognize or understand the meaning of sounds or words. Specific symptoms of agnosia vary, depending upon the location of the lesion within the cerebrum.

 WHAT DID YOU LEARN?

38 What is the definition of cognition?

13.8e Memory

 LEARNING OBJECTIVES

7. Compare and contrast short-term and long-term memory, and describe the parts of the brain involved with each.

8. Name the two regions of the limbic system involved in conversion of short-term memory to long-term memory.

Memory is a versatile element of human cognition involving different lengths of time and different storage capacities. Storing and retrieving information requires higher-order mental functions and depends upon complex interactions among different brain regions. On a broader scale, in addition to memory, information management by the brain entails both learning (the acquisition of new information) and forgetting (the elimination of trivial or nonuseful information).

Neuroscientists classify memory in various ways. For example, **sensory memory** occurs when we form important associations based on sensory input from the environment, such as the sounds coming from a crowded cafeteria, the smell from the food line, and the bright lights from the room. Sensory memory typically lasts for seconds at most.

Short-term memory (STM) follows sensory memory. It is generally characterized by limited capacity (approximately seven small segments of information) and brief duration (ranging from seconds to hours). Suppose that, in a Friday morning anatomy and physiology lecture, your instructor lists the general functions of the cerebral lobes on the board. Unless you study this information over the weekend, you will probably not recall it by Monday's lecture.

Some short-term memory, if adequately repeated and assessed, may be converted to long-term memory. Once information is placed into **long-term memory (LTM),** it may exist for limitless periods of time. So, for example, if over the weekend you read and recopy your lecture notes, review the text and figures in the book, and prepare note cards, you will have stored the information about the cerebral lobes as LTM. Not only will you be well prepared for your next examination, but you may even remember this information for years to come. (However, information in LTM needs to be retrieved occasionally or it can be "lost," and our ability to store and retrieve information declines with aging.)

It appears that our brain must organize complex information in short-term memory prior to storing it in long-term memory **(figure 13.30).** Conversion from STM to LTM is called **encoding,** or *memory consolidation.* Encoding requires the proper functioning of two components of the limbic system: the amygdaloid body and the hippocampus. The hippocampus is required for the formation of STM, whereas long-term memories are stored primarily in the association areas of the cerebral cortex. For example, voluntary motor activity memory is housed in the premotor cortex and the cerebellum, and memory of sounds is stored in the auditory association area.

Because STM and LTM involve different anatomic structures, loss of the ability to form STM does not affect the maintenance or accessibility of LTM.

 WHAT DID YOU LEARN?

39 What types of study habits best convert short-term memories into long-term memories?

13.8f Emotion

LEARNING OBJECTIVE

9. Explain the interactions of the prefrontal cortex and the limbic system in expression of emotions.

Emotional expression varies widely. For example, a horrible automobile accident may cause those involved and some observers to cry, scream,

Figure 13.30 Model of Information Processing.
Cognitive psychologists have proposed a model to show the relationships between sensory memory, short-term memory, and long-term memory, which develops later.

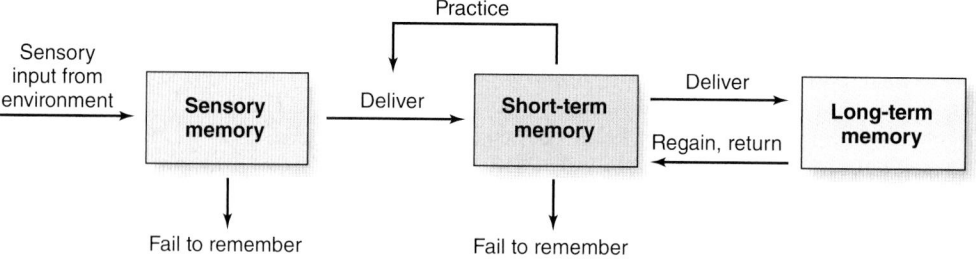

INTEGRATE

CLINICAL VIEW
Alzheimer Disease: The "Long Goodbye"

Alzheimer disease (AD) has become the leading cause of dementia in the developed world. The images illustrate the anatomic differences between a normal brain and an Alzheimer brain.

What are the classic symptoms of AD?
Typically, AD does not become clinically apparent until after the age of 65, although some individuals have earlier onset; its diagnosis is often delayed because of confusion with other forms of cognitive impairment. Symptoms include slow, progressive loss of higher intellectual functions and changes in mood and behavior. AD gradually causes language deterioration, impaired visuospatial skills, indifferent attitude, and poor judgment, while leaving motor function intact. Patients become confused and restless, often asking the same question repeatedly. AD progresses relentlessly over months and years, and thus has come to be known as "the long goodbye." Eventually, it robs its victims of their memory, their former personality, and even the capacity to speak.

What causes AD?
The underlying cause of AD remains a mystery, although both genetics and environment seem to play a role. Postmortem examinations of the brains of AD patients show marked and generalized cerebral atrophy. Microscopic examinations of brain tissue reveal a profound decrease in the number of cerebral cortical neurons, and a proliferation of two abnormal types of structures: **amyloid** (am×i-loyd) **plaques** and **neurofibrillary tangles**. Amyloid plaques are insoluble deposits of a protein termed beta amyloid as well as portions of neurons and microglial cells. The neurofibrillary tangles are formed from a protein called **tau** that is hyperphosphorylated. Researchers are not sure if the plaques and tangles are the cause of AD, or merely a by-product of the disease's manifestation. Biochemical alterations also occur, most significantly a decreased level of the neurotransmitter acetylcholine in the cerebrum.

Is there a cure or test for AD?
There is no cure at present for AD, although some medications help alleviate the symptoms and seem to slow the progress of the disease. In the meantime, researchers are trying to develop diagnostic tests that can better predict who may be at risk for AD. Until recently, the only way to definitively diagnose AD was at autopsy, when the brain could be macroscopically and microscopically examined. Now, positron emission tomography (PET) scans appear to be able to identify the early brain changes seen with AD.

Recent research has suggested that difficulty or loss in identifying common smells (e.g., lemon, cinnamon) is linked with an increased risk in developing AD. In fact, this loss of smell may be one of the first signs of developing the disease, presumably because the brain regions involved with smell are among the first regions to develop the amyloid plaques and neurofibrillary tangles of AD. So, in the near future researchers may develop a "scratch and sniff" test to help predict an individual's risk of developing AD.

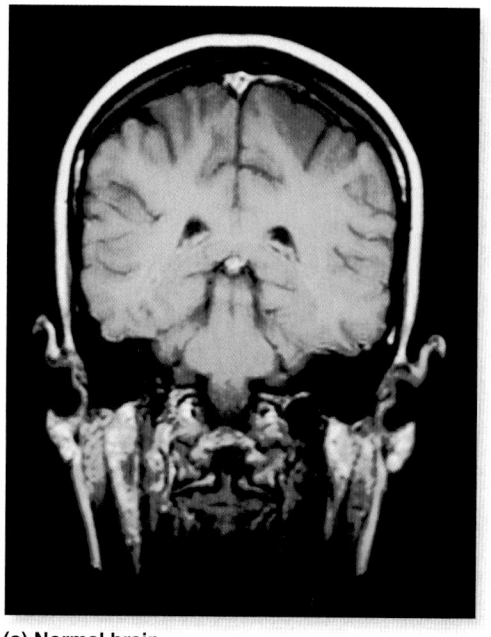

(a) Normal brain

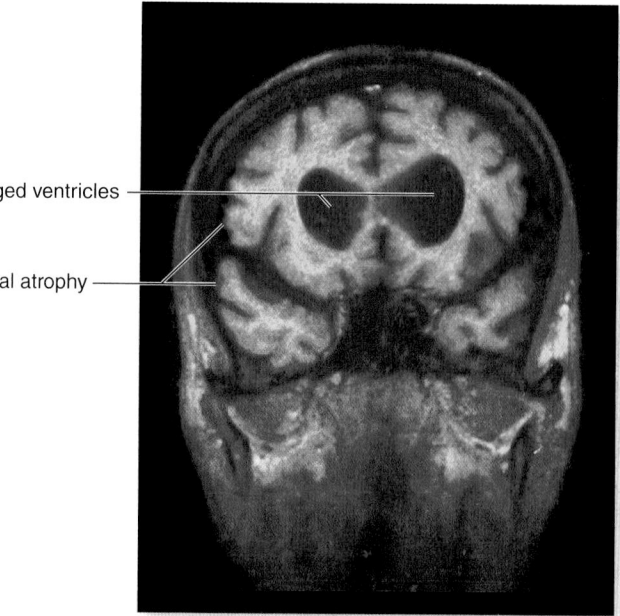

Enlarged ventricles

Cortical atrophy

(b) Alzheimer brain

Magnetic resonance imaging (MRI) scans show coronal sections of (a) a normal brain and (b) an Alzheimer brain. (Note the large ventricles and wide spaces between gyri in the Alzheimer brain.)

or totally lose "emotional control," whereas the responding emergency personnel generally appear stoic, wearing masked expressions as they go about their professional duties.

Expression of our emotions is interpreted by our limbic system but ultimately is controlled by the prefrontal cortex. Irrespective of how we feel, this cortical region decides the appropriate way to show our feelings. Researchers have identified the emotional control centers of the brain by using traditional techniques as well as by examining the behavior of experimental animals and individuals with brain lesions. Although interpreting the results is often difficult because of the complexities of both the brain and our behavior, researchers have learned that many important aspects of emotion depend upon an intact, functional amygdaloid body and hippocampus (components of the limbic system). If specific regions of either of these structures are damaged or artificially stimulated, we exhibit either deadened or exaggerated expressions of aggression, affection, anger, fear, love, pain, pleasure, or sexuality, as well as anomalies in learning and memory.

INTEGRATE

CLINICAL VIEW
Amnesia

Amnesia (am-nē'zē-ă) refers to complete or partial loss of memory. Most often, amnesia is temporary and affects only a portion of a person's experiences. Causes of amnesia range from psychological trauma to direct brain injury, such as a severe blow to the head or even a cerebrovascular accident (CVA).

Because memory processing and storage involve numerous regions of the brain, the type of memory loss that occurs in an episode of amnesia depends upon the area of the brain damaged. The most serious forms of amnesia result from damage to the thalamus and limbic structures, especially the hippocampus. If one or more of these structures is damaged, serious disruption or complete lack of memory storage and consolidation may follow. The nature of the underlying problem determines whether amnesia is complete or partial, and to what degree recovery, if any, is possible.

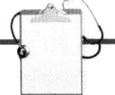

WHAT DID YOU LEARN?

40 What portions of the brain and limbic system are involved with modulation of emotion?

INTEGRATE

CLINICAL VIEW
Dyslexia

Dyslexia (dis-lek'sē-ă; *dys* = bad, *lexis* = word) is an inherited learning disability characterized by problems with single-word decoding. It often runs in families. Affected individuals not only have trouble reading but may also have problems writing and spelling accurately. These individuals may be able to recognize letters normally, but they demonstrate a level of reading competence far below that expected for their level of intelligence. Their writing may be disorganized and uneven, with the letters of words in incorrect order or even completely reversed. Some individuals appear to outgrow this condition, or at least develop improved reading ability over time. This improvement may reflect neural maturation or retraining of parts of the brain to better decode words and symbols. Some researchers have postulated that dyslexia is a form of **disconnect syndrome**, in which transfer of information between the cerebral hemispheres through the corpus callosum is impaired.

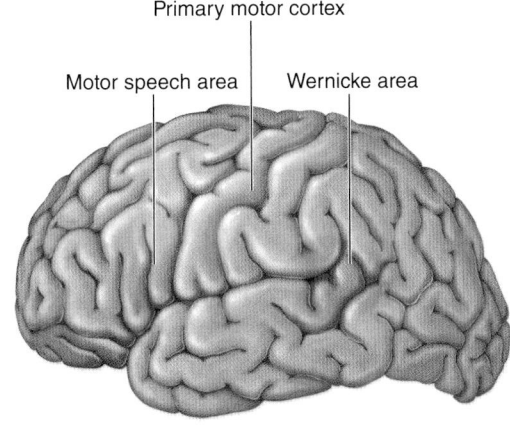

Primary motor cortex

Motor speech area | Wernicke area

(a) Lateral view

Figure 13.31 Functional Areas Within the Cerebral Cortex. (*a*) The left cerebral hemisphere in most people houses the Wernicke area, the motor speech area, and the prefrontal cortex. (*b*) A PET scan shows the areas of the brain that are most active during speech.

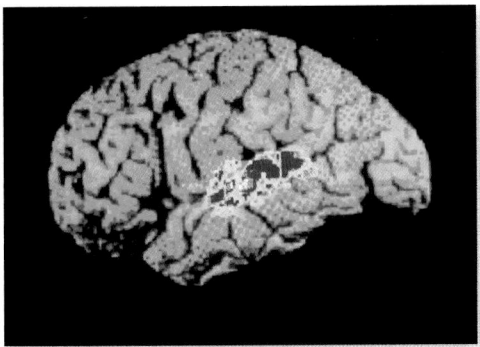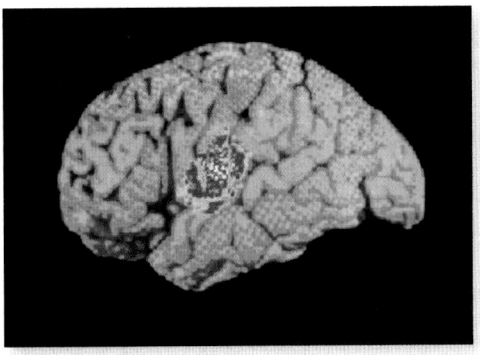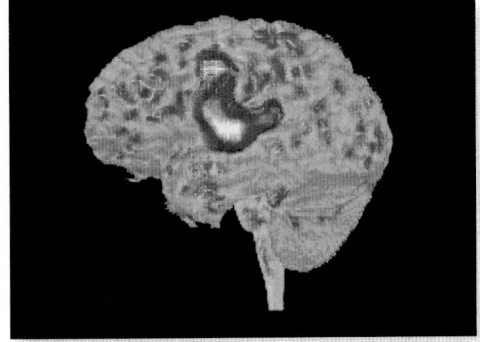

① Auditory information about a sentence travels to the primary auditory cortex. The Wernicke area then interprets the sentence.

② Information from the Wernicke area travels to the motor speech area.

③ Information travels from the motor speech area to the primary motor cortex, where motor commands involving muscles used for speech are given.

(b) PET scans

13.8g Language

LEARNING OBJECTIVE

10. List the cerebral centers involved in written and spoken language, and describe how these centers work together.

The higher-order processes involved in language include reading, writing, speaking, and understanding words. Recall that two important cortical areas involved in speech integration are the Wernicke area and the motor speech area (Broca area). The Wernicke area is involved in interpreting what we read or hear, whereas the motor speech area receives axons originating from the Wernicke area and then helps regulate the motor activities needed for us to speak. Thus, the Wernicke area is central to our ability to recognize written and spoken language. Immediately posterior to the Wernicke area is the angular gyrus, a region that processes the words we read into a form that we can speak (**figure 13.31**). First, the Wernicke area sends a speech plan to the motor speech area, which initiates a specific patterned motor program that is transmitted to the primary motor cortex. Next, neurons in the primary motor cortex signal other motor neurons, which then innervate the muscles of the cheeks, larynx, lips, and tongue to produce speech.

The Wernicke area is in the categorical hemisphere in most people. In the representational hemisphere, a cortical region opposite the Wernicke area recognizes the emotional content of speech. A lesion in this area of the cerebrum can make a person unable to

understand emotional nuances, such as bitterness or happiness, in spoken words. A lesion in the cortical region of the representational hemisphere opposite the motor speech area results in **aprosodia**, which causes dull, emotionless speech.

There are several speech disorders that affect the interpretation, processing, and execution of language (including sign language). For example, **apraxia of speech** is a motor function disorder. An individual is consciously aware of what she wants to say, but is unable to coordinate and execute the motor commands needed to produce the speech. As a result, these individuals may have difficulty in producing recognizable sounds and sequencing them properly for normal speech. In contrast, an individual with **aphasia** (a-fā′zē-ă; speechlessness) has difficulty understanding speech or writing, or is unable to produce comprehensible speech. Aphasic individuals may have consistent difficulties finding the correct words and may use nonsense words to describe certain objects. Often, they may not realize that others can't understand what they are saying. Many cases of aphasia are due to head injury or strokes that damage brain centers responsible for interpreting language. Speech-language pathologists (SLPs) work extensively with those suffering from either apraxia of speech or aphasia to help them improve language understanding and production.

WHAT DID YOU LEARN?

41 How is the Wernicke area involved in language processing?

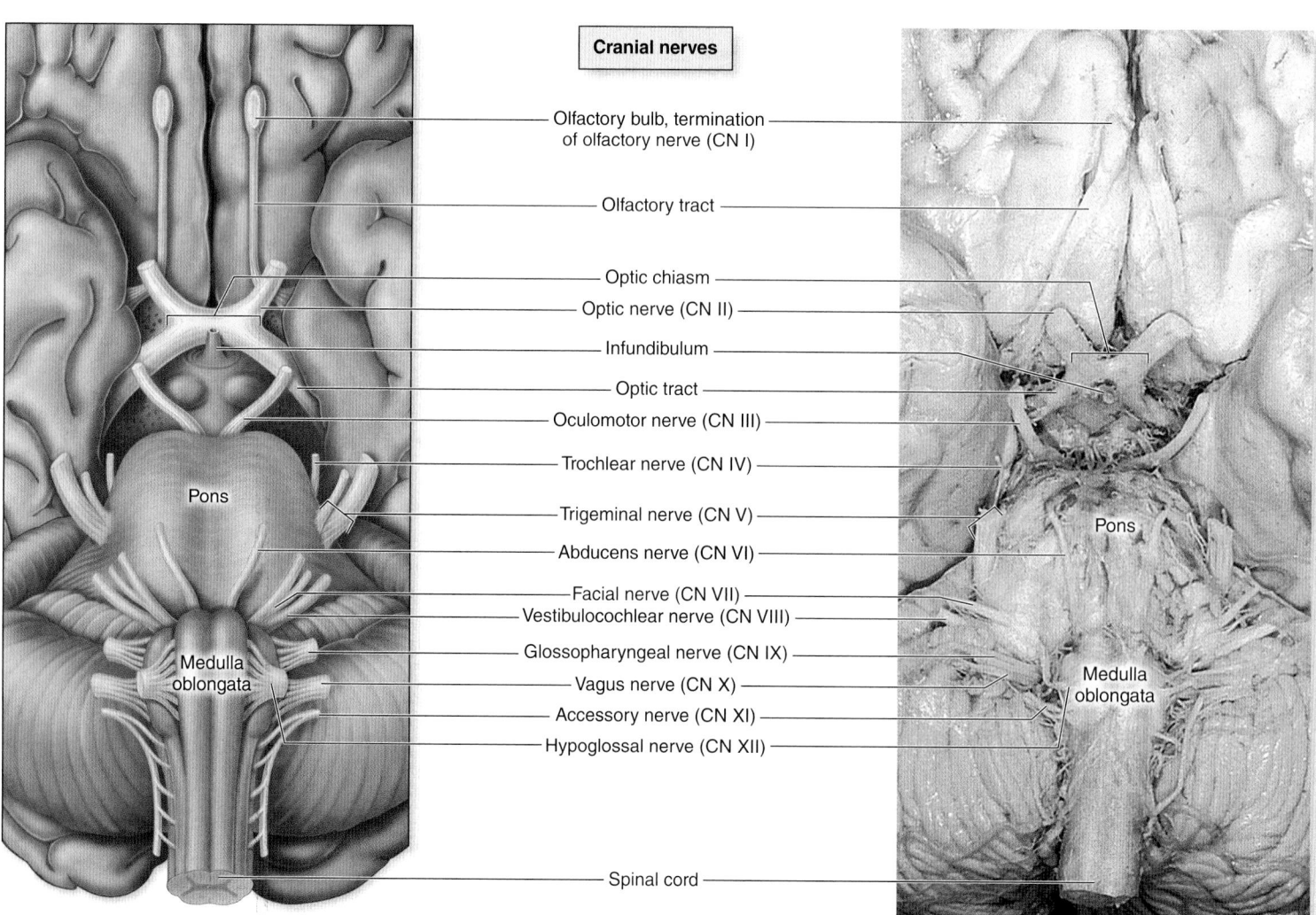

Cranial nerves

- Olfactory bulb, termination of olfactory nerve (CN I)
- Olfactory tract
- Optic chiasm
- Optic nerve (CN II)
- Infundibulum
- Optic tract
- Oculomotor nerve (CN III)
- Trochlear nerve (CN IV)
- Trigeminal nerve (CN V)
- Abducens nerve (CN VI)
- Facial nerve (CN VII)
- Vestibulocochlear nerve (CN VIII)
- Glossopharyngeal nerve (CN IX)
- Vagus nerve (CN X)
- Accessory nerve (CN XI)
- Hypoglossal nerve (CN XII)
- Spinal cord

Pons
Medulla oblongata

Figure 13.32 Cranial Nerves. A view of the inferior surface of the brain shows the 12 pairs of cranial nerves. AP|R

13.9 Cranial Nerves

LEARNING OBJECTIVES

1. List the names and locations of the 12 pairs of cranial nerves.
2. Compare the functions of each of the cranial nerves.

The 12 pairs of cranial nerves are part of the peripheral nervous system (PNS) and originate on the inferior surface of the brain. They are numbered with Roman numerals according to their positions, beginning with the most anteriorly placed nerve, and the number is sometimes preceded by the prefix *CN* (figure 13.32).

The name of each nerve generally has some relation to its function. The 12 pairs of cranial nerves are the olfactory (CN I), optic (CN II), oculomotor (CN III), trochlear (CN IV), trigeminal (CN V), abducens (CN VI), facial (CN VII), vestibulocochlear (CN VIII), glossopharyngeal (CN IX), vagus (CN X), accessory (CN XI), and hypoglossal (CN XII).

Table 13.4 summarizes the main motor and sensory functions of each cranial nerve. For easier reference, each main function of a nerve is color-coded. Pink represents a sensory function, blue stands for a somatic motor function (see section 12.1b), and green denotes a parasympathetic motor function. Table 13.5 lists the individual cranial nerves and discusses their functions, origins, and pathways. The color-coding in table 13.4 carries over to table 13.5, so you can easily determine whether a cranial nerve has motor and/or sensory components.

WHAT DID YOU LEARN?

 42 Which cranial nerves have sensory functions only?

INTEGRATE

LEARNING STRATEGY

Developing a code or phrase called a **mnemonic** (nē-mon′ik) may help you remember the cranial nerves. Here is a sample mnemonic for the cranial nerves:

Oh (*olfactory*)
once (*optic*)
one (*oculomotor*)
takes (*trochlear*)
the (*trigeminal*)
anatomy (*abducens*)
final (*facial*)
very (*vestibulocochlear*)
good (*glossopharyngeal*)
vacations (*vagus*)
are (*accessory*)
heavenly! (*hypoglossal*)

Table 13.4	Primary Functions of Cranial Nerves		
Cranial Nerve	**Sensory Function**	**Somatic Motor Function**	**Parasympathetic Motor (Autonomic) Function**[1]
I (olfactory)	Olfaction (smell)	*None*	*None*
II (optic)	Vision	*None*	*None*
III (oculomotor)	*None*[2]	Four extrinsic eye muscles (medial rectus, superior rectus, inferior rectus, inferior oblique); levator palpebrae superioris muscle (elevates eyelid)	Innervates sphincter pupillae muscle in eye to make pupil constrict; contracts ciliary muscles to make lens of eye more rounded (as needed for near vision)
IV (trochlear)	*None*[2]	Superior oblique eye muscle	*None*
V (trigeminal)	General sensory from anterior scalp, nasal cavity, nasopharynx, entire face, most of oral cavity, teeth, anterior two-thirds of tongue; part of auricle of ear; meninges	Muscles of mastication, mylohyoid, digastric (anterior belly), tensor tympani, tensor veli palatini	*None*
VI (abducens)	*None*[2]	Lateral rectus eye muscle	*None*
VII (facial)	Taste from anterior two-thirds of tongue	Muscles of facial expression, digastric (posterior belly), stylohyoid, stapedius	Increases secretion from lacrimal gland of eye, submandibular and sublingual salivary glands
VIII (vestibulocochlear)	Hearing (cochlear branch); equilibrium (vestibular branch)	*None*[3]	*None*
IX (glossopharyngeal)	General sensory and taste from posterior one-third of tongue, general sensory from part of pharynx, visceral sensory from carotid bodies	One pharyngeal muscle (stylopharyngeus)	Increases secretion from parotid salivary gland
X (vagus)	Visceral sensory information from heart, lungs, most abdominal organs General sensory information from external acoustic meatus, tympanic membrane, part of pharynx, laryngopharynx, and larynx	Most pharyngeal muscles; all laryngeal muscles	Innervates smooth muscle and glands of heart, lungs, larynx, trachea, most abdominal organs
XI (accessory)	*None*[2]	Trapezius muscle, sternocleidomastoid muscle	*None*
XII (hypoglossal)	*None*[2]	Intrinsic and extrinsic tongue muscles	*None*

1. The autonomic nervous system contains a parasympathetic division and sympathetic division. Some cranial nerves carry parasympathetic axons and are listed in this table. Detailed information about these divisions is found in section 15.2.
2. These nerves do contain some tiny proprioceptive sensory axons from the muscles, but in general, these nerves tend to be described as motor only.
3. A few motor axons travel with this nerve to the inner ear, but they are not considered a significant component of the nerve.

Table 13.5 Cranial Nerves

CN I OLFACTORY NERVE (ol-fak′tŏ-rē; *olfacio* = to smell)

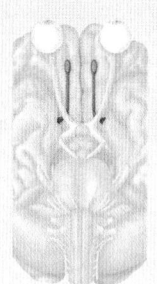

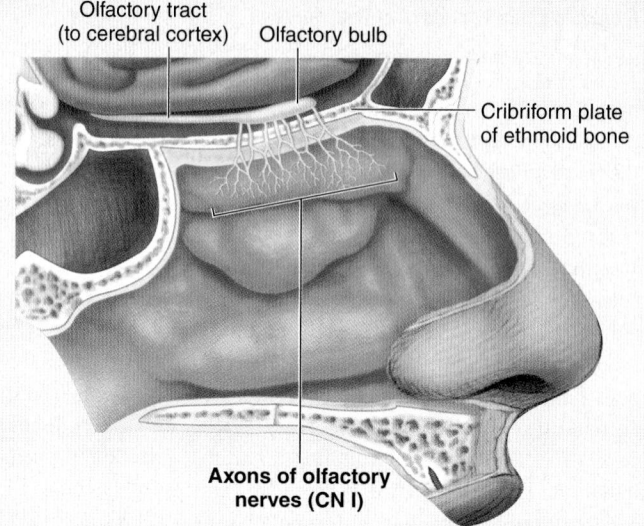

Olfactory tract (to cerebral cortex)

Olfactory bulb

Cribriform plate of ethmoid bone

Axons of olfactory nerves (CN I)

Description	Special sensory nerve that conducts olfactory (smell) sensation from the nose to the brain
Sensory function	Olfaction (smell)
Origin	Receptors (bipolar neurons) in olfactory epithelium of nasal cavity
Pathway	Travels through the cribriform foramina of ethmoid bone and synapses in the olfactory bulbs, which extend to various locations within the brain including the primary olfactory cortex of the temporal lobe
Conditions caused by nerve damage	Anosmia (partial or total loss of smell)
How to test for nerve damage	Test smell (have patient close eyes, close one nostril, and inhale an odor with the other nostril)

CN II OPTIC NERVE (op′tik; *ops* = eye)

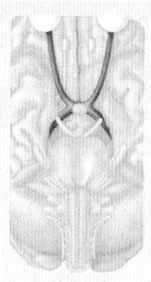

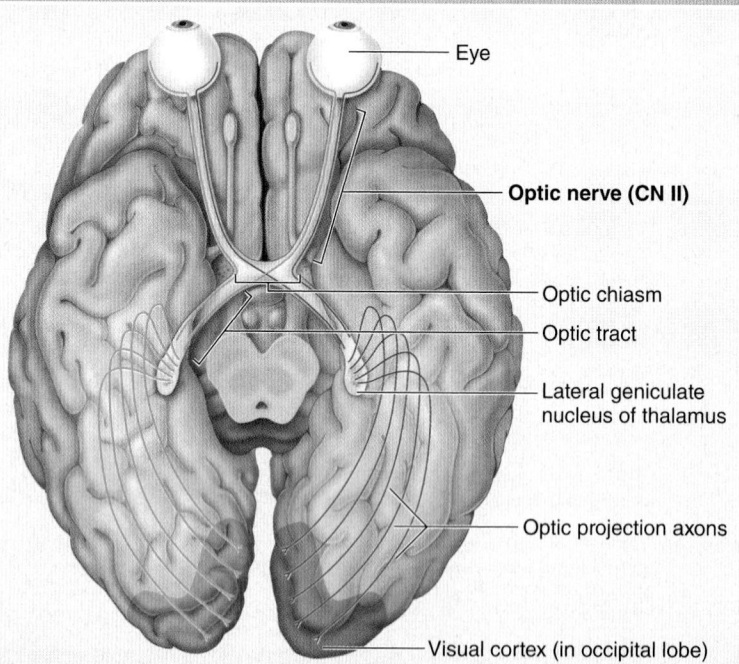

Eye

Optic nerve (CN II)

Optic chiasm

Optic tract

Lateral geniculate nucleus of thalamus

Optic projection axons

Visual cortex (in occipital lobe)

Description	Special sensory nerve that conducts visual information from the retina of the eye to the brain
Sensory function	Vision
Origin	Retina of the eye
Pathway	Enters cranium via optic canal of sphenoid bone; left and right optic nerves unite at optic chiasm; optic tract travels to lateral geniculate nucleus of thalamus; nerve fibers project to the occipital lobe
Conditions caused by nerve damage	Anopsia (visual defects)
How to test for nerve damage	Test vision (cover one eye and have patient view a visual acuity chart with the other eye)

| Table 13.5 | Cranial Nerves (continued) |

CN III OCULOMOTOR NERVE (ok'ū-lō-mō'tŏr; *oculus* = eye, *motorius* = moving)

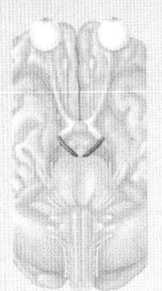

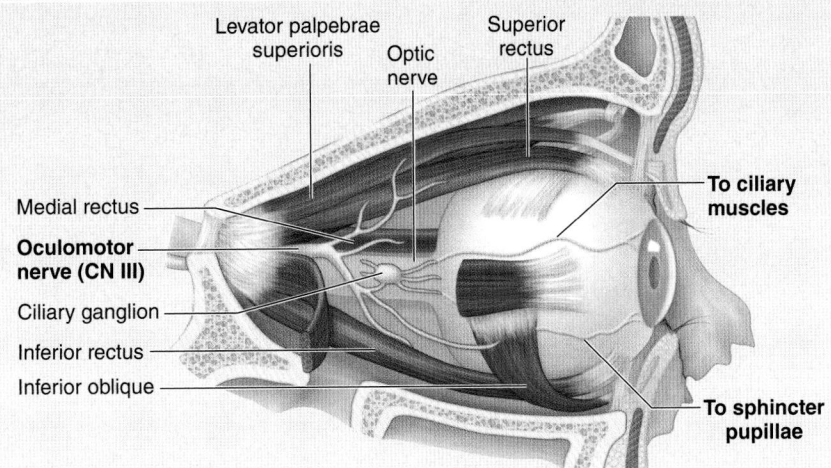

Description	Motor nerve that innervates four of the six extrinsic eye muscles, an upper eyelid muscle, and intrinsic eye muscles (smooth muscle within the eye)
Somatic motor function	Contracts four extrinsic eye muscles (superior rectus, inferior rectus, medial rectus, inferior oblique) to move eye Contracts levator palpebrae superioris muscle to elevate eyelid
Parasympathetic motor function	Contracts sphincter pupillae muscle of iris to make pupil constrict Contracts smooth muscle of ciliary body to make lens of eye more spherical (as needed for near vision)
Origin	Oculomotor and Edinger Westphal nuclei within the midbrain
Pathway	Leaves cranium via superior orbital fissure and travels to eye and eyelid (Parasympathetic axons travel to ciliary ganglion, and postganglionic parasympathetic axons than travel to iris and ciliary muscles)
Conditions caused by nerve damage	Ptosis (upper eyelid droop); paralysis of most eye muscles, leading to strabismus (eyes not in parallel/deviated improperly), diplopia (double vision), focusing difficulty, dilated pupil (mydriasis)
How to test for nerve damage	Determine if the upper eyelid droops, examine if pupil constricts in response to light, examine eye movement (have patient follow a moving object with eyes)

CN IV TROCHLEAR NERVE (trōk'lē-ar; *trochlea* = a pulley)

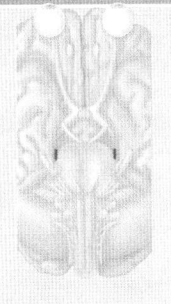

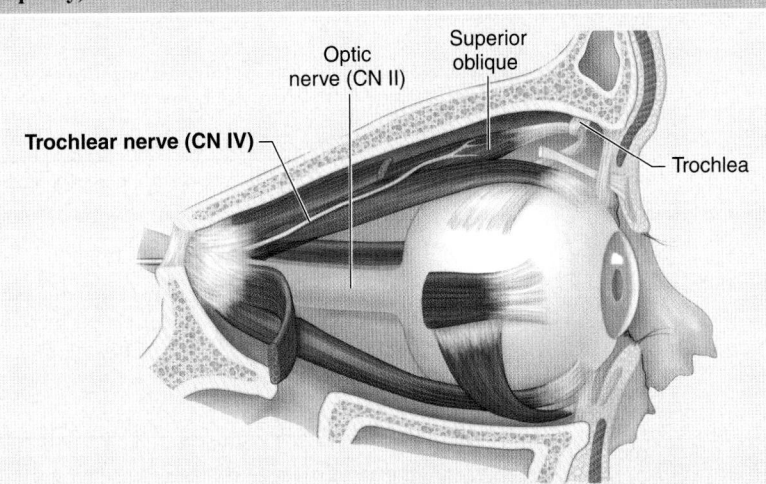

Description	Motor nerve that innervates one extrinsic eye muscle (superior oblique) that loops through a pulley-shaped ligament called a trochlea
Somatic motor function	Contracts one extrinsic eye muscle (superior oblique) to move eye inferiorly and laterally
Origin	Trochlear nucleus within the midbrain
Pathway	Leaves cranium via superior orbital fissure and travels to superior oblique muscle
Conditions caused by nerve damage	Paralysis of superior oblique, leading to strabismus (eyes not in parallel/deviated improperly), diplopia (double vision)
How to test for nerve damage	Examine eye movement (have patient follow a moving object with eyes)

(continued on next page)

Table 13.5   Cranial Nerves (continued)

CN V TRIGEMINAL NERVE (trī-jem′i-năl; *trigeminus* = threefold)

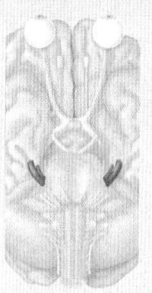

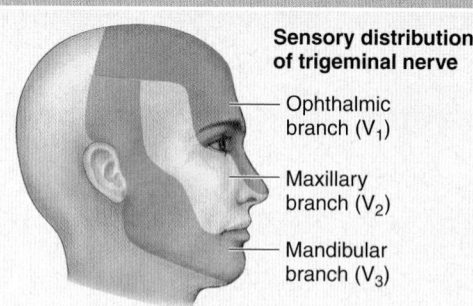

Sensory distribution of trigeminal nerve

— Ophthalmic branch (V_1)

— Maxillary branch (V_2)

— Mandibular branch (V_3)

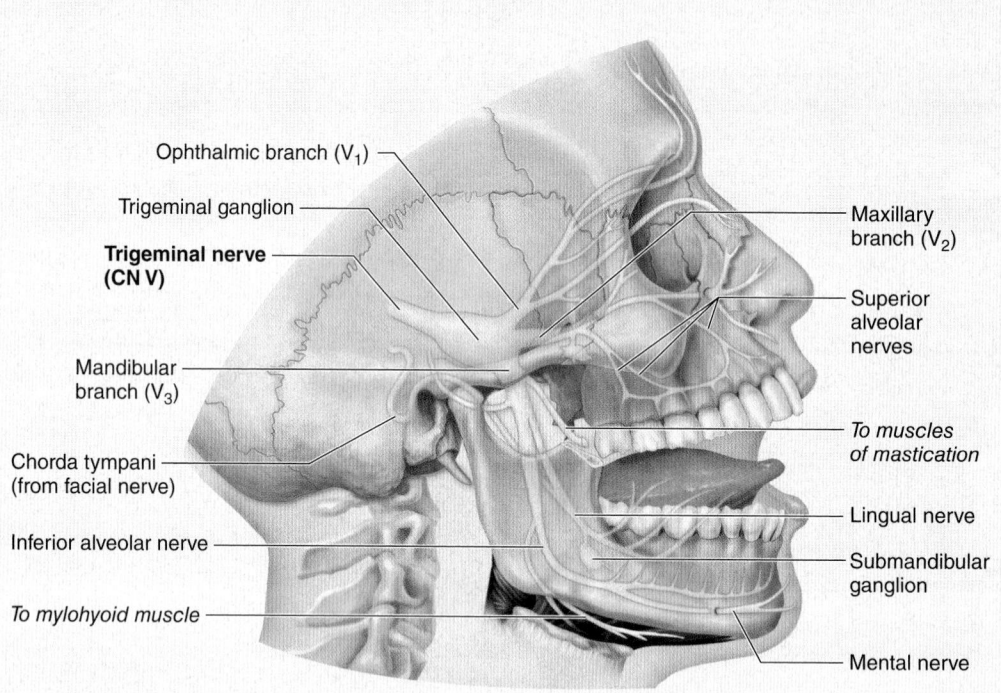

Ophthalmic branch (V_1)

Trigeminal ganglion

Trigeminal nerve (CN V)

Mandibular branch (V_3)

Chorda tympani (from facial nerve)

Inferior alveolar nerve

To mylohyoid muscle

Maxillary branch (V_2)

Superior alveolar nerves

To muscles of mastication

Lingual nerve

Submandibular ganglion

Mental nerve

Description	Mixed nerve that consists of three divisions: ophthalmic (V_1), maxillary (V_2), and mandibular (V_3); receives sensory nerve signals from face, oral cavity, nasal cavity, meninges, and anterior scalp, and innervates muscles of mastication
Sensory function	Sensory stimuli for this nerve are touch, temperature, and pain V_1: Conducts sensory nerve signals from cornea, nose, forehead, anterior scalp, meninges V_2: Conducts sensory nerve signals from nasal mucosa, palate, gums, cheek, meninges V_3: Conducts sensory nerve signals from anterior two-thirds of tongue, meninges, skin of chin, lower jaw, lower teeth; one-third from sensory axons of auricle of ear
Somatic motor function	Innervates muscles of mastication (temporalis, masseter, lateral and medial pterygoids), mylohyoid, anterior belly of digastric, tensor tympani muscle, and tensor veli palatini
Origin	Nuclei in the pons
Pathway	V_1: Sensory axons enter cranium via superior orbital fissure and travel to trigeminal ganglion before entering pons V_2: Sensory axons enter cranium via foramen rotundum and travel to trigeminal ganglion before entering pons V_3: Sensory axons travel through foramen ovale to trigeminal ganglion before entering pons. Motor axons leave pons and exit cranium via foramen ovale to supply muscles
Conditions caused by nerve damage	Trigeminal neuralgia (tic douloureux) is caused by inflammation of the sensory components of the trigeminal nerve and results in intense, pulsating pain lasting from minutes to several hours
How to test for nerve damage	Have patient close mouth against resistance; also have patient close eyes and then determine if an object moved along the face can be felt (such as a feather)

Table 13.5   Cranial Nerves (continued)

CN VI ABDUCENS NERVE (ab-dū′senz; to move away from)

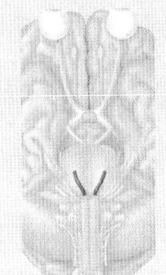

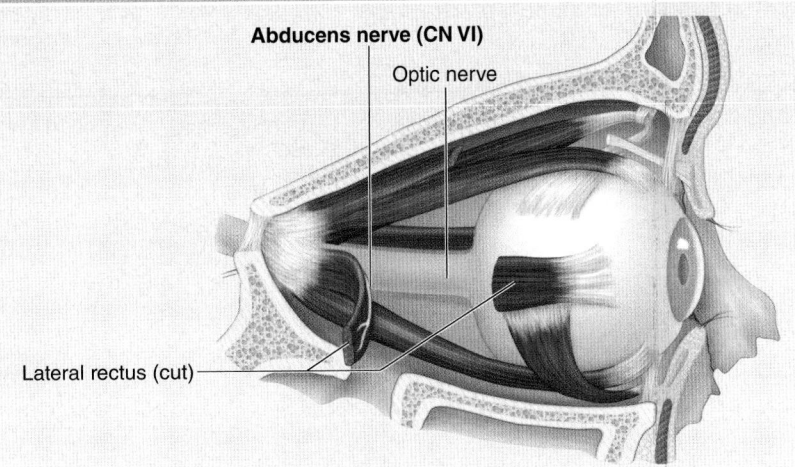

Abducens nerve (CN VI)

Optic nerve

Lateral rectus (cut)

Description	Motor nerve that innervates one extrinsic eye muscle (lateral rectus) to move the eye
Somatic motor function	Contracts lateral rectus for eye abduction
Origin	Pontine (abducens) nucleus in the pons
Pathway	Leaves cranium through superior orbital fissure and travels to lateral rectus muscle
Conditions caused by nerve damage	Paralysis of lateral rectus limits lateral movement of eye, diplopia (double vision)
How to test for nerve damage	Examine eye movement (have patient follow a moving object with eyes) and determine if the eye is able to be abducted

CN VII FACIAL NERVE (fā′shăl)

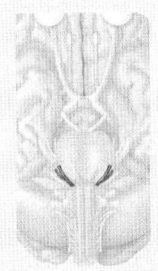

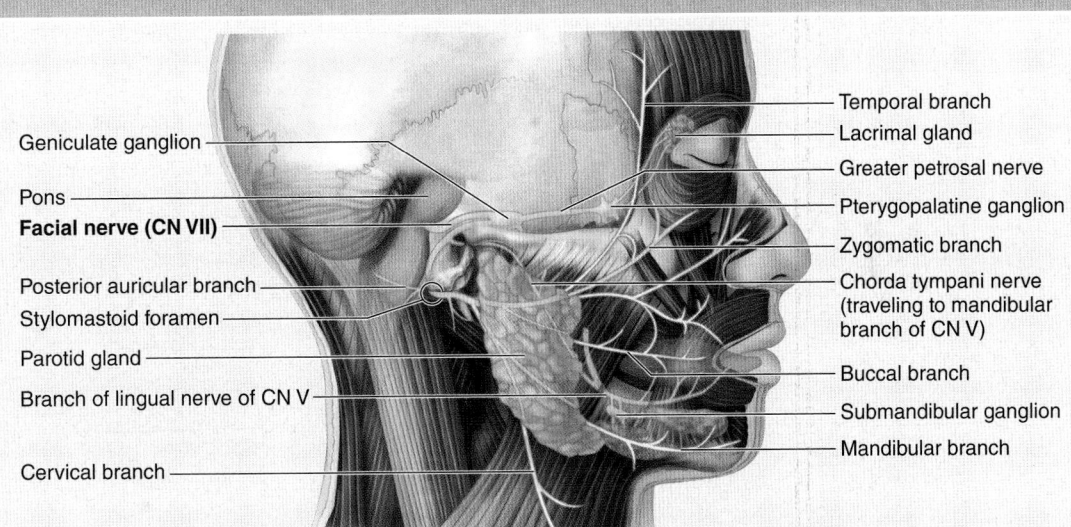

Geniculate ganglion

Pons

Facial nerve (CN VII)

Posterior auricular branch

Stylomastoid foramen

Parotid gland

Branch of lingual nerve of CN V

Cervical branch

Temporal branch

Lacrimal gland

Greater petrosal nerve

Pterygopalatine ganglion

Zygomatic branch

Chorda tympani nerve (traveling to mandibular branch of CN V)

Buccal branch

Submandibular ganglion

Mandibular branch

Description	Mixed nerve that conducts taste sensations from anterior two-thirds of tongue; relays motor output to muscles of facial expression; lacrimal (tear) gland, and most salivary glands
Sensory function	Conducts taste sensations from anterior two-thirds of tongue
Somatic motor function	The five major motor branches (temporal, zygomatic, buccal, mandibular, and cervical) innervate the muscles of facial expression, the posterior belly of the digastric muscle, and the stylohyoid and stapedius muscles
Parasympathetic motor function	Increases secretions of the lacrimal gland of the eye; increases secretions of the submandibular and sublingual salivary glands
Origin	Nuclei in the pons
Pathway	Sensory axons travel from the tongue via the chorda tympani branch of the facial nerve through a tiny foramen to enter the skull, and axons synapse at the geniculate ganglion of the facial nerve. Somatic motor axons leave the pons and enter the temporal bone through the internal acoustic meatus, project through temporal bone, and emerge through stylomastoid foramen to innervate the muscles of facial expression. Parasympathetic motor axons leave the pons, enter the internal acoustic meatus, leave with either the greater petrosal nerve or chorda tympani nerve, and travel to an autonomic ganglion before innervating their respective glands.
Conditions caused by nerve damage	Decreased tearing (dry eye) and decreased salivation (dry mouth); loss of taste sensation to anterior two-thirds of tongue, nerve paralysis (also known as Bell palsy; see Clinical View: "Idiopathic Facial Nerve Paralysis," section 11.3a) characterized by paralyzed facial muscles, lack of orbicularis oculi contraction, sagging at corner of mouth
How to test for nerve damage	Have patient smile, blink, and squint—inability to move facial muscles on one side indicates nerve damage

(continued on next page)

Table 13.5   Cranial Nerves (continued)

CN VIII VESTIBULOCOCHLEAR NERVE (ves-tib′ū-lō-kok′lē-ăr; relating to vestibule and cochlea of the ear)

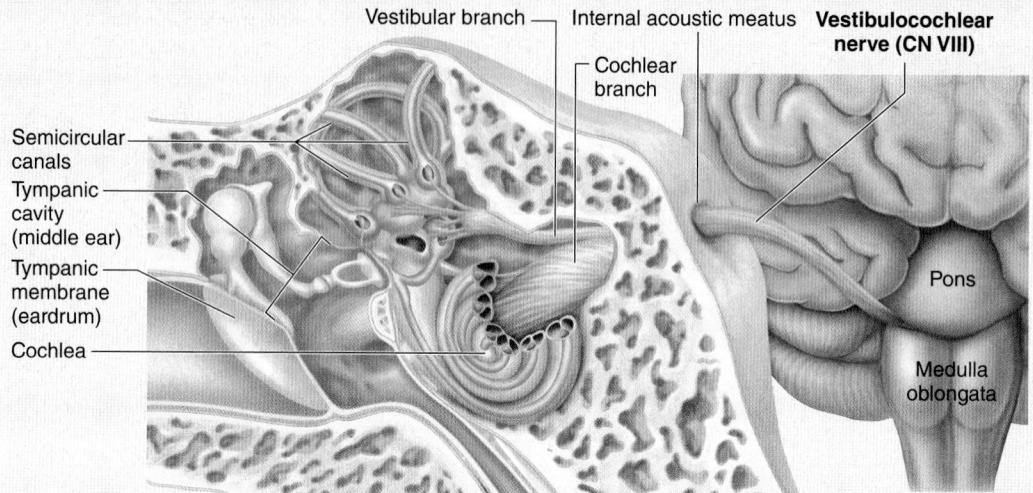

Description	Sensory nerve with two branches that conducts equilibrium and auditory (hearing) sensations from the inner ear to brain
Sensory function	Vestibular branch conducts nerve signals for equilibrium, while cochlear branch conducts nerve signals for hearing
Origin	Vestibular branch: Hair cells in the vestibule of the inner ear Cochlear branch: Cochlea of the inner ear
Pathway	Sensory cell bodies of the vestibular branch are located in the vestibular ganglion, whereas sensory cell bodies of the cochlear branch are located in the spiral ganglion near the cochlea. The vestibular and cochlear branches merge, and together enter cranial cavity through internal acoustic meatus and travel to junction of the pons and the medulla oblongata.
Conditions caused by nerve damage	Lesions in vestibular branch produce loss of balance, nausea, vomiting, and dizziness; lesions in cochlear branch result in deafness (loss of hearing)
How to test for nerve damage	Test hearing

CN IX GLOSSOPHARYNGEAL NERVE (glos′ō-fă-rin′jē-ăl; *glossa* = tongue, *pharynx* = throat)

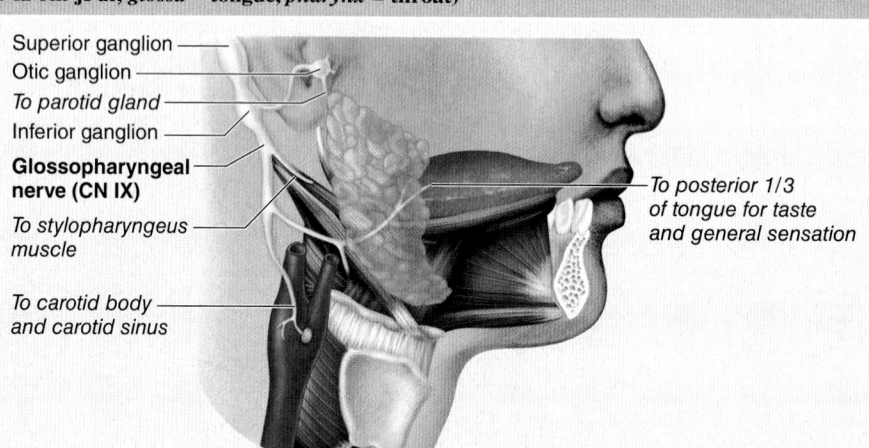

Description	Mixed nerve that receives taste and touch sensations from posterior one-third of the tongue, innervates one pharynx muscle and the parotid salivary gland
Sensory function	General sensation and taste from the posterior one-third of tongue; general sensation from most of pharynx; relay sensory information from the carotid arteries—from both chemoreceptors (which monitor blood levels of CO_2, H^+, and O_2) and baroreceptors (which monitor blood pressure)
Somatic motor function	Contracts stylopharyngeus (a pharynx muscle)
Parasympathetic motor function	Increases secretion of parotid salivary gland
Origin	Sensory axons originate on taste buds and mucosa of posterior one-third of tongue, as well as the carotid bodies Motor axons originate in nuclei in the medulla oblongata
Pathway	Sensory axons travel from posterior one-third of tongue and carotid bodies along nerve through the inferior or superior ganglion into the jugular foramen, and travel to pons. Somatic motor axons leave cranium via jugular foramen and travel to stylopharyngeus. Parasympathetic motor axons travel to otic ganglion and then to parotid gland.
Conditions caused by nerve damage	Reduced salivary secretion (dry mouth); loss of taste sensations to posterior one-third of tongue
How to test for nerve damage	Have patient open mouth and go "ahhh"—the soft palate should elevate and the uvula should remain in the midline under normal conditions

Table 13.5   Cranial Nerves (continued)

CN X VAGUS NERVE (vā′gŭs; wandering)

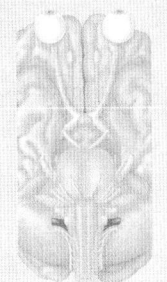

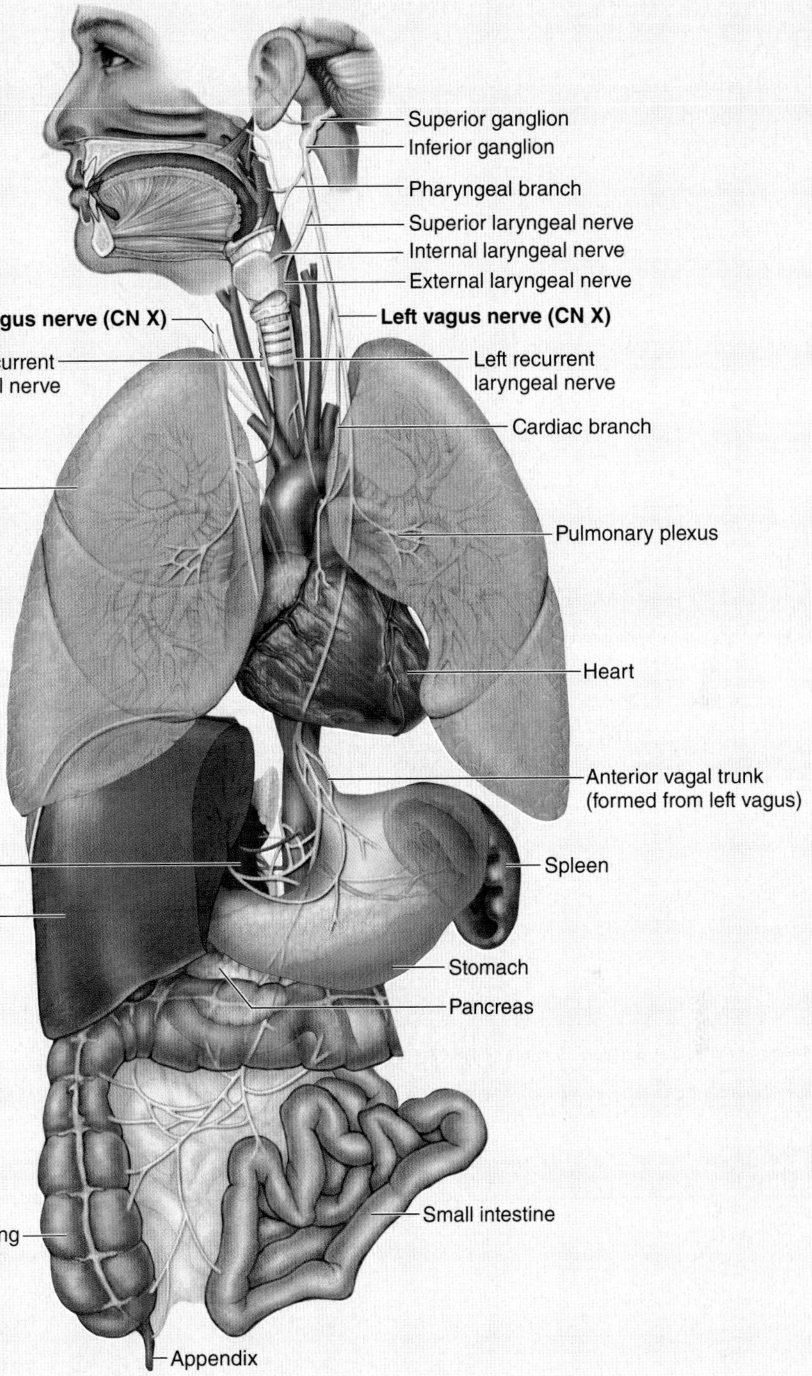

Superior ganglion
Inferior ganglion
Pharyngeal branch
Superior laryngeal nerve
Internal laryngeal nerve
External laryngeal nerve

Right vagus nerve (CN X)   **Left vagus nerve (CN X)**

Right recurrent laryngeal nerve

Left recurrent laryngeal nerve

Cardiac branch

Lung

Pulmonary plexus

Heart

Anterior vagal trunk (formed from left vagus)

Kidney
Liver

Spleen

Stomach
Pancreas

Small intestine

Ascending colon

Appendix

Description	Mixed nerve that innervates structures in the head and neck and in the thoracic and abdominal cavities
Sensory function	Visceral sensory information from heart, lungs, and most abdominal organs. General sensory information from external acoustic meatus, tympanic membrane (eardrum), laryngopharynx (inferior part of throat), and larynx (voice box)
Somatic motor function	Controls most pharynx muscles (for swallowing) and all larynx muscles (for production of speech)
Parasympathetic motor function	Innervates smooth muscle of thoracic and most abdominal organs, cardiac muscle, and glands of heart, lungs, pharynx, larynx, trachea, and most abdominal organs
Origin	Motor nuclei in medulla oblongata
Pathway	Leaves cranium via jugular foramen before traveling and branching extensively in neck, thorax, and abdomen; sensory neuron cell bodies are located in the superior and inferior ganglia associated with the nerve
Conditions caused by nerve damage	Paralysis leads to a variety of larynx problems, including hoarseness, monotone voice, or complete loss of voice. Other lesions may cause difficulty in swallowing or impaired gastrointestinal tract motility.
How to test for nerve damage	Ask patient if he has difficulties in swallowing. Determine if voice is hoarse or monotone. Have patient open mouth and go "ahhh"— the soft palate should elevate and the uvula should remain in the midline under normal conditions.

(continued on next page)

Table 13.5   Cranial Nerves (continued)

CN XI ACCESSORY NERVE (ak-ses'ō-rē)

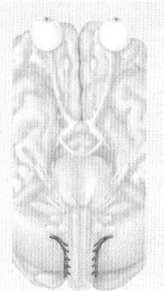

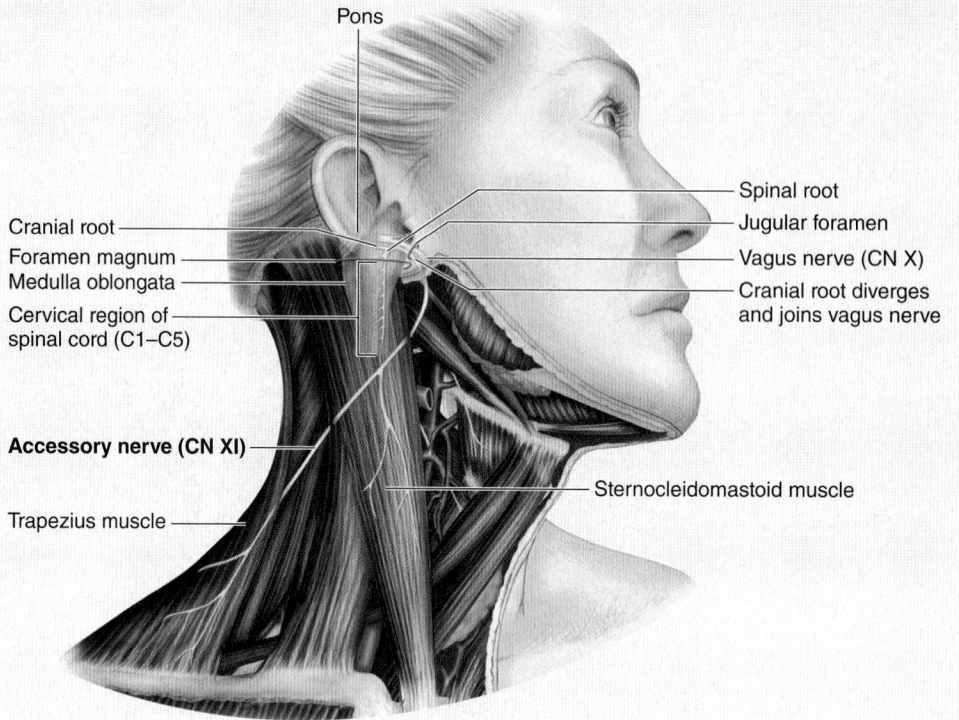

Description	Motor nerve that innervates trapezius and sternocleidomastoid muscles, also assists CN X to innervate pharynx muscles; formerly called the *spinal accessory nerve*
Somatic motor function	Cranial root: Travels with CN X to assist in innervating the pharynx muscles
	Spinal root: Contracts trapezius and sternocleidomastoid muscles
Origin	Cranial root: Nucleus in medulla oblongata
	Spinal root: Nucleus in spinal cord
Pathway	Spinal root travels superiorly to enter skull through foramen magnum; there, cranial and spinal roots merge and leave the skull via jugular foramen. Once outside the skull, cranial root splits to travel with CN X (vagus) to innervate pharynx muscles, and spinal root travels to sternocleidomastoid and trapezius.
Conditions caused by nerve damage	Paralysis of trapezius and sternocleidomastoid
How to test for nerve damage	Have patient elevate or shrug shoulders (tests trapezius function) or have patient turn head to opposite side (tests sternocleidomastoid function)

CN XII HYPOGLOSSAL NERVE (hī-pō-glos'ăl; *hypo* = below, *glossus* = tongue)

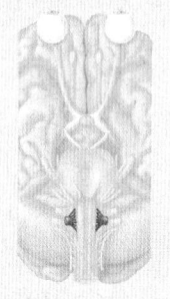

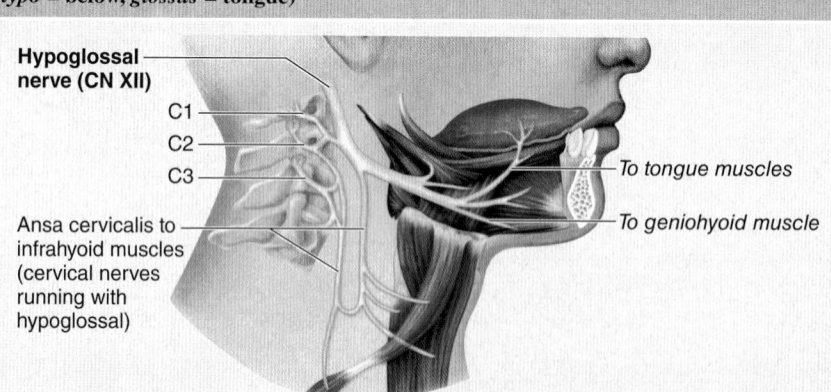

Description	Motor nerve that innervates both intrinsic and extrinsic tongue muscles; name means "under the tongue"
Somatic motor function	Contracts intrinsic and extrinsic tongue muscles to move tongue
Origin	Hypoglossal nucleus in medulla oblongata
Pathway	Leaves cranium via hypoglossal canal; travels inferior to mandible and to inferior surface of tongue
Conditions caused by nerve damage	Swallowing and speech difficulties due to impaired tongue movement
How to test for nerve damage	Have patient protrude (stick out) tongue: If a single hypoglossal nerve (either left or right) is paralyzed, a protruded (stuck-out) tongue deviates to the side of the damaged nerve

- The brain is composed of the cerebrum, diencephalon, brainstem, and cerebellum. Twelve cranial nerves are associated with the brain.

13.1 Brain Organization and Development 484

13.1a Overview of Brain Anatomy 484
- The brain has two cerebral hemispheres that exhibit folds called gyri with shallow sulci and deep fissures in between.

13.1b Development of Brain Divisions 484
- Three primary vesicles (prosencephalon, mesencephalon, and rhombencephalon) form from the neural tube by the late fourth week of development.
- Five secondary vesicles (telencephalon, diencephalon, mesencephalon, metencephalon, and myelencephalon) form from the primary vesicles by the fifth week of development.

13.1c Gray Matter and White Matter Distribution 491
- Most gray matter (composed of motor neuron and interneuron cell bodies, dendrites, and unmyelinated axons) is located on the periphery of the brain, whereas the white matter (composed primarily of myelinated axons) is located more internally.

13.2 Protection and Support of the Brain 493

- The brain is protected and isolated by the cranium, cranial meninges, cerebrospinal fluid, and a blood-brain barrier.

13.2a Cranial Meninges 493
- The cranial meninges are the pia mater, arachnoid mater, and dura mater.
- The cranial dural septa are folds of dura mater that project between the major parts of the brain and stabilize the brain's position.

13.2b Brain Ventricles 495
- Fluid-filled spaces in the brain are the paired lateral ventricles, the third ventricle, the cerebral aqueduct, and the fourth ventricle.

13.2c Cerebrospinal Fluid 495
- Cerebrospinal fluid (CSF) is a clear fluid that provides buoyancy, protection, and a stable environment for the brain and spinal cord.
- The choroid plexus (formed from ependymal cells and capillaries) produces CSF in the ventricles.
- CSF leaves the ventricles and enters the subarachnoid space, where it circulates around the brain and spinal cord. Excess CSF returns to the venous circulation through the arachnoid villi.

13.2d Blood-Brain Barrier 498
- The blood-brain barrier regulates movement of materials between the blood and the interstitial fluid of the brain.

13.3 Cerebrum 499

- The cerebrum is the center of our sensory perception, thought, memory, judgement, and voluntary motor actions.

13.3a Cerebral Hemispheres 499
- The left and right cerebral hemispheres are separated by a longitudinal fissure.

13.3b Lobes of the Cerebrum 499
- Each hemisphere contains five lobes: four superficial lobes (frontal, parietal, temporal, occipital lobes) and the insula, which is not visible from the surface.

13.3c Functional Areas of the Cerebrum 502
- The primary motor cortex in the frontal lobe directs voluntary movements.
- The primary somatosensory cortex in the parietal lobe collects somatic sensory information.
- Other primary cortices and association areas are housed in each of the five lobes.

13.3d Central White Matter 505
- The central white matter contains three major groups of axons: association tracts, commissural tracts, and projection tracts.

13.3e Cerebral Lateralization 506
- The left hemisphere is the categorical hemisphere in most individuals, and the right is the representational hemisphere.

13.3f Cerebral Nuclei 507
- The cerebral nuclei are masses of gray matter located within the central white matter.

13.4 Diencephalon 509

- The diencephalon is composed of the epithalamus, thalamus, and hypothalamus.

13.4a Epithalamus 509
- The epithalamus forms part of the posterior roof of the diencephalon; it contains the pineal gland (which secretes melatonin) and habenular nuclei (help relay signals from the limbic system to the midbrain).

13.4b Thalamus 510
- The thalamus is the main relay point for integrating, assimilating, and amplifying sensory signals sent to the cerebrum.

13.4c Hypothalamus 510
- The hypothalamus oversees the endocrine and autonomic nervous systems and houses many control and integrative centers.

(continued on next page)

13.5 Brainstem 512	• The brainstem is composed of the midbrain, pons, and medulla oblongata.
	13.5a Midbrain 512
	• The midbrain contains the cerebral peduncles, substantia nigra, tegmentum, tectal plate, and nuclei for two cranial nerves.
	13.5b Pons 514
	• The pons contains axon tracts, autonomic nuclei for control of respiration, and nuclei of four cranial nerves.
	13.5c Medulla Oblongata 514
	• The medulla oblongata connects the brain to the spinal cord. It contains sensory processing centers, autonomic reflex centers, and nuclei for four cranial nerves.
13.6 Cerebellum 516	**13.6a Structural Components of the Cerebellum 516**
	• The cerebellum is composed of left and right cerebellar hemispheres, with the vermis in between.
	• Cerebellar peduncles are thick axon tracts that connect the cerebellum to different parts of the brainstem.
	13.6b Functions of the Cerebellum 516
	• The cerebellum helps maintain posture and balance and fine-tunes skeletal muscle contractions leading to smooth, coordinated movement.
13.7 Functional Brain Systems 518	• A functional brain system consists of separate components of the brain that work together toward a common function. Examples include the limbic system and reticular formation.
	13.7a Limbic System 518
	• The limbic system includes a group of structures that surround the corpus callosum and thalamus. The limbic system functions in memory and emotional behavior.
	13.7b Reticular Formation 518
	• The reticular formation participates in cyclic activities such as arousing the cortex to consciousness and controlling the sleep-wake cycle.
13.8 Integrative Functions and Higher-Order Brain Functions 519	**13.8a Development of Higher-Order Brain Functions 519**
	• Higher-order functions mature and increase in complexity as development proceeds.
	13.8b Electroencephalogram 519
	• An electroencephalogram monitors brain activity by measuring brain waves through the use of electrodes.
	13.8c Sleep 520
	• Sleep is a period of rest for the brain and involves cycles of REM (rapid eye movement) and non-REM (non-rapid eye movement) activity.
	13.8d Cognition 521
	• Mental processes such as awareness, perception, thinking, and knowledge are collectively called cognition.
	13.8e Memory 521
	• Memory is a higher-order mental function involving the storage and retrieval of information gathered through previous activities.
	13.8f Emotion 521
	• Emotion is controlled by the limbic system and is regulated by the prefrontal cortex.
	13.8g Language 524
	• The motor speech area initiates a specific motor program for the movements involved in speech, whereas the Wernicke area is responsible for recognition of spoken and written language.
13.9 Cranial Nerves 525	• Twelve nerve pairs, called cranial nerves, project from the brain. Each nerve has a specific name and function and is designated by a Roman numeral.

CHALLENGE YOURSELF

Do You Know the Basics?

_____ 1. Which cranial nerve is responsible for innervating the intrinsic and extrinsic tongue muscles?

    a. accessory (CN XI)

    b. glossopharyngeal (CN IX)

    c. trigeminal (CN V)

    d. hypoglossal (CN XII)

_____ 2. The subdivision of the brain that does not initiate somatic motor movements, but rather coordinates and fine-tunes those movements, is the

    a. medulla oblongata.

    b. cerebrum.

    c. cerebellum.

    d. diencephalon.

3. Which of these is the least likely to affect information transfer from STM (short-term memory) to LTM (long-term memory)?

   a. emotional state

   b. repetition or rehearsal

   c. auditory association cortex

   d. cerebral nuclei

4. Which of the following is *not* a function of the hypothalamus?

   a. controls endocrine system

   b. regulates sleep-wake cycle

   c. controls autonomic nervous system

   d. initiates voluntary skeletal muscle movement

5. Which of the following statements is false about the choroid plexus?

   a. It is located within the ventricles of the brain.

   b. It is composed of ependymal cells and blood capillaries.

   c. It receives and filters all sensory information.

   d. It produces and circulates cerebrospinal fluid.

6. The _____ are descending motor tracts on the anterolateral surface of the midbrain.

   a. cerebral peduncles

   b. inferior colliculi

   c. pyramids

   d. tegmenta

7. Which cerebral lobe is located immediately posterior to the central sulcus and superior to the lateral sulcus?

   a. frontal lobe

   b. parietal lobe

   c. temporal lobe

   d. occipital lobe

8. The primary motor cortex is located in which cerebral structure?

   a. precentral gyrus

   b. postcentral gyrus

   c. motor speech area

   d. prefrontal cortex

9. The _____ are the isolated, innermost gray matter areas near the base of the cerebrum, inferior to the lateral ventricles.

   a. auditory association areas

   b. cerebral nuclei

   c. substantia nigra

   d. corpus callosum axons

10. Which structure contains some autonomic centers involved in regulating respiration?

    a. pons

    b. superior colliculi

    c. cerebellum

    d. thalamus

11. Describe (a) how and where the cerebrospinal fluid is formed, (b) its subsequent circulation, and (c) how and where it is reabsorbed into the vascular system.

12. Which specific area of the brain may be impaired if you cannot tell the difference between a smooth and a rough surface using your hands only?

13. What activities occur in the visual association area?

14. Describe the relationship between the cerebral nuclei and the cerebellum in motor activities.

15. List the functions of the hypothalamus.

16. Describe the pathway by which the pressure applied to the right hand during a handshake is transmitted and perceived in the left primary somatosensory cortex.

17. Identify the components of the limbic system.

18. During surgery to remove a tumor from the occipital lobe of the left cerebrum, a surgeon must cut into the brain to reach the tumor. List in order (starting with the covering skin) all of the layers that must be cut through to reach the tumor.

19. What is the difference between apraxia of speech and aphasia?

20. Which cranial nerves are associated with some aspect of eye movements or vision?

Can You Apply What You've Learned?

Use the following paragraph to answer questions 1 and 2.

Alex went to the dentist to get a cavity filled. The dentist used an anesthetic to numb the teeth prior to drilling.

1. Which nerve likely was anesthetized?

   a. CN IV (trochlear)

   b. CN V (trigeminal)

   c. CN VII (facial)

   d. CN IX (glossopharyngeal)

2. After the filling was inserted, Alex's gums on the same side of the tooth remained numb for a short while. What other problems may Alex temporarily experience until the anesthetic wears off?

   a. inability to close the mouth

   b. inability to protrude the tongue

   c. dry mouth

   d. numbness of the lips

Use the following paragraph to answer questions 3 and 4.

Shannon was the pitcher on her softball team. During one game, a batter hit the ball and it ricocheted off the left side of Shannon's head, knocking her temporarily unconscious. She eventually regained consciousness, but experienced severe pain near her left temple. Within a few hours, Shannon had trouble moving her right upper limb and became lethargic. Her team captain brought her to the emergency room, where the physician diagnosed her with an epidural hematoma.

3. An epidural hematoma causes an accumulation of blood to develop between what two structures?

   a. the skull and the periosteal layer of the dura mater

   b. the periosteal and meningeal layers of the dura mater

   c. the meningeal layer of the dura mater and the arachnoid mater

   d. the arachnoid mater and the pia mater

4. Why did Shannon experience the problems with her right upper limb? The hematoma likely was impinging on what brain structure?

   a. left precentral gyrus   c. left cerebral nuclei

   b. left postcentral gyrus   d. left cerebellum

5. A 25-year-old male named Carlos went to the optometrist with the complaint of double vision (diplopia). The optometrist performed various eye tests on Carlos. Carlos was able to read an eye chart with each individual eye but experienced the double vision when he tried to use both eyes to focus. The optometrist noticed that when he had Carlos look laterally with each eye, Carlos's right eye did not move as far laterally as his left could. Based on these tests, the optometrist suspected that the muscle innervated by _____ was not working properly.

   a. CN II (optic)   c. CN IV (trochlear)

   b. CN III (oculomotor)   d. CV VI (abducens)

Can You Synthesize What You've Learned?

1. Peyton felt strange when she awoke one morning. She could not hold a pen in her right hand when trying to write an entry in her diary, and her muscles were noticeably weaker on the right side of her body. Additionally, her husband noticed that she was slurring her speech, so he took her to the emergency room. What does the ER physician suspect has occurred? Where in the brain might the physician suspect that abnormal activity or perhaps a lesion is located, and why?

2. Parkinson disease is the result of decreased levels of the neurotransmitter dopamine in the brain. However, these patients can't take dopamine in drug form because the drug cannot reach the brain. What anatomic structure prevents the drug from reaching the brain? How could these same anatomic structures be beneficial to an individual under different circumstances?

3. During a robbery at his convenience store, Dustin was shot in the right cerebral hemisphere. He survived, although some specific functions were impaired. Would Dustin have been more likely or less likely to have survived if he had been shot in the medulla oblongata? Why?

INTEGRATE

ONLINE STUDY TOOLS      connect |ANATOMY & PHYSIOLOGY    LEARNSMART° AP|R

The following study aids may be accessed through Connect.

Clinical Case Study: A Case of the Shaking Palsy in an Elderly Woman

Interactive Questions: This chapter's content is served up in a number of multimedia question formats for student study.

LearnSmart: Topics and terminology include brain development and organization; protection and support of the brain; cerebrum;

diencephalon; brainstem; cerebellum; functional brain systems; integrative functions and higher-order brain functions; cranial nerves

Anatomy & Physiology Revealed: Topics include brain; divisions of the brain; meninges; CSF flow; MRI; thalamus; hypothalamus; brainstem; cerebral peduncle; cerebellum; hippocampus; inferior brain

Nervous System: Spinal Cord and Spinal Nerves

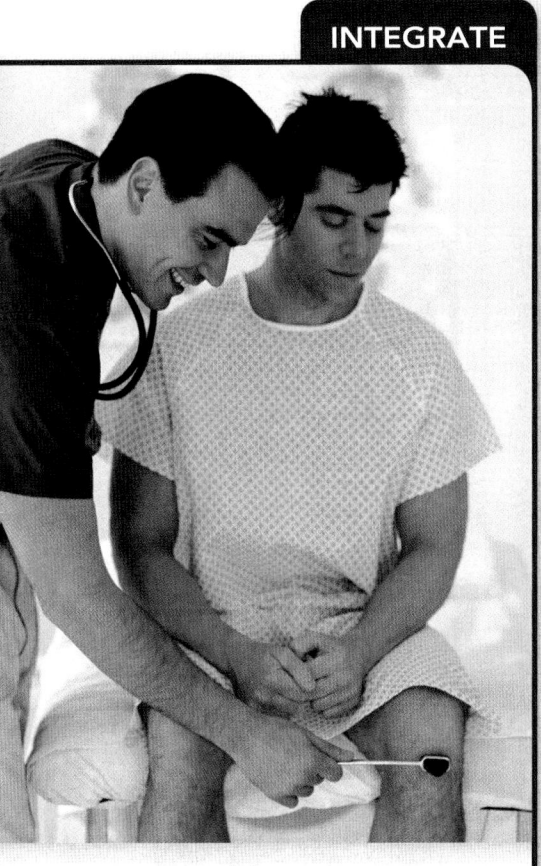

INTEGRATE

aprevealed.com

Module 7: Nervous System

CAREER PATH
Neurologist

A neurologist is a medical doctor who specializes in the diagnosis and treatment of nervous system disorders. A typical neurologic exam involves testing a patient's reflexes and detecting the patient's perception of various sensations. Hyperactive or hypoactive reflexes, or loss of sensation in specific dermatomes, may help pinpoint disorders in individual nerves, specific segments of the spinal cord, or regions of the brain. The neurologist has a thorough knowledge of the central nervous system, as well as motor and sensory components of the body's nerves, and uses this information to aid in an accurate diagnosis of the patient.

The spinal cord provides a vital link between the brain and the rest of the body, and yet it exhibits some functional independence from the brain. The spinal cord and its attached spinal nerves serve two important functions. First, they are a pathway for sensory and motor nerve signals. Second, the spinal cord and spinal nerves are responsible for some specific reflexes (spinal reflexes), which are our quickest reactions to a stimulus. This chapter describes the anatomy of the spinal cord and spinal nerves, then follows with a discussion of the integrative activities that occur there.

14.1 Spinal Cord Gross Anatomy

LEARNING OBJECTIVES

1. Describe the general composition of the spinal cord.

2. Identify the five anatomic subdivisions of the spinal cord and their associated spinal nerves.

3. Explain how the cauda equina arises in development.

A typical adult spinal cord is approximately ¾ of an inch in diameter and ranges between 42 and 45 centimeters (16 to 18 inches) in length (figure 14.1). It extends inferiorly from the medulla oblongata of the brain through the vertebral canal (see section 8.5) and ends at the inferior border of the L₁ vertebra. The tapering inferior end of the spinal cord is called the **conus medullaris** (kō′nŭs med-ū-lăr′is), which marks the official "end" of the spinal cord proper. Two enlargements are visible in the surface view of the spinal cord. These areas of the spinal cord are wider than other areas of the cord because of the greater number of neurons that extend from these enlargements to innervate the upper and lower limbs. The **cervical enlargement** is an enlarged region of the inferior cervical part of the spinal cord. (It contains the neurons that innervate the upper limbs.) The **lumbosacral enlargement** is an enlarged region of the mid-lumbar part of the spinal cord. (It contains the neurons that innervate the lower limbs.)

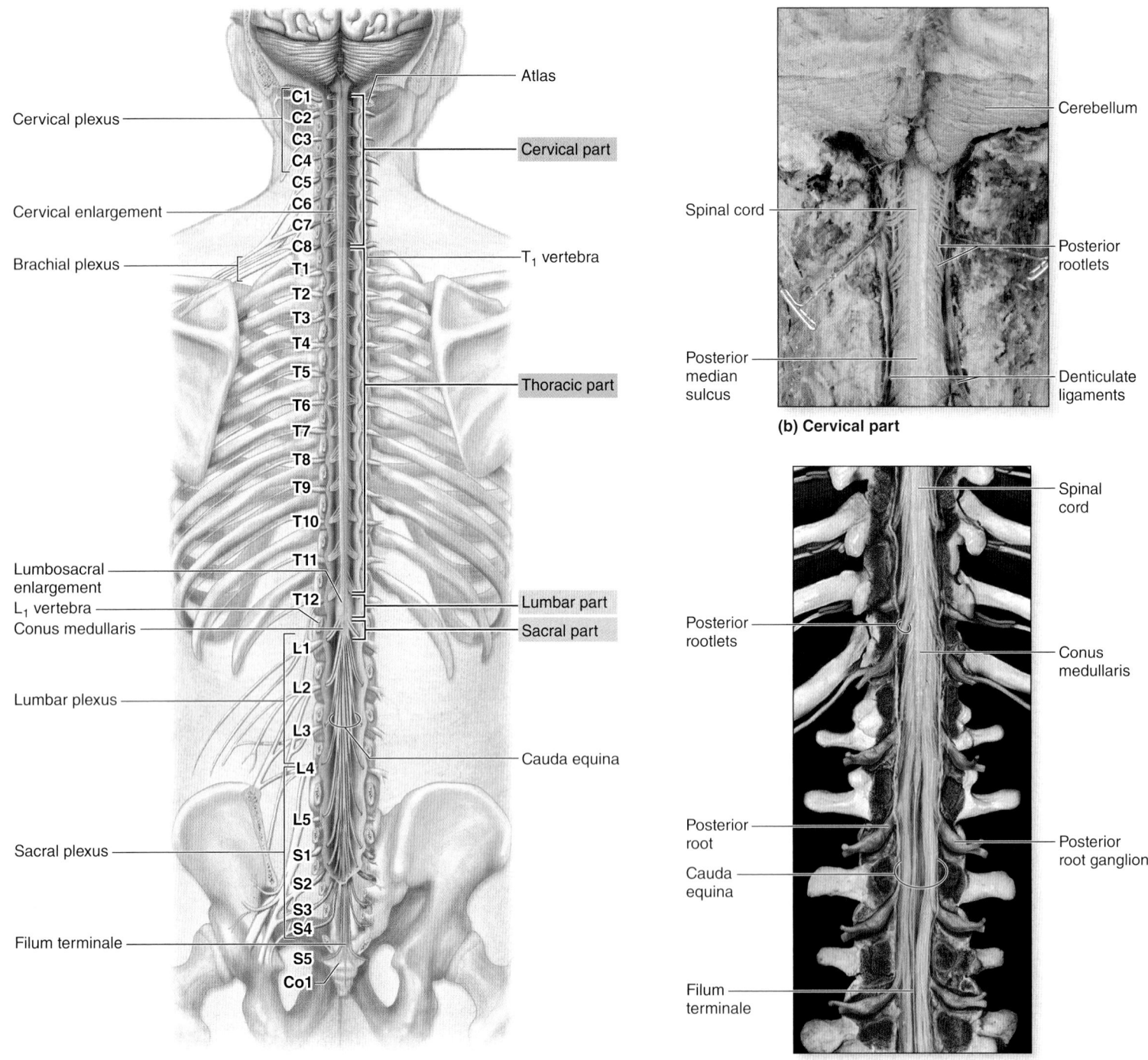

(a) Posterior view

(b) Cervical part

(c) Conus medullaris and cauda equina

Figure 14.1 Gross Anatomy of the Spinal Cord and Spinal Nerves. The spinal cord extends inferiorly from the medulla oblongata through the vertebral canal. (*a*) The vertebral arches have been removed to reveal the anatomy of the adult spinal cord and its spinal nerves. (*b*) Cadaver photo of the cervical part of the spinal cord. (*c*) Cadaver photo of the conus medullaris and the cauda equina. AP|R

The spinal cord is subdivided into five parts. Associated with each part are the rootlets that form the 31 pairs of spinal nerves. The spinal cord parts are as follows:

- The **cervical part** is the superiormost part of the spinal cord. It is continuous with the medulla oblongata. The cervical part contains neurons whose axons contribute to the 8 pairs of cervical spinal nerves (figure 14.1*b*).
- The **thoracic part** lies inferior to the cervical part. It contains the neurons for the 12 pairs of thoracic spinal nerves.
- The **lumbar part** is a shorter segment of the spinal cord that contains the neurons for the 5 pairs of lumbar spinal nerves.
- The **sacral part** lies inferior to the lumbar part and contains neurons for the 5 pairs of sacral spinal nerves.
- The **coccygeal** (kok-sij′ē-ăl) **part** is the most inferior tip of the spinal cord. (Some references consider this part a portion of the sacral part of the spinal cord.) One pair of coccygeal spinal nerves arises from this part.

Note that the different *parts* of the spinal cord do not match up exactly with the *vertebrae* of the same name. For example, the lumbar part of the spinal cord is actually closer to the inferior, or lower, thoracic vertebrae than to the lumbar vertebrae. This discrepancy is due to the fact that the vertebrae growth continued longer than the growth of the spinal cord itself. Thus, the spinal cord in an adult is shorter than the vertebral canal that houses it. Consequently, the rootlets of the more inferior spinal nerves including L2–L5, S1–S5, and Co1 extend inferiorly from the conus medullaris. Collectively, they are called the **cauda equina** (kaw′dă; *cauda* = tail; ē-kwī′nă; *equus* = horse). These nerve roots are so named because they resemble a horse's tail. Within the cauda equina is the **filum terminale** (fī′lŭm ter′mi-nă-lē; *terminus* = end). The filum terminale is a thin strand of pia mater (see section 14.2) that helps anchor the conus medullaris to the coccyx. Figure 14.1*c* shows the conus medullaris and the cauda equina.

The spinal cord is associated with 31 pairs of spinal nerves. Recall that the intervertebral foramina are lateral openings between vertebrae, which provide the passageway for each spinal nerve to extend from the spinal cord and exit the vertebral canal (see section 8.5). Each spinal nerve is identified by the first letter of the spinal cord part to which it attaches combined with a number. Thus, each side of the spinal cord contains 8 cervical nerves (called C1–C8), 12 thoracic nerves (T1–T12), 5 lumbar nerves (L1–L5), 5 sacral nerves (S1–S5), and 1 coccygeal nerve (Co1). Spinal nerve names can be distinguished from cranial nerve names (discussed in section 13.9) because cranial nerves are designated by CN followed by a roman numeral.

Viewed in cross section, the spinal cord is roughly cylindrical, but slightly flattened both posteriorly and anteriorly (**figure 14.2**). Its external surface has two longitudinal depressions: a narrow groove, the

INTEGRATE

LEARNING STRATEGY

With one exception, the number of spinal nerves matches the number of vertebrae in a particular body region. For example, the 12 pairs of thoracic spinal nerves correspond to the 12 thoracic vertebrae. The sacrum forms from 5 sacral vertebrae, and there are 5 pairs of sacral spinal nerves. The coccygeal vertebrae tend to fuse into one structure, and there is 1 pair of coccygeal nerves. The exception to this rule is that there are 8 pairs of cervical spinal nerves, but only 7 cervical vertebrae. This is because the first cervical pair emerges inferior to the occipital bone and superior to the atlas (the first cervical vertebra), and the eighth cervical nerve arises inferior to the seventh cervical vertebra.

posterior (or *dorsal*) **median sulcus,** dips internally on the posterior surface; and a slightly wider groove, the **anterior** (or *ventral*) **median fissure,** is seen on its anterior surface. Observe in figure 14.2 that both the size and shape of the spinal cord changes along its length because the amount of gray matter and white matter reflects the function of that part of the spinal cord. So for example, the spinal cord parts that control the upper and lower limbs are larger because more neuron cell bodies are located there, and more space is occupied by axons and dendrites.

WHAT DID YOU LEARN?

① What are the purposes of the cervical and lumbosacral enlargements?

② What are the total number of pairs of spinal nerves, and what are the specific types of spinal nerves?

14.2 Protection and Support of the Spinal Cord

LEARNING OBJECTIVES

1. Describe the locations and function of the spinal cord meninges.
2. Compare and contrast the three spaces associated with the spinal cord meninges.

The spinal cord is protected and encapsulated by **spinal cord meninges** (mĕ-nin′jes; mē′nin-jēz; sing., *meninx*, men′ingks; membrane), which are continuous with the cranial meninges described in section 13.2a. In addition, spaces between some of the meninges have clinical significance. The meninges and spaces listed from innermost

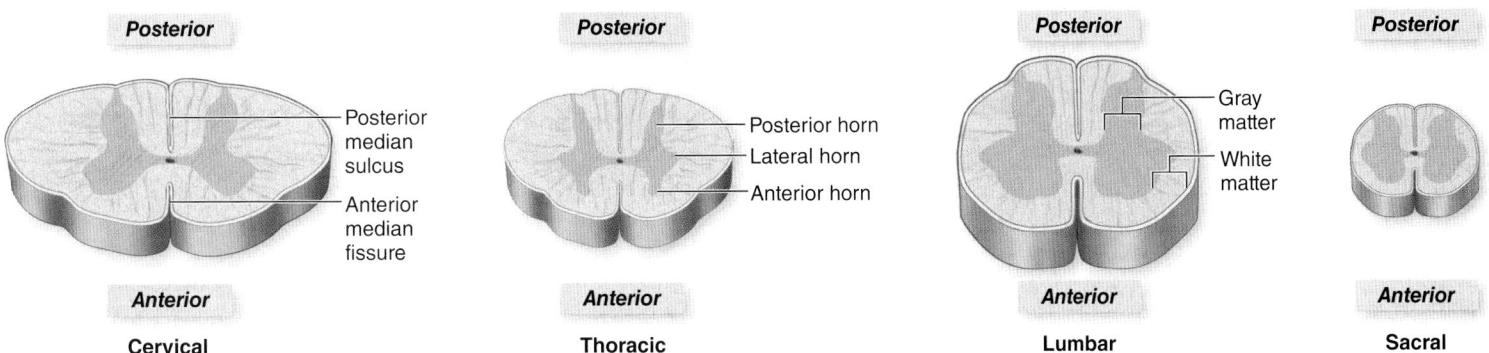

Cervical   Posterior / Posterior median sulcus / Anterior / Anterior median fissure

Thoracic   Posterior / Posterior horn / Lateral horn / Anterior horn / Anterior

Lumbar   Posterior / Gray matter / White matter / Anterior

Sacral   Posterior / Anterior

Figure 14.2 Cross Sections of the Spinal Cord. The cross sections of the cervical, thoracic, lumbar, and sacral parts of the spinal cord vary in shape and size.

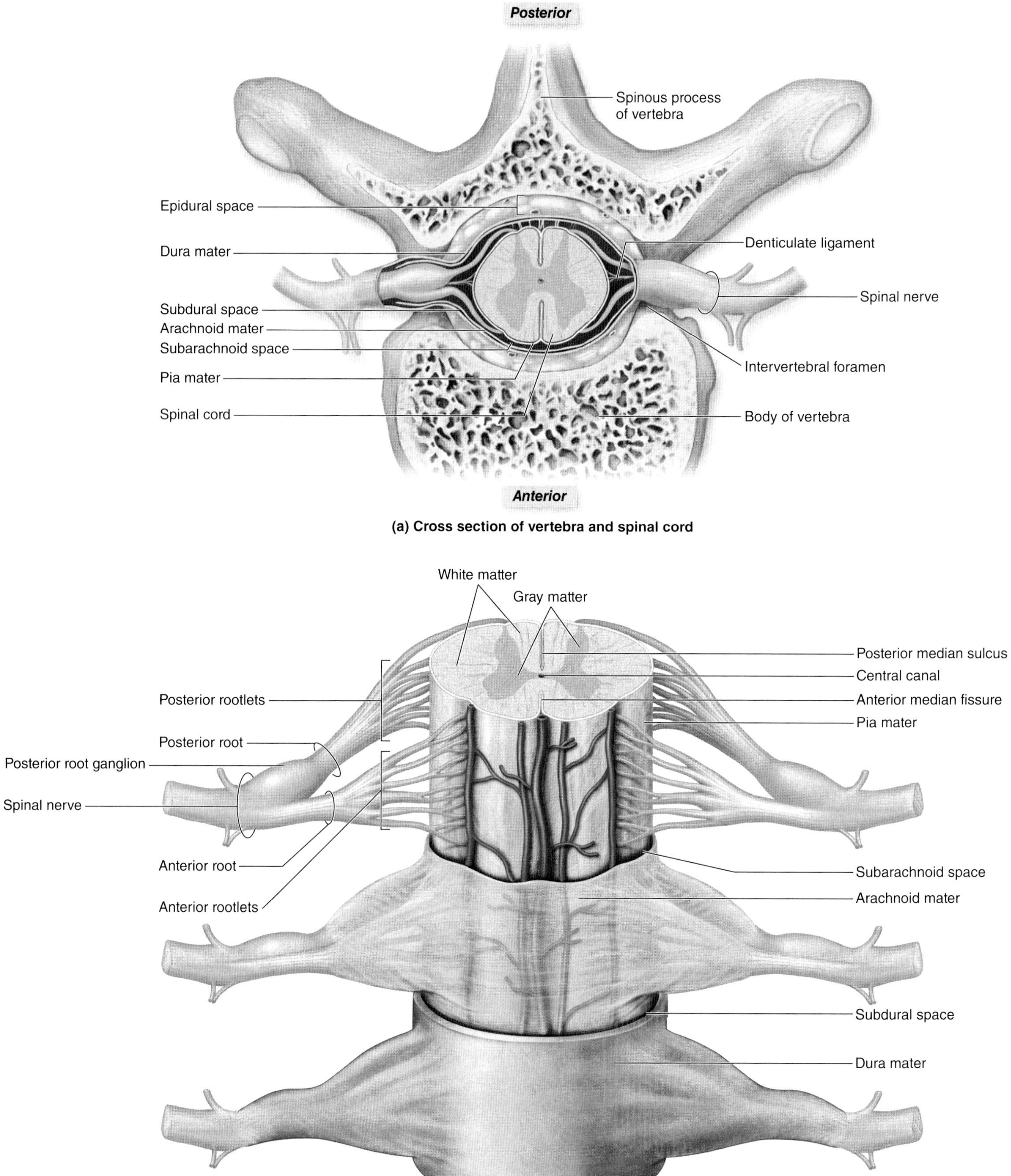

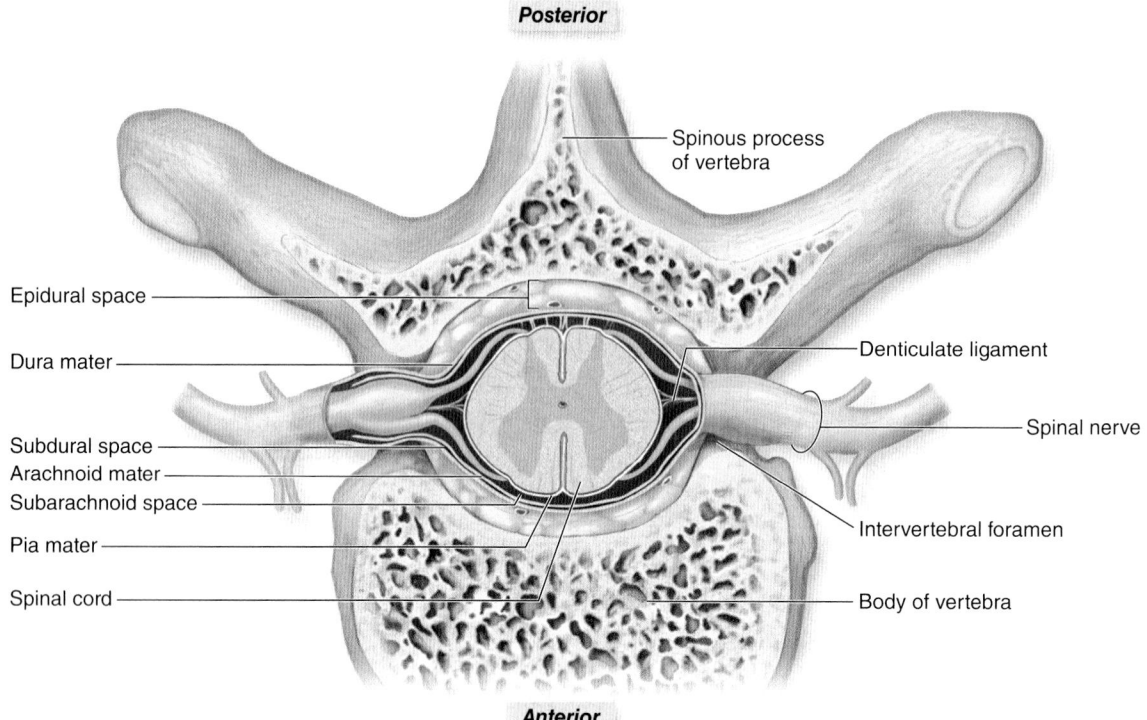

Posterior

Spinous process of vertebra

Epidural space

Dura mater

Denticulate ligament

Subdural space
Arachnoid mater
Subarachnoid space

Spinal nerve

Pia mater

Intervertebral foramen

Spinal cord

Body of vertebra

Anterior

(a) Cross section of vertebra and spinal cord

White matter

Gray matter

Posterior median sulcus
Central canal
Anterior median fissure
Pia mater

Posterior rootlets

Posterior root

Posterior root ganglion

Spinal nerve

Subarachnoid space

Arachnoid mater

Anterior root

Anterior rootlets

Subdural space

Dura mater

(b) Anterior view

Figure 14.3 Spinal Meninges and Structure of the Spinal Cord. (*a*) A cross section of the spinal cord shows the relationship between the meningeal layers and the superficial landmarks of the spinal cord and vertebral column. (*b*) Anterior view shows the spinal cord and meninges. **AP|R**

to outermost are as follows: pia mater, subarachnoid space, arachnoid mater, subdural space, dura mater, and epidural space (figure 14.3).

The **pia mater** directly adheres to the spinal cord. It is the delicate, innermost meningeal layer, which is composed of elastic and collagen fibers that support that some of the blood vessels supplying the spinal cord. **Denticulate** (den-tik′ū-lāt; *dentatus* = toothed) **ligaments** are paired, lateral triangular extensions of the spinal pia mater that attach to the dura mater (described shortly). These ligaments help suspend and anchor the spinal cord laterally (see figures 14.1*b* and 14.3*a*). Recall that the pia mater extends from the inferior end of the cord as the *filum terminale*, which serves to anchor the spinal cord inferiorly to the coccyx (see figure 14.1).

The **arachnoid mater** lies external to the pia mater. It is partially composed of a delicate web of collagen and elastic fibers termed the *arachnoid trabeculae*. Immediately deep to the arachnoid is the **subarachnoid space.** Cerebrospinal fluid circulates in this space (both around the spinal cord and around the brain). The **subdural space** is a potential space between the arachnoid and overlying dura mater.

The outermost layer of meninges is the **dura mater**, which is composed of dense irregular connective tissue. Although the cranial dura mater has an outer periosteal layer and an inner meningeal layer, the spinal dura mater consists of just one layer. The dura mater provides stability to the spinal cord. Additionally, at each intervertebral foramen, the dura mater extends between adjacent vertebrae and fuses with the connective tissue layers that surround the spinal nerves. The **epidural** (ep-i-dū′răl) **space** lies between the dura mater and the inner walls of the vertebra, and houses adipose and areolar connective tissue, and blood vessels. Epidural anesthetics, such as may be used to lessen pain during childbirth, are introduced into this space. (See Clinical View: "Lumbar Puncture.")

WHAT DO YOU THINK?

1. Why doesn't the spinal dura mater have two layers as the cranial dura mater does? What structures that are formed from cranial dura mater must be missing around the spinal cord?

WHAT DID YOU LEARN?

3. Where are the epidural, subdural, and subarachnoid spaces located? Which contains CSF?

14.3 Sectional Anatomy of the Spinal Cord

The spinal cord is partitioned into an inner gray matter region and outer white matter region (figure 14.4). The gray matter is dominated by neuron cell bodies, dendrites, glial cells, and unmyelinated axons, whereas the white matter is composed primarily of myelinated axons that extend to and from the brain (see section 13.1c).

14.3a Distribution of Gray Matter

LEARNING OBJECTIVES

1. Distinguish the four anatomic locations of gray matter in the spinal cord.
2. Name the types of neurons and functional groups (nuclei) found in each gray matter region.

The **gray matter** in the spinal cord is centrally located, and its shape resembles a letter H, or a butterfly. The gray matter may be subdivided into the following components: anterior horns, lateral horns, posterior horns, and the gray commissure.

Anterior horns are the left and right anterior masses of gray matter. The anterior horns primarily house the *cell bodies of somatic motor neurons,* which innervate skeletal muscle. **Lateral horns** are found in the T1–L2 parts of the spinal cord only. The lateral horns contain the *cell bodies of autonomic motor neurons,* which innervate cardiac muscle, smooth muscle, and glands. **Posterior horns** are the left and right posterior masses of gray matter. The *axons of sensory neurons* and the *cell bodies of interneurons* are located in the posterior horns. (Note that the cell bodies of these sensory neurons are not found in the posterior horns; rather, they are located in the posterior root ganglia, which are described in section 14.5a). The **gray commissure** (kom′i-shūr; *commissura* = a seam) is a horizontal bar of gray matter that surrounds a narrow **central canal.** The gray commissure primarily contains unmyelinated axons

INTEGRATE

CLINICAL VIEW
Lumbar Puncture

It is sometimes necessary to analyze the cerebrospinal fluid (CSF) to determine whether an infection or a disorder of the central nervous system is present. The clinical procedure for obtaining CSF is known as a **lumbar puncture** (commonly referred to as a *spinal tap*). The needle must be inserted through the skin, back muscles, and ligamentum flavum (between vertebrae). Then, the needle must pass through the epidural space, dura mater, arachnoid mater, and enter the subarachnoid space to obtain approximately 3 to 9 milliliters of CSF.

Since the adult spinal cord typically ends at the level of the L_1 vertebra, a lumbar puncture must be performed inferior to this level to ensure the spinal cord is not pierced by the needle. A lumbar puncture typically is made at the level of either the L_3 and L_4 vertebrae or the L_4 and L_5 vertebrae. To locate this level, the physician palpates the highest points of the iliac crests, which are at the same horizontal level as the spinous process of the L_4 vertebra. The physician can then insert the lumbar puncture needle either directly above or directly below the spinous process of L_4 when the vertebral column is flexed.

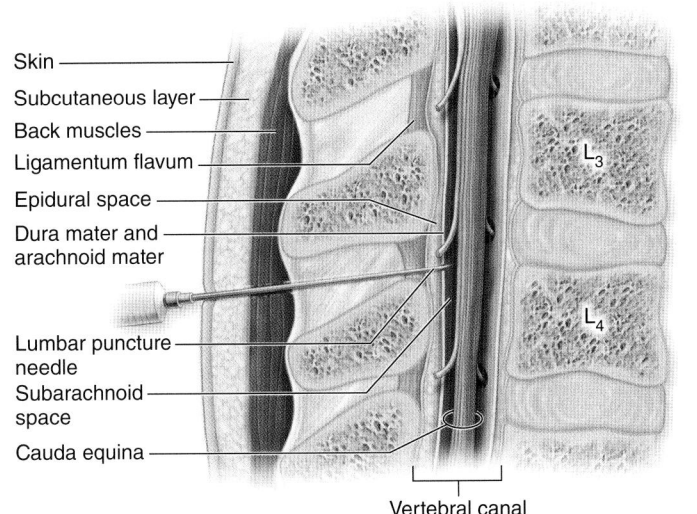

Skin
Subcutaneous layer
Back muscles
Ligamentum flavum
Epidural space
Dura mater and arachnoid mater
L_3
Lumbar puncture needle
Subarachnoid space
L_4
Cauda equina
Vertebral canal

Site of needle insertion for a lumbar puncture.

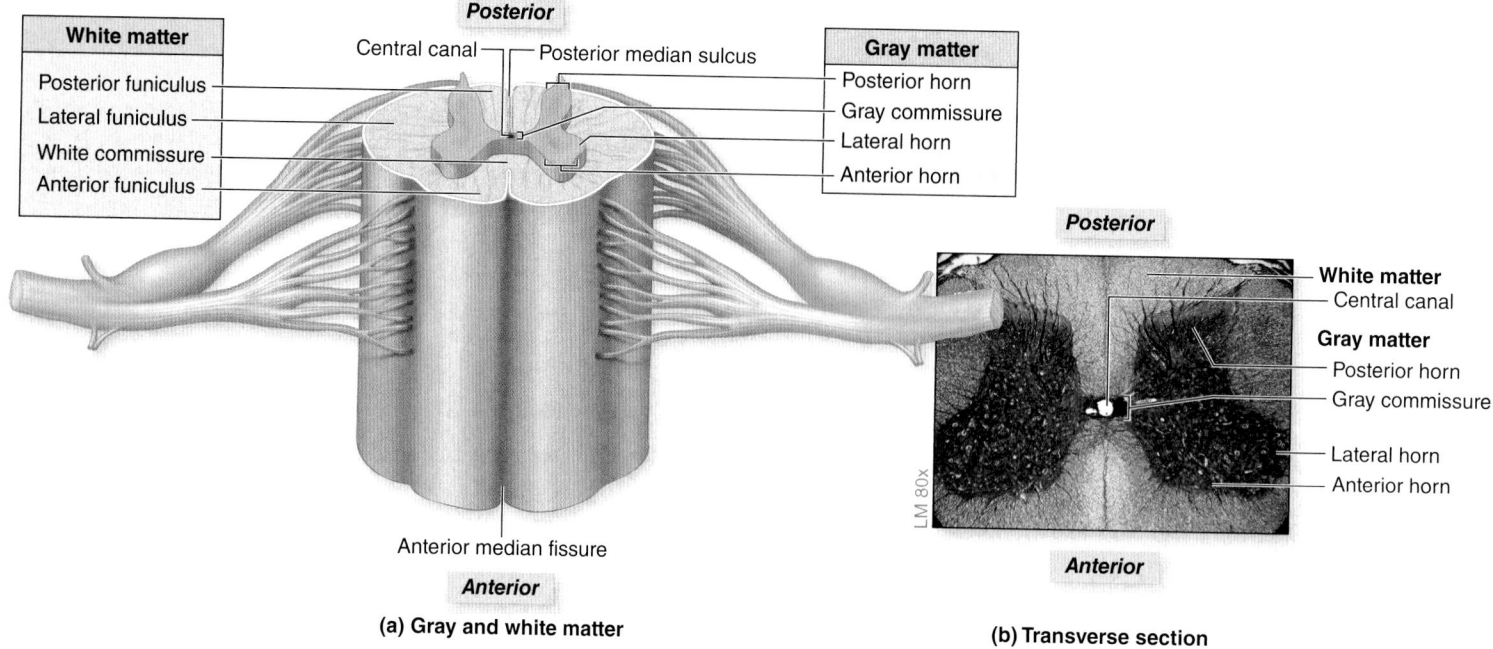

Figure 14.4 Gray Matter and White Matter Organization in the Spinal Cord. (*a*) The gray matter is centrally located, and the white matter is externally located. (*b*) Histology of a transverse section of the spinal cord. AP|R

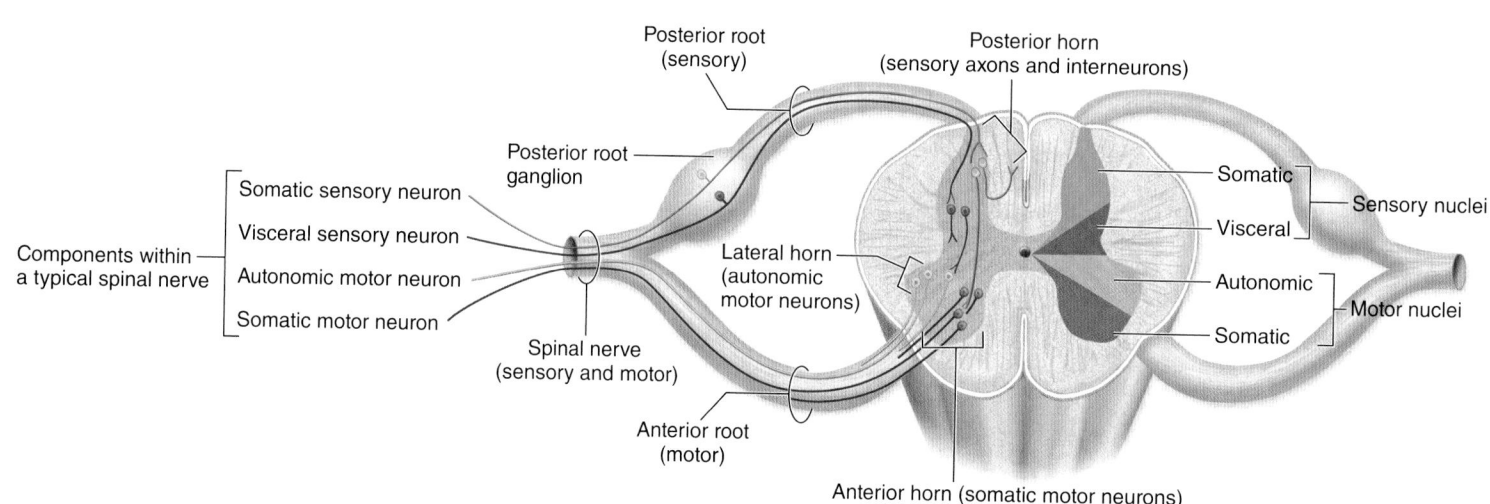

Figure 14.5 Neuron Pathways and Nuclei Locations. The collections of neuron cell bodies within the CNS form specific nuclei. Neurons are color-coded on the left side of the drawing; their respective nuclei are color-coded on the right side of the drawing.

and serves as a communication route between the right and left sides of the gray matter.

Within these parts of gray matter are various functional groups of neuron cell bodies called **nuclei (figure 14.5). Sensory nuclei** in the posterior horns contain interneuron cell bodies. **Somatic sensory nuclei** receive nerve signals from sensory receptors, (e.g., pain or pressure receptors in the skin), whereas **visceral sensory nuclei** receive nerve signals from blood vessels and viscera (e.g., from stretched smooth muscle within viscera). **Motor nuclei** in the anterior and lateral horns contain motor neuron cell bodies that send nerve signals to muscles and glands. The **somatic motor nuclei** in the anterior horn innervate skeletal muscle, whereas the **autonomic motor nuclei** in the lateral horns innervate smooth muscle, cardiac muscle, and glands.

 WHAT DID YOU LEARN?

④ Identify the nervous system structures found in the (a) anterior horn, (b) lateral horn, and (c) posterior horn.

14.3b Distribution of White Matter

LEARNING OBJECTIVES

3. Identify the location of white matter in the spinal cord.
4. Name the three anatomic divisions of the white matter, and explain their general composition.

The white matter of the spinal cord is external to the gray matter. White matter on each side of the cord is partitioned into three regions, each called a **funiculus** (fū-nik′ū-lŭs; pl. *funiculi,*[1] fū-nik′ū-lī; *funis* = cord) (figure 14.4). A **posterior funiculus** lies between the posterior gray horns on the posterior side of the cord and the posterior median sulcus. The **lateral funiculus** is the white matter on

---

[1]*Note:* Anterior and lateral funiculi were formerly called *columns.* The Federative Committee on Anatomical Terminology (FCAT) now states that the term "column" refers to structures within the gray matter of the spinal cord, whereas "funiculus" refers to the white matter regions.

INTEGRATE

CLINICAL VIEW

Poliomyelitis

Poliomyelitis (pō'lē-ō-mī'e-lī'tis; *polio* = gray, *myelos* = marrow, *itis* = infection) is an infection caused by one of the three strains of poliovirus. Infection is by oral-fecal or oral-oral route, and common routes of transmission are through contaminated food or water supplies. Most cases of polio are mild and may result in GI or flulike symptoms. However, in about 1% of the cases, the virus spreads to the nervous system and attacks somatic motor neurons in the anterior horn of the spinal cord. (The disease got its name by describing the inflammation of the gray matter of the spinal cord.) These cases are referred to as *paralytic polio*. Here, the motor neurons are damaged or destroyed, resulting in paralysis of the muscles innervated by those segments of the spinal cord. Paralysis may be temporary or permanent, depending on the extent of somatic motor neuron damage. Polio is rare in the Western world due to an active vaccination program, but is still endemic in Pakistan, Afghanistan, Nigeria, Syria, and Chad (where vaccination programs have been disrupted or incomplete).

INTEGRATE

CLINICAL VIEW

Treating Spinal Cord Injuries

Spinal cord injuries frequently leave individuals paralyzed and unable to perceive sensations to varying degrees, depending upon the location and extent of the injury. In recent years, advances have been made in the treatment of spinal cord injuries (although some of the findings are still preliminary). Prompt use of steroids immediately after the injury appears to preserve some muscular function that might otherwise be lost. Early use of antibiotics has substantially reduced the number of deaths caused by pulmonary and urinary tract infections that accompany spinal cord injuries. Recent research with rats has achieved reconnection and partial restoration of function of severed spinal cords. In addition, other research indicates that neural stem cells may be able to regenerate CNS axons.

each lateral side of the spinal cord. The **anterior funiculus** is composed of white matter that occupies the space on each anterior side of the cord between the anterior gray horns and the anterior median fissure; the anterior funiculi are interconnected by the **white commissure.**

The axons within each white matter funiculus are organized into smaller structural units called **tracts** (trakt; *tractus* = a drawing out) or **fasciculi** (fă-sik'ū-lī; *fascis* = bundle). Individual tracts conduct either sensory nerve signals (ascending tracts from the spinal cord to the brain) or motor nerve signals (descending tracts from the brain to the spinal cord) only. The lateral and anterior funiculi contain both ascending and descending tracts, and so they are composed of both motor and sensory axons. In contrast, the posterior funiculi contain sensory axons only, which extend as ascending tracts.

WHAT DID YOU LEARN?

 5 What are the three types of funiculi? List the specific tracts found in each.

14.4 Spinal Cord Conduction Pathways

The CNS communicates with peripheral body structures through **pathways.** These pathways conduct either sensory nerve signals from receptors to the CNS or motor nerve signals from the CNS to effectors; processing and integration occur along them. Here we describe the general features of these pathways and the details of both the sensory and motor pathways.

14.4a Overview of Conduction Pathways

LEARNING OBJECTIVES

1. Name the components of a conduction pathway and list the features common to all pathways.
2. Compare and contrast sensory and motor pathways.

Nervous system pathways are either sensory or motor. **Sensory pathways** are also called *ascending pathways* because the nerve signals transmitted from sensory receptors ascend through the spinal cord to the brain. **Motor pathways** are also called *descending pathways* because they transmit nerve signals that descend from the brain through the spinal cord to muscles or glands. Most nervous system pathways share several general characteristics:

- **Common location of neuron components:** The neuron cell bodies are located in one of three general places: ganglia within the peripheral nervous system, gray horns within the spinal cord, or nuclei within the brain along the pathway. In contrast, the axons of these neurons extend through the spinal cord and brain as tracts.

- **Composed of two or more neurons.** Most pathways are composed of a series of two or three neurons that work together. Sensory pathways have primary neurons, secondary neurons, and sometimes tertiary neurons that facilitate the pathway's functioning. In contrast, motor pathways use an upper motor neuron and a lower motor neuron.

- **Paired tracts.** All pathways are composed of paired tracts. A pathway on the left side of the CNS has a matching tract on the right side of the CNS.

- **Decussation.** Most pathways **decussate** (dē-kŭ-sāt'; *decusseo* = to make in the form of an X) (cross over) from one side of the body to the other side at some point along the pathway. This means that the left side of the brain processes sensory input from and motor output to the right side of the body, and vice versa. The term **contralateral** (kon-tră-lat'er-ăl; *contra* = opposite, *latus* = side) is used to indicate the relationship to the opposite side, whereas the term **ipsilateral** (ip-si-lat'er-ăl; *ipse* = same) means the same side. Over 90% of all pathways decussate; however, the point at which decussation occurs can vary among pathways.

INTEGRATE

LEARNING STRATEGY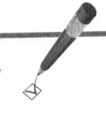

Pathways and tracts generally are named according to their origin and termination. For example, sensory pathways usually begin with the prefix *spino-*, indicating that they originate in the spinal cord. So the tract that originates in the spinal cord and terminates in the cerebellum is called the *spinocerebellar tract*. Motor pathways begin either with *cortico-*, indicating an origin in the cerebral cortex, or with the name of a brainstem nucleus, such as *rubro-*, indicating an origin within the red nucleus of the midbrain. Thus, both *corticospinal* and *rubrospinal* are motor tracts.

6 What are four characteristics common to most nervous system pathways?

14.4b Sensory Pathways

LEARNING OBJECTIVES

3. Define a sensory pathway, and describe its action.
4. List the neurons in the sensory pathway chain, and describe their roles.
5. Describe the three major somatosensory pathways.

Sensory pathways are ascending pathways that relay sensory input from receptors to the brain. Sensory input "informs" the brain about limb proprioception, touch, temperature, pressure, and pain. These pathways are organized into two categories that are dependent upon the type of receptor involved. *Somatosensory pathways* process stimuli received from receptors within the skin, muscles, and joints, whereas *viscerosensory pathways* process stimuli received from the viscera.

Sensory pathways use a series of two or three neurons to transmit nerve signals from the body to the brain:

- The first neuron in this chain is the **primary neuron** (or *first-order neuron*). The cell body of a primary neuron resides in the posterior root ganglion of a spinal nerve, and the axon extends to a secondary neuron.
- The **secondary neuron** (or *second-order neuron*) is an interneuron that extends from the primary neuron to either the tertiary neuron or to the cerebellum.
- The **tertiary neuron** (or *third-order neuron*) is also an interneuron. It extends from the secondary neuron to the cerebrum (specifically the primary somatosensory cortex of the parietal lobe (see section 13.3c). Pathways that lead to the cerebellum do not have a tertiary neuron.

There are three major types of **somatosensory pathways.** Each of these pathways has specific tracts that extend through the spinal cord **(figure 14.6)**. The posterior funiculus–medial lemniscal pathway relays sensory input through the fasciculus gracilis and fasciculus cuneatus. The anterolateral pathway, relays sensory input through the lateral and anterior spinothalamic tracts. The spinocerebellar pathway relays sensory input through the lateral and anterior spinocerebellar tracts. We now describe the details of each of these pathways.

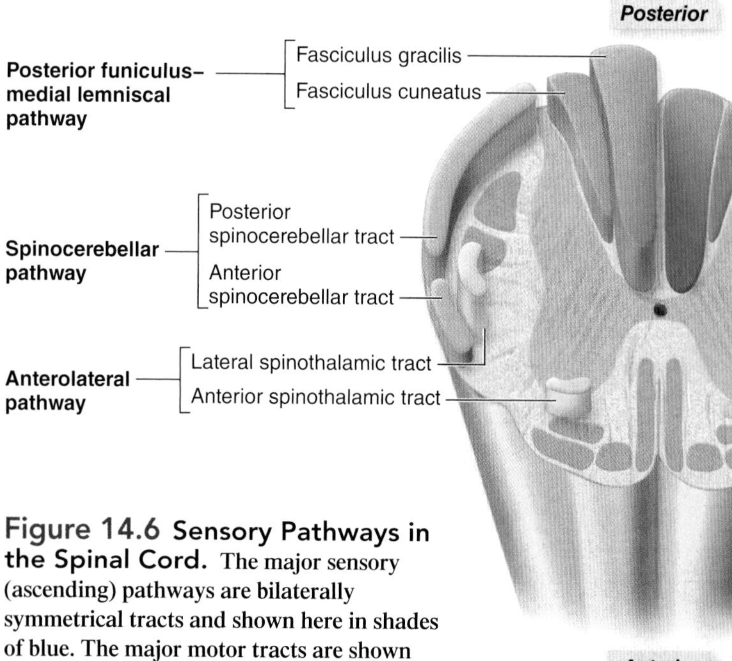

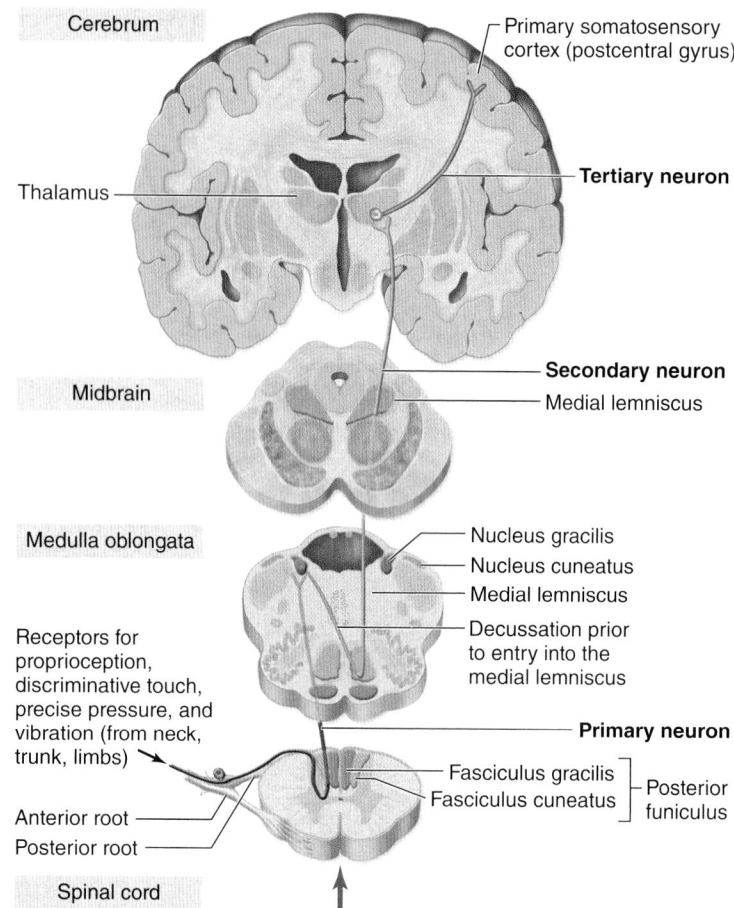

Right side of body   Left side of body

Cerebrum — Primary somatosensory cortex (postcentral gyrus)

Thalamus — **Tertiary neuron**

Midbrain — **Secondary neuron** — Medial lemniscus

Medulla oblongata — Nucleus gracilis / Nucleus cuneatus / Medial lemniscus / Decussation prior to entry into the medial lemniscus

Receptors for proprioception, discriminative touch, precise pressure, and vibration (from neck, trunk, limbs) — **Primary neuron** — Fasciculus gracilis / Fasciculus cuneatus — Posterior funiculus

Anterior root / Posterior root

Spinal cord

Pathway direction

Figure 14.7 Posterior Funiculus–Medial Lemniscal Pathway. This pathway transmits sensory information about limb position (proprioception), discriminative touch, precise pressure, and vibration. This pathway is bilaterally symmetrical—but to avoid confusion, only sensory input from the right side of the body is shown here. Decussation of axons occurs in the medulla oblongata. The primary neuron is purple, the secondary neuron is blue, and the tertiary neuron is green.

Posterior Funiculus–Medial Lemniscal Pathway

The **posterior funiculus–medial lemniscal pathway** projects from a somatic receptor to the primary somatosensory cortex of the cerebral cortex **(figure 14.7)**. Its name derives from two components: the tracts within the spinal cord, collectively called the **posterior funiculus,** and the tracts within the brainstem, collectively called the **medial lemniscus** (lem-nis′k ŭs; *lemniskos* = ribbon). This pathway transmits sensory input concerned with proprioceptive (posture and balance) information about limb position, as well as discriminative touch, precise pressure, and vibration sensations from the skin.

The posterior funiculus–medial lemniscal pathway uses a series of three neurons to signal the brain about a specific stimulus. Axons of the primary neurons reside in spinal nerves and reach the CNS through the posterior roots of spinal nerves. Upon entering the spinal cord, these axons ascend within a specific posterior funiculus, either the **fasciculus cuneatus** (kū′nē-ā-tŭs; *cuneus* = wedge) or the **fasciculus gracilis**. Sensory axons ascending within the posterior funiculi synapse on secondary neuron cell bodies housed within the nucleus cuneatus or nucleus gracilis, respectively. The axons of these secondary neurons then project to the thalamus on the opposite side of the brain through the medial

Figure 14.6 Sensory Pathways in the Spinal Cord.

Posterior

Posterior funiculus–medial lemniscal pathway — Fasciculus gracilis / Fasciculus cuneatus

Spinocerebellar pathway — Posterior spinocerebellar tract / Anterior spinocerebellar tract

Anterolateral pathway — Lateral spinothalamic tract / Anterior spinothalamic tract

Figure 14.6 Sensory Pathways in the Spinal Cord. The major sensory (ascending) pathways are bilaterally symmetrical tracts and shown here in shades of blue. The major motor tracts are shown for comparison in pale red.

Anterior

lemniscus. Decussation occurs after secondary neuron axons exit their specific nuclei within the medulla oblongata and before they enter the medial lemniscus.

The axons of the secondary neurons synapse on cell bodies of the tertiary neurons within the thalamus, where the sensory information is sorted according to the part of the body involved (somatotopically). Axons from these tertiary neurons transmit nerve signals to a specific location of the primary somatosensory cortex housed within the postcentral gyrus (see figure 13.13).

Anterolateral Pathway

The **anterolateral pathway** (or *spinothalamic pathway*) is located in the anterior and lateral white funiculi of the spinal cord **(figure 14.8)**. It is composed of the **anterior spinothalamic tract** (which extend through the anterior funiculi) and the **lateral spinothalamic tract** (which extend through the lateral funiculi). Axons within these pathways relay sensory input related to crude touch and pressure as well as pain and temperature. Typically, sensations that require us to act in response to the stimulus (such as an itch that makes us want to scratch, or tickling that makes us jerk away) are relayed through the anterolateral pathway. The anterolateral pathway also uses a chain of three

neurons to signal the brain about a specific stimulus. Axons of the primary neurons reside in spinal nerves and reach the CNS through the posterior roots of spinal nerves (like the previous pathway). However, these axons synapse on secondary neurons within the posterior horns of the spinal cord.

Axons of the secondary neurons in the anterolateral pathway decussate through the anterior white commissure and relay nerve signals to the opposite side of the spinal cord before ascending toward the brain and synapsing on tertiary neurons located within the thalamus. Axons from the tertiary neurons then transmit nerve signals to the appropriate part of the primary somatosensory cortex.

Spinocerebellar Pathway

The **spinocerebellar pathway** extends through the anterior and posterior white funiculi of the spinal cord **(figure 14.9)**. It is composed of the **anterior spinocerebellar tract** and the **posterior spinocerebellar tract.** Its name derives from the origin (spinal cord) and destination (cerebellum) of the secondary neurons of this pathway.

Axons within these pathways conduct nerve signals from proprioceptors related to postural input to the cerebellum. Information conducted in spinocerebellar pathways is integrated and acted on at a subconscious level.

The spinocerebellar pathway uses a chain of only two neurons to signal the cerebellum about a specific stimulus. (There is no tertiary neuron.) Axons of the primary neurons reside in spinal nerves and reach the CNS through the posterior roots of spinal nerves synapsing on secondary neurons within the posterior horns of the spinal cord (like the previous pathway).

Some of the secondary neuron axons in the spinocerebellar pathway decussate through the anterior white commissure and relay nerve signals to the opposite side of the spinal cord before ascending toward the brain. Secondary neuron axons transmit nerve signals to the cerebellum.

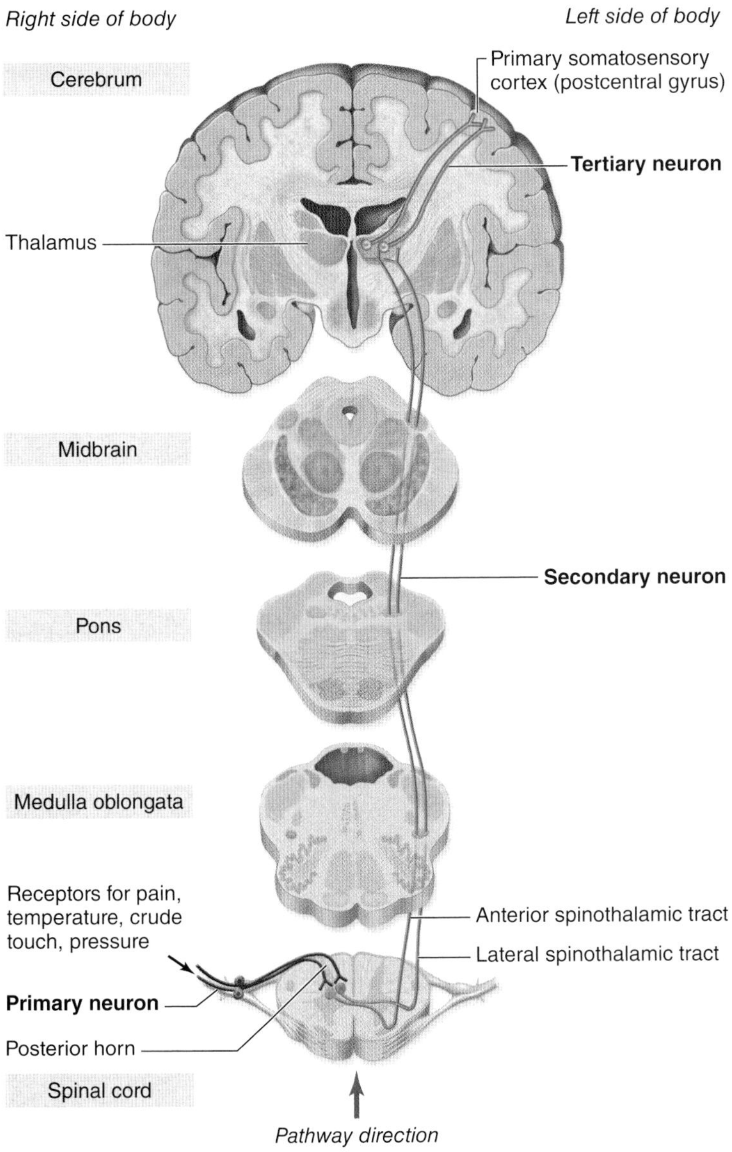

Figure 14.8 Anterolateral Pathway. This pathway conducts crude touch, pressure, pain, and temperature sensations toward the brain. Decussation of axons occurs at the level where the primary neuron axon enters the spinal cord. The primary neuron is purple, the secondary neuron is blue, and the tertiary neuron is green.

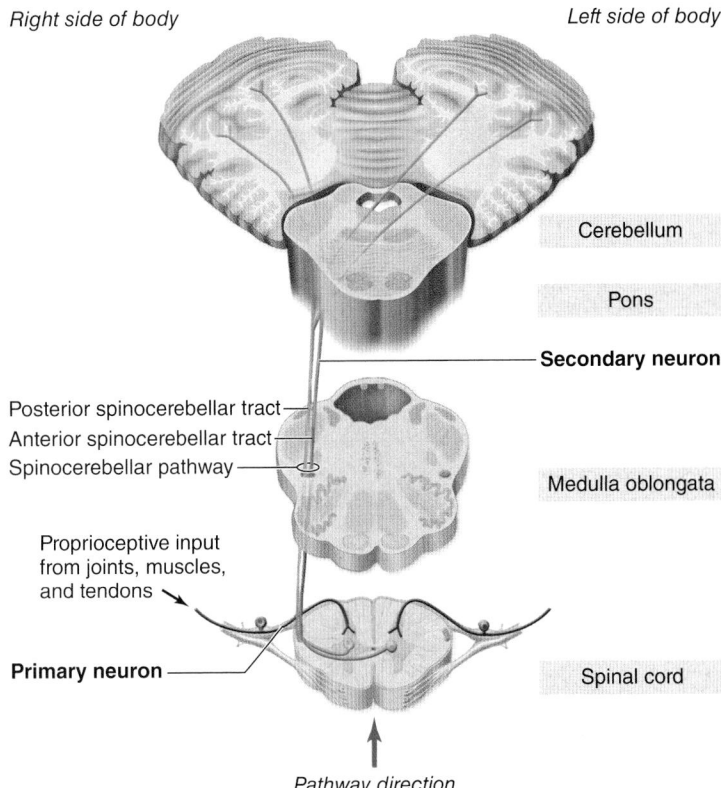

Figure 14.9 Spinocerebellar Pathway. This pathway conducts proprioceptive information to the cerebellum through both the anterior and posterior spinocerebellar tracts. Only some axons decussate, and these do so at the level where the primary neuron axon enters the spinal cord. Only primary (purple) and secondary (blue) neurons are found in this type of pathway.

Table 14.1 — Functions and Neuron Locations of Principal Sensory Spinal Cord Pathways

Pathway	Fasciculus Cuneatus	Fasciculus Gracilis	Anterior Spinothalamic	Lateral Spinothalamic	Anterior Spinocerebellar	Posterior Spinocerebellar
Function	Sensory input for limb position and discriminative touch, precise pressure, and vibration sensation		Sensory input for crude touch, pressure, pain, and temperature		Sensory input sent from proprioceptors to cerebellum for subconscious interpretation	
	Relays input from upper limb, superior trunk, neck, posterior head	Relays input from lower limb, inferior trunk	Relays input for crude touch and pressure	Relays input for pain and temperature	Relays input from inferior regions of trunk and lower limbs	Relays input from lower limbs, regions of trunk and upper limbs
Primary neuron	Extends from receptor to brainstem		Extends from receptor to spinal cord		Extends from receptor to spinal cord	
	Cell bodies within posterior root ganglion		Cell bodies within posterior root ganglion		Cell bodies within posterior root ganglion	
Secondary neuron	Extends from brainstem to thalamus		Extends from spinal cord to thalamus		Extends from spinal cord to cerebellum	
	Cell bodies within brainstem (nucleus cuneatus or nucleus gracilis)		Cell bodies within spinal cord (posterior horn)		Cell bodies within spinal cord (posterior horn)	
Tertiary neuron	Extends from thalamus to cerebral cortex		Extends from thalamus to cerebral cortex		N/A	
	Cell bodies within thalamus		Cell bodies within thalamus			
Structures involved in decussation	Axons of secondary neurons decussate prior to entry into medial lemniscus		Axons of secondary neurons decussate within spinal cord at level of entry		Some axons decussate in spinal cord and pons, whereas other axons do not decussate	Axons do not decussate

Table 14.1 summarizes the characteristics of the three major types of sensory pathways.

WHAT DID YOU LEARN?

7 What are the general locations and functions of primary, secondary, and tertiary neurons in sensory pathways?

8 What type of information does the posterior funiculus-medial lemniscal pathway transmit?

9 Compare the location of the secondary neurons between the anterior and lateral spinothalamic tracts.

14.4c Motor Pathways

LEARNING OBJECTIVES

6. Define a motor pathway, and describe its actions.

7. Distinguish between an upper motor neuron and a lower motor neuron, based on function and cell body location.

8. Compare and contrast the direct and indirect motor pathways.

Motor pathways, or *descending tracts,* are descending pathways in the brain and spinal cord that control effectors. Here we discuss the motor pathways that specifically control skeletal muscle. Motor pathways originate from the cerebral cortex, cerebral nuclei, or the brainstem (figure 14.10).

At least two motor neurons are present in the motor pathway: an upper motor neuron and a lower motor neuron. The cell body of the **upper motor neuron** is housed either within the cerebral cortex, cerebral nuclei, or a nucleus within the brainstem. An axon of the upper motor neuron synapses either directly on a lower motor neuron, or on an interneuron that ultimately synapses directly on a lower motor neuron. Upper motor neurons either excite or inhibit the activity of lower motor neurons. The cell body of a **lower motor neuron** is housed either within the anterior horn of the spinal cord or within a brainstem cranial nerve nucleus. An axon of the lower motor neuron exits the CNS and projects to the skeletal muscle to be innervated. The lower motor neuron is always excitatory because its axon connects directly to the skeletal muscle fibers.

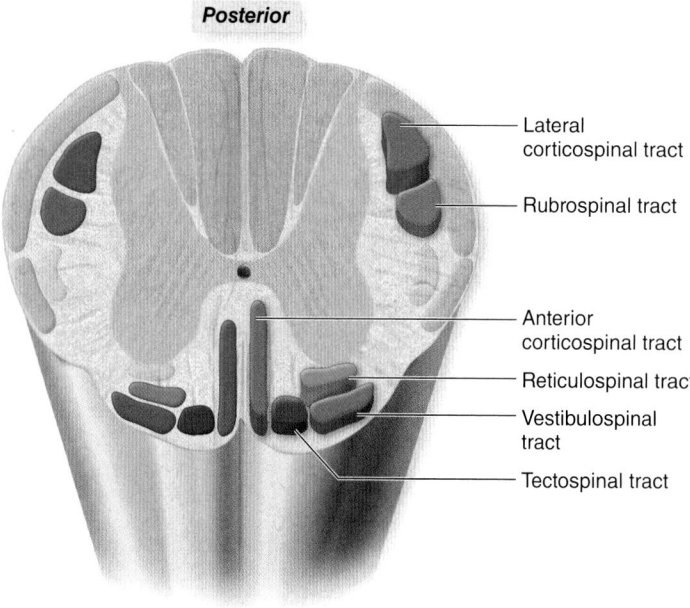

Posterior

Lateral corticospinal tract

Rubrospinal tract

Anterior corticospinal tract

Reticulospinal tract

Vestibulospinal tract

Tectospinal tract

Anterior

Figure 14.10 Motor Pathways in the Spinal Cord. Motor (descending) pathways are bilaterally symmetrical and shown here in red and orange. The major sensory (ascending) pathways are shown for comparison in light blue. **AP|R**

Motor neurons form two types of motor pathways: the direct pathway and the indirect pathway. The direct pathway is responsible for conscious control of skeletal muscle activity; the indirect pathway is responsible for subconscious (or reflexive) control of skeletal muscle.

Direct Pathway

The **direct pathway,** also called the **pyramidal** (pi-ram′i-dal) **pathway,** originates in the primary motor cortex in the frontal lobe (see section 13.3b). The name *pyramidal* is derived from the pyramid-like shape of the upper motor neuron cell bodies within this area.

The axons of pyramidal upper motor neurons descend through the internal capsule, enter the cerebral peduncles, and ultimately synapse in one of two locations: the brainstem or the spinal cord. Those tracts that extend from the cerebral cortex and synapse in the brainstem compose the corticobulbar tracts. In contrast, the tracts that extend from the cerebral cortex and synapse in the spinal cord compose the corticospinal tracts.

Corticobulbar Tracts The **corticobulbar** (kōr′ti-kō-bŭl′bar) **tracts** originate from the facial region of the motor homunculus within the primary motor cortex (see section 13.3c). The term *bulbar* means resembling a bulb and is used to indicate the embryonic rhombencephalon in the brainstem (see section 13.1b).

Axons of these upper motor neurons extend to the brainstem, where they synapse with lower motor neuron cell bodies that are housed within brainstem cranial nerve nuclei. Axons of these lower motor neurons help form some of the cranial nerves (see section 13.9). Thus, note that these tracts differ from the others that we are about to discuss in that (1) they do not pass through the spinal cord, and (2) they involve cranial nerves (and not spinal nerves). We discuss them here because they are considered one type of direct pathway.

Corticospinal Tracts The **corticospinal tracts** descend from the primary motor cortex of the cerebrum through the medulla oblongata and into the spinal cord where they synapse on lower motor neurons in the anterior horn of the spinal cord **(figure 14.11).** The corticospinal tracts are composed of two components: lateral and anterior corticospinal tracts. These two tracts differ in several significant ways including their point of crossing-over (decussation), their location within the spinal cord, and the specific muscles that they innervate and control.

The **lateral corticospinal tracts** include about 85% of the axons of the upper motor neurons that decussate within the pyramids of the medulla oblongata and then form the lateral corticospinal tracts in the lateral funiculi of the spinal cord. Axons of the lower motor neurons innervate skeletal muscles that control skilled movements in the *limbs,* such as playing a guitar, dribbling a soccer ball, or typing on your computer keyboard.

The **anterior corticospinal tracts** represent the remaining 15% of the axons of upper motor neurons that extend through the medulla oblongata. The axons of these neurons do not decussate at the level of the medulla oblongata; instead, they decussate at the level of a spinal cord segment through the anterior gray commissure. They subsequently synapse either with interneurons or with lower motor neurons in the anterior horn. Axons of the lower motor neurons innervate *axial* skeletal muscle.

Indirect Pathway

Several nuclei within the brainstem initiate motor commands for activities that occur at a subconscious or reflexive level. The **indirect pathway** is so named because upper motor neurons originate within brainstem nuclei and take a complex, circuitous route through the brain to the spinal cord. The indirect pathway modifies or helps control the pattern of somatic motor activity by exciting or inhibiting the lower motor neurons that innervate the muscles.

The different tracts of the indirect pathway are grouped according to their primary functions. The **lateral pathway** regulates and controls precise, discrete movements and tone in flexor muscles of the limbs—for example, the type of movement required to gently lay a baby in a crib. This pathway consists of the **rubrospinal** (rū′brō-spī′năl; *rubro* = red) **tracts** that originate in the red nucleus of the midbrain.

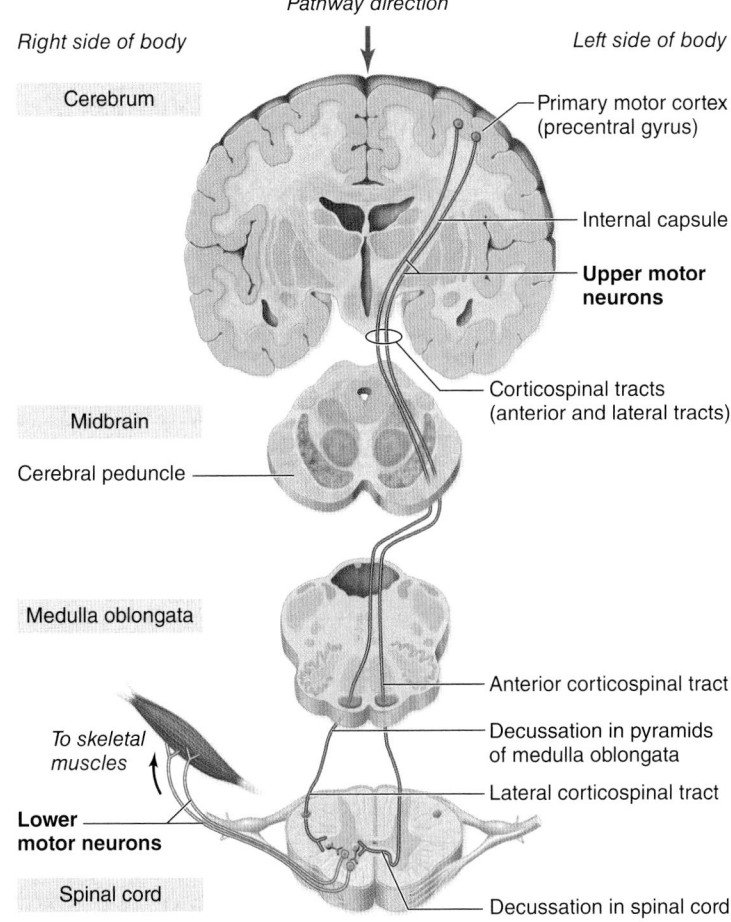

Pathway direction

Right side of body

Left side of body

Cerebrum

Primary motor cortex (precentral gyrus)

Internal capsule

Upper motor neurons

Corticospinal tracts (anterior and lateral tracts)

Midbrain

Cerebral peduncle

Medulla oblongata

Anterior corticospinal tract

To skeletal muscles

Decussation in pyramids of medulla oblongata

Lateral corticospinal tract

Lower motor neurons

Spinal cord

Decussation in spinal cord

Figure 14.11 Corticospinal Tracts. Corticospinal tracts originate within the cerebrum, decussate in the pyramids or the spinal cord and synapse on lower motor neurons within the anterior horns of the spinal cord. The upper motor neurons are red and the lower motor neurons are dark orange.

The **medial pathway** regulates reflexive muscle tone and gross movements of the muscles of the head, neck, proximal parts of the limbs, and trunk. The medial pathway consists of three groups of tracts—reticulospinal tracts, tectospinal tracts, and vestibulospinal tracts:

- The **reticulospinal** (re-tik-ū-lō-spī′năl) **tracts** originate from the reticular formation in the midbrain. They help control reflexive movements related to posture and maintaining balance.
- The **tectospinal** (tek-tō-spī′năl) **tracts** conduct motor output from the superior and inferior colliculi in the tectum of the midbrain to help regulate reflexive positional changes of the upper limbs, eyes, head, and neck as a consequence of visual and auditory stimuli.
- The **vestibulospinal** (ves-tib′ū-lō-spī′năl) **tracts** originate within vestibular nuclei of the brainstem. Nerve signals conducted within these tracts regulate reflexive muscular activity that helps maintain balance during sitting, standing, and walking.

Table 14.2 summarizes the characteristics of the principal types of motor pathways. Figure 14.12 summarizes the main differences between the sensory and motor pathways.

 WHAT DID YOU LEARN?

10 What are the locations and functions of upper and lower motor neurons in the motor pathways?

11 What are the differences between direct and indirect motor pathways?

Table 14.2	Principal Motor Spinal Cord Pathways			
Tract	**Manner of Decussation**	**Destination of Upper Motor Neurons**	**Termination Site**	**Function**
DIRECT PATHWAY				
Corticobulbar tracts	All cranial nerve motor nuclei receive bilateral (both ipsilateral and contra-lateral) input except CN VI, VII to the lower face, and XII (these nerves receive only contralateral input)	Brainstem only	Cranial nerve nuclei; reticular formation	Voluntary movement of cranial and facial muscles
Lateral corticospinal tracts	All decussate at the pyramids	Lateral funiculus	Gray matter region between posterior and anterior horns; anterior horn; all levels of spinal cord	Voluntary movement of appendicular muscles
Anterior corticospinal tracts	Decussation occurs in spinal cord at level of lower motor neuron cell body	Anterior funiculus	Gray matter region between posterior and anterior horns; anterior horn; cervical part of spinal cord	Voluntary movement of axial muscles
INDIRECT PATHWAY				
Lateral Pathway				
Rubrospinal tract	Decussate at ventral tegmentum	Lateral funiculus	Lateral region between posterior and anterior horns; anterior horn; cervical part of spinal cord	Regulates and controls precise discrete movements and tone in flexor muscles of the limbs
Medial Pathway				
Reticulospinal tract	No decussation (ipsilateral)	Anterior funiculus	Medial region between posterior and anterior horns; anterior horn; all parts of spinal cord	Controls reflexive movements related to posture and maintaining balance
Tectospinal tract	Decussate at dorsal tegmentum	Anterior funiculus	Medial region between posterior and anterior horns; anterior horn; cervical part of spinal cord	Regulates reflexive positional changes of the upper limbs, eyes, head, and neck due to visual and auditory stimuli
Vestibulospinal tract	Some decussate (contralateral) and some do not (ipsilateral)	Anterior funiculus	Medial region between posterior and anterior horns; anterior horn; medial tracts to cervical and superior thoracic parts of spinal cord; lateral tracts to all parts of spinal cord	Regulates reflexive muscular activity that helps maintain balance during sitting, standing, and walking

Figure 14.12 Differences Between Sensory and Motor Pathways.

(*a*) Sensory pathways transmit nerve signals from sensory receptors that ascend from the posterior and lateral funiculi of the spinal cord to the brain. These pathways use up to three neurons to transmit this information (primary, secondary, and tertiary neurons). (*b*) Motor pathways transmit nerve signals from the brain and descend to effectors. These pathways typically travel through the anterior and lateral funiculi of the spinal cord, and use two neurons (upper and lower motor neurons).

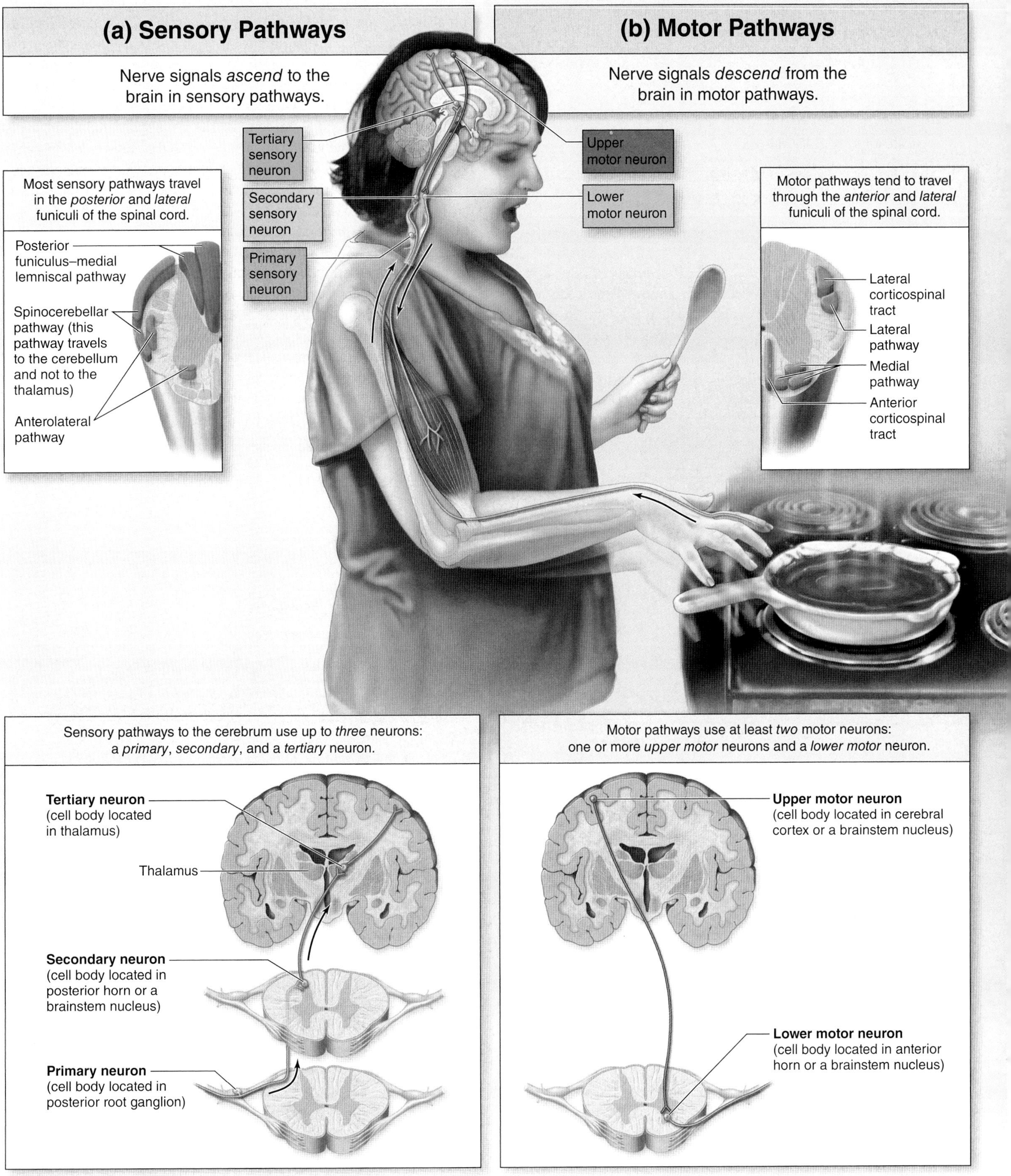

(a) Sensory Pathways

Nerve signals *ascend* to the brain in sensory pathways.

Tertiary sensory neuron

Secondary sensory neuron

Primary sensory neuron

Most sensory pathways travel in the *posterior* and *lateral* funiculi of the spinal cord.

Posterior funiculus–medial lemniscal pathway

Spinocerebellar pathway (this pathway travels to the cerebellum and not to the thalamus)

Anterolateral pathway

(b) Motor Pathways

Nerve signals *descend* from the brain in motor pathways.

Upper motor neuron

Lower motor neuron

Motor pathways tend to travel through the *anterior* and *lateral* funiculi of the spinal cord.

Lateral corticospinal tract

Lateral pathway

Medial pathway

Anterior corticospinal tract

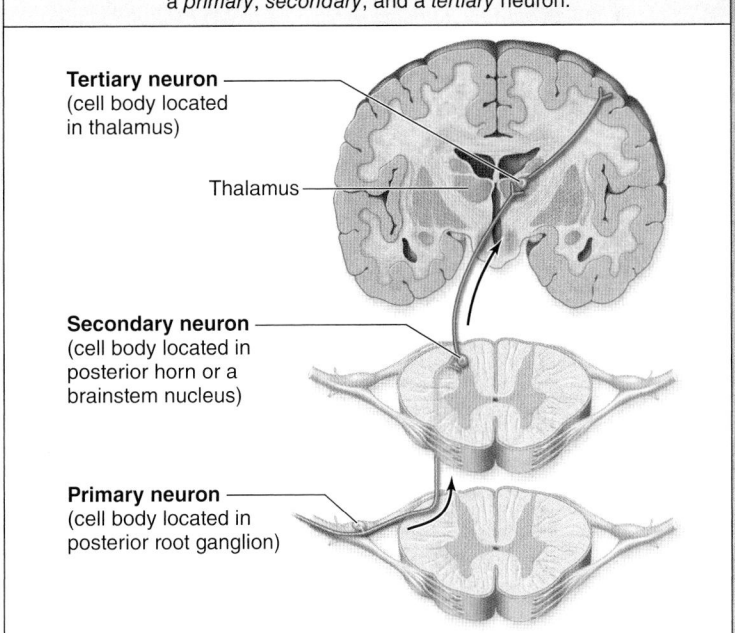

Sensory pathways to the cerebrum use up to *three* neurons: a *primary*, *secondary*, and a *tertiary* neuron.

Tertiary neuron (cell body located in thalamus)

Thalamus

Secondary neuron (cell body located in posterior horn or a brainstem nucleus)

Primary neuron (cell body located in posterior root ganglion)

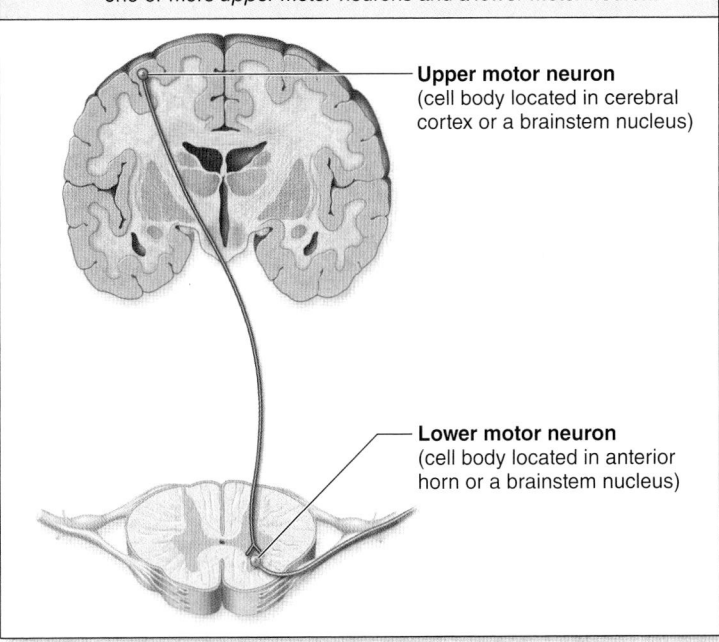

Motor pathways use at least *two* motor neurons: one or more *upper motor* neurons and a *lower motor* neuron.

Upper motor neuron (cell body located in cerebral cortex or a brainstem nucleus)

Lower motor neuron (cell body located in anterior horn or a brainstem nucleus)

14.5 Spinal Nerves

The 31 pairs of **spinal nerves** extend from the spinal cord to muscles and glands, and from sensory receptors to the spinal cord. Recall from section 12.2e that a nerve typically is formed from the union of thousands of motor and sensory axons and is enveloped in the three successive connective tissue wrappings: epineurium, perineurium, and endoneurium.

Here we describe the composition and anatomic pathways of the spinal nerves.

14.5a Overview of Spinal Nerves

LEARNING OBJECTIVES

1. Describe the components of a typical spinal nerve.
2. Compare and contrast the anterior and posterior rami of a spinal nerve.
3. Define a dermatome, and explain its clinical significance.

Recall that a spinal nerve is identified by the first letter of the spinal cord part from which it extends, combined with a number. Thus, each side of the spinal cord contains 8 cervical nerves (called C1–C8), 12 thoracic nerves (T1–T12), 5 lumbar nerves (L–L5), 5 sacral nerves (S1–S5), and 1 coccygeal nerve (Co1).

Each spinal nerve is formed by two roots—an anterior root and a posterior root. Each **anterior root** (or *ventral root*) contains motor axons only and is formed from the merging of multiple anterior rootlets. These motor axons arise from cell bodies in the anterior and lateral horns of the spinal cord (see figure 14.5). Motor axons conduct nerve signals from the CNS to effectors (muscles and glands).

The **posterior root** (or *dorsal root*), contains sensory axons only and is formed from the merging of multiple posterior rootlets. The cell bodies of these sensory neurons are located in a **posterior root ganglion,** (or *dorsal root ganglion*) which is within the posterior root. Sensory axons relay nerve signals from the receptors to the CNS.

Each anterior root and its corresponding posterior root unite within the intervertebral foramen to form a spinal nerve. Thus, all spinal nerves are mixed nerves because each spinal nerve contains both motor axons (from the anterior root) and sensory axons (from the posterior root). You can compare a spinal nerve to a cable composed of multiple wires. The "wires" within a spinal nerve are the motor and sensory axons, and each "wire'" transmits only in one direction.

Each cervical spinal nerve exits the vertebral canal and travels through an intervertebral foramen superior to the vertebra of the same number. For example, the second cervical spinal nerve exits the vertebral canal through the intervertebral foramen between the C_1 and the C_2 vertebrae. The eighth cervical spinal nerve is the exception; it leaves the intervertebral foramen between the C_7 and T_1 vertebrae. The spinal nerves inferior to the C8 nerve exit the vertebral canal and travel through an intervertebral foramen inferior to the vertebra of the same number. So, for example, the second thoracic spinal nerve exits the vertebral canal through the intervertebral foramen between the T_2 and T_3 vertebrae.

Because the spinal cord is shorter than the vertebral canal, the roots of the lumbar and sacral spinal nerves must extend inferiorly to reach their respective intervertebral foramina before they can merge and form a spinal nerve. Thus, the anterior and posterior roots of the lumbar, sacral, and coccygeal spinal nerves are much longer than the roots of the other spinal nerves, and collectively form the cauda equina (described in section 14.1).

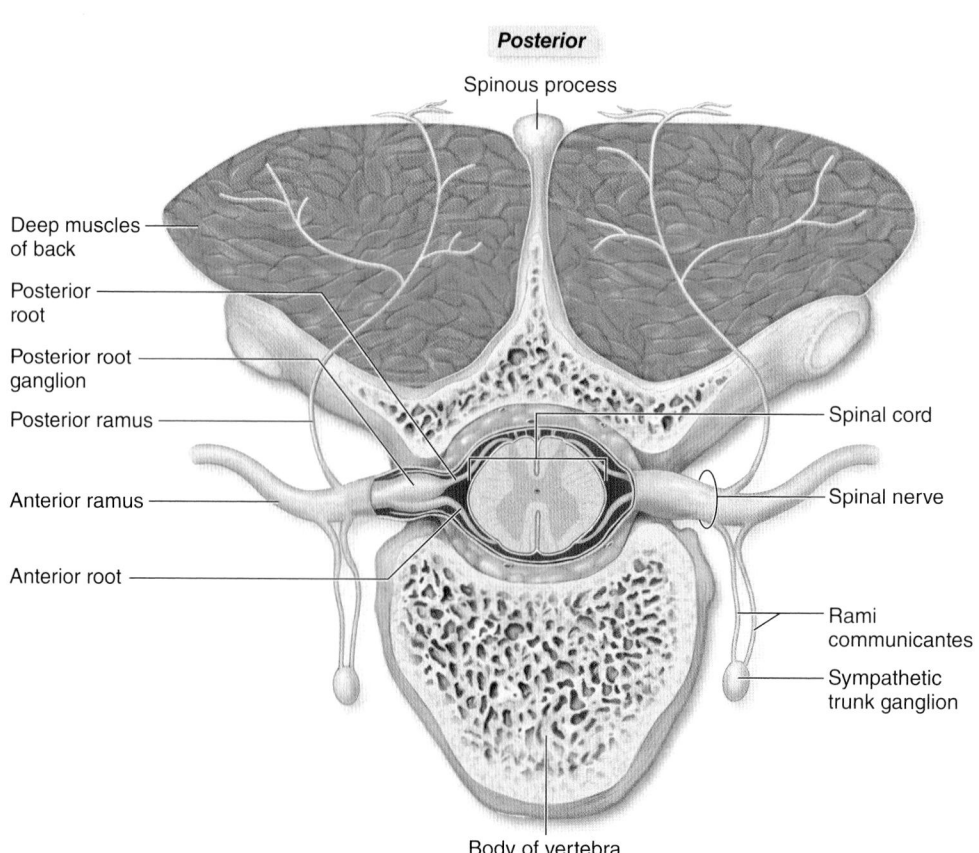

Figure 14.13 Spinal Nerve Branches. The major branches of a spinal nerve are the posterior ramus and the anterior ramus. AP|R

Distribution of Spinal Nerves

After exiting the intervertebral foramen, a typical spinal nerve almost immediately splits into branches, termed *rami* (figure 14.13). The **posterior** (*dorsal*) **ramus** (rā′mŭs; pl., *rami,* rā′mī; branch) is the smaller of the two main branches. It innervates the deep muscles of the back (e.g., erector spinae and transversospinalis; see section 11.4) and the skin of the back.

The **anterior** (*ventral*) **ramus** is the larger of the two main branches. The anterior ramus splits into multiple other branches, which innervate the anterior and lateral portions of the trunk, the upper limbs, and the lower limbs. Many of the anterior rami go on to form nerve plexuses, which are described in the following sections.

Additional rami, called the *rami communicantes,* are also associated with spinal nerves. These rami contain axons associated with the autonomic nervous system (ANS). Each set of rami communicantes extends between the spinal nerve and a ball-like structure called the sympathetic trunk ganglion. These ganglia are interconnected and form a beaded necklace-like structure called the sympathetic trunk that extends parallel and lateral to the vertebral column. The structures associated with the ANS are described in detail in section 15.4a.

WHAT DO YOU THINK?

2 Why is an anterior ramus so much larger than a posterior ramus?

Dermatomes

A **dermatome** (der′mă-tōm; *derma* = skin, *tome* = a cutting) is a specific segment of skin innervated by a single spinal nerve. All spinal nerves except for C1 innervate a segment of skin, and each area of the skin that is innervated by a specific spinal nerve has been mapped. Collectively, this map is called a dermatome map (figure 14.14). The dermatome map follows a segmental pattern along the body (although there is slight overlap between adjacent spinal nerves). For example, the horizontal segment of skin around the umbilicus (navel) region is supplied by the anterior ramus of the T10 spinal nerve.

Dermatomes are clinically important because they can indicate potential damage to one or more spinal nerves. For example, if a patient experiences **anesthesia** (an′es-thē′zē-ă; loss of sensation or numbness) along the medial side of the arm and forearm, the C8 spinal nerve may be damaged.

Dermatomes are also involved in **referred visceral pain,** a phenomenon in which pain or discomfort from one organ is mistakenly referred to a dermatome. For example, the appendix is innervated by axons from the T10 regions of the spinal cord, so appendicitis typically causes referred visceral pain to the T10 dermatome in the umbilicus region rather than in the abdominopelvic region of the appendix itself. Thus, pain in a dermatome typically arises from an organ nowhere near the dermatome. Referred visceral pain is explored further in section 16.2b.

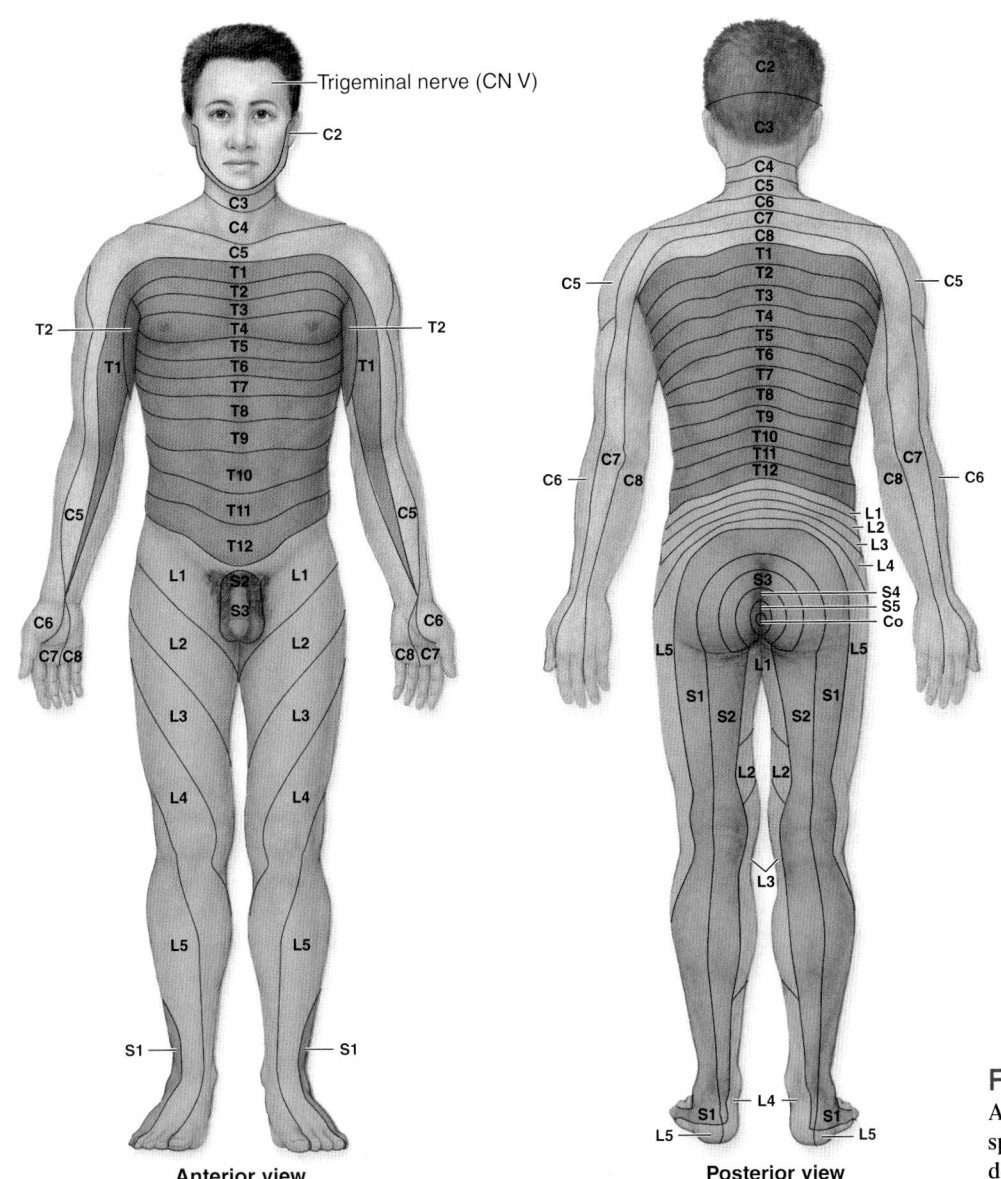

Anterior view **Posterior view**

Figure 14.14 Dermatome Maps.
A dermatome is an area of skin supplied by a single spinal nerve. These diagrams only approximate the dermatomal distribution. AP|R

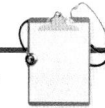
CLINICAL VIEW
Shingles

Some adults (usually after age 50) experience a reactivation of their childhood chickenpox infection, a condition termed **shingles** (shing'glz). Psychological stress, other infections (such as a cold or the flu), and even a sunburn can trigger the development of shingles.

During the initial infection, the chickenpox virus (varicella-zoster) sometimes leaves the skin and invades the posterior root ganglia. There, the virus remains latent until adulthood, when it becomes reactivated and proliferates, traveling through the sensory axons to the dermatome. (The word *shingles* is derived from the Latin word *cingulum,* meaning "girdle," reflecting the dermatomal pattern of its spread.) The virus gives rise to a rash and blisters along the dermatome, which are often accompanied by intense burning or tingling pain.

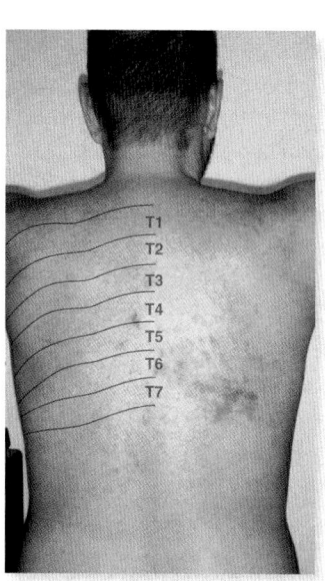

Antiviral medication (e.g., acyclovir) may reduce the severity and duration of symptoms of shingles. Additionally, older adults may receive a vaccine for shingles, which may help prevent or reduce the severity of the disease.

Typical dermatomal spread of a shingles rash in a 49-year-old man.

WHAT DID YOU LEARN?

  What are the differences between an anterior ramus and a posterior ramus of a typical spinal nerve?

  What is a dermatome, and why may a dermatome be clinically significant?

14.5b Nerve Plexuses
LEARNING OBJECTIVE
4. Define a nerve plexus.

A **nerve plexus** (plek's ŭs; a braid) is a network of interweaving anterior rami of spinal nerves. The anterior rami of most spinal nerves form nerve plexuses on both the right and left sides of the body. These nerve plexuses then split into multiple "named" nerves that innervate various body structures. The main plexuses are the cervical plexuses, brachial plexuses, lumbar plexuses, and sacral plexuses (see figure 14.1).

WHAT DO YOU THINK?
3 What is the benefit of having an intricate nerve plexus, rather than a single spinal nerve that innervates a structure?

Nerve plexuses are organized such that axons from each anterior ramus extend to body structures through several different branches. In addition, each terminal branch of the plexus houses axons from several different spinal nerves. Thus, damage to a single segment of the spinal cord or damage to a single spinal nerve generally does not result in complete loss of innervation to a particular muscle or region of skin.

Most of the thoracic spinal nerves, as well as nerves S5–Co1, do not form plexuses. We discuss the anterior rami of thoracic spinal nerves (called intercostal nerves) first, followed by the individual nerve plexuses.

WHAT DID YOU LEARN?
14 What is the composition of a typical nerve plexus?

14.5c Intercostal Nerves
LEARNING OBJECTIVE
5. Identify the distribution of the intercostal nerves.

The anterior rami of spinal nerves T1–T11 are called **intercostal nerves** because they travel in the intercostal space sandwiched between two adjacent ribs (**figure 14.15**). (T12 is called a **subcostal nerve,** because it arises inferior to the ribs, not between two ribs.) With the exception of T1, the intercostal nerves do not form plexuses. The intercostal nerves innervate much of the torso wall and portions of the upper limb (see the dermatomal map in figure 14.14). The specific innervation pattern of the T1–T12 nerves is as follows:

- A portion of the anterior ramus of T1 helps form the brachial plexus, but a branch of it travels within the first intercostal space.
- The anterior ramus of nerve T2 emerges from its intervertebral foramen and innervates the intercostal muscles of the second intercostal space. Additionally, a branch of T2 transmits sensory information from the skin covering the axilla and the medial surface of the arm.
- Anterior rami of nerves T3–T6 follow the costal grooves of the ribs to innervate the intercostal muscles and receive sensations from the anterior and lateral chest wall.
- Anterior rami of nerves T7–T12 innervate not only the inferior intercostal spaces, but also the abdominal muscles and their overlying skin.

WHAT DID YOU LEARN?
15 In general, what do the intercostal nerves innervate?

14.5d Cervical Plexuses
LEARNING OBJECTIVES
6. List the nerves of the cervical plexuses.
7. Explain the action of the phrenic nerve.

The left and right **cervical plexuses** are located deep on each side of the neck, immediately lateral to cervical vertebrae C_1–C_4 (**figure 14.16**). They are formed primarily by the anterior rami of spinal nerves C1–C4. The fifth cervical spinal nerve is not considered part of the cervical plexus, although it contributes some axons to one of the plexus branches. Branches of the cervical plexuses innervate anterior neck muscles (see section 11.3d) as well as the skin of the neck and portions of the head and shoulders. The branches of the cervical plexuses are described in detail in **table 14.3**. Note that in the tables, "Motor branches" indicates the portion that relays motor output to skeletal muscles and "Cutaneous branches" indicates the portion that relays sensory input from the skin.

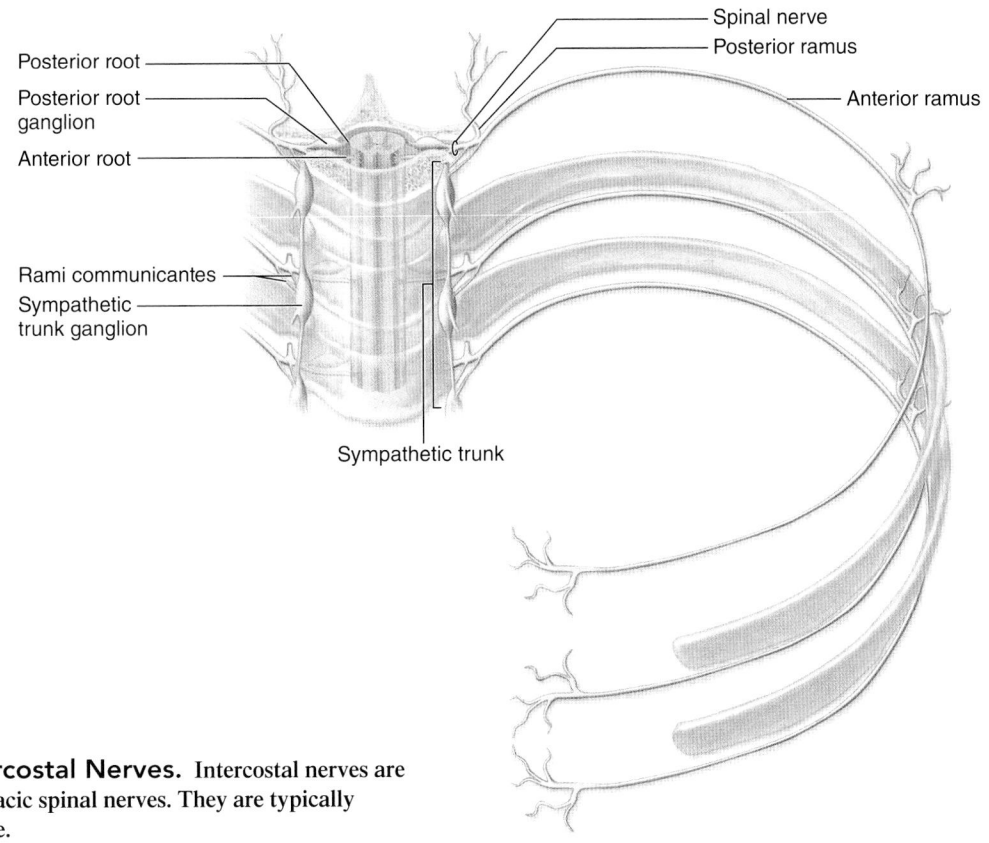

Figure 14.15 **Intercostal Nerves.** Intercostal nerves are the anterior rami of thoracic spinal nerves. They are typically distributed as shown here.

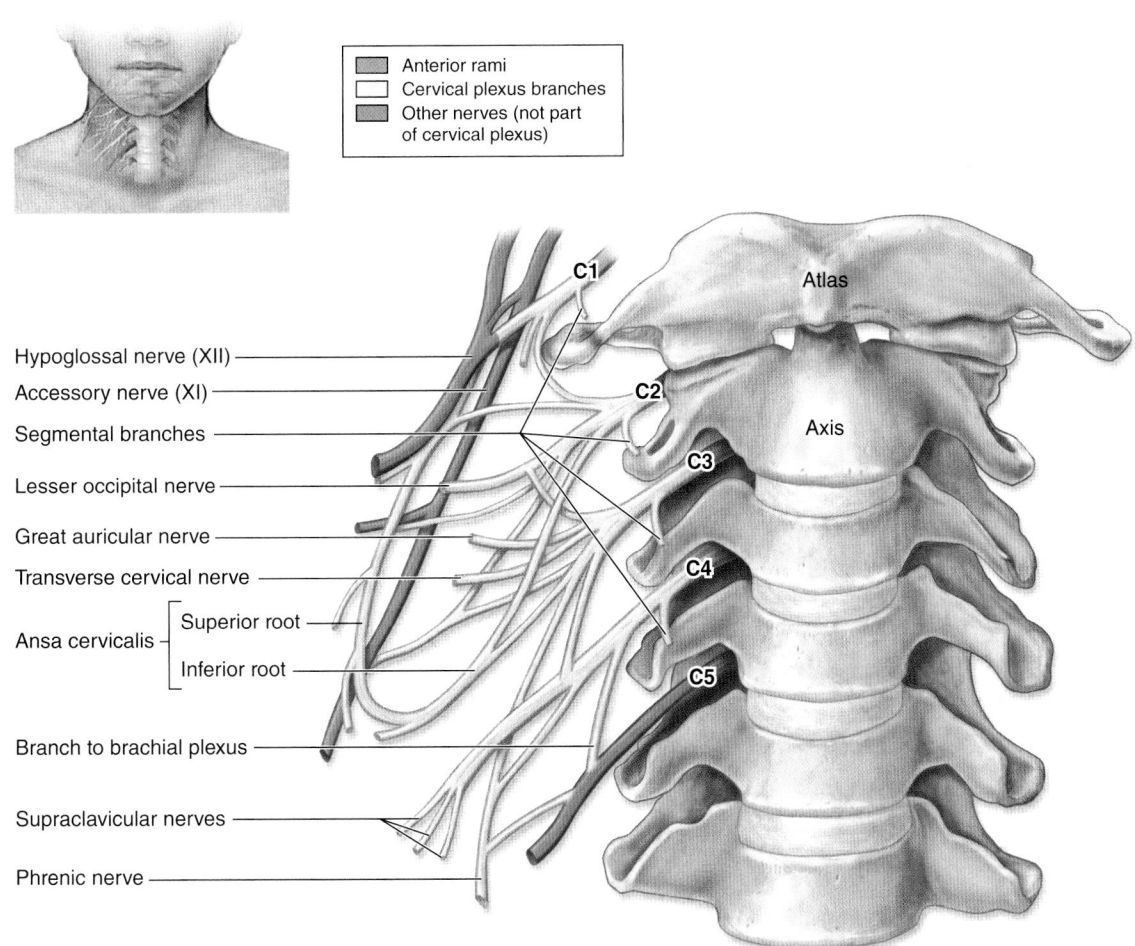

Figure 14.16 **Cervical Plexus.** Anterior rami of nerves C1–C4 form the cervical plexus, which innervates the skin and many muscles of the neck. AP|R

INTEGRATE

CONCEPT CONNECTION

Recall from Chapter 11 that the named skeletal muscles are innervated by specific spinal nerves. We are now learning the composition of these nerves in this chapter and which muscles they innervate. Should one or more of these spinal nerves become damaged, there will be a loss in sensory and motor function. Thus, if you recall the movements of a skeletal muscle and now you learn what nerve innervates that muscle, you may deduce the clinical symptoms a patient would have if one of these nerves is damaged.

Table 14.3	Branches of the Cervical Plexuses	
Nerves	**Anterior Rami**	**Innervation**
MOTOR BRANCHES		
Ansa cervicalis		Geniohyoid; infrahyoid muscles (omohyoid, sternohyoid, sternothyroid, and thyrohyoid)
Superior root	C1, C2	
Inferior root	C2, C3	
Segmental branches	C1–C4	Anterior and middle scalenes
CUTANEOUS BRANCHES		
Greater auricular	C2, C3	Skin on ear; connective tissue capsule covering parotid gland
Lesser occipital	C2	Skin of scalp superior and posterior to ear
Supraclavicular	C3, C4	Skin on superior part of chest and shoulder
Transverse cervical	C2, C3	Skin on anterior part of neck

Note: Although CN XII (hypoglossal) and the phrenic nerve (C3, C4, C5) travel with the nerves of the cervical plexus, hypoglossal and phrenic are not considered part of the plexus.

INTEGRATE

LEARNING STRATEGY

This mnemonic will help you remember the nerves that innervate the diaphragm: **C three, four, and five keep the diaphragm alive.**

One important branch of the cervical plexus is the **phrenic** (fren′ik; *phren* = diaphragm) **nerve**, which is formed primarily from the C4 nerve and some contributing axons from C3 and C5. The phrenic nerve travels through the thoracic cavity to innervate the thoracic diaphragm (see section 11.5).

WHAT DID YOU LEARN?

16 What is the action of the phrenic nerve?

14.5e Brachial Plexuses

LEARNING OBJECTIVES

8. Explain the structure of the brachial plexus, including the three trunks, two divisions, and three cords.

9. Describe the distribution of the five major nerve branches that arise from the three cords.

The left and right **brachial plexuses** are networks of nerves that supply the upper limb. Each brachial plexus is formed by the anterior rami of spinal nerves C5–T1 (**figure 14.17**). The components of the brachial plexus extend laterally from the neck, pass superior to the first rib, and then continue into the axilla. Each brachial plexus innervates the pectoral girdle and the entire upper limb of one side.

Structure of the Brachial Plexus

Structurally, each brachial plexus is more complex than a cervical plexus and is composed of anterior rami, trunks, divisions, and cords when examined from a medial to lateral perspective. The **anterior rami** (sometimes called *roots*) of the brachial plexus are simply the continuations of the anterior rami of spinal nerves C5–T1. These rami emerge through the intervertebral foramina and extend to the neck. The five rami unite in the posterior triangle of the neck to form the **superior, middle,** and **inferior trunks.** Nerves C5 and C6 unite to form the superior trunk; nerve C7 remains as the middle trunk; and nerves C8 and T1 unite to form the inferior trunk.

Portions of each trunk divide deep to the clavicle into an **anterior division** and a **posterior division** (shown in green and purple, respectively, in figure 14.7). These contain axons that primarily innervate the anterior and posterior parts of the upper limb, respectively.

Upon reaching the axilla, these anterior and posterior divisions converge to form three cords. They are named with respect to their position near the axillary artery:

- The **posterior cord** is posterior to the axillary artery and is formed by the posterior divisions of the superior, middle, and inferior trunks; therefore, it contains portions of C5–T1 nerves.

INTEGRATE

LEARNING STRATEGY

In general, nerves from the *anterior division* of the brachial plexus tend to innervate muscles that *flex* the parts of the upper limb. Nerves from the *posterior division* of the brachial plexus tend to innervate muscles that *extend* the parts of the upper limb.

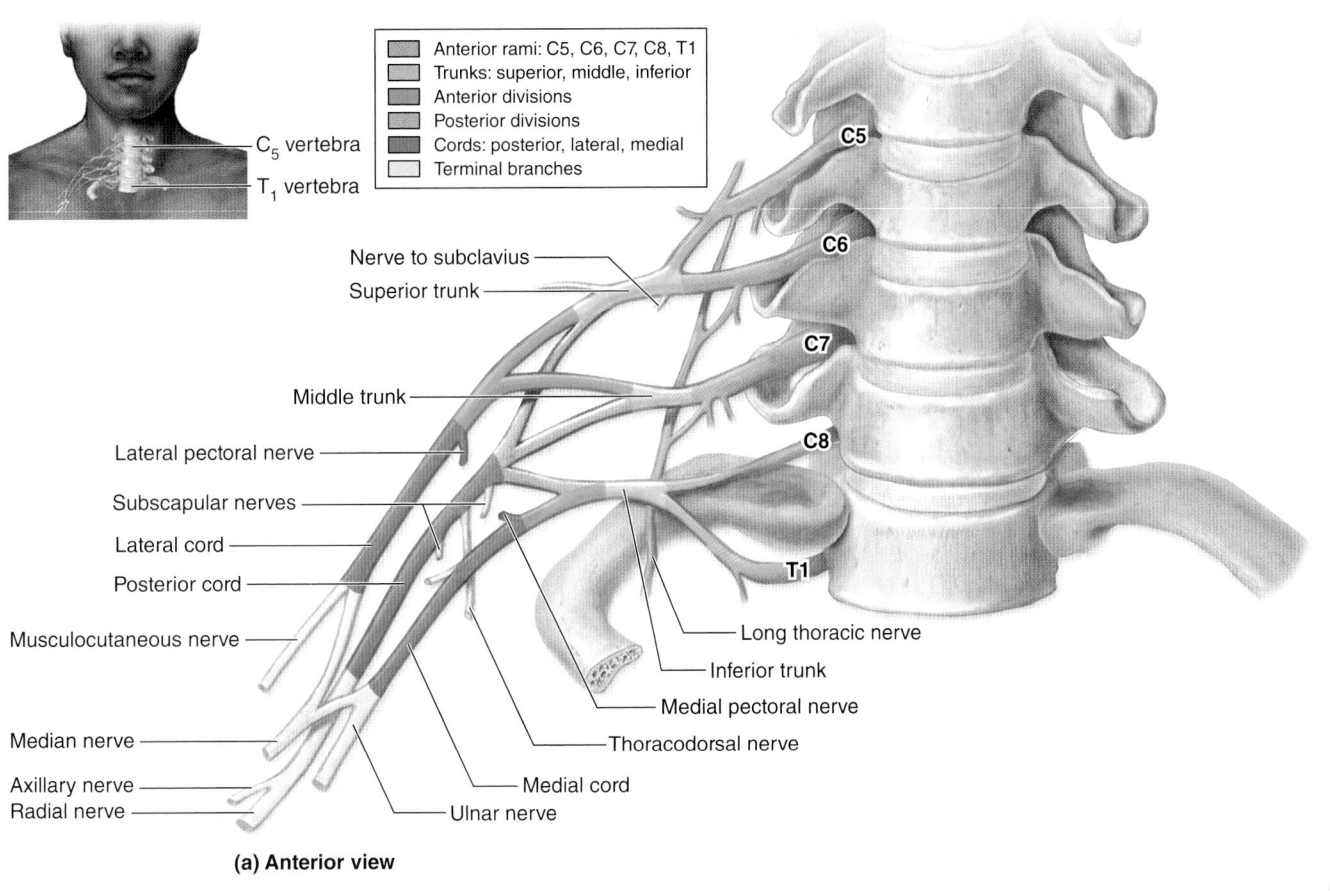

Anterior rami: C5, C6, C7, C8, T1
Trunks: superior, middle, inferior
Anterior divisions
Posterior divisions
Cords: posterior, lateral, medial
Terminal branches

C5 vertebra
T1 vertebra

C5
C6
C7
C8
T1

Nerve to subclavius
Superior trunk

Middle trunk

Lateral pectoral nerve
Subscapular nerves
Lateral cord
Posterior cord
Musculocutaneous nerve

Median nerve
Axillary nerve
Radial nerve

Long thoracic nerve
Inferior trunk
Medial pectoral nerve
Thoracodorsal nerve
Medial cord
Ulnar nerve

(a) Anterior view

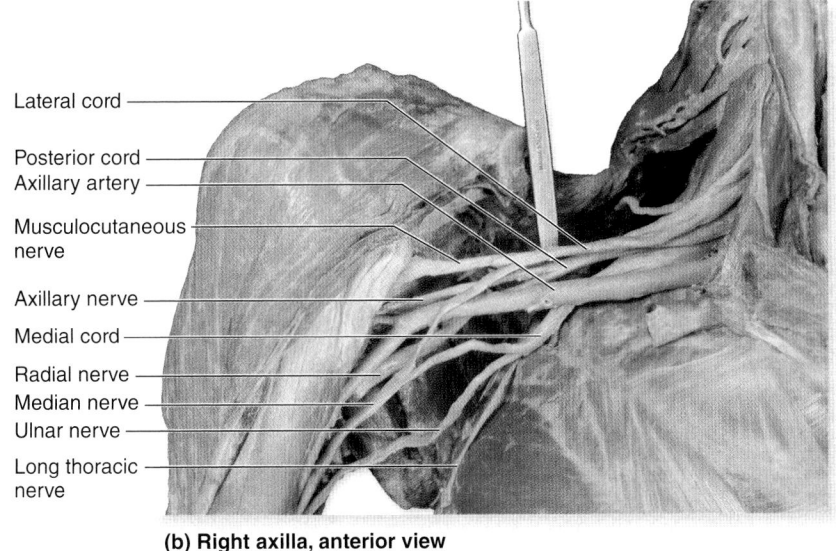

Lateral cord
Posterior cord
Axillary artery
Musculocutaneous nerve
Axillary nerve
Medial cord
Radial nerve
Median nerve
Ulnar nerve
Long thoracic nerve

(b) Right axilla, anterior view

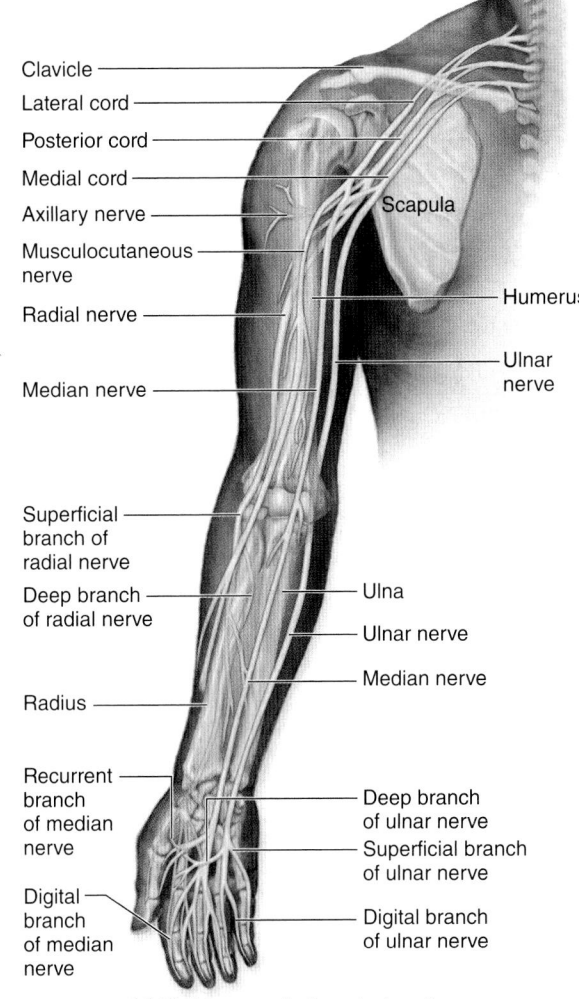

Clavicle
Lateral cord
Posterior cord
Medial cord
Axillary nerve
Musculocutaneous nerve
Radial nerve
Median nerve

Scapula
Humerus
Ulnar nerve

Superficial branch of radial nerve
Deep branch of radial nerve
Ulna
Ulnar nerve
Median nerve
Radius

Recurrent branch of median nerve
Digital branch of median nerve

Deep branch of ulnar nerve
Superficial branch of ulnar nerve
Digital branch of ulnar nerve

(c) Right upper limb, anterior view

Figure 14.17 Brachial Plexus. Anterior rami of nerves C5–T1 form the brachial plexus, which innervates the upper limb. (*a*) Rami, trunks, divisions, and cords form the subdivisions of this plexus. (*b*) A cadaver photo shows major nerves from the brachial plexus. (*c*) Complete pathways of main brachial plexus branches are shown in an anterior view of the right upper limb. **AP|R**

- The **medial cord** is medial to the axillary artery and is formed by the anterior division of the inferior trunk; it contains portions of nerves C8–T1.
- The **lateral cord** is lateral to the axillary artery and is formed from the anterior divisions of the superior and middle trunks; thus, it contains portions of nerves C5–C7.

Terminal Branches of the Brachial Plexus

Finally, five major **terminal branches** emerge from the three cords: the axillary nerve (from the posterior cord), median nerve (from the medial and lateral cords), musculocutaneous nerve (from the lateral cord), radial nerve (from the posterior cord), and ulnar nerve (from the medial cord). The nerves are compared in table 14.4.

The **axillary nerve** traverses through the axilla and posterior to the surgical neck of the humerus. The axillary nerve innervates both the deltoid and teres minor muscles (see section 11.8b). It receives sensory nerve signals from the superolateral part of the arm.

The **median nerve** travels along the midline of the arm and forearm, and deep to the carpal tunnel in the wrist. It innervates most of the anterior forearm muscles, the thenar muscles, and the lateral two lumbricals (see section 11.8e). It receives sensory nerve signals from the palmar side of the lateral 3½ fingers (thumb, index finger, middle finger, and the lateral one-half of the ring finger) and from the dorsal tips of these same fingers.

The **musculocutaneous** (mŭs′kū-lō-kū-tā′nē-ŭs) **nerve** innervates the anterior arm muscles (coracobrachialis, biceps brachii, and brachialis), which flex the humerus and/or flex the forearm (see sections 11.8b and c). It also receives sensory information from the lateral surface of the forearm.

The **radial nerve** travels along the posterior side of the arm and then along the radial side of the forearm. The radial nerve innervates the posterior arm muscles (forearm extensors) and the posterior forearm muscles (extensors of the wrist and digits, and the supinator of the forearm, see sections 11.8c and d). It receives sensory nerve signals from the posterior arm and forearm surface and the dorsolateral side of the hand.

The **ulnar nerve** descends along the medial side of the arm. It travels posterior to the medial epicondyle of the humerus and then extends along the ulnar side of the forearm. It innervates some of the anterior forearm muscles (the medial region of the flexor digitorum profundus and all of the flexor carpi ulnaris). It also innervates most of the intrinsic hand muscles, including the hypothenar muscles, the palmar and dorsal interossei, and the medial two lumbricals (see section 11.8e). It receives sensory input from the skin of the dorsal and palmar aspects of the medial 1½ fingers (the pinky finger and the medial half of the ring finger).

The brachial plexus also gives off numerous other nerves that innervate portions of the upper limb and pectoral girdle. These branches are not as large as the terminal branches (table 14.4).

WHAT DID YOU LEARN?

17 What spinal nerves typically compose the brachial plexus?

18 Which nerve might you have damaged if you have difficulty abducting your arm, and have anesthesia (lack of sensation) along the superolateral arm?

19 How do the ulnar nerve and radial nerve compare, with respect to motor and cutaneous innervation?

Table 14.4	Branches of the Brachial Plexus			
Terminal Branch		**Anterior Rami**	**Motor Innervation**	**Cutaneous Innervation**
Axillary Nerve Formed from posterior cord, posterior division of the brachial plexus		C5, C6	**Deltoid** (*arm abductor*) **Teres minor** (*lateral rotator of arm*)	Superolateral arm

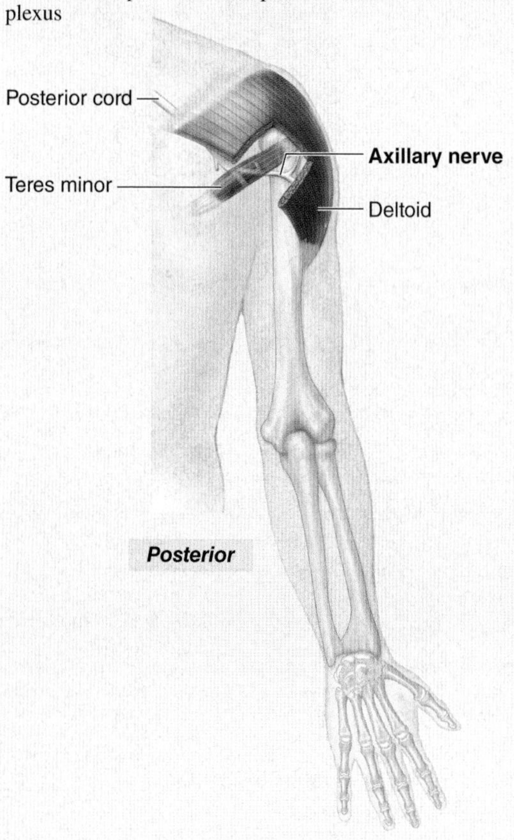

Posterior cord — Axillary nerve

Teres minor — Deltoid

Posterior

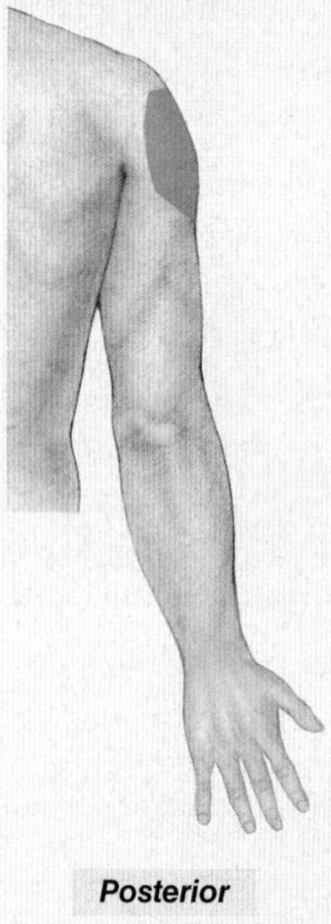

Posterior

Table 14.4	Branches of the Brachial Plexus (continued)

Terminal Branch	Anterior Rami	Motor Innervation	Cutaneous Innervation
Median Nerve Formed from medial and lateral cords, anterior division of the brachial plexus 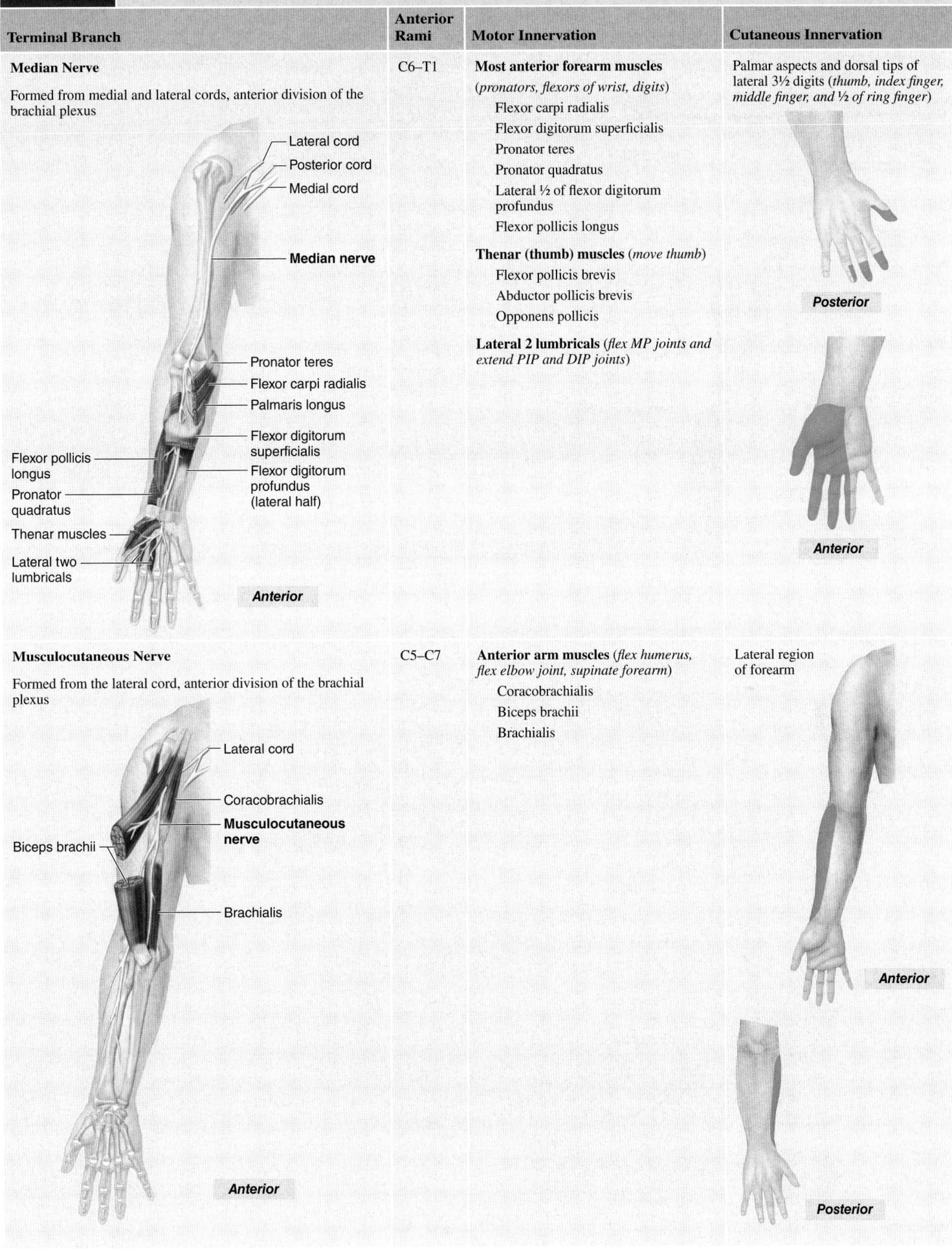	C6–T1	**Most anterior forearm muscles** (*pronators, flexors of wrist, digits*) 　Flexor carpi radialis 　Flexor digitorum superficialis 　Pronator teres 　Pronator quadratus 　Lateral ½ of flexor digitorum profundus 　Flexor pollicis longus **Thenar (thumb) muscles** (*move thumb*) 　Flexor pollicis brevis 　Abductor pollicis brevis 　Opponens pollicis **Lateral 2 lumbricals** (*flex MP joints and extend PIP and DIP joints*)	Palmar aspects and dorsal tips of lateral 3½ digits (*thumb, index finger, middle finger, and ½ of ring finger*)
Musculocutaneous Nerve Formed from the lateral cord, anterior division of the brachial plexus	C5–C7	**Anterior arm muscles** (*flex humerus, flex elbow joint, supinate forearm*) 　Coracobrachialis 　Biceps brachii 　Brachialis	Lateral region of forearm

(continued on next page)

Table 14.4   Branches of the Brachial Plexus (continued)

Terminal Branch	Anterior Rami	Motor Innervation	Cutaneous Innervation

Radial Nerve
 	
Formed from the posterior cord, posterior division of the brachial plexus
 	
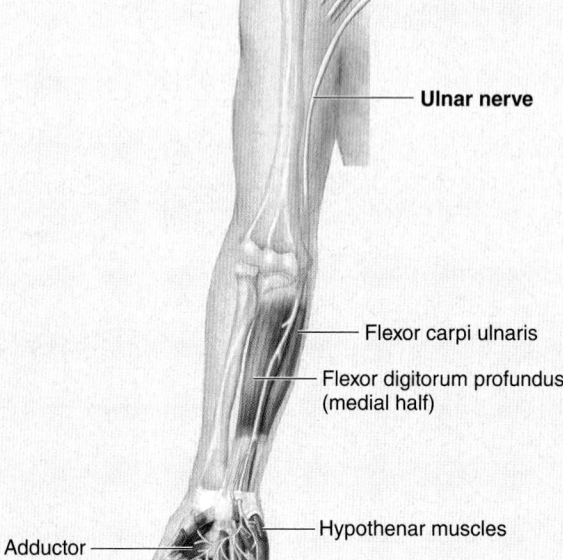
	C5–T1	**Posterior arm muscles** (*extend forearm*)
		  Triceps brachii
		  Anconeus
		Posterior forearm muscles (*supinate forearm, extend wrist, digits, 1 muscle that abducts thumb*)
		  Supinator
		  Extensor carpi radialis muscles
		  Extensor digitorum
		  Extensor carpi ulnaris
		  Extensor pollicis longus
		  Extensor pollicis brevis
		  Abductor pollicis longus
		  Extensor digiti minimi
		  Extensor indicis
		Brachioradialis (*flexes forearm*)
			Posterior region of arm
			Posterior region of forearm
			Dorsal aspect of lateral 3 digits (except their distal tips)
			
			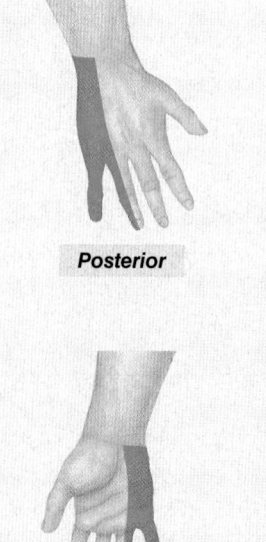

Ulnar Nerve
 	
Formed from the medial cord, anterior division of the brachial plexus
 	
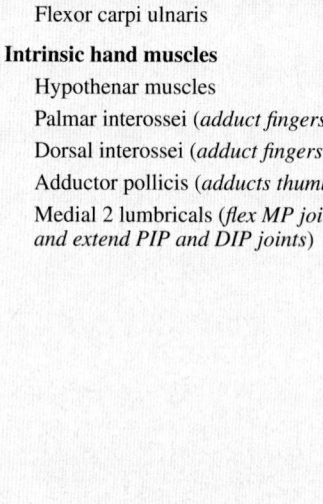
	C8–T1	**Anterior forearm muscles** (*flexors of wrist and digits*)
		  Medial ½ of flexor digitorum profundus
		  Flexor carpi ulnaris
		Intrinsic hand muscles
		  Hypothenar muscles
		  Palmar interossei (*adduct fingers*)
		  Dorsal interossei (*adduct fingers*)
		  Adductor pollicis (*adducts thumb*)
		  Medial 2 lumbricals (*flex MP joints and extend PIP and DIP joints*)
			Dorsal and palmar aspects of medial 1½ digits (*little finger, medial aspect of ring finger*)

Table 14.4	Branches of the Brachial Plexus (continued)			
Smaller Branches of the Brachial Plexus	**Anterior Rami**	**Motor Innervation**	**Cutaneous Innervation**	
Dorsal scapular	C5	Rhomboids, levator scapulae		
Long thoracic	C5–C7	Serratus anterior		
Lateral pectoral	C5–C7	Pectoralis major		
Medial pectoral	C8–T1	Pectoralis major Pectoralis minor		
Medial cutaneous nerve of arm	C8–T1		Medial side of arm	
Medial cutaneous nerve of forearm	C8–T1		Medial side of forearm	
Nerve to subclavius	C5–C6	Subclavius		
Suprascapular	C5–C6	Supraspinatus, infraspinatus		
Subscapular nerves	C5–C6	Subscapularis, teres major		
Thoracodorsal (nerve to latissimus dorsi)	C6–C8	Latissimus dorsi		

INTEGRATE

CLINICAL VIEW

Brachial Plexus Injuries

Injuries to parts of the brachial plexus are fairly common, especially in individuals aged 15–25. Minor plexus injuries may be treated by simply resting the limb. More severe brachial plexus injuries may require nerve grafts or nerve transfers, and for very severe injuries, no effective treatment exists.

Axillary Nerve Injury

The axillary nerve can be compressed within the axilla, or it can be damaged if the surgical neck of the humerus is broken (recall that the axillary nerve travels posterior to the surgical neck of the humerus). A patient whose axillary nerve is damaged has great difficulty abducting the arm due to paralysis of the deltoid muscle, as well as anesthesia (loss of sensation) along the superolateral skin of the arm.

Radial Nerve Injury

The radial nerve is especially subject to injury during humeral shaft fractures or in injuries to the lateral elbow. Nerve damage results in paralysis of the extensor muscles of the forearm, wrist, and fingers. A common clinical sign of radial nerve injury is *wrist drop*, where the patient is unable to extend his or her wrist. The patient also experiences anesthesia along the posterior arm, the forearm, and the part of the hand normally supplied by this nerve.

Posterior Cord Injury

The posterior cord of the brachial plexus (which includes the axillary and radial nerves) may be injured by improper use of crutches, a condition called *crutch palsy*. Similarly, the posterior cord can also be compressed if a person drapes the upper limb over the back of a chair for an extended period of time. Because this can happen if someone passes out in a drunken stupor, this condition is also referred to as *drunkard's paralysis*.

Median Nerve Injury

The median nerve may be impinged or compressed as a result of carpal tunnel syndrome (see Clinical View: "Carpal Tunnel Syndrome," in section 11.8d)

or by any deep laceration of the wrist. Median nerve injury often results in paralysis of the thenar group of muscles. The classic sign of median nerve injury is the *ape hand deformity*, which develops over time as the thenar eminence wastes away until the hand eventually resembles that of an ape (apes lack well-developed thumb muscles). The lateral two lumbricals are also paralyzed, and sensation is lost in the part of the hand supplied by the median nerve.

Ulnar Nerve Injury

The ulnar nerve may be injured by fractures or dislocations of the elbow because of this nerve's close proximity to the medial epicondyle of the humerus. When you "hit your funny bone," you actually have hit your ulnar nerve. Most of the intrinsic hand muscles are paralyzed, so the person is unable to adduct or abduct the fingers. In addition, the person experiences sensory loss along the medial side of the hand. A clinician can test for ulnar nerve injury by having a patient hold a piece of paper tightly between the fingers as the doctor tries to pull it away. If the person has weak or paralyzed interossei muscles, the paper can be easily extracted.

Superior Trunk Injury

The superior trunk of the brachial plexus can be injured by excessive separation of the neck and shoulder, as when a person riding a motorcycle is flipped from the bike and lands on the side of the head. A superior trunk injury affects the C5 and C6 anterior rami, so any brachial plexus branch that has these nerves is also affected to some degree.

Inferior Trunk Injury

The inferior trunk of the brachial plexus can be injured if the arm is excessively abducted, as when a neonate's arm is pulled too hard during delivery. In children and adults, inferior trunk injuries happen when grasping something above the head in order to break a fall—for example, grabbing a branch to keep from falling out of a tree. An inferior trunk injury involves the C8 and T1 anterior rami, so any brachial plexus branch that is formed from these nerves (such as the ulnar nerve) also is affected to some degree.

14.5f Lumbar Plexuses

LEARNING OBJECTIVES

10. Identify the spinal nerves that make up the lumbar plexus.

11. Compare and contrast the femoral and obturator nerve composition and distribution.

The left and right **lumbar plexuses** are formed from the anterior rami of spinal nerves L1–L4 located lateral to the L_1–L_4 vertebrae and along the psoas major muscle in the posterior abdominal wall (**figure 14.18**). This plexus innervates the inferior abdominal wall, anterior thigh, medial thigh, and the skin of the medial leg. The lumbar plexus is structurally less complex than the brachial plexus. However, like the brachial plexus, the lumbar plexus is also subdivided into an anterior division and a posterior division. The primary nerves of the lumbar plexus are listed in **table 14.5**.

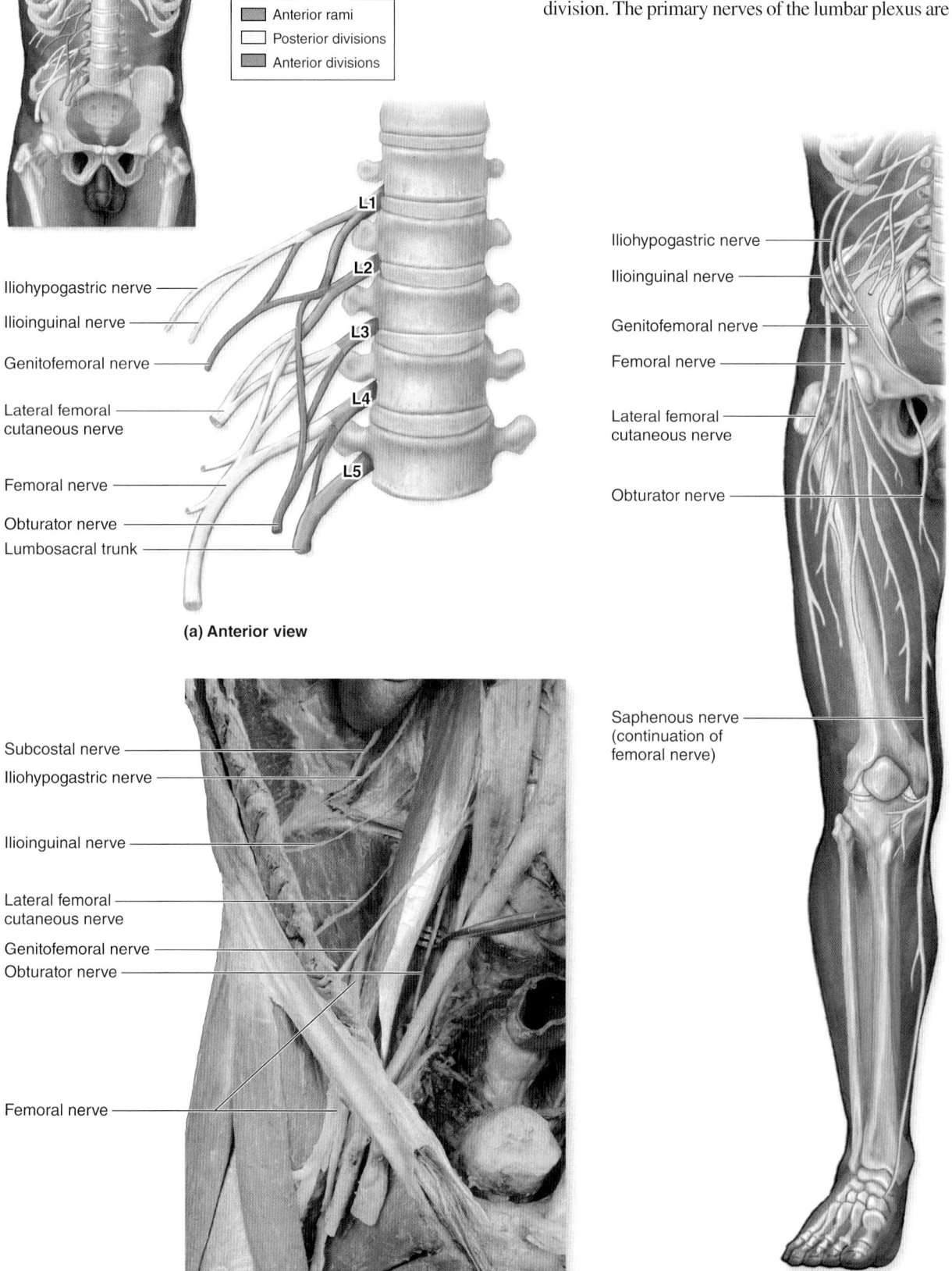

Anterior rami
Posterior divisions
Anterior divisions

Iliohypogastric nerve
Ilioinguinal nerve
Genitofemoral nerve
Lateral femoral cutaneous nerve
Femoral nerve
Obturator nerve
Lumbosacral trunk

L1
L2
L3
L4
L5

(a) Anterior view

Subcostal nerve
Iliohypogastric nerve
Ilioinguinal nerve
Lateral femoral cutaneous nerve
Genitofemoral nerve
Obturator nerve
Femoral nerve

(b) Right pelvic region, anterior view

Iliohypogastric nerve
Ilioinguinal nerve
Genitofemoral nerve
Femoral nerve
Lateral femoral cutaneous nerve
Obturator nerve

Saphenous nerve (continuation of femoral nerve)

(c) Right lower limb, anterior view

Figure 14.18 Lumbar Plexus. Anterior rami of nerves L1–L4 form the lumbar plexus. (*a*) The lumbar plexus may be subdivided into anterior divisions and posterior divisions, each of which form major nerves. (*b*) Cadaver photo shows the major nerves of the lumbar plexus. (*c*) An anterior view of the right lower limb shows the distribution of nerves of the lumbar plexus. AP|R

Table 14.5   Branches of the Lumbar Plexus

Main Branch		Anterior Rami	Motor Innervation	Cutaneous Innervation

Femoral Nerve 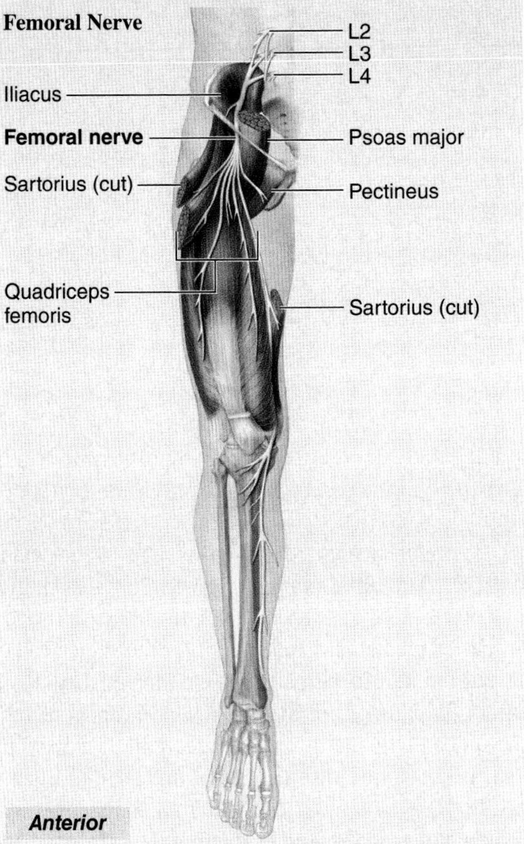
 	
Iliacus
Femoral nerve
Sartorius (cut)
Quadriceps femoris
 	
L2
L3
L4
Psoas major
Pectineus
Sartorius (cut)
 	
Anterior
		L2–L4	**Anterior thigh muscles**
			  Quadriceps femoris (*knee extensor*)
			  Iliopsoas (*hip flexor*)
			  Sartorius (*hip and knee flexor*)
			  Pectineus[1] (*hip flexor*)
				Anterior thigh
				Inferomedial thigh
				Medial side of leg
				Most medial aspect of foot 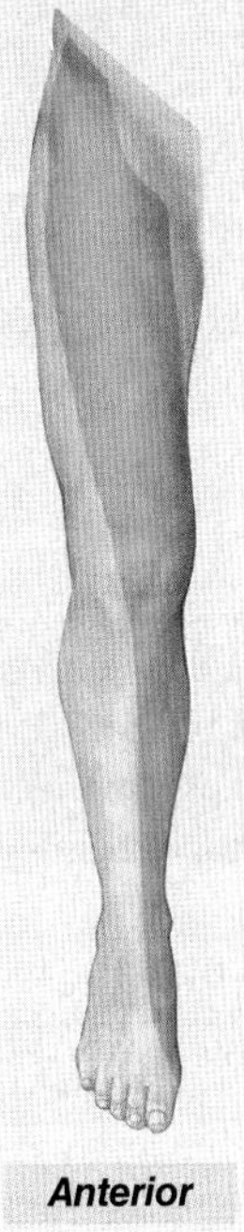
				
				Anterior

Obturator Nerve 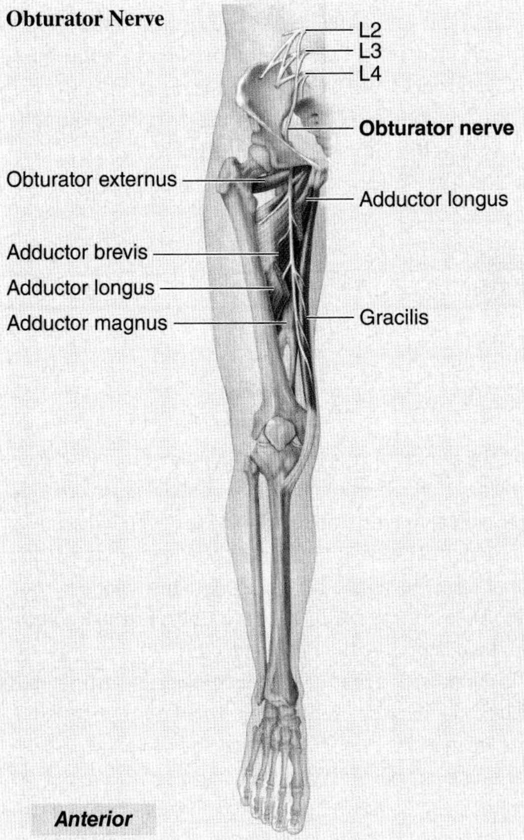
 	
Obturator externus
Adductor brevis
Adductor longus
Adductor magnus
 	
L2
L3
L4
Obturator nerve
Adductor longus
Gracilis
 	
Anterior
		L2–L4	**Medial thigh muscles** (*adductors of thigh*)
			  Adductors
			  Gracilis
			  Pectineus[1]
			  Obturator externus (*lateral rotator of thigh*)
				Superomedial thigh 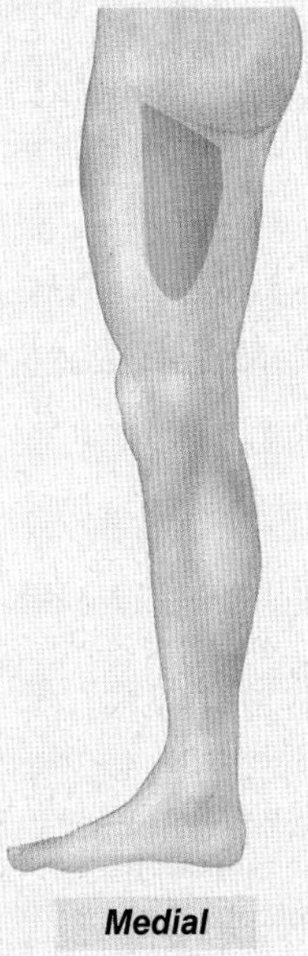
				
				Medial

1. Pectineus may be innervated by the femoral nerve, obturator nerve, or branches from both nerves.

(continued on next page)

Table 14.5	Branches of the Lumbar Plexus (continued)		
Smaller Branches of the Lumbar Plexus	Anterior Rami	Motor Innervation	Cutaneous Innervation
Iliohypogastric	L1	Partial innervation to abdominal muscles (*flex vertebral column*)	Superior lateral gluteal region Inferior abdominal wall
Ilioinguinal	L1	Partial innervation to abdominal muscles (*flex vertebral column*)	Inferior abdominal wall Scrotum (males) or labia majora (females)
Genitofemoral	L1, L2		Small area in anterior superior thigh Scrotum (males) or labia majora (females)
Lateral femoral cutaneous	L2, L3		Anterolateral thigh

The main nerve of the posterior division of the lumbar plexus is the **femoral nerve.** This nerve innervates the anterior thigh muscles, such as the quadriceps femoris (knee extensor) and the sartorius, psoas, and iliacus (hip flexors, see sections 11.9a, b). It also receives sensory input from the anterior and inferomedial thigh as well as the medial aspect of the leg.

The main nerve of the anterior division is the **obturator nerve,** which travels through the obturator foramen of the os coxae to the medial thigh. There, the nerve innervates the medial thigh muscles (which adduct the thigh, see section 11.9a) and receives sensory input from the superomedial skin of the thigh. Smaller branches of each lumbar plexus innervate the abdominal wall, portions of the external genitalia, and the inferior portions of the abdominal muscles (table 14.5; see also section 11.6).

 WHAT DID YOU LEARN?

 20 Which nerve of the lumbar plexus might you have damaged if you have difficulty extending your knee?

14.5g Sacral Plexuses

 LEARNING OBJECTIVES

12. List the spinal nerves that form the sacral plexus.
13. Describe the composition of the sciatic nerve and compare its branches.

The left and right **sacral plexuses** are formed from the anterior rami of spinal nerves L4–S4 and are located immediately inferior to the lumbar plexuses (**figure 14.19**). The lumbar and sacral plexuses are sometimes considered together as the *lumbosacral plexus.* The nerves emerging from a sacral plexus innervate the gluteal region, pelvis, perineum, posterior thigh, and almost all of the leg and foot.

The anterior rami of the sacral plexus are organized into an anterior division and a posterior division. The nerves that are formed from the anterior division tend to innervate muscles that flex (or plantar flex) parts of the lower limb, whereas the posterior division nerves tend to innervate muscles that extend (or dorsiflex) part of the lower limb. **Table 14.6** lists the main and smaller nerves of the sacral plexus.

The **sciatic** (sī-at′ik) **nerve,** also known as the *ischiadic* (is-kē-at′ik; hip joint) *nerve,* is the largest and longest nerve in the body. It is formed from portions of both the anterior and posterior divisions of the sacral plexus. This nerve projects from the pelvis through the greater sciatic notch of the os coxae and extends into the posterior region of the thigh. The sciatic nerve is actually composed of two divisions—the *tibial division* and the *common fibular division*—wrapped in a common sheath.

Just superior to the popliteal fossa, the two divisions of the sciatic nerve split into two nerves. The **tibial nerve** is formed from the anterior divisions of the sciatic nerve. In the posterior thigh, the tibial division of the sciatic nerve innervates the hamstrings (except for the short head of the biceps femoris) and the hamstring part of the adductor magnus. It travels in the posterior compartment of the leg, where it innervates the plantar flexors of the foot and the toe flexors (see sections 11.9c and d). In the foot, the tibial nerve splits into the lateral and medial plantar nerves, which innervate the plantar muscles of the foot and receive sensory input from the skin covering the sole of the foot.

The **common fibular** (*common peroneal*) **nerve** is formed from the posterior division of the sciatic nerve. As the common fibular division of the sciatic nerve, it innervates the short head of the biceps femoris muscle (see section 11.9b). Along the lateral knee, as it wraps around the neck of the fibula, this nerve splits into two main branches: the deep fibular nerve and the superficial fibular nerve.

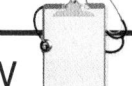

INTEGRATE

CLINICAL VIEW

Sacral Plexus Injuries

Some branches of the sacral plexus are readily subject to injury. For example, a poorly placed gluteal intramuscular injection can injure the superior or inferior gluteal nerves, and in some cases even the sciatic nerve. Additionally, a herniated intervertebral disc may impinge on the nerve branches that form the sciatic nerve. Injury to the sciatic nerve produces a condition known as **sciatica** (sī-at′i-kă), which is characterized by pain down the posterior of the thigh and leg.

The common fibular nerve is especially prone to injury due to fracture of the neck of the fibula or compression from a leg cast that is too tight. The anterior and lateral leg muscles may be paralyzed and leave the person unable to dorsiflex and evert the foot. One classic sign of fibular nerve injury is *foot drop.* Because the person cannot dorsiflex the foot to walk normally, he or she compensates by flexing the hip to lift the affected area and keep from tripping or stubbing the toes.

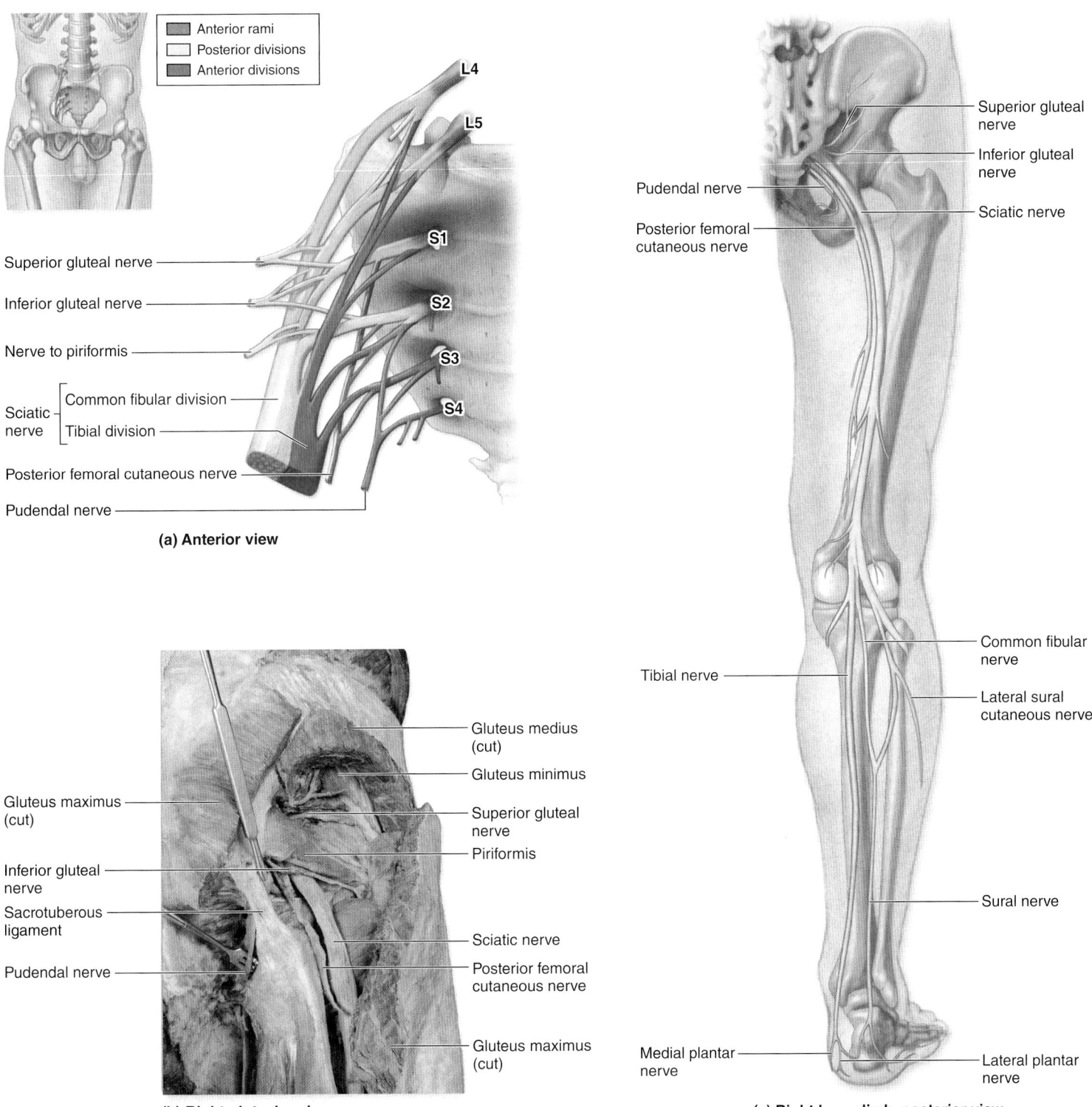

Figure 14.19 Sacral Plexus. Anterior rami of nerves L4, L5, and S1–S4 form the sacral plexus. (*a*) The sacral plexus has six roots and both anterior and posterior divisions. (*b*) Cadaver photo reveals the major sacral plexus nerves of the right gluteal region. (*c*) A posterior view shows the distribution of the major nerves of the sacral plexus.

The **deep fibular** (*deep peroneal*) **nerve** travels in the anterior compartment of the leg and terminates between the first and second toes. It innervates the anterior leg muscles (which dorsiflex the foot and extend the toes) and the muscles on the dorsum of the foot (which extend the toes, see sections 11.9c and d). In addition, this nerve receives sensory input from the skin between the first and second toes on the dorsum of the foot.

The **superficial fibular** (*superficial peroneal*) **nerve** travels in the lateral compartment of the leg. Just proximal to the ankle, this nerve

becomes superficial along the anterior part of the ankle and dorsum of the foot. The superficial fibular nerve innervates the lateral compartment muscles of the leg (foot evertors and weak plantar flexors, see sections 11.9c and d). It also receives sensory input from most of the dorsal surface of the foot and the anteroinferior part of the leg.

WHAT DID YOU LEARN?

21 What anterior rami form the sacral nerve plexus, and what general areas of the body does it innervate?

| Table 14.6 | Branches of the Sacral Plexus |

Main Branch	Anterior Rami	Motor Innervation	Cutaneous Innervation
Sciatic Nerve (Composed of tibial and common fibular divisions wrapped in a common sheath)	L4–S3	(See tibial and common fibular nerves)	(See tibial and common fibular nerves)

Tibial Nerve
 	
 L4
 L5
 S1
 S2
 S3
 Tibial division of sciatic nerve
 Biceps femoris (long head)
 Adductor magnus
 Semitendinosus
 Semimembranosus
 Tibial nerve
 Gastrocnemius
 Popliteus
 Soleus
 Tibialis posterior
 Flexor digitorum longus
 Flexor hallucis longus
 Medial plantar nerve
 Lateral plantar nerve
 Posterior
	L4–S3	**Posterior thigh muscles** (*extend thigh and flex leg*)
		   Long head of biceps femoris
		   Semimembranosus
		   Semitendinosus
		   Part of adductor magnus
		 Posterior leg muscles (*plantar flexors of foot, flexors of knee*)
		   Flexor digitorum longus
		   Flexor hallucis longus
		   Gastrocnemius
		   Soleus
		   Popliteus
		   Tibialis posterior (*inverts foot*)
		 Plantar foot muscles (*via medial and lateral plantar nerve branches*)
			Branches to the heel, and via its medial and lateral plantar nerve branches (which supply the sole of the foot)
			
			 Tibial nerve
			 Medial plantar nerve
			 Lateral plantar nerve
			 Plantar

Common Fibular Nerve (Divides into deep fibular and superficial fibular branches) L4 L5 S1 S2 S3 Common fibular division of sciatic nerve Biceps femoris short head **Common fibular nerve** Fibularis longus Fibularis brevis Tibialis anterior **Superficial fibular nerve** **Deep fibular nerve** Extensor digitorum longus Extensor hallucis longus Fibularis tertius Extensor digitorum brevis Extensor hallucis brevis **Anterior**	L4–S2	**Short head of biceps femoris** (*knee flexor*); see also deep fibular and superficial fibular nerves	(See deep fibular and superficial fibular nerves)

Table 14.6   Branches of the Sacral Plexus (continued)

Main Branch	Anterior Rami	Motor Innervation	Cutaneous Innervation

Deep Fibular Nerve
 	
Common fibular nerve
Superficial fibular nerve
Extensor digitorum longus
Fibularis tertius
Extensor digitorum brevis
Tibialis anterior
Deep fibular nerve
Extensor hallucis longus
Extensor hallucis brevis
Anterior
	L4–S1	**Anterior leg muscles** (*dorsiflex foot, extend toes*)
		   Tibialis anterior (*inverts foot*)
		   Extensor hallucis longus
		   Extensor digitorum longus
		   Fibularis tertius
		Dorsum foot muscles (*extend toes*)
		   Extensor hallucis brevis
		   Extensor digitorum brevis
			Dorsal interspace between first and second toes
			
			Anterior

Superficial Fibular Nerve
 	
Common fibular nerve
Fibularis longus
Fibularis brevis
Superficial fibular nerve
Anterior
	L5–S2	**Lateral leg muscles** (*foot evertors and plantar flexors*)
		   Fibularis longus
		   Fibularis brevis
			Anteroinferior part of leg; most of dorsum of foot
			
			Anterior

Smaller Branches of the Sacral Plexus	Anterior Rami	Motor Innervation	Cutaneous Innervation
Inferior gluteal nerve	L5–S2	Gluteus maximus (*thigh extensor*)	
Superior gluteal nerve	L4–S1	Gluteus medius, gluteus minimus, and tensor fasciae latae (*abductors of thigh*)	
Posterior femoral cutaneous nerve	S1–S3		Skin on posterior thigh
Pudendal nerve	S2–S4	Muscles of perineum, external anal sphincter, external urethral sphincter	Skin on external genitalia

14.6 Reflexes

Here we consider the characteristics of a reflex, the components of a reflex arc, the classification of reflexes, the different spinal reflexes, and how reflexes are tested in a clinical setting.

14.6a Characteristics of Reflexes

LEARNING OBJECTIVES

1. Describe the properties of a reflex.
2. Explain how reflexes are classified.

Reflexes are rapid, preprogrammed, involuntary responses of muscles or glands to a stimulus. An example of a reflex occurs when you accidentally touch a hot burner on a stove. Instantly and automatically, you remove your hand from the stimulus (the hot burner), even before you are completely aware that your hand was touching something extremely hot.

All reflexes have similar properties:

- A *stimulus* is required to initiate a reflex.
- A *rapid response* requires that few neurons are involved and synaptic delay is minimal.
- A *preprogrammed response* occurs the same way every time.
- An *involuntary response* requires no conscious intent or preawareness of the reflex activity. Thus, reflexes are usually not suppressed.

A reflex is a survival mechanism; it allows us to quickly respond to a stimulus that may be detrimental to our well-being without having to wait for the brain to process the information. Awareness of the stimulus occurs after the reflex action has been completed, in time to correct or avoid a potentially dangerous situation. (This is possible because sensory input has reached the cerebral cortex.)

WHAT DID YOU LEARN?

㉒ What are the four main properties of a reflex?

14.6b Components of a Reflex Arc

LEARNING OBJECTIVE

3. List the structures involved in a reflex arc and the steps in its action.

A **reflex arc** includes a sensory receptor, an effector, and the neural "wiring" between the two. It always begins at a receptor in the PNS, communicates with the CNS, and ends at a peripheral effector, either a muscle or gland. The number of intermediate steps varies, depending upon the complexity of the reflex. Generally, five steps are involved in a reflex, as illustrated in **figure 14.20** and described here:

① **A stimulus activates a sensory receptor.** A sensory receptor (dendritic endings of a sensory neuron or specialized receptor cells) responds to external and internal stimuli, such as temperature, pressure, or tactile changes. Proprioceptors are sensory receptors found in muscles and tendons, and a stimulus to a proprioceptor (such as the tapping of tendon) may initiate a reflex as well.

② **The sensory neuron transmits a nerve signal to the CNS.** A sensory neuron transmits a nerve signal from the receptor to the spinal cord (or brain).

③ **Information from the nerve signal is processed in the integration center by interneurons.** More complex reflexes may use a number of interneurons within the CNS to integrate and process incoming sensory nerve signals and transmit information to the motor neuron.

The simplest reflexes do not involve interneurons; rather, the sensory neuron synapses directly on a motor neuron in the CNS.

④ **The motor neuron transmits a nerve signal from the CNS to an effector.** The nerve signal is transmitted through the anterior root of a motor neuron and then the spinal nerve to the peripheral effector organ—a gland or muscle.

⑤ **The effector responds to the nerve signal from the motor neuron.** An effector is a muscle or a gland that responds to the nerve signal from the motor neuron. This response is intended to counteract or remove the original stimulus.

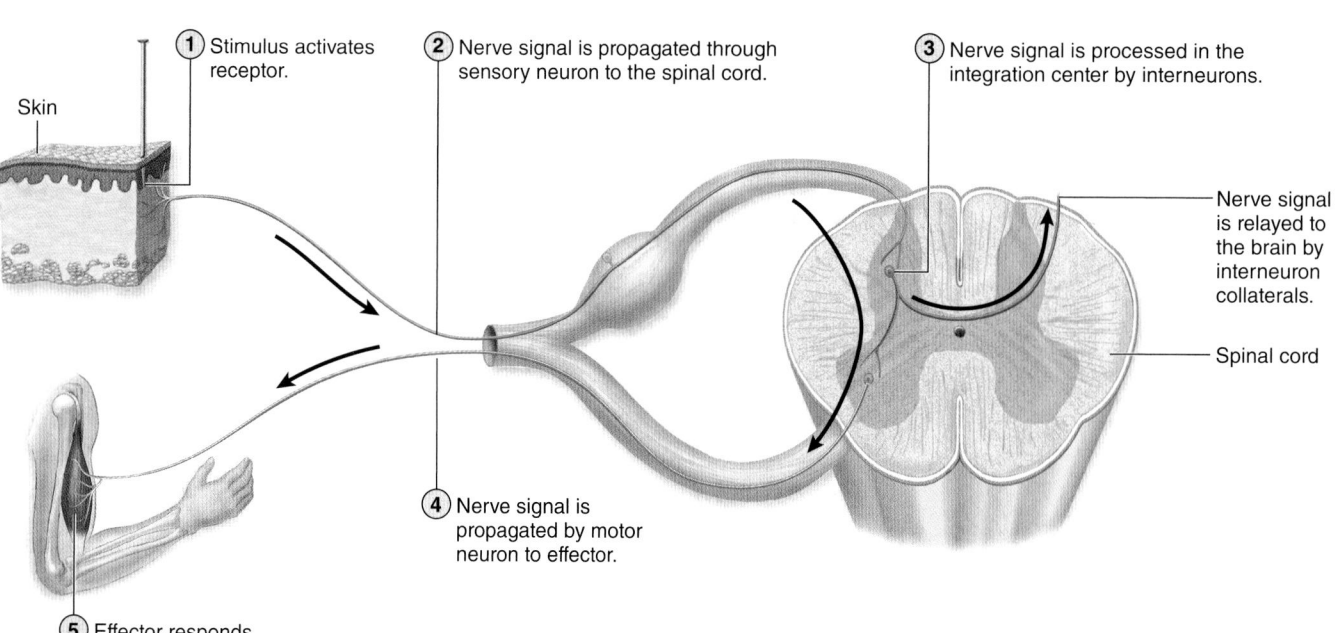

① Stimulus activates receptor.

Skin

② Nerve signal is propagated through sensory neuron to the spinal cord.

③ Nerve signal is processed in the integration center by interneurons.

Nerve signal is relayed to the brain by interneuron collaterals.

Spinal cord

④ Nerve signal is propagated by motor neuron to effector.

⑤ Effector responds.

Figure 14.20 Simple Reflex Arcs. A reflex arc is a neural pathway composed of neurons that control rapid, subconscious, preprogrammed responses to a stimulus. AP|R

4.6c Classifying Spinal Reflexes

LEARNING OBJECTIVE

4. Explain the five ways a reflex may be classified.

The specific components (or attributes) of a reflex can vary. Some reflexes involve the spinal cord, whereas others involve the brain. Some involve skeletal muscle and some involve other muscle types or glands. Some have only two neurons and some have more. We describe five different ways of classifying a reflex:

- **Spinal reflex or cranial reflex.** A reflex may be identified by the specific area of the central nervous system (integration center) that serves as the processing site. *Spinal reflexes* involve the spinal cord, whereas *cranial reflexes* involve the brain.

- **Somatic reflex or visceral reflex.** This classification criterion is determined by the type of effector that is stimulated by the motor neurons involved in the reflex. *Somatic reflexes* involve skeletal muscle as the effector. *Visceral (or autonomic) reflexes* involve cardiac muscle, smooth muscle, or a gland as the effector.

- **Monosynaptic reflex or polysynaptic reflex.** A reflex may also be classified by the number of neurons participating in the reflex. A *monosynaptic* (mon′ō-si-nap′tik; *monos* = single) *reflex* has only a sensory neuron and motor neuron (figure 14.21). The sensory axon synapses directly on the motor neuron, whose axon projects to the effector. Thus, there is only one synapse. Monosynaptic reflexes are the simplest, and they are also the most rapid. With only one synaptic delay, the response is very prompt. A *polysynaptic* (pol′ē-si-nap′tik; *polys* = many) *reflex* has one (or more) interneurons positioned between the sensory and motor neuron. These reflex arcs are more complicated and not as rapid.

- **Ipsilateral reflex or contralateral reflex.** The reflex may also be classified based upon whether it involves only one side of the body. An *ipsilateral reflex* is a reflex in which both the receptor and effector organs are on the same side of the spinal cord. A *contralateral reflex* is a reflex that involves an effector on the opposite side of the body of the receptor that detected the stimulus. Note that this terminology is only applicable to reflexes that involve the limbs. For example, an ipsilateral effect occurs when the muscles in your left arm contract to pull your left hand away from a hot object. In comparison, a contralateral effect occurs when you step on a sharp object with your left foot and then contract the muscles in your right leg to maintain balance as you withdraw your left leg from the damaging object.

- **Innate reflex or acquired reflex.** The reflex may be classified based upon whether you are born with it. An *innate reflex* is a reflex that you are born with, whereas an *acquired reflex* is one that is developed after birth.

As you read about the reflexes described in this section, see if you can determine how each would be classified based upon these five criteria. Keep in mind that there may be instances where you do not have enough information to classify the reflex in all five ways.

WHAT DID YOU LEARN?

23 What are the five steps involved in action of a reflex?

24 What is the major difference between monosynaptic and polysynaptic reflexes?

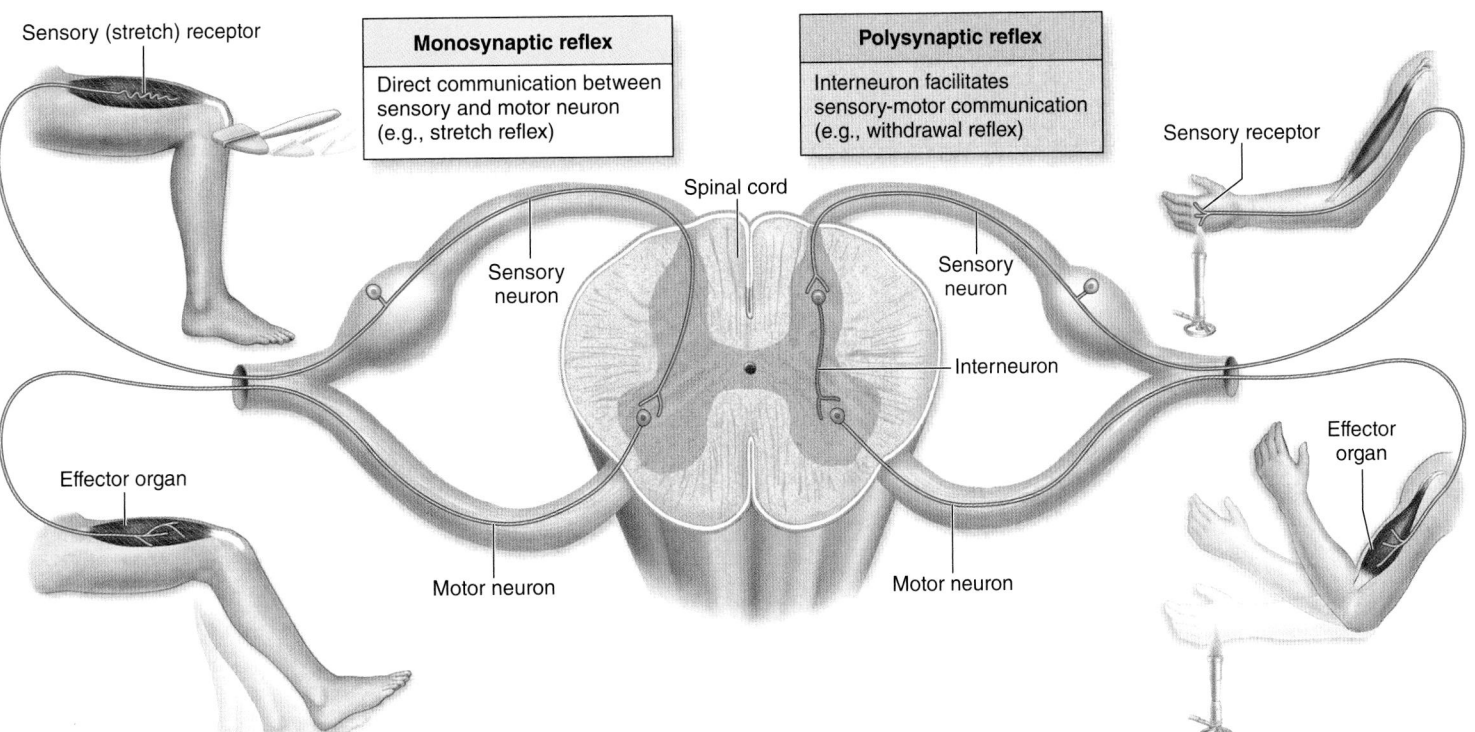

Figure 14.21 Monosynaptic and Polysynaptic Reflexes. The minimal number of neurons and the pathways of a monosynaptic reflex (left) are compared to those of a polysynaptic reflex (right).

14.6d Spinal Reflexes

LEARNING OBJECTIVE

5. Name and describe four common spinal reflexes.

Some common spinal reflexes are the stretch reflex, the Golgi tendon reflex, the withdrawal (flexor) reflex, and the crossed-extensor reflex.

Stretch reflexes and Golgi tendon reflexes are initiated by a stimulus that is detected by a specific sensory receptor called a proprioceptor. A **proprioceptor** resides in a muscle or tendon and detects any change to that structure, such as change in stretch or tension. The two principal proprioceptors are the muscle spindle and Golgi tendon organ.

A **muscle spindle** is a proprioceptor that detects changes in stretch of a muscle; for this reason, a muscle spindle is also known as a *stretch receptor* (figure 14.22). A muscle spindle is composed of **intrafusal muscle fibers** that are surrounded by a connective tissue capsule. These intrafusal muscle fibers lack myofilaments in their central regions and are contractile only at their distal regions. (Actin and myosin are found only at the ends of these fibers.) These muscle fibers are innervated by **gamma (γ) motor neurons,** so named because the term gamma refers to motor neurons with small diameter axons. These neurons detect changes *within* the muscle.

Around the muscle spindle are **extrafusal muscle fibers,** which are innervated by **alpha (α) motor neurons,** so named because these motor neurons have the largest diameter axons. (One study tip to remember the difference between the two neuron types is that **G**amma

Goes within the muscle, and **A**lpha wraps **A**round the muscle.) A muscle spindle is associated with a type of reflex called a stretch reflex.

A **Golgi tendon organ** is another type of proprioceptor that detects change in tension in a muscle tendon when a muscle contracts. A Golgi tendon organ is composed of sensory nerve endings within a tendon or near a muscle–tendon junction. This type of proprioceptor is associated with a Golgi tendon reflex.

Stretch Reflex

The **stretch reflex** is a muscle reflexively contracting in response to stretching of a muscle **(figure 14.22)**. Stretch in a muscle is monitored by a muscle spindle. A muscle spindle may be stretched if either the entire muscle is stretched or lengthened, or if a portion of the muscle that contains the muscle spindle is lengthened. When the muscle spindle is stretched, the sensation is detected by sensory neurons that are wrapped around the intrafusal muscle fibers of the muscle spindle. The sensory neurons transmit nerve signals to the spinal cord (CNS), where they synapse with the alpha motor neurons associated with that muscle. The alpha motor neurons then transmit nerve signals to the extrafusal muscle fibers, which causes the muscle to contract and thus resist the stretch.

The biceps reflex is an example of a stretch reflex. The biceps reflex may be classified as follows: it is a spinal reflex (it involves the spinal cord as the integration center), a somatic reflex (skeletal muscle is the effector), and a monosynaptic reflex (involves no interneurons), an ipsilateral reflex (the receptor and effector are located on the same

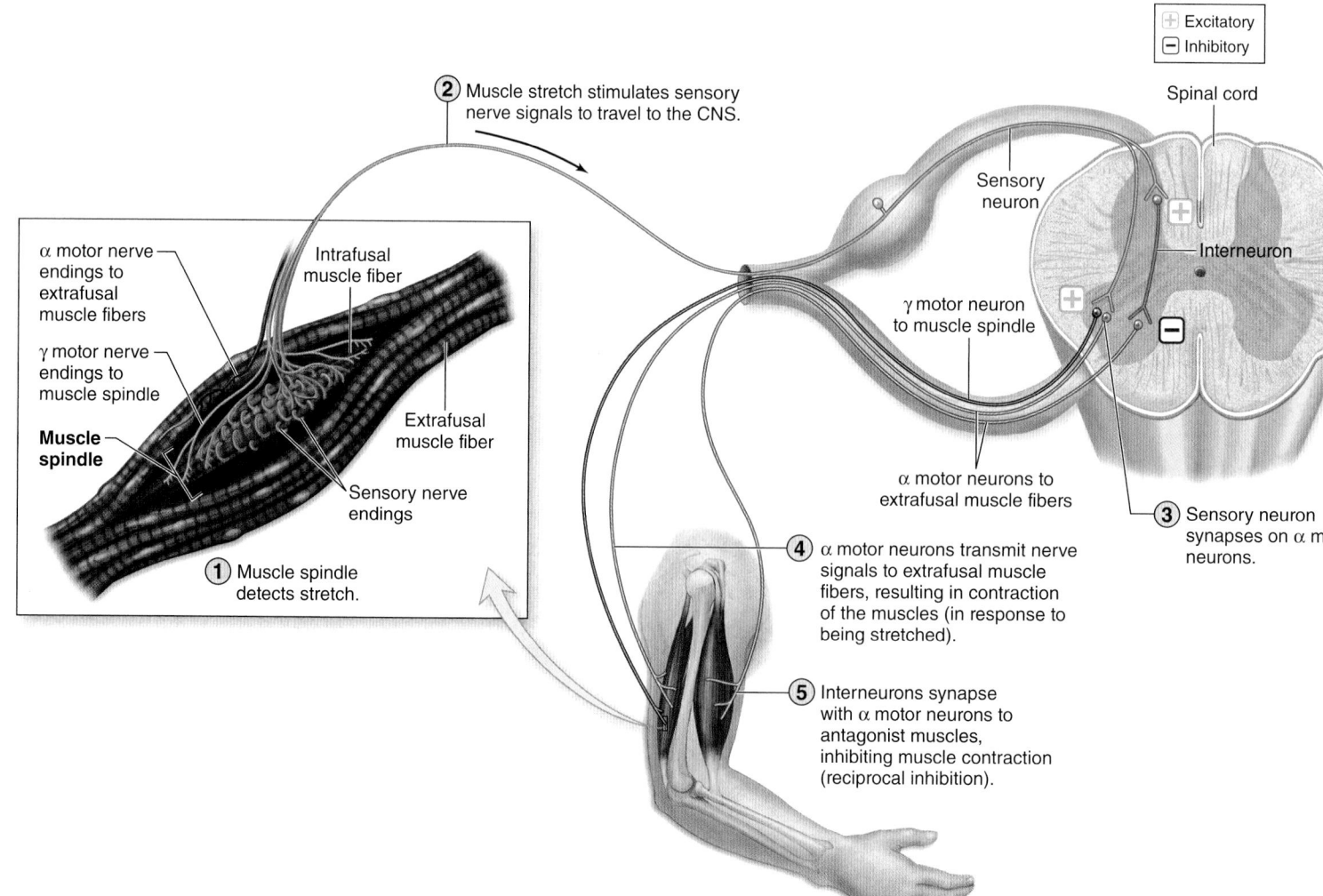

Figure 14.22 Stretch Reflex. A stretch reflex is a simple monosynaptic reflex. A stretching force detected by a muscle spindle results in the contraction of that muscle. Conversely, antagonistic muscle contraction is dampened, in a process called reciprocal inhibition.

side of the spinal cord) and is innate (we are born with this reflex). The stimulus (the tap on the biceps brachii tendon) stretches the muscle spindle in the biceps brachii muscle. Sensory neurons transmit nerve signals to the CNS, where they synapse with the alpha motor neurons. The alpha motor neurons transmit nerve signals to the extrafusal muscle fibers in biceps brachii, thereby initiating contraction of the muscle and flexing the elbow joint. (Note in figure 14.22 that the gamma motor neurons are also stimulated by the sensory nerve endings.)

Note in figure 14.22 that a stretch reflex also is indirectly involved in a process called **reciprocal inhibition.** When the sensory nerve signals reach the CNS, some of the sensory axons synapse with interneurons. These interneurons synapse with alpha motor neurons that inhibit antagonistic muscle contraction. In the case of the biceps reflex, the triceps brachii muscle is the inhibited antagonistic muscle. Thus, as the biceps brachii is stimulated, reciprocal inhibition results in the triceps brachii contraction being dampened, so the biceps movement will not be opposed by the triceps brachii.

The stretch reflex is a monosynaptic reflex, but the corresponding reciprocal inhibition is polysynaptic in nature, because it uses an interneuron within the circuit.

Golgi Tendon Reflex

Whereas the stretch reflex prevents muscles from stretching excessively, the Golgi tendon reflex prevents muscles from doing the opposite: tensing or contracting excessively. The **Golgi tendon reflex** is a polysynaptic reflex that results in muscle lengthening and relaxation in response to increased tension at a Golgi tendon organ **(figure 14.23)**.

As a muscle contracts, its associated tendon stretches, resulting in increased tension in the tendon and activation of the Golgi tendon organ. Sensory neurons in the Golgi tendon organ transmit nerve signals to interneurons in the spinal cord, which in turn inhibit the alpha motor neurons in the same muscle. When the motor neurons are inhibited, the associated muscle is allowed to relax, thus protecting the muscle and tendon from excessive tension damage.

Note that the sensory neurons also communicate with other interneurons in the spinal cord that stimulate alpha motor neurons for the *antagonistic* muscles. This process is called **reciprocal activation.** So, for example, if a Golgi tendon organ in the quadriceps femoris muscle detects excessive tension, then the Golgi tendon reflex ultimately relaxes the quadriceps femoris muscle, and reciprocal activation results in the hamstrings being stimulated to contract.

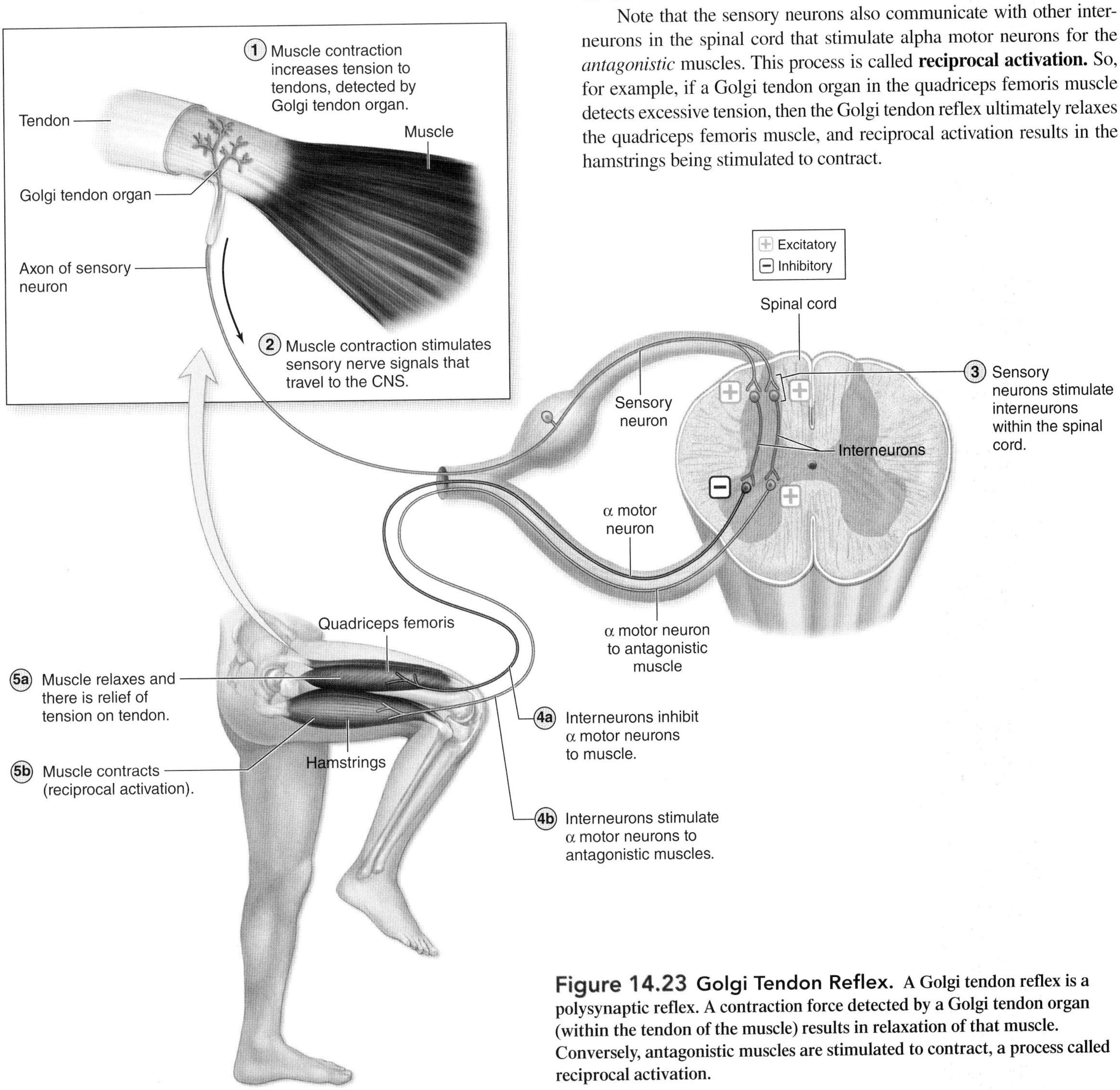

Figure 14.23 Golgi Tendon Reflex. A Golgi tendon reflex is a polysynaptic reflex. A contraction force detected by a Golgi tendon organ (within the tendon of the muscle) results in relaxation of that muscle. Conversely, antagonistic muscles are stimulated to contract, a process called reciprocal activation.

Golgi tendon reflexes help prevent a muscle or tendon from injury due to excessive tension and help ensure that the muscle contraction process occurs smoothly and efficiently. In cases where the muscle is under extreme tension (such as when one lifts a very heavy weight), a Golgi tendon reflex may nullify a stretch reflex (and thus the person drops the weight due to the excessive tension on the muscle). Thus, weightlifters are encouraged to have "spotters" who can help prevent the heavy weight from dropping unexpectedly.

Withdrawal Reflex

A **withdrawal** (*flexor*) **reflex** involves muscles contracting to withdraw the body part away from a painful stimulus. This reflex involves pain receptors, termed nociceptors (see section 16.1d). It is initiated by a painful stimulus, such as touching something very hot or painful (**figure 14.24**). This stimulation initiates a nerve signal that is transmitted by a sensory neuron to the spinal cord. Interneurons receive the sensory nerve signal and stimulate motor neurons to the flexor muscles. These flexor muscles contract in response.

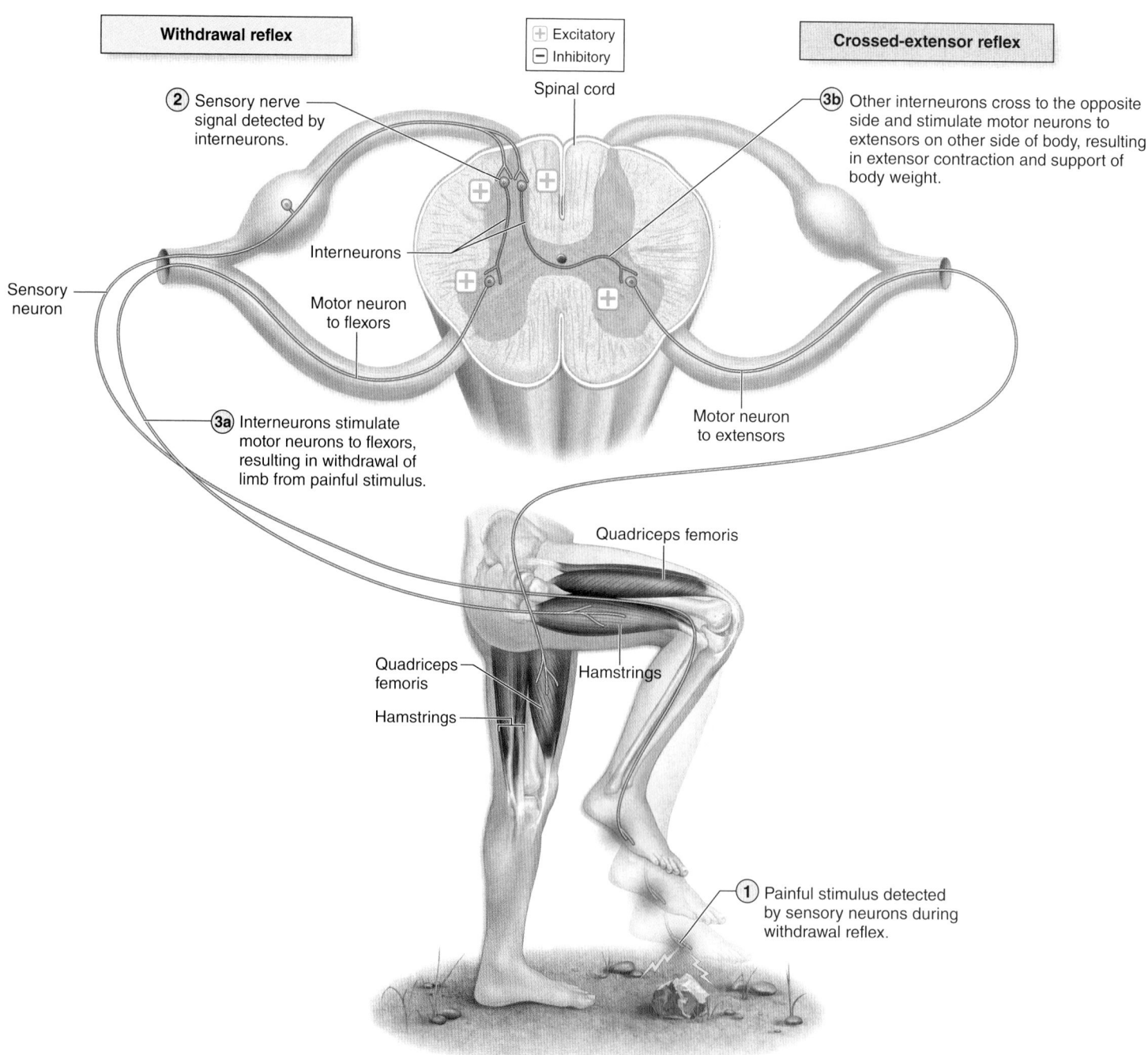

Figure 14.24 Withdrawal and Crossed-Extensor Reflexes. A withdrawal reflex is a polysynaptic reflex that is initiated by a painful stimulus. The crossed-extensor reflex occurs in response to the withdrawal reflex, by stimulating the extensor muscles in the opposite limb and thereby ensuring the opposite limb supports the body's weight.

For example, when you step on a sharp object, sensory neurons detect the sensation and transmit nerve signals to the spinal cord. They synapse with interneurons, which stimulate motor neurons to contract the flexor muscles (in this case, the hamstrings of the lower limb). You lift the lower limb up and away from the painful stimulus. In addition, reciprocal inhibition occurs with the extensor (quadriceps) muscles, so the hamstrings can contract unimpeded.

Crossed-Extensor Reflex

The **crossed-extensor reflex** often occurs in conjunction with the withdrawal reflex, usually in the lower (weight-bearing) limbs (figure 14.24). In essence, when the withdrawal reflex is occurring in one limb, the crossed-extensor reflex occurs in the other limb. So, when the sensory neurons transmit nerve signals to the spinal cord, some sensory branches synapse with interneurons involved in the stretch reflex, whereas other sensory branches synapse with interneurons involved in the crossed-extensor reflex. These latter interneurons cross to the other side of the spinal cord through the gray commissure and synapse with motor neurons that control antagonistic muscles in the opposite limb. These motor neurons are stimulated and cause these antagonistic muscles to contract.

Figure 14.24 illustrates how the withdrawal reflex causes the right lower limb to flex at the knee, due to contraction of the right hamstring muscles. In contrast, the crossed-extensor reflex stimulates the left quadriceps femoris muscle to contract, so the left limb remains extended and supports the body weight. Thus, the crossed-extensor reflex helps us maintain balance and shift body weight accordingly in these situations.

 WHAT DID YOU LEARN?

25 What are the four common spinal reflexes?

26 Identify the Golgi tendon reflex (which is an innate reflex) as (a) spinal or cranial, (b) somatic or visceral, (c) monosynaptic or polysynaptic, and (d) ipsilateral or contralateral.

14.6e Reflex Testing in a Clinical Setting

 LEARNING OBJECTIVE

6. Explain the indications of a hypoactive reflex versus those of a hyperactive reflex.

Reflexes can be an important diagnostic tool. Clinicians use them to test specific muscle groups and specific spinal nerves or segments of the spinal cord (table 14.7). Although some variation in reflexes is normal, a consistently abnormal reflex response may indicate damage to the nervous system or muscles.

A reflex response may be normal, hypoactive, or hyperactive. A **hypoactive reflex** refers to a reflex response that is diminished or absent. A hypoactive reflex may indicate damage to a segment of the spinal cord, or it may suggest muscle disease or damage to the neuromuscular junction.

A **hyperactive reflex** refers to an abnormally strong response. It may indicate damage somewhere in either the brain or spinal cord, especially if it is accompanied by **clonus** (klō′nŭs; *klonus* = tumult), rhythmic oscillations between flexion and extension, when the muscle reflex is tested.

 WHAT DID YOU LEARN?

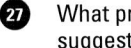

27 What problems could a hypoactive reflex suggest?

Table 14.7	Some Clinically Important Reflexes	
Reflex	**Spinal Nerve Segments Tested**	**Normal Action of Effector**
Biceps reflex	C5, C6	Flexes elbow when biceps brachii tendon is tapped
Triceps reflex	C6, C7	Extends elbow when triceps brachii tendon is tapped
Abdominal reflexes	T8–T12	Contract abdominal muscles when one side of the abdominal wall is briskly stroked
Cremasteric reflex	L1, L2	Elevates testis (due to contraction of cremaster muscle in scrotum) when medial side of thigh is briskly stroked
Patellar (knee-jerk) reflex	L2–L4	Extends knee when patellar ligament is tapped
Ankle (Achilles) reflex	S1	Plantar flexes ankle when calcaneal tendon is tapped
Plantar reflex	L5, S1	Flexes toes when plantar side of foot is briskly stroked[1]

1. This is the normal reflex response in adults; in adults with spinal cord damage and in normal infants, the **Babinski sign** occurs, which is extension of the great toe and fanning of the other toes.

14.7 Development of the Spinal Cord

LEARNING OBJECTIVE

1. Describe how the neural tube forms the gray matter structures in the spinal cord.

Recall from section 13.1b that the caudal (inferior) part of the neural tube forms the spinal cord. As the caudal part of the neural tube differentiates and specializes, the spinal cord begins to develop (**figure 14.25**). However, this developmental process is much less complex than that of the brain. A hollow **neural canal** in the neural tube develops into the central canal of the spinal cord. Note that the neural canal doesn't shrink in size; rather, the neural tube around it grows at a rapid rate. Thus, as the neural tube walls grow and expand, the neural canal in the newborn appears as a tiny hole called the central canal.

During the fourth and fifth weeks of embryonic development, the neural tube starts to grow rapidly and unevenly. Part of the neural tube forms the outer white matter of the spinal cord, whereas other components form the inner gray matter. By the sixth week of development, a horizontal groove called the **sulcus limitans** (lim′i-tanz; *limes* = boundary) forms in the lateral walls of the central canal (figure 14.25). The sulcus limitans also represents a dividing point in the neural tube as two specific regions become evident: the basal plates and alar plates.

The **basal plates** lie anterior to the sulcus limitans. The basal plates develop into the anterior and lateral horns, motor structures of the gray matter. They also form the anterior part of the gray commissure.

The **alar** (ā′lăr; *ala* = wing) **plates** lie posterior to the sulcus limitans. By about the ninth week of development, the alar plates develop into posterior horns, sensory structures of the gray matter. They also form the posterior part of the gray commissure.

During the embryonic period, the spinal cord extends the length of the vertebral canal. However, during the fetal period, growth of the vertebral column (and its vertebral canal) outpaces that of the spinal cord. By the sixth fetal month, the spinal cord is at the level of the S_1 vertebra; in contrast, a newborn's spinal cord ends at about the L_3 vertebra. By adulthood, the spinal cord length extends only to the level of the L_1 vertebra. This disproportionate growth explains why the lumbar, sacral, and coccygeal parts of the spinal cord and its associated nerve roots do not lie next to their respective vertebrae (as described in section 14.1).

WHAT DID YOU LEARN?

28 What structures develop from the alar plates, and what structures develop from the basal plates?

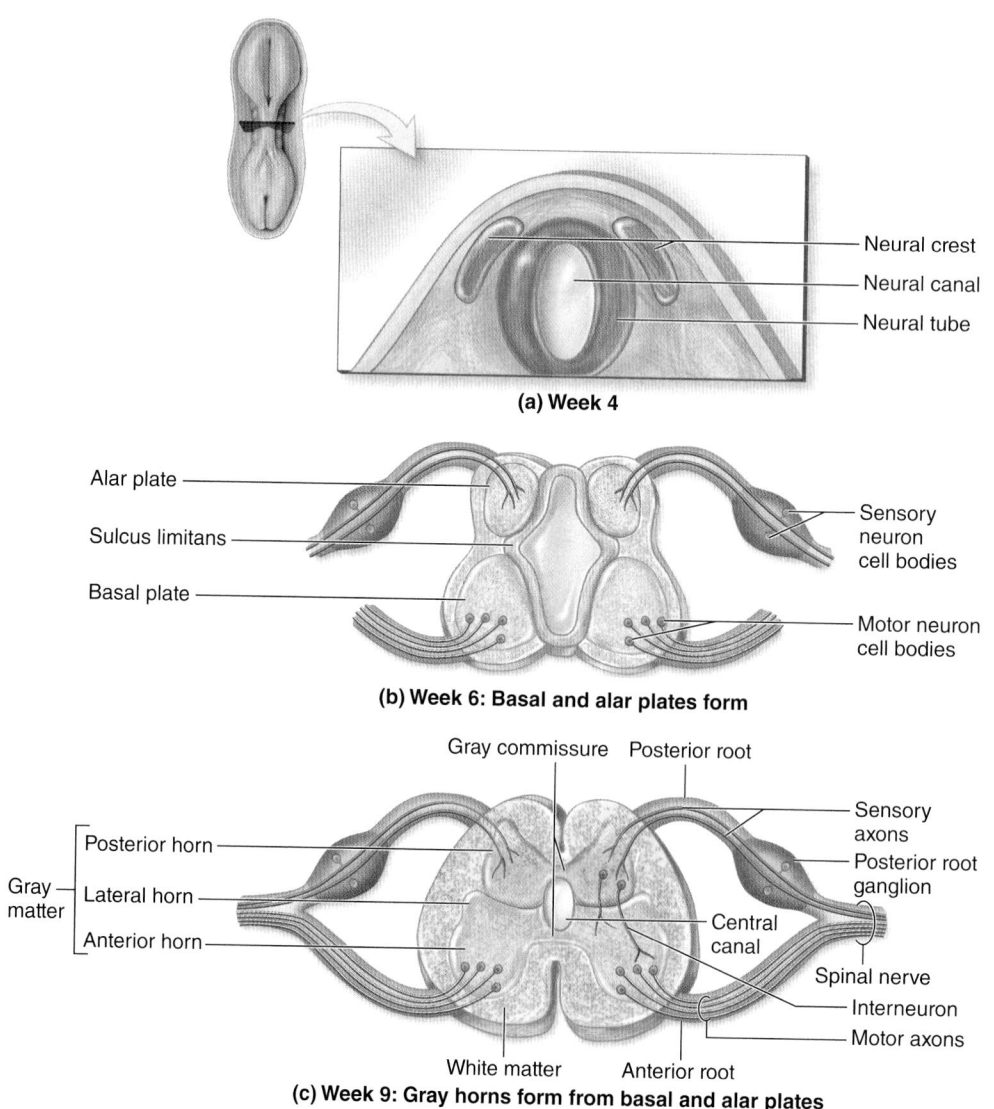

(a) Week 4

— Neural crest
— Neural canal
— Neural tube

(b) Week 6: Basal and alar plates form

Alar plate
Sulcus limitans
Basal plate

Sensory neuron cell bodies
Motor neuron cell bodies

(c) Week 9: Gray horns form from basal and alar plates

Gray commissure   Posterior root

Posterior horn
Gray matter   Lateral horn
Anterior horn

Sensory axons
Posterior root ganglion
Central canal
Spinal nerve
Interneuron
Motor axons

White matter   Anterior root

Figure 14.25 Spinal Cord Development. The spinal cord begins development as a tubular extension of the brain. (*a*) A cross section shows the structures of the neural tube of an embryo in week 4 of development. Transverse sections show (*b*) the formation of the alar and basal plates at week 6 and (*c*) the developing spinal cord at week 9.

CHAPTER SUMMARY

- The spinal cord and spinal nerves serve as a pathway for sensory and motor nerve signals and are responsible for reflexes.

- The adult spinal cord traverses the vertebral canal and typically ends at the level of the L₁ vertebra.
- There are 31 pairs of spinal nerves: 8 pairs of cervical nerves, 12 pairs of thoracic nerves, 5 pairs of lumbar nerves, 5 pairs of sacral nerves, and 1 pair of coccygeal nerves.

- The epidural space lies between the vertebral canal and the dura mater.
- A potential subdural space lies between the dura mater and arachnoid mater, whereas the subarachnoid space lies between the arachnoid mater and pia mater and contains cerebrospinal fluid (CSF).

- Gray matter is centrally located and composed of neuron cell bodies, dendrites, unmyelinated axons, and glial cells.
- White matter is peripherally located and composed of myelinated axons.

14.3a Distribution of Gray Matter 541

- Gray matter is composed of three horns: anterior (contains cell bodies of somatic motor neurons), lateral (contains cell bodies of autonomic motor neurons), and posterior (contains sensory axons and interneurons).

14.3b Distribution of White Matter 542

- White matter is organized into three pairs of funiculi, most of which contain sensory (ascending) tracts and motor (descending) tracts.

- The central nervous system (CNS) communicates with the body through conduction pathways that extend through the spinal cord.

14.4a Overview of Conduction Pathways 543

- Sensory pathways transmit ascending information from sensory receptors to the CNS, whereas motor pathways transmit descending information from the brain to muscles and glands.

14.4b Sensory Pathways 544

- Sensory pathways use primary, secondary, and tertiary neurons.
- The posterior funiculus–medial lemniscal pathway transmits stimuli of fine proprioception, discriminitive touch, precise pressure, and proprioception pressure.
- The anterolateral pathway transmits stimuli related to crude touch, pressure, pain, and temperature.
- The spinocerebellar pathway transmits stimuli from proprioceptors to the cerebellum.

14.4c Motor Pathways 546

- Motor pathways use upper and lower motor neurons.
- Somatic motor commands extend through either the direct system (for conscious control) or the indirect system (for unconscious control).
- The direct system consists of the corticobulbar and corticospinal tracts.
- The indirect system consists of the lateral pathway (rubrospinal tracts) and the medial pathway (reticulospinal, tectospinal, and vestibulospinal tracts).

- A spinal nerve is formed from the union of an anterior root and a posterior root.

14.5a Overview of Spinal Nerves 550

- Spinal nerves have two branches: A posterior ramus innervates the skin and deep muscles of the back, and an anterior ramus innervates the anterior and lateral portions of the trunk and the limbs.

14.5b Nerve Plexuses 552

- A nerve plexus is a network of interwoven anterior rami. Nerve plexuses occur in pairs.

14.5c Intercostal Nerves 552

- The anterior rami of spinal nerves T1–T11 do not form a plexus, but rather form the intercostal nerves. Nerve T12 is called a subcostal nerve.

14.5d Cervical Plexuses 552

- Each cervical plexus is formed from the anterior rami of C1–C4 spinal nerves. It innervates the anterior neck muscles and the skin along the neck and shoulders.

14.5e Brachial Plexuses 554

- Each brachial plexus is formed from the anterior rami of spinal nerves C5–T1. It innervates an upper limb.

14.5f Lumbar Plexuses 560

- Each lumbar plexus is formed from the anterior rami of spinal nerves L1–L4. It innervates the anterior and medial thigh, the lower abdominal wall, and the skin of the medial leg.

14.5g Sacral Plexuses 562

- Each sacral plexus is formed from the anterior rami of spinal nerves L4–S4. It innervates most of the lower limb as well as the perineum.

- A reflex is a rapid, preprogrammed motor response of muscles or glands to a stimulus.

14.6a Characteristics of Reflexes 566

- A reflex must be initiated by a stimulus.

14.6b Components of a Reflex Arc 566

- The five steps of a reflex are (1) activation of a receptor by a stimulus, (2) nerve signal propagation along a sensory neuron to the CNS, (3) integration and processing of information by interneurons, (4) nerve signal propagation along a motor neuron, and (5) effector response.

14.6c Classifying Spinal Reflexes 567

- Reflexes may be classified as to whether they are (a) spinal or cranial, (b) somatic or visceral, (c) monosynaptic or polysynaptic, (d) ipsilateral or contralateral, and (e) innate or acquired.

(continued on next page)

14.6 Reflexes (continued)	**14.6d Spinal Reflexes 568**
	• A stretch reflex is monosynaptic and contracts the muscle in response to increased stretch in a muscle spindle.
	• A Golgi tendon reflex is polysynaptic and prevents muscles from tensing excessively.
	• A withdrawal reflex is polysynaptic and activates flexor muscles to immediately remove a body part from a painful stimulus.
	• A crossed-extensor reflex stimulates the extensor muscles in the opposite limb, in response to the withdrawal reflex.
	14.6e Reflex Testing in a Clinical Setting 571
	• Reflex testing can help diagnose nervous system or muscular disorders.
	• Hypoactive reflexes, in which the response is diminished, may indicate spinal cord damage or muscle pathology.
	• Hyperactive reflexes, in which the response is abnormally strong, may indicate damage to the brain or spinal cord.
14.7 Development of the Spinal Cord 572	• The neural tube forms basal plates and alar plates.
	• Basal plates form the anterior horns, lateral horns, and anterior half of the gray commissure.
	• Alar plates form the posterior horns and the posterior half of the gray commissure.

CHALLENGE YOURSELF

Do You Know the Basics?

____ 1. Identify the meningeal layer immediately deep to the subdural space.
   a. arachnoid mater   c. dura mater
   b. pia mater   d. epidural space

____ 2. The anterior root of a spinal nerve contains
   a. axons of both motor and sensory neurons.
   b. axons of sensory neurons only.
   c. interneurons.
   d. axons of motor neurons only.

____ 3. Where are tertiary neurons found?
   a. extending between the posterior horn and anterior horn
   b. extending between the posterior horn and the brainstem
   c. extending between the thalamus and the primary somatosensory cortex
   d. extending between the primary motor cortex and the brainstem

____ 4. Which of the following is an example of a sensory pathway?
   a. reticulospinal tract
   b. spinocerebellar tract
   c. corticobulbar tract
   d. tectospinal tract

____ 5. The radial nerve originates from the _____ plexus.
   a. cervical   c. lumbar
   b. brachial   d. sacral

____ 6. Which structure sends motor nerve signals to the deep back muscles and receives sensory nerve signals from the skin of the back?
   a. posterior root
   b. posterior ramus
   c. anterior root
   d. anterior ramus

____ 7. Which statement is true about intercostal nerves?
   a. They are formed from the posterior rami of spinal nerves.
   b. They form a thoracic plexus of nerves.
   c. They originate from the thoracic part of the spinal cord.
   d. They innervate the lower limb.

____ 8. The _____ nerve innervates the anterior thigh muscles and the skin on the anterior thigh.
   a. femoral   c. sciatic
   b. obturator   d. tibial

____ 9. A _____ reflex is monosynaptic and responds to stretching in a muscle spindle.
   a. withdrawal   c. Golgi tendon
   b. crossed-extensor   d. stretch

____ 10. Which statement is correct about reflexes?
   a. The patellar reflex tests the S1-S3 spinal nerve segments.
   b. A hypoactive reflex may be indicative of damage to the neuromuscular junction.
   c. The more hyperactive the reflex, the healthier the individual.
   d. A normal biceps reflex response is extension of the elbow when the biceps tendon is tapped.

11. Identify the spinal cord parts, which spinal nerves are associated with them, and their relationship to the corresponding vertebrae.

12. List the three gray matter horns on each side of the spinal cord, and discuss the neuronal composition of each. In addition, list which types of nuclei (motor or sensory) are located in each horn.

13. Compare the main differences between the posterior funiculus–medial lemniscal pathway and the anterolateral pathway.

14. Describe the location and function of upper and lower motor neurons in the motor pathways.

15. What are the main terminal branches of the brachial plexus, and what muscles do these terminal branches innervate?

16. What anterior rami form the lumbar plexus, and what in general does this plexus innervate?

17. What muscles do the tibial and common fibular nerves innervate?

18. What are the five basic steps involved in a reflex arc?

19. What are the differences between a stretch and Golgi tendon reflex?

20. Where are the basal and alar plates of the neural tube located, and what do each form?

Can You Apply What You've Learned?

Use the following paragraph to answer questions 1–3.

Madeleine is an active 18-year-old who fell off her bike, fracturing the medial epicondyle of her humerus. In addition to experiencing severe pain in her elbow, she experienced numbness along the medial side of her hand. She went to the emergency room and the physician examined her.

1. After x-raying her elbow, the physician performed further tests on Madeline to determine what nerve damage she may have experienced. What nerve likely was injured when she fractured her medial epicondyle?
   a. radial
   a. ulnar
   c. musculocutaneous
   d. median

2. What diagnostic test would best help determine the potential damage to this nerve?
   a. Have Madeline extend her elbow against resistance.
   b. Have Madeline flex her wrist and fingers against resistance.
   c. Have Madeline hold a piece of paper between her fingers while the physician tries to pull the paper away.
   d. Have Madeline flex her elbow while the physician applies gentle pressure to the anterior forearm muscles.

3. What other muscle function could be impaired in this type of injury?
   a. adduction of the thumb
   b. flexion of the thumb
   c. extension of the thumb
   d. abduction of the thumb

4. George tried to lift a heavy box and in the process herniated one of his lumbar intervertebral discs. The herniated disc pinched on nerve roots and most likely caused which of the following ailments?
   a. pain down the back of the leg
   b. inability to flex the thigh
   c. inability to adduct the thigh
   d. pain down the medial side of the leg

5. Carlos experienced some muscle weakness in his right lower limb and went to see his physician. The physician noticed Carlos could not evert his foot and had anesthesia along most of the dorsum of his right foot. Carlos was still able to dorsiflex his foot and invert his foot. Based on these symptoms, what specific nerve likely was injured?
   a. common fibular
   b. tibial
   c. deep fibular
   d. superficial fibular

Can You Synthesize What You've Learned?

1. Arthur dove off a small cliff into water that was shallower than he expected and he hit his head. He is now a quadriplegic, which means that both his upper and lower limbs are paralyzed. Approximately where is the location of his injury? What is the likelihood that Arthur will recover from this injury?

2. Jessica fractured her fibula and had to wear a leg cast for several weeks. When the cast was removed, Jessica had trouble walking normally and experienced "foot drop." What structure likely was pinched by the leg cast, resulting in the foot drop?

3. Juanita was walking barefoot on the sidewalk when she stepped on a piece of glass with her right foot. Her right leg reflexively lifted up, away from the shard of glass. What is this type of reflex called? In addition, Juanita did not fall down when she lifted her leg. What action in her left lower limb helped stabilize her?

INTEGRATE

ONLINE STUDY TOOLS

 **connect** |ANATOMY & PHYSIOLOGY ▪ILEARNSMART° AP|R

The following study aids may be accessed through Connect.

Clinical Case Study: A Young Man with Rapidly Developing Muscle Weakness

Interactive Questions: This chapter's content is served up in a number of multimedia question formats for student study.

LearnSmart: Topics and terminology include spinal cord gross anatomy; protection and support of the spinal cord; sectional

anatomy of the spinal cord; spinal cord conduction pathways; spinal nerves; reflexes; development of the spinal cord

Anatomy & Physiology Revealed: Topics include cervical, thoracic, lumbar, and sacral regions of spinal cord; spinal nerves; brachial plexus; cervical plexus, lumbosacral plexus; reflex arc

chapter 15

Nervous System: Autonomic Nervous System

Anatomy & Physiology | REVEALED®
aprevealed.com

Module 7: Nervous System

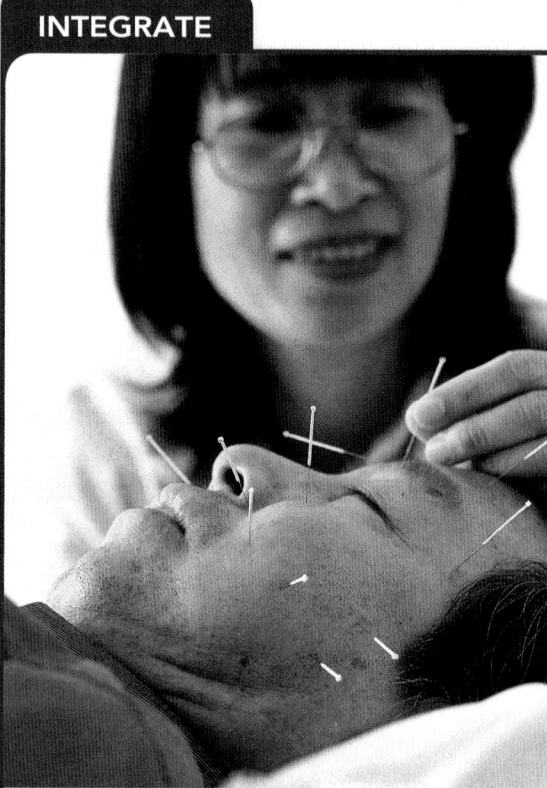

INTEGRATE

CAREER PATH
Acupuncturist

Acupuncture is one of the mainstays of traditional Chinese medicine (TCM) as well as complementary and alternative medicine (CAM). TCM is based on the premise that disease and pathology are due to a blockage of *qi* (chē; vital energy) at some location in the body. An acupuncturist inserts long, thin metallic needles into the skin at specific acupuncture sites to promote *qi* circulation. Acupuncture is effective at relieving some types of pain, and many studies suggest it can relieve nausea associated with chemotherapy or other medical treatments. Acupuncture has also been used to treat depression, anxiety, infertility and some neurologic disorders. Some researchers hypothesize that acupuncture manipulates our autonomic nervous system (ANS) in ways that are not yet understood. Because the ANS controls our internal environment and innervates our viscera, practices such as acupuncture may help the workings of these structures.

On a twisting downhill slope, an Olympic skier is concentrating on controlling her body to negotiate the course faster than anyone else in the world. Compared to the spectators in the viewing areas, her pupils are dilated, and her heart is beating faster and pumping more blood to her skeletal muscles. At the same time, organ system functions not needed in the race are practically shut down. Digestion, urination, and defecation can wait until the race is over. The skier exhibits a state of heightened readiness, called the "fight-or-flight" response, because the sympathetic division of the autonomic nervous system (ANS) is dominant. When the race is finished and she stops to rest and eat, the parasympathetic division of the ANS will dominate to meet the demands of digesting the meal. Thus, the ANS—both the sympathetic division and the parasympathetic division—functions to regulate the body's response to changing demands—whether they are the demands of exercise or the demands of supplying nutrients to the body.

15.1 Comparison of the Somatic and Autonomic Nervous Systems

We introduce this chapter on the autonomic nervous system by first describing how the nervous system is organized into a somatic nervous system and an autonomic nervous system. Here we compare the somatic and autonomic nervous systems' functional organization, their lower motor neurons, and the CNS regions that control the autonomic nervous system.

15.1a Functional Organization

LEARNING OBJECTIVE

1. List the similarities and differences between the SNS and the ANS.

Recall from section 12.1 that anatomists and physiologists have devised various ways to organize the nervous system both structurally and functionally. Here, we introduce a slightly different way of functionally organizing the nervous system into two systems that is based upon whether we are conscious of the process. The two components are the somatic nervous system and the autonomic nervous system (figure 15.1).

The **somatic nervous system (SNS)** includes processes that are perceived or controlled consciously (figure 15.1*a*). The **somatic sensory** portion includes detection of stimuli and transmission of nerve signals from the *special senses* (i.e., vision, hearing, equilibrium, smell and taste), *skin*, and *proprioceptors* (receptors in joints and muscles that detect body position) to the CNS. The sensory portion of the somatic nervous system, for example, allows us to see a beautiful mountain, smell a baby's skin, or taste a delicious dinner. The **somatic motor** portion involves initiation and transmission of nerve signals from the CNS to control *skeletal muscles*. This is exemplified by voluntary activities such as getting out of a chair, picking up a book, and throwing a ball for the dog to chase. Both the sensory input we consciously perceive and the motor output we voluntarily initiate to skeletal muscle involve the cerebrum (see section 13.3c). Reflexive skeletal muscle activity is controlled by the brainstem and spinal cord (see section 14.6).

The **autonomic nervous system (ANS)** (aw-tō-nom'ik; *auto* = self, *nomos* = law), also called the **autonomic motor** or *visceral motor system,* includes processes regulated below the conscious level (figure 15.1*b*). This system is a motor system only (see section 12.1b). These autonomic motor components initiate and transmit nerve signals from the CNS to cardiac muscle, smooth muscle, and glands. The autonomic nervous system often responds to input from **visceral sensory** components, such as receptors that detect stimuli associated with blood vessels and internal organs (viscera). Some of these sensory neurons, for example, monitor carbon dioxide concentration in the blood, whereas others detect pressure by measuring stretch in smooth muscles of visceral walls. Note that the visceral sensory structures are not part of the ANS per se, but rather, simply transmit input from sensory receptors that may result in motor output by the ANS.

The function of the ANS is to maintain **homeostasis**, or a constant internal environment (see section 1.5). Thus, the ANS regulates

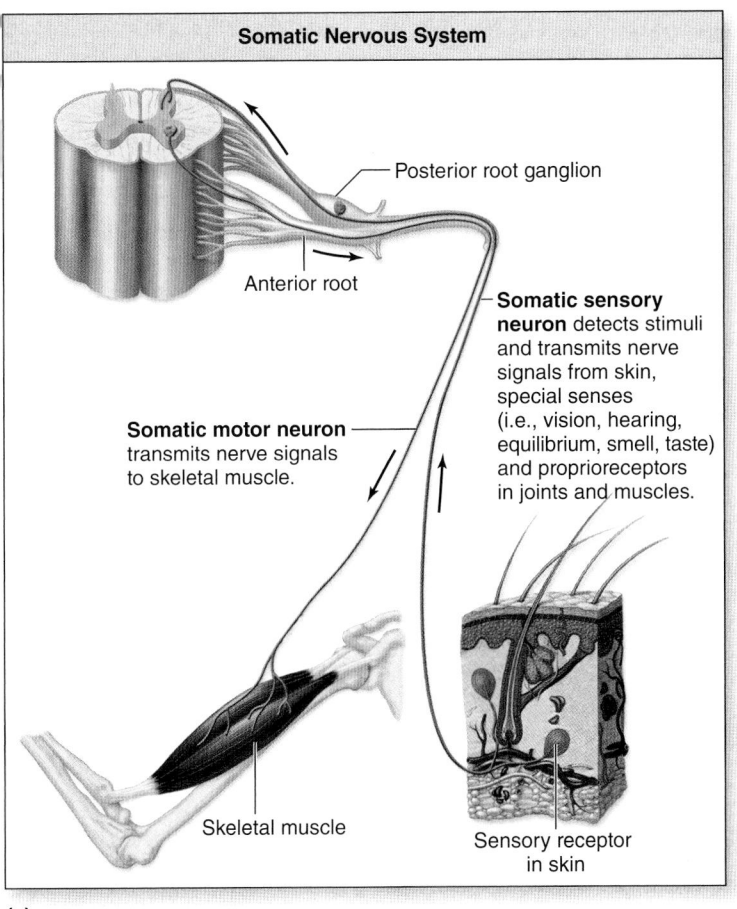

Somatic Nervous System

Posterior root ganglion

Anterior root

Somatic sensory neuron detects stimuli and transmits nerve signals from skin, special senses (i.e., vision, hearing, equilibrium, smell, taste) and proprioreceptors in joints and muscles.

Somatic motor neuron transmits nerve signals to skeletal muscle.

Skeletal muscle

Sensory receptor in skin

(a)

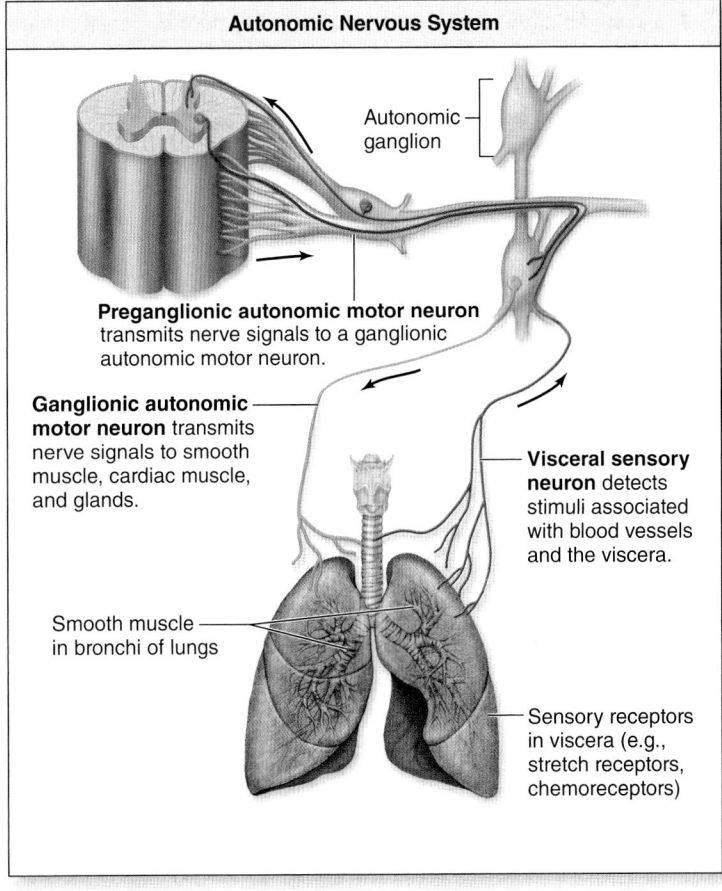

Autonomic Nervous System

Autonomic ganglion

Preganglionic autonomic motor neuron transmits nerve signals to a ganglionic autonomic motor neuron.

Ganglionic autonomic motor neuron transmits nerve signals to smooth muscle, cardiac muscle, and glands.

Visceral sensory neuron detects stimuli associated with blood vessels and the viscera.

Smooth muscle in bronchi of lungs

Sensory receptors in viscera (e.g., stretch receptors, chemoreceptors)

(b)

Figure 15.1 Comparison of Somatic and Autonomic Nervous Systems. The nervous system is functionally organized into the (*a*) somatic nervous system, which involves processes that we consciously perceive and control, and the (*b*) autonomic nervous system, which involves processes that occur below the conscious level. (Note that the visceral sensory structures are not part of the ANS per se.)

LEARNING STRATEGY

A good way to understand the two-neuron ANS chain is to compare it to the U.S. airline system, which uses connecting flights and "airport hubs" to transport the maximum number of people in the most cost-effective way.

Imagine that you are flying from Indianapolis to Miami for spring break: Your first flight from Indianapolis to Chicago is the **preganglionic axon.**

① "Preganglionic" flight from Indianapolis to Chicago (autonomic ganglion)

Chicago

Indianapolis

Miami

② "Postganglionic" flight from Chicago to Miami (effector organ)

Although flying north to Chicago is out of your way, the airline sends you to an airport hub because it is more efficient to send all Indianapolis passengers to this main location before they take different flights throughout the United States.

The airport hub in Chicago is the **autonomic ganglion,** the point where preganglionic and postganglionic "flights" meet up. Other preganglionic flights are arriving at the airport hub, and here all these passengers will connect with other flights.

Your connecting flight from Chicago to Miami is the **ganglionic neuron.** This flight will take you to your final destination, just as a postganglionic axon sends a nerve signal to an effector organ. On the plane with you are people from other preganglionic flights who all want to go to Miami as well.

Is using two different flights the most *direct* way for you to get from Indianapolis to Florida? Of course not. But it is the most *cost-efficient* (or *energy-efficient*) way for the airlines to transport and disperse many passengers with a limited number of planes.

The connecting-flight arrangement also allows for the extensive convergence of passengers to an airport "hub," and for the extensive divergence of passengers from the "hubs" to the final destinations. Thus, the two motor neurons used in the ANS allow for neuronal convergence and divergence.

The autonomic nervous system is similar to connecting airline flights and "airport hubs" in that both try to group and disperse many different items (nerve signals or passengers) with a limited number of neurons or flights.

---

all physiologic processes that must be maintained by the nervous system to keep the body alive including the regulation of heart rate, blood pressure, body temperature, sweating, and digestion. The ANS keeps these processes within optimal ranges and adjusts the variables to meet changing body needs.

WHAT DID YOU LEARN?

❶ What criterion is used to organize the nervous system into the SNS and the ANS? What sensory and motor components are associated with each of the two systems?

15.1b Lower Motor Neurons of the Somatic Versus Autonomic Nervous System

LEARNING OBJECTIVES

2. Compare and contrast lower motor neurons in the SNS and ANS.

3. Describe how the two-neuron chain in the ANS facilitates communication and control.

One significant anatomic difference between the somatic nervous system and autonomic nervous system is the number of lower motor neurons that extend from the CNS. A single lower motor neuron extends from the CNS to skeletal muscle fibers in the somatic nervous system (figure 15.1a). The cell body of a lower motor neuron lies within the brainstem or the spinal cord and its axon exits the CNS in either a cranial nerve or a spinal nerve and extends to a skeletal muscle, respectively (see sections 13.9 and 14.5). Additionally, the motor neurons of the somatic nervous system are (1) composed of myelinated axons with a large diameter allowing for fast propagation of a nerve signal (see section 12.9b), and (2) always release acetylcholine (ACh) neurotransmitter from its synaptic knobs to stimulate or excite the skeletal muscle fiber (see section 10.3).

In comparison, a chain of two lower motor neurons extend from the CNS to innervate cardiac muscle, smooth muscle, and glands in the ANS (figure 15.1b, **figure 15.2**). The first of the two ANS motor neurons is the **preganglionic** (prē′gang-lē-on′ik) **neuron.** Its cell body lies within the brainstem or the spinal cord. A **preganglionic axon** extends from this cell body and exits the CNS in either a cranial nerve or a spinal nerve. This axon projects to an autonomic ganglion in the peripheral nervous system. Preganglionic neurons have myelinated axons that typically are small in diameter, and nerve signals always results in the release of acetylcholine to excite the second neuron.

The second neuron in this pathway is called a **ganglionic neuron,** sometimes also referred to as a *postganglionic neuron.* (Note that the term *postganglionic* is not accurate, as the cell body of the neuron resides *within* the ganglion, and not *after* the ganglion.) Its cell body resides within an autonomic ganglion. A **postganglionic axon** extends from the cell body to an effector (cardiac muscle, smooth muscle, or a gland). (It is appropriate to call the axon *postganglionic* because it extends from the autonomic ganglion.) Ganglionic neurons have unmyelinated axons that are even smaller in diameter than preganglionic axons. The neurotransmitter released from the ganglionic neuron in response to a nerve signal is either ACh or norepinephrine (NE). Both neurotransmitters can either excite or inhibit an effector depending upon the type of receptors present within the effector (a concept described in section 15.5 in detail). Because motor neurons of the ANS are small and mostly unmyelinated, propagation of nerve signals is relatively slow in comparison to nerve signal propagation along somatic motor axons.

The two-neuron motor pathway in the ANS has a distinctive advantage over the one lower motor neuron of the somatic nervous system: It allows for increasing communication and control. This occurs because there is neuronal convergence and neuronal divergence (see section 12.11). **Neuronal convergence** (kon-ver′jens; *convergo* = to incline together) occurs because axons from numerous preganglionic neurons synapse with and influence a single ganglionic neuron. **Neuronal divergence** (di-ver′jens; *di* = apart)

Table 15.1	Comparison of Somatic and Autonomic Motor Nervous Systems	
Feature	**Somatic Nervous System**	**Autonomic Nervous Systems**
Functional Organization		
Sensory input	Special senses, skin; proprioceptors	Visceral senses (and some somatic senses)
Effectors	Skeletal muscle fibers	Cardiac muscle cells, smooth muscle cells, glands
CNS region of control	Cerebral cortex, cerebral nuclei, thalamus, cerebellum, brainstem, spinal cord	Hypothalamus, brainstem, spinal cord Cerebrum, thalamus, limbic system
Motor Neurons		
Number of neurons in pathway	One neuron from CNS: somatic motor neuron axon extends from CNS to effector	Two neurons from CNS: Preganglionic neuron has preganglionic axon that projects to ganglionic neuron; ganglionic neuron has postganglionic axon that projects to effector
Axon properties	Myelinated and thicker in diameter; have fast nerve signal propagation	Preganglionic axons are myelinated and thin in diameter Postganglionic axons are unmyelinated and are thinner in diameter; both have relatively slow nerve signal propagation
Neurotransmitter released	Acetylcholine (ACh)	Preganglionic axons release ACh Postganglionic axons release either ACh or norepinephrine (NE)
Response of effector	Excitation only	Either excitation or inhibition
Ganglia associated with motor neurons	None	Autonomic ganglia: parasympathetic division: terminal ganglia, intramural ganglia sympathetic division: sympathetic trunk ganglia, prevertebral ganglia

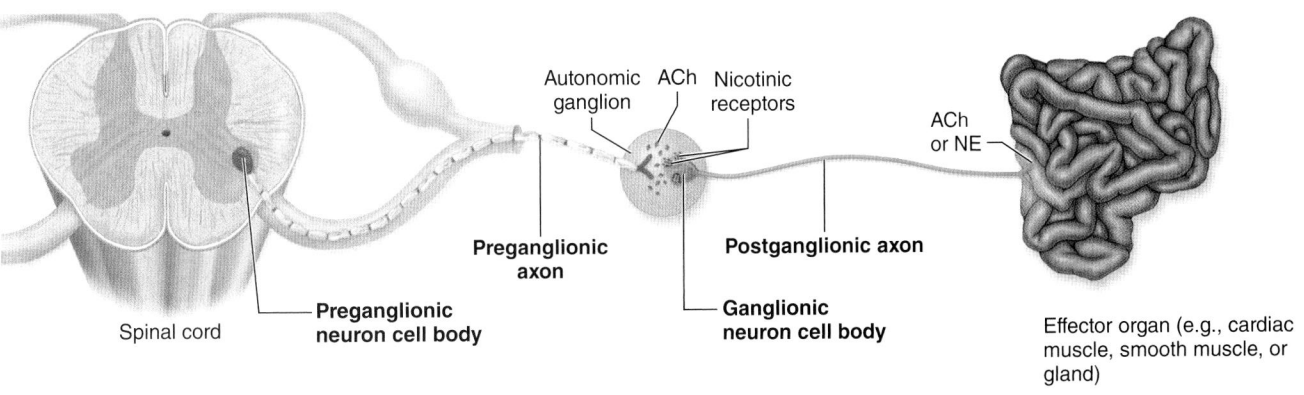

Figure 15.2 Lower Motor Neurons of the Autonomic Nervous System. The autonomic nervous system employs a chain of two lower motor neurons, a preganglionic neuron and a ganglionic neuron. The dendrites and cell body of a preganglionic neuron are housed within the CNS (brain or spinal cord). The preganglionic axon synapses with a ganglionic neuron within an autonomic ganglion.

occurs because axons from one preganglionic cell synapse with and influence numerous ganglionic neurons.

Table 15.1 compares the characteristics of the somatic and autonomic nervous systems.

WHAT DID YOU LEARN?

2 What are the anatomic features that distinguish the motor neurons in the SNS and ANS?

15.1c CNS Control of the Autonomic Nervous System

LEARNING OBJECTIVE

4. Describe the CNS hierarchy that controls the autonomic nervous system.

Several levels of CNS complexity are required to coordinate and regulate ANS function. Thus, despite the name "autonomic," the ANS is a regulated nervous system, not an independent one. Autonomic function is regulated by three CNS regions: the hypothalamus, brainstem, and spinal cord (figure 15.3). These CNS regions may be influenced by the cerebrum, thalamus, and limbic system.

The hypothalamus is the integration and command center for autonomic functions (see section 13.4c). It contains nuclei that control visceral functions in both divisions of the ANS, and it communicates with other CNS regions, including the brainstem and spinal cord. The hypothalamus is the central brain structure involved in emotions and physiologic processes, which are regulated through the ANS. For example, the sympathetic nervous system fight-or-flight response originates in the sympathetic nucleus in this brain region.

The brainstem nuclei mediate visceral reflexes (see section 13.5). These reflex centers control changes in blood pressure, blood vessel diameter, digestive activities, heart rate, pupil size, and eye lens shape for focusing on close-up objects (see section 15.7b).

Some autonomic responses, notably the parasympathetic activities associated with defecation and urination (in children), are processed and controlled at the level of the spinal cord without the involvement of the brain. However, the higher centers in the brain may consciously prevent defecation and urination by controlling the external sphincters.

ANS activities are affected by conscious activities in the cerebral cortex and subconscious communications between association areas in the cortex and the centers of parasympathetic and sympathetic control in the hypothalamus. Additionally, sensory processing in the thalamus and emotional states controlled in the limbic system directly affect the hypothalamus.

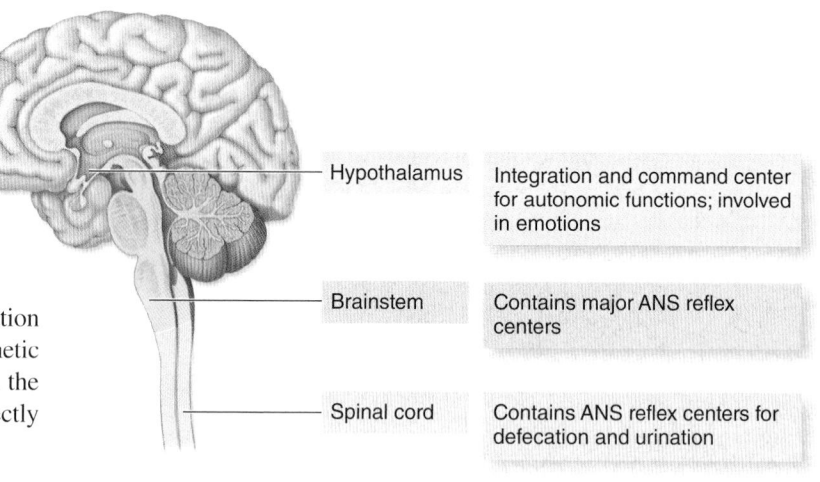

Hypothalamus	Integration and command center for autonomic functions; involved in emotions
Brainstem	Contains major ANS reflex centers
Spinal cord	Contains ANS reflex centers for defecation and urination

Figure 15.3 Control of Autonomic Functions by Higher Brain Centers. ANS functions are regulated by the hypothalamus, which in turn controls ANS regions in the brainstem and spinal cord.

WHAT DID YOU LEARN?

3 What CNS structure is the integration and command center for autonomic function?

INTEGRATE

LEARNING STRATEGY

The analogy of a corporation can help you understand the hierarchy of control of the ANS:

- The **hypothalamus** is the president of the Autonomic Nervous System corporation. It oversees all activity in this system.

- The **autonomic reflex centers** in the brainstem and spinal cord are the vice presidents of the corporation. They have a lot of control and power, but ultimately they must answer to the president (hypothalamus).

- The **preganglionic and ganglionic neurons** are the workers in the corporation. They are ultimately under the control of both the president and vice presidents. These workers are responsible for production.

15.2 Divisions of the Autonomic Nervous System

The motor component of the ANS is further subdivided into the parasympathetic division and the sympathetic division. Here we discuss both the general functional and anatomic differences between the two divisions, and also the degree of response (local or mass activation) that is possible when each division is activated.

15.2a Functional Differences

LEARNING OBJECTIVE

1. Describe the general functions of parasympathetic and sympathetic divisions of the autonomic nervous system.

The divisions perform dramatically different functions—but instead of being thought of as antagonistic, they should be considered complementary. The **parasympathetic** (par-ă-sim-pa-thet′ik; *para* = alongside, *sympatheo* = to feel with) **division** functions to maintain homeostasis when we are at rest. This division is primarily concerned with conserving energy and replenishing nutrient stores. Because it is most active when

the body is at rest or digesting a meal, the parasympathetic division has been nicknamed the "rest-and-digest" division.

The **sympathetic** (sim-pă-thet′ik) **division** functions to maintain homeostasis during exercise or times of stress or emergency, which includes the release of nutrients from stores (e.g., glucose released from the liver). Because of its function in regulating the more active states the sympathetic division has been nicknamed the "fight-or-flight" division. (To help you identify when the sympathetic division is active, remember the "three *E*'s": exercise, excitement, or emergency.)

WHAT DID YOU LEARN?

4 The parasympathetic division is responsible for what main functions?

15.2b Anatomic Differences in Lower Motor Neurons

LEARNING OBJECTIVE

2. Compare and contrast the anatomic differences in the lower motor neurons and associated ganglia of the parasympathetic and sympathetic divisions.

Anatomically, these two divisions are similar in that they typically both use a preganglionic neuron and a ganglionic neuron to innervate cardiac muscle, smooth muscle, or glands. Additionally, both divisions have autonomic ganglia that house the ganglionic neuron cell bodies. One of the major differences is where the preganglionic neuron cell bodies are housed in the CNS (**figure 15.4**). Parasympathetic preganglionic cell bodies are located in either the brainstem or the lateral gray matter of the S2–S4 spinal cord segments, and for this reason this division is also termed the *craniosacral* (krā′nē-ō-sā′krăl) *division*. In comparison, sympathetic preganglionic neuron cell bodies are located in the lateral horns of the T1–L2 spinal cord segments, and for this reason this division is also termed the *thoracolumbar* (thōr′ă-kō-lăm′bar) *division*.

Other anatomic differences between the parasympathetic and sympathetic nervous system include

- **Length of preganglionic and postganglionic axons.** Parasympathetic preganglionic axons are longer, and postganglionic axons are shorter, when compared to their counterparts in the sympathetic division. In the sympathetic division, preganglionic axons are shorter and postganglionic axons are longer.

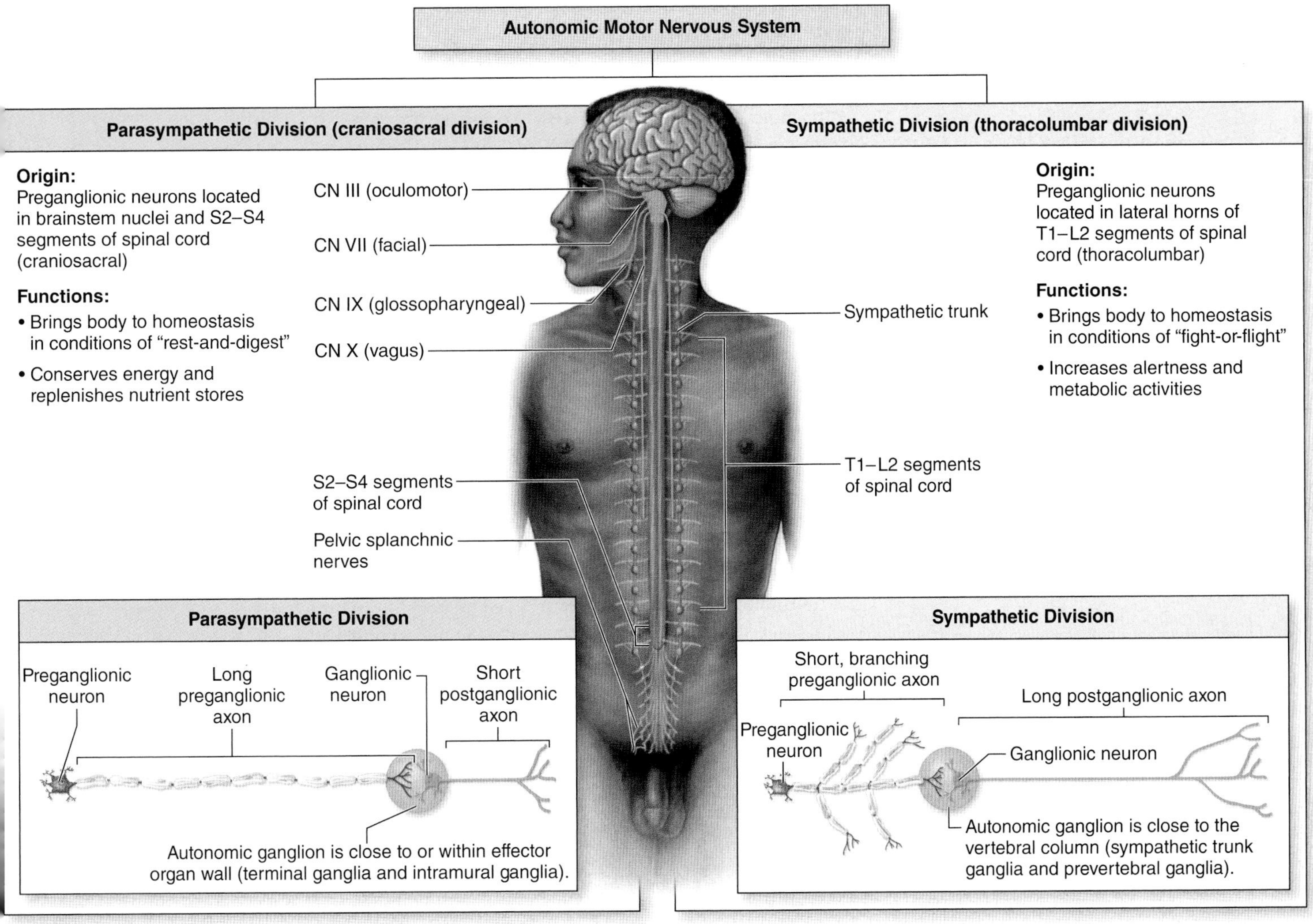

Autonomic Motor Nervous System

Parasympathetic Division (craniosacral division)

Origin:
Preganglionic neurons located in brainstem nuclei and S2–S4 segments of spinal cord (craniosacral)

Functions:
- Brings body to homeostasis in conditions of "rest-and-digest"
- Conserves energy and replenishes nutrient stores

CN III (oculomotor)
CN VII (facial)
CN IX (glossopharyngeal)
CN X (vagus)

S2–S4 segments of spinal cord
Pelvic splanchnic nerves

Sympathetic Division (thoracolumbar division)

Origin:
Preganglionic neurons located in lateral horns of T1–L2 segments of spinal cord (thoracolumbar)

Functions:
- Brings body to homeostasis in conditions of "fight-or-flight"
- Increases alertness and metabolic activities

Sympathetic trunk
T1–L2 segments of spinal cord

Parasympathetic Division

Preganglionic neuron | Long preganglionic axon | Ganglionic neuron | Short postganglionic axon

Autonomic ganglion is close to or within effector organ wall (terminal ganglia and intramural ganglia).

Sympathetic Division

Short, branching preganglionic axon
Long postganglionic axon
Preganglionic neuron
Ganglionic neuron
Autonomic ganglion is close to the vertebral column (sympathetic trunk ganglia and prevertebral ganglia).

Figure 15.4 Comparison of Parasympathetic and Sympathetic Divisions. The neurons of the parasympathetic division extend from the brainstem and sacral region, whereas the neurons of the sympathetic division extend from the thoracic and lumbar regions of the cord. The parasympathetic division axons exhibit very little branching, and autonomic ganglia lie close to or within the effector organ. The sympathetic division axons of both neurons show much branching, and autonomic ganglia lie close to the vertebral column. AP|R

- **Number of preganglionic axon branches.** Parasympathetic preganglionic axons tend to have few (less than 4) branches, whereas sympathetic preganglionic axons tend to have many branches (more than 20).
- **Location of ganglia.** Parasympathetic autonomic ganglia are either close to or within the effector (terminal ganglia and intramural ganglia, respectively). In comparison, sympathetic autonomic ganglia are relatively close to the spinal cord, and are on either side of the spinal cord or anterior to the spinal cord (sympathetic trunk ganglia and prevertebral ganglia, respectively). (These ganglia are described later in detail.)

WHAT DID YOU LEARN?

5 Describe the general anatomic differences in the parasympathetic and sympathetic divisions.

15.2c Degree of Response

LEARNING OBJECTIVE

3. Explain why parasympathetic activation is local and discrete, and sympathetic activation can result in mass activation.

It is the combination of long preganglionic axons with limited branches that results in a local response when the parasympathetic division is activated. Parasympathetic activity stimulates either one or only a few structures at the same time without necessarily having to "turn on" all the other organs.

In comparison, the combination of short preganglionic axons with more extensive branching within the sympathetic division facilitates the activation of many structures simultaneously; a process called **mass activation**. This process is facilitated when the adrenal medulla is stimulated by the sympathetic division, which causes this gland to release norepinephrine and epinephrine into the blood. Mass activation is especially important in response to stress, when it is necessary to coordinate rapid changes in activity with numerous structures at once. Think of all of the bodily changes that are initiated when you are exercising or scared: changes that include an increase in heart rate and blood pressure, increases in the amount of air that enters the lungs, dilation of the pupils, and mobilization of energy reserves from the liver. Keep in mind, however, that there are times when the sympathetic division may activate a single effector. For example, only a single effector is involved when the sympathetic division stimulates smooth muscle to increase the diameter of the pupil of the eye during low-light conditions.

WHAT DID YOU LEARN?

6 What is the difference between the degree of response for the parasympathetic division and the sympathetic division?

15.3 Parasympathetic Division

The parasympathetic division is primarily concerned with maintaining homeostasis at rest, and is functionally considered the "rest-and-digest" division. The parasympathetic division is also called the *craniosacral division* because of the anatomic origin of its preganglionic neuron from the brainstem and sacral region of the spinal cord. The two types of ganglia associated with the parasympathetic division are the **terminal** (ter'mi-năl; *terminus* = a boundary) **ganglia,** which are located close to the effector, and the **intramural** (in'tră-mū'răl; *intra* = within, *murus* = wall) **ganglia**, which are located within the wall of the target organ. Here we discuss the details of the structure and function of the cranial and sacral components of the parasympathetic division. Table 15.2 summarizes the different nerves associated with the parasympathetic division.

15.3a Cranial Components

LEARNING OBJECTIVE

1. Name the four cranial nerves associated with the parasympathetic division, and describe their actions.

The cranial nerves containing neurons from the parasympathetic division are the oculomotor (CN III), facial (CN VII), glossopharyngeal (CN IX), and vagus (CN X) nerves. (Recall that these cranial nerves are paired; they are found on the left and right side of the body.) Review table 13.5 for illustration of the cranial nerve pathways and the locations of their associated parasympathetic ganglia. The first three of these nerves transmit parasympathetic innervation to the head, whereas the vagus nerve is the source of parasympathetic stimulation for the thoracic and most abdominal organs (figure 15.5).

Oculomotor Nerve (CN III)

The **oculomotor nerve (CN III)** is formed by axons extending from cell bodies housed in nuclei in the midbrain. The preganglionic axons extend from CN III to the **ciliary** (sil'ē-ar-ē; *ciliaris* = eyelash) **ganglion** within the orbit. Postganglionic axons project from this ganglion to the ciliary muscle (within the eye) to control focusing of the lens to see close-up objects. Axons from this ganglion also extend to the sphincter pupillae muscle of the iris (which constricts the pupil) to allow less light into the eye, such as when we first walk outside on a bright sunny day (see section 16.4b).

Facial Nerve (VII)

The **facial nerve (CN VII)** contains parasympathetic preganglionic axons that exit the pons and control the production and secretion of tears, mucus, and saliva. Two branches of parasympathetic preganglionic axons exit the facial nerve and terminate at one of two ganglia. One branch (the greater petrosal nerve) terminates at the **pterygopalatine** (ter'i-gō-pal'ă-tīn; *pterygo* = wing shaped, *palatine* = of the palate) **ganglion** near the junction of the maxilla and palatine bones. Postganglionic axons project to the lacrimal glands and small glands of the nasal cavity, oral cavity, and palate to increase secretion by these glands.

Table 15.2	Parasympathetic Division Outflow		
Nerve(s)	**Origin of Preganglionic Neurons**	**Autonomic Ganglia**	**Effectors Innervated**
CN III (oculomotor)	Midbrain	Ciliary ganglion	Ciliary muscles to control lens for accommodation; sphincter pupillae muscle of eye to constrict pupil
CN VII (facial)	Pons	Pterygopalatine ganglion	Lacrimal glands; glands of nasal cavity, palate, oral cavity
		Submandibular ganglion	Submandibular and sublingual salivary glands
CN IX (glossopharyngeal)	Medulla oblongata	Otic ganglion	Parotid salivary glands
CN X (vagus)	Medulla oblongata	Multiple terminal and intramural ganglia	Thoracic viscera and most abdominal viscera
Pelvic splanchnic nerves	S2–S4 segments of spinal cord	Terminal and intramural ganglia	Some abdominal viscera and most pelvic viscera

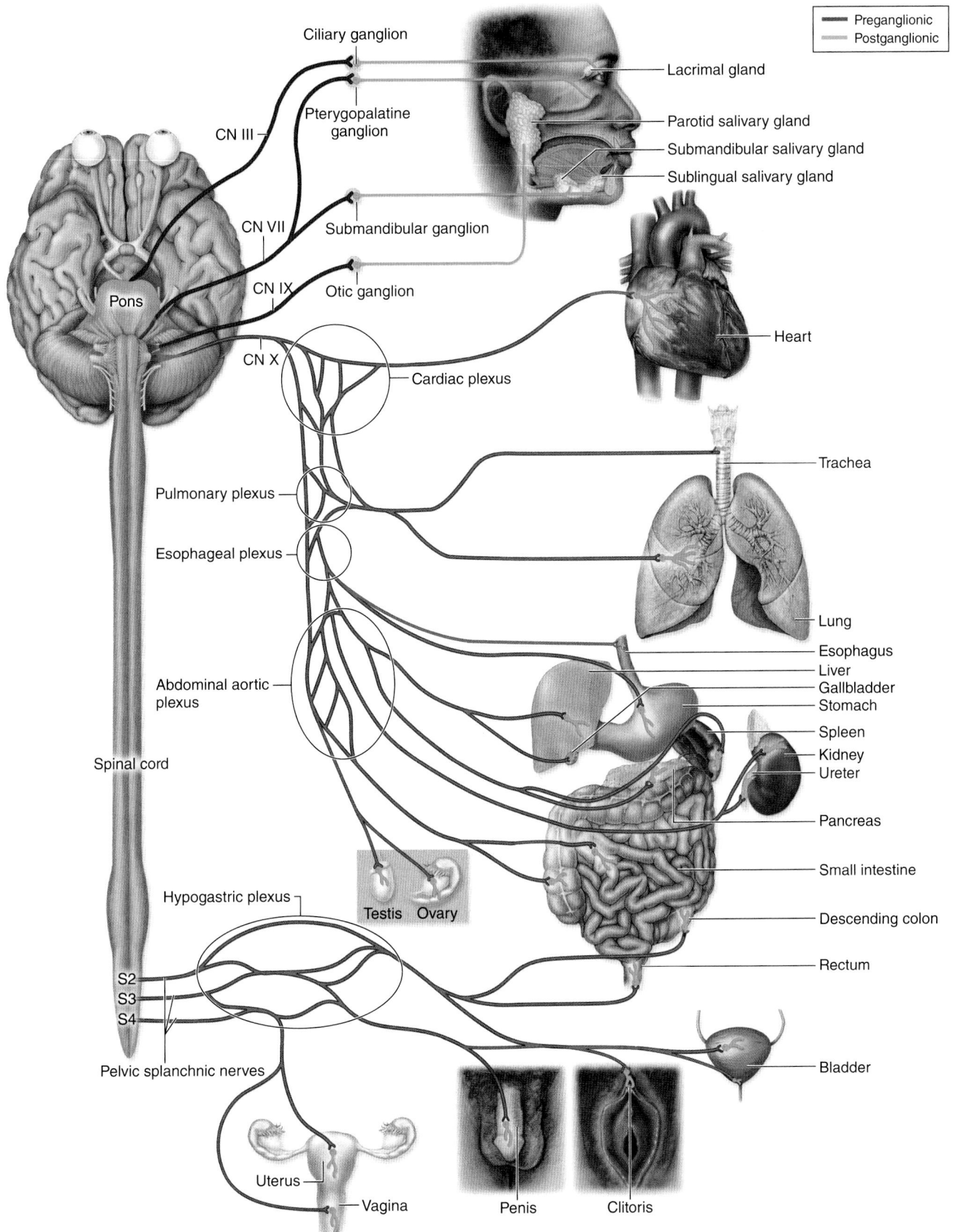

Figure 15.5 Overview of Parasympathetic Pathways. Preganglionic neurons of the parasympathetic division extend from the brain and sacral region of the spinal cord. Ganglionic neurons are located in terminal and intramural ganglia, and extend from these ganglia to innervate the viscera. AP|R

The other branch (the chorda tympani) terminates in the **submandibular** (sŭb-man-dib′ū-lăr; *sub* = under) **ganglion** near the angle of the mandible. Postganglionic axons projecting from this ganglion innervates the submandibular and sublingual salivary glands in the floor of the mouth, causing an increase in salivary gland secretions (see section 26.2b). Your mouth waters when you smell an aromatic meal due in part to these parasympathetic axons.

WHAT DO YOU THINK?

1 The pterygopalatine ganglion is sometimes nicknamed the "hay fever ganglion." Why is this nickname appropriate?

Glossopharyngeal Nerve (CN IX)

The **glossopharyngeal nerve (CN IX)** innervates the parotid salivary gland. Parasympathetic neurons extend from the brainstem in the glossopharyngeal nerve and synapse on ganglionic neurons in the **otic** (ō′tik; *ous* = ear) **ganglion,** which is positioned anterior to the ear. Postganglionic axons projecting from this ganglion stimulate the parotid salivary glands to increase their secretions.

Vagus Nerve (CN X)

The **vagus nerve (CN X)** innervates the thoracic organs and most of the abdominal organs, as well as the gonads (ovaries and testes).[1] The term vagus means "wanderer," which describes the wandering pathway the vagus nerve makes as it projects inferiorly through the neck and extends throughout the trunk.

In the thoracic cavity, branches of the vagus nerve extend to the heart to decrease heart rate (see section 19.9b) and to the bronchioles of the lung to cause bronchoconstriction (which allows less air into the lungs) (see section 23.5d). In the abdominal cavity branches of the vagus nerve extend to the GI tract to increase motility and release of secretions (see section 26.1e) and to the liver to stimulate the storage of glucose through glycogenesis (see section 27.6d).

WHAT DID YOU LEARN?

7 Which four cranial nerves have a parasympathetic component? What organs are innervated by each?

15.3b Pelvic Splanchnic Nerves

LEARNING OBJECTIVE

2. Explain the actions of the pelvic splanchnic nerves.

The remaining parasympathetic preganglionic axons originate from preganglionic neuron cell bodies housed within the lateral gray regions of the S2–S4 spinal cord segments. These preganglionic axons branch to form the **pelvic splanchnic** (splangk′nik; *splanchnic* = visceral) **nerves,** which contribute to a superior and inferior hypogastric plexus on each side of the body. The preganglionic axons that continue through each plexus project to the ganglionic neurons within either the terminal or intramural ganglia. The postganglionic axons extend to the effector.

The target organs innervated include the distal portion of the large intestine, the rectum, the urinary bladder, the distal part of the ureter, and most of the reproductive organs. This parasympathetic regulation of these target organs causes increased smooth muscle motility (muscle

---

[1]It is unclear what function, if any, these parasympathetic axons have on the gonads.

contraction) and secretory activity in these portions of digestive tract (see chapter 26), contraction of smooth muscle in the urinary bladder wall (see section 24.8c), and erection of the female clitoris and the male penis (see sections 28.3g and 28.4f).

WHAT DID YOU LEARN?

8 What organs are innervated by the pelvic splanchnic nerves?

15.4 Sympathetic Division

The sympathetic division is primarily concerned with preparing the body for exercise and emergencies, and is functionally considered the fight-or-flight division. Recall that the lower motor neurons of the sympathetic division extend only from the thoracic and lumbar regions of the spinal cord (T1–L2) which is why this division is also called the thoracolumbar division.

15.4a Organization and Anatomy of the Sympathetic Division

LEARNING OBJECTIVES

1. Give the location of the sympathetic preganglionic neuron cell bodies.

2. Describe the left and right sympathetic trunks and ganglia.

3. Compare and contrast white and gray rami regarding their location and composition.

4. Explain the differences between the sympathetic trunk ganglia and the prevertebral ganglia.

The sympathetic division is much more anatomically complex than the parasympathetic division so we first describe its anatomic components and then its pathways **(figure 15.6)**. The sympathetic preganglionic neuron cell bodies are housed in the **lateral horn** of the T1–L2 regions of the spinal cord. From there, the preganglionic sympathetic axons travel with somatic motor axons to exit the spinal cord and enter first the anterior roots and then the T1–L2 spinal nerves (see figures 14.3 and 14.13). However, these preganglionic sympathetic axons remain with the spinal nerve for only a short distance before they branch from the spinal nerve.

Sympathetic Trunks and Ganglia

Immediately lateral to the vertebral column and anterior to the paired spinal nerves are the left and right **sympathetic trunks (figure 15.7)**. A sympathetic trunk looks much like a pearl necklace. The "string" of the necklace is composed of bundles of axons, whereas the "pearls" are the **sympathetic trunk ganglia** (also known as *paravertebral,* or *sympathetic chain ganglia*), which house sympathetic ganglionic neuron cell bodies. Interestingly, the sympathetic division was so named because its sympathetic trunk ganglia "follows the spinal cord in sympathy."

One sympathetic trunk ganglion typically is associated with each spinal nerve. However, the cervical portion of each sympathetic trunk is partitioned into only three sympathetic trunk ganglia—the superior, middle, and inferior cervical ganglia—as opposed to the eight cervical spinal nerves. The **superior cervical ganglion** contains postganglionic sympathetic neuron cell bodies whose axons are distributed primarily to structures within the head and neck and to some thoracic viscera. These postganglionic axons innervate the sweat glands and smooth muscle in blood vessels of the head and neck, the dilator pupillae muscle of the eye, and the superior tarsal muscle of the eye

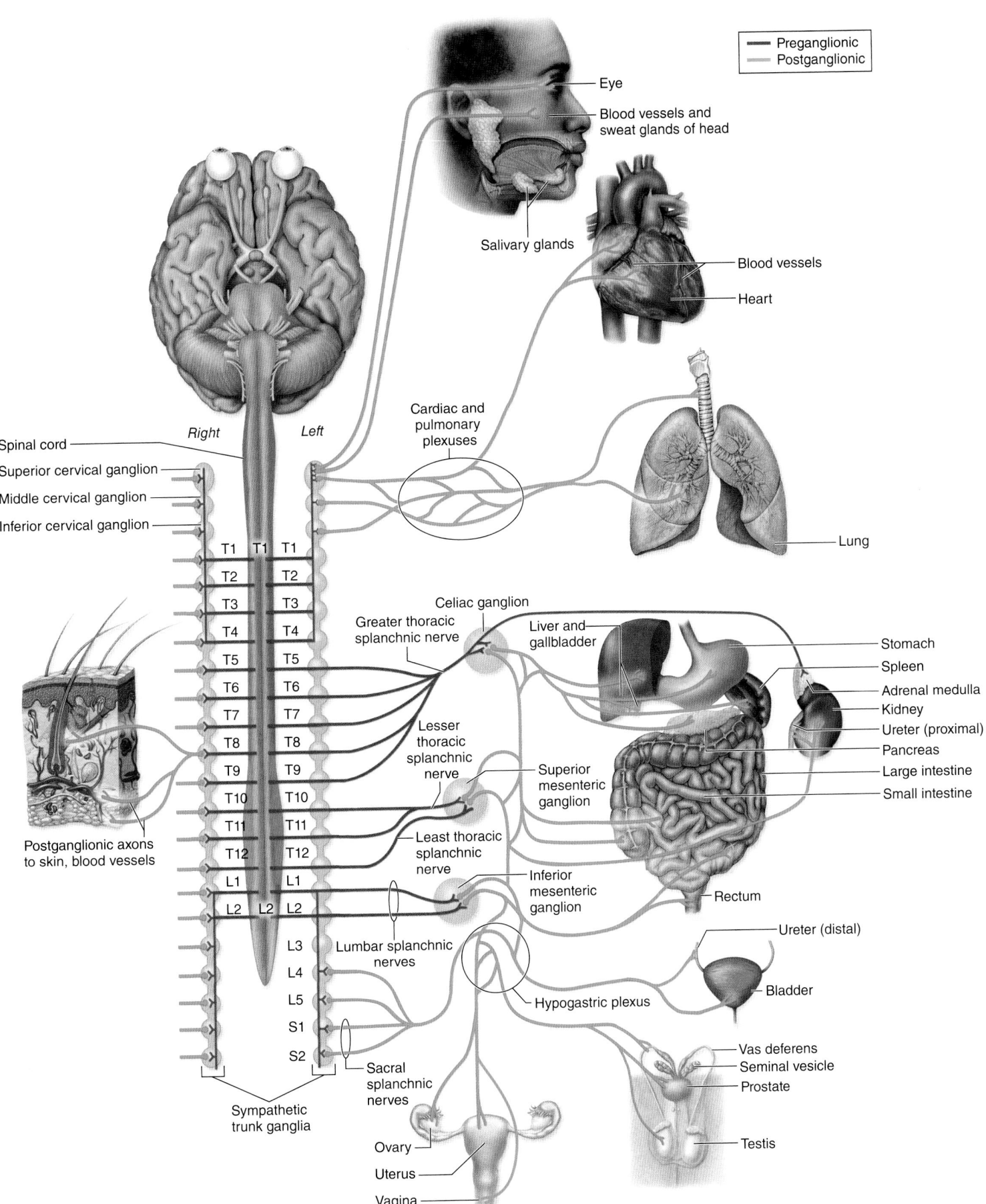

Figure 15.6 Overview of Sympathetic Pathways. Preganglionic axons of the sympathetic division extend from the T1–L2 regions of the spinal cord. Ganglionic axons are located in the sympathetic trunk and prevertebral ganglia, and extend to the target organs. (*Left*) The outflow of preganglionic axons and the distribution of postganglionic axons innervating the skin. (*Right*) Sympathetic postganglionic axon pathways through the gray rami, spinal nerves, and splanchnic nerves. AP|R

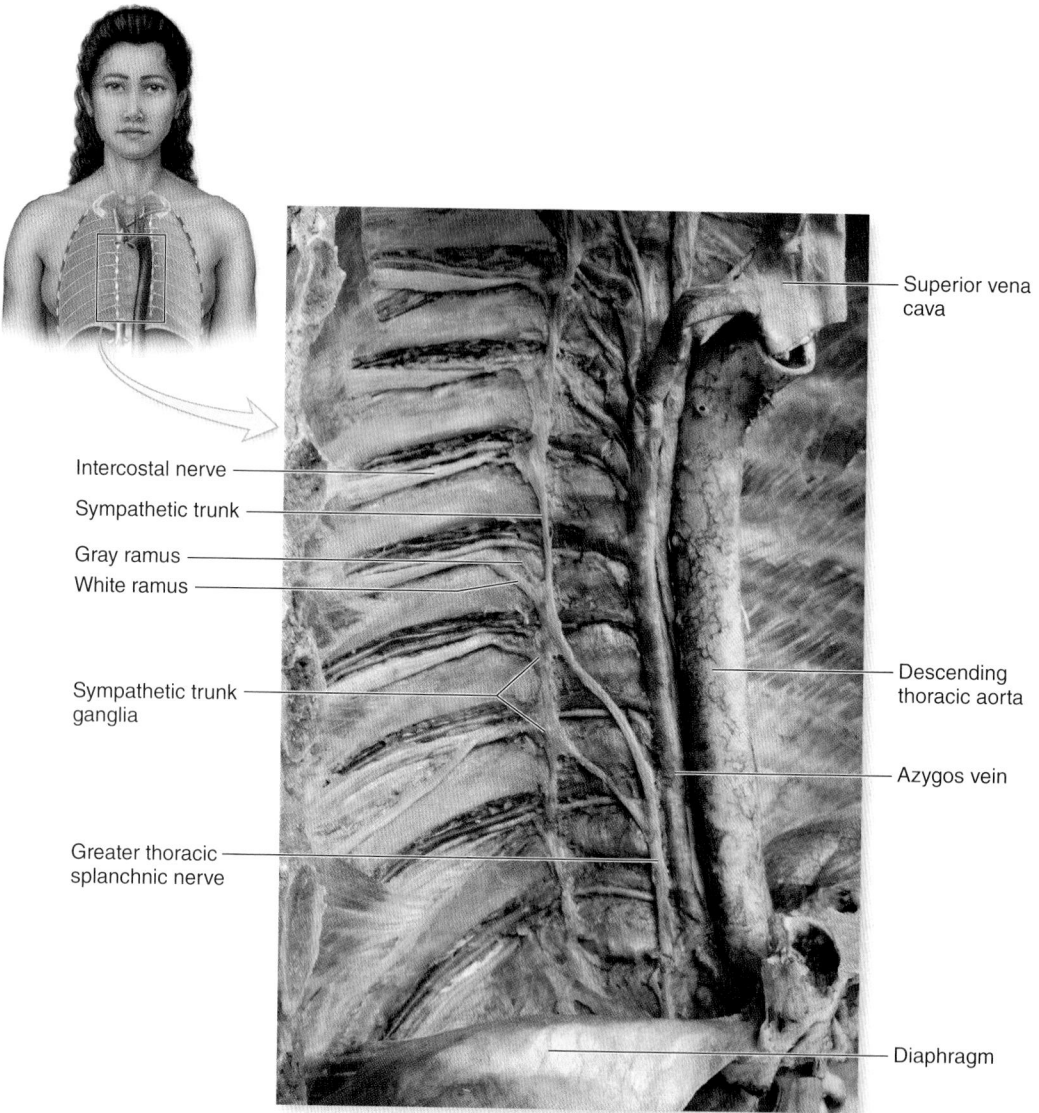

Intercostal nerve

Sympathetic trunk

Gray ramus

White ramus

Sympathetic trunk ganglia

Greater thoracic splanchnic nerve

Superior vena cava

Descending thoracic aorta

Azygos vein

Diaphragm

Figure 15.7 Sympathetic Trunk. An anterolateral photo of the right side of the thoracic cavity shows the sympathetic trunk, the white and gray rami, their attachment to the intercostal nerves, and the greater thoracic splanchnic nerve. **AP|R**

(which elevates the eyelid). The **middle** and **inferior cervical ganglia** also house neuron cell bodies that extend postganglionic axons to the thoracic viscera.

White and Gray Rami

Connecting the spinal nerves to each sympathetic trunk are rami communicantes (rā′mī kŏ-mū-ni-kan′tēz; sing., *ramus communicans; rami* = branches, *communico* = to share with someone) (see figures 14.13 and 15.7). **White rami communicantes** (or simply **white rami**) are composed of *pre*ganglionic sympathetic axons from the T1–L2 spinal nerves to the sympathetic trunk. Thus, white rami are associated only with the T1–L2 spinal nerves. Recall that preganglionic axons are myelinated, giving these rami a whitish appearance. White rami are similar to "entrance ramps" onto a highway.

    Gray rami communicantes (or simply **gray rami**) are composed of *post*ganglionic sympathetic axons that extend from the sympathetic trunk to the spinal nerve. Postganglionic axons are unmyelinated, so these rami have a grayish appearance. Gray rami connect to all spinal nerves, including the cervical, sacral, and coccygeal spinal nerves. By these routes, the sympathetic information that started out in the thoracolumbar region can be dispersed to all parts of the body. Gray rami are similar to "exit ramps" from a highway.

INTEGRATE

CLINICAL VIEW

Horner Syndrome

Horner syndrome is a condition caused by impingement of or injury to the cervical sympathetic trunk or the T1 sympathetic trunk ganglion, where postganglionic sympathetic axons traveling to the head originate. Symptoms typically are limited to the same side of the head where the original injury occurred. The patient presents with **ptosis** (tō'sis; a falling), in which the superior eyelid droops because the superior tarsal muscle is paralyzed. Paralysis of the pupil dilator muscle of the eye results in **miosis** (mī-ō'sis; *muein* = to close the eyes), which is a constricted pupil. **Anhydrosis** (an-hī-drō'sis; *an* = without, *hidros* = sweat) occurs because the sweat glands no longer receive sympathetic innervation. A fourth symptom is distinct facial flushing due to lack of sympathetic innervation to blood vessel walls, resulting in vasodilation.

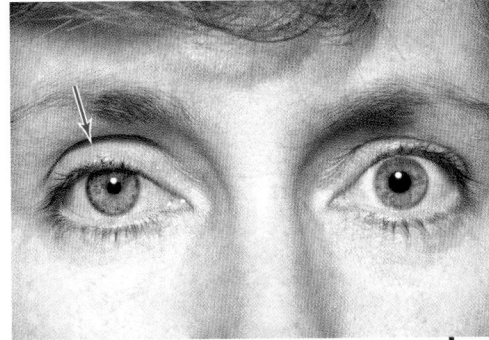

Horner syndrome in right eye. Note the ptosis (blue arrow) and constriction of the pupil.

Sympathetic Splanchnic Nerves

Sympathetic splanchnic nerves are composed of preganglionic sympathetic axons that did not synapse in a sympathetic trunk ganglion (figure 15.6). They run anteriorly from the sympathetic trunk to most of the abdominal and pelvic viscera. These splanchnic nerves should not be confused with the pelvic splanchnic nerves associated with the parasympathetic division described in section 15.3. Some of the larger splanchnic nerves have specific names, such as thoracic splanchnic nerves, lumbar splanchnic nerves, or sacral splanchnic nerves.

Prevertebral Ganglia

Splanchnic nerves typically terminate in **prevertebral** (*or collateral*) ganglia. The prevertebral ganglia differ from the sympathetic trunk ganglia in that (1) they are anterior to the vertebral column (prevertebral) on the anterolateral surface of the aorta; and (2) they are located only in the abdominopelvic cavity. Prevertebral ganglia include the celiac, superior mesenteric, and inferior mesenteric ganglia.

The **celiac ganglia** are adjacent to the origin of the celiac artery. The greater thoracic splanchnic nerves (composed of axons from T5–T9 segment of the spinal cord) synapse on ganglionic neurons within each celiac ganglion. Postganglionic axons from the celiac ganglia innervate the stomach, spleen, liver, gallbladder, and proximal part of the duodenum (first part of small intestine) and part of the pancreas.

The **superior mesenteric** (mez-en-ter'ik; *mesos* = middle, *enteron* = intestine) **ganglia** are adjacent to the origin of the superior mesenteric artery. The lesser and least thoracic splanchnic nerves project to and terminate within the superior mesenteric ganglia. Thus, these ganglia receives preganglionic sympathetic neurons from the T10–T12 segments of the spinal cord. Postganglionic axons extending from the superior mesenteric ganglia innervate the distal half of the duodenum, the remainder of the small intestine, the proximal part of the large intestine, part of the pancreas, the kidneys, and the proximal parts of the ureters.

The **inferior mesenteric ganglia** are adjacent to the origin of the inferior mesenteric artery. They receive sympathetic preganglionic axons via the lumbar splanchnic nerves, which originate in the L1–L2 segments of the spinal cord. The postganglionic axons project to and innervate the distal part of large intestine, rectum, urinary bladder, distal parts of the ureters, and most of the reproductive organs.

WHAT DID YOU LEARN?

9 What is the difference between sympathetic trunk ganglia and prevertebral ganglia?

10 What are the structural and functional differences between the white and gray rami communicantes?

15.4b Sympathetic Pathways

LEARNING OBJECTIVES

5. Describe the four pathways of sympathetic neurons.
6. Compare and contrast which general effector organs are innervated by each pathway.

All sympathetic preganglionic neurons originate in the lateral gray horns of the T1–L2 regions of the spinal cord. The axons of the preganglionic sympathetic neurons travel with somatic motor axons to exit the spinal cord within the anterior roots and then through the spinal nerves. However, these preganglionic sympathetic axons remain with the spinal nerve for only a short distance before they leave the spinal nerve within the white ramus. It is at this point where the major pathways of the sympathetic division differ. Each type of pathway is dependent upon the location of the effector organ being innervated. Axons exit the sympathetic trunk by one of four pathways.

Spinal Nerve Pathway

The **spinal nerve pathway** extends from the spinal cord to effectors of the skin of the neck, torso, and limbs. Skin effectors include sweat glands, smooth muscle forming arrector pili muscles (produce "goose bumps"), and smooth muscle cells within the walls of blood vessels (see chapter 6). In this pathway, a preganglionic neuron synapses with a ganglionic neuron in a sympathetic trunk ganglion (**figure 15.8a**). The postganglionic axon extends through a gray ramus that is at the same "level" as the ganglionic neuron. For example, if the preganglionic and ganglionic neurons synapse in the L4 ganglion, the postganglionic axon extends through the gray ramus at the level of the L4 spinal nerve. After the postganglionic axon extends through the gray ramus, it enters the spinal nerve and extends to its target organ.

Postganglionic Sympathetic Nerve Pathway

The **postganglionic sympathetic nerve pathway** extends from the spinal cord to the internal organs of the thoracic cavity (including the esophagus, heart, lungs, and thoracic blood vessels), effectors of the skin of the head and neck (sweat glands, arrector pili, and blood vessels of the skin), neck viscera as well as the superior tarsal and dilator pupillae muscles in the eye (for increasing the amount of light entering the eye) (see section 16.4b). In this pathway, the preganglionic neuron synapses with a ganglionic neuron in a sympathetic trunk ganglion, but the postganglionic axon does *not* leave the trunk via a gray ramus (figure 15.8b). Instead, the postganglionic axon extends away from the sympathetic trunk ganglion and projects directly to the effector organ.

Splanchnic Nerve Pathway

The **splanchnic nerve pathway** extends from the spinal cord to the abdominal and pelvic organs (e.g., stomach, small intestines, kidney). In this pathway, a preganglionic neuron does not synapse with a ganglionic neuron in a sympathetic trunk ganglion. Rather, the preganglionic axons pass through the sympathetic trunk ganglia without synapsing and extend to the prevertebral ganglia (figure 15.8c). There, the preganglionic axon synapses with a ganglionic neuron. The postganglionic axon then projects to the effector organs.

Adrenal Medulla Pathway

The final pathway is the **adrenal medulla pathway** (figure 15.8d). In this pathway, the internal region of the adrenal gland, called the **adrenal** (ă-drē′năl; *ad* = to, *ren* = kidney) **medulla**, is directly innervated by preganglionic sympathetic axons. (There is no ganglionic neuron.) The axons of the preganglionic neuron extend through both the sympathetic trunk and prevertebral ganglia and then synapse on neurosecretory cells within the adrenal medulla. Stimulation of these cells causes the release of **epinephrine** (ep′i-nef′rin) and **norepinephrine** (nōr′ep-i-nef′rin) into the blood. These hormones are transported

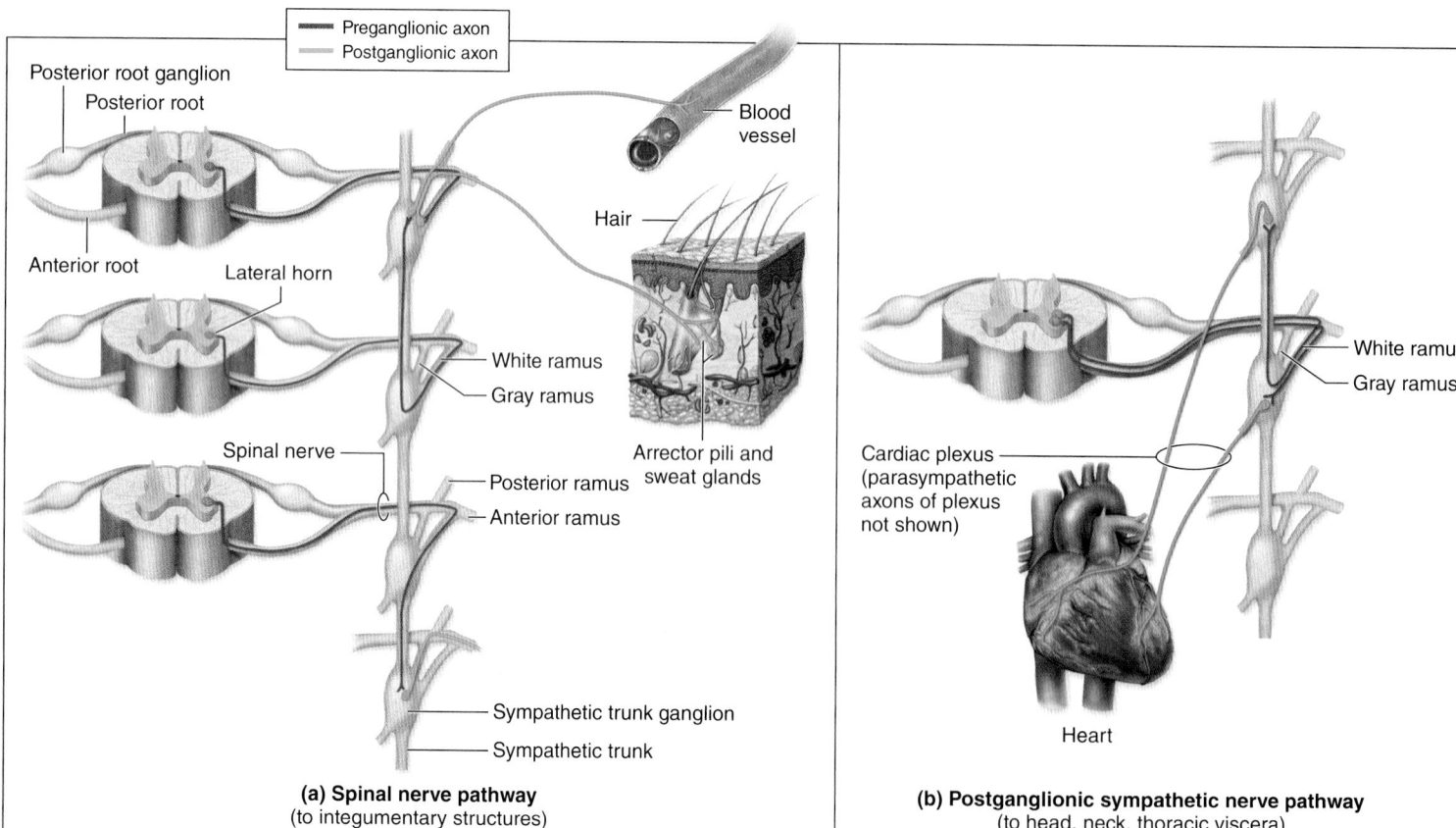

(a) Spinal nerve pathway
(to integumentary structures)

(b) Postganglionic sympathetic nerve pathway
(to head, neck, thoracic viscera)

Figure 15.8 Types of Sympathetic Pathways. Pathways of (*a*) a spinal nerve, (*b*) a postganglionic sympathetic nerve, (*c*) a splanchnic nerve, and (*d*) the adrenal medulla.

Table 15.3   Sympathetic Division Pathways

Pathway	Destination	Spinal Segment Origin	Postganglionic Axon Pathway from Sympathetic Trunk	Effectors Innervated[1]
Spinal nerve	Integumentary structures	T1–L2	Via cervical gray rami to all spinal nerves	Sweat glands, arrector pili muscles, blood vessels in skin of torso, limbs, and neck
Postganglionic sympathetic nerve	Head and neck viscera	T1–T2 (primarily from T1)	Via superior cervical ganglion and travel with blood vessels to the head and neck viscera	Sweat glands, arrector pili muscles, and blood vessels in skin of head; dilator pupillae muscle of eye; tarsal glands of eye; superior tarsal muscle of eye; neck viscera
	Thoracic organs	T1–T5	Via cervical and thoracic ganglia to autonomic nerve plexuses near organs	Esophagus, heart, lungs, blood vessels within thoracic cavity
Splanchnic nerve	Most abdominal organs	T5–T12	Via thoracic splanchnic nerves to prevertebral ganglia (e.g., celiac, superior mesenteric, and inferior mesenteric ganglia)	Abdominal portion of esophagus, stomach, liver, gallbladder, spleen, pancreas, small intestine, most of large intestine, kidneys, ureters, adrenal glands, blood vessels within abdominopelvic cavity
	Pelvic organs	T10–L2	Via lumbar and sacral splanchnic nerves to autonomic nerve plexuses that travel to effectors	Distal part of large intestine, anal canal, and rectum; distal part of ureters; urinary bladder; reproductive organs
Adrenal medulla	Adrenal gland	T8-T12	Via thoracic splanchnic nerves directly to adrenal medulla	Neurosecretory cells of adrenal medulla

1. Sympathetic axons innervate smooth muscle, cardiac muscle, and glands associated with the organs listed.

in the blood and bind to the same receptors as NE neurotransmitter, which prolongs the effects of sympathetic stimulation. Details for epinephrine and norepinephrine are listed in the summary table on regulating the stress response, which directly follows chapter 17 (Table R.5). (The relative amounts of these two hormones released are not equal. Typically, epinephrine accounts for approximately 80% and norepinephrine accounts for 20% of the hormone molecules that are released.) These hormones then circulate within the blood and help potentiate (prolong) the effects of the sympathetic stimulation. For example, if you narrowly miss getting into a car accident, your heart continues to beat quickly, you breathe rapidly, and you feel tense and alert well after the event because of prolonged effects of the sympathetic stimulation of the adrenal medulla. **Table 15.3** summarizes the sympathetic division pathways.

WHAT DO YOU THINK?

2   When a person is very stressed and tense, his or her blood pressure typically rises. What aspect of the sympathetic division causes this rise in blood pressure?

WHAT DID YOU LEARN?

11   How do the spinal nerve and splanchnic nerve sympathetic pathways differ, regarding both the pathway and the organs innervated?

12   In what ways does the adrenal medulla pathway help prolong the effects of the sympathetic stimulation?

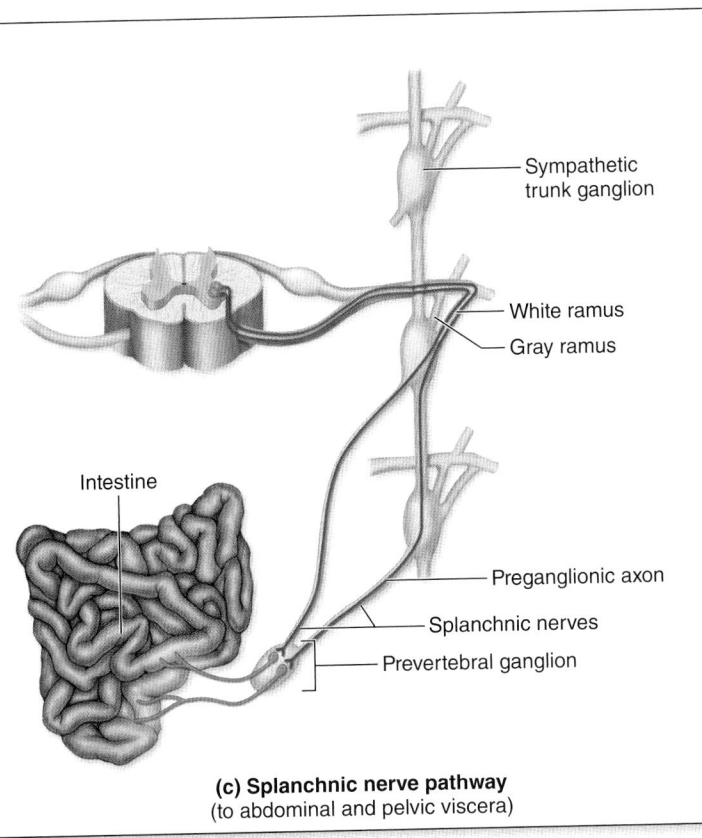

(c) Splanchnic nerve pathway
(to abdominal and pelvic viscera)

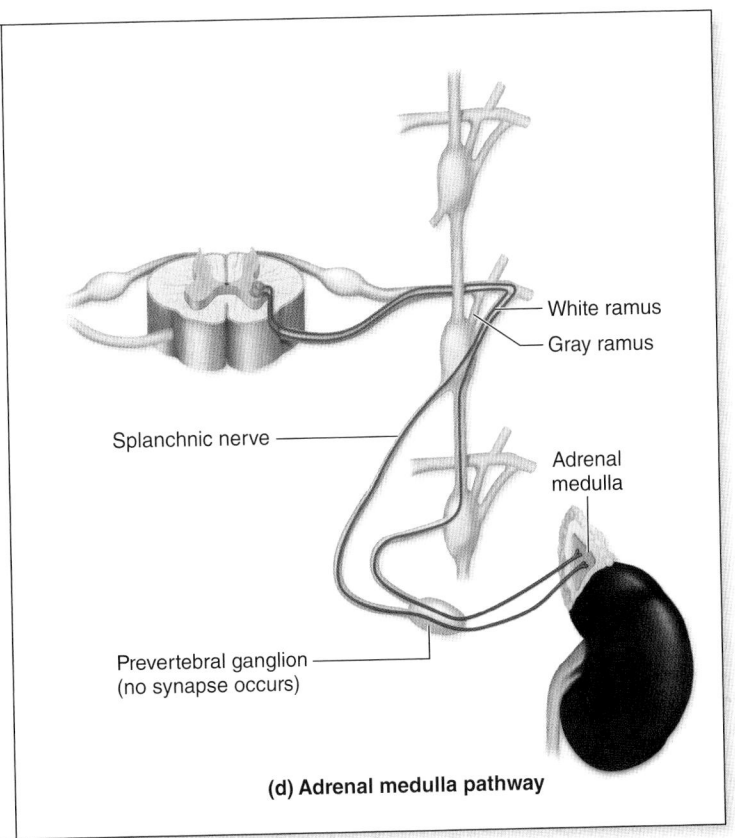

(d) Adrenal medulla pathway

15.5 Comparison of Neurotransmitters and Receptors of the Two Divisions

Transmission of a nerve signal to synaptic knobs causes the release of neurotransmitters into the synaptic cleft. The ANS utilizes several types of neurotransmitters, which are discussed here.

15.5a Overview of ANS Neurotransmitters

LEARNING OBJECTIVE

1. Identify the targets of the cholinergic and adrenergic neurotransmitters of the ANS.

Acetylcholine (ACh) and norepinephrine (NE) are the main neurotransmitters used in the ANS **(figure 15.9)**. These neurotransmitters will bind to specific receptors on the postsynaptic cell. Depending upon the receptor type, the neurotransmitter may cause either stimulation or inhibition.

Neurons that synthesize and release acetylcholine (ACh) are called **cholinergic** (kol-in-er′jik; *ergon* = work) **neurons.** Receptors that bind ACh are called **cholinergic receptors.** We will see in the next section that there are two types of cholinergic receptors (nicotinic and muscarinic), and each responds differently to binding of ACh. Cholinergic neurons include the following:

- All sympathetic and parasympathetic preganglionic neurons
- All parasympathetic ganglionic neurons
- The specific sympathetic ganglionic neurons that innervate sweat glands of the skin and blood vessels in skeletal muscle tissue

Neurons that synthesize and secrete norepinephrine (NE) are called **adrenergic** (ad-re-ner′jik) **neurons.** Most other sympathetic ganglionic neurons are adrenergic. Receptors that bind NE (or a related molecule, like epinephrine) are called **adrenergic receptors.** These receptors are subdivided into alpha (α) and beta (β) types, and are discussed in further detail in section 15.5c.

WHAT DID YOU LEARN?

13 Which ANS neurons are cholinergic? Which are adrenergic?

15.5b Cholinergic Receptors

LEARNING OBJECTIVE

2. Describe the two types of cholinergic receptors and the action of each when the neurotransmitter acetylcholine binds to them.

Two categories of cholinergic receptors, nicotinic and muscarinic, have been identified in the CNS and PNS. They were differentiated and named because molecules that are similar to ACh bind to them and cause their stimulation:

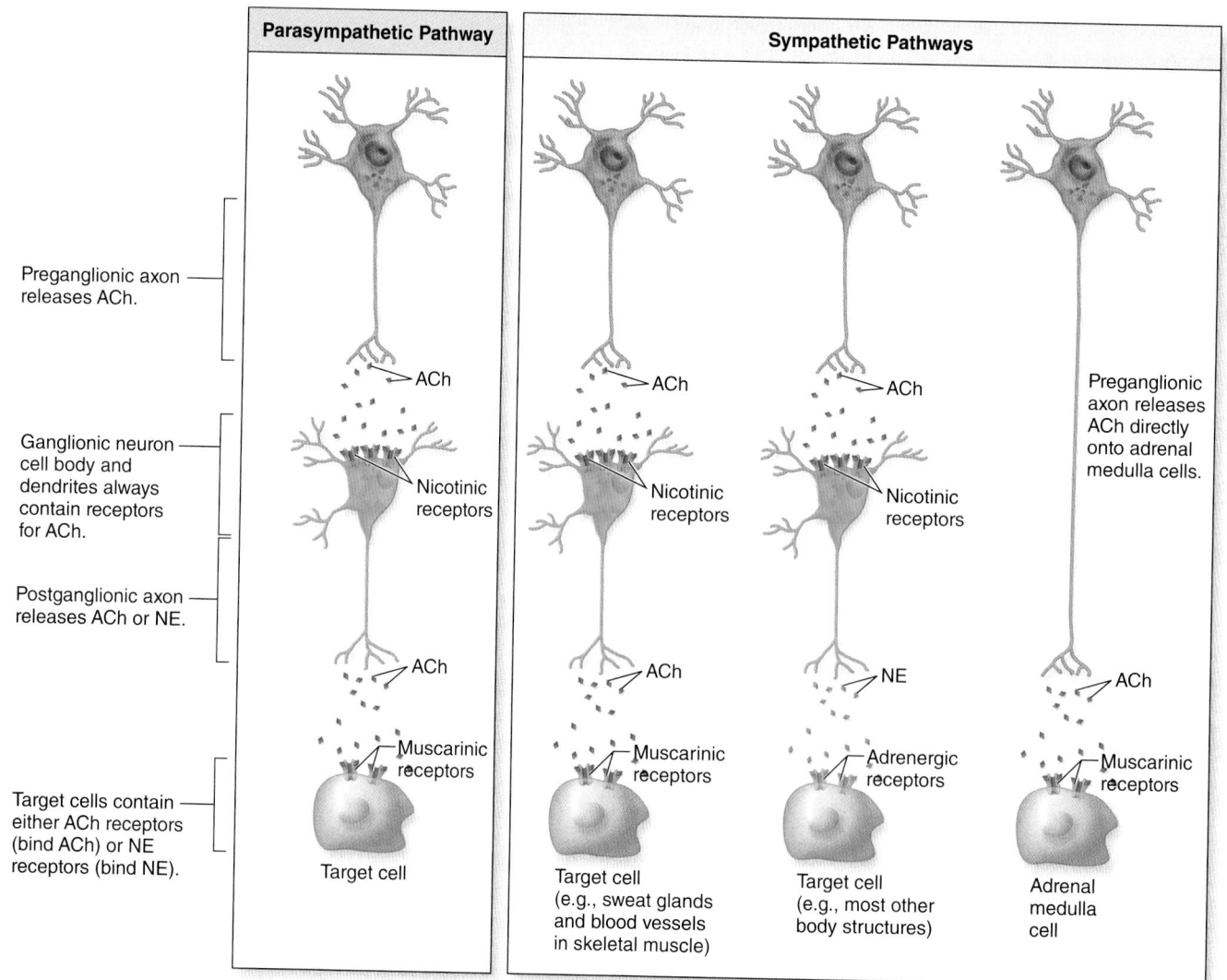

Preganglionic axon releases ACh.

Ganglionic neuron cell body and dendrites always contain receptors for ACh.

Postganglionic axon releases ACh or NE.

Target cells contain either ACh receptors (bind ACh) or NE receptors (bind NE).

Parasympathetic Pathway

ACh
Nicotinic receptors
ACh
Muscarinic receptors
Target cell

Sympathetic Pathways

ACh
Nicotinic receptors
ACh
Muscarinic receptors
Target cell (e.g., sweat glands and blood vessels in skeletal muscle)

ACh
Nicotinic receptors
NE
Adrenergic receptors
Target cell (e.g., most other body structures)

Preganglionic axon releases ACh directly onto adrenal medulla cells.
ACh
Muscarinic receptors
Adrenal medulla cell

Figure 15.9 Comparison of Neurotransmitters in the Autonomic Nervous System. In the parasympathetic pathway, both the preganglionic and postganglionic axons release acetylcholine (ACh). In the sympathetic pathways, all preganglionic axons and a few specific postganglionic axons release ACh. Most postganglionic sympathetic axons release norepinephrine (NE).

Nicotinic receptors were so named because they are all sensitive to the drug nicotine. These receptors are found on the cell bodies and dendrites of all ganglionic neurons (figure 15.9), as well as on the cells of the adrenal medulla. When nicotinic receptors bind ACh, they open ion channels to allow greater movement of sodium ions (Na^+) into the cell than potassium ions (K^+) from the cell. Thus, the membrane depolarizes and an excitatory postsynaptic potential (EPSP) is produced (see section 12.8a). In other words, ACh binding to nicotinic receptors always produces a stimulatory or excitatory response (This excitatory response to ACh also occurs in the somatic nervous system, and is described for the activation of skeletal muscle fibers in section 10.3b.)

Nicotinic receptors have various subtypes. This accounts for the difference in response that a given neurotransmitter may have at different locations. For example, the nicotinic receptor at the neuromuscular junction is blocked by the toxin curare, but it is not blocked by the cholinergic drugs hexamethonium or mecamylamine. The reverse conditions exist for nicotinic receptors on postganglionic neurons. Here, hexamethonium and mecamylamine bind, but curare does not.

WHAT DO YOU THINK?

3 What effect, if any, would smoking have on the nicotinic receptors of the ANS?

Muscarinic receptors were so named because they respond to muscarine, a mushroom toxin. They are found in all target organs stimulated by the parasympathetic division and in the few selected sympathetic target cells (e.g., sweat glands in the skin and blood vessels in skeletal muscle). There are different subtypes of muscarinic receptors, which have different effects on various body systems. These different subclasses of muscarinic receptors are either stimulated or inhibited by binding ACh. For example, the binding of ACh to smooth muscle in the gastrointestinal (GI) tract will result in contraction and increased motility of the muscle, whereas binding of ACh to muscarinic receptors on cardiac muscle pacemaker cells results in decreasing heartbeat rate. The drug pilocarpine (used to treat glaucoma) binds to muscarinic receptors to stimulate ciliary muscles to contract, which facilitates drainage of aqueous humor from the anterior chamber of the eye. All muscarinic receptors use second messenger systems (see section 17.5b).

Nicotinic and muscarinic receptors are compared in **table 15.4**.

WHAT DID YOU LEARN?

14 Where are nicotinic and muscarinic receptors each located?

15 When a neurotransmitter binds to a nicotinic effector, is the effect stimulatory or inhibitory?

15.5c Adrenergic Receptors

LEARNING OBJECTIVES

3. List the neurotransmitters categorized as catecholamines.
4. Name the four adrenergic receptors and give the locations of each.

Recall that signaling molecules (e.g., neurotransmitters, hormones) are called **ligands** when they specifically bind to receptors in the plasma membrane (see section 4.5b). The class of ligands that bind to adrenergic receptors in neurons are called **biogenic amines,** or *monoamines.* One category of biogenic amines is called **catecholamines** because of the presence of a catechol ring structure in the molecule. Catecholamines include dopamine, norepinephrine, and epinephrine (see section 17.3a).

As mentioned earlier, the two types of adrenergic receptors are **alpha (α)** and **beta (β) receptors,** which may be further divided into subclasses such as α_1, α_2, β_1, and β_2. Target cells with α receptors typically are stimulated, whereas those cells with β receptors may either be stimulated or inhibited in response to binding neurotransmitter (or hormone). (Note that there are exceptions to this rule.) These types of receptors may be further subdivided, as follows:

- **α_1 receptors** are located within plasma membranes of most smooth muscle cells and stimulate smooth muscle contraction. The specific organs with α_1 receptors include most blood vessels (including those going to the skin, GI tract, and kidneys), arrector pili muscles, uterus, ureters, internal urethral sphincter, and dilator pupillae of the eye. These receptors are involved with vasoconstriction of the above blood vessels, contraction of the uterine wall, contraction of the arrector pili muscles, closing of the internal urethral sphincter, and dilation of the pupil, respectively.

- **α_2 receptors** are located throughout the CNS and are in the pancreas to inhibit insulin secretion. In addition, these receptors are located in GI sphincters, and when they are stimulated they constrict the sphincter.

- **β_1 receptors** primarily have stimulatory effects. They are found within the heart, where they increase heart rate and force of contraction (see section 19.5b); and in the kidney, where they stimulate secretion of renin (see section 20.6b and 25.4a) to blood pressure.

- **β_2 receptors** primarily have inhibitory effects. They are present in smooth muscle plasma membranes of blood vessels supplying the heart wall, liver, and skeletal muscle. Unlike

Table 15.4	Cholinergic Receptors	
Characteristics	**Nicotinic**	**Muscarinic**
Primary locations	All postganglionic neurons in the ANS (both parasympathetic division and sympathetic division)	All effectors of parasympathetic target organs
	Adrenal medulla (gland innervated by preganglionic neuron of sympathetic division)	Limited sympathetic target organs that have muscarinic receptors (e.g., sweat glands)
	Neuromuscular junction of skeletal muscle	
	Some neurons of central nervous system (neurons involved in learning and memory)	
Excitatory or inhibitory	Always excitatory	Generally excitatory (except on cardiac muscle pacemaker cells)
Examples of drugs that interact with receptor	*Nicotine* binds with all nicotinic receptors	Muscarine binds with all muscarinic receptors
	Curare binds to specific subtype of nicotinic receptors at the neuromuscular junctions	*Pilocarpine* binds to specific subtype of muscarinic receptors of ciliary muscles within the eye
	Hexamethonium or *mecamylamine* binds to specific subtype of nicotinic receptors of postganglionic neurons in the ANS	

Table 15.5 — Adrenergic Receptors

Characteristics	Alpha (α) Receptors		Beta (β) Receptors		
	α₁	α₂	β₁	β₂	β₃
Primary locations and specific actions	Almost all effectors of sympathetic target organs (except cardiac muscle) Causes contraction of most smooth muscle including blood vessels of the skin, blood vessels of GI tract, blood vessels of kidneys, arrector pili, uterus, ureters, internal urethral sphincter, dilator pupillae muscle of eye	Pancreas (inhibits insulin release) CNS Platelets (facilitate blood clotting by promoting platelet aggregation)	Heart (both sinoatrial node and cardiac muscle; increases heart rate and force of contraction); kidney (stimulates release of renin) Increases blood pressure	Causes smooth muscle relaxation in blood vessels of the heart, liver, and skeletal muscle Causes smooth muscle relaxation within bronchioles, uterus, GI tract	Stimulates lipolysis in adipose connective tissue Increases blood pressure
General effect	Excitatory	Inhibitory or excitatory	Excitatory	Primarily inhibitory	Excitatory
Examples of drugs that interact with receptor	*Phenylephrine* causes vasoconstriction of nasal blood vessels, decreasing nasal secretions	*Clonidine* is used to treat high blood pressure by stimulating α₂ receptors in the vasomotor center of the brainstem	*Propranolol* is a nonselective beta blocker used to treat high blood pressure	*Albuterol* dilates bronchioles; used to treat asthma *Terbutaline* relaxes uterine wall to delay preterm labor	*Mirabegron* relaxes urinary bladder wall, used to treat overactive bladder

---

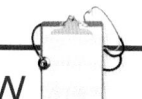

INTEGRATE

CLINICAL VIEW
Epinephrine for Treatment of Asthma

Asthma is a condition in which airflow into the lung is decreased due to the narrowing of the bronchioles. Bronchioles contain β₂ receptors, and epinephrine (hormone) binds more effectively to β₂ receptors, and causes greater relaxation of the smooth muscles of bronchioles than does norepinephrine. The greater the degree of bronchodilation, then the greater the rate of airflow moving into and out of the lungs. Consequently, epinephrine (not norepinephrine) is the active ingredient in medicines for treating asthma.

activation of α₁ receptors, activation of β₂ receptors causes smooth muscle relaxation, resulting in vasodilation of these specific blood vessels. Their stimulation also causes bronchodilation in the lung, relaxation of uterine and GI tract smooth muscle, and relaxation of the detrusor muscle of the urinary bladder.

- **β₃ receptors** may have stimulatory effects. They are found mainly in adipose connective tissue, where they stimulate lipolysis (breakdown of triglycerides into fuel molecules). They also cause smooth muscle relaxation in the urinary bladder.

Alpha and beta receptors are compared in **table 15.5**.

WHAT DID YOU LEARN?

16 What are the different types of catecholamines?

17 How is it possible for the stimulation of adrenergic receptors to result in either vasoconstriction or vasodilation of selected blood vessels?

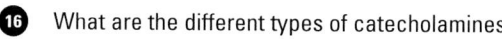

15.6 Interactions Between the Parasympathetic and Sympathetic Divisions

Most organs are innervated by both divisions of the autonomic nervous system, and these control the targets continuously to varying degrees. The parasympathetic and sympathetic division effects are compared in **table 15.6** and **figure 15.10**.

15.6a Autonomic Tone

LEARNING OBJECTIVE

1. Discuss the nature of autonomic tone and its effects.

The parasympathetic and sympathetic divisions both continuously release neurotransmitter to regulate specific target organs for either sustained stimulation or inhibition; a process referred to as the **autonomic tone.** Activity of an organ may be controlled merely by the change in tone within a single ANS division.

For example, the diameter of most blood vessels is maintained in a partially constricted state by the effects of sympathetic tone. A decrease in stimulation below the sympathetic tone causes vessel dilation, whereas an increase above sympathetic tone brings about vessel constriction. If

Table 15.6	Main Effects of the Parasympathetic and Sympathetic Divisions	
Effector	**Parasympathetic Innervation Effects**	**Sympathetic Innervation Effects**
CARDIOVASCULAR SYSTEM		
Heart	Decreases heart rate	Increases heart rate and force of contraction
Blood vessels		
Coronary	None	Vasodilation (β receptors) or vasoconstriction (α receptors)
To skeletal muscle	None	Vasodilation (β receptors)
Integumentary and most other blood vessels	None	Vasoconstriction (α receptors)
DIGESTIVE SYSTEM		
Salivary glands	Stimulate watery secretion	Stimulate more viscous secretion
GI tract gland secretion	Stimulates	Inhibits
Smooth muscle in GI tract	Stimulates peristalsis (motility)	Inhibits peristalsis (motility)
Sphincters	Relax (open, to allow passage of materials)	Contract (close to prevent passage of materials)
GI tract blood vessels	Vasodilation	Vasoconstriction
Liver	Stimulates glycogenesis (formation of glycogen from glucose)	Stimulates glycogenolysis (breakdown of glycogen into glucose)
RESPIRATORY SYSTEM		
Bronchi/bronchioles of lungs	Bronchoconstriction (airway narrows)	Bronchodilation (airway widens)
URINARY SYSTEM		
Kidneys	None	Stimulates release of renin (to raise blood pressure)
Bladder (muscle wall)	Contraction (to facilitate urination)	Relaxation
Internal urethral sphincter	Relax (open, to facilitate urination)	Contract (close, to inhibit urination)
REPRODUCTIVE SYSTEM		
Penis	Stimulates erection	Stimulates ejaculation
Clitoris	Stimulates erection	None
Uterine muscle	None	Contraction
Gland secretion	Stimulates	Inhibits
Vaginal muscular wall	None	Contraction
INTEGUMENTARY SYSTEM		
Arrector pili	None	Contraction to cause hair elevation (i.e., "goose bumps")
Sweat glands	None	Secretion
EYE		
Pupil size	Constriction (allows less light into the eye)	Dilation (allows more light into the eye)
Ciliary muscle	Contraction for near vision	None
Lacrimal gland	Stimulates secretion	None
ADRENAL MEDULLA		
	None	Stimulates release of epinephrine and norepinephrine
ADIPOSE CONNECTIVE TISSUE		
	None	Stimulates lipolysis (β_3 receptors)

the initial level of sympathetic tone were not present, then only vasoconstriction could occur as a result of sympathetic division activity.

WHAT DID YOU LEARN?

18 How does autonomic tone permit the control of blood vessel diameter by sympathetic innervation?

Figure 15.10 Comparison of the Parasympathetic and Sympathetic Divisions of the ANS. (*a*) The parasympathetic division, also known as the "rest-and-digest" division, has its preganglionic neurons located in the cranial region of the brainstem and sacral region of the spinal cord. (*b*) The sympathetic division, also known as the "fight-or-flight" division, contains preganglionic neurons located in the T1–L2 regions of the spinal cord.

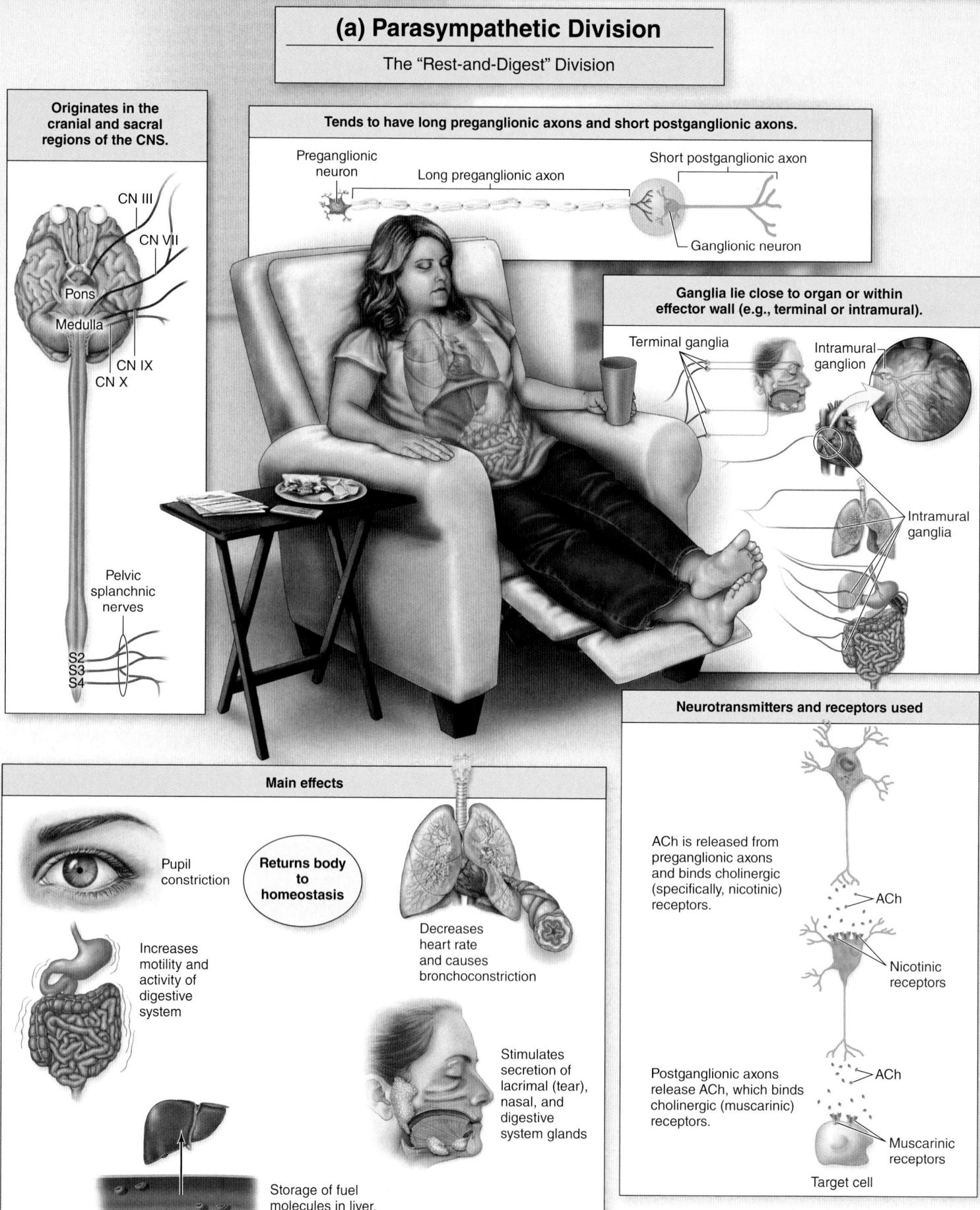

(a) Parasympathetic Division

The "Rest-and-Digest" Division

Originates in the cranial and sacral regions of the CNS.

CN III
CN VII
Pons
Medulla
CN IX
CN X

Pelvic splanchnic nerves

S2
S3
S4

Tends to have long preganglionic axons and short postganglionic axons.

Preganglionic neuron
Long preganglionic axon
Short postganglionic axon
Ganglionic neuron

Ganglia lie close to organ or within effector wall (e.g., terminal or intramural).

Terminal ganglia
Intramural ganglion
Intramural ganglia

Main effects

Pupil constriction

Returns body to homeostasis

Increases motility and activity of digestive system

Decreases heart rate and causes bronchoconstriction

Stimulates secretion of lacrimal (tear), nasal, and digestive system glands

Storage of fuel molecules in liver.

Neurotransmitters and receptors used

ACh is released from preganglionic axons and binds cholinergic (specifically, nicotinic) receptors.

ACh
Nicotinic receptors

Postganglionic axons release ACh, which binds cholinergic (muscarinic) receptors.

ACh
Muscarinic receptors

Target cell

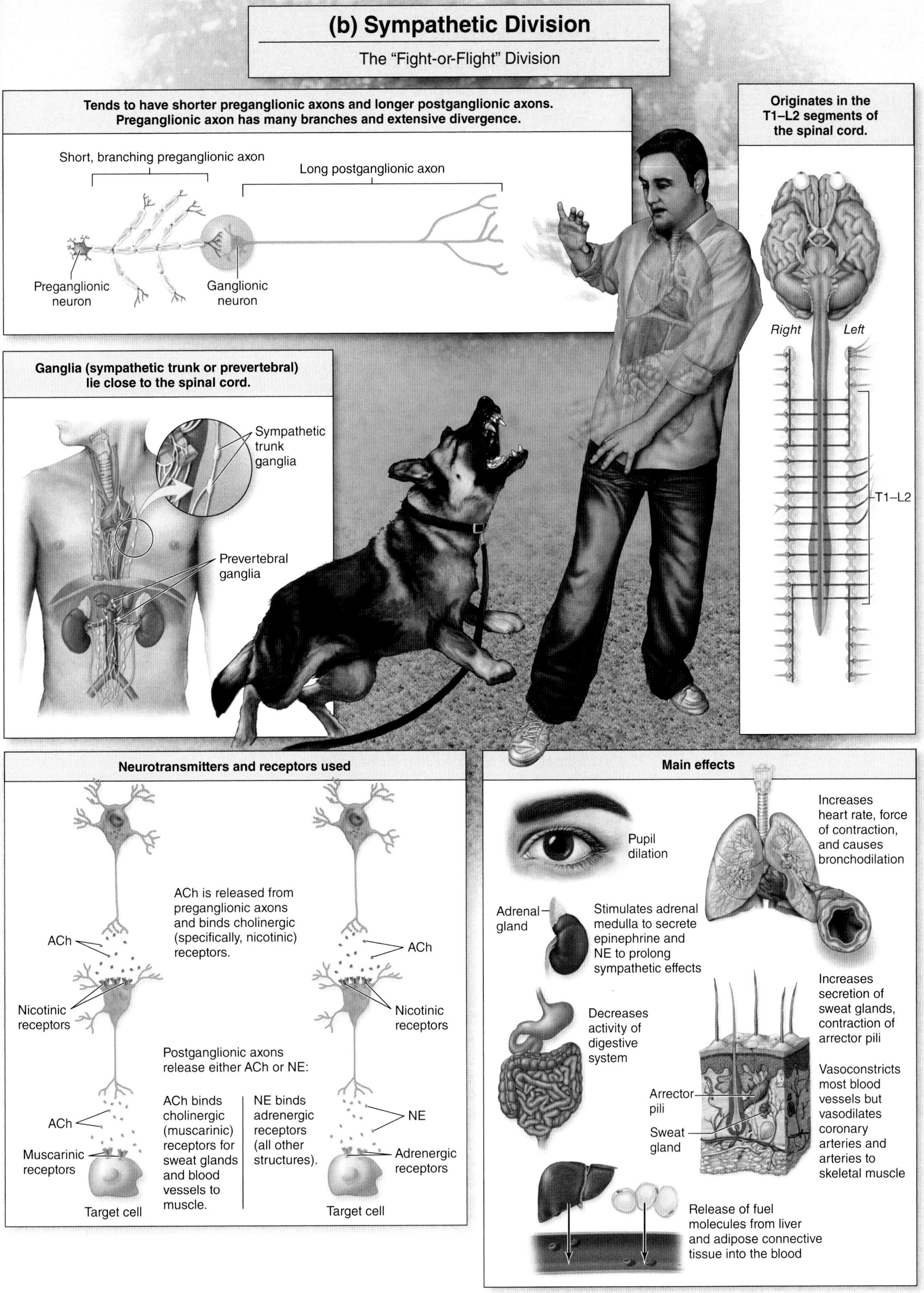

(b) Sympathetic Division

The "Fight-or-Flight" Division

Tends to have shorter preganglionic axons and longer postganglionic axons. Preganglionic axon has many branches and extensive divergence.

Short, branching preganglionic axon

Long postganglionic axon

Preganglionic neuron

Ganglionic neuron

Originates in the T1–L2 segments of the spinal cord.

Right   Left

T1–L2

Ganglia (sympathetic trunk or prevertebral) lie close to the spinal cord.

Sympathetic trunk ganglia

Prevertebral ganglia

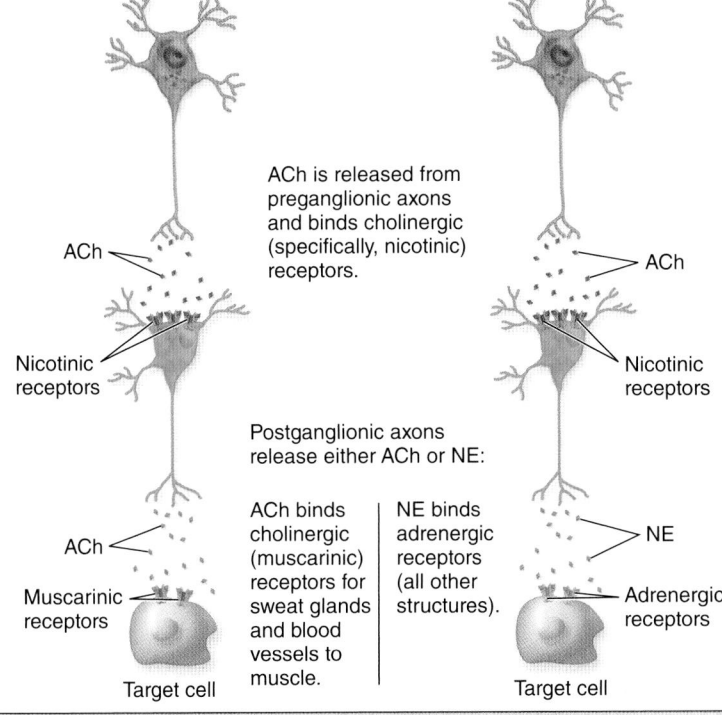

Neurotransmitters and receptors used

ACh

Nicotinic receptors

ACh is released from preganglionic axons and binds cholinergic (specifically, nicotinic) receptors.

ACh

Nicotinic receptors

ACh

Muscarinic receptors

Target cell

Postganglionic axons release either ACh or NE:

ACh binds cholinergic (muscarinic) receptors for sweat glands and blood vessels to muscle.

NE binds adrenergic receptors (all other structures).

NE

Adrenergic receptors

Target cell

Main effects

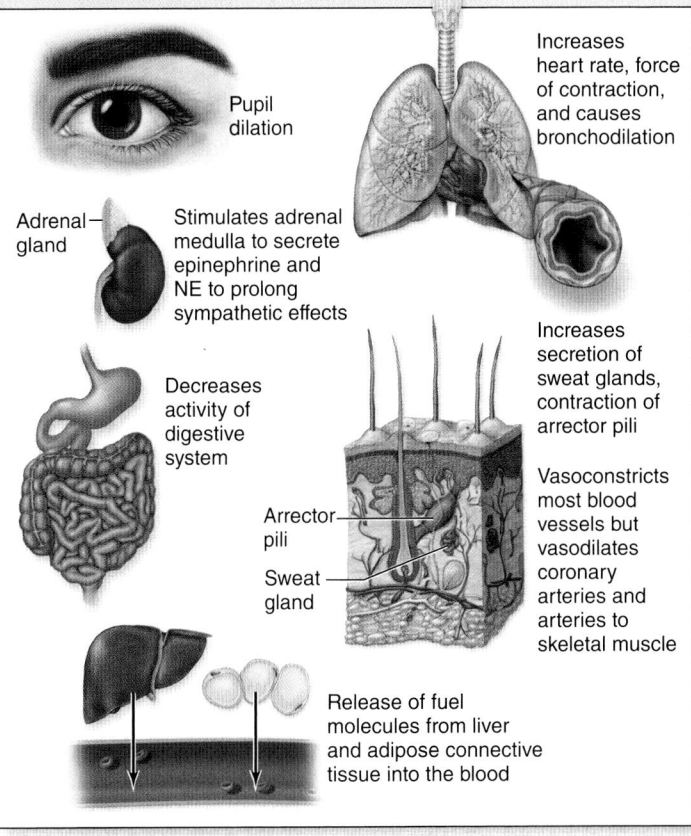

Pupil dilation

Adrenal gland

Stimulates adrenal medulla to secrete epinephrine and NE to prolong sympathetic effects

Decreases activity of digestive system

Arrector pili

Sweat gland

Increases heart rate, force of contraction, and causes bronchodilation

Increases secretion of sweat glands, contraction of arrector pili

Vasoconstricts most blood vessels but vasodilates coronary arteries and arteries to skeletal muscle

Release of fuel molecules from liver and adipose connective tissue into the blood

INTEGRATE

CLINICAL VIEW

Raynaud Syndrome

Raynaud syndrome, or *Raynaud phenomenon,* is a sudden constriction of the small arteries of the digits. The immediate decrease in blood flow results in blanching (loss of the normal hue) of the skin distal to the area of vascular constriction. The constriction is accompanied by pain, which may even continue for a while after the vessels have dilated and restored the local blood flow. Episodes are typically triggered by exposure to cold, although emotional stress has been known to precipitate a Raynaud attack. This condition is more common in women than in men, and is believed to result from an exaggerated local sympathetic response. The severity of this medical condition determines the frequency and the length of time of each occurrence. Most people affected with Raynaud syndrome must avoid the cold and other triggering circumstances.

15.6b Dual Innervation

LEARNING OBJECTIVES

2. Explain what is meant by dual innervation.
3. Describe the antagonistic and cooperative effects of dual innervation.

Many effectors of the ANS have **dual innervation,** meaning that they are innervated by postganglionic axons from both parasympathetic and sympathetic divisions. The actions caused by activities of both the divisions on the same organ usually result in effects that are antagonistic or cooperative.

Antagonistic Effects

Generally, the effects of parasympathetic and sympathetic innervation to the same organ are antagonistic—that is, they oppose each other. Some variations occur in how antagonistic effects are expressed, as follows:

- **Control of heart rate.** Parasympathetic stimulation slows down the heart rate, whereas sympathetic stimulation speeds up the heart rate. The same heart muscle effector cells receive this opposing stimulation. The two divisions are able to cause different responses because cardiac muscle cells contain more than one type of cellular receptor (e.g., muscarinic receptors and β_1 receptors).

- **Control of muscular activity in the gastrointestinal tract.** Parasympathetic stimulation of smooth muscle cells in the gastrointestinal tract wall increases their force of contraction and thus increases gastrointestinal tract motility. Conversely, sympathetic stimulation decreases the force of contraction and thus decreases motility. Again, both ANS divisions innervate the same effector cells, but house two different types of receptors (see section 26.1).

- **Control of pupil diameter in the iris of the eye.** Here, different effectors are innervated by the ANS divisions. Parasympathetic stimulation of the circular muscle layer in the iris causes pupil constriction, whereas sympathetic stimulation of the radial muscle layer in the iris causes pupil dilation (see section 16.4b).

Cooperative Effects

Cooperative effects are seen when both parasympathetic and sympathetic stimulation cause different responses that together produce a single, distinct result. The best example of cooperative effects occurs in the sexual function of the male reproductive system. The male penis becomes erect as a result of parasympathetic innervation, and ejaculation of semen from the penis is facilitated by stimulation from the sympathetic division (see section 28.4f). This synergistic effort of ANS divisions facilitates reproduction.

WHAT DID YOU LEARN?

19 What are the antagonistic effects of the sympathetic and parasympathetic divisions on the heart?

20 How do the sympathetic and parasympathetic divisions exhibit cooperative effects on the male reproductive system?

15.6c Systems Controlled Only by the Sympathetic Division

LEARNING OBJECTIVE

4. Describe the systems innervated only by the sympathetic division and how they function.

In some ANS effectors, opposing effects are achieved without dual innervation. For example, many blood vessels are innervated by sympathetic axons only. Increased sympathetic stimulation increases smooth muscle contraction, resulting in increased blood pressure. An analogy would be when you press down on a gas pedal to cause a car to accelerate. Decreasing the sympathetic stimulation below the autonomic tone will result in vasodilation, just as lifting your foot off of the gas pedal may slow a car down (because you are not supplying the gas). Thus, opposing effects are achieved merely by increasing or decreasing the autonomic tone in the sympathetic division.

Other examples of innervation by only the sympathetic division are seen in sweat glands in the trunk (stimulates sweating) and innervation of arrector pili muscles in the skin to cause "goose bumps" (see section 6.2b). The neurosecretory cells of the adrenal medulla also are innervated only by the sympathetic division. These cells have nicotinic receptors, so upon release of ACh by the postganglionic axon, the adrenal medulla cells are stimulated to release epinephrine and norepinephrine into the blood. There, these molecules act as hormones and prolong the fight-or-flight effects of the sympathetic division.

INTEGRATE

LEARNING STRATEGY

The mnemonic **ABE** may help you remember which structures receive only sympathetic innervation:

A = Adrenal medulla
B = Blood vessels
E = Effectors of the skin (arrector pili muscles and sweat glands)

🔎 **WHAT DID YOU LEARN?**

21 What are the body structures innervated by the sympathetic division only?

15.7 Integration of Autonomic System Function

Both divisions of the autonomic nervous system innervate organs through specific axon bundles called autonomic plexuses. These are networks of nerves that are located within the thoracic and abdominopelvic cavity. Homeostasis is maintained through autonomic reflexes that occur in the organs innervated by autonomic nerves via the plexuses. Overall, the autonomic functions are controlled by the CNS.

15.7a Autonomic Plexuses

LEARNING OBJECTIVE

1. Describe the structure and location of the five autonomic plexuses.

Autonomic plexuses are collections of sympathetic postganglionic axons and parasympathetic preganglionic axons, as well as some visceral sensory axons. These sympathetic and parasympathetic axons are close to one another, but they do not interact or synapse with each another. Although these plexuses look like disorganized masses of axons, they provide a complex innervation pattern to their target organs **(figure 15.11)**.

In the mediastinum of the thoracic cavity, the **cardiac plexus** consists of sympathetic postganglionic axons that originate in the cervical and thoracic sympathetic trunk ganglia, as well as parasympathetic preganglionic axons from the vagus nerve. Increased sympathetic activity increases heart rate and blood pressure, whereas increased parasympathetic activity decreases heart rate (see section 19.5b).

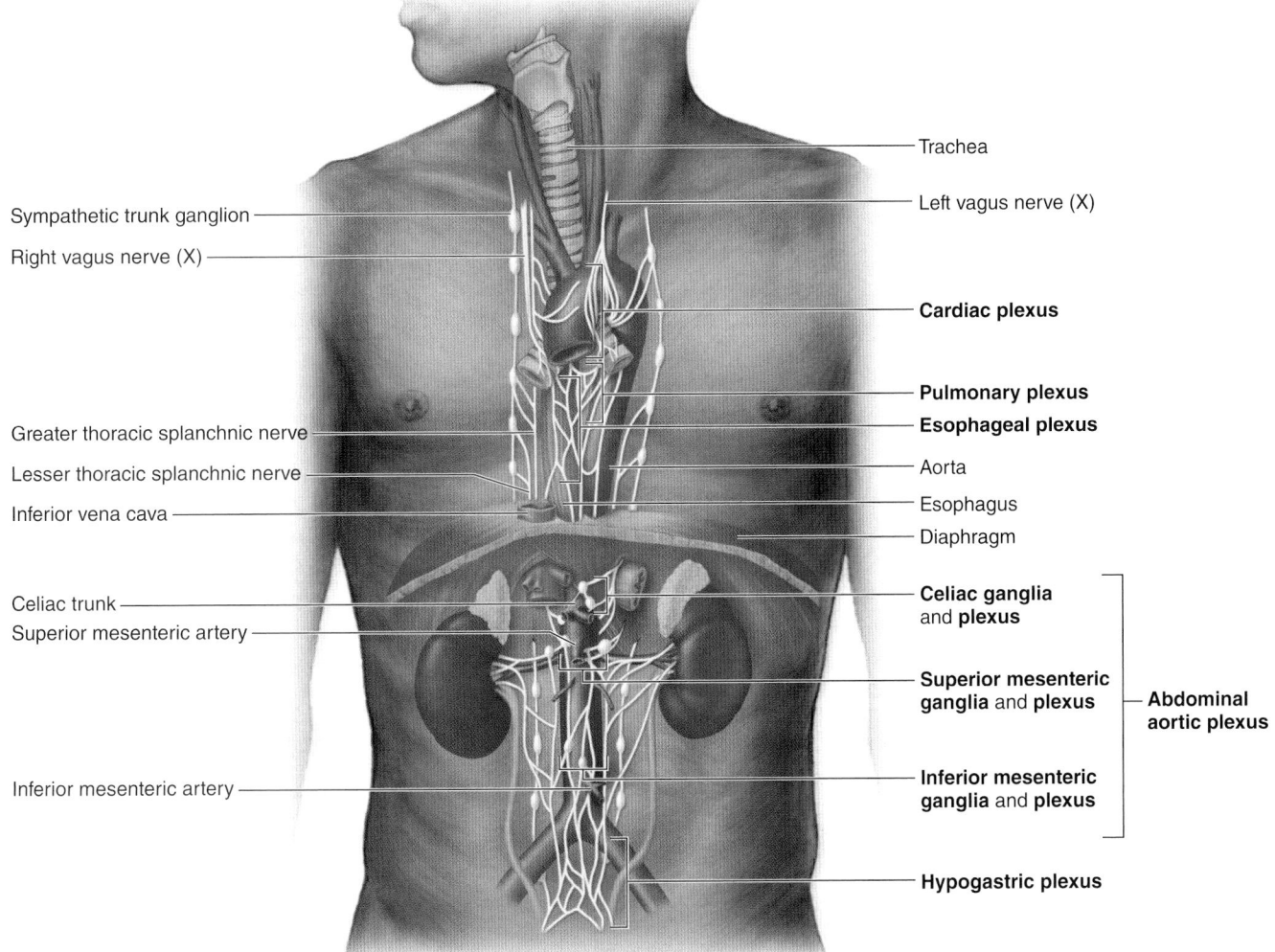

Figure 15.11 Autonomic Plexuses. Autonomic plexuses are located in the anterior body cavities in both the thoracic and abdominopelvic cavities. This anterior view shows the cardiac, pulmonary, and esophageal plexuses in the thoracic cavity and the abdominal aortic plexus (celiac, superior mesenteric, inferior mesenteric plexuses) in the abdominopelvic cavity. **AP|R**

The **pulmonary plexus** consists of sympathetic postganglionic axons from the cervical and thoracic sympathetic trunk ganglia and parasympathetic preganglionic axons from the vagus nerve. The axons project to the bronchi of the lungs. Sympathetic innervation causes bronchodilation (increase in the diameter of the bronchi), whereas stimulation of this parasympathetic pathway causes bronchoconstriction (a reduction in the diameter of the bronchi of the lungs, see section 23.3c).

The **esophageal plexus** consists of sympathetic postganglionic axons from the cervical and sympathetic trunk ganglia and parasympathetic preganglionic axons from the vagus nerve. Sympathetic innervation will inhibit muscle motility. Smooth muscle activity in the inferior esophageal wall is coordinated by parasympathetic axons that control the swallowing reflex in the inferior region of the esophagus by innervating the cardiac sphincter, a valve through which swallowed food and drink must pass (see section 26.2c).

The **abdominal aortic plexus** consists of the **celiac plexus, superior mesenteric plexus,** and **inferior mesenteric plexus.** It innervates all abdominal and some pelvic organs. The abdominal aortic plexus is composed of sympathetic postganglionic axons projecting from the prevertebral ganglia and parasympathetic preganglionic axons from either the vagus nerve or pelvic splanchnic nerves. Note that the celiac plexus also is known as the *solar plexus,* and it is partly responsible for the sensation of "getting the wind knocked out of you" when you are hit hard in the epigastric region.

The **hypogastric plexus** consists of a complex meshwork of sympathetic postganglionic axons (from the aortic plexus and the lumbar region of the sympathetic trunk) and preganglionic parasympathetic axons from the pelvic splanchnic nerves. Its axons innervate viscera within the pelvic region.

 WHAT DID YOU LEARN?

 22 What basic structures form an autonomic plexus?

INTEGRATE

CLINICAL VIEW
Autonomic Dysreflexia

Autonomic dysreflexia is a potentially dangerous vascular condition that causes blood pressure to rise profoundly, sometimes so high that blood vessels rupture. Specifically, it stimulates a sympathetic division reflex that causes systemic vasoconstriction and a marked increase in blood pressure. Autonomic dysreflexia is caused by hyperactivity of the autonomic nervous system in the weeks and months after a spinal cord injury. Often, the initial reaction to spinal cord injury is spinal shock, which is characterized by the loss of autonomic reflexes. However, paradoxically this decrease in reflex activities may cause certain viscera to respond abnormally to the lack of innervation, a phenomenon called **denervation hypersensitivity.** For example, when a person loses the ability to voluntarily evacuate the bladder, the bladder may continue to fill with urine to the point of overdistension. This induces a spinal cord reflex that causes involuntary relaxation of the internal urethral sphincter, allowing the bladder to empty.

15.7b Autonomic Reflexes

 **LEARNING OBJECTIVES**

2. Discuss how autonomic reflexes help maintain homeostasis.

3. Describe some major examples of autonomic reflexes.

Recall from section 14.6b that a reflex arc is a rapid, pre-programmed, response of a muscle or gland to a stimulus, which includes five components arranged in a reflex arc. These components include (1) a receptor (that detects the stimulus); (2) a sensory neuron that relays nerve signals from the receptor to the CNS; (3) an integration center (brain or spinal cord) that integrates the sensory input and initiates motor output; (4) motor neurons that relay nerve signals from the CNS to the effectors; and (5) effectors that bring about a response. Also note that whereas somatic reflexes involve skeletal muscles, autonomic reflexes involve cardiac muscle, smooth muscle, or glands. The autonomic nervous system helps maintain homeostasis through the involuntary activity of **autonomic reflexes,** also termed *visceral reflexes*. These reflexes enable the ANS to control visceral function. Autonomic reflexes consist of smooth muscle contractions (or relaxation), cardiac muscle contractions, or secretion by glands that are mediated by autonomic reflex arcs in response to a specific stimulus. Some autonomic reflexes are described here. See if you are able to determine the five components of the reflex arc for each of the examples:

- **Cardiovascular reflex.** A classic autonomic reflex involves the reduction of blood pressure. When blood pressure elevates, stretch receptors in the walls of large blood vessels are stimulated and nerve signals are propagated along visceral sensory neurons to the cardiac center in the medulla oblongata. These nerve signals inhibit sympathetic output and activate parasympathetic output to the heart to slow heart rate and decrease the volume of blood ejected, resulting in a decrease in blood pressure (see section 19.5).

- **Gastrointestinal reflex.** Autonomic reflexes control defecation. Fecal matter entering the rectum causes stretch of the rectal wall. Sensory neurons relay increased nerve signals to the spinal cord, which initiates a change in nerves signals along motor neurons to the rectum and anal sphincter. The rectum contracts and the internal anal sphincter relaxes (see section 26.3d).

- **Micturition reflex.** The mechanism leading to bladder emptying is similar to that described for fecal emptying of the colon. In a young child who is not yet toilet trained, stretch receptors send nerve signals to the sacral spinal cord when urine fills the bladder (**figure 15.12**). The reflex results in contraction of the smooth muscle in the bladder wall and relaxation of the urinary sphincters. (In a toilet-trained individual, sensory nerve signals end at the pons instead of the sacral region of the spinal cord, and urination occurs following voluntary relaxation of the external urethral sphincter. We discuss micturition in greater detail in section 24.8c).

Other autonomic reflexes include changing the size of bronchioles to regulate the amount of air entering the lungs, regulating digestive system activities, and changing pupil diameter.

 WHAT DID YOU LEARN?

 23 How does the cardiovascular reflex affect blood pressure?

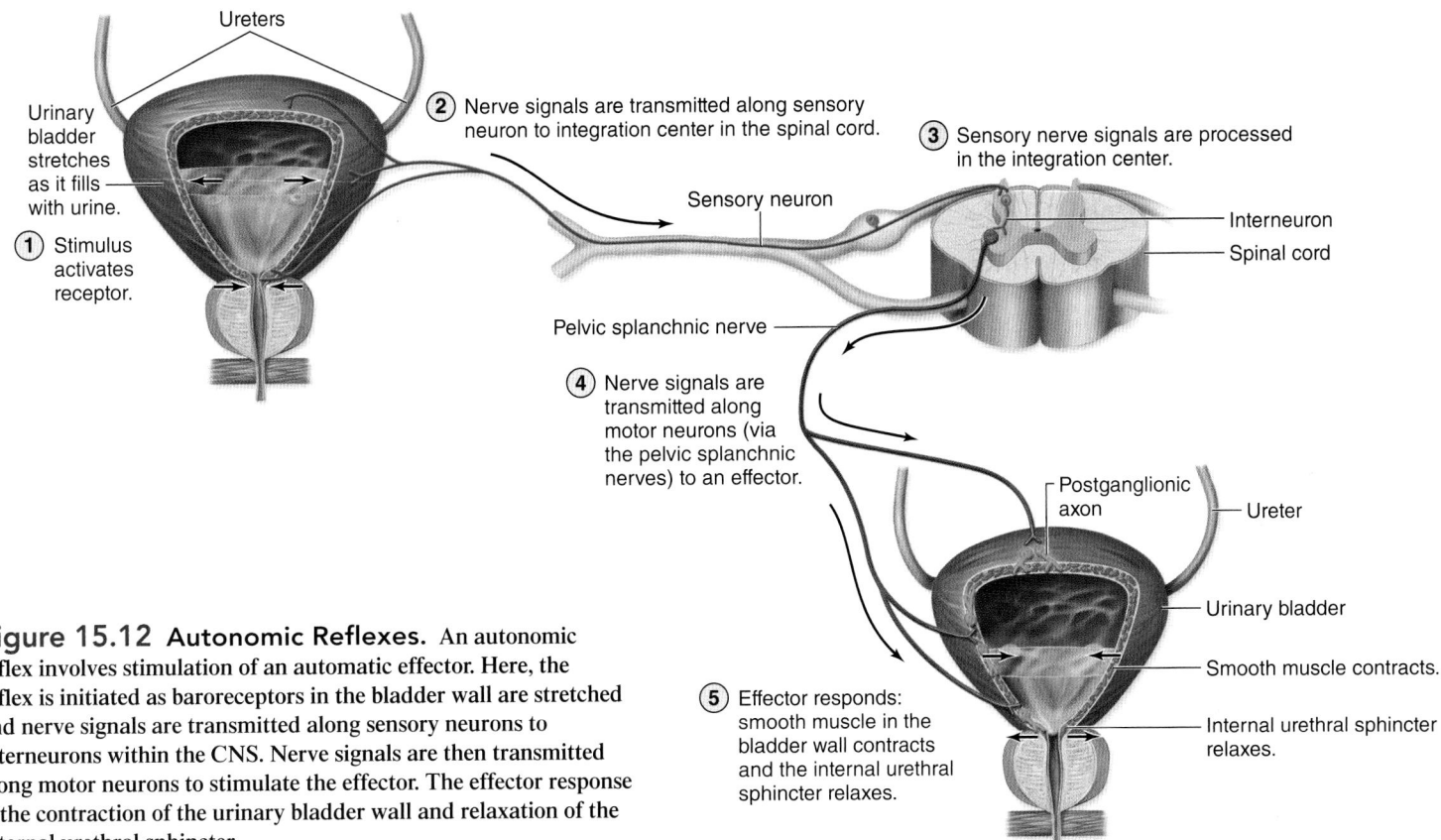

Figure 15.12 Autonomic Reflexes. An autonomic reflex involves stimulation of an automatic effector. Here, the reflex is initiated as baroreceptors in the bladder wall are stretched and nerve signals are transmitted along sensory neurons to interneurons within the CNS. Nerve signals are then transmitted along motor neurons to stimulate the effector. The effector response is the contraction of the urinary bladder wall and relaxation of the internal urethral sphincter.

(Figure labels:)

Ureters

Urinary bladder stretches as it fills with urine.

(1) Stimulus activates receptor.

(2) Nerve signals are transmitted along sensory neuron to integration center in the spinal cord.

Sensory neuron

(3) Sensory nerve signals are processed in the integration center.

Interneuron

Spinal cord

Pelvic splanchnic nerve

(4) Nerve signals are transmitted along motor neurons (via the pelvic splanchnic nerves) to an effector.

Postganglionic axon

Ureter

Urinary bladder

Smooth muscle contracts.

(5) Effector responds: smooth muscle in the bladder wall contracts and the internal urethral sphincter relaxes.

Internal urethral sphincter relaxes.

CHAPTER SUMMARY

- The autonomic nervous system controls the internal environment and helps maintain homeostasis.

15.1 Comparison of the Somatic and Autonomic Nervous Systems 577

- The nervous system can be functionally organized into the somatic nervous system and autonomic nervous system based upon whether the sensory input and motor output can be consciously regulated.

15.1a Functional Organization 577

- The somatic nervous system includes sensory input from the special senses, skin, muscles and joints, and motor output to control skeletal muscle.
- The autonomic nervous system includes involuntary motor output to cardiac muscle, smooth muscle, and glands and responds to sensory input from visceral sensory components.

15.1b Lower Motor Neurons of the Somatic Versus Autonomic Nervous System 578

- A single motor neuron innervates skeletal muscle fibers in the SNS, whereas the ANS has a two-neuron pathway consisting of preganglionic neurons in the CNS and ganglionic neurons in the PNS.

15.1c CNS Control of the Autonomic Nervous System 579

- Autonomic function is regulated by three CNS regions: hypothalamus, brainstem, and spinal cord.

15.2 Divisions of the Autonomic Nervous System 580

15.2a Functional Differences 580

- The parasympathetic division is primarily concerned with maintaining homeostasis when the body is at rest, which includes conserving energy and replenishing nutrient stores.
- The sympathetic division is primarily concerned with maintaining homeostasis in conditions of "fight-or-flight."

15.2b Anatomic Differences in Lower Motor Neurons 580

- Parasympathetic preganglionic neurons reside in the brainstem and sacral regions of the spinal cord, whereas sympathetic preganglionic axons reside in the thoracic and lumbar regions of the spinal cord.

15.2c Degree of Response 581

- The parasympathetic response tends to be discrete and localized, whereas the sympathetic response has the potential to produce a mass activation effect.

(continued on next page)

15.3 Parasympathetic Division 582	• The parasympathetic division is also known as the craniosacral system, because of the location of its preganglionic neurons.

15.3a Cranial Components   582
- Parasympathetic preganglionic axons extend through the oculomotor, facial, glossopharyngeal, and vagus cranial nerves.

15.3b Pelvic Splanchnic Nerves   584
- The remaining preganglionic parasympathetic cell bodies are housed within the S2–S4 segments of the spinal cord and form pelvic splanchnic nerves.

15.4 Sympathetic Division 584	• The sympathetic division is also known as the thoracolumbar division, because its preganglionic neurons reside in the T1–L2 segments of the spinal cord.

- A single effector may control one tissue, but many effectors often respond together, a phenomenon called mass activation.

15.4a Organization and Anatomy of the Sympathetic Division   584
- Preganglionic neuronal cell bodies are housed within the lateral gray horn of the spinal gray matter and their axons extend through white rami communicantes to the sympathetic trunk.
- Gray rami communicantes are composed of postganglionic sympathetic axons from the sympathetic trunk to the spinal nerve.
- Some preganglionic axons pass through the sympathetic trunk without synapsing and form splanchnic nerves that project to the prevertebral ganglia. Postganglionic axons extend from the prevertebral ganglia to the target organ.

15.4b Sympathetic Pathways   588
- In the spinal nerve pathway, the postganglionic axon enters the spinal nerve through the gray ramus and extends to target organs (blood vessels and glands of the skin).
- In the postganglionic sympathetic nerve pathway, the postganglionic axon extends from the sympathetic trunk and projects directly to the target organs (thoracic organs, neck viscera, and blood vessels and glands of the head and neck).
- In the splanchnic nerve pathway, the preganglionic axon passes through the sympathetic trunk and travels to the prevertebral ganglia, where it synapses with a ganglionic neuron, which extends to the target organs (most abdominal and pelvic viscera).
- In the adrenal medulla pathway, the preganglionic axon extends through both a sympathetic trunk ganglion and prevertebral ganglion without synapsing. It synapses on secretory cells in the adrenal medulla that release epinephrine and norepinephrine.

15.5 Comparison of Neuro-transmitters and Receptors of the Two Divisions 590	• The two main types of receptors are cholinergic and adrenergic receptors.

15.5a Overview of ANS Neurotransmitters   590
- Acetylcholine (ACh) is the neurotransmitter released by cholinergic neurons, which include all preganglionic neurons as well as all ganglionic parasympathetic neurons. ACh also is used by some ganglionic sympathetic neurons to sweat glands and blood vessels in skeletal muscle.
- Norepinephrine (NE) is the neurotransmitter released from adrenergic neurons, which include all other sympathetic postganglionic neurons.

15.5b Cholinergic Receptors   590
- Nicotinic receptors are located on all ganglionic neurons and cells in the adrenal medulla, and are always stimulatory.
- Muscarinic receptors are located on all parasympathetic target cells, on sweat glands in the skin, and on blood vessels in skeletal muscle. Their effect may be excitatory or inhibitory depending upon the receptor subtype.

15.5c Adrenergic Receptors   591
- Adrenergic receptors include α and β receptors. There are several subsets for both α and β types.

15.6 Interactions Between the Parasympathetic and Sympathetic Divisions 592	• Most organs are innervated by both divisions of the ANS.

15.6a Autonomic Tone   592
- Both ANS divisions maintain some continual activity, which is called the autonomic tone for that division.

15.6b Dual Innervation   596
- Many visceral effectors have dual innervation, meaning they are innervated by both ANS divisions. The actions of the divisions mostly are antagonistic (having opposite effects on a target organ), but a few are cooperative (working together to cause a single result).

15.6c Systems Controlled Only by the Sympathetic Division   596
- Most blood vessels, the adrenal medulla, and sweat glands of the skin are innervated by sympathetic axons only.

15.7 Integration of Autonomic System Function 597	• Autonomic functions ultimately are controlled by the CNS.

15.7a Autonomic Plexuses   597
- Autonomic plexuses are meshworks of sympathetic postganglionic axons and parasympathetic preganglionic axons, as well as some visceral sensory axons.

15.7b Autonomic Reflexes   598
- The autonomic nervous system helps maintain homeostasis through autonomic reflexes, which are also called visceral reflexes.

CHALLENGE YOURSELF

Do You Know the Basics?

_____ 1. A splanchnic nerve in the sympathetic division of the ANS

    a. connects neighboring sympathetic trunk ganglia.

    b. controls parasympathetic functions in the thoracic cavity.

    c. is formed by preganglionic axons that extend to prevertebral ganglia.

    d. travels through parasympathetic pathways in the head.

_____ 2. Some parasympathetic preganglionic neuron cell bodies are housed within the

    a. hypothalamus.

    b. sacral region of the spinal cord.

    c. cerebral cortex.

    d. thoracolumbar region of the spinal cord.

_____ 3. Which of the following is _not_ a function of the sympathetic division of the ANS?

    a. increases heart rate and breathing rate

    b. prepares for emergency

    c. increases digestive system motility and activity

    d. dilates pupils

_____ 4. Maintaining a resting level of ANS activity in a cell is called

    a. autonomic tone.

    b. cooperative effect.

    c. dual innervation.

    d. antagonistic effect.

_____ 5. Sympathetic division preganglionic axons travel to the _____ ganglia via the _____ rami.

    a. terminal; white

    b. sympathetic trunk; gray

    c. prevertebral; gray

    d. sympathetic trunk; white

_____ 6. All parasympathetic division synapses use _____ as a neurotransmitter.

    a. dopamine

    b. acetylcholine

    c. norepinephrine

    d. epinephrine

_____ 7. Which autonomic nerve plexus innervates the pelvic organs?

    a. cardiac plexus

    b. esophageal plexus

    c. hypogastric plexus

    d. inferior mesenteric plexus

_____ 8. A sympathetic postganglionic axon is

    a. long and unmyelinated.

    b. short and myelinated.

    c. short and unmyelinated.

    d. long and myelinated.

_____ 9. Nicotinic receptors are located on which of the following?

    a. plasma membranes of ganglionic neurons

    b. target cells that receive parasympathetic innervation

    c. blood vessels in skeletal muscles

    d. sweat glands

_____ 10. Which of the following is **true** about a beta receptor?

    a. It binds acetylcholine.

    b. Its effects are stimulatory only.

    c. It causes general vasoconstriction.

    d. It increases heart rate.

11. What are the three CNS regions that regulate autonomic function?

12. For the following ganglia, identify the location and the division of the ANS each is a part of sympathetic trunk ganglia, prevertebral ganglia, and terminal ganglia.

13. Compare and contrast the postganglionic axons of the parasympathetic and sympathetic divisions. Examine the axon length, myelination (or lack thereof), and the neurotransmitter used.

14. Compare and contrast sympathetic and parasympathetic innervation effects on digestive system structures.

15. Explain responses of nicotinic receptors and muscarinic receptors to simulation by ACh.

16. Describe the differences between cooperative effects and antagonistic effects in dual innervation of target organs.

17. Describe how the general functions of the sympathetic and parasympathetic divisions of the ANS differ.

18. What may occur with the mass activation of the sympathetic division of the ANS?

19. Describe the process of the micturition reflex.

20. How does sympathetic innervation regulate vasoconstriction or vasodilation in the same blood vessels?

Can You Apply What You've Learned?

Use the following paragraph to answer questions 1 and 2.

Arlene was crossing the street when a car ran a red light and nearly hit her. Arlene was not hurt, but she was very frightened and was in a heightened state of alertness well after the incident.

1. Arlene likely experienced all of the following physiologic effects _except_

    a. increased heart rate.

    b. pupil constriction.

    c. goose bumps.

    d. sweaty palms.

2. Arlene was in a heightened state of alertness well after the incident because
   a. the adrenal medulla secreted epinephrine and norepinephrine.
   b. the parasympathetic division stimulated regions of the brain.
   c. the sympathetic division decreased overall autonomic tone of blood vessels.
   d. All of these are correct.

3. George has hypertension (high blood pressure). His physician prescribed the drug propranolol, which is described as a beta-blocker, to reduce his blood pressure What do you suppose could be a side effect of propranolol?
   a. reduced heart rate
   b. increased blood clotting
   c. vasoconstriction of blood vessels to the skin
   d. bronchodilation

4. Albuterol is a drug designed to counteract the effects of asthma—namely, the medication, which may be used in an inhaler, facilitates bronchodilation. What receptors would you expect this drug to bind?
   a. α_1 receptors
   b. α_2 receptors
   c. β_1 receptors
   d. β_2 receptors

5. One surgical treatment for gastric ulcers is a selective vagotomy, where branches of the vagus nerve to the upper GI tract are cut. How would you suppose a vagotomy would help the treatment of a gastric ulcer?
   a. It would stimulate vasodilation of the blood vessels serving the stomach.
   b. It would reduce gastric gland secretion.
   c. It would promote faster movement of materials through the stomach.
   d. All of these are correct.

Can You Synthesize What You've Learned?

1. Our body expends a lot of energy activating and propagating the mass activation response of the sympathetic nervous system. Why is it necessary for us to have such an "expensive" mechanism at our disposal?

2. When you were younger, your parents may have told you to wait until 1 hour after you've eaten before you go swimming. Based on what you've learned about the ANS, can you hypothesize why swimming right after a meal may be problematic?

3. Some faculty dislike teaching lecture classes after lunch, complaining that the students do not pay attention at this time. From a physiologic viewpoint, what is happening to these students?

INTEGRATE

ONLINE STUDY TOOLS

McGraw Hill Education **connect** |ANATOMY & PHYSIOLOGY  LEARNSMART® AP|R

The following study aids may be accessed through Connect.

Clinical Case Study: A Secret Life Leading to an Unexpected Death

Interactive Questions: This chapter's content is served up in a number of multimedia question formats for student study.

LearnSmart: Topics and terminology include comparison of the somatic and autonomic nervous systems; divisions of the autonomic

nervous system; parasympathetic division; sympathetic division; comparison of neurotransmitters and receptors of the two divisions; interactions between the parasympathetic and sympathetic divisions; control and integration of autonomic system function

Anatomy & Physiology Revealed: Topics include sympathetic and parasympathetic overviews; thoracic region

Nervous System: Senses

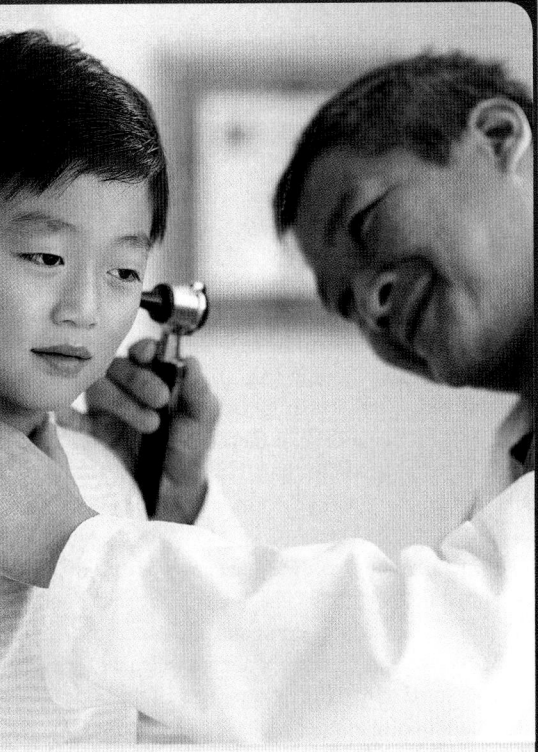

aprevealed.com

Module 7: Nervous System

CAREER PATH
Pediatrician

A pediatrician focuses on the health and well-being of children, from birth to early adulthood. One of the more common ailments of children is ear pain, which typically occurs as a result of a respiratory infection that has spread to the middle ear. Thus, knowledge of ear anatomy and physiology is essential for this health-care professional. A pediatrician uses a tool called an otoscope to examine the ear canal, eardrum, and middle ear for possible infection. This chapter provides examples of how the otoscope visualizes healthy versus infected middle ears, and why middle ear infections are more prevalent in children under 5.

We are barraged continuously with sensory information about the environment—both outside and inside our bodies. This information must be detected by various sensory receptors and transmitted to the brain and spinal cord so that it can be interpreted and the appropriate responses initiated. The sensory information comes in many different forms. Touch receptors react to physical contact, taste receptors detect chemicals in the food we eat, visual receptors respond to light, hearing receptors detect and react to sound waves, and pressure receptors within blood vessels respond to stretch. In this chapter on the senses, we first provide an introduction to sensory receptors and describe the structure and function of the general senses. We then examine in detail the special senses (smell, taste, vision, hearing, and equilibrium).

16.1 Introduction to Sensory Receptors

Sensory receptors (*recipio* = to receive) are components of the nervous system that provide us with information about both our external and internal environment. Here we describe the general function and structure of sensory receptors, the type of information they provide, and how the various types of sensory receptors are classified.

16.1a General Function of Sensory Receptors

LEARNING OBJECTIVE

1. Describe the general function of sensory receptors as transducers.

The general function of all sensory receptors is to respond to a **stimulus** (stim′ū-lus; pl., *stimuli* = a goad) and initiate sensory input to the central nervous system (CNS). This involves converting stimulus energy into an electrical signal. The original energy form detected is specific to the type of receptor (e.g., light energy is detected by the eye, sound energy by the ear, and mechanical energy by blood vessels). However, the form the energy is transduced or changed to is always electrical energy, and it is sent along a sensory neuron. This sensory information is propagated as an action potential (called a nerve signal; see section 12.9a) to the CNS for interpretation. It is because sensory receptors convert (or transduce) stimulus energy to electrical energy that sensory receptors are referred to as **transducers** (tranz-dū′sĕr; *trans* = across, *duco* = to lead).

Two features are critical to allow sensory receptors to function as transducers: (1) Sensory receptors, like neurons and muscle cells, establish and maintain a resting membrane potential (RMP) across their plasma membrane (see section 4.4). (2) Sensory receptors contain modality gated channels within their plasma membranes. A modality gated channel opens in response to a stimulus other than a neurotransmitter or a voltage change at the plasma membrane. (Recall that chemically gated channels open in response to a neurotransmitter and voltage-gated channels open in response to a voltage change; see section 12.6a). The details for the various types of receptors and the opening of their specialized modality gated channels are explained in detail throughout the later sections of this chapter.

WHAT DID YOU LEARN?

 How does a sensory receptor function as a transducer?

INTEGRATE

CONCEPT CONNECTION

Our sensory receptors initiate communication among most body systems. For example, sensory receptors in our integument signal our central nervous system to initiate cooling or heating mechanisms if the environmental temperature is either too hot or too cold (see sections 1.5b and 27.8b). Chemoreceptors in the blood vessels monitor carbon dioxide and hydrogen ion levels within the blood, and when their balance is changed both the cardiovascular system and respiratory system are stimulated to bring these levels back to homeostasis (see sections 20.6a and 23.5c). Receptors associated with our musculoskeletal system alert the nervous system about the body's position and balance (see section 13.6b). Finally, sensory receptors in the digestive tract allow us to perceive tastes in our food, detect textures and consistencies of foods, and let us know when our stomach has been extensively stretched when we have eaten too much (see sections 26.2 and 26.3).

16.1b General Structure of Sensory Receptors

LEARNING OBJECTIVE

2. Describe the general structure of a receptor, and explain the significance of a receptive field.

Sensory receptors range in complexity from the relatively simple, bare terminal endings of a single sensory neuron (e.g., some touch receptors; see figure 16.2) to specialized, complex structures called sense organs (e.g., the eye; see figure 16.12). Regardless of their anatomic complexity, however, all are functionally connected to the CNS by sensory neurons. This provides the means of relaying sensory information from the receptors to the brain and spinal cord.

The area that the terminal endings of a single sensory neuron is distributed is called its **receptive field.** The concept of a receptive field and its significance is most clearly shown with a comparison of receptive fields within the skin (**figure 16.1**). Note the relative amount of area that sensory neurons of the skin are distributed in two different regions of the body—the skin of the fingertips and the skin of the upper back. The size of the receptive field will determine the ability of the CNS to identify the exact location of a stimulus. A small receptive field provides us with the ability to identify the stimulus location more specifically. In contrast, a large receptive field allows us to determine only the general region of the stimulus.

Although it might seem advantageous for all sensory neurons to have small receptive fields (because it would provide us with enhanced

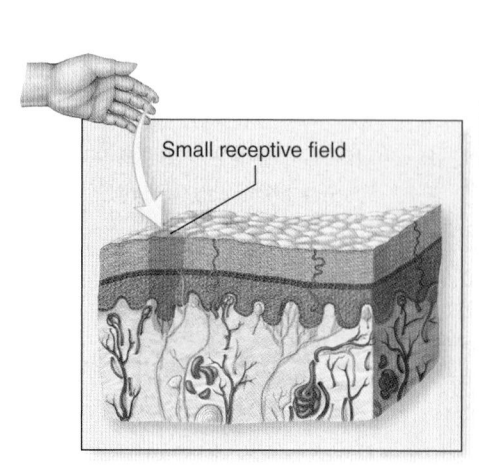
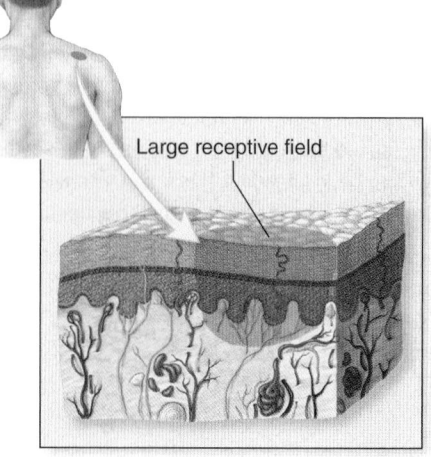

Figure 16.1 Receptive Fields. The specific area monitored by each sensory neuron is called its receptive field. A smaller receptive field offers greater specificity of localization than does a larger receptive field.

perceptive abilities), the number of sensory neurons in the body would have to markedly increase for us to gain this advantage. This greater number of sensory neurons would require a significant increase in body size, and the energy costs to maintain their activity would be enormous.

 WHAT DID YOU LEARN?

② Describe the general range in structural complexity of sensory receptors, and identify what is associated with all sensory receptors.

16.1c Sensory Information Provided by Sensory Receptors

LEARNING OBJECTIVES

3. Define a sensation.
4. Explain the characteristic of a stimulus that sensory receptors provide to the CNS.

Sensory input is relayed from sensory receptors to the CNS for interpretation. Whether we consciously perceive a stimulus is dependent upon which specific region of the CNS receives that sensory information. Only nerve signals that reach the *cerebral cortex* of the brain result in our conscious awareness. A stimulus that we are consciously aware of is called a **sensation.** A sensation occurs when we recognize a child's face or realize that the room is too warm. Although your body is constantly bombarded by numerous sensory stimuli, you are consciously aware of only a fraction of them. Much of the sensory input is relayed to other areas of the CNS (e.g., the hypothalamus, brainstem, spinal cord), where a response is initiated without your awareness. Sensory stimuli regarding your blood pressure, blood carbon dioxide levels, and chemical composition of material within the small intestine are examples of sensory information that is detected and responded to on a subconscious level.

A sensory receptor must be able to provide the CNS with several characteristics regarding a stimulus (whether it is consciously perceived or not). These characteristics include its modality, location, intensity, and duration. The **modality** (mō-dal′i-tē; *modus* = a mode) or form of a stimulus is provided by a given type of sensory receptor relaying sensory input along designated sensory neurons to specific regions of the CNS. For example, the receptors of the eye (the retina) initiate nerve signals along the optic nerve to the occipital lobe (visual cortex), and receptors of the ear (spiral organ) initiate nerve signals along the cochlear branch of the vestibulocochlear nerve to the temporal lobe (auditory cortex). In comparison, baroreceptors within the aorta send nerve signals along the vagus nerve to the cardiovascular center within the brainstem as part of blood pressure regulation. The brain is like a switchboard, and it interprets the source based upon which "line" the signal arrived.

The specific *location* of a stimulus is able to be determined by the CNS because sensory information is relayed either from different regions of a sensory receptor or from different locations within the body along designated sensory neurons within a given nerve that reaches specific regions of the CNS. For example, the optic nerve is composed of many sensory neurons that extend from different portions of the retina of the eye to communicate with designated regions within the visual cortex of the occipital lobe. The inner ear has a similar anatomic arrangement between the different regions of the spiral organ and designated regions within the auditory cortex of the temporal lobe. Recall that the specific location of receptors of the skin relay sensory input to designated regions of the postcentral gyrus of the parietal lobe for interpretation, a concept that was visually represented in the figure of the sensory homunculus (see figure 13.13).

The CNS is able to interpret the relative *intensity* of the stimulus because of the change in number of nerve signals that are arriving along a designated nerve. A greater stimulus results in both the most sensitive receptors initiating nerve signals more frequently and the less sensitive receptors (that are not typically active) initiating nerve signals. A lesser stimulus would result in fewer nerve signals being relayed by its associated sensory neurons. For example, a bright light results in a greater frequency of nerve signals relayed along the optic nerve to the visual cortex, whereas a softer sound results in a lesser frequency of nerve signals relayed along the vestibulocochlear nerve to the auditory cortex.

The CNS is able to determine the *duration* of stimulus because all sensory receptors become less sensitive to a constant stimulus and initiate a progressive decrease in nerve signals. This decrease in sensitivity to a continuous stimulus is called **adaptation.** However, the rate of decrease is different for the various types of sensory receptors. This difference in adaption is used to categorize sensory receptors as either tonic receptors or phasic receptors. **Tonic receptors** demonstrate limited adaptation. In response to a constant stimulus, tonic receptors continuously generate nerve signals and only *slowly* decrease the number relayed to the CNS. Examples of tonic receptors include receptors within the inner ear that determine head position, and proprioceptors in the joints and muscles that provide information of where your body is in space. In addition, all pain receptors are tonic receptors. This provides the motivation to address the cause of the pain and hopefully eliminate it so that the pain will stop. In comparison, **phasic receptors** exhibit rapid adaptation to a constant stimulus. Phasic receptors generate nerve signals only in response to a new (or changing) stimulus and *quickly* decrease the number of nerve signals relayed to the CNS. Examples include the deep pressure receptors that sense the increased pressure when we first sit down in a chair. We are immediately aware of the pressure increase wherever our body contacts the chair. But soon, we do not notice this pressure because adaptation has occurred in these receptors. You may have experienced adaptation after placing your glasses on the top of your head and then forgetting that they were there. It is advantageous for us to not be continuously bombarded by this type of sensory information.

 WHAT DID YOU LEARN?

❸ Explain how sensory receptors provide input regarding the modality, location, intensity, and duration of a stimulus.

16.1d Sensory Receptor Classification

LEARNING OBJECTIVE

5. Identify and describe the three criteria used to classify receptors.
6. Classify the various types of sensory receptors based upon each of the three criteria.

Three criteria are used to categorize sensory receptors—receptor distribution, stimulus origin, and modality of stimulus. These classification criteria are summarized in **table 16.1.**

Sensory Receptor Distribution

Sensory receptors may be classified based upon their distribution in the body. The two distribution types are the general senses and the special senses.

General sense receptors are distributed throughout the body and are simple in structure. The receptors for general senses are subdivided into two categories based upon their location in the body, and include somatic sensory receptors and visceral sensory receptors. **Somatic sensory** (or *somatosensory*) **receptors** are tactile receptors housed within both the skin and mucous membranes, which line the nasal cavity, oral cavity, vagina, and anal canal. These sensory receptors monitor several types of stimuli including texture of an object, pressure, vibration, temperature, and pain. Somatic sensory receptors are also within joints, muscles, and

Table 16.1	Criteria for Classifying Sensory Receptors		
Classification	**Description**		**Examples**
SENSORY RECEPTOR DISTRIBUTION (LOCATION OF RECEPTOR)			
General senses	Distributed throughout the body; structurally simple		
Somatic sensory receptors	Located in skin and mucous membranes		Tactile (touch) receptors
	Located in joints, muscles, and tendons		Joint receptors, muscle spindles, Golgi tendon organs
Visceral sensory receptors	Located within walls of viscera and blood vessels		Stretch receptors in stomach wall, chemoreceptors in blood vessels
Special senses	Located only in the head; structurally complex sense organs		Receptors for smell, taste, vision, hearing, and equilibrium
STIMULUS ORIGIN (LOCATION OF STIMULUS)			
Exteroceptors	Detect stimuli in the external environment		Receptors within skin or mucous membranes
			Receptors for smell, taste, vision, hearing, and equilibrium
Interoceptors	Detect stimuli within the body		Receptors within walls of viscera and blood vessels
Proprioceptors	Detect stimuli within joints, skeletal muscles, and tendons that sense body or limb movement		Joint receptors, muscle spindles, Golgi tendon organs
MODALITY OF STIMULUS (STIMULATING AGENT)			
Chemoreceptors	Detect chemicals (molecules or ions) dissolved in fluid		Taste receptors, receptors in blood vessels that monitor hydrogen ions (H^+)
Thermoreceptors	Detect changes in temperature		Receptors within skin and hypothalamus
Photoreceptors	Detect changes in light intensity, color, and movement		Eye
Mechanoreceptors	Detect physical deformation of the plasma membrane due to touch, pressure, vibration, and stretch; subtypes include baroreceptors, proprioceptors, tactile receptors, and other specialized cells such as the hair cells in the cochlea of the ear		Tactile receptors in the skin
Nociceptors	Detect painful stimuli		Pain receptors present in almost all organs

tendons, and include joint receptors, muscle spindles, and Golgi tendon organs, respectively. These detect stretch and pressure relative to position and movement of the skeleton and skeletal muscles. A Golgi tendon organ, for example, monitors stretch of a tendon when a muscle contracts. **Visceral sensory receptors** are located in the walls of the viscera (internal organs) and blood vessels. These receptors detect stretch in the smooth muscle within the walls of internal organs (e.g., stretch of the stomach wall), chemical changes in the contents within their lumen (e.g., a change in carbon dioxide levels in the blood), temperature, and pain.

In comparison, receptors of the **special senses** are located only within the head and are specialized, complex sense organs. The five special senses are olfaction (smell), gustation (taste), vision (sight), hearing (audition), and equilibrium (head position and acceleration).

Stimulus Origin

Sensory receptors can also be classified based upon where the stimulus originates. These classifications include exteroceptors, interoceptors, and proprioceptors.

Exteroceptors (eks′ter-ō-sep′ter, -tōr; *exterus* = external) detect stimuli from the external environment. Exteroceptors include the somatic sensory receptors of the skin and mucous membranes, as well as the receptors of the special senses. All of these types of receptors respond to a stimulus that is outside of the body.

Interoceptors (in′ter-ō-sep′ter; *inter* = between) detect stimuli from within our internal environment. Interoceptors include the visceral sensory receptors within the wall of internal organs and blood vessels.

Interoceptors keep our CNS informed about the changes that are occurring within our bodies.

Proprioceptors (prō′prē-ō-sep′ter; *proprius* = one's own) detect body and limb movements and include only the somatic sensory receptors within joints, muscles, and tendons.

Modality of Stimulus

Sensory receptors may also be classified according to the stimulus they respond to, which is called the **modality of stimulus,** or the *stimulating agent.* Thus, some sensory receptors respond only to temperature changes, whereas others respond to chemical changes. There are five groups of sensory receptors, based upon their modality of stimulus: chemoreceptors, thermoreceptors, photoreceptors, mechanoreceptors, and nociceptors:

- **Chemoreceptors** (kē′mō-rē-sep′tor, -ter) detect chemicals, either molecules or ions dissolved in fluid. These chemicals include the food and drink we have ingested, the composition of our body fluids, and the relative components of our inhaled air. The receptors in the taste buds on our tongue are chemoreceptors, because they respond to the specific molecules and ions in our ingested food to provide information to us about what is present in what we are eating. Likewise, chemoreceptors in some of our blood vessels monitor the concentration of carbon dioxide in our blood, to help influence our respiratory rate (see section 23.5c).

- **Thermoreceptors** (ther′mō-rē-sep′tōr; *therme* = heat) respond to changes in temperature. Thermoreceptors are present in both the skin and the hypothalamus. These receptors are components of reflexes that regulate and maintain body temperature.

- **Photoreceptors** (fō′tō-rē-sep′tōr; *phot* = light) are located in the eye, where they detect changes in light intensity, color, and movement.

- **Mechanoreceptors** (me-kăn′ō-rē-sep′tōr; *mechane* = machine) respond to distortion of the plasma membrane that occurs due to touch, pressure, vibration, and stretch. The various types of mechanoreceptors include baroreceptors, proprioceptors, tactile receptors, and other specialized cells such as the hair cells in the cochlea of the ear that move in response to sound waves. For example, **baroreceptors** (bar′ō-rē-sep′tōr; *baros* = weight) are a type of mechanoreceptor, which are stimulated by changes in stretch or distension within the wall of body structures. Baroreceptors located within blood vessel walls monitor stretch of these structures, as part of blood pressure regulation (see section 20.6a).

- **Nociceptors** (nō′si-sep′tōr; *noci* = pain) respond to painful stimuli. The purpose of nociceptors is to inform the body of injury or other damage so that an appropriate response may be made. **Somatic nociceptors** detect chemical, heat, or mechanical damage to the body surface or skeletal muscles. For example, our touching a hot pan or suffering a sprained ankle stimulates somatic nociceptors. **Visceral nociceptors** detect internal body damage within the viscera. When we feel discomfort in our internal organs, it often occurs because (1) a tissue has been deprived of oxygen—for example, as a result of a heart attack; (2) the smooth muscle in the wall of the organ has been stretched so much that we are uncomfortable; or (3) the tissue has suffered trauma, and damaged cells have released chemicals that stimulate specific nociceptors.

Classifying a Sensory Receptor

A given sensory receptor is described based upon each classification criterion: receptor distribution, stimulus origin, and modality of stimulus. The eyes, for example, are special senses because they are located in the head (receptor distribution); exteroceptors because they detect stimuli outside the body (stimulus origin); and photoreceptors because they detect light (modality of stimulus). In comparison, receptors that detect stretch of blood vessels are classified as general senses because they are distributed throughout the body, interoceptors because they detect stimuli within the body, and mechanoreceptors (specifically baroreceptors) because they detect changes in distension of the organ wall.

We use sensory receptor distribution (general senses versus special senses) as the criterion for organizing the remaining sections in this chapter on the senses. General senses are first described in section 16.2 followed by the details of the special senses in sections 16.3 to 16.5.

WHAT DID YOU LEARN?

Describe the following sensory receptors based upon the three classification criteria: the ear and the sense of hearing, the tongue and the sense of taste, and stretch receptors in the urinary bladder wall.

INTEGRATE

LEARNING STRATEGY

When you eat a very spicy meal, your mouth stings because the nociceptors in your mouth are somatic nociceptors. When you swallow and the food travels through your gastrointestinal (GI) tract, you may not experience a stinging, burning sensation because visceral nociceptors respond only to abnormal muscle stretch, oxygen deprivation, or chemical imbalance in the tissue. When the waste products from that spicy meal are expelled from the body, the anus may sting because the nociceptors around the anus and inferior anal canal are somatic nociceptors.

16.2 The General Senses

Receptors for general senses, as just described in section 16.1, are organized into somatic sensory receptors (tactile receptors of the skin and mucous membranes and proprioceptors) and visceral sensory receptors. Here we discuss somatic sensory receptors (tactile receptors only) and referred pain. (Proprioceptors are discussed in section 14.4b, and visceral sensory receptors are included throughout the text when discussing the physiology of the various body systems.)

16.2a Tactile Receptors

LEARNING OBJECTIVE

1. Compare and contrast unencapsulated and encapsulated tactile receptors.

Tactile (tak′til; *tango* = to touch) **receptors** are the most numerous type of sensory receptor (**figure 16.2**). They are mechanoreceptors located in the skin and mucous membranes. The dendritic endings that compose these sensory receptors are either unencapsulated or encapsulated (**table 16.2**).

Unencapsulated Tactile Receptors

Unencapsulated tactile receptors are simply terminal endings of sensory neurons with no protective covering. The three types of unencapsulated receptors are free nerve endings, root hair plexuses, and tactile discs.

Free nerve endings are the least complex of the tactile receptors and reside closest to the surface of the skin, usually in the papillary layer (superficial layer) of the dermis (see section 6.1b). Often, some branches extend into the deepest epidermal strata and terminate between the epithelial cells. Free nerve endings are also located in mucous membranes. These tactile receptors primarily detect temperature and pain stimuli, but some also detect light touch and pressure. They can be either tonic receptors (adapt slowly) or phasic receptors (adapt quickly).

Root hair plexuses are specialized terminal endings of sensory neurons that form a weblike sheath around hair follicles in the reticular layer (deeper layer) of the dermis. Any movement or displacement of the hair changes the arrangement of these terminal endings, initiating nerve signals. These phasic receptors quickly adapt; thus, although we feel the initial contact of a long-sleeved shirt on our arm hairs when we put on the garment, our conscious awareness subsides immediately until we move and the root hair plexuses are restimulated.

Figure 16.2 Tactile Receptors. Various types of tactile receptors in the skin receive information about touch, pressure, vibration, or temperature in our immediate environment. Unencapsulated tactile receptors are simply bare dendritic endings of sensory neurons, whereas encapsulated tactile receptors are dendritic endings of sensory neurons that are enclosed by either connective tissue or connective tissue and glial cells.

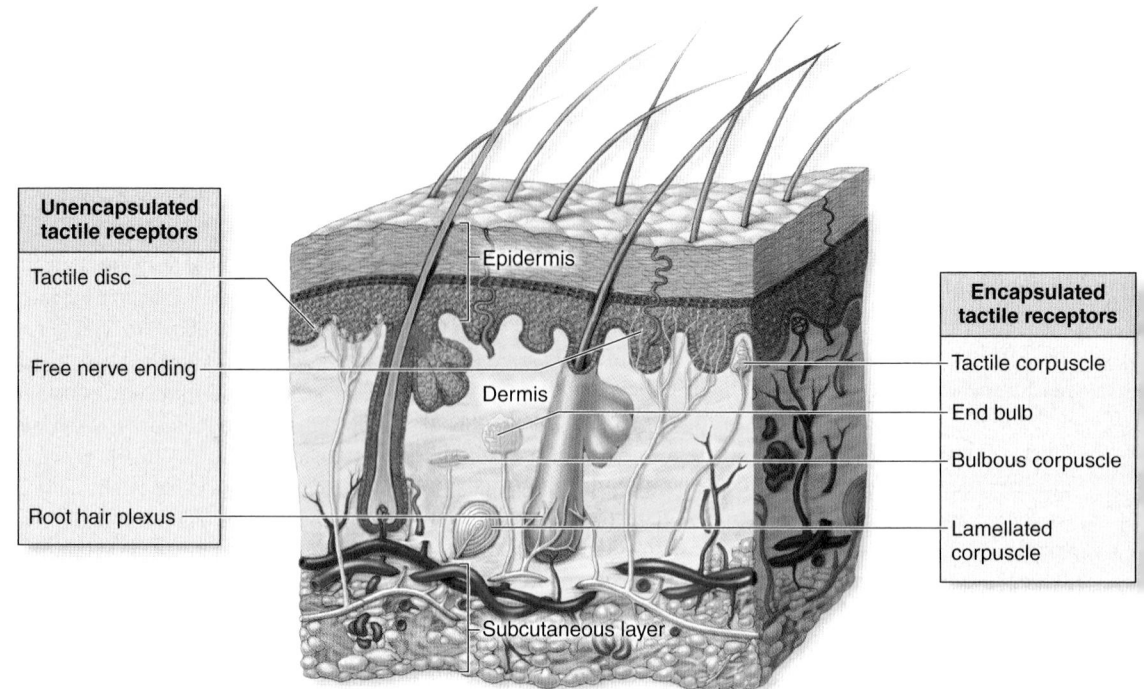

Tactile discs, previously called *Merkel discs,* are flattened terminal endings of sensory neurons that extend to **tactile cells** (*Merkel cells*), which are specialized epithelial cells located in the stratum basale (deepest layer) of the epidermis. These discs function as tonic receptors for light touch. (Note that tactile cells are the only specialized receptor cells; the remaining tactile receptors are simply the terminal endings of primary sensory neurons.)

Encapsulated Tactile Receptors

Encapsulated tactile receptors are terminal endings of sensory neurons that are wrapped either by connective tissue or by connective tissue and specialized glial cells called neurolemmocytes (previously called Schwann cells; see section 12.4). Encapsulated tactile receptors include end bulbs, lamellated corpuscles, bulbous corpuscles, and tactile corpuscles.

End bulbs, or *Krause bulbs,* are terminal endings of sensory neurons ensheathed in connective tissue. They are located both in the dermis of the skin, and in the mucous membranes of the oral cavity, nasal cavity, vagina, and anal canal. End bulbs are tonic receptors that detect light pressure stimuli and low-frequency vibration.

Lamellated (lam′e-lāt-ed; *lamina* = leaf) **corpuscles** (*corpus* = body), previously called *Pacinian corpuscles,* are large, leaf-shaped tactile receptors composed of several terminal endings ensheathed with an inner core of neurolemmocytes and outer concentric layers of connective tissue. They are phasic receptors found deep within the reticular layer of the dermis of the skin; in the hypodermis of the palms of the hands, soles of the feet, breasts, and external genitalia; and in the walls of some organs. The structure and location of lamellated corpuscles allow them to function in coarse touch, and sensing continuous deep-pressure and high-frequency vibration stimuli.

Bulbous corpuscles, or *Ruffini corpuscles,* are terminal endings of sensory neurons ensheathed within connective tissue that are housed within the dermis and subcutaneous layer. They are tonic receptors that detect both continuous deep pressure and distortion in the skin.

Tactile corpuscles, previously called *Meissner corpuscles,* are oval receptors formed from highly intertwined terminal endings of sensory neurons enclosed by modified neurolemmocytes, which are then covered with dense irregular connective tissue. They are housed within the dermal papillae (which are projections of the dermis; see section 6.1b), especially in the lips, palms, eyelids, nipples, and genitals. Tactile corpuscles are phasic receptors for discriminative touch to distinguish texture and shape of an object and for detecting light touch.

WHAT DID YOU LEARN?

5 What are the three types of unencapsulated tactile receptors, and where are they located within the integument?

Table 16.2	Types of Tactile Receptors

UNENCAPSULATED TACTILE RECEPTORS

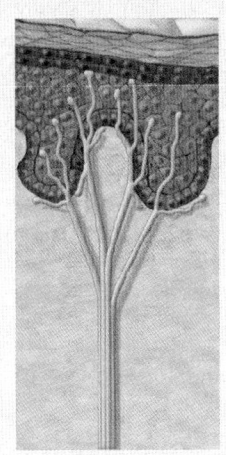 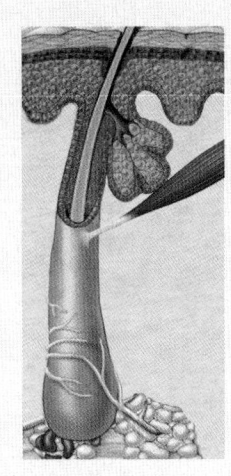 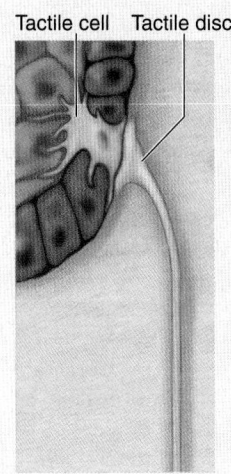

Tactile cell   Tactile disc

Receptor Type	**Free Nerve Ending**	**Root Hair Plexus**	**Tactile Disc**
Structure	Terminal endings of sensory neurons	Terminal endings of sensory neurons that surround hair follicles	Flattened terminal endings of sensory neurons that end adjacent to specialized tactile cells
Location	Closest to skin surface (papillary layer of the dermis and some terminal endings extend into the deepest layers of the epidermal strata); mucous membranes	Reticular layer of the dermis	Stratum basale of epidermis
Function	Detects temperature, pain, some detect light touch and pressure	Detects movement of the hair	Detects light touch
Rate of Adaptation	Phasic or tonic	Phasic	Tonic

ENCAPSULATED TACTILE RECEPTORS

 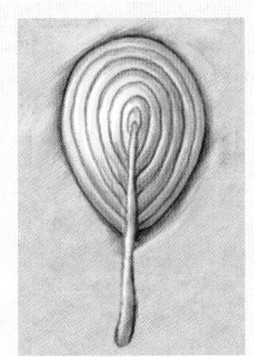 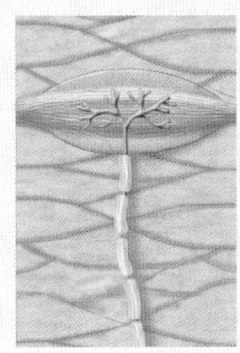 

Receptor Type	**End Bulb**	**Lamellated Corpuscle**	**Bulbous Corpuscle**	**Tactile Corpuscle**
Structure	Terminal endings of sensory neurons ensheathed in connective tissue	Terminal endings of sensory neurons ensheathed with an inner core of neurolemmocytes and outer concentric layers of connective tissue	Terminal endings of sensory neurons within connective tissue	Highly intertwined terminal endings of sensory neurons enclosed by modified neurolemmocytes and dense irregular connective tissue
Location	Dermis, mucous membranes of oral cavity, nasal cavity, vagina, and anal canal	Reticular layer of the dermis; hypodermis of the palms of the hands, soles of the feet, breasts, and external genitalia; and walls of some organs	Dermis and subcutaneous layer	Dermal papillae, especially in lips, palms, eyelids, nipples, genitals
Function	Detects light pressure and low-frequency vibration	Functions in coarse touch; detects continuous deep pressure and high-frequency vibration	Detects continuous deep pressure and skin distortion	Discriminative touch for distinguishing texture and shape of an object; light touch
Rate of Adaptation	Tonic	Phasic	Tonic	Phasic

16.2b Referred Pain

LEARNING OBJECTIVE

2. Define referred pain, and explain its significance in diagnosis.

Referred pain occurs when sensory nerve signals from certain viscera are perceived as originating not from the organ, but from somatic sensory receptors within the skin and skeletal muscle (see section 14.5a). Numerous somatic sensory neurons and visceral sensory neurons conduct nerve signals on the same ascending tracts within the spinal cord (**figure 16.3**). As a result, the somatosensory cortex in the brain (see section 13.3c) is unable to accurately determine the actual source of the stimulus, and thus the stimulus may be localized incorrectly.

Clinically, some common sites of referred pain are useful in medical diagnosis (**figure 16.4**). For example, cardiac problems are often a source of referred pain because the heart receives its sympathetic innervation from the T1–T5 segments of the spinal cord (see section 15.4). Pain associated with a myocardial infarction (heart attack) may be referred to the skin innervated by the T1–T5 spinal nerves, which lie along the pectoral region and the medial side of the arm. Thus, some individuals who are experiencing heart problems may perceive pain along the medial side of the left arm, where the T1 spinal nerve innervates (see Clinical View: "Angina Pectoris and Myocardial Infarction" in section 19.4). By the same token, kidney and ureter pain may be referred along the T10–L2 spinal nerves, which typically overlie the inferior abdominal wall

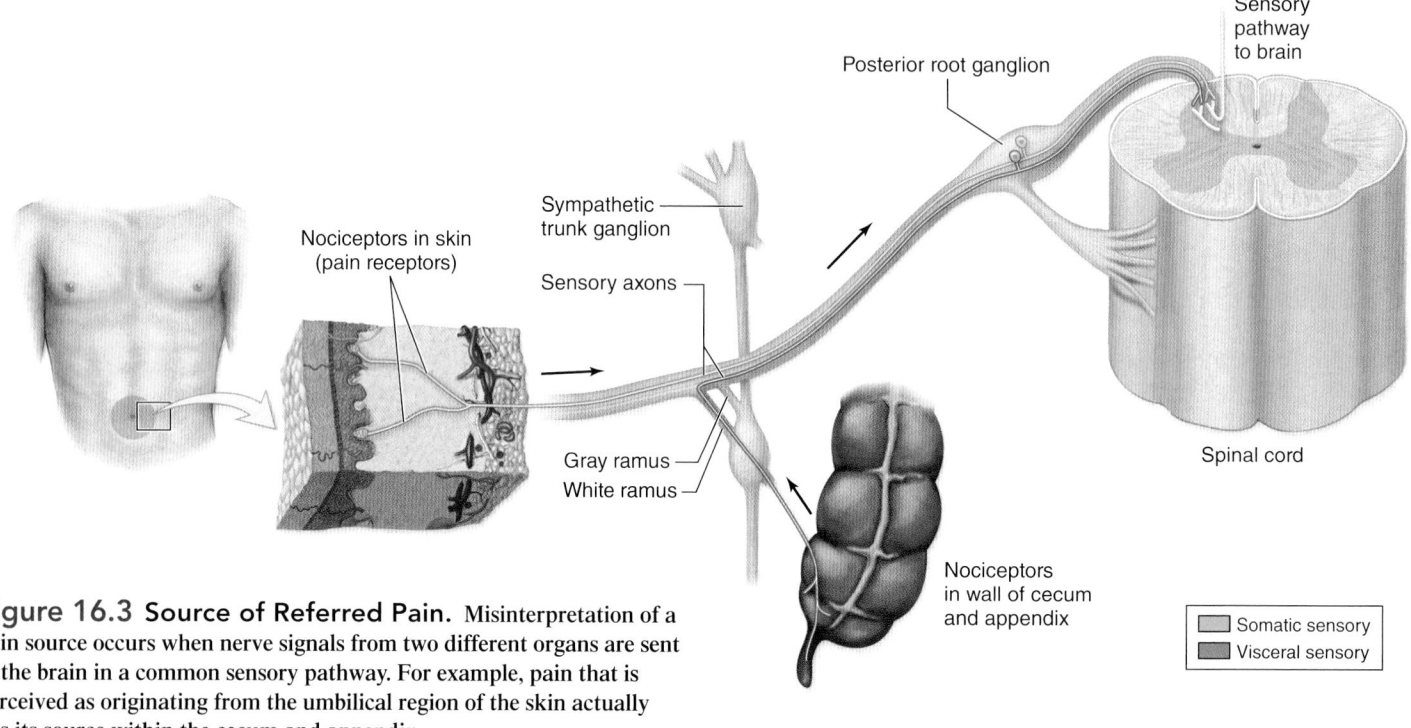

Figure 16.3 Source of Referred Pain. Misinterpretation of a pain source occurs when nerve signals from two different organs are sent to the brain in a common sensory pathway. For example, pain that is perceived as originating from the umbilical region of the skin actually has its source within the cecum and appendix.

Figure 16.4 Common Sites of Referred Pain Note: Appendicitis typically refers pain to the umbilical region. It is only in the later stages of appendicitis, when the pain becomes localized (due to the parietal peritoneum becoming inflamed as well), that the pain becomes felt in the lower right abdominal quadrant.

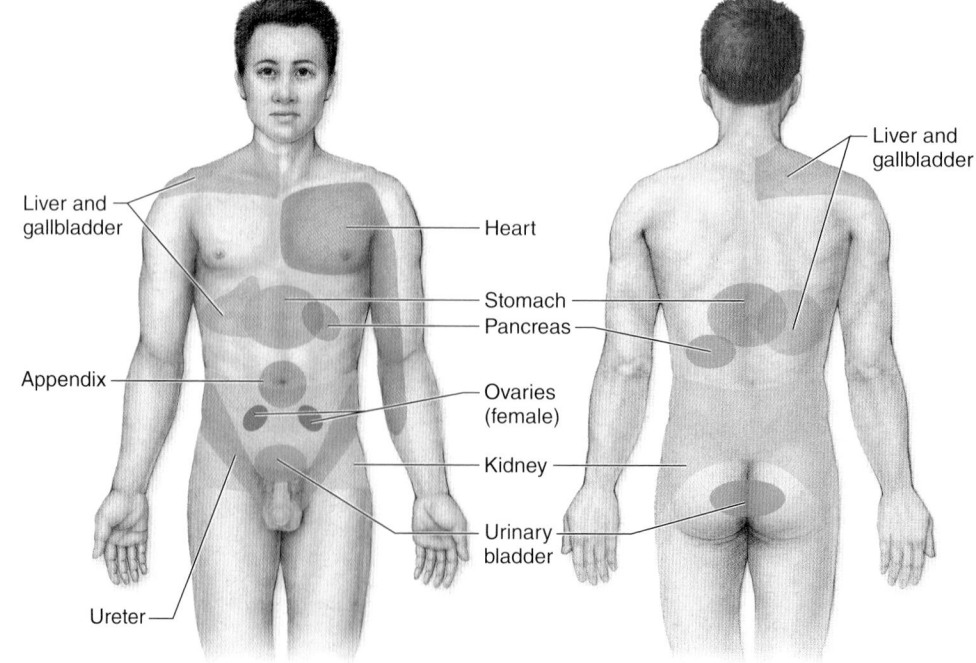

in the groin and loin regions (see Clinical View: "Renal Calculi" in section 24.8b).

Visceral pain is usually referred along the sympathetic nerve pathways, but sometimes it can follow parasympathetic pathways as well. Referred pain from the urinary bladder often can follow the parasympathetic pathways (via the pelvic splanchnic nerves; see section 15.3b). Because the pelvic splanchnic nerves lie in the S2–S4 region of the spinal cord, pain may be referred to the S2–S4 spinal nerves, which overlie the medial buttocks regions (see Clinical View: "Urinary Tract Infections" in section 24.8c).

WHAT DID YOU LEARN?

6 Explain the clinical significance of referred pain.

INTEGRATE

CLINICAL VIEW

Phantom Pain

Phantom (fan'tŏm; a ghost) **pain** is a sensation associated with a body part that has been removed. Following the amputation of an appendage, a patient often continues to experience pain from what is perceived as the removed part. For example, a patient may feel pain that seems localized in a foot that is no longer there. The stimulation of a sensory neuron pathway from the removed limb anywhere on the remaining intact portion of the pathway initiates nerve signals to the CNS, where they are interpreted as originating in the amputated limb. In other words, the cell bodies of the sensory neurons that had previously innervated the limb remain alive because they were not part of the limb but rather, reside in posterior root ganglia by the spinal cord. This so-called **phantom limb syndrome** can be quite debilitating. Some people experience extreme pain, whereas others have an insatiable desire to scratch a nonexistent itch.

16.3 Olfaction and Gustation

Both olfaction (the sense of smell) and gustation (the sense of taste) occur through receptors in the head that detect dissolved chemicals in the air and in our food, respectively. Thus, both of these receptors are classified as special senses, exteroceptors, and chemoreceptors. We cover these together because as you will see later, the appreciation of taste also involves our sense of smell.

16.3a Olfaction: The Sense of Smell

LEARNING OBJECTIVES

1. Name the components of the olfactory receptors, and discuss their mode of action.

2. Describe the olfactory pathways that relay sensory input to the brain.

Olfaction (ol-fak'shŭn; *olfacio* = to smell) is the sense of smell, whereby volatile molecules (called **odorants**) must be dissolved in the mucus in our nasal cavity to be detected by chemoreceptors. We use this sense to sample our environment for information about the food we will eat, the presence of other individuals in the room, or potential danger (e.g., spoiled food, smoke from a fire).

Compared to many other animals, our olfactory ability is much less sensitive and not as highly developed. Consequently, we do not rely as greatly on olfactory information to find food or communicate with others. Yet we do have the ability to distinguish one odor among thousands of different ones, a capability that we may appreciate as we walk through a garden of flowers.

Olfactory Epithelium

The sensory receptor organ for smell is the **olfactory epithelium.** This epithelium lines the superior region of the nasal cavity, covering both the inferior surface of the cribriform plate and superior nasal conchae of

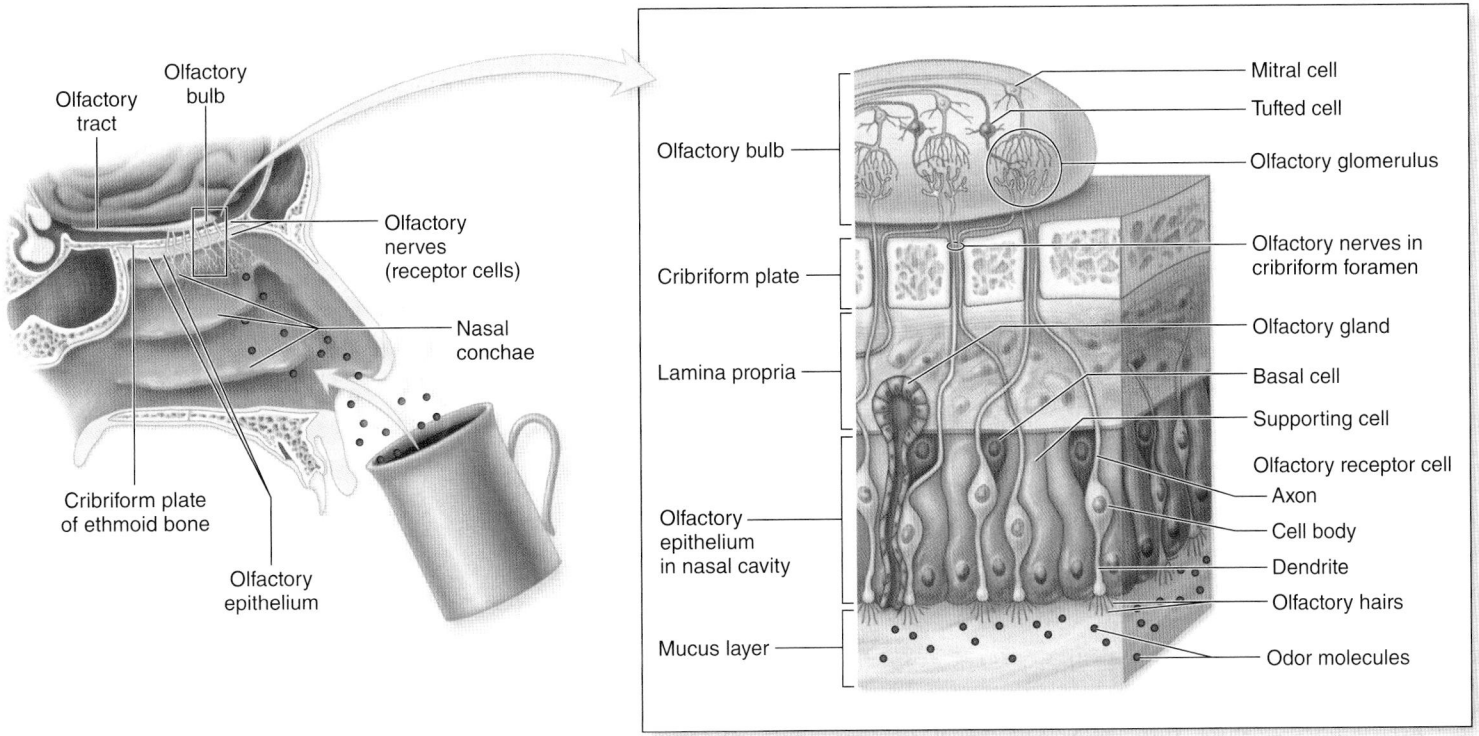

Figure 16.5 Olfactory Epithelium. The olfactory epithelium detects chemical stimuli in the air we breathe. When receptor cells within the olfactory epithelium are stimulated, they initiate nerve signals in axons that pass through the cribriform plate and synapse on neurons within the olfactory bulb. AP|R

the ethmoid bone (see section 8.2b). The olfactory epithelium is composed of three distinct cell types (figure 16.5):

- **Olfactory receptor cells** (also called *olfactory neurons*), which detect odors
- **Supporting cells** (also called *sustentacular cells*) that sustain the olfactory receptor cells
- **Basal cells**, which function as neural stem cells to continually replace olfactory receptor cells

Olfactory receptor cells are one of the few neuronal types that are replaced. Olfactory receptor neurons are regenerated every 40 to 60 days by basal cells within the olfactory epithelium. This process decreases with age, and the olfactory receptor cells that remain lose their sensitivity to odors. Thus, an elderly individual has a decreased ability to recognize odor molecules.

Internal to the olfactory epithelium is an areolar connective tissue layer called the **lamina propria** (lam′i-nă prō′prē-ă). Housed within the collagen fibers and ground substance of this layer are mucin-secreting **olfactory glands** (or *Bowman glands*) and many blood vessels and nerves. Secretions from both supporting cells and olfactory glands form the mucus that covers the exposed surface of the olfactory epithelium.

Olfactory Receptor Cells

Olfactory receptor cells are bipolar neurons that have undergone extensive differentiation and modification, and serve as the primary neuron in the sensory pathway for smell. Olfactory receptor cells have both a single dendrite and an unmyelinated axon. Projecting from the dendrites are numerous thin, nonmotile cilia called **olfactory hairs,** which extend into the layer of mucus. Olfactory hairs contain chemoreceptors within their plasma membrane that detect one specific odorant molecule. Depending upon which olfactory receptor cells are stimulated, different smells will be perceived. Axons of olfactory receptor cells form bundles (fascicles) of the **olfactory nerves (CN I)** (see section 13.9). These fascicles project through foramina (holes) in the cribriform plate of the ethmoid bone to enter an olfactory bulb.

Olfactory Nerve Structures and Pathways

The pair of **olfactory bulbs** are the terminal ends of olfactory tracts located inferior to the frontal lobes of the brain (see section 13.9). Axons of olfactory nerves synapse with both *mitral cells* and *tufted cells* (which are secondary neurons) within the olfactory bulbs. The resulting spherical structures are called **olfactory glomeruli** (glō-mar′yū-lī; sing., glomerulus; *glomerulus* = small ball). We have about 2000 glomeruli in our olfactory bulbs, and numerous olfactory receptor cells converge on each olfactory glomerulus. The convergence of signals within the glomerulus facilitates our ability to detect faint odors.

Axon bundles of the mitral and tufted cells form the paired **olfactory tracts** that project posteriorly along the inferior frontal lobe surface directly to the primary olfactory cortex in the temporal lobe (see table 13.5) and selected different regions of the brain, including the hypothalamus and amygdala. Unlike other sensory information, olfactory pathways do not project to the thalamus and therefore do not undergo any thalamic processing prior to reaching the cerebrum.

Detecting Smells

During normal relaxed breathing, most inhaled air does not pass across the olfactory epithelium. To be sure to detect different smells, we must sniff repeatedly or breathe deeply, which causes the inhaled air to mix and swirl in the superior region of the nasal cavity, so that odor molecules diffuse into the mucus layer covering the olfactory receptor cells.

INTEGRATE

CONCEPT CONNECTION

Stimulation of the hippocampus and amygdala, which are part of the limbic system involved with emotion and memory (see section 13.8f) links emotional reactions with smells like specific foods or perfumes encountered at the same time. This explains why scents are so strongly associated with either nostalgic memories or bad experiences.

Within the mucus, soluble proteins called **odorant-binding proteins** display an affinity for a variety of odorants.

Olfactory receptor cells are stimulated by contact of odorant-binding proteins with olfactory cell receptors. The olfactory pathway is so sensitive that only a few stimulating odorant-binding proteins with bound odorants are needed to bind to receptors and initiate olfactory sensation. Stimulation of olfactory receptor cells activates G proteins within these cells (see section 17.5b). In turn, activated G proteins stimulate adenylate cyclase enzymes, which convert ATP to cAMP (an intracellular second messenger). cAMP stimulates the opening of cation channels that allow the inflow of both Na^+ and Ca^{2+}. This net inflow of positive ions results in generation of local receptor potentials (graded potentials) within the olfactory hairs of the olfactory receptor cells. These local potentials will initiate an action potential that is propagated along the axon of olfactory receptor cells, causing the release of neurotransmitter from the terminal ends of the axon. This results in stimulation of different patterns of the approximately 2000 glomeruli. Consider for example that cooking a meal causes the release of numerous odorants, but the smell of the meal concoction has a unique "signature" recognized by the excitation pattern observed within the glomeruli.

Binding of neurotransmitter by secondary neurons results in propagation of nerve signals through the various olfactory pathways. Sensory information reaches different regions of the brain including the (1) cerebral cortex, which allows us to consciously perceive and identify the smell; (2) hypothalamus, which controls visceral reactions to smell, such as salivation, sneezing, or gagging; and (3) amygdala, which is a center for recognition of odors and often associating those odors to a particular emotion.

Note that once the receptors are stimulated, changes to the ion channels alter the flow of ions. This results in interfering with the subsequent generation of local receptor potentials within olfactory receptor cells, and adaptation of smell occurs rapidly. Thus, an initially strong smell (such as rotting food in a trash can) may seem to dissipate as your olfactory receptor cells quickly adapt to the foul odor.

WHAT DID YOU LEARN?

7 What is the role of the mucus in detection of smells?

8 Why do some smells stimulate an emotional reaction?

16.3b Gustation: The Sense of Taste

LEARNING OBJECTIVES

3. Describe the structure and function of papillae of the tongue.

4. Discuss the structure and location of gustatory receptors, and describe the gustatory pathways that relay sensory input to the brain.

5. Describe the five types of tastes, and explain the association of smell with taste.

Our sense of taste, called **gustation** (gŭs-tā′shŭn; *gusto* = to taste), occurs when we come in contact with the molecules or ions of what we eat and drink. Gustatory cells are chemoreceptors located within taste buds on the tongue and soft palate. The tongue and soft palate also house mechano-receptors and thermoreceptors to provide us with information about the texture and temperature of our food, respectively.

Papillae of the Tongue

On the dorsal surface of the tongue are epithelial and connective tissue elevations called **papillae** (pă-pil′ē; *papula* = a small nipple), which are of four types: filiform, fungiform, vallate, and foliate **(figure 16.6a, b)**:

Filiform (fil′i-fōrm; *filum* = thread) **papillae** are short and spiked; they are distributed on the anterior two-thirds of the tongue surface. These papillae do not house taste buds and thus, have no role in gustation. Instead, their bristlelike structure serves a mechanical function; they assist in detecting texture and manipulating food.

Fungiform (fŭn′ji-fōrm; = mushroom-shaped) **papillae** are blocklike projections primarily located on the tip and sides of the tongue. Each contains only a few taste buds.

Vallate (val′āt; *vallo* = to surround) **papillae**, or *circumvallate papillae*, are the least numerous (about 10 to 12) yet are the largest papillae on the tongue. They are arranged in an inverted V shape on the posterior dorsal surface of the tongue. Each papilla is surrounded by a deep, narrow depression. Most of our taste buds are housed within the walls of these papillae along the side facing the depression.

Foliate (fō′lē-āt; = leaflike) **papillae** are not well developed on the human tongue. They extend as ridges on the posterior lateral sides of the tongue and house only a few taste buds during infancy and early childhood.

Taste Buds

Taste buds are cylindrical sensory organs containing taste receptors and have the appearance of an onion (figure 16.6c, d). Each taste bud is composed of three distinct cell types:

- **Gustatory cells** (also called gustatory receptors), which detect tastants (taste-producing molecules and ions) in our food
- **Supporting cells** that sustain the gustatory cells
- **Basal cells**, which function as neural stem cells to continually replace the relatively short-lived gustatory cells

Gustatory cells are regenerated every 7 to 9 days by basal cells within the taste bud. This process decreases with age, and our sensitivity to taste also decreases. Beginning at about age 50, our ability to distinguish between different tastes declines.

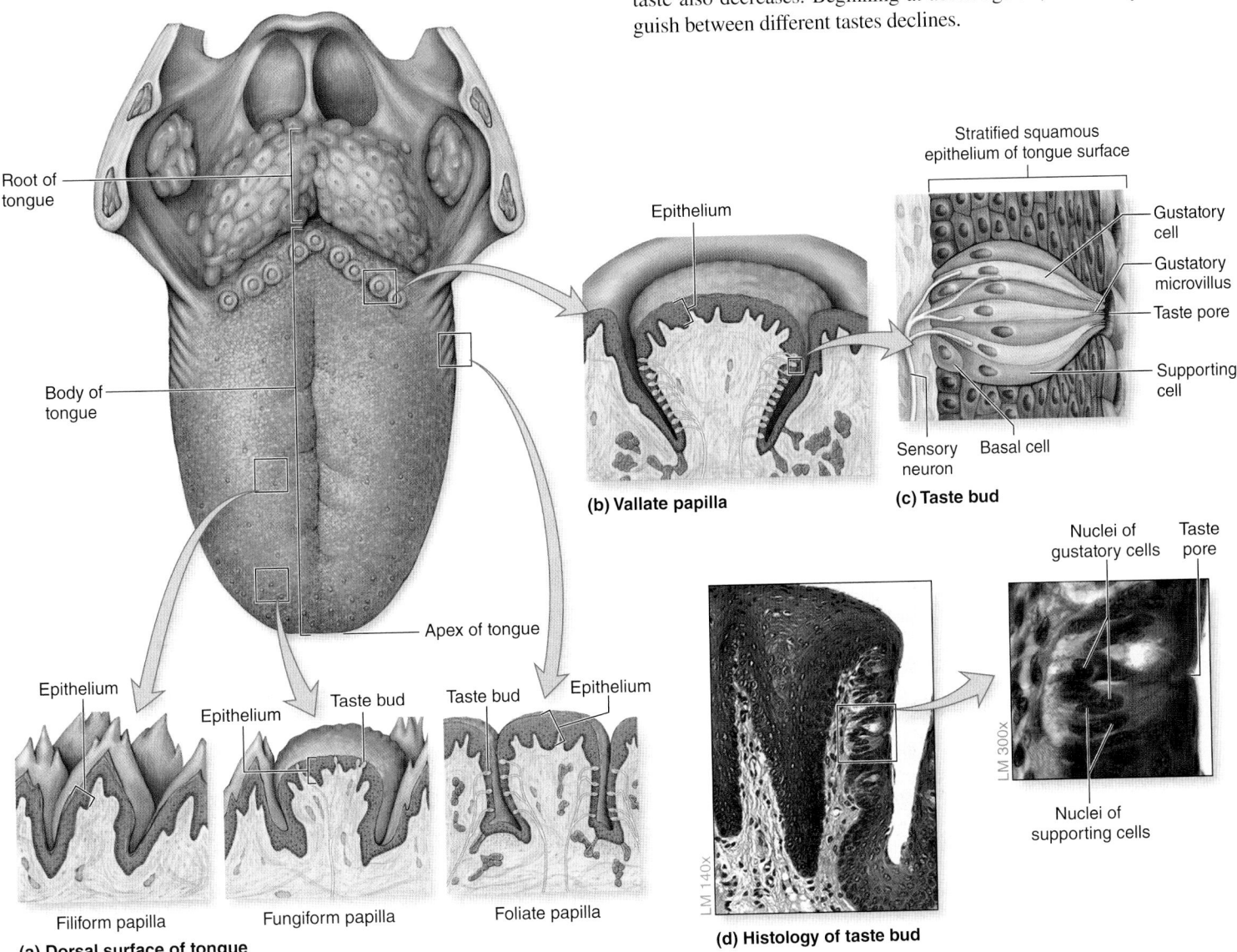

(a) Dorsal surface of tongue

(b) Vallate papilla

(c) Taste bud

(d) Histology of taste bud

Figure 16.6 Tongue Papillae and Taste Buds. (*a*) Papillae are small elevations on the tongue surface that exist in four types: filiform, fungiform, vallate, and foliate. (*b*) A vallate papilla exhibits many taste buds, one of which is shown in detail (*c*). (*d*) Photomicrographs show the histologic structure of a taste bud on a vallate papilla. AP|R

Gustatory Cells

Gustatory cells within the taste buds are specialized neuroepithelial cells. The dendritic ending of each gustatory cell is formed by a slender **gustatory microvillus,** sometimes called a *taste hair.* Many gustatory microvilli extend through an opening in the taste bud, called the **taste pore,** to the surface of the tongue. This is the receptive portion of the cell. Within the oral cavity, saliva keeps the environment moist; the tastants in our food dissolve in the saliva, and then they are able to contact and stimulate gustatory receptors.

Gustatory Pathways

Terminal endings of primary neurons are associated with gustatory cells, with each neuron contacting several gustatory cells. These sensory neurons are primarily components of the facial nerve (CN VII), which innervates taste buds from the anterior two-thirds of the tongue and the glossopharyngeal nerve (CN IX), which innervates taste buds from the posterior one-third of the tongue (figure 16.7). Axons within these nerves project to the medulla oblongata (specifically the nucleus solitarius) to synapse with secondary neurons. These secondary neurons project to the thalamus, and axons of tertiary neurons project to the primary gustatory cortex in the insula of the cerebrum.

Gustatory Discrimination and Physiology of Taste

In contrast to the large number of olfactory receptors we have in our nose, our tongue detects just five basic **taste sensations:** sweet, salty, sour, bitter, and umami:

- **Sweet** tastes are produced by organic compounds such as sugar or other molecules (e.g., artificial sweeteners).
- **Salt** tastes are produced by metal ions, such as sodium (Na^+) and potassium (K^+).
- **Sour** tastes are associated with acids in the ingested material, such as hydrogen ions (H^+) in vinegar.
- **Bitter** tastes are produced primarily by alkaloids such as quinine, unsweetened chocolate, nicotine, and caffeine.

- **Umami** stimuli: *Umami* (u'ma-mē) is a Japanese word meaning "delicious flavor." It is a taste related to amino acids, such as glutamate and aspartate, to produce a meaty flavor.

Researchers previously thought certain tastes were best interpreted along specific regions of the tongue; however, research has found these "taste maps" to be incorrect and has shown that taste sensations are spread over broader regions of the tongue than previously thought.

The tastants in our food bind to specific receptor plasma membrane proteins of gustatory cells. The initial binding of a specific tastant molecule to its receptor causes depolarization of that specific receptor cell, but the manner in which cell depolarization occurs varies. In general, sweet, bitter, and umami stimuli (which are tastant *molecules*) bind to receptors on the taste bud surface, which activate a G protein. The G protein activation causes the formation of a secondary messenger, which results in cell depolarization. In contrast, salt and sour stimuli (which are tastant *ions*) do not use a G protein and depolarize the cell directly.

Depolarization of a receptor taste cell generates local receptor potentials that initiate nerve signals that cause the release of neurotransmitter at its basal side. The neurotransmitter stimulates a primary neuron, which is within the facial or glossopharyngeal nerves that relay nerve signals to the brain.

Sensory information first reaches the medulla oblongata (specifically the nucleus solitarius) to trigger reflexes that increase salivation and release of stomach secretions in anticipation of the arrival of food. A gag or vomiting reflex may also occur in response to a nauseating substance. Sensory information is then relayed via secondary sensory neurons to the thalamus and then via tertiary sensory neurons to the primary gustatory cortex for the conscious perception of taste. This requires integrating taste sensations with those of temperature, texture, and smell.

WHAT DO YOU THINK?

1 When a person has a stuffy nose from a "cold" or hay fever, he or she typically can't detect tastes as well. Why?

Note that our ability to taste what we eat relies heavily on our olfactory sense. Recall the last time you had a severe cold or sinus infection. The aroma of the food could not reach your olfactory receptors, so the

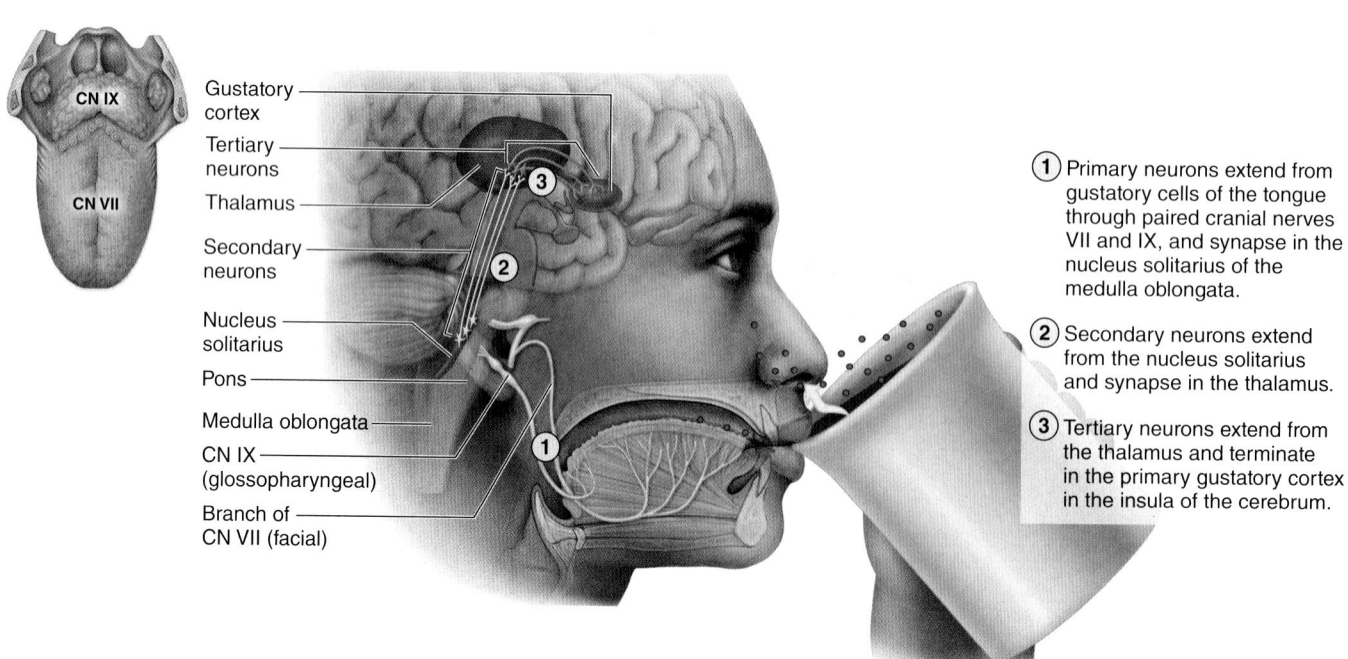

1 Primary neurons extend from gustatory cells of the tongue through paired cranial nerves VII and IX, and synapse in the nucleus solitarius of the medulla oblongata.

2 Secondary neurons extend from the nucleus solitarius and synapse in the thalamus.

3 Tertiary neurons extend from the thalamus and terminate in the primary gustatory cortex in the insula of the cerebrum.

Figure 16.7 Gustatory Pathway. Taste sensations are transmitted by the facial nerves (CN VII) from the anterior two-thirds of the tongue and by the glossopharyngeal nerves (CN IX) from the posterior one-third of the tongue. These taste sensations are transmitted to the nucleus solitarius of the medulla oblongata before being transmitted to the thalamus and finally entering the gustatory cortex of the cerebrum.

food was bland and almost tasteless. Together, taste and smell give our food its flavor. We then perceive the taste of the food when the brain interprets both the stimulation pattern from gustatory receptor sensory neurons and the aroma information from olfactory receptors.

WHAT DID YOU LEARN?

9 Which papillae of the tongue have taste buds, and what is the basic composition of a taste bud?

10 What are the five basic taste sensations, and what is the specific stimulus detected by each?

16.4 Visual Receptors

The sense of vision uses photoreceptors within the eyes to detect light, color, and movement to perceive detailed visual images of objects in our environment. We first describe the accessory structures of the eye before discussing the details of the anatomy of the eye and how it functions in vision.

16.4a Accessory Structures of the Eye

LEARNING OBJECTIVE

1. Describe the accessory structures of the eye, and list their functions.

The accessory structures of the eye are located either attached to the eye or around the eye (**figure 16.8**). These structures include the extrinsic eye muscles, eyebrows, eyelids, eyelashes, conjunctiva, and lacrimal glands.

The extrinsic eye muscles include six skeletal muscles attached externally to each eye and function in eye movement. These muscles are described in detail in section 11.3b.

Eyebrows and Eyelids

The **eyebrows** are slightly curved rows of thick, short hairs at the superior edge of the orbit along the supraorbital ridge. They function in both nonverbal communication associated with facial expressions and to prevent sweat from dripping into the eyes.

The **eyelids,** also called the *palpebrae* (pal-pē′brē; sing., *palpebra*), form the protective covering over the surface of the eye. Each eyelid is formed primarily by a fibrous core (the **tarsal plate**), the orbicularis oculi muscle (which closes the eyelid), and a thin covering of skin. A muscle associated only with the upper eyelid is the *levator palpebrae superioris muscle*, which pulls the upper eyelid to "open the eye." The space between the open eyelids is the **palpebral** (pal′pĕ-brăl) **fissure** (or eyeslit). The eyelids are joined at the **medial** and **lateral palpebral commissures** or *canthi*. At the medial commissure is a small, reddish body called the **lacrimal caruncle** (kar′ŭng-kl; *caro* = flesh).

Eyelashes extend from the free margins of the eyelids and help prevent particulate matter from entering the eye. Sensory receptors associated with the base of an eyelash trigger the blink reflex when the eyelash is touched.

Several glands are associated with the eyelids and eyelashes. **Tarsal glands**, which are sebaceous glands located within the tarsal plates of the eyelid, release an oily secretion at the edge of the eyelid. A *chalazion* (a cyst within the eyelid) forms from an infection of a tarsal gland. Both a sebaceous gland and a modified sweat gland are located at the base of each eyelash. A *stye* (which appears as a reddened area beneath the eyelid) is an infection of either of these two glands. These glands contribute to the gritty, particulate material often noticed around the eyelids after waking.

Conjunctiva

A specialized stratified columnar epithelium termed the **conjunctiva** (kon-jŭnk-tī′vă) forms a continuous transparent lining over the anterior surface of the sclera ("white") of the eye (the **ocular conjunctiva**) and the internal surface of the eyelid (the **palpebral conjunctiva**). The junction of the ocular conjunctiva and palpebral conjunctiva is called the **conjunctival fornix** (fōr′niks; = vault or arch). (This junction is what prevents a contact lens from moving behind the eye.)

The conjunctiva contains numerous goblet cells, which secrete mucin to lubricate and moisten the eye. Additionally, the conjunctiva contains many blood vessels, which supply oxygen and nutrients to the avascular sclera, as well as abundant nerve endings that detect foreign

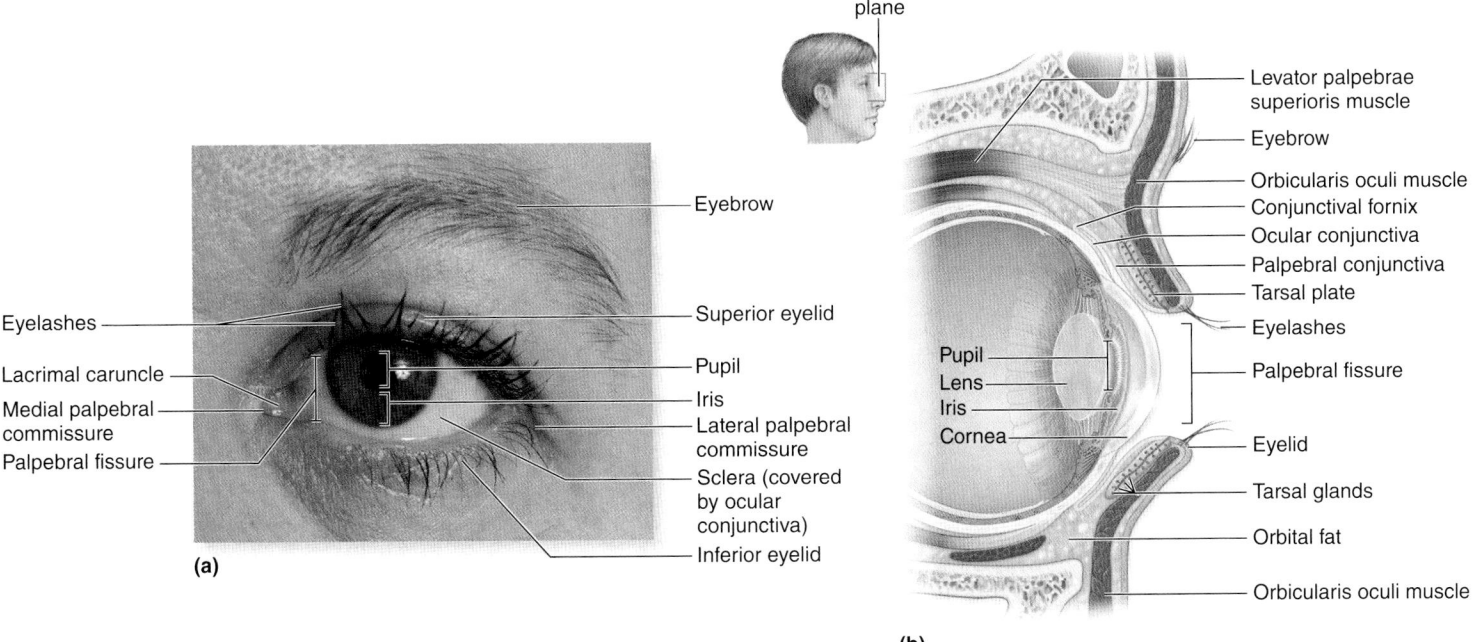

Figure 16.8 External Anatomy of the Eye and Surrounding Accessory Structures. (*a*) Photo of anterior view of the eye and accessory structures. (*b*) Diagram of a sagittal section shows the eye and its accessory structures. AP|R

INTEGRATE

CONCEPT CONNECTION

Five cranial nerves innervate the eye (see section 13.9). These include the CN II (optic nerve), which is a sensory nerve that relays input from the retina when it is stimulated by light, and CN V (trigeminal), which relays sensations from the cornea. Then there are three primarily motor nerves that relay motor output to the eye muscles. CN III (oculomotor nerve) innervates four of the six extrinsic eye muscles to control eye movement and the intrinsic eye muscles (i.e., the iris and ciliary muscles); CN IV (trochlear nerve) and VI (abducens nerve) each innervate an extrinsic eye muscle.

objects as they contact the eye. The conjunctiva does not cover the surface of the cornea (the transparent center of the anterior eye), so no blood vessels interfere with passage of light into the eye. **Conjunctivitis** (kon-jŭnk-ti-vī′tis) (or pink eye) is an inflammation of the conjunctiva caused either by infectious agents or irritants (e.g., pollen, contact lenses), and is the most common non-traumatic eye complaint seen by physicians.

Lacrimal Apparatus

A **lacrimal** (lak′ri-măl; *lacrima* = a tear) **apparatus** is associated with each eye; it produces, collects, and drains lacrimal fluid **(figure 16.9)**. Lacrimal fluid contains water, sodium ions, antibodies (see section 22.8), and an antibacterial enzyme called **lysozyme.** This fluid lubricates the anterior surface of the eye to reduce friction from eyelid movement; continuously cleanses and moistens the eye surface; helps prevent bacterial infection; and provides oxygen and nutrients to the corneal epithelium (described shortly).

The production and movement of lacrimal fluid occurs as follows (figure 16.9):

1. A **lacrimal gland,** which is about the size and shape of an almond and located within the superolateral depression of each orbit, continuously produces lacrimal fluid that drains through short ducts to the eye surface.
2. Blinking, which occurs about 15 to 20 times per minute, but less frequently when we are focusing on something such as reading a book, "washes" the lacrimal fluid over the eyes. Gradually, the lacrimal fluid is transferred to the lacrimal caruncle at the medial surface of the eye.
3. The lacrimal fluid drains into the **lacrimal puncta** (pungk′tă; sing., *punctum* = to prick), which are two small openings on the superior and inferior side of the lacrimal caruncle. (If you examine the lacrimal caruncle within your own eye, each punctum appears as a "hole."
4. Each lacrimal punctum has a **lacrimal canaliculus** (kan-ă-lik′yū-lūs; = small canal) that drains lacrimal fluid into a rounded **lacrimal sac.**
5. The fluid drains from the lacrimal sac into a **nasolacrimal duct.** This duct drains lacrimal fluid into the nasal cavity, where it mixes with mucus and then moves into the pharynx (throat) and is swallowed. Excess production of lacrimal fluid produces tears.

WHAT DID YOU LEARN?

11 Where is the conjunctiva located, and what are its functions?

12 How is lacrimal fluid spread across the eye surface and removed from the orbital region?

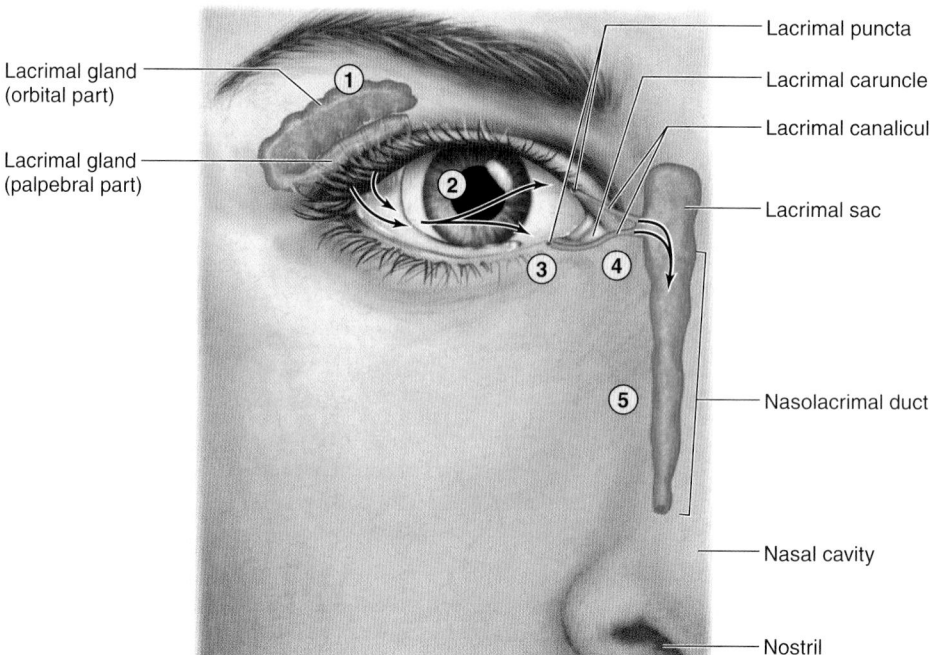

Lacrimal gland (orbital part)
Lacrimal gland (palpebral part)
Lacrimal puncta
Lacrimal caruncle
Lacrimal canaliculi
Lacrimal sac
Nasolacrimal duct
Nasal cavity
Nostril

Figure 16.9 Lacrimal Apparatus. The lacrimal apparatus continually produces lacrimal fluid (tears) that cleanses and maintains a moist condition on the anterior surface of the eye. Lacrimal fluid production and drainage occur in a series of steps.

16.4b Eye Structure

LEARNING OBJECTIVE

2. Describe the structures of the eye.

The eye is an almost spherical organ that measures about 2.5 centimeters (1 inch) in diameter. Most of the eye is receded into the orbit of the skull (see section 8.2d), a space also occupied by the lacrimal gland, extrinsic eye muscles, numerous blood vessels, and the cranial nerves that innervate the eye and other structures in the orbit. **Orbital fat** cushions the posterior and lateral sides of the eye (figure 16.8*b*), providing support, protection, and facilitating oxygen and nutrient delivery by the blood through its associated blood vessels.

The interior of the eye consists of two fluid-filled cavities. These two cavities are separated by the lens, which is a transparent biconvex structure enclosed in a fibrous capsule **(figure 16.10)**. The **posterior cavity,** which lies behind the lens, contains a permanent fluid called vitreous humor and the **anterior cavity,** which is in front of the lens, contains a circulating

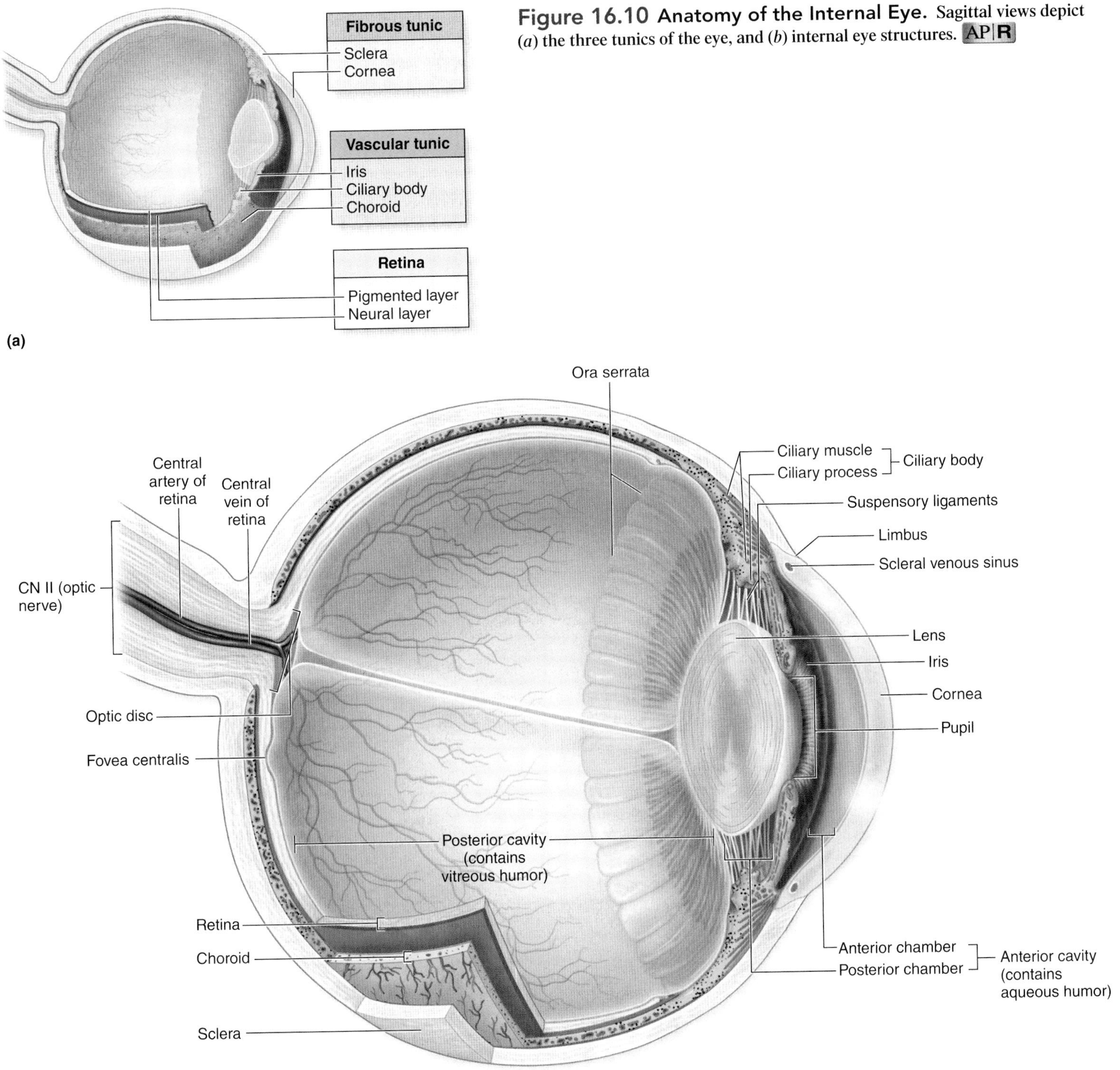

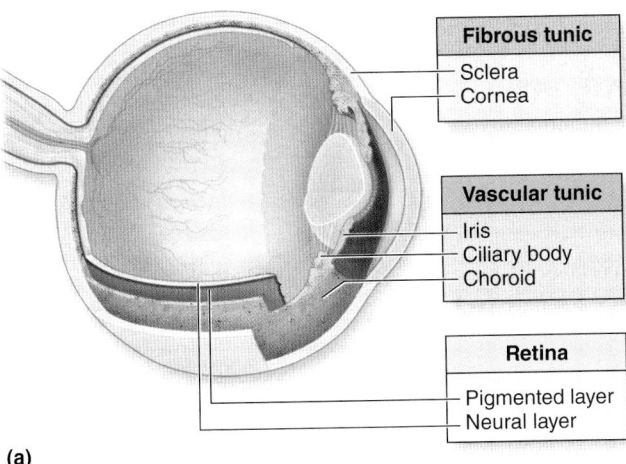

Fibrous tunic
Sclera
Cornea

Vascular tunic
Iris
Ciliary body
Choroid

Retina
Pigmented layer
Neural layer

(a)

Figure 16.10 Anatomy of the Internal Eye. Sagittal views depict (*a*) the three tunics of the eye, and (*b*) internal eye structures. **AP|R**

Ora serrata

Central artery of retina

Central vein of retina

CN II (optic nerve)

Optic disc

Fovea centralis

Retina

Choroid

Sclera

Ciliary muscle ⎤ Ciliary body
Ciliary process ⎦

Suspensory ligaments

Limbus

Scleral venous sinus

Lens

Iris

Cornea

Pupil

Posterior cavity (contains vitreous humor)

Anterior chamber ⎤ Anterior cavity
Posterior chamber ⎦ (contains aqueous humor)

(b)

fluid called aqueous humor. The lens and both fluids are described in detail later in this section.

The wall of the eye is formed by three principal tunics (or layers): the fibrous tunic (external layer), the vascular tunic (middle layer), and the retina (inner layer).

Fibrous Tunic

The external layer of the eye wall is called the **fibrous tunic,** or *external tunic.* It is composed of the posterior sclera and the anterior cornea.

Most of the fibrous tunic (the posterior five-sixths) is the tough **sclera** (sklĕr′ă; *skleros* = hard), a part of the outer layer that is called the "white" of the eye. It is composed of dense irregular connective tissue containing numerous blood vessels and nerves. The sclera provides for eye shape, protects the eye's delicate internal components, and serves as attachment site for extrinsic eye muscles. Posteriorly, the sclera is continuous with the dura mater that surrounds the optic nerve. ("Bloodshot" eyes occur with vasodilation of the scleral blood vessels, which become visible through the transparent conjunctiva.)

The **cornea** (kōr′nē-ă) is a convex, transparent structure that forms the anterior one-sixth of the fibrous tunic; its convex shape refracts (bends) light rays coming into the eye. The cornea is composed of an inner simple squamous epithelium, a middle layer of collagen fibers, and an outer stratified squamous epithelium, called the **corneal epithelium.** (Think of the cornea as a collagen protein sandwich with epithelial layers as the bread.) The cornea contains no blood vessels. Nutrients and oxygen are supplied to the internal epithelium of the cornea by aqueous humor within the anterior cavity of the eye, whereas the surface corneal epithelium receives its oxygen and nutrients from lacrimal fluid.

The cornea merges with the sclera at its outer edge; this region is called the **limbus** (lim′bŭs), or the *corneal scleral junction.* The corneal epithelium forming the external portion of the cornea is continuous with the ocular conjunctiva that covers the sclera. Thus, the entire eye is covered with an epithelium.

Vascular Tunic

The middle layer of the eye wall is the **vascular tunic,** also called the *uvea* (ū′vē-ă; *uva* = grape). The vascular tunic houses an extensive array of blood vessels, lymph vessels, and the intrinsic muscles of the eye. It is composed of three distinct regions; from posterior to anterior, they are the choroid, the ciliary body, and the iris.

The **choroid** (kōr′oyd) is the most extensive and posterior region of the vascular tunic, and it is composed of loose connective tissue that houses both an extensive network of capillaries and melanocytes. Two primary functions are associated with the choroid. Its vast network of blood vessels supplies oxygen and nutrients to the retina, and melanin produced by its melanocytes absorbs extraneous light to prevent it from scattering within the eye.

The **ciliary** (sil′ē-ar-ē; *cilium* = eyelid) **body** is located immediately anterior to the choroid, and is composed of both ciliary muscles and ciliary processes. **Ciliary muscles** are bands of smooth muscle. Extending from the ciliary muscle to the capsule surrounding the lens are suspensory ligaments, which anchor the lens. Relaxation and contraction of the ciliary muscles change the tension on the suspensory ligaments, thereby altering the shape of the lens. The **ciliary processes** contain capillaries that secrete aqueous humor (both functions of the ciliary body are discussed later in detail).

The most anterior region of the vascular tunic is the **iris** (ī′ris; = rainbow), which is the colored portion of the eye. The iris is composed of two layers of smooth muscle fibers, melanocytes, and an array of vascular and nervous structures. In the center of the iris is an opening called the **pupil** (pū′pil). The iris subdivides the

anterior cavity into two chambers (or rooms), including the **anterior chamber,** which is located between the iris and cornea, and the **posterior chamber,** which is located between the lens and the iris. The pupil is the opening between these two chambers. The iris controls pupil size or diameter—and thus the amount of light entering the eye—using its two smooth muscle layers **(figure 16.11).** The **sphincter pupillae** (pyū-pil′ē) **muscle** (or *pupillary constrictor*) is arranged in a pattern that resembles concentric circles around the pupil. This muscle contracts (and the pupil becomes *smaller*) when stimulated by visceral motor neurons of the parasympathetic division of the ANS that are within the oculomotor nerve (CN III). In comparison, the **dilator pupillae muscle** (or *pupillary dilator*) is organized in a radial pattern extending peripherally through the iris. This muscle contracts (and the pupil becomes *larger*) when stimulated by neurons of the sympathetic division of the ANS. Only one set of these smooth muscle layers can contract at a time. When stimulated by bright light the parasympathetic division causes the pupillary constrictors to contract and thus decrease pupil diameter, whereas low light levels activate the sympathetic division to cause pupil dilation. The **pupillary reflex** is the ability of the pupil to change in response to varying amounts of light. This reflex is tested if brain trauma is suspected (e.g., from a car accident or drug overdose).

Retina

The internal layer of the eye wall, called the **retina** (ret′i-nă; *rete* = a net) also is known as the *internal tunic* or *neural tunic.* It is composed of two layers: an outer pigmented layer and an inner neural layer **(figure 16.12).** The **pigmented layer** is immediately internal to the choroid and attached to it. Two primary functions are associated with the pigmented layer. It provides vitamin A for the photoreceptor (light-detecting) cells of the neural layer and absorbs extraneous light to prevent it from scattering within the eye (a function it shares with the choroid). The inner **neural layer** houses all of the photoreceptor cells and their associated neurons. This layer of the retina is responsible for

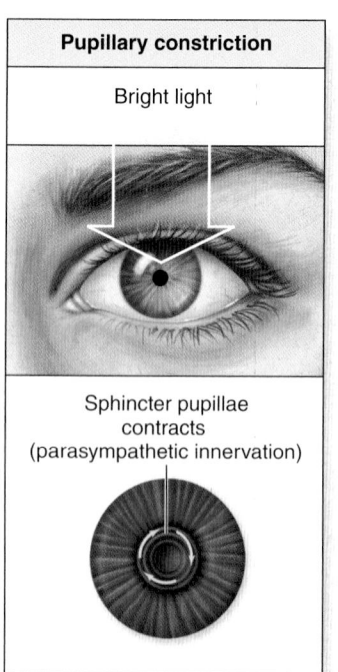
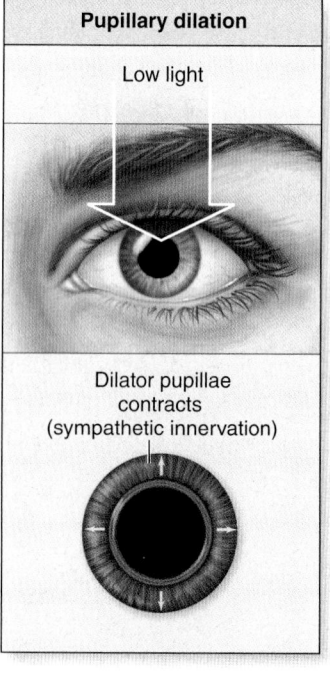

Pupillary constriction	Pupillary dilation
Bright light	Low light
Sphincter pupillae contracts (parasympathetic innervation)	Dilator pupillae contracts (sympathetic innervation)

Figure 16.11 Pupil Diameter. Pupillary constriction decreases the diameter of the pupil to reduce the amount of light entering the eye. Pupillary constriction is controlled by the parasympathetic division of the ANS. Pupillary dilation increases pupil diameter to increase light entry into the eye and is controlled by the sympathetic division.

Figure 16.12 Structure and Organization of the Retina. The retina is composed of two distinct layers: the outer pigmented layer and the inner neural layer, also called the neural retina. (*a*) The optic nerve is composed of ganglionic cell axons that originate in the neural retina. (*b, c*) The neural retina is composed of three primary cellular layers (in bold): the outer photoreceptor cells (rods and cones), the middle bipolar cells, and the inner ganglion cells. AP|R

(a)

Optic nerve
Retina
Sclera
Choroid
Optic disc
Fovea centralis

(b)

Retina
Pigmented layer
Neural layer

→ Incoming light
→ Nerve signal response to light through retina

Choroid
Photoreceptor cells:
Rods
Cones
Horizontal cell
Bipolar cells
Amacrine cell
Ganglion cells
Axons of ganglion cells to optic nerve
Posterior cavity

(c)

Pigmented layer
Neural layer
Retina
LM 250x

---

absorbing light rays and converting them into nerve signals that are transmitted to the brain.

The **ora serrata** (ōră sē-ră′tă; *serratus* = sawtooth) is a jagged margin between the photosensitive posterior region of the retina and the nonphotosensitive anterior region of the retina. This nonphotosensitive portion continues anteriorly to cover the ciliary body and the posterior aspect of the iris (see figure 16.10*b*).

Cells of the Neural Layer   Three distinct layers of neurons form the neural layer: photoreceptor cells, bipolar cells, and ganglion cells. The outermost layer of cells in the neural layer is composed of **photoreceptor** (*phot* = light) **cells,** which contain pigment molecules that react to light energy. The two types of photoreceptor cells are **rods,** which have a rod-shaped outer portion and function in dim light, and **cones,** which have a cone-shaped outer portion and function in high-intensity light and in color vision. These cells are described in more detail in section 16.4d.

Immediately internal to the photoreceptor cells is a layer of **bipolar cells.** Rods and cones form synapses on the dendrites of the bipolar cells.

Ganglion cells form the innermost layer in the neural layer and are adjacent to the posterior cavity. Axons of the ganglionic cells extend into and through the optic disc (described shortly) to form the optic nerve. Note that incoming light must pass through the ganglion cells and bipolar cells before reaching the photoreceptor cells.

Other cells that function in transmission of light stimuli include horizontal cells and amacrine cells. **Horizontal cells** are sandwiched between the photoreceptor and bipolar cells in a thin web. These horizontal cells regulate and integrate the stimuli sent from the photoreceptor cells to the other cell layers. **Amacrine** (am′ă-krin) **cells** are positioned between the bipolar and ganglion cells and help process and integrate visual information as it passes between bipolar and ganglion cells. Only the amacrine and ganglion cells in the retina produce action potentials; the other cells generate graded potentials (see section 12.9a).

INTEGRATE

CLINICAL VIEW
Detached Retina

A **detached retina** occurs when the outer pigmented and inner neural layers of the retina separate. Detachment may result from head trauma (soccer players and high divers are especially susceptible), or it may have no overt cause. Individuals who are nearsighted, due to a more elliptical eyeball, are at increased risk for detachment because their retina is typically thinned or stretched more than that of a normal eye. There is also increased risk for retinal detachment in diabetics and older individuals. A detached retina results in deprivation of nutrients for cells in the inner neural layer because it is pulled away from the vascularized choroid layer. Degeneration and death of the neural layer of the retina occur if the blood supply is not restored.

Symptoms of a detached retina include a large number of "floaters" (small, particle-like objects) in the vision; the appearance of a "curtain" in the affected eye; flashes of light; and decreased, watery, or wavy vision.

Pneumatic retinopexy is a treatment for upper retinal detachment. The physician inserts a needle into the anesthetized eye and injects a gas bubble into the vitreous humor. The gas bubble rises and pushes the neural layer back into its normal position. The gas bubble is absorbed and disappears over 1 to 2 weeks, and then a laser may be used to tack the two layers of the retina together. The **scleral buckle** is another treatment, and uses a silicone band to press inward on the sclera to hold the retina in place. A laser is then used to reattach the retina.

Components of the Retina   The distribution of rods and cones, the two types of photoreceptor cells, is not uniform throughout the retina. Three specific regions are identified: the optic disc, macula lutea, and peripheral retina (**figure 16.13**). The **optic disc** contains no photoreceptors. This is where axons of the ganglion cells extend from the back of the eye as the optic nerve (figure 16.12*a*). It is commonly called the **blind spot** because it lacks photoreceptor cells, and no image forms there. The **macula lutea** (mak'ū-lă lū'tē-a; *macula* = small spot; *lutea* = saffron-yellow) is a rounded, yellowish region just lateral to the optic disc (figure 16.13). Within the macula lutea is a depressed pit called the **fovea centralis** (fō'vē-ă sen'tră'lis; *fovea* = pit; *centralis* = central), which contains the highest proportion of cones and almost no rods. This pit is the area of sharpest vision; when you read the words in your text, they are precisely focused here. Although the other regions of the retina also receive and interpret light rays, no other region can focus as precisely as can the fovea centralis because of its high concentration of cones. The remaining most extensive region of the retina is called the **peripheral retina,** which contains primarily rods and functions most effectively in low light.

Lens

The **lens** is a strong yet deformable, transparent structure. It is composed of precisely arranged layers of cells that have lost their organelles and are filled completely by a protein called *crystallin,* which are enclosed by a dense, fibrous, elastic capsule. The lens focuses incoming light onto the retina, and its shape determines the degree of light refraction.

The **suspensory** (sŭs-pen'sŏ-rē; *suspendo* = to hang up) **ligaments** attach to the lens capsule at its periphery, where they transmit tension that enables the lens to change shape. The relative tension in the suspensory ligaments is altered by relaxation and contraction of the ciliary muscles in the ciliary body. When we view objects greater than 20 feet away, the ciliary muscles relax, the ciliary body moves away from the lens, and so the tension on the suspensory ligaments increases. This constant tension causes the lens to flatten (**figure 16.14***a*). This shape of the lens is the "resting" position of the lens.

In contrast, when we wish to view objects closer than 20 feet, the ciliary muscles contract, the ciliary body moves closer to the lens, and the tension on the suspensory ligaments decreases. This releases some of their pull on the lens so the lens can become more spherical or curved. The process of making the lens more spherical to view close-up objects is called **accommodation** (ă-kom'ŏ-dā'shŭn; *accommodo* = to adapt) (figure 16.14*b*). Accommodation is controlled by visceral motor neurons of the parasympathetic division that extend within the oculomotor nerve (CN III); see section 13.9.

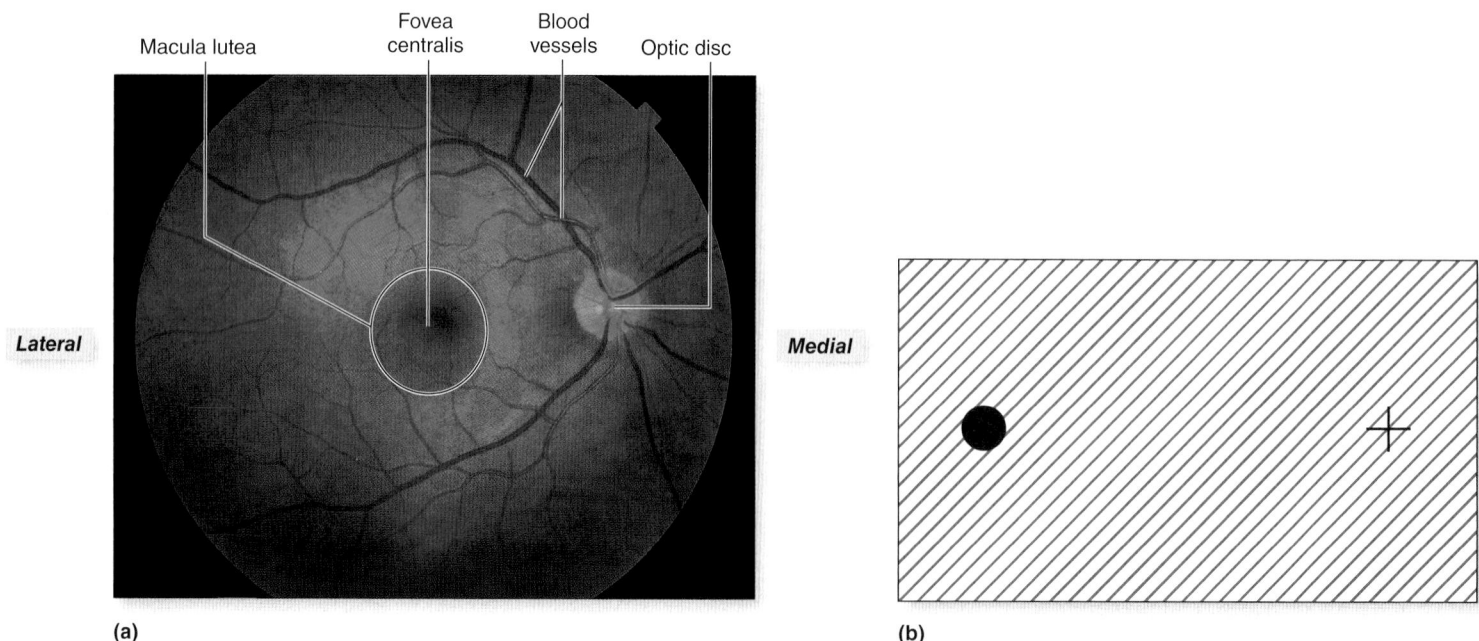

(a)　　　　　　　　　　　　　　　　(b)

Figure 16.13 Internal View of the Retina Showing the Optic Disc (Blind Spot). (*a*) An ophthalmoscope is used to view the retina through the pupil. Blood vessels travel through the optic nerve as it enters the eye at the optic disc. (*b*) Check your blind spot! Close your left eye. Hold this figure in front of your right eye, and stare at the black spot. Move the figure toward your open eye. At approximately 6 inches from your eye, the image of the plus sign is over the optic disc and the plus sign seems to disappear.

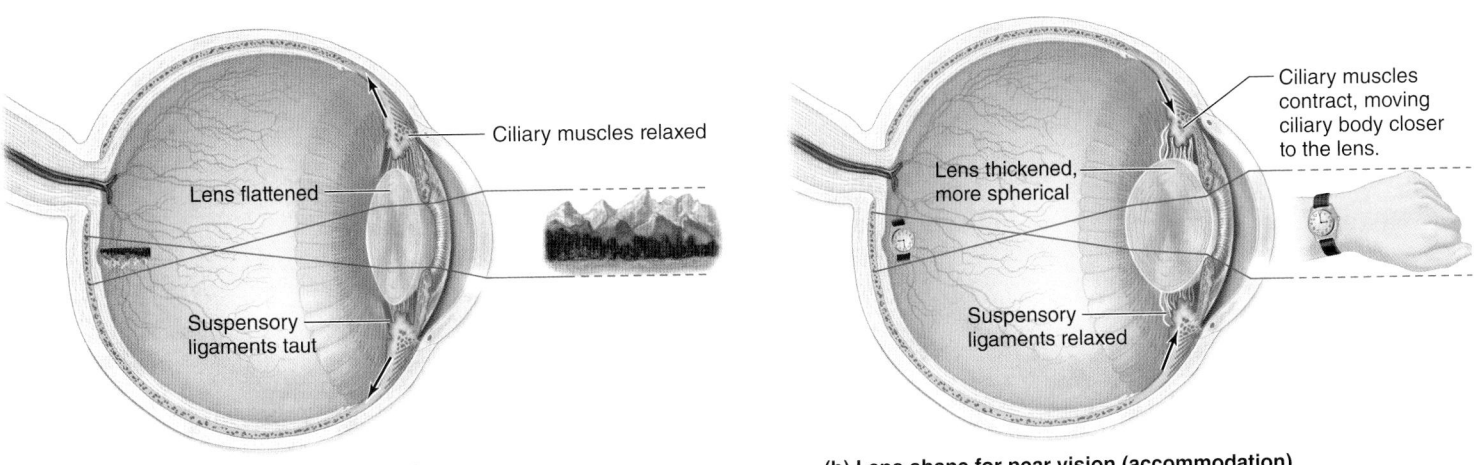

(a) Lens shape for distant vision

(b) Lens shape for near vision (accommodation)

Figure 16.14 Lens Shape in Far Vision and Near Vision. (*a*) To focus a distant object on the retina, the ciliary muscles within the ciliary body relax, which tenses the suspensory ligaments and flattens the lens. (*b*) To focus a near object on the retina, the ciliary muscles contract, causing release of tension on the suspensory ligaments and the lens to thicken (become more spherical or "puffy"). This process is called accommodation.

Vitreous Humor and Aqueous Humor

Vitreous humor (vit′rē-ŭs; *vitrum* = glassy), or *vitreous body,* is the transparent, gelatinous fluid that completely fills the posterior cavity. This permanent fluid is produced during embryonic development and both helps to maintain eye shape and support the retina to keep it flush against the back of the eye (see Clinical View: "Detached Retina").

Aqueous humor (ak′qwē-ŭs hū′mer; = watery fluid) is a transparent, watery fluid that circulates within the anterior cavity (**figure 16.15**). It is continuously produced by the ciliary processes. The circulation of aqueous humor provides nutrients and oxygen to both the avascular cornea (specifically, its inner epithelium) and lens.

INTEGRATE

CLINICAL VIEW
Macular Degeneration

Macular degeneration, which is the physical deterioration of the macula lutea, is the leading cause of blindness in developed countries. Although the majority of cases are reported in people over 55, the condition may occur in younger people as well. Most non-age-related cases are associated with conditions such as diabetes, an ocular infection, hypertension, or trauma to the eye.

Macular degeneration is typically associated with the loss of visual acuity in the center of the visual field, diminished color perception, and "floaters."

At present, there is no cure for macular degeneration. However, its progression may be slowed. Early detection has become an important element in treatment. To track the progression of the disease, doctors rely heavily on self-monitoring, in which the patient regularly performs a simple visual test using the *Amsler grid.* While staring at the dot, the patient looks for wavy lines, blurring, or missing parts of the grid, which would indicate degenerative changes in the macula.

Normal vision.

The same scene as viewed by a person with macular degeneration.

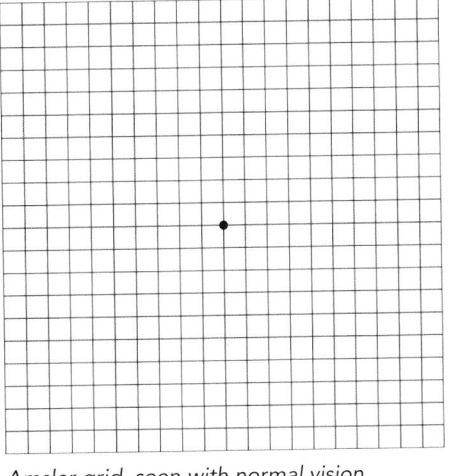

Amsler grid, seen with normal vision.

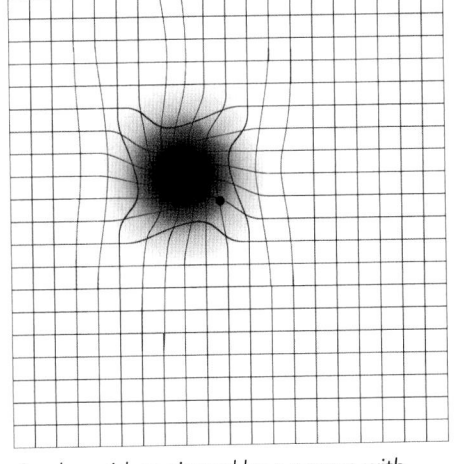

Amsler grid, as viewed by a person with macular degeneration.

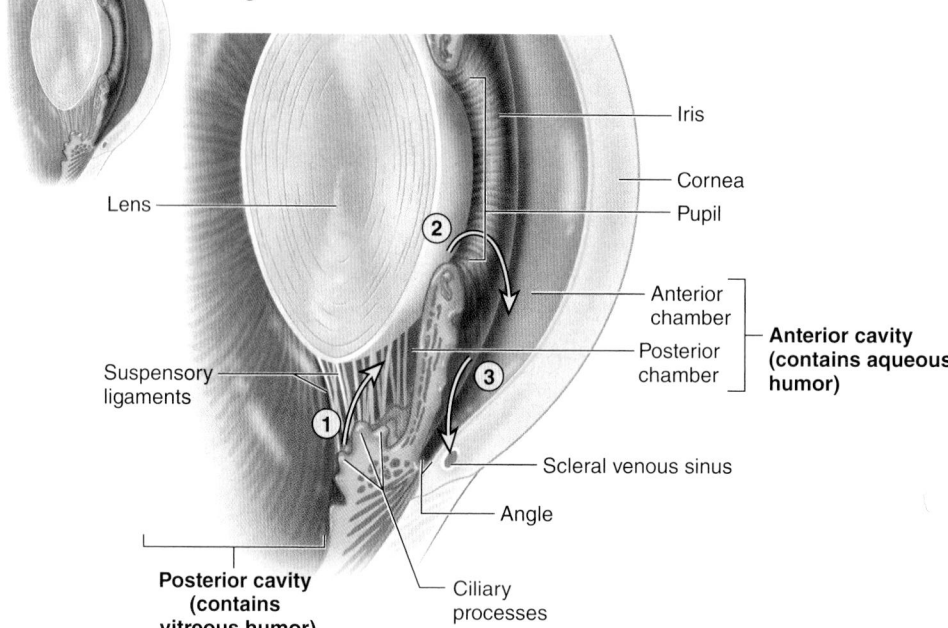

Anterior cavity
■ Anterior chamber
☐ Posterior chamber

▨ Posterior cavity

Figure 16.15 Aqueous Humor: Secretion and Resorption. Aqueous humor is a watery secretion that is continuously produced and circulated through the anterior cavity of the eye, to provide oxygen and nutrients to the inner portion of the cornea and the adjacent region of the lens.

Iris

Cornea

Pupil

Lens

Anterior chamber

Posterior chamber

Anterior cavity (contains aqueous humor)

Suspensory ligaments

Scleral venous sinus

Angle

Posterior cavity (contains vitreous humor)

Ciliary processes

(1) Aqueous humor is secreted by the ciliary processes into the posterior chamber.

(2) Aqueous humor moves from the posterior chamber, through the pupil, to the anterior chamber.

(3) Excess aqueous humor is resorbed into the scleral venous sinus.

INTEGRATE

CLINICAL VIEW
Cataracts

Cataracts (kat′ă-rakt) are small opacities within the lens that, over time, may coalesce to completely obscure the lens. Most cases occur as a result of aging, although other factors include diabetes, intraocular infections, excessive ultraviolet light exposure, and glaucoma. The resulting vision problems include difficulty focusing on close objects, reduced visual clarity due to clouding of the lens, "milky" vision, and reduced intensity of colors.

A cataract needs to be removed only when it interferes with normal daily activities. Newer surgical techniques include **phacoemulsification** (făk′ō-i-mul′se-fi-kā′shen), a process by which the opacified center of the lens is fragmented using ultrasonic sound waves, thus making it easier to remove. The destroyed lens is then replaced with an artificial intraocular lens, which becomes a permanent part of the eye.

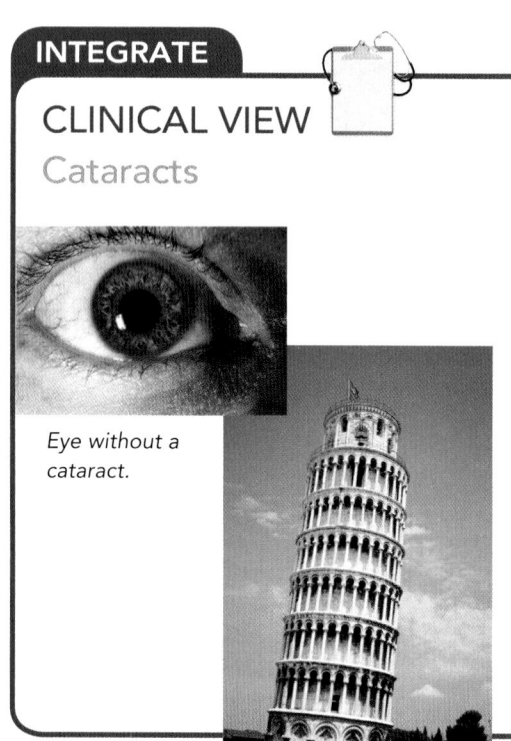

Eye without a cataract.

Normal vision.

Cataract

Eye with a cataract.

Image seen through cataract.

Blood plasma (see section 18.2) is filtered across the walls of capillaries of ciliary processes and enters the posterior chamber. (This process is similar to the formation of cerebrospinal fluid by the choroid plexuses within the ventricles of the brain; see section 13.2c.) Aqueous humor is filtered into the posterior chamber, circulates through the pupil, and moves into the anterior chamber. Aqueous fluid drains from the anterior chamber into a circular canal at the limbus called the **scleral venous sinus** (previously called the *canal of Schlemm*). Aqueous fluid then drains into nearby veins. Thus, as with cerebrospinal fluid, aqueous humor is produced from capillaries, circulates, and then enters into the venous circulation. Normally, the rate of formation by the ciliary processes is equal to the drainage into the scleral venous sinus; thus, a normal intraocular pressure is maintained. *Glaucoma* results from the blockage of aqueous humor drainage (see Clinical View: "Glaucoma").

WHAT DID YOU LEARN?

13 What are the three eye tunics; what is the primary function of each tunic?

14 Compare the anatomic structure of the cornea and the lens. Explain how oxygen and nutrients are supplied to these two structures.

15 What are the functions of the vitreous humor and the aqueous humor?

16.4c Physiology of Vision: Refraction and Focusing of Light

LEARNING OBJECTIVES

3. Describe refraction of light.

4. Discuss how light is focused on the retina.

The physiology of vision requires that light enters the eye and is transduced into electrical signals, which are then sent to the brain for integration and interpretation. The primary processes of vision include the refraction and focusing of light on the retina (which is discussed in this section), and phototransduction and relaying of sensory input along visual pathways (which are discussed in later sections).

Refraction of Light

Light rays are straight as they first enter the naked eye. However, the ability to see clearly requires refraction (or bending) of the light rays so that they hit on the retina—specifically, at the fovea centralis, which is the portion of the retina predominantly composed of cones, and provides the

INTEGRATE

CLINICAL VIEW

Glaucoma

Glaucoma (glaw-kō'mă) is a disease that exists in three forms, all characterized by increased intraocular pressure: primary angle-closure glaucoma, open-angle glaucoma, and congenital or juvenile glaucoma.

Angle-closure glaucoma and *open-angle glaucoma* both involve the angle formed in the anterior chamber of the eye by the union of the choroid and the corneal-scleral junction (figure 16.15). If this angle narrows, aqueous humor and pressure build up within the anterior chamber. Primary angle-closure glaucoma develops as a direct consequence of the narrowing of this angle, whereas open-angle glaucoma occurs if the drain angles are adequate but fluid transport out of the anterior chamber is impaired. *Congenital glaucoma* occurs only rarely and is due to hereditary factors or intrauterine infection.

Regardless of the cause, fluid buildup causes a posterior dislocation of the lens and a substantial increase in pressure in the posterior cavity. Compression of the choroid layer may occur, constricting the blood vessels that nourish the retina. Retinal cell death and increased pressure may distort the axons of ganglionic cells that form the optic nerve. Eventually, the patient may experience such symptoms as reduced field of vision, dim vision, and halos around lights. These symptoms are often unrecognized until it is too late and the damage is irreversible, so it is essential individuals get regular eye screenings where early stages may be detected by an optometrist or ophthalmologist.

CLINICAL VIEW
Functional Visual Impairments

Emmetropia (em-ĕ-trō′pē-ă; *emmetros* = according to measure, *ops* = eye) is the condition of normal vision, in which parallel rays of light are focused exactly on the retina. Any variation in the curvature of either the cornea or the lens, or in the overall shape of the eye, causes entering light rays to form an abnormal focal point. Conditions that can result include hyperopia, myopia, and astigmatism.

People with **hyperopia** (hī-per-ō′pē-ă) have trouble seeing close-up objects, and so are called *farsighted*. In this optical condition, only convergent rays (those that come together from distant points) can be brought to focus on the retina. The cause of hyperopia is a short eyeball; parallel light rays from objects close to the eye focus posterior to the retina. By contrast, people with **myopia** (mī-ō′pē-ă; *myo* = to shut) have trouble seeing faraway objects, and so are called *nearsighted*. In myopia, only rays relatively close to the eye focus on the retina. The cause of this condition is a long eyeball; parallel light rays from objects at some distance from the

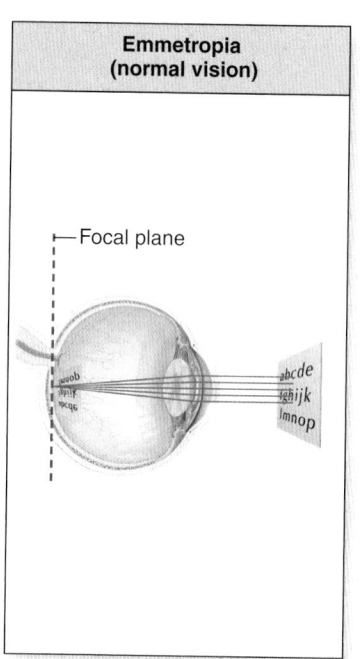

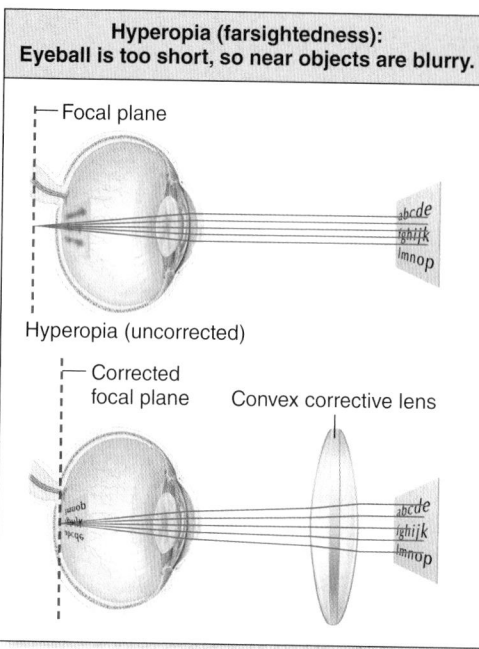

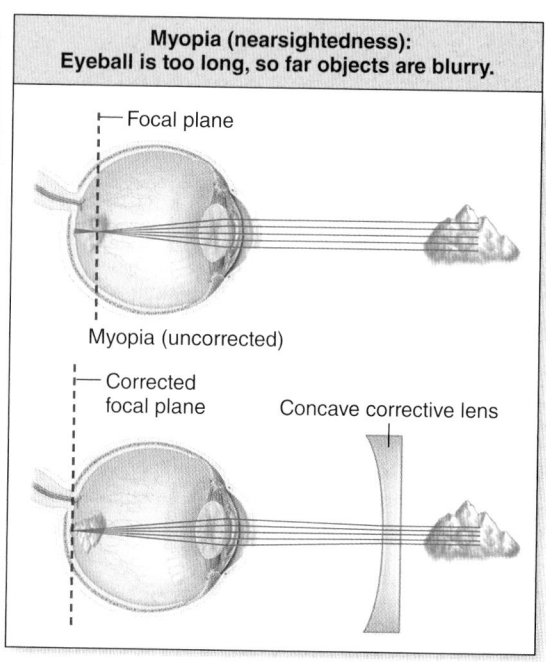

Vision correction using convex and concave lenses.

Figure 16.16 Refraction. Light waves change in speed as they pass between media of different densities. This results in the distortion of the image.

sharpest vision (figure 16.14). Light rays are refracted when (1) they pass between two media of different densities, and (2) these media meet at a *curved* surface. Each medium—such as air, water, and other clear fluids, and even clear solids such as glass—is assigned a *refractive index*, a number that represents its comparative density. The refraction of light rays is greater when there is a larger difference in the refractive index between adjacent media, such as between air and water (**figure 16.16**) and with increasing curvature of the media surface.

Before light can reach the photoreceptor cells, it must pass from the air, through the cornea, the aqueous humor, lens, and the vitreous humor, as well as through the cells forming the inner layers of the retina. Both the cornea and lens play a significant role in refraction of light for vision—the cornea because of the relatively large refraction of light that occurs as light passes from air into the cornea, and the lens because of its ability to change shape. Recall that the curvature of the lens (and thus the refraction of light) can be altered by relaxation and contraction of the ciliary muscles.

Focusing of Light

The processes that occur in the eye for us to see clearly are dependent upon how far away the object is that we are viewing. Let us first consider the three significant changes that occur when we view objects that are closer than 20 feet. These include convergence of the eyes, accommodation of the lens, and constriction of the pupil.

Convergence of the Eyes   **Convergence** of the eyes is the voluntary contraction of the extrinsic eye muscles to move the eyes medially. (The extreme case of this is when your eyes

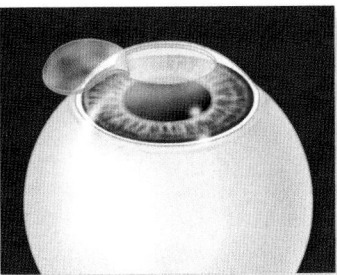

LASIK laser vision correction procedure.

① Cornea is sliced with a sharp knife. Flap of cornea is reflected, and deeper corneal layers are exposed.

② A laser removes microscopic portions of the deeper corneal layers, thereby changing the shape of the cornea.

③ Corneal flap is put back in place, and the edges of the flap start to fuse within 72 hours.

eye focus anterior to the retina within the vitreous body. Another variation is **astigmatism** (ă-stig′mă-tizm), which causes unequal focusing and blurred images due to unequal curvatures in one or more of the refractive surfaces (cornea, anterior surface, or posterior surface of the lens).

With increasing age, the lens becomes less resilient and less able to become spherical. Thus, even if the suspensory ligaments relax, the lens may not be able to spring out of the flattened position into the more spherical shape needed for near vision, and reading close-up words becomes difficult. This age-related change is called **presbyopia** (prez-bē-ō′pē-ă; *presbys* = old man, *ops* = eye).

The typical treatment for vision disturbances is eyeglasses. A concave eyeglass lens is used to treat myopia, because the concave lens bends the light rays to make the focused image appear directly on the retina, instead of too far in front of it. A convex lens is used to treat both hyperopia and presbyopia.

Corneal incision surgical techniques may help treat hyperopia, myopia, and astigmatism. These techniques involve cutting the cornea to change its shape, and thereby change its ability to refract light. One type of procedure, called a **radial keratotomy** (ker′ă-tot′ō-mē; *tome* = incision), or **RK,** treats nearsightedness. The ophthalmologist makes radial-oriented cuts in the cornea, which flatten the cornea and allow it to refract the light rays so that they focus on the retina.

Laser vision correction uses a laser to change the shape of the cornea. Two of the more popular types of laser vision correction are photorefractive keratectomy and laser-assisted in situ keratomileusis.

Photorefractive keratectomy (PRK) is called a *photoablation procedure* because the laser removes (ablates) tissue directly from the surface of the cornea. The removal of corneal tissue results in a newly shaped cornea that can focus better. This procedure is becoming less popular, because its regression rate is high—that is, the epithelial tissue that covers the surface of the cornea can regrow and regenerate, leading to partial return to uncorrected vision.

Laser-assisted in situ keratomileusis (ker′ă-tō-mī-lū′sis) (or **LASIK**) is rapidly becoming the most popular laser vision correction procedure. It can treat nearsightedness, farsightedness, and astigmatism. LASIK removes tissue from the inner, deeper layer of the cornea, which is less likely to regrow than surface tissue, so less vision regression occurs.

are "cross-eyed" such as occurs when you try to focus on your finger a few inches from your eyes.) This positions the eyes so that the image of the object being viewed is directed onto the fovea centralis. Individuals with extrinsic eye muscles that are weaker in one eye than in the other may be unable to converge the eyes and as a result will have **diplopia,** or *double vision.*

Accommodation of the Lens Recall that accommodation involves stimulation of the ciliary muscles by the parasympathetic division of the ANS when viewing objects closer than 20 feet. In response, the ciliary muscles contract to reduce tension in the suspensory ligaments, so the lens becomes more spherical or curved. Consequently, the light is refracted to a greater extent. Note that this change in light refraction is necessary because as objects become increasingly closer, the light rays reflecting from these objects must be bent to a greater degree so that the light rays hit on the retina.

Constriction of the Pupil The parasympathetic division also stimulates the sphincter pupillae muscle to contract to decrease the light rays passing through the edges of the lens. This is required when looking at objects closer than 20 feet because the lens must become more curved, but the edges of the lens are unable to curve to the extent that occurs at the center of the lens. Thus, light rays are not refracted at the edges of the lens to the same extent as the center of the lens. Consequently, light passing through the edges of the lens are not focused on the retina, and this portion of the object appears blurry. By constricting the pupil and allowing less light into the eye, light is passing only through the center of the lens. Collectively these changes associated with focusing on objects that are closer than 20 feet is called the **near response.**

In comparison, when viewing objects at a distance greater than 20 feet away, the near response is not occurring. Instead, the following is noted:

1. The eyes are facing forward and are not converging.
2. The ciliary muscles are relaxed and the lens is flatter so the light is refracted to a lesser extent (i.e., there is no accommodation).
3. The pupil is relatively dilated to allow a greater amount of light into the eye for maximizing visual input regarding the environment. Note that when viewing objects at any distance, it is an inverted image that hits the retina (see figure 16.14).

 WHAT DID YOU LEARN?

16 Describe how light is focused on the retina when viewing an object that is closer than 20 feet.

16.4d Physiology of Vision: Phototransduction

 **LEARNING OBJECTIVES**

5. Define phototransduction.
6. Compare and contrast the two general types of photoreceptors, including their photopigments.
7. Explain the bleaching reaction and how it relates to dark adaptation and light adaptation.

Phototransduction is converting (or transducing) light energy into an electrical signal. Photoreceptor cells (rods and cones) are the specific cells within the neural layer of the retina that engage in phototransduction. We first describe the anatomic details of photoreceptor cells (and other cells of the neural layer of the retina) and then discuss the process of phototransduction and the initiating of nerve signals that are sent to the brain.

Photoreceptors and Other Cells of the Neural Layer of the Retina

Both types of photoreceptor cells are composed of an outer segment, an inner segment, a cell body, and synaptic terminals. The **outer segment** extends into the pigmented layer of the retina (rod-shaped in rods and conical-shaped in cones) (**figure 16.17**). The outer segment is composed of hundreds of discs that are flattened membranous sacs, which are constantly being replaced. New discs are added at the base of the outer segment and begin to move externally toward the tip, where old, worn-out discs are removed by phagocytic cells within the pigmented layer. Usually it takes about 10 days for a disc to traverse this distance. The outer segment is connected to the **inner segment,** which contains the organelles for the cell, such as mitochondria. The inner segment connects to the **cell body,** which contain the nucleus. Synaptic terminals on the other side of the cell body house synaptic vesicles with glutamate neurotransmitter.

Rods and Cones   Rods are longer and narrower than cones. Each eye contains more than 100 million rods, and they are primarily located in the peripheral retina. Rods are activated by dim (low-intensity) light such as when you are in an unlit room at night, and provide no color recognition. Later we describe how rods produce limited sharpness of vision.

Cones occur at a density of less than 10 million per eye and are concentrated in the fovea centralis (the place of our most acute vision). Cones are activated by high-intensity light and provide color recognition and precise visual sharpness. Thus, when you notice the fine details in a colorful picture, the cones of your retina are responsible.

Why are there two visual pigments? One explanation is based upon both the arrangement of the three primary layers of neurons within the different regions of the retina and the varying degrees of sensitivity of rods and cones to light stimuli. In the peripheral retina, many rods converge on fewer bipolar cells, which synapse on one ganglion cell. In this arrangement, one ganglion cell is receiving input from many rods. In addition, recall that rods are more sensitive to light than are cones, and can be stimulated by low light. There is both an advantage and a disadvantage to this anatomic arrangement and characteristic of rods. The advantage to this arrangement can be seen in conditions of low light. Rods are stimulated and spatial summation occurs at both the bipolar cells and ganglion cells. The additive effect can cause sufficient neurotransmitter to be released from bipolar cells to stimulate ganglion cells to initiate an action potential that is propagated to the brain. However, because one ganglion is stimulated by numerous rods that

INTEGRATE

CLINICAL VIEW
Color Blindness

Color blindness is an X-linked recessive condition and occurs when an individual has an absence or deficit in one type of cone cell. The most common cone cells affected are the red and green cone cells, resulting in red-green color blindness. For these individuals, red and green colors appear similar and are difficult to distinguish. For example, in the adjacent image, a person with color blindness cannot distinguish the green number 74 from the rest of the speckled red background. The image instead would appear as a bunch of different sized dots without great differences in color. Color blindness is much more common in males (seen in about 8% of the male population) because it is an X-linked recessive trait.

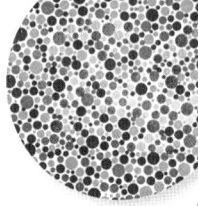

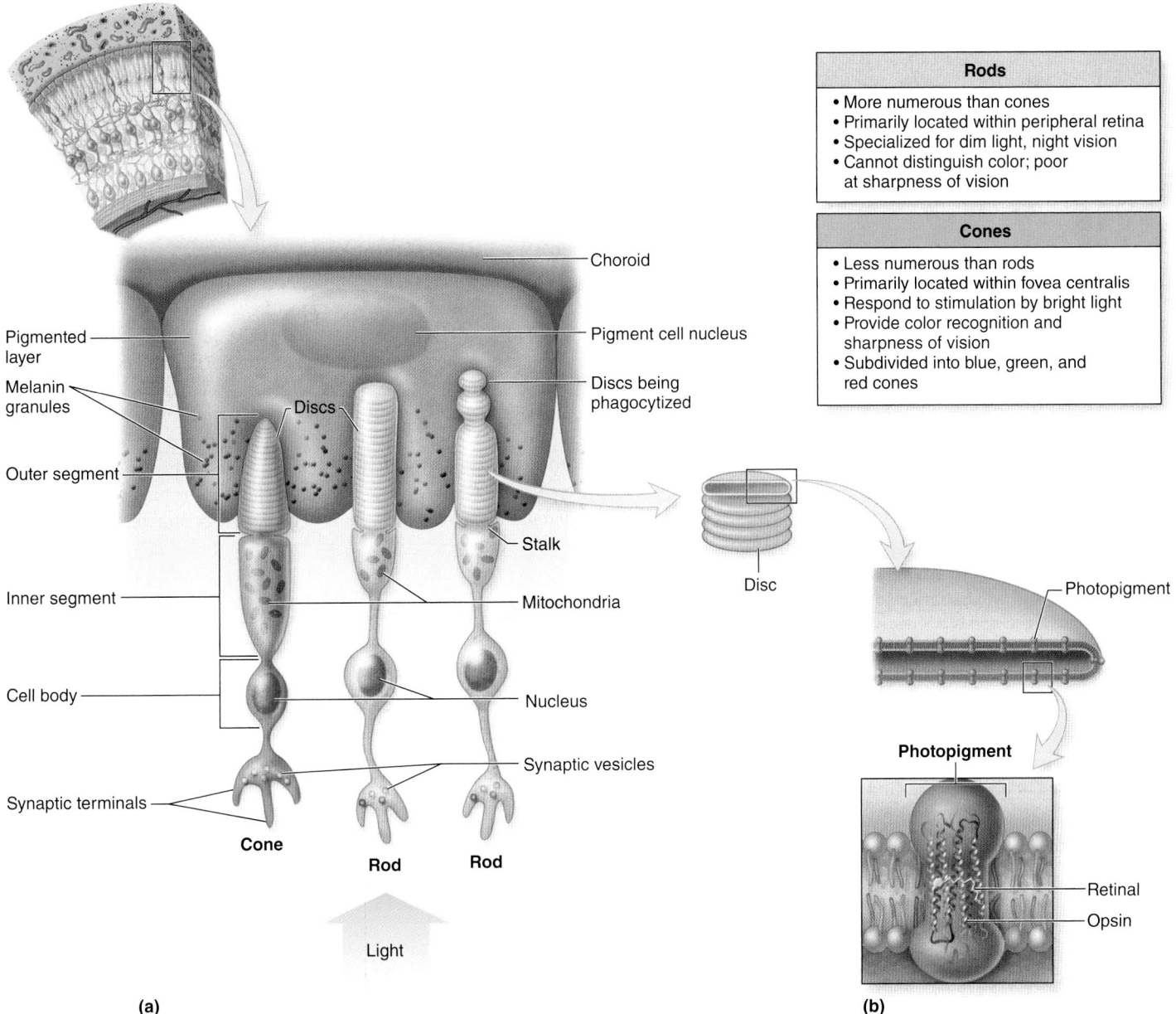

Rods
• More numerous than cones
• Primarily located within peripheral retina
• Specialized for dim light, night vision
• Cannot distinguish color; poor at sharpness of vision

Cones
• Less numerous than rods
• Primarily located within fovea centralis
• Respond to stimulation by bright light
• Provide color recognition and sharpness of vision
• Subdivided into blue, green, and red cones

Figure 16.17 Photoreceptors. (*a*) The outer segments of both rods and cones consist of stacks of discs embedded in the pigmented layer. (*b*) Membranes of the discs contain photopigments. Each photopigment is composed of an opsin and a retinal.

cover a relatively wide area (1 mm²), the perception by the brain of this sensory input is of a slightly blurry image.

In comparison, in the fovea centralis there is a one-to-one relationship between each cone and a bipolar cell, and between a bipolar cell and each ganglion cell. However, recall that cones are less sensitive to light than are rods, and require bright light to be stimulated. There is also both an advantage and a disadvantage to this anatomic arrangement and characteristic of cones. The advantage is having visual input from a very small area of the retina (about 1 μm²) that allows the brain to perceive a sharp image. However, cones are stimulated only in bright light, and sufficient light must be present for our cones to function. Thus, rods allow us to see in dim light, but produce a blurry image, whereas cones produce a sharp image, but require bright light.

Photopigments Photopigments are the specific molecules that absorb light and that are embedded within the plasma membrane of the outer segment of both rods and cones (figure 16.17*b*). A photopigment is composed of a protein called an **opsin** (op'sin; *opsis* = vision) and a light-absorbing molecule called a **retinal** (or a *retineme*), which is formed from vitamin A. There are several different types of photopigments that contain different opsins, and each type transduces different wavelengths of light. Thus, some photopigments may transduce light of longer wavelengths like reds, whereas other photopigments may transduce light of shorter wavelengths like blues. However, each photoreceptor expresses only one photopigment type.

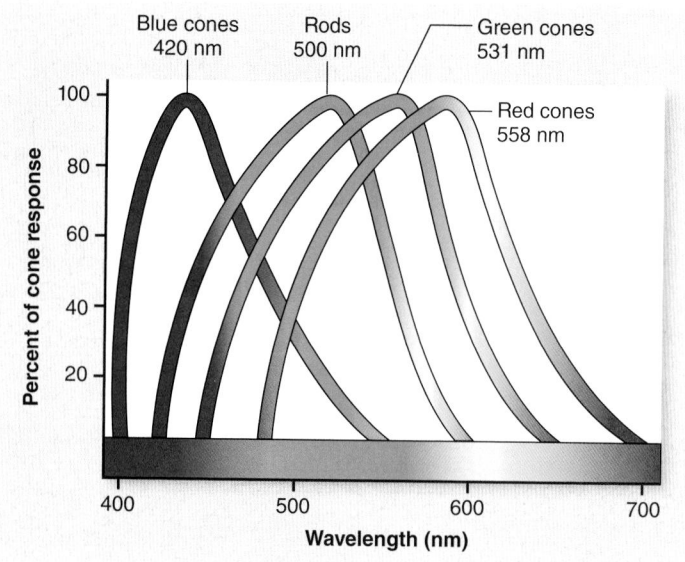

Figure 16.18 Absorption Wavelengths. Each photoreceptor detects a range of wavelengths, but not all wavelengths are detected to the same degree. For example, blue cones reach their peak stimulation at wavelengths of 420 nm—yet at 450 nm these blue cones respond at about 70% of their maximum. Mixtures of colors (such as a blue-green, at 470 nm) stimulate green and blue cones at about 50% of their maximum.

The photopigment in rods is called **rhodopsin** (rō-dop′sin; *rhodon = a rose*). Rhodopsin is involved in the transduction of dim light and is most sensitive to light at a 500-nm wavelength (**figure 16.18**). It is less sensitive to other light wavelengths.

The photopigment in cones is called **photopsin** (fō-top′sin; *phos = light*). There are three different photopsin proteins, and each type of photopsin protein maximally absorbs different wavelengths of light. Cones are categorized into three different types on the basis of the specific type of photopsin protein they contain and the wavelength to which it is most sensitive. **Blue cones** best detect wavelengths of light at about 420 nanometers (nm); **green cones** maximally absorb light at 531 nm; and **red cones** best detect light at 558 nm.

Thus, certain colors are best perceived by specific cones. But what about conditions where there is a mixture of colors (such as a greenish-blue hue)? In this case, we perceive intricate colors based on patterns of nerve signals arriving from a combination of these cones. Note in figure 16.18 that at about 470 nm, only the blue and green cones are stimulated; however, they are not stimulated at their peak. Instead, both types of cones are stimulated at about 50% capacity. When the blue and green cones initiate nerve signals, the brain interprets the color as a blue-green.

Phototransduction

Phototransduction occurs as light enters the eye and is transduced to an electrical signal by photoreceptor cells. We first discuss what happens with rods (**figure 16.19**). Prior to being activated by light, the retinal portion of the rhodopsin is in a bent, twisted shape called *cis*-**retinal.**

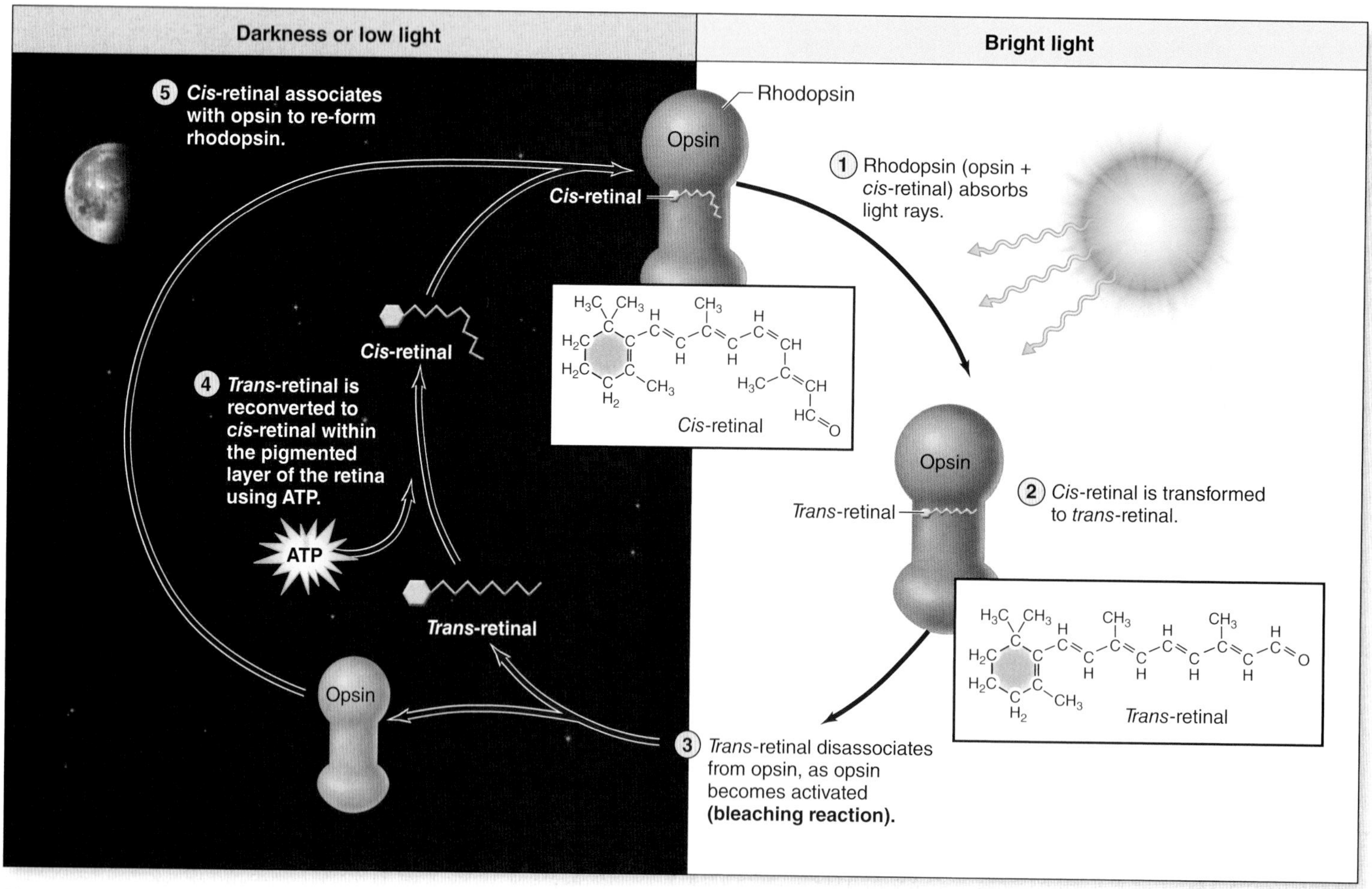

Darkness or low light	Bright light

5 *Cis*-retinal associates with opsin to re-form rhodopsin.

Rhodopsin

Opsin

Cis-retinal

1 Rhodopsin (opsin + *cis*-retinal) absorbs light rays.

Cis-retinal

4 *Trans*-retinal is reconverted to *cis*-retinal within the pigmented layer of the retina using ATP.

ATP

Trans-retinal

Opsin

H₃C CH₃ H CH₃ H CH₃ H
Cis-retinal

Opsin

Trans-retinal

2 *Cis*-retinal is transformed to *trans*-retinal.

Trans-retinal

3 *Trans*-retinal disassociates from opsin, as opsin becomes activated **(bleaching reaction).**

Figure 16.19 Bleaching Reaction and Regeneration of Rhodopsin. When light waves reach the rod, *cis*-retinal is transformed to *trans*-retinal, which then disassociates from opsin in a process called the bleaching reaction. Rhodopsin is re-formed when *trans*-retinal is converted back to *cis*-retinal, which then reunites with opsin.

Upon exposure to light, the retinal straightens out and reconfigures into a form called ***trans-retinal.*** *Trans*-retinal dissociates from the opsin (rhodopsin in rods) and phototransduction occurs. This dissociation of rhodopsin into its two components is a process termed a **bleaching reaction** because the rhodopsin goes from a bluish-purple color to colorless. Bleaching reduces rhodopsin amounts in rods and temporarily affects our ability to see in dim light conditions.

Rhodopsin must be regenerated for the rod cell to continue to function. Its regeneration occurs as follows: The disassociated *trans*-retinal is transported from the rod within the neural layer to the pigmented layer where it is converted back to its bent *cis*-retinal form—a process that requires ATP. The *cis*-retinal is then transported back to the rod where it associates with the opsin and re-forms the rhodopsin. This process is relatively slow; typically only half of your bleached rhodopsin is regenerated after about 5 minutes. Light will interfere with this process, and rhodopsin will bleach as fast as it is re-formed when we are in high-intensity light. For this reason, rods are essentially nonfunctional in bright light situations; the bleaching reaction in the neural retina occurs more quickly than rhodopsin re-formation in the pigmented layer.

A similar process occurs for the photopsin of cone cells. *Cis*-retinal transforms to *trans*-retinal, and a bleaching reaction occurs, but a more intense light is required. However, the regeneration of photopsin occurs much more quickly than the regeneration of rhodopsin; thus, cone cells are not as negatively affected by bright light as rods.

Recall a time when you went quickly from bright light conditions outside on a sunny day into a darkened movie theater. You probably remember the slow return of your sensitivity to low light levels. This phenomenon, called **dark adaptation,** occurs because initially our cones become nonfunctional in the low light (because they require a more intense stimulus), but our rods are still bleached from the bright light conditions from outside. It may take 20 to 30 minutes for your rhodopsin to be regenerated sufficiently that you can see well in low-light conditions.

In comparison, **light adaptation** is the process by which your eyes adjust from low light to bright light conditions, such as when you wake up at night and turn on the bright light in the bathroom. Even though your pupils constrict to reduce the amount of light entering your eyes, you are temporarily blinded as the rods become inactive and the cones, which initially were over-stimulated, gradually adjust to the brighter light. In about 5 to 10 minutes, the cones can produce sufficient visual acuity and color vision.

Initiating Nerve Signals

How does the splitting of photopigments within photoreceptor cells in the retina initiate nerve signals that are relayed to the brain? We first describe what is occurring in the rods, bipolar cells, and ganglion cells within the retina in the absence of light **(figure 16.20*a*)**. In the dark, the outer segment of rods continuously produces cyclic GMP (cGMP) from guanosine triphosphate (GTP) catalyzed by the enzyme guanylate cyclase. cGMP binds to cation channels in the plasma membrane of the outer segment allowing an influx of both Na^+ and Ca^{2+}. This influx of cations (called the dark current) depolarizes the photoreceptor cells to about -40 mV. These local currents of ions diffuse from the outer segment reaching voltage-gated Ca^{2+} channels at the synaptic terminals of the rods; this change in voltage triggers these channels to open. Calcium ion enters the synaptic terminals triggering the continuous (tonic) release of glutamate neurotransmitter. Glutamate binds with receptors of the bipolar cells to cause hyperpolarization of the bipolar cells. This inhibits the bipolar cells and prevents them from releasing glutamate neurotransmitter from their synaptic terminals. Thus, no nerve signals are generated by ganglion cells.

In the light (figure 16.20*b*), rhodopsin is split, and through a G protein second messenger pathway, the enzyme (phosphodiesterase) that breaks down cGMP is activated. Lower levels of cGMP result in the closing of the cation channels, preventing the influx of both Na^+ and Ca^{2+}. Thus, the dark current ceases, and the photoreceptor cells hyperpolarize. Voltage-gated Ca^{2+} channels at the synaptic terminals of the photoreceptor cells now close and release of glutamate neurotransmitter ceases. Bipolar cells are no longer inhibited and now release glutamate neurotransmitter that binds to receptors of ganglion cells. If sufficient neurotransmitter is released from the bipolar cells and the threshold is reached in a ganglion cell, a nerve signal is propagated along the axon of the ganglion cell into the brain.

WHAT DID YOU LEARN?

17 What are the differences between rods and cones with respect to their anatomy, their photopigments, and what light they process?

18 How does dark adaptation differ from light adaptation?

19 What occurs during phototransduction of light?

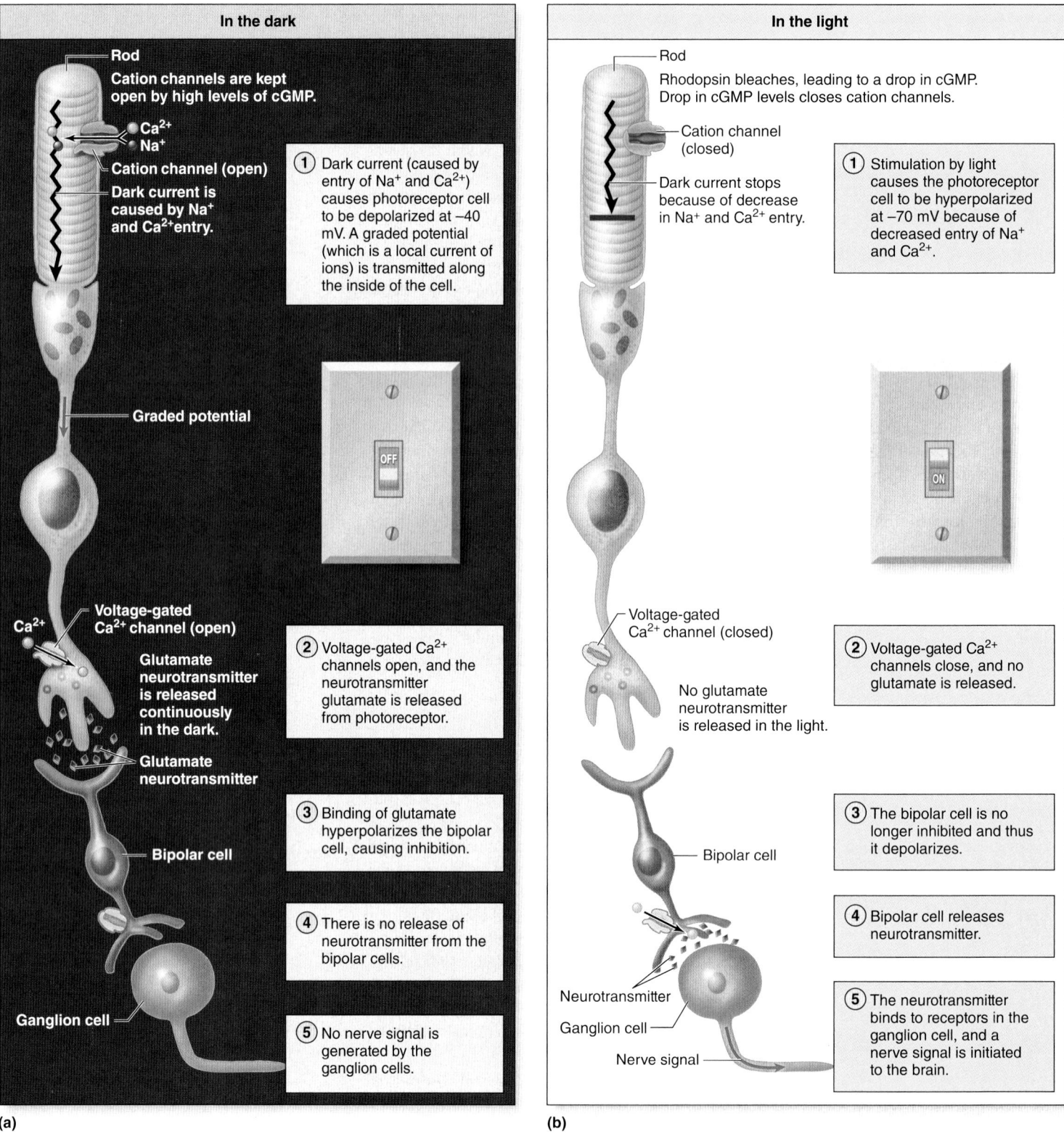

In the dark

Rod

Cation channels are kept open by high levels of cGMP.

Ca²⁺
Na⁺
Cation channel (open)

Dark current is caused by Na⁺ and Ca²⁺ entry.

Graded potential

OFF

Ca²⁺
Voltage-gated Ca²⁺ channel (open)

Glutamate neurotransmitter is released continuously in the dark.

Glutamate neurotransmitter

Bipolar cell

Ganglion cell

① Dark current (caused by entry of Na⁺ and Ca²⁺) causes photoreceptor cell to be depolarized at −40 mV. A graded potential (which is a local current of ions) is transmitted along the inside of the cell.

② Voltage-gated Ca²⁺ channels open, and the neurotransmitter glutamate is released from photoreceptor.

③ Binding of glutamate hyperpolarizes the bipolar cell, causing inhibition.

④ There is no release of neurotransmitter from the bipolar cells.

⑤ No nerve signal is generated by the ganglion cells.

(a)

In the light

Rod

Rhodopsin bleaches, leading to a drop in cGMP. Drop in cGMP levels closes cation channels.

Cation channel (closed)

Dark current stops because of decrease in Na⁺ and Ca²⁺ entry.

ON

Voltage-gated Ca²⁺ channel (closed)

No glutamate neurotransmitter is released in the light.

Bipolar cell

Neurotransmitter

Ganglion cell

Nerve signal

① Stimulation by light causes the photoreceptor cell to be hyperpolarized at −70 mV because of decreased entry of Na⁺ and Ca²⁺.

② Voltage-gated Ca²⁺ channels close, and no glutamate is released.

③ The bipolar cell is no longer inhibited and thus it depolarizes.

④ Bipolar cell releases neurotransmitter.

⑤ The neurotransmitter binds to receptors in the ganglion cell, and a nerve signal is initiated to the brain.

(b)

Figure 16.20 Phototransduction in Rod Photoreceptors. (*a*) Events associated with dim light that prevent initiation of nerve signals to the brain. (*b*) Events associated with bright light that initiate nerve signals to the brain.

16.4e Visual Pathways

LEARNING OBJECTIVES

8. Describe the visual pathway from the photoreceptors to the brain.

9. Explain how stereoscopic vision provides depth perception.

Figure 16.21 depicts the visual pathways. The visual pathways begin at the retina, where the photoreceptor cells transduce light stimuli to an electrical signal. The bipolar cells are the primary neurons, and the ganglion cells are the secondary neurons with the visual pathways (see sensory pathways from section 14.4b). The axons of the ganglion cells form the **optic nerve,** which exits the back of the eye at the optic disc (see figure 16.10). Optic nerves project from each eye and converge at the **optic chiasm** (kī′azm) (immediately anterior to the pituitary gland; not shown in figure). The optic chiasm is a flattened structure anterior to the infundibulum where many of the optic nerve axons decussate (cross) to the other side. The ganglionic axons originating from the medial region of each retina cross to the opposite side

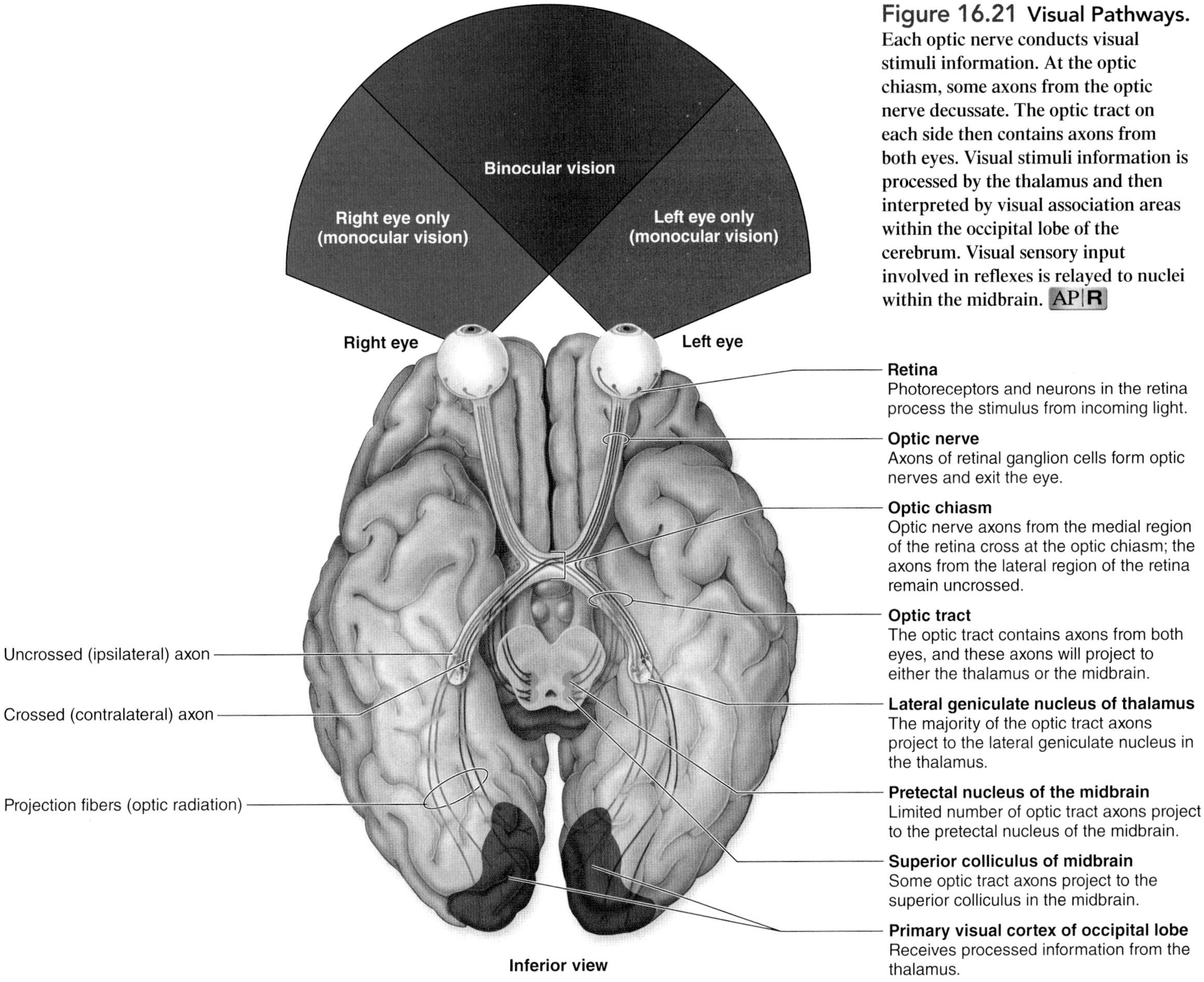

Figure 16.21 Visual Pathways. Each optic nerve conducts visual stimuli information. At the optic chiasm, some axons from the optic nerve decussate. The optic tract on each side then contains axons from both eyes. Visual stimuli information is processed by the thalamus and then interpreted by visual association areas within the occipital lobe of the cerebrum. Visual sensory input involved in reflexes is relayed to nuclei within the midbrain. **AP|R**

Binocular vision

Right eye only (monocular vision)

Left eye only (monocular vision)

Right eye

Left eye

Uncrossed (ipsilateral) axon

Crossed (contralateral) axon

Projection fibers (optic radiation)

Inferior view

Retina
Photoreceptors and neurons in the retina process the stimulus from incoming light.

Optic nerve
Axons of retinal ganglion cells form optic nerves and exit the eye.

Optic chiasm
Optic nerve axons from the medial region of the retina cross at the optic chiasm; the axons from the lateral region of the retina remain uncrossed.

Optic tract
The optic tract contains axons from both eyes, and these axons will project to either the thalamus or the midbrain.

Lateral geniculate nucleus of thalamus
The majority of the optic tract axons project to the lateral geniculate nucleus in the thalamus.

Pretectal nucleus of the midbrain
Limited number of optic tract axons project to the pretectal nucleus of the midbrain.

Superior colliculus of midbrain
Some optic tract axons project to the superior colliculus in the midbrain.

Primary visual cortex of occipital lobe
Receives processed information from the thalamus.

of the brain at the optic chiasm, whereas ganglionic axons originating from the lateral region of each retina remain on the same side of the brain and do not cross. **Optic tracts** extend laterally from the optic chiasm as a composite of ganglionic axons originating from the retina of each eye.

The majority of the optic tract axons extend to the thalamus, specifically to the **lateral geniculate** (je-nik′ū-lāt; *genu* = knee, referring to its appearance) **nucleus,** where visual information is processed within each thalamic body. Tertiary neurons project axons from the thalamus to the visual cortex of the occipital lobe for interpretation of incoming visual stimuli.

Note that the left and right eyes have somewhat overlapping visual fields. For the brain to interpret these two distinct visual images, it must process or unite them into one. These overlapping images then provide us with **stereoscopic vision,** or **depth perception,** which is the ability to determine how close or far away an object is. Animals that don't have overlapping visual fields (such as horses or deer) may have a greater visual range, but they cannot perceive visual depths.

In addition to the neural pathway that relays input to the visual cortex, a limited number of axons within each optic tract project to the midbrain. Those that extend to the **superior colliculi** (ko-lik′y ū-lī) coordinate the reflexive movements of the extrinsic eye muscles, whereas those that

extent to the **pretectal nuclei** function as the control centers in both the pupillary reflex to regulate the amount of the light that enters the eye and the accommodation reflex for focusing the lens. (Input sent to the pretectal nucleus is initiated by specialized ganglion cells that respond directly to light through the visual pigment melanopsin.)

WHAT DO YOU THINK?

2 Why would it be an advantage for a deer to have a wide visual field, as compared to overlapping, stereoscopic vision?

The physiology of vision, which involves the eyes, visual pathways, and the brain is integrated in **figure 16.22.** Vision loss may result from either disease or disorders in any structures involved with vision including the eye, optic nerve, optic tract, or brain.

WHAT DID YOU LEARN?

20 What areas of the brain consciously perceive visual stimuli, and which areas respond reflexively (below the conscious level)?

21 What is the significance of some ganglionic axons crossing to the opposite side of the brain?

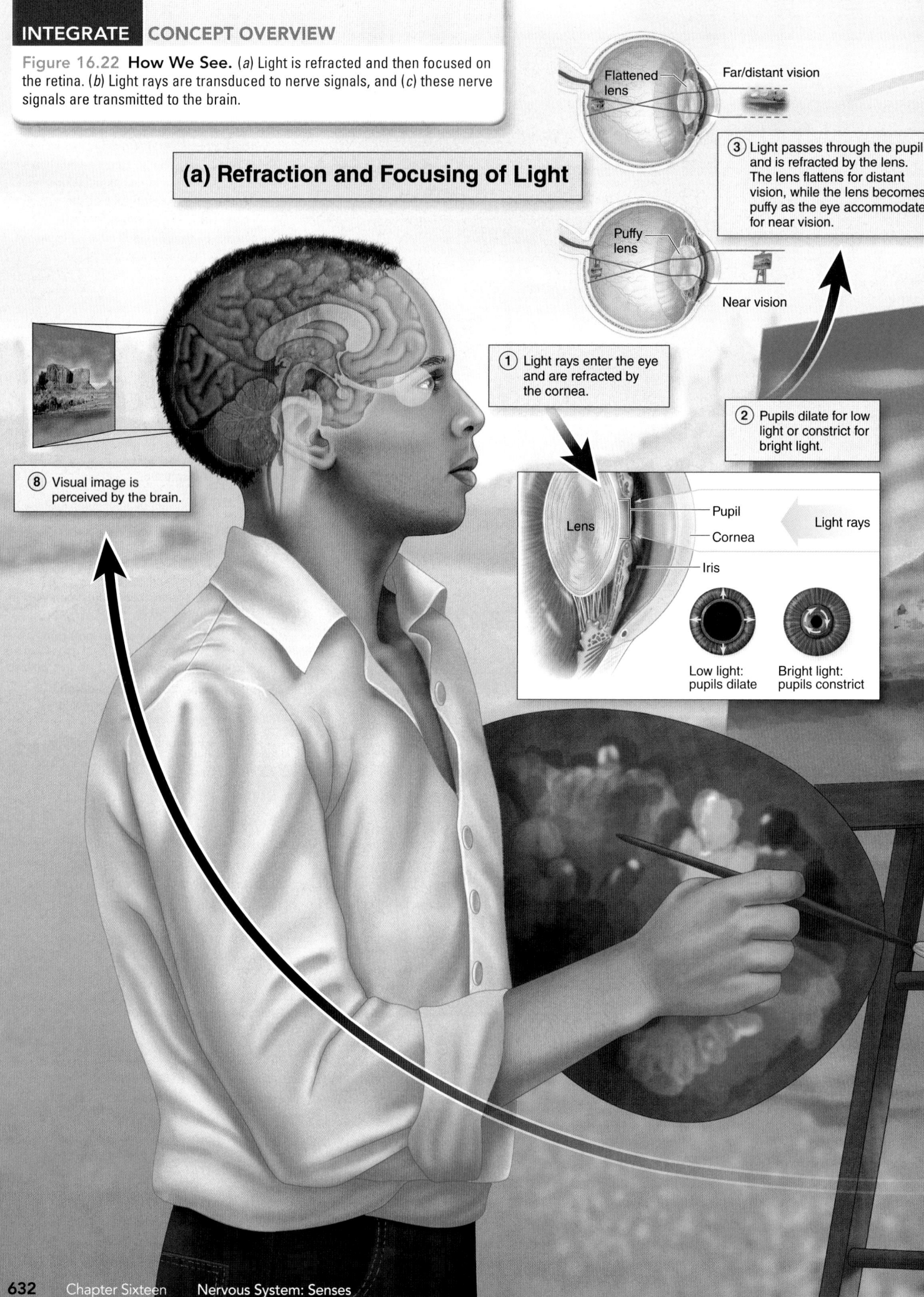

Figure 16.22 **How We See.** (*a*) Light is refracted and then focused on the retina. (*b*) Light rays are transduced to nerve signals, and (*c*) these nerve signals are transmitted to the brain.

(a) Refraction and Focusing of Light

Flattened lens

Far/distant vision

③ Light passes through the pupil and is refracted by the lens. The lens flattens for distant vision, while the lens becomes puffy as the eye accommodates for near vision.

Puffy lens

Near vision

① Light rays enter the eye and are refracted by the cornea.

② Pupils dilate for low light or constrict for bright light.

⑧ Visual image is perceived by the brain.

Lens

Pupil

Cornea

Light rays

Iris

Low light: pupils dilate

Bright light: pupils constrict

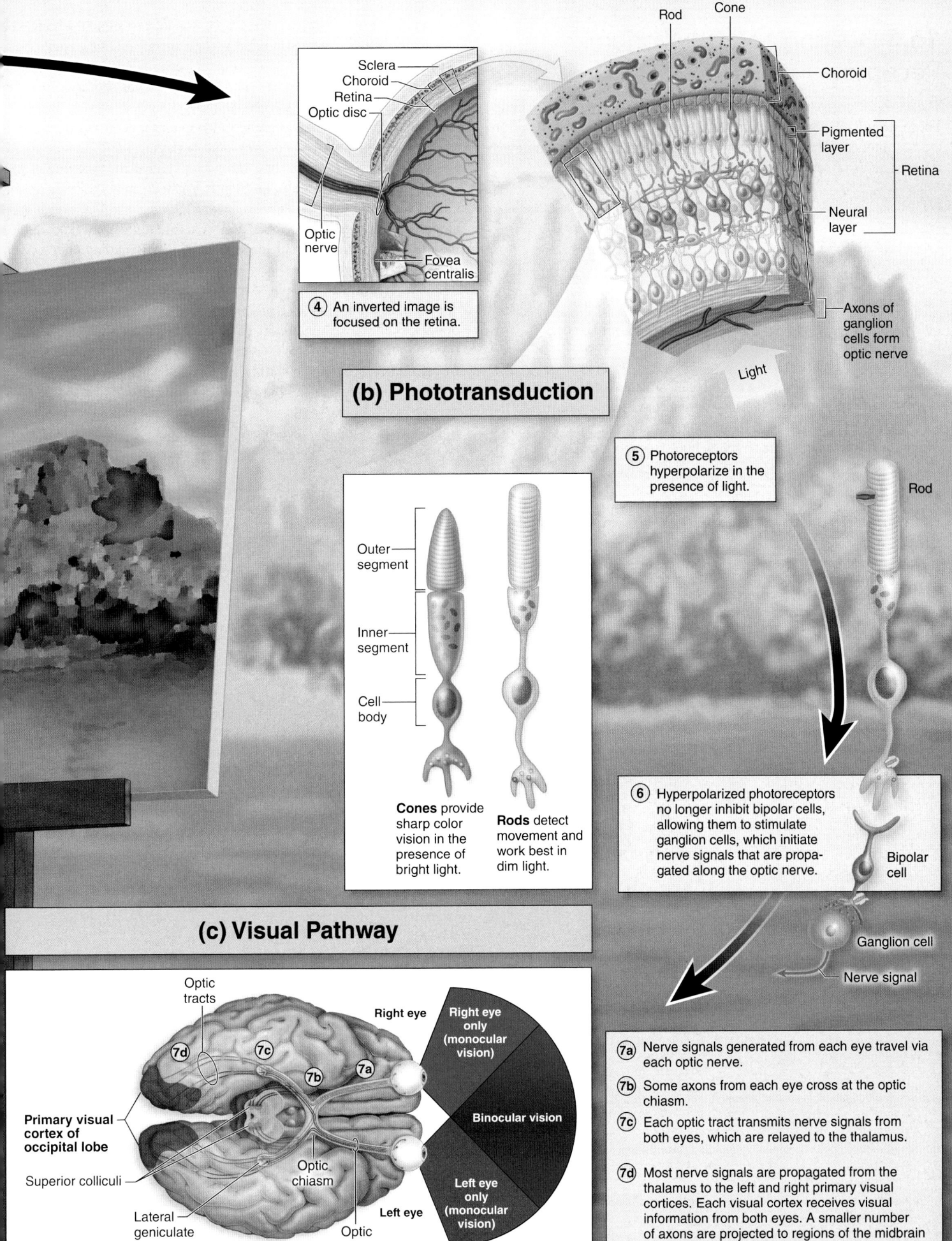

Sclera
Choroid
Retina
Optic disc

Optic nerve

Fovea centralis

④ An inverted image is focused on the retina.

Rod   Cone

Choroid

Pigmented layer

Neural layer

Retina

Axons of ganglion cells form optic nerve

Light

(b) Phototransduction

⑤ Photoreceptors hyperpolarize in the presence of light.

Rod

Outer segment

Inner segment

Cell body

Cones provide sharp color vision in the presence of bright light.

Rods detect movement and work best in dim light.

⑥ Hyperpolarized photoreceptors no longer inhibit bipolar cells, allowing them to stimulate ganglion cells, which initiate nerve signals that are propagated along the optic nerve.

Bipolar cell

Ganglion cell

Nerve signal

(c) Visual Pathway

Optic tracts

Right eye

Right eye only (monocular vision)

⑦d   ⑦c   ⑦b   ⑦a

Primary visual cortex of occipital lobe

Binocular vision

Superior colliculi

Optic chiasm

Left eye only (monocular vision)

Lateral geniculate nucleus

Optic nerve

Left eye

⑦a Nerve signals generated from each eye travel via each optic nerve.

⑦b Some axons from each eye cross at the optic chiasm.

⑦c Each optic tract transmits nerve signals from both eyes, which are relayed to the thalamus.

⑦d Most nerve signals are propagated from the thalamus to the left and right primary visual cortices. Each visual cortex receives visual information from both eyes. A smaller number of axons are projected to regions of the midbrain involved in visual reflexes.

16.5 Hearing and Equilibrium Receptors

The ear is the organ that detects both sound and movements of the head. These stimuli are transduced into nerve signals that are transmitted by the vestibulocochlear nerve (CN VIII) (see section 13.9), resulting in the sensations of hearing and equilibrium.

16.5a Ear Structure

✓ LEARNING OBJECTIVES

1. Describe the structures of the outer, middle, and inner ear.
2. Name the auditory ossicles and explain how they function in hearing.
3. Compare and contrast the bony labyrinth and the membranous labyrinth.

The ear is partitioned into three distinct anatomic regions: external, middle, and inner (figure 16.23). The external ear is located mostly on the outside of the body, and both the middle ear and inner ear are housed within the petrous part of the temporal bone (see section 8.2b).

External Ear

The most visible portion of the external ear is a skin-covered, elastic cartilage–supported structure called the **auricle** (aw'ri-kl; *auris* = ear), or *pinna* (pin'ă; wing). The auricle is funnel-shaped, and it serves both to protect the entry into the ear and to direct sound waves into the bony tube called the **external acoustic meatus** (or *external auditory meatus*, see section 8.2b). This canal, which is about 2.5 centimeters (1 inch) in length and 7.5 millimeters (0.3 inches) in diameter, extends slightly superiorly from the lateral surface of the head to the tympanic membrane.

The narrow external opening in the external acoustic meatus prevents large objects from entering and damaging the tympanic membrane. Near its entrance, fine hairs help guard the opening. Deep within the canal, ceruminous glands produce a waxlike secretion called **cerumen** (sĕ-rū-men; *cera* = wax), which combines with dead sloughed skin cells to form earwax. This material may help reduce infection within the external acoustic meatus by impeding microorganism growth.

The **tympanic** (tim-pan'ik; *tympanon* = drum) **membrane,** or *eardrum*, is a funnel-shaped membrane (approximately 1 centimeter in diameter) composed of fibrous connective tissue sandwiched between two epithelial sheets. It serves as the boundary between the external and middle ear. The tympanic membrane vibrates when sound waves hit it, and its vibrations provide the means for transmission of sound wave energy from the external ear to the middle ear. Pain associated with trauma to the tympanic membrane is relayed to the brain along sensory neurons within both the vagus and trigeminal nerves (see section 13.9).

Middle Ear

The middle ear contains an air-filled **tympanic cavity (figure 16.24)**. Medially, a bony wall separates the middle ear from the inner ear. Two membrane-covered openings are located within this wall: the oval

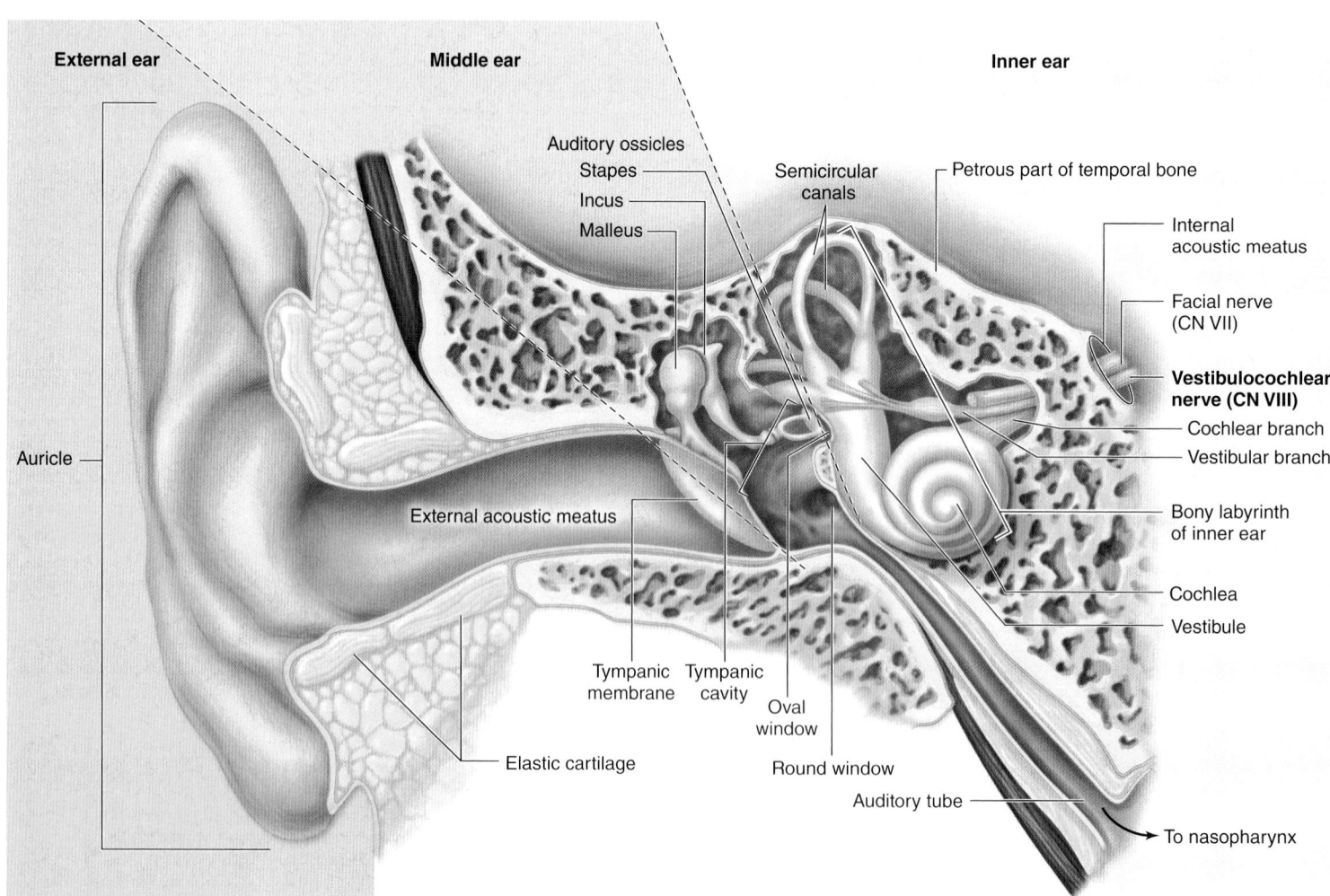

Figure 16.23 Anatomic Regions of the Right Ear. The ear is divided into external, middle, and inner regions. AP|R

CLINICAL VIEW
Otitis Media

Otitis (ō-tī′tis; *itis* = inflammation) **media** (mē′dē-ă; *medius* = middle) is an infection of the middle ear. It is most often experienced by young children, whose auditory tubes are horizontal, relatively short, and underdeveloped. If a young child has a respiratory infection, the causative agent may spread from the pharynx (throat) through the auditory tube. Fluid then accumulates in the middle ear cavity, resulting in pressure, pain, and sometimes impaired hearing. An **otoscope** (ō′tō-skōp; *skopeo* = to view) is an instrument used to examine the tympanic membrane, which normally appears white and pearly, but in cases of severe otitis media is red (due to inflammation and sometimes bleeding) and may even bulge due to fluid pressure in the middle ear.

Repeated ear infections, or a chronic ear infection that does not respond to antibiotic treatment, usually calls for a surgical procedure called a **myringotomy** (mir-ing-got′ō-mē; *myringa* = membrane), whereby a ventilation tube is inserted into the tympanic membrane. This procedure allows the infection to heal and the pus and mucus to drain from the middle ear into the external acoustic meatus and offers immediate relief from the pressure. Eventually, the inserted tube is sloughed, and the tympanic membrane heals.

When a child is about 5 years old, the auditory tube has become larger, more vertically angled, and better able to drain fluid and prevent infection from reaching the middle ear. Thus, the occurrence rate for ear infections drops dramatically at this time.

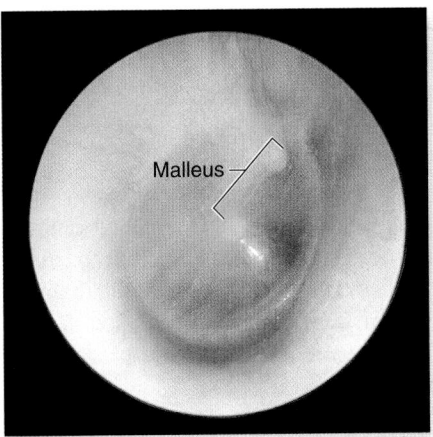

Normal tympanic membrane

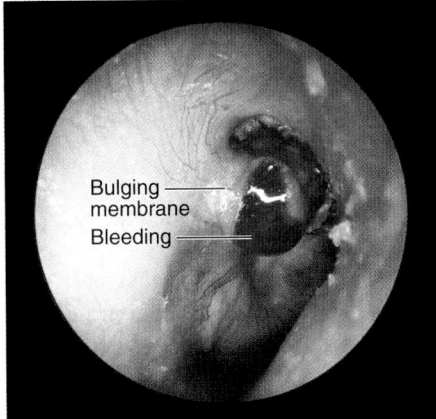

Otitis media

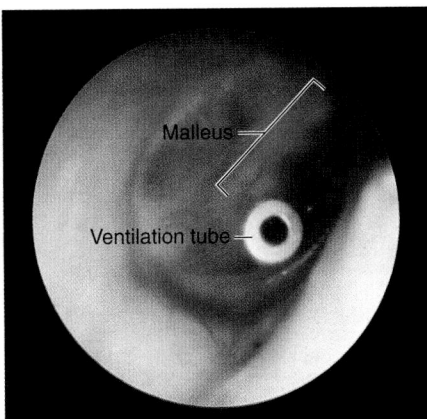

Myringotomy

Middle ear views as seen with an otoscope.

---

window and round window (both discussed later in detail). Inferiorly, the **auditory tube** (also called the *pharyngotympanic tube* or *Eustachian tube*), which is approximately 3.5 centimeters (1.5 inches) in length, serves as a passageway that extends from the middle ear into the nasopharynx (portion of upper throat posterior to the nasal cavity; see figure 23.5). This tube is normally closed. Air movement through this tube occurs as a result of chewing, yawning, and swallowing, which equalize pressure on either side of the tympanic membrane—allowing the tympanic membrane to vibrate freely. The tympanic cavity, auditory tube, and nasopharynx are lined with a continuous mucous membrane. Middle ear infections result when infectious agents (e.g., a cold virus) move from the nasopharynx through the auditory tube into the middle ear (see Clinical View: "Otitis Media").

WHAT DO YOU THINK?

3 When an airplane descends to a lower altitude, you may feel greater pressure in your ears, followed by a popping sensation—then more normal pressure resumes. What do you think has happened?

The tympanic cavity of each middle ear houses the three smallest bones of the body, the **auditory ossicles** (os′i-kl) (see section 8.3). These three bones, which are positioned between the tympanic membrane and the oval window, are, from lateral to medial, the malleus (hammer), the incus

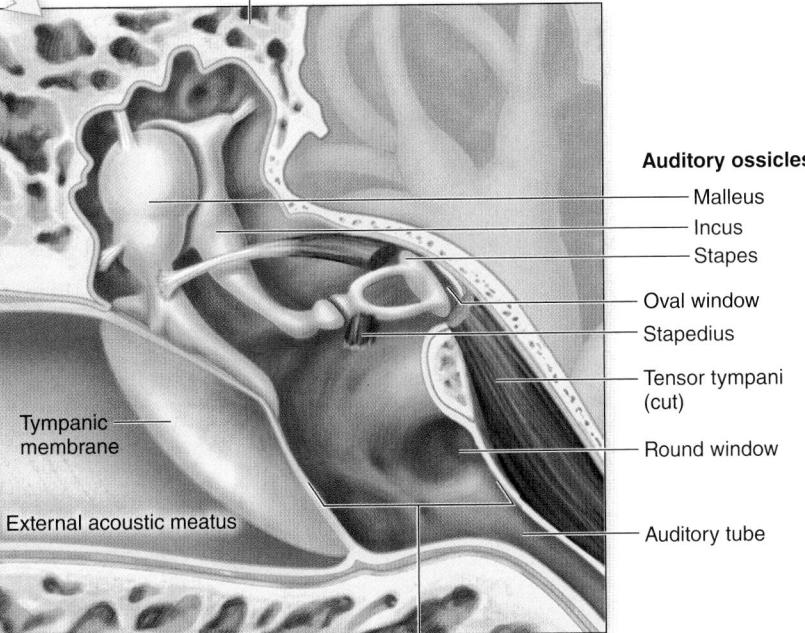

Figure 16.24 Middle Ear. The middle ear contains the auditory ossicles and associated structures within the tympanic cavity of the temporal bone.

(anvil), and the stapes (stirrup). (You can remember the order with the acronym MIS.) The **malleus** (mal′ē-ŭs), with its long handle and expanded end, resembles a large hammer in shape; the long handle has contact with the tympanic membrane. The **incus** (ing′kŭs), which is approximately triangular in shape and resembles an anvil, is the middle auditory ossicle. The **stapes** (stā′pēz), composed of an arch and plate of bone, resembles a stirrup on a saddle. Its cylindrical, disclike footplate fits into the oval window. The ossicles are anchored by various small ligaments to the surrounding structures.

The auditory ossicles are responsible for amplifying sound waves from the tympanic membrane to the oval window. When sound waves strike the tympanic membrane, the three middle ear ossicles vibrate along with the tympanic membrane, causing the footplate of the stapes to move in and out of the oval window. The movement of this ossicle initiates pressure waves in the fluid within the closed compartment of the inner ear. Two tiny skeletal muscles, the **tensor tympani** (attached to the malleus) and the **stapedius** (attached to the stapes), are located within the middle ear. These muscles reflexively restrict ossicle movement when loud sounds occur (including when we are speaking), and thus protect the sensitive sensory receptors within the inner ear. This reflexive response that involves contraction of the tensor tympani and stapedius to loud sounds takes approximately 40 milliseconds, and thus is not able to protect the sensory receptors from blasts of sounds, such as occurs with a gunshot. For this reason, loud blasts of sound are especially damaging to the sensory receptors within the inner ear.

Inner Ear

The inner ear is located within the petrous part of the temporal bone, where there are spaces or cavities called the **bony labyrinth** (lab′i-rinth; an intricate, mazelike passageway) **(figure 16.25).** Within the bony labyrinth are membranous, fluid-filled tubes and sacs called the **membranous labyrinth.** Receptors for both hearing and equilibrium are within the membranous labyrinth.

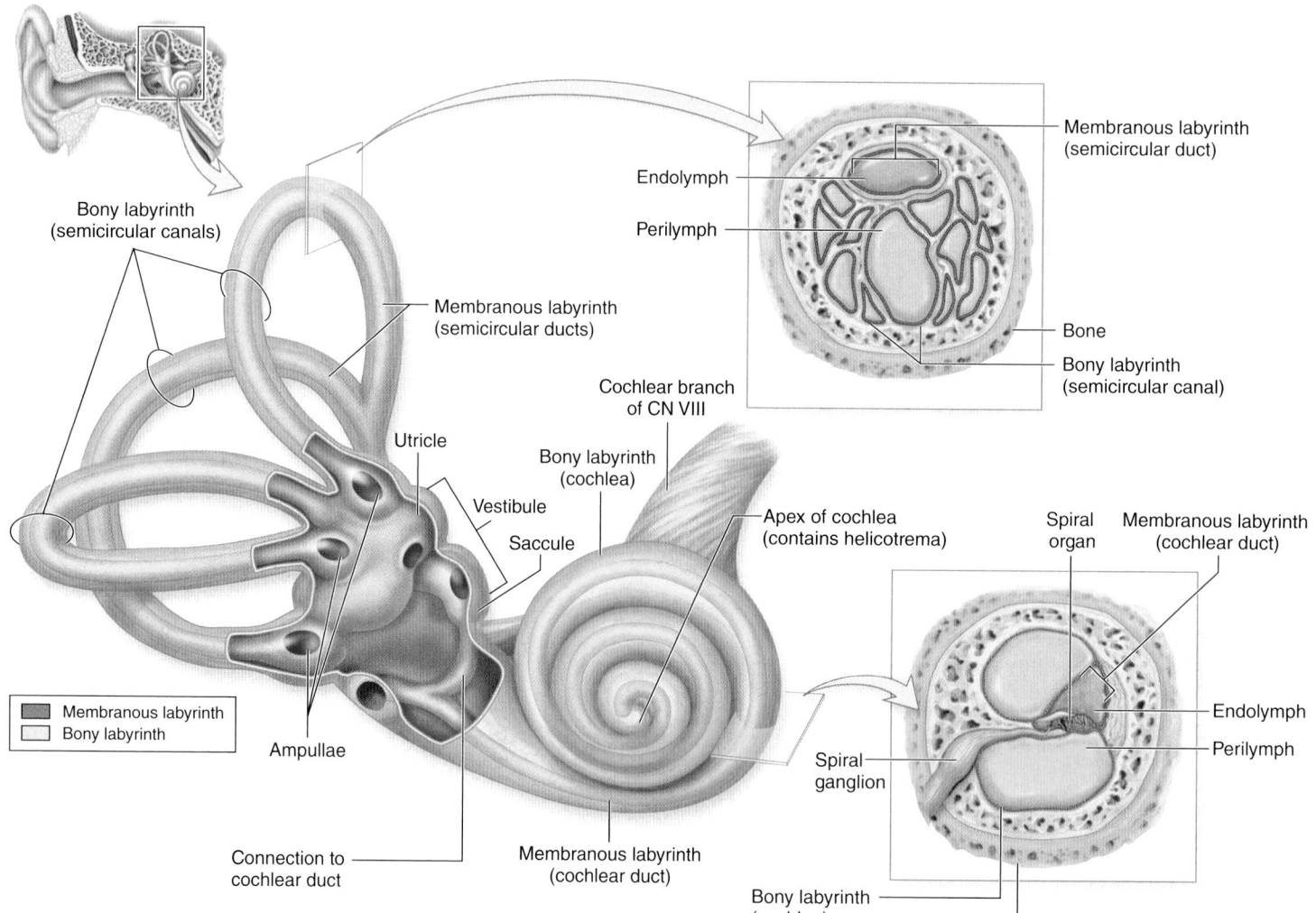

Figure 16.25 Inner Ear. The inner ear is composed of a bony labyrinth cavity that houses a fluid-filled membranous labyrinth. Within the bony labyrinth are the portions of the membranous labyrinth for hearing (the cochlear duct) and equilibrium and balance (saccule, utricle, and semicircular ducts). AP|R

The space between the outer walls of the bony labyrinth and the membranous labyrinth is filled with a fluid called **perilymph** (per'i-limf) that is similar in composition to interstitial fluid. The perilymph suspends, supports, and protects the membranous labyrinth from the wall of the bony labyrinth. The space within the membranous labyrinth contains a fluid called **endolymph** (en'dō-limf), which is similar in composition to intracellular fluid with relatively high levels of K+ (see section 4.4a).

The bony labyrinth is structurally and functionally partitioned into three distinct regions that include the cochlea, vestibule, and semicircular canals:

- The **cochlea** (kok'lē-ă = snail shell) houses a membranous labyrinth structure called the **cochlear duct** (or *scala media*).
- The **vestibule** (ves'ti-būl; vestibulum = entrance court) contains two saclike, membranous labyrinth structures—the **utricle** (ū'tri-kl; *uter* = leather bag) and the **saccule** (sak'ūl; *saccus* = sack).
- The **semicircular canals** each contain a membranous labyrinth structure called the **semicircular duct.**

The structures of the membranous labyrinth are continuous. The cochlear duct is continuous with the saccule, which is connected through a narrow passageway to the utricle, and the utricle is continuous with the semicircular ducts.

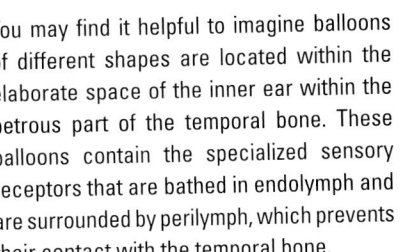

INTEGRATE

LEARNING STRATEGY

You may find it helpful to imagine balloons of different shapes are located within the elaborate space of the inner ear within the petrous part of the temporal bone. These balloons contain the specialized sensory receptors that are bathed in endolymph and are surrounded by perilymph, which prevents their contact with the temporal bone.

WHAT DID YOU LEARN?

22 What is the function of the external acoustic meatus?

23 Where are the auditory ossicles located, and what is their function?

24 What are the membranous labyrinth structures, and the specific bony labyrinth structure in which each resides?

16.5b Hearing

LEARNING OBJECTIVES

4. Explain the components of the cochlea and how they function in the sense of hearing.

5. Trace the path of a sound wave from outside the ear to stimulation of the vestibulocochlear nerve (CN VIII).

6. Distinguish between frequency and intensity of sound.

Hearing is the ability to detect and perceive sound. Here we follow the progression of how sound from the environment enters the ear and is transduced in the inner ear into electrical signals that are relayed to the brain. First, we consider the structures in the inner ear that are related to sound detection.

Structures for Hearing

Hearing organs are housed within the cochlea in both inner ears. View both **figure 16.26**, which is a cross section of the cochlea, and **figure 16.27**, which is a longitudinal section of a cochlea (partially "unrolled," so the structure can be viewed more easily), as you read through this section.

Cochlea The **cochlea** is a snail-shaped spiral chamber within the bone of the inner ear. Figure 16.26*a* depicts how this chamber "wraps" approximately 2.5 times around a spongy bone axis called the **modiolus** (mō-dī'ō-lŭs; = hub of a wheel), giving the cochlea its snail-shaped appearance. The membranous labyrinth called the cochlear duct is housed within the cochlea. The roof of the cochlear duct is formed by the **vestibular membrane,** and the floor is formed by the **basilar membrane** (figures 16.26*b* and 16.27). These membranes partition the bony labyrinth of the cochlea into two smaller chambers on either side of the cochlear duct; both are filled with perilymph. The chamber adjacent to the vestibular membrane is the **scala vestibuli** (*vestibular duct*), and the chamber adjacent to the basilar membrane is the **scala tympani** (*tympanic duct*). The scala vestibuli and scala tympani merge through a small channel called the **helicotrema** (hel'i-kō-trē'mă; *helix* = spiral, *trema* = hole) at the apex of the cochlea (figure 16.25).

Spiral Organ Protected within the membranous cochlear duct is the **spiral organ** (formerly called the *organ of Corti*), which is the sensory structure for hearing (figures 16.26*c, d* and 16.27). The cochlear duct contains endolymph. The spiral organ is a thick sensory epithelium

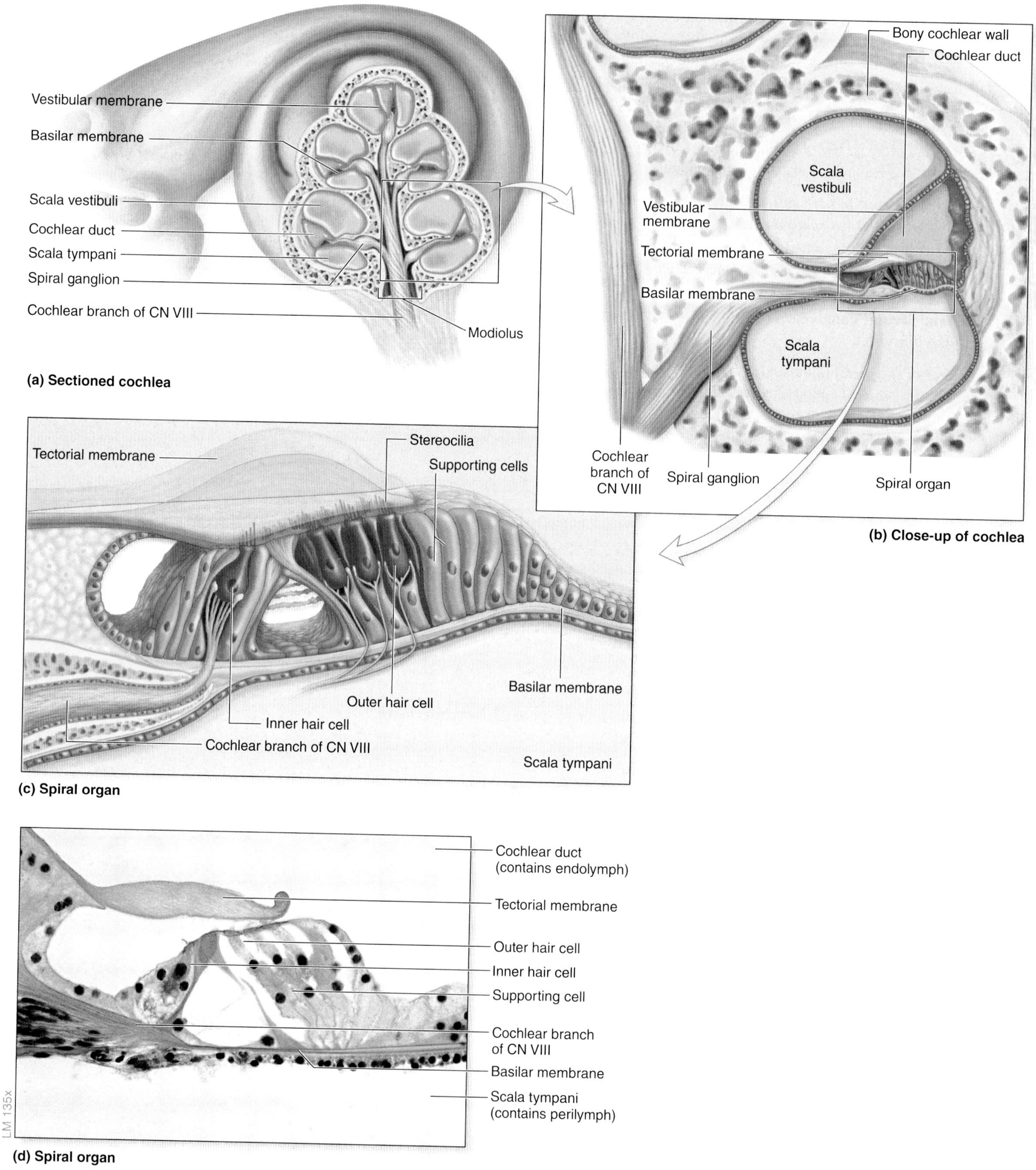

Vestibular membrane —

Basilar membrane —

Scala vestibuli —

Cochlear duct —

Scala tympani —

Spiral ganglion —

Cochlear branch of CN VIII —

Modiolus

(a) Sectioned cochlea

Bony cochlear wall
Cochlear duct

Scala vestibuli

Vestibular
membrane

Tectorial membrane

Basilar membrane

Scala tympani

Cochlear
branch of   Spiral ganglion
CN VIII

Spiral organ

(b) Close-up of cochlea

Tectorial membrane —

Stereocilia

Supporting cells

Outer hair cell

Inner hair cell

Cochlear branch of CN VIII

Basilar membrane

Scala tympani

(c) Spiral organ

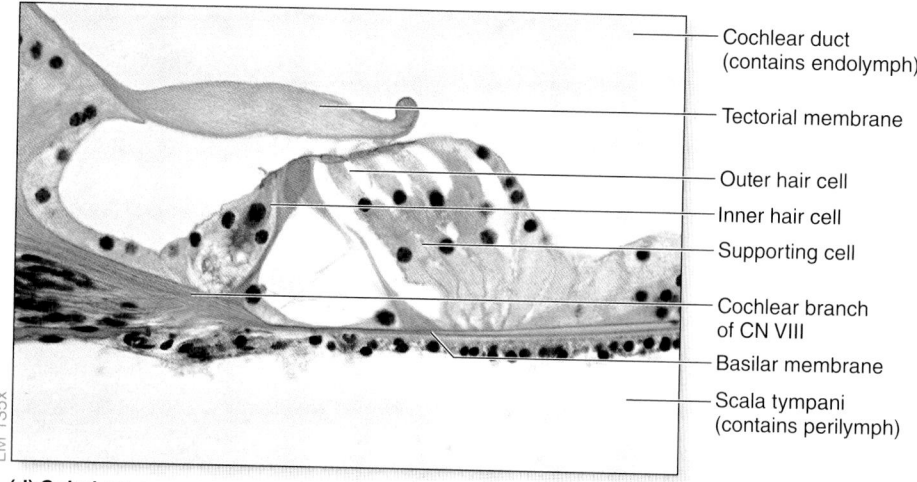

Cochlear duct
(contains endolymph)

Tectorial membrane

Outer hair cell

Inner hair cell

Supporting cell

Cochlear branch
of CN VIII

Basilar membrane

Scala tympani
(contains perilymph)

LM 135x

(d) Spiral organ

Figure 16.26 Structure of the Cochlea and Spiral Organ. The cochlea exhibits a snail-like spiral shape and is composed of three fluid-filled ducts. (*a*) A section through the cochlea details the relationship among the three ducts: the cochlear duct, scala vestibuli, and scala tympani. (*b*) A magnified view of the cochlea. (*c*) Hair cells rest on the basilar membrane of the spiral organ within the cochlear duct. (*d*) Light micrograph of the spiral organ. **AP|R**

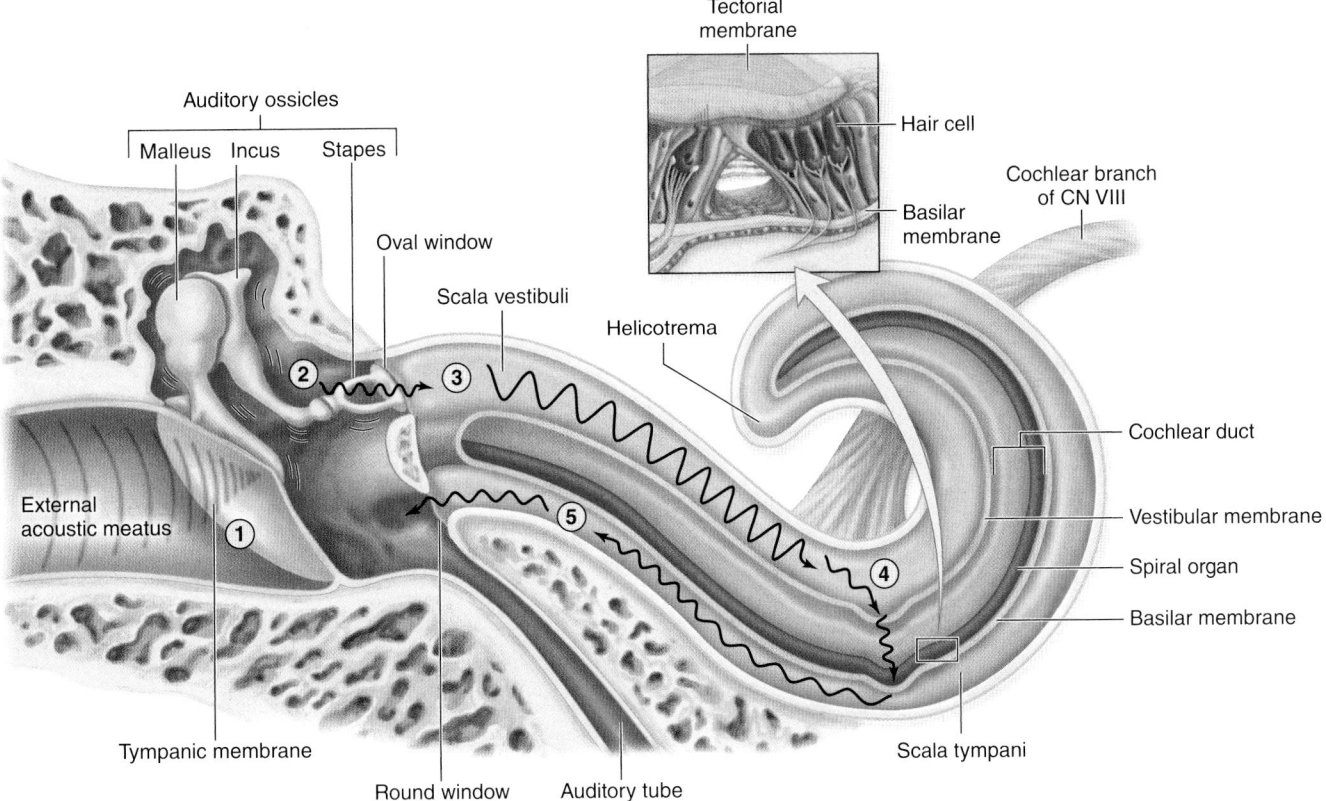

Auditory ossicles

Malleus | Incus | Stapes

Oval window

Scala vestibuli

External acoustic meatus

Tympanic membrane

Round window   Auditory tube

Tectorial membrane

Hair cell

Cochlear branch of CN VIII

Basilar membrane

Helicotrema

Cochlear duct

Vestibular membrane

Spiral organ

Basilar membrane

Scala tympani

(1) Sound waves are directed by the auricle into the external acoustic meatus causing the tympanic membrane to vibrate.

(2) Tympanic membrane vibration moves auditory ossicles (malleus, incus, and stapes); sound waves are amplified.

(3) The stapes at the oval window generates pressure waves in the perilymph within the scala vestibuli.

(4) Pressure waves cause the vestibular membrane to move, resulting in pressure wave formation in the endolymph within the cochlear duct and displacement of a specific region of the basilar membrane. Hair cells in the spiral organ are distorted, initiating nerve signals in the cochlear branch of the vestibulocochlear nerve (CN VIII).

(5) Remaining pressure waves are transferred to the perilymph within the scala tympani and are absorbed as the round window bulges slightly.

Figure 16.27 Sound Wave Pathways Through the Ear. Sound waves enter the external ear and vibrate the tympanic membrane to move the ossicles of the middle ear, which ultimately causes movement of a specific region of the spiral organ within the inner ear. AP|R

consisting of both **hair cells** and supporting cells that rests on the basilar membrane. Two categories of hair cells rest on the basilar membrane, including a single row of *inner hair cells* (which function as the sensory receptors for hearing) and three rows of *outer hair cells* (which alter the response of the spiral organ to sound). The apical surface of each hair cell has a covering of numerous (more than 50) long, stiff microvilli that are called **stereocilia** (ster′ē-ō-sil′ē-ă; *stereos* = solid) and one long cilium, called a **kinocilium** (kī-nō-sil′ē-ŭm; *kino* = movement). The stereocilia and kinocilium are embedded in an overlying gelatinous structure called the **tectorial** (tek-tōr′ē-ăl; *tectus* = to cover) **membrane.** Extending from the base of these hair cells are primary sensory neurons (about 90% from the inner hair cells and about 10% from the outer hair cells). The cell bodies of these sensory neurons are housed within the **spiral ganglia** in the modiolus, which is located medial to the cochlear duct (figure 16.26a, b).

Pathway from Sound Wave to Nerve Signal

How we detect sound waves, transduce sound energy to an electrical signal, and then transmit a nerve signal along the vestibulocochlear nerve to the brain is described here. See figure 16.27 as you read through this section.

Sound waves are collected and funneled by the auricle of the external ear to enter the external acoustic meatus, which make the tympanic membrane vibrate. The vibration of the tympanic membrane causes

movement of the auditory ossicles (malleus, incus, and stapes), which makes the oval window vibrate. Because the tympanic membrane is twenty times greater in diameter than the oval window, sounds transmitted across the middle ear are amplified twenty-fold. This a required step as the energy is transferred from air in the middle ear to fluid within the inner ear.

Vibration of the oval window causes pressure waves in the perilymph within the scala vestibuli. Pressure waves in the scala vestibuli cause deformation of the vestibular membrane, resulting in pressure waves in the endolymph within the cochlear duct. Depending upon the specific frequency of the sound waves, localized regions of the basilar membrane move (described shortly). Hair cells in the spiral organ of this region are distorted, initiating nerve signals in the cochlear branch of the vestibulocochlear nerve (CN VIII). Simultaneously, the pressure wave vibrations within the cochlear duct are transmitted to the perilymph within the scala tympani, and are absorbed at the round window, as the window bulges slightly.

Cochlear Hair Cell Stimulation

Recall that inner hair cells, which function in converting sound energy into an electrical signal, are components of the spiral organ within the cochlea. These cells reside on the basilar membrane and the tips of their stereocilia and kinocilium are embedded within the gelatinous tectorial membrane (figures 16.26 and 16.27). The details of the inner hair

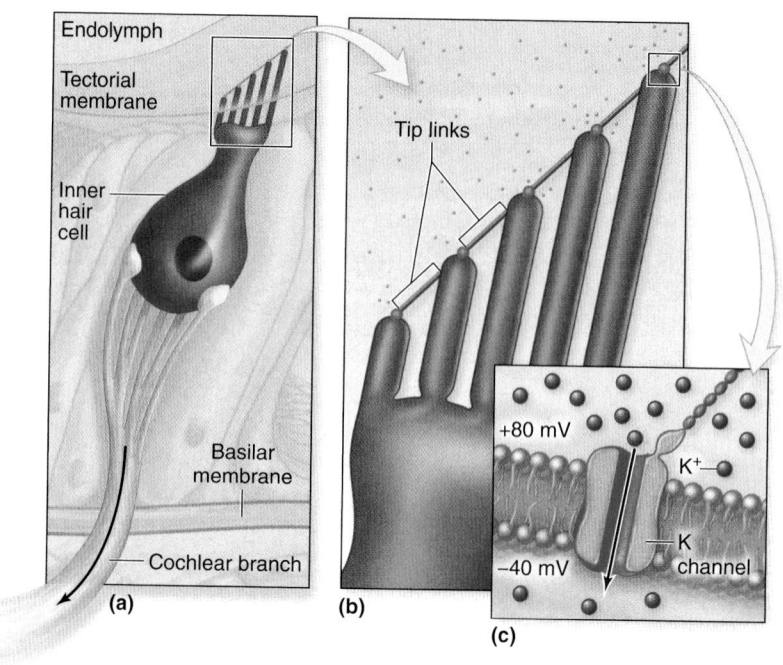

Figure 16.28 Inner Hair Cells. (*a*) The inner hair cells contain ion channels at their tips. (*b*) The ion channel is attached to its taller neighboring inner hair cell by a tip link protein. (*c*) Open ion channels allow K⁺ to move into the inner hair cell.

cells are shown in **figure 16.28**. Note that the inner hair cells are aligned, with each inner hair cell progressively taller. Also notice that the tips of each inner hair cell contain an ion channel, and that each ion channel is connected to the tip of the adjacent taller inner hair cell by a filamentous protein called a **tip link.** This arrangement of inner hair cells is bathed in endolymph. The endolymph is a fluid high in potassium ion (K⁺), contributing to the endolymph's electrical potential of +80 mV. In comparison, the cytosol of the inner hair cell has an electrical potential of −40 mV. This electrical difference between the endolymph and cytosol of the inner hair cells that exists when the inner hair cells are at rest represents potential energy (similar to the potential energy in resting neurons; see section 12.7b).

How does this arrangement result in the production of an electrical signal? Motion of the endolymph repeatedly moves the basilar membrane upward and downward. When the basilar membrane moves upward, the inner hair cells are pushed into the tectorial membrane, causing each inner hair cell to tilt toward its tall neighboring hair cell. This pulls on the tip link, and the movement of each tip link opens an ion channel on its shorter neighboring hair cell. Potassium ion (primarily) diffuses from the endolymph through the open ion channels into the inner hair cell. This movement of K⁺ causes depolarization of the inner hair cell, which triggers the release of neurotransmitter from the base of the inner hair cell. Neurotransmitter binds to dendrites of the primary neuron, causing graded potentials to be established, which move toward the axon. When the threshold is reached, an action potential is initiated along the axon of the primary neuron to the brain. In comparison, the downward movement of the basilar membrane pulls the inner hair cells away from the tectorial membrane, causing the inner cells to straighten. The pull by the tip links decreases, and the ion channels close. The inner hair cells temporarily hyperpolarize, and neurotransmitter is no longer released. Amazingly, this upward and downward movement of the basilar membrane can occur at a rate of up to 20,000 times per second.

Perception of Sound

Sound is the perception of pressure waves that are established by a vibrating object (e.g., a drum, guitar string, vocal cord). These waves of pressure move through any medium, including air, liquid, or a solid. The vibrating object pushes molecules within the medium. Those molecules push adjacent molecules, which in turn push adjacent molecules. However, no one molecule moves very far—it simply transfers the energy of its movement to other molecules. (This would be similar to a line of individuals, with each person pushing the next.)

Two properties of sound that we perceive are pitch and loudness. **Pitch** is ultimately dependent upon the frequency of the vibrating object. **Frequency** is the rate of back and forth motion of the vibrating object, and is measured in cycles per second and expressed in *Hertz* (*Hz*). The spiral organ of the human ear can perceive sounds with frequencies that range from about 20 Hz to 20,000 Hz, but we are most sensitive to sounds with frequencies between 1500 and 4000 Hz. We are able to perceive variations in pitch because there is a continuous change in relative "stiffness" of the basilar membrane, as it extends from being thick and short at the oval window to thin and long at the cochlear apex. Different regions of the basilar membrane move in response to the frequency of the sound waves. High-frequency sounds cause movement of the basilar membrane near the oval window (where the basilar membrane is relatively stiff), whereas low-frequency sounds move the basilar membrane far away from the oval window (where the basilar membrane is relatively flexible) (**figure 16.29**). (This is analogous to the difference in pitch of a piano that is produced from one end of the piano keys to the other.) We cannot perceive frequencies either lower or higher than this because the basilar membrane does not move in response to these rates of vibration.

Loudness is dependent upon the amount of back and forth motion of the vibrating object that establishes the degree of compression of the molecules (or the amplitude of the sound waves). Soft sounds cause relatively small movements of the basilar membrane in a relatively smaller area of the spiral organ, whereas loud sounds cause relatively larger movement of the basilar membrane in a wider area of the spiral organ. The greater movement of the basilar membrane associated with louder sounds increases both the rate of nerve signals that are initiated by the inner hair cells and the number of hair cells that are stimulated. The auditory cortex within the temporal lobe of the brain interprets this change in sensory input as a louder sound. The loudness of a sound is measured in decibels (dB). Zero decibels is the minimum sound (or threshold) that is heard by humans. Consider that the energy associated with sound increases ten times with every 10-decibel increase. This would mean that a sound of 20 dB has ten times the energy of a sound with 10 dB, and a sound of 30 dB has 100 times the energy of a sound of 10 dB. A normal conversation usually occurs at about 60 dB, and prolonged exposure to sounds of 90 dB or greater has sufficient energy to damage the sensory receptors that detect sound within the inner ear.

WHAT DID YOU LEARN?

25 What are the steps for detecting sounds?

26 Compare the difference in how we perceive pitch versus how we perceive loudness.

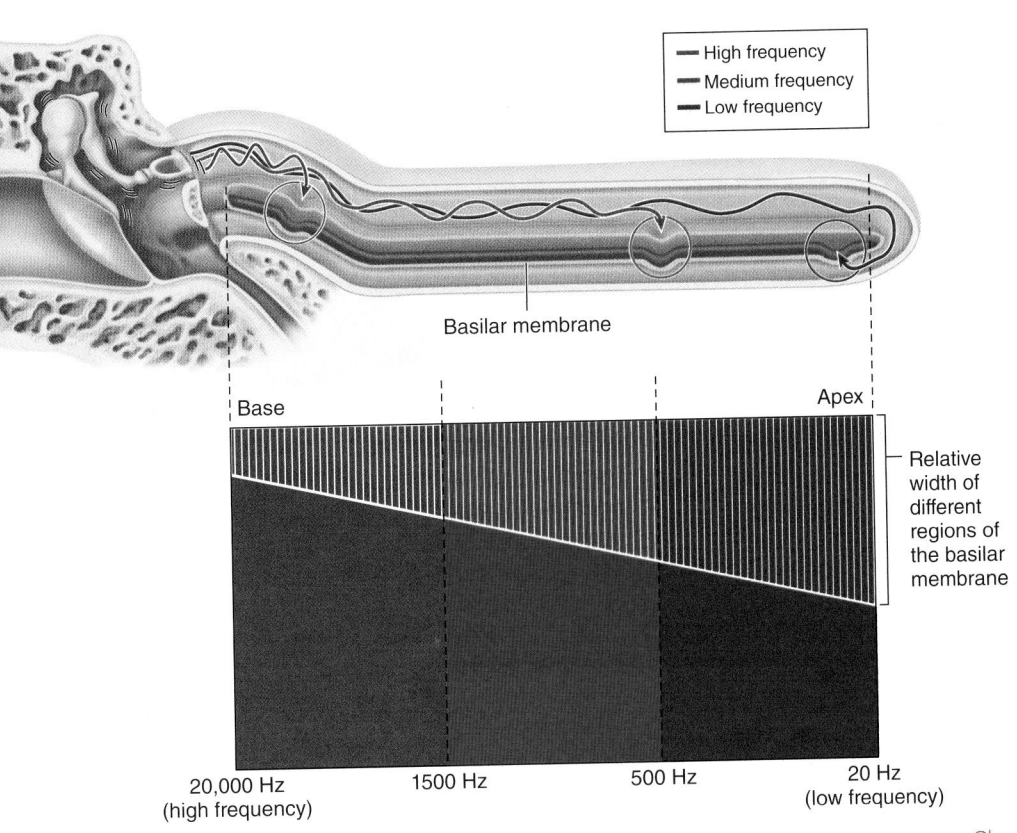

Figure legend:
— High frequency
— Medium frequency
— Low frequency

Basilar membrane

Base

Apex

Relative width of different regions of the basilar membrane

20,000 Hz (high frequency)   1500 Hz   500 Hz   20 Hz (low frequency)

Figure 16.29 Sound Wave Interpretation at the Basilar Membrane. Sound waves are interpreted at specific sites along the basilar membrane of the spiral organ (which is shown here straightened from its normal spiral shape). High-frequency sounds (red arrow) generate pressure waves that cause the basilar membrane to displace close to the base of the cochlea. Medium-frequency sounds (green arrow) generate pressure waves that cause the basilar membrane to displace near the center of the cochlea. Low-frequency sounds (blue arrow) generate pressure waves that cause the basilar membrane to displace near the helicotrema.

INTEGRATE

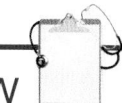

CLINICAL VIEW
Cochlear Implants

A **cochlear implant** is an electronic device that assists some hearing-impaired people. It is not a hearing aid and does not restore normal hearing; rather, it compensates for damaged or nonfunctioning parts of the inner ear. The components of a cochlear implant include (1) an external microphone to detect sound (typically worn behind one ear), (2) a speech processor to arrange the sounds from the microphone, and (3) a transmitter connected to a receiver/stimulator (placed within the cochlea) to convert the processed sound into electrical impulses.

A cochlear implant parallels normal hearing by selecting sounds, processing them into electrical signals, and then sending sound information to the brain for interpretation. However, this type of hearing is considerably different from normal hearing. Patients with implants report that voices sound squeaky and high-pitched, but this limitation does not prevent people with implants from having oral communication capabilities.

Although tens of thousands of people have received cochlear implants, some controversy accompanies their increasing use. First, any surgical installation procedure is always accompanied by a small risk for infection or complication. Second, cochlear implants do not always improve hearing quality. Cochlear implant opponents also have expressed concern that the implant represents a threat to the unique, sign-based culture of deaf

① Antenna, transmitter, and receiver are inserted in skin posterior and superior to the auricle. The transmitter lead is inserted into the inner ear.

Transmitter lead — Spiral organ

Transmitter
Receiver
Antenna

Cochlea cutaway

③ Electrical signals from the transmitter stimulate the cochlear nerve, which then transmits nerve signals to the brain.

Transmitter lead

② Sound waves are detected by the receiver and turned into electrical signals, which travel through the transmitter.

Cochlear implant.

people. Others question the use of implants in children, who may not have made the decision for themselves.

16.5c Auditory Pathways

LEARNING OBJECTIVE

7. Describe the auditory pathway from stimulation of the vestibulocochlear nerve (CN VIII) to the brain.

We discussed how sound waves travel through the inner ear and how an electrical signal is initiated by the inner hair cells within the cochlear duct in the previous section. But what is the nerve pathway that the initiated nerve signals travel to the brain? This auditory pathway, which is composed of a sequence of four sensory neurons instead of the normal two or three, is shown in **figure 16.30** and described here:

1. When the basilar membrane moves, the stereocilia of the spiral organ hair cells distort because they are anchored by the tectorial membrane. This distortion initiates nerve signals that are transmitted through the cochlear branch of the vestibulocochlear nerve (the primary neurons in this sensory pathway) to the **cochlear nucleus** within the medulla oblongata.

2. After integration and processing of the incoming information within the cochlear nucleus, nerve signals are transmitted along secondary neurons to the inferior colliculus in the midbrain. These nerve signals are relayed either (a) directly to the inferior colliculus, or (b) first to the superior olivary nucleus in the pons before transmission to the inferior colliculus. Nerve signals arriving at the **inferior colliculus** are involved in a reflex to loud sounds; here nerve signals are relayed along motor neurons to skeletal muscles of the body that cause us to jump and turn our head in response to loud sounds. Nerve signals arriving at the **superior olivary nucleus** function to (a) localize the sound and (b) respond reflexively to loud sounds by initiating nerve signals to the tensor tympani and stapedius to contract, which decreases the vibration of the ossicles.

3. Nerve signals are transmitted from the inferior colliculus to the **medial geniculate nucleus** of the thalamus (see section 13.4b) for initial processing and filtering of auditory sensory information.

4. Nerve signals are then relayed from the thalamus to the **primary auditory cortex** within the temporal lobe, where nerve signals are consciously perceived as sounds.

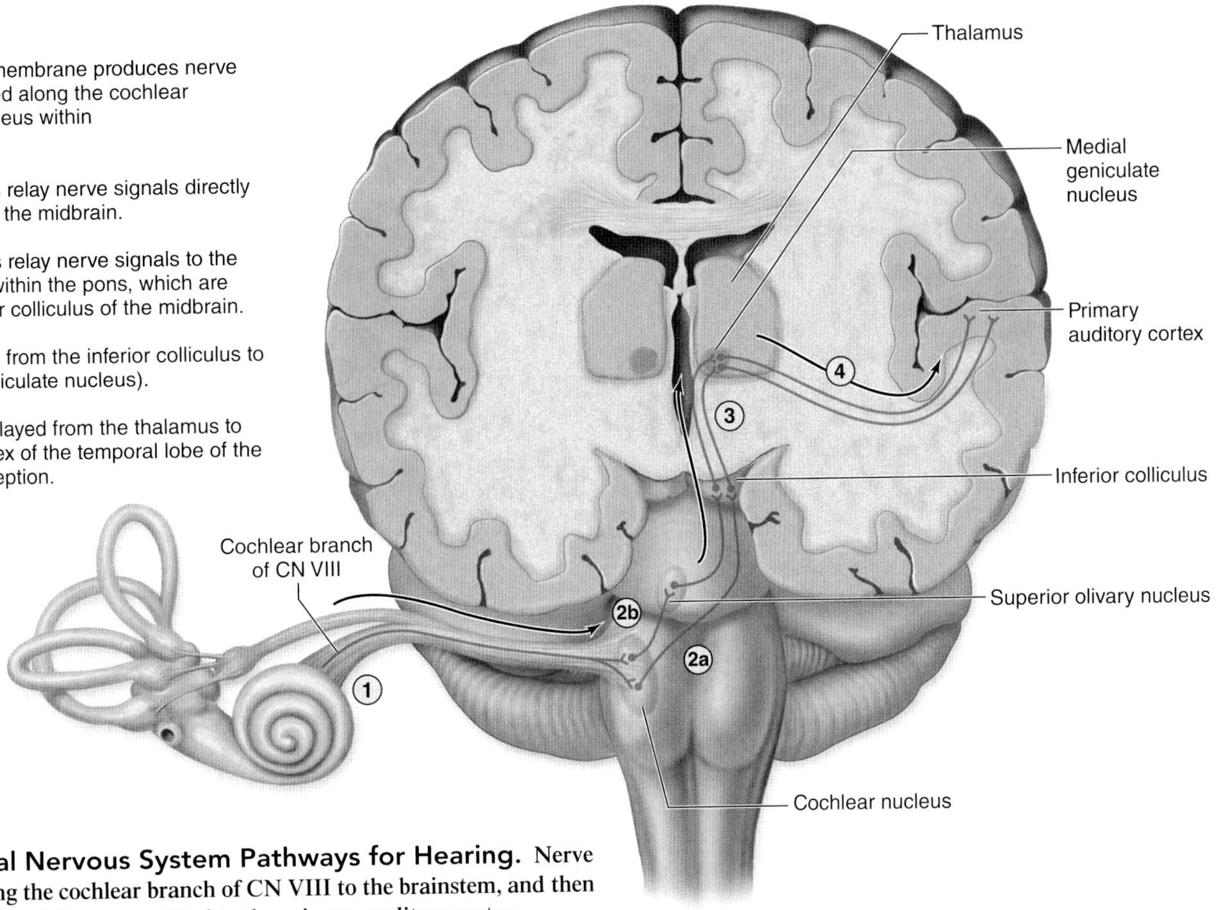

(1) Movement of the basilar membrane produces nerve signals that are propagated along the cochlear nerve to the cochlear nucleus within the medulla oblongata.

(2a) Some secondary neurons relay nerve signals directly to the inferior colliculus of the midbrain.

(2b) Some secondary neurons relay nerve signals to the superior olivary nucleus within the pons, which are then relayed to the inferior colliculus of the midbrain.

(3) Nerve signals are relayed from the inferior colliculus to the thalamus (medial geniculate nucleus).

(4) Nerve signals are then relayed from the thalamus to the primary auditory cortex of the temporal lobe of the cerebrum for sound perception.

Thalamus

Medial geniculate nucleus

Primary auditory cortex

Inferior colliculus

Superior olivary nucleus

Cochlear branch of CN VIII

Cochlear nucleus

Figure 16.30 Central Nervous System Pathways for Hearing. Nerve signals are propagated along the cochlear branch of CN VIII to the brainstem, and then relayed to the thalamus before being transmitted to the primary auditory cortex.

INTEGRATE

CLINICAL VIEW

Deafness

Deafness is defined as any hearing loss. Hearing loss is categorized as either conductive deafness or sensorineural deafness. **Conductive deafness** involves any interference with the transmission of sound waves from the pinna, through the external acoustic meatus, tympanic membrane, and ossicles. Examples of conductive deafness would include rupture of the tympanic membrane, fusion of the ossicles, or middle ear inflammation (otitis media). **Sensorineural deafness** involves malformation or damage to either the structures of the inner ear or the cochlear nerve. Examples would include damage to cilia of the inner ear from loud sounds (e.g., rock concerts or firing a weapon) or trauma to the cochlear nerve from a blow to the head.

To integrate what you've learned about how sounds are processed and interpreted by the brain, refer to **figure 16.31** to review the anatomy and physiology of hearing.

WHAT DID YOU LEARN?

27 What are the major brain structures involved in the auditory pathway, and what is the function of each?

INTEGRATE CONCEPT OVERVIEW

Figure 16.31 How We Hear. Sound waves enter the external ear and then (*a*) are transmitted to the middle ear. (*b*) Sound is amplified within the middle ear and transmitted to the inner ear, where (*c*) sound energy is transduced to nerve signals that (*d*) are transmitted along the auditory pathway to the brain.

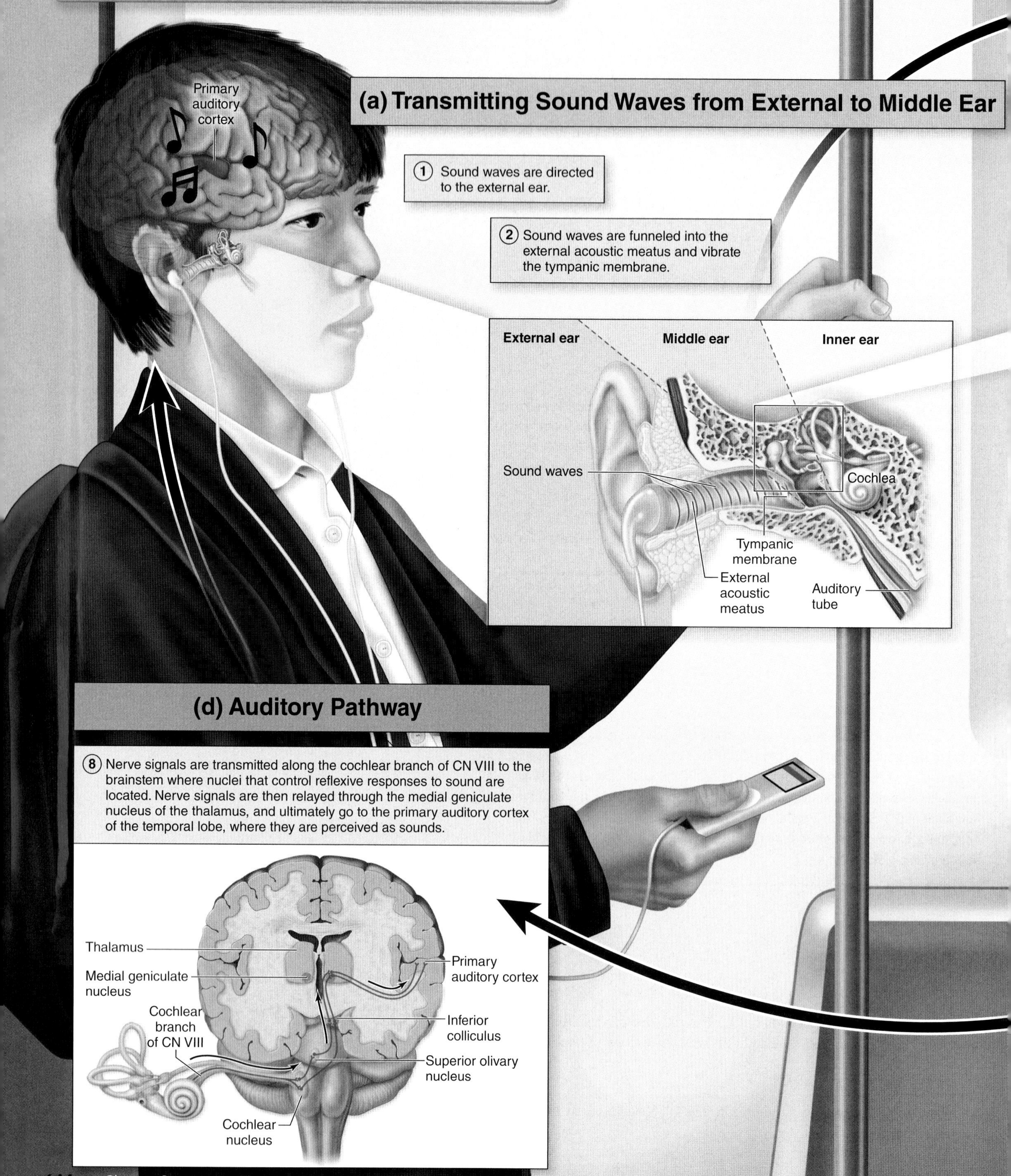

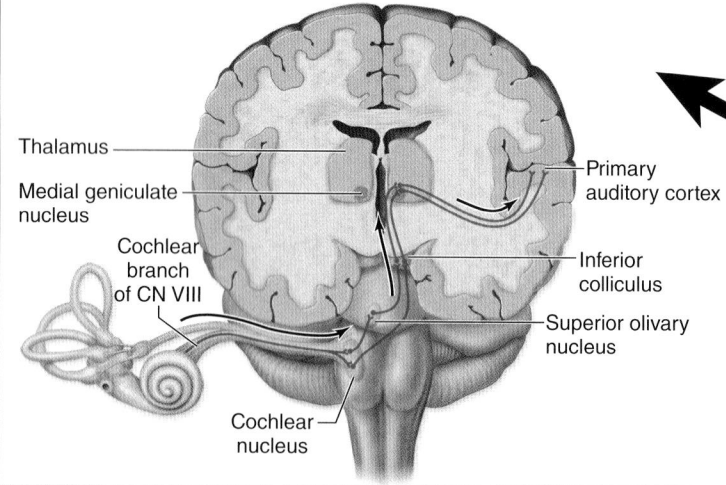

Primary auditory cortex
(a) Transmitting Sound Waves from External to Middle Ear
① Sound waves are directed to the external ear.
② Sound waves are funneled into the external acoustic meatus and vibrate the tympanic membrane.
External ear
Middle ear
Inner ear
Sound waves
Cochlea
Tympanic membrane
External acoustic meatus
Auditory tube
(d) Auditory Pathway
⑧ Nerve signals are transmitted along the cochlear branch of CN VIII to the brainstem where nuclei that control reflexive responses to sound are located. Nerve signals are then relayed through the medial geniculate nucleus of the thalamus, and ultimately go to the primary auditory cortex of the temporal lobe, where they are perceived as sounds.
Thalamus
Medial geniculate nucleus
Cochlear branch of CN VIII
Cochlear nucleus
Primary auditory cortex
Inferior colliculus
Superior olivary nucleus

(b) Amplification and Transmission of Sound from Middle to Inner Ear

③ The tympanic membrane vibration causes the auditory ossicles to vibrate, and they amplify the sound.

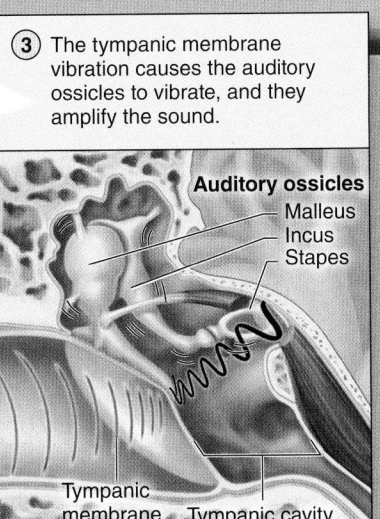

Auditory ossicles
- Malleus
- Incus
- Stapes

Tympanic membrane

Tympanic cavity

④ The stapes moves the oval window, generating pressure waves in the perilymph of the scala vestibuli.

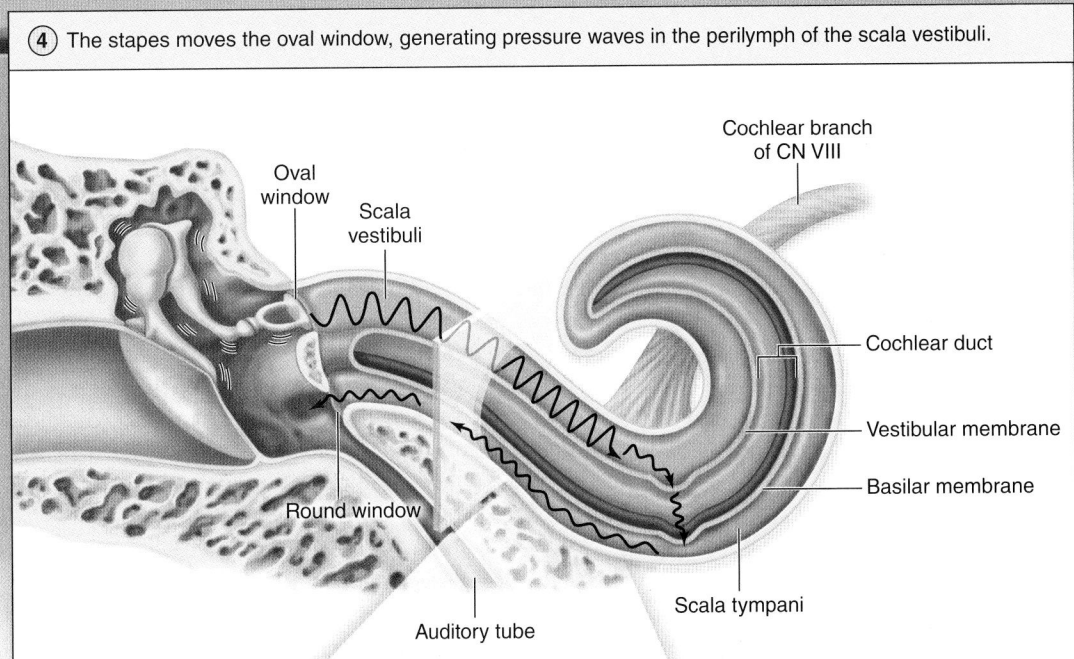

Cochlear branch of CN VIII

Oval window

Scala vestibuli

Cochlear duct

Vestibular membrane

Basilar membrane

Round window

Scala tympani

Auditory tube

(c) Transduction of Sound Energy to Nerve Signals

Cochlear duct

Scala vestibuli

Vestibular membrane

Tectorial membrane

Basilar membrane

Scala tympani

⑤ Pressure waves in the scala vestibuli cause vibration of the vestibular membrane and movement of endolymph within the cochlear duct.

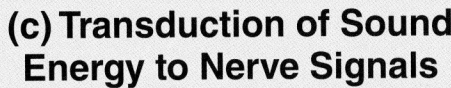

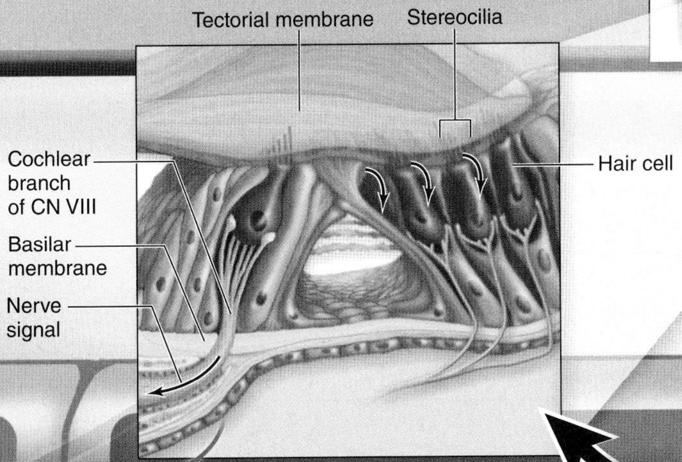

Tectorial membrane

Stereocilia

Cochlear branch of CN VIII

Basilar membrane

Nerve signal

Hair cell

⑥ The pressure waves displace specific regions of the basilar membrane, depending upon the frequency of the sounds.

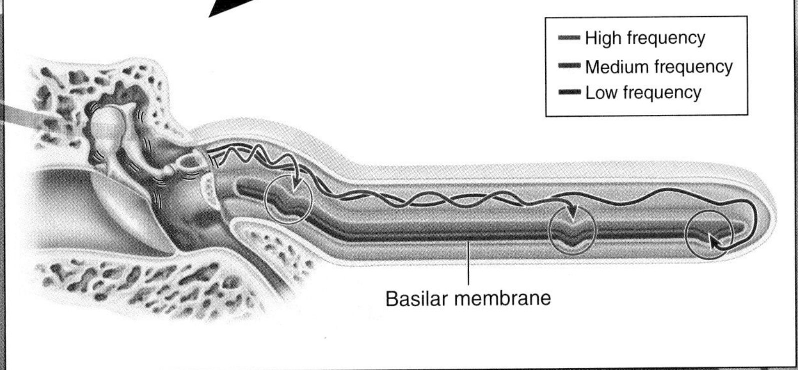

- High frequency
- Medium frequency
- Low frequency

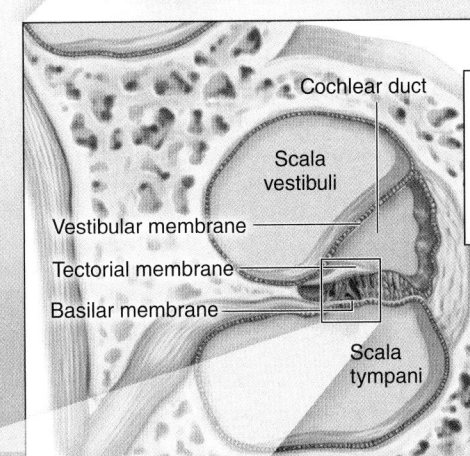

⑦ The displacement of the basilar membrane causes the stereocilia on the hair cells to bend, and nerve signals are propagated along the cochlear branch of CN VIII.

Basilar membrane

INTEGRATE

CLINICAL VIEW
Motion Sickness

Motion sickness is a sense of nausea, mild disorientation, and dizziness that some of us have felt while flying in an airplane or riding in an automobile. It develops when a person is subjected to acceleration and directional changes, but there is limited or discrepant visual contact with the outside horizon. In this situation, the vestibular complex of the inner ear is sending nerve signals to the brain that conflict with the visual reference. The eyes tell the brain we are standing still in an airplane or a ship's cabin, but the inner ear is saying something completely different.

Motion sickness may be alleviated by seeking a place of lesser movement and reestablishing the visual reference. Some people drink a carbonated beverage or eat soda crackers, although the reason this lessens the symptoms isn't clear. Antihistamines are effective in reducing symptoms, and a number of nonprescription oral or transdermal preparations are available as well, including dimenhydrinate (Dramamine), meclizine (Bonine), and cyclizine (Marezine).

16.5d Equilibrium and Head Movement

LEARNING OBJECTIVES

8. Describe the structures of the inner ear involved in equilibrium.
9. Explain how the utricle and saccule detect static equilibrium and linear movements of the head, and how the semicircular ducts function to detect rotational movements of the head.
10. Summarize the nerve pathways involved in equilibrium.

The term **equilibrium** refers to our awareness and monitoring of head position. Sensory receptors in the utricle, saccule, and semicircular ducts collectively called the **vestibular apparatus,** help monitor and adjust our equilibrium. Our brain receives this sensory input and, along with visual sensory input and input from our proprioceptors, integrates this information so we may keep our balance and make positional adjustments as necessary.

The utricle and saccule detect head position during **static equilibrium**—that is, when the head is stationary. So for example, when you are standing in the anatomic position, it is the utricle and saccule that inform your brain that your head is upright. The utricle and saccule also detect **linear acceleration** changes of the head. This occurs, for example, when you tilt your head downward to look at your shoes.

The semicircular ducts, in contrast, are responsible for detecting **angular acceleration,** or rotational movements of the head. The sensory receptors in the semicircular ducts are stimulated when you shake your head "no," or when a figure skater does a spin on the ice.

Static Equilibrium and Linear Acceleration

Both static equilibrium and linear acceleration are detected by the sensory receptors housed within the vestibule of the inner ear. The sensory receptor, called a **macula**, is located along the internal wall of both the membranous utricle and saccule. Each macula is composed of a mixed layer of hair cells and supporting cells **(figure 16.32)**. These hair cells not only have stereocilia, but each hair cell also has one long kinocilium on its apical surface. Stereocilia and kinocilia projecting from the hair cells embed within a gelatinous mass that completely covers the apical surface of the epithelium. This gelatinous layer is covered by a mass of small calcium carbonate crystals called **otoliths** (ō'tō-lith; *oto* = ear; *lithos* = stone), or *statoconia*. Together, the

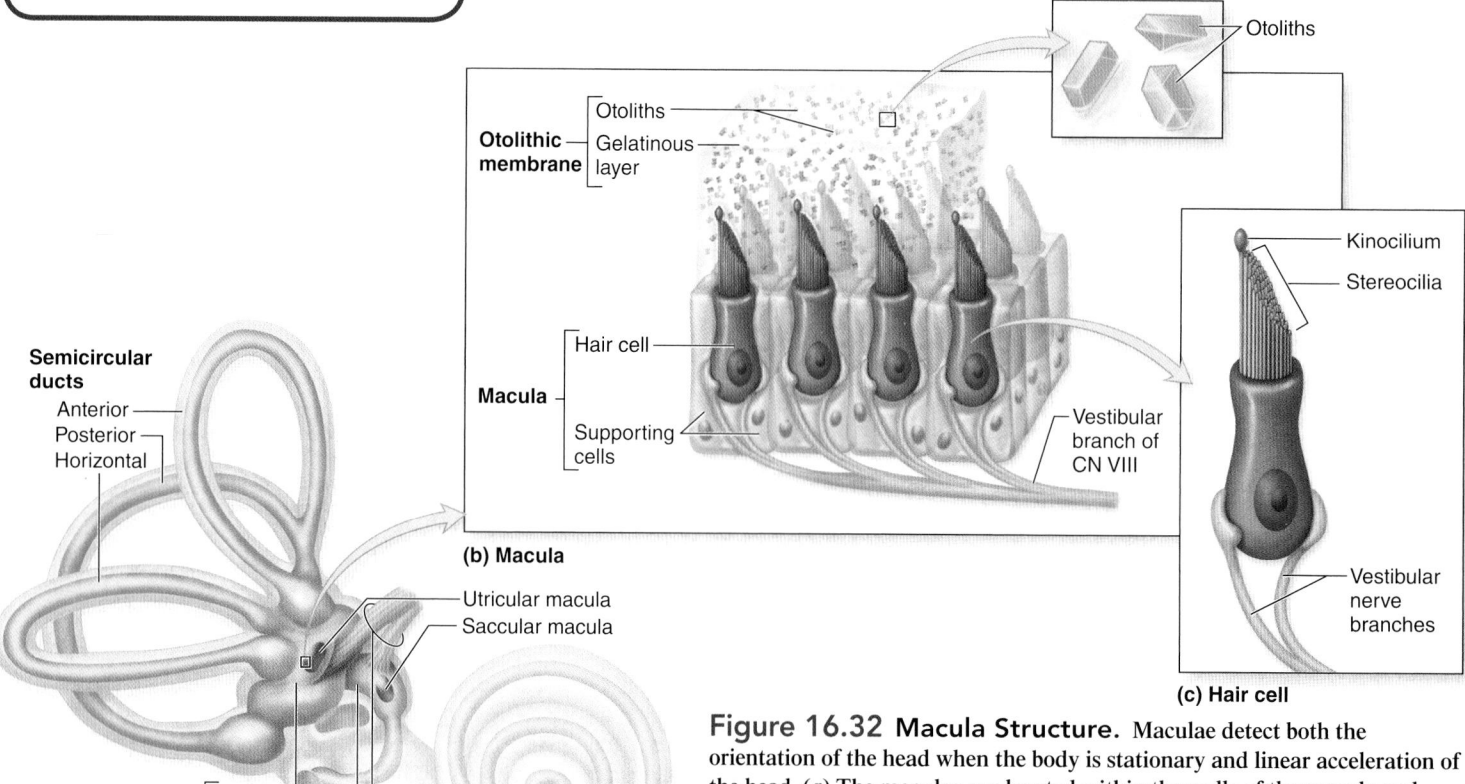

(b) Macula

(c) Hair cell

(a) Vestibular complex

Figure 16.32 Macula Structure. Maculae detect both the orientation of the head when the body is stationary and linear acceleration of the head. (*a*) The maculae are located within the walls of the saccule and utricle. (*b*) An enlarged view of a macula shows the apical surface of the hair cells covered by a gelatinous layer overlaid with otoliths, called the otolithic membrane. (*c*) An individual hair cell has numerous microvilli called stereocilia and a single, long kinocilium.

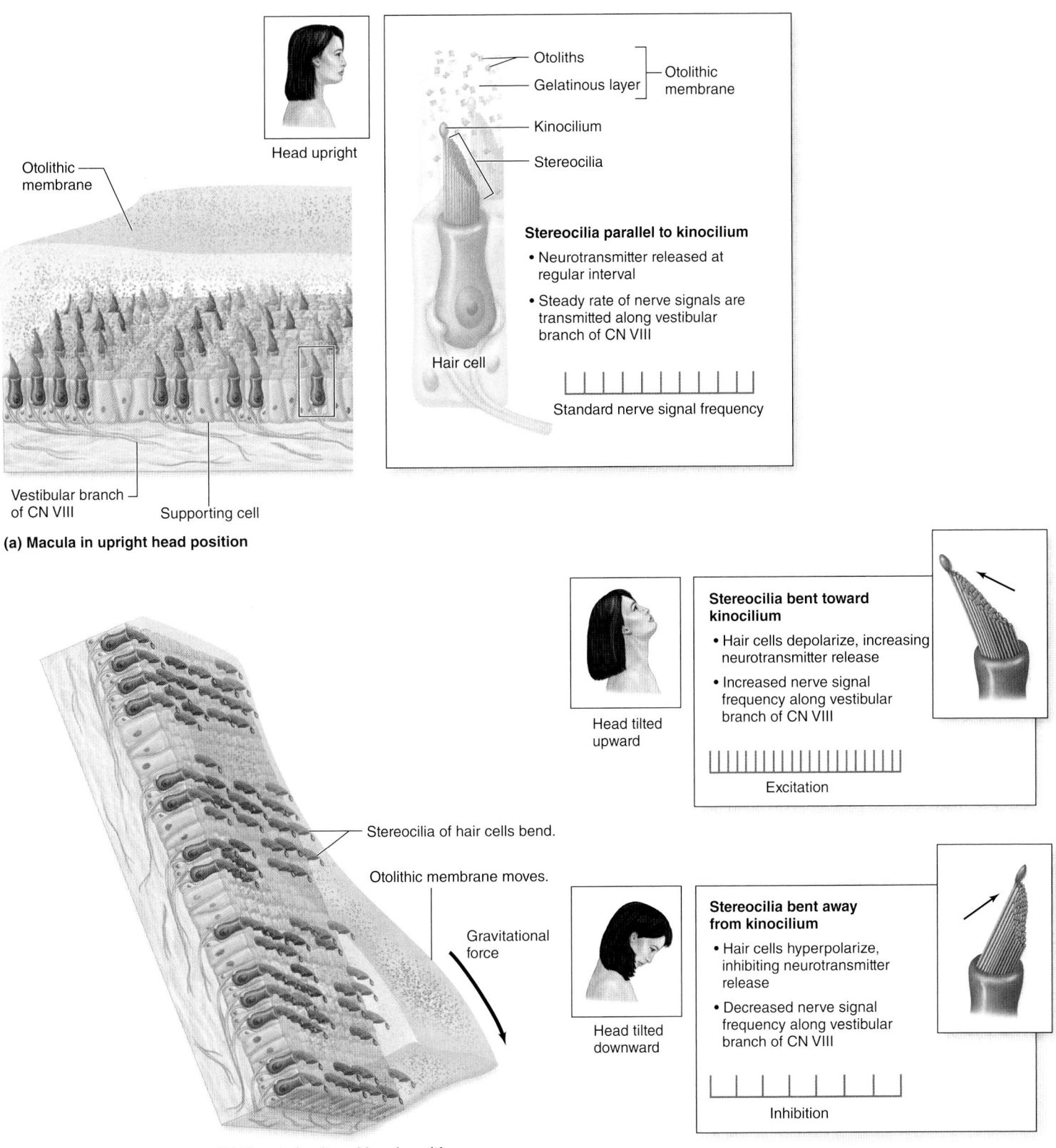

(a) Macula in upright head position

Otolithic membrane

Otoliths
Gelatinous layer
Otolithic membrane

Kinocilium
Stereocilia

Stereocilia parallel to kinocilium
- Neurotransmitter released at regular interval
- Steady rate of nerve signals are transmitted along vestibular branch of CN VIII

Hair cell

Standard nerve signal frequency

Head upright

Vestibular branch of CN VIII
Supporting cell

Stereocilia of hair cells bend.

Otolithic membrane moves.

Gravitational force

(b) Macula in altered head position

Head tilted upward

Stereocilia bent toward kinocilium
- Hair cells depolarize, increasing neurotransmitter release
- Increased nerve signal frequency along vestibular branch of CN VIII

Excitation

Head tilted downward

Stereocilia bent away from kinocilium
- Hair cells hyperpolarize, inhibiting neurotransmitter release
- Decreased nerve signal frequency along vestibular branch of CN VIII

Inhibition

Figure 16.33 How the Maculae Detect Head Position and Linear Acceleration of the Head. (*a*) When the head is in an upright position, hair cells and stereocilia are in parallel and supported by the otolithic membrane. (*b*) Tilting the head causes the otolithic membrane to move slightly, causing the stereocilia to bend and alter the frequency of propagated nerve signals, which reports the change in head position. When the stereocilia are bent toward the kinocilium, the hair cell depolarizes and thus nerve signal propagation increases in frequency. Conversely, when stereocilia are bent away from the kinocilium, the hair cell hyperpolarizes and thus nerve signal propagation decreases in frequency.

gelatinous layer and the crystals form the **otolithic membrane** (or *statoconic membrane*). The otoliths push on the underlying gelatinous layer, thereby increasing the weight of the otolithic membrane covering the hair cells.

The position of the head influences the position of the otolithic membrane (**figure 16.33**). When the head is held erect, the otolithic membrane applies pressure directly onto the hair cells, and minimal stimulation of the hair cells occurs. However, tilting the head causes the otolithic membrane to shift its position on the macula surface, thus distorting the stereocilia. Bending of the stereocilia results in a change in the amount of neurotransmitter released from the hair cells and a simultaneous change in the stimulation of the sensory neurons of the vestibular branch of the vestibulocochlear nerve (CN VIII).

Bending of the stereocilia toward the kinocilium causes the hair cells to depolarize and increase their rate of neurotransmitter release (figure 16.33b). A greater frequency (rate) of nerve signals is generated in the vestibular branch of CN VIII as a result. In contrast, bending of the stereocilia away from the kinocilium causes the hair cells to hyperpolarize and thus decrease their rate of neurotransmitter release. Ultimately, fewer and less frequent nerve signals are generated in the axons of the vestibular branch as time elapses. The brain interprets the change in nerve signals to determine the direction the head has tilted.

Angular Acceleration

Angular acceleration is detected by the sensory receptors housed within the semicircular ducts of the semicircular canals. At the base of each semicircular canal is an enlarged region called the **ampulla** (am-pul′lă; pl., *ampullae,* am-pul′lē; bottle shaped) **(figure 16.34)**. The ampulla contains an elevated region, called the **crista ampullaris** (or *ampullary crest*), which is covered by an epithelium of hair cells and supporting cells. Both the stereocilia and kinocilia are embedded into an overlying gelatinous dome called the **cupula** (kū′pū-lă).

The crista ampullaris in each semicircular duct detects rotational movements of the head **(figure 16.35)**. When the head first rotates, inertia causes endolymph to lag behind. This endolymph pushes against the cupula, causing bending of the stereocilia. Stereocilia bending results in altered neurotransmitter release from the hair cells and simultaneous stimulation of the sensory neurons. As previously described for sensory receptors within the vestibule, bending of the stereocilia in the direction of the kinocilium results in depolarization of the hair cells and increased frequency of nerve signals, whereas bending in the opposite direction results in hyperpolarization of the hair cells and decreased frequency of nerve signals (see figure 16.33b). Interestingly, these sensory receptors respond primarily to changes in velocity—meaning they respond to rotational acceleration or deceleration. If the head is rotating at a constant speed, eventually the movement of the endolymph catches up with the ampulla, and so the stereocilia on the hair cells are no longer bent, and the hair cells no longer are stimulated.

You may have done the following as a small child: You stood upright, closed your eyes, and then you spun in a clockwise direction for about 1 minute. Initially, you felt your head spinning, but after about 30 seconds that spinning sensation may have disappeared. The reason is that the endolymph movement caught up with the ampulla's rotational movement. When you stopped spinning, your head may still have felt like it was moving even though you were standing still. When you came to a stop (and decelerated), the endolymph's momentum resulted in endolymph movement after the body stopped.

INTEGRATE

CONCEPT CONNECTION

Multiple systems are involved in maintaining balance and equilibrium of the entire body. The vestibular apparatus detects motion of the head, our eyes provide the brain with visual information about body position (see section 16.4), and proprioceptors throughout the musculoskeletal system detect muscle and tendon tension (see section 14.6d).

Equilibrium Pathways

The nerve pathway that relays sensory input from the vestibular apparatus is shown in **figure 16.36** and described here:

1. When the stereocilia of either the maculae within the utricle and saccule housed within the vestibule or the crista ampullaris within the ampulla of a semicircular canal distort, nerve signals are initiated through the vestibular branch of the vestibulocochlear nerve (CN VIII).
2. The sensory axons of the vestibular branch (the primary neurons in this sensory pathway) terminate at either the medulla oblongata (vestibular nuclei) or cerebellum. (a) The paired **vestibular nuclei**

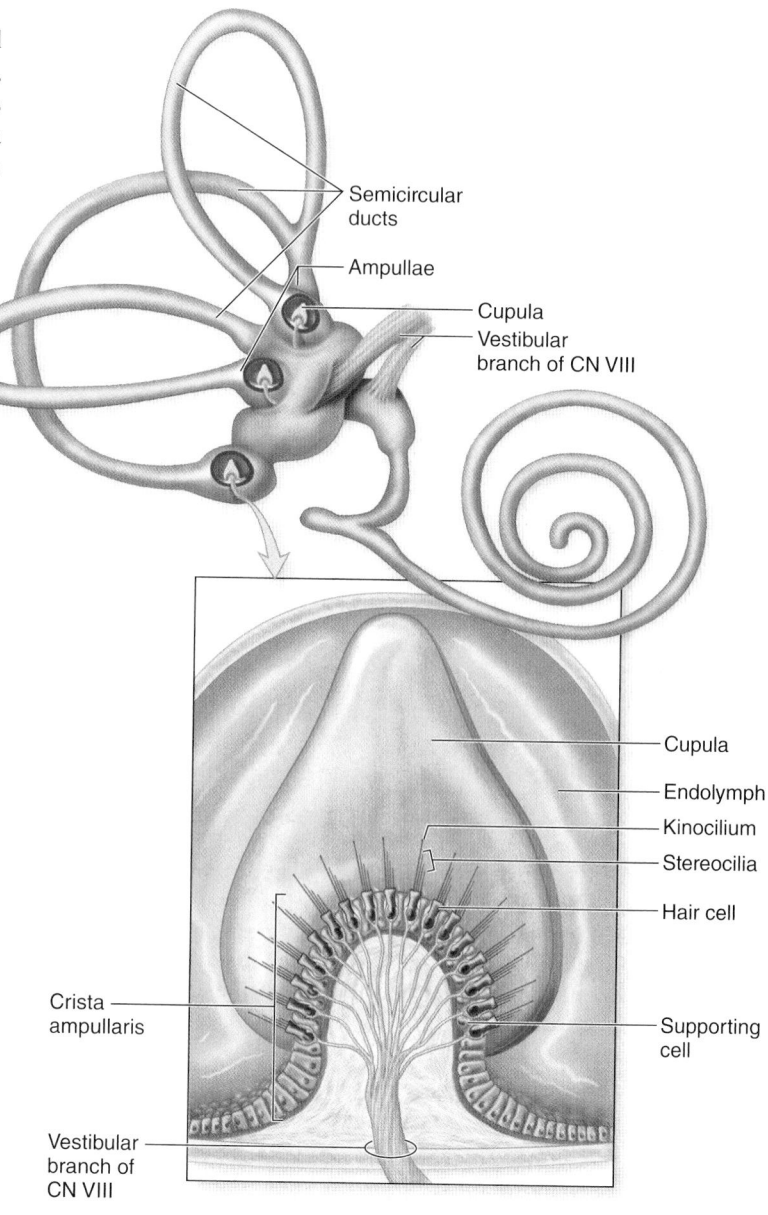

Figure 16.34 Ampulla. A diagrammatic section through an ampulla in a semicircular duct details the relationships among the hair cells and supporting cells, the cupula, and the endolymph in detecting rotational movement of the head.

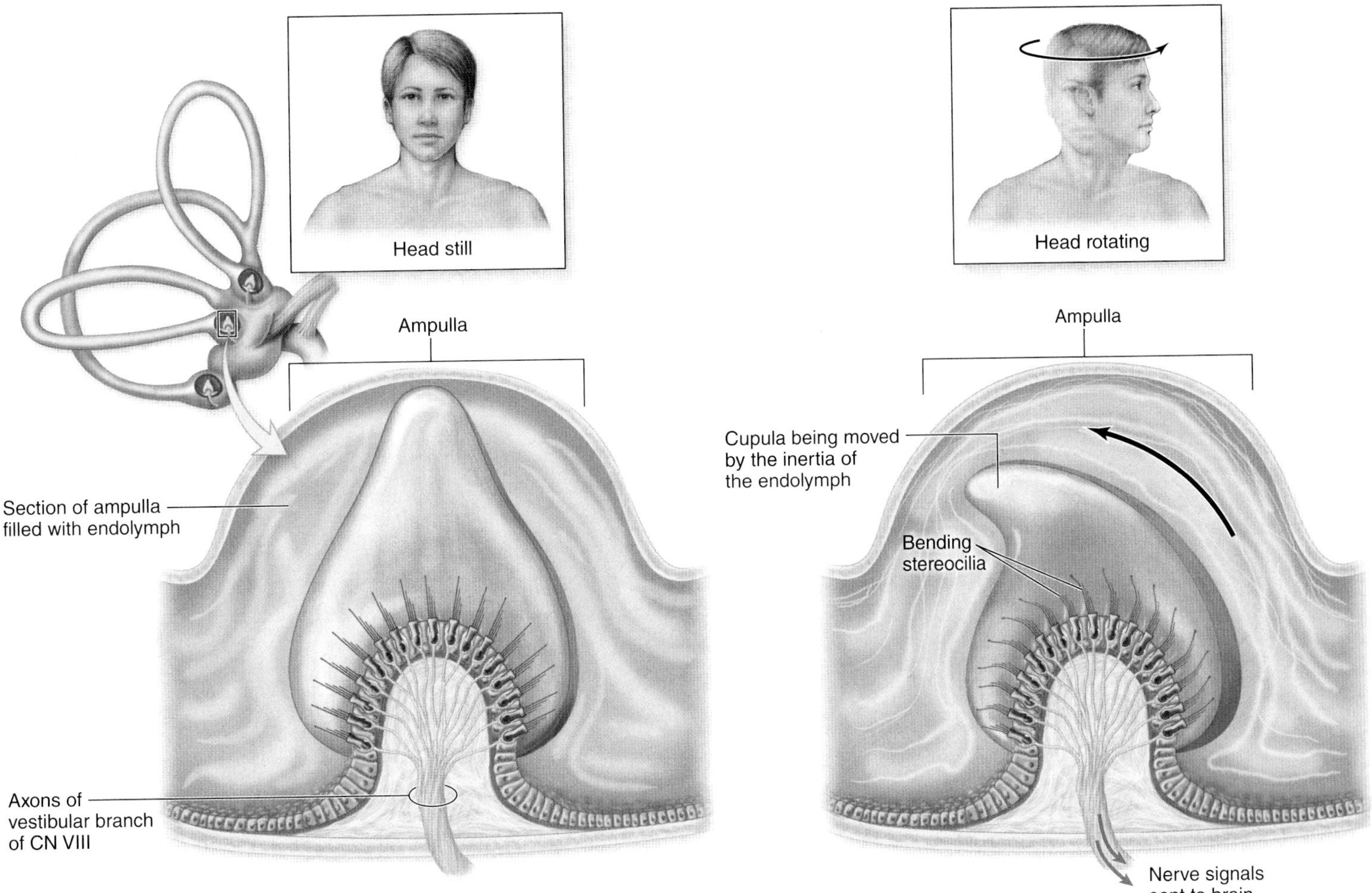

Head still

Ampulla

Section of ampulla filled with endolymph

Axons of vestibular branch of CN VIII

Head rotating

Ampulla

Cupula being moved by the inertia of the endolymph

Bending stereocilia

Nerve signals sent to brain

Figure 16.35 Function of the Crista Ampullaris. Rotation of the head causes endolymph within the semicircular duct to push against the cupula covering the hair cells, resulting in bending of their stereocilia and an alteration in the frequency of nerve signal propagation.

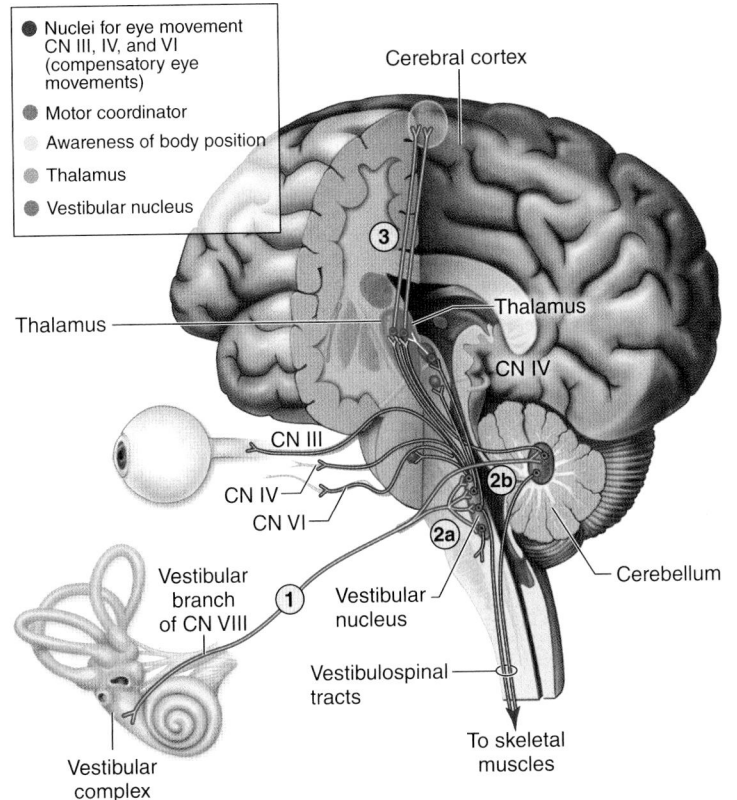

- ● Nuclei for eye movement CN III, IV, and VI (compensatory eye movements)
- ● Motor coordinator
- ○ Awareness of body position
- ● Thalamus
- ● Vestibular nucleus

Cerebral cortex

Thalamus

Thalamus

CN IV

CN III

CN IV

CN VI

3

2b

2a

Cerebellum

Vestibular branch of CN VIII

1

Vestibular nucleus

Vestibulospinal tracts

Vestibular complex

To skeletal muscles

Figure 16.36 Equilibrium Pathways. Information from the vestibular complex is sent to multiple parts of the brain, so posture and body movement may be adjusted accordingly.

within the superior region of the medulla oblongata integrate the stimuli from the vestibular apparatus (as well as the eyes and proprioceptors) to reflexively control movements of the eye (via the oculomotor, trochlear, and abducens cranial nerves) and to control skeletal muscle contraction for maintaining balance (via the descending vestibulospinal motor pathway; see section 14.4c). (b) The cerebellum integrates the sensory input from the vestibular apparatus (as well as input from the eyes, proprioceptors, and cerebrum) to coordinate skeletal muscle to maintain balance and muscle tone (via the descending vestibulospinal motor pathways).

3. Nerve signals are also sent from the vestibular nuclei and cerebellum to the thalamus and eventually the cerebral cortex for further processing and for our awareness of body position.

WHAT DID YOU LEARN?

28 What type of movement do the maculae detect, and how do they detect this movement?

29 What type of movement do the ampullae detect, and how do they detect this movement?

- Sensory receptors provide us with information about both our external and internal environment.

16.1 Introduction to Sensory Receptors 604

16.1a General Function of Sensory Receptors   604
- Sensory receptors function as transducers to change stimulus energy to an electrical signal that is transmitted to the CNS.

16.1b General Structure of Sensory Receptors   604
- Sensory receptors range in complexity from the bare terminal ending of a single sensory neuron to complex sense organs.

16.1c Sensory Information Provided by Sensory Receptors   605
- A sensation is a stimulus that is consciously perceived by the cerebral cortex. Most stimuli are not consciously perceived.
- A sensory receptor provides the CNS with several characteristics regarding a stimulus, including its modality, location, intensity, and duration.

16.1d Sensory Receptor Classification   605
- Sensory receptors are defined and characterized by receptor distribution (general senses and special senses); stimulus origin (exteroceptors, interoceptors, and proprioceptors); and modality of stimulus (chemoreceptors, thermoreceptors, photoreceptors, mechanoreceptors, and nociceptors).

16.2 The General Senses   607

16.2a Tactile Receptors   607
- Tactile receptors are the most numerous type of sensory receptors and are organized into unencapsulated and encapsulated receptors.
- Unencapsulated receptors include free nerve endings, root hair plexuses, and tactile discs.
- Encapsulated receptors include end bulbs, lamellated corpuscles, bulbous corpuscles, and tactile corpuscles.

16.2b Referred Pain   610
- Referred pain is pain that is perceived as if it originates in the skin and skeletal muscle but actually is caused by sensory nerve signals originating from an internal organ.

16.3 Olfaction and Gustation   611

- Both olfactory and gustatory receptors are classified as special senses, exteroceptors, and chemoreceptors.

16.3a Olfaction: The Sense of Smell   611
- The olfactory epithelium is the sensory receptor for the sense of smell and is located in the nasal cavity; it is composed of olfactory receptor cells, supporting cells, and basal cells.
- The sense of smell is transmitted by olfactory nerves to olfactory tracts, which project to the hypothalamus, amygdala, and directly to the olfactory cortex without first synapsing in the thalamus.

16.3b Gustation: The Sense of Taste   612
- Taste buds house the sensory receptors for the sense of taste, which are located primarily on the tongue; they are composed of gustatory cells, supporting cells, and basal cells.
- The five basic taste sensations are sweet, salt, sour, bitter, and umami.
- The sense of taste is transmitted by both the facial nerves (CN VII) and glossopharyngeal nerve (CN IX) through the gustatory pathway.

16.4 Visual Receptors   615

- The eyes house photoreceptors that detect light, color, and movement.

16.4a Accessory Structures of the Eye   615
- Accessory structures include extrinsic eye muscles, eyebrows, eyelashes, eyelids, conjunctiva, and lacrimal glands.
- The conjunctiva is a membrane that covers the anterior surface of the sclera of the eye (the ocular conjunctiva) and the internal surface of the eyelid (the palpebral conjunctiva); it does not cover the cornea.
- Each lacrimal apparatus includes a lacrimal gland that distributes lacrimal fluid to the eye surface. Lacrimal fluid is collected into the lacrimal puncta, which drain into the nasolacrimal duct and then the nasal cavity.

16.4b Eye Structure   617
- The eye is divided into two cavities by the lens: the posterior cavity, which contains the permanent gelatinous vitreous humor, and the anterior cavity, which contains the circulating watery aqueous humor.
- The fibrous tunic is composed of the sclera, which helps protect and maintain the shape of the eye and the cornea, which helps focus light on the retina.
- The vascular tunic of the eye wall has three regions: the choroid, which contains blood vessels to nourish the retina; the ciliary body, which assists lens shape changes with the suspensory ligaments and secretes aqueous humor; and the iris, which controls pupil diameter.
- The neural tunic, or retina, is composed of an outer pigmented layer (that absorbs stray light rays) and an inner neural layer that houses all of the photoreceptors and their associated neurons.
- The retina has a posterior, yellowish region called the macula lutea. Vision is sharpest at a depression within the center of the macula lutea called the fovea centralis.

16.4c Physiology of Vision: Refraction and Focusing of Light   623
- Refraction is the bending of light ray—a process that occurs primarily at the cornea and lens so that light hits the retina.
- Focusing of light when viewing objects closer than 20 feet involves convergence of the eyes, accommodation of the lens, and constriction of the pupil.

16.4d Physiology of Vision: Phototransduction   626
- Phototransduction involves photoreceptor cells (rods and cones), which are located in the neural layer of the retina, converting light energy into an electrical signal.
- Rods, which are primarily located in the peripheral retina, are activated by dim light and provide no color recognition. The image produced has limited visual sharpness.
- Cones, which are primarily located within the fovea centralis, are activated by bright light and provide color recognition. The image produced has visual sharpness. There are three types of cones, and each type maximally absorbs at different wavelengths of light.

16.4e Visual Pathways   630
- The optic nerves are formed by ganglionic axons that project from each eye and converge at the optic chiasm and extend into the brain as optic tracts.
- The optic tracts transmit nerve signals to the thalamus and superior colliculi. The thalamus then transmits visual information to the occipital lobe.

16.5 Hearing and Equilibrium Receptors   634
- Receptors in the ear provide for the senses of hearing and equilibrium.

16.5a Ear Structure   634
- The external acoustic meatus funnels sound waves to the tympanic membrane, which then directs sound waves to the middle ear.
- The middle ear is an air-filled space occupied by three auditory ossicles. The auditory ossicles (malleus, incus, and stapes) transmit and amplify the sound waves from the tympanic membrane to the oval window.
- The inner ear houses the structures for hearing and equilibrium. Specialized receptors are housed in the membranous labyrinth, which lies within a cavernous space in dense bone called the bony labyrinth.

16.5b Hearing   637
- The spiral organ (organ of hearing) is housed within the cochlea. Hair cells in the spiral organ of the cochlea rest on the basilar membrane and are anchored into the tectorial membrane.
- Sound is perceived when sound waves vibrate the tympanic membrane, resulting in auditory ossicle vibrations that lead to vibrations of the basilar membrane, which bend cilia to initiate nerve signals that are conducted along the cochlear branch of CN VIII.

16.5c Auditory Pathways   642
- Secondary axons from the cochlear nuclei project directly to the inferior colliculus, or to the superior olivary nuclei and then the inferior colliculus. Axons then project to the thalamus, where thalamic axons project to the primary auditory cortex so sound is consciously perceived.

16.5d Equilibrium and Head Movement   646
- The saccule and utricle detect both stationary position of the head and linear acceleration of the head.
- The semicircular canals are parts of the bony labyrinth that house the membranous semicircular ducts. Each semicircular duct is an expanded region called an ampulla that houses the hair cells for detecting rotational movements of the head.
- Nerve signals from the vestibular complex are transmitted via the vestibular branch of CN VIII either to the vestibular nuclei (in the medulla oblongata) or directly to the cerebellum. Nerve signals are then sent from the vestibular nuclei and cerebellum to the thalamus and eventually to the cerebral cortex for our conscious awareness of body position.

CHALLENGE YOURSELF

Do You Know the Basics?

_____ 1. Unencapsulated, terminal endings of dendrites of sensory neurons are called
- a. lamellated corpuscles.
- b. free nerve endings.
- c. bulbous corpuscles.
- d. end bulbs.

_____ 2. Each of these sensory receptors is accurately matched with its function *except*
- a. chemoreceptor; detects specific molecules and ions dissolved in fluid
- b. nociceptor; detects tissue damage or pain
- c. proprioceptor; detects change in muscle tension
- d. thermoreceptor; detects change in pressure

_____ 3. All of the following are accurate about the conjunctiva *except*

    a. it contains blood vessels that nourish the sclera.

    b. it covers the cornea.

    c. it contains sensory neurons that detect foreign objects on the eye.

    d. it is composed of stratified columnar epithelium.

_____ 4. Lacrimal fluid performs all of the following functions *except*

    a. cleansing the eye surface.

    b. preventing bacterial infection.

    c. humidifying the orbit.

    d. moistening the eye surface.

_____ 5. The arrangement of tunics in the eye, from the innermost to outermost aspect of the eye, is

    a. retina, vascular, fibrous.

    b. vascular, retina, fibrous.

    c. vascular, fibrous, retina.

    d. retina, fibrous, vascular.

_____ 6. When viewing an object closer than 20 feet, all of the following processes occur *except*

    a. the eyes accommodate to focus on the close-up object.

    b. the ciliary body contracts, so the suspensory ligaments become less taut.

    c. the lens becomes more rounded to better refract light rays.

    d. the pupils dilate to let in more light for the retina.

_____ 7. Which ear structure is correctly matched with its function?

    a. round window; transmits sound waves into the inner ear

    b. external acoustic meatus; directs sound waves to the tympanic membrane

    c. auditory ossicles; dampen sound waves before they reach the inner ear

    d. vestibular membrane; bends the stereocilia on hair cells to produce a nerve signal

_____ 8. Which statement is accurate about the cochlear duct?

    a. It detects linear acceleration of the head when the otolithic membrane bends the hair cells.

    b. It is filled with perilymph.

    c. It contains hair cells that convert sound waves into nerve signals.

    d. It contains a spiral organ that rests on a vestibular membrane.

_____ 9. Which of the following is accurate about the maculae of the vestibular apparatus?

    a. They detect rotational movements of the head.

    b. They are located in the semicircular canal.

    c. Nerve signals are generated when the otolithic membrane bends the stereocilia of the hair cells.

    d. They are the organs of hearing.

_____ 10. The only sensation to reach the cerebral cortex without first processing through the synapses in the thalamus is

    a. olfaction.

    b. vision.

    c. proprioception.

    d. touch.

11. What are the five classifications of receptors according to modality of stimulus? Give an anatomic example of each.

12. How are visceral nociceptors different from somatic nociceptors, and how do they relate to the phenomenon known as "referred pain"?

13. What is the pathway by which taste sensations reach the brain?

14. Describe the pathway by which olfactory stimuli travel from the nasal cavity to the brain.

15. What structures in the wall of the eye help control the amount of light entering the eye?

16. How is the lens able to focus images from a book that you are reading, and then immediately also focus the image of children playing in your backyard?

17. Compare lacrimal fluid and aqueous humor in terms of their formation, circulation, and function.

18. Briefly describe the structural relationship between the membranous labyrinth and the bony labyrinth.

19. Describe the pathway by which sound waves enter the ear, and how that sound is converted into a nerve signal.

20. Explain how the vestibule and semicircular canals detect equilibrium.

Can You Apply What You've Learned?

1. You are babysitting a 5-year-old child who does not want to eat his broccoli. He states that the broccoli tastes yucky. You tell him to hold his nose closed while he eats a piece of broccoli and see how it tastes. He says it doesn't taste as yucky when he holds his nose (although he still doesn't want to eat the broccoli). Why did the broccoli taste "less yucky" when the child held his nose?

    a. Cranial nerve IX was pinched when the child held his nose, so tastes could not be perceived as clearly.

    b. The activity of holding his nose took the child's mind off of eating the broccoli.

    c. Olfaction is responsible for a large component of taste, so by not smelling the broccoli, its taste is diminished.

    d. Holding the nose closed facilitated odiferous molecules to enter the mouth, thereby increasing the number of molecules that could be tasted by the taste buds.

2. Horner syndrome is a condition where sympathetic innervation to one side of the head and neck is damaged. What visual disturbances would you expect a person to have if she had Horner syndrome?

    a. inability to accommodate the eyes for near vision

    b. constricted pupil

    c. lens is permanently flattened

    d. abduction of the individual's affected eye

3. You may be familiar with "mosquito tones" for your cell phones—these are ringtones that are of a relatively high pitch that most children and teenagers can hear, but most adults cannot. Why do you think adults cannot hear these tones as well?

   a. Hair cells close to the oval window of the inner ear probably have become damaged as we age.

   b. The tympanic membrane loses its flexibility as we age, so it is not as able to transmit sound waves to the middle ear.

   c. Adults likely have damaged the spiral organ near the helicotrema, resulting in their inability to hear the tones.

   d. As we age, there is a gradual accumulation of cerumen that blocks the external auditory meatus and makes hearing sounds more difficult.

4. Individuals with macular degeneration may experience loss of vision in their central vision area, but their peripheral vision either is not affected or not as greatly affected. Why is this the case?

   a. Macular degeneration damages the optic disc, making central vision impaired but leaving peripheral vision intact.

   b. Only a single optic tract typically is damaged, so using your peripheral vision allows you to use the normal optic tract.

   c. Only rods are damaged in macular degeneration, so peripheral vision (where there are fewer rods) is not as affected.

   d. The central vision loss is due to damage to the fovea centralis; the peripheral retina is not greatly affected.

5. An elderly gentleman is brought in to the emergency room. He is complaining of pain radiating down his left arm and has shortness of breath. The ER physician suspects the gentleman is suffering a heart attack. Why would the man be experiencing pain down his left arm if the problem is with his heart?

   a. Chemoreceptors in the blood vessels of the left arm detected a change in oxygen levels in the blood and initiated the pain sensations.

   b. The cerebral cortex mistakenly located the source of the stimulus as being from skin and muscle in the arm instead of the heart.

   c. The patient was experiencing phantom pain in the upper limb.

   d. A heart attack stimulates baroceptors in the blood vessels of the upper limb.

Can You Synthesize What You've Learned?

1. Savannah is an active 3-year-old who began to cough and sniffle. Then she experienced an earache and a marked decrease in hearing acuity. During a physical examination, her pediatrician noted the following signs: elevated temperature, a reddened tympanic membrane with a slight bulge, and some inflammation in the pharynx. What might the pediatrician call this condition, and how would it be treated? How does Savannah's age relate to her illness?

2. After Alejandro quit smoking, he discovered that foods seemed much more flavorful than when he had smoked. How are smoking and taste perceptions linked?

3. George is unclear why nearsighted persons may need bifocal eyeglasses (which have two lenses—one for reading and one for nearsightedness) when they get older. He does not understand why the same set of glasses wouldn't work for both reading and for the nearsighted correction. Can you explain to George the difference between the two conditions, and why the treatment for both may appear to be similar?

INTEGRATE

ONLINE STUDY TOOLS

 connect |ANATOMY & PHYSIOLOGY   LEARNSMART® AP|R

The following study aids may be accessed through Connect.

Clinical Case Study: An Elderly Man Gets His Vision Back

Interactive Questions: This chapter's content is served up in a number of multimedia question formats for student study.

LearnSmart: Topics and terminology include introduction to sensory receptors; the general senses; olfaction and gustation; visual receptors; hearing and equilibrium receptors

Anatomy & Physiology Revealed: Topics include nasal cavity; tongue; eye; retina; vision; ear; cochlea; hearing

chapter 17

Endocrine System

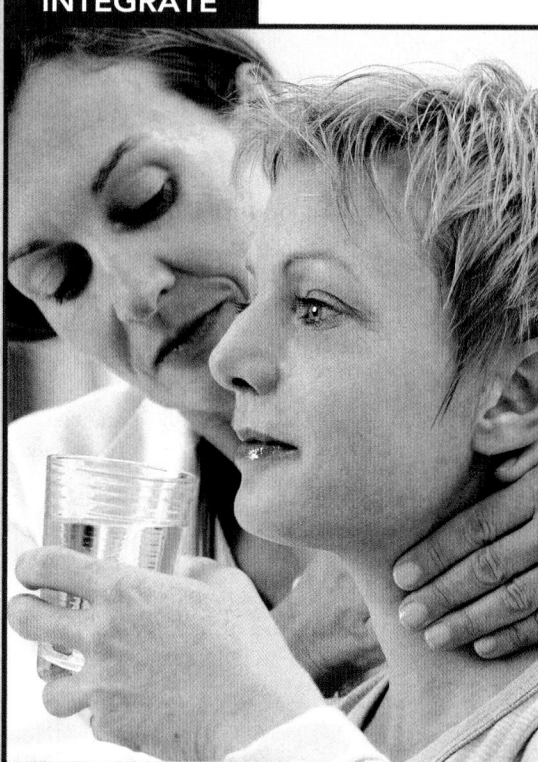

INTEGRATE

CAREER PATH
Endocrinologist

An endocrinologist is a physician that specializes
in the treating of endocrine disorders. The thyroid
gland is the most commonly treated endocrine
gland. This gland is located on the anterior side
of the neck. It produces and releases thyroid hormone to regulate and control metabolism. Palpation of a physical anomaly of the thyroid gland is
often the first indication of a thyroid gland disorder
(e.g., goiter or Graves disease).

Some days are just more hectic than others. We don't have time for breakfast and then
perhaps have a class that meets over the lunch hour, allowing us only enough time to grab
something from the snack machine before heading off to work or other classes in the
afternoon. Finally, upon returning home for the evening, we are so hungry we eat rapidly and
have a second helping. Fortunately, our endocrine system has come to our rescue, maintaining
blood glucose levels within normal homeostatic limits during this period of erratic food intake.
Through the release of hormones, the endocrine system provides the means to regulate and
control many diverse metabolic processes of the body, including regulating blood sugar.

This chapter has two primary purposes. The first is to provide an introduction and
general discussion of the endocrine system's central concepts, including endocrine
glands, hormone structure, hormone transport in the blood, and how ➡

Anatomy &
Physiology | **REVEALED**®
aprevealed.com

Module 8: Endocrine System

hormones interact with body cells. The second is to present the detailed processes by which selected representative hormones maintain body homeostasis. The remaining sections of this chapter provide both an overview of the functions of hormones released from other endocrine glands and the effects of aging on the endocrine system. Please note that each major hormone described in this text can be quickly accessed in the summary tables provided in the reference section immediately after this chapter, beginning on page 695.

17.1 Introduction to the Endocrine System

The **endocrine** (en′dō-krin) **system** is composed of endocrine glands located throughout the body that synthesize and secrete molecules called **hormones** (hōr′mōn; *hormao* = to rouse). Endocrine glands lack ducts and hormones are released into the blood and transported throughout the body (see section 5.1d). Cells with specific receptors for a hormone are called **target cells**. They bind the hormone, which initiates or inhibits selective metabolic activities within these cells. Thus, the endocrine system serves as one of the two major control systems that regulate the human body; the other is the nervous system, which was described in the preceding chapters.

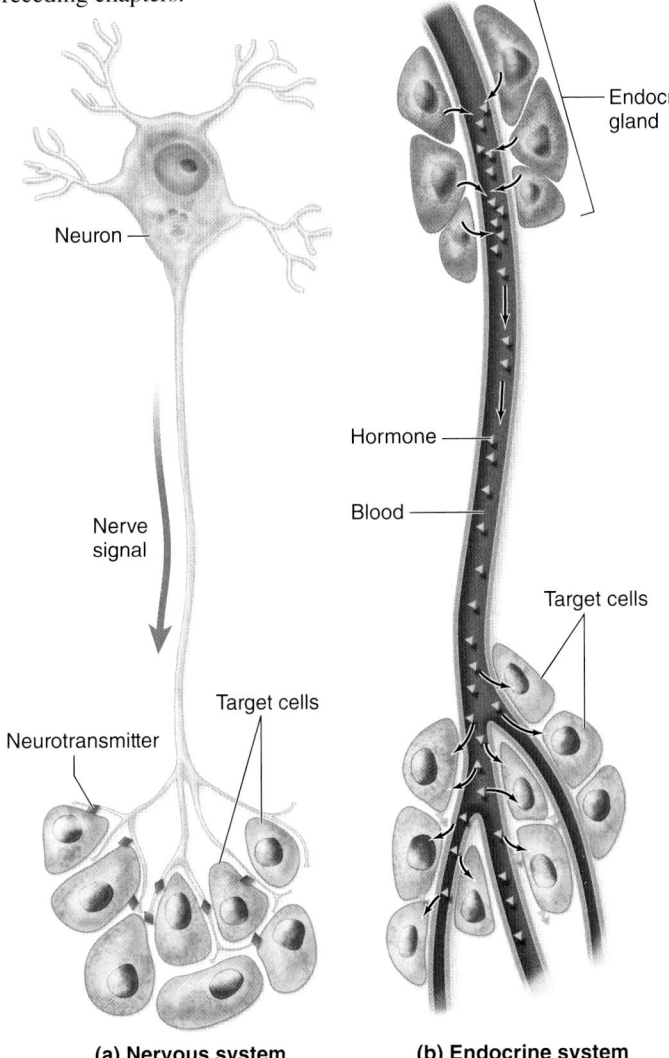

Neuron

Nerve signal

Neurotransmitter

Target cells

(a) Nervous system

Endocrine gland

Hormone

Blood

Target cells

(b) Endocrine system

Figure 17.1 Nervous and Endocrine System Communication Methods. (*a*) In the nervous system, neurons release neurotransmitters into a synaptic cleft to stimulate their target cells. (*b*) In the endocrine system, hormones are secreted by endocrine cells. The hormones enter the blood and travel throughout the body to reach their target cells.

17.1a Comparison of the Two Control Systems

LEARNING OBJECTIVE

1. Compare and contrast the actions of the nervous system and the endocrine system to control body function.

The nervous system and endocrine system, which serve as the two complementary systems of control, have two central features in common: (1) in response to stimuli, specialized cells of both systems release chemical substances (neurotransmitters or hormones) called **ligands** to communicate with particular target cells; and (2) the ligand binds to a cellular receptor in target cells to initiate a cellular change.

The methods and effects of the two control systems differ, however. The nervous system exercises control between two specific locations in the body by way of neurons **(figure 17.1a)**. Nerve signals trigger the release of neurotransmitter, which crosses the synaptic cleft and binds to another neuron, a muscle cell, or a gland cell to initiate a localized response of the target cell, such as muscle contraction or gland secretion. The neurotransmitter is quickly degraded or taken back into the neuron. The response induced by the nervous system is generally both rapid and short-lived.

The endocrine system, in comparison, exercises control between two specific locations in the body through secretion of hormones (figure 17.1b). Hormones are released from endocrine glands and transported within the blood to reach a target. This can be any cell in the body that contains a receptor for the hormone. The response induced by the endocrine system is generally widespread and involves both a relatively long response time and long-lasting effects. The general characteristics of these two complementary control systems are compared in **table 17.1**.

Table 17.1	Comparing the Nervous System and the Endocrine System	
Features	**Nervous System**	**Endocrine System**
Communication Method	A nerve signal causes neurotransmitter release from a neuron into a synaptic cleft	Secretes hormones into blood; hormones transported within the blood are distributed to target cells throughout body
Target of Stimulation	Other neurons, muscle cells, and gland cells	Any cell in the body with a receptor for the hormone
Response Time	Rapid reaction time: Typically milliseconds or seconds	Relatively slow reaction time: Seconds to minutes to hours
Range of Effect	Typically has localized, specific effects in the body	Typically has widespread effects throughout the body
Duration of Response	Short-term: Milliseconds; terminates with removal of stimulus	Long-lasting: Minutes to days to weeks; may continue after stimulus is removed

WHAT DID YOU LEARN?

❶ How does the endocrine system differ from the nervous system with respect to their target cells?

17.1b General Functions of the Endocrine System

LEARNING OBJECTIVE

2. Describe the general functions controlled by the endocrine system.

The endocrine system (through the release of hormones) can potentially communicate with any body cell. Consequently, its functions are very diverse. We have organized the functions of the endocrine system into the following four broad categories:

- **Regulating development, growth, and metabolism.** Hormones have regulatory roles in cell division and cell differentiation that occurs during both development and growth of the body. Hormones also control metabolic processes—both anabolic (synthesis) and catabolic (degradation) processes.

- **Maintaining homeostasis of blood composition and volume.** Hormones regulate the amount of specific substances dissolved within blood plasma, such as glucose, amino acids, and ions (e.g., Na^+, K^+, and Ca^{2+}). Additionally, hormones also regulate other characteristics of blood, including its volume, its cellular concentration (erythrocytes and leukocytes), and number of platelets.

- **Controlling digestive processes.** Several hormones influence both the secretory processes and the movement of materials through the gastrointestinal tract in our digestive system.

- **Controlling reproductive activities.** Hormones affect both development and function of the reproductive system as well as expression of sexual behaviors.

WHAT DID YOU LEARN?

❷ Diabetes mellitus is noted by sustained high blood glucose levels. Which of the four functions listed is the most directly affected?

Figure 17.2 Location of the Major Endocrine Glands and Organs Containing Endocrine Cells. AP|R
(Note: The placenta is not shown in this figure; see figure 29.7.)

17.2 Endocrine Glands

The next several sections provide a general overview of the four central features of the endocrine system. These include a discussion of (1) the location of the major endocrine glands and how they are stimulated to release their hormones; (2) the chemical structure of hormones; (3) how hormone molecules are transported within the blood; and (4) how hormones interact with their target cells. Here we describe the major endocrine glands.

17.2a Location of the Major Endocrine Glands

LEARNING OBJECTIVES

1. Distinguish between the two types of organization of endocrine cells.

2. Identify the major endocrine glands and their location within the body.

Endocrine glands are composed of epithelial tissue that produces and releases hormones, and these secretory cells are supported by a connective tissue framework. The secretory cells are organized in two general ways: either as a single organ with only an endocrine function or as cells housed in small clusters within organs or tissues that have some other primary function (**figure 17.2**).

An **endocrine organ** is a single organ that is entirely endocrine in function. Endocrine organs include the pituitary gland, pineal gland, thyroid gland, parathyroid glands, and adrenal glands.

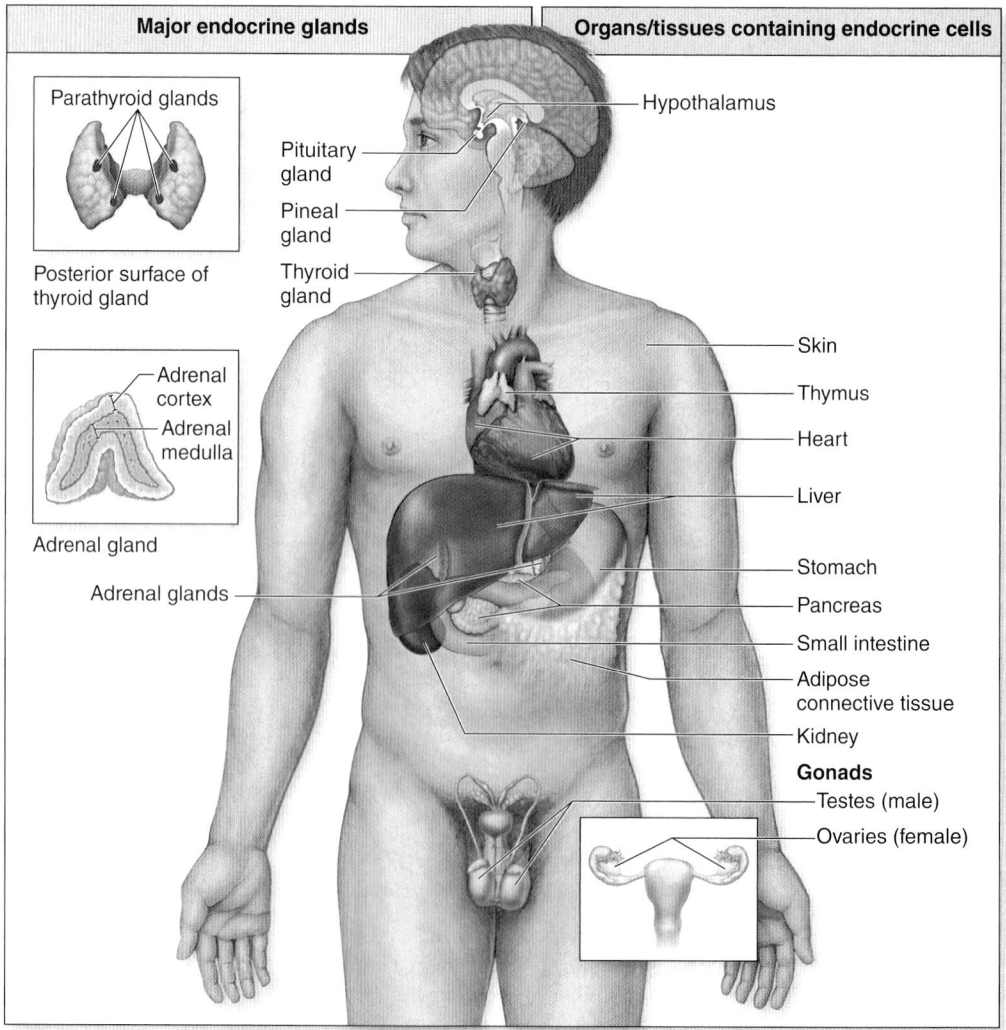

Major endocrine glands	Organs/tissues containing endocrine cells

Parathyroid glands
Posterior surface of thyroid gland

Adrenal cortex
Adrenal medulla
Adrenal gland
Adrenal glands

Pituitary gland
Pineal gland
Thyroid gland

Hypothalamus
Skin
Thymus
Heart
Liver
Stomach
Pancreas
Small intestine
Adipose connective tissue
Kidney
Gonads
Testes (male)
Ovaries (female)

Some endocrine cells are housed in tissue clusters within specific organs or tissues. These cells secrete one or more hormones, but the organs have some other primary function as well. These organs include the hypothalamus, skin, thymus, heart, liver, stomach, pancreas, small intestine, adipose connective tissue, kidneys, and gonads (testes and ovaries).

Note that the phrase "endocrine gland" will be used throughout the rest of this chapter when referring either to an endocrine organ or an organ or tissue containing endocrine cells. The hormones released from the major endocrine glands of the body and their general functions are listed in **table 17.2.**

Table 17.2	Endocrine Glands and Organs Containing Endocrine Cells	
Gland	**Hormone(s) Produced**	**Primary Function(s)**
Hypothalamus	Regulatory hormones	Control release of hormones from anterior pituitary
Hypothalamus (released from posterior pituitary)	Antidiuretic hormone (ADH)	Stimulates both the kidneys to decrease urine output and thirst center to increase fluid intake when the body is dehydrated; in high doses, ADH is a vasoconstrictor (thus, it is also called vasopressin)
	Oxytocin	Contraction of smooth muscle of uterus; ejection of milk; increases feelings of emotional bonding between individuals
Pituitary gland (anterior)	Thyroid-stimulating hormone	Stimulates thyroid gland to release thyroid hormone
	Prolactin (PRL)	Regulates mammary gland growth and breast milk production in females; may increase secretion of testosterone in males
	Follicle-stimulating hormone (FSH)	Controls development of both oocyte and ovarian follicle (spherical structure that houses an oocyte) within ovaries; controls development of sperm within testes
	Luteinizing hormone (LH)	Induces ovulation of secondary oocyte from the ovarian follicle
		Controls testosterone synthesis within testes
	Adrenocorticotropic hormone (ACTH)	Stimulates adrenal cortex to release corticosteroids (e.g., cortisol)
	Growth hormone (GH)	Release of insulin-like growth factors (IGFs) from liver; GH and IGFs function synergistically to induce growth
Pineal gland	Melatonin	Helps regulate the body's circadian rhythms (biological clock); functions in sexual maturation
Thyroid gland	Thyroid hormones: T_3 (triiodothyronine) and T_4 (tetraiodothyronine or thyroxine)	Increase metabolic rate of all cells; increase heat production (calorigenic effect)
	Calcitonin	Decreases blood calcium levels; most significant in children
Parathyroid glands	Parathyroid hormone (PTH)	Increases blood calcium levels by stimulating both release of calcium from bone tissue and decrease loss of calcium in urine; causes formation of calcitriol hormone (a hormone that increases calcium absorption from small intestine)
Thymus	Thymosin, thymulin, thymopoietin	Maturation of T-lymphocytes (a type of white blood cell or leukocyte)
Adrenal cortex	Mineralocorticoids (e.g., aldosterone)	Regulate blood Na^+ and K^+ levels by decreasing the Na^+ and increasing the K^+ excreted in urine
	Glucocorticoids (e.g., cortisol)	Participate in the stress response; increase nutrients (e.g., glucose) that are available in the blood
	Gonadocorticoids (e.g., dehydro-epiandrosterone [DHEA])	Stimulate maturation and functioning of reproductive system
Adrenal medulla	Epinephrine (EPI) and norepinephrine (NE)	Prolong effects of the sympathetic division of the autonomic nervous system
Pancreas	Insulin	Decreases blood glucose levels
	Glucagon	Increases blood glucose levels
Testes (gonads)	Testosterone	Stimulates maturation and function of male reproductive system
	Inhibin	Inhibits release of follicle-stimulating hormone (FSH) from anterior pituitary
Ovaries (gonads)	Estrogen and progesterone	Stimulates maturation and function of female reproductive system
	Inhibin	Inhibits release of follicle-stimulating hormone (FSH) from anterior pituitary
Heart	Atrial natriuretic peptide (ANP)	Functions primarily to decrease blood pressure by stimulating both the kidneys to increase urine output and the blood vessels to dilate
Kidneys	Erythropoietin (EPO)	Increases production of red blood cells (erythrocytes)
Liver	Angiotensinogen	Converted by enzymes released from the kidney and within the inner lining of blood vessels to angiotensin II; increases blood pressure by causing vasoconstriction and decreasing urine output; stimulates thirst center
	Insulin-like growth factors (IGFs)	Functions synergistically with growth hormone to regulate growth
	Erythropoietin (EPO)	Increases production of red blood cells (erythrocytes); note that kidneys are the major producers of EPO
Stomach	Gastrin	Facilitates digestion within stomach
Small intestine	Secretin	Regulates digestion within small intestine by helping to maintain normal pH within small intestine
	Cholecystokinin (CCK)	Regulates digestion within small intestine by facilitating digestion of nutrients within small intestine
Skin	Vitamin D_3	Converted by enzymes of liver and kidney to calcitriol; functions synergistically with PTH and increases calcium absorption from small intestine
Adipose connective tissue	Leptin	Helps regulate food intake
Placenta	Estrogen and progesterone	Stimulates development of fetus; stimulates physical changes within mother associated with pregnancy including those in the uterus and mammary glands

⚛ WHAT DID YOU LEARN?

❸ What are the major endocrine organs in the human body? What are the organs (or tissues) that have another primary function and contain endocrine cells?

⚛ WHAT DID YOU LEARN?

❹ Adrenocorticotropic hormone (ACTH) stimulates the adrenal cortex to release cortisol hormone. This is an example of what type of stimulation: (a) hormonal, (b) humoral, or (c) nervous system?

17.2b Stimulation of Hormone Synthesis and Release

⚛ LEARNING OBJECTIVE

3. Explain the three reflex mechanisms for regulating secretion of hormones.

The regulated secretion of a hormone from an endocrine gland is controlled through a reflex (see section 14.6a). Reflexes occur in both the nervous system and the endocrine system. Endocrine reflexes are initiated by one of three types of stimulation: hormonal stimulation, humoral stimulation, or nervous system stimulation (figure 17.3).

- **Hormonal stimulation.** The stimulus for the release of many hormones from its endocrine gland is the binding of another hormone. An example occurs when thyroid-stimulating hormone (which is released from the anterior pituitary) binds to the thyroid gland to cause release of thyroid hormone (figure 17.3a).
- **Humoral stimulation.** Some endocrine glands are stimulated to release their hormones in response to a changing level of nutrient molecules (e.g., glucose) or ions (e.g., Ca^{2+}) within the blood. (The term "humoral" is a historical term related to blood's being one of the four "humors," or fluids, of the body.) When either nutrient or ion levels decrease or increase within the blood, responsive cells of an endocrine gland release hormones. An example of humoral stimulation occurs when blood glucose increases and the pancreas releases insulin (figure 17.3b).
- **Nervous system stimulation.** A few endocrine glands are stimulated to release hormone(s) by direct stimulation from the nervous system. The classic example is the release of epinephrine and norepinephrine by the adrenal medulla in response to stimulation by the sympathetic division of the autonomic nervous system (figure 17.3c).

17.3 Hormones

All circulating hormones are synthesized within endocrine cells from either cholesterol or amino acids. Recall that cholesterol is a type of lipid with a four-ring structure (see section 2.7b) and amino acids are the building block components of proteins (see section 2.7e).

17.3a Categories of Circulating Hormones

⚛ LEARNING OBJECTIVES

1. Name the three structural categories of circulating hormones, and give examples within each category.

2. Distinguish the hormones that are lipid-soluble from those that are water-soluble.

Circulating hormones are grouped according to their chemical structure into three general categories: steroids, biogenic amines, and proteins. An example of each category is shown in figure 17.4. As you read about the chemical structure of these hormones, please note whether the molecules in each category are water-soluble or lipid-soluble. This difference in solubility influences both the transport of the hormone in the blood and how it interacts with its target cells.

Steroids

Steroids are lipid-soluble molecules synthesized from cholesterol (figure 17.4a). This category includes both the steroids produced in the gonads (e.g., estrogen and progesterone in the ovaries and testosterone in the testes) and the hormones synthesized by the adrenal cortex (e.g., corticosteroids such as cortisol, mineralocorticoids such as aldosterone).

Hormonal stimulation: Release of a hormone in response to another hormone	Humoral stimulation: Release of a hormone in response to changes in level of nutrient or ion in the blood	Nervous system stimulation: Release of a hormone in response to stimulation by the nervous system

① Anterior pituitary releases thyroid-stimulating hormone (TSH).

Anterior pituitary

② TSH stimulates thyroid gland to release thyroid hormone (TH).

TSH

Thyroid gland

Capillary

TH

① Blood glucose levels increase.

Pancreas

② Increased blood glucose stimulates pancreas to release insulin.

Insulin

Increased blood glucose

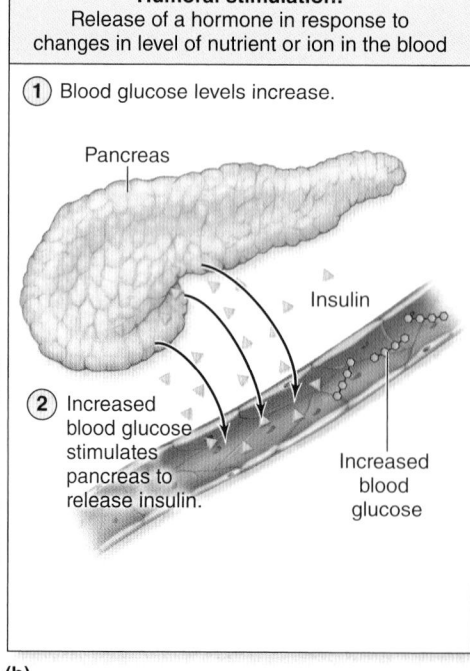

① Sympathetic division is activated.

Spinal cord

Nerve signal

Preganglionic axon

② Sympathetic preganglionic axons stimulate adrenal medulla to release epinephrine and norepinephrine.

Adrenal gland

Epinephrine and norepinephrine

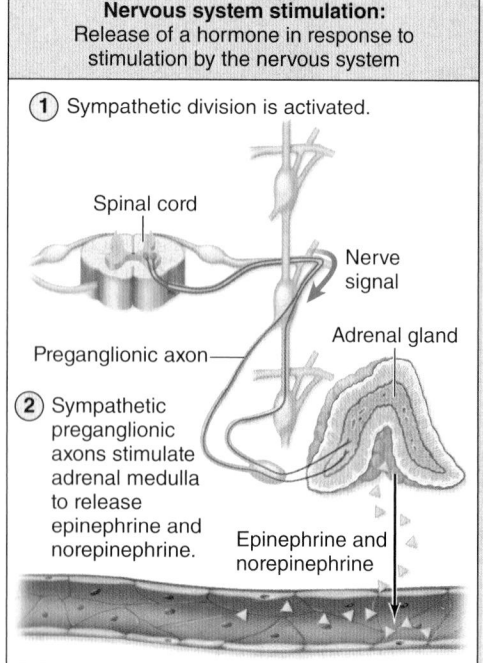

(a)　　　　　　　　　　(b)　　　　　　　　　　(c)

Figure 17.3 Types of Endocrine Stimulation. An endocrine gland can be stimulated to release its hormone in response to (*a*) hormonal stimulation, (*b*) humoral stimulation, or (*c*) nervous system stimulation. AP|R

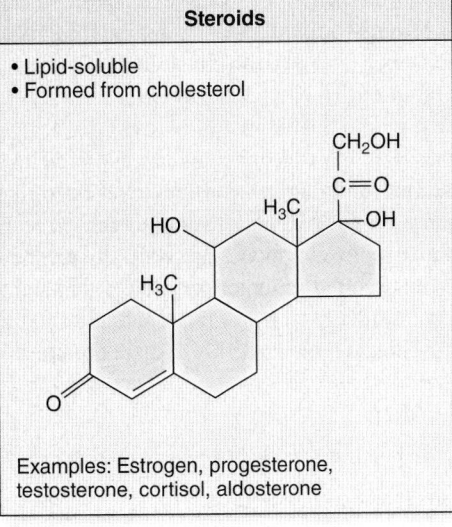

Steroids

• Lipid-soluble
• Formed from cholesterol

Examples: Estrogen, progesterone, testosterone, cortisol, aldosterone

(a)

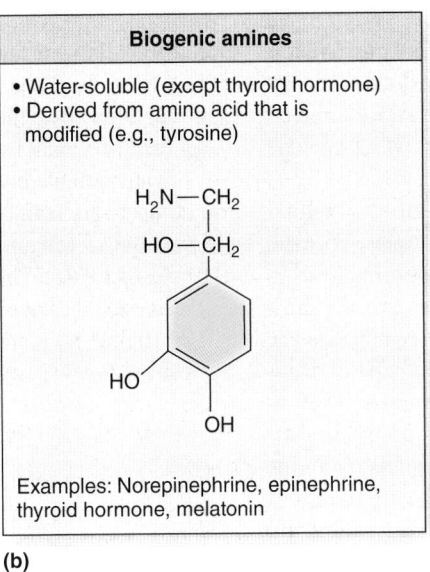

Biogenic amines

• Water-soluble (except thyroid hormone)
• Derived from amino acid that is modified (e.g., tyrosine)

Examples: Norepinephrine, epinephrine, thyroid hormone, melatonin

(b)

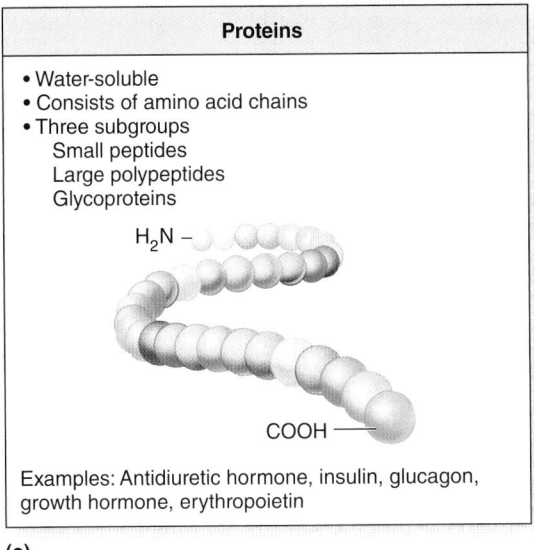

Proteins

• Water-soluble
• Consists of amino acid chains
• Three subgroups
  Small peptides
  Large polypeptides
  Glycoproteins

Examples: Antidiuretic hormone, insulin, glucagon, growth hormone, erythropoietin

(c)

Figure 17.4 Hormone Types. (*a*) Steroids (which include hormones released from both the gonads and adrenal cortex) are lipid-soluble and formed from cholesterol. In comparison, (*b*) biogenic amines (which include hormones released from the adrenal medulla, thyroid hormone, and melatonin) and (*c*) proteins (which includes most hormones) typically are water-soluble and formed from amino acids.

Calcitriol is the hormone produced from vitamin D (see section 7.6a), and it is sometimes classified as a steroid hormone. However, it is more accurately classified as a *sterol* hormone, a slightly different molecule but one that is also lipid-soluble.

Biogenic Amines

Biogenic amines are also called *monoamines*. They are modified amino acids (modification involves removal of a carboxyl functional group from an amino acid) (figure 17.4*b*). Biogenic amines include the catecholamines (e.g., epinephrine and norepinephrine) released from the adrenal medulla, thyroid hormone released from the thyroid gland, and melatonin from the pineal gland. Biogenic amines are water-soluble except for thyroid hormone. Thyroid hormone is lipid-soluble because it is produced from two tyrosine amino acids (see figure 2.25), which are amino acids that contain a nonpolar ring.

Proteins

Most hormones are **proteins.** These are molecules composed of small chains of amino acids and include small peptides, large polypeptides, and glycoproteins (figure 17.4*c*). All hormones in this category are water-soluble.

WHAT DID YOU LEARN?

5 Identify which of the following hormone categories is lipid-soluble: (a) reproductive hormones produced in the gonads, (b) adrenal cortex hormones, and (c) thyroid hormone.

6 What two events or processes associated with a hormone are influenced by whether a hormone is lipid-soluble or water-soluble?

17.3b Local Hormones

LEARNING OBJECTIVES

3. Describe the general structure, formation, and function of local hormones.
4. Compare autocrine and paracrine signaling that occurs through local hormones.

Local hormones are a large group of signaling molecules that do not circulate within the blood. Instead, these molecules are released from the cells that produce them and bind with either the same cell that produced them or neighboring cells. For this reason, some biologists do not classify these signaling molecules as hormones.

Eicosanoids (ī′kō-să-noydz; *eicosa* = twenty, *eidos* = formed) are a primary type of local hormones. The name "eicosanoids" is derived from the fact that this group of signaling molecules is formed from fatty acids that contain a chain of 20 carbon atoms. The source of these fatty acids is phospholipids within a cell's plasma membrane.

Eicosanoids are synthesized through a series of enzymes called an enzymatic cascade (**figure 17.5**). This process is

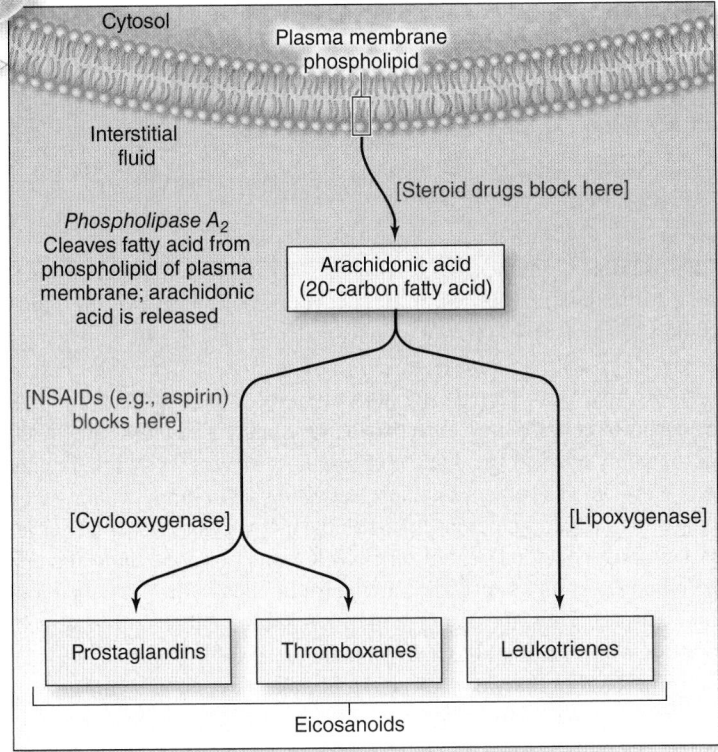

Figure 17.5 Eicosanoid Formation. Eicosanoids are synthesized by an enzymatic cascade that begins with phospholipids from within a plasma membrane. Subsequent to their formation, eicosanoids may stimulate the cell from which it is released (autocrine stimulation) or neighboring cells (paracrine stimulation).

initiated by an enzyme (phospholipase A$_2$) that removes a fatty acid from a phospholipid within a plasma membrane. The fatty acid that is released is called *arachidonic acid.* Other specific enzymes then act on arachidonic acid (e.g., cyclooxygenase, lipoxygenase) to convert its chemical structure to various types of eicosanoids, which include prostaglandins, thromboxanes, and leukotrienes.

Subsequent to their synthesis, eicosanoids initiate cellular changes in either (1) the same cell from which they were formed, a process called **autocrine stimulation,** or (2) neighboring cells, a process called **paracrine stimulation.** Prostaglandins are the most diverse category of eicosanoids and are thought to be synthesized in most tissues of the body. These local hormones stimulate pain receptors, reduce a fever, and increase the inflammatory response. Aspirin, ibuprofen, and other nonsteroidal anti-inflammatory drugs (NSAIDs) are medications that block the formation of prostaglandins, and thus inhibit these processes. In comparison, steroid medications inhibit the first step in this enzymatic cascade, and block the formation and action of prostaglandins, thromboxanes, and leukotrienes.

WHAT DID YOU LEARN?

7 Leukotrienes from damaged tissue cause smooth muscle in local blood vessels to vasodilate (increase the diameter of the vessel lumen). Is this an example of (a) autocrine stimulation or (b) paracrine stimulation? Explain.

17.4 Hormone Transport

Hormone molecules are transported within the blood and carried to target cells following their release from the endocrine gland that synthesized them. Here we consider how both lipid-soluble and water-soluble hormones are transported and the factors that influence the level of circulating hormone.

17.4a Transport in the Blood

LEARNING OBJECTIVE

1. Compare the transport of lipid-soluble hormones with that of water-soluble hormones.

Lipid-soluble hormones (e.g., steroids, calcitriol, and thyroid hormone) do not dissolve readily within the aqueous environment of the blood (blood plasma), and so they require carrier molecules. These molecules are water-soluble proteins synthesized by the liver. Think of these proteins as boats that "ferry" the hormone molecules within the blood. Some hormone carrier proteins may be very selective, binding and transporting only one specific lipid-soluble molecule (e.g., thyroxine-binding globulin), whereas others are nonselective (e.g., albumin), meaning they bind and transport numerous lipid-soluble molecules. A transport carrier protein has the added benefit of protecting the hormone molecule and helping to prevent its early destruction (or for small hormones their loss in the urine). Thus, the association of a hormone with a carrier protein often acts as a safeguard to help prolong the life of the hormone.

The binding between a lipid-soluble hormone and a carrier protein is temporary. Hormone molecules bind to the carrier molecule, detach from the carrier and float free within the blood, and then may later reattach to a different carrier molecule. Any hormone that is attached to a carrier is a **bound hormone,** whereas an unattached hormone is an **unbound (free) hormone.** Only unbound hormone, which represents a very small fraction of the hormone transported

within the blood (0.1% to 10%), is generally able to exit the blood and bind to cellular receptors of target organs. It is the maintaining of a given homeostatic level (or steady state) of the physiologically active, unbound hormone that is required. The bound hormone simply serves as a readily available source within the blood.

Water-soluble hormones (i.e., most biogenic amines and proteins), in comparison, readily dissolve within the aqueous environment of the blood, and so these hormones do not generally require carrier proteins (see section 2.4c). Thus, water-soluble hormones are generally released directly into the blood and transported to target cells. Note that some water-soluble hormones (e.g., insulin-like growth factor) are transported by carrier proteins, which function to prolong the life of these hormones.

WHAT DID YOU LEARN?

8 Why are carrier proteins necessary for lipid-soluble hormones?

9 What is the added benefit of a carrier protein?

17.4b Levels of Circulating Hormone

LEARNING OBJECTIVES

2. Describe the two primary factors that affect the concentration level of a circulating hormone.

3. Explain what is meant by the half-life of a hormone.

Hormones exert their physiologic effects primarily as a result of their blood concentration. Consequently, the amount of each hormone must be tightly regulated to prevent potential clinical consequences, such as organ or tissue malfunction and disease. For example, gigantism is due to high blood levels of growth hormone, whereas decreased metabolic rate is caused by low blood levels of thyroid hormone.

Two primary factors that interact to influence hormone concentration—hormone synthesis and hormone elimination:

- **Hormone synthesis.** Hormone synthesis occurs in an endocrine gland. If the rate of synthesis and release increases, then the concentration of the hormone within the blood is greater. In contrast, if synthesis and release of the hormone decreases, hormone concentration in the blood is lower.

- **Hormone elimination.** Hormones are typically eliminated through either (1) enzymatic degradation, which usually occurs in liver cells, or (2) removal of the hormone from the blood either by its excretion from the kidneys or by its uptake into target cells. The faster the rate of hormone elimination, the lower the hormone concentration within the blood, whereas the slower the rate of hormone elimination, the higher the hormone concentration within the blood.

To maintain homeostatic levels of each hormone, a balance is required between its rate of synthesis by its endocrine gland and its elimination from the blood by the activities of the liver, kidney, and target cells.

WHAT DO YOU THINK?

1 What effect would impaired function of the liver or kidneys potentially have on hormone concentration in the blood: increase, decrease, or no change? Explain.

Half-Life
The **half-life** is the amount of time necessary to reduce the hormone concentration within the blood to one-half of what had been secreted originally (or measured previously). Generally, water-soluble hormones

have a relatively short half-life that amounts to a few minutes or less for small peptides, and about an hour for larger proteins. Steroids generally have the longest half-life since their carrier protein protects them from early destruction or loss. The half-life of testosterone, for example, is 12 days. Note that the shorter the half-life of a hormone, the more frequently it must be replaced to maintain its normal concentration in the blood.

 WHAT DID YOU LEARN?

10 What is the relationship of hormone synthesis to the concentration of that hormone in the blood?

17.5 Target Cells: Interactions with Hormones

Hormones interact only with target cells (cells with receptors for that hormone) to initiate a specific cellular response. A specific hormone generally has different types of target cells. The hormone insulin, for example, binds with muscle cells, hepatocytes (liver cells), and adipose connective tissue cells. The greater the number of different target cells, the wider the potential influence that may be exhibited by a given hormone.

The specific process of how hormones interact with cell receptors, and the cellular changes that are initiated, is significantly different for lipid-soluble hormones and water-soluble hormones.

17.5a Lipid-Soluble Hormones

LEARNING OBJECTIVE

1. Describe how lipid-soluble hormones reach their target cell receptors and the type of cellular change they initiate.

Lipid-soluble hormones (e.g., steroids, calcitriol) are relatively small, nonpolar molecules that are **lipophilic** (lip'ō-fil'ik), or lipid-loving. Recall that the plasma membrane is not an effective barrier to small, nonpolar substances (see section 4.3a). Consequently, unbound lipid-soluble hormones such as steroids are able to diffuse across the plasma membrane. Upon entering the cell, the hormone binds to intracellular receptors located in either the cytosol or nucleus to form a **hormone-receptor complex (figure 17.6)**.

The hormone-receptor complex formed within the target cell then binds to a particular DNA sequence in regions of chromatin called **hormone-response elements (HREs)**. Binding to a specific DNA sequence results in transcription of messenger ribonucleic acid (mRNA). Subsequent translation of this mRNA by ribosomes synthesizes a specific protein (see section 4.8b). The change in protein

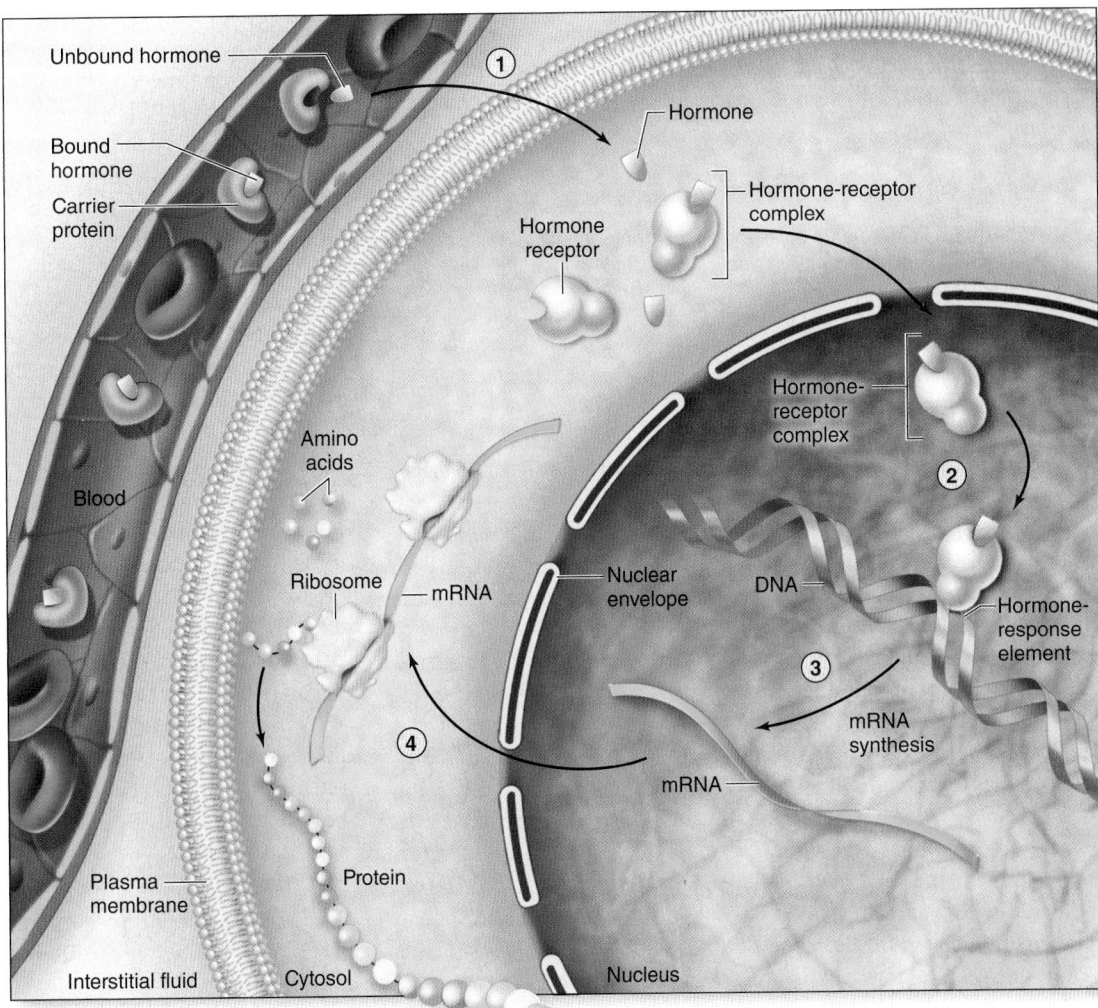

1 The unbound lipid-soluble hormone diffuses readily through the plasma membrane and binds with an intracellular receptor, either within the cytosol or the nucleus to form a hormone-receptor complex.

2 The hormone-receptor complex binds with a specific DNA sequence called a hormone-response element.

3 This binding stimulates mRNA synthesis.

4 mRNA exits the nucleus and is translated by a ribosome in the cytosol. A new protein is synthesized.

Figure 17.6 Lipid-Soluble Hormones and Intracellular Receptors. Lipid-soluble hormones enter a cell and ultimately cause the formation of new protein. AP|R

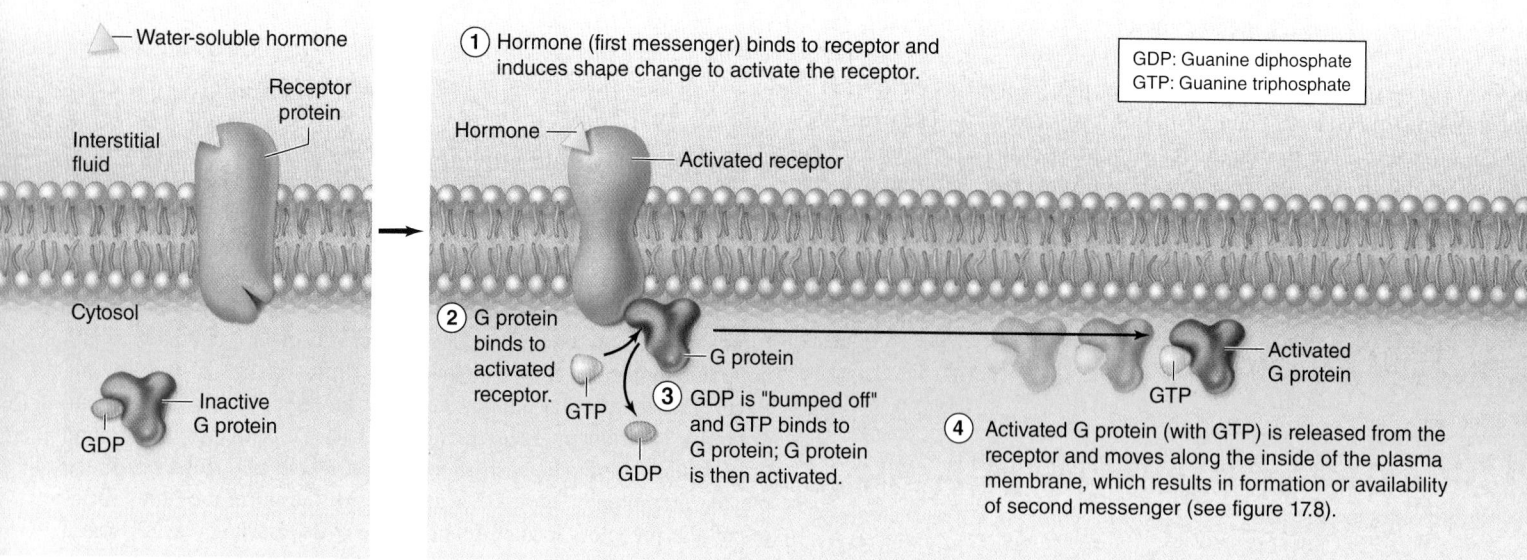

Figure 17.7 Activation of G Proteins. G proteins are activated in response to a water-soluble hormone binding to a plasma membrane receptor.

Within the figure:

- Water-soluble hormone
- Receptor protein
- Interstitial fluid
- Cytosol
- Inactive G protein
- GDP

① Hormone (first messenger) binds to receptor and induces shape change to activate the receptor.

- Hormone
- Activated receptor

② G protein binds to activated receptor.
- G protein
- GTP

③ GDP is "bumped off" and GTP binds to G protein; G protein is then activated.
- GDP

④ Activated G protein (with GTP) is released from the receptor and moves along the inside of the plasma membrane, which results in formation or availability of second messenger (see figure 17.8).
- Activated G protein
- GTP

> GDP: Guanine diphosphate
> GTP: Guanine triphosphate

---

synthesis pattern within the cell may result in either an alteration in cell structure (e.g., as occurs with sex hormones during puberty) or a shift in the target cells' metabolic activities if the newly synthesized proteins are enzymes.

Consider, for example, that testosterone results in larger muscles due to formation of contractile proteins, a deeper voice from longer and thicker vocal cords, and facial hair growth. These effects induced by testosterone reflect a cellular increase in protein synthesis.

WHAT DID YOU LEARN?

⑪ Where are lipid-soluble hormone receptors located? What is the general cellular change that occurs with binding of a lipid-soluble hormone?

17.5b Water-Soluble Hormones

LEARNING OBJECTIVE

2. Describe how water-soluble hormones induce cellular change in their target cells.

Water-soluble hormones (e.g., proteins and biogenic amines, except thyroid hormone) are polar molecules and are unable to cross the plasma membrane. Denied entry into the cell, water-soluble hormones instead must use an alternative, slightly more complex way to stimulate a target cell. This stimulation is initiated when the hormone binds to a plasma membrane receptor.

The binding of water-soluble hormones to a plasma membrane receptor initiates a series of biochemical events across the membrane called a **signal transduction pathway.** In this pathway, the hormone is the signaling molecule, or **first messenger.** Its binding to the receptor results in the formation of a different molecule within the cell called the **second messenger.** The second messenger then modifies some cellular activity. The mechanisms of initiating cellular changes are described here for the two most common signal transduction pathways involving adenylate cyclase activity and phospholipase C activity. We first describe the activation of the G protein complex, which is common to both of these pathways.

Hormone Binding and the Activation of G Protein

Both of the two common signal transduction pathways function through an internal plasma membrane protein complex called a **G protein** (first introduced in section 4.5b). This protein is named based upon its ability to bind guanine nucleotide (a nucleotide containing the base guanine; see section 2.7d). Guanine diphosphate (GDP) is bound when G protein is in the inactive state, and guanine triphosphate (GTP) is bound when G protein is in the activated state. The binding of hormone to a plasma membrane receptor causes a change in G protein from its inactive form to its activated form. The sequence of this occurrence is depicted and described in detail in **figure 17.7.**

Subsequent to its formation, activated G protein then generally activates (or inhibits) one of two plasma membrane enzymes associated with different intracellular enzymatic cascades: adenylate cyclase or phospholipase C. A cell may have either or both enzymatic cascades. Here we describe the activation of these enzymes.

Adenylate Cyclase Activity

Activated G protein moves along the inside of the plasma membrane, where it binds to the plasma membrane protein **adenylate** (a-den′i-lāt) **cyclase** (**figure 17.8a**). Activated adenylate cyclase increases the formation of the second messenger, cAMP (3′,5′-cyclic adenosine monophosphate) from ATP. The cAMP then activates a **protein kinase** (protein kinase A), a protein that phosphorylates (adds phosphate to) other molecules (see section 3.3g). Phosphorylation results in activation or inhibition of these molecules. Examples of hormones that function through the activation of adenylate cyclase include glucagon, antidiuretic hormone, epinephrine, thyroid-stimulating hormone, and follicle-stimulating hormone.

Phospholipase C Activity

The second possibility as a result of G protein activation occurs when it binds with a different plasma membrane protein called **phospholipase C** (fos′fō-lip′ās) (figure 17.8b). Activation of phospholipase C results in the splitting of **PIP₂** (phosphatidylinositol bisphosphate), a phospholipid molecule within the plasma membrane. The splitting of PIP₂ results in the formation of two secondary messenger molecules: **DAG (diacylglycerol;** dī′as-il′glis′ĕr-ōl) and **IP₃ (inositol;** in-ō′si-tōl, -tol) **triphosphate.**

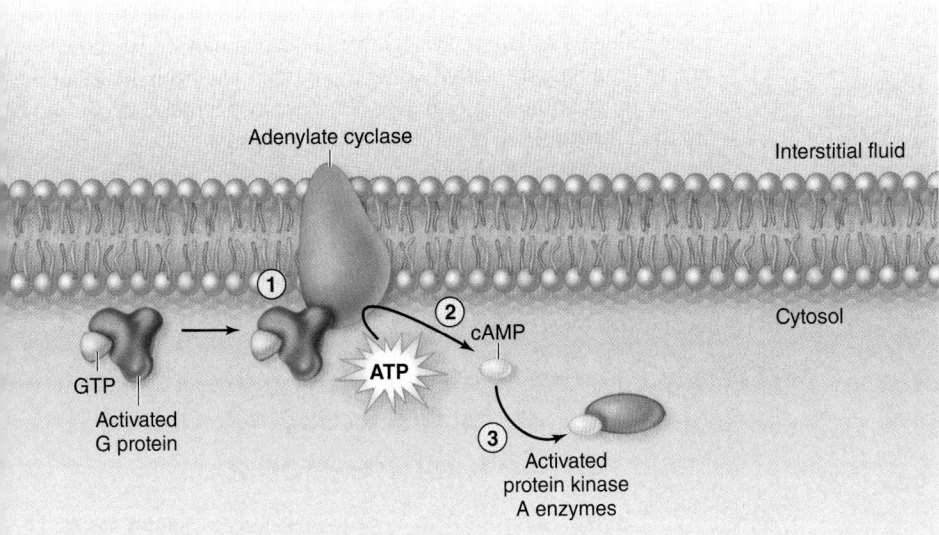

1 Activated G protein binds to and causes activation of the plasma membrane enzyme adenylate cyclase.

2 Adenylate cyclase converts ATP molecules to cAMP molecules.

3 cAMP serves as the "second messenger" by activating protein kinase A (a phosphorylating enzyme that adds phosphate to other molecules; these molecules may be activated or inhibited as a result).

a) **Activated G protein "turns on" adenylate cyclase.**

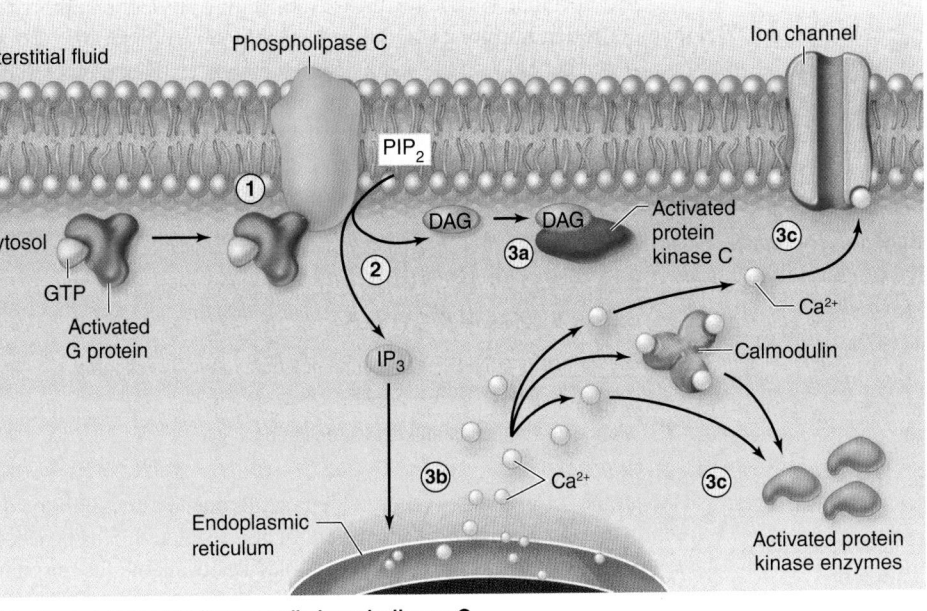

1 Activated G protein binds to and causes activation of the plasma membrane enzyme phospholipase C.

2 Phospholipase C splits PIP$_2$ into two second messengers: DAG (diacylglycerol) and IP$_3$ (inositol triphosphate).

3a DAG activates protein kinase C (a phosphorylating enzyme).

3b IP$_3$ increases Ca^{2+} in cytosol by stimulating Ca^{2+} release from the endoplasmic reticulum [ER] (and entry across the plasma membrane from the interstitial fluid, which is not shown in figure).

3c Ca^{2+} acts as a third messenger to activate protein kinase enzymes (Ca^{2+} does this directly or by first binding to calmodulin). Ca^{2+} may also alter activity of ion channels within the plasma membrane.

b) **Activated G protein "turns on" phospholipase C.**

Figure 17.8 Action of G Proteins. Following its activation, G protein is an intracellular molecule that moves along the inside of the plasma membrane and can stimulate other molecules. Two of the most common are (*a*) adenylate cyclase, which forms cyclic AMP second messenger, and (*b*) phospholipase C, which causes formation of DAG and IP$_3$ second messengers. Pathways involving G protein ultimately result in the activation of kinase enzymes that activate or inhibit other enzymes through phosphorylation, change cell permeability, or both. AP|R

Action of DAG DAG is a second messenger that remains in the plasma membrane. It is similar in action to cAMP in that it activates a protein kinase (here, protein kinase C). This enzyme in turn phosphorylates other molecules.

Action of IP$_3$ IP$_3$ is a second messenger that diffuses from the plasma membrane into the cytosol. It increases intracellular Ca^{2+} concentration by interacting with either the endoplasmic reticulum, causing the release of stored Ca^{2+}, or with Ca^{2+} channels in the plasma membrane (not shown in figure), permitting Ca^{2+} entry from the interstitial fluid. Increased intracellular Ca^{2+} acts as a third messenger within the cytosol to (1) activate protein kinase enzymes directly, or by first binding with an intracellular protein called calmodulin; or (2) alter plasma membrane permeability by binding to specific ion channels located within the plasma membrane and changing the flow of that specific ion either into or out of the cell. Examples of hormones that function through the

activation of phospholipase C include epinephrine, oxytocin, and antidiuretic hormone.

Action of Water-Soluble Hormones

Binding of water-soluble hormones activates G protein to alter transmembrane enzyme activity (adenylate cyclase or phospholipase C), which results in the formation of second and third messengers(e.g., cAMP, DAG, IP$_3$, Ca^{2+}). These messengers alter protein kinase activity, change a cell's permeability to ions, or both. Ultimately, the action may result in either activation or inhibition of enzymatic pathways, stimulation of growth through cellular division, release of cellular secretions, changes in membrane permeability, and muscle contraction or relaxation. The specific response in each cell type is dependent upon the hormone that binds to a cellular receptor, the types of second (and third) messengers that a given cell can produce, and the different enzymes within a cascade that are phosphorylated. For example:

- Glucagon is released from pancreatic cells in response to low blood glucose levels. Glucagon combines with receptors in plasma membranes of liver cells to cause an increase in cAMP synthesis and activation of kinase A enzymes. Kinase A enzymes phosphorylate specific enzyme pathways (e.g., glycogenolysis; see section 2.7c) that lead to the release of glucose from liver cells. Glucose enters the blood and helps to return blood glucose levels to within a normal homeostatic range.

- Oxytocin is released from the posterior pituitary during active labor for childbirth (see section 29.6c). It binds with membrane receptors of smooth muscle cells in the uterus to increase production of IP_3, thus increasing intracellular Ca^{2+} and causing stronger uterine muscle contractions that help expel the baby.

Intracellular Enzyme Cascade and Amplification of Response

Recall that the hormone molecule is not actually moved along a signaling pathway. The binding of the hormone results in specific "information" being passed in a signal conduction pathway called an *intracellular enzyme cascade*. The cascade includes G protein, transmembrane enzyme (adenylate cyclase or phospholipase C), second messenger, and protein kinase enzymes. The activated protein kinase enzymes may either stimulate or inhibit enzymatic pathways within the cell, alter cell permeability to an ion, or both. Signaling pathways have two advantages:

- The signaling pathway amplifies the signal at each enzymatic step to activate more molecules than were present in the previous step, which ultimately leads to a greater specific response. Consequently, the binding of a relatively small number of hormone molecules to their receptors in the plasma membrane of a target cell may result in the activation or inhibition of millions of molecules within that cell.

- Multistep signaling pathways provide more places and opportunities to fine-tune and regulate the pathway activities.

We have emphasized the amplification pathways in this discussion. However, metabolic cascade controls require precise regulation to be effective. Cells must have efficient mechanisms to quickly inactivate intermediates, including breakdown of second messenger molecules, and to terminate the activity of enzymes that had been activated during amplification (e.g., phosphodiesterase degrades cyclic AMP, limiting the response of those signaling pathways).

💡 **WHAT DID YOU LEARN?**

12 What is the specific role of the protein kinase enzymes in the signal transduction pathway initiated by water-soluble hormones?

17.6 Target Cells: Degree of Cellular Response

A cell's response to a hormone is not an isolated event. This is because a single target cell (1) displays differing numbers of receptors for the same hormone and can bind varying amounts of the same hormone simultaneously; and (2) may possess receptors for many different hormones, and thus it is able to respond to more than one type of hormone at the same time. Hepatocytes, for example, have changing numbers of receptors for both insulin and glucagon and respond to both of them. Consequently, the response of a given cell is dependent upon the net effect of both its displayed receptors and the amounts and kinds of hormone that it binds.

17.6a Number of Receptors

LEARNING OBJECTIVES

1. Describe the conditions that influence the number of receptors available for a specific hormone.

2. Define up-regulation and down-regulation.

The number of available receptors in a cell fluctuates, and both the direction and degree that it fluctuates are tightly regulated. This is necessary because the number of receptor molecules available for hormone binding directly influences the degree of cellular response. Cells may increase the number of receptors, thereby increasing cell sensitivity to a hormone, a process called **up-regulation.** In contrast, a cell may decrease its number of receptors and reduce the cell's sensitivity to a hormone through **down-regulation.**

Cells alter the number of receptors available in response to changes in hormone concentration within the blood. A cell may increase its number of receptors when there is a lower than normal hormone concentration level, or decrease its number of receptors in response to an elevated hormone level (**figure 17.9a**). The ability of a cell to change the number of its available receptors helps to maintain a normal level of cellular response, preventing either understimulation or overstimulation of the cell.

 WHAT DO YOU THINK?

2 What cellular response (up-regulation or down-regulation) would you predict occurs in response to high doses of a drug that binds to a specific hormone receptor? Will more or less of the drug be required for the same response over time? Explain.

Changes in receptor number also occur as a consequence of developmental maturity, the cell's state of activity, and the different stages of the cell cycle. For example, when a secretory cell becomes a mature cell, it may become less responsive to growth hormone (GH) because

INTEGRATE

LEARNING STRATEGY ✎

The interaction of water-soluble hormones and a target cell can be likened to a courier delivering a letter to a mansion.

⚠ The courier delivering the message plays the part of the **hormone.** He knocks on the mansion door (the **receptor**).

A butler (the **G protein**) answers the door and takes the message, but denies entry to the courier.

The message is passed along to various assistants before reaching the lady of the mansion (a series of events representing the **intracellular enzymatic cascade**). She then brings about a change in home activities based on the message in the letter.

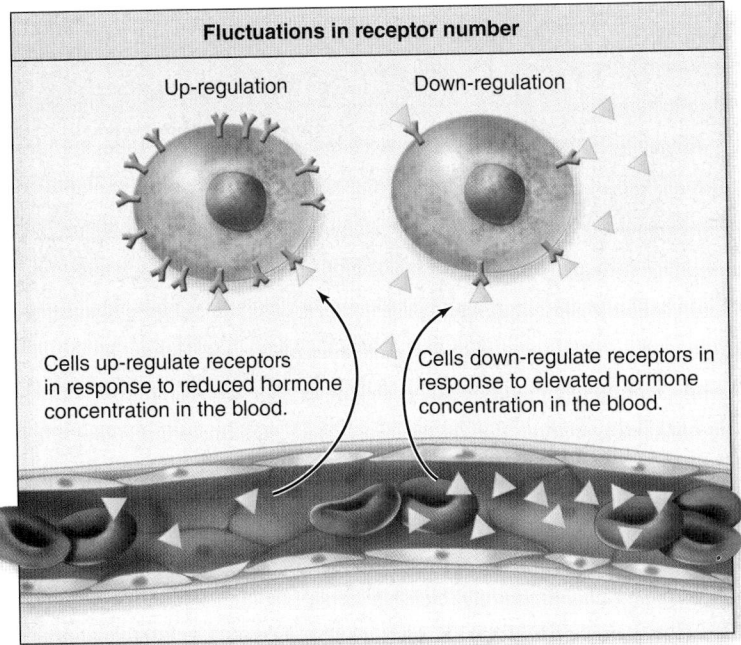

Fluctuations in receptor number

Up-regulation

Down-regulation

Cells up-regulate receptors in response to reduced hormone concentration in the blood.

Cells down-regulate receptors in response to elevated hormone concentration in the blood.

(a)

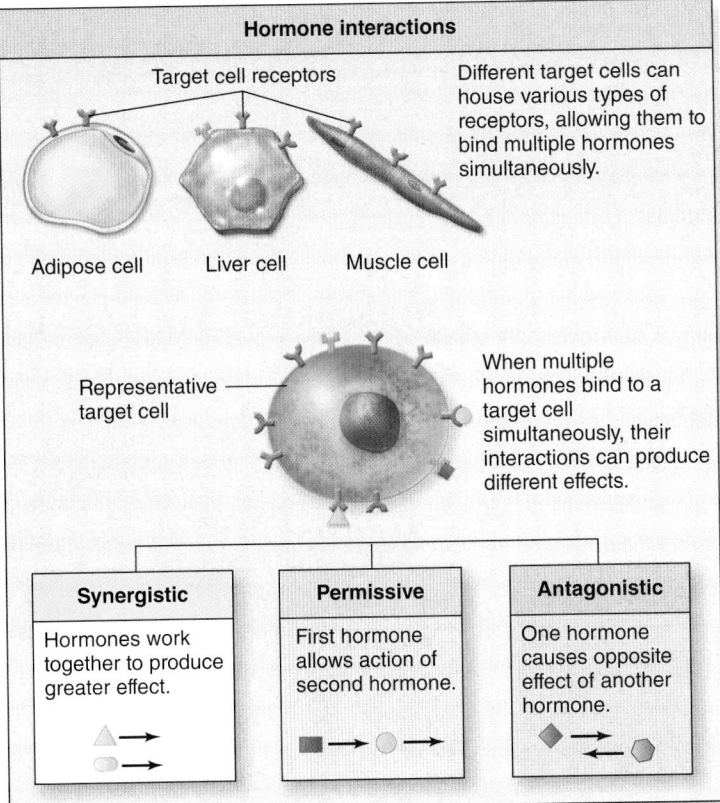

Hormone interactions

Target cell receptors

Different target cells can house various types of receptors, allowing them to bind multiple hormones simultaneously.

Adipose cell   Liver cell   Muscle cell

Representative target cell

When multiple hormones bind to a target cell simultaneously, their interactions can produce different effects.

Synergistic	Permissive	Antagonistic
Hormones work together to produce greater effect.	First hormone allows action of second hormone.	One hormone causes opposite effect of another hormone.

(b)

Figure 17.9 Cellular Response to Hormones.
A cell's response to hormones is dependent upon several factors, including the (a) number of cellular receptors available to bind a specific hormone and (b) possible interactions caused by more than one type of hormone binding to its different cellular receptors.

it no longer needs to continue growing at an accelerated rate. Thereafter, the mature secretory cell develops the ability to respond to hormones that stimulate it to secrete its product. This response occurs because the secretory cell now produces and displays receptors for different hormones.

WHAT DID YOU LEARN?

13. How does down-regulation of cellular receptors change responsiveness to a given hormone?

17.6b Receptor Interactions

LEARNING OBJECTIVE

3. Compare and contrast the three types of hormone interactions.

A target cell can simultaneously bind different hormones. The cellular response to binding more than one type of hormone is often such that some integration occurs within their signaling pathways. There are three principal ways in which hormones interact: synergistically, permissively, or antagonistically (figure 17.9b).

Synergistic interaction occurs when the activity of one hormone reinforces the activity of another hormone. Synergy means "working together." For example, female reproductive structures are more powerfully influenced by the presence of both estrogen and progesterone than by either hormone alone.

Permissive interaction takes place when the activity of one hormone requires a second hormone—as if one hormone "gives permission" for a different hormone to function. For example, prolactin is required to produce breast milk, and oxytocin is required for milk ejection from the breast. Oxytocin is unable to cause milk ejection without the previous, or accompanying, prolactin release to produce the milk.

Antagonistic interaction occurs when the effects of one hormone oppose the effects of another hormone. An example is the opposing effects on blood glucose levels exhibited by glucagon, which initiates cellular changes that increase blood glucose levels, and insulin, which initiates cellular changes that decrease blood glucose levels.

Figure 17.10 provides a visual overview of the general concepts of the endocrine system, as previously discussed.

WHAT DID YOU LEARN?

14. What effects are seen when hormones act synergistically?

INTEGRATE

CLINICAL VIEW
Hormone Analogs

Specificity of receptors allows the pharmaceutical industry to produce hormone analogs for use as medicines to inhibit or activate particular cellular functions. The drugs may activate a receptor and mimic the effect of the native hormone. One potential consequence of long-term or high-dose hormone analog use is that a patient's cells may down-regulate hormone receptors, an unintended effect. Consequently, following administration of the drug, the individual must be "weaned" from the medication (i.e., administered progressively smaller doses over several days). Weaning allows time for the body's cells to up-regulate the specific receptor to the normal level. You may be familiar with weaning an individual from steroids following treatment for inflammation (e.g., inflammation associated with asthma, allergic reactions, or rheumatoid arthritis).

Figure 17.10 **Endocrine System: Major Control System of the Body.** Hormones are produced by the glands of the endocrine system, released into the blood and transported to target organs. Lipid-soluble hormones enter target cells, whereas water-soluble hormones bind to receptors within the plasma membrane to induce cellular changes. A cell's response is determined both by its number of receptors and the different types of receptors that are available. Hormones are eliminated by the liver and kidney or taken uptake by target cells.

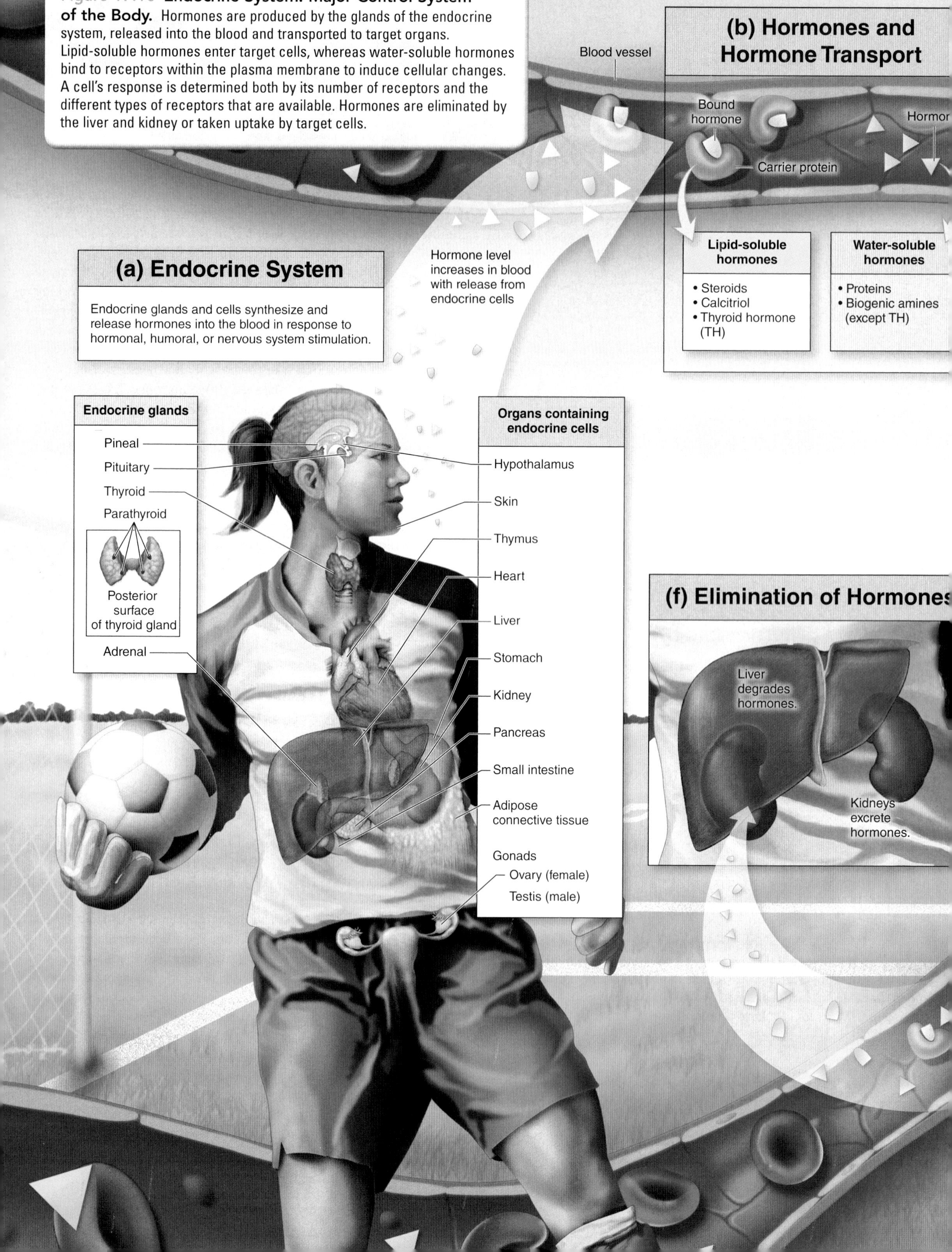

(b) Hormones and Hormone Transport

Blood vessel

Bound hormone

Hormor

Carrier protein

Lipid-soluble hormones	Water-soluble hormones
• Steroids • Calcitriol • Thyroid hormone (TH)	• Proteins • Biogenic amines (except TH)

(a) Endocrine System

Endocrine glands and cells synthesize and release hormones into the blood in response to hormonal, humoral, or nervous system stimulation.

Hormone level increases in blood with release from endocrine cells

Endocrine glands

Pineal
Pituitary
Thyroid
Parathyroid

Posterior surface of thyroid gland

Adrenal

Organs containing endocrine cells

Hypothalamus

Skin

Thymus

Heart

Liver

Stomach

Kidney

Pancreas

Small intestine

Adipose connective tissue

Gonads
Ovary (female)
Testis (male)

(f) Elimination of Hormones

Liver degrades hormones.

Kidneys excrete hormones.

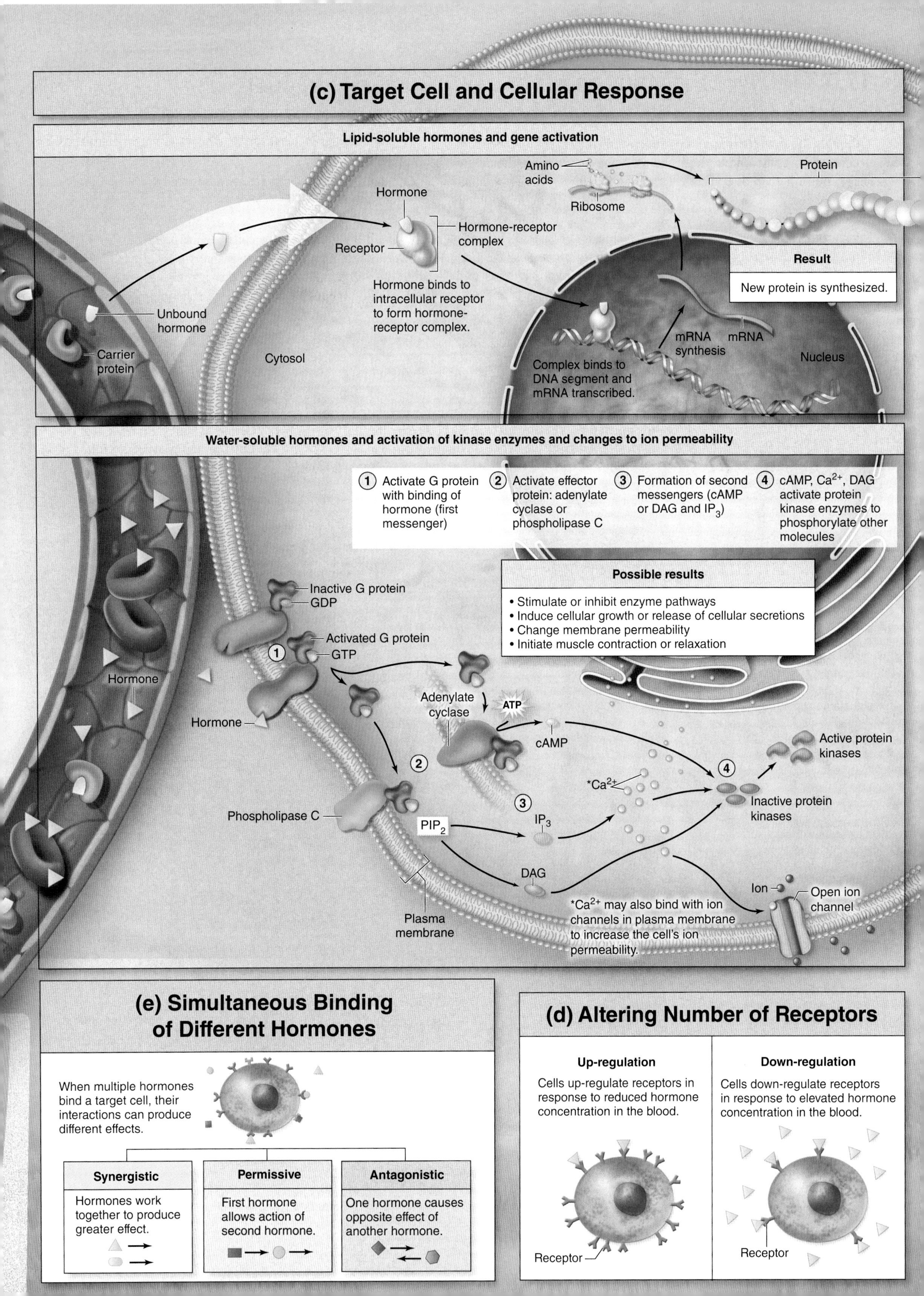

(c) Target Cell and Cellular Response

Lipid-soluble hormones and gene activation

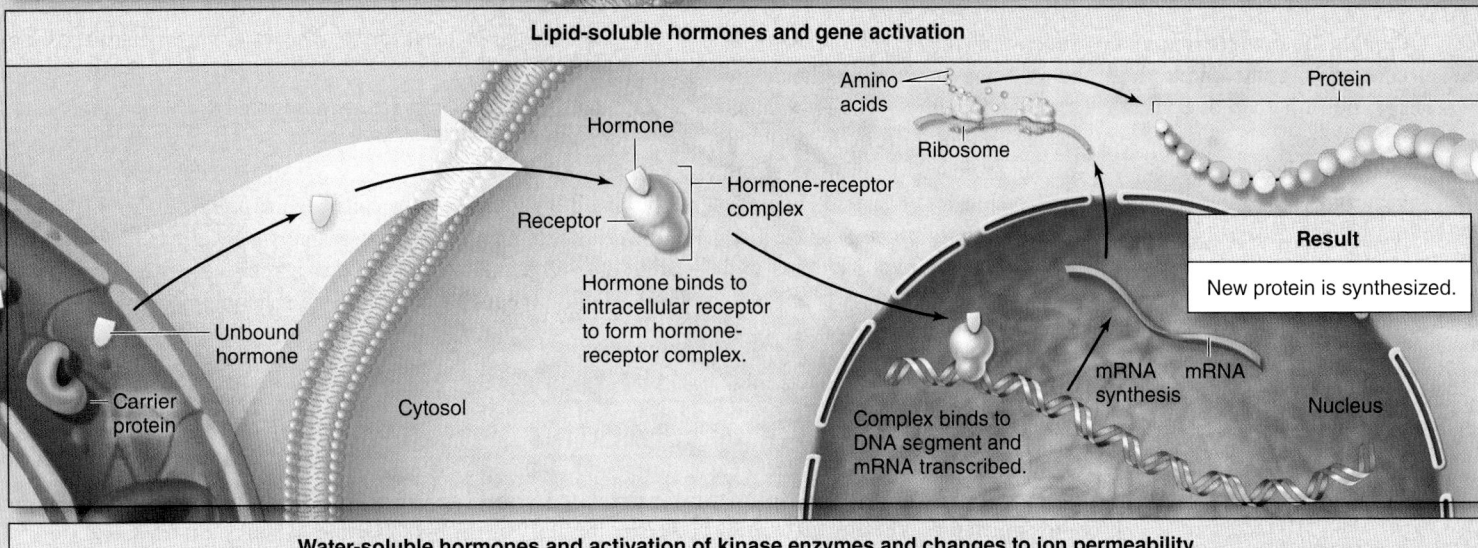

Hormone

Receptor

Hormone-receptor complex

Hormone binds to intracellular receptor to form hormone-receptor complex.

Unbound hormone

Carrier protein

Cytosol

Amino acids

Ribosome

Protein

Result

New protein is synthesized.

Complex binds to DNA segment and mRNA transcribed.

mRNA synthesis

mRNA

Nucleus

Water-soluble hormones and activation of kinase enzymes and changes to ion permeability

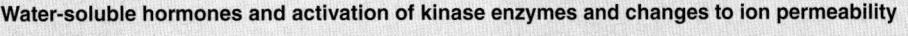

① Activate G protein with binding of hormone (first messenger)

② Activate effector protein: adenylate cyclase or phospholipase C

③ Formation of second messengers (cAMP or DAG and IP_3)

④ cAMP, Ca^{2+}, DAG activate protein kinase enzymes to phosphorylate other molecules

Possible results

- Stimulate or inhibit enzyme pathways
- Induce cellular growth or release of cellular secretions
- Change membrane permeability
- Initiate muscle contraction or relaxation

Inactive G protein
GDP

Activated G protein
GTP

①

Hormone

Hormone

Adenylate cyclase

ATP

cAMP

②

Phospholipase C

③

PIP_2

IP_3

DAG

$*Ca^{2+}$

④

Active protein kinases

Inactive protein kinases

Ion

Open ion channel

Plasma membrane

*Ca^{2+} may also bind with ion channels in plasma membrane to increase the cell's ion permeability.

(e) Simultaneous Binding of Different Hormones

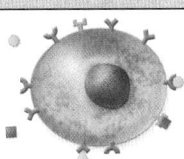

When multiple hormones bind a target cell, their interactions can produce different effects.

Synergistic	Permissive	Antagonistic
Hormones work together to produce greater effect.	First hormone allows action of second hormone.	One hormone causes opposite effect of another hormone.

(d) Altering Number of Receptors

Up-regulation

Cells up-regulate receptors in response to reduced hormone concentration in the blood.

Down-regulation

Cells down-regulate receptors in response to elevated hormone concentration in the blood.

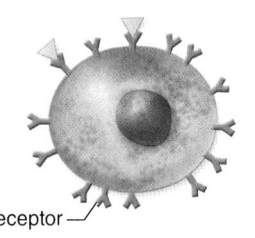

Receptor

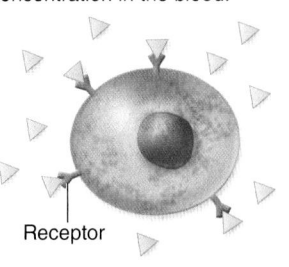

Receptor

17.7 The Hypothalamus and the Pituitary Gland

Although some endocrine glands function primarily as independent structures (e.g., the parathyroid glands), many endocrine glands are under the influence or control of the hypothalamus. The hypothalamus has direct control over the release of hormones from the pituitary gland, and it has indirect control over the release of hormones from the thyroid gland, adrenal gland, liver, testes, and ovaries. Thus, the hypothalamus controls a number of physiologic processes. We first describe the anatomic relationship between the hypothalamus and pituitary gland.

17.7a Anatomic Relationship of the Hypothalamus and the Pituitary Gland

LEARNING OBJECTIVES

1. Describe the anatomic relationship of the hypothalamus and the pituitary gland.

2. Identify the specific structures associated with the posterior pituitary and the anterior pituitary.

The **pituitary gland** is also called the *hypophysis* (hī-pof'i-sis; undergrowth). It lies inferior to the hypothalamus (**figure 17.11**). This small, slightly oval gland, which is approximately the size of a large pea, is housed within the sella turcica of the sphenoid bone (see section 8.2b).

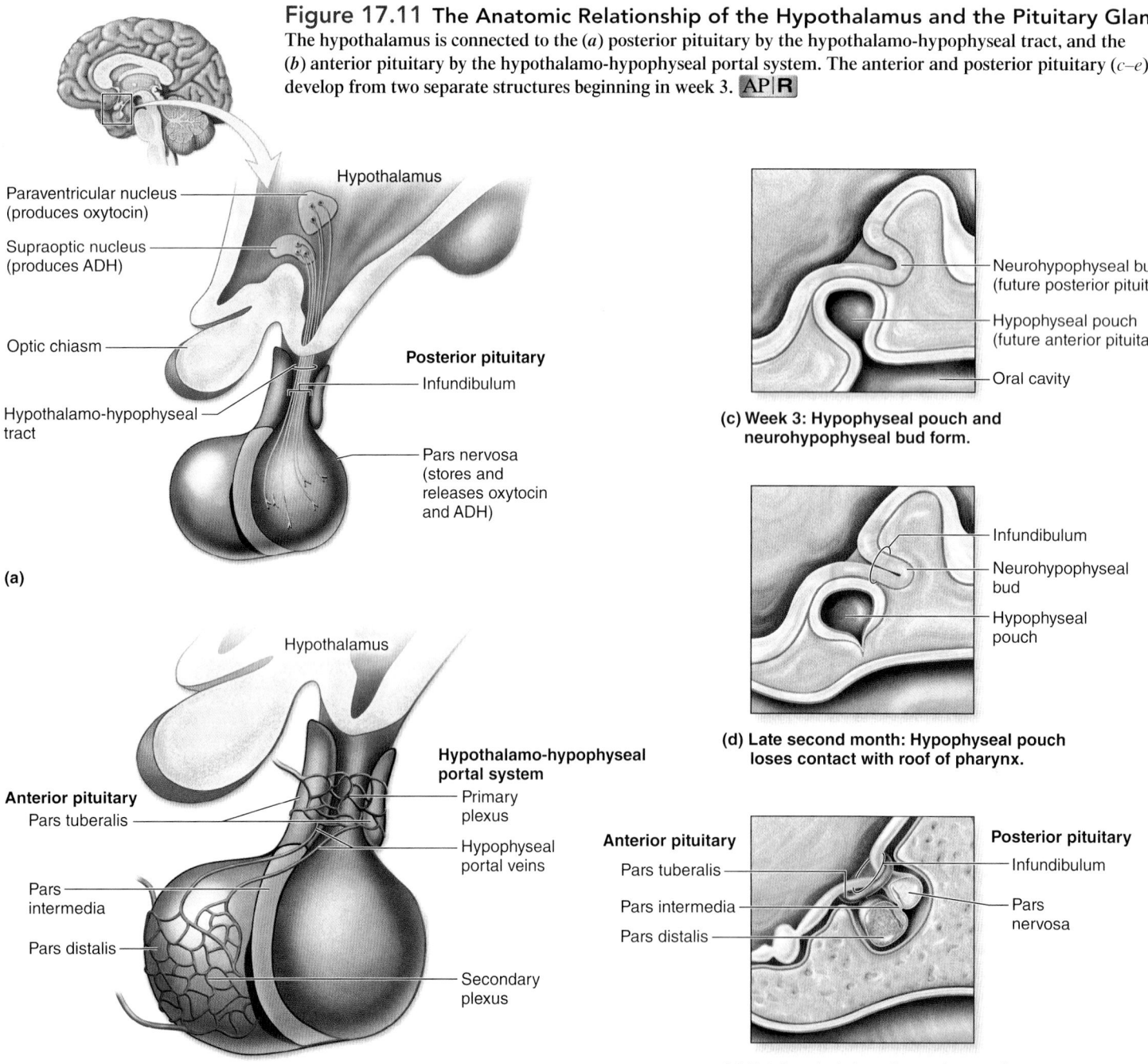

Figure 17.11 The Anatomic Relationship of the Hypothalamus and the Pituitary Gland. The hypothalamus is connected to the (*a*) posterior pituitary by the hypothalamo-hypophyseal tract, and the (*b*) anterior pituitary by the hypothalamo-hypophyseal portal system. The anterior and posterior pituitary (*c–e*) develop from two separate structures beginning in week 3. AP|R

Paraventricular nucleus (produces oxytocin)

Supraoptic nucleus (produces ADH)

Optic chiasm

Hypothalamo-hypophyseal tract

Hypothalamus

Posterior pituitary
Infundibulum

Pars nervosa (stores and releases oxytocin and ADH)

(a)

Neurohypophyseal bud (future posterior pituitary)

Hypophyseal pouch (future anterior pituitary)

Oral cavity

(c) Week 3: Hypophyseal pouch and neurohypophyseal bud form.

Infundibulum

Neurohypophyseal bud

Hypophyseal pouch

(d) Late second month: Hypophyseal pouch loses contact with roof of pharynx.

Hypothalamus

Anterior pituitary
Pars tuberalis

Pars intermedia

Pars distalis

Hypothalamo-hypophyseal portal system
Primary plexus

Hypophyseal portal veins

Secondary plexus

(b)

Anterior pituitary
Pars tuberalis

Pars intermedia

Pars distalis

Posterior pituitary
Infundibulum

Pars nervosa

(e) Fetal period: Anterior and posterior pituitary have formed.

The pituitary gland is connected to the hypothalamus by a very thin stalk, called either the **infundibulum** (in-fŭn′dib′ū-lŭm; a funnel), or *infundibular stalk*. The pituitary gland is partitioned both structurally and functionally into a posterior pituitary and an anterior pituitary and sometimes referred to as just the *posterior lobe* and *anterior lobe,* respectively. The characteristics of the posterior pituitary and anterior pituitary are summarized in table 17.3.

Posterior Pituitary

The **posterior pituitary** is also called *neurohypophysis* (nū′rō-hī-pof′i-sis). It is a neural part of the pituitary gland that develops beginning at about the third week of development as a bud that grows from the developing hypothalamus. The posterior pituitary is composed of a rounded lobe called the *pars nervosa* and the infundibulum; it makes up approximately one-quarter of the mass of the pituitary gland.

Axons from groups of neurons (approximately 10,000) extend from the hypothalamus to the pars nervosa of the posterior pituitary. The dendrites and cell bodies of these neurons are within the hypothalamus. Unmyelinated axons from these neurons extend through the infundibulum as the **hypothalamo-hypophyseal** (hī′pō-thal′ă-mō-hī-pō-fiz′ē-ăl) **tract.** The ends of the axons, including synaptic knobs, are within the pars nervosa. Two specific hypothalamic nuclei are recognized as being associated with the posterior pituitary: the **paraventricular** (par-ă′ven-trik′ū-lĕr) **nucleus** and the **supraoptic** (sū-pră-op′tik) **nucleus.**

Anterior Pituitary

Most of the pituitary gland (about three-quarters of the mass of the pituitary gland) is the **anterior pituitary,** or *adenohypophysis* (ad′ĕ-nō-hī-pof′i-sis; *adenos* = gland). It develops beginning at about the third week of development as an invagination of the ectoderm in the developing oral cavity to form the hypophyseal pouch (figure 17.11c–e). The anterior pituitary is partitioned into three distinct areas (figure 17.11b). The *pars distalis* is the large anterior rounded portion of the anterior pituitary. The *pars tuberalis* is a thin wrapping around the infundibulum.

A thin *pars intermedia* is a scant region between the pars distalis and the posterior pituitary.

The connection between the hypothalamus and the anterior pituitary involves two capillary plexuses interconnected by **portal veins** (a portal vessel is any vessel located between two capillary beds; see section 20.1e). The very porous capillary network associated with the hypothalamus is the **primary plexus,** or *primary capillary plexus.* The capillary network associated with the anterior pituitary is the **secondary plexus,** or *secondary capillary plexus.* Blood is drained from the primary plexus of the hypothalamus and transported to the secondary plexus of the anterior pituitary by the **hypophyseal portal veins.** This blood vessel network is called the **hypothalamo-hypophyseal portal system.** Thus, the hypothalamo-hypophyseal portal system provides a direct blood pathway between the hypothalamus and the anterior pituitary.

 WHAT DID YOU LEARN?

15 What is the anatomic connection between the hypothalamus and the posterior pituitary?

17.7b Interactions Between the Hypothalamus and the Posterior Pituitary Gland

 **LEARNING OBJECTIVE**

3. Identify the two hormones released from the posterior pituitary, and describe how the hypothalamus controls their release.

The posterior pituitary stores two hormones—oxytocin and antidiuretic hormone (figure 17.11a). Both hormones are synthesized in the hypothalamus: The paraventricular nucleus primarily produces oxytocin, and the supraoptic nucleus primarily forms ADH. For this reason, these neurons in the hypothalamus are called **neurosecretory** (nūr′ō-sē′krē-tōr-ē) **cells.** Following their synthesis in the hypothalamus, the hormones are packaged within secretory vesicles and transported by fast axonal transport (see section 12.2c) through the unmyelinated axons to their synaptic

Table 17.3	Comparison of Posterior and Anterior Pituitary	
Characteristic	**Posterior Pituitary**	**Anterior Pituitary**
Structure	¼ of pituitary gland; composed of nervous tissue	¾ of pituitary gland; composed of endocrine tissue
Mechanism of control	Nervous	Hormonal
Hormones synthesized by hypothalamus	Synthesized and transported to posterior pituitary Oxytocin (OT) Antidiuretic hormone (ADH)	Synthesized and released into hypothalamo-hypophyseal portal system Thyrotropin-releasing hormone (TRH) Prolactin-releasing hormone (PRH) Gonadotropin-releasing hormone (GnRH) Corticotropin-releasing hormone (CRH) Growth hormone–releasing hormone (GHRH) Prolactin-inhibiting hormone (PIH) Growth-inhibiting hormone (GIH)
Hormones released, or synthesized and released, by pituitary gland	*Released only* Oxytocin (OT) Antidiuretic hormone (ADH)	*Hormones synthesized and released* Thyroid-stimulating hormone (TSH) Prolactin (PRL) Follicle-stimulating hormone (FSH) Luteinizing hormone (LH) Adrenocorticotropic hormone (ACTH) Growth hormone (GH)

knobs within the posterior pituitary. Note that the posterior pituitary does not produce hormones; it is simply a storage site for these two hormones synthesized in the hypothalamus.

Hormone is released from the posterior pituitary when a nerve signal is sent from the hypothalamus along the hypothalamo-hypophyseal tract. Specifically, nerve signals from the paraventricular nucleus primarily stimulate release of oxytocin, and those from the supraoptic nucleus primarily cause release of ADH. These molecules are considered hormones, and not neurotransmitters, because they enter into the blood when released, even though they are released from synaptic knobs of neurons.

Oxytocin (OT) (ok-sē-tō′sin; *okytokos* = swift birth) stimulates contraction of smooth muscle of both the uterus during delivery and the breast to cause ejection of milk (see sections 29.6c and 29.8c). OT also interacts with the brain to increase feelings of emotional bonding between individuals.

Antidiuretic (an′tē-dī-yū-ret′-ik) **hormone (ADH)** (*anti* = against, *ouresis* = urination), also called *vasopressin*, stimulates both the kidneys to decrease urine output, the thirst center to increase fluid intake, and in high doses it causes vasoconstriction (see sections 24.6d and 25.4b). The functional details of both oxytocin and ADH are summarized in the reference tables following this chapter (see **table R.9** and **R.7**, respectively).

 WHAT DID YOU LEARN?

 16 How does the hypothalamus control the release of ADH from the posterior pituitary?

17.7c Interactions Between the Hypothalamus and the Anterior Pituitary Gland

 LEARNING OBJECTIVES

4. List the hormones released from the hypothalamus that control the anterior pituitary.

5. Explain how the hypothalamus controls the release of hormones from the anterior pituitary and the general function of each.

Hormonal stimulation triggers the release of hormones from the anterior pituitary. This occurs when **regulatory hormones** produced within the hypothalamus are released into the primary plexus and then transported via the hypophyseal portal veins to reach the secondary plexus within the anterior pituitary. The anterior pituitary then releases its hormones into the blood of the general circulation through which they reach target cells.

Hormones of the Hypothalamus

Regulatory hormones produced and released from the hypothalamus fall into one of two groups: **releasing hormones (RHs)** that stimulate the production and secretion of specific anterior pituitary hormones, and **inhibiting hormones (IHs)** that decrease the production and secretion of specific anterior pituitary hormones. The releasing hormones include: thyrotropin-releasing hormone (TRH), prolactin-releasing hormone (PRH), gonadotropin-releasing hormone (GnRH), corticotropin-releasing hormone (CRH), and growth hormone-releasing hormone (GHRH). The two inhibiting hormones are prolactin-inhibiting hormone (PIH) and growth-inhibiting hormone (GIH) (see table 17.3). The anterior pituitary also secretes melanocyte-stimulating hormone, but because it normally has little effect in humans and secretion ceases prior to adulthood, it will not be discussed further in this text.

Hormones of the Anterior Pituitary

The anterior pituitary secretes six major hormones: thyroid-stimulating hormone (TSH), prolactin (PRL), follicle-stimulating hormone (FSH), luteinizing hormone (LH), adrenocorticotropic hormone (ACTH), and

 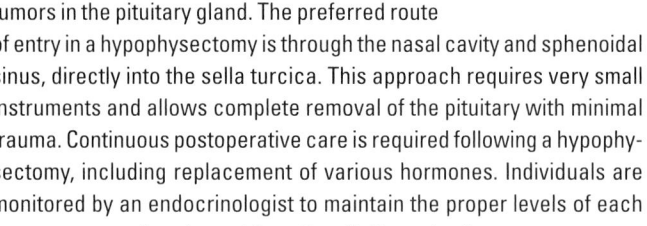

INTEGRATE

CLINICAL VIEW
Hypophysectomy

The surgical removal of the pituitary gland (the hypophysis) is called a **hypophysectomy** (hī′pof-i-sek′tō-mē), which is performed for tumors in the pituitary gland. The preferred route of entry in a hypophysectomy is through the nasal cavity and sphenoidal sinus, directly into the sella turcica. This approach requires very small instruments and allows complete removal of the pituitary with minimal trauma. Continuous postoperative care is required following a hypophysectomy, including replacement of various hormones. Individuals are monitored by an endocrinologist to maintain the proper levels of each hormone normally released from the pituitary gland.

growth hormone (GH). All of these hormones (except prolactin) are called **tropic** (trōp′ik) **hormones** because they stimulate other endocrine glands or cells to secrete other hormones.

Regulation of the Anterior Pituitary by the Hypothalamus

The hormones released from the hypothalamus into the hypothalamo-hypophyseal portal system, which control specific cells of the anterior pituitary to release their hormones into the general circulation, are shown in **figure 17.12**. Each hormone released from the hypothalamus is colorcoded with the specific hormone(s) that it triggers to release from the anterior pituitary:

-   **Thyrotropin-releasing hormone (TRH)** stimulates the anterior pituitary to release **thyroid-stimulating hormone (TSH)** (also called *thyrotropin*). TSH stimulates the thyroid gland to release **thyroid hormone (TH)**, which is the hormone that functions primarily to establish the body's metabolic rate. Thus, the general sequence of hormone release is TRH → TSH → TH (see section 17.8b).

-   **Prolactin-releasing hormone (PRH)** controls release of **prolactin** (prō-lak′tin; *lac* = milk) **(PRL)** from the anterior pituitary. Prolactin primarily regulates mammary gland growth and breast milk production in females (see section 29.8c). It also may increase secretion of testosterone in males. Prolactin release is inhibited by **prolactin-inhibiting hormone**.

-   **Gonadotropin-releasing hormone (GnRH)** regulates the release of both **follicle-stimulating hormone (FSH)** and **luteinizing** (lū′tēin-i-zing) **hormone (LH),** which are collectively called **gonadotropins.** These hormones act on the gonads in both females and males. In the female, FSH and LH act on the ovaries to control development of the oocyte and the follicle (the spherical structure that encloses the oocyte) and the release of estrogen and progesterone (see section 28.3b). In the male, FSH and LH act on the testes to regulate the development of sperm and the release of testosterone (see section 28.4b).

-   **Corticotropin-releasing hormone (CRH)** stimulates release of **adrenocorticotropic** (ă-drē′nō-kōr′ti-kō-trō′pik) **hormone (ACTH)** (also called *corticotropin*). ACTH stimulates the adrenal cortex to produce and secrete glucocorticoids (e.g., cortisol). Thus, the general sequence of hormone release is CRH → ACTH → glucocorticoids (see section 17.8c).

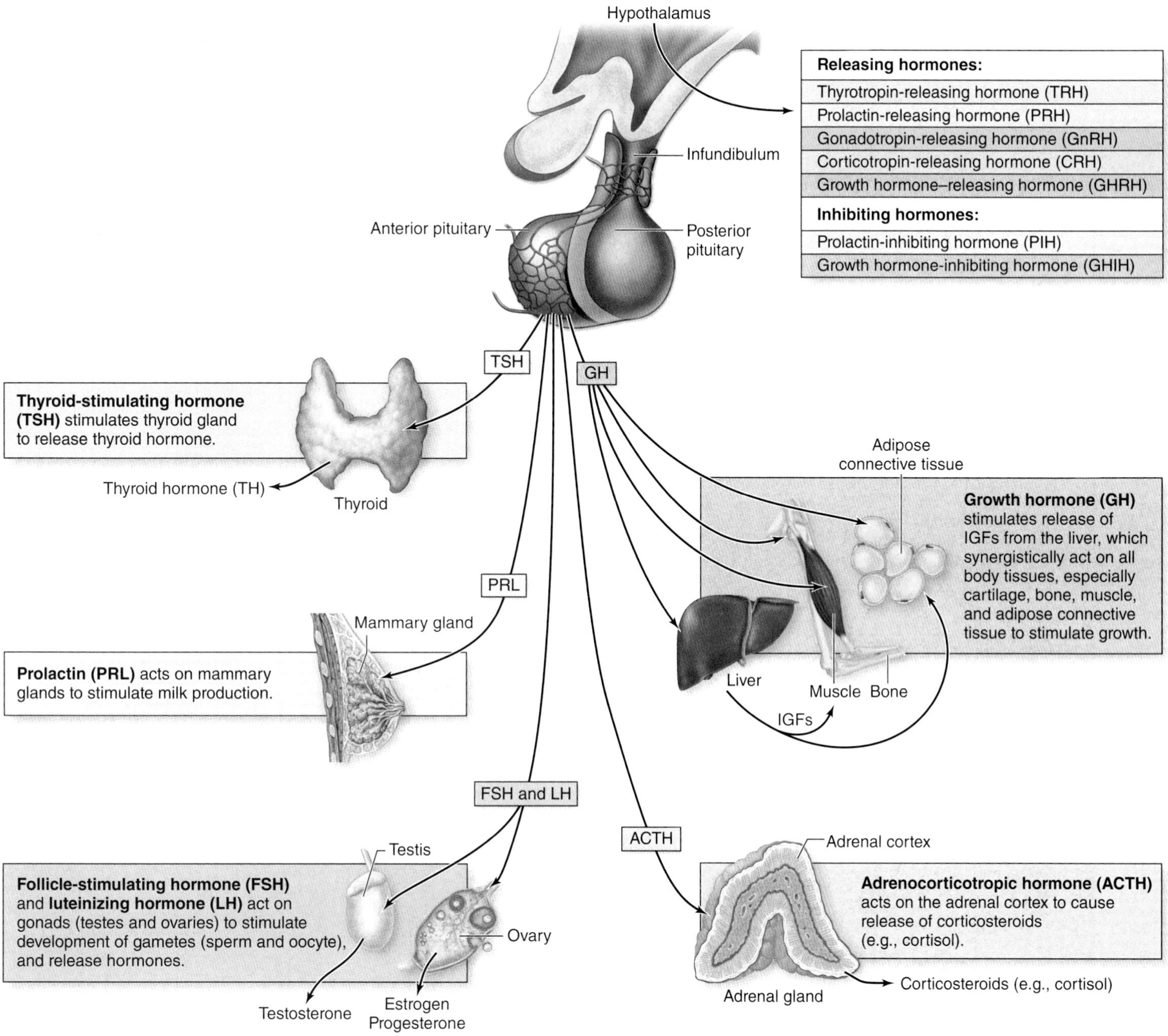

Hypothalamus

Infundibulum

Anterior pituitary

Posterior pituitary

Releasing hormones:

Thyrotropin-releasing hormone (TRH)
Prolactin-releasing hormone (PRH)
Gonadotropin-releasing hormone (GnRH)
Corticotropin-releasing hormone (CRH)
Growth hormone–releasing hormone (GHRH)

Inhibiting hormones:

Prolactin-inhibiting hormone (PIH)
Growth hormone-inhibiting hormone (GHIH)

TSH

GH

Thyroid-stimulating hormone (TSH) stimulates thyroid gland to release thyroid hormone.

Thyroid hormone (TH)

Thyroid

PRL

Mammary gland

Prolactin (PRL) acts on mammary glands to stimulate milk production.

Adipose connective tissue

Growth hormone (GH) stimulates release of IGFs from the liver, which synergistically act on all body tissues, especially cartilage, bone, muscle, and adipose connective tissue to stimulate growth.

Liver   Muscle Bone

IGFs

FSH and LH

Testis

Follicle-stimulating hormone (FSH) and **luteinizing hormone (LH)** act on gonads (testes and ovaries) to stimulate development of gametes (sperm and oocyte), and release hormones.

Ovary

Testosterone

Estrogen Progesterone

ACTH

Adrenal cortex

Adrenocorticotropic hormone (ACTH) acts on the adrenal cortex to cause release of corticosteroids (e.g., cortisol).

Adrenal gland

Corticosteroids (e.g., cortisol)

Figure 17.12 Anterior Pituitary Hormones. The hypothalamus releases regulatory hormones, to control the release of hormones from the anterior pituitary. Each regulatory hormone released from the hypothalamus is the same highlighted color as the specific hormone(s) from the anterior pituitary that it controls. AP|R

- **Growth hormone–releasing hormone (GHRH)** regulates the anterior pituitary to release **growth hormone (GH)** (or *somatotropin*). GH stimulates the liver to release both **insulin-like growth factor 1** and **2** (**IGF-1** and **IGF-2**) (also called *somatomedins*) (sō′mă-tō-mē′din). Both GH and IGFs function synergistically to stimulate cell growth and cell division, particularly within the skeletal and muscular systems. Thus, the general sequence of hormone release is GHRH → GH → IGFs (see section 17.8a). Release of GH is inhibited by **growth hormone–inhibiting hormone (GHIH).**

INTEGRATE

LEARNING STRATEGY

Remember the primary anterior pituitary hormones with the **TP-FLAG** mnemonic:

T = Thyroid-stimulating hormone

P = Prolactin

F = Follicle-stimulating hormone

L = Luteinizing hormone

A = Adrenocorticotropic hormone

G = Growth hormone

WHAT DID YOU LEARN?

17 What are the six primary hormones released from the anterior pituitary? How is the release of each of these hormones regulated by the hypothalamus?

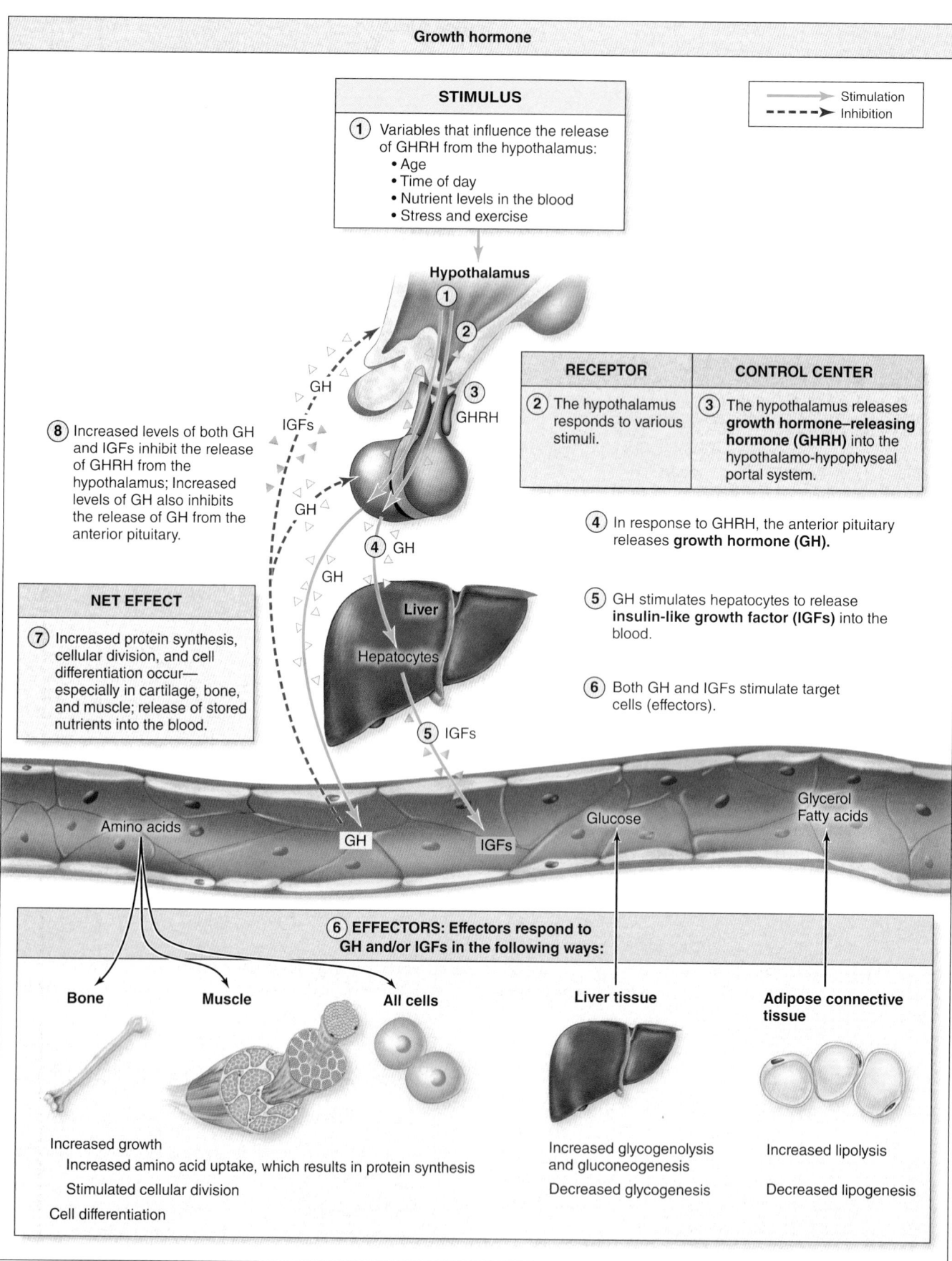

Growth hormone

Stimulation
- - - -> Inhibition

STIMULUS

① Variables that influence the release of GHRH from the hypothalamus:
- Age
- Time of day
- Nutrient levels in the blood
- Stress and exercise

Hypothalamus
① ②
GH
IGFs
③ GHRH

RECEPTOR	**CONTROL CENTER**
② The hypothalamus responds to various stimuli.	③ The hypothalamus releases **growth hormone–releasing hormone (GHRH)** into the hypothalamo-hypophyseal portal system.

⑧ Increased levels of both GH and IGFs inhibit the release of GHRH from the hypothalamus; Increased levels of GH also inhibits the release of GH from the anterior pituitary.

④ In response to GHRH, the anterior pituitary releases **growth hormone (GH).**

GH

④ GH

GH

Liver

NET EFFECT

⑦ Increased protein synthesis, cellular division, and cell differentiation occur— especially in cartilage, bone, and muscle; release of stored nutrients into the blood.

Hepatocytes

⑤ GH stimulates hepatocytes to release **insulin-like growth factor (IGFs)** into the blood.

⑥ Both GH and IGFs stimulate target cells (effectors).

⑤ IGFs

Glycerol
Fatty acids

Glucose

Amino acids

GH

IGFs

⑥ **EFFECTORS:** Effectors respond to GH and/or IGFs in the following ways:

Bone   **Muscle**   **All cells**   **Liver tissue**   **Adipose connective tissue**

Increased growth
Increased amino acid uptake, which results in protein synthesis
Stimulated cellular division
Cell differentiation

Increased glycogenolysis and gluconeogenesis
Decreased glycogenesis

Increased lipolysis
Decreased lipogenesis

Figure 17.13 Regulation and Action of Growth Hormone. The hypothalamus responds to particular stimuli by releasing growth hormone–releasing hormone (GHRH), which stimulates the anterior pituitary to release growth hormone (GH). GH stimulates the release of insulin-like growth factor (IGFs) from the liver. Together, GH and IGFs stimulate growth and alter the availability of nutrient molecules within the blood to provide additional energy for growth. (Direction of arrows between the blood and effectors indicates the net movement of the nutrients.)

17.8 Representative Hormones Regulated by the Hypothalamus

Here we examine the homeostatic system involving three of the six hormones regulated by the hypothalamus: growth hormone, thyroid hormone, and glucocorticoids (e.g., cortisol). All of these hormones have widespread effects on the body's metabolic processes. The other three—FSH, LH, and prolactin, which participate in regulation of the reproductive system—are described in detail in chapters 28 and 29.

17.8a Growth Hormone

LEARNING OBJECTIVE

1. Describe the homeostatic system involving growth hormone.

Growth hormone (GH) has numerous functions, including the stimulation of linear growth at the epiphyseal plate by causing an increase in cellular division and extracellular matrix formation within cartilage and bone, hypertrophy of muscle, and release of nutrients from storage into the blood to provide energy for growth.

The release of GH (**figure 17.13**) from the anterior pituitary is controlled by the release of growth hormone–releasing hormone (GHRH) from the hypothalamus. GHRH is produced and released from the hypothalamus in response to a number of factors, including a person's age, the time of day, nutrient levels in the blood, stress, and exercise (**figure 17.14**). GHRH enters the hypothalamo-hypophyseal portal system and is transported to the anterior pituitary (figure 17.13). GHRH binds to receptors in specific cells of the anterior pituitary (somatotropic cells) and stimulates the release of GH into the general circulation to be transported throughout the body.

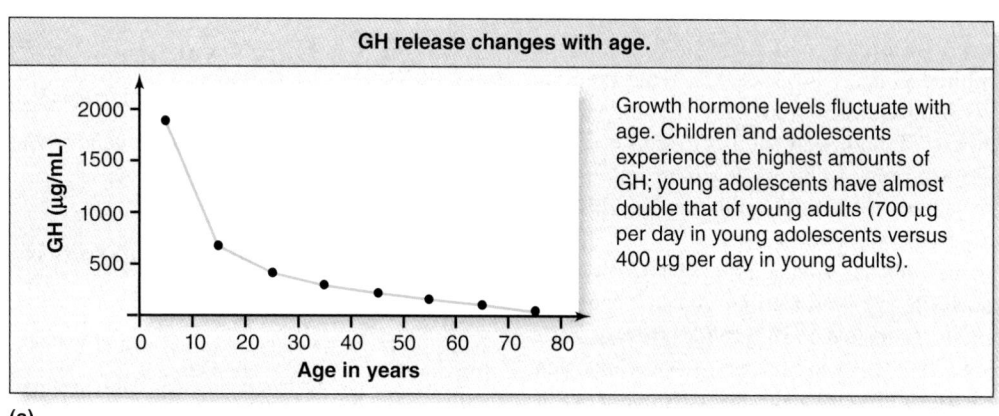

GH release changes with age.

Growth hormone levels fluctuate with age. Children and adolescents experience the highest amounts of GH; young adolescents have almost double that of young adults (700 μg per day in young adolescents versus 400 μg per day in young adults).

(a)

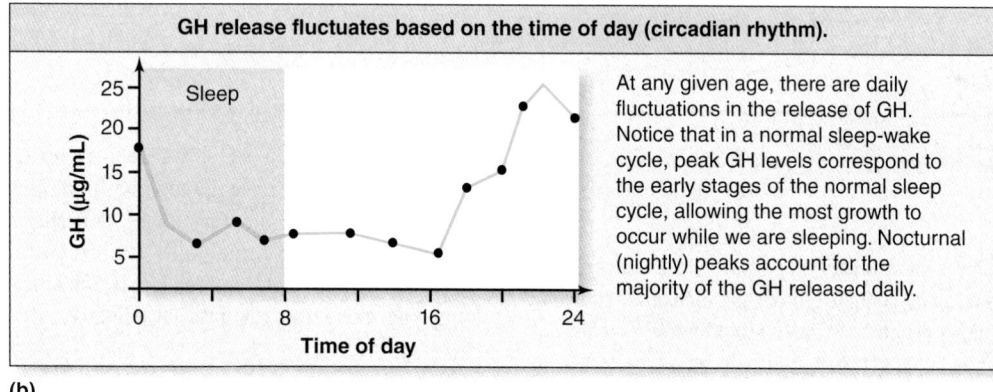

GH release fluctuates based on the time of day (circadian rhythm).

At any given age, there are daily fluctuations in the release of GH. Notice that in a normal sleep-wake cycle, peak GH levels correspond to the early stages of the normal sleep cycle, allowing the most growth to occur while we are sleeping. Nocturnal (nightly) peaks account for the majority of the GH released daily.

(b)

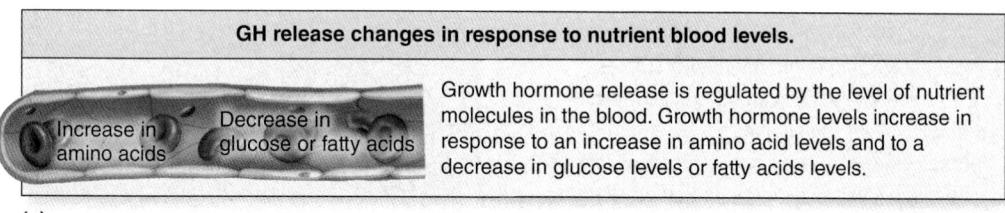

GH release changes in response to nutrient blood levels.

Growth hormone release is regulated by the level of nutrient molecules in the blood. Growth hormone levels increase in response to an increase in amino acid levels and to a decrease in glucose levels or fatty acids levels.

Increase in amino acids   Decrease in glucose or fatty acids

(c)

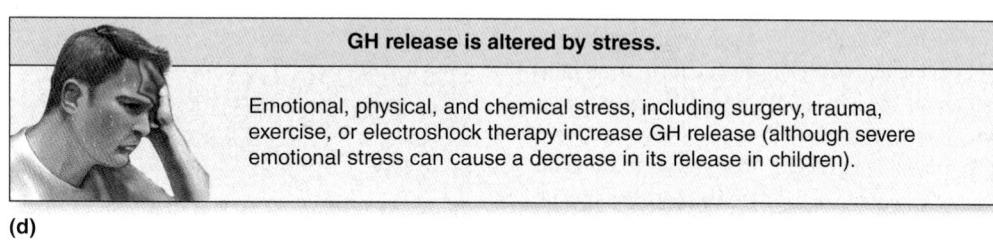

GH release is altered by stress.

Emotional, physical, and chemical stress, including surgery, trauma, exercise, or electroshock therapy increase GH release (although severe emotional stress can cause a decrease in its release in children).

(d)

Figure 17.14 Variables That Influence Growth Hormone Levels. Release of growth hormone is influenced by: (*a*) an individual's age, (*b*) time of day, (*c*) nutrient blood levels, and (*d*) stress.

One of the primary targets of GH is hepatocytes. Hepatocytes release the insulin-like growth factors (IGFs), into the general circulation when stimulated by GH. GH and IGFs have overlapping synergistic interactions (see section 17.6b), but IGFs are responsible for the greater response from the target cells. This results from the difference in their individual half-lives: GH (a protein hormone) has a half-life of 6 to 20 minutes, and IGFs (also protein hormones) have a half-life of approximately 20 hours because it is transported in the blood by carrier proteins that help protect IGFs from destruction (see section 17.4).

All cells of the body have receptors for GH, IGFs, or both. Binding of the hormones activates second messengers within the target cells, altering enzymatic pathways to increase protein synthesis, cellular division, cell differentiation, or a combination of these. Bone and muscle tissue in particular are affected by these hormones. In muscle fibers, uptake of amino acids increases and use of glucose decreases.

Hepatocytes are also stimulated by GH to increase both glycogenolysis (breakdown of glycogen into glucose molecules) and gluconeogenesis (formation of glucose from noncarbohydrate sources); at the same time, the glycogenesis (formation of glycogen from glucose molecules) pathway is inhibited. Glucose is released from the liver into the blood, thus blood glucose levels rise as a result. This increase in blood glucose is referred to as a **diabetogenic** (dī′ă-bet′ō-jen-ik) effect due to the similarity of elevated blood glucose levels in individuals with diabetes mellitus (see Clinical View: "Conditions Resulting in Abnormal Blood Glucose Levels" in section 17.9b).

Adipose connective tissue cells are stimulated by GH and IGFs to increase lipolysis (breakdown of triglycerides into glycerol and fatty acids) and decrease lipogenesis (formation of triglycerides from glycerol and fatty acids). Glycerol and fatty acids are released from adipocytes, causing an increase in the level of glycerol and fatty acids in the blood. Growth is an energy-requiring process. The additional glucose released from the liver, and the glycerol and fatty acids released from adipose connective tissue, together provide nutrient molecules required in cellular respiration for generating ATP (see section 3.4). Additional responses to the release of GH are listed in the summary table in the reference section, which directly follows this chapter (see **table R.3**).

The release of both GHRH from the hypothalamus and GH from the anterior pituitary are regulated by negative feedback in the following ways: In response to either increased levels of GH or increased levels of IGFs, the hypothalamus is stimulated to release **growth hormone–inhibiting hormone (GHIH),** which inhibits the release of GH from the anterior pituitary. GH also directly inhibits its own release from the anterior pituitary.

WHAT DO YOU THINK?

3 You may have heard someone describe a teenager going through puberty as having "thinned out." Which cellular process stimulated by GH directly accounts for the loss of fat tissue: glycogenolysis, lipolysis, or protein anabolism?

WHAT DID YOU LEARN?

18 How do GHRH, GH, and IGFs function together to regulate growth?

19 What are the primary target organs/tissues of GH and IGFs? Describe the effect on each.

17.8b Thyroid Gland and Thyroid Hormone

LEARNING OBJECTIVES

2. Describe thyroid gland location and anatomy.

3. Discuss how thyroid hormones are produced, stored, and secreted.

4. Explain the control of thyroid hormone by the hypothalamus and pituitary.

INTEGRATE

CLINICAL VIEW
Disorders of Growth Hormone Secretion

Growth hormone deficiency, also known as *pituitary dwarfism,* is a condition that exists at birth as a result of inadequate growth hormone production due to a hypothalamic or pituitary problem. Growth retardation is typically not evident until a child reaches 1 year of age, because the influence of growth hormone (GH) is minimal during the first 6 to 12 months of life. In addition to short stature, children with pituitary dwarfism often have periodic low blood sugar (hypoglycemia). Injections of growth hormone over a period of many years can bring about improvement, but not a normal state.

Oversecretion of growth hormone in childhood causes **pituitary gigantism.** Beyond extraordinary height (sometimes up to 8 feet), these people have enormous internal organs, a large and protruding tongue, and significant problems with blood glucose management. If untreated, a pituitary giant dies at a comparatively early age, often from complications of diabetes or heart failure.

Excessive growth hormone production in an adult results in **acromegaly** (ak′rō-meg′ă-lē) instead of gigantism because the epiphyseal plate is closed in an adult. The individual does not grow in height, but the bones of the face, hands, and feet enlarge and widen (appositional growth), along with growth in cartilage. An increase in mandible size leads to a protruding jaw (prognathism). Internal organs, especially the liver, increase in size, and increased release of glucose lead to the development of diabetes in all individuals with acromegaly. Acromegaly may result from loss of feedback control of growth hormone at either the hypothalamic or pituitary level, or it may develop because of a GH-secreting tumor of the pituitary. Current treatment includes using a growth-inhibiting hormone analog, which acts to inhibit the release of growth hormone from the anterior pituitary.

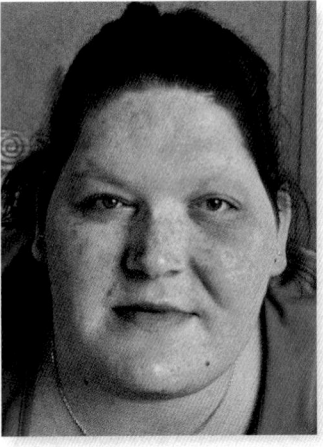

Acromegaly photos of a female in her 20s (left) and in her 40s illustrate the morphological changes associated with acromegaly.

The largest structure in the body entirely devoted to endocrine activities is the **thyroid** (thī′royd; *thyreos* = an oblong shield) **gland (figure 17.15).** Thyroid hormone produced by the thyroid gland increases the metabolic rate and body temperature. Release of thyroid hormone is under the control of the hypothalamus via the anterior pituitary.

Anatomy of the Thyroid Gland

The thyroid gland is a butterfly-shaped gland that is located immediately inferior to the thyroid cartilage of the larynx and anterior to the trachea. This gland is composed of **left** and **right lobes,** which are connected at the anterior midline by a narrow **isthmus** (is′mŭs). Both lobes of the thyroid gland are highly vascularized, giving the gland an intense reddish coloration. The entire gland is enclosed within a connective tissue capsule.

The thyroid gland at the histologic level is composed primarily of numerous microscopic, spherical structures called **thyroid follicles** (figure 17.15*b*). The wall of each follicle is formed by simple cuboidal epithelial cells, called **follicular cells,** which surround a central lumen. The lumen of each thyroid follicle houses a viscous, protein-rich fluid termed **colloid** (kol′oyd). (Think of thyroid follicles as microscopic gelatinous-filled balls.)

The follicular cells produce and later release **thyroid hormone (TH)** by first synthesizing a glycoprotein called **thyroglobulin (TGB)** and secreting it by exocytosis into the colloid-filled lumen. In brief, iodine molecules must be combined with the thyroglobulin in the colloid to produce hormone precursors, which are TGB molecules that contain immature thyroid hormone within their structure. The precursors are stored in the colloid until the secretion of thyroid hormone is needed.

When the thyroid gland is stimulated to secrete thyroid hormone, some of the colloid with thyroid hormone precursors is internalized by endocytosis into a follicular cell. It is transported to a lysosome, where an enzyme releases the immature thyroid hormone molecules from the precursor in preparation for its secretion from the follicular cells. The details of thyroid hormone synthesis and release are covered in **figure 17.16. Parafollicular** (par′ă′fo′lik′yū′lăr; *para* = alongside, near) **cells,** which are the other less numerous endocrine cells of the thyroid, are located around the follicular cells. These cells synthesize and release **calcitonin**—a hormone that functions to decrease blood calcium levels (see section 7.6c).

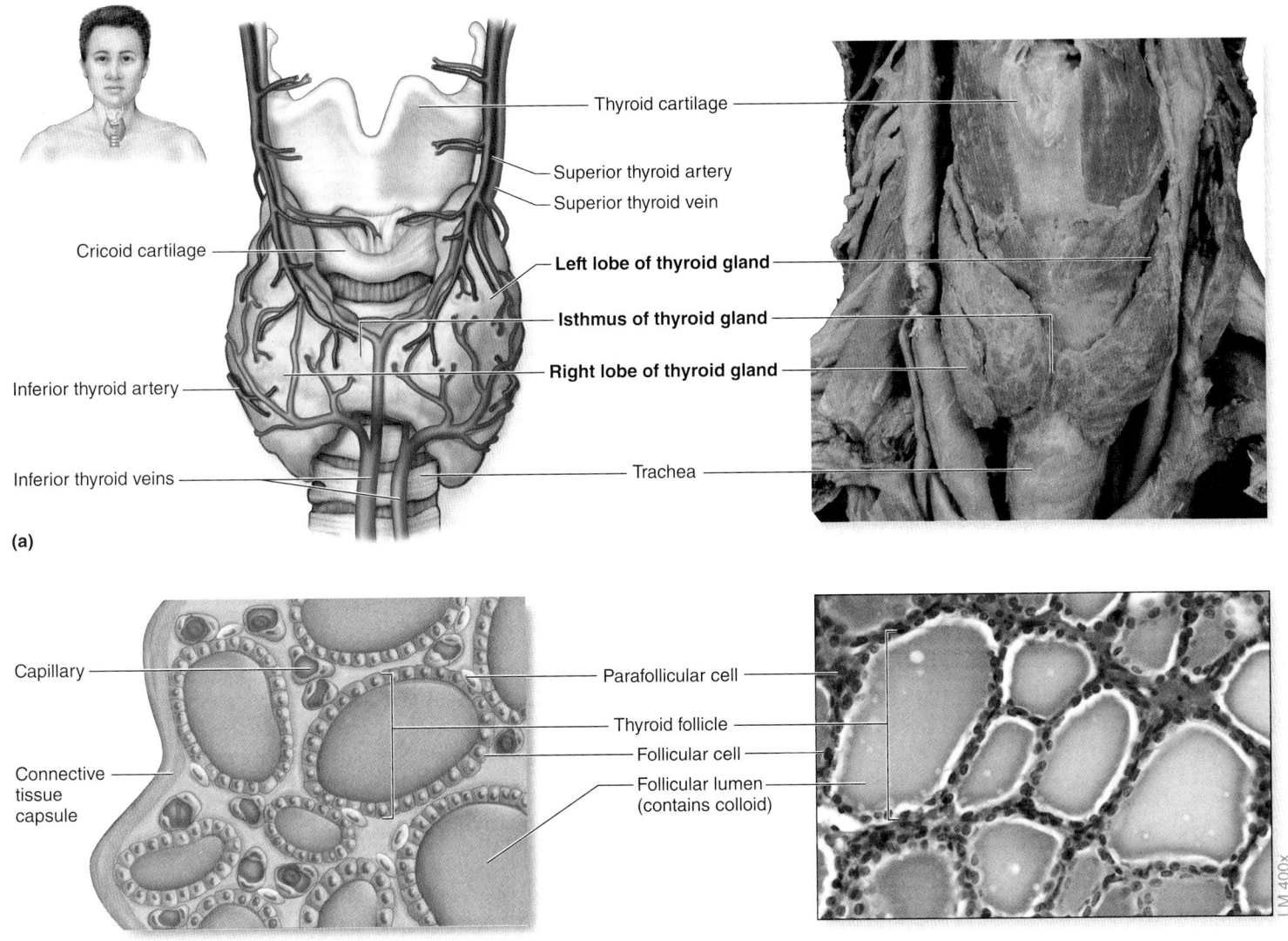

(a)

Thyroid cartilage
Superior thyroid artery
Superior thyroid vein
Left lobe of thyroid gland
Isthmus of thyroid gland
Right lobe of thyroid gland
Trachea
Cricoid cartilage
Inferior thyroid artery
Inferior thyroid veins

(b)

Capillary
Connective tissue capsule
Parafollicular cell
Thyroid follicle
Follicular cell
Follicular lumen (contains colloid)

LM 400x

Figure 17.15 The Thyroid Gland. (*a*) A drawing and cadaver photo illustrate that the thyroid gland is located anterior to the thyroid cartilage of the larynx (voicebox) and has a rich blood supply. (*b*) Microscopic anatomy of thyroid follicles, illustrating the simple cuboidal epithelium of the follicular cells, colloid within the follicle lumen, and the relationship of parafollicular cells to a follicle. Thyroid follicles produce thyroid hormone, and parafollicular cells produce the hormone calcitonin. AP|R

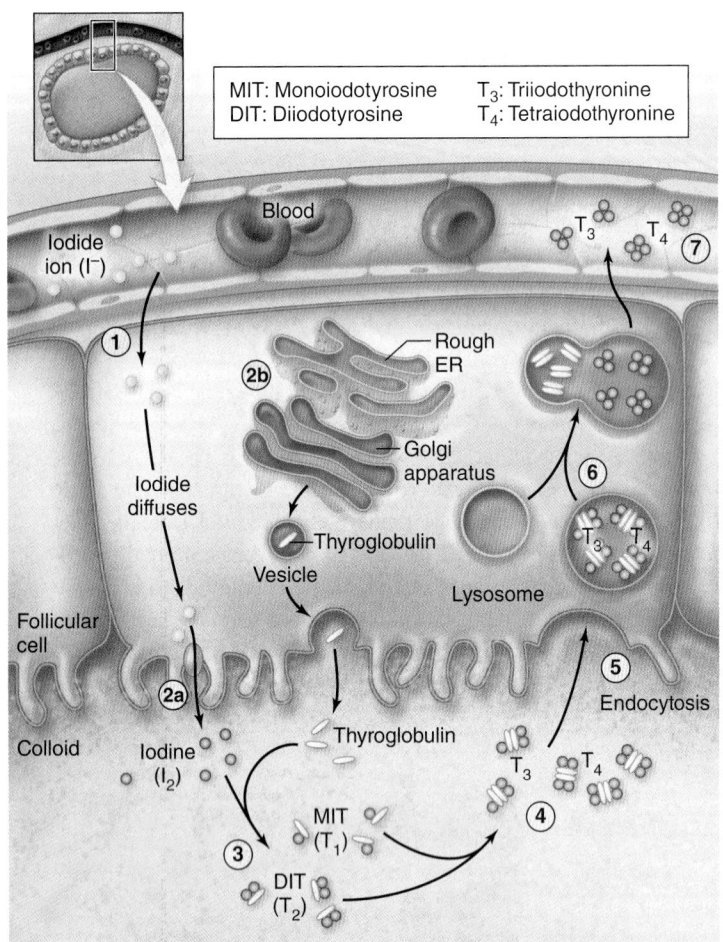

| MIT: Monoiodotyrosine | T_3: Triiodothyronine |
| DIT: Diiodotyrosine | T_4: Tetraiodothyronine |

Follicular cell

(1) Iodide ion uptake. Iodide ion (I^-) is moved by active transport from the blood into a follicular cell (because iodide concentration is higher within the cell than within the blood), diffuses through the cell, and is transported by facilitated diffusion into the colloid, which fills the follicle lumen.

(2a) Iodine molecule formation. Two I^- are joined to form molecular iodine (I_2) at the plasma membrane of the follicular cell.

(2b) Thyroglobulin synthesis. Synthesis of thyroglobulin includes production of the protein within the rough ER, and shipping it to the Golgi apparatus for addition of carbohydrate. Thyroglobulin (a glycoprotein) is incorporated into a vesicle and released from the follicular cell into the colloid by exocytosis.

Colloid

(3) MIT and DIT formation. I_2 are attached specifically to tyrosine amino acids of thyroglobulin by peroxidase enzymes, which facilitate removal of electrons; one I_2 is added to form MIT or T_1 (monoiodotyrosine) or two I_2 are added to form DIT or T_2 (diiodotyrosine).

(4) T_3 and T_4 formation. Enzymes within the colloid facilitate the joining of MIT and DIT. One MIT and one DIT are joined to form T_3, whereas two MITs are joined to form T_4. Both T_3 and T_4 remain attached to thyroglobulin.

(5) Endocytosis. Thyroglobulin with attached T_3 and T_4 is endocytosed into a follicular cell.

Follicular cell

(6) Release of T_3 and T_4 from thyroglobulin. A vesicle containing thyroglobulin with attached T_3 and T_4 fuses with a lysosome. Lysosomal enzymes cleave T_3 and T_4 from thyroglobulin.

(7) Release of T_3 and T_4 into the blood. T_3 and T_4, which are lipid-soluble molecules, move from the follicular cell into the blood by simple diffusion. More T_4 is released than T_3. T_3 and T_4 are collectively called **thyroxine.**

Figure 17.16 Thyroid Hormone: Synthesis, Storage, and Release. Molecular iodine (I_2) is attached to tyrosine amino acids of thyroglobulin within the colloid of thyroid follicles. Thyroglobulin with attached T_3 and T_4 is later transported into the follicular cells. Within lysosomes of follicular cells, T_3 and T_4 are cleaved from thyroglobulin and released into the blood.

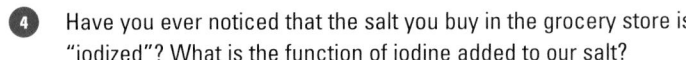 **WHAT DO YOU THINK?**

(4) Have you ever noticed that the salt you buy in the grocery store is "iodized"? What is the function of iodine added to our salt?

Action of Thyroid Hormone

Thyroid hormone is released from the thyroid gland as a result of the integrated activities of the hypothalamus and anterior pituitary. This physiologic relationship is referred to as the **hypothalamic-pituitary-thyroid axis (figure 17.17).**

The hypothalamus releases thyrotropin-releasing hormone (TRH), which enters into the hypothalamo-hypophyseal portal system in response to a decrease in blood levels of thyroid hormone. Other stimuli that may cause increased release of TRH include cold weather, pregnancy, high altitude, hypoglycemia—and in children, decreased body temperature.

TRH binds to receptors in cells of the anterior pituitary (thyrotropic cells) and stimulates the anterior pituitary to release thyroid-stimulating hormone (TSH) into the general circulation. TSH binds to receptors of the follicular cells of the thyroid gland and stimulates the release of thyroid hormone (TH). TH has two forms—triiodothyronine (T_3) and tetraiodothyronine (T_4) (thyroxine)—which are released into the circulation.

T_3 and T_4 are transported within the blood bound by carrier molecules (e.g., albumin, thyroxine-binding globulin, also called TBG). At any given time, a small percentage of T_3 and T_4 becomes unbound from the carrier proteins and can then exit from the blood.

A cellular transport system (carrier-mediated endocytosis) moves TH into target cells, where it binds to intracellular receptors. T_3 is the most active form of TH. Most cells, however, have an enzyme to remove one iodine to convert T_4 to T_3. This increases a cell's response to thyroid hormone because a much greater amount of T_4 (~90%) than T_3 (~10%) is produced and released from the thyroid gland.

TH is a lipid-soluble hormone that ultimately increases protein synthesis in all cells, especially neurons. TH specifically stimulates the synthesis of sodium-potassium (Na^+/K^+) pumps in neurons and the action of these additional ion pumps generates heat. The rise in temperature is referred to as the **calorigenic** (kă-lōr′i-jen′ik; *calor* = heat, *genesis* = production) **effect.** Increased amino acid uptake by cells provides the structural building blocks for these processes. TH also stimulates all cells to increase their glucose uptake. Concomitant with this increase is buildup in the number of cellular respiration enzymes within mitochondria.

TH stimulates other target cells to help meet the additional requirements for ATP due to the elevated metabolic rate. Hepatocytes are stimulated to increase both glycogenolysis and gluconeogenesis (see section 2.7c), while glycogenesis is inhibited, thus additional glucose is released into the blood. Adipose connective tissue cells respond to TH by both stimulating lipolysis and inhibiting lipogenesis (see section 2.7b). As a result, both glycerol and fatty acids are released into blood as alternative nutrient molecules for cellular respiration. This saves blood glucose for the brain and is called the **glucose-sparing effect.**

Respiration rate increases in response to TH to meet the additional demands for oxygen. Additionally, both heart rate and force of heart contraction increase blood flow to the tissues to deliver more nutrients and oxygen. This is a result of the cardiac muscle cells increasing their number of cellular receptors for epinephrine and norepinephrine, ensuring that the heart continues to be more responsive to these hormones

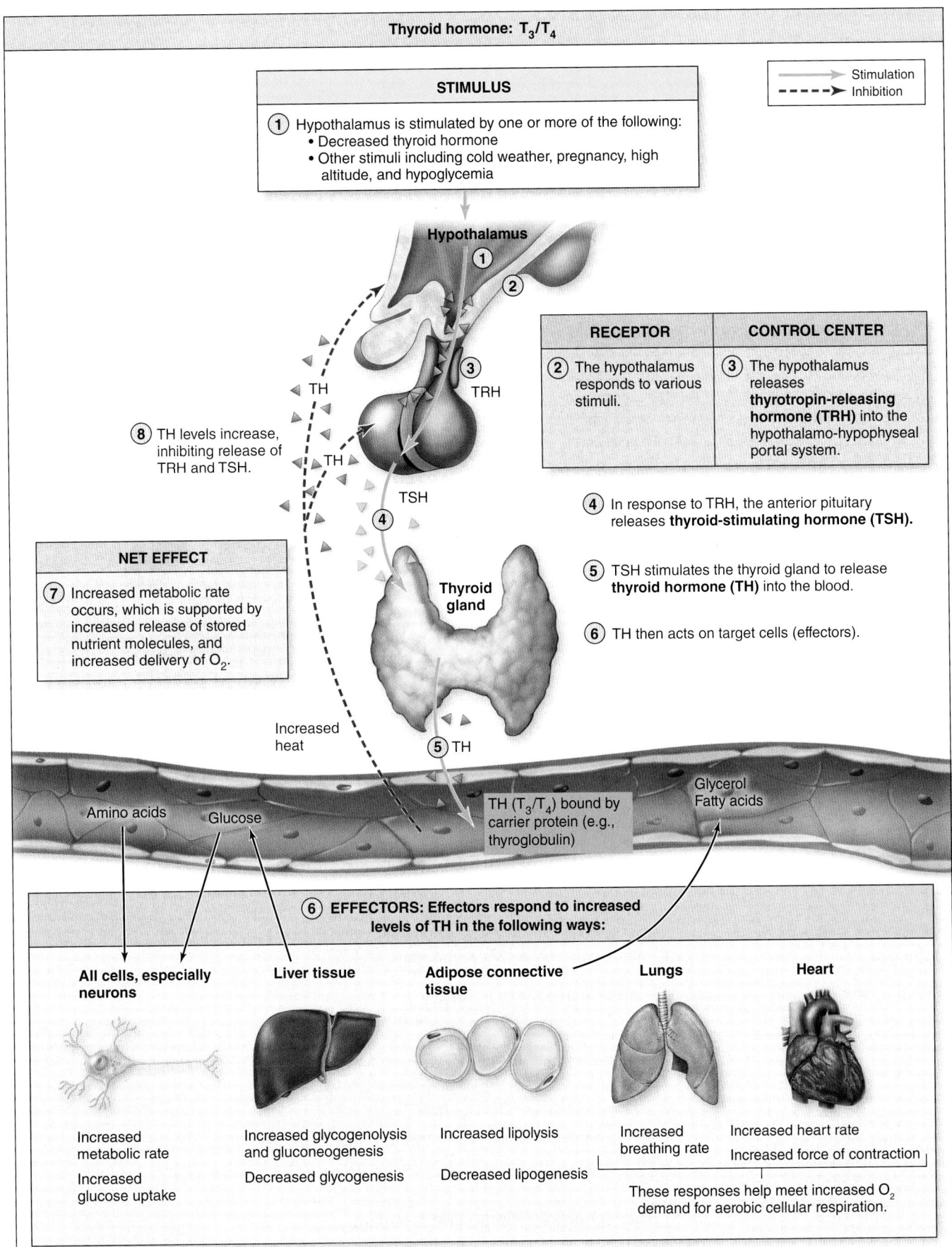

Thyroid hormone: T$_3$/T$_4$

Stimulation
Inhibition

STIMULUS

① Hypothalamus is stimulated by one or more of the following:
- Decreased thyroid hormone
- Other stimuli including cold weather, pregnancy, high altitude, and hypoglycemia

Hypothalamus
①
②
③ TRH

TH

TSH
④

⑧ TH levels increase, inhibiting release of TRH and TSH.

TH

RECEPTOR	CONTROL CENTER
② The hypothalamus responds to various stimuli.	③ The hypothalamus releases **thyrotropin-releasing hormone (TRH)** into the hypothalamo-hypophyseal portal system.

④ In response to TRH, the anterior pituitary releases **thyroid-stimulating hormone (TSH).**

⑤ TSH stimulates the thyroid gland to release **thyroid hormone (TH)** into the blood.

⑥ TH then acts on target cells (effectors).

Thyroid gland

NET EFFECT

⑦ Increased metabolic rate occurs, which is supported by increased release of stored nutrient molecules, and increased delivery of O$_2$.

Increased heat

⑤ TH

Amino acids   Glucose

TH (T$_3$/T$_4$) bound by carrier protein (e.g., thyroglobulin)

Glycerol Fatty acids

⑥ **EFFECTORS: Effectors respond to increased levels of TH in the following ways:**

All cells, especially neurons	**Liver tissue**	**Adipose connective tissue**	**Lungs**	**Heart**
Increased metabolic rate	Increased glycogenolysis and gluconeogenesis	Increased lipolysis	Increased breathing rate	Increased heart rate
Increased glucose uptake	Decreased glycogenesis	Decreased lipogenesis		Increased force of contraction

These responses help meet increased O$_2$ demand for aerobic cellular respiration.

Figure 17.17 Regulation and Action of Thyroid Hormone. The hypothalamus responds to particular stimuli by releasing thyrotropin-releasing hormone (TRH), which stimulates the anterior pituitary to release thyroid-stimulating hormone (TSH). TSH stimulates the thyroid gland to release thyroid hormone (TH). TH increases the body's metabolic rate and alters the availability of nutrient molecules within the blood to provide additional energy for the higher metabolic rate. (Direction of arrows between the blood and effectors indicates the net movement of the nutrients.)

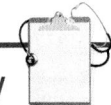
CLINICAL VIEW

Disorders of Thyroid Hormone Secretion

Thyroid hormone (TH) adjusts and maintains the basal metabolic rate of many cells in our body. TH release is very tightly controlled in the healthy state, but should the amount vary by even a little, a person could become hyperactive and heat-intolerant, or sluggish and overweight. Disorders of thyroid activity are among the most common metabolic problems clinicians see.

Hyperthyroidism results from excessive production of TH and is characterized by increased metabolic rate, weight loss, hyperactivity, and heat intolerance. Although there are a number of causes of hyperthyroidism, the more common ones are (1) ingestion of T_4 (weight control clinics sometimes use TH to increase metabolic activity); (2) excessive stimulation of the thyroid by the pituitary gland; and (3) loss of feedback control by the thyroid itself. This last condition, called **Graves disease,** is an autoimmune disorder involving the formation of autoantibodies that mimic TSH hormone. The autoantibodies bind to TSH receptors on the follicular cells of the thyroid, causing an abnormally high level of TH release. Graves disease includes all the symptoms of hyperthyroidism plus a peculiar change in the eyes known as *exophthalmos* (protruding and bulging eyeballs). Hyperthyroidism is treated by removing the thyroid gland, either by surgery or intravenous injections of radioactive iodine (I-131). In the procedure for treating hyperthyroidism, the thyroid literally "cooks itself" as it sequesters the I-131, but other organs are not damaged because they do not store iodine as does the thyroid. Patients whose thyroid glands have been removed or destroyed must take daily hormone supplements.

Hypothyroidism results from decreased production of TH. It is characterized by low metabolic rate, lethargy, a feeling of being cold, weight gain (in some patients), and photophobia (the disdain and avoidance of light). Hypothyroidism may be caused by decreased iodine intake, loss of pituitary stimulation of the thyroid, post-therapeutic hypothyroidism (resulting from either surgical removal or radioactive iodine treatments), or destruction of the thyroid by the person's own immune system (Hashimoto thyroiditis). Oral replacement of TH is the treatment for this type of hypothyroidism.

A **goiter** refers to enlargement of the thyroid, typically due to an insufficient amount of dietary iodine. Although the pituitary releases more TSH in an effort to stimulate the thyroid, the lack of dietary iodine prevents the thyroid from producing the needed TH. The long-term consequence of the excessive TSH stimulation is overgrowth of the thyroid follicles and the thyroid itself. Goiter was a relatively common deformity in the United States until food processors began adding iodine to table salt. It still occurs in parts of the world where iodine is lacking in the diet, and as such is referred to as endemic goiter. Unfortunately, goiters do not readily regress once iodine is restored to the diet, and surgical removal of the thyroid gland is often required.

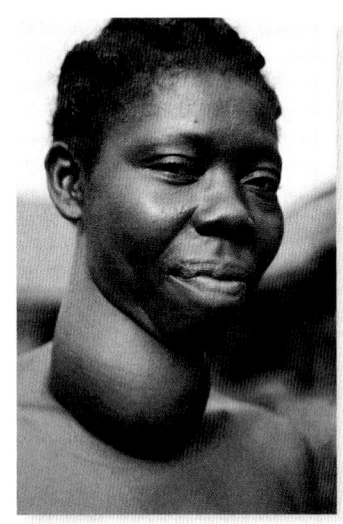

Goiter

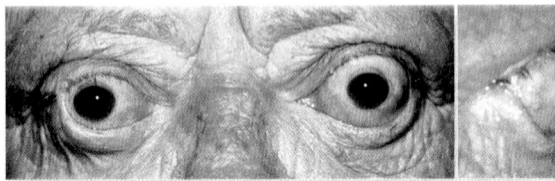

Exophthalmos seen in Graves disease.

(see section 19.9b). Additional responses to the release of thyroid hormone are listed in the summary table on thyroid hormone in the reference section, which directly follows this chapter (see **table R.4**).

The release of both TRH from the hypothalamus and TSH from the anterior pituitary are regulated by negative feedback. Increased thyroid hormone inhibits the release of both TRH from the hypothalamus and the release of TSH from the anterior pituitary. Increased TH also causes the release of growth hormone inhibiting hormone, which inhibits the release of TSH from the anterior pituitary (not shown on figure 17.17).

WHAT DO YOU THINK?

5 Predict the signs/symptoms that a person with hyperthyroidism would exhibit in each of the following circumstances: (a) high temperature or low temperature, (b) elevated pulse or decreased pulse, (c) elevated breathing or decreased breathing, and (d) plump or thin body.

WHAT DID YOU LEARN?

20 What is the relationship of TRH, TSH, and TH in regulating metabolism?

21 What are the primary target organs/tissues of TH? Describe the effect on each.

17.8c Adrenal Glands and Cortisol

LEARNING OBJECTIVES

5. Describe the structure and location of the adrenal glands.

6. Name the three zones of the adrenal cortex and the hormones produced in each zone.

7. Describe how the hypothalamus controls the release of glucocorticoid (cortisol) and the effects of cortisol.

The **adrenal** (ă-drē′năl; *ad* = to, *ren* = kidney) **glands,** or *suprarenal glands,* are paired, pyramid-shaped endocrine glands anchored on the superior surface of each kidney (**figure 17.18**).

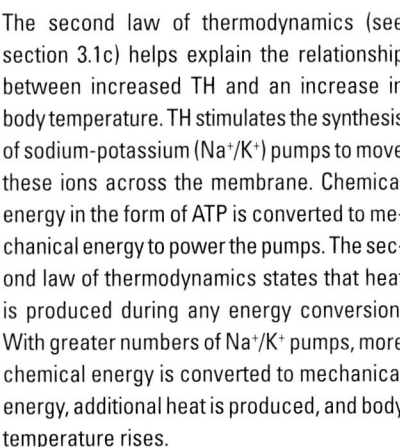

INTEGRATE

CONCEPT CONNECTION

The second law of thermodynamics (see section 3.1c) helps explain the relationship between increased TH and an increase in body temperature. TH stimulates the synthesis of sodium-potassium (Na^+/K^+) pumps to move these ions across the membrane. Chemical energy in the form of ATP is converted to mechanical energy to power the pumps. The second law of thermodynamics states that heat is produced during any energy conversion. With greater numbers of Na^+/K^+ pumps, more chemical energy is converted to mechanical energy, additional heat is produced, and body temperature rises.

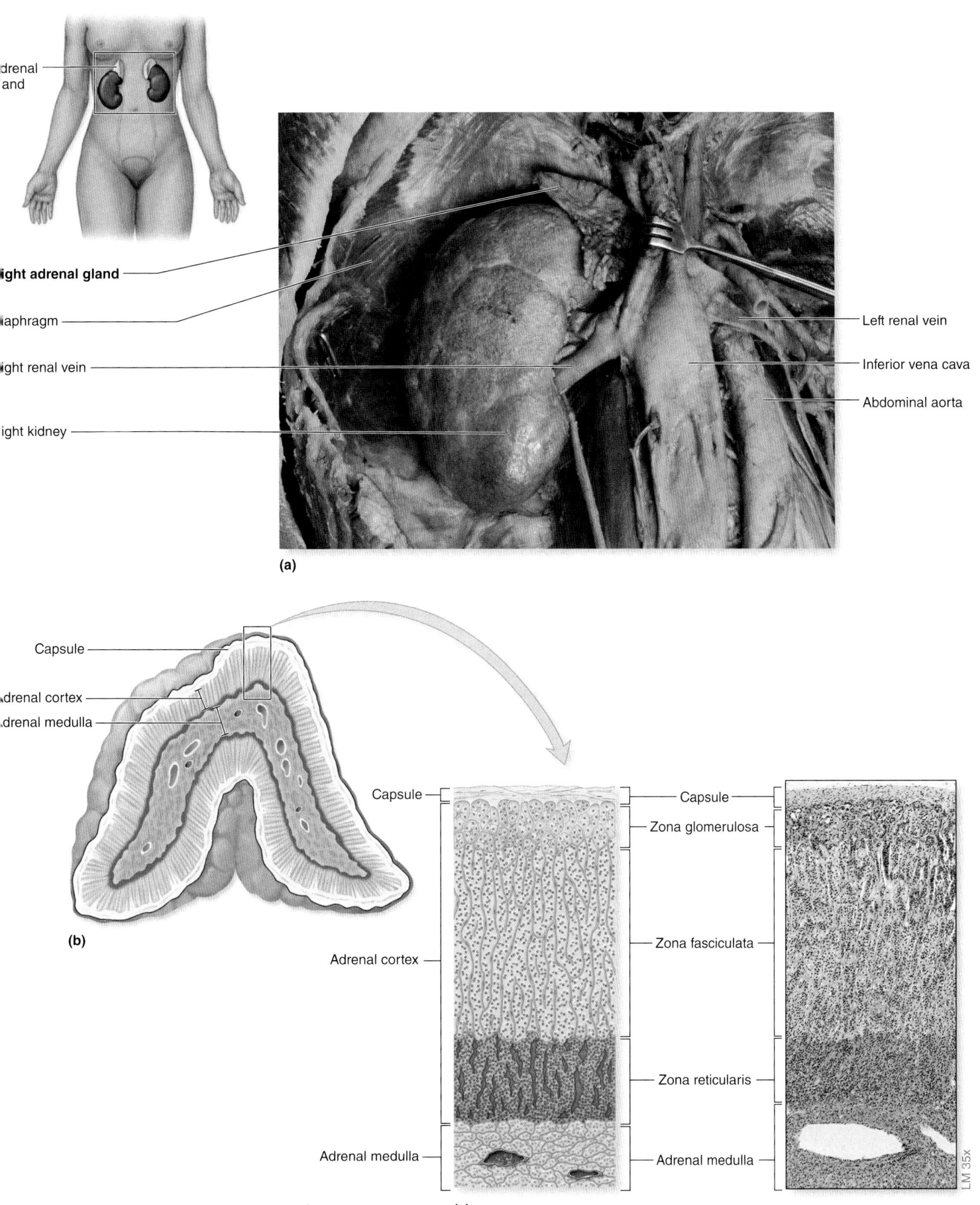

Adrenal
gland

Right adrenal gland

Diaphragm

Right renal vein

Right kidney

Left renal vein

Inferior vena cava

Abdominal aorta

(a)

Capsule

Adrenal cortex

Adrenal medulla

(b)

Capsule

Adrenal cortex

Adrenal medulla

Capsule

Zona glomerulosa

Zona fasciculata

Zona reticularis

Adrenal medulla

LM 35x

(c)

Figure 17.18 Adrenal Glands. Each adrenal gland is a two-part gland that secretes stress-related hormones. The adrenal cortex produces mineralocorticoids, glucocorticoids (e.g., cortisol), and gonadocorticoids from its different zones, and the adrenal medulla produces epinephrine and norepinephrine. (*a*) A cadaver photo show the relationships of the kidneys and adrenal glands. (*b*) A sectioned adrenal gland shows the outer cortex and the inner medulla. (*c*) A diagram and a micrograph illustrate the three zones of the adrenal cortex, as well as the relationship of the cortex to the external capsule and the internal medulla. AP|R

These glands are retroperitoneal (located posterior to the parietal peritoneum) and embedded within fat and fascia to minimize their movement. Two regions comprise an adrenal gland: the adrenal medulla and the adrenal cortex (figure 17.18b).

Anatomy of the Adrenal Glands

The **adrenal medulla** forms the inner core of each adrenal gland. It has a pronounced red-brown color due to its extensive vascularization. It releases the catecholamines epinephrine and norepinephrine (which are biogenic amines) in response to sympathetic nervous stimulation. Approximately 75% of the hormone released is epinephrine and about 25% is norepinephrine. Both hormones circulate within the blood and help prolong the fight-or-flight response, which is caused by activation of the sympathetic division (see section 15.4).

The **adrenal cortex** exhibits a distinctive yellow color as a consequence of the stored lipids within its cells. These cells synthesize more than 25 different lipid-soluble corticosteroids. The adrenal cortex is partitioned into three separate regions: the outer zona glomerulosa, the

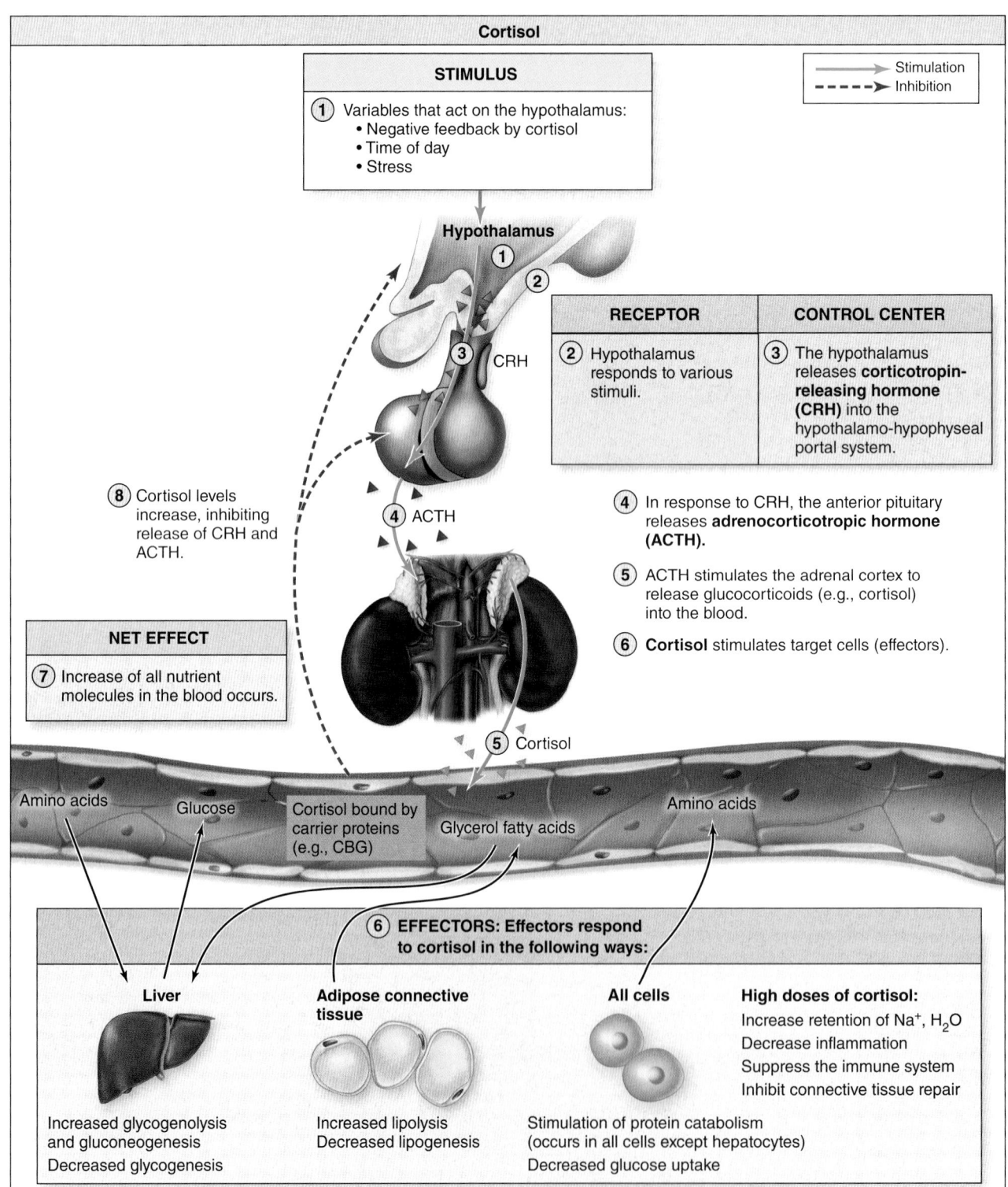

Figure 17.19 Regulation and Action of Cortisol Hormone. The hypothalamus responds to particular stimuli by releasing corticotropin-release hormone (CRH), which stimulates the anterior pituitary to release adrenocorticotropic hormone (ACTH). ACTH stimulates the adrenal cortex to release cortisol. Cortisol increases the availability of nutrient molecules to support the response to stress. (Direction of arrows between the blood and effectors indicates the net movement of the nutrients.)

iddle zona fasciculata, and the inner zona reticularis (figure 17.18c). ifferent functional categories of steroids are synthesized and secreted . the separate zones.

ormones of the Adrenal Cortex

he **zona glomerulosa** (zō′nă glō-měr-ū-lōs′ă; *glomerulus* = ball of arn) is the thin, outer cortical layer composed of dense, spherical usters of cells. These cells synthesize **mineralocorticoids** (min′er- -ō kōr′ti-koyd), a group of hormones that help regulate the com- osition and concentration of electrolytes (ions) in body fluids. The rincipal mineralocorticoid is **aldosterone** (al-dos′ter-ōn), which ·gulates the ratio of Na⁺ and K⁺ in our blood and body fluids by al- ·ring the amounts excreted by the kidney into the urine. Aldosterone imulates Na⁺ retention and K⁺ secretion. Severe imbalances in this tio can result in death (see section 25.4c for details of regulation). he functional details of aldosterone are described in sections 24.6d nd 25.4c and summarized in the reference tables following this chap- ·r (see **table R.7**).

The **zona fasciculata** (fă-sik′ū-la′tă; *fascicle* = bundle of parallel icks) is the middle layer and largest region of the adrenal cortex. It is omposed of parallel cords of lipid-rich cells that have a bubbly, almost ale appearance. The primary **glucocorticoids** (glū′kō-kōr-ti-koyd) ynthesized in this region are cortisol and corticosterone. Details of ortisol regulation and function are described shortly.

The innermost region of the cortex, the **zona reticularis** ě-tik′ū-lăr′is; *reticulum* = network), is a narrow band of small, branch- ig cells. They are capable of secreting minor amounts of sex hormones alled **gonadocorticoids.** The primary gonadocorticoids secreted are ale sex hormones called **androgens.** The androgens in females are onverted to estrogen. The amount of androgen secreted by the adrenal ortex is small compared to that secreted by the gonads; however, adre- al gland tumors in a female can result in elevated levels of testosterone nd varying degrees of masculinization.

INTEGRATE

LEARNING STRATEGY ✎

You can remember the general functions of the three layers of the adrenal cortex (from outermost to innermost) with the following: "salt, sugar, and sex."

Action of Cortisol

Cortisol (kōr′ti-sol) (or *hydrocortisone*) and **corticosterone** (kōr′ti-kos′ter-ōn) increase nutrient levels in the blood (glucose, fatty acids, and amino acids), especially in an attempt to resist stress and help repair injured or damaged tissues. Other physiologic changes are stimulated by glucocorticoids and become most evident when present in high doses.

Cortisol release from the adrenal cortex occurs through hypotha- lamic regulation by means of corticotropin-releasing hormone (CRH) and the subsequent release of adrenocorticotropic hormone (ACTH) from the anterior pituitary. This physiologic relationship is referred to as the **hypothalamic-pituitary-adrenal axis (figure 17.19).**

A decrease in cortisol level within the blood stimulates the hypo- thalamus to release CRH. The time of day and stress also influence the release of CRH and thus the release of cortisol **(figure 17.20).** CRH travels through the hypothalamo-hypophyseal portal system to the anterior pituitary, where it binds to receptors of the anterior pituitary (corticotropic cells) and stimulates the release of adrenocorticotropic hormone (ACTH) into general circulation (figure 17.19). ACTH then binds to receptors within adrenal cortex cells (zona fasciculata) and stimulates the release of cortisol and corticosterone. Cortisol accounts for 95% of the glucocorticoid activity. Cortisol is transported within the blood by carrier proteins (*corticosteroid-binding globulin,* or *CBG,*

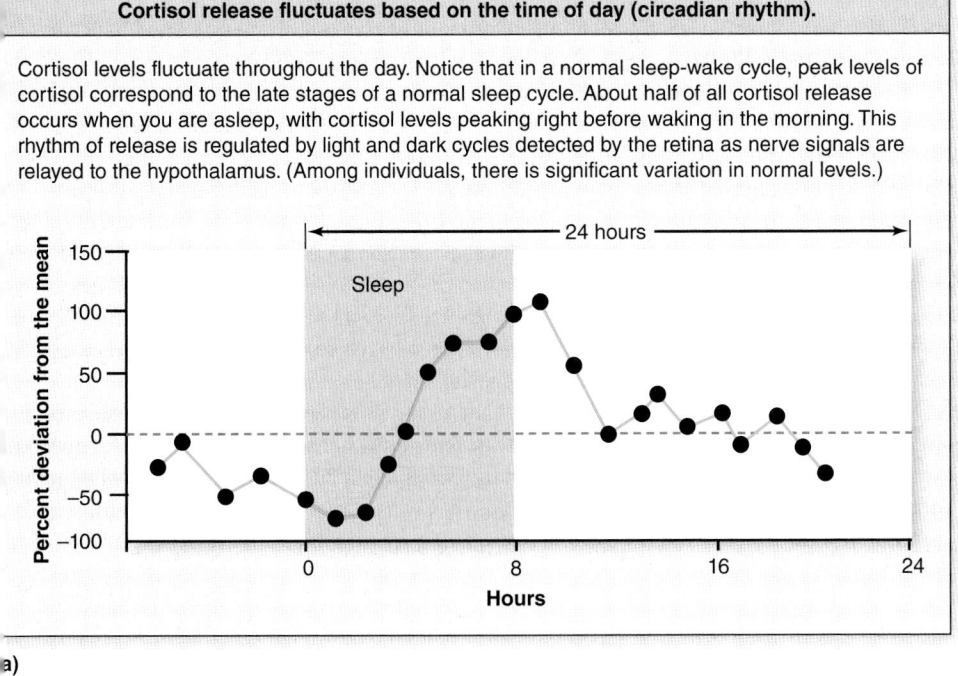

Cortisol release fluctuates based on the time of day (circadian rhythm).

Cortisol levels fluctuate throughout the day. Notice that in a normal sleep-wake cycle, peak levels of cortisol correspond to the late stages of a normal sleep cycle. About half of all cortisol release occurs when you are asleep, with cortisol levels peaking right before waking in the morning. This rhythm of release is regulated by light and dark cycles detected by the retina as nerve signals are relayed to the hypothalamus. (Among individuals, there is significant variation in normal levels.)

(a)

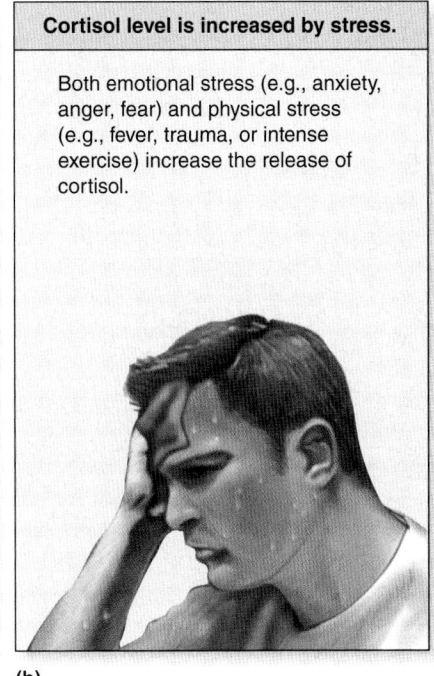

Cortisol level is increased by stress.

Both emotional stress (e.g., anxiety, anger, fear) and physical stress (e.g., fever, trauma, or intense exercise) increase the release of cortisol.

(b)

Figure 17.20 Variables That Influence Blood Levels of Cortisol. Blood levels of cortisol (*a*) fluctuate throughout the day, and) rise with increased stress, which is why cortisol is sometimes called the "stress hormone."

CLINICAL VIEW
Disorders in Adrenal Cortex Hormone Secretion

Three abnormal patterns of adrenal cortical function are Cushing syndrome, Addison disease, and androgenital syndrome.

Cushing syndrome results from the chronic exposure of the body's tissues to excessive levels of glucocorticoid hormones. This complex of symptoms is seen most frequently in people taking corticosteroids as therapy for autoimmune diseases such as rheumatoid arthritis, although

Photo prior to onset of Cushing syndrome.

Symptoms resulting from the excessive glucocorticoid secretion in Cushing syndrome include "moon face."

some cases result when the adrenal gland produces too much of its own glucocorticoid hormones. Corticosteroids are powerful immunosuppressant drugs, but they have serious side effects, such as osteoporosis, muscle weakness, redistribution of body fat, and salt retention (resulting in overall swelling of the tissues). Cushing syndrome is characterized by body obesity, especially in the face (called "moon face") and back ("buffalo hump"). Other symptoms include hypertension, excess hair growth, kidney stones, and menstrual irregularities.

Addison disease involves insufficient production of steroids (usually glucocorticoids and perhaps mineralocorticoids) from the adrenal cortex. Addison disease can result from (1) adrenal glands that were malformed during development, (2) impaired enzymatic pathways for steroid synthesis, and (3) destruction of the adrenal gland (typically by an autoimmune disorder that forms autoantibodies against the adrenal gland that results in its destruction). The symptoms include weight loss, general fatigue and weakness, hypotension, and darkening of the skin. Perhaps the most well-known person with Addison disease was John Fitzgerald Kennedy (JFK).

Adrenogenital syndrome, or **congenital adrenal hyperplasia,** first manifests in the embryo and fetus. It is characterized by the inability to synthesize corticosteroids. The anterior pituitary, sensing the deficiency of corticosteroids, releases massive amounts of ACTH in an unsuccessful effort to bring the glucocorticoid content of the blood up to a healthy level. This large amount of ACTH produces *hyperplasia* (increased size) of the adrenal cortex and causes the release of intermediary hormones that have a testosterone-like effect. The result is *virilization* (masculinization) in a newborn. Virilization in girls means the clitoris is enlarged, sometimes to the size of a male penis. The effect may be so profound that the sex of a newborn female is questioned or even mistaken. A virilized male may have an enlarged penis and exhibit signs of premature puberty as early as age 6 or 7. (See Clinical View: Intersex Conditions [Disorders of Sex Development] in section 28.5 for additional details.)

also called transcortin or serum albumin). Cortisol circulates in the blood and small amounts become unbound from their carrier protein and exits from the blood.

Cortisol readily passes through the plasma membrane and binds to intracellular receptor to form a hormone-receptor complex. This complex binds to DNA and stimulates cellular changes through the activation of genes. The cellular changes depend upon the tissue being stimulated. For example, hepatocytes are stimulated by cortisol to increase glycogenolysis and gluconeogenesis (see section 2.7c). At the same time, they are inhibited from using the glycogenesis pathway, so that blood glucose levels rise. Some cortisol is converted to cortisone in the liver.

Adipose connective tissue cells are stimulated by cortisol to increase lipolysis and decrease lipogenesis (see section 2.7b), resulting in the release of glycerol and fatty acids into blood. This provides alternative nutrients for gluconeogenesis.

Most cells, including muscle cells, lymphatic tissue cells, skin cells, and bone cells, increase protein catabolism (breakdown of protein into amino acids) in response to cortisol. An exception to this occurs in hepatocytes. In liver cells, additional amino acid released into the blood provides alternative nutrient molecules for gluconeogenesis. Cortisol additionally stimulates most cells to decrease glucose uptake. This is the *glucose-sparing effect*, which saves blood glucose for use in the brain.

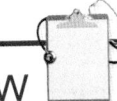

CLINICAL VIEW

The Stress Response (General Adaptation Syndrome)

Stressors may fall under the category of either *emotional stress* (e.g., anxiety, anger, fear, and excitement) or *physical stress* (e.g., fever, trauma, hemorrhage, surgery, and malnutrition). Stressors elicit the **stress response** (or *general adaptation syndrome*), which was defined by Hans Selye (a pioneer in the endocrinology of stress) as "the nonspecific response of the body to any demand made upon it." The body's response to stress is initiated by the hypothalamus and involves both the nervous system and the endocrine system. In 1936, Hans Selye described the response to stressors in three stages: the alarm reaction, the stage of resistance, and the stage of exhaustion.

The Alarm Reaction

The alarm reaction is the initial reaction to stress and is regulated primarily by the sympathetic division of the autonomic nervous system. The hypothalamus activates the sympathetic division with subsequent stimulation of the adrenal medulla to release epinephrine and norepinephrine into the blood. The following changes occur to the body (see table 15.6):

- Pupils dilate.
- Bronchioles dilate.
- Respiration rate increases.
- Blood pressure increases as both the heart rate and force of contraction increase.
- Blood vessels vasoconstrict.
- Sodium and water are retained, resulting in an increase in blood volume.
- Potassium and hydrogen ions are excreted.
- Both glucose and lipid levels in the blood increase.
- Sweating increases.
- Both digestion and urine production activities decrease.

The Stage of Resistance

The stage of resistance occurs after a few hours as glycogen stores in the liver are depleted. This stage is regulated primarily by the endocrine system. The major changes are induced by the release of glucocorticoids (e.g., cortisol). The principal function of this stage is to provide glucose to meet the increased energy demands. Glucose is especially important for nervous tissue because it is the main nutrient molecule for cell respiration that these tissues can use. To meet the increased demands for energy, gluconeogenesis is increased in the liver, and glucose is released into the blood. Glycerol and fatty acids increase in the blood due to increased lipolysis in adipose connective tissue cells. Amino acids increase as a result of increased protein catabolism (and decreased protein synthesis) in most cells. Glycerol and amino acids provide the liver with alternative nutrients for gluconeogenesis. Additionally, glucose uptake is inhibited in most cells. The net result is elevated blood glucose levels.

The Stage of Exhaustion

The stage of exhaustion occurs after weeks or months, as fat stores in adipose connective tissue are depleted. With fat stores depleted, and as structural proteins of the body's cells continue to be broken down for gluconeogenesis, the body becomes progressively weaker. Additionally, elevated levels of aldosterone may cause fluid, electrolyte, and pH imbalances. The combination of body weakness, electrolyte imbalances, and other factors may ultimately cause organ failure and death.

The release of both CRH from the hypothalamus and ACTH from the anterior pituitary is regulated by negative feedback. Increasing levels of cortisol inhibit the release of CRH from the hypothalamus and ACTH from the anterior pituitary. The details for cortisol are listed in the summary table, in the reference section, on regulating the stress response, which directly follows this chapter (see **table R.5**).

Note that corticosterone is used as a treatment for chronic inflammation. The side effects, especially when administered in high doses, include retention of Na⁺ and water; inhibited release of inflammatory agents (the anti-inflammatory effect); suppression of the immune system; and inhibited connective tissue repair. Note that suppression of the immune system increases an individual's susceptibility to infection and risk of cancer.

WHAT DID YOU LEARN?

22 What is the relationship of CRH, ACTH, and cortisol?

23 What are the primary target organs/tissues of cortisol? Describe the effect on each.

INTEGRATE

CONCEPT CONNECTION

The pancreas functions as both an endocrine gland and an exocrine gland. Endocrine glands release their secretions (hormones) into the blood. The pancreas releases both insulin and glucagon into the blood. In contrast, exocrine glands release their secretions into ducts. The pancreas also releases pancreatic juice into ducts that empty into the small intestine (duodenum), as described in section 26.3c.

17.9 Pancreatic Hormones

The pancreatic hormones, insulin and glucagon, function in regulating nutrient levels in the blood—both of which are released from the pancreas. We begin with a brief description of the anatomy of the pancreas before looking at the specifics of pancreatic hormone actions.

17.9a Anatomy of the Pancreas

LEARNING OBJECTIVES

1. Describe the gross anatomy and cellular structure of the pancreas.
2. Identify the primary types of pancreatic islet cells and the hormones they produce.

The **pancreas** (pan′krē-as; *pan* = all, *kreas* = flesh) is an elongated organ situated posterior to the stomach (**figure 17.21**). The pancreas performs both exocrine and endocrine activities; thus, it is considered a heterocrine, or mixed, gland. The endocrine cells of the pancreas are located within small clusters called **pancreatic islets** (ī′let), also known as *islets of Langerhans*. These endocrine cell clusters form only about 1% of the total pancreatic volume. (The other 99% are the pancreatic acini cells, which release secretions into ducts that lead into the small intestine; these secretions facilitate digestion; see section 26.3c.)

A pancreatic islet is composed of two primary types of cells: **alpha cells,** which secrete **glucagon** (glū′kă-gon), and **beta cells,** which secrete **insulin** (in′sū-lin; *insula* = island). Minor cells within the pancreatic islets include delta cells, which secrete somatostatin (also described a growth hormone–inhibiting hormone), and F cells, which secrete pancreatic polypeptide. These minor cells and their hormones will not be addressed.

WHAT DID YOU LEARN?

24 Why is the pancreas considered both an exocrine gland and an endocrine gland?

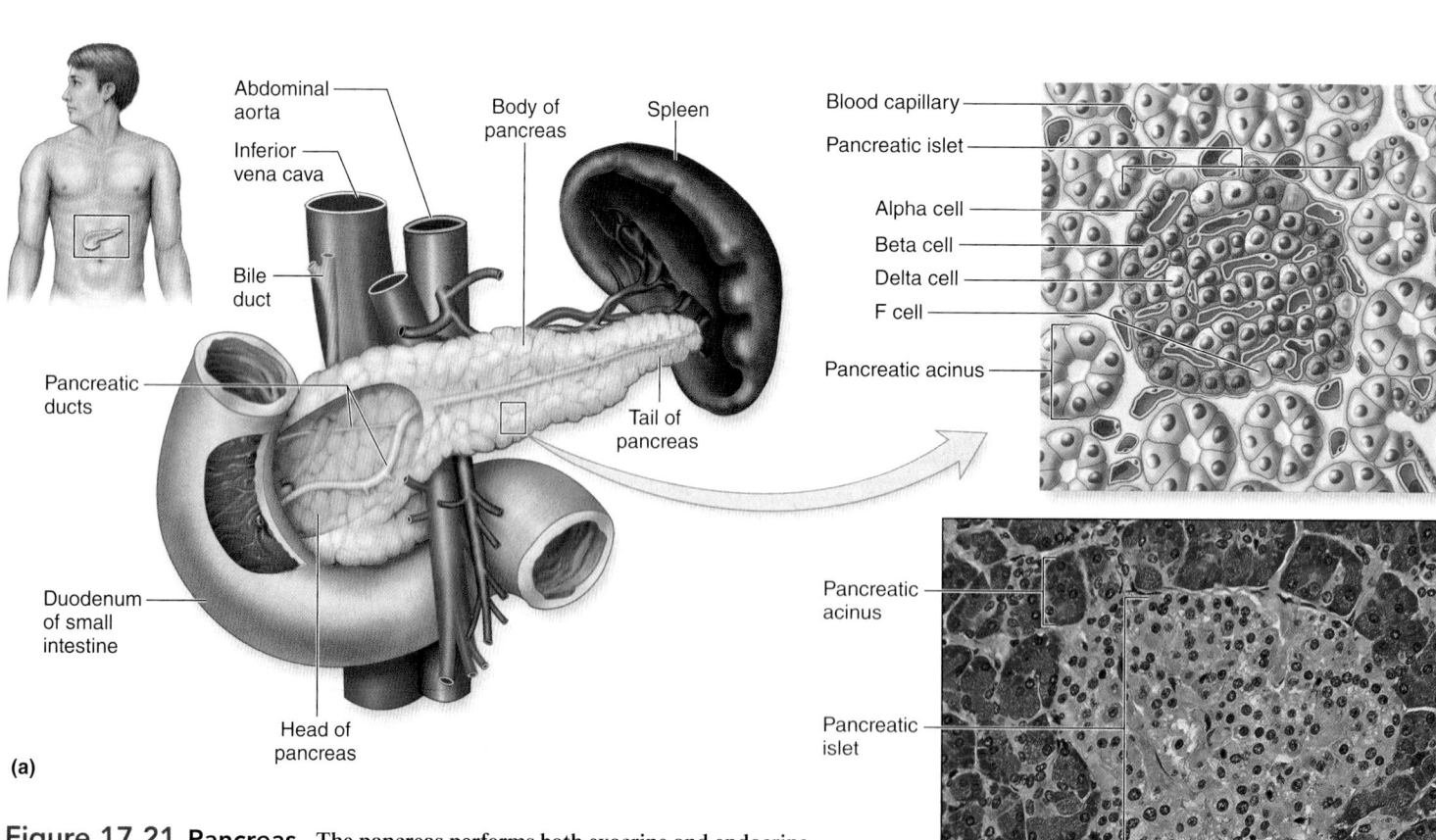

Figure 17.21 Pancreas. The pancreas performs both exocrine and endocrine activities. (*a*) An illustration shows the relationship between the pancreas and both the duodenum (first section of the small intestine) and spleen. (*b*) A diagram and micrograph reveal the histology of a pancreatic islet. Four types of islet cells are shown in the diagram, including alpha cells that release glucagon and beta cells that release insulin. **AP│R**

7.9b Effects of Pancreatic Hormones

LEARNING OBJECTIVES

3. Describe the action of insulin in lowering blood glucose concentration.

4. Explain the action of glucagon in raising blood glucose concentration.

The primary endocrine function of the pancreas is to maintain the concentration of glucose in the blood within a normal range of 70 of 110 milligrams of glucose per deciliter (mg/dL; a deciliter is an amount equivalent to 100 milliliters). Chronically high blood glucose levels can be very damaging to blood vessels and the kidneys, so this excess glucose must be transported into other body cells that can use or store this resource. Conversely, low blood glucose levels result in lethargy, impairment of mental and physical function, and death if glucose levels drop too low. Thus, blood glucose levels must be closely regulated. The homeostatic mechanisms for insulin and glucagon are described next and are summarized in the reference section table, which directly follows this chapter (see **table R.1**).

Lowering High Blood Glucose Levels with Insulin

Insulin is generally released from the pancreas following food intake **(figure 17.22)**. Chemoreceptors in the beta cells of the pancreas detect an increase in blood glucose (readings greater than the normal 70–110 mg/dL) and are stimulated to release insulin (a protein hormone). Insulin circulates in the blood and randomly exits from the blood into the interstitial fluid as it passes through capillaries.

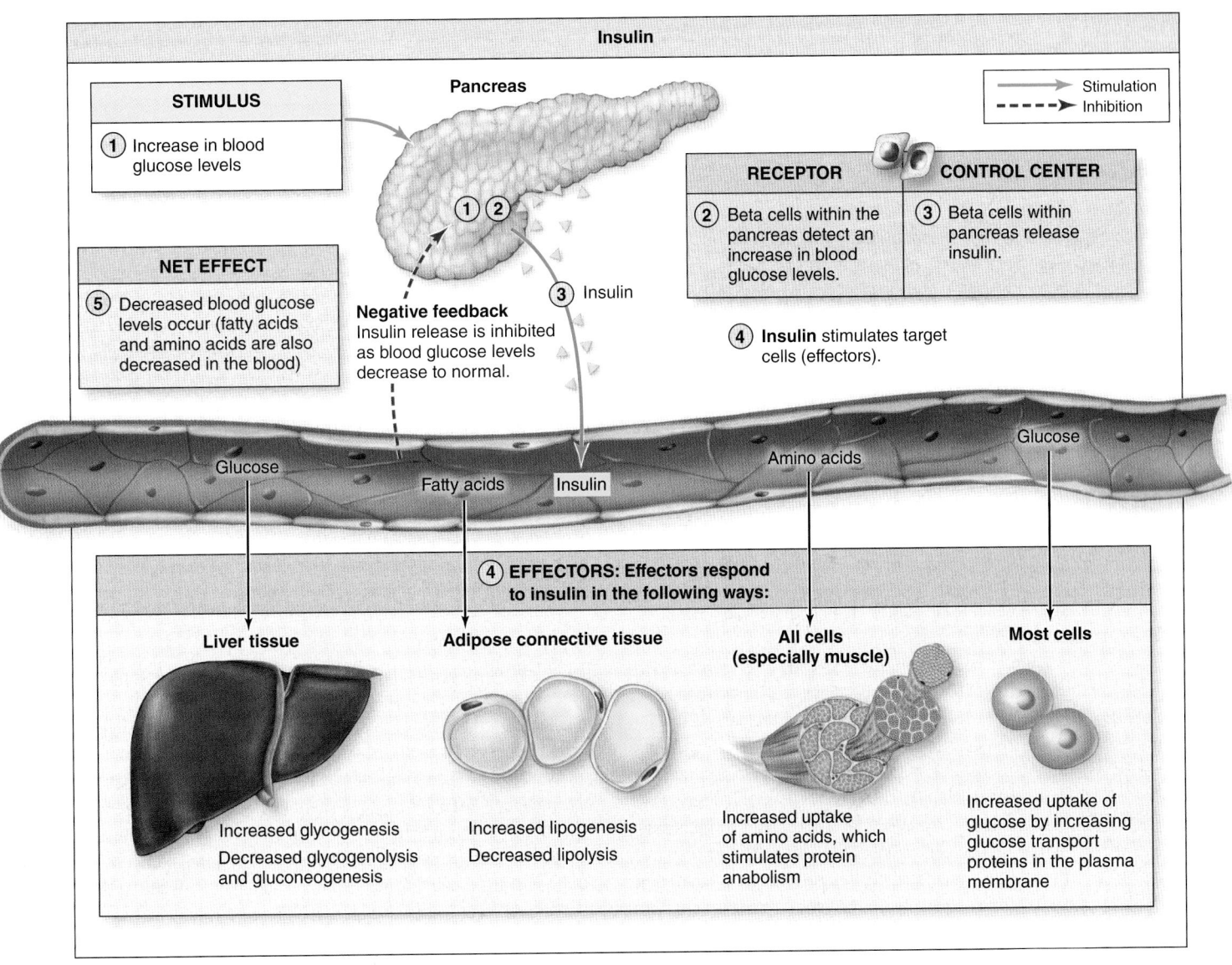

Figure 17.22 Regulation and Action of Insulin. Insulin is released from beta cells within pancreatic islets in response to high blood glucose. Insulin decreases the level of all nutrient molecules (glucose, fatty acids, and amino acids) within the blood. The uptake of fatty acids and amino acids from the blood limits their availability making it more likely that cells will use glucose available within the blood as their nutrient molecule for cell respiration. Thus, blood glucose more quickly returns to within normal homeostatic levels.

Target cells bind insulin, which activates second messengers within the target cell. (Additionally, recent evidence suggests that insulin may also enter the cell and directly bind to intracellular receptors.) Specific enzymatic pathways are altered as follows:

- Glycogenesis (see section 2.7c) in hepatocytes is stimulated, and both glycogenolysis and gluconeogenesis are inhibited, resulting in glucose molecules being removed from the blood and stored as glycogen within liver cells.
- Lipogenesis (see section 2.7b) in adipose connective tissue cells is stimulated, and lipolysis is inhibited. Fatty acid levels in the blood decrease, and the storage of fat is increased as a result.
- Most cells are stimulated to increase their cellular uptake of (1) amino acids (especially muscle cells), a change that induces cells to increase protein anabolism (synthesis of amino acids into protein); and (2) glucose, especially by the cells of muscle and adipose connective tissue. The uptake of glucose occurs as intracellular vesicles containing glucose transport proteins (specifically Glut4 transport proteins) fuse with the plasma membrane. Additional glucose transport molecules are placed

within the plasma membrane of cells, providing the glucose carrier molecules needed to bring more glucose into the cell. (These carriers are later removed as insulin levels decrease.)

In summary, the release of insulin results in both a decrease in *a* nutrients in the blood and in an increase in the synthesis of the storag form of these molecules within body tissues. By decreasing alternativ nutrients (fatty acids and amino acids), the cells of the body are mor likely to use the available glucose and help return blood glucose to normal level more quickly. The release of insulin is controlled by nega tive feedback; as blood glucose levels decrease, less insulin is release from the pancreas.

Note that some cells do not require insulin for glucose uptake These cells include neurons, kidney cells, hepatocytes, and erythrocyte (red blood cells). Each of these cells takes up glucose independentl without external stimulation.

 ### WHAT DO YOU THINK?

6 Body builders have been known to inject insulin to increase muscle bulk. Explain their reasoning. What is the risk of an insulin overdose?

INTEGRATE

CLINICAL VIEW

Conditions Resulting in Abnormal Blood Glucose Levels

Diabetes Mellitus

Diabetes mellitus (dí-ă-bē′tez = a siphon, me-lī′tŭs = sweetened with honey) is a metabolic condition marked by inadequate uptake of glucose from the blood. The name is derived from the phrase "sweet urine" because some of the excess glucose may be filtered into the urine, a condition called *glycosuria*. Chronically elevated blood glucose levels damage blood vessels, especially the smaller arterioles. Because of its damaging effects on the vascular system, diabetes is the leading cause of retinal blindness, kidney failure, and nontraumatic leg amputations in the United States. Diabetes is also associated with increased incidence of heart disease and stroke. In fact, heart disease or stroke is the cause of death in approximately 65% of individuals with diabetes.

Measuring the amount of glucose attached to hemoglobin molecules within erythrocytes (hemoglobin A1C test) is an accurate means for determining the degree of risk for an individual. The greater the amount of attached glucose, the higher the risk.

Three categories of diabetes mellitus are type 1 diabetes, type 2 diabetes, and gestational diabetes.

 **Type 1 diabetes** is also referred to as *insulin-dependent diabetes mellitus* (*IDDM*) or *juvenile diabetes*. It is characterized by absent or diminished production and release of insulin by the beta cells of the pancreatic islets. This type tends to occur in children and younger individuals, and is not directly associated with obesity. Type 1 diabetes may develop in a person who harbors a genetic predisposition, although some kind of triggering event is required to start the process. Often, the trigger is a viral infection leading to an autoimmune condition in which the beta cells of the pancreatic islets are destroyed. Treatment of type 1 diabetes requires daily injections of insulin. The recent use of stem cells shows promise as an effective means in treating type 1 diabetes.

Type 2 diabetes, also known as *insulin-independent diabetes mellitus* (*IIDM*), results from either decreased insulin release from the beta cells of the pancreatic islets or decreased insulin effectiveness at peripheral tissues. This type of diabetes was previously referred to as *adult-onset diabetes* because it tended to occur in people over the age of 30. However, type 2 diabetes is now often found in adolescents and young adults. Obesity plays a major role in the development of type 2 diabetes, and more young people are considered overweight than ever before. Most patients with type 2 diabetes can be successfully treated with a combination of diet, exercise, and medications that enhance insulin release or increase sensitivity to insulin at the tissue level. A person with type 2 diabetes must take insulin injections in more severe cases.

Gestational diabetes is seen in some pregnant women, typically in the latter half of the pregnancy. If untreated, gestational diabetes can pose a risk to the fetus as well as increase delivery complications. Most at risk for developing this condition are women who are overweight; those of African American, Native American, or Hispanic ancestry; or those who have a family history of diabetes. Although gestational diabetes usually resolves after giving birth, a woman having this condition has a 20% to 50% chance of developing type 2 diabetes within 10 years.

Hypoglycemia

Hypoglycemia occurs when blood glucose levels drop below 60 mg/dL. Hypoglycemia is not a disease; however, it may be a nonspecific indicator of some underlying homeostatic imbalance. The causes of hypoglycemia are numerous and include insulin overdose, prolonged and intense exercise, drinking alcohol on an empty stomach, liver or kidney dysfunction, deficiency of either glucocorticoids or growth hormone, and certain genetic conditions. Symptoms may include hunger, dizziness, nervousness, confusion, feeling anxious or weak, sweating, sleepiness, or any combination of these. The symptoms are thought to occur from insufficient glucose to the brain or from the activation of the sympathetic nervous system in response to low blood glucose levels.

If an individual is unable to eat or drink safely, such as if unconscious, unresponsive, or having convulsions, glucagon can be administered by injection. This provides a safe means to offset the low blood glucose level.

Raising Low Blood Glucose Levels with Glucagon

All nervous tissue depends almost exclusively upon glucose for cellular respiration. To prevent impairment of mental function, lethargy, and possibly death, blood glucose levels must be prevented from dropping too low. Glucagon is one of the important hormones released in response to low blood glucose levels (figure 17.23).

Chemoreceptors in the alpha cells of the pancreas detect decreased blood glucose levels and subsequently release glucagon (a polypeptide hormone) into the blood. Nutrients are stored in various body tissues, and glucagon facilitates the breakdown of these nutrients and their release into the blood. Glucagon binds to plasma membrane receptors to activate second messengers (cAMP) that cause the following:

- Glycogenolysis and gluconeogenesis (section 2.7c) in hepatocytes are stimulated, and glycogenesis is inhibited; glucose is released into the blood, thereby increasing blood glucose levels. (Glucose within muscle cells is not released but remains within muscle cells and is oxidized in cellular respiration.)
- Lipolysis (section 2.7b) in adipose connective tissue cells is stimulated, and lipogenesis is inhibited. Fatty acids and glycerol are released from fat storage and are increased within the blood.

In summary, the release of glucagon results in an increase in glucose, glycerol, and fatty acids in the blood and in a decrease in the storage form of these molecules within body tissues. An increase in blood glucose levels results in a decrease in glucagon release by negative feedback.

Note that glucagon has no effect on the structural and functional protein components of the body. The physiologic significance of this lack of effect is that the ongoing, and regular, release of this hormone (e.g., during periods between meals) does not tear down the muscles and other protein components of the body to maintain blood glucose levels in "nonemergency" situations.

Interestingly, paramedics may administer glucagon subcutaneously under certain conditions when low blood glucose is detected. This may be done if the individual is unconscious and is unable to be given sugar orally to directly raise blood glucose.

WHAT DID YOU LEARN?

25 Is the stimulus for insulin and glucagon release from the pancreas hormonal, humoral, or nervous?

26 What is the stimulus, receptor, control center, and effector response to the release of insulin? Indicate what happens to nutrient levels in the blood.

27 Which of these hormones causes release of glucose into the blood: growth hormone, thyroid hormone, cortisol, insulin, or glucagon?

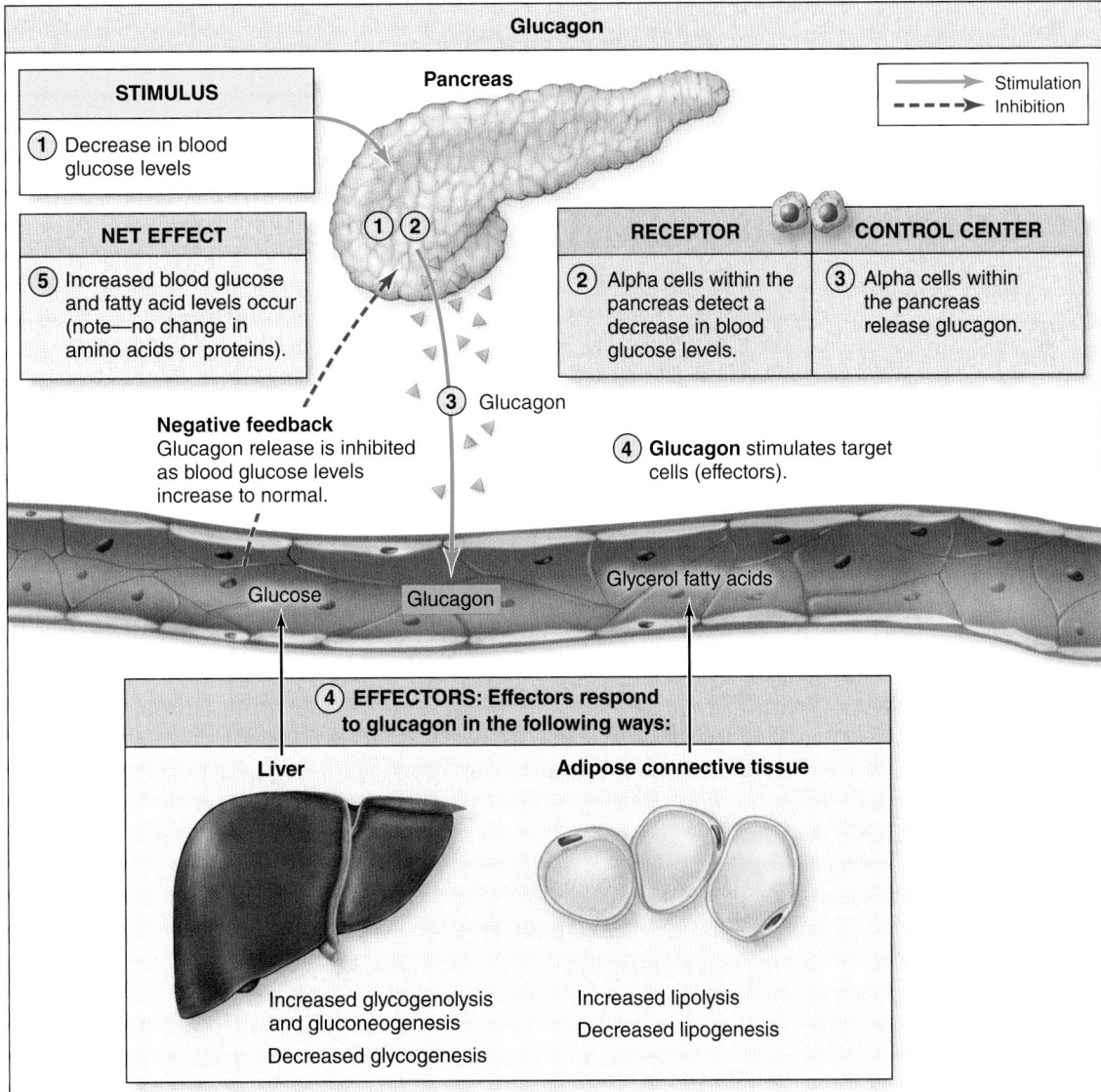

Figure 17.23 Regulation and Action of Glucagon. Glucagon is released from alpha cells within pancreatic islets in response to low blood glucose levels. Glucagon binds with target cells that increase both glucose, glycerol, and fatty acid levels within the blood.

17.10 Other Endocrine Glands

Here we describe other endocrine glands, including structures that have a primary function other than releasing hormones.

17.10a Pineal Gland

 LEARNING OBJECTIVE

1. Describe the general structure, location, and function of the pineal gland.

The **pineal** (pin′ē-ăl) **gland** (*or pineal body*) is a small, cone-shaped structure forming the posterior region of the epithalamus within the diencephalon (see figure 17.2 and section 13.4a). The pineal gland secretes **melatonin** (mel-ă-tōn′in; *melas* = dark hue, *ton as* = contraction), which makes us drowsy. Melatonin production tends to be cyclic; it increases at night, decreases during the day, and has the lowest levels around lunchtime. Melatonin helps regulate the circadian rhythm (24-hour body clock). Studies have linked low melatonin levels with mood (affective) disorders, such as seasonal affective disorder (SAD), a condition that may be treated with light therapy.

Melatonin also appears to affect the synthesis of a hypothalamic hormone (gonadotropin-releasing hormone). This hormone is responsible for regulating synthesis of two hormones from the anterior pituitary (follicle-stimulating hormone and luteinizing hormone), which in turn regulate the reproductive system. The role of melatonin in sexual maturation is not well understood. However, excessive melatonin secretion is known to delay puberty in humans.

WHAT DID YOU LEARN?

28 How do melatonin levels change throughout the day?

17.10b Parathyroid Glands

 LEARNING OBJECTIVE

2. Describe the general structure, location, and function of the parathyroid glands.

The small, brownish-red **parathyroid** (par-ă-thī′royd) **glands** are located on the posterior surface of the thyroid gland (see figure 17.2). These glands are usually four small nodules, but some individuals may have as few as two or as many as six of these glands. There are two different types of cells in the parathyroid glands: chief cells and oxyphil cells.

The more common **chief cells,** or *principal cells,* are the source of parathyroid hormone (PTH), which functions to increase blood calcium levels. It stimulates release of calcium from bone tissue, decreases loss of calcium in urine, and causes the kidney to release an enzyme to convert the inactive calcidiol hormone to the active calcitriol hormone (described shortly). The functional details of PTH and calcitriol are described in section 7.6b and summarized in the reference tables following this chapter (see **table R.2**).

The role of oxyphil cells is not known, although these cells are associated with a rare form of cancer called *oxyphil cell adenoma.*

WHAT DID YOU LEARN?

29 What is the primary hormone released from the parathyroid gland? What is its general function?

17.10c Structures with an Endocrine Function

LEARNING OBJECTIVE

3. Identify and provide a description of the general function of the hormone(s) released from each of the organs discussed in this section.

Thymus

The **thymus** (thī′mŭs) is a bilobed organ that is located anterior to the heart on its superior aspect (see figure 17.2). The thymus is relatively large in infants, continues to grow until puberty, and then begins to regress (decrease in size) after puberty. A connective tissue framework houses both the epithelial cells and maturing T-lymphocytes (a specific type of white blood cell). The T-lymphocytes migrate to the thymus following their formation in the bone marrow, and there epithelial cells secrete **thymic hormones** (thymosin, thymulin, and thymopoietin), which participate in the maturation of T-lymphocytes (see sections 21.3b and 22.5).

Heart

Endocrine tissue within the atria of the heart synthesizes and releases the hormone **atrial natriuretic** (nā′trē-yū-ret′ik; *natrium* = to carry, *ouron* = urine) **peptide (ANP).** It is a peptide hormone that functions in blood volume and blood pressure regulation. The primary function of ANP is to decrease blood pressure. This hormone stimulates both the kidneys to increase urine output and the blood vessels to dilate. Both of these actions facilitate blood pressure to decrease and return it to within normal homeostatic limits. The functional details of ANP are described in sections 24.5e, 24.6d, and 25.4d and summarized in the reference tables following this chapter (see **table R.7**).

Kidneys

Endocrine tissue within the kidneys release **erythropoietin (EPO)** (ĕ-rith′rō-poy′ĕ-tin) when specialized receptors (chemoreceptors) within the kidney detect low blood oxygen levels. EPO stimulates red bone marrow to increase the production rate of red blood cells (erythrocytes), which are the oxygen-carrying cells. The functional details of erythropoietin are described in section 18.3b and summarized in the reference tables following this chapter (see **table R.6**).

Liver

Recall that the liver releases insulin-like growth factors (IGFs) in response to growth hormone, as described in section 17.8a. The liver also releases **angiotensinogen** (an′jē-ō-ten-sin′ō-jen), an inactive hormone. The activation of angiotensinogen to **angiotensin II** (the active form of the hormone) requires both an enzyme released from the kidney (renin) and an enzyme anchored within the inner lining of blood vessels (angiotensin-converting enzyme or ACE) (see section 20.6b). The primary function of angiotensin II is to increase blood pressure. Angiotensin II has several effects. It (1) is a powerful blood vessel constrictor, (2) stimulates the kidney to decrease urine output, and (3) stimulates the thirst center. All of these processes function to keep blood pressure within normal homeostatic limits. (Note: The liver also secretes EPO; however, the primary producers of EPO are the kidneys.) The functional details of angiotensin II are described in sections 20.6b, 24.5e, and 25.4a and summarized in the reference tables following this chapter (see **table R.7**).

Stomach and Small Intestine

The stomach and small intestine are regions of the gastrointestinal (GI) tract. The stomach both synthesizes and releases gastrin. A primary function of **gastrin** (gas′trin; *gaster* = stomach) is to increase stomach activity (both its motility and its release of secretions) to facilitate digestion within the stomach (see section 26.2d). The small intestine is a long tube that is inferior to the stomach and located medially within the abdominal cavity. The small intestine releases both secretin and cholecystokinin, both of which function to facilitate digestion within the small intestine. A primary function of **secretin** (se-krē′tin) is to stimulate release of secretions from both the liver (bile) and pancreas (pancreatic juice) into the small intestine. A primary function of **cholecystokinin (CCK)** (ko′lē-sis-to-kī′nin; *chole* = bile, *cyst* = sac, *kinin* = to move), and what gives the hormone its name, is to stimulate release of bile from the gallbladder (a muscular sac on the inferior surface of the liver). The functional details of gastrin, secretin, and cholecystokinin are described in sections 26.2d and 26.3c and summarized in the reference tables following this chapter (see **table R.8**).

Skin

Ultraviolet light penetrates into surface skin cells (keratinocytes) to convert modified cholesterol molecules to vitamin D_3 (cholecalciferol), which is then released into the blood. Vitamin D_3 is converted to calcidiol by an enzyme within the liver and then by an enzyme within the kidney to **calcitriol,** the active hormone. Calcitriol is similar to parathyroid hormone because it increases blood calcium by stimulating release of calcium from bone tissue and decreases calcium loss in the urine. Additionally, calcitriol stimulates calcium absorption from the small intestine. The functional details of calcitriol are described in section 7.6b and summarized in the reference tables following this chapter (see **table R.2**).

Adipose Connective Tissue

Adipose connective tissue is located throughout the body, and it releases the hormone **leptin.** This hormone helps to regulate food intake by binding to the neurons within the hypothalamus that control appetite (see section 13.4c). Lower percentage of body fat is associated with lower blood levels of leptin, which stimulates the appetite. Thus, one of the functions of leptin is to regulate energy balance within the body. Clinicians and researchers have become more aware of other endocrine functions of adipose connective tissue by observing the outcomes of either excess or deficiency of this tissue. Excess adipose connective tissue has been linked with various types of cancers (e.g., colon, breast) and delay puberty in males, whereas abnormally low body fat can both delay the onset of puberty and interfere with a normal menstrual cycle in females.

The details of the major hormones discussed in this chapter (and throughout the text) are summarized in tables that are located directly following this chapter, and are organized based on their function (e.g., regulate blood glucose, regulate blood calcium). Each table includes the structure that produces it, the primary stimulus for its release, its primary target organs and cellular changes that it induces, and the section(s) in the text where it is described in detail.

 WHAT DID YOU LEARN?

30 What is the function of the kidney in the amount of erythrocytes circulating in the blood?

31 What organ releases angiotensinogen, and what is the function of angiotensinogen following its activation?

17.11 Aging and the Endocrine System

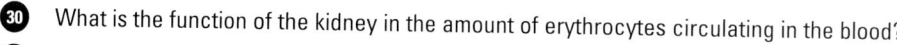 **LEARNING OBJECTIVE**

1. Describe how endocrine activity changes as people age.

The secretory activity of endocrine glands typically wanes as we age. Aging reduces the efficiency of endocrine system functions, and often normal levels of hormones decrease. Many conditions experienced after middle age, such as abdominal weight gain or muscle loss, are directly related to diminishing or reduced endocrine gland function.

One example is that the secretion of GH and sex hormones often decreases. Reduction in GH levels leads to loss of weight and body mass in the elderly, although continued exercise reduces this effect.

In addition, testosterone or estrogen levels decline as males and females age. Often hormone replacement therapy (HRT) attempts to supplement GH and sex hormone levels that have naturally diminished with age.

 WHAT DID YOU LEARN?

32 What general changes occur to the ability of endocrine glands to produce hormones as we age?

CHAPTER SUMMARY

- The endocrine system is composed of endocrine glands that produce chemical communication molecules called hormones.

17.1 Introduction to the Endocrine System 655

- The two control systems of the body include the nervous system and endocrine system.

17.1a Comparison of the Two Control Systems 655

- The nervous and endocrine systems complement each other to maintain homeostasis.

17.1b General Functions of the Endocrine System 656

- The primary processes controlled by hormones include regulating development, growth, and metabolism; maintaining homeostasis of blood composition and volume; controlling digestive processes; and controlling reproductive activities.

17.2 Endocrine Glands 656

- Endocrine glands are located throughout the body and are regulated to secrete their hormones into the blood.

17.2a Location of the Major Endocrine Glands 656

- Endocrine organs include the pituitary gland, pineal gland, thyroid gland, parathyroid glands, and adrenal glands.

- Endocrine tissue is housed in small clusters within the hypothalamus, skin, thymus, heart, liver, stomach, pancreas, small intestine, adipose connective tissue, kidneys, and gonads.

17.2b Stimulation of Hormone Synthesis and Release 658

- Release of hormones from endocrine cells is controlled through reflexes. There are three ways to stimulate these cells: (1) hormonal stimulation, (2) humoral stimulation by something in the blood other than another hormone, and (3) nervous system stimulation.

17.3 Hormones 658

- All hormones are synthesized within endocrine gland cells from either cholesterol or amino acids.

17.3a Categories of Circulating Hormones 658

- The three general categories of circulating hormones include steroids, biogenic amines, and proteins.

17.3b Local Hormones 659

- Eicosanoids are local hormones; they are synthesized from a fatty acid (arachidonic acid); following their formation, eicosanoids stimulate the cell that produced it (autocrine stimulation) or neighboring cells (paracrine stimulation).

17.4 Hormone Transport 660

- The mechanism of hormone transport is dependent upon whether the hormone is lipid-soluble or water-soluble.

17.4a Transport in the Blood 660

- Lipid-soluble hormones must attach to a carrier protein molecule to be transported within the blood.

- Water-soluble hormones readily dissolve in the aqueous environment of the blood and do not require a carrier protein.

17.4b Levels of Circulating Hormone 660

- The two primary factors that influence hormone concentration are hormone synthesis by endocrine glands and hormone elimination by the liver, kidneys, and target cells.

17.5 Target Cells: Interactions with Hormones 661

- Hormones bind with receptors of target cells; how this occurs is significantly different for lipid-soluble and water-soluble hormones.

17.5a Lipid-Soluble Hormones 661

- Hormones that are lipid-soluble (steroids, calcitriol, and thyroid hormone) stimulate cellular activity by binding to intracellular receptors: The hormone-receptor complex activates a region of DNA, resulting in the production of new proteins.

17.5b Water-Soluble Hormones 662

- Hormones that are water-soluble (proteins and biogenic amines, except thyroid hormone) bind with plasma membrane receptors; the hormone is the first messenger, and it causes the activation of G protein and the formation of a second messenger through an intracellular enzyme cascade.

- The cellular response may include activation or inhibition of enzymatic pathways, stimulation of growth through cellular division stimulation of cellular secretions, change membrane permeability, and muscle contraction or relaxation.

17.6 Target Cells: Degree of Cellular Response 664

- The degree of cellular response is a function of both its displayed receptors and the amounts and kinds of hormones that it binds.

17.6a Number of Receptors 664

- Up-regulation is the increase in the number of receptors, and down-regulation is a decrease in the number of receptors. Being able to change receptor number allows a target cell to modify its responsiveness to a hormone.

17.6b Receptor Interactions 665

- A single target cell may possess receptors for many different hormones.
- Hormones can interact with target cells to have one of three effects: synergistic, permissive, or antagonist.

(continued on next page)

17.7 The Hypothalamus and the Pituitary Gland 668	• The hypothalamus directly controls the release of hormones from the pituitary gland and indirectly controls the release of hormones from other endocrine glands.
	17.7a Anatomic Relationship of the Hypothalamus and the Pituitary Gland 668
	• The pituitary gland is inferior to the hypothalamus and connected to it by the infundibulum.
	• The hypothalamus communicates with the posterior pituitary via the hypothalamo-hypophyseal tract, which contains axons from two nuclei in the hypothalamus: the paraventricular nucleus and the supraoptic nucleus.
	• The hypothalamus communicates with the anterior pituitary via the hypothalamo-hypophyseal portal system, a vessel network that transports hormones from the hypothalamus to the anterior pituitary.
	17.7b Interactions Between the Hypothalamus and the Posterior Pituitary Gland 669
	• In response to nerve signals, the posterior pituitary releases antidiuretic hormone (ADH) or oxytocin (OT), which are hormones previously synthesized by the hypothalamus and stored in the posterior pituitary.
	17.7c Interactions Between the Hypothalamus and the Anterior Pituitary Gland 670
	• The regulatory hormones released from the hypothalamus include both "releasing" and "inhibiting" hormones.
	• Releasing hormones stimulate the release of specific hormones from the anterior pituitary, and inhibiting hormones decrease the release of hormones from the anterior pituitary.
	• The hormones synthesized and released from the anterior pituitary include thyroid-stimulating hormone (TSH), prolactin (PRL), follicle-stimulating hormone (FSH), luteinizing hormone (LH), adrenocorticotropic hormone (ACTH), and growth hormone (GH).
17.8 Representative Hormones Regulated by the Hypothalamus 673	• Regulatory hormones influence the secretion of growth hormone, thyroid hormone, and glucocorticoids (e.g., cortisol).
	17.8a Growth Hormone 673
	• Release of GH from the anterior pituitary is controlled by the release of growth hormone–releasing hormone (GHRH) and growth hormone–inhibiting hormone (GHIH) from the hypothalamus.
	• Growth hormone stimulates the release of insulin-like growth factors (IGFs) (somatomedins) from the liver; both GH and IGFs stimulate target cells (especially muscle) to increase protein synthesis, cell division, and cell differentiation; liver to increase both glycogenolysis and gluconeogenesis; and adipose connective tissue to increase lipolysis.
	17.8b Thyroid Gland and Thyroid Hormone 674
	• The thyroid gland is a butterfly-shaped gland anterior to the trachea and inferior to the larynx.
	• Decreased levels of thyroid hormone and certain stimuli cause the hypothalamus to secrete thyrotropin-releasing hormone (TRH), which causes release of thyroid-stimulating hormone (TSH) from the anterior pituitary. TSH reaches the thyroid gland and causes release of thyroid hormone (TH) from a stored precursor. This interactive sequence is referred to as the hypothalamic-pituitary-thyroid axis.
	• Thyroid hormone increases metabolism with a subsequent increase in body temperature.
	17.8c Adrenal Glands and Cortisol 678
	• The adrenal glands have both an inner medulla (which releases epinephrine and norepinephrine) and an outer cortex.
	• The adrenal cortex has three zones that produce mineralocorticoids, glucocorticoids (primarily cortisol), and gonadocorticoids.
	• Upon receiving certain stimulation, the hypothalamus secretes corticotropin-releasing hormone (CRH), which causes release of adrenocorticotropic hormone (ACTH) from the anterior pituitary. ACTH brings about release of cortisol by the adrenal cortex. This relationship is called the hypothalamic-pituitary-adrenal axis.
	• The net effect of cortisol is an increase in all nutrient molecules in the blood.
17.9 Pancreatic Hormones 684	• The pancreas releases both insulin and glucagon, which regulate nutrient blood levels.
	17.9a Anatomy of the Pancreas 684
	• The pancreas is both an exocrine and endocrine gland. The primary endocrine cells include alpha cells and beta cells.
	17.9b Effects of Pancreatic Hormones 685
	• The release of insulin into the blood from beta cells results in a decrease in all nutrients in the blood (including glucose) and an increase in the storage of these molecules within body tissues.
	• The release of glucagon from alpha cells results in an increase in glucose, glycerol, and fatty acids in the blood; it has no effect on structural and functional protein components of the body.
17.10 Other Endocrine Glands 688	• Hormones are also released from the pineal gland, parathyroid gland, and other structures.
	17.10a Pineal Gland 688
	• The pineal gland is a cone-shaped structure within the diencephalon that produces melatonin, which regulates circadian rhythms.
	17.10b Parathyroid Glands 688
	• The parathyroid glands produce parathyroid hormone, which increases blood calcium.
	17.10c Structures with an Endocrine Function 688
	• Structures with an endocrine function include the thymus, heart, kidneys, liver, stomach and small intestine, skin, and adipose connective tissue.
17.11 Aging and the Endocrine System 690	• The secretory activity of endocrine glands usually decreases with age, especially in regard to the production and activities of GH, testosterone, and estrogen.

CHALLENGE YOURSELF

Do You Know the Basics?

_____ 1. Which of the following is *not* a general process controlled by the endocrine system?

a. development, growth, and metabolism

b. control of reproductive activities

c. maintenance of homeostasis of blood composition

d. programmed cell death/destruction of aged cells

_____ 2. This hormone's primary function is to regulate metabolism.

a. calcitonin

b. thyroid hormone (TH)

c. growth hormone (GH)

d. glucagon

_____ 3. Which of the following are components of intracellular enzyme cascades initiated by water-soluble hormones?

a. G proteins

b. cAMP

c. protein kinase enzymes

d. All of these are correct.

_____ 4. A hormone released from the anterior pituitary is

a. glucagon.

b. growth hormone (GH).

c. melatonin.

d. epinephrine.

_____ 5. The action of water-soluble hormones may include all of the following *except*

a. activation or inhibition of enzymatic pathways.

b. bind with a hormone-responsive element.

c. muscle contraction or relaxation.

d. stimulation of cellular secretions.

_____ 6. Insulin increases _____ within hepatocytes to decrease blood glucose.

a. glycogenolysis

b. gluconeogenesis

c. glycogenesis

d. lipogenesis

_____ 7. Glucagon has an _____ effect to insulin on target cells.

a. antagonistic

b. synergistic

c. permissive

d. both permissive and synergistic

_____ 8. Glucocorticoids (e.g., cortisol) are produced in the adrenal cortex to help regulate

a. Na^+ and K^+ levels in body fluids.

b. blood pressure.

c. calcium levels in the blood.

d. glucose levels in the blood.

_____ 9. Thyroid-stimulating hormone stimulates the

a. anterior pituitary to release its hormones.

b. hypothalamus to release its hormones.

c. thyroid gland to release its hormones.

d. All of these are correct.

_____ 10. All of the following hormones are released from the hypothalamus to control the anterior pituitary gland *except*

a. growth-releasing hormone (GRH).

b. antidiuretic hormone (ADH).

c. prolactin-releasing hormone (PRH).

d. corticotropin-releasing hormone (CRH).

11. Describe similarities and differences between the endocrine system and the nervous system in their method of operation and effects.

12. List the four primary functions of the endocrine system.

13. Explain the three mechanisms used to stimulate hormone release from a target cell to initiate an endocrine reflex.

14. Identify the three chemical classes of hormones, and give an example of each. Most hormones belong to which class?

15. Describe how local hormones differ from circulating hormones.

16. Explain the function of carrier proteins in transporting lipid-soluble hormones in the blood, and how these hormones interact with cells.

17. Describe how water-soluble hormones interact with cells.

18. Explain how the hypothalamus oversees and controls endocrine system function of the posterior pituitary.

19. Explain how the hypothalamus oversees and controls endocrine system function of the anterior pituitary.

20. Discuss the homeostatic system involving insulin.

Can You Apply What You've Learned?

1. George is a 43-year-old construction worker who has developed a swelling in his neck that is painful and continues to grow. He visited the doctor and confided to his clinician that he has also lost weight and has become very hyperactive. What gland does the clinician suspect is functioning abnormally?

a. pituitary

b. thyroid

c. adrena

d. pancreas

2. What is the best diagnostic test to determine if this gland is not functioning normally?

   a. measuring the amount of radioactive iodine taken up by the thyroid

   b. doing body temperature scans every morning and evening at the same time

   c. watching weight fluctuation over a 1-month time period

   d. taking a blood sample to measure the amount of thyroid hormone (T_3 and T_4) present

3. Jelena is late for work and is rushing to get out the door. Her commute to work is slow due to rush-hour traffic, and she begins to become anxious and upset. As she is attempting to park, someone hits her car and she becomes angry. What specific hormones are released during this "emergency"?

   a. insulin/glucagon

   b. epinephrine/cortisol

   c. insulin/thyroid hormone

   d. melatonin/epinephrine

4. Blood samples from a young woman named Michelle indicate an elevated blood glucose level. This homeostatic imbalance is most likely caused by an insufficient amount, or decreased sensitivity, to which hormone?

   a. growth hormone

   b. glucagon

   c. insulin

   d. cortisol

5. Stephen is taking a new weight-loss supplement that is known to not only decrease the amount of adipose connective tissue (as advertised) but to also decrease glycogen stores in the liver and causes breakdown of muscle protein (protein catabolism). What substance in this weight-loss supplement is responsible for these changes?

   a. growth hormone

   b. glucagon

   c. insulin

   d. cortisol

Can You Synthesize What You've Learned?

1. After seeing a physician for a sudden weight loss, 19-year-old Harold is diagnosed with type 1 diabetes, which is a condition in which the beta cells of the pancreas are producing insufficient amounts of insulin or target cells not responding to insulin. Explain to Harold why results from his blood lab report indicate that he has an elevated blood glucose level.

2. Susan is a 35-year-old mother of two who works as an admissions officer at the university. She was recently diagnosed with a pituitary tumor. Consider the potential challenges she may experience by listing the hormones released by both the posterior pituitary and the anterior pituitary.

3. Henry is a well-informed patient who is interested in understanding how thyroid hormone is controlled by thyrotropin-releasing hormone (TRH) and thyroid-stimulating hormone (TSH). Briefly explain to him the hypothalamic-pituitary-thyroid axis.

INTEGRATE

ONLINE STUDY TOOLS

The following study aids may be accessed through Connect.

Clinical Case Study: A Giant of a Man is Defeated by a Young and Observant Boy

Interactive Questions: This chapter's content is served up in a number of multimedia question formats for student study.

LearnSmart: Topics and terminology include introduction to the endocrine system; endocrine glands; hormones; nutrient metabolism; the hypothalamus and the pituitary gland;

representative hormones regulated by the hypothalamus; pancreatic hormones; aging and the endocrine system

Anatomy & Physiology Revealed: Topics include pancreas; hypothalamus and pituitary glands; hormonal communication; intracellular receptor model; receptors and G proteins; thyroid; suprarenal (adrenal) glands

Animations: Topics include mechanism of lipid-soluble messengers; mechanism of thyroxine action

Major Regulatory Hormones of the Human Body

These tables provide a succinct reference for the major regulatory hormones of the human body. They are organized by the general variable or process regulated by each hormone or group of hormones. Each table includes the name of the hormone(s), its chemical structure (steroid, biogenic amine, or protein), the source of the hormone(s), primary stimulus for its release, its primary target organs and the cellular responses that it initiates, the net result or summary of its effect, an example of a related disease or condition, and the section reference(s) in the text where the hormone is discussed in detail.

Table R.1	Regulating Blood Glucose with Pancreatic Hormones	
Hormone	**Insulin**	**Glucagon**
Chemical structure	Protein (51 amino acids); water-soluble	Protein (29 amino acids); water-soluble
Source	Pancreas (beta cells)	Pancreas (alpha cells)
Primary stimulus for release	Increased blood glucose levels	Decreased blood glucose levels
Primary target organs and cellular changes	Liver: Increased glycogen storage (glycogen synthesized from glucose molecules, which are obtained from the blood) Adipose connective tissue: Increased triglyceride storage (triglycerides synthesized from glycerol and fatty acid, which are obtained from the blood) Skeletal muscle cells: Increased glycogen storage (glycogen synthesized from glucose molecules, which are obtained from the blood); Increased uptake of K^+ All target cells: Increased protein synthesis that results from increased uptake of amino acids from the blood; increased glucose uptake	Liver: Decreased glycogen storage (glycogen digested to glucose molecules); increased gluconeogenesis (glucose synthesis from noncarbohydrate sources—e.g., amino acids, lactate); glucose molecules are released into the blood Adipose connective tissue: Decreased triglyceride storage (triglycerides digested into glycerol and fatty acids, which are released into the blood) Skeletal muscle cells: Decreased glycogen storage (glycogen digested to glucose molecules, which are oxidized in cellular respiration to produce ATP within the muscle cells)
Net result	Increased synthesis/storage of fuel molecules (glycogen, triglycerides, and protein) Decreased blood levels of fuel molecules (glucose, glycerol, fatty acids, and amino acids)	Increased blood levels of glucose, glycerol, and fatty acids; decreased storage of glycogen and triglyceride molecules
Related diseases or conditions	Diabetes mellitus	Hypoglycemia
Text reference	Section 17.9b	Section 17.9b

Table R.2	Regulating Blood Calcium with Parathyroid Hormone and Calcitonin	
Hormone	**Parathyroid Hormone (PTH)**	**Calcitonin**
Chemical structure	Protein (84 amino acids); water-soluble	Protein (32 amino acids); water-soluble
Source	Parathyroid gland	Thyroid gland (parafollicular cells)
Primary stimulus for release	Decreased blood calcium levels	Increased blood calcium levels
Primary target organs and cellular changes	Bone: Increased osteoclast activity (Ca^{2+} released into the blood) Kidney: Decreased loss of Ca^{2+} and increased loss of phosphate (PO_4^{3-}) in urine Increased number of enzymes that convert calcidiol (inactive hormone in blood formed from vitamin D) to calcitriol, a hormone that functions synergistically with PTH and increases absorption of Ca^{2+} from small intestine	Bone: Primarily decreased activity of osteoclasts, especially in children (decreased release of Ca^{2+} into the blood) Kidney: Increased loss of Ca^{2+} in urine
Net result	Increased blood calcium levels	Decreased blood calcium levels
Related diseases or conditions	Hyperparathyroidism; hypoparathyroidism	Hyperthyroidism; hypothyroidism
Text reference	Section 7.6b	Section 7.6c

Table R.3	Regulating Growth with Growth Hormone and Insulin-like Growth Factor
Hormone	**Growth Hormone (GH)**
Chemical structure	Protein (191 amino acids); water-soluble
Source	Anterior pituitary
Primary stimulus for release	**Growth hormone–releasing hormone (GHRH)** is released from the hypothalamus and is transported in the hypophyseal portal veins to the anterior pituitary to stimulate the anterior pituitary to release **growth hormone (GH)**. GH stimulates the release of **insulin-like growth factors (IGFs)** from the liver.
Primary target organs and cellular changes[1]	Together growth hormone and IGFs interact with the following: • All cells, especially cartilage, bone, and muscle: Increased protein synthesis, cellular division, and cell differentiation • Liver: Decreased glycogen storage results from increased glycogenolysis (glycogen digested to glucose molecules); increased gluconeogenesis (glucose synthesis from noncarbohydrate molecules [e.g., amino acids, lactate]); glucose molecules are released into the blood • Adipose connective tissue: Decreased triglyceride storage results from increased lipolysis (triglycerides digested into glycerol and fatty acids), glycerol and fatty acid molecules are released into the blood
Net result	Cellular growth and protein synthesis; release of glucose, glycerol, and fatty acids into the blood provide necessary fuel molecules needed for growth
Related diseases or conditions	Pituitary dwarfism (in children), gigantism (in children), acromegaly (in adults)
Text reference	Section 17.8a

1. Other effects of growth hormone include increased hunger; skin development, including nails and hair; development of skeletal, muscular and nervous system; increased metabolic rate in mother and fetus during pregnancy; enhanced development of mammary glands during pregnancy; insulin antagonist; increased reabsorption of sodium ions (Na^+), potassium ions (K^+), and chlorine ions (Cl^-) by the kidneys; increased absorption of Ca^{2+} by the small intestine; and increased uptake of sulfur for synthesis of chondroitin sulfate by chondrocytes.

Table R.4	Regulating Metabolism with Thyroid Hormone
Hormone	**Thyroid Hormone (TH)**
Chemical structure	Biogenic amine (monoamine); water-insoluble (lipid-soluble)
Source	Thyroid gland
Primary stimulus for release	**Thyrotropin-releasing hormone (TRH)** is released from the hypothalamus and is transported in the hypophyseal portal veins to the anterior pituitary to stimulate the anterior pituitary to release **thyroid-stimulating hormone (TSH)** into the general circulation. TSH binds to cellular receptors of the thyroid gland, stimulating the release of **thyroid hormone (TH)—triiodothyronine (T_3) and tetraiodothyronine (T_4).**
Primary target organs and cellular changes[1]	All cells, especially neurons: Increased metabolic rate; increased amino acid uptake and protein synthesis; increased glucose uptake
	Liver: Decreased glycogen storage through increased glycogenolysis (glycogen digested to glucose molecules); increased gluconeogenesis (glucose synthesis from noncarbohydrate sources—e.g., amino acids, lactate); glucose molecules are released into the blood
	Adipose connective tissue: Decreased triglyceride storage results from increased lipolysis (triglycerides digested into glycerol and fatty acids); glycerol and fatty acid molecules are released into the blood
	Heart: Increased heart rate and force of contraction, which increases cardiac output
Net result	Increased metabolism (results in increased production of ATP, increased body temperature [calorigenic effect], increased blood P_{CO_2}, which stimulates respiratory center to increase resting breathing rate); increased oxygen consumption; increased release of glucose, glycerol, and fatty acids into the blood (provide necessary fuel molecules needed for increased metabolic rate)
Related diseases or conditions	Hyperthyroidism; hypothyroidism
Text reference	Section 17.8b

1. Other effects of thyroid hormone include increased appetite; increased alertness; bone growth and remodeling; skin development, including nails and hair; skeletal, muscular, and nervous system development; increased metabolic rate in mother and fetus during pregnancy; insulin antagonist; and release of GH from the anterior pituitary.

Table R.5	Regulating the Stress Response with Catecholamines and Glucocorticoids	
Hormone	**Catecholamines: Epinephrine and Norepinephrine**	**Glucocorticoids: Cortisol and Corticosterone**
Chemical structure	Biogenic amine (monoamine); water-soluble	Steroid hormones; water-insoluble (lipid-soluble)
Source	Adrenal medulla	Adrenal cortex (zona fasciculata)
Primary stimulus for release	Sympathetic division stimulation causes release of epinephrine (85%) and norepinephrine (15%)	**Corticotropin-releasing hormone (CRH)** is released from the hypothalamus and is transported in the hypophyseal portal veins to the anterior pituitary to stimulate the anterior pituitary to release **adrenocorticotropic hormone (ACTH)** into the general circulation. ACTH binds to cellular receptors of the adrenal cortex, stimulating the release of **glucocorticoids (e.g., cortisol).**
Primary target organs and cellular changes[1]	Increases all effects of the sympathetic division of the autonomic nervous system (see table 15.6)	Adipose connective tissue: Decreased triglyceride storage caused by increased lipolysis (triglycerides digested into glycerol and fatty acids); glycerol and fatty acid molecules are released into the blood
		All cells, except liver cells: Increased protein catabolism (proteins are digested into amino acids); amino acids released into the blood
		Liver cells: Increased gluconeogenesis (glucose synthesis from noncarbohydrate sources—e.g., amino acids, lactate); glucose molecules are released into the blood
		Immune system: Anti-inflammatory effect
Net result	Increased level of response and increased duration of response (about 30 minutes) that is initiated by the sympathetic division of the autonomic nervous system	Increased available fuel molecules in blood especially glucose (decreased storage of fuel molecules); anti-inflammatory
Related diseases or conditions	Chronic stress	Cushing syndrome, Addison disease
Text reference	Section 15.4	Section 17.8c

1. Other effects of glucocorticoids include impaired connective tissue repair (decreased fibroblasts and decreased protein synthesis of connective tissue matrix formation); atrophy of muscle, atrophy of skin, decrease in bone tissue; decreased lymphatic tissue (thymus, lymph nodes, spleen), decreased white blood cells (eosinophils, lymphocytes, and macrophages), decreased production of antibodies; decreased response by cell-mediated immunity and blocked production of fever; increased Na^+ and water retention; permissive effects for epinephrine and norepinephrine (catecholamines) that enhance vasoconstriction and increase cardiac output; increased gastric secretion; and inhibited release of luteinizing hormone, estrogen, and possibly testosterone.

Cortisol-like drugs are administered as treatment for inflammatory diseases (e.g., rheumatoid arthritis, eczema, asthma). High doses can result in side effects that include edema, muscle weakness, osteoporosis, thin skin, suppressed immune response, and infertility.

Table R.6	Regulating Erythrocyte Concentration in the Blood
Hormone	**Erythropoietin (EPO)**
Chemical structure	Glycoprotein; water-soluble
Source	Kidney (primarily) and liver
Primary stimulus for release	Decreased blood O_2; testosterone
Primary target organs and cellular changes	Red bone marrow: Increases rate of production of erythrocytes, cells that transport oxygen
Net result	Increased O_2-carrying capacity of blood
Related diseases or conditions	Anemia; polycythemia
Text reference	Section 18.3b

Table R.7	Regulating Fluid Balance, Blood Volume, and Blood Pressure			
Hormone	**Angiotensin II (Ang II)**	**Antidiuretic Hormone (ADH)**	**Aldosterone (ALDO)**	**Atrial Natriuretic Peptide (ANP)[1]**
Chemical structure	Protein (8 amino acids); water-soluble	Protein (9 amino acids); water-soluble	Steroid hormone; water-insoluble (lipid-soluble)	Protein (28 amino acids); water-soluble
Source	Liver (produces and releases angiotensinogen into the blood); activated in the blood in the presence of renin (released from kidney when stimulated) and angiotensin-converting enzymes (ACE) (endothelial layer of blood vessels, especially those in the lungs)	Produced by the hypothalamus and stored in the posterior pituitary (release is controlled by hypothalamus)	Adrenal cortex (zona glomerulosa)	Atrial chambers of the heart
Primary stimulus for release	Release of renin from juxtaglomerular apparatus of kidney stimulated by the following: • Decreased blood pressure • Sympathetic division stimulation	Hypothalamus sends a nerve signal along the hypothalamo-hypophyseal tract to the posterior pituitary in response to the following: • Angiotensin II • Increased blood osmolarity • Decreased nerve signals initiated by baroreceptors in atria, aorta, and carotid arteries in response to decrease in stretch	Angiotensin II Increased blood K^+ Decreased blood Na^+	Increased stretch of atrial wall (reflects increase in blood volume and blood pressure)
Primary target organs and cellular changes	Blood vessels: Potent vasoconstrictor, increases vascular resistance Kidney: Decreases urine output by decreasing glomerular filtration rate (GFR) to maintain blood volume Thirst center within the hypothalamus: Stimulation Hypothalamus: Releases ADH from posterior pituitary Adrenal cortex: Releases aldosterone	Kidney: Decreases H_2O excreted in urine Thirst center within the hypothalamus: Stimulation Blood vessels: Vasoconstrictor in high doses (why it is also called vasopressin)	Kidney: Decreases Na^+ and H_2O excreted in urine; increases K^+ excretion (except under conditions of low pH when H^+ is excreted instead)	Kidney: Increases GFR; increases Na^+ and H_2O excreted in urine; inhibits release of renin Blood vessels: Vasodilation, decreases vascular resistance Hypothalamus: Inhibits release of ADH from posterior pituitary Adrenal cortex: Inhibits release of aldosterone
Net result	Increased resistance, increased blood pressure; maintaining blood volume and blood pressure	Increased retention of water; maintaining blood volume and blood pressure	Maintenance of blood Na^+ and blood K^+ levels; maintaining blood volume and pressure by decreasing urine output	Increased urine output; decreased resistance, decreased blood volume and blood pressure
Related diseases or conditions	Blood pressure homeostasis	Diabetes insipidus	Hyperaldosteronism Hypoaldosteronism	Blood pressure homeostasis
Text reference	Sections 20.6b, 24.5e, and 25.4a	Sections 24.6d and 25.4b	Sections 24.6d and 25.4c	Sections 24.5e, 24.6d, and 25.4d

1. Other names include atrial natriuretic factor (ANF), atrial natriuretic hormone (ANH), and atriopeptin.

Table R.8	Regulating the Digestive System		
Hormone	Gastrin	Secretin	Cholecystokinin (CCK)
Chemical structure	Three forms of proteins (34, 17, 14 amino acids); water-soluble	Protein (27 amino acids); water-soluble	Protein (varying numbers of amino acids); water-soluble
Source	Stomach (enteroendocrine G cells)	Duodenal enteroendocrine cells	Duodenal enteroendocrine cells
Primary stimulus for release	Thought, smell, sight of food; parasympathetic division (vagal) stimulation; presence of partially digested protein in stomach; stretch of stomach wall, caffeine, increase in stomach pH	Acidic chyme entering small intestine	Chyme high in lipids and protein entering small intestine (duodenum)
Primary target organs and cellular changes	Stomach: Parietal cells release HCl and intrinsic factor, chief cells release pepsinogen; increased motility Gallbladder: Contraction causing release of bile into small intestine Pancreas: Secretion of pancreatic juice Small intestine: Contraction of wall Large intestine: Mass movements Pyloric sphincter: Relaxes Ileocecal valve: Relaxes	Stomach: Decreased secretions and motility Liver and pancreas: Increased secretion of bicarbonate ion (weak base) in bile and pancreatic juice	Stomach: Decreased secretions and motility Pancreas: Secretion of pancreatic juice containing pancreatic digestive enzymes Liver: Secretion of bile Gallbladder: Contraction causing release of bile into small intestine Hepatopancreatic sphincter: Relaxes
Net result	Increased secretions and motility of stomach; preparation of small intestine for arrival of chyme by relaxing pyloric sphincter and stimulating accessory gland secretions; movement of contents through the large intestine	Inhibited stomach activity; buffering of acidic chyme entering duodenum	Most important in increasing ability of small intestine to digest triglycerides by decreasing stomach motility, and increasing secretions from accessory glands (liver, gallbladder, and pancreas) into small intestine
Related diseases or conditions	Zollinger-Ellison syndrome	Low levels are associated with *Helicobacter pylori* infection	Change in release of CCK is associated with gallstones
Text reference	Section 26.2d	Sections 26.2d and 26.3c	Sections 26.2d and 26.3c

Table R.9	Regulating the Female Reproductive System			
Hormone	**Estrogen**	**Progesterone**	**Prolactin**	**Oxytocin**
Chemical structure	Steroid hormone; water-insoluble (lipid-soluble)	Steroid hormone; water-insoluble (lipid-soluble)	Protein (198 amino acids); water-soluble	Protein (9 amino acids); water-soluble
Source	Ovary (developing follicle)	Ovary (corpus luteum)	Anterior pituitary	Posterior pituitary
Primary stimulus for release	**Gonadotropin-releasing hormone (GnRH)** is released from the hypothalamus and is transported in the hypophyseal portal veins to stimulate the release of **follicle-stimulating hormone (FSH)** from the anterior pituitary into general circulation; FSH stimulates the development of follicles (in the ovary) and developing follicles produce **estrogen**	**Gonadotropin-releasing hormone (GnRH)** is released from the hypothalamus and is transported in the hypophyseal portal veins to stimulate the release of **luteinizing hormone (LH)** from the anterior pituitary into general circulation; LH triggers ovulation and the production of the corpus luteum; the corpus luteum produces **progesterone**	**Prolactin-releasing hormone (PRH)** is released from the hypothalamus and is transported in the hypophyseal portal veins to stimulate the release of **prolactin (PRL)** from the anterior pituitary into general circulation (Prolactin release is normally inhibited by the release of **prolactin-inhibiting hormone [PIH]** from the hypothalamus)	Hypothalamus sends nerve signals along the hypothalamo-hypophyseal tract to the posterior pituitary in response to sensory nerve signals from contractions of uterus or suckling of breast, to stimulate release of **oxytocin**
Primary target organs and cellular changes	Uterus: Builds endometrial lining Hypothalamus: Moderate levels inhibit release of GnRH and FSH; high levels stimulate release of GnRH and FSH Breast: Stimulates development of mammary gland Kidney: Retention of Na⁺ and water All cells: Increases protein anabolism	Uterus: Builds and maintains endometrial lining Mammary glands: Prepares for secretion of milk Kidney: Decreases retention of Na⁺ and water	Breast: Development of mammary ducts and glands	Uterus: Stimulates uterine contraction during delivery (and after delivery to expel the placenta, and continues to cause contractions to firm up the uterus) Breast: Stimulates ejection of milk from mammary glands Brain: Increases feelings of emotional bonding between individuals
Net result	Assists in ovarian follicle (and oocyte) development, regulation of female cycle; anabolic hormone; development of female characteristics; retention of Na⁺ and water	Regulates second half of uterine cycle; prepares mammary glands for milk production; eliminates Na⁺ and water	Stimulates development of mammary ducts and glands within the breast for milk production	Stimulates contraction of smooth muscle of uterus and breast; emotional bonding
Related diseases or conditions	Infertility	Infertility	Prolactinomas	Severe cases of autism may be associated with low levels of oxytocin
Text reference	Sections 28.3 and 29.8	Sections 28.3 and 29.8	Sections 28.3 and 29.8	Sections 28.3, 29.6, and 29.8

Table R.10	Regulating the Male Reproductive System	
Hormone	**Testosterone**	
Chemical structure	Steroid hormone; water-insoluble (lipid-soluble)	
Source[1]	Testes (interstitial cells)	
Primary stimulus for release	**Gonadotropin-releasing hormone (GnRH)** is released from hypothalamus into the hypophyseal portal veins; to stimulate the anterior pituitary to release both **luteinizing hormone (LH)** and **follicle-stimulating hormone (FSH)**; LH stimulates testes (interstitial cells) to release **testosterone**; FSH stimulates release of **androgen-binding protein (ABP)** and release of **inhibin** from sustentacular cells	
Primary target organs and cellular changes	Testes: With ABP, testosterone stimulates sperm production Hypothalamus: Inhibits GnRH release All cells with testosterone receptors: Increases protein anabolism Bone: Stimulates osteoblasts	Kidney: Increases production of erythropoietin Secondary sex characteristics Increases libido (sex drive)
Net result	Assists in sperm production, regulation of male cycle; anabolic hormone; increases erythrocytes; development of male characteristics	
Related diseases or conditions	Androgen insensitivity, infertility	
Text reference	Sections 18.3b and 28.4b	

1. In both males and females, **dehydroepiandrosterone (DHEA)** is produced from adrenal cortex and is converted to testosterone (this is the only source of testosterone in females).

Cardiovascular System: Blood

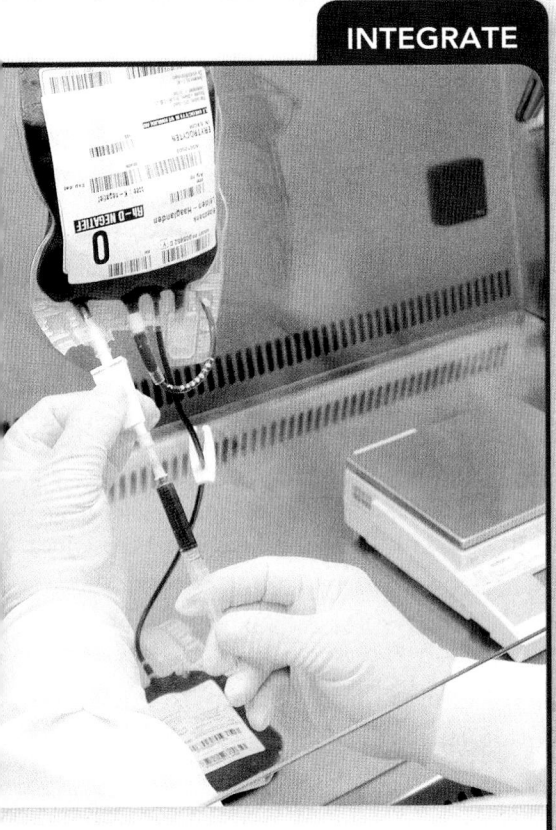

Module 9: Cardiovascular System

CAREER PATH
Blood Bank Technician

A blood bank technician is responsible for sampling and testing donated blood. This testing involves determining the blood type of the donated sample and whether the sample contains infectious agents that can cause serious diseases (such as hepatitis or HIV). The technician has a thorough understanding of blood composition and the importance of matching donor and recipient blood types prior to a transfusion.

A blood bank technician must practice *universal precautions* when working with blood. (Universal precautions refers to the practice of avoiding direct contact with a patient's bodily fluids by wearing gloves and wearing safety goggles if there is danger of fluid splashing.)

Within our bodies is a connective tissue so valuable that donating a portion of it to someone else can save that person's life. This tissue is regenerated continuously and is responsible for transporting the gases, nutrients, and hormones our bodies need for proper functioning. Losing too much of this tissue can be fatal and yet is something we frequently take for granted.

This valuable connective tissue is blood. Blood is considered a fluid connective tissue because it contains formed elements (red blood cells, white blood cells, platelets) and dissolved proteins in a liquid ground substance called plasma. Four to six liters of this warm, alkaline, viscous fluid is continuously pumped through our blood vessels. It may help to think of the circulation of blood as a "fluid conveyor belt" where cells, ions, and molecules are both continuously added to it and dropped off from it. As a result, the composition of blood is ever changing as it is pumped by our heart and makes its continuous journey through our vessels. Because of its intimate contact with the cells of the body, various blood tests can be performed, providing a physician with important information for an accurate diagnosis in assessing the state of our health.

In this chapter, we describe the function of blood and the various blood components. We then examine how these components are formed, how they function, and discuss hemostasis. We finish with a discussion of development and aging of blood.

18.1 Functions and General Composition of Blood

Blood is the specialized fluid that is transported through the **cardiovascular** (kar'dē-ō-vas'kū-lăr; *cardio* = heart, *vascular* = vessels) **system,** which is composed of the heart and blood vessels. Blood vessels form a circuit away from the heart and back to the heart that includes the arteries, capillaries, and veins. **Arteries** transport blood away from the heart, whereas **veins** transport blood toward the heart. **Capillaries** (kap'i-lār-ēz; *capillaries* = relating to hair) are permeable, microscopic vessels between arteries and veins. Capillaries serve as the sites of exchange between the blood and body tissues; it is from our capillaries that oxygen and nutrients exit the blood, and carbon dioxide and cellular wastes enter the blood.

   Blood is composed of **formed elements** (erythrocytes, leukocytes, and platelets) and plasma. **Erythrocytes** (ĕ-rith'rō-sīt; *erthyros* = red, *kytes* = cell) (or *red blood cells*) function to transport respiratory gases in the blood. **Leukocytes** (lu'kō-sit; *leuko* = white) (or *white blood cells*) contribute to defending the body against pathogens, and **platelets** help clot the blood and prevent blood loss from damaged vessels. **Plasma** is the fluid portion of blood containing plasma proteins and dissolved solutes. We begin by describing the general functions of blood, its physical characteristics, and its general components.

18.1a Functions of Blood

 **LEARNING OBJECTIVE**

1. Describe the general functions of blood.

Blood carries out a variety of important functions as it circulates throughout the body; these functions can be grouped as transportation, regulation, and protection.

Transportation

Blood transports formed elements and dissolved molecules and ions throughout the body. Consider that as blood is transported through the blood vessels it carries oxygen from and carbon dioxide to the lungs, nutrients absorbed from the gastrointestinal (GI) tract, hormones released by endocrine glands, and heat and waste products from the systemic cells. Even when you take a medication, it is the blood that delivers it to the cells of your body. Thus, it is the blood that serves as the "delivery system" for the body.

INTEGRATE

CONCEPT CONNECTION

Recall from sections 1.5b and 6.1d that the amount of heat absorbed and released is regulated by the hypothalamus. This is accomplished by both (1) stimulating muscle tissue contraction to increase the amount of heat generated (e.g., shivering and goose bumps), and (2) redistributing blood flow to the dermis through vasoconstriction (to retain heat) and vasodilation (to release heat).

Regulation

Blood participates in the regulation of body temperature, body pH, and fluid balance:

- **Body temperature.** Blood helps regulate body temperature. This is possible because blood absorbs heat from body cells, especially skeletal muscle, as it passes through blood vessels of body tissues. Heat is then released from blood at the body surface as blood is transported through blood vessels of the skin (see section 1.5).

- **Body pH.** Blood, because it absorbs acid and base from body cells, helps maintain the pH of cells. Blood contains chemical buffers (e.g., proteins, bicarbonate) that bind and release hydrogen ions (H^+) to maintain blood pH until the excess is eliminated from the body (see section 25.6).

- **Fluid balance.** Water is added to the blood from the gastrointestinal tract and lost in numerous ways (including in urine, sweat, and respired air). In addition, there is a constant exchange of fluid between the blood plasma in the capillaries and the interstitial fluid surrounding the cells of the body's tissues. Blood contains proteins and ions that exert osmotic pressure to pull fluid back into the capillaries to help maintain normal fluid balance (see section 20.3b).

Protection

Blood contains leukocytes, plasma proteins, and various molecules that help protect the body against potentially harmful substances. These substances are part of the immune system that is described in detail in chapter 22. Components of blood, including platelets and plasma proteins, also protect the body against blood loss, as described in section 18.4.

 WHAT DID YOU LEARN?

❶ What are some of the materials that blood transports?

❷ How does blood help regulate body temperature and fluid levels in the body?

18.1b Physical Characteristics of Blood

LEARNING OBJECTIVE

2. Name six characteristics that describe blood, and explain the significance of each to health and homeostasis.

Blood is a type of connective tissue (see section 5.2d) that can be described based on its physical characteristics including color, volume, viscosity, plasma concentration, temperature, and pH:

- **Color.** The color of blood depends upon whether it is oxygen-rich or oxygen-poor. Oxygen-rich blood is bright red or almost scarlet. Contrary to popular belief, oxygen-poor blood is not blue; rather, oxygen-poor blood is dark red. The bluish appearance of our veins can be attributed to both (1) the fact that we can see the blood traveling through the superficial veins in the skin; and (2) how light is reflected back to the eye from different colors. Lower-energy light wavelengths, like red, are absorbed by the skin and *not* reflected back to the eye, but higher-energy wavelengths like blue *are* reflected back to the eye, so the eyes can perceive only the blue coloration from the veins.

- **Volume.** The average volume of blood in an adult is 5 liters (L). Males tend to have, on average, 5 to 6 L, whereas females have, on average, 4 to 5 L. The greater amount of blood in males is due to their larger average size. Sustaining a normal blood volume is essential in maintaining blood pressure.

- **Viscosity.** Blood is about 4 to 5 times more *viscous* than water, meaning that it is thicker. Viscosity of blood depends upon the amount of dissolved substances in the blood relative to the amount of fluid; that is, viscosity is increased if the amount of substances—primarily erythrocytes—increases, or the amount of fluid decreases, or both.

- **Plasma concentration.** Although viscosity is a characteristic of whole blood, another important physical characteristic is the plasma concentration, which is the relative concentration of solutes (e.g., proteins and ions) in plasma. This is normally a 0.09% concentration, and it determines whether fluids move into or out of the plasma by osmosis as blood is transported through capillaries. For example, when an individual is dehydrated, the plasma becomes hypertonic (see section 4.3b), and fluid moves into the plasma from the surrounding tissues. Additionally, the plasma concentration is used to determine intravenous (IV) solution concentrations, which are usually isotonic to plasma.

- **Temperature.** The temperature of blood is about 1°C higher than measured body temperature; thus, if your body temperature is 37°C (98.6°F), your blood temperature is about 38°C (100.4°F). Therefore blood warms areas through which it travels.

- **Blood pH.** Blood plasma is slightly alkaline, with a pH between 7.35 and 7.45. Plasma proteins, like all proteins of the body, have a three-dimensional shape that is dependent upon H^+ concentration. If the pH is altered from the normal range, plasma proteins become denatured and are unable to carry out their functions (see section 2.8b).

The physical characteristics of blood are summarized in **table 18.1**.

WHAT DO YOU THINK?

1. If a woman has 5 L of blood and she donates 1 pint (about 0.5 L), what approximate percentage is she donating: 1%, 5%, 10%, or 15%? Why do you think individuals below a certain weight (i.e., less than 110 pounds) are not allowed to give blood?

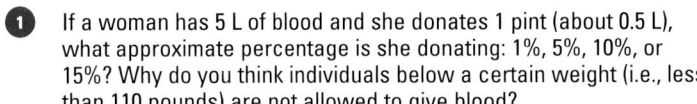

WHAT DID YOU LEARN?

3. Will blood be able to properly carry out its functions if blood pH is significantly altered? Why or why not?

18.1c Components of Blood

LEARNING OBJECTIVES

3. List the three components of a centrifuged blood sample.
4. Define hematocrit, and explain how the medical definition differs from the clinical usage.
5. Name the three formed elements of the blood, and compare their relative abundance.

Centrifuged Blood

Whole blood, which is both plasma and formed elements, can be separated into its liquid and cellular components by using a **centrifuge,** a device that spins the sample of blood in a tube so that heavier components collect at the bottom. **Figure 18.1** shows the resulting three components separated in the test tube. From bottom to top, these components are as follows:

- Erythrocytes form the lower layer of the centrifuged blood. They typically make up about 44% of a blood sample.
- A thin **buffy coat** makes up the middle layer. This slightly gray-white layer is composed of both leukocytes and platelets. The buffy coat forms less than 1% of a blood sample.
- Plasma is a straw-colored liquid that rises to the top in the test tube; it generally makes up about 55% of blood.

Table 18.1	Physical Characteristics of Blood	
Characteristics	**Normal Values**	
Color	Scarlet (oxygen-rich) to dark red (oxygen-poor)	
Volume	4–5 L (females) 5–6 L (males)	
Viscosity (relative to water)	4.5–5.5 × (whole blood)	
Plasma concentration	0.09%	
Temperature	38°C (100.4°F)	
pH	7.35–7.45	

INTEGRATE

CONCEPT CONNECTION

Recall from section 2.5c that an **acid** increases the concentration of H^+ by releasing it into the solution. Examples include hydrochloric acid (HCl) and carbonic acid (H_2CO_3). In contrast, a **base** decreases the concentration of H^+ in the solution. Examples include bicarbonate ions (HCO_3^-) and hydroxide ions (OH⁻). The **pH** is a measure of the relative amounts of H^+ in solution. A **buffer** helps prevent pH changes by binding or releasing excess H^+ to maintain the normal H^+ concentration in a solution.

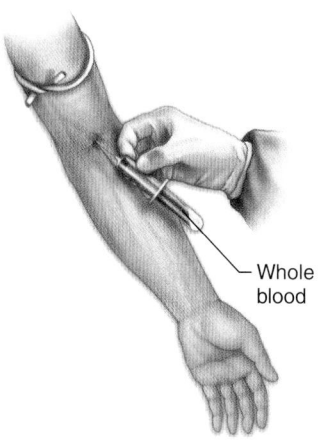

Whole blood

① Withdraw blood into a syringe and place it into a glass centrifuge tube.

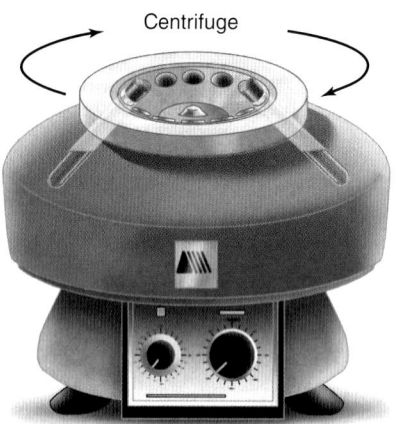

Centrifuge

② Place the tube into a centrifuge and spin for about 10 minutes.

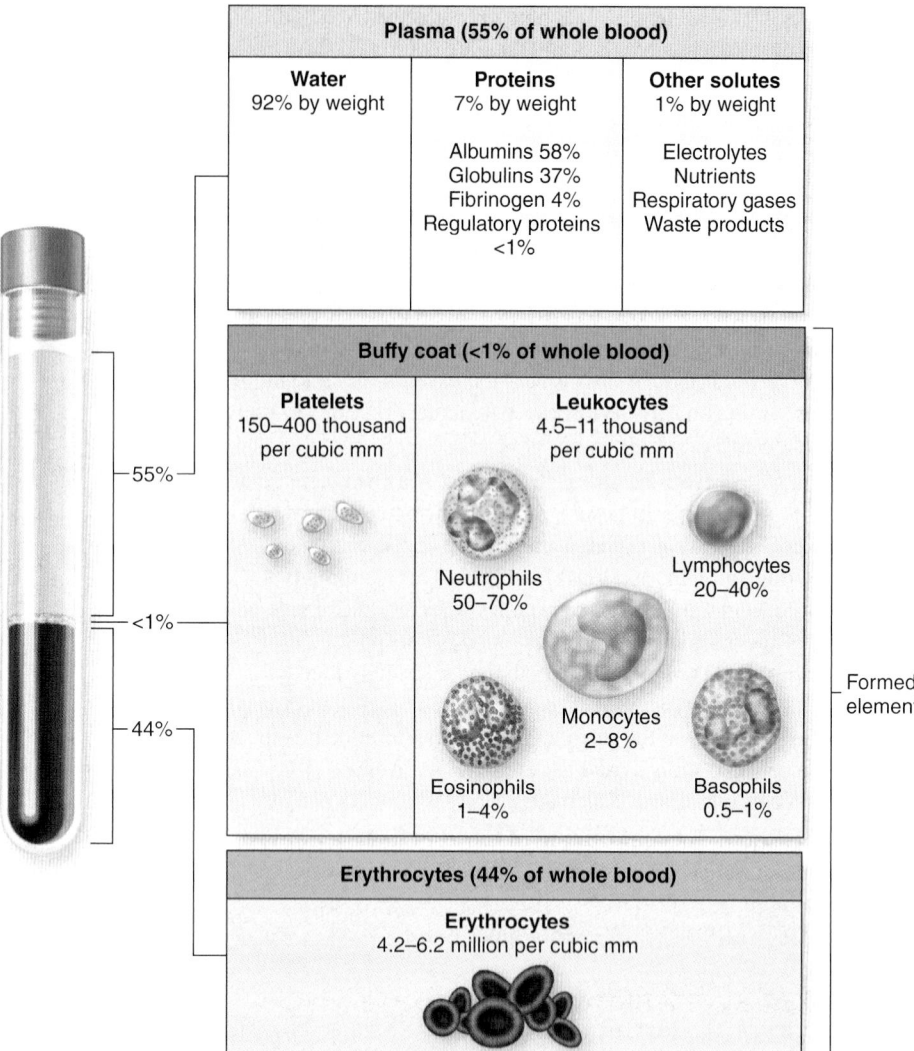

Plasma (55% of whole blood)		
Water 92% by weight	**Proteins** 7% by weight	**Other solutes** 1% by weight
	Albumins 58% Globulins 37% Fibrinogen 4% Regulatory proteins <1%	Electrolytes Nutrients Respiratory gases Waste products

Buffy coat (<1% of whole blood)

Platelets 150–400 thousand per cubic mm	Leukocytes 4.5–11 thousand per cubic mm
	Neutrophils 50–70%
	Lymphocytes 20–40%
	Monocytes 2–8%
	Eosinophils 1–4%
	Basophils 0.5–1%

Erythrocytes (44% of whole blood)

Erythrocytes
4.2–6.2 million per cubic mm

55%
<1%
44%

Formed elements

③ Components of blood separate during centrifugation to reveal plasma, buffy coat, and erythrocytes

Figure 18.1 Whole Blood Separation and Composition. Whole blood contains plasma (average is about 55%) and formed elements (average is about 45%). The percentages presented in this figure are average percentages of cells, and the values for components of the buffy coat represent average ranges. A cubic millimeter (mm³) of blood is equivalent to a microliter (μL). AP|R

INTEGRATE

CONCEPT CONNECTION

Not only does the blood contain buffers to help maintain the body's pH, but the urinary system and respiratory system also help to maintain this pH (see sections 25.5b, c). An increase in breathing rate can decrease blood carbon dioxide levels and H⁺ levels, thus increasing blood pH, whereas a decrease in breathing rate can cause an increase in carbon dioxide levels and H⁺ levels, thus decreasing blood pH. The urinary system helps maintain a normal blood pH by either producing HCO_3^- and eliminating H⁺ in the urine to increase blood pH, or eliminating HCO_3^- and retaining H⁺ to decrease blood pH.

The percentage of the volume of all formed elements (erythrocytes, leukocytes, and platelets) in the blood is called the **hematocrit** (hē'mă-tō-krit, hem'ă-; *hemato* = blood, *krino* = to separate). This medical dictionary definition of the true hematocrit differs slightly from the clinical definition, which equates the hematocrit to the percentage of only erythrocytes. (In practice, the true hematocrit and the clinical hematocrit are virtually the same.)

Hematocrit values vary somewhat and are dependent upon the age and sex of the individual. A very young child's hematocrit may vary from 30% to 60%, and that range will narrow to 35% to 50% as the child becomes older. Adult males tend to have a hematocrit ranging between 42% and 56%, whereas adult females' hematocrits range from 38% to 46%. Males typically have a higher hematocrit because testosterone stimulates the kidney to produce the hormone erythropoietin (EPO), which promotes erythrocyte production (see section 18.3b). An elevated hematocrit may indicate that the patient is either dehydrated or participating in blood doping, whereas a lowered hematocrit often suggests the patient is suffering from anemia.

Blood Smear

All of the components of the formed elements can be viewed by preparing a **blood smear.** A blood smear and the steps to produce a smear are shown in **figure 18.2**. Note the following:

• Erythrocytes are the most numerous of the formed elements. These are anucleate cells and appear as pink or pale purple, biconcave discs.

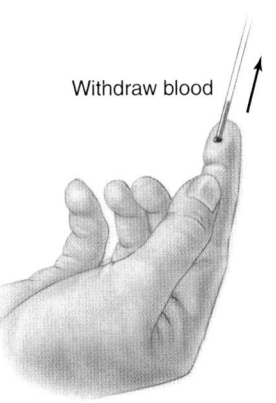

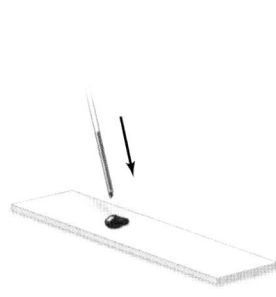

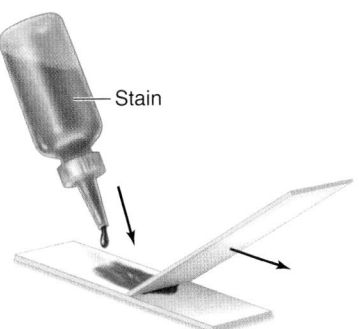

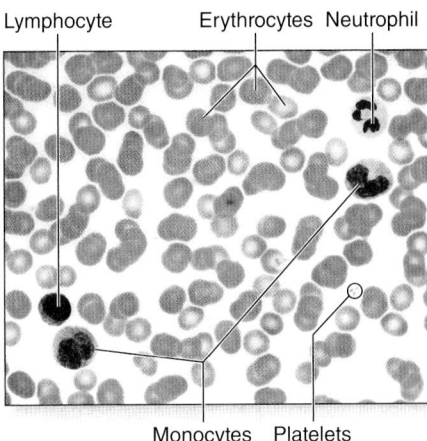

Lymphocyte   Erythrocytes Neutrophil

LM 640x

Monocytes   Platelets

(1) Prick finger and collect a small amount of blood using a microcapillary tube.

(2) Place a drop of blood on a slide.

(3a) Using a second slide, pull the drop of blood across the first slide's surface, leaving a thin layer of blood on the slide.

(3b) After the blood dries, apply a stain for contrast. Place a coverslip on top.

(4) When viewed under the microscope, blood smear reveals the components of the formed elements.

Figure 18.2 Preparing a Blood Smear.

- Leukocytes are larger than erythrocytes. The nucleus is very noticeable in leukocytes. Several leukocytes (a lymphocyte, neutrophil, and two monocytes) are shown in figure 18.2.
- Platelets appear as small fragments of cells.

WHAT DID YOU LEARN?

4 What are the three components visible in a centrifuged blood sample?

5 How does hematocrit vary among adult men and women, and how may dehydration affect hematocrit?

18.2 Composition of Blood Plasma

Plasma is composed primarily of water (about 92% of its volume), plasma proteins, and other solutes including electrolytes (e.g., Na^+), nutrients (e.g., glucose), respiratory gases (e.g., CO_2), and wastes (e.g., urea) **(table 18.2)**. Plasma is an extracellular fluid (ECF) because it is fluid found outside of cells. Plasma is similar to interstitial fluid, in that both have similar concentrations of electrolytes, nutrients, and waste products. However, one of the most significant differences is that protein concentration is higher in plasma than in the interstitial fluid (see section 25.1b). We first describe plasma proteins and then the specific substances transported in the plasma.

18.2a Plasma Proteins

LEARNING OBJECTIVES

1. Define colloid osmotic pressure.
2. Identify the various types of plasma proteins, and explain the general function of each.

Blood is considered a **colloid** (see section 2.6a) because it contains proteins in the plasma. Plasma proteins include albumin, globulins, fibrinogen and other clotting proteins, and regulatory proteins such as enzymes and some hormones. Most of these proteins are produced in the liver, including albumin, alpha- and beta-globulins, and both fibrinogen and other proteins involved with clotting. Plasma proteins, such as gamma-globulins and regulatory proteins, are produced by leukocytes and other organs, respectively.

Collectively, these plasma proteins exert osmotic pressure and prevent the loss of fluid from the blood as it moves through the capillaries. Osmotic pressure exerted by plasma proteins is called **colloid osmotic pressure.** This osmotic force is responsible for drawing fluids

Table 18.2	The Composition of Blood Plasma
Plasma Component (Percentage of Plasma)	**Functions**
Water (~92% of plasma)	The solvent in which formed elements are suspended and proteins and solutes are dissolved
PLASMA PROTEINS (~7% OF PLASMA): All proteins serve to buffer against pH changes	
Albumin (~58% of plasma proteins)	Exerts osmotic force to retain fluid within the blood Contributes to blood's viscosity Responsible for some fatty acid and hormone transport
Globulins (~37% of plasma proteins)	Alpha-globulins transport lipids and some metal ions Beta-globulins transport iron ions and lipids in blood Gamma-globulins are antibodies that immobilize pathogens
Fibrinogen (~4% of plasma proteins)	Participates in blood coagulation (clotting)
Regulatory proteins (<1% of plasma proteins)	Consists of enzymes and hormones
OTHER SOLUTES (~1% OF BLOOD PLASMA)	
Electrolytes (e.g., sodium, potassium, calcium, chloride, iron, bicarbonate, and hydrogen)	Help establish, maintain, and change membrane potentials, maintain pH balance, and regulate osmosis
Nutrients (e.g., amino acids, glucose, cholesterol, vitamins, fatty acids)	Energy source; precursor for synthesizing other molecules
Respiratory gases (e.g., oxygen: <2% dissolved in plasma, 98% bound to hemoglobin within erythrocytes, and carbon dioxide: ~7% dissolved in plasma, ~27% bound to hemoglobin within erythrocytes, ~66% converted to HCO_3^-)	Oxygen is needed for aerobic cellular respiration; carbon dioxide is a waste product produced by cells during this process
Wastes (breakdown products of metabolism) (e.g., lactate, creatinine, urea, bilirubin, ammonia)	Waste products serve no function in the blood plasma; rather, they merely are being transported to the liver and kidneys where they can be removed from the blood

into the blood and preventing excess fluid loss from blood capillaries into the interstitial fluid (see section 20.3b), thus helping to maintain blood volume and consequently blood pressure. If plasma protein levels decrease, such as might occur due to liver disease (resulting in decreased production of plasma proteins) or kidney damage (resulting in increased elimination of plasma proteins), colloid osmotic pressure also decreases. This decrease results in fluid loss from the blood and fluid retention in the interstitial space (i.e., edema).

Albumins (al-bū′min; *albumen* = white of egg) are the smallest and most abundant of the plasma proteins, making up approximately 58% of all plasma proteins. Because albumin is the most abundant type of plasma protein, it exerts the greatest colloid osmotic pressure to maintain blood volume and blood pressure. Secondarily, albumins act as transport proteins that carry ions, hormones, and some lipids in the blood.

Globulins (glob′ū-lin; *globules* = globule) are the second largest group of plasma proteins, forming about 37% of all plasma proteins. The smaller **alpha-globulins** and the larger **beta-globulins** primarily bind and transport certain water-insoluble molecules and hormones, some metals, and ions. **Gamma-globulins** are also called *immunoglobulins,* or *antibodies*, which play a part in the body's defenses (see section 22.8).

Fibrinogen (fī′brin-ō-jen; *fibra = fiber*) makes up about 4% of all plasma proteins. Fibrinogen as well as other clotting proteins are responsible for blood clot formation. Following trauma to the walls of blood vessels, fibrinogen is converted into long, insoluble strands of *fibrin,* which help form a blood clot. When the clotting proteins are removed from plasma, the remaining fluid is termed **serum** (ser′um; = whey). Blood clotting is described in more detail in section 18.4.

Regulatory proteins form a very minor class of plasma proteins (less than 1% of total plasma proteins). This group of proteins includes both enzymes to accelerate chemical reactions in the blood (see section 3.3a) and hormones being transported throughout the body to target cells (see section 17.3a).

 **WHAT DID YOU LEARN?**

6 How are plasma protein levels related to colloid osmotic pressure?

7 What is the most abundant type of plasma protein, and what are its two primary functions?

18.2b Other Solutes

 LEARNING OBJECTIVE

3. List dissolved substances in plasma by category.

Blood is also considered a *solution* because it contains dissolved ions as well as organic and inorganic molecules. These substances include electrolytes, nutrients, gases, and waste products. Recall from sections 2.3c and 17.4 that polar or charged substances (e.g., glucose and salts) dissolve readily in the blood, and nonpolar molecules (e.g., cholesterol, triglycerides, and fatty acids) do not readily dissolve in blood and require a carrier protein. Tables 18.3 and 18.4 list the normal ranges and functions of common solutes transported in blood plasma.

 WHAT DID YOU LEARN?

8 What are the main dissolved substances found in plasma?

Table 18.3 — Common Electrolytes in Arterial Plasma

Electrolytes (Ions)	Normal Ranges (Values)	Function	Substances and Structures That Regulate Electrolyte Blood Level
CATIONS			
Sodium (Na^+)	135–145 milliequivalents per liter (mEq/L)	Neuron and muscle function; fluid balance; cotransporter	Aldosterone, atrial natriuretic peptide (ANP), estrogen, progesterone, glucocorticoids
Potassium (K^+)	3.5–5.0 mEq/L	Neuron and muscle function	Aldosterone, ANP
Calcium (Ca^{2+})	8.4–10.2 milligrams per deciliter (mg/dL)	Hardens bone; release of neurotransmitter; muscle contraction; blood clotting; second messenger	Parathyroid hormone, calcitriol, calcitonin
Hydrogen (H^+)	pH 7.35–7.45	pH balance	Buffering systems—chemicals in blood, kidney, respiratory system
ANIONS			
Chloride (Cl^-)	96–106 mEq/L	Anion bound to sodium; component of gastric acid (HCl); chloride shift	Regulated indirectly through sodium
Bicarbonate (HCO_3^-)	23.1–26.7 mEq/L	pH balance	Dependent upon carbon dioxide and H^+ blood levels
Phosphate (PO_4^{3-})	2.5–4.1 mEq/L	Binds with calcium and deposited in bone	Parathyroid hormone

---

Table 18.4 — Common Molecules Found in Blood Plasma

Molecule	Normal Ranges (Values)	Function
Glucose	Fasting: 70–100 mg/dL; 2 hours after a meal: <145 mg/dL	Fuel molecule for cellular respiration (primary energy source for nervous tissue); tightly regulated by a number of hormones, including insulin and glucagon
Amino acids	*(Varies, based on specific amino acid being measured)*	Monomers for synthesizing protein; also regulated by some of the same hormones as glucose
Lactate	4.5–14.4 mg/dL	By-product of glycolysis
Lipids		Molecules that generally do not dissolve in water
Cholesterol	100–200 mg/dL	Plasma membrane component; synthesis of steroid hormones; bile salts
HDL	40–80 mg/dL	Transports lipids to the liver
VLDL/LDL	10–100 mg/dL	Transports lipids from the liver
Triglycerides	30–149 mg/dL	Fuel molecules
Phospholipids	6–12 mg/dL	Molecules that form plasma membrane bilayer

INTEGRATE

CONCEPT CONNECTION

The skeletal system is essential for hemopoeisis. Red bone marrow (the site for hemopoeisis) is found in the spongy bone of most children (see section 7.2d). However, as we age, much of that red bone marrow degenerates and is converted to fat (in the form of yellow bone marrow). As a result, adult red bone marrow (and thus, adult sites for hemopoiesis) are restricted to the flat bones of the skull, sternum, ribs, vertebrae, hip bones, and proximal epiphyses of the humerus and femur.

18.3 Formed Elements in the Blood

Collectively, erythrocytes, leukocytes, and platelets are called the formed elements and make up approximately 45% of whole blood. The term *formed elements* is a more appropriate description because mature erythrocytes contain neither nuclei nor organelles, and platelets are merely fragments broken off from a larger cell. **Table 18.5** summarizes the characteristics of the formed elements.

18.3a Hemopoiesis

LEARNING OBJECTIVES

1. Define hemopoiesis, and explain the role of colony-stimulating factors.
2. Describe the four cellular stages of erythropoiesis.
3. Compare the production of granulocytes, monocytes, and lymphocytes in leukopoiesis.
4. Summarize the process by which platelets are formed in thrombopoiesis.

Formed elements have a relatively short life span; new ones are continually produced by the process of **hemopoiesis** (hē′mō-poy-ē′sis; *poiesis* = a making), also called *hematopoiesis*. The red bone marrow (myeloid tissue) is responsible for hemopoiesis. Hemopoiesis occurs in most bones in young children, but as an individual reaches adulthood, hemopoiesis is restricted to selected bones primarily in the axial skeleton (see section 7.2d).

The process of hemopoiesis starts with hemopoietic stem cells called **hemocytoblasts** (hē′mō-sī′tō-blasts) **(figure 18.3)**. Hemocytoblasts are considered pluripotent cells, meaning that they can differentiate and develop into many different kinds of cells (see Clinical View: "Stem Cells" in section 5.6b). Hemocytoblasts produce two different lines for blood cell development: (1) The **myeloid** (mī′ĕ-loyd; *myelos* = marrow, *eidos* = appearance) **line** forms erythrocytes, all leukocytes except lymphocytes (this would include granulocytes and monocytes),

Table 18.5	Characteristics of the Formed Elements			
Formed Element	**Size (Diameter)**	**Function**	**Life Span**	**Density (Average Number per mm³ of Blood = μL)**
Erythrocytes	7.5 μm	Transport oxygen and carbon dioxide	~120 days	Females: ~4.8 million Males: ~5.4 million
Leukocytes (e.g., neutrophils, eosinophils, basophils, monocytes, and lymphocytes)	1.5 to 3 times larger than an erythrocyte; 11.25–22.5 μm	Initiate immune response; defend against potentially harmful substances	Varies from 12 hours (neutrophils) to years (lymphocytes)	4500–11,000
Platelets	<¼ the size of an erythrocyte; ~2 μm	Participate in hemostasis	~8–10 days	150,000–400,000

Table 18.6	Substances That Influence Hemopoiesis	
Substance	**Growth Factor or Hormone**	**Function**
Multi-colony-stimulating factor (multi-CSF)	Growth factor	Increases the formation of erythrocytes, granulocytes, monocytes, and platelets from myeloid stem cells
Granulocyte-macrophage colony-stimulating factor (GM-CSF)	Growth factor	Accelerates the formation of all granulocytes and monocytes from their progenitor cells
Granulocyte colony-stimulating factor (G-CSF)	Growth factor	Stimulates the formation of granulocytes from myeloblast cells
Macrophage colony-stimulating factor (M-CSF)	Growth factor	Stimulates the production of monocytes from monoblasts
Thrombopoietin	Growth factor	Stimulates both the production of megakaryocytes in the bone marrow and the subsequent formation of platelets
Erythropoietin (EPO)	Hormone (produced primarily by the kidneys)	Increase the rate of production and maturation of erythrocyte progenitor and erythroblast cells

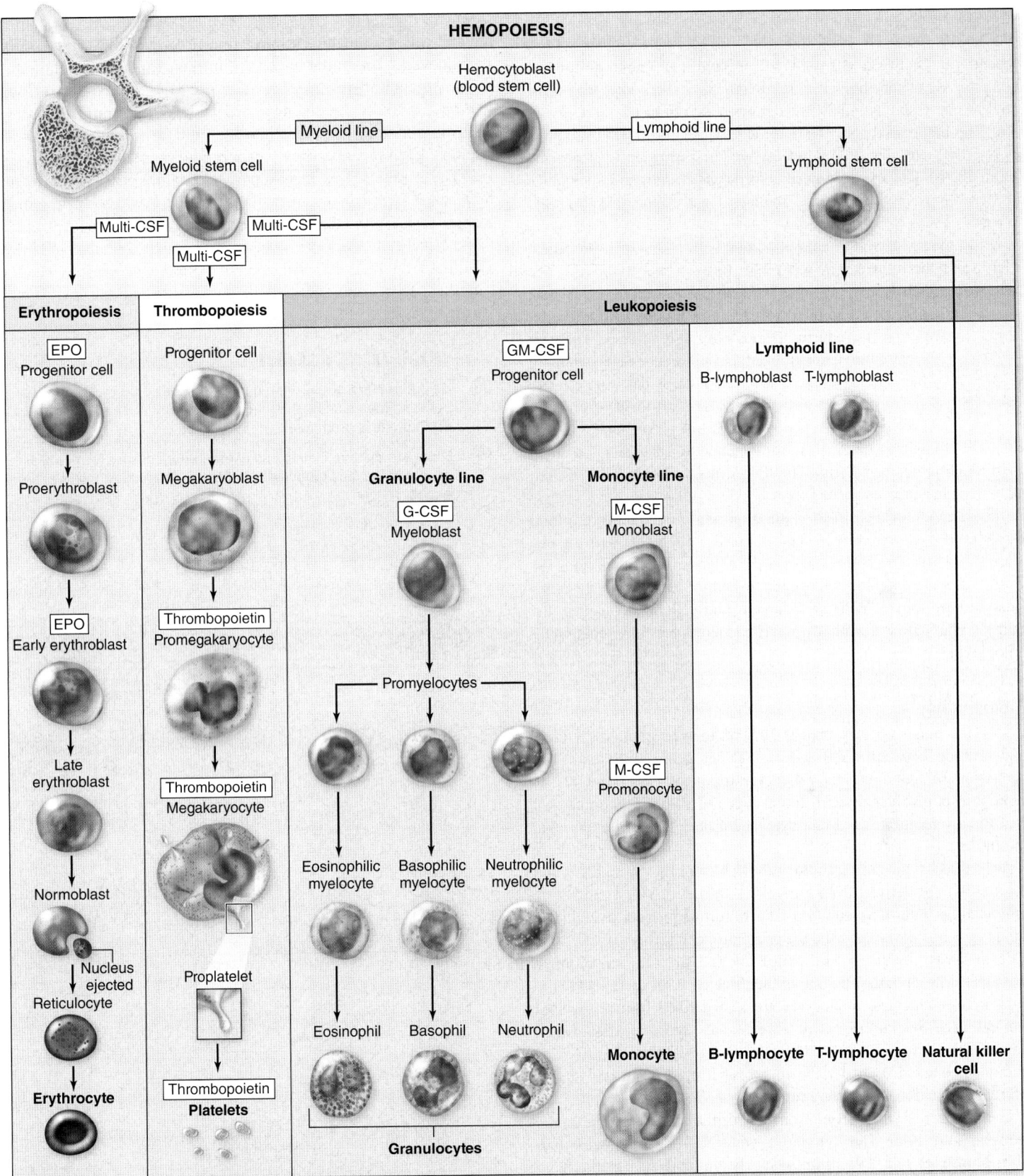

HEMOPOIESIS

Hemocytoblast
(blood stem cell)

Myeloid line — Lymphoid line

Myeloid stem cell

Lymphoid stem cell

Multi-CSF — Multi-CSF
Multi-CSF

Erythropoiesis	Thrombopoiesis	Leukopoiesis

Erythropoiesis

EPO
Progenitor cell

Proerythroblast

EPO
Early erythroblast

Late
erythroblast

Normoblast

Nucleus
ejected
Reticulocyte

Erythrocyte

Thrombopoiesis

Progenitor cell

Megakaryoblast

Thrombopoietin
Promegakaryocyte

Thrombopoietin
Megakaryocyte

Proplatelet

Thrombopoietin
Platelets

Leukopoiesis

GM-CSF
Progenitor cell

Granulocyte line — **Monocyte line**

G-CSF
Myeloblast

M-CSF
Monoblast

Promyelocytes

M-CSF
Promonocyte

Eosinophilic
myelocyte · Basophilic
myelocyte · Neutrophilic
myelocyte

Eosinophil · Basophil · Neutrophil

Monocyte

Granulocytes

Lymphoid line

B-lymphoblast · T-lymphoblast

B-lymphocyte · **T-lymphocyte** · **Natural killer cell**

Figure 18.3 Origin, Differentiation, and Maturation of Formed Elements. All formed elements are derived from common hemopoietic stem cells, called hemocytoblasts. Both myeloid stem cells and lymphoid stem cells are derived from hemocytoblasts. Myeloid stem cells give rise to erythrocytes, platelets, and to all leukocytes except lymphocytes. Lymphoid stem cells give rise to lymphocytes and natural killer (NK) cells. AP|R

and megakaryocytes (cells that produce platelets). (2) The **lymphoid** (lim′foyd) **line** forms only lymphocytes.

The maturation and division of hemopoietic stem cells is influenced by **colony-stimulating factors (CSFs),** or *colony-forming units* (*CFUs*). These molecules are all growth factors, except for erythropoietin, which is a hormone. These substances are described in **table 18.6** and shown in figure 18.3.

Erythropoiesis

Erythrocytes make up more than 99% of formed elements with a concentration between 4.2 and 6.2 million per cubic millimeter. The process of erythrocyte production is called **erythropoiesis** (ĕ-rith'rō-poy-ē'sis). Normally, erythrocytes are produced at the rate of about 3 million per second. The hormone erythropoietin (EPO) controls this rate by increasing the rate of erythrocyte formation (described in section 18.3b). Dietary requirements for normal erythropoiesis include iron, B vitamins (e.g., folic acid, riboflavin), and amino acids (to build proteins).

The process of erythropoiesis begins with a **myeloid stem cell,** which under the influence of multi-CSF forms a progenitor cell. The progenitor cell forms a **proerythroblast,** which is a large, nucleated cell. It then becomes an **erythroblast,** which is a slightly smaller cell that is producing hemoglobin in its cytosol. The next stage, called a **normoblast,** is a still smaller cell with more hemoglobin in the cytosol; its nucleus has been ejected. A cell called a **reticulocyte** (re-tik'ū-lō-sīt) eventually is formed. The reticulocyte has lost all organelles except some ribosomes, so it can continue to produce hemoglobin (through protein synthesis). The transformation from myeloid stem cell to reticulocyte takes about 5 days.

Some reticulocytes finish maturation while circulating in blood vessels (and in normal circumstances, make up 0.5–2.0% of the circulating blood). One to two days after entering the circulation, the ribosomes in the reticulocyte degenerate, and the reticulocyte becomes a mature erythrocyte. Without a nucleus and cellular organelles, the mature erythrocyte is essentially a plasma membrane "bag" containing hemoglobin.

Leukopoiesis

Leukocytes make up less than 0.01% of formed elements with a concentration between 4500 and 11,000 per cubic millimeter. The production of leukocytes is called **leukopoiesis** (lū'kō-poy-ē'sis). Leukopoiesis involves three different types of maturation processes: granulocyte maturation, monocyte maturation, and lymphocyte maturation.

All three types of granulocytes (neutrophils, basophils, and eosinophils) are derived from a myeloid stem cell. This stem cell is stimulated by multi-CSF and GM-CSF to form a progenitor cell. The granulocyte line develops when the progenitor cell forms a **myeloblast** (mī'ĕ-lō-blast) under the influence of G-CSF. The myeloblast ultimately differentiates into one of the three types of granulocytes.

Like granulocytes, monocytes are also derived from a myeloid stem cell. The myeloid stem cell differentiates into a progenitor cell, and under the influence of M-CSF this cell forms a **monoblast.** This is the monocyte line. Eventually, the monoblast forms a promonocyte that differentiates and matures into a **monocyte.**

Lymphocytes are derived from a **lymphoid stem cell** through the lymphoid line. The lymphoid stem cell differentiates into **B-lymphoblasts** and **T-lymphoblasts.** B-lymphoblasts mature into B-lymphocytes, whereas T-lymphoblasts mature into T-lymphocytes. Some lymphoid stem cells differentiate directly into natural killer (NK) cells. (Lymphocyte development and maturation are further described in section 22.5)

Thrombopoiesis

Platelets (or *thrombocytes*) make up less than 1% of formed elements with a concentration between 150,000 and 400,000 per cubic millimeter. The production of platelets is called **thrombopoiesis** (throm'bō-poy-ē'sis;

thrombos = clot). From the myeloid stem cell, a committed cell called a **megakaryoblast** (meg-ă-kar'ē-ō-blast; *mega* = big) is produced. It matures under the influence of thrombopoietin to form a **megakaryocyte** (meg-ă-kar'ē-ō-sīt) **(figure 18.4)**. Megakaryocytes are easily distinguished both by their large size (about 100 micrometers [μm] in diameter) and their dense, multilobed nucleus. Each megakaryocyte then produces thousands of platelets.

The process of how megakaryocytes produce platelets was in question until 2007, when researchers reported that megakaryocytes produce long extensions from themselves called *proplatelets*. While still attached to the megakaryocyte, these proplatelets extend through the blood vessel wall (between the endothelial cells) in the red bone marrow. The force from the blood flow "slices" these proplatelets into the fragments we know as platelets.

WHAT DID YOU LEARN?

9 Describe the process of erythropoiesis, beginning with the stem cell, and then placing the precursor cells in order, until a mature erythrocyte is produced.

10 What are the two main types of precursor cells for formed element development, and what mature formed elements are derived from each?

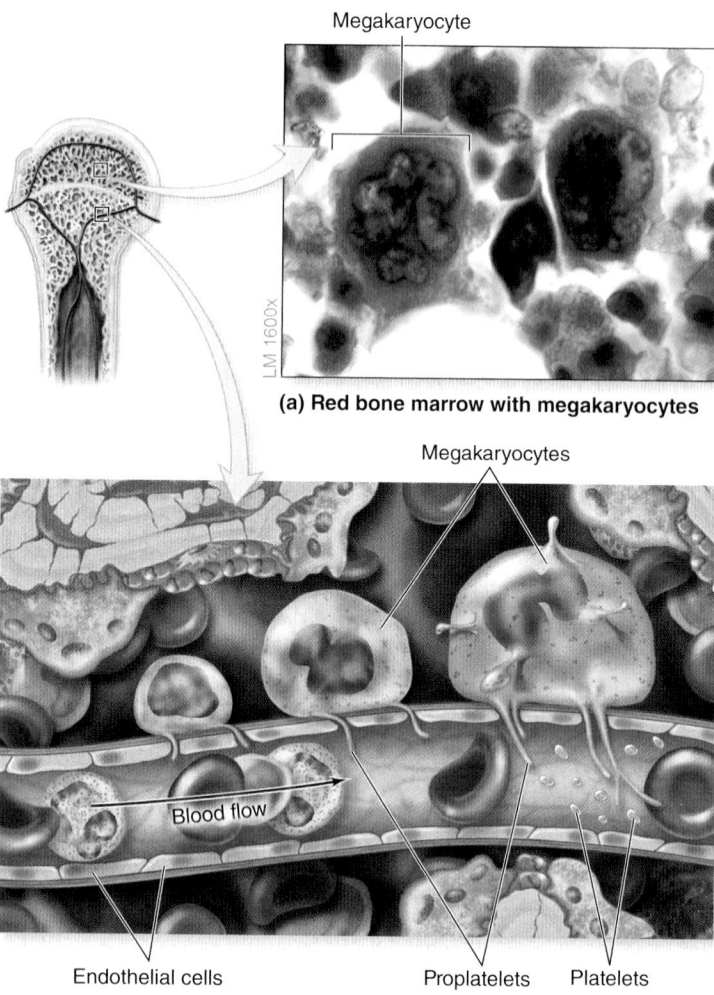

(a) Red bone marrow with megakaryocytes

(b) Platelets forming in blood vessel

Figure 18.4 Platelet Formation. Platelets are derived from megakaryocytes in the red bone marrow. (*a*) Photomicrograph of megakaryocytes in red bone marrow. (*b*) Megakaryocytes extend long processes (called proplatelets) through the blood vessel wall. These proplatelets are spliced by the force of the blood flow into platelets, which then circulate in the blood. AP|R

18.3b Erythrocytes

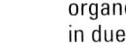 **LEARNING OBJECTIVES**

5. Describe the structure of erythrocytes.
6. List the events by which erythrocyte production is stimulated.
7. Explain the process by which erythrocyte components are recycled.
8. Compare and contrast the different blood types and their importance when transfusing blood.

Erythrocytes are very small, flexible cells, with a diameter of approximately 7.5 μm (figure 18.5). Although erythrocytes are commonly referred to as red blood cells, or RBCs, the term "red blood cell" is a misnomer because a mature erythrocyte lacks a nucleus and cellular organelles. As a result, it is more appropriate to call it a *formed element*. An erythrocyte has a unique biconcave disc structure (at its narrowest point about 0.75 μm and at its widest point about 2.6 μm). It is composed of a plasma membrane within which are housed about 280 million hemoglobin molecules.

Erythrocytes transport oxygen and carbon dioxide between the tissues and the lungs. The fact that erythrocytes lack a nucleus and organelles enables them to carry respiratory gases more efficiently. The biconcave shape and flexibility of erythrocytes allow them to stack and line up in single file, termed a **rouleau** (rū-lō′; pl., *rouleaux* = cylinder), as they pass through capillaries. A latticework of spectrin protein supports the plasma membrane of the erythrocyte on its internal surface and provides flexibility to the erythrocyte as it moves through the capillaries.

WHAT DO YOU THINK?

2 What does an erythrocyte gain by the loss of its nucleus and organelles? What three cellular processes can it no longer engage in due to the loss of its (a) nucleus and (b) mitochondria?

Hemoglobin

Hemoglobin (hē′mō-glō-bin) is a red-pigmented protein that transports oxygen and carbon dioxide. When blood is maximally loaded with oxygen, it is termed **oxygenated** (and appears bright red). Conversely, when some oxygen is lost and carbon dioxide is gained during systemic cellular gas exchange, blood is called **deoxygenated** (and appears dark red).

INTEGRATE

CLINICAL VIEW
Blood Doping

To enhance their performance in endurance events, some athletes may try to boost their bodies' ability to deliver oxygen to the muscles by increasing the number of erythrocytes in their blood. One way the number of erythrocytes can be increased naturally is by living and training at high altitude. The body compensates for the decreased oxygen concentration in the atmosphere by increasing the rate of erythrocyte production, thus increasing the number of erythrocytes per unit volume of blood.

An illegal procedure used by some athletes is called **blood doping**. There are two different methods for blood doping. In the first (and older) method the athlete essentially donates erythrocytes to himself or herself. Prior to competition, the athlete has a unit of blood removed and stored. As the kidneys detect the decreased blood oxygen, the hormone erythropoietin (EPO) is released and the bone marrow responds by increasing production of erythrocytes. This causes the body to increase erythrocyte production to make up for the ones just removed. A few days before the competition, the erythrocytes from the donated unit are transfused back into the athlete's body. The increased number of erythrocytes increases the amount of oxygen transported in the blood and is thought to favorably affect muscle performance, thereby improving athletic performance. The second method of blood doping has occurred with the development of pharmaceutical EPO, which is used to treat anemia. In this method of blood doping, an athlete is injected with pharmaceutical EPO to further increase erythrocyte levels

Potentially deadly dangers are inherent in blood doping. By increasing the number of erythrocytes per measured volume of blood, blood doping also increases the viscosity of the blood. Thus, the heart must work harder to pump this more viscous blood. Eventually, temporary athletic success may be overshadowed by permanent cardiovascular damage that can even lead to death, as has occurred with some athletes. Thus, blood doping has now been banned by athletic competition governing bodies.

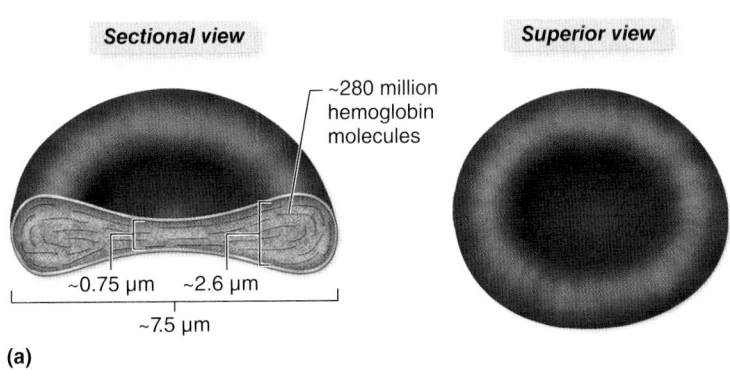

(a)

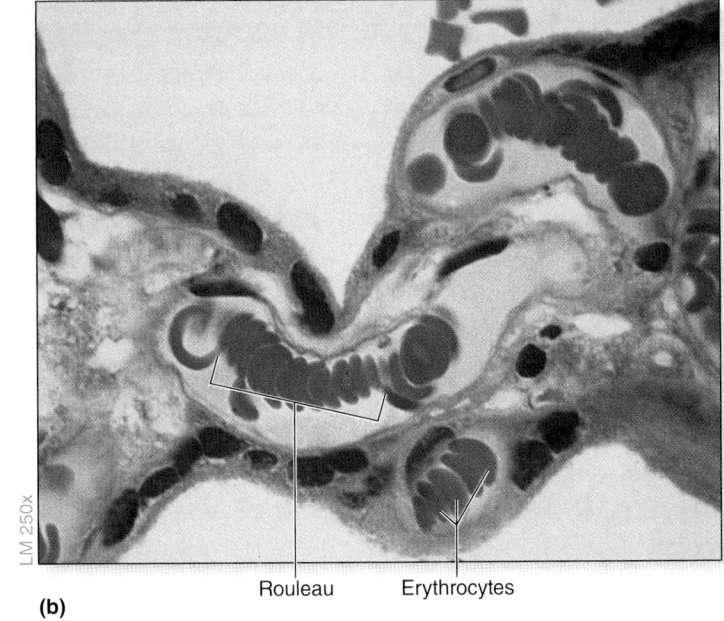

LM 250x

Rouleau   Erythrocytes

(b)

Figure 18.5 Erythrocyte Structure. (*a*) Erythrocytes have the gross structure of a biconcave disc, as shown here in sectional and superior views. (*b*) LM of erythrocytes shows their three-dimensional structure and a rouleau. AP|R

Each hemoglobin molecule consists of four protein molecules called **globins.** Two of these globins are called **alpha (α) chains,** and the other two, which are slightly different, are called **beta (β) chains (figure 18.6).** All globin chains contain a **heme** group, which is composed of a porphyrin (organic compound) ring, with an iron ion (Fe^{2+}) in its center. Oxygen binds to the Fe^{2+} in heme groups for transport in the blood.

Because each molecule of hemoglobin has four heme groups, it has four Fe^{2+} and is capable of binding four molecules of oxygen. The oxygen binding is fairly weak; this allows rapid attachment of oxygen with hemoglobin when erythrocytes pass through the blood capillaries of the lungs, and rapid detachment when erythrocytes pass through the systemic capillaries of body tissues.

Carbon dioxide and the globin molecule (not the Fe^{2+}) have a similar weak attachment relationship for the transport of carbon dioxide molecules. Carbon dioxide binds to the globin protein molecule as blood moves through systemic capillaries and is released as blood moves through the capillaries of the lungs.

The Role of EPO in Erythropoiesis

Erythropoiesis is controlled by the hormone **erythropoietin (EPO).** The kidneys are the primary producers of EPO, although the liver also secretes a small amount of EPO as well. The process by which EPO release is stimulated is shown in **figure 18.7.** The initial stimulus is a decrease in blood oxygen levels. This decrease may be caused by the continuous removal of aged erythrocytes, blood loss, or exposure to high altitudes (where atmospheric oxygen levels are lower). Chemoreceptors within the kidney detect low blood oxygen levels as the blood travels through blood vessels within the kidney. As a result, certain cells in the kidney release the hormone EPO into the blood. EPO is transported through the blood and reaches the red

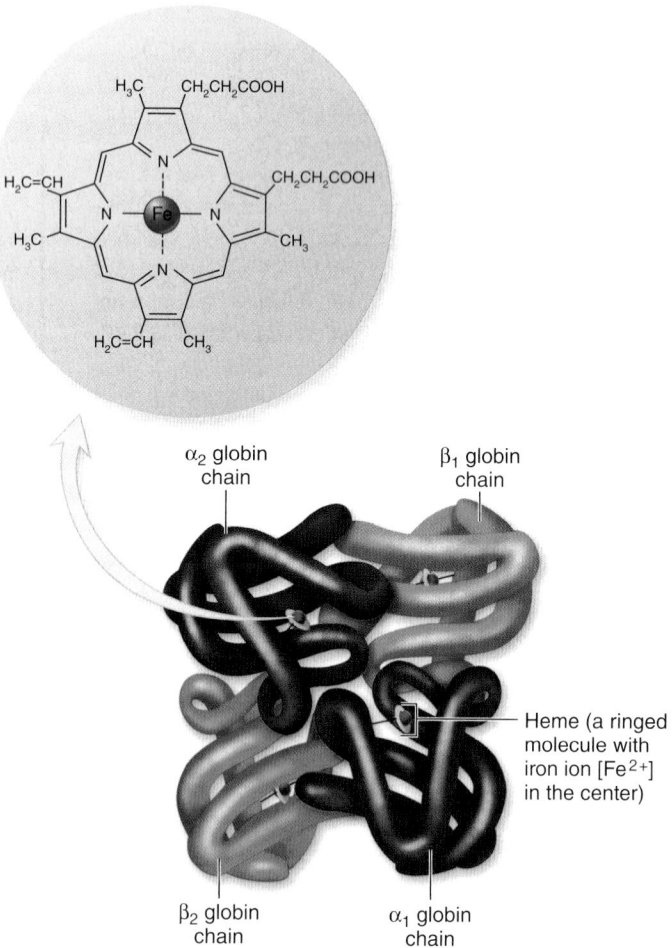

Figure 18.6 Molecular Structure of Hemoglobin. A single molecule of hemoglobin is composed of four protein subunits, called globins, each containing a lipid heme group that holds a single iron ion (Fe^{2+}) in its center. Each hemoglobin molecule transports four oxygen molecules that are weakly attracted to the Fe^{2+}. AP|R

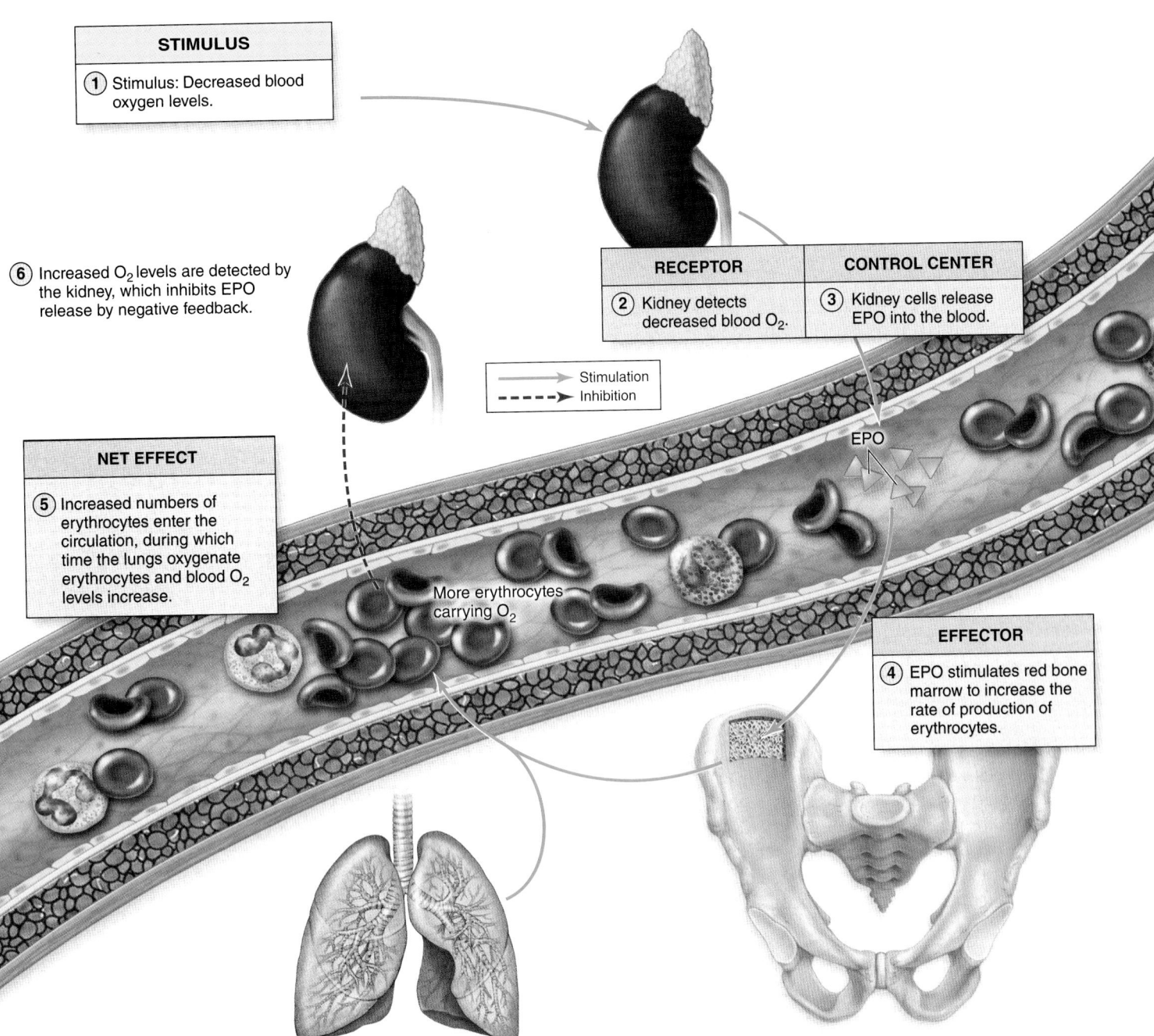

STIMULUS

1. Stimulus: Decreased blood oxygen levels.

6. Increased O_2 levels are detected by the kidney, which inhibits EPO release by negative feedback.

RECEPTOR	**CONTROL CENTER**
2. Kidney detects decreased blood O_2.	3. Kidney cells release EPO into the blood.

→ Stimulation
---→ Inhibition

EPO

NET EFFECT

5. Increased numbers of erythrocytes enter the circulation, during which time the lungs oxygenate erythrocytes and blood O_2 levels increase.

More erythrocytes carrying O_2

EFFECTOR

4. EPO stimulates red bone marrow to increase the rate of production of erythrocytes.

Figure 18.7 How EPO Regulates Erythrocyte Production. Low blood oxygen levels are detected by kidney chemoreceptors, which triggers the release of erythropoietin (EPO) into the blood. EPO stimulates the red bone marrow to produce more erythrocytes. EPO release is then inhibited when blood oxygen levels rise.

bone marrow. There, EPO stimulates myeloid cells in the red bone marrow to increase the rate of erythrocyte production. Additional erythrocytes are released into circulation (a process that takes a few days), so more oxygen can be transported from the lungs and delivered to the cells. Blood oxygen levels increase as a result. Increased oxygen levels inhibit release of EPO from kidney cells through negative feedback.

The adrenal gland secretes small amounts of testosterone in both sexes (see section 17.8c), and the testes secrete large amounts of testosterone in males (see section 28.4b and **table R.10** entitled Regulating the Male Reproductive System). In addition to its many other functions, testosterone stimulates the kidney to produce more EPO. Because males have higher levels of testosterone, they also usually have a higher erythrocyte count and a higher hematocrit.

Environmental factors, such as altitude, can affect EPO release and ultimately affect the hematocrit. Let us say a woman moves to a cabin high in the Rocky Mountains, where the atmospheric pressure is lower than it is at sea level. (As you'll learn in section 23.7c, a lower atmospheric pressure means less oxygen availability.) Each time she takes a breath at this altitude, she takes in relatively less oxygen than she would on an ocean beach. Her body compensates by releasing more EPO and making more erythrocytes over time; more erythrocytes in the blood can carry more oxygen to the tissues. However, having more erythrocytes increases the blood's viscosity, which could increase the chance of cardiovascular complications, such as major blood clots that lead to heart attacks or strokes. The details for erythropoietin are listed in the summary table Regulating Erythrocyte Concentration in the Blood, which directly follows chapter 17 (see table R.6).

Erythrocyte Destruction

The absence of both a nucleus and cellular organelles comes at a cost to the erythrocyte and affects its longevity. A mature erythrocyte cannot synthesize proteins either to repair itself or to replace damaged membrane regions. Aging and the wear-and-tear of circulation through blood vessels cause erythrocytes to become more fragile and less flexible. Therefore, the erythrocyte has a finite maximum life span of about 120 days. Every day, about 1% of the oldest circulating erythrocytes are removed from circulation. These old erythrocytes are phagocytized in both the spleen and liver by cells called macrophages.

Three molecular components must be accounted for in the destruction of hemoglobin: the globin protein, iron ion, and the heme group.

Two of the components are processed for recycling; the other component is metabolically altered and then excreted from the body, as shown in **figure 18.8**.

Globin proteins are broken down into free amino acids, most of which are used by the body for protein synthesis to make new erythrocytes or other body proteins.

The iron component in hemoglobin is removed and transported by a globulin protein called **transferrin** (trans-fer'in; *trans* = across, *ferrum* = iron) to the liver or spleen where the Fe^{2+} then is bound to storage proteins called **ferritin** (fer'i-tin) and **hemosiderin** (hē'mō-sid'ĕr-in). Ferritin is a large water-soluble protein that serves as the primary storage mechanism for iron. Iron is stored mainly in the liver and spleen, and it is transported by transferrin to the red bone marrow as needed for erythrocyte production. However, small amounts of iron, approximately 0.9 mg, are lost daily in sweat, urine, and feces. In females, additional iron is lost in those who have a monthly menstrual flow.

The heme group (minus the Fe^{2+}) released from hemoglobin is converted within macrophages first into a green pigment called **biliverdin** (bil-i-ver'din; *bilis* = bile, *verd* = green). Biliverdin is eventually converted into a yellowish pigment called **bilirubin** (bil-i-rū'bin; *rubin* = reddish), which is transported by albumin to the

INTEGRATE

CLINICAL VIEW
Anemia

Anemia (ă-nē'mē-ă; *an* = without, *haima* = blood) is any condition in which either the percentage of erythrocytes is lower than normal or the oxygen-carrying capacity of the blood is reduced (such as may occur if the hemoglobin is abnormal). In either case, there is decreased oxygen delivery to body tissues—and consequently, the heart must work harder to supply oxygen to the body. Symptoms of anemia include lethargy, shortness of breath, pallor of the skin and mucous membranes, fatigue, and heart palpitations. The types of anemia include the following:

- **Aplastic anemia** is characterized by significantly decreased formation of both erythrocytes and hemoglobin. This condition results from defective red bone marrow, perhaps as a result of poisons, toxins, or radiation exposure.

- **Congenital hemolytic anemia** occurs when destruction of erythrocytes is more rapid than normal. It is usually due to a genetic defect, which results in the production of abnormal membrane proteins that make the erythrocyte plasma membrane very fragile.

- **Erythroblastic anemia** is characterized by the presence of large numbers of immature, nucleated cells (called erythroblasts and normoblasts) in the circulating blood. An accelerated pace of cell maturation causes immature cells to be present in the blood. These cells cannot function normally and thus anemia results.

- **Hemorrhagic anemia** results from heavy blood loss. The hemorrhage may be caused, for example, by chronic ulcers or by heavy or prolonged menstrual flow.

- **Pernicious anemia** is a chronic progressive anemia of adults caused by failure of the body to absorb vitamin B_{12}. This vitamin is found in fish and meat, so most individuals receive enough B_{12} in their diet, unless they are vegans or strict vegetarians. (Thus, it is recommended that all vegetarians take a B_{12} vitamin supplement.)

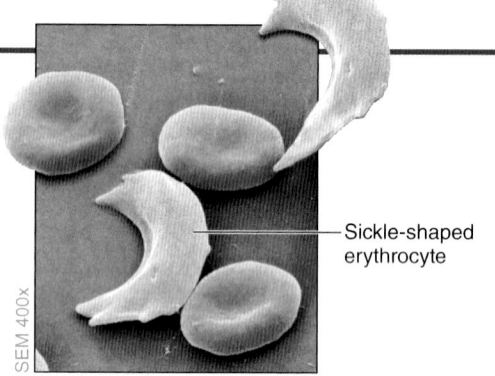

SEM of blood from a person with sickle-cell disease.

SEM 400x

Sickle-shaped erythrocyte

A defect in the production of **intrinsic factor,** a glycoprotein secreted by stomach lining cells to protect B_{12} in the stomach and enhance B_{12} absorption in the small intestine, leads to pernicious anemia. Individuals who have pernicious anemia due to defective intrinsic factor production must receive B_{12} intramuscular or subcutaneous injections, since they are unable to absorb oral B_{12} supplements.

- **Sickle-cell disease** is an autosomal recessive anemia that occurs when a person inherits two copies of the sickle-cell gene. Erythrocytes become sickle-shaped at lower blood oxygen concentrations, making them unable to flow efficiently through the blood vessels to body tissues and more prone to destruction (a process called *hemolysis*).

Most anemias are treated by letting the patient's own bone marrow replace the erythrocytes. This process may be facilitated through the use of pharmaceutical EPO. However, anemia is often a symptom of another disease or problem. For example, although many anemias are due to iron deficiency, the iron deficiency often is not because of diet, but rather the result of chronic blood loss, a process that depletes the body of its iron stores over months or years. The three most common causes of such chronic blood loss are excessive menstrual bleeding, undiagnosed stomach ulcer, and colon cancer. So, while restoring the patient's erythrocyte count, a physician should also look for any underlying cause of the anemia.

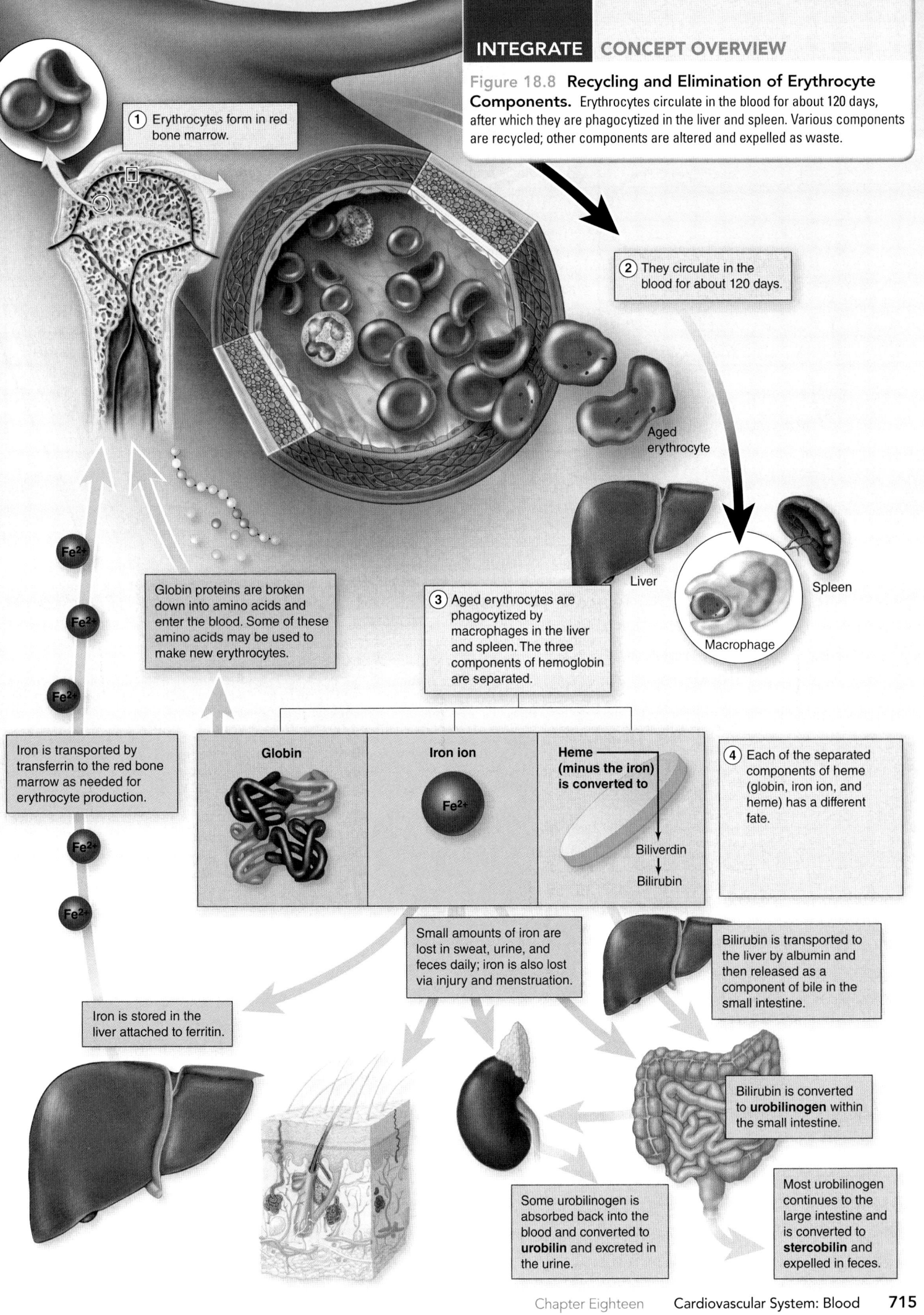

Figure 18.8 **Recycling and Elimination of Erythrocyte Components.** Erythrocytes circulate in the blood for about 120 days, after which they are phagocytized in the liver and spleen. Various components are recycled; other components are altered and expelled as waste.

(1) Erythrocytes form in red bone marrow.

(2) They circulate in the blood for about 120 days.

Aged erythrocyte

Liver

Spleen

Macrophage

Globin proteins are broken down into amino acids and enter the blood. Some of these amino acids may be used to make new erythrocytes.

(3) Aged erythrocytes are phagocytized by macrophages in the liver and spleen. The three components of hemoglobin are separated.

Iron is transported by transferrin to the red bone marrow as needed for erythrocyte production.

Fe^{2+}

Fe^{2+}

Fe^{2+}

Fe^{2+}

Fe^{2+}

Globin

Iron ion

Fe^{2+}

Heme (minus the iron) is converted to

Biliverdin

Bilirubin

(4) Each of the separated components of heme (globin, iron ion, and heme) has a different fate.

Small amounts of iron are lost in sweat, urine, and feces daily; iron is also lost via injury and menstruation.

Bilirubin is transported to the liver by albumin and then released as a component of bile in the small intestine.

Iron is stored in the liver attached to ferritin.

Bilirubin is converted to **urobilinogen** within the small intestine.

Some urobilinogen is absorbed back into the blood and converted to **urobilin** and excreted in the urine.

Most urobilinogen continues to the large intestine and is converted to **stercobilin** and expelled in feces.

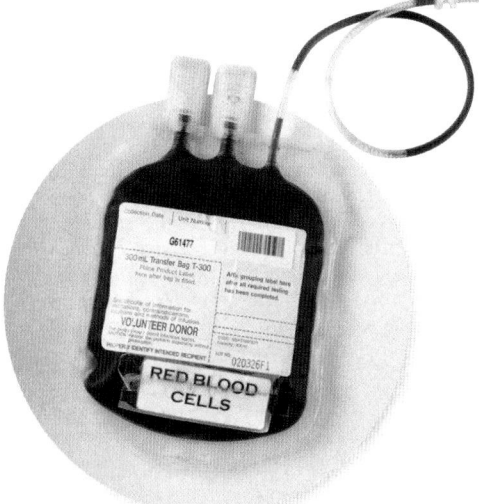

INTEGRATE

CLINICAL VIEW
Transfusions

A **transfusion** is the transfer of blood or blood components from a donor to a recipient. Whole blood is almost never transfused today, and it is not generally available. Rather, when you donate a unit of blood, it is almost immediately divided into three components: erythrocytes, clotting factors, and platelets. When a person needs one of these blood products, the physician administers only what is required, thus allowing one whole blood donation to serve up to three people.

To prevent health-related problems, donor blood must be collected under sterile conditions. The donated blood is first mixed with an anticoagulant to prevent clotting and is immediately refrigerated. Then the donated unit is tested for a variety of infectious agents that cause disease (e.g., hepatitis, AIDS). Next, the blood is separated into erythrocytes, clotting factors, and platelets. Should leukocytes be needed, they must be collected in a special apparatus that effectively filters the leukocytes from the blood and then returns the blood to the donor. (A donor with healthy red bone marrow can quickly replace the donated leukocytes and does not have to wait the full 8 weeks before being able to donate blood again.)

liver. Bilirubin is a component of a digestive secretion called bile, which is produced by the liver and released into the small intestine.

Bilirubin is converted to **urobilinogen** (yur-ō-bi-lin′ō-jen, *ouron* = urine) in the small intestine. Urobilinogen can either (1) continue through the large intestine and eventually be converted by the intestinal bacteria to **stercobilin,** a brown pigment that is expelled from the body as a component of feces; or (2) be absorbed back into the blood. In this latter case, it is converted to **urobilin,** a yellow pigment that is excreted by the kidneys.

Blood Types

The plasma membrane of an erythrocyte has numerous molecules called **surface antigens** (or *agglutinogens*), which project from the surface. These antigens have significant implications for blood transfusion, and in some cases, pregnancy. There are two groups of surface antigens that determine a person's blood type: the ABO blood group and the Rh protein.

ABO Blood Group The best-known antigens are those that form the **ABO blood group.** This group consists of two surface antigens (which are glycoproteins) called **A** and **B.** The presence or absence of the A antigen, the B antigen, or both is the criterion that determines your **ABO blood type,** as shown in **figure 18.9** and listed here:

- **Type A** blood has erythrocytes with surface antigen A only.
- **Type B** blood has erythrocytes with surface antigen B only.
- **Type AB** blood has erythrocytes having both surface antigens A and B.
- **Type O** blood has erythrocytes with neither surface antigen A nor B.

The ABO surface antigens on erythrocytes are accompanied by specific **antibodies** (or *agglutinins*) within the blood plasma. In general, an antibody is a Y-shaped protein that binds to a specific antigen that is perceived as foreign to the body (see section 22.4a). The ABO blood group has both **anti-A** and **anti-B antibodies** that react with the surface antigen A and the surface antigen B, respectively. You do not have antibodies in your blood plasma that bind to the surface antigens on your erythrocytes. Within the ABO blood group, the following blood types and antibodies are normally associated as follows:

- Type A blood has anti-B antibodies within its plasma.
- Type B blood has anti-A antibodies within its plasma.
- Type AB blood has *neither* anti-A nor anti-B antibodies within its plasma.
- Type O blood has *both* anti-A and anti-B antibodies in its blood plasma.

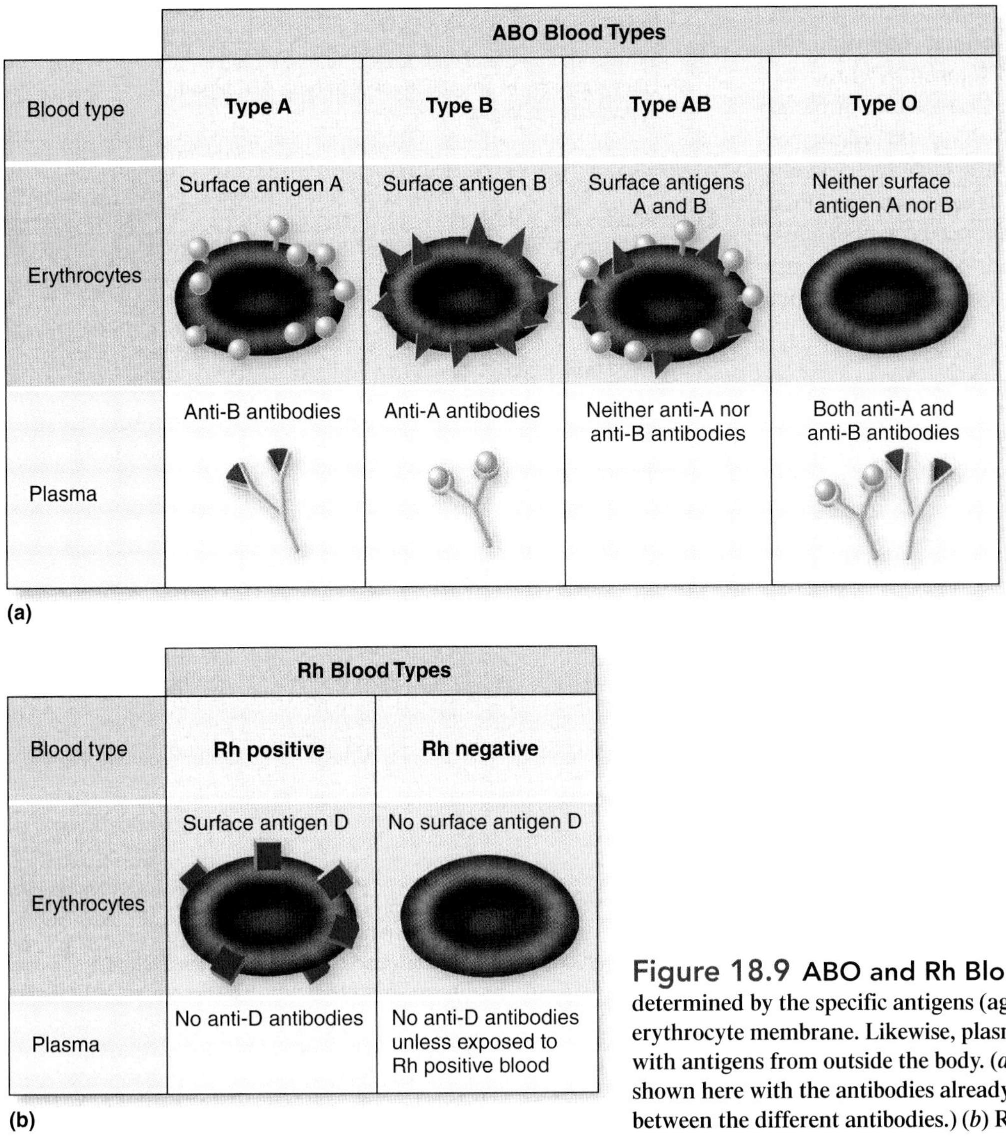

ABO Blood Types

Blood type	Type A	Type B	Type AB	Type O
Erythrocytes	Surface antigen A	Surface antigen B	Surface antigens A and B	Neither surface antigen A nor B
Plasma	Anti-B antibodies	Anti-A antibodies	Neither anti-A nor anti-B antibodies	Both anti-A and anti-B antibodies

(a)

Rh Blood Types

Blood type	Rh positive	Rh negative
Erythrocytes	Surface antigen D	No surface antigen D
Plasma	No anti-D antibodies	No anti-D antibodies unless exposed to Rh positive blood

(b)

Figure 18.9 ABO and Rh Blood Types. The blood type of an individual is determined by the specific antigens (agglutinogens) exposed on the surface of the erythrocyte membrane. Likewise, plasma contains antibodies (agglutinins) that react with antigens from outside the body. (*a*) ABO blood types. (Note that the antibodies are shown here with the antibodies already bound to the antigen, to clarify the differences between the different antibodies.) (*b*) Rh blood types.

Rh Factor Another common surface antigen on erythrocyte plasma membranes determines the Rh blood type. The **Rh blood type** is determined by the presence or absence of the Rh surface antigen, often called either **Rh factor** or **surface antigen D.** When the Rh factor is present, the individual is said to be **Rh positive (Rh$^+$).** Conversely, an individual is termed **Rh negative (Rh$^-$)** when the surface antigen is lacking (figure 18.9*b*).

In contrast to the antibodies of the ABO blood group, which may be found in the blood even without prior exposure to a foreign antigen, antibodies to the Rh factor (termed **anti-D antibodies**) appear in the blood only when an Rh negative individual is exposed to Rh positive blood. This most often occurs as a result of an inappropriate blood transfusion. Individuals who are Rh positive never exhibit anti-D antibodies, because they possess the Rh antigen on their erythrocytes. Only individuals who are Rh negative can exhibit anti-D antibodies, and that can occur only after exposure to Rh antigens.

The ABO and Rh blood types are usually reported together. For example, types AB and Rh$^+$ together are reported as AB$^+$. However, remember that ABO and Rh blood types are independent of each other, and neither of them interacts with or influences the presence or activities of the other group.

Clinical Considerations About Blood Types

Blood types become clinically important when a patient needs a blood transfusion. Compatibility between donor and recipient must be ascertained prior to blood transfusions. If a person is transfused with blood of an incompatible type, antibodies in the plasma bind to surface antigens of the transfused erythrocytes, and clumps of erythrocytes bind together in a process termed **agglutination** (ă-glū-tin-ā′shŭn; *ad* = to, *gluten* = glue). Clumped erythrocytes can block blood vessels and prevent the normal circulation of blood (**figure 18.10**). Eventually, some or all of the clumped erythrocytes may rupture, a process called **hemolysis** (hē-mol′i-sis; *lysis* = destruction). The release of erythrocyte contents and fragments into the blood often causes further hemolytic reactions and ultimately may damage organs. Therefore, compatibility between donor and recipient must be determined prior to blood donations and transfusions using an agglutination test.

WHAT DO YOU THINK?

3 Why is an individual with type O⁻ blood called a "universal donor"? Likewise, why is an individual with type AB⁺ blood called a "universal recipient"?

WHAT DID YOU LEARN?

11 What is the main function of an erythrocyte, and in what ways is an erythrocyte designed to efficiently carry out its function?

12 How do transferrin and ferritin participate in recycling erythrocyte components?

13 What are the structural and molecular differences between type A⁺ blood and type B⁻ blood?

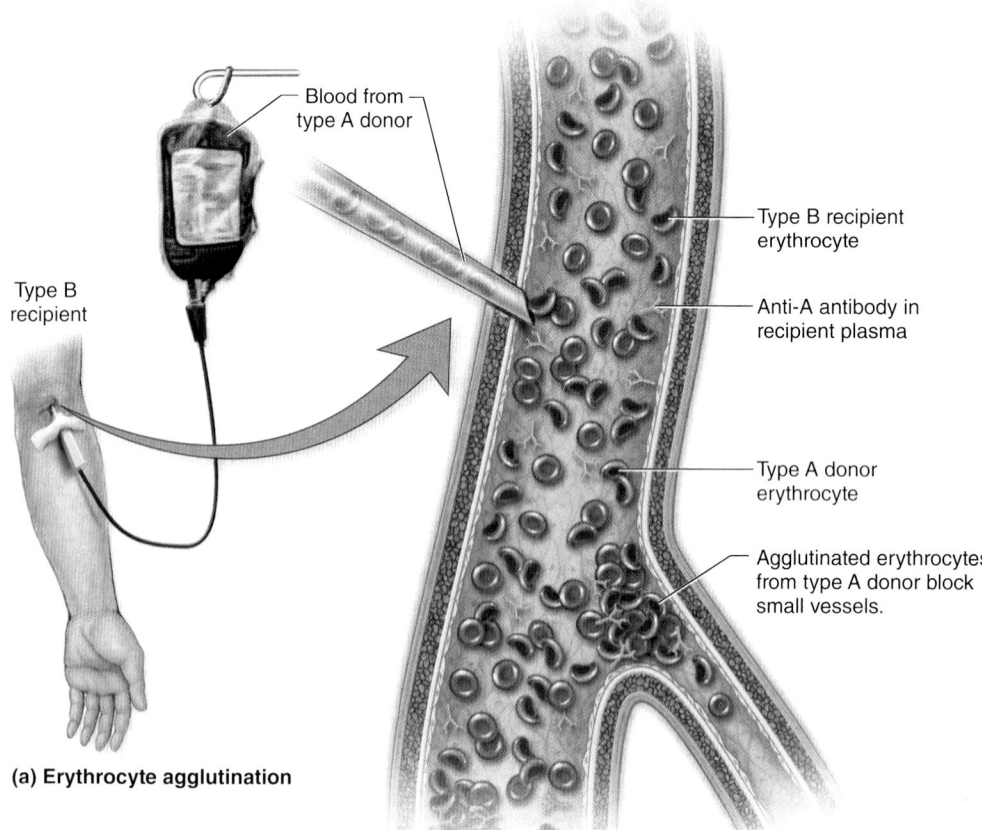

- Blood from type A donor
- Type B recipient
- Type B recipient erythrocyte
- Anti-A antibody in recipient plasma
- Type A donor erythrocyte
- Agglutinated erythrocytes from type A donor block small vessels.

(a) Erythrocyte agglutination

Figure 18.10 Agglutination Reaction. (*a*) If a person receives mismatched blood, antibodies in the blood plasma bind to their respective surface antigens within the erythrocyte plasma membranes; erythrocytes agglutinate (clump) and may block small blood vessels. (*b*) In a test between plasma and erythrocyte samples, a successful match (no clumping) is compared to an unsuccessful match (clumping).

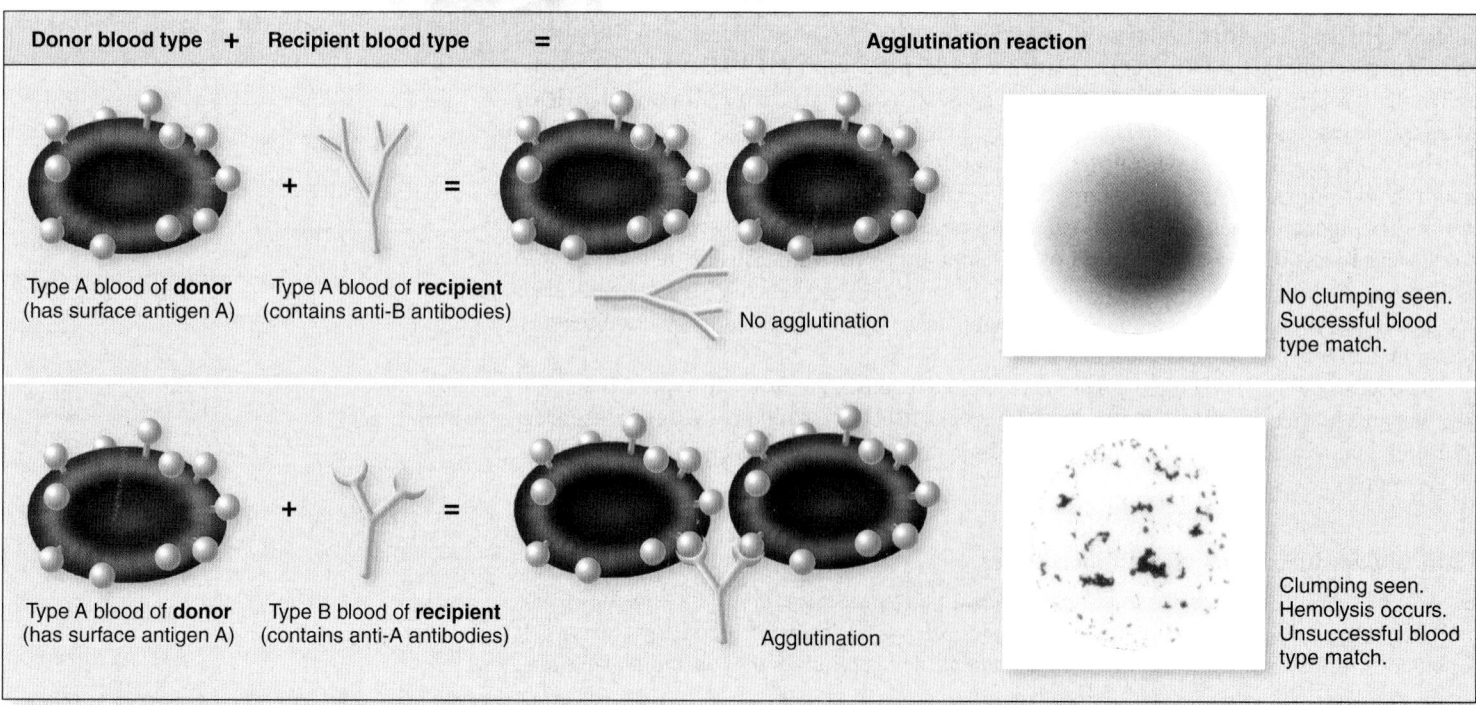

Donor blood type	+	Recipient blood type	=	Agglutination reaction	
Type A blood of **donor** (has surface antigen A)	+	Type A blood of **recipient** (contains anti-B antibodies)	=	No agglutination	No clumping seen. Successful blood type match.
Type A blood of **donor** (has surface antigen A)	+	Type B blood of **recipient** (contains anti-A antibodies)	=	Agglutination	Clumping seen. Hemolysis occurs. Unsuccessful blood type match.

(b) Agglutination test

CLINICAL VIEW

Rh Incompatibility and Pregnancy

The potential presence of anti-D antibodies is especially important in pregnant women who are Rh negative and have an Rh positive fetus. An Rh incompatibility may result during pregnancy if the mother has been previously exposed to Rh positive blood (such as can occur with a previously carried Rh positive fetus, typically at the time of childbirth). As a result of the prior exposure, the mother has anti-D antibodies that may cross the placenta and destroy the fetal erythrocytes, resulting in severe illness or death. The illness that occurs in the newborn is called **hemolytic disease of the newborn (HDN)**, or *erythroblastosis fetalis*. The newborn typically presents with anemia and *hyperbilirubinemia* (increased bilirubin in the blood) due to erythrocyte destruction. In severe cases, the infant may develop heart failure and must be given a blood transfusion to survive.

Giving a pregnant Rh negative woman special immunoglobulins (e.g., RhoGAM) between weeks 28 to 32 of her pregnancy and at birth prevents the mother from developing anti-D antibodies. Specifically, these immunoglobulins bind to fetal erythrocyte surface antigens—and in so doing, prevent the mother's immune system from recognizing Rh antigens and being stimulated to produce anti-D antibodies.

Hemolytic disease of the newborn.

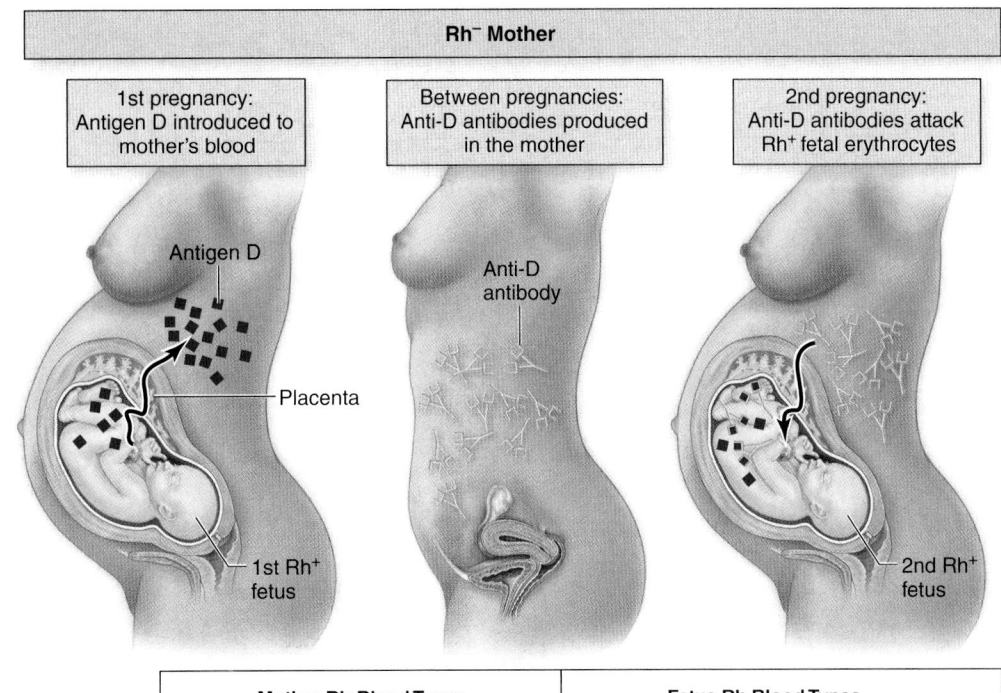

	Mother Rh Blood Types		Fetus Rh Blood Types	
	Pregnancy #1	**Pregnancy #2**	**Pregnancy #1**	**Pregnancy #2**
Blood type	**Rh negative**	**Rh negative**	**Rh positive**	**Rh positive**
Erythrocytes	No antigen D	No antigen D	Antigen D	Antigen D
Plasma	No anti-D antibodies	Anti-D antibodies (due to prior exposure)	No anti-D antibodies	No anti-D antibodies Anti-D antibodies from mother cross placenta and attack fetal erythrocytes causing **hemolytic disease of the newborn**

Rh⁻ Mother

- 1st pregnancy: Antigen D introduced to mother's blood — Antigen D, Placenta, 1st Rh⁺ fetus
- Between pregnancies: Anti-D antibodies produced in the mother — Anti-D antibody
- 2nd pregnancy: Anti-D antibodies attack Rh⁺ fetal erythrocytes — 2nd Rh⁺ fetus

18.3c Leukocytes

LEARNING OBJECTIVES

9. Explain the main function of leukocytes.
10. Distinguish between granulocytes and agranulocytes, and compare and contrast the various types.
11. Explain what is meant by a differential count and how it is clinically useful.

Leukocytes help defend the body against pathogens. Leukocytes differ from erythrocytes in that they are about 1.5 to 3 times larger in diameter, contain a nucleus and cellular organelles, and do not contain hemoglobin. The number of leukocytes in the blood normally ranges between 4500 and 11,000 per cubic millimeter (or microliter) of blood.

Leukocytes are motile (capable of movement within interstitial fluid) and remarkably flexible. In fact, most leukocytes are found within body tissues, as opposed to in the blood. Leukocytes enter the tissues from blood vessels by a process called **diapedesis** (dī′ă-pĕ-dē′sis; *dia* = through, *pedesis* = a leaping), whereby they squeeze between the endothelial cells of the blood vessel walls. **Chemotaxis** (kē-mō-tak′sis) is a process in which leukocytes are attracted to a site of infection by the presence of molecules released by damaged cells, dead cells, or invading pathogens (see figure 22.6). These processes are part of the inflammatory response discussed in depth in section 22.3d.

The five types of leukocytes are divided into two distinguishable classes—granulocytes and agranulocytes—based upon the visible presence or absence of secretory vesicles in the cytosol termed *specific granules*. The leukocytes are summarized in **table 18.7**.

Table 18.7   Leukocytes

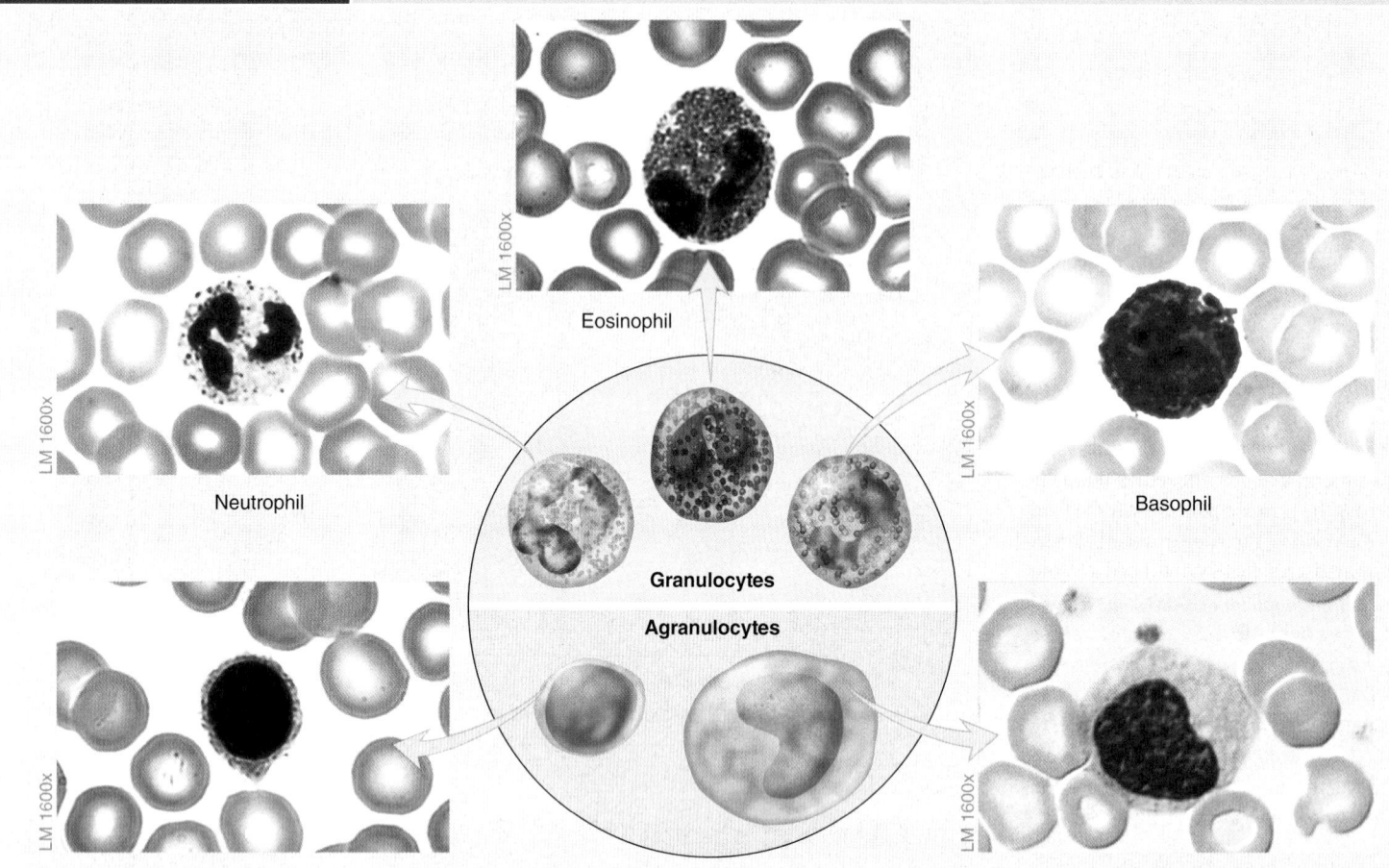

Eosinophil

Neutrophil

Basophil

Granulocytes

Agranulocytes

Lymphocyte

Monocyte

Type	Characteristics	Functions	Approximate %
GRANULOCYTES			
Neutrophils	Nucleus is multilobed (as many as 5 lobes) Cytosol contains neutral, or pale, specific granules (when stained)	Phagocytize pathogens, especially bacteria Release enzymes that target pathogens	50–70% of total leukocytes (1800–7800 cells per microliter)
Eosinophils	Nucleus is bilobed Cytosol contains reddish or pink-orange specific granules (when stained)	Phagocytize antigen-antibody complexes and allergens Release chemical mediators to destroy parasitic worms	1–4% of total leukocytes (100–400 cells per microliter)
Basophils	Nucleus is bilobed Cytosol contains deep blue-violet specific granules (when stained)	Release histamine (vasodilator and increases capillary permeability) and heparin (anticoagulant) during inflammatory reactions	0.5–1% of total leukocytes (20–50 cells per microliter)
AGRANULOCYTES			
Lymphocytes	Round or slightly indented nucleus (fills the cell in smaller lymphocytes) Nucleus is usually darkly stained Thin rim of cytosol surrounds nucleus	Coordinate immune cell activity Attack pathogens and abnormal and infected cells Produce antibodies	20–40% of total leukocytes (1000–4800 cells per microliter)
Monocytes	Kidney-shaped or C-shaped nucleus Nucleus is generally pale staining Abundant cytosol around nucleus	Exit blood vessels and become macrophages Phagocytize pathogens (e.g., bacteria, viruses), cellular fragments, dead cells, debris	2–8% of total leukocytes (100–700 cells per microliter)

Granulocytes

Granulocytes (gran′ū-lō-sīt) have specific granules in their cytosol that are clearly visible when viewed with a microscope. When a blood smear is stained to provide contrast, three types of granulocytes can be distinguished: neutrophils, eosinophils, and basophils. Their names refer to the granules' affinities for certain stains.

Neutrophils   The most numerous leukocyte in the blood is the **neutrophil** (nū′trō-fil; *neuter* = neither, *philos* = fond), constituting about 50–70% of the total number of leukocytes. The neutrophil is named for its neutral or pale-colored granules within a light lilac-colored cytosol. A neutrophil is about 1.5 times larger in diameter than an erythrocyte. Neutrophils exhibit a multilobed nucleus; as many as five lobes are interconnected by thin strands. For this reason, neutrophils also may be called *polymorphonuclear (PMN) leukocytes* because of the number of lobes and various shapes of their nuclei.

Neutrophils usually remain in circulation for about 10 to 12 hours before they exit the blood vessels and enter the tissue spaces, where they phagocytize infectious pathogens, especially bacteria. The mechanisms that help destroy bacteria are described in section 22.3b. The number of neutrophils in a person's blood rises dramatically in the presence of a chronic bacterial infection, as more neutrophils are produced that target the bacteria.

Eosinophils   **Eosinophils** (ē-ō-sin′ō-fil; *eos* = dawn) have reddish or pink-orange granules in their cytosol. Typically, eosinophils constitute about 1–4% of the total number of leukocytes. Their nucleus is bilobed, and the two lobes are connected by a thin strand. An eosinophil is about 1.5 times larger in diameter than an erythrocyte. Eosinophils phagocytize numerous antigen-antibody complexes or allergens (antigens that initiate a hypersensitive or allergic reaction). If the body is infected by parasitic worms, the eosinophils release chemical mediators that attack the worms.

Basophils   **Basophils** (bā′sō-fil; *basis* = base) are usually about 1.5 times larger than erythrocytes. They are the least numerous of the granulocytes; typically, basophils constitute about 0.5–1% of the total number of leukocytes. Basophils exhibit a bilobed nucleus and abundant deep blue-violet granules in the cytosol. The primary components of basophil granules are histamine and heparin. When histamine is released from these granules, it causes both an increase in the diameter of blood vessels (called **vasodilation**) and increased capillary permeability. The classic allergic symptoms of swollen nasal membranes, itchy and runny nose, and watery eyes are partially attributed to the release of histamine. The release of heparin from basophils inhibits blood clotting (a process called **anticoagulation**).

INTEGRATE

LEARNING STRATEGY

The mnemonic "Never let monkeys eat bananas" is a simple way to recall the leukocytes in order of their relative abundance:

Never = **N**eutrophil (most abundant)

Let = **L**ymphocyte

Monkeys = **M**onocyte

Eat = **E**osinophil

Bananas = **B**asophil (least abundant)

Agranulocytes

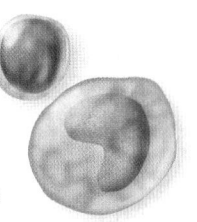

Agranulocytes are leukocytes that have such small specific granules in their cytosol that they are not clearly visible under the light microscope—hence the name agranulocyte (*a* = without). Agranulocytes include both lymphocytes and monocytes.

Lymphocytes   As their name implies, most **lymphocytes** (lim′fō-sīt) reside in lymphatic organs and structures (e.g., lymph nodes, spleen). Usually, lymphocytes constitute about 20–40% of the total number of leukocytes in the blood. Their dark-staining nucleus is typically rounded, and smaller lymphocytes exhibit only a thin rim of blue-gray cytosol around the nucleus. A lymphocyte usually is about the same size as or slightly larger than an erythrocyte. When activated, lymphocytes grow larger and have proportionally more cytosol. Thus, some of the smaller, nonactivated lymphocytes have a diameter less than that of an erythrocyte, whereas activated lymphocytes may be two times the diameter of an erythrocyte.

There are three categories of lymphocytes. **T-lymphocytes** (*T-cells*) manage and direct an immune response; some directly attack foreign cells and virus-infected cells. **B-lymphocytes** (*B-cells*) are stimulated to become plasma cells and produce antibodies. **NK cells** (*natural killer cells*) attack abnormal and infected tissue cells. Lymphocytes are examined in detail in chapter 22.

Monocytes   A **monocyte** (mon′ō-sīt) can be up to three times the diameter of an erythrocyte. Monocytes usually constitute about 2–8% of all leukocytes. The nucleus of a monocyte is kidney-shaped or C-shaped. After approximately 3 days in circulation, monocytes exit blood vessels and take up residence within the tissues, where they transform into large phagocytic cells called **macrophages** (mak′rō-fāj; *macros* = large, *phago* = to eat). Macrophages phagocytize bacteria, viruses, cell fragments, dead cells, and debris.

Differential Count and Changes in Leukocyte Profiles

Abnormal numbers of leukocytes result from various pathologic conditions. For example, a reduced number of leukocytes causes a serious disorder called **leukopenia** (lū-kō-pē′nē-ă; *penia* = poverty). This decreased number of leukocytes may increase the risk of a person developing an infection or decrease their ability to fight infection effectively. Conversely, **leukocytosis** (lū′kō-sī-tō′sis) results from a slightly elevated leukocyte count and may be caused by a variety of factors, such as a recent infection or stress.

The term **differential count,** or *white blood cell differential count,* measures the amount of each type of leukocyte in your blood, and determines whether any of the circulating leukocytes are immature. Infection, tissue necrosis, bone marrow failure, cancers, or some other stresses to the body can affect the total ranges or percentages of a specific type of leukocyte, so differential counts are useful for diagnosing ailments.

Acute bacterial infections, acute stress, and tissue necrosis typically are associated with an increase in neutrophils, called **neutrophilia.** Their numbers will be in the tens of thousands, and some ailments may cause neutrophil counts of up to 50,000. In addition, as the body tries to produce more and more neutrophils, some immature neutrophils (called *band neutrophils,* or *band cells*) enter the circulation and are detected by the differential count. Increased presence of these immature neutrophils is referred to as a **left-shifted differential.** This phrase reflects that historically a lab printout listed numbers of cells according to maturity from left to right, so that an increased number of immature neutrophils was listed on the left. Decreased neutrophil count, called **neutropenia,** may occur with certain anemias, drug or radiation therapy, and from other causes.

INTEGRATE

CLINICAL VIEW
Leukemia

Leukemia (lū-kē′mē-ă) is a malignancy (cancer) in the leukocyte-forming cells. There are several categories of leukemia, but all are marked by abnormal development and proliferation of leukocytes, both in the bone marrow and the circulating blood. Leukemias represent a malignant transformation of a leukocyte cell line, and as abnormal leukocytes increase in number, the erythrocytic and megakaryocytic lines typically decrease in numbers because the proliferating malignant cells overtake the marrow and leave no room for the normal cells. This decrease in erythrocyte and platelet production results in both anemia and bleeding, which are often the first signs of leukemia. Leukemias are classified based on their duration as either acute or chronic.

Acute leukemia progresses rapidly, and death typically occurs within a few months after the onset of symptoms (severe anemia, hemorrhages, and recurrent infections). Acute leukemias tend to occur in children and young adults. **Chronic leukemia** progresses more slowly; survival usually exceeds 1 year from the onset of symptoms. Symptoms include anemia and a tendency to bleed. Chronic leukemias usually occur in middle-aged and older individuals.

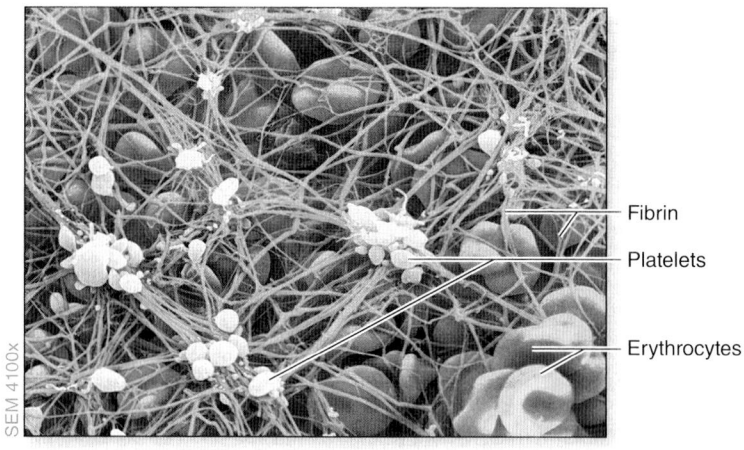

Figure 18.11 Blood Clotting. This SEM shows erythrocytes, fibrin, and platelets within a forming clot.

thrombos = clot); as with erythrocytes, that name is inappropriate because they are not true cells. Platelets are cell fragments, and unlike erythrocytes, they never had a nucleus. Recall that platelets are continually produced in the red bone marrow by megakaryocytes. Platelets serve an important function in hemostasis **(figure 18.11)**.

Normally, the concentration of platelets in an adult ranges from 150,000 to about 400,000 per cubic millimeter of blood, although the count may rise further during times of stress. Platelets can circulate in the blood for 8 to 10 days, unless they are needed earlier for hemostasis. Thereafter, they are broken down, and their contents are recycled. Approximately 30% of platelets are stored in the spleen (see section 21.4b).

 WHAT DID YOU LEARN?

16 What is the function of platelets, and what is their life span?

Viral infections, such as mumps, rubella, or mononucleosis, typically produce an increased number of lymphocytes. Lymphocyte values can increase to 20,000 in extreme cases. Additionally, the lymphocytes develop morphologic changes, in which their cytosol appears watery. Other conditions that can cause lymphocytosis include chronic bacterial infections, some leukemias, and multiple myeloma (a cancer of plasma cells, which are derived from B-lymphocytes). Decreased lymphocyte counts can occur with HIV infection, other leukemias, and sepsis, which is the presence of a pathogenic organism or substance in the blood.

Eosinophil numbers can increase in response to allergic reactions, parasitic infections, or some autoimmune diseases. Monocyte numbers may increase with chronic inflammatory disorders or tuberculosis, and may decrease due to prolonged prednisone (steroid) drug therapy. Finally, basophil counts can increase due to myeloproliferative disorders (which result from an overproduction of some formed elements in the bone marrow) and can decrease due to acute allergic and stress reactions.

 WHAT DID YOU LEARN?

14 What type of leukocyte may increase in number if you develop "strep throat" (an infection of the throat by *Streptococcus* bacteria)?

15 A person undergoes a routine blood test and is found to have a leukocyte count of 7000 cells per cubic millimeter, and 60% of the cells are neutrophils. Is the individual healthy? Explain.

18.3d Platelets

LEARNING OBJECTIVE

12. Explain the structure and function of platelets.

Platelets are irregular-shaped, membrane-enclosed cellular fragments that are about 2 μm in diameter (less than one-fourth the size of an erythrocyte). In stained preparations, they exhibit a dark central region. Platelets are sometimes called *thrombocytes* (throm′bō-sīt;

18.4 Hemostasis

When your blood vessels are healthy and functioning well, blood flows through them freely and does not clot unnecessarily. But if there is damage to a blood vessel, hemostasis is initiated. **Hemostasis** (hē′mō-stā′sĭs; *hemo* = blood, *stasis* = stability) is a stoppage of bleeding. It consists of three sequential phases, although there is some overlap between phases: vascular spasm, platelet plug formation, and coagulation phase **(figure 18.12)**.

18.4a Vascular Spasm

LEARNING OBJECTIVES

1. Describe vascular spasm, the first phase of hemostasis.

2. Name conditions that bring about vascular spasm.

When a blood vessel is injured, the first phase in hemostasis to occur is a **vascular spasm,** whereby the blood vessel constricts suddenly and, in so doing, limits the amount of blood that can leak from this damaged vessel. The spasm continues during the next phase, as both platelets and the endothelial cells of the blood vessel wall release an array of chemicals to further stimulate the vascular spasms.

The vascular spasm phase usually lasts from a few to many minutes. The more extensive the vessel and tissue damage, the greater the degree of vasoconstriction.

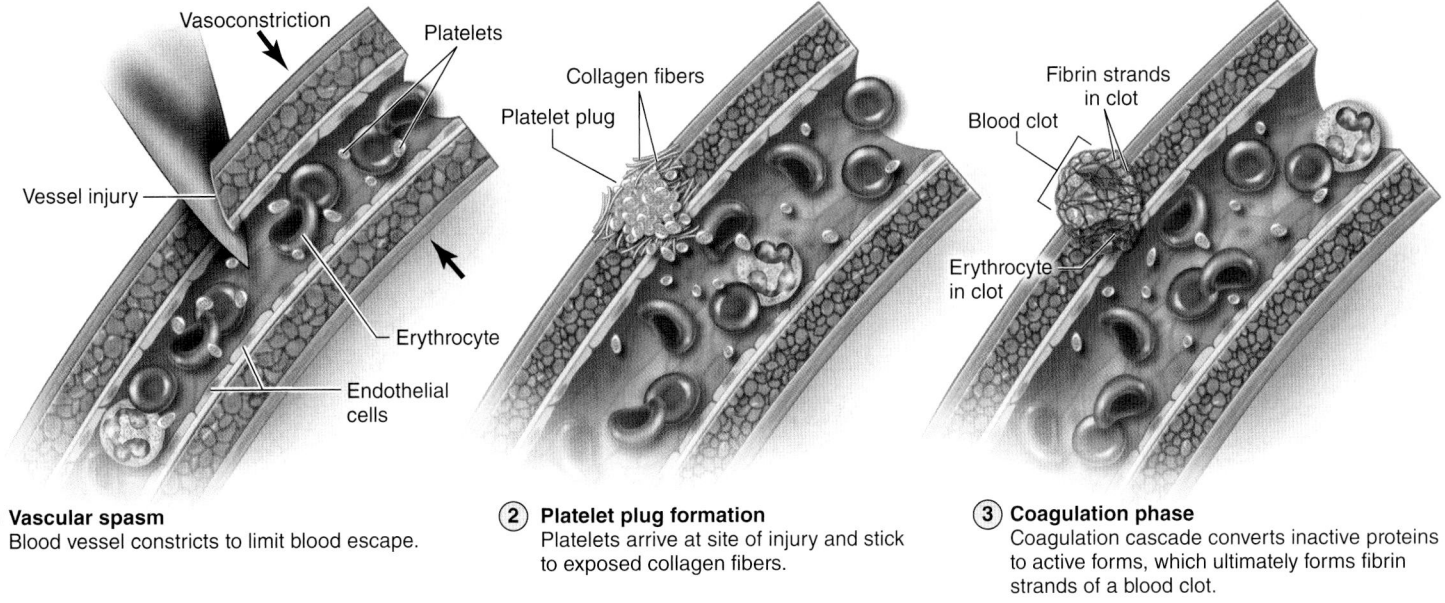

1 Vascular spasm
Blood vessel constricts to limit blood escape.

2 Platelet plug formation
Platelets arrive at site of injury and stick to exposed collagen fibers.

3 Coagulation phase
Coagulation cascade converts inactive proteins to active forms, which ultimately forms fibrin strands of a blood clot.

Figure 18.12 Hemostasis. Trauma to a blood vessel causes the blood to coagulate in a process called hemostasis.

WHAT DID YOU LEARN?

17 What occurs during a vascular spasm, and how long does this phase last?

18.4b Platelet Plug Formation

LEARNING OBJECTIVE

3. Describe what happens when platelets encounter damage in a blood vessel.

The next phase in hemostasis is the formation of a **platelet plug.** Normally, the endothelial wall (inner lining of a blood vessel) is smooth and is coated with an eicosanoid (see section 17.3b) called **prostacyclin,** which activates a pathway in both platelets and endothelial cells that involves production of cAMP to ultimately inhibit platelet activation. The end result is that prostacyclin serves as a platelet repellent.

Once a blood vessel is damaged, however, the collagen fibers within the connective tissue beneath the endothelial cells in the vessel wall become exposed. Platelets begin to stick to the exposed collagen fibers. Platelets adhere to the collagen fibers with the assistance of a plasma protein called *von Willebrand factor*, which serves as a bridge between platelets and collagen fibers.

As the platelets start to stick to the vessel wall, their morphology changes dramatically; they develop long processes that further adhere them to the blood vessel wall. As more and more platelets aggregate to the site, a platelet plug develops to close off the injury. The timing of this entire phase typically is less than a few minutes for a small- to medium-sized injury. Again, this is a temporary measure to block the flow of blood through the vessel wall where it is damaged.

Platelets undergo this morphologic change and become activated: their cytosol degranulates, releasing chemicals to assist with hemostasis. The following processes occur in response to these different chemicals:

- **Prolonged vascular spasms** with the release of *serotonin* and *thromboxane A_2* (an eicosanoid)
- **Attraction of other platelets** with the release of *adenosine diphosphate (ADP)* and *thromboxane A_2*, which facilitates the degranulation and release of these chemicals in other platelets

INTEGRATE

LEARNING STRATEGY

The three major events of hemostasis—vascular spasm, platelet plug formation, and blood coagulation—can be remembered with this succinct description of what happens—"squeeze, plug, clot."

- **Stimulation of coagulation** with the release of *procoagulants* that enhance blood clotting (the third phase)
- **Repair of the blood vessel** as platelets secrete substances to stimulate epithelial tissue, smooth muscle, and fibroblasts (cells of connective tissue) to replicate

WHAT DO YOU THINK?

4 Is the formation of the platelet plug an example of negative feedback or positive feedback?

Note that platelets are not only involved in the second phase of platelet plug formation, but they also increase events of the first and third phases, vascular spasm and coagulation phase, respectively. Thus, through the release of specific substances, platelets increase *all three* processes of hemostasis. (Thus, decreased hemostasis becomes a concern in individuals with a low platelet count, known as **thrombocytopenia.**)

The formation of the platelet plug is an example of positive feedback and typically is formed within 1 minute. But what is to prevent a platelet plug from becoming too big and growing out of control? As just mentioned, endothelial cells normally release prostacyclin. The healthy endothelial cells near the site of injury are still releasing their prostacyclin, so the plug does not grow larger than what is needed. The next phase, the coagulation phase, is beginning.

WHAT DID YOU LEARN?

18 What prevents platelets from forming plugs in healthy blood vessels?

19 How do platelets serve a central function in all three phases of hemostasis?

18.4c Coagulation Phase

LEARNING OBJECTIVES

4. Compare and contrast the intrinsic pathway and the extrinsic pathway for activating blood clotting.

5. Describe events in the common pathway.

6. Discuss the survival response that occurs when blood loss exceeds 10%.

Perhaps the most important and most complex component of hemostasis is **coagulation,** or blood clotting. A blood clot has an insoluble protein network composed of fibrin, which is derived from soluble fibrinogen. This meshwork of protein traps other elements of the blood, including erythrocytes, leukocytes, platelets, and plasma proteins, to form the clot (figure 18.11).

Substances Involved in Coagulation

Blood coagulation is a process that requires numerous substances, including calcium, clotting factors, platelets, and vitamin K. The specific clotting factors, including their numerical designation, name, function, pathway, and disorders associated with the factor are listed in **table 18.8.** Note that the clotting factor numbers are *in order of their discovery,* and not their position in the clotting pathway.

Most clotting factors are inactive enzymes, and most of these are produced by the liver. Vitamin K is a fat-soluble vitamin that is required for the synthesis of clotting factors II, VII,

Table 18.8	**Clotting Factors**				
Factor Designator[1]	**Name**	**Function**	**Extrinsic or Intrinsic Pathway**	**Clinical Syndrome (If Clotting Factor Is Deficient)**
I	Fibrinogen	Activated to fibrin	Both	Afibrinogemia (autosomal recessive disorder); during pregnancy can cause premature separation of placenta
II	Prothrombin	Protease; activated to thrombin	Both	Hypoprothrombinemia (autosomal recessive disorder); decreased synthesis in liver generally due to insufficient vitamin K (Note that mutation in the prothrombin gene causes hypercoagulation problems.[2])
III	Tissue factor (thromboplastin)	Cofactor; activates factor VII	Extrinsic	None known
IV	Calcium	Ion essential to both pathways	Both	None known
V	Proaccelerin	Cofactor; activates factor VII; combines with factor X to form prothrombin activator	Both	Parahemophilia (autosomal recessive) (Note that Leiden mutation causes hypercoagulation problems.[2])
VI	Accelerin	Redundant to activated factor V	Both	None known
VII	Proconvertin	Protease; activates factor X	Extrinsic	Hypoconvertinemia (autosomal recessive)
VIII	Antihemophilic factor A	Cofactor; activates factor X	Intrinsic	Hemophilia A (classical hemophilia); congenital X-linked trait
IX	Antihemophilic factor B (Christmas factor)	Protease; activates factor VIII	Intrinsic	Hemophilia B (Christmas disease[3]); congenital X-linked trait
X	Thrombokinase	Protease; combines with factor V to form prothrombin activator	Both	Stuart-Prower factor deficiency (autosomal recessive)
XI	Antihemophilic factor C	Protease; activates factor IX	Intrinsic	Hemophilia C, also known as plasma thromboplastin antecedent (PTA) deficiency (autosomal dominant)
XII	Hageman factor	Protease; activates factor XI and plasmin; converts prekallikrein to kallikrein	Intrinsic	Hageman trait (autosomal recessive disorder)
XIII	Fibrin-stabilizing factor	Cross-links fibrin	Both	The rarest of all of the clotting deficiencies; bleeding disorders apparent at birth (autosomal recessive disorder)

1. All proteins produced by the liver except: tissue factor (thromboplastin; factor III) formed by perivascular tissue; fibrin-stabilizing factor (factor XIII) produced by platelets and plasma; and factor IV, which is simply Ca^{2+} (not a protein). Hageman factor is produced by both the liver and platelets. Additional factors released from platelets: platelet factors 1, 2, 3, and 4.

2. This item is a hypercoagulation problem and not due to deficient clotting factor.

3. Named for the first person diagnosed with the disease.

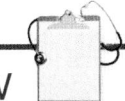

CLINICAL VIEW
Bleeding and Blood Clotting Disorders

If hemostasis fails to occur when needed, uncontrolled bleeding and death could result. On the other hand, if hypercoagulation develops, then unwanted blood clots could form in the blood and lead to the risk of deep vein thrombosis (blood clot in the leg), pulmonary embolism (blood clot in the lung), stroke (blood clot in the brain), or heart attack. Here we discuss several blood clotting disorders.

Bleeding Disorders

Bleeding disorders can be caused by several different conditions including hemophilia, a vitamin K deficiency, thrombocytopenia, or intake of various drugs.

Hemophilia is a group of bleeding disorders caused by specific genetic mutations. The two most common types of hemophilia are hemophilia A and hemophilia B, both of which are inherited in an X-linked recessive pattern. Females typically are carriers of the gene but may not experience symptoms because they have two X chromosomes, and one of the two X chromosomes may be normal. In contrast, males typically exhibit the full-blown disease because they have only one X chromosome.

Hemophilia A, also known as *classic hemophilia,* represents the vast majority of all hemophilias. It results in a deficiency or complete lack of normal factor VIII protein in the clotting cascade; the protein is abnormal and typically cannot participate in the proper clotting of the blood. This hemophilia occurs in approximately 1 in 5000 males in the United States. **Hemophilia B** is a deficiency of factor IX. It occurs in approximately 1 in 25,000 males in the United States. **Hemophilia C** is a relatively rare autosomal dominant deficiency of factor XI.

Vitamin K is a fat-soluble vitamin used by liver cells to produce many of the clotting factors. **Vitamin K deficiency** is more common in newborns than adults (because the newborn liver is immature and breast milk contains little vitamin K) and in individuals with liver or biliary diseases or chronic problems with fat absorption.

Thrombocytopenia is a deficiency in platelet count. This may be caused by an increased breakdown of platelets or decreased production of new platelets, as may occur with some bone marrow infections or cancers.

Various drugs, such as aspirin, ibuprofen (and other nonsteroidal anti-inflammatory drugs [NSAIDs]), and warfarin (Coumadin), and some herbal supplements (e.g., ginkgo biloba, garlic supplements) interfere with blood clotting and may cause bleeding if taken at sufficiently high doses. Physicians always should be monitoring their patients for bleeding disorders if they are taking any of these medications regularly. Additionally, they should ask their patients what herbal supplements they may be taking, so as to avoid any herbal supplement–drug interactions.

Hypercoagulation Problems

The term **hypercoagulation** refers to an increased tendency to clot blood. Hypercoagulation can lead to a **thrombus** (throm′bŭus; *thrombos* = clot), which is a clot within a blood vessel. If the thrombus dislodges and travels within the blood, it is called an **embolus** (em-bŏo′lŭus; *embolus* = wedge or stopper). An embolus is particularly dangerous because it can wedge within an artery and obstruct blood flow. A **pulmonary embolism** occurs in the pulmonary circulation of the lungs and can lead to breathing problems and perhaps death if not treated, whereas an embolus that travels to the blood vessels of the brain can cause a stroke. Treatment for a thrombus or embolus typically is with blood thinning medication (e.g., warfarin, heparin, or low-molecular-weight heparin).

Hypercoagulation can also have drug-related, environmental, and genetic causes. Certain medications such as birth control pills or hormone replacement therapy are associated with an increased risk of developing blood clots. Smoking greatly increases your risk, as nicotine is a vasoconstrictor, and it appears smoking increases levels of blood components related to clotting. Environmental causes include prolonged bedrest, surgery, pregnancy, or sitting in an airplane seat for a long time. In all of these cases, blood may pool in veins that are not being worked by exercise, and clotting can occur.

Genetic causes of hypercoagulation can be due to mutations of several genes, the most common of which is the **Leiden mutation,** which is a mutation of the gene for the synthesis of factor V. This mutation is present in 3% of the population and 3–15% of all Caucasians, and it accounts for 20–40% of all venous thromboses. Individuals with this mutation are unable to inactivate factor V in the clotting cascade, causing hypercoagulation and increased risk of thrombus formation. Young (under age 50), active individuals who present with a thrombus or embolus, and who don't have other significant risk factors for clotting, may need to be screened for these genetic mutations.

IX, and X; it acts as a coenzyme (see section 3.3c). Proteases such as factor VII and IX, when activated, act like scissors to convert another separate factor from its inactive to its active form. This active factor acts like scissors to convert yet another inactive factor to its active form; thus, the formation of a blood clot involves a cascade of changes.

Initiation of the Coagulation Cascade

The initiation of blood clotting can occur by two separate mechanisms: the intrinsic pathway (also known as the *contact activation pathway*) or the extrinsic pathway (or *tissue factor pathway*) **(figure 18.13)**. Both pathways converge, through a series of complicated steps, to the common pathway.

The **intrinsic pathway** is triggered by damage to the *inside* of the vessel wall and is initiated by platelets. This pathway typically takes approximately 3 to 6 minutes:

1. Platelets adhering to a damaged vessel wall release factor XII.
2. Factor XII converts the inactive factor XI to the active factor XI.
3. Factor XI converts inactive factor IX to active factor IX.
4. Factor IX binds with Ca^{2+} and platelet factor 3 to form a complex that converts inactive factor VIII to active factor VIII.
5. Factor VIII converts inactive factor X to active factor X.

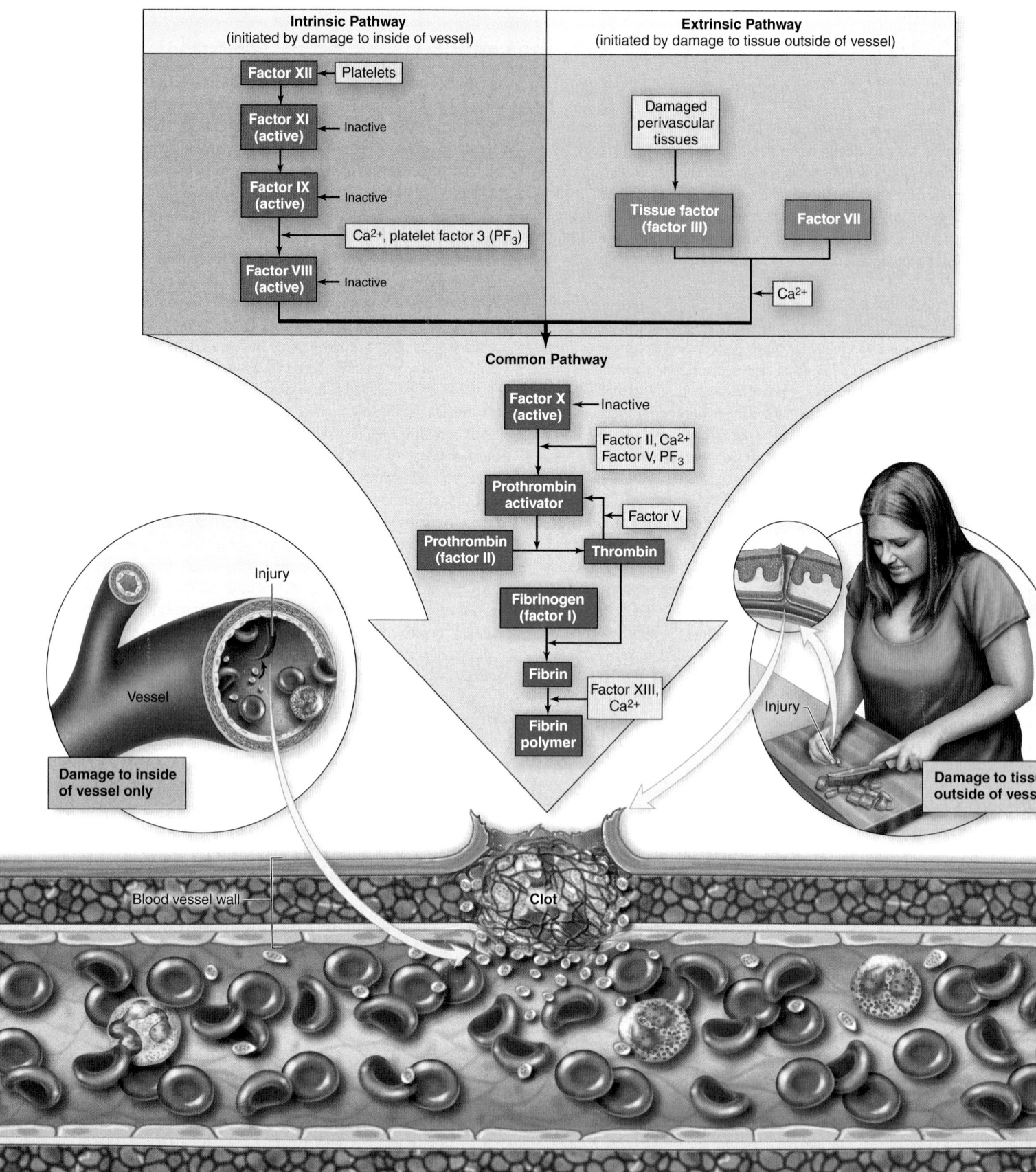

Figure 18.13 Coagulation Pathways. The process of coagulation typically involves both an intrinsic and an extrinsic pathway. Both pathways "merge" into a common pathway that ultimately converts fibrinogen (factor I) into fibrin.

In contrast, the **extrinsic pathway** is initiated by damage to the tissue that is *outside* the vessel, and this pathway usually takes approximately 15 seconds. This pathway occurs more quickly because there are fewer steps required. The steps include the following:

1. Tissue factor (thromboplastin; factor III) released from damaged tissues combines with factor VII and Ca^{2+} to form a complex.
2. This complex converts inactive factor X to active factor X.

Factor X, activated by either the intrinsic or extrinsic pathway, is the first step in the **common pathway:**

1. Active factor X combines with factors II and V, Ca^{2+}, and platelet factor 3 (PF_3) to form prothrombin activator.
2. Prothrombin activator activates prothrombin to thrombin.
3. Thrombin converts soluble fibrinogen into insoluble fibrin.
4. In the presence of Ca^{2+}, factor XIII is activated. Factor XIII cross-links and stabilizes the fibrin monomers into a fibrin polymer that serves as the "framework" of the clot.

Other components of blood become trapped in this spiderweb-like protein mesh. Like platelet plug formation, the clotting cascade is regulated by positive feedback. Once initiated by the intrinsic or extrinsic pathway, the events of the clotting cascade continue until a clot is formed (the climactic event). The size of the clot is limited because thrombin is either trapped within the clot or thrombin is quickly degraded by enzymes within the blood.

The Sympathetic Response to Blood Loss

In severe cases, when over 10% of the blood volume has been lost, a survival response is initiated.

As blood volume decreases, blood pressure decreases. If greater than 10% of the blood volume is lost from the blood vessels, the sympathetic division of the autonomic nervous system (ANS) is activated (see section 15.4), bringing about increased vasoconstriction of blood vessels, increased heart rate, and increased force of heart contraction in an attempt to maintain blood pressure. Blood flow is also redistributed to the heart and brain to keep these vital structures functioning. These processes are effective in maintaining blood pressure until approximately 40% of the blood is lost. Blood loss greater than 40% results in insufficient blood volume within the blood vessels, and blood pressure decreases to levels unable to support life.

WHAT DID YOU LEARN?

20 In what ways do the intrinsic and extrinsic pathways of the clotting cascade differ?

21 At what point in blood loss is the sympathetic division of the ANS typically activated, and what physiologic changes occur?

18.4d Elimination of the Clot

LEARNING OBJECTIVE

7. Explain the processes of clot retraction and fibrinolysis.

A blood clot is a temporary measure to stop blood loss through a damaged vessel wall. To return to normal, the blood vessel wall must be repaired and the clot eliminated. Elimination of the blood clot includes both clot retraction and fibrinolysis.

Clot retraction occurs as the clot is forming when actinomyosin, a contractile protein within platelets, contracts and squeezes the serum out of the developing clot. This makes the clot smaller as the sides of the vessel wall are pulled closer together.

To destroy the fibrin framework of the clot, **plasmin** (plaz′min) (or *fibrinolysin*) degrades the fibrin strands through **fibrinolysis** (fī′bri-nol′i-sis; *lysis* = dissolution). This is a process that begins within 2 days of the clot formation and occurs slowly over a number of days.

A "balancing act" is occurring continually within your blood between clot formation processes and those processes that prevent clot formation. The balance can be "tipped" so that blood clotting is initiated. A damaged blood vessel, impaired blood flow, atherosclerosis, or inflammation of the blood vessels can all potentially initiate blood clotting. Additionally, certain nutrients and vitamins must be present and available for blood

INTEGRATE

LEARNING STRATEGY

In both the intrinsic or extrinsic processes of clotting initiation, the final result is the production of factor X (factor 10). Use these tips to remember the sequence of factors:

- In the intrinsic pathway, a series of steps takes you to X by counting backward from 12 to 8—factors XII, XI, IX, and VIII (simply skip 10 because that is "the goal").
- The extrinsic pathway involves the coming together of factors VII and III, and if you add 7 and 3, the result is 10 (X).

clotting to occur normally. For example, calcium is used during the clotting process, and vitamin K is required for the synthesis of certain plasma proteins by the liver. Problems with the balance can lead to either bleeding or blood clotting disorders.

WHAT DO YOU THINK?

5 Predict why an individual confined to a wheelchair is more likely to develop unwanted clots.

WHAT DID YOU LEARN?

22 What is fibrinolysis, and what is its purpose?

18.5 Development and Aging of Blood

LEARNING OBJECTIVES

1. Describe when and how blood is formed in the embryo, fetus, childhood, and adulthood.
2. List some conditions that occur with the bone marrow and blood in the elderly.

The first primitive hemopoietic stem cells develop in the yolk sac wall of the embryo by the third week of development. The primitive hemopoietic stem cells go on to colonize other organs, such as the liver, spleen, and thymus. In these organs, these very primitive stem cells develop into the hemocytoblasts that produce all of the formed elements. Later in fetal development (perhaps beginning at 10 weeks), the hemocytoblasts begin to colonize red bone marrow, although the liver doesn't completely cease its blood cell production until close to birth.

Recall from section 7.2d that hemopoiesis occurs in most bones in young children, but as an individual reaches adulthood, hemopoiesis is restricted to selected bones in the axial skeleton. More red bone marrow is replaced with fat as individuals continue to age. Thus, older individuals have relatively less red bone marrow and may be more prone to developing anemia, which is a decrease in the number of circulating erythrocytes. In addition, older red bone marrow may be less able to meet any demands for an increased number of leukocytes. The leukocytes in the elderly may be less efficient and active than those in younger individuals, and the elderly also may have decreased numbers of leukocytes. Certain types of leukemias also are more prevalent among the elderly, probably due to the immune system (and its leukocytes) being less efficient.

WHAT DID YOU LEARN?

23 Where does hemopoiesis occur in (a) the fetus, (b) a child, and (c) an adult?

24 What are some ways that red bone marrow changes in the elderly?

CHAPTER SUMMARY

- Blood is a fluid connective tissue composed of formed elements, plasma, plasma proteins, and dissolved solutes that is transported through the cardiovascular system.

18.1 Functions and General Composition of Blood 702

18.1a Functions of Blood 702
- Blood transports nutrients, wastes, and respiratory gases; regulates body temperature, pH, and fluid levels; and protects the body against the activities of pathogens and the loss of blood.

18.1b Physical Characteristics of Blood 702
- The physical characteristics of blood include color, volume, viscosity, plasma concentration, temperature, and pH.

18.1c Components of Blood 703
- Centrifugation separates blood into three components: erythrocytes, a buffy coat composed of leukocytes and platelets, and plasma.
- Hematocrit represents the percentage of formed elements in blood. Males typically have higher hematocrits than females.
- Formed elements can be viewed in a blood smear.

18.2 Composition of Blood Plasma 705
- Blood plasma is a mixture of water, plasma proteins, and other solutes. It forms about 55% of whole blood.

18.2a Plasma Proteins 705
- Plasma is similar in composition to interstitial fluid, except the plasma contains proteins.
- Plasma proteins include albumin, globulins, fibrinogen and other clotting proteins, and regulatory proteins.

18.2b Other Solutes 707
- Other solutes include electrolytes, nutrients, respiratory gases, and waste products.

18.3 Formed Elements in the Blood 708
- Formed elements include erythrocytes that transport respiratory gases, leukocytes that serve some roles in protecting the body from harmful substances, and platelets that participate in hemostasis.

18.3a Hemopoiesis 708
- Hemopoiesis is the process by which formed elements develop.
- Formed elements develop in the red bone marrow via stem cells called hemocytoblasts.

18.3b Erythrocytes 711
- Erythrocytes exhibit a biconcave disc structure that facilitates the exchange of respiratory gases into and out of the erythrocytes.
- Hemoglobin is a pigmented protein within mature erythrocytes; it transports oxygen and carbon dioxide.
- Erythrocyte production is controlled by erythropoietin (and indirectly by testosterone).
- Aged erythrocytes are broken down and their components recycled after about 120 days in the blood.
- The ABO blood group consists of surface antigens (A or B) on the erythrocytes and plasma antibodies (anti-A or anti-B). Blood type is determined by the surface antigens on the erythrocytes (type O blood has no antigens). Individuals have plasma antibodies in their blood that are different from their surface antigens.
- The Rh blood group is determined by the presence or absence of the Rh surface antigen, called surface antigen D.

18.3c Leukocytes 719
- Leukocytes are white blood cells that defend against invading pathogens and remove damaged cells, debris, and antigen-antibody complexes.
- The types of leukocytes are granulocytes (neutrophils, eosinphils, and basophils) and agranulocytes (lymphocytes and monocytes).

18.3d Platelets 722
- Platelets are cellular fragments derived from megakaryocytes and function in hemostasis.

18.4 Hemostasis 722
- Hemostasis is a process to halt bleeding. There are three phases of hemostasis: vascular spasm, platelet plug formation, and the coagulation phase.

18.4a Vascular Spasm 722
- When a blood vessel wall is damaged, the vessel undergoes vascular spasms and constricts.

18.4b Platelet Plug Formation 723
- Platelets stick to the exposed collagen fibers within the damaged vessel wall, the platelets degranulate, and they release an array of chemicals to attract more platelets to the site.
- The platelets form a plug to close off the broken part of the blood vessel wall.

18.4c Coagulation Phase 724
- Coagulation is a clotting cascade involving the activation of numerous clotting factors that form a network of fibrin strands that trap formed elements and plasma proteins.
- The clotting cascade consists of an intrinsic pathway (which is initiated within a blood vessel by platelets), an extrinsic pathway (which is initiated by substances released outside the blood vessel), and a common pathway that produces fibrin.

18.4d Elimination of the Clot 727
- The blood clot is removed through clot contraction and fibrinolysis.

18.5 Development and Aging of Blood 727
- The first hemopoietic stem cells arise from the yolk sac wall in the embryo.
- In the fetus, hemopoiesis shifts to the liver, and then shifts to red bone marrow.
- Older individuals may have decreased production of formed elements.

Do You Know the Basics?

____ 1. Which individual is most likely to have the highest hematocrit level?

    a. a 10-year-old child

    b. a dehydrated adult male

    c. a healthy, nonmenstruating adult female

    d. a healthy, menstruating adult female

____ 2. Which type of leukocyte increases during allergic reactions and parasitic worm infections?

    a. basophil

    b. eosinophil

    c. lymphocyte

    d. neutrophil

____ 3. Which cell type forms platelets in the red bone marrow?

    a. lymphocyte

    b. megakaryocyte

    c. eosinophil

    d. reticulocyte

____ 4. Which of the following is *not* a function of blood?

    a. prevention of fluid loss

    b. transport of nutrients and waste

    c. maintenance of constant pH levels

    d. production of hormones

____ 5. A person with blood type A has

    a. anti-B antibodies in her blood plasma.

    b. anti-A antibodies in her blood plasma.

    c. both anti-A and anti-B antibodies in her blood plasma.

    d. no antibodies in her blood plasma.

____ 6. The hematocrit is a measure of

    a. water concentration in the plasma.

    b. the percentage of formed elements in the blood.

    c. the number of platelets in the blood.

    d. antibody concentration in the plasma.

____ 7. Oxygen attaches to a(n) _____ ion in hemoglobin.

    a. calcium

    b. sodium

    c. iron

    d. potassium

____ 8. During the recycling of components following the normal destruction of erythrocytes, globin is broken down, and its components are

    a. used to synthesize new proteins.

    b. stored as iron in the liver.

    c. eliminated from the body in the bile.

    d. removed in the urine.

____ 9. The extrinsic pathway of coagulation is initiated by

    a. platelets.

    b. fibrinogen.

    c. factor VIII.

    d. damage to tissue.

____ 10. A clot is best described as

    a. an aggregation of platelets.

    b. a fibrin network with trapped formed elements.

    c. agglutination of erythrocytes.

    d. All of these contribute to the formation of a clot.

11. How does blood help regulate body temperature?

12. What are alpha- and beta-globulins? What do they do?

13. When blood is centrifuged, a thin, whitish-gray layer called the buffy coat covers the packed erythrocytes. What are the components of the buffy coat?

14. What is the shape of an erythrocyte, and why is this shape advantageous to its function?

15. How are respiratory gases (oxygen and carbon dioxide) transported by erythrocytes?

16. What are the anatomic characteristics of each type of leukocyte? How can you tell these leukocytes apart when you view a blood smear under the microscope?

17. How do the functions of basophils differ from those of lymphocytes?

18. Briefly describe the origin, structure, and functions of platelets.

19. Compare and contrast the formation of lymphocytes versus granulocytes and monocytes. What precursor cells form each? What specific formed elements are formed from each? What is the common stem cell from which all leukocytes originate?

20. Describe the three phases of hemostasis, and list the major events that occur in each phase.

Can You Apply What You've Learned?

Use the following paragraph to answer questions 1–5.

Taylor is an active 25-year-old woman. She was brought into the emergency room after being hit by a car. Taylor suffered extensive blood loss, and the physicians felt she might be in need of a transfusion. They began the procedure of typing her blood.

1. Taylor's blood type was determined to be type A⁻. This meant she could be transfused with type

    a. A⁺.

    b. O⁺.

    c. AB⁻.

    d. O⁻.

2. Taylor was weak but conscious. The physicians asked her what medications, if any, she had taken. They asked her to list herbal supplements, pain relievers, and any multivitamins she took over the last few days. Her list included Advil (ibuprofen)—2 tablets that morning, 2 tablets that afternoon; a multivitamin—daily; birth control pill—daily; and an iron supplement—daily. Which of these medications, if any, would exacerbate and prolong Taylor's bleeding?

   a. Advil

   b. multivitamin

   c. birth control pills

   d. iron supplements

3. Which sequence or pathway best describes the process of hemostasis that must occur in Taylor to stop her bleeding?

   a. intrinsic pathway $\longrightarrow$ common pathway $\longrightarrow$ clotting

   b. common pathway $\longrightarrow$ extrinsic pathway $\longrightarrow$ clotting

   c. extrinsic pathway $\longrightarrow$ common pathway $\longrightarrow$ clotting

   d. intrinsic pathway $\longrightarrow$ extrinsic pathway $\longrightarrow$ clotting

4. Taylor was given the proper blood type and was admitted into the hospital to recoup from her injuries. The following evening, Taylor presented with a high fever, and the physicians suspected she had the complication of a bacterial infection. They sent her blood to the lab for a differential count. Assuming the physicians' guess was correct, the lab results would be returned with which of the following?

   a. increased number of eosinophils

   b. decreased number of basophils

   c. increased number of neutrophils

   d. increased number of lymphocytes

5. Taylor recovered from her injuries and left the hospital. Several months later, Taylor discovered that she was pregnant with her first child. Her OB/GYN told her she needed to take RhoGAM to ensure this pregnancy would culminate normally and there would be no problems for her child. If Taylor did *not* take RhoGAM, what complication could occur?

   a. The baby would develop anti-D antibodies to Taylor's A⁻ blood.

   b. The anti-D antibodies in Taylor's blood could attack the erythrocytes of all subsequent babies that Taylor carried.

   c. The child could survive only if it also had an A⁺ blood type.

   d. The anti-D antibodies in the baby's blood would attack Taylor's erythrocytes.

Can You Synthesize What You've Learned?

1. While taking a clinical laboratory class, Marilyn prepared and examined blood smears from several donors. One of the smears had an increased percentage (about 10% of observed leukocytes) of cells containing reddish-orange granules. Discuss the type of cell described and the condition that may have caused this increase in the donor.

2. Abby is a nurse on duty in a hospital emergency room when a critically injured patient is brought in. The physician calls for an immediate blood transfusion, but the patient's blood type is unknown. What blood type should the patient be given and why?

3. Your roommate and you are watching the track and field events at the Olympics. One of the sportscasters mentions that a particular athlete has been banned from competing because he was found guilty of blood doping. Your roommate asks you what blood doping is, and why an athlete would be banned for doping his blood. How would you answer your roommate?

INTEGRATE

ONLINE STUDY TOOLS

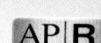 **connect** |ANATOMY & PHYSIOLOGY   ▮LEARNSMART® AP|R

The following study aids may be accessed through Connect.

Clinical Case Study: Leukemia in a Young Man, and a Special Treatment

Interactive Questions: This chapter's content is served up in a number of multimedia question formats for student study.

LearnSmart: Topics and terminology include functions and general composition of blood; composition of blood plasma;

formed elements in the blood; hemostasis; development and aging of blood

Anatomy & Physiology Revealed: Topics include blood; hemopoiesis; erythrocytes; hemoglobin breakdown; neutrophils, eosinophils, and basophils; lymphocytes; monocytes

Animations: Topics include hemoglobin breakdown

Cardiovascular System: Heart

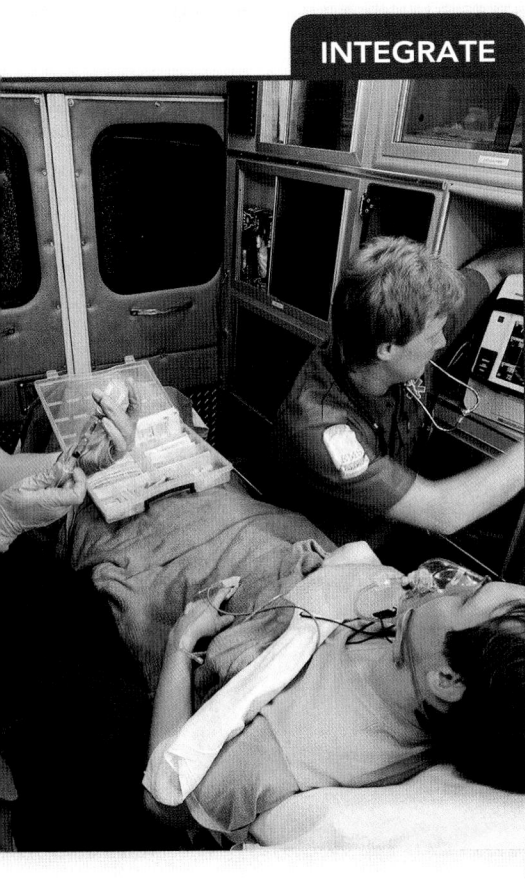

INTEGRATE

CAREER PATH
Paramedic

A paramedic is an allied health-care professional who administers initial treatment at the scene of a medical emergency or trauma. The central responsibilities of a paramedic include assessing cardiovascular health, administering certain preliminary treatments, and providing initial assessment to emergency room clinicians and personnel. In the United States, paramedic education may range from 6 months to 4 years (although most are trained at 2-year community colleges), and must pass a certification exam offered by the state in which they practice.

Anatomy & Physiology | REVEALED®
aprevealed.com

Module 9: Cardiovascular System

Who hasn't heard the saying "stay healthy, live longer"? When we think about staying healthy, many thoughts may come to mind: eating a diet that includes plenty of fruits and vegetables with limited amounts of fat and salt, participating in regular aerobic exercise, getting plenty of rest, and avoiding unhealthy behaviors such as smoking and heavy drinking. One significant reason for following this advice is to minimize our risk for developing cardiovascular disease. The importance of doing so becomes more obvious when we realize that cardiovascular disease—which is any disease that affects the heart or blood vessels—is the leading cause of death, both in the United States and worldwide.

➡️

We discuss the cardiovascular system in both this chapter on the heart and the next chapter on blood vessels. Our goal is to provide a clear and straightforward discussion on both a normal, healthy cardiovascular system and some of its more common malfunctions (e.g., myocardial infarction, heart murmurs, atherosclerosis, and circulatory shock). We hope to help you to develop both an understanding of, and an appreciation for, this system—a system so central to life that unfortunately when it fails, the outcome often proves fatal.

19.1 Introduction to the Cardiovascular System

Blood must circulate continuously to maintain homeostasis in the body: It must do so whether you are resting in bed or participating in intensive exercise. Circulation of blood is accomplished by the **cardiovascular** (kar′dē-ō-vas′kū-lǎr; *cardio* = heart, *vascular* = vessels) **system,** which is composed of both the heart and blood vessels **(figure 19.1)**. We first examine the

general function of this system and then include an overview of its components.

19.1a General Function

LEARNING OBJECTIVE

1. Describe the general function of the cardiovascular system.

The general function of the cardiovascular system is to transport blood throughout the body to allow the exchange of substances

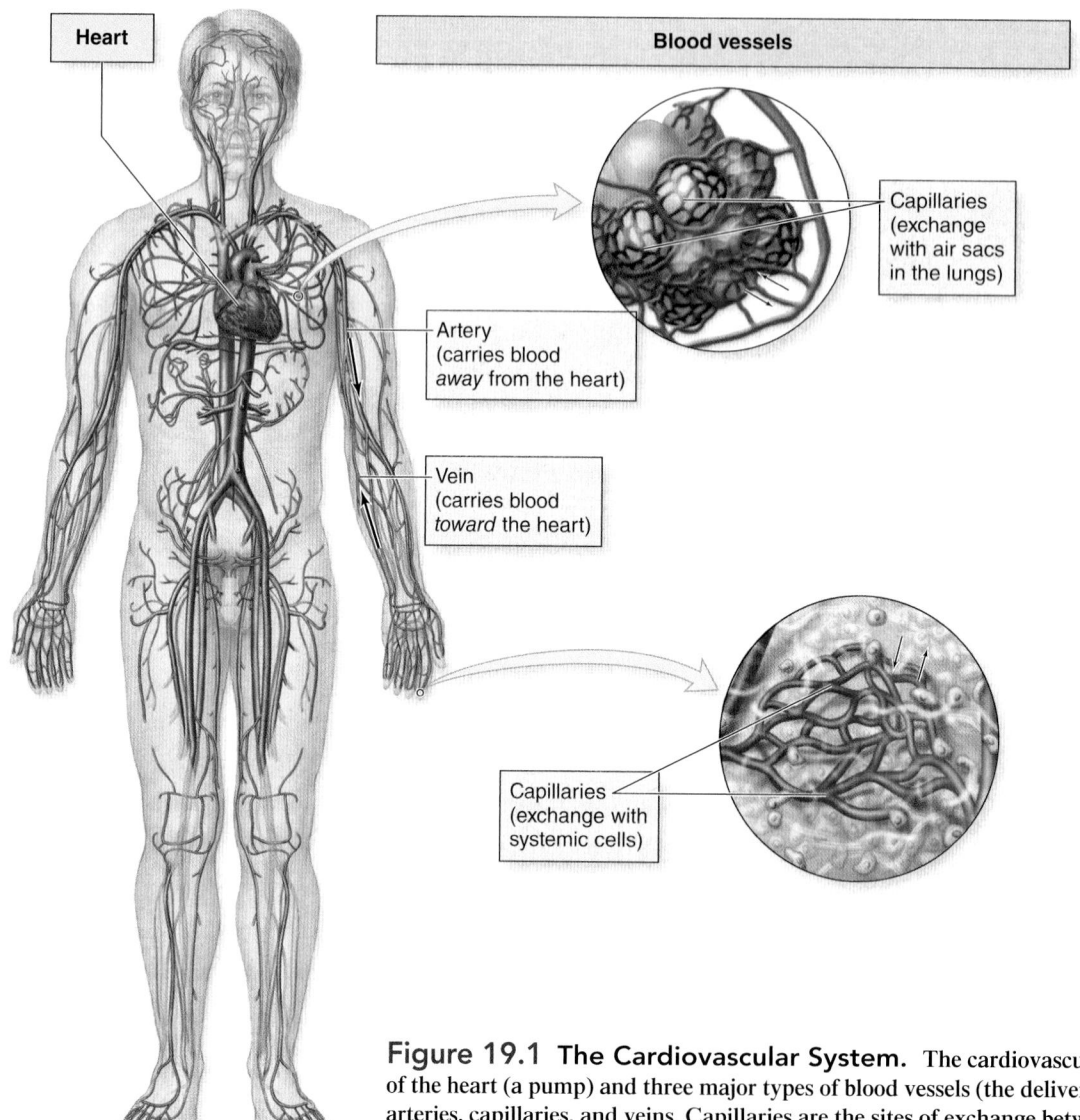

Figure 19.1 The Cardiovascular System. The cardiovascular system consists of the heart (a pump) and three major types of blood vessels (the delivery system), including arteries, capillaries, and veins. Capillaries are the sites of exchange between both the blood and air sacs in the lungs and blood and systemic cells in the body. AP|R

(e.g., respiratory gases, nutrients, and waste products) between the blood of capillaries and the body's cells. The "goal" of the cardiovascular system is to provide adequate perfusion of all body tissues. **Perfusion** (per-fyū′zhŭn; *perfusio* = pouring) is the delivery of blood per unit time per gram of tissue. It is typically expressed in milliliters per minute per gram (mL/min/g). *Adequate* perfusion involves delivering sufficient blood to maintain the health of all body cells.

The continual pumping action of the heart and healthy, patent (open and unblocked) vessels are essential to maintain good blood circulation and ample perfusion. If the heart fails to pump sufficient volumes of blood, or the vessels become hardened or occluded (blocked), then an adequate amount of blood may not reach the body's cells. Thus, tissues will be deprived of needed oxygen and nutrients, waste products accumulate, and cell death may occur.

WHAT DID YOU LEARN?

1 What are the potential consequences of a failing cardiovascular system?

19.1b Overview of Components

LEARNING OBJECTIVES

2. Differentiate between the three primary types of blood vessels.
3. Describe the general structure and function of the heart.
4. Compare and contrast pulmonary circulation and systemic circulation of the cardiovascular system. Trace blood flow through both circulations.

Here we provide an overview description of (1) the primary types of blood vessels, through which blood travels; (2) the heart, which pumps the blood; and (3) the two closed circulatory pathways of blood vessels that begin and end with the heart.

Blood Vessels

Blood vessels are the conduits or "soft pipes" of the cardiovascular system that transport blood throughout the body (figure 19.1). They are categorized into three primary types: **Arteries** (ar′ter-ēz) carry blood away from the heart; **veins** (vānz) carry blood back to the heart; and **capillaries** (kap′i-lār-ēz; capillaries = relating to hair) serve as the sites of exchange, either between the blood and the air sacs (called alveoli) of the lungs or the blood and systemic cells. (The details of blood vessel anatomy are described in section 20.1.)

A common error is to describe arteries as the vessels that always carry **oxygenated** (ok′si-je-nāt-ed) blood (blood high in O_2 and low in CO_2) and veins as the vessels that always transport **deoxygenated** blood (blood low in O_2 and high in CO_2). As you will see, this is *not* an accurate generalization; in some parts of the cardiovascular system, the reverse is true. The defining factor is whether the blood is moving away from the heart (as it does in arteries) or toward the heart (as it does in veins).

The Heart

The **heart** is the center of the cardiovascular system. It is a hollow, four-chambered organ, serving to pump blood throughout the body. Three anatomic features are significant in the normal function of the heart: (1) the two sides of the heart; (2) the great vessels attached to the heart; and (3) the two sets of valves that are located within the heart (**figure 19.2**).

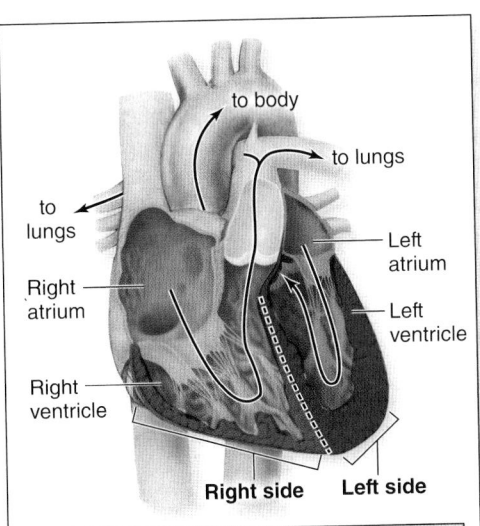

Two sides

Each side has a receiving chamber (atrium) and a pumping chamber (ventricle).

- **Right side:** Receives deoxygenated blood from the body and pumps it to the lungs.
- **Left side:** Receives oxygenated blood from the lungs and pumps it to the body.

(a)

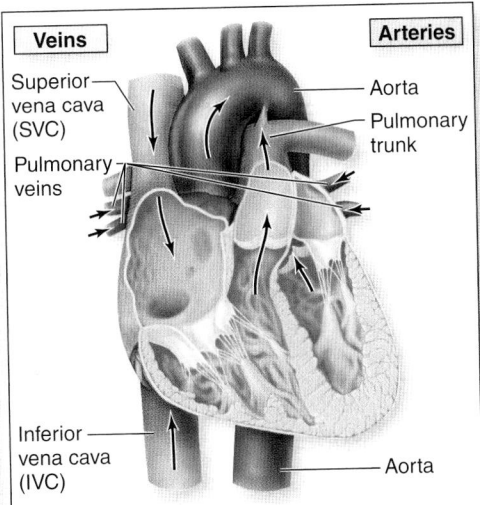

Great vessels

The great vessels are the large arteries and veins that are directly attached to the heart.

Arteries (arterial trunks) transport blood away from the heart.

- Pulmonary trunk transports from right ventricle.
- Aorta transports from left ventricle.

Veins transport blood toward the heart

- Venae cavae (SVC and IVC) drain into right atrium.
- Pulmonary veins drain into left atrium.

(b)

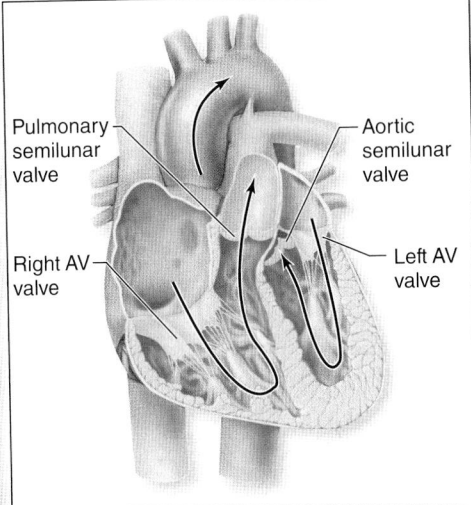

Valves

Heart valves prevent backflow to ensure one-way blood flow.

- **Atrioventricular (AV) valves** (i.e., right AV valve and left AV valve) are between an atrium and a ventricle.
- **Semilunar valves** (i.e., pulmonary semilunar valve and aortic semilunar valve) are between a ventricle and an arterial trunk.

(c)

Figure 19.2 Significant Anatomic Features of the Heart. The heart is composed of the following features: (*a*) a right side and a left side, (*b*) great vessels that are directly attached to it, and (*c*) two sets of valves that permit one-way flow of blood. **AP|R**

INTEGRATE

LEARNING STRATEGY

Arteries always carry blood **away** from the heart. **Veins** always carry blood **toward** the heart.

First, the heart is composed of two sides: the right side and the left side (figure 19.2a). Each side of the heart has two chambers: a superior chamber for receiving blood called an **atrium** (ā'trē-ŭm; entrance hall) and an inferior chamber for pumping blood away from the heart called a **ventricle** (ven'tri-kl; *venter* = belly). Thus, four chambers in the heart are identified: **right atrium** and **right ventricle** on the right side of the heart, and the **left atrium** and **left ventricle** on the left side of the heart. The two sides of the heart allow separation of circulating deoxygenated and oxygenated blood:

- The right side of the heart receives deoxygenated blood from the body and pumps it to the lungs.
- The left side of the heart receives oxygenated blood from the lungs and pumps it to the body.

Second, blood is transported directly to and from the chambers of the heart by the **great vessels** that are continuous with specific chambers of the heart (figure 19.2b). These vessels include both arteries and veins. There are two large arteries (or *arterial trunks*) attached to the superior border of the ventricles: They transport blood from a ventricle away from the heart. The **pulmonary trunk** (which splits into the pulmonary arteries) transports blood from the right ventricle, whereas the **aorta** (ā-ōr'tă) transports blood from the left ventricle. Large veins deliver blood to the heart into an atrium. These include the two venae cavae (the **superior vena cava [SVC]** and **inferior vena cava [IVC]),** which drain blood into the right atrium, and the **pulmonary veins,** which drain blood into the left atrium. Remember, ventricles pump blood into the arterial trunks, and atria receive blood from the veins.

 WHAT DO YOU THINK?

1 What vessels attached to the heart contain oxygenated blood? Are they both arteries? Explain.

Third, two sets of valves are located within the heart (figure 19.2c). The **atrioventricular** (ā'trē-ō-ven-trik'ū-lăr) **(AV) valves** are between the atrium and ventricle of each side of the heart. The **right AV valve,** also called the *tricuspid valve,* is located between the right atrium and the right ventricle. (Remember that the **TRI**cuspid is on the **RI**ght side.) The **left AV valve,** also called the *bicuspid valve,* or *mitral valve,* is located between the left atrium and left ventricle.

The other set of valves is the **semilunar valves**, which mark the boundary between a ventricle and its associated arterial trunk. The **pulmonary semilunar valve** is between the right ventricle and the pulmonary trunk, and the **aortic semilunar valve** is between the left ventricle and the aorta. The valves open to allow blood to flow through the heart and then close to prevent its backflow. This ensures one-way, or unidirectional, flow of blood through the heart.

Circulation Routes

The two sides of the heart and the blood vessels are arranged in two circuits: the pulmonary circulation and systemic circulation (**figure 19.3**).

The **pulmonary** (pŭl'mō-nār-ē; *pulmo* = lung) **circulation** includes the movement of deoxygenated blood through (1) the right side of the heart, (2) blood vessels to the lungs for the pickup of oxygen and the release of carbon dioxide, and (3) blood vessels that return blood to the left side of the heart.

The **systemic circulation** includes the movement of oxygenated blood through (1) the left side of the heart, (2) blood vessels to the systemic cells such as those of the liver, skin, muscle, and brain for the exchange of nutrients, respiratory gases, and wastes, and (3) blood vessels that return blood to the right side of the heart. Thus, the basic pattern of blood flow is the right side of the heart ⟶ lungs ⟶ the left side of the heart ⟶ systemic cells of the body ⟶ back to the right side.

One of the critical concepts in understanding the function of the cardiovascular system is the flow of blood through the heart and the two circulatory routes. This flow pattern is summarized in **figure 19.4**.

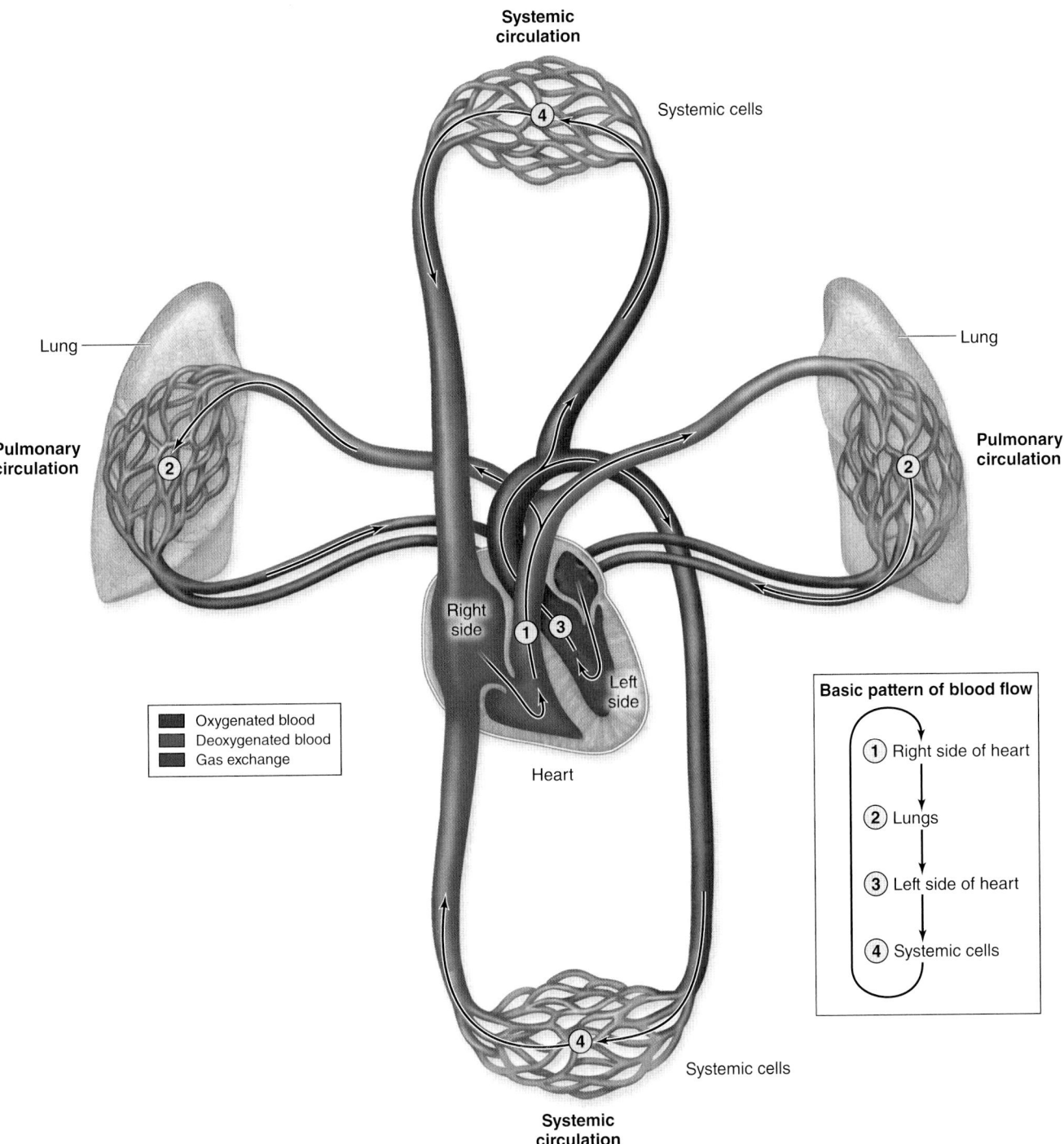

Figure 19.3 **Cardiovascular System Circulations.** The cardiovascular system is composed of the pulmonary circulation and the systemic circulation. AP|R

WHAT DID YOU LEARN?

2 What generalization can be made about all arteries? What generalization can be made about all veins?

3 What path does blood follow through the heart? Identify all structures that it passes through, including each chamber, valve, and great vessel. Begin at the right atrium.

4 Which of the great vessels is both an artery and transports deoxygenated blood? Which of the great vessels is both a vein and transports oxygenated blood?

Figure 19.4 Blood Flow Through the Heart and Circulatory Routes. The two circulatory routes for blood flow are (a) the pulmonary circulation, which includes the pumping of blood by the right side of the heart to the lungs and the return of blood to the left side of the heart, and (b) the systemic circulation, which includes the pumping of blood by the left side of the heart to the systemic cells and the return of blood to the right side of the heart.

(a) Pulmonary Circulation

Transports blood from the right side of the heart to the alveoli of the lungs for gas exchange and back to the left side of the heart

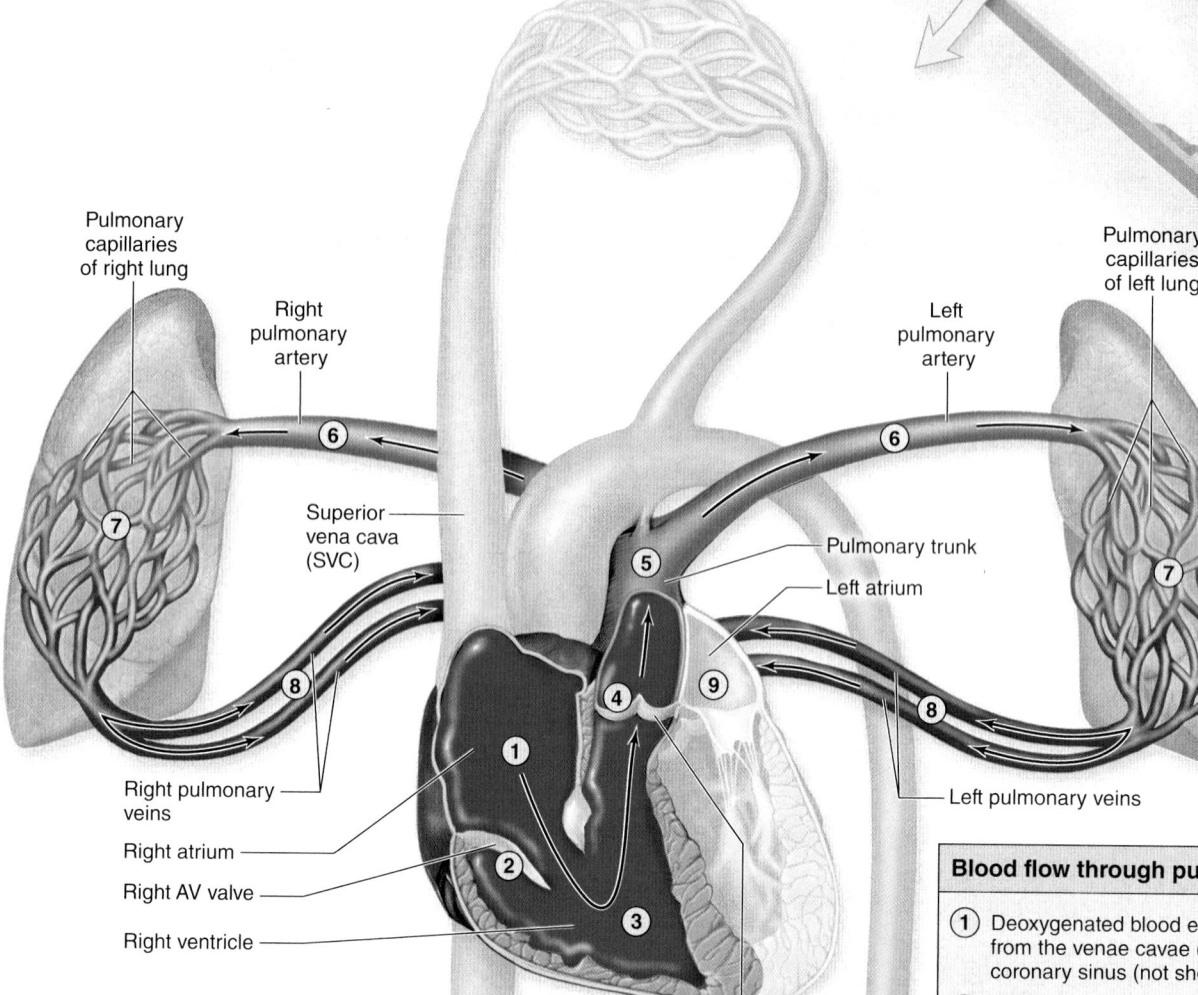

Pulmonary capillaries of right lung

Right pulmonary artery

Superior vena cava (SVC)

Right pulmonary veins

Right atrium

Right AV valve

Right ventricle

Inferior vena cava (IVC)

Pulmonary semilunar valve

Pulmonary capillaries of left lung

Left pulmonary artery

Pulmonary trunk

Left atrium

Left pulmonary veins

Blood flow through pulmonary circulation

1. Deoxygenated blood enters the **right atrium** from the venae cavae (SVC and IVC) and coronary sinus (not shown). This blood then

2. passes through the **right AV valve** (tricuspid valve),

3. enters the **right ventricle,**

4. passes through the **pulmonary semilunar valve,** and

5. enters the **pulmonary trunk.**

6. This blood continues through the **right and left pulmonary arteries** to both lungs, and

7. enters **pulmonary capillaries** of both lungs for gas exchange.

8. This blood, which is now oxygenated, enters **right and left pulmonary veins,** and is returned to

9. the **left atrium** of the heart.

(b) Systemic Circulation

Transports blood from the left side of the heart to the systemic cells of the body for nutrient and gas exchange, and back to the right side of the heart

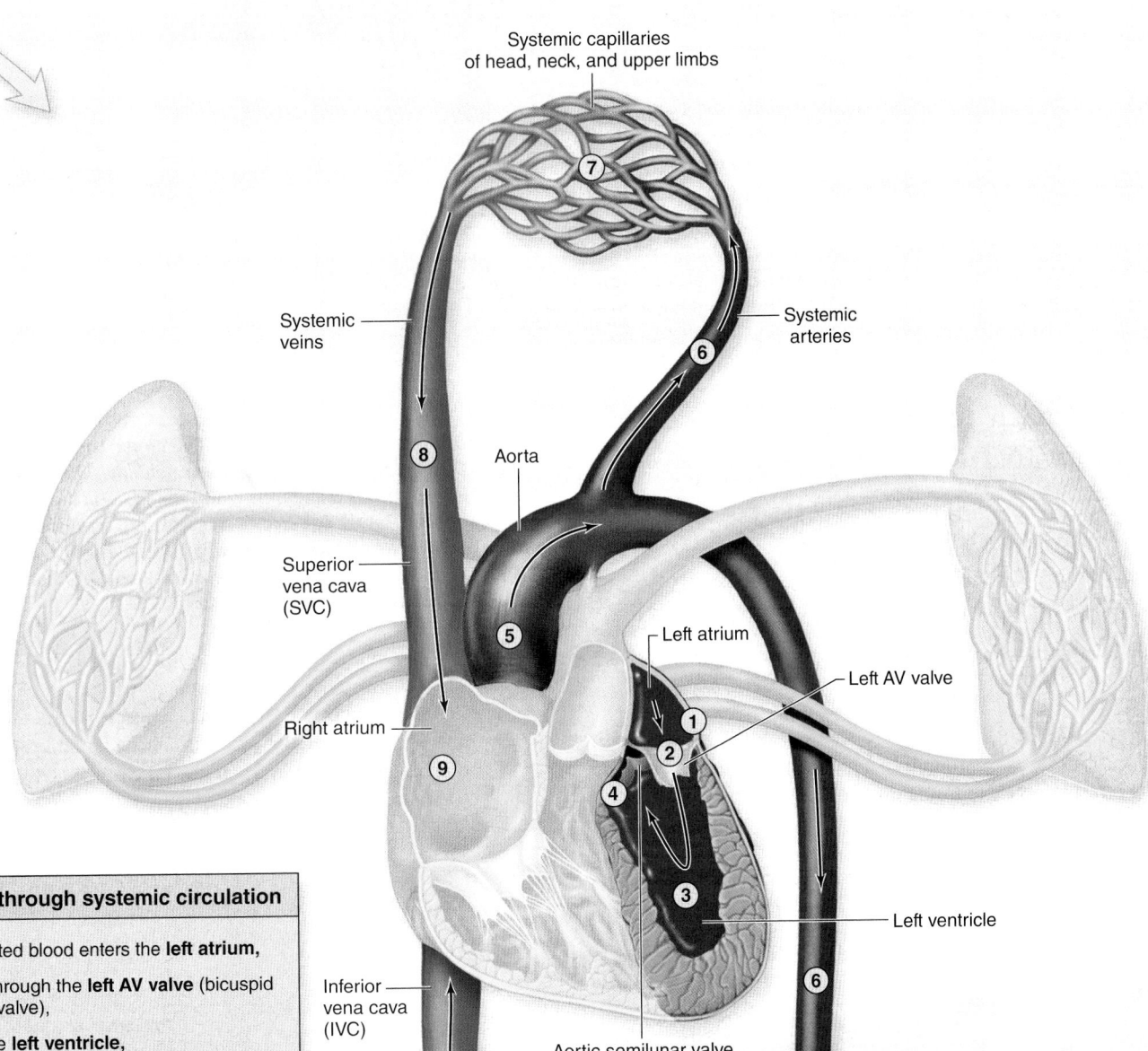

Systemic capillaries of head, neck, and upper limbs

Systemic veins

Systemic arteries

Aorta

Superior vena cava (SVC)

Left atrium

Left AV valve

Right atrium

Left ventricle

Inferior vena cava (IVC)

Aortic semilunar valve

Systemic veins

Systemic arteries

Systemic capillaries of trunk and lower limbs

Blood flow through systemic circulation

1. Oxygenated blood enters the **left atrium,**
2. passes through the **left AV valve** (bicuspid or mitral valve),
3. enters the **left ventricle,**
4. passes through **aortic semilunar valve,** and
5. enters the **aorta.**
6. This blood is distributed by the **systemic arteries,** and
7. enters **systemic capillaries** for nutrient and gas exchange.
8. This blood, which is now deoxygenated, ultimately drains into the **SVC, IVC,** and **coronary sinus** (not shown), and
9. enters the **right atrium.**

19.2 The Heart Within the Thoracic Cavity

The heart is located within the thoracic cavity and is enclosed within a fibroserous sac called the pericardium; both of these structures have roles in protection and support of the heart.

19.2a Location and Position of the Heart

LEARNING OBJECTIVE

1. Describe the location and position of the heart in the thoracic cavity.

The heart is located posterior to the sternum left of the body midline between the lungs within the **mediastinum** (me′dē-as-tī′nŭm; *medius* = middle) **(figure 19.5)**. The position of the heart is slightly rotated such that its right side or right border is located more anteriorly, whereas its left side or left border is located more posteriorly. The postero-superior surface of the heart is called the **base** (not labeled on figure 19.5). The inferior, conical end of the heart is called the **apex** (ā′peks; tip). It projects slightly anteroinferiorly toward the left side of the body with the right ventricle lying on the diaphragm. You may find it helpful to think of the heart's position like an "upside down" pyramid with the apex below the base.

WHAT DID YOU LEARN?

5 Which side of the heart is more visible in an anterior view?

19.2b Characteristics of the Pericardium

LEARNING OBJECTIVES

2. List the structural components of the pericardium.

3. Describe the function of the pericardium and the purpose of the serous fluid within the pericardial cavity.

The heart is enclosed in three layers, collectively called the **pericardium** (per-i-kar′dē-ŭm; *peri* = around, *kardia* = heart) **(figure 19.6)**. These layers are (from outside to inside) as follows:

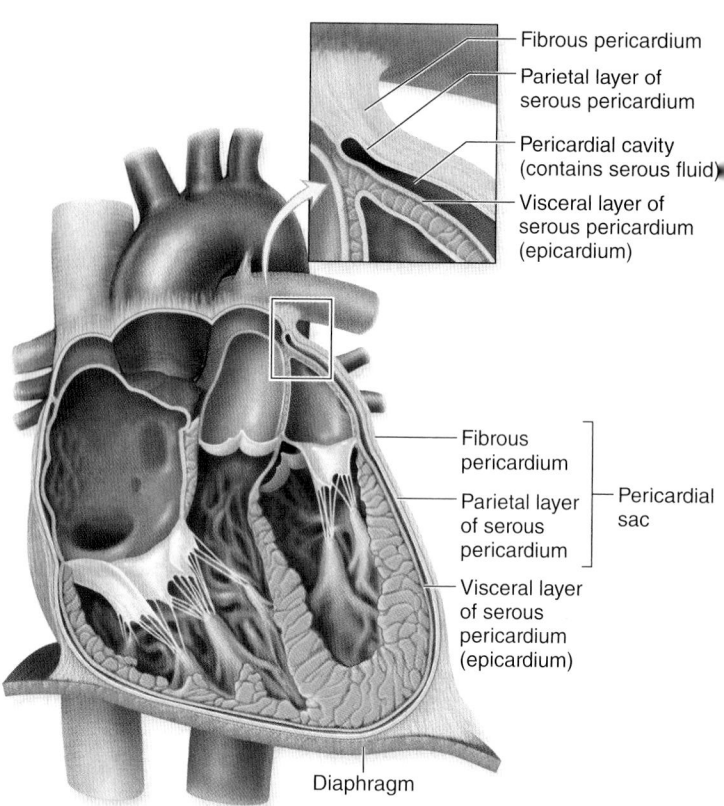

Figure 19.6 Pericardium. The protective layers of the heart include the pericardial sac composed of an outer fibrous pericardium and an inner serous membrane called the parietal layer of serous pericardium. Tightly adhered to the heart is a serous membrane called the visceral layer of serous pericardium. The space between the parietal and visceral layers is called the pericardial cavity, which contains serous fluid produced by both serous membranes. AP|R

- The **fibrous pericardium**, which is composed of tough, dense irregular connective tissue that encloses the heart, but does not attach to it. Rather, this layer is attached inferiorly to the diaphragm and superiorly to the base of the great arterial trunks (pulmonary trunk and aorta).

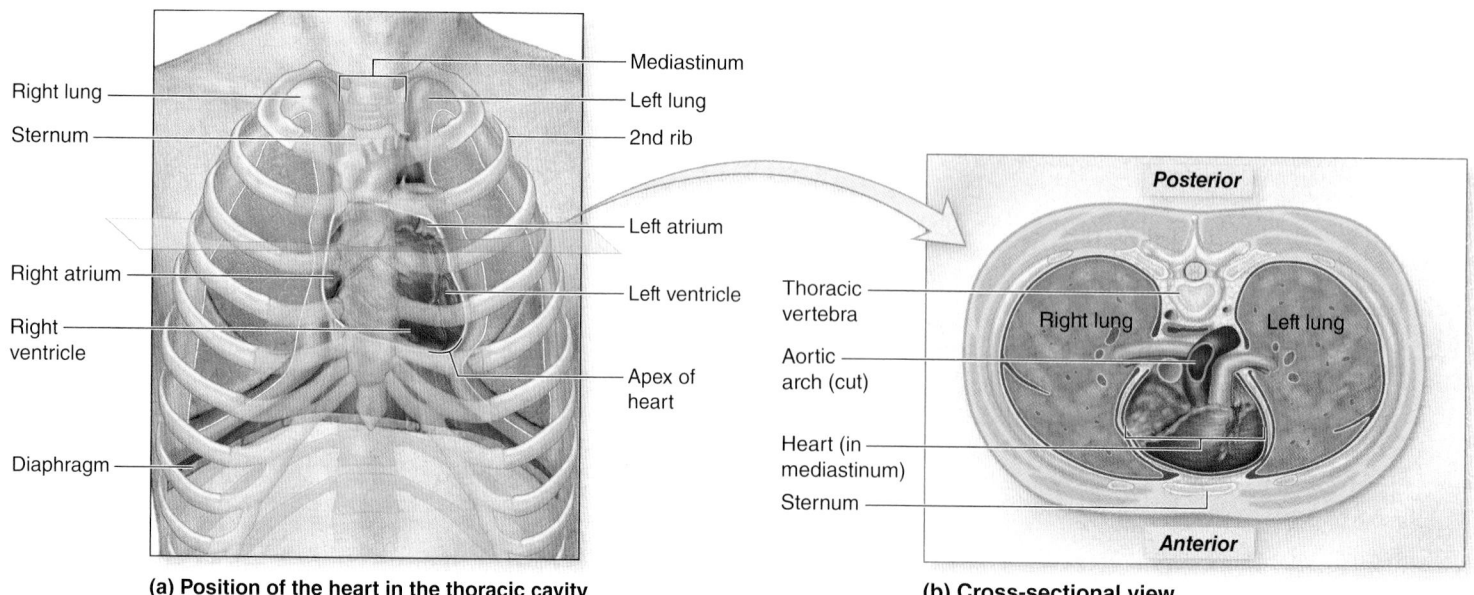

(a) Position of the heart in the thoracic cavity

(b) Cross-sectional view

Figure 19.5 Heart Position Within the Thoracic Cavity. (*a*) The heart is located within the mediastinum of the thoracic cavity between the lungs. (*b*) A cross-sectional view depicts the heart's relationship to the other organs in the thoracic cavity. AP|R

- The **parietal layer of the serous pericardium,** which is composed of simple squamous epithelium and an underlying delicate layer of areolar connective tissue, adheres to the inner surface of the fibrous pericardium.
- The **visceral layer of the serous pericardium** (also called the epicardium) is also composed of a simple squamous epithelium and an underlying delicate layer of areolar connective tissue. This serosal layer adheres directly to the heart. The two serosal layers are continuous with one another and separated by a potential space called the **pericardial cavity.**

The tough fibrous pericardium serves to both anchor the heart within the thoracic cavity and prevent the heart chambers from over-filling with blood. The two layers of the serous pericardium produce and release serous fluid into the pericardial cavity. This fluid has the consistency of an oily mixture, and it lubricates the serous membranes to decrease friction with every heartbeat.

 WHAT DID YOU LEARN?

 6 Describe the three layers that cover the heart. Where is the pericardial cavity relative to these layers?

INTEGRATE

LEARNING STRATEGY

Imagine your heart as a fist that is placed into a balloon (as shown). The two layers of the balloon are serous membranes. The portion adhered to your hand is the visceral layer, the outer portion is the parietal layer, and the space between them is the pericardial cavity. Now imagine a sandwich bag placed around your hand and balloon; this represents the fibrous pericardium.

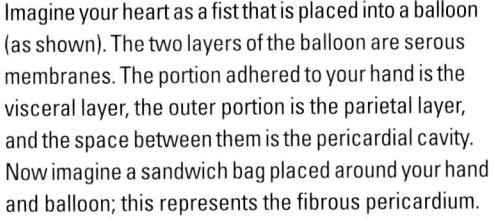
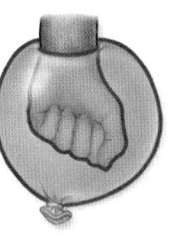

INTEGRATE

CLINICAL VIEW
Pericarditis

Pericarditis (per′i-kar-dī′tis; *itis* = inflammation) is an inflammation of the pericardium typically caused by viruses, bacteria, or fungi. The inflammation is associated with an increase in permeability of capillaries within the pericardium, which become more "leaky," resulting in excess fluid leaving the blood and accumulating within the pericardial cavity. At this point, the potential space of the pericardial cavity becomes an actual space as it fills with fluid. In severe cases, the excess fluid accumulation limits the heart's movement and keeps the heart chambers from filling with an adequate amount of blood. The heart is unable to pump blood, leading to a medical emergency called **cardiac tamponade** (tam′pŏ-nād′) and possibly resulting in heart failure and death.

A helpful physical finding in diagnosing pericarditis is **friction rub.** This is a crackling or scraping sound heard with a stethoscope that is caused by the movement of the inflamed pericardial layers against each other.

19.3 Heart Anatomy

When the pericardial sac is cut and reflected, the heart can be removed from the mediastinum and examined in greater detail. It is a relatively small, conical, muscular organ approximately the size of a person's clenched fist. In the average normal adult, it weighs about 300 grams (0.7 pound), but certain diseases may cause heart size to increase dramatically. Here we examine the heart's external and internal structures.

19.3a Superficial Features of the Heart

 LEARNING OBJECTIVE

1. Compare the superficial features of the anterior and posterior aspects of the heart.

The heart was previously described as being composed of four hollow chambers: two smaller atria and two larger ventricles (**figure 19.7**). The atria are separated from the ventricles externally by a relatively deep groove called the **coronary** (kōr′o-nār-ē; *corona* = crown) **sulcus** (or *atrioventricular sulcus*) that extends around the circumference of the heart. An **interventricular** (in-ter-ven-trik′ū-lăr; *inter* = between) **sulcus** is a groove between the ventricles that extends inferiorly from the coronary sulcus toward the heart apex, and delineates the superficial boundary between the right and left ventricles. The **anterior interventricular sulcus** is located on the anterior side of the heart, and the **posterior interventricular sulcus** is located on the posterior side of the heart. Located within all of these sulci are coronary vessels associated with supplying blood to the heart wall. Coronary vessels are described in detail in section 19.4.

Anterior View

The right atrium and right ventricle are prominent when observing the heart from an anterior view (figure 19.7*a*). The portion of the right atrium that is most noticeable is its wrinkled, flaplike extension called the **right auricle** (aw′ri-kl); *auris* = ear). You may find it helpful to think of an auricle as an "ear" of the atrium. Also seen in this view are both the aorta and pulmonary trunk, a small portion of the **left auricle** of the left atrium, the anterior interventricular sulcus, and part of the coronary sulcus.

Posterior View

The left atrium and left ventricle are prominent when observing the heart from a posterior view (figure 19.7*b*). The left atrium primarily forms the base on the posterosuperior surface of the heart. The pulmonary veins are attached to the left atrium. Also seen in this view are both the superior and inferior venae cavae, pulmonary arteries, the posterior interventricular sulcus, and part of the coronary sulcus that houses the coronary sinus (a modified vein that drains deoxygenated blood from the heart wall).

 WHAT DID YOU LEARN?

 7 Where is the coronary sulcus located? Is the coronary sinus visible from both the anterior view and posterior view?

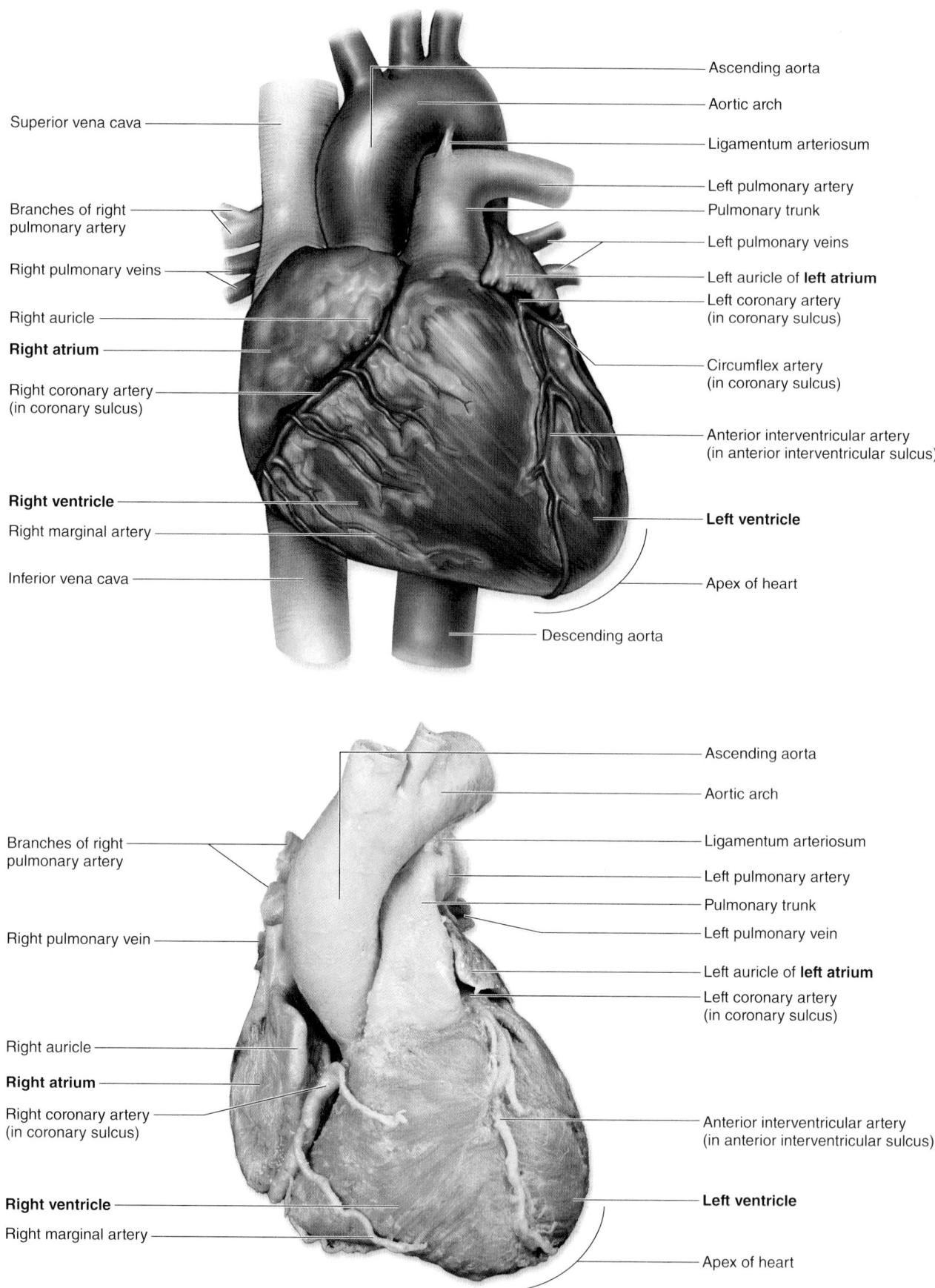

Figure 19.7 External Anatomy and Features of the Heart. A drawing and a cadaver photo show the superficial features of the heart. (*a*) Anterior view. AP|R

(a) Anterior view

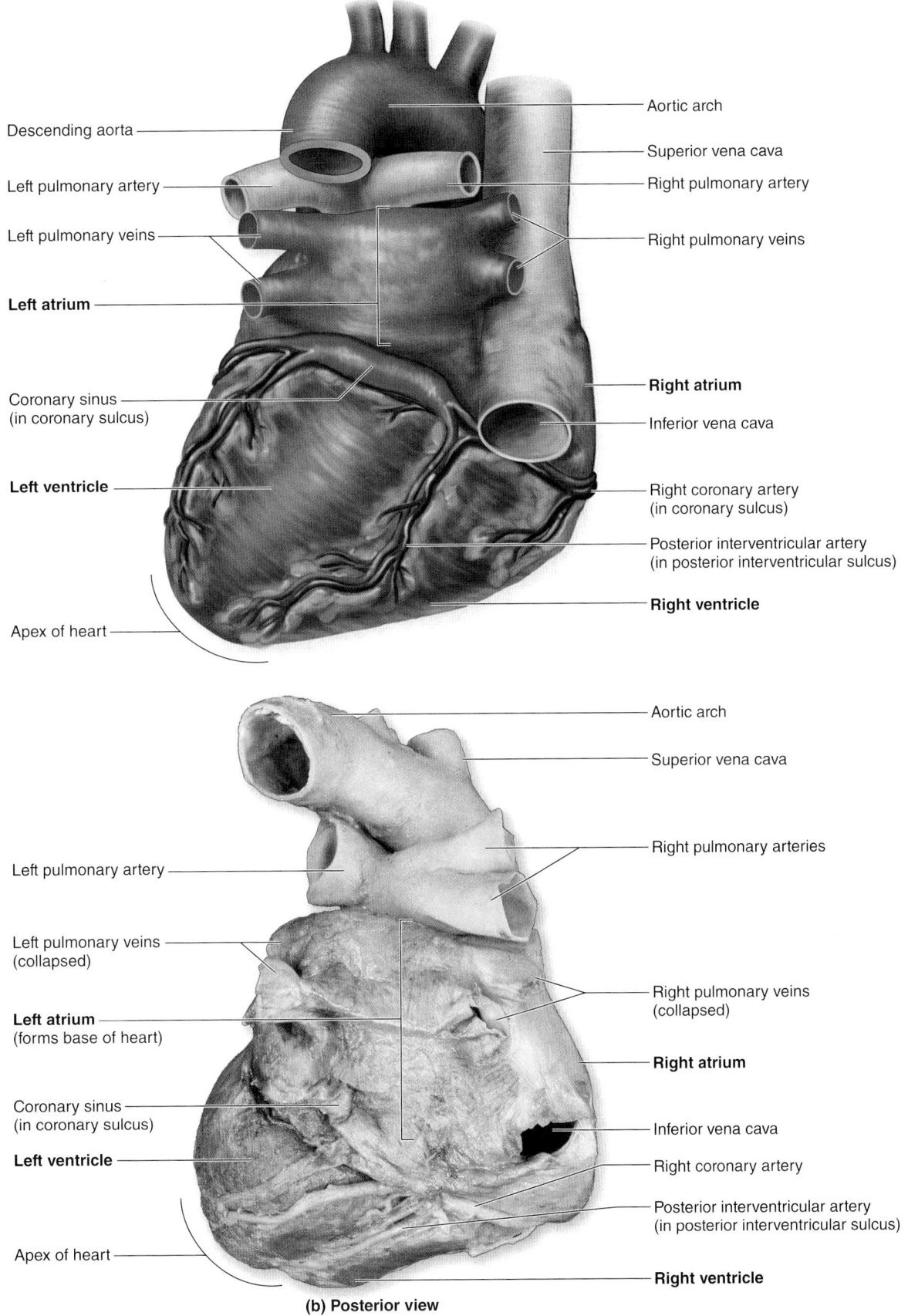

Descending aorta

Left pulmonary artery

Left pulmonary veins

Left atrium

Coronary sinus
(in coronary sulcus)

Left ventricle

Apex of heart

Aortic arch

Superior vena cava

Right pulmonary artery

Right pulmonary veins

Right atrium

Inferior vena cava

Right coronary artery
(in coronary sulcus)

Posterior interventricular artery
(in posterior interventricular sulcus)

Right ventricle

Left pulmonary artery

Left pulmonary veins
(collapsed)

Left atrium
(forms base of heart)

Coronary sinus
(in coronary sulcus)

Left ventricle

Apex of heart

Aortic arch

Superior vena cava

Right pulmonary arteries

Right pulmonary veins
(collapsed)

Right atrium

Inferior vena cava

Right coronary artery

Posterior interventricular artery
(in posterior interventricular sulcus)

Right ventricle

(b) Posterior view

Figure 19.7 External Anatomy and Features of the Heart *(continued).* (*b*) Posterior view. APR

19.3b Layers of the Heart Wall

LEARNING OBJECTIVE

2. Name the three layers of the heart wall and the tissue components of each.

If a coronal section is made through the heart, it is possible to view the internal structures of the heart (figure 19.8). First, observe the heart wall of the different heart chambers. Notice that the walls of the ventricles are thicker than the walls of the atria; this is because the ventricles are the "pumping chambers." Also note that the wall of the left ventricle is typically three times thicker than the right ventricular wall. This difference is because the left ventricle must generate enough pressure to force the blood through the entire systemic circulation. The right ventricle, in contrast, merely has to pump blood to the nearby lungs. This difference in thickness can also be seen when a transverse section is made through the heart (figure 19.9).

Three distinctive layers compose the wall of each chamber: an external epicardium, a thick middle myocardium, and an internal endocardium (figure 19.8b).

The **epicardium** (ep-i-kar′dē-ŭm; *epi* = upon) is the outermost heart layer and is also called the *visceral layer of serous pericardium*. This layer, as previously described, is composed of simple squamous epithelium and an underlying layer of areolar connective tissue. As we age, the epicardium thickens as it becomes more invested with adipose connective tissue.

The **myocardium** (mī-ō-kar′dē-ŭm; *mys* = muscle) is the middle layer of the heart wall. It is composed of cardiac muscle tissue (see section 5.3) and is the thickest of the three heart wall layers. Contraction of cardiac muscle composing the myocardium generates the force necessary to pump blood. The ventricular myocardium may change in thickness as we age. For example, it hypertrophies in response to narrowing of systemic arteries because the heart must work harder to pump the blood. We consider the microscopic anatomy of cardiac muscle in section 19.3e.

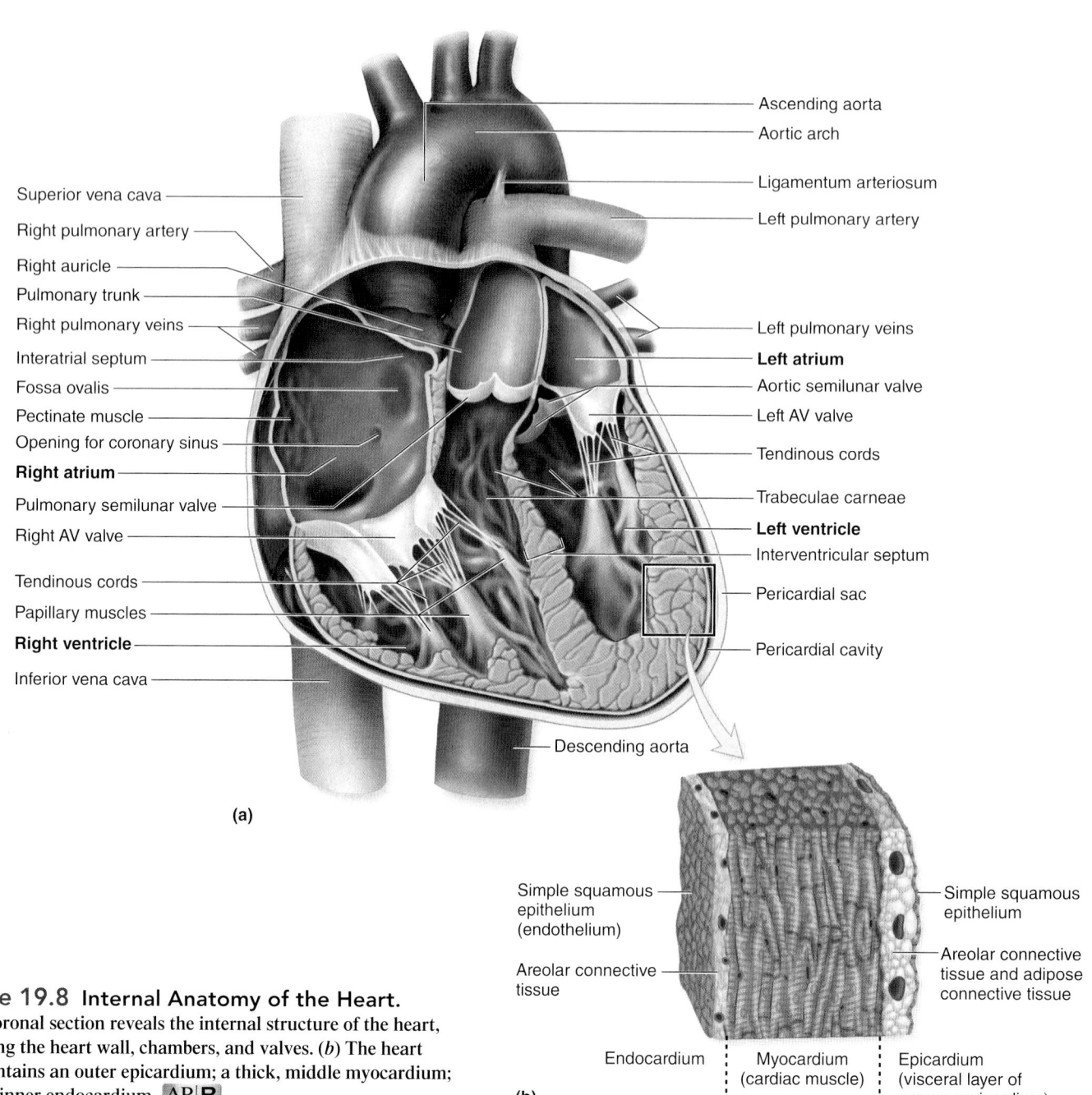

(a)

Figure 19.8 Internal Anatomy of the Heart.
(*a*) A coronal section reveals the internal structure of the heart, including the heart wall, chambers, and valves. (*b*) The heart wall contains an outer epicardium; a thick, middle myocardium; and an inner endocardium. AP|R

(b)

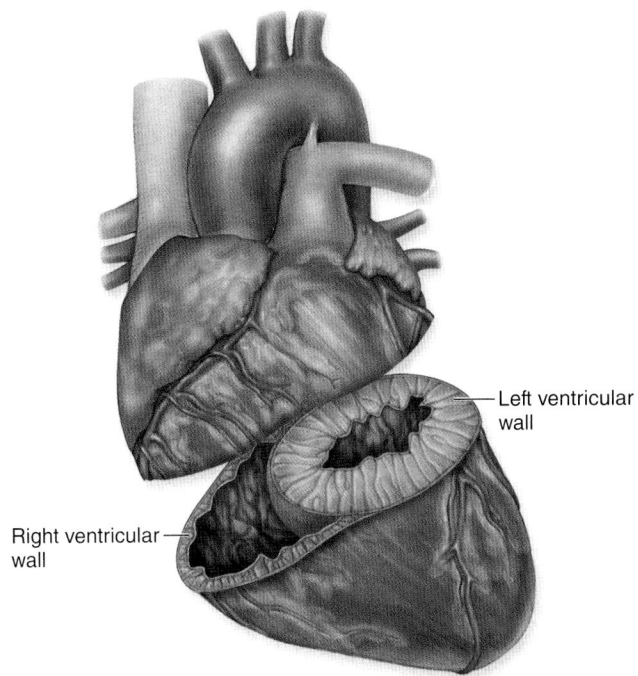

Left ventricular wall

Right ventricular wall

Figure 19.9 Comparison of Right and Left Ventricular Wall Thickness. The wall of the left ventricle is about three times thicker than that of the right ventricle, because the left ventricle must generate a force sufficient to push blood through the systemic circulation.

The internal surface of the heart and the external surfaces of the heart valves are covered by **endocardium** (en-dō-kar′dē-ŭm; *endon* = within). The endocardium, like the epicardium, is composed of a simple squamous epithelium and an underlying layer of areolar connective tissue. The epithelial layer of the endocardium is continuous with the epithelial layer called the endothelium, which lines the blood vessels (see figure 20.1).

 **WHAT DID YOU LEARN?**

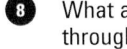 What are the layers of the heart (in order) that a scalpel would pass through during dissection? What are the two names given to the outer layer of the heart wall?

19.3c Heart Chambers

LEARNING OBJECTIVE

3. Characterize the four chambers of the heart and their functions.

Figure 19.8 also depicts the internal anatomy and structural organization of the four heart chambers. The right and left atrial chambers are separated by a thin wall called the **interatrial** (in-ter-ā′trē-ăl) **septum,** and the right and left ventricles are separated by a thick wall called the **interventricular septum.** The position of the interventricular septum is delineated on the heart's superficial surface with the interventricular sulci (see figure 19.7). (Thus, the interventricular "groove" on the surface is the landmark for the interventricular "wall" internal to it.)

Right Atrium

The internal wall of the right atrium is smooth on its posterior surface, but it exhibits muscular ridges, called **pectinate** (pek′ti-nāt; teeth of a comb) **muscles** on its anterior wall and within the auricle (figure 19.8*a*). Inspection of the interatrial septum reveals an oval depression called

the **fossa ovalis** (fos′ă; trench; ō-va′lis). It occupies the former location of the fetal **foramen ovale** (ō-val′ē), which shunted blood from the right atrium to the left atrium, bypassing the lungs during fetal life (fetal circulation is described in section 20.12). Immediately inferior to the fossa ovalis is the opening of the coronary sinus, which drains deoxygenated blood from the heart wall. Openings of the superior and inferior venae cavae are also visible. Separating the right atrium from the right ventricle is the right atrioventricular opening that contains the right AV valve.

Right Ventricle

The internal wall surface of the right ventricle displays characteristic large, smooth, irregular muscular ridges, called the **trabeculae carneae** (tră-bek′ū-lē; *trabs* = beam, kar′nē-ē; *carne* = flesh). Extending from the internal wall of the right ventricle are typically three cone-shaped, muscular projections called **papillary** (pap′i-lăr-ē; *papilla* = nipple) **muscles.** (The number of papillary muscles in the right ventricle can range from 2 to 9.) Papillary muscles anchor thin strands of collagen fibers called **tendinous cords** or **chordae tendineae** (kōr′dē ten′di-nē-ē) (or *heart strings*), which are attached to the right atrioventricular valve. The superior portion of the right ventricle narrows into a smooth-walled region leading into the pulmonary trunk. The pulmonary semilunar valve is positioned between the right ventricle and pulmonary trunk.

Left Atrium

Like the right atrium, the left atrium has pectinate muscles in its auricle. Openings of the pulmonary veins are visible. (Two are seen in figure 19.8.) Separating the left atrium from the left ventricle is the left atrioventricular opening that contains the left AV valve.

Left Ventricle

The internal wall surface of the left ventricle also displays characteristic trabeculae carneae. It has two papillary muscles that are anchored by tendinous cords. The entrance into the aorta is located at the superior aspect of the left ventricle. The aortic semilunar valve is positioned at the boundary of the left ventricle and ascending aorta.

 WHAT DID YOU LEARN?

 What is the structure that separates the two ventricles? What is the superficial landmark that identifies the location of this structure?

19.3d Heart Valves

LEARNING OBJECTIVE

4. Compare and contrast the structure and function of the two types of heart valves.

Effective blood flow requires valves to control blood flow and ensure it is "one-way." Recall that the two categories of heart valves are the atrioventricular (AV) valves and the semilunar valves. Each valve consists of endothelium-lined fibrous connective tissue flaps called cusps (**figure 19.10**).

Atrioventricular Valves

The right AV valve covers the right atrioventricular opening, and it has three cusps (which is why it is also called the tricuspid valve). The left AV valve covers the left atrioventricular opening, but it has only two cusps (which is why it is also called the bicuspid valve. The name "mitral" is

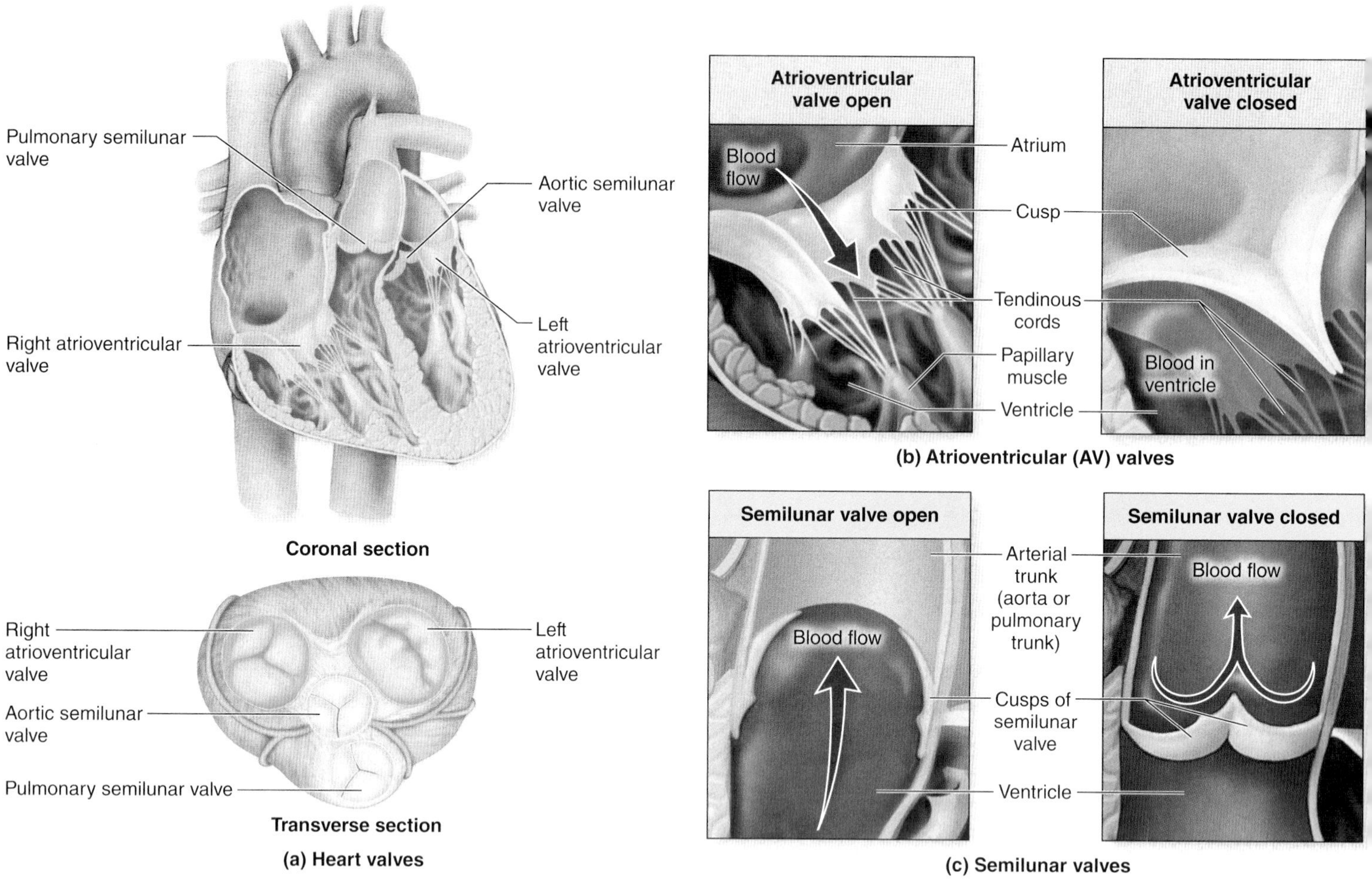

Heart Valves

Pulmonary semilunar valve

Aortic semilunar valve

Right atrioventricular valve

Left atrioventricular valve

Coronal section

Right atrioventricular valve

Left atrioventricular valve

Aortic semilunar valve

Pulmonary semilunar valve

Transverse section

(a) Heart valves

Atrioventricular valve open

Blood flow

Atrium

Cusp

Tendinous cords

Papillary muscle

Ventricle

Atrioventricular valve closed

Blood in ventricle

(b) Atrioventricular (AV) valves

Semilunar valve open

Blood flow

Semilunar valve closed

Blood flow

Arterial trunk (aorta or pulmonary trunk)

Cusps of semilunar valve

Ventricle

(c) Semilunar valves

Figure 19.10 Heart Valves. (*a*) Location of heart valves as viewed in coronal section and transverse section. (*b*) Atrioventricular valves in open and closed position. (*c*) Semilunar valves in open and closed position. AP|R

INTEGRATE

CLINICAL VIEW

Teenage Athletes and Sudden Cardiac Death

Sudden deaths of high school athletes have occurred during athletic events. These deaths were apparently caused by underlying, previously undetected cardiovascular disease. The *Journal of the American Medical Association* published a study of hundreds of cases that occurred between 1985 and 1995. This study indicated that no forewarning of a problem preceded these deaths. Autopsies revealed that most deaths were due to congenital heart defects and coronary artery anomalies. The result from these anomalies was cardiomegaly, which led to sudden death.

Cardiomegaly (kar′dē-ō-meg′ă-lē) is one term used to indicate an increase in the thickness of the heart muscle wall (hypertrophy) or an obvious increase or enlargement in heart size (dilation) due to stress applied to the heart. Many cardiovascular diseases can cause enlargement of the heart. However, the enlargement of the heart seen in young athletes is caused by a condition called **hypertrophic cardiomyopathy**. Other terms used are *hypertrophic heart, heart enlargement,* or *athlete's heart.* This condition involves both an enlargement of the heart walls (especially one or both ventricles) and a narrowing of the openings (outlets) for the blood

to pass through, which results in a decrease in cardiac output. Strenuous exercise exacerbates the condition, and patients may develop symptoms of shortness of breath, chest pain, and fainting. In some cases, sudden death may occur, often while the individual is participating in a physically demanding activity.

A standard x-ray that may accompany a physical examination often reveals potential cardiomegaly. The confirmation of the cardiomegaly being associated with hypertrophic cardiomyopathy (athlete's heart) depends upon an echocardiogram (a specialized ultrasound of the heart) that is used to evaluate the heart's function and its structure.

As a result, many school districts require student athletes to undergo a physical with an accompanying x-ray or echocardiogram prior to being approved to participate on athletic teams.

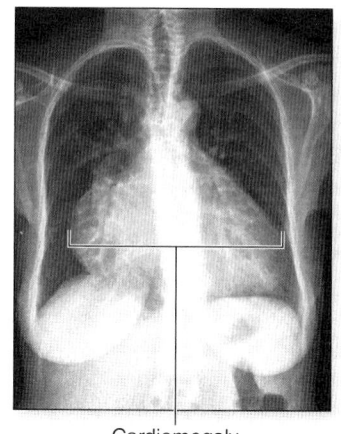

Cardiomegaly in an adult female. Note how the heart shadow encompasses most of the width of the thorax.

Cardiomegaly

INTEGRATE

CLINICAL VIEW

Heart Sounds and Heart Murmurs

There are two normal heart sounds associated with each heartbeat that collectively form the lubb-dupp sounds. The "lubb" sound is also known as the S1 sound and represents the closing of the atrioventricular valves. The "dupp" sound is also known as the S2 sound and is the closing of the semilunar valves. These heart sounds provide clinically important information about heart activity and the action of heart valves.

The place where sounds from each AV valve and each semilunar valve may best be heard does not correspond with the location of the valve, because some overlap of valve sounds occurs near their anatomic locations:

- The aortic semilunar valve is best heard in the second intercostal space to the right of the sternum.

- The pulmonary semilunar valve is best heard in the second intercostal space to the left of the sternum.

- The right AV valve is best heard by the right side of the sternum, inferior end.

- The left AV valve is best heard near the apex of the heart (at the level of the left fifth intercostal space, about 9 centimeters from the midline of the sternum).

An abnormal heart sound, generally called a **heart murmur,** is the first indication of heart valve problems. A heart murmur is usually the result of turbulence of the blood as it passes through the heart, and may be caused by valvular leakage, decreased valve flexibility, or a misshapen valve. Sometimes heart murmurs are of little consequence, but all of them need to be evaluated to rule out a more serious heart problem. Two types of heart murmurs are valvular insufficiency and valvular stenosis.

Valvular insufficiency, also termed *valvular incompetence,* occurs when one or more of the cardiac valves leaks because the valve cusps do not close tightly enough. Inflammation or disease may cause the free edges of the valve

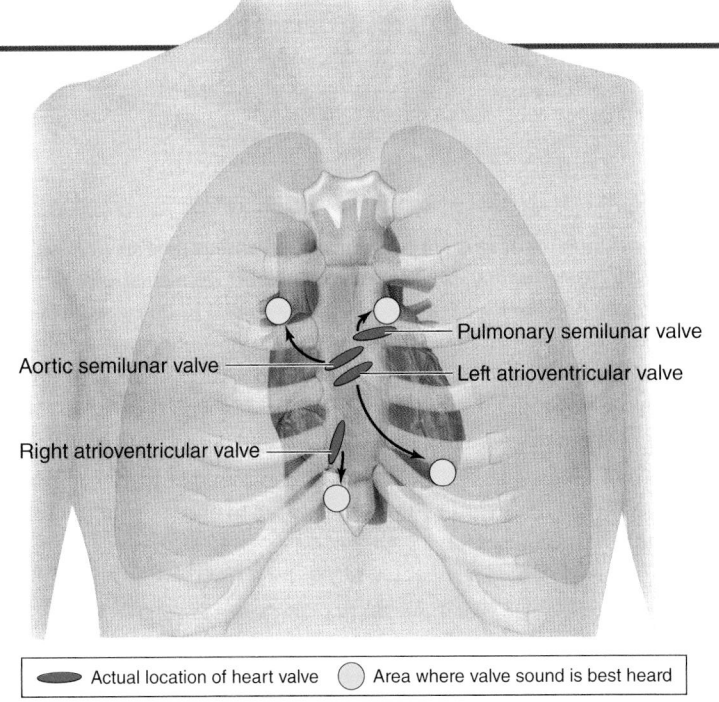

Locations of individual heart valves and the ideal listening sites for each valve are shown.

cusps to become scarred and constricted, allowing blood to regurgitate back through the valve and may cause further heart enlargement.

Valvular stenosis (ste-nō'sis; narrowing) is scarring of the valve cusps so that they become rigid or partially fused and cannot open completely. A stenotic valve is narrowed and presents resistance to the flow of blood, decreasing chamber output. Often the affected chamber undergoes hypertrophy and dilates—both conditions that may have dangerous consequences. A primary cause of valvular stenosis is **rheumatic** (rū-mat'ik) **heart disease,** which may follow a streptococcal infection of the throat.

---

also used for this valve because it resembles the headgear of bishops called a mitre.). The AV valves are shown in both the open and closed position in figure 19.10*b*. When open, the cusps of the valve extend into the ventricles. This allows blood to move from an atrium into the ventricle. When the ventricles contract, blood is forced superiorly as ventricular pressure rises. This causes the AV valves to close. The papillary muscles secure the tendinous cords that attach to the lower surface of each AV valve cusp. This prevents the valve from inverting and being "pushed open" into the atrium when the valve is closed. By being properly held in place, the cusps of the AV valves prevent blood flow back into the atrium.

Semilunar Valves

The pulmonary semilunar valve is located between the right ventricle and the pulmonary trunk, and the aortic semilunar valve is located between the left ventricle and the ascending aorta. Each valve is composed of three half-moon-shaped, pocketlike semilunar cusps (figure 19.10*a*, *c*). Neither papillary muscles nor tendinous cords are

associated with these valves. The semilunar valves are shown in both the open and closed position.

The semilunar valves open when the ventricles contract and the force of the blood pushes the semilunar valves open and blood enters the arterial trunks. The valves close when the ventricles relax and the pressure in the ventricle becomes less than the pressure in an arterial trunk. Blood in the arteries begins to move backward toward the ventricle and is caught in the cusps of the semilunar valves, and they close. The closure of the semilunar valves prevents blood flow back into the ventricle.

Both flexibility and elasticity of connective tissue composing heart valves decrease with aging (or disease). This may cause the heart valves to become inflexible. As a result, blood flow through the heart may be altered, and a heart murmur may be detected (see Clinical View: "Heart Sounds and Heart Murmurs").

 WHAT DID YOU LEARN?

10 What are the functions of the tendinous cords and papillary muscles?

19.3e Microscopic Structure of Cardiac Muscle

LEARNING OBJECTIVES

5. Describe the general structure of cardiac muscle.

6. Explain the intercellular structures of cardiac muscle.

7. Discuss how cardiac muscle meets its energy needs.

The myocardium is composed of cardiac muscle tissue (see section 5.3). This muscle tissue is made up of relatively short, branched cells that usually house one or two central nuclei **(figure 19.11)**. These cells are supported by areolar connective tissue, called an endomysium, that surrounds the cells. Other anatomic structures of cardiac muscle cells include the following:

- The sarcolemma (plasma membrane), which invaginates to form T-tubules that extend to the sarcoplasmic reticulum (SR). The T-tubules of cardiac muscle invaginate once per sarcomere and overlie Z discs.

- The SR surrounds bundles of myofilaments called myofibrils in cardiac muscle, but it is less extensive than the SR of skeletal muscle and lacks both terminal cisternae (the "end sacs" of sarcoplasmic reticulum) and a tight association with transverse tubules (figure 19.11c).

- Myofilaments are arranged in sarcomeres—thus, cardiac muscle appears striated when viewed under a microscope (figure 19.11d). Interestingly, maximum overlap of thin and thick filaments within the sarcomeres does *not* occur when cardiac muscle is at rest (as is the case within skeletal muscle; see length-tension relationship in section 10.7c). Instead, maximum overlap of thin and thick filaments occurs when cardiac muscle is stretched as blood is added to a heart chamber. As we will see, this provides a means of forming additional crossbridges between the thin and thick filaments and cardiac muscle contracting with increasingly greater degrees of force as additional blood enters the chamber.

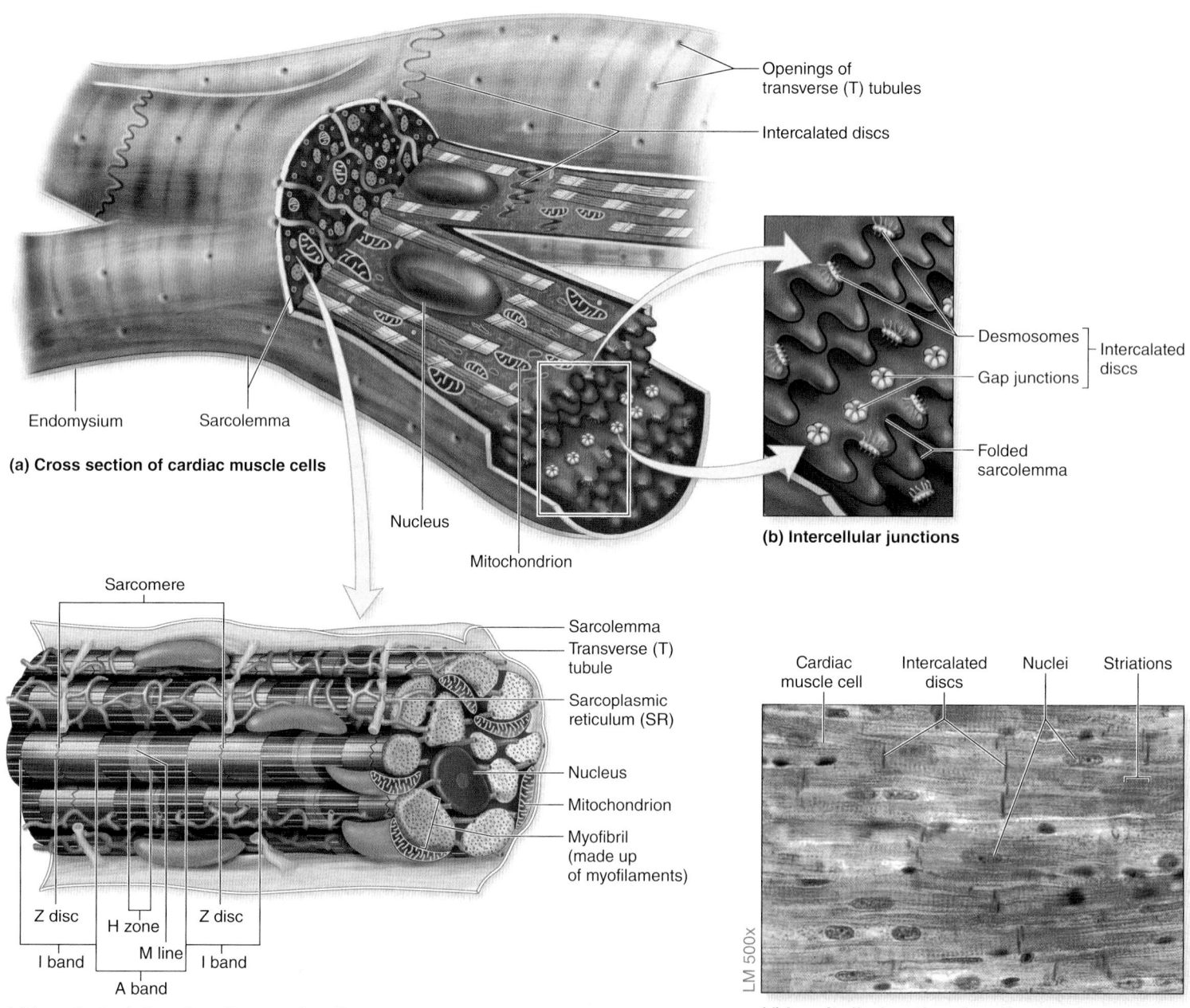

(a) Cross section of cardiac muscle cells

Endomysium   Sarcolemma   Nucleus   Mitochondrion

Openings of transverse (T) tubules

Intercalated discs

Desmosomes ⎤ Intercalated
Gap junctions ⎦ discs

Folded sarcolemma

(b) Intercellular junctions

(c) Longitudinal view of cardiac muscle cell

Sarcomere

Sarcolemma
Transverse (T) tubule
Sarcoplasmic reticulum (SR)
Nucleus
Mitochondrion
Myofibril (made up of myofilaments)

Z disc   H zone   Z disc
I band   M line   I band
A band

(d) Longitudinal section of cardiac muscle

Cardiac muscle cell   Intercalated discs   Nuclei   Striations

LM 500x

Figure 19.11 Histology of Cardiac Muscle. Cardiac muscle cells form the myocardium. (*a*) Individual cells are relatively short, branched, and striated. (*b*) They are connected to adjacent cells by intercalated discs composed of both desmosomes and gap junctions. (*c*) T-tubules are invaginations of the sarcolemma that extend internally to the sarcoplasmic reticulum. (*d*) Light micrograph of cardiac muscle in longitudinal section. Intercalated discs are visible as dark lines perpendicular to the cardiac muscle cells. AP|R

Intercellular Structures

Neighboring cardiac muscle cells have an extensively folded sarcolemma that permits adjoining membranes to interconnect, markedly increasing exposed surface areas between neighboring cells (figure 19.11b). This can be demonstrated by interlocking your fingers.) This increases structural stability of the myocardium and facilitates communication between cardiac muscle cells.

Unique structures called **intercalated** (in-ter′kă-lā-ted) **discs** are found at these cell-to-cell junctions. They link cardiac muscle cells together both mechanically and electrically and contain two distinctive structural features:

- **Desmosomes.** These are protein filaments that anchor into a protein plaque located on the internal surface of the sarcolemma (see section 4.6d). They act as mechanical junctions to prevent cardiac muscle cells from pulling apart.
- **Gap junctions.** These are protein pores between the sarcolemma of adjacent cardiac muscle cells. They provide a low-resistance pathway for the flow of ions between the cardiac cells. Gap junctions allow an action potential to move continuously along the sarcolemma of cardiac muscle cells, resulting in synchronous contraction of that chamber. Thus, a chamber functions as a single unit, or **functional syncytium.**

Metabolism of Cardiac Muscle

Cardiac muscle has features that support its great demand for energy, including an extensive blood supply, numerous mitochondria, and other structures such as myoglobin and creatine kinase (see section 10.4a).

Cardiac muscle relies almost exclusively on aerobic cellular respiration (see section 3.4). Its cellular structures and metabolic processes support this. Cardiac muscle has a large number of mitochondria (comprising approximately 25% of its volume compared to about 2% of the volume in skeletal muscle). It is also versatile in being able to use different types of fuel molecules, including fatty acids, glucose, lactate, amino acids, and ketone bodies. The relative amounts of these molecules that cardiac muscle cells use fluctuates depending upon conditions. For example, during intense exercise and increased production and release of lactate into the blood by skeletal muscle, cardiac muscle will absorb and use this resource.

Yet, as a consequence of its reliance on aerobic metabolism, cardiac muscle is quite susceptible to failure if ischemic (low-oxygen) conditions prevail. Cardiac muscle has limited capability in using glycolysis (see section 3.4b) or accruing an oxygen debt (see section 10.4b) in its activities. Therefore, any change that interferes with blood flow to the heart muscle, such as narrowing of the coronary arteries, can cause damage or death of the cardiac muscle cells composing the myocardium (see Clinical View: "Angina Pectoris and Myocardial Infarction").

WHAT DID YOU LEARN?

11 Which features of cardiac muscle support aerobic cellular respiration?

19.3f Fibrous Skeleton of the Heart

LEARNING OBJECTIVE

8. Describe the location and function of the fibrous skeleton.

The heart is supported internally by a **fibrous skeleton** composed of dense irregular connective tissue (**figure 19.12**). This fibrous skeleton performs the following functions:

- Provides structural support at the boundary between the atria and the ventricles
- Forms supportive fibrous rings to anchor the heart valves
- Provides a rigid framework for the attachment of cardiac muscle tissue
- Acts as an electric insulator because it does not conduct action potentials and thus prevents the ventricular chambers from contracting at the same time as the atrial chambers

Notice that cardiac muscle cells are arranged in spiral bundles around the heart chambers attached to the fibrous skeleton (figure 19.12b). When the atria contract, they compress the wall of the chambers inward to move the blood into the ventricles. When the ventricles contract, the action is similar to the wringing of a mop in that it begins at the apex of the heart and compresses superiorly, moving the blood into the great arterial trunks.

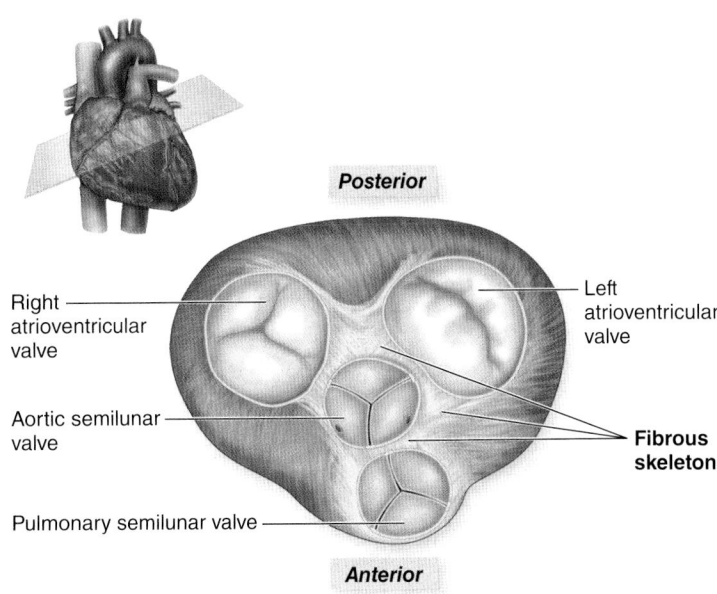

(a)

(b)

Figure 19.12 Fibrous Skeleton of the Heart.
(*a*) The fibrous skeleton is composed of dense irregular connective tissue that provides both mechanical support and electrical insulation within the heart. Shown in superior view, its structural support of heart valves is especially evident. (*b*) Diagram and photo of spiral pattern of cardiac muscle attached to the fibrous skeleton.

WHAT DID YOU LEARN?

12 Which function of the fibrous skeleton allows the atria to contract separately from the ventricles?

19.4 Coronary Vessels: Blood Supply Within the Heart Wall

Although the heart continuously pumps blood through its chambers, it cannot absorb the oxygen and nutrients it requires from this blood. The diffusion of oxygen and nutrients at the needed rates through the thick wall of the heart is not possible; instead, an intricate distribution system called the **coronary circulation** handles the task **(figure 19.13)**. The vessels that transport oxygenated blood to the wall of the heart are

called coronary arteries, whereas coronary veins transport deoxygenated blood away from the heart wall. Vessels on the posterior aspect of the heart are shown shaded in figures 19.13a and b.

19.4a Coronary Arteries

LEARNING OBJECTIVES

1. Identify the coronary arteries, and describe the specific areas of the heart supplied by their major branches.
2. Explain the significance of coronary arteries as functional end arteries.
3. Describe blood flow through the coronary arteries.

Right and **left coronary arteries** are positioned within the coronary sulcus of the heart to supply the heart wall (figure 19.13a). These

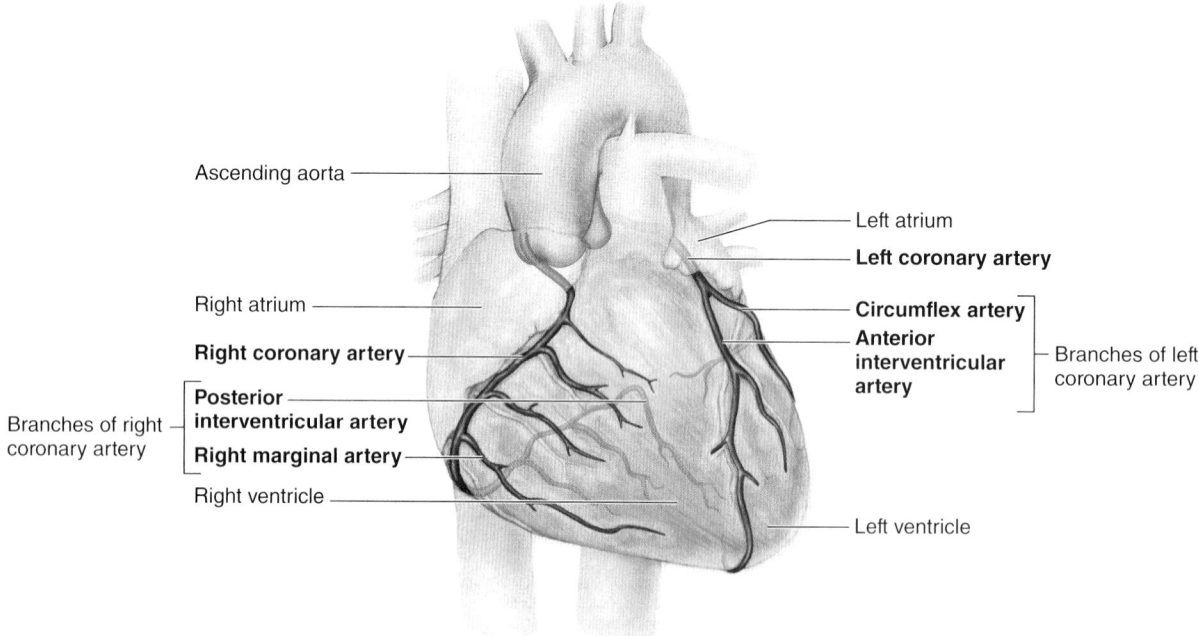

(a) Coronary arteries

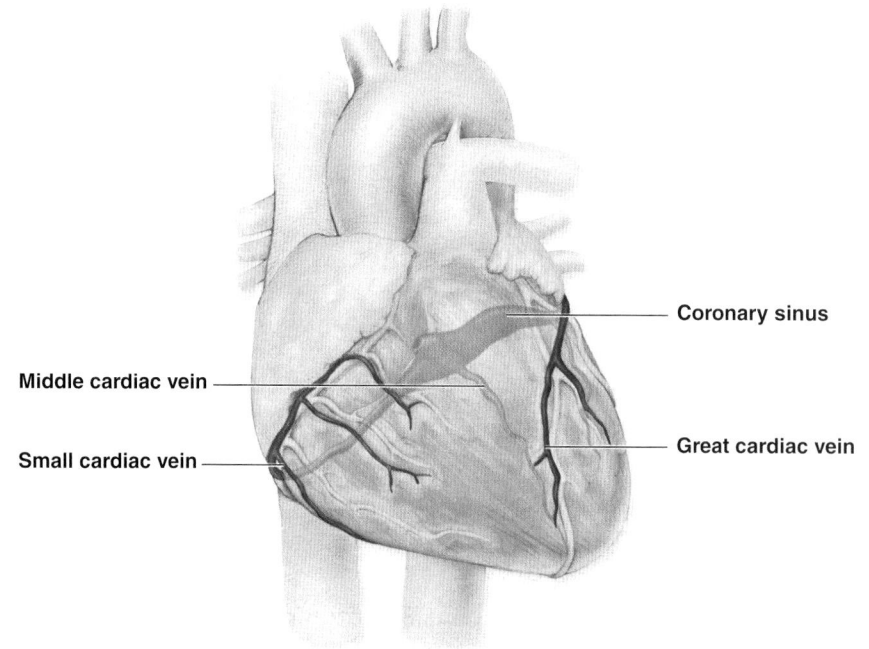

(b) Coronary veins

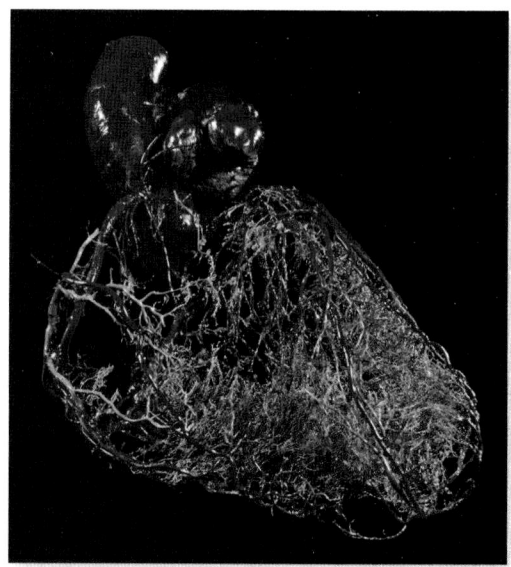

(c) Polymer cast of coronary vessels

Figure 19.13 Coronary Circulation. Diagram of the (*a*) anterior view of coronary arteries that transport blood to cardiac muscle tissue and (*b*) coronary veins that transport blood from cardiac muscle tissue. (Vessels on the posterior aspect of the heart are shown shaded.) (*c*) Photo of polymer cast of coronary vessels. AP|R

arteries are the first and only branches of the ascending aorta and originate immediately superior to the aortic semilunar valve.

The right coronary artery typically branches into the **right marginal artery** to supply the lateral wall of the right ventricle, and the **posterior interventricular artery** (or *posterior descending artery*) to supply the posterior wall of both the left and right ventricles. The left coronary artery typically branches into the **circumflex** (ser'kum-fleks; *circum* = around, *flexus* = to bend) **artery** to supply the lateral wall of the left ventricle, and the **anterior interventricular artery** (also called the *left anterior descending artery*) to supply both the anterior wall of the left ventricle and most of the interventricular septum. However, this arterial pattern can vary greatly among individuals. Note that the anterior interventricular artery is nicknamed the "widowmaker." This name reflects the fact that if this artery becomes occluded, there is a very high risk of a fatal heart attack, which makes a "widow" of the spouse. The coronary arteries and veins can also be viewed in figure 19.7.

Functional End Arteries

Body tissues are generally served by one artery; it is called an **end artery.** In comparison, some body tissues are served by two arteries; this is referred to as an arterial anastomosis (see section 20.1e). An arterial anastomosis provides two means of effectively delivering blood to a given region. The left and right coronary arteries both provide blood to the myocardium, and are described by some as an arterial anastomosis. However, if one of these arteries becomes blocked, this anastomosis is too small to shunt sufficient blood from one artery to the other, as may happen with coronary artery disease. As a result, the coronary arteries are more accurately called **functional end arteries.**

Blood Flow

Coronary arterial blood flow to the heart wall is intermittent. This occurs because coronary vessels are patent when the heart is relaxed and

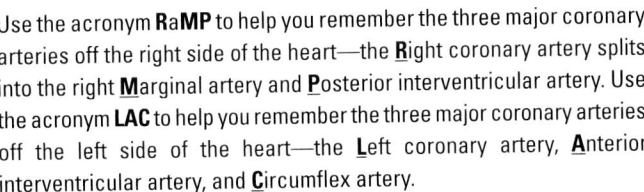

INTEGRATE

LEARNING STRATEGY

Use the acronym **RaMP** to help you remember the three major coronary arteries off the right side of the heart—the **R**ight coronary artery splits into the right **M**arginal artery and **P**osterior interventricular artery. Use the acronym **LAC** to help you remember the three major coronary arteries off the left side of the heart—the **L**eft coronary artery, **A**nterior interventricular artery, and **C**ircumflex artery.

blood flow is possible. However, coronary vessels are compressed when the heart contracts, temporarily interrupting blood flow. Thus, blood flow to the heart wall is not a steady stream; it is impeded and then flows, as the heart rhythmically contracts and relaxes.

WHAT DO YOU THINK?

2 What happens to the amount of time the coronary arteries are compressed when a heart beats at a faster rate, such as during exercise?

WHAT DID YOU LEARN?

13 What areas of the heart are deprived of blood when there is a blockage in the posterior interventricular artery?

19.4b Coronary Veins

LEARNING OBJECTIVE

4. Identify the coronary veins, and describe the specific areas of the heart drained by their major branches.

INTEGRATE

CLINICAL VIEW

Angina Pectoris and Myocardial Infarction

Unhealthy coronary arteries may not be able to supply sufficient blood flow to the heart wall. Coronary arteries may become narrowed and occluded with plaques in a condition called **atherosclerosis** (ath'er-ō-skler-ō'sis; *athere* = gruel, *sclerosis* = hardness), or *coronary artery disease*. In contrast, some individuals may experience coronary spasm, which is sudden narrowing of the vessels caused by smooth muscle contraction. Either atherosclerosis or coronary spasm can lead to angina pectoris or the more severe myocardial infarction.

Angina (an'ji-nă, an-jī'nă) **pectoris** is a poorly localized pain sensation in the left side of the chest, the left arm and shoulder, or sometimes the jaw and the back (although these symptoms may vary, especially in women). Generally it results from strenuous activity, when workload demands on the heart exceed the ability of the narrowed coronary vessels to supply blood. The pain from angina is typically referred (referred pain; see section 16.2b) along the sympathetic pathways (T1–T5 spinal cord segments, see section 15.4), so an individual may experience pain in the chest region or down the left arm, where the T1 dermatome (see section 14.5a) is located. The pain diminishes shortly after the person stops the exertion, and normal blood flow to the heart is restored. Treatment may include

medications that cause temporary vascular dilation, such as nitroglycerin. The prognosis and long-term therapy for angina depend upon the severity of the vascular narrowing or spasming.

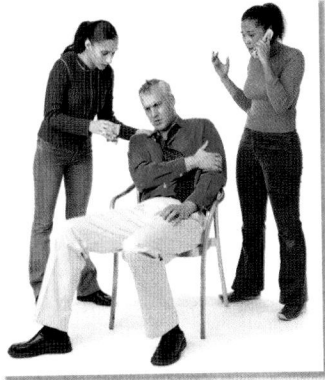

The term **infarction** refers to death of tissue due to lack of blood supply. **Myocardial infarction (MI)**, commonly called a *heart attack*, is a potentially lethal condition resulting from sudden and complete occlusion of a coronary artery. A region of the myocardium is deprived of oxygen, and some of this tissue may die (undergo necrosis). Some people experiencing an MI report the sudden development of excruciating and crushing substernal chest pain that typically radiates into the left arm or left side of the neck. Women may not experience these symptoms but instead be overcome with jaw pain, incredible fatigue, and flulike symptoms, and thus may be misdiagnosed. Other immediate symptoms include: weakness, shortness of breath, nausea, vomiting, anxiety, and marked sweating. Mature cardiac muscle cells have little or no capacity to regenerate (i.e., they cannot undergo cellular division), so when an MI results in cell death, scar tissue forms to fill the gap. If a large amount of tissue is lost, the person may die within a few hours or days because the heart has been profoundly and suddenly weakened.

Transport of deoxygenated blood from the myocardium occurs through one of several cardiac veins (figure 19.13b). These include the **great cardiac vein** within the anterior interventricular sulcus, positioned alongside the anterior interventricular artery; the **middle cardiac vein** within the posterior interventricular sulcus, positioned alongside the posterior interventricular artery; and a **small cardiac vein** positioned alongside the right marginal artery. These cardiac veins all drain into the **coronary sinus,** a large vein that lies within the posterior aspect of the coronary sulcus. The coronary sinus then returns this deoxygenated blood directly into the right atrium of the heart.

 WHAT DID YOU LEARN?

14 What is the function of the coronary sinus?

19.5 Anatomic Structures Controlling Heart Activity

The heart pumps blood continuously and depends upon rhythmic stimulation of cardiac muscle cells. Precise electrical events are orchestrated by the heart's conduction system, beginning with stimulation by the sinoatrial node and then transmission of an action potential by specialized conduction fibers of the heart to ensure that the atria are stimulated to contract prior to the ventricles. These events are influenced by the activity of the autonomic nervous system.

19.5a The Heart's Conduction System

 LEARNING OBJECTIVE

1. Identify and locate the components of the heart's conduction system.

Specialized cardiac muscle cells are found within the heart located internal to the endocardium; they are collectively called the heart's **conduction system** (figure 19.14a). These distinct cardiac cells do not contract, but rather they initiate and conduct electrical signals. The conduction system includes the following structures:

- The **sinoatrial** (sī′nō-ā′trē-ăl) **(SA) node** is located in the posterior wall of the right atrium, adjacent to the entrance of the superior vena cava. The cells here initiate the heartbeat and are commonly referred to as the *pacemaker* of the heart.
- The **atrioventricular (AV) node** is located in the floor of the right atrium between the right AV valve and the opening for the coronary sinus.
- The **atrioventricular (AV) bundle** (*bundle of His*) extends from the AV node into and through the interventricular septum. It divides into **left** and **right bundles.**
- The **Purkinje** (pŭr-kin′jē) **fibers** extend from the left and right bundles beginning at the apex of the heart and then continue through the walls of the ventricles.

 WHAT DID YOU LEARN?

15 Why is the SA node referred to as the pacemaker?

19.5b Innervation of the Heart

 LEARNING OBJECTIVE

2. Compare and contrast parasympathetic and sympathetic innervation to the heart.

While the heartbeat is initiated by the SA node, both the heart rate and the strength of contraction are regulated by the autonomic nervous system—specifically by the cardiac center within the medulla oblongata (see section 13.5c). Sensory input is relayed from receptors (baroreceptors and chemoreceptors within blood vessels and the right atrium) to the cardiac center. Parasympathetic and sympathetic axons extend from the **cardiac center** to the heart. This center houses both the cardioinhibitory and cardioacceleratory centers (figure 19.14b). The innervation from this center does not initiate the heartbeat; it merely modifies cardiac activity including both the heart rate and its force of contraction.

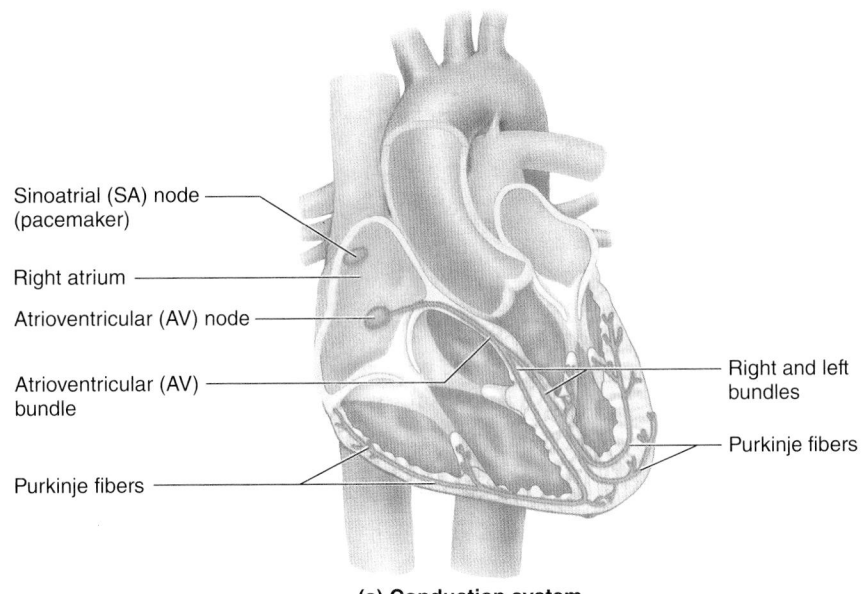

Sinoatrial (SA) node
(pacemaker)

Right atrium

Atrioventricular (AV) node

Atrioventricular (AV)
bundle

Purkinje fibers

Right and left
bundles

Purkinje fibers

(a) Conduction system

Cardioacceleratory center
Cardioinhibitory center
Cardiac center

Vagus nerve (left)

Glossopharyngeal nerve (CN IX)

Spinal cord

→ Sensory Input

Motor Output
→ Sympathetic axons
(preganglionic)
→ Sympathetic axons
(postganglionic)
→ Parasympathetic
axons (vagus)

Vagus nerve (right)

Cardiac nerve

Receptors
Baroreceptor
Chemoreceptor
Chemoreceptors
Baroreceptor
(Baroreceptors also within the right atrium)

Heart
SA node
AV node
Coronary arteries
Myocardium

(b) Innervation of the heart

Figure 19.14 Anatomic Structures Controlling Heart Activity. (*a*) The heart's conduction system, composed of the sinoatrial (SA) node, atrioventricular (AV) node, right and left bundles, and Purkinje fibers, initiates and conducts the electrical activity to cause heart contraction. (*b*) Heart rate and force of contraction are modified by autonomic centers (cardioacceleratory and cardioinhibitory centers) in the medulla oblongata. Parasympathetic axons extend from the cardioinhibitory center to both the SA node and AV node. Sympathetic axons extend from the cardioacceleratory center to the SA node, AV node, myocardium, and coronary vessels.

Parasympathetic innervation comes from the **cardioinhibitory center** via the right and left vagus nerves (CN X). As these nerves descend into the thoracic cavity, they give off branches that supply the heart (see section 15.3). Primarily, the right vagus nerve innervates the SA node, and the left vagus nerve innervates the AV node. Parasympathetic stimulation decreases heart rate, but generally has no direct effect on the force of contraction (because the vagus nerve does not innervate the myocardium).

Sympathetic innervation arises from the **cardioacceleratory center.** Neurons within the T1–T5 segments of the spinal cord extend to the SA node, AV node, and the myocardium (see section 15.4). Stimulation by the sympathetic division increases both heart rate and the force of heart contraction. There is also some sympathetic innervation to the coronary arteries, causing dilation of these vessels to support increased blood flow to the myocardium.

 WHAT DID YOU LEARN?

16 Which autonomic division is associated with the cardioacceleratory center in the brainstem, and how does it affect heart activity?

19.6 Stimulation of the Heart

The physiologic processes associated with heart contraction are organized into two major events. These are presented diagrammatically in figure 19.15 and include the following:

- **Conduction system.** Electrical activity is initiated at the SA node, and an action potential is then transmitted through the conduction system.
- **Cardiac muscle cells.** The action potential spreads across the sarcolemma of the cardiac muscle cells, causing sarcomeres within cardiac muscle cells to contract. These events occur *twice* in cardiac muscle cells during a heartbeat, first in the cells of the atria and then in the cells of the ventricles.

Here we discuss the events associated with the conduction system including the physiologic conditions of nodal cells at rest and how they serve as the heart's pacemaker. The processes associated with cardiac muscle cells are described in section 19.7.

 **19.6a Nodal Cells at Rest**

LEARNING OBJECTIVE

1. Describe a nodal cell at rest.

Nodal cells in the SA node are the pacemaker cells that initiate a heartbeat by spontaneously depolarizing to generate an action potential. These specialized cardiac cells exhibit several significant features. See figure 19.16a as your read through this section.

One essential feature of nodal cells is the electrical charge difference across the plasma membrane; the intracellular fluid (cytosol) just inside the plasma membrane is relatively negative in comparison to the fluid outside of the cell (interstitial fluid). This electrical charge difference when the nodal cell is at rest is called the **resting membrane potential (RMP).** Nodal cells have an RMP of about −60 millivolts (mV). An RMP, which is discussed in detail in section 4.4, is established and maintained by K⁺ leak channels, Na⁺ leak channels, and Na⁺/K⁺ pumps (not shown in figure 19.16a). The primary function of the Na⁺/K⁺ pumps is to maintain the concentration gradients for Na⁺ (with more Na⁺ outside the cell) and K⁺ (with more K⁺ inside the cell). Nodal cells also contain calcium ion (Ca²⁺) pumps that establish a Ca²⁺ concentration gradient with more Ca²⁺ outside the cell than inside. It is important to note that nodal cells (unlike other cells) do not have a stable RMP, as described shortly.

Additionally, nodal cells contain specific voltage-gated channels, including slow voltage-gated Na⁺ channels (which are open) and both fast voltage-gated Ca²⁺ channels and voltage-gated K⁺ channels (which are closed).

 WHAT DID YOU LEARN?

17 What is the resting membrane potential value of nodal cells?

19.6b Electrical Events at the SA Node: Initiation of the Action Potential

LEARNING OBJECTIVES

2. Define autorhythmicity.
3. Describe the steps for SA nodal cells to spontaneously depolarize and serve as the pacemaker cells.

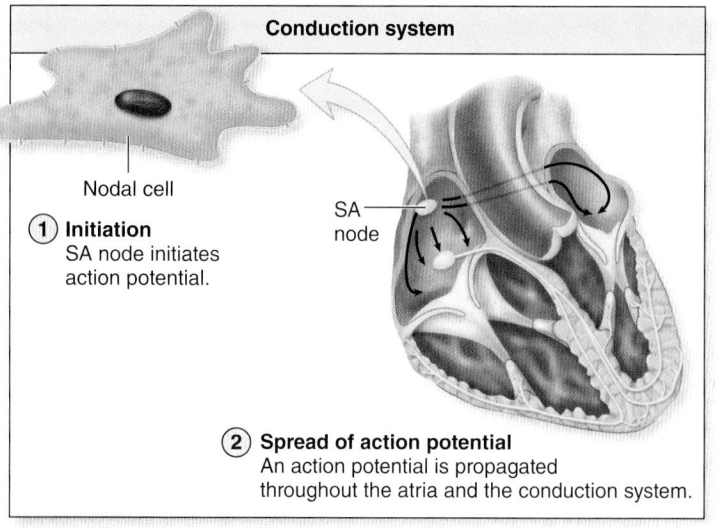

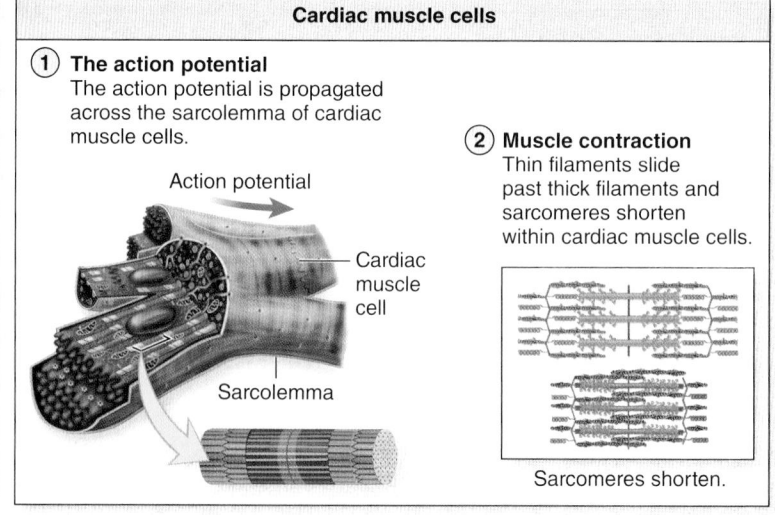

(a)   (b)

Figure 19.15 Physiologic Processes Associated with Heart Contraction. Both the conduction system and cardiac muscle cells have two significant physiologic processes that occur for heart contraction. (*a*) Initiation and spread of an action potential occurs in the conduction system. (*b*) Action potentials spread along the sarcolemma of cardiac muscle cells, triggering contraction of sarcomeres within these cardiac muscle cells. Events of cardiac muscle cells occur twice in one heartbeat; once in the cells of the atria and once in the cells of the ventricles.

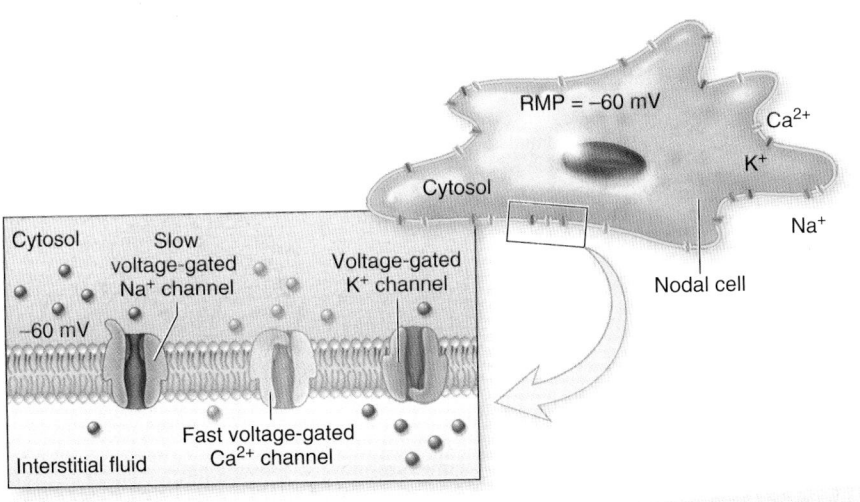

RMP = −60 mV

Ca²⁺

K⁺

Cytosol

Na⁺

Nodal cell

Cytosol | Slow voltage-gated Na⁺ channel | Voltage-gated K⁺ channel

−60 mV

Fast voltage-gated Ca²⁺ channel

Interstitial fluid

(a) Nodal cell at rest

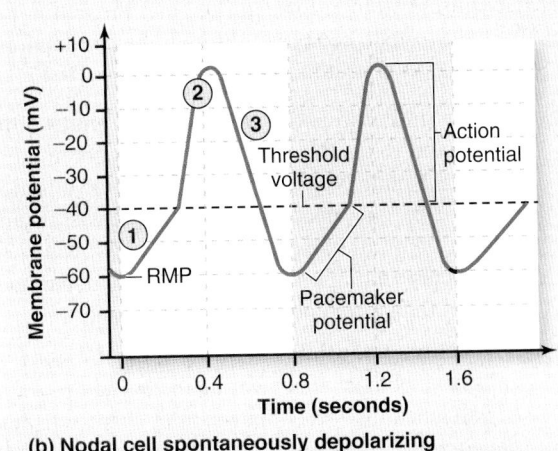

(b) Nodal cell spontaneously depolarizing

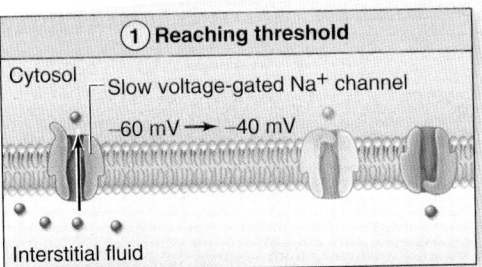

① Reaching threshold

Cytosol — Slow voltage-gated Na⁺ channel

−60 mV → −40 mV

Interstitial fluid

Slow voltage-gated Na⁺ channels open. Inflow of Na⁺ changes membrane potential from −60 mV to −40 mV.

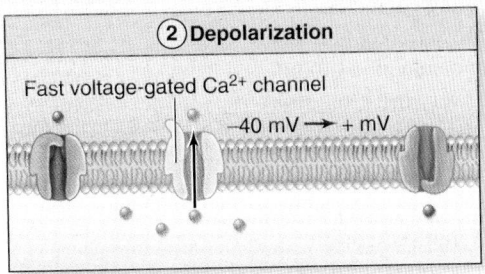

② Depolarization

Fast voltage-gated Ca²⁺ channel

−40 mV → + mV

Fast voltage-gated Ca²⁺ channels open. Inflow of Ca²⁺ changes membrane potential from −40 mV to just above 0 mV.

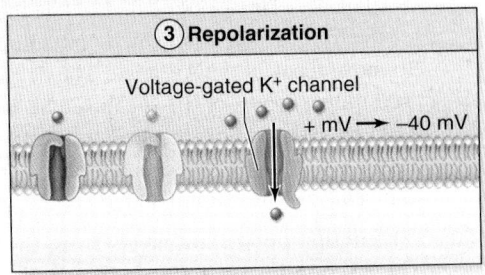

③ Repolarization

Voltage-gated K⁺ channel

+ mV → −40 mV

Fast voltage-gated Ca²⁺ channels close. Voltage-gated K⁺ channels open allowing K⁺ outflow. Membrane potential returns to RMP −60 mV, and K⁺ channels close.

Figure 19.16 SA Node Cellular Activity. (*a*) Nodal cells have an unstable resting membrane potential of −60 mV. Voltage-gated Na⁺ channels located in the plasma membrane of nodal cells open in response to changes in membrane potential. (*b*) Graph and accompanying diagrams of the sequential electrical changes occurring at the plasma membrane of SA nodal (pacemaker) cells to initiate stimulation of the heart.

SA nodal cells are unique in that they exhibit **autorhythmicity** (or *automaticity*), meaning that they are capable of depolarizing and firing an action potential spontaneously without any external influence. The following series of events occur within SA nodal cells, as depicted in figure 19.16*b*:

① **Reaching threshold.** Slow voltage-gated Na⁺ channels open (this is caused by repolarization from the previous cycle). The Na⁺ flows into the nodal cells, changing the resting membrane potential from −60 mV to −40 mV, which is the **threshold** value. Notice that the threshold is reached without outside stimulation.

② **Depolarization.** Changing of the membrane potential to the threshold triggers the opening of fast voltage-gated Ca²⁺ channels, and Ca²⁺ entry into the nodal cell causes a change in the membrane potential from −40 mV to a slightly positive membrane potential (just above 0 mV). This reversal of polarity is called **depolarization**.

③ **Repolarization.** Calcium channels close and voltage-gated K⁺ channels open; K⁺ flows out to change the membrane potential from a positive value to −60 mV, which is the RMP. The process of reestablishing the RMP is called **repolarization.** Repolarization triggers the reopening of slow voltage-gated Na⁺ channels, and the process begins again.

This process typically takes approximately 0.8 second, at rest; this results in a resting heart rate of 75 beats per minute. Interestingly, the inherent rhythm that SA nodal cells spontaneously depolarize is at a much faster rate of approximately 100 times per minute. (This inherent rhythm is determined when excised cardiac muscle cells are placed into cell culture, without autonomic nerve innervation.) The normal resting heart rate of 75 beats per minute is due to continuous parasympathetic stimulation of the SA node by the vagus nerve. This slowing of the heart rate is called **vagal tone.**

WHAT DO YOU THINK?

③ What specific type of channel is unique to cardiac nodal cells? What is the significance of this channel?

Comparison of Cardiac Nodal Cells and Neurons

Nodal cells are like neurons because they have an RMP. However, a significant difference between them is that neurons require stimulation (in the form of neurotransmitters or a modality; see section 12.6a), whereas nodal cells do not. This is because nodal cells do not have a stable resting membrane potential. Instead the RMP gradually increases to threshold when slow Na⁺ channels reopen. This ability to reach the threshold without stimulation is called a **pacemaker potential.** Another difference between neurons and nodal cells is that depolarization results from the entrance of Na⁺ into neurons, whereas it results from the entrance of Ca²⁺ into nodal cells.

WHAT DID YOU LEARN?

⑱ What is autorhythmicity? Describe how nodal cells function as autorhythmic cells to serve as the pacemaker of the heart.

19.6c Conduction System of the Heart: Spread of the Action Potential

LEARNING OBJECTIVE

4. Describe the spread of the action potential through the heart's conduction system.

Cardiac muscle stimulation requires that the action potential initiated by the SA node be spread through the conduction system. The sequence of events occurs as follows (figure 19.17):

① **Action potential is distributed throughout both atria and is relayed to the AV node.** The action potential initiated in the SA node is spread between cardiac muscle cells in the atria by gap junctions that allow almost instantaneous excitation of all muscle cells in the atrial walls. This electrical stimulation ultimately results in cardiac muscle cells of both atria contracting at the same time.

② **The action potential is delayed at the AV node.** AV nodal cells have both smaller fiber diameters and fewer numbers of gap junctions—thus, they exhibit characteristics that *slow the conduction rate* of the action potential by serving as "a bottleneck." This is facilitated by the insulating characteristics of the fibrous skeleton, which only allow the action potential to move through the AV node. The delay in conduction (about 100 milliseconds, or 0.1 second) may seem very brief, but it is long enough to allow the atria to finish contracting and force blood into the ventricles to complete ventricular filling before the ventricles are stimulated to contract.

③ **The action potential travels from the AV node through the AV bundle to Purkinje fibers.** The action potential spreads from the AV node along the AV bundle to the bundle branches to the Purkinje fibers.

④ **The action potential spreads throughout both ventricles via gap junctions.** The action potential is then spread between ventricular cardiac muscle cells by gap junctions that allow the almost simultaneous stimulation of all the cardiac muscle cells in the ventricular walls. Generally, these cells begin to contract within 120 to 200 milliseconds after the firing of the SA nodal cells.

Specialized Features Associated with Ventricles

The efficient functioning of the ventricles requires that there be a coordination of contraction that includes the following features:

- Purkinje fibers are larger in diameter than other cardiac conduction fibers, so action potential propagation is extremely rapid to the ventricular myocardium. Thus, the cardiac muscle cells of both ventricles contract at the same time.

- Papillary muscles within the ventricles are stimulated to contract immediately. These muscles anchor the tendinous cords to the AV valve cusps; they tighten the relaxed cords and cause them to start to pull on the AV valve cusps just prior to the increase in pressure within the ventricles. Thus, the valves are "braced" and better able to prevent backflow of blood into the atria.

- The stimulation of the ventricles begins at the apex of the heart. This feature ensures that blood is efficiently ejected superiorly toward the arterial trunks.

WHAT DID YOU LEARN?

❶⓽ What is the path of an action potential through the conduction system of the heart?

❷⓪ What anatomic features slow the conduction rate of the action potential as it passes through the AV node? What is the function of this delay?

① An action potential is generated at the sinoatrial (SA) node. It spreads via gap junctions between cardiac muscle cells throughout the atria to the atrioventricular (AV) node.

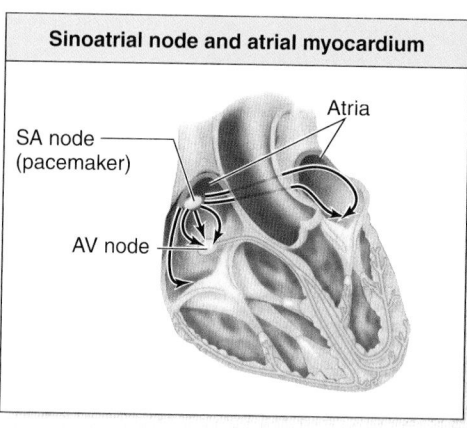

② The action potential is delayed at the AV node before it passes to the AV bundle within the interventricular septum.

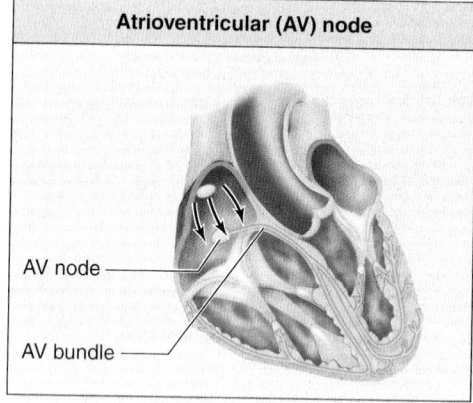

③ The AV bundle conducts the action potential to the left and right bundle branches and then to the Purkinje fibers.

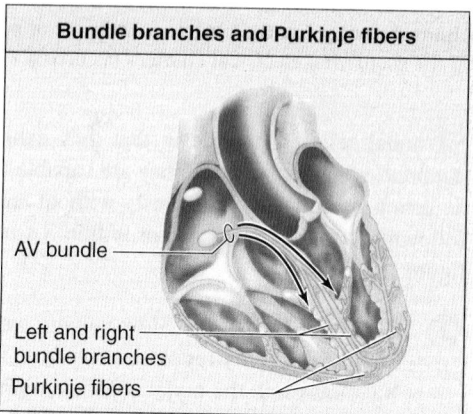

④ The action potential is spread via gap junctions between cardiac muscle cells throughout the ventricles.

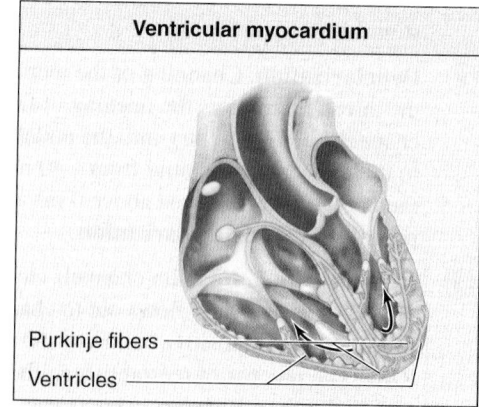

Figure 19.17 Initiation and Spread of an Action Potential Through the Cardiac Conduction System. The initiation and spread of an action potential begins at the SA node and is propagated through the cardiac conduction system. The average rate at rest is one per 0.8 second. AP|R

INTEGRATE

CLINICAL VIEW
Ectopic Pacemaker

A pacemaker other than the SA node is called an **ectopic** (ek-top′ik; *ektos* = outside, *topos* = place) **pacemaker**, or *ectopic focus*. Cells of the conduction system other than the SA node and cardiac muscle cells also have the ability to spontaneously depolarize and serve as the pacemaker. However, they depolarize at slower rates than the SA node. The AV node has an inherent rhythm of 40 to 50 beats per minute, and cardiac muscle cells have a rate of 20 to 40 beats per minute. If the SA node is not functioning, the AV node becomes the "default" pacemaker, and the AV node and then establishes the rhythm of the heart rate. Survival is possible because a rhythm of 40 to 50 beats per minute pumps sufficient blood to sustain life. However, if both the SA node and the AV node are not functioning, cardiac muscle cells will attempt to establish the rhythm. The heart rate will be only 20 to 40 beats per minute, which unfortunately is almost always too slow to support life.

19.7 Cardiac Muscle Cells

Two significant and interrelated events occur within cardiac muscle cells following their stimulation by the conduction system. These include propagation of an action potential at the sarcolemma and contraction of sarcomeres within the cardiac muscle cells. Remember that the events of cardiac muscle cells occur twice per heartbeat: first in the cardiac muscle cells of the atria, and then in the cardiac muscle cells of the ventricles.

19.7a Cardiac Muscle Cells at Rest

LEARNING OBJECTIVE

1. Describe the conditions at the sarcolemma of cardiac muscle cells at rest.

Cardiac muscle cells exhibit several significant features at their sarcolemma. See **figure 19.18a** as your read through this section.

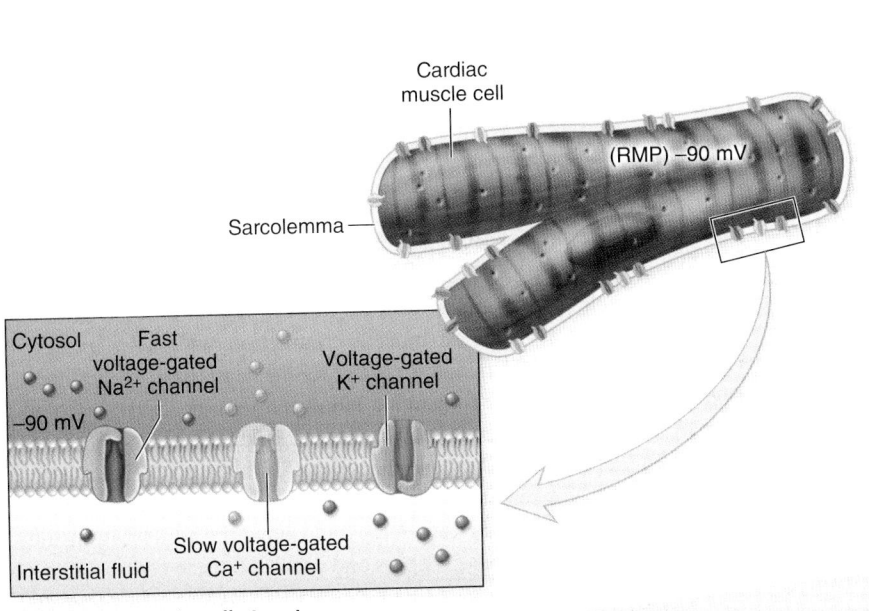

(a) Cardiac muscle cell at rest

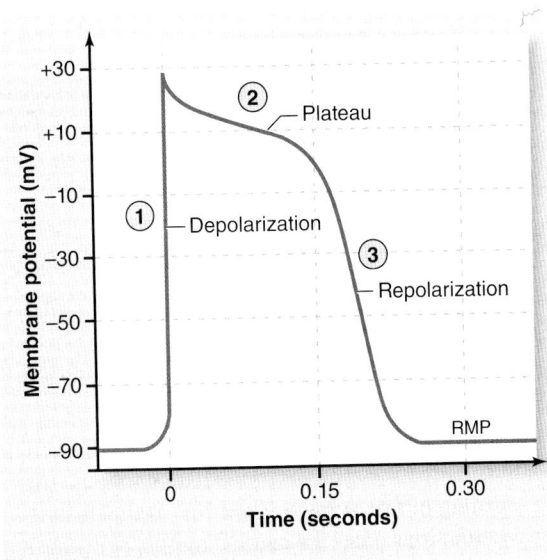

(b) Electrical events at the sarcolemma of a cardiac muscle cell

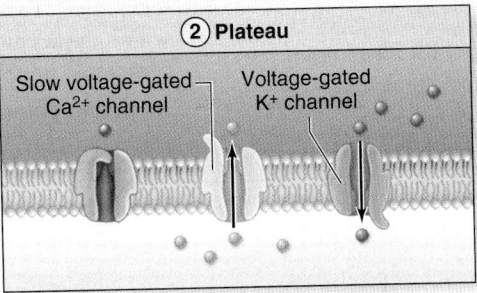

Fast voltage-gated Na⁺ channels open and Na⁺ rapidly enters the cell, reversing the polarity from negative to positive (−90 mV to +30 mV). These channels then close.

Voltage-gated K⁺ channels open and K⁺ flows out of cardiac muscle cells. Slow voltage-gated Ca^{2+} channels open and Ca^{2+} enters the cell, with no electrical change and the depolarized state is maintained.

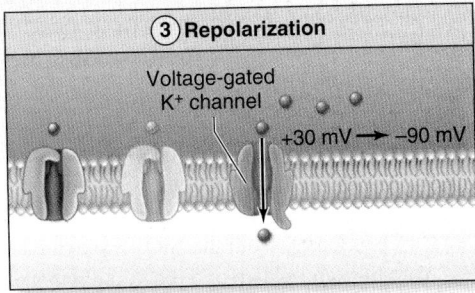

Voltage-gated Ca^{2+} channels close, voltage-gated K⁺ channels remain open, and K⁺ moves out of the cardiac muscle cell, and polarity is reversed from positive to negative (+30 mV to −90 mV).

Figure 19.18 Electrical Events of Cardiac Muscle Cells. (*a*) Cardiac muscle cells have a resting membrane potential of −90 mV. Voltage-gated channels in the sarcolemma are closed when the cardiac muscle cell is at rest. (*b*) Graph and accompanying diagrams of the sequential electrical events of an action potential at the sarcolemma of cardiac muscle cells.

The sarcolemma of cardiac muscle cells, like nodal cells, has K⁺ leak channels, Na⁺ leak channels, and Na⁺/K⁺ pumps, to establish and maintain a resting membrane potential with a greater concentration of Na⁺ outside the cardiac muscle cells and a greater concentration of K⁺ inside. The RMP value of cardiac muscle cells, however, is −90 millivolts (mV) (in comparison to −60 mV for nodal cells). Cardiac muscle cells also contain Ca^{2+} pumps that form a Ca^{2+} concentration gradient with more Ca^{2+} outside the cell than inside.

The sarcolemma of cardiac muscle cells has both fast voltage-gated Na⁺ channels that participate in depolarization at the membrane and K⁺ voltage-gated channels that participate in repolarization of the membrane. Additionally, cardiac muscle cells have slow voltage-gated Ca^{2+} channels within the sarcolemma. Slow voltage-gated Ca^{2+} channels, when open, allow Ca^{2+} into the cell. The movement of Ca^{2+} is significant in the normal function of cardiac muscle cells, as described shortly.

WHAT DID YOU LEARN?

21 In which direction does Ca^{2+} move in response to the opening of voltage-gated Ca^{2+} channels: into or out of the cardiac muscle cells?

19.7b Electrical and Mechanical Events of Cardiac Muscle Cells

LEARNING OBJECTIVES

2. List the electrical events of an action potential that occur at the sarcolemma.

3. Briefly summarize the mechanical events of muscle contraction.

As the heart contracts, we can distinguish certain electrical and mechanical events.

Electrical Events

Electrical events at the sarcolemma of cardiac muscle cells occur as follows (figure 19.18b):

① **Depolarization.** An action potential transmitted through the conduction system (or via gap junctions) triggers the opening of fast voltage-gated Na⁺ channels in the sarcolemma. Sodium ions enter the cardiac muscle cell, causing depolarization. The resting membrane potential changes from −90 mV to +30 mV. Voltage-gated Na⁺ channels close to the inactivated state (see section 12.6a).

② **Plateau.** Depolarization triggers the opening of voltage-gated K⁺ channels, and K⁺ leaves the cardiac muscle cells. There is a slight change in the membrane potential. Almost immediately, slow voltage-gated Ca^{2+} channels in the sarcolemma also open, and Ca^{2+} enters from the interstitial fluid into the cardiac muscle cells. The entering of Ca^{2+} stimulates the sarcoplasmic reticulum (SR) to release more Ca^{2+} (more than 80% of the Ca^{2+} is released from the SR). The exit of positively charged K⁺ from the sarcoplasm and the simultaneous entrance of positively charged Ca^{2+} into the sarcoplasm results in no electrical change at the sarcolemma. Thus, the sarcolemma of the cardiac muscle cell remains in a "depolarized state." This "leveling off" is referred to as the **plateau.**

③ **Repolarization.** Voltage-gated Ca^{2+} channels then close, and K⁺ channels remain open to complete repolarization as K⁺ exits the cardiac muscle cell. The membrane potential reverses, and the resting membrane potential of −90 mV is reestablished. This

allows a muscle cell to propagate a new action potential when the cardiac muscle is stimulated again.

Mechanical Events (Crossbridge Cycling)

The entry of Ca^{2+} into the sarcoplasm from both the interstitial fluid and SR (as described in step 2) initiates the internal mechanical events of muscle contraction. Calcium ions now bind to troponin to begin crossbridge cycling within a sarcomere, similar to the way in which skeletal muscle contracts (see section 10.3c).

The closing of voltage-gated Ca^{2+} channels (as described in step 3), reuptake of Ca^{2+} into the sarcoplasmic reticulum by Ca^{2+} pumps, and removal of Ca^{2+} from the cell by plasma membrane Ca^{2+} pumps decrease calcium levels in the sarcoplasm. Calcium is released from troponin with the subsequent decrease in crossbridges between the thin and thick filaments. Sarcomeres return to their resting length as the cardiac muscle cell now relaxes (see section 10.3d).

WHAT DO YOU THINK?

4 Calcium channel–blocking drugs are sometimes given to patients with heart problems. If a patient is given a Ca^{2+} channel blocker, how will this affect both heart rate and force of contraction? Explain.

WHAT DID YOU LEARN?

22 What three electrical events occur at the sarcolemma of cardiac muscle cells? Explain each event.

19.7c Repolarization and the Refractory Period

LEARNING OBJECTIVES

4. Define the refractory period.

5. Explain the significance of the plateau phase.

Cardiac muscle cells (unlike skeletal muscle fibers) *cannot* exhibit tetany, which is a sustained muscle contraction without relaxation (section 10.6c). This distinction is critical for the heart to be able to function as a mechanical pump to move blood through the cardiovascular system. Here we compare the characteristics of skeletal muscle fibers to cardiac muscle cells to understand the reason for this difference. Refer to **figure 19.19** as you read through this section.

Examine graphs in figure 19.19, and notice the following in the graphs of both skeletal muscle fibers and cardiac muscle cells:

- **Muscle tension,** which is depicted as the red line on each graph, includes both muscle contraction and muscle relaxation. (These mechanical events are due to sarcomeres within these muscle cells shortening and lengthening as contractile proteins slide past one another.)

- An **action potential,** which is shown as a blue line on the graph in figure 19.19a, c, includes both depolarization and repolarization. (These electrical events are occurring at the sarcolemma of these muscle cells.)

- The **refractory period,** which is the shaded green region on each graph, represents the time when the muscle cannot be re-stimulated to contract. (The refractory period is dependent upon how quickly depolarization and repolarization occur at the sarcolemma; the more quickly that repolarization occurs, the shorter the refractory period.)

Now observe just the two graphs that represent what is occurring in *skeletal muscle fibers* (figure 19.19a, b). Figure 19.19a represents the events of a single stimulation of skeletal muscle. Notice that muscle contraction and relaxation associated with sarcomeres within skeletal

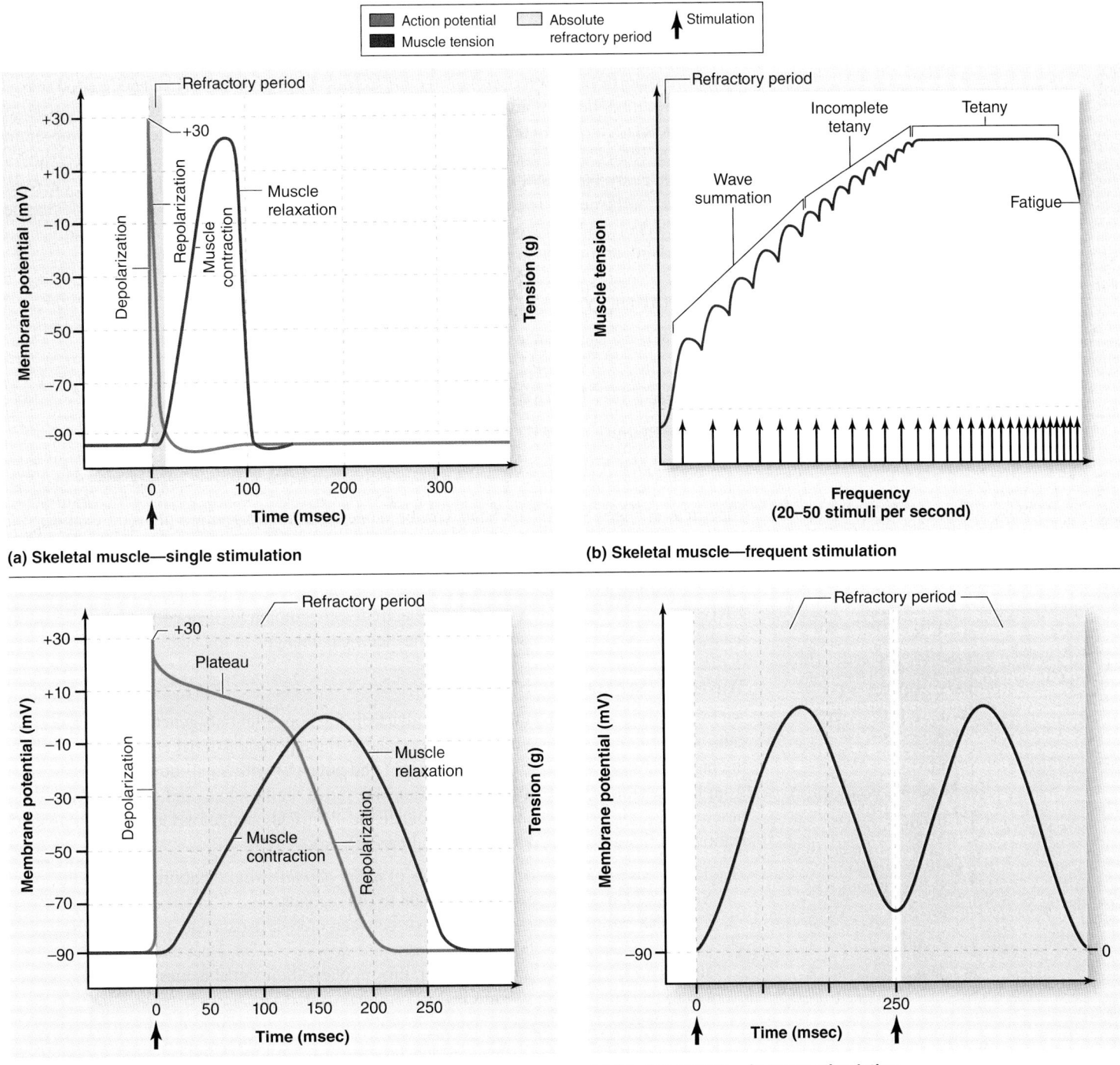

Legend:
- Action potential
- Muscle tension
- Absolute refractory period
- ↑ Stimulation

(a) Skeletal muscle—single stimulation

Refractory period
+30
Depolarization | Repolarization | Muscle contraction | Muscle relaxation

Membrane potential (mV): +30, +10, −10, −30, −50, −70, −90
Tension (g)
Time (msec): 0, 100, 200, 300

(b) Skeletal muscle—frequent stimulation

Refractory period
Incomplete tetany | Tetany
Wave summation
Fatigue
Muscle tension
Frequency (20–50 stimuli per second)

(c) Cardiac muscle—single stimulation

Refractory period
+30
Plateau
Depolarization | Muscle contraction | Repolarization | Muscle relaxation

Membrane potential (mV): +30, +10, −10, −30, −50, −70, −90
Tension (g)
Time (msec): 0, 50, 100, 150, 200, 250

(d) Cardiac muscle—frequent stimulation

Refractory period
Membrane potential (mV): −90
Time (msec): 0, 250
0

Figure 19.19 Comparison of Electrical and Mechanical Events in Skeletal Muscle Cells and Cardiac Muscle Cells. The electrical changes associated with an action potential propagated along the sarcolemma and the mechanical changes associated with generating muscle tension when a muscle contracts is shown for (*a*) skeletal muscle with a single stimulation, and (*b*) skeletal muscle with frequent stimulation, (*c*) cardiac muscle with a single stimulation, and (*d*) cardiac muscle with frequent stimulation. The extended refractory period in cardiac muscle cells (due to the plateau) allows time for cardiac muscle to contract and relax before being stimulated again. Thus, a sustained contraction (tetany) is prevented.

muscle fibers occurs over approximately 100 milliseconds. Observe also that repolarization at the sarcolemma occurs immediately following depolarization, which results in a refractory period that is relatively short (approximately 1 to 2 milliseconds). Consequently, while sarcomeres are still contracting within skeletal muscle fibers, the sarcolemma has already been repolarized to allow for a new stimulation.

Figure 19.19*b* represents the events of frequent stimulation of skeletal muscle. Notice that skeletal muscle, because it has a relatively short refractory period, can be restimulated at a frequency that does not allow the skeletal muscle sufficient time to relax completely. In fact, skeletal muscle can be stimulated at a rate that the muscle remains contracted

(without any relaxation). This condition of sustained muscle contraction is called **tetany.**

Now refer to the two graphs that represent what is occurring in *cardiac muscle cells* (figure 19.19*c, d*). Figure 19.19*c* represents the events of a single stimulation of cardiac muscle. Notice that muscle contraction and relaxation associated with sarcomeres within cardiac muscle cells occurs over approximately 250 milliseconds (a longer period than in skeletal muscle). Note also that repolarization at the sarcolemma does not occur immediately following depolarization, which results in a refractory period that is relatively long (almost 250 milliseconds). This relatively long refractory period is due to the plateau event at the sarcolemma, which delays repolarization. Consequently, while sarcomeres are still contracting and relaxing within cardiac muscle, the sarcolemma has *not* been repolarized to allow for a new stimulation.

Figure 19.19*d* represents the events of frequent stimulation of cardiac muscle. Notice that cardiac muscle, because it has a relatively long refractory period, has an extended period of time in which it cannot be restimulated. This delay in restimulation allows time for the sarcomeres of cardiac muscle cells within the heart chamber wall to fully contract and relax before being stimulated again. Thus, cardiac muscle cells composing the myocardium of the heart chamber walls do *not* exhibit tetany, but continue to both contract and relax following each stimulation—an essential feature for the heart to pump the blood. (Note that if tetany were possible in cardiac muscle, the chambers of the heart would experience a sustained contraction—this would be similar to a "pump ceasing or locking up.")

WHAT DID YOU LEARN?

23 What is the significance of the extended refractory period in cardiac muscle?

19.7d The ECG Recording

LEARNING OBJECTIVE

6. Identify the components of an ECG recording.

Electrical changes within the heart can be detected during a routine physical examination using monitoring electrodes attached to the skin—usually at the wrists, ankles, and six separate locations on the chest. The electrical signals are collected and charted as an **electrocardiogram** (ē-lek-trō-kar′dē-ō-gram; *gramma* = drawing), also called an **ECG** or **EKG.** When readings from the different electrodes are compared, they collectively provide an accurate, comprehensive assessment of the electrical changes of the heart.

Waves and Segments

An ECG provides a composite tracing of all cardiac action potentials generated by myocardial cells. A typical ECG tracing for one heart cycle has three principal deflections: a P wave above the baseline, a QRS complex that begins (Q) and ends (S) with small downward deflections from the baseline and has a large deflection (R) above the baseline, and a T wave above the baseline **(figure 19.20).** These waves indicate the electrical changes associated with depolarization and repolarization within specific heart regions:

1. The **P wave** reflects electrical changes of *atrial depolarization* that originates in the SA node. This event typically lasts 0.08 to 0.1 seconds.
2. The **QRS complex,** which usually lasts between 0.06 to 0.1 second, represents the electrical changes associated with *ventricular depolarization.* Note that the atria are simultaneously repolarizing; however, this repolarization signal is masked by the greater electrical activity of the ventricles.
3. The **T wave** is the electrical change associated with *ventricular repolarization.*

The two segments between the waves correspond with a plateau (where there is essentially no electrical change). During these time periods, the sarcomeres are shortening within the cardiac muscle cells. The **P-Q segment** is associated with the atrial plateau at the sarcolemma when the cardiac muscle cells within the atria are contracting, and the **S-T segment** is the ventricular plateau when the cardiac muscle cells within the ventricles are contracting.

Figure 19.20*b* shows two of the graphs of the electrical changes associated with the atria and ventricles (see figure 19.18*b*) superimposed on the ECG. Notice how the waves of an ECG reflect electrical changes—either depolarization or repolarization—and how the flat or level portions in the segments correspond to a plateau when there is no electrical change. The flat line between cycles represents when the heart is resting between beats.

Intervals

Additional characteristics of an ECG include two intervals: the P-R interval and the Q-T interval (figure 19.20*a*). Changes in length of an interval may reflect abnormal changes to the heart. The **P-R interval** represents the period of time from the beginning of the P wave (atrial depolarization) to the beginning of the QRS deflection (ventricular depolarization). This interval of time

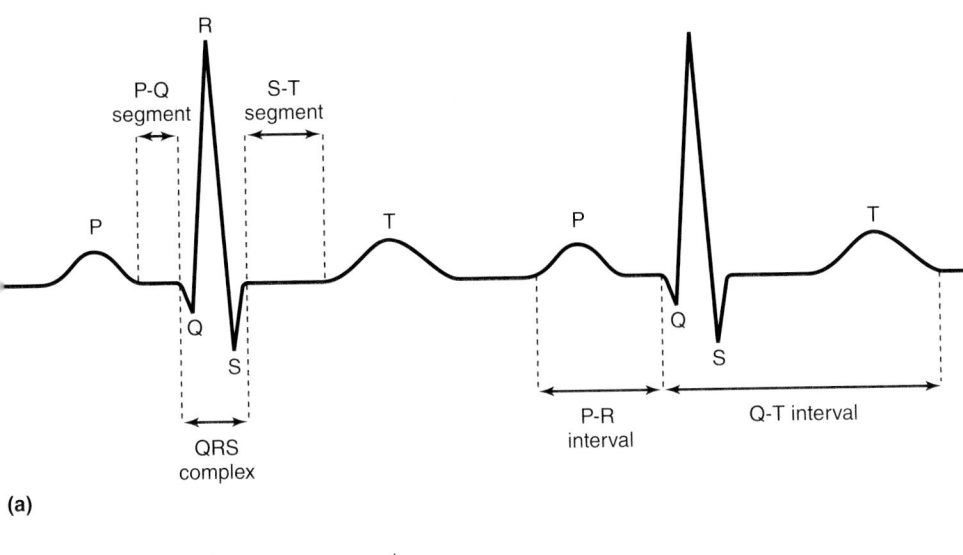

(a)

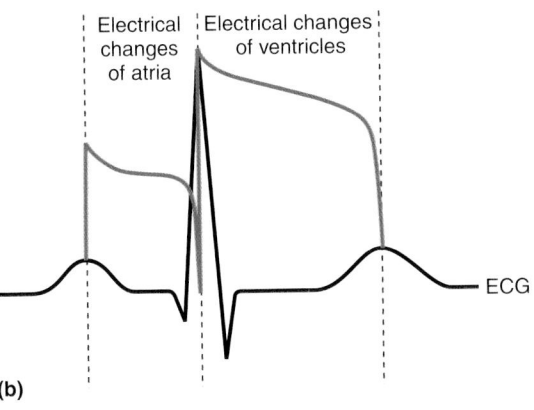

(b)

Figure 19.20 The Electrocardiogram. (*a*) A tracing of an electrocardiogram with the location of three waves, including the P wave, QRS complex, and T wave. Two segments are located between the waves and include the P-Q segment and the S-T segment. The P-R interval and Q-T interval are also shown. (*b*) A graph of electrical changes within atria and ventricles is superimposed on an ECG to help you to see the relationship of these electrical changes and their relationship to the waves of an ECG.

normally ranges from 0.12 to 0.20 second. It is the time required to transmit an action potential through the entire conduction system. P-R intervals that extend longer than 0.20 second generally indicate an impaired conduction within the ventricles, often indicating a *heart block*. (See Clinical View: "Cardiac Arrhythmias.")

The **Q-T interval** represents the time from the beginning of the QRS (ventricular depolarization) to the end of the T wave (ventricular repolarization). This is the time required for the action potential to occur within the ventricles. This interval depends upon heart rate and ranges from 0.2 to 0.4 second. Changes in the Q-T intervals may result in a fast, irregular heart rate, which is called **tachyarrhythmia** (tak'ē'ă'ridh'mē-ă; *tachys* = quick).

An ECG is a common diagnostic tool that is used in physical exams and after surgery to detect many changes to the heart, including abnormal heart rhythms (fast, slow, arrhythmia), hypertrophy of the heart, low blood flow to the heart (ischemia), and damage to the heart from myocardial infarction or high blood pressure. Although it is a very effective tool in diagnosing many conditions, it does not detect all abnormalities. Thus, it is one of several diagnostic tools used by cardiologists and other physicians in determining a diagnosis and development of a treatment plan.

 WHAT DID YOU LEARN?

24 What events in the heart are indicated by each of the following: P wave, QRS complex, and T wave? Identify the two segments of an ECG that reflects the plateau.

INTEGRATE

CLINICAL VIEW
Cardiac Arrhythmia

Cardiac arrhythmia (ă-rith′mē-ă), also called *dysrhythmia,* is any abnormality in the electrical activity of the heart.

Heart blocks involve impairment within the heart's conducting system. Heart blocks may result in a feeling of light-headedness, fainting, irregular heartbeat, and chest palpitations. The three different types of heart blocks differ in the extent to which electrical signals are not transmitted through the conduction system.

First-degree AV block is also called *PR prolongation* because of a lengthened PR interval. The action potentials are *slowed* between the atria and ventricles. This type of block is generally asymptomatic.

Second-degree AV block is the failure of *some* atrial action potentials to be conducted to the ventricles. The PR interval may be either normal or prolonged.

Third-degree AV block is a complete heart block, and is the failure of *all* action potentials to be conducted to the ventricles. This condition is life-threatening and requires medical intervention.

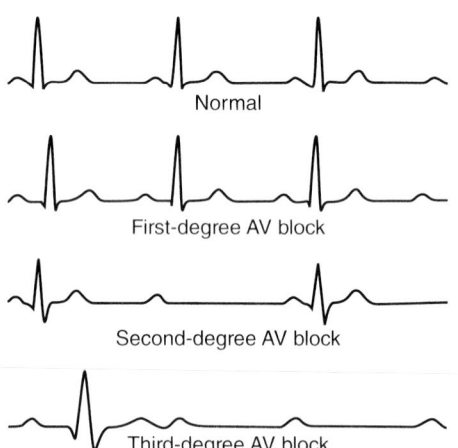

Normal

First-degree AV block

Second-degree AV block

Third-degree AV block

Atrial fibrillation (fĭ-bri-lā′shŭn) involves chaotic action potentials within the atria. The ventricles respond by increasing and decreasing contraction activities, which may lead to serious disturbances in cardiac rhythm.

Ventricular fibrillation is chaotic electrical activity within the ventricles. Muscle contraction of the cardiac muscle cells within the ventricles is uncoordinated, and the heart does not pump blood. Blood circulation stops (cardiac arrest), which leads to death of cardiac muscle (myocardial infarction) commonly called a "heart attack." To restore normal heart contractions, medical personnel apply a strong electrical shock to the skin of the chest using paddle electrodes in an attempt to synchronize the electrical events of cardiac muscle cells.

Premature ventricular contractions (PVCs) involve single or rapid bursts of abnormal action potentials initiated within the AV node or the ventricular conduction system instead of the SA node. PVCs often result from stress, stimulants such as caffeine, or sleep deprivation. They are not detrimental unless they occur in great numbers. Most PVCs go unnoticed, although occasionally one is perceived as the heart "skipping a beat" and then "jumping" in the chest.

PVC   PVC

19.8 The Cardiac Cycle

A **cardiac cycle** is the inclusive changes within the heart from the initiation of one heartbeat to the start of the next. One heartbeat involves the contraction and relaxation of the heart chambers. Note that the term **systole** (sis′tō-lē) refers to contraction of a heart chamber, whereas **diastole** (dī-as′tō-lē; dilation) refers to relaxation of a heart chamber.

19.8a Overview of the Cardiac Cycle

LEARNING OBJECTIVES

1. Identify the two processes within the heart that occur due to pressure changes associated with the cardiac cycle.

2. List the five phases of the cardiac cycle.

The alternating contraction and relaxation of both the atria and ventricles cause pressure changes within their chambers. Pressure increases during contraction and decreases during relaxation. These alternating pressure changes are responsible for two significant physiologic processes:

- Unidirectional movement of blood through the heart chambers, as blood moves along a pressure gradient (i.e., from an area of greater pressure to an area of lesser pressure)
- Opening and closing of heart valves to ensure that blood continues to move in a "forward" direction without backflow

INTEGRATE

LEARNING STRATEGY

You may find it helpful to think of how changing ventricular pressure has opposite effects on the two sets of valves by imagining yourself in a small room with two spring-loaded doors: One is open, and the other is closed. You apply pressure to the open door (AV valve) and it closes, and then you apply pressure to the closed door (semilunar valve) and it opens. You then release the pressure on the two spring-loaded doors in the reverse order. What happens? Each door returns to its original position—the one that was pushed open (semilunar valve) closes, and the door pushed closed (AV valve) opens.

The most important driving force to continually move blood through the heart, and to open and close heart valves during the cardiac cycle, is contraction and relaxation of the ventricles:

- **Ventricular contraction.** Ventricles contract and ventricular pressure rises. The one-way flow of blood occurs as both AV valves are pushed and kept closed, preventing backflow of blood into each atrium, and then the semilunar valves are pushed open, ejecting blood from a ventricle into an arterial trunk (aorta or pulmonary trunk).

- **Ventricular relaxation.** Ventricles relax and ventricular pressure decreases. Now, the semilunar valves close (because there is no longer pressure pushing them open), and then the AV valves open (because blood is entering the atria and there is no longer pressure pushing them closed).

Here we organize the cardiac cycle into five phases: atrial contraction and ventricular filling, isovolumetric contraction, ventricular ejection, isovolumetric relaxation, and atrial relaxation and ventricular filling. These events are summarized in **figure 19.21**. Consider the following as you read through the description of each of the phases of the cardiac cycle: (1) whether the atria and ventricles are contracted or relaxed, (2) whether the pressure in the ventricles is higher or lower than the pressure in the atria and higher or lower than the pressure in the arterial trunks (aorta and pulmonary trunk), and (3) if the AV valves and semilunar valves are opened or closed.

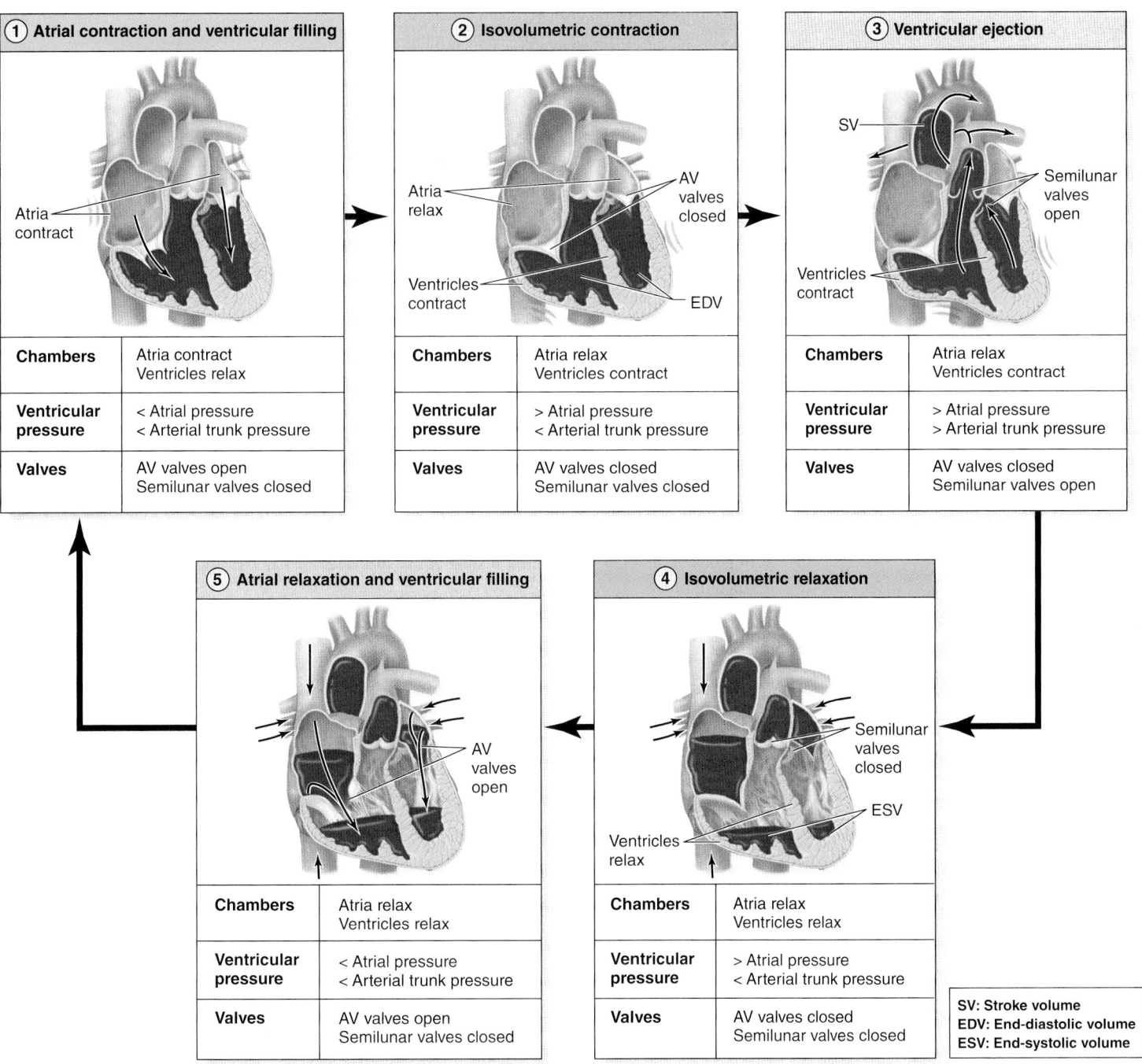

Figure 19.21 Phases of the Cardiac Cycle. Contraction and relaxation of heart chambers, changes in ventricular pressure, and opening and closing of heart valves occur during the phases of the cardiac cycle. AP|R

25 Pressure changes that occur during the cardiac cycle are responsible for what two physiologic processes within the heart?

19.8b Events of the Cardiac Cycle

LEARNING OBJECTIVES

3. List and describe what occurs during the five phases of the cardiac cycle.

4. Explain the significance of ventricular balance.

Because this is a cycle, we could start at any point and describe the events. However, for orientation and convenience we begin when all chambers are at rest and the atria are just initiating the contraction phase.

As the Cardiac Cycle Begins

We can note the following characteristics of the heart prior to atrial contraction as a new round of the cardiac cycle begins.

- All four chambers are at rest.
- Blood continues to return to the right atrium through the superior vena cava, inferior vena cava, and the coronary sinus, and to the left atrium through the pulmonary veins.
- Passive filling of the ventricles is under way. No contraction of the atria is needed at this time to assist with the continual ventricular filling.
- The AV valves are open because the pressure exerted by the blood filling the atria is greater than the pressure exerted by the blood remaining in the resting ventricles.
- The semilunar valves are closed because the pressure exerted by the blood remaining in the filling ventricles is lower than the pressure exerted by the blood in the arterial trunks.

Atrial Contraction and Ventricular Filling

This phase is distinguished from the resting conditions of the heart just described by two events: atrial contraction and completion of ventricular filling (phase 1 in figure 19.21). Atrial contraction is initiated by the SA node stimulating cardiac muscle cells of the atrial wall. Contraction of the atrial wall moves the remainder of blood that was within each atrium into the right or left ventricle.

Ventricular filling is complete upon termination of atrial contraction, and the ventricles now hold their maximum blood volume. This is appropriately called the **end-diastolic volume (EDV)** because it is the volume of blood within the ventricle at the end of diastole (or rest). EDV is labeled in phase 2 in figure 19.21. In a resting adult, the value of EDV is approximately 130 milliliters (mL) of blood.

Note that during atrial contraction, both the inflow of additional blood from the supplying veins into the atria (i.e., blood flow from the superior vena cava, inferior vena cava, and coronary sinus into the right atrium) and backflow of blood from the atria into these veins is prevented because contraction of the atria compresses the openings for the great veins.

You will find it helpful to remember that at the end of this phase, the atria relax and will remain relaxed until the beginning of the next cardiac cycle, as shown in phases 2 through 5 in figure 19.21. These four phases (2 through 5) involve the changes associated with contraction and relaxation of the ventricles.

Isovolumetric Contraction

Two distinctive changes occur in this phase: ventricular contraction and closure of the AV valves (phase 2 in figure 19.21). Ventricular contraction is initiated by the Purkinje fibers of the conduction system stimulating the cardiac muscle cells of the ventricular wall. As the ventricles begin to contract, the pressure within the ventricles increases and exceeds the pressure within the atria. Consequently, the AV valves are forced closed. Recall that they are braced by tendinous cords to the papillary muscles. The closure of these valves prevent backflow of blood from the ventricles into the atria. The semilunar valves remain closed because the pressure within the ventricle is still less than the pressure within the attached arterial trunk.

INTEGRATE

LEARNING STRATEGY

You may find this summary table of assistance in remembering what is occurring during each of the five phases of the cardiac cycle.

	1	2	3	4	5
Atria	C	R	R	R	R
Ventricles	R	C	C	R	R
AV valves	O	CL	CL	CL	O
Semilunar valves	CL	CL	O	CL	CL

R = Relaxation of a chamber
C = Contraction of a chamber
O = Valve open
CL = Valve closed

Note that all heart valves are closed at this time—thus, no blood may enter or leave the ventricles. Cardiac muscle cells are contracting, but the volume within the ventricular chambers remains the same. Thus, this phase is appropriately referred to as the time of **isovolumetric** (ī′sō-vol′yū-met′rik; *iso* = same) **contraction** of the ventricles.

Ventricular Ejection

Ventricular ejection, which is the movement of blood from the ventricles into the arterial trunks, occurs during this phase as ventricular contraction continues and the semilunar valves are forced open (phase 3 in figure 19.21). As the ventricles continue to contract, the pressure within the ventricles increases and exceeds the pressure within the arterial trunk. Consequently, the semilunar valves open, which allows blood to be ejected from the ventricles into their associated arterial trunk (e.g., blood is ejected from the left ventricle into the aorta). The AV valves remain closed because the pressure within the ventricles remains greater than the pressure within the atria. The amount of blood pumped out during ventricular contraction is termed the **stroke volume (SV).** Generally, this volume is approximately 70 mL of blood.

Not all of the blood in either ventricle is ejected. The blood remaining in a ventricle at the end of systole is appropriately called the **end-systolic volume (ESV)** because it is the volume of blood in the ventricle at the end of systole (or contraction), ESV is labeled in phase 4 in figure 19.21. ESV is determined by subtracting SV from EDV (EDV – SV = ESV): 130 mL – 70 mL = 60 mL.

Isovolumetric Relaxation

The ventricles are beginning to relax and the semilunar valves close during this phase (phase 4 in figure 19.21). As the ventricles start to relax and the ventricular chamber expands back to its resting size, pressure within the ventricles decreases below the pressure within the attached arterial trunks. Blood initially flows backward slightly within the arterial trunks but is caught in the semilunar valves, causing them to close. This closure of the semilunar valves prevents blood backflow from the arterial trunks into the ventricles. The AV valves remain closed because the pressure within the ventricles is still greater than the pressure within the atria.

Note that all heart valves are again closed simultaneously and the volume of blood within the ventricles remains the same. However, during this phase, cardiac muscle cells are relaxing. Consequently, this phase is referred to as the time of **isovolumetric relaxation** of the ventricles.

Atrial Relaxation and Ventricular Filling

The final phase of the cardiac cycle is distinguished by the continued relaxation of the ventricles and the opening of the AV valves (phase 5 in figure 19.21). As the ventricles continue to relax, the pressure within the ventricles decreases below the pressure within the atria, and the AV valves open. The opening of these valves allows blood to once again move from the atria into the ventricles for ventricular filling. The semilunar valves remain closed because the pressure within the ventricles remains lower than the pressure with the arterial trunks. Amazingly, these events of the cardiac cycle, as organized and described in these five phases, are repeated with each heartbeat!

All of the principal events of one heartbeat, including the: electrical events that control heart contraction, as depicted on an ECG; the contraction and relaxation of the heart chambers and position of the heart valves that occur during a cardiac cycle; the relative pressure within the left atrium, left ventricle, and aorta; and the changes in blood volume within the left ventricle are integrated in **figure 19.22.**

Ventricular Balance

It is important to realize that *equal* amounts of blood are normally pumped by the two ventricles through the two circulations, a condition called **ventricular balance.** However, the right ventricle has to pump the blood only to the adjacent lungs (a relatively short circuit), whereas the left ventricle has to pump the blood through the systemic circulation throughout the body (a relatively long circuit). Thus, the left ventricle must be larger and stronger than the right ventricle to pump the blood farther—but it is the same amount of blood pumped from both sides of the heart. Sustained pumping of *unequal* amounts of blood may result in **edema** (e-dē′mă; *aoidema* = swelling), which is excess fluid in the interstitial space or within cells. (See Clinical View: "Systemic and Pulmonary Edema.")

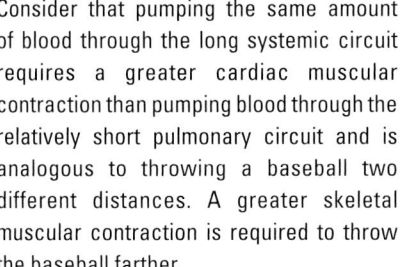

INTEGRATE

LEARNING STRATEGY

Consider that pumping the same amount of blood through the long systemic circuit requires a greater cardiac muscular contraction than pumping blood through the relatively short pulmonary circuit and is analogous to throwing a baseball two different distances. A greater skeletal muscular contraction is required to throw the baseball farther.

Figure 19.22 **Changes Associated with a Cardiac Cycle.** A cardiac cycle is the inclusive changes within the heart from the initiation of one heartbeat to the start of the next. The cardiac cycle is organized here into five phases, and includes (*a*) an ECG; (*b*) heart diagrams of the five phases; (*c*) graph of the relative pressure within the aorta, left atrium, and left ventricle, which is the driving force for both opening and closing the heart valves and moving blood through the heart (for the left side of the heart); and (*d*) a graph showing the changes in left ventricular blood volume.

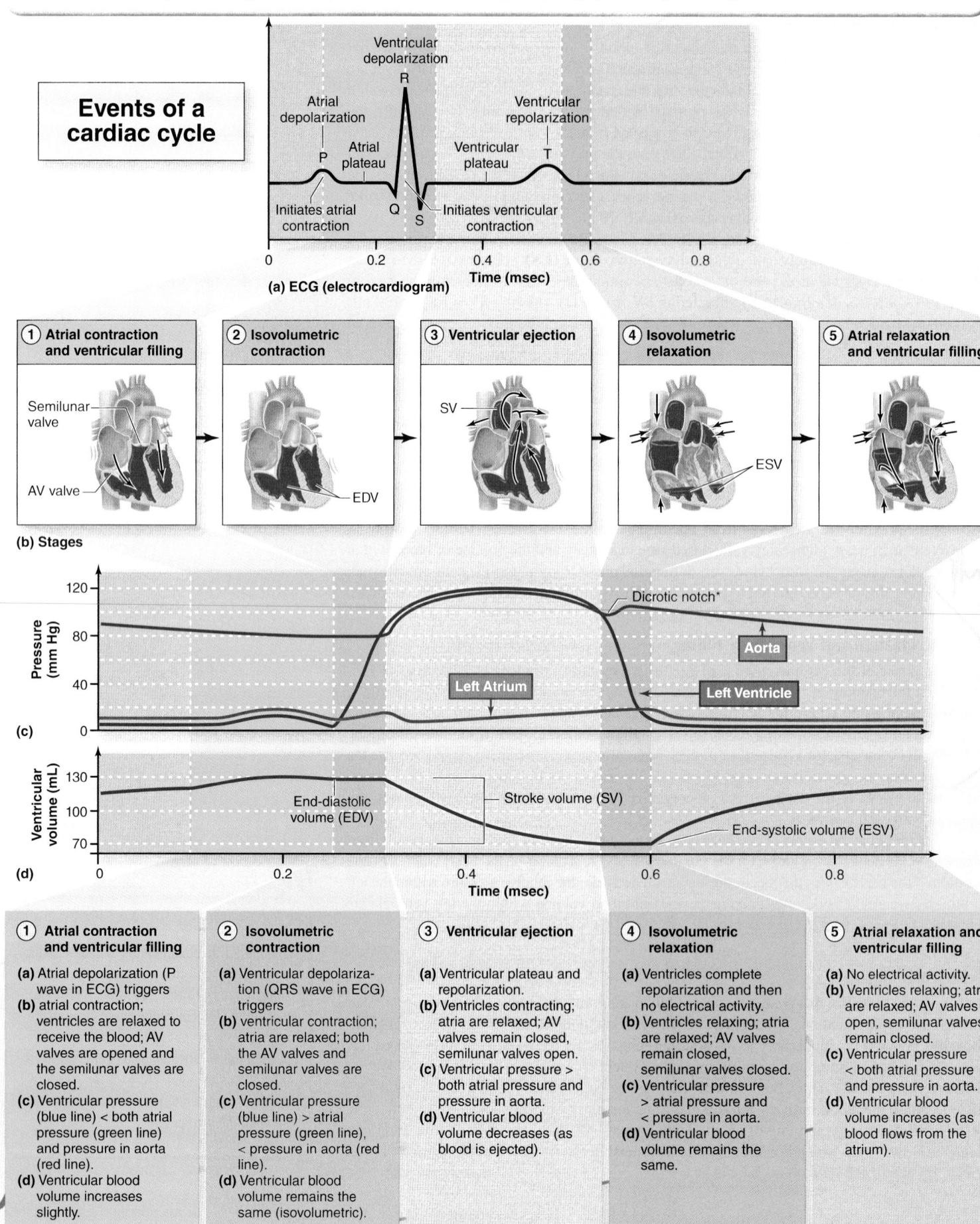

Events of a cardiac cycle

Ventricular depolarization
R

Atrial depolarization

Ventricular repolarization

P · Atrial plateau

Ventricular plateau · T

Initiates atrial contraction · Q · S · Initiates ventricular contraction

0 · 0.2 · 0.4 · 0.6 · 0.8
Time (msec)

(a) ECG (electrocardiogram)

(b) Stages

① **Atrial contraction and ventricular filling**
Semilunar valve
AV valve

② **Isovolumetric contraction**
EDV

③ **Ventricular ejection**
SV

④ **Isovolumetric relaxation**
ESV

⑤ **Atrial relaxation and ventricular filling**

(c)
Pressure (mm Hg): 120, 80, 40, 0
Dicrotic notch*
Aorta
Left Atrium
Left Ventricle

(d)
Ventricular volume (mL): 130, 100, 70
End-diastolic volume (EDV)
Stroke volume (SV)
End-systolic volume (ESV)
0 · 0.2 · 0.4 · 0.6 · 0.8
Time (msec)

① **Atrial contraction and ventricular filling**
(a) Atrial depolarization (P wave in ECG) triggers
(b) atrial contraction; ventricles are relaxed to receive the blood; AV valves are opened and the semilunar valves are closed.
(c) Ventricular pressure (blue line) < both atrial pressure (green line) and pressure in aorta (red line).
(d) Ventricular blood volume increases slightly.

② **Isovolumetric contraction**
(a) Ventricular depolarization (QRS wave in ECG) triggers
(b) ventricular contraction; atria are relaxed; both the AV valves and semilunar valves are closed.
(c) Ventricular pressure (blue line) > atrial pressure (green line), < pressure in aorta (red line).
(d) Ventricular blood volume remains the same (isovolumetric).

③ **Ventricular ejection**
(a) Ventricular plateau and repolarization.
(b) Ventricles contracting; atria are relaxed; AV valves remain closed, semilunar valves open.
(c) Ventricular pressure > both atrial pressure and pressure in aorta.
(d) Ventricular blood volume decreases (as blood is ejected).

④ **Isovolumetric relaxation**
(a) Ventricles complete repolarization and then no electrical activity.
(b) Ventricles relaxing; atria are relaxed; AV valves remain closed, semilunar valves closed.
(c) Ventricular pressure > atrial pressure and < pressure in aorta.
(d) Ventricular blood volume remains the same.

⑤ **Atrial relaxation and ventricular filling**
(a) No electrical activity.
(b) Ventricles relaxing; atria are relaxed; AV valves open, semilunar valves remain closed.
(c) Ventricular pressure < both atrial pressure and pressure in aorta.
(d) Ventricular blood volume increases (as blood flows from the atrium).

*Dicrotic notch: A temporary drop in pressure in the aorta that occurs when the aortic semilunar valve closes.

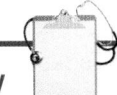

INTEGRATE

CLINICAL VIEW
Systemic and Pulmonary Edema

Edema (e-dē′mă) is accumulation of fluid in the interstitial space surrounding the cells. Edema may result from a decreased ability of the right ventricle, left ventricle, or both ventricles to pump blood effectively. If the *right* ventricle is impaired, it cannot keep up with the pumping action of the left ventricle. Consequently, blood returning from the systemic circulation into the right side of the heart accumulates within the systemic circulation. Additional fluid leaves the systemic capillaries to enter the interstitial space surrounding the body's cells, resulting in fluid accumulation within the body's tissue—a condition called **systemic edema**. Clinical signs of systemic edema include swelling of body tissues especially in the legs, ankles, and feet. If the *left* ventricle is impaired, it cannot keep up with the pumping action of the right ventricle, and blood returning from the pulmonary circulation into the left side of the heart accumulates within the pulmonary circulation. Additional amounts of fluid leave the pulmonary circulation to enter the interstitial space around the alveoli, resulting in swelling and fluid accumulation within the lungs—a condition called **pulmonary edema**. Clinical signs of pulmonary edema include pale skin, coughing up blood, and breathing difficulties. Pulmonary edema may lead to impaired gas exchange and may cause respiratory failure and death if not treated.

WHAT DID YOU LEARN?

26 What is occurring during ventricular ejection?

27 What pressure changes cause the closing of the AV valves and the opening of the semilunar valves?

28 Define end-diastolic volume, end-systolic volume, and stroke volume. How are they related?

19.9 Cardiac Output

The function of the cardiovascular system is to move blood throughout the body to transport respiratory gases, nutrients, and other substances. Cardiac output is a measure of how effective the cardiovascular system is in fulfilling its function. In a healthy individual, cardiac output increases during strenuous effort or exercise to meet the additional cellular demands for oxygen and nutrients and eliminating waste products. In contrast, individuals with impaired heart function may not experience an increase in cardiac output, thus limiting their ability to engage in physically demanding activities. Here we first define cardiac output and then discuss the variables that influence it.

19.9a Introduction to Cardiac Output

LEARNING OBJECTIVES

1. Define cardiac output.
2. Explain what is meant by cardiac reserve.

Cardiac output is defined as the amount of blood that is pumped by a *single* ventricle (left or right) in 1 minute and is typically expressed as liters per minute. As just described, both ventricles eject equal amounts of blood (i.e., balanced ventricular output), so either the left or right ventricle can be used.

Cardiac output is determined by heart rate and stroke volume. **Heart rate (HR)** is the number of beats per minute (bpm). **Stroke volume (SV)** is the volume of blood ejected during one beat and is expressed as milliliters per beat. Cardiac output (CO) is determined as follows:

HR	×	SV	=	CO
Beats per minute		mL ejected per beat		mL ejected per minute

So, for example, when a person is at rest and the HR is 75 beats per minute and SV is 70 mL per beat, then the CO is 5250 milliliters per minute (5250 mL/min) or 5.25 liters per minute (5.25 L/min) Note that 1000 mL = 1 liter.

The total volume of blood in the body is approximately 5 L (see section 18.1b). Thus, with a cardiac output of approximately 5 L per minute, the total blood volume is pumped through both the pulmonary circulation and systemic circulation every minute. This is equivalent to moving over 7000 L of blood daily!

Maintaining Resting Cardiac Output

The ability to maintain normal resting cardiac output is a function of both heart rate and stroke volume, and the value of one influences the value of the other. For example, individuals with smaller hearts produce a smaller stroke volume. Consequently, their resting heart rates must be higher to maintain normal resting cardiac output. This accounts for women's typically faster resting heart rate compared to men, and it also explains why infants and young children with their relatively small hearts have a faster resting heart rate than adults. Newborns, for example, typically have a resting heart rate between 120 to 160 beats per minute.

In comparison, highly trained athletes tend to have larger and stronger hearts. This is because the cardiac muscle cells of the heart wall hypertrophy in response to the additional cardiovascular demands made during consistent strenuous effort. The larger heart generates a greater stroke volume. Thus, an athlete's normal resting cardiac output is maintained with a larger stroke volume and a lower heart rate. Well-conditioned elite athletes, for example, may have resting heart rates below 40 beats per minute.

Cardiac Reserve

An increase in both heart rate and stroke volume results in an increase in cardiac output. During physical exertion, heart rate can be accelerated to more than 170 beats per minute. Likewise, stroke volume can be increased to more than 100 mL. **Cardiac reserve** is an increase in cardiac output above its level at rest. It can be determined by subtracting cardiac output at rest from the cardiac output during exercise.

Cardiac reserve is a measure of the level and duration of physical effort in which an individual can engage. Cardiac output may be increased by about four-fold in a healthy, nonathletic individual (to approximately 20 L/min), and up to seven-fold in a highly trained athlete (to approximately 35 L/min). In comparison, individuals with a weakened heart may have little cardiac reserve and thus experience limitations in exerting themselves.

WHAT DID YOU LEARN?

29 What are the two factors that determine cardiac output?

30 What is the cardiac output at rest and during exercise, and the cardiac reserve if (a) the heart rate is 75 beats per minute and stroke volume is 70 mL at rest, and (b) if the heart rate is 150 beats per minute and stroke volume is 100 mL during exercise?

19.9b Variables That Influence Heart Rate

LEARNING OBJECTIVE

3. Define chronotropic agents, and describe how they affect heart rate.

4. Discuss how autonomic reflexes alter heart rate.

Cardiac output is directly influenced by both heart rate and stroke volume. Both heart rate and stroke volume, in turn, have variables that influence each of them. Here we describe the variables that influence heart rate and then in section 19.9c discuss the variables that influence stroke volume.

The heart rate can be altered by external factors that act on the SA node (the pacemaker) and the AV node. The primary external factors to increase and decrease heart rate come from autonomic innervation and varying levels of some hormones. These factors that change heart rate are called **chronotropic** (kron′ō-trop′ik; *chrono* = time, *tropos* = change) **agents** and are classified as either positive chronotropic agents or negative chronotropic agents.

Positive chronotropic agents cause an increase in heart rate and include sympathetic nerve stimulation and certain types of hormonal stimulation. See **figure 19.23** as you read through this section. Sympathetic axons release the neurotransmitter norepinephrine (NE) to act directly on the SA nodal cells. The sympathetic division also causes release of both epinephrine (EPI) and NE from the adrenal medulla into the blood (step 1). Both NE and EPI bind to β_1-adrenergic receptors of the heart (step 2). This binding initiates an intracellular pathway involving G protein that results in the activation of adenylate cyclase enzyme with the accompanying production of the second messenger, cAMP (see section 17.5b). Ultimately, protein kinase enzymes phosphorylate Ca^{2+} channels, causing them to open. Positively charged Ca^{2+} enters the nodal cells, and nodal cells reach threshold more quickly, increasing the firing rate of the SA node (step 3). (Recall that an influx of Ca^{2+} is responsible for depolarization of nodal cells; see section 19.6b). Sympathetic stimulation of the AV node also increases calcium influx into these cells (not shown in figure 19.23). The delay in the AV node is decreased, and the conduction rate is increased. Thus, heart rate increases.

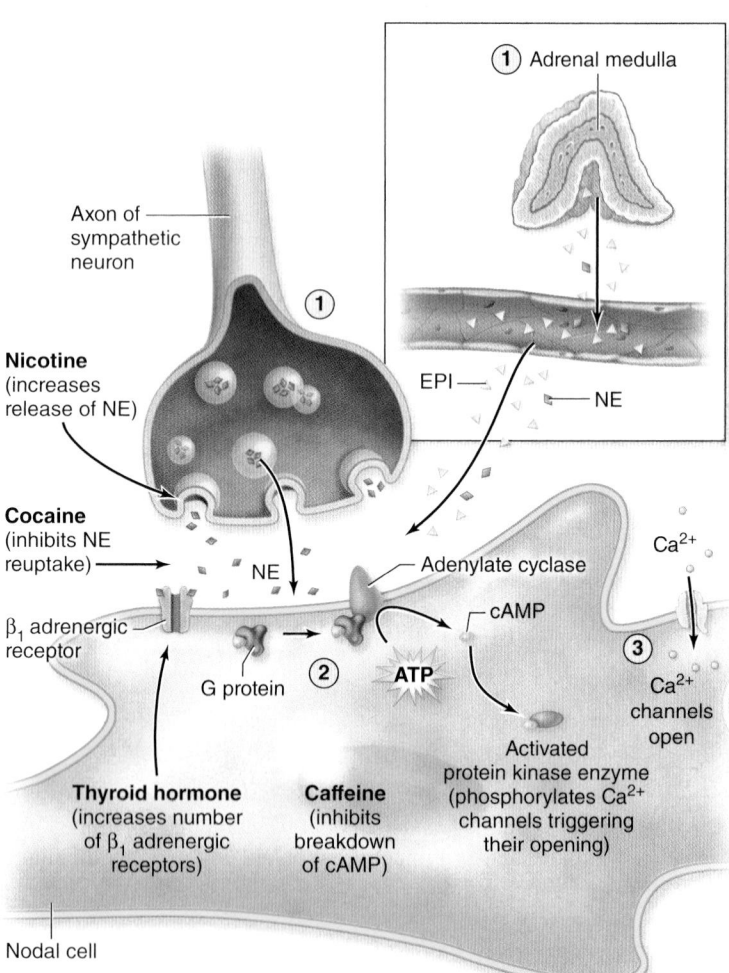

Figure 19.23 Sympathetic Innervation of Nodal Cells. Sympathetic axons release norepinephrine (NE), which acts as a positive chronotropic agent to increase heart rate. NE binds with receptors of nodal cells (step 1) to activate G protein and form a second messenger (step 2) that results in activation of a protein kinase enzyme and opening of Ca^{2+} channels (step 3). Both thyroid hormone (TH) and certain drugs increase heart rate at different steps within this pathway.

Thyroid hormone is also a positive chronotropic agent because it makes nodal cells more responsive to NE and EPI by increasing the number of β_1-adrenergic receptors (at step 1). Additionally, several drugs act on this pathway to increase heart rate. Nicotine and cocaine both increase the amount of NE present in the synaptic cleft (at step 1). Nicotine functions by increasing the release of NE and cocaine functions by inhibiting the reuptake of NE. Caffeine, in comparison, inhibits the breakdown of cAMP (at step 3). (Very high intake of caffeine in some "energy drinks" has resulted in several fatal heart attacks.)

In contrast, **negative chronotropic agents** decrease heart rate. Parasympathetic innervation is one of the most important. Parasympathetic axons release acetylcholine that binds to M2 muscarinic receptors, which are K^+ channels. These channels open, and K^+ moves down its concentration gradient to exit the cells. This loss of positive ion causes hyperpolarization (membrane potential becomes more negative) of nodal cells, and it takes a longer period of time for these cells to reach threshold—thus, the heart rate is slower. Beta-blocker drugs (which interfere with binding of norepinephrine and epinephrine to beta receptors) are another type of negative chronotropic agents and are used to treat high blood pressure.

Autonomic Reflexes

The ability of the autonomic nervous system to influence heart rate through the sympathetic and parasympathetic divisions is controlled through reflexes (see section 15.7b). The cardiac center receives and reflexively responds to sensory information from baroreceptors (blood pressure changes) and chemoreceptors (concerning carbon dioxide and hydrogen ion [H^+] levels) to alter parasympathetic and sympathetic nerve signals to the heart as needed to maintain homeostasis (see figure 19.14).

One specific reflex is the **atrial reflex,** also called the *Bainbridge reflex,* which protects the heart from overfilling. It is initiated when baroreceptors (stretch receptors) in the atrial walls are stimulated by an increase in venous return. Nerve signals are increased along sensory neurons to the cardioacceleratory center, resulting in an increase in nerve signals relayed along sympathetic axons to the heart. Heart rate increases so blood moves more quickly through the heart, thus decreasing atrial stretch. Arterial reflexes associated with regulating blood pressure that are initiated by baroreceptors and chemoreceptors in the aorta and carotid are described in section 20.6a.

WHAT DID YOU LEARN?

31 Distinguish the effect of a positive chronotropic agent and a negative chronotropic agent on heart rate, and give examples of each.

32 Describe the atrial reflex, which involves baroreceptors within the atria, the cardiac center, and the heart.

19.9c Variables That Influence Stroke Volume

LEARNING OBJECTIVES

5. List the three variables that may influence stroke volume.

6. Define each of the three variables, and describe the factors that influence each one and how it affects stroke volume.

Stroke volume (SV) is the volume of blood ejected per heartbeat **(figure 19.24).** Recall from the section on the cardiac cycle (see section 19.8b) that stroke volume is dependent upon the volume of blood that enters the heart at the end of heart relaxation (called the end-diastolic volume [EDV]). The typical EDV in a resting adult is approximately 130 mL. However, not all blood in either ventricle is normally ejected from the heart during ventricular contraction. The blood

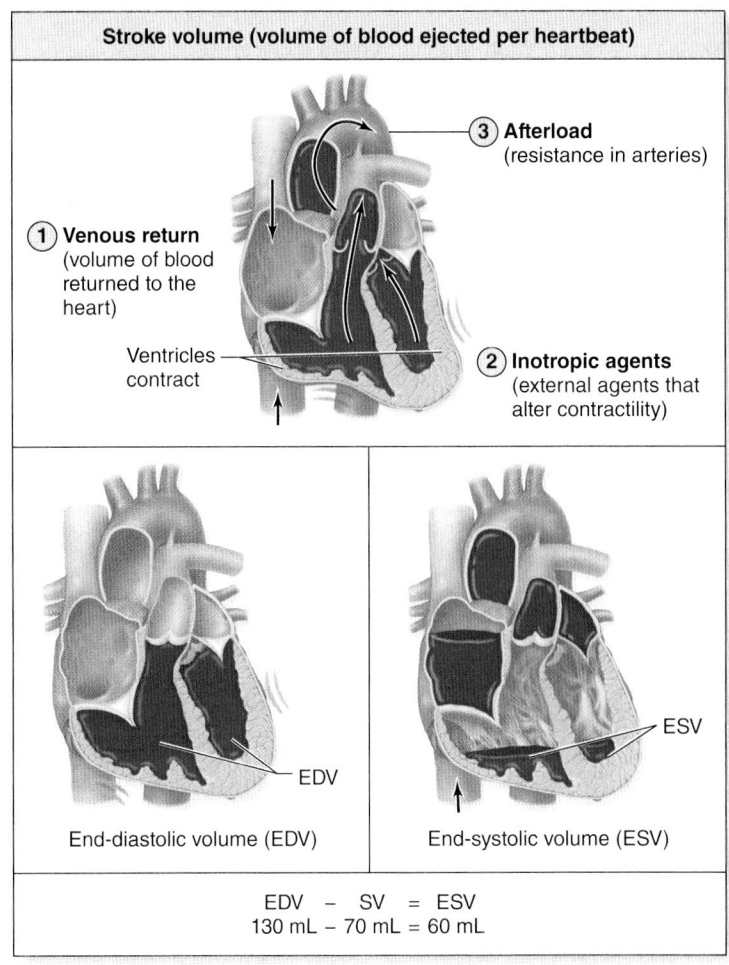

Stroke volume (volume of blood ejected per heartbeat)

① **Venous return** (volume of blood returned to the heart)

③ **Afterload** (resistance in arteries)

Ventricles contract

② **Inotropic agents** (external agents that alter contractility)

End-diastolic volume (EDV) — EDV

End-systolic volume (ESV) — ESV

$$EDV - SV = ESV$$
$$130 \text{ mL} - 70 \text{ mL} = 60 \text{ mL}$$

Figure 19.24 Relationship of EDV, ESV, and SV. Stroke volume is dependent upon venous return, presence of inotropic agents, and afterload. End-diastolic volume (EDV) is the volume of blood in a ventricle at the end of rest. End-systolic volume (ESV) is the volume of blood in a ventricle directly following a contraction. Stroke volume, which is the volume of blood pumped in one heartbeat, is equal to EDV–ESV.

remaining in a ventricle at the end of ventricular contraction is called the end-systolic volume (ESV). The typical ESV in an adult is 60 mL. Thus, stroke volume is the difference between EDV and ESV:

$$EDV - ESV = SV$$

$$130 \text{ mL} - 60 \text{ mL} = 70 \text{ mL}$$

These values for ESV, EDV, and SV are typical values for an adult. The specific volume of blood ejected as stroke volume varies and is influenced by several variables. These include (1) venous return, which is the amount of blood returned to the heart; (2) inotropic agents, which are external factors that alter the force of contraction of the myocardium; and (3) afterload, which is the resistance in the arteries to the ejection of blood from the heart.

Venous Return

Venous return is the volume of blood returned to the heart via the great veins and is directly related to stroke volume. Venous return determines the amount of blood in the ventricle at the end of rest immediately prior to contraction (i.e., end-diastolic volume, or EDV). This volume of blood, in turn, determines the preload on the heart. **Preload** is stretch of the heart wall due to the load to which a cardiac muscle is subjected before shortening.

The direct relationship of venous return and stroke volume can be explained by the **Frank-Starling law** of the heart (or simply *Starling's law*). This law essentially states that as the volume of blood entering the heart increases, there is greater stretch of the heart wall (or preload). This results in greater overlap of the thick and thin filaments in the sarcomeres of the cardiac muscle cells composing the myocardium, allowing formation of greater numbers of crossbridges. (Recall that cardiac muscle does not exhibit maximum overlap of thin and thick filaments at rest; see section 19.3e.) Consequently, a more forceful ventricular contraction is generated, and stroke volume increases.

As venous return decreases, there is less stretch of the heart wall (smaller preload), and this results in less overlap of the thick and thin filaments and fewer crossbridges form. Consequently, a less forceful ventricular contraction is generated, and stroke volume decreases.

What causes venous return, and thus preload, to either increase or decrease? Venous return is increased by either an increase in venous pressure or an increase in time to fill, and venous return may be decreased by a decrease in either of these two factors. You might find the analogy of filling a balloon helpful. The water pressure and the time to fill the balloon will determine how much water you can add to the balloon in a given period of time.

Venous return increases during exercise, for example, because of greater venous pressure. Veins are "squeezed" by skeletal muscles, which help return blood to the heart (see section 20.5a). Greater muscular movement increases the action of the skeletal muscle pump. During exercise, venous return can approximately double in comparison to its rate at rest. Venous return also increases with a slower heart rate. The slower heart rate allows a greater amount of time for blood to enter the heart. This is most noticeable in high-caliber athletes who have a very low resting heart rate.

In comparison, low blood volume (e.g., due to hemorrhage) or an abnormally high heart rate, decreases venous return. The result is a smaller end-diastolic volume and preload, and a smaller stroke volume.

Balanced ventricular output is primarily a function of the inherent ability of the heart to contract more forcefully in response to increased venous return. Consider, for example, that when you first start to exercise venous return into the right side of the heart increases, causing it to contract more forcefully and increasing stroke volume. Increased amount of blood then moves through the pulmonary circulation and returns to the left ventricle. The left ventricle, in turn, will experience greater stretch of its chamber wall and will contract more forcefully.

Inotropic Agents

Stroke volume is similar to heart rate in that it is altered by external factors. The primary external factors to increase and decrease heart rate come from autonomic innervation and varying levels of some hormones. These factors that change stroke volume are called **inotropic** (in′ō-trop′ik; *ino* = fiber) **agents.** Inotropic agents alter **contractility,** which is the force of contraction at a given stretch of the cardiac cells. An increase or decrease in the force of contraction is generally due to a change in the available Ca^{2+} in the sarcoplasm. Changes in Ca^{2+} concentration alter the number of crossbridges formed and thus the force of contraction generated.

A positive inotropic agent increases Ca^{2+} concentration, which results in formation of additional crossbridges. Positive inotropic agents include norepinephrine that is released from sympathetic axons and epinephrine and norepinephrine from the adrenal medulla. Thyroid hormone is also a positive inotropic agent because it increases the number of β_1-adrenergic receptors in the cardiac muscle cells. Certain drugs (e.g., digitalis) are positive inotropic agents used to treat abnormally low cardiac output that accompanies some heart conditions (e.g., congestive heart failure).

In comparison, a negative inotropic agent decreases contractility by decreasing available Ca^{2+} and fewer numbers of crossbridges are formed. Electrolyte imbalances (see table 25.2 and section 25.6) including an increase in either K^+ or H^+ act as negative inotropic agents. Certain drugs (e.g., nifedipine, a Ca^{2+} channel–blocking drug) are negative inotropic agents and are given to decrease cardiac output, typically in an effort to treat high blood pressure.

Afterload

Afterload is the resistance in arteries to the ejection of blood by the ventricles, and it represents the pressure that must be exceeded before blood is ejected from the chamber. Afterload generally becomes a consideration only in older individuals. Arteries typically develop atherosclerosis as we age. It is a condition in which plaque accumulates on the inner linings of a blood vessel. The smaller arterial lumen exerts greater resistance to the movement of blood into the arterial trunks, and stroke volume decreases. The relationship of these variables that influence stroke volume is integrated in **figure 19.25**.

 WHAT DID YOU LEARN?

② Which of the following increases stroke volume: (a) increased venous return, (b) increased Ca^{2+} in sarcoplasm, or (c) afterload? Explain.

19.9d Variables That Influence Cardiac Output

 **LEARNING OBJECTIVE**

7. Summarize the variables that influence cardiac output.

The factors that influence heart rate and stroke volume and ultimately cardiac output are integrated in a flowchart in **figure 19.26**. Note the following:

- **Heart rate.** An increase or decrease in heart rate is dependent upon chronotropic agents that influence the *conduction system.* These agents stimulate the SA node to change its firing rate or the AV node to alter the amount of delay.
- **Stroke volume.** An increase or decrease in stroke volume is generally due to changes in the *myocardium.* Venous return (which alters the stretch of the heart) and inotropic agents (which change the Ca^{2+} level in the sarcoplasm) influence the number of crossbridges, which alters the force of contraction.

INTEGRATE

CLINICAL VIEW

Bradycardia and Tachycardia

A persistently low resting heart rate in adults below 60 beats per minute is called **bradycardia.** Bradycardia is considered a normal change in highly trained athletes who engage in a sustained aerobic exercise program. Abnormal conditions that cause bradycardia include hypothyroidism, electrolyte imbalances, and congestive heart failure. In comparison, a persistently high resting heart rate above 100 beats per minute in adults is called **tachycardia.** Tachycardia is caused by abnormal conditions such as heart disease, fever, or anxiety.

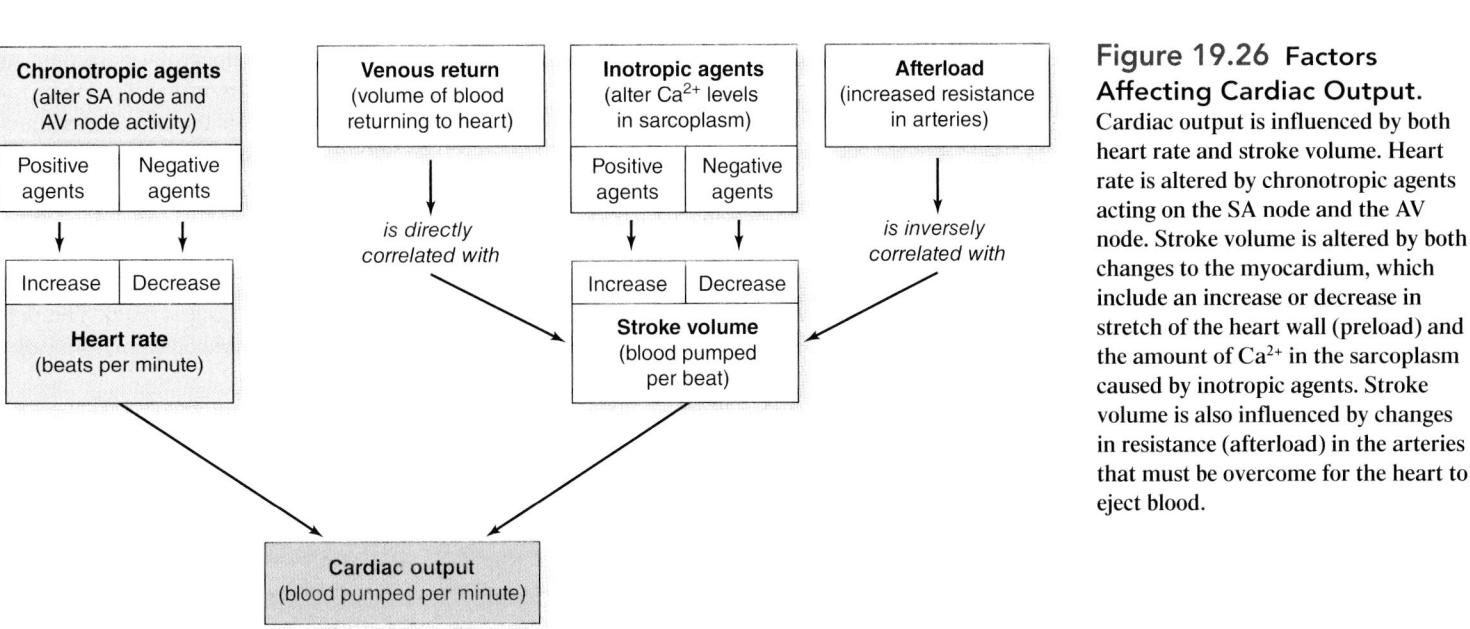

(a) Venous return

Volume of blood returned to the heart per unit time

Increased venous return (occurs with greater venous pressure or slower heart rate)

↓

Increases stretch of the heart wall (preload), which results in greater overlap of thick and thin filaments within the sarcomeres of the myocardium

↓

Additional crossbridges form, and ventricles contract with greater force

↓

Stroke volume increases

The opposite is seen with smaller venous return (e.g., occurs with hemorrhage or extremely rapid heart rate)

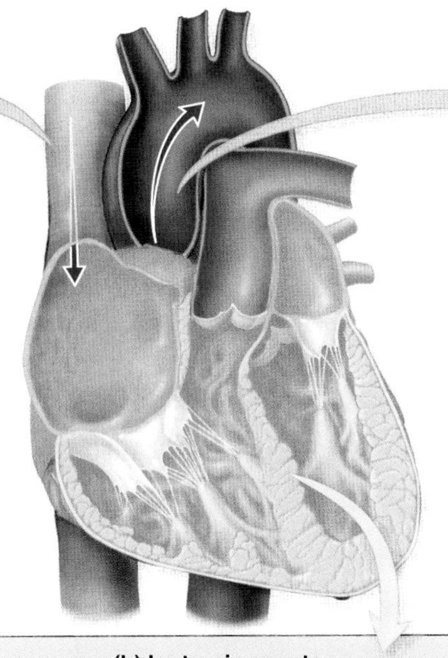

(c) Afterload

Resistance in arteries to ejection of blood

Atherosclerosis, which is deposition of plaque on the inner lining of arteries, is typically only a factor as we age

↓

Arteries become more narrow in diameter

↓

Increases the resistance to pump blood into the arteries

↓

Stroke volume decreases

(b) Inotropic agents

Substances that act on the myocardium to alter contractility

Positive inotropic agents (e.g., stimulation by sympathetic nervous system)

↓

Increased Ca^{2+} levels in the sarcoplasm results in greater binding of Ca^{2+} to troponin of thin filaments within sarcomeres of the myocardium

↓

Additional crossbridges form, and ventricles contract with greater force

↓

Stroke volume increases

The opposite is seen with negative inotropic agents (e.g., calcium channel blockers)

Figure 19.25 Variables That Influence Stroke Volume. Three variables influence stroke volume, which is the amount of blood ejected from the heart per beat. These include (*a*) venous return, (*b*) inotropic agents, and (*c*) afterload.

Chronotropic agents (alter SA node and AV node activity)

Positive agents	Negative agents
Increase	Decrease

Heart rate (beats per minute)

Venous return (volume of blood returning to heart)

is directly correlated with

Inotropic agents (alter Ca^{2+} levels in sarcoplasm)

Positive agents	Negative agents
Increase	Decrease

Stroke volume (blood pumped per beat)

Afterload (increased resistance in arteries)

is inversely correlated with

Cardiac output (blood pumped per minute)

Figure 19.26 Factors Affecting Cardiac Output. Cardiac output is influenced by both heart rate and stroke volume. Heart rate is altered by chronotropic agents acting on the SA node and the AV node. Stroke volume is altered by both changes to the myocardium, which include an increase or decrease in stretch of the heart wall (preload) and the amount of Ca^{2+} in the sarcoplasm caused by inotropic agents. Stroke volume is also influenced by changes in resistance (afterload) in the arteries that must be overcome for the heart to eject blood.

The only exception is afterload, which reflects increased resistance in arteries, making it more difficult for the heart to pump blood. Afterload is generally a factor only as we age.

- **Cardiac output.** Both heart rate and stroke volume are directly related to cardiac output. When both heart rate and stroke volume increase, cardiac output increases. In contrast, when both heart rate and stroke volume decrease, cardiac output decreases. It is not possible to predict the net effect on cardiac output (i.e., whether it will increase or decrease) if the heart rate and stroke volume change in the opposite direction to one another (e.g., stroke volume decreases from blood loss, and heart rate increases in an attempt to maintain cardiac output). The net effect will be determined by the relative change to both heart rate and stroke volume.

 WHAT DID YOU LEARN?

34 If both heart rate and stroke volume increase, cardiac output (a) stays the same, (b) increases, or (c) decreases? Thus, is the relationship between these two variables and cardiac output direct or inverse?

19.10 Development of the Heart

LEARNING OBJECTIVES

1. Explain how postnatal heart structures develop from the primitive heart tube.
2. Describe septal defects that may occur during development.

Development of the heart commences in the third week, when the embryo becomes too large to receive its nutrients through diffusion alone. The steps involved in heart development are complex, because the heart must begin working before its development is complete (**figure 19.27**).

By day 19 (middle of week 3), two **heart tubes** (*endocardial tubes*) form from mesoderm in the embryo. By day 21, these paired tubes fuse, forming a single primitive heart tube (figure 19.27*a*). By day 22, the heart begins to beat, and later in the fourth week, this single heart tube bends and folds upon itself to form the external heart shape (figure 19.27*b*). This tube develops the following named expansions that give rise to postnatal heart structures (listed from inferior to superior): **sinus venosus, primitive atrium, primitive ventricle,** and **bulbus cordis** (bŭl′bŭs kōr′dis). The sinus venosus and primitive atrium form parts of the left and right atria. The primitive ventricle forms most of the left ventricle. The bulbus cordis may be further subdivided into a **trabeculated part of the right ventricle,** which forms most of the right ventricle; the **conus cordis,** which forms the outflow tracts for the ventricles; and the **truncus arteriosus,** which forms the ascending aorta and pulmonary trunk (figure 19.27*c*).

The next major steps in heart development occur during weeks 5–8, when the single heart tube becomes partitioned into four chambers (two atria and two ventricles), and the great vessels form. This partitioning is complex, and errors in development lead to many of the more common congenital heart malformations.

The common atrium is subdivided into a left and right atrium by an interatrial septum, which consists of two parts (**septum primum** and **septum secundum**) that partially overlap (figure 19.27*d*). These two parts connect to tissue masses called endocardial cushions. An opening in the septum secundum (which is covered by the septum primum) is called the foramen ovale. Because the embryonic lungs are not functional, much of the blood is shunted from the right atrium to the left atrium by traveling through the foramen ovale and pushing the septum primum into the left atrium. Blood cannot flow back from the left atrium to the right atrium, because the septum primum's movement is stopped when it comes against the septum secundum. Thus, the septum primum acts as a unidirectional flutter valve.

When the baby is born and the lungs are fully functional, the blood from the left atrium pushes the septum primum and secundum together, closing the interatrial septum. The only remnant of the embryonic opening is an oval-shaped depression in the interatrial septum called the **fossa ovalis.**

Left and right ventricles are partitioned by an **interventricular septum** that grows superiorly from the floor of the ventricles. The AV valves, papillary muscles, and tendinous cords all form from portions of the ventricular walls as well. The superior part of the interventricular septum develops from superior structures (the aorticopulmonary septum, which is a spiral-shaped mass that also subdivides the truncus arteriosus into the pulmonary trunk and the ascending aorta).

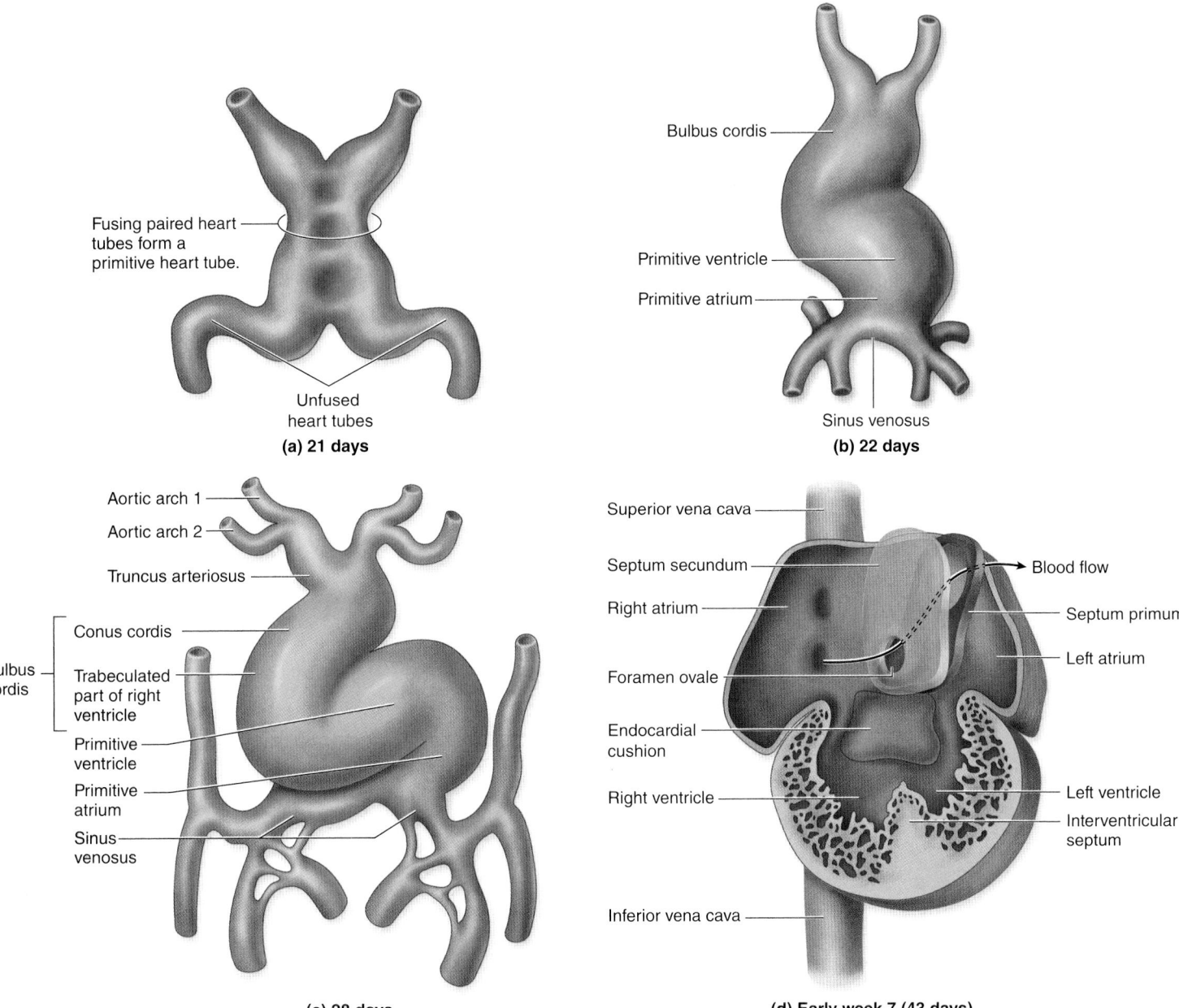

(a) 21 days

Fusing paired heart tubes form a primitive heart tube.

Unfused heart tubes

(b) 22 days

Bulbus cordis

Primitive ventricle

Primitive atrium

Sinus venosus

(c) 28 days

Aortic arch 1

Aortic arch 2

Truncus arteriosus

Conus cordis

Bulbus cordis

Trabeculated part of right ventricle

Primitive ventricle

Primitive atrium

Sinus venosus

(d) Early week 7 (43 days)

Superior vena cava

Septum secundum

Right atrium

Foramen ovale

Endocardial cushion

Right ventricle

Inferior vena cava

Blood flow

Septum primum

Left atrium

Left ventricle

Interventricular septum

Figure 19.27 Development of the Heart. The heart develops from mesoderm. By day 19, paired heart tubes are present in the cardiogenic region of the embryo. (*a*) These paired tubes fuse by day 21. (*b*) The single heart tube bends and folds upon itself, beginning on day 22. (*c*) By day 28, the heart tube is S-shaped. (*d*) By early week 7, the interatrial septum is formed by two overlapping septa. The foramen ovale is a passageway that detours blood away from the pulmonary circulation into the systemic circulation prior to birth.

Many congenital heart malformations result from incomplete or faulty development during the early weeks of development. For example, in an **atrial septal defect,** the postnatal heart still has an opening between the left and right atria. Thus, blood from the left atrium (the higher-pressure system) is shunted to the right atrium (the lower-pressure system). This can lead to enlargement of the right side of the heart.

Ventricular septal defects can occur if the interventricular septum is incompletely formed. A common malformation called **tetralogy of Fallot** occurs when the aorticopulmonary septum divides the truncus arteriosus unevenly. As a result, the patient has a ventricular septal defect, a very narrow pulmonary trunk (pulmonary stenosis), an aorta that overlaps both the left and right ventricles, and an enlargement of the right ventricle (right ventricular hypertrophy).

WHAT DID YOU LEARN?

35 What would be the path of blood flow through the heart if the foramen ovale did not close shortly after birth?

CHAPTER SUMMARY

- The cardiovascular system is composed of the heart and blood vessels organized into the pulmonary circulation and systemic circulation.

19.1a General Function 732

- The function of the cardiovascular system is to transport substances throughout the body and provide adequate perfusion to all tissues.

19.1b Overview of Components 733

- The primary types of blood vessels include arteries, capillaries, and veins.

- The heart acts as two "side-by-side" pumps with the right side composed of a right atrium and right ventricle and the left side composed of a left atrium and left ventricle. Valves within the heart help ensure a one-way flow of blood.

- The circulation routes include the pulmonary circulation and systemic circulation.

- The heart is a located in the thoracic cavity and is enclosed within a fibroserous sac.

19.2a Location and Position of the Heart 738

- The heart is located left of the body midline, posterior to the sternum in the mediastinum with its apex projecting anteroinferiorly.

19.2b Characteristics of the Pericardium 738

- The pericardium that encloses the heart includes the pericardial sac, which has an outer fibrous pericardium and an inner parietal layer of serous pericardium, and a visceral layer of serous pericardium (epicardium) that forms the outer layer of the heart wall.

- The pericardial cavity is a potential space between the layers of the serous pericardium that contain serous fluid, which is produced by the serous membranes and lubricates the surfaces to reduce friction.

- The heart is a relatively small, conical-shaped organ, which is approximately the size of a human fist.

19.3a Superficial Features of the Heart 739

- The coronary sulcus and interventricular sulci are visible from the external surface of the heart. These grooves house coronary vessels.

- The right side of the heart is more visible from the anterior view, and the left side of the heart is more visible from the posterior view.

19.3b Layers of the Heart Wall 742

- The heart wall has an epicardium (visceral layer of serous pericardium), myocardium, and endocardium.

19.3c Heart Chambers 743

- Atria are separated by an interatrial septum, and ventricles are separated by an interventricular septum.

- The four heart chambers include the right atrium, right ventricle, left atrium, and left ventricle.

19.3d Heart Valves 743

- The atrioventricular valves are located between an atrium and ventricle, and the semilunar valves are between a ventricle and an arterial trunk (pulmonary trunk or aorta).

19.3e Microscopic Structure of Cardiac Muscle 746

- Cardiac muscle cells are small, have one or two centrally located nuclei, and are branched.

- Intercalated discs, which are composed of desmosomes and gap junctions, tightly link the cardiac muscle cells together and permit the passage of action potentials, respectively.

- Cardiac muscle cells rely almost exclusively on aerobic cellular respiration for supplying ATP, which makes them susceptible to failure if the oxygen supply is inadequate.

19.3f Fibrous Skeleton of the Heart 747

- The fibrous skeleton provides an attachment site for heart valves and cardiac muscle, and prevents action potentials from spreading between the atria and ventricles except through the AV node.

- The coronary circulation is the circulation of blood to and from the heart wall.

19.4a Coronary Arteries 748

- Coronary arteries supply blood to the heart wall and include the left and right coronary arteries that branch off the ascending aorta.

19.4b Coronary Veins 749

- Venous return is through the cardiac veins into the coronary sinus, which drains into the right atrium of the heart.

- Heart activity is initiated by the conduction system and altered by the autonomic nervous system.

19.5a The Heart's Conduction System 750

- Stimulation of the heart involves initiation of an action potential at the SA node and its transmission through the conduction system.

- The conducting system includes sinoatrial (SA) node, atrioventricular (AV) node, AV bundle, and Purkinje fibers, which are composed of specialized cardiac cells that initiate and conduct action potentials, resulting in a heartbeat.

19.5b Innervation of the Heart 750

- Parasympathetic innervation comes from the cardioinhibitory center to decrease the heart rate. Sympathetic innervation comes from the cardioaccelerory center to increase both the heart rate and force of contraction.

CHALLENGE YOURSELF

Do You Know the Basics?

_____ 1. Which of the following is the correct circulatory sequence for blood to pass through part of the heart?

    a. right atrium ⟶ left AV valve ⟶ right ventricle ⟶ pulmonary semilunar valve

    b. right atrium ⟶ right AV valve ⟶ right ventricle ⟶ pulmonary semilunar valve

    c. left atrium ⟶ right AV valve ⟶ left ventricle ⟶ aortic semilunar valve

    d. left atrium ⟶ left AV valve ⟶ left ventricle ⟶ pulmonary semilunar valve

_____ 2. The pericardial cavity is located between the

    a. fibrous pericardium and the parietal layer of the serous pericardium.

    b. parietal and visceral layers of the serous pericardium.

    c. visceral layer of the serous pericardium and the epicardium.

    d. myocardium and the visceral layer of the serous pericardium.

_____ 3. How is blood prevented from backflowing from the pulmonary trunk into the right ventricle?

    a. closing of the right AV valves

    b. opening of the pulmonary semilunar valve

    c. contraction of the right atrium

    d. closing of the pulmonary semilunar valve

_____ 4. Venous blood draining from the heart wall enters the right atrium through the

    a. coronary sinus.

    b. inferior vena cava.

    c. pulmonary veins.

    d. superior vena cava.

_____ 5. Calcium channels in the nodal cells function to

    a. cause depolarization and initiate the cardiac action potential.

    b. assure excess calcium can leave the cell.

    c. bring the cell quickly to its resting membrane potential.

    d. sustain contraction of the cell.

_____ 6. Action potentials are spread rapidly between cardiac muscle cells by

    a. sarcomeres.

    b. intercalated discs.

    c. chemical neurotransmitters.

    d. the fibrous skeleton.

_____ 7. Why is it necessary to stimulate papillary muscles in the ventricle slightly earlier than the rest of the ventricular wall myocardium?

    a. to assure rapid conduction speed of the action potential

    b. to pull on AV valve cusps to prevent backflow

    c. to assure blood will surge toward the semilunar valves

    d. to assure coordinated contraction of the ventricular myocardium

_____ 8. Preload is a measure of

    a. stretch of heart chamber prior to contraction.

    b. contraction rate in cardiac muscle.

    c. reduced filling during exercise.

    d. autonomic nervous system stimulation of the heart.

_____ 9. All of the following occur when the ventricles contract _except_

    a. the AV valves close.

    b. blood is ejected into the aorta.

    c. the semilunar valves open.

    d. blood from the pulmonary trunk enters the atria.

_____ 10. What occurs during the atrial reflex?

    a. Atria slow their filling rate.

    b. There is a decrease in heart rate due to an increase in blood pressure.

    c. The SA node rhythm decreases.

    d. An increase in heart rate occurs in response to an increase in blood volume within the atria.

11. Describe and compare the differences between the pulmonary and systemic circulations.

12. Compare the structure, location, and function of the parietal and visceral layers of the serous pericardium.

13. Why are the tendinous cords required for the proper functioning of the AV valves?

14. Explain why the walls of the atria are thinner than those of the ventricles, and why the wall of the right ventricle are relatively thin when compared to the wall of the left ventricle.

15. Describe the structure and function of intercalated discs in cardiac muscle tissue.

16. Explain the general location and function of coronary vessels.

17. Describe the functional differences in the effects of the sympathetic and parasympathetic divisions of the autonomic nervous system on the activity of the heart.

18. Provide an overview of the two events for cardiac muscle contraction that include the conduction system and cardiac muscle cells.

19. List the five events of the cardiac cycle, and indicate for each if the atria are contracted or relaxed, if the ventricles are contracted or relaxed, if the AV valves are open or closed, and if the semilunar valves are open or closed.

20. Define cardiac output, and explain how it is influenced by both heart rate and stroke volume.

Can You Apply What You've Learned?

Use the following paragraph to answer questions 1 and 2.

A young man was doing some vigorous exercise when suddenly he felt chest pains just before he passed out. Upon being revived, he was taken to the hospital for examination. The medical personnel ran standard tests and ECG tracings, then said it would fine for him to resume his normal activities and workouts.

1. Why might the increase in heart rate associated with a vigorous exercise program be the cause of his becoming unconscious?

   a. increase in blood pressure throughout the body

   b. failure of the conduction system

   c. increase in return volume of blood to the heart

   d. coronary blood flow reduction due to tachycardia

2. Which of the following treatments would *not* be used to rule out that a myocardial infarction had occurred?

   a. doing an ECG

   b. performing constant blood pressure monitoring

   c. measuring blood levels of creatine kinase released from damaged heart muscle

   d. measuring blood levels of troponin released from damaged heart muscle

3. Calcium channel blockers are drugs that are given to cause

   a. an increase in the volume of blood being pumped.

   b. the afterload to be increased.

   c. contractility to decrease.

   d. preload to decrease.

4. A patient has been in a serious car accident and is hemorrhaging. Which of the following changes would be seen in this patient?

   a. a decrease in stroke volume

   b. an increase in heart rate

   c. a possible decrease in cardiac output

   d. All of these are correct.

5. During surgery, the right vagus nerve was accidently cut. Explain the effect to the heart rate.

   a. There is no change to the heart rate because the vagus nerve does not innervate the heart.

   b. The heart rate increases to the inherent rhythm of SA nodal cells.

   c. The heart stops beating, and heart rate becomes zero.

   d. The heart rate decreases to the inherent rhythm of SA nodal cells.

Can You Synthesize What You've Learned?

1. A young couple that you are friends with gave birth to a new baby. They were told by their physician that the foramen ovale did not close between the right atrium and left atrium (i.e., that their baby has "a hole in her heart"). They know that you are a nurse and have come to you to help them to understand what is going on. Explain both the normal flow of blood through the heart and the flow with the foramen ovale still open. Include in your description the concept of oxygenated and deoxygenated blood.

2. Josephine is a 55-year-old overweight woman who has a poor diet and does not exercise. One day while walking briskly, she experienced pain in her chest and down her left arm. Her doctor told her that she was experiencing angina due to heart problems. Josephine asks you to explain what causes angina, and why she was feeling pain in her arm even though the problem was with her heart. What do you tell her?

3. Your grandfather was told that his SA node (pacemaker) has stopped functioning. Explain how his heart is still beating at a rate of 40 to 50 times per minute. Are the atria stimulated to contract? Explain.

INTEGRATE

ONLINE STUDY TOOLS  **connect** |ANATOMY & PHYSIOLOGY  LEARNSMART AP|R

The following study aids may be accessed through Connect.

Concept Overview Interactive: Figure 19.22: Changes Associated with a Cardiac Cycle

Clinical Case Study: A Middle-Aged Woman with Back Pain

Interactive Questions: This chapter's content is served up in a number of multimedia question formats for student study.

LearnSmart: Topics and terminology include introduction to the cardiovascular system; location of the heart and the pericardium; heart anatomy; coronary vessels: blood supply of the heart wall; anatomic structures controlling heart activity; stimulation of the

heart; the cardiac cycle; cardiac output; development of the heart

Anatomy & Physiology Revealed: Topics include cardiovascular system overview; heart fly-through; blood flow through the heart; heart vasculature; internal features of the heart; cardiac muscle; cardiac cycle

Animations: Topics include action potentials in the sinoatrial node; cardiac cycle; mechanical events of the cardiac cycle; conduction system of the heart; cardiac excitation-contraction (EC) coupling

chapter
20

Cardiovascular System: Vessels and Circulation

INTEGRATE

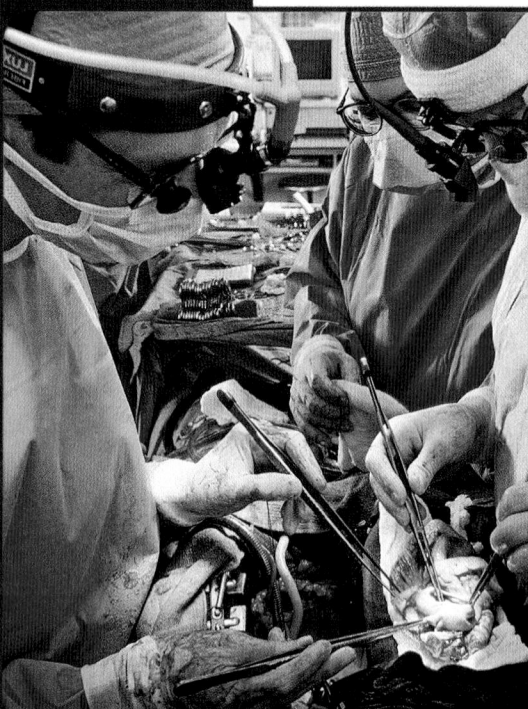

CAREER PATH
Cardiovascular Surgeon

A cardiovascular surgeon is a physician who specializes in diagnosing and treating disorders of the heart and blood vessels. One of the most common blood vessel disorders is atherosclerosis, where plaques develop in the vessels and cause lumen narrowing. If these vessels become completely blocked, the tissues they supply may die. Cardiovascular surgeons are trained to either remove blockages in vessels or add blood vessel bypasses to restore blood supply to an affected organ. They use medical imaging to help direct catheters through blood vessels and insert stents in blocked vessels.

Anatomy & Physiology | REVEALED®
aprevealed.com

Module 9: Cardiovascular System

Many of us are aware that high blood pressure (hypertension) is not healthy. Hypertension can damage blood vessels and lead to cardiovascular disease. But a minimal amount of blood pressure is needed to effectively pump blood (and the nutrients and respiratory gases it transports) throughout the body and deliver these materials to the body tissues. If blood pressure drops too low, the body will be deprived of nutrients and death may occur. Multiple body systems ➡

(including the endocrine, nervous, and urinary systems) participate in maintaining sufficient blood pressure to ensure that tissues are adequately perfused (supplied with blood) for the activity at hand.

We begin this chapter by describing the general structure and function of blood vessels, blood flow velocity, processes of capillary exchange, and factors that influence both blood flow and blood pressure in vessels. Pulmonary circulation and the major arteries and veins of systemic circulation are then described, and we conclude with a comparison of fetal versus adult circulation.

20.1 Structure and Function of Blood Vessels

Blood vessels are classified into three primary types based on function. **Arteries** (ar′ter-ē) convey blood away from the heart to the capillaries. **Capillaries** (kap′i-lar-ē; *capillaris* = relating to hair) are microscopic, relatively porous blood vessels for the exchange of substances between blood and tissues. **Veins** (vān) drain blood from the capillaries, transporting it back to the heart.

20.1a General Structure of Vessels

LEARNING OBJECTIVES

1. Describe the three tunics common to most vessels.
2. Explain the distinguishing features of the tunics found in arteries, capillaries, and veins.

Vessel walls are composed of layers called **tunics** (tū′nik; *tunica* = coat). The tunics surround the **lumen** (lū′men), or inside space of the vessel, through which blood flows. The three tunics are the tunica intima, tunica media, and tunica externa **(figure 20.1).**

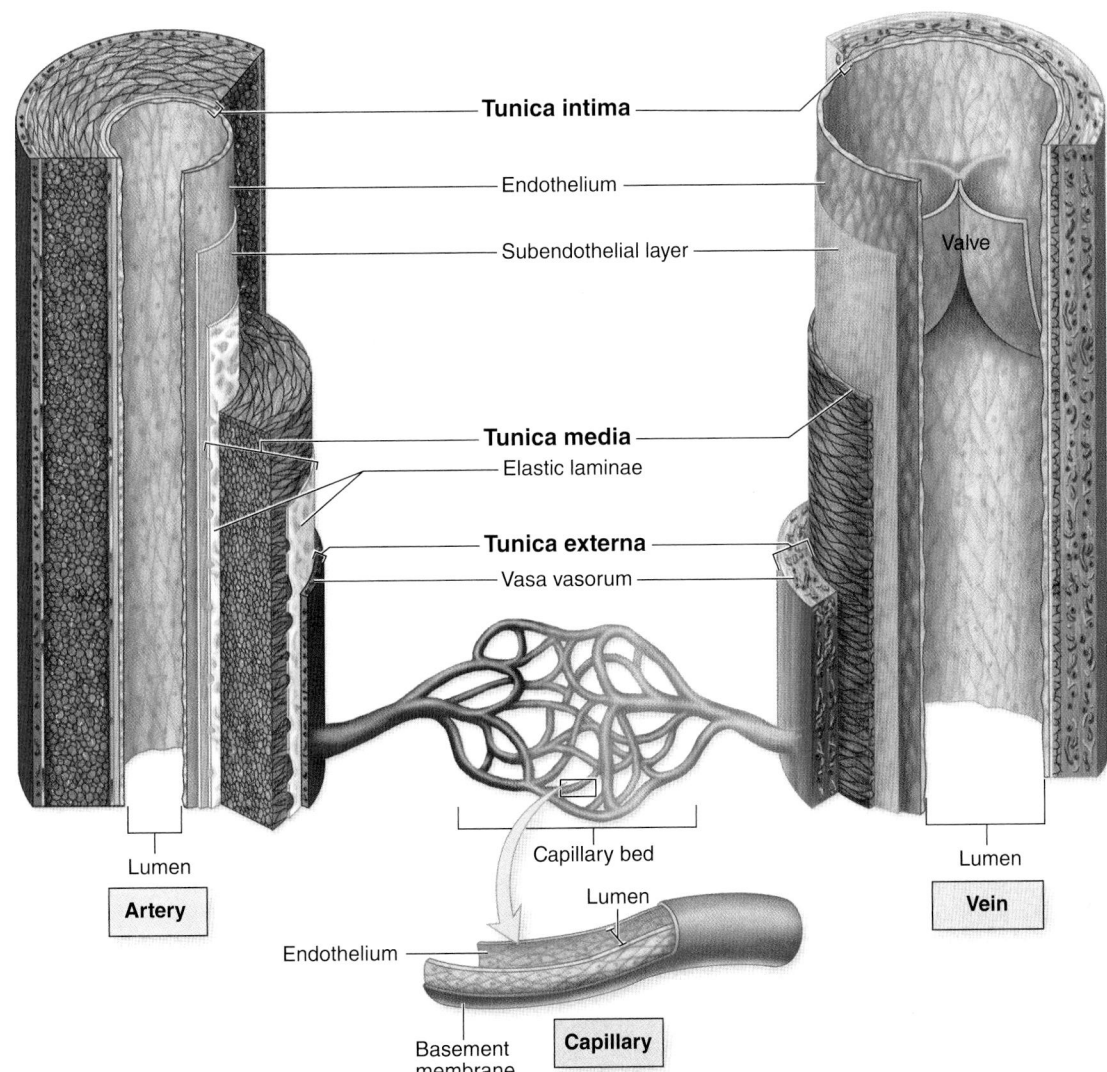

Figure 20.1 Walls of an Artery, a Capillary, and a Vein. Both arteries and veins have a tunica intima, tunica media, and tunica externa. However, an artery has a thicker tunica media and a relatively smaller lumen, whereas a vein's thickest layer is the tunica externa, and it has a larger lumen. Larger veins also have valves. Capillaries typically have only a tunica intima (basement membrane and endothelium), but they do not have a subendothelial layer. (Note: Different types of arteries vary in the distribution of elastic fibers within the tunica media. The array of elastic fibers depicted here, which are labeled elastic laminae, is the arrangement within a muscular artery.)

Arteries, capillaries, and veins differ in both the specific composition of their tunics and their functions.

Tunics

The innermost layer of a blood vessel wall is the **tunica intima** (tū′ni-kă in′tim-mă; *intimus* = inmost),or *tunica interna*. It is composed of an **endothelium** (a simple squamous epithelium, see section 5.1c)that faces the blood vessel lumen and a thin subendothelial layer of areolar connective tissue. The endothelium both provides a smooth surface as the blood moves through the lumen of the blood vessel and releases substances (e.g., nitric oxide, endothelin) to regulate contraction and relaxation of smooth muscle within the tunica media. Recall that the endothelium is continuous with the endocardium, which is the inner lining of the heart (see section 19.3b).

The **tunica media** (mē′dē-ă; *medius* = middle) is the middle layer of the vessel wall. It is composed predominantly of circularly arranged layers of smooth muscle cells that are supported by elastic fibers. Contraction of smooth muscle in the tunica media results in **vasoconstriction** (vā′sō-kon-strik′shŭn), or narrowing of the blood vessel lumen; relaxation of the smooth muscle causes **vasodilation** (vā′sō-dī-lā′shŭn), or widening of the blood vessel lumen.

The **tunica externa** (eks-ter′nă; *externe* = outside), or *tunica adventitia*, is the outermost layer of the blood vessel wall. It is composed of areolar connective tissue that contains elastic and collagen fibers. The tunica externa helps anchor the vessel to other structures. Very large blood vessels require their own blood supply to the tunica externa in the form of a network of small arteries called the **vasa vasorum** (vā′să vā-sōr′ŭm; vessels of vessels). The vasa vasorum extend through the tunica externa.

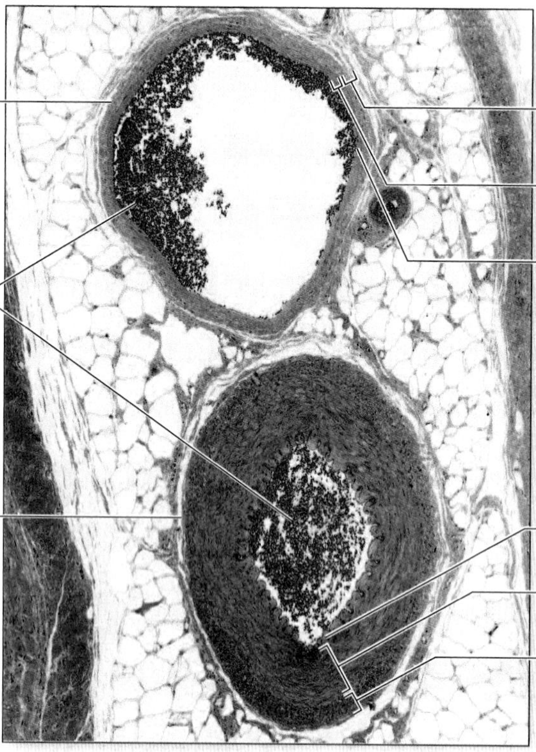

Comparison of the Different Vessel Types

Arteries and veins that supply the same body region and tend to lie next to one another are called **companion vessels. Figure 20.2** is a histologic image of a companion artery and vein. Compared to their venous companions,arteries have a thicker tunica media, a narrower lumen, and more elastic and collagen fibers. These differences mean that arterial walls can spring back to shape, and are more resilient and resistant to changes in blood pressure than are veins. In addition, an artery remains patent (open) even without blood in it.

In contrast, veins have a thicker tunica externa, a wider lumen, and less elastic and collagen fibers than a companion artery. The wall of a vein is typically collapsed if no blood is in the vessel. The characteristics of arteries and veins are summarized in **table 20.1**.

Figure 20.2 Microscopic Comparison of an Artery and Vein.
The artery generally maintains its shape in tissues as a result of its thicker wall. The wall of a companion vein often collapses when it is not filled with blood (although in this photo, the vein is not collapsed). AP|R

Table 20.1	Comparison of Companion Arteries and Veins	
Characteristic	**Artery**	**Vein**
Lumen Diameter	Narrower than vein lumen	Wider than artery lumen
General Wall Thickness	Thicker than vein	Thinner than artery
Cross-Sectional Shape	Cross-sectional shape retained without blood in vessel	Cross-sectional shape tends to flatten out (collapse) without blood in vessel
Thickest Tunic	Tunica media	Tunica externa
Elastic and Collagen Fibers in Tunics	More than in vein	Less than in artery
Valves	None	Present in most veins
Blood Pressure Range	Higher than in veins (100 mm Hg in larger arteries to 40 mm Hg in smaller arterioles)	Lower than in arteries (20 mm Hg in venules to 0 mm Hg in the inferior vena cava)
Blood Flow	Transports blood away from heart to the body	Transports blood from the body to the heart
Blood Oxygen Levels	Systemic arteries carry blood high in O_2 Pulmonary arteries carry blood low in O_2	Systemic veins carry blood low in O_2 Pulmonary veins carry blood high in O_2

Capillaries are unique in that they contain only the tunica intima composed of an endothelium and its underlying basement membrane; there is no subendothelial layer. Having this thin barrier allows for rapid gas and nutrient exchange between the blood in capillaries and the tissues.

WHAT DID YOU LEARN?

❶ What are three differences in anatomic structure between arteries and veins?

20.1b Arteries

LEARNING OBJECTIVE

3. Distinguish between elastic arteries, muscular arteries, and arterioles.

Arteries progressively branch into smaller vessels as they extend from the heart to the capillaries. There is both a corresponding decrease in lumen diameter and a change in the composition of the tunic wall that includes both a decrease in the relative amount of elastic fibers and an increase in the relative amount of smooth muscle. Arteries may be classified into three basic types: elastic arteries, muscular arteries, and arterioles (**figures 20.3** and **20.4**).

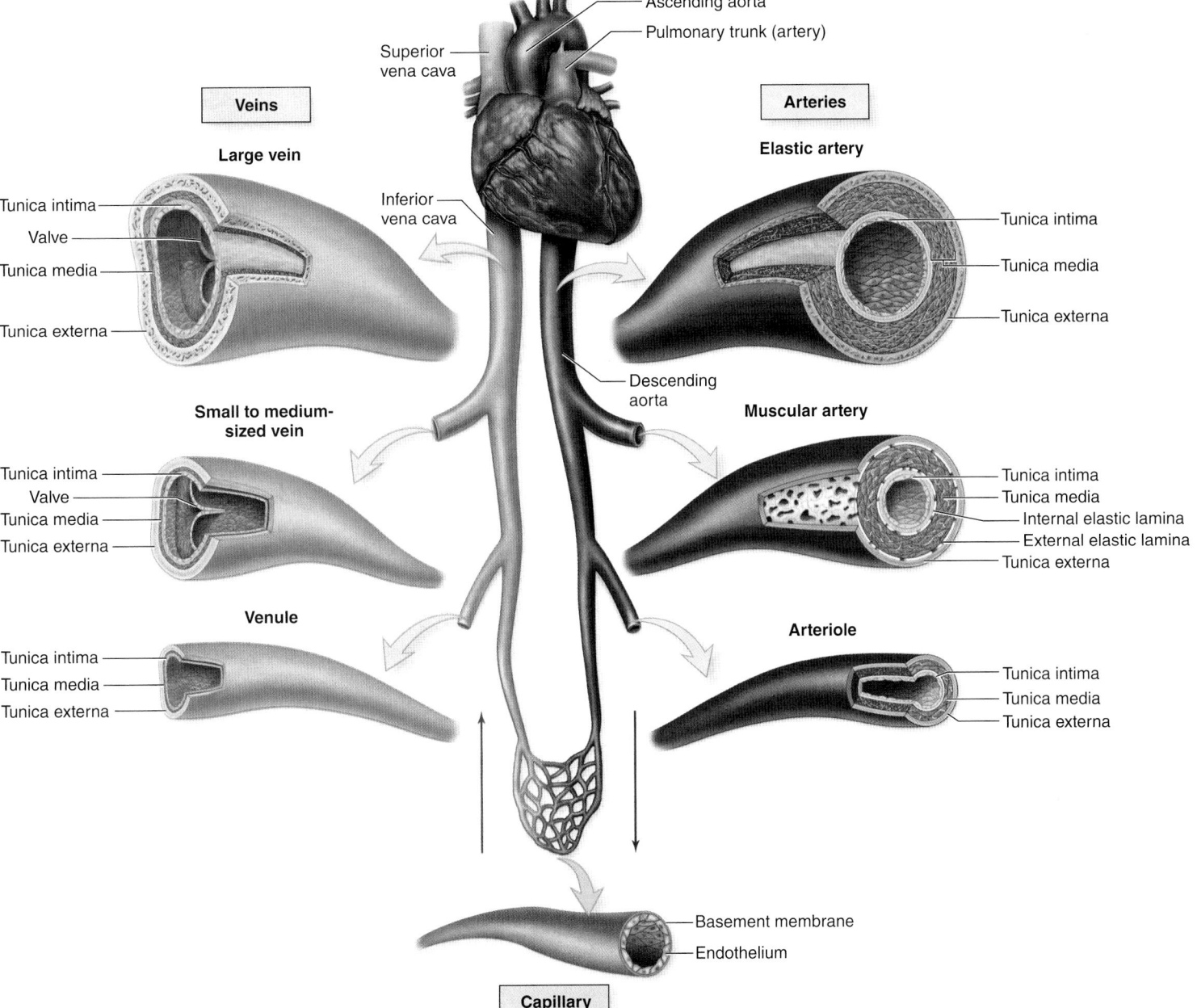

Figure 20.3 Comparison of Companion Vessels. The thickness of the tunics in companion arteries and veins differs, depending upon the size of the vessels.

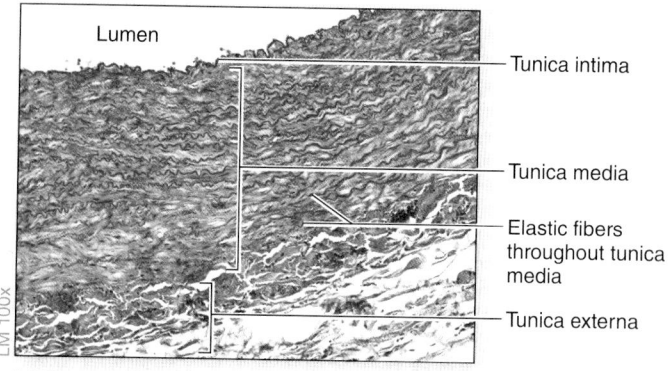

(a) Elastic artery

Lumen

Tunica intima

Tunica media

Elastic fibers throughout tunica media

Tunica externa

LM 100x

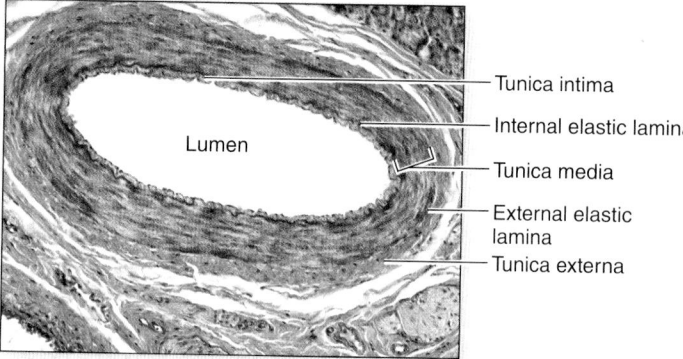

(b) Muscular artery

Lumen

Tunica intima

Internal elastic lamina

Tunica media

External elastic lamina

Tunica externa

LM 100x

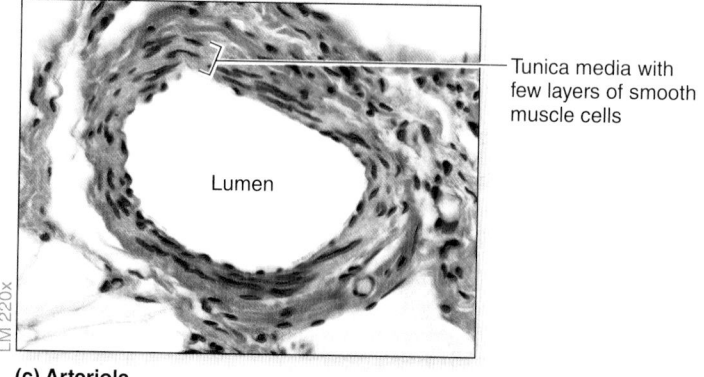

(c) Arteriole

Lumen

Tunica media with few layers of smooth muscle cells

LM 220x

Figure 20.4 Types of Arteries. Light micrograph images of three types of arteries. (*a*) Elastic arteries have vast arrays of elastic fibers in their tunica media. (*b*) Muscular arteries have a tunica media with numerous layers of smooth muscle flanked by elastic laminae. (*c*) Arterioles typically have a tunica media composed of six or fewer layers of smooth muscle cells.

Elastic Arteries

Elastic arteries are the largest arteries, with diameters ranging from 2.5 to 1 centimeter. They are also called *conducting arteries* because they conduct blood—from the heart to the smaller muscular arteries. As their name suggests, these arteries have a large proportion of elastic fibers; these are present throughout all three tunics, especially in the tunica media. The abundant elastic fibers allow the artery to stretch and accommodate the blood when a heart ventricle ejects blood into it during ventricular systole (contraction) and then recoil, which helps propel the blood through the arteries during ventricular diastole (relaxation).

The largest arteries close to the heart (e.g., aorta, pulmonary trunk, brachiocephalic, common carotid, subclavian) and the common iliac arteries are examples of elastic arteries (see figure 20.19*a*). Elastic arteries branch into muscular arteries.

INTEGRATE

CLINICAL VIEW
Atherosclerosis

Atherosclerosis (ath´er-ō-skler-ō´sis; *athere* = gruel, *sclerosis* = hardening) is a progressive disease of the elastic and muscular arteries. It is characterized by the presence of an **atheroma** (or *atheromatous plaque*), which leads to thickening of the tunica intima and narrowing of the arterial lumen.

Etiology

Although the etiology (cause) of atherosclerosis is not completely understood, the **response-to-injury hypothesis** is the most widely accepted. This proposal states that injury to the endothelium of an arterial wall, especially repeated injury caused by infection, trauma, or hypertension, results in an inflammatory reaction, eventually leading to the development of an atheroma. The injured endothelium becomes more permeable, which encourages leukocytes and platelets to adhere to the lesion and initiate an inflammatory response (see section 22.3d). Low-density lipoproteins and very-low-density lipoproteins

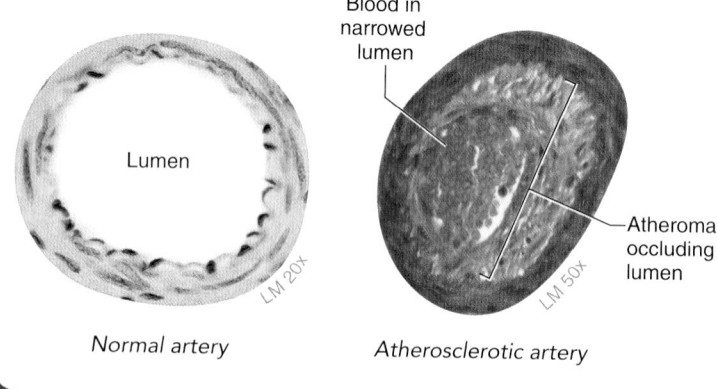

Blood in narrowed lumen

Lumen

LM 20x

Atheroma occluding lumen

LM 50x

Normal artery   *Atherosclerotic artery*

Muscular Arteries

Muscular arteries typically have diameters ranging from 1 centimeter to 0.3 millimeters. These medium-sized arteries are also called *distributing arteries* because they distribute blood to specific body regions and organs.

Muscular arteries have a proportionately thicker tunica media, with multiple layers of smooth muscle cells. Unlike in elastic arteries, the elastic fibers in muscular arteries are confined to two circumscribed sheets: The **internal elastic lamina** (lam´i-nă) separates the tunica media from the tunica intima, and the **external elastic lamina** separates the tunica media from the tunica externa. The relatively greater amount of muscle and lesser amount of elastic tissue result in a better ability to vasoconstrict and vasodilate, although with a lessened ability to stretch in comparison to elastic arteries.

Most named arteries (e.g., the brachial, anterior tibial, coronary, and inferior mesenteric arteries) are examples of muscular arteries (see figure 20.19*a*). Muscular arteries branch into arterioles.

Arterioles

Arterioles are the smallest arteries, with diameters ranging from 0.3 millimeters to 10 micrometers; these vessels are not named. In general, arterioles have fewer than six layers of smooth muscle in their tunica media. Larger arterioles have all three tunics, whereas the smallest arterioles may have a tunica intima surrounded by a single layer of

(LDLs and VLDLs) enter the tunica intima, combine with oxygen, and remain stuck to the vessel wall. This oxidation of these lipoproteins attracts monocytes, which adhere to the endothelium and migrate into the wall. As the monocytes migrate into the wall, they digest the lipids and develop into structures called **foam cells**. Eventually, smooth muscle cells from the tunica media migrate into the atheroma and proliferate, causing further enlargement of the atheroma and narrowing of the lumen of the blood vessel, thereby restricting blood flow to the regions the artery supplies.

Atherosclerosis is a progressive disease. The plaques typically begin to develop in early adulthood and grow and enlarge as we age. People are unaware of the plaques until they become large enough to restrict blood flow in an artery and cause vascular complications.

Risk Factors
Some individuals are genetically prone to atherosclerosis. **Hypercholesterolemia** (an increased amount of cholesterol in the blood), which also tends to run in families, has been positively associated with the rate of development and severity of atherosclerosis. Additionally, males tend to be affected more than females, and symptomatic atherosclerosis increases with age. Finally, smoking and hypertension both cause vascular injury, which increases the risk.

Treatment Options
If an artery is occluded (blocked) in one or just a few areas, a treatment called **angioplasty** (an'jē-ō-plas-tē; *angeion* = vessel, *plastos* = formed) is used. A physician inserts a balloon-tip catheter into an artery, and positions it at the site where the lumen is narrowed. Then the balloon is inflated, forcibly expanding the narrowed area, and a stent is placed in the vessel. A stent is a piece of wire-mesh that springs open to keep the vessel lumen open. For occluded coronary arteries, sometimes a much more invasive treatment

known as coronary bypass surgery may be needed. A vein (e.g., the great saphenous vein) or artery (e.g., the internal thoracic artery) is detached from its original location and grafted from the aorta to the coronary artery system, thus bypassing the area(s) of atherosclerotic narrowing.

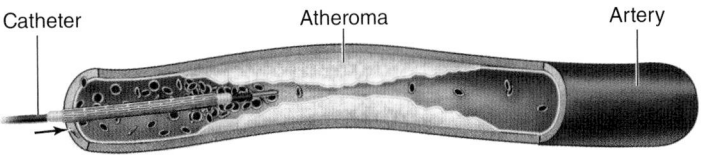

(1) An uninflated balloon and compressed stent are passed through a catheter to the area of the artery that is obstructed.

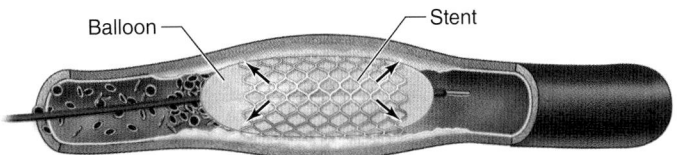

(2) Balloon inflates, which expands the stent and inserts it in place and also compresses the atheroma.

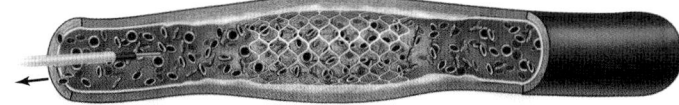

(3) The stent remains in the vessel as the balloon is deflated and the catheter is withdrawn.

Angioplasty is used to expand the narrowed region of an artery.

---

smooth muscle cells. Smooth muscle in the arterioles is slightly contracted (just as your skeletal muscles often are in a partial state of contraction; see section 10.7a). This contracted state is called **vasomotor tone** and is regulated by the vasomotor center in the brainstem. Vasomotor tone results in vasoconstriction, which allows for varying degrees of change from this slightly contracted state (see section 15.6a). Blood vessels can be either vasoconstricted to a greater degree to decrease

blood flow or vasodilated to allow more blood into an area. Arterioles have a significant role in regulating systemic blood pressure and blood flow to the different areas of the body.

WHAT DID YOU LEARN?

2 What changes are seen in the composition of the tunic wall of arteries as they branch into smaller and smaller vessels?

INTEGRATE

CLINICAL VIEW
Aneurysm AP|R

Elastic and muscular arteries are generally less able to withstand the forces from the pulsating blood as we age. Systolic blood pressure may increase with age, exacerbating this problem. As a result,

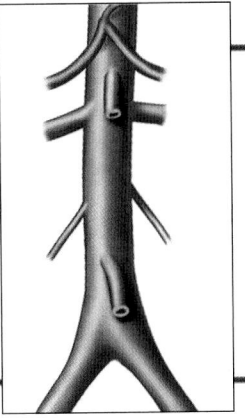
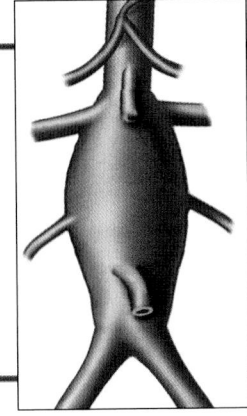

older individuals are more apt to develop an **aneurysm** (an'ū-riz-m), whereby part of the arterial wall thins and balloons out. This wall is more prone to rupture, which can cause massive bleeding and may lead to death. Aneurysms most commonly occur in the arteries at the base of the brain or in the aorta.

Normal aorta   *Large abdominal aortic aneurysm*

Table 20.2 — Types of Capillaries

Characteristics	(a) Continuous Capillary	(b) Fenestrated Capillary
Structure	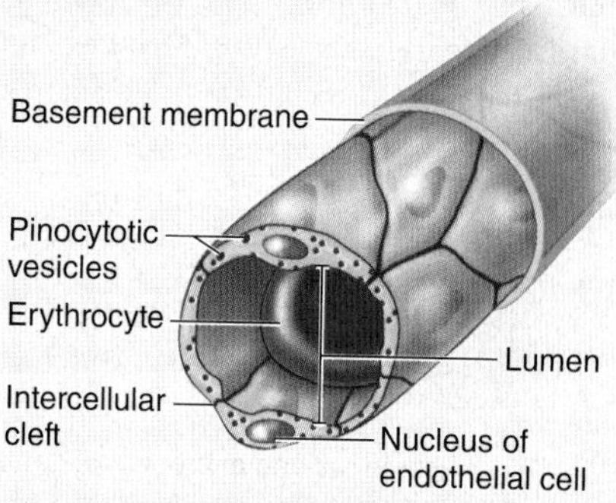 Basement membrane; Pinocytotic vesicles; Erythrocyte; Intercellular cleft; Lumen; Nucleus of endothelial cell	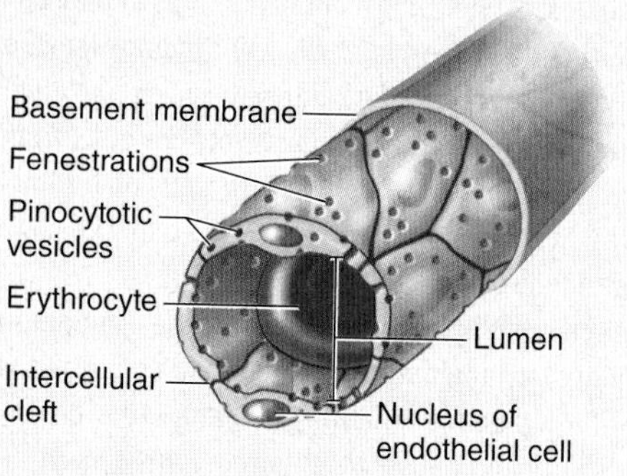 Basement membrane; Fenestrations; Pinocytotic vesicles; Erythrocyte; Intercellular cleft; Lumen; Nucleus of endothelial cell
Description	Lining of endothelial cells is complete around lumen; basement membrane is complete; intercellular clefts between endothelial cells	Same as continuous capillaries, except also contain fenestrations (range from 10 to 100 nanometers in diameter)
Materials That Pass Through Vessel Wall	Plasma and its contents (except most proteins); some leukocytes	Large amounts of materials are filtered, released, or absorbed; some smaller proteins
Locations	Most capillaries (e.g., capillaries within muscles, skin, thymus, lungs, and central nervous system [CNS])	Small intestine; for absorbing nutrients Ciliary process; to produce aqueous humor in the eye Choroid plexus; to produce cerebrospinal fluid (CSF) in the brain Most endocrine glands; for release of hormones into the blood Kidneys; for filtering blood

20.1c Capillaries

LEARNING OBJECTIVES

4. Describe the general anatomic structure and function of capillaries.
5. Compare the anatomic structure, function, and location of continuous capillaries, fenestrated capillaries, and sinusoids.
6. Trace the movement of blood through a capillary bed.

Capillaries are the smallest blood vessels. They connect arterioles to venules (the smallest veins). The average capillary is approximately 1 mm in length with a diameter of 8 to 10 micrometers, just slightly larger than the diameter of a single erythrocyte. The narrow vessel diameter means erythrocytes must travel in single file (termed **rouleau**) through each capillary (see section 18.3b). Capillaries consist solely of an endothelial layer (of simple squamous cells) resting on a basement membrane. The narrow vessel diameter and the thin wall are optimal for exchange of substances between blood and body tissues.

Types of Capillaries

Capillaries are differentiated based on their relative degree of permeability and include continuous capillaries, fenestrated capillaries, and sinusoids (table 20.2).

Continuous capillaries are the most common type of capillary. The endothelial cells form a complete, *continuous* lining around the lumen that rests on a complete basement membrane. Tight junctions (see section 4.6d) secure endothelial cells to one another; however, they do not form a complete "seal." The gaps between the endothelial cells are called **intercellular clefts.** Materials can move into or out of the blood either through endothelial cells by membrane transport processes (e.g., diffusion, pinocytosis; see section 4.3), or between endothelial cells through intercellular clefts by diffusion and bulk flow (see section 20.3).

The size of intercellular clefts prevents movement of large substances, including formed elements and plasma proteins, while allowing the movement of fluid containing small substances (smaller than about 5 nanometers), such as glucose, amino acids, and ions. Continuous capillaries are found, for example, in muscle, the skin, lungs, and central nervous system.

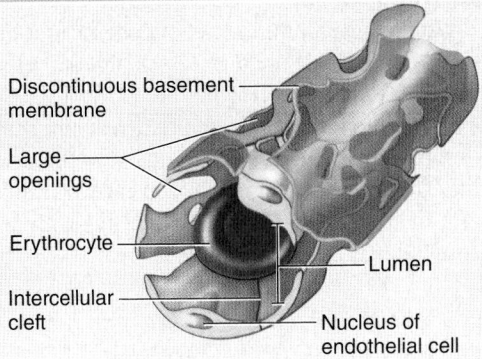

(c) Sinusoid

Discontinuous basement membrane

Large openings

Erythrocyte

Intercellular cleft

Lumen

Nucleus of endothelial cell

Lining of endothelial cells is incomplete around lumen; basement membrane is incomplete or absent

Large substances (formed elements, large plasma proteins) and plasma

Red bone marrow (where formed elements enter the blood)
Liver and spleen (where old erythrocytes are phagocytized by macrophages and taken out of circulation)
Some endocrine glands (anterior pituitary, adrenal, and parathyroid)

Fenestrated (fen′es-trā′ted; *fenestra* = window) **capillaries** are also composed of a complete, continuous lining of endothelial cells, and a complete basement membrane. However, small regions of the endothelial cells (typically 10 to 100 nanometers in diameter) are extremely thin; these thin areas are called **fenestrations** (or *pores*). Fenestrations are small enough to prevent formed elements from passing through the wall, yet large enough to allow movement of some smaller plasma proteins. Fenestrated capillaries are seen where a great deal of fluid transport between the blood and interstitial tissue occurs. Examples of structures that contain fenestrated capillaries include the small intestine for the absorption of nutrients, the ciliary process of the eye in the production of aqueous humor, choroid plexus of the brain in the production of cerebrospinal fluid, most of the endocrine glands to facilitate the absorption of hormones into the blood, and the kidney for the filtering of blood.

Sinusoids (si′nŭ-soyd; *sinus* = cavity, *eidos* = appearance), or *discontinuous capillaries,* have an incomplete lining of the endothelial cells with large openings or gaps, and the basement membrane is either discontinuous or absent. These openings allow for transport of large substances (formed elements, large plasma proteins), as well as plasma between the blood and tissues. Sinusoids are found in red bone marrow for entrance of formed elements into the circulation, the liver and spleen for removing aged erythrocytes from circulation, and some endocrine glands (e.g., anterior pituitary, parathyroid glands) for facilitating the movement of hormone molecules into the blood. A common feature of structures with sinusoids is their reddish color.

INTEGRATE

CONCEPT CONNECTION

The blood-brain barrier (BBB) (see section 13.2d) is formed by modified continuous capillaries that have thickened basement membranes and no intercellular clefts. Substances can pass through endothelial cells only by regulated membrane transport processes (e.g., facilitated diffusion, active transport). However, movement of nonpolar substances is not regulated by cells (see section 4.3a) and so these nonpolar substances (e.g., nicotine, alcohol) may pass through the cells by simple diffusion and enter the brain cells.

Capillary Beds

Capillaries do not function independently; rather, a group of capillaries (10 to 100) function together and form a **capillary bed (figure 20.5)**. A capillary bed is fed by a **metarteriole** (met′ar-tēr′ē-ōl; *meta* = between),which is a vessel branch of an arteriole. The proximal part of the metarteriole is encircled by scattered smooth muscle cells, whereas the distal part of the metarteriole (called the **thoroughfare channel**) has no smooth muscle cells. The thoroughfare channel connects to a **postcapillary venule** (ven′ūl, vē′nūl), which drains the capillary bed.

Vessels called **true capillaries** branch from the metarteriole and make up the bulk of the capillary bed. At the origin of each true capillary, a smooth muscle ring called the **precapillary sphincter** controls blood flow into the true capillaries. Sphincter relaxation permits blood to flow into the true capillaries, whereas sphincter contraction causes blood to flow directly from the metarteriole and thoroughfare channel

into the postcapillary venule with blood bypassing the capillary bed. The precapillary sphincters go through cycles of contracting and relaxing at a rate of about 5 to 10 cycles per minute. This cyclical process is called **vasomotion.**

At any given time only about one-quarter of the capillary beds are open, because there are over 60,000 miles of capillaries and only about 250 to 300 milliliters (mL) of blood (about 5% of the total blood volume) moving through the capillaries at any particular moment. There is simply not enough blood available to fill all capillaries at the same time. The specific amount of blood entering capillaries per unit time per gram of tissue is called **perfusion,** typically expressed in milliliters per minute per gram (mL/min/g).

WHAT DID YOU LEARN?

3 What type of capillary is the most permeable, and where in the body are they found?

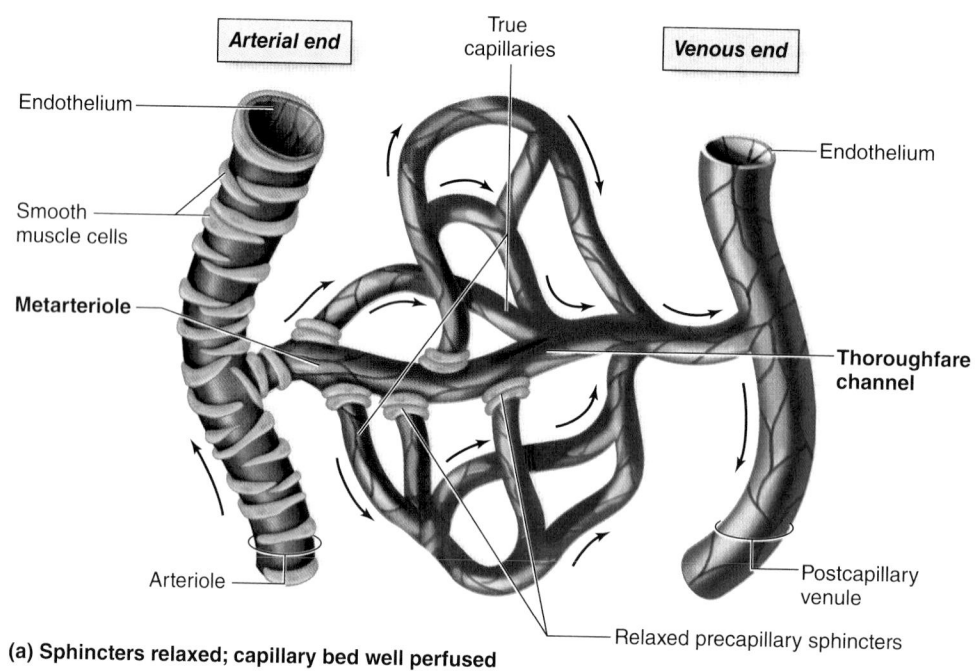

(a) Sphincters relaxed; capillary bed well perfused

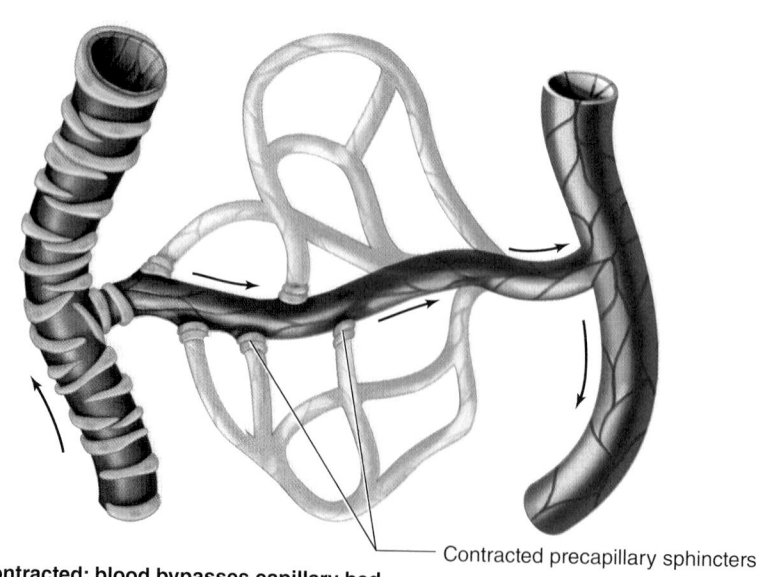

(b) Sphincters contracted; blood bypasses capillary bed

Figure 20.5 Capillary Bed Structure and Perfusion Through the Bed. A capillary bed originates from a metarteriole. The metarteriole continues as a thoroughfare channel that merges with the postcapillary venule. True capillaries branch from the thoroughfare channel, and blood flow into these true capillaries is regulated by the precapillary sphincters. (*a*) A well-perfused capillary bed, with all of the precapillary sphincters relaxed, and (*b*) a capillary bed where most blood bypasses the capillary bed due to contracted precapillary sphincters.

20.1d Veins

LEARNING OBJECTIVES

7. Describe the structure and general function of veins.

8. Explain how veins serve as a blood reservoir for the cardiovascular system.

Veins merge into larger and larger vessels with a corresponding increase in lumen diameter as they extend from the capillaries to the heart (see figure 20.3).

Venules

Venules are the smallest veins, measuring from 8 to 100 micrometers in diameter. Venules are companion vessels with arterioles. The smallest venules, called postcapillary venules, drain capillaries. Smaller venules merge to form larger venules. The largest venules have all three tunics. Venules merge to form veins.

Small, Medium-Sized, and Large Veins

A venule becomes a vein when its diameter is greater than 100 micrometers. Small and medium-sized veins are companion vessels with muscular arteries, whereas the largest veins travel with elastic arteries. Most veins contain numerous **valves,** so as to prevent blood from pooling in the limbs. The valves are formed primarily of tunica intima and strengthened by elastic and collagen fibers. Valves have an anatomic structure similar to semilunar valves of the heart (see section 19.3d).

Systemic Veins as Blood Reservoirs

The percentage of the total blood that is moving through each of the different components of the cardiovascular system while *at rest* is

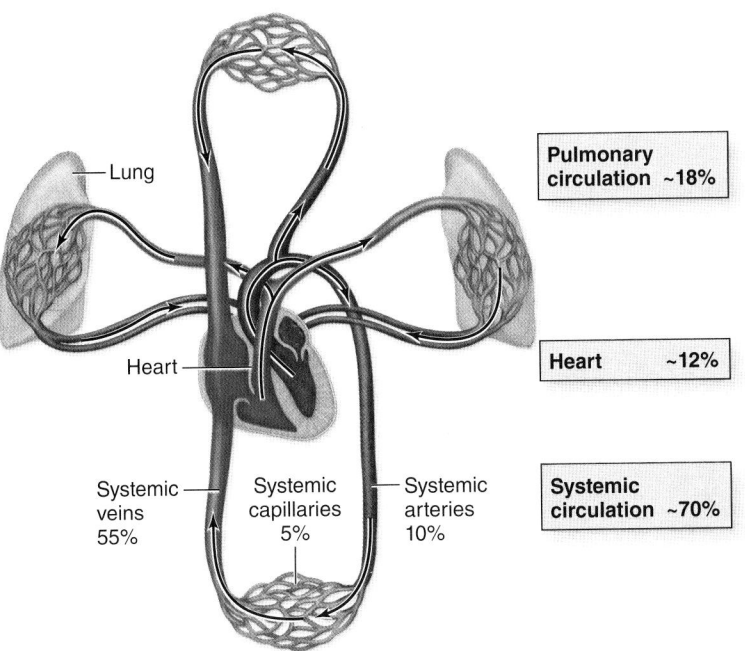

Figure 20.6 Blood Distribution at Rest. Different components of the cardiovascular system vary in the amounts of blood moving through them, with systemic veins containing the largest percentage.

INTEGRATE

CONCEPT CONNECTION

A modified vein that has very thin walls and no smooth muscle is referred to as a **sinus.** A sinus is supported by surrounding tissue. Two examples include the coronary sinus located in the coronary sulcus of the heart, supported by adipose connective tissue (see figure 19.7), and the dural venous sinuses of the brain, supported by the dense connective tissue of the dura mater (see figures 13.5 and 13.6).

Coronary sinus

illustrated in **figure 20.6**. Relatively small amounts of blood are within the pulmonary circulation (about 18%) and the heart (about 12%). The largest percentage of blood is within the systemic circulation (about 70%), with the greatest amount (about 55%) within the body's systemic veins. The relatively large amount of blood within veins allows veins to function as **blood reservoirs.** Blood may be shifted from venous reservoirs into circulation through vasoconstriction of veins, when more blood is needed with increased physical exertion—and shifted back into venous reservoirs through vasodilation of veins, when less blood is needed at rest.

Figure 20.7 integrates how each blood vessel structure is adapted for the specific functions of that vessel.

WHAT DID YOU LEARN?

4. How does a vein serve as a blood reservoir?

20.1e Pathways of Blood Vessels

LEARNING OBJECTIVE

9. Compare and contrast the simple and alternative pathways of blood vessels.

Blood vessels are arranged in either simple or alternative pathways **(figure 20.8)**. Each pathway is described in detail.

Simple Pathway

In the simple pathway, one major artery delivers blood to the organ or body region and then branches into smaller and smaller arteries to become arterioles. Each arteriole feeds into a single capillary bed. A venule drains blood from the capillaries and merges with other venules to form one major vein that drains blood from the organ or body region. Thus, the simple pathway includes one artery, capillary bed, and vein associated with an organ or body region.

Blood transported to and from the spleen is an example of a simple blood flow pathway. A single splenic artery delivers oxygenated blood to the spleen with the exchange made in a capillary bed of the spleen, and a single splenic vein drains deoxygenated blood from the spleen (see figure 21.7). Arteries that provide only one pathway through which blood can reach an organ are referred to as **end arteries.**

Alternative Pathways

Several alternative circulatory pathways are possible, and these include various types of anastomoses and the portal systems. They differ from the simple pathway in the number of arteries, capillary beds, or veins that serve an organ or body region.

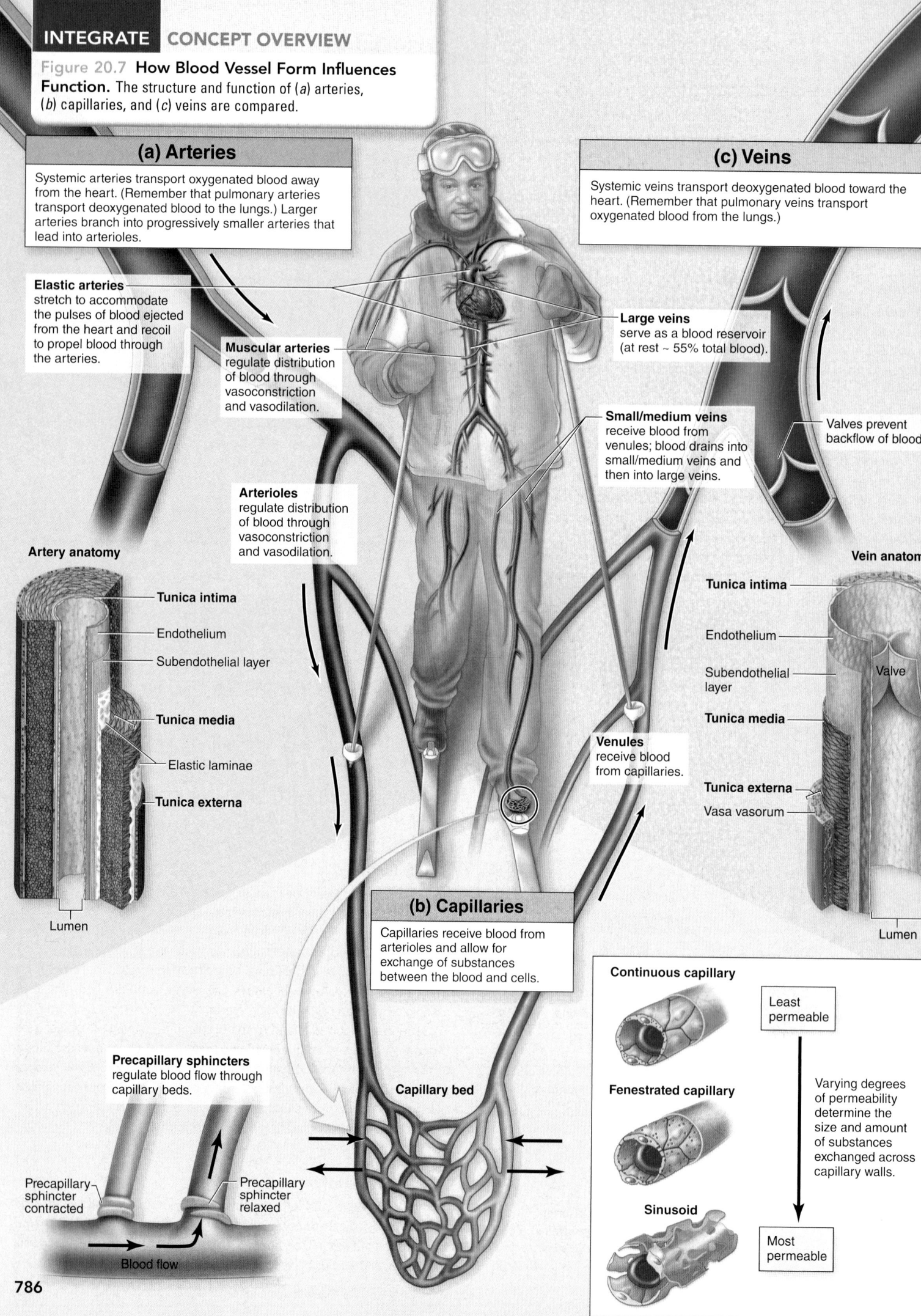

Figure 20.7 **How Blood Vessel Form Influences Function.** The structure and function of (a) arteries, (b) capillaries, and (c) veins are compared.

(a) Arteries

Systemic arteries transport oxygenated blood away from the heart. (Remember that pulmonary arteries transport deoxygenated blood to the lungs.) Larger arteries branch into progressively smaller arteries that lead into arterioles.

Elastic arteries stretch to accommodate the pulses of blood ejected from the heart and recoil to propel blood through the arteries.

Muscular arteries regulate distribution of blood through vasoconstriction and vasodilation.

Arterioles regulate distribution of blood through vasoconstriction and vasodilation.

Artery anatomy

- Tunica intima
  - Endothelium
  - Subendothelial layer
- Tunica media
  - Elastic laminae
- Tunica externa

Lumen

Precapillary sphincters regulate blood flow through capillary beds.

Precapillary sphincter contracted

Precapillary sphincter relaxed

Blood flow

(c) Veins

Systemic veins transport deoxygenated blood toward the heart. (Remember that pulmonary veins transport oxygenated blood from the lungs.)

Large veins serve as a blood reservoir (at rest ~ 55% total blood).

Small/medium veins receive blood from venules; blood drains into small/medium veins and then into large veins.

Valves prevent backflow of blood.

Venules receive blood from capillaries.

Vein anatomy

- Tunica intima
  - Endothelium
  - Subendothelial layer
- Tunica media
- Tunica externa
  - Vasa vasorum

Valve

Lumen

(b) Capillaries

Capillaries receive blood from arterioles and allow for exchange of substances between the blood and cells.

Capillary bed

Continuous capillary

Least permeable

Fenestrated capillary

Varying degrees of permeability determine the size and amount of substances exchanged across capillary walls.

Sinusoid

Most permeable

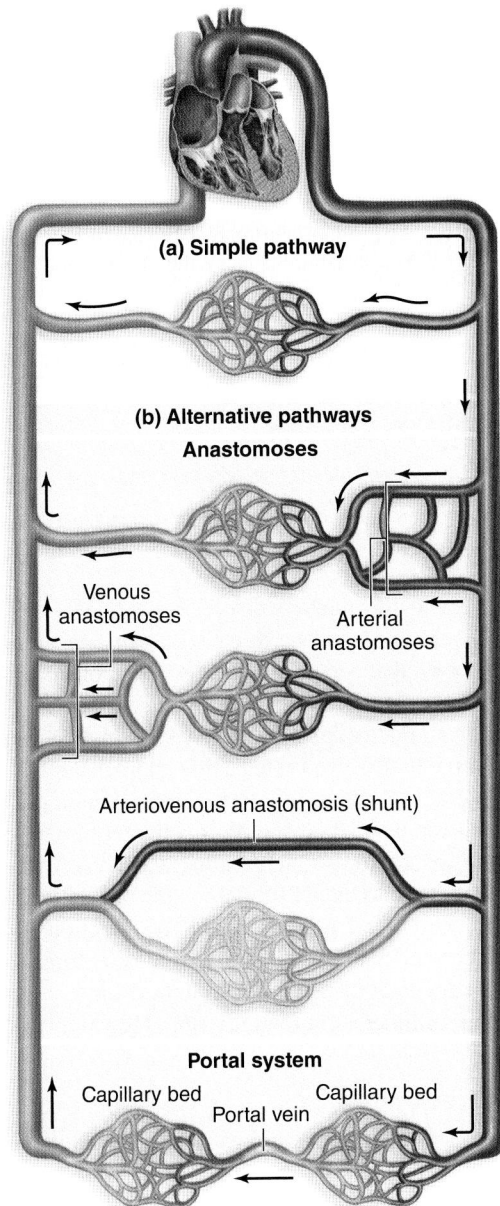

Figure 20.8 Comparison of Simple and Alternative Blood Flow Pathways. (*a*) Simple pathways use one major artery, one capillary bed, and one major vein to deliver blood to a body region. (*b*) Alternative pathways include anastomoses and portal systems. Anastomoses differ in the number of either arteries, capillary beds, or veins that serve a body region. In a portal system, venous blood is directed through the capillary bed of another organ (like the liver) first before going back to the heart.

Three of the alternative pathways are designated as anastomoses. An **anastomosis** (a-nas'tō-mō'sis; pl., *anastomoses*) is the joining together of blood vessels. An **arterial anastomosis** includes two or more *arteries* converging to supply the same body region. (For example, figure 20.22*a* shows anastomoses among the superior and inferior epigastric arteries that serve the abdominal wall.) Other vessels, such as the coronary arteries, may have anastomoses that are so tiny that the function of the arteries may almost be considered end arteries; these arteries are called **functional end arteries** (see section 19.4a).

A **venous anastomosis** includes two or more *veins* draining the same body region. Veins tend to form many more anastomoses than do arteries. Veins that drain the upper limb including the basilic, brachial, and cephalic veins provide an example of a venous anastomosis (see figure 20.19*b*).

An **arteriovenous anastomosis,** or *shunt,* transports blood from an artery directly into a vein, bypassing the capillary bed. These shunts are present in the fingers, toes, palms, and ears, and they allow these areas to be bypassed if the body is becoming hypothermic (cold). It is the bypassing of these body structures, as blood is shunted through an arteriovenous anastomosis, that makes them particularly vulnerable to frostbite (see Clinical View: "Frostbite and Dry Gangrene" in section 27.8b).

A different type of blood vessel arrangement is a **portal system.** Blood flows through *two* capillary beds, with the two capillary beds separated by a portal vein. A **portal vein** delivers blood to another organ first, before the blood is sent back to the heart. Thus, the sequence of blood vessels is as follows: an artery, capillary bed, portal vein, capillary bed, and a vein. One example is the hypothalamo-hypophyseal portal system that extends between the hypothalamus and the anterior pituitary (see section 17.7a). This portal system provides a more direct means of delivering hypothalamic hormones to the anterior pituitary. Another portal system is the hepatic portal system, which is described in section 20.10d.

WHAT DID YOU LEARN?

5 How does each of the various alternative blood vessel pathways differ from the simple pathway?

20.2 Total Cross-Sectional Area and Blood Flow Velocity

LEARNING OBJECTIVES

1. Describe the relationship of the total cross-sectional area and velocity of blood flow.
2. Predict the significance of slow blood flow in the capillaries.

The **cross-sectional area** of a vessel is the diameter of the vessel's lumen. The **total cross-sectional area** is estimated as the aggregate lumen diameter across the total number of a given type of vessel (artery, capillary, or vein) if they were all positioned side by side. You may be surprised to learn that, although the cross-sectional area of an individual artery is relatively large, the total cross-sectional area of arteries is relatively small. This observation is also accurate for veins, which have a relatively small total cross-sectional area. See the blue line on the graph in **figure 20.9.** In comparison, an individual capillary has a very small

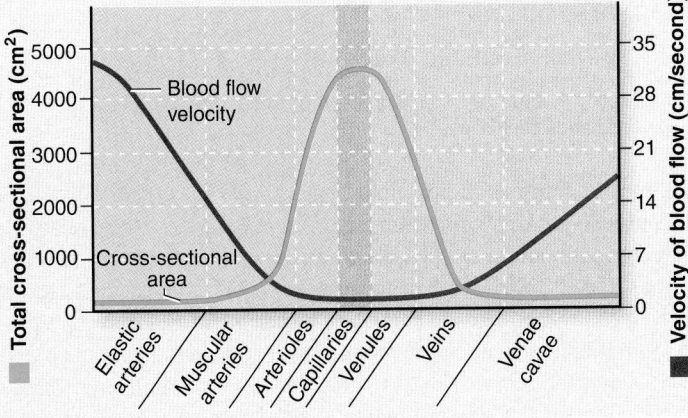

Figure 20.9 Relationship of Total Cross-Sectional Area and Velocity of Blood Flow. The greater the total cross-sectional area, the slower the blood flow. Notice that capillaries have the greatest cross-sectional area and the slowest blood flow—conditions that facilitate the exchange of substances between the blood and tissues.

cross-sectional area; however, the total cross-sectional area of capillaries—of which there are approximately 60,000 miles—is the largest with a value of approximately 4500 square centimeters (cm^2). The total cross-sectional area is physiologically significant because of its influence on blood flow velocity.

Blood flow velocity is the rate of blood transported per unit time and typically measured in centimeters per second. Observe the graph in figure 20.9 and notice that there is an *inverse* relationship between the total cross-sectional area and blood flow velocity (red line on the graph). Blood flow velocity in both arteries and veins, with their relatively small total cross-sectional area, is relatively fast. In comparison, blood flow velocity in capillaries, with their relatively large total cross-sectional area, is relatively slow. Thus, blood flow velocity changes as it moves through the different types of vessels: Velocity of blood flow is relatively fast in the arteries, slowest in the capillaries, and moves relatively fast again as it moves through the veins. Consider the analogous phenomenon of water flow in a river. In regions where the river is narrow, the river moves more quickly, and where the river is wider, the river moves more slowly. Of course, the amount of water flow is the same in these different regions and always moves toward the ocean. Likewise, blood flow velocity is altered as it moves through the different portions of the vasculature, but it always moves along a blood pressure gradient (described in section 20.5a) as it is transported through the vasculature of the cardiovascular system.

What is the significance of the much slower rate of blood flow through the capillaries? The slower blood flow rate allows sufficient time for efficient capillary exchange of respiratory gases, nutrients, wastes, and hormones between the tissues and the blood (see section 20.3).

 WHAT DID YOU LEARN?

6 In which type of vessel is blood flow the slowest? Explain the anatomic structure that accounts for this and the physiologic significance.

20.3 Capillary Exchange

The function of capillaries is to allow for the exchange of substances (e.g., respiratory gases, nutrients, wastes, and hormones) between the blood and the surrounding tissues. Exchange processes include diffusion, vesicular transport, and bulk flow.

20.3a Diffusion and Vesicular Transport

 LEARNING OBJECTIVE

1. Explain the process of diffusion and vesicular transport between capillaries and tissues.

Within systemic capillaries, substances such as oxygen, hormones, and nutrients move by **diffusion** (see section 4.3a) from their relatively high concentration in the blood into the interstitial fluid and then into the tissue cells, where the concentration of these materials is lower. Conversely, carbon dioxide and waste products diffuse from the higher concentration in the tissue cells to the lower concentration in the blood. Very small solutes (e.g., O_2, CO_2, glucose, ions) and fluids may diffuse via the endothelial cells or intercellular clefts, while larger solutes, such as small proteins, must pass through the fenestrations in fenestrated capillaries or gaps in sinusoids.

Vesicular transport occurs when endothelial cells use pinocytosis (see section 4.3c) to form fluid-filled vesicles, which are then transported to the other side of the cell and released by exocytosis.

Substances can be moved either from the blood into the interstitial fluid or from the interstitial fluid into the blood. Solutes such as certain hormones (e.g., insulin) and fatty acids are transported across the endothelial cells by this method.

 WHAT DID YOU LEARN?

7 What substances are transported by diffusion through the capillaries? Which substances leave the capillaries by vesicular transport?

20.3b Bulk Flow

LEARNING OBJECTIVES

2. Explain the processes of bulk flow, filtration, and reabsorption.

3. Compare and contrast hydrostatic pressure and colloid osmotic pressure in the capillaries.

Bulk flow refers to the movement of large amounts of fluids and their dissolved substances in one direction down a pressure gradient. **Filtration,** a process that occurs on the arterial end of a capillary, is the movement of fluid by bulk flow out of the blood through the openings in the capillaries (e.g., intercellular clefts, fenestrations). During this process, fluids and small, dissolved solutes flow through easily, while large solutes are generally blocked. In contrast, reabsorption occurs on the venous end of a capillary. **Reabsorption** is the movement of fluid by bulk flow in the opposite direction, back into the blood (**figure 20.10**).

How is it possible to have filtration on the arterial end, and reabsorption on the venous end, of a capillary? The direction of movement is dependent upon the net pressure of two opposing forces at the capillary level—hydrostatic pressure and colloid osmotic pressure; both are measured in millimeters of mercury (mm Hg).

Hydrostatic Pressure

Hydrostatic pressure (HP) is the physical force exerted by a fluid on a structure. For example, **blood hydrostatic pressure (HP_b)** (or simply *blood pressure*) is the force exerted per unit area by the blood as it presses against the vessel wall. Blood hydrostatic pressure promotes filtration from the capillary.

The interstitial fluid also has its own hydrostatic pressure, called **interstitial fluid hydrostatic pressure (HP_{if}),** which is the force of the interstitial fluid on the external surface of the blood vessel. For most tissues, the interstitial fluid hydrostatic pressure is very small and for simplicity's sake is assumed to be close to zero. Thus, for our discussion, the main hydrostatic pressure is the blood hydrostatic pressure, which pushes materials *out* of the capillary.

Colloid Osmotic Pressure

The other main force regulating filtration and reabsorption is osmotic pressure, which refers to the "pull" of water into an area by osmosis due to the higher relative concentration of solutes. **Colloid osmotic pressure (COP)** refers to the pull of water back into a tissue by the tissue's concentration of proteins (colloid).

The **blood colloid osmotic pressure (COP_b)** is the force that draws fluid back into the blood due to the proteins in blood, such as albumin. Blood colloid osmotic pressure opposes hydrostatic pressure, and thus promotes reabsorption. Clinicians also use the term **oncotic** (*onkosis* = swelling) **pressure** to describe the blood colloid osmotic pressure.

An **interstitial fluid colloid osmotic pressure (COP_{if})** also exists, but its value is relatively low because few proteins are present in the

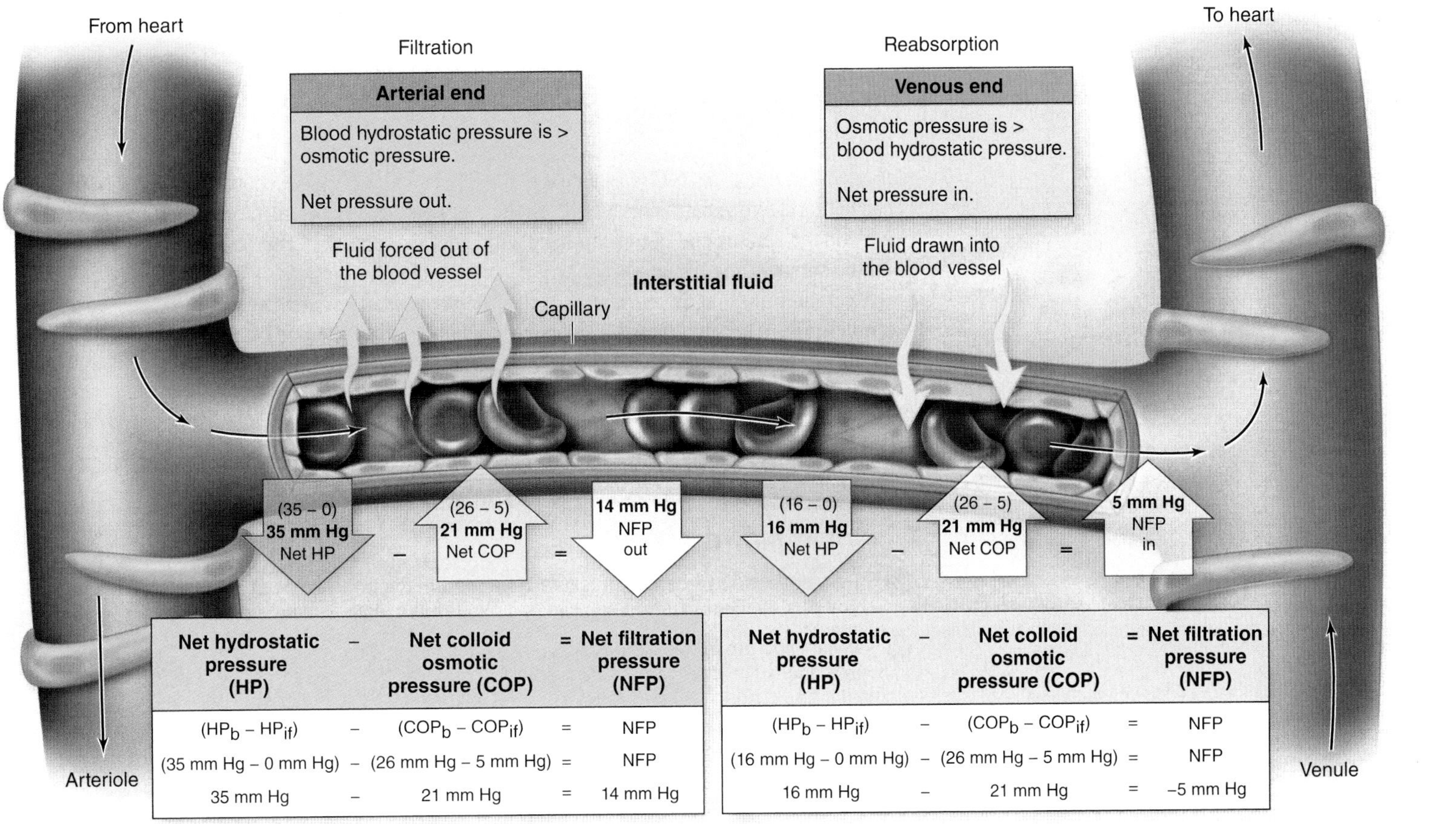

Figure caption area with diagram labels:

Filtration

Arterial end
Blood hydrostatic pressure is > osmotic pressure.

Net pressure out.

Reabsorption

Venous end
Osmotic pressure is > blood hydrostatic pressure.

Net pressure in.

From heart

To heart

Fluid forced out of the blood vessel

Interstitial fluid

Fluid drawn into the blood vessel

Capillary

(35 – 0) 35 mm Hg Net HP

(26 – 5) 21 mm Hg Net COP

14 mm Hg NFP out

(16 – 0) 16 mm Hg Net HP

(26 – 5) 21 mm Hg Net COP

5 mm Hg NFP in

Arteriole

Venule

Net hydrostatic pressure (HP)	–	Net colloid osmotic pressure (COP)	= Net filtration pressure (NFP)
$(HP_b - HP_{if})$	–	$(COP_b - COP_{if})$	= NFP
(35 mm Hg – 0 mm Hg)	–	(26 mm Hg – 5 mm Hg) =	NFP
35 mm Hg	–	21 mm Hg	= 14 mm Hg

Net hydrostatic pressure (HP)	–	Net colloid osmotic pressure (COP)	= Net filtration pressure (NFP)
$(HP_b - HP_{if})$	–	$(COP_b - COP_{if})$	= NFP
(16 mm Hg – 0 mm Hg)	–	(26 mm Hg – 5 mm Hg) =	NFP
16 mm Hg	–	21 mm Hg	= –5 mm Hg

Figure 20.10 Bulk Flow at Capillaries. Blood hydrostatic pressure is greater at the arterial end of the capillary and less at the venous end of the capillary. The osmotic pressures remain relatively constant, resulting in a net filtration out of the arterial end of the capillary, and a net reabsorption into the venous end of the capillary. AP|R

interstitial fluid. COP_{if} may range from 0 to 5 mm Hg. By knowing the specific values for the hydrostatic and osmotic pressures, the direction of bulk flow can be determined through the calculation of net filtration pressure.

WHAT DID YOU LEARN?

❽ What is the difference between hydrostatic pressure and osmotic pressure?

20.3c Net Filtration Pressure

LEARNING OBJECTIVES

4. Define net filtration pressure (NFP).
5. Calculate net filtration pressure for both the arterial end and the venous end of a capillary.

Net filtration pressure (NFP) is the difference between the net hydrostatic pressure (difference between the blood and interstitial fluid hydrostatic pressures) and the net colloid osmotic pressure (difference between the blood and the interstitial fluid colloid osmotic pressures) and is shown in figure 20.10. The net filtration pressure may be determined by the following equation:

$$NFP = (HP_b - HP_{if}) - (COP_b - COP_{if})$$

where HP_b = the hydrostatic pressure of blood, HP_{if} = the hydrostatic pressure of the interstitial fluid, COP_b = the colloid osmotic pressure of blood, and COP_{if} = the colloid osmotic pressure of the interstitial fluid. This equation is a variation of **Starling's law,** developed by the physiologist Ernest Starling. Starling was among the first to discover

that hydrostatic and osmotic forces work against one another to drive the filtration and reabsorption of materials across a capillary wall.

Net filtration pressure changes as the blood moves from the arterial end of the capillary to the venous end of the capillary. On the arterial end of a capillary, HP_b is typically around 35 mm Hg, HP_{if} is assumed to be 0 mm Hg, COP_b is around 26 mm Hg, and COP_{if} is around 5 mm Hg. The net filtration pressure is calculated as follows:

$$(35 \text{ mm Hg} - 0 \text{ mm Hg}) - (26 \text{ mm Hg} - 5 \text{ mm Hg}) =$$
$$35 \text{ mm Hg} - 21 \text{ mm Hg} = 14 \text{ mm Hg}$$

Notice that NFP has a positive value of 14 mm Hg at the arterial end. A positive value indicates that the hydrostatic pressure pushing fluids out of the blood is greater than the net colloid osmotic pressure pulling the fluid back into the capillary. Consequently, *filtration,* or the net movement of fluid out of the blood vessel into the surrounding tissue, occurs on the arterial end of a capillary.

Now consider the events on the venous end of a capillary. Blood hydrostatic pressure decreases continuously as it moves through the capillary to reach the venule end, because there is a net movement of fluids out of the capillary as blood moves through the length of the capillary. This smaller amount of blood means the blood hydrostatic pressure at the venous end of a capillary is lower—usually around 16 to 20 mm Hg.

In contrast, the colloid osmotic pressures of both the blood and the interstitial fluid remain relatively constant throughout the capillary and have values similar to those on the arterial end of the capillary: COP_b is around 26 mm Hg and COP_{if} is around 5 mm Hg. Again, HP_{if} is assumed to be 0 mm Hg. Net filtration pressure is calculated on the venous end as follows:

$$(16 \text{ mm Hg} - 0 \text{ mm Hg}) - (26 \text{ mm Hg} - 5 \text{ mm Hg}) = \text{NFP}$$
$$16 \text{ mm Hg} \quad - \quad 21 \text{ mm Hg} \quad = -5 \text{ mm Hg}$$

Notice that in this case, NFP has a negative value (–5 mm Hg). A negative value for NFP occurs because blood hydrostatic pressure is less than net osmotic pressure. Consequently, reabsorption occurs on the venous end of the capillary with a net movement of fluid back into the blood vessel from the surrounding tissue.

Keep in mind that the numbers for the pressures used to calculate net filtration pressure are examples only. The specific values depend upon the part of the body, the amount of blood entering a specific capillary bed, and the general health of the individual.

WHAT DO YOU THINK?

1 Is it possible for blood pressure to decrease to such a degree that capillary exchange ceases? Explain.

WHAT DID YOU LEARN?

9 How does the hydrostatic pressure change from the arterial end of a capillary to the venous end of a capillary? Do you see similar changes in colloid osmotic pressure?

10 Which two pressures have the largest values? Explain how each of these specifically influences filtration and reabsorption.

20.3d Role of the Lymphatic System

LEARNING OBJECTIVE

6. Explain the lymphatic system's role at the capillary bed.

Although net filtration occurs at the arterial end of a capillary and net reabsorption at its venous end, not all of the fluid is reabsorbed at the venous end of the capillary. The capillary typically reabsorbs only about 85% of the fluid that has passed into the interstitial fluid. What happens to the excess 15% of fluid that wasn't reabsorbed?

Another body system, the **lymphatic system,** is responsible for picking up this excess fluid and returning it to the blood. Lymph vessels reabsorb this excess fluid, filter it, and return it to the venous circulation (see section 21.1). The lymphatic system is described in detail in chapter 21.

WHAT DID YOU LEARN?

11 If these lymph vessels were nonfunctional, what would happen to the amount of interstitial fluid around the capillary bed?

20.4 Local Blood Flow

Recall that there is simply not enough blood in the body to fill all capillaries at the same time. Blood must therefore be directed to organs and tissues where it is most needed and away from areas where it is not. **Local blood flow** is the blood delivered locally to the capillaries of a specific tissue and is measured in milliliters per minute. Recall that the specific amount of blood entering capillaries per unit time per gram of tissue is called perfusion. The ultimate function of the cardiovascular system is for *adequate perfusion of all tissues* (see section 19.1a).

The amount of blood delivered to a specific organ or tissue is dependent upon several factors, including (1) the degree of vascularization of the tissue, (2) the myogenic response, (3) local regulatory factors that alter blood flow, and (4) total blood flow.

20.4a Degree of Vascularization and Angiogenesis

LEARNING OBJECTIVES

1. Describe what is meant by degree of vascularization.

2. Explain the process of angiogenesis and how it aids perfusion.

The **degree of vascularization,** or the extent of blood vessel distribution within a tissue, determines the potential ability of blood delivery. Organs that are very active metabolically, such as the brain, skeletal muscle, the heart, and the liver, generally are highly vascularized. In comparison, some structures, such as tendons and ligaments, have little vascularization; blood delivery to these tissues is limited. Additionally, some structures contain no capillaries (or are avascularized); these include epithelial tissue, cartilage, and the cornea and lens of the eye.

The amount of vascularization in a given tissue may change over time through the process of angiogenesis. **Angiogenesis** (an′jē-ō-jen′ĕ-sis; *angio* = vessel, *genesis* = production) is the formation of new blood vessels in tissues that require them. This process helps provide adequate perfusion through long-term anatomic changes that occur over several weeks to months. For example, angiogenesis is stimulated in skeletal muscle in response to aerobic training; in adipose tissue, angiogenesis occurs in adipose connective tissue when an individual gains weight in the form of fat deposits. Angiogenesis also occurs in response to a slow, gradual occlusion (blockage) of coronary vessels, thus potentially providing alternative routes to deliver blood to the heart wall.

Regression (or return to previous state) of vessels is also possible. For example, some skeletal muscle blood vessels regress when an individual who was physically active becomes sedentary, or blood vessels in adipose tissue regress when the amount of adipose tissue is decreased through restriction of food and increased physical activity.

 WHAT DID YOU LEARN?

12 In what ways is angiogenesis stimulated in skeletal muscle? In adipose tissue?

20.4b Myogenic Response

 LEARNING OBJECTIVE

3. Describe the myogenic response that maintains normal blood flow through a tissue.

Systemic blood pressure, which is the driving force to move blood through blood vessels, changes under different conditions (e.g., when we get excited or become relaxed). However, blood flow into a tissue may remain relatively constant because of the **myogenic** (mī′-ō-jen′ik; *mys* = muscle) **response,** which is contraction and relaxation of smooth muscle in response to changes in stretch of the blood vessel wall.

An increase in systemic blood pressure causes an additional volume of blood to enter the blood vessel, which stretches the smooth muscle cells within the blood vessel wall. This stimulates the smooth muscle cells to contract, resulting in vasoconstriction. Thus, although systemic blood pressure is higher, which would drive additional blood into the blood vessel, the resulting vasoconstriction decreases the size of the blood vessel lumen offsetting the change, and blood flow into a tissue remains constant. In contrast, a decrease in systemic blood pressure causes a lower volume of blood to enter the blood vessel, with decreased stretch of the smooth muscle cells within the blood vessel wall. This stimulates the smooth muscle cells to relax, resulting in vasodilation. Thus, although systemic blood pressure is lower, which would drive less blood into the blood vessel, the resulting vasodilation increases the size of the blood vessel lumen offsetting the change, and blood flow into a tissue remains constant.

 WHAT DID YOU LEARN?

13 Explain the myogenic response to an increase in system blood pressure.

20.4c Local, Short-Term Regulation

 LEARNING OBJECTIVES

4. Compare and contrast a vasodilator and a vasoconstrictor.

5. Explain how a tissue autoregulates local blood flow based on metabolic needs.

6. Describe how local blood flow is altered by tissue damage and as part of the body's defense.

Local regulation of blood flow occurs on a continual basis in response to changes in metabolic activity of tissues. Local blood flow is also altered in response to tissue damage or as part of the body's defense systems. The stimulus is changing concentrations of certain chemicals, collectively called **vasoactive chemicals.** They are classified according to their action as either vasodilators or vasoconstrictors. **Vasodilators** are substances that cause smooth muscle relaxation, which results in both vasodilation of arterioles and opening of precapillary sphincters. Consequently, blood

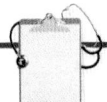 **INTEGRATE**

CLINICAL VIEW
Tumor Angiogenesis

Like normal cells of the body, abnormal cells of cancerous tumors require the delivery of oxygen and nutrients and the removal of their wastes. Thus, a critical event in development of a cancerous tumor is the formation of blood vessels within the tumor, a process called **tumor angiogenesis.** The process is initiated when cancerous cells release molecules that cause normal host cells to release growth factors that stimulate angiogenesis. One active area of cancer research is inhibiting tumor angiogenesis, with the intent of reducing or eliminating the tumor by depriving it of the oxygen and nutrients needed for growth.

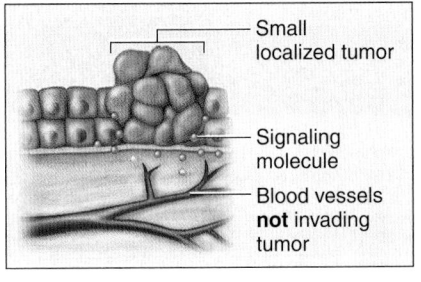

- Small localized tumor
- Signaling molecule
- Blood vessels **not** invading tumor

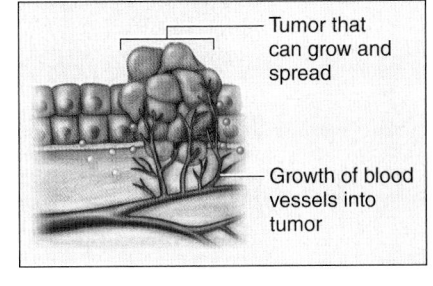

- Tumor that can grow and spread
- Growth of blood vessels into tumor

flow increases into a capillary bed. **Vasoconstrictors** are substances that cause smooth muscle contraction, which results in both arterioles vasoconstricting and precapillary sphincters closing. Thus, blood flow decreases into a capillary bed (see figure 20.5).

Autoregulation and Changing Metabolic Activity

Autoregulation is the process by which a tissue itself regulates or controls its local blood flow in response to its changing metabolic needs. The initial stimulus typically is inadequate perfusion due to increased metabolic activity of the tissue. If the tissue is not receiving adequate blood perfusion, then the oxygen and nutrient levels decline, while there is an increase in carbon dioxide, lactate, hydrogen ion (H^+), and potassium ion (K^+) levels. These altered levels act as local vasodilators, and as a result additional blood enters the capillaries serving the tissue. As perfusion increases in the tissue and these levels adjust back to homeostatic values, the vessels constrict. Thus, there is a negative feedback loop between elevated levels of these molecules and the degree of vasodilation.

Autoregulation is most noticeable when blood supply is temporarily disrupted and then restored. When the blood flow is temporarily disrupted, the tissue is deprived of needed oxygen and nutrients, and metabolic wastes accumulate. When the local blood flow is restored, there is a marked increase in blood flow to the affected tissue—a condition called **reactive hyperemia.** The additional blood is required to resupply the oxygen and nutrients and eliminate the accumulated wastes.

An example of reactive hyperemia occurs when you enter a warm room after being outside in the cold, and your cheeks turn bright red. When you were outside in the cold, the blood vessels in the dermis constricted, forcing more blood away from the dermis through an arteriovenous shunt to conserve heat. When you leave the cold, the reddish color of your cheeks is due to increased blood flow in the dermis. After a short period of time, the reddish color subsides, as normal blood flow in the skin resumes.

Short-Term Regulation Due to Damaged Tissue or as Part of the Body's Defense System

Local blood flow is also regulated when vasoactive chemicals are released from damaged tissue, leukocytes, and platelets in response to tissue damage or as part of the body's defense. This process is referred to as *inflammation* and is discussed in detail in section 22.3d. For example, **histamine** and **bradykinin** are released in response to a trauma, allergic reaction, infection, or even exercise. These chemicals cause vasodilation by either directly stimulating arterioles or indirectly by stimulating endothelial cells of the vessel to release nitric oxide. **Nitric oxide** is a very powerful, but short-lived, vasodilator.

Other vasoactive substances, such as **prostaglandins** and **thromboxanes,** which are local hormones released with tissue injury, can cause vasoconstriction (see description of local hormones in section 17.3b). Recall from section 18.4a that if endothelial cells are damaged, they release an array of chemicals that are powerful vasoconstrictors to help prevent blood loss from the damaged vessel. Systemic hormones also alter blood flow, and their effects are described in section 20.6b. See table 20.3 for a list of vasodilators and vasoconstrictors.

 WHAT DID YOU LEARN?

14 What relationship exists between metabolic activity and local blood flow?

20.4d Relationship of Local and Total Blood Flow

 LEARNING OBJECTIVE

7. Explain the general relationship of total blood flow to local blood flow.

Maintaining sufficient local blood flow transported throughout the body (to ensure adequate perfusion of all tissues) is ultimately dependent on total blood flow. **Total blood flow** is the amount of blood transported throughout the entire vasculature in a given period of time (usually expressed in liters per minute). Total blood flow equals cardiac output. The average cardiac output at rest is 5.25 liters per minute (L/min), and may increase substantially during exercise (see section 19.9). If cardiac output increases, total blood flow increases, and additional blood is available to body tissues. If cardiac output decreases, total blood flow decreases, and less blood is available to the tissues. The factors that regulate total blood flow are dependent upon the two components of the cardiovascular system (both the heart and blood vessels), as well the blood that is transported through them. These factors are described in detail in the next section.

 WHAT DID YOU LEARN?

15 How is local blood flow dependent on total blood flow?

Table 20.3		Substances and Systems That Affect Blood Pressure and Flow
Effect	**Local Substances**	**Hormones and Neurotransmitters**
Vasodilators Vasodilation Blood flow 	Decreased oxygen levels Decreased nutrient levels Increased CO_2, H^+, K^+, lactate levels Histamine Bradykinin Nitric oxide	Atrial natriuretic peptide (ANP) Epinephrine (bound to β-adrenergic receptors within coronary and skeletal muscle blood vessels)
Vasoconstrictors Vasoconstriction Blood flow	Increased oxygen levels Increased nutrient levels Decreased CO_2, H^+, K^+, lactate levels Endothelins Prostaglandins Thromboxanes	Angiotensin II Aldosterone Antidiuretic hormone (ADH) Norepinephrine (bound to α-adrenergic receptors of most blood vessels, including the skin and abdominal organs)[1]

1. A *decrease* in sympathetic stimulation will result in a decrease in the listed effect, much like taking the foot off of the gas pedal will slow down a car.

20.5 Blood Pressure, Resistance, and Total Blood Flow

Here we integrate concepts on the heart, blood vessels, and blood to describe: first, blood pressure and how the pumping action of the heart establishes a pressure gradient to drive the blood through the vasculature; second, how blood experiences resistance as it is transported through blood vessels; and third, how together blood pressure and resistance determine total blood flow.

20.5a Blood Pressure

LEARNING OBJECTIVES

1. Define blood pressure and a blood pressure gradient.
2. Compare and contrast blood pressure and blood pressure gradients in the arteries, capillaries, and veins.
3. Calculate pulse pressure and mean arterial pressure (MAP) in the arteries.
4. Explain the mechanisms that help overcome the small pressure gradient in veins to return blood to the heart.

Blood pressure is the force per unit area that blood exerts against the inside wall of a vessel (as described earlier in the section on bulk flow). A **blood pressure gradient** is the change in blood pressure from one end of a blood vessel to its other end. A blood pressure gradient exists in the vasculature because blood pressure is highest in the arteries as the heart rhythmically contracts, and it is lowest in the veins. Blood pressure gradients are both clinically and physiologically significant because they are the driving force that propels blood through the vessels. Please refer to **figure 20.11** as you read through this section.

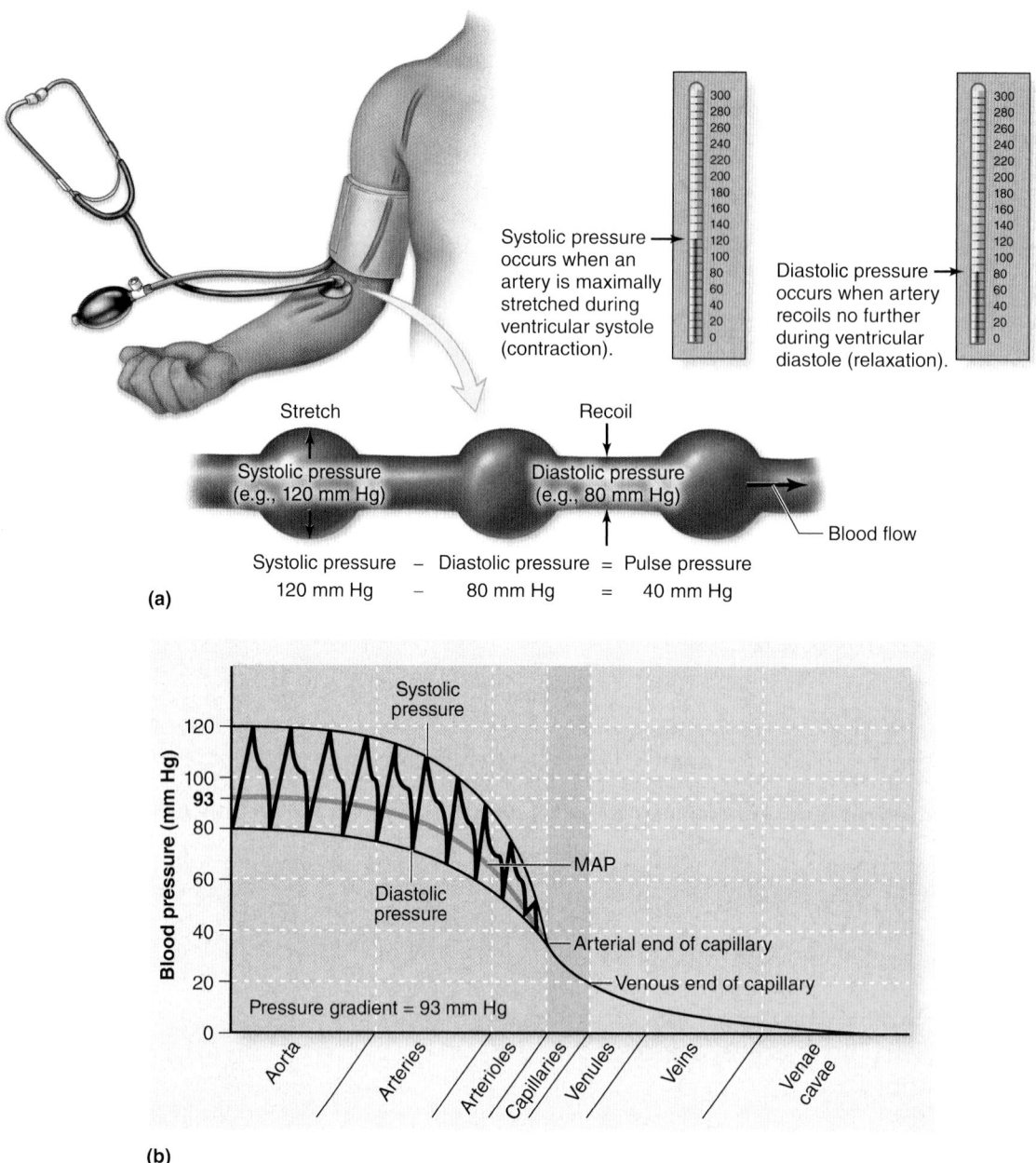

(a)

Systolic pressure occurs when an artery is maximally stretched during ventricular systole (contraction).

Diastolic pressure occurs when artery recoils no further during ventricular diastole (relaxation).

Stretch

Recoil

Systolic pressure (e.g., 120 mm Hg)

Diastolic pressure (e.g., 80 mm Hg)

Blood flow

Systolic pressure − Diastolic pressure = Pulse pressure
120 mm Hg − 80 mm Hg = 40 mm Hg

(b)

Figure 20.11 **Blood Pressure.** (*a*) Blood pressure within the arteries is pulsatile. Systolic pressure is when the artery is maximally stretched during ventricular contraction. Diastolic pressure is when the vessel recoils no further, which occurs during ventricular relaxation. Pulse pressure is the additional pressure placed on the artery from when the heart is relaxing to when it is contracting. (*b*) A tracing compares the blood pressure changes as blood flows through the different types of vessels in the cardiovascular system. (MAP is mean arterial pressure, which is discussed on page 795.)

Arterial Blood Pressure

Blood flow is pulsing, or *pulsatile,* in arteries as a consequence of the ventricles contracting and relaxing. The highest blood pressure generated in arteries is during ventricular systole when the artery is maximally stretched; this value is recorded as the **systolic pressure.** The lowest pressure is during ventricular diastole when the artery recoils no further; this value is recorded as the **diastolic pressure.** Arterial blood pressure is expressed as a ratio, in which the numerator (upper number) is the systolic pressure and the denominator (lower number) is the diastolic pressure. The average adult has a blood pressure of 120/80 mm Hg, but blood pressure can vary greatly among individuals. Two values can be calculated from systolic pressure and diastolic pressure: pulse pressure and mean arterial pressure.

Pulse Pressure   **Pulse pressure** is the additional pressure placed on the arteries from when the heart is resting (diastolic blood pressure) to when the heart is contracting (systolic blood pressure). Pulse pressure can be calculated by taking the difference between the systolic and the diastolic blood pressure. So, for an individual with a blood pressure of 120/80 mm Hg, the pulse pressure would be 40 mm Hg (120 mm Hg − 80 mm Hg = 40 mm Hg).

Pulse pressure is significant because it is a measure of the elasticity and recoil of arteries. Healthy elastic arteries expand and recoil easily, assisting in the movement of blood through the cardiovascular system. However, as vessels age or become diseased (e.g., with atherosclerosis), arteries lose their elasticity and expand and recoil less readily, making it more difficult for the heart to pump blood. Thus, although temporary changes in pulse pressure are associated

CLINICAL VIEW
Detecting a Pulse Point

A **pulse** refers to the rhythmic throbbing of an arterial wall as blood is being pumped through it. A pulse reflects the pulse pressure. Measuring the pulse is clinically significant for the following reasons:

1. It allows us to indirectly determine the rate of our heartbeat.
2. The force of the pulse can inform us about blood pressure, as high blood pressure will produce a more forceful pulse, and low blood pressure will produce a weaker pulse.

3. The absence of a pulse can allow us to determine if blood flow to a body part is lacking.

Pulse points can be found throughout the body and typically are located where an artery may be compressed against a bone or other solid structure. Two fingers are used to palpate the pulse, but the thumb is typically not used because it has a weak pulse of its own, which may interfere with pulse detection. Some common pulse locations are shown and listed in **table 20.4**. See if you can palpate these pulse points on yourself.

Table 20.4	Common Pulse Points	
Artery	**Best Location to Detect Pulse**	
Superficial temporal	Anterior to the ear superior to the root (origin) of the zygomatic process of the temporal bone	Superficial temporal artery — Facial artery — Common carotid artery — Brachial artery — Radial artery — Femoral artery — Popliteal artery — **Posterior** — Posterior tibial artery — Dorsalis pedis artery —
Facial	Immediately anterior to the angle of the mandible and the masseter muscle	
Common carotid	Anterior to the sternocleidomastoid muscle and lateral to the larynx and trachea	
Brachial	Along the medial surface of the arm, midway between the axilla and antecubital regions	
Radial	Radial (lateral) side of the wrist, between the brachioradialis and flexor carpi radialis tendons	
Femoral	Immediately inferior to the inguinal ligament (groin), approximately halfway between the pubis and the anterior superior iliac spine	
Popliteal	Popliteal fossa, with the knee flexed slightly	
Posterior tibial	Posteroinferior to the medial malleolus of the tibia	
Dorsalis pedis	Either over the navicular tarsal (on the dorsal medial side of the foot) or between the dorsal interspace between the first and second toes	

with increased cardiac output, such as would occur during exercise, permanent changes in pulse pressure may be an indication of unhealthy arteries.

The rhythmic throbbing sensations associated with pulse pressure are palpated in superficial elastic and muscular arteries to determine one's "pulse" (see table 20.4).

Mean Arterial Pressure **Mean arterial pressure (MAP)** is the average (or mean) measure of the blood pressure forces on the arteries. Because diastolic pressure usually lasts slightly longer than systolic

pressure, MAP is not simply an average of these two pressures. Rather, MAP may be estimated as follows:

$$\text{MAP} = \text{Diastolic pressure} + \frac{1}{3} \text{ Pulse pressure}$$

So for a person with average blood pressure of 120/80 mm Hg, his or her MAP would be approximately 93 mm Hg (80 + 40/3 = 93).

Mean arterial pressure is clinically significant because it provides a numerical value for how well body tissues and organs are perfused. A MAP of 70 to 110 mm Hg typically indicates good perfusion. A MAP lower than 60 mm Hg may indicate insufficient blood flow, and a very

INTEGRATE

CLINICAL VIEW
Cerebral Edema

Maintaining a normal mean arterial pressure (MAP) to the brain is critical. A possible consequence of an elevated MAP is **cerebral edema,** which is excess interstitial fluid in the brain. It can occur if MAP is greater than 160 mm Hg, a pressure that is reached, for example, with a blood pressure reading of 240/140 mm Hg. The high MAP substantially increases filtration in the capillaries of the brain, and because there are no lymph vessels associated with the central nervous system (see section 21.1a), excess fluid remains in the interstitial space.

high MAP could indicate the delivery of too large of blood flow to body tissues with the possibility of causing edema (swelling) in the tissues (see Clinical View: "Cerebral Edema").

Pulse pressure and mean arterial pressure are highest in the arteries closest to the heart, such as the aorta. As the arteries branch into smaller vessels and are greater distances from the ventricles, both of these pressures decrease. The pressure gradient in arteries is relatively steep and facilitates the movement of blood through the arteries (see figure 20.11*b*).

Capillary Blood Pressure

By the time the blood reaches the capillaries, fluctuations between systolic and diastolic blood pressure disappear, so the pulse pressure disappears. Blood flow is smooth and even as it enters the capillaries.

Capillary blood pressure must be sufficient for exchange of substances between the blood and surrounding tissue, but not be so high that it would damage the relatively fragile capillaries. Blood pressure on the arterial end of the capillary is about 40 mm Hg and drops quickly (along the approximately 1-millimeter length of a capillary) to below 20 mm Hg on the venous end of the capillary. These blood pressure values are used to determine net filtration pressure for capillary exchange as previously described in section 20.3. Recall that the

relatively high blood pressure on the arteriole end of the capillary accounts for filtration, and the relatively low blood pressure on the venous end allows for reabsorption as colloid osmotic pressure pulls fluids back into the blood.

Venous Blood Pressure

The movement of blood from the capillaries back to the heart via the veins is called **venous return.** Blood pressure in the venules and veins is not pulsatile because the blood is far removed from (and thus not influenced by) the pumping action of the heart. Therefore, veins also have no demonstrable pulse pressure. Blood pressure is 20 mm Hg in the venules and almost 0 mm Hg by the time blood travels through the inferior vena cava to the right atrium of the heart. Thus, the pressure gradient in the veins is only 20 mm Hg. This relatively small blood pressure gradient is generally insufficient to move blood through the veins under given conditions without assistance, such as when an individual is standing. Thus, venous return must be facilitated by valves within veins and two "pumps"—the skeletal muscle pump and the respiratory pump (figure 20.12).

The **skeletal muscle pump** assists the movement of blood primarily within the limbs. As skeletal muscles contract, veins are squeezed to help propel the blood toward the heart, and valves prevent blood backflow. When skeletal muscles are more active—for example, when a person is walking—blood is pumped more quickly and efficiently toward the heart by the skeletal muscle pump. Conversely, extended inactivity leads to blood pooling in the leg veins, which increases an individual's risk for development of deep vein thrombosis (see Clinical View: "Deep Vein Thrombosis").

The **respiratory pump** assists the movement of blood within the thoracic cavity. The diaphragm contracts and flattens as we inspire (inhale). Intra-abdominal pressure increases and places pressure on the vessels within the abdominal cavity. Concomitantly, thoracic cavity volume increases and intrathoracic pressure decreases. Blood is propelled from the vessels in the abdominopelvic cavity into the vessels in the thoracic cavity. When we expire (exhale), the diaphragm

INTEGRATE

CLINICAL VIEW
Deep Vein Thrombosis

Deep vein thrombosis (throm-bō'sis; a clotting) **(DVT)** refers to a **thrombus** (blood clot) in a vein. The most common site for the thrombus is a vein in the calf (sural) region. DVT typically occurs in individuals with heart disease or those who are inactive or immobile for a long period of time, such as bedridden patients. Even healthy individuals who have been on a long airline trip may develop DVT.

Initial signs of DVT include fever, tenderness and redness in the affected area, severe pain and swelling in the areas drained by the affected vein, and rapid heartbeat. The most serious complication of DVT is a **pulmonary embolus** (em'bō-lŭs; a plug), in which a blood clot breaks free and is transported to the lung, eventually blocking a branch of the pulmonary artery and potentially causing respiratory failure and death. If a DVT is diagnosed, the patient is given anticoagulation medication, such as low-molecular-weight heparin, to help prevent further clotting and break up the existing clot.

INTEGRATE

CLINICAL VIEW
Varicose Veins

Varicose (var'i-kōs; *varix* = dilated vein) **veins** are dilated and tortuous (having many curves or twists). The valves in these veins have become nonfunctional, causing blood to pool in one area and the vein to swell and bulge. Varicose veins are most common in the superficial veins of the lower limbs. They may be a result of genetic predisposition, aging, or some form of stress that inhibits venous return (such as standing for long periods of time, obesity, or pregnancy). Varicose veins in the anorectal region are called **hemorrhoids** (hem'ŏ-rōydz). Hemorrhoids occur due to increased intra-abdominal pressure, as when a person strains to have a bowel movement or is in labor during childbirth.

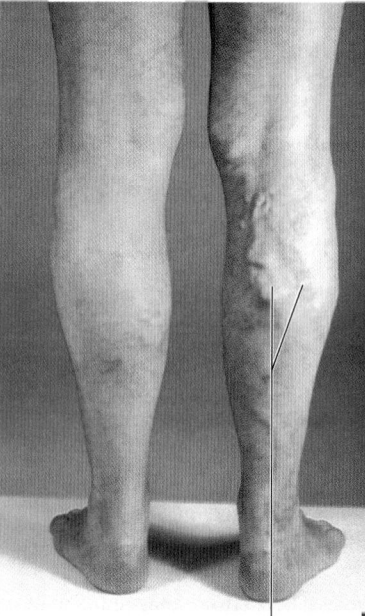

Varicose veins

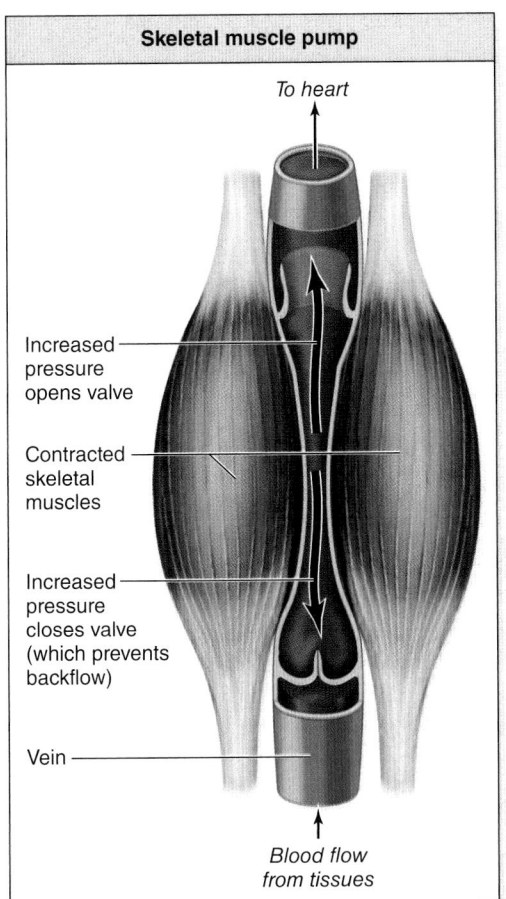

INTEGRATE

CLINICAL VIEW
Circulatory Shock

Circulatory shock is any state in which there is insufficient blood flow for adequate perfusion of the body's tissues, typically due either to impaired pumping of the heart (e.g., caused by congestive heart failure or a malfunctioning pacemaker) or to low venous return. Circulatory shock associated with low venous return can result from the following:

- Decreased blood volume due to hemorrhage, dehydration, or systemic release of histamine in an allergic reaction that increases capillary permeability
- An obstructed vein
- Venous pooling caused by extended immobility or by extensive vasodilation (resulting from certain bacterial toxins or from brainstem trauma that results in loss of vasomotor tone)

Blood Pressure Gradient in the Systemic Circulation

We now know the normal range of blood pressure values in the various portions of the vasculature. We can use the average blood pressure in the arteries close to the heart and the blood pressure in the inferior vena cava to calculate the total blood pressure gradient in the systemic circulation. This can be determined by viewing figure 20.11. The average or mean arterial blood pressure (MAP) in the arteries is 93 mm Hg. The blood pressure in the inferior vena cava is 0 mm Hg. Thus, the blood pressure gradient established by the pumping action of the heart is 93 mm Hg.

Most importantly, this blood pressure gradient is the driving force to move blood through the vasculature. Changes in the blood pressure gradient are directly correlated with changes in total blood flow. An increase in the blood pressure gradient increases total blood flow, whereas a decrease in the blood pressure gradient decreases total blood flow.

INTEGRATE

LEARNING STRATEGY

A blood pressure gradient is established by the heart when it pumps blood through the vessels, much like a pressure gradient is created by a water pump to force water through pipes. In both cases, the stronger the pump, the greater the pressure gradient that can be established, and the greater the flow that is possible.

relaxes and returns to its dome shape. Thoracic cavity volume decreases and intrathoracic pressure increases, which places pressure on vessels within the thoracic cavity. Blood moves from the vessels in the thoracic cavity back into the heart. In addition, intra-abdominal pressure decreases, allowing blood to move from the lower limbs into the abdominal vessels. When breathing rate increases—for example, when a person is exercising—blood is moved more quickly back to the heart by the respiratory pump.

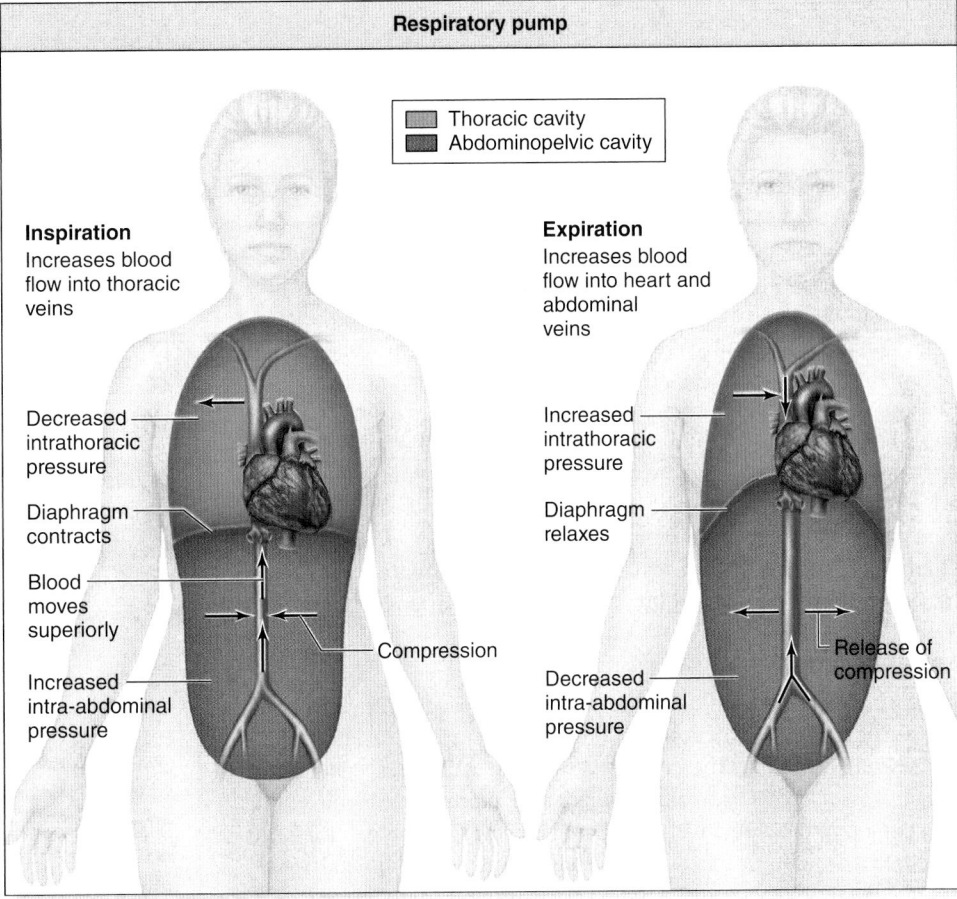

Figure 20.12 Factors That Influence Venous Return. To help overcome the small blood pressure gradient within veins, blood return to the heart via veins is facilitated by both (*a*) a skeletal muscle pump within the limbs, and (*b*) the respiratory pump within the torso.

How is it possible to change the steepness of the pressure gradient? The blood pressure gradient is altered by changes in cardiac output. An increase in cardiac output will increase the pressure gradient. Conversely, a decrease in cardiac output will decrease the pressure gradient (see section 19.9).

 WHAT DID YOU LEARN?

16 A 55-year-old female has an arterial blood pressure reading of 155/95 mm Hg. What are her pulse pressure and mean arterial pressure?

17 What is the physiologic significance of capillary blood pressure?

18 How is the small pressure gradient in veins overcome?

19 How is the pressure gradient to move blood through the systemic circulation calculated? What is the significance of this pressure gradient?

20.5b Resistance

LEARNING OBJECTIVE

5. Define resistance, and explain how it is influenced by blood viscosity, vessel length, and vessel radius.

Resistance also influences total blood flow. **Resistance** is defined as the amount of friction the blood experiences as it is transported through the blood vessels. Blood flow is always opposed by resistance. This friction is due to the contact between blood and the blood vessel wall. The term **peripheral resistance** is typically used when discussing the resistance of blood in the blood vessels (as opposed to the resistance of blood in the heart). Several factors affect peripheral resistance, including blood viscosity, blood vessel length, and the size of the lumen of blood vessels (as indicated by vessel radius).

Blood Viscosity

Viscosity refers to the resistance of a fluid to its flow. More generally, it is the "thickness" of a fluid. The thicker the fluid, the more viscous it is, and the greater its resistance to flow. The thickness is dependent upon the relative percentage of particles in the fluid and their interactions with one another. Because blood contains formed elements (erythrocytes, leukocytes, platelets) and plasma proteins, it is approximately 4.5 to 5.5 times more viscous than water. As a result, blood exhibits greater resistance to flow than water.

A change in blood's viscosity causes a change in resistance of blood flow through a vessel. For example, if someone is anemic (and has a lower than normal number of erythrocytes), then that person's blood viscosity would be lower, and the blood would have less resistance to flow. Conversely, if someone has a greater than normal amount of formed elements (e.g., if an individual increases the number of erythrocytes in his or her blood; see Clinical View: "Blood Doping" in section 18.3b), or the individual is dehydrated, then the blood would be more viscous and have greater resistance to flow.

Vessel Length

Increasing vessel length increases resistance, because the longer the vessel, the greater the friction the fluid experiences as it travels through the vessel. Consequently, shorter vessels offer less resistance than longer vessels with comparable diameters. Normally, vessel length in a person remains relatively constant. However, if one gains a large amount of weight, the body must produce miles of additional vessel length by angiogenesis for blood to be transported through the extra fat. Thus, vessel resistance increases if one gains weight, and decreases

if someone loses a lot of weight (because those vessels are no longer needed and regress).

Vessel Radius

Blood viscosity and vessel length remain relatively constant in a typical healthy individual. The major way resistance may be regulated is by altering vessel lumen radius (and thus changing the vessel diameter.)

How specifically does vessel radius influence resistance? Blood tends to flow fastest in the center of the vessel lumen, while blood near the sides of the vessels slows, because it encounters resistance from the nearby vessel wall. This difference in flow rate within a blood vessel (or in any conduit) is called **laminar flow.** You can see evidence of laminar flow by studying a river: The water flow near the banks or edges of the river is a bit slower or sluggish, while the water flow near the center of the river is quite fast in comparison. So, if vessel diameter increases, relatively less blood flows near the edges and overall blood flow increases. In contrast, if vessel radius decreases, relatively more blood flows near the edges and overall blood flow decreases.

The relationship between blood flow and the radius of blood vessel lumen may be stated as follows (where the symbol $\propto$ means "is proportional to"):

$$F \propto r^4$$

where F = flow and r = radius of the lumen of a vessel (The radius is 1/2 the diameter of the lumen.). This mathematical expression reflects that flow is directly proportional to the fourth power of a radius. If a vessel vasodilates and its radius increases from 1 millimeter (mm) to 2 mm, the overall change in flow is 16 times greater: If $r = 1$ mm, then $r^4 = 1$, and $F = 1$ mm per second; and when $r = 2$ mm, then $r^4 = 16$, and $F = 16$ mm per second. In contrast, if a vessel vasoconstricts and its radius decreases from 2 mm to 1 mm, the overall change in flow is 16 times less.

Any vessel may vasoconstrict or vasodilate; however, resistance usually is regulated specifically by vasoconstriction and vasodilation of muscular arteries and arterioles, a change systemically controlled by the sympathetic division (see section 15.6c). Thus, because there are so many muscular arteries and arterioles (and the overall length of all these vessels is so great), even small changes

---

INTEGRATE

LEARNING STRATEGY

Drinking through a straw provides an easy way to demonstrate the variables that influence resistance:

- **Viscosity.** Water or soda moves through a straw more easily than does a thick vanilla milkshake.
- **Vessel length.** It is easier to drink through a relatively short straw compared with a very long, twisty straw.
- **Vessel radius.** Water can be sipped through a normal-sized straw more easily than through a hollow coffee stirrer with its smaller diameter.

Greater resistance is experienced when drinking thick liquids, using a longer straw, or using a more narrow straw. In the same way, an increase in resistance is exhibited with more viscous blood, increased total length of blood vessels, and vasoconstriction of blood vessels.

in vessel radius significantly change resistance, and therefore can have dramatic effects on blood flow.

Note that atherosclerosis (see Clinical View: "Atherosclerosis" in section 20.1b) is a disease process where a plaque narrows the lumen of blood vessels. Thus, resistance is increased as blood moves through the atherosclerotic blood vessels.

 **WHAT DID YOU LEARN?**

20 How is resistance defined?

21 What are the three factors that alter resistance? How does each affect blood flow in vessels?

20.5c Relationship of Blood Flow to Blood Pressure Gradients and Resistance

LEARNING OBJECTIVES

6. Explain the relationship of both the blood pressure gradient and resistance to total blood flow.

7. Discuss why blood pressure increases with increased resistance in the systemic circulation.

Total blood flow is the amount of blood that moves through the cardiovascular system per unit time and is influenced by both blood pressure gradients and resistance, as previously described. This relationship is expressed mathematically as follows:

$$F \propto \frac{\Delta P}{R}$$

where F = blood flow, ΔP = pressure gradient ($P_1 - P_2$), and R = resistance.

Systemic Blood Pressure Gradient

The preceding formula demonstrates that blood flow is *directly* related to the pressure gradient. Thus, as the blood pressure gradient increases, total blood flow is greater, and as the blood pressure gradient decreases, total blood flow lessens (assuming resistance remains the same). Recall that an increase in cardiac output will increase the pressure gradient, whereas a decrease in cardiac output will decrease the pressure gradient. Consequently, increasing blood flow is possible with a steeper blood pressure gradient—however, it is a change exhibited only with greater effort exerted by the heart.

Resistance

The preceding mathematical expression also demonstrates that blood flow is inversely related to resistance. If resistance increases, blood flow lessens, whereas if resistance decreases, blood flow increases (assuming the pressure gradient remains the same).

Recall that resistance is increased with (1) an increase in blood viscosity (e.g., as occurs with greater concentration of erythrocytes); (2) an increase in blood vessel length (e.g., a change that accompanies weight gain); and (3) a decrease in vessel lumen diameter (e.g., a change that occurs with a net increase in vasoconstriction and in individuals with atherosclerosis).

Relationship of Systemic Blood Pressure and Resistance

Individuals with *sustained* increased resistance (e.g., that accompanies significant weight gain, or with atherosclerosis) generally exhibit elevated arterial blood pressure readings. This condition is clinically significant; it occurs because a greater pressure gradient must be produced to overcome the higher resistance and ensure normal blood flow and adequate perfusion of all tissues. The relationship of blood flow, blood pressure gradients, and resistance is summarized in **figure 20.13**.

 WHAT DID YOU LEARN?

22 In general, would you predict a higher blood pressure, lower blood pressure, or a normal blood pressure in individuals with sustained increased resistance? Explain.

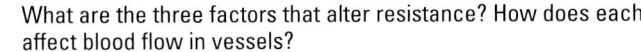

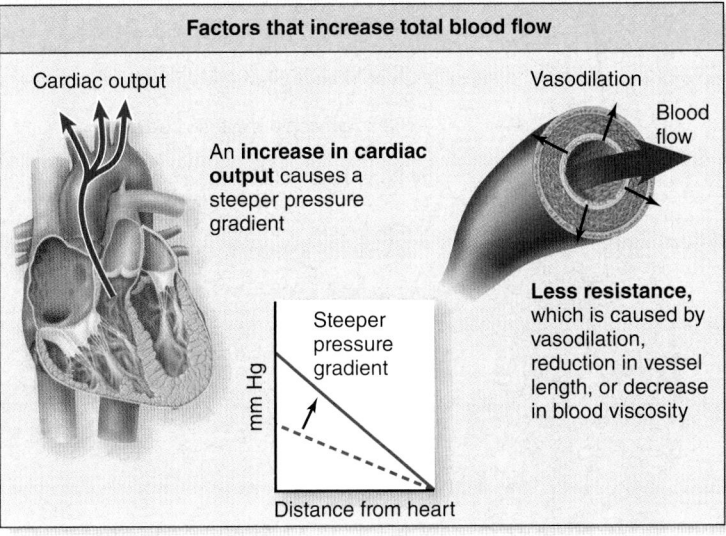

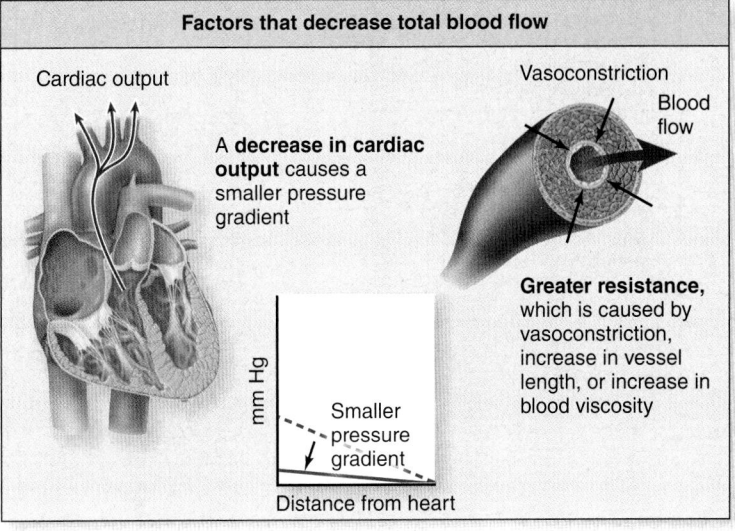

(a) (b)

Figure 20.13 Factors That Influence Total Blood Flow. Maintaining adequate perfusion of systemic tissues is dependent upon maintaining total blood flow. Total blood flow is a function of the pressure gradient and resistance and can be either (*a*) increased or (*b*) decreased.

20.6 Regulation of Blood Pressure and Blood Flow

Blood pressure must be high enough to propel blood through the cardiovascular system to maintain adequate perfusion of all tissues, yet not so high that it causes damage to the blood vessels. Blood pressure is dependent upon three primary variables: cardiac output, resistance, and blood volume. Regulation of these variables is critical in maintaining homeostasis and occurs through short-term mechanisms of the nervous system, long-term mechanisms of the endocrine system, or both.

20.6a Neural Regulation of Blood Pressure

LEARNING OBJECTIVES

1. Describe the anatomic components associated with regulating blood pressure through short-term mechanisms.

2. Explain the autonomic reflexes that alter blood pressure.

Short-term regulation of blood pressure occurs through autonomic reflexes involving nuclei within the medulla oblongata (see section 13.5c). These reflexes adjust blood pressure quickly, such as occurs when you arise from a sitting to a standing position, by altering cardiac output, resistance, or both. We first describe the anatomic structures involved in the autonomic reflexes and then discuss how they function to maintain a normal blood pressure. Please refer to **figure 20.14** as you read through this section.

Cardiovascular Center

Two distinct groups of autonomic nuclei in the medulla oblongata participate in the regulation of blood pressure, specifically the cardiac center and the vasomotor center. Collectively, these two centers are called the **cardiovascular center.** The **cardiac center** regulates heart activity (and thus cardiac output), and the **vasomotor center** controls the degree of vasoconstriction and vasodilation of blood vessels (and thus, regulates resistance).

Cardiac Center Recall that two regulatory nuclei are housed within the cardiac center: the **cardioacceleratory center** and the **cardioinhibitory center** (see section 19.5b). Sympathetic division pathways extend from the cardioacceleratory center to both the sinoatrial (SA) node and the atrioventricular (AV) node. Nerve signals relayed along these sympathetic pathways cause release of norepinephrine from their ganglionic neurons, which both (1) increases the firing rate of the SA node and (2) shortens the delay of nerve signals at the AV node as they are relayed along the heart's conduction pathway. Thus, sympathetic stimulation increases heart rate. In addition, sympathetic division pathways extend to the myocardium that when stimulated cause a more forceful contraction. Thus, sympathetic stimulation also increases stroke volume. The resulting increase in both heart rate and stroke volume produce a greater cardiac output.

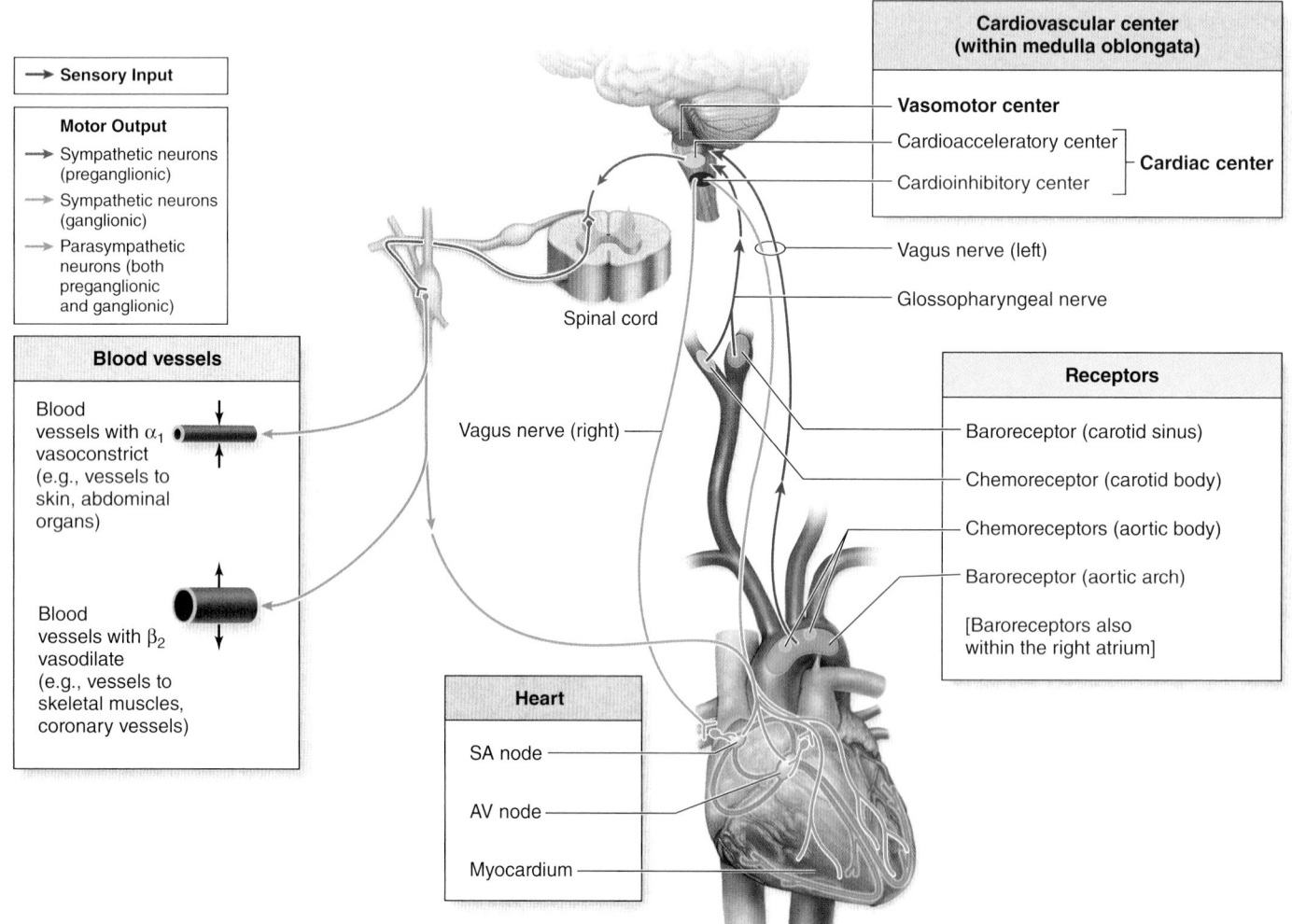

Figure 20.14 Cardiovascular Center. The cardiovascular center regulates blood pressure via a negative feedback loop. It receives sensory input from baroreceptors and chemoreceptors within the carotid arteries and aortic arch. This center then regulates blood pressure (through motor output along sympathetic nerves and vagus nerves) by adjusting cardiac output and peripheral resistance experienced by blood in blood vessels.
AP|R

Parasympathetic division pathways extend from the cardioinhibitory center to also innervate both the SA node and AV node. Nerve signals relayed along these parasympathetic pathways cause release of acetylcholine from their ganglionic neurons, which both decreases the firing rate of the SA node and lengthens the delay of nerve signals at the AV node as they are relayed along the heart's conduction pathway. Thus, parasympathetic stimulation decreases heart rate. The resulting decrease in heart rate produces a smaller cardiac output. (Note that parasympathetic pathways do not extend to the myocardium and so do not alter the force of contraction and stroke volume.)

Vasomotor Center   The vasomotor center, in comparison, regulates the degree of vasoconstriction and vasodilation of blood vessels. Blood vessels, unlike the heart, do not have dual innervation (see section 15.6b). Instead, they are typically innervated by only the sympathetic division with no innervation by the parasympathetic division. Thus, only sympathetic division pathways extend from the vasomotor center to most blood vessels; the neurotransmitter norepinephrine is also released from these ganglionic neurons.

Blood Vessels   Variations in response of blood vessels (vasoconstriction and vasodilation) is dependent upon the subtypes of receptors in the smooth muscle cells within the tunica media of the blood vessel wall. The receptor subtype is specific to the location of the blood vessels within the different vascularized tissues of the body (e.g., skeletal muscle, skin). These smooth muscle cells typically contain one of several subtypes that bind different neurotransmitters and hormones. Two of the most common subtypes are **alpha1 (α_1) receptors** and **beta2 (β_2) receptors** (see section 15.5c). Smooth muscle cells that contain α_1 receptors contract in response to norepinephrine, which causes vasoconstriction of these blood vessels. In contrast, smooth muscle cells that contain β_2 receptors relax in response to epinephrine, which causes vasodilation of these blood vessels. Note that the source of epinephrine hormone (and some norepinephrine) is the adrenal medulla, which releases both of these hormones in response to stimulation by the sympathetic division; see section 15.4b.

Activation of the sympathetic division and subsequent stimulation of the adrenal medulla cause the following:

- **Increased peripheral resistance.** More blood vessels are stimulated to vasoconstrict than vasodilate because there are more blood vessels that house α_1 receptors than β_2 receptors; consequently, peripheral resistance is increased, raising blood pressure.
- **Larger circulating blood volume.** Vasoconstriction of veins shifts blood from venous reservoirs and circulating blood volume increases, thus increasing blood pressure.
- **Redistribution of blood flow.** More blood flow reaches skeletal muscles and the heart, and less blood flow goes to most other structures. Thus, organs requiring additional nutrients and oxygen have adequate perfusion.

Inhibition of the sympathetic division reverses these changes: Peripheral resistance decreases, blood shifts to venous reservoirs, and blood flow distribution returns toward its previous levels.

Baroreceptors

Baroreceptors are specialized sensory nerve endings that respond to stretch (see section 16.1d). Here the baroreceptors detect stretch of the blood vessel wall that occurs as the blood volume within them changes. The two main baroreceptors for the cardiovascular system are those located within the aortic arch and carotid sinuses, specifically within the tunica externa.

The aortic arch baroreceptors transmit nerve signals to the cardiovascular center through the vagus nerve (CN X). The aortic arch baroreceptors are important in regulating systemic blood pressure.

The carotid sinus baroreceptors, which are located within each internal carotid artery, near the artery's initial bifurcation (branching) from the common carotid artery, transmit nerve signals to the cardiovascular center through the glossopharyngeal nerve (CN IX). The baroreceptors within the carotid sinuses monitor blood pressure changes in the head and neck—thus, it is important in monitoring blood pressure affecting the brain. Baroreceptors in the carotid sinuses are more sensitive to blood pressure changes than baroreceptors in the aortic arch—a condition that is not surprising given the importance of delivering sufficient blood to the brain.

Baroreceptors continuously transmit nerve signals along sensory neurons within the vagus nerve and glossopharyngeal nerves to the cardiovascular center at a particular rate. Their firing rate changes when the stretch in the blood vessel wall changes.

Autonomic Reflexes

Baroreceptors are activated in response to changes in stretch of the blood vessel wall to initiate autonomic reflexes that help regulate blood pressure. These reflexes are appropriately called **baroreceptor reflexes.** These reflexes are initiated by either a decrease or an increase in blood pressure.

If blood pressure decreases:

1. Decreased stretch in the blood vessel wall (reflecting a decrease in blood pressure, such as might occur when you arise from bed in the morning) is detected by baroreceptors in the aortic arch, carotid sinuses, or both.
2. The baroreceptors *decrease* the frequency of nerve signals relayed along sensory neurons within the vagus and glossopharyngeal nerves to both the cardiac center and vasomotor center.
3. In response, the cardioacceleratory center within the cardiac center *increases* nerve signals relayed along sympathetic pathways extending to the heart, including the SA node, AV node, and myocardium. Concomitantly, the cardioinhibitory center of the cardiac center *decreases* nerve signals relayed along parasympathetic pathways that extend to the SA node and AV node. Consequently, both heart rate and stroke volume increase, resulting in a greater cardiac output. (This would be analogous to both putting your foot on the gas and taking your foot off the brake when driving a car.)
4. Simultaneously, the vasomotor center *increases* nerve signals along sympathetic pathways that extend to blood vessels, resulting in net vasoconstriction and an increase in peripheral resistance, along with shifting of blood from venous reservoirs.

   The resulting increase in cardiac output, increase in resistance, and larger circulating blood volume quickly elevate blood pressure to maintain sufficient blood pressure to move blood through the vasculature.

If blood pressure increases:

1. Increased stretch in the blood vessel wall (reflecting an increase in blood pressure) is detected by baroreceptors in the aortic arch, carotid sinuses, or both.
2. The baroreceptors *increase* the frequency of nerve signals relayed along sensory neurons within the vagus nerve and glossopharyngeal nerve to both the cardiac center and vasomotor center.

3. In response, the cardioacceleratory center within the cardiac center *decreases* nerve signals relayed along sympathetic pathways extending to the heart, including the SA node, AV node, and myocardium. Concomitantly, the cardioinhibitory center of the cardiac center *increases* nerve signals relayed along parasympathetic pathways that extend to the SA node and AV node. Consequently, both heart rate and stroke volume decrease, resulting in a smaller cardiac output. (This would be analogous to both taking your foot off the gas and putting your foot on the brake when driving a car.)

4. Simultaneously, the vasomotor center *decreases* nerve signals along sympathetic pathways to blood vessels, resulting in net vasodilation and a decrease in resistance, along with shifting of blood to venous reservoirs.

The resulting decrease in cardiac output, decrease in resistance, and smaller circulating blood volume lower blood pressure, and blood flow returns to its resting levels.

Baroreceptors respond best to sudden, short-term changes in blood pressure, but they are not effective long-term or chronic blood pressure regulators. If someone has chronically high blood pressure, eventually the baroreceptors will adapt to the change in blood pressure and thus adjust their normal "set point" (see section 1.5).

 WHAT DO YOU THINK?

2 Nicotine stimulates both the SA node to increase its firing rate and the myocardium to contract more forcefully, which raises cardiac output. It also causes arteriole vasoconstriction. Given this information, would you expect the blood pressure of a smoker to be relatively higher or lower than that of a nonsmoker? Explain.

Chemoreceptor Reflexes

Chemoreceptors, although more important in regulating respiration, are also secondarily involved in regulating blood pressure. When chemoreceptors are stimulated, they initiate **chemoreceptor reflexes,** which are negative feedback loops that ultimately bring blood chemistry levels back to normal.

The two main peripheral chemoreceptors are the **aortic bodies** and **carotid body.** The aortic bodies are located in the arch of the aorta whereas a carotid body is located at the bifurcation of each common carotid as it splits into an external carotid and an internal carotid artery. Both aortic and carotid bodies send sensory input to the cardiovascular center; the aortic bodies send nerve signals via the vagus nerve, and the carotid body along the glossopharyngeal nerve.

High carbon dioxide levels, low pH, and very low oxygen levels stimulate the chemoreceptors, and their increased firing primarily stimulates the vasomotor center. The vasomotor center responds by increasing nerve signals along sympathetic pathways to blood vessels, which increases resistance and shifts blood from venous reservoirs to increase venous return. The changes raise blood pressure and increase blood flow, including blood flow to the lungs, which allows for an accompanying change in respiratory gas exchange. As a result, blood gas levels return to normal. (The reason high carbon dioxide, low blood pH, and very low oxygen levels function in stimulating chemoreceptors is discussed in detail in section 23.5c.)

Higher Brain Centers

Blood pressure is also influenced by higher brain centers. For example, increases in body temperature, or experiencing the fight-or-flight response associated with exercise or an emergency, result in the hypothalamus increasing cardiac output and resistance. Even anxiety over a new or dangerous experience, such as skydiving, can increase blood pressure. In addition, the limbic system alters blood pressure in response to emotions or emotional memories (see section 13.8f).

 WHAT DID YOU LEARN?

23 When are the short-term mechanisms for regulation of blood pressure important?

24 What is the initial change to blood pressure when you arise in the morning? Describe the autonomic reflex to maintain your blood pressure when you arise.

20.6b Hormonal Regulation of Blood Pressure

 **LEARNING OBJECTIVES**

3. Describe the hormones that regulate blood pressure.

4. Explain the renin-angiotensin system and its influence on blood pressure.

5. Contrast the effects of aldosterone, antidiuretic hormone, and angiotensin II on blood pressure with those of atrial natriuretic peptide.

Several hormones, in addition to epinephrine and norepinephrine, participate in regulating blood pressure. These include angiotensin II, antidiuretic hormone, aldosterone, and atrial natriuretic peptide. These typically regulate blood pressure by altering resistance, blood volume, or both. Blood volume is regulated by stimulating fluid intake (assuming fluid intake occurs) or altering urine output. A general description of each hormone is presented next. A more detailed discussion of these hormones is included in chapters 24 and 25.

Renin-Angiotensin System

The **renin-angiotensin system** "straddles" short-term neural regulation and long-term hormonal regulation because the synthesis of the angiotensin II is initiated by the nervous system (short-term mechanisms), and angiotensin II causes the release of other hormones (long-term mechanisms).

The liver produces a plasma protein called angiotensinogen (an inactive hormone) and continuously releases it into the blood. The kidney releases the enzyme **renin** into the blood in response to either low blood pressure or stimulation by the sympathetic division **(figure 20.15).** Within the blood, renin converts **angiotensinogen** into **angiotensin I.** Angiotensin I is then converted to **angiotensin II** by **angiotensin-converting enzyme (ACE),** an enzyme associated with the capillary endothelium. ACE is found in very high concentrations on the pulmonary capillary endothelium, so most (but not all) angiotensin conversion occurs in the lungs.

Having most of the ACE enzyme in pulmonary capillary endothelium helps ensure that sufficient angiotensin I is converted to angiotensin II. This is because all blood moves through the pulmonary circulation to be oxygenated, so contact between angiotensin I and ACE is maximized.

Angiotensin II has several important effects: It is a powerful vasoconstrictor—much more powerful than comparable hormones, such as norepinephrine—and thus it increases peripheral resistance and raises blood pressure to a greater extent. Angiotensin II stimulates the thirst center; fluid intake increases blood volume, which increases blood pressure. Angiotensin II also regulates blood volume by direct action in the kidneys to decrease urine formation, and by indirect action through stimulating the release of other hormones (aldosterone

Figure 20.15 Renin-Angiotensin System.
The enzyme renin, released by the kidney in response to low blood pressure, initiates a series of enzymatic chemical reactions within the blood that ultimately help raise blood pressure.

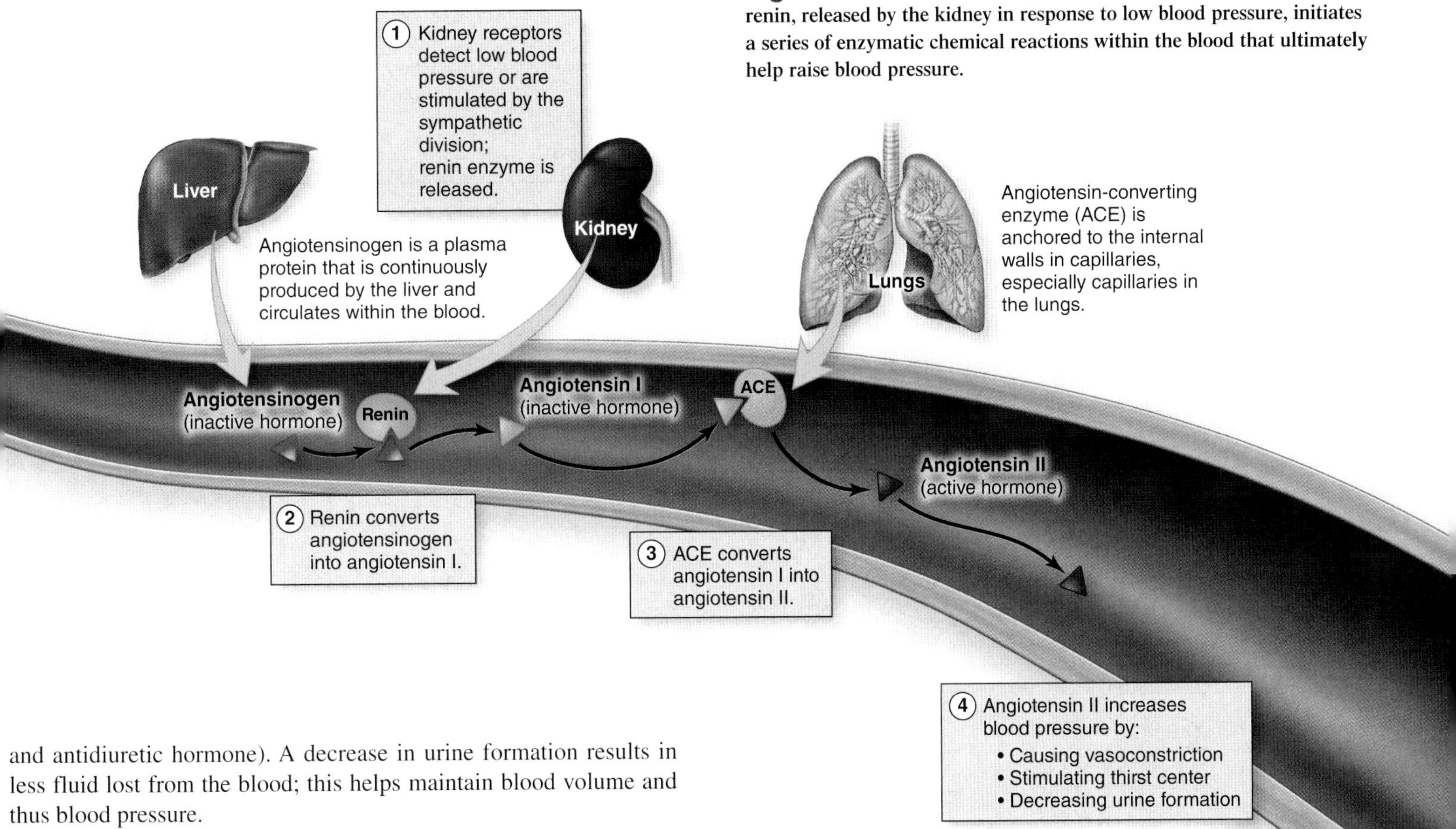

① Kidney receptors detect low blood pressure or are stimulated by the sympathetic division; renin enzyme is released.

Angiotensinogen is a plasma protein that is continuously produced by the liver and circulates within the blood.

Liver

Kidney

Angiotensin-converting enzyme (ACE) is anchored to the internal walls in capillaries, especially capillaries in the lungs.

Lungs

Angiotensinogen (inactive hormone) Renin

Angiotensin I (inactive hormone)

ACE

Angiotensin II (active hormone)

② Renin converts angiotensinogen into angiotensin I.

③ ACE converts angiotensin I into angiotensin II.

④ Angiotensin II increases blood pressure by:
- Causing vasoconstriction
- Stimulating thirst center
- Decreasing urine formation

and antidiuretic hormone). A decrease in urine formation results in less fluid lost from the blood; this helps maintain blood volume and thus blood pressure.

WHAT DO YOU THINK?

3 Do you think an ACE inhibitor drug would be used to treat hypertension or hypotension? Explain.

Aldosterone and Antidiuretic Hormone

Aldosterone is released from the adrenal cortex in response to several stimuli, including angiotensin II. Aldosterone increases the absorption of sodium ion (Na^+) and water in the kidney, decreasing their loss in the urine; this helps maintain blood volume and blood pressure.

Antidiuretic hormone (ADH) is released from the posterior pituitary in response to nerve signals from the hypothalamus (see section 17.7b). The hypothalamus stimulates the posterior pituitary following either detection of increased concentration of blood (typically correlated with low blood volume) or stimulation of the hypothalamus by angiotensin II. ADH increases the absorption of water in the kidney, decreasing its loss in the urine; this helps maintain blood volume and blood pressure. ADH also stimulates the thirst center so there is fluid intake, and blood volume increases. During extreme cases of low blood volume, such as might occur with hemorrhaging, extensive release of ADH occurs, which causes vasoconstriction. This vasoconstriction increases peripheral resistance and blood pressure. This is why ADH is also referred to as *vasopressin*.

In summary, angiotensin II, aldosterone, and ADH decrease urine output to help maintain blood volume and blood pressure. Angiotensin II and ADH (in high doses) increase peripheral resistance and blood pressure, and further increase blood pressure if there is fluid intake.

Atrial Natriuretic Peptide

Atrial natriuretic peptide (ANP) is released from the atrium of the heart in response to an increase in stretch of the atrial walls due to increased blood volume and increased venous return. ANP both (1) stimulates vasodilation, which decreases peripheral resistance and (2) increases urine output, which decreases blood volume. The net effect is a decrease in blood pressure. The details for these four hormones are listed in the summary table on regulating fluid balance, blood volume, and blood pressure, which directly follows chapter 17 (table R.7).

Integration of Variables That Influence Blood Pressure

Homeostatic mechanisms to maintain a normal blood pressure are dependent on three primary variables: cardiac output, resistance, and blood volume. All three of these variables are *directly* related to blood pressure. An increase in any of the three increases blood pressure, and a decrease in any of the three decreases blood pressure. The relationship of these primary variables is summarized in **figure 20.16**.

WHAT DID YOU LEARN?

25 How is angiotensinogen activated to become angiotensin II? How does angiotensin II influence blood pressure?

26 Which hormone decreases blood pressure?

Figure 20.16 Factors That Regulate Blood Pressure. Three primary factors influence blood pressure: (*a*) cardiac output, (*b*) peripheral resistance, and (*c*) blood volume.

(a) Cardiac Output (CO)

Cardiac output is the volume of blood pumped per minute. CO is a function of heart rate (HR) and stroke volume (SV): **CO = HR × SV.**

Heart rate (HR)

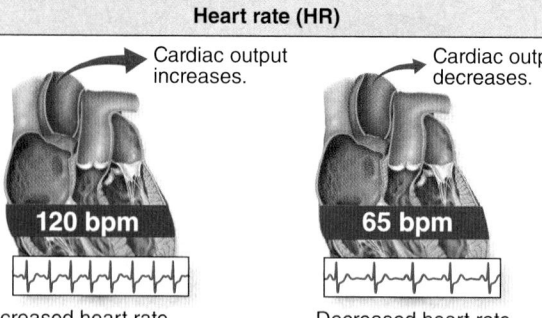

Cardiac output increases.

Cardiac output decreases.

120 bpm

65 bpm

Increased heart rate increases cardiac output and blood pressure.

Decreased heart rate decreases cardiac output and blood pressure.

Stroke volume (SV)

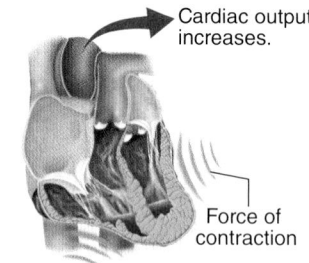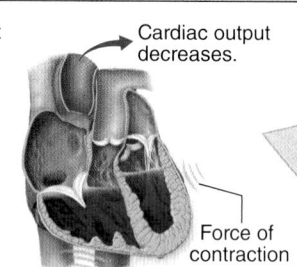

Cardiac output increases.

Cardiac output decreases.

Force of contraction

Force of contraction

Increased stroke volume increases cardiac output and blood pressure.

Decreased stroke volume decreases cardiac output and blood pressure.

INTEGRATE

CLINICAL VIEW

Measuring Blood Pressure

Arterial blood pressure is measured indirectly using a **sphygmomanometer** (sfig´mō-mă-nom´ĕ-ter; *phygmos* = pulse, *metron* = measure). A cuff is wrapped around the arm, and a stethoscope is placed just distal to the compressed artery, allowing a practitioner to listen for pulse sounds. The cuff is inflated until the brachial artery is completely compressed, temporarily stopping blood flow. The pressure in the cuff is then decreased as air is gradually released. Two values are recorded (e.g., 120/80), and the unit of measurement is millimeters of mercury (mm Hg).

Brachial artery
Stethoscope

Blood pressure cuff

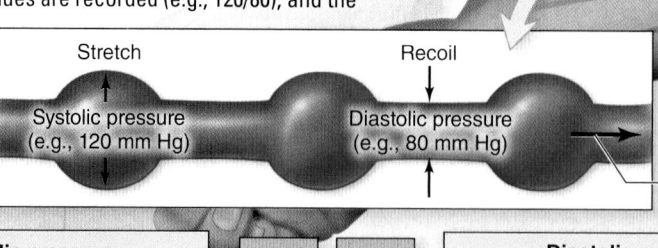

Stretch

Recoil

Systolic pressure (e.g., 120 mm Hg)

Diastolic pressure (e.g., 80 mm Hg)

Blood flow

Systolic pressure

- "Top" number of a blood pressure reading
- Pressure in the arteries when the ventricles contract
- Recorded when the first pulsating sound is heard (when pressure in the brachial artery is sufficient to overcome pressure in the cuff, reestablishing blood flow)

300	300
280	280
260	260
240	240
220	220
200	200
180	180
160	160
140	140
120	120
100	100
80	80
60	60
40	40
20	20
0	0

Diastolic pressure

- "Bottom" number of a blood pressure reading
- Pressure in the arteries when the ventricles relax
- Recorded when sounds are no longer heard (because pressure in the cuff is no longer compressing the artery)

(b) Peripheral Resistance

Peripheral resistance is the opposition to flow of blood in vessels, and is a function of vessel radius, vessel length, and blood viscosity.

Vessel radius	Vessel length	Blood viscosity
Vasoconstriction narrows vessel and forces blood through a narrower lumen, increasing peripheral resistance and blood pressure. 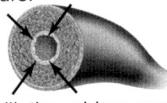 Vasodilation widens vessel, decreasing peripheral resistance and blood pressure. 	Longer vessels increase peripheral resistance, which raises blood pressure.  Shorter vessels decrease peripheral resistance, which lowers blood pressure.	Increased blood viscosity increases peripheral resistance and blood pressure.  Decreased blood viscosity decreases peripheral resistance and blood pressure. 

(c) Blood Volume

Fluid intake increases blood volume and blood pressure.
Fluid output decreases blood volume and blood pressure.

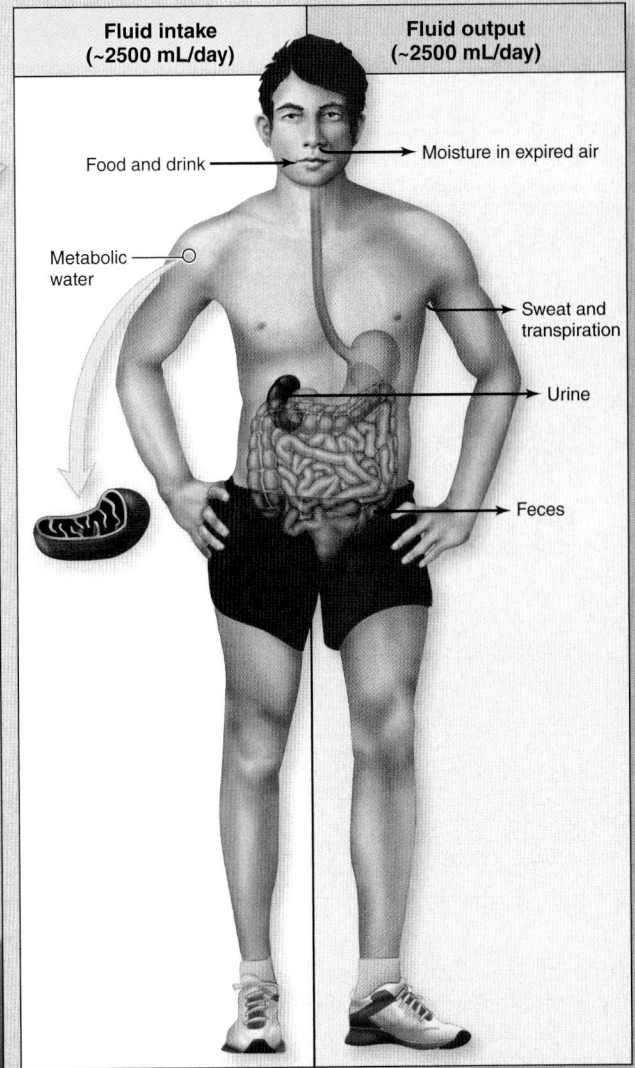

Fluid intake (~2500 mL/day)	Fluid output (~2500 mL/day)
Food and drink Metabolic water	Moisture in expired air Sweat and transpiration Urine Feces

INTEGRATE

CLINICAL VIEW
Hypertension and Hypotension

Hypertension is chronically elevated blood pressure, defined as a systolic pressure greater than 140 mm Hg or a diastolic pressure greater than 90 mm Hg. Hypertension may damage blood vessel walls, making arteries more likely to develop atherosclerosis, or it may thicken the arteriole walls and reduce their luminal diameter, a condition called **arteriolosclerosis.** Hypertension also is a major cause of heart failure owing to the extra workload placed on the heart.

In contrast, **hypotension** is a chronically low blood pressure that results in symptoms such as fatigue, dizziness, and fainting. Some physicians define hypotension as a systolic pressure below 90 mm Hg or a diastolic pressure below 60 mm Hg, whereas other physicians state that "normal" low blood pressure for one individual may be hypotensive for another.

Orthostatic hypotension, or *postural hypotension,* is a drop in blood pressure when an individual suddenly changes position, such as when a person stands up after lying down. As a result, the person may experience dizziness, light-headedness, and fainting after this positional change. Essentially, orthostatic hypotension occurs when the nervous system responses that help regulate blood pressure do not function quickly enough, and MAP temporarily decreases below 60 mm Hg. Thus, the blood remains pooled in the veins and not enough reaches the cerebral vasculature, resulting in the above symptoms.

20.7 Blood Flow Distribution During Exercise

1. Compare total blood flow and distribution at rest and during exercise.

During exercise, there is both an increase in total blood flow due to a faster and stronger heartbeat and because blood is removed from the "reservoirs" of the veins to the active circulation. There is also a redistribution of blood. Both of these changes help ensure that the most metabolically active tissues are receiving adequate blood flow to meet the needs of the tissue cells.

Figure 20.17 provides an example in which blood flow changes from 5.25 L/min (5250 mL/min) at rest to 17.5 L/min (17,500 mL/min) during exercise. The following increases in blood flow to specific regions or organs can be noted:

- Blood flow to the coronary vessels of the heart increases approximately three-fold (from 250 mL/min to 750 mL/min), a change that helps to ensure that sufficient oxygen reaches the cardiac muscle within the heart wall.
- Skeletal muscle blood flow increases an amazing 11-fold (from 1100 mL/min to 12,500 mL/min)—which is approximately 70% of the total cardiac output—a change needed to meet the high metabolic demands experienced by skeletal muscle during exercise.
- The percentage of blood flow to the skin increases to almost five times its resting level (from 400 mL/min to 1900 mL/min) to dissipate heat.

In contrast, relatively less total blood flow is distributed to the abdominal organs slowing digestive processes; less is transported to the kidneys, which decreases urine output to maintain blood volume and blood pressure. Smaller amounts reach other structures that are not as metabolically active during exercise.

⌇⌇ **WHAT DID YOU LEARN?**

27 Which organs have an increased proportion of cardiac output during exercise? Which receive a decreased proportion of cardiac output during exercise?

20.8 Pulmonary Circulation

The pulmonary circulation is responsible for carrying deoxygenated blood from the right side of the heart to the lungs and then returning the newly oxygenated blood to the left side of the heart.

20.8a Blood Flow Through the Pulmonary Circulation

⌇⌇ **LEARNING OBJECTIVE**

1. Trace the pathway of vessels from the right ventricle to the lungs and back to the left atrium.

In the pulmonary circulation, deoxygenated blood is pumped out of the right ventricle into the **pulmonary trunk** (figure 20.18). This vessel bifurcates into a **left** and **right pulmonary artery** that go to the corresponding lungs. The pulmonary arteries divide into smaller arteries that continue to subdivide to form arterioles. These arterioles finally branch into pulmonary capillaries, where gas exchange occurs. Carbon dioxide diffuses from the blood and enters the tiny air sacs

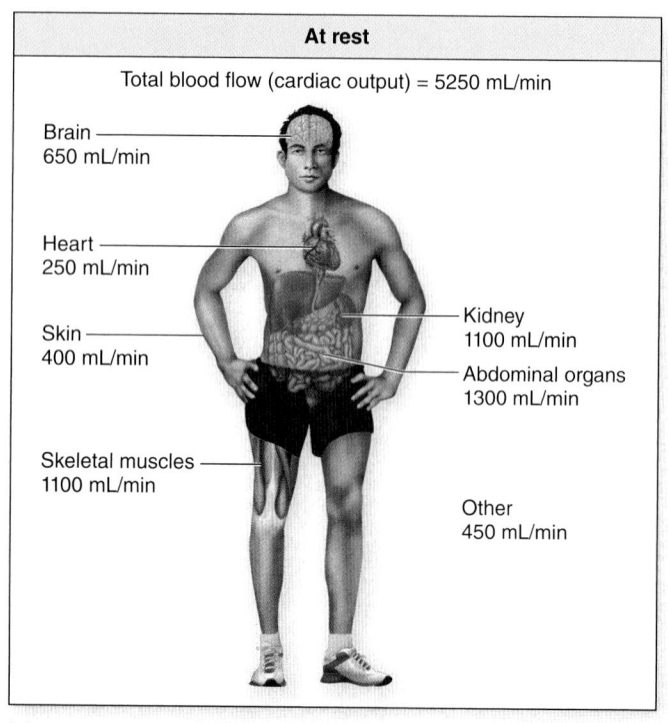

(a)

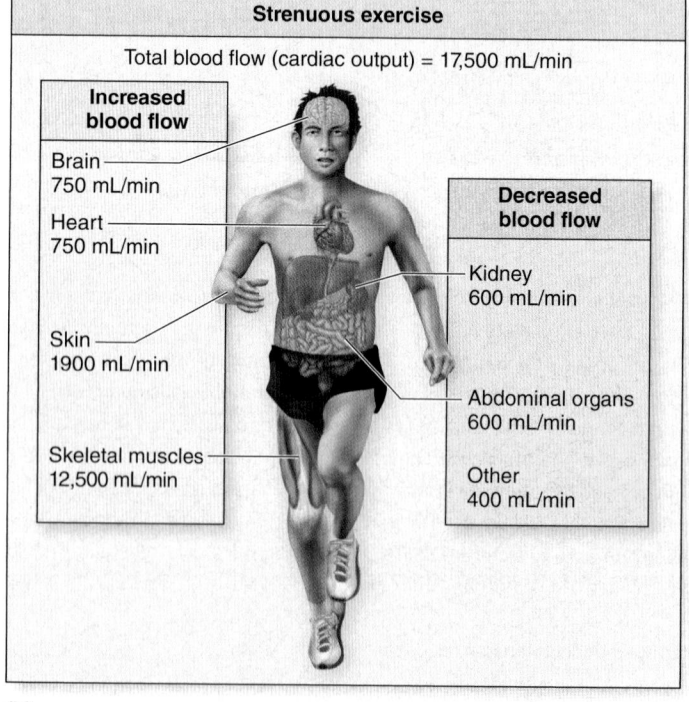

(b)

Figure 20.17 Comparison of Systemic Circulation Blood Flow During Rest and Strenuous Exercise. (*a*) At rest, cardiac output is approximately 5.25 L/min (5250 mL/min). (*b*) Cardiac output increases to 17.5 L/min (1750 mL/min) during strenuous exercise in this example. A greater percentage of blood is directed to the coronary vessels of the heart, skeletal muscles, and skin, and relatively less blood is sent to the abdominal organs, kidneys, and other less metabolically active tissues.

(alveoli) of the lungs, while oxygen moves in the opposite direction, from the air sacs into the blood.

The capillaries merge to form venules and then the **pulmonary veins.** Typically, two left and two right pulmonary veins carry the newly oxygenated blood to the left atrium of the heart.

WHAT DID YOU LEARN?

28 What is the percentage of blood returning to the right side of the heart that is then pumped to the lungs?

20.8b Characteristics of the Pulmonary Circulation

LEARNING OBJECTIVE

2. Identify features of the pulmonary circulation that distinguish it from systemic circulation.

Pulmonary arteries have less elastic connective tissue and wider lumens than systemic arteries. Compared to the systemic circulation, pulmonary vessels are relatively short, because the lungs are close to the heart. As a result, blood pressure is lower throughout the pulmonary circulation in comparison to the systemic circulation. The pressure changes associated with the pulmonary circulation are as follows:

- Blood leaves the right ventricle with a systolic pressure of about 15 to 25 mm Hg, depending upon whether the body is resting or active.
- Blood pressure drops as the blood passes through the pulmonary trunk and right and left pulmonary arteries, reaching an overall pressure of about 10 mm Hg in the pulmonary capillaries of the alveoli. This lower pressure means that the blood moves more slowly in pulmonary capillaries than in systemic capillaries, facilitating gas exchange within the lungs.
- Blood exits the pulmonary capillaries into progressively larger veins that become the pulmonary veins; blood pressure is almost 0 mm Hg as these veins empty into the left atrium.

WHAT DID YOU LEARN?

29 How does blood pressure in the pulmonary circulation compare to the pressure in the systemic circulation?

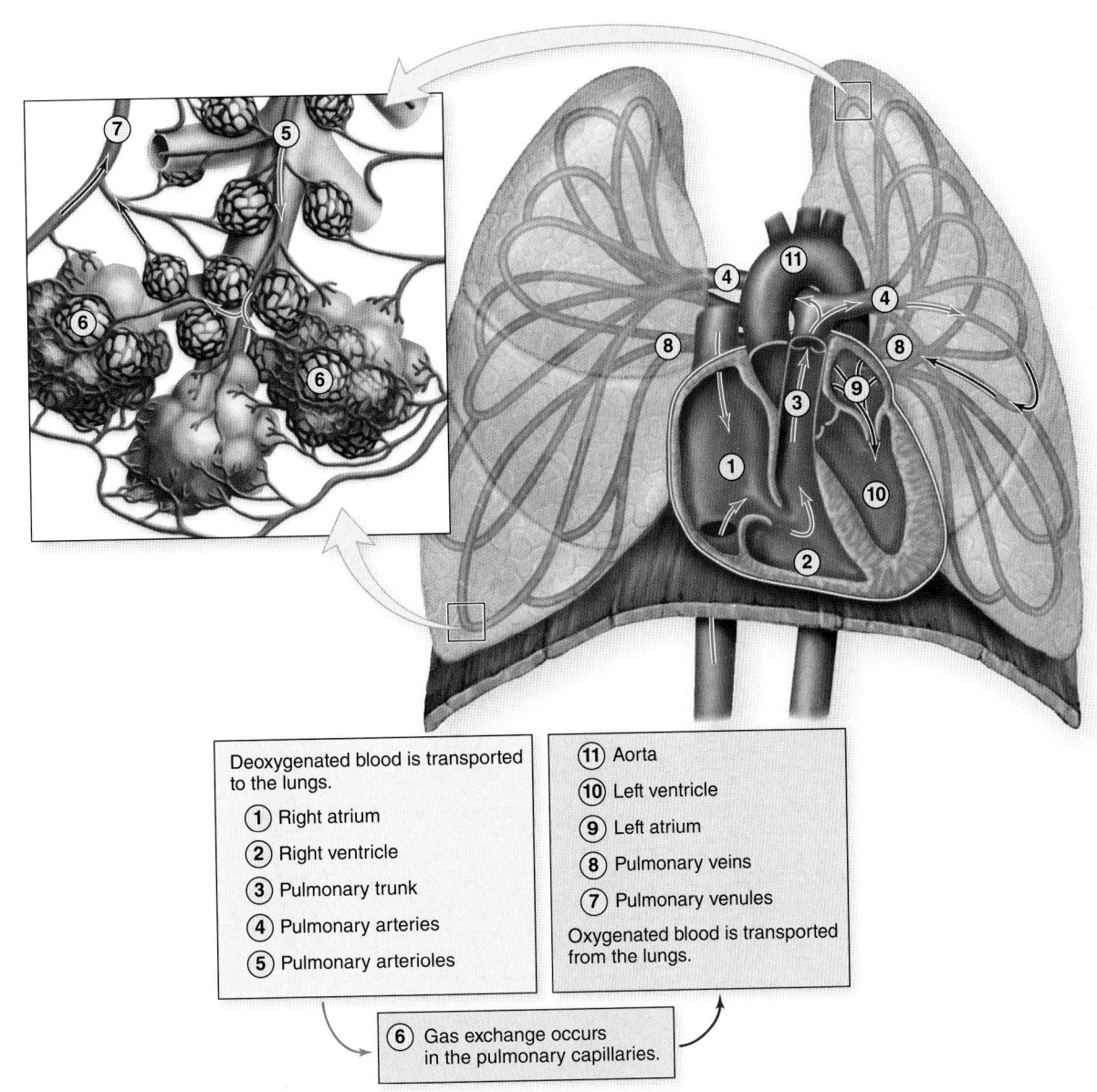

Deoxygenated blood is transported to the lungs.
1. Right atrium
2. Right ventricle
3. Pulmonary trunk
4. Pulmonary arteries
5. Pulmonary arterioles

11. Aorta
10. Left ventricle
9. Left atrium
8. Pulmonary veins
7. Pulmonary venules
Oxygenated blood is transported from the lungs.

6. Gas exchange occurs in the pulmonary capillaries.

Figure 20.18 Pulmonary Circulation. The pulmonary circulation transports blood to and from the gas exchange surfaces of the lungs. Blood circulation through the heart is indicated by colored arrows: deoxygenated blood (blue arrows), oxygenated blood (red arrows). AP|R

20.9 Systemic Circulation: Vessels from and to the Heart

We discuss the vessels from and to the heart and then look at the circulatory routes for each body region: the head and trunk and the upper and lower limbs. As you read these descriptions, refer to **figure 20.19**, which shows the locations of the major arteries and veins.

20.9a General Arterial Flow Out of the Heart

LEARNING OBJECTIVE

1. List the arteries that transport blood away from the left ventricle of the heart to the major areas of the body.

Oxygenated blood is pumped out of the left ventricle of the heart and enters the ascending aorta. The **left** and **right coronary arteries** emerge immediately from the wall of the ascending aorta and supply the heart wall (figure 20.19 *inset*).

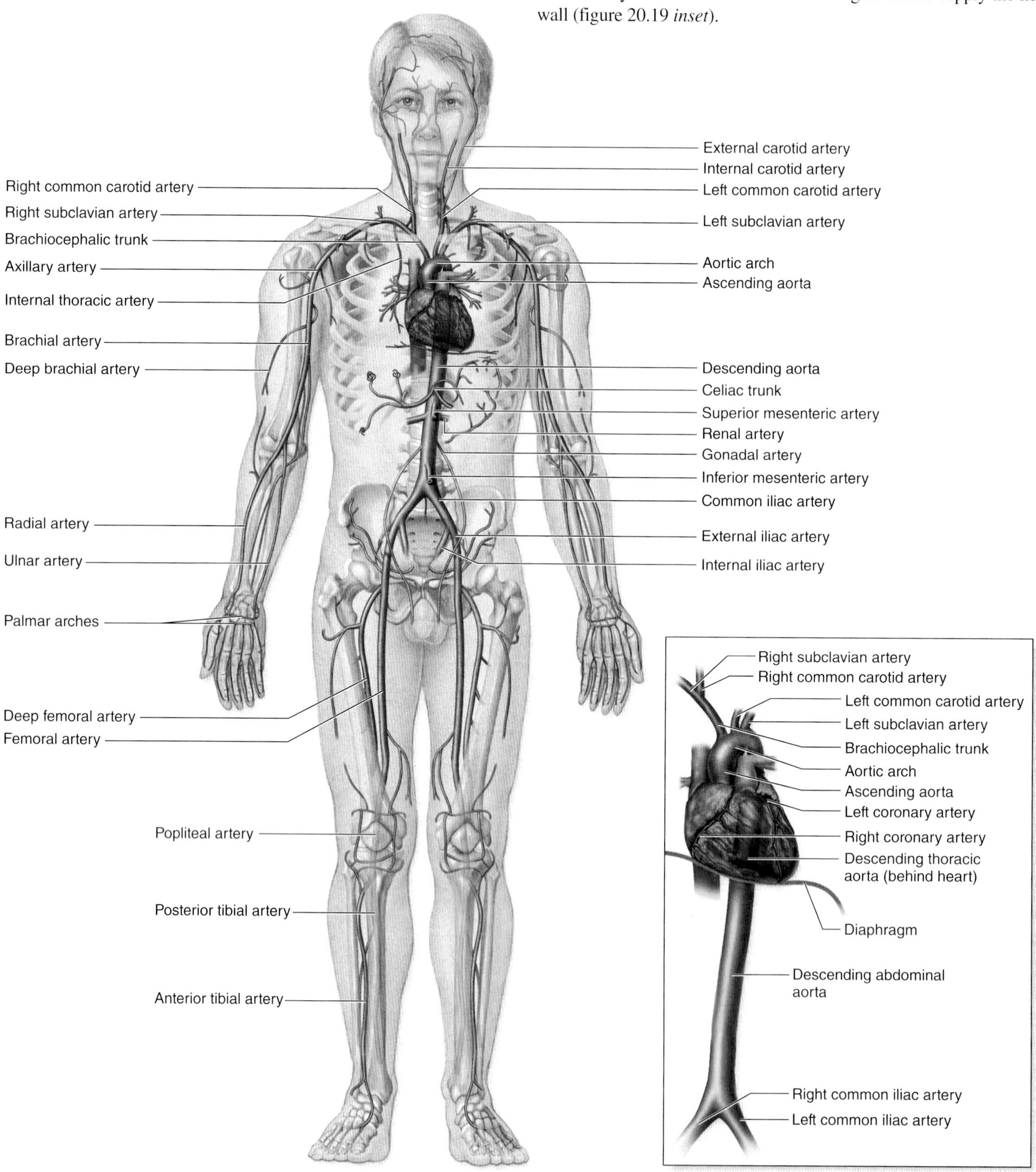

(a) Arteries, anterior view

Figure 20.19 General Vascular Distribution. Arteries in the systemic circulation transport blood from the heart to systemic capillary beds; systemic veins return this blood to the heart. (*a*) Anterior view of the systemic arteries, with an inset of the heart and aorta. AP|R

The ascending aorta curves toward the left side of the body and becomes the **aortic arch** (also called the *arch of the aorta*). Recall that the aortic bodies for regulating blood pressure are within the tunica externa of the aortic arch. Three main arterial branches emerge from the aortic arch:

1. The **brachiocephalic** (brā-kē-ō-se-fal′ik) **trunk,** which bifurcates into the **right common carotid** (ka-rot′id) **artery,** supplying arterial blood to right side of the head and neck, and

the **right subclavian** (sŭb-klā′vē-an; *sub* = beneath + clavicle) **artery,** supplying the right upper limb and some thoracic structures
2. The **left common carotid artery,** supplying the left side of the head and neck
3. The **left subclavian artery,** supplying the left upper limb and some thoracic structures

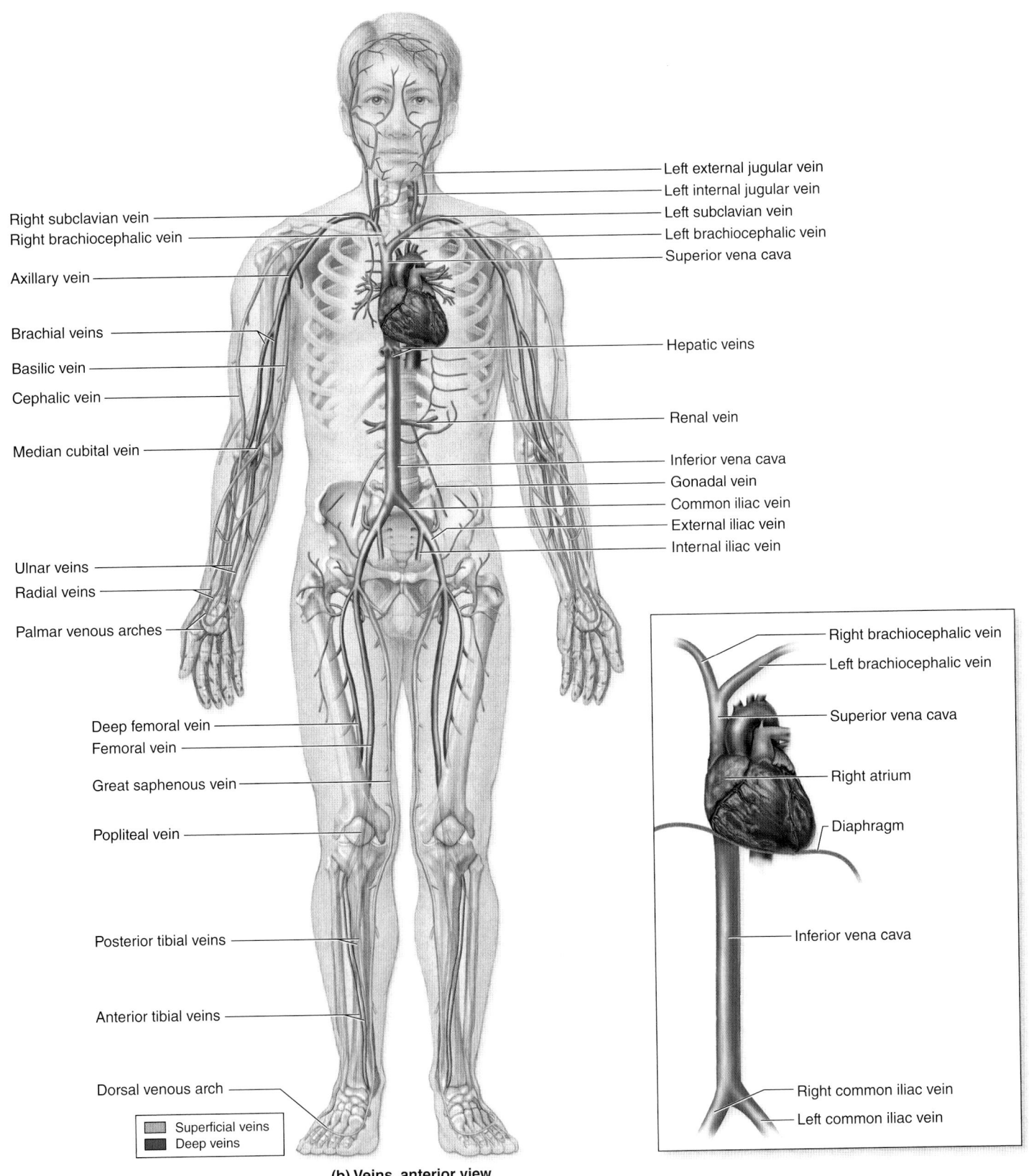

(b) Veins, anterior view

Figure 20.19 General Vascular Distribution (*continued*). (*b*) Anterior view of the systemic veins, with an inset of the major veins that drain into the heart.

INTEGRATE

LEARNING STRATEGY

Here are some tips to help you remember the names of blood vessels:

1. Blood vessels often share names with either the body region through which they traverse or the bone next to them. For example, the axillary artery is in the axillary region and the radial artery travels near the radius.

2. Some blood vessels are named for the structure they supply. For example, the renal arteries supply the kidneys, the gonadal arteries supply the gonads, and the facial arteries supply the face.

3. Arteries and veins that travel together (called companion vessels) often share the same name. For example, the femoral artery is accompanied by the femoral vein.

4. Writing out your own simplified flowchart of blood vessels for each body region is a good way to reinforce your memory of the vessels.

The aortic arch curves and projects inferiorly as the **descending thoracic aorta,** several branches of which supply the thoracic wall and internal organs. As this artery extends inferiorly through the aortic opening (hiatus) in the diaphragm, it is renamed the **descending abdominal aorta,** where it supplies the abdominal wall and internal organs.

At the level of the fourth lumbar vertebra, the descending abdominal aorta bifurcates into **left** and **right common iliac** (il'ē-ak; *ileum* = groin) **arteries.** Each of these arteries further divides into an internal iliac artery (to supply pelvic and perineal structures) and an external iliac artery (to supply the lower limb).

 WHAT DID YOU LEARN?

30 What are the three main branches off of the aortic arch? What areas in general do these branches serve?

20.9b General Venous Return to the Heart

 LEARNING OBJECTIVE

2. Name the veins that return blood from the systemic circulation to the right atrium of the heart.

Blood is returned to the right atrium of the heart by three vessels: the superior vena cava, the inferior vena cava, and the coronary sinus (figure 20.19b). The veins that drain the head, neck, upper limbs, and thoracic and abdominal walls from each side of the body merge to form the **left** and **right brachiocephalic veins.** The two brachiocephalic veins merge to form the **superior vena cava.** The veins that drain blood inferior to the diaphragm merge to collectively form the **inferior vena cava.** The inferior vena cava is responsible for transporting venous blood toward the heart from the lower limbs, pelvis and perineum, and abdominal structures. It lies to the right side of the descending abdominal aorta and extends through an opening (caval opening) in the diaphragm.

Recall from section 19.4b that the **coronary sinus** also drains into the right atrium, delivering deoxygenated blood from the heart myocardium.

 WHAT DID YOU LEARN?

31 Which body regions are drained by (a) the superior vena cava and (b) the inferior vena cava?

20.10 Systemic Circulation: Head and Trunk

Blood flow to the head and to the thoracic and abdominal organs is critical to survival, so it is not surprising that these regions receive blood flow soon after it leaves the heart.

20.10a Head and Neck

 LEARNING OBJECTIVES

1. Name the arteries and veins associated with the head and neck structures.

2. Diagram and explain the cerebral arterial circle and its function.

3. Describe the general structure and function of dural venous sinuses.

Arterial supply to the head and neck comes from the branches of the aortic arch. Venous drainage of the head and neck is through the jugular veins, which then drain into the brachiocephalic veins. We discuss the arterial supply first.

Arterial Supply

Most of the blood to the head and neck is supplied by the **common carotid arteries** (**figure 20.20a**). The common carotid arteries travel parallel immediately lateral to either side of the trachea. At the superior border of the thyroid cartilage of the larynx, each artery divides into an **external carotid artery** that supplies structures external to the skull, and an **internal carotid artery** that supplies internal skull structures. Recall that the carotid sinus, a receptor that helps regulate blood pressure, is within the internal carotid artery near its bifurcation from the common carotid artery (see section 19.5b).

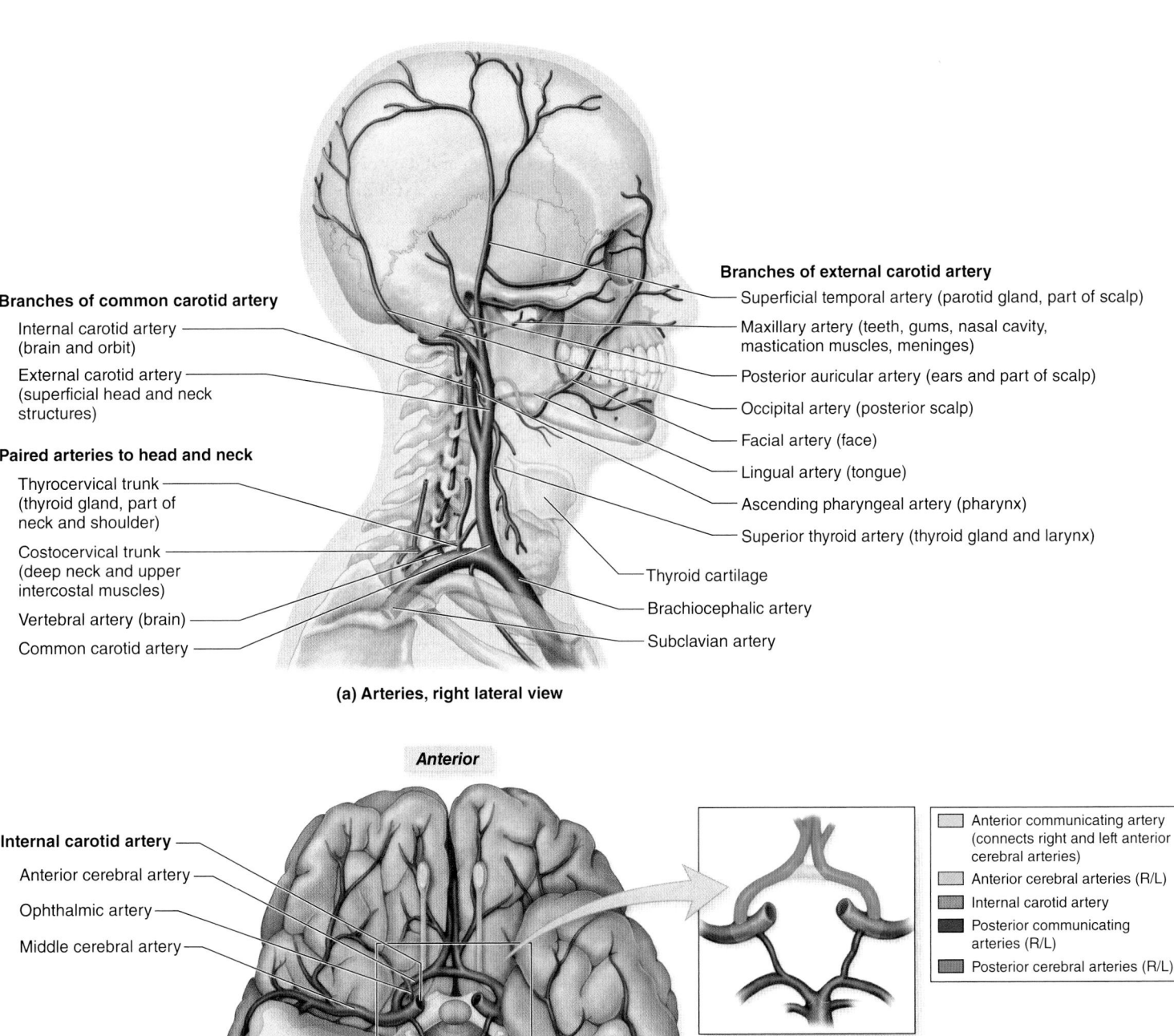

Branches of common carotid artery

Internal carotid artery (brain and orbit)

External carotid artery (superficial head and neck structures)

Paired arteries to head and neck

Thyrocervical trunk (thyroid gland, part of neck and shoulder)

Costocervical trunk (deep neck and upper intercostal muscles)

Vertebral artery (brain)

Common carotid artery

Branches of external carotid artery

Superficial temporal artery (parotid gland, part of scalp)

Maxillary artery (teeth, gums, nasal cavity, mastication muscles, meninges)

Posterior auricular artery (ears and part of scalp)

Occipital artery (posterior scalp)

Facial artery (face)

Lingual artery (tongue)

Ascending pharyngeal artery (pharynx)

Superior thyroid artery (thyroid gland and larynx)

Thyroid cartilage

Brachiocephalic artery

Subclavian artery

(a) Arteries, right lateral view

Anterior

Internal carotid artery

Anterior cerebral artery

Ophthalmic artery

Middle cerebral artery

Anterior communicating artery (connects right and left anterior cerebral arteries)

Anterior cerebral arteries (R/L)

Internal carotid artery

Posterior communicating arteries (R/L)

Posterior cerebral arteries (R/L)

Cerebral arterial circle (circle of Willis)

Posterior cerebral artery

Basilar artery

Vertebral artery

Posterior

(b) Arteries of the brain, inferior view

Figure 20.20 Arterial Blood Flow to the Head and Neck. (*a*) Right lateral view shows the branches that supply blood to the head and neck. (*b*) An inferior view of the brain shows the branches of the internal carotid and vertebral arteries that supply the brain. An inset shows an enlarged view of the cerebral arterial circle, which is an anastomosis of blood vessels that surrounds the sella turcica of the sphenoid bone. AP|R

Additional blood to the head and neck comes from the **vertebral artery, thyrocervical** (thy′rō-ser′vi-kal) **trunk,** and **costocervical** (kos′tō-ser′vi-kal) **trunk.** These are all branches of the subclavian artery.

External Carotid   The external carotid artery supplies blood to several branches that include the **superior thyroid artery, ascending pharyngeal** (fă-rin′jē-ăl; *pharynx* = throat) **artery, lingual** (lin′gwăl) **artery, facial artery, occipital artery,** and **posterior auricular artery.** Thereafter, the external carotid artery divides into the **maxillary artery** and the **superficial temporal artery.** The specific regions supplied by these arteries are listed in figure 20.20*a*.

Internal Carotid   The internal carotid artery branches only after it enters the skull through the carotid canal. Once inside the skull, it forms multiple branches, including the **anterior** and **middle cerebral arteries,** which supply the brain, and the **ophthalmic** (op-thal′mik; *ophthalmos* = eye) **arteries,** which supply the eyes and some of the surrounding structures (figure 20.20*b*).

Vertebral Arteries   The **vertebral arteries** emerge from the subclavian arteries and travel through the transverse foramina of the cervical vertebrae before entering the skull through the foramen magnum, where they merge to form the **basilar** (bas′i-lăr; *basis* = base) **artery.** The basilar artery travels immediately anterior to the pons and extends many branches prior to subdividing into the **posterior cerebral arteries,** which supply the posterior portion of the cerebrum.

Cerebral Arterial Circle   The **cerebral arterial circle** (*circle of Willis*) is an important arterial anastomosis around the sella turcica (inset, figure 20.20*b*). The circle is formed from posterior cerebral arteries, **posterior communicating arteries** (branches of the posterior cerebral arteries), internal carotid arteries, anterior cerebral arteries, and an **anterior communicating artery** (which connects the two anterior cerebral arteries). This arterial circle equalizes blood pressure in the brain and can provide collateral channels should one vessel become blocked.

Venous Drainage

Three primary pairs of veins drain the neck and head (**figure 20.21**). On each side of the head is a **vertebral vein** and an **external jugular vein,** both of which empty into the subclavian vein. The third vein is an **internal jugular vein,** which joins with the subclavian vein to form the brachiocephalic vein. The external jugular primarily drains superficial head and neck structures, while the internal jugular drains blood from the cranial cavity. The right and left brachiocephalic veins join to form the superior vena cava.

Cranial Cavity   Some cranial venous blood is drained by the vertebral veins that extend through the transverse foramina of the cervical vertebrae and drain into the brachiocephalic veins. However, most of the venous blood of the cranium drains through several large veins collectively known as the **dural** (dū′răl) **venous sinuses.** Recall that these large modified veins are formed between the two layers of dura mater and also receive excess cerebrospinal fluid (see section 13.2a). Blood from the dural venous sinus system is drained primarily into the internal jugular vein.

 WHAT DID YOU LEARN?

32   What major arteries supply the head and neck? What primary veins drain these regions?

33   What is the function of the dural venous sinuses?

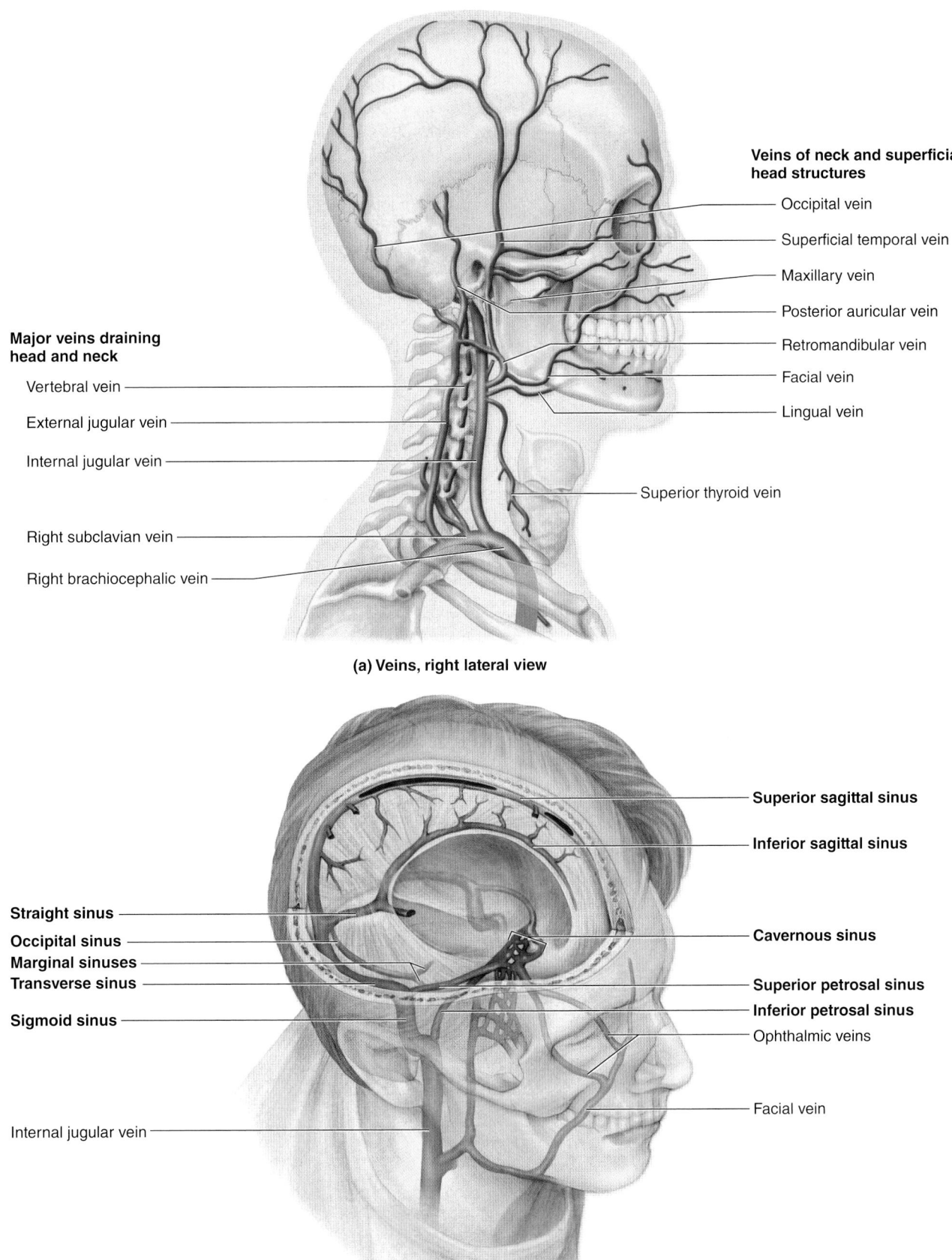

Veins of neck and superficial head structures

Occipital vein

Superficial temporal vein

Maxillary vein

Posterior auricular vein

Retromandibular vein

Facial vein

Lingual vein

Superior thyroid vein

Major veins draining head and neck

Vertebral vein

External jugular vein

Internal jugular vein

Right subclavian vein

Right brachiocephalic vein

(a) Veins, right lateral view

Superior sagittal sinus

Inferior sagittal sinus

Straight sinus

Occipital sinus

Marginal sinuses

Transverse sinus

Sigmoid sinus

Cavernous sinus

Superior petrosal sinus

Inferior petrosal sinus

Ophthalmic veins

Facial vein

Internal jugular vein

(b) Dural venous sinuses, right superior anterolateral view

Figure 20.21 Venous Blood Flow from the Head and Neck. (*a*) Right lateral view shows the major veins and their tributaries that drain blood from the head and neck. (*b*) Venous drainage of the cranium from a superior anterolateral view. The dural venous sinuses are labeled in bold. **AP|R**

20.10b Thoracic and Abdominal Walls

LEARNING OBJECTIVES

4. Describe the pairs of arteries that supply the thoracic wall.

5. Discuss the arteries that supply the abdominal wall.

6. List veins that drain the thoracic and abdominal walls and delineate their pathways.

The systemic circulation to the walls of the trunk region is primarily via paired vessels that have extensive anastomoses. The venous drainage is more complex than the arterial supply.

Arterial Supply

The thoracic and abdominal walls are both supplied by several pairs of arteries (figure 20.22a). A left and right **internal thoracic artery** emerge from each subclavian artery to supply the anterior thoracic wall and mammary gland. Each internal thoracic artery is located lateral to the sternum and has the following branches: the first six **anterior intercostal arteries** and a **musculophrenic** (mŭs′kū-lō-fren′ik; *phren* = diaphragm) **artery** that divides into anterior intercostal arteries 7–9. The intercostal arteries supply the contents of the intercostal spaces. The internal thoracic artery then becomes the **superior epigastric** (ep-i-gas′trik; *epi* = upon, *gastric* = stomach) **artery,** which carries blood to the superior abdominal wall.

The **inferior epigastric artery,** a branch of the external iliac artery, supplies the inferior abdominal wall. This artery anastomoses extensively with the superior epigastric artery.

The **supreme intercostal artery** is a branch of the costocervical trunk; it branches into the first and second **posterior intercostal arteries.** Posterior intercostal arteries 3–11 are branches of the descending thoracic aorta. The posterior and anterior intercostal arteries anastomose,

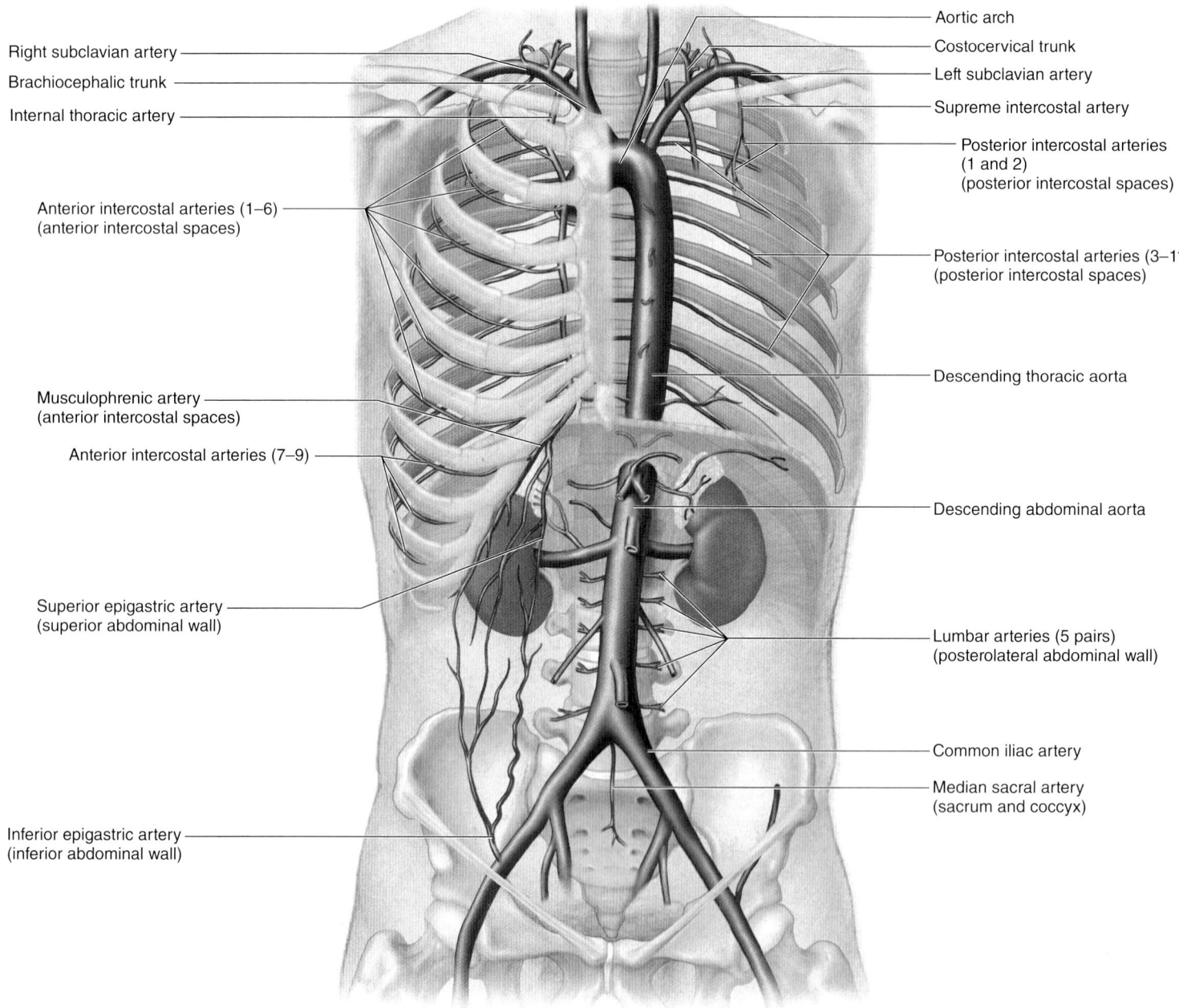

(a) Arteries, anterior view

Figure 20.22 Circulation to the Thoracic and Abdominal Body Walls. (*a*) The arteries that supply the thoracic and abdominal wall are shown. Notice the anastomoses of the superior and inferior epigastric arteries. AP|R

and each pair forms a horizontal vessel arc that spans a segment of the thoracic wall.

Finally, five pairs of **lumbar arteries** branch from the descending abdominal aorta to supply the posterolateral abdominal wall. In addition, a single **median sacral artery** arises at the bifurcation of the aorta in the pelvic region to supply the sacrum and coccyx.

Venous Drainage

Venous drainage of the thoracic and abdominal walls is a bit more complex than the arterial pathways (figure 20.22*b*). **Anterior intercostal veins,** a **musculophrenic vein,** and the **superior epigastric vein** all merge into the **internal thoracic vein,** which drains into the brachiocephalic vein.

The **inferior epigastric vein** merges with the **external iliac vein.** The first and second **posterior intercostal veins** merge with the **supreme intercostal vein** that drains into the brachiocephalic vein.

The **lumbar veins** and **posterior intercostal veins** drain into the azygos system of veins along the posterior thoracic wall. The **hemiazygos** (hem′ē-az′ī-gos) and **accessory hemiazygos veins** on the left side of the vertebrae drain the left-side veins. The **azygos vein** drains the right-side veins and also receives blood from the hemiazygos veins. The azygos vein drains into the superior vena cava. A schematic for venous return can be viewed in figure 20.22*c*.

 WHAT DID YOU LEARN?

34 Does the azygos system, which receives venous blood from the thoracic and abdominal wall, drain into the superior vena cava or the inferior vena cava?

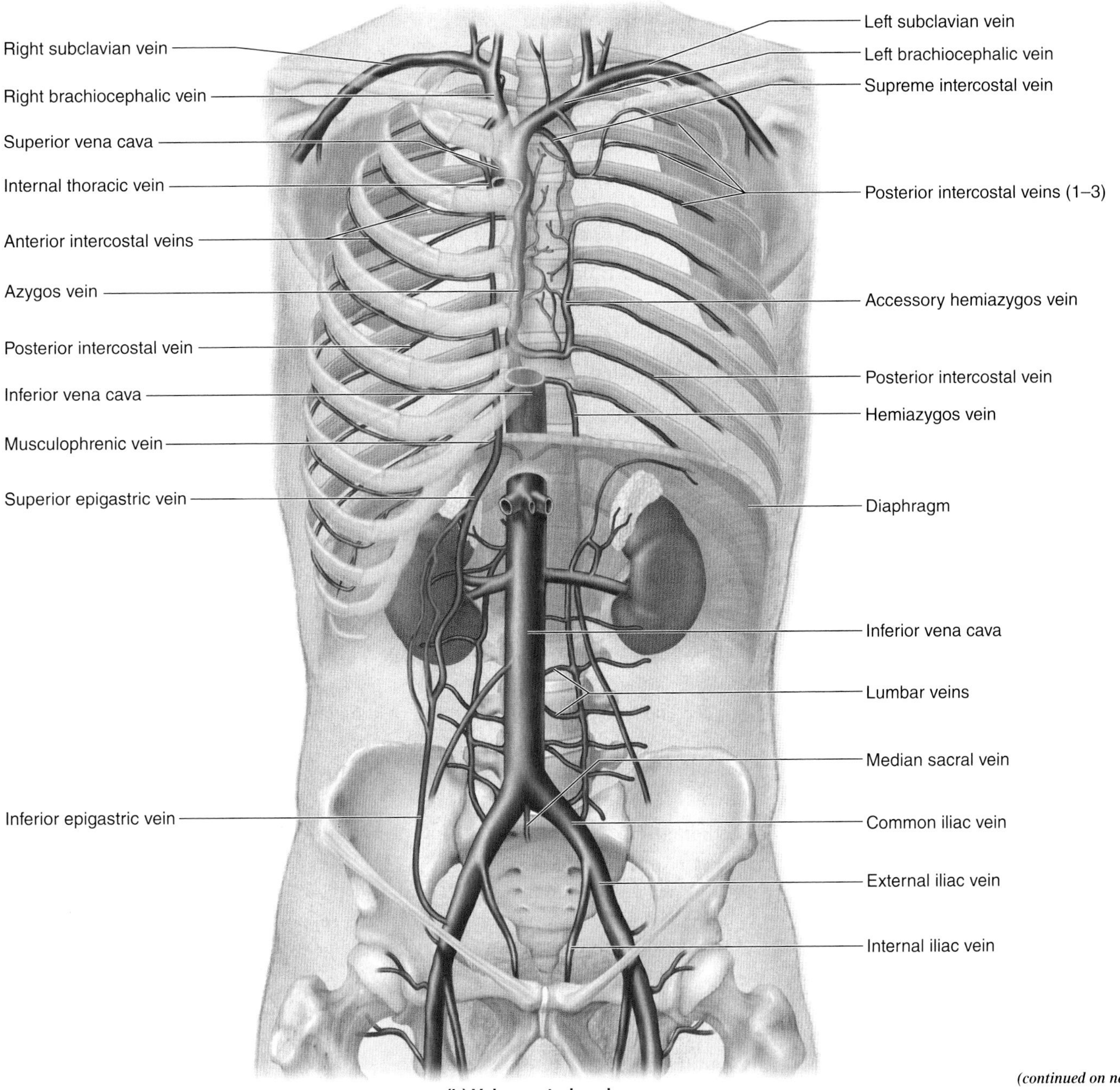

Right subclavian vein

Right brachiocephalic vein

Superior vena cava

Internal thoracic vein

Anterior intercostal veins

Azygos vein

Posterior intercostal vein

Inferior vena cava

Musculophrenic vein

Superior epigastric vein

Inferior epigastric vein

Left subclavian vein

Left brachiocephalic vein

Supreme intercostal vein

Posterior intercostal veins (1–3)

Accessory hemiazygos vein

Posterior intercostal vein

Hemiazygos vein

Diaphragm

Inferior vena cava

Lumbar veins

Median sacral vein

Common iliac vein

External iliac vein

Internal iliac vein

(b) Veins, anterior view

(continued on next page)

Figure 20.22 Circulation to the Thoracic and Abdominal Body Walls (*continued*). (*b*) The venous drainage of the thoracic and abdominal wall is more complex than its arterial supply. AP|R

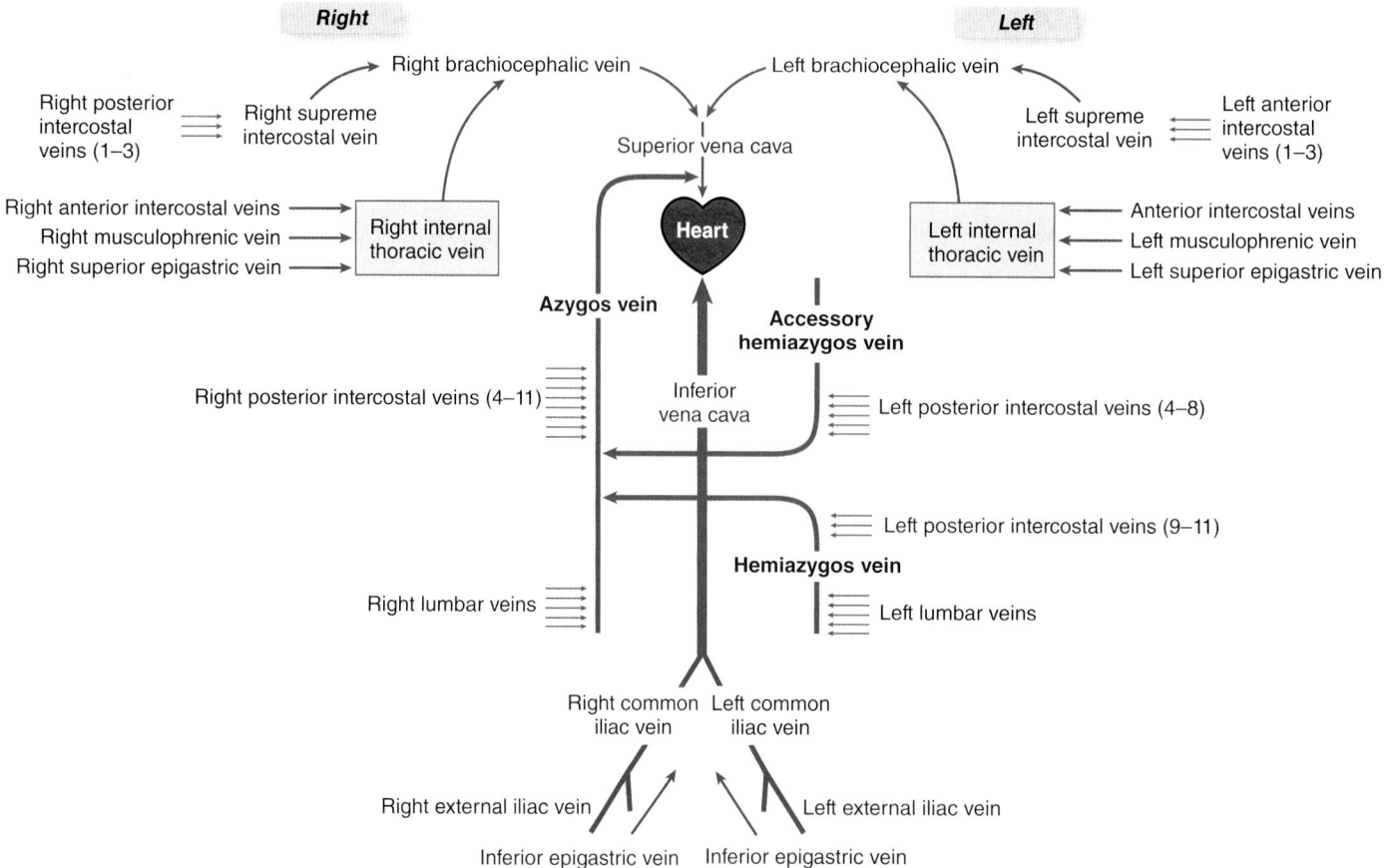

Figure 20.22 Circulation to the Thoracic and Abdominal Body Walls (*continued*). (*c*) A schematic of the venous drainage.

20.10c Thoracic Organs

LEARNING OBJECTIVE

7. Describe the vessels that supply and drain the lungs, esophagus, and diaphragm.

The main thoracic organs include the heart, lungs, esophagus, and diaphragm. The vessels supplying and draining the heart were described in section 19.4. The vessels of the other thoracic organs are discussed here (**figure 20.23**).

Lungs

The bronchi and bronchioles (airways of the lung) and connective tissue of the lungs are supplied by three or four small **bronchial arteries** that emerge as tiny branches from the anterior wall of the descending thoracic aorta. Left and right **bronchial veins** (not shown) drain into the azygos system of veins and the pulmonary veins. The rest of the lung receives its oxygen via diffusion directly from the tiny air sacs (alveoli) of the lungs.

Esophagus

Several small **esophageal** (ē-sof′a-jē′ăl, ē′sŏ-faj′ē-ăl) **arteries** emerge from the anterior wall of the descending thoracic aorta and supply the esophagus. Additionally, the left gastric artery forms several **esophageal branches** that supply the abdominal portion of the esophagus. **Esophageal veins** drain the esophageal wall and may take either of two routes: into the azygos vein or into the left gastric vein (not shown). The latter merges with the hepatic portal vein (described in the next section).

Diaphragm

Arterial blood is supplied to the diaphragm by paired vessels. **Superior phrenic** (fren′ik; *phren* = diaphragm) **arteries** emerge from the descending thoracic aorta; **inferior phrenic arteries** emerge

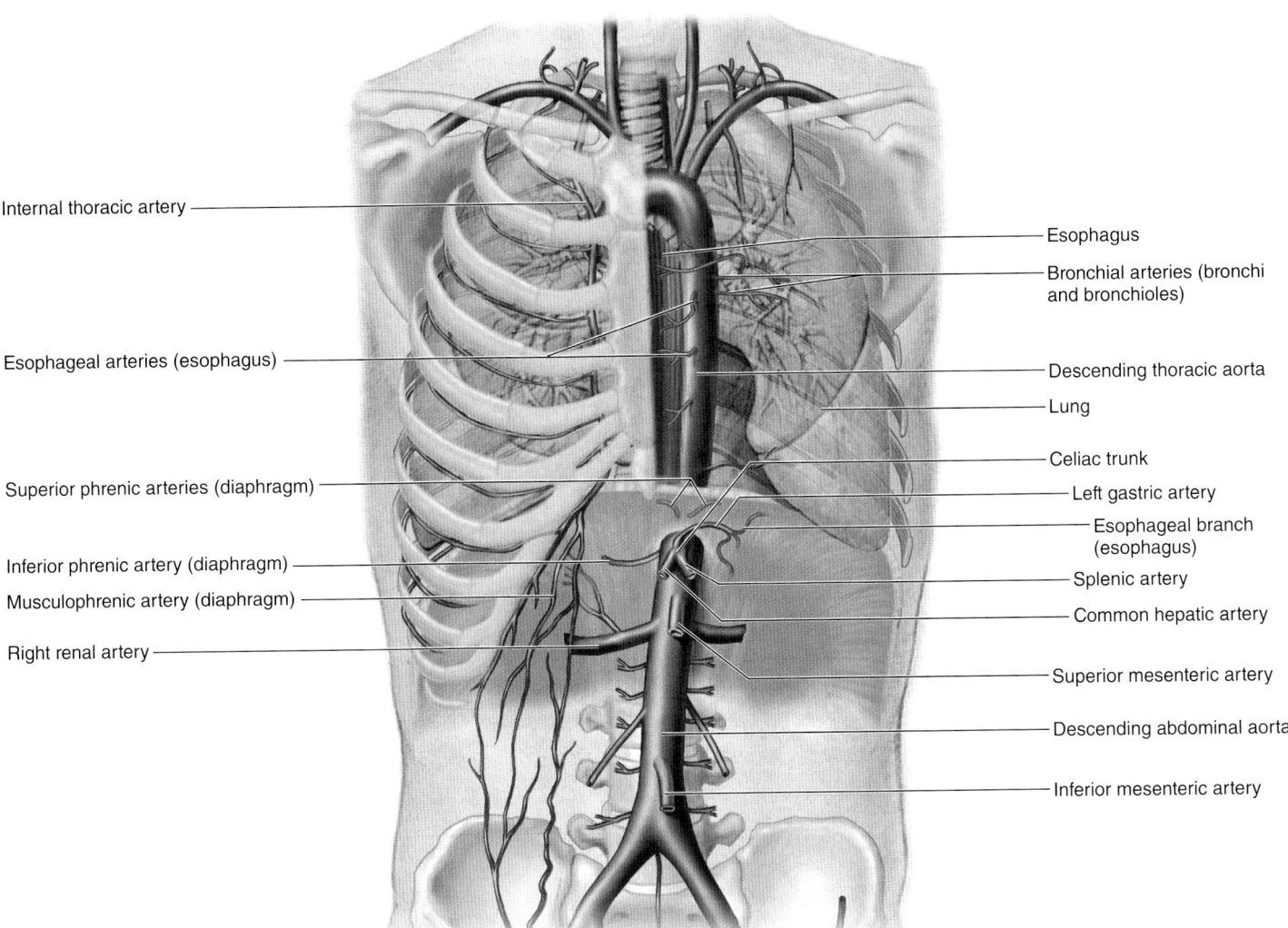

Internal thoracic artery

Esophageal arteries (esophagus)

Superior phrenic arteries (diaphragm)

Inferior phrenic artery (diaphragm)

Musculophrenic artery (diaphragm)

Right renal artery

Esophagus

Bronchial arteries (bronchi and bronchioles)

Descending thoracic aorta

Lung

Celiac trunk

Left gastric artery

Esophageal branch (esophagus)

Splenic artery

Common hepatic artery

Superior mesenteric artery

Descending abdominal aorta

Inferior mesenteric artery

Figure 20.23 Arterial Blood Supply of Organs of the Thoracic Cavity and Diaphragm. The blood vessels to the bronchi (airways), esophagus, and diaphragm are shown. The ribs and superficial vessels have been removed from the left side of the body so as to see the deeper vessels. AP|R

from the descending abdominal aorta to supply the diaphragm; and both musculophrenic arteries and *pericardiacophrenic arteries* (not shown) arise from the internal thoracic artery. Superior phrenic and inferior phrenic veins drain with the inferior vena cava, while the musculophrenic drains into the internal thoracic veins that merge with the brachiocephalic veins. These veins are not illustrated.

WHAT DID YOU LEARN?

35 What are the systemic arteries that supply oxygenated blood to the lungs? What specific structures of the lung receive the blood?

20.10d Gastrointestinal Tract

LEARNING OBJECTIVES

8. Name the three major arteries that branch from the descending aorta to supply the gastrointestinal tract, and list their major branches.

9. Explain the function of the hepatic portal system.

10. Trace the route of blood from the gastrointestinal tract to the inferior vena cava.

The gastrointestinal (GI) tract receives arterial blood from unpaired arteries that arise from the abdominal aorta. Venous blood travels through a hepatic portal system before draining into the inferior vena cava.

Arterial Supply to the Abdomen

Three unpaired arteries emerge from the anterior wall of the descending abdominal aorta to supply the GI tract. From superior to inferior, these arteries are the celiac trunk, superior mesenteric artery, and inferior mesenteric artery (see figure 20.22a).

Celiac Trunk   The **celiac** (sē′lē-ak; *koilia* = belly) **trunk** is located immediately inferior to the aortic opening (hiatus) of the diaphragm. Three branches emerge from this arterial trunk **(figure 20.24)**: (1) the **left gastric artery,** (2) the **splenic artery,** and (3) the **common hepatic artery.** The common hepatic artery divides into the

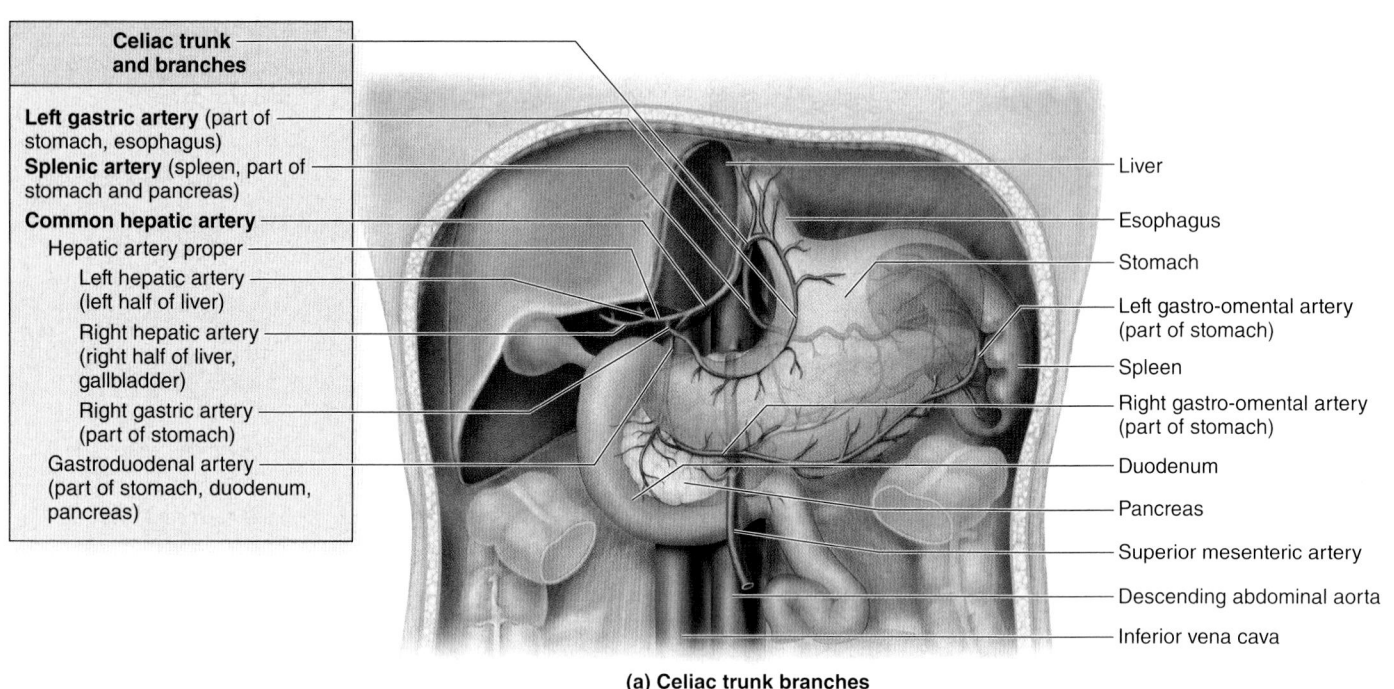

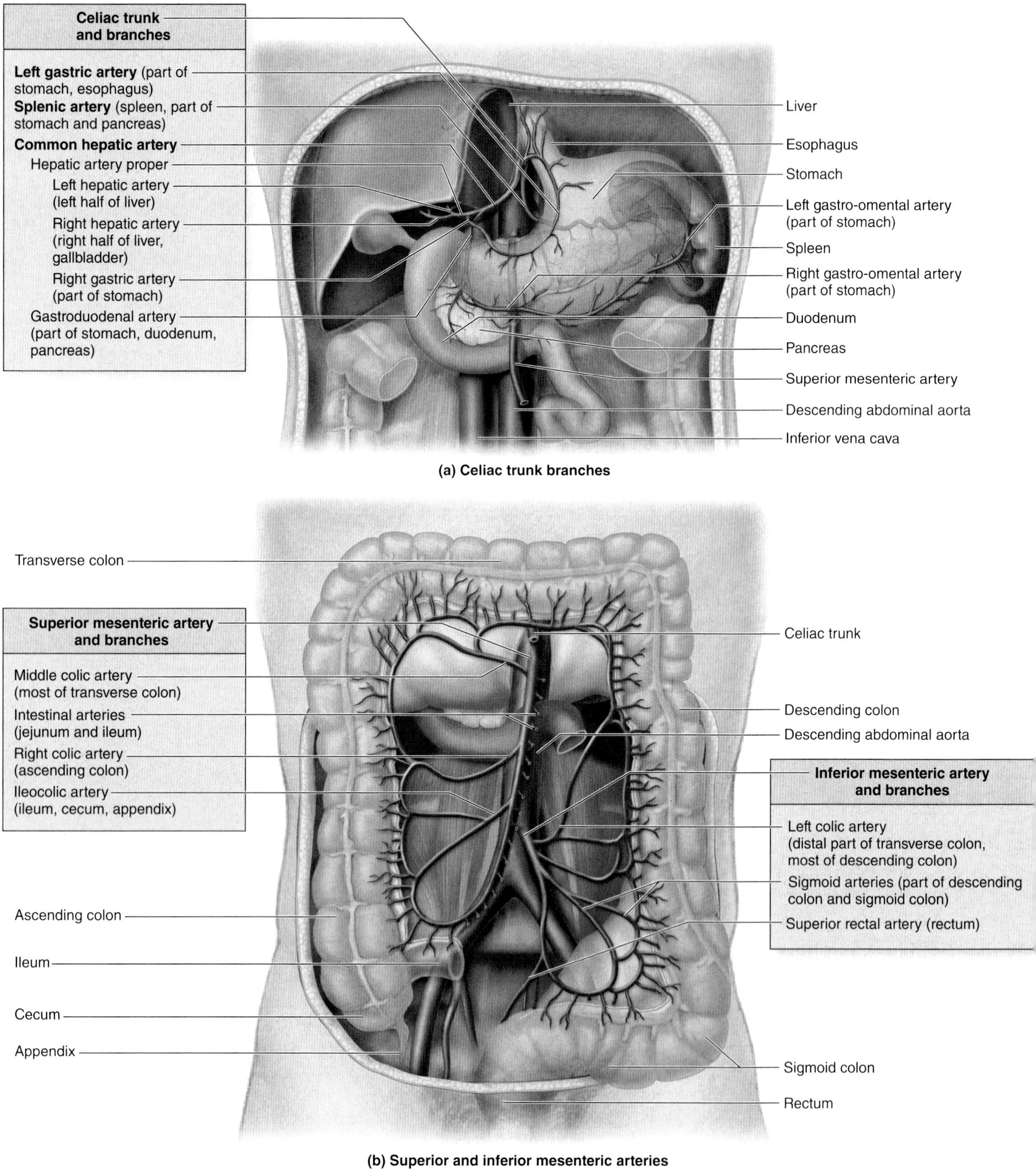

Celiac trunk and branches

Left gastric artery (part of stomach, esophagus)
Splenic artery (spleen, part of stomach and pancreas)
Common hepatic artery
  Hepatic artery proper
    Left hepatic artery (left half of liver)
    Right hepatic artery (right half of liver, gallbladder)
    Right gastric artery (part of stomach)
  Gastroduodenal artery (part of stomach, duodenum, pancreas)

Liver
Esophagus
Stomach
Left gastro-omental artery (part of stomach)
Spleen
Right gastro-omental artery (part of stomach)
Duodenum
Pancreas
Superior mesenteric artery
Descending abdominal aorta
Inferior vena cava

(a) Celiac trunk branches

Transverse colon

Superior mesenteric artery and branches

Middle colic artery (most of transverse colon)
Intestinal arteries (jejunum and ileum)
Right colic artery (ascending colon)
Ileocolic artery (ileum, cecum, appendix)

Ascending colon

Ileum

Cecum

Appendix

Celiac trunk

Descending colon
Descending abdominal aorta

Inferior mesenteric artery and branches

Left colic artery (distal part of transverse colon, most of descending colon)
Sigmoid arteries (part of descending colon and sigmoid colon)
Superior rectal artery (rectum)

Sigmoid colon

Rectum

(b) Superior and inferior mesenteric arteries

Figure 20.24 Arterial Supply to the Gastrointestinal Tract and Abdominal Organs. The celiac trunk, superior mesenteric artery, and inferior mesenteric artery supply most of the abdominal organs. (*a*) Branches of the celiac trunk supply part of the esophagus, stomach, spleen, pancreas, liver, and gallbladder. (*b*) Branches of the superior mesenteric and inferior mesenteric arteries primarily supply the intestines.

hepatic artery proper and the **gastroduodenal** (gas'trō-dū'ō-dē'năl) **artery.**

Superior Mesenteric Artery   The **superior mesenteric** (mez-en-ter'ĭk; *mesos* = middle, *enteron* = intestine) **artery** is located immediately inferior to the celiac trunk. Its branches include 18–20 **intestinal arteries,** the **middle colic artery,** the **right colic artery,** and the **ileocolic** (il'ē-ō-kol-ik) **artery.**

Inferior Mesenteric Artery   The **inferior mesenteric artery** emerges approximately 5 centimeters superior to bifurcation of the aorta at about the level of the L₃ vertebra. Its branches include the **left colic artery,** the **sigmoid arteries,** and the **superior rectal** (rek'tăl; *rectus* = straight) **artery.** The specific regions that each arterial branch supplies blood to are included in figure 20.24.

Venous Return from the Abdomen and the Hepatic Portal System

Blood that has passed through the capillaries of digestive organs and spleen that is then drained by its respective veins is not directly returned to the inferior vena cava and returned to the heart **(figure 20.25).** Instead, the blood is transported from the veins of the digestive organs and spleen into a **hepatic portal system** that drains the blood into the liver before this blood drains to the inferior vena cava.

The hepatic portal system provides the means for the liver to process blood that has passed through the blood vessels of the digestive organs before it is returned to the heart and redistributed throughout the body. This blood is nutrient-rich, deoxygenated, and may potentially contain harmful substances (e.g., alcohol, toxins) that were absorbed from the digestive organs. The hepatic portal system allows for the most efficient route for handling these absorbed substances. The hepatic portal system also receives products of erythrocyte destruction from the spleen, so that the liver can recycle some of these components.

Within the hepatic portal system, blood from the digestive organs drains into three main venous branches:

1. The **splenic vein,** a horizontally positioned vein
2. The **inferior mesenteric vein,** a vertically positioned vein
3. The **superior mesenteric vein,** another vertically positioned vein on the right side of the body

Blood from all three of these drain into the **hepatic portal vein,** which drains blood to the liver. Some small veins, such as the left and right gastric veins, drain directly into the hepatic portal vein. The venous blood in the hepatic portal vein flows through the sinusoids of the liver. In these sinusoids, the venous blood mixes with arterial oxygenated blood entering the liver via the hepatic arteries. Thus, deoxygenated but high-nutrient-filled blood from digestive organs and oxygenated blood from the hepatic artery flow within the liver sinusoids.

Blood leaves the liver through **hepatic veins** that merge with the inferior vena cava.

WHAT DID YOU LEARN?

36 What are the three branches off of the celiac trunk, and which organs do these vessels supply?

37 What are the three primary veins that drain into the hepatic portal vein of the hepatic portal system? What is the function of the hepatic portal system?

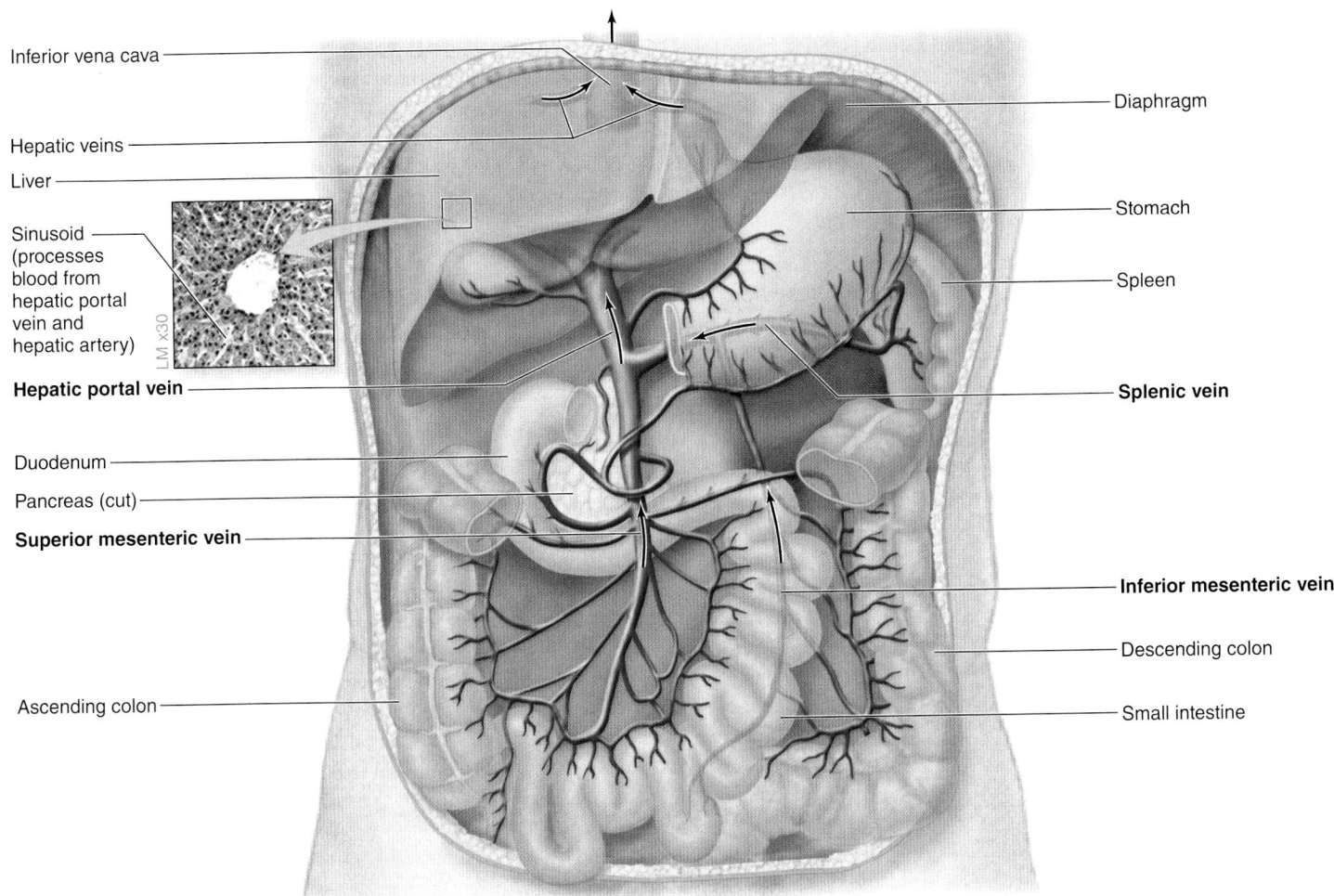

Figure 20.25 Hepatic Portal System. The hepatic portal system is a network of veins that transports venous blood from the digestive organs and spleen to the liver for nutrient processing. Black arrows indicate the direction of blood flow. AP|R

INTEGRATE

LEARNING STRATEGY

Although the pattern of the veins of the hepatic portal system can vary, together they typically resemble the side view of a chair. The front leg of the chair represents the inferior mesenteric vein, while the back leg of the chair represents the superior mesenteric vein. The seat of the chair is the splenic vein, while the chair back represents the hepatic portal vein.

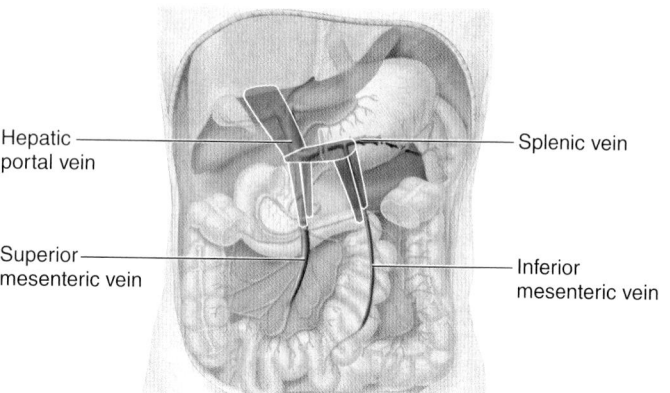

Hepatic portal vein — Splenic vein

Superior mesenteric vein — Inferior mesenteric vein

The configuration of the veins of the hepatic portal system resembles the side view of a chair.

20.10e Posterior Abdominal Organs, Pelvis, and Perineum

LEARNING OBJECTIVES

11. Describe the arteries and veins that supply and drain the adrenal glands, kidneys, and gonads.
12. Name the main vessels associated with the pelvis and perineum.

Branches from the descending abdominal aorta and the internal iliac arteries supply the posterior abdominal organs and pelvis, and veins having the same names drain them **(figure 20.26)**. Several paired arterial branches emerge from the sides of the descending abdominal aorta in addition to the arteries already mentioned.

Posterior Abdominal Organs

Each **middle suprarenal artery** supplies an adrenal gland; each **renal artery** supplies a kidney; and each **gonadal artery** supplies a gonad. Subsequently, these organs are drained by veins having the same name as the arteries.

Pelvis and Perineum

The aorta divides at its inferior end into the right and left common iliacs. Each common iliac further divides into an internal iliac artery and an external iliac artery. The **internal iliac artery** is the primary arterial supply to the pelvis and perineum. Some branches of the internal iliac artery include the **superior** and **inferior gluteal** (glū′tē-ăl) **arteries,** the **superior vesical artery,** the **middle rectal artery,** the **vaginal artery** and **uterine artery** (in females), the **internal pudendal** (pū-den′dăl; *pudeo* = to feel ashamed) **artery,** and the **obturator** (ob′tū-rā-tŏr) **artery.** Remnant vessels of fetal circulation are the **medial umbilical ligaments**—previously, umbilical arteries that transported blood from the fetus to the placenta.

The pelvis and perineum are drained by veins with the same name as the supplying arteries (see figure 20.22*b*). The veins merge with the internal iliac vein that merges with the common iliac vein, which subsequently drains into the inferior vena cava.

WHAT DID YOU LEARN?

 ㉘ What arteries supply the kidneys? The adrenal glands? The female uterus?

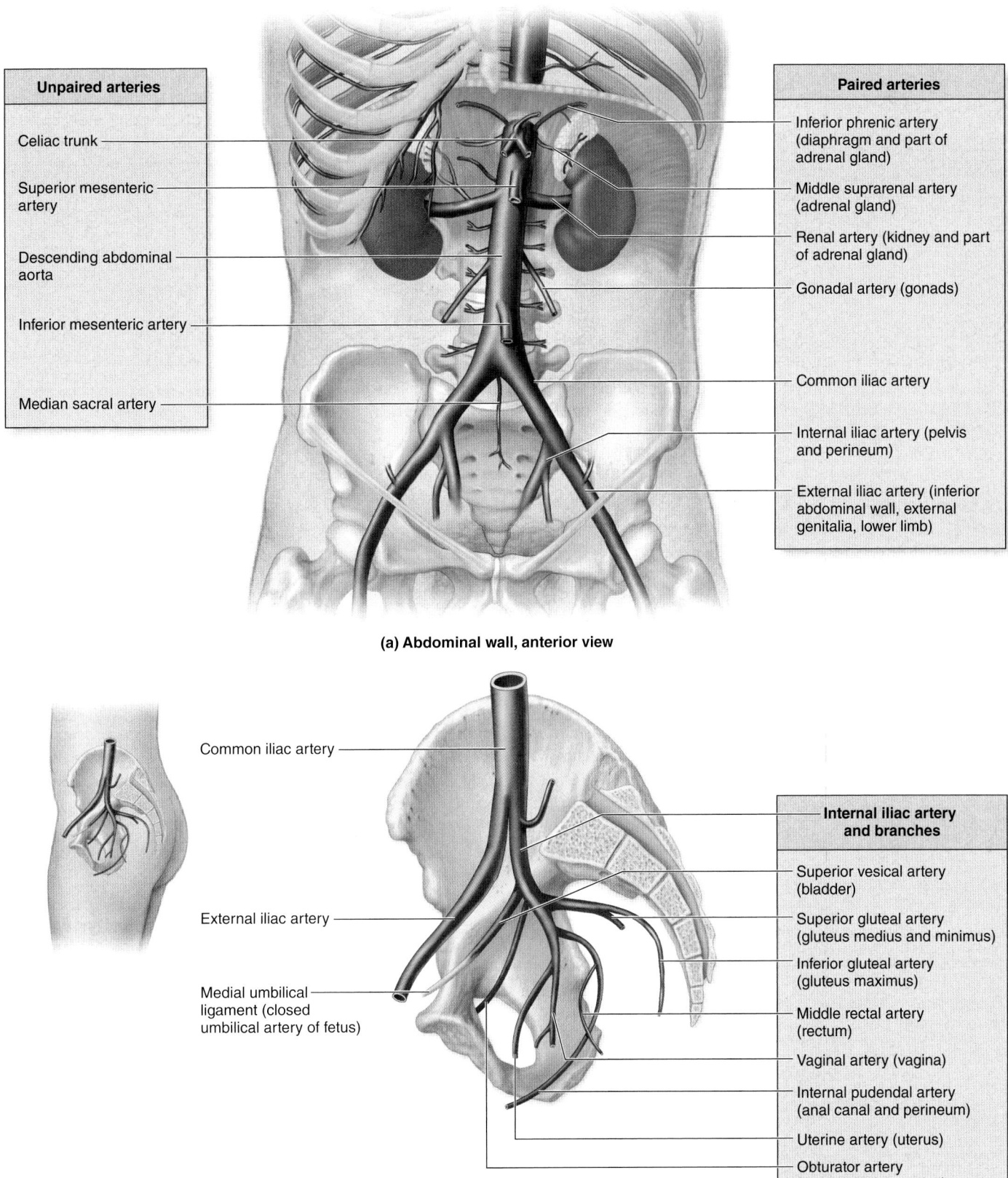

Unpaired arteries

Celiac trunk

Superior mesenteric artery

Descending abdominal aorta

Inferior mesenteric artery

Median sacral artery

Paired arteries

Inferior phrenic artery (diaphragm and part of adrenal gland)

Middle suprarenal artery (adrenal gland)

Renal artery (kidney and part of adrenal gland)

Gonadal artery (gonads)

Common iliac artery

Internal iliac artery (pelvis and perineum)

External iliac artery (inferior abdominal wall, external genitalia, lower limb)

(a) Abdominal wall, anterior view

Common iliac artery

External iliac artery

Medial umbilical ligament (closed umbilical artery of fetus)

Internal iliac artery and branches

Superior vesical artery (bladder)

Superior gluteal artery (gluteus medius and minimus)

Inferior gluteal artery (gluteus maximus)

Middle rectal artery (rectum)

Vaginal artery (vagina)

Internal pudendal artery (anal canal and perineum)

Uterine artery (uterus)

Obturator artery (medial thigh muscles)

(b) Female pelvis, medial view

Figure 20.26 Arterial Supply to the Abdominal Organs, Pelvis, and Perineum. (*a*) Paired arteries directly off the descending abdominal aorta supply blood to the adrenal glands, kidneys, gonads, and pelvis. (*b*) Branches of the right internal iliac artery distribute blood to the pelvic organs on the right side of the body. Shown is a female pelvis; a male pelvis would have no uterine artery, and instead of a vaginal artery, would have an inferior vesical artery (not all internal iliac artery branches are labeled).

20.11 Systemic Circulation: Upper and Lower Limbs

Blood flow through the upper and the lower limbs mirror one another. Both the upper and lower limbs (1) are supplied by a main arterial vessel: the subclavian artery for the upper limb, and the external iliac artery for the lower limb; (2) have their main artery bifurcating at either the elbow or the knee; (3) have arterial and venous arches; and (4) have superficial and deep networks of veins.

20.11a Upper Limb

LEARNING OBJECTIVES

1. Trace the arteries of the upper limb from the subclavian artery to the fingers.
2. Compare and contrast the superficial venous drainage and the deep venous drainage of the upper limb.

Each upper limb is supplied by a subclavian artery and drained by a subclavian vein. The venous system includes both deep and superficial drainage.

Arterial Flow to the Upper Limb

A **subclavian artery** supplies blood to each upper limb. The left subclavian artery emerges directly from the aortic arch, while the right subclavian artery is a division of the brachiocephalic trunk (see figure 20.19*a*).

Before extending into the arm, the subclavian artery extends multiple branches to supply parts of the body's upper region: the vertebral artery, the thyrocervical trunks, the costocervical trunk, and the internal thoracic artery, as described earlier.

After the subclavian artery passes over the lateral border of the first rib, it is renamed the **axillary** (ak′sil-ār-ē) **artery** (**figure 20.27**). The axillary artery extends many branches to the shoulder and thoracic region. When the axillary artery passes the inferior border of the teres major muscle, it is renamed the **brachial** (brā′kē-ăl) **artery.** One of its branches is the **deep brachial artery,** which supplies blood to most brachial (arm) muscles. In the cubital fossa, the brachial artery divides into a **radial artery** and an **ulnar artery.** Both arteries supply the forearm and wrist before they anastomose and form two arterial arches in the palm: the **deep palmar arch** (formed primarily from the radial artery), and the **superficial palmar arch** (formed primarily from the ulnar artery). **Digital arteries** emerge from the arches to supply the fingers.

WHAT DO YOU THINK?

 If the left ulnar artery is cut, would any blood be able to reach the left hand and fingers? Why or why not?

Venous Drainage of the Upper Limb

Venous drainage of the upper limb, which converges on the axillary vein and into the subclavian vein, is accomplished through two groups of veins: superficial and deep.

Superficial Venous Drainage On the dorsum of the hand, a dorsal venous network (or arch) of veins drains into both the medially located **basilic** (ba-sil′ik) **vein** and the laterally located **cephalic** (se-fal′ik) **vein.** These veins drain into the axillary vein, and they also have perforating branches that allow them to connect to the deeper veins.

In the cubital region, an obliquely positioned **median cubital vein** connects the cephalic and basilic veins. The median cubital vein is a common site for venipuncture, in which a vein is punctured with a hollow needle to draw blood or inject fluids and medications. All of these superficial veins are highly variable among individuals and have multiple superficial tributaries draining into them.

Deep Venous Drainage The **digital veins** and **deep** and **superficial palmar venous arches** drain into *pairs* of **radial veins** and **ulnar veins** that run parallel to arteries of the same name. At the level of the cubital fossa, the radial and ulnar veins merge to form a pair of **brachial veins** that travel with the brachial artery. Brachial veins and the basilic vein merge to form the **axillary vein.**

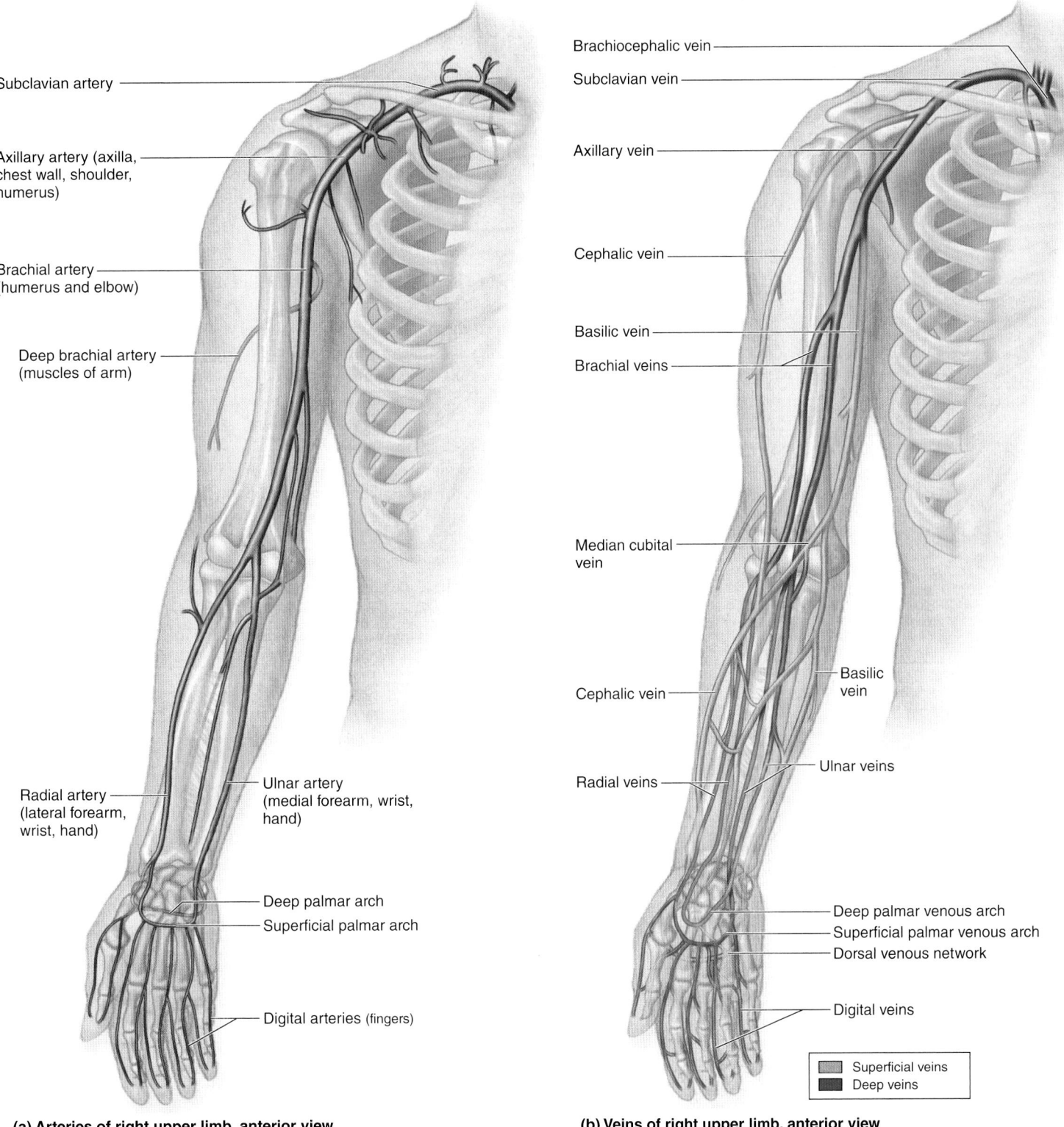

(a) Arteries of right upper limb, anterior view

(b) Veins of right upper limb, anterior view

Figure 20.27 Vascular Supply to the Upper Limb. The subclavian artery transports oxygenated blood to the upper limb; veins merge to return deoxygenated blood to the heart. (*a*) Arteries that supply the upper limb. (*b*) Superficial and deep veins that return blood from the upper limb. **AP|R**

Superior to the lateral border of the first rib, the axillary vein is renamed the **subclavian vein** (see figure 20.19*b*). When the subclavian vein and internal jugular veins of the neck merge, they form the brachiocephalic vein. As we have seen, the left and right brachiocephalic veins form the superior vena cava.

WHAT DID YOU LEARN?

39 What is the sequence of vessels from the subclavian artery to the digital artery of the thumb?

40 What are the primary superficial veins of the upper limb?

20.11b Lower Limb

LEARNING OBJECTIVES

3. Trace the arteries of the lower limb from the external iliac artery to the toes.

4. Compare and contrast the superficial venous drainage and the deep venous drainage of the lower limb.

The arterial and venous blood flow of the lower limb is very similar to that of the upper limb. As we discuss lower limb blood flow, compare it with that of the upper limb.

Arterial Flow to the Lower Limb

The main arterial supply for the lower limb is the **external iliac artery,** which is a branch of the common iliac artery **(figure 20.28a).** The external iliac artery travels inferior to the inguinal ligament, where it is renamed the **femoral** (fem′ŏ-răl) **artery.** The **deep femoral artery** (or *deep artery of the thigh*) emerges from the femoral artery to supply the hip joint (via medial and lateral femoral circumflex arteries) and many of the thigh muscles, before traversing posteromedially along the thigh. When the femoral artery enters the popliteal fossa, it is renamed the **popliteal** (pop-lit′ē-ăl, pop-li-tē′ăl) **artery.** The popliteal artery supplies the knee joint and muscles in this region.

The popliteal artery divides into an **anterior tibial artery** that supplies the anterior compartment of the leg, and a **posterior tibial artery** that supplies the posterior compartment of the leg. The posterior tibial artery extends a branch called the **fibular artery,** which supplies the lateral compartment leg muscles.

The posterior tibial artery continues to the plantar side of the foot, where it branches into **medial** and **lateral plantar arteries.** The anterior tibial artery crosses over the anterior surface of the ankle, where it is renamed the **dorsalis pedis artery** (*dorsal artery of the foot*). The dorsalis pedis artery and a branch of the lateral plantar artery unite to form the **plantar arterial arch** of the foot. **Digital arteries** extend from the plantar arch and supply the toes.

Venous Drainage of the Lower Limb

Venous drainage of the lower limb is also through superficial and deep veins (figure 20.28b).

Superficial Venous Drainage   On the dorsum of the foot, a dorsal venous arch drains into the **great saphenous** (săf-ĕ′nŭs, să-fē′nŭs) **vein** and the **small saphenous vein.** The great saphenous vein originates in the medial ankle and extends adjacent to the medial surface of the entire lower limb before it drains into the femoral vein. The small saphenous vein extends adjacent to the lateral ankle and then travels along the posterior calf, before draining into the popliteal vein. These superficial veins have perforating branches that connect to the deeper veins. If the valves in these veins (or the perforating branches) become incompetent, varicose veins develop. (See Clinical View: "Varicose Veins" in section 20.5a.)

Deep Venous Drainage   The **digital veins** and deep veins of the foot drain into pairs of **medial** and **lateral plantar veins.** These veins and a pair of **fibular veins** drain into a pair of **posterior tibial veins.** On the dorsum of the foot and ankle, deep veins drain into a pair of **anterior tibial veins,** which traverse alongside the anterior tibial artery. The **anterior** and **posterior tibial veins** merge to form a **popliteal vein** that curves to the anterior portion of the thigh and is renamed the **femoral vein.** Once this vein passes superior to the inguinal ligament, it is renamed as the **external iliac vein.** The external and internal iliac veins merge in the pelvis, forming the common iliac vein. Left and right common iliac veins then merge to form the inferior vena cava.

WHAT DID YOU LEARN?

41 Can you trace one pathway that blood may travel through from the external iliac artery to the digital arteries?

42 How do the great saphenous vein and small saphenous vein compare in terms of their anatomic position and length?

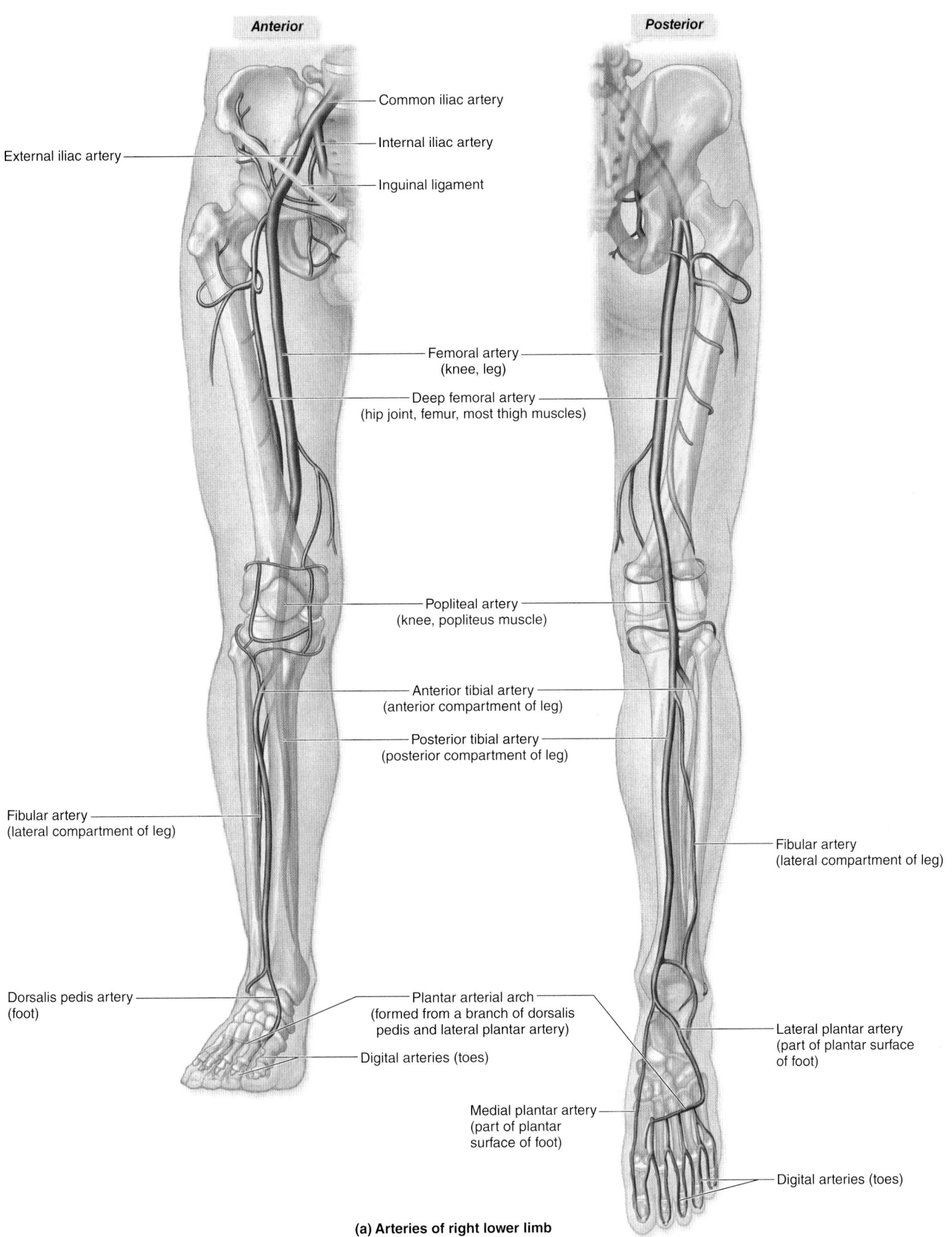

Anterior

Posterior

External iliac artery

Common iliac artery

Internal iliac artery

Inguinal ligament

Femoral artery
(knee, leg)

Deep femoral artery
(hip joint, femur, most thigh muscles)

Popliteal artery
(knee, popliteus muscle)

Anterior tibial artery
(anterior compartment of leg)

Posterior tibial artery
(posterior compartment of leg)

Fibular artery
(lateral compartment of leg)

Fibular artery
(lateral compartment of leg)

Dorsalis pedis artery
(foot)

Plantar arterial arch
(formed from a branch of dorsalis
pedis and lateral plantar artery)

Digital arteries (toes)

Lateral plantar artery
(part of plantar surface
of foot)

Medial plantar artery
(part of plantar
surface of foot)

Digital arteries (toes)

(a) Arteries of right lower limb

Figure 20.28 Vascular Supply to the Lower Limb. The external iliac artery transports oxygenated blood to the lower limb; veins merge to return deoxygenated blood to the heart. (*a*) Anterior and posterior views of arteries distributed throughout the lower limb. **AP|R**

(continued on next page)

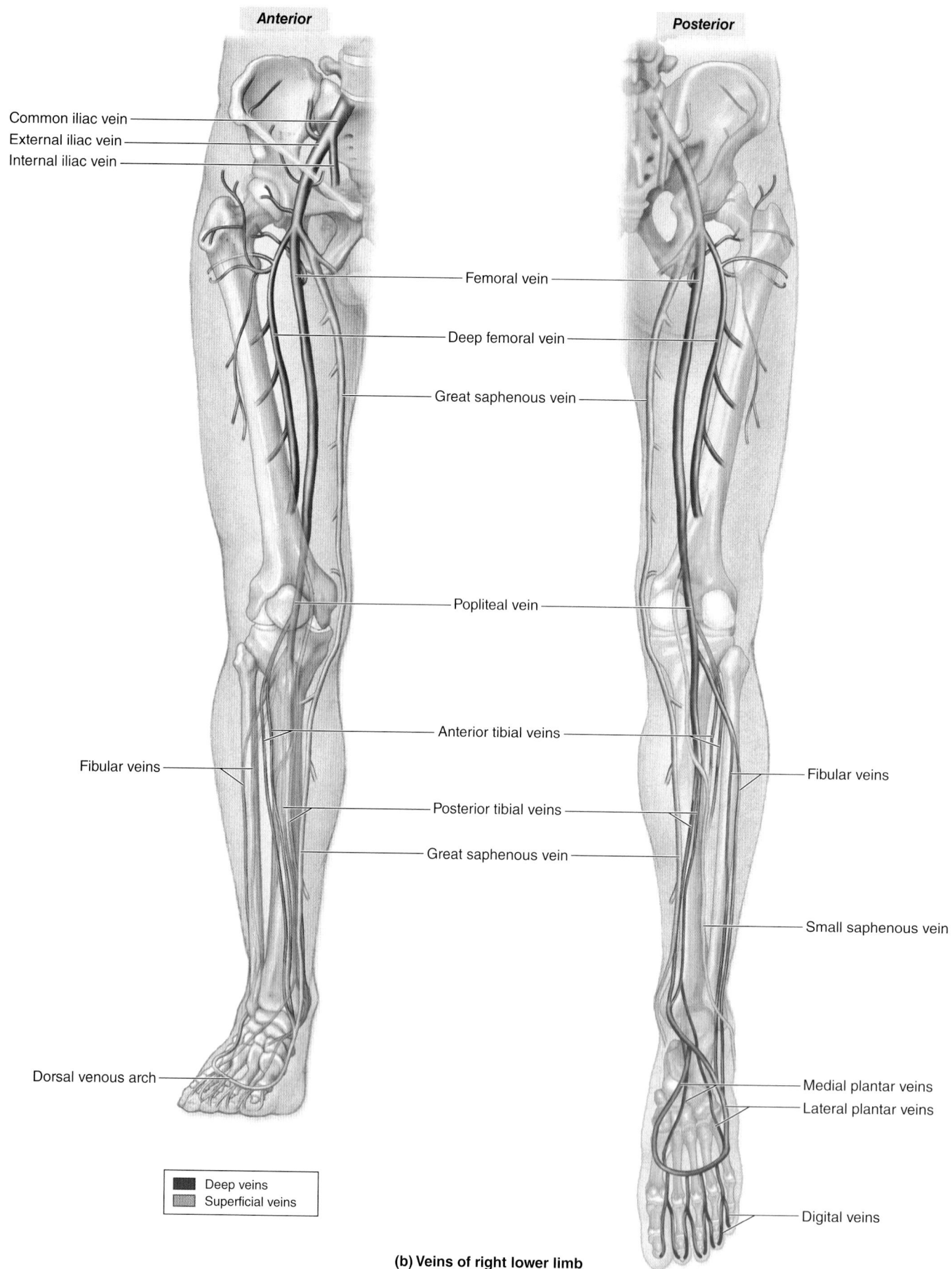

Anterior

Common iliac vein
External iliac vein
Internal iliac vein

Femoral vein

Deep femoral vein

Great saphenous vein

Popliteal vein

Anterior tibial veins

Fibular veins

Posterior tibial veins

Great saphenous vein

Dorsal venous arch

Deep veins
Superficial veins

Posterior

Great saphenous vein

Fibular veins

Small saphenous vein

Medial plantar veins
Lateral plantar veins

Digital veins

(b) Veins of right lower limb

Figure 20.28 Vascular Supply to the Lower Limb (*continued*). (*b*) Anterior and posterior views of the superficial and deep veins that return blood from the lower limb.

20.12 Comparison of Fetal and Postnatal Circulation

The cardiovascular system of the fetus is structurally and functionally different from that of the newborn. Whereas the fetus receives oxygen and nutrients directly from the mother through the placenta, the newborn's postnatal cardiovascular system must be independent. Additionally, because the fetal lungs are not functional, the blood pressure in the pulmonary arteries and right side of the heart is greater than the pressure in the left side of the heart. Finally, several fetal vessels help shunt blood directly to the organs in need and away from the organs that are not yet functional. As a result, the fetal cardiovascular system has some structures that are modified or that cease to function once the baby is born.

20.12a Fetal Circulation

LEARNING OBJECTIVE

1. Trace the pathway of blood circulation in the fetus.

The fetal circulatory route is listed here and shown in **figure 20.29**:

1. Oxygenated blood from the placenta enters the body of the fetus through the **umbilical vein.**

2. The blood from the umbilical vein is shunted away from the liver and directly toward the inferior vena cava through the **ductus venosus** (dŭk′tŭs vē-nō′sŭs).

3. Oxygenated blood in the ductus venosus mixes with deoxygenated blood in the inferior vena cava.

4. Blood from the superior and inferior venae cavae empties into the right atrium.

5. Because pressure is greater on the right side of the heart as compared to the left side, most of the blood is shunted from the right atrium to the left atrium via the **foramen ovale.** This blood flows into the left ventricle and then is pumped out through the aorta.

6. A small amount of blood enters the right ventricle and pulmonary trunk, but much of this blood is shunted from the pulmonary trunk to the aorta through a vessel detour called the **ductus arteriosus** (ar-tēr′ē-ō′sŭs).

7. Blood travels to the rest of the body, and the deoxygenated blood returns to the placenta through a pair of **umbilical arteries.**

8. Nutrient and gas exchange occurs at the **placenta,** and the cycle repeats.

At birth, the fetal circulation begins to change into the postnatal pattern. When the baby takes its first breath, pulmonary resistance drops, and the pulmonary arteries dilate. As a result, pressure on the right side of the heart decreases so that the pressure is greater on the left side of the heart, which handles the systemic circulation.

WHAT DID YOU LEARN?

43 List the five structures of fetal circulation, and identify the function of each.

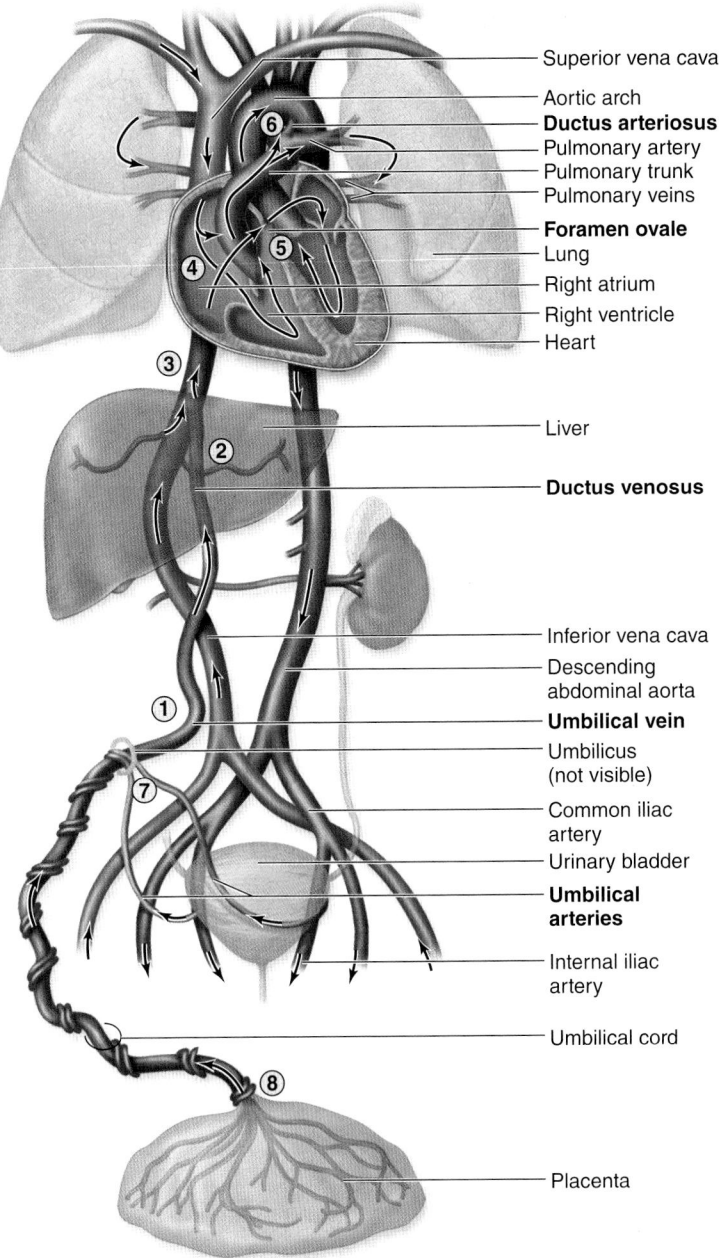

Fetal cardiovascular structure	Postnatal structure
Ductus arteriosus	Ligamentum arteriosum
Ductus venosus	Ligamentum venosum
Foramen ovale	Fossa ovalis
Umbilical arteries	Medial umbilical ligaments
Umbilical vein	Round ligament of liver (ligamentum teres)

Figure 20.29 Fetal Circulation. Structural changes in both the heart and blood vessels accommodate the different needs of the fetus and the newborn. The pathway of blood flow is indicated by black arrows. The chart at the bottom of the drawing summarizes the fate of each fetal cardiovascular structure following birth.

20.12b Postnatal Changes

LEARNING OBJECTIVE

2. Describe the changes that occur after the baby is born and must utilize the pulmonary circulation.

The postnatal changes occur as follows:

- The umbilical vein and umbilical arteries constrict and become nonfunctional. They turn into the **round ligament of the liver** (or *ligamentum teres*) and the **medial umbilical ligaments,** respectively.

- The ductus venosus ceases to be functional and constricts, becoming the **ligamentum venosum.**

- Because pressure is now greater on the left side of the heart, the two flaps of the interatrial septum close off the foramen ovale. The only remnant of the foramen ovale is a thin, oval depression in the wall of the septum called the **fossa ovalis.**

- Within 10 to 15 hours after birth, the ductus arteriosus constricts and becomes a fibrous structure called the **ligamentum arteriosum.**

INTEGRATE

CLINICAL VIEW
Patent Ductus Arteriosus

In some infants, especially those born prematurely, the ductus arteriosus fails to close after birth. A **patent** (open) **ductus arteriosus** serves as a conduit through which blood from the aorta can enter the pulmonary system; it can result in high blood pressure in the pulmonary circulation and in mixing of oxygenated blood from the aorta with deoxygenated blood from the pulmonary trunk. Because prostaglandins help keep the ductus arteriosus open during fetal life, a patent ductus arteriosus may be treated with prostaglandin-inhibiting medication. Surgery to remove the ductus arteriosus is a second treatment option.

WHAT DID YOU LEARN?

44 Why is it important for the ductus arteriosus and foramen ovale to both close after birth?

CHAPTER SUMMARY

20.1 Structure and Function of Blood Vessels 777	• Arteries take blood away from the heart, capillaries are responsible for gas and nutrient exchange, and veins take blood to the heart.

20.1a General Structure of Vessels   777

- Arteries and veins have a tunica intima (innermost layer), a tunica media (middle layer), and a tunica externa (outermost layer).
- Capillaries have a tunica intima, composed of an endothelial layer and a basement membrane only.

20.1b Arteries   779

- Arteries progressively branch into elastic arteries, muscular arteries, and arterioles.

20.1c Capillaries   782

- Capillaries, the smallest blood vessels, connect arterioles with venules. Gas and nutrient exchange occurs in the capillaries.
- The three types of capillaries are continuous capillaries, fenestrated capillaries, and sinusoids, which have varying degrees of permeability and are arranged in capillary beds.

20.1d Veins   785

- Venules are small veins that merge into larger veins. One-way valves prevent blood backflow in veins.
- Blood pressure is low in the veins, which act as reservoirs and hold about 55% of the body's blood at rest.

20.1e Pathways of Blood Vessels   785

- Blood vessels may be arranged in simple pathways (artery ⟶ capillary bed ⟶ veins ⟶ heart) or in alternative pathways, which include anastomoses and portal venous systems.

20.2 Total Cross-Sectional Area and Blood Flow Velocity 787	• The velocity of blood flow is inversely related to the total cross-sectional area of the blood vessels. • Blood flows slowest in the capillaries, increasing efficiency of nutrient and gas exchange.

20.3 Capillary Exchange 788	• Materials may pass through the capillary walls via diffusion, vesicular transport, and bulk flow.

20.3a Diffusion and Vesicular Transport   788

- Small solutes (e.g., oxygen, carbon dioxide, glucose, ions) move between blood and interstitial fluid via diffusion.
- Certain hormones, such as insulin, and fatty acids are transported via vesicular transport.

20.3b Bulk Flow   788

- Bulk flow refers to the net filtration and reabsorption that occurs at the capillary level due to hydrostatic pressure and osmotic pressure.

20.3c Net Filtration Pressure   789

- The net filtration pressure is the difference between the net hydrostatic pressure and the net colloid osmotic pressure.
- The net filtration pressure is positive at the arterial end of the capillary, resulting in filtration, and is negative at the venous end of the capillary, resulting in reabsorption.

(continued on next page)

20.8 Pulmonary Circulation 806	**20.8a Blood Flow Through the Pulmonary Circulation 806** • Blood is pumped from the right ventricle to the lungs and back to the left side of the heart.
	20.8b Characteristics of the Pulmonary Circulation 807 • The pulmonary circulation is a smaller circuit and has lower blood pressure than the systemic circulation.
20.9 Systemic Circulation: Vessels from and to the Heart 808	**20.9a General Arterial Flow Out of the Heart 808** • Oxygenated blood is pumped from the left ventricle through the ascending aorta; the coronary arteries branch from the ascending aorta. • Three branches of the aortic arch are the brachiocephalic trunk, the left common carotid, and the left subclavian. • The descending abdominal aorta bifurcates into the left and right common iliac arteries.
	20.9b General Venous Return to the Heart 810 • Veins from the head, neck, upper limbs, and thoracic and abdominal walls merge to form left and right brachiocephalic veins, which merge to form the superior vena cava.
20.10 Systemic Circulation: Head and Trunk 810	**20.10a Head and Neck 810** • The common carotid artery supplies most of the blood to the head and neck and divides into the external carotid artery and internal carotid artery. • The external carotid artery supplies superficial regions of the head and organs of the neck while the internal carotid artery supplies the brain and orbits. • Blood from the cranium drains via the vertebral veins and the dural venous sinuses. Venous return is via the internal and external jugular veins.
	20.10b Thoracic and Abdominal Walls 814 • Two internal thoracic arteries as well as other arteries supply the thoracic and abdominal walls. • Venous return from the thoracic and abdominal walls is complex and involves the azygos system of veins.
	20.10c Thoracic Organs 816 • The lungs are supplied by bronchial arteries branching from the descending thoracic aorta; they are drained by bronchial veins. • The esophagus receives blood from the esophageal arteries that branch from the aorta and from the esophageal branches of the left gastric artery; it is drained via the esophageal veins. • The diaphragm is supplied by phrenic arteries and drained by the phrenic veins.
	20.10d Gastrointestinal Tract 817 • The three major unpaired arteries from the descending abdominal aorta that supply the abdominal organs are the celiac trunk, the superior mesenteric artery, and the inferior mesenteric artery. • The hepatic portal vein receives oxygen-poor but nutrient-rich blood from the gastrointestinal (GI) organs and spleen; this vessel takes this blood to the liver for processing. Blood leaves the liver via the hepatic veins.
	20.10e Posterior Abdominal Organs, Pelvis, and Perineum 820 • Paired branches of the descending abdominal aorta supply the posterior abdominal organs, pelvis, and perineum. Venous drainage is by veins of the same name as the arteries.
20.11 Systemic Circulation: Upper and Lower Limbs 822	• The upper and lower limbs are each supplied by a single artery that branches into multiple vessels. Both superficial and deep veins drain the limbs. **20.11a Upper Limb 822** • The upper limb is supplied by the subclavian artery and its branches and is drained into the subclavian vein.
	20.11b Lower Limb 824 • The lower limbs are supplied by the external iliac arteries and veins.
20.12 Comparison of Fetal and Postnatal Circulation 827	**20.12a Fetal Circulation 827** • In the fetus, oxygenated blood is supplied by the umbilical vein from the placenta, and deoxygenated blood leaves through a pair of umbilical arteries. • The fetal cardiovascular system bypasses the liver via the ductus venosus and bypasses the pulmonary circulation via the foramen ovale and ductus arteriosus.
	20.12b Postnatal Changes 828 • In the newborn, the umbilical vein and arteries degenerate, and the ductus venosus, foramen ovale, and ductus arteriosus close as the infant is able to breathe on its own.

CHALLENGE YOURSELF

Do You Know the Basics?

_____ 1. Which of the following is *not* a characteristic of a capillary?

    a. Fenestrated capillaries have pores that allow for relatively large materials to slip out of the capillary.

    b. Sinusoid capillaries are the main type of capillary around the brain.

    c. Capillaries often are arranged in a capillary bed that is supplied by an arteriole.

    d. The capillary wall consists of an endothelium and basement membrane only—there is no subendothelial layer.

_____ 2. Which statement is accurate about veins?

    a. Veins always carry deoxygenated blood.

    b. Veins drain into smaller vessels called venules.

    c. The largest tunic in a vein is the tunica externa.

    d. The lumen of a vein tends to be smaller than that of a comparably sized artery.

_____ 3. Vasa vasorum are found in the tunica _____ of a large blood vessel.

    a. intima

    b. media

    c. externa

    d. All of these are correct.

_____ 4. Which of the following decreases perfusion of a tissue?

    a. decreased total blood flow

    b. vasodilators

    c. angiogenesis

    d. increased carbon dioxide, increased H^+, and low O_2

_____ 5. The _____ is a type of vessel with the smallest pressure gradient that must be overcome by contraction of skeletal muscles and breathing.

    a. artery

    b. vein

    c. capillary

    d. sinusoid

_____ 6. An increase in _____ will result in an increase in blood flow to an area.

    a. vessel length

    b. vessel diameter

    c. blood viscosity

    d. the number of formed elements

_____ 7. Which of the following is accurate?

    a. Total blood flow increases with a steeper pressure gradient (assuming resistance stays the same).

    b. Total blood flow decreases with increased resistance (assuming cardiac output stays the same).

    c. Total blood flow is significant in maintaining adequate perfusion of all tissues.

    d. All of these are correct.

_____ 8. Velocity of blood flow is the slowest in

    a. muscular arteries.

    b. capillaries.

    c. veins.

    d. elastic arteries.

_____ 9. Blood pressure is regulated by the

    a. cardiac center.

    b. vasomotor center.

    c. hormones.

    d. All of these are correct.

_____ 10. Name the correct pathway that blood flows through in the upper limb arteries.

    a. subclavian ⟶ axillary ⟶ ulnar ⟶ radial ⟶ brachial

    b. subclavian ⟶ axillary ⟶ brachial ⟶ cephalic ⟶ basilic

    c. subclavian ⟶ ulnar ⟶ brachial ⟶ radial

    d. subclavian ⟶ axillary ⟶ brachial ⟶ radial and ulnar

11. List and describe the three tunics found in most blood vessels.

12. Compare and contrast arteries and veins with respect to function, tunic size, lumen size, and blood pressure.

13. Explain the difference between hydrostatic and osmotic pressure. How do these pressures change from the arteriole end of a capillary to the venule end of a capillary?

14. Write the formula for determining net filtration pressure in the capillaries.

15. Describe the relationship of vessel radius, vessel length, blood viscosity, and blood pressure to blood flow.

16. How is it possible for the sympathetic division to cause vasoconstriction in most blood vessels, but vasodilation in coronary and skeletal muscle blood vessels? Explain.

17. Briefly explain how changes in cardiac output, resistance, and blood volume influence blood pressure.

18. Compare how the cardiac center and vasomotor center regulate blood pressure and blood flow.

19. Compare the systemic circulation and pulmonary circulation. Discuss the function of arteries and veins in each system.

20. What postnatal changes occur in the heart and blood vessels? Why do these occur?

Can You Apply What You've Learned?

1. If a patient has cirrhosis of the liver and is unable to produce sufficient albumin and other plasma proteins, then what variable is changed in capillary exchange, and what is the effect?

   a. There is no change in the capillary exchange process.

   b. Hydrostatic pressure in the blood decreases, and fluid remains in the blood.

   c. Colloid osmotic pressure in the capillary increases, and blood volume increases in the blood vessels.

   d. Colloid osmotic pressure in the capillary decreases, and fluid remains in the interstitial fluid, potentially causing edema.

2. Arlene heads to the gym and initiates a vigorous exercise regimen. Which hormone will *not* be released?

   a. epinephrine

   b. atrial natriuretic peptide

   c. angiotensin II

   d. norepinephrine

3. After her workout, Arlene lies down on a mat to stretch out her leg muscles. She stands up suddenly after stretching and feels a bit light-headed. However, the light-headed feeling passes quickly. What physiologic process occurred that resulted in the passing of her light-headed feeling?

   a. The aortic bodies detected an increase in blood pressure in the head, and initiated a chemoreceptor reflex that decreased blood pressure.

   b. The carotid bodies detected low CO_2 and high O_2 levels from exercising, and stimulated the vasomotor center to vasoconstrict the vessels of the head and neck.

   c. Only α receptors in the head and neck initially were being stimulated, resulting in the light-headed feeling.

   d. The baroreceptors within the carotid detected decreased stretch in the carotid vessel wall and initiated a baroreceptor reflex to increase blood flow to the head and neck.

4. Over a period of 6 months, Harold loses 40 pounds. When Harold next visits his doctor, he discovers that his blood pressure levels have fallen too. Which of the following best explains why Harold's blood pressure dropped?

   a. When fat is lost, the extra length in blood vessels supplying the fat is lost too. Decreased blood vessel length resulted in decreased blood pressure.

   b. Most of the fat that Harold lost was around his thoracic organs, which compressed these organs. By losing the weight, he removed this compressive force.

   c. The resistance in the blood vessels increased due to the fat loss, resulting in decreased blood pressure.

   d. The fat previously constricted the blood vessels. When Harold lost the fat, the blood vessels were allowed to dilate, and thus blood pressure was reduced.

5. Alejandro was hit by a car, which resulted in massive injury and bleeding to his right foot. You realize that you have to compress a lower limb artery so that Alejandro will not lose too much blood. Which vessel, if compressed, will completely stop the flow of blood to the foot?

   a. anterior tibial artery

   b. posterior tibial artery

   c. deep femoral artery

   d. popliteal artery

Can You Synthesize What You've Learned?

1. Your patient is Thomas, a 50-year-old man that rarely exercises, is overweight, and only occasionally eats healthy meals, which puts him at risk for atherosclerosis. Explain to him the relationship between atherosclerosis and high blood pressure.

2. Arteries tend to have a lot of vascular anastomoses around body joints, such as the elbow and knee. Can you think of a reason why this would be beneficial?

3. Explain why an overweight individual with high blood pressure is advised to lose weight. Include in your response the relationship of blood vessel length, resistance, and blood pressure.

INTEGRATE

ONLINE STUDY TOOLS

The following study aids may be accessed through Connect.

Clinical Case Study: A Man with Shortness of Breath Following an Airline Flight

Interactive Questions: This chapter's content is served up in a number of multimedia question formats for student study.

LearnSmart: Topics and terminology include structure and function of blood vessels; velocity of blood flow; capillary exchange; local blood flow; blood pressure, resistance, and total blood flow; regulation of blood pressure and blood flow; blood flow distribution during exercise; pulmonary circulation; systemic circulation: vessels from and to the heart; systemic circulation: head and trunk; systemic circulation: upper and lower limbs; comparison of fetal and postnatal circulation

Anatomy & Physiology Revealed: Topics include muscular artery and medium-sized vein; aneurysm; relationship between cross section and velocity of flow; capillary exchange; baroreceptor reflex; pulmonary and systemic circulation; cardiovascular system overview; brain arteries and veins; celiac trunk; inferior vena cava; abdominal aorta; hepatic portal system; shoulder and arm vasculature; hip and thigh vasculature

Animations: Topics include fluid exchange across walls of capillaries; arteriolar radius and blood flow; arteriolar resistance and blood pressure; baroreceptor reflex control of blood pressure; chemoreceptor reflex control of blood pressure

Lymphatic System

chapter 21

INTEGRATE

ap001revealed.com

Module 10: Lymphatic System

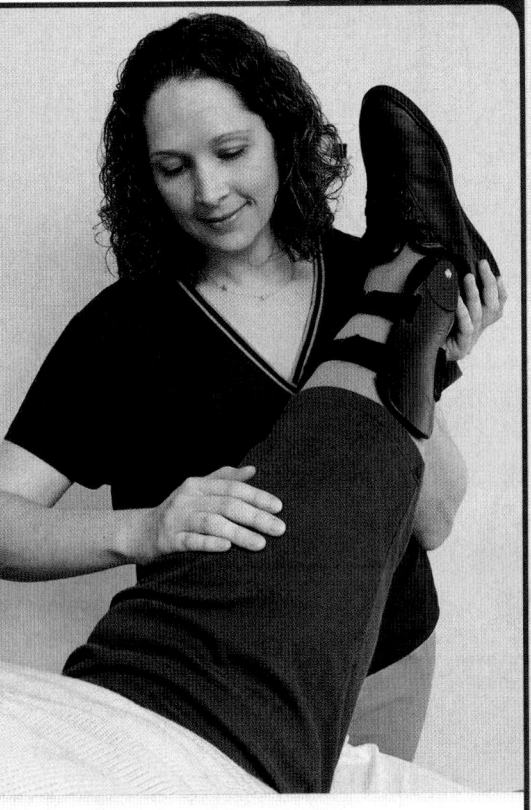

CAREER PATH
Lymphedema Therapist

A lymphedema therapist uses massage, compression wraps/garments, and exercise with patients to alleviate edema, a common condition if lymph vessels and lymph nodes are removed as a part of the treatment for cancer. Massage assists the movement of lymph in the lymph vessels and its return to the venous circulation. The certification to become a lymphedema therapist is typically earned by nurses, physical therapists, occupational therapists, and massage therapists.

It is not uncommon for young children to develop tonsillitis—a condition in which the tonsils become inflamed and enlarged. Often, the lymph nodes in the neck also are swollen, and the spleen may become enlarged. You might initially be curious about why the lymph nodes and the spleen are also affected when a child has tonsillitis. All of these structures—the tonsils, lymph nodes, and the spleen—are components of the lymphatic system, and these organs function in helping the immune system defend the body against infectious agents. The inflammation and enlargement that occur in lymphatic structures are signs that these organs are actively engaged in defending the body against potentially harmful substances.

The lymphatic system has two significant functions, and both provide support to other body systems: (1) The lymphatic system transports and houses lymphocytes and other immune cells that help the immune system defend against foreign substances. (2) The lymphatic system aids the cardiovascular system by returning excess fluid to venous blood within the cardiovascular system to maintain fluid balance, blood volume, and blood pressure. Interestingly, the lymphatic system has no unique function of its own.

Here, we first describe lymph vessels and the absorption of excess fluid from the interstitial space and then discuss lymphatic organs, including lymph nodes and the spleen. Integrated within the chapter are brief descriptions of some of the common malfunctions of the lymphatic system. The details of the immune system are described in chapter 22.

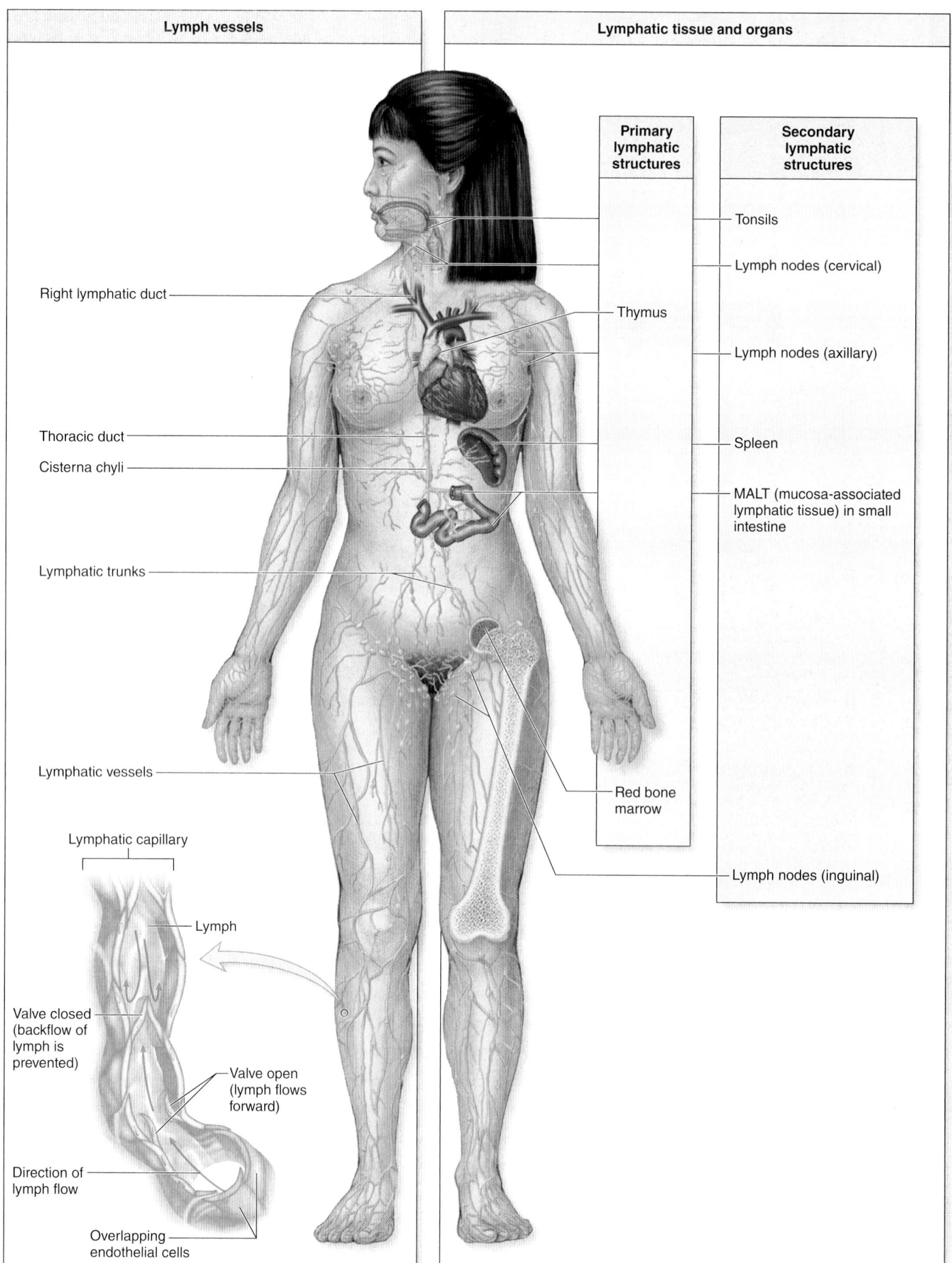

Lymph vessels

Right lymphatic duct

Thoracic duct

Cisterna chyli

Lymphatic trunks

Lymphatic vessels

Lymphatic capillary

Lymph

Valve closed (backflow of lymph is prevented)

Valve open (lymph flows forward)

Direction of lymph flow

Overlapping endothelial cells

Lymphatic tissue and organs

Primary lymphatic structures	Secondary lymphatic structures

Tonsils

Lymph nodes (cervical)

Thymus

Lymph nodes (axillary)

Spleen

MALT (mucosa-associated lymphatic tissue) in small intestine

Red bone marrow

Lymph nodes (inguinal)

Figure 21.1 Lymphatic System. The lymphatic system is composed of both lymph vessels and lymphatic tissue and organs. Lymphatic tissue and organs are organized into primary and secondary lymphatic structures.

21.1 Lymph and Lymph Vessels

The **lymphatic** (lim-fat′ik) **system** is composed of lymph vessels and lymphatic tissues and organs (**figure 21.1**). Lymph is the fluid transported within lymph vessels. We begin by describing the formation and characteristics of lymph. We then discuss its transport through progressively larger lymph vessels until the lymph is returned to the cardiovascular system.

21.1a Lymph and Lymphatic Capillaries

LEARNING OBJECTIVES

1. Describe lymph and its contents.
2. Discuss the location and anatomic structure of lymphatic capillaries.
3. Explain how fluid enters lymphatic capillaries.

Lymph originates as interstitial fluid surrounding tissue cells; it moves passively into the lymphatic capillaries due to a hydrostatic pressure gradient. Lymphatic capillaries merge to form larger lymph vessels.

Characteristics of Lymph

Approximately 15% of the fluid that enters the interstitial space surrounding the cells is not reabsorbed back into the blood capillaries during capillary exchange (see section 20.3d). This interstitial fluid amounts to about 3 liters daily and is normally absorbed into lymphatic capillaries.

Once inside the lymph vessels, the interstitial fluid is called **lymph** (limf; *lympha* = clear spring water). The components of lymph include water, dissolved solutes (e.g., ions), a small amount of protein (approximately 100 to 200 grams that leaked into the interstitial space during capillary exchange), sometimes foreign material that includes both cell debris and pathogens, and perhaps metastasized cancer cells (see Clinical View: "Metastasis").

INTEGRATE

CLINICAL VIEW
Metastasis

Although the lymph vessels provide an essential function by rerouting excess interstitial fluid back into the blood, these vessels sometimes can participate in the spread of pathogens or disease. For example, cancerous cells can break free from a primary tumor and be transported in the lymph. These cancerous cells may establish secondary tumors that develop in other locations within the body, a process referred to as **metastasis** (mĕ-tas′tă-sis). For example, breast cancer may metastasize to the lungs, and this cancer in the lungs is referred to as metastasized breast cancer, not lung cancer. A biopsy of a lymph node that is "positive" for the presence of cancer cells from another body organ verifies that the cancer has metastasized.

Lymphatic Capillaries

The lymph vessel network begins with **lymphatic capillaries,** which are the smallest lymph vessels (**figure 21.2**). Lymphatic capillaries are microscopic, closed-ended vessels that absorb interstitial fluid. They are interspersed throughout areolar connective tissue among blood capillary networks, except those within the red bone marrow, spleen, and the central nervous system. Note that lymphatic capillaries are absent within avascular tissues such as epithelia and cartilage.

A lymphatic capillary resembles the anatomic structure of a blood capillary in that its wall is composed of an endothelium (figure 21.2b). However, lymphatic capillaries are typically larger in diameter than blood capillaries, lack a basement membrane, and have overlapping endothelial cells. These overlapping endothelial cells act as one-way flaps to allow fluid to enter the lymphatic capillary, but prevent its loss. **Anchoring filaments** help hold these endothelial cells to the nearby

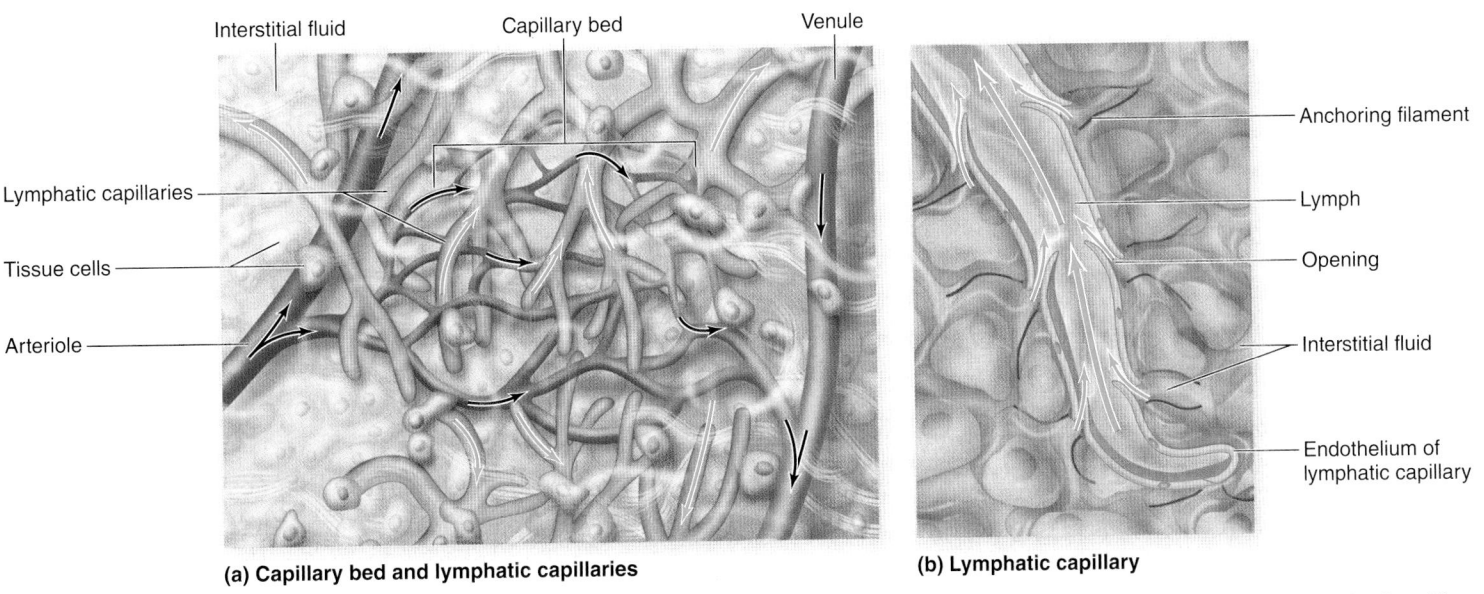

(a) Capillary bed and lymphatic capillaries

(b) Lymphatic capillary

Figure 21.2 Lymphatic Capillaries. (*a*) Lymphatic capillaries begin as blind-end vessels within connective tissue among most blood capillary networks and absorb excess fluid left during capillary exchange. (*b*) A lymphatic capillary takes up excess interstitial fluid through overlapping endothelial cells. The fluid is then called lymph. Here, the black arrows show blood flow and the green arrows show lymph flow. AP|R

INTEGRATE

CONCEPT CONNECTION

We described the formation of cerebrospinal fluid (CSF) in section 13.2c. Approximately 40% of the CSF is contributed from fluid that drains from the interstitial spaces of the brain. This excess fluid moves from the brain across the pia mater into the subarachnoid space. Its drainage into the subarachnoid space circumvents the need for lymph vessels within the central nervous system.

structures. Lymphatic capillaries located within the gastrointestinal (GI) tract, called **lacteals** (lak′tē-ăl; *lactis* = milk), allow for the absorption of lipid-soluble substances from the GI tract, a concept detailed in section 26.4c.

Movement of Lymph into Lymphatic Capillaries

The driving force to move fluid into the lymphatic capillaries is an increase in hydrostatic pressure within the interstitial space. Interstitial hydrostatic pressure rises as additional fluid is filtered from the blood capillaries (see section 20.3b). An increase in pressure at the margins of the lymphatic capillary endothelial cells "pushes" interstitial fluid into the lymphatic capillary lumen. The higher the interstitial fluid pressure, the greater the amount of fluid that enters the lymphatic capillary. The anchoring filaments extending between lymphatic capillary cells and the surrounding tissue prevent the collapse of the lymphatic capillaries as pressure exerted by the interstitial fluid increases.

The pressure exerted by lymph after it enters the lymphatic capillary forces the endothelial cells of these vessels to close. Thus,

INTEGRATE

CONCEPT CONNECTION

Lacteals are lymphatic capillaries within the small intestine that absorb dietary lipids and lipid-soluble vitamins that are unable to enter the blood directly from the gastrointestinal (GI) tract (see section 26.4c). The lipids are absorbed as part of chylomicrons (lipid droplets enveloped within protein). The lymph from the GI tract has a milky color due to the chylomicrons, and for this reason the GI tract lymph is also called chyle.

INTEGRATE

LEARNING STRATEGY

Fluid entry into a lymphatic capillary is analogous to the movement of the entryway door to your house or apartment. Imagine that the door is unlocked and the knob is turned. Putting pressure on the outside of the door (like the pressure of interstitial fluid on the outside of the lymphatic capillary wall) causes it to open to the inside so you can enter. Once inside, pressure applied to the inside surface of the door (or fluid pressure against the inside lymphatic capillary surface) causes it to close.

lymph becomes "trapped" within the lymphatic vessel and cannot be released back into the interstitial space. Lymph is then transported through a network of increasingly larger vessels that include (in order) lymphatic capillaries, lymphatic vessels, lymphatic trunks, and lymphatic ducts.

 WHAT DID YOU LEARN?

❶ What substances typically are absorbed from the interstitial space into lymphatic capillaries?

❷ How does fluid enter and become "trapped" in the lymphatic capillaries?

21.1b Lymphatic Vessels, Trunks, and Ducts

LEARNING OBJECTIVES

4. Explain the mechanisms that move lymph through lymphatic vessels, trunks, and ducts.

5. Name the five types of lymphatic trunks and the regions of the body from which they drain lymph.

6. Describe the regions that are drained by the right lymphatic duct and by the thoracic duct.

After lymph enters the lymphatic capillaries, it continues to flow into increasingly larger lymphatic vessels, trunks, and ducts. Ultimately, the lymph empties into the blood circulation.

Lymphatic Vessels

Lymphatic capillaries merge to form larger structures that are called **lymphatic vessels** (figure 21.1). Superficial lymphatic vessels are generally positioned adjacent to the superficial veins of the body; in contrast, deep lymphatic vessels are next to deep arteries and veins. Lymphatic vessels resemble small veins because both contain all three vessel tunics (intima, media, and externa) and have **valves** within their lumen. Valves are required to prevent lymph from pooling in these vessels and help prevent lymph backflow because the lymphatic vessel network is a low-pressure system. These valves are especially important in areas where lymph flow is against the direction of gravity, such as in the lower limbs.

The lymphatic system lacks a pump. It relies on several mechanisms to move lymph through its vessels: (1) contraction of nearby skeletal muscles in the limbs (skeletal muscle pump) and the respiratory pump in the torso (as described in section 20.5a), (2) the pulsatile movement of blood in nearby arteries, and (3) rhythmic contraction of smooth muscle in walls of larger lymph vessels (trunks and ducts).

Some lymphatic vessels connect directly to lymphatic organs called lymph nodes. Foreign or pathogenic material is filtered as lymph passes through lymph nodes. They are arranged in a series along lymph vessels, are described in more detail in section 21.4a.

Lymphatic Trunks

Lymphatic vessels feed into **lymphatic trunks** on both the right and left side of the body (**figure 21.3**). Each lymphatic trunk removes lymph from a specific major body region:

- *Jugular trunks* drain lymph from both the head and neck.
- *Subclavian trunks* remove lymph from the upper limbs, breasts, and superficial thoracic wall.
- *Bronchomediastinal trunks* drain lymph from deep thoracic structures.
- *Intestinal trunks* drain lymph from most abdominal structures.

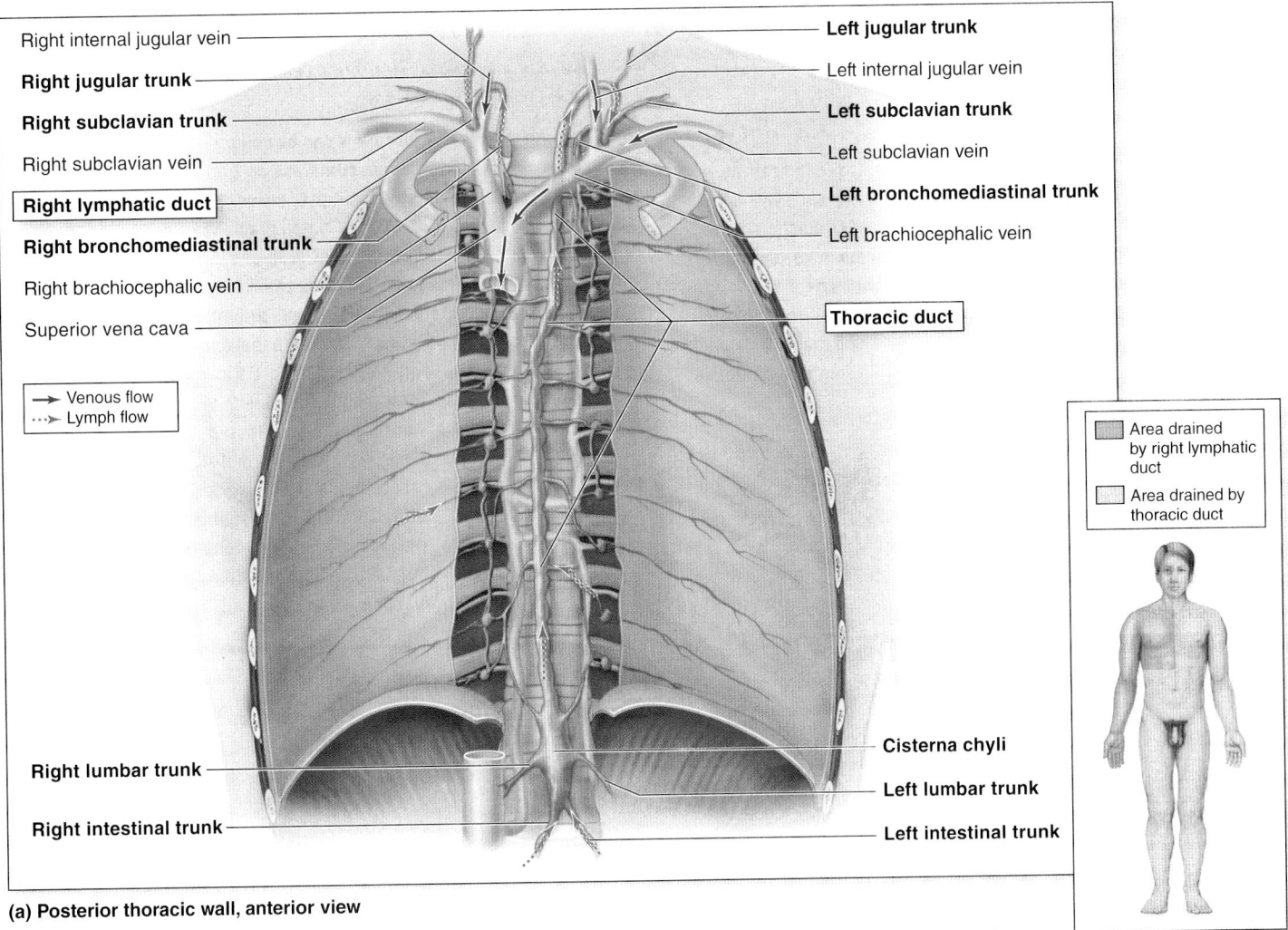

Right internal jugular vein
Right jugular trunk
Right subclavian trunk
Right subclavian vein
Right lymphatic duct
Right bronchomediastinal trunk
Right brachiocephalic vein
Superior vena cava

→ Venous flow
··▸ Lymph flow

Left jugular trunk
Left internal jugular vein
Left subclavian trunk
Left subclavian vein
Left bronchomediastinal trunk
Left brachiocephalic vein

Thoracic duct

Right lumbar trunk
Right intestinal trunk

Cisterna chyli
Left lumbar trunk
Left intestinal trunk

Area drained by right lymphatic duct
Area drained by thoracic duct

(a) Posterior thoracic wall, anterior view

(b) Lymph drainage pattern

Figure 21.3 Lymphatic Trunks and Ducts. Lymph drains from lymphatic trunks into two lymphatic ducts (right lymphatic duct and thoracic duct) that empty into the junctions of the right jugular and right subclavian veins and left jugular and left subclavian veins, respectively. (*a*) An anterior view of the posterior thoracic wall illustrates the major lymphatic trunks and ducts and the site of lymph drainage into the venous circulation of the cardiovascular system. (*b*) Areas of lymph drainage into the right lymphatic duct and the thoracic duct. AP|R

- *Lumbar trunks* remove lymph from the lower limbs, abdominopelvic wall, and pelvic organs.

WHAT DO YOU THINK?

1 Predict what may occur with respect to lymphatic drainage if a group of lymph nodes and their associated lymph vessels are surgically removed, such as might occur when breast cancer has metastasized.

Lymphatic Ducts

Lymphatic trunks drain into the largest lymph vessels called **lymphatic ducts.** There are two lymphatic ducts: the right lymphatic duct and the thoracic duct. Both of these convey lymph back into the venous blood circulation.

Right Lymphatic Duct   The **right lymphatic duct** is located near the right clavicle. It receives lymph from the lymphatic trunks that drain the following areas: (1) the right side of the head and neck, (2) the right upper limb, and (3) the right side of the thorax. It returns the lymph into the junction of the right subclavian vein and the right internal jugular vein. Thus, the right lymphatic duct drains lymph from the upper right quadrant of the body.

Thoracic Duct   The larger of the two lymphatic ducts is the **thoracic duct.** It has a length of about 37.5 to 45 centimeters (15 to 18 inches) and extends from the diaphragm to the junction of the left subclavian and left jugular veins. The thoracic duct drains lymph from the remaining areas of the body (left side of head and neck, left upper limb, left thorax, all of the abdomen, and both lower limbs).

At the base of the thoracic duct and anterior to the L_2 vertebra is a rounded, saclike structure called the **cisterna chyli** (sis-ter′nă kī′lī). The cisterna chyli gets its name from the milky, lipid-rich lymph called **chyle** (kīl; *chylos* = juice) it receives from vessels that drain the small intestine of the gastrointestinal (GI) tract. Both left and right intestinal and lumbar trunks drain into the cisterna chyli. The thoracic duct extends superiorly from the cisterna chyli and lies directly anterior to the vertebral bodies. It passes through the aortic opening of the diaphragm then ascends to the left of the vertebral body midline.

WHAT DID YOU LEARN?

3 What mechanisms are used to assist lymph movement through lymph vessels?

4 Which major body regions drain lymph to the right lymphatic duct?

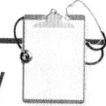

CLINICAL VIEW
Lymphedema

Lymphedema (limf'e-dē'mă; *oidema* = a swelling) is recognized as an accumulation of interstitial fluid that occurs due to interference with lymph drainage in a part of the body. Interstitial fluid increases, and the affected area both swells and becomes painful. If lymphedema is left untreated, the protein-rich interstitial fluid may interfere with wound healing and can even contribute to an infection by acting as a growth medium for bacteria.

Most cases of lymphedema are *obstructive*, meaning they are caused by blockage of lymph vessels. There are several causes of obstructive lymphedema:

- Trauma or infection of the lymph vessels
- The spread of malignant tumors within the lymph nodes and lymph vessels
- Radiation therapy that results in scar formation in lymph vessels or nodes
- Any surgery that requires removal of a group of lymph nodes (e.g., breast cancer surgery when the axillary lymph nodes are removed)

Although lymphedema has no cure, it can be controlled. Patients may wear compression stockings or other compression garments, which reduce swelling and assist interstitial fluid return to the circulation. Certain exercise and massage regimens may improve lymph drainage as well.

Millions of individuals in Southeast Asia and Africa have developed lymphedema as a result of infection by threadlike, parasitic filarial worms. In **lymphatic filariasis** (fil-ă-rī'ă-sis; *filum* = thread), filarial worms lodge in the lymphatic system, live and reproduce there for years, and eventually obstruct lymph drainage. Mosquitoes are the most common vector for transmitting filariasis, although some filarial worms gain entrance through cracks in the skin of the foot. An affected body part can swell to many times its normal size. In these extreme cases, the condition is known as **elephantiasis** (el'ĕ-fan-tī'ă-sis). Patients are treated with antiparasitic medications to kill the filarial worms, although the damage to the lymphatic system may be irreversible.

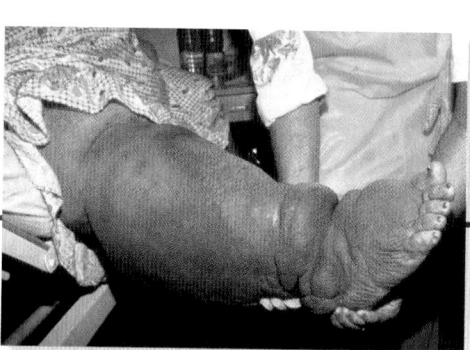

Elephantiasis caused by lymphatic filariasis.

21.2 Overview of Lymphatic Tissue and Organs

LEARNING OBJECTIVE

1. Name the two categories of lymphatic tissue and organs, and identify components of the body that belong to each category.

The lymphatic system also is made up of specialized lymphatic (lymphoid) tissue and organs. These tissues and organs include red bone marrow, the thymus, lymph nodes, the spleen, tonsils, lymphatic nodules, and MALT (mucosa-associated lymphatic tissue) (see figure 21.1; diffuse lymphatic nodules not shown).

Lymphatic system tissues and organs are categorized as either primary or secondary structures:

- **Primary lymphatic structures** are involved in the formation and maturation of lymphocytes. Both the red bone marrow and thymus are considered primary lymphatic structures.
- **Secondary lymphatic structures** are not involved in lymphocyte formation, but instead serve to house both lymphocytes and other immune cells following their formation. Secondary lymphatic structures are the sites where an immune response is initiated (see section 22.6). The major secondary lymphatic structures include the lymph nodes, spleen, tonsils, lymphatic nodules, and MALT.

Each component is described in more detail in the next several sections and is summarized in **table 21.1**.

WHAT DID YOU LEARN?

❺ How are primary lymphatic structures and secondary lymphatic structures differentiated? What are examples of each?

	Table 21.1	Lymphatic Structures		
Component	**Primary or Secondary Lymphatic Structures**	**Location**	**Function**	
Red bone marrow	Primary	Located within spongy bone of certain bones	Formation of all formed elements; site of B-lymphocyte maturation	
Thymus	Primary	Superior mediastinum (in adults); anterior and superior mediastinum (in children)	Site of T-lymphocyte maturation and differentiation	
Lymph nodes	Secondary	Along length of lymphatic vessels; clusters are present in axillary, inguinal, and cervical regions	Filter lymph; where immune response is initiated against a substance in the lymph	
Spleen	Secondary	Left upper quadrant of abdomen, near 9th–11th ribs, "wraps" partially around stomach	Filters blood; where immune response is initiated against a substance in the blood; removes aged erythrocytes and platelets; serves as erythrocyte and platelet reservoir	
Tonsils	Secondary	Within oral cavity and pharynx (throat)	Protect against inhaled and ingested substances	
Lymphatic nodules	Secondary	Every body organ and wall of appendix	Protect body organs	
MALT (mucosa-associated lymphatic tissue)	Secondary	Clusters of lymphatic nodules within walls of gastrointestinal (GI), respiratory, urinary, and reproductive tracts	Protects mucosal membranes against foreign substances	

21.3 Primary Lymphatic Structures

The general structure and function of the primary lymphatic structures are described in this section. These include the red bone marrow and the thymus.

21.3a Red Bone Marrow

LEARNING OBJECTIVES

1. Describe the location and general function of red bone marrow.
2. Identify the two major types of lymphocytes.

Red bone marrow is located within trabeculae in selected portions of spongy bone within the skeleton. In adults, these include the flat bones of the skull, the vertebrae, the ribs, the sternum, the ossa coxae, and proximal epiphyses of each humerus and femur (see section 7.2d).

Red bone marrow is responsible for hemopoiesis (production of formed elements), which was described in section 18.3. Recall that the formed elements include erythrocytes, platelets, granulocytes (neutrophils, eosinophils, and basophils), and agranulocytes (monocytes and lymphocytes) (figure 21.4). The two major types of lymphocytes are **T-lymphocytes** (also called *T-cells*) and **B-lymphocytes** (also called *B-cells*). Most formed elements move from the red bone marrow into the blood following hemopoiesis. Unlike the other formed elements, T-lymphocytes must migrate to the thymus to complete their maturation. The functions of both T-lymphocytes and B-lymphocytes are described in detail in chapter 22. The "T" in the name "T-lymphocytes" originates from the requirement of these cells to complete their maturation in the thymus.

WHAT DID YOU LEARN?

6 Why is red bone marrow considered a primary lymphatic structure?

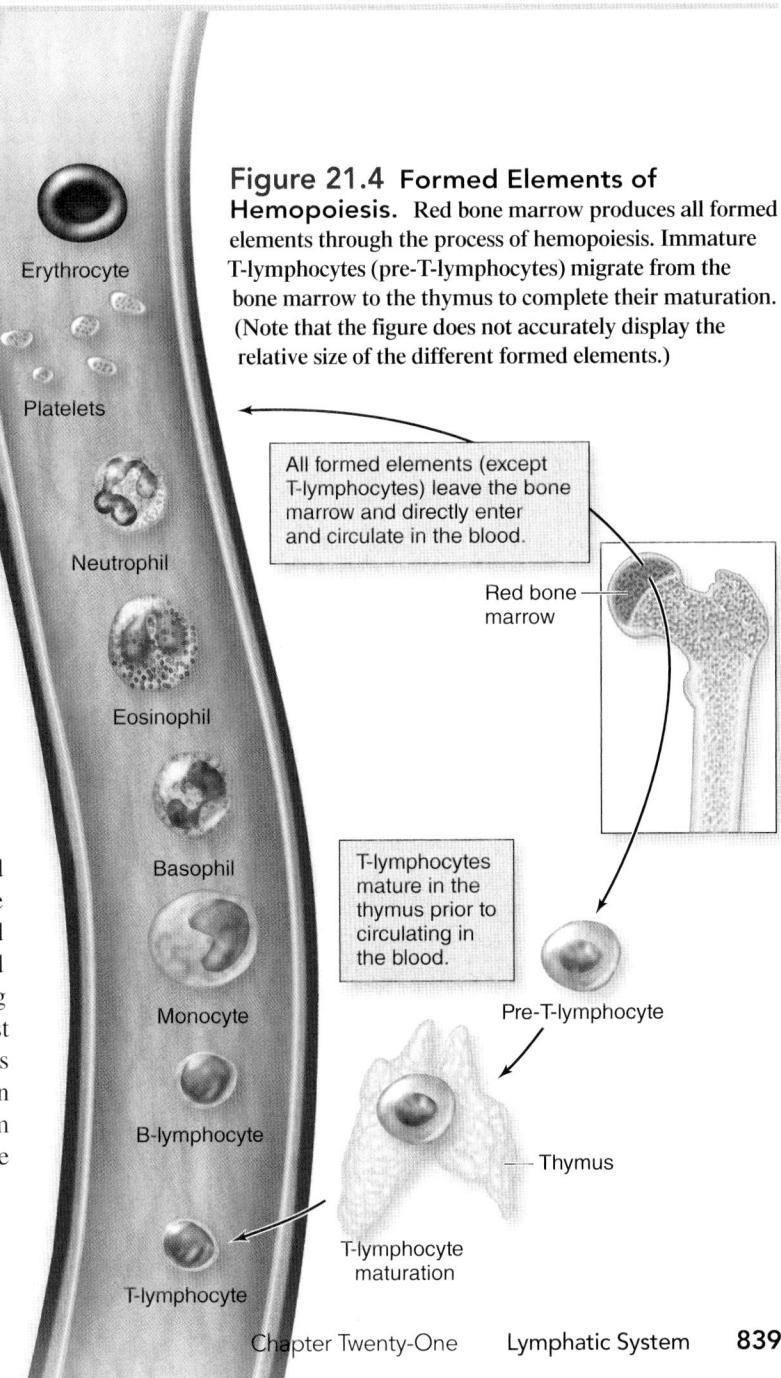

Figure 21.4 Formed Elements of Hemopoiesis. Red bone marrow produces all formed elements through the process of hemopoiesis. Immature T-lymphocytes (pre-T-lymphocytes) migrate from the bone marrow to the thymus to complete their maturation. (Note that the figure does not accurately display the relative size of the different formed elements.)

Erythrocyte

Platelets

Neutrophil

Eosinophil

Basophil

Monocyte

B-lymphocyte

T-lymphocyte

All formed elements (except T-lymphocytes) leave the bone marrow and directly enter and circulate in the blood.

Red bone marrow

T-lymphocytes mature in the thymus prior to circulating in the blood.

Pre-T-lymphocyte

Thymus

T-lymphocyte maturation

INTEGRATE

LEARNING STRATEGY

Lymphocytes can be identified according to the tissue or organ where they mature: **T**-lymphocytes mature in the **T**hymus. **B**-lymphocytes mature in the **B**one.

21.3b Thymus

LEARNING OBJECTIVE

3. Describe the structure and general function of the thymus.

The **thymus** (thī′mŭs) is a bilobed organ that is located in the superior mediastinum and functions in T-lymphocyte maturation (**figure 21.5**). In infants and young children, the thymus is quite large and extends into the anterior mediastinum as well. The thymus continues to grow until puberty, when it reaches a maximum weight of 30 to 50 grams. Cells within the thymus begin to regress after it reaches this size. Thereafter, much of the thymic tissue is replaced by adipose connective tissue.

The thymus in a child, consists of two fused **thymic lobes,** each surrounded by a connective tissue **capsule**. Fibrous extensions of the capsule, called **trabeculae** (tră-bek′ū-lē), or *septa,* subdivide the thymic lobes into **lobules.** Each lobule is arranged into an outer **cortex** and inner **medulla.** Both parts are composed primarily of epithelial

INTEGRATE

CONCEPT CONNECTION

Recall from chapter 17 how some organs, in addition to their primary function, also house clusters of endocrine tissue that produce hormones. The thymus, for example, contains (1) lymphatic tissue and is considered part of the lymphatic system, and (2) epithelial tissue that produces and secretes hormones (e.g., thymosin, thymulin, thymopoietin) and is considered part of the endocrine system.

tissue infiltrated with T-lymphocytes in varying stages of maturation. The cortex contains immature T-lymphocytes (pre-T-lymphocytes) and the medulla contains mature T-lymphocytes. The epithelial cells secrete thymic hormones (e.g., thymulin) that participate in the maturation of T-lymphocytes (see section 17.10c). Because the thymus contains both lymphatic cells and epithelial tissue, it is described as a *lymphoepithelial organ.* The details of T-lymphocyte maturation are described in section 22.5.

WHAT DID YOU LEARN?

❼ How are the two types of T-lymphocytes arranged in the cortex and medulla?

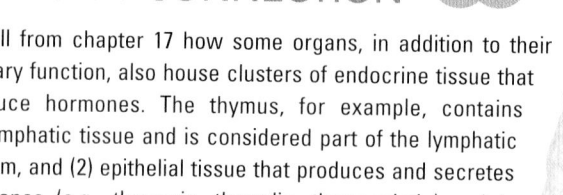

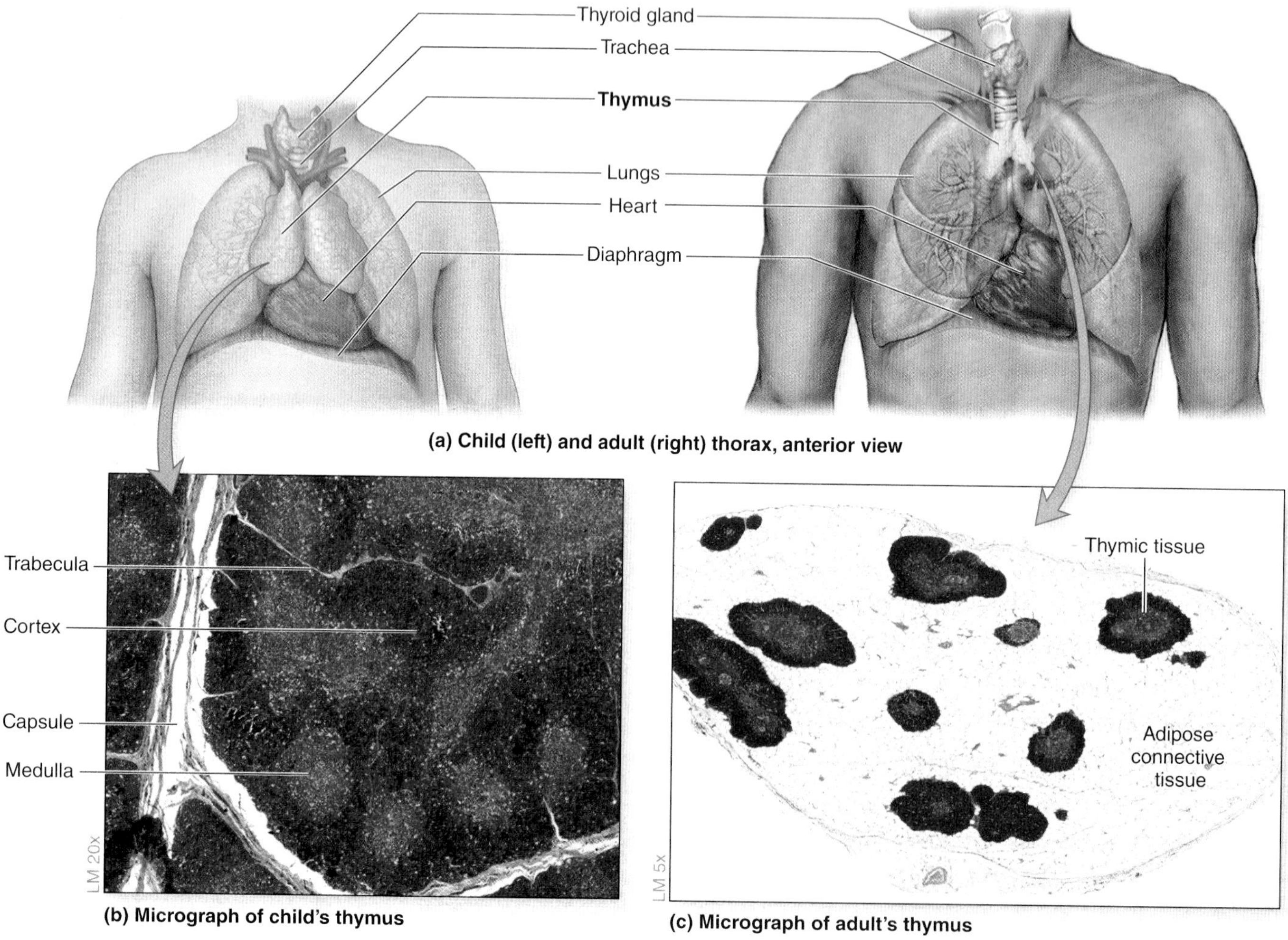

Thyroid gland
Trachea
Thymus
Lungs
Heart
Diaphragm

(a) Child (left) and adult (right) thorax, anterior view

Trabecula
Cortex
Capsule
Medulla

LM 20x

(b) Micrograph of child's thymus

Thymic tissue

Adipose connective tissue

LM 5x

(c) Micrograph of adult's thymus

Figure 21.5 Thymus. (*a*) The thymus is a bilobed lymphatic organ that is most prominent in children. (*b*) A micrograph of a child's thymus reveals the arrangement of the cortex and the central medulla within a lobule. (*c*) A micrograph of an adult's thymus showing loss of thymic tissue and gain of adipose connective tissue. AP|R

21.4 Secondary Lymphatic Structures

Structures that house both lymphocytes and other immune cells are called secondary lymphatic structures. These structures are composed of lymphatic cells that are enmeshed within a reticular extracellular connective tissue matrix.

Secondary lymphatic structures are organized into both **lymphatic organs** and aggregates of **lymphatic nodules.** They are differentiated by the presence or absence of a capsule composed of dense irregular connective tissue that encloses the lymphatic structure. A *complete* capsule is present in lymphatic organs, which include lymph nodes and the spleen. A capsule is either *incomplete* or *absent* in other lymphatic structures, which include tonsils, MALT, and diffuse lymphatic nodules.

21.4a Lymph Nodes

LEARNING OBJECTIVES

1. Describe the structure of lymph nodes.
2. Explain the function of lymph nodes.

Lymph nodes are small, round or oval encapsulated structures, which are located along the pathways of lymph vessels where they serve as the main lymphatic organ. Lymph nodes function in the filtering of lymph and removal of unwanted substances.

Lymph nodes vary in both their size (from 1 to 25 millimeters) and their number (estimated between 500 and 700 throughout the entire body). Lymph nodes are located both superficially and deep within the body and typically occur in clusters that receive lymph from selected body regions. Some examples of clustered lymph nodes include the *cervical lymph nodes* that receive lymph from the head and neck; the *axillary lymph nodes* in the armpit that receive lymph from the breast, axilla, and upper limb; and *inguinal lymph nodes* in the groin that receive lymph from the lower limb and pelvis (see figure 21.1). In addition to clusters, lymph nodes are found individually distributed throughout the body.

Numerous **afferent lymphatic vessels** bring lymph into a lymph node (figure 21.6). There is typically only one **efferent lymphatic vessel,** which originates at the involuted portion of the lymph node called the **hilum** (hī′lŭm), or *hilus.* Lymph is drained via the efferent lymphatic vessel from this region of the lymph node.

The capsule of a lymph node is composed of dense irregular connective tissue that both encapsulates the node and sends internal extensions into it that are called **trabeculae.** They subdivide the node into compartments. The connective tissue provides a pathway through which blood vessels and nerves may enter the lymph node.

The lymph node regions deep to the capsule are subdivided into an outer **cortex** and an inner **medulla.** The cortex is composed in part of multiple **lymphatic nodules.** Each lymphatic nodule within the lymph node is composed of reticular fibers, which support an inner **germinal center** that houses both proliferating B-lymphocytes and

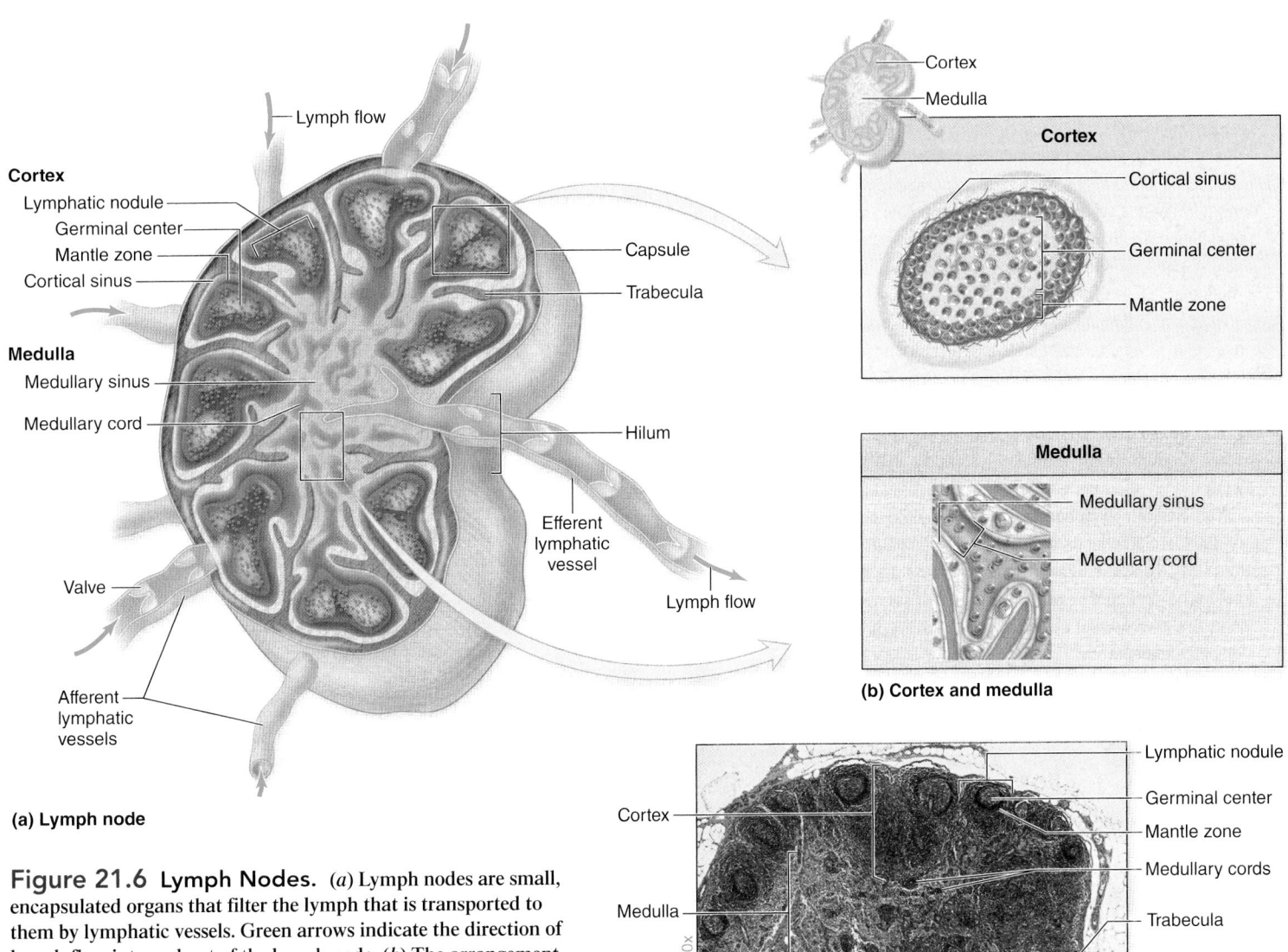

(a) Lymph node

(b) Cortex and medulla

(c) Lymph node section

Figure 21.6 Lymph Nodes. (*a*) Lymph nodes are small, encapsulated organs that filter the lymph that is transported to them by lymphatic vessels. Green arrows indicate the direction of lymph flow into and out of the lymph node. (*b*) The arrangement of regions within the cortex and medulla of a lymph node. (*c*) A micrograph of a lymph node shows the cortex and medulla.

AP|R

some macrophages. The germinal center is surrounded by an outer region called a **mantle zone,** which contains T-lymphocytes, macrophages, and dendritic cells (described in section 22.2a). The medulla differs because it has strands of connective tissue fibers that support B-lymphocytes, T-lymphocytes, and macrophages. These structures are called **medullary cords.**

Both the cortex and medulla of a lymph node also contain tiny open channels called lymphatic sinuses (**cortical sinuses** and **medullary sinuses,** respectively). These spaces are lined with macrophages.

Lymph Flow Through Lymph Nodes

Lymph enters a lymph node through numerous afferent lymphatic vessels, then it flows through the lymph node sinuses, before typically draining through one efferent lymphatic vessel. Numerous afferent lymphatic vessels and generally one efferent lymphatic vessel are significant in the normal function of a lymph node. Consider that in this anatomic arrangement the collective diameter of the "inflow pipes" (afferent lymphatic vessels) is larger than the diameter of the "outflow pipe" (efferent lymphatic vessel). This creates a higher fluid pressure in the intervening structures (the lymph node sinuses), which helps to force the lymph through the lymph node.

INTEGRATE

CLINICAL VIEW
Lymphoma

A **lymphoma** (lim-fō'mă; *oma* = tumor) is a malignant neoplasm that develops within lymphatic structures. These tumors develop most often from abnormal B-lymphocytes, and less commonly from abnormal T-lymphocytes. Usually (but not always) a lymphoma presents as a nontender, enlarged lymph node, often in the neck or axillary region. Some patients have no further symptoms, whereas others may experience night sweats, fever, and unexplained weight loss in addition to the nodal enlargement. Lymphomas are grouped into two categories: Hodgkin lymphoma and non-Hodgkin lymphoma.

Hodgkin lymphoma (or *Hodgkin disease*) is characterized by the presence of the **Reed-Sternberg cell,** a large cell whose two nuclei resemble owl eyes, surrounded by lymphocytes within the affected lymph node. Hodgkin lymphoma affects young adults (ages 16–35) and people over 60. It arises in a lymph node and then spreads to other nearby lymph nodes. If caught early, Hodgkin lymphoma can be treated and cured by excision of the tumor, followed by radiation, chemotherapy, or both.

Non-Hodgkin lymphomas are much more common than Hodgkin lymphomas. Some kinds of non-Hodgkin lymphoma are aggressive and often fatal, whereas others are slow-growing and more responsive to treatment. Treatment depends upon the type of non-Hodgkin lymphoma, the extent of its spread at the time of discovery, and the rate of progression of the malignancy.

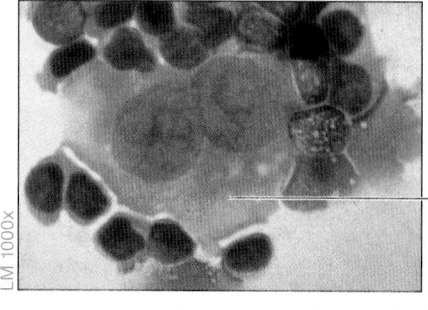

LM 1000×

Reed-Sternberg cell

Reed-Sternberg cell, a characteristic of Hodgkin lymphoma.

INTEGRATE

CONCEPT CONNECTION

Dendritic cells are specialized phagocytic cells that are formed in bone marrow and are housed in both epithelial and connective tissue of the skin (see sections 6.1a and 6.1b) and mucosal membranes. Those cells within the epidermis of the skin are specifically called epidermal dendritic cells or Langerhans cells (see section 6.1a). Dendritic cells migrate from the skin and mucosal membranes to a lymph node following endocytosis of foreign substances (see section 22.2a).

Lymph is continuously monitored for the presence of foreign or pathogenic material as it passes through nodes. Macrophages residing in the lymph node remove foreign debris from the lymph. Lymph then exits the lymph node through an efferent lymphatic vessel. Recall that lymph nodes are often found in clusters, so after one lymph node receives and filters lymph, the lymph is then transported to another lymph node in the cluster, then to another, and so on. Thus, lymph is repeatedly screened for unwanted substances.

Lymphocytes housed within the lymph node also come into contact with foreign substances. An immune response may be initiated after this contact, during which lymphocytes undergo cellular division, especially in the germinal centers. Some of these new lymphocytes remain within the lymph node, whereas others are transported within the lymph and then enter into the blood, to ultimately reach areas of infections (see section 22.6).

When a person has an infection, often some lymph nodes are swollen and tender to the touch. This is a condition erroneously termed *swollen glands.* These enlarged nodes are a sign that lymphocytes are proliferating and attempting to fight an infection. Swollen superficial lymph nodes, such as those in the neck and axilla, can generally be palpated (felt).

 WHAT DID YOU LEARN?

 8 How does lymph flow through a lymph node, and how is it monitored by macrophages and lymphocytes?

21.4b Spleen

 LEARNING OBJECTIVES

3. Describe the spleen and its location.
4. Distinguish between white pulp and red pulp.
5. List the functions of the spleen.

The **spleen** (splēn) is the largest lymphatic organ in the human body. It is located in the left upper quadrant of the abdomen, inferior to the diaphragm and adjacent to ribs 9–11 (**figure 21.7**). This deep red organ lies lateral to the left kidney and posterolateral to the stomach. The spleen can vary considerably in size and weight, but typically is about 12 centimeters (5 inches) long and 7 centimeters (3 inches) wide.

The spleen's posterolateral aspect (called the *diaphragmatic surface*) is convex and rounded; the concave anteromedial border (the *visceral surface*) contains the **hilum** (or *hilus*), where blood vessels and nerves enter and leave the spleen. A **splenic** (splen'ik) **artery** delivers blood to the spleen, whereas blood is drained by a **splenic vein.**

The spleen is surrounded by a connective tissue capsule from which trabeculae extend into the organ. The spleen lacks a cortex and medulla. Rather, the trabeculae subdivide the spleen into regions of white pulp and red pulp. **White pulp** consists of spherical clusters of T-lymphocytes, B-lymphocytes, and macrophages, which surround a **central artery.**

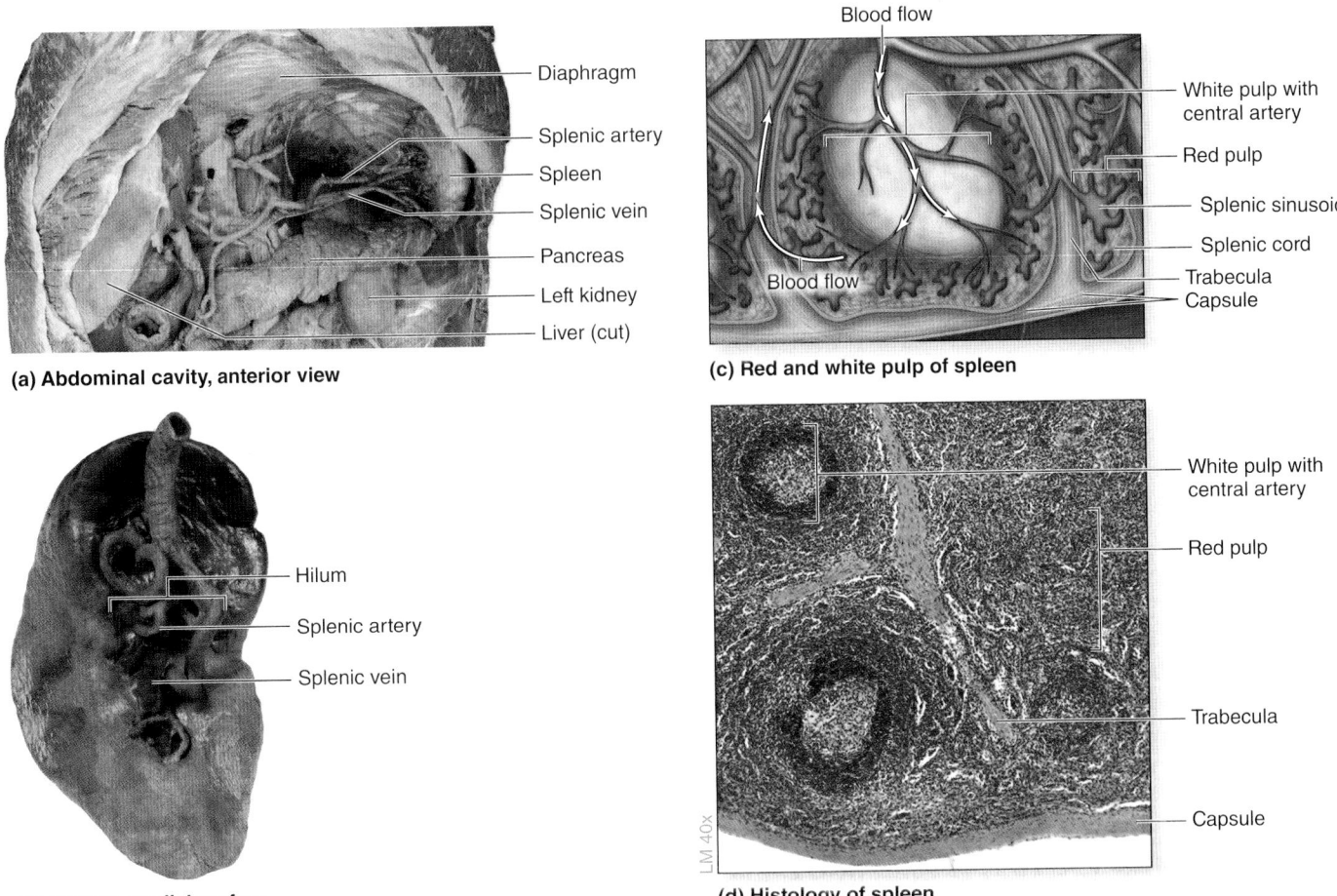

(a) Abdominal cavity, anterior view

Diaphragm
Splenic artery
Spleen
Splenic vein
Pancreas
Left kidney
Liver (cut)

(b) Spleen, medial surface

Hilum
Splenic artery
Splenic vein

(c) Red and white pulp of spleen

Blood flow
White pulp with central artery
Red pulp
Splenic sinusoid
Splenic cord
Trabecula
Capsule
Blood flow

(d) Histology of spleen

White pulp with central artery
Red pulp
Trabecula
Capsule
LM 40x

Figure 21.7 Spleen. (*a*) A cadaver photo shows the position of the spleen within the body. The pancreas has been moved inferiorly in this photo to show more clearly the splenic vessels. (*b*) A second photo shows the medial surface of the spleen, where the splenic artery and splenic vein extend from the hilum. A diagram (*c*) and micrograph (*d*) of spleen histology depict the microscopic arrangement of red pulp and white pulp. AP|R

The remaining splenic tissue, called **red pulp,** contains erythrocytes, platelets, macrophages, and B-lymphocytes. The cells in red pulp are housed in reticular connective tissue and form structures called **splenic cords** (*cords of Bilroth*). **Splenic sinusoids** are associated with red pulp. (Recall from section 20.1c that sinusoids are very permeable capillaries that have a discontinuous basal lamina, so blood cells can easily enter and exit across the vessel wall.) The sinusoids drain into small venules that ultimately lead into a **splenic vein.**

Red pulp of the spleen serves as a blood reservoir, including a storage site for both erythrocytes and platelets (about 30% of all platelets are stored in the spleen). In situations where more erythrocytes and platelets are needed, such as during hemorrhage, these stored formed elements reenter the blood (see section 18.3d).

WHAT DO YOU THINK?

2 Predict the consequences to the spleen during an accident if an individual wears his or her seatbelt more superiorly across the abdomen instead of more inferiorly across the pelvis.

Monitoring Blood as It Flows Through the Spleen

The spleen functions to filter *blood* (not lymph). As blood enters the spleen and flows through the central arteries, the white pulp monitors the blood for foreign materials, bacteria, and other potentially harmful substances.

After passing through a central artery, blood travels through sinusoids of red pulp. As blood moves through the sinusoids, macrophages lining the sinusoids phagocytize bacteria and foreign debris from the blood, as well as both old and defective erythrocytes and platelets. Thus, the general flow of blood through the spleen is the

splenic artery, the central artery (of white pulp), the splenic sinusoid (of red pulp), venules (that drain sinusoids), and ultimately the splenic vein (figure 21.7*c*).

In summary, the spleen serves several functions, including: (1) phagocytosis of bacteria and other foreign materials in the blood as part of the body's defense (red and white pulp); (2) phagocytosis of old, defective erythrocytes and platelets from circulating blood (red pulp); and (3) acting as a blood reservoir and storage site for both erythrocytes and platelets (red pulp).

INTEGRATE

CLINICAL VIEW

Splenectomy

Splenectomy (splē-nek'tō-mē) is surgical removal of the spleen. This procedure may be performed for several reasons including a severe splenic infection or cyst within the spleen, Hodgkin lymphoma or other types of cancers, and certain blood disorders (e.g., sickle cell anemia). The most common reason, however, is a ruptured spleen resulting from abdominal injury.

If the spleen is removed, other lymphatic organs may take over many of the spleen's duties, but the person may be more prone to contracting a serious or life-threatening infection. For this reason, individuals who have had their spleens removed are encouraged to receive vaccines against the flu and pneumonia and are placed on antibiotic therapy for an extended period of time, sometimes for their entire life.

During fetal development through the fifth month, the spleen also engages in the formation of blood cells, a function performed after birth by the bone marrow. This function remains latent in the spleen and may be reactivated under certain conditions (e.g., some hematologic disorders).

WHAT DID YOU LEARN?

9 What are the general functions of the spleen? Indicate whether red pulp or white pulp is responsible for each function.

10 Which lymphatic structures filter lymph? Which filters blood?

21.4c Tonsils

LEARNING OBJECTIVE

6. Identify the main groups of tonsils and their location and function.

Tonsils (ton'sillz; *tonsilla* = a stake) are secondary lymphatic structures that are not completely surrounded by a connective tissue capsule. They are found in the pharynx (throat) and oral cavity. A **pharyngeal tonsil** is found in the posterior wall of the nasopharynx; when this tonsil becomes enlarged, it is called **adenoids** (ad'ĕ-noydz; *aden* = gland). **Palatine tonsils** are in the posterolateral region of the oral cavity, and **lingual tonsils** are along the posterior one-third of the tongue (**figure 21.8**). Tonsils help protect against foreign substances that may be either inhaled or ingested.

Invaginated outer edges called *tonsillar crypts* increase the tonsil's surface area to help trap material. Lymphatic nodules, some containing germinal centers, are housed with the tonsils.

WHAT DID YOU LEARN?

11 What are the three main groups of tonsils and their function?

21.4d Lymphatic Nodules and MALT

LEARNING OBJECTIVES

7. Describe the composition of individual lymphatic nodules.

8. Compare the locations of MALT and Peyer patches.

Lymphatic nodules and MALT make up the last category of secondary lymphatic structures. These relatively small masses of lymphatic tissue are located throughout the body.

Lymphatic Nodules

Lymphatic nodules (nod'ūlz), or *lymphatic follicles,* are small, oval clusters of lymphatic cells (e.g., B-lymphocytes, T-lymphocytes, macrophages) with some extracellular matrix that are *not* completely surrounded by a connective tissue capsule. Scattered lymphatic nodules are referred to as *diffuse lymphatic tissue.* This tissue can be found in every body organ and within the wall of the appendix where it helps to defend against infections in these structures. However, in some areas of the body, many lymphatic nodules group together to form larger structures, such as MALT.

INTEGRATE

CLINICAL VIEW
Tonsillitis and Tonsillectomy

The tonsils help to protect the pharynx from infection. They frequently become inflamed and infected, a condition called **acute tonsillitis** (ton'si-lī'tis). The palatine tonsils are most commonly affected. The tonsils redden and enlarge; in severe cases, they may partially obstruct the pharynx and cause respiratory distress.

Tonsils may be infected by viruses such as adenoviruses or bacteria such as *Streptococcus*. Streptococcal tonsillitis often results in very red tonsils that have whitish specks (called whitish exudate). The symptoms of tonsillitis include fever, chills, sore throat, and difficulty swallowing.

Persistent or recurrent infections may result in permanent enlargement of the tonsils and development of a condition that is called **chronic tonsillitis.** If medical treatment does not help the chronic tonsillitis, surgical removal of the tonsils **(tonsillectomy)** may be indicated. Medical guidelines suggest performing a tonsillectomy only if the person has had 7 throat infections (e.g., tonsillitis, strep throat) in 1 year, 5 infections in 2 years, or 3 infections per year for 3 years. Research indicates that tonsillectomy does not significantly affect the body's response to new infections.

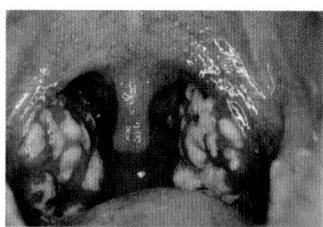

Acute tonsillitis

MALT

MALT (mucosa-associated lymphatic tissue) is located in the lamina propria of the mucosa (see section 5.5b) of the gastrointestinal, respiratory, urinary, and reproductive tracts. The lymphatic cells in the MALT help defend against foreign substances that come in contact with mucosal membranes.

MALT is very prominent in the mucosa of the small intestine, primarily in the ileum. There, collections of lymphatic nodules called **Peyer patches** can become quite large and bulge into the gastrointestinal tract lumen (see figure 26.16).

Figure 21.9 is a visual representation of the role of the lymphatic system, which assists the cardiovascular system in maintaining fluid balance by returning excess fluid from the interstitial space to the blood, and assists the immune system by participating in defending the body against potentially harmful substances.

WHAT DID YOU LEARN?

12 What is the function of MALT in the mucosal linings of the gastrointestinal, respiratory, urinary, and reproductive tracts?

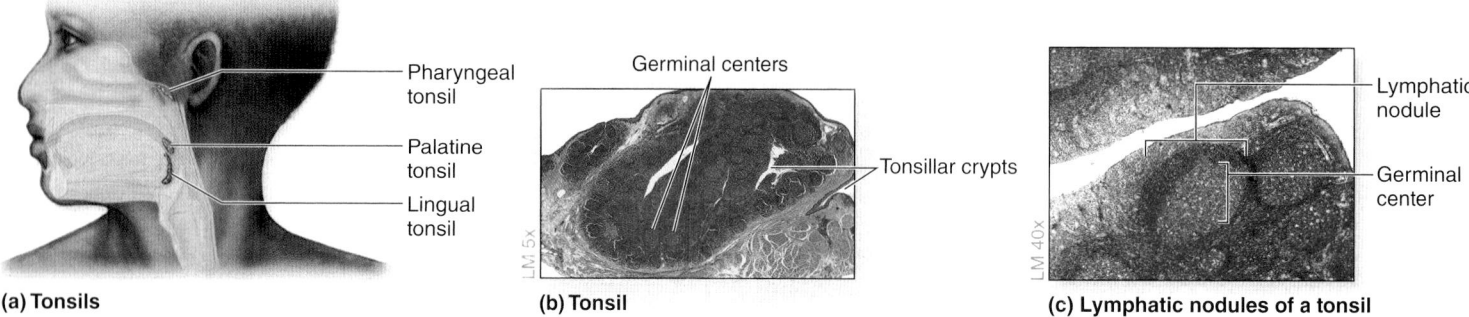

(a) Tonsils — Pharyngeal tonsil, Palatine tonsil, Lingual tonsil

(b) Tonsil — Germinal centers, Tonsillar crypts, LM 5x

(c) Lymphatic nodules of a tonsil — Lymphatic nodule, Germinal center, LM 40x

Figure 21.8 Tonsils. (*a*) Tonsils reside in the wall of the oral cavity and pharynx. A micrograph shows (*b*) tonsillar crypts that help trap foreign substances and (*c*) the germinal centers within the lymphatic nodules of the tonsils. AP|R

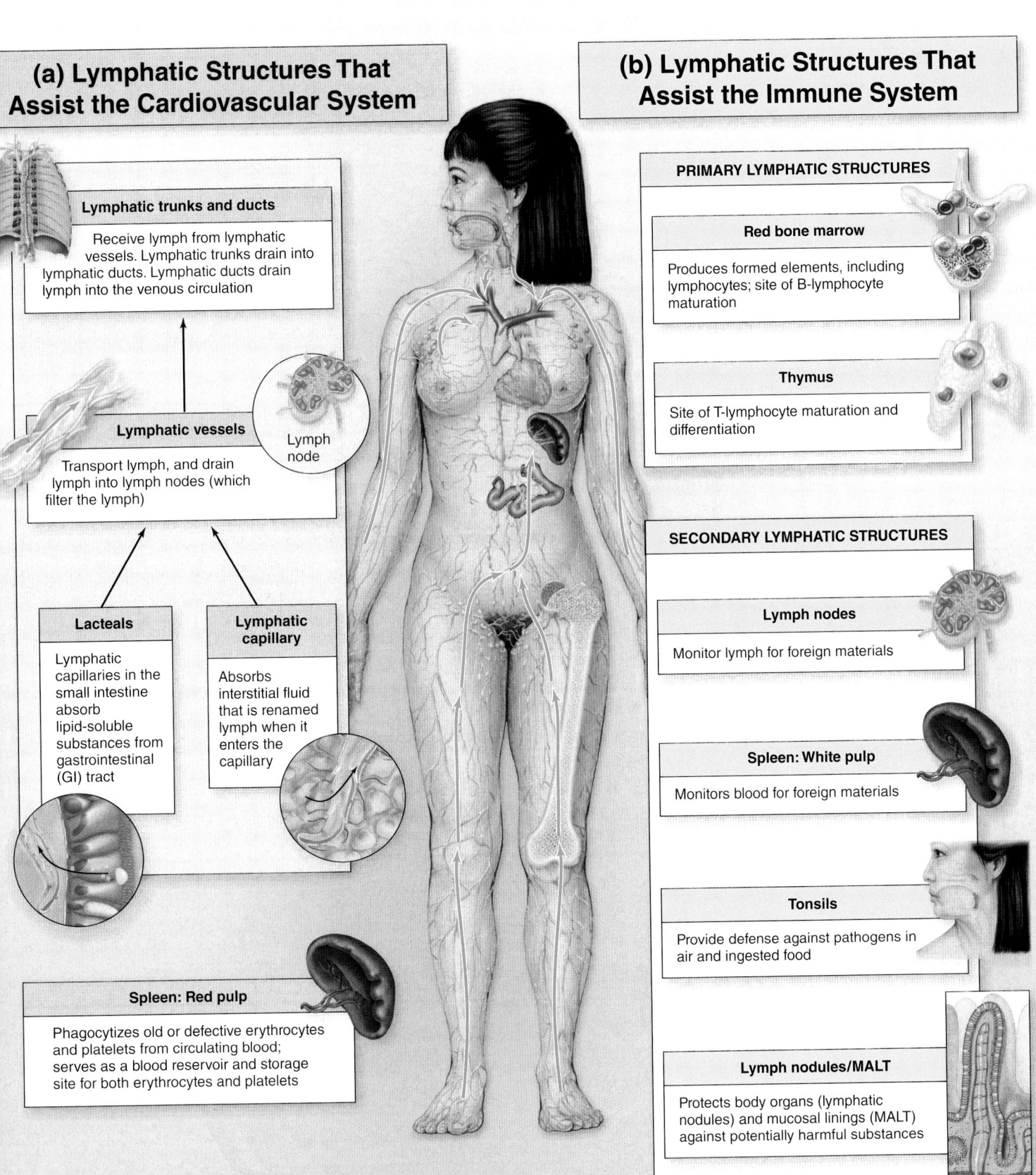

INTEGRATE CONCEPT OVERVIEW

Figure 21.9 Relationship of the Lymphatic System to Both the Cardiovascular System and Immune System. The lymphatic system assists both (a) the cardiovascular system by returning fluid from the interstitial space back into the blood to help maintain fluid balance, blood volume, and blood pressure; and (b) the immune system in the body's defense.

(a) Lymphatic Structures That Assist the Cardiovascular System

Lymphatic trunks and ducts

Receive lymph from lymphatic vessels. Lymphatic trunks drain into lymphatic ducts. Lymphatic ducts drain lymph into the venous circulation

Lymphatic vessels

Transport lymph, and drain lymph into lymph nodes (which filter the lymph)

Lymph node

Lacteals

Lymphatic capillaries in the small intestine absorb lipid-soluble substances from gastrointestinal (GI) tract

Lymphatic capillary

Absorbs interstitial fluid that is renamed lymph when it enters the capillary

Spleen: Red pulp

Phagocytizes old or defective erythrocytes and platelets from circulating blood; serves as a blood reservoir and storage site for both erythrocytes and platelets

(b) Lymphatic Structures That Assist the Immune System

PRIMARY LYMPHATIC STRUCTURES

Red bone marrow

Produces formed elements, including lymphocytes; site of B-lymphocyte maturation

Thymus

Site of T-lymphocyte maturation and differentiation

SECONDARY LYMPHATIC STRUCTURES

Lymph nodes

Monitor lymph for foreign materials

Spleen: White pulp

Monitors blood for foreign materials

Tonsils

Provide defense against pathogens in air and ingested food

Lymph nodules/MALT

Protects body organs (lymphatic nodules) and mucosal linings (MALT) against potentially harmful substances

- The lymphatic system, which is composed of lymph vessels and lymphatic tissues and organs, supports the functions of the cardiovascular and immune system.

21.1 Lymph and Lymph Vessels 835

- Lymph is transported in lymph vessels.

21.1a Lymph and Lymphatic Capillaries 835

- Lymph is interstitial fluid containing solutes and sometimes foreign material that is absorbed into lymphatic capillaries and transported through lymph vessels and returned to the blood.
- Lymphatic capillaries are endothelium-lined vessels with overlapping endothelial cells where interstitial fluid enters lymphatic vessels to become lymph.

21.1b Lymphatic Vessels, Trunks, and Ducts 836

- Lymph is transported through a network of increasing larger vessels that include lymphatic capillaries, lymphatic vessels, lymphatic trunks, and lymphatic ducts.
- Movement of lymph through lymph vessels is assisted by the presence of valves within the larger lymphatic vessels and trunks and several mechanisms that help propel the lymph.

21.2 Overview of Lymphatic Tissue and Organs 838

- Primary lymphatic structures are involved in the formation and maturation of lymphocytes.
- Secondary lymphatic structures house lymphocytes and other immune cells, serving as sites to initiate the immune response.

21.3 Primary Lymphatic Structures 839

21.3a Red Bone Marrow 839

- Red bone marrow produces all formed elements including lymphocytes.

21.3b Thymus 840

- The thymus is a bilobed organ found in the mediastinum. It is most active in childhood until puberty, and then it declines in size and function.
- The thymus is the site of T-lymphocyte maturation, which occurs in the presence of thymic hormones.

21.4 Secondary Lymphatic Structures 841

- Secondary lymphatic structures may be organized into lymphatic organs, which are enclosed within a complete capsule, and lymphatic nodules that have an incomplete or absent capsule.
- Both lymphatic organs and lymphatic nodules are composed of a reticular connective tissue matrix that houses lymphatic cells.

21.4a Lymph Nodes 841

- Lymph nodes are numerous, small, encapsulated lymphatic organs that filter lymph.
- Each lymph node is organized into an outer cortex and inner medulla that house immune cells (e.g., B-lymphocytes).

21.4b Spleen 842

- The spleen is the largest lymphatic organ and it is partitioned into white pulp and red pulp that filters blood.
- The spleen functions in the removal of bacteria and other foreign material from blood, the removal of old erythrocytes and platelets, and as a blood reservoir and storage site for both erythrocytes and platelets.

21.4c Tonsils 844

- Tonsils are large clusters of partially encapsulated lymphatic cells that are located in the pharynx and oral cavity to protect against foreign substances that may be either inhaled or ingested.
- Tonsils are named based on their specific location and include the pharyngeal tonsil, palatine tonsils, and lingual tonsils.

21.4d Lymphatic Nodules and MALT 844

- Lymphatic nodules can be found in every organ throughout the body.
- MALT (mucosa-associated lymphatic tissue) is composed of large groups of lymphatic nodules housed in the mucosal-lined walls of the gastrointestinal tract, respiratory tract, reproductive tract, and urinary tract.

CHALLENGE YOURSELF

Do You Know the Basics?

_____ 1. What body systems are supported by the lymphatic system?

    a. cardiovascular and urinary

    b. cardiovascular and immune

    c. respiratory and cardiovascular

    d. respiratory and urinary

_____ 2. Lymph is drained into the thoracic duct from which of the following body regions?

    a. right lower limb

    b. right upper limb

    c. right side of the head

    d. right side of the thorax

____ 3. The spleen is a secondary lymphatic structure that is located

    a. in the oral cavity.

    b. along lymphatic vessels.

    c. attached to the gastrointestinal tract.

    d. inferior to the diaphragm adjacent to the stomach.

____ 4. What is the function of the thymus?

    a. It is the site of T-lymphocyte maturation.

    b. It filters lymph.

    c. It filters blood.

    d. It produces the formed elements of the blood.

____ 5. Which type of lymph vessel consists solely of an endothelium and has one-way flaps that allow interstitial fluid to enter but not exit?

    a. lymphatic vessel

    b. lymphatic capillary

    c. lymphatic duct

    d. lymphatic trunk

____ 6. Which statement is accurate about lymph nodes?

    a. Lymph nodes do not become swollen and tender.

    b. Lymph nodes filter blood.

    c. Lymph enters the lymph node through afferent lymphatic vessels.

    d. Lymphatic sinuses are located in the cortex of a lymph node only.

____ 7. In an early *Streptococcus* infection of the throat, all of the following structures may swell *except* the

    a. pharyngeal tonsil.

    b. spleen.

    c. cervical lymph node.

    d. palatine tonsil.

____ 8. The lymphatic trunk that drains lymph from the upper limb, breasts, and superficial thoracic wall is the

    a. lumbar trunk.

    b. jugular trunk.

    c. subclavian trunk.

    d. bronchomediastinal trunk.

____ 9. Aged erythrocytes are removed from circulation by the

    a. mucosa-associated lymphatic tissue.

    b. lymph nodes.

    c. thymus.

    d. spleen.

____ 10. Interstitial fluid that is absorbed into lymph vessels will be monitored by the _____ before the fluid is dumped into venous blood.

    a. mucosa-associated lymphatic tissue (MALT)

    b. spleen

    c. thymus

    d. lymph nodes

11. List the anatomic structures of the lymphatic system, including lymph vessels, primary lymphatic structures, and secondary lymphatic structures.

12. Explain what distinguishes a primary lymphatic structure from a secondary lymphatic structure.

13. Describe what lymph is, and draw a flowchart that illustrates what structures the lymph travels through to return to the blood.

14. Which body regions have their lymph drained to the thoracic duct?

15. Describe how the thymus's anatomy changes as we age.

16. Describe the basic anatomy of a lymph node, how lymph enters and leaves the node, and the functions of this organ.

17. Compare and contrast the red and white pulp of the spleen with respect to the anatomy and functions of each.

18. Describe the specific locations of the tonsils.

19. Describe the location and function of diffuse lymphatic nodules and mucosa-associated lymphatic tissue (MALT).

20. Explain how the lymphatic system supports the functions of both the cardiovascular system and the immune system.

Can You Apply What You've Learned?

1. A tick has embedded itself in the scalp of a young boy. His mother is most likely to find which lymph nodes to be swollen?

    a. cervical

    b. inguinal

    c. axillary

    d. femoral

2. A child born without his thymus would not have mature

    a. macrophages.

    b. B-lymphocytes.

    c. dendritic cells.

    d. T-lymphocytes.

3. A young woman was in a car accident and had to have her spleen removed as a consequence of its rupture during the accident. She now has a greater risk of

   a. an overactive immune response.

   b. bacterial infections.

   c. low blood pressure.

   d. a pathogen surviving in her lymph.

4. One of the postoperative complications from the removal of lymph nodes during a mastectomy would be

   a. edema in the upper limb.

   b. regrowth of the lymph nodes.

   c. an overactive immune system.

   d. tender and swollen lymph nodes in other regions of the body.

5. All of the following can result in lymphedema (the accumulation of interstitial fluid due to interference with lymphatic drainage) *except*

   a. surgical removal of a group of lymph nodes.

   b. obstruction of lymph vessels that drain a lymph node, such as might occur with a tumor or infection.

   c. radiation therapy, which may cause scar formation of lymph vessels.

   d. exercise that increases the flow of lymph in the lymph vessels.

Can You Synthesize What You've Learned?

1. Arianna was diagnosed with mononucleosis, an infectious disease that targets B-lymphocytes. The doctor palpated her left side, just below the rib cage, and told Arianna she was checking to see if a certain organ was enlarged, a complication that can occur with mononucleosis. What lymphatic organ was the doctor checking, and why would it become enlarged? Include some explanation of the anatomy and histology of this organ in your answer.

2. Jordan has an enlarged lymph node along the side of his neck, and he is worried that the structure may be a lymphoma. Explain how malignant cells may have reached the lymph node.

3. Mark has come to the emergency care facility complaining of a sore throat. Upon examination, his tonsils appear swollen. When questioned further, he reports that although he has no history of his tonsils being swollen, his throat has been sore for almost a week. He is concerned that he will need to go to the hospital to have his tonsils removed. Explain the conditions that would generally require a tonsillectomy.

INTEGRATE

ONLINE STUDY TOOLS

The following study aids may be accessed through Connect.

Clinical Case Study: A Young Woman with a Neck Mass

Interactive Questions: This chapter's content is served up in a number of multimedia question formats for student study.

LearnSmart: Topics and terminology include lymph and lymph vessels; overview of lymphatic tissue and organs; primary lymphatic structures; secondary lymphatic structures

Anatomy & Physiology Revealed: Topics include lymphatic system overview; thoracic duct; thymus; lymph node; spleen; tonsils

Animations: Topics include lymphatic system overview

Immune System and the Body's Defense

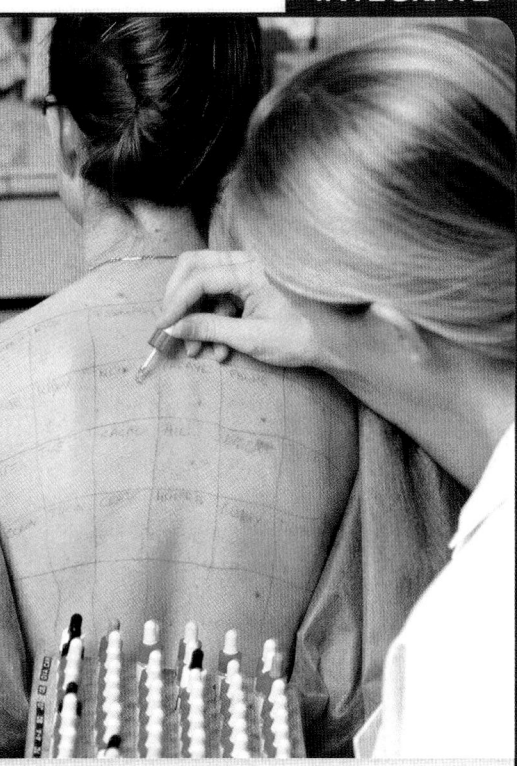

CAREER PATH

Allergist

An allergist is a physician who specializes in allergies, asthma, and other disorders of the immune system. Practicing specialists and research scientists who study the immune system analyze the normal activities, malfunctions, and the best means of treating disorders involving this system. Hypersensitivity disorders including allergies are one common malfunction. Skin tests for allergies are performed on an area of the back by placing allergens into the epidermis and noting the reaction that occurs. The formation of a rash where the substance is tested indicates that the individual is allergic to that substance.

Module 10: Lymphatic System

There are children born without an **immune system**—a condition called severe combined immune deficiency syndrome (SCIDS). The most famous of these children was a boy named David Vetter, born in 1971. He was dubbed by the media as "The Boy in the Bubble" because without an immune system, David literally had to live in a sterile plastic bubble. At the age of 12, David received a bone marrow transplant from his sister with the hope of his being able to live outside the bubble. This new bone marrow would have the ability to produce the immune cells that David's bone marrow was unable to form. Sadly, residing in his sister's bone marrow was a dormant (not active) ➡️ virus called Epstein-Barr. This virus is generally kept under control by a healthy immune system. However, because David did not have a functioning immune system, the bone marrow with the Epstein-Barr virus induced formation of cancerous tumors throughout his body, and he died several months later.

Most of us may take for granted that we have a functioning immune system. While each of us may catch a cold or the flu on occasion, our immune system is—without our typically being aware of it—protecting us from infectious agents and other harmful substances. This system is unique because, unlike all other body systems, the immune system is not made up of organs. Rather, it is composed of numerous cellular and molecular structures located throughout the body that function together in the body's defense to provide us with **immunity** (i-myū'ni-tē).

Our body's immunity is organized into two categories: innate immunity and adaptive immunity. Here we present the details of both types of immunity. Visual summaries that integrate the many different means the body has to defend itself are included for both. Our coverage of the immune system in this chapter is not comprehensive. Rather, our discussion is tailored to provide a general overview of how the immune system functions, as well as a description of some of the more common immune system malfunctions (e.g., hypersensitivities, acquired immunodeficiency syndrome [AIDS], and autoimmune disorders). We hope to help you to develop both an understanding of, and an appreciation for, this system that so diligently functions to protect us. We first provide a brief overview of the different infectious agents. This is a necessary preliminary step because immune system function is often dependent upon the specific type of infectious agent against which it must defend.

22.1 Overview of Diseases Caused by Infectious Agents

LEARNING OBJECTIVES

1. Compare and contrast the five major classes of infectious agents.
2. Describe prions, and name a disease they cause.

Infectious agents are organisms that cause damage, or possibly death, to the host organism that they invade. Infectious agents that cause harm to the host are said to be **pathogenic.** The five major categories of infectious agents include bacteria, viruses, fungi, protozoans, and multicellular parasites. These categories, including examples of each, are summarized and compared in **table 22.1** and described here:

- **Bacteria** are microscopic, single-celled organisms 1 to 2 micrometers in size that are enclosed by both a plasma membrane and a cell wall. Bacterial cells are fundamentally different from the cells of humans and other living things so are called *prokaryotic cells.* The cells of humans and other living organisms are termed *eukaryotic cells.* Different species of bacteria are spherical, rodlike, or coiled in shape, and thus are referred to as cocci, bacilli, and spirilla, respectively. Certain bacteria can cause disease, but the vast majority of bacterial species do not.

  Some bacteria have increased **virulence** (vir'yū-lĕns)— the ability to cause serious illness—due to the presence of an external, sticky polysaccharide capsule. Additionally, some pathogenic bacteria cause disease by releasing enzymes or toxins that interfere with the function of cells; an example is the bacterium *Clostridium tetani* (see Clinical View: "Muscular Paralysis and Neurotoxins" in section 10.3c).

- **Viruses** are *not* cells. Viruses are much smaller than a bacterial cell at about one-hundredth of a micrometer, and they are composed of DNA or RNA within a protein capsid, or shell. The protein capsid of some viruses may also be enclosed

within a membrane. Viruses are **obligate intracellular parasites;** that is, they must enter a cell to replicate. The process of viral reproduction includes directing the infected cell to synthesize copies of both the viral DNA or RNA molecule and its capsid protein. New *viral particles* are then formed within the infected cells and released from them to enter surrounding cells. A virus, or the immune system's response to it, ultimately kills the cells it invades.

Table 22.1	Major Categories of Infectious Agents
Characteristic	**Bacteria**
Structure	
Cellular Characteristic	Prokaryotic
Significant Features	Intracellular and extracellular microbes, some of which produce enzymes, toxins
Selected Diseases Caused by the Agent	Streptococcal infections (e.g., strep throat), staphylococcal infections, tuberculosis, syphilis, diphtheria, tetanus, Lyme disease, salmonella, and anthrax

Viruses cause different diseases depending upon the type of cell they infect. Examples of viral diseases include the common cold, chickenpox, ebola, and HIV.

- **Fungi** (fun'jī) are eukaryotic cells that have a cell wall external to the plasma membrane. This group includes molds, yeasts, and multicellular fungi that produce spores. Proteolytic enzymes released from fungi induce inflammation that causes redness and swelling of the infected area.

  Fungal diseases (*mycoses*) in healthy individuals in the United States are usually limited to superficial infections of the skin, scalp, and nails (e.g., ringworm, "athlete's foot"). Other fungal diseases involve infections of mucosal linings (e.g., vaginal yeast infections) or may cause internal fungal infections (e.g., histoplasmosis, which affects the respiratory system).

- **Protozoans** are also eukaryotic cells, but they lack a cell wall. Protozoan infections include malaria and trichomoniasis.

- **Multicellular parasites** are nonmicroscopic organisms (larger than a centimeter in size) that reside within a host from which they take nourishment. Parasitic worms such as tapeworms, for example, infect the intestinal tract of humans and other mammals.

Prions (prī'on) are small fragments of infectious proteins that cause disease in nervous tissue. Variant Creutzfeldt-Jakob disease (also known as *bovine spongiform encephalopathy*, or "mad cow disease") is an example. This prion disease can be spread from cows to humans by consuming infected meat because nerves within the muscle are contaminated with prions. Prions are neither cells nor viruses, and research into how they cause disease is ongoing.

 WHAT DID YOU LEARN?

1 Which pathogen must enter a cell to replicate? Which type of pathogen is composed of prokaryotic cells?

22.2 Overview of the Immune System

Cells, plasma proteins, and secreted products of the immune system protect the body from infectious agents and other potentially harmful substances. Here we provide a brief introduction to the immune cells, as well as the specialized hormonelike chemicals called cytokines that regulate them. We then give an overview of how the immune system is organized into two overlapping and complementary components: innate immunity and adaptive immunity.

22.2a Immune Cells and Their Locations

 **LEARNING OBJECTIVE**

1. List the types of leukocytes of the immune system, and describe where they may be found.

Leukocytes are formed in the red bone marrow (section 18.3c) prior to circulating in the blood. They include (1) the three types of granulocytes (neutrophils, eosinophils, and basophils); (2) monocytes that become macrophages or dendritic cells when they exit blood vessels and take up residence in the tissues; and (3) the three types of lymphocytes, which include T-lymphocytes (or T-cells), B-lymphocytes (or B-cells), and NK (natural killer) cells.

Structures That House Immune System Cells

Most leukocytes are found in body tissues (as opposed to in the blood). The primary locations that house immune cells include lymphatic tissue, select organs, epithelial layers of the skin and mucosal membranes, and connective tissues of the body (**figure 22.1**):

- **Lymphatic tissue.** T-lymphocytes, B-lymphocytes, macrophages, dendritic cells, and NK cells are housed in secondary

Viruses	Fungi	Protozoans	Multicellular Parasites
Not a cell; DNA or RNA within a capsid protein	Eukaryotic	Eukaryotic	Eukaryotic
Obligate intracellular parasites; must enter cell to replicate	Produce spores; release proteolytic enzymes	Intracellular and extracellular parasites that interfere with normal cellular functions	Live within a host; grow in size with nutrients provided by the host
Common cold, influenza, polio, mumps, measles, hepatitis, rubella, chickenpox, ebola, herpes, and HIV (which leads to AIDS)	Ringworm, diaper rash, jock itch, athlete's foot, yeast infections, and histoplasmosis	Malaria, toxoplasmosis, giardiasis, amoebiasis, leishmaniasis, trichomoniasis, and African sleeping sickness	Parasitic infection from tapeworms, lung flukes, liver flukes, blood flukes, hookworms, *Trichinella, Ascaris,* whipworms, and pinworms

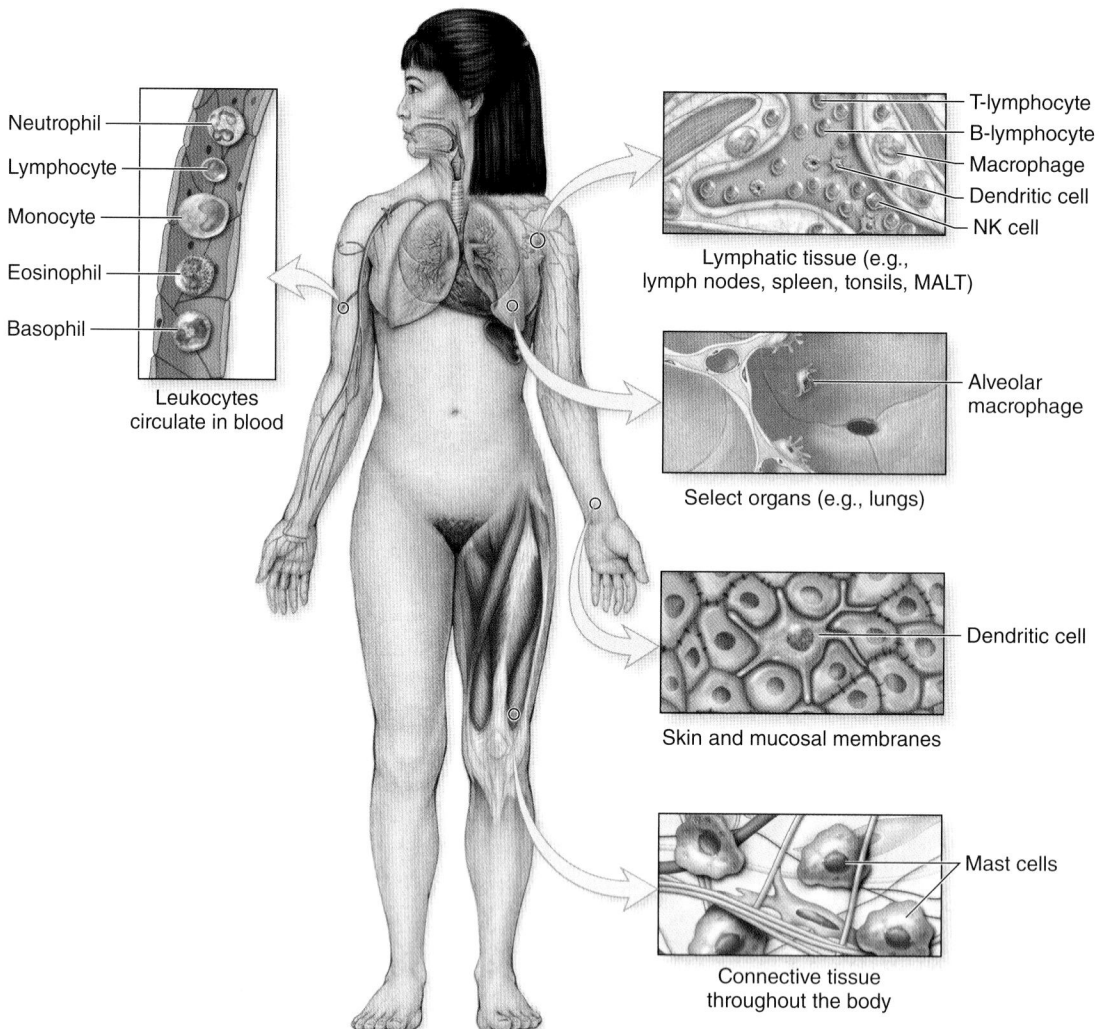

Figure 22.1 Primary Location of Immune Cells. Immune cells circulate in the blood, but are primarily located in lymphatic tissue, select organs, epithelial tissue of the skin and mucosal membranes, and connective tissue throughout the body.

lymphatic structures of lymph nodes, the spleen, tonsils, MALT (mucosa-associated lymphatic tissue), and lymphatic nodules (see section 21.4).

- **Select organs.** Macrophages are also housed in other organs; some are specifically named based on their location, such as alveolar macrophages of the lungs and microglia of the brain. Macrophages may be permanent residents, referred to as fixed macrophages, or migrate through tissues and are called wandering macrophages.

- **Epithelial layers of the skin and mucosal membranes.** Dendritic cells are located in the skin and mucosal membranes, and are typically derived from monocytes. These cells in the epidermis of the skin are more specifically called *epidermal dendritic cells* (see section 6.1a). Dendritic cells engulf pathogens in the skin and mucosal membranes and subsequently migrate to a lymph node through lymph vessels that drain the tissue (see section 21.4a).

- **Connective tissue.** Mast cells (cells similar to basophils) are located within the connective tissue throughout the body, typically in close proximity to small blood vessels (see section 5.2a). They are especially abundant in the dermis of the skin and the mucosal linings of the respiratory, digestive, urinary, and reproductive tracts. However, they are also housed in connective tissue of organs, such as the endomysium that ensheathes muscle fibers (see section 10.2a).

WHAT DID YOU LEARN?

❷ What types of immune cells does lymph contact as it is filtered through a lymph node?

22.2b Cytokines

LEARNING OBJECTIVES

2. Define cytokines and describe their similarities to hormones.
3. List the general categories of cytokines.

Cytokines (sī′tō-kīn; *cyto* = cell, *kinesis* = movement) are small, soluble proteins produced by cells of both the innate and adaptive immune system to regulate and facilitate immune system activity. These soluble proteins: (1) serve as a means of communication between the cells; (2) control the development and behavior of effector cells of immunity; (3) regulate the inflammatory response of innate immunity; and (4) function as weapons to destroy cells. Cytokines have also recently been shown to influence other, non-immune cells such as those of the nervous system.

A cytokine is released from one cell and binds to a specific receptor of a target cell, where its action is similar to that of a hormone. Cytokines can act on the cell that released it (autocrine stimulation), local neighboring cells (paracrine stimulation), or circulate in the blood to cause systemic effects (endocrine stimulation). To prevent continuous stimulation, cytokines have a short half-life (see section 17.4b).

Table 22.2 Major Categories of Cytokines

Category	Primary Function	Source	Designation	Examples
Interleukin (IL)	Regulates immune cells	T-lymphocytes, macrophages, endothelial cells, and other various cells	IL followed by number	IL-1 IL-2
Tumor necrosis factor (TNF)	Destroys tumor cells; may have other functions as well	T-lymphocytes, macrophages, mast cells, dendritic cells	TNF followed by Greek letter	TNF-α
Colony-stimulating factor (CSF)	Stimulates leukopoiesis in bone marrow to increase synthesis of a specific type (colony) of leukocytes (see table 18.6 in section 18.3a)	T-lymphocytes, monocytes	First letter of cell(s) it is regulating, followed by CSF	G-CSF (granulocyte CSF) GM-CSF (granulocyte-macrophage CSF)
Interferon (IFN)	Three classes: IFN-α and IFN-β are antiviral agents, and IFN-γ is a pro-inflammatory agent	Leukocytes, fibroblasts	IFN followed by Greek letter	IFN-α

Although the terminology of cytokines continues to evolve within the discipline of immunology, a current method of organizing cytokines identifies these different categories: interleukin (IL), tumor necrosis factor (TNF), colony-stimulating factor (CSF), and interferon (IFN). **Table 22.2** summarizes the cytokine categories.

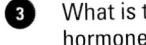 **WHAT DID YOU LEARN?**

❸ What is the definition of a cytokine? How are cytokines similar to hormones?

INTEGRATE

LEARNING STRATEGY

A military analogy can help describe how the cells of the immune system function in the body's defense. The cells are the "troops," the pathogen is the "opposition," and cytokines serve as both a means of communication and as weapons available to the troops to fight infection.

22.2c Comparison of Innate Immunity and Adaptive Immunity

LEARNING OBJECTIVE

4. Compare and contrast the primary features of innate and adaptive immunity.

The cells, cytokines, and processes of the immune system are organized into two categories based on the type of immunity that is provided. The two categories are innate immunity and adaptive immunity (**figure 22.2**). Although both work to protect us from potentially harmful agents, the two differ in several respects including the participating cells that are involved, the specificity with which the cells respond, the mechanisms involved in eliminating harmful substances, and the amount of time required for a response.

Innate Immunity

Some defense mechanisms of the immune system serve to protect us against numerous different substances, and because we are born with these defenses, this type of immunity is referred to as **innate immunity** (or *nonspecific immunity*). Innate immunity includes the barriers of the skin and mucosal membranes that prevent entry, as well as nonspecific cellular and molecular internal defenses. The structures and

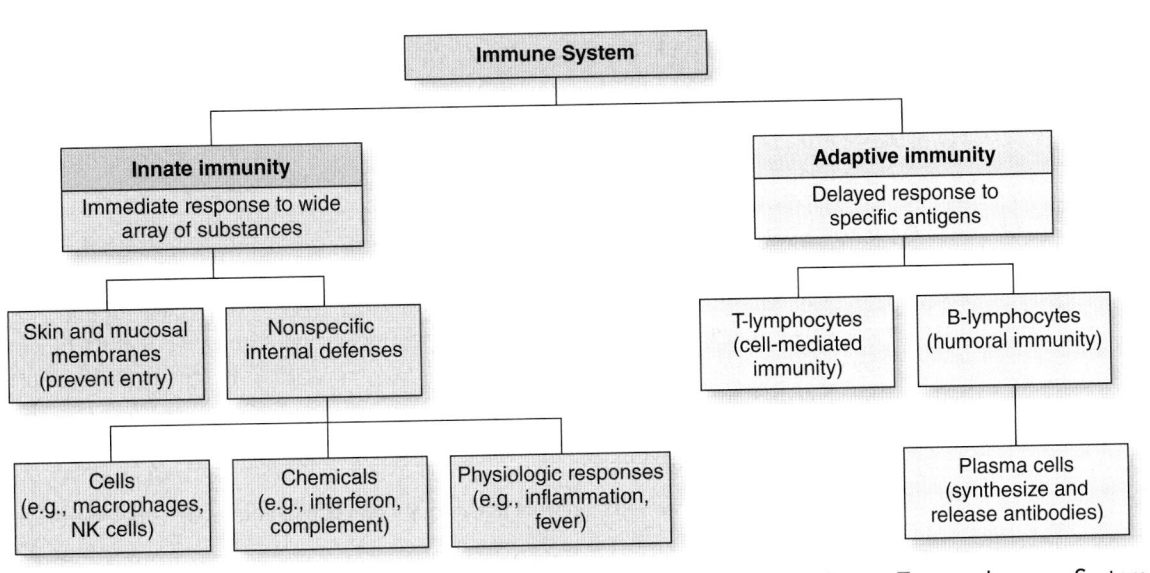

Figure 22.2 Overview of the Immune System. The immune system is composed of two overlapping and complementary components: innate immunity, which is initiated immediately against a wide array of substances; and adaptive immunity, which involves a delayed response to a specific antigen.

mechanisms associated with innate immunity do not require previous exposure to a foreign substance, and they respond immediately to any potentially harmful agent.

Adaptive Immunity

Adaptive immunity (or *acquired immunity*) involves specific T-lymphocytes and B-lymphocytes, which respond to different foreign substances (or antigens) to which we are exposed during our lifetime. For example, one lymphocyte may respond to the virus that causes chickenpox, but this same lymphocyte does not respond if it encounters the bacterium that causes strep throat. Lymphocytes provide a powerful means of eliminating foreign substances. However, although the process begins immediately, adaptive immunity typically takes several days to be effective.

In the past, the immune system was defined as the functional system composed only of lymphocytes and their response to specific foreign substances. That is, the immune system was synonymous with adaptive immunity. The definition of the immune system has since changed because of a greater understanding of the interdependency of adaptive immunity and innate immunity, and it now includes the structures and processes of both. Although we consider these two components of immunity separately, you will see that innate and adaptive immunity work together to defend the body.

  WHAT DID YOU LEARN?

4 What are the cells of adaptive immunity?

22.3 Innate Immunity

Innate immunity includes all body structures and processes that (1) prevent entry of potentially harmful substances and (2) respond nonspecifically to a wide range of potentially harmful substances following their entry into the body—that is, nonspecific internal defenses (figure 22.2). If you imagine your body to be like a fortress, the skin and mucosal membranes provide it with its *first line of defense*. The internal processes of innate immunity are considered the body's *second line of defense*. These internal defenses include (1) activities of various types of cells including neutrophils, macrophages, dendritic cells, basophils, mast cells, NK cells, and eosinophils; (2) chemicals such as interferon and complement; and (3) physiologic processes that include the inflammatory response and development of a fever. Each of these is discussed in detail in this section.

22.3a Preventing Entry

 LEARNING OBJECTIVE

1. Describe the physical, chemical, and biological barriers to entry of harmful agents into the body.

The epithelial tissues of the epidermis and connective tissue of the dermis of the **skin** provide a physical barrier that very few microbes can penetrate, if the skin is intact. The cells of the skin also release a number of antimicrobial substances, including sebum, lysozyme, defensins, and dermicidin. Additionally, nonpathogenic microorganisms, termed the *normal flora,* reside on the skin and help prevent the growth of pathogenic microorganisms.

Mucosal membranes that line the openings of the body produce mucin that when hydrated forms mucus and also release lysozyme, defensins, and immunoglobulin A (IgA). In addition, harmless bacteria also live in the linings of the various tracts of the body and suppress the growth of other potentially more virulent types. The various mechanisms used by the skin, mucosal membranes, and other structures to help prevent entry are summarized in **table 22.3**.

These mechanisms generally are very successful. However, if microbes are present in sufficient numbers, or the barrier is compromised, such as occurs with a puncture or burns, microbes may be able to penetrate into the underlying connective tissue and establish an infection. At this point, the internal processes of both innate immunity (the *second line of defense*) and adaptive immunity (the *third line of defense*) are both set in motion to eliminate the infectious agent.

  WHAT DID YOU LEARN?

5 What is the role of the skin and mucosal membranes in the body's defenses?

Table 22.3	First Line of Defense: Preventing Entry of Pathogens	
Structure, Substance, or Process	**Description**	**Function**
SKIN		
Epidermis; dermis	Stratified squamous epithelium forms epidermis; areolar and dense irregular connective tissue forms dermis	Provide a physical, chemical, and biological barrier for body surface
Normal flora	Commensal flora, including nonpathogenic bacteria	Help prevent growth of pathogenic microbes
Exfoliation	Sloughing off of epidermal cells	Removes potential pathogens from skin surface
Hyaluronic acid	Mucopolysaccharide with a gel-like consistency that is located in areolar tissue of the dermis	Slows migration of microbes that have penetrated the epidermis
Sebaceous (oil) gland secretions	Secretions called sebum that contain lactate and fatty acids	Create a low pH (3–5) that interferes with the growth of microbes
Sweat gland secretions	Secretions that contain lysozyme, defensins, and dermicidin	Help wash away microbes; contain antibacterial and antifungal substances
MUCOUS MEMBRANES		
Epithelial and connective tissue	Lining of respiratory, gastrointestinal, urinary, and reproductive tracts; contain hyaluronic acid	Provide a physical, chemical, and biological barrier of body structures exposed to the external environment
Normal flora	Commensal flora, including nonpathogenic bacteria	Help prevent growth of pathogenic microbes
Mucus	Formed from hydrated mucin; contains lysozyme, defensins, and IgA	Thick secretion that helps trap microbes; contains antimicrobial substances
RESPIRATORY TRACT		
Nasal secretions	Secretions that contain lysozyme, defensins, and IgA	Contain antimicrobial substances
Vibrissae	Hairs in nasal cavity	Trap microbes in the nose
Cilia	Extensions of plasma membranes	Sweep mucus in the respiratory tract so that it can be expectorated or swallowed
Coughing and sneezing	Blasts of expired air	Mechanical elimination of microbes or other foreign substances from the respiratory tract
GASTROINTESTINAL TRACT		
Saliva	Secretions released into the mouth from the salivary glands; contains lysozyme and IgA	Helps wash away microbes; contains antimicrobial substances
Hydrochloric acid (HCl)	Strong acid produced within the stomach	Creates very low pH (pH ~2) that destroys many bacteria, bacterial toxins, and other microbes that enter the stomach
Defecation and vomiting	Removal of waste from the digestive tract	Eliminate microbes before they can be absorbed into the blood
UROGENITAL TRACT		
Urine	Urine formed in kidneys is transported out of the body through the urinary tract	Flow of urine flushes microbes from urinary tract
Lactate	Weak acid	Produced by the vagina; creates a low pH that slows or prevents the growth of microbes
SECRETIONS PRODUCED BY THE SKIN AND MUCOUS MEMBRANES		
Lysozyme	Antibacterial enzyme	Attacks the cell wall of some bacteria (gram positive bacteria)
Defensins	Small proteins	Form pores in the plasma membrane of microbes, compromising their structural integrity
Dermicidin	Small proteins produced by the skin	Antibacterial agent against both gram positive and gram negative bacteria; antifungal agent
Immunoglobulin A (IgA)	Specific type of antibody present in areas exposed to the environment	Binds with a specific foreign substance (antigen)
OTHER SECRETIONS		
Lacrimal fluid	Fluid produced by lacrimal glands; contains lysozyme and IgA	Washes microbes away from surface of eyes; contains antimicrobial agents
Cerumen	Waxy secretions within external auditory meatus	Waterproofs external auditory meatus; may trap microbes in external ear

22.3b Cellular Defenses

LEARNING OBJECTIVE

2. Describe the cells that function in innate immunity.

Cells of innate immunity include neutrophils, macrophages, dendritic cells, basophils, mast cells, NK cells, and eosinophils. The structure and function of each are described here. Please refer to **figure 22.3** as you read through this section.

Neutrophils, Macrophages, and Dendritic Cells

Neutrophils, macrophages, and dendritic cells engulf unwanted substances such as infectious agents and cellular debris through phagocytosis (see section 4.3c). What follows phagocytosis is dependent upon the cell type. Both **neutrophils** and **macrophages** function to destroy infectious agents through a process that involves a lysosome and a respiratory burst (figure 22.3a). The vesicle containing the unwanted substance (a *phagosome*) merges with a lysosome to form a *phagolysosome*. Within the phagolysosome, digestive enzymes contributed from the lysosome chemically digest the unwanted substances. Destruction of microbes and viruses is facilitated by the production of reactive oxygen-containing molecules, such as nitric oxide, hydrogen peroxide, and superoxide; the release of these molecules is called a **respiratory burst** (or *oxidative burst*). Degraded residues of the engulfed substance are then released from the cell by exocytosis.

Dendritic cells function to destroy infectious agents and then present fragments of the microbe on its cell surface to T-lymphocytes—a process called antigen presentation, which is necessary for initiating adaptive immunity (see sections 6.1a and 22.4c). This role of antigen presentation is shared with macrophages. (This role for macrophages is not depicted in figure 22.3a.)

Basophils and Mast Cells

Basophils and **mast cells** are both proinflammatory chemical-secreting cells (figure 22.3b). Recall that basophils circulate in the blood, and mast cells reside in connective tissue of the skin, mucosal linings, and various internal organs. Substances secreted by basophils and mast cells increase fluid movement from the blood to an injured tissue. They also serve as *chemotactic chemicals*, which are chemicals that attract immune cells as part of the inflammatory response (see section 22.3d).

Basophils and mast cells release granules during the inflammatory response. These granules contain various substances including **histamine,** which increases both vasodilation and capillary permeability, and **heparin,** an anticoagulant. They also release **eicosanoids** (ī′kō-să-noydz; *ecosa* = twenty; *eidos* = form) from their plasma membrane (see section 17.3b), which increase inflammation.

Natural Killer Cells

NK (natural killer) cells destroy a wide variety of unwanted cells, including virus-infected cells, bacteria-infected cells, tumor cells, and cells of transplanted tissue (figure 22.3c). NK cells are formed in the bone marrow, circulate in the blood, and accumulate in secondary lymphatic structures of the lymph node, spleen, and tonsils.

NK cells patrol the body in an effort to detect unhealthy cells, a process referred to as **immune surveillance.** NK cells make physical contact with unhealthy cells and destroy them by release of cytotoxic chemicals. These cytotoxic chemicals include *perforin,* which forms a transmembrane pore in the unwanted cells, and *granzymes,* which then enter the cell through the transmembrane pore initiating apoptosis. Apoptosis (see section 4.10) is a form of cellular death in which the cell does not lyse, but rather "shrivels"; this helps limit the spread of the infectious agent.

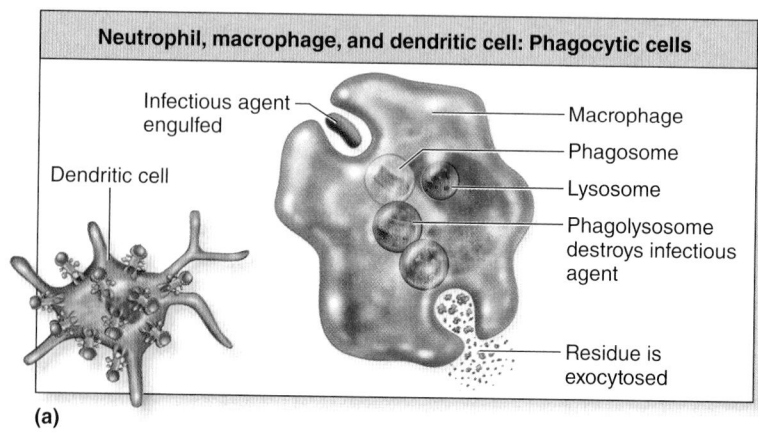

Neutrophil, macrophage, and dendritic cell: Phagocytic cells

- Infectious agent engulfed
- Dendritic cell
- Macrophage
- Phagosome
- Lysosome
- Phagolysosome destroys infectious agent
- Residue is exocytosed

(a)

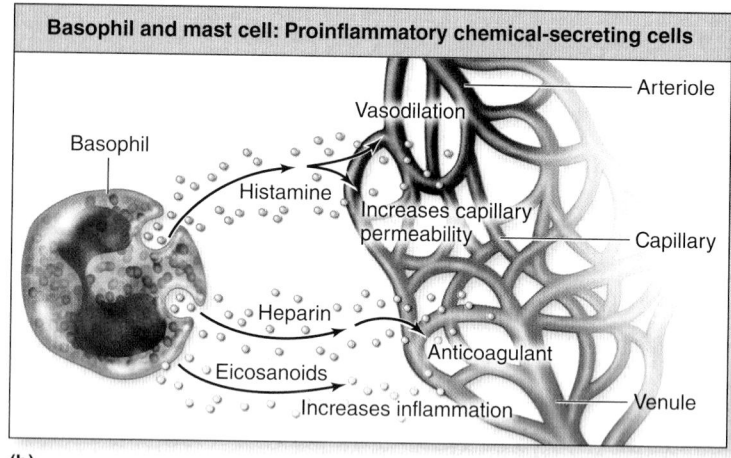

Basophil and mast cell: Proinflammatory chemical-secreting cells

- Basophil
- Vasodilation
- Arteriole
- Histamine
- Increases capillary permeability
- Capillary
- Heparin
- Anticoagulant
- Eicosanoids
- Increases inflammation
- Venule

(b)

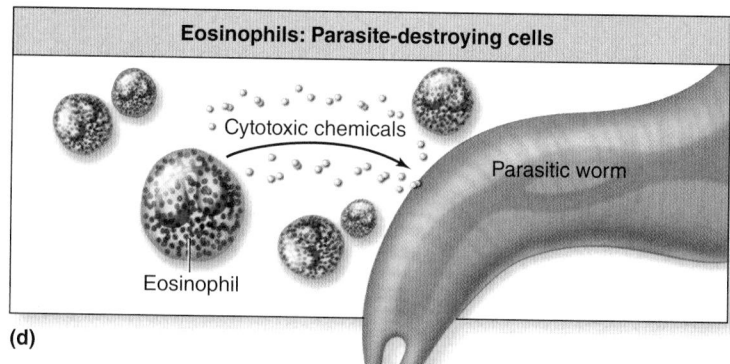

NK cell: Apoptosis-initiating cells

- Perforin and granzyme
- Perforin forms a transmembrane pore
- Granzymes enter pore, causing apoptosis of cell
- NK cell
- Unhealthy or unwanted cell
- Apoptosis

(c)

Eosinophils: Parasite-destroying cells

- Cytotoxic chemicals
- Parasitic worm
- Eosinophil

(d)

Figure 22.3 Cells of Innate Immunity. The cells of innate immunity use multiple tactics to combat pathogens, including (*a*) phagocytosis (example shown is a macrophage), (*b*) chemical secretion that increases inflammation (example shown is a basophil), (*c*) chemical secretion that destroys unhealthy cells by inducing apoptosis, and (*d*) chemical secretion that helps eliminate parasites. AP|R

6 What distinguishes neutrophils from dendritic cells? How do basophils differ from mast cells?

7 How do NK cells accomplish the task of eliminating unwanted cells?

INTEGRATE

LEARNING STRATEGY

The cells associated with innate immunity have a "military-like" function:

- **Neutrophils** are the "foot soldiers" that are the first to arrive at the site of infection.
- **Macrophages** are the "big eaters"—the cleanup crew that arrives at the injured or infected scene late and stays longer.
- **Basophils/mast cells** engage in chemical warfare that causes inflammation.
- **NK cells** serve as "security guards" that "search for and destroy" unwanted cells.
- **Eosinophils** are the "heavy artillery" to take on the "big guys" (parasites).

Eosinophils

Eosinophils (ē′ō-sin′ō-fil) target parasites (figure 22.3d). Mechanisms of destruction include degranulation and release of enzymes and other substances (e.g., reactive oxygen-containing compounds, neurotoxins) that are lethal to the parasite. Like NK cells, eosinophils can release proteins that form a transmembrane pore to destroy cells of the multicellular organism.

Eosinophils also participate in the immune response associated with allergy and asthma (see Clinical View: "Hypersensitivities" in section 22.9c) and engage in phagocytosis of antigen-antibody complexes (see section 22.8b).

Cells of the innate immune system are able to recognize foreign microbes because they possess pattern recognition receptors (e.g., toll-like receptors or TLRs) on their cell surface. These receptors bind to common molecular patterns (or motifs) of microbes including those of bacteria and viruses. TLRs are actually a class (or family) of receptors, with each class recognizing a specific microbial component.

INTEGRATE

CONCEPT CONNECTION

Earlier we discussed the differential count of white blood cells (see section 18.3c), a process that measures the amount of each type of leukocyte in your blood. Differential counts are useful in diagnosing different types of infections. For example:

- An increase in neutrophils is associated with an acute bacterial infection.
- Monocytes may increase with chronic inflammatory disorders or tuberculosis.
- An increase in eosinophils occurs in response to a parasitic infection.
- An elevated number of lymphocytes generally is associated with viral infections or with chronic bacterial infections.

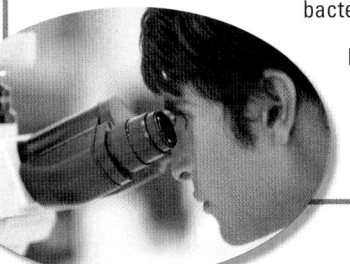

In contrast, decreased lymphocyte counts can occur with HIV infection and sepsis (presence of a large number of pathogens in the blood).

22.3c Antimicrobial Proteins

LEARNING OBJECTIVES

3. Explain the general function of interferons.
4. Define the complement system and describe how it is activated.
5. Describe the four major means by which complement participates in innate immunity.

Antimicrobial proteins are specific types of molecules of the innate immune system that function against microbes. Two of those, interferons and complement, are described in this section.

Interferons

Interferons (IFNs) are a category of cytokines that include (1) IFN-α and IFN-β produced by leukocytes and virus-infected cells and (2) IFN-γ produced by T-lymphocytes and NK cells. IFNs serve as a nonspecific defense mechanism against the spread of any viral infection. A virus-infected cell, although "doomed" because either the virus or the immune cells will destroy it, helps prevent further spread of the virus by releasing IFNs. Following their release, IFNs function as follows (**figure 22.4**):

- IFN-α and IFN-β bind to receptors of neighboring cells, preventing them from becoming infected by triggering synthesis of enzymes that both destroy viral RNA or DNA and inhibit synthesis of viral proteins. IFN-α and IFN-β also stimulate NK cells to destroy virus-infected cells.

- IFN-γ is released from NK cells to stimulate macrophages to also destroy virus-infected cells.

Complement System

One of the most important antimicrobial groups of substances of innate immunity is the **complement system.** It is composed of at least

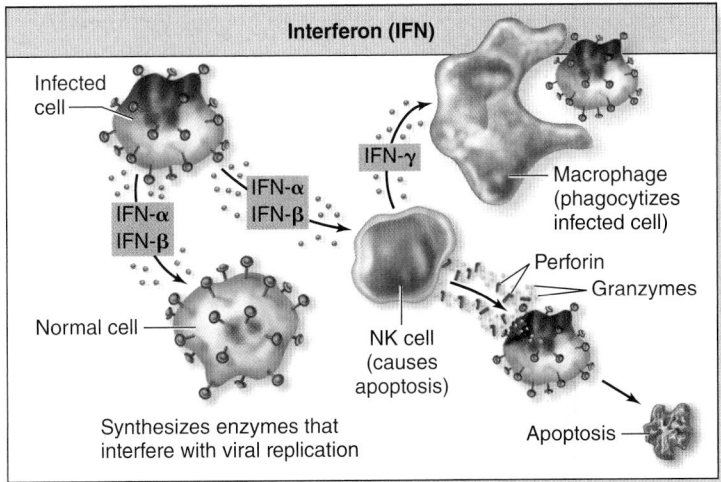

Figure 22.4 Effects of Interferon. Interferon (IFN) is one category of cytokines released from a variety of cells. A virus-infected cell releases IFN-α and IFN-β, which stimulate antiviral changes to neighboring cells to prevent their infection and induce NK cells to both destroy virus-infected cells and release IFN-γ to activate macrophages to destroy virus-infected cells.

30 plasma proteins that make up approximately 10% of the blood serum proteins. These proteins are collectively referred to as **complement.** The name is derived from how they complement, or work along with, antibodies (proteins produced by differentiated B-lymphocytes, which are described in section 22.8). Individual complement proteins are generally identified with the letter "C" followed by a number (e.g., C1, C2).

The liver continuously synthesizes and releases inactive complement proteins into the blood. Once in the blood, inactive complement proteins are activated by an enzyme cascade. (Recall that a similar process of an enzyme cascade of inactive proteins produced by the liver is also involved in blood clotting; see section 18.4c.)

Activation of complement occurs following entry of a pathogen into the body. Two of the major means of activation include the **classical pathway,** in which a complement protein binds to an antibody that has previously attached to a foreign substance (e.g., a portion of a bacterium); and the **alternative pathway,** in which surface polysaccharides of certain bacterial and fungal cell walls bind directly with a complement protein. Note that antibody is required for the activation by the classical pathway, but not for activation by the alternative pathway.

Following its activation, the complement system mediates several important defense mechanisms, and it is especially potent against bacterial infections **(figure 22.5).**

- **Opsonization** (op′sŏn-ī-zā′shŭn) is the binding of a protein (in this case, complement) to a portion of bacteria or other cell type that enhances phagocytosis. The binding protein is called an **opsonin** (op′sŏ-nin). The binding of complement makes it more likely that a substance is identified and engulfed by a phagocytic cell (e.g., macrophage).

- **Inflammation.** Complement increases the inflammatory response through the activation of mast cells and basophils and by attracting neutrophils and macrophages (see section 22.3d).

- **Cytolysis.** Various complement components (e.g., C5–C9) trigger direct killing of a target by forming a protein channel in the plasma membrane of a target cell called a **membrane attack complex (MAC).** The MAC protein channel compromises the cell's integrity, allowing an influx of fluid that causes lysis of the cell.

- **Elimination of immune complexes.** Complement links immune (antigen-antibody) complexes to erythrocytes so they may be transported to the liver and spleen. Erythrocytes are stripped of

INTEGRATE

LEARNING STRATEGY

You can remember the various actions of complement with the acronym **O-ICE: O**psonization, **I**nflammation, **C**ytolysis, and **E**limination of immune complexes.

these complexes by macrophages within these organs, and the erythrocytes then continue circulating in the blood.

WHAT DID YOU LEARN?

8 How is the complement system defined? What are the four major means by which complement participates in innate immunity?

22.3d Inflammation

LEARNING OBJECTIVES

6. Define inflammation, and discuss the basic steps involved, including the formation of exudate and its role in removing harmful substances.
7. Describe the benefits of inflammation.
8. List the cardinal signs of inflammation, and explain why each occurs.

Inflammation, or the **inflammatory response,** is an immediate, local, nonspecific event that occurs in vascularized tissue against a great variety of injury-causing stimuli. Inflammation occurs, for example, in response to a scratch of your skin, a bee sting, overuse of a body structure (e.g., pitching arm), or from proteolytic enzymes released by fungi. This physiologic process is the *major* effector response of innate immunity and is successful in helping to eliminate most infectious agents and other unwanted substances from the body!

Events of Inflammation

Inflammation involves several steps **(figure 22.6).** The first step is the *release of various chemicals.* Damaged cells of injured tissue, basophils, dendritic cells, macrophages, mast cells, and infectious organisms release numerous chemicals, including histamine, leukotrienes, prostaglandins, interleukins, TNFs, and chemotactic factors. **Table 22.4** lists various chemicals of inflammation, describes their function, and identifies their source.

Opsonization	Increases inflammation	Cytolysis	Elimination of immune complexes

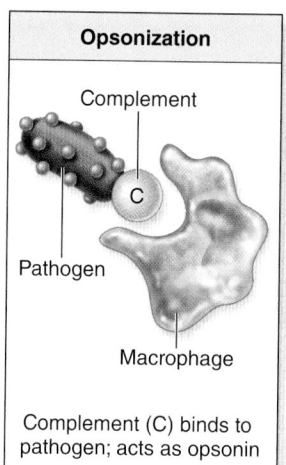
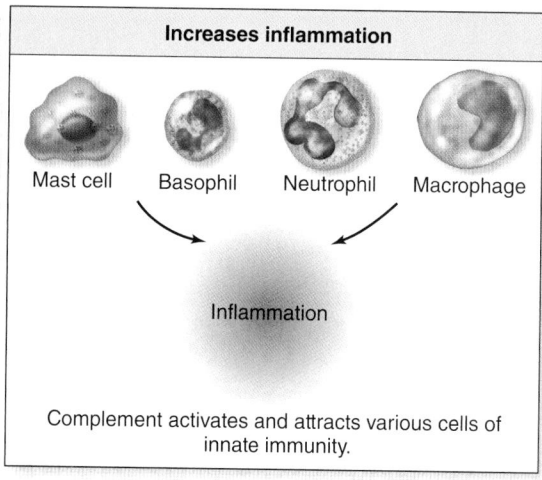
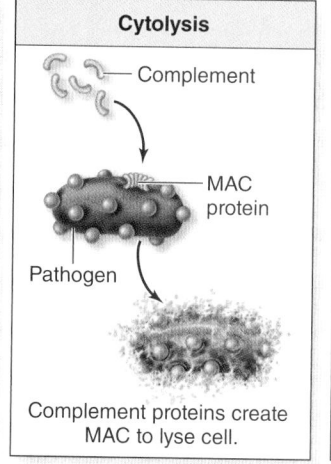
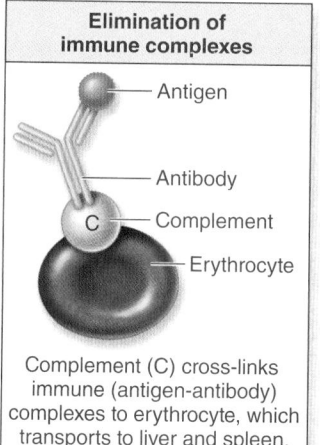

Complement (C) binds to pathogen; acts as opsonin

Complement activates and attracts various cells of innate immunity.

Complement proteins create MAC to lyse cell.

Complement (C) cross-links immune (antigen-antibody) complexes to erythrocyte, which transports to liver and spleen.

Figure 22.5 Complement System. Upon activation, complement (C) proteins serve to protect the body through various mechanisms, including opsonization, increasing inflammation, cytolysis of target cells, and elimination of immune complexes.

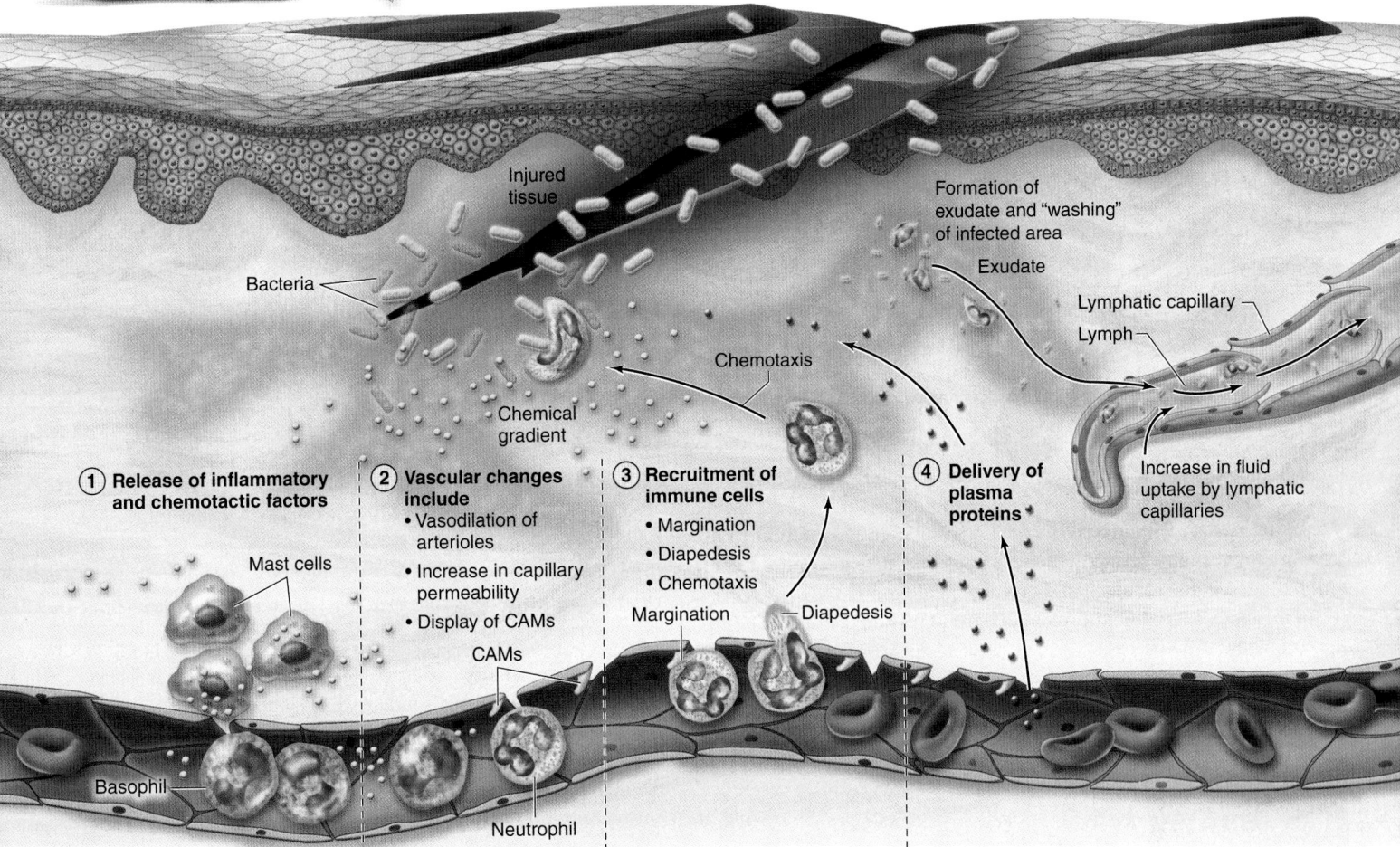

The second step encompasses *vascular changes*. Released chemicals cause a variety of responses in local blood vessels, including vasodilation, increase in capillary permeability, and stimulation of the capillary endothelium to provide molecules for leukocyte adhesion (**cell-adhesion molecules,** or **CAMs**).

The third step involves the *recruitment of leukocytes*. Leukocytes make their way from the blood to the infected tissue through the following processes:

- **Margination** is the process by which CAMs on leukocytes adhere to CAMs on the endothelial cells of capillaries within the injured tissue. The result is similar to "cellular Velcro." Neutrophils are generally the first to arrive and are short-lived, followed later by the longer-lived macrophages.
- **Diapedesis** (dī′ă-pĕ-dē-sis) is the process by which cells exit the blood by "squeezing out" between vessel wall cells, usually in the postcapillary venules, and then migrate to the site of infection (see section 18.3c).
- **Chemotaxis** is migration of cells along a chemical gradient (see section 18.3c). Chemicals released from damaged cells, dead cells, or invading pathogens diffuse outward and form a chemical gradient that attracts immune cells. Recruited cells also participate in the inflammatory response through the release of specific cytokines, such as granulocyte-macrophage

colony-stimulating factor (GM-CSF), that stimulate leukopoiesis within red bone marrow (see section 18.3a). This helps account for the increase in leukocyte count that occurs during an active infection. Macrophages may also release pyrogens, such as interleukin 1 (IL-1), that induce a fever (see section 22.3e).

Delivery of plasma proteins also occurs as shown in the fourth step. Selective plasma proteins are brought into the injured or infected site, including immunoglobulins (described in section 22.8), complement (just described), clotting proteins, and kinins. **Clotting proteins** lead to formation of a clot that walls off microbes and prevents them from spreading into blood and other tissues (see section 18.4c). However, some bacterial species can dissolve clots. **Kinins** (kī′nin) are produced from kininogens, which are inactive plasma proteins produced by the liver (and released into and transported by the blood) and locally by numerous other cells. Kinins, including bradykinin, are activated by tissue injury and have similar effects to histamine; they increase capillary permeability and the production of CAMs by the capillary endothelium. Kinins also stimulate sensory pain receptors and are the most significant stimulus for causing the pain associated with inflammation.

Effects of Inflammation

One of the most important consequences produced by the inflammatory response is a net movement of additional fluid from the blood

Table 22.4	Chemicals of Inflammation	
Substance	**Function in Inflammation**	**Source**
Histamine	Vasodilation; increased permeability of capillaries; conversion of an inactive plasma protein (kininogen) into active peptides called kinins; released early in inflammation	Mast cells, basophils, platelets
Kinins (e.g., bradykinin)	Vasodilation and increased permeability of capillaries; increase production of CAMs; stimulate sensory pain receptors	Plasma protein produced by the liver and other cells as kininogen; activated by tissue injury
Leukotrienes (slow-reacting substance of anaphylaxis [SRS-A])	Effects similar to histamine; released later in the inflammatory response than histamine and longer lasting	Eicosanoids produced from arachidonic acid molecules of mast cell and basophil plasma membranes
Prostaglandins	Vasodilation, fever, stimulate sensory pain receptors (categories include E, D, A, F, and B)	Eicosanoids produced from arachidonic acid molecules of mast cell and basophil plasma membranes
Chemotactic factor	Attracts immune cells; release of specific chemotactic factors attract a specific type of cell (e.g., neutrophil chemotactic factor attracts neutrophils early in the inflammatory response; with a parasitic infection, eosinophil chemotactic factor attracts eosinophils)	Mast cells and basophils
Serotonin	Effects similar to histamine	Platelets
Nitric oxide	Vasodilation; may inhibit mast cells and platelets	Endothelium of blood vessels
Alpha-1 antitrypsin	Inhibits damage to connective tissue by enzymes released from destroyed phagocytes	Plasma protein formed by the liver
C-reactive protein	Activates complement by binding to polysaccharides on bacteria surface	Liver
IL-1 and TNF-α	Increase cell-adhesion molecules (CAMs) to cause margination; cause endothelial cell contraction to facilitate diapedesis	Dendritic cells, macrophages

through the infected or injured area and then into the lymph. Increased fluid, protein, and immune cells leave the capillaries and then enter the interstitial space of the tissue; this fluid is collectively referred to as **exudate** (eks′ū-dāt). Exudate delivers cells and substances needed to eliminate the injurious agent and promote healing.

This increase in fluid movement is due to several factors including the following:

- **Vasodilation,** which allows more blood into the infected area
- **Increased capillary permeability** as endothelial cells lining the blood vessel wall contract, which causes larger openings

INTEGRATE

CLINICAL VIEW

Pus and Abscesses

Pus may form in severe infections. Pus is exudate (excess fluid, protein, and immune cells that leave the capillaries and then enter the interstitial space of the tissue) that contains destroyed pathogens, dead leukocytes, macrophages, and cellular debris. Pus may be removed by the lymphatic system or through the skin (for surface injuries). If the pus is not completely cleared, an **abscess** may form in the area, whereby the pus is walled off with collagen fibers. If an abscess forms, it usually requires surgical intervention to remove.

INTEGRATE

CLINICAL VIEW

Applying Ice for Acute Inflammation

The advice typically given for acute inflammation is to apply ice. Ice serves to vasoconstrict blood vessels (decreasing the inflammatory response) and numb the area so that it seems less painful.

between the endothelial cells and allows more fluid to move from the blood into the interstitial fluid

- **Loss of plasma protein,** which decreases capillary osmotic pressure, resulting in less fluid being retained in the blood and reabsorbed back into the blood during capillary exchange (see section 20.3b)

Increased hydrostatic pressure exerted by the interstitial fluid causes additional fluid uptake by lymphatic capillaries (see section 21.1a). The newly formed lymph carries with it unwanted substances that include infectious agents, dead cells, and cellular debris. The contents of lymph can then be monitored as it passes through a series of lymph nodes. You may find it helpful to think of the inflammatory response as "washing" the infected or injured area.

The inflammatory response typically slows down and tissue healing begins within 72 hours. Monocytes exit the blood and become macrophages to begin the cleanup of the affected area. Bacteria, damaged host cells, and dying neutrophils are engulfed and destroyed by macrophages. Tissue repair begins as fibroblasts multiply and synthesize collagen, forming new connective tissue. (Connective tissue formation may lead to the formation of scar tissue in the case of an extensive injury.)

Several benefits are associated with the inflammatory response, including helping to eliminate pathogens by limiting their spread; destroying infectious agents and removing cellular debris; and producing the conditions for tissue repair and healing.

WHAT DO YOU THINK?

❶ How can you tell by looking at an injured area of the skin that inflammation has occurred?

Cardinal Signs of Inflammation

Inflammation is accompanied by certain cardinal signs that may include the following:

- **Redness,** due to increased blood flow
- **Heat,** due to increased blood flow and increased metabolic activity within the area
- **Swelling,** resulting from increase in fluid loss from capillaries into the interstitial space

- **Pain,** which is caused by stimulation of pain receptors from compression due to accumulation of interstitial fluid, and chemical irritation by kinins, prostaglandins, and substances released by microbes
- **Loss of function** (which may occur in more severe cases of inflammation due to pain and swelling)

The inflammatory response typically lasts no longer than 8 to 10 days under normal conditions. The ending of the normal **acute inflammatory response** (the process just described) is necessary to prevent the unwanted detrimental effects of **chronic inflammation** (see Clinical View: "Chronic Inflammation").

 WHAT DID YOU LEARN?

9 What is inflammation and what are the basic steps involved in the inflammatory response?

10 In what ways does exudate assist in the body's defense?

22.3e Fever

 **LEARNING OBJECTIVES**

9. Define a fever, and describe how it occurs.

10. List the benefits and risks of a fever.

A fever may accompany the inflammatory response. A **fever** is defined as an abnormal elevation of body temperature **(pyrexia)** of at least 1°C (1.8°F) from the typically accepted core body temperature of 37°C (98.6°F). It results from release of fever-inducing molecules called **pyrogens** (e.g., IL-1, IL-6, TNF-α) that are released from either infectious agents (e.g., bacterial toxins) or immune cells in response to infection, trauma, drug reactions, and brain tumors.

Events of Fever

Pyrogens are released and circulate in the blood; they target the hypothalamus (see section 13.4c) and cause release of prostaglandin E_2 (PGE_2). It raises the temperature set point of the hypothalamus from the normal 37°C. The following stages of a fever occur in response: onset, stadium, and defervescence. (Keep in mind that these stages can be cyclical until the pathogen is eliminated or at least brought under control.)

During the **onset** of a fever, the hypothalamus stimulates blood vessels in the dermis of the skin to vasoconstrict to decrease heat loss through the skin, and a person shivers to increase heat production through muscle contraction (see section 1.5b). Consequently, body temperature rises. The person may experience chills during this stage, which leads to the shivering.

The period of time where the elevated temperature is maintained is referred to as **stadium.** The metabolic rate increases to promote physiologic processes involved in eliminating the harmful substance. The liver and spleen bind zinc and iron (minerals needed by microbes) to slow microbial reproduction.

Defervescence (def'ĕr-ves'ents) occurs when the temperature returns to its normal set point. This happens when the hypothalamus is no longer stimulated by pyrogens, prostaglandin release decreases, and the temperature set point reverts to its normal value. The hypothalamus then stimulates the mechanisms to release heat from the body, including vasodilation of blood vessels in the skin and sweating. The person may appear flushed and the skin warm to the touch. An increase in fluid intake should occur during a fever to prevent dehydration caused by an increased loss of body fluid.

Benefits of Fever

Fever actually has numerous benefits. A fever inhibits replication of bacteria and viruses, promotes interferon activity, increases activity of adaptive immunity, and accelerates tissue repair. Most recently, it has been demonstrated that a fever also increases CAMs on the endothelium of capillaries in the lymph nodes, resulting in additional immune cells migrating out of the blood and into the lymphatic tissue. Thus, it is not necessary (and may be detrimental) to treat a mild fever. Most physicians now recommend letting a fever "run its course" and give fever-reducing medication only if the fever becomes very high or if the patient is in significant discomfort from the fever.

Risks of a High Fever

A fever is significant when it is above 100°F. High fevers (103°F in children, and slightly lower in an adult) are potentially dangerous because of the changes in metabolic pathways and denaturation of body proteins (see section 2.8b). Seizures may occur at sustained body temperature above 102°F (although generally they occur at much higher temperatures), irreversible brain damage may occur at body temperatures that are sustained at greater than 106°F, and death is likely when body temperature reaches 109°F.

Figure 22.7 is a visual summary of innate immunity. Both the structures and processes that prevent entry (first line of defense) and the internal nonspecific structures and processes (second line of defense) are included. Remember that the processes of internal nonspecific defenses of innate immunity are generally effective in eliminating most pathogens.

 WHAT DID YOU LEARN?

11 What is a fever and what are the three stages of a fever?

12 What are the benefits and risks of a fever?

INTEGRATE

CLINICAL VIEW
Chronic Inflammation

Chronic inflammation is the condition in which inflammation continues for longer than 2 weeks. Chronic inflammation elicits all the discomfort of the inflammatory response without necessarily ridding the body of the foreign substance. Whereas the acute inflammatory response is generally characterized by neutrophils, chronic inflammation is generally characterized by the presence of cells that arrive later in the inflammatory response, such as macrophages and lymphocytes.

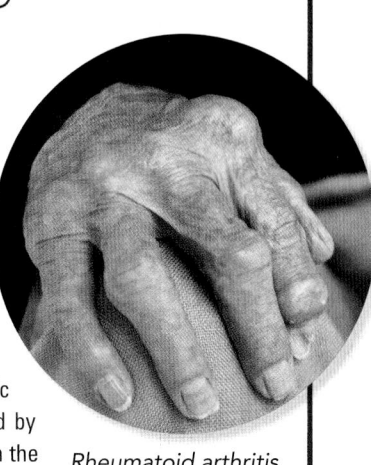

Rheumatoid arthritis

One primary cause of chronic inflammation is overuse injuries, which occur as a result of repetitive minute trauma to a given area of the body. These include tennis elbow, swimmer's shoulder, or shin splints.

Some forms of chronic inflammation occur when the processes of acute inflammation do not eliminate the pathogen or injurious agent such as with tuberculosis, allergens, a splinter that is not removed, injury to a blood vessel, or from an autoimmune disorder (e.g., rheumatoid arthritis, shown in the image). Unfortunately, chronic inflammation can lead to tissue destruction and formation of scar tissue (fibrosis).

Figure 22.7 Innate Immunity. Innate immunity is the nonspecific means of defending the body against potentially harmful substances. These include the (*a*) first line of defense (cellular and molecular structures that help prevent entry), and (*b*) second line of defense (all internal, nonspecific means of eliminating a foreign substance).

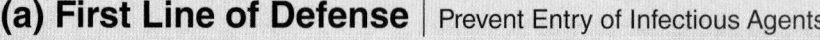

(a) First Line of Defense | Prevent Entry of Infectious Agents

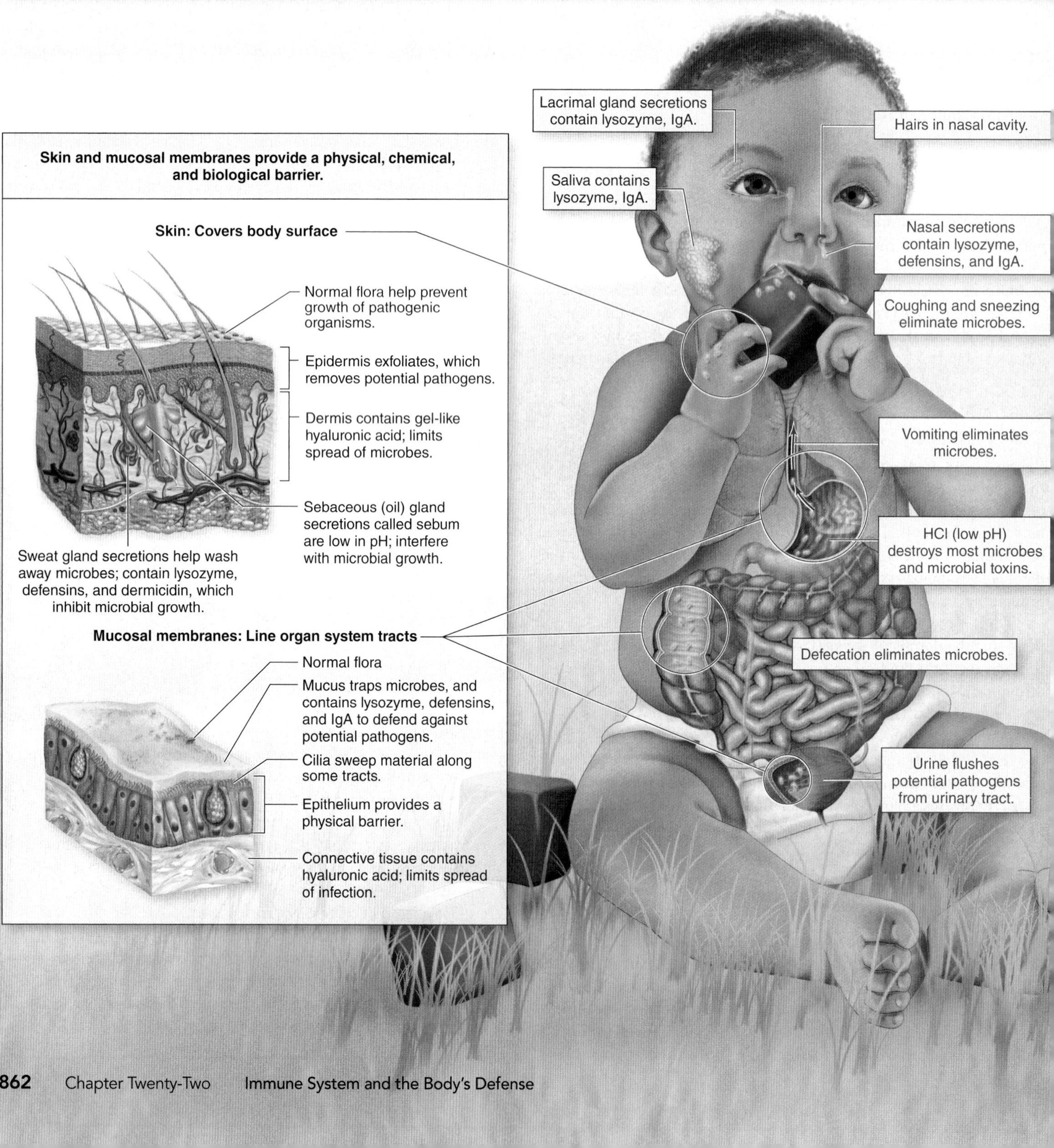

Skin and mucosal membranes provide a physical, chemical, and biological barrier.

Skin: Covers body surface

Normal flora help prevent growth of pathogenic organisms.

Epidermis exfoliates, which removes potential pathogens.

Dermis contains gel-like hyaluronic acid; limits spread of microbes.

Sebaceous (oil) gland secretions called sebum are low in pH; interfere with microbial growth.

Sweat gland secretions help wash away microbes; contain lysozyme, defensins, and dermicidin, which inhibit microbial growth.

Mucosal membranes: Line organ system tracts

Normal flora

Mucus traps microbes, and contains lysozyme, defensins, and IgA to defend against potential pathogens.

Cilia sweep material along some tracts.

Epithelium provides a physical barrier.

Connective tissue contains hyaluronic acid; limits spread of infection.

Lacrimal gland secretions contain lysozyme, IgA.

Saliva contains lysozyme, IgA.

Hairs in nasal cavity.

Nasal secretions contain lysozyme, defensins, and IgA.

Coughing and sneezing eliminate microbes.

Vomiting eliminates microbes.

HCl (low pH) destroys most microbes and microbial toxins.

Defecation eliminates microbes.

Urine flushes potential pathogens from urinary tract.

FEVER

Hypothalamus regulates body temperature including a fever. Benefits of fever include inhibition of microbe reproduction, enhanced immune response, and accelerated tissue repair.

INFLAMMATION

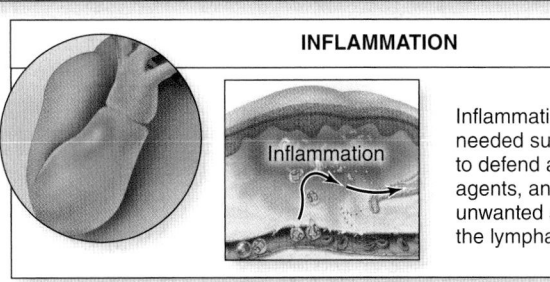

Inflammation

Inflammation delivers needed substances to defend against injurious agents, and flushes unwanted substances into the lymphatic capillaries.

CELLULAR DEFENSES

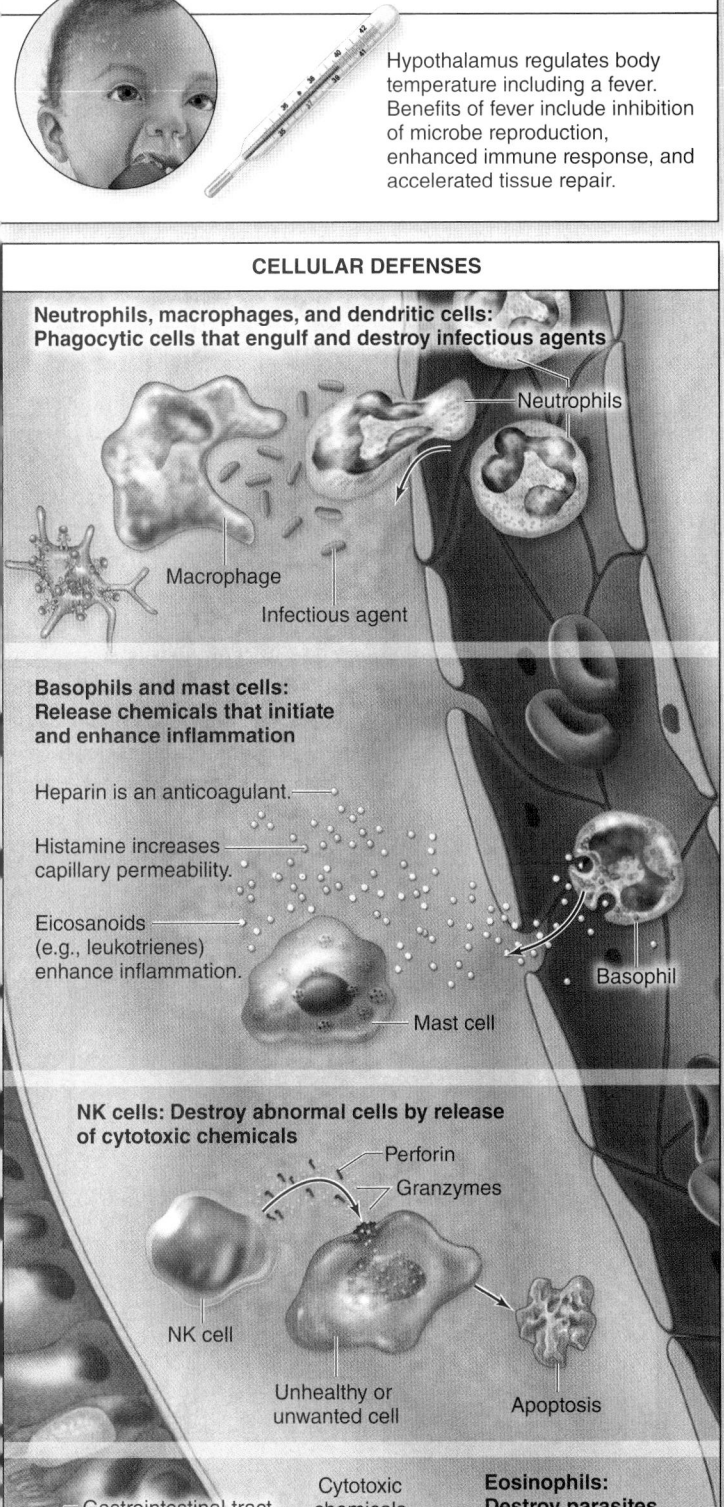

Neutrophils, macrophages, and dendritic cells: Phagocytic cells that engulf and destroy infectious agents

Neutrophils

Macrophage

Infectious agent

Basophils and mast cells: Release chemicals that initiate and enhance inflammation

Heparin is an anticoagulant.

Histamine increases capillary permeability.

Eicosanoids (e.g., leukotrienes) enhance inflammation.

Basophil

Mast cell

NK cells: Destroy abnormal cells by release of cytotoxic chemicals

Perforin

Granzymes

NK cell

Unhealthy or unwanted cell

Apoptosis

Gastrointestinal tract

Cytotoxic chemicals

Eosinophils: Destroy parasites by release of cytotoxic chemicals

Eosinophil

Parasitic worm

ANTIMICROBIAL PROTEINS AND CHEMICALS

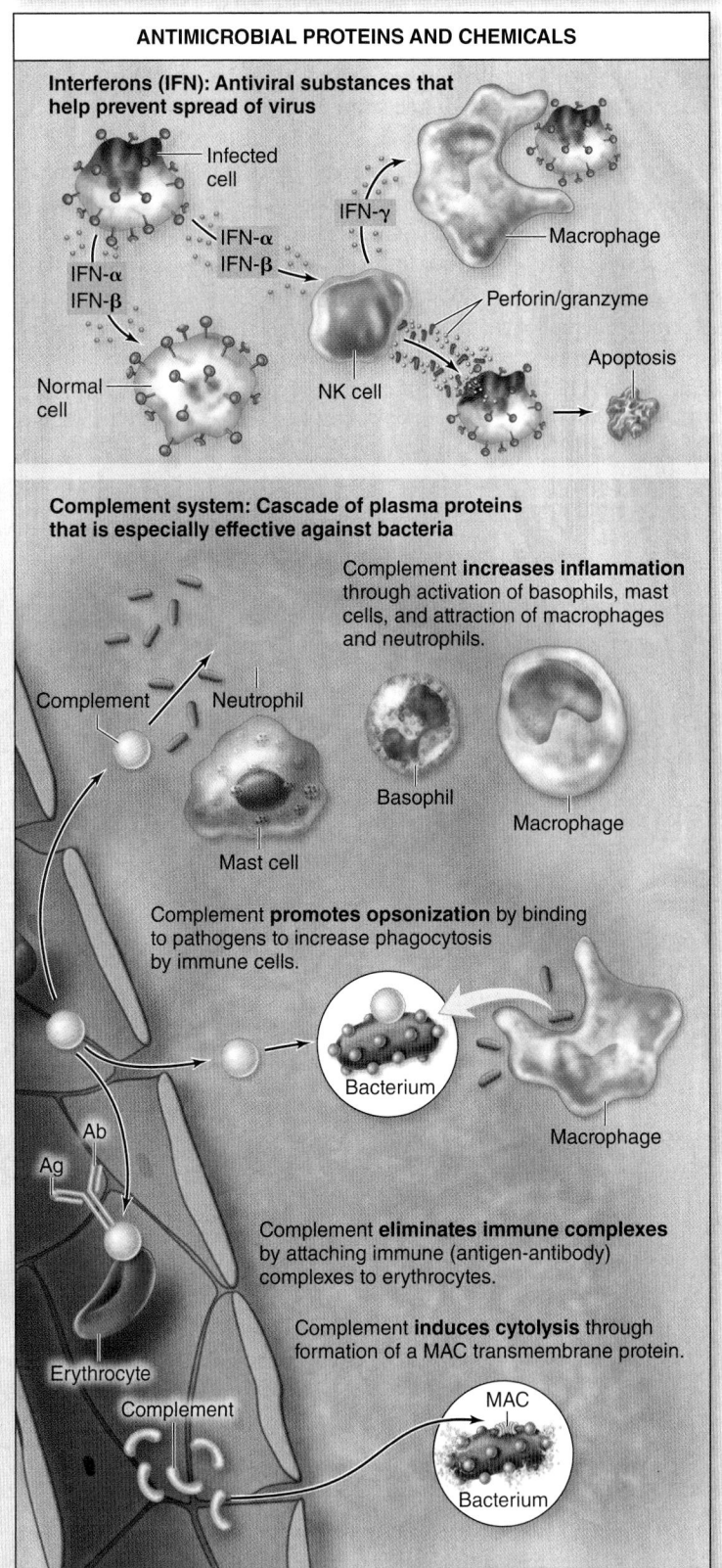

Interferons (IFN): Antiviral substances that help prevent spread of virus

Infected cell

IFN-α
IFN-β

IFN-γ

Macrophage

IFN-α
IFN-β

Perforin/granzyme

Apoptosis

Normal cell

NK cell

Complement system: Cascade of plasma proteins that is especially effective against bacteria

Complement **increases inflammation** through activation of basophils, mast cells, and attraction of macrophages and neutrophils.

Complement

Neutrophil

Basophil

Macrophage

Mast cell

Complement **promotes opsonization** by binding to pathogens to increase phagocytosis by immune cells.

Bacterium

Macrophage

Ab

Ag

Complement **eliminates immune complexes** by attaching immune (antigen-antibody) complexes to erythrocytes.

Erythrocyte

Complement **induces cytolysis** through formation of a MAC transmembrane protein.

Complement

MAC

Bacterium

863

22.4 Adaptive Immunity: An Introduction

Adaptive immunity is also initiated upon entry of a foreign substance (or antigen); however, it takes longer to respond than innate immunity. Contact with an antigen causes a lymphocyte to proliferate and differentiate to form a specialized clone, or "army," of lymphocytes against that antigen. The lymphocytes formed and the products they secrete in the body's defense are collectively referred to as the **immune response.** At least several days are generally required to develop an immune response upon the first exposure to the antigen, and it is for this reason that adaptive immunity is considered the *third line of defense.*

The immune response involving T-lymphocytes, which differentiate into helper T-lymphocytes and cytotoxic T-lymphocytes, is more specifically called **cell-mediated immunity** (or *cellular immunity*). In comparison, the immune response involving B-lymphocytes that develop into plasma cells to synthesize and release antibodies is termed **humoral immunity** (or *antibody-mediated immunity*). A general overview of the two branches of adaptive immunity is shown in **figure 22.8**.

In this section, we introduce adaptive immunity by describing several central concepts, including: a description of antigens, the general structure of lymphocytes, antigen-presenting cells and MHC molecules (structures that interact with T-lymphocytes), and an overview of the critical events in the life cycle of a lymphocyte.

22.4a Antigens

LEARNING OBJECTIVES

1. Describe the features of an antigen, and explain what is meant by antigenic determinant.
2. Explain immunogenicity, and list attributes that affect it.
3. Discuss how haptens stimulate immune responses.

Pathogenic organisms and other foreign substances are detected by T-lymphocytes and B-lymphocytes because they contain antigens. An **antigen** (an'ti-gen; *anti[body]* + *gen* = producing) is a substance that binds to a component of adaptive immunity (T-lymphocyte or an antibody). An antigen is usually a protein or a large polysaccharide (see section 2.7). Examples of antigens include parts of infectious agents such as the protein capsid of viruses, cell wall of bacteria or fungi, and bacterial toxins. Tumor cells also contain antigens. In the case of cancerous cells, mutations occur that generally result in the production of unique (abnormal) proteins designated as *tumor antigens.*

Foreign antigens or *nonself-antigens* bind with the body's immune components because they are different enough in structure from the human body's molecules. In contrast, the body's molecules (structures called **self-antigens**) typically do not bind with the body's immune components. The immune system is generally very effective in distinguishing a self-antigen from foreign antigen. One malfunction of the immune system, however, involves the immune system reacting to self-antigens as if they were foreign. These conditions are collectively called **autoimmune disorders** (see Clinical View: "Autoimmune Disorders"). (Note that plastic and some metals are generally not antigens. This accounts for why plastics and metals, including titanium, stainless steel, and cobalt chrome, are used in artificial implants such as for a hip replacement.)

Lymphocytes normally have contact with only a portion of the antigen. The specific site on the antigen molecule that is recognized by components of the immune system is referred to as the **antigenic determinant,** or *epitope.* Each type of antigenic determinant has a different shape, and a pathogenic organism can have numerous different antigenic determinants. **Figure 22.9** shows an antigen and several antigenic determinants.

An antigen that induces an immune response is more specifically called an **immunogen,** and its ability to cause an immune response is termed its **immunogenicity.** Important attributes that affect an

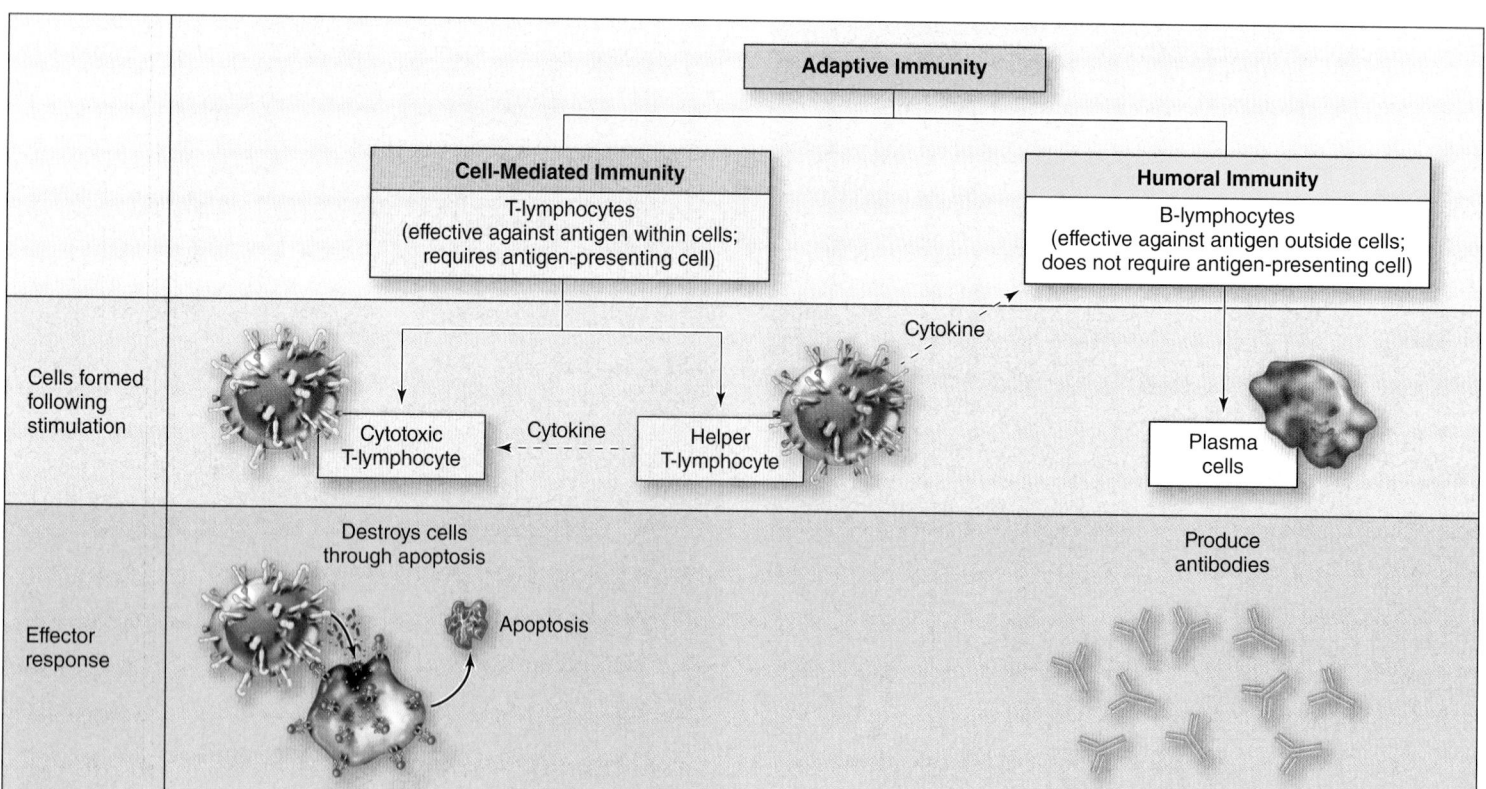

Figure 22.8 Two Branches of Adaptive Immunity. The two branches of adaptive immunity include cell-mediated immunity, which involves T-lymphocytes, and humoral immunity, which involves B-lymphocytes.

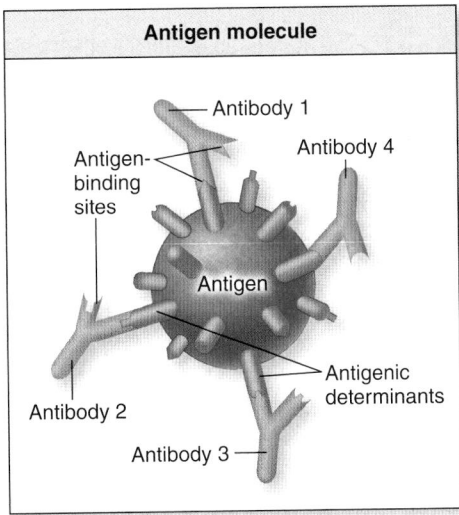

Antigen molecule

Figure 22.9 Antigens and Antigenic Determinants. An antigenic determinant is the specific portion of an antigen to which the components of the adaptive immune system bind. Typically each antigen has several antigenic determinants. AP|R

antigen's immunogenicity include degree of foreignness, size, complexity, and the quantity of the antigen. An increase in one or more of these attributes increases the antigen's ability to elicit an immune response, and thus its immunogenicity.

Some substances are too small to function as an antigen alone, but when attached to a carrier molecule in the host, become antigenic and trigger an immune response; these small molecules are called **haptens** (hap'ten; *hapto* = to grasp). An example is the lipid toxin in poison ivy, which penetrates the skin and triggers an immune response after combining with a body protein. Hapten-stimulating immune responses account for hypersensitivity reactions to drugs, such as penicillin, and to chemicals in the environment, such as pollen, animal dander, mold, and snake or bee venom (see Clinical View: "Hypersensitivities" in section 22.9c).

WHAT DID YOU LEARN?

13 How is an antigenic determinant related to an antigen?

14 What distinguishes a hapten from an antigen?

22.4b General Structure of Lymphocytes

LEARNING OBJECTIVE

4. Describe receptors of both T-lymphocytes and B-lymphocytes.

T-lymphocytes and B-lymphocytes differ from other immune cells because each lymphocyte has a unique **receptor complex,** which are composed of several different and separate proteins. There are typically about 100,000 receptor complexes per cell. A receptor complex will bind one specific antigen. The antigen receptor (which is a portion of a receptor complex) of a T-lymphocyte is referred to as the **TCR** (or *T-cell receptor*), and the antigen receptor of a B-lymphocyte is called a **BCR** (or *B-cell receptor*) (**figure 22.10**).

WHAT DO YOU THINK?

2 If an antigen mutates, will the same lymphocytes recognize it?

The initial contact made between a BCR or TCR of a lymphocyte and the antigen it recognizes is different in B-lymphocytes and T-lymphocytes. B-lymphocytes can make direct contact with an antigen. In contrast, T-lymphocytes must first have the antigen processed and presented in the plasma membrane of another type of cell. T-lymphocytes simply are not able to recognize the antigen without this preliminary step.

T-lymphocytes have additional receptor molecules (called coreceptors) that facilitate T-lymphocyte physical interaction with a cell presenting antigen. One significant category of coreceptors is the CD molecules. In fact, the two major types of T-lymphocytes—helper T-lymphocytes and cytotoxic T-lymphocytes—can be distinguished based on the specific CD protein associated with the TCR (figure 22.10a). The plasma membranes of helper T-lymphocytes contain the CD4 protein, and the plasma membranes of cytotoxic T-lymphocytes contain the CD8 protein.

INTEGRATE

CLINICAL VIEW

Autoimmune Disorders

Autoimmune disorders occur when the immune system does not have tolerance for a specific self-antigen and subsequently initiates an immune response to these self-antigens as if they were foreign. Lack of tolerance and the development of autoimmune disorders can result from cross reactivity, altered self-antigens, or when immune cells enter an area of immune privilege.

Cross-reactivity occurs when a foreign antigen is similar in structure to a self-antigen, and the immune system is unable to distinguish between the two. For example, antigens of *Streptococcus* bacteria are similar to certain heart proteins, and immune cells damage the bicuspid (mitral) and aortic valves, resulting in *rheumatic heart disease.*

Altered self-antigens occur when a microbe induces changes in a specific protein in the body (self-antigen) and the immune cells now respond to it as if it were foreign. Examples of diseases include the following:

- *Type 1 (insulin-dependent) diabetes,* which is thought to result from a microbe's inducing changes in the proteins of the beta cells of the

pancreatic islets in the pancreas; the immune system then destroys these cells (see Clinical View: "Conditions Resulting in Abnormal Blood Glucose Levels" in section 17.9b).

- *Multiple sclerosis,* which results from the destruction of the myelin sheath formed by oligodendrocytes; this destruction occurs by T-lymphocytes (see Clinical View: "Nervous System Disorders Affecting Myelin" in section 12.4c).

Areas of **immune privilege** are structures that prevent or limit access of immune cells (e.g., the brain, eye, testes, ovaries, and placenta). Previously, it was thought that these areas were protected passively by the blood-tissue barrier and lack of lymphatic drainage. However, recent studies have shown that these areas actively participate in maintaining immune privilege by producing various molecules, including specific immunosuppressive cytokines and special plasma proteins that actively destroy T-lymphocytes that infiltrate the area. If large numbers of immune cells enter an immune privileged area, they can destroy structures that are "perceived" as foreign. For example, if a male takes a forceful blow to the scrotum that destroys the blood-testis barrier, immune cells may destroy developing sperm cells and cause sterility.

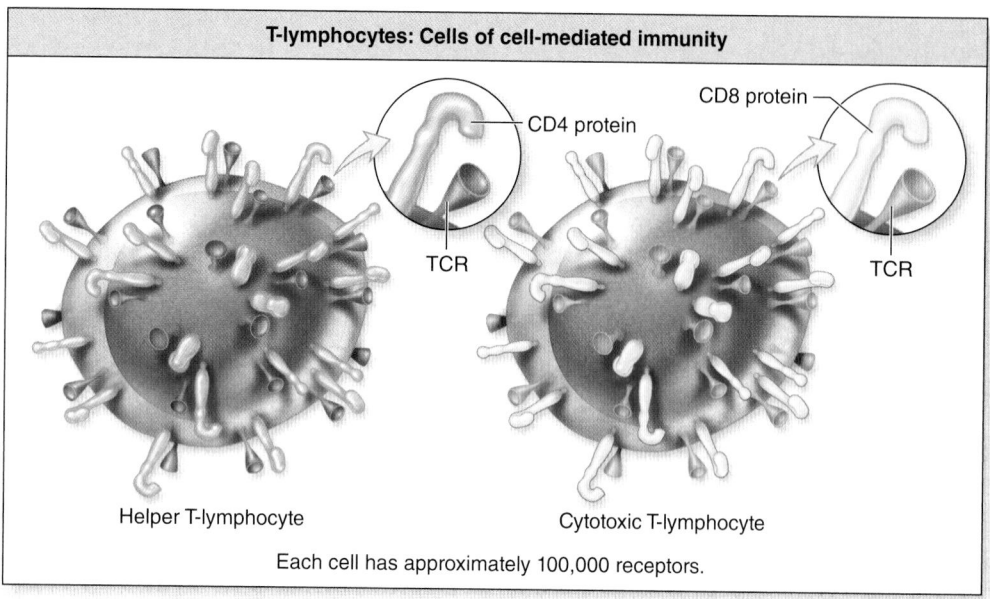

T-lymphocytes: Cells of cell-mediated immunity

CD4 protein

CD8 protein

TCR

TCR

Helper T-lymphocyte

Cytotoxic T-lymphocyte

Each cell has approximately 100,000 receptors.

(a)

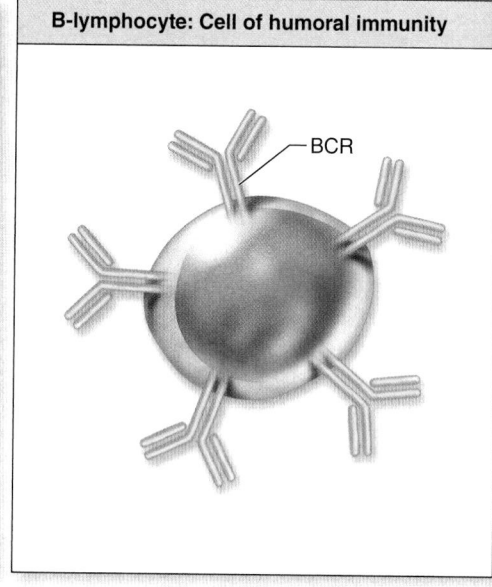

B-lymphocyte: Cell of humoral immunity

BCR

(b)

Figure 22.10 T-Lymphocytes and B-Lymphocytes. The receptors of T-lymphocytes and B-lymphocytes are plasma membrane molecules. (*a*) Helper T-lymphocytes contain TCRs (T-cell receptors) and CD4 proteins, whereas cytotoxic T-lymphocytes contain TCRs and CD8 protein. (*b*) B-lymphocytes contain BCRs (B-cell receptors). Note: There are many other receptors embedded within both T-lymphocytes and B-lymphocytes. This figure depicts only the TCR, CD4, or CD8 receptors of T-lymphocytes and the BCR of B-lymphocytes. AP|R

Note that the terminology associated with T-lymphocytes can be confusing because they may have several designations. Keep in mind that their names reflect either the lymphocyte's function or the type of membrane protein receptor associated with the TCR:

- **Helper T-lymphocytes** (T$_H$) function to coordinate the immune response—helping both cell-mediated immunity and humoral immunity, as well as enhancing certain aspects of innate immunity (e.g., activate NK cells and macrophages); it is for this reason they are called "helper" T-lymphocytes. Structurally, helper T-lymphocytes contain the CD4 plasma membrane protein and are also classified as **CD4** (or CD4+) **cells.**

- **Cytotoxic T-lymphocytes** (T$_C$) release chemicals that are toxic to cells, resulting in their destruction. Because cytotoxic T-lymphocytes contain the CD8 plasma membrane protein, they are also classified as **CD8** (or CD8+) **cells.**

Various other types of T-lymphocytes are also formed, including (1) memory T-lymphocytes (both T$_C$ and T$_H$), which cause a more rapid response to an antigen when future encounters of the same antigen occur, and (2) regulatory T-lymphocytes (Tregs), which function in suppressing the immune response.

WHAT DID YOU LEARN?

15 What features distinguish the receptors of helper T-lymphocytes, cytotoxic T-lymphocytes, and B-lymphocytes?

22.4c Antigen-Presenting Cells and MHC Molecules

LEARNING OBJECTIVES

5. Define antigen presentation.
6. Describe antigen-presenting cells, and list cells that serve this function.
7. Explain the process of formation of MHC class I molecules in nucleated cells and MHC class II molecules in professional antigen-presenting cells.

8. Diagram the general interaction of TCR and CD receptors of a T-lymphocyte with antigen associated with the MHC molecules of other cells.

Antigen presentation is the display of antigen on a cell's plasma membrane surface. This is a necessary process performed by other cells so that T-lymphocytes can recognize an antigen. Generally, two categories of cells present antigen to T-lymphocytes: all nucleated cells of the body (i.e., all cells except erythrocytes) and a category of cells called antigen-presenting cells. The term **antigen-presenting cell (APC)** is used to describe any immune cell that functions specifically to communicate the presence of antigen to *both* helper T-lymphocytes and cytotoxic T-lymphocytes. Dendritic cells, macrophages, and B-lymphocytes function as APCs.

Antigen presentation requires the physical attachment of antigen to a specialized transmembrane protein called **MHC.** MHC is an abbreviation for **major histocompatibility** (his′tō-kom-pat′i-bil′i-tē; *histo* = tissue) **complex.** This name refers to the group of genes that code for MHC molecules embedded within plasma membranes. There are two primary categories of MHC molecules: MHC class I molecules and MHC class II molecules. All nucleated cells present antigen with MHC class I molecules, whereas APCs display antigen with both MHC class I molecules and with MHC class II molecules (a molecule displayed only by APCs).

Synthesis and Display of MHC Class I Molecules on Nucleated Cells

MHC class I molecules are glycoproteins; they are genetically determined and are unique to each individual (see Clinical View: "Organ Transplants and MHC Molecules").

MHC class I molecules are continuously synthesized by the rough endoplasmic reticulum (RER), inserted into the ER, shipped within and modified by the endomembrane system (see section 4.6a), and then embedded within the plasma membrane for the purposes of displaying peptide fragments of **endogenous proteins** (proteins within the cell). This process is referred to as the **endogenous pathway** (figure 22.11*a*).

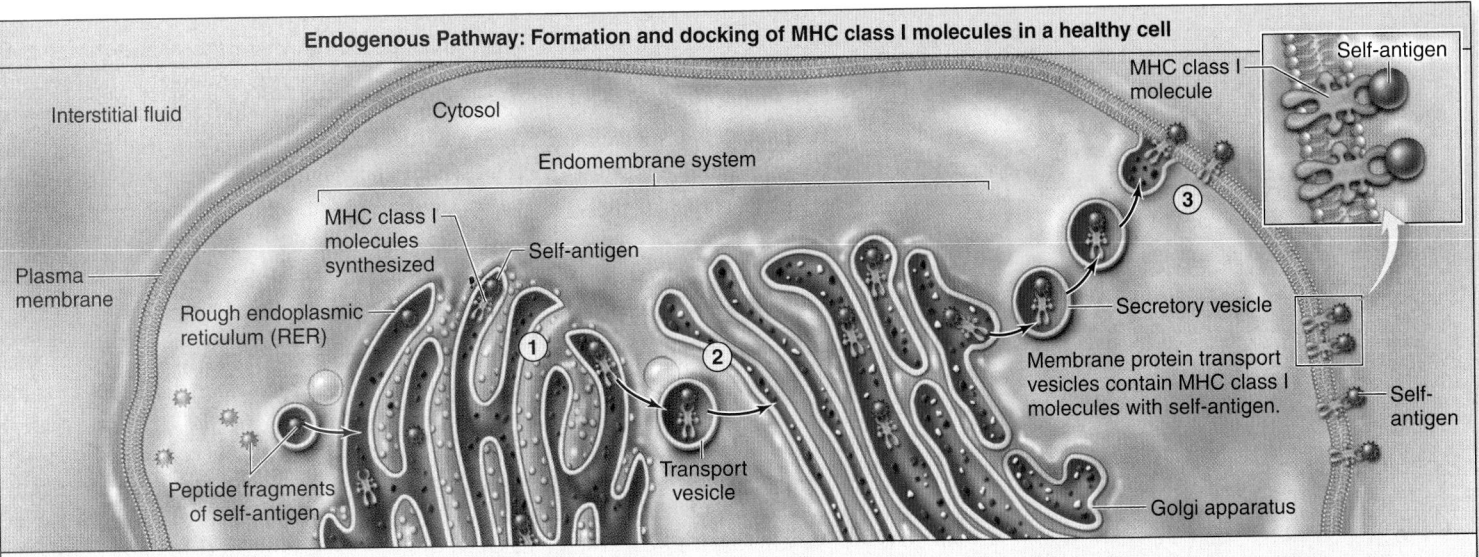

Endogenous Pathway: Formation and docking of MHC class I molecules in a healthy cell

Interstitial fluid

Cytosol

Endomembrane system

MHC class I
molecule

Self-antigen

Plasma
membrane

MHC class I
molecules
synthesized

Self-antigen

Rough endoplasmic
reticulum (RER)

Secretory vesicle

Membrane protein transport
vesicles contain MHC class I
molecules with self-antigen.

Self-antigen

Transport
vesicle

Peptide fragments
of self-antigen

Golgi apparatus

① MHC class I molecules are continuously synthesized by the RER. During production, peptide fragments of the cell (self-antigens) bind with the MHC class I molecules.

② Transport vesicles are produced from the RER that contain MHC class I molecules with bound self-antigen. They are shipped by the endomembrane system through the Golgi apparatus to the plasma membrane.

③ MHC class I molecules with bound self-antigen are displayed within the plasma membrane following fusion of the secretory vesicles with the plasma membrane.

(a)

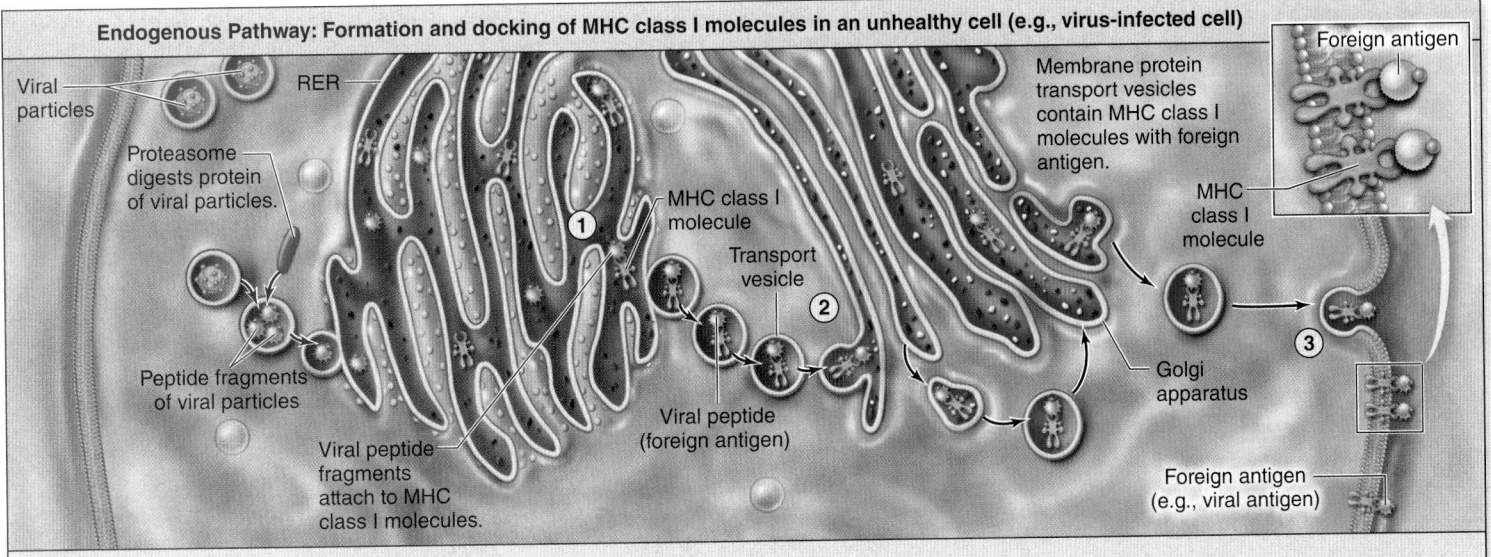

Endogenous Pathway: Formation and docking of MHC class I molecules in an unhealthy cell (e.g., virus-infected cell)

Viral particles

RER

Foreign antigen

Membrane protein
transport vesicles
contain MHC class I
molecules with foreign
antigen.

Proteasome
digests protein
of viral particles.

MHC class I
molecule

MHC
class I
molecule

Transport
vesicle

Peptide fragments
of viral particles

Golgi
apparatus

Viral peptide
(foreign antigen)

Viral peptide
fragments
attach to MHC
class I molecules.

Foreign antigen
(e.g., viral antigen)

Proteins of viral particles (or other microbes) are digested by proteasomes into peptide fragments; peptide fragments are taken up into the RER.

① As MHC class I molecules are synthesized by the RER, peptide fragments of the viral particle (foreign antigen) become attached to MHC class I molecules.

② Transport vesicles are produced from the RER that contain MHC class I molecules with viral peptide fragments. They are shipped by the endomembrane system through the Golgi apparatus to the plasma membrane.

③ MHC class I molecules with bound foreign antigen are displayed within the plasma membrane following fusion of the secretory vesicles with the plasma membrane.

(b)

Figure 22.11 Formation and "Docking" of MHC Class I Molecules in Nucleated Cells. MHC class I molecules are produced and shipped to the plasma membrane of all nucleated cells. (*a*) A normal, healthy cell only displays self-antigens with the MHC class I molecules. (*b*) An infected or unhealthy cell displays foreign antigen with the MHC class I molecules. This "alerts" cytotoxic T-lymphocytes that this cell is infected or unhealthy and should be destroyed.

A significant event occurs during the synthesis and transport of MHC class I molecules to the cell surface involving the endogenous pathway: Peptide fragments in the cell randomly bind with the MHC class I molecules. This occurs within the RER. These peptide fragments in uninfected, healthy cells are simply partially degraded proteins of the cell and are considered "self." Consequently, in uninfected, healthy cells MHC

class I molecules are displaying self-antigens on their surface. These self-antigens are ignored or tolerated by the immune system cells.

However, if the cell is infected, the antigens presented are foreign antigens (figure 22.11*b*). Proteins of an intracellular infectious agent (e.g., viral particle) are cleaved by a proteasome in the cytosol into peptide fragments of 3 to 15 amino acids; these degraded peptide

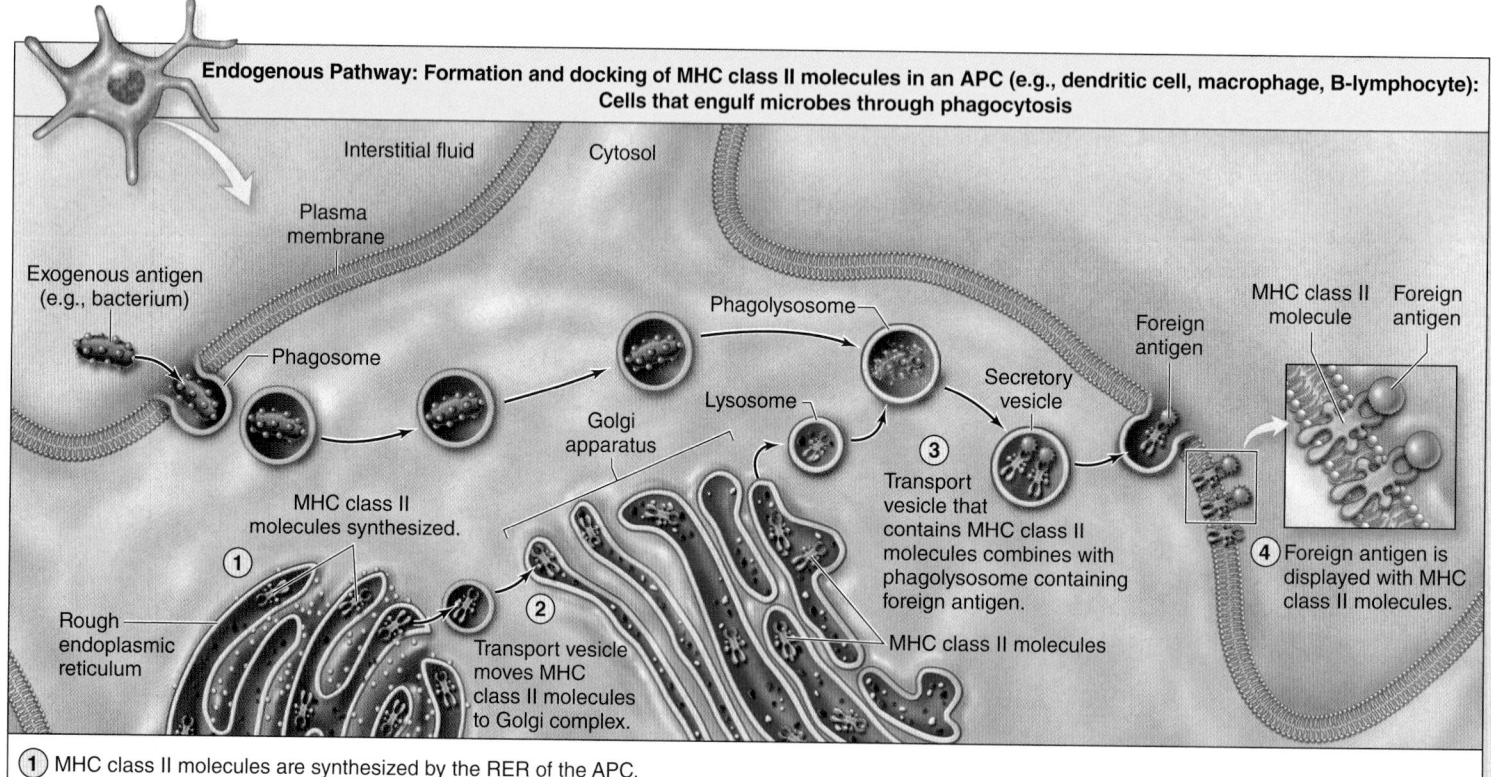

INTEGRATE

LEARNING STRATEGY ✐

You may find it helpful to consider the military analogy of an antigen-presenting cell (APC) as a "sentinel," a cell that keeps watch for potentially harmful substances and reports their presence to T-lymphocytes.

fragments of the infectious agent are considered "nonself." The peptide fragments of the infectious agent that are in the cytosol are shipped into the RER, where the peptide fragments combine with MHC class I molecules within the RER. Through the endomembrane system, the MHC class I molecules carrying the foreign antigens are shipped to the plasma membrane where they are displayed at the cell surface. We will see that the display of foreign antigens with an MHC class I molecule provides the means of communicating specifically with *cytotoxic* T-lymphocytes and will result in the destruction of these infected cells.

Synthesis and Display of MHC Class II Molecules on Professional Antigen-Presenting Cells

Recall that APCs display both MHC class I and MHC class II molecules. MHC class I molecules are synthesized in an APC in a manner similar to other nucleated cells. Here we describe the synthesis and display of MHC class II molecules.

The **MHC class II molecule,** like the MHC class I molecule, is a glycoprotein continuously synthesized by the rough endoplasmic reticulum (RER), modified by the endomembrane system, and then embedded within the plasma membrane **(figure 22.12)**. However, antigens are presented with MHC molecules only after an APC engulfs **exogenous antigens** (pathogens, cellular debris, or other potentially harmful substances located *outside* of cells). The process involving proteins that are engulfed from outside of a cell is referred to as the **exogenous pathway.** Recall from section 22.3b that cells of innate immunity, including dendritic cells and macrophages, recognize microbes through pattern recognition receptors (e.g., toll-like receptors) that are displayed on their cell surface.

Exogenous antigen, through the process of endocytosis, is brought into the cell. A phagosome (vesicle) is formed. The phagosome containing foreign antigen merges with a lysosome to form a phagolysosome, where the substance is digested into peptide fragments. The vesicle containing peptide fragments (antigens) then merges with vesicles containing newly synthesized MHC class II molecules. The peptide fragments are then "loaded" into the MHC class II molecules. These vesicles in turn then merge with the plasma membrane of the APC, with exogenous

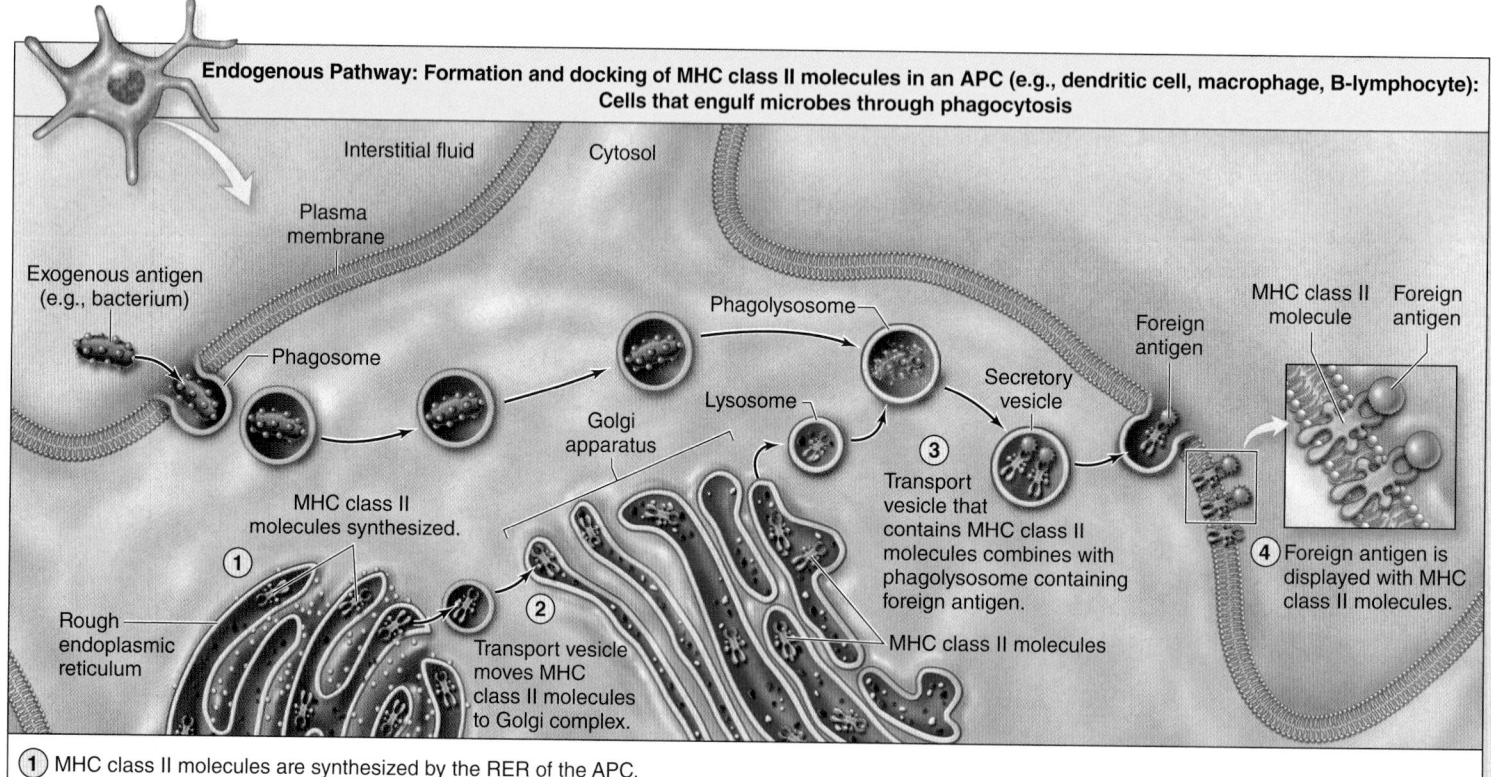

Endogenous Pathway: Formation and docking of MHC class II molecules in an APC (e.g., dendritic cell, macrophage, B-lymphocyte): Cells that engulf microbes through phagocytosis

Interstitial fluid — Cytosol

Plasma membrane

Exogenous antigen (e.g., bacterium)

Phagosome

Phagolysosome

Lysosome

Golgi apparatus

Secretory vesicle

Foreign antigen

MHC class II molecule · Foreign antigen

MHC class II molecules synthesized.

(1)

Rough endoplasmic reticulum

Transport vesicle moves MHC class II molecules to Golgi complex.

(2)

(3) Transport vesicle that contains MHC class II molecules combines with phagolysosome containing foreign antigen.

MHC class II molecules

(4) Foreign antigen is displayed with MHC class II molecules.

(1) MHC class II molecules are synthesized by the RER of the APC.

(2) MHC class II molecules are shipped by the endomembrane system through the Golgi apparatus to the plasma membrane.

(3) During the process of phagocytosis and destruction of an exogenous antigen, vesicles containing digested peptide fragments merge with vesicles containing MHC class II molecules; the foreign antigen binds with MHC class II molecules within the vesicles.

(4) MHC class II molecules and foreign antigen are displayed within the plasma membrane.

Figure 22.12 Formation and "Docking" of MHC Class II Molecules in Antigen-Presenting Cells. Antigen-presenting cells (APCs), which include dendritic cells, macrophages, and B-lymphocytes, display both MHC class I and MHC class II molecules. Here we see the engulfment of exogenous antigen (Ag) (antigen from outside of a cell) and its presentation with MHC class II molecules in the plasma membrane.

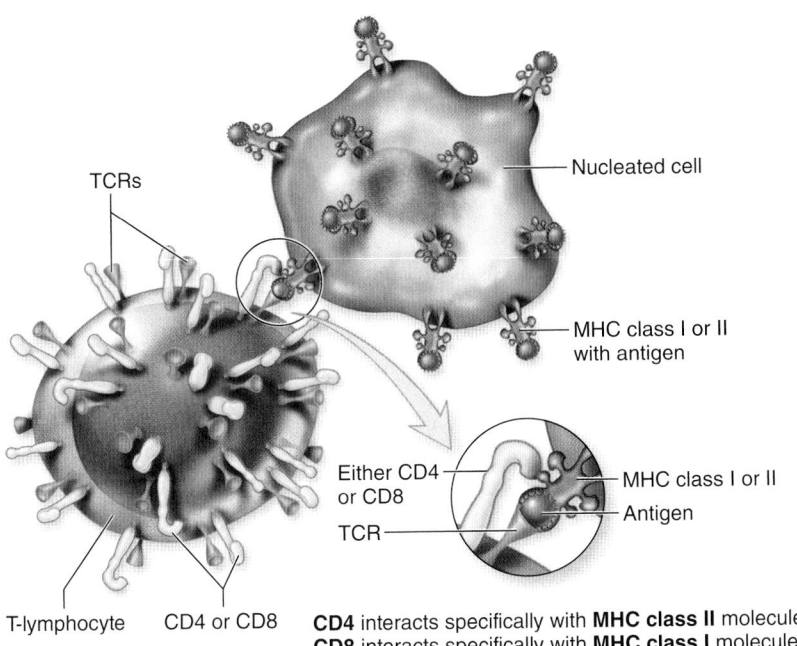

Figure 22.13 Interaction of Receptors of T-Lymphocytes with MHC Molecules of Other Cells. The CD4 or CD8 plasma protein of a T-lymphocyte directly attaches to the MHC molecules of the other cells. This helps keep the two cells together as the TCR of the T-lymphocyte examines the peptide fragment to determine if it is self-antigen or a foreign antigen.

TCRs

Nucleated cell

MHC class I or II with antigen

Either CD4 or CD8

MHC class I or II

TCR

Antigen

T-lymphocyte   CD4 or CD8   **CD4** interacts specifically with **MHC class II** molecules.
CD8 interacts specifically with **MHC class I** molecules.

antigen now displayed bound to MHC class II molecules. This display of foreign antigen with an MHC class II molecule provides the means of communicating specifically with *helper* T-lymphocytes. Degraded components of the engulfed exogenous antigen are also removed from the cell by exocytosis. A similar process occurs by APCs to display antigen with MHC class I molecules. However, this display of foreign antigen with an MHC class I molecule provides the means of communicating specifically with *cytotoxic* T-lymphocytes. In either case, the communication between the APCs and T-lymphocyte will trigger their activation (as described in detail in section 22.6).

Figure 22.13 illustrates a general example of how the receptors of a T-lymphocyte (TCR and CD4 or CD8) interact with an MHC class I or II molecule that contains antigen. Details of these interactions are described in section 22.6.

 WHAT DID YOU LEARN?

 Which type of MHC class molecules is found on all nucleated cells and is used to communicate with cytotoxic T-lymphocytes? Which classes are displayed on APCs, and which class is used specifically to communicate with (a) helper T-lymphocytes and (b) cytotoxic T-lymphocytes?

CLINICAL VIEW

Organ Transplants and MHC Molecules

An **organ transplant** involves the transfer of an organ from one individual to another (see Clinical View: "Tissue Transplants" in section 5.6c). Examples of transplanted organs include the kidney, liver, heart, and lungs. Prior to the transplant, the donor and recipient are tested for the major histocompatibility complex (MHC) antigens and the ABO blood group antigens. No two individuals (except identical twins) have the same MHC genes, so no two individuals have exactly the same MHC molecules.

Organ transplants are risky because the MHC molecules of the cells of the transplanted tissue or organ may be detected by the immune system as foreign. Consequently, components of both the adaptive and innate immune system attempt to destroy it. Thus, the recipient's immune system is suppressed with drugs that make it less likely that it will detect the foreign antigens and cause rejection. However, these immunosuppressive drugs increase the risk of infectious disease and tumor cells going undetected.

An important exception to these required precautions involves corneal transplants. The cornea of the eye is in an area of immune privilege (i.e., an area where immune cells are generally prevented from entering). Thus, a cornea can be transplanted from one individual to another without the need for matching of the tissue or for the administration of immunosuppressive drugs.

22.4d Overview of Life Events of Lymphocytes

LEARNING OBJECTIVE

9. Identify the three significant events that occur in the lifetime of a lymphocyte.

Participation of lymphocytes in the body's defense involves three significant events (figure 22.14):

- **Formation of lymphocytes.** The formation and maturation of lymphocytes occur within primary lymphatic structures (red bone marrow and the thymus; see

Figure 22.14 Overview of Life Events of Lymphocytes.
(a) Lymphocyte formation occurs in primary lymphatic structures (bone marrow and thymus), and mature (immunocompetent) cells migrate to secondary lymphatic structures. (b) Activation occurs in secondary lymphatic structures. (c) The effector response occurs at the site of infection.

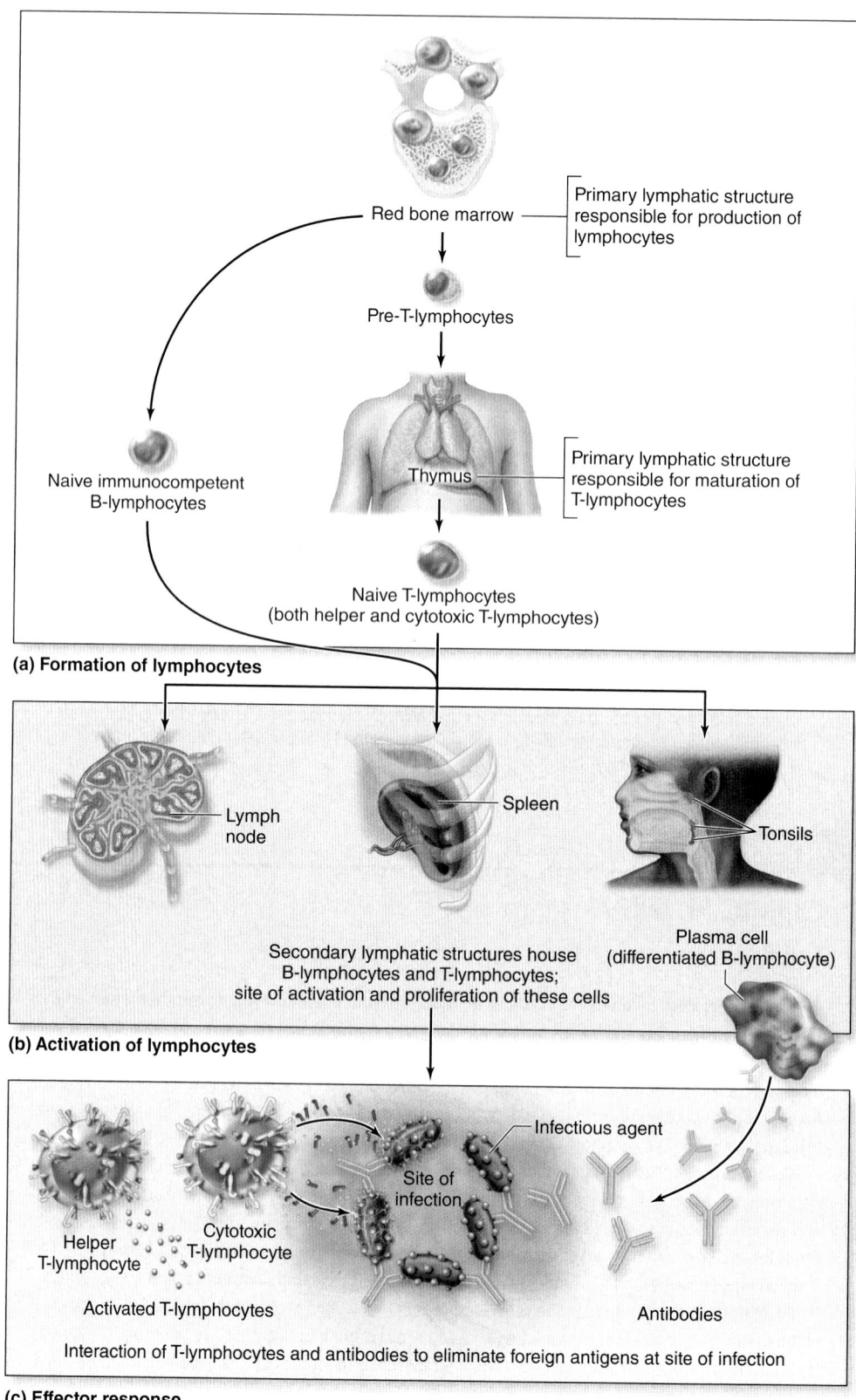

Red bone marrow — Primary lymphatic structure responsible for production of lymphocytes

Pre-T-lymphocytes

Naive immunocompetent B-lymphocytes

Thymus — Primary lymphatic structure responsible for maturation of T-lymphocytes

Naive T-lymphocytes (both helper and cytotoxic T-lymphocytes)

(a) Formation of lymphocytes

Lymph node

Spleen

Tonsils

Plasma cell (differentiated B-lymphocyte)

Secondary lymphatic structures house B-lymphocytes and T-lymphocytes; site of activation and proliferation of these cells

(b) Activation of lymphocytes

Infectious agent

Site of infection

Helper T-lymphocyte

Cytotoxic T-lymphocyte

Activated T-lymphocytes

Antibodies

Interaction of T-lymphocytes and antibodies to eliminate foreign antigens at site of infection

(c) Effector response

section 21.3). Here T-lymphocytes and B-lymphocytes become able to recognize only one specific foreign antigen.

- **Activation of lymphocytes.** Following their formation, lymphocytes then migrate to secondary lymphatic structures (e.g., lymph nodes, the spleen, tonsils, MALT) where they are housed (see section 21.4). Typically, these locations are where lymphocytes have their first exposure to the antigen that they bind, and thus become activated. In response to activation, lymphocytes replicate to form many identical lymphocytes.

- **Effector response.** The effector response is the specific action of the T-lymphocytes and B-lymphocytes to help eliminate the antigen at the site of infection. T-lymphocytes leave the secondary lymphatic structures, migrating to the site of infection. B-lymphocytes (as differentiated plasma cells) primarily remain within the secondary lymphatic structures, synthesizing and releasing large quantities of antibodies against the antigen. The antibodies enter the blood and lymph and are transported to the site of infection.

Please note that these processes are sequential but generally differ in where they take place in the body: lymphocyte development in primary lymphatic structures, activation of lymphocytes in secondary lymphatic structures, and effector response at the site of infection.

  WHAT DID YOU LEARN?

17 Where does a lymphocyte typically encounter an antigen for the first time: primary lymphatic structure, secondary lymphatic structure, or site of infection?

22.5 Formation and Selection of Lymphocytes

Lymphocytes originate in red bone marrow (see section 18.3a). Following their formation, lymphocytes are "tested" to see whether they are able to bind antigen and respond to it—that is, whether they are **immunocompetent.** This process occurs primarily during development and shortly after birth in primary lymphatic structures (bone marrow and thymus). We describe how this occurs in T-lymphocytes.

22.5a Formation of T-Lymphocytes
LEARNING OBJECTIVE

1. Explain how T-lymphocytes mature.

T-lymphocytes originate in red bone marrow and then migrate to the thymus to complete their maturation. (The "T" of T-lymphocytes reflects the role of the thymus in its maturation.) Millions of pre-T-lymphocytes migrate from the red bone marrow to the thymus; they possess a unique TCR receptor *and* initially both the CD4 and CD8 proteins (referred to as "double positive"). These cells are immature T-lymphocytes with a TCR that was produced randomly through "gene shuffling," a concept beyond the scope of this text.

Each T-lymphocyte must have its TCR "tested" to determine not only whether it is able to bind to the MHC molecule with presented antigen, but also whether it binds only to antigen that is foreign or "nonself." This testing results in T-lymphocyte selection.

  WHAT DID YOU LEARN?

18 Where does the maturation of T-lymphocytes take place?

22.5b Selection of T-Lymphocytes

≈✓≤ LEARNING OBJECTIVE

2. Compare and contrast positive and negative selection of T-lymphocytes.

T-lymphocytes go through two selection processes in the thymus, collectively called *thymic selection* (see **figure 22.15** as you read through this section):

1. **Positive selection.** The TCR embedded in the plasma membrane of a T-lymphocyte must be able to recognize and bind an MHC molecule. This is "tested" by having T-lymphocytes bind with thymic epithelial cells that have MHC molecules. Those

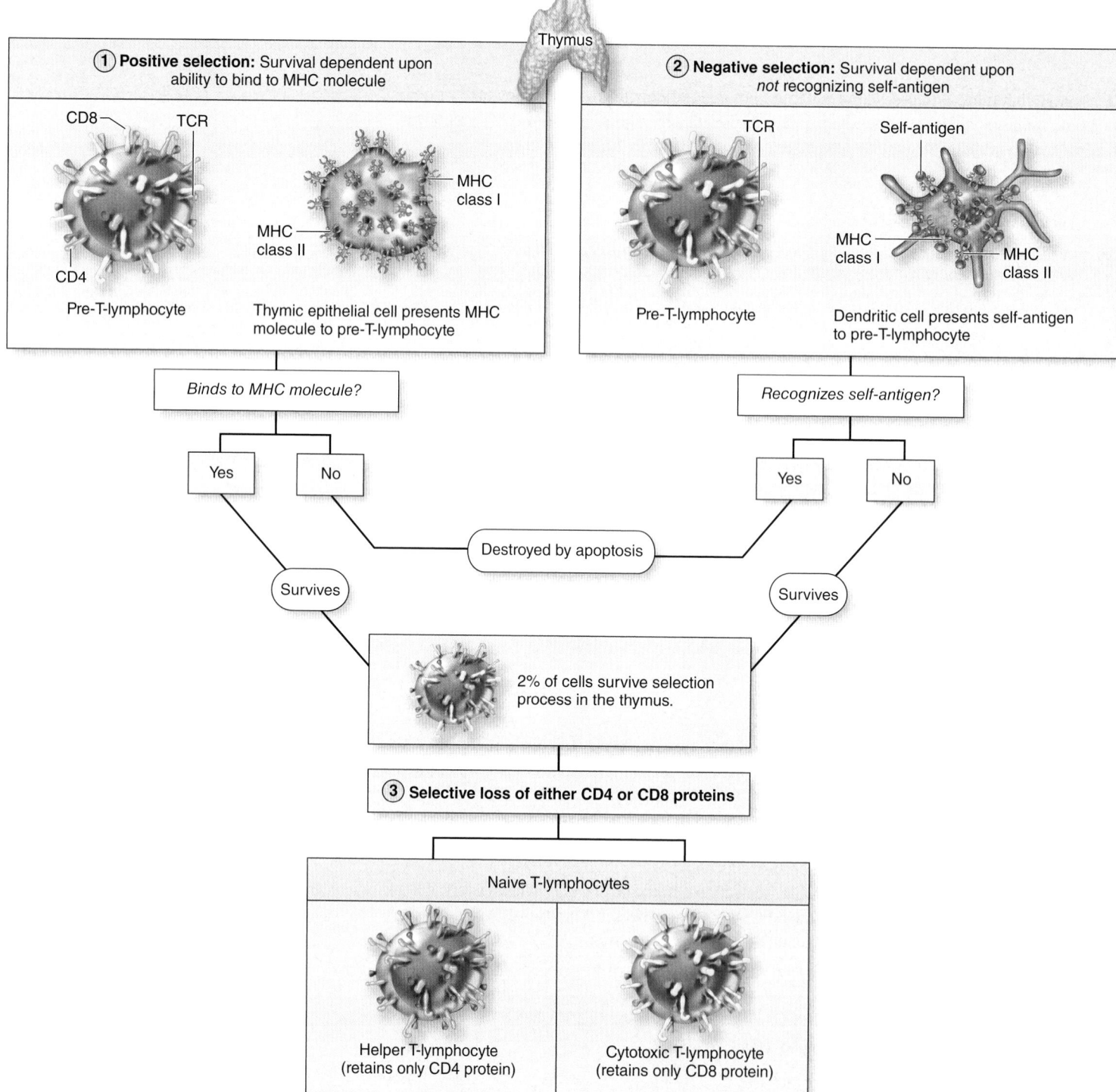

Figure 22.15 Thymic Selection. T-lymphocytes complete their maturation in the thymus to form naive T-lymphocytes that are immunocompetent. The process of thymic selection of T-lymphocytes includes (1) positive selection, (2) negative selection, and (3) the selective loss of either CD4 or CD8 proteins. (Note that positive selection occurs prior to negative selection.)

T-lymphocytes that can bind MHC survive, and those that cannot are eliminated. Thus, those T-lymphocytes that perform this function are selected *for,* a process referred to as **positive selection** (figure 22.15, step 1).

2. **Negative selection.** The newly formed T-lymphocyte must also *not* bind to any self-antigens that are presented within an MHC molecule. This is "tested" by thymic dendritic cells presenting self-antigens with MHC class I and II molecules. If the T-lymphocyte does bind to the self-antigen, then it is destroyed. Thus, those T-lymphocytes that perform this function are selected *against,* a process referred to as **negative selection** (figure 22.15, step 2). This is the process by which cells generally learn to "ignore" molecules of the body or self-antigens, a state referred to as **self-tolerance.** This process, which occurs in the primary lymphatic structures, is more specifically called **central tolerance.**

Consequently, T-lymphocytes that survive both positive and negative selection can bind an MHC molecule and recognize foreign antigen. Only approximately 2% of the originally formed T-lymphocytes survive both selection processes; the remaining 98% are eliminated in the thymus by apoptosis (see section 4.10).

  WHAT DID YOU LEARN?

19 What would happen if a T-lymphocyte that failed the negative selection test was not destroyed?

22.5c Differentiation and Migration of T-Lymphocytes

 **LEARNING OBJECTIVE**

3. Explain the additional changes to T-lymphocytes after selection.

The final step in T-lymphocyte selection is the differentiation of each T-lymphocyte into either a helper T-lymphocyte (CD4+ cell) by the selective loss of the CD8 protein, or a cytotoxic T-lymphocyte (CD8+ cell) by the selective loss of CD4 protein (figure 22.15, step 3). Consequently, two primary types of T-lymphocytes leave the thymus: helper T-lymphocytes (that are CD4+) and cytotoxic T-lymphocytes (that are CD8+).

The T-lymphocytes that leave the thymus are immunocompetent cells. However, each of these T-lymphocytes is also classified as a **naive T-lymphocyte** because it has not yet been exposed to the specific foreign antigen it recognizes. Naive helper T-lymphocytes and naive cytotoxic T-lymphocytes migrate from the thymus to secondary lymphatic structures, where they are housed (figure 22.14b).

It should be noted, however, that a subclass of CD4+ cells is also formed called **regulatory T-lymphocytes (Tregs).** (These cells were previously called *suppressor T-lymphocytes.*) Tregs are formed from T-lymphocytes that bind self-antigens to a moderate extent compared to other CD4+ cells. Tregs migrate to the periphery (body structures outside the primary lymphatic structures), where they release inhibitory chemicals that turn off both the cell-mediated immune response and the humoral immune response. Tregs function in self-tolerance outside of the primary lymphatic structures—a process that is more specifically called **peripheral tolerance.** Current studies have focused extensively on the role of Tregs in disease (e.g., microbial infections, autoimmune disorders, allergies, tumors; see Clinical View: "Regulatory T-Lymphocytes and Tumors").

A similar process to form and select B-lymphocytes occurs in the red bone marrow. (However, MHC is not involved in selection of B-lymphocytes.) Naive B-lymphocytes also then migrate to secondary lymphatic structures where they are housed, and come in contact with foreign antigen that stimulates them to proliferate and differentiate. New T- and B-lymphocytes continue to form throughout one's lifetime, and these cells participate in the immune response when exposed to a particular foreign antigen.

 WHAT DID YOU LEARN?

20 In general, how does central tolerance differ from peripheral tolerance?

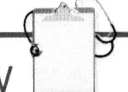

INTEGRATE

CLINICAL VIEW

Regulatory T-Lymphocytes and Tumors

Tumors have been shown to induce Tregs (specifically called tumor Tregs) to proliferate. The increased production of these cells results in greater than normal suppression of the immune system. Consequently, immune cells are less likely to detect the tumor cells, and the tumor continues to grow. Potential cancer treatments are in development that involve diminishing the inhibitory influences of tumor Tregs.

22.6 Activation and Clonal Selection of Lymphocytes

Lymphocyte activation involves physical contact between a lymphocyte and the antigen it recognizes and the subsequent proliferation and differentiation of lymphocytes into a clone of identical cells that have the same TCR or BCR that matches that specific antigen. This process of forming a clone in response to a specific antigen is called **clonal selection.**

The first encounter between an antigen and lymphocyte is called an **antigen challenge.** It typically occurs in secondary lymphatic structures. The specific secondary lymphatic structure in the body in which the antigen challenge occurs generally depends upon the point of entry of the antigen. Antigen in the blood is taken to the spleen; antigen that penetrates the skin is engulfed and transported by epidermal dendritic cells to a lymph node; and antigen that enters through the mucosal membrane of the respiratory, gastrointestinal, urinary, or reproductive tracts comes into contact with the tonsils or MALT (mucosa-associated lymphatic tissue).

22.6a Activation of T-Lymphocytes

LEARNING OBJECTIVES

1. Describe how both helper T-lymphocytes and cytotoxic T-lymphocytes are activated.

2. Explain the role played by IL-2 in both activations.

Both types of T-lymphocytes must undergo activation before they can carry out immune system functions. Activation of both types of T-lymphocytes requires two signals; however, the specific process differs between the two types.

Activation of Helper T-Lymphocytes

The specifics of activation of helper T-lymphocytes is shown in **figure 22.16b.** The **first signal** is direct physical contact between the MHC class II molecule of an antigen-presenting cell (APC) and the TCR of a helper T-lymphocyte. Exogenous antigen previously engulfed by an APC is presented on its surface with MHC class II molecules (as described in section 22.4c). The APC is either housed in the secondary lymphatic structure (e.g., macrophage) or migrates there from the skin (e.g., dendritic cells) to make contact with the helper T-lymphocyte.

A helper T-lymphocyte binds to the APC to inspect the antigen: The specific TCR of a T-lymphocyte binds with the peptide fragment presented with an MHC *class II* molecule of the APC. This interaction is stabilized by the CD4 molecule of the helper T-lymphocyte binding to other regions of the MHC class II molecule. If the TCR does not recognize the presented antigen, it disengages from the APC. If it does recognize the antigen, contact between the two cells lasts several hours.

The **second signal** takes place when other receptors of the APC (e.g., B7) interact with receptors of the helper T-lymphocyte (e.g., CD28). Ultimately, helper T-lymphocytes are induced to synthesize and release the cytokine interleukin 2 (IL-2), which occurs within about 24 hours. IL-2 acts as an autocrine hormone to further stimulate the helper T-lymphocyte from which it was released.

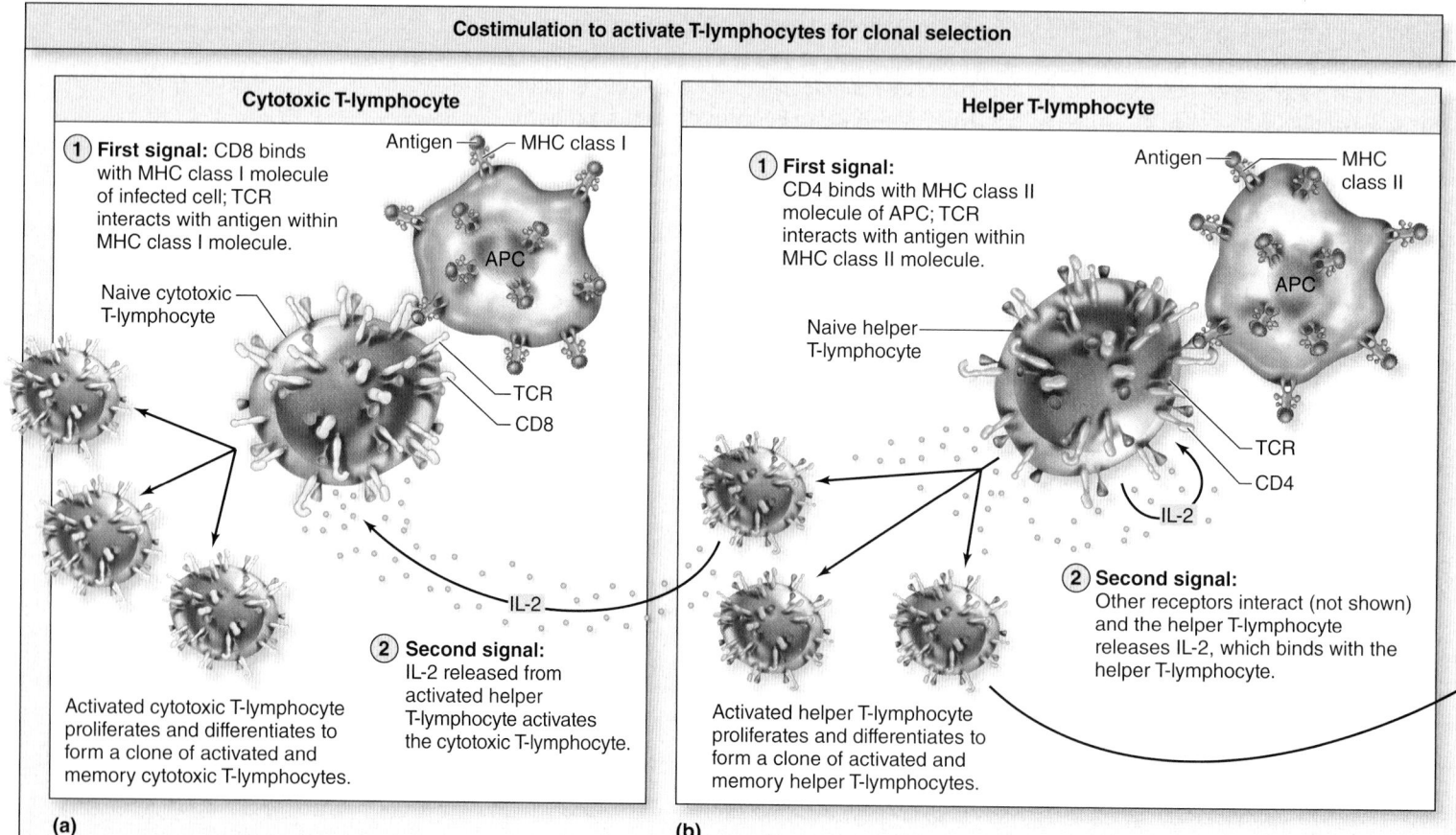

Costimulation to activate T-lymphocytes for clonal selection

Cytotoxic T-lymphocyte

(1) **First signal:** CD8 binds with MHC class I molecule of infected cell; TCR interacts with antigen within MHC class I molecule.

Antigen — MHC class I

Naive cytotoxic T-lymphocyte

APC

TCR
CD8

IL-2

(2) **Second signal:** IL-2 released from activated helper T-lymphocyte activates the cytotoxic T-lymphocyte.

Activated cytotoxic T-lymphocyte proliferates and differentiates to form a clone of activated and memory cytotoxic T-lymphocytes.

(a)

Helper T-lymphocyte

(1) **First signal:** CD4 binds with MHC class II molecule of APC; TCR interacts with antigen within MHC class II molecule.

Antigen — MHC class II

Naive helper T-lymphocyte

APC

TCR
CD4

IL-2

(2) **Second signal:** Other receptors interact (not shown) and the helper T-lymphocyte releases IL-2, which binds with the helper T-lymphocyte.

Activated helper T-lymphocyte proliferates and differentiates to form a clone of activated and memory helper T-lymphocytes.

(b)

Figure 22.16 Activation of Lymphocytes. Activation of lymphocytes occurs in secondary lymphatic structures, usually the lymph nodes or spleen. Activation results in lymphocyte proliferation and differentiation to form a clone of identical cells that includes memory cells. Two signals (costimulation) are required to activate each type of lymphocyte: (*a*) cytotoxic T-lymphocyte, (*b*) helper T-lymphocyte, and (*c*) B-lymphocyte. **AP|R**

T-lymphocytes are activated and proliferate to form a "clone" of helper T-lymphocytes (T-lymphocytes that possess TCRs that bind that specific antigen). Some of the cells produced are *activated helper T-lymphocytes* that continue to produce IL-2, and some are *memory helper T-lymphocytes,* cells available for subsequent encounters with the specific antigen. (Note that lack of second signal is thought to result in helper T-lymphocytes becoming Tregs. Recall that Tregs can also be formed in the thymus when CD4+ cells bind self-antigen with moderate affinity.)

Activation of Cytotoxic T-Lymphocytes

The **first signal** for a cytotoxic T-lymphocyte is similar to the first stimulation for a naive helper T-lymphocyte (figure 22.16*a*). However, direct physical contact is made between the TCR of a cytotoxic T-lymphocyte and a peptide fragment presented with an MHC *class I* molecule of either an APC or an infected cell. This interaction is stabilized by the CD8 of the cytotoxic T-lymphocyte binding to other regions of the MHC class I molecule. The **second signal** is the binding of IL-2 that is released from *helper* T-lymphocytes. IL-2 acts as a paracrine hormone to stimulate the cytotoxic T-lymphocyte. IL-2 is required for cytotoxic T-lymphocytes to become fully activated. (Note that only APCs [e.g., dendritic cells] are able to activate naive cytotoxic T-lymphocytes—that is, when cytotoxic T-lymphocytes are first exposed to the antigen they recognize.)

Upon activation, cytotoxic T-lymphocytes proliferate and differentiate into clones, some becoming *activated cytotoxic T-lymphocytes*, and others developing into *memory cytotoxic T-lymphocytes* that are activated upon reexposure to the same antigen.

WHAT DID YOU LEARN?

21 What type of cell is required to activate both helper T-lymphocytes and naive cytotoxic T-lymphocytes?

22 How do cytokines released by helper T-lymphocytes participate in activation of both helper and cytotoxic T-lymphocytes?

22.6b Activation of B-Lymphocytes

LEARNING OBJECTIVE

3. Compare the activation of B-lymphocytes with that of T-lymphocytes.

Immunocompetent but naive B-lymphocytes are also activated by a specific antigen in secondary lymphatic structures. As with T-lymphocytes, two signals are required. However, B-lymphocytes do not require antigen to be presented by other nonlymphocyte cells. B-lymphocytes recognize and respond to antigens *outside of cells,* such as antigens of viral particles, bacteria, bacterial toxins, or yeast spores.

The **first signal** occurs when intact antigen binds to the BCR, and the antigen cross-links BCRs (figure 22.16*c*). The stimulated B-lymphocyte engulfs, processes, and presents the antigen to the helper T-lymphocyte that recognizes that antigen. (This is similar to the action of other APCs.) The **second signal** occurs when an activated helper T-lymphocyte releases IL-4 to stimulate the B-lymphocyte.

Activation of B-lymphocytes causes the B-lymphocytes to proliferate and differentiate. Most of the *activated B-lymphocytes* differentiate into **plasma cells** that produce antibodies, and the

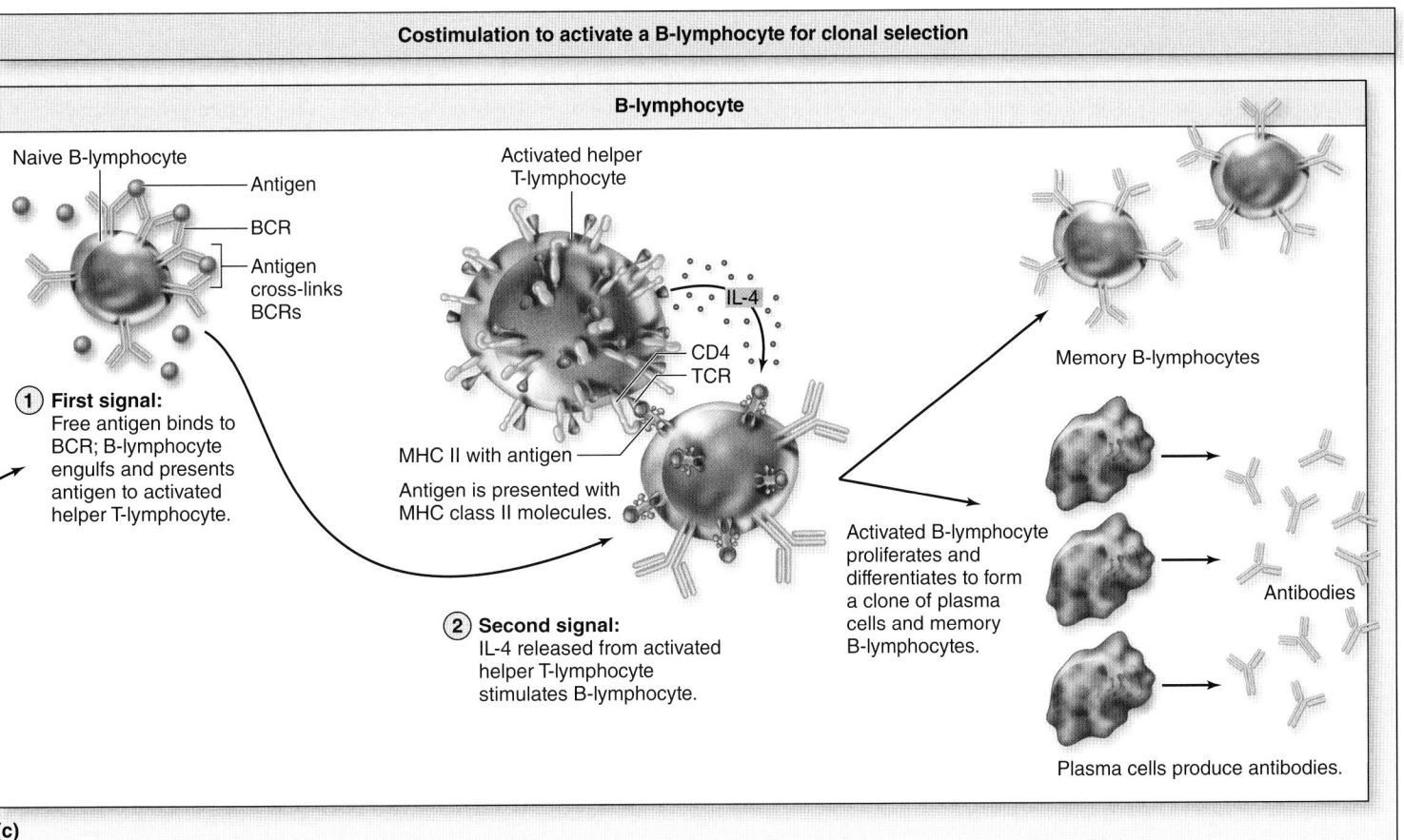

Costimulation to activate a B-lymphocyte for clonal selection

B-lymphocyte

Naive B-lymphocyte

Antigen

BCR

Antigen cross-links BCRs

1 First signal: Free antigen binds to BCR; B-lymphocyte engulfs and presents antigen to activated helper T-lymphocyte.

Activated helper T-lymphocyte

IL-4

CD4

TCR

MHC II with antigen

Antigen is presented with MHC class II molecules.

2 Second signal: IL-4 released from activated helper T-lymphocyte stimulates B-lymphocyte.

Activated B-lymphocyte proliferates and differentiates to form a clone of plasma cells and memory B-lymphocytes.

Memory B-lymphocytes

Antibodies

Plasma cells produce antibodies.

(c)

remainder become *memory B-lymphocytes* that are activated upon reexposure of the same antigen. Memory B-lymphocytes differ from plasma cells in some respects: (1) The memory B-lymphocytes retain their BCRs, and (2) memory B-lymphocytes have a much longer life span (months to years) than plasma cells (typically 5 to 7 days). Note that B-lymphocytes can be stimulated by antigen without direct contact between a B-lymphocyte and helper T-lymphocyte under certain conditions. However, the production of memory B-lymphocytes and the various forms of antibodies (described in section 22.8) requires helper T-lymphocyte participation during B-lymphocyte activation. Observe figure 22.16 and notice the central role that helper T-lymphocytes play in activating both cytotoxic T-lymphocytes (cell-mediated branch of immunity) and **B-lymphocytes** (humoral branch of immunity).

WHAT DID YOU LEARN?

23 Is a separate APC required for B-lymphocyte activation, or is a B-lymphocyte able to serve the role of an APC?

24 Explain the role of cytokines that are released from helper T-lymphocytes in the activation of B-lymphocytes.

22.6c Lymphocyte Recirculation

LEARNING OBJECTIVE

4. Describe lymphocyte recirculation and explain its general function.

One of the hurdles facing adaptive immunity is the requirement of direct physical contact between antigen and the specific lymphocyte with the unique receptor that recognizes the antigen. It is estimated that only 1 in every 100,000 to 1,000,000 T-lymphocytes or B-lymphocytes can bind with a specific antigen on the first exposure to that antigen—that is, during the antigen challenge. The "odds" for contact are increased because lymphocytes reside only temporarily in any given secondary lymphatic structure, and after a period of time they exit and then circulate through blood and lymph every several days. This process is referred to as **lymphocyte recirculation.** Lymphocyte recirculation provides a means of delivering different lymphocytes to secondary lymphatic structures, making it more likely that a lymphocyte will encounter its antigen, if present.

WHAT DID YOU LEARN?

25 What advantage is provided by lymphocyte recirculation?

22.7 Effector Response at Infection Site

The **effector response** comprises the specific mechanisms that activated lymphocytes use to help eliminate the antigen. Each type of lymphocyte has a unique function. Helper T-lymphocytes release IL-2, IL-4, and other cytokines that regulate (or stimulate) cells of both adaptive and innate immunity. Cytotoxic T-lymphocytes destroy unhealthy cells by apoptosis. Plasma cells (differentiated B-lymphocytes) produce antibodies (see figure 22.8).

22.7a Effector Response of T-Lymphocytes

LEARNING OBJECTIVES

1. Explain the effector response of helper T-lymphocytes.
2. Explain how an unhealthy cell is destroyed by cytotoxic T-lymphocytes.
3. Explain why the processes of T-lymphocytes are collectively called the cell-mediated branch of adaptive immunity.

Just as with activation, the effector response of helper T-lymphocytes and cytotoxic T-lymphocytes differs, as is reflected in their names.

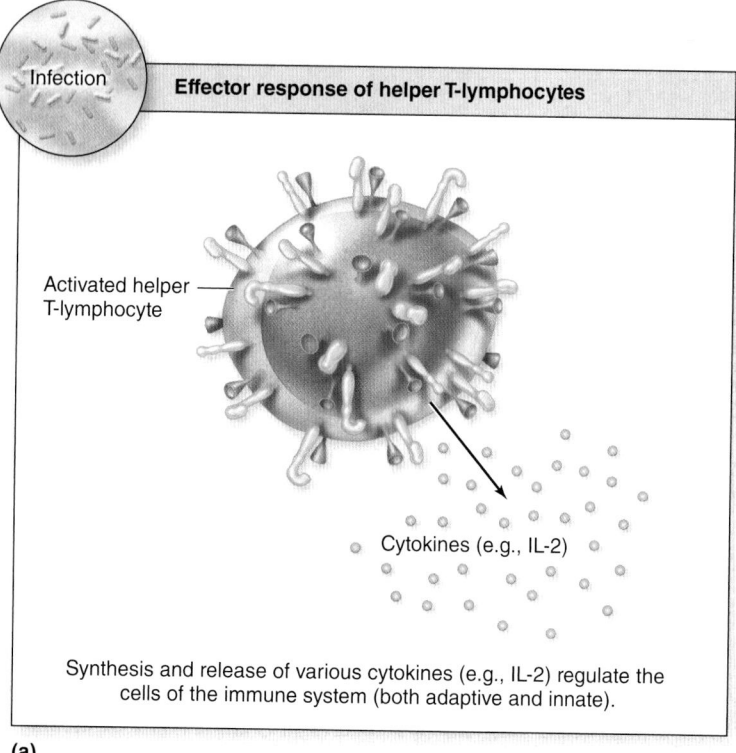

Effector response of helper T-lymphocytes

Infection

Activated helper T-lymphocyte

Cytokines (e.g., IL-2)

Synthesis and release of various cytokines (e.g., IL-2) regulate the cells of the immune system (both adaptive and innate).

(a)

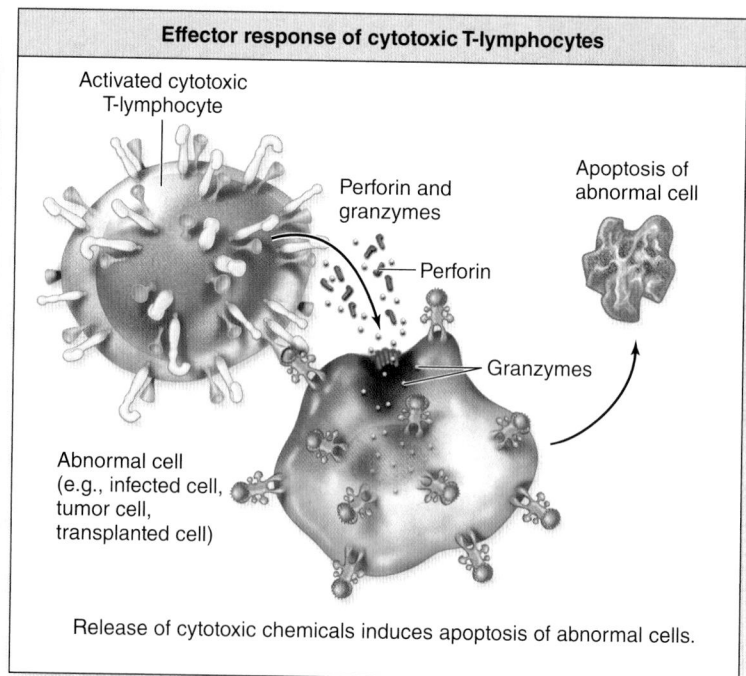

Effector response of cytotoxic T-lymphocytes

Activated cytotoxic T-lymphocyte

Perforin and granzymes

Apoptosis of abnormal cell

Perforin

Granzymes

Abnormal cell (e.g., infected cell, tumor cell, transplanted cell)

Release of cytotoxic chemicals induces apoptosis of abnormal cells.

(b)

Figure 22.17 Effector Response of T-Lymphocytes. The effector response of helper T-lymphocytes and cytotoxic T-lymphocytes occurs at the site of infection. (*a*) Helper T-lymphocytes release various cytokines, and (*b*) cytotoxic T-lymphocytes destroy abnormal cells through the release of the cytotoxic chemicals perforin and granzymes, which induces apoptosis.

INTEGRATE

LEARNING STRATEGY ✎

The "military" contribution of the different lymphocytes of adaptive immunity can be thought of as follows:

- **Helper T-lymphocytes** are generals of the army; they recruit and regulate other immune cells.
- **Cytotoxic T-lymphocytes** are elite military forces; they engage in "cell-to-cell" combat against a specific foe.
- **B-lymphocytes** are the elite military forces that release "munitions" (antibodies) that are typically released from a distance.

Effector Response of Helper T-Lymphocytes

Activated and memory helper T-lymphocytes leave the secondary lymphatic structure after several days of exposure to antigen. They migrate to the site of infection, where they continue to release the cytokines to regulate other immune cells (**figure 22.17a**).

Although helper T-lymphocytes were named based on their function in helping activate B-lymphocytes, their contributions are much more encompassing. Helper T-lymphocytes activate cytotoxic T-lymphocytes, as described previously, through the release of cytokines (e.g., IL-2); they also enhance formation and activity of cells of the innate immune system, including macrophages and NK cells. Thus, healthy helper T-lymphocytes play a central role in a normal functioning immune system (see Clinical View: "HIV and AIDS" in section 22.9c).

💡 **WHAT DO YOU THINK?**

3 HIV (the virus that causes AIDS) specifically targets the helper T-lymphocytes, causing the destruction of these cells. Given the role of helper T-lymphocytes, why does this disease increase a person's susceptibility to infectious diseases?

Effector Response of Cytotoxic T-Lymphocytes

Like helper T-lymphocytes, activated and memory cytotoxic T-lymphocytes also leave the secondary lymphatic structure after several days and migrate to the site of infection in the body's tissue. Cytotoxic T-lymphocytes destroy infected cells that display the antigen. The effector response of cytotoxic T-lymphocytes is initiated when physical contact is made between a cytotoxic T-lymphocyte and the specific foreign antigen displayed on an unhealthy or foreign *cell* (e.g., a virus-infected cell, bacteria-infected cell, tumor cell, or foreign transplanted cell) (figure 22.17b).

If the cytotoxic T-lymphocyte recognizes the antigen presented by the infected cell (with MHC class I molecules), it destroys the cell by releasing granules containing the cytotoxic chemicals perforin and granzymes (the same substances released from NK cells described in section 22.3b). Perforin forms a channel in the target plasma membrane that increases the cell's permeability; granzymes enter the cell through the perforin channels. Granzymes induce cell death by apoptosis, which helps to limit spread of the infectious agent. It is because the immune response of T-lymphocytes is effective against antigens associated with cells that it is referred to as cell-mediated immunity.

💡 **WHAT DID YOU LEARN?**

26 Are cells of both innate immunity and adaptive immunity regulated by cytokines released by helper T-lymphocytes?

27 Cell-mediated immunity is specifically effective against what type of targets? Provide examples.

22.7b Effector Response of B-Lymphocytes

✔ **LEARNING OBJECTIVES**

4. Describe the function of plasma cells in the effector response of B-lymphocytes.

5. Define antibody titer.

Antibodies are the effectors of humoral immunity. Antibodies are formed primarily by plasma cells (although limited amounts are produced by B-lymphocytes). Plasma cells typically remain in the lymph nodes, continuing to synthesize and release antibodies. Antibodies circulate throughout the body in the lymph and blood, ultimately coming in contact with antigen at the site of infection. Plasma cells, over their life span of about 5 days, produce hundreds of millions of antibodies against the specific antigen.

The circulating blood concentration of antibody against a specific antigen is referred to as the **antibody titer.** This can be a measure of immune response. The details of antibody structure and function are described next.

💡 **WHAT DID YOU LEARN?**

28 What is the specific role of plasma cells?

22.8 Immunoglobulins

An **antibody** is an **immunoglobulin** (im′ū-nō-glob′ū-lin) (**Ig**) protein produced against a particular antigen (**figure 22.18**). The structure of antibodies reflects their ability to target specific antigens that they may encounter. Antibodies do not destroy pathogens directly but facilitate the destruction by other immune cells. You may find it helpful to think of the function of antibodies as "tagging" a specific antigen so that it can be eliminated. Here we consider both the structure of immunoglobulins and their actions.

22.8a Structure of Immunoglobulins

✔ **LEARNING OBJECTIVE**

1. Describe the general structure of an immunoglobulin molecule, including its two functional regions.

An immunoglobulin molecule is a Y-shaped, soluble protein composed of four polypeptide chains: two identical heavy chains and two identical light chains, with flexibility at the hinge region of the two heavy chains. These four polypeptide chains are held together by disulfide bonds to form an **antibody monomer** (i.e., a single Y-shaped protein). Two important functional regions of the antibody monomer are the variable regions and the constant region.

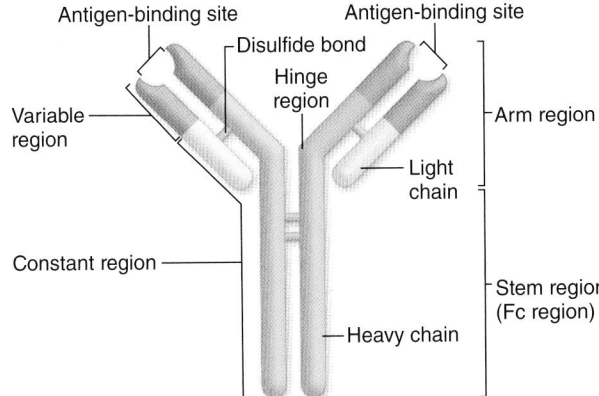

Figure 22.18 Antibody Structure. The structure of an antibody is a Y-shaped protein that includes two variable regions that bind antigen and one constant region that determines its biological activity.

INTEGRATE

CONCEPT CONNECTION

Recall from section 18.2a that most plasma proteins, including albumin, clotting proteins, and alpha- and beta-globulins, are produced by the liver. In contrast, antibodies (immunoglobulins), which are the most significant gamma-globulins, are produced by plasma cells.

Variable Regions

The **variable regions** located at the ends of the "arms" of the antibody contain the antigen-binding site, which attach to a specific antigenic determinant of an antigen. Most antibodies have two antigen-binding sites, which allow each antibody to bind to two antigenic determinants. The variable region binds the antigen through weak intermolecular forces, including hydrogen bonds, ionic bonds, and hydrophobic interactions (see section 2.3d).

Constant Region

The **constant region** contains the **Fc region**, which is the portion of the antibody that determines the biological functions of the antibody. The constant region is the same or nearly the same in structure for antibody molecules of a given class; there are five major classes of immunoglobulins: IgG, IgM, IgA, IgD, and IgE. These classes are described in greater detail in section 22.8c.

WHAT DID YOU LEARN?

29 What is the significance of the variable regions of an immunoglobulin molecule?

22.8b Actions of Antibodies

LEARNING OBJECTIVE

2. List the functions of the antigen-binding site and Fc region of antibodies, and briefly describe how each occurs.

Antibodies are effective against antigens through the binding of the antigen-binding site with the antigen to cause the following (**figure 22.19a**):

- **Neutralization.** An antibody physically covers an antigenic determinant of a pathogen to make it ineffective in establishing an infection or causing harm. For example, neutralization occurs when an antibody covers the region of a virus used to bind to a cell receptor, preventing entry of the virus into a cell. A similar process neutralizes toxins.

- **Agglutination.** Antibody cross-links antigens of foreign cells, causing them to agglutinate or "clump." This is especially effective against bacterial cells and mismatched erythrocytes in a blood transfusion (see section 18.3b).

- **Precipitation.** Antibody can cross-link soluble, circulating antigens such as viral particles (not whole cells) to form an *antigen-antibody complex*. These complexes become insoluble and precipitate out of body fluids. The precipitated complexes are then engulfed and eliminated by phagocytic cells such as macrophages.

The Fc region of the antibody projects externally after the variable region of the antibody binds to the antigen. The exposed Fc region may participate in several important actions including the following (figure 22.19b):

- **Complement fixation.** The Fc region of certain classes of antibodies (IgG and IgM) can bind specific complement proteins to cause activation of complement by the classical pathway. The functions of complement are described in section 22.3c and include opsonization, increasing inflammation, inducing cytolysis, and elimination of immune complexes.

- **Opsonization.** The Fc region of certain classes of antibodies (e.g., IgG) can also cause opsonization (making it more likely that a target is "seen" by phagocytic calls). Phagocytic cells such as neutrophils and macrophages have receptors for the Fc region of certain antibody classes. The phagocytic receptors bind in a "zipperlike" fashion to the Fc region of the antibodies to engulf both the antigen and antibody.

- **Activation of NK cells.** The Fc region of certain classes of antibodies (IgG) bind to specific receptors on NK cells (much like phagocytes). This induces NK cells to destroy abnormal cells by the release of cytotoxic chemicals that cause apoptosis of the cell. This process is called **antibody-dependent cell-mediated cytotoxicity (ADCC).**

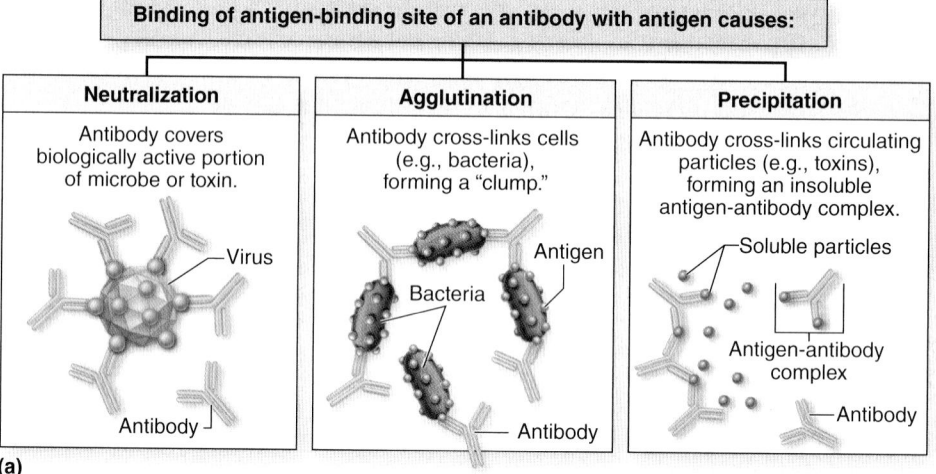

(a)

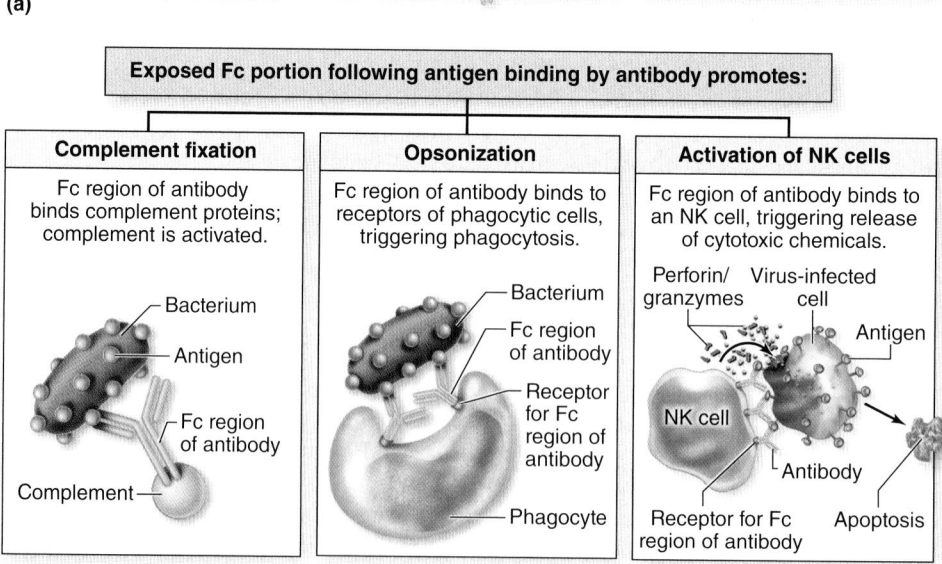

(b)

Figure 22.19 Antibody Functions. Antibody (Ab) functions to "tie up" the antigen (Ag) until it can be eliminated. (*a*) Three functions of antibodies involve the binding of the antigen-binding site to an antigen to cause neutralization, agglutination, or precipitation. (*b*) Three other functions first require the binding of antibody to an antigen; the Fc region of the antibody projects externally. The Fc region can then bind complement, bind to phagocytic cells to enhance phagocytosis of the unwanted cell, or bind to NK cells to trigger apoptosis of the unwanted cell.

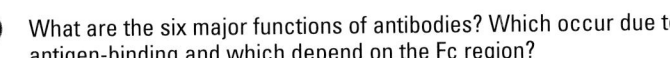

INTEGRATE

LEARNING STRATEGY

Generally, the role of antibodies as weapons is to "tie up the prisoner" until other help arrives. You can remember the six functions of an antibody with the acronyms NAP and CON: Neutralization, Agglutination, and Precipitation (NAP), as well as Complement, Opsonization, and NK cells (CON). Remember—a NAP can help you CONcentrate.

The antibodies immobilize specific antigens and ultimately cause their elimination by other immune cells. Antibodies are especially effective in binding to viral particles, bacteria, toxins, and yeast spores. It is because the immune response of B-lymphocytes is effective against soluble antigens (antigens dissolved in the body's "humors") that it is referred to as humoral immunity.

WHAT DID YOU LEARN?

30 What are the six major functions of antibodies? Which occur due to antigen-binding and which depend on the Fc region?

22.8c Classes of Immunoglobulins

LEARNING OBJECTIVE

3. Describe the structure, location, and specific function of the five major classes of immunoglobulins.

The five classes of immunoglobulins are IgG, IgM, IgA, IgD, and IgE. These may be remembered with the acronym G-MADE. Each class of immunoglobulin is unique in a number of aspects. **Table 22.5** summarizes the characteristics of the five immunoglobulin classes.

IgG is the major class of immunoglobulins. It makes up 75–85% of antibody in the blood and is the predominant antibody in the lymph, cerebrospinal fluid, serous fluid, and peritoneal fluid. IgG can participate in all of the functions previously listed for actions of antibodies including the neutralization of toxins such as snake venom. Additionally, IgG antibodies cross the placenta and can be responsible for hemolytic disease of the newborn (see Clinical View: "Rh Incompatibility and Pregnancy" in section 18.3b).

IgM is normally a pentamer (composed of five monomers) found mostly in the blood. IgM is not as versatile in its biological functions as IgG. For example, IgM is not efficient at virus neutralization. However, IgM is the most effective at causing agglutination of cells and binding complement. Additionally, naturally occurring IgM antibodies are responsible for rejection of mismatched blood transfusions (see Clinical View: "Transfusions" in section 18.3b).

IgA is found in areas exposed to the environment, such as mucosal membranes and tonsils, and it is produced in various secretions, including mucus, saliva, tears, and breast milk. IgA plays a significant role in protecting the respiratory and gastrointestinal tract. In secretions, IgA is a dimer (i.e., composed of two antibody molecules). It helps to prevent pathogens (e.g., viruses, bacteria) from adhering to epithelial tissue and penetrating underlying tissue through the process of neutralization. IgA is especially effective at agglutination.

IgD (along with a monomer form of IgM) functions as the antigen-specific B-lymphocyte receptor. It also functions to identify when immature B-lymphocytes may be ready for activation to participate in adaptive immunity.

IgE has a very low rate of synthesis and is generally formed in response to allergic reactions and to parasitic infections. IgE causes release of histamine and other mediators of inflammation from basophils and mast cells, and it attracts eosinophils. The formation of IgE and its response to allergens is described in detail (see Clinical View: "Hypersensitivities" in section 22.9c).

Class Switching

Each plasma cell has the potential to produce different classes of antibodies. This process of changing the class of antibody produced from IgM to IgG, IgE, and IgA is called *class switching*. Direct contact between the helper T-lymphocytes (e.g., CD40 surface protein) and plasma calls (e.g., CD154) is required, along with the release of various cytokines from the helper T-lymphocytes. Specific cytokines determine the type of antibody class formed.

Figure 22.20 is a visual summary of adaptive immunity. Adaptive immunity is considered the third line of defense because, although its response is initiated immediately, on the first exposure to an antigen there is a lag time that occurs from the time of exposure to the development of the immune response. This process may take several days.

WHAT DID YOU LEARN?

31 Which subclass of antibodies is most prevalent? Which specific functions of antibodies can it engage in?

Table 22.5 Major Classes of Immunoglobulins

Characteristic	IgG	IgM	IgA	IgD	IgE
Primary Locations	Body fluids including blood, lymph, cerebrospinal fluid, serous fluid, peritoneal fluid, breast milk	Blood (monomer is B-lymphocyte receptor); breast milk	External secretions (mucus, saliva, tears, breast milk, and colostrum)	B-lymphocyte receptor	Blood
Actions	Neutralization (viruses, bacteria, toxins); agglutination; precipitation; complement activation; opsonization; natural killer cell activation (antibody-dependent cell-mediated cytotoxicity)	Neutralization, agglutination (great potential), complement binding (great potential)	Neutralization, agglutination (great potential)	BCR (B-cell receptor)	Produced during allergic reactions or as a result of a parasitic infection; activation of mast cells and basophils
Half-Life (Days) in Blood	23	5	5.5	2.8	2.0
Unique Characteristics	Used for passive immunity; crosses placenta	First produced antibody; only antibody produced in fetus	Associated with mucosal membranes; helps protect against local respiratory or gastrointestinal (GI) infections	Identifies when immature B-lymphocytes may be ready for activation	Causes release of products from basophils and mast cells; attracts eosinophils

INTEGRATE CONCEPT OVERVIEW

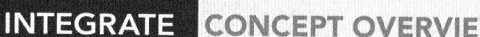

Figure 22.20 Adaptive Immunity. Adaptive immunity is considered the third line of defense. It is the specific means by which lymphocytes defend the body against specific antigens. Adaptive immunity requires three events in the life of lymphocytes: (*a*) formation of lymphocytes within primary lymphatic structures, (*b*) activation and clonal selection of lymphocytes within secondary lymphatic structures, and (*c*) effector response of T-lymphocytes and antibodies at the site of infection.

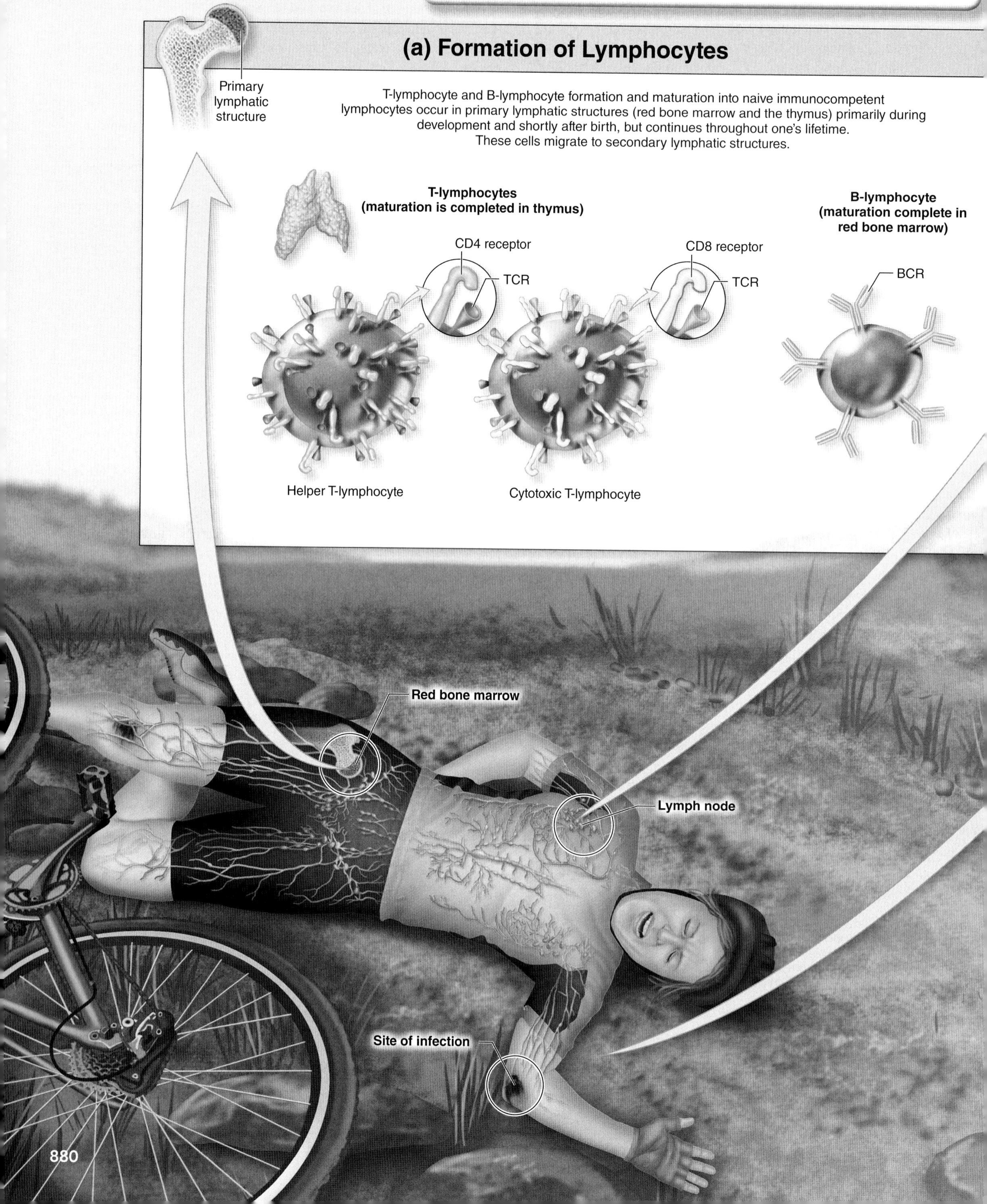

(a) Formation of Lymphocytes
Primary lymphatic structure
T-lymphocyte and B-lymphocyte formation and maturation into naive immunocompetent lymphocytes occur in primary lymphatic structures (red bone marrow and the thymus) primarily during development and shortly after birth, but continues throughout one's lifetime. These cells migrate to secondary lymphatic structures.
T-lymphocytes (maturation is completed in thymus)
B-lymphocyte (maturation complete in red bone marrow)
CD4 receptor
TCR
CD8 receptor
TCR
BCR
Helper T-lymphocyte
Cytotoxic T-lymphocyte
Red bone marrow
Lymph node
Site of infection

(b) Activation and Clonal Selection of Lymphocytes

The first exposure to antigen (the antigen challenge) typically occurs in secondary lymphatic structures (e.g., lymph nodes, spleen, tonsils, MALT). Clones of activated and memory helper and cytotoxic T-lymphocytes, and plasma cells and memory B-lymphocytes are formed.

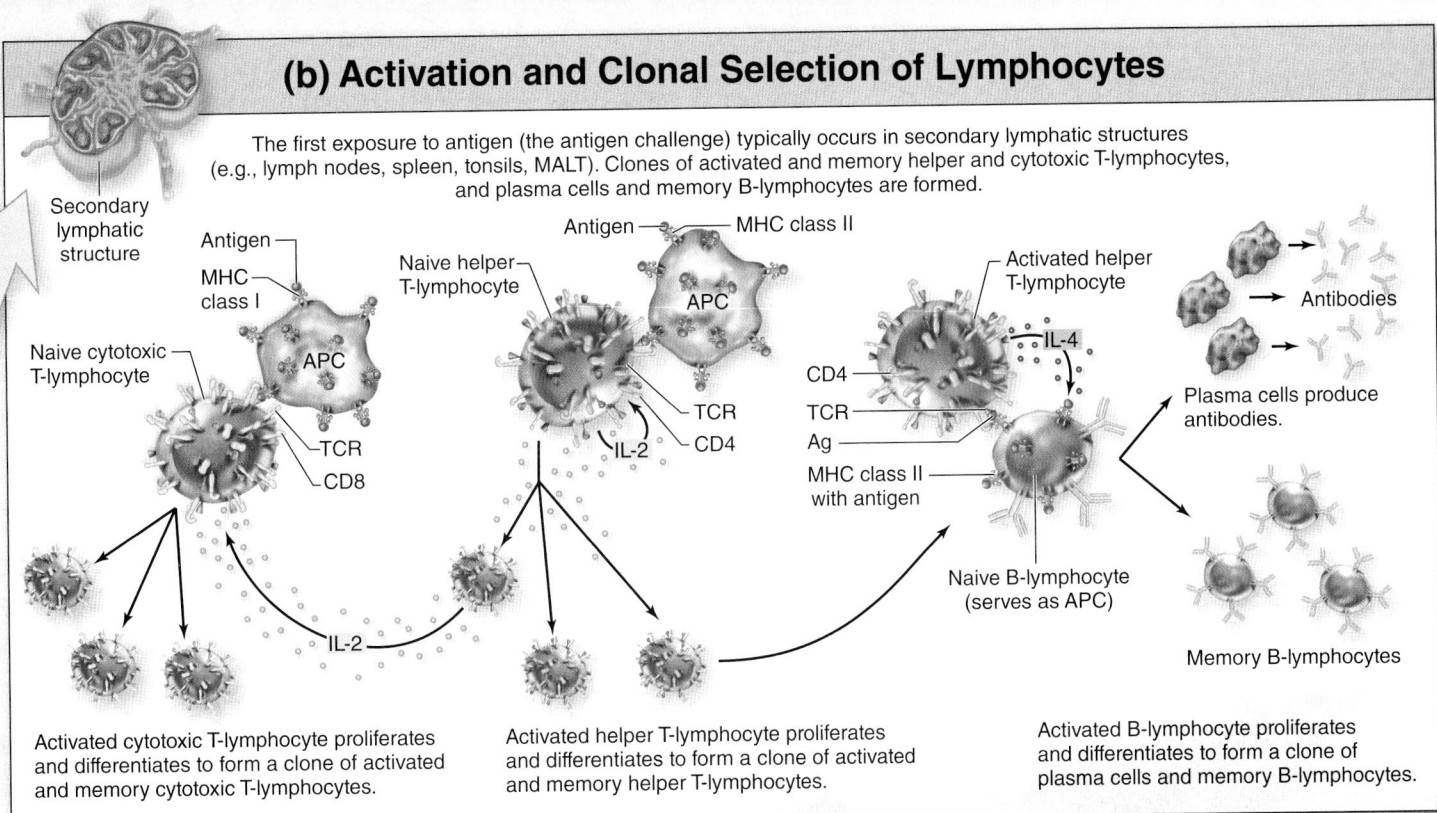

Secondary lymphatic structure

Naive cytotoxic T-lymphocyte

Antigen
MHC class I
APC
TCR
CD8

Naive helper T-lymphocyte
Antigen — MHC class II
APC
TCR
CD4
IL-2

Activated helper T-lymphocyte
CD4
TCR
Ag
MHC class II with antigen
IL-4

Antibodies
Plasma cells produce antibodies.

Naive B-lymphocyte (serves as APC)

Memory B-lymphocytes

IL-2

Activated cytotoxic T-lymphocyte proliferates and differentiates to form a clone of activated and memory cytotoxic T-lymphocytes.

Activated helper T-lymphocyte proliferates and differentiates to form a clone of activated and memory helper T-lymphocytes.

Activated B-lymphocyte proliferates and differentiates to form a clone of plasma cells and memory B-lymphocytes.

(c) Effector Response

Site of infection

CELL-MEDIATED IMMUNITY

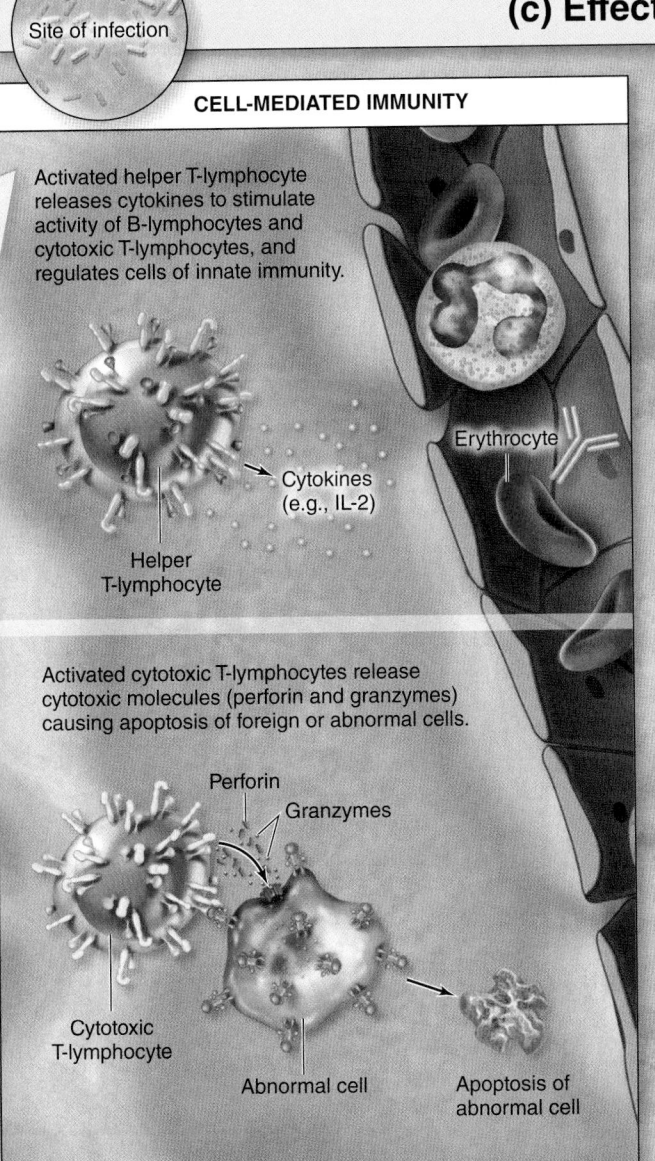

Activated helper T-lymphocyte releases cytokines to stimulate activity of B-lymphocytes and cytotoxic T-lymphocytes, and regulates cells of innate immunity.

Erythrocyte

Cytokines (e.g., IL-2)

Helper T-lymphocyte

Activated cytotoxic T-lymphocytes release cytotoxic molecules (perforin and granzymes) causing apoptosis of foreign or abnormal cells.

Perforin
Granzymes

Cytotoxic T-lymphocyte

Abnormal cell

Apoptosis of abnormal cell

HUMORAL IMMUNITY

Fab region of antibody binds to antigen to cause several consequences including neutralization of microbial cells (e.g., bacteria) and particles (e.g., virus, toxins); agglutination of cells; and precipitation of particles (e.g., toxins).

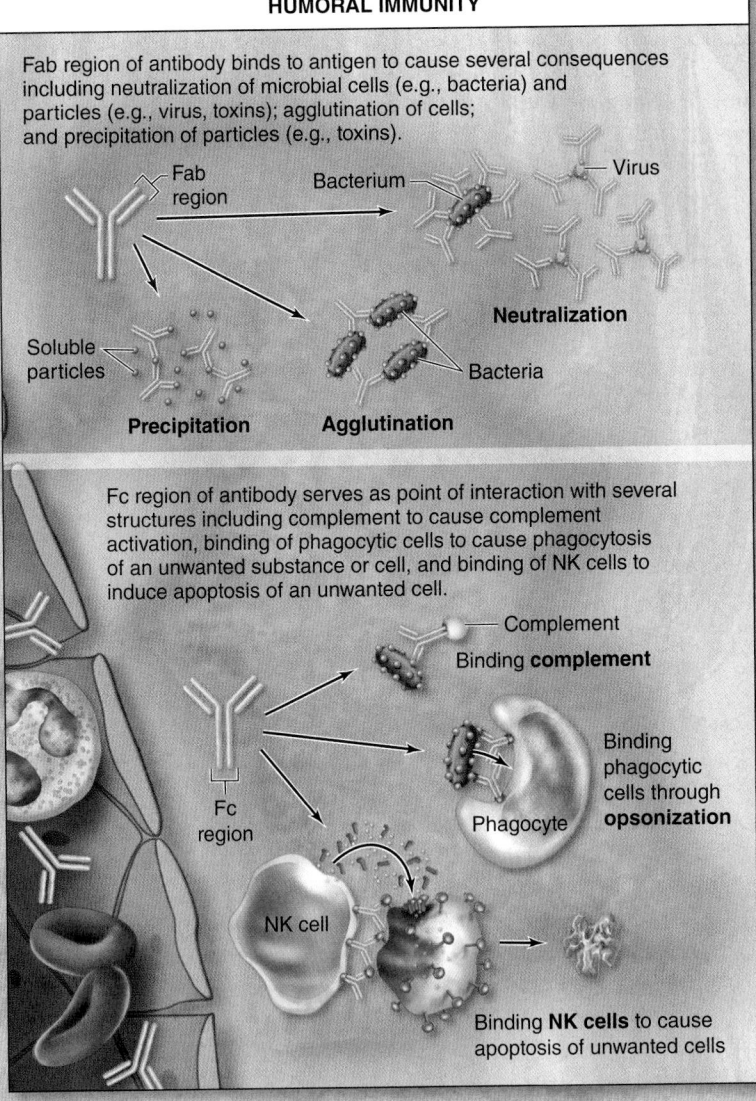

Fab region
Bacterium
Virus

Neutralization

Soluble particles
Bacteria

Precipitation
Agglutination

Fc region of antibody serves as point of interaction with several structures including complement to cause complement activation, binding of phagocytic cells to cause phagocytosis of an unwanted substance or cell, and binding of NK cells to induce apoptosis of an unwanted cell.

Complement
Binding **complement**

Binding phagocytic cells through **opsonization**

Fc region
Phagocyte

NK cell

Binding **NK cells** to cause apoptosis of unwanted cells

22.9 Immunologic Memory and Immunity

One of the central features of adaptive immunity is the development of "memory" that provides immunity against a specific antigen. Here we discuss immunologic memory and how immunity is obtained through both active and passive means.

22.9a Immunologic Memory

 LEARNING OBJECTIVE

1. Define immunologic memory and explain how it occurs.

Activation of adaptive immunity requires direct physical contact between a lymphocyte and an antigen. On the first exposure to an antigen (the antigen challenge), limited numbers of helper T-lymphocytes, cytotoxic T-lymphocytes, and B-lymphocytes recognize the antigen (about 1 in 100,000 to 1,000,000). Generally a lag time occurs between the body's initial exposure to the antigen and the physical contact with lymphocytes required to develop an immune response.

The antigen challenge, however, causes the formation of memory cells in response to the activation of T-lymphocytes and B-lymphocytes, as described in earlier sections. These long-lived lymphocytes represent an "army" of thousands against specific antigens and are responsible for immunologic memory.

On subsequent exposures to an antigen, these vast number of memory cells make contact with the antigen more rapidly and produce a more powerful response, which is referred to as the **secondary response,** *memory response,* or *anamnestic* (an′am-nes′tik; *an* = not, *amnesia* = forgetfulness) *response.* On each subsequent exposure to a specific pathogen, the pathogen is typically eliminated even before disease symptoms develop. For example, a person who develops measles will not develop measles again, even if reexposed to that virus. The virus is eliminated by activated memory T-lymphocytes, memory B-lymphocytes, and antibodies before it causes harm. This feature of immunologic memory makes adaptive immunity a highly potent protector. Vaccines are typically effective in developing memory, and a secondary response is seen on exposure to the substance vaccinated against (see Clinical View: "Vaccinations").

 WHAT DID YOU LEARN?

32 Briefly describe immunologic memory, and explain its significance.

22.9b Measure of Immunologic Memory

LEARNING OBJECTIVE

2. Discuss the difference between the primary response and the secondary response to antigen exposure.

Antibody titer (concentration) in blood serum is one measure of immunologic memory. The graphs shown in **figure 22.21** reflect the changes in serum antibody titer (specifically, the amount of IgM and IgG in the blood) over time in response to both the initial exposure (the antigen challenge) and subsequent exposures to an antigen. The *degree* of protection is indicated by levels of circulating IgG.

Initial Exposure and the Primary Response

The initial exposure to a specific antigen can be in the form of an active infection or a vaccine. The measurable response of antibody production to the first exposure is called the **primary response:**

- **Lag or latent phase.** There is initially a period of no detectable antibody in the blood. This period may extend 3 to 6 days. Antigen detection, activation, proliferation, and differentiation of lymphocytes, including development of memory lymphocytes, occur during the lag phase.
- **Production of antibody.** Within 1 to 2 weeks, plasma cells produce IgM and then IgG. Antibody titer levels peak and then generally decrease over time.

Subsequent Exposures and the Secondary Response

Subsequent exposures to an antigen can occur after varying lengths of time following the initial exposure, and the measurable response to subsequent exposure is called the **secondary response:**

INTEGRATE

CLINICAL VIEW
Vaccinations

A **vaccine** is an attenuated (weakened) or dead microorganism, or some component of the microorganism (may be bioengineered) that is administered through one of several routes: oral, intradermal, intravenous (IV), intraperitoneal, or intranasal.

The function of a vaccine is to stimulate the immune system to develop memory B-lymphocytes (predominantly) while providing a relatively safe means for the initial exposure to a microorganism. The risk is relatively low because the microorganism (or its components) has no ability (or limited ability) to establish an infection. If an individual is later exposed to the same antigens, the secondary response is triggered and it will be swift and powerful; the individual is generally not even aware of contact with the microbe.

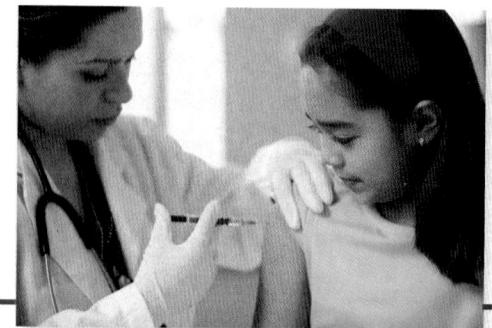

However, a vaccine is different from having an active infection in these ways:

- The immune response to a vaccine is predominantly from the humoral-mediated branch (e.g., oral polio vaccine). The reason is that B-lymphocytes bind unattached microbes to induce humoral-mediated immunity, but few, if any, of the weakened or dead microbes in the vaccine can infect cells to induce T-lymphocytes to establish cell-mediated immunity. (The B-lymphocytes present the antigen to T-lymphocytes in some instances.)
- Depending upon the life span of the particular memory B-lymphocytes, the vaccine may provide lifelong immunity, or periodic **booster shots** may be needed to ensure continued protection against the antigen. For example, booster shots are needed every 10 years for protection against tetanus.

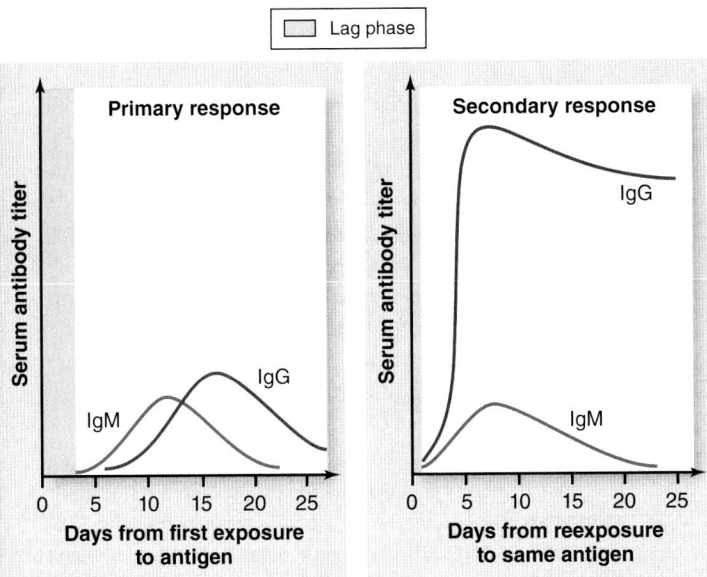

Primary response

Serum antibody titer

IgM

IgG

0   5   10 15 20 25

Days from first exposure to antigen

Secondary response

Serum antibody titer

IgG

IgM

0   5   10 15 20 25

Days from reexposure to same antigen

Figure 22.21 Primary and Secondary (Anamnestic) Response in Humoral Immunity. Graph of the primary response that shows the level of IgM and IgG antibody produced by plasma cells on the first exposure to a given antigen. Graph of the secondary response that depicts the level of these antibodies on all subsequent exposures to the same antigen.

- **Lag or latent phase.** A much shorter lag phase occurs with subsequent exposures to the same antigen. This difference is due to the presence of memory lymphocytes.
- **Production of antibody.** Antibody levels rise more rapidly, with a greater proportion of the IgG class of antibodies. This higher level of IgG production may continue for longer periods, perhaps even years.

WHAT DID YOU LEARN?

33 How does the secondary response differ from the primary response? What is the advantage of the secondary response?

22.9c Active and Passive Immunity

LEARNING OBJECTIVES

3. Define active immunity and passive immunity.

4. Describe how both active and passive immunity can be acquired naturally and artificially.

Immunity can be active or passive **(figure 22.22)**. **Active immunity** results from a direct encounter with a pathogen or foreign substance that results in the production of memory cells and can be obtained either naturally or artificially. *Naturally* acquired active immunity occurs when an individual is directly exposed to the antigen of an infectious agent. *Artificially* acquired active immunity takes place when the exposure occurs through a vaccine. In both cases, memory cells against that specific antigen are formed.

In contrast to active immunity, **passive immunity** is obtained from another individual or animal, and it can also be obtained naturally or artificially. Naturally acquired passive immunity occurs from the transfer of antibodies from the mother to the fetus across the placenta (IgG) or to the baby in the mother's breast milk (IgA, IgM, and IgG). In contrast, when serum containing antibodies against a specific antigen is transferred from one individual to another, this process is referred to as artificially acquired passive immunity. For example, serum containing antibodies against the toxins associated with tetanus and botulism, can be transferred to an individual who is at risk from one of these toxins. Antibodies to a poisonous snake venom (antivenom) can also be transferred to an individual who has been bitten by that species of snake. The antibodies neutralize the toxin or venom to prevent it from doing harm until the body is able to eliminate it.

In both types of passive immunity, the individual has not had an antigenic challenge and has not produced memory cells. Passive immunity lasts only as long as the antibody proteins are present in the individual. For example, the half-life of IgG in the blood is 23 days; IgM is 5 days (table 22.5).

WHAT DID YOU LEARN?

34 Which type of immunity—active or passive—results in the production of memory cells and generally provides lifelong protection from that antigen?

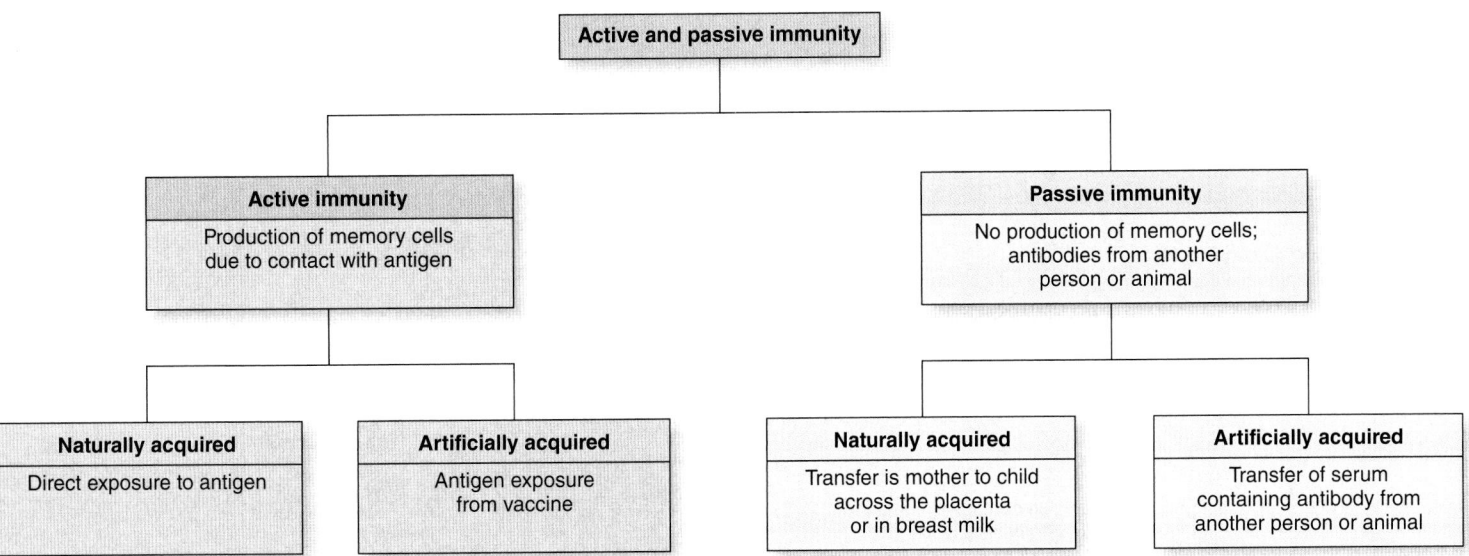

Figure 22.22 Active and Passive Immunity. Active and passive immunity are distinguished by the production of memory cells. Memory cells are produced in active immunity, whereas in passive immunity they are not.

CLINICAL VIEW
Hypersensitivities

Hypersensitivity is an abnormal and exaggerated response of the immune system to an antigen. The various types of hypersensitivities are categorized based on the amount of time required for the immune response to occur following exposure to an antigen: **Acute hypersensitivities** occur within seconds; **subacute hypersensitivities** within 1 to 3 hours, and **delayed hypersensitivities** occur within 1 to 3 days. The difference in time of onset reflects the components of the immune system that are involved. Both acute and subacute hypersensitivity reactions involve humoral immunity. Immunoglobulin E (IgE) antibodies are involved in acute hypersensitivities, whereas subacute hypersensitivity reactions are triggered by immunoglobulin G (IgG) and immunoglobulin M (IgM). In contrast, delayed hypersensitivity reactions involve cell-mediated immunity. We limit our discussion to acute hypersensitivities.

Acute Hypersensitivity (Allergies)

An acute hypersensitivity is more commonly referred to as an **allergy** (or *type I hypersensitivity*), and it is an overreaction of the immune system to a noninfectious substance, or **allergen.** Examples of different allergens include pollen, latex, peanuts, and bee venom. Following exposure to the allergen, an allergic reaction occurs within seconds and continues for about a half hour.

There are three major phases associated with an allergic reaction:

1. **Sensitization phase.** An individual is exposed to an allergen. The allergen is engulfed by an antigen-presenting cell (APC) and presented to a helper T-lymphocyte (not shown in figure). The helper T-lymphocytes release cytokines that cause the B-lymphocytes to mature into plasma cells that produce IgE antibodies against the allergen. The IgE antibodies bind to basophils and mast cells (by the Fc region of the antibody), and may remain bound to these cells for several weeks or longer.

2. **Activation phase.** If the individual is reexposed to the same allergen, the allergen binds to the IgE antibodies that are bound to the basophils and mast cells, cross-linking the receptors.

3. **Effector phase.** The mast cells or basophils release chemicals (histamine, heparin, and eicosanoids) that cause an inflammatory response. The inflammatory response is responsible for the symptoms associated with allergies. The symptoms an individual experiences depend upon where the inflammatory response occurs in the body:

- Contact with the mucous membranes of the nasal passage and conjunctiva of the eye result in a runny nose and watery eyes (**allergic rhinitis**, or "*hay fever*"). This is the most common site with approximately 20% of the general population experiencing allergic rhinitis.
- Exposure of the skin surface can result in red welts and itchy skin (**hives**).
- Entrance into the respiratory passageway causes bronchoconstriction and increases secretion of mucus, causing labored breathing and coughing (**allergic asthma**).
- Entrance into the digestive tract causes increased fluid secretions and peristalsis that may cause vomiting and diarrhea (not shown in figure).
- Circulation in the blood through a bee sting or injection by needle causes systemic vasodilation and inflammation. In extreme cases, extensive loss of fluid from the blood into the interstitial space results in a marked decrease in blood volume and blood pressure. Consequently, the individual may have insufficient blood pressure to maintain adequate perfusion (**anaphylactic shock**).

Acute hypersensitivity reactions involve the IgE antibody. The three stages of acute hypersensitivity include the sensitization phase, activation phase, and effector phase.

Allergies (or Type I hypersensitivity)

Initial exposure to allergen

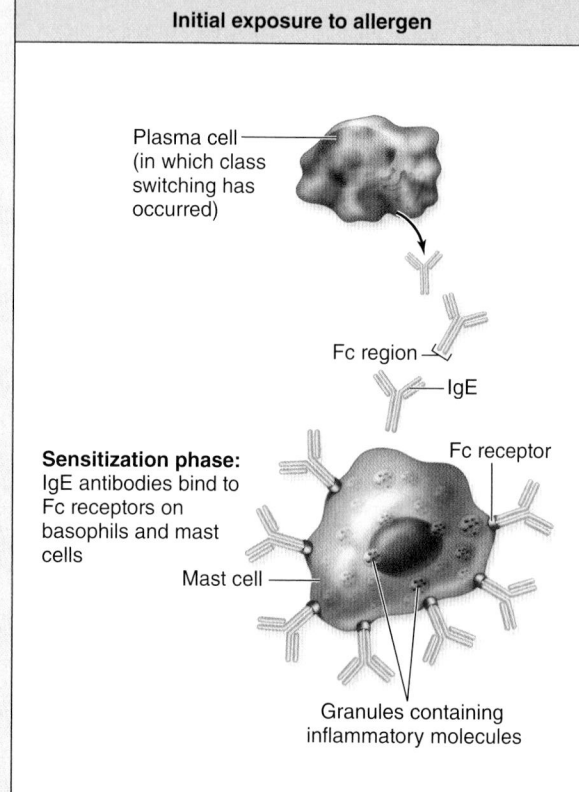

Plasma cell (in which class switching has occurred)

Fc region —

IgE

Sensitization phase:
IgE antibodies bind to Fc receptors on basophils and mast cells

Fc receptor

Mast cell —

Granules containing inflammatory molecules

Reexposure to same allergen

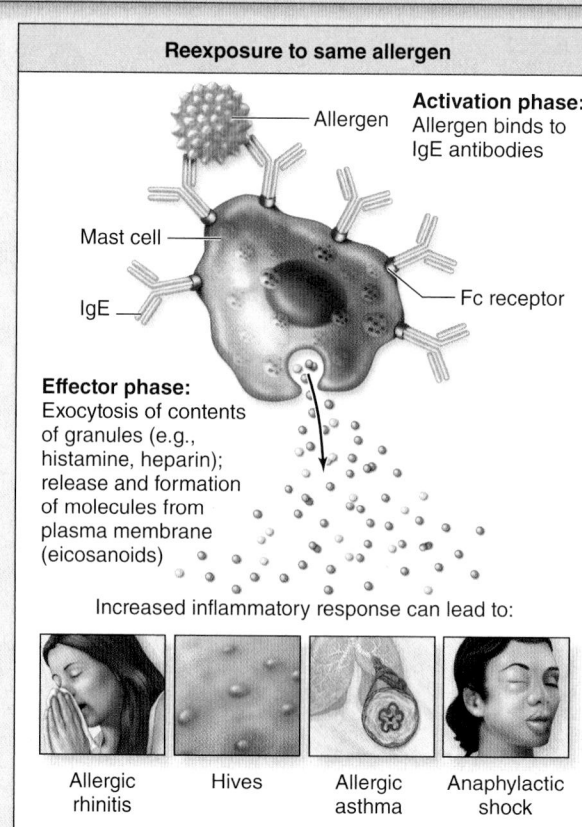

Allergen

Activation phase:
Allergen binds to IgE antibodies

Mast cell —

IgE —

Fc receptor

Effector phase:
Exocytosis of contents of granules (e.g., histamine, heparin); release and formation of molecules from plasma membrane (eicosanoids)

Increased inflammatory response can lead to:

| Allergic rhinitis | Hives | Allergic asthma | Anaphylactic shock |

INTEGRATE

CLINICAL VIEW
HIV and AIDS

AIDS (acquired immunodeficiency syndrome) is a life-threatening condition that is the result of the **human immunodeficiency virus (HIV).** An HIV infection targets the immune system—in particular, the helper T-lymphocyte (CD4+ T-cell). HIV infects and destroys these helper T-lymphocytes over a period of time (months to years). Prolonged HIV infection leads to the devastating effects of AIDS.

Process of the human immunodeficiency virus (HIV) infecting a cell.

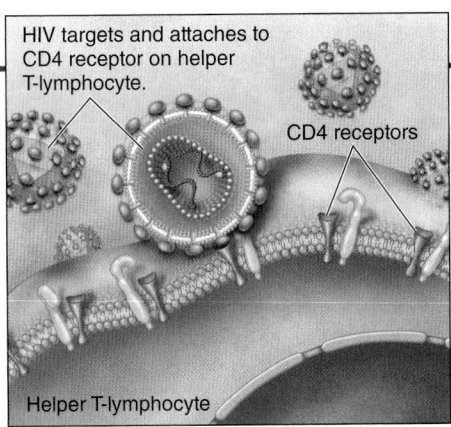

Epidemiology

HIV resides in the body fluids of infected individuals, including their blood, semen, vaginal secretions, or even breast milk. HIV can be transmitted in the following ways: during unprotected sexual (vaginal or anal) intercourse, sharing hypodermic needles with other intravenous drug users, during development (because HIV crosses the placenta), or breastfeeding an infant. Current evidence indicates HIV is *not* spread by casual kissing, sharing eating utensils, using a public toilet, or casual physical contact. First seen in the United States in the early 1980s among the homosexual male population, it is now a prominent disease among heterosexual populations. The United Nations program on AIDS (UNAIDS) estimates that 90% of all HIV infections are currently transmitted heterosexually. Prior to 1985, before HIV and AIDS were well known, HIV could be transmitted through the donated blood supply. Individuals who received blood transfusions sometimes received HIV-infected blood, thereby infecting them as well. This discovery has led to more stringent screening of blood donors.

Since the early 1980s, over 60 million people have become infected with HIV, and over 27 million people have died. The incidence of AIDS is increasing throughout the world, but the disease is particularly rampant on the continents of Africa and Asia. The AIDS epidemic in Africa has led to massive numbers of deaths, and children are frequently orphaned as both parents succumb to the disease. Asian countries are seeing a surge of new HIV and AIDS cases, especially among sex workers and their clients.

Prevention

The key to limiting the spread of HIV infection is to refrain from behaviors that allow the virus's transmission. Unprotected sexual intercourse (especially anal intercourse) can spread HIV, so individuals should either practice abstinence or protect themselves with the use of condoms. HIV can also be spread through oral sex. (Other contraceptives like birth control pills do *not* protect an individual from HIV infection.) Both partners in a monogamous relationship should be tested for presence of the HIV virus (done via a simple blood test) before engaging in sexual intercourse. Intravenous drug users should not share needles. Health-care workers should wear gloves and be careful around patient bodily fluids. HIV-infected pregnant women need special prenatal care to help prevent the transmission of the virus to their fetus. In addition, HIV-infected mothers are discouraged from breastfeeding their infants, as the virus is present in breast milk.

HIV cannot survive for long periods of time outside the human body. The virus can be eliminated from medical equipment or personal care items by cleaning them with a common disinfectant (such as bleach or hydrogen peroxide) or by heating them to temperatures above 135°F.

How HIV Causes Its Damage

Helper T-lymphocytes are destroyed by HIV infection. This destruction occurs in several ways. Some helper T-lymphocytes are programmed to produce HIV RNA at such a fast rate that the cell undergoes lysis, or bursts. Other helper T-lymphocytes are targeted and destroyed by other immune cells, such as macrophages or cytotoxic T-lymphocytes. Over a period of months to years, the helper T-lymphocyte population declines to a dangerously low level. Helper

T-lymphocytes initiate and oversee the body's immune response; therefore, a decrease in helper T-lymphocytes results in a loss of normal immune function.

Early Symptoms

Approximately several weeks to several months after initial HIV infection, many individuals will experience flulike symptoms, including sore throat, fever, fatigue, headache, and swollen lymph nodes. Some people also may experience night sweats as well, whereas still others may be completely asymptomatic. Often, these symptoms disappear after a few weeks when the body's other immune cells target and destroy HIV infected cells. Healthy helper T-lymphocytes divide to replace those cells that were destroyed. However, HIV continues to replicate at a faster rate than the immune system can rid itself of infected cells; in addition, the virus mutates to avoid detection. Over a period of years, the helper T-lymphocyte population drops to very low numbers, setting the stage for AIDS.

What Do HIV Blood Tests Look For?

HIV blood tests look for the evidence of HIV antibodies in the blood. These antibodies are produced by plasma cells about a month after initial infection. These antibodies indicate the body is responding to HIV infection. It can take up to 6 months for antibody levels in the blood to rise to a point where they can be detected by the blood test. Thus, individuals who have been exposed to HIV but get tested within the first 6 months may receive false negative results and are still at risk for infecting others.

When Does HIV Become AIDS?

HIV infection is diagnosed as AIDS when either a person's helper T-lymphocyte count drops to below 200 cells per cubic milliliter (in comparison to 800 to 1200 cells per cubic millimeter for a healthy individual) or a person develops an opportunistic infection or illness.

Opportunistic infections are those that thrive due to the compromised immune system. Some examples of opportunistic illness include protozoan infections (e.g., toxoplasmosis and pneumonia caused by *Pneumocystis jiroveci*); fungal infections (e.g., candidiasis, histoplasmosis), some bacterial infections; and neoplasms (cancers; e.g., Kaposi sarcoma, aggressive non-Hodgkin lymphoma, and cervical cancer). Opportunistic infections account for up to 80% of all AIDS-related deaths. Additionally, many AIDS patients have some form of central nervous system (CNS) complications, including meningitis, encephalitis, neurologic deficits, and neuropathies.

Treatment Options

There is no cure for HIV, so HIV infection is a lifelong illness. Current pharmaceutical treatments are "cocktails" of multiple drugs that alleviate symptoms or help prevent the spread of HIV in the body, but they cannot eradicate HIV from an infected individual. Most of these drugs also have unpleasant side effects.

Unfortunately, HIV drugs are expensive and not widely available in developing countries, where their need is greatest. One hopeful sign is that pharmaceutical companies are negotiating with these governments to make cheaper forms of the drug available. In the meantime, education about preventing HIV infection continues throughout the world.

CHAPTER SUMMARY

- The immune system is a functional system composed of cells, plasma proteins, and other substances that protect the body from harmful agents.

22.1 Overview of Diseases Caused by Infectious Agents 850

- The five major classes of infectious agents are bacteria, viruses, fungi, protozoans, and multicellular parasites.
- Bacteria are composed of prokaryotic cells; viruses are composed of either DNA or RNA in a protein capsid; and fungi, protozoans, and parasites are composed of eukaryotic cells.

22.2 Overview of the Immune System 851

- The immune system is composed of immune cells and cytokines, and it is organized into innate immunity and adaptive immunity.

22.1a Immune Cells and Their Locations 851

- Immune cells circulate in the blood and are also located in body tissues, including lymphatic tissues, select organs, epithelial tissue of the skin and mucosal membranes, and connective tissue throughout the body.

22.2b Cytokines 852

- Cytokines are small, soluble proteins produced by immune cells that function similarly to hormones; cytokines include interleukins, tumor-necrosis factors, colony-stimulating factors, and interferons.

22.2c Comparison of Innate Immunity and Adaptive Immunity 853

- Innate immunity encompasses defenses we are born with and includes barriers to prevent entry and nonspecific internal defenses.
- Adaptive immunity encompasses defenses developed in response to exposure to specific antigens and includes both cell-mediated immunity (T-lymphocytes) and humoral immunity (B-lymphocytes).

22.3 Innate Immunity 854

- The advantages of innate immunity are that the components develop an immediate response against a wide array of potentially harmful substances; however, the responses do not result in memory.

22.3a Preventing Entry 854

- The skin and mucosal membranes provide a physical, chemical, and biological barrier that is usually successful in preventing entry of harmful substances. These are considered the body's first line of defense.

22.3b Cellular Defenses 856

- Cells of innate immunity include neutrophils, macrophages, dendritic cells, basophils and mast cells, NK (natural killer) cells, and eosinophils.

22.3c Antimicrobial Proteins 857

- Antimicrobial proteins include interferon and the complement system.

22.3d Inflammation 858

- Inflammation is an immediate, local, nonspecific response that occurs in vascularized tissue against a great variety of injury-causing stimuli. Inflammation is the major effector response of innate immunity.

22.3e Fever 861

- A fever is an abnormal elevation of body temperature of at least 1°C (1.8°F), and three phases associated with a fever are onset, stadium, and defervescence. Mild fevers are beneficial, whereas high fevers may be detrimental.

22.4 Adaptive Immunity: An Introduction 864

- Adaptive immunity is developed when T-lymphocytes and B-lymphocytes are stimulated by antigen.

22.4a Antigens 864

- An antigen is a substance that binds to a component of adaptive immunity (i.e., T-lymphocytes or antibodies).

22.4b General Structure of Lymphocytes 865

- Helper T-lymphocytes contain TCRs (T-cell receptors) and CD4 proteins, cytotoxic T-lymphocytes contain TCRs and CD8 proteins, and B-lymphocytes contain BCRs (B-cell receptors).
- TCRs bind with presented antigen, and BCRs bind with free antigen (e.g., viral particles).

22.4c Antigen-Presenting Cells and MHC Molecules 866

- Major histocompatibility complex (MHC) molecules, which are plasma membrane proteins, display antigen on a cell's surface so the antigen can be encountered by T-lymphocytes.
- All nucleated cells present antigen with MHC class I molecules, and antigen-presenting cells (APCs) present antigen with both MHC class I and MHC class II molecules.

22.4d Overview of Life Events of Lymphocytes 870

- Three significant events occur in the lifetime of a lymphocyte: (a) formation, which occurs in primary lymphatic structures; (b) activation, which occurs within secondary lymphatic structures; and (c) participation in an effector response, which occurs at the site of infection.

22.5 Formation and Selection of Lymphocytes 871

22.5a Formation of T-Lymphocytes 871

- Formation of T-lymphocytes begins in the red bone marrow and is completed in the thymus to produce immunocompetent but naive lymphocytes, a process that contributes to central self-tolerance.

22.5b Selection of T-Lymphocytes 872

- T-lymphocytes undergo positive selection, in which their ability to recognize foreign antigen attached to MHC is determined, and negative selection, in which their tolerance for self-antigens attached to MHC is tested. Only those lymphocytes that successfully pass both types of selection are allowed to survive to become either helper T-lymphocytes or cytotoxic T-lymphocytes. (Tregs are formed from CD4+ cells that moderately bind self-antigen.)

CHALLENGE YOURSELF

Do You Know the Basics?

_____ 1. All of the following are phagocytic cells *except*

    a. neutrophils.

    b. T-lymphocytes.

    c. macrophages.

    d. eosinophils.

_____ 2. This cell releases cytokines to activate B-lymphocytes, increase the activity of macrophages, and in general regulates the overall immune response.

    a. cytotoxic T-lymphocyte

    b. helper T-lymphocyte

    c. natural killer cell

    d. basophil

_____ 3. This cell is activated by binding antigen, and then engulfing and presenting the antigen with MHC class II molecules to helper T-lymphocytes. The helper T-lymphocytes release cytokines as the second form of stimulation.

    a. NK (natural killer) cell

    b. macrophage

    c. B-lymphocyte

    d. cytotoxic T-lymphocyte

_____ 4. These two cells destroy an infected cell by releasing chemicals that cause apoptosis.

    a. NK cell and cytotoxic T-lymphocyte

    b. macrophage and NK cell

    c. helper T-lymphocyte and cytotoxic T-lymphocyte

    d. B-lymphocyte and T-lymphocyte

_____ 5. All of the following are functions of antibodies *except*

    a. neutralization of pathogen.

    b. destruction of antigen.

    c. agglutination of antigen.

    d. opsonization.

_____ 6. The four characteristics of adaptive immunity include all of the following *except*

    a. activation by a specific antigen.

    b. memory.

    c. production of clones of cells that have the same TCR or BCR.

    d. each lymphocyte is effective against a wide array of pathogens.

_____ 7. During which process does additional fluid enter an injured or infected area from the blood and additional fluid is removed by the lymph vessels?

    a. fever

    b. clonal selection

    c. inflammation

    d. activation of helper T-lymphocytes

_____ 8. This chemical is released by virus-infected cells to decrease the spread of virus to nearby cells.

    a. interferon

    b. bradykinin

    c. perforin

    d. complement

_____ 9. The correct sequence of the major events in the life of a lymphocyte includes

    a. effector response, formation, and activation.

    b. activation, formation, and effector response.

    c. activation, effector response, and formation.

    d. formation, activation, and effector response.

_____ 10. Two of the major actions of complement include

    a. increasing the inflammatory response and cytolysis.

    b. recognizing and destroying a specific antigen and cytolysis.

    c. producing antibody and increasing the inflammatory response.

    d. releasing cytokines to increase the immune response and production of antibody.

11. Compare the general characteristics of innate immunity and adaptive immunity in terms of the cells involved, specificity, general mechanisms, and time required.

12. Define the inflammatory response, and explain its benefits.

13. Describe an antigen.

14. Describe class I and class II MHC molecules, and explain how they function in assisting T-lymphocytes in recognizing an antigen.

15. Explain positive and negative selection that occurs during selection of T-lymphocytes.

16. Describe how helper T-lymphocytes play a pivotal role in a healthy, normal functioning immune system.

17. Explain the general function of cytotoxic T-lymphocytes.

18. Describe both the function of antibodies and complement in defending the body.

19. There are two branches of adaptive immunity: cell-mediated and humoral immunity. Distinguish the types of antigens they are each effective against.

20. Explain the difference between the primary and secondary immune response.

Can You Apply What You've Learned?

1. Maria, who is 3 years old, was stung by a bee. The area where the stinger entered the skin became red, warm, and swollen. This is a normal response to the foreign venom called
   a. a fever.
   b. the complement cascade.
   c. an inflammatory response.
   d. an antigen challenge.

2. Jay, a young dad, takes his baby to the pediatrician several times in the first year of the child's life. These visits will stimulate the baby to make memory cells against specific antigens. Why are these visits necessary?
   a. The baby is being vaccinated.
   b. This is to verify that the baby has a normal inflammatory response.
   c. The baby must have an unusual immune deficit that must be monitored.
   d. The baby's blood is filtered to remove foreign antigen.

3. A young woman has just been diagnosed with the human immunodeficiency virus (HIV). This virus is especially devastating because it infects the cell that regulates the immune response. The cells HIV infects are
   a. B-lymphocytes.
   b. helper T-lymphocytes.
   c. NK cells.
   d. cytotoxic T-lymphocytes.

4. One-year-old Matthew always seems to be sick. When his blood is tested, there are no antibodies. The physician concludes that the child is lacking
   a. the ability to develop an inflammatory response.
   b. helper T-lymphocytes.
   c. cytotoxic T-lymphocytes.
   d. humoral immunity.

5. Upon further testing, it is found that Matthew does have normal cellular immunity. However, without antibodies Matthew will be less able to
   a. destroy cancer cells.
   b. destroy virus-infected cells.
   c. bind viral particles.
   d. destroy intracellular pathogens.

Can You Synthesize What You've Learned?

1. Dianne is an avid tennis player but has recently been complaining of tendonitis in her elbow. She knows that you work in health care and asks you to explain what caused this flare-up.

2. Stephanie is in her first year of college and has recently come down with a cold. She is running a slight fever of 100°F. Explain to her why a fever is beneficial.

3. Describe the events that occur in an individual who has an allergy to ragweed.

INTEGRATE

ONLINE STUDY TOOLS

■ connect
|ANATOMY & PHYSIOLOGY

■ LEARNSMART® **AP|R**

The following study aids may be accessed through Connect.

Concept Overview Interactive: Figure 22.20: Adaptive Immunity

Clinical Case Study: A Young Man with Memory Loss

Interactive Questions: This chapter's content is served up in a number of multimedia question formats for student study.

LearnSmart: Topics and terminology include overview of diseases caused by infectious agents; overview of immune system; innate immunity; adaptive immunity: an introduction; formation and selection of lymphocytes; activation and clonal selection of lymphocytes; effector response at infection site; immunoglobulins; immunologic memory and immunity

Anatomy & Physiology Revealed: Topics include immune response; antigen processing; cytotoxic T cells; helper T cells

Animations: Topics include antiviral activity of interferon; activation of complement; inflammatory response; antigenic (epitopes) determinants; antigen processing

chapter
23

Respiratory System

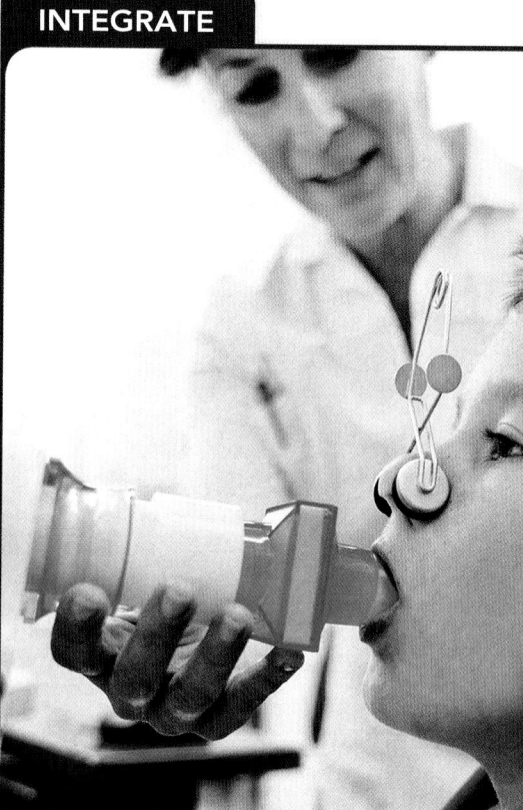

INTEGRATE

CAREER PATH
Respiratory Therapist

A respiratory therapist is an allied health-care professional who assesses, manages, and provides treatment to individuals with breathing or other cardiopulmonary disorders. These include both acute conditions that often accompany a stroke or a heart attack and chronic conditions such as bronchitis, asthma, and emphysema.

Anatomy & Physiology **REVEALED**
aprevealed.com

Module 11: Respiratory System

The **respiratory** (res′pi-ră-tōr′ē; *respire* = to breathe) **system** provides the means for gas exchange necessary for living cells. As cells engage in aerobic cellular respiration they need both an uninterrupted supply of oxygen and the removal of carbon dioxide waste that is produced. The continuous exchange of oxygen and carbon dioxide between the atmosphere and body cells occurs through collective processes called **respiration.** Respiration requires both coordinated and integrated physiologic processes of a number of systems, including the respiratory, skeletal, muscular, nervous, and cardiovascular systems. The respiratory system is responsible for the exchange of gases between the atmosphere and the lungs. The skeletal and muscular systems alter ➡️ the volume and pressure within the thoracic cavity to facilitate movement of air into and out of the lungs, whereas the nervous system stimulates and coordinates the contraction of skeletal muscles associated with breathing. The cardiovascular system transports oxygen and carbon dioxide between the lungs and the cells.

We begin by discussing the general functions and anatomic structures of the respiratory system components. Next, we consider the processes involved in respiration, including (1) how the respiratory, skeletal, muscular, and nervous systems function together during breathing (pulmonary ventilation); (2) the exchange of respiratory gases between the lungs and blood (alveolar gas exchange) and between the blood and systemic cells (systemic gas exchange); and (3) gas transport by the cardiovascular system. We conclude by considering the influence of breathing rate on homeostasis.

23.1 Introduction to the Respiratory System

The respiratory system consists of the respiratory passageways extending through the head, neck, and trunk, and the lungs themselves. Here we examine how the respiratory system serves numerous functions in the body, is organized both structurally and functionally, and is lined internally and protected with a mucous membrane.

23.1a General Functions of the Respiratory System

LEARNING OBJECTIVE

1. State the functions of the respiratory system.

The primary, and perhaps only, function most of us associate with the respiratory system is breathing. However, the respiratory system has several purposes that include the following:

- **Air passageway.** The respiratory tract is a passageway for air between the external environment and the alveoli (air sacs) of the lungs. Air is moved from the atmosphere to alveoli as we breathe in and then expelled into the atmosphere as we breathe out.
- **Site for the exchange of oxygen and carbon dioxide.** A thin barrier between the alveoli and the pulmonary capillaries provides the site for exchange of oxygen and carbon dioxide. Oxygen diffuses from the alveoli into the blood, and carbon dioxide diffuses from the blood into the alveoli.
- **Detection of odors.** Olfactory receptors located in the superior regions of the nasal cavity detect odors as air moves past them. Sensory input from these receptors is then relayed to various regions of the brain for interpretation.
- **Sound production.** The vocal cords of the larynx (voice box) vibrate as air moves across them to produce sounds; these sounds then resonate in upper respiratory structures.

Additionally, the rate and depth of breathing influence (1) the blood levels of oxygen (O_2), carbon dioxide (CO_2), and hydrogen ion (H^+), and (2) the venous return of blood and lymph due to changes in compression within the thoracic cavity that occur during breathing. How breathing rate and depth influence both blood composition and blood and lymph transport is described in detail at the end of the chapter.

WHAT DID YOU LEARN?

1. Which respiratory structure is associated with the exchange of respiratory gases?

23.1b General Organization of the Respiratory System

LEARNING OBJECTIVE

2. Distinguish between the structural organization and the functional organization of the respiratory system.

The respiratory system is organized into two structural regions: an upper respiratory tract and a lower respiratory tract (figure 23.1). The nose, nasal cavity, and pharynx form the **upper respiratory tract.** The larynx, trachea, bronchi, bronchioles (including terminal and respiratory bronchioles), alveolar ducts, and alveoli are the components of the **lower respiratory tract.**

The structures of the respiratory system are also categorized based on function. Passageways that serve to transport or conduct air are part of the **conducting zone;** these structures include the passageways from the nose to the end of the terminal bronchioles. Structures that participate in gas exchange with the blood—including the respiratory bronchioles, alveolar ducts, and alveoli—are part of the **respiratory zone.**

WHAT DID YOU LEARN?

2. What structures compose the upper respiratory tract? What structures compose the respiratory zone?

23.1c Respiratory Mucosa

LEARNING OBJECTIVES

3. Describe the structure of the mucosa that lines the respiratory tract and the structural changes observed along its length.

4. Explain the function of mucus produced by the mucosa.

The respiratory passageway is exposed to the external environment and is lined internally by a **mucosa,** also called a **mucous membrane.** In general, the mucosa is composed of an epithelium resting upon a basement membrane, and an underlying lamina propria composed of areolar connective tissue. The epithelium is ciliated in most portions of the respiratory tract conducting zone.

A general pattern of structural change is observed in the epithelium along the length of the respiratory tract. The epithelium becomes progressively thinner from the nasal cavity to the alveoli; it changes

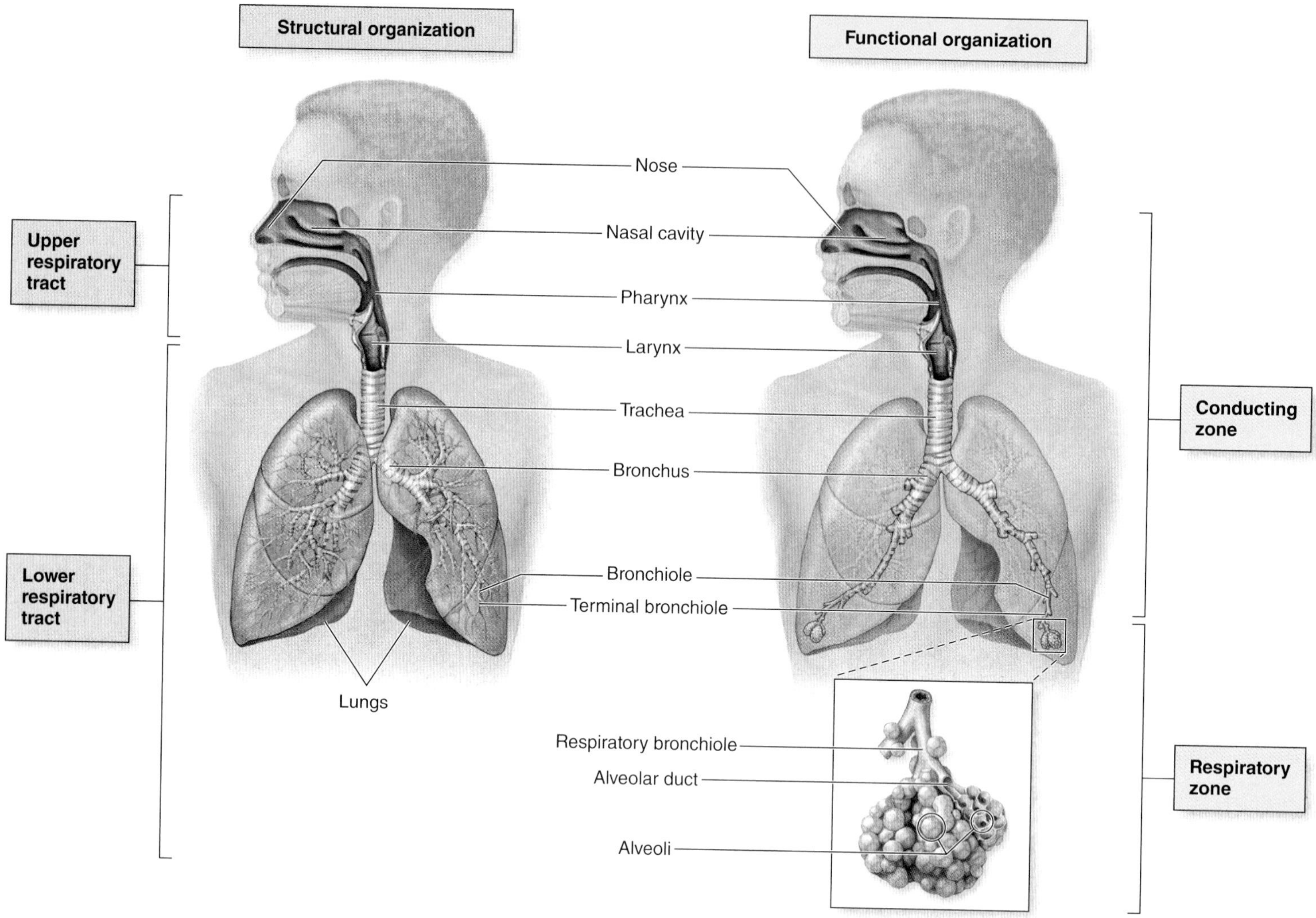

Structural organization

Functional organization

Upper respiratory tract

Lower respiratory tract

Conducting zone

Respiratory zone

Nose

Nasal cavity

Pharynx

Larynx

Trachea

Bronchus

Bronchiole

Terminal bronchiole

Lungs

Respiratory bronchiole

Alveolar duct

Alveoli

Figure 23.1 General Anatomy of the Respiratory System. Structurally, the respiratory system is organized into the upper respiratory tract and lower respiratory tract. Functionally, the respiratory system is divided into the conducting zone and respiratory zone.

from pseudostratified ciliated columnar to simple ciliated columnar to simple cuboidal to simple squamous (see section 5.1c). Exceptions to this general pattern occur in selected regions of the respiratory tract. These exceptions include (1) portions of the pharynx that serve as a passageway for both air and food (i.e., oropharynx and laryngopharynx), and (2) components of the larynx that include the vocal cords and the area immediately superior to them. These areas are lined by a nonkeratinized stratified squamous epithelium (instead of ciliated pseudostratified columnar epithelium) to withstand abrasion. **Figure 23.2** summarizes the types of epithelia found in the mucosa along the respiratory tract.

The epithelium lining most of the respiratory tract contains goblet cells, and the underlying lamina propria houses both mucous and serous glands. Mucus is produced from the combined secretions of these cells and glands. The amount of mucus produced daily is approximately 1 to 7 tablespoons, but that amount increases with exposure to irritants. Mucous secretions contain **mucin,** a protein that increases viscosity of mucus to more effectively trap inhaled dust, dirt particles, microorganisms, and pollen. The secretions also contain specific substances to help defend the body against microbes, including lysozyme (an antibacterial enzyme), defensins (antimicrobial proteins), and immunoglobulin A (antibodies).

Both mucus and saliva entrap materials, which may be coughed up as a viscous substance called **sputum.** Physicians may request sputum samples from their patients to diagnose potential respiratory infections.

 WHAT DID YOU LEARN?

❸ In what ways does the epithelium of the upper respiratory tract differ from the epithelium in the alveoli?

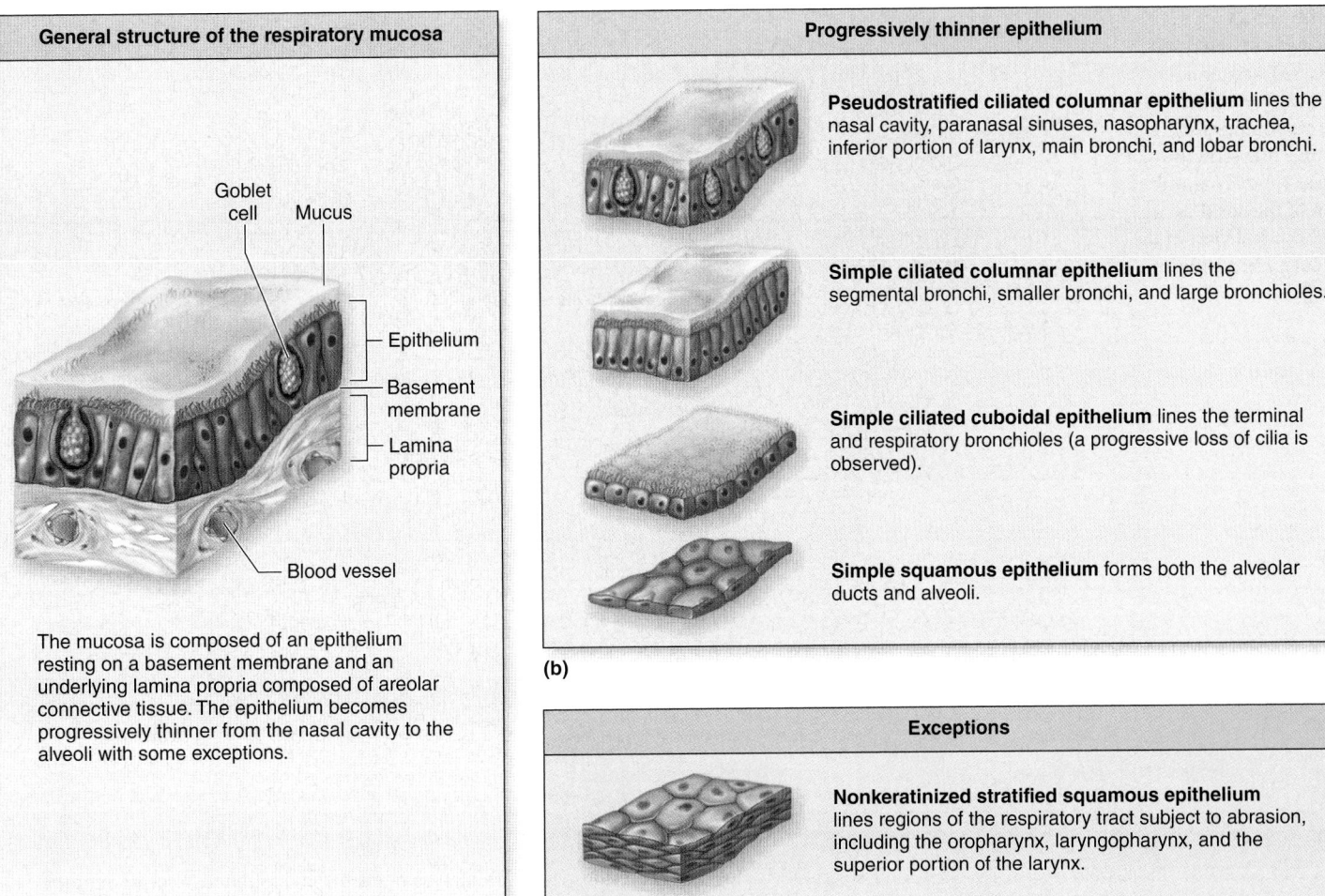

General structure of the respiratory mucosa

Goblet cell

Mucus

Epithelium

Basement membrane

Lamina propria

Blood vessel

The mucosa is composed of an epithelium resting on a basement membrane and an underlying lamina propria composed of areolar connective tissue. The epithelium becomes progressively thinner from the nasal cavity to the alveoli with some exceptions.

(a)

Progressively thinner epithelium

Pseudostratified ciliated columnar epithelium lines the nasal cavity, paranasal sinuses, nasopharynx, trachea, inferior portion of larynx, main bronchi, and lobar bronchi.

Simple ciliated columnar epithelium lines the segmental bronchi, smaller bronchi, and large bronchioles.

Simple ciliated cuboidal epithelium lines the terminal and respiratory bronchioles (a progressive loss of cilia is observed).

Simple squamous epithelium forms both the alveolar ducts and alveoli.

(b)

Exceptions

Nonkeratinized stratified squamous epithelium lines regions of the respiratory tract subject to abrasion, including the oropharynx, laryngopharynx, and the superior portion of the larynx.

(c)

Figure 23.2 Respiratory Mucosa. A mucosa forms the inner lining of the respiratory tract and is called the respiratory mucosa. (*a*) The general structure of the mucosa includes three major layers: an epithelium, basement membrane, and an underlying lamina propria. (*b*) The types of epithelia in the mucosa become progressively thinner along the length of the respiratory tract. (*c*) Exceptions to the general thinning pattern occur in places subject to abrasion. AP|R

23.2 Upper Respiratory Tract

The upper respiratory tract, as described previously, includes the nose, nasal cavity, and pharynx (**figure 23.3a**).

23.2a Nose and Nasal Cavity

LEARNING OBJECTIVES

1. Describe the structure and function of the nose.

2. Provide a general description of the structure and function of the nasal cavity.

The **nose** is the first structure of the conducting passageway for inhaled air (figure 23.3b); it is formed by bone, hyaline cartilage, and dense irregular connective tissue covered with skin externally. Paired nasal bones form the bridge of the nose and support it superiorly. Antero-inferiorly from the bridge, there is one pair of **lateral cartilages** and there are two pairs of **alar cartilages.** The flared components of the paired **nostrils,** or *nares* (nā´res; sing., *naris*) are composed of dense irregular connective tissue. The nostrils open into the inferior surface of the nose that leads into the nasal cavity.

The **nasal cavity** is oblong-shaped, and it extends from the nostrils to paired openings called **choanae** (kō´an-ē; sing., *choana*) or *posterior nasal apertures* (figure 23.3c, d). The choanae lead into the pharynx. The floor of the nasal cavity is formed by the hard and soft palate, and the

Figure 23.3
Upper Respiratory Tract. (*a*) Anatomic regions that compose the upper respiratory tract, (*b*) supporting structures of the nose, (*c*) midsagittal section of the nasal cavity, and (*d*) coronal view of the nasal cavity in a cadaver. AP|R

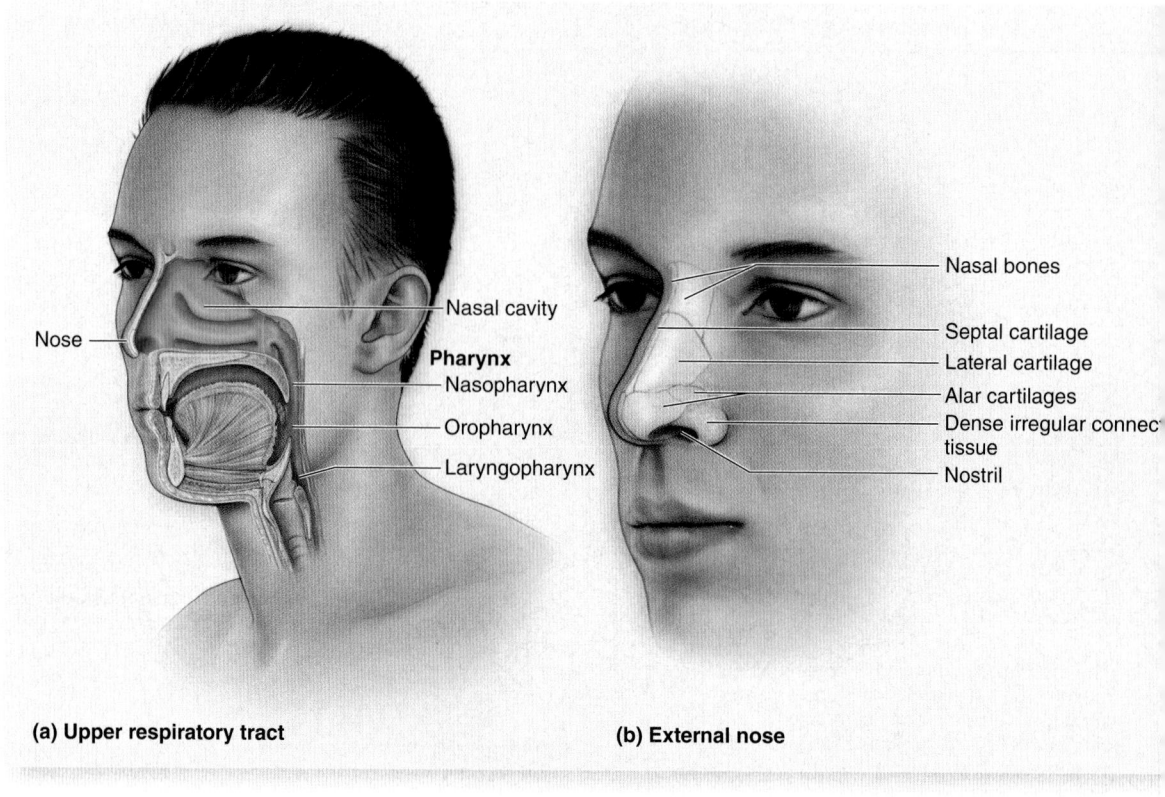

(a) Upper respiratory tract

(b) External nose

roof is composed of the nasal, frontal, ethmoid, and sphenoid bones, and some cartilage of the nose. The **nasal septum** divides the nasal cavity into left and right portions. The septum is formed anteriorly by the **septal nasal cartilage** and posteriorly by a thin, bony sheet composed of the perpendicular plate of the ethmoid superiorly and the vomer bone inferiorly (see section 8.2d).

WHAT DO YOU THINK?

❶ What is a deviated nasal septum, and how would it affect breathing?

Three paired, bony projections are located along the lateral walls of the nasal cavity: the **superior, middle,** and **inferior nasal conchae** (kon'kē; sing., *concha;* a shell). Because the conchae help produce turbulence in the inhaled air, they are sometimes called the *turbinate* bones. The conchae partition the nasal cavity into separate air passages (or "valleys"), each called a **nasal meatus** (mē-ā'tŭs). A meatus is located immediately inferior to its corresponding nasal concha.

The nasal cavity is divided into three parts (figure 23.3*c*): the nasal vestibule, olfactory region, and respiratory region. The **nasal vestibule** is immediately internal to the nostrils and is lined by skin and coarse hairs called **vibrissae** (vi-bris'ē; sing., *vibrissa; vibro* = to quiver) to trap large particulates.

The **olfactory region** is the superior portion of the nasal cavity. It contains the olfactory epithelium (pseudostratified ciliated columnar epithelium and olfactory receptors). Airborne molecules that dissolve in the mucus covering the olfactory epithelium stimulate olfactory receptors to detect different odors (see section 16.3a).

The **respiratory region** of the nasal cavity is lined by a mucosa composed of pseudostratified ciliated columnar epithelium. The lamina propria of this mucosal lining has an extensive vascular network. Nosebleeds (*epistaxis*) are especially common because of both the vast distribution of blood vessels and their superficial location (just deep to the epithelium). Additionally, paired **nasolacrimal ducts** drain lacrimal

secretions from the surface of each eye into the respiratory region of the nasal cavity (see figure 16.9 in section 16.4a).

A primary function of the nasal cavity is to condition the air (which means to warm, cleanse, and humidify the air) as it enters the respiratory tract. The air is warmed to body temperature by the extensive array of blood vessels within the nasal cavity lining. These vessels dilate in response to cold air, resulting in increased blood flow that helps to more effectively warm the inhaled air. The air is cleansed as inhaled microbes, dust, and other foreign material become trapped in the mucus covering the inner lining of the respiratory tract. Cilia then "sweep" the mucus and its trapped contents towards the pharynx to be swallowed. The air is also humidified as it passes through the moist environment of the nasal passageway. Conditioning of air is enhanced by conchae, which cause air turbulence that increases the amount of contact between the inhaled air and the mucosa (figure 23.3*c, d*).

WHAT DID YOU LEARN?

❹ What changes occur to inhaled air as it passes through the nasal cavity?

❺ What is the function of nasal conchae?

23.2b Paranasal Sinuses

LEARNING OBJECTIVE

3. Describe the structure and function of the four paired paranasal sinuses.

The **paranasal** (par-ă-nā'săl; *para* = alongside) **sinuses,** first described in section 8.2d, are associated with the nasal cavity (**figure 23.4**). These sinuses are spaces within the skull bones and named for the specific skull bones in which they are located. Thus, from a superior to inferior direction, they are the paired **frontal, ethmoidal,** and

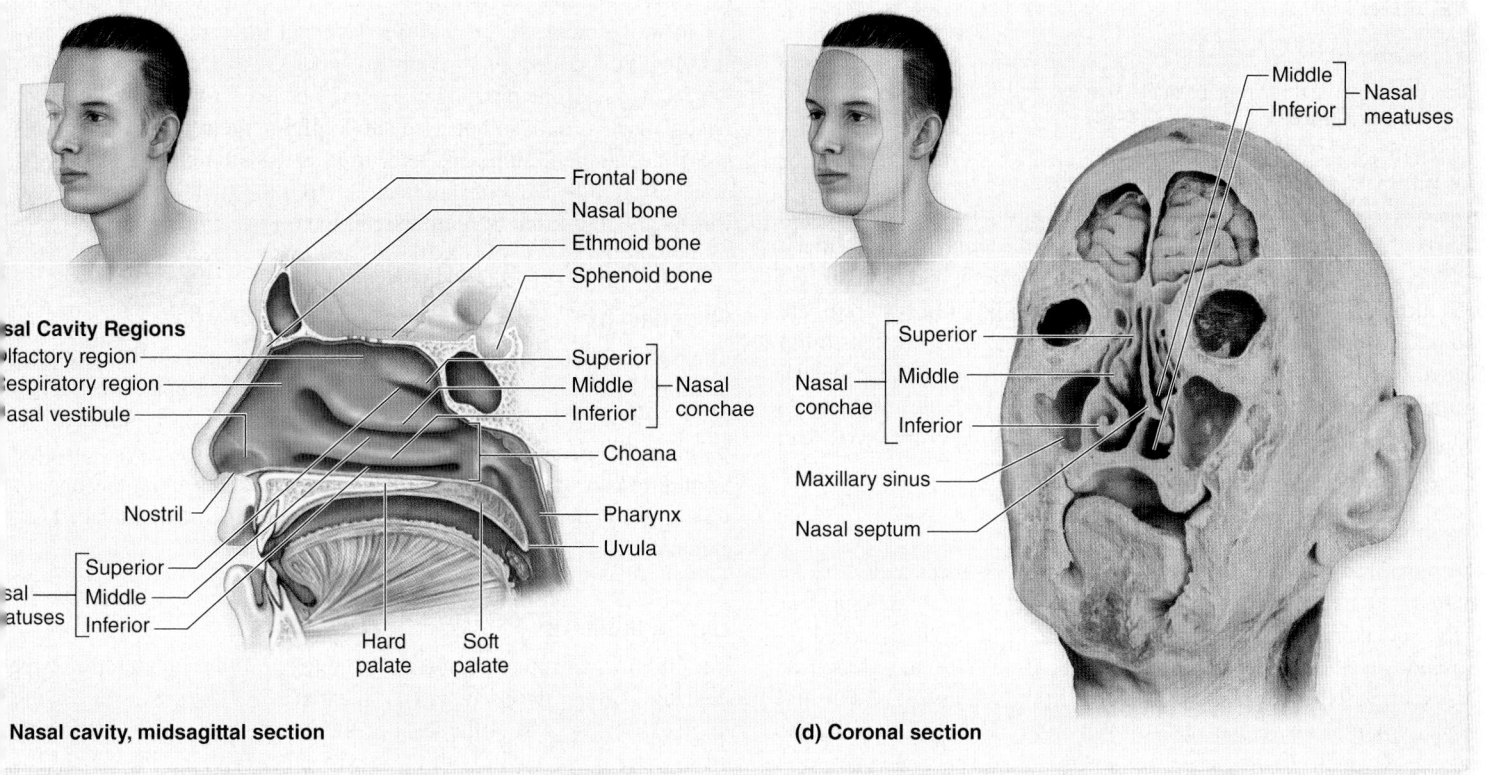

Nasal Cavity Regions
- Olfactory region
- Respiratory region
- Nasal vestibule

- Frontal bone
- Nasal bone
- Ethmoid bone
- Sphenoid bone

- Superior
- Middle — Nasal conchae
- Inferior

- Choana
- Pharynx
- Uvula

- Nostril

Nasal — Superior / Middle / Inferior — meatuses

Hard palate   Soft palate

Nasal cavity, midsagittal section

Middle / Inferior — Nasal meatuses

Nasal conchae — Superior / Middle / Inferior

Maxillary sinus

Nasal septum

(d) Coronal section

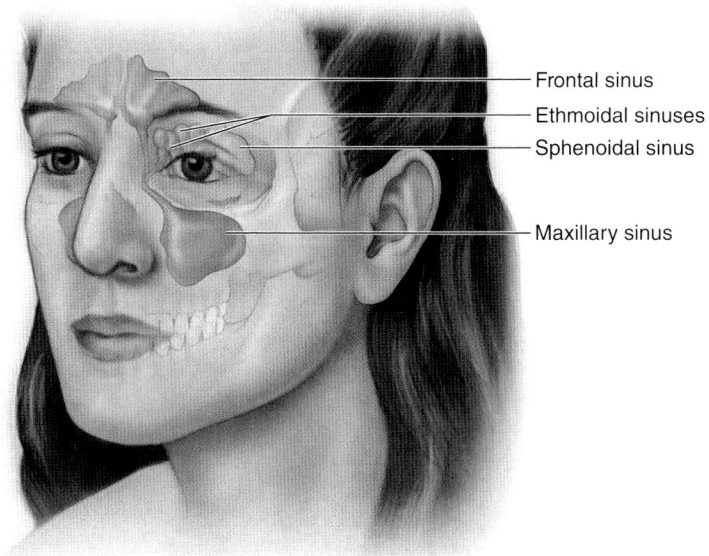

- Frontal sinus
- Ethmoidal sinuses
- Sphenoidal sinus
- Maxillary sinus

Anterolateral view

Figure 23.4 Paranasal Sinuses. The paranasal sinuses are air-filled cavities named for the bones in which they occur: frontal, ethmoidal, sphenoidal, and maxillary.

maxillary sinuses; the **sphenoidal sinuses** are located posterior to the ethmoidal sinuses. Ducts connect all sinuses to the nasal cavity. Both the sinuses and their ducts are lined by a pseudostratified ciliated columnar epithelium that is continuous with the mucosa of the nasal cavity. The mucus, with its trapped particulate matter, is swept by cilia from each paranasal sinus into the nasal cavity and then into the pharynx, where it is swallowed.

WHAT DID YOU LEARN?

6 How are the paranasal sinuses connected to the nasal cavity?

INTEGRATE

CLINICAL VIEW
Runny Nose

A **runny nose,** or **rhinorrhea** (rī′nō-rē-ă; *rhin* = nose, *rhoia* = flow), can occur as a result of (1) an increased production of mucus, such as occurs in response to a cold virus or allergy; (2) crying, due to increased secretions from the lacrimal glands that drain into the nasal cavity; or (3) exposure to cold air.

When cold air enters the nasal cavity and is exposed to the warm temperatures, water condensation occurs and the water mixes with mucus. In addition, the watery mucus is more likely to remain within the nasal cavity because the "chilled" cilia on epithelial cells are less able to sweep the mucus into the nasopharynx.

INTEGRATE

CLINICAL VIEW
Sinus Infections and Sinus Headaches

The mucosa of the ducts that drain from the paranasal sinuses into the nasal cavity can become inflamed in response to a respiratory infection or an allergy. Drainage of mucus is decreased and mucus accumulates in the paranasal sinuses as a result. A **sinus infection** can occur due to the lack of proper drainage of mucus. **Sinus headaches** result from increased pressure in the paranasal sinuses, due to the swelling of the mucosa. They also can result from pressure changes associated with swimming or high altitudes.

23.2c Pharynx

LEARNING OBJECTIVE

4. Compare the three regions of the pharynx, and describe their associated structures.

The **pharynx** (far'ingks), commonly called the *throat,* is a funnel-shaped passageway that averages 13 centimeters (5.1 inches) in length (**figure 23.5**). It is located posterior to the nasal cavity, oral cavity, and larynx. Air is conducted along its entire length, and both air and food along its inferior portions. The lateral walls of the pharynx are composed of skeletal muscles that both contribute to distensibility needed to accommodate swallowed food and help force these materials into the esophagus. The pharynx is partitioned into three regions—from superior to inferior, they are the nasopharynx, oropharynx, and laryngopharynx.

Nasopharynx

The **nasopharynx** (nā'zō-far'inks) is the superiormost region of the pharynx. Located directly posterior to the nasal cavity and superior to the soft palate, the nasopharynx, like the nasal cavity, is lined by a pseudostratified ciliated columnar epithelium. Normally, only air passes through the nasopharynx. Material from both the oral cavity and oropharynx typically is blocked from entering the nasopharynx by the soft palate, which elevates when we swallow. However, sometimes food or drink enters the nasopharynx and the nasal cavity, as when a person tries to swallow and then laughs at the same time. The soft palate cannot form a good seal for the nasopharynx, and the force from the laugh may propel some of the material into the nasal cavity. If the laugh is forceful enough, the material may come out the nostrils.

The nasopharynx lateral walls have paired openings into **auditory tubes** (*eustachian tubes,* or *pharyngotympanic tubes*) that connect the nasopharynx to the middle ear. These tubes equalize air pressure on either side of the tympanic membrane (eardrum) by allowing air to move between the nasopharynx and the middle ear (see section 16.5a). A collection of lymphatic nodules, called the tubal tonsils, are located near the pharyngeal opening of these tubes. The posterior nasopharynx wall also houses a single **pharyngeal tonsil.** When this tonsil is enlarged, clinicians refer to it as the **adenoids** (ad'ĕ-noydz; *aden* = gland, *eidos* = resemblance). Both the tubal and pharyngeal tonsils are composed of lymphatic tissue (see section 21.4c) and help to prevent the spread of infections.

Oropharynx

The middle pharyngeal region, called the **oropharynx** (ōr'ō-far'ingks), is immediately posterior to the oral cavity. The oropharynx extends from the level of soft palate superiorly to the hyoid bone inferiorly. The **palatine tonsils** are located on lateral walls of the oropharynx, and the **lingual tonsils** are at the base of the tongue (and thus are in the anterior region of oropharynx), providing defense against ingested or inhaled foreign materials (see figure 21.8).

Laryngopharynx

The inferior, narrowed region of the pharynx is the **laryngopharynx** (lă-ring'gō-far'ingks), which is located directly posterior to the larynx. It extends from the level of the hyoid bone and is continuous on its inferior end with both the larynx anteriorly and the esophagus posteriorly. Both the oropharynx and laryngopharynx serve as a common passageway for food and air. They are lined by a nonkeratinized stratified squamous epithelium (see figure 23.2c).

WHAT DID YOU LEARN?

❼ What two regions of the pharynx contain tonsils? What is their purpose?

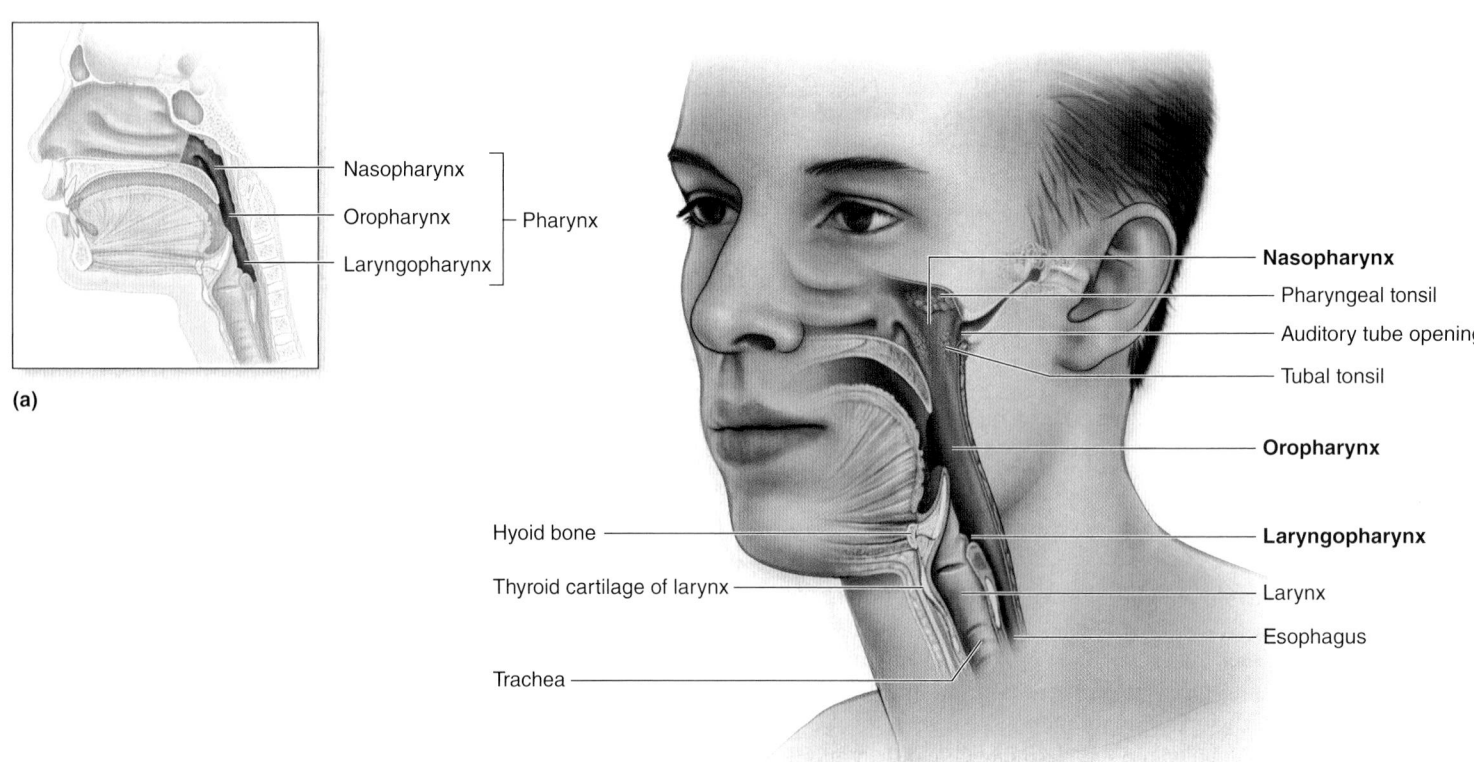

(a)

Nasopharynx
Oropharynx — Pharynx
Laryngopharynx

(b)

Hyoid bone
Thyroid cartilage of larynx
Trachea

Nasopharynx
Pharyngeal tonsil
Auditory tube opening
Tubal tonsil
Oropharynx
Laryngopharynx
Larynx
Esophagus

Figure 23.5 Pharynx. (*a*) The three specific regions of the pharynx (nasopharynx, oropharynx, and laryngopharynx) are highlighted in a sagittal section. (*b*) The pharynx is shown in relationship to the larynx, trachea, and esophagus in an anterolateral view. AP|R

23.3 Lower Respiratory Tract

The structures of the **lower respiratory tract** include both conducting pathways (the larynx, trachea, bronchi, and bronchioles) and those involved in gas exchange (respiratory bronchioles, alveolar ducts, and alveoli) (see figure 23.1).

23.3a Larynx

LEARNING OBJECTIVES

1. Describe the general functions and structure of the larynx.
2. Explain how the larynx functions in sound production.

The **larynx** (lar′ingks), also called the *voice box,* is a somewhat cylindrical structure that averages about 4 centimeters (1.6 inches) in length (figure 23.5). It is continuous superiorly with the laryngopharynx and inferiorly with the trachea.

Functions of the Larynx

The larynx has several major functions:

- **Serves as a passageway for air.** The larynx is normally open to allow the passage of air.
- **Prevents ingested materials from entering the respiratory tract.** During swallowing, the superior opening of the larynx is covered by the epiglottis to prevent ingested materials from entering the lower respiratory passageway.
- **Produces sound for speech.** Ligaments within the larynx, called vocal cords, vibrate as air is passed over them during an expiration.
- **Assists in increasing pressure in the abdominal cavity.** The epiglottis of the larynx closes over the opening of the larynx so air cannot escape, and simultaneously abdominal muscles

INTEGRATE

CONCEPT CONNECTION

The Valsalva maneuver facilitates several physiologic processes including: elimination of both urine from the urinary bladder (see section 24.8c) and feces from the gastrointestinal tract (see section 26.3d), and the expulsion of a baby during childbirth (see section 29.6d).

contract to increase abdominal pressure. This action is referred to as the **Valsalva maneuver** (see Integrate: Concept Connection). You can experience the increase in abdominal pressure associated with the Valsalva maneuver by holding your breath while forcefully contracting your abdominal muscles.

- **Participates in both a sneeze and cough reflex.** Both a sneeze and a cough result in an explosive blast of exhaled air. This occurs when the abdominal muscles contract forcefully and the vocal cords are initially closed and then open abruptly as the pressure increases in the thoracic cavity. Sneezing is initiated by irritants in the nasal cavity and coughing by irritants in the trachea and bronchi. Both help remove irritants from the respiratory tract.

Larynx Anatomy

The opening that connects the pharynx and larynx is called the **laryngeal** (lă-rin′jē-ăl) **inlet,** *laryngeal aperture,* or *laryngeal aditus.* The larynx is formed and supported by a framework of nine pieces of cartilage that are held in place by ligaments and muscles. The nine cartilages include: the single thyroid, cricoid, and epiglottis cartilages, and the paired arytenoid, corniculate, and cuneiform cartilages as shown in **figure 23.6.**

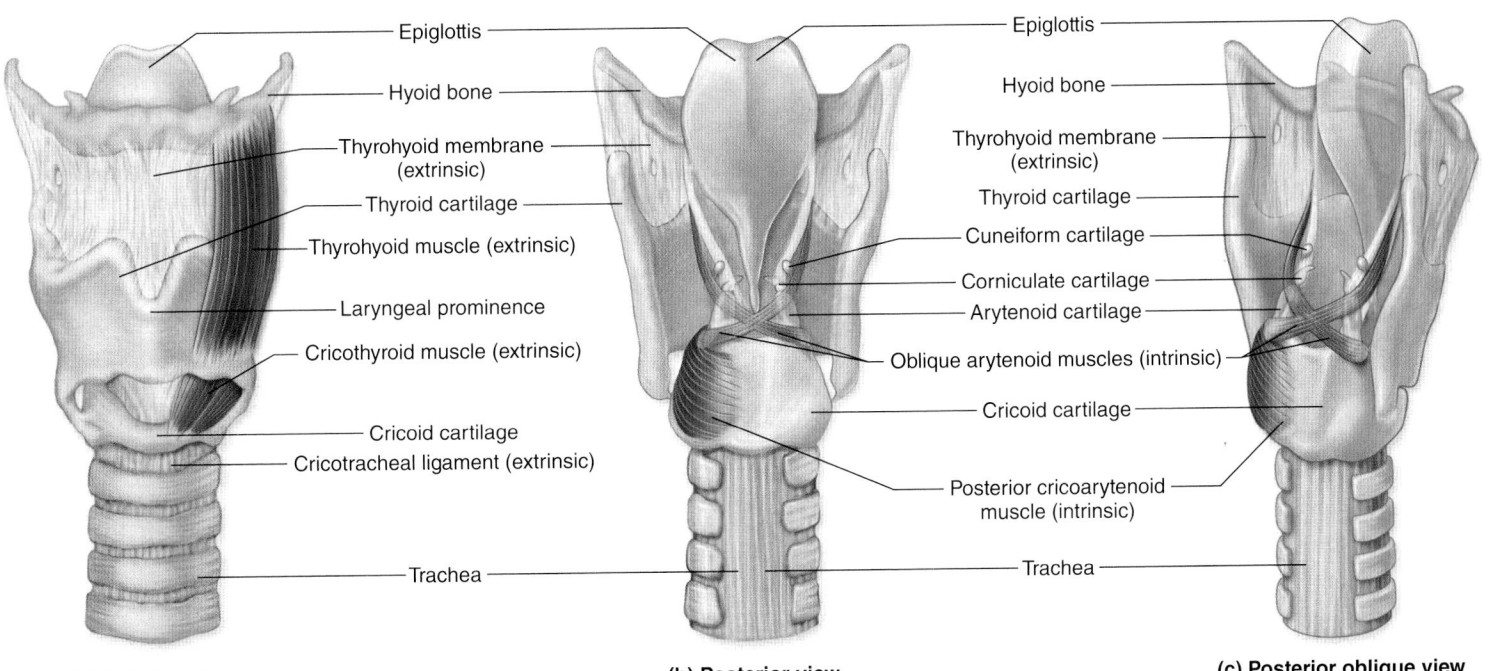

(a) Anterior view   **(b) Posterior view**   **(c) Posterior oblique view**

Figure 23.6 Larynx. The larynx shown in an (*a*) anterior view, (*b*) posterior view, and (*c*) posterior oblique view. It is composed of nine cartilages, various ligaments, and skeletal muscle. The nine cartilages, as well as extrinsic and intrinsic ligaments, form the flexible support of the larynx. Extrinsic muscles participate in the elevation of the larynx, and intrinsic muscles function in sound production. **AP|R**

The **thyroid cartilage** is the largest laryngeal cartilage. Shaped like a shield, it forms the anterior and lateral walls of the larynx. The almost V-shaped anterior projection of the thyroid cartilage is called the **laryngeal prominence** (commonly referred to as the *Adam's apple*). This protuberance is generally larger in males because (1) the laryngeal inlet is narrower in males (90 degrees) than in females (120 degrees), and (2) it enlarges at puberty due to testosterone-induced growth.

The thyroid cartilage is attached to the lateral surface of the ring-shaped **cricoid** (krī'koyd; *kridos* = a ring) **cartilage** located inferior to the thyroid cartilage. The large, spoon- or leaf-shaped **epiglottis** (ep-i-glot'is; *epi* = on, *glottis* = mouth of windpipe) is anchored to the inner aspect of the thyroid cartilage and projects posterosuperiorly into the pharynx. It closes over the laryngeal inlet during swallowing. The three smaller, paired cartilages, the **arytenoid** (ar-i-tē'noyd), **corniculate** (kōr'-ni-kū-lāt; *corniculatus* = horned), and **cuneiform** (kū'nē-i-fōrm; *cuneus* = wedge) **cartilages** are located internally. All cartilages of the larynx, except the epiglottis, are composed of hyaline cartilage. The epiglottis, which opens and closes over the laryngeal inlet, is composed of the more flexible elastic cartilage (see section 5.2d).

Laryngeal ligaments are classified as either extrinsic ligaments or intrinsic ligaments. **Extrinsic ligaments** attach to the external surface of laryngeal cartilages and extend to other structures that include the superiorly located hyoid bone and inferiorly located trachea.

The **intrinsic ligaments** are located within the larynx and include both vocal ligaments and vestibular ligaments (**figure 23.7**). The **vocal ligaments** are composed primarily of elastic connective tissue and extend anterior to posterior between the thyroid cartilage and the arytenoid cartilages. These ligaments are covered with a mucosa to form the **vocal folds.** Vocal folds also are called the *true vocal cords*

because they produce sound when air passes between them. They are distinctive from the surrounding tissue because they are avascular and white in color. The opening between these folds is called the **rima glottidis** (rī'mă; slit; glo-tī'dis; *rima* = slit). Together the vocal folds and the rima glottidis form the **glottis.**

The *vestibular ligaments* form the other intrinsic ligaments. These extend between the thyroid cartilage to the arytenoid and corniculate cartilages. Together with the mucosa covering them, they form the **vestibular folds** located superior to the vocal folds. These folds also are called the *false vocal cords* because they have no function in sound production, but serve to protect the vocal folds. The opening between the vestibular folds is called the *rima vestibuli.*

Skeletal muscles compose part of the larynx wall and are classified as either extrinsic or intrinsic muscles. The **extrinsic muscles** originate on either the hyoid bone or the sternum and insert on the thyroid cartilage. Extrinsic muscles elevate the larynx during swallowing (see section 26.2c).

The **intrinsic muscles** are located within the larynx and attach to both the arytenoid and corniculate cartilages. Contraction of the intrinsic muscles causes the arytenoid cartilages to pivot resulting in a change in the dimension of the rima glottidis. The opening becomes narrower as the vocal folds are adducted and becomes wider if the vocal folds are abducted. They function in voice production and help close off the larynx when swallowing.

Sound Production

Sound production originates as the vocal folds begin to vibrate. This occurs when the intrinsic laryngeal muscles narrow the opening of the rima glottidis and air is forced past the vocal cords during an expiration.

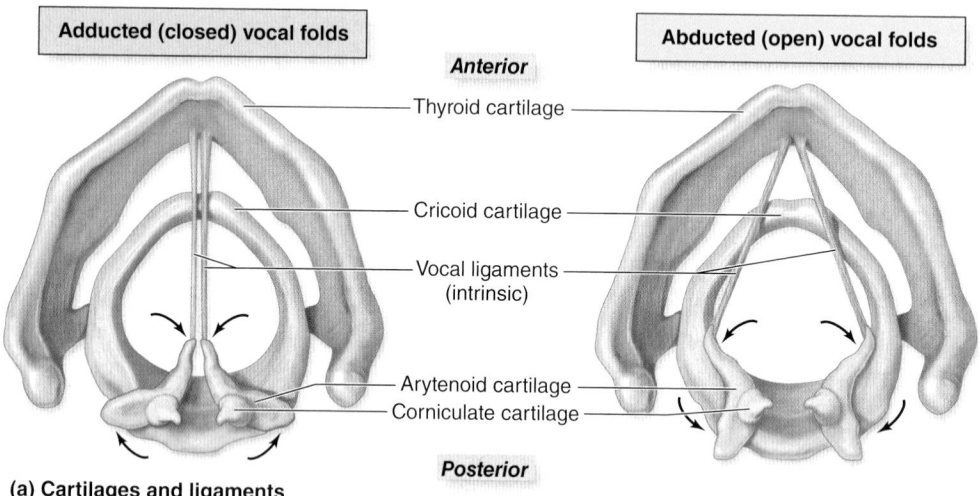

(a) Cartilages and ligaments

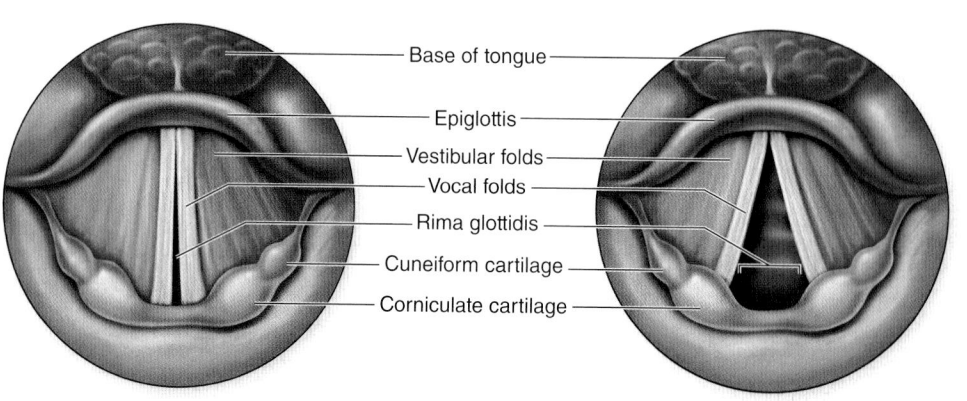

(b) Laryngoscopic view

Figure 23.7 Vocal Folds. The vocal folds (true vocal cords) are elastic ligaments covered with a mucosa that extend between the thyroid and arytenoid cartilages. These folds surround the rima glottidis and are involved in sound production. Adducted (closed) and abducted (open) vocal folds are shown in (*a*) a superior view of the cartilages and ligaments only, and (*b*) a diagrammatic laryngoscopic view of the coverings around these cartilages and ligaments. (*c*) A photo of a superolateral laryngoscopic view, showing the vestibular folds, the vocal folds, and the rima glottidis opening into the trachea.

(c) Larynx, superolateral view

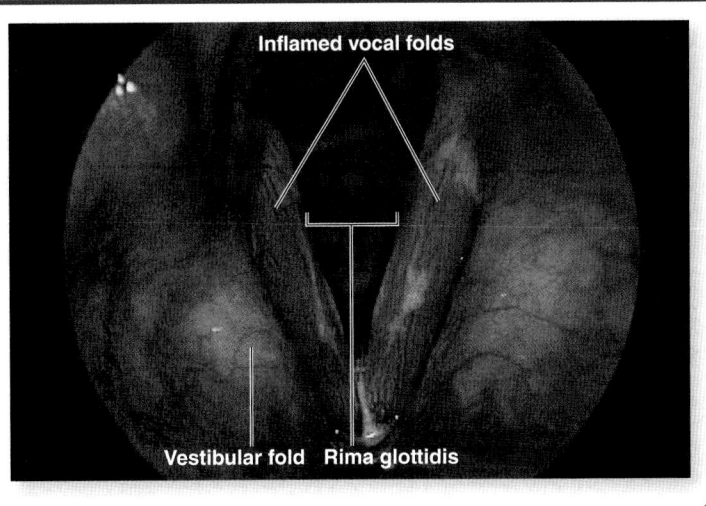

INTEGRATE

CLINICAL VIEW
Laryngitis

Laryngitis (lar-in-jī'tis) is an inflammation of the larynx that may extend to its surrounding structures. Viral or bacterial infection is the number one cause of laryngitis. Less frequently, laryngitis follows overuse of the voice, such as yelling for several hours at a football game. Symptoms include hoarse voice, sore throat, and sometimes fever. Severe cases can produce inflammation and swelling extending to the epiglottis. Children's airways are proportionately smaller, and a swollen and inflamed epiglottis (called *epiglottitis*) may lead to sudden airway obstruction and become a medical emergency.

A laryngoscopic view shows the inflamed, reddened vocal folds characteristic of laryngitis.

Inflamed vocal folds

Vestibular fold   Rima glottidis

The characteristics of the sound include range, pitch, and loudness. The **range** of a voice (be it soprano or bass) is determined by the length and thickness of the vocal folds. Males generally have longer and thicker folds than do females, and thus males produce sounds that are in a lower range. Our vocal folds increase in length as we grow, which is why our voices become lower or deeper as we mature to adulthood.

Pitch refers to the frequency of sound waves. Pitch is determined by the amount of tension or tautness on the vocal folds as regulated primarily by the intrinsic laryngeal muscles. Increasing the tension on the vocal folds causes the vocal folds to vibrate more when air passes by them and thus to produce a higher sound. Conversely, the less taut the vocal folds, the less they vibrate, and the lower the pitch of the sound.

Loudness depends on the force of the air passing across the vocal cords. A lot of air forced through the rima glottidis produces a loud sound, whereas a little air produces a soft sound. When you whisper, only the most posterior portion of the rima glottidis is open, and the vocal folds do not vibrate. Because the vocal folds are not vibrating, the whispered sounds are all of the same pitch.

Speech also requires the participation of the spaces in the pharynx, nasal and oral cavities, and the paranasal sinuses, which serve as resonating chambers for sound, and the structures of the lips, teeth, and tongue that help form different sounds. Young children tend to have high, nasal-like voices because their sinuses are not yet well developed, so they lack large resonant chambers. When you hold your nose and speak, your voice sounds quite different because air doesn't pass through the nasal cavity.

WHAT DID YOU LEARN?

8 How does the larynx assist in increasing abdominal pressure?

9 What are the three unpaired cartilages in the larynx?

10 What are the structural and functional differences between the vocal folds and vestibular folds?

23.3b Trachea

LEARNING OBJECTIVES

3. Describe the structure of the trachea.

4. Explain the structure and function of the tracheal cartilages.

The **trachea** (trā'kē-ă; = rough) is a flexible, slightly rigid, tubular organ often referred to as the "windpipe" **(figure 23.8)**. It extends inferiorly through the neck into the mediastinum from the larynx to the main bronchi. The trachea lies immediately anterior to the esophagus and posterior to part of the sternum.

Gross Anatomy of the Trachea

The trachea averages approximately 13 centimeters (5.1 inches) in length and 2.5 centimeters (1 inch) in diameter. The anterior and lateral walls of the trachea are supported by 15 to 20 C-shaped rings of hyaline cartilage called **tracheal cartilages.** The tracheal cartilages are connected superiorly and inferiorly with one another by elastic connective tissue sheets called **anular** (an'ū-lăr; *anulus* = ring) **ligaments.** An internal ridge of mucosal covered cartilage called the **carina** (kă-rī'nă) is located at the split of the trachea into the main bronchi (figure 23.8c). The sensory receptors of the carina are extremely sensitive and can induce a forceful cough when stimulated by irritants.

Each C-shaped tracheal cartilage is ensheathed in a perichondrium and a dense fibrous membrane. The open ends of the cartilage rings are attached posteriorly by both the **trachealis muscle** and an elastic ligamentous membrane. These rings reinforce and provide structural support to the tracheal wall to ensure that the trachea remains open (patent) at all times, whereas the more flexible trachealis muscle and ligamentous membrane on the posterior aspect of the trachea allows for distension during swallowing of food through the esophagus. The trachealis contracts during coughing to reduce the diameter of the trachea, thus facilitating the more rapid expulsion of air, helping to dislodge material (foreign objects or food) from the air passageway.

Tracheotomy (trā-kē-ot'ō-mē; *tome* = incision) is one of the oldest surgical procedures: it involves making an incision into the trachea to facilitate breathing when a patient's airway is blocked or respiratory ventilation is compromised by disease or injury. Although a tracheotomy is a potentially life-saving procedure, it is not without risks and should be undertaken only by trained medical personnel.

Histology of the Tracheal Wall

The innermost to outermost layers that form the wall of the trachea are (1) the **mucosa,** which is composed of a pseudostratified ciliated columnar epithelium with goblet cells and a lamina propria; (2) the

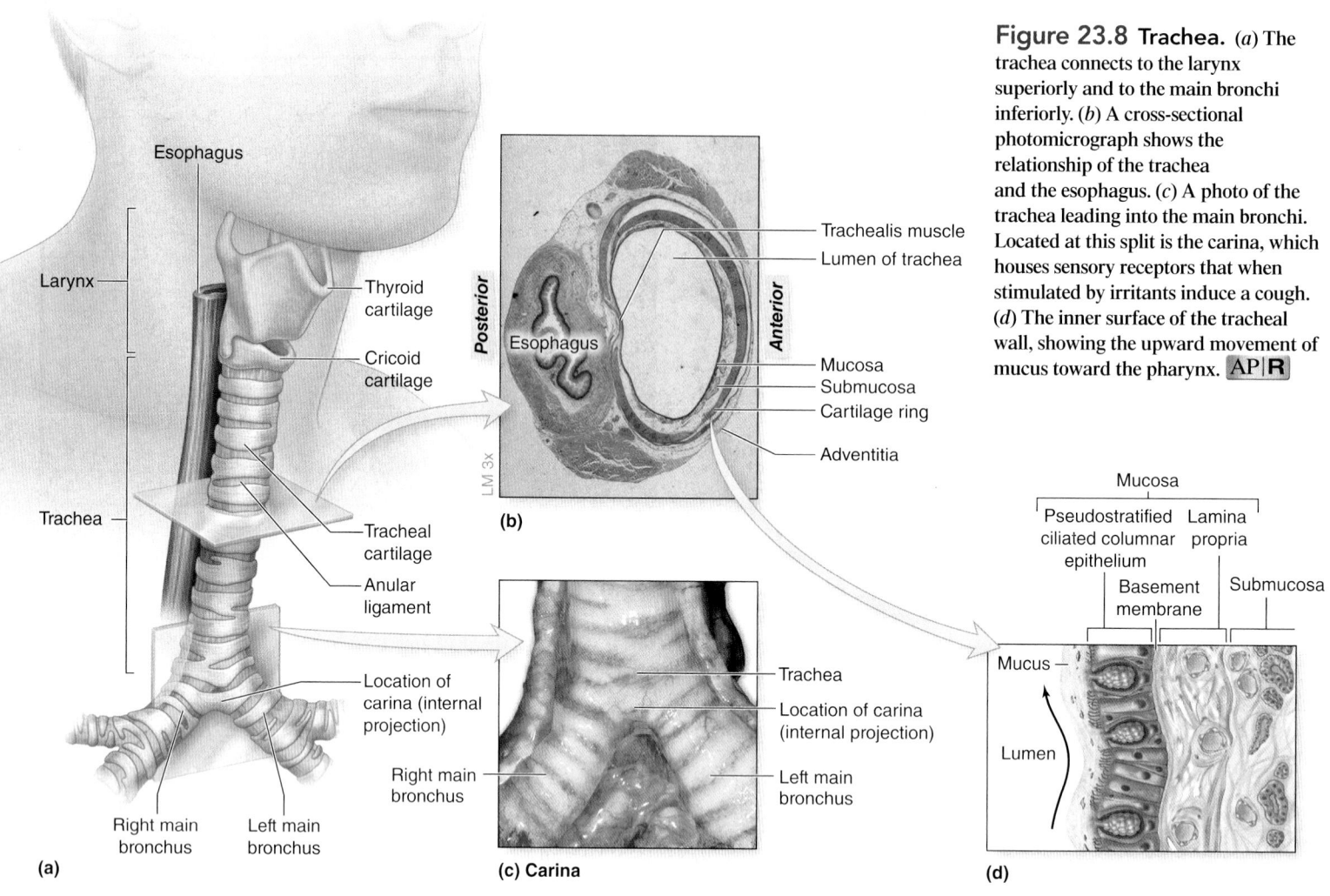

Figure 23.8 Trachea. (*a*) The trachea connects to the larynx superiorly and to the main bronchi inferiorly. (*b*) A cross-sectional photomicrograph shows the relationship of the trachea and the esophagus. (*c*) A photo of the trachea leading into the main bronchi. Located at this split is the carina, which houses sensory receptors that when stimulated by irritants induce a cough. (*d*) The inner surface of the tracheal wall, showing the upward movement of mucus toward the pharynx. AP|R

submucosa, consisting of areolar connective tissue that houses larger blood vessels, nerve endings, serous and mucous glands, and lymphatic tissue; (3) **tracheal cartilage** (described earlier); and (4) the **adventitia,** composed of elastic connective tissue.

The movement of cilia in the mucosal epithelium propels mucus laden with dust, microbes, and other particles superiorly toward the larynx and the pharynx, where it may be swallowed or expelled (figure 23.8*d*).

WHAT DO YOU THINK?

2 The lining of the trachea and bronchi in chronic smokers changes from pseudostratified ciliated columnar epithelium to stratified squamous epithelium. Why do you think this change occurs, and what are some consequences of this change?

WHAT DID YOU LEARN?

11 What is the function of the C-shaped tracheal cartilages? How do the trachealis muscle and elastic ligamentous membrane that complete each ring posteriorly function?

23.3c Bronchial Tree

LEARNING OBJECTIVES

5. Describe the structural subdivisions of the bronchial tree.

6. Explain the processes of bronchoconstriction and bronchodilation.

The **bronchial** (brong′kē-ăl) **tree** is a highly branched system of air-conducting passages that originates at the main bronchi and progressively branches into narrower tubes that diverge throughout the lungs before ending in the smallest bronchiole passageway (**figure 23.9**).

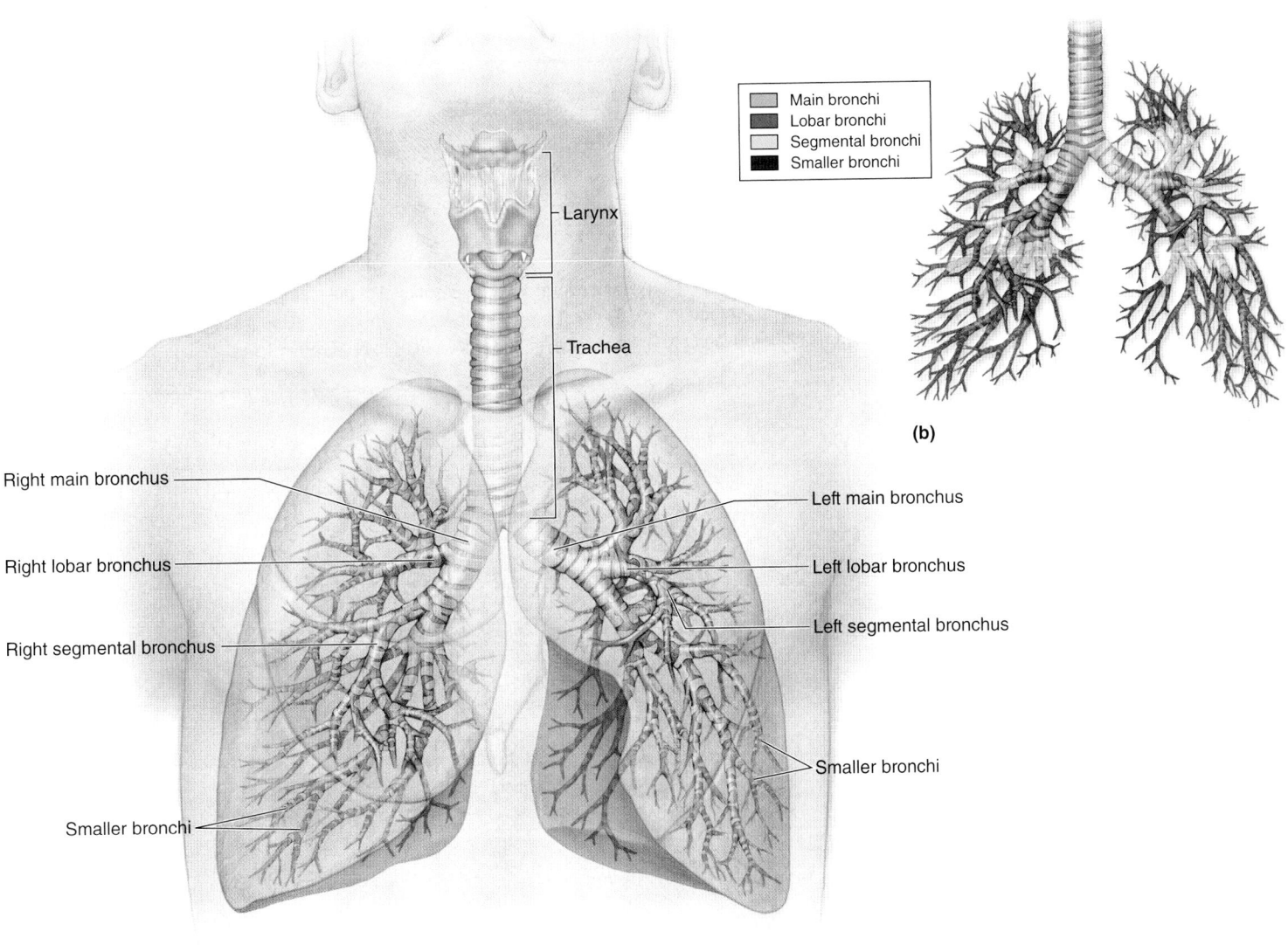

- Larynx
- Trachea

Main bronchi
Lobar bronchi
Segmental bronchi
Smaller bronchi

(b)

Right main bronchus

Right lobar bronchus

Right segmental bronchus

Smaller bronchi

Left main bronchus

Left lobar bronchus

Left segmental bronchus

Smaller bronchi

(a) Anterior view

Figure 23.9 Bronchial Tree. The bronchial tree is composed of conducting passageways that originate at the two main bronchi and end at the terminal bronchioles. (*a*) Larynx, trachea, and bronchi are shown. (*b*) The major divisions of the bronchial tree are color-coded. AP|R

Gross Anatomy of the Bronchial Tree

The trachea splits at the level of the sternal angle (where the manubrium and body of the sternum articulate see section 8.6a) into the right and left **main bronchi** (brong′kī; sing., *bronchus*) also known as *primary bronchi*. Each main bronchus projects inferiorly and laterally into a lung. The right main bronchus is shorter, wider, and more vertically oriented than the left main bronchus—thus, foreign particles are more likely to lodge in the right main bronchus. Both main bronchi, along with all associated pulmonary vessels, lymph vessels, and nerves, enter a lung on its medial surface.

Each main bronchus then branches into **lobar bronchi** (or *secondary bronchi*), that extend to each lobe of the lung. The right lung with three lobes has three lobar bronchi, and the left lung with two lobes has two lobar bronchi. Lobar bronchi are smaller in diameter than main bronchi. They further divide into **segmental bronchi** (or *tertiary bronchi*) that serve a division of the lung called a bronchopulmonary segment (described later). The right lung is supplied with 10 segmental bronchi, and the left lung is supplied by 8 to 10 segmental bronchi. The bronchial tree continues to divide into more numerous and smaller bronchi and then bronchioles. There are approximately 9 to 12 different levels or generations of bronchial branch divisions; the main, lobar, and segmental bronchi are the first, second, and third generations of bronchi, respectively.

Bronchi lead into tubes that have a diameter of less than 1 millimeter called **bronchioles** (brong′kē-ōl). **Terminal bronchioles** are the last portion of the conducting pathway. They lead into respiratory bronchioles, the first segments of the respiratory zone.

INTEGRATE

CLINICAL VIEW
Bronchitis

Bronchitis (brong-kī'tis) is inflammation of the bronchi caused by a viral or bacterial infection, or by inhaling irritants such as vaporized chemicals, particulate matter, or cigarette smoke. Clinically, bronchitis is divided into two categories: acute and chronic.

Acute bronchitis develops rapidly either during or after an infection, such as a cold. Symptoms include coughing, sneezing, pain upon inhalation, and fever. Most cases of acute bronchitis resolve completely within 10 to 14 days.

Chronic bronchitis results from long-term exposure to irritants. Chronic bronchitis is defined medically as the production of large amounts of mucus, associated with a cough lasting 3 continuous months. If exposure to the irritant persists, permanent changes to the bronchi occur, including thickened bronchial walls with subsequent narrowing of their lumens and overgrowth (hyperplasia) of the mucin-secreting cells of the bronchi. These long-term changes in the bronchi increase the likelihood of future bacterial infections within the respiratory tract..

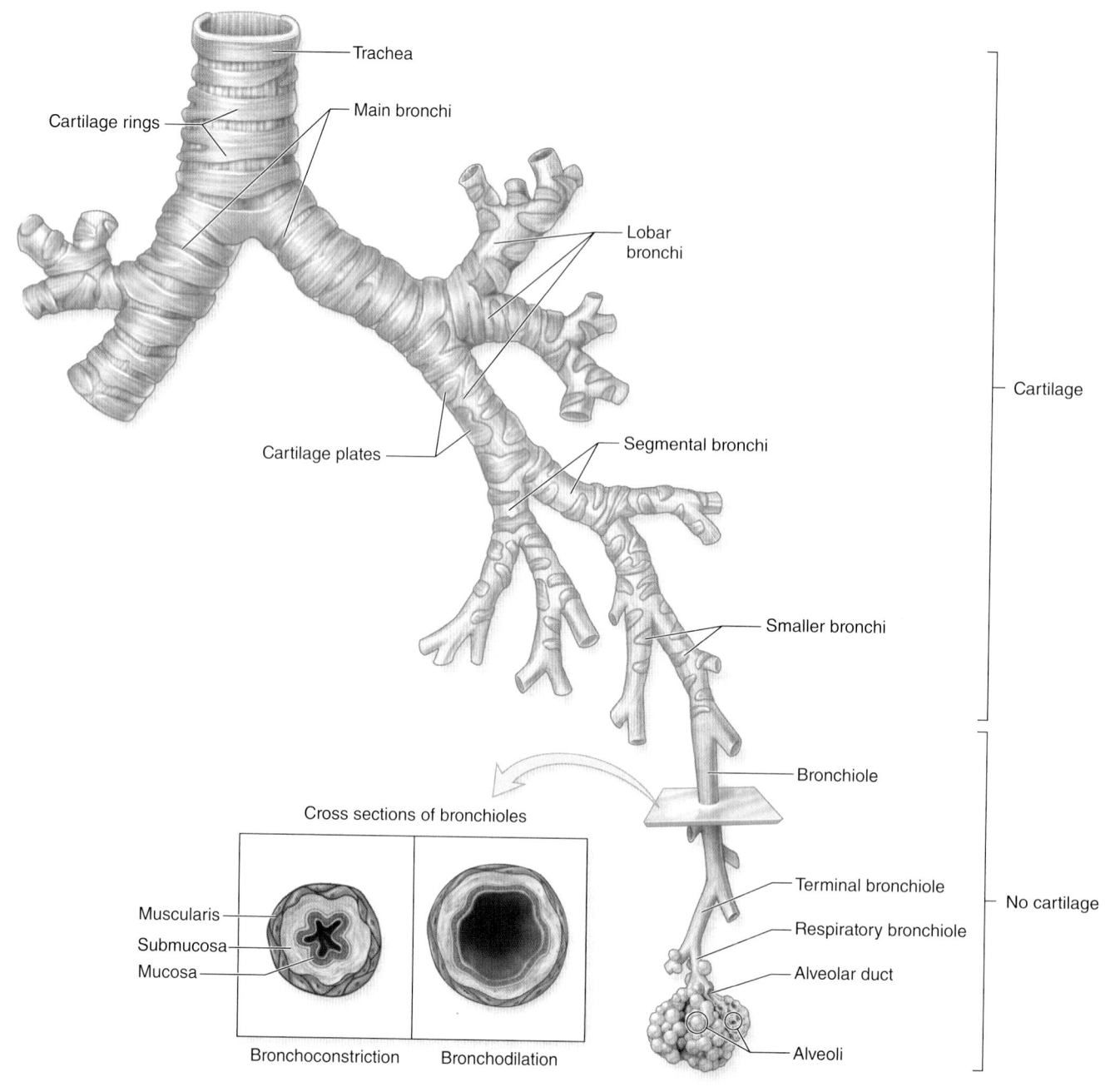

Figure 23.10 Structure of the Bronchial Wall. Irregular cartilage plates of decreasing size support the branching bronchi. In contrast, bronchioles do not contain cartilage, but rather have a proportionately thicker layer of smooth muscle. This smooth muscle layer allows for bronchoconstriction and bronchodilation that change the size of the lumen, regulating the amount of air reaching the alveoli.

CLINICAL VIEW

Asthma (az'mă) is a chronic condition characterized by episodes of bronchoconstriction coupled with wheezing, coughing, shortness of breath, and excess pulmonary mucus. Typically, the affected person develops sensitivity to an airborne agent, such as pollen, smoke, mold spores, dust mites, or particulate matter. Upon reexposure to this triggering substance, a localized immune reaction occurs in the bronchi and bronchioles, resulting in bronchoconstriction, swollen submucosa, and increased production of mucus. Episodes typically last an hour or two. Continual exposure to the triggering agent increases the severity and frequency of asthma attacks. The walls of the bronchi and bronchioles eventually may become permanently thickened, leading to chronic and unremitting

airway narrowing and shortness of breath. If airway narrowing is extreme during a severe asthma attack, death could occur.

The primary treatment for asthma consists of administering inhaled steroids (cortisone-related compounds) to reduce the inflammatory reaction, combined with bronchodilators to alleviate the bronchoconstriction. Allergy shots have proven helpful for some patients. Individuals with severe asthma may need oral doses of steroids to help control the allergic hyper-response and reduce the inflammation. A new treatment called bronchial thermoplasty uses heat to remove some of the outer layers of smooth muscle. This decreases the muscle contractions associated with bronchoconstriction to lessen the severity of asthma.

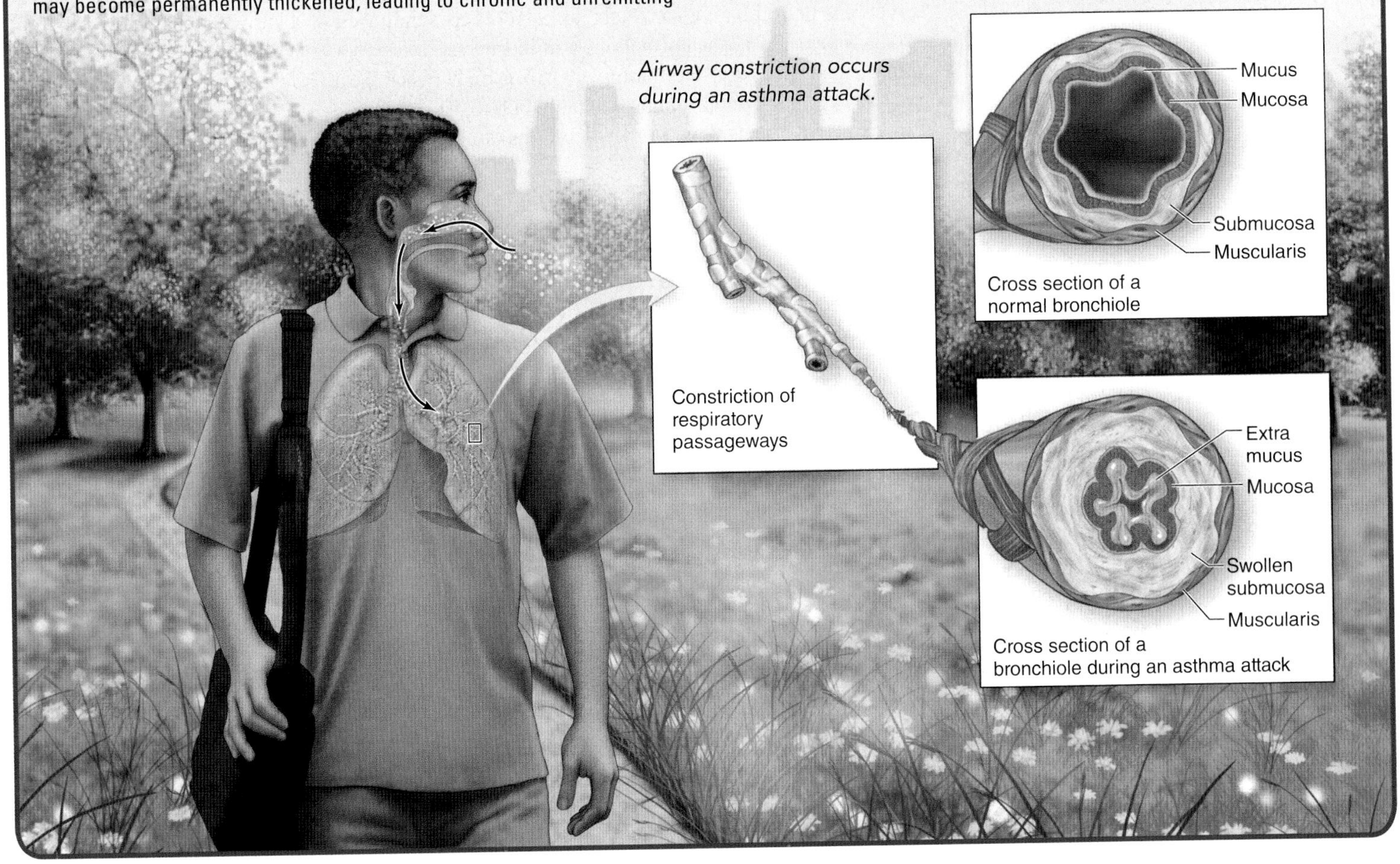

Airway constriction occurs during an asthma attack.

Constriction of respiratory passageways

Mucus
Mucosa

Submucosa
Muscularis

Cross section of a normal bronchiole

Extra mucus
Mucosa

Swollen submucosa
Muscularis

Cross section of a bronchiole during an asthma attack

Histology of the Bronchial Tree

Incomplete rings or plates of hyaline cartilage support the walls of the main bronchi to ensure that they remain open (**figure 23.10**). The extent of wall support lessens as bronchi divide and their diameter decreases. It first appears as various-sized irregular plates of cartilage in their walls that continue to decrease in size and number as the branching of smaller air passageways continues. Unlike bronchi, bronchioles have no cartilage in their walls, because their small diameter alone normally prevents collapse. Instead, bronchioles have a proportionately thicker layer of smooth muscle than do bronchi. Contraction of this smooth muscle narrows the diameter of the bronchiole, referred to as **bronchoconstriction,** and decreases the amount of air passing through the bronchial tree. In comparison, relaxation of smooth muscle widens the diameter of the bronchiole, referred to as **bronchodilation,** and increases the amount of air passing through the bronchial tree. Note that bronchoconstriction lessens the amount of potentially harmful substances that may be inhaled into the alveoli (e.g., smoke, toxins, allergens), whereas bronchodilation maximizes the amount of oxygen that is delivered to the alveoli and the amount of carbon dioxide that is removed. The changes in the epithelium of the bronchial tree are summarized in figure 23.2.

WHAT DID YOU LEARN?

12 What are the significant structural differences between bronchi and bronchioles?

23.3d Respiratory Zone: Respiratory Bronchioles, Alveolar Ducts, and Alveoli

LEARNING OBJECTIVES

7. Describe the structure and function of the components of the respiratory zone.

8. List three types of cells found in alveoli, and describe the function of each.

The respiratory zone was described earlier as being composed of respiratory bronchioles, alveolar ducts, and alveoli. These are all microscopic structures. The smallest **respiratory bronchioles** subdivide into thin airways called **alveolar ducts** that lead into **alveolar sacs**, a grape-like cluster of alveoli at the end of an alveolar duct **(figure 23.11)**. Both respiratory bronchioles and alveolar ducts are airways that contain small (about 0.25 to 0.5 millimeters in diameter) saccular outpocketings called **alveoli** (al-vē′ō-lī; sing., al-vē′ō-lŭs; *alveus* = hollow sac). Respiratory bronchioles typically are composed of a simple cuboidal epithelium, whereas both the alveolar ducts and alveoli are composed of a simple squamous epithelium (see figure 23.2). The epithelium within the respiratory zone is much thinner than in the conducting portion, thus facilitating gas diffusion between the alveolus and pulmonary capillaries.

Each lung contains approximately 300 to 400 million alveoli by the time a person is about 8 years old. The packing of these millions of air-filled alveoli give the lung its spongy nature.

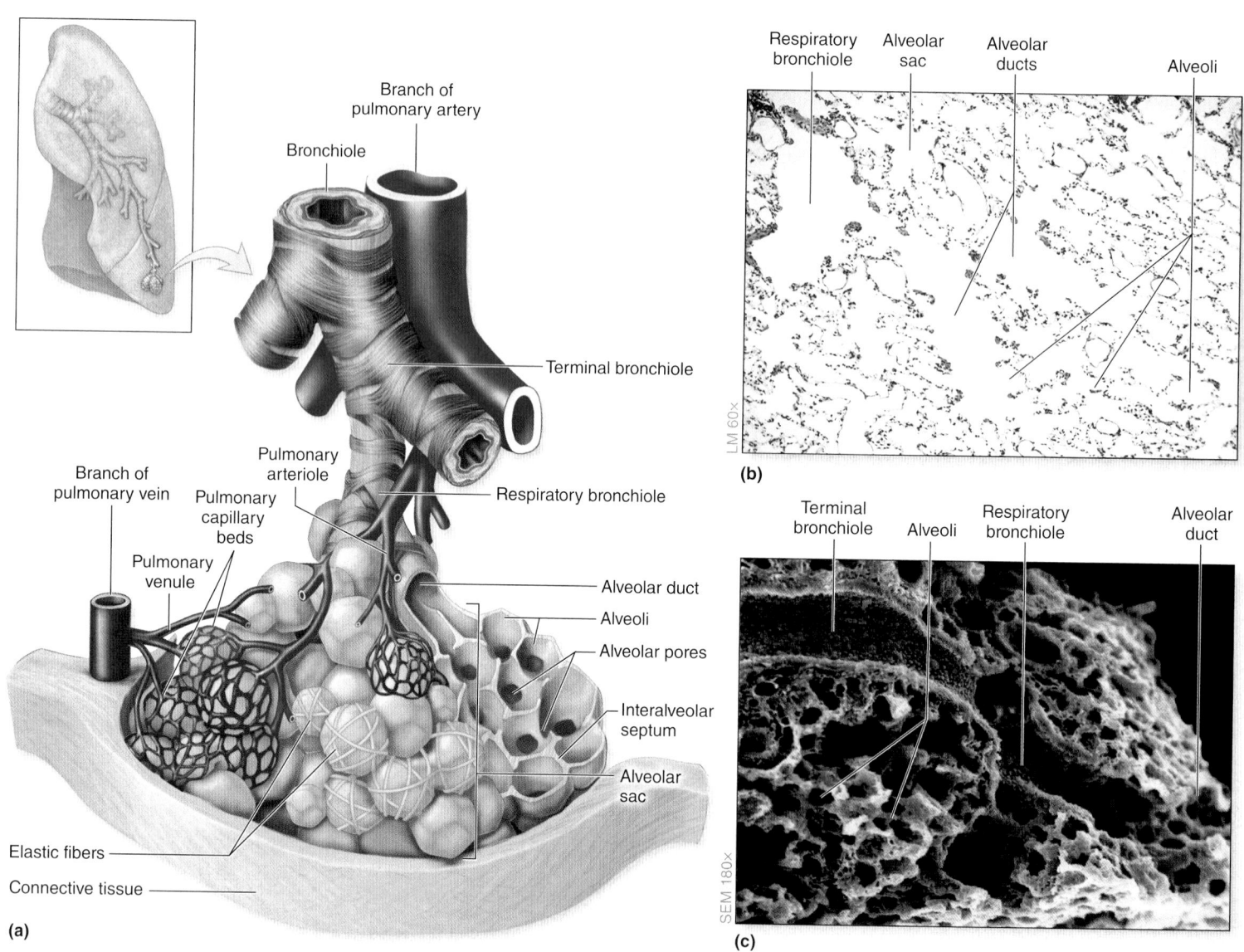

Figure 23.11 Bronchioles and Alveoli. Bronchioles and alveoli form the terminal ends of the respiratory passageway. (*a*) Terminal bronchioles branch into respiratory bronchioles in the respiratory zone, which then branch into alveolar ducts and alveoli. Pulmonary vessels travel with the bronchioles, and the pulmonary capillaries wrap around the alveoli for gas exchange. Elastic tissue is also wrapped around alveoli. (*b*) A photomicrograph shows the relationship of respiratory bronchiole, alveolar ducts, and alveoli. (*c*) SEM of a terminal bronchiole, a respiratory bronchiole, alveolar duct, and alveoli reveals the honeycomb appearance of alveoli. AP|R

Alveoli abut one another, so their sides become slightly flattened. Thus, an alveolus in cross section actually looks more hexagonal or polygonal in shape than circular. Small openings in the walls, called **alveolar pores,** occur between some adjacent alveoli; these openings provide for collateral ventilation of alveoli. Pulmonary capillaries surround each alveolus to facilitate gas exchange. The **interalveolar septum** contains elastic fibers that contribute to the ability of the lungs to stretch during inspiration and recoil during expiration.

Two cell types form the alveolar wall: the simple squamous alveolar type I cell, and the almost cuboidal-shaped alveolar type II cell **(figure 23.12a)**. The predominant **alveolar type I cell** is also called a *squamous alveolar cell.* It makes up approximately 95% of the alveolar surface area and forms part of a thin barrier that separates the air in the alveoli from the blood in pulmonary capillaries. The internal surface of alveoli formed by type I cells is moist, thus making alveoli prone to collapse because the moist inner surface causes a high surface tension. The smaller population of **alveolar type II cells** (or *septal cells*) secrete an oily fluid, called **pulmonary surfactant** (ser-fak′tănt). This oily fluid (which contains a mixture of lipid and protein molecules) coats the inner alveolar surface. If an alveolus begins to collapse, such as would occur during expiration, the pulmonary surfactant molecules become more tightly packed together and tend to collectively oppose the collapse of the alveolus.

A third type of cell that is part of the alveolus is the **alveolar macrophage,** also called a *dust cell.* This cell is a leukocyte that may be either fixed or free. Fixed alveolar macrophages remain within the connective tissue of the alveolar walls, whereas free alveolar macrophages are migratory cells that continually move across the alveolar surface within the alveoli. Both types of alveolar macrophages engulf any microorganisms or particulate material that reach the alveoli. The alveolar macrophages are able to leave the lungs either by entering the lymph vessels or by being coughed up in sputum and then expectorated from the mouth. **Table 23.1** summarizes the structures of the lower respiratory tract. A lower respiratory tract infection affects some or all of these structures.

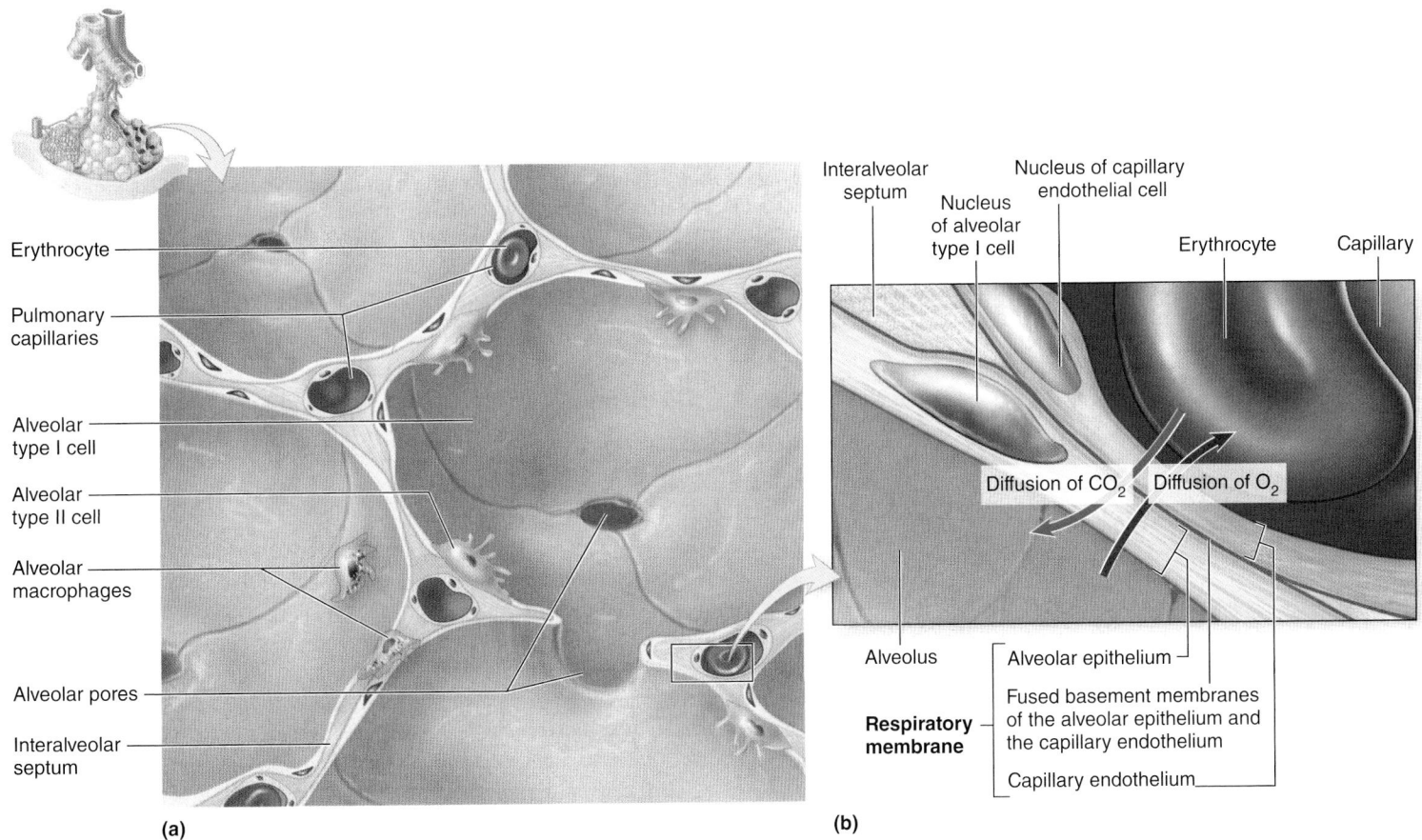

Figure 23.12 Alveoli and the Respiratory Membrane. (*a*) Microscopic alveoli form the terminal end of the air passageway. (*b*) Gas exchange between the alveoli and the pulmonary capillaries occurs across a thin respiratory membrane. The respiratory membrane consists of an alveolar type I cell, an endothelial cell of a capillary, and their fused basement membranes. Oxygen diffuses from an alveolus into the blood within the capillary, and carbon dioxide diffuses in the opposite direction. (Note that the pulmonary surfactant covering layer is not shown here.) AP|R

Table 23.1	Structures of the Lower Respiratory Tract			
Structure[1]	Anatomic Description	Wall Support	Epithelial Lining	Function
Larynx	A somewhat cylindrical structure between the pharynx and trachea	Nine pieces of cartilage; supported by ligaments and skeletal muscle	Nonkeratinized stratified squamous epithelium superior to vocal folds; pseudostratified ciliated columnar epithelium inferior to vocal folds	Conducts air; prevents ingested material from entering trachea; produces sound; assists in increasing pressure in abdominal cavity; participates in both a sneeze and cough reflex
Trachea	Flexible, semirigid tubular organ connecting larynx to main bronchi	C-shaped cartilage rings keep trachea patent (open)	Pseudostratified ciliated columnar epithelium	Conducts air
Bronchi	Largest airways of bronchial tree; consist of main, lobar, segmental, and smaller bronchi	Incomplete rings and irregular plates of cartilage; some smooth muscle	Larger bronchi lined by pseudostratified ciliated columnar epithelium; smaller bronchi lined by simple ciliated columnar epithelium	Conduct air
Bronchioles	Smaller conducting airways of bronchial tree; larger bronchioles branch into smaller bronchioles; terminal bronchioles are last part of conducting zone	No cartilage; proportionately greater amounts of smooth muscle in walls	Ranges from simple ciliated columnar epithelium (for largest bronchioles) to simple cuboidal epithelium (for smaller bronchioles)	Conduct air; smooth muscle in walls allows bronchoconstriction and bronchodilation
Respiratory bronchioles	First structures of respiratory zone	No cartilage; smooth muscle is scarce in walls	Simple cuboidal epithelium	Gas exchange
Alveolar ducts	Small airways that branch off respiratory bronchioles; multiple alveoli found along walls of alveolar duct	No cartilage, no smooth muscle	Simple squamous epithelium	Gas exchange
Alveoli	Small air sacs	No cartilage, no smooth muscle	Simple squamous epithelium	Gas exchange

1. Structures are listed in the order that air passes through them during inspiration.

 WHAT DID YOU LEARN?

13 Which of the following respiratory structures are supported by cartilage: nose, larynx, trachea, bronchi, bronchioles, and alveolar sacs?

14 The respiratory tract can be damaged from desiccation (drying out), cold air, microbes, or exposure to chemicals or particulate matter. Which of the following help protect the respiratory tract: nasal hairs, mucus, tonsils, cilia, macrophages, sneezing, or coughing? Explain how for each.

15 List the conducting and respiratory structures (in order) that air passes through from the atmosphere to the alveoli.

23.3e Respiratory Membrane

 LEARNING OBJECTIVE

9. Explain the structure of the respiratory membrane.

The **respiratory membrane** (figure 23.12b) is the thin barrier (only 0.5 micrometer thick) that oxygen and carbon dioxide diffuse across during gas exchange between the alveoli and the blood in the pulmonary capillaries. It consists of an alveolar epithelium and its basement membrane, and a capillary endothelium and its basement membrane. The two basement membranes are fused. Oxygen diffuses from the alveolus across the respiratory membrane into the pulmonary capillary, thereby allowing the erythrocytes in the blood to become oxygenated. Oxygen is then transported by the blood to systemic cells. Conversely, carbon dioxide diffuses from the blood within the pulmonary capillary through the respiratory membrane to enter each alveolus. Once in the alveoli, carbon dioxide is expired from the respiratory system into the external environment.

 WHAT DID YOU LEARN?

 16 List, in order, the structures of the respiratory membrane that oxygen must cross to move from an alveolus into the blood.

23.4 Lungs

The paired lungs house both the bronchial tree and all the respiratory portions of the respiratory tract. Here we examine the position of the lungs in the thoracic cavity and their anatomic structure. We also discuss circulation to and innervation of the lungs, pleural membranes, and how lung inflation is maintained.

23.4a Gross Anatomy of the Lung

LEARNING OBJECTIVES

1. Describe the location and general structure of the lungs.
2. Compare and contrast the right versus left lung.

The paired lungs are located within the thoracic cavity on either side of the mediastinum, the median region that houses the heart. The lungs are enclosed and protected by the thoracic cage (**figure 23.13**). Each lung has a wide, concave **base** that rests inferiorly upon the muscular diaphragm, and an **apex** (or *cupula*) that is slightly superior and posterior to the clavicle. The lung surfaces are adjacent to the ribs, mediastinum, and diaphragm, and are respectively referred to as the **costal surface, mediastinal surface,** and **diaphragmatic surface** of the lungs.

Each lung has a conical shape with an indented region on its mediastinal surface called the **hilum,** through which pass the bronchi, pulmonary vessels, lymph vessels, and autonomic nerves (**figure 23.14**). Collectively, these structures that extend from the hilum are termed the **root** of the lung.

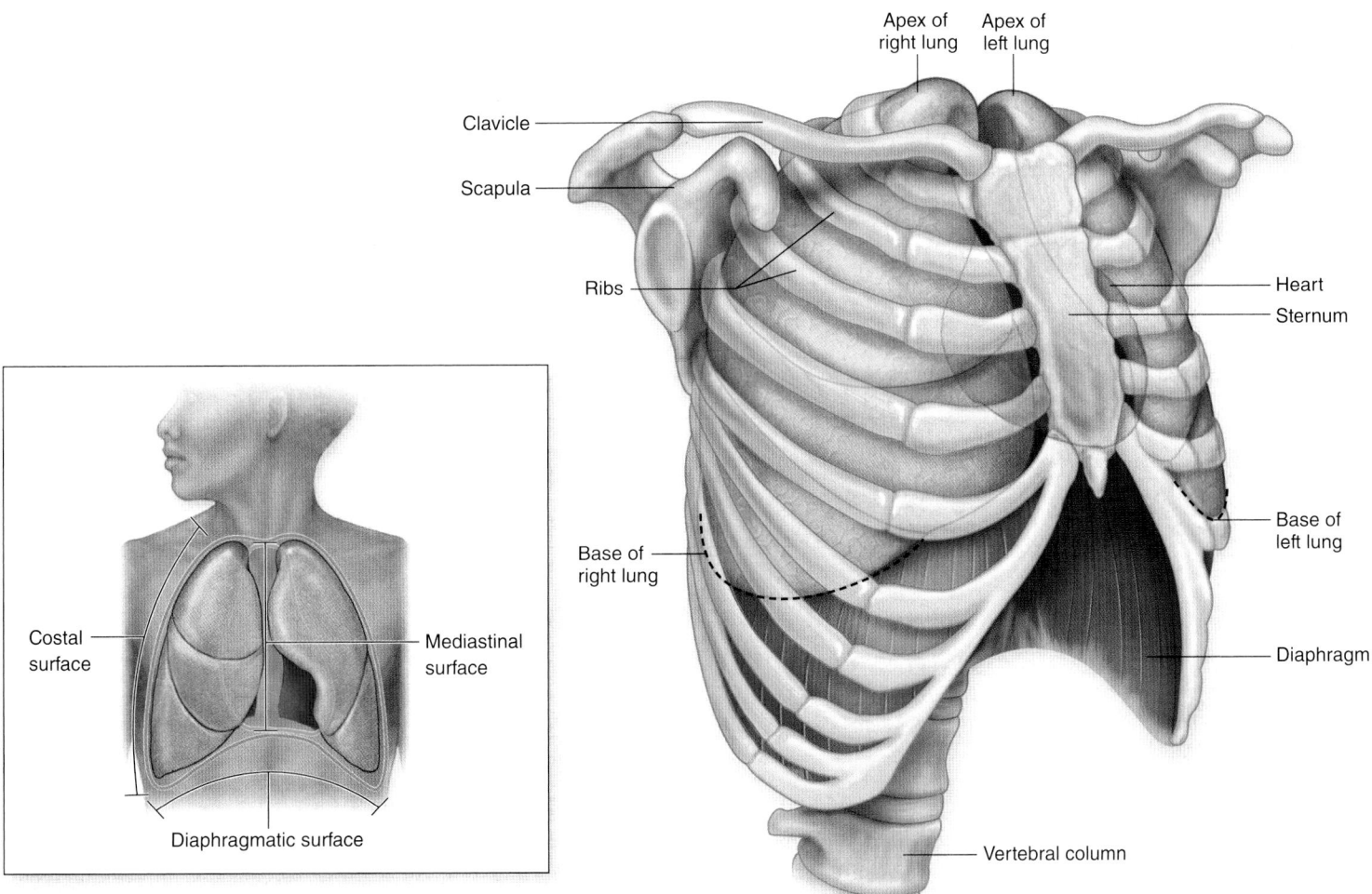

Figure 23.13 Position of the Lungs. Within the thoracic cavity, the lungs are bordered and protected by the thoracic cage. They are lateral to the mediastinum. The base of each lung rests on the diaphragm, and its apex is slightly superior and posterior to the clavicle. AP|R

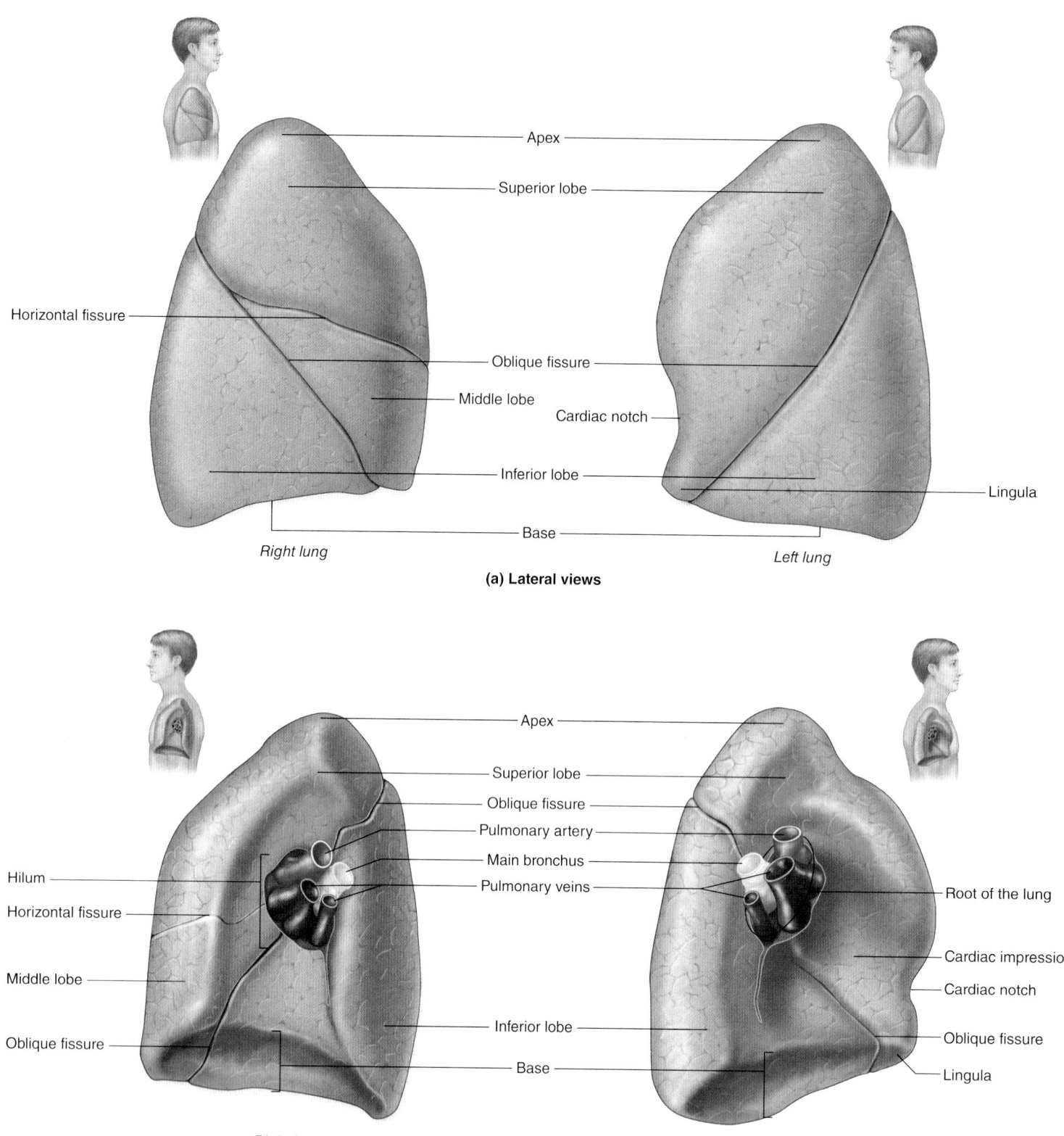

Figure 23.14 Lungs. The lungs are composed of lobes separated by distinct depressions called fissures. (*a*) Lateral views show the three lobes of the right lung and the two lobes of the left lung. (*b*) Medial views show the hilum of each lung, where the pulmonary vessels and bronchi extend as part of the root of the lung. AP|R

The right and left lungs exhibit some obvious structural differences. The right lung is larger and wider than the left lung, and is subdivided by two fissures into three lobes. The **horizontal fissure** separates the **superior** (*upper*) **lobe** from the **middle lobe,** whereas the **oblique fissure** separates the middle lobe from the **inferior** (*lower*) **lobe.** In contrast, the left lung—which is slightly smaller than the right lung because the heart projects into the left side of the thoracic cavity—has only two lobes. The left lung has an oblique fissure that separates the superior lobe from the inferior lobe. The **lingula** of the left lung is a projection from the superior lobe that is homologous to the middle lobe of the right lung. The left lung also has two surface indentations

CLINICAL VIEW
Smoking

Smoking results in the inhalation of over 200 chemicals into the respiratory passageways of the lungs. These chemicals blacken the respiratory passageways and cause respiratory changes that increase the risk of (1) respiratory infections, including the common cold, influenza, pneumonia, and tuberculosis; and (2) cellular and genetic damage to the lungs that may lead to emphysema or lung cancer.

The deleterious effects of smoking are not limited to the respiratory system. Nicotine causes vasoconstriction in the cardiovascular system, carbon monoxide interferes with oxygen binding to hemoglobin, and the risk and severity of atherosclerosis are increased. These changes decrease blood flow, thus resulting in decreased delivery of nutrients and oxygen to cells of systemic tissues. Women who smoke during pregnancy typically have babies with lower birth weight. This condition occurs in part because the umbilical arteries vasoconstrict, decreasing blood flow to the placenta.

Smoking increases the risk of both stomach ulcers caused by *Helicobacter pylori* infection and the risk of cancer of the esophagus, stomach, and pancreas. In the reproductive system, smoking increases the risks associated with human papillomavirus (HPV) infection. (HPV is linked with increased risk of cervical cancer.) Smoking also increases the risk of Alzheimer disease.

Studies indicate an association between secondhand smoke exposure and increased risk of bronchitis, asthma, and ear infections in children. New evidence also shows that thirdhand smoke, the toxins that are present in clothes, furniture, carpet, and other materials even after the cigarette is extinguished, poses a health risk— especially to infants and young children.

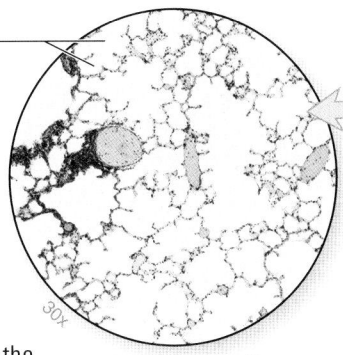

Alveoli

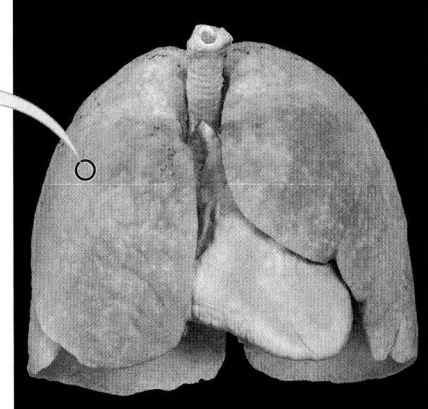

Alveoli are small, numerous, and well formed.

Nonsmoker's lungs: lungs are pink

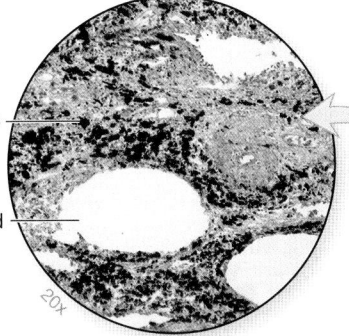

Deposits

Enlarged alveolus

Alveoli are enlarged, less numerous, and contain black deposits.

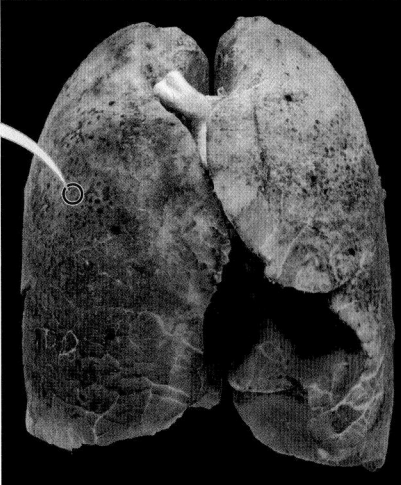

Smoker's lungs: lungs are blackened

---

to accommodate the heart: the **cardiac impression** on its medial surface and a **cardiac notch** on its anterior surface. In addition, a groovelike impression on the medial surface of the left lung accommodates the descending thoracic aorta.

Each lung is partitioned into **bronchopulmonary** (brong′kō-pul′mō-nār′ē) **segments;** there are 10 segments in the right lung and typically 8 to 10 in the left lung (**figure 23.15**). (The discrepancy in segment number for the left lung comes from the merging of some left lung segments into combined ones that occurs during development.) Each bronchopulmonary segment is an autonomous unit, encapsulated within connective tissue, and supplied by its own segmental bronchus, a branch of both the pulmonary artery and vein, and lymph vessels. Consequently, if a portion of a lung is diseased, a surgeon can remove the entire bronchopulmonary segment that is affected, and the remaining healthy segments continue to function.

Within each segment, the lung is organized into marble-sized **lobules,** each of which is surrounded by connective tissue and supplied by a terminal bronchiole, arteriole, venule, and lymph vessel.

 WHAT DID YOU LEARN?

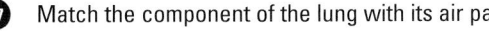

17 Match the component of the lung with its air passageway.

main bronchus	bronchopulmonary segment
lobar bronchus	lobe of lung
segmental bronchus	lobule of lung
terminal bronchiole	lung (right and left)

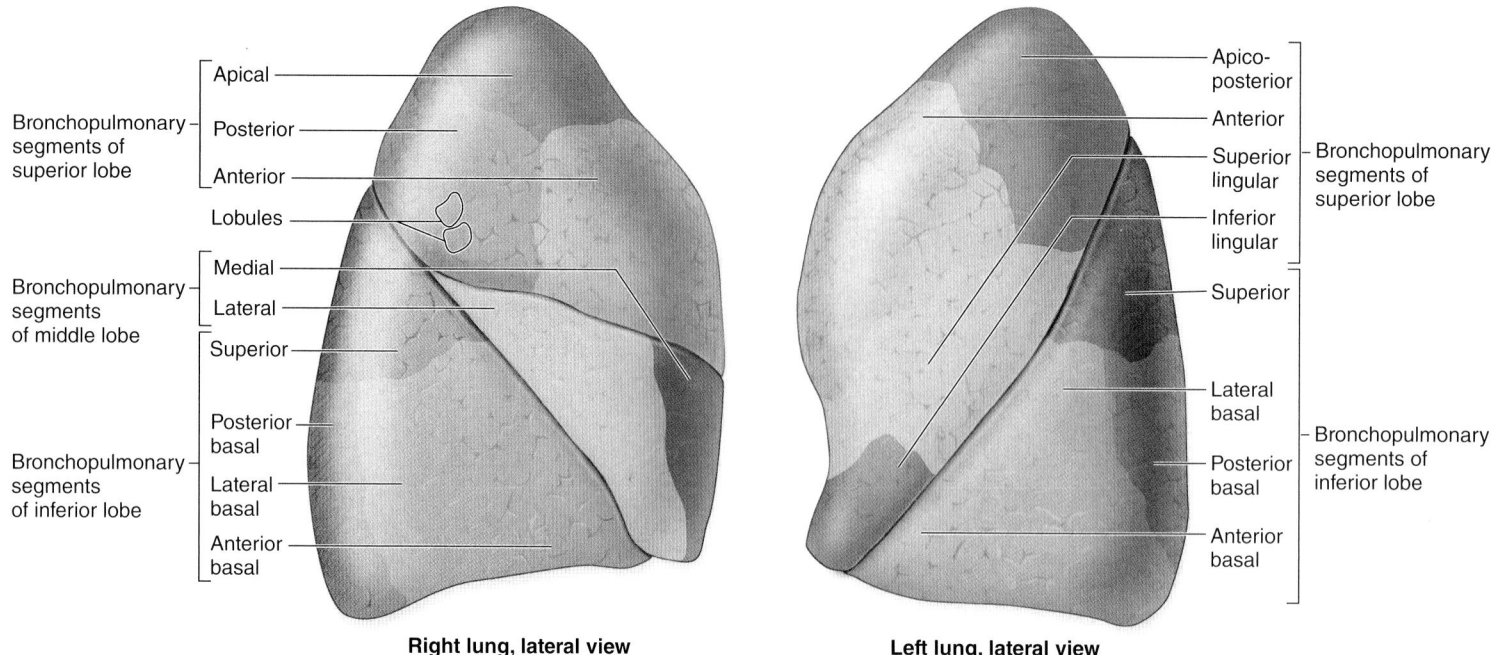

Figure 23.15 Bronchopulmonary Segments and Lobules of the Lungs. Both lungs are partitioned into self-contained bronchopulmonary segments (represented by different colors), each supplied by a segmental bronchus. Lobules are visible in each bronchopulmonary segment. (Note: The medial bronchopulmonary segment is not visible on the left lung.

23.4b Circulation to and Innervation of the Lungs

✔ LEARNING OBJECTIVES

3. Distinguish between the two types of blood circulation through the lungs.

4. Describe the innervation of lung structures by the autonomic nervous system.

The circulation of blood to and from the lungs, lymph circulation from the lungs, and innervation of the lungs by autonomic axons are described here.

Blood Supply

Two types of blood circulation are associated with the lungs: the pulmonary circulation and the bronchial circulation. Recall from our examination of the heart and blood vessels that the **pulmonary circulation (figure 23.16)** conducts blood to and from the gas exchange surfaces of the lungs to replenish its depleted oxygen levels and get rid of excess carbon dioxide. Pulmonary arteries carry deoxygenated blood to pulmonary capillaries within the lungs. The deoxygenated blood that enters these capillaries is reoxygenated here before it returns through a series of pulmonary venules and veins to the left atrium.

In comparison, the **bronchial circulation** is a component of the systemic circulation and transports oxygenated blood to the tissues of the lungs (see figure 20.23). It consists of both small bronchial arteries and veins that supply the walls of bronchi and bronchioles. The cells of the smallest respiratory structures (such as alveoli and alveolar ducts) exchange their respiratory gases directly with the inhaled

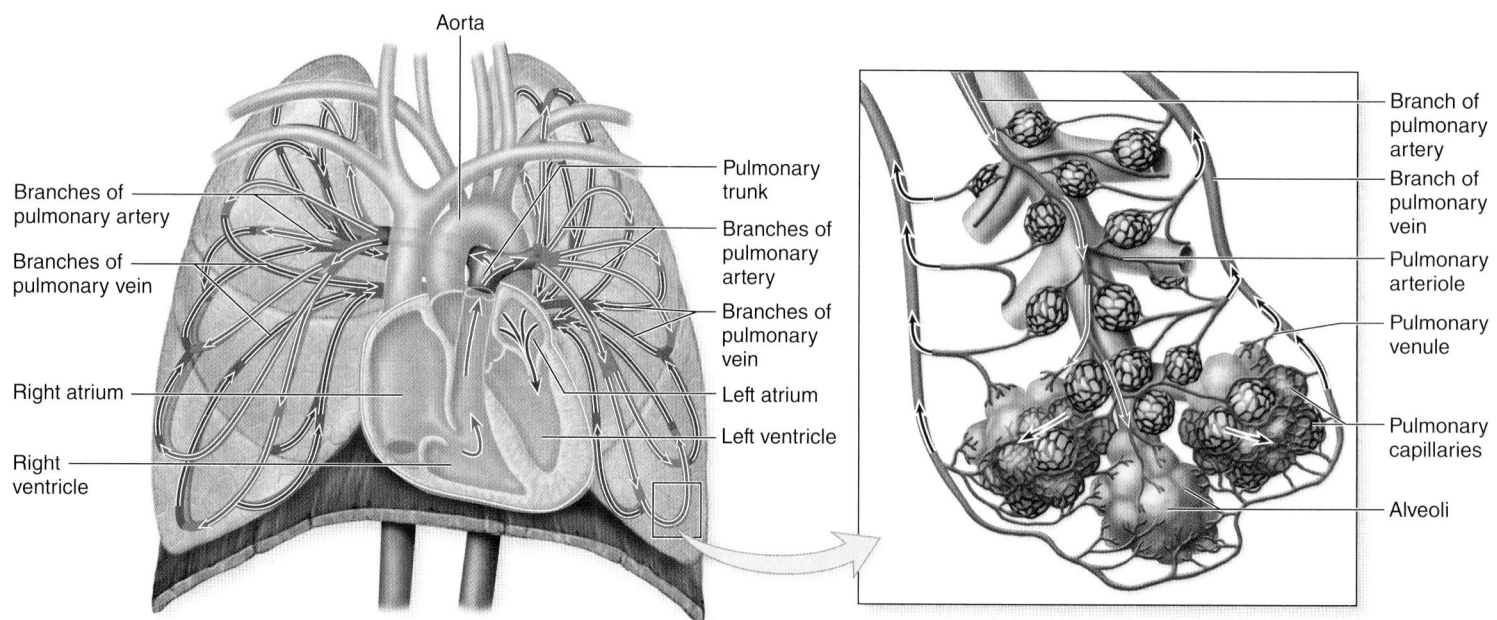

Figure 23.16 Pulmonary Circulation of the Lungs. Pulmonary circulation delivers blood to the lungs for reoxygenation and removal of carbon dioxide. **AP|R**

air. Approximately three or four **bronchial arteries** branch from the anterior wall of the descending thoracic aorta (or its branches). Thereafter, bronchial arteries divide to form capillary beds supplying structures in the bronchial tree. **Bronchial veins** collect venous blood from these capillary beds. Some of this deoxygenated blood drains into pulmonary veins, where it mixes with the freshly oxygenated blood. Consequently, blood exiting the lungs within the pulmonary veins, which will be returned to the left side of the heart and circulated throughout the body, is slightly less oxygenated than the blood immediately leaving the pulmonary capillaries following gas exchange.

Lymph Drainage

Lymph vessels and lymph nodes are located within the connective tissue of the lung, around the bronchi and in the pleura. Lymph vessels are important in removing excess fluid from the lungs. Lymph, absorbed by lymph vessels, is filtered through lymph nodes, which collect carbon, dust particles, and pollutants that were not "swept out" by cilia lining the respiratory tract. Lymph nodes may become dark in color as they accumulate the microscopic matter we inhale over a lifetime.

Innervation of the Respiratory System

The smooth muscle and glands of the larynx, trachea, bronchial tree, and lungs are innervated by the autonomic nervous system. The bronchioles receive both sympathetic and parasympathetic innervation. Sympathetic innervation to the lungs originates generally from the T1–T5 segments of the spinal cord (see section 15.4a). This innervation primarily causes bronchodilation.

Parasympathetic innervation to the lungs is from vagus nerves (CN X), and it stimulates bronchoconstriction (see section 15.3a). The vagus nerves are also the primary source of innervation to the larynx. Thus, damage to one of the vagus nerves going to the larynx can cause a person to develop a monotone or a permanently hoarse voice.

  **WHAT DID YOU LEARN?**

18 Which arteries deliver oxygenated blood to tissues of the lungs and which veins drain deoxygenated blood from the lungs? What vessels receive some of this deoxygenated blood?

23.4c Pleural Membranes and Pleural Cavity

LEARNING OBJECTIVES

5. Describe the pleural membranes and pleural cavity.
6. Explain the function of serous fluid in the pleural cavity.

The outer lung surfaces and the adjacent internal thoracic wall are lined by a serous membrane called **pleura** (plūr′ă). It is composed of simple squamous epithelium and areolar connective tissue. The **visceral pleura** tightly adheres to the lung surface, whereas the **parietal pleura** lines the internal thoracic walls, the lateral surfaces of the mediastinum, and the superior surface of the diaphragm **(figure 23.17)**. The parietal pleural layer is continuous with the visceral pleural layer at the hilum of each lung (see figure 1.9c). Each lung is enclosed in a separate visceral pleural membrane, and the heart in a visceral pericardial membrane (see section 19.2b); thus, these organs are compartmentalized, which helps limit spread of infections.

The **pleural cavity** is located between the visceral and parietal serous membrane layers. When the lungs are fully inflated, the pleural cavity is considered a *potential space* because the visceral and parietal pleurae are almost in contact with each other. An oily,

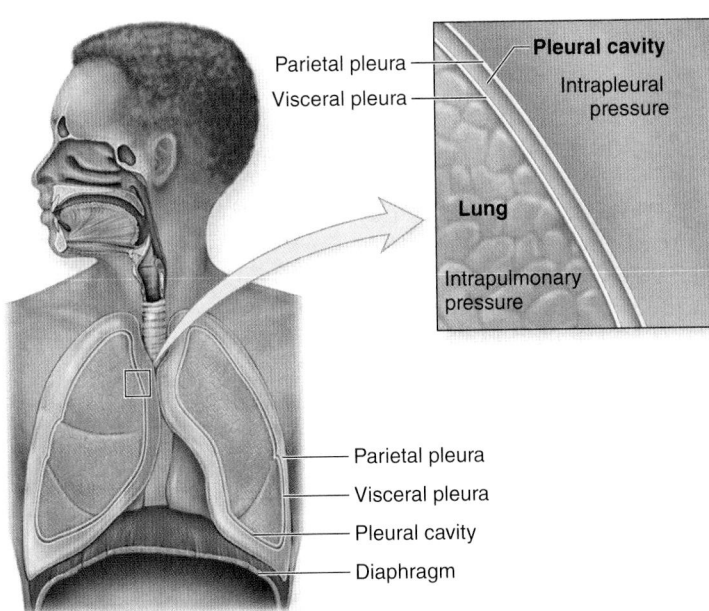

Figure 23.17 Pleural Membranes and Pressures Associated with the Lungs. The two pleural membranes include both the visceral pleura, which covers the outer surface of the lungs and the parietal pleura, which lines the inner surface of the thoracic wall. Between the two pleural membranes is a potential space called the pleural cavity. Two pressures associated with the lungs include the intrapulmonary pressure (the pressure within the lungs) and the intrapleural pressure (pressure within the pleural cavity). **AP|R**

INTEGRATE

CLINICAL VIEW

Lung Cancer

Lung cancer is a highly aggressive and frequently fatal malignancy that originates in the epithelium of the respiratory system. It claims over 150,000 lives annually in the United States. Smoking causes about 85% of all lung cancers. Symptoms include chronic cough, coughing up blood, excess pulmonary mucus, and increased likelihood of pulmonary infections.

Metastasis, the spread of cancerous cells to other tissues (most commonly the adrenal glands, liver, brain, and bone), occurs early in the course of the disease, making a surgical cure unlikely for most patients. Some people are diagnosed based upon symptoms that develop after the cancer has already metastasized to a distant site. For example, in some cases lung cancer is not discovered until the patient seeks treatment for a seizure disorder related to cancer in the brain.

Lung cancers are classified by their histologic appearance into three basic patterns: squamous cell carcinoma, adenocarcinoma, and small-cell carcinoma. **Squamous cell carcinoma** (kar-si-nō′mă; *karkinos* = cancer, *oma* = tumor) is the most common form of lung cancer. (These affected squamous cells are the cells that are formed as part of the stratified squamous epithelium that replaces the pseudostratified ciliated columnar epithelium lining air passageways in the lungs to withstand the chronic inflammation and injury caused by tobacco smoke.) **Adenocarcinoma** (ad′ĕ-nō-kar′si-nō′mă) is a less common form of lung cancer. An adenocarcinoma of the lung arises from the mucin-producing glands in the respiratory epithelium. **Small-cell carcinoma** occurs less frequently and originates from the small neuroendocrine cells in the larger bronchi.

INTEGRATE

CLINICAL VIEW
Pleurisy and Pleural Effusion

Pleurisy (plŭr'ĭ-sē) is inflammation of the pleural membranes. Usually only one lung is affected because the lungs are located within distinct compartments. The inflamed membranes increase the friction between the visceral and parietal layers, causing the pleural layers to rub against one another. Patients with this condition report severe chest pain associated with breathing.

Pleural effusion is excess fluid in the pleural cavity, which occurs when the capacity of the lymph vessels to remove fluid from the pleural cavity is exceeded by its formation. This accumulation may be caused by (1) systemic factors, such as failure of the left side of the heart, pulmonary embolism, or cirrhosis of the liver; (2) viral or bacterial infections of the lung; or (3) lung cancer that triggers the inflammatory response of the immune system within the lungs.

INTEGRATE

CLINICAL VIEW
Pneumothorax and Atelectasis

Pneumothorax (nū-mō-thōr'aks; *pneuma* = air) occurs when air gets into the pleural cavity. Pneumothorax may develop in one of two ways. Air may be introduced externally from a penetrating injury to the chest, such as a knife wound or gunshot, or it may originate internally either when a broken rib lacerates the surface of the lung or an alveolus ruptures.

Lung expansion is dependent upon intrapleural pressure being lower than the intrapulmonary pressure. However, pneumothorax sometimes causes intrapleural and intrapulmonary pressures to become equal, so the lungs are released from the "outward pull" of the chest wall and collapse, or deflate. A collapsed lung is termed **atelectasis** (at-ĕ-lek'tă-sis; *ateles* = incomplete, *ektasis* = extension). The collapsed portion of the lung remains down until the air has been removed from the pleural space. If the amount of air is small, it exits naturally within a few days. However, a large introduction of air is a medical emergency and requires insertion of a tube into the pleural space to suction out the air.

serous fluid is produced by the serous membranes and released into the pleural cavity. Serous fluid acts as a lubricant, ensuring the pleural surfaces slide by each other with minimal friction during breathing. Each pleural cavity normally contains less than 15 mL of serous fluid and is drained continuously by lymph vessels within the visceral pleura. A balance normally exists between formation and removal of serous fluid in the pleural cavity.

 WHAT DID YOU LEARN?

19 What is the function of serous fluid within the pleural cavity?

23.4d How Lungs Remain Inflated
LEARNING OBJECTIVE

7. Explain the anatomic properties that keep lungs inflated.

Lung inflation occurs due to the expanding properties of the chest wall, the recoiling properties of the lungs, and the anatomic arrangement of the pleural cavity between chest wall and lungs.

The chest wall is anatomically configured to expand outwardly. This is readily observed when the thoracic cage is opened surgically, because cutting through the chest wall causes it to "spring" open. The lungs cling to the internal surface of the chest wall as it expands outward because of the surface tension caused by the serous fluid within the pleural cavity. However, the lungs are composed of vast amounts of elastic connective tissue that is "stretched" as the lungs expand. The natural tendency of elastic connective tissue is to recoil, which causes the lungs to exhibit a noticeable inward pull.

The contrasting outward pull of the chest wall and the opposing inward pull of the lungs results in a vacuum or "suction" within the pleural cavity. Consequently, the pressure generated in the pleural cavity, called the **intrapleural pressure,** is lower than the pressure inside the lungs, called the **intrapulmonary** (or *intra-alveolar*) **pressure** (figure 23.17). This difference in pressure keeps the lungs inflated. The lungs remain inflated similar to the way in which a balloon remains inflated—namely, the pressure inside is greater than the pressure outside. Note that when intrapleural pressure becomes equal to intrapulmonary pressure the lungs deflate, like the deflation of a balloon when the stem is released.

 WHAT DID YOU LEARN?

20 Why is the intrapleural pressure normally lower than intrapulmonary pressure? What is the function of this difference in pressure?

23.5 Respiration: Pulmonary Ventilation

Respiration is a general term for the exchange of respiratory gases (oxygen and carbon dioxide) between the atmosphere and the systemic cells of the body—it is organized into four continuous and simultaneously occurring processes:

- **Pulmonary ventilation**—movement of respiratory gases between the atmosphere and the alveoli of the lungs
- **Alveolar gas exchange** (or *external respiration*)—exchange of respiratory gases between the alveoli and the blood in the pulmonary capillaries
- **Gas transport**—transport of respiratory gases within the blood between the lungs and systemic cells of the body
- **Systemic gas exchange** (or *internal respiration*)—exchange of respiratory gases between the blood in the systemic capillaries and systemic cells of the body

Figure 23.18 and **table 23.2** provide a summary for the movement of respiratory gases involving these four processes. Notice that the net movement of oxygen is from the atmosphere to the systemic cells, and involves the following, which occur simultaneously and continuously:

1. Air containing oxygen is inhaled into the alveoli during the inspiratory phase of pulmonary ventilation.

2. Oxygen diffuses from alveoli into the blood of pulmonary capillaries during alveolar gas exchange.

3. Oxygen is transported within the blood from the lungs to the systemic cells of the body.

4. Oxygen diffuses from the blood within the systemic capillaries into the systemic cells during systemic gas exchange.

Table 23.2 Respiration Processes

Process	Description	Body Systems
Pulmonary ventilation	Movement of air between atmosphere and the alveoli • Net movement of oxygen from atmosphere to alveoli during inspiration (step 1) • Net movement of carbon dioxide from alveoli to atmosphere during expiration (step 8)	Respiratory, skeletal, muscular, and nervous
Alveolar gas exchange	Exchange of respiratory gases between alveoli of the lungs and the blood • Oxygen diffuses from alveoli into blood (step 2) • Carbon dioxide diffuses from blood into alveoli (step 7)	Respiratory and cardiovascular
Gas transport	Blood transport of respiratory gases between lungs and systemic cells of the body • Oxygen is transported from lungs to systemic cells (step 3) • Carbon dioxide is transported from systemic cells to lungs (step 6)	Cardiovascular
Systemic gas exchange	Exchange of respiratory gases between blood and systemic cells • Oxygen diffuses from blood into systemic cells (step 4) • Carbon dioxide diffuses from systemic cells into blood (step 5)	Cardiovascular

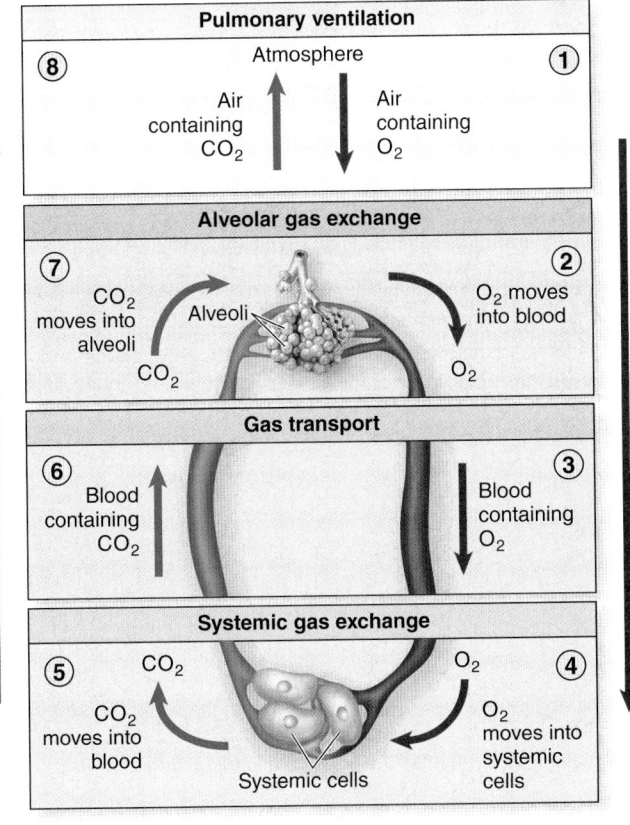

Figure 23.18 Overview of Respiration. Respiration involves four processes that include pulmonary ventilation, alveolar gas exchange, gas transport, and systemic gas exchange.

(5) Carbon dioxide diffuses from systemic cells into the blood within systemic capillaries during systemic gas exchange.

(6) Carbon dioxide is transported within the blood from systemic cells to the lungs.

(7) Carbon dioxide diffuses from the blood within the pulmonary capillaries into the alveoli during alveolar gas exchange.

(8) Air containing carbon dioxide is then exhaled from the alveoli into the atmosphere during the expiratory phase of pulmonary ventilation.

Pulmonary ventilation is discussed in this section. The other three processes of respiration are covered later in sections 23.6 and 23.7.

23.5a Introduction to Pulmonary Ventilation

LEARNING OBJECTIVE

1. Give an overview of the process of pulmonary ventilation.

Pulmonary ventilation is known as *breathing,* which is the movement of air between the atmosphere and the alveoli. It consists of two cyclic phases: **inspiration** (also called *inhalation*), which brings air into the lungs and **expiration** (also called *exhalation*), which forces air out of the lungs. Breathing may be either quiet or forced. **Quiet breathing** is the rhythmic breathing that occurs at rest; **forced breathing** is vigorous breathing that accompanies exercise or hard exertion.

The principles involved in breathing, whether quiet or forced, use the same physiologic processes. Autonomic nuclei in the brainstem (see sections 13.5b and c) regulate the skeletal muscles involved with breathing to cyclically contract and relax, resulting in thoracic cavity volume changes. Dimensional changes within the thoracic cavity during breathing result in pressure changes, establishing a changing pressure gradient between the lungs and the atmosphere. Air moves down the pressure gradient either to enter the lungs during inspiration or to exit the lungs during expiration. Three major topics are addressed in this section: how pressure gradients are established during breathing (mechanics of breathing), how breathing is controlled by the central nervous system, and the different pulmonary measurements associated with breathing.

WHAT DID YOU LEARN?

21 What are the general steps of pulmonary ventilation beginning with the autonomic nuclei?

23.5b Mechanics of Breathing

LEARNING OBJECTIVES

2. Explain how pressure gradients are established and result in pulmonary ventilation.

3. State the relationship between pressure and volume as described by Boyle's law.

4. Distinguish between quiet and forced breathing.

The mechanics of breathing involve several integrated aspects: (1) the specific actions of the skeletal muscles of breathing; (2) dimensional (volume) changes within the thoracic cavity; (3) pressure changes resulting from volume changes (based on Boyle's gas law); (4) pressure gradients; and (5) volumes and pressures associated with breathing.

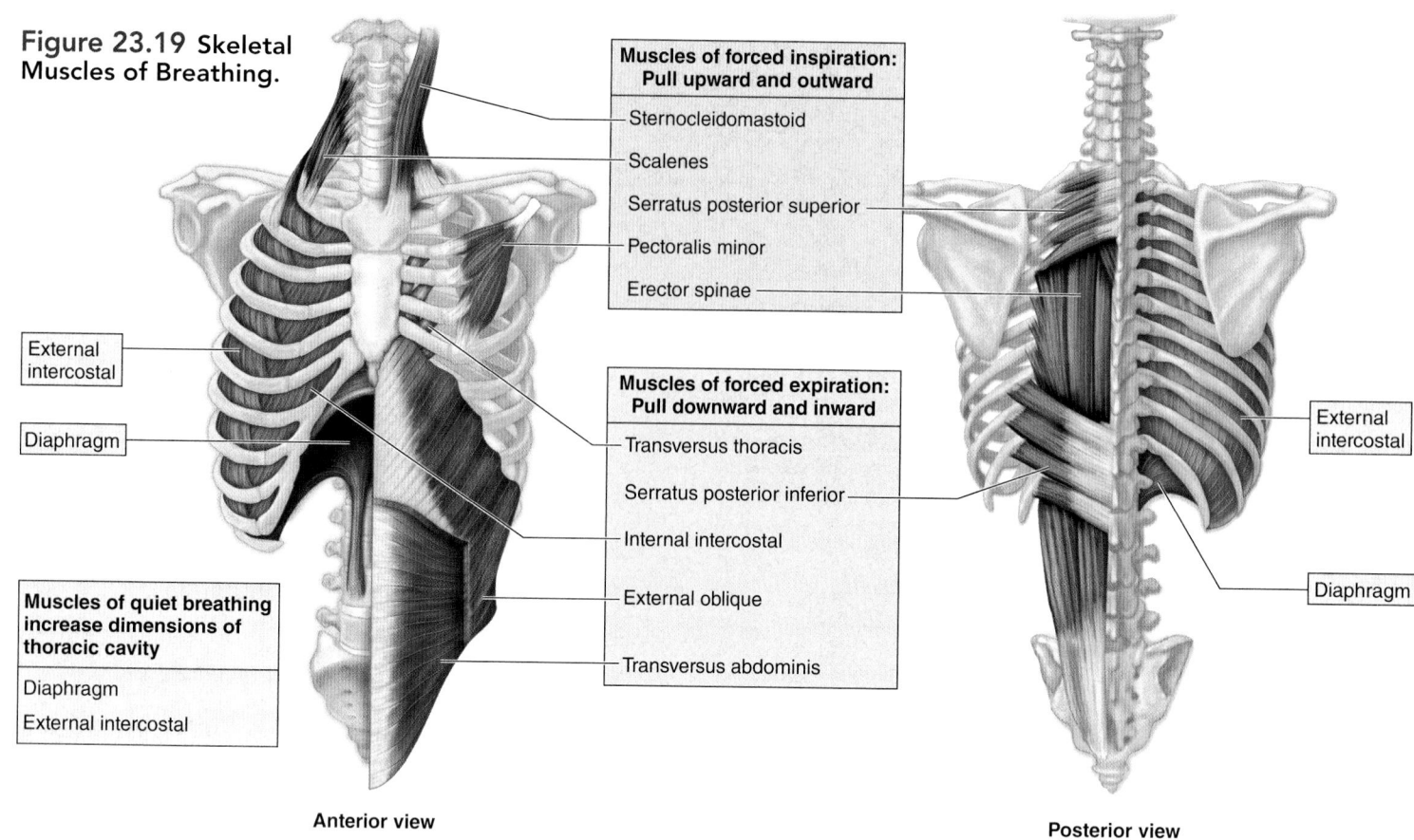

Figure 23.19 Skeletal Muscles of Breathing.

Muscles of forced inspiration: Pull upward and outward
- Sternocleidomastoid
- Scalenes
- Serratus posterior superior
- Pectoralis minor
- Erector spinae

Muscles of forced expiration: Pull downward and inward
- Transversus thoracis
- Serratus posterior inferior
- Internal intercostal
- External oblique
- Transversus abdominis

External intercostal

Diaphragm

Muscles of quiet breathing increase dimensions of thoracic cavity

Diaphragm

External intercostal

External intercostal

Diaphragm

Anterior view

Posterior view

Muscles of Breathing	
Muscles of quiet breathing (increase dimensions of thoracic cavity)	The **diaphragm** forms the rounded "floor" of the thoracic cavity and is dome-shaped when relaxed. It alternates between the relaxed domed position and the contracted flattened position and changes the vertical dimensions of the thoracic cavity.
	The **external intercostals** extend from a superior rib inferomedially to the adjacent inferior rib. These elevate the ribs and increase the transverse dimensions of the thoracic cavity.
Muscles of forced inspiration (pull upward and outward)	The **sternocleidomastoid** attaches to sternum and clavicle; lifts rib cage.
	The **scalenes** attach to ribs 1 and 2; elevate ribs 1 and 2.
	The **pectoralis minor** attaches to ribs 3–5; elevates ribs 3–5.
	The **serratus posterior superior** attaches to ribs 2–5 on its anterior surface; lifts ribs 2–5.
	The **erector spinae** is a group of deep muscles along the length of the vertebral column; extend or straighten the vertebral column.
Muscles of forced expiration (pull downward and inward)	The **internal intercostals** lie deep and at right angles to the external intercostals; depress the ribs and decrease the transverse dimensions of the thoracic cavity.
	The **abdominal muscles** (primarily the external obliques and transversus abdominis) compress the abdominal contents, forcing the diaphragm into a higher domed position and the rectus abdominis pulls the sternum and rib cage inferiorly.
	The **transversus thoracis** extends across the inner surface of the thoracic cage and attaches to ribs 2–6; depresses ribs 2–6.
	The **serratus posterior inferior** extends between the ligamentum nuchae and the lower border of ribs 9–12; depresses ribs 9–12.

Each of these aspects of breathing are discussed and then this information is integrated to describe how breathing occurs.

Skeletal Muscles of Breathing

The skeletal muscles of breathing (see section 11.5) are classified into the following three categories: muscles of quiet breathing, muscles of forced inspiration, and muscles of forced expiration (**figure 23.19**):

- **Muscles of quiet breathing** are the skeletal muscles involved in normal rhythmic breathing that occurs at rest. They are the **diaphragm** and the **external intercostals.** These muscles

alternately contract and relax, resulting in movement of air into and out of the lungs. The diaphragm forms the rounded "floor" of the thoracic cavity and is dome-shaped when relaxed. When it contracts, its central portion flattens and moves inferiorly to press against the abdominal viscera. The external intercostal muscles extend from a superior rib inferomedially to the adjacent inferior rib; contraction of these muscles elevate the ribs.

- **Muscles of forced inspiration** are used during a deep inspiration, such as occurs during heavy exercise or prior to "holding a long note" while singing. These muscles include the **sternocleidomastoid, scalenes, pectoralis minor,**

INTEGRATE

CONCEPT CONNECTION

Cellular respiration is the metabolic pathway whereby glucose and other fuel molecules, such as fatty acids, are oxidized and their chemical energy is transferred to form adenosine triphosphate (ATP) as described in section 3.4. The carbon atoms of the oxidized fuel molecules are released as carbon dioxide, a waste product. Oxygen eventually "catches electrons" and joins with hydrogen ions to form water as the final step in the electron transport chain. The respiratory system and cardiovascular system support the process of cellular respiration by delivering the oxygen and removing the carbon dioxide waste.

serratus posterior superior, and **erector spinae.** These muscles, except for the erector spinae, are located in a more superior location relative to the thoracic cavity and can effectively move the rib cage superiorly, laterally, and anteriorly, resulting in a greater increase in the volume within the thoracic cavity than occurs during quiet inspiration. Located along the length of the vertebral column, the erector spinae muscles also aid in lifting the rib cage, but do so by extending the vertebral column as occurs when you "sit up straight."

- **Muscles of forced expiration** contract during a hard expiration, for example, when one blows up a balloon or coughs. The muscles of forced expiration include the **internal intercostals, abdominal muscles, transversus thoracis,** and **serratus posterior inferior.** In general, these muscles either pull the rib cage inferiorly, medially, and posteriorly or compress the abdominal contents to move the diaphragm superiorly into the thoracic cavity, resulting in a greater decrease in volume within the thoracic cavity than occurs during quiet expiration.

The muscles of forced inspiration and forced expiration are collectively referred to as the *accessory muscles of breathing.*

Volume Changes in the Thoracic Cavity

The cyclic activity of breathing muscles causes thoracic cavity volume changes that occur in three dimensions: vertically, laterally, and in an anterior-posterior direction **(figure 23.20)**.

Vertical dimension changes of the thoracic cavity result from the movement of the diaphragm. When it contracts, its central portion flattens and moves inferiorly, which increases the vertical dimensions of the thoracic cavity. When the diaphragm relaxes and returns to its original position, the vertical dimensions decrease.

Only small movements of the diaphragm are required for breathing, and usually the changes in vertical dimension measure only a few millimeters during quiet breathing. Greater changes in the superior movement of the diaphragm occur during forced expiration because of contraction of the abdominal muscles.

Lateral dimension changes occur either as the rib cage is elevated and the thoracic cavity widens, or as the rib cage depresses and thoracic cavity narrows. This action can be mimicked by placing your hands at the sides of your ribs and then abducting and adducting your hands relative to your ribs.

Anterior-posterior dimension changes occur as the inferior portion of the sternum moves anteriorly and then posteriorly. This action can be visualized by placing one hand on the front of your lower chest and lifting it outwardly away from the chest and then back. In general, lateral and anterior-posterior dimensional changes both occur as a result of the contraction and relaxation of all muscles of breathing shown in figure 23.19, except for the diaphragm.

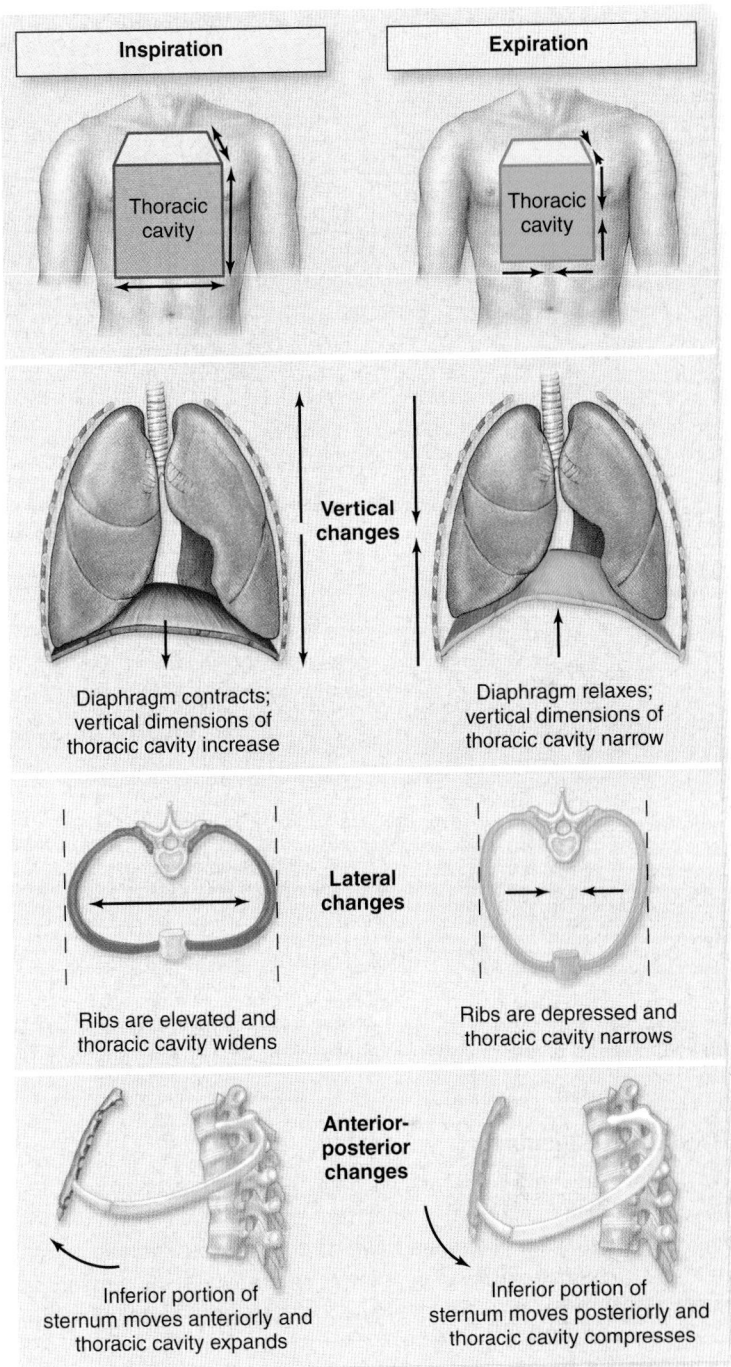

Figure 23.20 Thoracic Cavity Dimensional Changes Associated with Breathing. The box-like thoracic cavity changes size during inspiration and expiration. The box increases in vertical, lateral, and anterior-posterior dimensions during inspiration due to movement of the diaphragm, ribs, and sternum. These dimensions decrease upon expiration.

Boyle's Gas Law: Relationship of Volume and Pressure

Volume changes in the thoracic cavity cause gas pressure changes in the thoracic cavity. **Boyle's law** states that at a constant temperature, the pressure (P) of a gas decreases if the volume (V) of the container increases, and vice versa. The law may be expressed by the following formula:

$$P_1V_1 = P_2V_2$$

where P_1 and V_1 represent the initial conditions, and P_2 and V_2 represent the changed conditions for pressure and volume, respectively. This formula expresses an inverse relationship that exists between gas pressure and volume. **Figure 23.21a** visually helps you to understand this relationship.

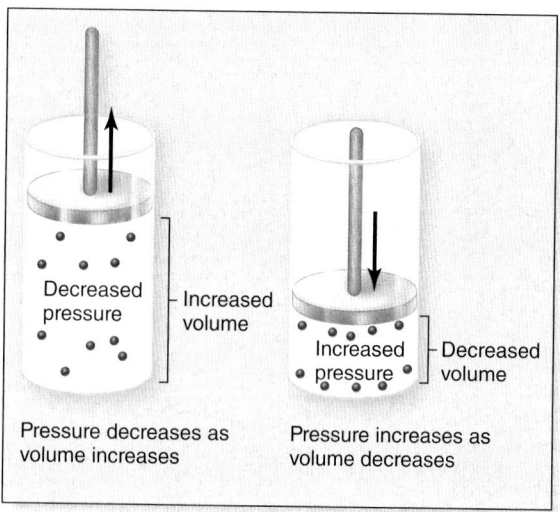

(a) Boyle's law

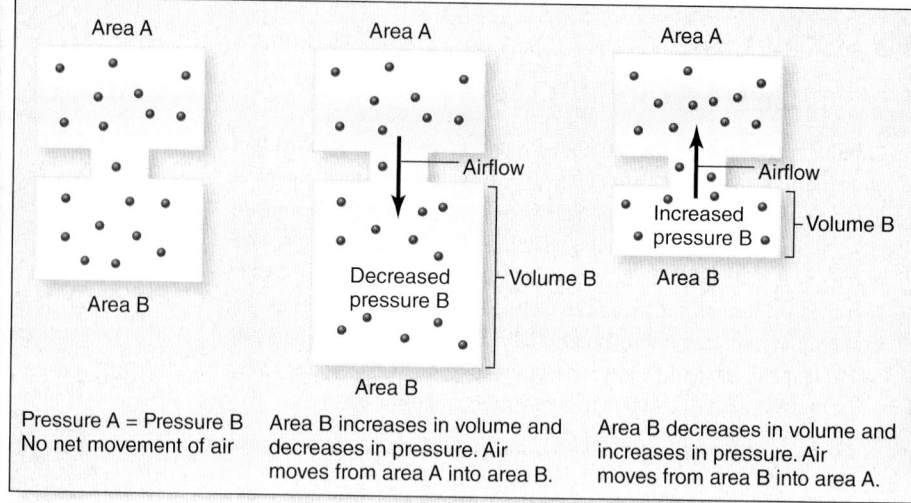

(b) Pressure gradients

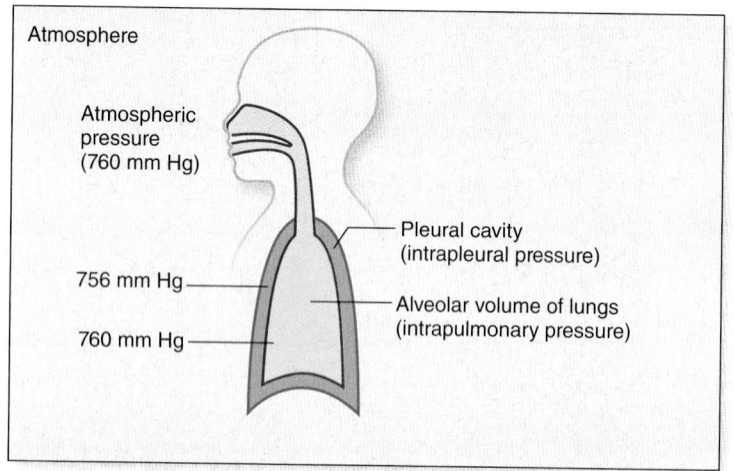

(c) Volumes and pressures with breathing (at the end of an expiration)

Figure 23.21 Boyle's Law and Pressure Gradients.
(*a*) Boyle's law states that volume and pressure are inversely related.
(*b*) The figure shows that air does not move if the pressure is equal between two areas. It also shows how air moves from an area of high pressure to an area of low pressure when pressure gradients are established by changes in volume. (*c*) Three significant pressures are associated with breathing. Atmospheric pressure is a set value of 760 mm Hg at sea level. The other two are the intrapulmonary pressure and the intrapleural pressure, which both change during breathing as a result of changing volumes. **AP|R**

Pressure Gradients

An air pressure gradient occurs when the force per unit area is greater in one area than in another. If a pressure gradient exists between two regions, and they are interconnected, then air moves from the region of higher pressure to the region of lower pressure until the pressure in the two regions becomes equal. Figure 23.21*b* shows this relationship.

Volumes and Pressures Associated with Breathing

A similar relationship exists with respect to the atmosphere and the lungs, which are interconnected by the respiratory passageway (figure 23.21*c*).

The **atmosphere** is the air in the environment that surrounds us. **Atmospheric pressure** is the pressure (weight) gases in the air exert in the environment. Its value changes with altitude. The standard value is given at sea level because air becomes less dense or "thins" with increased altitude. The "thinner air" exerts a correspondingly lower atmospheric pressure. The value for atmospheric pressure at sea level can be expressed in several ways: 14.7 pounds per square inch = 1 atmosphere (atm) = 760 mm Hg. The value used in this text is 760 mm Hg. (Note that 1 millimeter of mercury is expressed as 1 mm Hg, and it is the pressure exerted by a column of mercury that is 1 millimeter high in a glass tube.) Atmospheric pressure does not change in the context of breathing.

The thoracic cavity contains the lungs. The collective volume of the alveoli within the lungs is called the **alveolar volume,** and its associated pressure is the intrapulmonary pressure (described in section 23.4d). This pressure fluctuates with breathing and may be higher, lower, or the same as atmospheric pressure. Intrapulmonary

pressure is equal to atmospheric pressure (at sea level = 760 mm Hg) at the end of both inspiration and expiration.

Recall that the lungs are separated from the wall of the thoracic cavity by the pleural cavity. The pressure exerted within the pleural cavity is called the intrapleural pressure. Intrapleural pressure also fluctuates with breathing and is always lower than intrapulmonary pressure so that the lungs remain inflated. Prior to inspiration, it is generally about 4 mm Hg lower than intrapulmonary pressure (756 mm Hg).

Note that a volume change in the thoracic cavity occurring during inspiration and expiration establishes a pressure gradient between the atmosphere and the thoracic cavity that determines the direction of airflow. Thus, an increase in volume of the thoracic cavity, with an accompanying decrease in pressure, results in air moving into the lungs during inspiration. In contrast, a decrease in volume of the thoracic cavity, with an accompanying increase in pressure, results in air moving out of the lungs during expiration.

Quiet Breathing

Recall that quiet breathing is the normal breathing that occurs when you are relaxed. Refer to **figure 23.22** as you read through this description of the stepwise sequence of events that occurs during quiet breathing:

Inspiration

1. Initially, both intrapulmonary pressure and atmospheric pressure are 760 mm Hg at sea level prior to inspiration. Intrapleural

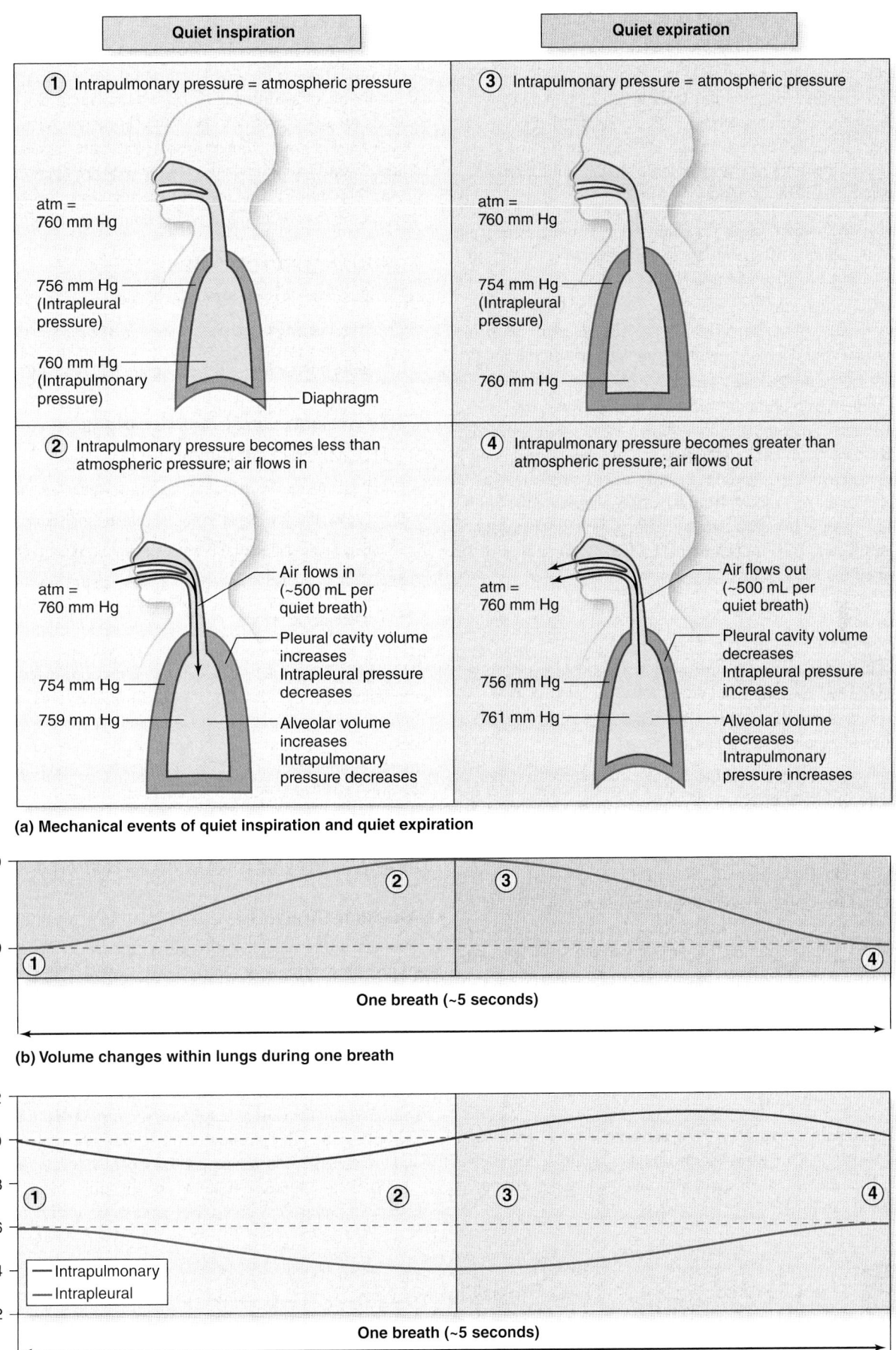

(a) Mechanical events of quiet inspiration and quiet expiration

(b) Volume changes within lungs during one breath

(c) Pressure changes within lungs and pleural cavity

Figure 23.22 Volume and Pressure Changes Associated with the Mechanics of Quiet Breathing. Circled numbers correspond to the steps described in the text. (*a*) Schematic of changes in volume, pressure, and air flow that are associated with breathing. (*b*) Approximately 500 mL of air is inspired and then expired during quiet breathing; these volumes are associated with (*c*) relatively small changes in intrapulmonary and intrapleural pressures.

pressure is slightly lower at approximately 756 mm Hg, or about 4 mm Hg lower than intrapulmonary pressure.

(2) The diaphragm contracts, increasing the thoracic cavity vertical dimensions, and the external intercostals contract, increasing both lateral and anterior-posterior dimensions. During quiet breathing, diaphragmatic movement may account for about two-thirds of the thoracic cavity volume change, and external intercostal movement for about one-third.

- Muscle contraction results in pleural cavity volume increasing with an accompanying decrease in intrapleural pressure from approximately 756 to 754 mm Hg.
- Simultaneously, the lungs expand because of the surface tension caused by serous fluid in the pleural cavity with an accompanying decrease in intrapulmonary pressure from 760 to 759 mm Hg.
- When the intrapulmonary pressure decreases below atmospheric pressure, air moves down the pressure gradient from the environment into the alveoli, until the intrapulmonary pressure is once again equal to atmospheric pressure. The volume of air that moves from the atmosphere into the lungs during a single breath in quiet breathing is approximately 500 milliliters (mL) or 0.5 liters (L). This volume of air is called the tidal volume.

Expiration

(3) Initially, the intrapulmonary pressure has equalized with atmospheric pressure (both at 760 mm Hg) prior to quiet expiration, and intrapleural pressure is about 754 mm Hg, still lower than intrapulmonary pressure.

(4) The diaphragm and external intercostals relax, decreasing the dimensions of the thoracic cavity.

- The pleural cavity volume decreases as the diaphragm relaxes and the thoracic wall recoils. Intrapleural pressure now increases from 754 mm Hg to return to 756 mm Hg.
- Simultaneously, the alveolar volume decreases because the lungs are pulled inward by the recoil of elastic connective tissue in the lungs. Intrapulmonary pressure now increases from 760 to 761 mm Hg.
- When intrapulmonary pressure exceeds atmospheric pressure, air is forced out of the alveoli into the atmosphere. This continues until the intrapulmonary pressure is once again equal to atmospheric pressure. Approximately 500 mL of air moves out of the lungs.

The changes in thoracic cavity volume and both intrapulmonary and intrapleural pressures that are associated with quiet breathing are shown in figure 23.22b, c. The events of quiet inspiration and quiet expiration are summarized in **table 23.3**.

Forced Breathing

Forced breathing involves steps similar to quiet breathing. However, both forced inspiration and expiration are active processes, requiring contraction of additional muscles (see figure 23.19). Their activity causes greater changes in both the thoracic cavity volume and intrapulmonary pressure. Consequently, more air moves into and out of the lungs. Significant chest volume changes are apparent that accompany forced breathing, unlike the barely perceptible changes in the chest cavity volume that occur during quiet breathing.

 WHAT DID YOU LEARN?

22 Describe the sequence of events of quiet inspiration.

23 How are larger amounts of air moved between the lungs and atmosphere during forced inspiration and forced expiration? Is more energy expended during forced breathing? Why?

23.5c Nervous Control of Breathing

 LEARNING OBJECTIVES

5. Describe the anatomic structures involved in regulating breathing.

6. Explain the physiologic events associated with controlling quiet breathing.

7. Explain the different reflexes that alter breathing rate and depth.

8. Distinguish between nervous system control of structures of the respiratory system and nervous system control of structures involved in breathing.

The skeletal muscles of breathing are coordinated by nuclei within the brainstem, and the rate and depth of breathing are altered through either reflexes or consciously controlled by the cerebrum. Here we first describe the anatomic components of the (1) respiratory center, which regulates breathing; (2) skeletal muscles of breathing and their innervation; and (3) receptors that detect stimuli and relay sensory input to the respiratory center to alter breathing rate and depth. We then integrate these structures to discuss how the nervous system controls breathing. Please refer to **figure 23.23** as you read through this section.

Anatomic Structures

The autonomic nuclei within the central nervous system (CNS) that control breathing are collectively called the **respiratory center** (figure 23.23a). One portion of this center, which is located within the medulla oblongata, is called the **medullary respiratory center.** Two groups of nuclei compose the medullary respiratory center. These are the **ventral respiratory group (VRG),** located within the anterior region of the medulla (which contains both inspiratory neurons and expiratory neurons), and the **dorsal respiratory group (DRG),** located posterior to the VRG. The other portion of the respiratory center is within the pons and is called the **pontine respiratory center** (or *pneumotaxic center*).

Skeletal muscles of breathing include the diaphragm, external intercostal muscles, and other accessory muscles of breathing, as

Table 23.3	Changes Associated with Quiet Breathing	
Variable	**Inspiration**	**Expiration**
Diaphragm and external intercostals	Contracting (active)	Relaxing (passive)
Pleural cavity	Volume increases; pressure decreases	Volume decreases; pressure increases
Lungs	Volume increases; pressure decreases	Volume decreases; pressure increases
Air movement	Into lungs	Out of lungs

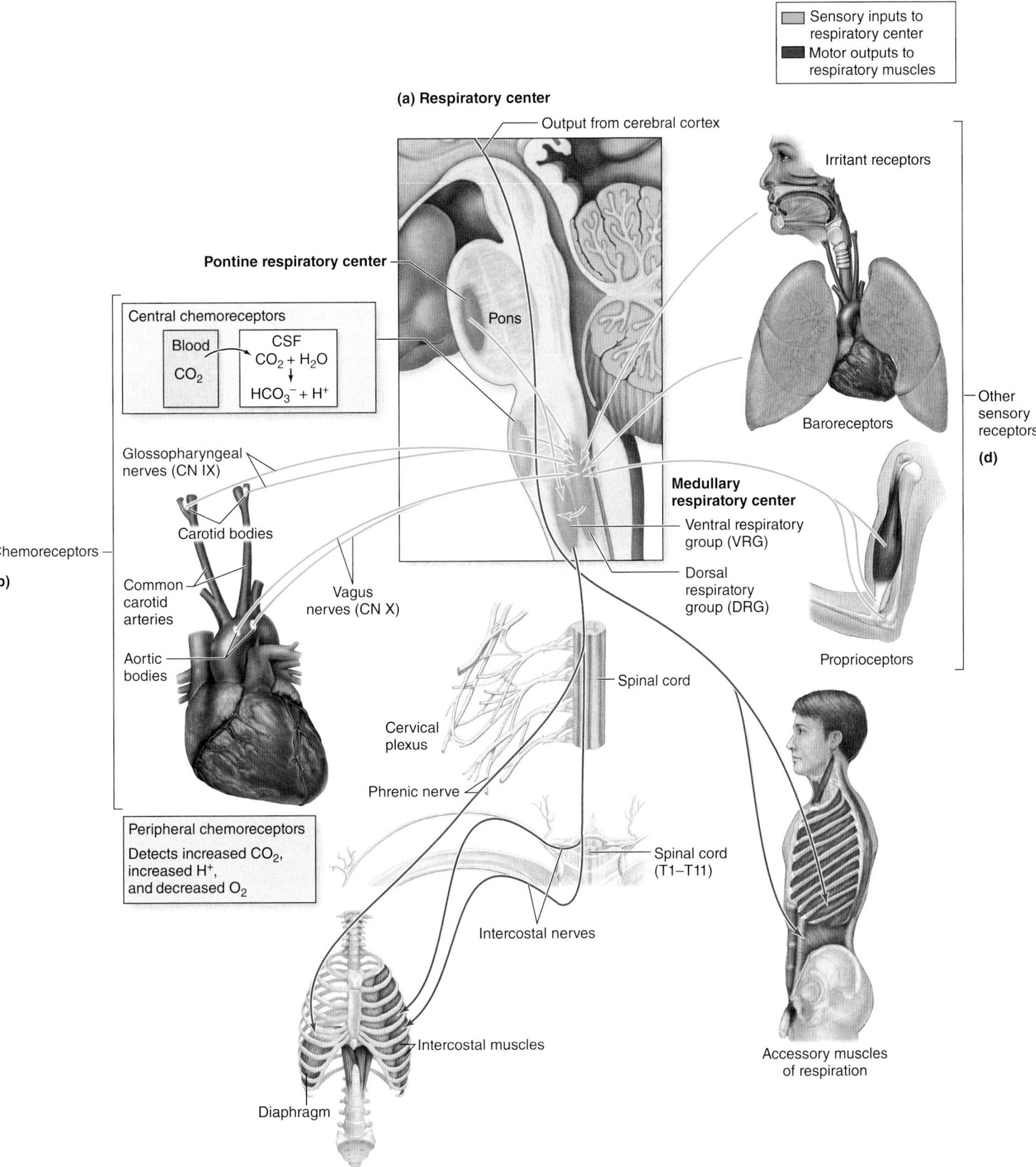

(a) Respiratory center

Output from cerebral cortex

Irritant receptors

Pontine respiratory center

Central chemoreceptors

Blood	CSF
CO_2	$CO_2 + H_2O$
	$\downarrow$
	$HCO_3^- + H^+$

Pons

Baroreceptors

Other sensory receptors

(d)

Glossopharyngeal nerves (CN IX)

Medullary respiratory center

Ventral respiratory group (VRG)

Carotid bodies

Dorsal respiratory group (DRG)

Chemoreceptors

Vagus nerves (CN X)

(b)

Common carotid arteries

Proprioceptors

Aortic bodies

Spinal cord

Cervical plexus

Peripheral chemoreceptors

Detects increased CO_2, increased H^+, and decreased O_2

Phrenic nerve

Spinal cord (T1–T11)

Intercostal nerves

Accessory muscles of respiration

Intercostal muscles

Diaphragm

(c) Skeletal muscles of breathing

Sensory inputs to respiratory center
Motor outputs to respiratory muscles

Figure 23.23 Respiratory Center. (*a*) The respiratory center rhythmically sends motor output to the diaphragm and external intercostals to regulate quiet breathing. The respiratory center is stimulated to alter the breathing rate and depth primarily by (*b*) sensory input from central chemoreceptors in the brain and peripheral chemoreceptors within the carotid and aortic bodies. Based on this sensory input, nerve signals are relayed to (*c*) the skeletal muscles of breathing to alter the breathing rate or depth. Sensory input is also relayed to the respiratory center from (*d*) other sensory receptors, including irritant receptors of the mucosal lining of the respiratory tract, baroreceptors (stretch receptors) of the lungs and visceral pleura, and proprioceptors of muscles, tendons, and joints. Note that breathing can be consciously controlled by the cerebral cortex; its motor output bypasses the respiratory center to directly stimulate lower motor neurons innervating skeletal muscles of breathing. AP|R

previously described (see figure 23.19). Upper motor neurons from the VRG synapse with lower motor neurons that extend from the spinal cord to the skeletal muscles of breathing (figure 23.23c). These lower motor neurons are within either the **phrenic nerves** that innervate the diaphragm or the **intercostal nerves** that innervate the intercostal muscles. Accessory muscles of respiration are innervated by other individually named somatic nerves.

Chemoreceptors are the primary sensory receptors involved in altering breathing (figure 23.23b). Chemoreceptors monitor fluctuations in concentration of both hydrogen ions (H⁺) and respiratory gases (P_{CO_2} and P_{O_2}) within both the cerebrospinal fluid (CSF) and the blood. Note that the concentrations of respiratory gases are expressed as the partial pressure of carbon dioxide (P_{CO_2}) and the partial pressure of oxygen (P_{O_2}). Partial pressures are discussed later in section 23.6a. For now, just remember that the higher the partial pressure for a gas, the greater its concentration. Chemoreceptors are housed both within the brain (central chemoreceptors) and in specific blood vessels (peripheral chemoreceptors):

- **Central chemoreceptors** are within the medulla oblongata in close proximity to the medullary respiratory center (see section 13.5c). Central chemoreceptors monitor only pH changes of CSF induced by changes in blood P_{CO_2}. Carbon dioxide diffuses from the blood into the CSF. In the CSF, carbonic anhydrase catalyzes the formation of carbonic acid from carbon dioxide and water. The carbonic acid then dissociates into bicarbonate (HCO_3^-) and hydrogen ions (H⁺). Thus, central chemoreceptors are monitoring H⁺ concentration within the CSF, which is formed from CO_2 that diffuses there from the blood.

- **Peripheral chemoreceptors** are located both within the aortic arch (called the **aortic bodies**) and at the split of each common carotid artery into the external and internal carotid arteries (called **carotid bodies**). Peripheral chemoreceptors normally detect changes in the concentration of both H⁺ and P_{CO_2} within arterial blood. The peripheral chemoreceptors differ from central chemoreceptors because they are stimulated by changes in H⁺ produced independently of P_{CO_2}. This occurs, for example, when H⁺ accumulates either during metabolic acidosis, as a result of either kidney failure (kidneys normally eliminate H⁺, see section 24.6d), or when ketoacidosis develops, in the case of diabetes mellitus (ketoacids are a by-product from metabolism of fatty acids; see section 25.6c). Peripheral chemoreceptors can also be stimulated by relatively large changes in blood P_{O_2}. When stimulated either by changes in blood pH or blood respiratory gases, the carotid bodies alter nerve signals relayed along the glossopharyngeal nerves and the aortic bodies alter nerve signals relayed along the vagus nerves to the respiratory center.

INTEGRATE

CLINICAL VIEW
Apnea

Apnea (ap'nē-ă; a = absence, pnea = breathing) is the absence of breathing. It can occur voluntarily (during swallowing or holding your breath), be drug-induced (e.g., anesthesia), or result from neurologic disease or trauma. **Sleep apnea** is the temporary cessation of breathing during sleep.

Other receptors include **proprioceptors** located within joints and muscles that are stimulated by body movement, **baroreceptors** located within both the visceral pleura and bronchiole smooth muscle that are stretch receptors, and **irritant receptors** located within the respiratory passageways that are stimulated by dust and other particulate matter (figure 23.23d).

Physiology of Breathing

Although there is much still unknown about how breathing is completely controlled, the following description provides some insight into what is generally accepted.

Quiet Breathing *Quiet inspiration* is initiated when the inspiratory neurons of the VRG within the medullary respiratory center spontaneously depolarize or are "turned on." Nerve signals initiated within the inspiratory center of the VRG are sent through the nerve pathways to the skeletal muscles of quiet breathing for approximately 2 seconds. The intensity of the nerve signals increases over these 2 seconds. This stimulates both the diaphragm and external intercostal muscles to contract, resulting in an increase in thoracic cavity volume. Thus, a pressure gradient is established and air moves from the atmosphere into the alveoli.

Quiet expiration occurs when the VRG is inhibited or "turned off." This inhibition results because during quiet inspiration, nerve signals from the VRG inspiratory neurons are also relayed to the VRG expiratory neurons. The VRG expiratory neurons in response will then send inhibitory signals back to the VRG inspiratory neurons (like an inhibitory feedback loop). This inhibition causes the VRG inspiratory neurons to "turn off." Consequently, nerve signals are no longer sent through the nerve pathways to the skeletal muscles of quiet breathing; this lasts typically for approximately 3 seconds. Lack

INTEGRATE

LEARNING STRATEGY

This summary can help you to remember what is monitored by the two different types of chemoreceptors.

- **Central chemoreceptors** *monitor CSF:* H⁺ levels in CSF (which is produced from blood CO_2).

- **Peripheral chemoreceptors** *monitor blood:* (1) CO_2 levels; (2) H⁺ levels that are produced through metabolic processes (e.g., lactic acid, ketoacids); and (3) O_2.

INTEGRATE

CLINICAL VIEW
Hypoxic Drive

Certain respiratory disorders (such as emphysema) result in decreased ability to expire carbon dioxide, and blood P_{O_2} levels can become the stimulus for breathing. This is called **hypoxic** (hī-pok'sik) **drive.** It occurs as carbon dioxide levels in the blood become elevated and remain elevated over a long period of time. Chemoreceptors become less sensitive to P_{CO_2} and by default, decreased P_{O_2} levels stimulate them. Since a low P_{O_2} is the stimulus for breathing, administering oxygen would elevate P_{O_2} and thus interfere with the person's ability to breathe on his or her own.

of nerve stimulation causes both the diaphragm and external intercostal muscles to relax, resulting in a decrease in thoracic cavity volume. Thus, a pressure gradient is established and air moves from the alveoli into the atmosphere.

A breathing rhythm that involves 2 seconds of inspiration followed by 3 seconds of expiration results in an average respiratory rate of 12 times per minute. The average range for the rate of quiet breathing is generally between 12 and 15 times per minute—a rate referred to as **eupnea** (yūp-nē-ă). (Note: Physiologists previously thought the roles of the VRG and DRG were reversed from what is explained here.)

The specific role of the pontine respiratory center in breathing is to relay nerve signals to the medullary respiratory center to facilitate a smooth transition between inspiration and expiration. Individuals that have experienced damage to this center are able to still breathe. However, the breathing is erratic and involves long gasping inspirations followed by occasional expirations.

WHAT DO YOU THINK?

3 The phrenic nerves extend from the cervical plexus formed by the rami of spinal nerves C3–C5, whereas the intercostal nerves are the anterior rami of spinal nerves T1–T11. Predict the consequences to breathing for each condition if a spinal cord injury is (a) at C2 or above, (b) between C6 and T11, or (c) T12 or below.

Reflexes That Can Alter Breathing Rate and Depth

Altering Breathing Rate and Depth Through Reflexes Involving Chemoreceptors   Breathing rate and depth are primarily altered by reflexes that respond to sensory input from chemoreceptors. Nerve signals from these receptors are sent along sensory neurons to the DRG. When the DRG is activated, nerve signals are subsequently relayed to the VRG, resulting in a change in the rate and depth of breathing. Change in the *rate* of breathing is accomplished by altering the amount of time spent in both inspiration and expiration, whereas altering the *depth* of breathing is accomplished through stimulation of accessory muscles, which results in greater thoracic volume changes.

Breathing rate and depth can be reflexively *increased* if either the central chemoreceptors detect an increase in H^+ concentration in the CSF or the peripheral chemoreceptors detect an increase in blood H^+ concentration, an increase in blood P_{CO_2}, H^+, or both. Central chemoreceptors relay additional nerve signals directly to the closely located DRG, and peripheral chemoreceptors relay additional nerve signals through both the glossopharyngeal nerves from the carotid bodies and vagus nerves from the aortic bodies to the DRG. The DRG then relays this information to the VRG. Respiration rate and depth are increased and additional CO_2 is expelled, ultimately returning blood P_{CO_2} and CSF pH to normal levels. Conversely, a decrease in either H^+ or P_{CO_2} initiates fewer nerve signals relayed to the DRG, and respiration rate and depth are decreased.

The most important stimulus affecting breathing rate and depth is blood P_{CO_2}. The respiratory center is very sensitive to changes in blood P_{CO_2} levels; increases in P_{CO_2} levels as small as 5 mm Hg will double the breathing rate. The changes in blood P_{CO_2} can stimulate respiratory rate changes most powerfully when the carbon dioxide is joined to water, forming carbonic acid in the CSF. This is because, unlike the blood, the CSF *lacks proteins* to buffer the gain or loss of H^+. (See chemical buffers in section 25.5d.) Consequently, pH changes in the CSF most accurately reflect the changes in P_{CO_2}.

Generally, changes in blood P_{O_2} are *not* an independent means of regulating breathing. Note that the arterial oxygen level in the blood must decrease substantially from its normal P_{O_2} level of 95 mm Hg to an abnormally low level of 60 mm Hg before it can stimulate the chemoreceptors independently of P_{CO_2}. This relationship can have deadly consequences to swimmers who hyperventilate before swimming underwater. This hyperventilation causes the blood P_{CO_2} levels to decrease to the point where chemoreceptors are not stimulated. The swimmer's blood P_{O_2} level is decreased at the same time by exertion, but not enough to stimulate chemoreceptors. A swimmer may lose consciousness and drown as a result.

P_{O_2} levels normally influence breathing rate by causing the chemoreceptors to be more sensitive to changes in blood P_{CO_2}. This relationship has a synergistic effect: The combination of decreased P_{O_2} and increased P_{CO_2}, along with the subsequent production of H^+, causes greater stimulation of the respiratory center.

Altering Breathing Patterns Through Reflexes Involving Other Receptors   Receptors other than chemoreceptors alter breathing patterns. (1) Proprioceptors within joints and muscles, when stimulated by body movement, increase nerve signals to the respiratory center with a subsequent increase in breathing depth. (2) Baroreceptors within both the visceral pleura and bronchiole smooth muscle are stimulated by stretch. These sensory receptors initiate a reflex to prevent overstretching of the lungs by inhibiting inspiration activities. This reflex is referred to as the **inhalation reflex,** or *Hering-Breuer reflex*. It effectively protects the lungs from damage due to overinflation. When overstretched, these baroreceptors send nerve signals through the vagus nerves to the respiratory center to shut off inspiration activity, thus resulting in expiration. This may be a normal means of controlling respiration in infants, but it is thought to serve only as a protective reflex after infancy. (3) Irritant receptors, when stimulated, initiate either a sneezing or coughing reflex. Sneezing and coughing involve an exaggerated intake of breath, the closure of the larynx and contraction of abdominal muscles, followed by an explosive blast of exhaled air when the vocal cords open abruptly. A sneeze is initiated when irritant receptors are stimulated within the nasal cavity, and a cough is initiated when irritant receptors within lower regions of the respiratory passageway are stimulated.

Action of Higher Brain Centers   Higher brain centers, including the hypothalamus, limbic system, and cerebral cortex, can influence breathing rate. The hypothalamus increases the breathing rate if the body is warm and decreases it if the body is cold. The limbic system alters the breathing rate in response to emotions and emotional memories. The frontal lobe of the cerebral cortex controls voluntary changes in our breathing pattern for various activities, such as talking, singing, breath-holding, performing the Valsalva maneuver, and other actions. Unlike the other higher areas of the brain that relay impulses to the respiratory center, nerve signals from the cerebral cortex bypass the respiratory center to directly stimulate lower motor neurons in the spinal cord (see figure 23.23).

Nervous Control of Anatomic Structures of the Respiratory System and Anatomic Structures of Breathing

We distinguish between the innervation to the anatomic structures of the respiratory system and the anatomic structures that function in breathing. Anatomic structures of the respiratory system, which are composed of both smooth muscle and glands, are innervated by the axons of lower motor neurons of the autonomic nervous system and controlled by autonomic nuclei within the brainstem. In contrast, breathing muscles, which are composed of skeletal muscle tissue, are innervated by axons of the lower motor neurons of the somatic nervous system. However, the control of the breathing muscles comes from both autonomic nuclei in the brainstem, as well as somatic nuclei in the cerebral cortex. The autonomic nuclei forming the respiratory center regulate normal breathing with their rhythmic output along the lower motor neurons of the phrenic and intercostal nerves, and this center alters breathing rate and depth in response to various sensory input, as described. The cerebral cortex consciously regulates breathing by directly stimulating lower motor neurons that extend to the skeletal muscles of breathing. This diverse input from the nervous system allows breathing to be controlled both reflexively and consciously.

WHAT DID YOU LEARN?

24 What are the functions of the VRG and DRG within the respiratory centers?

25 Which of the following stimuli will cause an increase in the respiratory rate: (a) increase in blood P_{CO_2}, (b) increase in blood H^+, (c) increase in H^+ within the CSF, and (d) increase in blood P_{O_2}?

26 Are the skeletal muscles of breathing innervated by somatic nerves or autonomic nerves? Explain.

23.5d Airflow, Pressure Gradients, and Resistance

LEARNING OBJECTIVES

9. Define airflow.

10. Explain how pressure gradients and resistance determine airflow.

Airflow is the amount of air that moves into and out of the respiratory tract with each breath. Sufficient exchange of air must be maintained for normal body functions; this exchange is in part determined by airflow and variables that affect it. Airflow is a function of two factors: (1) the *pressure gradient* established between atmospheric pressure and intrapulmonary pressure, and (2) the *resistance* that occurs due to conditions within the respiratory tract, lungs, and chest wall. The formula for airflow is expressed here:

$$F = \frac{\Delta P}{R} \qquad \text{or} \qquad F = \frac{P_{atm} - P_{alv}}{R}$$

The parameters are F = flow, ΔP = difference in pressure between atmosphere (atm) and the intrapulmonary pressure within the alveoli (alv), and R = resistance.

This mathematical expression demonstrates that flow is *directly* related to the pressure gradient between atmosphere and lungs and *inversely* related to resistance. If the pressure gradient increases, then airflow into the lungs increases, but if the pressure gradient decreases, airflow into the lungs lessens (assuming resistance remains the same). In contrast, if resistance increases, then airflow lessens, whereas if resistance decreases, airflow increases (assuming the pressure gradient remains the same).

The **pressure gradient** (ΔP) is the difference between atmospheric pressure and intrapulmonary pressure ($P_{atm} - P_{alv}$). It can be changed by altering the volume of the thoracic cavity. The contraction of both the diaphragm and external intercostals during quiet breathing cause small volume changes that allow approximately 500 mL of air to enter into the lungs. The thoracic cavity volume is further increased if accessory muscles of forced inspiration are stimulated, causing a larger decrease in intrapulmonary pressure. Airflow into the lungs increases because a steeper pressure gradient is established between atmospheric pressure and intrapulmonary pressure.

Airflow is always opposed by resistance. **Resistance** includes all the factors that make it more difficult to move air from the atmosphere through the respiratory passageway into the alveoli. Resistance may be altered in three possible ways: (1) a decrease in elasticity of the chest wall and lungs, (2) a change in the bronchiole diameter or the size of the passageway through which air moves, and (3) the collapse of alveoli.

A decrease in elasticity of the chest wall and lungs results in an increase in resistance. Young, healthy chest walls and lung tissue are naturally elastic. However, as we age, elastic connective tissue decreases in both the chest wall and the lungs. Elasticity of these structures is decreased when (1) an individual has vertebral column malformation, such as scoliosis; (2) arthritis develops within the thoracic cage; or (3) elastic connective tissue in the lungs is replaced with inflexible scar tissue, which occurs with pulmonary fibrosis.

Bronchiole diameter also influences resistance. Resistance increases with bronchoconstriction caused by parasympathetic stimulation, histamine release, or exposure to cold. In addition, decreasing lumen size through accumulation of mucus or inflammation in bronchioles also increases resistance. Resistance decreases with bronchodilation caused by sympathetic stimulation and the subsequent release of epinephrine from the adrenal medulla, or with external administration of epinephrine.

Increase in surface tension occurs if alveolar type II cells are not producing sufficient pulmonary surfactant because this condition increases resistance. This variable is generally important only with premature infants that are unable to produce sufficient pulmonary surfactant (healthy lungs continue to produce pulmonary surfactant beginning approximately 2 months prior to full-term birth). Without pulmonary surfactant, the alveoli in the lungs of these premature infants collapse with each expiration. With each inspiration, the high surface tension caused by the wet inner surface of the alveoli must be overcome for air to enter and reinflate the alveoli for gas exchange. These infants, therefore, experience greater resistance to airflow. This condition is referred to as **respiratory distress syndrome (RDS).**

Both surface tension and elasticity of the chest wall and lung determine compliance. **Compliance** is the ease with which the lungs and chest wall expand. Thus, the easier the lung expansion, the greater the compliance. In contrast, the more difficult it is for the lungs to expand, the lower the compliance.

WHAT DO YOU THINK?

4 Epinephrine is administered in the treatment of asthma. Does epinephrine increase or decrease airway resistance? Does epinephrine increase or decrease airflow?

Respiratory diseases and anatomic abnormalities that increase resistance to airflow are associated with either a decrease in the size of the lumen of bronchioles (e.g., asthma) or a decrease in compliance (e.g., pulmonary fibrosis), or both. In either case, it produces an increase in resistance. If adequate airflow is to be maintained, the increased resistance must be met with more forceful inspirations to establish a steeper pressure gradient.

The muscles of inspiration must work harder, and a greater amount of the body's metabolic energy must be spent on breathing, for more forceful inspirations to occur. Approximately 5% of the total energy expenditure of the body normally is spent for quiet breathing.

INTEGRATE

CONCEPT CONNECTION

A similar mathematical relationship expresses blood flow as a function of a pressure gradient and resistance. The blood pressure gradient is established by the heart, and the resistance is experienced by the blood as it is transported through the blood vessels (see section 20.5c).

This value increases as airway resistance increases, and it can reach 20–30% of energy expenditure. This four-fold to six-fold increase in energy is so demanding that individuals with these disorders can become exhausted simply from breathing.

 WHAT DID YOU LEARN?

27 The two factors that determine airflow are the pressure gradient and resistance. What are the three major factors that increase resistance to airflow? What changes to breathing must occur to maintain adequate airflow if resistance is increased?

23.5e Pulmonary and Alveolar Ventilation

 **LEARNING OBJECTIVES**

11. Distinguish between pulmonary ventilation and alveolar ventilation, and discuss the significance of each.

12. Explain the relationship between anatomic dead space and physiologic dead space.

Previously we described the process of moving air into and out of the lungs and referred to it as pulmonary ventilation. The term **pulmonary ventilation** may also refer to the amount of air that is inhaled in 1 minute. The normal adult breathes approximately 500 mL per breath (tidal volume), and this occurs about 12 times per minute. The amount of air taken in during 1 minute (pulmonary ventilation) is calculated using the following formula:

$$\underset{\substack{\text{(amount of air} \\ \text{per breath)}}}{\text{Tidal volume}} \times \underset{\substack{\text{(number of breaths} \\ \text{per minute)}}}{\text{Respiration rate}} = \text{Pulmonary ventilation}$$

$$500\ \text{mL} \times 12\ \text{breaths/min} = 6000\ \text{mL/min} = 6\ \text{L/min}$$

 WHAT DO YOU THINK?

5 Is *all* of the air taken in during pulmonary ventilation available for gas exchange? Why or why not?

Note that only the air reaching the alveoli is available for gas exchange with the blood. When air is moved from the atmosphere into the respiratory tract, a portion of it remains in the conducting zone. This collective space, where there is no exchange of respiratory gases, is referred to as the **anatomic dead space,** and it has an average volume of approximately 150 mL. The amount of air that reaches the alveoli and is available for gas exchange per minute is termed **alveolar ventilation.** This volume is less than pulmonary ventilation because the volume of air in the anatomic dead space must be subtracted from the volume of air inhaled with each breath. Thus, alveolar ventilation is calculated using the following mathematical formula:

$$(\text{Tidal volume} - \text{anatomic dead space}) \times \text{Respiration rate}$$
$$= \text{Alveolar ventilation}$$
$$(500\ \text{mL} - 150\ \text{mL}) \times 12 =$$
$$350\ \text{mL} \times 12 = 4200\ \text{mL/min} = 4.2\ \text{L/min}$$

Deeper breathing is more effective for maximizing alveolar ventilation than faster, shallower breathing. Assuming you take one deep breath, you have to overcome the dead air space only one time. All the additional air inhaled in that breath is available for gas exchange. If you take two quick breaths, you have to fill the dead air space twice.

Some respiratory disorders result in a decreased number of alveoli participating in gas exchange. This decrease can be due either to damage to the alveoli or a change in the respiratory membrane, such as when fluid accumulates in the lungs with pneumonia. The difference in volume of air available for gas exchange is accounted for by the more inclusive term **physiologic dead space,** which is the normal anatomic dead space plus any loss of alveoli. The anatomic dead space is equivalent to the physiologic dead space in a healthy individual, because the usual loss of alveoli should be minimal.

 WHAT DID YOU LEARN?

28 A person in yoga class is encouraged to take long, slow, deep breaths. Would this person have greater or lesser alveolar ventilation than an individual with a more shallow breathing? Explain.

23.5f Volume and Capacity

 LEARNING OBJECTIVES

13. Define the four different respiratory volume measurements.

14. Explain the four respiratory capacities that are calculated from the volume measurements.

15. Give the meaning of forced expiratory volume (FEV) and maximum voluntary ventilation (MVV).

The volume of air that enters and leaves the lungs can be measured with an instrument called a **spirometer.** Respiratory volumes vary throughout a 24-hour period and during different stages of your life. They also vary from individual to individual. The variation is significant enough to be used as a diagnostic tool for determining the health of an individual's respiratory system. Values for an individual are compared to standard values of a reference population. Respiratory measurements are often used to diagnose respiratory disease, monitor changes in respiratory impairment over time, and assess effectiveness of treatment.

Four major respiratory volumes are typically measured **(figure 23.24** and **table 23.4). Tidal volume (TV)** is the amount of air inhaled or exhaled per breath during quiet breathing. **Inspiratory reserve volume (IRV)** is the amount of air that can be forcibly inhaled beyond the tidal volume (after a normal inspiration). IRV is a measure of lung compliance. **Expiratory reserve volume (ERV)** is the amount that can be forcibly exhaled beyond the tidal volume (after a normal expiration). ERV is a measure of lung and chest wall elasticity. Finally, **residual volume (RV)** is the amount of air left in the lungs even after the most forceful expiration.

There are four major respiratory capacities that can be calculated from the summation of two or more of these respiratory volumes. **Inspiratory capacity (IC)** is the sum of the tidal volume plus the inspiratory reserve volume. **Functional residual capacity (FRC)** is the sum of the expiratory reserve volume plus the residual volume. It is the volume of air that is normally left in the lungs after a quiet expiration. **Vital capacity (VC)**

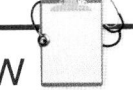

INTEGRATE

CLINICAL VIEW

Minimal Volume

Prior to birth, the fetus's respiratory system is nonfunctional because gas exchange occurs between fetal blood and maternal blood at the placenta. The lungs and pulmonary vessels of the fetus are collapsed; most of the blood is shunted to the systemic circulation.

When a newborn takes its first breath, the alveoli inflate. Thereafter, a small amount of air, termed **minimal volume,** remains in the lungs even if the lungs later collapse. In contrast, a stillborn infant never takes a breath. The test at autopsy of whether an infant has been stillborn is to remove the lungs and then to immerse the lungs in water. If the infant was stillborn, the lungs sink because they contain no minimal volume of air.

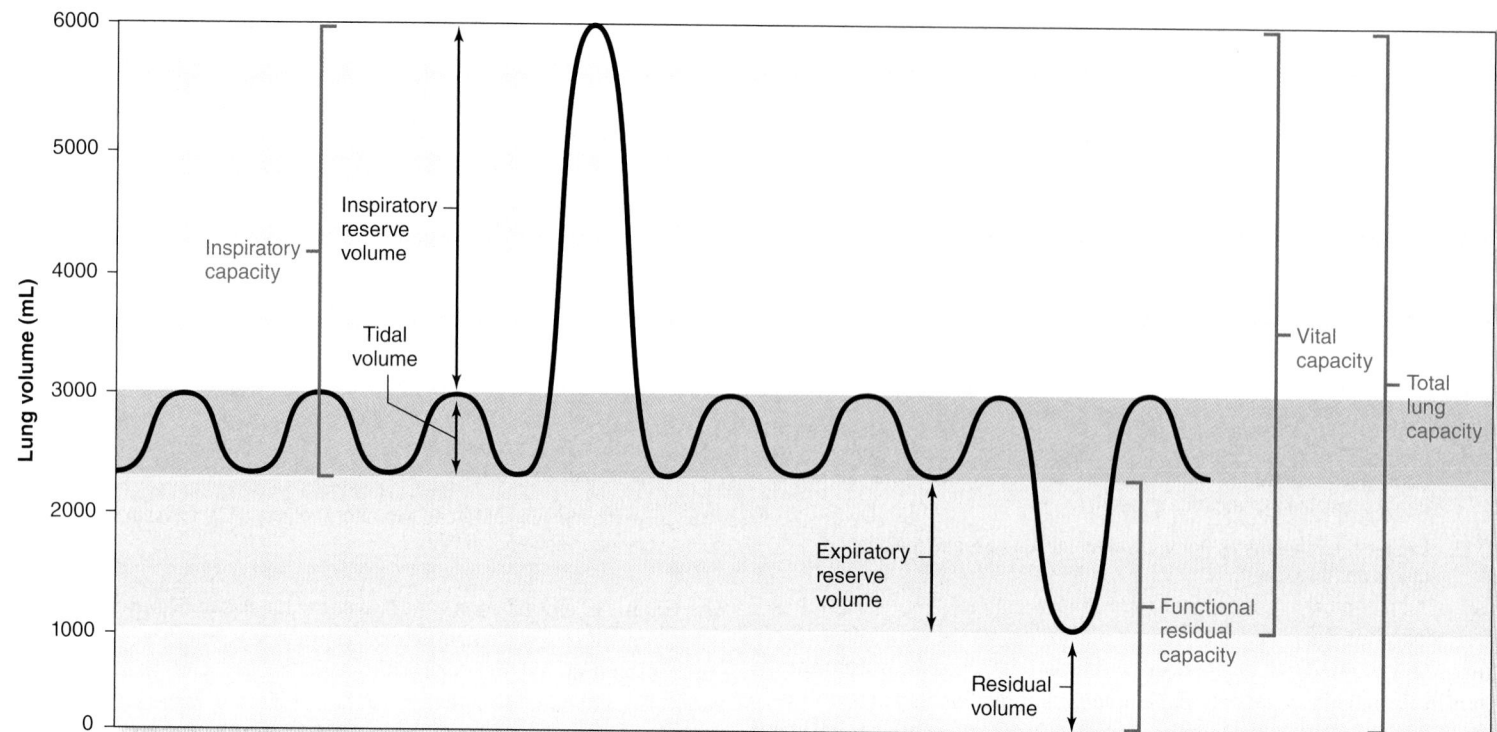

Figure 23.24 Respiratory Volumes and Capacities. Pulmonary volumes include tidal volume, inspiratory reserve volume, expiratory reserve volume, and residual volume. Capacities are the sum of two or more volumes. Inspiratory capacity includes tidal volume and inspiratory reserve volume. Functional residual capacity includes expiratory reserve volume and residual volume. Vital capacity includes tidal volume, inspiratory reserve volume and expiratory reserve volume. Total lung capacity is the sum of all four volumes.

Table 23.4	Respiratory Volumes and Capacities			
VOLUMES				
Volume	**Definition**		**Normal Values (Male)**	**Normal Values (Female)**
Tidal volume (TV)	The amount of air taken into or expelled out of lungs during a quiet breath		500 mL	500 mL
Inspiratory reserve volume (IRV)	The amount of air taken into the lungs during a forced inspiration, following a quiet inspiration; IRV is a measure of lung compliance		3100 mL	1900 mL
Expiratory reserve volume (ERV)	The amount of air expelled from the lungs during a forced expiration, following a quiet expiration; ERV is a measure of lung and chest wall elasticity		1200 mL	700 mL
Residual volume (RV)	The amount of air left (residual) in lungs following a forced expiration		1200 mL	1100 mL
CAPACITIES				
Capacity	**Formula**	**Definition**	**Normal Values (Male)**	**Normal Values (Female)**
Inspiratory capacity	TV + IRV	Total ability to inspire	3600 mL	2400 mL
Functional residual capacity	ERV + RV	Amount of air normally left (residual) in lungs after you expire quietly	2400 mL	1800 mL
Vital capacity	TV + IRV + ERV	Measure of the amount of air the lungs are capable of holding	4800 mL	3100 mL
Total lung capacity	TV + IRV + ERV + RV	Total amount of air that can be in lungs	6000 mL	4200 mL

is the sum of the tidal volume plus both the inspiratory reserve volume and expiratory reserve volume. The residual volume is not part of the vital capacity. VC is significant because it is a measure of the total amount of air that a person can exchange through forced breathing. Finally, **total lung capacity (TLC)** is the sum of all the volumes, including the residual volume, and is the maximum volume of air that the lungs can hold.

Two additional common respiratory measurements of note are forced expiratory volume and maximum voluntary ventilation. **Forced expiratory volume (FEV)** is the percentage of the vital capacity that

can be expelled in a specific period of time. For example, FEV_1 is the vital capacity percentage that is expired in 1 second. This value is obtained by inspiring as much air as possible and then expelling the air from the lungs as quickly as possible. A healthy person should be able to expel 75–85% of the vital capacity in 1 second. Individuals with decreased ability to expire (e.g., individuals with emphysema) typically exhibit decreased FEV values.

Maximum voluntary ventilation (MVV) is the greatest amount of air that can be taken into, and then expelled from, the lungs in

1 minute. MVV levels are obtained by breathing as quickly and as deeply as possible for 1 minute. Maximum voluntary ventilation can be as high 30 L/min (compared to the pulmonary ventilation at rest of 6 L/min). All individuals with respiratory disorders have an impaired ability to inspire, expire, or both—thus, they will exhibit lower than normal MVV values. Note that both the FEV and MVV amounts are timed and reflect *rates* of air movement.

 WHAT DID YOU LEARN?

 29 How are capacities calculated?

23.6 Respiration: Alveolar and Systemic Gas Exchange

Gas exchange is the movement of respiratory gases between blood and either alveoli or cells of systemic tissues. The movement of these gases between blood in pulmonary capillaries and the alveoli of the lungs is called alveolar gas exchange, whereas the movement of respiratory gases between blood in systemic capillaries and systemic cells is called systemic gas exchange. Here we describe the general chemical principles of gas exchange; then we describe the specifics of alveolar gas exchange in the lungs and systemic gas exchange at systemic cells.

23.6a Chemical Principles of Gas Exchange

LEARNING OBJECTIVES

1. Define partial pressure and the movement of gases relative to a partial pressure gradient.
2. Describe the partial pressures that are relevant to gas exchange.
3. Explain the laws that govern gas solubility.

The gases in air move collectively down a (total) pressure gradient during breathing, as previously described in the section on pulmonary ventilation. In contrast, each gas moves independently down its own partial pressure gradient during gas exchange. Additionally, each gas moves between air (in the alveoli) and a liquid (blood).

Partial Pressure and Dalton's Law

Partial pressure is the pressure exerted by each gas within a mixture of gases and is measured in mm Hg; it is written with a P followed by the symbol for the gas. For example, the partial pressure for oxygen is written as P_{O_2}. We can use atmospheric pressure and the mixture of gases in air to more fully explain partial pressure.

Atmospheric pressure is the total pressure all gases collectively exert in the environment. These molecules include nitrogen (N_2), oxygen (O_2), carbon dioxide (CO_2), water vapor (H_2O), and a number of other minor gases. We have seen that at sea level atmospheric pressure has a value of 760 mm Hg. The amount each gas in the atmosphere contributes to the total pressure—that is, its partial pressure—is determined by multiplying the total pressure exerted by the gas mixture by the percent of the specific gas of interest:

Total pressure × % of gas = Partial pressure of that gas

Thus, the partial pressure exerted for each gas in the atmosphere can be calculated from the total pressure (760 mm Hg at sea level) and

the percentage of each of the most common gases: nitrogen (78.6%), oxygen (20.9%), carbon dioxide (0.04%), and water vapor (0.46%) as follows:

P_{N_2}	760 mm Hg × 78.6%	= 597 mm Hg
P_{O_2}	760 mm Hg × 20.9%	= 159 mm Hg
P_{CO_2}	760 mm Hg × 0.04%	= 0.3 mm Hg
P_{H_2O}	760 mm Hg × 0.46%	= 3.5 mm Hg
	Total pressure	= 760 mm Hg

When these partial pressures are added together, their sum must equal the total atmospheric pressure. As just shown,

$$P_{N_2} + P_{O_2} + P_{CO_2} + P_{H_2O} = 760 \text{ mm Hg}$$

The relationship of partial pressure to total pressure is summarized by **Dalton's law,** which states that the total pressure in a mixture of gases is equal to the sum of all of the individual partial pressures.

Partial Pressure Gradients

A **partial pressure gradient** exists when the partial pressure for a specific gas is higher in one region than in another. If a partial pressure gradient exists between two regions for a given gas, then the gas moves from the region of its higher partial pressure to the region of its lower partial pressure, and it may continue to move until the partial pressures in the two regions become equal. The exchange of respiratory gases in both alveolar gas exchange and systemic gas exchange is dependent upon partial pressure gradients.

Relevant Partial Pressures in the Body

The partial pressures that are relevant for understanding respiratory gas exchange are both the P_{O_2} and P_{CO_2} within alveoli in the lungs, systemic cells, and in circulating blood. Refer to **figure 23.25** as you read through this section.

P_{O_2} and P_{CO_2} in the Alveoli   Although air from the environment is inhaled directly into the lungs, the partial pressures of the gases within the alveoli are different from the respective atmospheric partial pressures measured for several reasons: (1) Air from the environment mixes with the air remaining in the anatomic dead space in the respiratory tract; (2) oxygen diffuses out of the alveoli into the blood, and carbon dioxide diffuses from the blood into the alveoli; and (3) more water vapor is present within the alveoli because of the higher humidity there. Consequently, within the alveoli the percentage of oxygen is lower (13.7%) and the percentage of carbon dioxide is higher (5.2%) than in the atmosphere. The calculation of the partial pressure of the respiratory gases in the alveoli is as follows (figure 23.25*b*):

Alveolar Partial Pressures (at Sea Level)

P_{O_2}	760 mm Hg × 13.7% = 104 mm Hg
P_{CO_2}	760 mm Hg × 5.2% = 40 mm Hg

The P_{O_2} is *lower* in the alveoli (P_{O_2} = 104 mm Hg) than it is in the atmosphere (P_{O_2} = 159 mm Hg), and the P_{CO_2} is *higher* in the alveoli (P_{CO_2} = 40 mm Hg) than it is in the atmosphere (P_{CO_2} = 0.3 mm Hg). Note that under normal circumstances, the alveolar partial pressures remain constant because we are rhythmically exchanging air between the atmosphere and the alveoli of the lungs as we breathe.

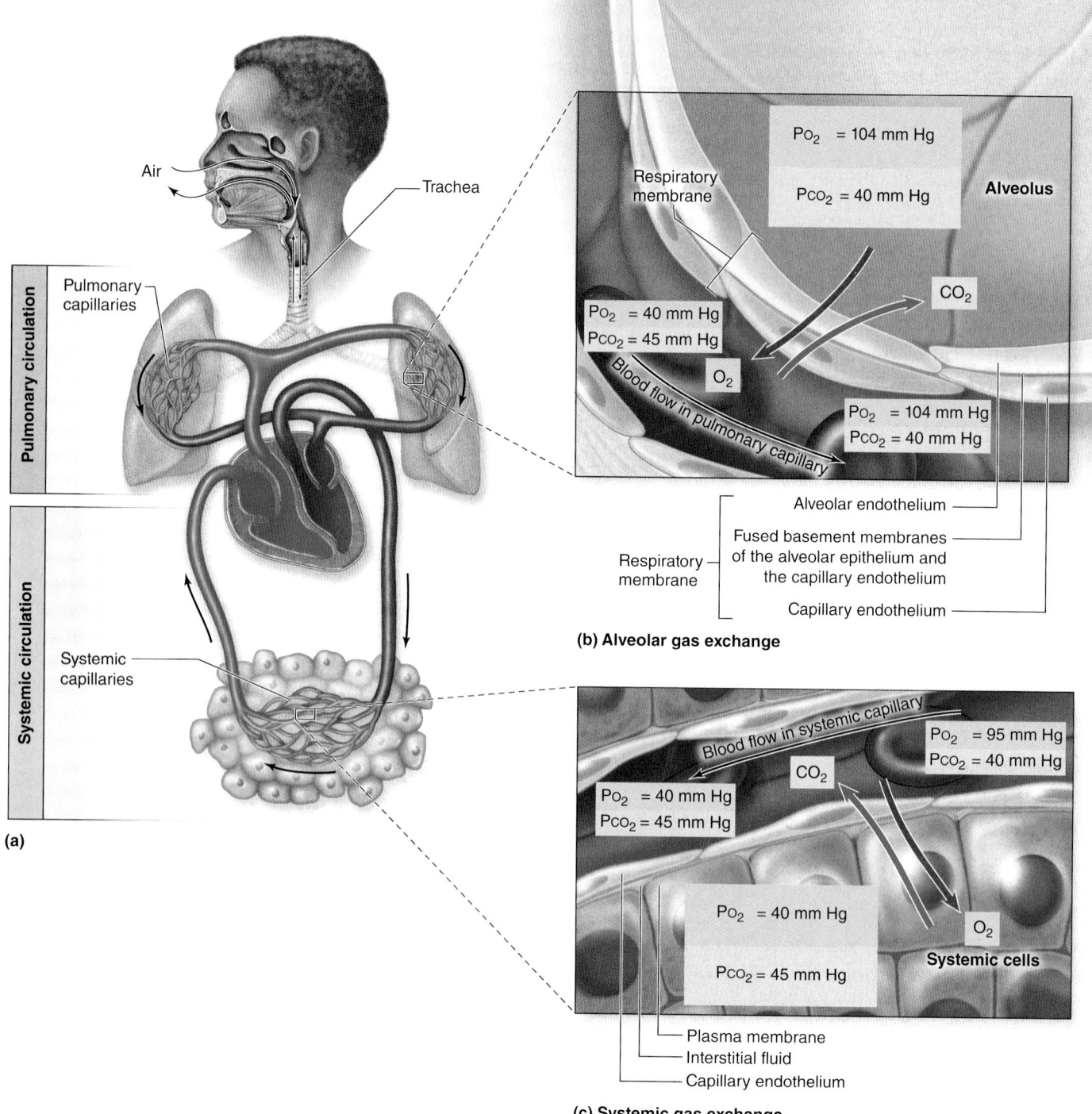

(b) Alveolar gas exchange

(c) Systemic gas exchange

Figure 23.25 Alveolar and Systemic Gas Exchange. (*a*) Pulmonary circulation delivers blood to and from the lungs, and systemic circulation delivers blood to and from systemic cells. (*b*) Alveolar gas exchange occurs when respiratory gases are exchanged between the alveoli and the pulmonary capillaries across the respiratory membrane. Oxygen diffuses from the alveoli into the blood of the pulmonary capillaries. Carbon dioxide simultaneously diffuses in the reverse direction. (*c*) Oxygen diffuses from the blood in the systemic capillaries to systemic cells during systemic gas exchange, and carbon dioxide simultaneously diffuses in the reverse direction. (Values for P_{O_2} are shown in pink; values for P_{CO_2} are shown in blue.) AP|R

P_{O_2} and P_{CO_2} in Systemic Cells   The partial pressures of both O_2 and CO_2 in systemic cells reflect the activities of cellular respiration. Cells use oxygen during cellular respiration and produce carbon dioxide as a waste product. So in comparison to the percentages in alveoli, the percentage of oxygen in the systemic cells is lower and the percentage of carbon dioxide is higher. The partial pressures of the respiratory gases in systemic cells at rest usually exhibit the measured values as shown here (figure 23.25*c*):

Systemic cells partial pressures
(Resting conditions)
P_{O_2} 40 mm Hg
P_{CO_2} 45 mm Hg

Note that under normal, resting conditions, the partial pressures in systemic cells remain constant because the oxygen is continuously used and carbon dioxide is continuously produced by cells during cellular respiration.

P_{O_2} and P_{CO_2} in the Circulating Blood   Both P_{O_2} and P_{CO_2} levels in the blood (figure 23.25*b, c*) are not set values, unlike the relatively constant values in the alveoli and systemic cells. The P_{O_2} and P_{CO_2} in the blood change continuously as the blood flows through the pulmonary capillaries, where oxygen enters the blood and carbon dioxide leaves the blood. In addition, as blood flows through systemic capillaries, the reverse occurs: Oxygen leaves the blood and carbon dioxide enters the blood. We explain in sections 23.6b and c how the values for blood P_{O_2} and blood P_{CO_2} change during alveolar and systemic gas exchange.

INTEGRATE

CLINICAL VIEW
Decompression Sickness and Hyperbaric Oxygen Chambers

Decompression sickness, also known as *the bends,* or *Caisson disease,* occurs when a diver is submerged in water beyond a certain depth and returns too quickly to the surface. Divers are breathing air under higher pressure while diving; the further they submerge, the higher the pressure. Nitrogen, although it has low solubility, is forced into the blood. If a diver rises to the surface too quickly, there is not enough time to expel all of the nitrogen through expiration. Instead, the dissolved nitrogen comes out of solution while still in the blood and tissues, and nitrogen gas bubbles develop in these tissues, including the joints. This is similar to how carbon dioxide comes out of a soda when a can is opened. The nickname *the bends* comes from the fact that divers bend their joints to try to relieve the pain.

Hyperbaric (hī'pĕr-bar'ik; *hyper* = over; *baros* = pressure) **oxygen chambers** are used to treat individuals with decompression sickness. The partial pressure gradient for oxygen is increased when individuals are placed in a chamber with oxygen pressures higher than atmospheric pressure (i.e., greater than 1 atm). The high partial pressure of oxygen forces additional oxygen to dissolve in the blood plasma.

Hyperbaric oxygen chambers also can be used to treat certain disorders such as carbon monoxide poisoning, foot ulcers associated with diabetes, traumatic crush injuries, severe anemia, and gas gangrene. The additional oxygen that enters the tissue accelerates the tissue healing rate.

Gas Solubility and Henry's Law

Specific additional chemical principles govern the exchange of gas between air (a gas) and blood (a liquid). These principles are summarized by **Henry's law,** which states that at a given temperature, the solubility of a gas in a liquid (i.e., how much gas can either enter or leave the liquid) is dependent upon (1) the partial pressure of the gas in the air and (2) the solubility coefficient of the gas in the liquid.

The partial pressure of a gas is the driving force to move it into a liquid. Recall that a given partial pressure is determined by the total pressure and the specific percentage of the gas of interest within the mixture—if either of these changes, the amount of the gas that enters the liquid changes. Carbon dioxide gas in a soft drink is an example of forcing more gas into a liquid by increasing its partial pressure. The CO_2 is forced into the soda under high pressure, and then the container is sealed. When the can is opened and the pressure released, carbon dioxide leaves the soda because of the lower P_{CO_2} of the atmosphere; over time, the soda becomes "flat."

The **solubility coefficient** is the volume of gas that dissolves in a specified volume of liquid at a given temperature and pressure. This coefficient is a constant that depends upon the interactions between molecules of both the gas and the liquid. The more favorable these molecular interactions are for a given partial pressure, the greater the amount of gas that dissolves in the liquid.

Gases vary in their solubility in water. For example, oxygen has a very low solubility in water (solubility coefficient = 0.024), whereas carbon dioxide has solubility about 24 times that of oxygen (solubility coefficient = 0.57). Nitrogen is the least soluble of these three major gases in the atmosphere and is about half as soluble as oxygen. The order of solubility for the three gases dissolved in water would be

$$CO_2 > O_2 > N_2$$

Because the amount of a gas that dissolves in a liquid is dependent upon both its partial pressure and its solubility coefficient, gases with low solubility coefficients require larger pressure gradients to "push" the gas into the liquid. You can see this relationship when comparing the partial pressure gradients for oxygen and carbon dioxide. Nitrogen, with its very low solubility, does not dissolve in the blood in significant amounts under conditions at or above sea level. However, as mentioned in the Clinical View: "Decompression Sickness and Hyperbaric Oxygen Chambers," divers using compressed air may be subject to dangerous effects of increased nitrogen forced into the blood because of the greater pressures experienced when submerged.

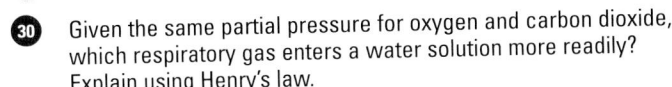

WHAT DID YOU LEARN?

30 Given the same partial pressure for oxygen and carbon dioxide, which respiratory gas enters a water solution more readily? Explain using Henry's law.

23.6b Alveolar Gas Exchange (External Respiration)

LEARNING OBJECTIVES

4. Describe alveolar gas exchange and the partial pressure gradients responsible.
5. Name the two anatomic features of the respiratory membrane that contribute to efficient alveolar gas exchange.
6. Explain ventilation-perfusion coupling and how it maximizes alveolar gas exchange.

The partial pressure gradients between the alveoli and blood, and the subsequent movement of respiratory gases across the respiratory membrane during alveolar gas exchange, are described in this section. The factors that contribute to the efficiency of alveolar gas exchange follow.

Figure 23.25*b* depicts events that occur in alveolar gas exchange. Notice that the P_{O_2} in the alveoli is 104 mm Hg, and the blood entering the pulmonary capillaries has a P_{O_2} of 40 mm Hg. Oxygen diffuses across the respiratory membrane from the alveoli into the capillaries because of the P_{O_2} partial pressure gradient, until the P_{O_2} in the blood is equal to that of the alveoli at 104 mm Hg. Thus, blood P_{O_2} has increased from 40 to 104 mm Hg as blood moves through the pulmonary capillaries. However, the P_{O_2} in the alveoli remains constant because oxygen is continuously entering the alveoli through the respiratory passageways.

Simultaneously, CO_2 is diffusing in the opposite direction; the P_{CO_2} in the alveoli is 40 mm Hg and that of the blood entering the pulmonary capillaries is 45 mm Hg. Carbon dioxide diffuses down its partial pressure gradient from the blood into the alveoli until the P_{CO_2} in the blood is equal to that of the alveoli at 40 mm Hg. Thus, blood P_{CO_2} has decreased from 45 to 40 mm Hg as blood moves through the pulmonary capillaries. As with oxygen, the P_{CO_2} in the alveoli also remains constant because carbon dioxide is continuously leaving the alveoli through the respiratory passageways.

Efficiency of Gas Exchange at the Respiratory Membrane

The efficiency of both O_2 and CO_2 diffusion during alveolar gas exchange is dependent upon anatomic features of the respiratory membrane. These features include both its large surface area and minimal thickness. The aggregate surface area of the respiratory membrane in a healthy lung measures approximately 70 square meters—a little less than half the size of a tennis court. The minimal thickness of this barrier measures approximately 0.5 micrometers.

Physiologic adjustments also contribute to maximizing gas exchange at the alveoli. Some alveoli, at any given time, are well ventilated and some are not; similarly, some regions of the lung have ample blood moving through pulmonary capillaries, and some do not. The smooth muscle of both the bronchioles that lead into the alveoli and the arterioles that carry blood to pulmonary capillaries can contract and relax to maximize gas exchange. This inherent ability of bronchioles to regulate airflow and arterioles to regulate blood flow simultaneously is called **ventilation-perfusion coupling** (figure 23.26).

Ventilation is altered by changes in bronchodilation and bronchoconstriction. Bronchioles dilate in response to an increase in P_{CO_2}, whereas they constrict in response to a decrease in P_{CO_2}.

Perfusion is altered by changes in pulmonary arteriole dilation and constriction. These arterioles dilate in response to either an increase in P_{O_2} or a decrease in P_{CO_2}, whereas they constrict in response to either a decrease in P_{O_2} or an increase in P_{CO_2}. Note these changes in bronchioles and pulmonary arterioles occur independently of one another.

WHAT DID YOU LEARN?

31 How do the partial pressures of oxygen and carbon dioxide in blood change during alveolar gas exchange?

32 Which of the following would decrease alveolar gas exchange: loss of alveoli, fluid accumulation in the lungs, arteriole vasoconstriction, and bronchiole dilation? Explain.

23.6c Systemic Gas Exchange (Internal Respiration)

LEARNING OBJECTIVES

7. Explain the partial pressure gradients between systemic cells and the blood in capillaries.

8. Differentiate between alveolar and systemic gas exchange.

Gases move between the blood and systemic cells of the body during systemic gas exchange (figure 23.25c). Oxygen diffuses out of systemic capillaries into the interstitial fluid around systemic cells, prior to crossing the plasma membrane to enter a cell. At the same time, carbon dioxide exits the cell moving in the opposite direction to enter the blood in systemic capillaries. The driving force is the same as in alveolar gas exchange—the partial pressure gradient that exists for both O_2 and CO_2.

Notice that the P_{O_2} in the systemic cells is 40 mm Hg, and the blood as it enters the surrounding systemic capillaries has a P_{O_2} of 95 mm Hg. Therefore, oxygen diffuses out of the systemic capillaries down its partial pressure gradient into the cells until the blood P_{O_2} is equal to the partial pressure in the cells at 40 mm Hg. Thus, blood

INTEGRATE

CLINICAL VIEW

Emphysema

Emphysema (em-fi-zē'mă; *en* = in, *physema* = blowing) is an irreversible loss of pulmonary gas exchange surface area due to inflammation of air passageways distal to the terminal bronchioles, in conjunction with widespread destruction of pulmonary elastic connective tissue. These combined events lead to dilation (an increase in the diameter) of individual alveoli, as well as merging of individual alveoli with others. The result is a decrease in the total number of alveoli, and the subsequent loss of gas exchange surface area.

A person with advanced emphysema is unable to expire effectively, so that stagnant, deoxygenated air builds up within the abnormally large (but numerically diminished) alveoli. Most cases of emphysema result from damage caused by smoking. Once the tissue in the lung has been destroyed, it cannot regenerate, and thus there is no cure for emphysema. The best therapy for an emphysema patient is to stop smoking and try to get optimal use from the remaining lung tissue by using a bronchodilator, seeking prompt treatment for pulmonary infections, and taking oxygen supplementation if necessary.

Cases of emphysema not linked to smoking are generally caused by a genetic disorder that involves deficiency of alpha-1 antitrypsin. Alpha-1 antitrypsin is a plasma protein normally produced by the liver. It is transported in the blood to the lungs where it destroys neutrophil elastase, which is an enzyme released by neutrophils in the lungs. Normally, neutrophils release this enzyme as they fight infections. Although this enzyme can cause damage to alveoli, it is usually kept under control by alpha-1 antitrypsin. Individuals with this genetic disorder produce decreased amounts of alpha-1 antitrypsin. Neutrophil elastase therefore increases, causing elevated levels of damage to alveoli. The rate of damage is accelerated if the individual also smokes.

Emphysema causes dilation of the alveoli and loss of elastic tissue, decreasing respiratory membrane surface area and resulting in decreased alveolar gas exchange. (a) A section of an emphysemic lung shows the dilated alveoli. (b) Microscopically, the alveoli are abnormally large and nonfunctional.

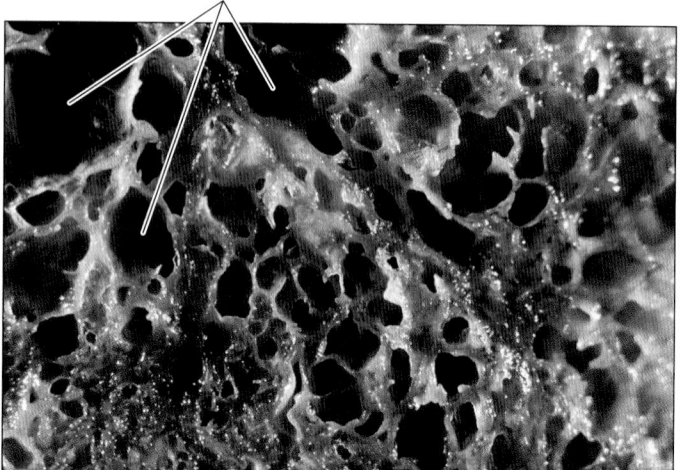

Dilated, nonfunctional alveoli

(a) Gross section of an emphysemic lung

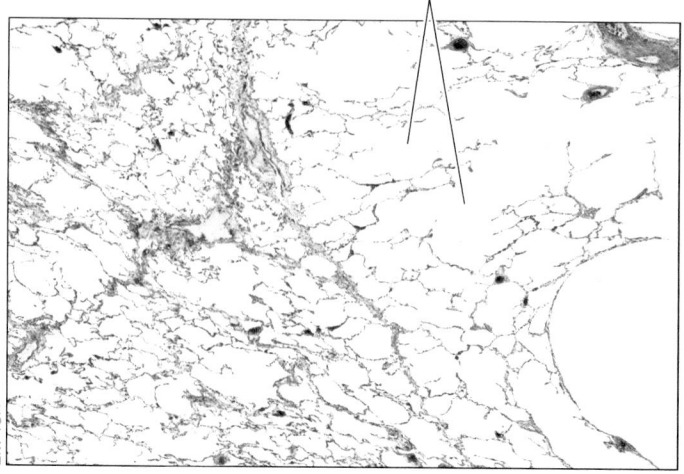

Dilated, nonfunctional alveoli

LM 15×

(b) Microscopic view of an emphysemic lung

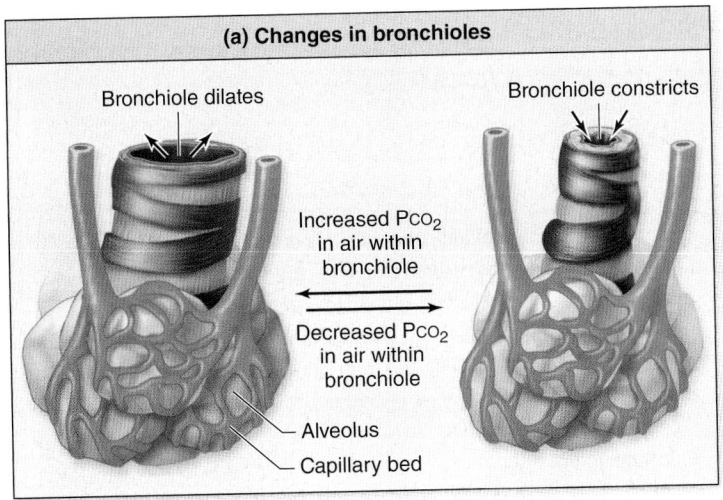

(a) Changes in bronchioles

Bronchiole dilates

Bronchiole constricts

Increased Pco₂ in air within bronchiole

Decreased Pco₂ in air within bronchiole

Alveolus

Capillary bed

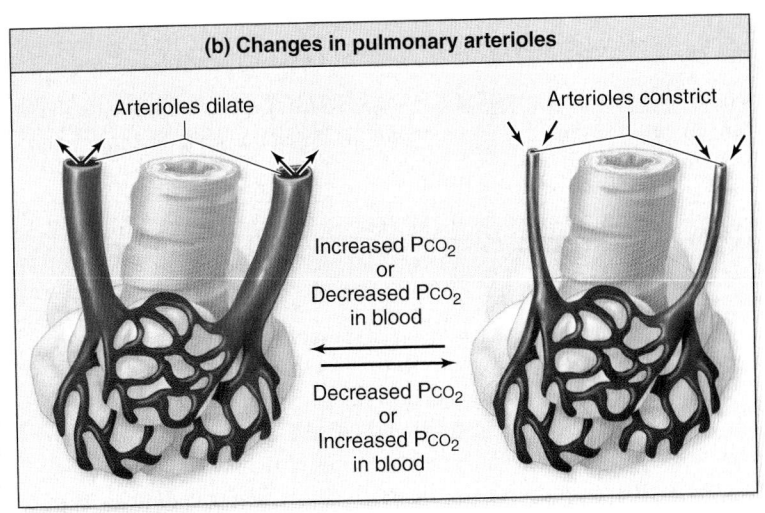

(b) Changes in pulmonary arterioles

Arterioles dilate

Arterioles constrict

Increased Pco₂ or Decreased Pco₂ in blood

Decreased Pco₂ or Increased Pco₂ in blood

Figure 23.26 Ventilation-Perfusion Coupling. (*a*) Bronchioles dilate or constrict in response to changes in CO_2 in air within the bronchioles. (*b*) Pulmonary arterioles dilate or constrict in response to changes in either blood P_{O_2} or blood P_{CO_2}.

P_{O_2} has decreased from 95 to 40 mm Hg as blood moves through the systemic capillaries.

Simultaneously, carbon dioxide is diffusing in the opposite direction. The P_{CO_2} in systemic cells is 45 mm Hg, and the blood entering the systemic capillaries is 40 mm Hg. Carbon dioxide diffuses down its partial pressure gradient from the cells into the blood until blood P_{CO_2} is 45 mm Hg. Thus, blood P_{CO_2} has increased from 40 to 45 mm Hg as blood moves through the systemic capillaries.

Unless conditions change, such as when engaging in strenuous activity, the partial pressure of each gas in the systemic cells remains relatively constant because the continuous delivery of oxygen and removal of carbon dioxide correspond with the amounts associated with cellular respiration. The results of both alveolar and systemic gas exchange are summarized in **table 23.5**.

Integration of Alveolar and Systemic Gas Exchange

During alveolar gas exchange, blood P_{CO_2} decreases from 45 to 40 mm Hg; during systemic gas exchange, blood P_{CO_2} increases from 40 to 45 mm Hg (figure 23.25*b*, *c*). Notice that the P_{CO_2} values reverse as the blood makes its way through the two cardiovascular circuits: 45 to 40 mm Hg in the pulmonary capillaries, 40 to 45 mm Hg in the systemic capillaries.

WHAT DO YOU THINK?

6 What can account for the blood P_{O_2} arriving at systemic capillaries with a lower P_{O_2} value than when it left the pulmonary capillaries?

Blood P_{O_2} increases from 40 to 104 mm Hg during alveolar gas exchange, and blood P_{O_2} decreases from 95 to 40 mm Hg during systemic gas exchange. When we discussed pulmonary blood circulation earlier, we noted that bronchial veins dump small amounts of deoxygenated blood into the pulmonary veins prior to the blood returning to the heart, where it is subsequently pumped by the left ventricle through the systemic circulation. This input of deoxygenated blood accounts for the decrease in P_{O_2} from 104 to 95 mm Hg.

WHAT DID YOU LEARN?

33 How do the partial pressures of oxygen and carbon dioxide in blood change during systemic gas exchange?

Table 23.5	Gas Exchange[1]	
Characteristic	**Alveolar Gas Exchange**	**Systemic Gas Exchange**
Definition	Exchange of respiratory gases between alveoli in lungs and blood in pulmonary capillaries	Exchange of respiratory gases between systemic cells and blood in systemic capillaries
Changes in Blood P_{O_2}	Blood P_{O_2} increases from 40 to 104 mm Hg	Blood P_{O_2} decreases from 95 to 40 mm Hg
Changes in Blood P_{CO_2}	Blood P_{CO_2} decreases from 45 to 40 mm Hg	Blood P_{CO_2} increases from 40 to 45 mm Hg

1. General conditions: Location is at sea level and individual is at rest.

INTEGRATE

CLINICAL VIEW
Respiratory Diseases and Efficiency of Alveolar Gas Exchange

Certain diseases can decrease the efficiency of oxygen and carbon dioxide exchange due to changes in the anatomic structure of the respiratory membrane. For example, individuals with emphysema, lung cancer, or tuberculosis, or those who have survived surgical removal of a lung, have decreased numbers of alveoli and therefore *decreased surface area for gas exchange.* Individuals with pneumonia or those with congestive heart failure of the left side of the heart are at risk for fluid buildup, resulting in a *thickened respiratory membrane.*

Changes in ventilation-perfusion coupling can also decrease efficiency of gas exchange. Individuals with narrowing of air passageways from bronchitis or asthma experience decreased air reaching the alveoli, or those with obstructed blood flow from a pulmonary embolism, have decreased blood flow into the pulmonary capillaries.

These disease conditions result in a decrease in blood P_{O_2} as less oxygen enters the blood and an increase in blood P_{CO_2} as more carbon dioxide remains within the blood.

23.7 Respiration: Gas Transport

The fourth process of respiration is gas transport. Gas transport is the movement of respiratory gases within the blood between the lungs and systemic cells.

23.7a Oxygen Transport

LEARNING OBJECTIVE

1. Explain why hemoglobin is essential to oxygen transport.

Oxygen is transported within blood from the alveoli through pulmonary veins of the pulmonary circulation to the left side of the heart. Then blood is pumped into the aorta and through arteries of the systemic circulation to enter systemic capillaries (see figure 19.4). Oxygen then diffuses from the blood in systemic capillaries into systemic cells. The ability of blood to transport oxygen is dependent upon two factors: the solubility coefficient of oxygen in blood plasma and the presence of hemoglobin (Hb).

Our discussion of gas solubility and Henry's law pointed out that the solubility coefficient of oxygen is very low (0.024). This means that only small amounts of oxygen (less than 2%) are dissolved in the plasma. Consequently, about 98% of the oxygen in the blood must be transported within erythrocytes where it attaches to the iron within hemoglobin molecules (see figure 18.6). Oxygen bound to hemoglobin is referred to as **oxyhemoglobin** (abbreviated HbO_2). Hemoglobin without bound oxygen is called **deoxyhemoglobin** (abbreviated as HHb):

$$HHb \text{ (deoxyhemoglobin)} + O_2 \rightleftharpoons HbO_2 \text{ (oxyhemoglobin)}$$

WHAT DID YOU LEARN?

34 Why is such a small percentage (about 2%) of oxygen dissolved in plasma and most transported on hemoglobin?

23.7b Carbon Dioxide Transport

LEARNING OBJECTIVES

2. Describe the three ways carbon dioxide is transported in the blood.

3. Explain the conversion of CO_2 to and from HCO_3^- within erythrocytes.

Cells typically produce about 200 mL/min of carbon dioxide as a waste product during cellular respiration. Carbon dioxide is transported from systemic cells within deoxygenated blood through veins of the systemic circulation to the right side of the heart, then pumped into the pulmonary trunk and pulmonary arteries to enter pulmonary capillaries (see figure 19.4). Carbon dioxide then diffuses from the blood within capillaries into the alveoli.

Whereas hemoglobin is the major means of transporting oxygen, carbon dioxide has three means of being transported in the blood from the systemic cells to the alveoli: (1) CO_2 dissolved in plasma, (2) CO_2 attached to the globin portion of hemoglobin, and (3) as bicarbonate (HCO_3^-) dissolved in plasma.

We previously noted that the solubility coefficient of carbon dioxide is 0.57. Due to both this value and the small partial pressure gradient for CO_2, approximately 7% of carbon dioxide is transported to the alveoli as a dissolved gas within the plasma of blood.

Hemoglobin is capable of transporting about 23% of the CO_2 as a carbaminohemoglobin compound. The CO_2 is attached to amine groups ($-NH_2$) in the globin protein:

$$CO_2 + Hb \rightleftharpoons HbCO_2 \text{ (carbaminohemoglobin)}$$

CLINICAL VIEW
Measuring Blood Oxygen Levels with a Pulse Oximeter

One way to measure blood oxygen levels is to take a blood sample. A noninvasive and indirect way to measure blood oxygen levels is to use a **pulse oximeter**. This device is applied to a translucent area of the body, usually a finger or earlobe. Two wavelengths of light, both red (660 nm) and infrared (940 nm), are emitted from light-emitting diodes (LEDs) and directed through the finger or earlobe. The device contains a photodiode on the other side that can measure the hemoglobin saturation by determining the ratio of oxyhemoglobin to deoxyhemoglobin based on absorption of the two wavelengths. Normal readings include hemoglobin saturation that is greater than 95%.

The remaining 70% of the CO_2 diffuses into erythrocytes and combines with water to form bicarbonate (HCO_3^-) and H^+:

$$CO_2 + H_2O \rightleftharpoons H_2CO_3 \rightleftharpoons HCO_3^- + H^+$$

HCO_3^- then diffuses into the plasma. Thus, the largest percentage of CO_2 is carried from the systemic cells to the lungs in plasma as dissolved HCO_3^-. Carbon dioxide is regenerated when blood moves through pulmonary capillaries and this process is reversed. **Figure 23.27** provides a synopsis of the details of this conversion and its reversal within erythrocytes.

WHAT DID YOU LEARN?

35 How is the majority of carbon dioxide transported within the blood?

23.7c Hemoglobin as a Transport Molecule

LEARNING OBJECTIVES

4. Name the three substances carried by hemoglobin.

5. Explain the significance of the oxygen-hemoglobin saturation curve for both alveolar and systemic gas exchange.

Hemoglobin (see figure 18.6) transports three substances relative to respiration activities: (1) oxygen attached to iron, (2) carbon dioxide bound to the globin, and (3) hydrogen ions bound to the globin. A critical aspect of this transport is that the binding or release of one substance causes a conformational change that temporarily alters the shape of the hemoglobin molecule. This change influences the ability of hemoglobin to bind or release the other two substances.

Oxygen-Hemoglobin Saturation Curve

Because one hemoglobin molecule has four iron atoms, each hemoglobin molecule may bind a maximum of four O_2 molecules. The amount of oxygen bound to hemoglobin is expressed as the *percent O_2 saturation of hemoglobin*. For example, when oxygen is bound to one-quarter of the available iron binding sites, the hemoglobin is said to be 25% saturated, and if all iron sites are occupied by oxygen, hemoglobin is 100% saturated.

Hemoglobin saturation is determined by several variables. The most important variable is the P_{O_2}. Predictably, as the P_{O_2} increases, hemoglobin saturation increases. The binding of each O_2 molecule

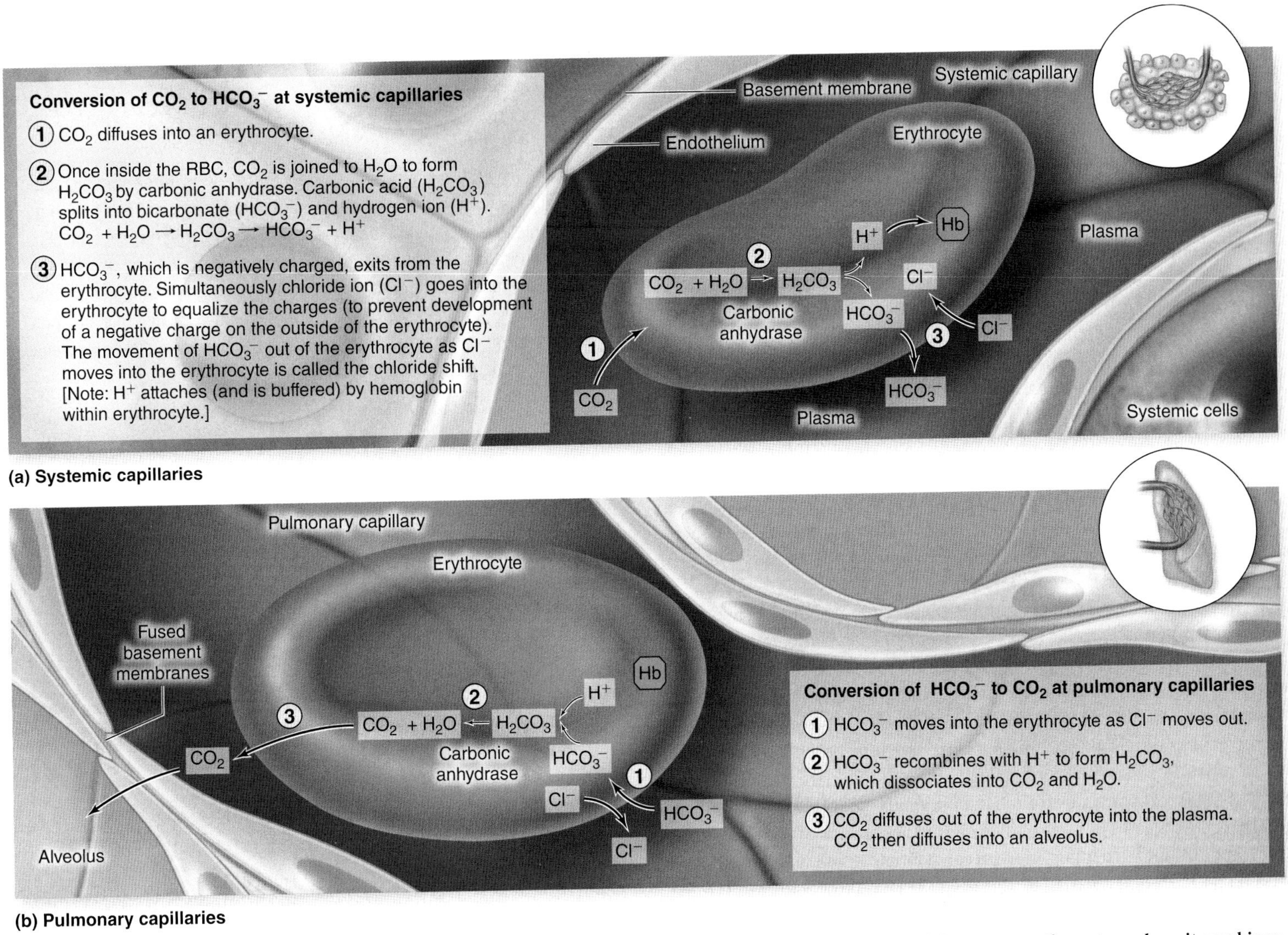

Conversion of CO₂ to HCO₃⁻ at systemic capillaries

① CO_2 diffuses into an erythrocyte.

② Once inside the RBC, CO_2 is joined to H_2O to form H_2CO_3 by carbonic anhydrase. Carbonic acid (H_2CO_3) splits into bicarbonate (HCO_3^-) and hydrogen ion (H^+).
$$CO_2 + H_2O \rightarrow H_2CO_3 \rightarrow HCO_3^- + H^+$$

③ HCO_3^-, which is negatively charged, exits from the erythrocyte. Simultaneously chloride ion (Cl^-) goes into the erythrocyte to equalize the charges (to prevent development of a negative charge on the outside of the erythrocyte). The movement of HCO_3^- out of the erythrocyte as Cl^- moves into the erythrocyte is called the chloride shift. [Note: H^+ attaches (and is buffered) by hemoglobin within erythrocyte.]

(a) Systemic capillaries

Conversion of HCO₃⁻ to CO₂ at pulmonary capillaries

① HCO_3^- moves into the erythrocyte as Cl^- moves out.

② HCO_3^- recombines with H^+ to form H_2CO_3, which dissociates into CO_2 and H_2O.

③ CO_2 diffuses out of the erythrocyte into the plasma. CO_2 then diffuses into an alveolus.

(b) Pulmonary capillaries

Figure 23.27 Conversion of Carbon Dioxide to Bicarbonate. (*a*) In systemic capillaries, CO_2 enters erythrocytes, where it combines with H_2O to form carbonic acid (H_2CO_3): This compound is then converted to bicarbonate ion (HCO_3^-) and hydrogen ion (H^+). HCO_3^- leaves the erythrocyte and is replaced by Cl^-. HCO_3^- is transported in plasma. H^+ produced during the chloride shift binds with and is buffered by hemoglobin, thus helping to prevent a decrease in pH. (*b*) The process is reversed in pulmonary capillaries. Cl^- leaves the erythrocyte and HCO_3^- enters. HCO_3^- recombines with H^+ to reform CO_2 and H_2O. CO_2 exits the erythrocyte. It then diffuses into the alveoli and is expelled.

causes a conformational change in hemoglobin that makes it progressively easier for each additional O_2 molecule to bind to an available iron. This increase in the ease of oxygen binding is termed the **cooperative binding effect** of O_2 loading.

The graph shown in **figure 23.28** relates the Po_2 and percent O_2 saturation of hemoglobin called the **oxygen-hemoglobin saturation curve** (or *oxyhemoglobin dissociation curve*). Notice that the relationship between Po_2 and hemoglobin saturation is not linear (a straight line). The plotted points on the graph produce an S-shaped, or sigmoidal curve.

Relatively large changes occur in the hemoglobin saturation as the Po_2 increases initially. For example, a change in Po_2 from 20 to 40 mm Hg, a relatively small difference, results in the hemoglobin saturation increasing from about 35% to 75% saturated (a change of 40%) as a result of the cooperative binding effect. The curve is very steep in this part of the graph. A Po_2 of at least 60 mm Hg causes hemoglobin to become over 90% saturated. When Po_2 values are higher than 60 mm Hg, only relatively small changes in hemoglobin saturation occur. A small change in Po_2 from 80 to 100 mm Hg changes the hemoglobin saturation from 95% to 98% saturated—a very small change of only 3%. Figure 23.28 has application to the physiologic processes of oxygen loading that occurs during both alveolar gas exchange in the lungs and oxygen unloading that occurs during systemic gas exchange at the systemic cells as described next.

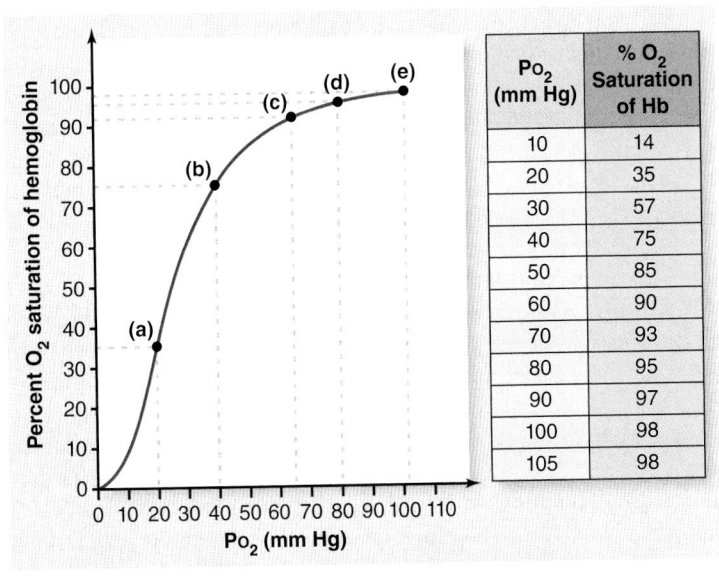

Po_2 (mm Hg)	% O_2 Saturation of Hb
10	14
20	35
30	57
40	75
50	85
60	90
70	93
80	95
90	97
100	98
105	98

Figure 23.28 Oxygen-Hemoglobin Saturation Curve. The percent O_2 saturation of hemoglobin increases as Po_2 increases. Relatively large changes in percent O_2 saturation of hemoglobin occur as Po_2 changes from 0 to 60 mm Hg. At Po_2 greater than 60 mm Hg changes in percent O_2 saturation of hemoglobin are much smaller. (The letter labels are discussed in text.)

The Oxygen-Hemoglobin Saturation Curve and Alveolar Gas Exchange

Hemoglobin loads with oxygen as blood moves through the pulmonary capillaries in the lungs. The alveolar P_{O_2} is 104 mm Hg at sea level. What would hemoglobin saturation be after the blood is transported through the pulmonary capillaries? The graph in figure 23.28 (label e) helps us determine that hemoglobin would be approximately 98% saturated.

If we ascend a high mountain, the air thins and environmental P_{O_2} decreases; this is accompanied by a decrease in alveolar P_{O_2}. How does this influence the hemoglobin saturation that occurs during alveolar gas exchange? Figure 23.28 can be used to determine approximate oxygen percent saturation values by selecting several specific alveolar P_{O_2} values that are correlated with given altitudes. For example, if the alveolar P_{O_2} value is 81 mm Hg (which corresponds to an elevation of about 5000 feet), hemoglobin saturation would be approximately 95% (figure 23.28, label d); and if the P_{O_2} value is 65 mm Hg (altitude 9000 feet), the hemoglobin saturation would be about 91% (figure 23.28, label c). In comparison, if alveolar P_{O_2} is 40 mm Hg (altitude 17,000 feet), hemoglobin saturation would be only 75% (figure 23.28, label b).

Increases in altitude from sea level (with accompanying decreases in alveolar P_{O_2}) initially result in only small changes in hemoglobin saturation; thus, changes in oxygen delivery are minimal. However, ascending to a very high altitude, with its accompanying large decrease in alveolar P_{O_2}, results in large decreases in hemoglobin saturation. The adverse physiologic effects from a decrease in alveolar P_{O_2}, referred to as **altitude sickness,** can occur in some individuals at altitudes as low as 6600 feet but occur for most individuals at altitudes greater than 8200 feet. Milder symptoms of altitude sickness include headache, nausea, and difficulty sleeping; more severe symptoms include pulmonary edema and cerebral edema.

 WHAT DO YOU THINK?

7 If a professional athlete is on the side of the field breathing pure oxygen, do you think this will make a difference in his or her athletic performance? Explain.

One additional observation can be made concerning oxygen loading in the alveoli and hemoglobin saturation. Because a person's hemoglobin is 98% saturated at sea level, if the partial pressure of oxygen is raised higher and higher (say, through the administration of pure oxygen), little would be gained—a potential gain of only the saturation of the remaining 2%. However, the binding of oxygen to reach 100% hemoglobin saturation can be reached only if atmospheric pressure is increased from 1 atm to 3 atm (760 to 2280 mm Hg), pressures that generally can be simulated only in hyperbaric oxygen chambers.

The Oxygen-Hemoglobin Saturation Curve and Systemic Gas Exchange

Oxygen is released from hemoglobin during its transport through systemic capillaries. The partial pressure of oxygen in systemic cells during resting conditions is approximately 40 mm Hg, and hemoglobin saturation is therefore 75% (figure 23.28, label b). Hemoglobin in the blood is 98% saturated with oxygen as it leaves the lungs, and then after it flows through the systemic capillaries during resting conditions it is still relatively saturated with oxygen at approximately 75%. Therefore, under resting conditions only a small percentage of the oxygen (approximately 20–25%) transported by the hemoglobin is released as it passes through systemic capillaries.

The amount of oxygen that remains bound to the hemoglobin after passing through the systemic capillaries is referred to as the **oxygen reserve.** The oxygen reserve provides a means for additional oxygen to be delivered to systemic cells under increased metabolic demands, such as occurs during exercise. If systemic

INTEGRATE

CONCEPT CONNECTION

Hemoglobin is a protein, and like all proteins, its three-dimensional structure is maintained by weak intramolecular interactions between the amino acid monomers that compose it, as described in section 2.8b. These interactions include hydrogen bonding, electrostatic attractions between oppositely charged groups, and association of nonpolar groups. Increases in temperature, changes in pH, and the binding of some ions weaken or break these bonds, ultimately resulting in conformational changes in the shape of a protein.

cell P_{O_2} decreases to 20 mm Hg, such as occurs during vigorous exercise, then what is the hemoglobin saturation? Vigorous exercise produces a large decrease in the hemoglobin saturation, meaning that more oxygen is unloaded. The hemoglobin saturation in the blood leaving systemic capillaries would be only 35% (figure 23.28, label a).

 WHAT DO YOU THINK?

8 Is more or less oxygen delivered to the systemic cells under the conditions of the lower cellular P_{O_2} (such as during exercise)? Is the oxygen reserve higher or lower?

Other Variables That Influence Oxygen Release from Hemoglobin During Systemic Exchange

Blood P_{O_2} is the most significant factor in hemoglobin's ability to bind and release oxygen during systemic gas exchange (**figure 23.29a**). However, other factors—such as changes in temperature, changes in pH [H^+], synthesis of the molecule 2,3-BPG (2,3-bisphosphoglycerate) or 2,3-DPG (2,3-diphosphoglycerate) in an alternative biochemical pathway in erythrocytes, and CO_2 binding to hemoglobin—can influence the amount of oxygen binding to hemoglobin (figure 23.29b-e). Each of these variables causes a conformational change in hemoglobin that decreases the ability of iron within the heme to bind O_2. Consequently, additional oxygen is released. We discuss how these variables influence the release of oxygen within systemic capillaries (figure 23.29):

- **Temperature.** Metabolic activities increase body temperature. This elevated temperature interferes with hemoglobin's ability to bind and hold oxygen, so additional oxygen is released from hemoglobin (figure 23.29b).

- **H^+ binding to hemoglobin.** H^+ produced when CO_2 enters erythrocytes (carbonic anhydrase reaction) binds to the globin protein in hemoglobin, causing a conformational change in hemoglobin and additional oxygen to be released. This H^+-induced decrease in affinity of O_2 for hemoglobin is known as the **Bohr effect** (figure 23.29c).

- **Presence of 2,3-BPG.** 2,3-BPG binds to hemoglobin, causing the release of additional oxygen as blood moves through systemic capillaries (figure 23.29d). 2,3-BPG is produced in an alternative (glycolytic) metabolic pathway within erythrocytes. The enzymatic process for 2,3-BPG production is stimulated by certain hormones that bind to erythrocyte receptors including thyroid hormone, epinephrine, growth hormone, and testosterone.

- **CO_2 binding to hemoglobin.** The binding of carbon dioxide to globin also causes the release of more oxygen from hemoglobin (figure 23.29e).

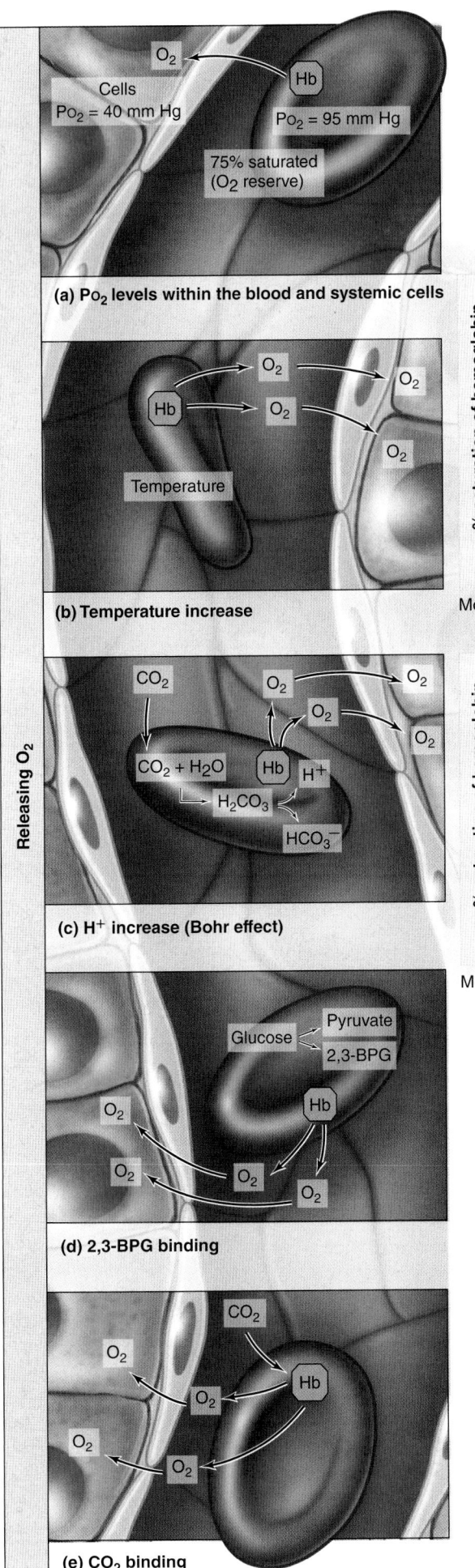

(a) Po₂ levels within the blood and systemic cells

Cells Po₂ = 40 mm Hg

O₂

Hb

Po₂ = 95 mm Hg

75% saturated (O₂ reserve)

(b) Temperature increase

Temperature

(c) H⁺ increase (Bohr effect)

CO_2

$CO_2 + H_2O$

Hb H⁺

H_2CO_3

HCO_3^-

(d) 2,3-BPG binding

Glucose ⇄ Pyruvate

2,3-BPG

Hb

(e) CO_2 binding

CO_2

Hb

Releasing O₂

Figure 23.29 Hemoglobin and Oxygen Release.

(*a*) The most important variable that influences oxygen release from hemoglobin during systemic gas exchange is blood Po₂. Other variables cause a conformational change in hemoglobin that increases the release of oxygen, including (*b*) an increase in temperature; (*c*) an increase in H⁺ (referred to as the Bohr effect); (*d*) the binding of 2,3-BPG; and (*e*) the binding of carbon dioxide. The influence of temperature and H⁺ levels (i.e., pH) on the oxygen-hemoglobin saturation curve is shown in figure (*b*) and figure (*c*), respectively.

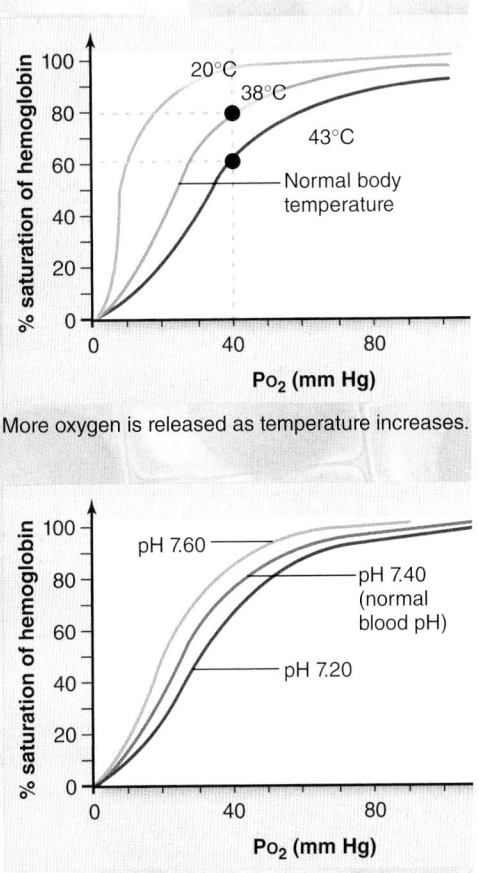

More oxygen is released as temperature increases.

More oxygen is released as pH decreases.

Interestingly, the release of oxygen during systemic gas exchange also causes a conformational change in hemoglobin that enhances its ability to bind CO_2. Consequently, the more oxygen that is released from hemoglobin, the more carbon dioxide that binds to it—this is called the **Haldane effect.**

Examination of the oxygen-hemoglobin saturation curve reveals the effects two of these variables (temperature and pH) have on oxygen release (figure 23.29, graphs). When temperature increases from 38°C to 43°C, at any given Po₂, hemoglobin saturation is lower than when the temperature is 38°C (figure 23.29*b*, graph). If temperature decreases, hemoglobin saturation increases. Similar changes in hemoglobin saturation are observed with changes in pH (figure 23.29*c*, graph). Factors that bring about a decrease in oxygen affinity to hemoglobin (e.g., increase in temperature, increase in H⁺) and the additional release of oxygen are said to cause a **shift right** in the saturation curve. In contrast, the variables that bring about an increase in oxygen affinity to hemoglobin (e.g., decrease in temperature, decrease in H⁺) result in release of less oxygen and are said to cause a **shift left.**

Summary of Respiration

Figure 23.30 is a visual overview of the four events of respiration. It depicts the seemingly effortless task of moving oxygen and carbon dioxide between the environment and our systemic cells. Remember that all of the processes are occurring simultaneously and continuously.

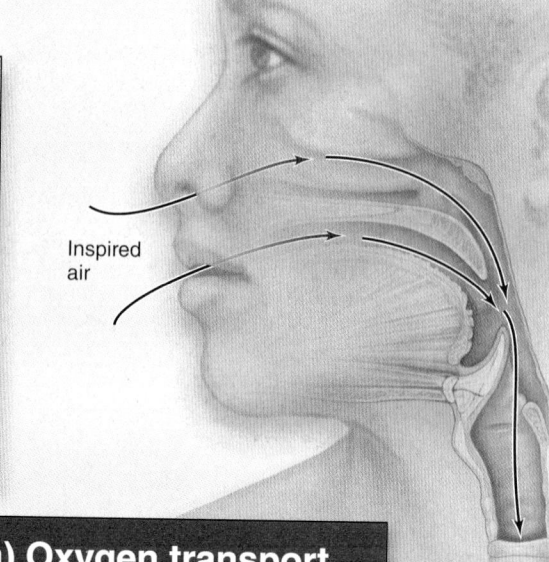

(a) Oxygen transport from the atmosphere into the cells. (Steps 1–4)

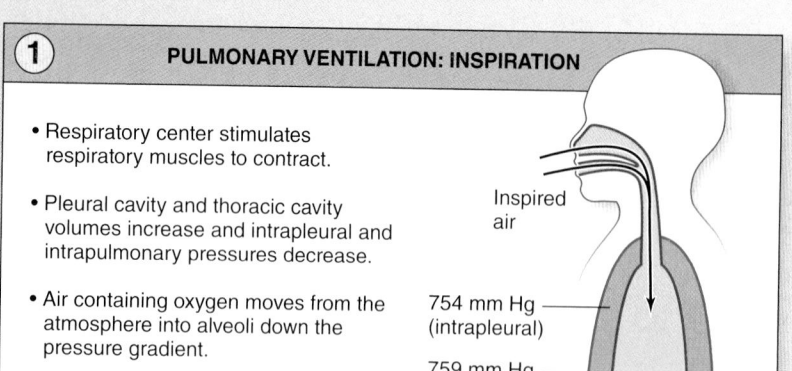

① PULMONARY VENTILATION: INSPIRATION

- Respiratory center stimulates respiratory muscles to contract.

- Pleural cavity and thoracic cavity volumes increase and intrapleural and intrapulmonary pressures decrease.

- Air containing oxygen moves from the atmosphere into alveoli down the pressure gradient.

Inspired air

754 mm Hg (intrapleural)

759 mm Hg (intrapulmonary)

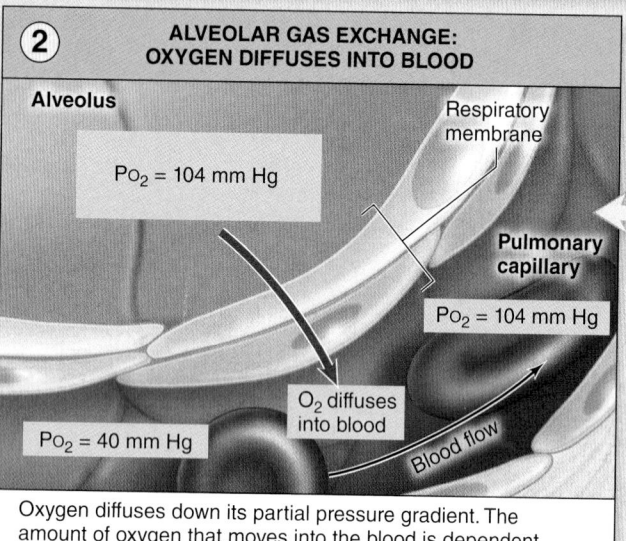

② ALVEOLAR GAS EXCHANGE: OXYGEN DIFFUSES INTO BLOOD

Alveolus

Respiratory membrane

P_{O_2} = 104 mm Hg

Pulmonary capillary

P_{O_2} = 104 mm Hg

O_2 diffuses into blood

P_{O_2} = 40 mm Hg

Blood flow

Oxygen diffuses down its partial pressure gradient. The amount of oxygen that moves into the blood is dependent upon surface area and thickness of the respiratory membrane, and ventilation-perfusion coupling.

③ GAS TRANSPORT OF OXYGEN WITHIN THE BLOOD

>98% of O_2 binds to iron of hemoglobin

<2% of O_2 is dissolved in plasma

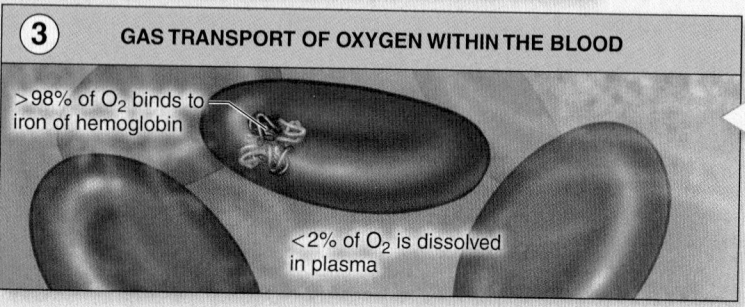

④ SYSTEMIC GAS EXCHANGE: OXYGEN DIFFUSES INTO SYSTEMIC CELLS

P_{O_2} = 40 mm Hg

O_2 diffuses into systemic cells

Systemic cells

P_{O_2} = 95 mm Hg

Systemic capillary

Blood flow

P_{O_2} = 40 mm Hg

Additional oxygen is released from hemoglobin with: an increase in temperature, H^+, 2,3-BPG, and CO_2 binding.

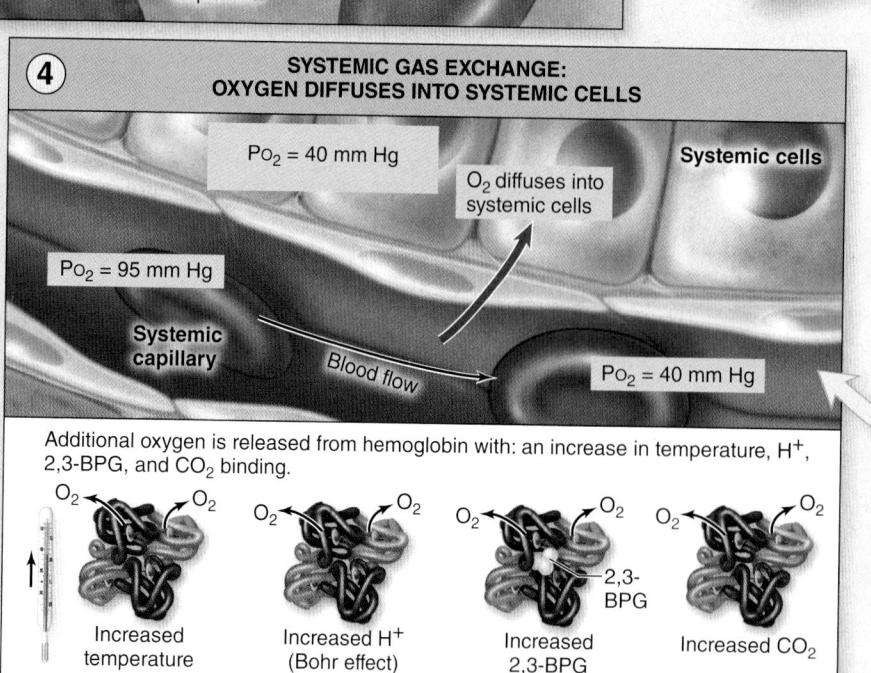

O_2 ← → O_2

Increased temperature

O_2 ← → O_2

Increased H^+ (Bohr effect)

O_2 ← → O_2

2,3-BPG

Increased 2,3-BPG

O_2 ← → O_2

Increased CO_2

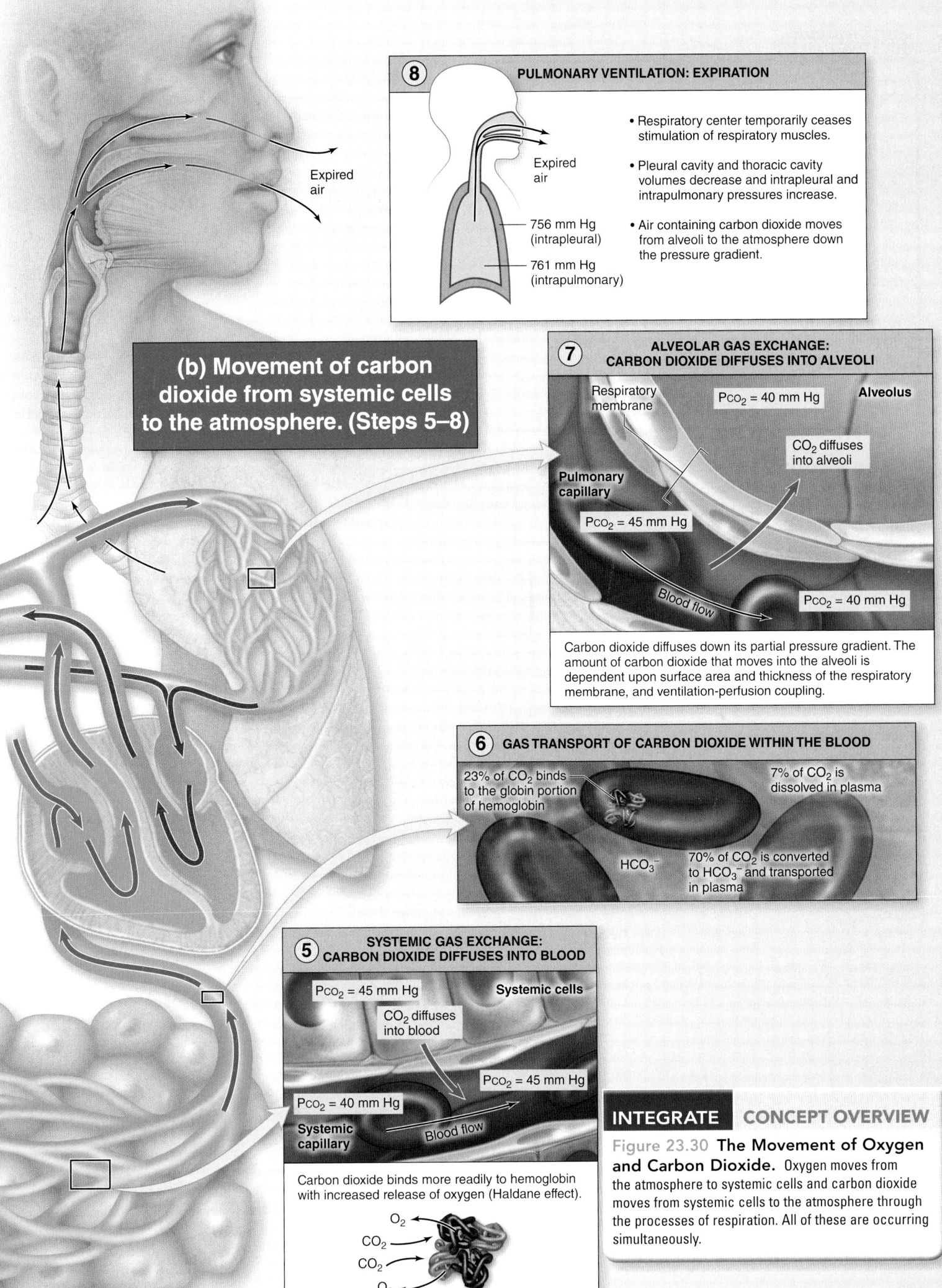

(b) Movement of carbon dioxide from systemic cells to the atmosphere. (Steps 5–8)

⑧ PULMONARY VENTILATION: EXPIRATION

Expired air

Expired air

756 mm Hg (intrapleural)

761 mm Hg (intrapulmonary)

- Respiratory center temporarily ceases stimulation of respiratory muscles.
- Pleural cavity and thoracic cavity volumes decrease and intrapleural and intrapulmonary pressures increase.
- Air containing carbon dioxide moves from alveoli to the atmosphere down the pressure gradient.

⑦ ALVEOLAR GAS EXCHANGE: CARBON DIOXIDE DIFFUSES INTO ALVEOLI

Respiratory membrane

P_{CO_2} = 40 mm Hg   Alveolus

CO_2 diffuses into alveoli

Pulmonary capillary

P_{CO_2} = 45 mm Hg

Blood flow

P_{CO_2} = 40 mm Hg

Carbon dioxide diffuses down its partial pressure gradient. The amount of carbon dioxide that moves into the alveoli is dependent upon surface area and thickness of the respiratory membrane, and ventilation-perfusion coupling.

⑥ GAS TRANSPORT OF CARBON DIOXIDE WITHIN THE BLOOD

23% of CO_2 binds to the globin portion of hemoglobin

7% of CO_2 is dissolved in plasma

HCO_3^-

70% of CO_2 is converted to HCO_3^- and transported in plasma

⑤ SYSTEMIC GAS EXCHANGE: CARBON DIOXIDE DIFFUSES INTO BLOOD

P_{CO_2} = 45 mm Hg   Systemic cells

CO_2 diffuses into blood

P_{CO_2} = 45 mm Hg

P_{CO_2} = 40 mm Hg

Systemic capillary   Blood flow

Carbon dioxide binds more readily to hemoglobin with increased release of oxygen (Haldane effect).

O_2
CO_2
CO_2
O_2

INTEGRATE   CONCEPT OVERVIEW

Figure 23.30 **The Movement of Oxygen and Carbon Dioxide.** Oxygen moves from the atmosphere to systemic cells and carbon dioxide moves from systemic cells to the atmosphere through the processes of respiration. All of these are occurring simultaneously.

935

INTEGRATE

CLINICAL VIEW

Fetal Hemoglobin and Physiologic Jaundice

A different type of hemoglobin molecule, called **fetal hemoglobin**, is found in the erythrocytes of unborn babies. Fetal hemoglobin has a greater affinity for binding oxygen than adult hemoglobin, ensuring a net movement of oxygen from the mother's blood to the blood of the fetus. After the infant is born, the erythrocytes with fetal hemoglobin are replaced with erythrocytes containing adult hemoglobin. The infant may exhibit **physiologic jaundice**—a yellowish tinge to the skin due to elevated levels of bilirubin— as the fetal hemoglobin breaks down. This condition lasts about a week, and sometimes 2 weeks in premature infants. To facilitate and accelerate the breakdown of bilirubin, an infant with jaundice may be placed under a Bili-Lite or partially "wrapped" in a BiliBlanket.

If any of the systems involved in respiration—respiratory, skeletal, muscular, nervous, or cardiovascular—are unable to function normally, a homeostatic imbalance in respiratory gas exchange occurs. **Table 23.6** summarizes many of the causes of these homeostatic imbalances.

WHAT DID YOU LEARN?

36 How does oxygen movement occur during alveolar gas exchange, gas transport, and systemic gas exchange?

37 How does carbon dioxide movement occur during systemic gas exchange, gas transport, and alveolar gas exchange?

38 Does hemoglobin saturation increase or decrease during alveolar gas exchange?

39 How is oxygen release from hemoglobin during systemic gas exchange altered by: P_{O_2}, temperature, H^+ concentration, 2,3-BPG, and CO_2?

23.8 Breathing Rate and Homeostasis

The respiratory center stimulates breathing to occur at a rate of 12 to 15 times per minute and a tidal volume depth of 500 mL at rest. The respiratory center adjusts the rate and depth of breathing to maintain homeostasis in response to various stimuli, including changes in blood P_{CO_2}, levels of blood H^+, and changes in blood P_{O_2}. Recall that blood P_{CO_2} level is the most important stimulus.

Changes in breathing rate generally help to maintain homeostatic levels of respiratory gases and pH in the blood. However, changes in breathing rate may also result in homeostatic imbalances. Here we cover the changes that occur with hyperventilation, hypoventilation, and during exercise, including how homeostasis may be affected.

23.8a Effects of Hyperventilation and Hypoventilation on Cardiovascular Function

LEARNING OBJECTIVES

1. Explain how hyperventilation and hypoventilation influence the chemical composition of blood.

2. Describe how breathing rate and depth affect venous return of blood and lymph.

Hyperventilation is a breathing rate or depth that is increased above the body's demand. Hyperventilation may be caused by anxiety, panic,

Table 23.6	Causes of Respiratory Homeostatic Imbalances	
System	**Physiologic Consequences**	**Clinical Example**
RESPIRATORY SYSTEM		
Airway obstruction	Decreased airflow into alveoli	Asthma, bronchitis
Thickened respiratory membrane	Decreased alveolar gas exchange	Pulmonary edema, pneumonia
Loss of respiratory membrane surface	Decreased alveolar gas exchange	Emphysema, lung cancer
SKELETAL SYSTEM		
Arthritis or deformities of thoracic cage or vertebral column	Impaired ability to cause dimensional volume changes in thoracic cavity	Rheumatoid arthritis, congenital deformities
MUSCULAR SYSTEM		
Paralysis of respiratory muscles	Impaired ability to cause dimensional volume changes in thoracic cavity	Polio, muscular dystrophy
NERVOUS SYSTEM		
Brainstem injury Oversedation of respiratory center	Decreased ability to stimulate muscles of breathing	Trauma, drug use
Spinal cord injuries	Decreased ability to stimulate muscles of breathing	Trauma (e.g., diving or motorcycle accident)
CARDIOVASCULAR SYSTEM		
Pulmonary embolism	Blockage in pulmonary artery, and blood does not reach lung capillaries for gas exchange	Slowed blood flow from immobilization (e.g., prolonged bedrest, long plane flight, confined to wheelchair)
Anemia	Decreased erythrocytes or hemoglobin concentration with decreased gas transport	Low iron, pernicious anemia (inability to absorb B_{12})
Impeded blood flow	Decreased gas transport and gas exchange	Atherosclerosis, congestive heart failure, hemorrhage

or ascending to a high altitude (as the individual breathes faster to compensate for the lower oxygen levels). Hyperventilation may also be performed voluntarily when you consciously inspire and expire excessively at an accelerated rate. P_{O_2} levels increase and P_{CO_2} levels decrease in the alveoli. This increases the partial pressure gradients between alveoli and blood for both P_{O_2} and P_{CO_2}. These changes affect the blood as follows: (1) Additional oxygen does not enter the blood despite the steeper P_{O_2} gradient because hemoglobin is generally 98% saturated even during quiet breathing. (2) However, additional carbon dioxide leaves the blood to enter the alveoli due to a steeper P_{CO_2} gradient. Consequently, blood P_{CO_2} decreases below normal levels, a condition called **hypocapnia** (hī′pō-kap′nē-ă).

Low blood P_{CO_2} causes vasoconstriction of blood vessels. The blood vessels within the brain are especially vulnerable to these changes. Ironically, one result of hyperventilation is *decreased* oxygen delivery to the brain due to this generalized vasoconstriction. Low blood P_{CO_2} may also result in a decrease in blood H^+ concentration. If the body's buffering capacity is exceeded, this may result in **respiratory alkalosis,** a condition discussed in more detail in section 25.6b.

Symptoms that accompany hyperventilation may include feeling faint or dizzy, numbness, tingling of the mouth and fingertips, muscular cramps, and tetany. If hyperventilation is prolonged, it can cause disorientation, loss of consciousness, coma, and possibly death. If the cause is panic, breathing rate typically returns to a normal level once the individual loses consciousness. A person who is hyperventilating is sometimes directed to breathe into and out of a paper bag, because this action is thought to slow the loss of CO_2.

Hypoventilation is breathing that is either too slow (called **bradypnea,** brad-ip-nē′-ă) or too shallow (called **hypopnea,** hī-pop′nē-ă) to adequately meet the metabolic needs of the body. Causes of hypoventilation are varied and include airway obstruction, pneumonia, brainstem injury, obesity (which restricts lung expansion), and any other condition that interferes with pulmonary ventilation or alveolar gas exchange. Oxygen levels decrease and carbon dioxide levels increase in the alveoli. This results in smaller partial pressure gradients between the alveoli and blood for both O_2 and CO_2. Thus, the diffusion of the respiratory gases is altered during alveolar gas exchange as follows:

- Lower amounts of oxygen diffuse from the alveoli into the blood, and blood P_{O_2} decreases. This is a condition called **hypoxemia** (hī-pok-sē-mē-ă), which may lead to low oxygen in the tissue called **hypoxia** (hī-pok-sē-ă).
- Lower amounts of carbon dioxide diffuse from the blood into the alveoli. This causes blood P_{CO_2} to increase, a condition called **hypercapnia** (hī′pĕr-kap′nē-ă).

Low blood oxygen levels may result in insufficient oxygen delivery to systemic cells, with a subsequent decrease in aerobic cellular respiration (see section 3.4).

INTEGRATE

CONCEPT CONNECTION

Breathing rate also influences venous return of blood and lymph, as is described, respectively, in the cardiovascular system (see section 20.5a) and lymphatic system (see section 21.1b). The contraction and relaxation of the skeletal muscles of breathing cause rhythmic pressure changes during breathing, and this cycle is referred to as the *respiratory pump*. The action of the respiratory pump increases during hyperventilation, increasing venous return of blood and lymph, whereas during hypoventilation, decreased action of the respiratory pump decreases venous return of blood and lymph.

High blood P_{CO_2} may result in a decrease in pH because as CO_2 increases, this equation is driven to the right and H^+ levels increase ($CO_2 + H_2O \rightarrow H_2CO_3 \rightarrow HCO_3^- + H^+$). If the body's buffering capacity is exceeded, this shift may lead to **respiratory acidosis,** described in more detail in section 25.6b.

Symptoms resulting from either insufficient blood P_{O_2}, increased blood P_{CO_2}, or both include lethargy, sleepiness, headache, polycythemia (low oxygen triggers release of erythropoietin), and cyanotic tissues in which skin color appears blue as a consequence of low oxygen saturation of hemoglobin. If hypoventilation is prolonged, it may lead to convulsions, unconsciousness, and possibly death.

Hypoventilation, or even the cessation of breathing, can be performed voluntarily. However, you cannot hold your breath long enough to die. The accumulation of CO_2 in the blood stimulates chemoreceptors to relay nerve signals to the respiratory center to initiate inspiration before or after loss of consciousness—but always before the brain suffers damage from lack of oxygen.

 WHAT DID YOU LEARN?

40 How does blood P_{O_2} and P_{CO_2} change if an individual is hyperventilating?

23.8b Breathing and Exercise

LEARNING OBJECTIVE

3. Explain the changes in breathing that accompany exercise.

A number of observations have been made regarding breathing and exercise, although not all aspects are fully understood.

While participating in vigorous exercise, a person's breathing *depth* increases while the breathing *rate* remains the same. This type of breathing, which is deeper, but not faster, is referred to as **hyperpnea** (hī′pĕr-nē-ă). Hyperpnea can be differentiated from hyperventilation in that hyperventilation is a condition where both breathing depth and rate are increased.

Both oxygen consumption and carbon dioxide production increase in response to elevated rates of cellular respiration during exercise. However, blood P_{O_2} and P_{CO_2} levels remain relatively the same. This occurs because deeper breathing, increased cardiac output, and increased blood flow are able to deliver the additional oxygen needed and eliminate the greater amount of carbon dioxide produced (i.e., supply increases to meet demand). Thus, blood P_{O_2} and P_{CO_2} levels are not thought to be the stimuli that cause breathing to change because they remain relatively constant during exercise. Although questions remain as to the exact cause, stimulation of the respiratory center generally is thought to occur from one or more of the following causes, including

- Sensory signals relayed from proprioceptors in joints, muscles, and tendons in response to movement
- Motor output originating in the cerebral cortex that initiate muscular movement during exercise, simultaneously relaying signals to the respiratory center
- The conscious anticipation of participating in exercise

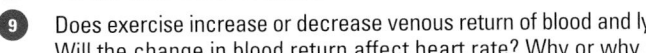 **WHAT DO YOU THINK?**

9 Does exercise increase or decrease venous return of blood and lymph? Will the change in blood return affect heart rate? Why or why not?

 WHAT DID YOU LEARN?

41 How does blood P_{O_2} and P_{CO_2} change during exercise? Explain.

42 What are three possible reasons that breathing changes during exercise?

- The respiratory system consists of the lungs and the respiratory passageways extending through the head, neck, and trunk.

23.1 Introduction to the Respiratory System 891

- The functions, organization, and mucosal lining are described for the respiratory system.

23.1a General Functions of the Respiratory System 891

- The primary functions of the respiratory system include serving as a passageway for air between the atmosphere and the alveoli, site for gas exchange between the alveoli and blood, detection of odors, and sound production.
- Breathing rate influences blood levels of oxygen, carbon dioxide, and hydrogen ion (pH), as well as venous return of blood and lymph.

23.1b General Organization of the Respiratory System 891

- Structurally, the respiratory tract is composed of the upper respiratory tract and the lower respiratory tract.
- Functionally, the respiratory tract is composed of the conducting zone and respiratory zone.

23.1c Respiratory Mucosa 891

- The respiratory tract is lined by a mucosa that is composed of an epithelium, basement membrane, and lamina propria.
- The epithelium of the mucosa becomes progressively thinner along its length, except for areas that must withstand abrasion.

23.2 Upper Respiratory Tract 893

- The upper respiratory tract includes the nose, nasal cavity, and pharynx.

23.2a Nose and Nasal Cavity 893

- The nose leads into the nasal cavity, a space formed by the nose and skull that can be divided into the nasal vestibule, olfactory region, and respiratory region.

23.2b Paranasal Sinuses 894

- The four paranasal sinuses—frontal, ethmoidal, maxillary, and sphenoidal—have ducts that lead into the nasal cavity.

23.2c Pharynx 896

- The pharynx is portioned into three regions: the nasopharynx, oropharynx, and laryngopharynx.

23.3 Lower Respiratory Tract 897

- The lower respiratory tract is composed of the structures from the larynx to the alveoli.

23.3a Larynx 897

- The larynx is a passageway for air, closes off the respiratory tract during swallowing, is involved in sound production, assists in increasing pressure in the abdominal cavity, and participates in both a sneeze and cough reflex.

23.3b Trachea 899

- The trachea is a patent, flexible tube with C-shaped cartilage "rings" that extends from the larynx to the main bronchi.

23.3c Bronchial Tree 900

- The bronchial tree is a highly branched system of air-conducting passages that originate from the left and right main bronchi and end in the terminal bronchioles.
- Airflow is regulated through the bronchial tree by contraction and relaxation of smooth muscle to cause bronchoconstriction and bronchodilation, respectively, especially within bronchioles.

23.3d Respiratory Zone: Respiratory Bronchioles, Alveolar Ducts, and Alveoli 904

- The respiratory zone includes the regions of the respiratory tract with a wall thin enough for gas exchange; it extends from respiratory bronchioles, alveolar ducts, and alveoli.
- Alveoli are composed of alveolar type I cells, which are the primary cell that forms the alveoli; alveolar type II cells, which produce surfactant to decrease surface tension; and alveolar macrophages, which engulf microorganisms and particulate matter.

23.3e Respiratory Membrane 906

- The respiratory membrane is the boundary between the alveolar lumen and the lumen of the pulmonary capillary.
- Respiratory gases cross this barrier in alveolar gas exchange between the alveoli of the lungs and the blood in the pulmonary capillaries.

23.4 Lungs 907

- Lungs are located in the thoracic cavity lateral to the mediastinum and enclosed in the thoracic cage.

23.4a Gross Anatomy of the Lung 907

- Each lung is conical in shape and divided into lobes.

23.4b Circulation to and Innervation of the Lungs 910

- Pulmonary circulation transports deoxygenated blood to the gas exchange surface of the lung for re-oxygenation, and bronchial circulation delivers oxygenated blood to the bronchi and bronchioles.
- The smooth muscle and glands of the larynx, trachea, bronchial tree, and lungs are innervated by the autonomic nervous system.

23.4c Pleural Membranes and Pleural Cavity 911

- The lung surface and the internal thoracic walls are lined by serous pleural membranes.
- The pleural cavity is between the pleural layers and contains serous fluid.

23.4d How Lungs Remain Inflated 912

- The intrapleural pressure (the pressure in the pleural cavity) is lower than the intrapulmonary pressure (the pressure in the lungs), and this pressure difference keeps the lungs inflated.

Do You Know the Basics?

_____ 1. Respiration normally requires the

    a. respiratory system.

    b. muscular system.

    c. nervous system.

    d. All of these are correct.

_____ 2. Arrange the following from largest to smallest for the divisions of the lung.

    a. lobes, bronchopulmonary segments, lobules, alveoli

    b. lobes, alveoli, lobules, bronchopulmonary segments

    c. alveoli, lobules, bronchopulmonary segments, lobes

    d. lobules, bronchopulmonary segments, lobes, alveoli

_____ 3. The lungs do not normally collapse because

    a. they are attached to the thoracic wall with parietal ligaments.

    b. they are attached to the thoracic wall with the visceral ligaments.

    c. the pressure in the intrapleural cavity is lower than the pressure in the intrapulmonary space (lungs).

    d. the pressure in the intrapleural cavity is greater than the pressure in the intrapulmonary space (lungs).

_____ 4. Which of the following correctly represents the sequence of events associated with the thoracic cavity to produce inspiration?

    a. muscle contraction, increase in volume, decrease in pressure

    b. decrease in pressure, increase in volume, muscle contraction

    c. muscle contraction, decrease in pressure, increase in volume

    d. increase in volume, muscle contraction, decrease in pressure

_____ 5. The diaphragm and the external intercostals relax. Volume decreases and pressure increases in the thoracic cavity. This description (by itself) describes

    a. quiet inspiration.

    b. quiet expiration.

    c. forced inspiration.

    d. forced expiration.

_____ 6. Which areas of the brain contain the respiratory center?

    a. medulla oblongata and hypothalamus

    b. hypothalamus and pons

    c. medulla oblongata and pons

    d. medulla oblongata and cerebrum

_____ 7. The amount of which substance in the blood normally increases the respiratory rate?

    a. oxygen

    b. carbon dioxide

    c. hydrogen gas (H_2)

    d. bicarbonate

_____ 8. The movement of oxygen from the blood into the systemic cells is referred to as

    a. pulmonary ventilation.

    b. alveolar gas exchange.

    c. systemic gas exchange.

    d. gas transport.

_____ 9. Most carbon dioxide is transported in the blood

    a. dissolved in the plasma as carbon dioxide.

    b. in association with hemoglobin.

    c. as bicarbonate ion.

    d. in combination with oxygen.

_____ 10. All of the following are accurate statements about hemoglobin except

    a. hemoglobin carries oxygen on the Fe ion.

    b. hemoglobin carries carbon dioxide on the globin.

    c. hemoglobin carries only a small portion of the total carbon dioxide in the blood (less than 25%).

    d. hemoglobin releases oxygen at the level of the cell, making hemoglobin more saturated.

11. Explain how the respiratory tract is organized both structurally and functionally.

12. Describe the relationship of the visceral pleura, parietal pleura, pleural cavity, and serous fluid to keep the lungs inflated.

13. List the four processes of respiration, in order, for moving oxygen from the atmosphere to the body's tissues. List, in order, the processes for the movement of carbon dioxide.

14. Describe the muscles, volume changes, and pressure changes involved with quiet inspiration and expiration.

15. Explain how additional air is moved during a forced inspiration or expiration.

16. Describe how quiet breathing is controlled by the respiratory center.

17. Explain alveolar and systemic gas exchange.

18. List the two means by which oxygen is transported in the blood and the three means by which carbon dioxide is transported.

19. Describe the relationship of P_{O_2} and hemoglobin percent saturation.

20. List the variables that increase the release of oxygen (by decreasing the affinity of oxygen to hemoglobin) as blood passes through systemic capillaries.

Can You Apply What You've Learned?

1. Paramedics arrived at a car accident to find an elderly gentleman breathing erratically. He was not wearing his seat belt and had hit his head on the windshield. Given his erratic breathing, they were concerned that he had damage to the

    a. medulla oblongata.

    b. cerebral cortex.

    c. pons.

    d. spinal cord.

Use the following to answer questions 2–4.

Michelle, age 45, has been smoking for over 30 years. She now is beginning to have trouble breathing. Her physician sent her to have her respiratory function tested. The results showed that she has emphysema, which is characterized by a decreased surface area for alveolar gas exchange.

2. What component of the respiratory system is most affected in emphysema, causing her to struggle with breathing?

    a. Nasal passageways are inflamed, and less air can move into and out of the respiratory tract.

    b. Bronchi are inflamed.

    c. Bronchioles are dilated.

    d. Damage to the alveoli has occurred.

3. Because physiologic dead space is increased, she would be exhibiting

    a. increased resistance during inhaling.

    b. smaller amounts of air entering her lungs.

    c. harder and deeper breathing that leaves her feeling tired.

    d. decreased blood P_{CO_2}.

4. Blood samples were also taken. Which results might be expected for her blood respiratory gases and blood pH?

    a. decrease in P_{O_2}, increase in P_{CO_2}, decrease in blood pH

    b. decrease in P_{O_2}, decrease in P_{CO_2}, decrease in blood pH

    c. decrease in P_{O_2}, increase in P_{CO_2}, increase in blood pH

    d. increase in P_{O_2}, decrease in P_{CO_2}, increase in blood pH

5. A 10-year-old boy named Michael is playing soccer and begins to have trouble breathing. His mother takes him to the pediatrician, who suspects that he has asthma. The pediatrician recommends that Michael see a respiratory specialist, and results from his pulmonary function tests indicate that Michael does have asthma. Michael's difficulty in breathing, as a result of the asthma, is caused by

    a. inflammation of his nasal passageways.

    b. inflammation and bronchoconstriction of his bronchioles.

    c. spastic closure of his epiglottis, temporarily closing off his trachea.

    d. damage to the alveoli, resulting in decreased surface area of the respiratory membrane.

Can You Synthesize What You've Learned?

1. Tiffany had returned to her college dorm and was having difficulty breathing. She knew she was having an asthma attack. What changes are occurring in her respiratory tract to cause her difficulty in breathing? What potential changes would you predict for her blood P_{O_2} and P_{CO_2}? What substance is generally given as a treatment, which will result in her bronchioles dilating?

2. The nerve to the sternocleidomastoid muscle was damaged during surgery. What type of breathing will be impaired—quiet inspiration, quiet expiration, forced inspiration, or forced expiration?

3. Mark is attempting his first serious mountain climb above 8000 feet. He begins to breathe harder during his ascent, feels light-headed, and has difficulty thinking clearly. What has stimulated him to breathe harder? What effect does this have on P_{CO_2}? What potential change in blood pH can occur? Why? Does more or less oxygen reach the brain? Explain.

INTEGRATE

ONLINE STUDY TOOLS

  **LEARNSMART**

The following study aids may be accessed through Connect.

Concept Overview Interactive: Figure 23.30: The Movement of Oxygen and Carbon Dioxide

Clinical Case Study: How the Smallest Things Become Critically Important

Interactive Questions: This chapter's content is served up in a number of multimedia question formats for student study.

LearnSmart: Topics and terminology include upper and lower respiratory tracts, lungs, pulmonary ventilation, alveolar and systemic gas exchange, gas transport, and breathing rate and homeostasis

Anatomy & Physiology Revealed: Topics include pharynx and larynx, trachea and bronchi, alveolar pressure changes, partial pressure

chapter 24

Urinary System

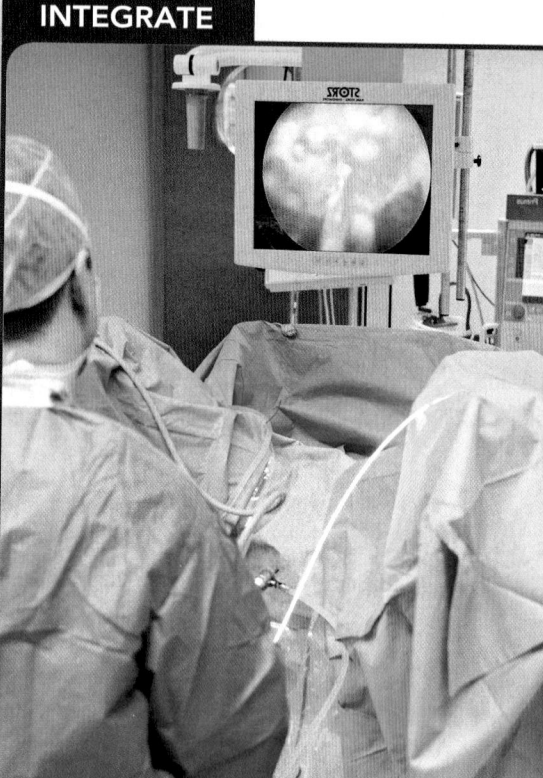

INTEGRATE

CAREER PATH
Urologist

Urologists are medical and surgical professionals who diagnose and treat individuals who have disorders of the urinary system and the male reproductive system. Conditions treated by urologists include kidney stones, stress incontinence, urinary tract infections, congenital abnormalities, benign prostatic hyperplasia, and various types of cancers. This photo shows a urologist viewing a video monitor during prostate surgery.

Imagine a river that supplies drinking water to a town. The water in the river may become polluted with sediment, animal waste, and motorboat fuel, but the town has a water treatment plant that removes these wastes. In an analogous way, the cells of all body systems produce waste products and they end up in the blood. These unwanted substances are filtered from the blood by the kidneys to form urine, and the urine is then eliminated from the body by the ureters, urinary bladder, and urethra. Thus, the urinary system serves as the body's "water treatment plant."

Anatomy & Physiology REVEALED
aprevealed.com

Module 13: Urinary System

Here we first introduce the urinary system and then discuss the details of kidney structure and its function in filtering blood and forming urine. The chapter ends with a description of the anatomic structures of the ureters, urinary bladder, and urethra, and their roles in transporting, storing, and eliminating urine.

24.1 Introduction to the Urinary System

LEARNING OBJECTIVES

1. Identify the structures that compose the urinary system, and provide a description of the general function of each.

2. List the functions of the kidneys.

The urinary (ūr′i-nār-ē) system is collectively composed of the kidneys, ureters, urinary bladder, and urethra (figure 24.1). One of the primary functions of the kidneys is to filter blood and convert the filtrate into urine. This liquid waste is then transported by the ureters from the kidneys to the urinary bladder, which is an expandable muscular sac that stores as much as 1 liter (L) of urine until it is eliminated from the body through the urethra.

Several essential physiologic processes occur to the filtrate when it is within the kidney being converted to urine. These include the following:

- **Elimination of metabolic wastes.** The kidneys remove waste that is present in the filtrate (e.g., urea, uric acids) to prevent these substances from reaching toxic levels within the blood.

- **Regulation of ion levels.** The kidneys help control the blood's ion balance, such as sodium ions (Na^+), potassium ions (K^+), calcium ions (Ca^{2+}), and phosphate ions (PO_4^{3-}) by eliminating varying amounts of these substances in the urine, depending upon dietary intake.

- **Regulation of acid-base balance.** The kidneys aid in maintaining acid-base balance by altering blood levels of both hydrogen ions (H^+) and bicarbonate ions (HCO_3^-) by also eliminating varying amounts of these ions.

- **Regulation of blood pressure.** The kidneys also help regulate blood pressure by excreting fluid in the urine. Regulating fluid lost in the urine helps to regulate blood volume. The kidneys also release the enzyme renin, which is required for production of angiotensin II, a hormone that increases blood pressure (see section 20.6b and table R.7). Arguably, blood pressure regulation is one of the kidney's most important functions and is described in detail in section 25.4.

- **Elimination of biologically active molecules.** Loss of small molecules that are biologically active (e.g., hormones, drugs) also occurs as these substances are filtered, not reclaimed, and then become part of urine.

The kidneys perform other functions, however, in addition to their role in filtering blood and processing filtrate to form urine. As you read through these functions, you may remember that many of these have been described in previous chapters. The other functions of the kidney include the following:

- **Formation of calcitriol.** The kidneys synthesize the final enzyme in calcitriol hormone formation. Calcitriol increases the absorption of calcium from the small intestine to increase blood calcium concentration (see section 7.6).

- **Production and release of erythropoietin.** As the kidneys are filtering the blood, they are also indirectly measuring the oxygen level of blood. In response to low blood oxygen levels, cells within the kidney secrete erythropoietin (EPO) hormone. EPO stimulates red bone marrow to increase its rate of erythrocyte formation (see section 18.3b and table R.6). The greater numbers of erythrocytes transport additional oxygen from the lungs to systemic cells (see section 23.7a).

- **Potential to engage in gluconeogenesis.** During prolonged fasting or starvation, the kidneys may engage in gluconeogenesis to produce glucose from noncarbohydrate sources; this helps to maintain normal blood glucose levels during periods of extreme nutrient deprivation (see section 2.7c).

A general observation can be made about the various functions of the kidneys. They take care of our blood. The kidneys remove unwanted materials from the blood, maintain blood plasma concentration of ions (e.g., Ca^{2+}, Na^+, and K^+), help regulate blood pH (i.e., H^+ and HCO_3^- concentration), alter blood volume, regulate the number of erythrocytes, and also help maintain blood glucose with severe limitation on nutrient intake. Healthy blood is intimately dependent upon healthy kidneys.

WHAT DO YOU THINK?

1. Which of the following would accompany a loss of kidney function: (a) accumulation of wastes, (b) anemia, (c) changes in blood pressure, or (d) acid-base imbalance?

WHAT DID YOU LEARN?

1. Which structure of the urinary system forms urine, and which structure stores urine?

2. What are the two means by which the kidney helps to regulate blood pressure?

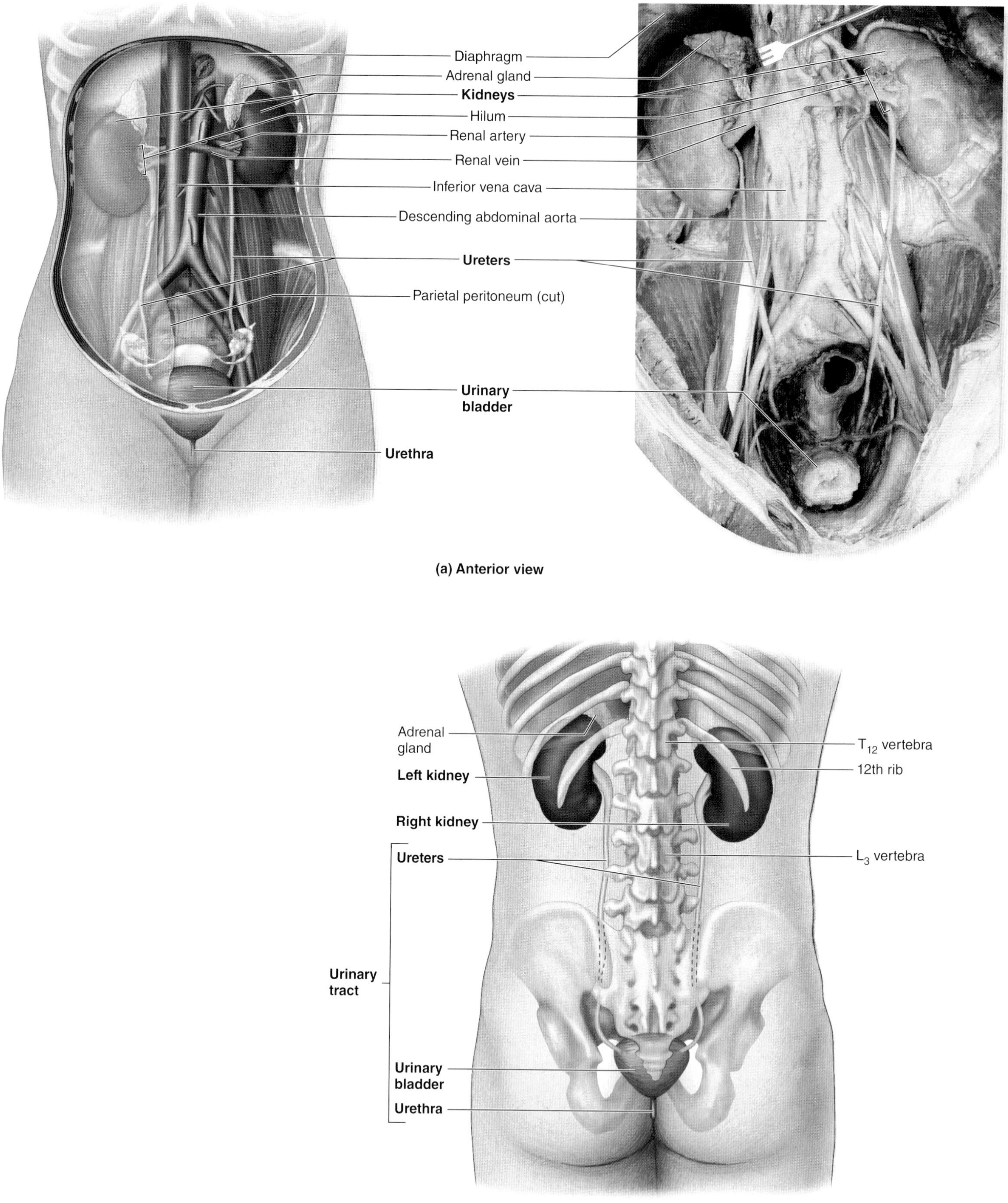

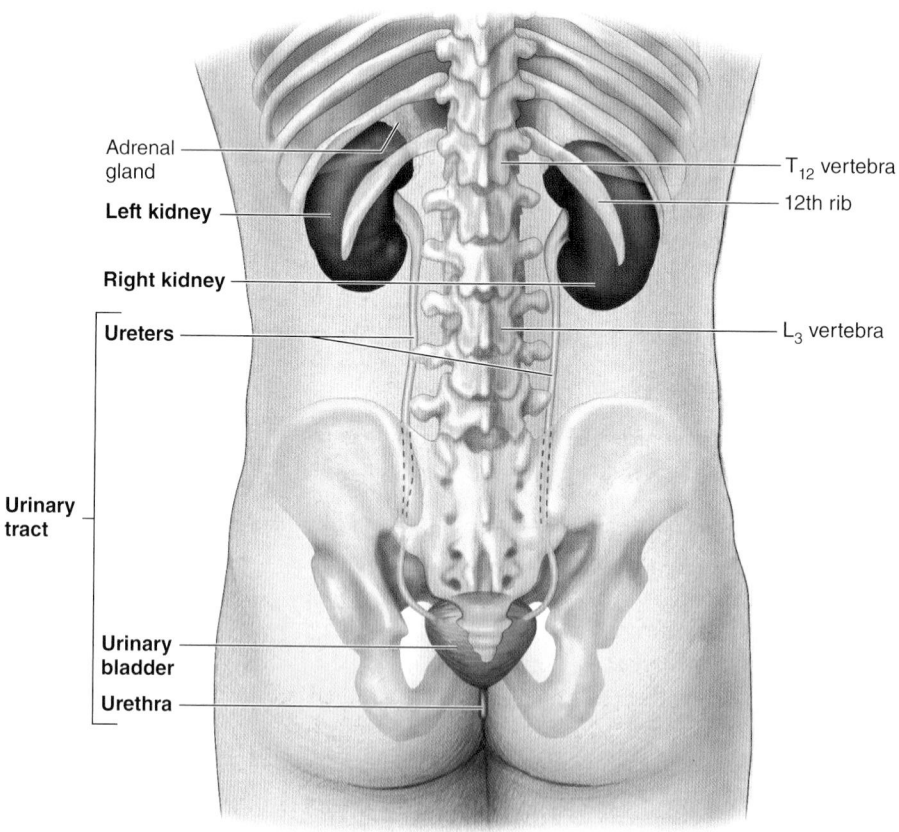

Diaphragm
Adrenal gland
Kidneys
Hilum
Renal artery
Renal vein
Inferior vena cava
Descending abdominal aorta
Ureters
Parietal peritoneum (cut)
Urinary bladder
Urethra

(a) Anterior view

Adrenal gland
Left kidney
Right kidney
Ureters
Urinary tract
Urinary bladder
Urethra

T₁₂ vertebra
12th rib
L₃ vertebra

(b) Posterior view

Figure 24.1 **Urinary System.** The urinary system is composed of two kidneys, two ureters, a single urinary bladder, and a single urethra, shown here in (*a*) anterior and (*b*) posterior views. (Components of urinary system are labeled in bold font.) AP|R

24.2 Gross Anatomy of the Kidney

The **kidneys** are two symmetrical, bean-shaped, reddish-brown organs (figure 24.1). Each kidney measures about 12 centimeters (4.7 inches) in length, 6.5 centimeters (2.5 inches) in width, and 2.5 centimeters (1 inch) in thickness, and is about the same size as your hand to the second knuckle (between the proximal and middle phalanges). A kidney weighs approximately 100 grams (g). Each has a concave medial border called the **hilum** (hī′lŭm = a small bit), where vessels, nerves, and the ureter connect to the kidney. The kidney's lateral border is convex. An adrenal gland rests on the superior aspect of each kidney.

24.2a Location and Support

LEARNING OBJECTIVES

1. Describe the location of the kidneys in the body.
2. Name and describe the four tissue layers that surround and support the kidneys.

Kidneys are located along the posterior abdominal wall, lateral to the vertebral column. The left kidney is between the level of the T_{12} and L_3 vertebrae, and the right kidney is about 2 centimeters inferior to the left kidney to accommodate the large size of the liver. Both kidneys are only partially protected by the rib cage, making them vulnerable to forceful blows to the inferior region of the back.

The kidneys are positioned posterior to the parietal peritoneum in the **retroperitoneal** (ret′rō-per′i-tō-nē′ăl; *retro* = back) **space**

INTEGRATE

LEARNING STRATEGY

To understand the retroperitoneal position of the kidneys, imagine placing an eraser against a blackboard, which represents the posterior abdominal wall. Then hang a cloth that represents the parietal peritoneum so that the eraser is between the blackboard and the sheet. The eraser, which is located posterior to the sheet (the *parietal peritoneum*), is in a region called retroperitoneal. Structures that would be in front of (and enclosed by) the sheet are described as being *intraperitoneal*.

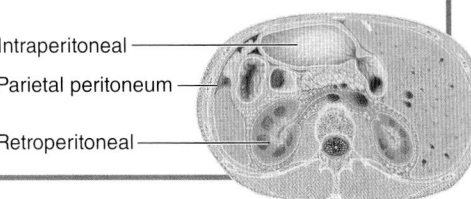

Intraperitoneal
Parietal peritoneum
Retroperitoneal

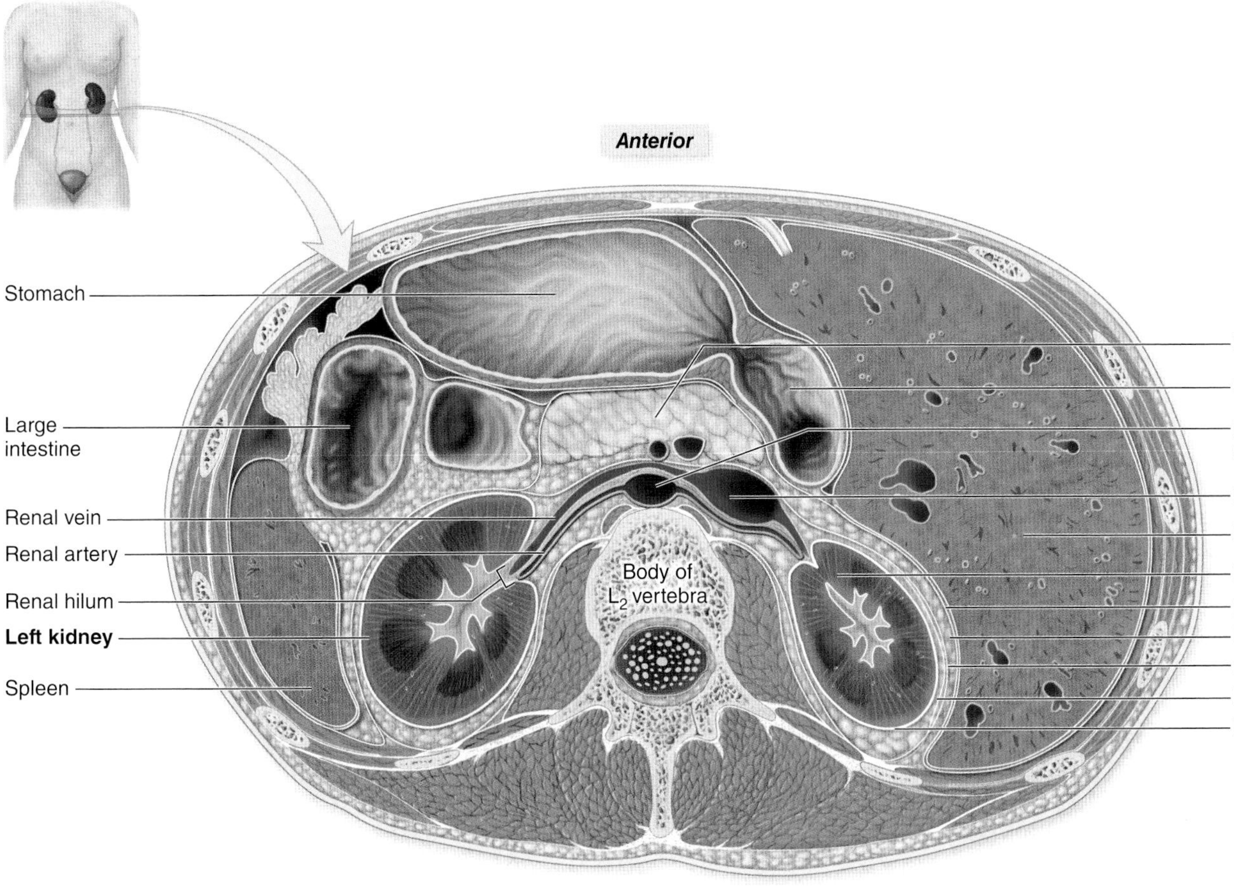

(figure 24.2). Thus, only the anterior surface of the kidneys is covered with parietal peritoneum.

Each kidney is surrounded and supported by several tissue layers. From innermost (closest to the kidney) to outermost, these layers are the fibrous capsule, perinephric fat, renal fascia, and paranephric fat:

- The **fibrous capsule** (*capsa* = box) (or *renal capsule*) is directly adhered to the external surface of the kidney. It is composed of dense irregular connective tissue and maintains

Anterior

Stomach

Large intestine

Renal vein
Renal artery
Renal hilum
Left kidney
Spleen

Body of L_2 vertebra

Pancreas
Small intestine
Descending abdominal aorta
Inferior vena cava
Liver
Right kidney
Parietal peritoneum
Paranephric fat
Renal fascia
Perinephric fat
Fibrous capsule

Posterior

Figure 24.2 Position and Support of the Kidneys. A cross-sectional view shows that the kidneys are located against the posterior abdominal wall and are covered on their anterior surface by the parietal peritoneum and surrounded by four concentric tissue layers (from innermost to outermost): fibrous capsule, perinephric fat, renal fascia, and paranephric fat. AP|R

INTEGRATE

CLINICAL VIEW
Renal Ptosis

The loss of adipose connective tissue in very thin elderly people or individuals with anorexia nervosa may result in **renal ptosis** (tō′sēz), which is the "dropping" or inferior movement of the kidney within the abdominal cavity. Consequently, the ureter may kink, resulting in a decrease or blockage of urine flow from the kidney to the urinary bladder. Urine backs up into the kidney which will result in a swelling of the kidney, called **hydronephrosis.** If the cause of hydronephrosis is not treated, renal failure may occur.

the kidney's shape, protects it from trauma, and helps prevent infectious pathogens from penetrating the kidney.

- The **perinephric** (*peri* = around) **fat** is also called *perirenal fat* or *adipose capsule*, is external to the fibrous capsule, and contains adipose connective tissue. It provides cushioning and support for the kidney.
- The **renal fascia** (rē′năl fash′ē-ă; *ren* = kidney) is external to the perinephric fat and is composed of dense irregular connective tissue. It anchors the kidney to surrounding structures.
- The **paranephric** (*para* = next to) **fat** (also called *pararenal fat* or *paranephric body*) is the outermost layer surrounding the kidney. It is composed of adipose connective tissue and also provides cushioning and support for the kidney.

WHAT DID YOU LEARN?

3 What tissue composes the fibrous capsule that directly adheres to the kidney, and what are its functions?

24.2b Sectional Anatomy of the Kidney

LEARNING OBJECTIVES

3. Identify the two distinct regions of the kidney and the components of each.
4. Explain the relationship between minor calyces, major calyces, and renal pelvis.

When a kidney is sectioned along a coronal plane, the *parenchyma* or "functioning tissue" is visible. The two distinct regions of the parenchyma include an outer **renal cortex** and an inner **renal medulla** (figure 24.3).

Extensions of the cortex, called **renal columns,** project into the medulla and subdivide it into **renal pyramids** (also termed *medullary pyramids*) that appear striated or striped. An adult kidney typically contains 8 to 15 renal pyramids. The wide base of a renal pyramid lies at the external edge of the medulla, where it meets the cortex: This is called the **corticomedullary junction,** or *corticomedullary border*. The medially directed apex (or tip) of the renal pyramid is called the **renal papilla.**

The parenchyma of a human kidney can also be divided into 8 to 15 **renal lobes.** A renal lobe consists of a renal pyramid, portions of renal columns adjacent to either side of the renal pyramid, and the renal cortex external to the pyramid base.

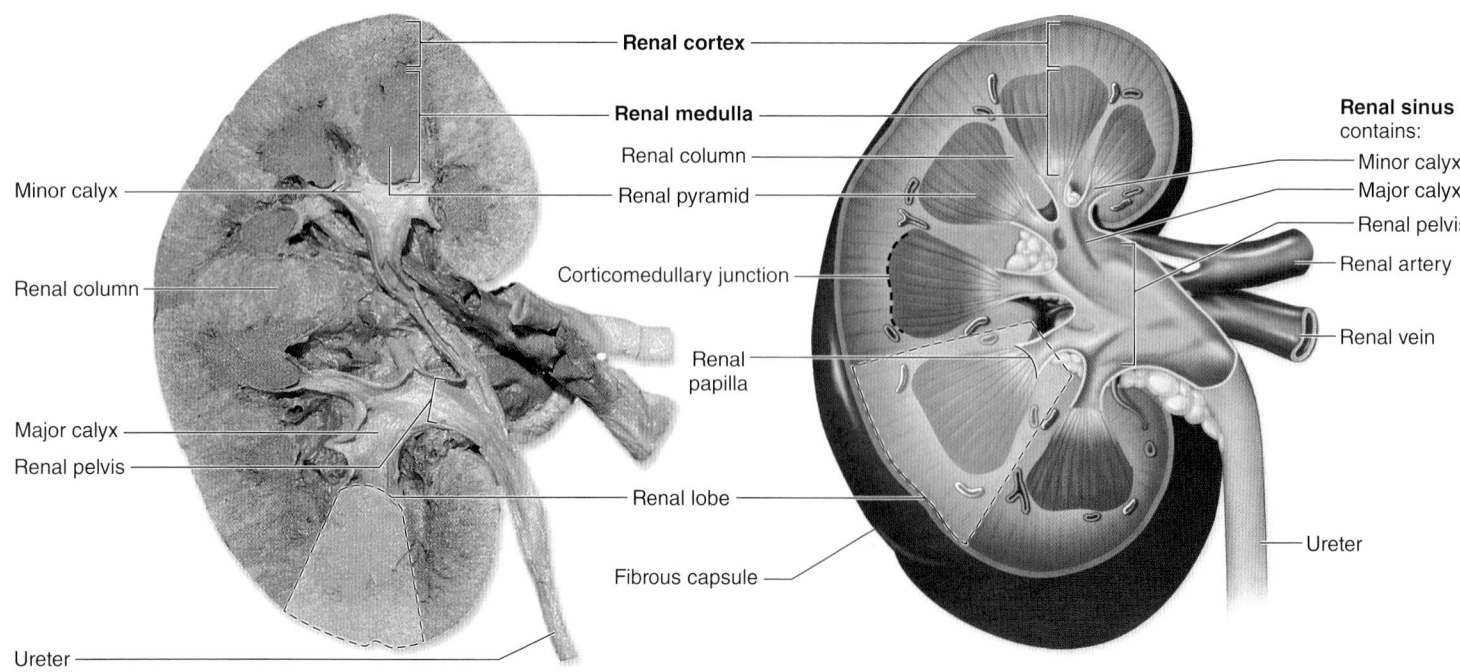

Right kidney, coronal section

Figure 24.3 Kidney. A coronal cut through the right kidney reveals the parenchyma and urine drainage areas of the kidney. AP|R

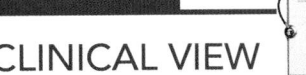

INTEGRATE

CLINICAL VIEW
Kidney Variations and Anomalies

Anatomic variations of the kidneys can occur during development. **Renal agenesis** (ā-jĕn'e-sis) is the failure of a kidney to develop. If one kidney fails to develop, it is called unilateral renal agenesis, and it occurs in about 1 per 1000 births. Bilateral renal agenesis is the failure of both kidneys to develop, and it occurs in about 1 per 3000 births. Unilateral renal agenesis is often asymptomatic, whereas bilateral renal agenesis is invariably fatal. A **pelvic kidney** may occur if the developing kidney fails to migrate from

the pelvic cavity to the abdominal cavity. A **horseshoe kidney** develops when the inferior parts of the left and right kidneys fuse as they ascend from the pelvic cavity into the abdominal cavity. Horseshoe kidneys are fairly common, occurring in about 1 per 600 births. Both pelvic kidneys and horseshoe kidneys typically are asymptomatic and function normally. **Supernumerary** (sū-per-nū'mer-ār-ē) **kidneys** are extra kidneys that develop, but they are rare and usually have no clinical significance. In fact, some individuals find out that their kidneys have anatomic variations in number or structure only when seeing a physician regarding an unrelated health issue.

Each kidney contains a medially located space called the **renal sinus,** in addition to the parenchyma. This space serves as the urine drainage area. It is organized into minor calyces, major calyces, and a renal pelvis. Each of the 8 to 15 funnel-shaped **minor calyces** (kāl'i-sēz, sing. *calyx* = cup of a flower) is associated with a renal pyramid. Several minor calyces merge to form a larger **major calyx.** Each kidney typically contains two or three major calyces. The major calyces merge to form a large, funnel-shaped **renal pelvis.** The renal pelvis merges at the medial edge of the kidney with the ureter. Housed within the space around the renal pelvis are the renal artery, renal vein, lymph vessels, nerves, and a variable amount of fat. (Lymph vessels and nerves of the kidney are not shown in figure 24.3.)

 WHAT DID YOU LEARN?

4 What are the regions of the kidney that drain urine?

24.2c Innervation of the Kidney
 LEARNING OBJECTIVE

5. List the structures of the kidney innervated by the sympathetic division.

Each kidney is innervated by both divisions of the autonomic nervous system. Sympathetic nerves extend from the T10–T12 segments of the spinal cord to the blood vessels of the kidney (see section 15.4), including the afferent and efferent arterioles, and also innervate the juxtaglomerular apparatus (these structures are described in section 24.3). The general effect of sympathetic stimulation of the kidneys is to decrease urine production. Parasympathetic nerves extend from the brain within the vagus nerve (CN X), but the specific effects of parasympathetic innervation to the kidney are not known.

 WHAT DID YOU LEARN?

5 What three anatomic structures of the kidney are innervated by the sympathetic division of the autonomic nervous system?

24.3 Functional Anatomy of the Kidney

The functional anatomy of the kidneys encompasses structures that include nephrons, collecting tubules, collecting ducts, and their associated structures.

24.3a Nephron
 LEARNING OBJECTIVES

1. Describe a renal corpuscle and its components.
2. Identify the location, and describe the structure, of the three components of a renal tubule.
3. Name and compare the two types of nephrons and the functional differences between them.

The **nephron** (nef'ron: *nephros* = kidney) is the microscopic, functional filtration unit of the kidney. Each nephron consists of two major structures: a renal corpuscle and a renal tubule (**figure 24.4**). All of the renal corpuscles and almost all of the renal tubules reside in the cortex (the exception being the nephron loop, which also extends into the medulla).

Renal Corpuscle

The **renal corpuscle** (kōr'pus-l; *corpus* = body, *cle* = tiny) is an enlarged, bulbous portion of a nephron housed within the renal cortex. It is composed of two structures: the glomerulus and the glomerular capsule.

The **glomerulus** (glō-mer'yū-lŭs; *glomus* = ball of yarn, *ulus* = small) is a thick tangle of capillary loops called the *glomerular capillaries.* Blood enters the glomerulus by an **afferent** (af'er-ent; *aferrens* = to bring to) **arteriole** and exits by an **efferent** (ef'er-ent; *efferens* = to bring out) **arteriole.**

The **glomerular** (glō-mer'yū-lăr) **capsule** (*Bowman's capsule*) is formed by two layers: an internal permeable **visceral layer** that directly overlies the glomerular capillaries (described later in more detail), and an external impermeable **parietal layer** composed of simple squamous epithelium. Between these two layers is a **capsular space** that receives the filtrate (described later), which is then modified to form urine.

The renal corpuscle has two opposing poles: a **vascular pole,** where afferent and efferent arterioles attach to the glomerulus, and a **tubular pole,** where the renal tubule originates.

Renal Tubule

The **renal tubule** makes up the remaining part of a nephron and is composed of a simple (single layer) epithelium resting on a basement membrane. It consists of three continuous sections: the proximal convoluted tubule, the nephron loop, and the distal convoluted tubule. The convoluted tubules reside in the cortex, whereas the nephron loop typically extends from the cortex into the medulla.

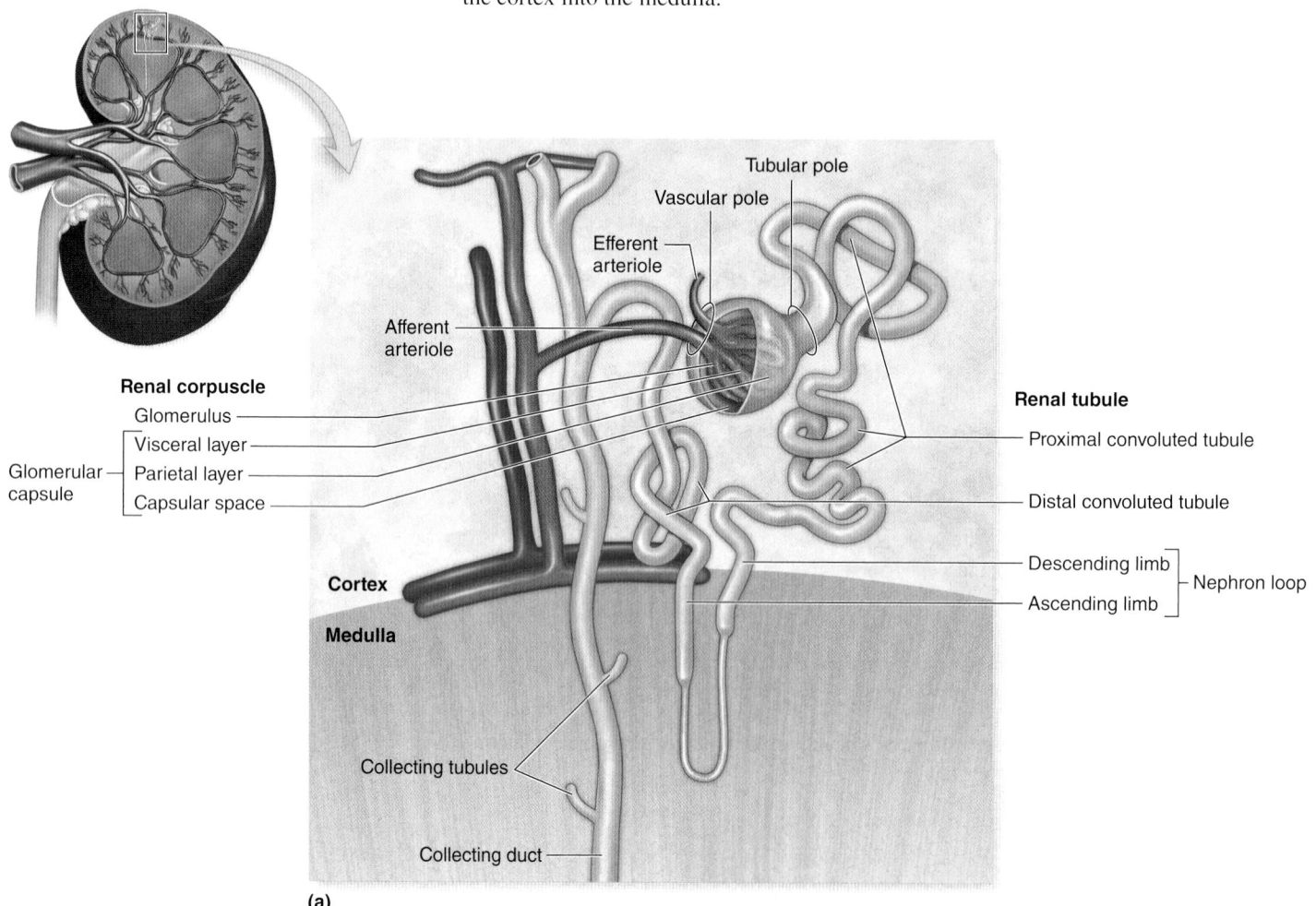

(a)

Figure 24.4 Nephron Structure. The nephron is composed of a renal corpuscle and renal tubule. (*a*) The nephron is shown in its anatomic orientation relative to its position in the cortex and medulla of the kidney. AP|R

The **proximal convoluted tubule (PCT)** is the first region of the renal tubule. It originates at the tubular pole of the renal corpuscle and is composed of a simple cuboidal epithelium with tall, apical microvilli that markedly increase its surface area and thus its reabsorption capacity. When viewed with a light microscope, the lumen of the proximal convoluted tubule looks fuzzy due to the brush border formed by these tall microvilli (see section 5.1c).

The **nephron loop** (*loop of Henle*) originates at a sharp bend in the proximal convoluted tubule. Each nephron loop has two limbs: a descending limb and an ascending limb that connect at a "hairpin turn" within the medulla. The **descending limb** extends medially from the proximal convoluted tubule to the tip of the nephron loop; conversely, the **ascending limb** of the nephron loop returns to the renal cortex and terminates at the distal convoluted tubule. Portions of both limbs are classified as either thick or thin according to their lining epithelia. The **thick segments** of each limb have a simple cuboidal epithelium lining, whereas the **thin segments** of each limb have a simple squamous epithelium.

The **distal convoluted tubule (DCT)** originates in the renal cortex at the end of the nephron loop's thick ascending limb and extends to a collecting tubule. Like the proximal convoluted tubule, the distal convoluted tubule is lined with simple cuboidal epithelium. However, the distal convoluted tubule epithelial lining cells are smaller and have only sparse, short, apical microvilli. Thus, the distal convoluted tubule lumen appears clear and not fuzzy when viewed with a light microscope.

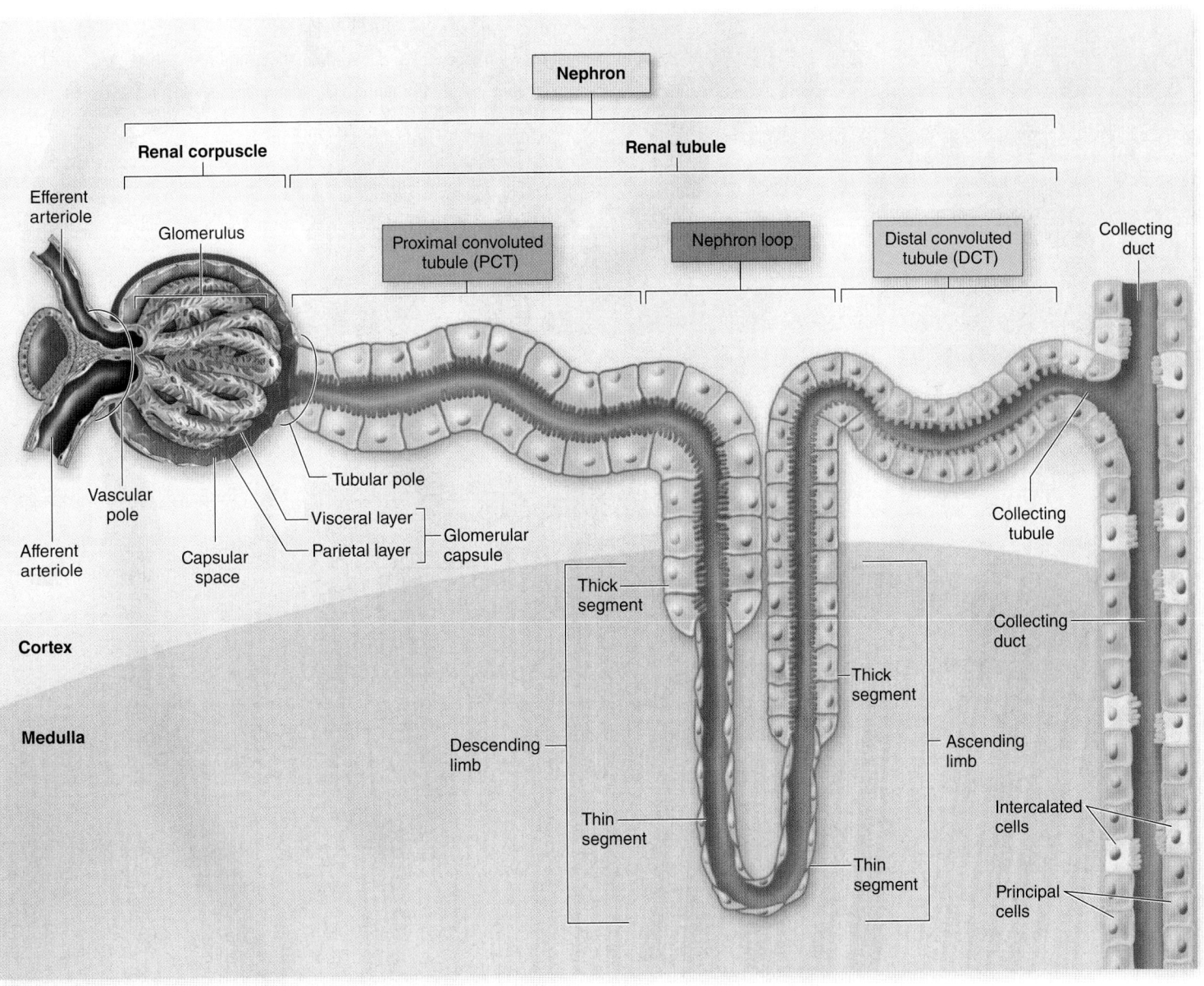

(b)

Figure 24.4 Nephron Structure *(continued)*. (*b*) In this diagrammatic representation, the nephron has been straightened out, with each component color coded for clarity. Note that neither the collecting tubule nor the collecting duct is part of a nephron.

Two Types of Nephrons

The renal corpuscle and both the proximal and distal convoluted tubules are housed within the cortex; the nephron loops extend from the cortex medially toward or into the medulla, as previously described. The relative position of the renal corpuscle in the cortex and the length of the nephron loop are used to classify nephrons into two categories: cortical nephrons and juxtamedullary nephrons (figure 24.5).

Cortical nephrons are oriented with their renal corpuscles near the peripheral edge of the cortex and have a relatively short nephron loop that barely penetrates the medulla. Thus, the bulk of a cortical nephron resides within in the cortex. Approximately 85% of nephrons are cortical nephrons.

The remaining 15% of nephrons are called **juxtamedullary** (jŭks′tă-med′yŭ-lār-ē; *juxta* = near) **nephrons.** Their renal corpuscles lie adjacent to the corticomedullary junction, and they have relatively long nephron loops that extend deep into the medulla. Juxtamedullary nephrons are important in establishing a salt concentration gradient within the interstitial space that lies outside of the nephron loop, the collecting tubules, and the collecting ducts—thus allowing for the regulation of urine concentration by antidiuretic hormone (ADH) (a process described in section 24.6f).

WHAT DID YOU LEARN?

6 What two structures compose the renal corpuscle? Provide a brief description of each.

7 What is the order of the components of a renal tubule?

8 What differences exist between cortical and juxtamedullary nephrons?

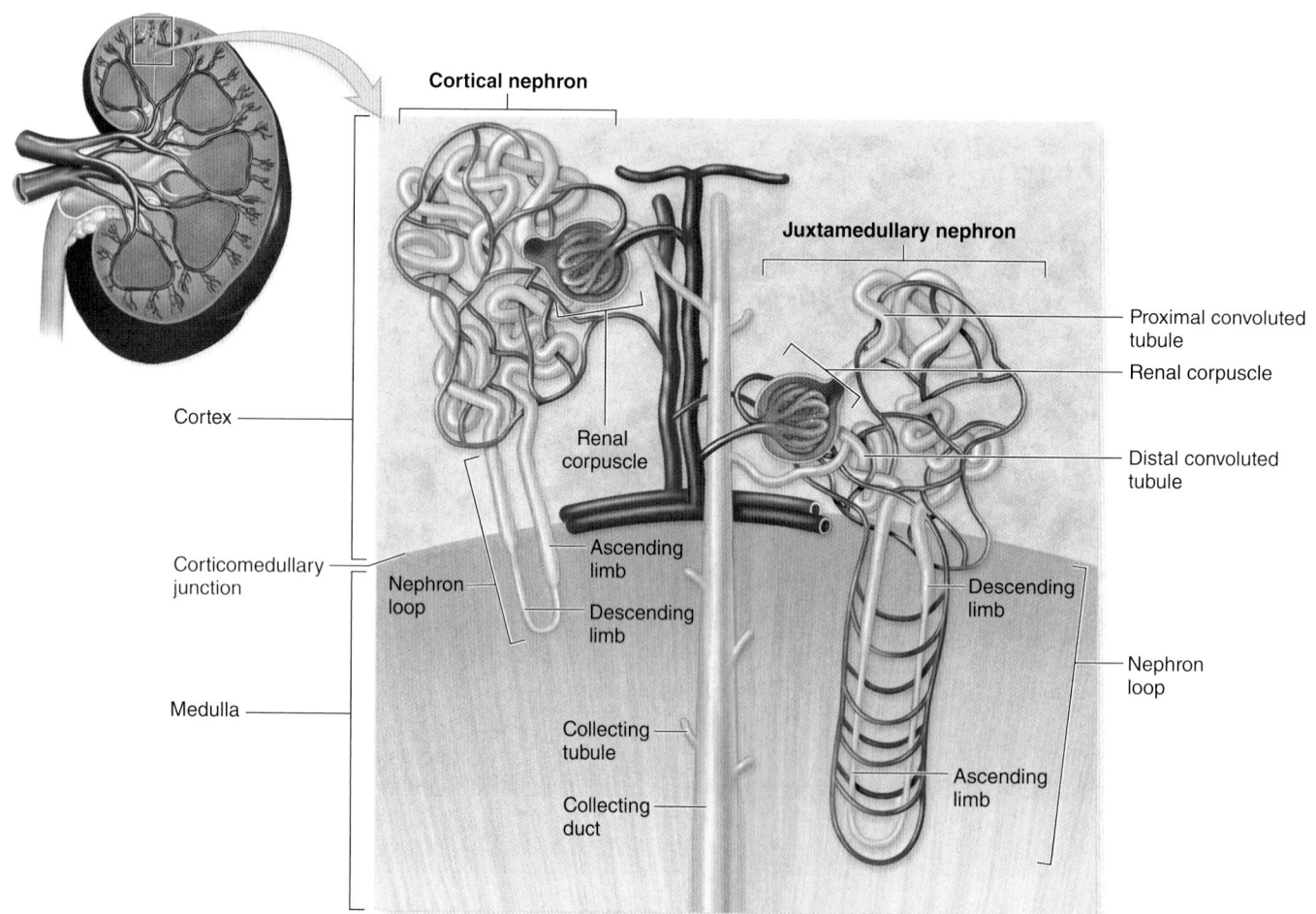

Figure 24.5 Two Types of Nephrons. Cortical nephrons are located almost completely in the cortex and have short nephron loops that barely penetrate into the medulla. Juxtamedullary nephrons, which lie close to the corticomedullary junction, have relatively long nephron loops that extend deeper into the medulla.

24.3b Collecting Tubules and Collecting Ducts

LEARNING OBJECTIVES

4. State the relationship between collecting tubules and collecting ducts.

5. Identify the two types of specialized epithelial cells found within distal convoluted tubules and collecting tubules and ducts.

Several nephrons drain into each **collecting tubule.** Each kidney contains thousands of collecting tubules, and a series of collecting tubules empty into larger **collecting ducts.** Both collecting tubules and collecting ducts project through the renal medulla toward the renal papilla. Numerous collecting ducts then empty into a **papillary duct** located within the renal papilla (see figure 24.9). (Note: The nephrons that drain into a single collecting duct compose a *lobule,* which is a microscopic structure within the cortex.)

The epithelial cells of the collecting tubules are cuboidal-shaped, but the cells become very tall columnar cells in the collecting ducts near the renal papilla. The striations seen within the renal pyramid are formed collectively by the straight ascending and descending tubules of nephron loops, the collecting tubules, and collecting ducts (see figures 24.3 and 24.4).

The distal convoluted tubule, collecting tubules, and collecting ducts contain two types of specialized epithelial cells called principal cells and intercalated cells (see figure 24.4b). **Principal cells** have cellular receptors to bind both aldosterone (released from the adrenal cortex) and antidiuretic hormone (released from the posterior pituitary). **Intercalated cells** (types **A** and **B**) are specialized epithelial cells that help regulate urine pH and blood pH. Both types of cells are described in section 24.6d, but for now keep in mind that type A cells eliminate acid (H^+) and type B cells eliminate base (HCO_3^-). Histologic images of the kidney are provided in figure 24.6.

WHAT DID YOU LEARN?

9 Differentiate between the function of principal cells and the intercalated cells within the kidney.

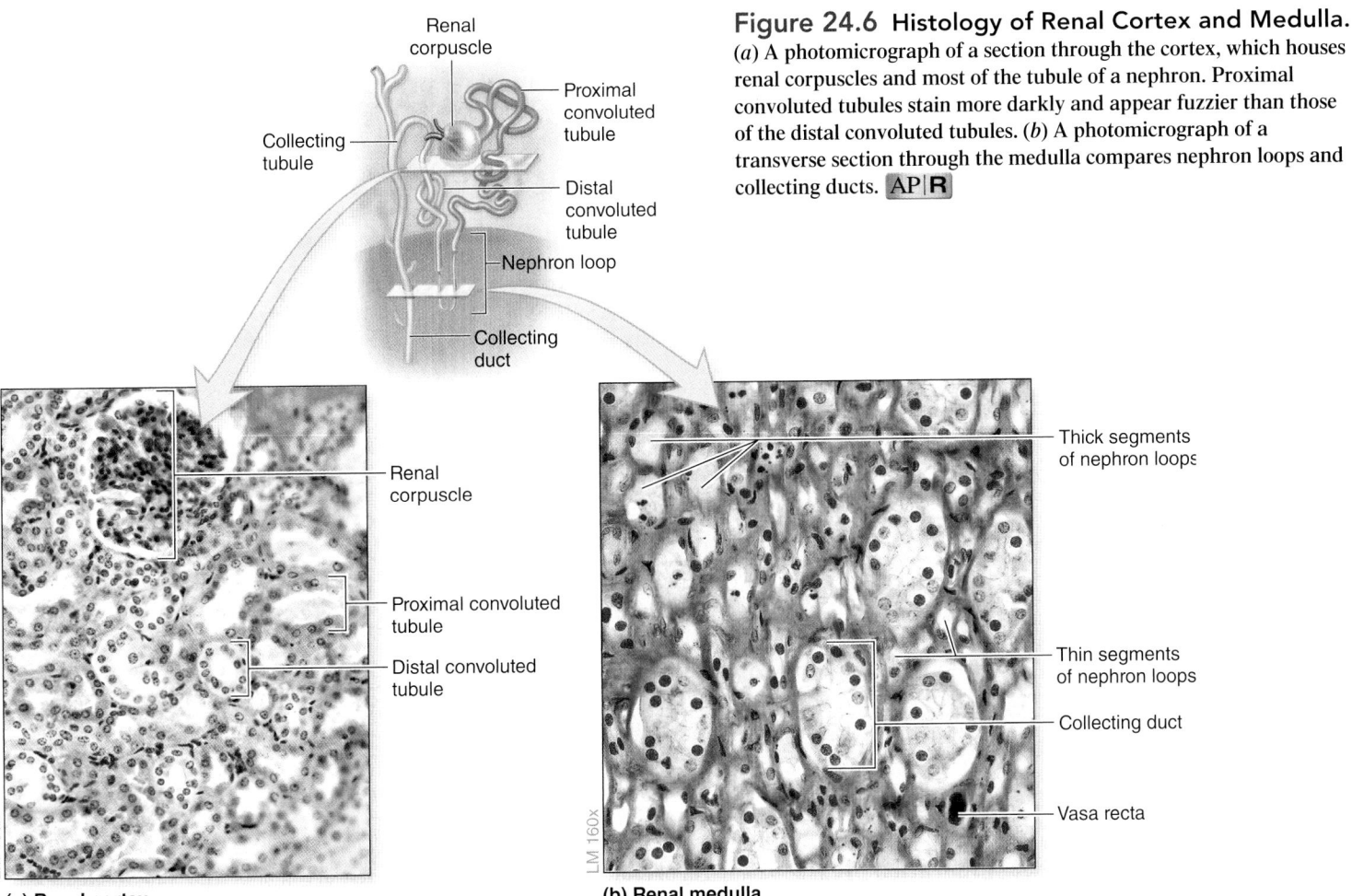

Figure 24.6 Histology of Renal Cortex and Medulla.
(*a*) A photomicrograph of a section through the cortex, which houses renal corpuscles and most of the tubule of a nephron. Proximal convoluted tubules stain more darkly and appear fuzzier than those of the distal convoluted tubules. (*b*) A photomicrograph of a transverse section through the medulla compares nephron loops and collecting ducts. AP|R

Renal corpuscle
Collecting tubule
Proximal convoluted tubule
Distal convoluted tubule
Nephron loop
Collecting duct

Renal corpuscle
Proximal convoluted tubule
Distal convoluted tubule

LM 160x

(a) Renal cortex

Thick segments of nephron loops
Thin segments of nephron loops
Collecting duct
Vasa recta

LM 160x

(b) Renal medulla

24.3c Juxtaglomerular Apparatus

LEARNING OBJECTIVES

6. Describe the location and structure of the juxtaglomerular apparatus.

7. Explain the two actions of granular cells.

8. Describe the function of cells of the macula densa.

The nephron was "stretched out" in figure 24.4b to more clearly view its components. A representation of the normal orientation of a nephron would show the distal convoluted tubule of the nephron contacting the afferent arteriole of that same nephron (figure 24.7). This accurate physical arrangement of the nephron allows us to understand the anatomic features of a specialized region of the nephron called the **juxtaglomerular (JG)** (jŭks'tă-glō-mer'ū-lăr) **apparatus,** an important structure in regulating filtrate formation and systemic blood pressure.

The primary components of the JG apparatus include both granular cells and macula densa cells. **Granular cells** (or *juxtaglomerular cells*) are modified smooth muscle cells of the afferent arteriole located near its entrance into the renal corpuscle. Granular cells have two functions: (1) They contract when stimulated either by stretch or by the sympathetic division; and (2) they synthesize, store, and release the enzyme **renin.** Renin is required in the production of angiotensin I, which is then converted by angiotensin converting enzyme (ACE) to angiotensin II, as shown in figure 20.15; see section 20.6b.

The **macula densa** (mak'ū-lă = spot, den'să = dense) is a group of modified epithelial cells in the wall of the distal convoluted tubule where it contacts the granular cells. The cells of the macula densa are located only on the tubule side next to the afferent arteriole, and they are narrower and taller than other distal convoluted tubule epithelial cells. The macula densa cells detect changes in the sodium chloride (NaCl) concentration of fluid within the lumen of the distal convoluted tubule. Macula densa cells signal granular cells to release renin through paracrine stimulation (see section 17.3b).

Extraglomerular mesangial cells, located just outside the glomerulus within the gap between the afferent arteriole and efferent arteriole, are also components of the JG apparatus. These cells communicate with the other cells of the JG apparatus via gap junctions and the release of paracrine hormones. Their specific function is not well understood.

WHAT DID YOU LEARN?

10 What are the two primary cellular components of the juxtaglomerular apparatus, and how is each stimulated?

24.4 Blood Flow and Filtered Fluid Flow

At least 20% to 25% of the resting cardiac output (or about 1 liter per minute [L/min]) normally flows through the kidneys for the purpose of removing unwanted substances. A filtrate is formed when blood flows through the glomerulus, and some components of the plasma enter the capsular space. As a result, two fluid flow patterns must be noted.

- The flow of blood into and out of the kidney
- The flow of filtrate, tubular fluid, and urine through the nephron and other urinary system structures

24.4a Blood Flow Through the Kidney

LEARNING OBJECTIVES

1. Name the arteries that supply the kidney, in sequence from largest to smallest.

2. Describe the two capillary beds through which blood must pass in the kidney.

3. List the veins through which blood leaves the kidney, in sequence from smallest to largest.

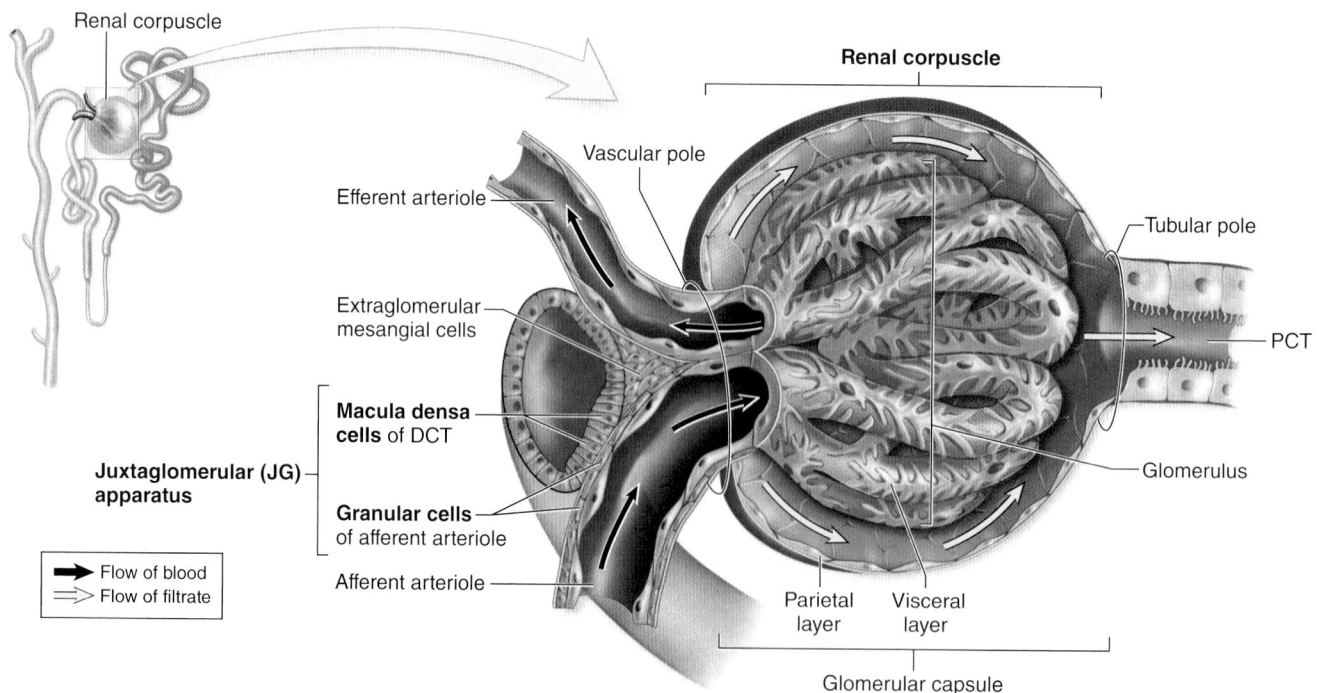

Figure 24.7 Juxtaglomerular Apparatus. With the nephron in the normal orientation, there is physical contact between the afferent arteriole and the adjacent distal convoluted tubule (DCT) forming the juxtaglomerular apparatus. The juxtaglomerular apparatus is composed of granular cells of the afferent arteriole and macula densa cells of the distal convoluted tubule. The juxtaglomerular apparatus monitors blood pressure and releases renin into the blood in response to either low blood pressure or stimulation by the sympathetic division of the ANS. AP|R

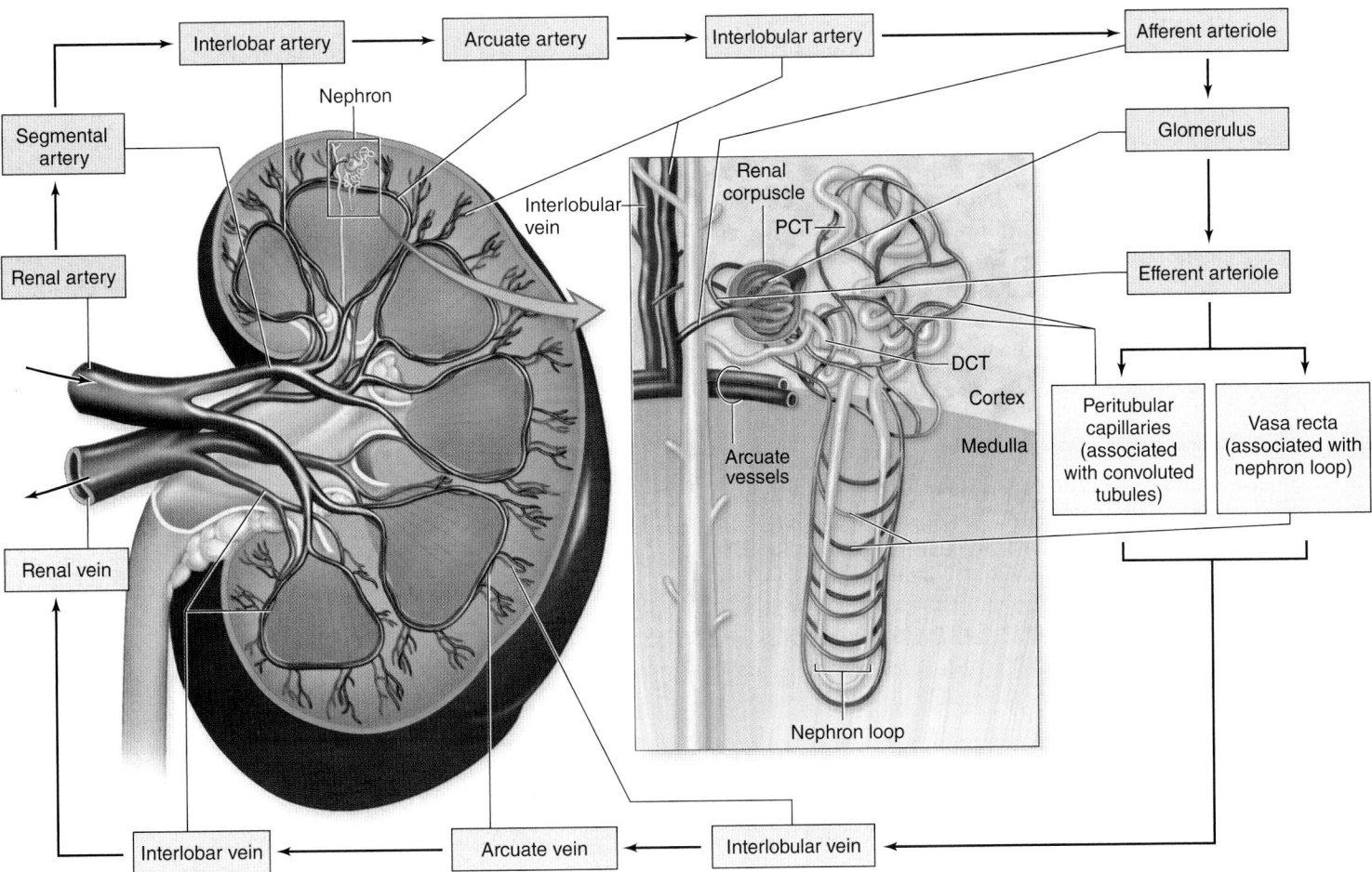

Figure 24.8 Blood Supply to the Kidneys. A coronal view depicts kidney circulation. An expanded view shows circulation to a nephron. Pink boxes indicate vessels with arterial blood; lavender boxes indicate vessels where reabsorbed materials reenter the blood; blue boxes indicate vessels returning blood to the general circulation.

The blood flow pathway is shown in **figure 24.8**. Be sure to refer back to this figure often as you read this section on blood flow through the kidney.

Arteries

Blood is delivered to each kidney by way of a **renal artery** that arises from the abdominal aorta. The renal artery splits as it enters the renal sinus into **segmental arteries.** While still in the renal sinus, the segmental arteries branch to form the **interlobar** (in-ter-lō′bar) **arteries**, which are located between the renal lobes (within the renal columns). The interlobar arteries extend to the corticomedullary junction, where they branch to form **arcuate** (ar′kū-āt; *arcuatus* = bowed) **arteries.**

Arcuate arteries are arch-shaped and project parallel to the base of the medullary pyramid at the corticomedullary junction. The arcuate arteries extend branches called **interlobular** (in-ter-lob′ū-lăr) **arteries** (or *cortical radiate arteries*) that project peripherally into the cortex (where renal lobules are located). As the interlobular arteries extend into the cortex, numerous small afferent arterioles branch from them.

Arterioles and Capillaries

Each **afferent arteriole** supplies blood to a **glomerulus.** Some blood plasma is filtered at the glomerulus (and this fluid enters the glomerular capsule of the renal corpuscle). After this filtration event, blood still remaining within the glomerulus exits the glomerulus through an **efferent arteriole.** Each efferent arteriole now branches into a second capillary network, either the **peritubular**

INTEGRATE

LEARNING STRATEGY

The names of the blood vessels in the kidney give you clues as to their location or appearance:

- Interlobar vessels are positioned "inter" = *between* the lobes of the kidney.
- Arcuate vessels form "arcs" as these vessels run parallel to the corticomedullary junction.
- Interlobular vessels are located between the lobules of the kidney cortex.

- Afferent arterioles carry blood *to* the glomerulus ("afferent" means carrying toward).
- Efferent arterioles take blood *away from* the glomerulus ("efferent" means to carry away, or exit).
- Peritubular capillaries are "peri" = *around* the convoluted tubules in the cortex.
- Vasa recta means "straight vessels," and these vessels run parallel to the long, straight components of the nephron loop.

(per′ē-tū′byū-lăr; *peri* = around, *tubule* = small tube) **capillaries** or **vasa recta** (vā′să rek′tă; straight vessel). The peritubular capillaries are associated with, and intertwined around, both the proximal and distal convoluted tubules; thus, they primarily reside in the cortex of the kidney. In comparison, vasa recta capillaries are "straight vessels" associated with the nephron loop; thus, they primarily reside in the medulla of the kidney.

Notice that all blood moves through two capillary beds as it flows through the kidney. Blood first enters from the afferent arteriole into the glomerular capillaries, where it is filtered. When the blood reaches the second capillary bed of either the peritubular capillaries or the vasa recta, the exchange of gases, nutrients, and wastes occurs between the tissues of the kidney and the blood. The peritubular capillaries and vasa recta then drain into the network of veins.

Veins

Blood drains from both peritubular and vasa recta capillary beds into small veins. The smallest of these veins are the **interlobular veins** (or *cortical radiate veins*). Interlobular veins merge to form **arcuate veins** at the base of the medullary pyramids, and these merge to form **interlobar veins**, which extend through the renal columns. Interlobar veins merge in the renal sinus to form the **renal vein.** Note that there are no segmental veins; rather, the interlobar veins directly form the renal vein. The renal vein leaves the kidney at its hilum and drains into the inferior vena cava.

WHAT DID YOU LEARN?

11 What is the pathway that blood follows as it enters via the renal artery and later leaves via the renal vein?

12 What are the three major types of capillaries associated with the nephron? Describe the location and general function of each.

24.4b Filtrate, Tubular Fluid, and Urine Flow

LEARNING OBJECTIVES

4. Distinguish between filtrate, tubular fluid, and urine.

5. Trace the fluid from its formation at the renal corpuscle until it exits the body through the urethra.

The flow of fluid through a nephron, other renal tubules, and then the rest of the urinary system is integrated in **figure 24.9**. Be sure to refer to this figure as you read through this section.

When blood flows through the glomerulus, both water and solutes are filtered from the blood plasma, moving across the wall of the glomerular capillaries and into the capsular space to form **filtrate.** This filtrate then enters the proximal convoluted tubule, where it is now called **tubular fluid.** It flows through the proximal convoluted tubule, nephron loop, and distal convoluted tubule. Tubular fluid from several distal convoluted tubules enters into small collecting tubules that empty into larger collecting ducts.

Tubular fluid is not changed further after leaving the collecting ducts, and it is now called **urine.** It enters a papillary duct located within a renal papilla and then flows progressively through spaces

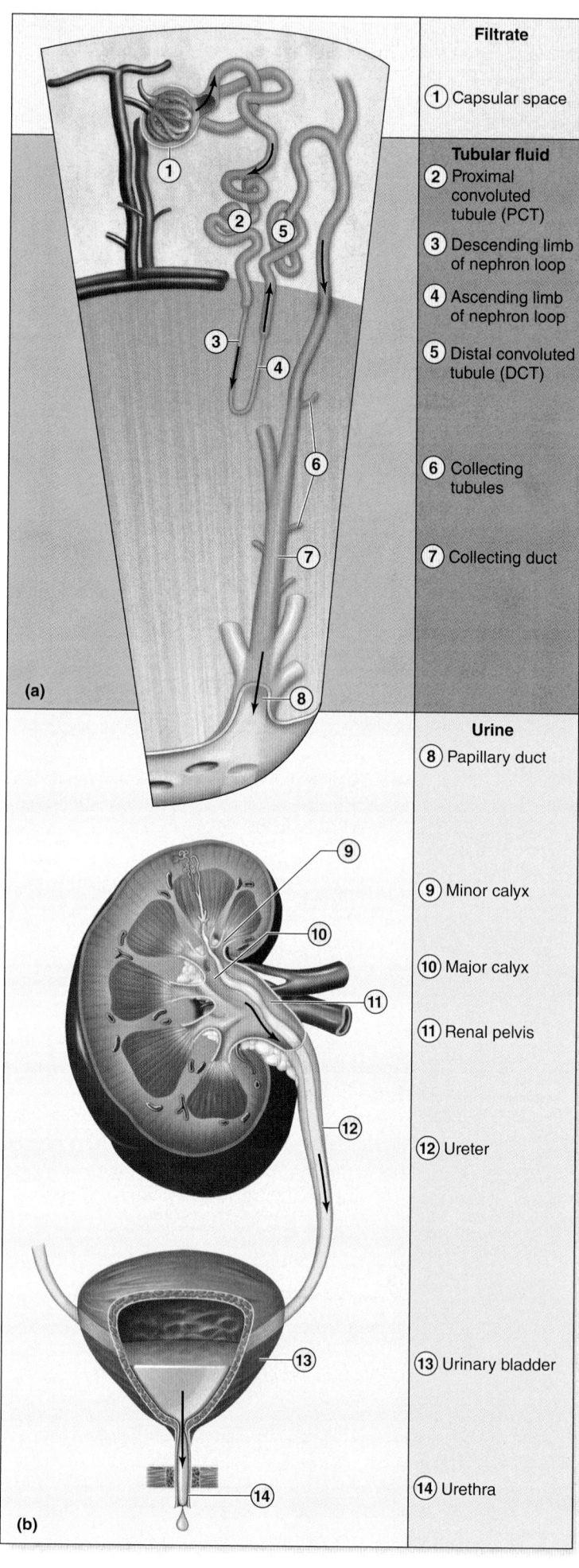

Filtrate

(1) Capsular space

Tubular fluid
(2) Proximal convoluted tubule (PCT)
(3) Descending limb of nephron loop
(4) Ascending limb of nephron loop
(5) Distal convoluted tubule (DCT)

(6) Collecting tubules

(7) Collecting duct

(a)

Urine
(8) Papillary duct

(9) Minor calyx

(10) Major calyx

(11) Renal pelvis

(12) Ureter

(13) Urinary bladder

(14) Urethra

(b)

Figure 24.9 Structures That Transport Fluids Through the Urinary System. (*a*) Microscopic view depicts flow of fluid through the nephron, and collecting tubules and ducts. (*b*) Macroscopic view depicts urine flow through the renal calyces and renal pelvis of the kidney, ureters, and then into the urinary bladder and expelled through the urethra.

within the renal sinus of the kidney. These spaces occur in the following order: minor calyx, major calyx, and renal pelvis. The renal pelvis transports urine from the kidney into a ureter, and a ureter from each kidney transports urine into the urinary bladder, where it is stored until it is excreted from the body through the urethra.

 WHAT DID YOU LEARN?

13 What is the pathway of fluid filtered by the kidney from the glomerulus to its eventual excretion?

24.5 Production of Filtrate Within the Renal Corpuscle

Filtrate production occurs within the renal corpuscle. Water and solutes cross the filtration membrane and enter into the capsular space. We first present a brief overview of the entire process of urine production and then consider details of filtrate production at the renal corpuscle. The processes that occur within the renal tubules are discussed in the next section.

24.5a Overview of Urine Formation

 LEARNING OBJECTIVE

1. Compare and contrast the renal processes of filtration, reabsorption, and secretion.

Urine is formed in the kidneys through three interrelated processes: filtration, reabsorption, and secretion (**figure 24.10**):

- **Glomerular filtration** passively separates some water and dissolved solutes from the blood plasma within the glomerular capillaries. Water and solutes enter into the capsular space of the renal corpuscle due to pressure differences across the filtration membrane. Collectively, this separated fluid is called filtrate, which is essentially protein-free plasma.

- **Tubular reabsorption** occurs when components within the tubular fluid move by membrane transport processes (e.g., diffusion, osmosis, active transport) from the lumen of the renal tubules, collecting tubules, and collecting ducts across their walls and return to the blood within the peritubular capillaries and vasa recta. Generally, all vital solutes and most water that was in the filtrate are reabsorbed, whereas excess solutes, some water, and waste products remain within the tubular fluid.

- **Tubular secretion** is the movement of solutes, usually by active transport, out of the blood within the peritubular and vasa recta capillaries into the tubular fluid. Materials are moved selectively into the tubules to be eliminated or excreted from the body. Remember, *secretion results in excretion.*

Notice that the processes of tubular reabsorption and tubular secretion involve the movement of materials in opposite directions. In tubular reabsorption, substances are moved from the tubular fluid into the blood, whereas in tubular secretion, substances are moved from the blood into the tubular fluid. In addition, there is generally more tubular reabsorption than tubular secretion.

 WHAT DID YOU LEARN?

14 How does tubular reabsorption differ from tubular secretion?

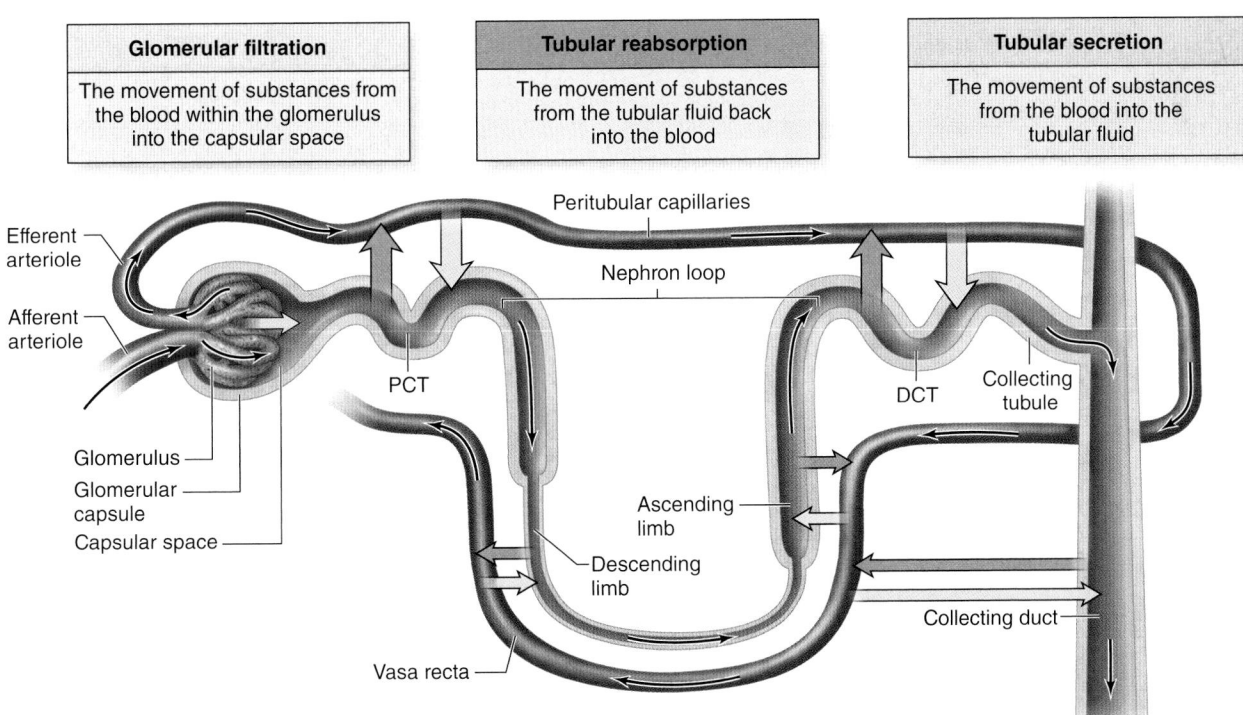

Glomerular filtration	Tubular reabsorption	Tubular secretion
The movement of substances from the blood within the glomerulus into the capsular space	The movement of substances from the tubular fluid back into the blood	The movement of substances from the blood into the tubular fluid

Efferent arteriole
Afferent arteriole
Peritubular capillaries
Nephron loop
PCT
DCT
Collecting tubule
Glomerulus
Glomerular capsule
Capsular space
Ascending limb
Descending limb
Collecting duct
Vasa recta

Figure 24.10 Overview of Processes of Urine Formation. The three major processes of urine formation include glomerular filtration, tubular reabsorption, and tubular secretion. AP|R

24.5b Filtration Membrane

LEARNING OBJECTIVE

2. Describe the three layers that make up the glomerular filtration membrane.

The **filtration membrane** is a porous, thin (0.1 micrometer wide), and negatively charged structure that is formed by the glomerulus and visceral layer of the glomerular capsule (**figure 24.11**). It is composed of three sandwiched layers. For a substance in the blood to become part of the filtrate, it must be able to pass through these three layers of the "filter," which are listed from innermost (closest to the lumen of the glomerulus) to outermost:

1. **Endothelium of glomerulus.** The endothelium of the glomerulus is fenestrated (see types of capillaries in section 20.1c). It allows plasma and its dissolved substances to be filtered, while restricting the passage of large structures, such as the formed elements (erythrocytes, leukocytes, and platelets).

2. **Basement membrane of glomerulus.** The porous basement membrane is composed of glycoprotein and proteoglycan molecules. It restricts the passage of large plasma proteins such as albumin, while allowing smaller structures to pass through.

3. **Visceral layer of glomerular capsule.** The visceral layer of the glomerular capsule is composed of specialized cells called **podocytes** (pod′ō-sīt; *podos* = foot). Podocytes are octopus-like cells that have long, "foot-like" processes called **pedicels** (ped′ĭ-sel; *pedicellus* = little foot) that wrap around the glomerular capillaries to support the capillary wall but do not completely ensheathe it. The pedicels are separated by thin spaces called **filtration slits** that are covered with membrane. Pedicels of one podocyte interlock with pedicels of a different podocyte, similar to how the fingers of your hands can interlock.

INTEGRATE

LEARNING STRATEGY

The filtration membrane is similar to an elaborate colander or sieve. Use the following to imagine the three layers that form this "sieve" that materials must pass through to become filtrate: (1) The glomerular capillary is a drinking straw with small holes in it (the holes are the fenestrations of the endothelium). (2) The straw is wrapped in modeling clay (representing the basement membrane). (3) The third layer is represented by your hand wrapped around the clay-covered straw. Your hand represents a podocyte and the fingers of your hand, the pedicels. The spaces between your fingers are the filtration slits. If your fingers were thinly webbed, the webbing would represent the membrane-covered filtration slits.

The membrane-covered filtration slits restrict the passage of most small proteins.

Mesangial (mes-an′jē-ăl) **cells** (or more specifically called *intraglomerular mesangial cells*), are specialized cells positioned within and between the capillary loops of the glomerulus (not shown in figure 24.11). These cells have both phagocytic and contractile properties. These functions are described in detail in section 24.5e.

WHAT DO YOU THINK?

2 If a substance within the blood does not become part of the filtrate, what happens to it? By what structure do these substances exit the glomerulus?

WHAT DID YOU LEARN?

15 How are the components of the filtration membrane of the glomerulus arranged?

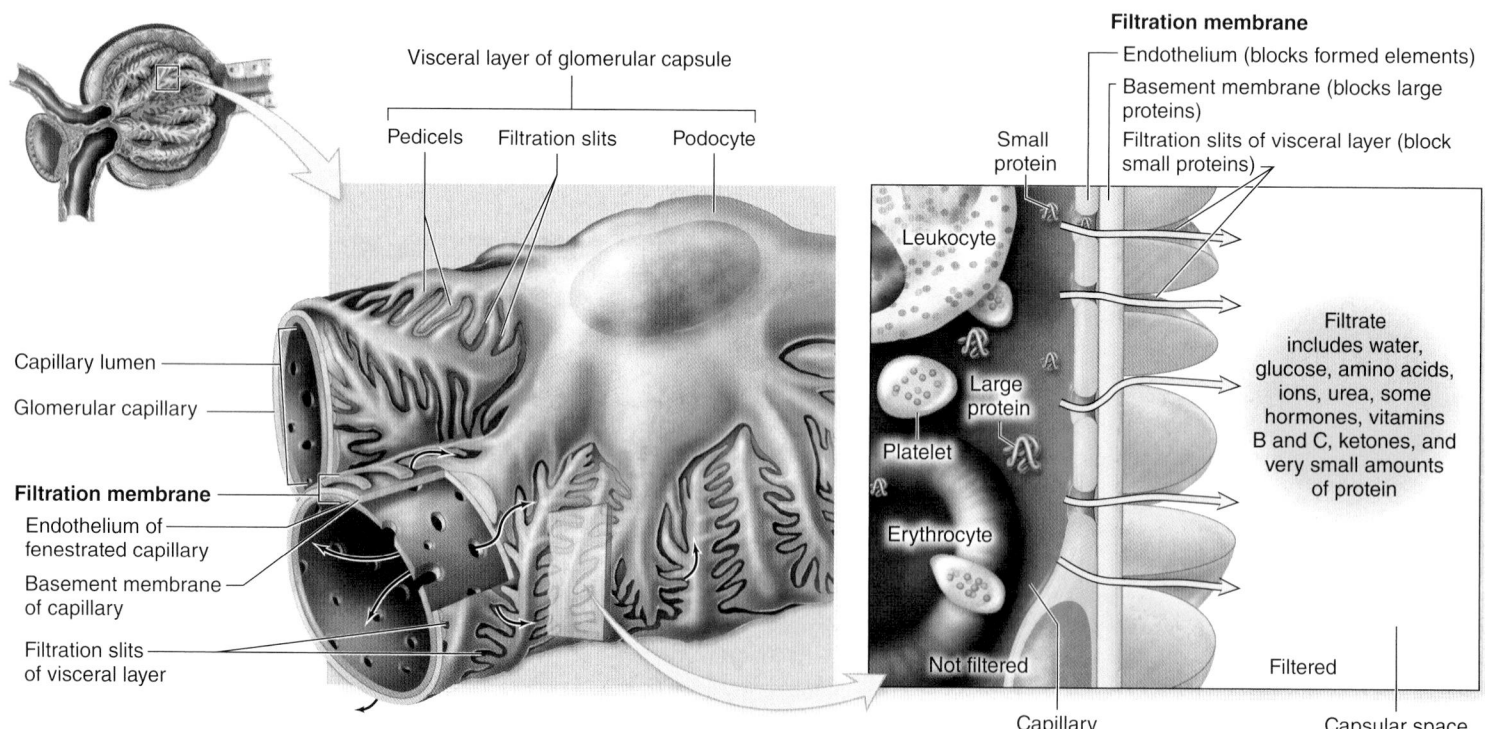

Capillary lumen
Glomerular capillary
Filtration membrane
Endothelium of fenestrated capillary
Basement membrane of capillary
Filtration slits of visceral layer

Visceral layer of glomerular capsule
Pedicels   Filtration slits   Podocyte

(a) Filtration membrane

Filtration membrane
Endothelium (blocks formed elements)
Basement membrane (blocks large proteins)
Filtration slits of visceral layer (block small proteins)

Leukocyte
Small protein
Large protein
Platelet
Erythrocyte

Filtrate includes water, glucose, amino acids, ions, urea, some hormones, vitamins B and C, ketones, and very small amounts of protein

Not filtered   Filtered
Capillary   Capsular space

(b) Substances filtered by filtration membrane

Figure 24.11 Filtration Membrane. Filtrate is produced in the renal corpuscle when blood plasma is forced across the filtration membrane under pressure. (*a*) The filtration membrane is composed of three layers, including a fenestrated capillary endothelium, basement membrane of the capillary, and membrane covered filtration slits formed by podocytes of the visceral layer of the glomerular capsule. (*b*) Diagrammatic view of the three layers of the filtration membrane, including materials not filtered (left of membrane) and filtered (right of membrane). AP|R

24.5c Formation of Filtrate and Its Composition

LEARNING OBJECTIVES

3. Give examples of substances that are freely filtered, that are not filtered, and that are filtered in a limited way.
4. Describe the phagocytic function of mesangial cells.

An average of 180 L of filtrate is produced daily by the kidneys through the glomerular filtration membrane. However, because of the size of the openings in the membrane and the overall negative charge across the membrane, not all substances within the blood plasma are filtered equally (figure 24.11b). Consequently, substances in the blood can be placed into one of three categories based on the degree to which the substance is filtered.

- **Freely filtered.** Small substances such as water, glucose, amino acids, ions, urea, some hormones, water-soluble vitamins (i.e., vitamins B and C), and ketones can pass easily through the filtration membrane and become part of the filtrate. These substances have the same concentration of ions, molecules, and wastes in filtrate as in the plasma.
- **Not filtered.** Formed elements (i.e., erythrocytes, leukocytes, and platelets) and large proteins are structures that cannot normally pass through the filtration membrane. These substances are usually restricted from becoming part of the filtrate.
- **Limited filtration.** Proteins that are of intermediate size are generally not filtered. They are blocked from filtration either because their size prevents movement through the openings of the filtration membrane or because they are negatively charged and repelled by the membrane's negative charge. Only limited amounts of these substances would become part of the filtrate.

Filtrate is characterized as filtered plasma with certain solutes and *minimal* amounts of protein. Filtrate is caught within the capsular space and then funneled into the proximal convoluted tubule. Components of blood that are not filtered exit the renal corpuscle through the efferent arteriole and then continue through either the peritubular or vasa recta capillaries (see figure 24.8).

Some of the material being filtered becomes trapped within the basement membrane. One of the functions of the mesangial cells is to phagocytize macromolecules (e.g., immunoglobulins) that become caught within the basement membrane, thus helping to keep the filtration membrane clean.

WHAT DID YOU LEARN?

16 What is normally filtered across the glomerular membrane? What is not normally filtered?

17 Certain diseases, kidney trauma, heavy metals, and some bacterial toxins can damage the filtration membrane. What effect would this have on relative permeability of the membrane and the substances that are filtered?

24.5d Pressures Associated with Glomerular Filtration

LEARNING OBJECTIVES

5. Define glomerular hydrostatic pressure (HP$_g$), and explain why it is higher than the pressure in other capillaries.
6. Name two pressures that oppose HP$_g$.
7. Explain how to calculate the net filtration pressure.
8. Define glomerular filtration rate and the factors that influence it.

Filtrate is produced due to the difference between hydrostatic pressure of the blood in the glomerulus and the opposing pressures of both the osmotic blood pressure and fluid pressure in the capsular space of the renal corpuscle. This difference is termed the net filtration pressure.

Glomerular Hydrostatic (Blood) Pressure

Blood pressure in the glomerulus is called the **glomerular hydrostatic (blood) pressure (HP$_g$).** It is the driving force that "pushes" water and some dissolved solutes out of the glomerulus and into the capsular space of the renal corpuscle (figure 24.12). It is the HP$_g$ that promotes filtration.

HP$_g$ has a higher value than the blood pressure of other systemic capillaries (60 mm Hg versus about 20 to 40 mm Hg; see section 20.3c). This higher pressure is required for filtration to occur, and it is due to the relative diameter size difference in the afferent and efferent arterioles. The afferent arteriole has a larger internal diameter than the efferent arteriole. Consider that in this anatomic arrangement the "inflow pipe" is larger than the "outflow pipe." This results in a higher pressure in the intervening structures (i.e., the glomerular capillaries). As a consequence, these capillaries are more vulnerable to damage than other capillaries due to this relatively high blood pressure.

Pressures That Oppose Glomerular Hydrostatic Pressure

Two pressures oppose glomerular hydrostatic pressure and thus oppose filtration. They are the blood colloid osmotic pressure and capsular hydrostatic pressure.

Blood colloid osmotic pressure (OP$_g$) is the osmotic pressure exerted by the blood due to the dissolved solutes it contains. The most important of these solutes are the plasma proteins (colloid). Blood colloid osmotic pressure opposes filtration because it tends to pull or draw

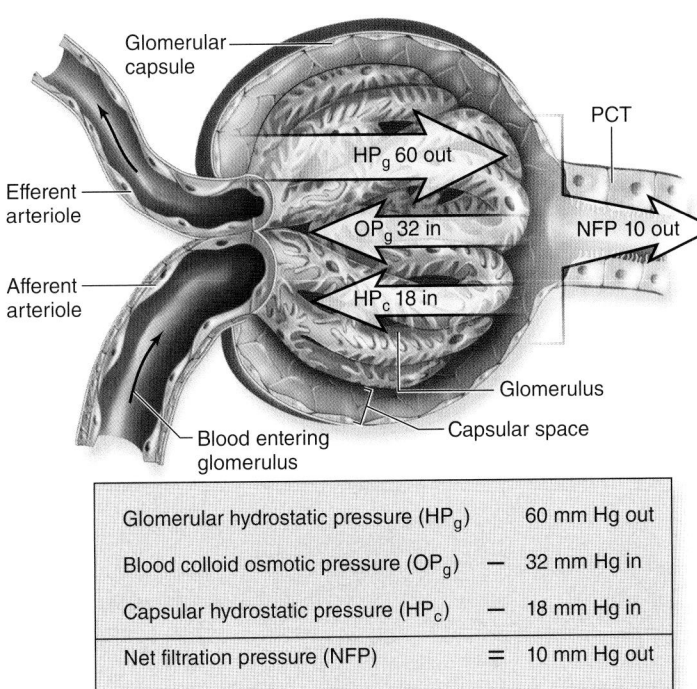

Glomerular hydrostatic pressure (HP$_g$)	60 mm Hg out
Blood colloid osmotic pressure (OP$_g$) −	32 mm Hg in
Capsular hydrostatic pressure (HP$_c$) −	18 mm Hg in
Net filtration pressure (NFP) =	10 mm Hg out

Figure 24.12 Pressures That Determine Net Filtration Pressure. Blood colloid osmotic pressure and capsular hydrostatic pressure are subtracted from glomerular hydrostatic pressure to determine net filtration pressure. **AP|R**

fluids into the glomerulus. The typical value is 32 mm Hg. This is very similar to the colloid osmotic pressure of 26 mm Hg within other systemic capillaries.

Capsular hydrostatic pressure (HP$_c$) is the pressure in the glomerular capsule due to the amount of filtrate already within the capsular space. The presence of this filtrate impedes the movement of additional fluid from the blood into the capsular space and thus it also opposes filtration. A typical value is 18 mm Hg.

Determining Net Filtration Pressure

Filtration occurs if the pressure that promotes filtration, HP$_g$, is greater than the sum of the pressures that oppose filtration (OP$_g$ and HP$_c$). The difference in these pressures is the **net filtration pressure (NFP).** To determine NFP, use the following calculation:

$$HP_g - (OP_g + HP_c) = NFP$$
$$60 \text{ mm Hg} - (32 \text{ mm Hg} + 18 \text{ mm Hg}) = NFP$$
$$60 \text{ mm Hg} - 50 \text{ mm Hg} = 10 \text{ mm Hg}$$

 WHAT DO YOU THINK?

3 An individual with cirrhosis of the liver produces lower-than-normal amounts of plasma proteins. Which pressure is affected: HP$_g$, OP$_g$, or HP$_c$? What do you predict regarding net filtration pressure: Will it increase or decrease as a result? Will more or less filtrate be formed? Explain.

Variables Influenced by Net Filtration Pressure

The **glomerular filtration rate (GFR)** is an important variable influenced by net filtration pressure. It is defined as the rate at which the volume of filtrate is formed, and it is expressed as volume per unit time (usually 1 minute).

The net filtration pressure directly influences the GFR. As NFP increases, usually as the consequence of HP$_g$, the GFR also increases. Likewise, as NFP decreases, GFR decreases.

Net filtration pressure also influences other variables. As the NFP increases and GFR increases more filtrate is produced. The increase in amount of filtrate results in increased fluid volume moving more rapidly through the tubules, so there is less time to reabsorb substances from the tubular fluid. Consequently, more substances remain in the tubular fluid and are excreted in the urine.

The relationship of these variables to an increase in NFP is shown here:

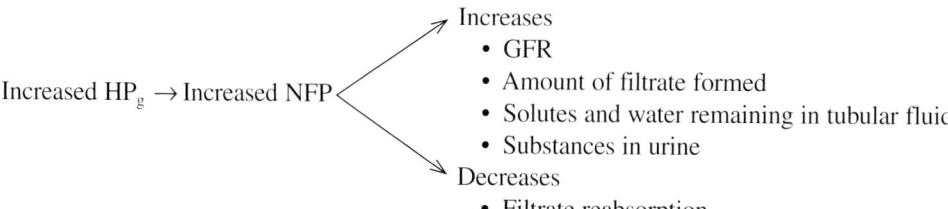

Increased HP$_g$ → Increased NFP

Increases
- GFR
- Amount of filtrate formed
- Solutes and water remaining in tubular fluid
- Substances in urine

Decreases
- Filtrate reabsorption

The significance of these relationships will be referred to throughout different sections of the chapter. Specifically, notice the direct relationship between HP$_g$ and the amount of solutes and water remaining in the tubular fluid. As HP$_g$ increases, substances in the filtrate, including NaCl, increase. As HP$_g$ decreases, substances in the filtrate, including NaCl, decrease. The level of NaCl is monitored by the macula densa of the JG apparatus to indirectly monitor blood pressure in the glomerulus. This concept is discussed in more detail in the following section.

 WHAT DID YOU LEARN?

18 What is the value of the NFP if the glomerular hydrostatic pressure (HP$_g$) is 65 mm Hg, OP$_g$ is 30 mm Hg, and HP$_c$ is 20 mm Hg?

19 What happens to the value of the NFP in question 18 if the HP$_g$ increases from 65 mm Hg to 75 mm Hg?

20 If HP$_g$ increases, what is the effect on NFP? Is the relationship between HP$_g$ and NFP direct or inverse?

24.5e Regulation of Glomerular Filtration Rate

LEARNING OBJECTIVES

9. Describe what is meant by intrinsic and extrinsic controls, and give examples of each.
10. Compare and contrast the myogenic response and the tubuloglomerular feedback mechanism, which are involved in renal autoregulation.
11. Explain the effects of sympathetic division stimulation on the glomerular filtration rate.
12. Describe the effects of atrial natriuretic peptide on the glomerular filtration rate.

The glomerular filtration rate is tightly regulated because of its influence on the amount of substances reabsorbed into the blood and the amount excreted in the urine. By being able to control the glomerular filtration rate, the kidney is able to control urine production based on physiologic conditions, such as state of hydration.

Glomerular filtration rate is primarily influenced both by changing the luminal diameter of the afferent arteriole ("the inflow pipe") and by altering the surface area of the filtration membrane. The processes involved include (1) **intrinsic control** (within the kidney itself), which consists of *renal autoregulation* that maintains GFR at a normal level; and (2) **extrinsic controls** (external to the kidney) that involve nervous system or hormonal regulation, which may decrease or increase GFR, respectively.

Renal Autoregulation: Intrinsic Controls

Renal autoregulation is the intrinsic ability of the kidney to maintain a constant blood pressure and glomerular filtration rate despite changes in systemic arterial pressure. This allows the kidney to produce urine at a constant rate despite fluctuations in systemic blood pressure. Renal autoregulation functions by two mechanisms: the myogenic response and the tubuloglomerular feedback mechanism. Refer to **figure 24.13** as you read this section on renal autoregulation.

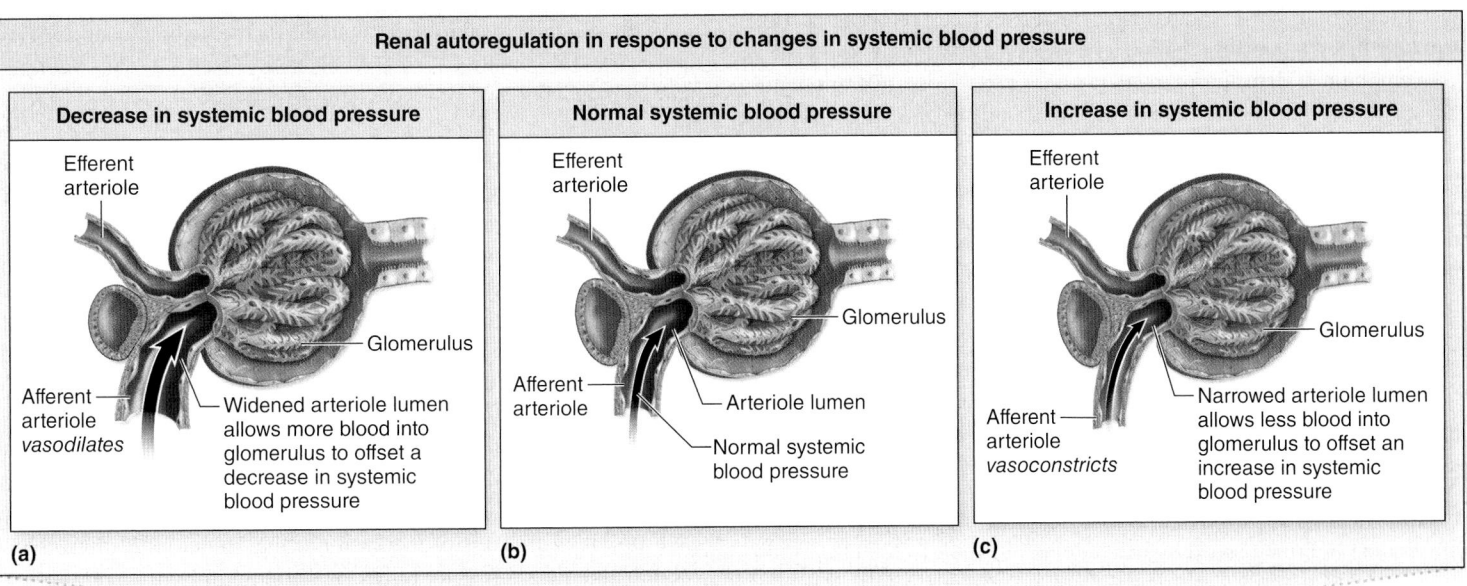

Renal autoregulation in response to changes in systemic blood pressure

| Decrease in systemic blood pressure | Normal systemic blood pressure | Increase in systemic blood pressure |

(a) Efferent arteriole / Glomerulus / Afferent arteriole *vasodilates* / Widened arteriole lumen allows more blood into glomerulus to offset a decrease in systemic blood pressure

(b) Efferent arteriole / Glomerulus / Afferent arteriole / Arteriole lumen / Normal systemic blood pressure

(c) Efferent arteriole / Glomerulus / Afferent arteriole *vasoconstricts* / Narrowed arteriole lumen allows less blood into glomerulus to offset an increase in systemic blood pressure

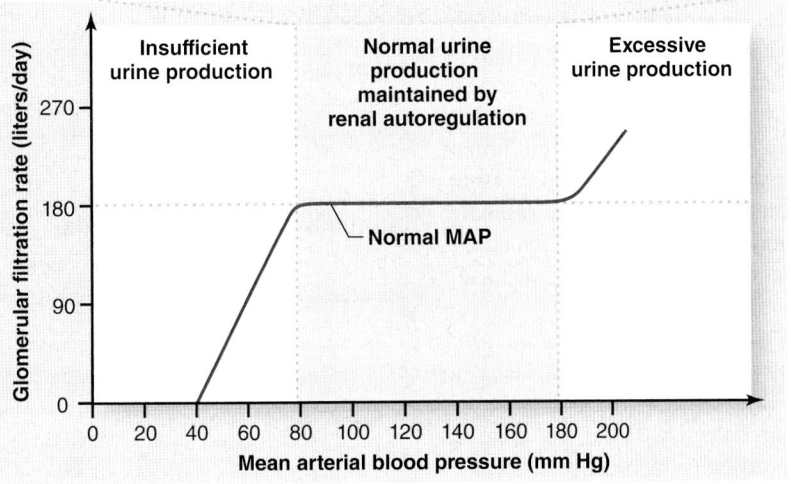

(d)

Figure 24.13 Renal Autoregulation. Renal autoregulation is the intrinsic ability of the kidney to maintain a normal glomerular filtration rate despite changes in systemic blood pressure. The figures depict the response to (*a*) a decrease in systemic blood pressure, (*b*) normal resting systemic blood pressure, and (*c*) an increase in systemic blood pressure. (*d*) Renal autoregulation is effective in maintaining normal glomerular filtration for mean arterial pressure (MAP) typically between 80 mm Hg and 180 mm Hg.

Myogenic Response The **myogenic** (mī′ō-jen′ik; *mys* = muscle, *genesis* = origin) **response** involves contraction and relaxation of smooth muscle in the wall of the afferent arteriole in response to changes in stretch. (The myogenic response is introduced in section 20.4b.) A decrease in systemic blood pressure (e.g., such as occurs when you are taking a nap) results in a lower volume of blood entering the afferent arteriole, reducing the stretch of the smooth muscle in the arteriole wall. The smooth muscle cells in the vessel relax, resulting in vasodilation. The wider vessel lumen of the afferent arteriole allows more blood into the glomerulus, which compensates for the lower systemic blood pressure. Glomerular blood pressure and GFR remain normal (figure 24.13a).

In contrast, an increase in systemic blood pressure (e.g., such as occurs when you exercise) causes an additional volume of blood to enter the afferent arteriole, stretching the smooth muscle in the arteriole wall. The smooth muscle cells in the vessel contract, resulting in vasoconstriction. The narrower vessel lumen of the afferent arteriole allows less blood into the glomerulus, which compensates for the greater systemic blood pressure. Glomerular blood pressure and GFR remain normal (figure 24.13c).

Tubuloglomerular Feedback Mechanism The juxtaglomerular apparatus also helps maintain a normal glomerular blood pressure through the **tubuloglomerular feedback mechanism,** which is based on detection of NaCl levels in the tubular fluid.

If the myogenic response is not sufficient to maintain normal glomerular blood pressure in response to an increase in systemic blood pressure, then glomerular blood pressure increases and the amount of NaCl remaining in the tubular fluid increases (see preceding section showing this relationship). Ultimately, an increase in tubular fluid NaCl concentration is detected by macula densa cells in the juxtaglomerular apparatus. The macula densa cells respond by releasing a signaling molecule (most likely ATP is the signaling molecule) that binds to, and stimulates contraction of smooth muscle cells in the afferent arteriole wall. This paracrine stimulation results in further vasoconstriction of the afferent arteriole and a decreased volume of blood entering the glomerulus. Mesangial cells are also stimulated to contract, which decreases filtration membrane surface area. Both vasoconstriction of the afferent arteriole and smaller membrane surface area cause GFR and the amount of filtrate formed to return to normal levels.

You might find it helpful to think of the tubuloglomerular feedback as a "backup mechanism" to the myogenic response in response to increased systemic blood pressure.

Limitations to Maintaining GFR It is possible to control GFR and the amount of urine formed by maintaining blood pressure in the glomerulus, as just described. However, there are limitations to this regulation. Renal autoregulation is effective in maintaining a normal glomerular blood pressure only if the systemic mean arterial blood pressure (MAP; see section 20.5a) remains between 80 and 180 mm Hg (figure 24.13d).

A decrease in systemic blood pressure results in vasodilation of the afferent arteriole, with maximal arteriole dilation being reached at a MAP of 80 mm Hg. Further decreases in MAP can bring no further arteriole dilation, resulting in a decrease in glomerular blood pressure and glomerular filtration rate. If systemic blood pressure is extremely low (e.g., resulting from a severe hemorrhage), filtration and the elimination of wastes in urine cease, resulting in accumulation of toxic metabolic wastes in the blood.

If systemic blood pressure increases, then the afferent arteriole vasoconstricts. The afferent arteriole vessels are maximally constricted at a MAP value of approximately 180 mm Hg. Further increases in systemic blood pressure can bring no further arteriole constriction, resulting in an increase in both glomerular blood pressure and GFR. Ultimately, the amount of urine increases.

Neural and Hormonal Control: Extrinsic Controls

The intrinsic controls of renal autoregulation, which were just described, help to maintain the GFR within normal homeostatic limits. In comparison, extrinsic controls, which are described here, involve physiologic processes to *change* GFR to adjust urine output based on physiologic need. GFR can be decreased when the kidney is stimulated by the sympathetic division, and it can be increased with atrial natriuretic peptide stimulation.

Decreasing GFR Through Sympathetic Division Stimulation Activation of the sympathetic division as part of the "fight-or-flight" response (see section 15.4) results in a decrease in GFR through both vasoconstriction of the afferent arteriole and decreasing the surface area of the filtration membrane.

The sympathetic division sends increased nerve signals to the kidneys during exercise or in an emergency (**figure 24.14a**). Both afferent and efferent arterioles vasoconstrict as a result. Severe vasoconstriction of the afferent arteriole greatly reduces blood flow into the glomerulus. Glomerular blood pressure and GFR decrease.

Sympathetic stimulation also causes granular cells of the JG apparatus to release renin, with the subsequent production of angiotensin II (see section 20.6b). Angiotensin II stimulates myofilaments within mesangial cells to contract. This results in a decrease in the surface area of the filtration membrane, and subsequently the GFR decreases.

Severe vasoconstriction of the afferent arteriole and contraction of mesangial cells both contribute to a decrease in GFR with an accompanying decrease in urine production. Fluid is retained within the blood, maintaining blood volume.

This is a critical adjustment that allows the body to conserve fluid under stressful conditions, such as extensive exertion (e.g., running a marathon on a hot day) or severe hemorrhaging.

Increasing GFR Through Atrial Natriuretic Peptide **Atrial natriuretic** (nā′trē-yū-ret′ik) **peptide (ANP)** increases GFR to eliminate fluid. This peptide hormone is released from atrial cardiac muscle cells into the blood in response to distension of these chambers; this occurs when there is either an increase in blood volume return or an increase in blood pressure. ANP is transported in the blood to the kidney following its release from the heart. Atrial natriuretic peptide relaxes the afferent arteriole and inhibits the release of renin from the granular cells to ultimately cause relaxation of the mesangial cells to increase filtration membrane surface area (figure 24.14b). The net result is an increase in GFR that is accompanied by both an increase in urine volume and a decrease in blood volume. Thus, the release of ANP provides a means of increasing urine output to decrease blood volume and blood pressure to within normal limits.

The complex interrelationship of the urinary, nervous, and endocrine systems in controlling systemic blood volume and blood pressure is discussed in detail in chapter 25. Note: The details for both angiotensin II and ANP are included in the summary table "Regulating Fluid Balance, Blood Volume, and Blood Pressure," which directly follows chapter 17 (see **table R.7**).

Summary of GFR Regulation

The glomerular filtration rate can be maintained, decreased, or increased by the following mechanisms:

- Renal autoregulation *maintains GFR* by altering the size of the afferent arteriole in response to changes in systemic blood pressure. The afferent arteriole vasoconstricts if systemic blood pressure increases, whereas the afferent arteriole vasodilates if systemic blood pressure decreases. This occurs with assistance from macula densa cells and mesangial cells. Renal

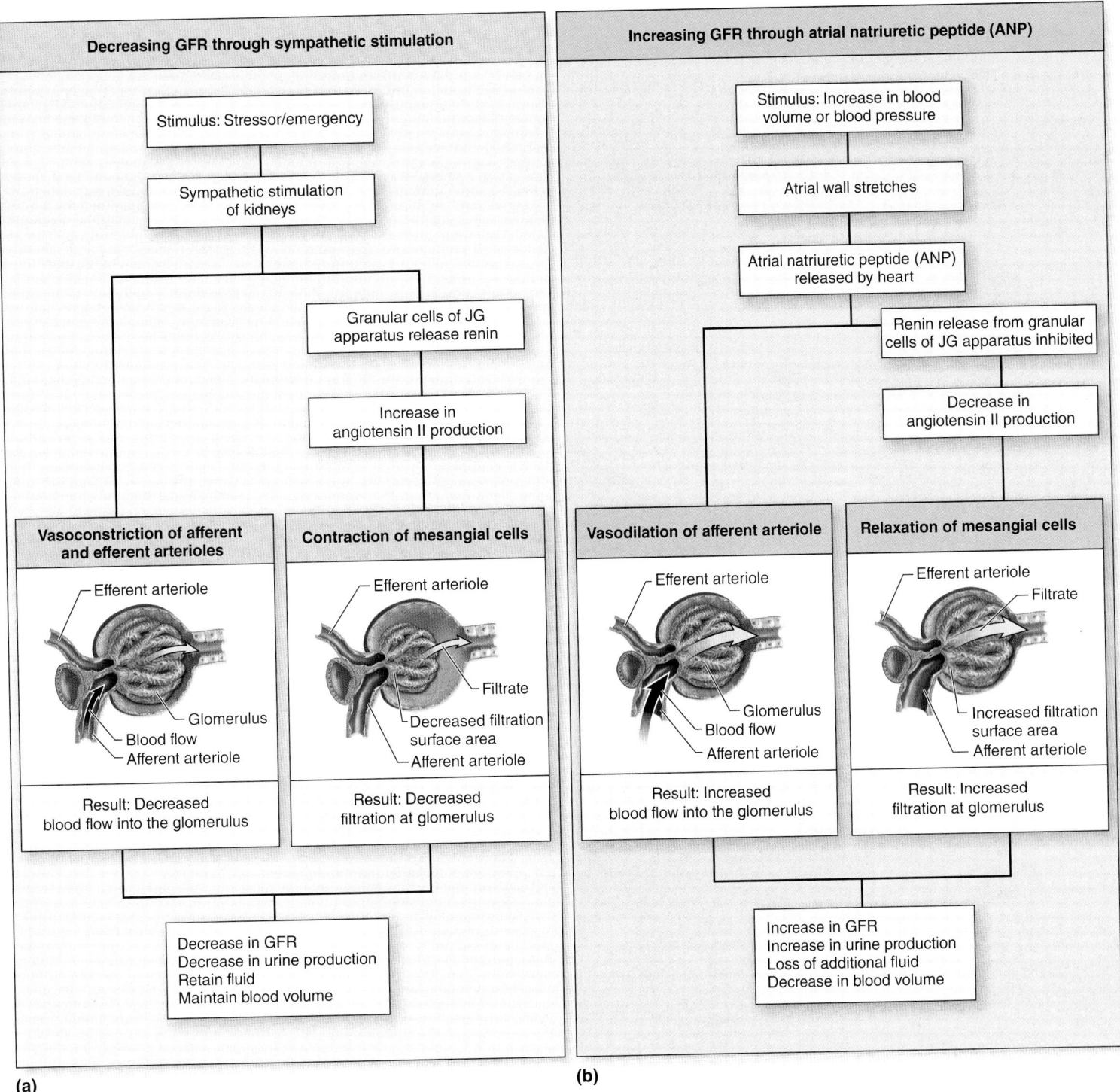

Figure 24.14 Extrinsic Mechanisms for Changing the Glomerular Filtration Rate (GFR) Through Nervous and Hormonal Control. Glomerular filtration can be (a) decreased by extensive sympathetic stimulation and (b) increased by atrial natriuretic peptide (ANP) stimulation.

autoregulation is effective in maintaining GFR if MAP is within the range of 80 to 180 mm Hg (figure 24.13d).

- Direct stimulation by the sympathetic division causes (1) vasoconstriction of afferent arterioles to decrease blood flow into the glomerulus and (2) renin release with the subsequent production of angiotensin II and contraction of mesangial cells that decrease the surface area of the glomerulus. Both GFR and urine production are decreased.

- Atrial natriuretic peptide (ANP) *increases GFR* through (1) vasodilation of the afferent arteriole to increase blood flow into the glomerulus and (2) inhibition of renin release and the subsequent relaxation of mesangial cells that increase the surface area of the glomerulus. Both GFR and urine production are increased.

Figure 24.15 provides a visual summary of the structures and processes involved in glomerular filtration.

WHAT DID YOU LEARN?

21 Does urine production increase, decrease, or stay the same in response to an increase in glomerular filtration rate?

22 What are the three factors that regulate glomerular filtration rate? Does each of these increase, decrease, or maintain GFR?

23 Renal autoregulation is effective with a MAP between 80 and 180 mm Hg. Would renal autoregulation be effective in an individual with a blood pressure of 300/150 mm Hg? A pressure of 70/55 mm Hg? Explain.

Figure 24.15 **Glomerular Filtration and Its Regulation.**
A visual summary of glomerular filtration, including (*a*) the filtration membrane and components of filtrate, (*b*) determining net filtration pressure, and (*c*) the mechanisms for controlling glomerular filtration.

(a) Filtration Membrane and Components of Filtrate

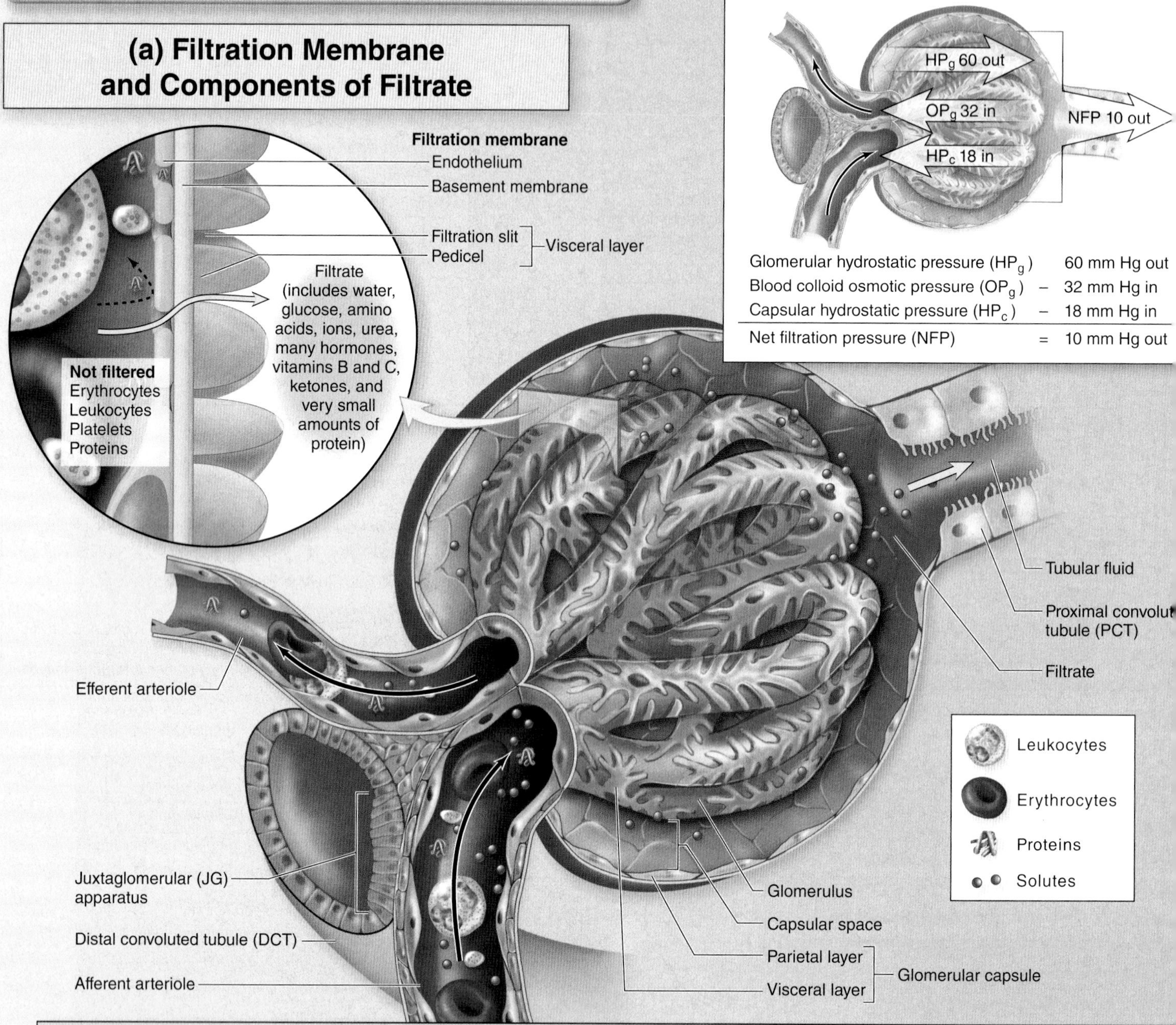

Filtration membrane
- Endothelium
- Basement membrane

- Filtration slit — Visceral layer
- Pedicel

Filtrate (includes water, glucose, amino acids, ions, urea, many hormones, vitamins B and C, ketones, and very small amounts of protein)

Not filtered
Erythrocytes
Leukocytes
Platelets
Proteins

Efferent arteriole

Juxtaglomerular (JG) apparatus

Distal convoluted tubule (DCT)

Afferent arteriole

Tubular fluid

Proximal convolut tubule (PCT)

Filtrate

Leukocytes

Erythrocytes

Proteins

Solutes

Glomerulus

Capsular space

Parietal layer
Visceral layer — Glomerular capsule

(b) Determining Net Filtration Pressure in the Renal Corpuscl

HP$_g$ 60 out

OP$_g$ 32 in — NFP 10 out

HP$_c$ 18 in

Glomerular hydrostatic pressure (HP$_g$)		60 mm Hg out
Blood colloid osmotic pressure (OP$_g$)	–	32 mm Hg in
Capsular hydrostatic pressure (HP$_c$)	–	18 mm Hg in
Net filtration pressure (NFP)	=	10 mm Hg out

(c) Control of Glomerular Filtration Rate (GFR)

MAINTAIN GFR

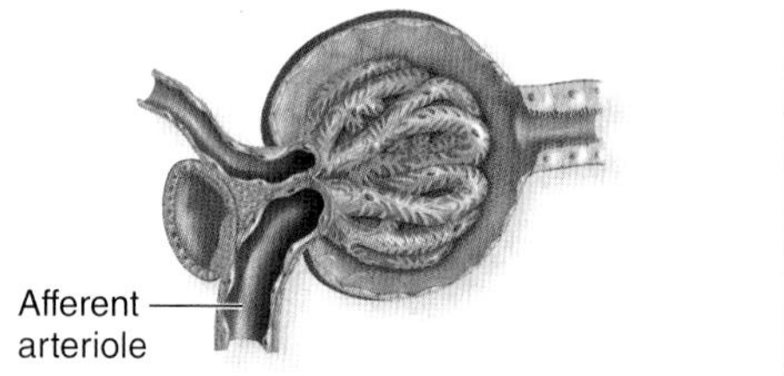

Afferent arteriole

Renal autoregulation maintains GFR despite changes in systemic blood pressure (BP):
- Decreased systemic BP results in vasodilation of afferent arteriole
- Increased systemic BP results in vasoconstriction of afferent arteriole

DECREASE GFR

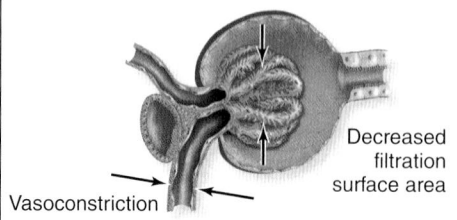

Decreased filtration surface area

Vasoconstriction

The sympathetic division decreases GFR by:
- Afferent arteriole vasoconstriction
- Triggering mesangial cells to contract, which decreases filtration surface area

Urine production is decreased, which helps maintain blood volume.

INCREASE GFR

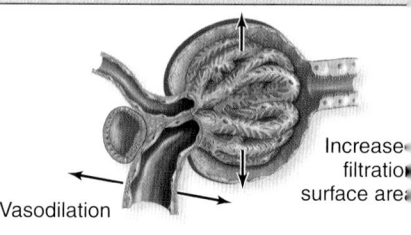

Increase filtratio surface are

Vasodilation

Atrial natriuretic peptide (ANP) increases GFR b
- Afferent arteriole vasodilation
- Triggering mesangial cells to relax, which increases filtration surface area

Urine production is increased, which decrease blood volume.

24.6 Reabsorption and Secretion in Tubules and Collecting Ducts

Tubular fluid flows through the proximal convoluted tubule, nephron loop, distal convoluted tubule, and then into collecting tubules and collecting ducts, as described earlier. Substances are reabsorbed when they move from the tubular fluid back into the blood. Additionally, some substances that were not initially filtered at the glomerulus, but must be eliminated from the blood, become a component of urine through tubular secretion.

Here we provide an overview of transport processes, and consider transport maximum and renal threshold. We then consider the details of the substances reabsorbed and secreted. These are organized as follows:

- Substances reabsorbed completely
- Substances with regulated reabsorption (meaning some, but not all, of substance reabsorbed)
- Substances eliminated as waste products

Finally, we describe the establishment of a concentration gradient in the interstitial fluid, which drives the reabsorption of water controlled by ADH.

24.6a Overview of Transport Processes

LEARNING OBJECTIVE

1. Describe five characteristics and conditions that affect tubular reabsorption and secretion.

This section provides a general overview of important anatomic structures and physiologic conditions in the kidney that influence reabsorption and secretion of substances. Please refer to **figure 24.16** as you read through the following description:

1. The barrier that a substance must cross is the simple epithelium of the tubule wall.
2. Substances can either pass between the epithelial cells of the tubular wall by **paracellular** (păr′a-sel′yū-lăr; *para* = alongside) **transport,** or more commonly, move through the epithelial cells by **transcellular** (trănz′sel′yū-lăr; *trans* = across) **transport.**
3. During transcellular transport, a substance must cross two plasma membranes: the **luminal membrane** that is in contact with tubular fluid and the **basolateral membrane** that rests on the basement membrane. The order in which the substance crosses these membranes depends upon whether it is being reabsorbed or secreted.

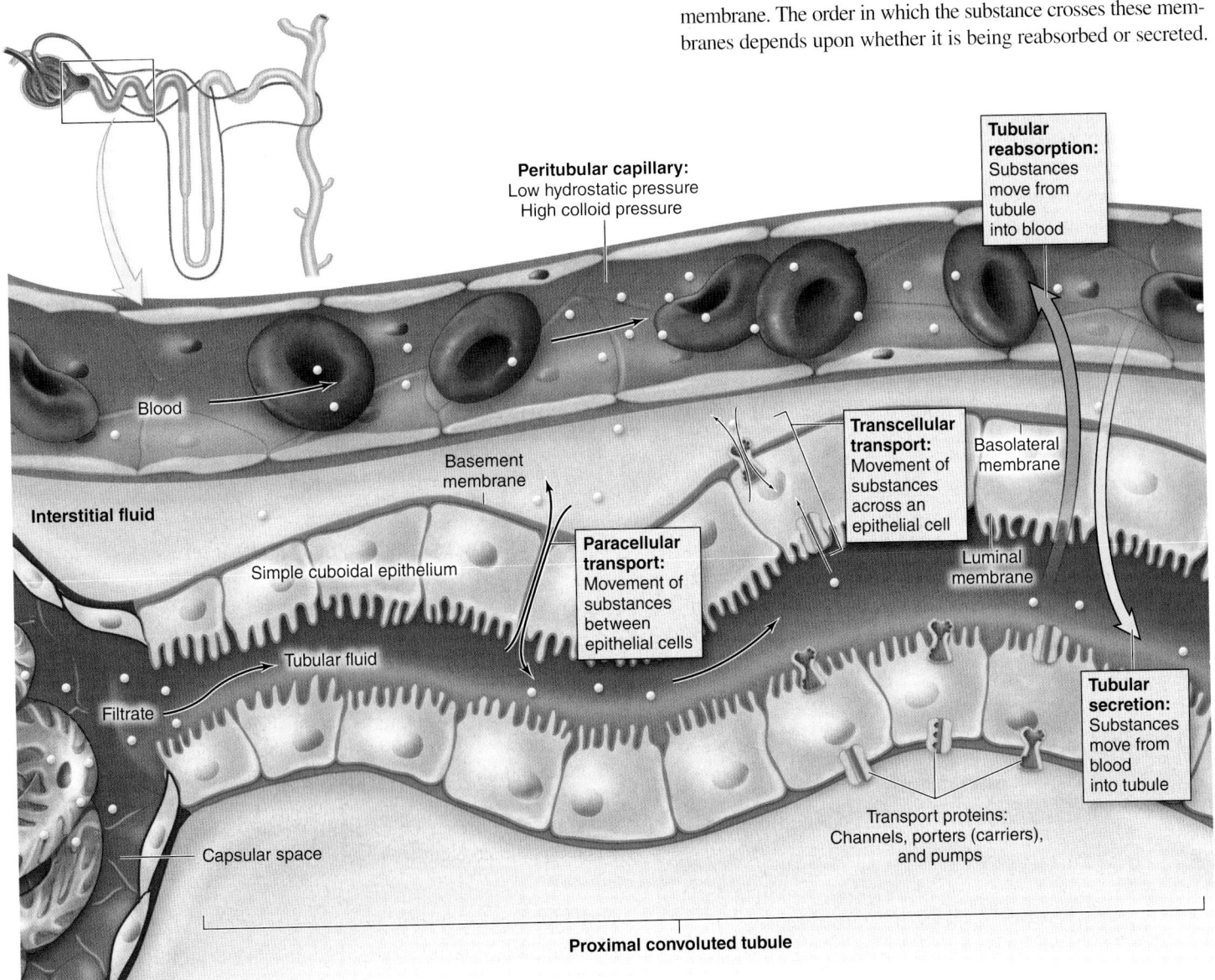

Peritubular capillary:
Low hydrostatic pressure
High colloid pressure

Tubular reabsorption:
Substances move from tubule into blood

Blood

Transcellular transport:
Movement of substances across an epithelial cell

Basolateral membrane

Basement membrane

Interstitial fluid

Paracellular transport:
Movement of substances between epithelial cells

Simple cuboidal epithelium

Luminal membrane

Tubular fluid

Filtrate

Tubular secretion:
Substances move from blood into tubule

Capsular space

Transport proteins:
Channels, porters (carriers), and pumps

Proximal convoluted tubule

Figure 24.16 Convoluted Tubules and Peritubular Capillaries. The anatomic structures and physiologic conditions that influence tubular reabsorption and tubular secretion.

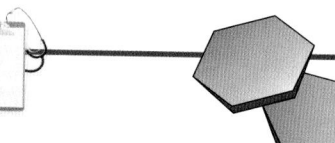

LEARNING STRATEGY

Movement of filtrate and tubular fluid through the nephron is analogous to movement of items along a conveyor belt. The fluid's path is like the conveyer belt, and substances are first loaded onto the beginning of the conveyor belt (filtration). Then, some of these substances are retrieved from the conveyor belt (tubular reabsorption). Additional substances not originally filtered can be added to the conveyor belt along its length (tubular secretion). Whatever is on the conveyor belt at the end represents the substances that make up urine.

CLINICAL VIEW
Glucosuria

Glucosuria (or *glycosuria*) is the excretion of glucose in urine. This occurs when the plasma glucose level exceeds approximately 300 milligrams per deciliter (mg/dL) and the transport maximum (T_m) rate for glucose of 375 milligrams per minute (mg/min) has been exceeded. Glucose molecules act as an osmotic diuretic, pulling water into the tubular fluid and causing additional loss of fluid in the urine. Glucosuria, along with frequent urination and thirst, are classical signs of diabetes mellitus (see Clinical View: "Conditions Resulting in Abnormal Blood Glucose Levels" in section 17.9b).

4. Different transport proteins (table 24.1) are embedded within the two membranes. They control the movement of various substances using membrane transport processes that include simple or facilitated diffusion, osmosis, primary and secondary active transport, and vesicular transport (see section 4.3).
5. Peritubular capillaries have both low hydrostatic pressure (8 mm Hg), because of the loss of fluid during filtration in the glomerulus, and high colloid (oncotic) pressure (>30 mm Hg) exerted by protein, because most proteins remain in the blood during filtration. These two important properties facilitate reabsorption of substances into the peritubular capillaries through bulk flow (see section 20.3b).

Although reabsorption and secretion of substances occur along the entire length of the nephron tubule, collecting tubules, and collecting ducts, most reabsorption occurs in the proximal convoluted tubule, where the uptake processes are aided by the extensive microvilli on the cells' luminal surfaces that increase their surface area.

WHAT DID YOU LEARN?

24 What are the significant anatomic and physiologic factors that influence tubular reabsorption and tubular secretion?

24.6b Transport Maximum and Renal Threshold
LEARNING OBJECTIVES

2. Define the transport maximum of a substance.
3. Explain what is meant by renal threshold.

The **transport maximum (T_m)** is the maximum amount of a substance that can be reabsorbed (or secreted) across the tubule epithelium in a given period of time (i.e., its rate of movement). This maximum is dependent upon the number of the transport proteins in the epithelial cell membrane specific for the substance.

Table 24.1	Transport Proteins in Cells of the Renal Tubule	
Pumps	**Carriers (Porters)**	**Channels**
Na$^+$/K$^+$ ATPase	Glucose uniporter	Na$^+$ channels
H$^+$ ATPase	Na$^+$/glucose symporter	K$^+$ channels
Ca^{2+} ATPase	Na$^+$/HCO$_3^-$ symporter	Aquaporins (H$_2$O channels)
	Na$^+$/H$^+$ antiporter	
	Cl$^-$/HCO$_3^-$ antiporter	
	Na$^+$/K$^+$/2 Cl$^-$ symporter	

For example, the T_m for glucose reabsorption by the glucose transport proteins is approximately 375 milligrams per minute (mg/min). As long as the tubular fluid contains no more than of this amount of glucose passing through a region of the renal tubule every minute, all of it will be reabsorbed. If the tubular fluid contains more than this amount, then the transport proteins are saturated, and the excess glucose is excreted in the urine.

The maximum *plasma concentration* of a substance that can be transported in the blood without eventually appearing in the urine is called the **renal threshold.** For example, the renal threshold for glucose is 300 milligrams per deciliter (mg/dL). Above this concentration, the substance is so concentrated in the plasma, and thus in the tubular fluid following filtration, that the transport proteins cannot reabsorb all of the substance from the fluid. The T_m is exceeded, and the substance remains within the tubule fluid and is excreted in the urine.

WHAT DO YOU THINK?

4 Explain why increased urine production and dehydration are symptoms of diabetes mellitus.

WHAT DID YOU LEARN?

25 What is the transport maximum of a substance? How is it different from the renal threshold of the substance?

24.6c Substances Reabsorbed Completely
LEARNING OBJECTIVES

4. Explain the reabsorption of nutrients such as glucose.
5. Describe the process by which protein is transported out of the filtrate and into the blood.

Some substances are not normal components of urine because 100% of the substance is reabsorbed from the tubular fluid back into the blood. The location for this reabsorption typically is in the proximal convoluted tubule. Two major types of substances fall into this category: nutrients (e.g., glucose, amino acids, lactate), and the small amount of filtered plasma proteins.

Nutrient Reabsorption

Nutrients are normally reabsorbed completely in the proximal convoluted tubule where each nutrient has its own specific transport proteins. We use glucose as an example to describe how nutrients may be reabsorbed completely (figure 24.17).

Glucose concentration is relatively high inside the tubule cell and relatively low within both the tubular fluid and interstitial fluid. Glucose is first transported into the tubule cell across the luminal membrane by Na^+/glucose symporter proteins. Energy from Na^+ moving down its concentration gradient into the tubule cell is used to move glucose up its concentration gradient into the tubule cell by secondary active transport (see section 4.3c). Glucose is then moved by glucose uniporters (carriers) out of the tubule cell down its concentration gradient via facilitated diffusion across the basolateral membrane.

Glucose ultimately is returned to the blood in the peritubular capillaries. As with many other substances that rely on membrane transport proteins, there is a maximum amount of glucose that can be reabsorbed per unit time.

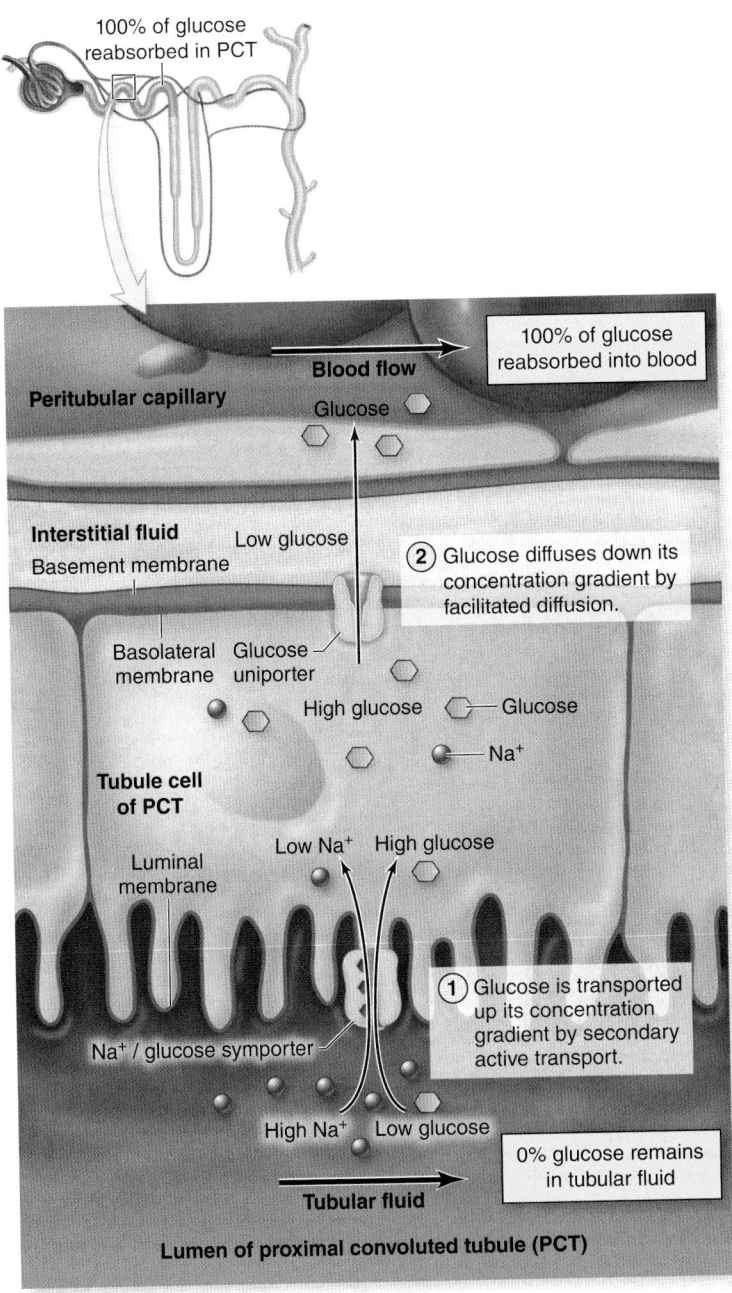

Figure 24.17 Glucose Reabsorption. Glucose reabsorption occurs in the proximal convoluted tubule. In a healthy individual, 100% of the glucose is reabsorbed. Glucose is transported (1) across the luminal membrane against its concentration gradient—this occurs by secondary active transport via Na^+/glucose symporters, and (2) across the basolateral membrane down its concentration gradient by facilitated diffusion via glucose uniporters.

Transport of Proteins

Although most proteins are not freely filtered in the glomerulus because of their size and negative charge, some small and medium-sized peptides, such as insulin and angiotensin, and limited amounts of large proteins (e.g., about 0.02% of albumin) may appear in the filtrate. Protein is transported from the tubular fluid in the proximal convoluted tubule back into the blood so as not to be excreted in the urine.

We use the general term "transported" here (instead of reabsorbed) because the proteins actually undergo transformational changes while being reabsorbed. Protein is moved across the luminal membrane by pinocytosis or receptor-mediated endocytosis (see section 4.3c). Lysosomes in these tubule cells then digest the proteins into their amino acid building blocks. These amino acids are moved by facilitated diffusion across the basolateral membrane back into the blood. Very small peptides, such as angiotensin II, are degraded by peptidases within the luminal membrane and the amino acids are absorbed directly into the tubule cell. Thus, proteins and small peptides are first degraded into amino acids, which are then absorbed into the blood.

WHAT DO YOU THINK?

5 What happens to blood concentration of plasma proteins in an individual with renal disease that results in either (a) damage to the filtration membrane or (b) decreased filtration at the filtration membrane?

Given the kidney's normal function in not filtering most plasma proteins and degrading the plasma proteins that are filtered, what are the clinical implications of renal disease that alter these processes?

First, consider the consequences if the renal disease results in higher than normal filtration of plasma proteins (e.g., damage to the glomerular membrane or high systemic blood pressure). These abnormal changes result in saturation of the tubule cellular structures involved in protein uptake, and protein is lost in the urine. Consequently, blood concentration of all plasma proteins decreases.

Second, consider the consequences if the renal disease results in lower than normal filtration (e.g., changes that accompany chronic kidney disease to varying degrees, depending upon the state of the disease). These abnormal changes result in a decrease in the filtration of small plasma proteins and their degradation within the kidney tubules. Consequently, blood concentration of these plasma proteins increases. Plasma protein levels that are either too low or too high result in unwanted physiologic outcomes.

WHAT DID YOU LEARN?

26 How is glucose reabsorbed across the two membranes of the tubule cells?

27 Why are proteins said to be transported rather than simply reabsorbed in the proximal convoluted tubule?

24.6d Substances with Regulated Reabsorption

LEARNING OBJECTIVES

6. List substances for which reabsorption is regulated.
7. Describe how the reabsorption of sodium, potassium, calcium, and phosphate occurs.
8. Describe the reabsorption of water, and compare how it is regulated by the actions of aldosterone and antidiuretic hormone.
9. Describe how pH is regulated in the collecting tubules.

Some substances are recovered completely from the tubular fluid, whereas others are not completely reabsorbed, resulting in a variable and often small percentage of that substance being excreted into the urine. By varying the amount of a substance excreted, the nephron has a substantial role in regulating the blood level of that substance. A number of substances fall into the category of undergoing regulated reabsorption, including Na^+, water, K^+, HCO_3^-, and Ca^{2+}. We will see that the reabsorption of Na^+ plays a pivotal role in the reabsorption of many of these other substances.

WHAT DO YOU THINK?

6 If a substance in the blood is filtered, and not all of it is completely reabsorbed back into the blood, does its level in the blood increase, decrease, or stay the same? Explain.

Sodium (Na⁺) Reabsorption

The amount of Na^+ reabsorbed from the tubular fluid can vary from 98% to 100%. Unlike glucose and other nutrients, Na^+ is reabsorbed along the entire length of the nephron tubule, and collecting tubules and ducts, with the majority of the Na^+ (about 65%) reabsorbed in the proximal convoluted tubule (**figure 24.18a**). Approximately 25% is reabsorbed

in the nephron loop, and a varying (and regulated) amount in the distal convoluted tubule and collecting tubules and ducts.

Sodium concentration is relatively low inside the tubule cell and relatively high within both the tubular fluid and interstitial fluid (figure 24.18b). (The Na^+ concentration gradient is established by Na^+/K^+ pumps, as described shortly.) Thus, Na^+ moves passively by facilitated diffusion down its concentration gradient across the luminal membrane into the tubular cell. The specific type of transport protein involved in the movement across the luminal membrane varies with the different sections of the renal tubule.

Embedded in the basolateral membrane are Na^+/K^+ pumps. These pumps move Na^+ from within the tubular cell into the interstitial fluid, while K^+ is moved from the interstitial fluid into the tubular cell. This keeps the Na^+ concentration relatively low within tubule cells. The energy consumption of these pumps is substantial; they use approximately 80% of all energy invested in active transport within nephrons. The Na^+ then enters into the peritubular and vasa recta capillaries through intercellular clefts.

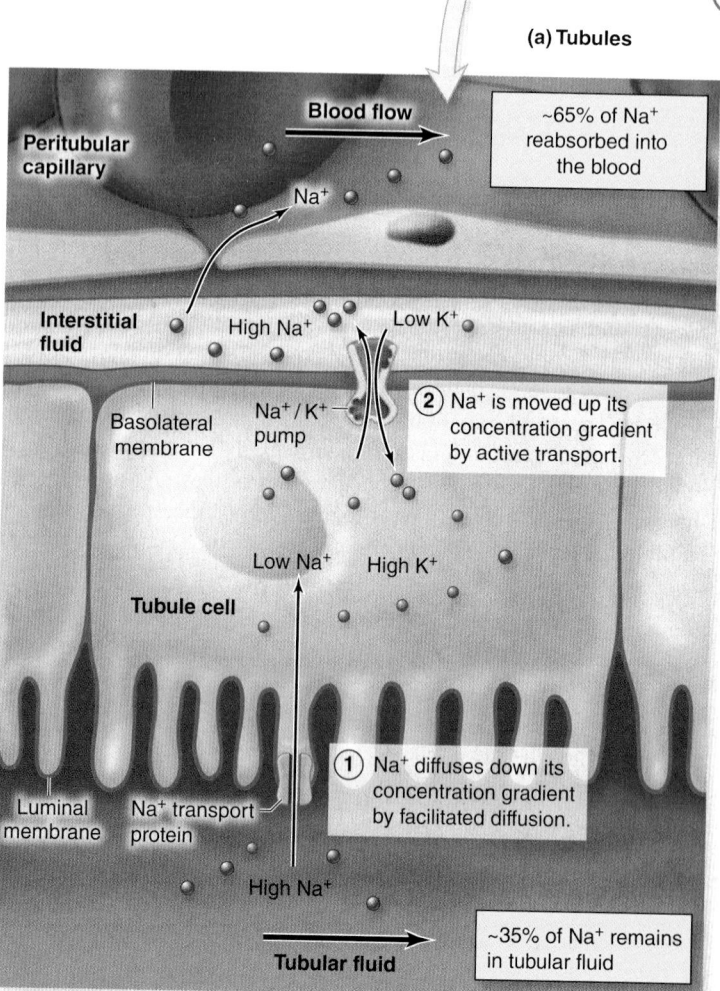

~65% of Na⁺ reabsorbed in PCT

~25% of Na⁺ reabsorbed in nephron loop

Regulated percent of Na⁺ reabsorbed in DCT, CT, and CD

(a) Tubules

(b) Lumen of proximal convoluted tubule (PCT)

- Peritubular capillary
- **Blood flow**
- ~65% of Na⁺ reabsorbed into the blood
- Interstitial fluid
- High Na⁺ — Low K⁺
- Na⁺
- Basolateral membrane
- Na⁺/K⁺ pump
- ② Na⁺ is moved up its concentration gradient by active transport.
- Low Na⁺ — High K⁺
- Tubule cell
- Luminal membrane — Na⁺ transport protein
- ① Na⁺ diffuses down its concentration gradient by facilitated diffusion.
- High Na⁺
- **Tubular fluid**
- ~35% of Na⁺ remains in tubular fluid

(c) Lumen of the distal convoluted tubule (DCT), collecting tubule (CT), or collecting duct (CD)

- **Tubular fluid**
- Luminal membrane
- Na⁺ channels
- High Na⁺
- H₂O
- Aquaporin
- Principal cell
- High K⁺ — K⁺
- Low Na⁺
- Basolateral membrane
- Na⁺/K⁺ pumps
- **Interstitial fluid**
- High Na⁺
- Osmosis
- **Blood flow**
- Na⁺
- H₂O
- Aldosterone increases both the number of Na⁺ channels and Na⁺/K⁺ pumps, resulting in an increase in Na⁺ reabsorption.
- % of Na⁺ varies from 0–2%
- Peritubular capillary

Figure 24.18 Sodium (Na⁺) Reabsorption. (*a*) Na⁺ reabsorption occurs along most of the length of the renal tubule. (*b*) The majority of Na⁺ is reabsorbed in the proximal convoluted tubule. Na⁺ is transported across the luminal membrane down its concentration gradient by facilitated diffusion via various types of Na⁺ transport proteins, and across the basolateral membrane against its concentration gradient by Na⁺/K⁺ pumps. (*c*) The amount of Na⁺ excreted in the urine is regulated in the distal convoluted tubule, collecting tubules, and collecting ducts by hormones. Aldosterone binds to principal cells, increasing both the number of Na⁺ channels and Na⁺/K⁺ pumps. The net effect is that additional Na⁺ is reabsorbed, water follows by osmosis, and additional K⁺ is secreted.

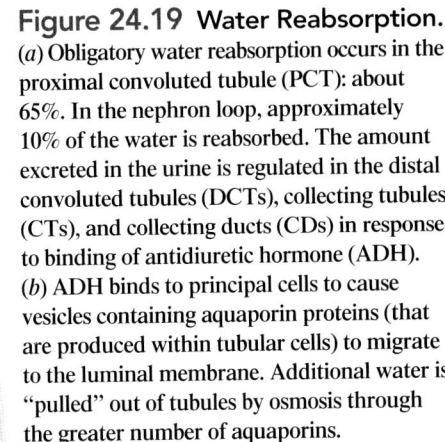

INTEGRATE

LEARNING STRATEGY

Use the phrase "Al likes salt" to remember that *al*dosterone helps our body retain sodium. Also, note the word roots of atrial **natriuretic** peptide, which causes sodium (**Na⁺**, from Latin **natrium**) to be excreted in the urine (**-uretic**).

Sodium reabsorption is regulated by hormones that bind to tubular cells near the end of the tubule (figure 24.18c). If we consider that Na⁺ intake can vary from as little as 0.05 gram per day (g/day) if an individual is on a low-sodium diet to perhaps as high as 25 g/day, regulation of Na⁺ output is critical if blood Na⁺ levels are to be maintained within a normal homeostatic range.

The amount of Na⁺ that is excreted in urine ranges from 0% to 2% of the total filtered Na⁺. This amount is controlled by aldosterone and atrial natriuretic peptide hormones in the DCT, CT, and CD. **Aldosterone** (al-dos'tĕr-ōn) is a steroid hormone (see section 17.3a) produced by the adrenal cortex. It enters principal cells to bind with intracellular receptors to form a hormone-receptor complex that stimulates protein synthesis of Na⁺ channels and Na⁺/K⁺ pumps. These additional transport proteins become embedded in the plasma membranes of principal cells and Na⁺ reabsorption increases. Water follows the Na⁺ by osmosis, resulting in reabsorption of an isotonic fluid. Note that K⁺ is secreted into the tubular fluid during this process.

Atrial natriuretic peptide (ANP) inhibits both the reabsorption of Na⁺ in the proximal convoluted tubule and collecting tubules and the release of aldosterone (not shown in figure 24.18). Consequently, more Na⁺ and water (because water follows Na⁺ by osmosis) are excreted in urine. Recall that ANP increases GFR, a process that also increases urine output (see section 24.5e). Thus, the release of ANP provides a means of increasing urine output to decrease blood volume and blood pressure to within normal limits. Note that the details for both aldosterone and ANP are included in the summary table "Regulating Fluid Balance, Blood Volume, and Blood Pressure," which directly follows chapter 17 (see **table R.7**). Hormones that influence Na⁺ reabsorption are included in **table 24.2**.

Water (H₂O) Reabsorption

Water movement occurs by osmosis. It is reabsorbed by paracellular transport between cells, or by transcellular transport through specific water transport proteins called **aquaporins**. Of the approximately 180 L of water filtered daily, all except an average of approximately 1.5 L is reabsorbed. The exact amount of water reabsorbed depends upon both fluid intake and fluid excreted through other routes (e.g., sweating). The tubule varies in its permeability to water along its length, and the concentration of the tubular fluid changes as it moves through the different sections of the tubule.

Approximately 65% of the water in the tubular fluid is reabsorbed in the proximal convoluted tubule (**figure 24.19a**). The aquaporins here are permanent components of the luminal membrane and are relatively constant in number. The movement of water out of the proximal convoluted tubule follows Na⁺ by osmosis, and it is referred to as **obligatory water reabsorption**. This water movement is dependent upon the movement of Na⁺, so the water is "obliged" to follow Na⁺. The movement of Na⁺ and water is equivalent, and an isotonic tubular fluid is maintained as it moves through the proximal convoluted tubule.

In the nephron loop, only about 10% of the filtered water in the tubular fluid is reabsorbed. Water moves from the descending limb of the nephron loop into the vasa recta. Both the ascending limb of the nephron loop and distal convoluted tubule are impermeable to water, and no water reabsorption occurs there.

As tubular fluid moves through the collecting tubules and ducts, water reabsorption is controlled primarily by aldosterone and antidiuretic hormone. Recall that aldosterone increases the number of Na⁺/K⁺ pumps

Figure 24.19 Water Reabsorption.
(a) Obligatory water reabsorption occurs in the proximal convoluted tubule (PCT): about 65%. In the nephron loop, approximately 10% of the water is reabsorbed. The amount excreted in the urine is regulated in the distal convoluted tubules (DCTs), collecting tubules (CTs), and collecting ducts (CDs) in response to binding of antidiuretic hormone (ADH). (b) ADH binds to principal cells to cause vesicles containing aquaporin proteins (that are produced within tubular cells) to migrate to the luminal membrane. Additional water is "pulled" out of tubules by osmosis through the greater number of aquaporins.

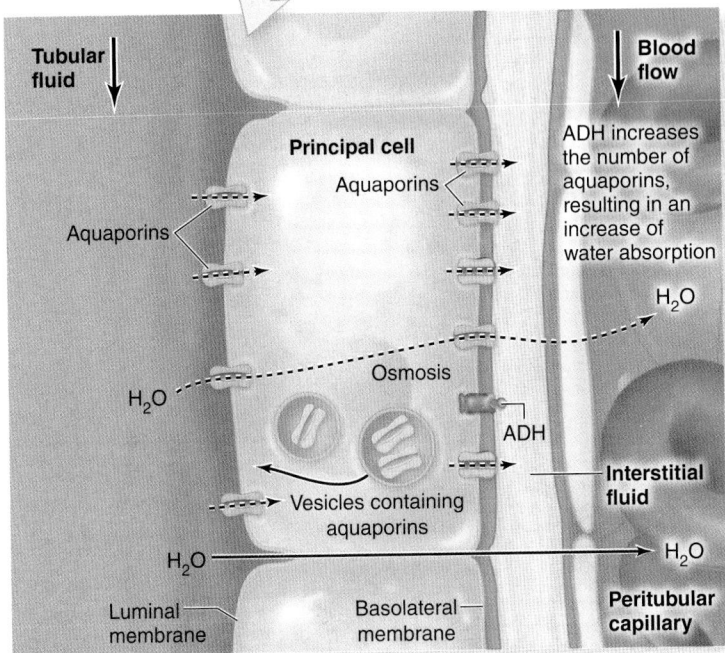

(a) Tubules

(b) Lumen of CT and CD

Table 24.2	Hormones That Influence Na⁺ Reabsorption	
Increase	**Decrease**
Aldosterone	Atrial natriuretic peptide
Cortisol	Progesterone
Estrogen	Parathyroid hormone
Growth hormone	Glucagon
Thyroid hormone	
Insulin	

and Na+ channels in principal cells, thus increasing both Na+ and water reabsorption. Consequently, the concentration of tubular fluid is maintained (figure 24.18). In contrast, **antidiuretic** (an′tē-dī′yū′ret′ik; *anti* = against, *ouresis* = urination) **hormone (ADH)** that is released from the posterior pituitary gland when we are dehydrated binds to receptors of principal cells to increase migration of vesicles containing aquaporins to the luminal membrane. This action provides additional channels for water reabsorption (figure 24.19*b*).

The osmotic force caused by the concentration gradient within the interstitial fluid (described in section 24.6f) pulls water from the tubule. Thus, water reabsorption regulated by ADH near the end of the tubule is *independent* of Na+ reabsorption, and as a result solute concentration of the tubular fluid increases. Tubular reabsorption in response to ADH is referred to as **facultative water reabsorption.**

ADH and Urine Production How concentrated the urine will become is dependent upon the relative amounts of ADH bound to the principal cells. Greater amounts of ADH result in larger amounts of water reabsorption by principal cells and less water remaining in the tubular fluid; thus, a smaller volume of a more concentrated urine is produced. Conditions of extreme dehydration result in subsequent high concentrations of ADH—thus, urine volume can be as little as about 0.5 L/day. Urine can be as concentrated as 1200 milliosmoles (mOsm), which is the same concentration as the interstitial fluid outside the tubules. Under these conditions, the urine generally appears noticeably darker yellow in color. In comparison, lesser amounts of ADH result in smaller amounts of water reabsorption by principal cells and more water remaining in the tubular fluid—which produces a larger volume of a less concentrated urine. Conditions of overhydration (e.g., drinking a liter of soda) result in lower concentrations of ADH—thus urine output that can be several liters per day, and (with the aid of pumps that move salt out of the tubular fluid) can be as dilute as 50 mOsm. Under these conditions, the urine will appear a light yellow to very pale yellow color.

Chapter 25 presents a detailed discussion of the effects of these hormones (aldosterone, ANP, and ADH, as well as angiotensin II) on electrolyte, fluid, and acid-base balance. The details for these four hormones are included in the summary table "Regulating Fluid Balance, Blood Volume, and Blood Pressure," which directly follows chapter 17 (see **table R.7**).

Reabsorption and Secretion of Potassium

Potassium is unlike other substances previously covered because it is both reabsorbed and secreted (**figure 24.20**). The result may be a net reabsorption of K+, with little being lost in urine or a net secretion, with larger losses in the urine.

INTEGRATE

CONCEPT CONNECTION

Blood pressure is dependent upon three major variables, which include cardiac output (see section 19.9), resistance (see section 20.5b), and blood volume (see section 0.6). These variables and their influence on blood pressure are integrated in figure 20.16. The kidneys influence blood pressure by altering urine output. An increase in urine output decreases blood volume (and thus blood pressure), whereas a decrease in urine output helps to maintain blood volume (and thus blood pressure).

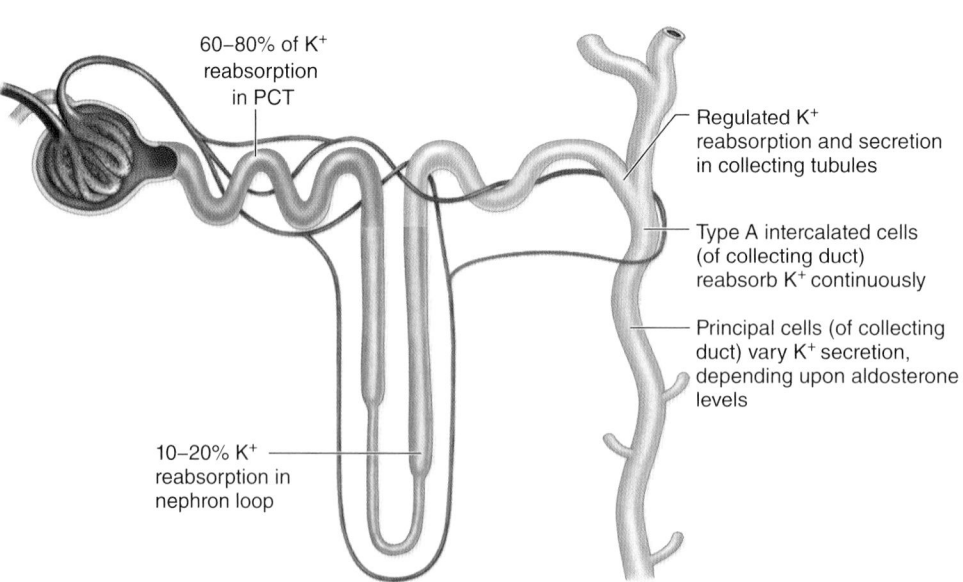

60–80% of K+ reabsorption in PCT

Regulated K+ reabsorption and secretion in collecting tubules

Type A intercalated cells (of collecting duct) reabsorb K+ continuously

Principal cells (of collecting duct) vary K+ secretion, depending upon aldosterone levels

10–20% K+ reabsorption in nephron loop

Figure 24.20 Potassium Movement. Potassium is both reabsorbed and secreted along different segments of the nephron tubule, collecting tubule, and collecting duct. The amount of K+ lost in the urine is dependent upon the activity of principal cells in the collecting tubules and ducts in response to aldosterone.

In the proximal convoluted tubule, 60% to 80% of the K^+ in the tubular fluid is reabsorbed by paracellular transport; it is dependent upon the movement of Na^+, as follows:

1. Sodium is reabsorbed across the luminal membrane.
2. Water follows the Na^+.
3. The concentration of the remaining solutes in the tubular fluid increases as water follows the movement of Na^+.
4. Consequently, the solute concentration of tubular fluid is greater than in the interstitial fluid, creating a gradient between the tubular fluid and interstitial fluid.
5. Potassium moves down its concentration gradient from the tubular fluid by the paracellular route.
6. These conditions also allow the passive reabsorption of other solutes, including other cations (Mg^{2+}, Ca^{2+}), phosphate ion (PO_4^{3-}), fatty acids, and urea.

Approximately 10% to 20% of the K^+ in tubular fluid is reabsorbed in the thick segment of the nephron loop ascending limb by both transcellular and paracellular transport.

Either a net reabsorption or net secretion of K^+ may occur in the collecting tubule. The two specialized cells in the collecting tubule, intercalated cells and principal cells, have opposite effects on K^+ movement. Type A intercalated cells reabsorb K^+ continuously, whereas principal cells secrete K^+ at varying rates based on aldosterone levels.

We have seen that aldosterone stimulates principal cells to secrete K^+, while Na^+ and water are reabsorbed. The most powerful stimulant to cause release of aldosterone from the adrenal cortex is an elevated K^+ blood level. Thus, this mechanism involves negative feedback to keep blood K^+ levels close to optimum values.

Calcium and Phosphate Balance

Calcium and phosphate are two substances generally considered together because 99% of the body calcium is stored in bone, and the majority is stored as calcium phosphate. Approximately 50% of the Ca^{2+} in blood becomes part of the filtrate and then the tubular fluid. The remainder of the Ca^{2+} is bound to protein in the blood and is prevented from being filtered. In comparison, 90% to 95% of the PO_4^{3-} is filtered as blood passes through glomerular capillaries.

The amount of Ca^{2+} and PO_4^{3-} excreted in the urine is regulated by parathyroid hormone (PTH), and thus it influences blood levels of both Ca^{2+} and PO_4^{3-}. Recall from section 7.6b that PTH is released from the parathyroid gland in response to decreased blood Ca^{2+}. One target of PTH is the kidney tubules (figure 24.21). PTH inhibits PO_4^{3-} reabsorption in the proximal convoluted tubule and it stimulates Ca^{2+} reabsorption in the distal convoluted tubule. As additional PO_4^{3-} is eliminated via the urine, less PO_4^{3-} is available to form calcium phosphate, the major calcium salt deposited in bone. Thus, Ca^{2+} redeposited in the bone decreases, and blood Ca^{2+} increases.

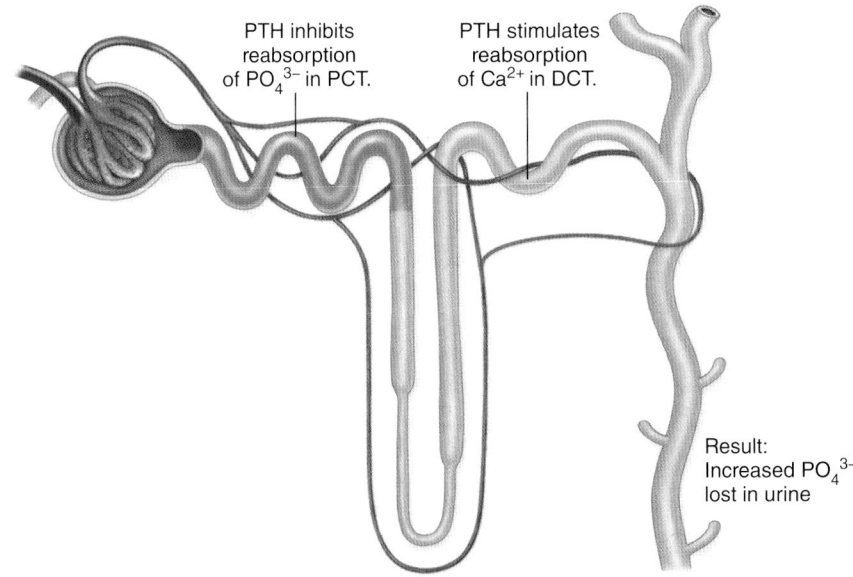

PTH inhibits reabsorption of PO_4^{3-} in PCT.

PTH stimulates reabsorption of Ca^{2+} in DCT.

Result: Increased PO_4^{3-} lost in urine

Figure 24.21 PTH Regulation of Calcium Ion and Phosphate Ion Reabsorption. Calcium and phosphate ions are both reabsorbed in various segments of the tubule. The amounts reabsorbed are regulated by parathyroid hormone (PTH), which both inhibits the reabsorption of PO_4^{3-} in the proximal convoluted tubule (PCT) and stimulates Ca^{2+} reabsorption in the distal convoluted tubule (DCT).

Bicarbonate Ions, Hydrogen Ions, and pH

The movement of bicarbonate ions (HCO_3^-, a weak base) and hydrogen ions (H^+, an acid) play a significant role in regulating pH of both urine and the blood. Bicarbonate ions move freely across the filtration membrane, whereas only small amounts of H^+ are filtered with most remaining in the blood. (H^+ is not generally filtered because it is attached to chemical structures in the blood, such as proteins.) The filtered HCO_3^- must be reabsorbed to ensure blood pH does not become too acidic (**figure 24.22a**). Approximately 80% to 90% of HCO_3^- is "reclaimed" from the tubular fluid, primarily in the proximal convoluted tubule. The remaining 10% to 20% is taken up from the thick segment of the ascending limb of the nephron loop. Thus, as the tubular fluid enters the distal convoluted tubule, usually 100% of the HCO_3^- originally in the filtrate has been reabsorbed. Note that filtered HCO_3^- is not specifically reabsorbed but rather is "replaced" by a process that is described in figure 24.22b.

The pH of urine, and consequently the pH of blood, are regulated in the collecting tubules. Exactly how this occurs depends upon the blood concentration of H^+, which is expressed as $[H^+]$. Increased $[H^+]$ typically occurs, for example, in individuals consuming a more acidic diet, which is a diet that includes animal protein and wheat (a typical American diet). As a result, newly synthesized HCO_3^- is reabsorbed into the blood, and H^+ is secreted into the tubular fluid, by *type A intercalated cells*. The result is an increase in blood pH (more alkaline) and a decrease in urine pH (more acidic), which averages a pH of about 6.0. The details of this process are described in figure 24.22c.

Decreased blood $[H^+]$ is more typical of individuals consuming a more alkaline diet, which is a diet high in fruits and vegetables and little or no animal protein. In this case, *type B intercalated cells* are active (not shown in the figure). The action of type B cells is the reverse of that of type A cells. Think of type B intercalated cells as "flipped" type A cells: Type B cells reabsorb H^+ and secrete HCO_3^- to lower blood pH and increase urine pH.

The actions of the type A and type B intercalated cells in regulating blood pH are a significant component of how the body maintains pH balance (see section 25.5b).

WHAT DID YOU LEARN?

28 How does Na^+ reabsorption occur? Which two hormones are involved?

29 What is the effect of parathyroid hormone on the reabsorption of both PO_4^{3-} and Ca^{2+}?

30 How is the movement of H^+ and HCO_3^- regulated by type A and type B intercalated cells?

24.6e Substances Eliminated as Waste Products

LEARNING OBJECTIVES

10. Identify the three nitrogenous waste products, and describe the fate of each.

11. Give examples of other materials eliminated by kidneys.

The urinary system prevents the accumulation of metabolic waste, various hormones and their metabolites, and foreign substances (e.g., drugs and chemicals) by eliminating them in the urine. Materials to be excreted in the urine are filtered at the glomerulus, secreted along the tubule pathway to ensure their elimination, or both filtered and secreted.

Elimination of Nitrogenous Waste

Nitrogenous waste is metabolic waste that contains nitrogen. The body's main nitrogenous waste products are (1) **urea,** a small, water-soluble molecule produced from protein breakdown in the liver; (2) **uric acid,** produced from nucleic acid breakdown primarily in the liver and (3) **creatinine,** (krē-at′i-nēn, -nin) produced from metabolism of creatine in muscle tissue.

Urea and uric acid are both reabsorbed and secreted, whereas creatinine is only secreted. Here we describe the processing of urea. Urea levels in the blood range from 3 milliequivalents per liter (mEq/L) to 9 mEq/L and depend upon protein intake. Urea is freely filtered and is both reabsorbed and secreted in different areas of the tubules. Approximately half is reabsorbed at the proximal convoluted tubule by the paracellular movement but is then secreted back into the tubule at the nephron loop by urea uniporters. Thus, 100% of the filtered urea is present in the tubular fluid as it enters the distal convoluted tubules.

However, about 50% of the urea is then reabsorbed at the collecting tubules, leaving about 50% of the filtered urea to be excreted in the urine. In addition to being eliminated as a waste product, urea serves a functional role in the kidneys by helping to establish the concentration gradient in the interstitial fluid of the kidney (described in section 24.6f).

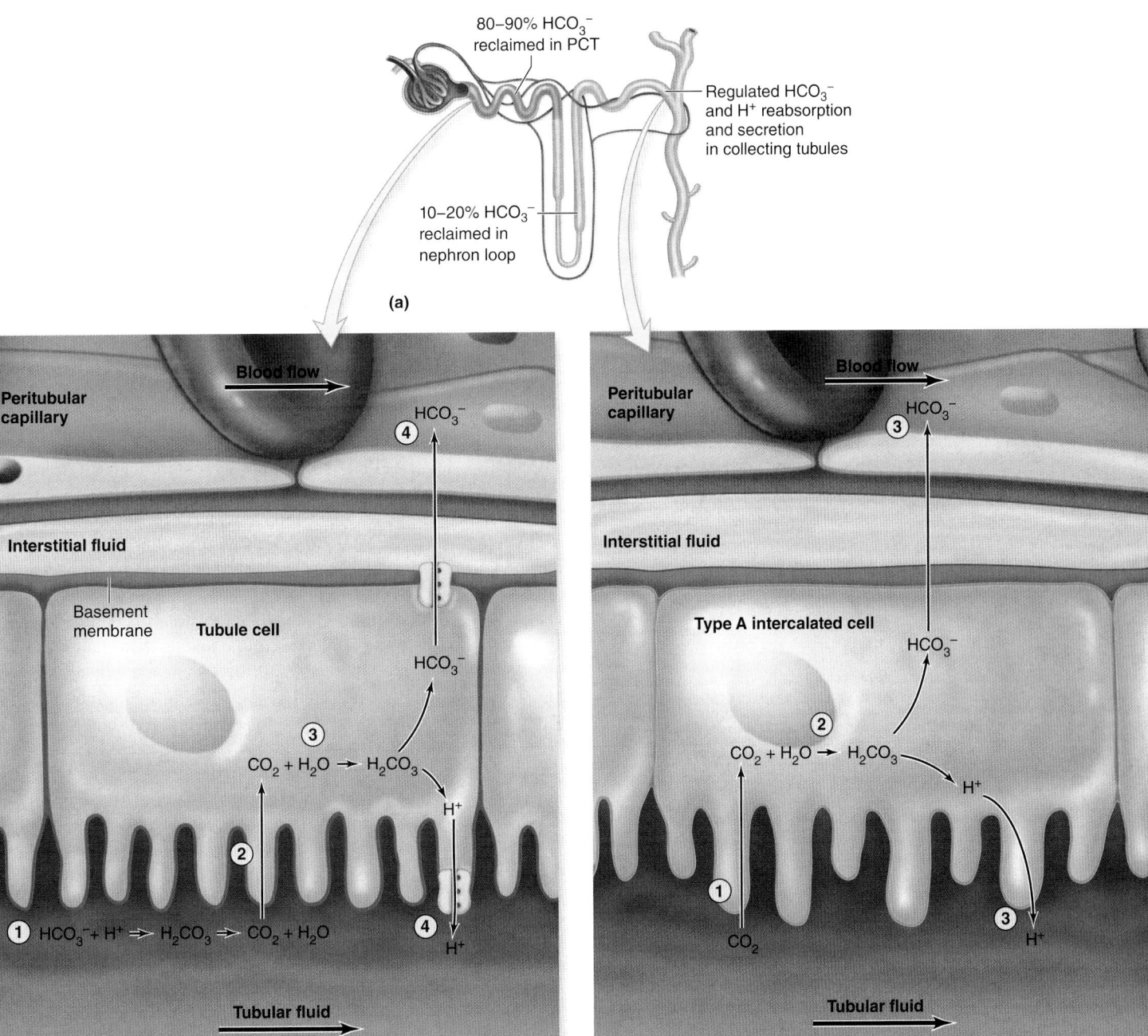

(a)

(b) **Lumen of proximal convoluted tubule and nephron loop**

① Within filtrate, HCO_3^- combines with H^+ to form carbonic acid (H_2CO_3), which is unstable and dissociates to form CO_2 and H_2O.

② CO_2 enters tubule cell by simple diffusion.

③ Within tubule cell, CO_2 combines with H_2O and the process is reversed to form HCO_3^- and H^+.

④ HCO_3^- is absorbed into the blood, and H^+ is secreted into the tubular fluid.

Net effect is "reclaiming" the HCO_3^- that was just filtered and "losing" H^+ in urine (this always occurs).

(c) **Lumen of collecting tubules: Type A intercalated cells**

① CO_2 in tubular fluid enters tubule cell.

② CO_2 combines with H_2O to form HCO_3^- and H^+.

③ HCO_3^- is absorbed into the blood and H^+ is secreted into the tubular fluid.

Net effect is the "capturing" of newly formed HCO_3^- and "losing" H^+ in urine (occurs with an increase in blood H^+ levels).

Figure 24.22 Bicarbonate Ion and Hydrogen Ion Movement. (*a*) Bicarbonate movement occurs along the entire length of the nephron. (*b*) In the proximal convoluted tubule (PCT), approximately 80–90% of the filtered HCO_3^- is "replaced" by an indirect mechanism. The remainder of filtered HCO_3^- is absorbed in the nephron loop (thick segment of ascending limb). (*c*) In the collecting tubules, when $[H^+]$ is elevated, type A intercalated cells synthesize a new molecule of HCO_3^- and an H^+. Bicarbonate ion is absorbed into blood, and H^+ is secreted into tubular fluid. Type B cells (not shown) are essentially "flipped" type A cells. Type B cells function when $[H^+]$ levels are decreased by secreting HCO_3^- and absorbing H^+.

Elimination of Drugs and Bioactive Substances

The kidneys eliminate potentially harmful substances through both filtration and secretion in addition to eliminating nitrogenous wastes. Most secretion of these and other substances (except K^+) occur in the proximal convoluted tubule and include the following:

- **Certain drugs.** Antibiotics including penicillin and sulfonamides, aspirin (salicylate), morphine, chemotherapy drugs, saccharin, and chemicals in marijuana are just a few examples of the drugs that are eliminated in the urine.

- **Other metabolic wastes.** Urobilin (bilirubin breakdown product; see section 18.3b) and hormone metabolites are examples of metabolic wastes that are eliminated in urine.

- **Some hormones.** Human chorionic gonadotropin (hCG) is a hormone produced when a woman is pregnant: It is an example of a hormone that is eliminated in the urine (see section 29.2c). The presence of hCG in the urine can be used to verify the pregnancy with a pregnancy test completed at home or in the doctor's office. Other examples of excreted hormones include epinephrine and prostaglandins.

WHAT DID YOU LEARN?

31 What are some substances eliminated in the urine?

24.6f Establishing the Concentration Gradient

LEARNING OBJECTIVES

12. Explain what is meant by the countercurrent multiplier that occurs within the nephron loop.
13. Describe the countercurrent exchange system that maintains the concentration gradient.
14. Discuss the contribution of urea cycling to the concentration gradient.

You previously learned that the action of antidiuretic hormone (ADH) is dependent upon the concentration gradient in the interstitial fluid surrounding renal tubules. This concentration gradient is established by various solutes, including Na^+ and chloride ion (Cl^-), which progressively increase in concentration from the cortex into the medulla. The concentration gradient exerts osmotic pull to move more water from the tubular fluid into the interstitial fluid when ADH is present because ADH increases the number of aquaporins in the collecting tubules and ducts. The water then moves into the capillaries.

This section describes how the concentration gradient in the interstitial fluid is established and maintained by (1) the nephron loop via the countercurrent multiplier, (2) the vasa recta via the countercurrent exchange system, and (3) urea recycling.

The Nephron Loop

A positive feedback mechanism called the **countercurrent multiplier** involves the nephron loop and is partially responsible for establishing the salt concentration gradient within the interstitial fluid **(figure 24.23a)**. "Countercurrent" refers to the tubular fluid's "reversing" its relative direction as it moves first through the descending limb and then through the ascending limb of the loop. "Multiplier" refers to the positive feedback loop that increases the concentration of salts (e.g., Na^+ and Cl^-) within the interstitial fluid. The juxtamedullary nephrons with their long nephron loops are primarily important in this process.

The *descending* limb is permeable to water. It is also impermeable to the movement of salts out of the tubule. Water is moved from the tubular fluid into the interstitial fluid as a result, whereas salts are retained within the tubular fluid. (Note that solute concentration in the tubule increases from 300 mOsm to as much as 1200 mOsm—by the loss of water and retention of salt.)

In contrast to the descending limb, the *ascending* limb is impermeable to water, and it actively pumps salt out of the tubular fluid into the interstitial fluid. As a result, water is retained in the tubular fluid, and salt is moved from the tubular fluid into the interstitial fluid. (Note that the solute concentration becomes diluted from 1200 mOsm to 100 mOsm as it moves through the ascending limb.) Most importantly,

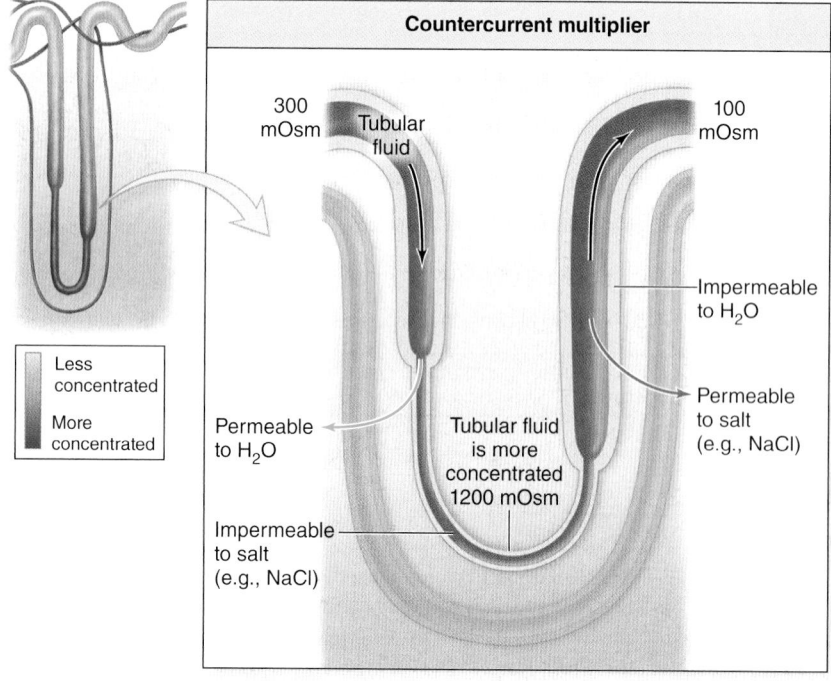

Countercurrent multiplier

300 mOsm — Tubular fluid

100 mOsm

Impermeable to H_2O

Permeable to salt (e.g., NaCl)

Less concentrated

More concentrated

Permeable to H_2O

Tubular fluid is more concentrated 1200 mOsm

Impermeable to salt (e.g., NaCl)

(a) Nephron loop

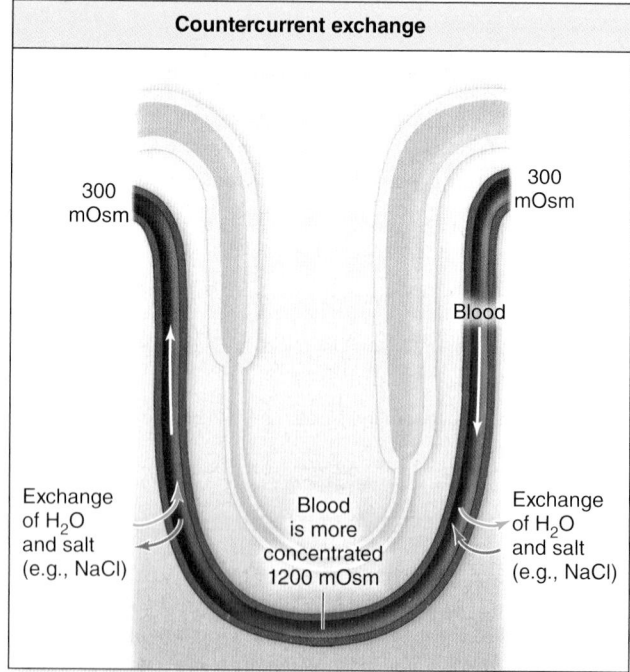

Countercurrent exchange

300 mOsm

300 mOsm

Blood

Exchange of H_2O and salt (e.g., NaCl)

Blood is more concentrated 1200 mOsm

Exchange of H_2O and salt (e.g., NaCl)

(b) Vasa recta

Figure 24.23 Interstitial Fluid Concentration Gradient. A visual representation of how the concentration gradient is established in the interstitial fluid surrounding the nephron loop. (*a*) Countercurrent multiplier refers to the actions of the nephron loop. (*b*) Countercurrent exchange system involves the processes of the vasa recta. These processes occur simultaneously.

INTEGRATE

CLINICAL VIEW
Diuretics

Diuretics (dī-ū-ret′ik) are substances that increase excretion of urine. They are a common treatment for hypertension because by increasing urine output, both blood volume and blood pressure are decreased.

Diuretic drugs are grouped into different classes depending upon their mechanism of action. **Loop diuretics** such as furosemide (Lasix) and bumetanide target the transport protein (Na$^+$/K$^+$/2 Cl$^-$ transporters) that pump the Na$^+$ salts within the nephron loop and lower the concentration gradient of the interstitial fluid, so less osmotic force is present to pull fluid out of the collecting tubule and collecting ducts, and more fluid is excreted in urine. **Thiazides** such as chlorothiazide and hydrochlorothiazide interfere with Na$^+$ and water reabsorption in the distal convoluted tubule. Thus, less Na$^+$ and water are reabsorbed, increasing urine output.

however, is that the salt concentration in the interstitial fluid surrounding the tubules increases, which sets up a positive feedback loop.

Consider that as the salt concentration in the interstitial fluid increases (1) an even greater amount of water moves out of the descending limb, (2) this increases the salt concentration in the tubular fluid that flows into the ascending limb, which (3) allows even more salts to be pumped out of the ascending limb. Consequently, the concentration of salts in the interstitial fluid is increased or "multiplied" through this positive feedback loop.

Vasa Recta

The concentration gradient established in the interstitial fluid by the nephron loop is maintained or preserved by the unique anatomic arrangement and physiologic conditions of the vasa recta (figure 24.23*b*). The blood within the vasa recta travels in the *opposite* direction to the tubular fluid flow within the adjacent nephron loop. So, the blood flows deep into the medulla alongside the *ascending* limb of the nephron loop, and it then flows toward the cortex alongside the *descending* limb of the nephron loop.

The process by which the vasa recta help maintain the concentration gradient is termed the **countercurrent exchange system.** Recall that "countercurrent" refers to flow in opposite directions—in this case, blood flows in opposite directions within neighboring, parallel capillaries. The term "exchange" refers to the passive exchange of both salts and water between the blood and the interstitial fluid around the nephron loop.

The countercurrent exchange process occurs as follows:

- As the blood flows through the vasa recta deep into the renal medulla alongside the ascending limb, water moves by osmosis out of these capillaries into the more concentrated interstitial fluid. At the same time, salts in the interstitial fluid enter the vasa recta by diffusion down their concentration gradients. Thus, the blood in the vasa recta is losing water and gaining salts, and the concentration of total salt in the blood increases. Thus, as blood within the vasa recta travels into the deepest part of the medulla, it becomes more and more concentrated.

- If the vasa recta were to continue deep into the medulla, these salts would be transported away in the blood. However, the path of the blood flow in the vasa recta makes a 180-degree turn and is positioned alongside the descending limb of the nephron loop

toward the cortex. Salts are now transported into a region in which osmotic and solute gradients reverse. Here, the salts diffuse back out of the blood into the interstitial fluid, while water moves into the vasa recta.

Thus, as blood within the vasa recta travels toward the cortex, it becomes less and less concentrated. In fact, the blood within the vasa recta returning to the renal cortex has approximately the same concentration (or is slightly less concentrated) from when it first left the cortex. Most importantly, it leaves the salts responsible for the concentration gradient in the interstitial fluid.

Urea Recycling

An important and additional contribution to maintaining the concentration gradient in the interstitial fluid occurs with **urea recycling.** Recycled urea in fact makes up approximately one-half of the solutes of the interstitial fluid concentration gradient. Urea is removed from the tubular fluid in the collecting duct by urea uniporters as mentioned previously; it diffuses back into the tubular fluid in the thin segment of the ascending limb. Because both the thick segment of the ascending limb and DCT are not permeable to urea, urea remains within the tubular fluid until it reaches the collecting duct, where it is removed from the tubular fluid. Thus, urea is "cycled" between the collecting tubule and nephron loop. Some of this urea remains in the interstitial fluid, contributing to its concentration gradient.

Summary of Reabsorption and Secretion

The processes by which the nephron tubules, collecting tubules, and collecting ducts transform filtrate into urine in the different components of the nephron can be summarized and integrated, as follows **(figure 24.24):**

1. After filtration of plasma occurs, a majority of most other substances are either reabsorbed or secreted into the proximal convoluted tubules. Additional amounts of substances are reabsorbed as the tubular fluid moves through the nephron loop.

2. The nephron loop and the surrounding vasa recta are, along with the recycling of urea between the collecting duct and nephron loop, responsible for establishing the concentration gradient of the interstitial fluid. This gradient is necessary for the normal function of ADH.

3. The regulation of the amount of specific substances excreted in urine is controlled mainly by principal cells and by type A and type B intercalated cells within the distal convoluted tubule, collecting tubules, and collecting ducts. The fluid leaving the collecting ducts at the renal papilla is called urine.

4. Urine is composed of water and various dissolved substances including ions (e.g., Na$^+$, K$^+$, and Cl$^-$), various waste products, and perhaps drugs. Urine is drained into the renal sinus of the kidney to be excreted by the urinary tract. Under normal circumstances, urine does not contain either formed elements of the blood (erythrocytes, leukocytes, platelets) because they are not filtered or nutrients because they are 100% reabsorbed in the PCT.

WHAT DID YOU LEARN?

32 How is the concentration gradient that is essential for normal function of ADH in water reabsorption established and maintained?

33 Which substances are reabsorbed in tubular processing in the different regions of the nephron? Which are secreted? What are the general processes that occur in the different regions of the tubules: (a) PCT, (b) nephron loop, and (c) DCT, CT, and CD?

Figure 24.24 **Tubular Reabsorption and Tubular Secretion.** Various reabsorption and secretion processes occur along the (*a*) proximal convoluted tubule, (*b*) nephron loop, (*c*) distal convoluted tubule, collecting tubules, and collecting ducts.

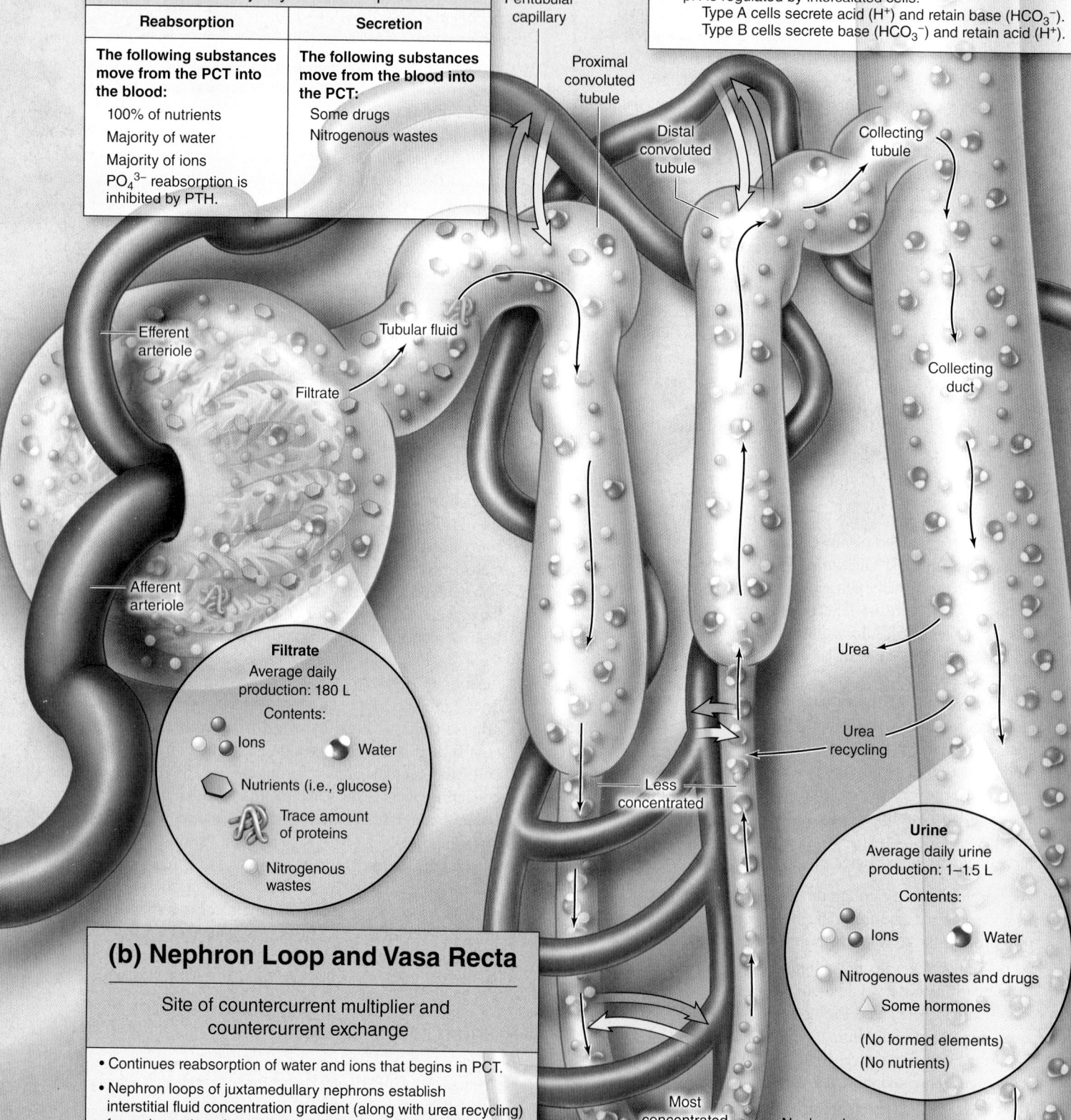

(a) Proximal Convoluted Tubule

Site for majority of reabsorption

Reabsorption	Secretion
The following substances move from the PCT into the blood:	The following substances move from the blood into the PCT:
100% of nutrients	Some drugs
Majority of water	Nitrogenous wastes
Majority of ions	
PO_4^{3-} reabsorption is inhibited by PTH.	

(c) Distal Convoluted Tubule, Collec Tubule, and Collecting Duct

Sites of regulation

- Na^+ reabsorption is regulated by aldosterone and ANP.
- Water reabsorption is regulated by aldosterone and ADH.
- Amount of K^+ secreted into the tubular fluid is dependent upon both intercalated cells and principal cells.
- Ca^{2+} reabsorption is increased by parathyroid hormone (PTH)
- pH is regulated by intercalated cells:
  Type A cells secrete acid (H^+) and retain base (HCO_3^-).
  Type B cells secrete base (HCO_3^-) and retain acid (H^+).

Peritubular capillary

Proximal convoluted tubule

Distal convoluted tubule

Collecting tubule

Collecting duct

Efferent arteriole

Tubular fluid

Filtrate

Afferent arteriole

Filtrate
Average daily production: 180 L
Contents:
Ions
Water
Nutrients (i.e., glucose)
Trace amount of proteins
Nitrogenous wastes

Urea

Urea recycling

Less concentrated

Urine
Average daily urine production: 1–1.5 L
Contents:
Ions
Water
Nitrogenous wastes and drugs
Some hormones
(No formed elements)
(No nutrients)

(b) Nephron Loop and Vasa Recta

Site of countercurrent multiplier and countercurrent exchange

- Continues reabsorption of water and ions that begins in PCT.
- Nephron loops of juxtamedullary nephrons establish interstitial fluid concentration gradient (along with urea recycling) for reabsorption of water induced by ADH.

Most concentrated

Nephron loop

Vasa recta

24.7 Evaluating Kidney Function

Some means of determining the effectiveness of kidney function is necessary, given the critical role the kidneys play in regulating concentrations of various substances within the blood and in eliminating wastes and foreign substances. This testing is especially crucial while monitoring and treating kidney disease.

24.7a Measuring Glomerular Filtration Rate

LEARNING OBJECTIVES

1. Describe the procedure for measuring the glomerular filtration rate.
2. Explain the formula for calculating the glomerular filtration rate.

One way to assess kidney function is to measure the rate filtrate is formed per unit time—that is, the glomerular filtration rate (GFR). To conduct this test, an individual is injected with *inulin,* a polysaccharide derived from plants that is freely filtered and neither reabsorbed nor secreted in the kidney, so the amount in the urine is equal to the amount that is filtered. (Inulin should not be confused with the hormone insulin, which regulates blood glucose levels.)

Enough inulin is injected into a subject to achieve a blood plasma concentration of 1 mg/mL. Urine is collected and measured for volume and concentration of the inulin. Additionally, blood is drawn and the plasma concentration of inulin is measured at given time intervals. Glomerular filtration rate is determined by the following formula:

$$GFR = \frac{UV}{P}$$

where U = concentration of inulin in urine, V = volume of urine produced per minute, and P = concentration of inulin in plasma. For example, if the concentration of inulin in urine = 125 mg/mL, the volume of urine = 1 mL/min, and the concentration of inulin in plasma = 1 mg/mL, then

$$GFR = \frac{\frac{125 \text{ mg}}{mL} \times \frac{1 \text{ mL}}{min}}{\frac{1 \text{ mg}}{mL}} = 125 \text{ mL/min}$$

Normal adult GFR is 125 mL/min. A lower glomerular filtration rate indicates a decrease in kidney function, and thus is more likely that nitrogenous wastes and other unwanted substances are accumulating in the blood.

WHAT DID YOU LEARN?

34 What is the purpose of measuring the glomerular filtration rate?

24.7b Measuring Renal Plasma Clearance

LEARNING OBJECTIVES

3. Define renal plasma clearance and its importance.
4. Identify the substance that may be measured to estimate the glomerular filtration rate.

Another means of assessing kidney function is by measuring renal plasma clearance. The **renal plasma clearance** test measures the volume of plasma that can be completely cleared of a substance in a given period of time—usually in 1 minute. We may infer from this test whether a substance is reabsorbed or secreted. If a substance is neither reabsorbed nor filtered (like inulin), its renal plasma clearance is equal to the GFR (125 mL/min). However, if a substance is reabsorbed, its renal plasma clearance is lower than GFR because less of the substance is excreted, or "cleared," in the urine.

For example, the renal plasma clearance of urea is 70 mL/min. If urea is filtered at a rate of 125 mL/min (the normal GFR), and only 70 mL/min is cleared, the rest (55 mL/min) is reabsorbed. In contrast, the renal plasma clearance of glucose is normally 0 mL/min because in a healthy individual 100% of the glucose is reabsorbed and none is excreted in the urine.

Substances that are filtered and secreted have renal plasma clearance values higher than the GFR. This occurs because additional amounts of the substance are secreted into the tubular fluid

INTEGRATE

CLINICAL VIEW
Renal Failure, Dialysis, and Kidney Transplant

Renal Failure

Renal failure refers to greatly diminished or absent renal functions caused by the destruction of about 90% of the kidney. Renal failure often results from a chronic disease that affects the glomerulus or the small blood vessels of the kidney, as a result of autoimmune conditions, high blood pressure, or diabetes. Once the kidney's structures have been destroyed, there is no chance they will regenerate or begin functioning again. Thus, the two main treatments are dialysis or kidney transplant.

Dialysis

The term **dialysis** (dī-al'i-sis) comes from a Greek word meaning "to separate agents or particles on the basis of their size." Two forms of dialysis are commonly used today: peritoneal dialysis and hemodialysis. In **peritoneal dialysis,** a catheter is permanently placed into the peritoneal cavity, to which a bag of dialysis fluid may be attached externally. A volume of a special dialysis fluid is introduced, and the harmful waste products in the blood are transferred, or dialyzed, into the fluid across the peritoneal membrane. After several hours, the fluid is drained from the peritoneal cavity and replaced with fresh fluid.

In **hemodialysis,** the patient's blood is cycled through a machine that filters the waste products across a specially designed membrane. A vascular connection is made between a conveniently located superficial artery and vein. This connection is called a shunt and is externally accessible. The patient must remain stationary for the time it takes to cycle the blood through the dialysis unit while metabolic waste products are removed. Hemodialysis must be performed three to four times a week and each treatment takes about 4 hours.

Kidney Transplant

A **kidney transplant** from a genetically similar person ("matched" for major histocompatability complex; see section 22.4c) may successfully

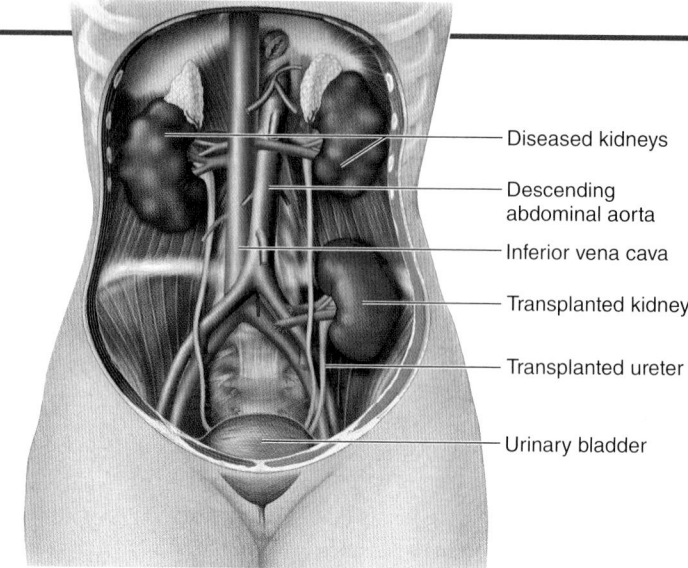

Location of a transplanted kidney in the abdominopelvic cavity.

Diseased kidneys
Descending abdominal aorta
Inferior vena cava
Transplanted kidney
Transplanted ureter
Urinary bladder

restore renal function. The kidney is generally removed from the donor by a laparoscopic procedure. A recent development for obtaining the kidney is to remove the kidney through the umbilicus (navel, or belly button) after making a single small incision. This decreases the donor's recovery time from approximately 3 months to just under a month.

The replacement kidney is attached to an artery and vein in the inferior abdominopelvic region, where it is relatively easy to establish a vascular connection. The new kidney rests either on the superior surface of, or immediately lateral to, the urinary bladder. Because a pelvic artery and vein connect the donor kidney to the patient's blood supply, having the kidney near the bladder means only a short segment of ureter is needed for the bladder connection. The diseased kidney is not removed.

The transplanted kidney is a foreign tissue. Therefore, immunosuppressant drugs are regularly administered to suppress the immune system's activity. The recipient may need to take these drugs for the remainder of his or her life.

and excreted in the urine. For example, creatinine has a renal plasma clearance of 140 mL/min, indicating that the substance is both filtered and secreted.

Many medications also have renal plasma clearances higher than the GFR because they are secreted. For this reason, it is important to determine renal plasma clearance to know the amount and timing of drug dosage. The higher the renal plasma clearance, the more often the medications must be given to maintain therapeutic levels.

Note that in clinical practice, renal plasma clearance of creatinine can be used to approximate glomerular filtration rate because its clearance is only slightly higher than GFR. Measuring creatinine clearance avoids the need to inject inulin into the patient's blood.

 WHAT DID YOU LEARN?

35 What information is gained by measuring the renal plasma clearance for a specific substance (e.g., medication)?

24.8 Urine Characteristics, Transport, Storage, and Elimination

Urine is the fluid that exits the kidneys and passes into the ureters. Thereafter, it is transported through the ureters into the urinary bladder, where it is stored until it can be eliminated from the body through the urethra. The process of eliminating urine is referred to as micturition or urination. Here we describe the characteristics of urine, the components of the urinary tract, and the process of micturition.

24.8a Characteristics of Urine

 LEARNING OBJECTIVES

1. Describe the composition of urine and its characteristics.
2. Explain what is meant by specific gravity.

Table 24.3　Abnormal Urinalysis

Abnormal Urine Constituent	Clinical Term	Possible Cause(s)
Glucose	Glucosuria or glycosuria	Diabetes mellitus
Ketones (acetone, acetoacetic acid, beta-hydroxybutyric acid)	Ketonuria	Diabetes mellitus, starvation (including anorexia), low-carbohydrate diet (e.g., Atkins, South Beach); ketones produced as by-products of fatty acid metabolism
Protein (Note that trace amounts of protein [5–10 mg/dL] are not clinically significant)	Proteinuria	Increased glomerular permeability caused by kidney trauma, heavy metals, bacterial toxins, exertion (e.g., running a marathon), hypertension, cold exposure, glomerular nephritis
Bile pigment	Bilirubinuria	Liver pathology (e.g., hepatitis or cirrhosis), obstructed bile duct due to gallstones
Erythrocytes	Hematuria	Glomerular damage, kidney trauma, pathology along the urinary tract (such as occurs with urinary tract stones), contamination from menstrual flow
Hemoglobin (Hb)	Hemoglobinuria	Accelerated rate of erythrocyte destruction (e.g., hemolytic anemia or a transfusion reaction), burns, renal damage, contamination from menstrual flow
Leukocytes	Pyuria	Urinary tract infection, acute glomerulonephritis, contamination from female reproductive tract
Nitrites	Nitrituria	Urinary tract infection because bacteria convert nitrates (NO_3^-) to nitrites (NO_2^-)
Myoglobin	Myoglobinuria	Rhabdomyolysis (a condition where skeletal muscle breaks down; caused by excessive exercise and is an uncommon side effect of statins)

Urine is the product of filtered and processed blood plasma. It typically is a sterile excretion unless contaminated with microbes in the kidney or urinary tract. Urine characteristics include its composition, volume, pH, specific gravity, color and turbidity, and smell.

Typical urine composition is approximately 95% water and 5% solutes. These solutes include salts (Na^+, Cl^-, K^+, Mg^{2+}, Ca^{2+}, SO_4^{2-}, $H_2PO_4^-$, and NH_4^+), nitrogenous wastes (e.g., urea, uric acid, and creatinine), some hormones, drugs, and small amounts of ketone bodies (a waste product of digesting fatty acids). Table 24.3 lists some substances that should not normally be found within urine (i.e., they are abnormal constituents) and possible causes for their presence.

The average daily urine volume is normally 1–2 L. Variations occur due to many variables, including fluid intake, blood pressure, diet, body temperature, use of diuretics, diabetes, and the amount of fluid excreted by other means (e.g., heavy sweating, vomiting). A minimum of about 0.5 L of urine per day is required to eliminate the wastes from the body. If urine production drops below 0.40 L, wastes will accumulate in the blood. Table 24.4 list factors that influence urine volume output.

The normal pH for urine ranges between 4.5 and 8.0; the average value is 6.0, which is slightly acidic. The pH of urine may be affected by diet. It generally decreases to become more acidic when we consume proteins and wheat, and it tends to increase and become more basic with a diet high in fruits and vegetables. Urine pH can also be influenced by other factors, such as metabolism and bacterial infections.

Specific gravity is the density (g/mL) of a substance compared to the density (g/mL) of water. For example, if your urine were composed only of pure water, it would have a specific gravity of 1.000. The average specific gravity of urine is slightly higher, with levels ranging from 1.003 to 1.035 because solutes *are* normal components of urine. Levels vary due to time of day, amount of food and liquids consumed, and amount of exercise. Generally, if you are well hydrated, urine output increases and specific gravity values decrease. A specific gravity value below 1.010 indicates relative hydration, whereas a specific gravity value above 1.020 indicates relative dehydration.

The color of urine ranges from almost clear to dark yellow, depending upon the concentration of pigment from urobilin (bilirubin breakdown product; see section 18.3b). An increased volume of urine generally is lighter in color, and a decreased volume of urine is darker in color. However, intake of some substances (e.g., beets, certain vitamins) can change the color of urine. Normal vaginal secretions, excessive substances in urine (cellular material, protein), crystallization or precipitation of salts if collected and left standing, and bacteria will increase the turbidity (cloudiness) of the urine.

Urinoid is the term used for the normal smell of fresh urine. Urine may develop an ammonia smell if allowed to stand because bacteria convert the nitrogen in urea into ammonia (NH_3). Asparagus and certain other foods can alter the smell of urine. One of the indications that a person has diabetes mellitus is that the urine smells fruity (like acetone, which is found in fingernail polish) from the accelerated production of ketoacids from fatty acid metabolism (see section 3.4h).

Table 24.4　Factors That Influence Urine Volume

Decrease Urine Volume	Increase Urine Volume
Increase in ADH	Decrease in ADH
Increase in aldosterone	Decrease in aldosterone
Decrease in ANP	Increase in ANP
Decrease in fluid intake	Increase in fluid intake
Decrease in blood pressure	Increase in blood pressure
Increase in other fluid output (e.g., sweating, vomiting, diarrhea, hemorrhage)	Diabetes mellitus / Diuretics (e.g., medications, alcohol)

WHAT DID YOU LEARN?

36 What characteristics are used to describe urine? What conditions cause variation in pH of urine?

INTEGRATE

CONCEPT CONNECTION

The inner epithelial lining of the urinary tract (except for distal portions of the male urethra) is composed of transitional epithelium (see section 5.1c). When stretched, the cells forming the transitional epithelium change from cuboidal or polyhedral to become almost squamous in shape. In addition, the transitional epithelium contains tight junctions between the cells (see section 4.6d). This combination of the "stretchable" transitional epithelium and tight junctions function to prevent leakage of urine through the bladder wall.

24.8b Urinary Tract (Ureters, Urinary Bladder, Urethra)

LEARNING OBJECTIVES

3. Describe the structure and function of the ureters.

4. Explain the structure of the urinary bladder.

5. List distinguishing characteristics of the female and male urethra.

The urinary tract consists of the ureters, urinary bladder, and urethra. The urinary tract is responsible for transporting the urine produced in the kidneys, storing the urine until the time and place are right for its elimination from the body, and transporting the urine from storage to the exterior of the body.

Ureters

The **ureters** (ū-rē′ter) are long, epithelial-lined, fibromuscular tubes that transport urine from the kidneys to the urinary bladder **(figure 24.25)**. Each tube averages about 25 centimeters (about 10 inches) in length and is retroperitoneal. The ureters originate from the renal pelvis as it exits the hilum of the kidney and then extend inferiorly to enter the posterolateral wall of the base of the urinary bladder. The wall of the ureter is composed of three concentric tunics. From innermost to outermost, these tunics are the mucosa, muscularis, and adventitia. (Note that the ureter does not have a submucosa layer.)

The **mucosa** is formed by a transitional epithelium (see section 5.1c) that is both distensible and impermeable to the passage of urine. External to the transitional epithelium of the mucosa is the lamina propria, composed of a fairly thick layer of dense irregular connective tissue.

The middle **muscularis** consists of an inner longitudinal and an outer circular layer of smooth muscle cells. The presence of urine within the renal pelvis causes these muscle layers to contract rhythmically to propel the urine through the ureters into the urinary bladder. In areas where the ureter is not filled with urine, the mucosa folds to fill the lumen.

 WHAT DO YOU THINK?

7 Why do the ureters use smooth muscle contraction to actively move urine to the urinary bladder? Why don't they rely only on gravity to move the urine to the inferiorly located bladder?

The external layer of the ureter wall is the **adventitia.** It is composed of a dispersed array of collagen and elastic fibers within areolar connective tissue. Some extensions of this areolar connective tissue layer also anchor the ureter to the posterior abdominal wall.

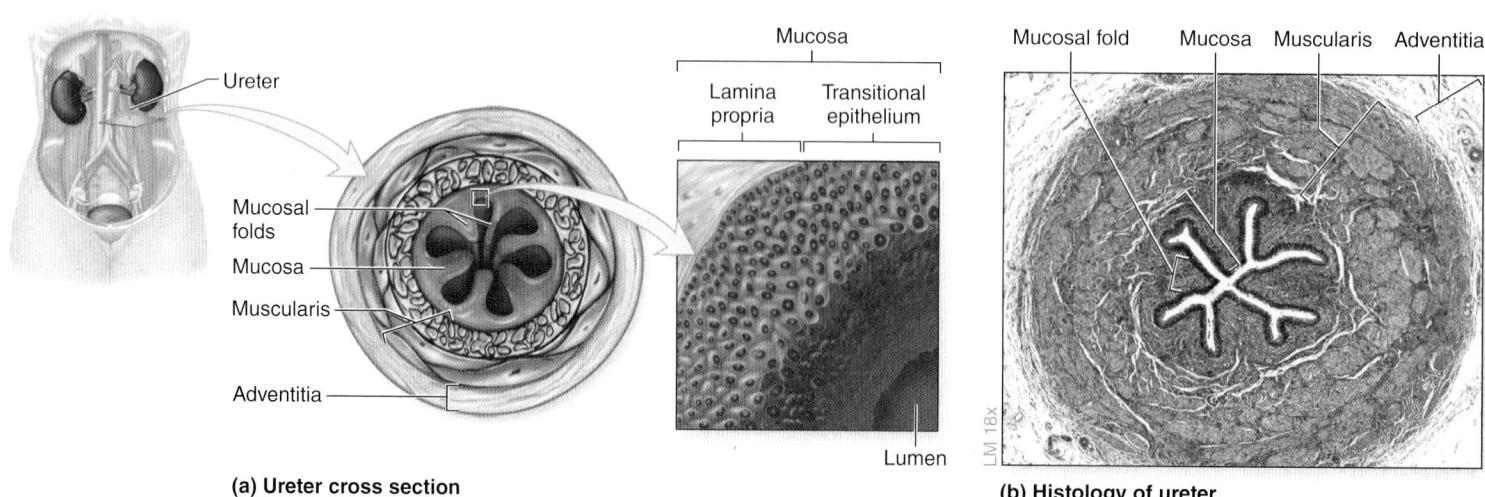

(a) Ureter cross section

(b) Histology of ureter

Figure 24.25 Ureters. The ureters transport urine from the kidneys to the urinary bladder for storage prior to eliminating from the body. (*a*) Features of a ureter in cross-sectional view including its transitional epithelium. (*b*) A photomicrograph of a ureter in cross section shows its mucosal folds and thick muscularis. **AP|R**

INTEGRATE

CLINICAL VIEW
Intravenous Pyelogram

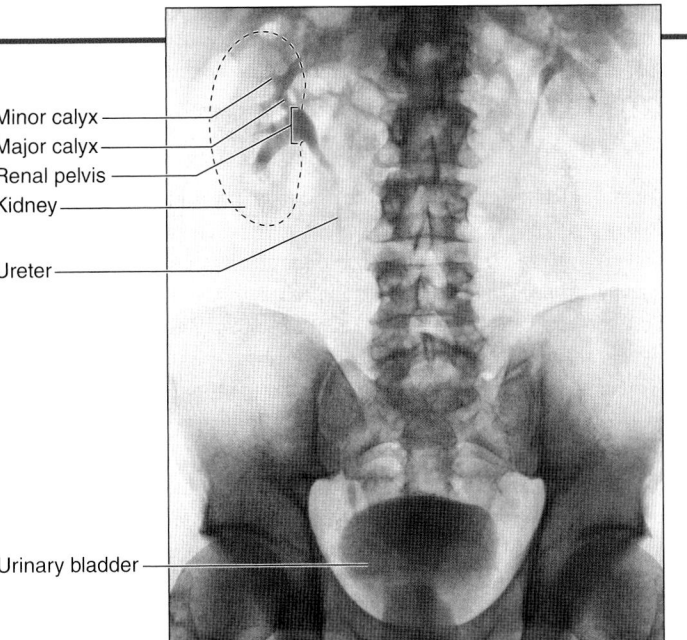

Minor calyx
Major calyx
Renal pelvis
Kidney
Ureter
Urinary bladder

Sometimes a physician needs to visualize the kidneys, ureters, and urinary bladder, especially when the flow of urine from one or both of the kidneys into the bladder becomes blocked. A physician may ask for an x-ray study known as an **intravenous pyelogram** (pī′el-ō-gram; *pyelos* = tub, *gram* = recording), which is produced by injecting a small amount of radiopaque dye into a vein. As the dye passes through the kidneys and is cleared into the urine, a series of sequential abdominal x-rays provide a "time lapse" view of urinary system flow.

An IV pyelogram is useful in diagnosing renal calculi (kidney stones), and may determine if there is abnormal dilation in any parts of the drainage system (including the calyces). If urine flow is normal, the entire pathway of the urinary tract should appear dark on the x-ray. However, if there is a blockage (as is the case with a renal calculus), the dark coloration will suddenly 'stop' at the area of the blockage.

A pyelogram enables physicians to visualize the urinary system organs and identify the location of any blockages within the urinary system.

The ureters project through the bladder wall obliquely. Because of the oblique course of the ureters through the bladder wall, the ureter walls are compressed as the bladder distends, decreasing the likelihood of urine refluxing (backflowing) into the ureters from the bladder when it is emptying.

The ureters are innervated by the autonomic nervous system (see chapter 15). Sympathetic axons extend from the T11–L2 segments of the spinal cord. Pain from the ureter (e.g., caused by a kidney stone lodged in the ureter) is referred to the T11–L2 dermatomes (see figure 16.4). These dermatomes are distributed along a "loin-to-groin" region, so "loin-to-groin" pain typically means ureter or kidney discomfort. The vagus nerve (CN X) innervates the superior region of the ureter, and the pelvic splanchnic nerves innervate the inferior region of the ureter. The effects of this parasympathetic stimulation are unknown.

Urinary Bladder

The **urinary bladder** is an expandable, muscular organ that serves as a reservoir for urine. The bladder is positioned immediately posterior to the pubic symphysis. In females, the urinary bladder is anteroinferior to the uterus and directly anterior to the vagina (see figure 28.3); in males, the bladder is anterior to the rectum and superior to the prostate gland (see figure 28.14). The urinary bladder is a retroperitoneal organ, and only its superior surface is covered with the parietal peritoneum.

When it is empty, the urinary bladder exhibits an inverted pyramidal shape (**figure 24.26**). As it fills with urine, the bladder distends superiorly until it assumes an oval shape.

A posteroinferior triangular area of the urinary bladder wall, called the **trigone** (trī′gōn; *trigonum* = triangle), is formed by imaginary lines connecting the two ureter openings and the urethral opening. The trigone remains immobile as the urinary bladder fills and evacuates. It functions as a funnel to direct urine into the urethra as the bladder wall contracts to evacuate the stored urine. Because the ureters and urethra form the three points of this triangular region, infections are more common within this area.

The four tunics that form the wall of the bladder are the mucosa, submucosa, muscularis, and adventitia. The innermost **mucosa** is adjacent to the bladder lumen; it is formed by a transitional epithelium that accommodates the shape changes occurring with distension, and by a highly vascularized lamina propria that supports the mucosa. Additionally, **mucosal folds** or *rugae* (rū′jē) allow for even greater distension. Within the trigone region, the mucosa is smooth, thick, and lacking mucosal folds. The **submucosa** lies immediately external to the mucosa and is formed by dense irregular connective tissue that supports the urinary bladder wall.

The **muscularis** is formed by three layers of smooth muscle, collectively called the **detrusor** (dē-trū′ser, -sōr; *detrudo* = to drive away) **muscle.** These smooth muscle bundles exhibit such

INTEGRATE

CLINICAL VIEW
Renal Calculi

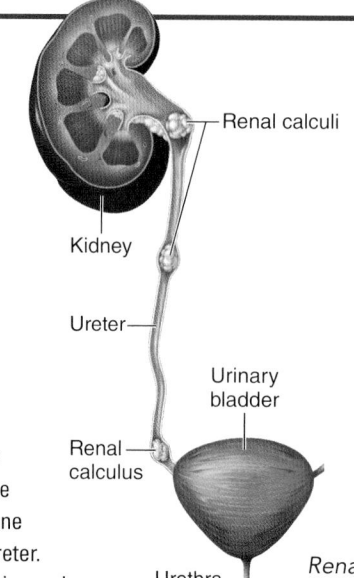

A **renal calculus** (pl., *calculi*), or *kidney stone,* is formed from crystalline minerals that build up in the kidney. Causes and risk factors for kidney stone formation include inadequate fluid intake and dehydration, reduced urinary flow and volume, frequent urinary tract infections, and certain abnormal chemical or mineral levels in the urine. **Hypercalcuria,** a high level of calcium in the urine, can lead to kidney stones composed of calcium. In general, males tend to develop stones more often than females.

Very small renal calculi may be asymptomatic, and the person may excrete them without their being aware of the stones being eliminated in the urine. However, a larger stone can become obstructed in the kidney, renal pelvis, or ureter. Symptoms include severe pain along the "loin-to-groin" region and possibly nausea and vomiting. Following the intake of plenty of water (2 to 3 quarts per day) to assist movement, most stones smaller than about 4 millimeters in diameter eventually pass through the urinary tract on their own. Administration of intravenous fluids may also assist this process.

If the stone is too large to pass on its own, one common treatment is **lithotripsy** (lith′ō-trip-sē; *lithos* = stone, *tresis* = boring), in which sonic shock waves are directed toward the stones to pulverize them into smaller particles that can be expelled in the urine. Alternatively, using **ureteroscopy,** a scope is inserted from the urethra into the urinary bladder and ureter to break up and remove the stone. If these treatments aren't viable, traditional surgery may be required.

Renal calculi may become lodged at various sites along the urinary tract.

---

complex orientations that it is difficult to delineate individual layers in random histology sections. At the neck of the urinary bladder, an involuntary internal urethral sphincter is formed by the smooth muscle that encircles the urethral opening.

The **adventitia** is the outer layer of areolar connective tissue covering the urinary bladder. A peritoneal membrane covers only the superior surface of the urinary bladder.

Urethra

The **urethra** (ū-rē′thră) is an epithelial-lined fibromuscular tube that extends from the urinary bladder from its anteroinferior surface to the urethral opening. The urethra transports urine to the exterior of the body (**figure 24.27**).

Two urethral sphincters restrict the release of urine until the pressure within the urinary bladder is high enough and both involuntary and voluntary activities are activated to release the urine. The

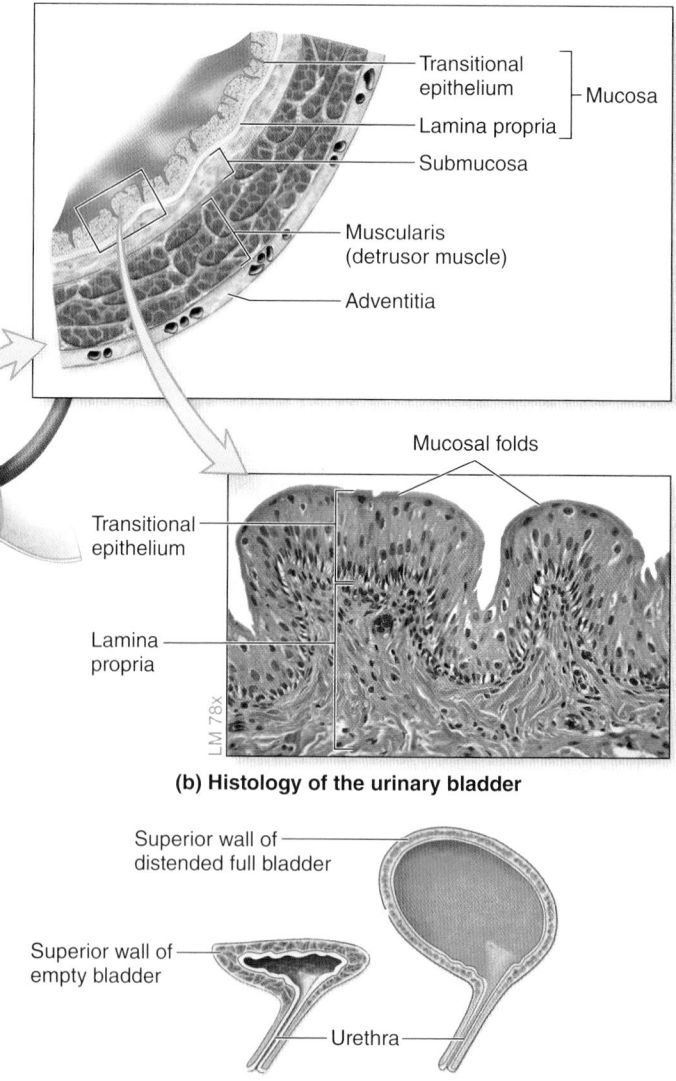

(a) Urinary bladder, anterior view

(b) Histology of the urinary bladder

(c) Urinary bladder, sagittal view

Figure 24.26 Urinary Bladder. (*a*) The urinary bladder is an expandable, muscular sac. This view depicts a female bladder. (*b*) Illustration and photomicrograph of the urinary bladder wall show its tunics. (*c*) Sagittal views show that the urinary bladder expands superiorly and becomes more oval in shape as it fills with urine. **AP|R**

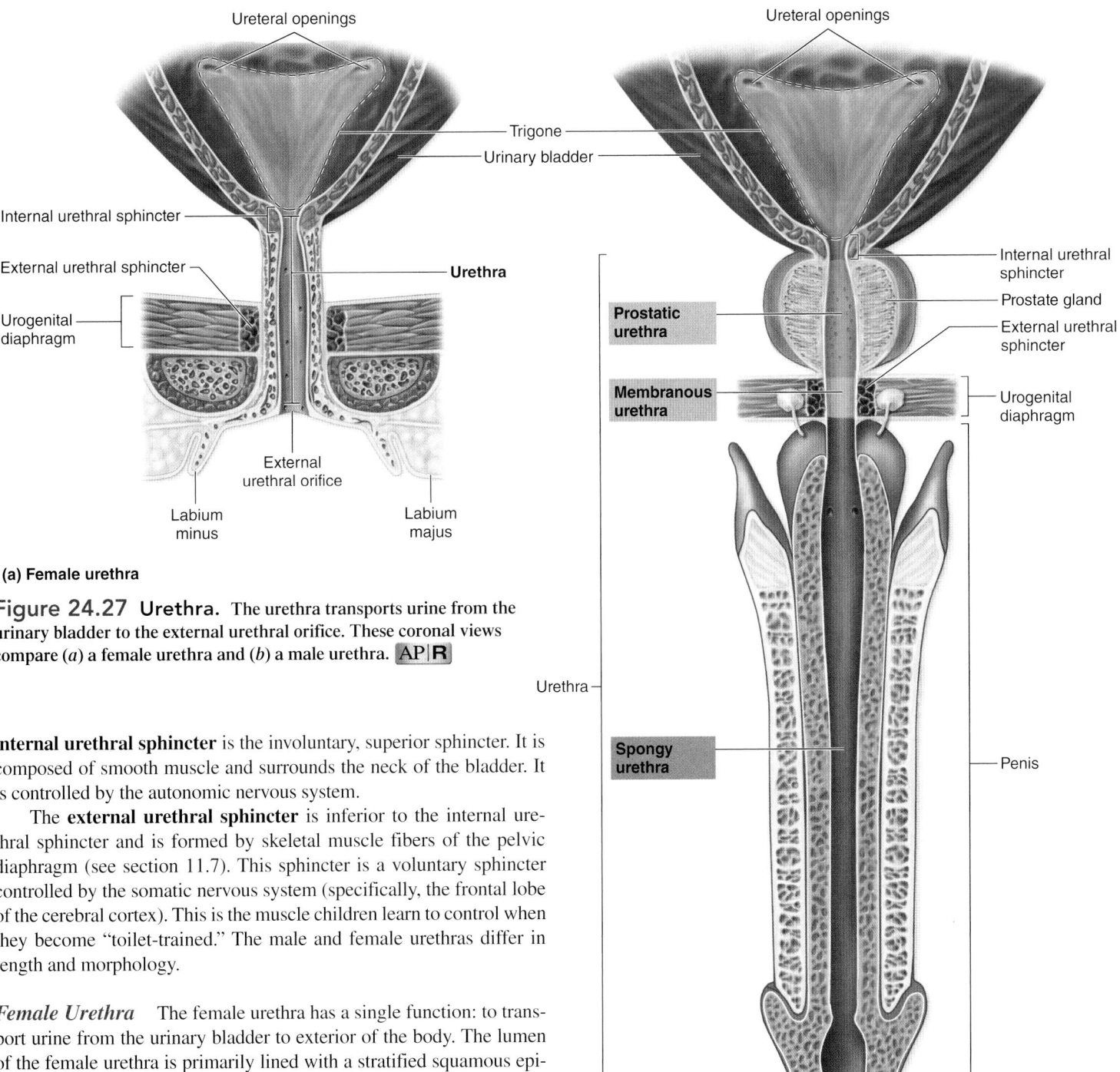

(a) Female urethra

Figure 24.27 Urethra. The urethra transports urine from the urinary bladder to the external urethral orifice. These coronal views compare (*a*) a female urethra and (*b*) a male urethra. AP|R

(b) Male urethra

internal urethral sphincter is the involuntary, superior sphincter. It is composed of smooth muscle and surrounds the neck of the bladder. It is controlled by the autonomic nervous system.

The **external urethral sphincter** is inferior to the internal urethral sphincter and is formed by skeletal muscle fibers of the pelvic diaphragm (see section 11.7). This sphincter is a voluntary sphincter controlled by the somatic nervous system (specifically, the frontal lobe of the cerebral cortex). This is the muscle children learn to control when they become "toilet-trained." The male and female urethras differ in length and morphology.

Female Urethra   The female urethra has a single function: to transport urine from the urinary bladder to exterior of the body. The lumen of the female urethra is primarily lined with a stratified squamous epithelium. The urethra is approximately 4 centimeters (about 1.6 inches) long, and it opens to the outside of the body at the external urethral orifice located in the female perineum.

Male Urethra   The male urethra has both urinary and reproductive functions because it serves as a passageway for both urine and semen (but not at the same time). It is approximately 19 centimeters long (7.5 inches) and is partitioned into three segments: the prostatic urethra, the membranous urethra, and the spongy urethra.

The **prostatic** (pros-tat′ik) **urethra** is approximately 3.5 centimeters (about 1.5 inches) in length and is the most dilatable portion of the urethra. It extends through the prostate gland immediately inferior to the male bladder, where multiple small prostatic ducts enter it. The urethra in this region is lined by a transitional epithelium. Two smooth muscle bundles surround the mucosa: an internal longitudinal bundle and an external circular bundle. The external circular muscular bundles are a continuation of the thickened circular region of the internal urethral sphincter at the bladder outlet.

The **membranous** (mem′bră-nus) **urethra** is the shortest and least dilatable portion of the male urethra. It extends from the inferior surface of the prostate gland through the urogenital diaphragm. As a result, it is surrounded by skeletal muscle fibers that form the external urethral sphincter of the urinary bladder. The epithelium in this region is often either stratified columnar or pseudostratified columnar.

The **spongy urethra** (or *penile urethra*) is the longest part of the male urethra at approximately 15 centimeters long (about 6 inches). It is encased within a cylinder of erectile tissue in the penis called the corpus spongiosum, and it extends to the **external urethral orifice.** The proximal part of the spongy urethra is lined by a pseudostratified columnar epithelium, whereas the distal part has a stratified squamous epithelial lining.

WHAT DID YOU LEARN?

37 What are the major components of the urinary tract? What tunics are found in each of these?

38 How does the urethra of a male and female differ?

24.8c Micturition

LEARNING OBJECTIVES

6. Define micturition.
7. Compare and contrast the storage reflex and the micturition reflex.
8. Explain conscious control over micturition.

The expulsion of urine from the bladder is called **micturition** (mik-chū-rish′ŭn; *micturio* = to desire to make water), *urination,* or *voiding.* Two reflexes are associated with the process of micturition: the storage reflex and the micturition reflex. These reflexes are regulated by the sympathetic and parasympathetic divisions of the autonomic nervous system, respectively.

Innervation of the Urinary Bladder and Urethral Sphincters

The urinary bladder has sympathetic, parasympathetic, and somatic innervation. Sympathetic axons extend from T11 to L2 segments of the spinal cord. Sympathetic stimulation causes contraction of the internal urethral sphincter and inhibits contraction of the detrusor muscle. Thus, stimulation by the sympathetic division inhibits micturition. Innervation by the parasympathetic division is from the micturition center located

in the pons, and extends from the spinal cord segments S2–S4 in the pelvic splanchnic nerves. The pelvic splanchnic nerves cause contraction of the detrusor muscle and relaxation of the internal urethral sphincter, so that urine can be expelled. Thus, the parasympathetic division stimulates micturition. The external urethral sphincter is composed of skeletal muscle. It is innervated by the somatic nervous system via the pudendal nerve (which is formed from the S2–S4 segments of the spinal cord). The pudendal nerve voluntarily contracts this sphincter to prevent urination.

Storage Reflex

The storage of urine in the urinary bladder is controlled by both the autonomic and somatic nervous system **(figure 24.28a).** During the filling of the urinary bladder, urine moves through the ureters from the kidneys. Varying nerve signals conducted by sympathetic axons cause smooth muscle cells of (1) the detrusor muscle of the urinary bladder wall to relax, which allows the bladder to accommodate the urine; and (2) internal urethral sphincter to contract, so that urine is retained within the bladder.

This process is referred to as the **storage reflex.** The skeletal muscle of the external urethral sphincter is also continuously stimulated by nerve signals along the pudendal nerve so it remains contracted.

Micturition Reflex

The process of micturition is also controlled by both the autonomic and somatic nervous systems in a toilet-trained individual, and it usually proceeds as follows (figure 24.28b):

1. When the volume of urine retained within the bladder reaches approximately 200 to 300 mL, the bladder becomes distended, and baroceptors in the bladder wall are activated.

2. These baroceptors send nerve signals through visceral sensory neurons to stimulate the micturition center within the pons.

3. The micturition center alters nerve signals propagated down the spinal cord and through the pelvic splanchnic nerves (which are parasympathetic nerves).

4. Parasympathetic stimulation causes the smooth muscle cells composing the detrusor muscle to contract and internal urethral sphincter to relax. In infants, urination occurs at this point because they lack voluntary control of the external urethral sphincter.

The sensation of having to urinate is relayed along sensory axons to the cerebral cortex.

Conscious Control of Urination

An individual's conscious decision to urinate is due to altering nerve signals relayed from the cerebral cortex through the spinal cord and along the pudendal nerve to cause relaxation of the external urethral sphincter. Expulsion of urine is facilitated by both the voluntary contraction of muscles in the abdominal wall and expiratory muscles as part of the Valsalva maneuver (see section 23.3a). Upon emptying of the urinary bladder (all but about 10 mL), the detrusor muscle relaxes, and the neurons of the micturition reflex center are inactivated, while those of the storage reflex are activated.

If an individual does not urinate at the time of the first micturition reflex, the detrusor muscle relaxes as a consequence of the "stress-relaxation response" of smooth muscle (see section 10.10d). The bladder continues to fill with urine, and the micturition reflex is

INTEGRATE

CLINICAL VIEW
Urinary Tract Infections

A **urinary tract infection (UTI)** occurs when either bacteria (most commonly *E. coli*) or fungi (e.g., yeast) enter and multiply within the urinary tract. Women are more prone to UTIs because they have a short urethra that is close to the anus, allowing bacteria from the gastrointestinal (GI) tract to more readily enter the urethra. Sexual intercourse or medical use of a urinary catheter also increase the risk of UTIs.

A UTI often develops first in the urethra, an inflammation called **urethritis** (ū-rē-thrī′tis). If the infection spreads to the urinary bladder, **cystitis** (sis-tī′tis) results. Occasionally, bacteria from an untreated UTI can spread up the ureters to the kidneys, a condition termed **pyelonephritis** (pī′ĕ-lō-ne-frī′tis), which is inflammation of the kidney that is potentially fatal. Symptoms of a UTI include difficult and painful urination, called **dysuria** (dis-ū′rē-ă; *dys* = bad, difficult), frequent urination, and a feeling of uncomfortable pressure in the pubic region. If the infection spreads to the kidneys, sharp back and flank pain, fever, and occasionally nausea and vomiting may occur. A UTI can be diagnosed through **urinalysis** (ū-ri-nal′i-sis), a test of the urine that can reveal the presence of leukocytes, blood, and bacteria or fungi, as well as nitrituria (see table 24.3 in section 24.8a). Appropriate antibiotic therapy cures most UTIs caused by bacteria.

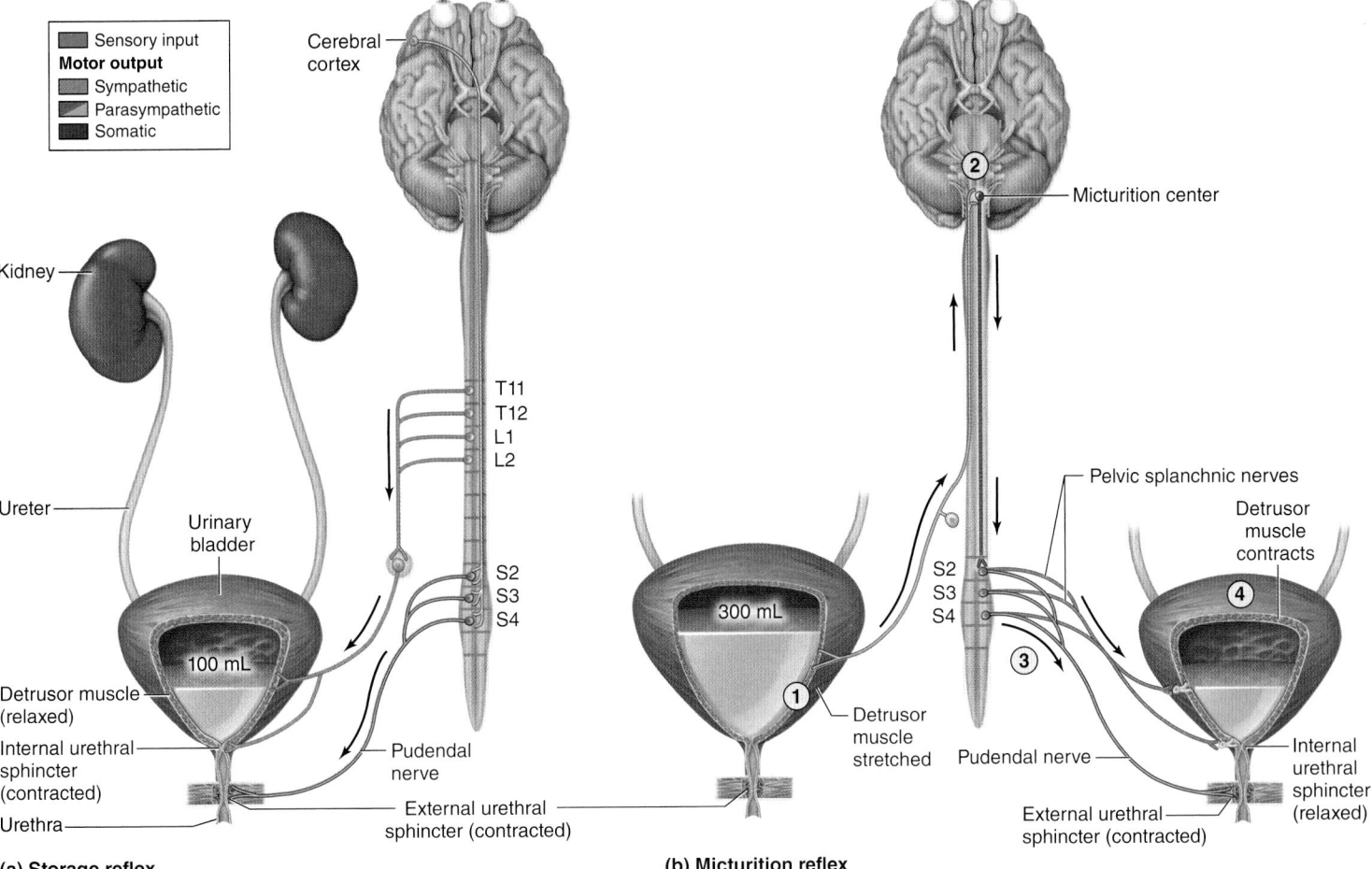

Figure 24.28 Micturition. (*a*) The storage reflex allows urine to be held in the urinary bladder until it can be eliminated from the body. (*b*) The micturition reflex is the involuntary portion of eliminating urine. The relaxation of the external urethral sphincter permits voluntary control of micturition. AP|R

INTEGRATE

CLINICAL VIEW

Impaired Urination

Two common terms are associated with impaired urination: incontinence and retention. **Incontinence** (in-kon'ti-nens; *in* = not, *contineo* = to hold together) is the inability to voluntarily control urination. Incontinence may occur as a result of childbirth (stress incontinence), stronger than normal detrusor muscle contractions (urge incontinence), or the secondary result of medications. Uncontrolled urination may also occur when an individual is frightened, as nerve signals initiated in the limbic system are sent to the micturition center, causing urination.

Retention (rē-ten'shŭn) involves failure to eliminate urine normally. Causes include side effects following general anesthesia and an enlarged prostate in a male that impedes urine flow through the prostatic urethra. If necessary, a small tube called a *catheter* may be placed into the urethra to allow urine to flow out of the bladder.

initiated again after another 200–300 mL of urine is added. This cycle continues until there is between 500 mL and 600 mL. At this point, urination occurs without conscious control.

Additionally, an individual may choose to empty the bladder prior to the initiation of the micturition reflex by contracting the abdominal muscles as part of the Valsalva maneuver and compressing the bladder. This compression of the bladder stimulates baroreceptors within the bladder wall and the micturition reflex is initiated.

The urinary system, which is composed of the kidneys, ureters, urinary bladder, and urethra is responsible for filtering our blood and forming, storing, and eliminating urine. By adjusting the amount of urine that is formed, the kidneys play a critical role in regulating blood volume and blood pressure. These processes of the urinary system also function to maintain blood ion levels within normal homeostatic levels and eliminate both metabolic wastes and biologically active molecules—all of which function to keep our blood healthy.

WHAT DID YOU LEARN?

39 What steps lead to micturition? At what point does the reflex overcome conscious control?

- The urinary system is composed of the kidneys and the urinary tract (ureters, urinary bladder, and the urethra).

24.1 Introduction to the Urinary System 943

- The kidneys filter blood to produce urine, which is subsequently eliminated from the body through the ureters, urinary bladder, and urethra.
- The kidneys perform numerous functions, including eliminating metabolic wastes from the blood, regulating ion and acid-base balance, regulating blood pressure, eliminating biologically active molecules, producing the enzyme that forms calcitriol, producing and releasing erythropoietin, and having the potential to engage in gluconeogenesis to regulate blood glucose.

24.2 Gross Anatomy of the Kidney 945

- The kidneys are two bean-shaped reddish organs.

24.2a Location and Support 945

- The kidneys are located along the posterior abdominal wall within the retroperitoneal space and are partially protected by the ribs.
- The kidneys are surrounded by a fibrous capsule, perinephric fat, renal fascia, and paranephric fat.

24.2b Sectional Anatomy of the Kidney 946

- Each kidney is composed of an outer cortex and an inner medulla, and the medulla is subdivided into 8 to 15 renal pyramids (or 8 to 15 renal lobes).
- Renal sinuses are spaces within the kidney and include the minor calyces, major calyces, and a single renal pelvis.

24.2c Innervation of the Kidney 947

- Sympathetic axons innervate the afferent and efferent arterioles and the juxtaglomerular apparatus.

24.3 Functional Anatomy of the Kidney 947

- Nephrons, collecting tubules, collecting ducts, and associated structures form the functional anatomy of the kidney.

24.3a Nephron 947

- A nephron consists of a renal corpuscle, which is composed of a glomerulus plus glomerular capsule, and a renal tubule (proximal convoluted tubule, nephron loop, and distal convoluted tubule).

24.3b Collecting Tubules and Collecting Ducts 951

- Numerous nephrons drain into a collecting tubule, and numerous collecting tubules drain into a collecting duct.

24.3c Juxtaglomerular Apparatus 952

- The juxtaglomerular (JG) apparatus is composed of granular cells of the afferent arteriole and a macula densa of the distal convoluted tubule. It functions in the regulation of filtrate formation and systemic blood pressure.

24.4 Blood Flow and Filtered Fluid Flow 952

- Filtrate is formed from blood as it flows through the glomerulus.

24.4a Blood Flow Through the Kidney 952

- Blood is transported to the kidneys in renal arteries, which divide into segmental arteries and then interlobar arteries. Branching off the interlobar arteries are arcuate arteries (at the corticomedullary junction) and then interlobular arteries, which branch into afferent arterioles that supply a glomerulus.
- Blood leaving the glomerulus follows this path: efferent arteriole, peritubular capillaries or vasa recta, interlobular veins, arcuate veins, interlobar veins, and a renal vein.

24.4b Filtrate, Tubular Fluid, and Urine Flow 954

- Filtrate is collected in the capsular space of the glomerular capsule.
- Tubular fluid flows through the proximal convoluted tubule, nephron loop, distal convoluted tubule, collecting tubules and ducts.
- Urine flows from numerous collecting ducts into a papillary duct, then into the minor calyx, major calyx, renal pelvis, ureter, and urinary bladder (where it is stored), and excreted via the urethra.

24.5 Production of Filtrate Within the Renal Corpuscle 955

- The renal corpuscle is the site of filtrate production.

24.5a Overview of Urine Formation 955

- Filtration, reabsorption, and secretion are the three processes involved in urine formation.

24.5b Filtration Membrane 956

- The filtration membrane is composed of the fenestrated endothelium and basement membrane of the glomerulus and the visceral layer of the glomerular capsule.

24.5c Formation of Filtrate and Its Composition 957

- Filtrate is (essentially) protein-free filtered plasma.

24.5d Pressures Associated with Glomerular Filtration 957

- The three pressures that determine net filtration pressure are glomerular (blood) hydrostatic pressure (HP_g), which promotes filtration; and both blood colloid osmotic pressure (OP_g) and capsular hydrostatic pressure (HP_c), which oppose filtration.
- Net filtration pressure is determined by subtracting OP_g and HP_c from HP_g.

24.5e Regulation of Glomerular Filtration Rate 959

- Glomerular filtration rate (GFR) is maintained through renal autoregulation (myogenic and tubuloglomerular mechanisms), decreased by sympathetic stimulation, and increased by atrial natriuretic peptide hormone.

CHALLENGE YOURSELF

Do You Know the Basics?

_____ 1. All of following are functions of the kidney *except*

    a. gluconeogenesis, the formation of glucose from non-carbohydrate sources.

    b. release of erythropoietin to control erythrocyte production.

    c. control of blood pressure through the release of renin.

    d. production of plasma proteins to control blood volume.

_____ 2. When the kidneys are described as being retroperitoneal, this refers to the fact that the kidneys

    a. are within the parietal peritoneal lining of the abdominal cavity.

    b. are posterior to the parietal peritoneal lining of the abdominal cavity.

    c. are superior to the peritoneal lining of the abdominal cavity.

    d. have no protective covering.

_____ 3. Which of the following is located within the renal medulla?

    a. collecting duct

    b. glomerulus

    c. renal corpuscle

    d. distal convoluted tubule

_____ 4. All of the following are capillaries *except*

    a. the glomerulus.

    b. the minor calyx.

    c. the vasa recta.

    d. peritubular capillaries.

_____ 5. Which of the following is a component of filtrate but not normally a component of urine?

    a. water

    b. erythrocytes

    c. nitrogenous waste

    d. glucose

_____ 6. If blood pressure in the glomerulus increases, then

    a. net filtration pressure decreases.

    b. the percentage of substances reabsorbed increases.

    c. urine production increases.

    d. renin is released.

_____ 7. Which hormone increases Na^+ and water reabsorption and K^+ secretion?

    a. antidiuretic hormone

    b. angiotensin II

    c. atrial natriuretic peptide

    d. aldosterone

_____ 8. If the tubular maximum is exceeded, then

    a. the carrier mechanism will work harder.

    b. the excess will be reabsorbed in the bladder.

    c. the excess will appear in the urine.

    d. its blood concentration will increase.

_____ 9. The function unique to the nephron loop is to

    a. regulate pH.

    b. excrete water.

    c. establish a concentration gradient in the medulla interstitial fluid.

    d. regulate the concentration of blood Ca^{2+}.

_____ 10. If antidiuretic hormone (ADH) concentration increases,

    a. urine volume increases.

    b. urine concentration decreases.

    c. urine volume decreases, and urine concentration increases.

    d. urine volume increases, and urine concentration decreases.

11. Trace blood flow into and out of the kidney. Identify the appropriate vessels.

12. Describe where filtrate, tubular fluid, and urine are found in the functioning kidney.

13. Describe the anatomic components of the juxtaglomerular apparatus.

14. Describe the filtration membrane, and explain the structures that do not normally pass through it.

15. Explain how glomerular filtration rate (GFR) is maintained by renal autoregulation, decreased by sympathetic stimulation, and increased by atrial natriuretic peptide.

16. Discuss the affect of aldosterone and antidiuretic hormone (ADH) on principal cells and their effect on urine production.

17. Explain how antidiuretic hormone (ADH) is dependent upon the concentration gradient in the medulla of the kidney.

18. Describe the significant differences between blood plasma, filtrate, and urine.

19. Identify all of the following that are functions of the kidney: (a) maintain blood pH; (b) regulate blood ion concentrations; (c) regulate blood volume and blood pressure; (d) eliminate wastes, some hormones, and certain drugs from the blood; (e) release renin; (f) release erythropoietin; and (g) stimulate the final step in calcitriol formation.

20. Explain the process of micturition.

Can You Apply What You've Learned?

Use the following paragraph to answer questions 1–3.

Maria is 8 months pregnant. When she goes to her doctor for her checkup, the nurse takes her blood pressure. It is unusually high: 200/100 mm Hg. The nurse also collects a urine sample and detects elevated levels of protein in her urine.

1. What can account for the protein in her urine?

    a. Protein is a normal component of urine and is no cause for concern.

    b. Excessive amounts of plasma proteins are filtered at the renal corpuscle.

    c. Tubular cells produce additional protein in response to the elevated blood pressure.

    d. The tubular cells have exocytosed protein into the tubular fluid.

2. What would you predict has happened to her plasma protein concentration?

    a. It has no effect.

    b. Plasma protein levels will have increased.

    c. Plasma protein levels will have decreased.

    d. Plasma proteins levels will have remained the same or increase.

3. What do you think has happened to her urine production? (Consider the pressures involved in determining net filtration pressure and osmotic force generated by solutes in the filtrate.)

    a. Urine production stays the same.

    b. Urine production increases.

    c. Urine production decreases.

    d. It is not possible to predict the effect on urine production.

4. Martin, a young man of 20, was in a car accident and is hemorrhaging. When he arrives by ambulance at the hospital, he is stabilized. However, they notice that over the next day he does not urinate. Which of the three variables for determining net filtration pressure has changed and can account for the lack of urine formation?

    a. Glomerular hydrostatic (blood) pressure has decreased.

    b. Capsular hydrostatic pressure has decreased.

    c. Blood colloidal osmotic pressure has decreased.

    d. Glomerular hydrostatic (blood) pressure has increased.

5. A 19-year-old male named Paul was in a diving accident and severed his spinal cord at the T1 segment. Which of the following describes one of the physiologic changes that he will experience?

    a. His kidneys will no longer be able to filter his blood and produce urine.

    b. His urinary bladder will no longer be able to contract to expel urine.

    c. His filtration membrane will contract, and filtration will no longer occur.

    d. He will no longer be able to consciously control urination.

Can You Synthesize What You've Learned?

1. A patient with cancer is treated with chemotherapy. This specific medication is eliminated by the kidney. Results from a renal plasma clearance test are higher than average. Will the medication need to be given more or less often? Explain.

2. Theoretically, which one of the following hormones could be administered to decrease blood volume: antidiuretic hormone, aldosterone, parathyroid hormone, or atrial natriuretic peptide? Explain.

3. Males who suffer from either benign prostatic hyperplasia (noncancerous prostate gland enlargement) or prostate cancer often have problems with urination. Based on your knowledge of the male urethra, hypothesize why these urination problems occur.

INTEGRATE

ONLINE STUDY TOOLS
  connect | ANATOMY & PHYSIOLOGY   LEARNSMART   AP|R

The following study aids may be accessed through Connect.

Concept Overview Interactive: Figure 24.15: Glomerular Filtration

Clinical Case Study: A Young Child with a Fever and Foamy Urine

Interactive Questions: This chapter's content is served up in a number of multimedia question formats for student study.

LearnSmart: Topics and terminology include introduction to the urinary system; gross anatomy of the kidney; functional anatomy of the kidney; blood flow and filtered fluid flow; production of filtrate within the renal corpuscle; reabsorption and secretion in tubules and collecting ducts; evaluating kidney function; urine characteristics, transport, storage, and elimination

Anatomy & Physiology Revealed: Topics include urinary system overview; kidney gross anatomy; kidney microscopic anatomy; renal cortex; podocyte; urine formation; upper and lower urinary system; micturition reflex

Animations: Topics include urinary system overview; kidney gross anatomy; kidney microscopic anatomy; basic renal processes; urine formation; renal clearance; diffusion, facilitated diffusion; cotransport; sodium-potassium exchange pump

Fluid and Electrolytes

INTEGRATE

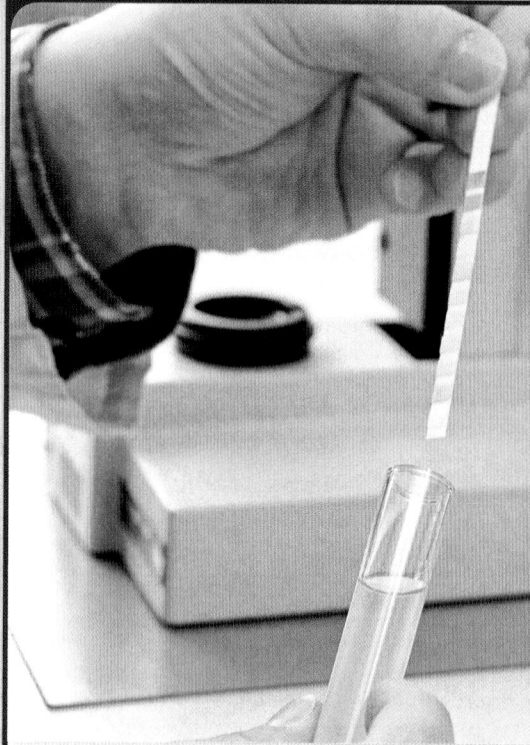

CAREER PATH
Medical Laboratory Technologist

A medical laboratory technologist is a health-care professional who examines and performs biological, chemical, and hematological tests on body fluid samples or specimens including urine, blood, stool, and pericardial fluid. Results are analyzed for accuracy and reported to physicians. These medical care specialists are employed at hospitals, doctor's offices, and specialized medical laboratories.

Anatomy &
Physiology **REVEALED**
aprevealed.com

Module 13: Urinary System

Our body's ability to maintain fluid balance is something most of us take for granted despite our sometimes erratic fluid intake. For example, we may drink several cups of coffee first thing in the morning, go the rest of the day without drinking anything, perhaps run several miles in the late afternoon, and then drink large quantities of fluid at the end of the day before retiring for the night. Fortunately, our body compensates for our inconsistent fluid intake by making physiologic adjustments throughout the day so that fluids are maintained within homeostatic limits. In a similar way, electrolyte levels, including those that determine pH, are being regulated to maintain their balance.

Fluid and electrolyte imbalances, however, can occasionally occur. Insufficient fluid intake, excessive sweating, vomiting, diarrhea, or abnormal changes in the function of an organ (e.g., kidney disease, congestive heart failure) all may tax the body to such an extent that fluid and electrolyte levels are driven out of balance. Here we describe the major concepts and mechanisms involved in fluid and electrolyte balance. These include the percentage, distribution, chemical composition, and movement of body fluid; maintenance of electrolyte levels; hormonal regulation of fluid balance and its influence on blood pressure; acid-base control and balance; and acid-base disturbances.

25.1 Body Fluids

We begin the discussion of fluid balance by describing the percentage of body fluid, its distribution into two major fluid compartments, and how fluid moves between the fluid compartments within the body.

25.1a Percentage of Body Fluid

LEARNING OBJECTIVES

1. List the factors that influence the percentage of body fluid.
2. Explain the significance of percentage of body fluid relative to fluid balance.

The human body contains between 45% and 75% fluid by weight, with an average of about 65% (figure 25.1). The fluid percentage depends upon two variables: the age of an individual, and the relative amounts of adipose connective tissue and skeletal muscle tissue:

- **Age.** Infants have the highest percentage of fluid, at approximately 75% fluid by weight. In contrast, elderly individuals have the lowest percentage of fluid at 45%. Children and young and middle-aged adults are usually somewhere in between these two extremes, with a general trend of decreasing percentage of body fluid seen with increasing age.

- **Relative amounts of adipose connective tissue to skeletal muscle tissue.** The percentage of fluid in the body at each age depends upon the ratio of adipose connective tissue and skeletal muscle tissue because of the difference in water content of these tissues. Adipose connective tissue is approximately 20% water, whereas skeletal muscle tissue is approximately 75% water. This accounts for the general differences in the percentage of body fluid that are noted between females and males of the same age after puberty.

    Lean adult females are, on average, typically composed of 55% body fluid, whereas lean adult males are, on average, typically composed of 60% body fluid. This difference reflects the relatively lower amounts of skeletal muscle and relatively higher amounts of adipose connective tissue in a lean adult female than in a lean adult male.

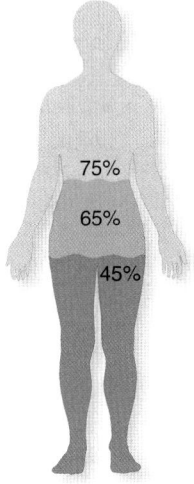

Figure 25.1 Water Content of the Body. The percentage of water that composes the body varies from 45–75%, with an average of 65%.

Most importantly, your body's water content influences how susceptible you are to a fluid imbalance; individuals that have a lower percentage of body fluid are more likely to experience a fluid imbalance. For example, one reason elderly individuals are more prone as a group to fluid imbalances than are young adults is because of their relatively smaller percentage of body fluid.

WHAT DID YOU LEARN?

❶ If you have an increase in muscle mass as a result of weight training, will your percentage of body fluid increase, decrease, or stay the same? Explain.

25.1b Fluid Compartments

LEARNING OBJECTIVES

3. Describe the two major body fluid compartments, and compare their compositions.
4. Explain how fluid moves between the major body fluid compartments.

The fluid in our body is partitioned into intracellular fluid and extracellular fluid compartments. **Intracellular fluid (ICF)** is the fluid within our cells. A majority or approximately two-thirds of the total fluid is within our cells (figure 25.2). The barrier enclosing this fluid is the

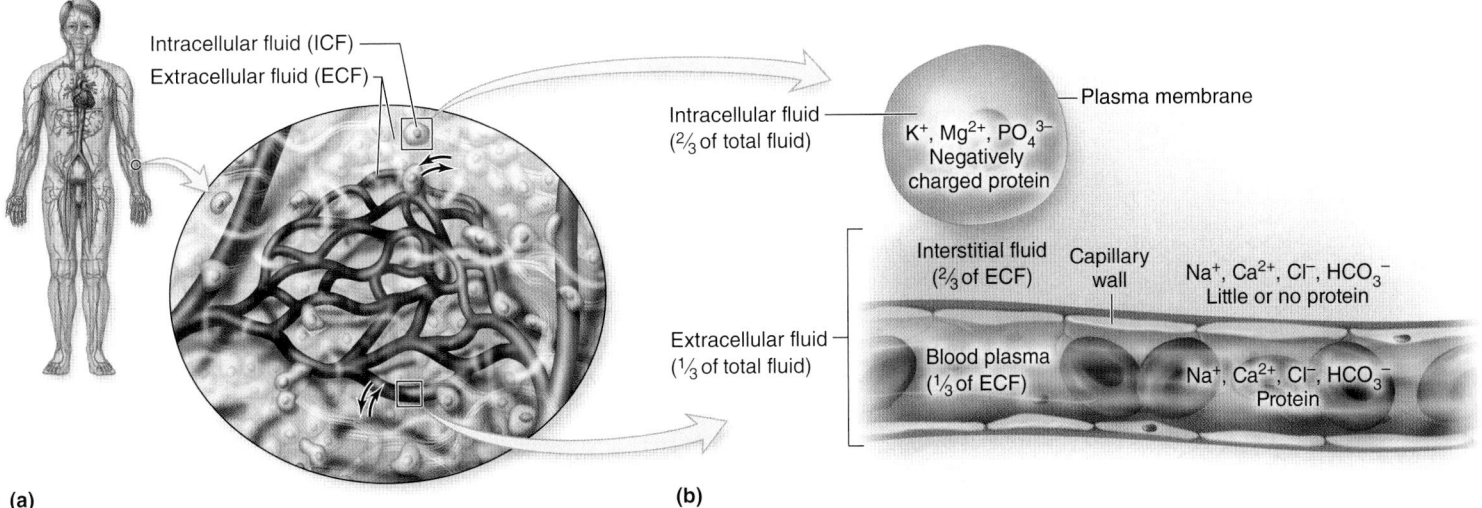

(a)   (b)

Figure 25.2 Fluid Compartments. (*a*) Fluid is contained within two major compartments: intracellular and extracellular fluid compartments. (*b*) K^+, Mg^{2+}, PO_4^{3-}, and proteins are more common in the intracellular fluid. The extracellular fluid is composed of both interstitial fluid and blood plasma, where Na^+, Ca^{2+}, Cl^-, and HCO_3^- are more prevalent than they are in the intracellular fluid. Within the extracellular fluid, significant amounts of protein are present within the blood plasma, with little or no protein in the interstitial fluid.

INTEGRATE

LEARNING STRATEGY

The "**rule of thirds**" can help you to remember the distribution of fluid: intracellular fluid (ICF) = ⅔ and extracellular fluid (ECF) = ⅓. When only the extracellular fluid is considered, the distribution within the ECF is two-thirds interstitial fluid and one-third blood plasma.

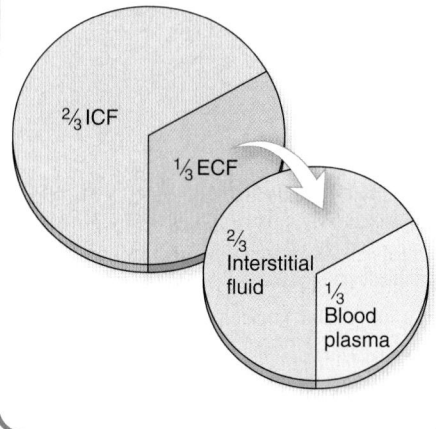

plasma membrane (see section 4.2). The selectively permeable plasma membrane allows some, but not all, substances through it.

The fluid outside of cells is referred to collectively as **extracellular fluid (ECF).** Extracellular fluid includes both **interstitial fluid (IF),** the fluid that surrounds and "bathes" the cells, and **blood plasma,** which is the fluid within the blood vessels (see section 18.2).

Interstitial fluid composes approximately two-thirds of the extracellular fluid, and blood plasma about one-third. The barrier separating blood plasma from the interstitial fluid is the capillary vessel wall that consists of a single layer of simple squamous epithelium resting on a basement membrane (see section 20.1c). A capillary wall typically is more permeable than a plasma membrane. Consequently, the interstitial fluid and the blood plasma are more chemically similar to each other than they are to the intracellular fluid (as described shortly).

Examples of specific extracellular fluids include: cerebrospinal fluid of the brain and spinal cord, synovial fluid of joints, aqueous and vitreous humor of the eye, fluids of the inner ear (peri-lymph and endolymph), and serous fluid within body cavities (pleural, pericardial, and peritoneal). These specific extracellular fluids are not typically subject to significant daily fluid gains and losses, and for this reason they are essentially ignored in this chapter. Only fluid in the two major compartments—and the factors that influence them, such as lymph return to blood—is addressed.

Composition of Body Fluids in the Two Compartments

Each fluid compartment is chemically distinct based on the relative amounts of substances dissolved in it (figure 25.2b). Intracellular fluid is the most distinct compartment; it contains more potassium (K^+) and magnesium (Mg^{2+}) cations, phosphate anion (PO_4^{3-}), and negatively charged proteins than the extracellular fluid.

These differences in chemical composition reflect both processes within the cell and the regulatory activity of transport proteins within the plasma membrane that move substances into and out of the cell. For example, the relatively large amount of protein within intracellular fluid is a result of protein synthesis, whereas the relatively high concentration of intracellular K^+ occurs as a result of sodium-potassium (Na^+/K^+) pumps embedded within the plasma membrane that transport K^+ into a cell as Na^+ is transported out (see section 4.3c).

The two fluids composing the extracellular fluid—interstitial fluid and blood plasma—are, in comparison, both distinct chemically from intracellular fluid and similar in chemical composition to one another. Both interstitial fluid and blood plasma have a high concentration of these ions: sodium (Na^+) and calcium (Ca^{2+}) cations, and chloride (Cl^-) and bicarbonate (HCO_3^-) anions.

Interstitial fluid and blood plasma exhibit one significant difference—namely, that protein is present in blood plasma—but very little protein is within the interstitial fluid. The similarity in ionic composition and the difference in protein composition reflect the relative permeability of the capillary wall: Proteins are generally too large to move out of the blood through the openings in the capillary wall to enter the interstitial fluid, whereas fluids and ions move freely. Therefore, during capillary exchange, blood plasma and all of its dissolved substances—except for most proteins—are filtered to become part of the interstitial fluid (see section 20.3). Specific percentages of the different substances in the intracellular fluid and extracellular fluid (both IF and blood plasma) are given in **table 25.1**.

Fluid Movement Between Compartments

Fluid movement between compartments occurs continuously in response to changes in relative osmolarity (concentration). This happens when the fluid concentration in one fluid compartment becomes either hypotonic or hypertonic (see section 4.3b), with respect to another compartment; water immediately moves by osmosis between the two compartments until the water concentration is once again equal **(figure 25.3)**. You may recall from section 4.3b that water always moves by osmosis from the hypotonic solution to the hypertonic solution. This movement of water between the compartments is possible because the plasma membranes and the capillary wall are both permeable to water.

When you drink water, water enters your blood from the gastrointestinal (GI) tract and becomes part of the blood plasma. Consequently, the plasma osmolarity decreases and blood plasma becomes hypotonic to the ICF. Thereafter, as blood moves through the capillaries, water first moves out of the blood plasma to become part of the interstitial fluid, and then moves from the interstitial fluid into cells (figure 25.3a). This means there is a net movement of water from the blood plasma into the cells. In contrast, if water is lost from the body without being replaced in a timely manner, then dehydration occurs. Blood plasma osmolarity

Table 25.1	Percentages of Solutes in Body Fluids[1]	Extracellular Fluid %	
Solutes	Intracellular Fluid %	Interstitial Fluid %	Blood Plasma %
CATIONS			
Potassium (K⁺)	75	3	3
Magnesium (Mg²⁺)	17	1	1
Sodium (Na⁺)	6	94	94
Calcium (Ca²⁺)	2[2]	2	2
PROTEINS AND ANIONS			
Proteins	27	Trace	10
Phosphate (PO₄³⁻)	20	1	1
Bicarbonate (HCO₃⁻)	6	18	16
Chloride (Cl⁻)	2	77	69
Other	45	4	4

1. Values given are approximate percentages in skeletal muscle tissue.
2. Calcium levels in the cytosol of skeletal muscle fluctuate between when a muscle is at rest and when it is contracting.

increases, and blood plasma becomes hypertonic to both the interstitial fluid and the cells. Consequently, a net movement of water occurs from the cells into the blood plasma. It moves first into the interstitial fluid and then into the blood plasma (figure 25.3b).

WHAT DID YOU LEARN?

2 Which ions are more prevalent in the intracellular fluid? Which are more prevalent in the extracellular fluid?

3 What is the major distinction in the chemical composition of blood plasma and interstitial fluid?

4 When you are dehydrated, is the net movement of fluid from the blood plasma into the cells or from the cells into the blood plasma?

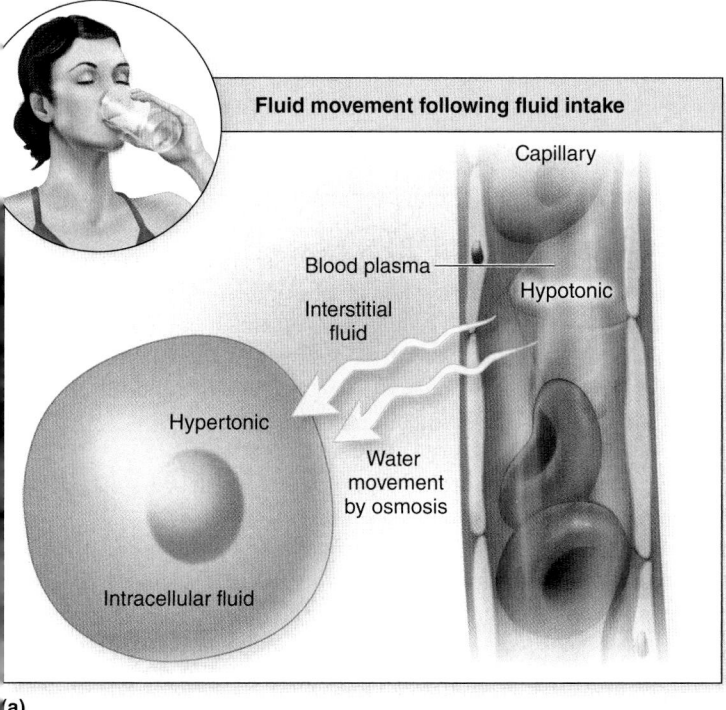

(a)

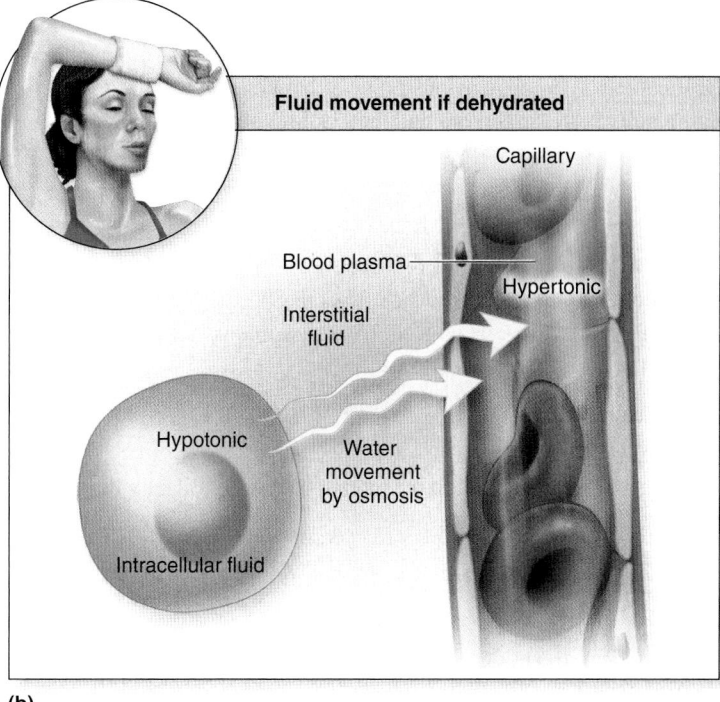

(b)

Figure 25.3 Fluid Movement Between Compartments. (a) A net movement of water from the blood plasma into the intracellular compartment occurs following fluid intake. (b) A net movement of water from the intracellular compartment into the blood plasma occurs when an individual is dehydrated.

25.2 Fluid Balance

Many systems influence fluid balance, including the digestive, cardiovascular, lymphatic, integumentary, respiratory, urinary, nervous, and endocrine systems. Here we discuss how these systems help bring fluid into the body, are involved with movement of fluid within the body, participate in the loss of fluid from the body, and regulate the processes of fluid balance.

25.2a Fluid Intake and Fluid Output

 **LEARNING OBJECTIVES**

1. Define fluid balance.
2. List the sources of fluid intake.
3. Distinguish between the categories of water loss.

Fluid balance exists when fluid intake is equal to fluid output, and a normal distribution of water and solutes is present in the two major fluid compartments. Our discussion begins by describing and comparing how fluid intake and fluid output occur.

Fluid Intake

Fluid intake is the addition of water to the body. It is divided into two categories, preformed water and metabolic water (**figure 25.4**):

- **Preformed water** includes the water absorbed from food and drink taken into the GI tract. On average, this is approximately 2300 milliliters (mL) of fluid intake per day.
- **Metabolic water** includes the water produced daily from aerobic cellular respiration (see section 3.4e) and dehydration synthesis (see section 2.7a). It is approximately 200 mL of fluid per day.

The sum of the average preformed water intake and metabolic water produced is 2500 mL. Given that fluid intake from both food and drink is approximately 2300 mL of the total 2500 mL, then the fluid absorbed from the digestive tract accounts for about 92% of daily fluid intake. A significant and perhaps obvious point is that normally the only way to significantly increase fluid in the body is through fluid intake from the food and drink that we consume (see Clinical View: "Intravenous [IV] Solution").

Fluid Output

Fluid output is the loss of water from the body. Fluid output must equal fluid intake to maintain fluid balance. Fluid is lost from the body through the normal mechanisms of

- Breathing
- Sweating
- Cutaneous transpiration (evaporation of water directly through the skin)
- Defecation
- Urination

The amount of water lost through each of these processes depends upon physical activity, environmental conditions, and internal conditions of the body. Average amounts for each type of fluid output are shown in figure 25.4. Notice that of the average fluid output, 1500 mL out of the 2500 mL (or approximately 60%) is lost in urine, and the remaining 40% of fluid is lost in expired air, through the skin by sweat and cutaneous transpiration, and in feces.

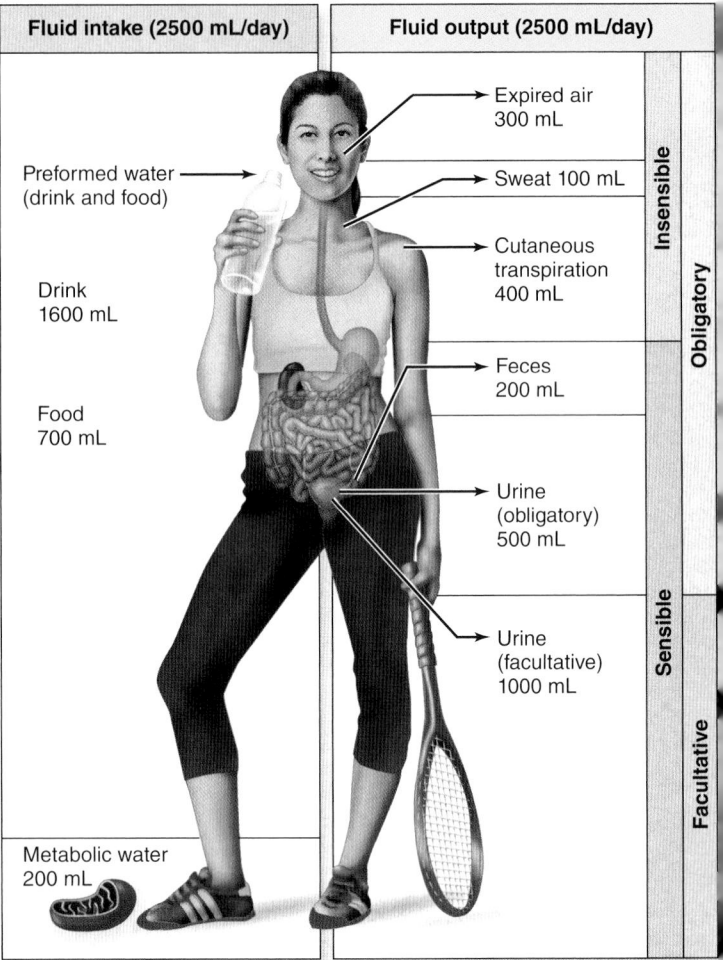

Figure 25.4 Fluid Intake and Fluid Output. Fluid balance is determined by relative amounts of fluid intake and fluid output. Fluid intake results from GI tract absorption of ingested food and drink, as well as water produced by metabolic processes. Fluid output includes fluid lost from the body as a component of expired air, through the skin by both sweat and cutaneous transpiration, and in the feces and urine. Fluid output is called sensible if it is measurable or insensible if it is not. It may also be categorized as obligatory if it is always lost, or facultative if the loss in the urine is regulated. (Values given are approximate averages.)

Water loss can be described in two ways: either as sensible and insensible water loss or as obligatory and facultative water loss.

Sensible and Insensible Water Loss Sensible and insensible fluid loss reflects whether the fluid loss is measurable.

Sensible water loss is measurable, and it includes fluid lost through feces and urine. In contrast, **insensible water loss** is not measurable. It includes both fluid lost in expired air and fluid lost from the skin through sweat and cutaneous transpiration.

Obligatory and Facultative Water Loss Water loss may also be described as either obligatory or facultative.

Obligatory water loss is a loss of water that always occurs, regardless of the state of hydration of the body. It includes water lost through breathing and through the skin (insensible water loss), as well as fluid lost in the feces and in the minimal amount of urine produced to eliminate wastes from the body, approximately 0.5 L (500 mL) per day.

Facultative (fak′ŭl-tā-tiv) **water loss** is controlled water loss through regulation of the amount of urine expelled from the body. It is dependent upon the degree of hydration of the body and is hormonally regulated in the distal convoluted tubule, collecting tubules, and collecting ducts in nephrons of the kidney (see sections 24.6d and 25.4).

INTEGRATE

CLINICAL VIEW
Intravenous (IV) Solution

A patient may be administered an **intravenous** (in'tră-vē'nŭs) **(IV) solution** as a clinical means of increasing fluid input directly into the blood. In this procedure, a needle is inserted into the individual's vein either in the hand or the antecubital region, with a tube extending from the needle to a bag.

The two most common IV solutions are a 0.9% saline solution and a 5% dextrose, or D5W, solution.

A **0.9% saline solution** is a solution containing 0.9 grams of sodium chloride to every 100 mL of sterile water. This solution is the standard IV solution for fluid replacement because it has a similar osmolarity to blood plasma. Thus, it will not cause a net shift of fluid between compartments following the addition of this fluid.

A **5% dextrose**, or **D5W**, solution is composed of 5 grams of dextrose (glucose) sugar to every 100 mL of sterile water. However, this solution is a hypertonic solution (to blood plasma) when in the holding bag—but when infused into the body, the dextrose is metabolized, and the solution becomes hypotonic. This solution is used to supply the body with both water and an energy source (glucose).

Note that the only physiologic mechanism to control fluid output is through the hormonal regulation of urine output. When the body is overhydrated, hormonally controlled facultative water loss normally plays a significant role in eliminating excess fluid. In comparison, obligatory fluid output always occurs, regardless of the hydrated state of the body. Therefore, hormonal regulation of urine output can decrease fluid loss when the body is dehydrated, but not inhibit it completely. Over time, the body continues to become more and more dehydrated unless fluids are replaced. Death may result in severe cases of dehydration.

WHAT DID YOU LEARN?

 5 What are the two major sources of fluid intake? What are two ways that fluid output is categorized, and which one is based on the hydrated state of the body?

25.2b Fluid Imbalance

LEARNING OBJECTIVES

4. Name the different causes of fluid imbalance.

5. Compare and contrast the different types of fluid imbalances.

6. Explain what is meant by fluid sequestration.

A fluid imbalance occurs either when fluid output does not equal fluid intake or when fluid is distributed abnormally. Fluid imbalances can be organized into five categories that include volume depletion, volume excess, dehydration, hypotonic hydration, and fluid sequestration. The first four of these categories (which occur when fluid output does not equal fluid intake) can be differentiated using two criteria:

1. Does the fluid imbalance change the osmolarity (concentration) of body fluid?

2. Is the fluid imbalance caused by an excess or deficiency of body fluid?

Fluid Imbalance with Constant Osmolarity

Fluid imbalances with constant osmolarity occur when isotonic fluid is lost or gained. **Volume depletion** occurs when isotonic fluid loss is greater than isotonic fluid gain. Examples of conditions that result in volume depletion include hemorrhage, severe burns, chronic vomiting, diarrhea, or the hyposecretion of aldosterone (a hormone that stimulates both Na+ and water reabsorption in the kidney; see section 24.6d).

Volume excess occurs when isotonic fluid gain is greater than isotonic fluid loss. This typically results when fluid intake is normal, but there is decreased fluid loss through the kidneys (e.g., from either renal failure or aldosterone hypersecretion).

In both volume depletion and volume excess, there is no change in osmolarity. Consequently, there is no net movement of water between fluid compartments.

Fluid Imbalance with Changes in Osmolarity

Certain types of fluid imbalance involve fluid loss or gain that is not isotonic. **Dehydration** can result from profuse sweating, diabetes mellitus, intake of alcohol, hyposecretion of antidiuretic hormone (ADH—a hormone that stimulates water reabsorption in the kidney; see section 24.6d), insufficient water intake, or overexposure to cold weather. In each case, the water loss is greater than the loss of solutes, and the blood plasma becomes hypertonic. Consequently, water shifts between fluid compartments with a net movement of water from the cells into the interstitial fluid and then into blood plasma. Body cells may become dehydrated as a result (figure 25.3b).

Hypotonic hydration is also called *water intoxication,* or *positive water balance.* It can result from ADH hypersecretion, but it is generally caused from drinking a large amount of plain water following excessive sweating. An example would be an amateur athlete who runs a marathon, and drinks excessive amounts of plain water instead of using an electrolyte-enhanced solution. Both Na+ and water are lost during sweating, and drinking water replaces only the water, but not the solutes. The blood plasma then becomes hypotonic to the other fluid compartments. Fluid moves from blood plasma into the interstitial fluid, and then into the cells (figure 25.3a). Cells may become swollen with fluid.

One of the consequences of extreme hypotonic hydration is cerebral edema. Brain cells become impaired as they swell with excess fluid. The person may experience headaches, nausea, or both. Convulsions, coma, or death may result in severe cases. In addition, some individuals have died after having been forced or enticed to drink excessive amounts of water (e.g., fraternity hazings and water-drinking contests).

Fluid Sequestration

Fluid sequestration (sē'kwes-trā'shŭn; *sequestro* = to lay aside) differs from the other fluid imbalances because total body fluid may be normal, but it is distributed abnormally. Fluid accumulates in a particular location, and it is not available for use elsewhere.

Edema is an example of fluid sequestration in which fluid accumulates in the interstitial space around cells, and is characterized by puffiness or swelling. Anatomic or physiologic changes that can result in edema are depicted in **figure 25.5**. Notice as you review this figure that edema is generally a result of abnormal changes in the cardiovascular system (heart or blood vessels), blood composition, or changes to lymph vessels (the vessels that return fluid to the cardiovascular system). These changes alter the net filtration pressure (NFP) at systemic capillaries (see section 20.3c), causing additional fluid to either leave the capillaries or remain in the interstitial space.

Other examples of fluid sequestration include hemorrhage, ascites, pericardial effusion, and pleural effusion. *Ascites* (ă-sī'tēz) is the accumulation of fluid within the peritoneal cavity, *pericardial effusion* is fluid within the pericardial cavity, and *pleural effusion* is the

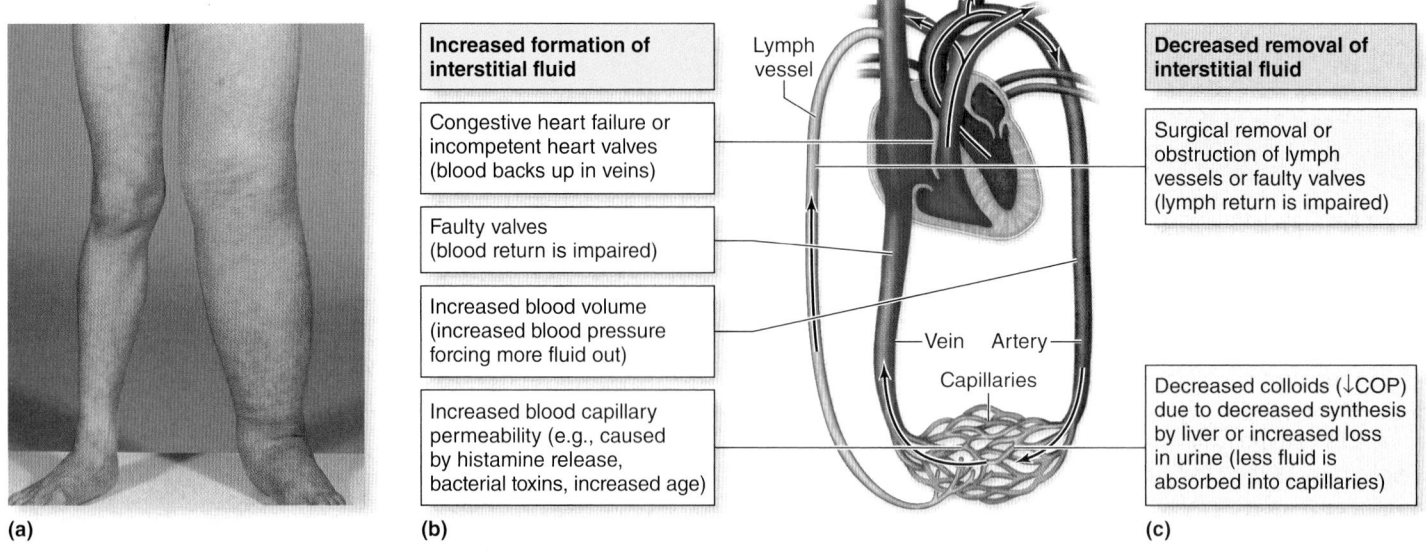

Figure 25.5 Edema. (*a*) Edema is excess fluid within the interstitial space around cells that is characterized by swelling. The individual shown has edema in the left lower limb. Edema may result from either (*b*) increased formation of interstitial fluid or (*c*) decreased removal of interstitial fluid.

accumulation of fluid, sometimes up to several liters, in the pleural cavity as a consequence of lung infections.

 WHAT DID YOU LEARN?

 ⁶ How would you distinguish fluid deficiency from dehydration in terms of changes in total body fluid, changes in blood osmolarity, and fluid movement between compartments?

25.2c Regulation of Fluid Balance

 LEARNING OBJECTIVES

7. Describe the stimuli that increase fluid intake.
8. Explain the conditions and stimuli that decrease fluid intake.
9. Name the four hormones that are involved in regulating fluid output.

Maintaining fluid balance involves the regulation of both fluid intake and fluid output to prevent fluid imbalances. Because no receptors directly monitor either water volume or grams of dissolved solutes, such as Na^+ or K^+, the mechanisms for monitoring fluid are indirect. Fluid balance is regulated by monitoring blood volume, blood pressure, and blood plasma osmolarity (concentration of solutes in blood plasma). The relationships between these variables can be described as follows:

- **Fluid intake increases blood volume.** The additional volume of blood has two consequences: Blood pressure increases, and if water gain exceeds solute gain, then blood osmolarity decreases.
- **Fluid output decreases blood volume.** Therefore, blood pressure is lower, and if more water is lost than solutes, blood osmolarity increases.

INTEGRATE

CLINICAL VIEW
Dehydration in Infants and the Elderly

Both infants and the elderly are more vulnerable to dehydration than are young and middle-aged adults.

Infants are especially vulnerable to dehydration for the following reasons:

- Infants have a greater ratio of skin surface area to volume in comparison to adults, and thus lose more fluid via sweating and transpiration relative to their body size.
- They have immature kidneys that cannot effectively concentrate urine; therefore, more water is required to eliminate wastes.
- They have a higher metabolic rate, which produces more metabolic waste to be eliminated.
- They do not have a completely developed homeostatic mechanism for temperature regulation (see section 1.5); thus, they develop fevers

that are both higher and longer lasting. As a result, additional water is required to cool the body.

In comparison, the elderly are also susceptible to dehydration, but for different reasons:

- Due to the loss of skeletal muscle tissue, elderly individuals are on average composed of 45% water by weight. Smaller losses in water thus may result in a fluid imbalance.
- The kidneys become less effective at concentrating and diluting urine as we age, so more fluid is lost in the urine.
- Cutaneous loss of fluid increases due to decreased thickness of the skin and loss of subcutaneous tissue.
- Fluid replacement may not occur with regularity because the thirst mechanism is less effective. Some elderly people, especially nonambulatory individuals, may also voluntarily restrict fluid intake due to concerns of incontinence.
- Elders are also at greater risk for hyperglycemia, with the elevated blood glucose causing the loss of additional water from the body.

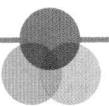

INTEGRATE

CONCEPT CONNECTION

You learned in section 20.3c that capillary dynamics are determined by the **net filtration pressure (NFP)**. NFP is determined by subtracting the net colloid osmotic pressure from the net hydrostatic pressure. On the arterial end of the capillary, capillary hydrostatic pressure is sufficient to push fluid out of the capillary with an NFP of approximately 14 mm Hg. Fluid exits the blood into the interstitial space, but most plasma proteins remain within the blood.

In contrast, at the venous end of capillaries, the NFP is approximately −5 mm Hg. This relatively negative pressure causes fluid to be pulled in the opposite direction from the interstitial space back into the blood. However, most but not all of the fluid returns to the blood on the venous end. The excess fluid remaining in the interstitial space, which is approximately 15% of what was originally filtered, is taken up by the lymphatic system (see section 21.1a). Changes to either NFP or lymph transport can result in edema (figure 25.5).

Fluid intake and fluid output regulation generally occur in response to the fluctuation of these variables, which are discussed in the following sections.

Regulating Fluid Intake

Fluid intake is controlled by various stimuli that either activate or inhibit the **thirst center** located within the hypothalamus (see section 13.4c).

Stimuli to Turn on the Thirst Center Stimuli for activating the thirst center, which occurs when fluid intake is less than fluid output, include the following:

- **Decreased salivary secretions.** Saliva production decreases, and mucous membranes are not as moist, when less fluid is available. Sensory input is relayed from sensory receptors in the mucous membranes of the mouth and throat to the thirst center.
- **Increased blood osmolarity.** This occurs most commonly from insufficient water intake and dehydration. The increase in blood osmolarity stimulates sensory receptors in the thirst center directly, and also stimulates the hypothalamus to initiate nerve signals to the posterior pituitary to release antidiuretic hormone (ADH) (see section 17.7b). ADH also stimulates the thirst center. This stimulation of the thirst center occurs with as little as a 2–3% increase in ADH.
- **Decreased blood pressure.** When fluid intake is less than fluid output, blood volume decreases with an accompanying decrease in blood pressure. Renin is released from the kidney in response to a lower blood pressure (see section 20.6b). Renin initiates the conversion of angiotensinogen to angiotensin II. An increase of 10–15% in the concentration of angiotensin II within the blood stimulates the thirst center. This mechanism is especially important when extreme volume depletion occurs; for example, when an individual is hemorrhaging.

When the thirst center is activated, nerve signals are relayed to the cerebral cortex, and we then become conscious of our thirst. If we take fluid into the body by drinking or eating, water is absorbed from the GI tract into the blood, and the water then moves into the interstitial space and ultimately into the cells (figure 25.3a).

Stimuli to Turn off the Thirst Center Stimuli for inhibiting the thirst center are produced when fluid intake is greater than fluid output. All of these stimuli (except distension of the stomach, described here) oppose stimuli that activate the thirst center. These include the following:

- **Increased salivary secretions.** When body fluid level is high, salivary secretions increase, and the mucous membranes of the mouth and throat become moist. Sensory input to the thirst center decreases.
- **Distension of the stomach.** Fluid entering the stomach causes it to stretch, and nerve signals are relayed to the hypothalamus to inhibit the thirst center. (Note that an empty stomach does not stimulate the thirst center; rather, only a stretched stomach wall will inhibit the thirst center.)
- **Decreased blood osmolarity.** Blood osmolarity decreases when additional fluid enters the blood. In response, the thirst center is no longer stimulated directly, and the hypothalamus decreases stimulation of ADH release from the posterior pituitary.
- **Increased blood pressure.** Blood volume and blood pressure increase with the addition of fluid. This rise in blood pressure inhibits the kidney from releasing renin, and the subsequent production of angiotensin II decreases. A decrease in angiotensin II results in a reduced stimulation of the thirst center.

The stimuli that inhibit the thirst center can be divided into two categories, depending upon both the time required to inhibit the thirst center and their level of accuracy in reflecting the hydrated state of the body. Stimuli that immediately inhibit the thirst center, but are less accurate concerning the hydrated state, include both the moistening of the mucous membranes and distension of the stomach. Signals from these stimuli will inhibit the thirst center for approximately 30 to 45 minutes, which is long enough for the absorption of fluids from the GI tract into the blood plasma.

Once fluids are absorbed, blood osmolarity decreases (with an accompanying decrease in ADH release), and both blood volume and blood pressure increase (with an accompanying decrease in renin release and production of angiotensin II). These are less immediate changes but more accurately reflect the body's state of hydration.

 WHAT DO YOU THINK?

1 Many times during a long-distance race, water stations are positioned along the side of the road so that runners may rehydrate during the race. Sometimes a runner will take a drink, swirl it around in the mouth, and then spit the water back out instead of swallowing it. Do you think this practice should be encouraged? Explain in terms of the effect on the thirst center and the hydrated state of the body.

Regulating Fluid Output

Recall that fluid output is regulated through the kidneys by controlling urine output. Four major hormones are involved in regulating urine output: angiotensin II, antidiuretic hormone (ADH), aldosterone, and atrial natriuretic peptide (ANP).

Angiotensin II, ADH, and aldosterone help decrease urine output. These three hormones function to maintain both blood volume and blood pressure. In contrast, ANP increases urine output to decrease both blood volume and blood pressure. The specific mechanisms employed by each of these hormones in regulating fluid output in the kidneys also function in regulating some electrolytes (e.g., Na$^+$). Thus, it is more appropriate to describe the details of these hormones at the end of the next section on electrolytes.

💡 **WHAT DID YOU LEARN?**

7 What stimuli activate the thirst center?

8 Which of these four hormones—angiotensin II, antidiuretic hormone, aldosterone, and atrial natriuretic peptide—increases urine output?

INTEGRATE

LEARNING STRATEGY ✎

The body can be thought of as a "leaky bucket" that we must continuously refill.

Fluid input. The only significant way to increase total water in the "bucket" is from food and drink, with fluid intake regulated through the thirst center. Fluid input can also be increased therapeutically through administration of an IV solution.

Fluid output (obligatory). Water continuously "leaks" from the body even when we are dehydrated. Obligatory water loss includes expired air, sweat, fluid loss through the skin (sweat and cutaneous transpiration), and defecation. Even when we are dehydrated, a minimum of urine production is also required to eliminate nitrogenous wastes and other substances. Abnormal fluid loss may also occur through vomiting or blood loss.

Fluid output (regulated). The primary means of regulating water loss is through hormonal stimulation of the kidney to control water loss. Three hormones decrease water loss (antidiuretic hormone [ADH], aldosterone, and angiotensin II), whereas atrial natriuretic peptide (ANP) increases water loss. This is like having a small tap in the "bucket" that can be opened or closed. However, even with this regulation, the bucket is still leaky; thus fluids must be replaced daily.

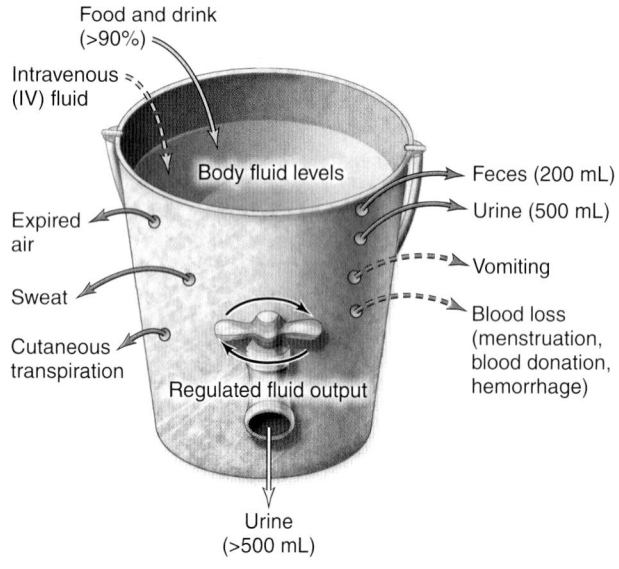

Food and drink (>90%)
Intravenous (IV) fluid
Body fluid levels
Feces (200 mL)
Urine (500 mL)
Expired air
Vomiting
Sweat
Cutaneous transpiration
Regulated fluid output
Blood loss (menstruation, blood donation, hemorrhage)
Urine (>500 mL)

Fluid input
⟹ Normal fluid input
==⟹ Clinical fluid input
Fluid output
⟹ Normal obligatory fluid loss
==⟹ Abnormal fluid loss
⟹ Hormonally controlled fluid loss

Regulated fluid output
Decrease urine output
• Angiotensin II
• Antidiuretic hormone
• Aldosterone
Increase urine output
• Atrial natriuretic peptide

25.3 Electrolyte Balance

Water movement between fluid compartments is caused by a relative difference in the concentration of solutes in those compartments. The greater the difference in the number of solutes, the greater the osmotic pressure (see section 4.3b). Because it is the relative number of solutes that determines the osmotic pressure, solutes are differentiated into nonelectrolytes and electrolytes. Nonelectrolytes and electrolytes were first introduced in section 2.4c.

25.3a Nonelectrolytes and Electrolytes

👁 **LEARNING OBJECTIVES**

1. Describe the difference between a nonelectrolyte and an electrolyte.
2. Explain the general role of electrolytes in fluid balance.

Molecules that do not dissociate (or come apart) in solution are called **nonelectrolytes.** Most of these substances are covalently bonded organic molecules (e.g., glucose, urea, and creatinine). In contrast, an **electrolyte** is any substance that dissociates in solution to form cations and anions. The term electrolyte refers directly to the ability of these substances, when dissolved and dissociated in solution, to conduct an electric current. Electrolytes include salts, acids, bases, and some negatively charged proteins. Examples of electrolytes and the dissociated ions they produce when placed into an aqueous solution (aq) include the following:

Salts: $NaCl \; (aq) \longrightarrow Na^+ + Cl^-$ or $CaCl_2 \; (aq) \longrightarrow Ca^{2+} + 2 \; Cl^-$
Acid: $HCl \; (aq) \longrightarrow H^+ + Cl^-$
Base: $NaOH \; (aq) \longrightarrow Na^+ + OH^-$

NaCl dissociates into two ions, Na^+ and Cl^-. Because osmotic pressure is dependent upon the *number* of solutes, NaCl exerts twice the osmotic pressure of the same concentration of a nonelectrolyte, such as glucose, which does not dissociate into ionic forms. Further, $CaCl_2$ in solution dissociates into three components, Ca^{2+} and two Cl^-, and exerts three times the osmotic pressure compared to that of glucose.

To account for this difference in exerting osmotic pressure, the concentration of electrolytes in solution is commonly expressed in milliequivalents per liter (mEq/L), which reflects the amount or equivalent number of electrical charges in 1 liter of solution. (Milliequivalents is a measure of either the amount of H^+ an anion can bind, or the amount of bicarbonate ion (HCO_3^-) a cation can bind, in 1 liter of solution.)

💡 **WHAT DID YOU LEARN?**

9 Why do electrolytes exert a greater osmotic pressure than nonelectrolytes?

25.3b Major Electrolytes: Location, Functions, and Regulation

👁 **LEARNING OBJECTIVES**

3. List the six major electrolytes found in body fluids, other than H^+ and HCO_3^-.
4. Explain why Na^+ is a critical electrolyte in the body.
5. Describe the variables that influence K^+ distribution.
6. Identify the main location, functions, and the means of regulation for each electrolyte.

The human body fluids contain common electrolytes. The common electrolytes include Na^+, K^+, Cl^-, Ca^{2+}, PO_4^{3-}, Mg^{2+} (as well as H^+

and HCO_3^-). Each electrolyte has unique functions in the body in addition to its general function of contributing to the exertion of osmotic pressure. To carry out these functions effectively, each electrolyte must be maintained within a normal concentration range in the blood plasma (see table 18.3).

Regulating normal concentration of the different electrolytes in the blood plasma requires maintaining input equal to output, similar to what we discussed with respect to water balance. Here we present a description of the location, functions, and regulation of the major electrolytes, with the exception of H^+ and HCO_3^-, which are covered in section 25.5 on acid-base balance.

Sodium Ion (Na⁺)

Approximately 99% of Na^+ is in the ECF and only 1% in the ICF, a gradient that is maintained by Na^+/K^+ pumps. Sodium is the principal cation in the ECF. It is usually present in the form of either sodium bicarbonate ($NaHCO_3$) or sodium chloride ($NaCl$) and exerts the greatest osmotic pressure in the ECF.

Sodium functions in a number of physiologic processes, many of which have been presented in previous chapters (e.g., neuromuscular functions and cotransport in kidney tubules). We now describe how Na^+ concentration is regulated to maintain its balance, and how it functions in determining plasma osmolarity and regulating fluid balance.

Sodium Balance Our normal blood plasma Na^+ concentration is 135–145 mEq/L (**figure 25.6a**). We obtain Na^+ from our diet. Although the daily dietary requirement for Na^+ is only 1.5 to 2.3 grams per day (g/day), our intake may vary from as little as 0.5 g on a low-sodium diet to as high as 20–25 g with high Na^+ intake. A typical American diet averages 3–7 g of Na^+ per day. Most of our Na^+ comes from table salt and processed foods such as canned soups, lunch meats, and crackers. Sodium loss is through urine, feces, and sweat. The blood concentration of Na^+ is regulated by three hormones, namely aldosterone, ADH, and ANP. These hormones also indirectly help in regulating fluid balance and are described in detail later in this chapter.

 WHAT DO YOU THINK?

2 Discuss why physicians might recommend a diet low in Na^+ for individuals with high blood pressure.

Sodium's Role in Blood Plasma Osmolarity As the most common cation in the ECF—composing approximately 90% of the solute concentration in the ECF—Na^+ is therefore the most important electrolyte in determining blood plasma osmolarity and in regulating fluid balance. The ECF becomes temporarily hypertonic if Na^+ concentration increases from either elevated Na^+ intake or decreased water content. Consequently, water moves from the other compartments into the blood plasma in an attempt to reestablish the normal Na^+ concentration (figure 25.6b).

The ECF becomes temporarily hypotonic if Na^+ concentration decreases either as a consequence of decreased Na^+ intake or increased water content. Water then moves from the plasma into the cells until the normal Na^+ concentration is reestablished.

Note that changes in plasma volume directly influence both blood volume and blood pressure. Thus, retention of Na^+ and water increases both blood volume and blood pressure, whereas the loss of Na^+ and water causes both a decrease in blood volume and blood pressure. For this reason, individuals with high blood pressure may be instructed by their physician to restrict their sodium intake. Although high blood pressure has many causes, and individuals retain Na^+ to varying degrees, studies have shown that decreasing dietary intake of Na^+ is effective in lowering blood pressure in some individuals.

Sodium imbalance is one of the most common types of electrolyte imbalances. A Na^+ imbalance occurs when the Na^+ concentration is either above the normal levels (**hypernatremia;** hī′per-nă-trē′mē-ă) or below normal levels (**hyponatremia;** hī′pō-nă-trē′mē-ă). Any condition that alters Na^+ intake or output, or any condition that alters water intake or output, may cause a change in Na^+ plasma concentration. Most occurrences result from changes in water content. The causes, effects, and symptoms of electrolyte imbalances are included in **table 25.2**.

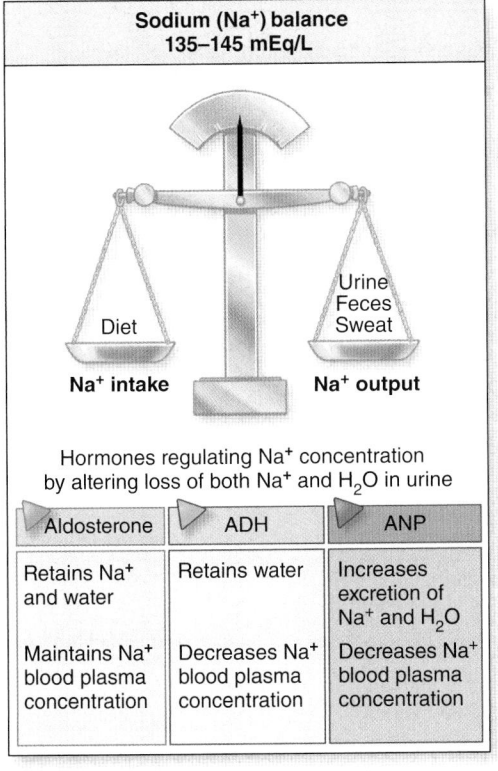

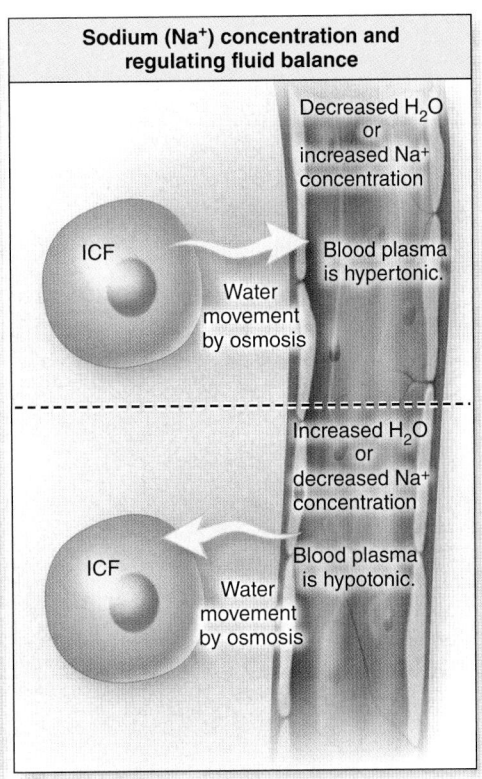

Figure 25.6 Sodium Balance.
Sodium is the principal cation in the extracellular fluid (ECF). (*a*) Normal blood plasma Na^+ concentration is between 135 mEq/L and 145 mEq/L and is important in maintaining fluid balance. Sodium level increases through the diet and decreases through urine, feces, and sweating. Sodium content and concentration are regulated by aldosterone, antidiuretic hormone (ADH), and atrial natriuretic peptide (ANP). (*b*) Changes in Na^+ concentration cause movement of water between fluid compartments.

Potassium Ion (K⁺)

In contrast to Na⁺, approximately 98% of **potassium** is in the ICF and only 2% is in the ECF. This gradient is also maintained by the Na⁺/K⁺ pump. Potassium is the principal cation to exert intracellular osmotic pressure because it is the most abundant cation in the ICF. Potassium is required for normal neuromuscular physiologic activities, and it has a significant role in controlling heart rhythm.

Potassium Balance　Although the vast majority of the total body K⁺ is located within the ICF, it is only the 2% of the K⁺ in the ECF that is continually regulated. The normal plasma value for K⁺ in the ECF is between 3.5 mEq/L and 5.0 mEq/L (**figure 25.7a**). Small changes in blood plasma K⁺ levels can readily lead to a K⁺ imbalance—the *most potentially lethal* of the electrolyte imbalances. Neuromuscular changes (e.g., cardiac arrhythmia, muscle weakness) are the most significant effects of a K⁺ imbalance. These can lead to either cardiac or respiratory arrest.

Both the total body K⁺ and the distribution of K⁺ must be regulated in order to maintain K⁺ balance in our body fluids. **Total body potassium** is regulated by K⁺ intake and output. The daily dietary requirement for K⁺ is 40 mEq/L, although it may vary from 40–150 mEq/L. Potassium is generally obtained from fruits and vegetables, but input is increased with salt substitutes that contain K⁺.

Only small amounts of K⁺ are lost from the body through sweat and in feces. Most K⁺ (approximately 80–90%) is lost in urine. Some K⁺ is always being lost because the body has no means to conserve all of its K⁺. The amount of K⁺ lost in the urine fluctuates, and greater amounts are lost during conditions of high blood plasma K⁺, increased aldosterone secretion, and high blood pH.

Potassium distribution is the ratio of K⁺ in the ICF to that in the ECF. This ratio is dependent upon both the level of activity of Na⁺/K⁺ pumps that actively pump K⁺ into the cell, and the K⁺ leak channels that allow K⁺ to flow back out of the cell. Variables that influence K⁺ distribution are described next.

Potassium Shifts　Potassium can redistribute between compartments under certain conditions that include changing blood plasma K⁺ concentration, changing blood plasma H⁺ concentration, and the presence of specific hormones in the blood (figure 25.7b).

Changes in blood plasma K⁺ concentration cause shifts in K⁺ between the ECF and ICF. An increase in blood plasma K⁺ results in K⁺ moving from the ECF into the ICF, whereas a decrease in blood plasma K⁺ results in the movement of K⁺ from the ICF into the ECF. Thus, the ICF serves as a reservoir for blood plasma K⁺ levels, taking on excess K⁺ in response to elevated blood plasma K⁺ concentration and releasing K⁺ when blood plasma K⁺ concentration begins to decline.

Changes in blood plasma H⁺ concentration also cause a shift in K⁺ between the ECF and ICF. When H⁺ increases (pH decreases) in the ECF, excess H⁺ moves from the ECF into the ICF in an attempt to reestablish acid-base balance (described later in this chapter). At the same time, K⁺ moves in the opposite direction, from the ICF into the ECF, to prevent the development of an electrochemical gradient (see section 4.4b). The reverse movement of both H⁺ and K⁺ occurs if H⁺ decreases (pH increases). A primary reason an acid-base imbalance can be lethal is because of these shifts in K⁺ that occur in response to pH changes.

Certain hormones also induce shifts in K⁺ between the ECF and ICF; insulin is one important example. Insulin is typically released following a meal and decreases not only blood glucose (as described in section 17.9b) but also blood K⁺. Insulin decreases blood plasma K⁺ by stimulating the activity of Na⁺/K⁺ pumps in cells, thus increasing the transport of K⁺ from the ECF into the ICF. Physiologists reason that this helps prevent elevated K⁺ in the blood plasma (a condition termed **hyperkalemia;** hī′per-kă-lē′mē-ă) that could potentially occur as K⁺ is absorbed into the blood following a meal. Interestingly, individuals exhibiting hyperkalemia may be treated with the administration of insulin (along with glucose).

WHAT DO YOU THINK?

❸ Body builders have been known to inject insulin to increase muscle mass because it stimulates protein anabolism. What is one of the risks associated with this practice in terms of K⁺ levels: elevated blood plasma K⁺ (hyperkalemia) or decreased blood plasma K⁺ (hypokalemia)? Could the practice of injecting insulin to increase muscle mass be fatal?

Figure 25.7 Potassium Balance. Potassium is the principal cation in the intracellular fluid (ICF). (*a*) Normal blood plasma K⁺ concentration is 3.5–5.0 mEq/L. Potassium balance is dependent upon both total K⁺ and its distribution in the fluid compartments. Total K⁺ levels are a function of intake from the diet and loss through urine, feces, and sweat. (*b*) Potassium distribution changes in response to changing levels of K⁺ in the extracellular fluid (ECF), changes in H⁺ blood plasma concentration, and to the presence of specific hormones (e.g., insulin).

Total potassium (K⁺) and its distribution

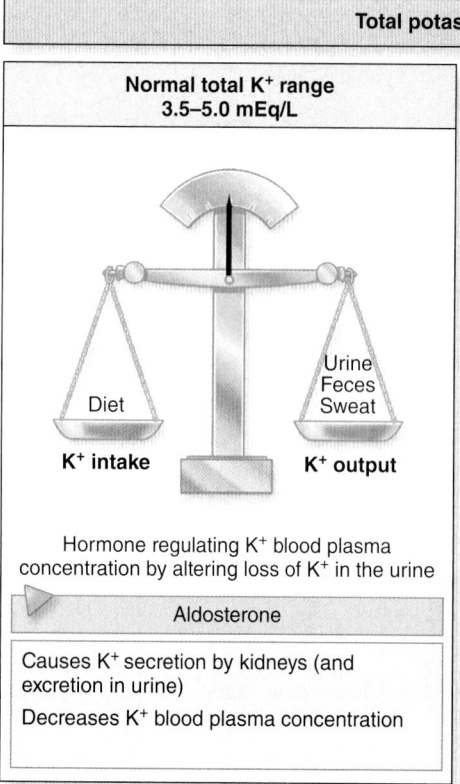

(a) Normal total K⁺ range 3.5–5.0 mEq/L

Diet — K⁺ intake

Urine Feces Sweat — K⁺ output

Hormone regulating K⁺ blood plasma concentration by altering loss of K⁺ in the urine

Aldosterone

Causes K⁺ secretion by kidneys (and excretion in urine)

Decreases K⁺ blood plasma concentration

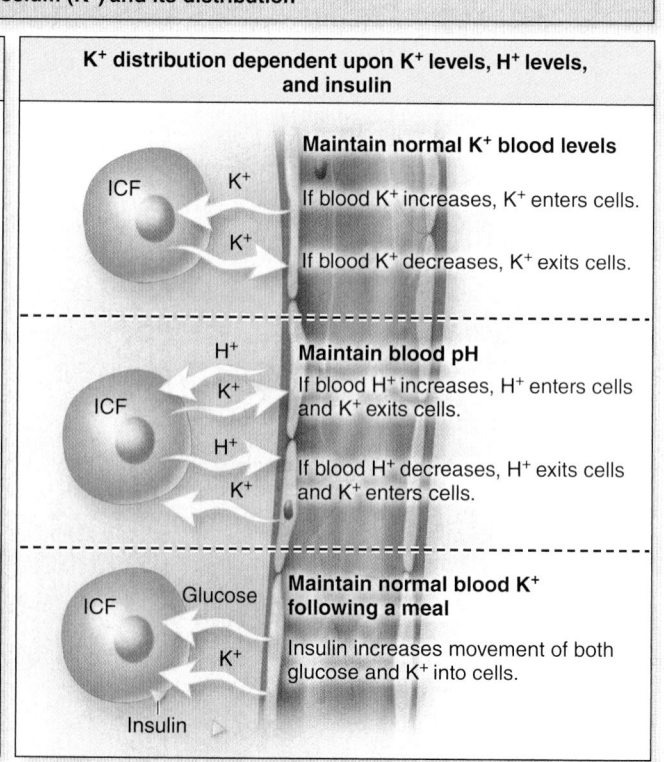

(b) K⁺ distribution dependent upon K⁺ levels, H⁺ levels, and insulin

Maintain normal K⁺ blood levels
ICF — K⁺
If blood K⁺ increases, K⁺ enters cells.
If blood K⁺ decreases, K⁺ exits cells.

Maintain blood pH
ICF — H⁺, K⁺
If blood H⁺ increases, H⁺ enters cells and K⁺ exits cells.
If blood H⁺ decreases, H⁺ exits cells and K⁺ enters cells.

Maintain normal blood K⁺ following a meal
ICF — Glucose, K⁺
Insulin increases movement of both glucose and K⁺ into cells.
Insulin

Table 25.2	Blood Electrolyte Imbalances		
Condition	**Causes**	**Effects**	**Symptoms**
Hypernatremia ($Na^+ > 145$ mEq/L)	Dehydration, diabetes insipidus (hyposecretion of ADH), administration of intravenous (IV) saline	Cell shrinkage, resulting in neurologic impairment; may result in excess fluid causing high blood pressure or edema	Confusion, coma, paralysis of breathing muscles, death
Hyponatremia ($Na^+ < 135$ mEq/L)	Excessive water intake, hypersecretion of antidiuretic hormone (ADH), diuretic abuse, severe diarrhea, burns, hyposecretion of aldosterone, excessive sweating	Cell swelling and decrease in blood volume and blood pressure	Nausea, lethargy, confusion, headache, muscle cramps, seizures, coma, death
Hyperkalemia ($K^+ > 5.0$ mEq/L)	Aldosterone hyposecretion, renal failure, acidosis, decreased insulin, extensive cellular trauma from a crushing injury or burn, transfusion of outdated blood, hemolytic anemia	A fast rise in K^+ results in K^+ diffusing into cells, altering the resting membrane potential; interferes with an action potential A slow rise in K^+ inactivates Na^+ channels; action potentials impaired, with muscles and nerves being less excitable	Nausea, vomiting, diarrhea, skeletal muscle weakness, tingling of skin, numbness of hands or feet, and irregular heartbeat that can lead to cardiac arrest[1] Can be asymptomatic, or an individual may have nausea, fatigue, muscle weakness
Hypokalemia ($K^+ < 3.5$ mEq/L)	Diuretic abuse, aldosterone hypersecretion, chronic vomiting, diarrhea, excessive laxative or enema use, heavy sweating, alkalosis, increased insulin, or formation of new tissue because K^+ is the predominant intracellular fluid (ICF) cation	Cells become hyperpolarized, interfering with neuron and muscle function	Nausea, vomiting; nervous and muscle tissue are less excitable, resulting in numbness, muscle weakness, decreased tone of smooth muscle, flaccid paralysis; decreased and irregular heart rate that may lead to cardiac arrest; muscle weakness can be severe enough to cause paralysis of diaphragm, leading to respiratory arrest
Hyperchloremia ($Cl^- > 107$ mEq/L)	Excess Cl^- in diet, intravenous (IV) solution containing NaCl, diarrhea, hyperaldosteronism, hyperparathyroidism, kidney disease; usually associated with metabolic acidosis	Decreased nerve excitability; causes changes in concentration of other ions	May be asymptomatic or the individual may have irregular breathing, weakness, and intense thirst; symptoms generally arise from other conditions associated with hyperchloremia
Hypochloremia ($Cl^- < 97$ mEq/L)	Heavy sweating, vomiting, gastrointestinal (GI) suction, kidney disease, metabolic alkalosis	Increased nerve excitability; causes changes in concentration of other ions	Many individuals are asymptomatic, but may experience tetany, hyperactive reflexes, muscle cramps; if severe— arrhythmias, seizures, coma, or respiratory arrest
Hypercalcemia ($Ca^{2+} > 5.2$ mEq/L)	Hyperparathyroidism, hypothyroidism (decreased calcitonin), bone loss from cancer, immobilization, excessive intake of vitamin D or calcium antacids, renal failure	Interferes with normal muscle and nerve function and causes underexcitability of muscles and nerves	May be asymptomatic; muscle weakness, fatigue, depression, confusion, loss of appetite, nausea, vomiting, constipation, and abnormal heart rhythms; severe calcification of soft tissue
Hypocalcemia ($Ca^{2+} < 4.5$ mEq/L)	Vitamin D deficiency, diarrhea, pregnancy, lactation; less calcium released from bone due to hypoparathyroidism, or hyperthyroidism (increased calcitonin)	Overexcitability of muscles and nerves	Numbness, tingling, muscle spasms, increased heart rate; skeletal muscle also more excitable and can go into tetany, causing laryngospasms and suffocation
Hyperphosphatemia ($PO_4^{3-} > 2.9$ mEq/L)	Kidney impairment; increased dietary intake of supplements; laxatives or enemas; respiratory acidosis, cellular damage	Precipitation of Ca^{2+}, resulting in hypocalcemia	Generally asymptomatic by itself; however, may lead to hypocalcemia as PO_4^{3-} binds with Ca^{2+}; symptoms affect muscles, nerves, and bone calcification
Hypophosphatemia ($PO_4^{3-} < 1.8$ mEq/L)	Insufficient PO_4^{3-} intake; hyperparathyroidism; respiratory alkalosis	Many effects due to decreased production of ATP	Weakness and dysfunction of nervous system and both skeletal and cardiac muscle
Hypermagnesemia ($Mg^{2+} > 2.1$ mEq/L)	Kidney impairment, high intake of Mg^{2+}, such as in antacids or enemas	Changes in neuromuscular, cardiovascular, and central nervous system	Depressed neuromuscular activity, hypotension, cardiac arrest, weakness, lethargy, and respiratory depression
Hypomagnesemia ($Mg^{2+} < 1.4$ mEq/L)	Impaired absorption from the GI tract; increased loss in kidneys, or redistribution of Mg^{2+}; chronic alcoholism	Changes in neuromuscular, cardiovascular, and central nervous system	Muscle twitches, tremors, hyperactive reflexes, mental confusion and depression

1. Lethal injections for euthanizing animals and for capital punishment include high doses of K^+, causing cardiac arrest.

Chloride Ion (Cl⁻)

Chloride is a common anion that is normally associated with Na⁺ (as NaCl). It is the most abundant anion in the ECF. Chloride ions are found in the lumen of stomach as hydrochloric acid (see section 26.2d), and they also participate in the chloride shift within erythrocytes for transport of carbon dioxide in the form of HCO_3^- (see figure 23.27).

Normal blood plasma Cl⁻ concentration is 96–106 mEq/L. Chloride is obtained in the diet, primarily from table salt and processed foods. Chloride is normally lost in the sweat, stomach secretions, and urine. The amount lost in urine is primarily dependent upon blood plasma Na⁺. Chloride levels are directly correlated with Na⁺ levels, because Cl⁻ follows Na⁺ by electrostatic interactions and is regulated by the same mechanisms as Na⁺.

Calcium Ion (Ca²⁺)

Calcium is the most abundant electrolyte in bone and teeth. Approximately 99% of all body Ca²⁺ (most commonly as calcium phosphate, $Ca_3[PO_4]_2$), is stored within the extracellular matrix of these structures, causing them to harden (see section 7.2e). To prevent hardening or calcification of other tissue, Ca²⁺ is normally moved by Ca²⁺ pumps either out of cells or into the sarcoplasmic reticulum within muscle cells. This prevents Ca²⁺ from binding to the abundant PO_4^{3-} within cells. In addition to hardening of both bones and teeth, Ca²⁺ is needed to initiate muscle contraction and release neurotransmitters; it also serves as a second messenger, and participates in blood clotting.

Normal blood plasma Ca²⁺ concentration is approximately 5 mEq/L. In blood plasma, Ca²⁺ can exist either bound to protein such as albumin (about 35%), associated with anions such as PO_4^{3-} (about 15%), or in an ionized or unbound state (about 50%). Only the ionized form is physiologically active. Calcium is obtained in the diet from yogurt, milk, soy products, cheese, sardines, and green leafy vegetables (such as broccoli and collard greens). Calcium is lost from the body in urine, feces, and sweat. (The regulation of Ca²⁺ by parathyroid hormone, calcitriol, and calcitonin is discussed in sections 7.6 and 24.6d.)

Phosphate Ion (PO₄³⁻)

Phosphate ions (PO_4^{3-}) occur as both hydrogen phosphate (HPO_4^{2-}) and dihydrogen phosphate ($H_2PO_4^-$). Phosphate is the most abundant anion in the ICF. Approximately 85% is stored in the extracellular matrix of bone and teeth as calcium phosphate, $Ca_3(PO_4)_2$. Phosphate is also a component of nucleotides of DNA and RNA and of phospholipid molecules within plasma membranes. It serves as an intracellular buffer against pH changes (described later) and is a common buffer in urine.

Normal blood plasma concentration of PO_4^{3-} ranges from 1.8–2.9 mEq/L. Most PO_4^{3-} is ionized (about 90%) in blood plasma and the rest is bound to plasma proteins such as albumin. Phosphorus is obtained in the diet from milk, meat, and fish. In addition, many food additives (used to prevent spoilage) and most soft drinks contain PO_4^{3-}. Phosphate is regulated by many of the same mechanisms as Ca²⁺; this is because as just described 99% of body calcium is stored in bone, with the majority stored as calcium phosphate.

Magnesium Ion (Mg²⁺)

Magnesium is primarily located within bone or within cells. After K⁺, Mg²⁺ is the most abundant cation in the ICF. Magnesium participates in over 300 enzymatic reactions including ATP synthesis, the synthesis of protein, and enzymatic reactions involving carbohydrate metabolism. It also assists in the movement of Na⁺ and K⁺ across the plasma membrane by the Na⁺/K⁺ pump, and it is important in muscle relaxation.

Normal blood plasma concentration of Mg²⁺ ranges from 1.4–2.1 mEq/L. Because most Mg²⁺ is located within cells, this represents only a small fraction of total body Mg²⁺. In blood plasma, Mg²⁺ is either in a free, ionized form or bound to protein. The ionized form is physiologically active. Magnesium is obtained in the diet from beans and peas, leafy green vegetables, and other sources, and is lost from the body in sweat and urine. Magnesium blood plasma levels are regulated through the kidney.

 WHAT DID YOU LEARN?

 10 What is the net direction of K⁺ movement in response to a decrease in pH? Explain.

25.4 Hormonal Regulation

Four hormones play a major role in homeostatic regulation of both fluid and electrolytes: angiotensin II, antidiuretic hormone, aldosterone, and atrial natriuretic peptide. We look closely at the details of regulatory mechanisms for each hormone, including the stimulus for its release and its subsequent effects on numerous variables, such as fluid intake, fluid output, blood volume, blood pressure, and blood osmolarity.

Before reading this section, recall that an increase in fluid intake increases blood volume and systemic blood pressure, and may decrease blood osmolarity if more water than solutes are taken in. The reverse occurs with a decrease in fluid intake.

25.4a Angiotensin II

 LEARNING OBJECTIVES

1. Explain the means by which angiotensin II formation can be triggered.
2. List the four primary effects of angiotensin II.

Angiotensin II is a peptide hormone that is unique among the four hormones that affect fluid balance because (1) it has two significantly different means of stimulation for its formation, and (2) when produced, it stimulates the most diverse effects **(figure 25.8)**.

We previously described angiotensin II as a potent vasoconstrictor that helps regulate blood pressure (see sections 17.10c, 20.6b, and 24.5e). Recall that angiotensinogen is an inactive hormone synthesized and released continuously from the liver. Its activation, which occurs within the blood, is initiated by the enzyme renin. Renin is released from the juxtaglomerular (JG) apparatus of the kidneys in response to either (1) low blood pressure (as detected by decreased stretch of baroreceptors within granular cells, or by decreased NaCl detected by chemoreceptors within macula densa cells; see section 24.3c); or (2) stimulation by the sympathetic division. The sequential action of renin and angiotensin converting enzyme (ACE) (which is bound to the endothelial lining of blood vessels) cause the formation of angiotensin II (the active form of the hormone).

Angiotensin II has a number of effectors (target organs), and it initiates the following changes when it binds to these structures:

- **Blood vessels.** Stimulates vasoconstriction of systemic blood vessels to increase total peripheral resistance, which increases systemic blood pressure (see section 20.6b).

- **Kidneys.** Decreases urine output from the kidneys as a result of decreased glomerular filtration rate (GFR) in the nephrons by stimulating vasoconstriction of afferent arterioles and contraction of the mesangial cells within the glomerulus (see section 24.5e). This decreases urine output and helps to maintain systemic blood volume, and thus blood pressure.

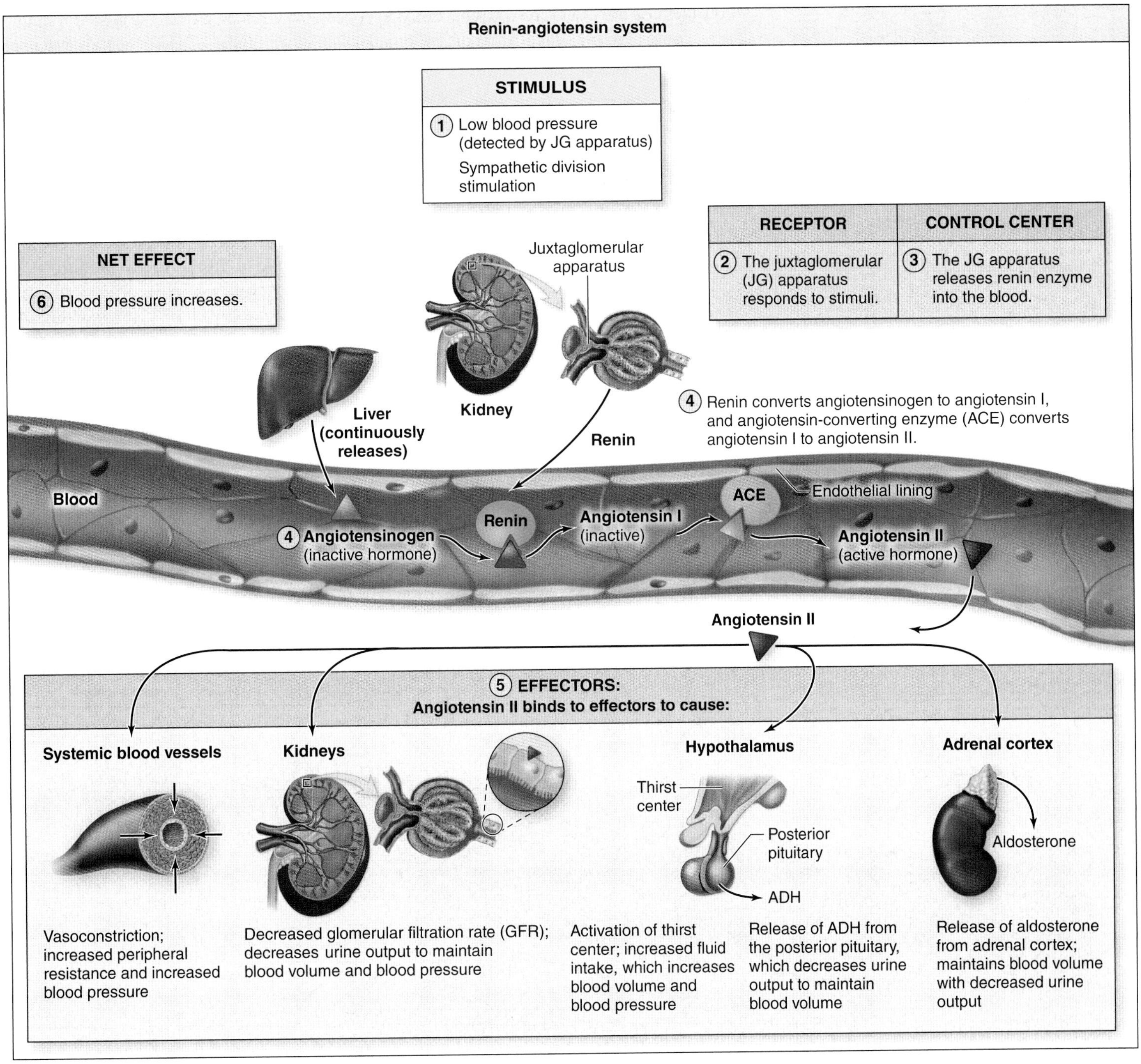

Renin-angiotensin system

STIMULUS

(1) Low blood pressure (detected by JG apparatus)

Sympathetic division stimulation

RECEPTOR	CONTROL CENTER
(2) The juxtaglomerular (JG) apparatus responds to stimuli.	(3) The JG apparatus releases renin enzyme into the blood.

NET EFFECT

(6) Blood pressure increases.

Juxtaglomerular apparatus

Liver (continuously releases)

Kidney

Renin

(4) Renin converts angiotensinogen to angiotensin I, and angiotensin-converting enzyme (ACE) converts angiotensin I to angiotensin II.

Blood

ACE — Endothelial lining

(4) Angiotensinogen (inactive hormone)   Renin   Angiotensin I (inactive)   Angiotensin II (active hormone)

Angiotensin II

(5) EFFECTORS:
Angiotensin II binds to effectors to cause:

Systemic blood vessels	Kidneys	Hypothalamus		Adrenal cortex
		Thirst center	Posterior pituitary / ADH	Aldosterone
Vasoconstriction; increased peripheral resistance and increased blood pressure	Decreased glomerular filtration rate (GFR); decreases urine output to maintain blood volume and blood pressure	Activation of thirst center; increased fluid intake, which increases blood volume and blood pressure	Release of ADH from the posterior pituitary, which decreases urine output to maintain blood volume	Release of aldosterone from adrenal cortex; maintains blood volume with decreased urine output

Figure 25.8 Renin-Angiotensin System. Angiotensin II production is initiated when renin enzyme is released into the blood from the kidney in response to either decreased blood pressure or stimulation by the sympathetic division. The effect of angiotensin II is to increase peripheral resistance, and maintain and increase blood volume and blood pressure to within normal homeostatic limits. AP|R

- **Thirst center.** Stimulates the thirst center in the hypothalamus. If fluid intake occurs, this increases blood volume, which increases systemic blood pressure.

- **Hypothalamus and adrenal cortex.** Stimulates both the hypothalamus to activate the posterior pituitary to release ADH and adrenal cortex to release aldosterone (both described shortly).

The formation and action of angiotensin II can be summarized as follows: It is synthesized either when blood pressure is low or the sympathetic division is activated. It causes an increase in resistance, decrease in fluid output (which helps to maintain blood volume and blood pressure), and an increase in blood volume (if fluid intake occurs). Consequently, blood pressure increases. Increasing blood pressure is aided by the release of both ADH and aldosterone. As blood pressure returns to within normal homeostatic levels, both renin release and angiotensin II synthesis are decreased by negative feedback.

WHAT DID YOU LEARN?

(11) How does angiotensin II alter fluid output and potentially alter fluid intake?

25.4b Antidiuretic Hormone

LEARNING OBJECTIVES

3. Explain how release of antidiuretic hormone (ADH) occurs from the posterior pituitary.

4. Describe the three actions of antidiuretic hormone.

Antidiuretic hormone (ADH) (also called **vasopressin**) is released from the posterior pituitary in response to nerve signals from the hypothalamus **(figure 25.9)**. ADH is a peptide that is synthesized by the hypothalamus and then transported to the posterior pituitary, where it is stored and released (see sections 17.7b and 24.6d).

Three primary types of stimuli—low blood pressure, low blood volume, or an increase in blood osmolarity—signal the need to retain fluid:

- Angiotensin II binds with receptors of cells of the hypothalamus, having been released in response to low blood pressure, as just described.
- Sensory input from baroreceptors within the atria of the heart and aorta and carotid blood vessels stimulate the hypothalamus. A decrease in this nerve stimulation is initiated in response to decreased stretch of baroreceptors within the atria, aorta, and the carotid arteries; the decrease in stretch is caused by low blood volume in the

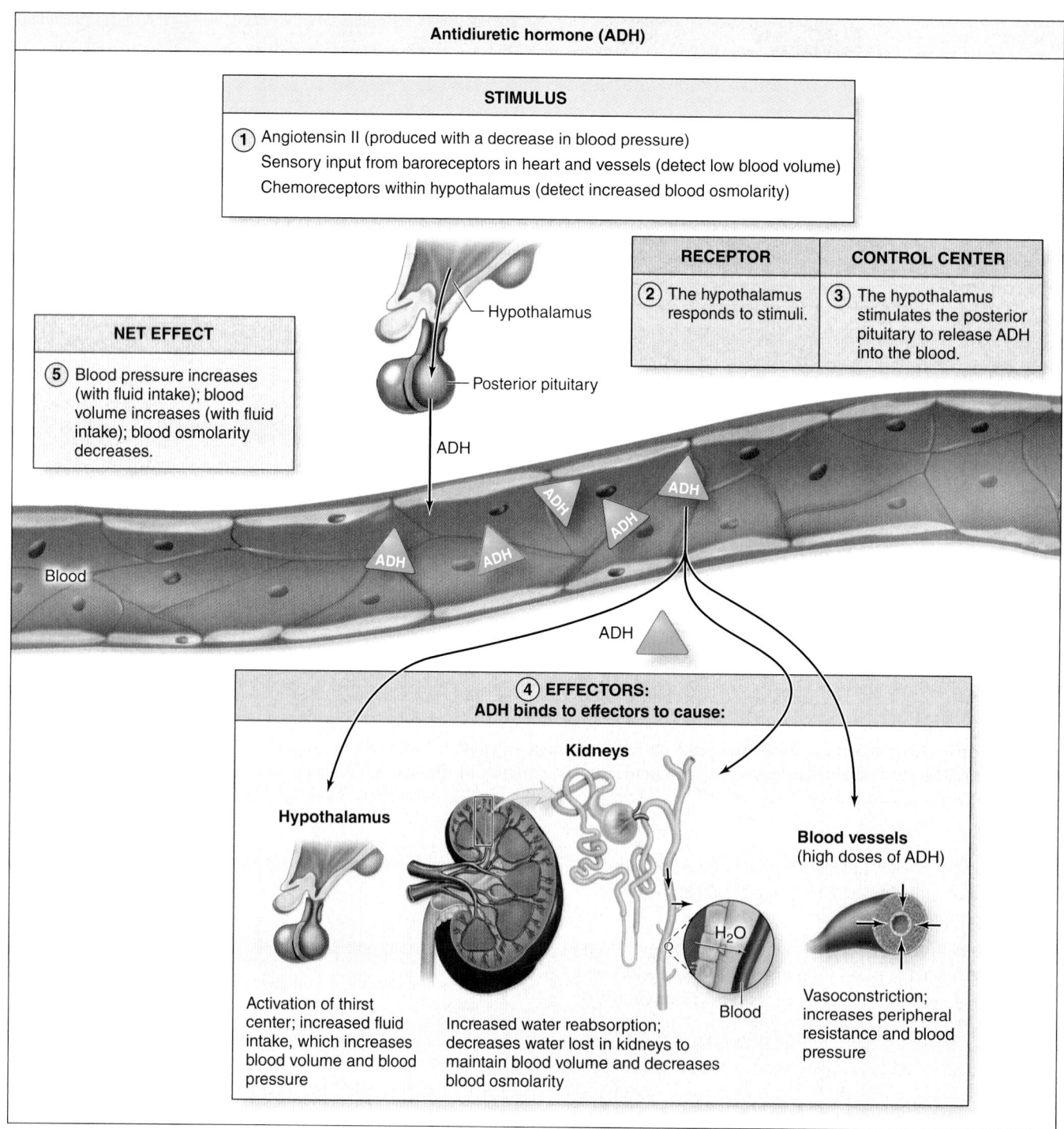

Figure 25.9 Actions and Effects of Antidiuretic Hormone. Antidiuretic hormone (ADH) is released in response to various stimuli, including angiotensin II, low blood volume, and high blood osmolarity. The effect of ADH on target organs is to maintain and possibly increase blood volume and blood pressure, and decrease blood osmolarity to within normal homeostatic limits.

vasculature (see section 20.6a). Please note that this decrease in sensory nerve input from the heart and blood vessels to the hypothalamus is critical under conditions of severe blood loss.

- Chemoreceptors within the hypothalamus detect an increase in blood osmolarity as blood is transported through the capillaries of the hypothalamus. Increased blood osmolarity is the primary stimulus for release of ADH.

In response to any of these stimuli, the hypothalamus increases release of ADH from the posterior pituitary: The larger the change in these stimuli, the greater the amount of ADH released.

ADH is transported in the blood and initiates the following changes when it binds to these various effectors:

- **Thirst center.** ADH stimulates the thirst center in the hypothalamus. If fluid intake occurs, blood volume and blood pressure increase, and blood osmolarity decreases.
- **Kidneys.** ADH increases water reabsorption in the kidneys. ADH binds to principal cells of the collecting tubules and ducts, stimulating these cells to increase the number of aquaporins in the tubular membrane. Additional water is reabsorbed through these aquaporins in response to the osmotic gradient of the interstitial fluid (as described in section 24.6d). This helps to both maintain blood volume by decreasing fluid loss in urine and decrease blood osmolarity.
- **Blood vessels.** High doses of ADH (which occur, for example, with severe hemorrhage) cause vasoconstriction of systemic blood vessels, which increases peripheral resistance. This action is the reason that ADH is also referred to as vasopressin. Systemic blood pressure increases as a result.

The release and action of ADH can be summarized as follows: ADH is released under conditions of low blood pressure (action of angiotensin II), low blood volume (detected by stretch receptors in the heart and blood vessels), and high blood osmolarity (dehydration) to stimulate both fluid intake and water reabsorption in the kidneys. If fluid intake occurs, then blood pressure increases, blood osmolarity is further decreased, and blood volume increases. As the stimuli return to within normal homeostatic levels, ADH release is decreased through negative feedback.

 WHAT DID YOU LEARN?

12 How does the homeostatic system involving ADH function? Include how it is released and its actions.

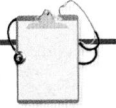
INTEGRATE

CLINICAL VIEW
Diabetes Insipidus

Diabetes insipidus (which affects approximately 3 per 100,000 in the general population) results from either hyposecretion of ADH from the posterior pituitary or the inability of the kidney to respond to ADH. The result is a decrease in fluid retention, with increased urine production. Individuals with diabetes insipidus, if left untreated, can lose up to 20 L of fluid daily!

25.4c Aldosterone

 LEARNING OBJECTIVES

5. List three conditions that lead to aldosterone release.
6. Describe the changes that occur in response to binding of aldosterone by kidney cells.

Aldosterone (ALDO) is normally released from the adrenal cortex in response to angiotensin II, decreased blood plasma Na^+ levels, or most importantly, increased blood plasma K^+ levels (**figure 25.10**).

Aldosterone is a steroid hormone that is transported within the blood plasma and eventually binds to receptors within principal cells of the kidney, as described in sections 17.8c and 24.6d. The binding of aldosterone to these cells causes increased reabsorption and retention of both Na^+ and water, and increased secretion and then excretion of K^+. Aldosterone increases the number of Na^+/K^+ pumps and Na^+ channels, so more Na^+ is reabsorbed from the filtrate back into the blood. Water follows the Na^+ movement by osmosis. Fluid retention results in decreased urine output. Because equal amounts of Na^+ and water are reabsorbed, blood osmolarity remains constant.

Note that K^+ excretion is normally increased except under conditions of low pH. In acidic conditions, as Na^+ and water are reabsorbed from the tubule into the blood, H^+ (instead of K^+) is secreted from the blood into the tubule. This loss of excess H^+ assists in returning blood pH to within normal homeostatic limits.

Both low blood pressure and changes in Na^+ and K^+ blood plasma levels cause aldosterone release. Blood volume and blood pressure are maintained through the reabsorption of both Na^+ and water in the kidneys, with no change to osmolarity. K^+ secretion is increased (unless there is an increase in H^+). As blood volume, blood pressure, and both Na^+ and K^+ blood plasma levels return to

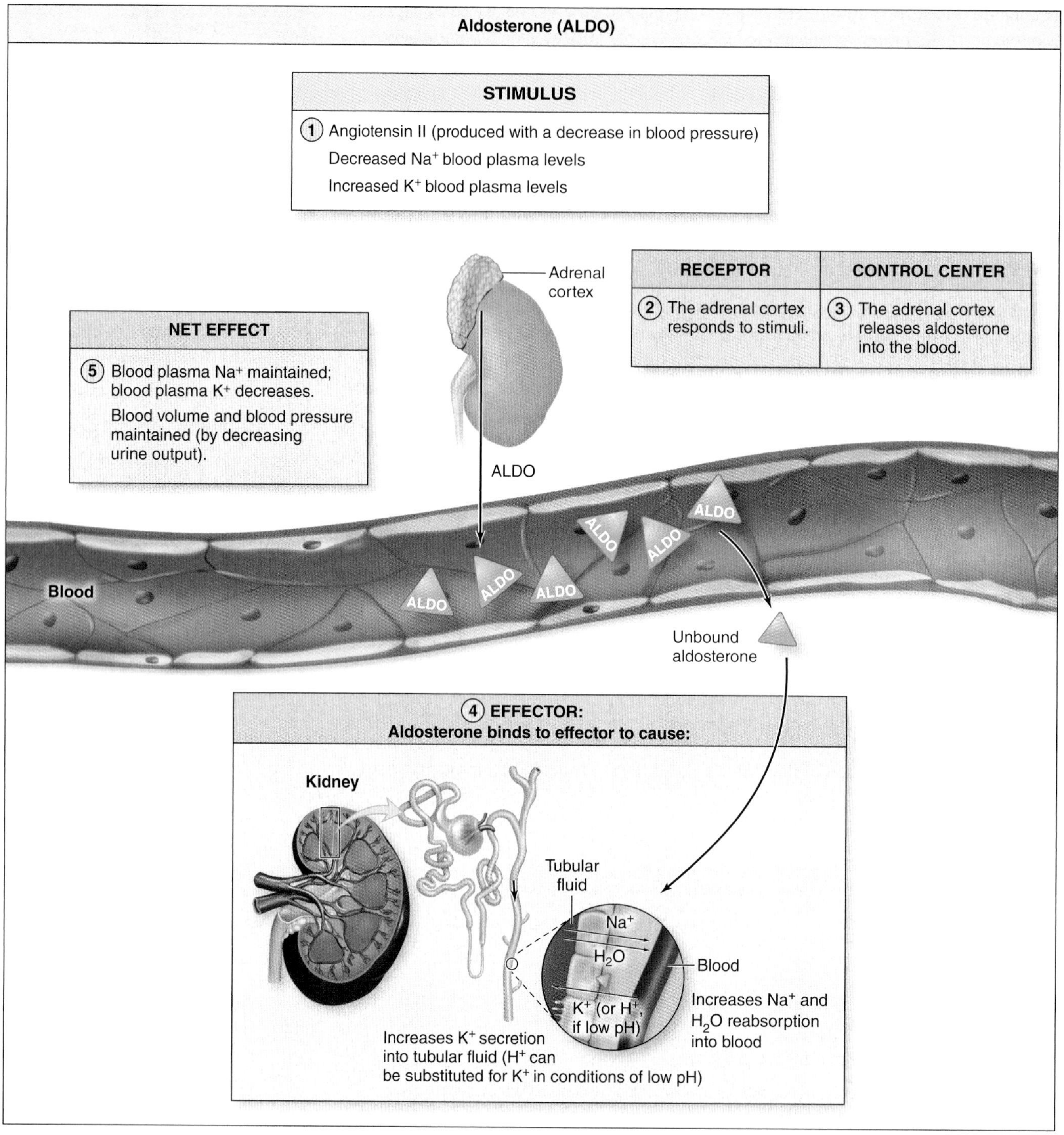

Aldosterone (ALDO)

STIMULUS

① Angiotensin II (produced with a decrease in blood pressure)
Decreased Na^+ blood plasma levels
Increased K^+ blood plasma levels

Adrenal cortex

RECEPTOR	CONTROL CENTER
② The adrenal cortex responds to stimuli.	③ The adrenal cortex releases aldosterone into the blood.

NET EFFECT

⑤ Blood plasma Na^+ maintained; blood plasma K^+ decreases.
Blood volume and blood pressure maintained (by decreasing urine output).

ALDO

Blood

ALDO ALDO ALDO ALDO ALDO ALDO

Unbound aldosterone

④ **EFFECTOR:**
Aldosterone binds to effector to cause:

Kidney

Tubular fluid

Na^+
H_2O ——Blood
K^+ (or H^+, if low pH)

Increases Na^+ and H_2O reabsorption into blood

Increases K^+ secretion into tubular fluid (H^+ can be substituted for K^+ in conditions of low pH)

Figure 25.10 Actions and Effects of Aldosterone. Aldosterone is released from the adrenal cortex in response to various stimuli, including angiotensin II, decreased blood plasma levels of Na^+, or increased blood plasma levels of K^+. The effect of aldosterone is to decrease urine output to maintain blood volume, and alter blood plasma concentration of Na^+ and K^+ to within normal homeostatic limits.

normal ranges, aldosterone release is decreased. Thus, as with angiotensin II and ADH, aldosterone release and its effects are regulated by negative feedback.

WHAT DID YOU LEARN?

⑬ How does aldosterone influence the contents and volume of fluid output?

25.4d Atrial Natriuretic Peptide

LEARNING OBJECTIVES

7. Describe the stimulus for the release of atrial natriuretic peptide (ANP) and its three actions.

8. Explain the ways in which the effects of atrial natriuretic peptide differ from the effects of angiotensin II, ADH, and aldosterone.

Atrial natriuretic (nā′trē-yū-ret′ik; *natrium* = sodium, *uresis* = urination) **peptide (ANP)** is a peptide hormone that opposes the actions of the three hormones just discussed (figure 25.11). ANP is released into the blood from cells in the heart atria. The stimulus for its release is increased stretch of these chambers, which is an indication of both increased blood volume and blood pressure (see sections 17.10c, 20.6b, 24.5e, and 24.6d).

ANP decreases both blood volume and blood pressure by binding to these target organs and causing the following responses:

- **Blood vessels.** Dilates systemic blood vessels, resulting in decreased total peripheral resistance. Systemic blood pressure decreases as a result.
- **Kidneys.** Causes vasodilation of the afferent arterioles in the kidneys and relaxation of mesangial cells; both increase the glomerular filtration rate (see section 24.5e).

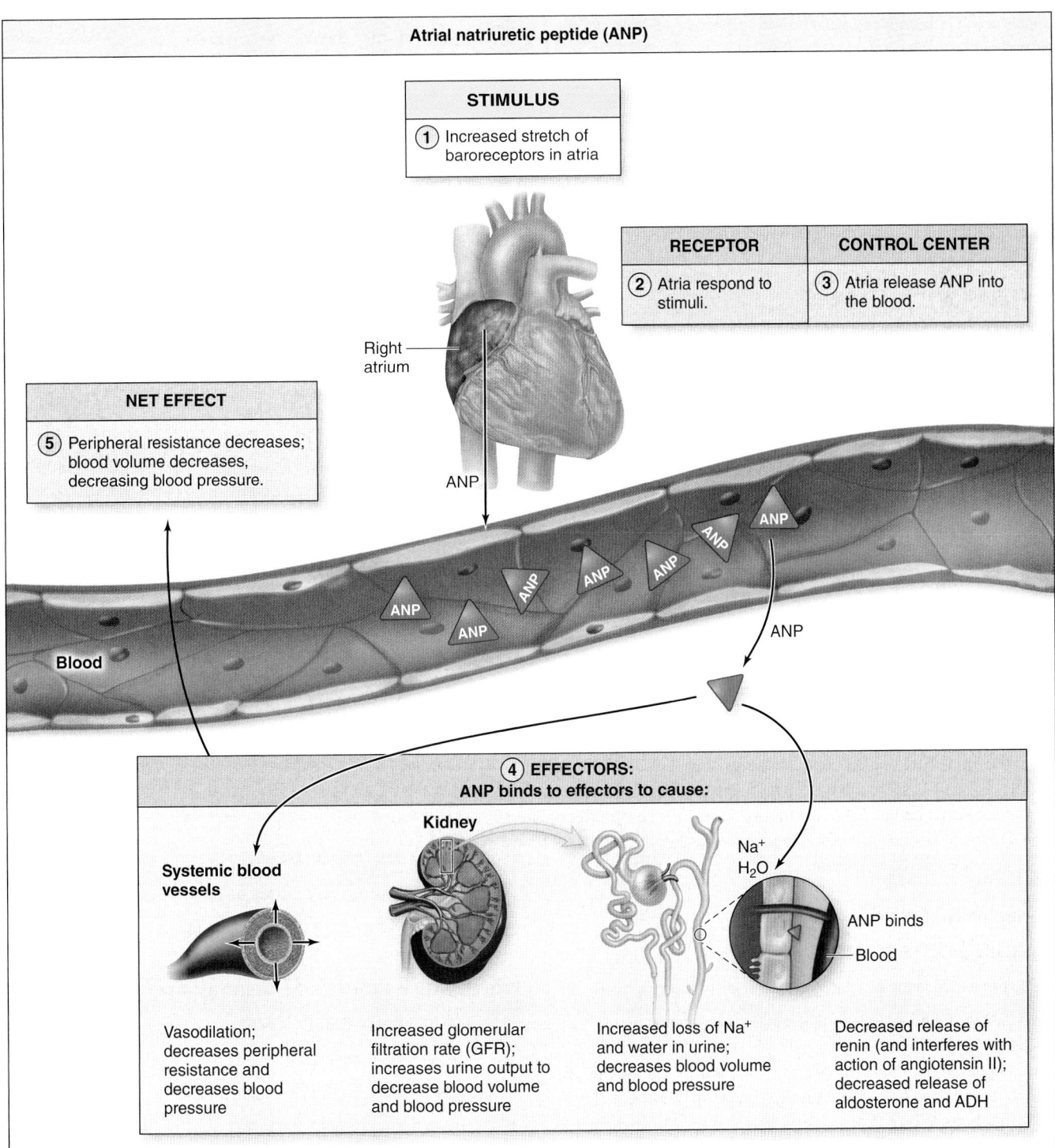

Figure 25.11 Actions and Effects of Atrial Natriuretic Peptide. Atrial natriuretic peptide (ANP) is released in response to increased stretch in atria as a consequence of high blood volume and high blood pressure. The effect of ANP is to decrease both peripheral resistance and blood volume with a resultant decrease in blood pressure to within normal homeostatic limits.

CONCEPT CONNECTION

Blood pressure is one of the most fascinating and complex physiologic variables of the body. Blood pressure is similar to body temperature because it has a set point that is maintained through a variety of mechanisms. Three major variables influence blood pressure, as described in other chapters:

- Cardiac output (see section 19.9)
- Resistance within blood vessels (see section 20.5)
- Blood volume (see section 20.6b)

In this chapter, we have discussed and integrated how various hormones regulate both blood volume and resistance to alter blood pressure.

CONCEPT CONNECTION

Recall from section 2.8b that normal three-dimensional folding of protein is pH dependent. If pH changes occur, the protein may permanently unfold or denature. Under acidic conditions, H^+ binds to the protein; under alkaline conditions, H^+ is released from the protein. In either case, the weak intermolecular bonds that hold the shape of a protein are broken, and it loses its three-dimensional shape and ability to function. Because most enzymes are proteins, one of the most serious consequences of an acid-base imbalance is decreased enzymatic activity in metabolic pathways. If severe enough, these disruptions can be fatal.

Additionally, ANP inhibits Na^+ and water reabsorption by nephron tubules, resulting in additional loss of Na^+ and water (see section 24.6d). These changes increase urine output. Blood volume and systemic blood pressure decreases.

In addition, atrial natriuretic peptide inhibits the release of renin, the action of angiotensin II, and the release of ADH and aldosterone, thus preventing the actions of these hormones.

Note: The details for angiotensin II, ADH, aldosterone, and ANP are included in the summary table "Regulating Fluid Balance, Blood Volume, and Blood Pressure," which directly follows chapter 17 (see **table R.7**).

 WHAT DID YOU LEARN?

14 How does ANP influence fluid output, blood volume, and systemic blood pressure?

25.5 Acid-Base Balance

Acid-base balance is also called *pH balance.* It requires the regulation of hydrogen ion (H^+) concentration in body fluids in order to maintain a slightly alkaline arterial blood pH that is between 7.35 and 7.45. The regulation of H^+ concentration is under the same constraints as other electrolytes; that is, the input must equal output. However, there are some unique considerations to acid-base regulation. Hydrogen ion concentration is altered by the input and output of both acids and bases, by the respiratory rate, and by chemical buffers. We describe here the details of each of the factors that influence acid-base balance.

25.5a Categories of Acid

 LEARNING OBJECTIVES

1. Distinguish between the two categories of acids in the body.
2. Name the two buffering systems that regulate each category.

You first encountered the concepts of acidity, alkalinity, and pH in section 2.5, and you've seen how maintaining a proper pH balance is critical to body functions. The pH of a solution was shown to be inversely related to H^+ concentration. Adding an acid increases the H^+ concentration, resulting in a lower pH; adding a base reduces the amount of H^+ in solution and results in a higher pH.

Two major categories of acid are present in the body; fixed acid and volatile acid. **Fixed acid** (or *metabolic acid*) is the wastes produced from metabolic processes (other than from carbon dioxide). Examples of fixed acids include lactic acid from glycolysis, phosphoric acid from nucleic acid metabolism, and ketoacids from metabolism of fat.

Volatile (vol′ă-til; to evaporate quickly) **acid** is *carbonic acid* produced when carbon dioxide combines with water, as shown:

$$CO_2 + H_2O \rightleftharpoons H_2CO_3 \text{ (carbonic acid)}$$

This reaction occurs readily in the presence of the enzyme carbonic anhydrase; H_2CO_3 then dissociates into H^+ and HCO_3^-. The entire chemical reaction is as follows:

$$CO_2 + H_2O \rightleftharpoons H_2CO_3 \rightleftharpoons H^+ + HCO_3^-$$

The term "volatile acid" refers to the fact that carbonic acid is produced from a gas that is normally expired (or "evaporated"). Because CO_2 is readily converted to carbonic acid in the presence of carbonic anhydrase, CO_2 itself is often referred to as the volatile acid.

The level of each acid is regulated by separate *physiologic* buffering systems:

- Fixed acid is regulated by the kidney through the reabsorption and elimination of HCO_3^- and H^+.
- Volatile acid is regulated by the respiratory system through the regulation of the respiratory rate.

We first consider the sources of fixed acids along with the regulation of acid-base balance by the kidney, and later we explore the regulation of volatile acids through the respiratory system.

WHAT DID YOU LEARN?

15 What is meant by acid-base balance?
16 How are fixed acids distinguished from volatile acids?

25.5b The Kidneys and Regulation of Fixed Acids

LEARNING OBJECTIVES

3. List the various sources of fixed acid.
4. Describe how the kidneys counteract excess blood H^+.
5. Explain how the kidneys function with decreasing blood H^+.

Daily input of fixed acid and base varies, and the net effect can tend to either decrease or increase the body's pH. As a general rule, the daily processes that influence acid-base balance tend to increase blood H^+ concentration, thus decreasing blood pH. Maintaining acid-base balance requires that the normal excretion processes eliminate the excess H^+. We examine these typical conditions looking first at input and then how pH is maintained.

Typical Conditions That Cause Blood H^+ Concentration to Increase

The input of acid into the blood, except that produced from CO_2, occurs from two major sources: nutrients absorbed by the gastrointestinal (GI) tract and body cells (figure 25.12a). Typically, more acid is absorbed from the GI tract because most individuals, at least in the United States, consume a diet rich in animal protein and wheat. These various ingested items contribute H^+ to the blood. Cells also contribute acid as waste products from metabolic processes. These products include lactic acid, phosphoric acid, and ketoacids. Acidic conditions may also result from the excessive loss of HCO_3^- (a weak base) as a consequence of diarrhea. (HCO_3^- is normally lost in the feces, but excess amounts can be lost when an individual has diarrhea; see Clinical View: "How Does Vomiting or Diarrhea Alter Blood H^+ Concentration?")

The additional H^+ must be eliminated to maintain acid-base balance. Recall from section 24.6d that kidney tubules respond to increased blood H^+ concentration, as follows:

- Reabsorb all filtered HCO_3^- along the length of the nephron tubule
- Synthesize and absorb new HCO_3^- while excreting H^+ into the filtrate through type A intercalated cells in the distal convoluted tubule and collecting tubules

Thus, under conditions when blood H^+ concentration is increasing, kidneys help to maintain a normal blood pH by both eliminating excess H^+ and synthesizing and reabsorbing HCO_3^-.

Less Typical Conditions That Cause Blood H^+ Concentration to Decrease

Although the general daily input of acid tends to decrease pH, some conditions tend to increase pH as a result of excess retention of base or loss of acid (figure 25.12b). Substances that increase blood pH also enter the blood from the GI tract. For example, this may occur if an individual regularly eats a vegetarian diet (that is rich in fruits and vegetables and

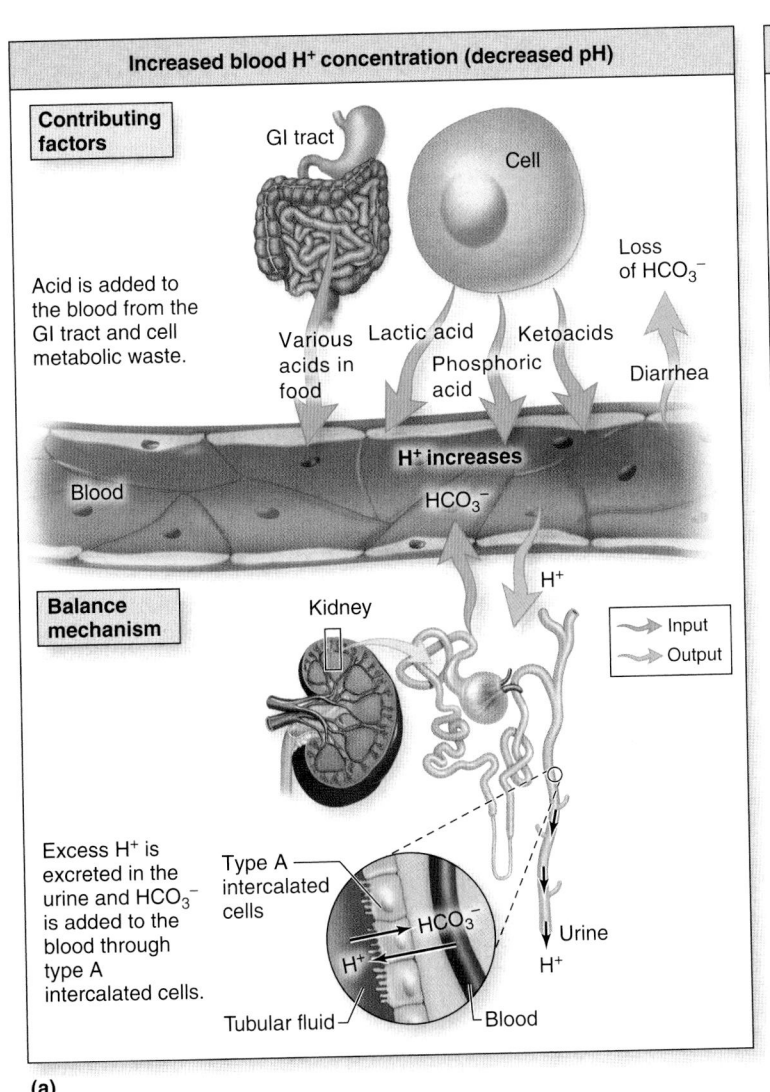

(a)

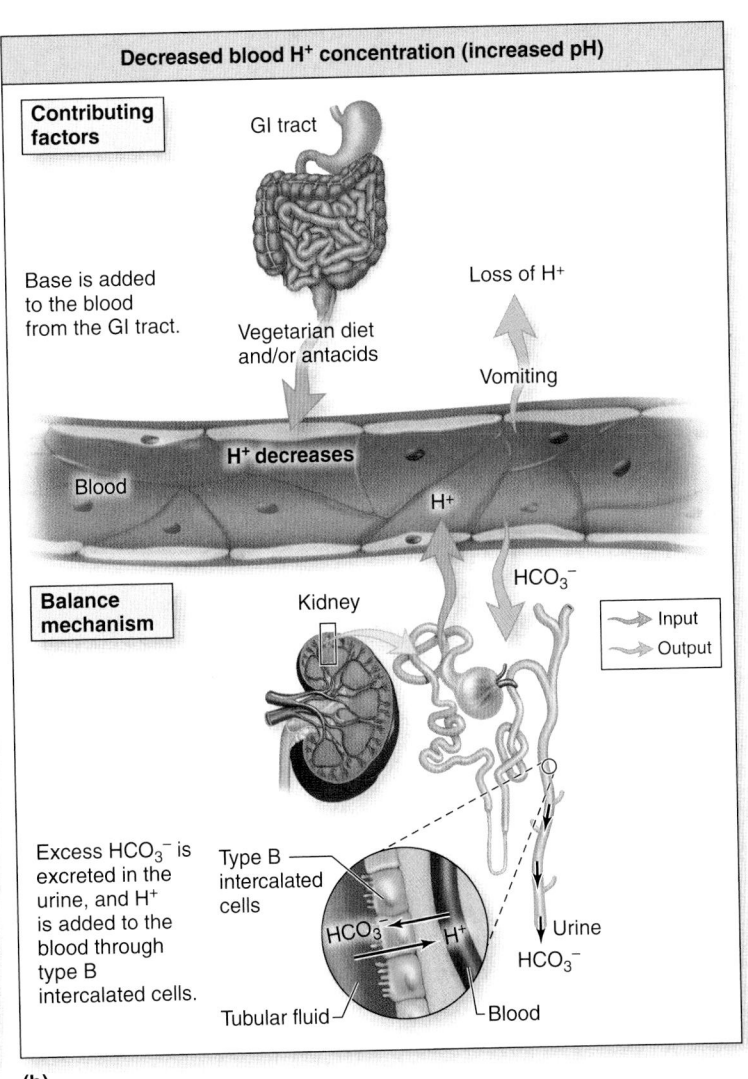

(b)

Figure 25.12 Altered Blood H^+ Concentration and Adjustments by the Kidneys. (a) Acid is normally added to the blood from the gastrointestinal (GI) tract as a result of diets rich in animal protein and wheat, and cellular metabolic wastes that include lactic acid from glycolysis, phosphoric acid from nucleic acid metabolism, and ketoacids from fat metabolism. Excessive loss of HCO_3^- as a consequence of diarrhea further increases blood H^+ concentration. The kidney adjusts blood plasma H^+ concentration as it eliminates excess H^+ and synthesizes and reabsorbs HCO_3^- through action of type A cells. (b) A vegetarian diet, high intake of antacids, or the loss of acid from vomiting decreases blood H^+ concentration. The kidney adjusts blood H^+ concentration as it eliminates HCO_3^- and reabsorbs H^+ through action of type B cells. AP|R

INTEGRATE

CLINICAL VIEW

How Does Vomiting or Diarrhea Alter Blood H⁺ Concentration?

Although different regions of the gastrointestinal (GI) tract may contribute either acid or base to the blood, the overall effect of these secretions is generally neutral in pH. This balance changes, however, with vomiting and diarrhea.

Vomiting, or *emesis* (em'e'sis; *emeo* = to vomit), is the ejection of the contents of the stomach through the mouth that occurs by involuntary spasms of the stomach and forceful contraction of the abdominal muscles. The ejected material is referred to as *vomitus* and contains hydrochloric acid (HCl). The loss of HCl triggers cells of the epithelial lining of the stomach to replace the

lost H^+. These cells are similar to the type A intercalated cells of the kidney; they excrete H^+ into the lumen of the stomach and simultaneously transport HCO_3^- into the blood. The addition of HCO_3^- to the blood increases blood pH. If severe enough, it can result in metabolic alkalosis.

Diarrhea is an excessive amount of watery stool. Severe cases of diarrhea are potentially life-threatening because of the excessive loss of HCO_3^- in the feces. The HCO_3^- loss triggers epithelial lining cells of the intestine to replace the lost HCO_3^-. These cells are similar to the type B intercalated cells of the kidney, and they both excrete HCO_3^- into the lumen of the intestine and simultaneously transport H^+ into the blood. The addition of H^+ into the blood decreases blood pH. Severe cases can result in metabolic acidosis.

Metabolic alkalosis and acidosis are described in section 25.6c.

---

low in animal protein), or regularly ingests antacids. These increase the absorption of alkaline substances into the blood, and blood H^+ concentration decreases. Alkaline conditions may also be caused by the abnormal loss of HCl as a result of vomiting.

When alkaline conditions are present, renal tubules respond in the following way:

- Do not reabsorb all of the filtered HCO_3^- along the length of the nephron tubule
- Secrete HCO_3^- from the blood into the filtrate, while reabsorbing H^+ in exchange through type B intercalated cells

Some of the filtered HCO_3^- (which is not reabsorbed) and secreted HCO_3^- is lost in the urine, while H^+ is reabsorbed. Blood plasma H^+ concentration increases, and blood pH decreases to normal.

The kidneys therefore act as a physiologic buffering system to eliminate either excess acid or base from the body. The process is relatively slow, taking from several hours to days; however, this is the only way to eliminate fixed acid or base, and it provides the most powerful means of maintaining blood H^+ concentration and preventing pH changes.

 WHAT DID YOU LEARN?

17 How do the kidneys regulate fixed acids to help maintain blood pH in response to increased blood H^+ concentration?

25.5c Respiration and Regulation of Volatile Acid

 **LEARNING OBJECTIVE**

6. Explain the relationship between breathing rate and blood pH.

Like the kidneys, the respiratory system serves as a physiologic buffering system to maintain acid-base balance, but does so by regulating the level of the volatile carbonic acid (H_2CO_3).

The respiratory system normally eliminates CO_2 from the lungs at rates that are equivalent to the production of CO_2. Recall that chemoreceptors detect changes in H^+, CO_2, and O_2 and relay to the respiratory center to alter breathing rate, and the most significant variable is changes in CO_2 (see section 23.5c). Blood levels of CO_2 are normally maintained such that the partial pressure of CO_2 (Pco_2) is between 35 and 45 mm Hg. Given that H_2CO_3 production is dependent upon CO_2 levels:

$$CO_2 + H_2O \rightleftarrows H_2CO_3 \rightleftarrows H^+ + HCO_3^-$$

the respiratory system also maintains the volatile acid H_2CO_3 within a normal homeostatic range ($H_2CO_3 = 3$ micromoles per deciliter [μmol/dL]).

Consequently, H_2CO_3 does not usually have daily fluctuations that affect acid-base balance.

An abnormal increase or decrease in the respiratory rate, however, that occurs independently of changes in the Pco_2 may affect acid-base balance because it drives the reversible reaction involving carbonic anhydrase either to the left or to the right (see section 3.2b). The following changes occur within several minutes **(figure 25.13)**:

- **An abnormal increase of the respiratory rate** causes elevated levels of CO_2 to be expired, resulting in a decrease in blood CO_2 concentration (blood Pco_2 decreases). The decreased level of blood CO_2 drives the chemical reaction to the left (figure 25.13*a*). Blood H^+ concentration decreases, and blood pH increases (becomes more alkaline).
- **An abnormal decrease of the respiratory rate** (or exchange of respiratory gases) results in an increase in the amount of CO_2 retained, thus elevating blood CO_2 (blood Pco_2 increases). The increased levels of blood CO_2 drive the equation to the right (figure 25.13*b*). Blood H^+ concentration increases, and blood pH decreases (becomes more acidic).

 WHAT DID YOU LEARN?

18 Describe the change (increase, decrease, or stays the same) for each of the following variables if an individual hyperventilates (breathes at a higher than normal breathing rate): (a) blood CO_2, (b) blood H^+ concentration, and (c) blood pH.

25.5d Chemical Buffers

 LEARNING OBJECTIVES

7. Describe the components of the protein buffering system.

8. Explain the reactions of the phosphate buffering system.

9. Discuss how the bicarbonate buffering system maintains acid-base balance in the ECF.

The physiologic buffering systems previously described are extremely effective in helping to maintain body fluid pH. The physiologic processes within the kidneys occur over several hours to several days, whereas those of the respiratory system require only a few minutes to respond.

In contrast, **chemical buffering systems** act quickly and temporarily to help prevent pH changes that would occur in response to the addition of acid or base, such as happens shortly after a meal, or as a result of abnormal loss of acid or base.

Chemical buffering systems are composed of one or two types of molecules that can bind and release H^+ within a fraction of a second to

Abnormal increases in respiratory rate	Abnormal decreases in respiratory rate

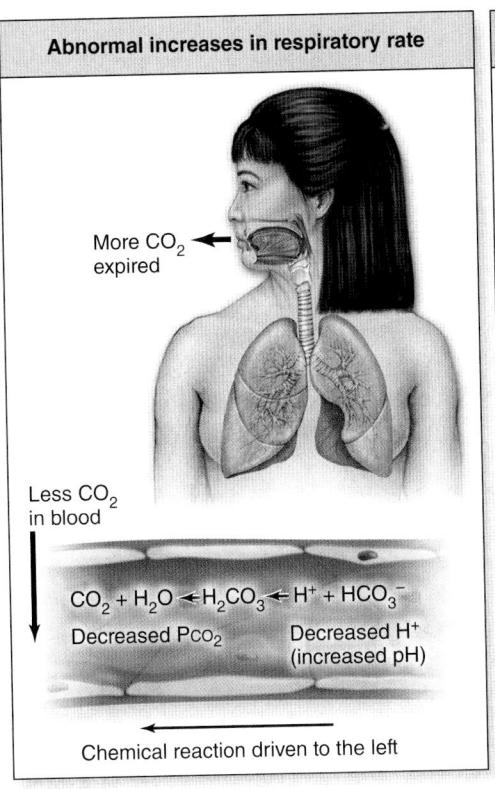

More CO$_2$ expired

Less CO$_2$ in blood

$CO_2 + H_2O \leftarrow H_2CO_3 \leftarrow H^+ + HCO_3^-$

Decreased Pco$_2$ Decreased H$^+$ (increased pH)

Chemical reaction driven to the left

(a)

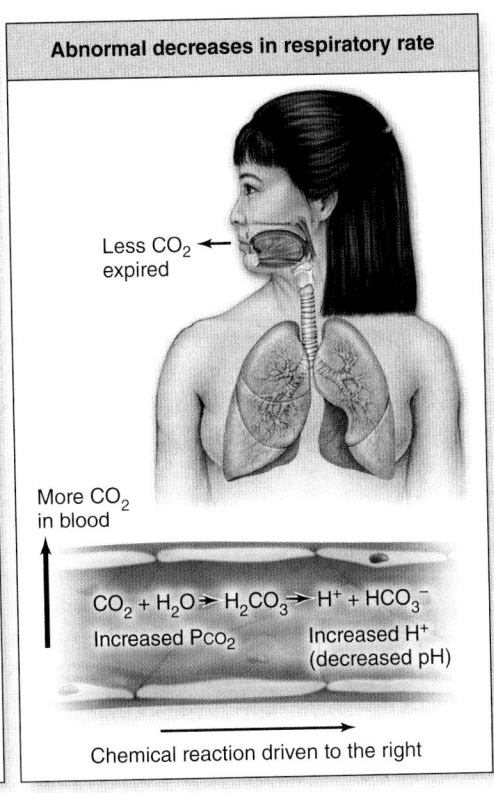

Less CO$_2$ expired

More CO$_2$ in blood

$CO_2 + H_2O \rightarrow H_2CO_3 \rightarrow H^+ + HCO_3^-$

Increased Pco$_2$ Increased H$^+$ (decreased pH)

Chemical reaction driven to the right

(b)

Figure 25.13 Respiration Rate and Blood pH. The respiratory system normally eliminates CO$_2$ at the same rate that cells produce it; thus, blood Pco$_2$ of 35–45 mm Hg is maintained under normal conditions. (*a*) An abnormal increase in respiratory rate decreases blood Pco$_2$, driving the chemical reaction to the left. Subsequently, blood H$^+$ concentration decreases, increasing blood pH. (*b*) An abnormal decrease in respiratory rate increases blood Pco$_2$, driving the chemical reaction to the right. Subsequently blood H$^+$ concentration increases, decreasing blood pH.

INTEGRATE

CONCEPT CONNECTION

The respiratory center receives sensory nerve signals from both central and peripheral chemoreceptors (see section 23.5c). These chemoreceptors are stimulated by increased H$^+$, and they initiate sensory input to the medullary respiratory center. Respiration rate and depth are increased to rid the body of more CO$_2$. Conversely, a decrease in H$^+$ reduces stimulation of chemoreceptors, causing a reduction in breathing rate and depth resulting in more CO$_2$ being retained. Under normal conditions, adjustments in the respiratory rate maintain the partial pressure of carbon dioxide (Pco$_2$) within a range of 35 to 45 mm Hg.

prevent large pH changes. The molecule, or molecules, of chemical buffering systems are composed of both a weak base that can bind excess H$^+$ and a weak acid that can release H$^+$. Remember that chemical buffers cause both temporary and limited adjustments until the body can eliminate the excess acid or base through physiologic buffering systems (namely, the kidneys or the respiratory system).

The three most important chemical buffering systems include the following:

- The protein buffering system within both cells and the blood
- The phosphate (PO$_4^{3-}$) buffering system within cells
- The bicarbonate (HCO$_3^-$) buffering system within the ECF, particularly the blood

Although similar in the mechanism to buffer both acid and base, the three chemical buffering systems differ in their locations and in the molecules that compose them.

Protein Buffering System

The **protein buffering system** is a chemical buffering system that is composed of proteins within cells and in blood plasma. It accounts for about three-quarters of the chemical buffering in body fluids. The **amine group (—NH$_2$)** of amino acids acts as a weak base to buffer acid, whereas the **carboxylic acid (—COOH)** of amino acids acts as a weak acid to buffer base (see section 2.8 for a review of amino acid and protein structures). The chemical change to each group is shown here:

$$\underset{\text{(weak base)}}{-NH_2} + \underset{\text{(strong acid)}}{H^+} \longrightarrow \underset{\text{(weak acid)}}{NH_3^+} \qquad (1)$$

$$\underset{\text{(weak acid)}}{-COOH} + \underset{\text{(strong base)}}{OH^-} \longrightarrow \underset{\text{(weak base)}}{COO^-} + H_2O \qquad (2)$$

With the addition of strong acid, shown in equation (1), the weak base (—NH$_2$) of the protein buffering system binds the H$^+$ that was added to the solution. This weak base becomes a weak acid (NH$_3^+$) as a result. The net effect is elimination of a strong acid (H$^+$) and the production of a weak acid (NH$_3^+$). In comparison, the addition of strong base, as in equation (2), causes the weak acid (—COOH) of the protein buffering system to release H$^+$, and as it does so, it becomes a weak base (COO$^-$). The net effect is removal of a strong base (OH$^-$) and the production of a weak base (COO$^-$).

Proteins are components within both cells and blood, and their buffering systems help minimize pH changes throughout the body. The most notable exception is in the cerebrospinal fluid (CSF), where there are no proteins. Proteins that help buffer pH changes include both intracellular proteins, plasma proteins, and hemoglobin in erythrocytes.

Phosphate Buffering System

The **phosphate buffering system** is found in intracellular fluid (ICF). It is especially effective in buffering metabolic acid produced by cells because phosphate (PO$_4^{3-}$) is the most common anion within cells. The phosphate buffering system is also composed of both a weak base and a weak acid. Here **hydrogen phosphate (HPO$_4^{2-}$)** is the weak base and **dihydrogen phosphate (H$_2$PO$_4^-$)** is the weak acid. The chemical change to HPO$_4^{2-}$ and H$_2$PO$_4^-$ is shown here:

$$\underset{\text{(weak base)}}{HPO_4^{2-}} + \underset{\text{(strong acid)}}{H^+} \longrightarrow \underset{\text{(weak acid)}}{H_2PO_4^-} \qquad (3)$$

You can think of chemical buffering systems as "sponges" for H^+. When pH decreases, the chemical buffering system soaks up H^+, and when pH increases, the chemical buffering system releases H^+. In this way, the H^+ concentration in solution may be kept within normal range, helping to prevent pH changes.

$$\underset{\text{(weak acid)}}{H_2PO_4^-} + \underset{\text{(strong base)}}{OH^-} \longrightarrow \underset{\text{(weak base)}}{HPO_4^{2-}} + H_2O \quad \textbf{(4)}$$

The addition of acid, shown in equation (3), is buffered by the weak base (HPO_4^{2-}) that binds the H^+ to become a weak acid ($H_2PO_4^-$). In contrast, the addition of base, as in equation (4), causes the weak acid $H_2PO_4^-$ to release H^+. As it does so, it becomes the weak base HPO_4^{2-}, and water is formed. As with the protein buffering system, the net result is either a strong acid buffered to produce a weak acid or a strong base buffered to produce a weak base.

Bicarbonate Buffer System

The **bicarbonate buffering system** in the blood is the most important buffering system in the ECF. Bicarbonate ion and carbonic acid are the key components of this buffering system. **Bicarbonate (HCO_3^-)** serves as a weak base, whereas **carbonic acid (H_2CO_3)** acts as a weak acid. The chemical change to HCO_3^- and H_2CO_3 is shown here:

$$\underset{\text{(weak base)}}{HCO_3^-} + \underset{\text{(strong acid)}}{H^+} \longrightarrow \underset{\text{(weak acid)}}{H_2CO_3} \quad \textbf{(5)}$$

$$\underset{\text{(weak acid)}}{H_2CO_3} + \underset{\text{(strong base)}}{OH^-} \longrightarrow \underset{\text{(weak base)}}{HCO_3^-} + H_2O \quad \textbf{(6)}$$

The addition of acid is buffered in equation (5) by the weak base HCO_3^-. Bicarbonate binds the excess H^+ and becomes the weak acid $H_2CO_3^-$. In comparison, with the addition of base as in equation (6), the weak acid H_2CO_3 releases H^+. As it does so, it becomes a weak base (HCO_3^-), and water is formed. The net result again is that a strong acid is buffered to produce a weak acid, or a strong base is buffered to produce a weak base.

In summary, the phosphate buffering system is important in buffering against pH changes within cells, whereas the bicarbonate buffering system buffers against pH changes in the blood. Proteins buffer in both cells (intracellular proteins) and within the blood plasma (plasma proteins, hemoglobin).

The moderation of pH changes by chemical buffers usually allows time for the kidneys to alter the excretion of H^+ or HCO_3^-, or for the respiratory system to adjust the expiration of CO_2. All the chemical buffering systems, however, are limited in the amount of acid or base that they can buffer. This is called the **buffering capacity.** If the buffering capacity is exceeded, then pH levels may decrease or increase beyond normal limits. These changes in pH are referred to as acid-base disturbances and are described in the next section.

The body's buffering systems—both the chemical buffering systems and physiologic buffering systems—that are involved in maintaining acid-base balance are integrated in **figure 25.14**.

 WHAT DID YOU LEARN?

19 What are the three chemical buffering systems, and where do they function?

20 What is the general amount of time required to maintain pH by (a) the chemical buffering systems, (b) the respiratory system, and (c) the kidneys?

The addition of either a strong acid or strong base causes greater changes in pH than the addition of a weak acid or a weak base (see section 2.5b).

A **strong acid,** such as HCl, dissociates completely in solution, whereas a **weak acid,** such as H_2CO_3, only partially dissociates in solution. As a result, a strong acid releases more H^+ into solution, causing greater decreases in pH.

In a similar way, a **strong base,** such as NaOH, dissociates completely in solution, and a **weak base,** such as $NaHCO_3$, only partially dissociates. As a result, a strong base binds more H^+ from the solution, resulting in larger increases in pH.

By independently binding and releasing H^+, **buffers** help limit pH changes by converting strong acids to weak acids and strong bases to weak bases.

Recall that following systemic gas exchange, CO_2 enters erythrocytes and binds to water, and then it dissociates from the water to produce H^+ and HCO_3^-. The HCO_3^- shifts out of the erythrocyte to be carried in the blood plasma, whereas the H^+ attaches to hemoglobin protein within the erythrocyte (see section 23.7b). This process then reverses when the blood reaches the lungs for alveolar gas exchange. During this process, hemoglobin acts as a buffer (first binding and then releasing H^+) to help maintain the concentration of free H^+ within both erythrocytes and blood plasma (see figure 23.27).

Figure 25.14 Maintaining Acid-Base Balance.
Buffering systems help maintain normal body pH. (*a*) Chemical buffering systems are located within cells and the blood, and they function within seconds to correct imbalances. (*b*) Physiologic buffering systems include both the respiratory system, which buffers within minutes, and the kidneys, which buffer within hours to days.

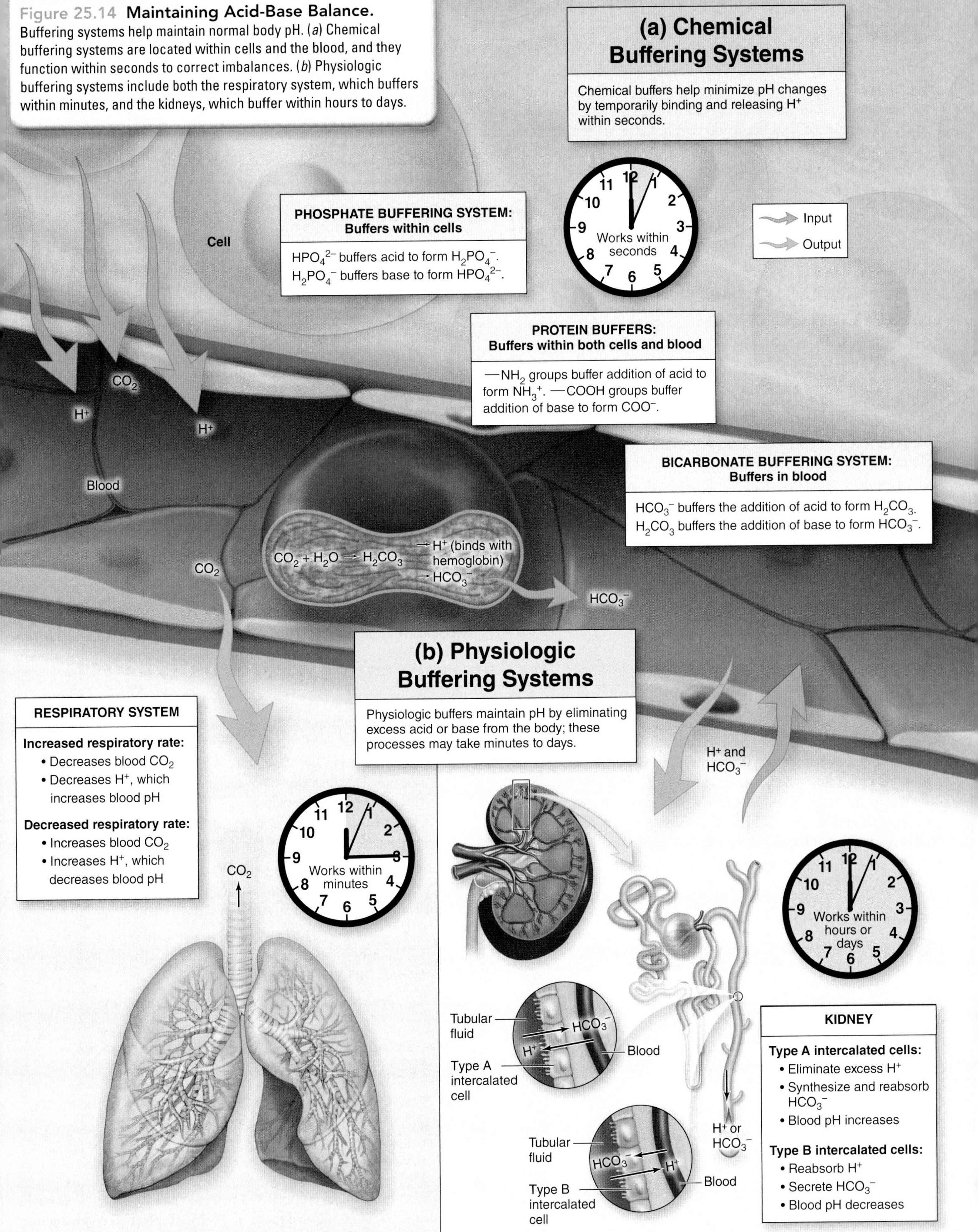

(a) Chemical Buffering Systems

Chemical buffers help minimize pH changes by temporarily binding and releasing H^+ within seconds.

Works within seconds

→ Input
→ Output

Cell

PHOSPHATE BUFFERING SYSTEM:
Buffers within cells

HPO_4^{2-} buffers acid to form $H_2PO_4^-$.
$H_2PO_4^-$ buffers base to form HPO_4^{2-}.

CO_2

H^+

H^+

PROTEIN BUFFERS:
Buffers within both cells and blood

—NH_2 groups buffer addition of acid to form NH_3^+. —COOH groups buffer addition of base to form COO^-.

Blood

BICARBONATE BUFFERING SYSTEM:
Buffers in blood

HCO_3^- buffers the addition of acid to form H_2CO_3.
H_2CO_3 buffers the addition of base to form HCO_3^-.

CO_2

$CO_2 + H_2O \rightarrow H_2CO_3$ → H^+ (binds with hemoglobin)
   → HCO_3^-

HCO_3^-

(b) Physiologic Buffering Systems

Physiologic buffers maintain pH by eliminating excess acid or base from the body; these processes may take minutes to days.

H^+ and HCO_3^-

RESPIRATORY SYSTEM

Increased respiratory rate:
- Decreases blood CO_2
- Decreases H^+, which increases blood pH

Decreased respiratory rate:
- Increases blood CO_2
- Increases H^+, which decreases blood pH

CO_2

Works within minutes

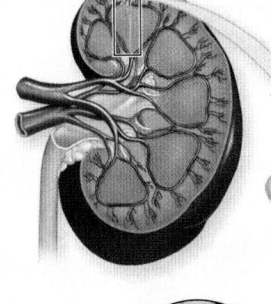

Tubular fluid
HCO_3^-
H^+
Blood
Type A intercalated cell

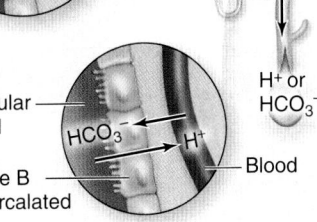

Tubular fluid
HCO_3^-
H^+
Blood
Type B intercalated cell

H^+ or HCO_3^-

Works within hours or days

KIDNEY

Type A intercalated cells:
- Eliminate excess H^+
- Synthesize and reabsorb HCO_3^-
- Blood pH increases

Type B intercalated cells:
- Reabsorb H^+
- Secrete HCO_3^-
- Blood pH decreases

25.6 Disturbances to Acid-Base Balance

Severe vomiting, diarrhea, uncontrolled diabetes, emphysema, pneumonia, brainstem trauma, renal failure, congestive heart failure, climbing to high altitude, and diuretic abuse all have something in common: All of these conditions can cause an acid-base disturbance. **Acidosis,** or *acidemia,* is an arterial blood pH reading below 7.35, whereas **alkalosis,** or *alkalemia,* is an arterial blood pH reading above 7.45.

Before describing the different types of acid-base disturbances, we begin by describing some relevant terminology, including acid-base disturbance, compensation, and acid-base imbalance.

25.6a Overview of Acid-Base Imbalances

 LEARNING OBJECTIVE

1. Explain acid-base disturbance, compensation, and acid-base imbalance.

An **acid-base disturbance** occurs when the buffering capacity of chemical buffering systems is exceeded, and there is a transient or temporary change in blood H^+ concentration, resulting in a change in blood pH beyond the normal range of 7.35 to 7.45. In response to a transient acid-base disturbance, the physiologic buffering system of the kidneys, the respiratory system, or both function to offset the disturbance. The response of physiologic buffering systems to acid-base disturbances that results in the return of blood pH to normal is called **compensation.**

If these physiologic buffering systems are not effective in returning the pH to normal, then pH disturbance is referred to as **uncompensated** (or *partially uncompensated*). When an uncompensated, temporary pH disturbance results in a persistent pH change, it is referred to as an **acid-base imbalance.** An acid-base imbalance for any extended period of time is life-threatening. Without intervention, an acid-base imbalance continues to worsen, and if the pH reaches values below 7.0 or above 7.7, it is fatal within a few hours.

Four major types of acid-base disturbances are distinguished based on two criteria: whether the primary disturbance is respiratory or metabolic in nature; and whether the pH change is acidic or alkaline. The four categories are respiratory acidosis, respiratory alkalosis, metabolic acidosis, and metabolic alkalosis. We examine each of these in terms of the cause of the primary disturbance, and then we describe the compensation made by the physiologic buffering system that helps prevent an acid-base imbalance.

 WHAT DID YOU LEARN?

 21 How does a compensated acid-base imbalance differ from an uncompensated acid-base imbalance?

25.6b Respiratory-Induced Acid-Base Disturbances

 LEARNING OBJECTIVES

2. Define respiratory acidosis, and identify some of the causes of this type of acid-base disturbance.

3. Explain why infants are more susceptible to respiratory acidosis.

4. Define respiratory alkalosis, and identify some of the causes of this type of acid-base disturbance.

Respiratory acid-base disturbances occur when changes in respiratory function result in either increased or decreased blood P_{CO_2} that

is outside the normal range of 35 to 45 mm Hg, with the subsequent change in H_2CO_3 levels:

$$CO_2 + H_2O \rightleftarrows H_2CO_3 \rightleftarrows H^+ + HCO_3^-$$

Respiratory Acidosis

The most common acid-base disturbance occurs because of impaired elimination of CO_2 by the respiratory system (caused by either a decrease in breathing rate or impaired gas exchange at the respiratory membrane). **Respiratory acidosis** is defined as occurring when the P_{CO_2} in the arterial blood becomes elevated above 45 mm Hg. Causes include the following:

- Injury to the respiratory center, perhaps caused by trauma or by poliovirus infection
- Disorders of the nerves or muscles involved with breathing, such as the loss of muscle strength associated with muscular dystrophy
- Airway obstruction (e.g., chronic obstructive pulmonary disease)
- Decreased gas exchange due to reduced respiratory surface area or thickened width of the respiratory membrane (these two conditions are associated with emphysema or pulmonary edema, respectively)

Continued impairment results in the accumulation of CO_2 in the blood that ultimately causes an increase in blood H_2CO_3 and a subsequent increase in H^+ concentration as the chemical reaction shifts to the right:

$$CO_2 + H_2O \rightarrow H_2CO_3 \rightarrow H^+ + HCO_3^-$$

Infants are more susceptible to respiratory acidosis because their smaller lungs and lower residual volume (see section 23.5f) do not eliminate CO_2 as effectively. CO_2 accumulates in the blood, with a subsequent increase in carbonic acid (H_2CO_3).

Respiratory Alkalosis

Respiratory alkalosis occurs when the P_{CO_2} decreases to levels below 35 mm Hg due to an increase in respiration. Faster breathing (hyperventilation) may occur in response to the following:

- Severe anxiety
- Any condition in which an individual is not receiving sufficient oxygen (e.g., as might occur when climbing to a high altitude where there is a decrease in the partial pressure of oxygen [P_{O_2}]; during congestive heart failure; as a result of severe anemia; or due to low blood pressure)
- Aspirin overdose (a condition that stimulates the respiratory center)

Continued elimination of CO_2 results in decreased levels of blood CO_2 that in turn cause a decrease in blood H_2CO_3 and lowered H^+ concentration as the chemical reaction shifts to the left:

$$CO_2 + H_2O \leftarrow H_2CO_3 \leftarrow H^+ + HCO_3^-$$

 WHAT DID YOU LEARN?

 22 How does hyperventilation bring about respiratory alkalosis?

25.6c Metabolic-Induced Acid-Base Disturbances

 LEARNING OBJECTIVES

5. Explain how metabolic acid-base disturbances differ from respiratory acid-base disturbances.

6. Define both metabolic acidosis and metabolic alkalosis, and identify some of the causes of each type of acid-base disturbance.

Metabolic disturbances are reflected by blood plasma HCO_3^- concentration that is outside its normal range of 22–26 mEq/L. This condition may occur either from the loss or gain of H⁺ (except for H_2CO_3 produced from CO_2), or the loss or gain of HCO_3^-. These disturbances involve physiologic processes other than changes associated with the respiratory system.

Metabolic Acidosis

The most common metabolic acid-base disturbance occurs as a result of a decrease in HCO_3^-. This decrease may result from an excessive loss of HCO_3^-, but more generally it occurs when there is an accumulation of fixed acid. The excess H⁺ binds with HCO_3^- to form H_2CO_3; thus, HCO_3^- levels in the blood decrease. **Metabolic acidosis** occurs when arterial blood levels of HCO_3^- fall below 22 mEq/L. This condition is usually caused by unhealthy changes in physiologic processes that were described earlier in the chapter (see figure 25.12a). These include the following:

- Increased production of metabolic acids, such as ketoacidosis from diabetes mellitus, increased lactic acid from glycolysis, or excessive production of acetic acid from excessive intake of alcohol
- Decreased elimination of acid due to renal dysfunction
- Increased elimination of HCO_3^- as a result of severe diarrhea

Infants are especially vulnerable to this condition because they produce larger amounts of acidic metabolic wastes due to a higher metabolic rate.

Metabolic Alkalosis

Metabolic alkalosis is defined as arterial blood levels of HCO_3^- that exceed 26 mEq/L. A loss of H⁺ or an increase in HCO_3^- can bring about this condition (see figure 25.12b). Causes include the following:

- Vomiting (the most common cause) or prolonged nasogastric suction
- Increased loss of acids by the kidneys with overuse of diuretics (medications that increase urine output)
- Increased alkaline input from consuming large amounts of antacids

 WHAT DID YOU LEARN?

 ❷❸ What is the primary cause of metabolic alkalosis?

25.6d Compensation

 LEARNING OBJECTIVE

7. Describe renal and respiratory compensation.

It is possible for both the kidneys and the respiratory system to make adjustments to compensate for changes in H⁺ and HCO_3^- in an attempt to return blood pH to within its normal range. However, it is important to remember that both the initial disturbance *and* the system compensation are reflected in abnormal (low or high) concentrations of CO_2 and HCO_3^- in lab results that measure these variables. These changes in blood levels may be monitored with arterial blood gas (ABG) lab results (see Clinical View: "Arterial Blood Gas [ABG] and Diagnosing Different Types of Acid-Base Disturbances").

Renal Compensation

Renal compensation (physiologic adjustments of the kidney to changes in pH) occurs in response to elevated blood H⁺ concentration due to a cause other than a renal dysfunction. We know that the normal physiologic activities of type A intercalated cells result in excretion of H⁺ and

reabsorption of HCO_3^- (see figure 25.12a). However, during renal compensation, this occurs to a greater degree than normal. Consequently, higher than normal levels of H⁺ are excreted, and higher than normal amounts of HCO_3^- are synthesized and reabsorbed into the blood. Renal compensation results in elevated values for blood HCO_3^-. As expected, urine pH is lower than normal.

In comparison, renal compensation in response to a decrease in blood H⁺ concentration involves the normal response of type B intercalated cells to excrete HCO_3^- and reabsorb H⁺ (see figure 25.12b). During renal compensation, this activity occurs to a greater degree than normal; higher than normal levels of H⁺ are reabsorbed into the blood, and higher amounts of HCO_3^- are excreted. Renal compensation results in lower than normal values for blood HCO_3^-. Urine pH in this case is higher than normal.

Renal compensation is generally effective in neutralizing acid-base disturbances that are not due to renal dysfunction, and in returning the pH back to normal. However, impaired renal function places an individual at higher risk for an acid-base imbalance. Renal function decreases as we age, and because of this, the elderly are at greater risk for pH imbalances.

Respiratory Compensation

The respiratory system attempts to compensate for metabolic acidosis by increasing the respiratory rate as a result of increased H⁺ concentration (figure 25.13a). During respiratory compensation, higher than normal amounts of CO_2 are expired, as evidenced by a lower than normal blood P_{CO_2} value.

Changes in the respiratory rate also may compensate for metabolic alkalosis, such as occurs with vomiting. The normal response of decreasing respiratory rate in response to decreased blood H⁺ concentration occurs during respiratory compensation (figure 25.13b). However, during respiratory compensation, this occurs to a greater degree than normal, resulting in lower than normal amounts of CO_2 expired. The evidence is in the higher than normal blood CO_2 values that are measured.

INTEGRATE

CLINICAL VIEW

Lactic Acidosis and Ketoacidosis

When insufficient oxygen is available (called *ischemia*), cells are forced to shift from aerobic cellular respiration to glycolysis (with increased production of lactic acid). This occurs during prolonged intense exercise or when the cardiovascular system is impaired, such as in congestive heart failure. Blood pH drops as a result. If substantial amounts of lactic acid are produced, acid-base balance is disturbed, resulting in the condition called **lactic acidosis.**

During conditions of insufficient cellular glucose uptake, cells are forced to shift from glucose metabolism to fat metabolism for their ATP production (ketoacids are a by-product). Individuals who have diabetes mellitus type 1 experience this condition because insufficient insulin is produced (see Clinical View: "Conditions Resulting in Abnormal Blood Glucose Levels" in section 17.9b). Glucose is unable to enter most cells without insulin. This forces cells to use fat for ATP production. The greater the deficiency of insulin, the greater the degree of fat metabolism, and the larger the amount of ketoacids formed. If substantial amounts of ketoacids are produced, a disturbance in the acid-base balance may occur that is called **ketoacidosis.**

Ketoacidosis may also occur if blood glucose levels are too low, and the body cells must metabolize fat for ATP production. Ketoacidosis is one of the risks of very-low-carbohydrate diets that are not medically monitored.

CLINICAL VIEW

Arterial Blood Gas (ABG) and Diagnosing Different Types of Acid-Base Disturbances

Measurements of **arterial blood gas (ABG)** are used to diagnose and monitor an acid-base disturbance and its compensation. These measurements include pH, $Paco_2$ (arterial Pco_2), and HCO_3^-. Changes in these values can help formulate a diagnosis, differentiate between the four types of acid-base disturbances, determine whether compensation is occurring—and if so, whether it is complete or partial.

Normal Conditions

First, consider the ABG results if an individual is in acid-base balance (no disturbance or compensation):

- Blood pH is within the normal range of 7.35 to 7.45.
- $Paco_2$ is 35 to 45 mm Hg, which reflects a normal respiratory rate and elimination of CO_2. By maintaining $Paco_2$, an H_2CO_3 concentration of approximately 3 μmol/dL is also regulated.
- Blood HCO_3^- concentration is 22–26 mEq/L, reflecting the effectiveness of the kidney in retaining or excreting HCO_3^-.

When an individual is in acid-base balance, a normal ratio between H_2CO_3 concentration and HCO_3^- concentration of 1:20 exists. A "balance" of acid and base with a normal ratio of 1:20 can be represented with a scale.

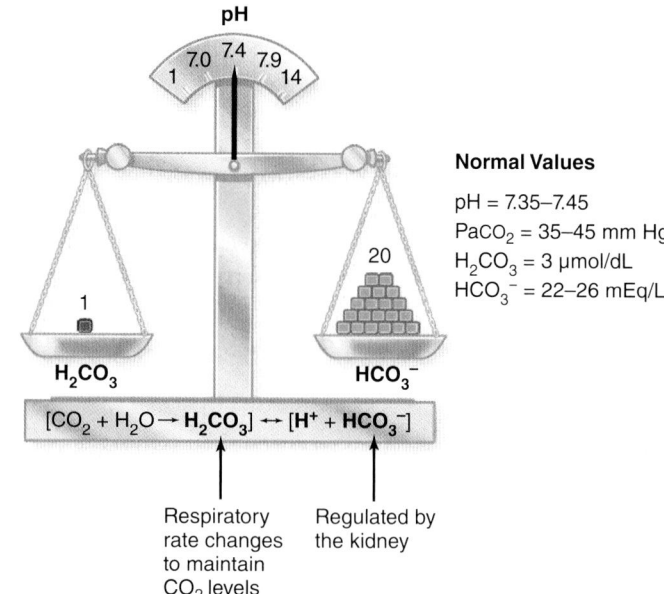

Normal Values

pH = 7.35–7.45
$PaCO_2$ = 35–45 mm Hg
H_2CO_3 = 3 μmol/dL
HCO_3^- = 22–26 mEq/L

Acid-Base Disturbances and Compensation

Now consider what occurs to both ABG values and the ratio of H_2CO_3 to HCO_3^- in each of the four types of acid-base disturbances and the resulting compensation. When compensation is complete, pH returns to normal, but altered blood levels of HCO_3^- and CO_2 are still observed in the ABG values. These changes and their compensation are visually represented with scales. Each primary disturbance and the accompanying compensation are presented in a separate box. Each box contains two figures; the left figure represents the primary disturbance and the right figure represents compensation.

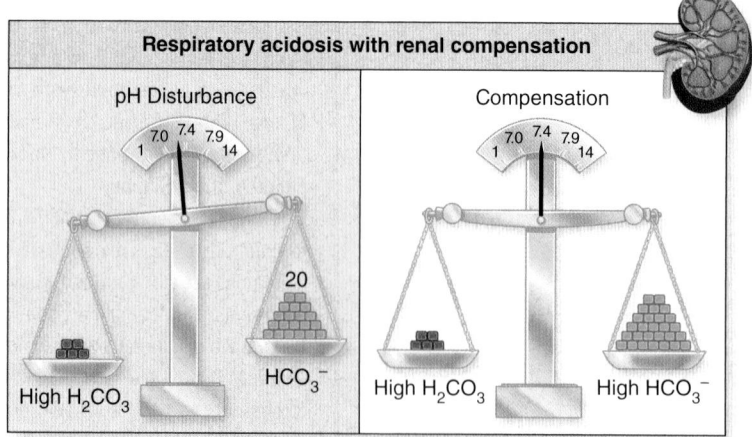

Respiratory acidosis with renal compensation.

In respiratory acidosis, respiratory rate decreases (hypoventilation) or breathing is too shallow (hypopnea) and blood CO_2 increases. This drives the chemical equation to the right ($CO_2 + H_2O \longrightarrow H_2CO_3$), and H_2CO_3 increases. This increase in H_2CO_3 is represented with a "tipped" scale. The kidneys compensate with increased secretion of H^+ and increased reabsorption of HCO_3^-. The increase in blood HCO_3^- from renal compensation reestablishes acid-base balance. However, although blood pH is returned to normal and the ratio of H_2CO_3 to HCO_3^- is returned to 1:20, notice that blood levels of H_2CO_3 and HCO_3^- are both greater than normal.

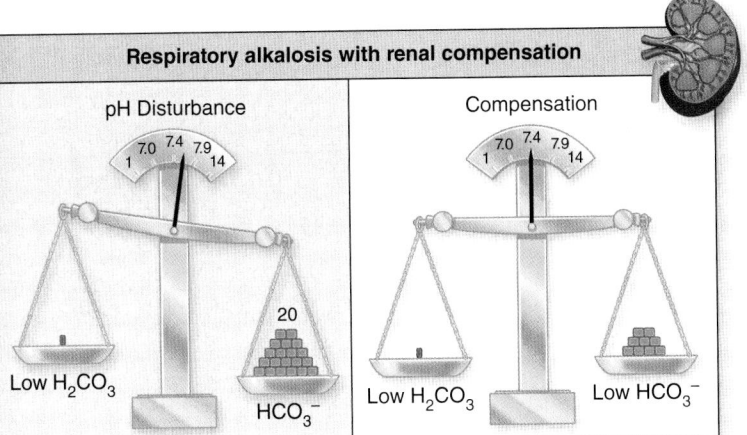

Respiratory alkalosis with renal compensation.

In respiratory alkalosis, respiratory rate increases (hyperventilation) and blood CO_2 decreases. This drives the chemical equation to the left ($CO_2 + H_2O \longleftarrow H_2CO_3$), and H_2CO_3 decreases. Again, this decrease in H_2CO_3 is shown with a "tipped" scale. The kidneys compensate with increased secretion of HCO_3^- and increased reabsorption of H^+. The decrease in blood HCO_3^- from renal compensation reestablishes acid-base balance. However, although blood pH is returned to normal and the ratio of H_2CO_3 to HCO_3^- is returned to 1:20, both H_2CO_3 and HCO_3^- are lower than normal.

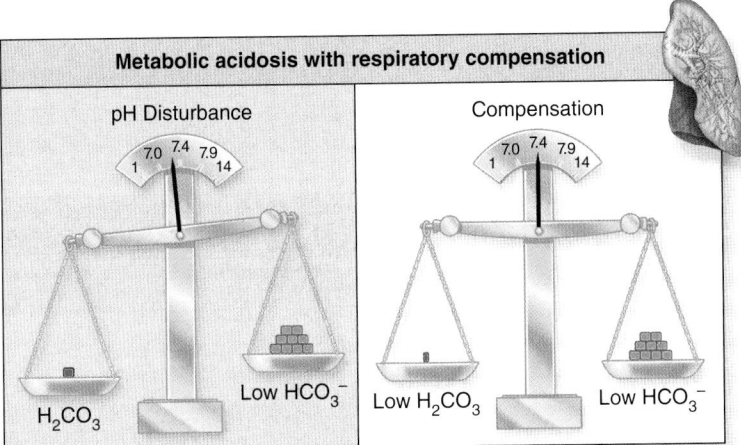

Metabolic acidosis with respiratory compensation.

In metabolic acidosis, blood H^+ increases, driving the chemical equation to the left ($CO_2 + H_2O \longleftarrow H_2CO_3 \longleftarrow H^+ + HCO_3^-$), and HCO_3^- decreases. The respiratory system compensates with an increased respiratory rate, and CO_2 (and H_2CO_3) decrease. The decrease in blood H_2CO_3 from respiratory compensation reestablishes acid-base balance. However, although blood pH is returned to normal and the ratio of H_2CO_3 to HCO_3^- is returned to 1:20, again notice that both H_2CO_3 and HCO_3^- are lower than normal.

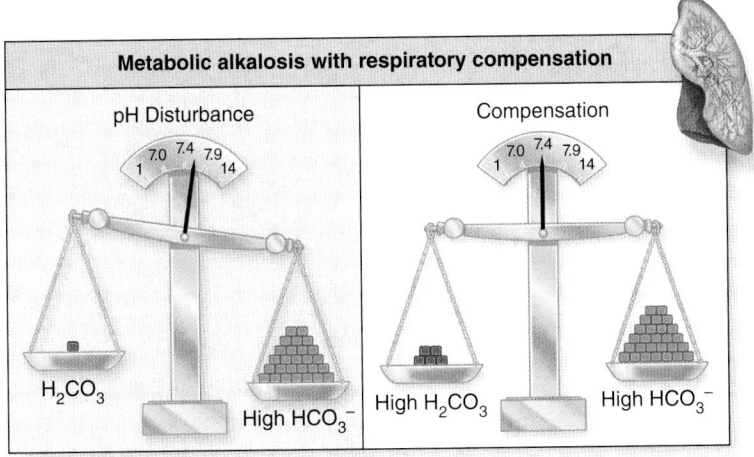

Metabolic alkalosis with respiratory compensation.

In metabolic alkalosis, blood H^+ has decreased, driving the chemical equation to the right ($CO_2 + H_2O \longrightarrow H_2CO_3 \longrightarrow H^+ + HCO_3^-$), and HCO_3^- increases. The respiratory system compensates with a decreased respiratory rate, and CO_2 (and H_2CO_3) increase. The increase in blood H_2CO_3 from respiratory compensation reestablishes acid-base balance. However, although blood pH is returned to normal and the ratio of H_2CO_3 to HCO_3^- is returned to 1:20, again notice that both H_2CO_3 and HCO_3^- are greater than normal.

Notice that if compensation occurs, both respiratory acidosis and metabolic alkalosis have similar outcomes: an increase in H_2CO_3 and an increase in HCO_3^-. In a like manner, both respiratory alkalosis and metabolic acidosis have similar results if compensation occurs: a decrease in H_2CO_3 and a decrease in HCO_3^-. To differentiate between the pH imbalances that have similar lab results, you would need further information.

If compensation is not complete, you will be able to differentiate between the acid-base imbalances based on pH. For example, to distinguish respiratory acidosis from metabolic alkalosis, a low pH would occur if the primary disturbance was caused by respiratory acidosis, and a high pH would occur from metabolic alkalosis. It would also be possible to differentiate between these based on the diagnosis of the individual's condition. For example, you would most likely conclude that the pH imbalance was respiratory acidosis if the individual had emphysema, or that it was metabolic alkalosis if the person was taking diuretics to treat high blood pressure. Similar analysis could be used to differentiate respiratory alkalosis and metabolic acidosis.

Table 25.3 — Acid-Base Imbalances

Variable	Normal Values	Respiratory Acidosis	Respiratory Alkalosis	Metabolic Acidosis	Metabolic Alkalosis
PRIMARY DISTURBANCE					
pH	7.35–7.45	< 7.35	> 7.45	< 7.35	> 7.45
P_{CO_2}	35–45 mm Hg	> 45 mm Hg (high)	< 35 mm Hg (low)	Normal	Normal
HCO_3^-	22–26 mEq/L	Normal	Normal	< 22 mEq/L (low)	> 26 mEq/L (high)
COMPENSATION[1]					
pH	7.35–7.45	Normal	Normal	Normal	Normal
P_{CO_2}	35–45 mm Hg	> 45 mm Hg (high)	< 35 mm Hg (low)	< 35 mm Hg (low)	> 45 mm Hg (high)
HCO_3^-	22–26 mEq/L	> 26 mEq/L (high)	< 22 mEq/L (low)	< 22 mEq/L (low)	> 26 mEq/L (high)

1. If completely compensated, pH returns to normal. The ratio of H_2CO_3 to HCO_3^- is once again 1:20.

Generally, respiratory compensation is less effective in addressing metabolic acid-base disturbances than renal compensation. The ability to decrease respiratory rate in response to metabolic alkalosis is limited by the development of hypoxemia. As the respiratory rate decreases, P_{O_2} levels also decrease. When P_{O_2} levels decrease below critical values, the respiratory rate is stimulated to increase. This may prevent complete compensation for metabolic alkalosis.

Acid-base imbalances are summarized in **table 25.3**.

 WHAT DID YOU LEARN?

 How does renal compensation affect blood plasma levels of HCO_3^-? How does respiratory compensation affect blood P_{CO_2}?

CHAPTER SUMMARY

25.1 Body Fluids 989

- Normal homeostatic levels of fluid and electrolytes are usually maintained despite sometimes erratic fluid intake.
- The percentage of fluid that composes our body varies between individuals.

25.1a Percentage of Body Fluid 989
- Your total body water content is dependent upon your age and the relative amounts of adipose connective tissue to skeletal muscle tissue.

25.1b Fluid Compartments 989
- The two major fluid compartments are intracellular fluid (fluid inside of a cell) and extracellular fluid (fluid outside of a cell).
- Extracellular fluid is composed of interstitial fluid (fluid around cells) and fluid within the blood plasma.
- Fluid movement between compartments occurs continuously in response to change in the relative fluid concentrations between the compartments.

25.2 Fluid Balance 992

- Maintaining fluid balance involves the interaction of many systems of the body.

25.2a Fluid Intake and Fluid Output 992
- Fluid intake includes preformed water in our food and drink, and metabolic water that is produced within the body.
- Fluid output includes losses when we expire air, through our skin, feces, and urine.
- Fluid output is categorized as sensible or insensible, depending upon if it is measurable; or as obligatory or facultative, depending upon if the loss is regulated based on state of hydration.

25.2b Fluid Imbalance 993
- A fluid imbalance occurs when fluid output does not equal fluid intake or when fluid is distributed abnormally.
- Fluid imbalances can be organized into five categories including volume depletion, volume excess, dehydration, hypotonic hydration, and fluid sequestration.

25.2c Regulation of Fluid Balance 994
- Fluid intake is regulated through stimuli that turn on and turn off the thirst center.
- Fluid output is hormonally regulated by altering the amount of fluid lost in urine (facultative water loss); is decreased by the renin-angiotensin system, aldosterone, and antidiuretic hormone; and is increased by atrial natriuretic peptide.

25.3 Electrolyte Balance 996

- Solutes are differentiated into nonelectrolytes and electrolytes.

25.3a Nonelectrolytes and Electrolytes 996
- Nonelectrolytes do not dissociate in solution. Electrolytes are substances that conduct an electric current following their dissociation in solution, and include salts, acids, bases, and some negatively charged proteins.

25.3b Major Electrolytes: Location, Functions, and Regulation 996
- Major electrolytes include sodium ion (Na^+), potassium ion (K^+), chloride ion (Cl^-), calcium ion (Ca^{2+}), phosphate ion (PO_4^{3-}), and magnesium ion (Mg^{2+}). Each is normally maintained within homeostatic levels.

CHALLENGE YOURSELF

Do You Know the Basics?

_____ 1. Which of the following individuals is most likely to contain the largest percentage of water?

    a. a frail 76-year-old woman

    b. a chunky 52-year-old male athlete

    c. a healthy 88-year-old man

    d. a lean 35-year-old male athlete

_____ 2. The fluid compartment with the largest percentage of fluid is

    a. intracellular.

    b. interstitial.

    c. plasma.

    d. extracellular.

_____ 3. Which of the following would result in fluid moving from the blood into the interstitial space?

    a. dehydration

    b. increased blood pressure

    c. burns

    d. vomiting

_____ 4. If an individual has decreased saliva production, increased blood osmotic pressure, decreased blood volume, and decreased blood pressure, he or she is experiencing

    a. edema.

    b. dehydration.

    c. hypotonic hydration.

    d. acidosis.

_____ 5. Which hormone decreases total body fluid, blood volume, and blood pressure?

    a. antidiuretic hormone (ADH)

    b. aldosterone

    c. atrial natriuretic peptide (ANP)

    d. angiotensin II

_____ 6. Which of the following describes an electrolyte? An electrolyte

    a. does not conduct an electric current.

    b. is covalently bonded and usually contains carbon.

    c. does not dissociate in solution.

    d. plays a significant role in regulating fluid balance.

_____ 7. The major role of this electrolyte is to maintain osmotic pressure in the extracellular fluid (ECF).

    a. HCO_3^-

    b. Ca^{2+}

    c. Na^+

    d. K^+

_____ 8. An increase in blood CO_2 levels is followed by a(n) _____ in blood H^+ levels and a(n) _____ in blood pH.

    a. decrease; decrease

    b. increase; increase

    c. decrease; increase

    d. increase; decrease

_____ 9. Which of the following is _not_ a chemical buffer in the body?

    a. carbohydrate

    b. protein

    c. phosphate

    d. hemoglobin

_____ 10. The kidney can act to buffer the blood by

    a. secreting H^+.

    b. reabsorbing H^+.

    c. producing HCO_3^-.

    d. All of these are correct.

11. List the three variables that determine the percentage of total water in the body.

12. Describe the movement of water between the compartments after taking a drink of water.

13. List the stimuli that turn on, and turn off, the thirst center.

14. Explain the homeostatic system involving the renin angiotensin system, ADH, and aldosterone.

15. Describe how ANP is regulated and how it opposes the action of the other three hormones (angiotensin II, ADH, and aldosterone).

16. Describe the functions of Na^+ and how it is regulated.

17. Describe what occurs in the kidney to maintain acid-base balance when there is an increase in blood plasma H^+.

18. Explain how increasing breathing rate increases blood pH.

19. List the three chemical buffers, and describe how and where they buffer pH.

20. Describe respiratory acidosis and its compensation.

Can You Apply What You've Learned?

1. Maria brings her baby to the emergency room. She reports that her daughter Sophia has had a fever for 2 days. She has also been vomiting and had diarrhea, and refuses to drink any fluid. Sophia is at risk for

    a. volume excess.

    b. volume depletion.

    c. dehydration.

    d. hypotonic hydration.

2. A young man has burns over 25% of his body, destroying his cells. This electrolyte is released from the damaged cells, resulting in elevated blood plasma levels. Results from the lab would indicate elevated levels of this common intracellular electrolyte.

    a. Ca^{2+}

    b. K^+

    c. Na^+

    d. Cl^-

3. Harriet, a young, poor, single mother, brings her baby to the emergency room because the baby is crying constantly. She reports that she is unable to buy formula and has been unable to breastfeed. Instead she has been giving the baby a bottle with only water. What is the potential fluid imbalance this infant might be experiencing?

    a. volume excess

    b. dehydration

    c. volume depletion

    d. hypotonic hydration

4. Harold has been suffering from diabetes mellitus for approximately 15 years. As a result of having recently lost his job and being under a great deal of stress, and he has not been adequately managing his diabetes. His breath smells sweet, a sign of producing ketoacids. Harold is most likely suffering from which type of pH imbalance?

    a. respiratory acidosis

    b. metabolic acidosis

    c. respiratory alkalosis

    d. metabolic alkalosis

5. Harold's blood lab results would most likely show _____ caused by the primary disturbance and _____ caused by compensation.

    a. decreased HCO_3^-; decreased P_{CO_2}

    b. decreased HCO_3^-; increased P_{CO_2}

    c. increased HCO_3^-; decreased P_{CO_2}

    d. increased HCO_3^-; increased P_{CO_2}

Can You Synthesize What You've Learned?

1. Morgan is a nurse at the local hospital. She received the lab results back from a patient that said the patient has hyperaldosteronism, a high level of aldosterone. Based on this diagnosis, explain what you would expect in regards to blood levels of Na^+ and K^+ as well as blood pressure. Explain why.

2. Ms. Taylor, 68 years old, has been vomiting for 2 days. On the day of admission to the hospital, she feels agitated and anxious. Her lab results, and normal lab values, are given in the following table. What is the primary disturbance: respiratory acidosis, respiratory alkalosis, metabolic acidosis, or metabolic alkalosis? How is her body attempting to compensate, and is it successful? Explain.

Variable	Patient's Lab Results	Normal Values
Blood pH	7.38	7.35–7.45
HCO_3^-	30 mEq/L	22–26 mEq/L
P_{CO_2}	60 mm Hg	35–45 mm Hg

INTEGRATE

ONLINE STUDY TOOLS

 connect |ANATOMY & PHYSIOLOGY LEARNSMART® AP|R

The following study aids may be accessed through Connect.

Clinical Case Study: An Elderly Woman Who Suddenly Develops Severe Shortness of Breath

Interactive Questions: This chapter's content is served up in a number of multimedia question formats for student study.

LearnSmart: Topics and terminology include body fluids; fluid balance; electrolyte balance; hormonal regulation; acid-base balance; disturbances to acid-base balance

Anatomy & Physiology Revealed: Topics include renal corpuscle: macula densa; renal corpuscle: distal convoluted tubule

chapter 26

Digestive System

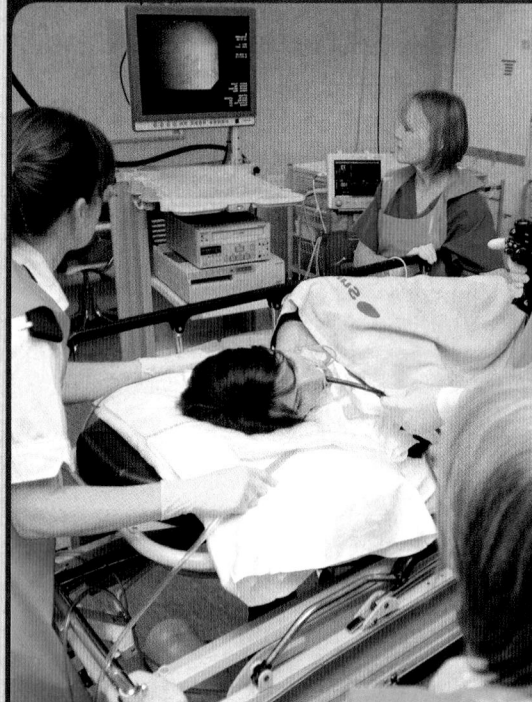

INTEGRATE

CAREER PATH
Gastroenterologist

A gastroenterologist specializes in both the diagnosis and treatment of gastrointestinal (GI) tract disorders, as well as disorders of the pancreas, liver, and gallbladder. These medical specialists treat malfunctions of the digestive system, including: gastroesophageal reflux (heartburn), peptic ulcers, irritable bowel syndrome, hepatitis, colitis, colon polyps, and cancer. This photo shows a gastroenterologist viewing a video monitor during a colonoscopy, which shows the inner lining of the large intestine.

Module 12: Digestive System

Each time we eat a meal and drink fluids, our body takes in the nutrients required for our survival. However, these nutrients are typically unusable by us in their original forms. The **digestive system** provides the means to break down ingested nutrients and then absorb them.

Here we present the many organs and processes associated with the digestive system. We begin with an introduction to the system and its organs, and then explore each component in more detail: the upper gastrointestinal (GI) tract, lower GI tract, and the accessory organs associated with them. The chapter concludes with a review of the major nutrients and the details of their digestion and absorption, including the portions of the digestive system in which these functions occur. This will provide you with an opportunity to view the "big picture" of enzyme pathways that occur in the different organs for each type of molecule—an approach that integrates form and function.

26.1 Introduction to the Digestive System

The **digestive system** includes the organs that ingest the food, mix and move the ingested materials, add secretions to facilitate digestion of these materials into smaller usable components, absorb the necessary nutrients into the blood or lymph, and expel the waste products from the body.

26.1a General Functions of the Digestive System

LEARNING OBJECTIVE

1. List and describe the six general functions of the digestive system.

The digestive system performs six main functions: ingestion, motility, secretion, digestion, absorption, and elimination of wastes.

- **Ingestion** (in-jes'chŭn; *ingero* = to carry in) is the introduction of solid and liquid nutrients into the oral cavity (*mouth*). It is the first step in the process of digesting and absorbing nutrients.
- **Motility** (mō-til'i-tē) is a general term describing both voluntary and involuntary muscular contractions for mixing and moving materials through the gastrointestinal tract.
- **Secretion** (se-krē'shŭn) is the process of producing and releasing substances such as digestive enzymes, acid, and bile into the gastrointestinal tract. These secretions facilitate digestion.
- **Digestion** is the breakdown of ingested food into smaller structures that may be absorbed from the gastrointestinal tract. **Mechanical digestion** occurs when ingested material is physically broken down into smaller units by chewing and mixing without changing their chemical structure. **Chemical digestion** involves the activity of specific enzymes to break chemical bonds to change larger complex molecules into smaller molecules that can then be absorbed.
- **Absorption** (ab-sōrp'shŭn) involves membrane transport of digested molecules, electrolytes, vitamins, and water from the gastrointestinal tract into the blood or lymph.
- **Elimination** is the expulsion of indigestible components that are not absorbed.

WHAT DID YOU LEARN?

❶ What is the primary difference between mechanical digestion and chemical digestion?

26.1b Organization of the Digestive System

LEARNING OBJECTIVES

2. Identify the six organs that make up the gastrointestinal (GI) tract.

3. List the accessory organs and structures involved in the digestive process.

The digestive system has two separate categories of organs: those composing the gastrointestinal tract, and the accessory digestive organs **(figure 26.1)**. The **gastrointestinal** (gas'trō-in-tes'tin-ăl; *gastro* = the stomach) **(GI) tract** is also called either the *digestive tract* or *alimentary canal*. The GI tract organs essentially form a continuous tube that includes the oral cavity (mouth), pharynx (throat), esophagus, stomach, small intestine, and large intestine, and ends at

the anus. It is typically about 30 feet in length in an adult cadaver—although, due to smooth muscle tone, it is significantly shorter in a living individual. Within the confines of the GI tract, ingested food is broken down into smaller components that can then be absorbed along its length. Note that materials within the lumen of the GI tract are not considered part of the body until they are absorbed.

Accessory digestive organs assist in the breakdown of food. Accessory digestive glands produce secretions that empty into the GI tract and include the salivary glands, liver, and pancreas. Other accessory digestive organs are not glands. They include the teeth and tongue, which participate in the chewing and swallowing of food, and the gallbladder, which concentrates and stores the secretions of the liver.

WHAT DID YOU LEARN?

❷ How is the gastrointestinal (GI) tract distinguished from accessory digestive organs?

26.1c Gastrointestinal Tract Wall

LEARNING OBJECTIVES

4. List and describe the four tunics (layers) that make up the gastrointestinal wall.

5. Briefly describe the general process of absorption.

6. Explain the action of the muscularis tunic.

The GI tract from the esophagus through the large intestine is a hollow tube composed of four concentric layers, called **tunics.** From innermost (adjacent to the lumen) to outermost, the four general tunics are the mucosa, submucosa, muscularis, and adventitia or serosa **(figure 26.2)**.

Mucosa

The **mucosa** (mū-kō'sa) is the inner-lining mucous membrane. It typically consists of an epithelium, an underlying lamina propria, and a thin layer of muscularis mucosae.

The **epithelium** is in contact with the contents within the lumen. It is a simple columnar epithelium for most of the GI tract (stomach, small intestine, large intestine). Recall from section 5.1c that this type of epithelium allows for secretion and absorption. The portions of the GI tract that must withstand abrasion (such as the esophagus) are lined by a nonkeratinized, stratified squamous epithelium.

The underlying **lamina propria** consists of areolar connective tissue that contains small blood vessels and nerves. Absorption occurs when substances are moved through the epithelial cells that line the GI tract wall, and are absorbed into blood or lymphatic capillaries located within the lamina propria (figure 26.2*b*).

The **muscularis** (mŭs-kū-lā'ris) **mucosae** is a thin layer of smooth muscle deep to the lamina propria. Contractions of this layer cause slight movements in the mucosa, which (1) facilitate the release of secretions from the mucosa into the lumen, and (2) increase contact of materials in the lumen with the mucosa. In other words, the muscularis mucosae gently "shakes things up."

Submucosa

The **submucosa** is composed of areolar and dense irregular connective tissue, the relative amounts of each vary depending upon the specific region of the GI tract. Many large blood vessels, lymph vessels, nerves, and glands are within the submucosa. Fine branches of the nerves extend into the mucosa and along with their associated autonomic ganglia are collectively referred to as the **submucosal nerve plexus,** or *Meissner plexus.* These nerves innervate both the smooth muscle and glands of the mucosa and submucosa.

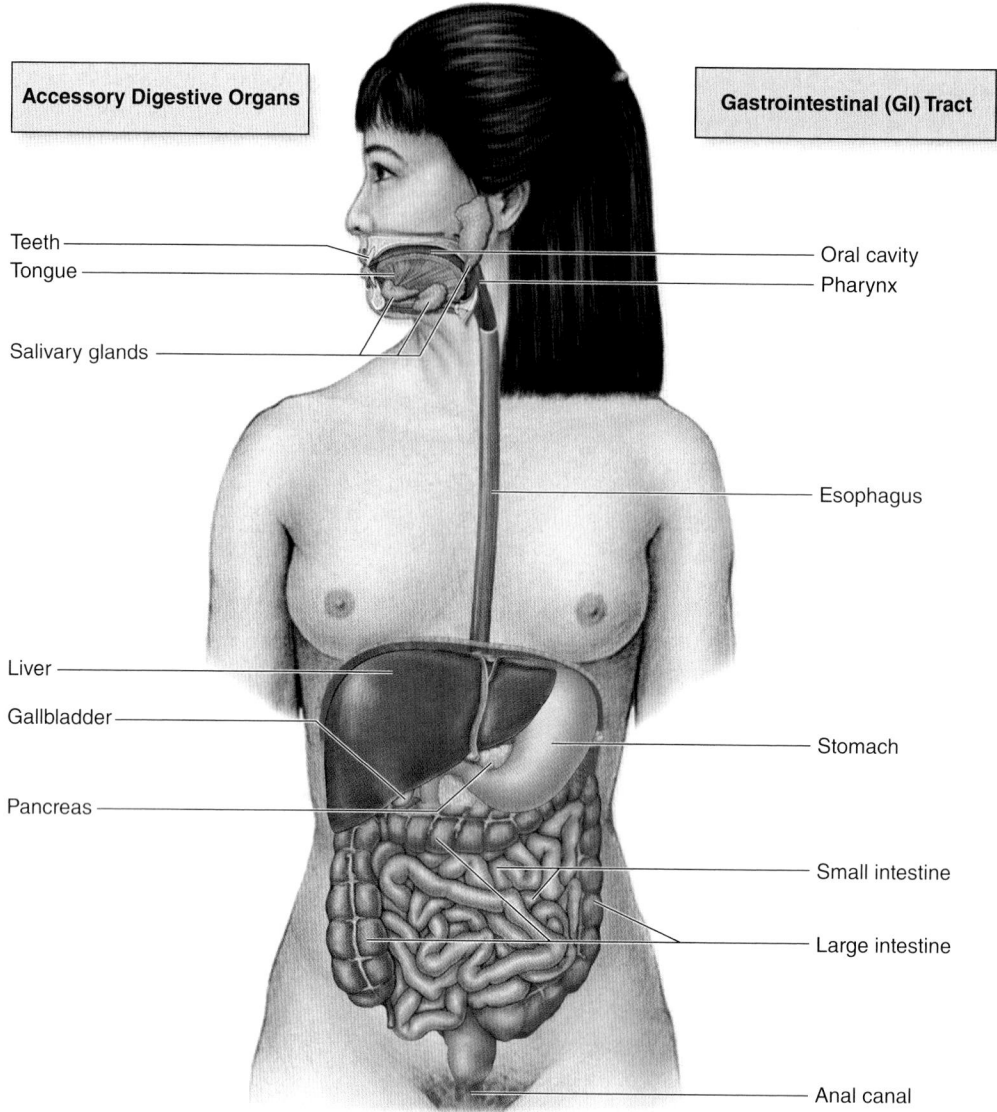

Accessory Digestive Organs

Teeth
Tongue
Salivary glands
Liver
Gallbladder
Pancreas

Gastrointestinal (GI) Tract

Oral cavity
Pharynx
Esophagus
Stomach
Small intestine
Large intestine
Anal canal

Figure 26.1 Digestive System. The digestive system is composed of the gastrointestinal (GI) tract and accessory digestive organs that assist the GI tract in the process of digestion. **AP|R**

The areolar connective tissue of both the lamina propria of the mucosa and the submucosa house mucosa-associated lymphatic tissue (MALT). In the small intestine (and especially in its last portion, the ileum), larger aggregates of lymphatic nodules in the submucosa are called **Peyer patches.** The presence of MALT helps prevent ingested microbes from crossing the GI tract wall and entering the body (see section 21.4d).

Muscularis

The **muscularis** usually contains two layers of smooth muscle. The smooth muscle cells of the inner layer are oriented circumferentially within the GI tract wall and are called the **inner circular layer.** The cells of the outer layer are oriented lengthwise and are called the **outer longitudinal layer.** Fine branches of nerves and their associated autonomic ganglia are located between these two layers of smooth muscle; these nerve branches control muscle contractions and are collectively referred to as the **myenteric** (mī-en-ter'ik; *mys* = muscle, *enteron* = intestine) **nerve plexus,** or *Auerbach plexus.*

Both the submucosal nerve plexus and the myenteric nerve plexus together compose the **enteric nervous system.** Sensory neurons within these plexuses detect both changes in the GI tract wall (e.g., stretch) and chemical makeup of the contents of the lumen. Thus, this system is composed of both motor neurons of the autonomic nervous system and

visceral sensory neurons (see chapter 15). It is interesting to note that some researchers consider the enteric nervous system a completely separate nervous system because of both the number of neurons that compose it and the fact that it can function independently of the central nervous system.

The inner circular muscle layer is greatly thickened at several locations along the GI tract to form a **sphincter.** A sphincter closes off the lumen at some point, and in so doing it can control the movement of materials into the next section of the GI tract.

The function of the muscularis is to mix and propel the contents within the GI tract. If you think of it as a hollow tube, then contractions of the circular layer constrict the lumen of the tube, whereas contractions of the longitudinal layer shorten the tube. Contractions of these smooth muscle layers are associated with two primary types of motility: peristalsis and mixing (figure 26.2c):

- **Peristalsis** (per'i-stal'sis; *stalsis* = constriction) is the alternating contraction sequence of both the inner circular and outer longitudinal muscle layers for the purpose of propelling ingested materials through the GI tract.
- **Mixing** is the "backward-and-forward," or kneading, motion that occurs at any point in time within different regions but lacks directional movement. Mixing is for the purpose of blending ingested materials with the secretions within the GI tract.

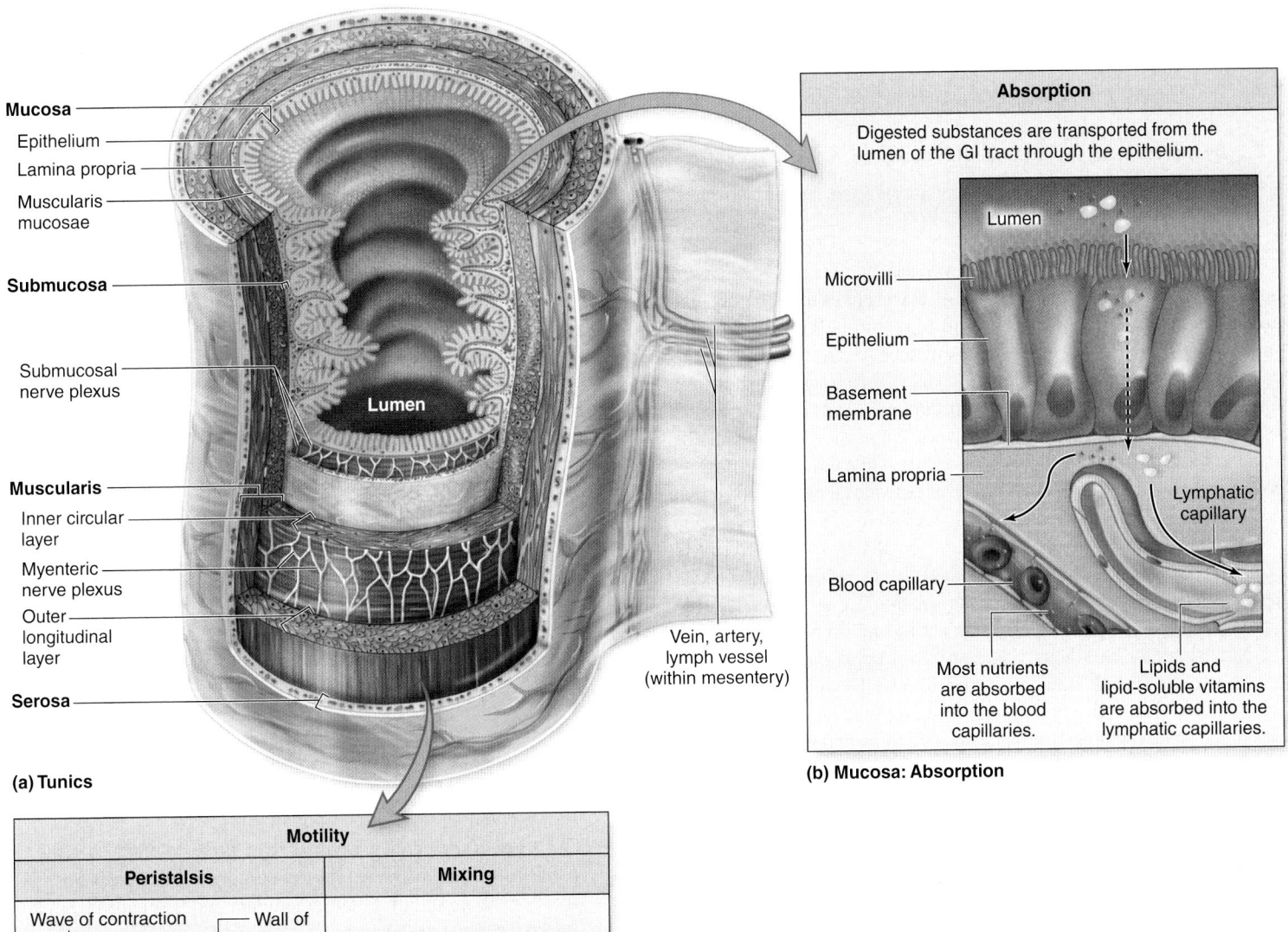

Mucosa
- Epithelium
- Lamina propria
- Muscularis mucosae

Submucosa

- Submucosal nerve plexus

Lumen

Muscularis
- Inner circular layer
- Myenteric nerve plexus
- Outer longitudinal layer

Serosa

(a) Tunics

Vein, artery, lymph vessel (within mesentery)

Absorption

Digested substances are transported from the lumen of the GI tract through the epithelium.

Lumen

- Microvilli
- Epithelium
- Basement membrane
- Lamina propria
- Blood capillary
- Lymphatic capillary

Most nutrients are absorbed into the blood capillaries.

Lipids and lipid-soluble vitamins are absorbed into the lymphatic capillaries.

(b) Mucosa: Absorption

Motility

Peristalsis	Mixing

Peristalsis
- Wave of contraction
- Wall of GI tract
- Lumen
- Relaxation
- Bolus

Mixing
- Mixing
- Further mixing

(c) Muscularis: Motility

Figure 26.2 Tunics of the Abdominal GI Tract. (*a*) The wall of the abdominal GI tract has four tunics: the mucosa, submucosa, muscularis, and adventitia or serosa. (*b*) Substances need only to cross the epithelium of the mucosa through membrane transport processes to be absorbed into the blood capillaries or lymphatic capillaries. (*c*) Motility includes both peristalsis, which propels material through the lumen of the GI tract, and mixing, which is a type of muscular contraction that facilitates the blending of materials within the GI tract. AP|R

Adventitia or Serosa

The outermost tunic may be either an adventitia or a serosa. An **adventitia** (ad-ven-tish′ă) is composed of areolar connective tissue with dispersed collagen and elastic fibers and is associated with portions of the GI tract that are located outside the peritoneal cavity (described shortly). A **serosa** (se-rō′să) has the same composition as the adventitia, but it is completely covered by a serous membrane called the visceral peritoneum and is associated within portions of the GI tract within the peritoneal cavity.

Some GI organs deviate from the typical pattern of the tunics just described. For example, the esophagus has a nonkeratinized stratified squamous epithelium in its mucosa to protect its lining, and the

stomach has three layers of smooth muscle in the muscularis. Being familiar with the basic tunic pattern, and then discovering how an organ may deviate from this pattern, provides clues as to the organ's function.

WHAT DID YOU LEARN?

3 What specific layer(s) must substances cross to enter the blood or lymphatic capillaries during their absorption?

4 What purpose is served by muscular sphincters at various locations along the length of the GI tract?

5 How does peristalsis differ from mixing?

26.1d Serous Membranes of the Abdominal Cavity

LEARNING OBJECTIVES

7. Describe the structure of the serous membranes associated with the GI tract.

8. Distinguish between intraperitoneal and retroperitoneal organs.

9. Explain the function of the mesentery, and describe the five individual mesenteries of the abdominopelvic cavity.

Two serous membranes are associated with the abdominal cavity (see section 1.4e). The serous membrane that lines the inner surface of the abdominal wall is called the **parietal peritoneum** (per′i-tō-nē′ŭm; *periteino* = to stretch over). The portion of the serous membrane that reflects over and covers the surface of internal organs is called the **visceral peritoneum.** Between these two layers is the thin **peritoneal cavity.** It is a potential space in which both peritoneal layers secrete a lubricating serous fluid. This small volume of fluid in the peritoneal cavity lubricates both the internal abdominal wall and the external organ surfaces. It allows the abdominal organs to move freely and reduces friction resulting from this movement.

Organs within the abdomen that are completely surrounded by visceral peritoneum are called **intraperitoneal** (in′tră-per′i-tō-nē′ăl) **organs.** They include the stomach, most of the small intestine, and parts of the large intestine. **Retroperitoneal** (re-trō-per′i-tō-nē′ăl) **organs** lie outside the parietal peritoneum directly against the posterior abdominal wall, so only their anterolateral portions are covered with the parietal peritoneum. Retroperitoneal digestive organs include most of the duodenum (the first part of the small intestine), the pancreas, ascending and descending colon (parts of the large intestine), and the rectum (see figure 1.9).

Mesentery

The general term **mesentery** (mes′en-ter-ē) refers to the double layer of peritoneum that supports, suspends, and stabilizes the intraperitoneal GI tract organs. Blood vessels, lymph vessels, and nerves that supply

INTEGRATE

CONCEPT CONNECTION

The serous peritoneum is very similar to the serous pleura (see section 23.4c) and the serous pericardium (see section 19.2b). These membranes have an outer parietal layer lining the inside of the cavity and an inner visceral layer that covers the external surface of the organ. The space between the parietal and visceral layers is where serous fluid is secreted, and this fluid acts as a lubricant to prevent friction between the layers as the organs move.

the GI tract are sandwiched between the two folds. Several terms are used to identify the individual mesenteries associated with specific organs (**figure 26.3;** see also figure 1.9):

- The **greater omentum** (ō-men′tŭm) extends inferiorly like an apron from the inferolateral surface of the stomach (greater curvature) and covers most of the abdominal organs. It often accumulates large amounts of adipose connective tissue, thus it is referred to as the "fatty apron" and serves to both insulate the abdominal organs and as a storage for excess fat.

- The **lesser omentum** connects the superomedial surface of the stomach (lesser curvature) and the proximal end of the duodenum to the liver.

- The **falciform** (fal′si-fōrm; *falx* = sickle) **ligament** is a flat, thin, crescent-shaped peritoneal fold that attaches the liver to the internal surface of the anterior abdominal wall.

- The **mesentery proper,** which is sometimes referred to as the *mesentery,* is a fan-shaped fold of peritoneum that suspends most of the small intestine (the jejunum and the ileum) from the internal surface of the posterior abdominal wall.

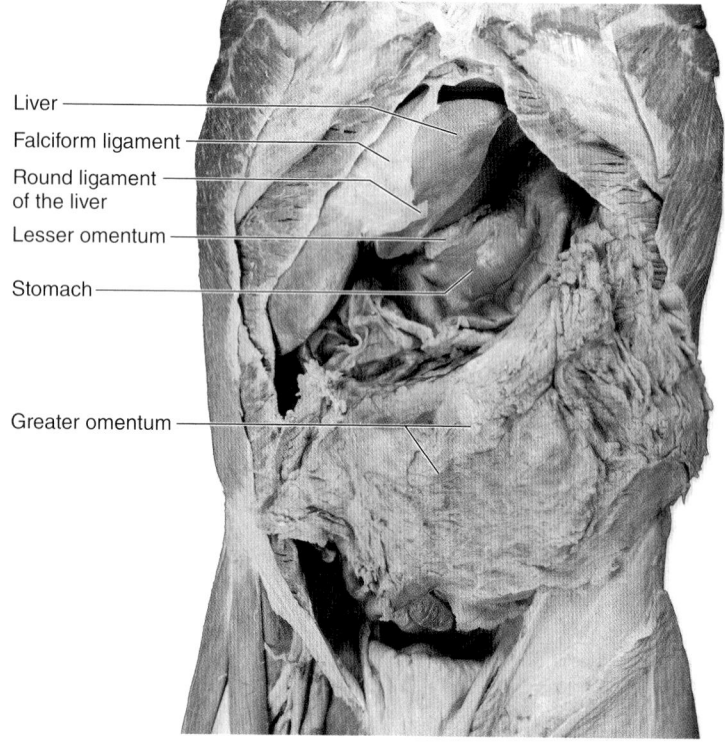

Liver
Falciform ligament
Round ligament of the liver
Lesser omentum
Stomach
Greater omentum

(a) Omenta

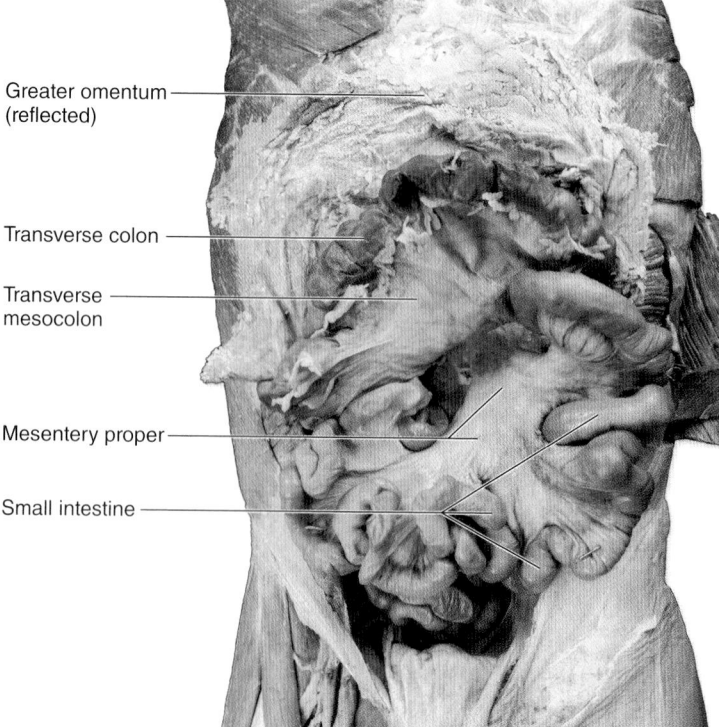

Greater omentum (reflected)
Transverse colon
Transverse mesocolon
Mesentery proper
Small intestine

(b) Mesentery proper and mesocolon

Figure 26.3 Serous Membranes. Many abdominal organs are held in place by double-layered serous membranes called mesenteries, which include (*a*) the greater and lesser omenta, and (*b*) the mesentery proper and the mesocolon.

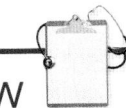

INTEGRATE

CLINICAL VIEW

Peritonitis

Peritonitis is an inflammation of the peritoneum and is associated with abdominal pain. The most common cause is perforation of the GI tract, which allows the contents of the GI tract to leak out and come in contact with the peritoneum. Perforations may result from within—for example, ulcers of the stomach or duodenum that "eat through" the lining of the GI tract, or a sharp object that has been ingested and pierces through the GI tract. Perforations may also result from without—for example, from a gunshot wound or during abdominal surgery.

- The **mesocolon** (mez'ō-kō'lon) is a fold of the peritoneum that attaches parts of the large intestine to the posterior abdominal wall. The mesocolon has several distinct sections, each named for the portion of the colon it suspends. For example, transverse mesocolon is associated with the transverse colon, whereas sigmoid mesocolon is associated with the sigmoid colon.

 WHAT DID YOU LEARN?

6 What is the difference between intraperitoneal and retroperitoneal organs? List the digestive organs that are intraperitoneal organs.

7 Where is the greater omentum located?

26.1e Regulation of Digestive System Processes

 LEARNING OBJECTIVES

10. Identify the type and location of the two main receptors of the digestive system.

11. Describe and compare long reflexes and short reflexes.

12. List the major hormones that regulate the processes of digestion.

INTEGRATE

LEARNING STRATEGY

Organs that are retroperitoneal are positioned against the posterior wall outside of the parietal peritoneum. These organs are associated with the digestive system, urinary system, and cardiovascular system. A common medical mnemonic to remember these organs is SAD PUCKER:

S = **S**uprarenal glands (adrenal glands)

A = **A**orta and inferior vena cava

D = **D**uodenum (most of)

P = **P**ancreas

U = **U**rinary bladder and ureters

C = **C**olon (ascending and descending)

K = **K**idneys

E = **E**sophagus (abdominal portion)

R = **R**ectum

Regulation of digestion involves receptors that monitor changes associated with the GI tract and its contents. Stimulation of these receptors brings about both nervous system and hormonal responses.

Receptors

Two major types of general sense receptors (see section 16.1d) are embedded throughout the mucosa and submucosa of the GI tract length: **baroreceptors** within the wall that detect either stretch or pressure of a particular region of the GI tract as the contents move through, and **chemoreceptors** that detect the presence of specific substances (e.g., H^+, fatty acids) of the passing contents within the lumen.

Nervous Control

Sensory input from both mechanoreceptors and chemoreceptors is relayed to the central nervous system (CNS) in response to stimulation. Autonomic motor output is then relayed through three cranial nerves (the facial, glossopharyngeal, and vagus nerves; see sections 13.9 and 15.3a) to different digestive system effectors, such as salivary glands, the pancreas, and the muscularis layers in the GI tract wall. The result is coordinated secretory and smooth muscle contractions involved in digestive responses. These autonomic interactions of the CNS are described as **long reflexes**.

Some digestive reflexes do not involve the CNS. Instead, they are local and occur only within the neurons of the enteric nervous system that are housed within the GI tract wall. These are called **short reflexes.** The details of reflexes are described in later sections.

Hormonal Control

Three primary hormones participate in the regulation of the processes of digestion: gastrin that is released from the stomach; and secretin and cholecystokinin that are released from the small intestine. The specific functions of these hormones are described in sections 26.2d and 26.3c.

Now that an overview of the main functions and features of the GI tract has been presented, we are ready to proceed with a more detailed exploration through the system, beginning with the upper GI tract.

 WHAT DID YOU LEARN?

8 Does a long reflex involve the central nervous system?

9 What three primary hormones regulate digestive processes?

INTEGRATE

CONCEPT CONNECTION

Three of the four cranial nerves (see sections 13.9 and 15.3a) containing parasympathetic axons are involved in regulating digestive activities:

The **facial nerve (CN VII)** extends to the sublingual and submandibular glands to stimulate salivary secretions.

The **glossopharyngeal nerve (CN IX)** extends to the parotid gland to stimulate salivary secretions.

The **vagus nerve (CN X)** innervates most of the organs of the digestive system (e.g., stomach, small intestine, most of the large intestine, pancreas, liver) and stimulates their digestive activities.

26.2 Upper Gastrointestinal Tract

The upper gastrointestinal tract is where initial processing by both mechanical and chemical means takes place. It consists of the oral cavity and salivary glands, pharynx, esophagus, stomach, and duodenum.

26.2a Overview of the Upper Gastrointestinal Tract Organs

LEARNING OBJECTIVE

1. Describe the components of the upper gastrointestinal tract.

A superficial view of the upper GI tract organs and accessory structures helps integrate their general structures with their digestive activities and functions (see figure 26.1):

- **Oral cavity and salivary glands.** Mechanical digestion (mastication) begins in the oral cavity. Saliva is secreted from the salivary glands in response to food being present within the oral cavity. It is mixed with the ingested materials to form a wet mass called a bolus. One component of saliva is salivary amylase, an enzyme that initiates the chemical digestion of starch (amylose).
- **Pharynx** (far′ingks). The bolus is moved into the pharynx, where swallowing occurs. Mucus secreted in saliva and in the superior part of the esophagus provides lubrication to facilitate swallowing.
- **Esophagus.** The bolus is transported from the pharynx through the esophagus into the stomach. Mucus secretion by the esophagus lubricates the passage of the bolus.
- **Stomach.** The bolus is mixed with gastric secretions as smooth muscle in the stomach wall contracts. These secretions are produced by epithelial cells of the stomach mucosa and include acid (hydrochloric acid [HCl]), digestive enzymes, and mucin. The mixing continues as an acidic "purée" called chyme is formed.

Note that the first part of the small intestine is the duodenum. It is also included in the upper GI tract; however, it will be described with the rest of the small intestine later in the chapter.

WHAT DID YOU LEARN?

10. What structures are considered part of the upper gastrointestinal tract?

26.2b Oral Cavity and Salivary Glands

LEARNING OBJECTIVES

2. Identify the anatomic structures of the oral cavity.
3. Describe the structure and function of salivary glands and how the release of saliva is regulated.
4. Explain the process of mastication.
5. Discuss the structure and development of the teeth.

The Oral Cavity

The **oral cavity,** or *mouth,* is the entrance to the GI tract (**figure 26.4**). Food is ingested into the oral cavity, where it undergoes the initial processes of mechanical and chemical digestion.

Gross Anatomy The oral cavity has two distinct spatial regions: (1) the vestibule (or buccal cavity), which is the space between the gums, lips, and cheeks; and (2) the oral cavity proper, which lies central to the teeth. The oral cavity is bounded laterally by the cheeks, anteriorly by the teeth and lips, and it leads posteriorly into the oropharynx.

The cheeks are covered externally by the integument and contain the buccinator muscles (see figure 11.5). These muscles compress the cheeks against the teeth to hold solid materials in place during mastication (chewing). The cheeks terminate at the fleshy **lips** (or *labia*) that are formed primarily by the orbicularis oris muscle. Lips have a reddish hue because of their abundant supply of superficial blood vessels and the reduced amount of keratin within their outer skin. The internal surfaces of both the superior and inferior lips each are attached to the gingivae (gums) by a thin mucosa fold in the midline, called the **labial frenulum** (lā′bē-ăl; *labium* = lip, fren′ū-lŭm; *frenum* = bridle).

The **palate** (pal′ăt) forms the superior boundary, or "roof," of the oral cavity and acts as a barrier to separate it from the nasal cavity. The anterior two-thirds of the palate is hard and bony (called the hard palate), whereas the posterior one-third is soft and muscular (called the soft palate). The **hard palate** is formed by the palatine processes of the maxillae and the horizontal plates of the palatine bones (see section 8.2b). It exhibits prominent **transverse palatine folds,** or *friction ridges,* that assist the tongue in manipulating ingested materials prior to swallowing. Extending inferiorly from the posterior part of the soft

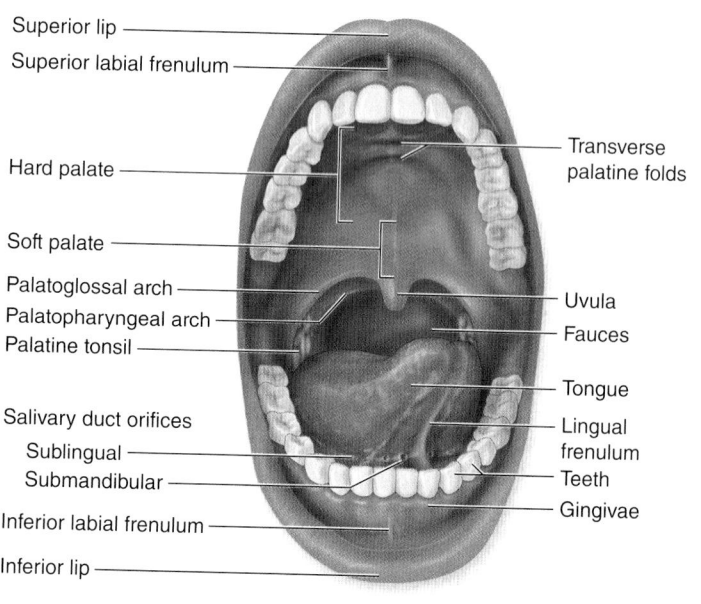

Superior lip
Superior labial frenulum
Hard palate
Soft palate
Palatoglossal arch
Palatopharyngeal arch
Palatine tonsil
Salivary duct orifices
Sublingual
Submandibular
Inferior labial frenulum
Inferior lip
Transverse palatine folds
Uvula
Fauces
Tongue
Lingual frenulum
Teeth
Gingivae

(a) Oral cavity, anterior view

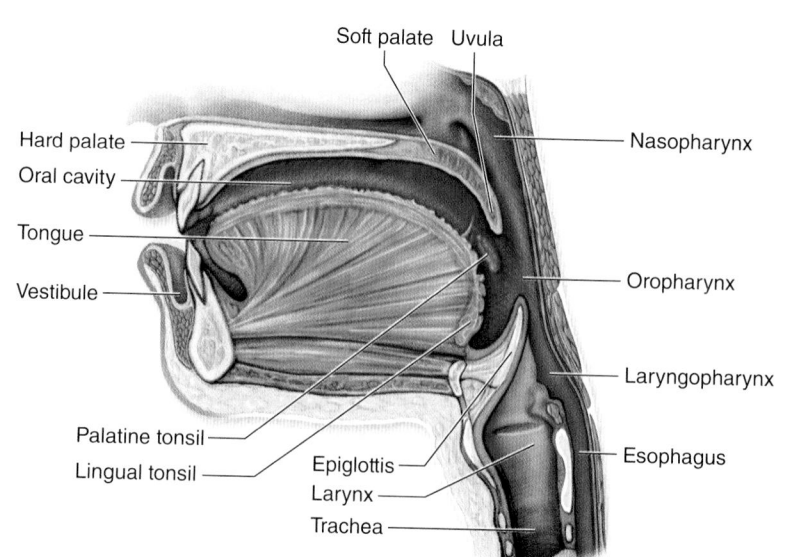

Soft palate   Uvula
Hard palate
Oral cavity
Tongue
Vestibule
Palatine tonsil
Lingual tonsil
Epiglottis
Larynx
Trachea
Nasopharynx
Oropharynx
Laryngopharynx
Esophagus

(b) Oral cavity and pharynx, sagittal section

Figure 26.4 Oral Cavity. Ingested food and drink enter the GI tract through the oral cavity and move into the pharynx. (*a*) An anterior view shows the structures of the oral cavity. (*b*) A sagittal section shows the structures of both the oral cavity and the pharynx. AP|R

palate is a cone-shaped medial projection called the **uvula** (ū′vū-lă). When you swallow, the soft palate and the uvula elevate to close off the posterior entrance into the nasopharynx and prevent ingested materials from entering the nasal region.

The **fauces** (faw′sēz) represent the opening between the oral cavity and the oropharynx. The fauces are bounded by paired muscular folds: the **palatoglossal** (păl′a-tō-glos′ăl) **arch** (anterior fold) and the **palatopharyngeal** (păl′a-tō-fa-rin′jē-ăl) **arch** (posterior fold). The palatine tonsils are housed between the arches. These tonsils serve as an "early line of defense" as they monitor ingested food and drink for foreign antigens and initiate an immune response when necessary (see section 21.4c).

The inferior surface, or floor, of the oral cavity houses the tongue. The **tongue** is formed primarily from skeletal muscle. Both extrinsic and intrinsic muscles move the tongue (see section 11.3c). Numerous small projections called **papillae** (pă-pil′ē; sing., *papilla, papula* = pimple) cover the superior (dorsal) surface of the tongue and are involved in the sense of taste (see section 16.3b). The posteroinferior region of the tongue also contains round masses of lymphatic tissue called the lingual tonsils. The inferior surface of the tongue attaches to the floor of the oral cavity by a thin vertical mucous membrane, the **lingual frenulum.** The tongue manipulates and mixes ingested

materials during chewing and helps compress the partially digested materials against the palate to assist in mechanical digestion. The tongue also performs important functions in both swallowing and in speech production. The teeth and gums are described in detail at the end of this section.

Histology   The epithelial lining of the oral cavity is a stratified squamous epithelium that protects against the abrasive activities associated with mechanical digestion. The nonkeratinized type of epithelium lines most of the oral cavity; the keratinized type lines the lips, portions of the tongue, and a small region of the hard palate.

Salivary Glands

Salivary glands, which produce **saliva** (sa-li′va), are located both within the oral cavity (intrinsic salivary glands) and outside the oral cavity (extrinsic salivary glands). **Intrinsic salivary glands** are unicellular glands that continuously release relatively small amounts of secretions independent of the presence of food. Only the secretions from the intrinsic salivary glands contain **lingual lipase,** an enzyme that begins the digestion of triglycerides. Most saliva, however, is produced from multicellular exocrine glands outside the oral cavity called **extrinsic salivary glands.**

Gross Anatomy   Three pairs of multicellular salivary glands are located external to the oral cavity: the parotid, submandibular, and sublingual glands (**figure 26.5a**).

The **parotid** (pă-rot′id; *para* = beside, *ot* = ear) **salivary glands** are the largest salivary glands. Each parotid gland is located anterior and inferior to the ear, partially overlying the masseter muscle. The parotid

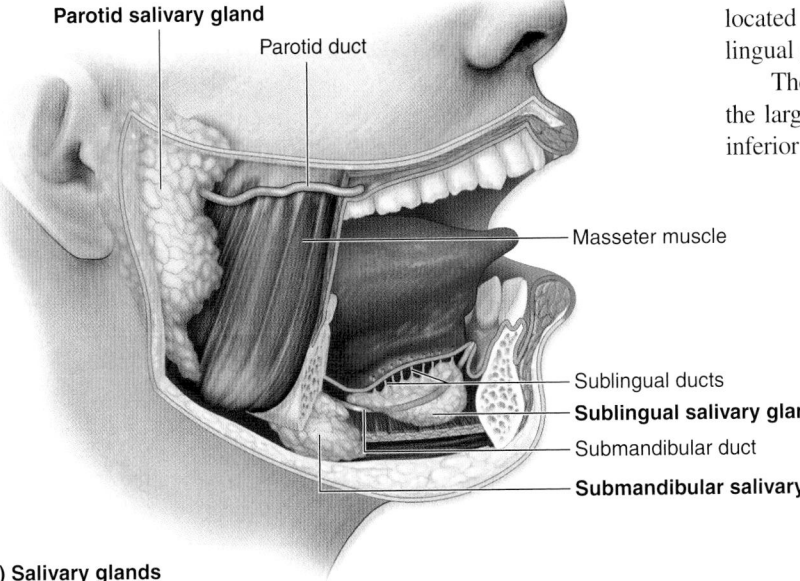

(a) Salivary glands

Figure 26.5 Salivary Glands. Saliva is produced primarily by three paired extrinsic salivary glands. (*a*) The relative locations of the parotid, submandibular, and sublingual salivary glands are shown in a side view. (*b*) Serous and mucous alveoli are shown in a diagrammatic representation of salivary gland histology. (*c*) Both mucous and serous cells may be seen in a micrograph of the submandibular salivary gland. AP|R

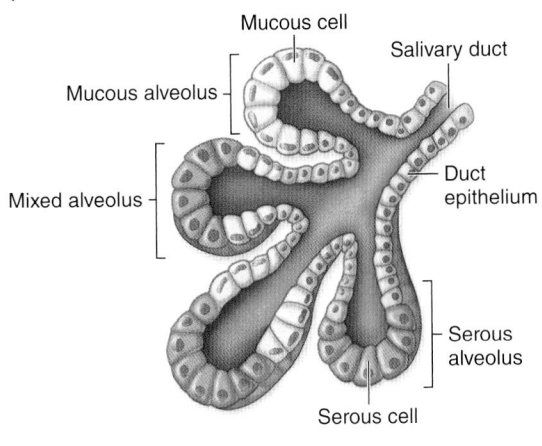

(b) Salivary gland histology

LM 200x

(c) Submandibular salivary gland

salivary glands produce about 25–30% of the saliva, which is transported through the **parotid duct** to the oral cavity. The parotid duct extends from the gland, across the external surface of the masseter muscle, before penetrating the buccinator muscle and opening into the vestibule of the oral cavity near the second upper molar. Infection of the parotid glands by a myxovirus causes *mumps.* Children are protected against mumps when immunized with the MMR (measles, mumps, and rubella) vaccine.

The **submandibular salivary glands** are both inferior to the floor of the oral cavity and medial to the body of the mandible as their name suggests. The submandibular salivary glands produce most of the saliva (about 60–70%). A **submandibular duct** opens from each gland through a papilla in the floor of the oral cavity on either side of the lingual frenulum.

The **sublingual salivary glands** are inferior to the tongue, and medial and anterior to the submandibular salivary glands. Each sublingual salivary gland extends multiple tiny sublingual ducts that open onto the inferior surface of the oral cavity, posterior to the submandibular duct papilla. These tiny glands contribute only about 3–5% of the total saliva.

Histology   Two types of secretory cells are housed within the large paired salivary glands and collectively produce the components of saliva: mucous cells and serous cells (figure 26.5*b, c*). Mucous cells secrete mucin, which forms mucus upon hydration, whereas serous cells secrete a watery fluid containing electrolytes and salivary amylase. The proportion of mucous cells to serous cells varies among the three types of salivary glands. The parotid glands produce only serous secretions, whereas the submandibular and sublingual glands produce both mucus and serous secretions. The structure and secretions of salivary glands are summarized in table 26.1.

Saliva   The volume of saliva secreted daily ranges between 1 and 1.5 liters. Most saliva is produced during mealtime, but smaller amounts are produced continuously to ensure that the oral cavity mucous membrane remains moist. Saliva is composed of 99.5% water and a mixture of solutes. Saliva is formed as water and electrolytes are filtered from plasma within capillaries then through cells (acini) of a salivary gland. Other components are added by cells of the salivary glands, including salivary amylase, mucin, and lysozyme. These components permit saliva to participate in various functions:

- Moistens ingested food as it is formed into a **bolus** (bō′lŭs; *bolos* = lump), a globular wet mass of partially digested material that is more easily swallowed

- Initiates the chemical breakdown of starch in the oral cavity because of the salivary amylase it contains
- Acts as a watery medium into which food molecules are dissolved so taste receptors may be stimulated (see section 16.3b)
- Cleanses the oral cavity structures
- Helps inhibit bacterial growth in the oral cavity because it contains antibacterial substances including lysozyme and antibodies (IgA) (IgA is formed by plasma cells in the lamina propria and transported across the epithelial cells.)

Regulation of Salivary Secretions   The salivary nuclei within the brainstem (see section 13.5c) regulate salivation. A basal level of salivation in response to parasympathetic stimulation ensures that the oral cavity remains moist. Input to the salivary nuclei is received from chemoreceptors or mechanoreceptors in the upper GI tract. These receptors detect various types of stimuli, including the introduction of substances into the oral cavity, especially those that are acidic, such as a lemon; and arrival of foods into the stomach lumen, especially foods that are spicy or acidic. If one eats spoiled food, bacterial toxins within the stomach stimulate receptors that initiate sensory nerve signals to the salivary nuclei. Input is also received by the salivary nuclei from the higher brain centers in response to the thought, smell, or sight of food. Stimulation of the salivary nuclei results in increased nerve signals relayed along parasympathetic neurons within both the facial nerve, which innervates the submandibular and sublingual salivary glands, and glossopharyngeal nerve, which innervates the parotid salivary glands, and additional saliva is released.

Sympathetic stimulation, which occurs during exercise or when an individual is excited or anxious (see section 15.4), results in a more viscous saliva by decreasing the water content of saliva. (This occurs because sympathetic stimulation constricts the capillaries of the salivary gland and decreases the fluid added to saliva.)

 WHAT DO YOU THINK?

1 Research suggests that a dry mouth (inadequate production of saliva) is correlated with an increase in dental problems, such as cavities. What are the possible reasons for this correlation?

Mechanical Digestion: Mastication

Mechanical digestion in the oral cavity is called **mastication** (mas′ti-kā′shŭn; *mastico* = to chew) or *chewing.* It requires the coordinated activities of teeth, skeletal muscles in lips, tongue, cheeks, and jaws that are controlled by nuclei within the medulla oblongata and pons, collectively called the **mastication center.**

Table 26.1	Salivary Glands and Secretions			
SALIVARY GLANDS				
Gland	**Structure and Location**		**Types of Secretion**	**Percentage of Saliva Produced**
Parotid	Largest salivary gland; located anterior and inferior to ears; parotid duct opens into vestibule near second upper molar		Only serous secretions	25–30%
Submandibular	Located inferior to the floor of the oral cavity; duct opens lateral to lingual frenulum		Both serous and mucus secretions	60–70%
Sublingual	Smallest salivary gland; located inferior to tongue; ducts open into floor of oral cavity		Both serous and mucus secretions	3–5%
SALIVA CHARACTERISTICS				
Production Rate	**Solute Components**		**pH Range and Composition**	
1–1.5 liters per day	Salivary amylase, lingual lipase (from intrinsic salivary glands), mucin, ions (Na^+, K^+, Cl^-, HCO_3^-), lysozyme (antibacterial enzyme), immunoglobulin A (from plasma cells)		Slightly acidic (pH 6.4–6.8): 99.5% water, 0.5% solutes	

The primary function in chewing the food is to mechanically reduce its bulk into smaller particles to facilitate swallowing. Chemical digestion and absorption are affected very little by chewing, except that the surface area of the food is increased, which facilitates exposure to and action by digestive enzymes. Mastication also promotes salivation to help soften and moisten the food to form a bolus.

Note that medications composed of small, nonpolar molecules (e.g., nitroglycerin, to treat angina pectoris) may be absorbed directly into the blood from the mouth. When these medications are placed under the tongue, they pass through the oral cavity epithelium by simple diffusion (see section 4.3a) and are absorbed into the blood.

Teeth

The **teeth** are collectively known as the **dentition** (den-tish′ŭn; *dentition* = teething). A tooth has an exposed **crown,** a constricted **neck,** and one or more **roots** that anchor it to the jaw (figure 26.6a). The roots of the teeth fit tightly into dental alveoli, which are sockets within the alveolar processes of both the maxillae and the mandible. Collectively, the roots, the dental alveoli, and the periodontal ligament

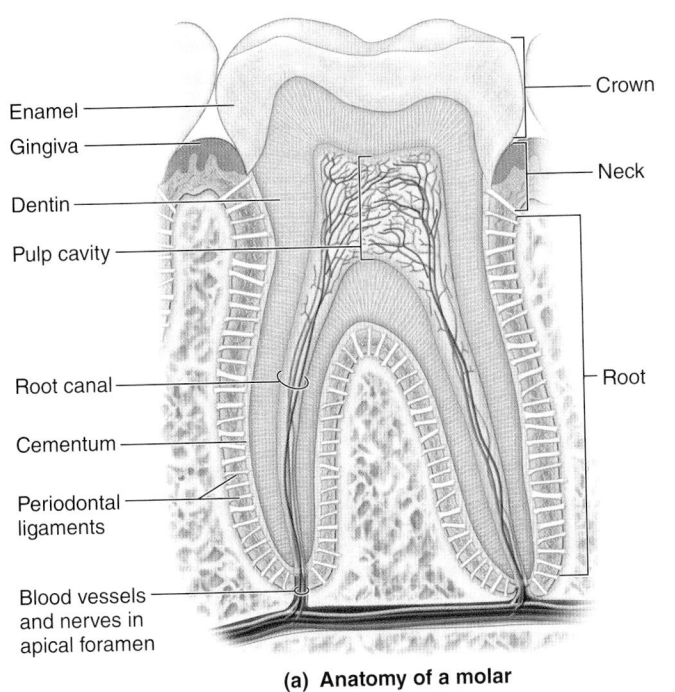

Enamel
Gingiva
Dentin
Pulp cavity
Root canal
Cementum
Periodontal ligaments
Blood vessels and nerves in apical foramen
Crown
Neck
Root

(a) Anatomy of a molar

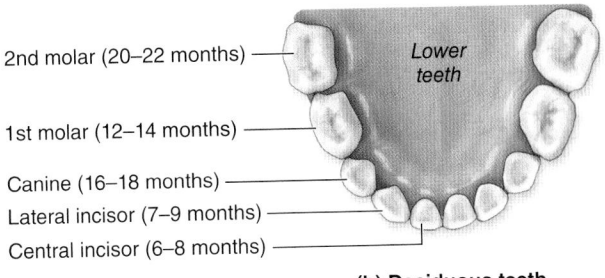

Central incisor (7–9 months)
Lateral incisor (9–11 months)
Canine (18–20 months)
1st molar (14–16 months)
2nd molar (24–30 months)
Upper teeth

2nd molar (20–22 months)
Lower teeth
1st molar (12–14 months)
Canine (16–18 months)
Lateral incisor (7–9 months)
Central incisor (6–8 months)

(b) Deciduous teeth

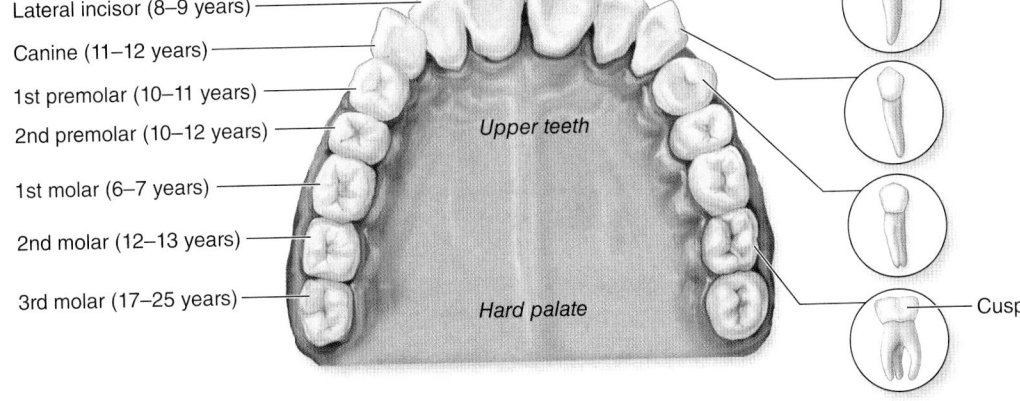

Central incisor (7–8 years)
Lateral incisor (8–9 years)
Canine (11–12 years)
1st premolar (10–11 years)
2nd premolar (10–12 years)
1st molar (6–7 years)
2nd molar (12–13 years)
3rd molar (17–25 years)
Upper teeth
Hard palate
Cusp

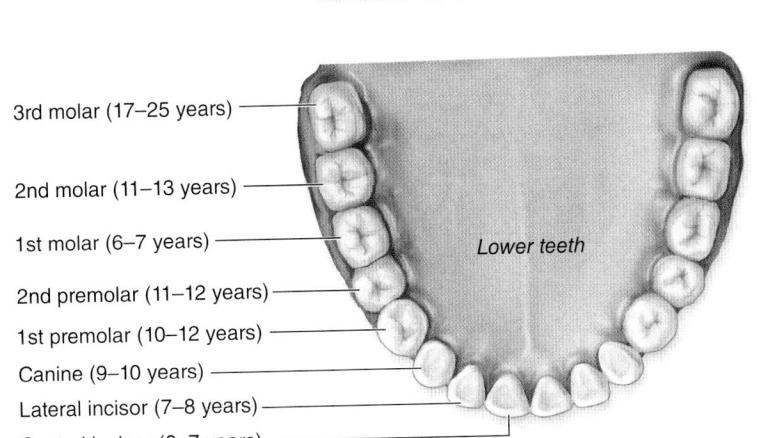

3rd molar (17–25 years)
2nd molar (11–13 years)
1st molar (6–7 years)
2nd premolar (11–12 years)
1st premolar (10–12 years)
Canine (9–10 years)
Lateral incisor (7–8 years)
Central incisor (6–7 years)
Lower teeth

(c) Permanent teeth

Figure 26.6 Teeth. Ingested food is chewed by the teeth in the oral cavity. (*a*) Anatomy of a molar. (*b, c*) Comparison of the average dentition of deciduous and permanent teeth, including the approximate age at eruption for each tooth. AP|R

that binds the roots to the alveolar processes form a gomphosis joint (described in section 9.2a).

Dentin (den'tin; *dens* = tooth) forms the primary mass of a tooth. Dentin is comparable to bone but harder. On the external surface of the dentin, a tough, durable layer of **enamel** forms the crown of the tooth. Enamel is the hardest substance in the body and is primarily composed of calcium phosphate crystals. The center of the tooth is a **pulp cavity** that contains a connective tissue called **pulp.**

A **root canal** is continuous with the pulp cavity and opens into the connective tissue surrounding the root through an opening called the apical foramen. Blood vessels and nerves housed in the pulp pass through the apical foramen. Each root of a tooth is ensheathed within hardened material called **cementum** (se-men'tŭm).

Deciduous and Permanent Teeth

Two sets of teeth develop and erupt during a normal lifetime (figure 26.6*b, c*). In an infant, 20 **deciduous** (dē-sid'ū-ŭs; *deciduus* = falling off) **teeth,** also called *milk teeth,* erupt between 6 months and 30 months after birth. These teeth are eventually lost and replaced by 32 **permanent teeth.** As figure 26.6*b* shows, the more anteriorly placed permanent teeth tend to appear first, followed by the posteriorly placed teeth. (The major exception to this rule is the first molars, which appear at about age 6 and sometimes are referred to as the *6-year molars.*) The last teeth to erupt are the third molars, often called *wisdom teeth,* in the late teens or early 20s. The jaw often lacks space to accommodate these final molars, and they may either emerge only partially or grow at an angle and become impacted (wedged against another structure). Impacted teeth cannot erupt properly because of the angle of their growth.

The most anteriorly placed permanent teeth are called **incisors** (in-sī'zōr; *incido* = to cut into). They are shaped like a chisel and have a single root. They are designed for slicing or cutting into food. Immediately posterolateral to the incisors are the **canines** (kā'nīn; *canis* = dog) that have a pointed tip for puncturing and tearing food. **Premolars** are located posterolateral to the canines and anterior to the molars. They have flat crowns with prominent ridges called

cusps that are used to crush and grind ingested materials. Premolars may have one or two roots. The **molars** are the thickest and most posteriorly placed teeth. They have large, broad, flat crowns with distinctive cusps, and three or more roots. Molars are also adapted for grinding and crushing ingested materials. If the oral cavity is divided into quadrants, each quadrant contains the following number of permanent teeth: two incisors, one canine, two premolars, and three molars.

The **gingivae** (jin'ji-vă, -vē) are the *gums.* They are composed of dense irregular connective tissue, with an overlying nonkeratinized stratified squamous epithelium that covers the alveolar processes of the upper and lower jaws and surrounds the neck of the teeth.

WHAT DID YOU LEARN?

11 What are the roles of the tongue, teeth, and salivary glands in forming a bolus?

26.2c Pharynx and Esophagus

LEARNING OBJECTIVE

6. Discuss the anatomy of the pharynx and esophagus and their complementary activities in the process of swallowing.

The pharynx and the esophagus connect the oral cavity to the stomach (**figure 26.7a**).

Gross Anatomy of the Pharynx

The pharynx was previously described in detail in section 23.2c. It is a funnel-shaped, muscular passageway with distensible lateral walls that serves as the passageway for both air and food. Three skeletal muscle pairs called the superior, middle, and inferior **pharyngeal constrictors** form the wall of the pharynx (see figure 11.10). The oropharynx and laryngopharynx are lined with nonkeratinized stratified squamous epithelium that provides protection against abrasion associated with swallowing ingested materials.

Gross Anatomy of the Esophagus

The **esophagus** (ĕ-sof'ă-gŭs) is a normally collapsed, tubular passageway. It is about 25 centimeters (10 inches) long in an adult and begins

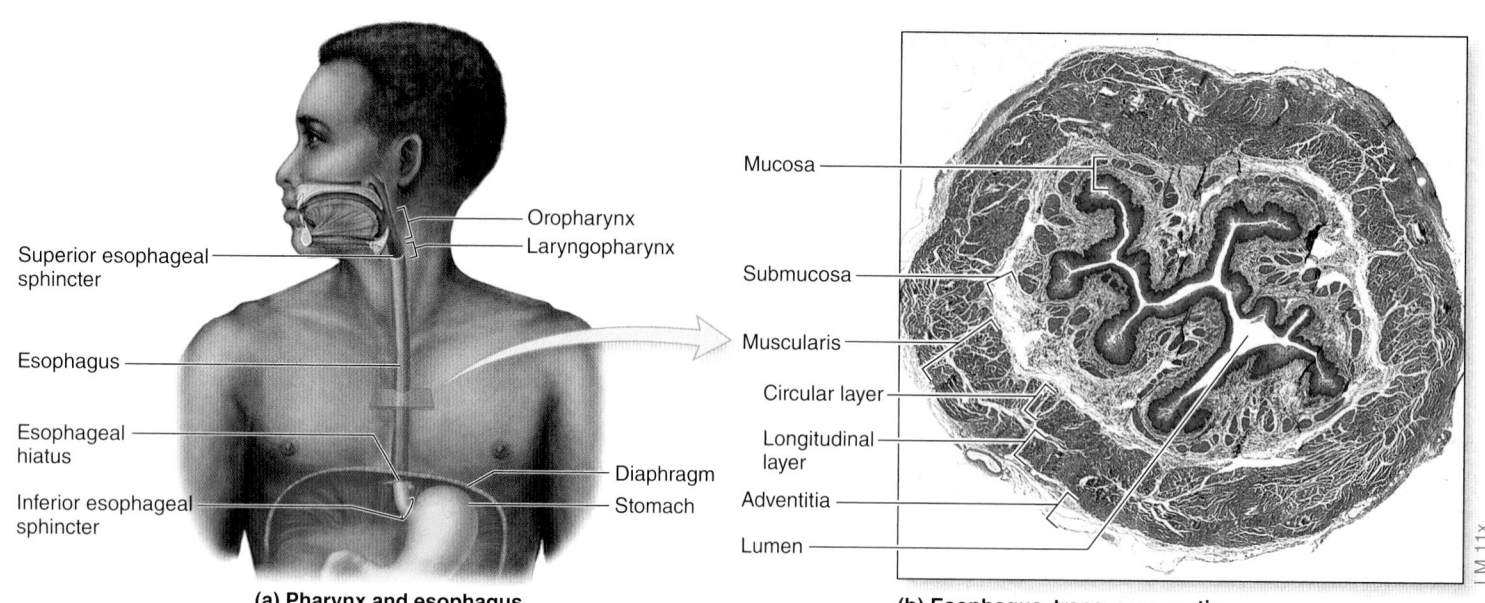

(a) Pharynx and esophagus

(b) Esophagus, transverse section

Figure 26.7 The Esophagus. (*a*) The esophagus extends inferiorly from the pharynx to the stomach and functions in the passage of food and drink. (*b*) A photomicrograph of a transverse section through the esophagus identifies the tunics in its wall. AP|R

INTEGRATE

CLINICAL VIEW
Reflux Esophagitis and Gastroesophageal Reflux Disease

Sometimes acidic chyme refluxes into the esophagus, causing the burning pain and irritation of **reflux esophagitis.** Because the pain is felt posterior to the sternum and may be so intense that it is mistaken for a heart attack, this condition is commonly known as *heartburn.*

Unlike the stomach epithelium, the esophageal epithelium is poorly protected against acidic contents and easily becomes inflamed and irritated. Reflux esophagitis is seen most frequently in overweight individuals, smokers, those who have eaten a very large meal (especially just before bedtime), and people with **hiatal hernias** (hī-ā′tăl her′nē-ă; rupture), in which a portion of the stomach protrudes through the diaphragm into the thoracic cavity. Eating spicy foods, or ingesting too much caffeine, may exacerbate the symptoms in people affected by reflux esophagitis. Preventive treatment includes lifestyle changes such as losing weight, quitting smoking, limiting meal size, and not lying down until 2 hours after eating. Sleeping with the head of the bed elevated 4 to 6 inches, so that the body lies at an angle rather than flat, also appears to alleviate symptoms.

Chronic reflux esophagitis may lead to **gastroesophageal reflux disease (GERD).** Frequent gastric reflux erodes the esophageal tissue in this condition, so over a period of time, scar tissue builds up in the esophagus,

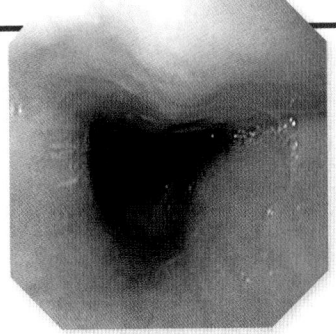

Endoscopic view of a normal esophagus.

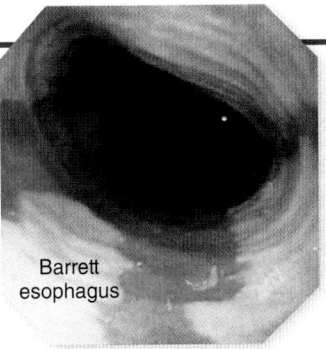

An endoscopic view of the esophagus shows the signs of Barrett esophagus.

leading to narrowing of the esophageal lumen. In more advanced cases, the esophageal epithelium may change from stratified squamous to columnar secretory epithelium, a condition known as **Barrett esophagus.** The secretions of secretory columnar epithelium may provide protection from the erosive gastric secretions. Unfortunately, this metaplasia increases the risk of cancerous growths.

GERD may be treated with a series of medications. Proton pump inhibitors (e.g., omeprazole [Prilosec], esomeprazole [Nexium]) limit acid secretion in the stomach by acting on the proton (hydrogen ion [H^+]) pumps that help produce acid. Histamine (H_2) blockers (e.g., famotidine [Pepcid], nizatidine [Axid], ranitidine [Zantac]) also help limit acid secretion in the stomach. Antacids also help neutralize stomach acid.

---

at approximately the level of the cricoid cartilage of the larynx, with most of its length within the thoracic cavity. This tube is directly anterior to the vertebral bodies, until it passes through the diaphragm. The inferior region of the esophagus connects to the stomach, where it passes through an opening in the diaphragm called the **esophageal hiatus** (hī-ā′tŭs; to yawn). Only the last 1.5 centimeters (slightly more than one-half inch) of the esophagus is located within the abdominal cavity.

The **superior esophageal sphincter** (or *pharyngoesophageal sphincter*) is a contracted ring of circular skeletal muscle at the superior end of the esophagus. It is the area where the esophagus and the pharynx meet. This sphincter is closed during inhalation of air, so air does not enter the esophagus and instead enters the larynx and trachea.

The **inferior esophageal sphincter** (*gastroesophageal,* or *cardiac sphincter*) is a contracted ring of circular smooth muscle at the inferior end of the esophagus. This sphincter isn't strong enough alone to prevent materials from refluxing back into the esophagus; instead, the muscles of the diaphragm at the esophageal opening contract to help prevent materials from regurgitating from the stomach into the esophagus.

Histology

The mucosa of the esophagus is lined with a nonkeratinized stratified squamous epithelium (figure 26.7*b*). It serves to protect this region from abrasion as food is swallowed.

The submucosa of the esophagus is thick and composed of abundant elastic fibers that permit distension during swallowing. It also houses numerous mucous glands that provide thick, lubricating mucus for the epithelium. The ducts of these glands project through the mucosa and open into the lumen.

The muscularis of the esophagus is unique in that it contains a blend of both skeletal and smooth muscle. The two layers of muscle in the superior one-third of the esophageal muscularis are skeletal, rather than smooth, to ensure that the swallowed material moves rapidly out

of the pharynx and into the esophagus before the next respiratory cycle begins. (Remember that smooth muscle contracts more slowly than does skeletal muscle; see section 10.10c.) Skeletal muscle and smooth muscle cells intermingle in the middle one-third of the esophageal muscularis, and only smooth muscle is found within the wall of the inferior one-third of this muscularis. This transition marks the beginning of a continuous smooth muscle muscularis that extends throughout the stomach and the small and large intestines to the anus. The outermost layer of the esophagus is an adventitia.

Motility: The Swallowing Process

Swallowing is called *deglutition* (dē-glū-tish′ŭn). It is the process of moving ingested materials from the oral cavity to the stomach. Swallowing has three phases: the voluntary phase, the pharyngeal phase, and the esophageal phase (**figure 26.8**).

The **voluntary phase** occurs after ingestion. It is controlled by the cerebral cortex (primarily the temporal lobes and motor cortex of the frontal lobe). Ingested materials and saliva mix in the oral cavity. Chewing forms a bolus that is mixed and manipulated by the tongue and then pushed superiorly against the hard palate. Transverse palatine folds in the hard palate help direct the bolus posteriorly toward the oropharynx.

The arrival of the bolus at the entryway to the oropharynx initiates the swallowing reflex of the **pharyngeal phase.** The pharyngeal phase is involuntary. Tactile sensory receptors around the fauces are stimulated by the bolus and initiate nerve signals along sensory neurons to the **swallowing center** (or *deglutition center*) in the medulla oblongata (see section 13.5c). Nerve signals are then relayed along motor neurons to effectors to cause the following response:

- Entry of the bolus into the oropharynx
- Elevation of the soft palate and uvula to block the passageway between the oropharynx and nasopharynx

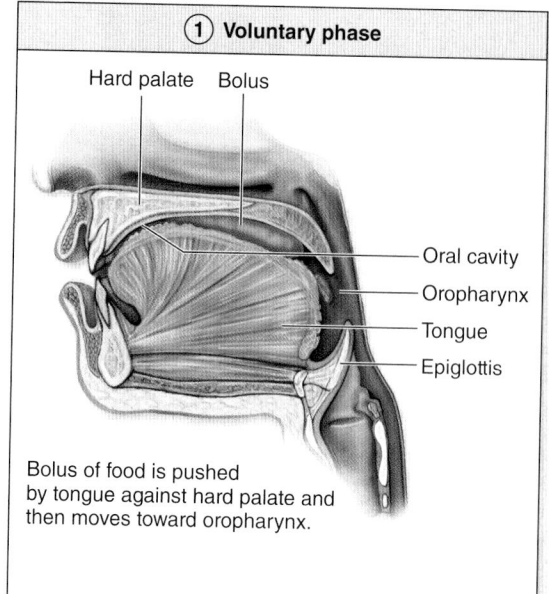

① Voluntary phase

Hard palate   Bolus

Oral cavity
Oropharynx
Tongue
Epiglottis

Bolus of food is pushed by tongue against hard palate and then moves toward oropharynx.

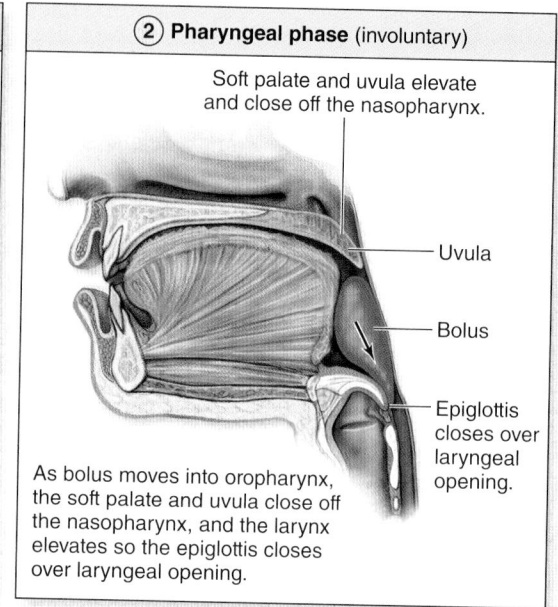

② Pharyngeal phase (involuntary)

Soft palate and uvula elevate and close off the nasopharynx.

Uvula

Bolus

Epiglottis closes over laryngeal opening.

As bolus moves into oropharynx, the soft palate and uvula close off the nasopharynx, and the larynx elevates so the epiglottis closes over laryngeal opening.

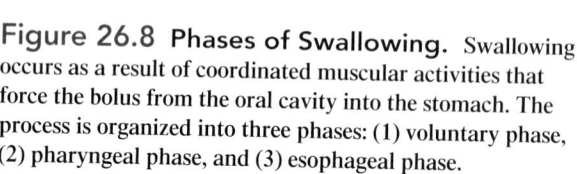

① **Voluntary phase**

② **Pharyngeal phase** (involuntary)

③ **Esophageal phase** (involuntary)

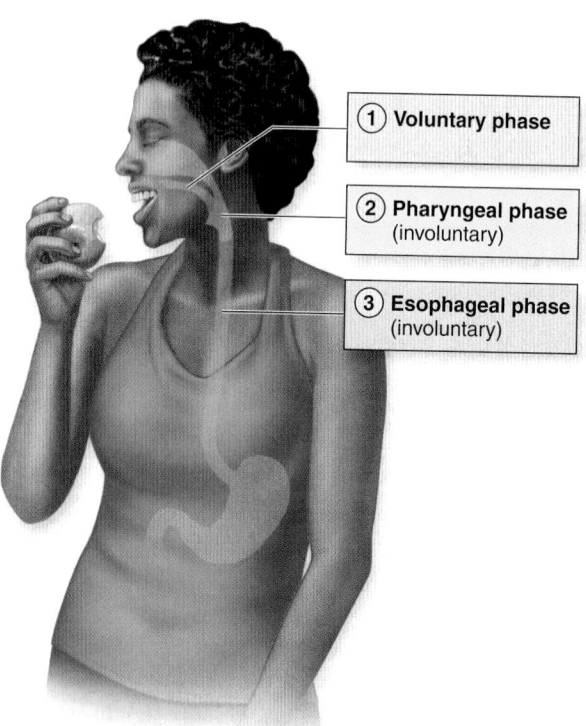

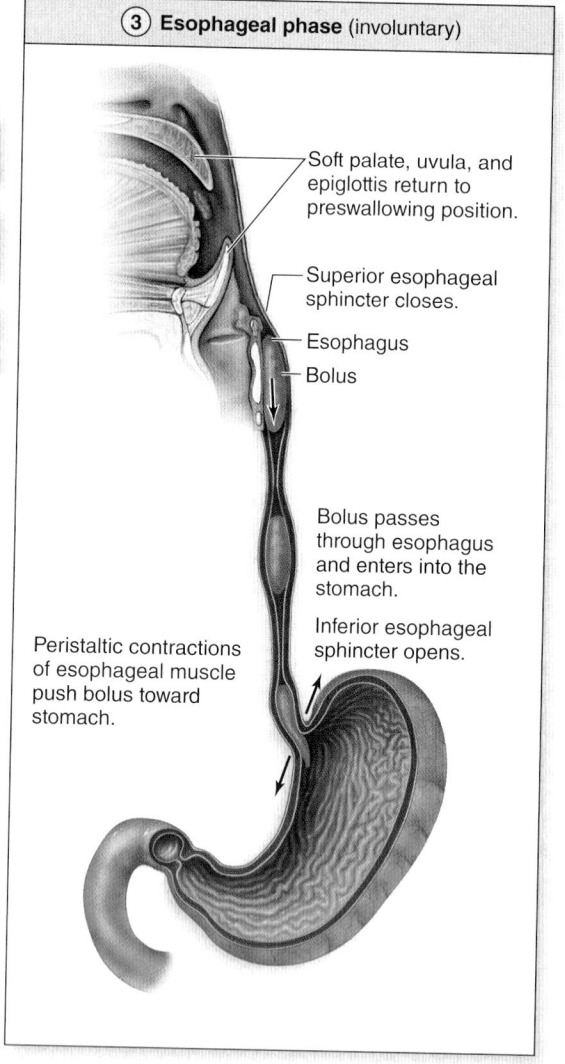

③ Esophageal phase (involuntary)

Soft palate, uvula, and epiglottis return to preswallowing position.

Superior esophageal sphincter closes.

Esophagus
Bolus

Bolus passes through esophagus and enters into the stomach.

Inferior esophageal sphincter opens.

Peristaltic contractions of esophageal muscle push bolus toward stomach.

Figure 26.8 Phases of Swallowing. Swallowing occurs as a result of coordinated muscular activities that force the bolus from the oral cavity into the stomach. The process is organized into three phases: (1) voluntary phase, (2) pharyngeal phase, and (3) esophageal phase.

- Elevation of the larynx by the extrinsic muscles (see section 23.3a) move the larynx anteriorly and superiorly, resulting in the epiglottis covering the laryngeal opening; this prevents ingested material from entering the trachea

In addition, nerve signals are relayed to the respiratory center within the medulla oblongata to assure that a breath is not taken during swallowing. During this time, the bolus passes quickly and involuntarily through the pharynx to the esophagus—about 1 second elapses in this phase. Sequential contraction of the pharyngeal constrictors decreases the diameter of the pharynx, beginning at its superior end and moving toward its inferior end. This creates a pressure difference, forcing swallowed material from the pharynx into the esophagus.

The **esophageal phase** is also involuntary. It is the time during which the bolus passes through the esophagus and into the stomach—about 5 to 8 seconds. The presence of the bolus within the lumen of the esophagus stimulates sequential peristaltic waves of muscular contraction that assist in propelling the bolus toward the stomach.

Higher pressure occurs in the superior region of the esophagus relative to the inferior region.

The superior and inferior esophageal sphincters are normally closed at rest. When the bolus is swallowed, these sphincters relax to allow it to pass through the esophagus. The inferior esophageal sphincter contracts after passage of the bolus, helping to prevent reflux of materials and fluids from the stomach into the esophagus.

WHAT DID YOU LEARN?

12 How do the tunics of the esophagus differ from the "default" tunic pattern in both the mucosa and muscularis?

13 How is the bolus moved from the oral cavity into the stomach, as described in the three phases of swallowing?

26.2d Stomach

LEARNING OBJECTIVES

7. Describe the gross anatomy and histology of the stomach.

8. Explain the two general functional activities of the stomach.

9. Describe the phases that regulate motility and secretion in the stomach.

The **stomach** (stŭm'ŭk) is a holding sac in the superior left quadrant of the abdomen immediately inferior to the diaphragm. Under normal conditions, between 3 and 4 liters of food, drink, and saliva enter the stomach daily and generally spend between 2 and 6 hours there, depending upon the amount and composition of the ingested material. It mixes the ingested food with secretions released from the stomach wall and mechanically digests the contents into a semifluid mass called chyme. Chemical digestion of both protein and fat begins in the stomach, but absorption from it is limited to small, nonpolar substances that are in contact with the mucosa of the stomach. Both alcohol and aspirin are examples of substances that are absorbed in the stomach. One significant function of the stomach is to serve as a "holding bag" for controlled release of partially digested materials into the small intestine, where most chemical digestion and absorption occurs. However, the only essential function performed by the stomach is the release of intrinsic factor (a substance required for the absorption of vitamin B_{12}, which occurs within the small intestine).

INTEGRATE

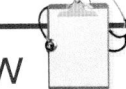

CLINICAL VIEW
Gastric Bypass

Gastric bypass is a treatment used in obese individuals to assist in weight loss. It is a surgical procedure that involves sectioning off a small part of the stomach (so the individual eats smaller portions) and attaching it to a lower part of the small intestine (so less nutrients are absorbed). Several changes are noted following surgery including a decrease in appetite and altered response to hormones including insulin. One of the most surprising changes is that the surgery can induce type 2 diabetes into remission many times within a few days of the surgery. The International Diabetes Foundation now endorses gastric bypass surgery for treatment of type 2 diabetes.

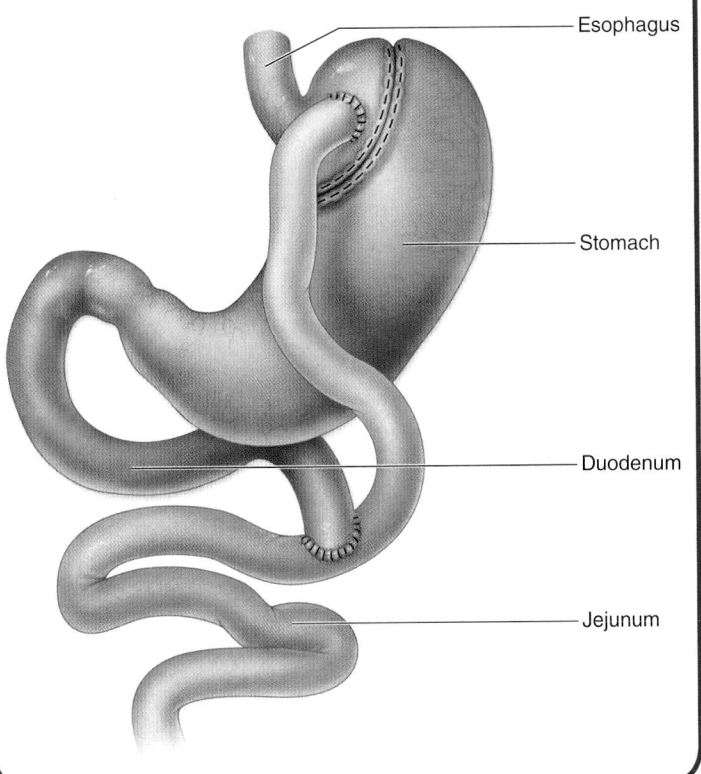

Gross Anatomy of the Stomach

The stomach is a muscular J-shaped organ (**figure 26.9**). It has both a larger convex inferolateral surface called the **greater curvature** and a smaller concave superomedial surface called the **lesser curvature.** This organ is composed of four regions:

- The **cardia** (kar'dē-ă) is a small, narrow, superior entryway into the stomach lumen from the esophagus. The internal opening where the cardia meets the esophagus is called the **cardiac orifice,** which is the location of the cardiac sphincter.
- The **fundus** (fŭn'dŭs; bottom) is the dome-shaped region lateral and superior to the esophageal connection with the stomach. Its superior surface contacts the diaphragm. The fundus has both weaker contractions and a higher pH than other regions of the stomach.
- The **body** is the largest region of the stomach; it is inferior to the cardiac orifice and the fundus and extends to the pylorus.
- The **pylorus** (pī-lōr'ŭs; *pylorus* = gatekeeper) is the narrow, funnel-shaped terminal region of the stomach. Its opening into

the duodenum of the small intestine is called the **pyloric orifice.** Surrounding this pyloric orifice is a thick ring of circular smooth muscle called the **pyloric sphincter.** The pyloric sphincter regulates the entry of material into the small intestine.

The internal stomach lining is composed of numerous **gastric folds,** or *rugae* (rū'jē; *ruga* = wrinkle). These gastric folds are seen only when it is empty. They allow the stomach to expand greatly when it fills with food and drink and then return to its normal J-shape when it empties.

Two serous membranes are associated with the stomach: the *greater omentum* and the *lesser omentum,* which were described previously. The greater omentum extends inferiorly from the greater curvature of the stomach, forming the fatty apron that covers the anterior surface of abdominal organs. The lesser omentum extends superiorly from the lesser curvature of the stomach and duodenum to the liver.

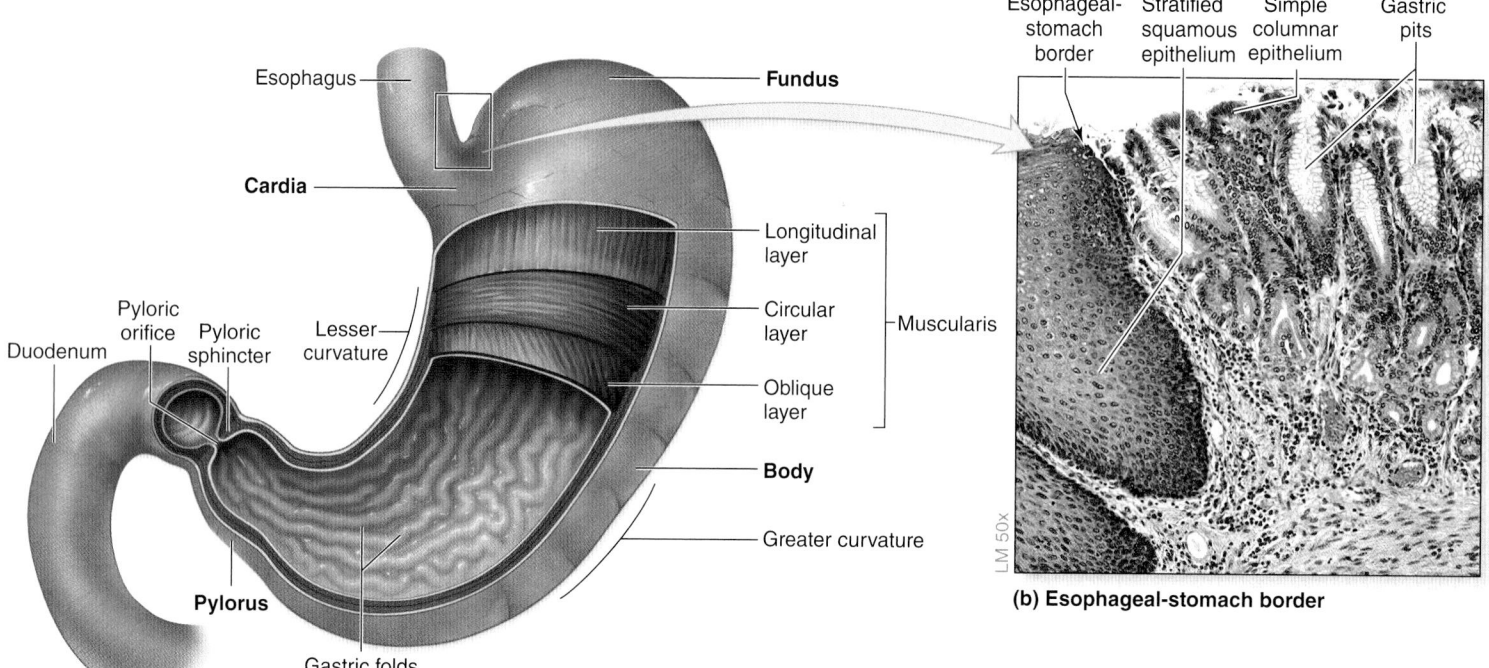

Esophageal-stomach border
Stratified squamous epithelium
Simple columnar epithelium
Gastric pits

Esophagus
Fundus
Cardia
Longitudinal layer
Circular layer
Oblique layer
Muscularis
Duodenum
Pyloric orifice
Pyloric sphincter
Lesser curvature
Body
Greater curvature
Pylorus
Gastric folds

LM 50x

(a) Stomach regions, anterior view

(b) Esophageal-stomach border

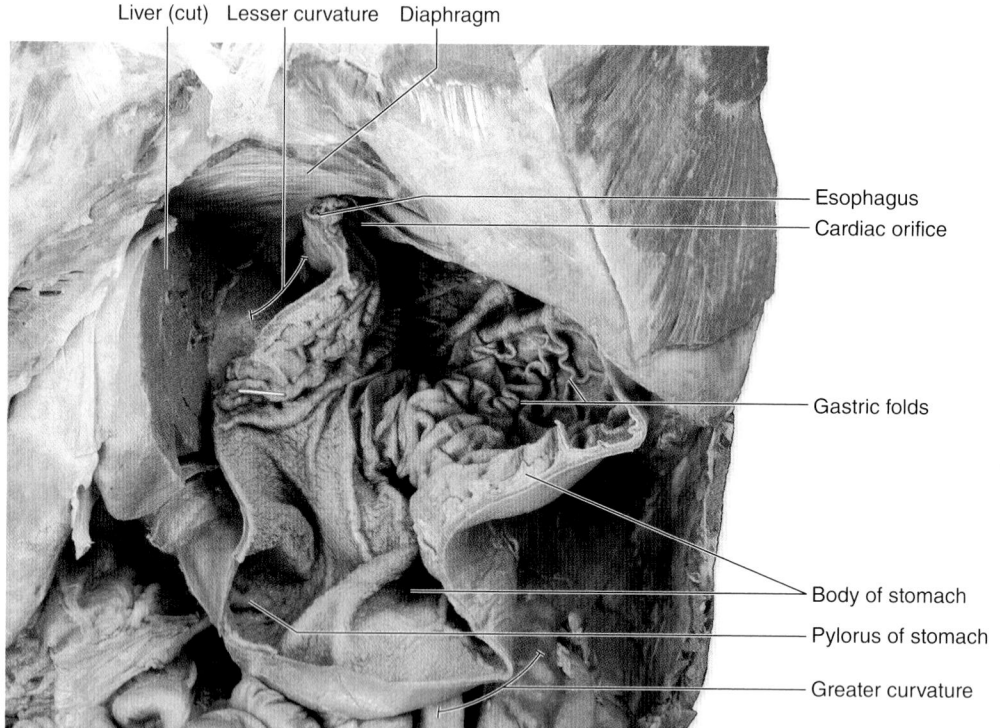

Liver (cut)   Lesser curvature   Diaphragm

Esophagus
Cardiac orifice

Gastric folds

Body of stomach
Pylorus of stomach
Greater curvature

(c) Gross anatomy of stomach (cut open)

Figure 26.9 Gross Anatomy of the Stomach. The stomach is a muscular sac where mechanical and chemical digestion of the bolus occurs. (*a*) The major regions of the stomach are the cardia, the fundus, the body, and the pylorus. Three layers of smooth muscle make up the muscularis tunic. (*b*) A photomicrograph of the abrupt transition from stratified squamous epithelium in the esophagus to simple columnar in the stomach. (*c*) A cadaver photo shows an anterior, open section of the stomach, revealing the gastric folds and the cardiac orifice (opening of stomach connected to esophagus). AP|R

Histology

The mucosa of the stomach is only 1.5 millimeters at its thickest region (about the thickness of a nickel). This inner lining has three significant features (**figure 26.10**):

- It is lined by a simple columnar epithelium supported by lamina propria. The transition from stratified squamous in the esophagus to simple columnar epithelium in the stomach is abrupt (figure 26.9*b*).

- The lining is indented by numerous depressions called **gastric pits.**

- Several **gastric glands** extend deep into the mucosa from the base of each gastric pit. The muscularis mucosae partially surround the gastric glands and helps expel gastric gland secretions when it contracts.

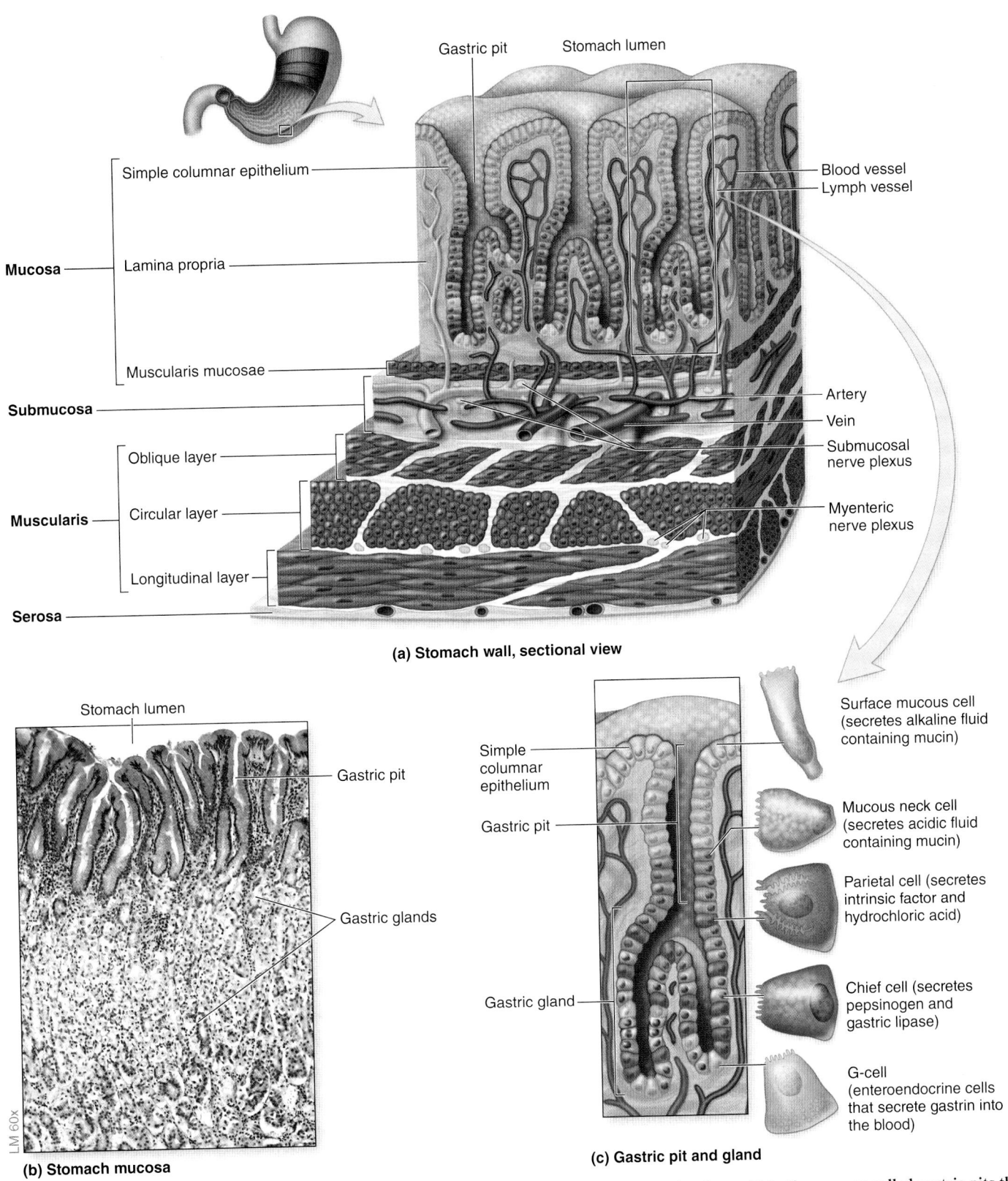

Gastric pit   Stomach lumen

Mucosa
- Simple columnar epithelium
- Lamina propria
- Muscularis mucosae

Submucosa

Muscularis
- Oblique layer
- Circular layer
- Longitudinal layer

Serosa

Blood vessel
Lymph vessel

Artery
Vein
Submucosal nerve plexus

Myenteric nerve plexus

(a) Stomach wall, sectional view

Stomach lumen

Gastric pit

Gastric glands

LM 60x

(b) Stomach mucosa

Simple columnar epithelium

Gastric pit

Gastric gland

Surface mucous cell (secretes alkaline fluid containing mucin)

Mucous neck cell (secretes acidic fluid containing mucin)

Parietal cell (secretes intrinsic factor and hydrochloric acid)

Chief cell (secretes pepsinogen and gastric lipase)

G-cell (enteroendocrine cells that secrete gastrin into the blood)

(c) Gastric pit and gland

Figure 26.10 Histology of the Stomach Wall. (*a*) The stomach wall contains invaginations within the mucosa called gastric pits that lead into gastric glands. (*b*) A photomicrograph shows the cells lining the gastric pit and gastric glands. (*c*) A diagrammatic section of a gastric gland shows its structure and the distribution of different secretory cells. AP|R

The muscularis of the stomach varies from the general GI tract pattern in that it is composed of three smooth muscle layers instead of two: an inner oblique layer, a middle circular layer, and an outer longitudinal layer. The presence of a third (oblique) layer of smooth muscle assists the continued churning and blending of the swallowed bolus to help mechanically digest the food. The muscularis becomes increasingly thicker as it progresses from the body to the pylorus. The outermost layer of the stomach is a serosa called the visceral peritoneum because the stomach is intraperitoneal.

WHAT DO YOU THINK?

2 The stomach secretes highly acidic gastric juices that facilitate the breakdown of food. What prevents the gastric juices from eating away at the stomach itself?

Gastric Secretions

Five types of secretory cells of the gastric epithelium are integral contributors to the process of digestion (figure 26.10c). Four of these cell types produce the approximate 3 liters per day of **gastric juice** associated with the stomach. The fifth type of cell secretes a hormone into the blood.

Surface Mucous Cells **Surface mucous cells** line the stomach lumen and extend into the gastric pits. They continuously secrete an alkaline product containing mucin onto the gastric surface. Mucin becomes hydrated, producing a 1- to 3-millimeter mucus layer. This mucus layer, along with a high rate of cell turnover in the mucosa, helps to prevent ulceration of the stomach lining upon exposure to both the high acidity of the gastric fluid and gastric enzymes.

Mucous Neck Cells **Mucous neck cells** are located immediately deep to the base of the gastric pit and are interspersed among the parietal cells (described next). These cells produce an acidic mucin that differs structurally and functionally from the alkaline mucin secreted by the surface mucous cells. The acidic mucin helps maintain the acidic conditions resulting from the secretion of hydrochloric acid by parietal cells. The mucus produced from mucin secretion by both types of mucous cells has lubricating properties to protect the stomach lining from abrasion or mechanical injury.

Parietal Cells **Parietal cells** (also called *oxyntic cells*) are responsible for the addition of two substances into the lumen of the stomach:

- **Intrinsic factor.** The production of this glycoprotein is the *only* essential function performed by the stomach. Intrinsic factor is required for absorption of vitamin B_{12} in the ileum (the final portion of the small intestine). B_{12} is necessary for production of normal erythrocytes. A critical decrease or absence of B_{12} results in pernicious anemia (see Clinical View: "Anemia" in section 18.3b).

- **Hydrochloric acid (HCl).** HCl is not formed within the parietal cell; it would destroy the cell. Instead, it forms from the H^+ and Cl^- secreted from the parietal cells' surface. The details of this process are described in **figure 26.11**. HCl is responsible for the low pH of between 1.5 and 2.5 within the stomach. (Vomiting causes increased formation of HCl; this results in increased HCO_3^- in the blood, raising blood pH. Extensive vomiting can lead to metabolic alkalosis; see Clinical View: "How Does Vomiting or Diarrhea Alter Blood H^+ Concentration?" in section 25.5b.)

Hydrochloric acid has several functions in the digestive processes of the stomach:

- It converts the inactive enzyme pepsinogen into active pepsin and provides the optimal pH environment for pepsin activity.
- It kills most microorganisms that enter the stomach (most cannot survive in the extremely low pH).
- It contributes to the breakdown of plant cell walls and animal connective tissue.

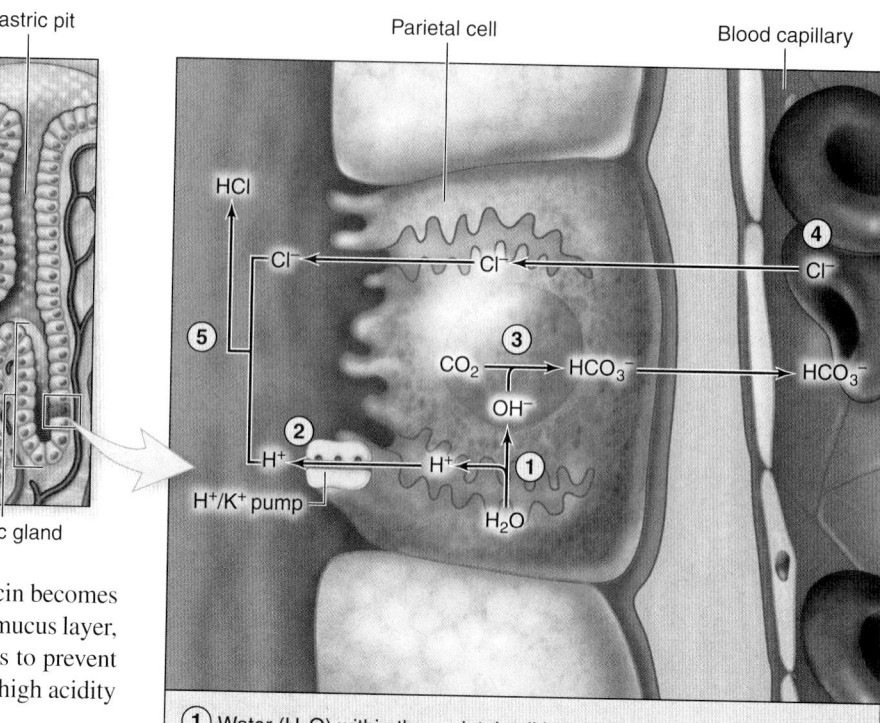

1. Water (H_2O) within the parietal cell is split into a hydrogen ion (H^+) and hydroxide ion (OH^-).
2. H^+ is pumped into the lumen of the gastric gland by an H^+/K^+ pump.
3. OH^- bonds with carbon dioxide (CO_2) to form bicarbonate ion (HCO_3^-).
4. An exchange occurs as HCO_3^- is transported out of the parietal cell (HCO_3^- then enters the blood), while chloride ion (Cl^-) is transported into the parietal cell; Cl^- then enters the lumen of the gastric gland.
5. Within the lumen of the gastric gland, Cl^- combines with H^+ to form hydrochloric acid (HCl).

Figure 26.11 Formation of HCl from Parietal Cells.
H^+ and HCO_3^- are produced within epithelial cells of the stomach. The net movement is H^+ into the lumen of the stomach and HCO_3^- into the blood. Additionally, Cl^- is transferred from the blood into the stomach lumen, where it combines with H^+ to form hydrochloric acid (HCl). **AP|R**

- It denatures proteins (see section 2.8b) by causing them to unfold, thus facilitating chemical digestion by enzymes.

Chief Cells **Chief cells** (also called *zymogenic cells*, or *peptic cells*) are the most numerous secretory cells within the gastric glands, hence the name "chief" cells. These cells produce and secrete packets of zymogen granules primarily containing pepsinogen. Pepsinogen is the inactive precursor of the proteolytic enzyme pepsin. Pepsin must be produced in this inactive form to prevent the destruction of chief cell proteins.

Pepsinogen is activated following its release into the stomach. It is activated by both HCl and other active pepsin molecules already present in the stomach. The pepsin chemically digests denatured proteins in the stomach into smaller peptide fragments (oligopeptides). Chief cells also produce **gastric lipase,** an enzyme that has a limited role in fat digestion (digests about 10–15% of the ingested fat).

G-Cells **G-cells** are **enteroendocrine** (en′ter-ō-en′dō-krin; *enteron* = gut, intestine) **cells** that are widely distributed in the gastric glands of the stomach. G-cells secrete gastrin hormone into the blood. **Gastrin** stimulates stomach motility and secretions. Other enteroendocrine cells (there are at least eight types) produce other hormones, such as somatostatin, a peptide hormone that modulates the function of nearby enteroendocrine and exocrine cells.

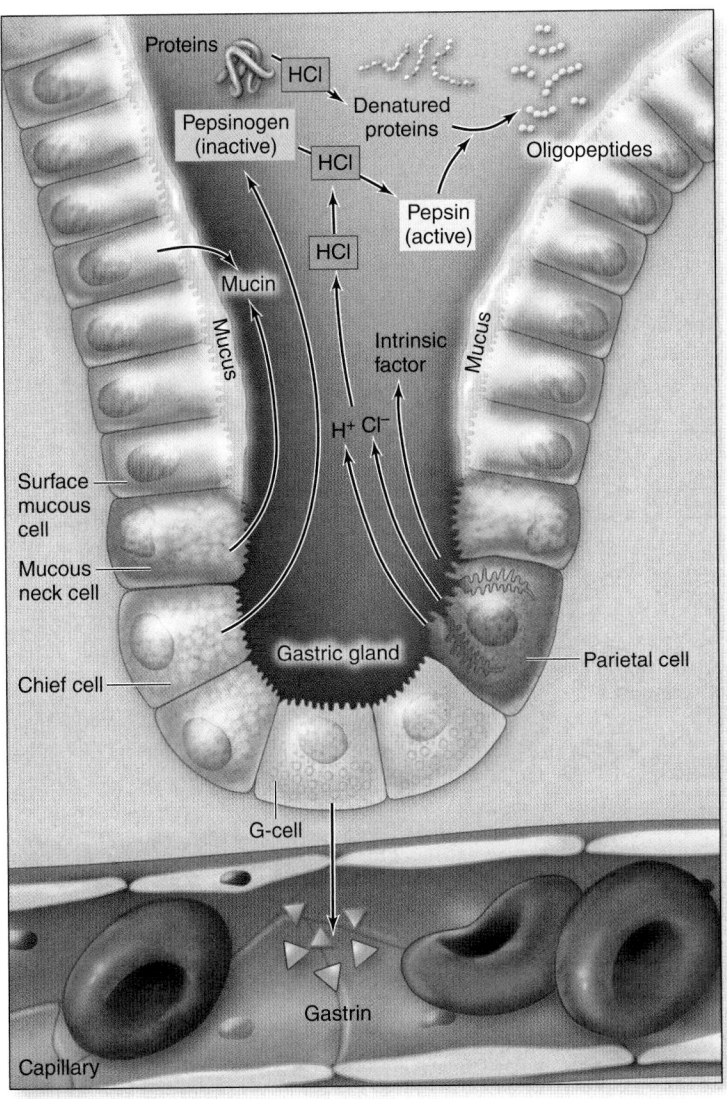

Figure 26.12 Gastric Secretions. Secretory cells and their products are identified. Secretions enter the lumen of the stomach, except the hormone gastrin, which enters the blood.

The five types of secretory cells and their products are summarized in **figure 26.12**.

Motility in the Stomach

Smooth muscle activity in the stomach wall has two primary functions: (1) mixing the bolus with gastric juice to form chyme, and (2) emptying chyme from the stomach into the small intestine.

Gastric mixing is a form of mechanical digestion that changes the semidigested bolus into **chyme** (kīm; *chymos* = juice). Chyme has the consistency of a pastelike soup. Contractions of the stomach's thick muscularis layer churn and mix the bolus with the gastric secretions, leading to a reduction in the size of swallowed particles.

Gastric emptying is the movement of acidic chyme from the stomach through the pyloric sphincter into the duodenum, which is facilitated by the progressive thickening of the muscularis layer in the pyloric region. As a wave of peristaltic muscular contraction moves through the pylorus toward the pyloric sphincter, a pressure gradient is established that drives the stomach contents toward the small intestine.

The interaction here is unique: The peristaltic wave establishes a greater pressure on the contents in the pylorus than the pressure exerted by the pyloric sphincter to stay closed and prevent movement. Consequently, a few milliliters (about 3 milliliters) of chyme are emptied into the small intestine. After the peristaltic wave has moved past the pyloric sphincter, the pressure of the sphincter is once again greater than the pressure on the contents, and the pyloric sphincter closes. As this sphincter closes, stomach contents are squeezed back toward the stomach body. This reverse flow event is called **retropulsion.** Retropulsion not only results in the prevention of further chyme moving into the small intestine, but it also contributes to additional mixing of the stomach contents to further reduce size of food particles. The processes of both gastric mixing and gastric emptying are summarized in **figure 26.13**.

Regulation of the Digestive Processes in the Stomach

The stomach is essentially a holding bag for partially digested food until it enters the small intestine, where its digestion will be completed. This process is highly regulated.

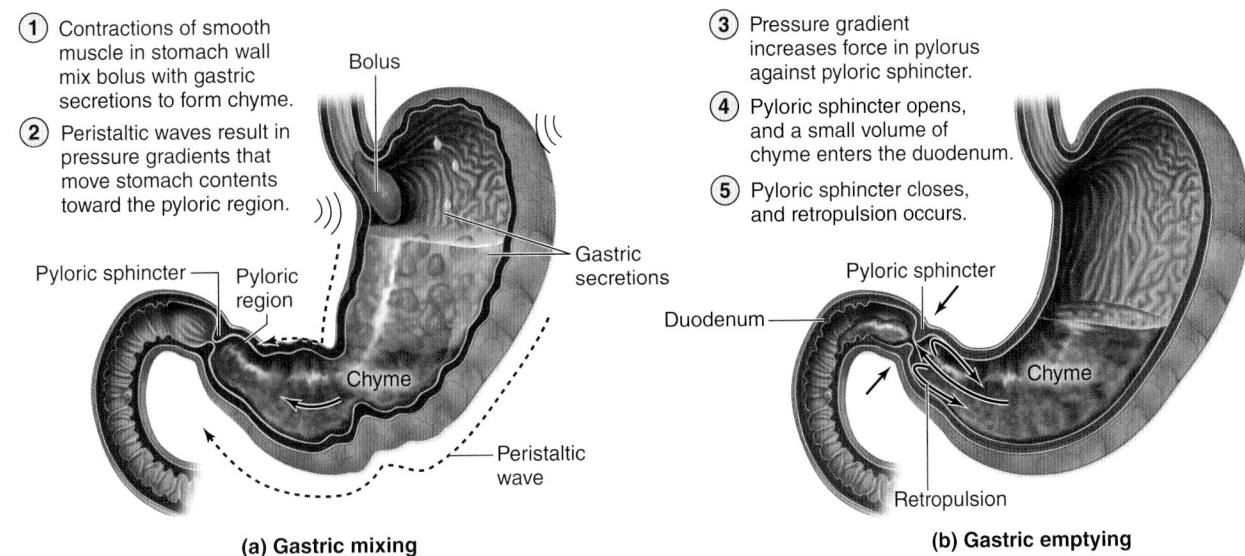

① Contractions of smooth muscle in stomach wall mix bolus with gastric secretions to form chyme.

② Peristaltic waves result in pressure gradients that move stomach contents toward the pyloric region.

③ Pressure gradient increases force in pylorus against pyloric sphincter.

④ Pyloric sphincter opens, and a small volume of chyme enters the duodenum.

⑤ Pyloric sphincter closes, and retropulsion occurs.

(a) Gastric mixing

(b) Gastric emptying

Figure 26.13 Gastric Mixing and Emptying. (*a*) Chyme is formed in the stomach as the bolus is mixed with gastric secretions. (*b*) A small volume is then forced into the duodenum through the partially open pyloric sphincter and then the sphincter closes and retropulsion occurs.

INTEGRATE

CLINICAL VIEW

Peptic Ulcers

Normally a balance exists in the stomach between the acidic gastric juices and the protective, regenerative nature of the mucosa lining. When this balance is thrown off, the stage is set for the development of a **peptic ulcer**—a chronic, solitary erosion of a portion of the lining of either the stomach or the duodenum. Annually, over 4 million people in the United States are diagnosed with an ulcer.

Gastric ulcers are peptic ulcers that occur in the stomach, whereas **duodenal ulcers** are peptic ulcers that occur in the superior part of the duodenum, the first segment of the small intestine. Duodenal ulcers are common because the first part of the duodenum receives the acidic chyme from the stomach but has yet to receive the alkaline bile and pancreatic juice that may neutralize chyme's acidic content.

Symptoms of an ulcer include a gnawing, burning pain in the epigastric region, which may be worse after eating a meal; nausea; vomiting; and extreme belching. Bleeding also may occur, and the partially digested blood results in dark, tarlike stools. If left untreated, an ulcer may erode the entire organ wall and cause **perforation**, which is a medical emergency.

Irritation of the gastric mucosa (**gastritis**) has been linked to many cases of peptic ulcer. Nonsteroidal anti-inflammatory drugs (NSAIDs), such as ibuprofen and aspirin, are a common cause of gastritis, and these drugs also impair healing of the gastric lining. However, the major player in peptic ulcer formation is a bacterium called *Helicobacter pylori,* which is present in over 70% of gastric ulcer cases and well over 90% of duodenal ulcer cases. *H. pylori* resides in the stomach and produces enzymes that break down the components in the gastric mucus, lessening its protective effects. As leukocytes enter the stomach to destroy the bacteria, they also destroy the mucous neck cells. The decreased ability to produce mucus both irritates the stomach causing possible erosion of the layers of the stomach wall and allows for a favorable environment for continued proliferation of *H. pylori.* This sets up a positive feedback cycle that can potentially lead to perforation of the stomach wall. Thus, the bacteria initiate a cascade of events that may lead to erosion of the gastric lining and eventual perforation of the stomach wall, if not treated.

Categories of medications that help are an antibiotic taken for 2 weeks to eradicate *H. pylori,* and treatments that are similar for gastric reflux, including proton-pump inhibitors, a histamine (H_2) blocker, and antacids.

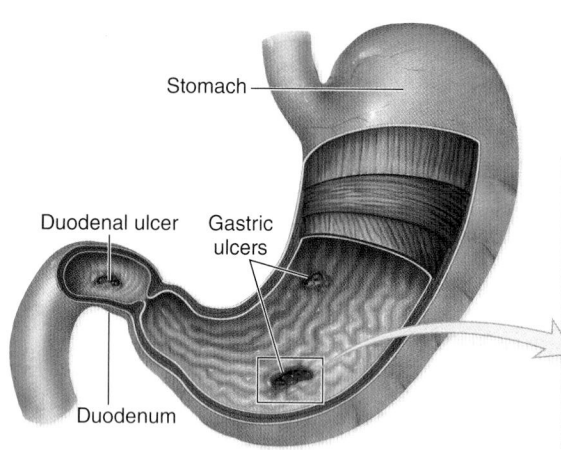

Common locations of gastric and duodenal ulcers.

Stomach

Duodenal ulcer   Gastric ulcers

Duodenum

Mucosa
Submucosa
Muscularis
Serosa

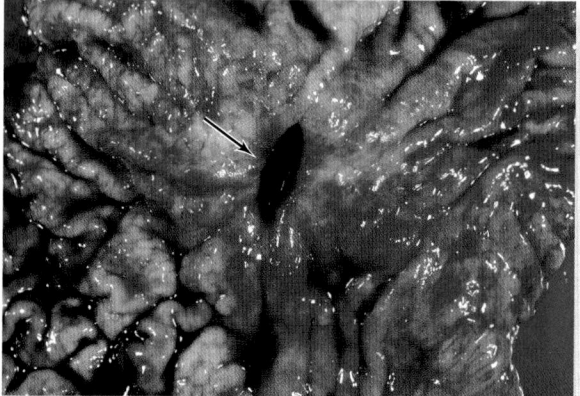

Perforated gastric ulcer

Pacemaker cells *(interstitial cells of Cajal)* in the stomach wall spontaneously depolarize less than four times per minute and establish its basic rhythm of muscular contraction in this organ. Electrical signals spread through the smooth muscle cells in the muscularis layer of the stomach via gap junctions. Muscular contractions by the stomach wall are regulated by both nervous reflexes and hormones, which alter the force but not rate of contraction, which is constant. Secretory activity of gastric glands is also altered. How this occurs is organized into three phases: cephalic phase, gastric phase, and intestinal phase. The cephalic and gastric phases involve the events before and during a meal, whereas the intestinal phase involves the events that occur after a meal, as the ingested materials are being digested. See **figure 26.14** as you read through this section.

Cephalic Phase   The **cephalic phase** primarily involves the *cephalic reflex,* which is a nervous system reflex initiated by the thought, smell, sight, or taste of food. Nerve signals from the higher regions of the brain are sent to the hypothalamus, which then relays nerve signals to the medulla oblongata. The medulla oblongata increases parasympathetic stimulation of the stomach via the vagus nerve (vagal stimulation),

which causes both an increase in contractile force in the gastric wall (increases motility) and secretory activity of the gastric glands. Sometimes you become aware of these processes when your stomach "growls."

Gastric Phase   The **gastric phase** involves the processes following the bolus reaching the stomach. This phase is regulated by both the nervous system via the gastric reflex and the endocrine system through the release of gastrin hormone. The *gastric reflex* is initiated as food enters the stomach. Baroreceptors in the wall of the stomach detect increased distension in the stomach wall, and chemoreceptors detect both protein and an increase in pH of gastric contents. (Proteins buffer H^+ and increase pH; see section 25.5d.) Nerve signals are relayed along sensory neurons to the medulla oblongata, resulting in the same effects as described in the cephalic reflex—an increase in both stomach motility and secretory activity of gastric cells.

The presence of food (especially protein) in the stomach also causes release of gastrin from enteroendocrine cells. Gastrin enters the blood and circulates back to the stomach to further stimulate the contractile activity of muscle in the gastric wall and to primarily increase

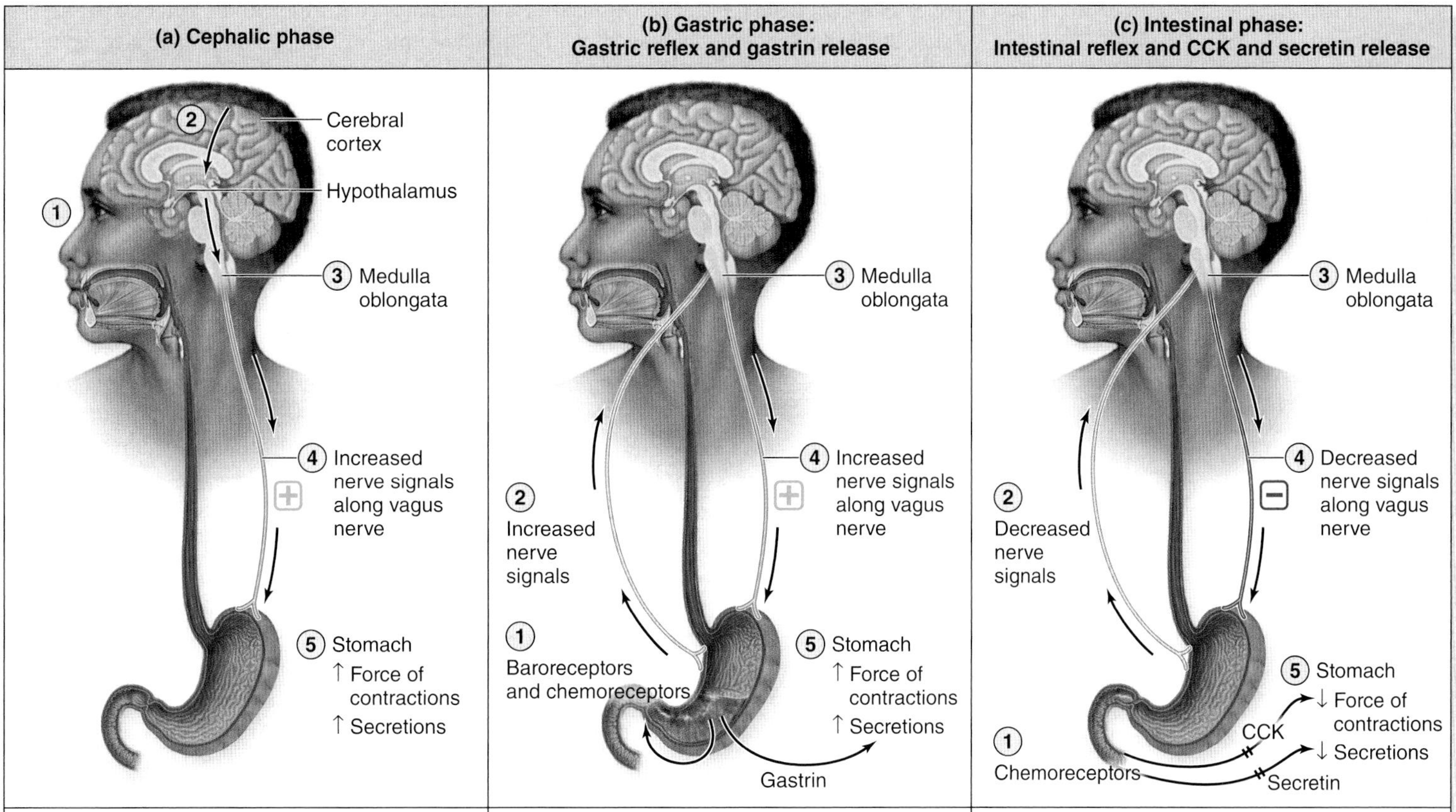

(a) Cephalic phase	(b) Gastric phase: Gastric reflex and gastrin release	(c) Intestinal phase: Intestinal reflex and CCK and secretin release

The cephalic reflex:
Initiated by thought, smell, sight, or taste of food (or even sounds of food preparation)

1. *Receptors:* Special senses (e.g., nose, eyes)
2. *Sensory input:* Increased nerve signals relayed from the cerebral cortex and hypothalamus to the medulla oblongata
3. Medulla oblongata integrates input from higher brain centers
4. *Motor output:* Increased nerve signals relayed from medulla oblongata to stomach
5. *Effector:* Stomach stimulated to both increase force of contraction and release of secretions

The gastric reflex:
Initiated by presence of food in stomach

1. *Receptors:* Baroreceptors in stomach wall detect stretch; chemoreceptors detect protein or high pH in stomach contents
2. *Sensory input:* Increased nerve signals relayed to medulla oblongata
3. Medulla oblongata integrates sensory input
4. *Motor output:* Increased nerve signals relayed from medulla oblongata to stomach
5. *Effector:* Stomach stimulated to both increase force of contraction and release of secretions

In addition, the presence of food in the stomach causes release of **gastrin,** which targets both the stomach to increase force of contraction and release of secretions (especially HCl) and also stimulates contraction of the pyloric sphincter.

The intestinal reflex:
Initiated by presence of acidic chyme in duodenum

1. *Receptors:* Chemoreceptors in intestinal wall detect acidic chyme or low pH in stomach contents
2. *Sensory input:* Decreased nerve signals relayed to medulla oblongata
3. Medulla oblongata integrates sensory input
4. *Motor output:* Decreased nerve signals relayed from medulla oblongata to stomach
5. *Effector:* Stomach stimulated to both decrease force of contraction and release of secretions

In addition, the presence of fatty chyme in the duodenum causes release of **cholecystokinin (CCK),** which decreases the force of contraction in the stomach. The presence of acidic chyme causes release of **secretin,** which inhibits release of stomach secretions.

Figure 26.14 Regulation of Digestive Processes in the Stomach. Both muscular contractions of the stomach wall and secretions released by the stomach are regulated by nervous reflexes and hormones. These processes are organized into three phases: cephalic phase, gastric phase, and intestinal phase.

INTEGRATE

CLINICAL VIEW

Vomiting

Vomiting is the rapid expulsion of gastric contents through the oral cavity. Prior to vomiting, heart rate and sweating increase, nausea is felt, and a noticeable increase in saliva production occurs. The vomiting reflex is a complicated act that is controlled by the vomiting center in the medulla oblongata. This brainstem region responds to head injury, motion sickness, infection, toxicity, or food irritation in the stomach and intestines.

Vomiting is initiated following a deep inspiration and the closure of both nasal passages and the glottis. Skeletal muscle contraction (abdominal muscles and diaphragm) increase pressure within the stomach and thus supply the primary force for expulsion of digestive tract contents. As the pressure increases in the stomach, the acidic gastric contents are forced into and through the esophagus and out of the oral cavity.

Care must be taken that vomit is not aspirated into the respiratory tract; a risk for a semiconscious or unconscious individual. Thus, it is critical that individuals undergoing surgical procedures have an empty stomach and small intestine because general anesthesia has an associated risk of inducing nausea and vomiting.

release of HCl from parietal cells. Gastrin also stimulates contraction of the pyloric sphincter to slow stomach emptying, thereby allowing sufficient time for completion of digestive activities associated with the stomach before the chyme is moved into the small intestine.

Intestinal Phase   The **intestinal phase** involves the processes following the chyme reaching the small intestine, a phase that is also regulated by both the nervous system and the endocrine system. The intestinal phase involves both the intestinal reflex and the release of two significant hormones: cholecystokinin (CCK) and secretin. The *intestinal reflex* opposes the other two reflexes (cephalic reflex and gastric reflex). It protects the small intestine from being overloaded with chyme. The intestinal reflex is initiated with entry of acidic chyme into the duodenum, which causes a decrease in nerve signals relayed to the medulla oblongata. Consequently, vagal stimulation to the stomach is decreased with a concomitant decrease in both motility and secretory activity of the stomach.

Cholecystokinin (released primarily in response to fatty chyme within the small intestine) causes a decrease in stomach motility. Secretin (released primarily in response to acidic chyme in the small intestine) decreases secretory activity of the stomach. Both cholecystokinin and secretin inhibit the release of gastrin. This slows down the emptying of the stomach, thus allowing small intestine to continue its digestive processes before additional chyme is added. Both hormones also influence the digestive processes in the lower GI tract and are discussed in more detail later in section 26.3c. In addition, the details for gastrin, secretin, and CCK hormones are included in the summary table "Regulating the Digestive System," which directly follows chapter 17 (see **table R.8**). (Note that researchers initially believed an additional hormone released from the small intestine named *gastric inhibitory peptide* regulated stomach activity. However, this hormone is now thought to primarily regulate the release of insulin in response to increased glucose concentration in the contents of the small intestine. To better reflect its role, it has been renamed **glucose-dependent insulinotropic peptide (GIP).**

WHAT DID YOU LEARN?

14 List the secretory cell types in the stomach, their products, and the function of the products.

15 Which neural reflex is initiated by food in the stomach and what does it control?

26.3 Lower Gastrointestinal Tract

The lower gastrointestinal (GI) tract continues the processes of digestion and importantly functions in the absorption of nutrients. Material that cannot be digested and absorbed is then eliminated.

26.3a Overview of the Lower Gastrointestinal Tract Organs

LEARNING OBJECTIVE

1. Describe the three components of the lower gastrointestinal tract.

First, we take a superficial view of the lower GI tract organs and accessory organs to help integrate their structures with their digestive activities and functions (**figure 26.15**):

- **Small intestine.** The small intestine is divided into three continuous regions (duodenum, jejunum, and ileum). Recall that the duodenum is considered part of the upper GI tract, but it is described in this section on the lower GI tract organs. The small intestine receives chyme from the stomach that is then mixed with accessory organ secretions. Most chemical digestion and absorption occurs within the small intestine.

- **Accessory organs.** Accessory organ secretions include bile and pancreatic juice. Bile is produced by the liver and then stored, concentrated, and released by the gallbladder. Pancreatic juice contains numerous digestive enzymes and is produced and released by the pancreas.

- **Large intestine.** The large intestine continues absorption, primarily of water, electrolytes, and vitamins (including vitamins B and K produced by

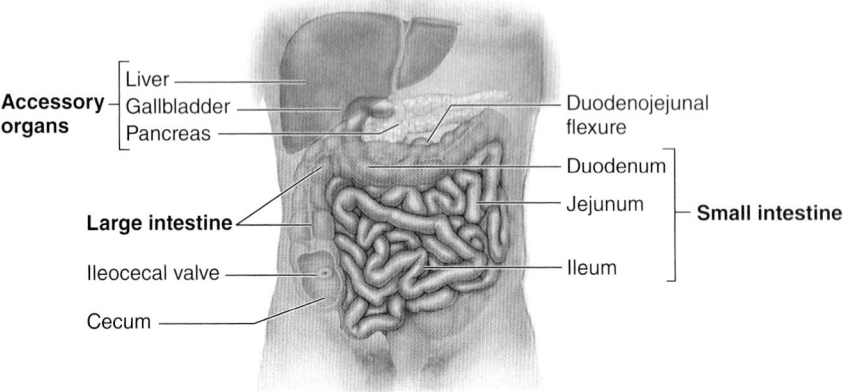

Figure 26.15 Gross Anatomy of the Lower GI Tract Organs and Accessory Organs. The three regions of the small intestine—duodenum, jejunum, and ileum—are continuous and framed within the abdominal cavity by the large intestine. Accessory organs release secretions into the duodenum.

INTEGRATE
CLINICAL VIEW
Intestinal Disorders

The term **inflammatory bowel disease (IBD)** applies to two autoimmune disorders, Crohn disease and ulcerative colitis. In both of these disorders, selective regions of the intestine become inflamed.

Crohn disease is a condition of young adults characterized by intermittent and relapsing episodes of intense abdominal cramping and diarrhea. Although any region of the GI tract may be involved, the distal ileum is the most frequently and severely affected site. Inflammation involves the entire thickness of the intestinal wall, extending from the mucosa to the serosa. For reasons that are not clear, lengthy regions of the intestine having no trace of injury or inflammation may be followed abruptly by several inches of markedly diseased intestine.

The age distribution and symptoms of **ulcerative colitis** are similar to those of Crohn disease, but ulcerative colitis involves only the large intestine. The rectum and descending colon are the first to show signs of inflammation and are generally the most severely affected. Also, in ulcerative colitis the inflammation is confined to the mucosa, instead of the full thickness of the intestinal wall. Finally, unlike Crohn disease, ulcerative colitis is associated with a profoundly increased risk of colon cancer.

Treatment of inflammatory bowel disease is complex. Anti-inflammatory drugs, as well as stress reduction and possibly nutritional supplementation, help control symptoms. Surgery may be necessary.

Crohn disease and ulcerative colitis are distinctly different from a much more common disorder called **irritable bowel syndrome (IBS)**. IBS is characterized by abnormal function of the colon with symptoms of crampy abdominal pain, bloating, constipation, and diarrhea. It occurs in about one in every five people in the United States, and is more common in women than men. Irritable bowel syndrome may be diagnosed if a medical evaluation has ruled out Crohn disease and ulcerative colitis. Although neither a cause nor a cure for IBS is known, most people can control their symptoms by reducing stress, changing their diet, and using certain medications.

bacteria within the large intestine). The digestive process is completed as feces is produced and then eliminated through the anus.

WHAT DID YOU LEARN?

16 What organs are considered part of the lower GI tract?

26.3b Small Intestine

LEARNING OBJECTIVES

2. Describe the anatomy of the small intestine.

3. List the glands found in the small intestine and their secretions.

4. Explain motility within the small intestine.

The **small intestine** is a long tube that is inferior to the stomach and located medially within the abdominal cavity. Generally, about 9 to 10 liters of ingested food, water, and digestive system secretions enter the small intestine daily. Ingested nutrients typically spend at least 12 hours in the small intestine. The small intestine finishes chemical digestion and is responsible for absorbing almost all of the nutrients and a large percentage of the water and electrolytes. Vitamins ingested in the diet are also absorbed by the small intestine. Fat-soluble vitamins are absorbed with lipids, and water-soluble vitamins are absorbed through various membrane transport mechanisms. (Recall from our discussion on parietal cells that absorption of vitamin B$_{12}$ requires the presence of intrinsic factor, which is produced by the stomach.)

Gross Anatomy of the Small Intestine

The **small intestine** is also called the *small bowel*. It is a coiled, thin-walled tube about 1 inch in diameter and approximately 6 meters (20 feet) in length in the unembalmed cadaver. (It is much shorter in a living individual due to smooth muscle tone.) It extends from the pylorus of the stomach to the cecum of the large intestine; thus, it occupies a significant portion of the abdominal cavity. The small intestine consists of three specific segments: the duodenum, jejunum, and ileum.

The **duodenum** (dū-ō-dē′nŭm, dū-od′ĕ-nŭm; breadth of twelve fingers) forms the first segment of the small intestine. It is approximately 25 centimeters (10 inches) long and originates at the pyloric sphincter (see figure 26.9). The duodenum is arched into a C-shape around the head of the pancreas and becomes continuous with the jejunum at the **duodenojejunal flexure** (flek′sher; *fleksura* = bend). Most of the duodenum is retroperitoneal, although the very initial portion is intraperitoneal. The duodenum is attached to the liver by the lesser omentum. The duodenum receives accessory organ secretions from the liver, gallbladder, and pancreas, as well as chyme from the stomach.

The **jejunum** (jĕ-jū'nŭm; *jejunus* = empty) is the middle region of the small intestine (figure 26.15). Extending approximately 2.5 meters (7.5 feet), it makes up about two-fifths of the small intestine's total length. The jejunum is the primary region within the small intestine for chemical digestion and nutrient absorption.

The **ileum** (il'ē-ŭm; *eiles* = twisted) is the last region of the small intestine. At about 3.6 meters (10.8 feet) in length, the ileum forms approximately three-fifths of the small intestine. Its distal end terminates at the **ileocecal** (il'ē-ō-sē'kăl) **valve,** a sphincter that controls the entry of materials into the large intestine. Absorption of digested materials continues in the ileum. Both the jejunum and the ileum are intraperitoneal and suspended within the abdomen by the mesentery proper (see figure 26.3).

The mucosal and submucosal tunics of the small intestine are thrown into internal **circular folds** (also called *plicae* [plī'sē] *circulares*) that extend inward toward the lumen **(figure 26.16)**. Circular folds are easily seen by the naked eye: They help increase the surface area through which nutrients are absorbed. They also act like "speed bumps" to slow down the movement of chyme and ensure that it remains within the small intestine for maximal nutrient absorption. Circular folds are more numerous in the duodenum and jejunum and least numerous in the ileum.

WHAT DO YOU THINK?

3 Why are the circular folds much more numerous in the duodenum and least numerous in the ileum? How does the abundance of circular folds relate to the main functions of the duodenum?

Histology

In the small intestine, the length of the muscularis mucosae is shorter than the two layers internal to it, forcing the two inner layers of the mucosa (i.e., the epithelium, and lamina propria) into folds to form small fingerlike projections of the mucosa called **villi.** Villi are larger and most numerous in the jejunum. The epithelium and lamina propria of each

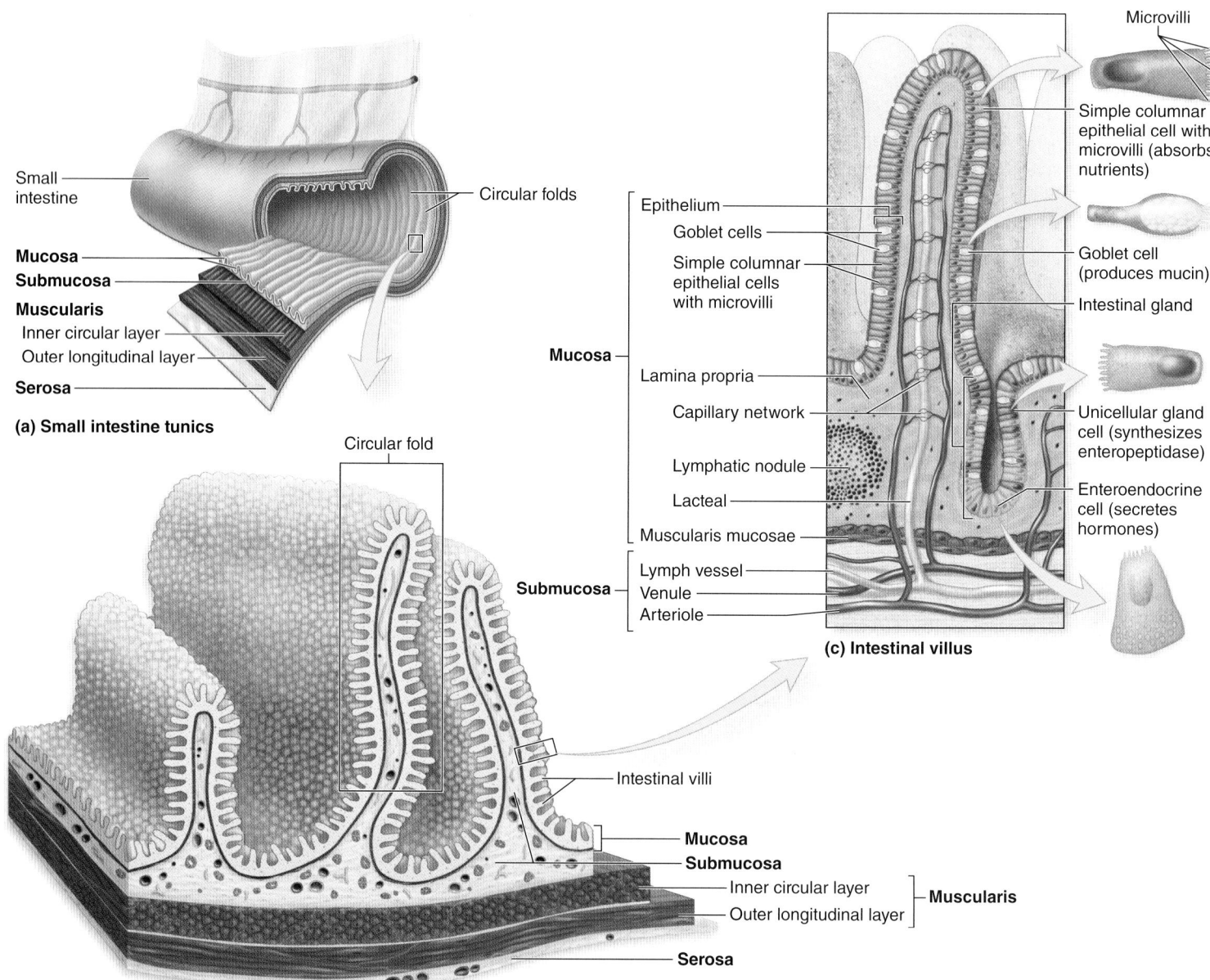

(a) Small intestine tunics

(b) Section of small intestine

(c) Intestinal villus

Figure 26.16 Histology of the Small Intestine. (*a*) The wall of the small intestine is formed by four tunics: mucosa, submucosa, muscularis, and serosa. (*b*) Circular folds are inward projections of both the mucosa and submucosa, whereas villi are fingerlike projections of only the mucosa. Both circular folds and villi increase the surface area of the small intestine. (*c*) The epithelial cells covering the surface of each villus have microvilli to further increase the surface area. The lamina propria within each villus houses both blood capillaries and lymphatic capillaries (lacteals). AP|R villus appears analogous to a glove (epithelium) covering a finger (lamina propria). Each villus contains an arteriole, a rich capillary network, and a venule. The capillaries absorb most nutrients. A **lacteal** is a type of lymphatic capillary also within the villus (described in section 21.1a). A lacteal is responsible for absorbing lipids and lipid-soluble vitamins that are too large to be absorbed by the blood capillaries. Both the circular folds and villi help increase the surface area for absorption and secretion.

Microvilli (mī-krō-vil'ī; *micros* = small) are extensions of the plasma membrane of the simple columnar epithelial cells lining the small intestine. Microvilli further increase the surface area of the small intestine. Individual microvilli are not clearly visible in light micrographs of the small intestine; instead, they appear as a fuzzy edge of the simple columnar cells called the **brush border.** Embedded within this brush border are various enzymes that complete the chemical digestion of most nutrients immediately before absorption. Collectively these are called **brush border enzymes.** Located in close proximity and also embedded within the plasma membrane are the required proteins for membrane transport of digested molecules. Histologic images of both villi and microvilli are shown in figure 26.17.

Small Intestine Secretions

Between the intestinal villi are invaginations of the mucosa called **intestinal glands** (also known as *intestinal crypts,* or *crypts of Lieberkühn*); these mucosal cells secrete **intestinal juice.** These glands extend to the base of the mucosa and slightly resemble the anatomy of the gastric glands of the stomach (figure 26.16c).

- **Goblet cells** produce mucin that when hydrated form mucus, which lubricates and protects the intestinal lining. These cells increase in number from the duodenum to the ileum, because more lubrication is needed as digested materials are absorbed and undigested materials are left behind.

- **Unicellular gland cells** synthesize **enteropeptidase,** an enzyme described in section 26.4b.
- The **enteroendocrine cells** release hormones such as CCK and secretin. We have already discussed the stomach-associated functions of these hormones. Several other functions are described in section 26.3c.

Another type of gland housed within the submucosal layer and found only in the proximal duodenum is called a **submucosal** or **duodenal gland** (or *Brunner gland*). This gland produces a viscous, alkaline mucus secretion that protects the duodenum from the acidic chyme.

Motility of the Small Intestine

Smooth muscle activity in the small intestine wall has three primary functions: (1) mixing chyme with accessory gland secretions, (2) moving the chyme continually against new areas of the brush border, and (3) propelling the contents through the small intestine toward the large intestine.

All these functions facilitate chemical digestion and absorption, employing the processes of segmentation and peristalsis. When chyme first enters the small intestine, segmentation is more prevalent than peristalsis. **Segmentation** mixes chyme with accessory gland secretions through a "backward-and-forward" motion (see figure 26.2c).

Peristalsis then propels material within the GI lumen by alternating contraction of the circular and longitudinal muscle layers in small regions. The rhythm of muscular contractions is more frequent in the duodenum than the ileum, thus the net movement of intestinal contents is toward the large intestine.

The **gastroileal** (gas'trō-il'ē-ăl) **reflex** (which is thought to have both short reflexes and long reflexes that involve the medulla oblongata) is initiated by food entering the stomach. As part of this reflex, the ileum contracts, ileocecal sphincter relaxes, and the cecum (the first part of the large intestine) relaxes. Thus, contents within the GI tract are moved from the ileum through the open ileocecal sphincter into the cecum. Then, the ileocecal sphincter contracts to prevent backflow from the cecum into the ileum.

WHAT DID YOU LEARN?

17 What are the three anatomic structures that increase the surface area of the small intestine? Describe each.

18 Which type of motility is primarily responsible for mixing the chyme and accessory gland secretions within the small intestine—segmentation or peristalsis?

26.3c Accessory Organs and Ducts

LEARNING OBJECTIVES

5. Describe the accessory digestive organs associated with the small intestine and the contributions of each to digestive processes.

6. Explain how blood and bile flow through the liver.

7. Discuss the regulation of the accessory glands associated with the small intestine.

Three accessory digestive organs release secretions into the duodenum: the liver, the gallbladder, and the pancreas. Here we examine the ducts from each organ to the duodenum, the anatomy and histology of each organ, and their products conveyed to the small intestine that contribute to the digestion of the chyme arriving from the stomach. The accessory organs are summarized in **table 26.2.**

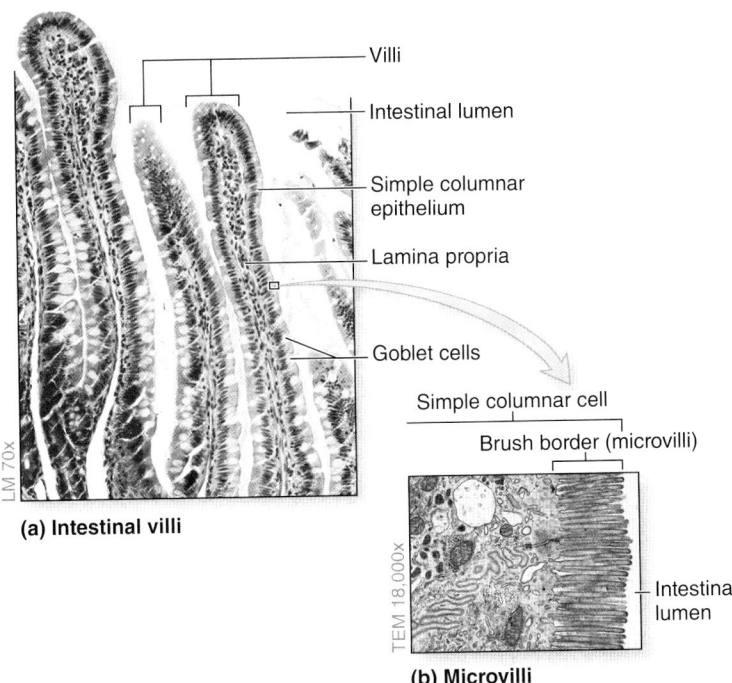

LM 70x

(a) Intestinal villi

TEM 18,000x

(b) Microvilli

Villi

Intestinal lumen

Simple columnar epithelium

Lamina propria

Goblet cells

Simple columnar cell

Brush border (microvilli)

Intestinal lumen

Figure 26.17 Intestinal Villi and Microvilli. Photomicrographs show (*a*) the internal structure of villi projecting into the intestinal lumen, and (*b*) microvilli epithelium's apical surface.

Table 26.2	Accessory Organs Associated with the Duodenum	
Organ	**Description**	**Digestive functions**
Liver	Largest internal organ	Primary digestive function is to produce and release bile into the duodenum
Gallbladder	Small muscular sac on inferior side of liver	Stores, concentrates, and releases bile into the duodenum
Pancreas	Mixed gland with endocrine and exocrine functions	Exocrine cells (acinar cells) function to produce and release numerous digestive enzymes, and HCO_3^- is released from pancreatic ducts to form a mixture called "pancreatic juice" into the duodenum

Accessory Organ Ducts

A series of ducts deliver secretions to the duodenum of the small intestine (**figure 26.18**). These ducts include the biliary apparatus from the liver and gallbladder, and the pancreatic ducts from the pancreas.

The **biliary** (bil′ē-ăr-ē) **apparatus** is a network of thin ducts that include the left and right hepatic ducts, which drain the left and right lobes of the liver, respectively. The **left** and **right hepatic ducts** merge to form a single **common hepatic duct.** The union of the **cystic duct** from the gallbladder and the common hepatic duct forms the **common bile duct** that extends inferiorly through the pancreas to open into the duodenum.

The common bile duct becomes the **hepatopancreatic ampulla** (or *ampulla of vater*), which is a swelling either adjacent to or within the posterior duodenal wall. The main pancreatic duct joins with the hepatopancreatic ampulla. Associated with the hepatopancreatic ampulla is the **hepatopancreatic sphincter** that regulates movement of bile (from the liver and gallbladder) and pancreatic juice (from the pancreas) into the duodenum. Within the duodenum, the **major duodenal papilla** is a projection where the hepatopancreatic ampulla penetrates the wall, and both bile and pancreatic juice enter the duodenum. An alternative way that a small amount of pancreatic juice may also enter the duodenum is through an **accessory pancreatic duct.** This duct penetrates the duodenal wall, forming the **minor duodenal papilla.**

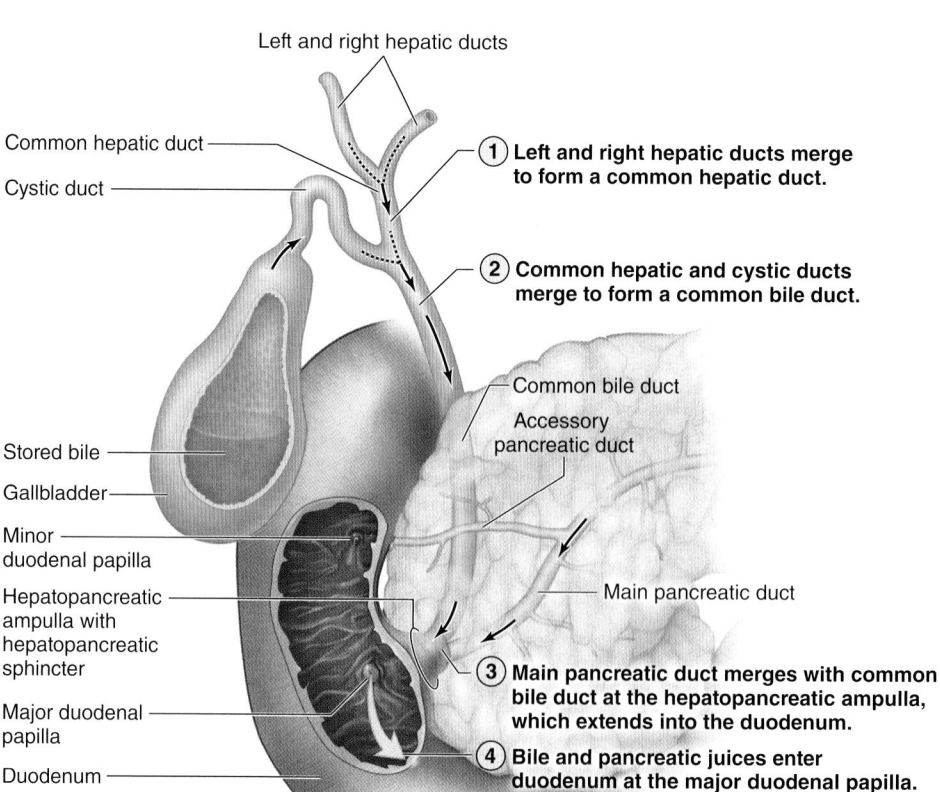

Left and right hepatic ducts

Common hepatic duct

Cystic duct

(1) Left and right hepatic ducts merge to form a common hepatic duct.

(2) Common hepatic and cystic ducts merge to form a common bile duct.

Common bile duct

Accessory pancreatic duct

Stored bile

Gallbladder

Minor duodenal papilla

Hepatopancreatic ampulla with hepatopancreatic sphincter

Major duodenal papilla

Duodenum

Main pancreatic duct

(3) Main pancreatic duct merges with common bile duct at the hepatopancreatic ampulla, which extends into the duodenum.

(4) Bile and pancreatic juices enter duodenum at the major duodenal papilla.

Figure 26.18 Ducts of Accessory Organs. Various ducts merge to transport bile and pancreatic juice from accessory organs to the duodenum.

Liver

The **liver** is an accessory digestive organ. It is located in the right upper quadrant of the abdomen, immediately inferior to the diaphragm (see figure 26.1). It has numerous functions (described in section 27.6), but its main function in digestion is the production of bile.

Gross Anatomy of the Liver   The liver is the largest internal organ, weighing 1 to 2 kilograms (2 to 4 pounds), and it constitutes approximately 2% of an adult's body weight. The liver is covered by a connective tissue capsule except at the porta hepatis (described shortly). Covering the connective tissue capsule is a layer of visceral peritoneum, except for a small region on its diaphragmatic surface called the bare area (**figure 26.19**).

The liver is composed of four partially separated lobes and is supported by two ligaments. The major lobes are the **right lobe** and the **left lobe.** The right lobe is separated from the smaller left lobe by the falciform ligament, a peritoneal fold that secures the liver to the anterior abdominal wall as discussed in section 26.1d. In the inferior free edge of the falciform ligament lies the **round ligament of the liver** (*ligamentum teres*), which represents the remnant of the fetal umbilical vein. Within the right lobe are the **caudate** (kaw′dāt; *cauda* = tail) **lobe** and the **quadrate** (kwah′drāt; *quadrates* = square) **lobe.** The caudate lobe is adjacent to the inferior vena cava, and the quadrate lobe is adjacent to the gallbladder.

Along the inferior surface of the liver are several structures that collectively resemble the letter H. The gallbladder and the round ligament of the liver form the vertical superior parts of the H; the **inferior vena cava** and the **ligamentum venosum** form the vertical inferior parts. (Recall from section 20.12 that the ligamentum venosum is a remnant of the ductus venosus in the embryo. This vessel shunted blood from the umbilical vein to the inferior vena cava.) Finally, the **porta hepatis** (pōr′tă, = gate; hep′ă-tis, *hepatikos* = liver), the horizontal crossbar of the H, is the site at which blood and lymph vessels, bile ducts, and nerves (not shown) extend from the liver. In particular, the hepatic portal vein and branches of the hepatic artery proper enter at the porta hepatis.

The cells of the liver receive blood from two sources; one is oxygenated and the other is deoxygenated. The **hepatic artery** is a branch of the celiac trunk that extends off of the aorta and transports oxygenated blood to the liver. The **hepatic portal vein,** as part of the hepatic portal system (see section 20.10d), transports deoxygenated and nutrient-rich blood from the capillary beds of the GI tract, spleen, and pancreas. It brings approximately 75% of the blood volume to the liver. Blood from branches of these two vessels mixes as it passes into and through the hepatic lobules (described next).

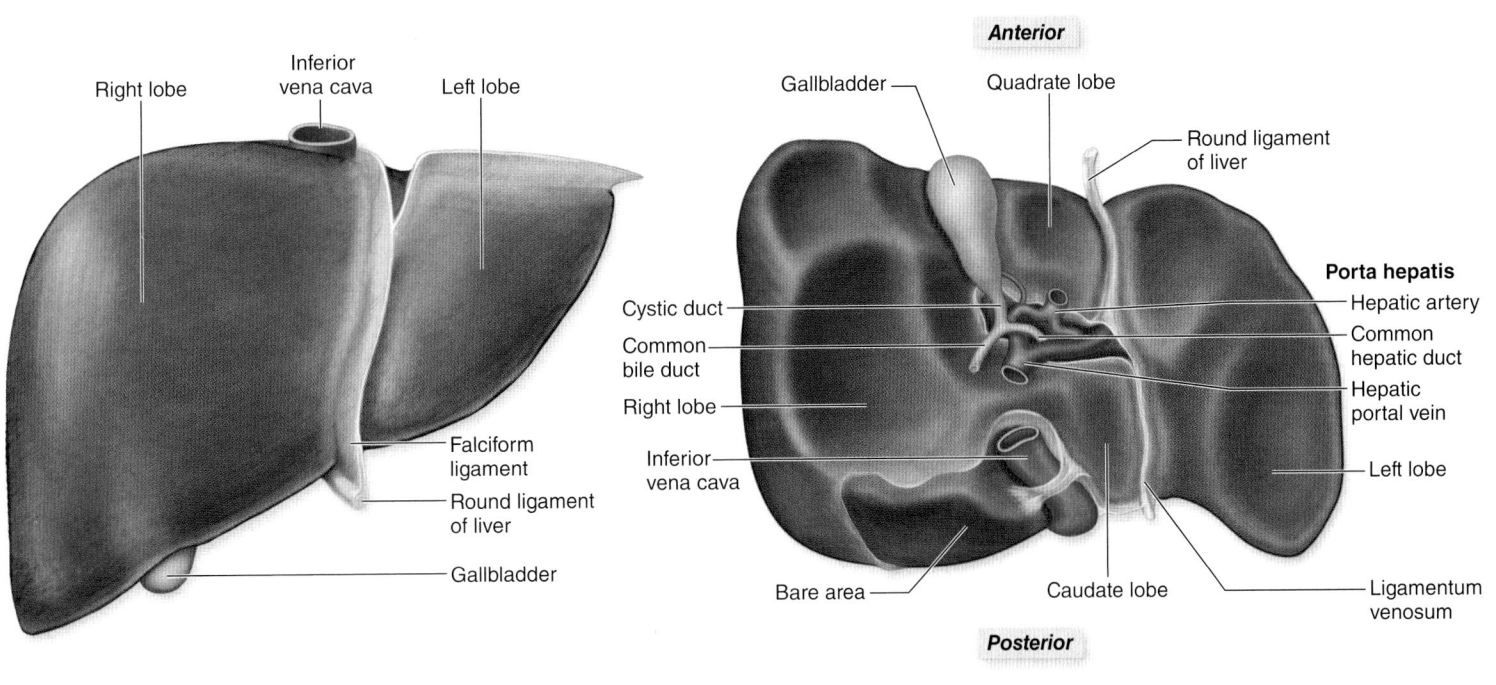

(a) Anterior view

(b) Posteroinferior view

Figure 26.19 Gross Anatomy of the Liver. The liver is in the upper right quadrant of the abdomen. (*a*) Anterior and (*b*) posteroinferior views show the four lobes of the liver. AP|R

INTEGRATE

CONCEPT CONNECTION

The hepatic portal system is a venous network that drains the digestive organs and shunts the blood to the liver (see section 20.10d). The blood exits the liver through hepatic veins that merge with the inferior vena cava.

The GI tract absorbs digested nutrients. These nutrients must be transported to and processed within the liver. The liver also detoxifies any potentially harmful agents that have been absorbed into the GI blood vessels. It is much more efficient to transport blood containing these substances first to the liver via the hepatic portal system for processing, before it is distributed throughout the body.

Histology The liver's connective tissue capsule branches throughout the organ and forms septa that partition the liver into thousands of microscopic polyhedral **hepatic lobules** that are the structural and functional units of the liver **(figure 26.20)**. Within hepatic lobules are liver cells called **hepatocytes** (hep'ă-tō-sīt). At the periphery of each lobule are several **portal triads,** composed of a bile ductule, and *branches* of both the hepatic portal vein and the hepatic artery. At the center of each lobule is a **central vein** that drains the blood flow from the lobule. Central veins collect the blood and merge throughout the liver to form numerous **hepatic veins** that eventually empty into the inferior vena cava.

In cross section, a hepatic lobule looks like a side view of a bicycle wheel. The central vein is like the hub of the wheel. At the circumference of the wheel (where the tire would be) are the portal triads that are usually equidistant apart. Cords of hepatocytes make up the numerous spokes of the wheel, and they are bordered by **hepatic sinusoids** (see section 20.1c).

INTEGRATE

CLINICAL VIEW

Cirrhosis of the Liver

Liver cirrhosis (sir-rō'sis; *kirrhos* = yellow) results when hepatocytes have been destroyed and are replaced by fibrous scar tissue. This scar tissue, which results from collagen synthesis by reticuloendothelial cells in the liver, often surrounds isolated nodules of regenerating hepatocytes. The fibrous scar tissue also compresses (1) the blood vessels resulting in **hepatic portal hypertension** (high blood pressure in the hepatic portal venous system) and (2) bile ducts in the liver, which impedes bile flow.

Liver cirrhosis is caused by injury to the hepatocytes, as may result from chronic alcoholism, liver diseases, or certain drugs or toxins. Chronic hepatitis is a long-term inflammation of the liver that leads to necrosis of liver tissue. Most frequently, viral infections from either **hepatitis B** or **hepatitis C** produce chronic hepatitis. Other disorders that result in liver cirrhosis include some inherited diseases, chronic biliary obstruction, and biliary cirrhosis.

Early stages of liver cirrhosis may be asymptomatic. However, once liver function begins to falter, the individual complains of fatigue, weight loss, and nausea, and may have pain in the right upper quadrant. During an exam, a doctor may palpate an abnormally small and hard liver. To confirm

the diagnosis, a liver biopsy is done to obtain a small portion of liver tissue through a needle passed into the liver; the cells are then examined microscopically. The fibrosis and scarring of liver cirrhosis are irreversible. However, further scarring may be slowed or prevented by treating the cause of the cirrhosis (e.g., hepatitis, alcoholism).

Advanced liver cirrhosis may have a variety of complications :

- **Jaundice** (yellowing of the skin and sclerae of the eyes) occurs when the liver's ability to eliminate bilirubin is impaired.
- **Edema** (e-dē'mă), the accumulation of fluid in body tissues, is evident due to reduced formation and release of albumin.
- **Ascites** (ă-sī'tēz; fluid accumulation in the abdomen) develops because of decreased albumin production.
- Intense itching occurs when bile products are deposited in the skin.
- Toxins in the blood and brain accumulate because the liver cannot effectively process them.
- Hepatic portal hypertension may lead to dilated veins of the inferior esophagus (esophageal varices).

End-stage liver cirrhosis may be treated only with a liver transplant. Otherwise, death results either from progressive liver failure or from the complications.

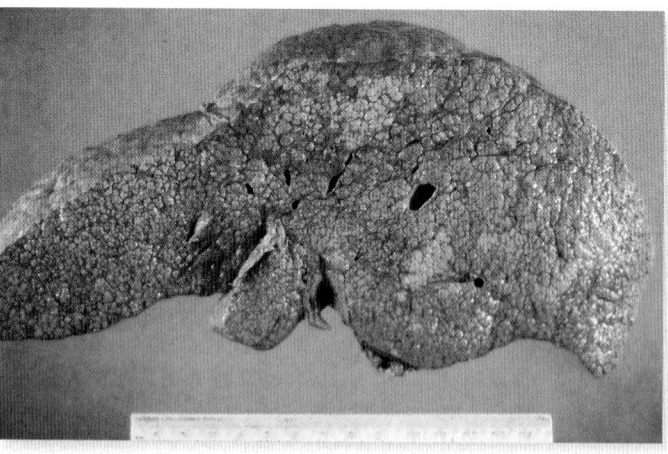

This gross specimen depicts a type of nodular cirrhosis of the liver.

Fibrous scar tissue

LM 100x

Normal hepatocytes

A photomicrograph shows how fibrous scar tissue infiltrates and replaces hepatocytes.

Hepatic sinusoids are thin-walled capillaries with large gaps between these cells, which make the sinusoids more permeable than other capillaries (see section 20.1c). Venous blood of the hepatic portal system and arterial blood are mixed within hepatic sinusoids and then flow slowly toward the central vein (figure 26.20b). Nutrients are absorbed from the sinusoids and enter the hepatocytes. Additionally, the sinusoids are lined with stellate (star-shaped) cells called **reticuloendothelial cells** (*Kupffer cells*). These cells are macrophages that are responsible for engulfing potentially harmful substances (e.g., microbes) as the blood is transported through the liver sinusoids.

Sandwiched between each cord of hepatocytes is a **bile canaliculus** (kan'ă-lik-yū-lŭs). This is a small channel that transports bile produced by hepatocytes to the bile ductule in the portal triad.

Bile The liver secretes bile, a yellowish-green alkaline fluid containing mostly water, bicarbonate ions (HCO_3^-), bile salts (formed from cholesterol), bile pigments (e.g., bilirubin), cholesterol, lecithin (a phospholipid) and mucin. The liver produces bile at a rate of 0.5 to 1 liter per day. Bile salts and lecithin function in the mechanical digestion of lipids, allowing more efficient chemical digestion of triglycerides.

Figure 26.20 Histology of the Liver.
(*a*) The functional units of the liver are called hepatic lobules.
(*b*) A central vein projects through the center of a hepatic lobule, and several portal triads are positioned at its periphery.
(*c*) A photomicrograph depicts a liver lobule and portal triad.
(*d*) Diagram of the hepatic portal system. AP|R

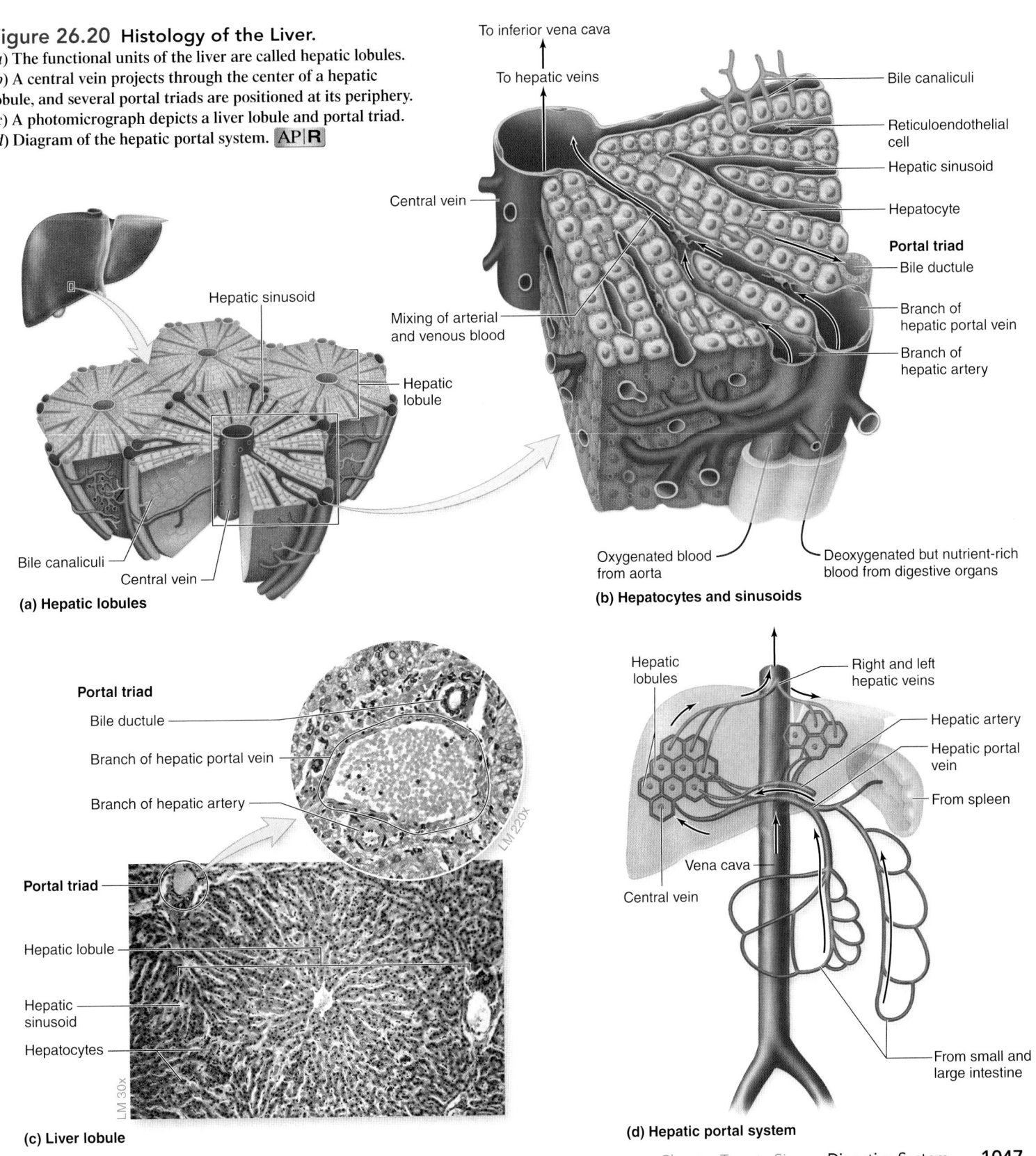

(a) Hepatic lobules

(b) Hepatocytes and sinusoids

(c) Liver lobule

(d) Hepatic portal system

INTEGRATE

CLINICAL VIEW
Gallstones (Cholelithiasis)

Concentration of materials in the bile may lead to the eventual formation of **gallstones**. Gallstones occur twice as frequently in women as in men and are more prevalent in developed countries. Obesity, increasing age, female sex hormones, Caucasian ethnicity, and lack of physical activity are all risk factors for developing gallstones.

The term **cholelithiasis** (kō'lē-li-thī'ă-sis; *chol* = bile, *lithos* = stone, *iasis* = condition) refers to the presence of gallstones in either the gallbladder or the biliary apparatus. Gallstones are typically formed from condensations of either cholesterol or calcium and bile salts. These stones may vary from the tiniest grains to almost golf-ball-sized. The majority of gallstones are asymptomatic until a gallstone becomes lodged in the neck of the cystic duct, causing the gallbladder to become inflamed (**cholecystitis**) and dilated. The most common symptom is severe pain (called biliary colic) perceived in the right hypochondriac region or sometimes between the shoulder blades

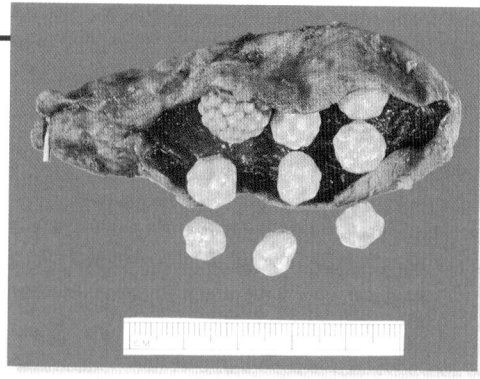

A photograph of gallstones in a gallbladder.

or in the area of the right shoulder. Nausea and vomiting may occur, along with indigestion and bloating. Symptoms are typically worse after eating a fatty meal. Treatment consists of surgical removal of the gallbladder, called **cholecystectomy** (kō'lē-sis-tek'tō-mē; *kystis* = bladder, *ektome* = excision). Following surgery, the liver continues to produce bile, even in the absence of the gallbladder, but concentrating bile (which is a function of the gallbladder) no longer occurs.

Gallbladder

Attached to the inferior surface of the liver (figure 26.19), the **gallbladder** (or *cholecyst*) is a saclike organ that stores, concentrates, and releases bile that the liver produces. The gallbladder has three tunics: an inner mucosa, a middle muscularis, and an external serosa. The mucosa is thrown into folds that permit distension of the wall as the gallbladder fills with bile. Earlier we learned that the **cystic** (sis'tik; *cysto* = bladder) **duct** connects the gallbladder to the common bile duct.

At the neck of the gallbladder, a sphincter valve controls the flow of bile into and out of the gallbladder. Bile enters the gallbladder when the hepatopancreatic sphincter associated with the hepatopancreatic ampulla is closed. It backs up through both the common bile duct and cystic duct into the gallbladder. The gallbladder can hold approximately 40 to 60 milliliters of concentrated bile. Concentrated bile is transported from the gallbladder through the cystic duct and then the common bile duct through the hepatopancreatic ampulla into the duodenum.

WHAT DO YOU THINK?

4 If your gallbladder were surgically removed, how would this affect your digestion of fatty meals? What diet alterations might you have to make after this surgery?

Pancreas

The **pancreas** has both endocrine and exocrine functions. Endocrine cells produce and secrete hormones such as insulin and glucagon (see section 17.9). Exocrine cells (also called acinar cells) produce pancreatic juice to assist with digestive activities.

Gross Anatomy of the Pancreas The pancreas is approximately 5 to 6 inches in length and about 1 inch thick. It is a retroperitoneal organ that extends horizontally from the duodenum toward the left side of the abdominal cavity, where it has contact with the spleen. The pancreas exhibits a wide **head** adjacent to the curvature of the duodenum; a central, elongated **body** projecting toward the left lateral abdominal wall; and a **tail** that tapers as it approaches the spleen (**figure 26.21**).

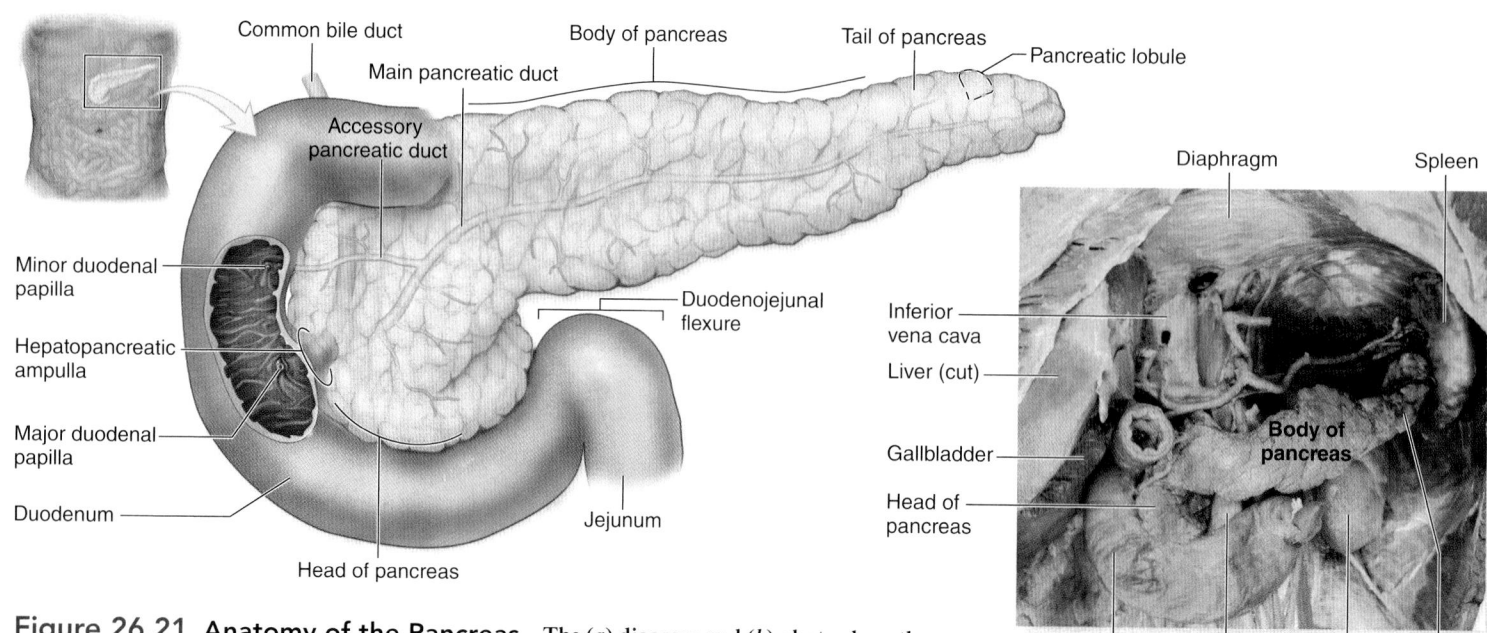

Figure 26.21 Anatomy of the Pancreas. The (*a*) diagram and (*b*) photo show the components of the pancreas in relationship to the duodenum. **AP|R**

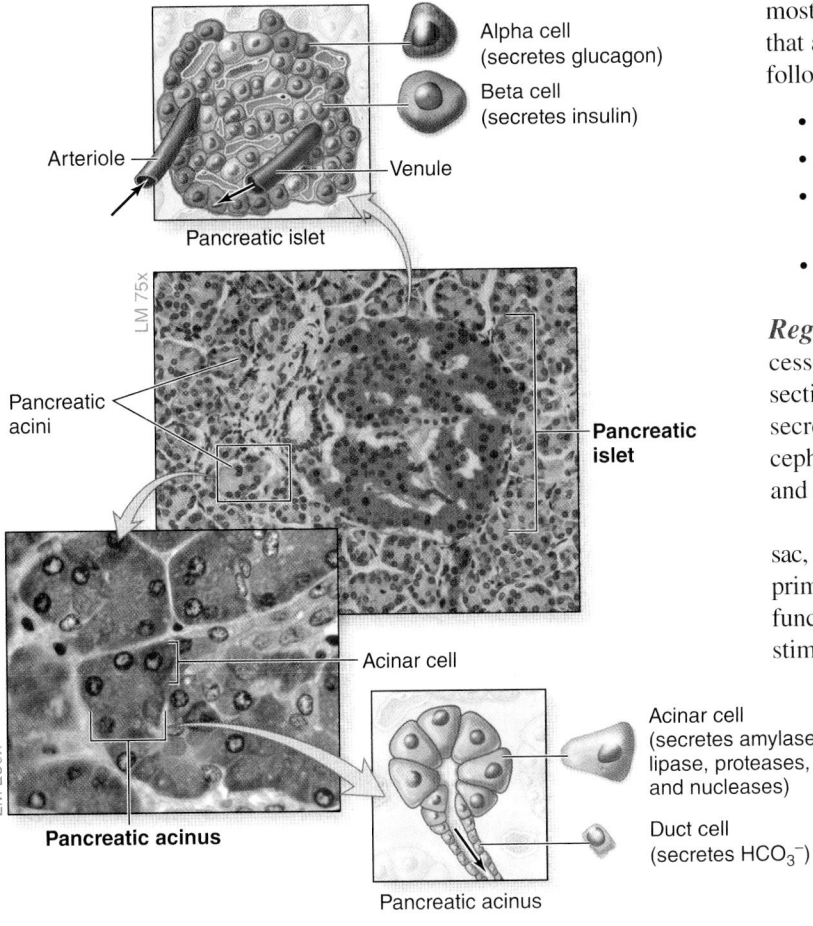

Alpha cell (secretes glucagon)

Beta cell (secretes insulin)

Arteriole — Venule

Pancreatic islet

LM 75x

Pancreatic acini

Pancreatic islet

Acinar cell

LM 200x

Pancreatic acinus

Acinar cell (secretes amylase, lipase, proteases, and nucleases)

Duct cell (secretes HCO_3^-)

Pancreatic acinus

Figure 26.22 Histology of the Pancreas. Photomicrographs and diagrams of the histology of pancreatic islets and pancreatic acini.

Histology The pancreas contains modified simple cuboidal epithelial cells called **acinar** cells that are arranged in saclike **acini** (sing., *acinus* = grape) **(figure 26.22)**. These cells are organized into large clusters termed **lobules.** Acinar cells produce and release digestive enzymes. Small ducts lead from each acinus into larger ducts that empty into larger pancreatic ducts that lead to the duodenum (as described earlier). The simple cuboidal epithelial cells lining the pancreatic ducts have the important function of secreting alkaline HCO_3^- fluid.

Pancreatic Secretions Together, secretions of acinar cells and cells that line the pancreatic ducts form **pancreatic juice.** Pancreatic juice (approximately 1 to 1.5 liters per day) is an alkaline fluid containing

mostly water, HCO_3^-, and a versatile mixture of digestive enzymes that are described in detail in section 26.4. These enzymes include the following:

- Pancreatic amylase to digest starch
- Pancreatic lipase for the digestion of triglycerides
- Inactive proteases (trypsinogen, chymotrypsinogen, and procarboxypeptidase) that, when activated, digest protein
- Nucleases for the digestion of nucleic acids (DNA and RNA)

Regulation of Accessory Structures Several of the same processes that regulate stomach motility and secretions described in section 26.2d and depicted in figure 26.14 also control the release of secretions from the accessory organs. Vagal stimulation during the cephalic and gastric phase, in addition to stimulating stomach motility and secretion, also activates the pancreas to release pancreatic juice.

Cholecystokinin (CCK) (ko′lē-sis-to-kī′nin; *chole* = bile; *cyst* = sac, *kinin* = to move) is a hormone released from the small intestine primarily in response to free fatty acids in chyme. One of the primary functions of cholecystokinin (and from which its name is derived) is stimulation of the smooth muscle in the gallbladder wall to strongly contract, causing the release of concentrated bile. Other functions of CCK include stimulating the pancreas to release enzyme-rich pancreatic juice, and relaxing the smooth muscle within the hepatopancreatic ampulla, allowing entry of the bile and pancreatic juice into the small intestine. CCK also inhibits both stomach motility and release of gastric secretions.

Secretin (se-krē′tin) is released from the small intestine primarily in response to an increase in chyme acidity. Secretin primarily causes the release of an alkaline solution that contains HCO_3^- from both the liver and ducts of the pancreas. Upon entering the small intestine, this alkaline fluid helps neutralize the acidic chyme. Secretin also inhibits gastric motility and secretions. Hormones that regulate digestion are summarized in **table 26.3.**

WHAT DID YOU LEARN?

19 Where do deoxygenated, nutrient-rich blood and oxygenated blood first come together within a liver lobule?

20 Does the liver produce digestive enzymes? If not, what substance does it produce that assists in digestion?

21 What are the primary functions of pancreatic juice?

Table 26.3	Hormones That Control Digestion		
Hormone	**Secreted by**	**Stimulus for Release**	**Primary Target(s) and Effects**
Gastrin	G-cells in stomach	Bolus in the stomach—especially if the bolus contains proteins	Parietal cells (primarily): Stimulates secretion of hydrochloric acid (HCl)
			Chief cells: Stimulates secretion of pepsinogen
			Pyloric sphincter: Stimulates contraction
Cholecystokinin (CCK)	Enteroendocrine cells of small intestine	Chyme containing amino acids and fatty acids entering small intestine	Stomach: Inhibits stomach motility and gastric secretion
			Gallbladder: Stimulates release of bile
			Pancreas: Stimulates release of enzyme-rich pancreatic juice
			Hepatopancreatic sphincter: Causes relaxation
Secretin	Enteroendocrine cells of small intestine	Primarily with increase in acidity of chyme entering the small intestine	Stomach: Inhibits gastric secretion and stomach motility
			Pancreas: Stimulates secretion of alkaline solution from pancreatic ducts
			Liver: Stimulates secretion of alkaline solution

CLINICAL VIEW
Appendicitis

Inflammation of the appendix is called **appendicitis** (ă-pen-di-sī'tis). Most cases of appendicitis occur because fecal matter obstructs the appendix, although sometimes an appendix becomes inflamed without any obstruction. As the tissue becomes inflamed, the appendix swells, the blood supply is compromised, and bacteria may proliferate in the wall. Untreated, the appendix may burst and release its contents into the peritoneum, causing a massive and potentially deadly infection called peritonitis.

During the early stages of acute appendicitis, the smooth muscle wall contracts and goes into spasms. Because this smooth muscle is innervated by the autonomic nervous system, pain is referred to the T10 dermatome around the umbilicus (see section 16.2b).

As the inflammation worsens and the parietal peritoneum becomes inflamed as well, the pain becomes sharp and localized to the right lower quadrant of the abdomen. Individuals with appendicitis typically experience nausea or vomiting, abdominal tenderness in the inferior right quadrant, a low fever, and an elevated leukocyte count. An inflamed appendix is surgically removed in a procedure called an *appendectomy*.

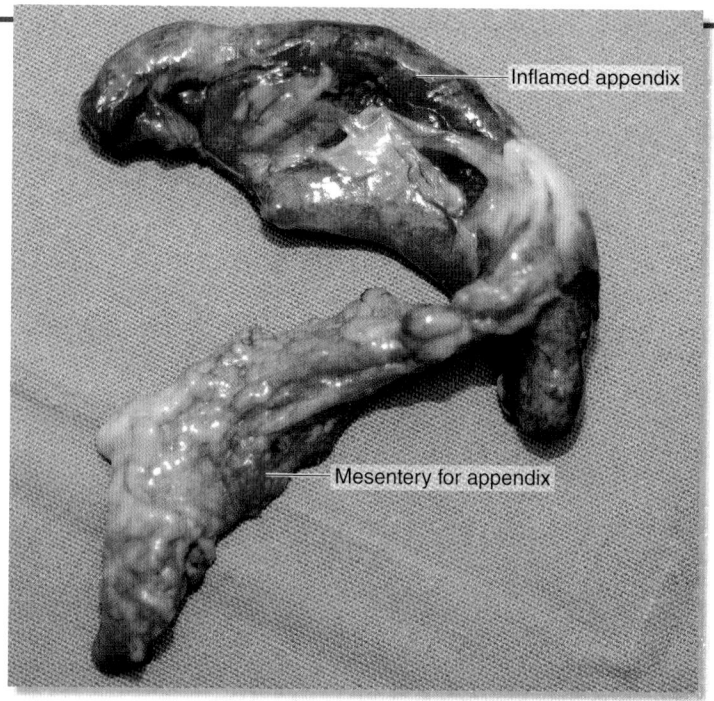

Inflamed appendix

26.3d Large Intestine

LEARNING OBJECTIVES

8. Name the three major regions of the large intestine and four segments of the colon of the large intestine.

9. Explain the distinguishing histologic features of the large intestine.

10. Describe the bacterial action that takes place in the large intestine.

The **large intestine** is called the *large bowel* and it is a relatively wide tube that is significantly shorter than the small intestine. It is called the "large" intestine because its diameter is greater than that of the small intestine. Approximately 1 liter of digested material passes from the small intestine to the large intestine daily. Most nutrients have been absorbed within the small intestine. The large intestine absorbs water and electrolytes (primarily sodium [Na$^+$] and chloride [Cl$^-$] ions) from the remaining digested material that enters into it from the small intestine (as well as vitamins B and K that are synthesized by bacteria within the large intestine). It is estimated that only 100 milliliters of the water entering the colon daily is lost in feces. Thus, the watery chyme that first enters the large intestine (including all of the undigested materials as well as the waste products secreted by the accessory organs (e.g., bilirubin by the liver) soon solidifies and is compacted into **feces** (fē'cēz; *faex* = dregs), or fecal material. The large intestine then stores this fecal material until it is eliminated through defecation.

Gross Anatomy of the Large Intestine

The large intestine has a diameter of 6.5 centimeters (2.5 inches) and an approximate length of 1.5 meters (5 feet) from its origin at the ileocecal junction to its termination at the anus (figure 26.23). Three major regions comprise the large intestine: the cecum, the colon, and the rectum.

Cecum The **cecum** (sē'kŭm; *caecus* = blind) is a blind sac. It is the first portion of the large intestine and located in the right lower abdominal quadrant. This pouch extends inferiorly from the ileocecal valve. Chyme enters the cecum from the ileum. Projecting inferiorly from the posteromedial region of the cecum is the **vermiform appendix** (ver'mi-fōrm; *vermis* = worm; ă-pen'diks; appendage), a thin, hollow, fingerlike sac lined by lymphocyte-filled lymphatic nodules (see section 21.4d). Both the cecum and the vermiform appendix are intraperitoneal organs. Research studies suggest that the appendix may harbor bacteria that are beneficial to the function of the colon.

Colon At the level of the ileocecal valve, the second region of the large intestine, the **colon,** begins and forms an inverted U-shaped arch. The colon is partitioned into four segments: the ascending colon, transverse colon, descending colon, and sigmoid colon.

The **ascending colon** originates at the ileocecal valve and extends superiorly from the superior edge of the cecum along the right lateral border of the abdominal cavity. The ascending colon is retroperitoneal, since its posterior wall directly adheres to the posterior abdominal wall, and only its anterior surface is covered with peritoneum. As it approaches the inferior surface of the liver, the ascending colon makes a 90-degree turn toward the left side and anterior region of the abdominal cavity. This bend in the colon is called the **right colic** (kol'ik) **flexure,** or the *hepatic flexure.*

The **transverse colon** originates at the right colic flexure and curves slightly anteriorly as it projects horizontally to the left across the anterior region of the abdominal cavity. The transverse colon is intraperitoneal. As the transverse colon approaches the spleen in the left upper quadrant of the abdomen, it makes a 90-degree turn inferiorly and posteriorly. The resulting bend in the colon is called the **left colic flexure,** or the *splenic flexure.*

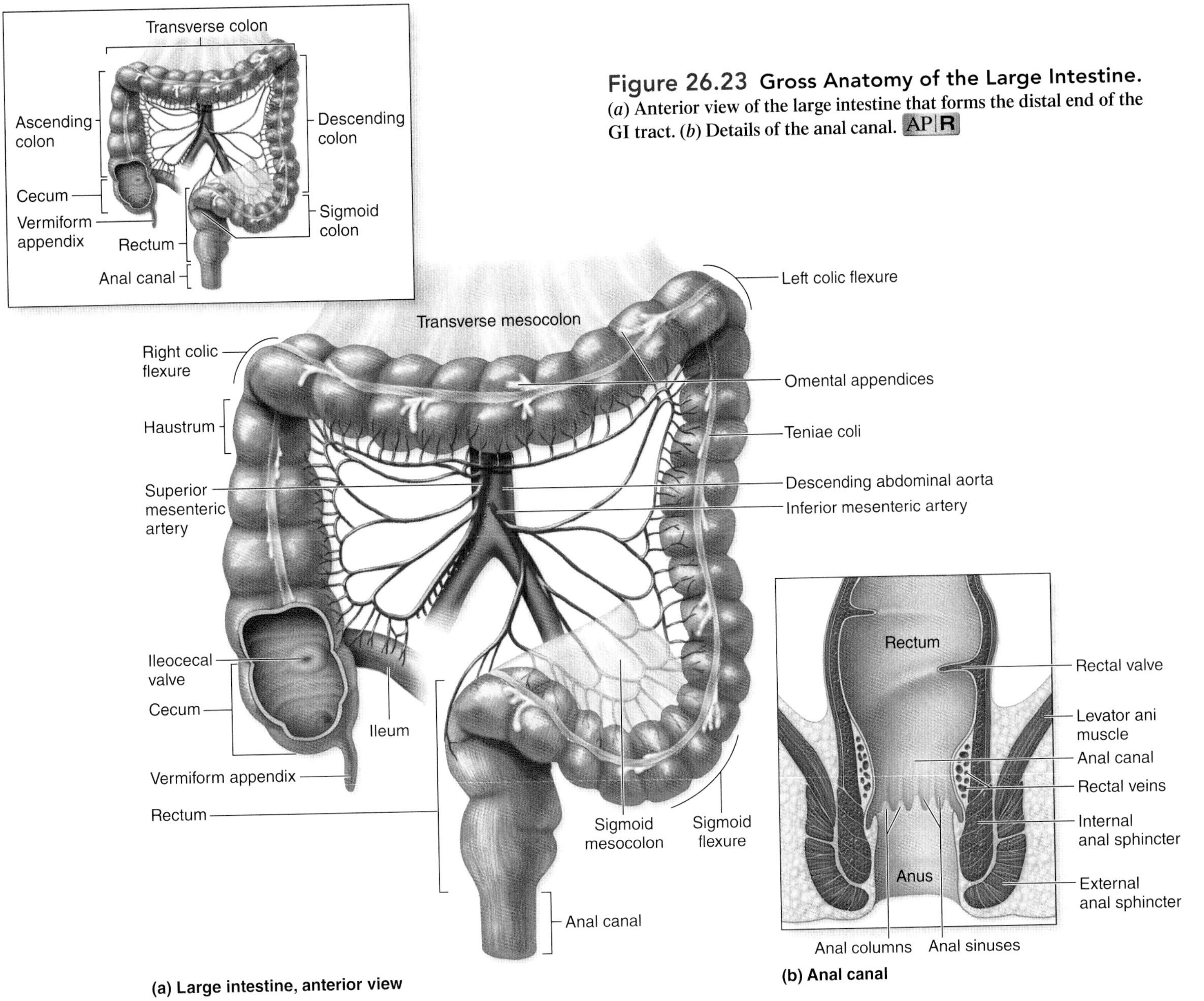

Figure 26.23 Gross Anatomy of the Large Intestine.
(*a*) Anterior view of the large intestine that forms the distal end of the GI tract. (*b*) Details of the anal canal. AP|R

(a) Large intestine, anterior view

(b) Anal canal

The **descending colon** is retroperitoneal and located along the left side of the abdominal cavity and slightly posterior. It originates at the left colic flexure and descends vertically to the sigmoid colon.

The **sigmoid** (sig′moyd; resembling letter S) **colon** originates at the **sigmoid flexure** and turns inferomedially into the pelvic cavity. The sigmoid colon, like the transverse colon, is intraperitoneal. The sigmoid colon terminates at the rectum. Recall that a type of mesentery, called the mesocolon, attaches each section of the colon to the posterior abdominal wall, with the mesocolon of each region specifically named (e.g., ascending mesocolon, transverse mesocolon).

Rectum The **rectum** (rek′tŭm; *rectus* = straight) is the third major region of the large intestine. It is a retroperitoneal structure that extends from the sigmoid colon. The rectum is a muscular tube that readily expands to store accumulated fecal material prior to defecation. Three thick transverse folds of the rectum, called **rectal valves**, ensure that fecal material is retained during the passing of gas.

The **anal canal** makes up the terminal few centimeters of the large intestine. The anal canal is lined by a stratified squamous epithelium, and it passes through an opening in the levator ani muscles of the pelvic floor

(see section 11.7) and terminates at the anus. The internal lining of the anal canal contains relatively thin longitudinal ridges, called **anal columns,** between which are small depressions termed **anal sinuses.** As fecal material passes through the anal canal during defecation, pressure exerted on the anal sinuses causes their cells to release mucin to form mucus. The extra mucus lubricates the anal canal during defecation. At the base of the anal canal are the involuntary smooth muscle **internal anal sphincter** and voluntary skeletal muscle **external anal sphincter,** which generally close off the opening to the anal canal. The muscles composing these sphincters relax and allow the sphincter to open during defecation.

Specialized Structures of the Large Intestine

Several features are unique to the large intestine including teniae coli, haustra, and epiploic appendages. **Teniae coli** (tē′nē-ē; ribbons, band; kō′lī) are thin, distinct, longitudinal bundles of smooth muscle. The teniae coli act like elastic in a waistband—they help bunch up the large intestine into many sacs, collectively called **haustra** (haw′strӑ; sing., *haustrum; haustus* = to drink up). Hanging off the external surface of the haustra are lobules of fat called **omental appendices,** or *epiploic* (e-pi-plō′ik; membrane-covered) *appendages.*

INTEGRATE

CLINICAL VIEW
Colorectal Cancer

Colorectal cancer is the second most common type of cancer in the United States, with over 140,000 cases occurring annually and 60,000 of those resulting in death. The term **colorectal cancer** refers to a malignant growth anywhere along the large intestine (colon or rectum). The majority of colorectal cancers appear in the distal descending colon, sigmoid colon, and rectum, which are the segments of the large intestine that have the longest contact with fecal matter before it is expelled from the body. Most colorectal cancers arise from **polyps** (pol'ip), which are outgrowths from the colon mucosa. Note, however, that colon polyps are very common, and most of them never become cancerous. Low-fiber diets have also been implicated in increasing the risk of colon cancer, because decreased dietary fiber leads to decreased stool bulk and longer time for stools to remain in the large intestine. Theoretically, this condition exposes the large intestine mucosa to toxins in the stools for longer periods of time. Other risk factors include a family history of colorectal cancer, personal history of ulcerative colitis, and increased age (most patients are over the age of 40).

Initially, most patients are asymptomatic. Later, they may notice rectal bleeding (often evidenced as blood in the stool or on the toilet paper) and a persistent change in bowel habits (typically constipation). Eventually, the person may experience abdominal pain, fatigue, unexplained weight loss, and anemia.

The cancerous growth must be removed surgically, and sometimes radiation or chemotherapy are used as well. Colorectal cancers that are limited to the mucosa have a 5-year survival rate, but the prognosis is poor for cancers that have spread into deeper colon wall tunics or metastasized to the lymph nodes.

The key to an increased survival rate for colorectal cancer is early detection. If caught early, colorectal cancer is very treatable. An individual should see their doctor if rectal bleeding or any persistent change in bowel habits is experienced. By age 50 (or earlier, for those that have symptoms or a family history of colorectal cancer), the following screening methods are recommended:

1. A yearly **fecal occult** (ō-kŭlt', ok'ŭlt) **blood test,** which checks for the presence of blood in the stools.
2. A **sigmoidoscopy** (sig'moy-dos'kŏ-pē; *skopeo* = to view) every 5 years. In this procedure, which is done in the doctor's office, an endoscope is inserted into the rectum and sigmoid colon to check for polyps or cancer.
3. A **colonoscopy** (kō-lon-os'kŏ-pē) every 10 years. A colonoscopy is more extensive than a sigmoidoscopy. The endoscope is inserted into the large intestine at least up to the right colic flexure of the colon, and sometimes as far proximally as the ileocecal valve.

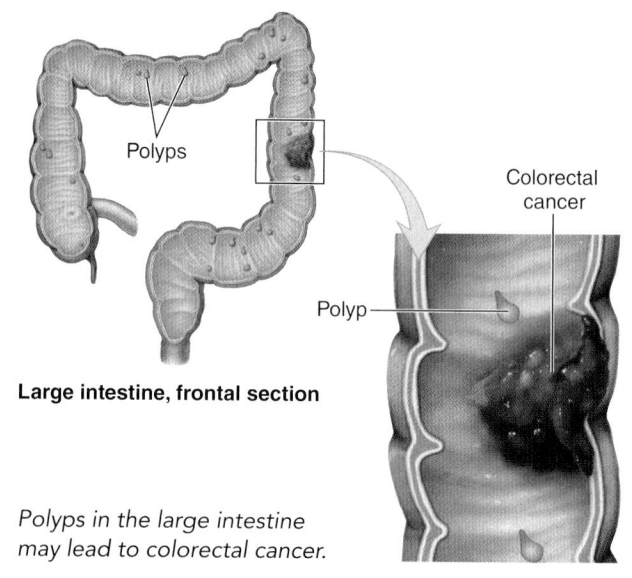

Large intestine, frontal section

Polyps in the large intestine may lead to colorectal cancer.

INTEGRATE

CONCEPT CONNECTION

Recall that heme is broken down into bilirubin and absorbed from the blood in the liver (see section 18.3b). Bilirubin is then released as a component of bile into the small intestine. It is either reabsorbed back into the blood and eliminated as urobilin in the urine, or it continues through the GI tract into the large intestine, where bacteria convert it to stercobilin. Stercobilin is a brown pigment that accounts for the normal color of feces.

Histology

The mucosa of the large intestine is lined by a simple columnar epithelium with numerous goblet cells (**figure 26.24**). Unlike the small intestine, the large intestine mucosa is smooth and lacks intestinal villi, yet it resembles the small intestine in that it has epithelial cells and numerous **intestinal glands** (or crypts) that extend inward toward the muscularis mucosae. The glands' mucous cells secrete mucin to lubricate the undigested material and facilitate its passage through the large intestine. Many lymphatic nodules and lymphatic cells occupy the lamina propria of the large intestine.

The muscularis of the cecum and colon has two layers of smooth muscle, but the outer longitudinal layer is discontinuous and does not completely surround the colon and cecum. Instead, these longitudinal smooth muscle fibers form the teniae coli, just described.

Bacterial Action in the Large Intestine

Numerous normal bacterial flora inhabit the large intestine; they are termed the *indigenous microbiota*. These bacteria are responsible for the chemical breakdown of complex carbohydrates, proteins, and lipids that remain in the chyme after it has passed through the small intestine. Bacterial actions produce carbon dioxide, H^+, hydrogen sulfide, methane, indoles, and skatoles. Some of these substances account for the odor of feces. Additionally, B vitamins and vitamin K are produced by the bacterial flora, which are then absorbed from the large intestine into the blood. (Note that these vitamins are also absorbed in the small intestine from the foods that we eat.) Feces is the final product formed and then eliminated from the GI tract. Feces is composed of water, salts, epithelial cells (sloughed from GI lining), bacteria, and undigested material.

Figure 26.24 Histology of the Large Intestine. (*a*) The luminal wall of the large intestine is composed of four tunics: mucosa, submucosa, muscularis, and serosa. (*b*) A photomicrograph shows the histology of the mucosa and submucosa of the large intestine wall.

Opening to intestinal gland

Goblet cells

Lumen

Simple columnar epithelium

Mucosa

Intestinal gland

Lamina propria

Lymphatic nodule

Muscularis mucosae

Submucosa

Muscularis
Circular layer

Longitudinal layer (tenia coli)

Nerves   Arteriole   Venule

Serosa

(a) Large intestine tunics

Lumen

Opening to intestinal gland

Goblet cells

Simple columnar epithelium

Intestinal gland

Muscularis mucosae

LM 80x

(b) Large intestine mucosa and submucosa

Motility and Regulation of the Large Intestine

Several types of movements are noted in the large intestine:

- **Peristalsis** of the large intestine is usually weak and sluggish, but otherwise it resembles the peristalsis that occurs in the wall of the small intestine.
- **Haustral churning** occurs after a relaxed haustrum fills with digested or fecal material until its distension stimulates reflex contractions in the muscularis. These contractions increase churning and move the material to more distal haustra.
- **Mass movements** are powerful, peristaltic-like contractions involving the teniae coli, which propel fecal material toward the rectum. A wave of contraction begins in the middle of the transverse colon, forcing a large amount of fecal matter into the descending colon, sigmoid colon, and the rectum. Generally, mass movements occur two or three times a day, often during or immediately after a meal.

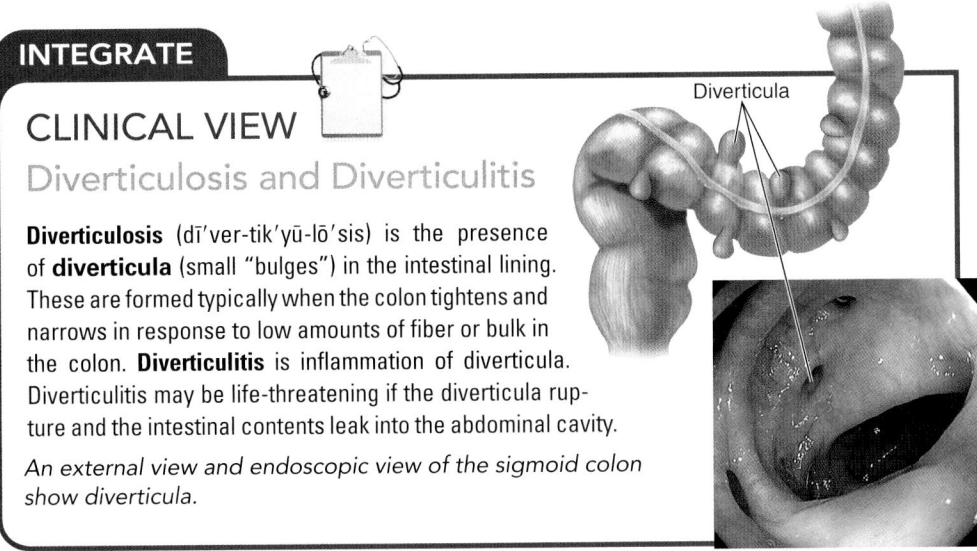

INTEGRATE

CLINICAL VIEW

Diverticulosis and Diverticulitis

Diverticulosis (dī'ver-tik'yū-lō'sis) is the presence of **diverticula** (small "bulges") in the intestinal lining. These are formed typically when the colon tightens and narrows in response to low amounts of fiber or bulk in the colon. **Diverticulitis** is inflammation of diverticula. Diverticulitis may be life-threatening if the diverticula rupture and the intestinal contents leak into the abdominal cavity.

An external view and endoscopic view of the sigmoid colon show diverticula.

Diverticula

INTEGRATE

CLINICAL VIEW
Constipation and Diarrhea

Constipation is typically a temporary, impaired ability to defecate and generally results in compacted feces that is difficult to eliminate. Constipation may result from a combination of factors, including a diet low in fiber (insufficient bulk), dehydration, lack of exercise, and improper bowel habits (not defecating when the urge arises, allowing additional water to be absorbed from the feces). Constipation is also a common side effect of general anesthesia.

In contrast, **diarrhea** may result as a consequence of disruption in normal mechanisms to absorb intestinal water, or excessive amounts of osmotically active solutes that keep water in the intestinal lumen and prevent its reabsorption. One example of such an osmotically active solute is sorbitol, a sugar substitute often found in sugar-free ice cream, diet sodas, cough syrups, and sugar-free gum. Sorbitol is considered a nonstimulant laxative because it draws water into the large intestine and stimulates bowel movements.

Two major reflexes are associated with motility in the large intestine:

- The **gastrocolic** (gas′trō-kol′ik) **reflex** is initiated by stomach distension to cause a mass movement (motility in the large intestine just described).
- The **defecation reflex.** The elimination of feces from the GI tract by the process of **defecation** (def-ĕ-kā′shŭn; *defaeco* = to remove the dregs) **(figure 26.25).** Filling of the rectum initiates the urge to defecate. This stimulus results in transmission of nerve signals from the receptors to the spinal cord. In response, increased nerve signals are along cells of the salivary glands called parasympathetic motor neurons, which causes both the sigmoid colon and rectum to contract and the internal (involuntary) anal sphincter to relax. (The defecation reflex is an example of a monosynaptic reflex; see section 14.6c.)

Voluntary elimination of feces from the body typically is learned sometime after age 3 years. The conscious decision (initiated by the cerebral cortex) to defecate involves both the Valsalva maneuver (see section 23.3a) and relaxation of the external anal sphincter.

WHAT DID YOU LEARN?

22 What is the pathway of chyme from its entry into the large intestine until feces is eliminated?

23 What are the general functions of bacteria within the large intestine?

24 Which substances are typically absorbed by the large intestine?

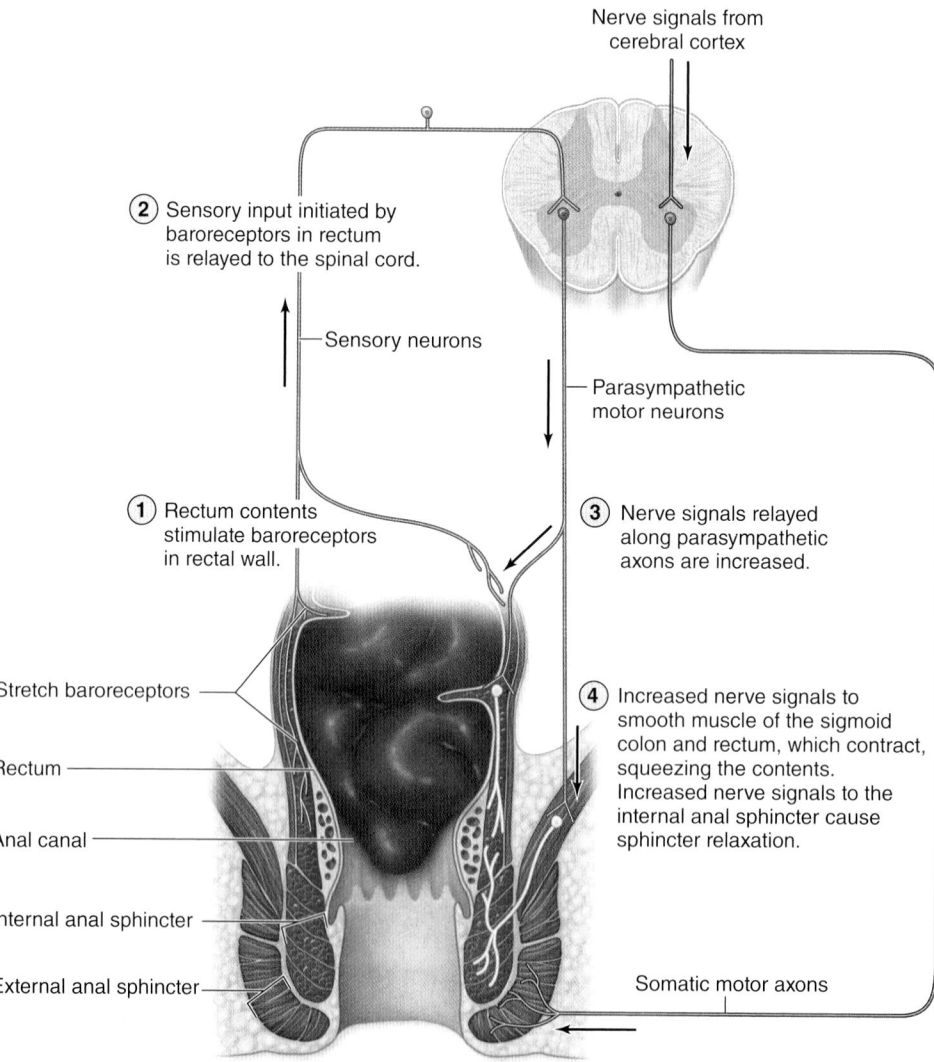

Nerve signals from cerebral cortex

2 Sensory input initiated by baroreceptors in rectum is relayed to the spinal cord.

Sensory neurons

Parasympathetic motor neurons

1 Rectum contents stimulate baroreceptors in rectal wall.

3 Nerve signals relayed along parasympathetic axons are increased.

Stretch baroreceptors

Rectum

Anal canal

Internal anal sphincter

External anal sphincter

4 Increased nerve signals to smooth muscle of the sigmoid colon and rectum, which contract, squeezing the contents. Increased nerve signals to the internal anal sphincter cause sphincter relaxation.

Somatic motor axons

5 The conscious decision to defecate is controlled by the cerebral cortex. External anal sphincter relaxes and Valsalva maneuver is initiated, eliminating the feces.

Figure 26.25 Defecation. The defecation reflex involves sensory perception of stretch in the rectum that initiates a spinal reflex that stimulates muscles in the rectal wall to contract and the internal anal sphincter to relax. Conscious regulation of defecation requires the relaxation of the external anal sphincter and the Valsalva maneuver.

26.4 Nutrients and Their Digestion

The term **essential nutrients** indicates substances that must constitute part of the diet for survival. The six essential nutrients are carbohydrates, proteins, lipids, minerals, vitamins, and water. These nutrients and their functions in the body are described in depth in sections 27.1–3.

This section discusses the chemical digestion of the foods that we eat. In addition to the mechanism of their breakdown, we describe their absorption and general use by the body. We describe the breakdown of carbohydrates, lipids, and proteins, as well as nucleic acids, the last of which is not an essential nutrient. All are digested by the process of *hydrolysis,* the breaking of chemical bonds with an accompanying addition of a water molecule (see section 2.7a).

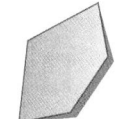

26.4a Carbohydrate Digestion

LEARNING OBJECTIVES

1. Name the three classes of carbohydrates.
2. Explain the processing in the oral cavity that initiates carbohydrate digestion.
3. Describe the chemical digestion of carbohydrates that occurs in the small intestine.

Carbohydrates are organized based upon the number of repeating units of simple sugars. You learned in section 2.7c that carbohydrates may be classified as **monosaccharides** (e.g., glucose, fructose, and galactose), **disaccharides** (e.g., sucrose, maltose, and lactose), and **polysaccharides** (e.g., starch and cellulose). Chemical digestion of carbohydrates consists of (1) the breakdown of starch into individual glucose molecules, and (2) the breakdown of disaccharides into the individual monosaccharides that compose them. The oral cavity and small intestine are the main sites of carbohydrate digestion.

Carbohydrate Breakdown in the Oral Cavity

Digestion of starch begins in the oral cavity. It is catalyzed by **salivary amylase** that is synthesized and released from the salivary glands. Salivary amylase breaks the chemical bonds between glucose molecules, within the starch molecule, to partially digest the starch molecule. The extent of starch digestion is dependent upon the length of time the salivary amylase is allowed to act on the starch.

Salivary amylase is inactivated by the low pH of the stomach when the bolus is swallowed. This inactivation typically occurs within 15 to 20 minutes after the bolus enters the stomach. The larger the meal, the longer salivary amylase remains active. This extended activity occurs because it takes longer for the swallowed bolus to be mixed with the low pH of the gastric juices that inactivate the salivary amylase. This is more likely when the bolus is within the fundus of the stomach, where smooth muscle contractions are the weakest and the pH is the highest. No new enzymes for carbohydrate digestion are introduced in the stomach.

Carbohydrate Breakdown in the Small Intestine

Pancreatic amylase is synthesized and released by the pancreas as a component of pancreatic juice into the small intestine through the main or accessory pancreatic duct **(figure 26.26)**. This enzyme continues the digestion of starch into shorter strands of glucose (5 to 25 glucose molecules, oligosaccharides), maltose (disaccharide of two glucose molecules), and individual glucose molecules.

The completion of starch breakdown is accomplished by brush border enzymes embedded within the epithelial lining of the small intestine. These enzymes include **dextrinase** and **glucoamylase,** which break the bonds between glucose subunits of oligosaccharides, and **maltase** that breaks the bond between the two glucose molecules that compose maltose.

The digestion of other ingested disaccharides, such as lactose (milk sugar) and sucrose (table sugar), requires only one enzyme each. Each enzyme is specifically named for the substrate it digests. **Lactase** digests lactose to glucose and galactose, and **sucrase** digests sucrose to glucose and fructose. These enzymes are brush border enzymes. Individuals with either a reduced amount or lack of lactase enzyme are referred to as being **lactose intolerant** (see Clinical View: "Lactose Intolerance" in section 3.4a).

The monosaccharides released from these enzymatic reactions include glucose, fructose, and galactose. They are absorbed across the small intestinal epithelial lining into the blood. All venous blood from the small intestine is transported through the hepatic portal vein to the liver, where fructose and galactose will be converted into glucose. Glucose has different fates. It can become

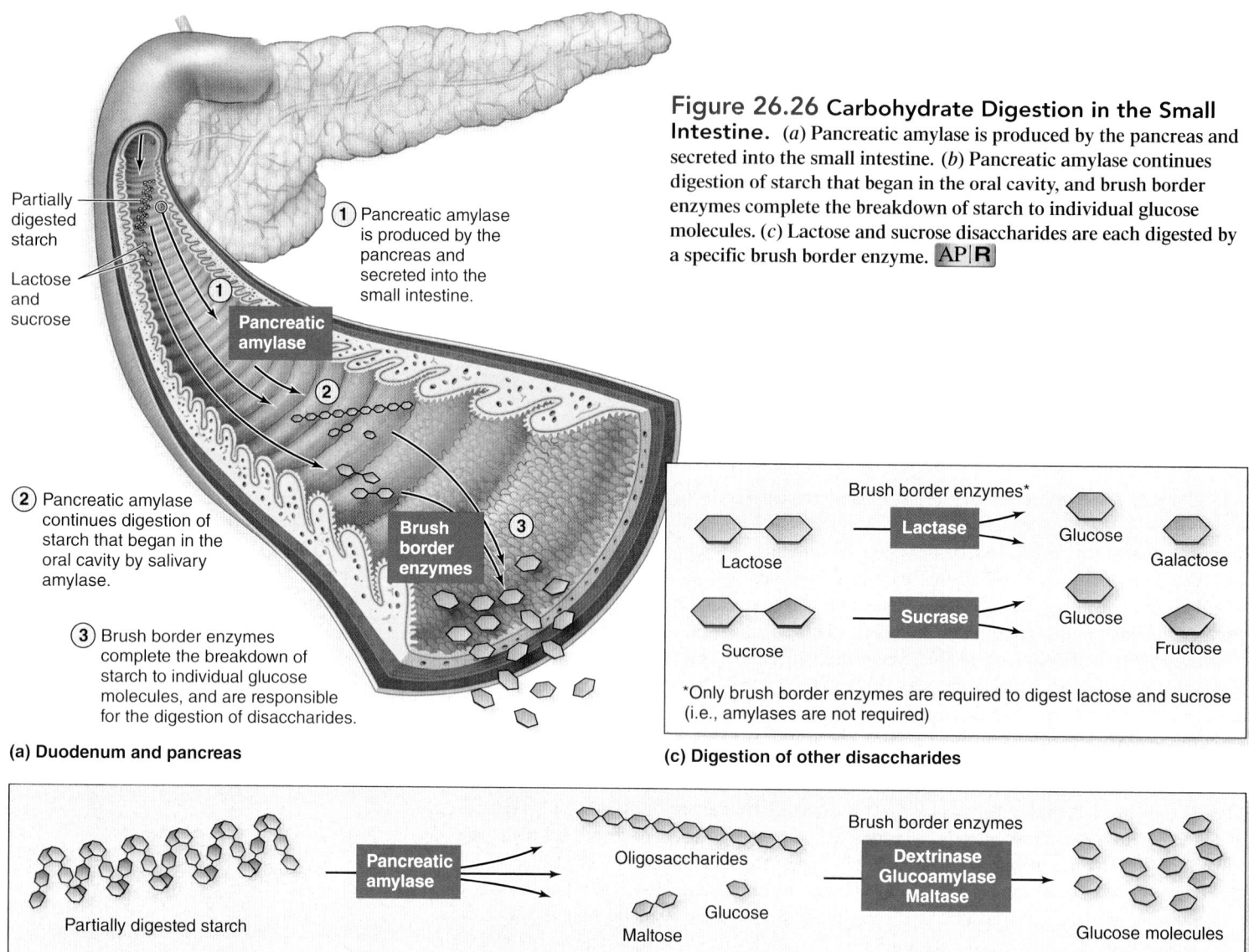

Figure 26.26 Carbohydrate Digestion in the Small Intestine. (*a*) Pancreatic amylase is produced by the pancreas and secreted into the small intestine. (*b*) Pancreatic amylase continues digestion of starch that began in the oral cavity, and brush border enzymes complete the breakdown of starch to individual glucose molecules. (*c*) Lactose and sucrose disaccharides are each digested by a specific brush border enzyme. AP|R

Partially digested starch

Lactose and sucrose

(1) Pancreatic amylase is produced by the pancreas and secreted into the small intestine.

Pancreatic amylase

(2) Pancreatic amylase continues digestion of starch that began in the oral cavity by salivary amylase.

Brush border enzymes

(3) Brush border enzymes complete the breakdown of starch to individual glucose molecules, and are responsible for the digestion of disaccharides.

(a) Duodenum and pancreas

Brush border enzymes*

Lactose → Lactase → Glucose + Galactose

Sucrose → Sucrase → Glucose + Fructose

*Only brush border enzymes are required to digest lactose and sucrose (i.e., amylases are not required)

(c) Digestion of other disaccharides

Partially digested starch → Pancreatic amylase → Oligosaccharides + Maltose + Glucose → Brush border enzymes Dextrinase Glucoamylase Maltase → Glucose molecules

(b) Digestion of starch

part of the blood glucose, be taken up by any cell to be oxidized through cellular respiration (see section 3.4), be taken up by liver cells and muscle cells and synthesized into glycogen and stored, or converted into fat (triglycerides) and stored in adipose connective tissue.

Cellulose is a carbohydrate that is a component of plant cell walls. Cellulose is not chemically digested because we lack the enzymes required to break the bonds between its glucose molecules. Thus, cellulose, along with other indigestible substances, is fiber that adds "bulk" to the contents of the lumen and facilitates its moving through the GI tract. A summary of carbohydrate digestion is included in table 26.4.

WHAT DID YOU LEARN?

25 What enzymes are required to completely break down starch? Name the specific source of each of these enzymes.

26.4b Protein Digestion

LEARNING OBJECTIVES

4. Identify the enzyme that initiates protein digestion in the stomach, and explain its activation and action.

5. Explain why the proteolytic enzymes of the stomach and pancreas are synthesized in inactive forms.

6. Describe the chemical digestion of proteins that occurs in the small intestine.

Proteins are polymers composed of amino acid subunits linked by peptide bonds (see section 2.7e). Digestion of protein releases individual amino acids so that the amino acids may be absorbed into the blood and transported to cells for the synthesis of new proteins (see section 4.8).

Table 26.4	Digestion of Macromolecules			
Location	**Carbohydrates**	**Proteins**	**Lipids**	**Nucleic Acids**
Oral cavity (saliva)	Starch—*salivary amylase*—→ partially digested starch	No protein digestion	*Lingual lipase* added but activated in low pH of stomach	No nucleic acid digestion
Stomach (gastric juice)	No additional enzymes added	Protein—*pepsin*—→ polypeptide and peptide fragments	Triglyceride—*lingual lipase*—→ monoglyceride and fatty acids (limited amounts) Triglyceride—*gastric lipase*—→ monoglyceride and fatty acids (limited amounts)	No nucleic acid digestion
Small intestine (pancreatic juice secreted into duodenum)	Partially digested starch— *pancreatic amylase*—→ oligosaccharides, maltose, and glucose	Protein—*trypsin*—→ polypeptide and peptide fragments Protein—*chymotrypsin*—→ polypeptide and peptide fragments Protein—*carboxypeptidase* —→ amino acids from carboxy-end of peptides	Triglyceride—*pancreatic lipase*—→ monoglyceride and fatty acids (within micelles)	DNA— *deoxyribonuclease*—→ deoxyribonucleotides RNA—*ribonuclease*—→ ribonucleotides
Small intestine (brush border enzymes)	Oligosaccharides—*dextrinase* and *glucoamylase*—→ maltose, glucose Maltose—*maltase*—→ glucose Lactose—*lactase*—→ glucose, galactose Sucrose—*sucrase*—→ glucose, fructose	Dipeptides—*dipeptidase*—→ amino acids Peptides—*aminopeptidase*—→ amino acids from amino end of peptides	No brush border enzymes are required for completing lipid digestion	Nucleotides—*phosphatase*—→ nucleosides and phosphate Nucleosides— *nucleosidase*—→ nitrogenous base and sugar (ribose or deoxyribose)

Proteins are broken down into amino acids by enzymes that target peptide bonds either between specific adjacent amino acids within the protein or nonspecifically release amino acids from either end of a protein. All enzymes that digest protein are released from both the stomach and pancreas as inactive enzymes. These enzymes must be activated (e.g., pepsinogen is activated to pepsin within the low pH of the stomach). This is because the proteolytic enzymes would destroy the proteins within the cells that produce them or in the case of protein-digesting enzymes produced in the pancreas would destroy the cells lining the main and accessory pancreatic ducts as they pass through those ducts.

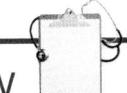

INTEGRATE

CLINICAL VIEW

Celiac Disease (Gluten-Sensitive Enteropathy)

Celiac disease (also known as *gluten-sensitive enteropathy*) is an autoimmune disorder that affects up to 1% of the population in the United States. **Gluten**—a protein common in wheat, rye, and barley (but not rice or corn)—stimulates an immune response in affected individuals that damages the villi of the small intestine, which interferes with absorption. Common foods that contain gluten include breads, pizza, pasta, and many processed foods. Symptoms include abdominal pain and chronic diarrhea, which leads to nutrient deficiencies because of malabsorption. Although there is no cure for celiac disease, it can be effectively managed by following a gluten-free diet.

Protein Breakdown in the Stomach

Protein digestion begins within the stomach lumen with the enzyme pepsin. **Pepsin** is formed from pepsinogen, an inactive precursor released by chief cells. Hydrochloric acid that is released from parietal cells causes a low pH within the stomach that both activates pepsinogen to pepsin and denatures proteins to facilitate their chemical breakdown (as described in section 26.2d).

Protein Breakdown in the Small Intestine

The high pH of the small intestine inhibits further action by pepsin on protein shortly following the entry of chyme into the small intestine. Three of the enzymes that continue the digestion of protein are synthesized and released from the pancreas into the small intestine in inactive forms—trypsinogen, chymotrypsinogen, and procarboxypeptidase (**figure 26.27**). Once these inactive forms of the enzymes reach the small intestine, they are activated by the enzyme **enteropeptidase** (en'tĕr-ō-pep'ti-dās; previously called *enterokinase*), an enzyme previously synthesized by the small intestine and released into the lumen of the small intestine (see section 26.3b). Enteropeptidase activates trypsinogen to **trypsin.** Trypsin in turn activates additional molecules of trypsinogen to trypsin, as well as chymotrypsinogen to **chymotrypsin** (kī-mō-trip'sin) and procarboxypeptidase to **carboxypeptidase** (kar-bok'sē-pep'ti-dās).

Trypsin and chymotrypsin break the bonds between specific amino acids within the protein to produce smaller strands of amino acids called peptides. (Trypsin cleaves a protein specifically at positively charged amino acids of arginine and lysine, whereas chymotrypsin

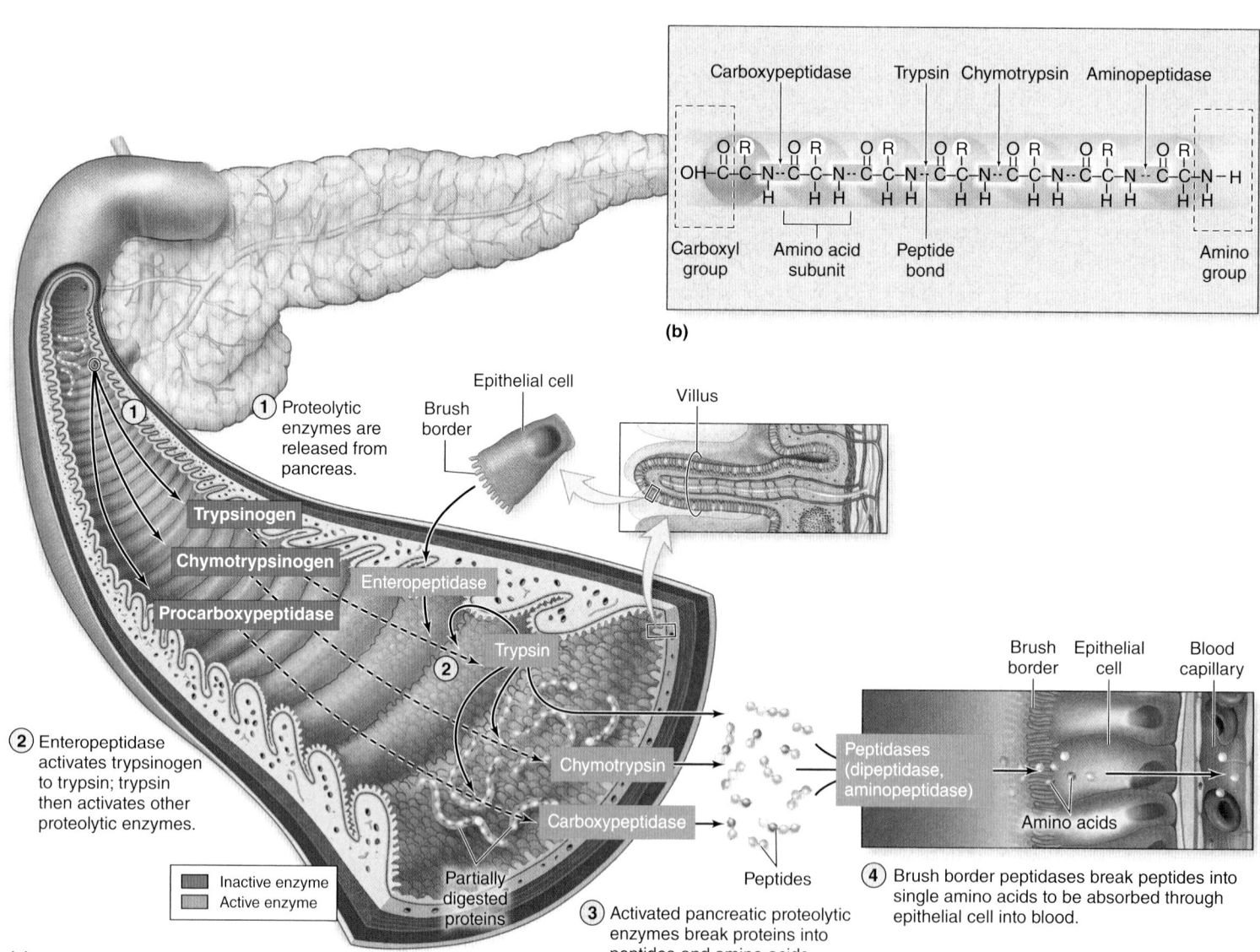

Figure 26.27 Protein Digestion in the Small Intestine. (*a*) Pancreatic enzymes for protein digestion are produced and secreted into the small intestine in an inactive form. After reaching the small intestine, these enzymes are activated. (*b*) The locations where different protein-digesting enzymes break peptide bonds are highlighted in purple.

cleaves specifically at hydrophobic amino acids including phenylalanine, tryptophan, and tyrosine; see figure 2.25, which shows the 20 different amino acids.) Carboxypeptidase is restricted to breaking the bond only between an amino acid on the carboxyl end and the remaining protein (it releases one amino acid at a time). Dipeptides and free amino acids are the breakdown products of carboxypeptidase.

The brush border enzyme **dipeptidase** breaks the final bond between the two amino acids of a dipeptide so that both may be absorbed. The brush border enzyme **aminopeptidase** generates free amino acids from the amino end of peptides.

Free amino acids are absorbed across the small intestine epithelial lining and enter into the blood. Amino acids can be used as building blocks of new proteins by cells, or if excess amino acids are absorbed, they are converted either into glucose (by gluconeogenesis in the liver primarily or kidney) or deaminated (amine [—NH₂] group is removed in the liver) and used as fuel for cellular respiration (see section 3.4h). A summary of protein digestion is included in table 26.4.

 WHAT DID YOU LEARN?

26 How are proteolytic enzymes activated in the stomach and in the small intestine? Explain why this is necessary.

26.4c Lipid Digestion

 LEARNING OBJECTIVES

7. Explain the role of bile salts in lipid digestion.

8. Discuss the process by which lipids are absorbed.

Lipids are highly variable structures that contain different arrangements of their building blocks (see section 2.7b). Lipids have one unifying property, which is that they are not water-soluble. Two major ingested lipids are triglycerides (or neutral fats) and cholesterol. Triglycerides are composed of a glycerol molecule and three fatty acids bonded to the glycerol, and enzymes are required to break the bonds between glycerol and fatty acids. Digestion of cholesterol is not required for its absorption.

Lipid Breakdown in the Stomach

Lingual lipase (produced by intrinsic salivary glands in the mouth) is a component of saliva in the oral cavity. The optimal pH of this enzyme (4.5 to 5.4), however, means it is not activated until it reaches the stomach. Once in the stomach, triglycerides undergo limited digestion by both lingual lipase and gastric lipase (an enzyme produced by chief cells of the stomach). These "acidic lipases" digest approximately 30% of the triglycerides to diglyceride and a fatty acid. Neither of these lipase enzymes requires the participation of bile salts (see next section).

Lipid Breakdown in the Small Intestine

The majority of triglyceride digestion occurs in the small intestine and is facilitated by **pancreatic lipase,** an enzyme produced by the pancreas and released into the duodenum. Pancreatic lipase digests each triglyceride into a monoglyceride and two free fatty acids. However, because triglycerides are lipids and do not dissolve in the luminal fluids of the digestive system, triglycerides form relatively large lipid masses. For example, when butter is added to water, the butter does not stay mixed with the water but remains separate.

For pancreatic lipase to effectively digest the fat, the large lipid droplets must first be mechanically separated into smaller droplets; this process is called **emulsification** (ē-mŭl-si-fĭ-kā′shun; **figure 26.28**). (This is similar to breaking an ice cube into ice chips.) Emulsification occurs by the action of **bile salts,** which are part of bile. Bile is produced by the liver and stored, concentrated, and released from the gallbladder. Bile salts are amphipathic molecules composed of a polar head and a nonpolar tail. The nonpolar tails position themselves around the fat with the polar heads next to the aqueous fluid in the lumen (like an inverted spiked ball with the fat in the center; see section 2.4c). This structure is called a **micelle** (mi-sel′ or mī-sel; *micella* = small morsel). Thus, the function of bile salts is to emulsify fats so that pancreatic lipase has greater "access" to the triglyceride molecules and may more effectively chemically digest the fat molecules. (Note that the process of emulsification is facilitated by *lecithin*, which are phospholipid molecules within bile.) Cholesterol is also

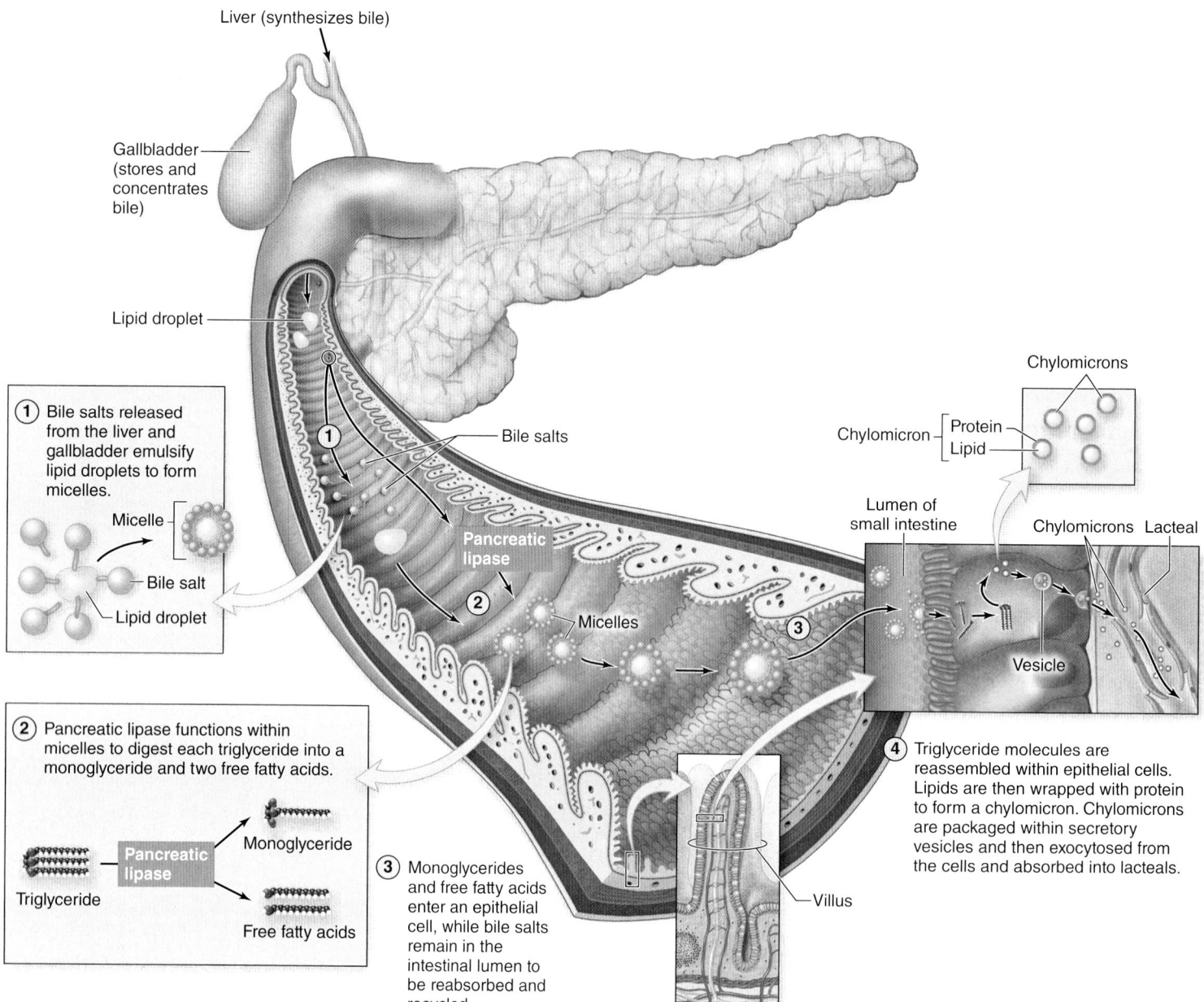

Liver (synthesizes bile)

Gallbladder
(stores and
concentrates
bile)

Lipid droplet

1 Bile salts released
from the liver and
gallbladder emulsify
lipid droplets to form
micelles.

Micelle

Bile salt

Lipid droplet

Bile salts

**Pancreatic
lipase**

2

Micelles

3

2 Pancreatic lipase functions within
micelles to digest each triglyceride into a
monoglyceride and two free fatty acids.

Monoglyceride

**Pancreatic
lipase**

Triglyceride

Free fatty acids

3 Monoglycerides
and free fatty acids
enter an epithelial
cell, while bile salts
remain in the
intestinal lumen to
be reabsorbed and
recycled.

Chylomicrons

Chylomicron ⎡ Protein
            ⎣ Lipid

Lumen of
small intestine

Chylomicrons   Lacteal

Vesicle

4 Triglyceride molecules are
reassembled within epithelial cells.
Lipids are then wrapped with protein
to form a chylomicron. Chylomicrons
are packaged within secretory
vesicles and then exocytosed from
the cells and absorbed into lacteals.

Villus

Figure 26.28 Lipid Digestion and Absorption in the Small Intestine. Bile salts emulsify lipids within the small intestine, forming micelles to facilitate fat digestion by pancreatic lipase. After the pancreatic lipase has digested the triglyceride, the free fatty acids and the monoglycerides are taken into an epithelial cell where triglycerides are re-formed, then wrapped along with other lipids within protein to form a chylomicron. Chylomicrons are released from the epithelial cells by exocytosis and then enter a lacteal within the lamina propria of the small intestine wall.

within the micelle, but it is not chemically digested. No brush border enzymes are required in the breakdown of triglycerides.

Eventually, in the last portion of the ileum, bile salts are recovered from the GI tract back into the blood by active transport and ultimately recycled to the liver for reuse. New bile salts are synthesized by hepatocytes to replace those inevitably lost during elimination of feces. Table 26.4 summarizes lipid digestion.

Lipid Absorption

Digested triglycerides (monoglycerides and free fatty acids), cholesterol, other lipids (e.g., lecithin), and fat-soluble vitamins are contained within micelles. Micelles transport lipids to the simple columnar epithelial lining of the small intestine. Here, the lipids enter the epithelial cells, whereas the bile salts remain in the small intestine lumen to be recycled and reused.

Once inside the epithelial cells, the fatty acids are reattached to the monoglyceride to re-form triglycerides. Triglycerides, cholesterol, and other lipid molecules are then "wrapped" with protein to form a **chylomicron** (kī′lō-mī′kron; *chylos* = juice, *micros* = small). The Golgi apparatus (see section 4.6a) packages chylomicrons into secretory vesicles. Vesicles containing chylomicrons merge with the plasma membrane of epithelial cells to release chylomicrons by exocytosis. Chylomicrons are too large to pass through blood capillary walls but instead enter the lacteals, the lymphatic capillaries of the small intestine. The overlapping endothelial cells of the lacteals act like one-way valves to permit entry of chylomicrons.

Recall that all lymph enters the blood, via either the right lymphatic duct or thoracic duct, at the junction of the internal jugular vein and the subclavian vein (see section 21.1b). Chylomicrons enter the blood and deliver lipids to the liver and other tissues (e.g., adipose connective tissue, skeletal muscle tissue, and cardiac muscle).

 WHAT DID YOU LEARN?

 27 What is the function of bile salts in lipid digestion? Is this considered chemical digestion or mechanical digestion?

26.4d Nucleic Acid Digestion

 **LEARNING OBJECTIVE**

9. Describe the digestion of nucleic acids.

Nucleic acids are polymers of nucleotides. The two types of nucleic acid polymers are deoxyribonucleic acid (DNA) and ribonucleic acid (RNA). A nucleotide monomer is composed of three components: sugar (either deoxyribose or ribose), a phosphate group, and a nitrogenous base (see section 2.7d). Recall that nucleic acids are not an essential nutrient, but like essential nutrients, they are digested by specific enzymes of the digestive system.

Nucleic Acid Breakdown in the Small Intestine

Nucleic acid digestion occurs in the small intestine. The nucleases (**deoxyribonuclease** and **ribonuclease**), synthesized and released by the pancreas, begin the digestion of nucleic acids. Each breaks the phosphodiester bond between the individual nucleotides of DNA and RNA, respectively.

The breakdown of the nucleotides is further accomplished by brush border enzymes embedded in the epithelial lining of the small intestine. These enzymes include (1) **phosphatase,** which breaks the bond holding the phosphate to the rest of the nucleotide (without the phosphate, this molecule is called a *nucleoside*); and (2) **nucleosidase,** which breaks the bond between the sugar and the nitrogenous base of the nucleoside, releasing the sugar and nitrogenous base.

All nucleic acid component building blocks are absorbed across the epithelium of the small intestine into the blood. These include phosphate, the sugar (ribose or deoxyribose), and the nitrogenous bases (thymine, adenine, guanine, cytosine, and uracil). Table 26.4 summarizes nucleic acid digestion. A summary of the digestive processes is presented in figure 26.29.

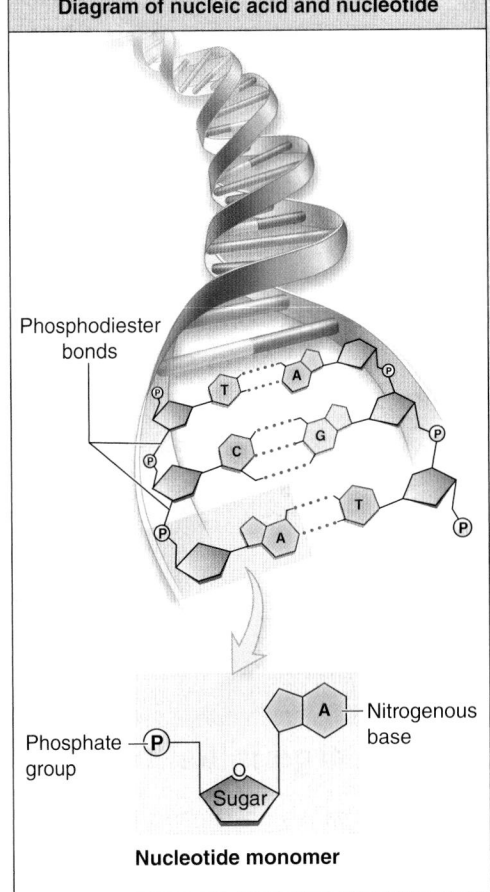

Diagram of nucleic acid and nucleotide

Phosphodiester bonds

Phosphate group

Nitrogenous base

Sugar

Nucleotide monomer

 WHAT DID YOU LEARN?

 28 Where does nucleic acid digestion occur?

INTEGRATE CONCEPT OVERVIEW

Figure 26.29 **Structures and Functions of the Digestive System.** Chemical digestion of nutrients occurs in the (*a*) oral cavity, (*b*) stomach, and (*c*) small intestine. Undigested substances are modified by bacteria within the *(d)* large intestine before being eliminated as feces. (Digestion of nucleic acids is not shown.)

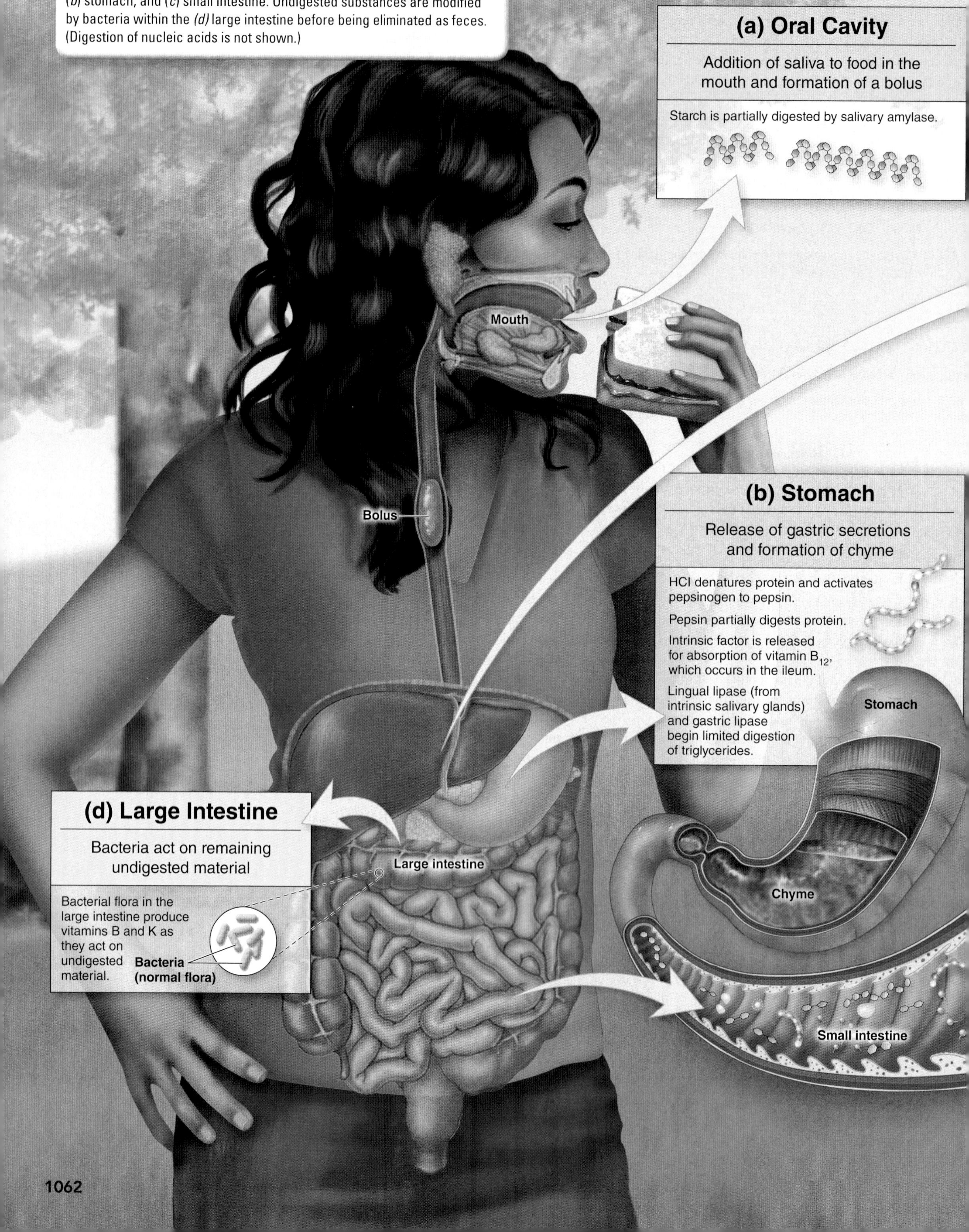

(a) Oral Cavity
Addition of saliva to food in the mouth and formation of a bolus
Starch is partially digested by salivary amylase.
Mouth
Bolus
(b) Stomach
Release of gastric secretions and formation of chyme
HCl denatures protein and activates pepsinogen to pepsin.
Pepsin partially digests protein.
Intrinsic factor is released for absorption of vitamin B12, which occurs in the ileum.
Lingual lipase (from intrinsic salivary glands) and gastric lipase begin limited digestion of triglycerides.
Stomach
Chyme
Small intestine
(d) Large Intestine
Bacteria act on remaining undigested material
Bacterial flora in the large intestine produce vitamins B and K as they act on undigested material.
Bacteria (normal flora)
Large intestine

(c) Small Intestine and Accessory Organs

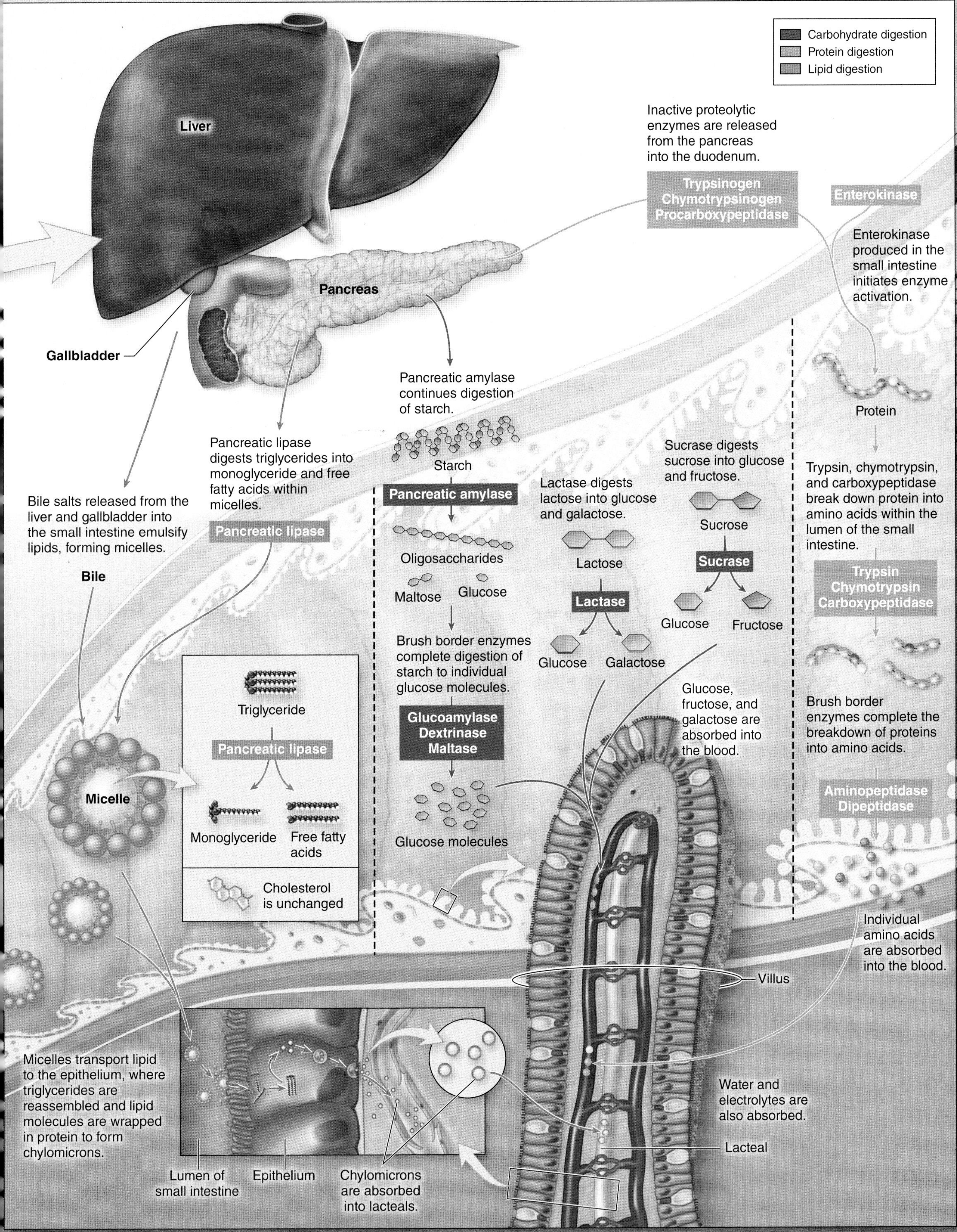

Legend:
- Carbohydrate digestion
- Protein digestion
- Lipid digestion

Liver

Pancreas

Gallbladder

Inactive proteolytic enzymes are released from the pancreas into the duodenum.

Trypsinogen
Chymotrypsinogen
Procarboxypeptidase

Enterokinase

Enterokinase produced in the small intestine initiates enzyme activation.

Bile salts released from the liver and gallbladder into the small intestine emulsify lipids, forming micelles.

Bile

Pancreatic lipase digests triglycerides into monoglyceride and free fatty acids within micelles.

Pancreatic lipase

Pancreatic amylase continues digestion of starch.

Starch

Pancreatic amylase

Lactase digests lactose into glucose and galactose.

Sucrase digests sucrose into glucose and fructose.

Sucrose

Sucrase

Protein

Trypsin, chymotrypsin, and carboxypeptidase break down protein into amino acids within the lumen of the small intestine.

Trypsin
Chymotrypsin
Carboxypeptidase

Oligosaccharides

Maltose   Glucose

Lactose

Lactase

Glucose   Fructose

Glucose   Galactose

Brush border enzymes complete digestion of starch to individual glucose molecules.

Glucoamylase
Dextrinase
Maltase

Glucose, fructose, and galactose are absorbed into the blood.

Brush border enzymes complete the breakdown of proteins into amino acids.

Aminopeptidase
Dipeptidase

Glucose molecules

Triglyceride

Pancreatic lipase

Monoglyceride   Free fatty acids

Cholesterol is unchanged

Micelle

Individual amino acids are absorbed into the blood.

Villus

Water and electrolytes are also absorbed.

Micelles transport lipid to the epithelium, where triglycerides are reassembled and lipid molecules are wrapped in protein to form chylomicrons.

Lacteal

Lumen of small intestine   Epithelium   Chylomicrons are absorbed into lacteals.

CHAPTER SUMMARY

- The digestive system provides the means to process, break down, and absorb nutrients needed for our survival.

26.1a General Functions of the Digestive System 1021

- The digestive system performs six major functions: ingestion, motility, secretion, digestion, absorption, and elimination.

26.1b Organization of the Digestive System 1021

- The gastrointestinal (GI) tract includes the oral cavity, pharynx, esophagus, stomach, small intestine, and large intestine.

- Accessory digestive organs are connected to the GI tract: salivary glands, liver, gallbladder, and pancreas. Teeth and tongue also assist in digestion activities.

26.1c Gastrointestinal Tract Wall 1021

- The GI tract tunics—from adjacent to the tube lumen to the outer wall—are the mucosa, submucosa, muscularis, and serosa or adventitia.

26.1d Serous Membranes of the Abdominal Cavity 1024

- Components of serous membranes are the parietal peritoneum (lining the inside wall of the body cavity) and visceral peritoneum (layer covering the organ); a peritoneal cavity is between them with a lubricating serous fluid. Organs enclosed within the parietal peritoneum are intraperitoneal; those posterior to the parietal peritoneum are retroperitoneal.

- Serous membranes suspend and stabilize some GI tract organs and include the greater omentum, lesser omentum, falciform ligament, mesentery proper, and mesocolon.

26.1e Regulation of Digestive System Processes 1025

- Major receptors throughout the GI tract include mechanoreceptors and chemoreceptors. Nervous control of digestive processes is associated with either long reflexes (involving the CNS) or short reflexes among neurons in the GI tract wall.

- Hormones that participate in regulating digestive processes include gastrin, which is released from the stomach, and secretin and cholecystokinin, which are released from the small intestine.

- Digestion begins in upper GI tract organs.

26.2a Overview of the Upper Gastrointestinal Tract Organs 1026

- Upper GI tract organs include the oral cavity, pharynx, esophagus, stomach, and the initial part of small intestine (duodenum).

26.2b Oral Cavity and Salivary Glands 1026

- The oral cavity is where salivary gland secretions are mixed with ingested food to form a bolus and where both mechanical digestion and chemical digestion of food begin.

- Saliva is produced by both intrinsic salivary glands and three sets of extrinsic salivary glands: parotid, submandibular, and sublingual.

26.2c Pharynx and Esophagus 1030

- The pharynx and esophagus connect the oral cavity to the stomach.

- Swallowing occurs in three phases: (1) voluntary phase, (2) pharyngeal phase, and (3) esophageal phase.

26.2d Stomach 1033

- Both mechanical and chemical digestion activities continue in the stomach.

- The tunica mucosa has a simple columnar epithelium, indentations (gastric pits) to increase its surface area, and deep penetrations (gastric glands) that house unique secretory cells.

- Gastric juice is composed of secretions from mucous cells (mucin) to protect the stomach lining. Parietal cells release intrinsic factor (to help absorb vitamin B_{12}) and hydrochloric acid. Chief cells release pepsinogen.

- G-cells in the stomach secrete gastrin, a hormone that stimulates both stomach motility and release of secretions.

- Pacemaker cells of the stomach control the rate of muscular contraction.

- Three phases are associated with stomach activity and include the cephalic phase, gastric phase, and intestinal phase.

- The lower GI tract continues the processes of digestion that are begun in the upper GI tract and is responsible for absorption of all nutrients.

26.3a Overview of the Lower Gastrointestinal Tract Organs 1040

- The lower GI tract organs include most of the small intestine (ileum and jejunum), large intestine, and accessory organs (liver, gallbladder, and pancreas), which release their secretions into the duodenum.

26.3b Small Intestine 1041

- The small intestine has three parts: duodenum, jejunum, and ileum.

- Small intestine motility has three primary functions: (1) mixing chyme with accessory gland secretions, (2) exposing the chyme to different regions of the brush border, and (3) moving small intestine contents toward the large intestine.

- The secretions from the cells of the small intestine include mucin, enteropeptidase, and hormones (cholecystokinin and secretin).

26.3c Accessory Organs and Ducts   1043

- Secretions from the liver, gallbladder, and pancreas enter the small intestine through a series of ducts that include the hepatic ducts, cystic ducts, and pancreatic ducts.

- The liver produces bile to facilitate mechanical digestion of triglycerides.

- The gallbladder stores and concentrates bile produced by the liver.

- Pancreatic cells (acinar cells) and pancreatic duct cells produce pancreatic juice, which is an alkaline "cocktail" of numerous digestive enzymes.

- Nervous system reflexes and endocrine secretions (gastrin, cholecystokinin, and secretin) control stomach activity, small intestine motility, and secretions from accessory organs.

26.3d Large Intestine   1050

- The function of the large intestine (which is composed of the cecum, colon, and rectum) is primarily the absorption of water and electrolytes and the formation of feces.

- Movements in the large intestine include peristalsis, haustral churning, and mass movements. Reflexes associated with the large intestine include the gastrocolic reflex, and the defecation reflex.

26.4 Nutrients and Their Digestion   1055

- The six essential nutrients are carbohydrates, proteins, lipids, minerals, vitamins, and water.

26.4a Carbohydrate Digestion   1055

- Only monosaccharides (e.g., glucose, fructose, and galactose) may be directly absorbed.

- Starch must first be digested into smaller units by amylase (from the oral cavity) and then by specific pancreatic and brush border enzymes into smaller structures.

- Disaccharides (lactose and sucrose) only require one brush border enzyme each to be digested.

26.4b Protein Digestion   1056

- All enzymes that digest protein are released in an inactive state to prevent digestion of our body proteins.

- Protein digestion begins in the stomach by pepsin, and continues in the small intestine by various enzymes released from the pancreas (trypsin, chymotrypsin, and carboxypeptidase) and then completed by brush border enzymes.

26.4c Lipid Digestion   1059

- Triglyceride digestion is facilitated by an interaction with bile salts that emulsify lipids to form micelles.

- Within micelles, pancreatic lipase digests triglyceride molecules into a monoglyceride and free fatty acids.

- Digested triglycerides are reassembled into triglycerides within epithelial cells and then along with other lipids are wrapped in proteins to form a chylomicron; chylomicrons then are absorbed into lacteals (lymphatic capillaries of the small intestine).

26.4d Nucleic Acid Digestion   1061

- Deoxyribonucleases and ribonucleases break apart DNA and RNA, respectively. Brush border enzymes break the bonds within the nucleotide, and the sugar, phosphate, and nitrogenous base components are subsequently absorbed into the blood.

CHALLENGE YOURSELF

Do You Know the Basics?

_____ 1. Which organ is located in the right upper quadrant of the abdomen?

    a. liver

    b. spleen

    c. descending colon

    d. appendix

_____ 2. The _____ cells of the stomach are responsible for the formation of hydrochloric acid (HCI).

    a. chief

    b. parietal

    c. mucous

    d. G-

_____ 3. Which of the following is an unregulated process in the digestive tract?

    a. secretion of cholecystokinin (CCK) by the duodenal mucosa

    b. absorption of amino acids across epithelium of the small intestine

    c. release of bicarbonate ion (HCO_3^-) by pancreatic duct cells

    d. peristalsis in the stomach

_____ 4. Which organ (or part of an organ) is retroperitoneal?

    a. stomach

    b. transverse colon

    c. descending colon

    d. ileum

_____ 5. Pancreatic juice contains

    a. HCO_3^- and digestive enzymes.

    b. bile.

    c. bile and digestive enzymes.

    d. gastrin hormone.

_____ 6. Bile is transported through the

    a. lacteals.

    b. accessory pancreatic duct.

    c. cystic duct.

    d. teniae coli.

_____ 7. Digestion of proteins begins in the

    a. oral cavity.

    b. stomach.

    c. small intestine.

    d. large intestine.

_____ 8. Micelles help promote

    a. denaturation of proteins.

    b. digestion and absorption of triglycerides.

    c. storage of digested carbohydrates.

    d. increased degradation of carbohydrates.

_____ 9. Digestive enzymes that chemically digest most of our ingested food are secreted by the

    a. liver.

    b. small intestine.

    c. pancreas.

    d. salivary glands.

_____ 10. Most of the absorption of our digested food occurs within the

    a. large intestine.

    b. pancreas.

    c. small intestine.

    d. esophagus.

11. The GI tract from the esophagus to the anal canal is composed of four tunics. Describe the general histology of the tunics and the specific features of the esophageal tunics.

12. Discuss the reason why the involuntary sequence of contraction in the pharyngeal phase of swallowing involves skeletal muscle.

13. Explain why intrinsic factor release is the only essential function of the stomach.

14. Compare the structure of the circular folds, villi, and microvilli (the three structures that increase the surface area of the small intestine).

15. Discuss why the tunica mucosa in the colon has a high percentage of goblet cells to produce mucin.

16. Why is it necessary for enzymes that digest protein to be secreted in an inactive form?

17. What is the role of the gallbladder in digestion?

18. Describe the different forms of mechanical digestion in the oral cavity, stomach, and small intestine.

19. Describe the chemical digestive processes that occur in the stomach.

20. How are lipids absorbed in the GI tract?

Can You Apply What You've Learned?

1. Maria was diagnosed with stomach cancer and had to have her stomach surgically removed. Following the removal of her stomach, Maria must obtain shots of which essential substance?

    a. pepsinogen

    b. B_{12}

    c. HCl

    d. gastrin

2. Harold had his gallbladder removed (cholecystectomy). What advice is given concerning his diet that directly relates to the removal of his gallbladder?

    a. Eat a diet low in protein.

    b. Avoid meals high in fat.

    c. Consume a diet low in carbohydrates.

    d. Eat large meals once a day.

3. What component of the digestive tract can you not live without because almost all absorption of nutrients occurs here?

    a. small intestine

    b. large intestine

    c. stomach

    d. gallbladder

4. The pancreatic ducts are blocked with a thick, sticky mucus in individuals with the genetic condition called cystic fibrosis. What difficulties will these individuals experience?

    a. inability to produce mucin

    b. inability to produce HCl

    c. insufficient digestive enzymes

    d. excess production of bile

5. Mark was rushed to the hospital when his appendix ruptured. This condition is life-threatening due to

    a. entrance of bacteria into the abdominal cavity.

    b. serous fluid entering the GI tract.

    c. leaking of intestinal enzymes into the abdominal cavity.

    d. leaking of HCl into the abdominal cavity.

Can You Synthesize What You've Learned?

1. Alexandra experienced vomiting and diarrhea and was diagnosed with gastroenteritis (stomach flu). What specific digestive system organs were affected by the illness?

2. A key event in the chemical digestion processes that occur within the stomach is the release of H^+ and Cl^- into the lumen of the stomach. Explain how the parietal cells of the gastric glands are able to produce HCl that is a million times more acidic than blood without destroying the cells themselves.

3. Most cases of colorectal cancer occur in the most distal part of the large intestine (the rectum, sigmoid colon, and descending colon). Why does the proximal part of the large intestine tend to have fewer instances of colon cancer? Include the anatomy and function of the colon structures in your explanation.

INTEGRATE

ONLINE STUDY TOOLS

The following study aids may be accessed through Connect.

Clinical Case Study: A Teenage Boy with Abdominal Pain and Weight Loss

Interactive Questions: This chapter's content is served up in a number of multimedia question formats for student study.

LearnSmart: Topics and terminology include introduction to the digestive system; upper gastrointestinal tract; lower gastrointestinal tract; nutrients and their digestion

Anatomy & Physiology Revealed: Topics include digestive system overview; oral cavity and pharynx; salivary glands; teeth; esophagus; stomach; HCl production; small intestine; duodenum; liver; biliary ducts; abdominal cavity; hydrolysis of sucrose

Animations: Topics include organs of digestion; digestive system overview; stomach; hydrochloric acid production of the stomach; hormones and gastric secretion; three phases of gastric secretions; liver; reflexes in the colon

chapter
27

Nutrition and Metabolism

INTEGRATE

CAREER PATH
Dietitians and Nutritionists

Dietitians and nutritionists are health-care professionals who are concerned with proper diet to maintain good health. They evaluate and assess the needs of individual patients and help develop appropriate nutritional programs. Often they recommend modifications of the diet to help prevent or treat illnesses. Dietitians manage and provide dietary services in nursing care facilities, schools, and hospitals.

Module 12: Digestive System

Many individuals express the desire to "eat healthy." Given the amount of media coverage regarding dietary intake, most of us are aware that eating healthy includes eating a diet with plenty of fruits, vegetables, and whole grains, which also includes small portions of protein, and is low in fat and sodium. By eating healthy we provide our bodies with the nutrients required for both the synthesis of new molecules and the energy for maintenance, growth, and repair, and we avoid foods that may cause us harm.

Here we provide an overview to nutrition. Our discussion includes a description of nutrients, various guidelines for obtaining adequate nutrition, and how the body regulates nutrient levels in the blood during and following a meal. We also examine the functions of the liver including its essential role in regulating nutrient blood levels, the central role of cellular respiration in nutrient metabolism, the production of heat from metabolic processes, and the regulation of body temperature.

27.1 Introduction to Nutrition

LEARNING OBJECTIVES

1. Define both nutrition and nutrients.
2. Distinguish macronutrients from micronutrients and essential from nonessential nutrients.
3. Explain the meaning of recommended daily allowance (RDA).

Nutrition is the study of the means by which living organisms both obtain and utilize the nutrients they need to grow and sustain life. **Nutrients** include most organic biomolecules (carbohydrates, lipids, proteins), vitamins, and minerals that the body needs for development and growth, maintaining anatomic structures and physiologic processes, and repair of damaged tissues. Water is also considered to be a nutrient because the body requires 2 to 3 liters of it daily. Water and its properties are described in section 2.4, and water balance is discussed in section 25.2.

Nutrients can be described in two ways: either as (1) macronutrients and micronutrients or (2) essential nutrients and nonessential nutrients. Macronutrients and micronutrients reflect the daily amounts that are required. **Macronutrients** must be consumed in relatively large quantities. All macronutrients are organic biomolecules (see section 2.7): They include carbohydrates, lipids, and proteins. **Micronutrients** must be consumed in relatively small quantities, and include both vitamins and minerals.

Nutrients are also organized into two categories, depending upon whether it is vital that you obtain them in your diet. **Essential nutrients** (see section 26.4) must be obtained and absorbed by the processes of the digestive system, and thus it is required (essential) that these nutrients be part of your dietary intake. **Nonessential nutrients** can be adequately provided by biochemical processes within the body, and for this reason they are not required to be part of your dietary intake.

Recommended Daily Allowance

Federal government agencies have established values for the amount of each nutrient that must be obtained each day called the **recommended daily allowances (RDAs)**. These were originally established by the Food and Nutrition Board (FNB) as directed by the National Academy of Sciences (NAS), and they are reviewed and updated periodically based on additional data from nutrition studies. RDA information may be accessed online at the USDA's food and nutrition information center (http://www.health.gov/dietaryguidelines/). These government-established values are used for food planning, food labeling, clinical dietetics, food programs, and educational programs on nutrition. Although RDA values are currently based on population studies, in the future these RDA levels could be established for each individual based on one's specific genetic makeup.

WHAT DID YOU LEARN?

1. List the six nutrients required by the body.
2. Distinguish between macronutrients and micronutrients.
3. What is meant by the RDA of nutrients?

27.2 Macronutrients

All macronutrients (carbohydrates, lipids, proteins) have a common function; they provide fuel for the processes of cellular respiration to form ATP (see section 3.4); thus, these molecules provide energy to the body. The amount of energy is measured in calories. A **calorie** is defined as the amount of heat required to raise the temperature of 1 gram of water by 1 degree Celsius. A kilocalorie is equal to 1000 calories or 1 Calorie (with a capital C). Body weight is maintained when there is balance between calories consumed in the diet and those expended through metabolic processes and physical activity. **Table 27.1** provides a description and the general functions for the macronutrients.

27.2a Carbohydrates

LEARNING OBJECTIVE

1. Identify the categories that are dietary sources of carbohydrates, and give examples of each category.

We described carbohydrates in section 2.7c. Recall that **carbohydrates** are structurally classified as **monosaccharides** (the monomer); **disaccharides,** which have two monosaccharides; and **polysaccharides** formed from chains of monosaccharides. However, when describing dietary sources of carbohydrates, three different categories are recognized, as follows:

- **Sugars.** These include both the monosaccharides glucose, fructose, and galactose and the disaccharides sucrose (table sugar, maple syrup, and fruits), lactose (milk sugar), and maltose (found in cereals). Other sugars include dextrose, brown sugar, honey, malt syrup, corn syrup, corn sweetener, high fructose corn syrup, invert sugar, molasses, raw sugar, turbinado sugar, and trehalose.
- **Starch.** This is a polysaccharide polymer of glucose molecules found within certain types of foods including tubers (e.g., potatoes, carrots), grains (e.g., wheat, barley, rice, corn), and beans and peas (kidney beans, garbanzo beans, lentils). Refined starches are sometimes added as thickeners and stabilizers. Cornstarch is an example of a refined starch.
- **Fiber.** It includes the fibrous molecules (e.g., cellulose) of both plants and animals that cannot be chemically digested and absorbed by the gastrointestinal (GI) tract. Sources of fiber include lentils, peas, beans, whole grains, oatmeal, berries, and nuts.

Table 27.1	Macronutrients and Their Functions	
Nutrient	**Description**	**Functions**
Carbohydrates	Carbohydrates are classified as monosaccharides, disaccharides, and polysaccharides	Glucose is broken down to release energy for life processes; glycogen is a polymer of glucose molecules, which are stored in the liver and skeletal muscle tissue
Lipids	Lipids are a diverse group of hydrophobic biomolecules that include triglycerides, steroids, phospholipids, and eicosanoids	Triglycerides serve as a form of energy storage (gram for gram, triglycerides provide more energy than carbohydrates); phospholipids compose cell membranes; steroids are components of plasma membranes, serve as hormones, and are precursors for synthesis of bile salts; eicosanoids are local hormones
Proteins	Proteins are polymers composed of amino acids	Proteins are both structural and functional components of the body; provide energy

Sugars and starch are usually converted to glucose, which is one of the primary nutrients supplying energy to cells. We have seen that each glucose molecule is oxidized during the process of cellular respiration to net approximately 30 ATP molecules (see section 3.4f). Stated in terms of calories, this is the equivalent of approximately 4 kilocalories of energy per gram of glucose. Interestingly, glucose is not considered an essential nutrient because it can be synthesized within hepatocytes from other monosaccharides (e.g., fructose, galactose) and noncarbohydrate molecules (e.g., amino acids, fatty acids) by the process of gluconeogenesis (see section 2.7c).

The fiber in our diet serves a different purpose. Fiber is neither chemically digested nor is it absorbed. Rather it remains within the lumen of the GI tract and adds bulk. This bulk stimulates peristalsis of the large intestine, which facilitates the movement of the GI tract contents and their ultimate elimination as a component of feces. In other words, fiber helps to "keep you regular." In addition, some types of fiber (water-soluble fiber including beta-glucan, psyllium, pectin, guar gum) decrease low-density lipoprotein (LDL) cholesterol levels (see section 27.6c). It is thought that water-soluble fiber interferes with the reabsorption of bile acids (which are formed from cholesterol) and the bile acids are then eliminated in the feces. Consequently, the liver must absorb additional cholesterol from the blood to replace the bile acids that have been lost.

WHAT DID YOU LEARN?

4 What is the general function of both sugars and starch in our diet?

5 How does dietary fiber differ from other dietary carbohydrates?

27.2b Lipids

LEARNING OBJECTIVES

2. Identify the types and dietary sources of triglycerides, and describe their functions.

3. Describe the sources and functions of cholesterol.

Lipids are a diverse group of biologically active, hydrophobic molecules that include triglycerides, phospholipids, steroids (including cholesterol), and eicosanoids (see section 2.7b). Here we describe both triglycerides and cholesterol.

Triglycerides (or fats) are composed of glycerol and fatty acids. Fatty acids are organized into three categories, which depend upon their degree of saturation (see figure 2.18):

- **Saturated fatty acids** have no double bonds (each carbon in the fatty acid chain is completely saturated with hydrogen atoms). Sources of saturated fats are generally solid at room temperature and dietary sources include the fat in meat, milk, cheese, coconut oil, and palm oil.

- **Unsaturated** (also called *monounsaturated*) **fatty acids** have one double bond. Unsaturated fats are typically liquid at room temperature. Dietary unsaturated fats include nuts and certain vegetable oils (e.g., canola oil, olive oil, sunflower oil).

- **Polyunsaturated fatty acids** have two or more double bonds. Sources of polyunsaturated fats are also liquid at room temperature and dietary sources include some vegetable oils (e.g., soybean oil, corn oil, safflower oil).

Triglycerides are also a primary nutrient supplying energy to cells. Oxidation of triglyceride molecules yields approximately 9 kilocalories of energy per gram of fat—more than twice that of glucose. Fats are also necessary for the absorption of fat-soluble vitamins (vitamins A, D, E, and K).

In general, fat molecules and the fatty acids that compose them can be synthesized by the body. The only two fatty acids that the body needs, but cannot normally synthesize, are alpha-linoleic acid and linoleic acid. However, other fatty acids (e.g., docosahexaenoic acid) may be required in the diet in certain disease conditions.

Cholesterol is required as a component of the plasma membrane of our cells. It is also the precursor molecule for the formation of steroid hormones, bile salts, and vitamin D. Cholesterol is made available either through our diet (a component of animal-based products, including meat, eggs, and milk) or is synthesized by metabolic pathways within the liver. Both cholesterol synthesis and transport of lipids are described in section 27.6.

WHAT DID YOU LEARN?

6 What are two primary functions of fats?

INTEGRATE

CLINICAL VIEW

High Fructose Corn Syrup

The manufacture of **high fructose corn syrup** begins with cornstarch that is liquefied, broken down into glucose, and some of the glucose is converted to fructose. High fructose corn syrups vary in the relative amounts of glucose and fructose they contain.

Individuals in the United States have a much higher intake of high fructose corn syrup today as compared to 30 years ago. The reason is that many manufacturers of food products now sweeten their products with high fructose corn syrup, which is sweeter and typically less expensive than sucrose. Many processed foods contain some high fructose corn syrup, including soft drinks, cereals, breads, soups, breakfast bars, yogurt, lunch meats, cookies, and condiments (e.g., ketchup, mustard).

The intake of large amounts of high fructose corn syrup, and its potential effects on an individual's health, remains controversial. On the one hand, some claim that high fructose corn syrup leads to higher blood levels of triglycerides, is partially responsible for the obesity epidemic in the United States, and exacerbates the condition of diabetes mellitus. On the other hand, the manufacturers of high fructose corn syrup claim high fructose corn syrup is natural and is fine in moderation. Research continues to investigate whether health issues are associated with a large intake of this sweetener.

27.2c Proteins

LEARNING OBJECTIVES

4. Describe why protein is required in our diet and the general amount that is needed.

5. Explain the difference between a complete protein and an incomplete protein.

6. Discuss nitrogen balance and include the difference between a positive and negative nitrogen balance.

Proteins are the most structurally and functionally diverse molecules. (There are an estimated 50,000 to 100,000 different proteins in the body; see table 2.6.) Proteins are synthesized from 20 different amino acids (see figure 2.25). Adequate dietary intake of protein provides the amino acids to synthesize new proteins to replace body protein structures. The specific amount of protein required in the diet is generally dependent upon one's age and sex, but ranges between approximately 45 and 60 grams (about 1/10 pound) daily. Required protein is increased when fighting an infection, following an injury, under conditions of stress, and during pregnancy. It is also higher in infants and children to synthesize protein molecules needed for growth.

Eight of the twenty amino acids (isoleucine, leucine, lysine, methionine, phenylalanine, threonine, tryptophan, and valine) are essential in adults and therefore must be obtained from the diet. The other twelve can be synthesized within the body. (Histidine is also essential in infants because they are not able to synthesize it.) Dietary protein sources vary in whether they provide all of the essential amino acids, and are identified as either complete proteins or incomplete proteins. **Complete proteins** contain all of the essential amino acids, whereas **incomplete proteins** do not. Generally, animal proteins (meats, poultry, fish, eggs, milk, cheese, yogurt) are complete proteins, and plant proteins (legumes, vegetables, grains) tend to be lacking in one or more of the essential amino acids and thus they are incomplete proteins. For example, the protein in corn has low amounts of both lysine and tryptophan. Combinations of dishes containing plant proteins (such as legumes and grains) can provide all of the essential amino acids (e.g., beans and corn, soybeans and rice, red beans and rice).

Note that there is no storage of excess amino acids. Consequently, all of the amino acids required for protein synthesis must be supplied regularly either through animal proteins or through plant protein combinations. A **vegetarian** is an individual who does not eat meat, fish, or poultry. **Lacto-ovo vegetarians** do not eat animal flesh, but do eat milk, eggs, and cheese. **Vegans** do not eat any animal products. A concern for vegetarians is that plant-based protein sources are often individually incomplete. These individuals must be sure to obtain the essential amino acids from a variety of complementary protein sources. These balanced food combinations do not need to be eaten in the same meal, but when eaten regularly they may supply the necessary essential amino acids.

Nitrogen Balance   Proteins are also a source of **nitrogen** (recall that each amino acid contains a —NH_2 functional group; see section 2.7e). Nitrogen is a chemical element that is needed for synthesizing other nitrogen-containing molecules, such as nitrogenous bases of nucleic acids (DNA, RNA), and porphyrin (a component of heme in hemoglobin). Adequate availability of nitrogen is expressed as nitrogen balance. **Nitrogen balance** occurs when equilibrium exists between its dietary intake and its loss in urine and feces. Input of nitrogen must equal output of nitrogen to maintain a nitrogen balance. A **positive nitrogen balance** occurs when an individual absorbs more nitrogen than is excreted, such as occurs in individuals who are growing, pregnant, or recovering from injury. A **negative nitrogen balance** occurs when more nitrogen is excreted than is absorbed; this is a condition that might result from malnutrition

or blood loss, for example. An individual in sustained negative nitrogen balance has insufficient nitrogen to synthesize nitrogen-containing molecules, and a decrease of as little as a few hundred grams of total nitrogen may be fatal.

WHAT DID YOU LEARN?

⑦ Explain the general difference in animal proteins and plant proteins in terms of obtaining essential amino acids.

⑧ How may a vegetarian obtain all of the essential amino acids?

27.3 Micronutrients

Micronutrients, which include both vitamins and minerals, are primarily obtained in the foods that we eat, and plants tend to be very good dietary sources. Each type of food may contain a variety of vitamins and minerals, but no one food contains all that are required. Consequently, it is necessary to have variety in the diet to meet the body's needs. Some individuals may also obtain vitamins and minerals through supplements.

27.3a Vitamins

LEARNING OBJECTIVES

1. Distinguish between water-soluble and fat-soluble vitamins.

2. List examples of how both water-soluble vitamins and fat-soluble vitamins function in the body.

3. Describe the difference between essential and nonessential vitamins.

Vitamins are organic molecules required for normal metabolism, yet they are present in only small amounts in food. Vitamins are categorized in two ways: (1) water-soluble vitamins or fat-soluble vitamins and (2) essential vitamins or nonessential vitamins. **Table 27.2** lists the water-soluble and fat-soluble vitamins and their sources, the recommended daily allowance (RDA), and deficiency symptoms.

WHAT DO YOU THINK?

① Predict what happens to the level of fat-soluble vitamins stored in adipose connective tissue as a result of taking megadoses of vitamins.

Water-Soluble and Fat-Soluble Vitamins

Water-soluble vitamins dissolve in water: They include both the B vitamins and vitamin C. These vitamins are easily absorbed into the blood from the digestive tract. If dietary intake of water-soluble vitamins exceeds what is needed by the body, the excess is excreted into the urine. There are several different types of B vitamins, each of which is designated with a number and with a name (e.g., B_1 is thiamine; B_2 is riboflavin). B vitamins serve as coenzymes in various enzymatic chemical reactions. For example, vitamin B_3, also called niacin, is a necessary hydrogen carrier in mitochondria during adenosine triphosphate (ATP) synthesis (see section 3.4).

Vitamin C (or ascorbic acid) is required for the synthesis of collagen, which is an important protein in connective tissue. This vitamin, along with vitamins A and K, functions as an antioxidant by removing free radicals (damaging chemical structures that contain unpaired electrons).

Fat-soluble vitamins dissolve in fat (not in water) and include vitamins A, D, E, and K. They are absorbed from the gastrointestinal tract within the lipid of micelles and ultimately enter into the lymphatic capillaries (lacteals) (see section 26.4c). If dietary intake of fat-soluble vitamins exceeds body requirements (generally caused by excessive intake through supplements), the excess is stored within the body fat and may reach toxic levels (a condition termed *hypervitaminosis*).

Table 27.2

Vitamins Required by Adults

Vitamin	Source	RDA (mg)	Deficiency Symptoms
WATER-SOLUBLE VITAMINS			
B-complex vitamins			
Vitamin B_1 (thiamine)	Meat, nuts, whole grains, eggs, seeds, legumes	1.1–1.2	Weakened heart, beriberi, edema
Vitamin B_2 (riboflavin)	Meat, whole grains, eggs, greens	1.1–1.3	Ariboflavinosis—bright pink tongue, cracked lips, throat swelling, bloodshot eyes, anemia
Vitamin B_3 (niacin)	Meat, liver, whole grains	14–16	Nerve inflammation, pellagra
Vitamin B_5 (pantothenic acid)	Meat, eggs, whole grains, legumes		Fatigue, decreased coordination; paraesthesia—prickling and burning sensations in hands and feet
Vitamin B_6 (pyridoxine)	Meat, whole grains, legumes	1.3–1.7	Anemia, convulsions
Vitamin B_{12} (cyanocobalamin)	Dairy products, meat, eggs	2.4	Pernicious anemia; hypoalbuminemia; damage to nervous tissue
Biotin (vitamin B_7)	Meat, eggs, cheese, legumes, nuts	0.03–0.10	Skin rash, hair loss, anemia, mental conditions including hallucinations
Folate (folic acid; B_9)	Many green vegetables, liver, eggs, whole grains, seeds, legumes	400	Risk of neural tube defects to unborn child if mother is deficient during pregnancy
Vitamin C	Leafy green vegetables, fruit, citrus, broccoli, tomatoes, cabbage	75–90	Scurvy—skin spots, bleeding gums, loss of teeth, fever, lethargy, and death
FAT-SOLUBLE VITAMINS			
Vitamin A (retinol)	Liver, milk products, green vegetables, fish oil, eggs, margarine	700–900	Night blindness, dry flaky skin, child blindness
Vitamin D (calciferol)	Cod liver oil, dairy products; UV rays (when synthesized by keratinocytes of epidermis)	15	Rickets, bone abnormalities
Vitamin E (tocopherol)	Leafy green vegetables, seeds, margarine, butter, fish oil, wheat germ	15	Rare
Vitamin K	Leafy green vegetables, liver (synthesized by bacteria in colon)	90–120	Abnormal bleeding

The functions of fat-soluble vitamins vary.

- Vitamin A (retinol) is a precursor molecule for the formation of the visual pigment retinal (see section 16.4d).
- Vitamin D (calciferol) is modified to form calcitriol: This is a hormone that increases calcium absorption from the gastrointestinal tract (see section 7.6).
- Vitamin E (tocopherol) helps stabilize and prevent damage to cell membranes that are described in section 4.2.
- Vitamin K is required for synthesis of specific blood clotting proteins (see section 18.4c).

Essential and Nonessential Vitamins

Vitamins are also categorized based on their requirement in the diet. Essential vitamins must be provided in the diet. If an essential vitamin is lacking in our intake, or its absorption is impaired, then a vitamin deficiency disease results (*avitaminosis*). The vitamins in table 27.2 are all essential vitamins.

Nonessential vitamins are cofactors that the body is able to produce and recycle as needed. Examples are the NADH and $FADH_2$ that occur in the citric acid cycle (see section 3.4d).

 WHAT DID YOU LEARN?

 9 Which vitamins are water-soluble? Which are fat-soluble? Which of these groups may be dangerous in excess?

27.3b Minerals

LEARNING OBJECTIVES

4. Define minerals, and list examples of how minerals absorbed in the small intestine function in the body.

5. Distinguish between major minerals and trace minerals.

Minerals are inorganic ions such as iron, calcium, sodium, potassium, iodine, zinc, magnesium, and phosphorus. Minerals have diverse functions in the body. For example:

- **Iron** is present both in hemoglobin within erythrocytes, where it binds oxygen (see section 18.3b), and within the mitochondria in the electron transport system to bind electrons (see section 3.4e).
- **Calcium** is required for formation and maintenance of the skeleton (see section 7.2e), muscle contraction (see section 10.3c), exocytosis of neurotransmitters (see section 12.8d), and blood clotting (see section 18.4c).
- **Sodium** and **potassium** function to maintain a resting membrane potential in excitable cells, and are required in the generation of an action potential (see sections 10.3b and 12.8c).
- **Iodine** is needed to produce thyroid hormone (see section 17.8b).
- **Zinc** has roles in both protein synthesis (see section 4.8) and wound healing (see section 6.3).

All minerals are essential and must be obtained from the diet. We have a daily requirement for all minerals, and the amount required places minerals into one of two categories: **major minerals,** which are needed at levels greater than 100 milligrams (mg) per day, and **trace minerals,** which are required at less than 100 mg per day. Major minerals include calcium, chloride, magnesium, phosphorus, potassium, sodium, and sulfur; trace minerals include chromium, cobalt, copper, fluoride, iodine, iron, manganese, molybdenum, selenium, and zinc.

Minerals absorbed from the GI tract are stored to varying degrees within the body. These reserves are "hedges" against unexpected decreases in availability; however, if stored minerals are depleted, clinical problems may occur. **Table 27.3** lists the minerals, their sources, recommended daily allowance, and deficiency symptoms.

Table 27.3	Minerals Required by Adults		
Mineral	**Source**	**RDA (mg)**	**Deficiency Symptoms**
MAJOR MINERALS			
Calcium	Green vegetables, dairy products	1000–1300	Loss of bone mass, muscle weakness, depressed nerve activity
Chloride	Cured meat, table salt (NaCl)	2399	Muscle cramps
Magnesium	Green leafy vegetables, grains (whole)	310–420	Muscle weakness, nervous system disturbances
Phosphorus	Meat, nuts, dairy products, grains	700	Loss of appetite; anxiety, fatigue
Potassium	Meat, fruits, grains, vegetables, dairy products, salt substitutes	4700	Muscle weakness, abnormal cardiac function
Sodium	Table salt (NaCl), canned foods	< 2300	Neuromuscular disturbances, hypertension
Sulfur	Kale, cauliflower, cabbage	Unknown	Symptoms associated with protein deficiency
TRACE MINERALS			
Chromium	Whole grains, meat, cereals	0.05–0.25	Impaired glucose tolerance associated with insulin resistance
Cobalt	Fish, green vegetables, nuts	Unknown	Anemia
Copper	Seafood, legumes, liver, nuts	900	Anemia
Fluoride	Fluorinated water, fish, tea	1.5–4.0	Increased risk of dental caries (cavities)
Iodine	Iodized table salt	0.15	Goiter
Iron	Red meat, poultry, fish, dark green leafy vegetables, nuts, whole grains	8–18	Anemia, weakness, fatigue, pale skin, headaches, sensitivity to cold
Manganese	Vegetables, nuts, tea, fruits	2.5–5.0	Behavioral changes, abnormal bone formation
Molybdenum	Whole grains, nuts, beans	0.005–0.25	Rare
Selenium	Cereals, nuts, meat	55	Rare
Zinc	Meat, seeds	8–11	Changes in skin, eyes; hair loss; diarrhea

INTEGRATE

CLINICAL VIEW

Iron Deficiency

Iron (Fe) has several critical functions within the body: (1) It is required in the structure of hemoglobin in erythrocytes and myoglobin in muscle cells; (2) Iron is a component of specific electron transport proteins (cytochromes) within mitochondria; (3) It is required for some enzymes in the synthesis of certain hormones, neurotransmitters, and amino acids.

Iron is obtained in the diet from diverse sources including red meat, poultry, fish, dark green leafy vegetables, nuts, and whole grains. Although it is an essential mineral, iron is toxic in its free or unattached form. The toxic effects of iron are avoided by having it bound to specific proteins. **Mucosal ferritin** binds iron within epithelial cells in the small intestine lining following its absorption from the gastrointestinal (GI) tract. When iron is needed in the body, mucosal ferritin transfers iron to **mucosal transferrin** (of epithelial cells of the small intestine) that then transfers it to blood transferrin. **Blood transferrin** (a plasma protein) delivers iron to tissues throughout the body, including red bone marrow for hemoglobin synthesis, or to the liver for storage. Within the liver, iron is bound to **ferritin**.

Iron deficiency is one of the most common nutritional deficiencies. Symptoms of iron deficiency typically include fatigue, weakness, pale skin, headaches, and sensitivity to cold temperatures. It results from insufficient intake of iron based on loss of iron from the body. Iron is obtained in the diet. How much is obtained is dependent upon the quantities present in the consumed food and the amount absorbed from the GI tract. Heme iron (iron obtained in eating animal flesh of meats, poultry, and fish) is absorbed much more readily than nonheme iron (iron not associated with heme, which is obtained by eating iron-rich plants such as spinach and beans). Absorption may be altered by other food substances that are also present in the GI tract. For example, vitamin C increases the absorption of iron, whereas tea, which contains tannic acid, interferes with absorption of iron. In fact, high intake of tea may lead to iron deficiency. Changes to the intestinal lining that accompany certain diseases, such as Crohn disease (an inflammatory disease) or celiac sprue (an autoimmune disorder) also interfere with iron absorption.

Iron is lost in feces (from the cells of the intestinal lining), urine, and sweat. Excessive loss of iron from bleeding (e.g., heavy or frequent menses, internal bleeding, trauma, blood donation) also may lead to an iron deficiency. Women are more prone to iron deficiencies because of the loss of blood with each menses. Pregnant women and women that are breast-feeding have an even higher risk because of the extra demands for iron placed on them.

Note that some foods are **fortified** by adding one or more essential nutrients, which may or may not be normally present in the food to increase its nutritional value. **Fortification** may include, for example, vitamins (e.g., vitamin D in milk to increase calcium absorption, folic acid [B_9] in flour to prevent neural tube defects), or minerals (e.g., iodine in salt to prevent goiter).

 WHAT DID YOU LEARN?

10 What are the major minerals? The trace minerals? How are these two groups distinguished?

27.4 Guidelines for Adequate Nutrition

LEARNING OBJECTIVES

1. Describe MyPlate, which was developed by the USDA to help people eat healthy.

2. Identify the items that are included on a food label.

The amount of each type of nutrient biomolecule, vitamin, and mineral required by the body is not the same for all individuals; rather, it is dependent upon a number of factors, including an individual's age, sex, body mass, level of physical activity, health, and whether an individual is pregnant. A number of guidelines (recommended daily allowance, the USDA MyPlate, and required nutritional labels on packaged foods) have been provided or are required by scientific governmental agencies in the United States.

MyPlate

Not only may malnutrition be avoided, but an individual's state of health may be maximized when all of the nutrients are obtained at appropriate levels in the diet. Previously, food groups were organized by the United States Department of Agriculture (USDA) into a food guide pyramid that consisted of breads/grains, vegetables, fruits, dairy, proteins, and fats/oils/sweets. This organization provided a general approach for the number of servings that should be consumed daily from each category.

In 2011, the USDA revised the pyramid to **MyPlate** (figure 27.1). This visual of a divided plate helps us to see at a glance the relative proportions of the types of foods we need to stay healthy. One half of the plate is vegetables and fruits and the other half is protein and grains, with dairy off to the side. Examples of each category of food include the following:

- **Vegetables**—broccoli, lettuce, peppers, and carrots
- **Fruits**—apples, blueberries, bananas, and oranges

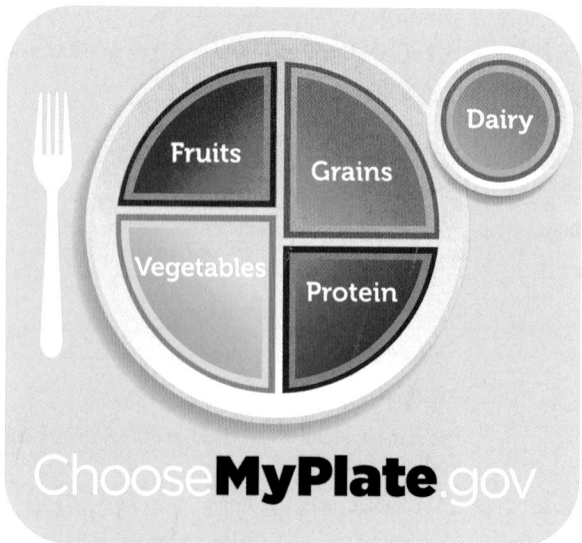

Figure 27.1 MyPlate. MyPlate shows the relative proportions of fruits, vegetables, grains, protein, and dairy for eating healthy.

- **Protein**—poultry, fish, beans, eggs, and nuts
- **Grains**—bread, cereals, rice, and pasta
- **Dairy**—skim milk, yogurt, and cheese

You will notice several additional recommendations listed if you go to the USDA website (www.choosemyplate.gov). These include the suggestion to balance calories by eating less, switch from whole milk to low-fat or no-fat milk, reduce the intake of foods that are high in salt (e.g., canned soup), and cut back on sugary drinks by drinking water instead.

Nutritional Food Labels

Specific details on the composition of prepackaged items are made available on food labels **(figure 27.2)**. This information is helpful for individuals who are (1) interested in eating a healthy diet, (2) meal planning for weight-loss programs, and (3) restricting intake of nutrients such as sugar or sodium. These labels are required by law for most (but not all) products, and it is mandatory that the following information be included, as follows:

- Servings per container and calories per serving, which allows you to determine if there is more than one serving per container and how many calories are being consumed
- Total fat and the different types of fat (e.g., unsaturated fat, saturated fat, trans fat) and cholesterol
- Carbohydrates including grams of dietary fiber
- Protein
- Vitamins
- Some minerals (e.g., sodium)

The label also provides both the Percent Daily Value, which is based on a diet of 2000 or 2500 calories, and the product ingredients. Product ingredients are listed in order of product weight—those having the greatest weight listed first. The list of product ingredients helps us to determine if the product contains healthy ingredients (e.g., 100% whole grains), or unhealthy ingredients. Unhealthy ingredients include both relatively large amounts of sugar (which, recall from section 27.2a, comes in many forms) and trans fats, which are produced during the process of hydrogenation of unsaturated fats to saturated fats. Product ingredients are also helpful to individuals with allergies (e.g., peanut allergies) to make sure that they avoid these products.

Several changes to nutrition labels will be required by the Food and Drug Administration (FDA), including information to reflect the actual serving size that an individual would eat—not what they should eat—how much sugar is added, and the amount of potassium and vitamin D that the food contains. Both the original food label that was introduced over 20 years ago and the recently revised nutrition label are shown in figure 27.2.

Foods that do not typically require food labels include (1) those prepared on site (e.g., bakery, deli, candy store) or (2) those with limited nutrients (e.g., coffee, tea, some spices). In addition, some businesses are exempt from having to provide food labels because of either their small size or low annual sales income. In addition, dietary supplements (e.g., multivitamins) are required to have a supplemental label instead of a food label.

WHAT DID YOU LEARN?

11. What categories of food are shown on the USDA MyPlate? Which food category takes up the largest portion of the plate?

12. What is the purpose of the requirement for nutritional labeling on packaged foods?

Nutrition Facts
Serving Size 2/3 cup (55g)
Servings Per Container About 8

Amount Per Serving	
Calories 230	Calories from Fat 72

	% Daily Value*
Total Fat 8g	**12%**
Saturated Fat 1g	**5%**
Trans Fat 0g	
Cholesterol 0mg	**0%**
Sodium 160mg	**7%**
Total Carbohydrate 37g	**12%**
Dietary Fiber 4g	**16%**
Sugars 1g	
Protein 3g	

Vitamin A	10%
Vitamin C	8%
Calcium	20%
Iron	45%

* Percent Daily Values are based on a 2,000 calorie diet. Your daily value may be higher or lower depending on your calorie needs

	Calories:	2,000	2,500
Total Fat	Less than	65g	80g
Sat Fat	Less than	20g	25g
Cholesterol	Less than	300mg	300mg
Sodium	Less than	2,400mg	2,400mg
Total Carbohydrate		300g	375g
Dietary Fiber		25g	30g

(a) Current nutrition label

Nutrition Facts
8 servings per container
Serving size 2/3 cup (55g)

Amount per 2/3 cup
Calories 230

% DV*	
12%	**Total Fat** 8g
5%	Saturated Fat 1g
	Trans Fat 0g
0%	**Cholesterol** 0mg
7%	**Sodium** 160mg
12%	**Total Carbs** 37g
14%	Dietary Fiber 4g
	Sugars 1g
	Added Sugars 0g
	Protein 3g
10%	Vitamin D 2mcg
20%	Calcium 260mcg
45%	Iron 8mg
5%	Potassium 235mg

* Footnote on Daily Values (DV) and calories reference to be inserted here.

(b) Proposed nutrition label

Figure 27.2 Food Label. A food label is included on prepackaged foods and provides serving size, calories, and nutrient information. (*a*) A current nutrition label introduced over 20 years ago and (*b*) a recently proposed nutrition label are shown.

27.5 Regulating Blood Levels of Nutrients

Recall that blood is composed of plasma and formed elements. Plasma contains water, proteins, electrolytes, nutrients, respiratory gases, and waste products (see section 18.2). The percentage of these substances is regulated by several processes, many of which have been described in the preceding chapters. Here we limit our discussion to regulation of levels of nutrients as they vary during the day relative to the intake of meals. Generally, regulation of blood nutrients is dependent upon the length of time since you have eaten. These conditions are referred to as the absorptive state and the postabsorptive state.

27.5a Absorptive State

LEARNING OBJECTIVE

1. Explain when the absorptive state occurs and how nutrient levels are regulated during this time.

The **absorptive state** includes the time you are eating, digesting, and absorbing nutrients. It usually lasts approximately 4 hours after a given meal. If you eat three meals spread throughout the day, you typically spend about 12 hours daily in the absorptive state. During the absorptive state, the concentrations of glucose, triglycerides, and amino acids are increasing within the blood as they are absorbed from the GI tract **(figure 27.3)**. The body's challenge is to maintain

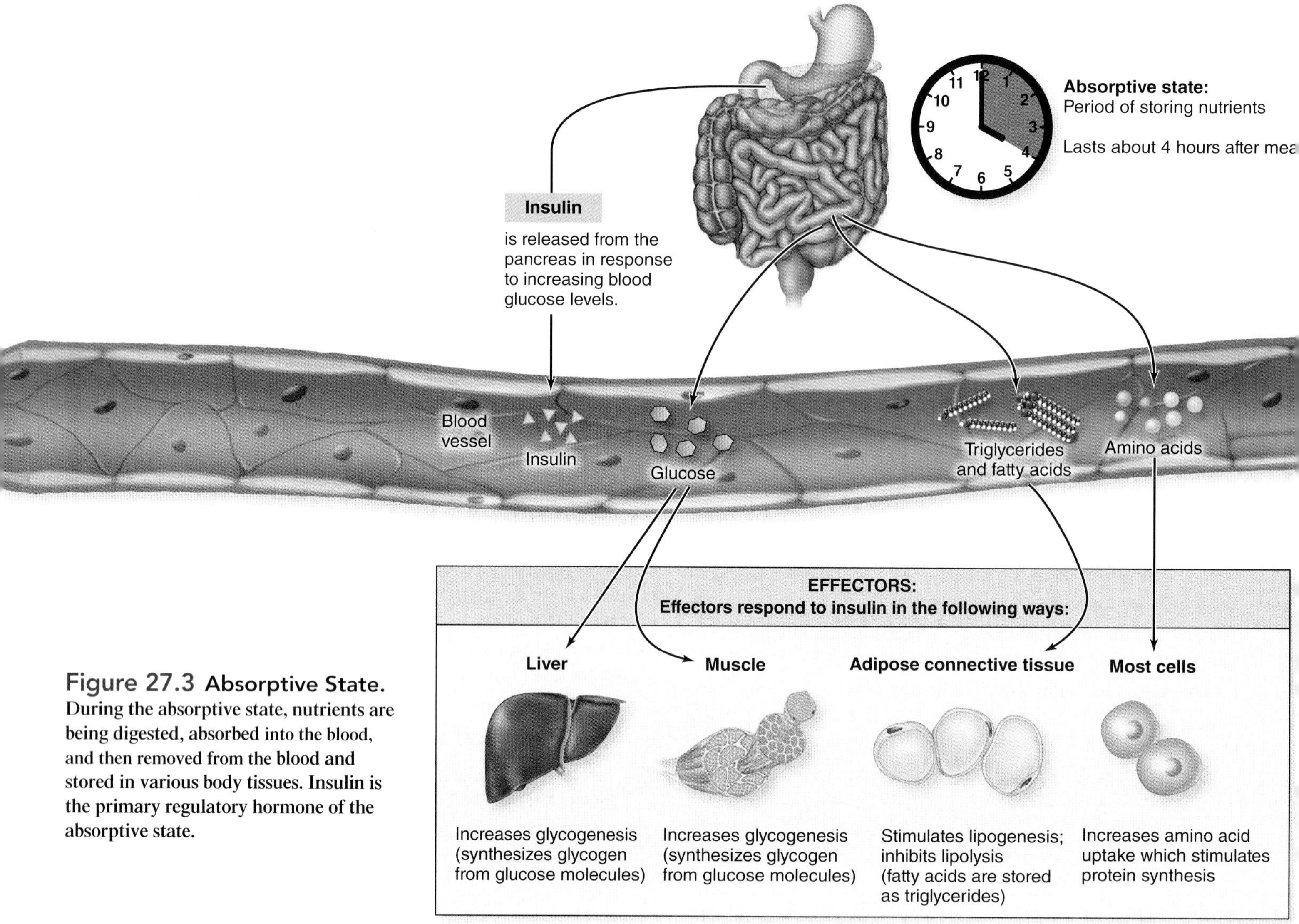

Insulin

is released from the pancreas in response to increasing blood glucose levels.

Absorptive state:
Period of storing nutrients

Lasts about 4 hours after mea[l]

Blood vessel

Insulin

Glucose

Triglycerides and fatty acids

Amino acids

EFFECTORS:
Effectors respond to insulin in the following ways:

Liver	Muscle	Adipose connective tissue	Most cells
Increases glycogenesis (synthesizes glycogen from glucose molecules)	Increases glycogenesis (synthesizes glycogen from glucose molecules)	Stimulates lipogenesis; inhibits lipolysis (fatty acids are stored as triglycerides)	Increases amino acid uptake which stimulates protein synthesis

Figure 27.3 Absorptive State.
During the absorptive state, nutrients are being digested, absorbed into the blood, and then removed from the blood and stored in various body tissues. Insulin is the primary regulatory hormone of the absorptive state.

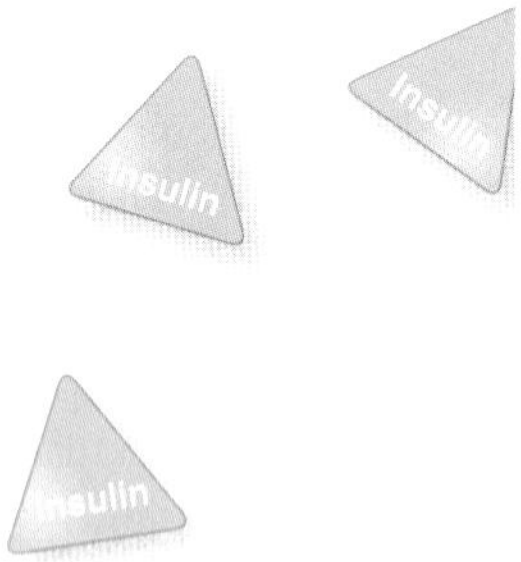

homeostatic levels of these nutrients during this time—primarily while maintaining homeostatic blood glucose levels to within the range of 70 to 110 milligrams per deciliter (mg/dL).

Insulin (discussed in detail in section 17.9b and summarized in table R.1 Regulating Blood Glucose with Pancreatic Hormones) is the major regulatory hormone that is released during the absorptive state. Its release from the pancreas occurs in response to an increase in blood glucose levels. A few of the responses initiated by insulin that alter blood nutrient levels are listed here:

- Stimulates both liver cells and muscle cells to form the polysaccharide glycogen from glucose by increasing glycogenesis
- Causes adipose connective tissue to increase uptake of triglycerides from the blood and decreases the breakdown of triglycerides by stimulating lipogenesis and inhibiting lipolysis
- Stimulates most cells (especially muscle cells) to increase amino acid uptake that causes an accelerated rate in protein synthesis

Consequently, the release of insulin results in a decrease in all energy-releasing molecules (glucose, triglycerides, and amino acids) in the blood, an increase in the storage of glycogen and triglycerides, and the formation of protein within body tissues. Note that excess amino acids are not "stored" as protein. The amino acids become structural and functional components of the body.

WHAT DID YOU LEARN?

 How does the storage of nutrient molecules change during the absorptive state?

INTEGRATE

CONCEPT CONNECTION

Insulin and glucagon, as well as cortisol, thyroid hormone, and growth hormone participate in regulating nutrient levels in the blood. These five hormones are described in detail in the discussion and examination of the endocrine system (see sections 17.8 and 17.9).

27.5b Postabsorptive State

⚕ LEARNING OBJECTIVE

2. Explain when the postabsorptive state occurs, and how nutrient levels are regulated during this time.

The **postabsorptive state** is the time between meals when the body relies on its stores of nutrients because no further absorption of nutrients is occurring **(figure 27.4)**. Assuming that an individual eats three meals spread out through the day, and spends 12 hours in the absorptive state, the other 12 hours are spent in the postabsorptive state. The challenge is to maintain homeostatic levels of many nutrients (e.g., monosaccharides, triglycerides, and amino acids) as these substances are decreasing in the blood.

Glucagon (discussed in detail in section 17.9b and summarized in table R.1 Regulating Blood Glucose with Pancreatic Hormones) is the major regulatory hormone that is released during the postabsorptive state. The pancreas releases glucagon in response to decreasing blood glucose levels. Glucagon has several effects, including the following:

- Stimulates liver cells to engage in catabolism of glycogen to glucose by increasing glycogenolysis. Glucagon may also increase the formation of glucose from noncarbohydrate sources by stimulating gluconeogenesis
- Causes adipose connective tissue to break down triglycerides to glycerol and fatty acids by stimulating lipolysis

Glucose is released from the liver, and fatty acids (and glycerol) are released from fat storage in response to glucagon stimulation. The levels of these molecules increase in the blood.

There is no storage form of either amino acids or proteins in cells; thus, glucagon has no effect on body proteins. The physiologic significance of this becomes apparent when you realize that a hormone released on a regular basis (i.e., during periods between meals) is not tearing down the structural and functional protein components of the body to maintain blood glucose levels.

💡 WHAT DID YOU LEARN?

14 What is the major regulatory hormone in the postabsorptive state, and what are its functions?

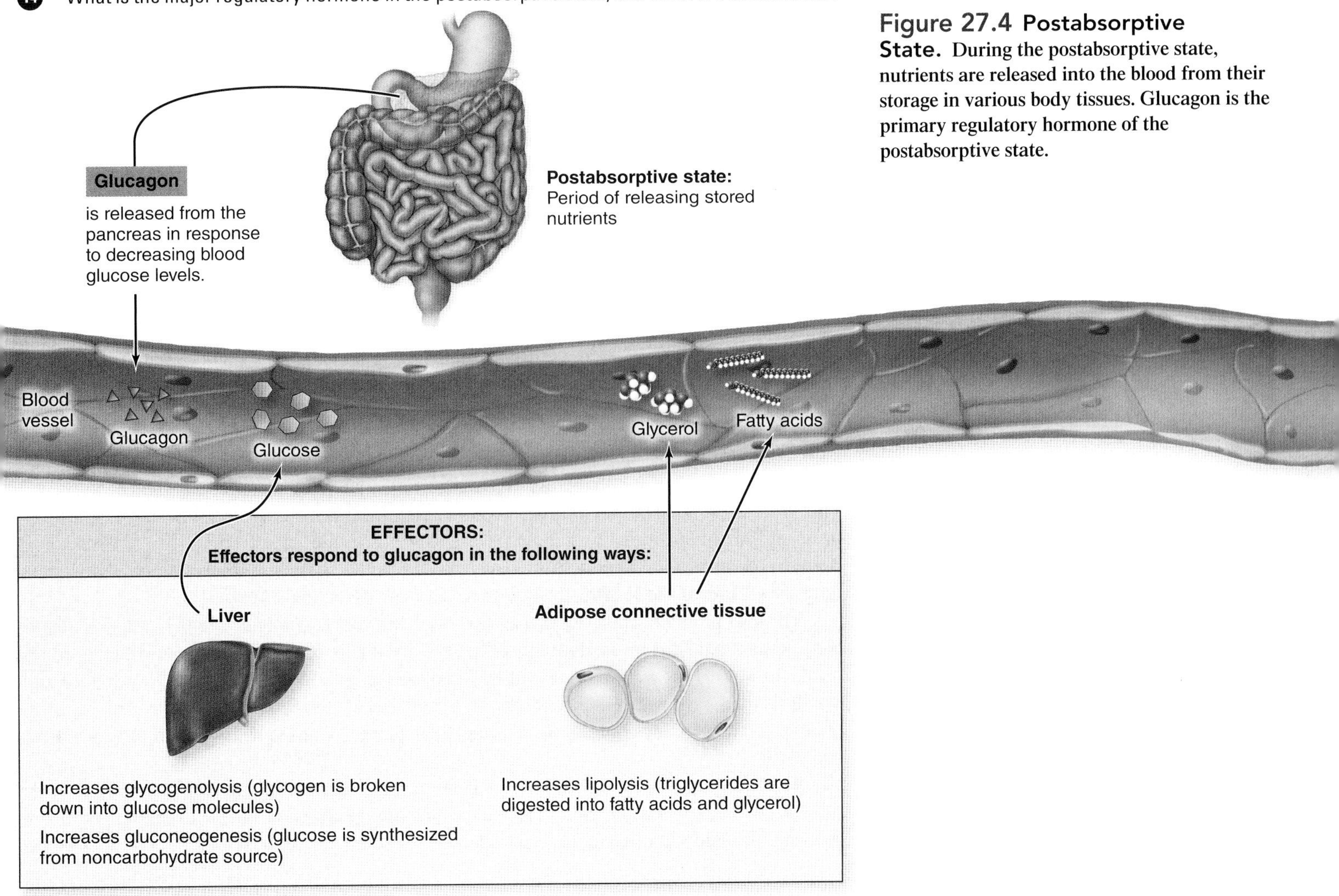

Figure 27.4 Postabsorptive State. During the postabsorptive state, nutrients are released into the blood from their storage in various body tissues. Glucagon is the primary regulatory hormone of the postabsorptive state.

Glucagon is released from the pancreas in response to decreasing blood glucose levels.

Postabsorptive state: Period of releasing stored nutrients

Blood vessel — Glucagon — Glucose — Glycerol — Fatty acids

EFFECTORS:
Effectors respond to glucagon in the following ways:

Liver
Increases glycogenolysis (glycogen is broken down into glucose molecules)
Increases gluconeogenesis (glucose is synthesized from noncarbohydrate source)

Adipose connective tissue
Increases lipolysis (triglycerides are digested into fatty acids and glycerol)

27.6 Functions of the Liver

The **liver** is the largest internal organ. We described the details of both its anatomic and histologic structure in section 26.3c. This amazing organ is estimated to have several hundred different functions. Several of these have been described in earlier chapters, including (1) the formation of plasma proteins (e.g., albumin and clotting proteins) (see section 18.2a); (2) removal of bilirubin from blood, following breakdown of hemoglobin as part of erythrocyte destruction (see section 18.3b); and (3) formation and release of bile into the small intestine for emulsification of lipids (see section 26.4c). We just described its role in regulating blood glucose in the absorptive and postabsorptive states by serving as a "glucose bank."

Here we first review the histologic structure of liver lobules and then examine more closely liver function in both the synthesis of cholesterol and the transport of lipids in the blood. We conclude this section by integrating the structure and the various functions of the liver that have been described throughout the text.

27.6a Anatomy of Liver Lobules

 **LEARNING OBJECTIVE**

1. Describe the anatomic arrangement of liver lobules, including the central vein, hepatocytes, sinusoids, and bile canaliculi.

Liver lobules are the functional unit of the liver (as described in section 26.3c). These microscopic structures are composed of cords of hepatocytes that radiate out from a central vein **(figure 27.5a)**. Sinusoids (modified capillaries) are located between the cords of hepatocytes.

Both oxygenated blood from branches of the hepatic artery and deoxygenated (nutrient-rich) blood delivered from branches of the hepatic portal vein combine and are transported through the liver sinusoids toward the central vein (sections 20.10d and 26.3c). This blood then flows out of the liver via several hepatic veins into the inferior vena cava (IVC). The bile canaliculi are sandwiched between each cord of hepatocytes and receive bile synthesized by hepatocytes. Bile then drains through a series of progressively larger ducts until it enters the small intestine.

 **WHAT DID YOU LEARN?**

15 What is the path of blood flow through a liver lobule?

27.6b Cholesterol Synthesis

 **LEARNING OBJECTIVE**

2. Explain the relationship of dietary intake of cholesterol and level of cholesterol synthesis in the liver.

We previously stated that cholesterol is obtained from animal products as dietary sources, such as meat, milk, and eggs. Hepatocytes also contain metabolic pathways that synthesize cholesterol (figure 27.5b).

Cholesterol Synthesis

Fatty acids within the blood are transported from a liver sinusoid to enter hepatocytes where they are broken down into numerous two-carbon units; each is formed into acetyl CoA. This process is called **beta oxidation**. Acetyl CoA molecules are used to synthesize cholesterol in an enzymatic pathway that includes a specific enzyme called **HMG-CoA (3-hydroxy-3-methylglutaryl CoA) reductase.** (This enzyme is the target for cholesterol-lowering statin drugs; see Clinical View: "Blood Cholesterol Levels" in section 27.6c.)

The liver produces cholesterol at a basal level that varies among individuals. An individual's basal level is adjusted inversely to his or her dietary intake of cholesterol. A low dietary intake results in lower blood cholesterol and less cholesterol entering hepatocytes. Thus, cholesterol synthesis by the liver increases. In contrast, a high dietary intake of cholesterol increases blood cholesterol with more cholesterol entering hepatocytes. Consequently, cholesterol synthesis decreases.

Following its formation, cholesterol is either (1) released into the blood as a component of very-low-density lipoproteins (VLDLs), which are described in the next section, or (2) synthesized into bile salts (bile acids) and released as a component of bile into the small intestine

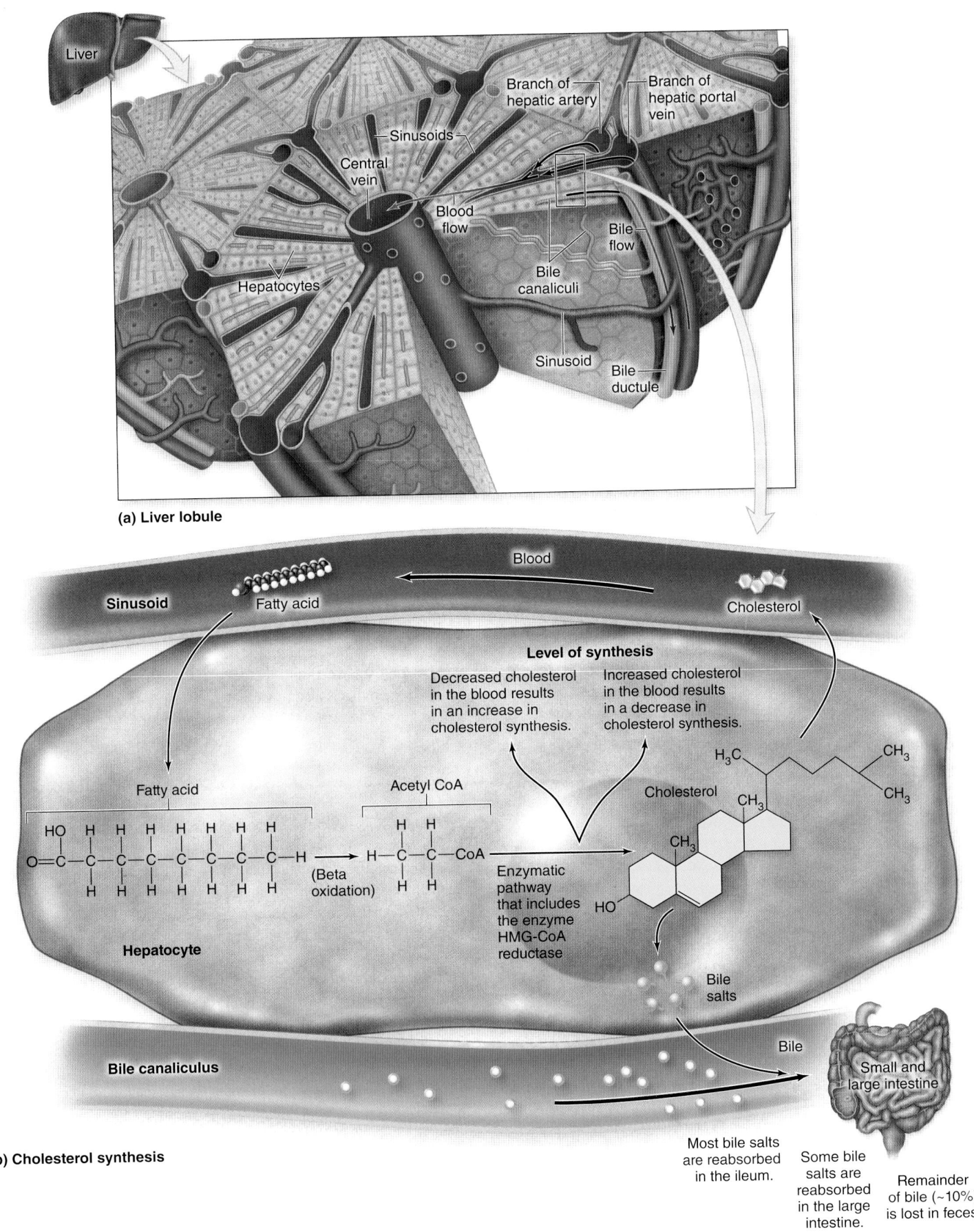

(a) Liver lobule

Liver

Branch of hepatic artery

Branch of hepatic portal vein

Sinusoids

Central vein

Blood flow

Bile flow

Hepatocytes

Bile canaliculi

Sinusoid

Bile ductule

Blood

Sinusoid — Fatty acid

Cholesterol

Level of synthesis

Decreased cholesterol in the blood results in an increase in cholesterol synthesis.

Increased cholesterol in the blood results in a decrease in cholesterol synthesis.

Fatty acid

Acetyl CoA

Cholesterol

(Beta oxidation)

Enzymatic pathway that includes the enzyme HMG-CoA reductase

Hepatocyte

Bile salts

Bile

Small and large intestine

Bile canaliculus

(b) Cholesterol synthesis

Most bile salts are reabsorbed in the ileum.

Some bile salts are reabsorbed in the large intestine.

Remainder of bile (~10%) is lost in feces.

Figure 27.5 The Liver and Cholesterol Synthesis. (*a*) Both oxygenated and deoxygenated blood (which is nutrient-rich) flow through the sinusoids of liver lobules toward the central vein. Hepatocytes produce bile that is released into bile canaliculi that moves toward microscopic bile ductules. These ductules merge to form larger ducts that ultimately release bile into the small intestine. (*b*) Cholesterol is synthesized within individual hepatocytes that compose liver lobules. Following their synthesis, cholesterol molecules are released either into the blood of the liver sinusoids or synthesized into bile salts and released as a component of bile into bile canaliculi. **AP|R**

(see section 26.4c). A majority of the bile salts are reabsorbed back into the blood primarily while moving through the ileum (and to a limited extent while moving through the large intestine). A small amount of bile salts (about 10%) continue into the large intestine and are removed from the body as a component of feces. This provides a means of eliminating excess cholesterol from the body and lowering blood cholesterol levels.

 WHAT DID YOU LEARN?

16 Is cholesterol synthesis increased or decreased in response to high dietary intake of cholesterol? What are the two fates of cholesterol following its synthesis?

27.6c Transport of Lipids

 LEARNING OBJECTIVE

3. Describe the transport of lipids within the blood.

Lipids are hydrophobic molecules and are insoluble in blood. Their transportation within the blood requires that they are first wrapped in a water-soluble protein. The lipid and the protein "wrap" are collectively called a **lipoprotein:** These are a general category of structures that contain triglycerides, cholesterol, and phospholipids within the "confines" of a protein.

We previously discussed one group of lipoproteins—chylomicrons (see section 26.4c). Recall that the absorption of lipids from the small intestine requires the formation of chylomicrons within epithelial cells lining the small intestine. A chylomicron is mostly composed of triglycerides and some cholesterol enveloped in protein. After its formation, it is absorbed into a lacteal and transported within the lymph until it enters into venous blood at the junction of the jugular and subclavian vein (see section 21.1b). Chylomicrons deliver lipids to the liver and other tissues

(e.g., adipose connective tissue for storage and skeletal muscle and cardiac muscle for energy use). Chylomicron remnants are then taken up by the liver.

Various other lipoproteins are formed within the liver. The relative density of these structures is used to classify these lipoproteins. The three broad categories of lipoproteins are (1) very-low-density lipoproteins (VLDLs), which contain the most lipid; (2) low-density lipoproteins (LDLs) with somewhat less lipid; and (3) high-density lipoproteins (HDLs) with the least amount of lipid. These function in transport of lipids between the liver and peripheral tissues.

Transport from the Liver to Peripheral Tissues
Both very-low-density lipoproteins and low-density lipoproteins are associated with transport of lipids from the liver to the peripheral tissues (figure 27.6):

- **Very-low-density lipoproteins (VLDLs)** contain various types of lipids (e.g., triglycerides and cholesterol) packaged with protein. VLDLs are assembled within the liver and then released into the blood. These "lipid delivery vehicles" circulate in the blood to release triglycerides to all cells of peripheral tissues, but primarily to adipose connective tissue. A change in density accompanies the release of triglycerides from these structures, and the lipoprotein is then called a low-density lipoprotein.

- **Low-density lipoproteins (LDLs)** contain relatively high amounts of cholesterol. LDLs deliver cholesterol to cells. LDLs bind to LDL receptors displayed within the plasma membrane of a cell and are subsequently engulfed into the cell by receptor-mediated endocytosis (see section 4.3c). Cholesterol is incorporated into the plasma membrane of all cells or is used by certain tissues (e.g., testes, ovaries, and the adrenal cortex) to produce steroid hormones (see section 17.5a).

INTEGRATE

CLINICAL VIEW
Blood Cholesterol Levels

Blood cholesterol levels are tested to determine blood levels for total cholesterol, low density lipoproteins (LDLs), and high density lipoproteins (HDLs). An important comparison also includes the relative levels of HDLs and LDLs. The results are clinically significant because high blood levels for either total cholesterol or LDLs, as well as low blood levels for HDLs are considered risk factors for cardiovascular disease.

A total cholesterol value above 200 mg/dL is considered "high." In addition, individual LDL and HDL levels are reported to provide a more specific measure of cholesterol levels. A normal value for LDL blood levels for a healthy adult is less than 100 mg/dL, and for those at risk for cardiovascular disease an acceptable value is below 70 mg/dL. In comparison, a normal value for HDL blood levels is between 40 and 50 mg/dL for adult males and between 50 and 60 mg/dL for adult females. A healthy ratio of HDL to LDL is considered to be above 0.3, or even better above 0.4.

A healthy ratio of HDL to LDL is a more accurate reading than total cholesterol because of the difference in the function of LDLs and HDLs. LDLs are considered to be the "bad cholesterol" because excess cholesterol not needed by peripheral tissue is deposited on the inner arterial walls, an event that may lead to development of plaque and arteriosclerosis (a risk factor for heart attacks and strokes). HDLs are considered the "good cholesterol" because they transport lipid (e.g., cholesterol, triglycerides) from the arterial wall to the liver for eventual elimination from the body through the GI tract.

Dietary intake of various types of fats influences blood cholesterol levels. The omega-3 fatty acids found in some cold-water fish, for example, lower total blood cholesterol levels. In comparison, high intake of saturated fats (especially the trans fats) has been shown to decrease HDL blood levels and increase LDL blood levels. Other risks factors correlated with elevated blood cholesterol include cigarette smoking, caffeine intake, and stress.

Statin drugs (e.g., Zocor, Lipitor, Crestor) were developed for the purpose of lowering blood cholesterol. They are typically prescribed for individuals with LDL blood levels higher than 130 mg/dL. Statin drugs act as a competitive inhibitor for HMG-CoA (3-hydroxy-3-methylglutaryl CoA) reductase, a specific enzyme within the metabolic pathway for cholesterol synthesis (figure 27.5). Thus, these drugs help decrease cholesterol synthesis. Subsequently, liver cells increase their uptake of cholesterol, which lowers blood cholesterol levels. Statin drugs do have side effects (e.g., muscle weakness, nausea, difficulty sleeping), so they should be taken only by individuals with high blood cholesterol or those who are at risk for cardiovascular disease.

Note that elevated blood cholesterol is only one of many risk factors for developing cardiovascular disease. Given that cardiovascular disease is the cause of death in approximately 50% of disease-related deaths in the United States, research continues in determining the most accurate risk factors and how to help individuals to lower their risk.

Transport from Peripheral Tissues to the Liver

High-density lipoproteins (HDLs) are associated with the transport of lipid from the peripheral tissues to the liver. These structures are produced in stages and have a function that opposes both VLDLs and LDLs. The proteins are formed in the liver and are released into the blood without the addition of lipid (i.e., these structures are "empty" HDLs). HDL molecules circulate throughout the blood and "fill" with lipids (e.g., cholesterol and triglycerides) from peripheral tissue and the inner lining of arterial walls. HDLs then transport these lipids to the liver. Excess cholesterol is converted to bile salts within the liver and released into the small intestine as a component of bile (recall that this is the means of eliminating cholesterol from the body). Note that HDLs, like LDLs, make cholesterol available to steroid-producing tissues. However, in contrast to LDLs, HDLs have cholesterol removed without being engulfed by the cell (not shown in figure).

WHAT DO YOU THINK?

2 Which lab test result is more informative: (a) one that gives total cholesterol or (b) one that provides the levels of both LDL and HDL? Explain.

WHAT DID YOU LEARN?

17 Which of these molecular structures (VLDLs, LDLs, or HDLs) is responsible for transporting cholesterol from peripheral tissues to the liver?

27.6d Integration of Liver Structure and Function

LEARNING OBJECTIVE

4. Identify and briefly describe the numerous roles of the liver in metabolism.

The liver is responsible for many different metabolic processes. Its structure and some of its many functions are integrated, visually represented, and described in **figure 27.7**. A summary of liver functional categories include the following:

- Carbohydrate metabolism
- Protein metabolism
- Lipid metabolism
- Transport of lipids
- Other functions (e.g., storage and drug detoxification)

WHAT DID YOU LEARN?

18 Provide an example for each of the five categories of liver functions integrated in figure 27.7.

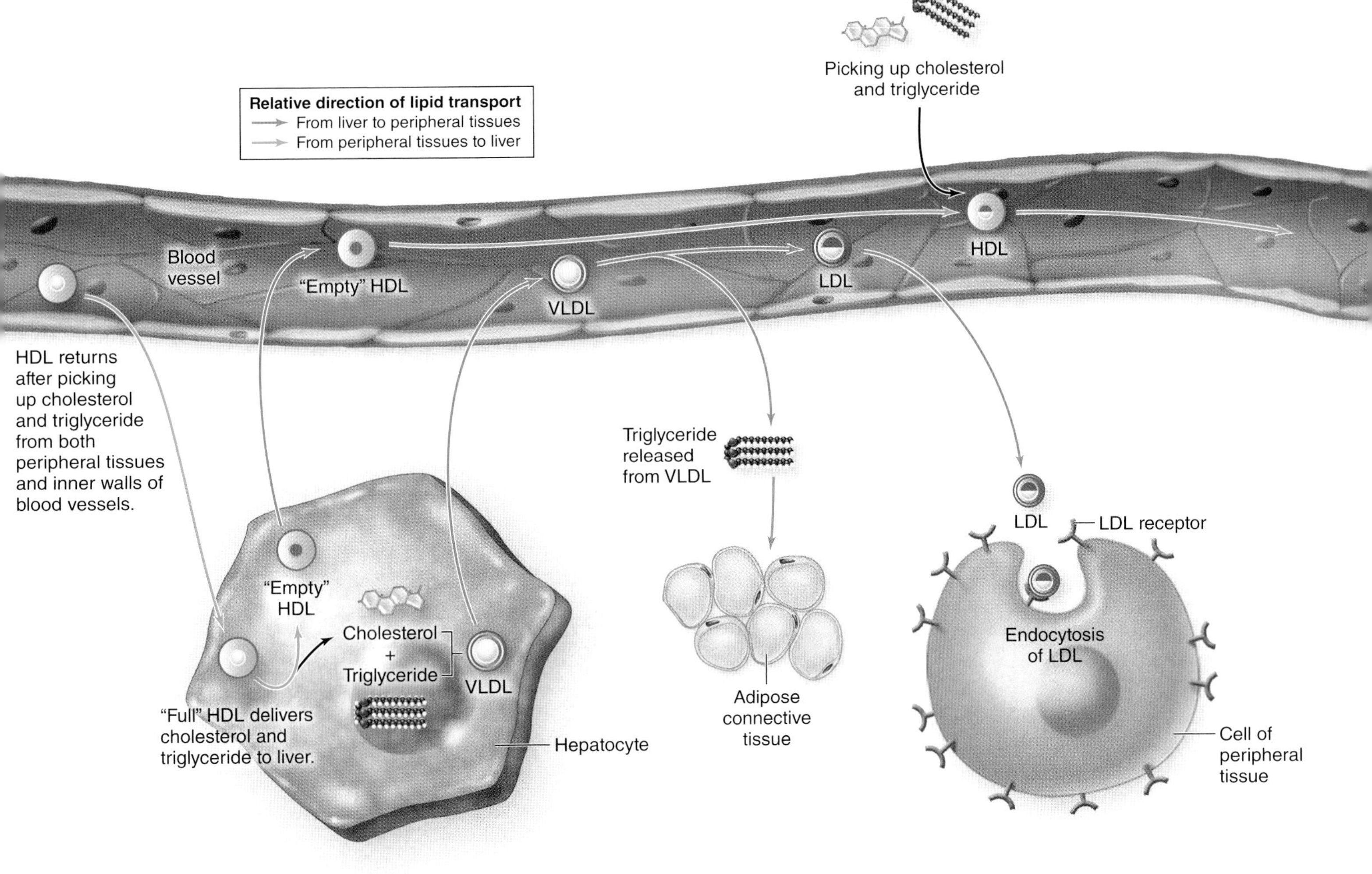

Figure 27.6 Lipid Transport. Very-low-density lipoproteins (VLDLs), which become low-density lipoproteins (LDLs), participate in transporting lipid from the liver to the peripheral tissues. In contrast, high-density lipoproteins (HDLs) transport lipid from peripheral tissues to the liver.

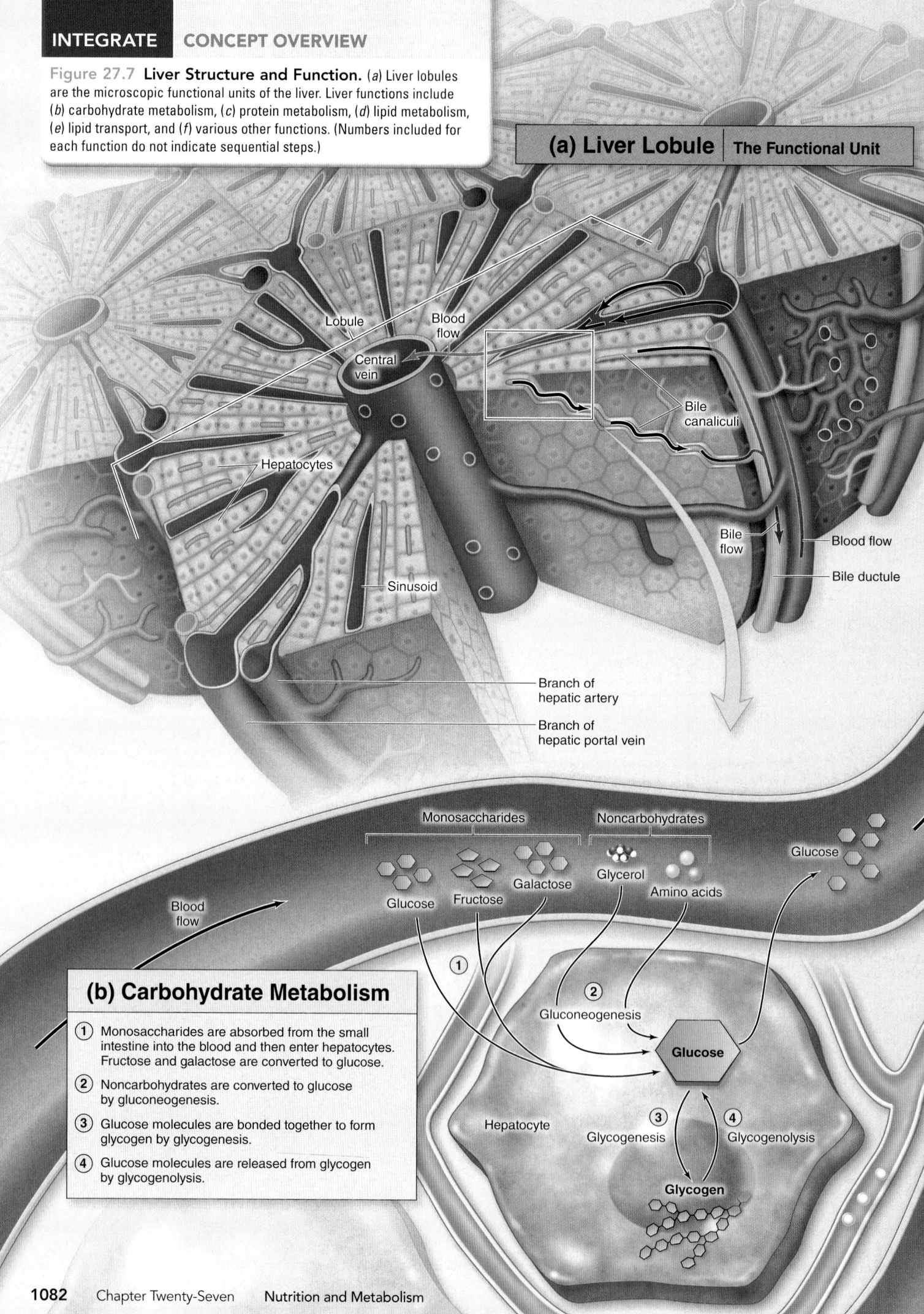

Figure 27.7 **Liver Structure and Function.** (*a*) Liver lobules are the microscopic functional units of the liver. Liver functions include (*b*) carbohydrate metabolism, (*c*) protein metabolism, (*d*) lipid metabolism, (*e*) lipid transport, and (*f*) various other functions. (Numbers included for each function do not indicate sequential steps.)

(a) Liver Lobule | The Functional Unit

Lobule

Blood flow

Central vein

Hepatocytes

Sinusoid

Bile canaliculi

Bile flow

Blood flow

Bile ductule

Branch of hepatic artery

Branch of hepatic portal vein

Monosaccharides

Noncarbohydrates

Glucose

Glucose   Fructose   Galactose   Glycerol   Amino acids

Blood flow

(b) Carbohydrate Metabolism

(1) Monosaccharides are absorbed from the small intestine into the blood and then enter hepatocytes. Fructose and galactose are converted to glucose.

(2) Noncarbohydrates are converted to glucose by gluconeogenesis.

(3) Glucose molecules are bonded together to form glycogen by glycogenesis.

(4) Glucose molecules are released from glycogen by glycogenolysis.

(1)

(2) Gluconeogenesis

Glucose

Hepatocyte

(3) Glycogenesis   (4) Glycogenolysis

Glycogen

(c) Protein Metabolism

1. Deamination: Amine group removed from amino acids

   NH_2 is converted to urea and urea enters blood (urea eliminated by kidney)

   Remaining components oxidized in cellular respiration to generate ATP from the liver

2. Amino acids used to form proteins, including plasma proteins

3. Transamination: Amino acids converted from one form to another

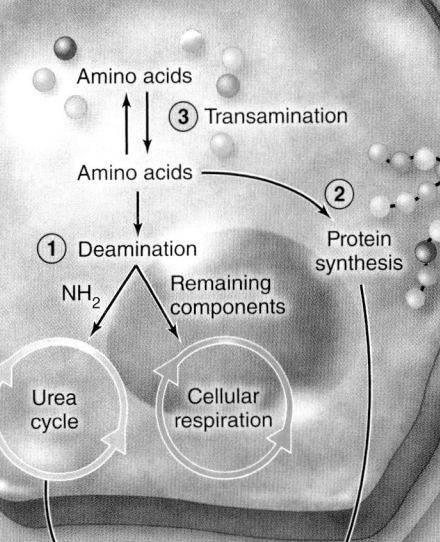

Amino acids

3 Transamination

Amino acids

1 Deamination

NH_2

Remaining components

2 Protein synthesis

Urea cycle

Cellular respiration

Urea transported to kidney and excreted

Plasma proteins

Blood flow

VLDLs

Ketone bodies

Bile

VLDLs

Bile

Bile salts ← Cholesterol ← Acetyl CoA → Ketone bodies

5

4

Glycerol

3 Beta oxidation

1 Lipogenesis

Triglycerides

2 Lipolysis

Fatty acids

(d) Lipid Metabolism

1. Fatty acids joined with glycerol to form triglycerides (lipogenesis)

2. Fatty acids released from triglycerides (lipolysis)

3. Fatty acids broken down into acetyl CoA (beta oxidation)

4. Acetyl CoA changed to ketone bodies (water-soluble molecules); ketone bodies released into blood, transported to other cells, where they can be oxidized in cell respiration pathways

5. Acetyl CoA used in cholesterol synthesis; cholesterol released into blood within VLDLs, and some used to form bile salts and released as a component of bile

Blood flow

(e) Lipid Transport

Transport both triglycerides and cholesterol (within VLDLs and LDLs) from the liver to peripheral tissues

"Empty" HDLs released to pick up lipids (e.g., cholesterol) from peripheral tissues and blood vessels; return as "full" HDL to the liver

(f) Other Liver Functions

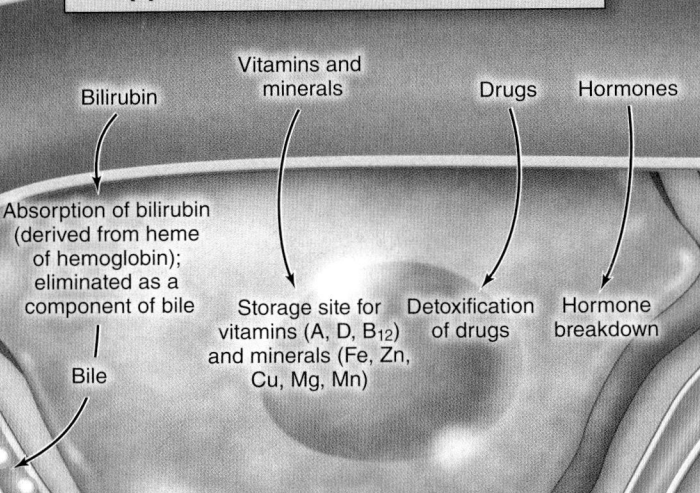

Bilirubin

Vitamins and minerals

Drugs

Hormones

Absorption of bilirubin (derived from heme of hemoglobin); eliminated as a component of bile

Bile

Storage site for vitamins (A, D, B_{12}) and minerals (Fe, Zn, Cu, Mg, Mn)

Detoxification of drugs

Hormone breakdown

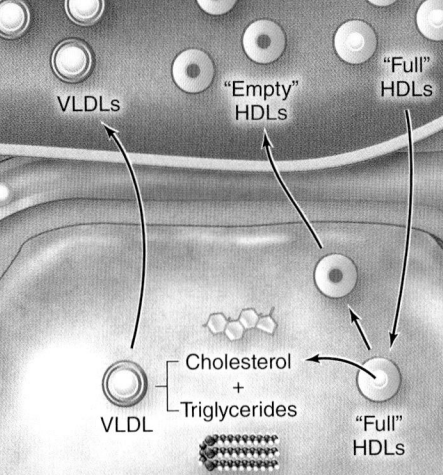

VLDLs

"Empty" HDLs

"Full" HDLs

Cholesterol + Triglycerides

VLDL

"Full" HDLs

INTEGRATE

CONCEPT CONNECTION

We first introduced the details of cellular respiration early in the text in section 3.4. This information provided the necessary background to understand later sections including muscle metabolism in section 10.4a, muscle fiber types in section 10.5a, and metabolism of cardiac muscle in section 19.3e. We also discussed how cellular respiration is supported by the general processes of respiration, and how an individual's ability to produce sufficient energy (ATP) is dependent upon both a healthy respiratory system and cardiovascular system.

27.7 Central Role of Cellular Respiration

The biochemical pathways of cellular respiration were discussed in detail in section 3.4 for the oxidation of glucose to generate ATP. Here we first review ATP generation through the metabolic pathways of cellular respiration. We then discuss how these pathways are also used to oxidize other nutrient molecules (e.g., fatty acids and amino acids) both to generate ATP and to interconvert between various biomolecules and their building blocks.

27.7a ATP Generation

LEARNING OBJECTIVES

1. Describe where the following nutrient molecules enter the metabolic pathway of cellular respiration: glucose, the breakdown products of triglycerides, and amino acids.

2. Explain deamination of proteins.

Recall from our discussion of cellular respiration that it involves four stages (**figure 27.8**):

① **Glycolysis** is a metabolic pathway that occurs in the cytoplasm of a cell that does not require oxygen. Glucose is oxidized to two pyruvate molecules. Two ATP molecules are formed (by substrate-level phosphorylation), and hydrogen is transferred to 2 NAD$^+$ molecules to form 2 NADH molecules during this process (see section 3.4b). If insufficient oxygen is available to the cell, then the pyruvate is converted to lactate (see section 3.4g). The other steps in cell respiration occur within the mitochondria and require oxygen.

② The **intermediate stage** is a multienzyme step during which pyruvate is converted to acetyl CoA, and a molecule of carbon dioxide (CO_2) is formed by decarboxylation. One molecule of NADH is produced per pyruvate (see section 3.4c).

③ The **citric acid cycle** is the metabolic pathway in which acetyl CoA binds to oxaloacetic acid (OAA) to form citric acid. Thereafter, two molecules of carbon dioxide (CO_2) are formed by decarboxylation. One ATP molecule, 3 NADH molecules, and 1 FADH$_2$ molecule are produced per "turn" of the cycle. Remember that both the intermediate stage and the citric acid cycle occur twice in the breakdown of a glucose molecule because there are two pyruvate molecules formed from oxidation of each glucose molecule entering glycolysis (see section 3.4d).

④ The **electron transport system** involves the transfer of hydrogen and an electron from the coenzymes NADH and FADH$_2$. ATP formation subsequently occurs through oxidative phosphorylation (see section 3.4e).

Glycerol and Fatty Acids

The building blocks of triglycerides are glycerol and fatty acids. They may enter into the pathway at certain stages of cellular respiration to release their chemical energy to generate ATP. Glycerol specifically enters the pathway of glycolysis. Glycerol is converted to glucose through gluconeogenesis within the liver. The carbons of fatty acids are removed two at a time to form acetyl CoA (through beta oxidation). Acetyl CoA molecules enter the citric acid cycle.

Amino Acids

The amino acids that compose a protein may be used at various stages of cellular respiration to generate ATP. However, recall that amino acids contain nitrogen in an amine group (—NH$_2$). The amine group is removed by a process called **deamination** (dē-am′i-nā′shŭn; *de* = remove) within hepatocytes of the liver. You might find it helpful to think of the amine group as a "contaminant" that must be removed in order for amino acids to be used in cellular respiration to generate ATP.

The remaining portion of the amino acid following deamination enters the metabolic pathway of cellular respiration at different steps, depending upon the specific amino acid. They may enter (1) into the pathway of glycolysis, (2) at the intermediate stage, or (3) at specific points within the citric acid cycle. The amine group (—NH$_2$) is converted to urea (urea cycle) and is eliminated through the kidney as a component of urine (see section 24.8a).

WHAT DID YOU LEARN?

⑲ Where in the biochemical pathway of cellular respiration (glycolysis, intermediate stage, or citric acid cycle) does each of the following enter: (a) glycerol, (b) fatty acids, and (c) deaminated amino acids?

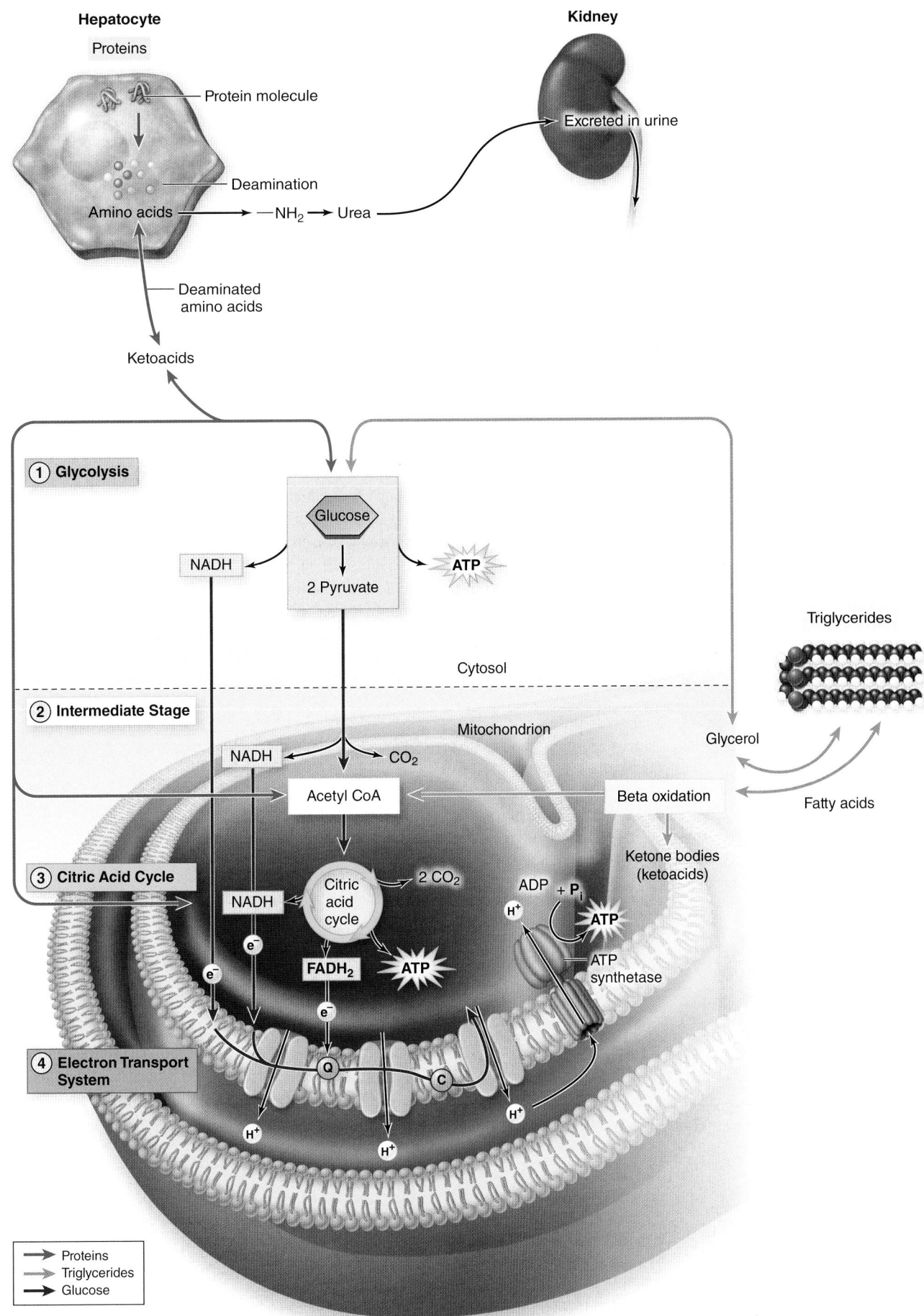

Figure 27.8 Cellular Respiration: Generation of ATP Molecules and Interconversion of Nutrient Molecules. The metabolic pathways of cellular respiration serve to both generate ATP molecules through the oxidation of glucose, triglycerides, and proteins, and provide a means of converting one type of nutrient biomolecule to another.

27.7b Interconversion of Nutrient Biomolecules and Their Building Blocks

LEARNING OBJECTIVE

3. Describe the physiologic advantages of the ability to interconvert nutrient biomolecules.

Individuals who eat excess amounts of carbohydrates, including sweets, can deposit additional triglycerides into their adipose connective tissue and gain weight. How is consumed sugar turned into fat? Some weight-loss plans advocate following a low-carbohydrate diet (e.g., the Atkins diet). Considering that nervous tissue primarily uses glucose, how is it possible that individuals on low-carbohydrate diets are able to provide the cells of the nervous tissue with the needed glucose?

Interconversion of nutrient biomolecules, which is the changing of one nutrient biomolecule to another, is possible because of the biochemical pathways that are associated with cellular respiration. Notice in figure 27.8 how the three nutrient biomolecules can be converted to each other through pathways that involve cellular respiration. This is indicated with arrows extending in both directions between the different pathways. For example, if energy is not needed, glucose can be broken down to acetyl CoA that is then synthesized into triglycerides and stored, instead of entering the citric acid cycle.

 WHAT DID YOU LEARN?

20 How is excess sugar (glucose) converted to fat (triglycerides)?

27.8 Energy and Heat

How energy may be converted from one form to another was first discussed in section 3.1. For example, the chemical energy in food is first converted into ATP (another form of chemical energy) that is used to generate the mechanical energy of muscle contraction that occurs as we walk. We have previously stated that energy conversions result in the production of heat (according to the second law of thermodynamics). Here we describe how heat is related to metabolic rate and how we regulate body temperature, which has normal variations among individuals.

27.8a Metabolic Rate

 LEARNING OBJECTIVES

1. Define metabolic rate.

2. Explain how both basal metabolic rate and total metabolic rate are measured, and the variables that influence each.

The **metabolic rate** is the measure of energy used in a given period of time. There are two means of expressing metabolic rate: basal metabolic rate and total metabolic rate.

Basal Metabolic Rate

Basal metabolic rate (BMR) is the amount of energy required when an individual is at rest (and not eating). Resting conditions are determined as follows: The individual has not eaten for 12 hours, is reclining and relaxed, and is exposed to specific environmental conditions, including a room temperature between 20°C and 25°C.

BMR may be measured by either of two methods:

- A **calorimeter,** which is a water-filled chamber into which an individual is placed. Heat released from the person's body alters the temperature of the water, and the change in temperature is measured. This is considered a direct method because heat is directly measured.

- A **respirometer,** which is an instrument to measure oxygen consumption. It is used to indirectly measure BMR because a relationship exists between oxygen consumption and heat production. Oxygen is used to produce ATP in aerobic cellular respiration, and ATP is utilized in metabolic processes that produce heat. Typically, for every liter of oxygen consumed, 4.8 kilocalories of heat are produced. This is considered an indirect method because the amount of heat produced is inferred from oxygen consumption.

The BMR of individuals varies because of their age, lean body mass, sex, and levels of various hormones in the blood. The BMR decreases with age with about 3% decrease each decade beginning at age 30. Individuals with greater lean body mass (greater amounts of skeletal muscle relative to adipose connective tissue) tend to have a higher BMR. This is why males tend to have a higher BMR than females (average about 5–10% higher). Thyroid hormone increases BMR with an accompanying increase in lipolysis occurring within adipose connective tissue (see section 17.8b and table R.4 Regulating Metabolism with Thyroid Hormone). Individuals with hypothyroidism have a lower than normal BMR and tend to gain weight, whereas those with hyperthyroidism have a higher than normal BMR and tend to lose weight.

Another important variable in BMR, however, is body surface area. The reason is that heat is lost through the surface of the skin. The greater the surface area of the skin, the more heat that is lost. Heat lost through the surface is offset by the metabolic rate of cells, within limits. The more heat that is being lost, the more metabolically active body cells must be to maintain body temperature.

 WHAT DO YOU THINK?

3 Which individual would you predict to have a higher basal metabolic rate, a man who is 6 feet tall weighing 200 pounds, or a man who is 5 feet, 9 inches tall weighing 160 pounds? Explain your answer.

Total Metabolic Rate

Total metabolic rate (TMR) is the amount of energy used by the body, including energy needed for physical activity. Thus, TMR is the BMR plus metabolism associated with physical activity. The TMR varies widely, depending upon several factors:

- **The amount of skeletal muscle and its activity.** For example, a rapid elevation in TMR occurs during vigorous exercise and stays elevated for several hours after exercise.

- **Food intake.** Metabolic rate increases following ingestion of a meal but decreases after the absorption of nutrients has been completed.

- **Changing environmental conditions.** Metabolic rate increases, for example, when one is exposed to cold temperatures.

 WHAT DID YOU LEARN?

21 Would your total metabolic rate increase when you exit your home on a cold winter day? Explain.

27.8b Temperature Regulation

 LEARNING OBJECTIVES

3. Define core body temperature, and explain why it must be maintained.

4. Explain the neural and hormonal controls of temperature regulation.

We have just seen that a direct relationship exists between the metabolic rate of an individual and heat production—as the metabolic rate increases, more heat is produced. Because metabolic rate fluctuations are influenced by level of activity, food intake, and environmental temperature, the amount of heat produced also varies. Consequently, it is imperative that body temperature be maintained within certain physiologic limits in response to these changes. Homeostatic mechanisms usually keep our body temperature near the normal value of 98.6°F (37°C). This temperature regulation is accomplished through a balance of heat production and heat loss, and it occurs through both neural and hormonal controls.

One of the critical aspects of regulating body temperature is maintaining **core body temperature,** which is the temperature of the vital portions of the body, or *core,* that consists of the head and torso. The temperature of these regions is kept relatively constant, or stable, to assure that life is maintained. This generally occurs by allowing fluctuations in the temperature of peripheral regions, such as the limbs. (Regulation of body temperature was first discussed in section 1.5b.)

Nervous System Control

Nervous system control of body temperature is mediated primarily through the hypothalamus (see section 13.4c). Motor pathways extend from the hypothalamus to the sweat glands in the skin, skeletal muscles, and peripheral blood vessels **(figure 27.9)**. The hypothalamus detects changes in body temperature either by monitoring the temperature of blood as it passes through the hypothalamus or nerve signals received from the skin.

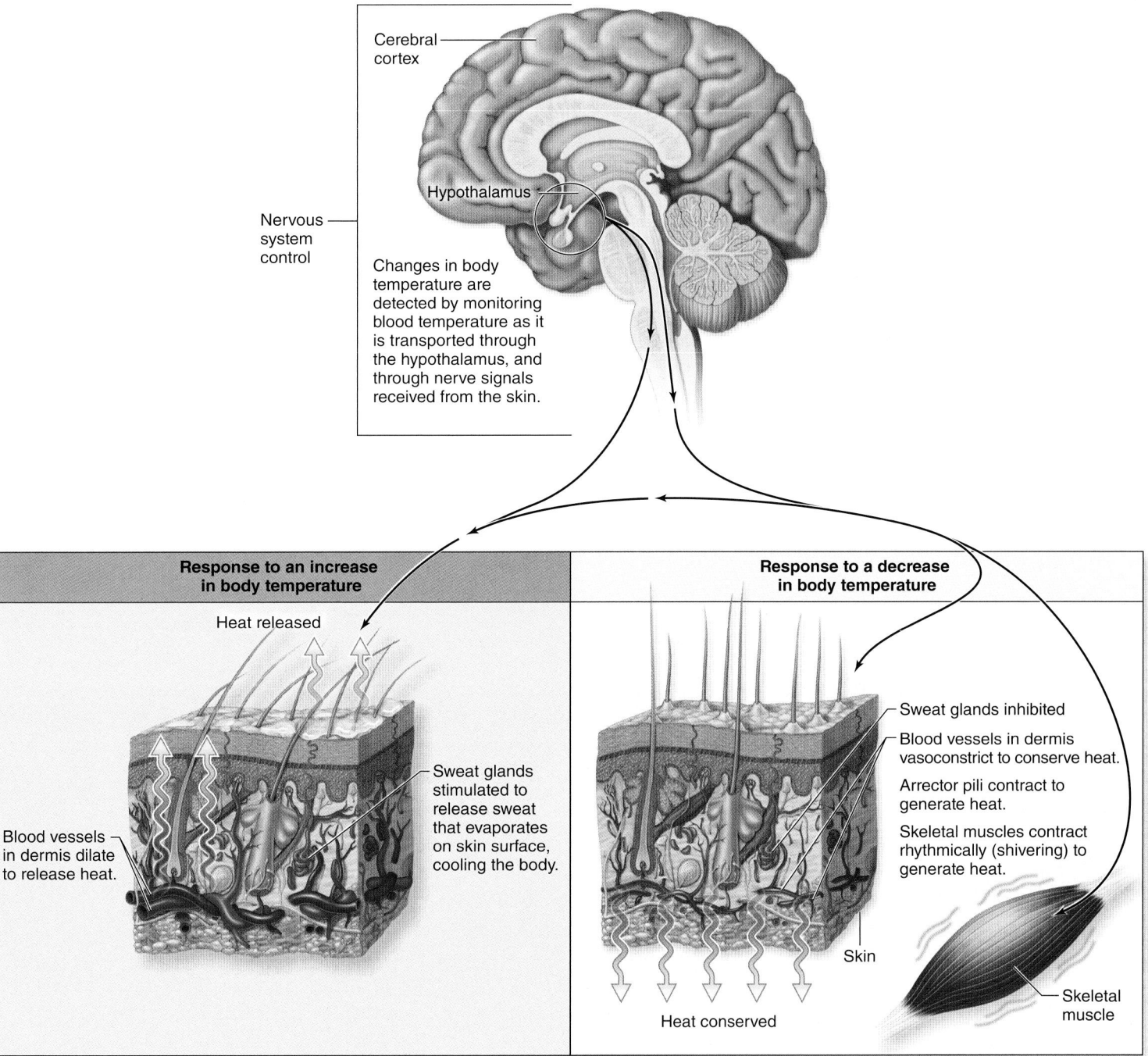

Figure 27.9 Nervous System Control of Body Temperature. The hypothalamus regulates body temperature by controlling sweat glands, vasoconstriction and vasodilation of blood vessels, contraction of smooth muscle of arrector pili, and the involuntary skeletal muscle contractions of shivering.

An increase in metabolic rate causes a subsequent increase in body temperature, and heat must be released. The hypothalamus responds by stimulating sweat glands to release sweat onto the surface of the body to draw heat away by both evaporation and transpiration (see section 25.2a) and stimulating vasodilation of peripheral blood vessels to bring heat to the skin surface.

In contrast, when metabolic rate decreases, it causes a subsequent decrease in body temperature, and additional heat must be generated. Now the hypothalamus inhibits sweat gland activity, stimulates constriction of peripheral blood vessels, thereby reducing blood circulation and heat loss at the periphery, and induces both smooth muscle contraction of arrector pili (to cause "goosebumps") and skeletal muscle contraction through shivering to generate heat. With extreme cold, the result of vasoconstriction to the peripheral regions may be frostbite, leading to dry gangrene (see Clinical View: "Frostbite and Dry Gangrene").

Conscious changes in behavior that are initiated by the cerebral cortex can help regulate body temperature. When we are hot, heat release is expedited by the removal of clothing, stopping work to rest, going inside where it is air-conditioned, or jumping into a pool; but when cold we attempt to both conserve heat by putting on additional clothing or curling up under a blanket, and we generate additional heat by becoming more active.

Hormonal Control

Temperature regulation is also mediated by hormone secretion, including thyroid hormone, epinephrine and norepinephrine, growth hormone, and testosterone. The most significant is thyroid hormone. Recall from section 17.8b, that thyroid hormone establishes the metabolic rate, causing us to produce a certain amount of heat to keep the body warm. As your body temperature begins to drop, the hypothalamus releases thyrotropin-releasing hormone (TRH); TRH stimulates the anterior pituitary to release thyroid-stimulating hormone (TSH); and TSH stimulates the thyroid gland to release the thyroid hormones (T_3 and T_4). Thyroid hormone is able to help maintain body temperature by increasing the metabolic rate of almost all cells, especially neurons.

Neurons are specifically stimulated to increase their number of sodium-potassium (Na^+/K^+) pumps. Because there are more Na^+/K^+ pumps, more energy is converted as the pumps use ATP to move the ions, then more heat is produced, and body temperature is maintained.

WHAT DID YOU LEARN?

22 What is the core body temperature, and why must it be maintained?

23 Explain the role of the hypothalamus in regulating the response to an increase in body temperature.

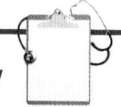
INTEGRATE

CLINICAL VIEW
Frostbite and Dry Gangrene

Frostbite is damage to superficial cells due to exposure to extreme cold. The skin appears white and may be blistered. Loss of sensation may also occur. It may lead to dry gangrene in severe cases. **Dry gangrene** (necrosis or pathologic death) is a form of gangrene in which the involved body part is desiccated (loss of water), sharply demarcated (distinct change in coloration), and shriveled. These changes are usually due to extensive and prolonged vasoconstriction of blood vessels as a result of exposure to extreme cold. This results in oxygen deprivation and death of the tissue.

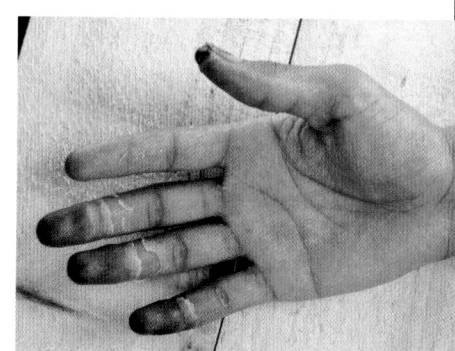
Photo of frostbite and dry gangrene.

CHAPTER SUMMARY

	• Nutrition is the study of the way living organisms acquire and utilize the nutrients necessary to grow and maintain life.
27.1 Introduction to Nutrition 1069	• Nutrients include biomolecules, vitamins, minerals, and water; they are obtained through the diet via the digestive system.
	• Nutrients are classified as both macronutrients or micronutrients and essential or nonessential.
	• Recommended daily allowances (RDAs) of nutrients have been established by the government and are updated to reflect new research.
27.2 Macronutrients 1069	• All macronutrients (carbohydrates, lipids, proteins) provide energy to the body through cellular respiration.
	27.2a Carbohydrates 1069
	• Dietary sources of carbohydrates include three different categories: sugars, starch, and fiber.
	27.2b Lipids 1070
	• Lipids are a diverse group of biologically active, hydrophobic molecules that include both triglycerides and cholesterol.
	27.2c Proteins 1071
	• Proteins are the most structurally and functionally diverse molecules and are synthesized from 20 different amino acids.
	• Complete proteins contain all of the essential amino acids, whereas incomplete proteins do not.

(continued on next page)

27.8 Energy and Heat   1086

- The amount of heat produced by the body is related to metabolic rate.
- Physiologic mechanisms maintain normal body temperature in response to changes in heat produced and environmental conditions.

27.8a Metabolic Rate   1086

- The metabolic rate is the rate that energy is used in a given period of time.
- Basal metabolic rate (BMR) is the amount of energy used when an individual is at rest and can be measured directly with a calorimeter and indirectly with a respirometer. The BMR of individuals varies, depending upon their age, lean body mass, sex, hormone levels, and body surface area.
- Total metabolic rate (TMR) is the total amount of energy used by an individual and varies, depending upon amount of skeletal muscle, level of physical activity, food intake, and changing environmental conditions.

27.8b Temperature Regulation   1086

- A direct relationship exists between an individual's metabolic rate and heat production.
- Temperature regulation is accomplished through a balance of heat production and heat loss, and occurs through both nervous system and hormonal controls.

CHALLENGE YOURSELF

Do You Know the Basics?

_____ 1. Which of the following is a nutrient?

    a. proteins

    b. minerals

    c. vitamins

    d. All of these are correct.

_____ 2. Which of these vitamins typically serves as a coenzyme?

    a. vitamin D

    b. vitamin E

    c. vitamin B

    d. vitamin A

_____ 3. What is the major macronutrient in a potato?

    a. carbohydrate

    b. protein

    c. triglyceride

    d. vitamin A

_____ 4. During the absorptive state

    a. blood glucose is initially decreasing.

    b. glucagon is released.

    c. insulin is released.

    d. several hours (more than 4 hours) have passed since the last meal.

_____ 5. When the pancreas releases insulin, it causes

    a. amino acid uptake in muscle.

    b. hepatocytes to form glycogen from glucose.

    c. lipogenesis in adipocytes.

    d. All of these are correct.

_____ 6. Which of the following nutrient biomolecules can be oxidized in biochemical pathways of cellular respiration to generate ATP?

    a. glucose

    b. amino acids

    c. fatty acids

    d. All of these are correct.

_____ 7. Which of the following conditions is _not_ applicable to measuring basal metabolic rate?

    a. reclining

    b. the time immediately after a meal

    c. room temperature

    d. resting

_____ 8. Total metabolic rate increases under which condition?

    a. warm weather

    b. starvation

    c. being confined to bed

    d. eating a meal

_____ 9. All of the following are functions of the liver _except_ the

    a. release of insulin.

    b. formation of plasma proteins (e.g., albumin).

    c. formation of LDL particles.

    d. elimination of bilirubin.

_____ 10. Thyroid hormone is the primary hormone that regulates body temperature. It does so by

    a. stimulating muscle contraction.

    b. increasing the number of Na^+/K^+ pumps in neurons.

    c. vasoconstriction of peripheral blood vessels.

    d. sweating.

11. Define nutrition.

12. Describe the general functions of vitamin C.

13. Distinguish between nonessential and essential amino acids.

14. Explain the risk of excessive intake of fat-soluble vitamins.

15. Define minerals, and give examples of their functions in the body.

16. Define nitrogen balance, and compare positive nitrogen balance and negative nitrogen balance.

17. Define the postabsorptive state. What is the major hormone that regulates blood nutrients during this time, and what is its role?

18. Explain the function of the liver in the transport of lipid in the blood.

19. Briefly describe the role of the liver in carbohydrate metabolism, protein metabolism, and lipid metabolism.

20. Explain neural and hormonal induced changes when you are cold.

Can You Apply What You've Learned?

1. An individual has recently adopted a vegetarian lifestyle. She has been on this type of diet for approximately 6 months and admits that she has not been conscientious about making healthy food choices. She is feeling weak, appears gaunt, and is losing body mass. What type of nutrient biomolecule is most likely missing from her diet?

    a. nonessential amino acids

    b. essential amino acids

    c. fats (triglycerides)

    d. carbohydrates

2. A young man explains to his doctor that he has been having headaches and fluctuations in body temperature, going from shivering to sweating, that are not induced by changes in his environment. After a great deal of diagnostic work, the physician orders a magnetic resonance imaging (MRI) scan. A mass is most likely detected in which area of his brain?

    a. hypothalamus

    b. pons

    c. thalamus

    d. medulla oblongata

3. Derrick is a young man who lifts weights for several hours a day. He explains that remaining physically fit is very important to him and has been taking megadoses of vitamins as part of maintaining his health. You are concerned about

    a. reduced carbohydrate intake that may result in ketoacidosis.

    b. toxic buildup of water-soluble vitamins.

    c. loss of protein mass.

    d. toxic buildup of fat-soluble vitamins.

4. An elderly gentleman has recently been diagnosed with cirrhosis of the liver. He admits to being a heavy drinker. This man is your patient. Given the damage to his liver, you would have concerns about all of the following *except*

    a. regulating his ability to maintain body temperature, because of decreased thyroid hormone levels.

    b. internal bleeding, because he has impaired ability to synthesize clotting proteins.

    c. edema, because of impaired synthesis of plasma proteins (e.g., albumin).

    d. jaundice, because his liver is less efficient at removing bilirubin from the blood.

5. Howard is a middle-aged man and has been slowly gaining weight after reaching the age of 30. He has decided to lose weight by following a low-carbohydrate diet. What physiologic process allows his brain to still have sufficient glucose to function normally?

    a. maintaining nitrogen balance

    b. hypothalamus regulation

    c. gluconeogenesis

    d. transamination

Can You Synthesize What You've Learned?

1. As a dietician in the hospital, explain to a patient the new MyPlate.

2. Explain the challenge in meeting dietary requirements to an individual who has adopted a vegetarian lifestyle.

3. Explain to a patient who is being admitted to the hospital with frostbite what has occurred.

INTEGRATE

ONLINE STUDY TOOLS

 **connect** |ANATOMY & PHYSIOLOGY | **LEARNSMART** AP|R

The following study aids may be accessed through Connect.

Clinical Case Study: A Young Man with Lung Disease and Bruising

Interactive Questions: This chapter's content is served up in a number of multimedia question formats for student study.

LearnSmart: Topics and terminology include nutrients; obtaining nutrients from food; regulating blood levels of nutrients; functions of the liver; central role of cellular respiration; energy and heat

Anatomy & Physiology Revealed: Topic includes liver

Animations: Topics include B vitamins; complementary proteins; anatomy of a food label

chapter
28

Reproductive System

Anatomy & Physiology **REVEALED**®
aprevealed.com

Module 14: Reproductive System

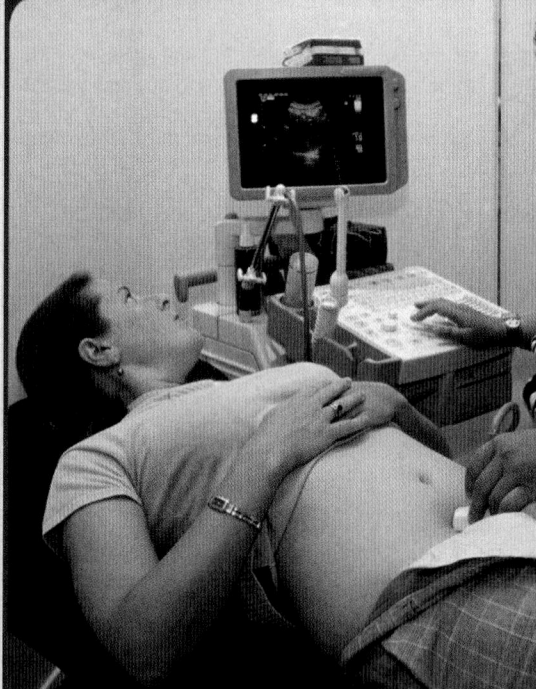

INTEGRATE

CAREER PATH
Ultrasound Technician

An ultrasound technician is trained in the medical field of sonography, which uses sound waves to visualize internal organs. In addition to a general knowledge of anatomy and physiology, these health-care professionals receive additional training in the use of the ultrasound equipment. This individual must be able to accurately locate internal organs and recognize potential abnormalities. Here, an ultrasound technician is examining the female reproductive organs and determining the viability of an early pregnancy.

The female and male reproductive systems provide the means for both the sexual maturation of each individual and the production of special cells necessary to propagate the next generation. Here we first discuss the general similarities between the two reproductive systems. Gametogenesis (the formation of gametes) and the process of meiosis (sex cell division) are discussed and examined before we describe the structures and functions of both the female and male reproductive systems. Our discussion concludes with a description of reproductive system development.

28.1 Overview of Female and Male Reproductive Systems

The female and male reproductive systems have obvious differences, and yet they share several general characteristics. For example, some mature reproductive system structures in both sexes are derived from common developmental structures called primordia and serve a common function in adults. Such structures are called **homologues** (hōm'ō-log; *homo* = same, alike, *logos* = relation). They are compared in **table 28.1** and are described in detail later in the chapter.

28.1a Common Elements of the Two Systems

LEARNING OBJECTIVE

1. List the similarities between the female and male reproductive systems.

Both reproductive systems have elements in common, including the general processes that produce the gametes, the production of hormones, and the maturation of reproductive capability:

- Both females and males have primary reproductive organs called **gonads** (gō'nad; *gone* = seed): These are ovaries in females and testes in males. The gonads produce sex cells called **gametes** (gam'ēt; husband or wife), which unite at fertilization to initiate the formation of a new individual.
- The gonads also produce relatively large amounts of **sex hormones,** which affect maturation, development, and changes in the activity of the reproductive system organs.
- Both sexes also exhibit **accessory reproductive organs,** including ducts to carry gametes away from the gonads toward the site of fertilization (in females) or to the outside of the body (in males).

The sexual union between a female and a male is known as **copulation** (kop-ū-lā'shŭn; *copulatio* = a joining), **coitus** (kōi'-tŭs; to come together), or **sexual intercourse.** If fertilization occurs, then the support, protection, and nourishment of the developing human occurs within the female reproductive tract.

WHAT DID YOU LEARN?

1 What general components make up the reproductive systems in both females and males?

Table 28.1	Reproductive System Homologues	
Female Organ	**Male Organ Homologue**	**Common Function**
Ovary	Testis	Produces gametes and sex hormones
Clitoris	Glans of penis	Contains erectile tissue that stimulates sexual arousal and climax
Labia minora	Body of penis	Contains erectile tissue that stimulates sexual arousal and climax
Labia majora	Scrotum	Protect and cover some reproductive structures
Greater vestibular gland	Bulbourethral gland	Secretes mucin for lubrication

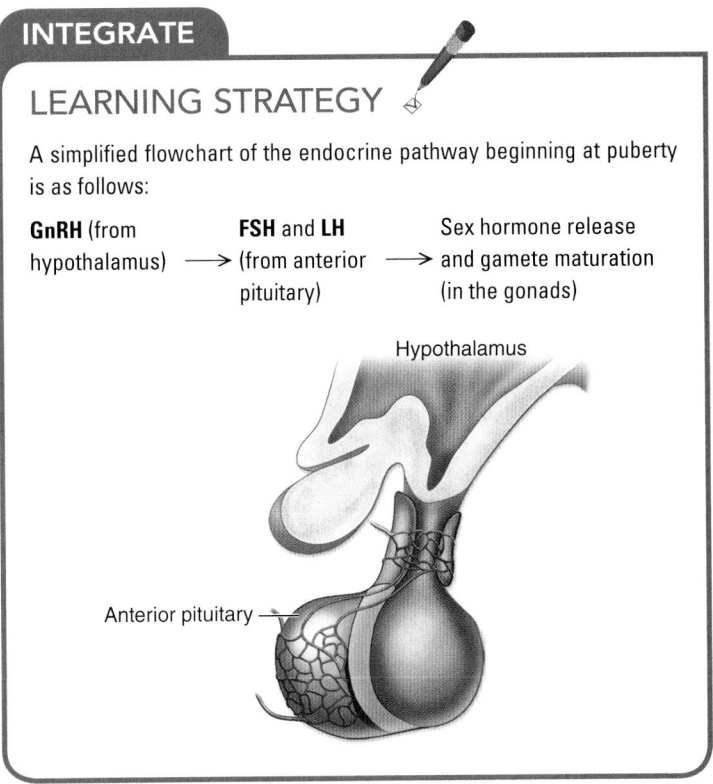

INTEGRATE

LEARNING STRATEGY

A simplified flowchart of the endocrine pathway beginning at puberty is as follows:

GnRH (from hypothalamus) ⟶ **FSH** and **LH** (from anterior pituitary) ⟶ Sex hormone release and gamete maturation (in the gonads)

Hypothalamus

Anterior pituitary

28.1b Sexual Maturation in Females and Males

LEARNING OBJECTIVE

2. Identify the hormones responsible for initiating puberty in females and males.

Both the female and male reproductive systems are primarily nonfunctional and dormant until a time in adolescence known as puberty. At **puberty** (pū'ber-tē; *puber* = grown up), external sex characteristics become more prominent, such as breast enlargement in females, pubic hair growth in both sexes, and fully functional reproductive organs in both sexes. Gametes begin to mature, and the gonads start to secrete their sex hormones.

Puberty is initiated when the hypothalamus begins secreting **gonadotropin-releasing hormone (GnRH)** (see section 17.7). GnRH acts on specific endocrine cells in the anterior pituitary and stimulates them to release the gonadotropins **follicle-stimulating hormone (FSH)** and **luteinizing hormone (LH).** (Prior to puberty, FSH and LH are virtually nonexistent in girls and boys.) As levels of FSH and LH increase, the gonads produce significant levels of sex hormones and start the processes of both gamete and sexual maturation.

Although both reproductive systems produce gametes, the female typically produces and releases only a single gamete, termed an oocyte (ō'ō-sīt'; *oon* = egg), monthly, whereas the male produces large numbers of gametes, or sperm, daily—about 100 million per day. These male gametes are stored for a short time, and if they are not expelled from the body within that period, they are resorbed.

WHAT DID YOU LEARN?

2 What hormones begin to be secreted at puberty, and what are their general functions?

28.1c Anatomy of the Perineum

LEARNING OBJECTIVE

3. Compare the components of the perineum in females and males.

The **perineum** (per'i-nē'ŭm) is a diamond-shaped area between the thighs in both females and males that is bounded anteriorly by the pubic

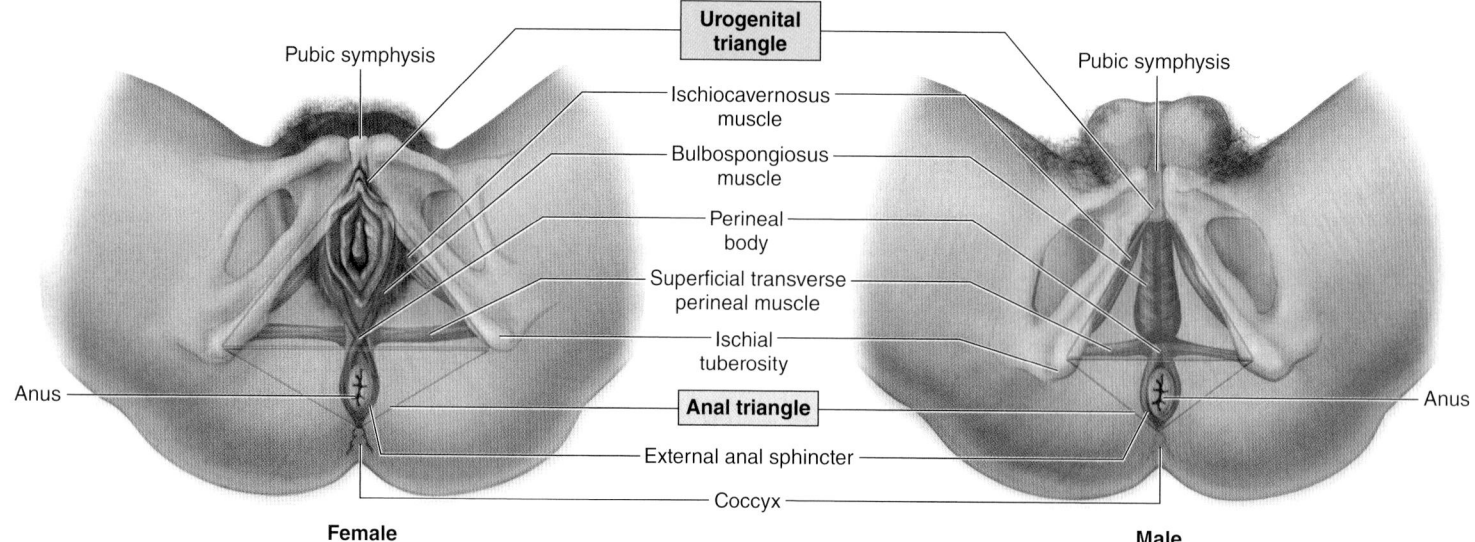

Urogenital
triangle

Pubic symphysis

Pubic symphysis

Ischiocavernosus
muscle

Bulbospongiosus
muscle

Perineal
body

Superficial transverse
perineal muscle

Ischial
tuberosity

Anal triangle

Anus

Anus

External anal sphincter

Coccyx

Female

Male

Figure 28.1 Perineum. In both females and males, the perineum is the diamond-shaped area between the thighs, extending from the pubis anteriorly to the coccyx posteriorly, and bordered laterally by the ischial tuberosities. An imaginary horizontal line extending from the ischial tuberosities subdivides the perineum into a urogenital triangle anteriorly and an anal triangle posteriorly. AP|R

symphysis, laterally by the ischial tuberosities, and posteriorly by the coccyx (figure 28.1). It is an area that may be partially torn during childbirth, due to the extensive stretching needed to expel the baby. (Sometimes, this region may have to be surgically cut during birth if there is difficulty expelling the baby, see section 29.6d.) Two distinct triangle bases are formed by an imaginary horizontal line extending between the ischial tuberosities of the ossa coxae. Both triangles house specific structures in the floor of the pelvis:

- The anterior triangle, called the **urogenital triangle,** contains the urethral and vaginal orifices in females and the base of the penis and the scrotum in males. Within the urogenital triangle are the muscles that surround the external genitalia, called the ischiocavernosus, bulbospongiosus, and superficial transverse perineal muscles.

- The posterior triangle, called the **anal triangle,** is the location of the anus in both sexes. Surrounding the anus is the external anal sphincter.

The external anal sphincter, bulbospongiosus, and superficial transverse perineal muscles are partly anchored by a dense connective tissue structure called the perineal body. Review table 11.12 and figure 11.17 in section 11.7 when learning these structures.

WHAT DID YOU LEARN?

3 What are the components of the urogenital triangles in females and males?

28.2 Gametogenesis

Gametogenesis (gam′ĕ-tō-jen′ĕ-sis; *gameto* = gamete, *genesis* = beginning) is the process of forming human sex cells, called gametes. Female gametes are secondary oocytes (commonly referred to as "eggs"), whereas male gametes are sperm. The process of gametogenesis begins with a specific type of cell division called meiosis. The events of meiosis generally are similar in both females and males, with a few differences. Here we describe the basics of heredity and meiosis; sex-specific differences are described later in this chapter.

28.2a A Brief Review of Heredity

LEARNING OBJECTIVES

1. Distinguish between autosomes and sex chromosomes.

2. Explain why somatic cells contain 2*n* chromosomes, but gametes must contain *n* chromosomes.

Humans pass on their traits to a new individual when they reproduce. This hereditary information is carried on chromosomes. Human somatic (body) cells contain 23 pairs of chromosomes—22 pairs of autosomes and one pair of sex chromosomes, for a total of 46 chromosomes:

- **Autosomes** contain genes that code for cellular functions. These genes also help determine most human characteristics, such as eye color, hair color, height, and skin pigmentation. A pair of matching autosomes are called **homologous chromosomes** (hō-mol′ō-gŭs; *homos* = same, *logos* = relation).
- The pair of **sex chromosomes** consists of either two X chromosomes or an X and a Y chromosome. These chromosomes primarily determine whether an individual is female (two X chromosomes) or male (one X chromosome and one Y chromosome), although they also contain genes that code for cellular functions.

One member of each pair of chromosomes is inherited from each parent. In other words, if you examined one of your body cells, you would discover that 23 of the chromosomes came from your mother, and the other 23 chromosomes in this same cell came from your father. A cell that contains 23 pairs of chromosomes is said to be **diploid** (dip′loyd; *diploos* = double) and to have $2n$ chromosomes, where n is the unpaired chromosome number.

If the gametes were diploid, then each new individual would receive two sets of paired chromosomes, or $4n$. This situation is not what happens in nature, and it is not compatible in life. The gametes from either sex are **haploid** (hap′loyd; *haplos* = simple, *eidos* = appearance) because they contain 23 chromosomes only, not 23 pairs, and thus their chromosome number is designated as $1n$, or just n. (Heredity is discussed in more detail in section 29.9.)

 WHAT DID YOU LEARN?

4 How do sex chromosomes differ from autosomes?

5 Why do gametes have to be haploid instead of diploid?

28.2b An Overview of Meiosis

LEARNING OBJECTIVES

3. Compare and contrast meiosis and mitosis.
4. Describe events during interphase, before cell division begins.
5. Explain the difference between homologous chromosome pairs and sister chromatids.

Meiosis (mī-ō′sis; *meiosis* = lessening) is sex cell division that starts off with a diploid parent cell and produces haploid daughter cells called gametes. Mitosis (somatic cell division, described in section 4.9) and meiosis (sex cell division) differ in the following ways:

- Mitosis produces *two* daughter cells that are genetically *identical* to the parent cell. In contrast, meiosis produces *four* daughter cells that are genetically *different* from the parent cell.
- Mitosis produces daughter cells that are *diploid,* whereas meiosis produces daughter cells that are *haploid.*
- Meiosis includes a process called **crossing over,** whereby genetic material is exchanged between homologous chromosomes.

Crossing over helps "shuffle the genetic deck of cards," so to speak. Thus, crossing over is a means of combining different genes from both parents on one of the homologous chromosomes. Crossing over does not occur in mitosis.

Meiosis begins with a diploid parent cell located in the gonad (ovary or testis). In this cell, 23 chromosomes came from the organism's mother (23 maternal chromosomes), and 23 from the father (23 paternal chromosomes). To produce haploid gametes, this parent cell must undergo meiosis.

The cell phase called interphase occurs prior to meiosis. During interphase, the DNA in each chromosome is replicated, or duplicated exactly, in the parent cell. These **replicated chromosomes** (also known as *double-stranded chromosomes*) are composed of two identical structures called **sister chromatids** (krō′mă-tid; *chromo* = color, *id* = resembling). Each sister chromatid contains an identical copy of DNA at this point. The sister chromatids are attached at a specialized region termed the **centromere.**

Note that sister chromatids are *not* the same as a "pair" of chromosomes. A chromosome composed of sister chromatids resembles a written letter X, whereas a homologous *pair* of

LEARNING STRATEGY

Meiosis I (the first meiotic division) randomly separates maternal and paternal *pairs* of replicated (double-stranded) chromosomes, whereas meiosis II (the second meiotic division) separates replicated chromosomes into single chromosomes. Also, meiosis II is very similar to mitosis. Thus, if you remember the steps of mitosis, you can figure out the steps of meiosis II.

Table 28.2 — Cell Division Terminology

Term	Definition	Image
Replicated chromosome (also known as a *double-stranded chromosome, duplicated chromosome*)	A chromosome that initially has two identical sister chromatids joined at the centromere (Note: Once crossing over occurs, the sister chromatids are no longer identical.)	Sister chromatids / Centromere
Pair of chromosomes	A homologous maternal chromosome and paternal chromosome	
Single chromosome (also known as a *single-stranded chromosome*)	A chromosome consisting of a single chromatid and a centromere	

chromosomes is composed of a matching maternal chromosome and paternal chromosome, which are not attached at a centromere. Therefore, after interphase, there are 23 pairs of replicated chromosomes. To assist you with this terminology, please refer to **table 28.2**.

Once the DNA is replicated in interphase, the phases of meiosis begin (**figure 28.2**). Meiosis is comprised of two separate sequential events: meiosis I and meiosis II.

WHAT DID YOU LEARN?

6 In what ways do mitosis and meiosis differ?

7 What is the difference between sister chromatids and paired homologous chromosomes?

28.2c Meiosis I: Reduction Division

LEARNING OBJECTIVES

6. Discuss why meiosis I is termed reduction division.

During **meiosis I** (also known as the *first meiotic division*), the homologous pairs of replicated chromosomes are separated when the cell divides. The result is two cells, each of which contains 23 chromosomes only (not 23 pairs) that consist of replicated sister chromatids held together at a centromere. Because meiosis I results in the reduction of the number of chromosomes in the daughter cells, it is also called *reduction division*. Meiosis I consists of four phases plus cytokinesis, described next.

MEIOSIS I

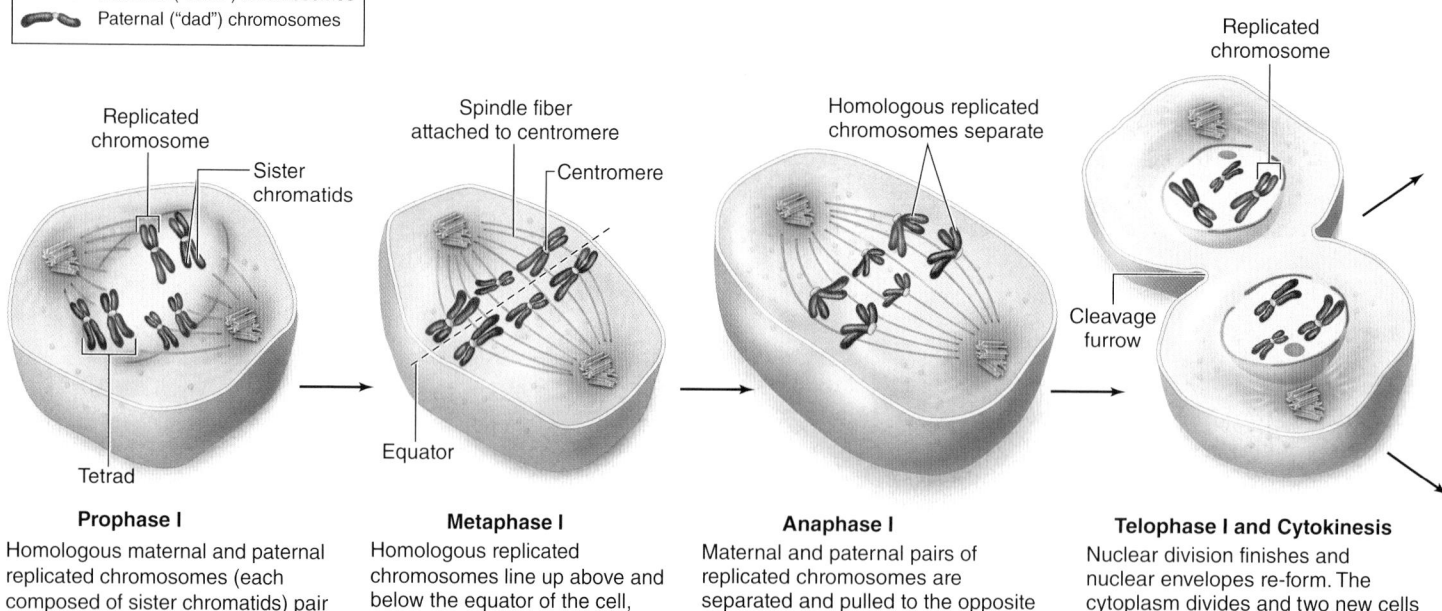

| Maternal ("mom") chromosomes |
| Paternal ("dad") chromosomes |

Prophase I
Homologous maternal and paternal replicated chromosomes (each composed of sister chromatids) pair up (synapsis), and the pair forms a tetrad. Crossing over occurs between homologous maternal and paternal chromosomes, increasing genetic diversity.

Metaphase I
Homologous replicated chromosomes line up above and below the equator of the cell, forming a double line of chromosomes. Spindle fibers attach to the centromeres.

Anaphase I
Maternal and paternal pairs of replicated chromosomes are separated and pulled to the opposite ends of the cell, a process called reduction division. Note that the sister chromatids remain attached in each replicated chromosome.

Telophase I and Cytokinesis
Nuclear division finishes and nuclear envelopes re-form. The cytoplasm divides and two new cells are produced, each containing 23 replicated chromosomes only. The replicated chromosomes are still composed of sister chromatids.

Figure 28.2 Meiosis. Meiosis is a type of cell division that results in the formation of gametes (sex cells). In meiosis I, homologous pairs of chromosomes are separated after synapsis and crossing over occurs. In meiosis II, sister chromatids are separated in a sequence of phases that resembles the steps of mitosis. AP|R

Prophase I

Homologous replicated chromosomes in the parent cell pair up to form a **tetrad.** (Keep in mind that each chromosome here consists of two sister chromatids.) The process by which homologous chromosomes pair up is called **synapsis** (si-nap′sis; *syn* = together).

As the maternal and paternal chromosomes come close together, crossing over occurs. The homologous chromosomes within each tetrad exchange genetic material at this time. A portion of the genetic material in a sister chromatid of a maternal chromosome is exchanged with the same portion of genetic material in a sister chromatid of a paternal chromosome. This shuffling of the genetic material ensures continued genetic diversity in new organisms. (Note that after crossing over, the sister chromatids in a replicated chromosome are no longer identical.) Prophase I ends with the breakdown of the nuclear envelope.

Metaphase I

The homologous pairs of each tetrad line up on either side of the midline or equator of the cell, forming a double line of chromosomes. This alignment of paired chromosomes is random with respect to whether the original maternal or paternal chromosome of a pair is on one side of the equator or the other. For example, some maternal chromosomes may be to the left of the equator, and other maternal chromosomes may be to the right. This random alignment of homologous chromosomes is called **independent assortment.**

Spindle fibers formed by microtubules extend from centrioles at opposite ends of the cell and attach to the centromere of each homologous replicated chromosome.

Anaphase I

The homologous pairs of chromosomes separate and are pulled to the opposite ends of the cell. Thus, a maternal chromosome consisting of two sister chromatids is pulled to one side of the cell, while the homologous paternal chromosome is pulled to the opposite side. The process by which maternal and paternal chromosome pairs are separated and move to opposite ends of the cell is referred to as **reduction division,** because each daughter cell receives only one-half the starting number of chromosomes (only 23 chromosomes of the original 23 pairs). Note that the pairs of chromosomes are no longer together; however, each replicated chromosome still consists of two sister chromatids. In addition, recognize that not all maternal chromosomes are pulled to the same side of the cell; because of independent assortment in metaphase I, there is a random alignment of maternal and paternal chromosomes.

Telophase I and Cytokinesis

The replicated chromosomes arrive at opposite ends of the cell, and then a nuclear membrane may re-form around these chromosomes. A cleavage furrow forms in the cell, and the cell cytoplasm divides (cytokinesis) to produce two new cells. Each daughter cell now contains 23 replicated chromosomes only, but each replicated chromosome is still composed of two sister chromatids bound together.

The two cells formed in this stage must undergo another round of cell division, meiosis II, to separate the sister chromatids.

WHAT DID YOU LEARN?

8 What is independent assortment and during what phase of meiosis does it occur?

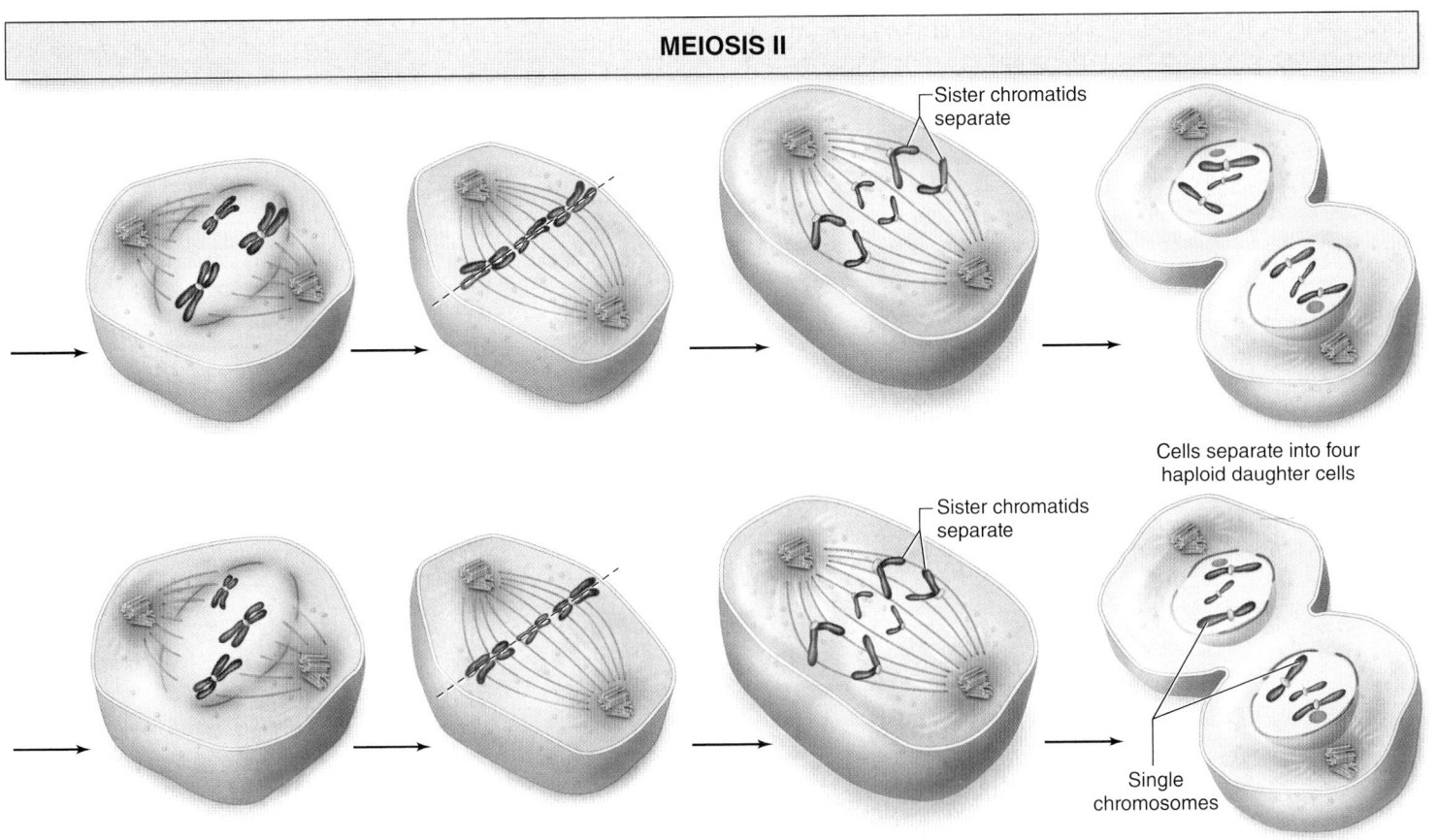

MEIOSIS II

Sister chromatids separate

Cells separate into four haploid daughter cells

Sister chromatids separate

Single chromosomes

Prophase II	Metaphase II	Anaphase II	Telophase II and Cytokinesis
Nuclear envelope breaks down, and replicated chromosomes cluster together. (There is no crossing over in prophase II.)	Spindle fibers extend from the centrioles to each sister chromatid in each chromosome and align replicated chromosomes along the equator of the cell.	Sister chromatids of each replicated chromosome are pulled apart at the centromere. Sister chromatids (now called single chromosomes) migrate to opposite ends of the cell.	Nuclear division finishes, and the nuclear envelopes re-form. The four new daughter cells that are produced each contain 23 single chromosomes only.

9 What is reduction division and why is it necessary in production of gametes?

28.2d Meiosis II: Separation of Sister Chromatids

LEARNING OBJECTIVE

7. Describe the events of meiosis II and the final outcome of the two stages of meiosis.

After meiosis I, the two daughter cells formed each contain 23 replicated chromosomes consisting of two connected sister chromatids. In **meiosis II** (also known as the *second meiotic division*), these chromatids separate and become single chromosomes in haploid cells.

Prophase II

The second prophase event of meiosis resembles the prophase stage of mitosis. In each of the two new cells, the nuclear envelope breaks down, and the chromosomes collect together. However, crossing over does not occur in this phase because no homologous chromosome pairs are present; they were separated in anaphase I.

Metaphase II

Spindle fibers extend from the centrioles to the centromere of each chromatid. The replicated chromosomes (composed of sister chromatids) are aligned to form a single line along the equator in the middle of the cell.

Anaphase II

The sister chromatids of each replicated chromosome are pulled apart at the centromere and are now separated. Each sister chromatid is now called a single chromosome (or *single-stranded chromosome*) and is pulled to the opposite pole of the cell.

Telophase II and Cytokinesis

The single chromosomes arrive at opposite ends of the cell. Nuclear membranes re-form, a cleavage furrow forms, and the cytoplasm in both cells divides.

These meiotic processes produce four daughter cells from the original single cell. These daughter cells are haploid, because they contain 23 chromosomes only, not 23 pairs. Twenty-two of these are autosomes, and one is a sex chromosome.

These daughter cells mature into secondary oocytes in females, and they mature into sperm in males. These processes are described in more detail later in this chapter.

WHAT DID YOU LEARN?

10 At what point during meiosis do the daughter cells become haploid?

INTEGRATE

CLINICAL VIEW
Nondisjunction

Normally, the two members of a homologous chromosome pair separate during meiosis I, and paired sister chromatids separate during meiosis II. In **nondisjunction,** either homologous chromosomes or sister chromatids fail to separate properly, and both components move into one cell. As a result of nondisjunction, one gamete may receive two copies of a single chromosome and thus have 24 chromosomes, whereas the other gamete would receive no copy of a specific chromosome and would have only 22 chromosomes. If either of these cells unites with a normal gamete at fertilization, the resulting individual will have either 47 chromosomes **(trisomy)** or 45 chromosomes **(monosomy).** Trisomy means the individual has three copies of a chromosome, whereas monosomy means an individual has only one copy of a chromosome.

Most trisomies and monosomies are lethal to the embryo, but some that involve smaller chromosomes allow survival. A trisomy disorder is named according to the specific chromosome that has three copies. The most well-known of these is **trisomy 21,** also known as **Down syndrome.** A person with trisomy 21 typically has some form of intellectual disability, slanting epicanthal folds (eye creases), heart defects, poor muscle tone, and short stature. Many, but not all, cases of Down syndrome occur due to nondisjunction in the formation of the oocyte. The incidence of Down syndrome increases with the mother's age; however, cases are known in which the nondisjunction occurs in formation of sperm. (A less common cause of Down syndrome is the result of part of chromosome 21 translocating [breaking off and attaching] to chromosome 14.)

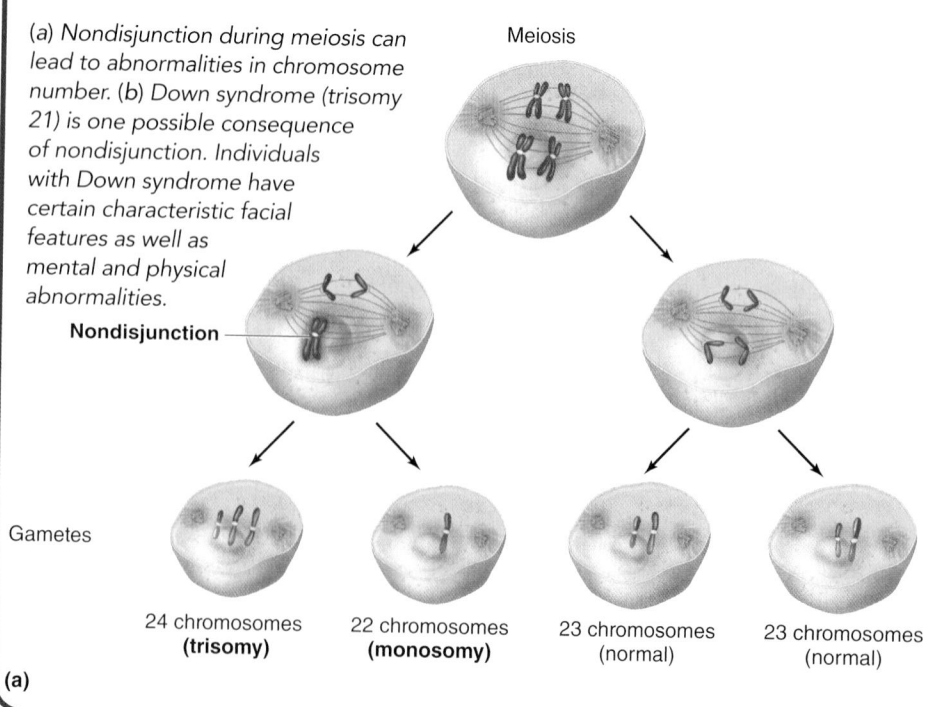

(a) Nondisjunction during meiosis can lead to abnormalities in chromosome number. (b) Down syndrome (trisomy 21) is one possible consequence of nondisjunction. Individuals with Down syndrome have certain characteristic facial features as well as mental and physical abnormalities.

Meiosis

Nondisjunction

Gametes

24 chromosomes **(trisomy)**   22 chromosomes **(monosomy)**   23 chromosomes (normal)   23 chromosomes (normal)

(a)

(b)

28.3 Female Reproductive System

A sagittal section through the female pelvis illustrates the internal reproductive structures and their relationships to the urinary bladder and rectum (figure 28.3). The peritoneum folds over selected pelvic organs and produces two major dead-end recesses, or pouches. The anterior **vesicouterine** (ves'i-kō-yū'ter-in; *vesica* = bladder, *utero* = uterus) **pouch** forms the space between the urinary bladder and the uterus, and the posterior **rectouterine** (rek'tō-yū'ter-in) **pouch** forms the space between the rectum posteriorly and the uterus anteriorly.

The primary reproductive organs of the female are the ovaries. The accessory reproductive organs include the uterine tubes, uterus, vagina, external genitalia, and mammary glands.

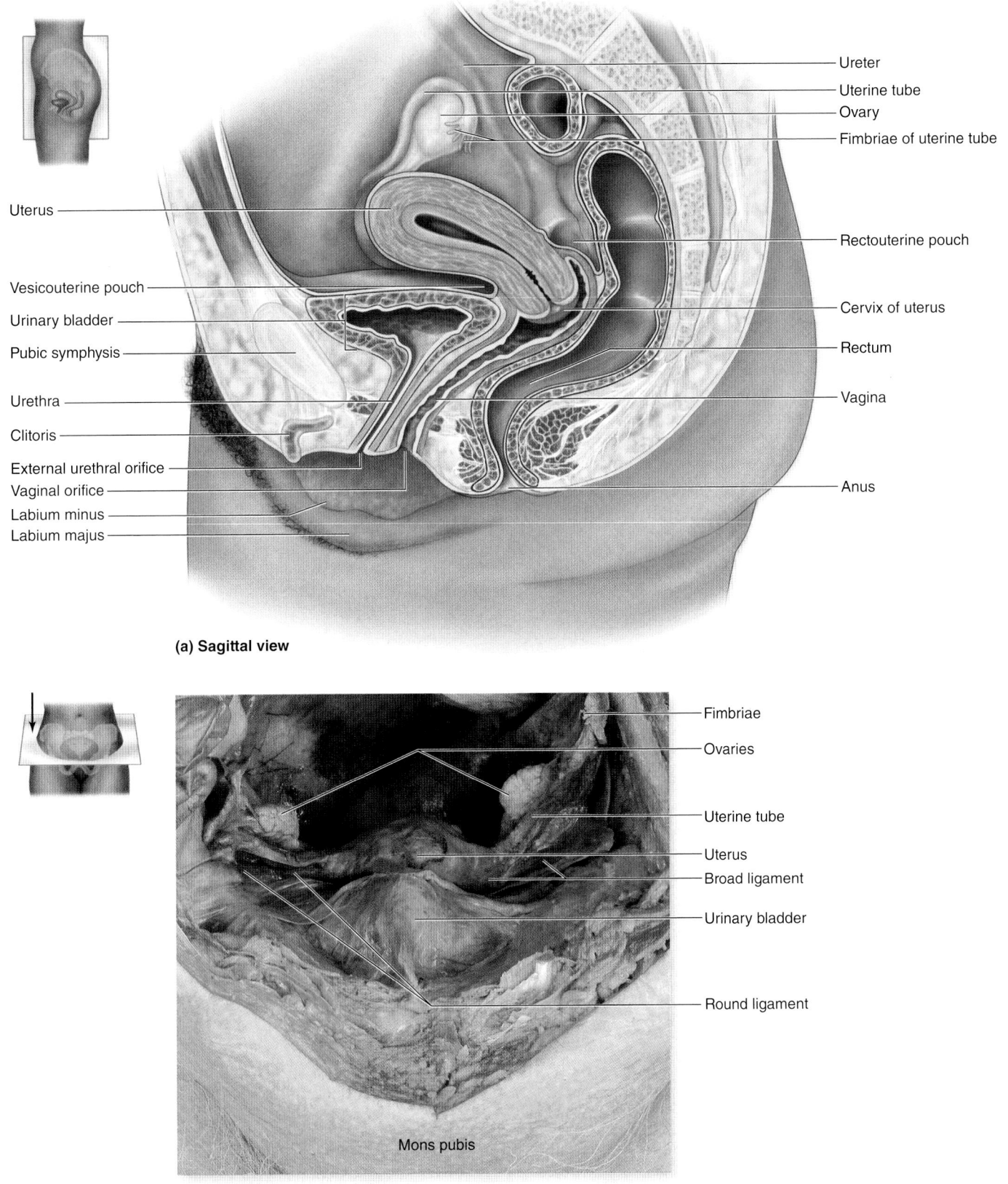

(a) Sagittal view

(b) Anterosuperior view

Figure 28.3 Female Pelvic Region. (*a*) A sagittal section of the female pelvis illustrates the position of the uterus with respect to the rectum and urinary bladder. (*b*) A cadaver photo provides an anterosuperior view of the female pelvis and reproductive organs. AP|R

28.3a Ovaries

LEARNING OBJECTIVES

1. Describe the gross and microscopic anatomy of the ovary.

2. Compare the different types of ovarian follicles that form in the ovary.

Here we consider the anatomy of the ovaries, and more specifically, the types of ovarian follicles seen in a mature ovary. The ovarian follicles are the site of oocyte production and release of the female sex hormones estrogen and progesterone. (Note: The details for both estrogen and progesterone are included in the summary table "Regulating the Female Reproductive System," which directly follows chapter 17 (see **table R.9**).

Anatomy of Ovaries

The **ovaries** are paired, oval organs located within the pelvic cavity lateral to the uterus (**figure 28.4**). In an adult, the ovaries are slightly larger than an almond—about 2 to 3 centimeters (cm) long, 2 cm wide, and 1 to 1.5 cm thick. Their size usually varies during each menstrual cycle as well as during pregnancy.

The ovaries are anchored within the pelvic cavity by specific cords and sheets of connective tissue. A double fold of peritoneum, called the **mesovarium** (mez′ō-vā′rē-ŭm), attaches to each ovary at its **hilum**, which is the anterior surface of the ovary where its blood vessels and nerves enter and leave the organ. The mesovarium secures each ovary to a **broad ligament**, which is a drape of peritoneum that hangs over the uterus. Each ovary is anchored to the posterior aspect of the broad ligament by an **ovarian ligament**, which is the superior portion of the round ligament of the uterus. Finally, a **suspensory ligament** attaches to the lateral edge of each ovary and projects superolaterally to the pelvic wall. The ovarian blood vessels and nerves are housed within each suspensory ligament, and they join the ovary at its hilum.

Each ovary is supplied by both an **ovarian artery** and an **ovarian vein.** The ovarian arteries are direct branches off the aorta, immediately inferior to the renal vessels. The ovarian veins exit the ovary and drain into either the inferior vena cava or one of the renal veins. Traveling with the ovarian artery and vein are autonomic nerves (see sections 15.3 and 15.4). Sympathetic axons extend from the T10 segments of the spinal cord, whereas parasympathetic axons extend from CN X (vagus).

When an ovary is sectioned and viewed microscopically, many features are visible (figure 28.4b). Surrounding the ovary is a thin, simple cuboidal epithelial layer called the **germinal epithelium,** so named because early anatomists erroneously thought it was the origin of the female germ (sex) cells. Deep to the germinal epithelium is a dense connective tissue capsule called the **tunica albuginea** (al-bū-jin′ē-ă; *albugo* = white spot). Deep to the tunica albuginea, the ovary can be partitioned into an outer **cortex** and an inner **medulla**. The cortex contains highly cellular connective tissue and ovarian follicles (described next), while the medulla is composed of areolar connective tissue and contains branches of the ovarian blood vessels, lymph vessels, and nerves.

Ovarian Follicles

Within the cortex are thousands of **ovarian follicles.** Ovarian follicles consist of an **oocyte** surrounded by **follicle cells** (or *granulosa cells*), which support the oocyte. There are several different types of ovarian follicles, each representing a different stage of development; here we describe the six main types (**figure 28.5**):

1. A **primordial follicle** is the most primitive type of ovarian follicle. Each primordial follicle consists of a **primary oocyte**

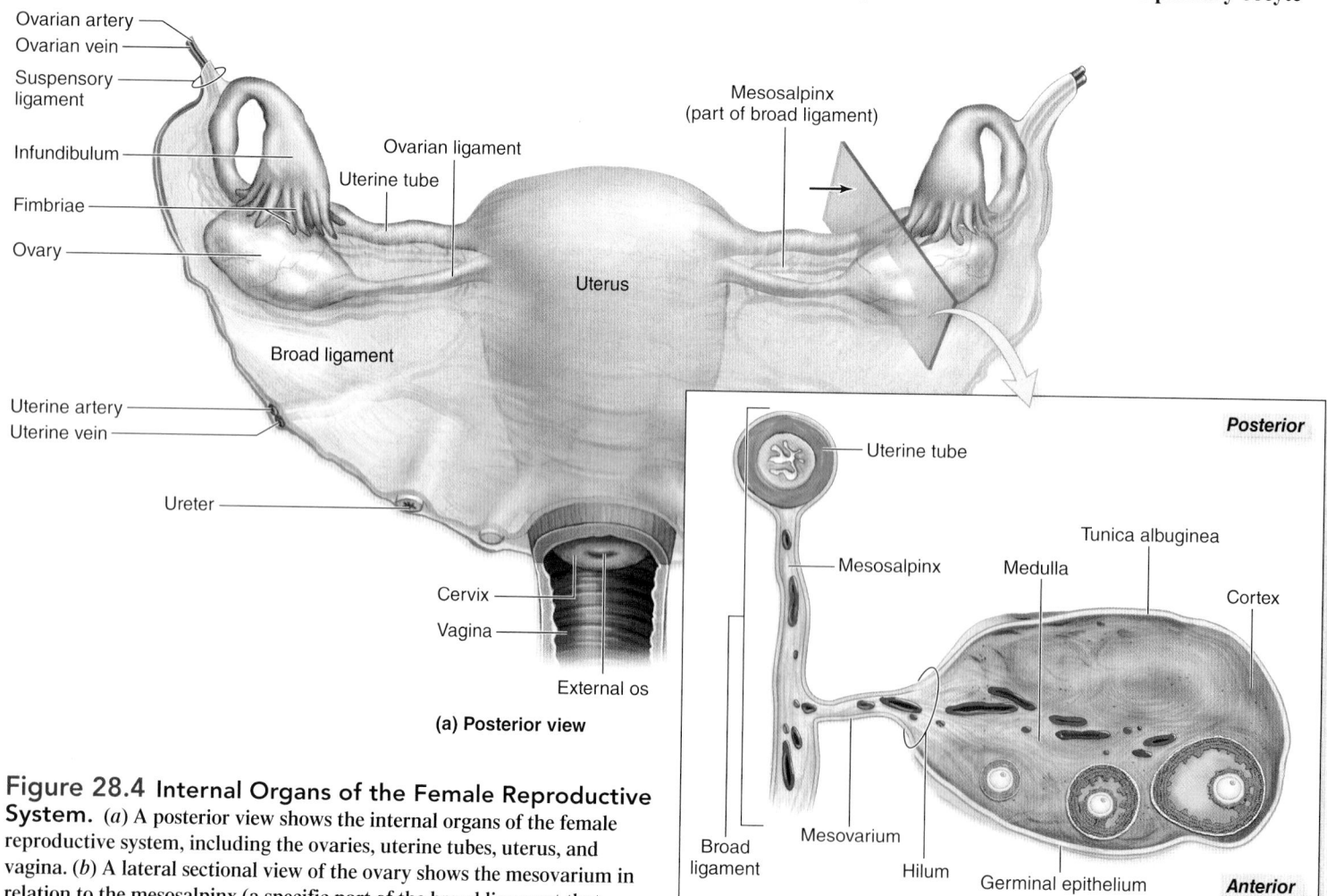

(a) Posterior view

Figure 28.4 Internal Organs of the Female Reproductive System. (*a*) A posterior view shows the internal organs of the female reproductive system, including the ovaries, uterine tubes, uterus, and vagina. (*b*) A lateral sectional view of the ovary shows the mesovarium in relation to the mesosalpinx (a specific part of the broad ligament that overlies the uterine tubes). AP|R

(b) Lateral sectional view

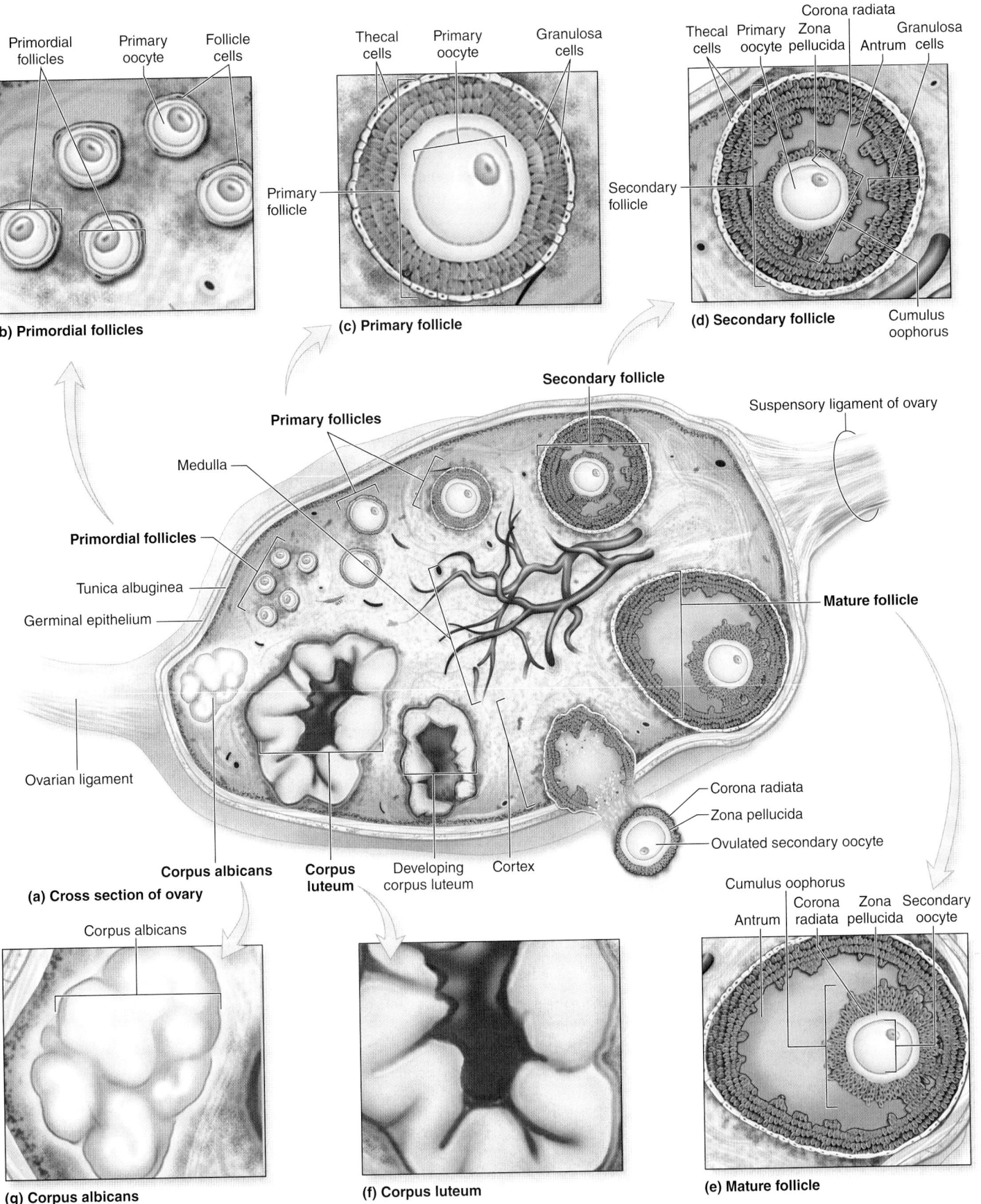

(b) Primordial follicles

Primordial follicles Primary oocyte Follicle cells

(c) Primary follicle

Thecal cells Primary oocyte Granulosa cells Primary follicle

(d) Secondary follicle

Corona radiata
Thecal cells Primary oocyte Zona pellucida Antrum Granulosa cells
Secondary follicle
Cumulus oophorus

(a) Cross section of ovary

Primary follicles
Secondary follicle
Suspensory ligament of ovary
Medulla
Primordial follicles
Tunica albuginea
Germinal epithelium
Mature follicle
Ovarian ligament
Corona radiata
Zona pellucida
Ovulated secondary oocyte
Corpus albicans Corpus luteum Developing corpus luteum Cortex

(g) Corpus albicans

Corpus albicans

(f) Corpus luteum

(e) Mature follicle

Cumulus oophorus
Antrum Corona radiata Zona pellucida Secondary oocyte

Figure 28.5 Ovary. The ovary produces and releases both female gametes (secondary oocytes) and sex hormones. (*a*) A coronal view of the ovary depicts the different stages of follicle maturation, ovulation, and corpus luteum development and degeneration. Note that all of the follicles and structures shown in this image would appear at different times during the ovarian cycle—they do not occur simultaneously. Further, the follicles do not migrate through the ovary; rather, all follicles are shown together merely for comparative purposes. (*b–e*) The stages of follicle development, ending with the mature follicle, which ruptures during ovulation. After ovulation, the remnant of the mature follicle forms (*f*) the corpus luteum, which then degenerates into (*g*) the corpus albicans. AP|R

INTEGRATE

LEARNING STRATEGY

To distinguish a primary oocyte from a secondary oocyte, remember that:

- A **p**rimary oocyte is arrested in prophase **I** (the term "primary" also means "one").

- A **s**econdary oocyte is arrested in metaphase **II** (the term "secondary" means "two").

Additionally, remember that the *only* ovarian follicle containing a secondary oocyte is a mature follicle—all other ovarian follicles have primary oocytes only.

surrounded by a single layer of *flattened* follicle cells. A primary oocyte is an oocyte that is arrested in the first meiotic prophase. About 1.5 million of these types of follicles are present in both ovaries at birth.

2. A **primary follicle** forms from a maturing primordial follicle. Each primary follicle consists of a primary oocyte surrounded by one or more layers of *cuboidal* follicular cells, which are now called **granulosa cells.** Each primary follicle secretes **estrogen** (es′trō-jen; *oistrus* = estrus, *gen* = producing) as it continues to mature, which stimulates changes in the uterine lining. Some connective tissue derived cells, called **thecal cells,** are located around and on the periphery of the ovarian follicle. These cells help initiate and control follicle development by influencing hormone production. Specifically, the thecal cells secrete androgens, which are converted to estrogen by the granulosa cells.

3. A **secondary follicle** forms from a primary follicle. Each secondary follicle contains a primary oocyte, many layers of granulosa cells, and a fluid-filled space called an **antrum.** Within the antrum, serous fluid accumulates and increases in volume as ovulation nears. The oocyte is forced toward one side of the follicle, where it is surrounded by a cluster of follicle cells termed the **cumulus oophorus** (kū′mū-lŭs; heap) (ō-of′ōr-ŭs; *phorus* = bearing).

   Surrounding the primary oocyte are two protective structures, the zona pellucida and the corona radiata. The **zona pellucida** (zō′nă; zone) (pe-lū′sĭd-ă; *pellucidus* = allowing the passage of light) is a translucent structure that contains glycoproteins. External to the zona pellucida is the **corona radiata** (kō-rō′nă; crown) (rā-dē-ā′tă; radiating), which is the innermost layer of cumulus oophorus cells.

4. A **mature follicle** (also called a *vesicular follicle,* or *Graafian follicle*) forms from a secondary follicle. A mature follicle becomes quite large, and contains a **secondary oocyte** (surrounded by a zona pellucida and the corona radiata), numerous layers of granulosa cells, and a large, fluid-filled, crescent-shaped antrum. A secondary oocyte has completed meiosis I and is arrested in the second meiotic metaphase. Typically, only one mature follicle forms each month.

5. When a mature follicle ruptures and expels its oocyte (a process called ovulation), the remnants of the follicle in the ovary turn into a yellowish structure called the **corpus luteum** (lū-tē′ŭm; *luteus* = saffron yellow). The corpus luteum secretes the sex hormones **progesterone** (prō-jes′ter-ōn; *pro* = before; gestation) and estrogen. These hormones stimulate and support the continuing buildup of the uterine lining and prepare the uterus for possible implantation of a fertilized oocyte.

6. When a corpus luteum undergoes regression (breaks down), it turns into a white connective tissue scar called the **corpus albicans** (al′bi-kanz; white). Most corpus albicans structures are completely resorbed, and only a few may remain within an ovary.

Table 28.3 summarizes the different structures that develop during a female's monthly cycle.

 WHAT DID YOU LEARN?

11 What are the broad ligament, ovarian ligament, and suspensory ligament of an ovary?

12 How are the primordial, primary, secondary, and mature follicles similar? How are they different?

Table 28.3	Ovarian Follicles and Structures That Develop in the Ovary		
Ovarian Structure	**Type of Oocyte**	**Anatomic Characteristics**	**Time of First Appearance**
Primordial follicle	Primary oocyte	Single layer of flattened follicular cells surround an oocyte	Fetal period
Primary follicle	Primary oocyte	Single or multiple layers of cuboidal granulosa cells surround an oocyte	Puberty
Secondary follicle	Primary oocyte	Multiple layers of granulosa cells surround the oocyte and a small, fluid-filled antrum	Puberty
Mature follicle	Secondary oocyte	Many layers of granulosa cells surround the oocyte and a very large antrum	Puberty
Corpus luteum	No oocyte	Yellowish, collapsed folds of granulosa cells	Puberty
Corpus albicans	No oocyte	Whitish connective tissue scar, remnant of a degenerated corpus luteum	Puberty

28.3b Oogenesis and the Ovarian Cycle

LEARNING OBJECTIVES

3. List the hormones responsible for oogenesis, and explain each hormone's effect on oogenesis.
4. Identify the ovarian follicles that develop before birth and those that form after puberty.
5. Describe the three phases of the ovarian cycle.

Oogenesis (ō′ō-jen′ĕ-sis; *oon* = egg, *genesis* = origin) is the maturation of a primary oocyte to a secondary oocyte. We describe this process as it relates to the life stages of a female; it is illustrated in **figure 28.6**.

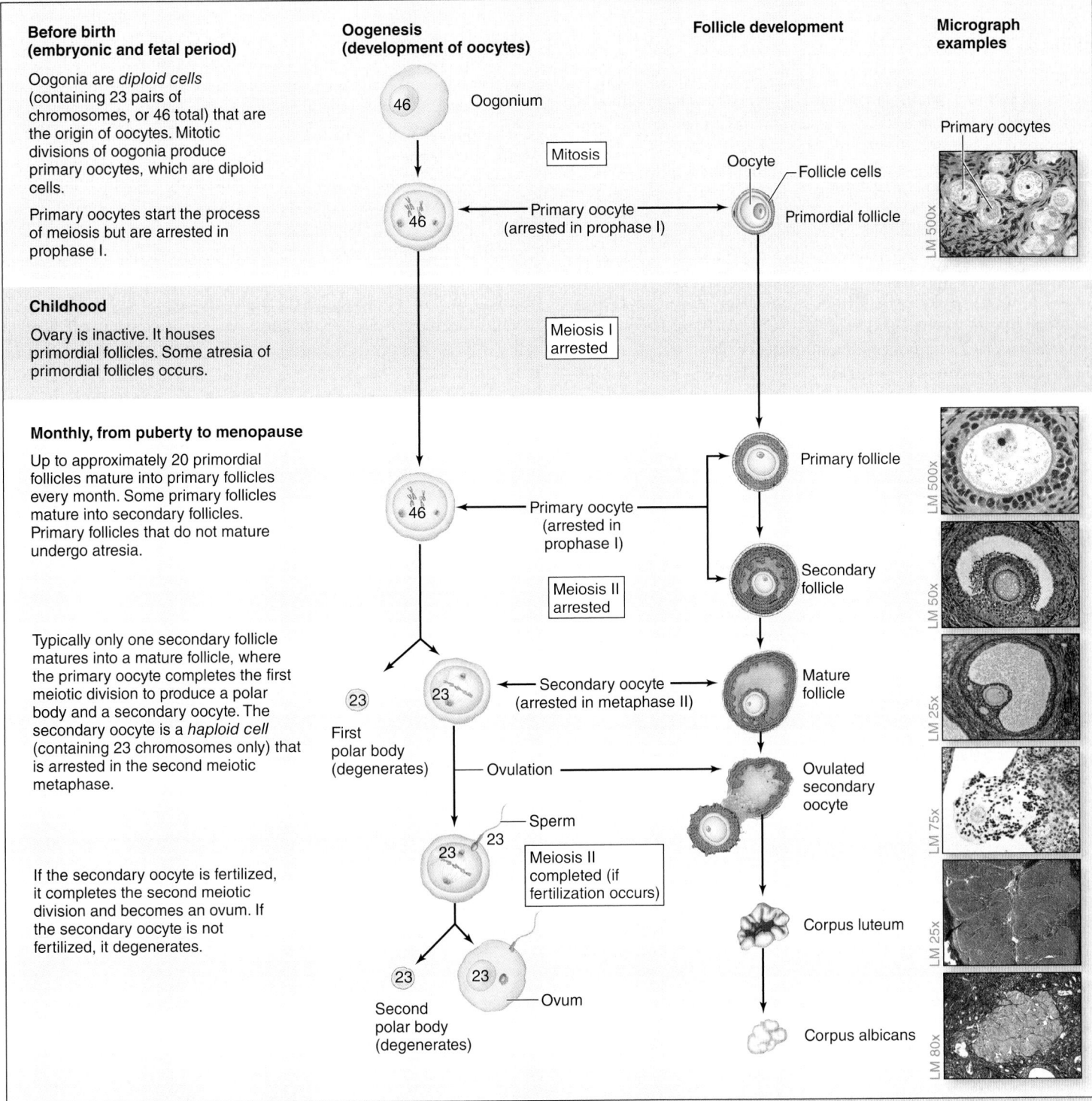

Figure 28.6 Oogenesis. Oogenesis begins in a female fetus, when primary oocytes form in primordial follicles. The ovary and these follicles remain inactive in childhood. At puberty, a select number of primordial follicles each month undergoes maturation and produces a female gamete (secondary oocyte).

Before Birth

The process of oogenesis begins in a female fetus before birth. At this time, the ovary contains primordial germ cells called **oogonia** (ō-ō-gō′nē-ă; sing., *oogonium*), which are diploid cells (they have 23 pairs of chromosomes). During the fetal period, the oogonia divide by mitosis to produce primary oocytes. Primary oocytes are the structures that start the process of meiosis, but they are arrested in prophase I. At this point, the cells are the primary oocytes described in the preceding section. At birth, the ovaries of a female child are estimated to contain a total of approximately 1.5 million primordial follicles within its cortex. The primary oocytes in the primordial follicles remain arrested in prophase I until after puberty.

Childhood

During childhood, a female's ovaries are inactive, and no follicles develop. In fact, the continuing event that occurs during childhood is **atresia** (ă-trē′zē-ă; *a* = not, *tresis* = a perforation), in which some primordial follicles regress. By the time a female child reaches puberty, only about 400,000 primordial follicles remain in the ovaries.

From Puberty to Menopause

When a female reaches puberty, the hypothalamus releases gonadotropin-releasing hormone (GnRH). GnRH stimulates the anterior pituitary to release follicle-stimulating hormone (FSH) and luteinizing hormone (LH). The levels of FSH and LH vary in a cyclical pattern and produce a monthly sequence of events in follicle development called the **ovarian cycle.**[1] The three phases of the ovarian cycle are the follicular phase, ovulation, and the luteal phase (**figure 28.7**).

Follicular Phase The **follicular phase** occurs during days 1–13 of an approximate 28-day ovarian cycle. At the beginning of the follicular phase, FSH and LH stimulate up to about 20 primordial follicles to mature into primary follicles. It is unclear why some of the primordial follicles are stimulated to mature while the rest remain

[1]Although we simplistically discuss the ovarian cycle as happening within a single 28-day time frame, individual follicle maturation takes almost a year.

unaffected. As the follicles develop, their follicular cells release the hormone **inhibin,** which helps inhibit further FSH production, thus preventing excessive ovarian follicle development and allowing the current primary follicles to mature.

Shortly thereafter, a few of these primary follicles mature and become secondary follicles. The primary follicles that do not mature undergo atresia. Late in the follicular phase, typically only one secondary follicle in an ovary matures into a mature follicle. Under the influence of LH, the volume of fluid increases within the antrum of this follicle, and the oocyte is forced toward one side of the follicle, where it is surrounded by the cumulus oophorus. The innermost layer of the cumulus oophorus cells is the corona radiata.

As the secondary follicle matures into a mature follicle, its primary oocyte finishes meiosis I (where pairs of replicated chromosomes are separated), and two cells form (figure 28.6). One of these cells receives a minimal amount of cytoplasm and forms a **polar body,** which is a nonfunctional cell that later regresses. The other cell receives the bulk of the cytoplasm and becomes the secondary oocyte, which continues to develop and reaches metaphase II of meiosis before it is arrested again. This secondary oocyte does not complete meiosis II (where sister chromatids are separated) unless it is fertilized by a sperm. If the oocyte is not fertilized, it breaks down and regresses about 24 hours later.

Ovulation **Ovulation** (ov′yū-lā′shŭn) occurs on day 14 of a 28-day ovarian cycle and is defined as the release of the secondary oocyte from a mature follicle (figure 28.7). Typically, only one ovary ovulates each month—that is, typically the left ovary ovulates one month, and the right ovary ovulates the next. Ovulation is induced only when there is a peak in LH secretion. As the time of ovulation approaches, the follicle cells in the mature follicle increase their rate of fluid secretion, forming a larger antrum and causing further swelling within the follicle. The edge of the follicle that continues to expand at the ovarian surface becomes quite thin and eventually ruptures, expelling the secondary oocyte.

Luteal Phase The **luteal phase** occurs during days 15–28 of the ovarian cycle, when the remaining follicle cells in the ruptured mature follicle become the corpus luteum.

The corpus luteum is essentially a temporary endocrine gland. It secretes progesterone and estrogen that stabilize and build up the uterine lining, and prepare for possible implantation of a fertilized oocyte.

INTEGRATE

CLINICAL VIEW

Ovarian Cancer

Ovarian cancer refers to a primary malignancy in the ovaries. The cancer may originate from the oocytes, the connective tissue of the ovary, or the surface epithelium. Not all tumors that develop in the ovaries are malignant (or cancerous), but when the tumors are malignant, a diagnosis of ovarian cancer is made. Ovarian cancer is the fifth most common cancer in women. Women that have the inherited mutation in the *BRCA1* and *BRCA2* genes (which normally produce tumor suppressor proteins) have a lifetime risk of about 10–40% of developing ovarian cancer (compared to a 1–2% risk for those without the mutation in these genes). Other risk factors for the disease include increasing age, obesity, infertility, and a family history of ovarian cancer, breast cancer, or colorectal cancer. Birth control medication (in premenopausal women) reduces the risk of ovarian cancer, whereas postmenopausal estrogen therapy increases the risk.

Ovarian cancer is difficult to diagnose early because initial symptoms may be nonspecific and because there is no reliable test for the disease. Initial symptoms may include constipation, increased need and urgency to urinate, abdominal swelling and bloating, pelvic pain, indigestion, nausea, unexplained back pain, and changes in weight. As all of these symptoms may be nonspecific, the causes of the symptoms may not be correctly attributed to ovarian cancer. A blood test for CA-125 (a cancer antigen protein found on ovarian cancer cells) may indicate the likelihood of ovarian cancer, but because CA-125 levels also may be elevated in numerous other noncancerous conditions (such as endometriosis), a high CA-125 test alone is not considered reliable for a diagnosis of cancer.

Once a diagnosis of cancer is made, treatment typically involves a combination of surgery and chemotherapy. If it is diagnosed early, 5-year-survival-rate prognoses are very good. Unfortunately, many ovarian cancers are not detected until after the cancer has metastasized to other organs, which dramatically reduces the chances for good long-term survival rates. Therefore, women who experience the initial symptoms described above should seek medical consultation promptly.

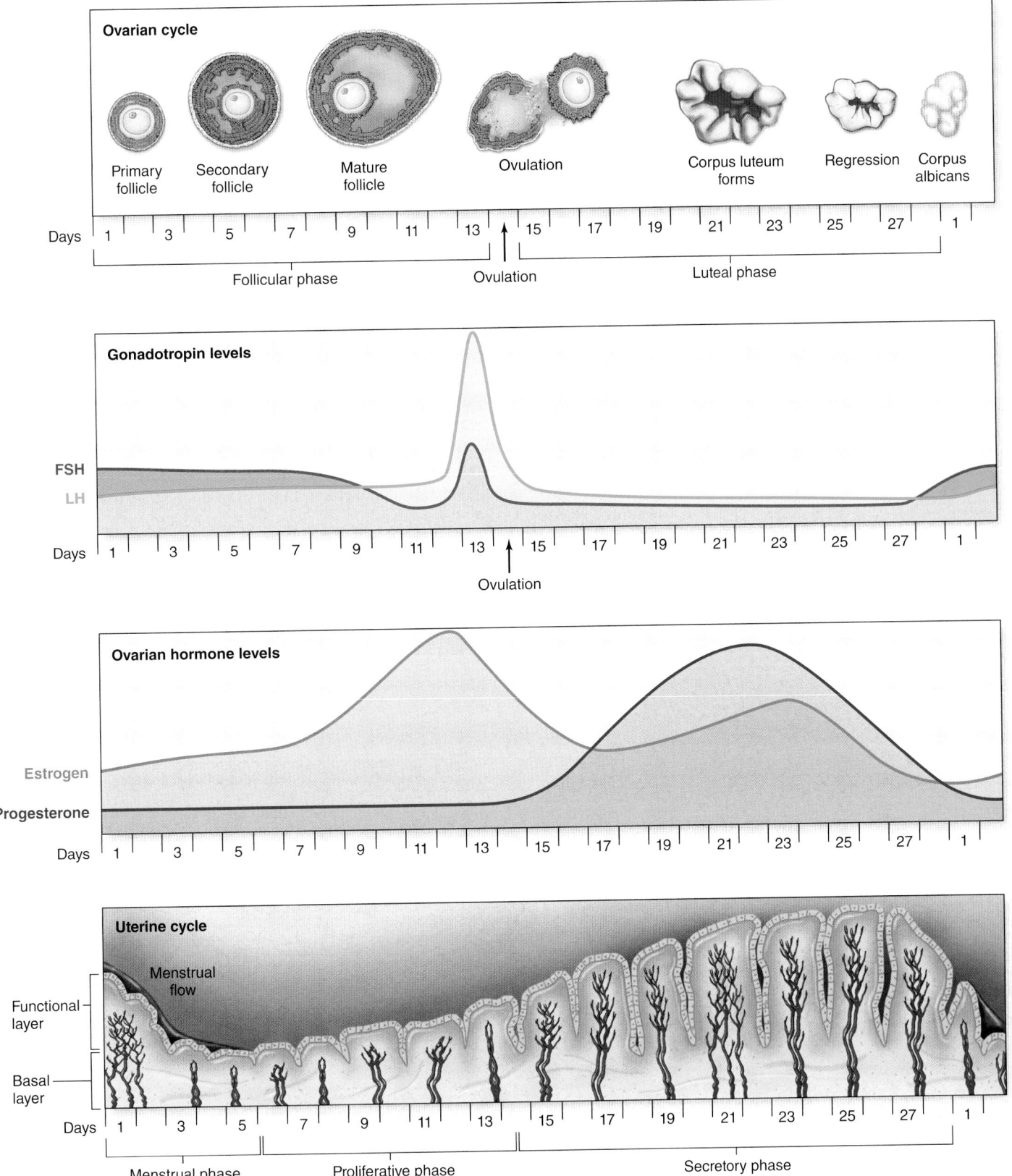

Figure 28.7 Hormonal Changes in the Female Reproductive System. Follicle-stimulating hormone (FSH) stimulates development of ovarian follicles during the follicular phase of the ovarian cycle. Estrogen (secreted from ovarian follicles) stimulates the proliferative phase in the uterine cycle, whereas a peak in luteinizing hormone (LH) promotes ovulation and the development of the corpus luteum, which produces both progesterone and estrogen to promote uterine lining development. AP|R

It also secretes inhibin, which along with the high levels of estrogen and progesterone, inhibits the hypothalamus and anterior pituitary from secreting their reproductive hormones. The corpus luteum has a life span of about 10 to 13 days if the secondary oocyte is not fertilized. After this time, the corpus luteum regresses and becomes a corpus albicans. As the corpus luteum regresses, its levels of secreted

progesterone and estrogen drop, causing the uterine lining to be shed in a process called **menstruation** (men-strū-ā′shŭn). The shed lining is called **menses** or a *period*. This marks the end of the luteal phase. A female's first menstrual cycle, called **menarche** (me-nar′kē; *men* = month, *arche* = beginning), indicates the culmination of female puberty and typically occurs around age 11–12.

☀ **WHAT DO YOU THINK?**

① If a woman has one ovary surgically removed, can she still become pregnant? Why or why not?

After Menopause

When a woman has stopped having monthly menstrual cycles for 1 year and is not pregnant, or has a medical condition that leads to a cessation of menstruation—such as extremely low body fat or in cases of anorexia nervosa—she is said to be in **menopause** (men'ō-pawz; *pauses* = cessation).

The age at normal onset of menopause varies considerably, but typically it is between 45 and 55 years. The reason a woman reaches menopause is that either there are no more ovarian follicles remaining, or ovarian follicle maturation has halted. As a result, significant amounts of estrogen and progesterone are no longer being secreted. Thus, a woman's uterine lining does not grow, and she no longer has a menstrual period. Menopause is discussed further in section 28.5f.

Regulation of the Ovarian Cycle in Depth

We have given you an overview of the hormones involved in regulating the ovarian cycle in the previous sections. However, the interplay among the hormones is much more intricate and involves both negative and positive feedback mechanisms to complete a single ovarian cycle. A stepwise description of these hormonal effects in the ovarian cycle are listed here and shown in **figure 28.8**:

① **The hypothalamus initiates the ovarian cycle by secreting GnRH,** which stimulates the anterior pituitary to secrete FSH and LH.

② **FSH and LH target the ovaries and stimulate follicular development.** These hormones cause maturation of follicles, as described earlier, and both hormones also affect the ovarian follicle's secretion of other hormones.

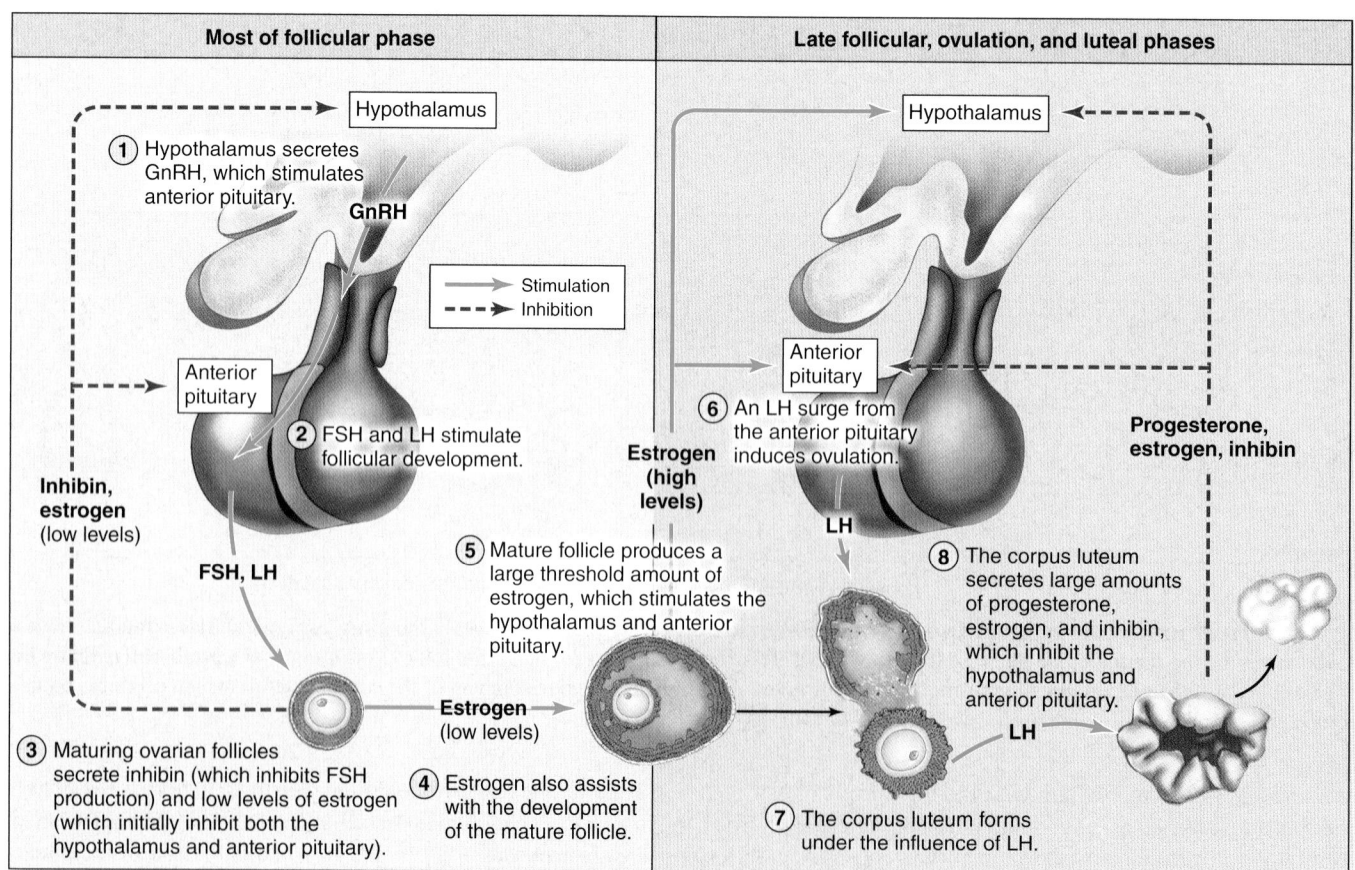

Figure 28.8 Hormonal Interactions in the Ovarian Cycle. The ovarian cycle is initiated when the hypothalamus secretes gonadotropin-releasing hormone (GnRH), which stimulates the anterior pituitary to secrete follicle-stimulating hormone (FSH) and luteinizing hormone (LH). The cascade of events that occur during the cycle is illustrated.

(3) **The maturing ovarian follicles secrete inhibin and estrogen.** These hormones have a negative feedback effect on the hypothalamus and anterior pituitary. Specifically, inhibin helps inhibit FSH release by the anterior pituitary; low levels of estrogen inhibit production of GnRH, FSH, and LH.

(4) **Estrogen also assists with the development of the dominant (mature) ovarian follicle.** A secondary follicle matures into a mature follicle under the influence of estrogen.

(5) **The mature follicle, once it develops, produces a larger threshold amount of estrogen.** Paradoxically, higher levels of estrogen have a stimulatory effect on the hypothalamus and anterior pituitary, and thus a positive feedback loop is initiated.

(6) **The positive feedback loop results in an LH surge from the anterior pituitary, which induces ovulation.** Without this surge in LH, the mature follicle would not ovulate (expel its secondary oocyte). Most oral contraceptives regulate estrogen and progesterone levels, which in turn prevent this LH surge (see Clinical View: "Contraception Methods" in section 28.3f). Estrogen levels decline slightly right after ovulation, presumably because ovulation damaged some of the estrogen-secreting follicular cells in the mature follicle.

(7) **A corpus luteum forms from the ovulated follicle.** After ovulation, LH induces the remaining follicular cells in the ovary to develop into the corpus luteum—hence the name "luteinizing."

(8) **The corpus luteum secretes large amounts of progesterone, estrogen, and inhibin.** This combination inhibits both the hypothalamus and anterior pituitary and builds the uterine lining. The corpus luteum degenerates in 10 to 13 days (if the oocyte is not fertilized), which results in decreases in some hormone levels.

(9) **The ovarian cycle repeats.** (Step 9 not shown in figure 28.8.) Unless the oocyte is fertilized, the negative feedback events described in step 8 serve to reduce LH levels, resulting in the degeneration of the corpus luteum. Consequently, the levels of progesterone, estrogen, and inhibin drop, and the hypothalamus again is able to secrete GnRH to initiate the cycle.

If the secondary oocyte is fertilized, and if it successfully implants in the uterine lining, this fertilized structure, now called a pre-embryo, starts secreting **human chorionic gonadotropin (hCG),** a hormone that enters the mother's blood and acts on the corpus luteum (see section 29.2c). Essentially, hCG mimics the effects of LH and continues to stimulate the corpus luteum. As a result, the corpus luteum continues producing progesterone and estrogen, which maintains and continues building the uterine lining. After 3 months, the placenta of the developing fetus starts producing its own progesterone and estrogen. Thus, typically by the end of the third month the corpus luteum has usually regressed into a corpus albicans.

WHAT DID YOU LEARN?

13 What follicles are present at birth? What follicles are present at puberty?

14 What are the specific effects of FSH and LH on the ovarian cycle?

15 What are the three phases of the ovarian cycle, and what main events occur in each phase?

28.3c Uterine Tubes, Uterus, and Vagina

LEARNING OBJECTIVES

6. Describe the anatomy and function of the uterine tubes.

7. List the functions of the uterus, and compare its three tunics.

8. Explain the gross anatomy of the vagina.

The uterine tubes, uterus, and vagina are accessory reproductive organs responsible for the transport and implantation of the fertilized oocyte, and the eventual expulsion of the fetus.

Uterine Tubes

The **uterine tubes,** also called the *fallopian* (fa-lō′pē-an) *tubes* or *oviducts* (ō′vi-dŭkt; *duco* = to lead), extend laterally from both sides of the uterus toward the ovaries **(figure 28.9)**. They function to transport the ovulated oocyte to the uterus, and are the site of fertilization of an oocyte. The uterine tubes are small in diameter, and reach their maximum length of between 10 and 12 centimeters after puberty. These tubes are covered and suspended by the **mesosalpinx** (mez′ō-sal′pinks; *salpinx* = trumpet), a specific superior part of the broad ligament of the uterus (see figure 28.4a). Each uterine tube is composed of the following contiguous segments:

- The **infundibulum** (in-fŭn-dib′ū-lŭm; funnel) is the free, funnel-shaped, lateral margin of the uterine tube. Its numerous fingerlike folds are called **fimbriae** (fim′brē-ē; fringes). The

fimbriae of the infundibulum envelop the ovary only at the time of ovulation.

- The **ampulla** (am-pul′lă; two-handled bottle) is the expanded region medial to the infundibulum. Fertilization of an oocyte typically occurs there.

- The **isthmus** (is′mus) is the constricted region that extends medially from the ampulla toward the lateral wall of the uterus. It forms about one-third of the length of the uterine tube.

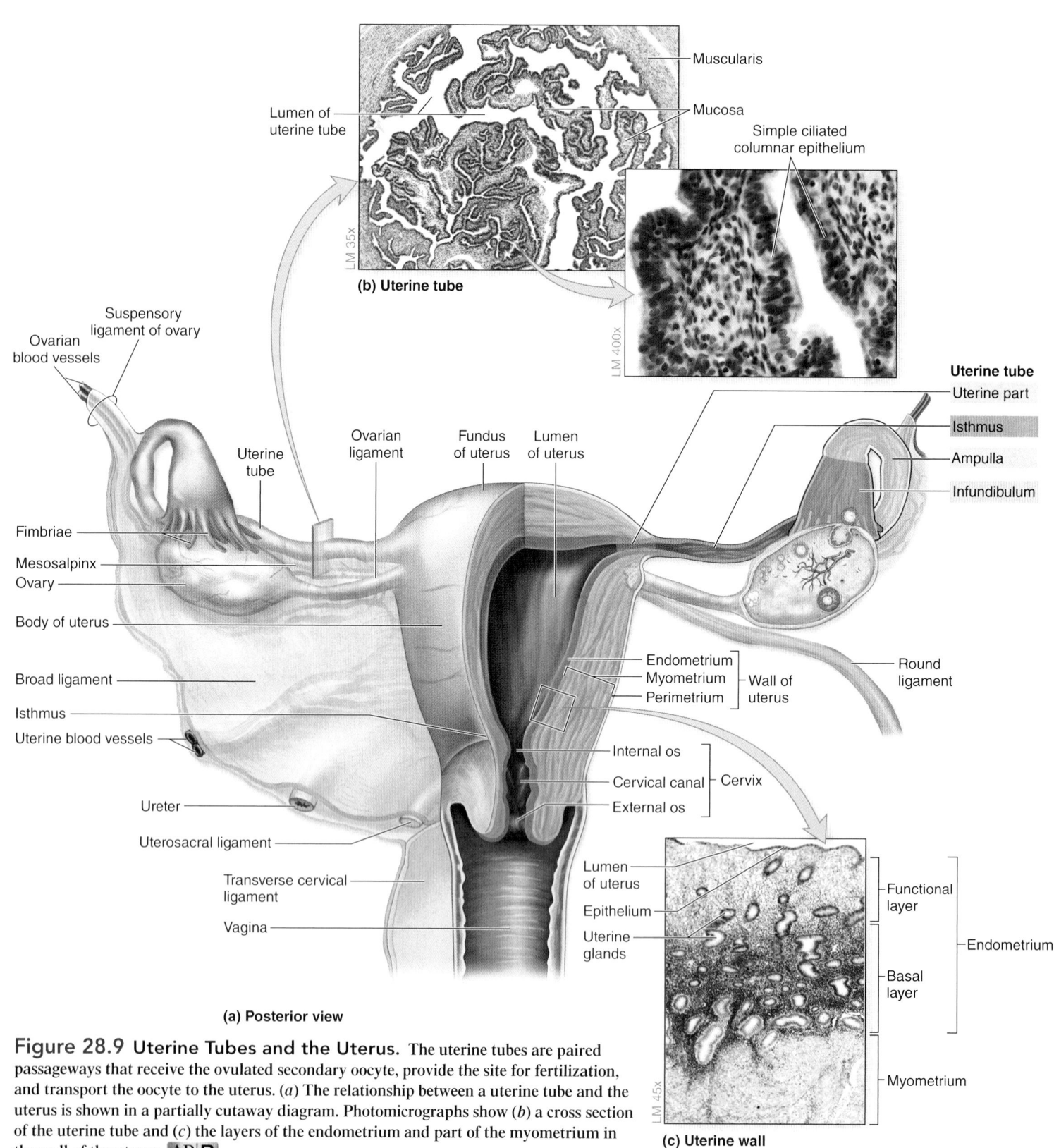

(b) Uterine tube

(a) Posterior view

(c) Uterine wall

Figure 28.9 Uterine Tubes and the Uterus. The uterine tubes are paired passageways that receive the ovulated secondary oocyte, provide the site for fertilization, and transport the oocyte to the uterus. (a) The relationship between a uterine tube and the uterus is shown in a partially cutaway diagram. Photomicrographs show (b) a cross section of the uterine tube and (c) the layers of the endometrium and part of the myometrium in the wall of the uterus. AP|R

- The **uterine part** is also called the *intramural part,* or *interstitial segment*, and it extends medially from the isthmus and penetrates the wall of the uterus.

The wall of the uterine tube consists of a mucosa, a muscularis, and a serosa. The **mucosa** is formed from a simple ciliated columnar epithelium and an underlying layer of areolar connective tissue. The mucosa is thrown into linear folds, which reduce the size of the uterine tube lumen. After ovulation, the cilia on the apical surface of the epithelial cells of both the infundibulum and the ampulla begin to beat in the direction of the uterus. This causes a slight current in the fluid within the uterine tube lumen, drawing the oocyte into the uterine tube and moving it toward the uterus.

The **muscularis** is composed of an inner circular layer and an outer longitudinal layer of smooth muscle. Some peristaltic contractions in the muscularis help propel the oocyte, or pre-embryo if fertilization has occurred, through the uterine tube toward the uterus. The **serosa** is the external serous membrane covering the uterine tube.

Uterus

The **uterus** (ū′ter-ŭs; womb) is a hollow, pear-shaped, thick-walled muscular organ within the pelvic cavity. It has a lumen (internal space) that connects to the uterine tubes superolaterally and to the vagina inferiorly (figure 28.9*a*). Normally, the uterus is angled anterosuperiorly across the superior surface of the urinary bladder, a position referred to as **anteverted** (an-te-vert′ed; *ante* = before, *versio* = a turning). In figure 28.3, the uterus is shown in an anteverted position. If the uterus is positioned posterosuperiorly, so that it is projecting toward the rectum, this position is called **retroverted** (re′trō-ver-ted). The uterus may shift from an anteverted to a retroverted position in older women.

The uterus serves many functions. Following fertilization, the pre-embryo makes contact with the uterine lining and implants in the inner uterine wall. The uterus then supports, protects, and nourishes the developing embryo/fetus by forming a vascular connection that later develops into the placenta. The uterus ejects the fetus at birth after maternal oxytocin levels increase to initiate the uterine contractions of labor (see section 29.6). If an oocyte is not fertilized, the muscular wall of the uterus contracts and sheds its inner lining as menstruation.

The uterus and uterine tubes receive their blood supply from the uterine arteries (branches of the internal iliac arteries; see section 20.10e), which branch into smaller vessels that supply the inner lining of the uterus.

The uterus is partitioned into the following regions:

- The **fundus** (fŭn′dŭs) is the broad, curved superior region extending between the lateral attachments of the uterine tubes.
- The middle region, called the **body**, is the major part of the organ and is composed of a thick wall of smooth muscle.
- The **isthmus** is the narrow, constricted inferior region of the body that is superior to the cervix.
- The **cervix** is the narrow inferiormost portion of the uterus that projects into the vagina.

A narrow channel within the cervix that is called the **cervical canal** connects to the vagina inferiorly. The superior opening of this canal is the **internal os** (*os* = mouth). The inferior opening of the cervix into the lumen of the vagina is the **external os.** The external os is covered with nonkeratinized stratified squamous epithelium. The cervix contains mucin-secreting glands that help form a thick mucus plug at the external os. This mucus plug is suspected to be a physical barrier that prevents pathogens from invading the uterus from the vagina. The mucus plug thins considerably around the time of ovulation, so that sperm may more easily enter the uterus.

Several structures support the uterus and help hold it in place:

- The muscles of the pelvic floor (**pelvic diaphragm** and **urogenital diaphragm;** see figure 11.17) hold the uterus and vagina in place and help resist intra-abdominal pressure exerted inferiorly on the pelvis.
- The **round ligaments** (figure 28.9*a*) extend from the lateral sides of the uterus, traverse through the inguinal canal and attach to the labia majora.
- The **transverse cervical ligaments,** or *cardinal ligaments,* run from the sides of the cervix and superior vagina laterally to the pelvic wall.
- The **uterosacral ligaments,** or *sacrocervical ligaments,* connect the inferior portion of the uterus posteriorly to the sacrum.

Many of these ligaments are located between the folds of the broad ligament. Weakness in either the pelvic floor muscles or these ligaments can lead to **prolapse** (prō-laps′; *prolapsus* = a failing) of the uterus, in which the uterus starts to protrude through the vagina. Despite its name, the broad ligament is not a strong support for the uterus, but rather is a peritoneal drape over the uterus.

The uterine wall is composed of three concentric tunics: the perimetrium, myometrium, and endometrium (figure 28.9). The outer tunic is a serosa called the **perimetrium** (per-i-mē′trē-ŭm; *metra* = uterus).

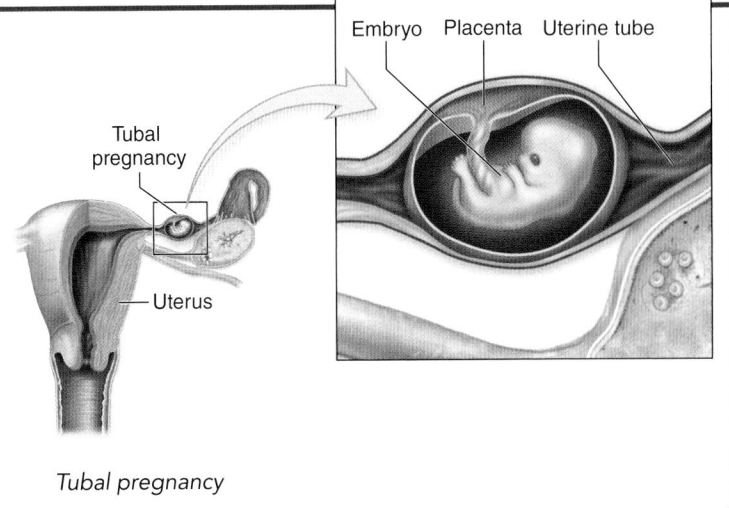

INTEGRATE

CLINICAL VIEW

Tubal Pregnancy

An *ectopic pregnancy* refers to a fertilized oocyte implanting in a location other than the uterine endometrium. In a **tubal pregnancy** (one specific type of ectopic pregnancy), the fertilized oocyte implants in the uterine tube rather than the uterus. The uterine tube is unable to expand as the embryo grows, so the embryo can remain viable no later than week 8, at which time it has become too large for the confines of the uterine tube. The woman experiences severe cramping, and the uterine tube will rupture if the embryo is not surgically removed. Rupture of the uterine tube results in a massive hemorrhage into the abdominopelvic cavity that may endanger the life of the mother. No current method is available to treat a tubal pregnancy that can spare the developing embryo.

Embryo   Placenta   Uterine tube

Tubal pregnancy

Uterus

Tubal pregnancy

INTEGRATE

CLINICAL VIEW
Endometriosis

Endometriosis (en'dō-mē-trē-ō'sis) occurs when part of the endometrium is displaced onto the external surface of organs within the abdominopelvic cavity. Scientists think that during the regular uterine (menstrual) cycle of some women, some endometrial tissue may be expelled from the uterine tubes and become implanted on the surface of various abdominal and pelvic organs. This displaced endometrium may grow under the influence of hormones, but it cannot be expelled through the vagina during menses. Thus, the ensuing hemorrhage and breakdown of the displaced endometrium may cause considerable pain and scarring that often leads to deformities of the uterine tubes. Treatments include the use of hormones designed to retard the growth and cycling of the displaced endometrial tissue, as well as surgical removal of the ectopic endometrium.

The pelvic organs show extensive endometrial tissue attachments.

The perimetrium is continuous with the broad ligament. The **myometrium** (mī'ō-mē'trē-ŭm; *mys* = muscle) is the thick, middle tunic of the uterine wall formed from three intertwining layers of smooth muscle. The innermost tunic of the uterus, called the **endometrium** (en'dō-mē'trē-ŭm), is an intricate mucosa composed of a simple columnar epithelium and an underlying lamina propria. The lamina propria is filled with compound tubular glands (also called **uterine glands**), which enlarge during the uterine cycle.

Two distinct layers form the endometrium. The deeper layer is the **basal layer,** also called the *stratum basalis* (bā-sā'lis). The basal layer is immediately adjacent to the myometrium and is a permanent layer that undergoes few changes during each uterine cycle. The more superficial of the two endometrial layers is the **functional layer,** or *stratum functionalis* (fŭnk'shŭn-ăl'is). Beginning at puberty, the functional layer grows from the basal layer under the influence of estrogen and progesterone secreted from the ovarian follicles. If fertilization and implantation do not occur, the functional layer is shed as menses. The basal layer gives rise to a new functional layer after the end of each menses.

Vagina

The **vagina** (vă-jī'nă) is a thick-walled, fibromuscular tube that forms the inferiormost region of the female reproductive tract and measures about 10 centimeters in length in an adult female (see figure 28.3a). The vagina connects the uterus with the outside of the body and functions as the birth canal. The vagina is also the copulatory organ of the female, as it receives the penis during intercourse, and it serves as the passageway for menstruation.

Arterial supply comes from the vaginal arteries (branches of the internal iliac arteries), and venous drainage is via vaginal veins. The lumen of the vagina is flattened anteroposteriorly. The vagina's relatively thin, distensible wall consists of three tunics: an inner mucosa, a middle muscularis, and an outer adventitia **(figure 28.10)**. The **mucosa** consists of a nonkeratinized stratified squamous epithelium (after puberty; prior to puberty the epithelium is relatively thin) and a highly vascularized lamina propria. Vaginal epithelial cells produce an acidic secretion that helps prevent bacterial and other pathogenic infection. The inferior region of the vaginal mucosa contains numerous transverse folds, or rugae. Near the external opening of the vagina, called the **vaginal orifice,** these mucosal folds project into the lumen to form a vascularized, membranous barrier called the **hymen** (hī'men; membrane). The hymen typically is perforated during the first instance of sexual intercourse, but also may be perforated by tampon use, medical exams, or very strenuous physical activity. The **muscularis** of the vagina has both outer and inner layers of smooth muscle. The **adventitia** contains some inner elastic fibers and an outer layer of areolar connective tissue.

INTEGRATE

CLINICAL VIEW
Cervical Cancer

Cervical cancer is one of the most common malignancies of the female reproductive system. The most important risk factor is previous human papillomavirus (HPV) infection. Researchers have developed a vaccine (Gardasil) for the four most common types of HPV that cause cervical cancer. However, for the vaccine to be effective, it must be given to girls who are not yet sexually active, so the recommended age range for the vaccine is 11 to 13 years.

The **Papanicolaou (Pap) smear** has become a very effective method of detecting cervical cancer in its early and curable stage. A health-care professional inserts a metal or plastic instrument called a **speculum** (spek'ū-lŭm; mirror) into the vagina to keep the vagina open to examine the cervix. Epithelial cells are scraped from the edge of the cervix and examined for any abnormal cellular development (termed **dysplasia**).

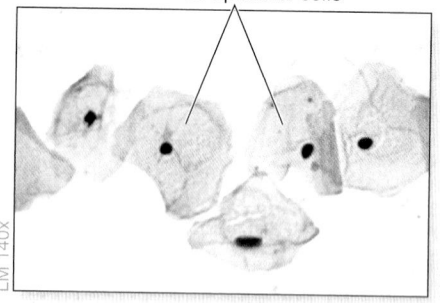

Normal Pap smear

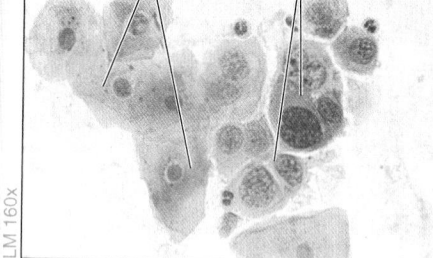

Abnormal Pap smear

If dysplastic cells are detected, the health-care professional will likely request a follow-up Pap smear and possibly even a biopsy. Sometimes, dysplastic cells are a result of irritation, infection, or some undetermined cause, and are not cancerous. If cervical cancer is present, a portion of the cervix may be removed, a procedure known as a **cone biopsy**. Invasive cancer is treated with the removal of the entire uterus, called a **hysterectomy** (his-ter-ek'tō-mē; *hystera* = womb).

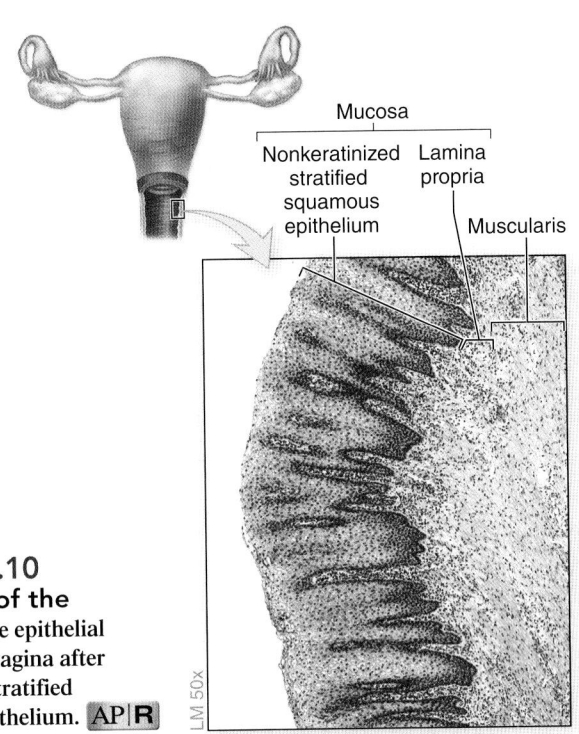

WHAT DID YOU LEARN?

16 What are the four segments of the uterine tubes, and in which segment does fertilization occur?

17 How does each of the three tunics of the uterus contribute to its function?

18 What are the main functions of the vagina, and how does its structure support these functions?

Mucosa

Nonkeratinized stratified squamous epithelium

Lamina propria

Muscularis

LM 50x

Figure 28.10
Histology of the Vagina. The epithelial lining of the vagina after puberty is a stratified squamous epithelium. AP|R

28.3d Uterine (Menstrual) Cycle and Menstruation

LEARNING OBJECTIVES

9. Compare the three phases of the uterine cycle.

10. List and explain what hormones influence the events in each part of the ovarian cycle.

11. Explain how the ovarian cycle and the uterine cycle are interrelated.

The cyclical changes in the endometrial lining occur under the influence of estrogen and progesterone secreted by the developing follicles and the corpus luteum. The **uterine cycle** (or *menstrual cycle*) consists of three distinct phases of endometrium development: the menstrual phase, proliferative phase, and secretory phase (see figure 28.7, *bottom*). The length of the uterine cycle may vary greatly among females; the timeline listed in the following paragraphs represents a typical 28-day uterine cycle. Note that the uterine cycle could be shorter (21 days or less) or longer (35 days or more).

WHAT DO YOU THINK?

2 What factors could influence the length and timing of a woman's monthly uterine cycle?

The **menstrual** (men'stru-ăl; *menstrualis* = monthly) **phase** occurs approximately during days 1–5 of the cycle. This phase is marked by sloughing of the functional layer of the endometrium and lasts through the period of menstrual bleeding. The **proliferative** (prō-lif'er-ă-tiv; *proles* = offspring, *fero* = to bear) **phase** follows, spanning approximately days 6–14. The initial development of the new functional layer of the endometrium overlaps the time of follicle growth and estrogen secretion by the ovary. The last phase is the **secretory** (se-krēt'ĕ-rē, sē'krĕ-tōr-ē) **phase,** which occurs at approximately days 15–28. During the secretory phase, increased vascularization and development of uterine glands occurs primarily in response to progesterone secretion from the corpus luteum.

The corpus luteum degenerates and progesterone level drops dramatically if fertilization does not occur. Without significant levels of progesterone, the functional layer lining sloughs off, and the next uterine cycle begins with the menstrual phase.

Table 28.4 and figure 28.7 compare the uterine cycle with the ovarian cycle discussed previously, and table 28.5 summarizes the hormones that influence the ovarian and uterine cycles. **Figure 28.11** illustrates how the hormones, ovarian cycle, and uterine cycle are interrelated and follow a cyclical process. The menstrual and proliferative phases of the menstrual cycle occur

Table 28.4	Comparison of Ovarian and Uterine Cycle Phases	
Day[1]	**Ovarian Cycle Phase**	**Uterine Cycle Phase**
1–5	Follicular phase	Menstrual phase
6–13	Follicular phase	Proliferative phase
14	Ovulation	Proliferative phase
15–28	Luteal phase	Secretory phase

1. This table assumes a 28-day cycle between menstrual periods. If a woman has a longer or shorter cycle, the day ranges for the follicular, menstrual, and proliferative phases will vary.

Table 28.5	Influence of Hormones on the Ovarian and Uterine Cycles	
Hormone	**Primary Source of Hormone**	**Effects**
Gonadotropin-releasing hormone (GnRH)	Hypothalamus	Stimulates anterior pituitary to produce and secrete FSH and LH
Follicle-stimulating hormone (FSH)	Anterior pituitary	Stimulates development and maturation of ovarian follicles
Luteinizing hormone (LH)	Anterior pituitary	Stimulates ovulation (when there is a peak in LH)
Estrogen	Ovarian follicles (before ovulation), corpus luteum (after ovulation), or placenta (during pregnancy)	Initiates and maintains growth of the functional layer of the endometrium
Progesterone	Corpus luteum or placenta (during pregnancy)	Primary hormone responsible for functional layer growth after ovulation; causes increase in blood vessel distribution, uterine gland size, and nutrient production
Inhibin	Ovarian follicles	Inhibits FSH secretion so as to prevent excessive follicular development

Figure 28.11 The Interrelationships Between Hormones, the Ovarian Cycle, and the Uterine (Menstrual) Cycle. (*a*) Days 1–5 include menstruation and new follicular development. (*b*) Days 6–12 begin the proliferative phase of the uterine cycle and the development of a mature follicle. (*c*) Days 13–14 begin with a peak in LH and end with ovulation. (*d*) Days 15–28 include development of the corpus luteum and continued buildup of the uterine lining.

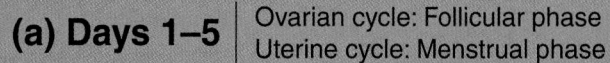

(a) Days 1–5 | Ovarian cycle: Follicular phase / Uterine cycle: Menstrual phase

GnRH stimulates FSH and LH secretion: Some ovarian follicles develop and produce estrogen. Functional layer of the endometrium is shed.

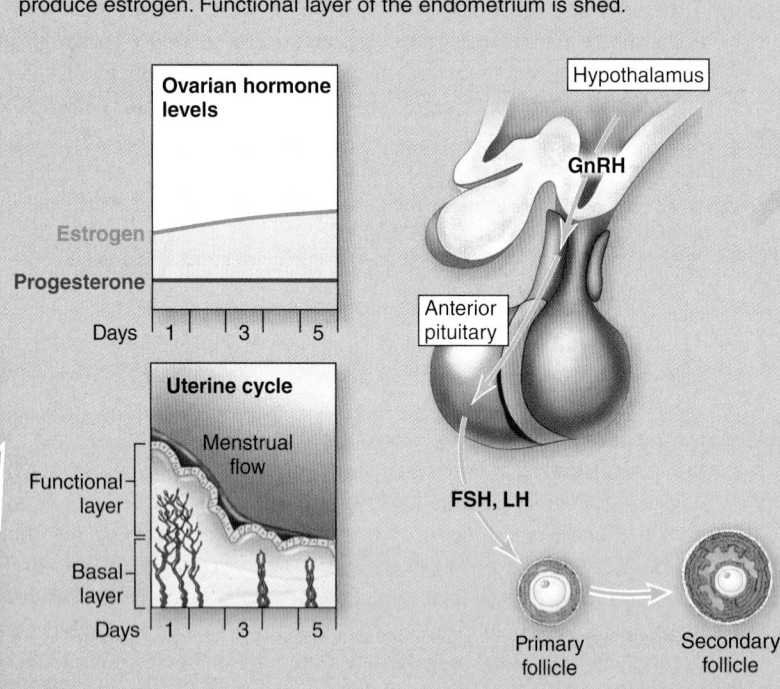

Ovarian hormone levels

Estrogen
Progesterone

Days 1 3 5

Hypothalamus

GnRH

Anterior pituitary

FSH, LH

Primary follicle → Secondary follicle

Uterine cycle

Menstrual flow

Functional layer

Basal layer

Days 1 3 5

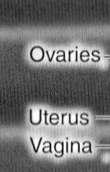

Ovaries
Uterus
Vagina

(d) Days 15–28 | Ovarian cycle: Luteal phase / Uterine cycle: Secretory phase

Corpus luteum forms and secretes large amounts of estrogen, progesterone, and inhibin. Combined, these inhibit GnRH, FSH, and LH secretion. Progesterone stimulates uterine lining growth. If the oocyte is not fertilized, the corpus luteum regresses and hormone levels drop.

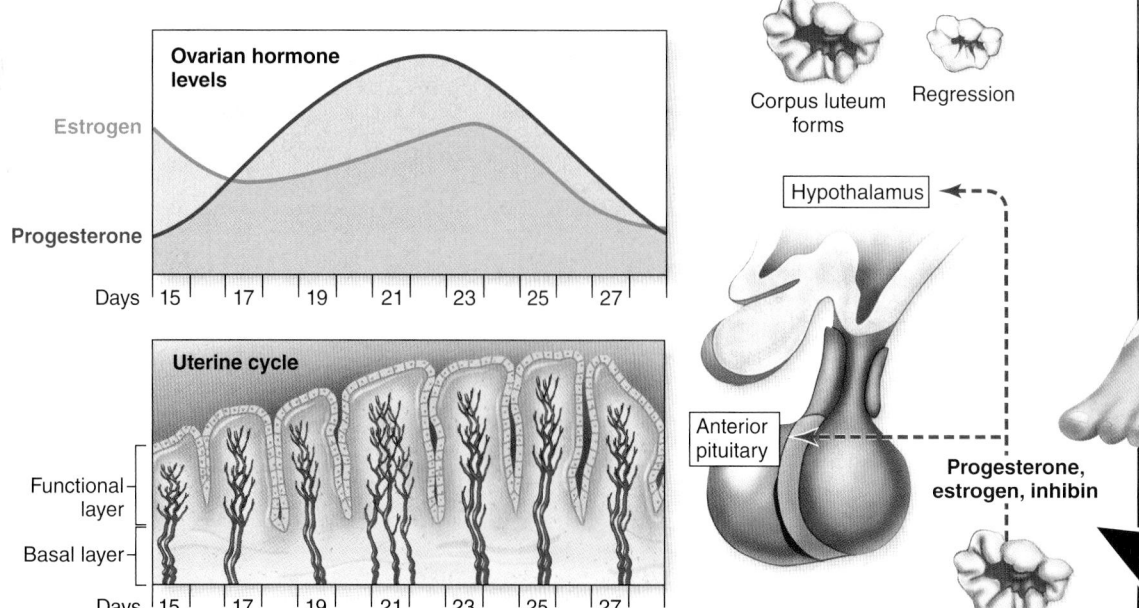

Ovarian hormone levels

Estrogen
Progesterone

Days 15 17 19 21 23 25 27

Corpus luteum forms   Regression

Hypothalamus

Anterior pituitary

Progesterone, estrogen, inhibin

Uterine cycle

Functional layer
Basal layer

Days 15 17 19 21 23 25 27

(b) Days 6–12 | Ovarian cycle: Follicular phase
Uterine cycle: Proliferative phase

Estrogen and inhibin inhibit the hypothalamus and anterior pituitary, causing a drop in FSH. One follicle continues to mature and produce estrogen. Functional layer of the endometrium is rebuilding.

Gonadotropin levels

FSH
LH

Days | 7 | 9 | 11

Ovarian hormone levels

Estrogen

Progesterone

Days | 7 | 9 | 11

Uterine cycle

Functional layer
Basal layer

Days | 7 | 9 | 11

Hypothalamus

Anterior pituitary

Inhibin, estrogen (low levels)

Estrogen (low levels)

Secondary follicle

Mature follicle

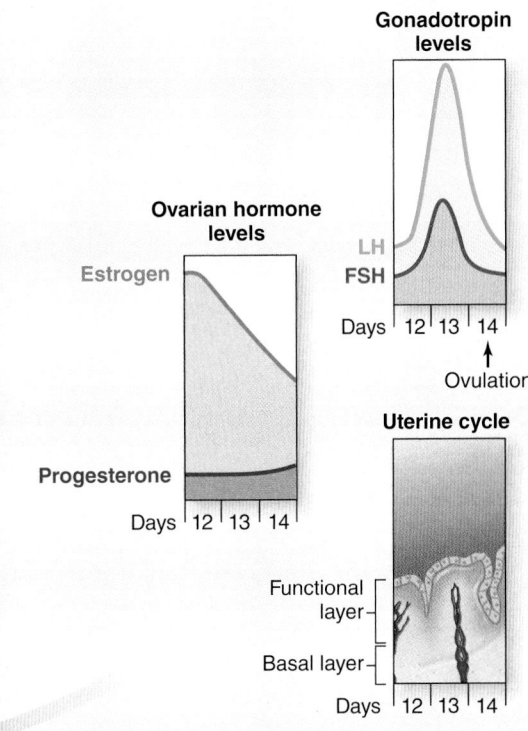

(c) Days 13–14 | Ovarian cycle: Follicular phase, ovulation
Uterine cycle: Proliferative phase

The increase in estrogen above threshold stimulates the hypothalamus and anterior pituitary, causing an LH surge. The LH surge induces ovulation.

Gonadotropin levels

LH
FSH

Days | 12 | 13 | 14

Ovulation

Ovarian hormone levels

Estrogen

Progesterone

Days | 12 | 13 | 14

Uterine cycle

Functional layer
Basal layer

Days | 12 | 13 | 14

Hypothalamus

Anterior pituitary

Estrogen (high levels)

LH

Mature follicle

Ovulation

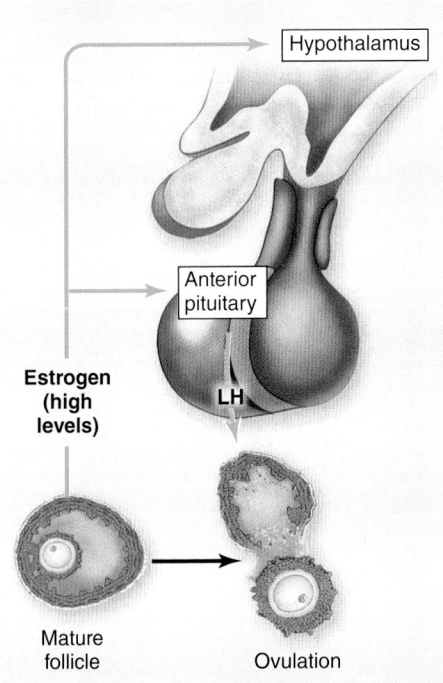

in conjunction with the follicular and ovulation phases of the ovarian cycle. The secretory phase of the menstrual cycle occurs at the same time as the luteal phase of the ovarian cycle.

The day ranges listed in the figures assume that the woman has a 28-day cycle, meaning she ovulates at day 14 and has a menstrual period about every 28 days. If a woman has a longer cycle, her menstrual phase or her proliferative phase is longer than average. Typically, a woman ovulates 14 days before menstruation (regardless of the length of her cycle), so there is little variation in the number of days for the secretory phase.

WHAT DID YOU LEARN?

19 What are the three phases of the uterine cycle, and what major events occur in each?

20 Compare and contrast the ovarian and uterine cycles. How are they interrelated?

28.3e External Genitalia

LEARNING OBJECTIVE

12. Describe the components of the female external genitalia.

The external reproductive organs of the female are termed the **external genitalia,** or **vulva** (vŭl′vă; a covering) **(figure 28.12).** The **mons pubis** (monz; mountain) is an expanse of skin and subcutaneous connective tissue immediately anterior to the pubic symphysis. The mons pubis is covered with pubic hair in postpubescent females. The **labia majora** (lā′bē-ă mă-jŏr′ă; sing., *labium majus; labium* = lip, *majus* = larger) are paired, thickened folds of skin and connective tissue. The labia majora are homologous to the scrotum of the male. In adulthood, their outer surface is covered with coarse pubic hair; they contain numerous sweat and sebaceous glands. The **labia minora** (mī-nŏr′ă; sing., *labium minus; minus* = smaller) are paired folds immediately internal to the labia majora. They are devoid of hair and contain a highly vascular layer of areolar connective tissue. Sebaceous glands are located within these folds, as are numerous melanocytes, resulting in enhanced pigmentation of the folds.

The space between the labia minora is called the **vestibule.** Within the vestibule are the **urethral opening** and the vaginal orifice. On either side of the vaginal orifice is an erectile body called the **bulb of the vestibule,** which engorges with blood and increases in sensitivity during sexual intercourse. A pair of **greater vestibular glands** (previously called *glands of Bartholin*) are housed within the posterolateral walls of the vestibule and secrete mucin, which forms mucus to act as a lubricant for the vagina. Secretion increases during sexual intercourse, when additional lubrication is needed. These secretory structures are homologous to the male bulbourethral glands (which are discussed with the male reproductive system).

The **clitoris** (klit′ō-ris) is a small erectile body, usually less than 2 centimeters in length, located at the anterior regions of the labia minora. It is homologous to the penis of the male. Two small erectile bodies called **corpora cavernosa** form the **body** of the clitoris. The corpora cavernosa become engorged with blood and increase in sensitivity during sexual intercourse. Extending from each of these bodies posteriorly are elongated masses, each called the **crus** (krūs) of the clitoris, which attach to the pubic arch. Capping the body of the clitoris is the **glans** (glanz; acorn). The many specialized sensory nerve receptors housed in the clitoris provide pleasure to the female during sexual intercourse. The **prepuce** (prē′pūs; foreskin) is an external fold of the labia minora that forms a hoodlike covering over the clitoris.

WHAT DID YOU LEARN?

21 What are the individual components of the female external genitalia and their functions?

28.3f Mammary Glands

LEARNING OBJECTIVES

13. Explain the gross anatomy of the mammary glands.

14. Compare the hormones responsible for milk production and milk ejection.

Each **mammary gland,** or *breast,* is located within the anterior thoracic wall and is composed of a compound tubuloalveolar exocrine

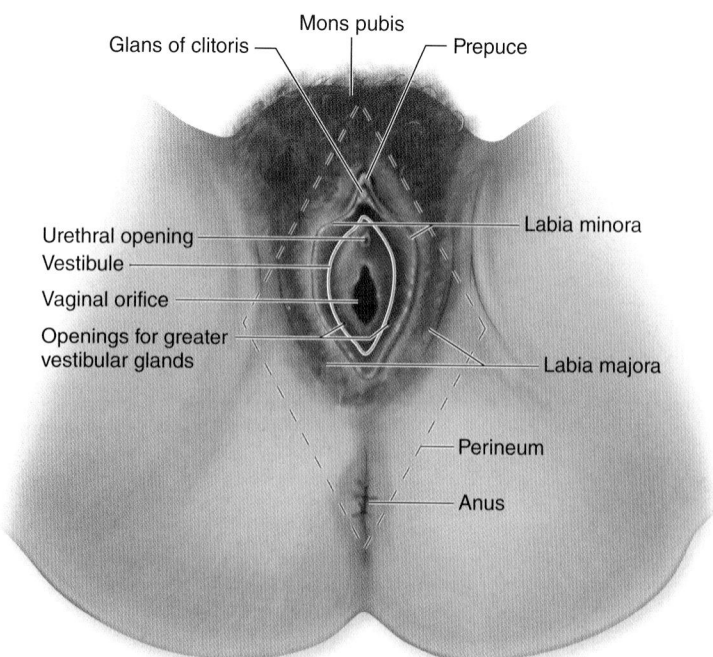

(a) Superficial structures of external genitalia

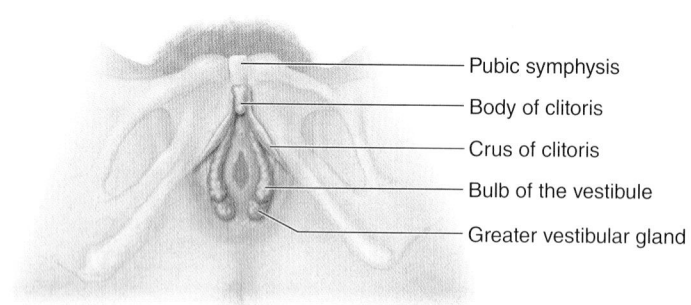

(b) Deep structures of external genitalia

Figure 28.12 Female External Genitalia. (*a*) Inferior view of the external genitalia, illustrating the urethral opening and the vaginal opening, which are within the vestibule and bounded by the labia minora. (*b*) Deep structures associated with the external genitalia.

INTEGRATE

CLINICAL VIEW
Breast Cancer

Breast cancer affects approximately 1 in every 8 women in the United States, and it also occurs infrequently in males. Some well-documented risk factors for breast cancer include maternal relatives with breast cancer, longer reproductive span (early menarche coupled with delayed menopause), obesity, nulliparity, late age at first pregnancy, and having mutations in specific genes (*BRCA1* and *BRCA2*) that normally produce tumor suppressor proteins. Most of these risk factors are related

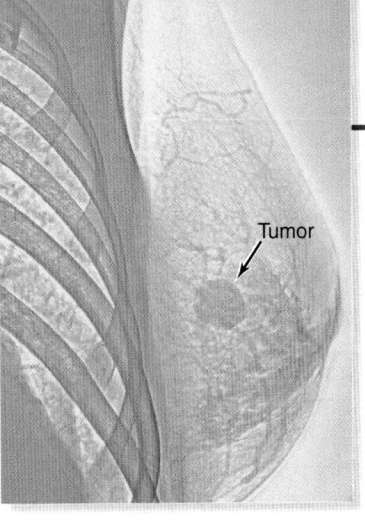

Mammogram showing a tumor.

to increased exposure to estrogen over a long period of time. Breast cancers arise from the duct epithelium, not the alveoli. Monthly self-examination is an important means of early detection of breast malignancies. Mammography, which is an x-ray of the breast that can detect small areas of increased tissue density, can identify many small malignancies that are not yet palpable in a self-examination. Recommendations vary, but most physicians agree that women over the age of 40 should have a mammogram done every 1 to 2 years. Women with a family history of breast cancer should consider regular mammography before the age of 40.

gland (**figure 28.13**). The gland's secretory product is called breast milk and it contains proteins, fats, and a lactose sugar to provide nutrition to infants.

The **nipple** is a cylindrical projection on the center of the mammary gland. It contains multiple tiny openings of the excretory ducts that transport breast milk. The **areola** (ă-rē′ō-lă; small area) is the pigmented, rosy or brownish ring of skin around the nipple. Its surface often appears uneven and grainy due to the numerous sebaceous glands, called **areolar glands,** immediately internal to the surface. The color of the areola may vary, depending upon whether or not a woman has given birth. In a **nulliparous** (nŭl-ip′ă-rŭs; *nullus* = none, *pario* = to bear) woman, one who has never given birth, the areola is rosy or light brown in color. In a **parous** (par′ŭs) woman, one who has given birth, the areola may change to a darker color.

Internally, the mammary glands are supported by fibrous connective bands called **suspensory ligaments.** These thin bands extend

from the skin and attach to the deep fascia overlying the pectoralis major muscle. Thus, the mammary gland and the pectoralis major muscle are structurally linked.

The mammary glands are subdivided into **lobes,** which are further subdivided into smaller compartments called **lobules.** Lobules contain secretory units termed **alveoli** that produce milk in the lactating female. Alveoli become more numerous and larger during pregnancy. Tiny ducts drain milk from the alveoli and lobules. The tiny ducts of the lobules merge and form 10 to 20 larger channels called **lactiferous** (lak-tif′er-ŭs; *lact* = milk, *fero* = to bear) **ducts.** A lactiferous duct drains breast milk from a single lobe. As each lactiferous duct approaches the nipple, its lumen expands to form a **lactiferous sinus,** a space where milk is stored prior to release from the nipple.

Breast milk is released by a process called **lactation** (lak-tā′shŭn; *lactatio* = to suckle), which occurs in response to a complex sequence

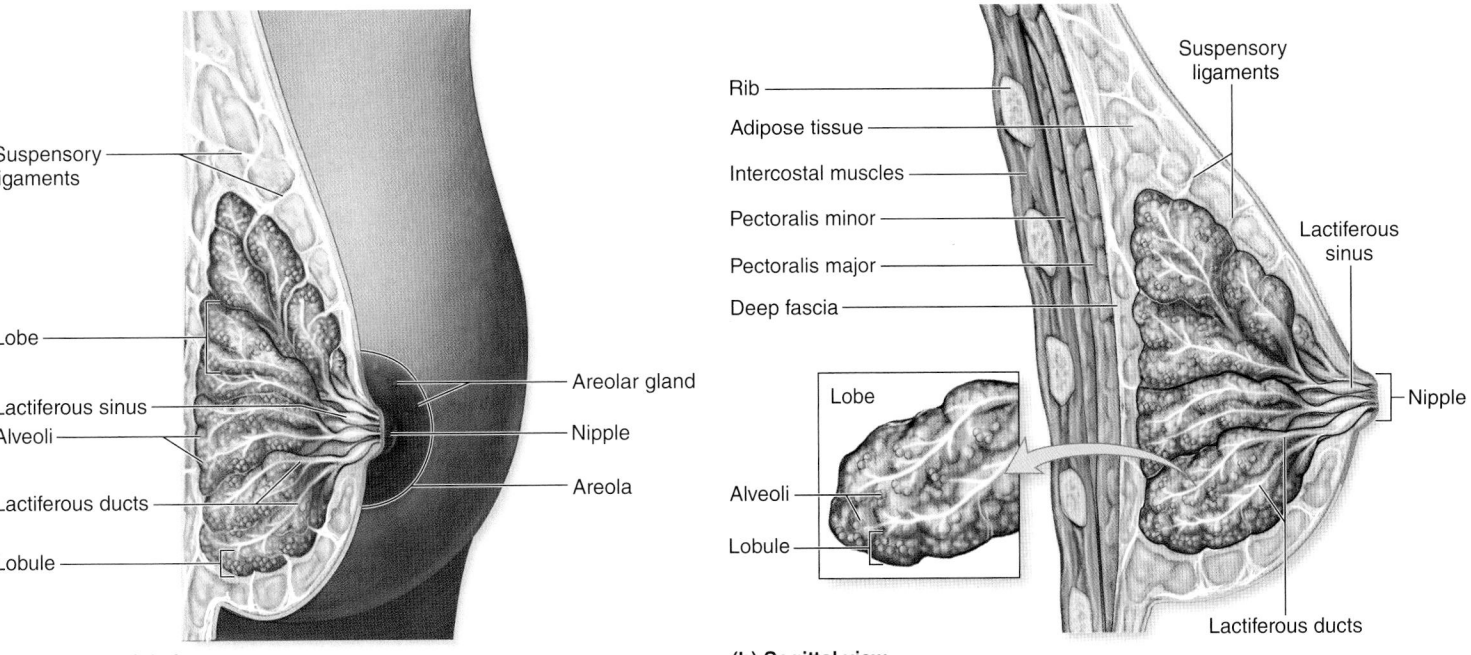

(a) Anteromedial view

(b) Sagittal view

Figure 28.13 Mammary Glands. The mammary glands are composed of glandular tissue and a variable amount of fat. (*a*) An anterior view is partially cut away to reveal internal structures. (*b*) A diagrammatic sagittal section of a mammary gland shows the distribution of alveoli within lobules and the extension of ducts to the nipple. AP|R

CLINICAL VIEW
Contraception Methods

The term **contraception** (kon-tră-sep'shun) refers to birth control, or the prevention of pregnancy. A wide range of birth control methods are available, and they have varying degrees of effectiveness.

Abstinence (ab'sti-nens; *abstineo* = to hold back) means refraining from sexual intercourse. Abstinence is the only 100% proven way to prevent pregnancy.

The **rhythm method** requires avoiding sexual intercourse during the time when a woman is ovulating. Because sperm can live for several days in the female reproductive tract, it is best to avoid intercourse both a few days prior and a few days after ovulating. The rhythm method requires that a woman know when she is ovulating, which may be difficult to determine because there is great variation in ovarian cycles. As a result, this method has a high failure rate (~25%).

The **withdrawal method,** also known as the *pull-out method,* is where the male removes his penis from the vagina before he ejaculates. This method has a high failure rate, in part because it may be difficult to predict the point prior to ejaculation. Additionally, it may be possible for pre-ejaculate (emissions secreted prior to ejaculation) to collect enough sperm in the urethra from a prior ejaculation to impregnate a female. The failure rate for the withdrawal method is about 19%.

Barrier methods of birth control use a physical barrier to prevent sperm from reaching the uterine tubes. Barrier methods include the following:

- **Condoms,** when used properly, collect the sperm and prevent them from entering the female reproductive tract. They are also the only birth control method that helps protect against sexually transmitted infections (STIs), such as human papillomavirus (HPV), herpes, and HIV. Condoms for males fit snugly on the erect penis, whereas **vaginal condoms** are placed in the vagina prior to sexual intercourse. The typical failure rate for condoms is about 15%.

- **Spermicidal foams and gels** are chemical barrier methods that kill sperm before they travel to the uterine tubes. They are inserted into the vagina or placed on the penis prior to sexual intercourse. Foams and gels are not the most effective method of birth control (when used alone, the failure rate is about 25%); rather, they should be used in conjunction with a physical barrier method.

- **Diaphragms** and **cervical caps** are structures made of rubber or silicone that are inserted into the vagina and placed over the cervix prior to sexual intercourse. Spermicidal gel is used around the edges to help prevent sperm from entering the cervix. Some women find it difficult to correctly place the diaphragm or cervical cap, and incorrect placement can result in sperm entering the uterus. The failure rate may reach as high as 20% if the structure is not inserted properly.

Lactation is the production of milk for nursing a baby, and can prevent ovulation and menstruation for many months after childbirth *if* a woman nurses her child constantly (i.e., much more than five times a day). Frequent lactation sends signals to the hypothalamus to prevent follicle-stimulating hormone (FSH) and luteinizing hormone (LH) from being secreted, thus preventing ovulation. If a woman is lactating, she should always use another form of birth control as well, because she will not know when her ovulation cycle begins again until many weeks later, when she begins menstruating again.

Intrauterine devices (IUDs) are T-shaped, flexible plastic structures inserted into the uterus by a health-care provider. Once in place, the IUD prevents fertilization from occurring, although researchers aren't sure specifically how. The IUD may contain copper or levonorgestrel (a synthetic progesterone). IUDs containing copper are effective for up to 10 years, and

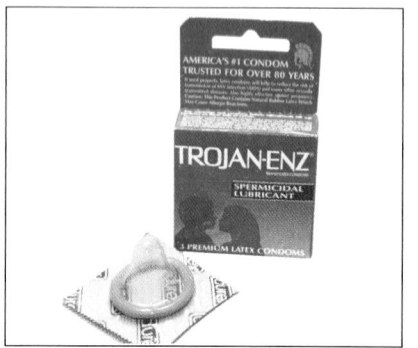

(a) Condoms

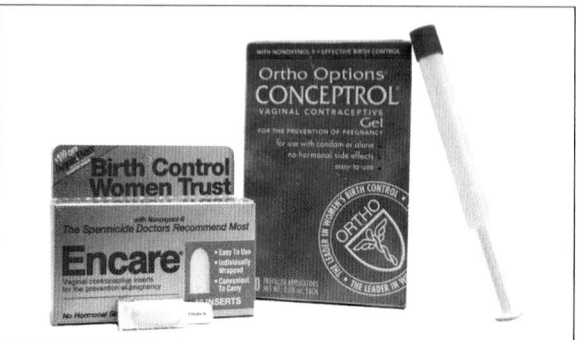

(b) Spermicidal foams

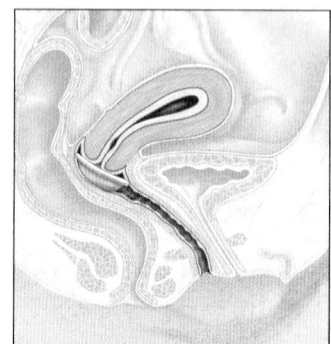

(c) Diaphragm

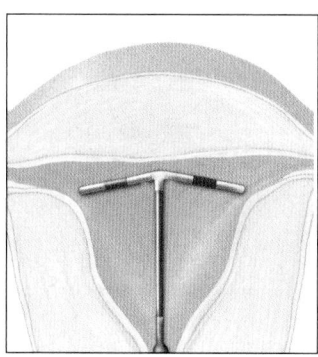

(d) Intrauterine device (IUD)

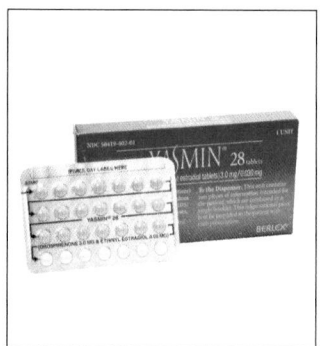

(e) Oral contraceptive

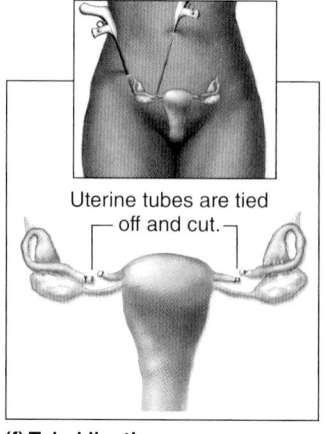

Uterine tubes are tied off and cut.

(f) Tubal ligation

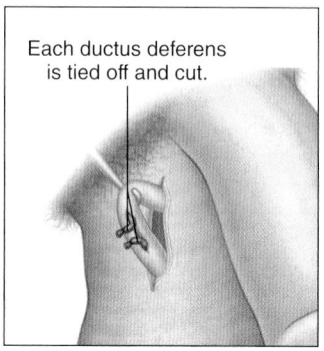

Each ductus deferens is tied off and cut.

(g) Vasectomy

Contraception includes barrier, chemical, and surgical methods.

those containing levonorgestrel (e.g., Skyla or Mirena) are effective for 3 to 5 years. Their failure rate typically is low (1–2%).

Chemical methods of birth control are very effective if used properly. They include the following:

- **Oral contraceptives** are commonly called *birth control pills* and typically come in 28-day packets. The first 21 days of pills contain low levels of estrogen and progestins, and the last 7 days are sugar pills. (Progesterone is one type of progestin.) The low levels of estrogen and progestins prevent the LH "spike" needed for ovulation. Thus, oral contraceptives prevent ovulation. During the 7 days of sugar pills, the circulating levels of estrogen and progestins drop, and menstruation occurs. Menstrual flow typically is much lighter when a woman takes oral contraceptives because the circulating levels of hormones were low to begin with, so the uterine lining is not as thick. Oral contraceptives require a woman to take a pill a day, at about the same time each day. If she misses one or more days of pills, ovulation may occur. In addition, some medications (such as antibiotics) may interfere with their effectiveness, so a backup method of contraception should be used when taking these medications. When used properly, the failure rate is very low (0.1%) but may increase to 7% if the timing of taking the pills varies or if the woman is on medication that may interfere with the pill's effectiveness.

- **Estrogen/progestin** (prō-jes′tin; *pro* = before; gestation) **patches** are alternatives to the daily oral contraceptive. A patch placed on the body delivers a regular amount of estrogen and progestin through the skin (transdermally). The patch is replaced each week and failure rate is 1%.

- **Injected/implanted progestins** help prevent pregnancy by preventing ovulation and thickening the mucus around the cervix (thus creating a slight physical barrier to the sperm). Medroxy-progesterone (Depo-Provera) is an injectable contraceptive given once every 3 months, whereas etonogestrel (Implanon) is an implantable contraceptive that lasts for up to 3 years. The drawback is that ovulation may not occur for many months after stopping the injections. Typical use failure rate is 0.3%.

- **Morning-after pills** containing levonorgestrel (e.g., Plan B One-Step) can be taken within 72 hours after having unprotected intercourse. These pills work by inhibiting ovulation, altering the menstrual cycle to delay ovulation, or irritating the uterine lining to prevent implantation. Many of these brands are available over the counter (without a prescription).

- **Mifepristone** (Mifeprex in the United States; RU-486 in Europe) may be used during the first 7 weeks of pregnancy. Mifepristone blocks progesterone receptors, so progesterone cannot attach to these receptors and thereby maintain a pregnancy. When taken with a prostaglandin drug, mifepristone induces a miscarriage. Mifepristone's availability and use is very politically charged, with each side of the abortion debate arguing for or against it.

Surgical methods of contraception are both **tubal ligation** for women and **vasectomy** (va-sek′tō-mē) for men. A tubal ligation requires that uterine tubes are cut and the ends are tied off or cauterized shut to prevent both sperm from reaching the oocyte and the oocyte from reaching the uterus. The uterine tubes are ligated either laparoscopically (shown in the adjacent figure) or via entry through the uterus (which is a much more difficult procedure for the surgeon.) A vasectomy is an outpatient procedure whereby each ductus deferens is cut, a short segment is removed, and then the ends are tied off. Sperm cannot leave the testis and thus are broken down and resorbed. Both surgeries are effective birth control methods, but they are usually permanent and irreversible, so they are not options for people who wish to have more children.

of internal and external stimuli. Normally, a woman starts to produce breast milk when she has recently given birth. Recall from section 17.7c that the hormone **prolactin** is produced in the anterior pituitary and is responsible for milk production. Thus, when the amount of prolactin increases, the mammary gland grows and forms more expanded and numerous alveoli. The hormone **oxytocin,** produced by the hypothalamus and released from the posterior pituitary, is responsible for milk ejection.

WHAT DID YOU LEARN?

22 What is the relationship between lobes, suspensory ligaments, and lactiferous ducts in the mammary gland?

23 What are the effects of prolactin and oxytocin on the female mammary gland?

28.3g Female Sexual Response

LEARNING OBJECTIVE

15. Explain how the female sexual response and orgasm is elicited.

The female sexual response refers to a series of physiologic events that occur during stimulation of the female reproductive organs. The response begins with the **excitement phase,** where reproductive structures such as the mammary glands, clitoris, vaginal wall, bulbs of the vestibule, and labia become engorged with blood. The nipples become erect as a result of the blood engorgement in the mammary glands. The vestibular glands and the glands within the vaginal wall both produce mucin for lubrication. The uterus shifts from an anteverted position to a more erect position within the pelvic cavity.

As the excitement phase continues, the erectile tissue of the clitoris swells as it engorges with blood and becomes very sensitive to tactile stimuli. The inferior part of the vaginal wall constricts slightly. The woman's heart rate, blood pressure, and respiratory rate increase during this time as orgasm nears. Both divisions of the autonomic nervous system control many of these physiologic responses, and the excitement phase also is facilitated by somatosensory signals, such as the sensation of the penis in the vagina or the caressing of parts of the body.

Orgasm refers to the time period where there are intense feelings of pleasure, a feeling of a release of tension, perhaps a feeling of warmth, and some pelvic throbbing. The vagina and uterus contract rhythmically for a period of many seconds.

INTEGRATE

CONCEPT CONNECTION

The male and female sexual response each requires an intricate interplay among various body systems beyond the reproductive system. The nervous system (both autonomic and somatic components) is involved with coordinating the entire event and controlling the numerous physiologic processes. The cardiovascular system is responsible for the engorgement of erectile tissue with blood, and heart rate and blood pressure are increased just prior to orgasm. In addition, the respiratory system is involved as breathing rate increases prior to orgasm. For the male, there is an additional step as the internal urethral sphincter (part of the urinary system) is contracted so no urine enters the urethra during ejaculation.

The **resolution phase** follows orgasm. The uterus returns to its original position and the vaginal wall relaxes at this time. The excess blood leaves the other reproductive organs. The cycle may begin again. Unlike men, women do not have a refractory period, which is a period of time where the body cannot have another orgasm; thus, women have the potential to have multiple orgasms during a single sexual experience.

WHAT DID YOU LEARN?

24 What are the three phases of female sexual response?

28.4 Male Reproductive System

The primary reproductive organs in the male are the **testes** (tes'tēz; sing., *testis*). The accessory reproductive organs include a complex set of ducts and tubules leading from the testes to the penis, a group of male accessory glands, and the penis, which is the organ of copulation **(figure 28.14)**.

28.4a Scrotum

LEARNING OBJECTIVE

1. Describe the gross anatomy and function of the scrotum.

The ideal temperature for producing and storing sperm is about 2–3°C lower than internal body temperature. The **scrotum** (skrō'tŭm), a skin-covered sac between the thighs, provides the cooler environment needed for normal sperm development and maturation **(figure 28.15)**. The scrotum is homologous to the labia majora in the female.

Externally, the scrotum contains a distinct, ridgelike seam at its midline, called the **raphe** (rā'fē; *rhaphe* = seam). The raphe extends in an anterior direction along the inferior surface of the penis and in a posterior direction to the anus. The wall of the scrotum is composed of an external layer of skin, a thin layer of superficial fascia immediately internal to the skin, and a layer of smooth muscle, the **dartos** (dar'tos; skinned) **muscle,** immediately internal to the fascia.

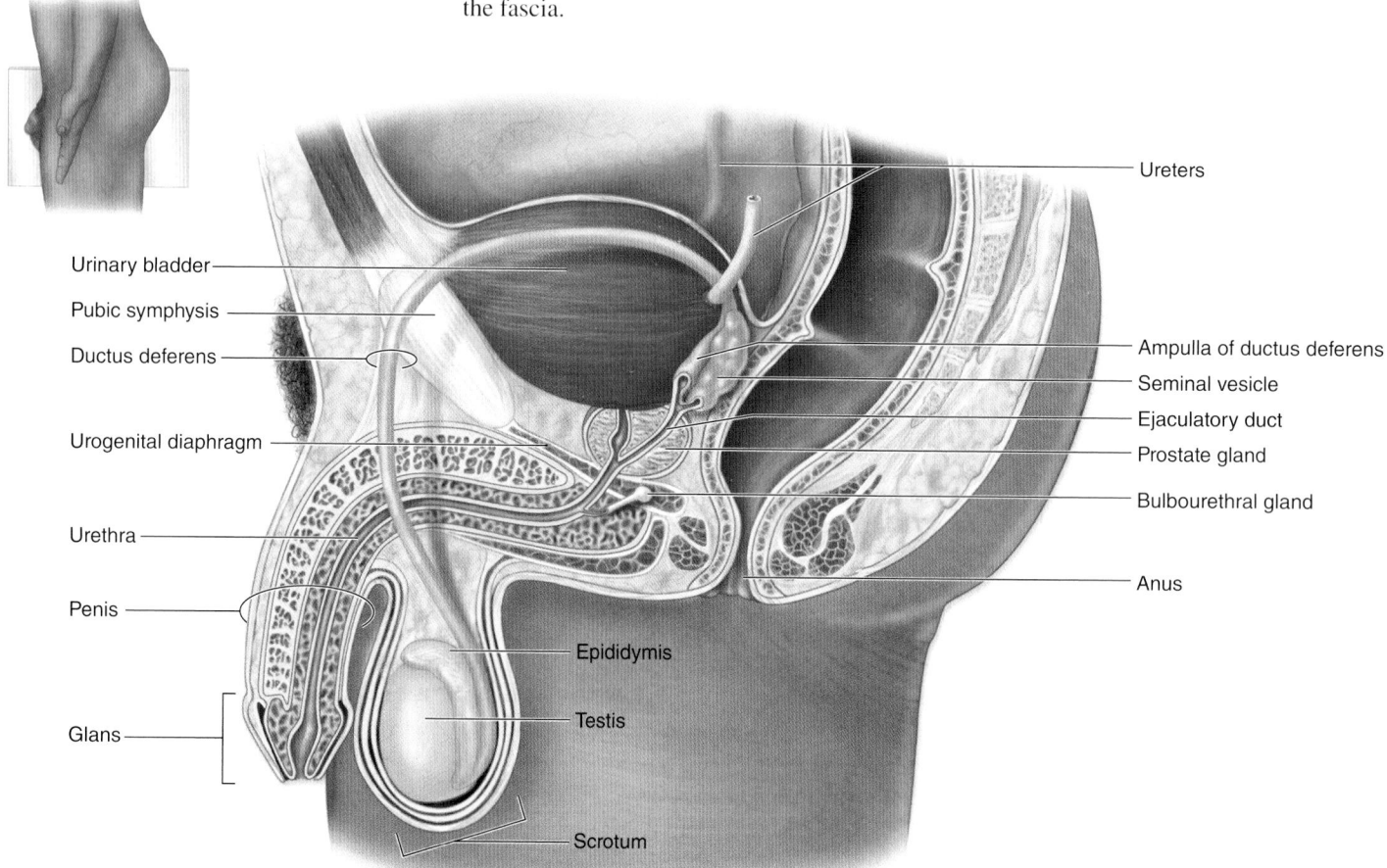

Labels (left): Urinary bladder, Pubic symphysis, Ductus deferens, Urogenital diaphragm, Urethra, Penis, Glans, Epididymis, Testis, Scrotum

Labels (right): Ureters, Ampulla of ductus deferens, Seminal vesicle, Ejaculatory duct, Prostate gland, Bulbourethral gland, Anus

Figure 28.14 Male Pelvic Region. A sagittal diagrammatic section shows the location and relationship of the male pelvic structures. AP|R

The blood vessels and nerves supplying each testis extend from within the abdomen to the scrotum in a multilayered structure called the **spermatic cord**. The spermatic cord originates in the **inguinal canal,** a tubelike passageway through the inferior abdominal wall. The spermatic cord wall consists of three layers:

- An **external spermatic fascia** is formed from the aponeurosis of the external oblique muscle.
- The **cremaster** (krē-mas′ter; a suspender) **muscle** and **cremasteric fascia** are formed from muscle fiber extensions of the internal oblique muscle and its aponeurosis, respectively.
- An **internal spermatic fascia** is formed from fascia deep to the abdominal muscles.

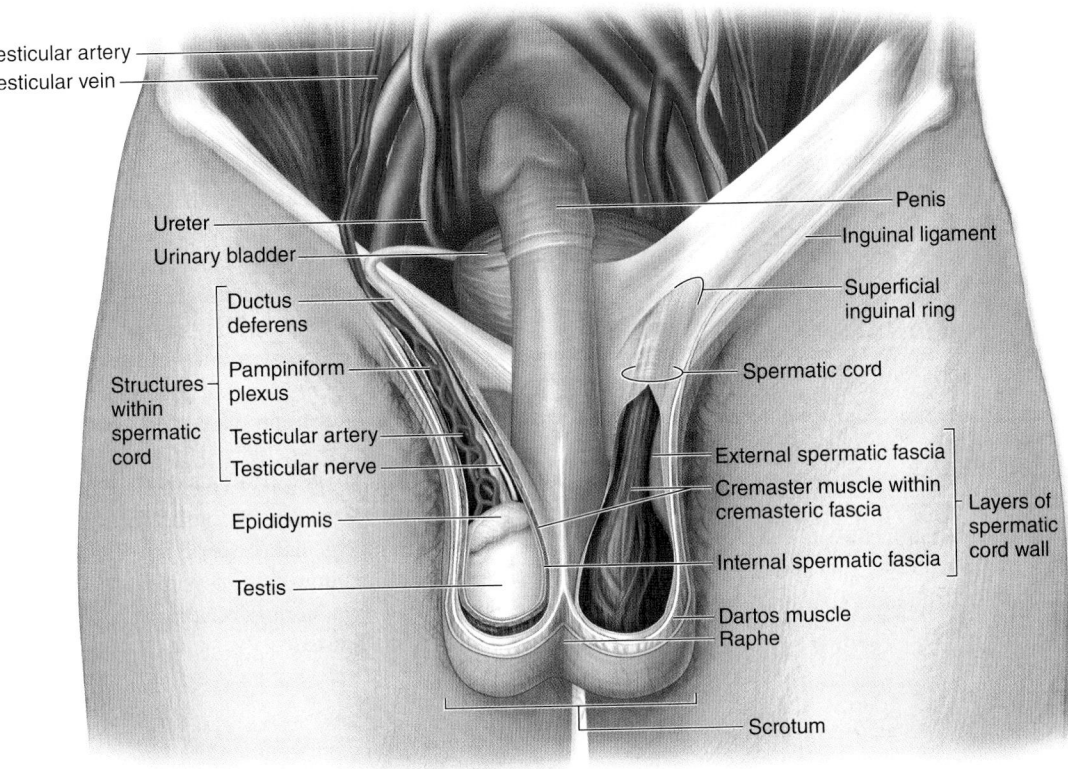

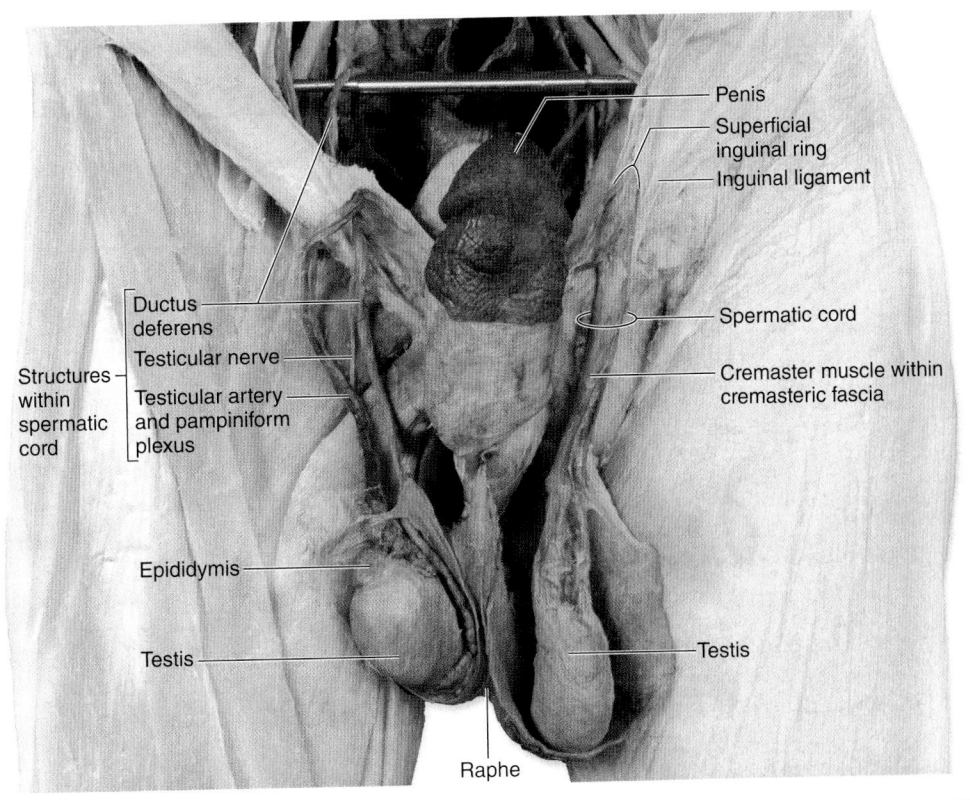

Figure 28.15 Scrotum and Testes. Anterior view and cadaver photo show the scrotum, a skin-covered sac that houses the testes. A multi-layered spermatic cord houses the blood vessels, nerves, and the ductus deferens for each testis. **AP|R**

Within the spermatic cord is a singular **testicular artery** that is a direct branch from the abdominal aorta. The testicular artery is surrounded by a plexus of veins called the **pampiniform** (pam-pin′i-form; *pampinus* = tendril, *forma* = form) **plexus.** This venous plexus is a means to provide thermoregulation by pre-cooling arterial blood prior to its reaching the testes. Autonomic nerves travel with these vessels and innervate each testis.

When the testes are exposed to elevated temperatures, the dartos muscle relaxes, which unwrinkles the skin of the scrotum and allows the testes to move inferiorly further away from the body. This movement away from the body cools the testes. At the same time, the cremaster muscle also relaxes to allow the testes to move inferiorly away from the body. The opposite occurs if the testes are exposed to cold. In this case, both the dartos and cremaster muscles contract, pulling the testes and scrotum closer to the body to conserve heat.

WHAT DID YOU LEARN?

25 How does the scrotum help regulate the temperature of the testes?

28.4b Testes and Spermatogenesis

LEARNING OBJECTIVES

2. Describe the gross anatomy and microscopic anatomy of the testes.

3. Explain the process of spermatogenesis and spermiogenesis.

4. Compare and contrast spermatogenesis with oogenesis.

In the adult human male, the testes are relatively small, oval organs housed within the scrotum (figures 28.14 and 28.15). Each weighs approximately 10 to 12 grams, and displays average dimensions of 4 centimeters (cm) in length, 2 cm in width, and 2.5 cm in anteroposterior diameter. The testes produce sperm and androgens (male sex hormones), the most common of which is testosterone. Note: The details for testosterone are included in the summary table "Regulating the Male Reproductive System," which directly follows chapter 17 (see **table R.10**).

WHAT DO YOU THINK?

3 If a man's testes were removed, would he still be able to produce androgens?

Each testis is covered anteriorly and laterally by a serous membrane, the **tunica vaginalis** (văj-in-ăl′ĭs; ensheathing) **(figure 28.16a).** This membrane is derived from the peritoneum of the abdominal cavity. The tunica vaginalis has an outer **parietal layer** and an inner **visceral layer** that are separated by a cavity filled with serous fluid. A thick, whitish, fibrous capsule called the **tunica albuginea** covers the testis and lies immediately deep to the visceral layer of the tunica vaginalis. At the posterior margin of the testis, the tunica albuginea thickens and projects into the interior of the organ as the **mediastinum testis.** Blood vessels, a system of ducts, lymph vessels, and some nerves enter or leave each testis through the mediastinum testis.

The tunica albuginea projects internally into the testis and forms delicate connective tissue **septa,** which subdivide the internal space into about 250 separate **lobules.** Each lobule contains up to four extremely convoluted, thin and elongated **seminiferous** (sem′in-if′er-ŭs; *semen* = seed, *fero* = to carry) **tubules.** The seminiferous tubules contain two types of cells: (1) a group of nondividing support cells called the **sustentacular** (sŭs-ten-tak′ū-lăr; *sustento* = to hold upright) **cells,** which also are termed *Sertoli cells,* or *nurse cells*; and (2) a population of dividing germ cells that continuously produce sperm beginning at puberty.

The sustentacular cells provide a protective environment for the developing sperm, and their cytoplasm helps nourish the developing sperm (figure 28.16b). In addition, sustentacular cells release the hormone **inhibin** when sperm count is high. Inhibin inhibits FSH secretion and thus regulates sperm production. (Conversely, when sperm count declines, inhibin secretion decreases.)

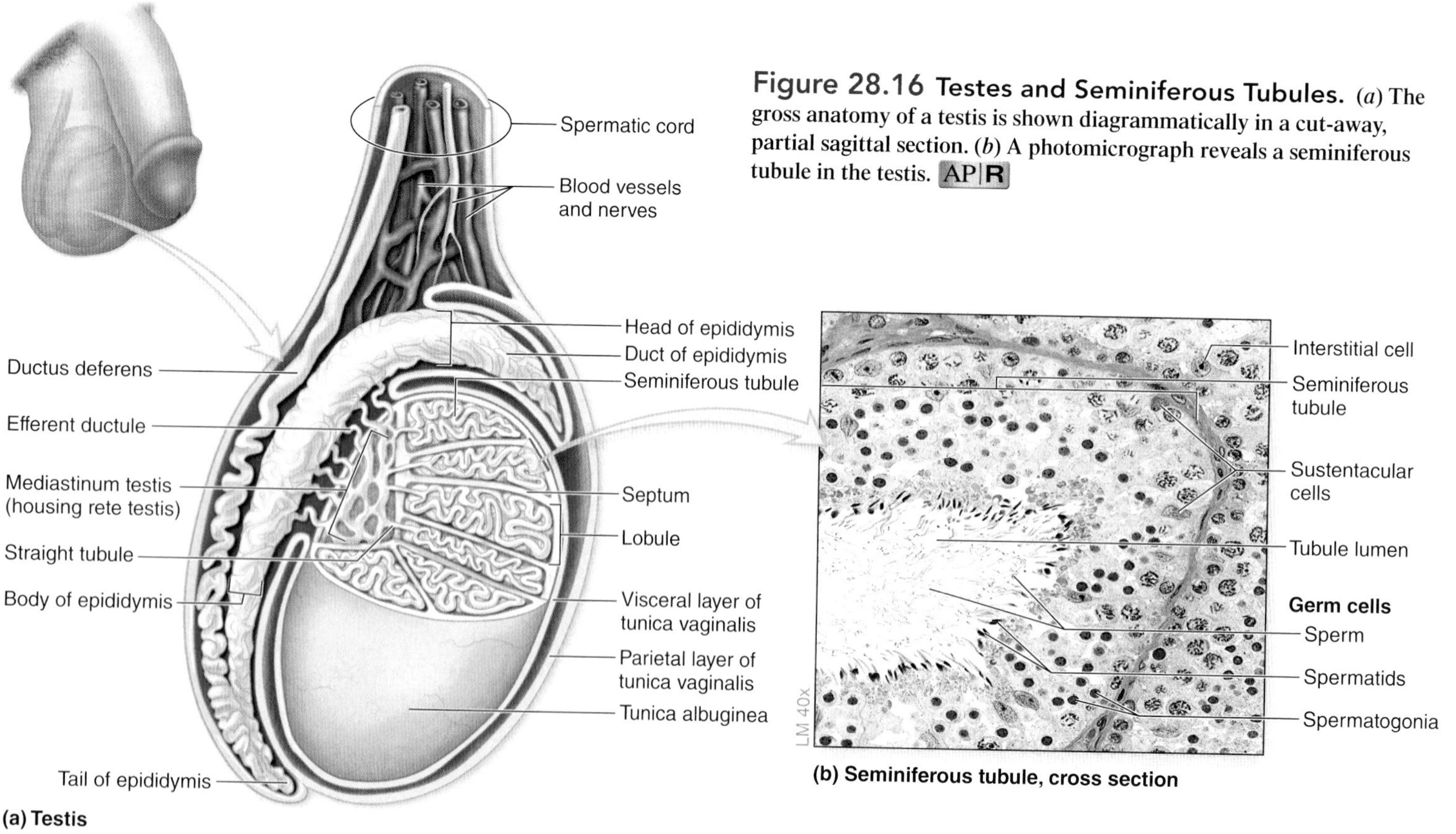

Figure 28.16 Testes and Seminiferous Tubules. (*a*) The gross anatomy of a testis is shown diagrammatically in a cut-away, partial sagittal section. (*b*) A photomicrograph reveals a seminiferous tubule in the testis.

Spermatic cord

Blood vessels and nerves

Ductus deferens

Efferent ductule

Mediastinum testis (housing rete testis)

Straight tubule

Body of epididymis

Tail of epididymis

Head of epididymis
Duct of epididymis
Seminiferous tubule

Septum

Lobule

Visceral layer of tunica vaginalis

Parietal layer of tunica vaginalis

Tunica albuginea

(a) Testis

Interstitial cell

Seminiferous tubule

Sustentacular cells

Tubule lumen

Germ cells
Sperm

Spermatids

Spermatogonia

LM 40x

(b) Seminiferous tubule, cross section

The sustentacular cells are bound together by tight junctions, which form a **blood-testis barrier** that is similar to the blood-brain barrier. The blood-testis barrier helps protect developing sperm from materials in the blood. It also protects the sperm from the body's leukocytes, which may perceive the sperm as foreign because they have different chromosome numbers and proteins. (In contrast, the female oocyte likely is protected from materials in the blood by the ovarian follicle structures that surround the oocyte.)

The spaces surrounding the seminiferous tubules are called **interstitial spaces.** Within these spaces reside the **interstitial cells** (or *Leydig cells*). Luteinizing hormone stimulates the interstitial cells to produce hormones called **androgens** (an'drō-jen; *andros* = male human). There are several types of androgens; the most common one is testosterone. Although the adrenal cortex secretes a small amount of androgens in both sexes, the vast majority of androgen release is via the interstitial cells in the testis in males, beginning at puberty.

Hormonal Regulation of Androgen Production and Sperm Development

Like the hormone interactions involved in the ovarian cycle, the interplay among the hormones involved in regulating the spermatogenesis and androgen production is intricate and involves several negative feedback mechanisms. A stepwise description of these hormonal effects are listed here and shown in **figure 28.17**:

① **The hypothalamus initiates spermatogenesis by secreting GnRH,** which stimulates the anterior pituitary to secrete FSH and LH.

② **FSH and LH target the testes and stimulate both spermatogenesis and androgen production.** Specifically,

LH stimulates interstitial cells in the testis to secrete testosterone, and FSH stimulates the sustentacular cells to secrete androgen-binding protein (ABP). ABP binds to testosterone (and other androgens) to ensure that testosterone levels remain high in the testes. Therefore, the testis and its cells are the effectors, and they release stimuli (testosterone and ABP) in this cycle.

③ **The increased levels of testosterone have various immediate effects on the body.** The high testosterone levels facilitate spermatogenesis. However, this same circulating testosterone inhibits GnRH secretion and reduces anterior pituitary sensitivity to GnRH. Thus, rising levels of testosterone have a negative feedback effect on the entire cycle.

④ **Sustentacular cells respond to rising sperm count levels and secrete inhibin.** Inhibin primarily causes inhibition of FSH secretion from the anterior pituitary, and it serves as an additional negative feedback mechanism on the cycle.

⑤ **Circulating testosterone stimulates libido (sex drive) and development of secondary sex characteristics.** Testosterone acts on the brain so there is an increased desire for and sensitivity to sexual stimulation. Secondary sex characteristics include growth and development of hair in the pubic and axillary regions, a deeper voice (because testosterone affects larynx development), and the growth of facial hair.

Note that sperm development and androgen production are controlled by negative feedback only, whereas the ovarian cycle uses both positive and negative feedback mechanisms.

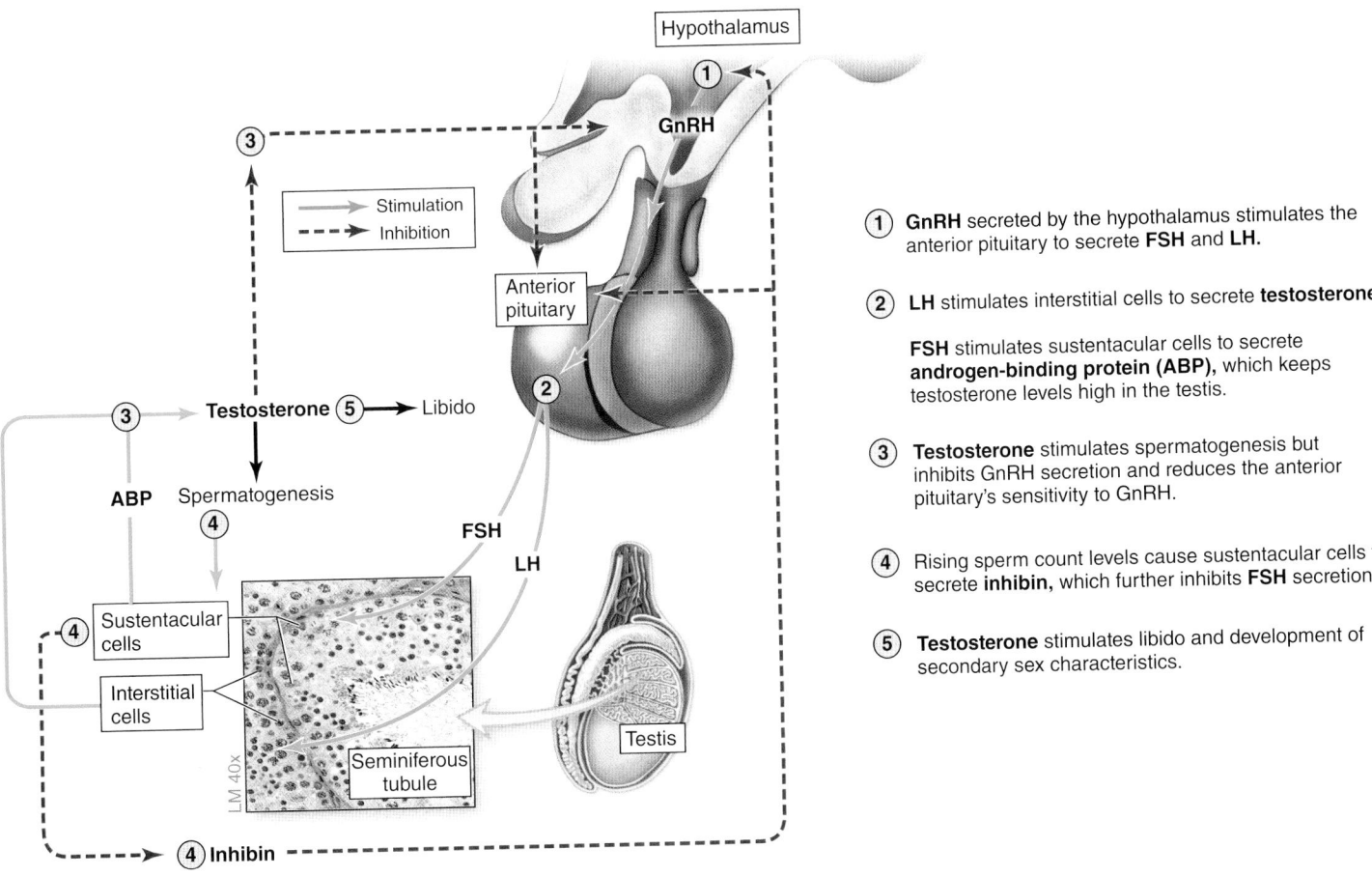

① **GnRH** secreted by the hypothalamus stimulates the anterior pituitary to secrete **FSH** and **LH.**

② **LH** stimulates interstitial cells to secrete **testosterone.**

FSH stimulates sustentacular cells to secrete **androgen-binding protein (ABP),** which keeps testosterone levels high in the testis.

③ **Testosterone** stimulates spermatogenesis but inhibits GnRH secretion and reduces the anterior pituitary's sensitivity to GnRH.

④ Rising sperm count levels cause sustentacular cells to secrete **inhibin,** which further inhibits **FSH** secretion.

⑤ **Testosterone** stimulates libido and development of secondary sex characteristics.

Figure 28.17 Hormonal Regulation of Spermatogenesis and Androgen Production. The hypothalamus secretes gonadotropin-releasing hormone (GnRH), which stimulates the anterior pituitary to release follicle-stimulating hormone (FSH) and luteinizing hormone (LH). FSH and LH stimulate spermatogenesis and testosterone (and other androgen) production, as shown in this image.

INTEGRATE

CLINICAL VIEW
Sexually Transmitted Infections

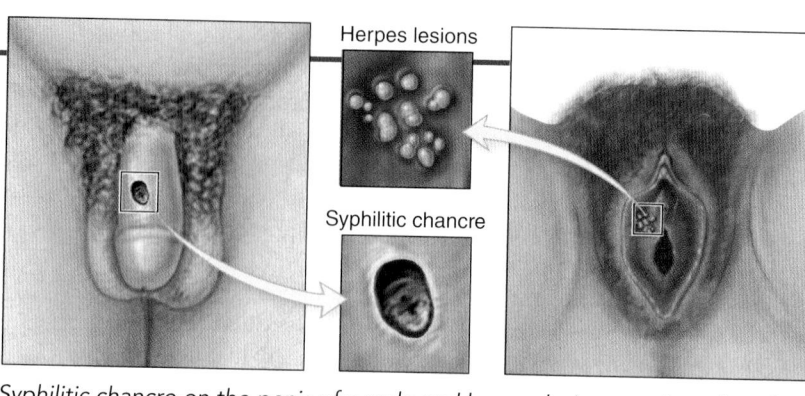

Syphilitic chancre on the penis of a male and herpes lesions on the vulva of a female.

Sexually transmitted infections (STIs), also known as *sexually transmitted diseases,* or *venereal diseases,* are a group of infectious diseases that are usually transmitted via sexual contact. Often the symptoms of STIs are not immediately noticeable, so infected individuals may infect someone else without realizing it. Mothers also may transmit STIs to their newborns, either directly across the placenta or at the time of delivery. Condoms have been shown to help prevent the spread of STIs, but they are not 100% effective.

STIs are a leading cause of **pelvic inflammatory disease** in women, in which the reproductive organs become infected. Should bacteria from an STI infect the uterus and uterine tubes, scarring is likely to follow, leading to blockage of the tubes and infertility. We've already discussed two types of STIs: human papillomavirus infections (see Clinical View: "Cervical Cancer" earlier in this chapter) and AIDS (see Clinical View: "HIV and AIDS" in section 22.9c). We now describe other common STIs.

Chlamydia (kla-mid′ē-ă) is the most frequently reported bacterial STI in the United States. The responsible agent is the bacterium *Chlamydia trachomatis.* Most infected people are asymptomatic; the rest develop symptoms within 1 to 3 weeks after exposure. These symptoms include abnormal vaginal discharge, painful urination (in both men and women), and low back pain. Chlamydia is treated with antibiotics.

Genital herpes (her′pēz; *herpo* = to creep) is caused by herpes simplex virus type 1 (HSV-1) or type 2 (HSV-2). Infected individuals undergo cyclic outbreaks of blister formation in the genital and anal regions; the blisters are filled with fluid containing millions of infectious viral particles. The blisters then break and turn into tender sores that remain for 2 to 4 weeks. Typically, future cycles of blistering are less severe and shorter in duration than the initial episode. There is no cure for herpes, but antiviral medications can lessen the severity and length of an outbreak.

Gonorrhea (gon-ō-rē′ă) is caused by the bacterium *Neisseria gonorrhoeae,* and it is spread either by sexual contact or from mother to newborn at the time of delivery. Symptoms include painful urination and a yellowish discharge from the penis or vagina. Gonorrhea is treated with antibiotics, although in recent years many gonorrhea strains have become resistant to some antibiotics. If untreated, women may develop pelvic inflammatory disease, and men may develop epididymitis, a painful condition of the epididymis that can lead to infertility. If a newborn acquires the disease, then blindness, joint problems, or a life-threatening blood infection may result.

Syphilis (sif′i-lis) is caused by the corkscrew-shaped bacterium *Treponema pallidum.* The bacterium is spread sexually via contact with a syphilitic sore called a **chancre** (shan′ker), or a newborn may acquire it *in utero.* Babies can acquire congenital syphilis from their mothers and are often stillborn, but if they live, they have a high incidence of skeletal malformations and neurologic problems. Syphilis can be treated with antibiotics.

Development of Sperm: Spermatogenesis and Spermiogenesis

Spermatogenesis (sper′mă-tō-jen′ĕ-sis; *genesis* = origin) is the process of sperm development that occurs within the seminiferous tubule of the testis. Spermatogenesis does not begin until puberty, when significant levels of FSH and LH stimulate the testis to begin gamete development.

The process of spermatogenesis is shown in **figure 28.18a.** All sperm develop from primordial germ (stem) cells called **spermatogonia** (sper′mă-tō-gō′nē-ă; sing., *spermatogonium; sperma* = seed, *gone* = generation). Spermatogonia are diploid cells (meaning they have 23 pairs of chromosomes for a total of 46). These cells lie near the base of the seminiferous tubule, surrounded by the cytoplasm of a sustentacular cell. To produce sperm, spermatogonia first divide by mitosis. One of the cells produced is a new spermatogonium (a new germ cell), to ensure that the numbers of spermatogonia never become depleted, and the other cell is a committed cell called **primary spermatocyte.** Primary spermatocytes are diploid and are the cells that undergo meiosis.

When a primary spermatocyte undergoes meiosis I, the two cells produced are called **secondary spermatocytes.** Secondary spermatocytes are haploid cells, meaning they each have 23 chromosomes only. These cells are still surrounded by the sustentacular cell cytoplasm, but they are relatively closer to the lumen of the seminiferous tubule as opposed to the base of the seminiferous tubule.

Secondary spermatocytes complete meiosis II to form **spermatids** (sper′mă-tid). A spermatid is a haploid cell and is surrounded by the sustentacular cell cytoplasm, very near to the lumen of the seminiferous tubule. The spermatids still have a circular appearance, rather than the sleek shape of mature sperm.

In the final stage of spermatogenesis, a process called **spermiogenesis** (sper′mē-ō-jen′ĕ-sis), the newly formed spermatids differentiate to become anatomically mature **spermatozoa** (sing., *spermatozoon;* sper′mă-to-zo′on) or **sperm** (figure 28.18b). During spermiogenesis, the

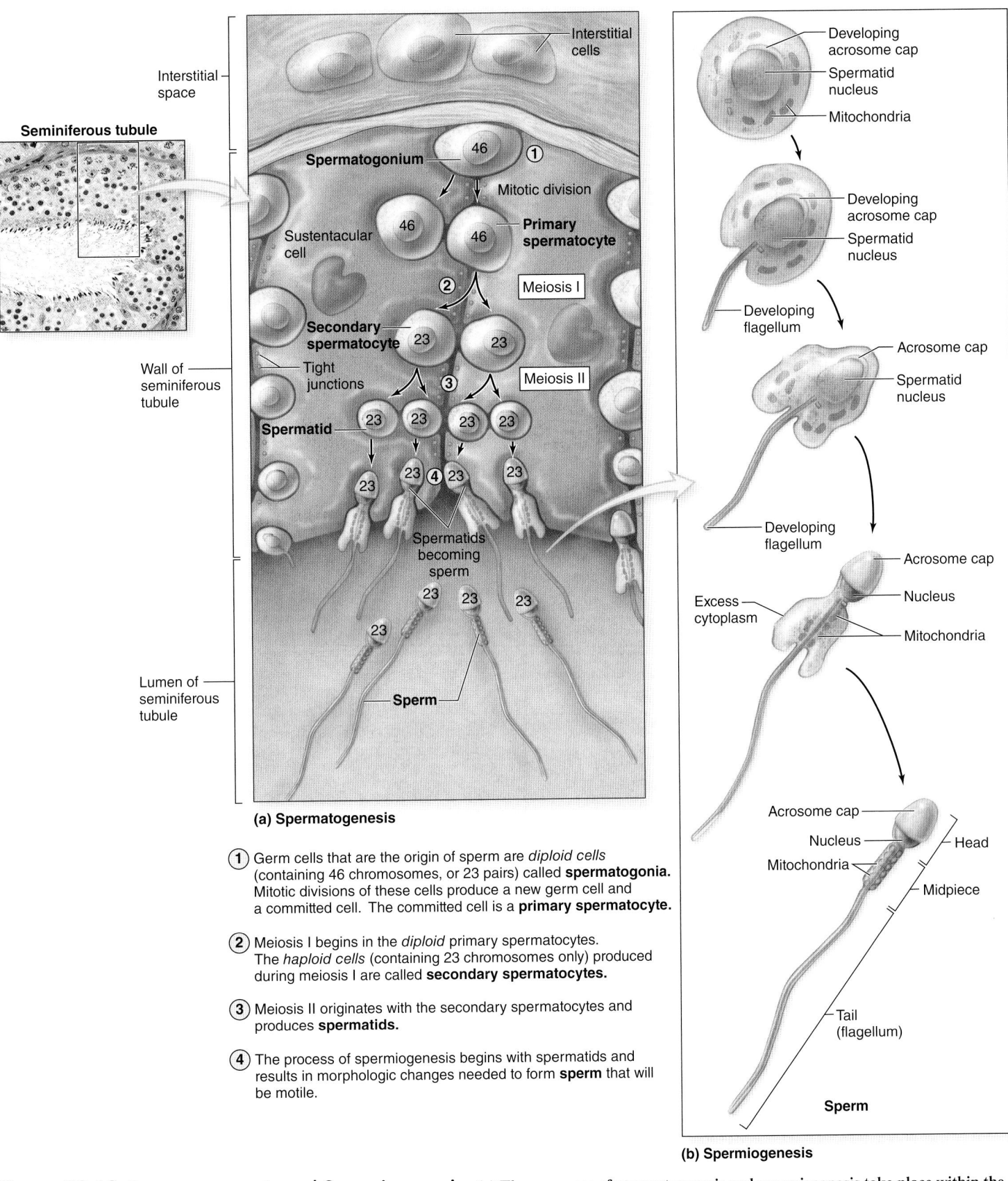

(a) Spermatogenesis

(1) Germ cells that are the origin of sperm are *diploid cells* (containing 46 chromosomes, or 23 pairs) called **spermatogonia.** Mitotic divisions of these cells produce a new germ cell and a committed cell. The committed cell is a **primary spermatocyte.**

(2) Meiosis I begins in the *diploid* primary spermatocytes. The *haploid cells* (containing 23 chromosomes only) produced during meiosis I are called **secondary spermatocytes.**

(3) Meiosis II originates with the secondary spermatocytes and produces **spermatids.**

(4) The process of spermiogenesis begins with spermatids and results in morphologic changes needed to form **sperm** that will be motile.

(b) Spermiogenesis

Figure 28.18 Spermatogenesis and Spermiogenesis. (*a*) The processes of spermatogenesis and spermiogenesis take place within the wall of the seminiferous tubule. (*b*) Structural changes occur during spermiogenesis as a sperm forms from a spermatid. AP|R

spermatid sheds excess cytoplasm and its nucleus elongates. A structure called the **acrosome cap** (ak′rō-sōm; *akros* = tip, *soma* = body) forms over the nucleus. This structure contains digestive enzymes that help penetrate the secondary oocyte for fertilization. As the spermatid elongates, a **tail,** also called the *flagellum,* forms from the organized microtubules within the cell. The tail is attached to a **midpiece** or *neck* region containing mitochondria and a centriole. These mitochondria provide the energy to move the tail.

Table 28.6	Stages of Spermatogenesis		
Cell Type	**Number of Chromosomes**	**Haploid or Diploid**	**Action**
Spermatogonium	23 pairs (46)	Diploid	Divides by mitosis to produce a new spermatogonium and a primary spermatocyte
Primary spermatocyte	23 pairs (46)	Diploid	Completes meiosis I to produce secondary spermatocytes
Secondary spermatocyte	23 only	Haploid	Completes meiosis II to produce spermatids
Spermatid	23 only	Haploid	Undergoes spermiogenesis, where most of its cytoplasm is shed and a midpiece, tail, and head form
Spermatozoon (sperm)	23 only	Haploid	Leaves seminiferous tubule and matures in epididymis

INTEGRATE

CLINICAL VIEW

Paternal Age Risks for Disorders in the Offspring

Although an increase in certain developmental disorders (such as trisomy 21s) have been well documented in older mothers, recent research indicates that increased paternal age (age at when a male becomes a father) also may be related to an increased risk of selected genetic disorders in the offspring. Because spermatocytes undergo cell division frequently throughout a male's lifetime, then the older the male becomes, the more likely the spermatocytes may acquire mutations as a result of these frequent cell divisions. Males over 45 are four times more likely to pass along genetic mutations to their offspring than men in their 20s, and this risk increases as the male continues to age.

Several large population-based cohort studies have demonstrated that children born to fathers older than 45 years were at increased risk of developing various psychiatric disorders. Specifically, these children were at increased risk of developing bipolar disorder, schizophrenia, autism spectrum disorder, major depressive disorder, and attention deficit hyperactivity disorder (ADHD). Interestingly, increased maternal age is not correlated with an increase in these disorders, further supporting the view that it is the more frequent mutations in older sperm that cause these particular disorders. Thus, both advanced maternal age and advanced paternal age in individuals carry increased risks in developmental problems in their children.

Although the sperm look mature, they do not yet have all of the characteristics needed to successfully travel through the female reproductive tract and fertilize an oocyte. The sperm must leave the seminiferous tubule through a network of ducts (described next) and reside in the epididymis for a period of time in order to become fully motile. **Table 28.6** summarizes the stages of spermatogenesis.

Note that both female and male gametogenesis events share some similarities and have some notable differences. Both female and male gametes undergo meiosis, but only a single viable secondary oocyte is produced, whereas four sperm are produced. All female oocytes have initiated (and then become arrested in) meiosis prior to the female being born. In contrast, male spermatogonia do not undergo spermatogenesis until puberty, but after this time they can divide and produce spermatocytes throughout a male's adult lifetime.

WHAT DID YOU LEARN?

26 What are the major cell types in the seminiferous tubules?

27 What hormones are produced by the interstitial cells and the sustentacular cells, and how do these hormones affect the hypothalamus and anterior pituitary?

28 How does a spermatogonium divide to produce spermatozoa? Give specific steps.

29 What main events occur in spermiogenesis?

28.4c Duct System in the Male Reproductive Tract

LEARNING OBJECTIVES

5. Explain the function of each component of the ducts associated with the male reproductive system.

6. Trace the pathway that sperm travel through the testes and duct system.

The left and right testes each have their own set of ducts. These ducts store and transport sperm as they mature and pass out of the male body **(figure 28.19)**.

Ducts Within the Testis

The **rete** (rē'tē; net) **testis** is a meshwork of interconnected channels in the mediastinum testis that receive sperm from the seminiferous tubules via straight tubules. The rete testis is lined by simple cuboidal epithelium with short microvilli covering its luminal surface. The channels of the rete testis merge to form the efferent ductules (see figure 28.16a).

Approximately 12 to 15 **efferent ductules** (duk'tūl) connect the rete testis to the epididymis. They are lined with both ciliated columnar epithelia that propel the sperm toward the epididymis and nonciliated columnar epithelia that absorb excess fluid secreted by the seminiferous tubules. The efferent ductules drain into the epididymis.

Epididymis

The **epididymis** (ep-i-did'i-mis; pl., *epididymides*; *epi* = upon, *didymis* = twin) is a comma-shaped structure composed of an internal duct and an external covering of connective tissue. Its **head** lies on the superior surface of the testis, whereas the **body** and **tail** are on the posterior surface of the testis (see figure 28.16a). Internally, the epididymis contains a long, convoluted **duct of the epididymis,** which is approximately 4 to 5 meters in length and lined with pseudostratified columnar epithelium that contains stereocilia (long microvilli) (figure 28.19d).

The epididymis stores sperm until they are fully mature and capable of being motile. Just as a newborn has the superficial anatomic characteristics of an adult, but cannot move as an adult, the sperm that first enter the epididymis look like mature sperm but can't move like mature sperm. If they are expelled too soon, they lack the ability to be motile,

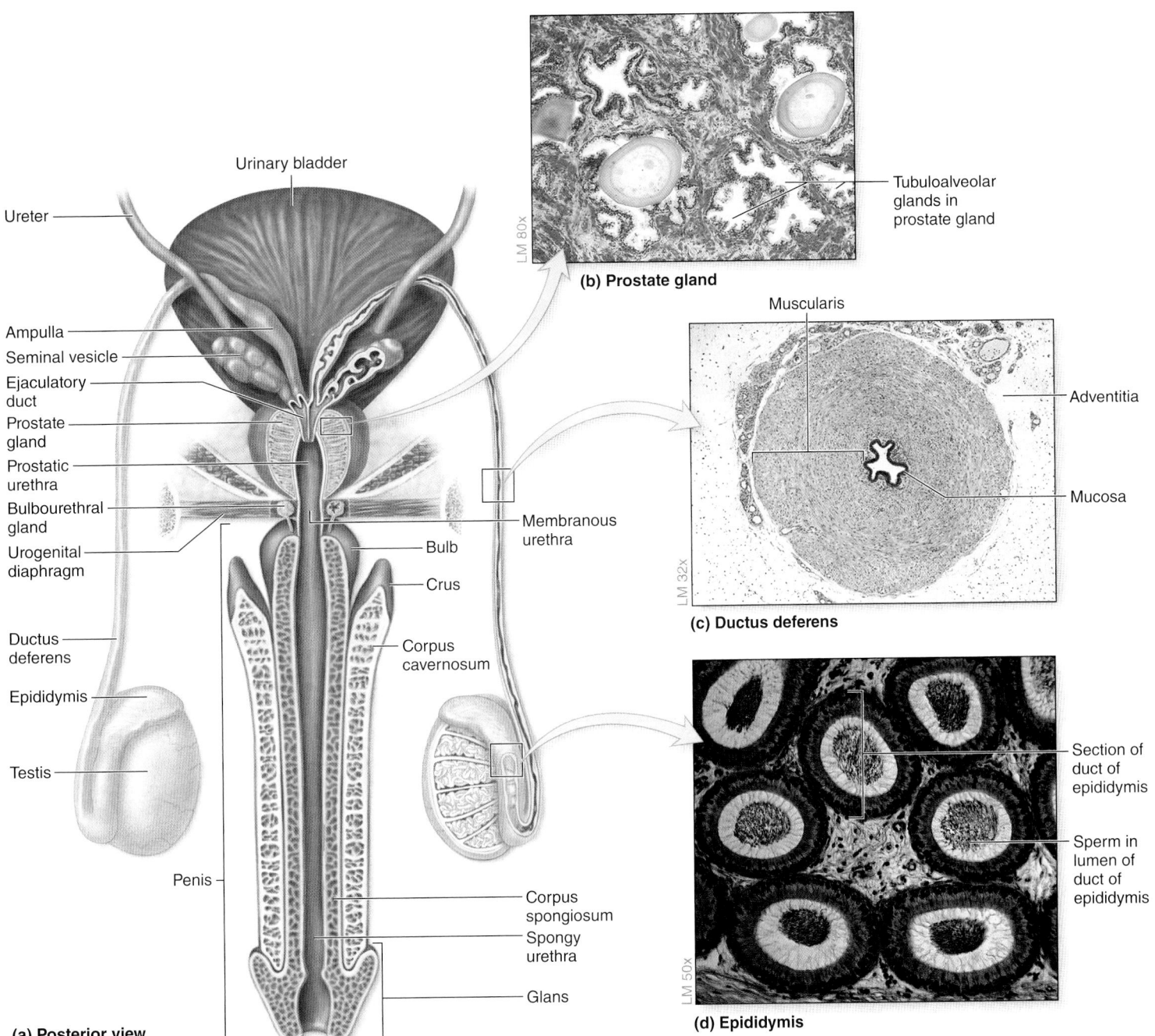

Urinary bladder

Ureter

Ampulla

Seminal vesicle

Ejaculatory duct

Prostate gland

Prostatic urethra

Bulbourethral gland

Urogenital diaphragm

Ductus deferens

Epididymis

Testis

Penis

(a) Posterior view

Tubuloalveolar glands in prostate gland

(b) Prostate gland

Muscularis

Adventitia

Mucosa

(c) Ductus deferens

Membranous urethra

Bulb

Crus

Corpus cavernosum

Corpus spongiosum

Spongy urethra

Glans

Section of duct of epididymis

Sperm in lumen of duct of epididymis

(d) Epididymis

Figure 28.19 Duct System and Accessory Glands in Male Reproductive Tract. (*a*) A posterior view depicts the structural components of the male reproductive ducts and accessory glands. Micrographs show cross sections through the (*b*) prostate gland, (*c*) ductus deferens, and (*d*) epididymis. AP|R

which is necessary to travel through the female reproductive tract and fertilize a secondary oocyte. If sperm are not ejected from the male reproductive system in a timely manner, the old sperm degenerate and are resorbed by cells lining the duct of the epididymis.

Ductus Deferens

When sperm leave the epididymis during ejaculation, they enter the **ductus deferens** (dĕf′er-ens; carry away), also called the *vas deferens*. The ductus deferens is a thick-walled tube that is located within the spermatic cord, and extends through the inguinal canal (see figure 28.15), and into the pelvic cavity (figure 28.19*a*). When the ductus deferens travels through the inguinal canal and enters the pelvic cavity, it separates from the other spermatic cord components and extends posteriorly along the superolateral surface of the bladder. It then

travels inferiorly and terminates close to the region where the bladder and prostate gland meet. As the ductus deferens approaches the superoposterior edge of the prostate gland, it enlarges and forms the **ampulla** of the ductus deferens (figure 28.19*a*). The ampulla of the ductus deferens unites with the proximal region of the seminal vesicle to form the terminal portion of the reproductive duct system, called the ejaculatory duct.

The ductus deferens wall is composed of an inner **mucosa** (lined by a pseudostratified ciliated columnar epithelium), a middle **muscularis,** and an outer **adventitia** (figure 28.19*c*). The muscularis contains three layers of smooth muscle: an inner longitudinal, middle circular, and outer longitudinal layer. Contraction of the muscularis is necessary to move sperm through the ductus deferens, because sperm do not exhibit motility until they are ejaculated from the penis.

Ejaculatory Duct

Each **ejaculatory duct** is between 1 and 2 centimeters long. The epithelium of the ejaculatory duct is a pseudostratified ciliated columnar epithelium. The ejaculatory duct conducts sperm (from the ductus deferens) and a component of seminal fluid (from the seminal vesicle) toward the urethra. Each ejaculatory duct opens into the prostatic urethra.

Urethra

The **urethra** transports semen from both ejaculatory ducts to the outside of the body. Recall from figure 24.27*b* that the urethra is subdivided into a **prostatic** (pros-tat′ik) **urethra** that extends from the bladder through the prostate gland, a **membranous urethra** that continues through the urogenital diaphragm, and a **spongy urethra** that extends through the penis.

 WHAT DID YOU LEARN?

 What is the sequence of ducts through which male gametes pass?

28.4d Accessory Glands and Semen Production

LEARNING OBJECTIVES

7. Describe the anatomy and function of the accessory glands.
8. Compare and contrast sperm, seminal fluid, and semen.
9. List the major components of semen and which accessory gland produces each component.

Recall from earlier in this chapter that the vagina has a highly acidic environment to prevent bacterial growth. Sperm cannot survive in this type of environment, so an alkaline secretion called **seminal** (sem′i-nal) **fluid** is needed to neutralize the acidity of the vagina. In addition, as the sperm travel through the female reproductive tract (a process that can take hours to several days), they are nourished by nutrients within the seminal fluid. The components of seminal fluid are produced by the accessory glands that include the seminal vesicles, the prostate gland, and the bulbourethral glands.

Seminal Vesicles

The paired **seminal vesicles** are located on the posterior surface of the urinary bladder lateral to the ampulla of the ductus deferens. Each seminal vesicle is an elongated, hollow organ approximately 5 to 8 centimeters long. The wall of each vesicle contains mucosal folds of pseudostratified columnar epithelium. The medial (proximal) portion of the seminal vesicle merges with a ductus deferens to form the ejaculatory duct.

The seminal vesicles secrete a viscous, whitish-yellow, alkaline fluid containing both fructose and prostaglandins. Fructose nourishes the sperm as they travel through the female reproductive tract. Prostaglandins are hormonelike substances that promote the widening and slight dilation of the external os of the cervix, which facilitates sperm entry into the uterus.

Prostate Gland

The **prostate** (pros′tāt; one who stands before) **gland** is a compact, encapsulated organ that weighs about 20 grams and is shaped like a walnut, measuring approximately 2 cm by 3 cm by 4 cm. It is located immediately inferior to the bladder. The prostate gland includes submucosal glands that produce mucin and more than 30 tubuloalveolar glands that open directly through numerous ducts into the prostatic

urethra (figure 28.19*b*). Together, this aggregate of secretory structures contributes a component to the seminal fluid.

The prostate gland secretes a slightly milky fluid that is weakly acidic and rich in citric acid, seminalplasmin, and **prostate-specific antigen (PSA).** The citric acid is a nutrient for sperm health, the seminalplasmin is an antibiotic that combats urinary tract infections in the male, and the PSA acts as an enzyme to help liquify semen following ejaculation. (Note that the slightly acidic prostate secretion of the prostate is not sufficient to cause seminal fluid to be acidic, and thus seminal fluid is still alkaline and functions to neutralize the acidity of the vagina.)

INTEGRATE

CLINICAL VIEW

Benign Prostatic Hyperplasia and Prostate Cancer

Benign prostatic hyperplasia (BPH) is a noncancerous enlargement of the prostate gland. BPH is a common disorder in older men, and hormonal changes in aging males are the cause of the enlargement.

In BPH, large nodules form within the prostate and compress the prostatic urethra. Thus, the patient has difficulty starting and stopping a stream of urine, and often complains of **nocturia** (frequent urinating at night), **polyuria** (more frequent urination), and **dysuria** (painful urination). Some drug regimens help inhibit hormones that cause prostate enlargement, but when medications are no longer effective, surgical removal of the prostatic enlargement is indicated. The most commonly performed surgical procedure is called a **TURP (transurethral resection of the prostate),** in which an instrument called a **resectoscope** (rē-sek′tō-skōp) is inserted into the urethra to cut away the problematic enlargement.

Prostate cancer is one of the most common malignancies among men over 50, and the risk of developing it increases with age. Prostate cancer forms hard, solid nodules, most often in the posterior part of the prostate gland. Early stages of the cancer are generally asymptomatic, but as it progresses, urinary symptoms may develop.

A very effective screening tool is a **digital rectal exam,** whereby a physician inserts a gloved finger into the rectum and palpates adjacent structures (including the prostate gland). Additionally, most physicals for men over the age of 50 now include a test for prostate-specific antigen (PSA) in the blood. An elevated PSA level may indicate either benign prostatic hyperplasia or prostate cancer.

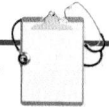

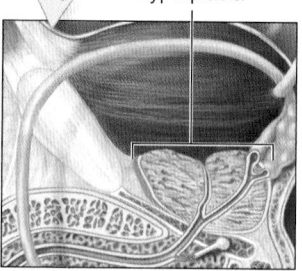

Benign prostatic hyperplasia

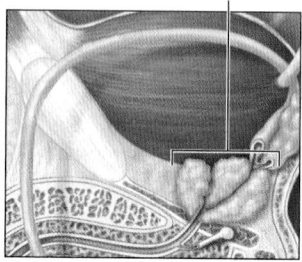

Prostate cancer

For earlier stages of prostate cancer, radiation therapy may be beneficial—either traditional external-beam radiation or **interstitial radiotherapy,** in which radioactive palladium or iodine "seeds" are permanently implanted in the prostate. For patients with a more aggressive cancer, the entire prostate and some surrounding structures are surgically removed, a procedure called a **radical prostatectomy.**

Bulbourethral Glands

Paired, pea-shaped **bulbourethral** (bŭl′bō-ū-rē′thrăl) **glands** (or *Cowper glands*) are located within the urogenital diaphragm on each side of the membranous urethra (figure 28.19a; see also figure 28.14). Each gland has a short duct that projects into the bulb (base) of the penis and enters the spongy urethra. Bulbourethral glands are tubuloalveolar glands that have a simple columnar and pseudostratified columnar epithelium. Their secretory product is a clear, viscous mucin that forms mucus. This mucus coats and lubricates the urethra for the passage of sperm during sexual intercourse.

Semen

Seminal fluid from the accessory glands combines with sperm from the testes to make up **semen** (sē′men; seed). When released during intercourse, semen is called **ejaculate** (ē-jak′ū-lāt; *eiaculatus* = to shoot out), and it normally measures about 3 to 5 milliliters in volume and contains approximately 200 to 500 million spermatozoa. In a sexually active male, the average transit time of human spermatozoa—from their release into the lumen of the seminiferous tubules, passage through the duct system, and presence in the ejaculate—is about 2 weeks. Because semen is composed primarily of seminal fluid, a

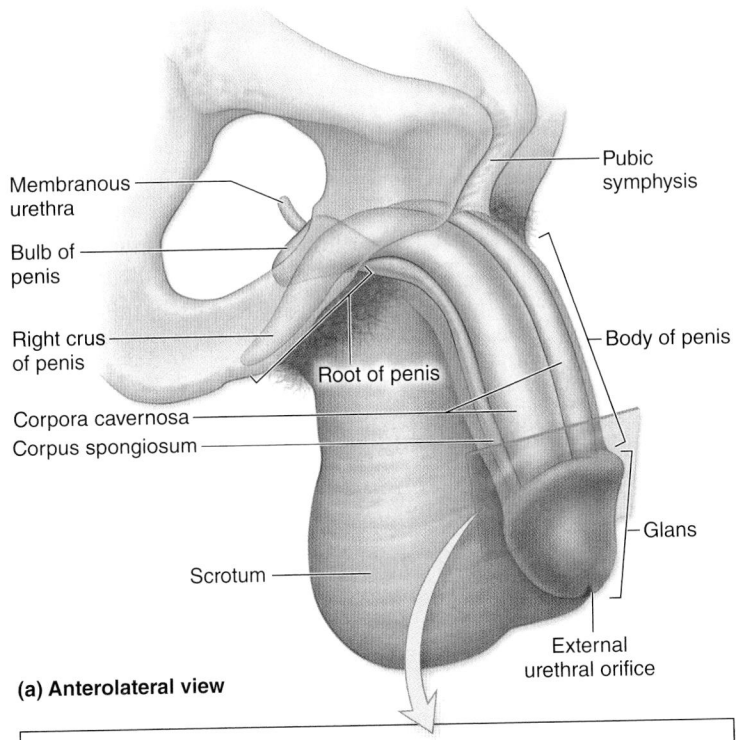

(a) Anterolateral view

Membranous urethra
Bulb of penis
Right crus of penis
Corpora cavernosa
Corpus spongiosum
Scrotum
Root of penis
Pubic symphysis
Body of penis
Glans
External urethral orifice

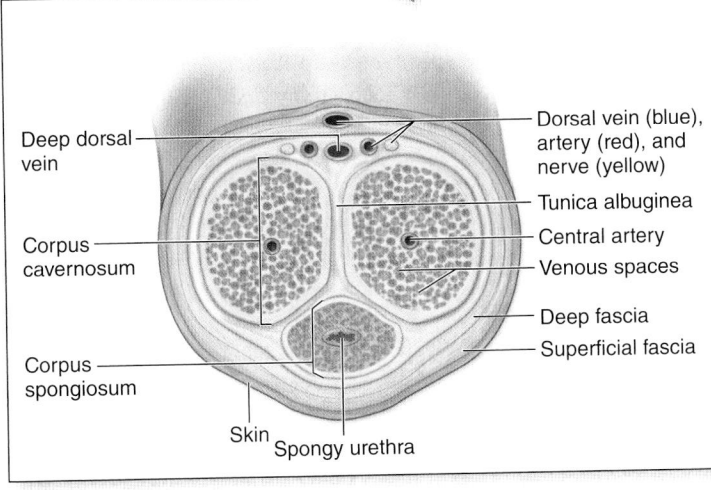

(b) Cross section

Deep dorsal vein
Corpus cavernosum
Corpus spongiosum
Skin
Spongy urethra
Dorsal vein (blue), artery (red), and nerve (yellow)
Tunica albuginea
Central artery
Venous spaces
Deep fascia
Superficial fascia

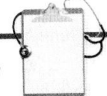

INTEGRATE

CLINICAL VIEW
Circumcision

Circumcision (ser-kŭm-sizh′ŭn; *circum* = around, *caedo* = to cut) is the surgical removal of the prepuce (foreskin) of the penis (see figure 28.20 for a comparison of circumcised versus uncircumcised penises). It is practiced among many cultures. Circumcision has several health benefits: Circumcised males appear less likely to develop urinary tract infections because the bacteria that cause these infections tend to stick to the foreskin. Circumcision may also protect against both penile inflammation (because the glans of a circumcised penis can be kept clean more easily) and penile cancer. Finally, some research has suggested that circumcised males have a reduced risk of acquiring and passing on sexually transmitted infections (STIs), including HIV. Circumcised men were approximately 50% less likely to become infected than uncircumcised men.

The drawbacks to circumcision include the following: Infants are sometimes circumcised without anesthesia, subjecting them to pain and elevated stress levels. Circumcision also carries a risk of complications, including infection, excessive bleeding, and in rare cases, subsequent surgery. Some individuals have suggested that circumcision may affect sensation during sexual intercourse, although this hypothesis has not been systematically tested or proven.

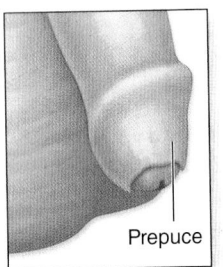

Prepuce

(c) Uncircumcised penis

male who is very active sexually may have a reduced sperm count because there are fewer sperm to be released from the epididymis; however, the total semen volume remains close to normal for that individual.

WHAT DO YOU THINK?

④ If a man has a vasectomy, is he still able to produce sperm? If so, what happens to those sperm? How is the composition of semen changed in an individual who has had a vasectomy?

WHAT DID YOU LEARN?

㉛ What are the specific functions of the accessory glands?

㉜ How is seminal fluid different from semen?

28.4e Penis

LEARNING OBJECTIVE

10. Describe the structure and functions of the penis.

The **penis** (pē′nis; tail) and the scrotum form the external genitalia in males (**figure 28.20**). Internally, the attached portion of the penis is the **root,** which is dilated internal to the body surface, forming both the **bulb** and the **crus** of the penis. The bulb attaches the penis to the bulbospongiosus muscle in the urogenital triangle, and the crus attaches the penis to the pubic arch. The **body,** or *shaft,* of the penis is the elongated,

Figure 28.20 Anatomy of the Penis. (*a*) An anterolateral view of the penis. The dorsum of the penis faces anteriorly, whereas the ventral surface is adjacent to the scrotum. (*b*) A diagrammatic, cross section shows the arrangement of the erectile bodies. (*c*) An uncircumcised penis.

movable portion. The tip of the penis is called the **glans,** and it contains the **external urethral orifice.** The skin of the penis is thin and elastic. At the distal end of the penis, the skin is attached to the raised edge of the glans and forms a circular fold called the **prepuce** (prē′pūs), or *foreskin.*

Within the shaft of the penis are three cylindrical erectile bodies. The paired **corpora cavernosa** (kav′er-nō-sa′; sing., *corpus cavernosum; caverna* = grotto) are located dorsolaterally. Ventral to them in the midline is the single **corpus spongiosum** (spŭn′jē-ō-sŭm), which contains the spongy urethra. Each corpus cavernosum terminates in the shaft of the penis, whereas the corpus spongiosum continues within the glans. The erectile bodies are ensheathed by the **tunica albuginea,** which also provides an attachment to the skin over the shaft of the penis.

 WHAT DID YOU LEARN?

33 What are the similarities and differences between the corpora cavernosa and the corpus spongiosum of the penis?

28.4f Male Sexual Response

 **LEARNING OBJECTIVES**

11. Compare and contrast the processes of erection and ejaculation.

12. Explain how the male sexual response (and ejaculation) is elicited.

The male sexual response begins with an excitement phase. The erectile bodies of the penis are composed of a complex network of **venous spaces** surrounding a central artery. During sexual excitement, blood enters the erectile bodies and fills the venous spaces. As these venous spaces become engorged with blood, the erectile bodies become rigid, a process called **erection** (ē-rek′shŭn; *erecto* = to set up). The rigid erectile bodies compress the veins that drain blood away from the venous spaces. Thus, the spaces fill with blood, but the blood cannot leave the erectile bodies until the sexual excitement ceases. Parasympathetic innervation (through the pelvic splanchnic nerves; see section 15.3b) is responsible for increased blood flow and thus the erection of the penis. Specifically, the parasympathetic innervation facilitates local release of nitric oxide (NO) into the tissues, which assists with erection. (Recall from section 20.4c that NO is a vasodilator.) Near the end of the excitement phase but before orgasm, increases occur in heart rate, blood pressure, and respiratory rate.

Orgasm refers to the time period during which there are intense feelings of pleasure, a feeling of a release of tension, and expulsion of semen. In the beginning phases of orgasm, the ductus deferens undergoes peristalsis and moves the sperm towards the urethra. Later in this phase, the accessory glands secrete their components of seminal fluid, which combine with the sperm to form semen. The internal urethral sphincter of the urinary bladder contracts to ensure that no urine enters the urethra at this time. **Ejaculation** (ē-jak-ū-lā′shŭn) typically occurs at the ending stage of an orgasm and is the process by which semen is expelled from the penis with the help of rhythmic contractions of the smooth muscle in the wall of the urethra. Sympathetic innervation (from the lumbar splanchnic nerves; see section 15.4a) is responsible for ejaculation.

 **INTEGRATE**

LEARNING STRATEGY

One way to remember the autonomic innervation for the penis is the phrase "Point and Shoot." The **p** in **p**oint (erection) also stands for **p**arasympathetic innervation, while the **s** in shoot (ejaculation) stands for **s**ympathetic innervation.

Although in most body systems sympathetic and parasympathetic innervation tend to perform opposite functions, the male reproductive system is an exception. Here, parasympathetic innervation is necessary to achieve an erection, while sympathetic innervation promotes ejaculation. Reduction of autonomic activity after sexual excitement reduces blood flow to the erectile bodies and shunts most of the blood to other veins, thereby returning the penis to its flaccid condition.

Following an orgasm, there is a resolution phase, which is marked by feelings of intense relaxation. The sympathetic division is stimulated to contract the central artery of the penis and contract small muscles around the erectile tissue, which serves to expel the engorged blood. Gradually, the penis becomes soft and flaccid again. Resolution in men is followed by a **refractory period,** during which the man cannot attain another erection. This period may last for many minutes or for hours. This refractory period becomes longer as men age. (Women, in contrast, have no refractory period and thus have the potential to experience multiple orgasms in a single sexual experience.)

  WHAT DID YOU LEARN?

34 How do erection and ejaculation differ?

35 How do both parasympathetic and sympathetic stimulation contribute to penile function during sexual arousal?

28.5 Development and Aging of the Female and Male Reproductive Systems

The female and male reproductive structures originate from the same basic embryonic primordia, but they differentiate into either female or male structures, depending upon the molecular signals the primordia receive. To better explain this process, we must first distinguish between the genetic and phenotypic sex of an individual.

28.5a Genetic Versus Phenotypic Sex

LEARNING OBJECTIVES

1. Compare and contrast genetic versus phenotypic sex.

2. List the gene(s) responsible for producing a phenotypic male.

Genetic sex is also called *genotypic sex,* and it refers to the sex of an individual based on the sex chromosomes inherited. An individual with two X chromosomes is a genetic female, whereas a person with one X and one Y chromosome is a genetic male. Genetic sex is determined at fertilization.

In contrast, **phenotypic** (fē′nō-tip-ik, fen′ō-) **sex** refers to the appearance of an individual's internal and external genitalia. A person with ovaries and female external genitalia (labia) is a phenotypic female, whereas a person with testes and male external genitalia (penis, scrotum) is a phenotypic male. Phenotypic sex starts to become apparent after the seventh week of development.

What determines the development of the primordial tissue into female reproductive organs or male reproductive organs? In males, the **sex-determining region Y (SRY) gene** is located within the larger testis-determining factor (TDF) region on the Y chromosome. If the Y chromosome is present and the *SRY* gene is appropriately expressed, this gene produces proteins to stimulate the production of androgens that initiate male phenotypic development. If a Y chromosome is absent, or if the Y chromosome is either lacking or has an abnormal *SRY* gene, a female phenotypic sex results.

⟨icon⟩ **WHAT DID YOU LEARN?**

36 What gene normally present on the Y chromosome is responsible for a male phenotypic sex?

28.5b Formation of Indifferent Gonads and Genital Ducts

⟨icon⟩ **LEARNING OBJECTIVE**

3. Describe what anatomic structures are formed from the mesonephric and paramesonephric ducts.

Early in the fifth week of embryonic development, paired **genital ridges,** also called *gonadal ridges,* form from **intermediate mesoderm** (see section 5.6 and figure 29.12). The genital ridges will form the gonads. These longitudinal ridges are medial to the developing kidneys at about the level of the tenth thoracic vertebra (**figure 28.21,** *top*). Between weeks 5 and 6, primordial **germ cells** migrate from the yolk sac to the genital ridges. These germ cells will form the future gametes (either oocytes or sperm). Shortly thereafter, two sets of duct systems are formed.

- The **mesonephric** (mez-ō-nef′rik) **ducts,** also known as *Wolffian ducts,* form most of the male duct system. The mesonephric ducts also connect the mesonephros to the developing urinary bladder.

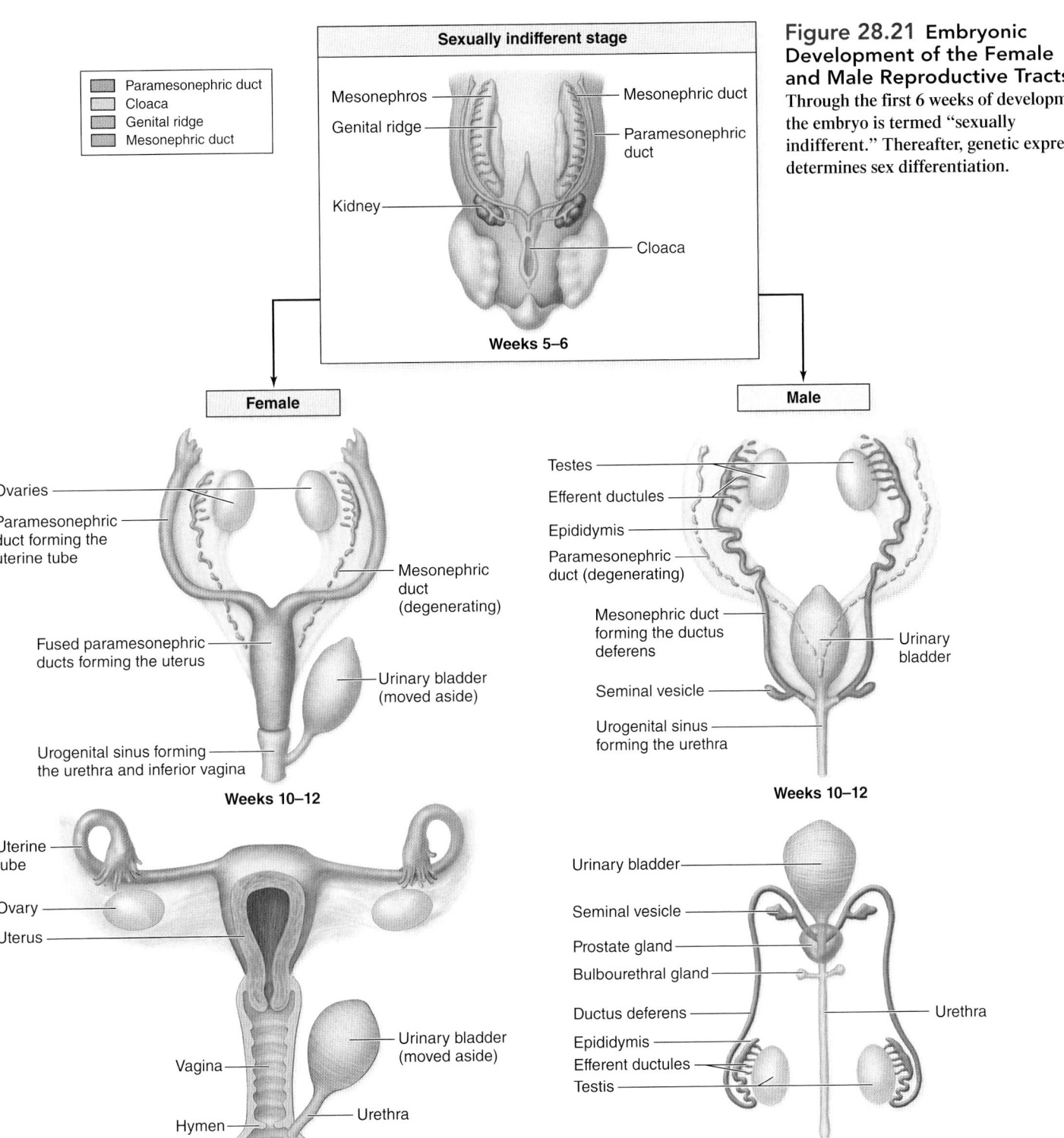

Figure 28.21 Embryonic Development of the Female and Male Reproductive Tracts. Through the first 6 weeks of development, the embryo is termed "sexually indifferent." Thereafter, genetic expression determines sex differentiation.

- The **paramesonephric ducts,** or *Müllerian ducts,* form most of the female duct system, including the uterine tubes, uterus, and superior part of the vagina. These ducts appear lateral to the mesonephric ducts.

All human embryos develop both duct systems, but only one of the duct systems remains in the fetus. If the embryo is female, the paramesonephric ducts develop, and the mesonephric ducts degenerate. If the embryo is male, the mesonephric ducts grow and differentiate into male reproductive structures, while the paramesonephric ducts degenerate.

 WHAT DID YOU LEARN?

37 Which duct system persists in female embryos? In male embryos?

28.5c Internal Genitalia Development

 LEARNING OBJECTIVES

4. Describe the events that cause the female internal reproductive organs to develop.

5. Identify the hormone responsible for inducing paramesonephric duct degeneration, and identify the cells responsible for secreting this hormone.

The development of the female internal reproductive structures is traced in figure 28.21, *left.* Because no SRY proteins are produced in the developing female, the mesonephric ducts degenerate. Between weeks 8 and 20 of development, the paramesonephric ducts develop and differentiate. The caudal (inferior) ends of the paramesonephric ducts fuse, forming the uterus and the superior part of the vagina. The cranial (superior) parts of the paramesonephric ducts remain separate and form two uterine tubes. The remaining inferior part of the vagina is formed from the urogenital sinus, which also forms the urinary bladder and urethra.

By about week 7 of development in the male, the *SRY* gene on the Y chromosome begins influencing the indifferent gonad to become a testis, which then forms sustentacular cells and interstitial cells. Once the sustentacular cells form, they begin secreting **anti-Müllerian hormone (AMH)** (also known as *Müllerian inhibiting substance*), which inhibits the development of the paramesonephric ducts (figure 28.21, *right*). These paramesonephric ducts degenerate, and between weeks 8 and 12, the mesonephric ducts begin to form the male duct system—efferent ductules, epididymides, vasa deferens, seminal vesicles, and ejaculatory ducts.

The prostate and bulbourethral glands do not form from the mesonephric ducts. Instead, they begin to form as endodermal "buds" or outgrowths of the developing urethra between weeks 10 and 13. As the prostate gland and bulbourethral glands develop, they incorporate mesoderm into their structures as well.

Finally, note that the indifferent gonad originates near the level of the T_{10} vertebra. Throughout prenatal development, the developing testis descends inferiorly from the abdominal region toward the developing scrotum. A thin band of connective tissue called the **gubernaculum** (gū′ber-nak′ū-lŭm; helm) attaches to the testis and assists the testis descent from the abdomen, through the developing inguinal canal, to its placement in the scrotum (not shown in figure 28.21). As the embryo grows (but the gubernaculums remains the same length), the testis is passively pulled into the scrotum. This process is slow, beginning in the third month and not completed until the ninth month. It is common for premature male babies to have undescended testes because they were born before the testes had fully descended into the scrotum. Their testes usually descend shortly after birth.

 WHAT DID YOU LEARN?

38 What is the fate of the paramesonephric ducts in a female embryo?

39 What hormone do the sustentacular cells secrete in the male embryo, and how does this affect internal genitalia development?

28.5d External Genitalia Development

 LEARNING OBJECTIVE

6. List the common primordial external genitalia structures, and compare their development in females and males.

As with the internal genitalia, both female and male external genitalia develop from the same primordial structures (**figure 28.22**). By the sixth week of development, the following external structures are seen:

- The **urogenital folds** (or *urethral folds*) are paired, elevated structures on either side of the urogenital membrane, a thin partition separating the urogenital sinus from the outside of the body.
- The **genital tubercle** is a rounded structure anterior to the urogenital folds.
- The **labioscrotal swellings** (or *genital swellings*) are paired elevated structures lateral to the urethral folds.

INTEGRATE

CLINICAL VIEW

Intersex Conditions (Disorders of Sex Development)

Intersex conditions, or *disorders of sex development,* refer to a series of disorders where there is a discrepancy between a person's genotype (and gonad development) and their external genitalia. (The previous term to describe this condition is *hermaphrodite* (her-maf′rō-dīt), which is derived from the Greek name Hermaphroditus, the mythological son of the Greek god Hermes and the goddess Aphrodite.) **True gonadal intersex,** also known as *true hermaphroditism,* refers to an individual with both ovarian and testicular structures (which typically does not have the potential for fertility) and ambiguous (or female) external genitalia. The person may be a genetic male (XY) or a genetic female (XX). True hermaphroditism is very rare.

Pseudohermaphroditism (sū′dō-her-maf′rō-dī-tizm; *pseudes* = false) refers to an individual whose genetic sex and phenotypic sex do not match. A **46 XY intersex** individual, previously known as a *male pseudohermaphrodite,* is a genetic male (XY) whose external genitalia resemble those of a female (female phenotypic sex). The 46 XY intersex condition most commonly is caused by **androgen insensitivity syndrome (AIS),** whereby the body cells cannot respond to androgens, because the androgen receptors do not form or function properly. Another cause may be from either a reduction or lack of male hormones (e.g., testosterone) during development.

A **46 XX intersex** individual, previously known as a *female pseudohermaphrodite,* is a genetic female (XX) with external genitalia that resemble those of a male (male phenotypic sex). The clitoris enlarges to look like a small penis, and the two labia may become partially fused to resemble a scrotum. The 46 XX intersex condition may result if the female fetus is exposed to excessive androgens (e.g., if the pregnant mother was given certain medications to help prevent miscarriage). More commonly, it is caused by **congenital adrenal hyperplasia,** in which the fetus's adrenal glands produce excessive amounts of androgens.

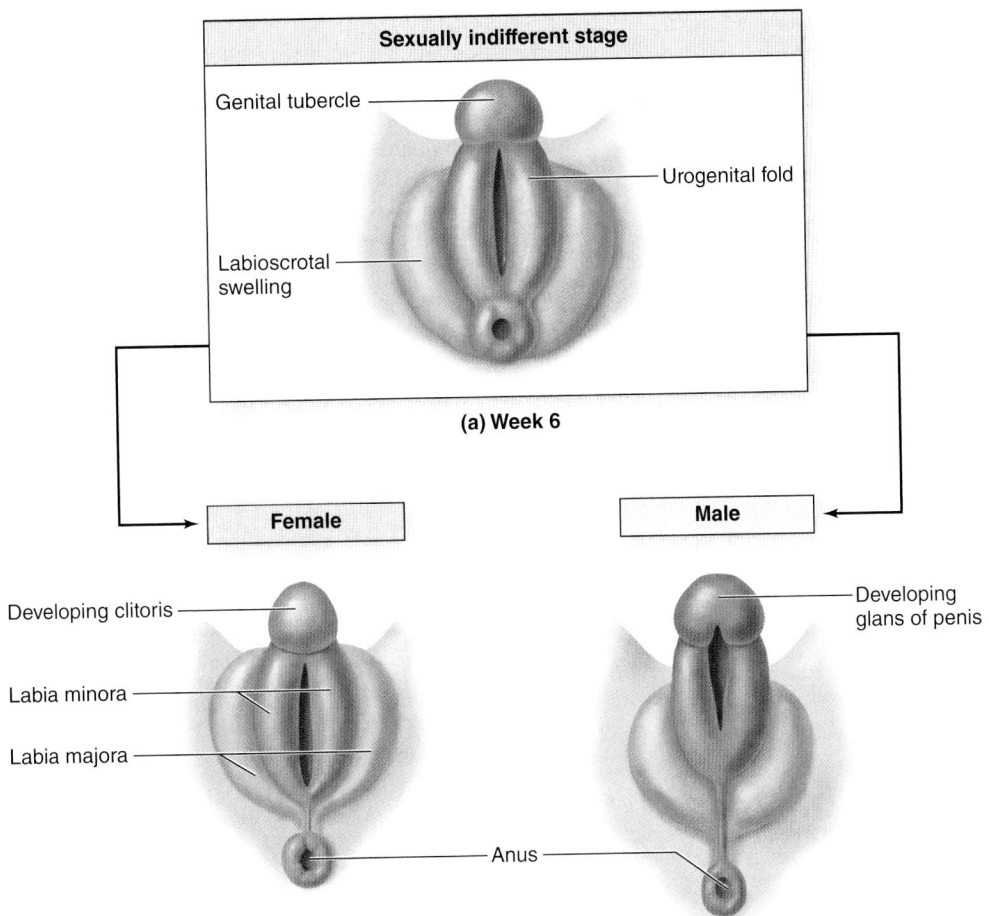

Sexually indifferent stage

Genital tubercle

Urogenital fold

Labioscrotal swelling

(a) Week 6

Female

Male

Developing clitoris

Developing glans of penis

Labia minora

Labia majora

Anus

(b) Week 12: Urogenital folds begin to fuse in the male

Glans of clitoris

Urethral orifice

Glans of penis

Body of penis

Urethral orifice

Vaginal orifice

Labia minora

Labia majora

Scrotum

Anus

(c) Week 20: External genitalia well differentiated

Figure 28.22 Development of External Genitalia. (*a*) At 6 weeks of development, the external genitalia are undifferentiated. (*b*) By 12 weeks, the urogenital folds begin to fuse in the male and remain open in the female. (*c*) By 20 weeks, external genitalia are well differentiated.

The external genitalia appear very similar in both females and males until about week 12 of development: They do not become clearly differentiated until about week 20. In the absence of testosterone, female external genitalia develop. The genital tubercle becomes the clitoris. The urogenital folds do not fuse, but become the labia minora. Finally, the labioscrotal folds also remain unfused and become the labia majora.

Production and circulation of testosterone in the male cause the primitive external structures to differentiate. The genital tubercle enlarges and elongates, forming the glans of the penis and part of the dorsal side of the penis. The urogenital folds grow and fuse around the developing urethra and form most of the body of the penis. Finally, the labioscrotal swellings fuse at the midline, forming the scrotum.

 WHAT DID YOU LEARN?

40 What causes the differentiation of tissues into male reproductive structures?

28.5e Puberty

LEARNING OBJECTIVES

7. Give the common definition for puberty, and list the age range during which it commonly occurs.

8. List some of the common developmental events that occur during puberty.

Recall that both the female and male reproductive systems are primarily nonfunctional and "dormant" until puberty. **Puberty** (pū′ber-tē; *puber* = grown up) is a period in adolescence where the reproductive organs become fully functional and the external sex characteristics become more prominent, such as breast enlargement in females and pubic hair in both sexes. The timing of puberty may be affected by genetics, environmental factors, and overall health of the individual.

Puberty is initiated when the hypothalamus begins secreting GnRH, which stimulates the anterior pituitary to release the gonadotropins FSH and LH). Prior to puberty, FSH and LH are virtually nonexistent in girls and boys. As levels of FSH and LH increase, the gonads produce significant levels of sex hormones and start the processes of gamete maturation and sexual maturation.

The earliest signs of puberty are the development of breast buds in girls and the appearance of pubic and axillary hair in both boys and girls. Menarche is one of the later events of puberty, and tends to occur about 2 years after the first signs appear. Boys experience growth in the testicles and penis, and they may begin to get erections at this time and experience ejaculations during the night. Male voices start to change and become lower in pitch, as the increase in testosterone causes rapid growth of the laryngeal structures.

The timing of puberty differs between females and males and varies among different populations. Girls generally reach puberty about 2 years prior to boys. African American girls tend to experience puberty about 1 year earlier than their Caucasian counterparts. Additionally, children in better economic conditions experience puberty earlier than children who are underprivileged, presumably because the former group experience better nutrition and heath care. Over the past 100 years, the age of puberty has dropped, so current-day children are starting puberty about 2 years earlier than children in the past. Currently, puberty typically begins between ages 8 and 12 for girls and between ages 9 and 14 for boys (although puberty may occur much later in either population).

Precocious puberty is apparent when the signs of puberty develop much earlier than normal—such as at the ages of 7 to 8 for girls, or age 9 for boys. Precocious puberty may be idiopathic (no known cause) or due to brain injury or infection, or a tumor in the pituitary or gonads. If a child experiences signs of precocious puberty, he or she should be taken to a pediatrician for an evaluation to see if any treatment is needed.

WHAT DID YOU LEARN?

41 What factors affect the age that menarche first occurs?

28.5f Menopause and Male Climacteric

LEARNING OBJECTIVES

9. Give the common definition and symptoms for menopause.

10. Describe events of the male climacteric.

After reaching sexual maturity, the female and male reproductive systems exhibit marked differences in their response to aging.

Female Menopause

Gametes typically stop maturing in females by their 40s or 50s, and menopause occurs. The time when a woman is nearing menopause is called **perimenopause.** During perimenopause, estrogen levels begin to drop, and a woman may experience irregular periods, skip some periods, or have very light periods. Recall that when a woman has stopped having monthly menstrual cycles for 1 year and is not pregnant, she is said to be in **menopause** (men′ō-pawz; *pauses* = cessation). The age at normal onset of menopause varies considerably, but typically is between 45 and 55 years. A reduction in hormone production that accompanies menopause causes some atrophy of the reproductive organs and the breasts. The vaginal wall thins and there is a reduction in glandular secretions for maintaining a moist, lubricated lining. The uterus shrinks and atrophies, becoming much smaller than it was before puberty. The woman's endometrial lining does not grow, and she no longer has a menstrual period.

The lack of significant amounts of estrogen and progesterone in a menopausal woman also affects other organs and body systems. Women may experience "hot flashes," in which their bodies perceive periodic elevations in body temperature, and they may develop thinning scalp hair and an increase in facial hair. Menopausal women are at greater risk for osteoporosis and heart disease due to the drop in estrogen and progesterone levels.

Formerly, **hormone replacement therapy (HRT)** in the form of estrogen and progesterone supplements was routinely offered to menopausal women to help diminish these symptoms and risks. However, recent studies indicate that the risks associated with HRT (e.g., increased risk of breast cancer and lack of protection against heart disease) may outweigh the benefits in older women. As a result, many physicians have stopped prescribing HRT for menopausal symptoms, although some women still choose to receive HRT if their symptoms are severe.

Male Climacteric

In contrast to females, males do not experience the relatively abrupt change in reproductive system function that females do. A slight decrease in the size of the testes parallels a reduction in the size of the seminiferous tubules and the number of interstitial cells. As a consequence of the reduced number of interstitial cells, decreased testosterone levels in men in their 50s signal a change called the **male climacteric** (klī-mak′ter-ik, klī-mak-ter′ik). Most men experience few symptoms, but some may experience mood swings, decreased sex drive, and hot flashes and sweating episodes. However, men generally do not stop producing gametes as women do following menopause. Additionally, while men experience a reduction in testosterone levels, this reduction is gradual and not as steep or sudden as the estrogen and progesterone drop seen in menopausal women.

Most men experience prostate enlargement (either benign or cancerous) as they age. This prostate enlargement can interfere with sexual and urinary functions. Also associated with aging are **erectile dysfunction** and **impotence,** which refer to the inability to achieve or maintain an erection. Besides aging, other risk factors for this condition include heart disease, diabetes, smoking, and prior prostate surgery. Many drugs (e.g., sildenafil [Viagra]) have entered the market that treat erectile dysfunction by prolonging vasodilation of the penile arteries and thus inhibit relaxation of the erectile bodies. (Patients who have heart disease and take nitrates for chest pain are discouraged from taking sildenafil, because the combination of drugs can cause an unsafe drop in blood pressure.)

WHAT DID YOU LEARN?

42 How do the female and male reproductive systems differ with respect to aging?

CHAPTER SUMMARY

28.1 Overview of Female and Male Reproductive Systems 1093

- The female and male reproductive systems function to propagate the next generation.

28.1a Common Elements of the Two Systems 1093

- Both reproductive systems have gonads that produce gametes (sex cells) and sex hormones and accessory reproductive organs (that transport or sustain the gametes).

28.1b Sexual Maturation in Females and Males 1093

- Sexual maturation begins at puberty when the hypothalamus begins secreting gonadotropin-releasing hormone (GnRH). This hormone stimulates the anterior pituitary to release follicle-stimulating hormone (FSH) and luteinizing hormone (LH).

- FSH and LH stimulate the gonads to release sex hormones and produce gametes.

28.1c Anatomy of the Perineum 1093

- The perineum is a diamond-shaped area between the thighs that houses the urogenital and anal triangles.

28.2 Gametogenesis 1094

- Gametogenesis is the process of forming sex cells.

28.2a A Brief Review of Heredity 1094

- Humans have 22 pairs of autosomes and 1 pair of sex chromosomes in the body's diploid cells. The sex chromosomes, termed X and Y, determine gender: XX is female, and XY is male.

- The gametes, or sex cells, are haploid cells; they contain 22 chromosomes and 1 sex chromosome, not pairs of chromosomes. When gametes unite in fertilization, the diploid condition is restored.

28.2b An Overview of Meiosis 1095

- Meiosis is sex cell division that produces haploid gametes from diploid parent cells.

- Meiosis involves two rounds of division: meiosis I and meiosis II. As a result of these two rounds of division, one diploid cell gives rise to four haploid cells.

28.2c Meiosis I: Reduction Division 1096

- Chromosomes become replicated prior to meiosis I so that each chromosome consists of two sister chromatids held together by a single centromere.

- The pairs of replicated homologous chromosomes separate during meiosis I. The stages are prophase I, metaphase I, anaphase I, and telophase I and cytokinesis.

- Crossing over and exchange between homologous chromosomes occurs in prophase I and helps produce gametes that are genetically different from the parent cell.

28.2d Meiosis II: Separation of Sister Chromatids 1098

- The replicated sister chromatids separate in meiosis II. The stages are prophase II, metaphase II, anaphase II, and telophase II and cytokinesis.

- The result of the two sequences of meiosis is that four haploid cells are produced that may go on to form gametes.

28.3 Female Reproductive System 1099

- Female reproductive organs include paired ovaries and uterine tubes, a uterus, a vagina, external genitalia, and the mammary glands.

28.3a Ovaries 1100

- The cortex of the ovary houses ovarian follicles that consist of an oocyte surrounded by follicle (or granulosa) cells.

28.3b Oogenesis and the Ovarian Cycle 1103

- GnRH stimulates release of FSH and LH, which bring about follicle maturation. As the follicles mature, they release estrogen, and when these levels reach a threshold, the estrogen positively stimulates the hypothalamus and anterior pituitary to release more of their hormones.

- Changing levels of FSH and LH cause a primordial follicle to mature into a primary follicle. A secondary follicle matures from a primary follicle, and a mature follicle matures from a secondary follicle.

- A peak in LH causes the secondary oocyte to be released from the mature follicle at ovulation; remaining follicular cells become the hormone-producing corpus luteum.

- The ovarian cycle consists of the follicular phase, ovulation, and the luteal phase.

28.3c Uterine Tubes, Uterus, and Vagina 1107

- The uterine tubes extend from the uterus to each ovary; the uterine tube is the site of fertilization.

- The uterus is a thick-walled muscular organ that functions as the site of pre-embryo implantation, supports and nourishes the embryo/fetus, and is the site of menstruation.

- The uterine wall consists of an inner mucosa, the endometrium; a thick-walled middle muscular layer, the myometrium; and an outer serosa, the perimetrium.

- The vagina is an epithelial-lined fibromuscular tube that serves as the birth canal for the fetus, the organ of copulation during intercourse, and the passageway for menstrual discharge.

28.3d Uterine (Menstrual) Cycle and Menstruation 1111

- The endometrium has a functional layer that is sloughed off as menses and a deeper basal layer that regenerates a new functional layer during the next uterine cycle.

- The menstrual and proliferative phases of the menstrual cycle occur in conjunction with the follicular and ovulation phases of the ovarian cycle. The secretory phase of the menstrual cycle occurs at the same time as the luteal phase of the ovarian cycle.

(continued on next page)

28.3 Female Reproductive System (continued)	**28.3e External Genitalia 1114** • The external genitalia, collectively called the vulva, include the mons pubis, labia majora and minora, and the clitoris. **28.3f Mammary Glands 1114** • The female mammary glands produce breast milk. • Prolactin is responsible for milk production; oxytocin is responsible for milk ejection. **28.3g Female Sexual Response 1117** • The female sexual response includes three phases: excitement phase, orgasm, and resolution phase. • Orgasm refers to the time period where there are intense feelings of pleasure, a feeling of a release of tension, perhaps a feeling of warmth, and some pelvic throbbing.
28.4 Male Reproductive System 1118	• The primary male reproductive system organs are the testes; accessory reproductive organs include ducts, accessory glands, and the penis. **28.4a Scrotum 1118** • The scrotum houses the testes outside the body, where the lower temperature is needed to form functional sperm. **28.4b Testes and Spermatogenesis 1120** • The testes contain seminiferous tubules (coiled tubes that house developing sperm and sustentacular cells) and interstitial cells, which produce androgens. • Spermatogenesis is the meiotic process that forms haploid spermatids, whereas spermiogenesis is the process by which spermatids differentiate into sperm. • The hypothalamus secretes GnRH, which stimulates the anterior pituitary to secrete FSH and LH, which stimulates spermatogenesis and testosterone production. • Rising levels of testosterone and inhibin inhibit both GnRH production and anterior pituitary sensitivity to GnRH, resulting in a negative feedback loop. **28.4c Duct System in the Male Reproductive Tract 1124** • The ducts store and transport sperm and include the rete testis, efferent ductules, epididymis, ductus deferens, and the ejaculatory duct. • The male urethra transports either urine or semen (but not both) at any one time. **28.4d Accessory Glands and Semen Production 1126** • The accessory glands (seminal vesicles, prostate gland, and bulbourethral glands) produce seminal fluid, which is a nutrient-rich fluid that supports sperm. • Semen is a mixture of seminal fluid and sperm. **28.4e Penis 1127** • The penis is the male copulatory organ, and contains three parallel erectile bodies plus the urethra. **28.4f Male Sexual Response 1128** • Erection is where the erectile tissue fills with blood and the penis enlarges. This process is primarily controlled by the parasympathetic division of the ANS. • Ejaculation is where the penis expels semen and is controlled by the sympathetic division of the ANS.
28.5 Development and Aging of the Female and Male Reproductive Systems 1128	• Both female and male reproductive structures originate from the same basic primordia. Gene expression determines how they differentiate. **28.5a Genetic Versus Phenotypic Sex 1128** • Genetic sex is based on chromosome type; phenotypic sex refers to the appearance of the internal and external genitalia. **28.5b Formation of Indifferent Gonads and Genital Ducts 1129** • Genital ridges in embryonic development form from intermediate mesoderm. Primordial germ cells migrate from the yolk sac to the genital ridges and form the future gametes. **28.5c Internal Genitalia Development 1130** • In the absence of a Y-chromosome (and thus the absence of the sex-determining region [*SRY*] gene), the female reproductive pattern develops. The male pattern develops as a result of the SRY protein. **28.5d External Genitalia Development 1130** • The external genitalia appear very similar until about week 12; external genitalia become fully differentiated by about week 20. **28.5e Puberty 1132** • Puberty is the time when the reproductive system starts producing gametes and sex hormones. • The timing for puberty varies, but it typically occurs earlier in girls than boys. **28.5f Menopause and Male Climacteric 1132** • Menopause is when a woman stops ovulating and having menstrual periods for 1 year. Some women may experience additional menopausal symptoms such as hot flashes, thinner scalp hair, growth of darker facial hair, and increased risk for osteoporosis. • Men may experience a male climacteric, which is the result of decreased testosterone levels. However, men continue to produce gametes throughout their lifetime.

Do You Know the Basics?

_____ 1. The female homologue to the glans of the penis is/are the

    a. labia majora.

    b. labia minora.

    c. clitoris.

    d. vagina.

_____ 2. Ovulation occurs due to a dramatic "peak" of which hormone?

    a. progesterone

    b. LH (luteinizing hormone)

    c. FSH (follicle-stimulating hormone)

    d. prolactin

_____ 3. Which statement is accurate about the uterus?

    a. The endometrium basal layer is shed each month as menses.

    b. The myometrium has several layers of skeletal muscle.

    c. The cervix projects into the vagina.

    d. The round ligament is peritoneum that drapes over the uterus.

_____ 4. Which structure contains a primary oocyte, several layers of granulosa cells, and an antrum?

    a. primordial follicle

    b. primary follicle

    c. secondary follicle

    d. mature follicle

_____ 5. In the male, what cells produce androgens?

    a. spermatogonia

    b. interstitial cells

    c. sustentacular cells

    d. All of these are correct.

_____ 6. All of the following organs produce a component of seminal fluid _except_ the

    a. bulbourethral glands.

    b. testes.

    c. seminal vesicles.

    d. prostate gland.

_____ 7. Spermatogonia divide by mitosis to form a new spermatogonium and

    a. a sperm.

    b. spermatids.

    c. a primary spermatocyte.

    d. zygotes.

_____ 8. Sperm are stored in the _____, where they remain until they are fully mature and capable of motility.

    a. epididymis

    b. seminiferous tubule

    c. ductus deferens

    d. rete testis

_____ 9. Which statement is accurate about the ovarian cycle?

    a. GnRH inhibits the anterior pituitary from secreting FSH and LH.

    b. Progesterone stimulates the follicle cells to mature.

    c. Estrogen production ceases when the secondary oocyte is ovulated.

    d. High levels of estrogen stimulate the anterior pituitary to secrete FSH and LH.

_____ 10. The paramesonephric ducts in the embryo form which of the following?

    a. uterine tubes and uterus

    b. ovary

    c. ductus deferens

    d. seminal vesicle

11. What are some anatomic similarities between the male and female reproductive systems? What are the anatomic homologues between these systems?

12. What hormones are associated with the female reproductive system, and what is the function of each hormone?

13. Describe the differences among a primary follicle, secondary follicle, and mature follicle.

14. List the uterine wall layers, and describe the basic anatomy of each layer.

15. Compare and contrast the ovarian cycle phases and the uterine cycle phases. When do they occur? What specific events are associated with each phase?

16. Describe the relationship among the hypothalamus, anterior pituitary, and the testis with regard to sperm and androgen production. What hormones initiate the cycle? What hormones inhibit the feedback loop?

17. What is the function of sustentacular cells in the production of spermatozoa?

18. Describe the process of spermatogenesis, including which cells are diploid and which are haploid.

19. How do erection and ejaculation occur in the male?

20. What structures are formed from the paramesonephric ducts? From the mesonephric ducts?

Can You Apply What You've Learned?

Use the following paragraph to answer questions 1–5.

Luisa and Victor are a young married couple trying to start a family but are having trouble conceiving. They make an appointment with a fertility clinic to determine what the problem may be. Blood samples are taken from Luisa and Victor and they are asked a series of questions. The physician learns that Victor likes to take long hot baths each evening, as a way to unwind, and that Luisa has had an irregular menstrual cycle since her early 20s.

1. The physician tells Victor that the long hot baths he takes may be interfering with them conceiving. Based on your knowledge of the reproductive system, why do you think this would be the case?
   a. Soaking the penis in the hot water prevents the penis from becoming fully erect.
   b. The autonomic axons responsible for erection and ejaculation become singed in the bath.
   c. The testes' temperature rises in the hot water, preventing successful spermatogenesis.
   d. The long hot baths prevent the hypothalamus from secreting GnRH to begin the hormonal cycle for sperm development.

2. Luisa tells the physician that her periods are irregular, but most recently she had a 35-day cycle (meaning 35 days elapsed from day 1 of menses to the next menstrual period). Given this particular menstrual cycle, what day of this cycle would be the most likely time Luisa and Victor could conceive?
   a. day 5   c. day 21
   b. day 14   d. day 28

3. The physician reviews Luisa's and Victor's blood work. They discover that Victor has low levels of circulating luteinizing hormone (LH). How could this specifically affect Victor's fertility?
   a. The interstitial cells would not be producing enough testosterone for spermatogenesis.
   b. The sustentacular cells are secreting too much inhibin.
   c. Spermatids are degenerating into primary spermatocytes.
   d. The sustentacular cells are unable to produce enough androgen-binding protein (ABP) for sperm development.

4. Victor and Luisa are questioned about their past sexual history. Specifically, they are asked if they ever tested positive for any sexually transmitted infection (STI) or experienced symptoms of an STI before. Why would this information be helpful?
   a. This information is not necessary for analysis of their fertility—useful only if they carry a child to term.
   b. Some STIs cause pelvic inflammatory disease, which can affect fertility.
   c. Syphilitic chancres are associated with a drop in sperm count.
   d. Women with herpes sores have difficulty ovulating.

5. The physician prescribes a fertility medication for Luisa that mimics the effects of follicle-stimulating hormone (FSH). This medication would be expected to
   a. facilitate sperm traveling to the uterine tube for easy fertilization.
   b. cause a greater number of ovarian follicles to mature.
   c. build up the uterine lining for successful implantation.
   d. extend the time of ovulation.

Can You Synthesize What You've Learned?

1. Jennifer is a 44-year-old woman who just gave birth to a son. The son has poor muscle tone, slanting eye creases, and a heart defect. What is the likely cause of the son's characteristics? Does Jennifer's age have anything to do with her son's condition? Why or why not?

2. Caitlyn had unprotected sex with her fiancé approximately 2 weeks after her last period and is worried that she might have become pregnant. She asks her physician if there are times during her monthly menstrual cycle when she might be more likely to become pregnant. She also asks how birth control pills prevent a woman from becoming pregnant. What will the physician tell Caitlyn?

3. If parents wish to know the sex of their unborn baby, they usually have to wait until weeks 18–22 of development before a sonogram determining the sex can be performed. Based on your knowledge of reproductive system development, explain why the sex of the unborn baby can't be determined easily with a sonogram before this time.

INTEGRATE

ONLINE STUDY TOOLS

connect
|ANATOMY & PHYSIOLOGY

LEARNSMART   AP|R

The following study aids may be accessed through Connect.

Clinical Case Study: A Young Man with a Scrotal Mass

Interactive Questions: This chapter's content is served up in a number of multimedia question formats for student study.

LearnSmart: Topics and terminology include overview of female and male reproductive systems; gametogenesis; female reproductive system; male reproductive system; development and aging of the female and male reproductive systems

Anatomy & Physiology Revealed: Topics include female perineum; meiosis; female reproductive system overview; female pelvis; ovary; female reproductive cycles; vagina; breast; male reproductive system overview; penis and scrotum; seminiferous tubule; spermatogenesis; testes and spermatic cord

Animations: Topics include comparison of meiosis and mitosis; meiosis; meiosis with crossing over; female reproductive cycles; female reproductive system overview; ovulation through implantation; male reproductive system overview; spermatogenesis

chapter 29

Development, Pregnancy, and Heredity

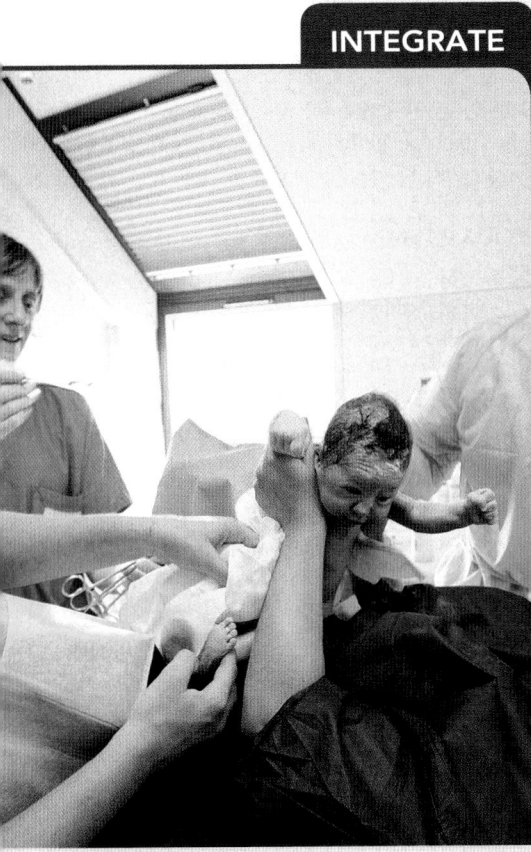

CAREER PATH
Obstetrician/Gynecologist

An obstetrician/gynecologist (OB/GYN) is a physician whose medical training is focused on the female reproductive system, pregnancy, labor, and the postpartum period (the period following birth). An OB/GYN must be aware of the anatomic and physiologic changes that occur to the mother during pregnancy, and have a thorough understanding of key developmental events for the fetus. The OB/GYN not only assists the woman during birth, but also will see the new mother in the weeks after giving birth, as the mother's body returns to its pre-pregnancy state.

Anatomy & Physiology | **REVEALED®**
aprevealed.com

Module 14: Reproductive System

Humans are like all other organisms in that they undergo **development,** a series of progressive changes that leads to the formation and organization of the diverse cell types in the body. In this chapter we focus briefly on the developmental events that occur prior to birth, a discipline known as **embryology** (em-brē-ol'ō-jē; *embryon* = a young one, *logos* = study). Embryology encompasses the developmental events that occur during the prenatal period, the first 38 weeks of human development culminating at birth.

As the embryo, and later the fetus, is developing, the body of the expectant mother undergoes a series of remarkable changes. We describe the anatomic and physiologic changes of pregnancy and explain the processes that must occur during parturition, or childbirth. We then explore postpartum (post-birth) changes and feeding of the newborn. This chapter concludes with an overview of heredity, the transmission of genetic traits from parent to newborn.

29.1 Overview of the Prenatal Period

LEARNING OBJECTIVE

1. Define the prenatal period, and identify the three shorter periods that occur during the prenatal period.

The **prenatal period** begins with fertilization, when a secondary oocyte (see section 28.3b) and sperm unite, and it ends approximately 38 weeks later with birth.[1] The general term for the new organism is a conceptus, but you will see that different terms are used during specific parts of the prenatal period.

1. Some physicians refer to pregnancy as a 40-week gestation period. This time frame is measured from a woman's last period to the birth of the newborn, so fertilization does not occur until week 2 (when a woman ovulates). Physicians may use this reference because a woman knows the time of her last period, but she may not know the day on which she ovulated and had a secondary oocyte fertilized. Even when the date of conception is known, normal pregnancy length may vary up to 37 days from this average.

The prenatal period is broken down into three shorter periods (figure 29.1):

- The **pre-embryonic period** is the first 2 weeks of development (the first 2 weeks after fertilization), when the single cell produced by fertilization (the zygote) becomes a spherical, multicellular structure (a blastocyst). This period ends when the blastocyst implants in the lining of the uterus.
- The **embryonic period** includes the third through eighth weeks of development. It is a remarkably active time during which rudimentary versions of the major organ systems appear in the body, which is now called an embryo.
- The **fetal period** includes the remaining 30 weeks of development prior to birth, when the organism is called a **fetus** (fē′tus; offspring). During the fetal period, the fetus continues to grow, and its organs increase in complexity.

In sections 29.2 and 29.3, we describe the developmental processes that occur in the pre-embryonic and embryonic periods, which are known collectively as **embryogenesis.** The fetal period is discussed in section 29.4.

WHAT DID YOU LEARN?

❶ What distinguishes the pre-embryonic period from the embryonic period? The embryonic period from the fetal period?

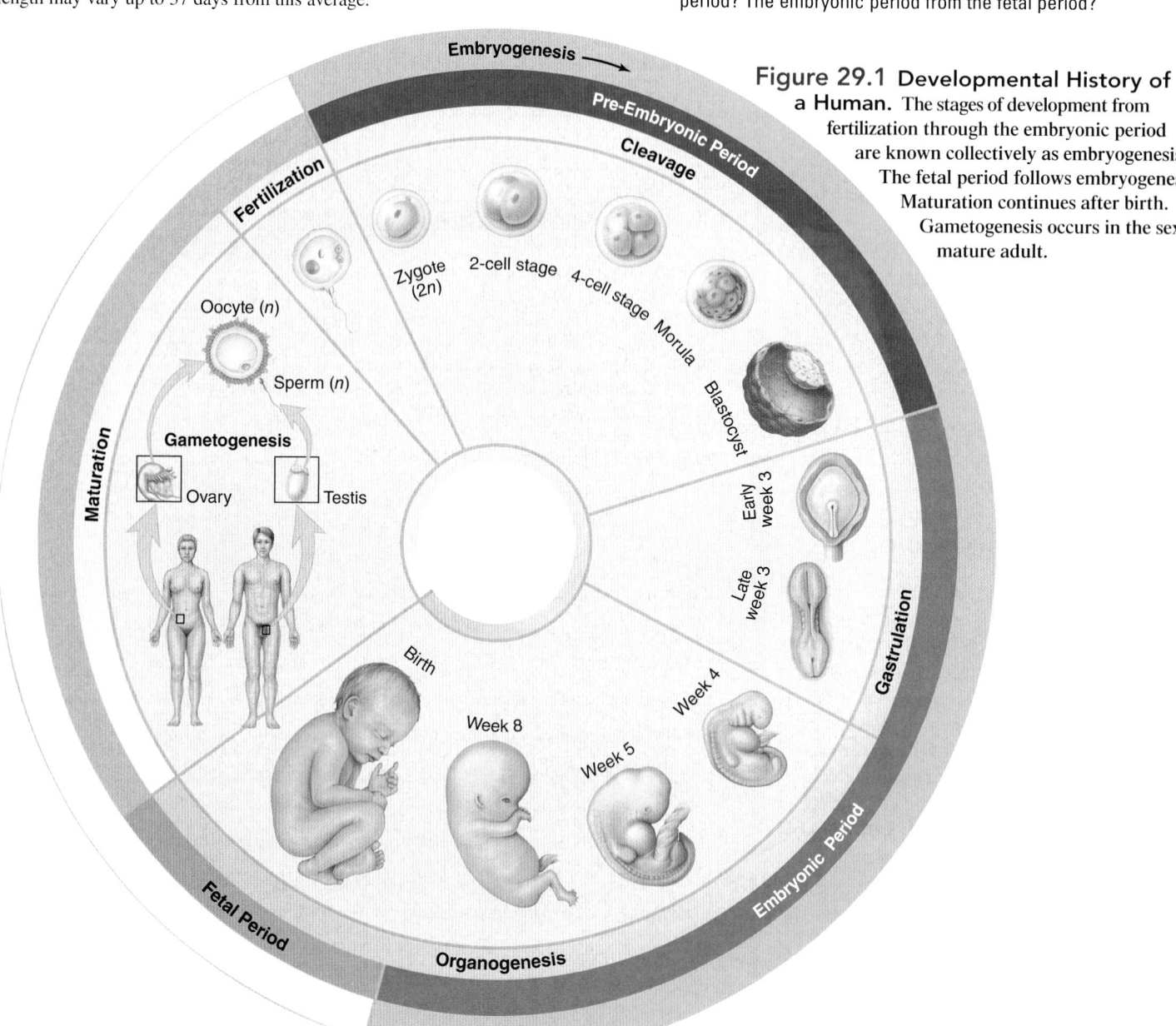

Figure 29.1 Developmental History of a Human. The stages of development from fertilization through the embryonic period are known collectively as embryogenesis. The fetal period follows embryogenesis. Maturation continues after birth. Gametogenesis occurs in the sexually mature adult.

29.2 Pre-Embryonic Period

The pre-embryonic period in human development begins with fertilization, when the male's sperm and the female's secondary oocyte unite to form a single diploid cell called the **zygote** (zī′gōt; *zygotes* = yoked). The zygote is the same size as the secondary oocyte, which typically is between 100 and 120 micrometers (μm) in diameter. During the first 2 weeks, the zygote undergoes mitotic cell divisions, and the number of cells increases, forming a **pre-embryo.** The pre-embryonic stage of development spans the time from fertilization in the uterine tube through completion of implantation (burrowing and embedding) into the wall of the mother's uterus. **Table 29.1** traces the sequence of these events.

Table 29.1	Chronology of Events in Pre-Embryonic Development		
Developmental Stage or Process	**Time of Occurrence**	**Location**	**Events**
Fertilization 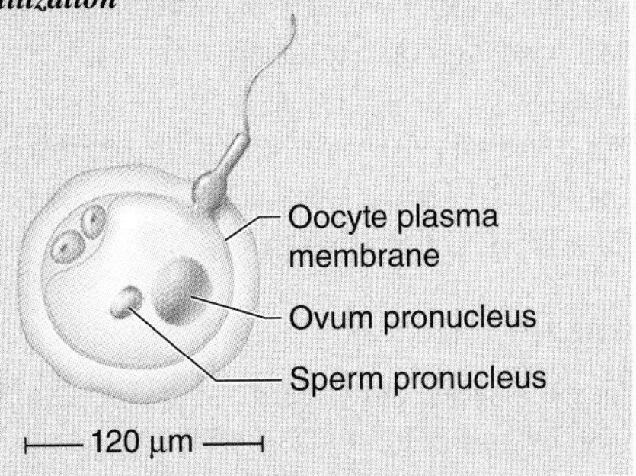 Oocyte plasma membrane — Ovum pronucleus — Sperm pronucleus — 120 μm —	Within 12–24 hours after ovulation	Ampulla of uterine tube	Sperm penetrates secondary oocyte; secondary oocyte completes meiosis II and becomes an ovum; ovum and sperm plasma membranes fuse
Zygote 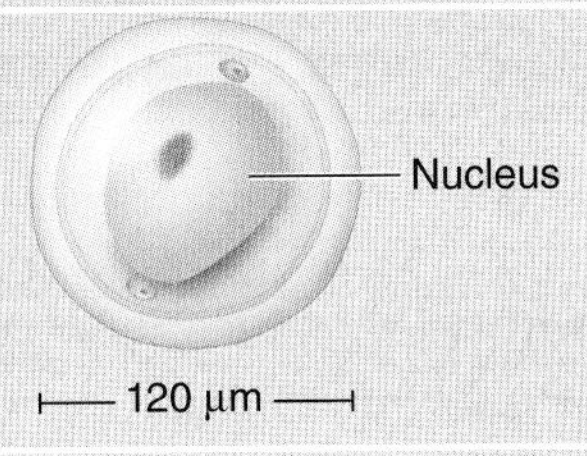 — Nucleus — 120 μm —	At the end of fertilization	Ampulla of uterine tube	Diploid cell produced when ovum and sperm pronuclei fuse
Cleavage 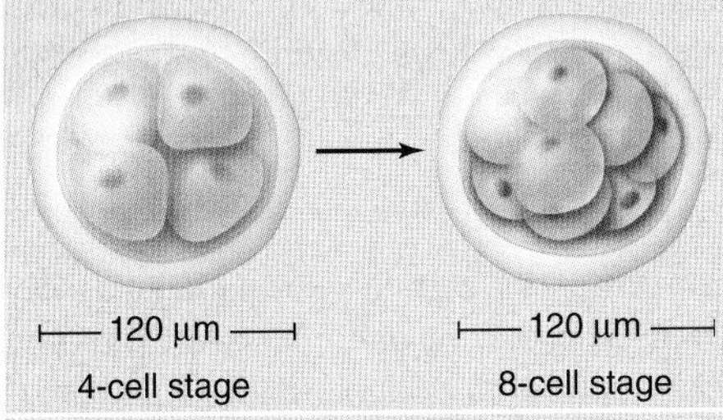 — 120 μm — 4-cell stage — 120 μm — 8-cell stage	30 hours to day 3 post-fertilization	Uterine tube	Zygote undergoes cell division by mitosis to increase cell number, but overall size of structure remains constant
Morula — 120 μm — Morula 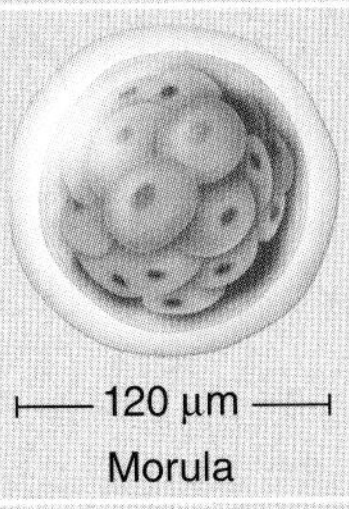	Days 3–4 post-fertilization	Uterine tube	Structure formed resembles a solid ball of cells; 16 or more cells are present, but there is no change in diameter from original zygote
Blastocyst 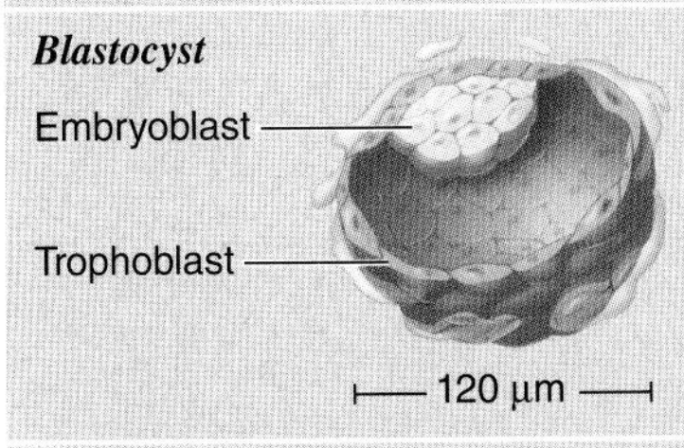 Embryoblast — Trophoblast — — 120 μm —	Days 5–6	Uterus	Hollow ball of cells; outer ring formed by trophoblast; inner cell mass (embryoblast) is a cell cluster inside blastocyst
Implantation 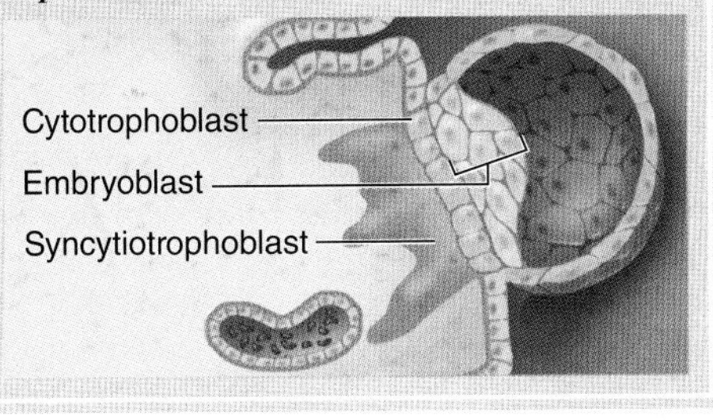 Cytotrophoblast — Embryoblast — Syncytiotrophoblast —	Begins late first week and is complete by end of second week	Functional layer of endometrium of uterus	Blastocyst adheres to functional layer of uterus; trophoblast cells and functional layer together begin to form the placenta

29.2a Fertilization

LEARNING OBJECTIVES

1. Describe the events of fertilization.
2. Explain capacitation of sperm and its relationship to fertilization.

Fertilization is the process by which two gametes (sex cells) fuse to form a new diploid cell containing genetic material derived from both parents. Besides combining the male and female genetic material, fertilization restores the diploid number of chromosomes, determines the sex of the organism, and initiates cleavage (discussed later in this section). Fertilization typically occurs in the widest part of the uterine tube, called the ampulla. Following ovulation, the secondary oocyte remains viable in the female reproductive tract for no more than 24 hours, whereas sperm remain viable for an average of 3 to 4 days after ejaculation from the male.

Upon arrival in the female reproductive tract, sperm are not yet capable of fertilizing the secondary oocyte. Sperm must undergo **capacitation** (kă-pas′i-tā′shun; *capacitas* = capable of), which is a physiologic conditioning, before they can accomplish fertilization. Capacitation occurs in the female reproductive tract, and typically takes several hours. During this time, a glycoprotein coat and some proteins are removed from the sperm plasma membrane that overlies the acrosomal region of the sperm.

Normally, millions of sperm are deposited in the vagina of the female reproductive tract during intercourse. However, many sperm leak out of the vagina, and some are not completely motile (able to swim). Other sperm do not survive the acidic environment of the vagina, and still more lose direction as they move through the uterus and get "churned" by its muscular contractions (which may occur as a result of sexual intercourse). Thus, while the male releases millions of sperm during sexual intercourse, only a few hundred have a chance at fertilization.

The oocyte cytosol and the cumulus cells release chemotaxic signals (chemicals) to attract the sperm to its location. Specifically, the cumulus cells around the oocyte release progesterone, which binds to specific channels found on the flagella of sperm and causes an influx of calcium ions (Ca^{2+}). The influx of Ca^{2+} is necessary for calcium-dependent actions such as capacitation, the acrosome reaction (discussed next), and fertilization.

When sperm reach the secondary oocyte, they attempt fertilization. Only one sperm typically is able to fertilize the secondary oocyte; the remaining sperm are prevented from penetrating the oocyte.

WHAT DO YOU THINK?

1. Rarely, two sperm may penetrate a secondary oocyte. Do you think this fertilized cell will survive for long? Why or why not?

The phases of fertilization are corona radiata penetration, zona pellucida penetration, and fusion of the sperm and oocyte plasma membranes **(figure 29.2)**.

Corona Radiata Penetration

The sperm that successfully reach the secondary oocyte are initially prevented entry by both the corona radiata and the zona pellucida. When sperm reach the corona radiata, their motility allows the sperm to push through the cell layers. Once the sperm have penetrated the corona radiata, they then encounter the more solid zona pellucida.

Zona Pellucida Penetration

After the sperm have made a pathway through the corona radiata, they release digestive enzymes from their acrosomes to penetrate the zona pellucida. This release of enzymes (primarily hyaluronidase and acrosin) from the acrosome is known as the **acrosome reaction.** When one sperm successfully penetrates the zona pellucida and its nucleus enters

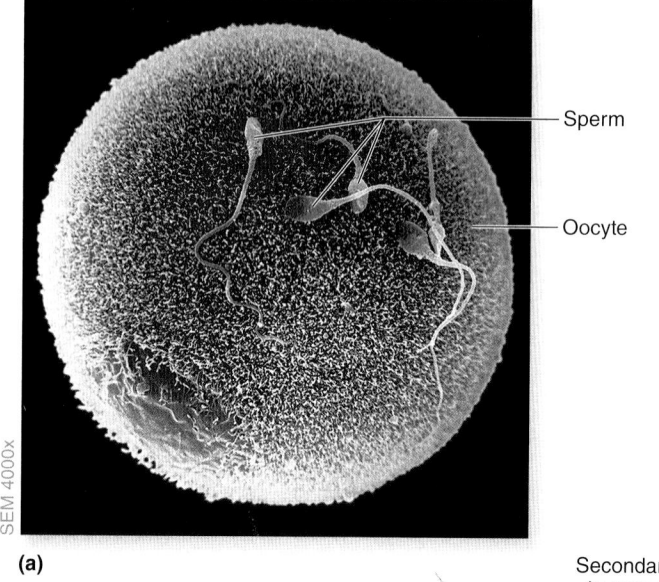

(a)

SEM 4000x

Sperm

Oocyte

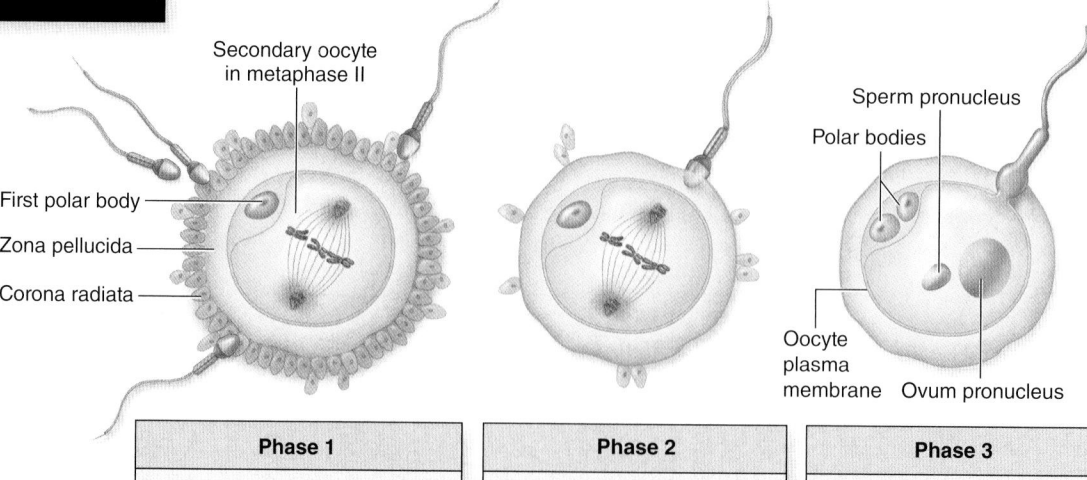

Figure 29.2 Fertilization of a Secondary Oocyte in Humans. A secondary oocyte is ovulated in a "developmentally arrested" state at metaphase II in meiosis. (*a*) Scanning electron micrograph of sperm surrounding the secondary oocyte. (*b*) Schematic representation of the three phases of fertilization.

Secondary oocyte in metaphase II

First polar body
Zona pellucida
Corona radiata

Sperm pronucleus
Polar bodies

Oocyte plasma membrane   Ovum pronucleus

Phase 1	Phase 2	Phase 3
Sperm penetrates corona radiata.	Sperm undergoes acrosome reaction and penetrates zona pellucida.	Sperm and oocyte plasma membranes fuse.

(b) Phases of fertilization

the secondary oocyte, immediate changes occur to both the zona pellucida and the oocyte, so that no other sperm can enter. In essence, the zona pellucida hardens, preventing other sperm from binding to and ultimately digesting their way through this layer. This process is necessary to ensure that only one sperm fertilizes the oocyte.

On very rare occasions, two or more sperm cell nuclei simultaneously enter the secondary oocyte, a phenomenon called **polyspermy** (pol′ē-sper-mē; *poly* = many). Polyspermy is immediately fatal because it causes the fertilized oocyte to have 23 triplets (if two sperm enter) or 23 quadruplets (if three sperm enter) of chromosomes, instead of the normal 23 pairs of chromosomes.

Fusion of Sperm and Oocyte Plasma Membranes and Nuclei

When the sperm and oocyte plasma membranes come into contact, they immediately fuse. Only the nucleus of the sperm enters the cytosol of the secondary oocyte. The midpiece and flagellum of the sperm degenerate shortly thereafter, and typically never enter the fertilized cell. When the nucleus of the sperm enters the secondary oocyte, the secondary oocyte completes the second meiotic division and forms an **ovum** (see section 28.3b). The nucleus of the sperm and the nucleus of the ovum are called **pronuclei** (*pro* = before, precursor of) because they have a haploid number of chromosomes. These pronuclei come together and fuse, forming a single nucleus that contains a diploid number (23 pairs) of chromosomes. The single diploid cell formed is the zygote.

💡 **WHAT DID YOU LEARN?**

2 What are some factors or events that can prevent sperm from reaching the secondary oocyte?

3 What are the three events that occur during fertilization?

29.2b Cleavage

LEARNING OBJECTIVES

3. Define cleavage, and explain when it occurs.

4. Compare and contrast the structures of the zygote, morula, and blastocyst.

Following fertilization, the zygote begins the process of becoming a multicellular organism. After the zygote divides once and reaches the 2-cell stage, a series of mitotic divisions called **cleavage** (klēv′ij) results in an increase in cell number, but not an increase in the overall size of the structure. The diameter of the structure remains about 120 μm, so the mitotic divisions produce greater numbers of smaller cells to fit in this structure. The structure will not increase in *size* until it implants in the uterine wall and derives a source of nourishment from the mother (**figure 29.3**).

Before the 8-cell stage, cells are not tightly bound together, but after the third cleavage division, the cells become tightly compacted into a ball. The process by which contact between cells is increased to the maximum is called **compaction.** These cells now divide again,

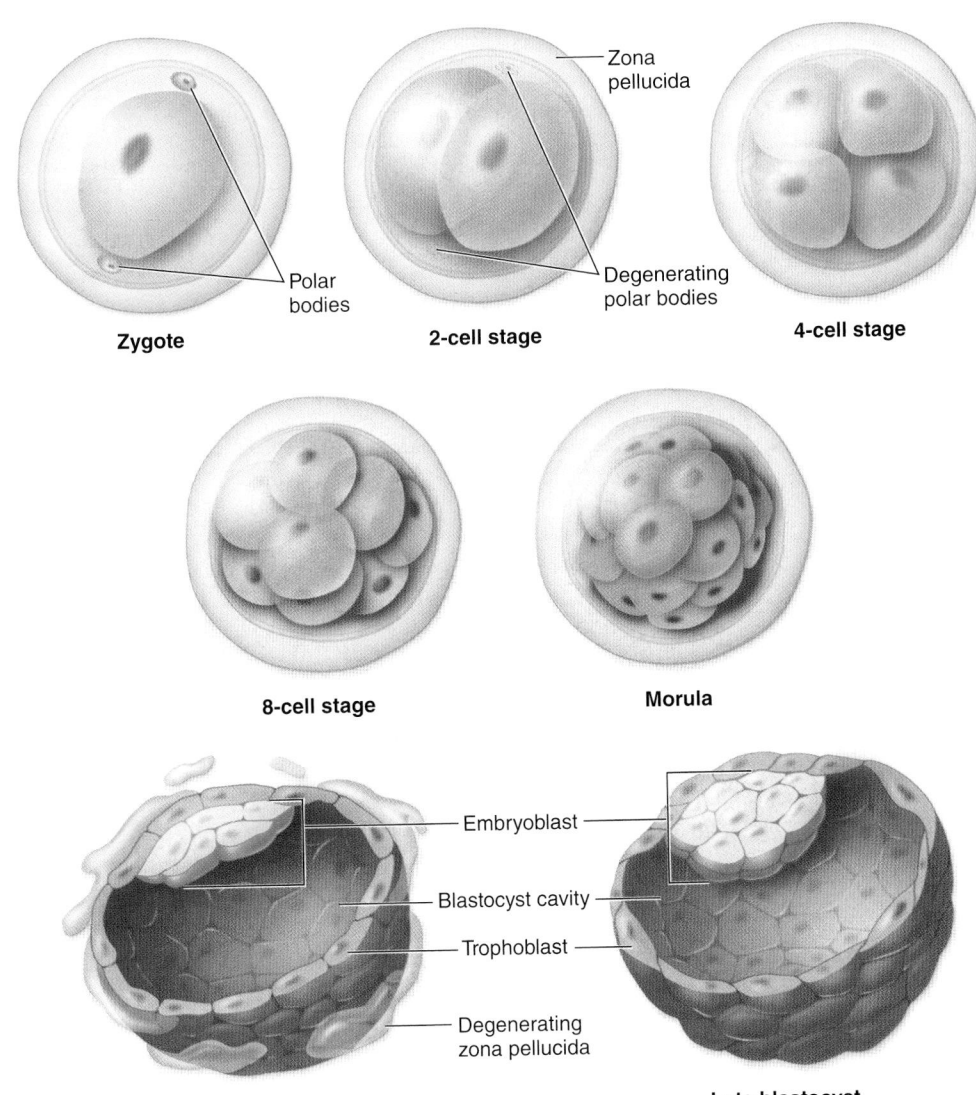

Zygote

Polar bodies

2-cell stage

Zona pellucida

Degenerating polar bodies

4-cell stage

8-cell stage

Morula

Embryoblast

Blastocyst cavity

Trophoblast

Degenerating zona pellucida

Early blastocyst

Late blastocyst

Figure 29.3 Cleavage in the Pre-Embryo. Shortly after fertilization, the zygote begins to undergo a series of cell divisions, termed cleavage. Divisions increase the number of cells, but the pre-embryo remains the same size. During each succeeding division, the cells are smaller than those in the previous generation, until they reach the size of most cells of the body. (Polar bodies are discussed in section 28.3b.)

forming a 16-cell stage, the **morula** (mōr'ū-lă; *morus* = mulberry). The cells of the morula continue to divide further.

Shortly after the morula enters the lumen of the uterus, fluid begins to leak through the degenerating zona pellucida into the morula. As a result, a fluid-filled cavity, called the **blastocyst cavity,** develops within the morula. The pre-embryo at this stage of development is known as a **blastocyst** (blas'tō-sist; *blastos* = germ), and it has two distinct components:

- The **trophoblast** (trof'ō-blast; *trophe* = nourishment) is an outer ring of cells surrounding the fluid-filled cavity. These cells will form the chorion, one of the extraembryonic membranes discussed later in this section.
- The **embryoblast,** or *inner cell mass,* is a tightly packed group of cells located only within one side of the blastocyst. The embryoblast will form the embryo proper. These early cells are **pluripotent** (plū-rip'ō-tent; *pluris* = multi, *potentia* = power), which means they have the power to differentiate into any cell or tissue type in the body. (For more information about pluripotent cells, see Clinical View: "Stem Cells" in section 5.6.)

An overview of fertilization and cleavage, including the movement of the pre-embryo from the uterine tube into the uterus, is given in **figure 29.4**.

WHAT DID YOU LEARN?

4 How many cells are present initially in the morula?

5 What are the two cell layers of the blastocyst?

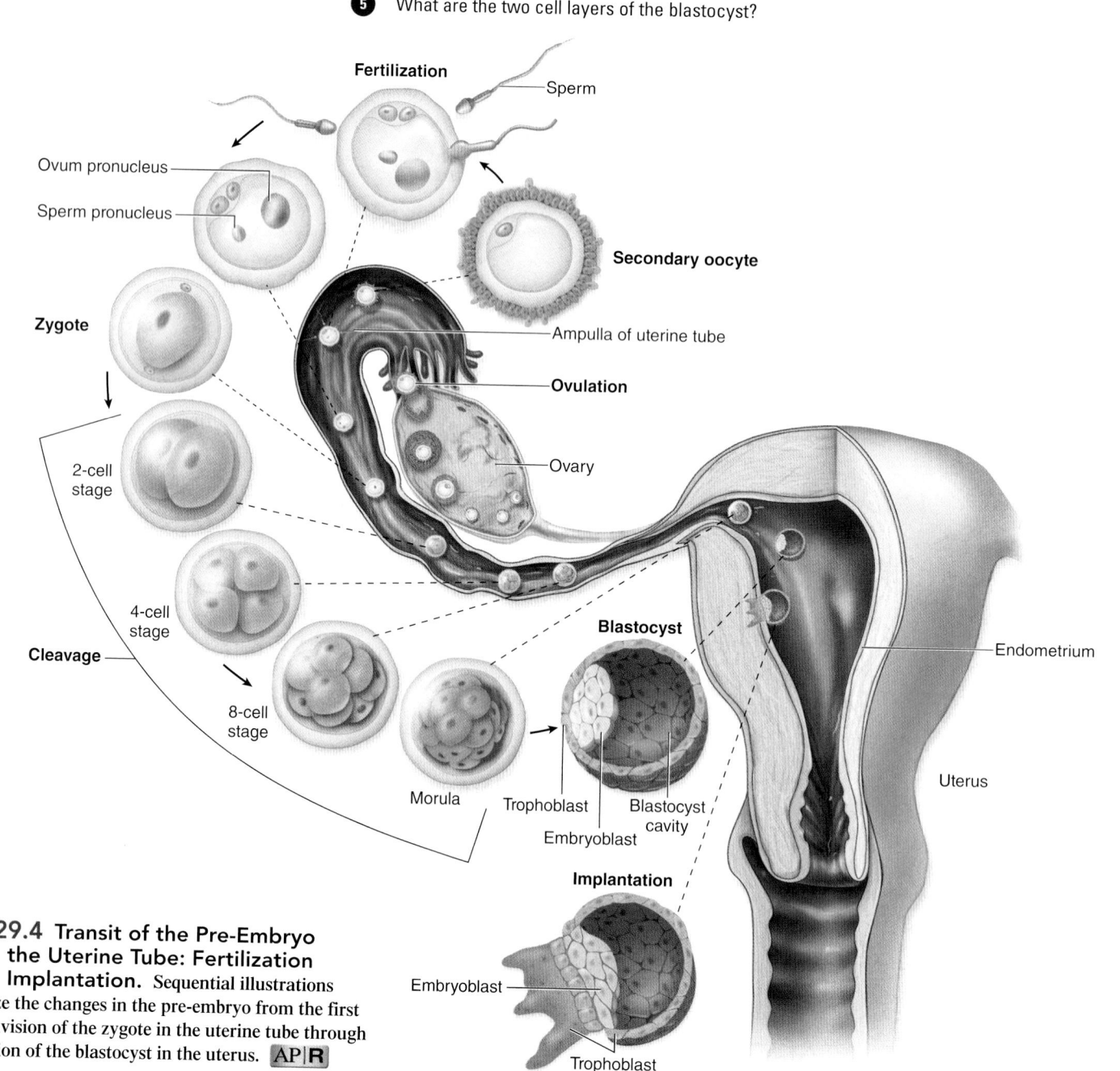

Figure 29.4 Transit of the Pre-Embryo Through the Uterine Tube: Fertilization Through Implantation. Sequential illustrations characterize the changes in the pre-embryo from the first cleavage division of the zygote in the uterine tube through the formation of the blastocyst in the uterus. AP|R

29.2c Implantation

LEARNING OBJECTIVES

5. Define implantation, and explain when it occurs.

6. Explain the physiologic significance of the syncytiotrophoblast's production of hCG.

By the end of the first week after fertilization, the blastocyst enters the lumen of the uterus. The zona pellucida around the blastocyst begins to break down as the blastocyst prepares to invade the functional layer of the uterus. **Implantation** is the process by which the blastocyst burrows into and embeds within the endometrium.

The blastocyst begins the implantation process by about day 7 (the end of the first week of development), when trophoblast cells begin to invade the functional layer of the endometrium (**figure 29.5**). Simultaneously, the trophoblast subdivides into two layers: a **cytotrophoblast** (sī′-tō-trō′fō-blast; *kytos* = cell), which is the inner cellular layer of the trophoblast, and a **syncytiotrophoblast** (sin-sish′ē-ō-trō′fō-blast), which is the outer, thick layer of the trophoblast where no plasma membranes are visible. Over the next few days, the syncytiotrophoblast cells burrow into the functional layer of the endometrium and bring with them the rest of the blastocyst.

By day 9, the blastocyst has completely burrowed into the uterine wall and makes contact with the pools of nutrients in the uterine glands. Thus, implantation begins during the first week of development and is not complete until the second week.

The syncytiotrophoblast is responsible for producing a hormone called **human chorionic gonadotropin (hCG).** Recall from section 28.3b that hCG signals the corpus luteum that fertilization and implantation have occurred. Thus, the corpus luteum does not degenerate but rather persists for another 3 months, producing large amounts of progesterone and estrogen that thicken and maintain the uterine lining.

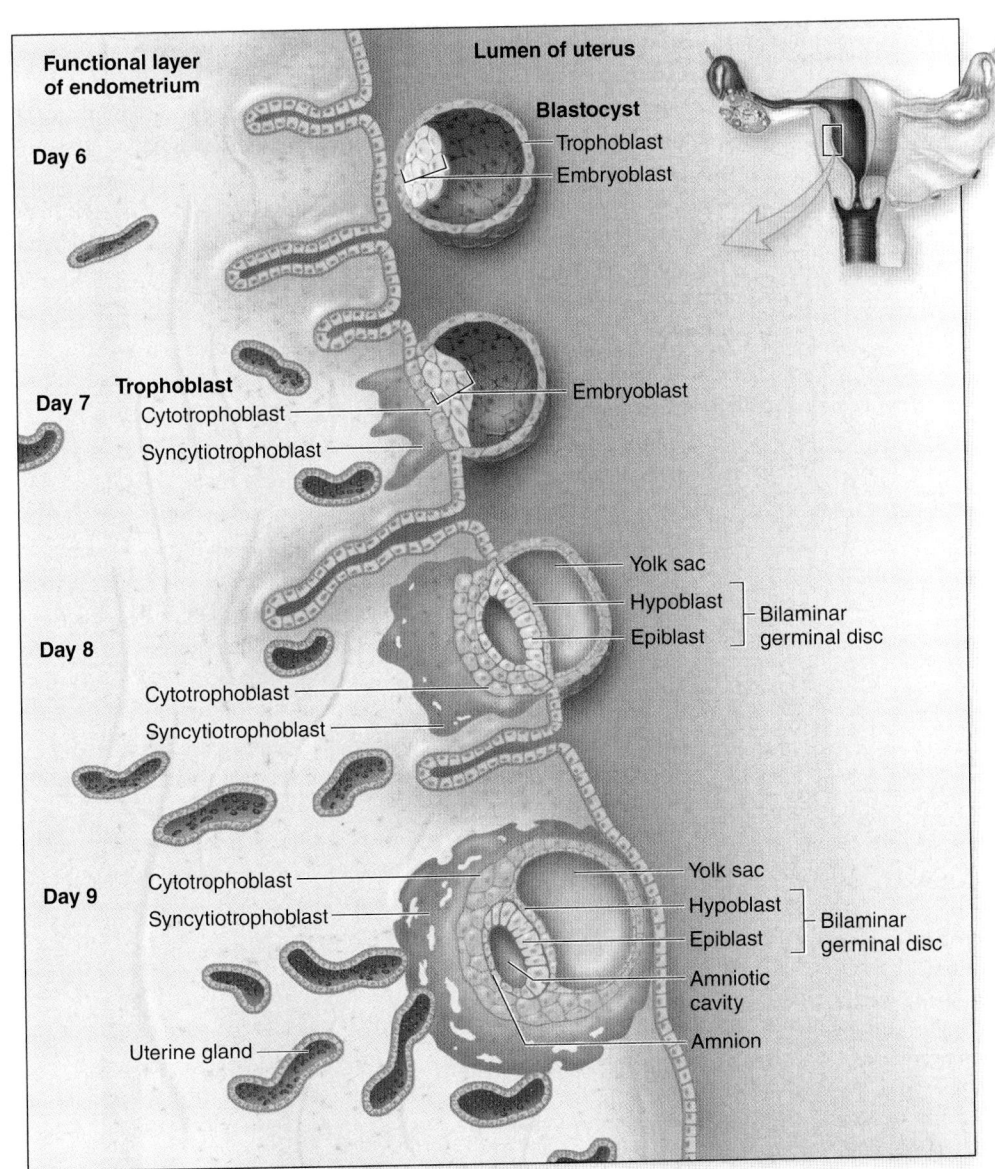

Figure 29.5 Implantation of the Blastocyst. The pre-embryo becomes a blastocyst in the uterine lumen prior to implantation. Contact between the blastocyst and the uterine wall begins the process of implantation about day 7. The implanting blastocyst makes contact with the maternal blood supply (via uterine glands) about day 9.

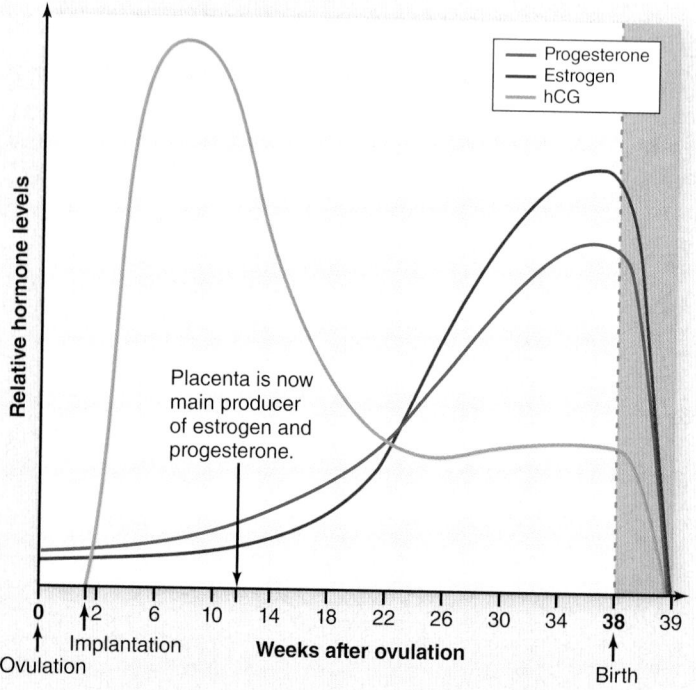

INTEGRATE

CLINICAL VIEW

Chromosomal Abnormalities and Spontaneous Abortion

Abnormalities in chromosome number or structure occur regularly during gametogenesis, fertilization, or cleavage. If chromosomal abnormalities are severe enough, they result in the spontaneous abortion (miscarriage) of the blastocyst or embryo. Many of these spontaneous abortions occur early in pregnancy (within 2 to 3 weeks after fertilization), so a woman often spontaneously aborts without ever realizing she was pregnant.

Some estimates propose that approximately 50% of all pregnancies terminate as a result of spontaneous abortion; perhaps half of these are caused by chromosomal abnormalities in the developing organism. Thus, although 2–3% of all infants are born with some type of birth defect, this percentage would be much higher if not for the high frequency of spontaneous abortions very early in pregnancy.

By the end of the second week of development, sufficient quantities of hCG are produced to be detected in a woman's urine (when it was filtered from the blood within the kidneys) **(figure 29.6)**. The presence of hCG in urine indicates a woman is pregnant, and thus hCG is the basis for most modern-day pregnancy tests. For the first 3 months of pregnancy, hCG levels remain high, but after that they decline. When

hCG declines, the corpus luteum degenerates as well. However, by this time, the corpus luteum is no longer needed because the placenta is producing its own estrogen to maintain the pregnancy.

WHAT DID YOU LEARN?

6 What would happen if a mutation prevented development of the syncytiotrophoblast cell layer?

7 Why is detection of significant levels of hCG a reliable test for pregnancy?

29.2d Formation of the Bilaminar Germinal Disc and Extraembryonic Membranes

LEARNING OBJECTIVES

7. Describe the development of the bilaminar germinal disc.

8. Name the three extraembryonic membranes, and summarize their functions.

During the second week of development, as the blastocyst is undergoing implantation, changes also occur to the embryoblast portion of the blastocyst (the inner cell mass described earlier). By day 8, the cells of the embryoblast begin to differentiate into two layers. A layer of small, cuboidal cells adjacent to the blastocyst cavity is termed the **hypoblast** layer, and a layer of columnar cells adjacent to the amniotic cavity is called the **epiblast** layer (figure 29.5). Together, these layers form a flat disc termed a **bilaminar germinal disc,** or *blastodisc.*

The bilaminar germinal disc and trophoblast also produce **extraembryonic membranes** to mediate between them and the environment **(figure 29.7)**. They first appear during the second week of development and continue to develop during the embryonic and fetal periods. They protect the embryo and assist in vital functions such as nutrition, gas exchange, and removal and storage of waste materials. These extraembryonic membranes are as follows:

- The **yolk sac** is the first extraembryonic membrane to develop. It is formed from and continuous with the hypoblast layer. In humans, it does not store yolk as it does in eggs of birds and reptiles, but it is an important site for early blood cell and blood vessel formation.

- The **amnion** (am'nē-on; *amnios* = lamb) is a thin membrane that is formed from and continuous with the epiblast layer. The amnion eventually encloses the entire embryo in a fluid-filled sac called the **amniotic cavity** to protect the embryo from desiccation (drying out). The amniotic membrane is specialized to secrete the amniotic fluid that bathes the embryo.

- The **chorion** (kō'rē-on; membrane covering the fetus) is the outermost extraembryonic membrane and is formed from both the rapidly growing cytotrophoblast cells and syncytiotrophoblast. These cells blend with the functional layer of the endometrium and eventually form the placenta, the site of gas and nutrient exchange between the embryo and the mother.

WHAT DID YOU LEARN?

8 What are the two cell layers of the bilaminar germinal disc?

9 Which cell layers give rise to each of the three extraembryonic membranes?

Relative hormone levels

— Progesterone
— Estrogen
— hCG

Placenta is now main producer of estrogen and progesterone.

0 2 6 10 14 18 22 26 30 34 38 39
Ovulation Implantation **Weeks after ovulation** Birth

Figure 29.6 Hormone Levels During Pregnancy.
When the pre-embryo implants, it secretes human chorionic gonadotropin (hCG), which sustains the corpus luteum for 12 weeks (3 months). After the twelfth week, levels of hCG drop and the corpus luteum degenerates, and the placenta becomes the main producer of the estrogen and progesterone.

Figure 29.7 Formation of Extraembryonic Membranes. The extraembryonic membranes (yolk sac, amnion, and chorion) first appear during the second week of development. Their changes in growth and form are shown at (*a*) week 3, (*b*) early week 4, and (*c*) late week 4 of development.

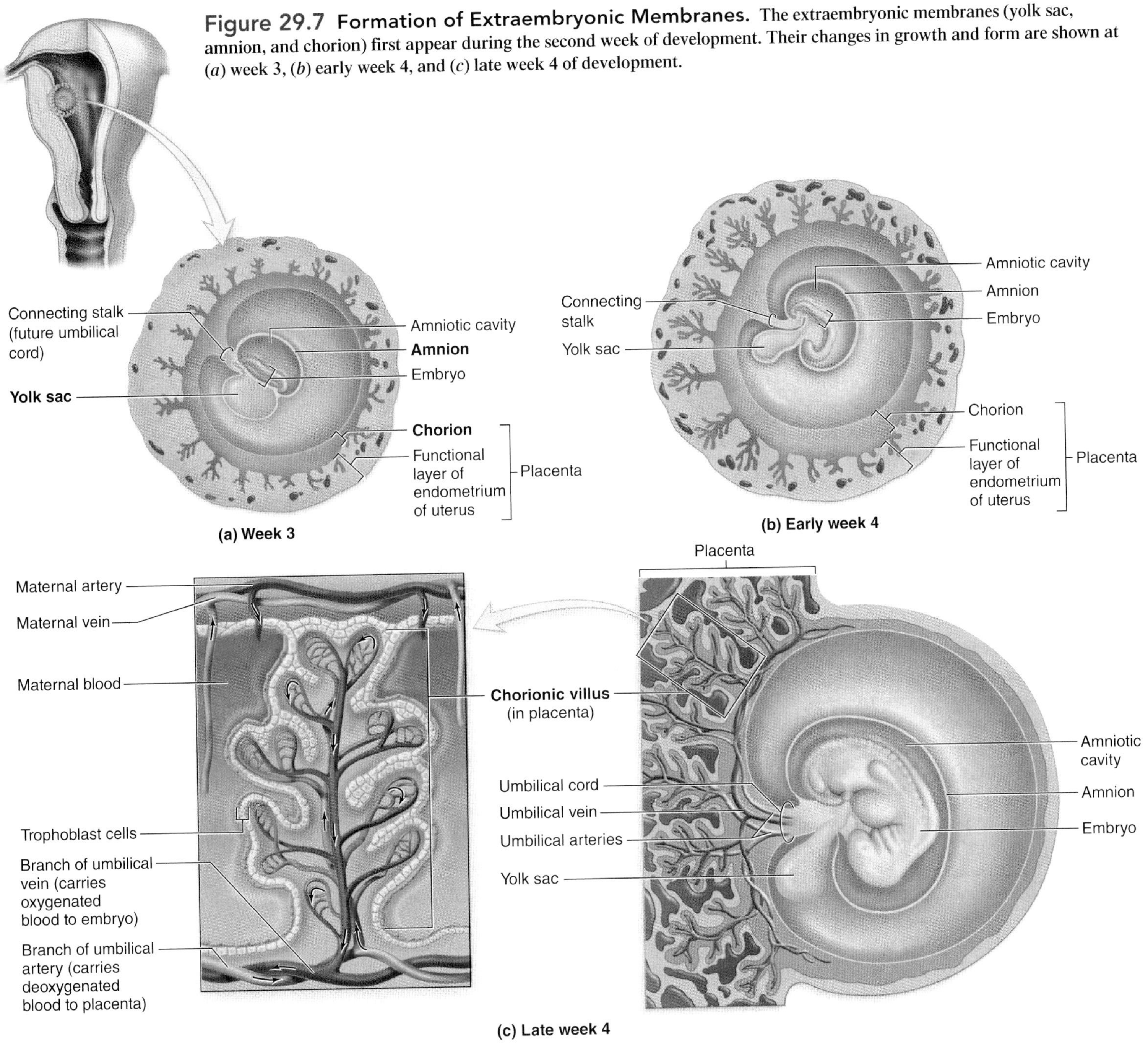

Connecting stalk (future umbilical cord)

Yolk sac

Amniotic cavity
Amnion
Embryo

Chorion
Functional layer of endometrium of uterus
— Placenta

(a) Week 3

Connecting stalk
Yolk sac

Amniotic cavity
Amnion
Embryo

Chorion
Functional layer of endometrium of uterus
— Placenta

(b) Early week 4

Placenta

Maternal artery
Maternal vein

Maternal blood

Trophoblast cells

Branch of umbilical vein (carries oxygenated blood to embryo)

Branch of umbilical artery (carries deoxygenated blood to placenta)

Chorionic villus (in placenta)

Umbilical cord
Umbilical vein
Umbilical arteries

Yolk sac

Amniotic cavity

Amnion

Embryo

(c) Late week 4

LEARNING STRATEGY

The *second week* of development may be thought of as the "period of twos," because many paired structures develop.

- A *two-layered* (epiblast and hypoblast) germinal disc forms.
- *Two membranes* (the yolk sac and the amnion) develop on either side of the bilaminar germinal disc.
- The placenta develops from *two* components that merge (the chorion and the functional layer of the endometrium of the uterus).

29.2e Development of the Placenta

LEARNING OBJECTIVES

9. Compare the maternal and fetal portions of the placenta.
10. Describe the main functions of the placenta, and name the hormones that promote its development.

Recall that the blastocyst is approximately the same size as the initial zygote, but the blastocyst contains many more cells than the zygote. To develop into an embryo and fetus, the blastocyst must receive nutrients and respiratory gases from the maternal blood supply. The connection between the embryo or fetus and the mother is the richly vascular

placenta (plă-cen′tă; a cake). The main functions of the placenta are as follows:

- Exchange of nutrients, waste products, and respiratory gases between the maternal and fetal blood
- Transmission of maternal antibodies to the developing embryo or fetus (see section 22.8c)
- Production of estrogen and progesterone to maintain and build the uterine lining

The placenta begins to form during the second week of development. The fetal portion of the placenta develops from the chorion, whereas the maternal portion of the placenta forms from the functional layer of the uterus. The early organism is connected to the placenta via a structure called the **connecting stalk** (figure 29.7a, b). This connecting stalk eventually contains the umbilical arteries and umbilical vein that distribute blood through the embryo or fetus. The connecting stalk is the precursor to the future **umbilical cord** (figure 29.7c).

Figure 29.7 illustrates how the components of the placenta become better defined during the embryonic period. Fingerlike structures called **chorionic villi** form from the chorion. The chorionic villi contain branches of the umbilical vessels. Adjacent to the chorionic villi is the functional layer of the endometrium, which contains maternal blood. Note that fetal blood and maternal blood do not mix; however, the bloodstreams are so close to one another that exchange of gases and nutrients can occur. The concentration of O_2 and nutrients is higher in the maternal blood, and therefore these diffuse into the fetal blood. Conversely, the concentration of CO_2 and waste products is higher in the fetal blood, so these materials diffuse from the fetal blood into the maternal circulation.

Although the placenta first forms during the pre-embryonic period, most of its growth and development occur during the fetal period. When the placenta matures, it is disc shaped and adheres firmly to the wall of the uterus. Immediately after the baby is born, the placenta is also expelled from the uterus. The expelled placenta is often called the *afterbirth*.

The placenta may be thought of as a selectively permeable structure. Certain materials enter freely through the placenta into the fetal blood, whereas other substances are effectively blocked. For example, respiratory gases and nutrients may freely cross the placental barrier, but certain microorganisms and high levels of maternal hormones are prevented from crossing this barrier into the developing fetus. Unfortunately, a number of undesirable substances, such as many viruses (e.g., HIV and rubella) and bacteria (e.g., *Treponema*, the bacterium that causes syphilis) *can* cross the placental barrier, infecting the fetus and sometimes causing birth defects or death. Most drugs, alcohol, and the toxins from smoking can pass through the placental barrier as well.

Some fetuses may be more susceptible to materials that cross the placental barrier than other fetuses. Additionally, the *dose* of the material crossing the placental barrier and the timing of this crossing both affect fetus susceptibility. These facts help explain why some newborns are strongly affected by materials that cross the placental barrier, whereas other newborns are relatively unaffected.

Prior to implantation, the blastocyst is not harmed by undesirable substances because it does not yet have a connection with the mother's uterine lining. However, once implantation begins and the placenta starts to form, the developing organism is exposed to most of the chemicals and other substances to which the mother is exposed. For these reasons, pregnant women are strongly urged to quit smoking and to refrain from taking drugs and drinking alcohol during their pregnancies.

WHAT DID YOU LEARN?

10 What are the main functions of the placenta?

29.3 Embryonic Period

The embryonic period begins in week 3 with the establishment of the three primary germ layers through the process of gastrulation. Subsequent interactions and rearrangements among the cells of the three layers prepare for the formation of specific tissues and organs, a process called organogenesis. By week 4 the embryo has a beating heart, and by the end of the embryonic period (week 8), the main organ systems have been established, and the major features of the external body form are recognizable. **Table 29.2** summarizes the events that occur during the embryonic period.

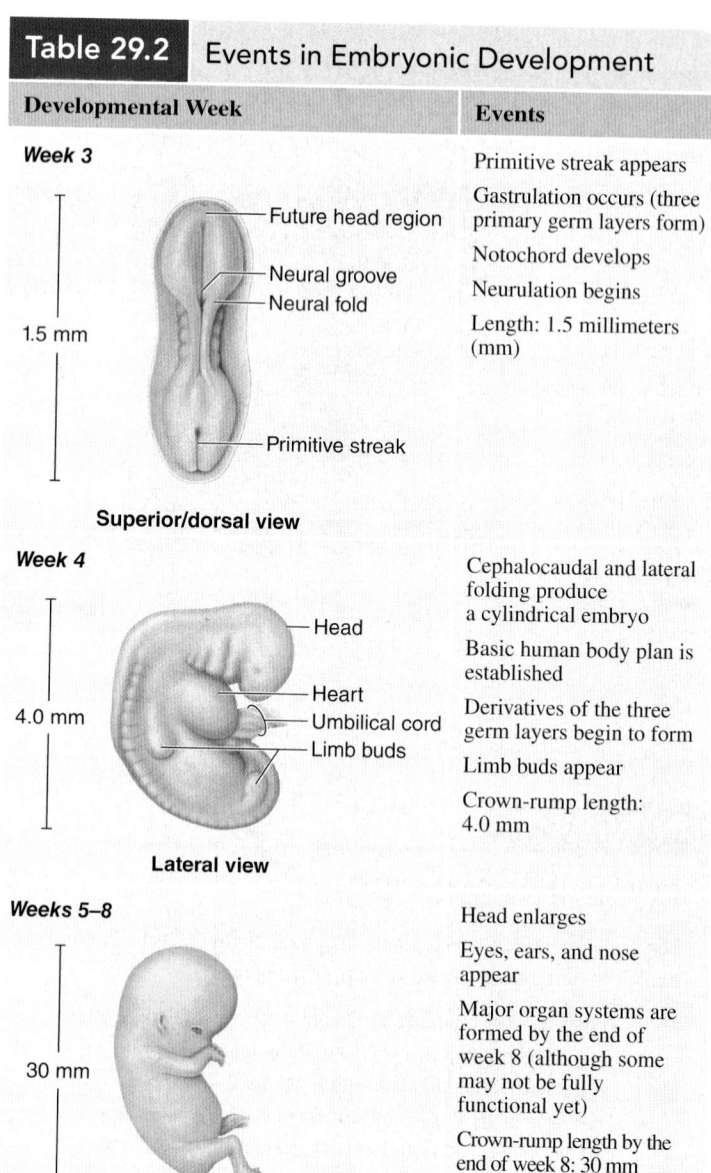

Table 29.2	Events in Embryonic Development
Developmental Week	**Events**
Week 3 1.5 mm Future head region Neural groove Neural fold Primitive streak *Superior/dorsal view*	Primitive streak appears Gastrulation occurs (three primary germ layers form) Notochord develops Neurulation begins Length: 1.5 millimeters (mm)
Week 4 4.0 mm Head Heart Umbilical cord Limb buds *Lateral view*	Cephalocaudal and lateral folding produce a cylindrical embryo Basic human body plan is established Derivatives of the three germ layers begin to form Limb buds appear Crown-rump length: 4.0 mm
Weeks 5–8 30 mm *Lateral view*	Head enlarges Eyes, ears, and nose appear Major organ systems are formed by the end of week 8 (although some may not be fully functional yet) Crown-rump length by the end of week 8: 30 mm

INTEGRATE

LEARNING STRATEGY

The *third week* of development produces an embryo with *three* primary germ layers: ectoderm, mesoderm, and endoderm.

29.3a Gastrulation and Formation of the Primary Germ Layers

⇲ LEARNING OBJECTIVES

1. Describe the process of gastrulation.
2. List the three primary germ layers that compose the embryo.

Gastrulation (gas-trū-lā′shŭn; *gaster* = belly) occurs during the third week of development immediately after implantation, and is one of the most critical periods in the development of the embryo. Gastrulation is a process by which the cells of the epiblast migrate and form the three **primary germ layers,** which are the cells from which all body tissues develop. The three primary germ layers are called ectoderm, mesoderm, and endoderm. Once these three layers have formed, the developing trilaminar (three-layered) structure may be called an **embryo.**

Gastrulation begins with formation of the **primitive streak,** a thin depression on the surface of the epiblast (**figure 29.8a, b**). The cephalic (head) end of the streak, known as the **primitive node,** consists of a slightly elevated area surrounding a small **primitive pit.**

Cells detach from the epiblast layer and migrate through the primitive streak between the epiblast and hypoblast layers. This inward movement of cells is known as **invagination.** Migrating cells first displace the hypoblast and form the **endoderm** (en′dō-derm; *endo* = inner). Next, more epiblast cells invaginate and form a new primary germ layer known as **mesoderm** (mez′ō-derm; *meso* = middle, *derma* = skin). Cells remaining in the epiblast then form the **ectoderm** (ek′tō-derm; *ektos* = outside). Thus, the epiblast, through the process of gastrulation, is the source of the three primary germ layers from which all body tissues and organs eventually derive (figure 29.8c).

💡 WHAT DID YOU LEARN?

11 What events occur during gastrulation? What are the locations of the three germ layers?

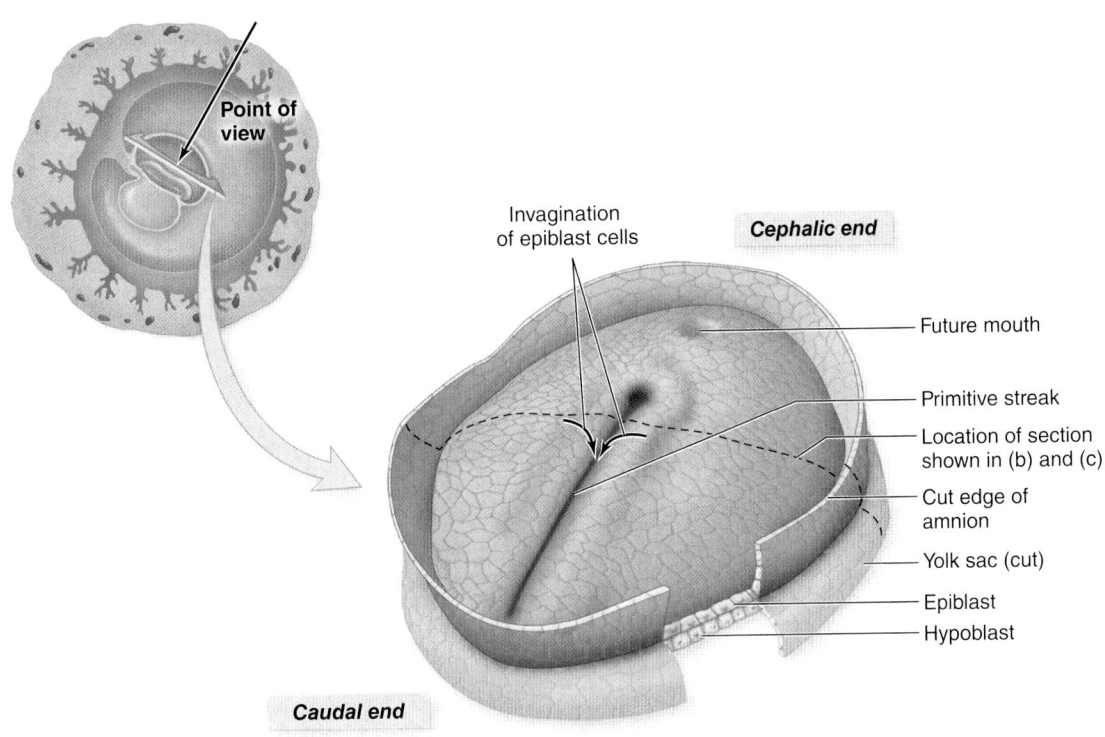

(a) Early week 3, superolateral view

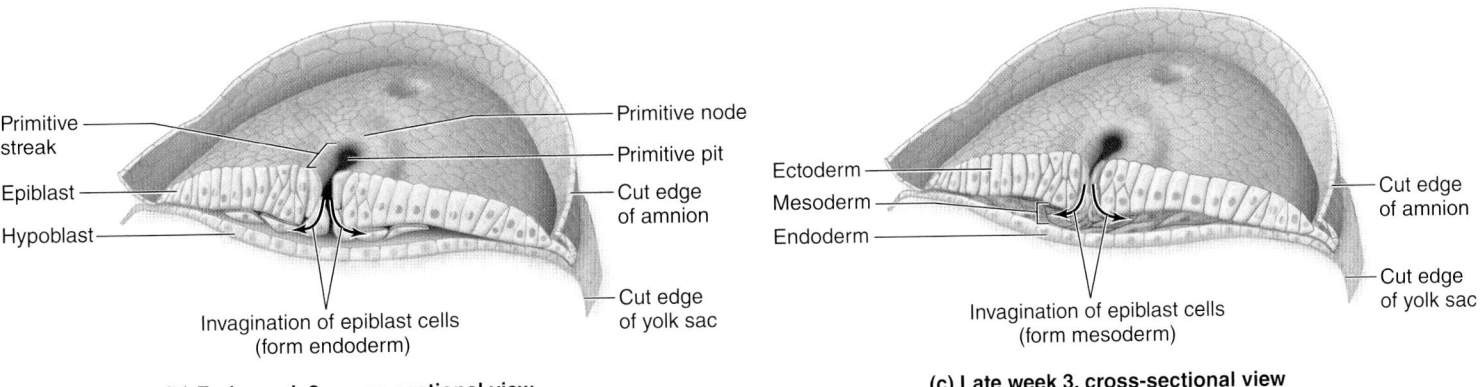

(b) Early week 3, cross-sectional view

(c) Late week 3, cross-sectional view

Figure 29.8 The Role of the Primitive Streak in Gastrulation. (*a*) The primitive streak is a raised groove on the epiblast surface of the bilaminar germinal disc that appears early in the third week. (*b, c*) During gastrulation, epiblast cells migrate toward the primitive streak, where some become embryonic endoderm, and others form mesoderm between the epiblast and the new endoderm. Cells remaining in the epiblast become the ectoderm. Thus, the epiblast forms all three primary germ layers.

29.3b Folding of the Embryonic Disc

LEARNING OBJECTIVES

3. Explain the process and the purpose of the folding of the embryonic disc.

4. Describe how the three primary germ layers differentiate.

The 3-week embryo is a flattened, disc-shaped structure. For this reason, the structure is also referred to as an **embryonic disc (figure 29.9)**. So how does this flattened structure develop into a three-dimensional human?

The shape transformation begins during the late third and fourth weeks of development, when cells are rapidly dividing, and certain regions of the embryo grow faster than others. This rapid division and growth of cells causes folding; thus, the embryonic disc starts to fold on itself and become more cylindrical. **Figure 29.10** illustrates the two types of folding that occur: cephalocaudal folding and transverse folding.

Cephalocaudal (sef'ă-lō-kaw'dăl) **folding** occurs in the cephalic (head) and caudal (tail) regions of the embryo. Essentially, the embryonic disc and amnion grow very rapidly, but the yolk sac does not grow at all. This differential growth causes the head and tail regions to fold on themselves.

Transverse folding (or *lateral folding*) occurs when the left and right sides of the embryo curve and migrate toward the midline. As these sides come together, they restrict and start to pinch off the yolk sac. Eventually, the sides of the embryonic disc fuse in the midline and create a cylindrical embryo. Thus, the ectoderm is now solely along the entire exterior of the embryo, whereas the endoderm is confined to the internal region of the embryo. As this midline fusion occurs, the yolk sac pinches off from most of the endoderm (with the exception of one small region of communication called the *vitelline duct*).

Thus, cephalocaudal folding helps form the future head and buttocks region of the embryo, whereas transverse folding creates a cylindrical trunk or torso region of the embryo. We now examine the specific derivatives of these primary germ layers (which were previously introduced in section 5.6a).

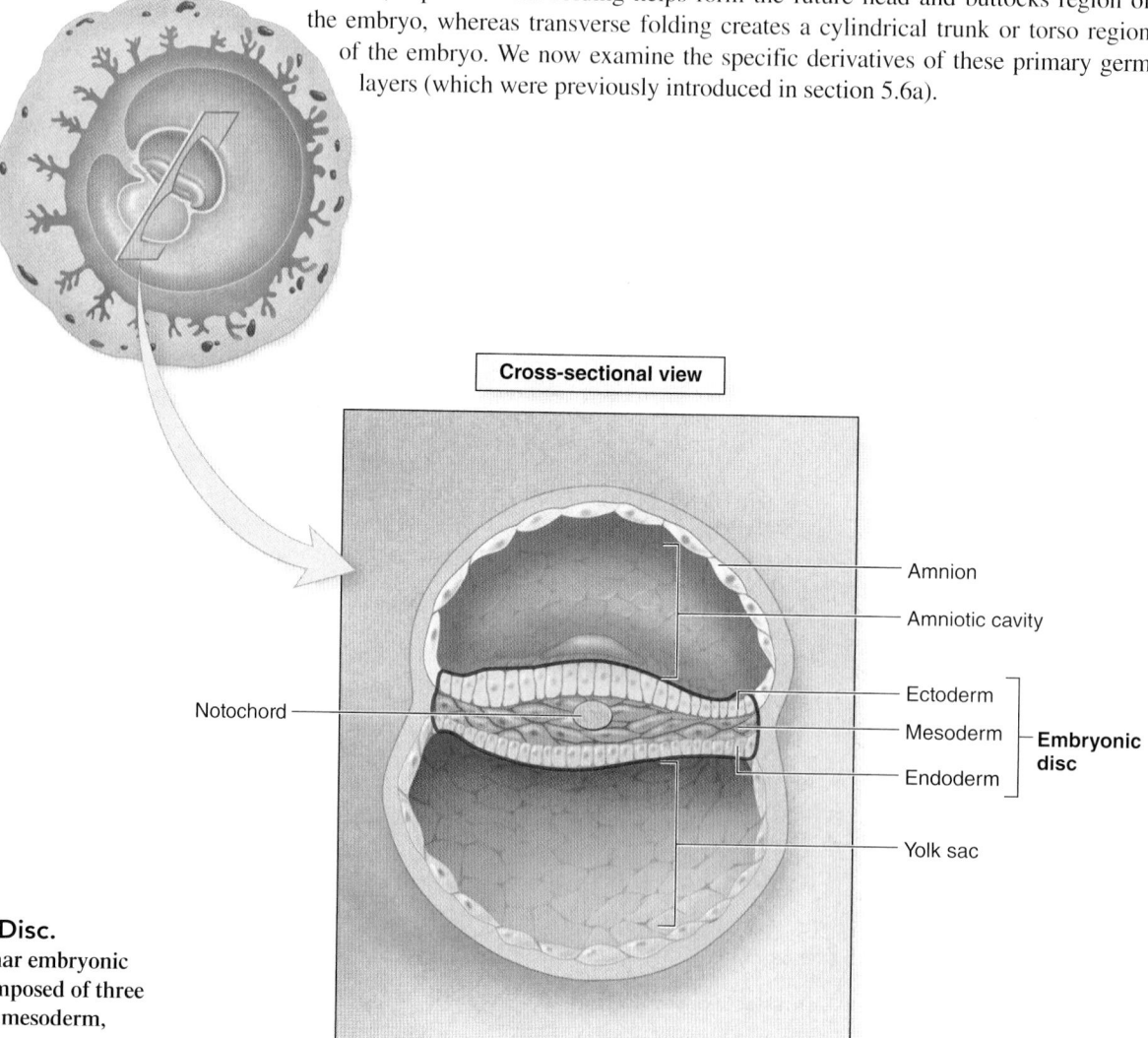

Cross-sectional view

- Amnion
- Amniotic cavity
- Notochord
- Ectoderm
- Mesoderm — **Embryonic disc**
- Endoderm
- Yolk sac

Figure 29.9 Embryonic Disc.
Gastrulation produces a trilaminar embryonic disc (outlined in pink) that is composed of three primary germ layers: endoderm, mesoderm, and ectoderm.

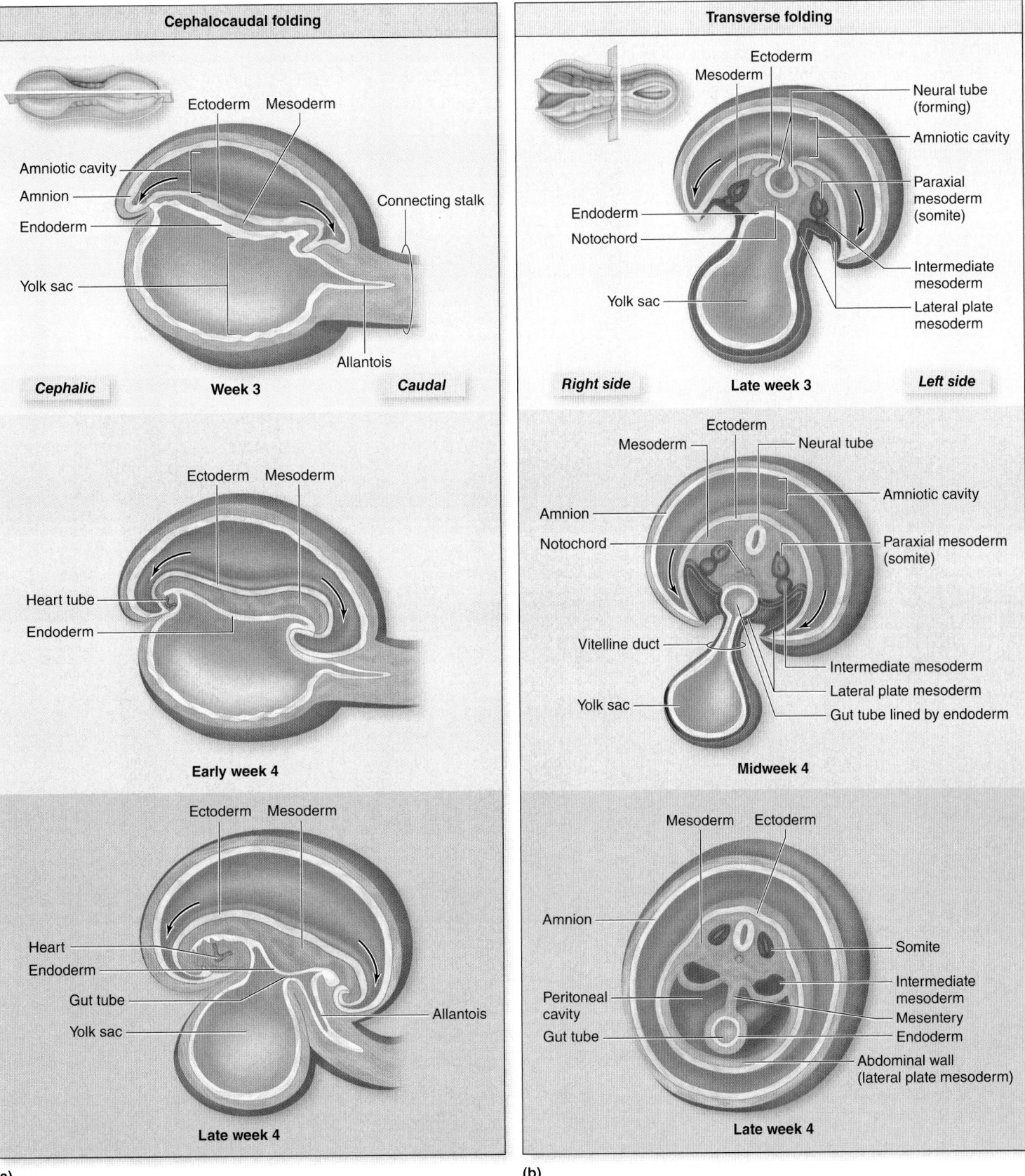

Figure 29.10 Folding of the Embryonic Disc. During the third and fourth weeks of development, the flat embryo undergoes both (*a*) cephalocaudal folding and (*b*) transverse folding.

Differentiation of Ectoderm

After the embryo undergoes cephalocaudal and transverse folding, the ectoderm is located on the external surface of the now-cylindrical embryo. The ectoderm is responsible for forming nervous system tissue in a process called *neurulation* (see section 13.1b). The ectodermal cells covering the embryo after neurulation form the epidermis. Ectoderm also forms the epidermis and epidermal derivatives, sense organs, and pituitary gland. It forms the adrenal medulla, enamel of teeth, and lens of the eye. With a few exceptions, ectoderm gives rise to those organs and structures that maintain contact with the outside world **(figure 29.11)**.

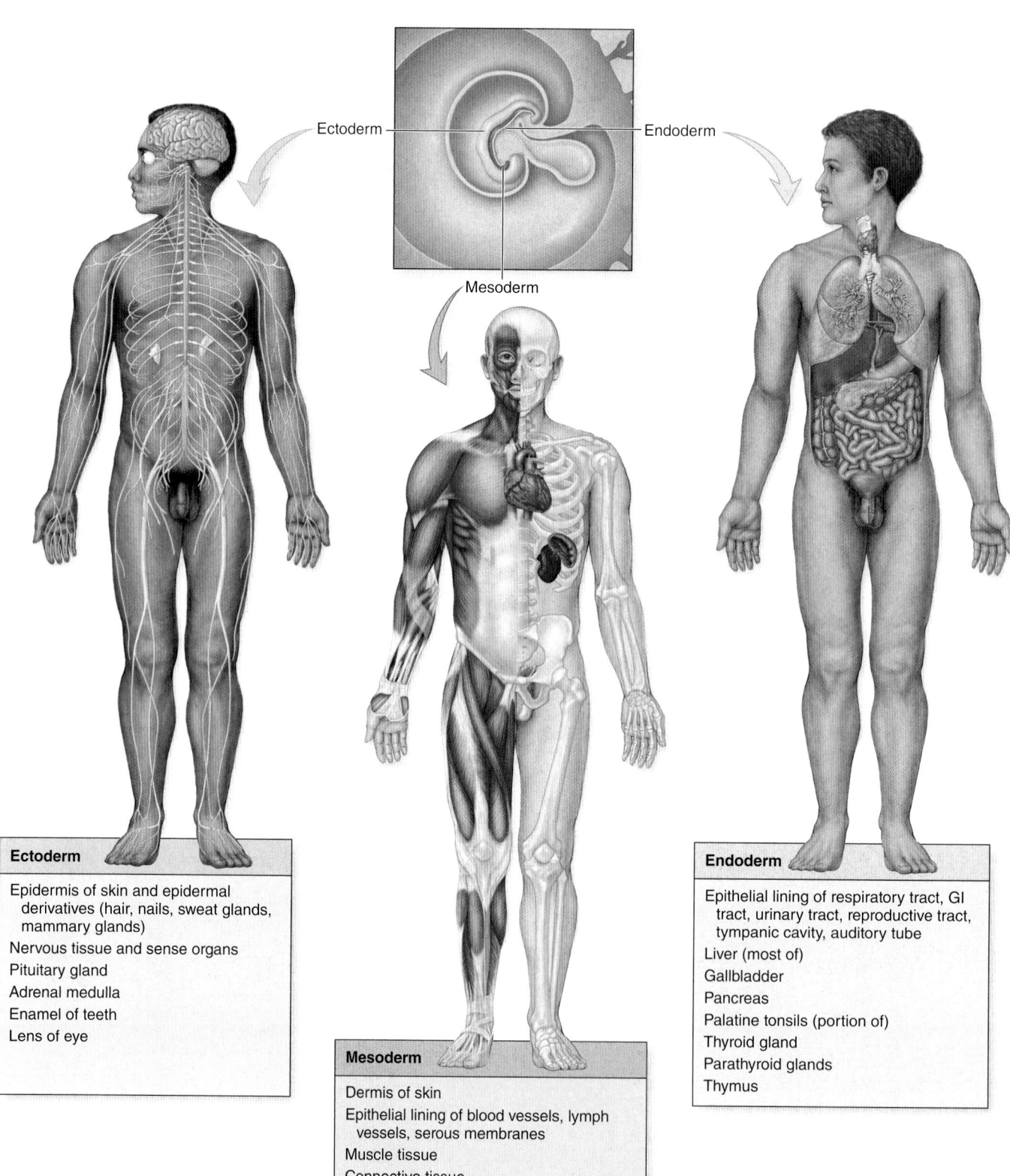

Ectoderm

Endoderm

Mesoderm

Ectoderm

Epidermis of skin and epidermal
   derivatives (hair, nails, sweat glands,
   mammary glands)
Nervous tissue and sense organs
Pituitary gland
Adrenal medulla
Enamel of teeth
Lens of eye

Mesoderm

Dermis of skin
Epithelial lining of blood vessels, lymph
   vessels, serous membranes
Muscle tissue
Connective tissue
Adrenal cortex
Heart
Kidneys and ureters
Internal reproductive organs
Spleen

Endoderm

Epithelial lining of respiratory tract, GI
   tract, urinary tract, reproductive tract,
   tympanic cavity, auditory tube
Liver (most of)
Gallbladder
Pancreas
Palatine tonsils (portion of)
Thyroid gland
Parathyroid glands
Thymus

Figure 29.11 The Three Primary Germ Layers and Their Derivatives.
Ectoderm, mesoderm, and endoderm give rise to all of the tissues in the body.

Differentiation of Mesoderm

Mesoderm subdivides into the following five categories (**figure 29.12**):

- The tightly packed midline group of mesodermal cells, also called *chordamesoderm,* forms the **notochord.** The notochord serves as the basis for the central body axis and the axial skeleton, and induces the formation of the neural tube, as previously described in section 13.1b.

- **Paraxial mesoderm** is found on both sides of the neural tube. The paraxial mesoderm then forms **somites** (sō′mīt; *soma* = body), which are blocklike masses responsible for the formation of the axial skeleton, most muscle (including limb musculature), and most of the cartilage, dermis, and connective tissues of the body.

- Lateral to the paraxial mesoderm are cords of **intermediate mesoderm,** which forms most of the kidneys, ureters, and the reproductive system.

- The most lateral layers of mesoderm on both sides of the neural tube remain thin and are called the **lateral plate mesoderm.** These give rise to the spleen, adrenal cortex, most of the components of the cardiovascular system, the serous membranes of the body cavities, and all the connective tissue components of the limbs.

- The last region of mesoderm, called the **head mesenchyme** (mez′en-kīm), forms connective tissues and musculature of the face (not shown in figure).

The derivatives of the mesoderm are listed and illustrated in figure 29.11.

Differentiation of Endoderm

Endoderm becomes the innermost tissue when the embryo undergoes transverse folding. Among the structures formed by embryonic endoderm are the linings of the gastrointestinal (GI), respiratory, urinary, and reproductive tracts (figure 29.11). It forms the tympanic cavity (middle ear) and the auditory tube. Endoderm also forms most of the liver, gallbladder, pancreas, portions of the palatine tonsils, the thyroid gland, parathyroid glands, and thymus.

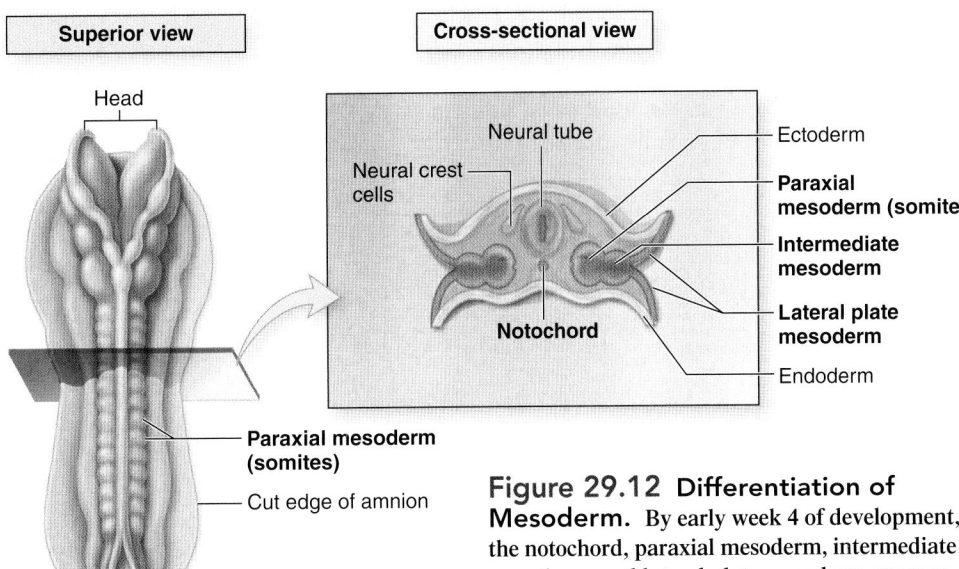

Figure 29.12 Differentiation of Mesoderm. By early week 4 of development, the notochord, paraxial mesoderm, intermediate mesoderm, and lateral plate mesoderm are seen.

 WHAT DID YOU LEARN?

12 How are the results of cephalocaudal folding and transverse folding different?

13 What structures form from each of the three germ layers? Give specific examples.

29.3c Organogenesis

LEARNING OBJECTIVE

5. Define organogenesis and explain the risk of teratogens during this period.

Once the three primary germ layers have formed and the embryo has undergone cephalocaudal and transverse folding, the process of **organogenesis** (ōr′gă-nō-jen-ĕ-sis), or organ development, begins. The upper and lower limbs attain their adult shapes, and the rudimentary forms of most organ systems have developed by week 8 of development.

By the end of the embryonic period, the embryo is slightly longer than 2.5 centimeters (1 inch), and yet it already has the outward appearance of a human (see figure 29.1). During the embryonic period, the embryo is particularly sensitive to **teratogens** (ter′ă-tō-jen, te-ră′tō-jen; *teras* = monster, *gen* = producing), substances that can cause birth defects or the death of the embryo. Teratogens include alcohol, tobacco smoke, drugs, some viruses, and even some seemingly benign medications, such as aspirin. Because the embryonic period includes organogenesis, exposure to teratogens at this time can result in the malformation of some or all organ systems.

Although rudimentary versions of most organ systems have formed during the embryonic period, different organ systems undergo "peak development" periods at different times. For example, the peak development for limb maturation is weeks 4–8, whereas peak development of the external genitalia begins in the late embryonic period and continues through the early fetal period. Teratogens cause the most harm to an organ system during its peak development period. So, a drug such as thalidomide (which causes limb defects) causes the most limb development damage if taken by the mother during weeks 4–8 (see Clinical View: "Limb Malformations" in section 8.12).

 WHAT DID YOU LEARN?

14 Why is it important for a pregnant woman to consult with her physician before using any medication?

29.4 Fetal Period

LEARNING OBJECTIVE

1. Describe the major events that occur during the fetal stage of development.

The fetal period extends from the beginning of the third month of development (week 9) to birth. It is characterized by maturation of tissues and organs, and rapid growth of the body. Fetal length increases dramatically in months 3 to 5. The length of the fetus is usually measured in centimeters, either as the crown-rump length (CRL) or the crown-heel length (CHL). The 2.5-centimeter (1-inch) embryo will grow in the fetal period to an average length of 53 centimeters (21 inches).

Fetal weight increases steadily as well, although the weight increase is most striking in the last 2 months of pregnancy. The average weight of a full-term fetus ranges from 2.5 to 4.5 kilograms (5.5–9.9 pounds). The major events that occur during the fetal period are listed in table 29.3.

 WHAT DID YOU LEARN?

15 The fetal period is characterized by what changes?

Table 29.3	Fetal Stage of Development
Time Period	**Major Events**
Weeks 9–12 CRL: 9 cm	Primary ossification centers appear in most bones Reproductive organs begin to develop Coordination between nerves and muscles for movement of limbs occurs Brain enlarges Body elongates Epidermis and dermis of the skin become more fully developed Permanent kidneys develop Palate (roof of mouth) develops Average crown-rump length at 12 weeks: 9 centimeters (cm) Average weight: 28 grams (g)
Weeks 13–16 CRL: 14 cm	Body grows rapidly Ossification in the skeleton continues Limbs become more proportionate in length to body Brain and skull continue to enlarge Average crown-rump length at 16 weeks: 14 cm Average weight: 170 g
Weeks 17–20 CRL: 19 cm	Muscle movements become stronger and more frequent Lanugo covers skin Vernix caseosa covers skin Limbs near final proportions Brain and skull continue to enlarge Average crown-rump length at 20 weeks: 19 cm Average weight: 454 g
Weeks 21–38 CRL: 36 cm	Body gains major amount of weight Subcutaneous fat is deposited Eyebrows and eyelashes appear Eyelids open Testes descend into scrotum (month 9) Blood cells form in marrow only Average crown-rump length at 38 weeks: 36 cm Average *total* length at 38 weeks: 53 cm Average weight: 2.5–4.5 kilograms (kg)

CLINICAL VIEW
Infertility and Infertility Treatments

Infertility refers to the inability to conceive and maintain a pregnancy. Typically, it is defined medically as the inability to conceive after at least 1 year of regular sexual intercourse without protection.

Contrary to popular belief, the causes of infertility are equally split between females and males. One of the most common causes of infertility in women is blocked uterine tubes due to pelvic inflammatory disease or complications from endometriosis (see Clinical View: "Endometriosis" in section 28.3c and Clinical View: "Sexually Transmitted Infections" in section 28.4b). Ovulation disorders (e.g., abnormal follicle-stimulating hormone [FSH] and luteinizing hormone [LH]) secretion or polycystic ovarian syndrome) are another common cause of female infertility. Some men (and more rarely, some women) develop **anti-sperm antibodies,** molecules that mark and target the sperm for destruction by the immune system. Other male-related causes of infertility include low sperm count, morphologically abnormal sperm, or impaired sperm delivery. Obesity, alcohol intake, drug use, smoking, exposure to certain environmental chemicals, and higher levels of stress have been associated with an increased risk of infertility in both sexes. In some cases, the cause for infertility remains unknown.

Several potential treatments are available for infertility, depending upon the initial cause:

- **Intrauterine insemination,** also known as *artificial insemination,* consists of injection of specially prepared sperm directly into the uterus through the vagina. Prior to injection, the sperm are prepared in the lab and stimulated to undergo capacitation.
- **Clomiphene citrate** (Clomid) is an oral medication that competes for estrogen receptors in the hypothalamus, pituitary, and ovaries—and in so doing, it reduces the negative feedback effects of the body's own estrogens on the ovarian cycle. Thus, follicular development and ovulation is stimulated. Clomid sometimes is associated with multiple pregnancies and ovarian hyperstimulation.
- **Human menopausal gonadotropins** (e.g., Repronex, Menopur, and Pergonal) and pure **follicle-stimulating hormone (FSH)** (Gonal-F) are injected medications that stimulate follicular development. These medications may have multiple adverse effects, including multiple pregnancy (from too many follicles being stimulated), ectopic pregnancy, miscarriages, and ovarian hyperstimulation syndrome.
- **Bromocriptine** (Parlodel) is an oral medication used to treat high prolactin levels seen in some infertile women. When prolactin levels drop, normal follicular development may ensue.

If pregnancy does not ensue after about 6 months of treatment with medications, or with multiple intrauterine insemination attempts, a couple may wish to explore the more complex **assisted reproductive technologies (ARTs).** These technologies may cost tens of thousands of dollars and have variable success rates. Among the ARTs are the following:

- **In vitro fertilization (IVF)** is a procedure by which pre-ovulatory oocytes are surgically extracted from the ovaries and each is injected with a prepared sperm in the lab. Prior to this process, the woman often takes medications (such as the ones previously listed) to stimulate the ovaries to produce multiple mature follicles.

Some infertility treatments have a greater likelihood of multiple births, which can be risky for both mother and the developing fetuses.

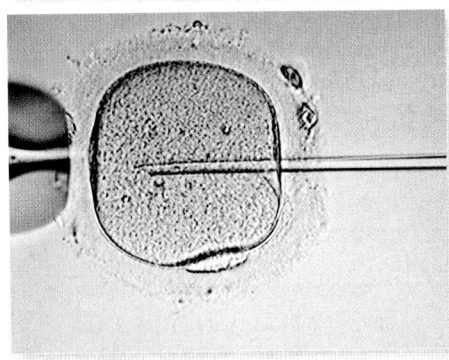

In vitro fertilization (IVF) is a procedure where a sperm is injected directly into the secondary oocyte.

One or more fertilized oocytes (now multicellular pre-embryos) are then surgically implanted in the uterus, and any remaining fertilized pre-embryos are cryopreserved (frozen) for future use. This procedure may have a success rate ranging from 10–35%, depending upon the age of the mother and other factors. One complication of IVF is multiple pregnancies, due to the fact that usually two or more pre-embryos are transferred during any one procedure.

- **Donor oocytes** may be retrieved from a female donor surgically. The oocytes may be fertilized with sperm in vitro and then placed in the future mother's uterus. Alternatively, the donor oocytes may be injected into the future mother's uterine tube with sperm (a process referred to as **gamete intrafallopian transfer**), in the hopes that the sperm will fertilize one or more of the donor oocytes.
- Likewise, **donor embryos** may be given to, and used by, an infertile couple. These embryos often were cryopreserved from another couple who may have undergone IVF.

Some infertile couples have chosen to have a **surrogate mother.** This surrogate may be impregnated with a pre-embryo from the infertile couple, or she may use her own oocyte and the male's donated sperm. However, the use of surrogates is fraught with ethical and legal complications; for example, the surrogate and the couple may fight for legal rights to the baby, or the couple may have issues with respect to how the surrogate is caring for herself during the pregnancy.

Many of these infertility treatments have come under ethical and legal scrutiny. For example, is it ethical to impregnate a woman with multiple pre-embryos? Who has legal rights to cryopreserved embryos, and who decides about their fate if the mother and father disagree? If a couple uses a surrogate mother, and that same couple later divorces, could the father claim sole custody because he is the only parent biologically related to the child? The medical advances of infertility treatments have outpaced their regulation and oversight, and medical ethicists will continue to debate these issues for years to come.

29.5 Effects of Pregnancy on the Mother

Pregnancy is an approximately nine-month process that, if all goes well, leads to the birth of a healthy baby. In the previous sections we discussed how the developing embryo/fetus undergoes a variety of changes within these 9 months. However, pregnancy also has dramatic anatomic and physiologic effects on the mother. Some of these changes can be very demanding, as the mother's body adapts to caring for itself and that of the developing fetus.

Here we discuss some of the typical anatomic and physiologic changes a woman experiences throughout her pregnancy. However, one very important thing to keep in mind is that great variation exists in how women experience pregnancy.

29.5a The Course of Pregnancy

LEARNING OBJECTIVE

1. Compare and contrast the first, second, and third trimesters of pregnancy.

As mentioned earlier, the length of the pregnancy is subdivided into trimesters:

- The **first trimester** encompasses the first 3 months of pregnancy (or the first 12 weeks of development of the embryo and fetus). During this time period, the zygote develops into an embryo and then into an early fetus.

- The **second trimester** includes months 4 to 6 of pregnancy and is marked by growth of the fetus and expansion of maternal tissues.

- The **third trimester** encompasses months 7 to 9 of pregnancy. During this time period, the fetus grows most rapidly and gains weight, and the mother's body prepares for the eventual labor and delivery.

Women's experience of pregnancy varies greatly. For example, some women may have little or no "morning sickness" during pregnancy, whereas others may have to be hospitalized due to extreme nausea and vomiting that lasts well past the first trimester. Some women may have little weight gain, whereas others may gain a significant amount of weight. Even the *length* of the pregnancy isn't the same for everyone—some women may deliver well before their projected due dates, whereas some may need to have labor induced because they are well past the due date. In addition, one woman may have different pregnancy experiences for each of her children. One pregnancy may be uneventful, whereas another may have medical complications; or one birth may be long and difficult, and another is relatively quick. As you read this section about the typical anatomic and physiologic changes that occur, please keep in mind that every pregnancy, and how it is experienced by the mother, is unique.

WHAT DID YOU LEARN?

16 How would you describe the differences between the first trimester and the second trimester? Between the second trimester and the third?

INTEGRATE

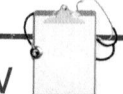

CLINICAL VIEW
Gestational Diabetes

Gestational (jes-tā′shŭn-ăl) **diabetes** is diabetes that develops for the first time during pregnancy. It is marked by increasing insulin resistance and high blood glucose levels, which may develop due to the effects of human placental lactogen (HPL), estrogen, and progesterone (all of which increase insulin resistance). It also may be due to the increased aldosterone and cortisol levels produced by pregnancy.

Gestational diabetes occurs in 3 to 8 per 100 pregnancies and typically is tested for in the second trimester (weeks 24–28 of pregnancy), when symptoms usually appear. Risk factors for gestational diabetes include maternal obesity, existence of pre-diabetes or a family history of diabetes, or a previous birth of a baby weighing over 9 pounds. African American, Native American, Hispanic, and Asian women have a greater risk of developing gestational diabetes compared to Caucasian women. Mothers with gestational diabetes may develop high blood pressure during pregnancy and may have complications with childbirth. The baby may grow exceedingly large due to the elevated glucose, increasing the risk of cesarean section and birth complications, and may develop breathing problems and hypoglycemia immediately after birth. Although gestational diabetes typically resolves after giving birth, both mother and child are at increased risk for developing type 2 diabetes later on in life.

If a woman is diagnosed with gestational diabetes, she is put on a special diet to regulate her blood glucose levels. She monitors her blood glucose levels on a regular basis, is encouraged to increase her physical activity, and in some cases may have to take insulin. After giving birth, the mother should take steps to reduce her risk factors for type 2 diabetes and be tested for the disease at regular intervals throughout her life.

29.5b Hormonal Changes

LEARNING OBJECTIVES

2. Discuss the critical effects of estrogen and progesterone during pregnancy.

3. Identify other hormones whose levels are altered during pregnancy.

A wide array of hormonal changes occurs during pregnancy. We do not go into detail about all hormonal changes here; instead, we summarize the major alterations.

Recall that when the blastocyst implants in the uterus, its syncytiotrophoblast begins secreting large amounts of human chorionic gonadotropin (hCG), which signals the female reproductive system to maintain and build the uterine lining. During the first trimester of pregnancy, hCG levels remain high to maintain the corpus luteum (so it continues to produce progesterone and estrogen), but then they decline (see figure 29.6). When hCG declines, the corpus luteum degenerates as well—but by this time, the corpus luteum is no longer needed because the placenta is producing its own estrogen and progesterone to maintain the pregnancy.

In the second and third trimesters of pregnancy, the placenta is the major producer of progesterone and estrogen. These hormones are responsible for the majority of effects we discuss later in this section. The high levels of progesterone and estrogen suppress FSH and LH secretion, so the ovarian cycle and additional follicular development are arrested during the pregnancy. They facilitate uterine enlargement, mammary gland enlargement, and fetal growth. Note: The details for both progesterone and estrogen are included in the summary table "Regulating the Female Reproductive System," which directly follows chapter 17 (see **table R.9**).

Estrogen and progesterone have a dramatic effect on the integumentary system; many pregnant women report faster-growing and stronger nails (likely due to increased levels of estrogen and progesterone), and their hair tends to be fuller and thicker in response to these hormones (because the hormones prevent normal cyclical hair loss, and a greater percentage of hair follicles remain in the resting stage). In addition, estrogen is primarily responsible for relaxation of many ligamentous joints, such as the sacroiliac joints and pubic symphysis, in preparation for labor. Progesterone is responsible for functional layer growth and the prevention of menstruation during pregnancy.

Another hormone called **relaxin** is secreted by the corpus luteum and placenta. Despite its name, research suggests that this hormone is not responsible for the ligament relaxation seen in pregnancy. Rather, relaxin appears to promote blood vessel growth in the uterus. The placenta also becomes a major secretor of **corticotropin-releasing hormone (CRH)**. Recall from section 17.8c that small amounts of CRH are produced by the hypothalamus and stimulate the anterior pituitary to secrete adrenocorticotropic hormone (ACTH), which acts on the adrenal cortex. In comparison to the hypothalamus, the placenta secretes large amounts of CRH during pregnancy, and this hormone is believed to play a role in the length of pregnancy and the timing of childbirth. CRH also is responsible for the rise in aldosterone in the mother, which promotes fluid retention (see sections 24.6d and 25.4c) and an overall increase in blood volume during pregnancy. (Recall that aldosterone regulates sodium and potassium levels.) Thus, a woman may experience **edema** (e-dē′mă; swelling) during the later stages of pregnancy.

The placenta secretes **human chorionic thyrotropin (HCT),** which is similar to thyroid-stimulating hormone and thus stimulates the thyroid gland. As a result, a pregnant woman's metabolic rate increases.

Human placental lactogen (HPL), also known as *human chorionic somatomammotropin (hCS)*, is secreted by the developing placenta—specifically, by the syncytiotrophoblast cells. The levels of this hormone rise linearly beginning the fifth week to maximum levels by the thirty-sixth week of development. Although this hormone is named for its effects on inducing lactation in mammals generally, its effects on human lactation have not been demonstrated. Specifically, HPL does affect how the pregnant woman metabolizes certain nutrients—the mother metabolizes more fatty acids instead of glucose, leaving greater glucose reserves for the fetus. HPL also inhibits the effects of insulin (see section 17.9b), so there are greater circulating levels of glucose in the blood (again, for the use of the fetus).

Other hormones whose levels increase include prolactin and oxytocin. You learned in sections 17.7b and 28.3f that oxytocin is produced by the hypothalamus and is involved in uterine contractions as well as milk expulsion from the mammary glands. Oxytocin levels increase in the second and third trimesters, in response to rising estrogen levels, and peak during labor. Prolactin is produced by the anterior pituitary and is responsible for milk production. Prolactin levels increase tenfold during pregnancy to ensure that lactation occurs after giving birth.

WHAT DID YOU LEARN?

17 How do estrogen and progesterone act to sustain pregnancy?

18 What are the actions of CRH, HPL, oxytocin, and prolactin during pregnancy?

29.5c Uterine and Mammary Gland Changes

LEARNING OBJECTIVES

4. Explain the changes to the uterus in a pregnant woman.

5. Describe the hormones that affect mammary gland development during pregnancy.

The uterus undergoes a dramatic metamorphosis during pregnancy. **Figure 29.13** illustrates how both the uterus and the mammary glands change shape throughout the entire pregnancy.

Prior to pregnancy, the uterus is approximately 8 cm by 5 cm and situated within the pelvic cavity. Once implantation occurs, the uterus begins to enlarge and expand as its muscle cells hypertrophy and undergo hyperplasia. Uterine enlargement can be detected during a vaginal exam by 4 weeks after fertilization, and by 12 weeks (the end of the first trimester) the uterus is just superior to the level of the pubic symphysis and is about the size and shape of a large grapefruit. As the uterus expands, it impinges on the

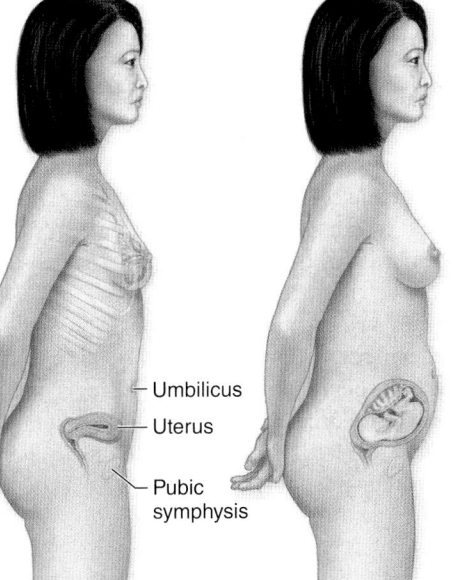

Diaphragm

Xiphoid process

Umbilicus

Uterus

Pubic symphysis

Pubic symphysis

(a) Before conception
Uterus the size of a fist and resides in pelvis

(b) 4 months
Fundus of uterus between pubic symphysis and umbilicus

(c) 7 months
Fundus superior to the umbilicus

(d) 9 months
Fundus at level of xiphoid process of sternum

Figure 29.13 Uterine and Mammary Gland Changes During Pregnancy. The uterus enlarges dramatically, reaching the xiphoid process of the sternum by the ninth month of pregnancy. The mammary glands increase in size as well, as they prepare to produce breast milk.

space occupied by the urinary bladder, so more frequent urination during this trimester is a common complaint. Recall that the developing fetus has a crown-to-rump length of only 9 cm at this time, so the bulk of the enlargement is due to myometrium muscle hypertrophy and hyperplasia, placental growth, and amniotic fluid production.

By 16 weeks (4 months) (figure 29.13b) the uterus has expanded into the abdominal cavity and its fundus typically is at a midpoint level between the pubic symphysis and umbilicus. The uterus continues to expand and reaches the level of the umbilicus by about week 22. During this second trimester, the superior growth of the uterus temporarily decreases pressure on the urinary bladder, so the need to frequently urinate may lessen during this trimester.

The expansion continues during the third trimester, as now the fetus is preparing for its own rapid growth sequence. By about 28 weeks (7 months) the fundus of the uterus is superior to the umbilicus, and by the ninth month the fundus of the uterus is at the level of the xiphoid process of the sternum (figure 29.13d). The enlarged uterus pushes against the diaphragm and compresses many of the abdominopelvic organs, resulting in certain gastrointestinal (GI) ailments (described in the next section). In addition, now the uterus is so large that it impinges on the bladder, so frequent urination becomes a common complaint again.

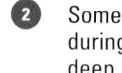 **WHAT DO YOU THINK?**

2 Some pregnant women experience difficulty with deep breathing during the last trimester. Can you think of any reasons as to why deep breathing (versus shallow breathing) is more difficult during this time?

The mammary glands typically are tender and sore during the first trimester due to the increasing levels of estrogen and progesterone. The placenta secretes **melanocyte-stimulating hormone (MSH),** which is in part responsible for the darkening of the areolae and nipples during this time (figure 29.13d). MSH also is responsible for a darkening of the linea alba, the ligamentous connection between left and right rectus sheaths, turning it into a temporary vertical dark line called the **linea nigra** (nī′grä; *niger* = black). Mammary glandular tissue grows and additional acini develop, especially under the influence of prolactin. It is typical for a pregnant woman to increase one bra cup size during this time due to the mammary gland growth. Further discussion of lactation occurs later in this chapter.

WHAT DID YOU LEARN?

19 What impact does the growth of the uterus during pregnancy have on other organs of the abdomen?

20 What hormones affect the mammary glands of a pregnant woman?

29.5d Digestive System, Nutrient, and Metabolic Changes

LEARNING OBJECTIVES

6. Describe the effects of HPL and other hormones on the pregnant woman's ability to utilize glucose.

7. List some common GI complaints and conditions that occur during pregnancy and their causes.

You read earlier that human placental lactogen (HPL) affects how the mother metabolizes certain nutrients, so that she metabolizes more fatty acids instead of glucose, leaving the glucose for the developing fetus. The increased levels of corticosteroids, estrogen, progesterone, and human placental lactogen also result in increased insulin resistance in

INTEGRATE

CLINICAL VIEW
Hyperemesis Gravidarum

Hyperemesis gravidarum (HG) (hī-per-em-ē′sis; *emesis* = vomiting; grav′ĭ-dăr-ŭm; *gravid* = a pregnant woman) is severe, continuous nausea and vomiting during pregnancy that results in dehydration, electrolyte imbalance, and weight loss in the mother. In severe cases, the woman may have to be hospitalized and given intravenous fluids or be fed with a nasogastric tube until either symptoms subside or she is ready to give birth. Fortunately, hyperemesis gravidarum is relatively rare, compared to the milder forms of morning sickness.

the pregnant mother. This increased insulin resistance in some cases can lead to gestational diabetes in the mother (see Clinical View: "Gestational Diabetes" in section 29.5a).

Many expectant mothers may experience **morning sickness** during the first trimester of their pregnancy. Contrary to its name, morning sickness does not occur just in the morning, although typically symptoms tend to be most severe then. Symptoms may vary dramatically among women—some may experience a little nausea for a few months only, whereas others may have dehydrating vomiting that lasts throughout their pregnancy and may be debilitating. In the most severe cases, a woman may need to be hospitalized and replenished with IV fluids until the pregnancy comes to term.

There are several hypotheses about the cause of morning sickness, but none have been definitively proven. One suggests that the high levels of circulating hormones, especially hCG, are the cause. However, several research studies have not been able to demonstrate this. Others have suggested that morning sickness is an evolutionary adaptation to protect the developing fetus from harmful toxins in the food. Women with morning sickness tend to prefer bland carbohydrates (which may be less likely to be spoiled or have toxins) and may have aversions to meat and eggs (which may be more likely to have toxins or be spoiled with pathogens).

The higher circulating levels of progesterone result in relaxed smooth muscle and thus slowed intestinal motility, so digested materials remain in the GI tract for longer periods of time. In the later stages of pregnancy, the expanding uterus compresses the abdominal organs and may impinge on part of the intestines. All of these factors may result in heartburn and indigestion; mothers also may experience constipation (see section 26.3d). Chronic constipation, as well as problems with venous circulation (caused by compression of the lower veins by the fetus) also can lead to hemorrhoids (see Clinical View: "Varicose Veins" in section 20.5a).

A pregnant woman typically needs about 300 extra calories daily to supply both herself and her fetus. Adequate levels of folic acid, calcium, protein, and iron are especially important during pregnancy. A woman experiences weight gain during pregnancy, but only about 20 pounds of this weight gain is due to the fetus, placenta, breast and uterine enlargement, and fluid retention. Any additional weight gain typically is additional adipose tissue, fluid retention, or both.

WHAT DID YOU LEARN?

21 Why is it important to check a woman's blood glucose levels during pregnancy?

22 What conditions other than overeating might explain weight gain during pregnancy?

INTEGRATE

CLINICAL VIEW

Preeclampsia

Preeclampsia (prē-ē-klamp'sē-ă), also known as *toxemia* or *pregnancy-induced hypertension,* is high blood pressure that occurs by the second half of pregnancy and is marked by protein in the urine (proteinuria). It occurs in about 2–6% of all pregnancies, and risk factors include maternal obesity, diabetes, advanced age, and having a previous preeclampsic birth. The cause is unknown, but researchers suspect that endothelial cell injury, due to a poorly perfused placenta, initiates the disorder. The high blood pressure carries with it the risks of general hypertension for the mother, and the poor perfusion of the placenta means the fetus may not get an adequate supply of nutrients. The only cure for preeclampsia is giving birth, so depending upon when symptoms occur, a woman may be put on bedrest and medications (to keep blood pressure levels lower), or labor may be induced if the baby is close to term.

Eclampsia is high blood pressure that causes seizures or convulsions in the mother and is a medical emergency. The concern for every preeclampsia patient is to make sure her symptoms do not develop into full-blown eclampsia.

29.5e Cardiovascular and Respiratory System Changes

LEARNING OBJECTIVES

8. List the cardiovascular changes a woman typically exhibits during pregnancy.
9. Explain the changes to the respiratory system during pregnancy.

Because of the needs of both the mother and the embryo/fetus, the cardiovascular system undergoes dramatic changes throughout pregnancy. Respiratory system function also becomes altered to meet increased requirements for gas exchange.

Cardiovascular System

The cardiovascular system must distribute respiratory gases and nutrients to both the mother and the growing fetus. More blood is needed, so plasma volume increases by about 50% throughout pregnancy. The increased blood volume needs an increased cardiac output to pump this blood through the circulation. Cardiac output increases 30–50% beginning at week 6 of pregnancy and peaks about weeks 24–28 of pregnancy, before dropping slightly. To increase cardiac output, the body increases both heart rate (on average from 70 to 90 beats per minute) and stroke volume.

The increase in blood volume may initially cause an increase in blood pressure during the first trimester. However, by the second trimester the blood pressure drops because of a decrease in peripheral vascular resistance that results from both a decrease in blood viscosity (a pregnant woman has a lower hematocrit than a nonpregnant female) and a decreased sensitivity to the hormone angiotensin (see sections 20.6b and 25.4a).

By the third trimester, the uterus and fetus compress the abdominal blood vessels, so venous return from the lower part of the body may be impaired. As a result, some pregnant women may develop varicose veins, hemorrhoids, and edema in the lower limbs.

Respiratory System

Earlier we mentioned that in the later part of pregnancy, the expanding uterus prevents the diaphragm from fully descending and the lungs from fully expanding with air. **Dyspnea** (disp'nē-a) is the uncomfortable awareness of breathing and may occur during periods of exertion. Increased estrogen levels may cause the nasal mucosa to become fluid-filled, have increased blood circulation, and congestion. The expectant mother also may experience **epistaxis** (nosebleeds) due to the increased circulation.

Progesterone increases the sensitivity of brainstem chemoreceptors to blood CO_2 levels, ultimately functioning to lower the blood CO_2 levels. These lower blood CO_2 levels facilitate the diffusion of gases across the placenta. To lower the blood CO_2 levels, the tidal volume and pulmonary ventilation levels (see section 23.5a) increase by 30–40% as breathing depth increases. Breathing rate increases slightly as well. Additionally, the mother's oxygen consumption increases about 20–30% to meet the oxygen demands of both mother and fetus. Thus, all of these alterations serve to provide enough oxygen to mother and fetus, as well as facilitate gas exchange between the placenta and the maternal blood.

WHAT DID YOU LEARN?

23. How is a woman's cardiovascular system altered during pregnancy?
24. Why is it beneficial for CO_2 receptor sensitivity to be increased when a woman is pregnant?

29.5f Urinary System Changes

LEARNING OBJECTIVE

10. Describe the effects of pregnancy on the mother's urinary system.

The pregnant woman's urinary system is responsible for eliminating not only her own metabolic waste products but also the waste products of her fetus. In addition, up to 50% more plasma volume must be filtered by the kidneys. Glomerular filtration rate (GFR; see section 24.5d) thus increases about 30–50% during pregnancy and urine output increases slightly.

Recall that compression by the expanding uterus can lead to frequent urination in both the first and third trimesters. In contrast, the uterus places relatively less pressure on the bladder during the second trimester.

Progesterone causes smooth muscle relaxation in the ureters, which may cause uretal and renal pelvic dilation. This dilation and the increased urine volume may result in urine stasis (slowing or stopping) anywhere along its path. In addition, compression of the ureter or kidney by the uterus can result in urine drainage issues. All of these factors put pregnant women at much greater risk for urinary tract infections (UTIs), which are perhaps the most common type of bacterial infection during pregnancy.

WHAT DID YOU LEARN?

25. What factors make pregnant women more likely to get urinary tract infections?

INTEGRATE

CONCEPT CONNECTION

As you've seen, pregnancy has an effect on virtually every body system in the pregnant woman. The reproductive, endocrine, digestive, cardiovascular, respiratory, urinary, and even the integumentary system respond and adapt to the changes brought about during pregnancy. The skeletal and muscular systems adjust to her body's increasing demand for calcium, and several skeletomuscular changes develop as well (e.g., **lordosis**, or accentuated lumbar curve seen in late-term pregnancies, see section 8.5b). The immune system typically is suppressed slightly so the woman's body will not mistakenly attack the fetus as something foreign. In fact, pregnancy is a wonderful example of how all body systems may adapt and develop their own forms of homeostasis in response to such a physiologic challenge.

29.6 Labor (Parturition) and Delivery

Labor is also known as *parturition* (par-tūr-ĭsh'ŭn; *parturitio* = to be in labor) and is the physical expulsion of the fetus and placenta from the uterus. True labor typically occurs at 38 weeks for a full-term pregnancy, but not all uterine contractions lead to true labor. We first consider factors that initiate labor, and then compare the contractions of false labor with those of true labor; we conclude with a description of the three stages of labor.

29.6a Factors That Lead to Labor

 LEARNING OBJECTIVE

1. Explain the physiologic processes that initiate labor.

Throughout the pregnancy, as the uterus enlarges and stretches, the uterine myometrium prepares itself for uterine contractions. In the later stages of pregnancy, the increasing levels of estrogen counteract the calming influence of progesterone on the uterine myometrium, and increase the uterine myometrium sensitivity. In addition, the rising levels of estrogen stimulate the production of oxytocin receptors on the uterine myometrium, so as the levels of oxytocin also rise, more receptors are available on the uterus for binding this hormone.

All of these factors result in the uterine myometrium becoming more sensitive and "irritable" in the later stages of pregnancy, and contractions begin to occur. These contractions typically are weak and irregular, but as levels of estrogen and oxytocin continue to rise in the later stages, they become more intense and frequent. Thus, weak contractions may occur and be noticed as soon as the second trimester of pregnancy.

Premature labor refers to labor that occurs prior to 38 weeks. Premature labor (and giving birth to a premature infant) is not desirable because the infant's body systems, especially the lungs, may not be fully developed. Very premature infants are at greater risk for morbidity and mortality due to their underdeveloped organ systems. Thus, the ideal outcome is for labor to begin as close as possible to term.

WHAT DID YOU LEARN?

26 How do progesterone, estrogen, and oxytocin interact to eventually bring about labor?

29.6b False Labor

 LEARNING OBJECTIVE

2. List the signs and characteristics of false labor.

False labor is defined as uterine contractions that do *not* result in the three stages of labor and the expulsion of the fetus. The contractions of false labor are known as **Braxton-Hicks contractions,** named for a nineteenth-century gynecologist. It may be difficult for a woman to know whether she is experiencing Braxton-Hicks contractions or true labor contractions, and it is not uncommon for a woman to mistakenly think she is about to give birth.

In general, Braxton-Hicks contractions have the following characteristics:

- They tend to be irregularly spaced and do not become more frequent as time passes.
- They tend to be relatively weak, do not increase in intensity, and may stop entirely if the woman changes position or activity (e.g., goes for a walk).
- The pain from these contractions is usually limited to the lower abdomen and pelvic region, instead of radiating through the entire abdominal region and back (as with true labor contractions).
- The pain from the contractions may stop or change in response to movement.
- They do not lead to the cervical changes seen in the three stages of labor (described in section 29.6d).

WHAT DID YOU LEARN?

27 What are five signs of false labor?

INTEGRATE

CLINICAL VIEW
Inducing Labor

If a woman has not undergone labor within 2 weeks after her due date, her physician may recommend inducing labor. The woman typically is admitted to the hospital the night before labor is to progress, and a prostaglandin gel is smoothed onto her cervix. The prostaglandin gel assists with cervical dilation (thinning and widening of the cervix) and may even "kick start" the true labor process. If true labor has not begun the following morning, the physician may administer an intravenous (IV) drip of synthetic oxytocin (called Pitocin) to initiate true labor.

29.6c Initiation of True Labor

LEARNING OBJECTIVES

3. Explain the signs and characteristics of true labor.

4. Describe the positive feedback mechanism of true labor.

True labor is defined as uterine contractions that increase in intensity and regularity, and that result in changes to the cervix. The mother and the fetus both have an active role in initiating true labor.

As the pregnancy nears term, the mother's hypothalamus secretes increasing levels of oxytocin, as described earlier. (This increase in oxytocin levels is in response to a cascade of changes in both fetal and maternal hormones responsible for maintaining pregnancy.) In addition, the fetus's hypothalamus secretes its own oxytocin near the beginning of true labor. Both sources of oxytocin stimulate the placenta to secrete **prostaglandins** (pros-tă-glan′dĭn), which are fatty acids and hormonelike substances that also stimulate smooth muscle contraction, most notably uterine muscle contraction (see section 17.3b for a review of prostaglandins and other types of eicosanoids). Prostaglandins are also responsible for the softening and dilating of the cervix. The combined action of maternal oxytocin, fetal oxytocin, and the rising levels of prostaglandins initiate the rhythmic contractions of true labor.

In comparison to Braxton-Hicks contractions, true labor contractions have the following characteristics:

- They tend to increase in frequency over time, so a woman who starts with contractions that occur roughly every 15 minutes eventually will have contractions that occur about every 5 minutes.
- Contractions increase in intensity as labor progresses.
- The pain from the contractions tends to radiate from the upper abdomen inferiorly to the lower back (or vice versa), instead of being localized in the lower abdomen or groin (as with Braxton-Hicks contractions).
- The pain from the contractions does not go away or change in response to movement.
- The contractions facilitate cervical dilation and expulsion of the fetus and placenta.

True labor also initiates a positive feedback mechanism **(figure 29.14)**. The more intense uterine contractions result in the fetus's head pushing against the cervix, stimulating the

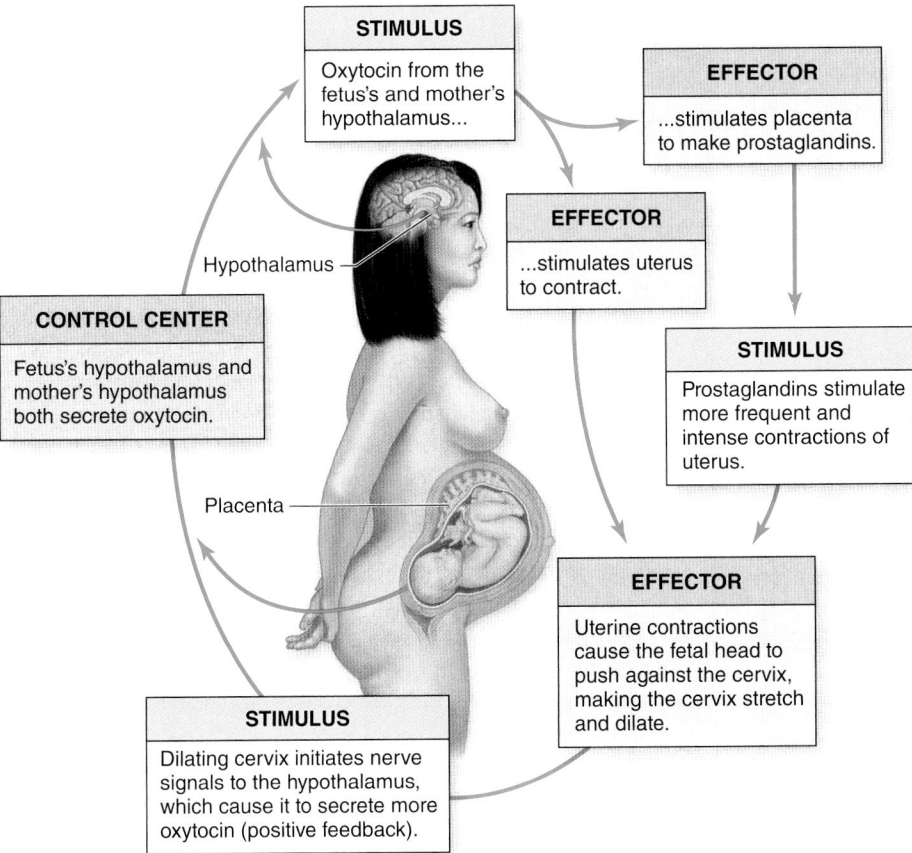

STIMULUS
Oxytocin from the fetus's and mother's hypothalamus...

EFFECTOR
...stimulates placenta to make prostaglandins.

EFFECTOR
...stimulates uterus to contract.

Hypothalamus

CONTROL CENTER
Fetus's hypothalamus and mother's hypothalamus both secrete oxytocin.

STIMULUS
Prostaglandins stimulate more frequent and intense contractions of uterus.

Placenta

EFFECTOR
Uterine contractions cause the fetal head to push against the cervix, making the cervix stretch and dilate.

STIMULUS
Dilating cervix initiates nerve signals to the hypothalamus, which cause it to secrete more oxytocin (positive feedback).

Figure 29.14 Positive Feedback Mechanism of True Labor. Rising levels of maternal and fetal oxytocin together initiate the positive feedback mechanism of true labor. Oxytocin stimulates uterine contractions as well as the release of prostaglandins, which promote cervical stretching and facilitate uterine contractions. Sensory input relayed from the uterus to the hypothalmus stimulates it to release more oxytocin from the posterior pituitary. The process escalates until the fetus is expelled from the uterus.

INTEGRATE

CLINICAL VIEW

Anesthetic Procedures to Facilitate True Labor

Uterine contractions can be quite painful and discomforting for the expectant mother. Although some women may choose not to have any pain medication and have what is referred to as a *natural childbirth,* others may select oral pain medication or a variety of anesthetic procedures.

A **pudendal nerve block** numbs the pudendal nerve, which is the main sensory nerve of the perineum and also innervates the perineal muscles. This type of anesthesia numbs the lower part of the vagina as well as the skin and muscles of the perineum, but the woman can still feel uterine contractions and some vaginal stretching. The obstetrician places the needle in the vagina and puts his or her finger above the needle and below the fetus's head, so the needle will not injure the fetus. A pudendal nerve block may be given during the second stage of labor, in conjunction with an episiotomy (described later in this section), or after labor if repairs to torn tissue are needed.

An **epidural nerve block** for labor is placed in the epidural space, usually somewhere between the L_1 and L_4 vertebrae. As the anesthetic traverses the lower part of the epidural space, it bathes the outer covering of the lower thoracic, lumbar, and sacral nerves. Thus, all of the nerves involved with the uterus, vagina, and perineum are numbed, as well as the lower limbs to some extent (depending upon the type of epidural anesthetic used.) It relieves the pain associated with uterine labor contractions, but usually does not interfere with the contractions themselves. However, some women who receive an epidural may also need an IV of synthetic oxytocin (Pitocin) to ensure uterine contractions progress well.

A **spinal nerve block** typically is reserved for cesarean sections. In this case, the needle is placed in the subarachnoid space, typically between the T_{10} and L_1 vertebrae. The placement of the anesthetic here ensures that the anterior abdominal wall is numbed, prior to its incision. The lower limbs and pelvis are completely numbed as well as a consequence of the procedure. Side effects from spinal anesthetic include nausea and a headache, perhaps from the loss of some CSF from the injection site.

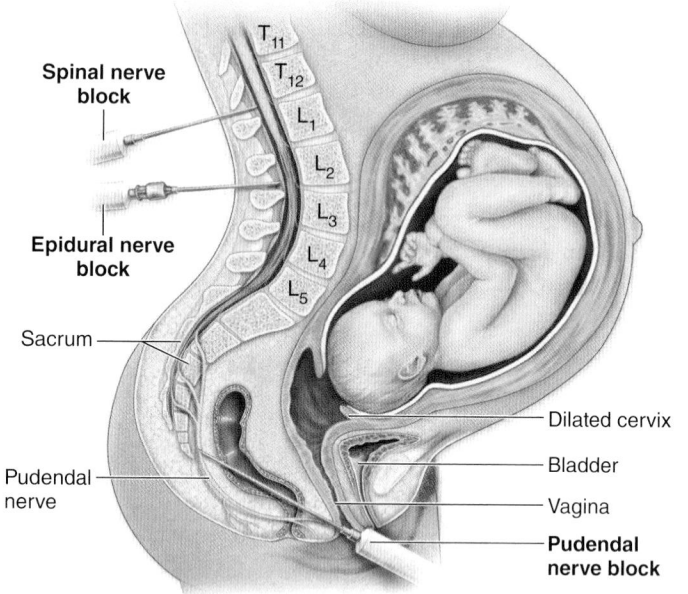

Comparisons are shown for the placement of pudendal anesthesia, epidural anesthesia, and spinal anesthesia.

---

stretching and dilation of the cervix. This manual stretching of the cervix and uterine contractions initiate nerve signals to the hypothalamus, causing it to stimulate the posterior pituitary to secrete more oxytocin. In addition, uterine contractions stimulate the placenta to secrete more prostaglandins, which also result in more intensive uterine contractions. Thus, true labor intensifies until the fetus is expelled from the uterus. Once the fetus and placenta are expelled (and thus the major source of prostaglandins is removed from the body) and the uterus and cervix are no longer fully stretched, oxytocin levels drop, and labor ceases.

 WHAT DID YOU LEARN?

28 What are the five signs of true labor?

29 How does the positive feedback mechanism of true labor lead to birth of the fetus?

29.6d Stages of True Labor

 **LEARNING OBJECTIVE**

5. List the three stages of true labor and events of each stage.

True labor involves three stages: the dilation stage, the expulsion stage, and the placental stage **(figure 29.15).** The lengths for each stage are discussed next, but note that the timing may vary greatly among different women.

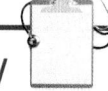
INTEGRATE

CLINICAL VIEW

Fetal Positioning and the Dilation Stage

Normally, the fetus is in a **vertex** (head down) position in the cervix, and its face is toward the sacrum. This position is the ideal position for both dilating the cervix and pushing the fetus through the vagina. If a fetus is in **breech** (buttocks first) position, the head is not in a position to dilate the cervix, and there may be a delay or problems with cervical dilation. Likewise, if the fetus is head-down but the face is facing the pubic symphysis instead of the sacrum, the head is positioned such that cervical dilation is not optimal. Instead, each uterine contraction causes the fetal head to push against the mother's sacrum more than the cervix, which can result in a delay in cervical dilation and cause more intense labor pain for the mother. For both of these variant positions, there also may be greater difficulties in expelling the fetus through the vagina and alternative means (e.g., forceps, vacuum) may be needed to extract the baby through the vagina. Alternatively, the fetus may have to be delivered via **cesarean section** (se-zā′rē-ăn; *C-section*), where a midline horizontal incision is made above the pubic symphysis, through the abdominal and uterine walls, and the fetus is delivered through this incision.

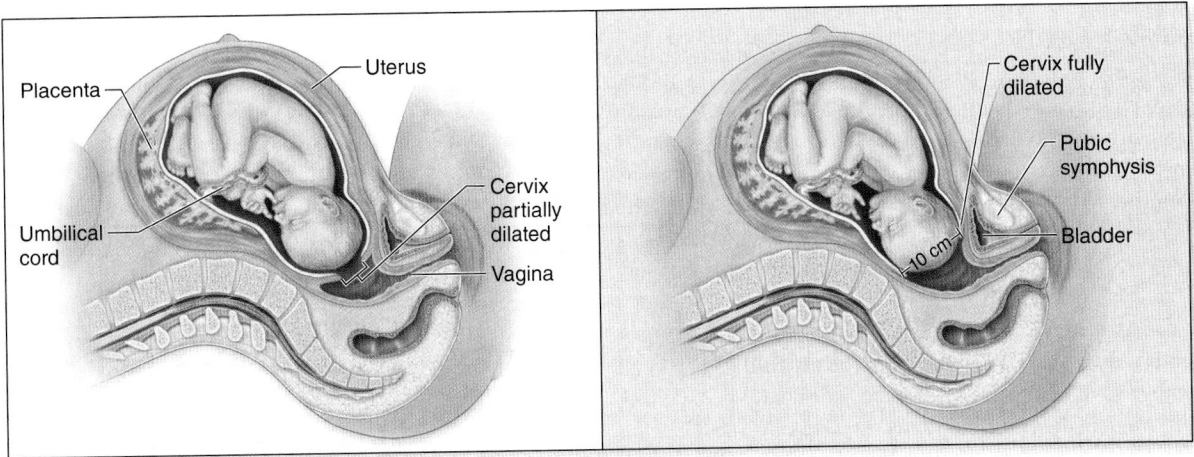

(a) Early dilation stage

(b) Late dilation stage

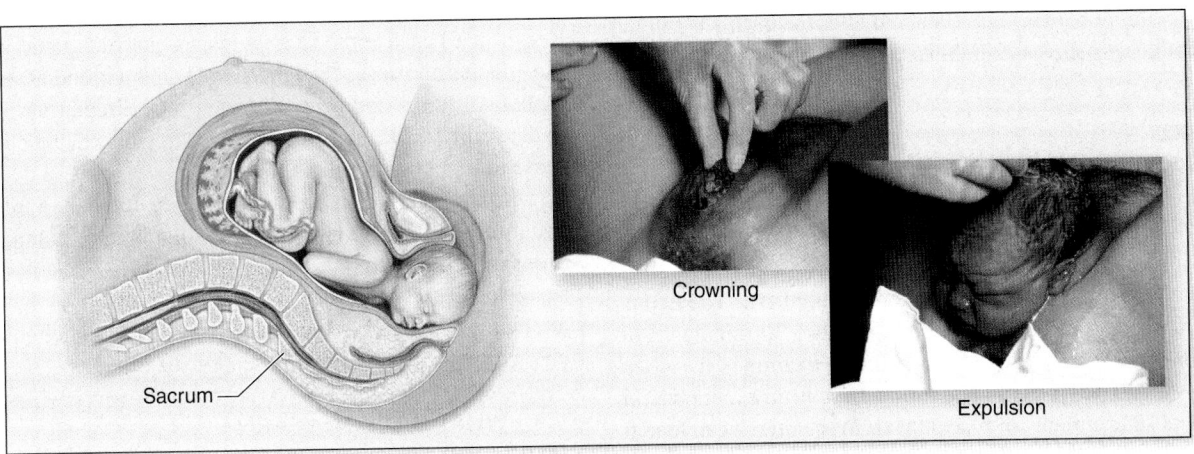

(c) Expulsion stage

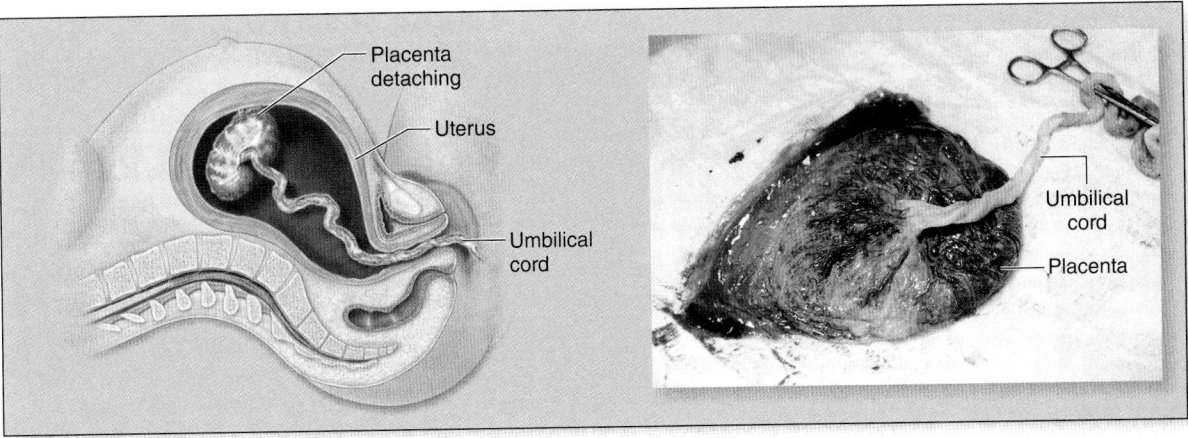

(d) Placental stage

Figure 29.15 Stages of True Labor and Childbirth. The dilation stage begins (*a*) with a distension of the cervix, due to the manual forces exerted by the fetal head. (*b*) Late in the dilation stage, the cervix will dilate to 10 cm in diameter. The expulsion stage (*c*) is where the fetus is pushed out of the uterus, and is followed by (*d*) the placental stage, where the placenta is expelled.

Dilation Stage

The **dilation stage** (figure 29.15*a, b*) begins with the onset of regular uterine contractions and ends when the cervix is **effaced** (thinned) and **dilated** (expanded) to 10 centimeters in diameter. This is the longest of the three stages and is the stage marked by the greatest variability. **Nulliparous women** (those who have never given birth before) generally experience a longer dilation stage—on average, 8 to 24 hours—than do **parous women** (those who have given birth previously); the dilation stage in parous women may range from 4 to about 12 hours.

The dilation stage starts with regularly spaced uterine contractions that increase in intensity and frequency. Each time the uterus contracts, the baby's head pushes against the cervix and causes it to efface and dilate slightly, until the cervix dilates from 1 cm to 10 cm in diameter.

This stage is also marked by the rupture of the amniotic sac and the release of amniotic fluid, also commonly known as the "water breaking." If the amniotic sac doesn't rupture on its own by the end of this stage, the physician or other birth practitioner will manually rupture the sac in preparation for the expulsion stage.

Expulsion Stage

The **expulsion stage** (figure 29.15c) begins with the complete dilation of the cervix and ends with the expulsion of the fetus from the mother's body. This stage may last as little as several minutes but typically takes 30 minutes to several hours. As with the dilation stage, nulliparous women typically have a longer expulsion stage than parous women. The uterine contractions help push the fetus through the vagina and may be facilitated if the woman "bears down" (i.e., uses the Valsalva maneuver to increase abdominal pressure as she pushes) with each contraction.

When the first part of the baby's calvarium (i.e., skullcap; see section 8.2a) distends the vagina, this is referred to as **crowning.** The head is then followed by the rest of the body. If there is difficulty in expelling the baby from the vagina, then an **episiotomy** (e-piz-ē-ot′ō-mē; *episeion* = pundenda, *tome* = incision) may be performed, which is where the perineal muscles are surgically incised to create a wider opening for the baby to pass through. (This cut is sutured after the birth.) When the baby's body is fully expelled, the umbilical cord is clamped and tied off.

Placental Stage

The **placental stage** (figure 29.15d) occurs after the baby is expelled. The uterus continues to contract, and these contractions help compress uterine blood vessels and also help displace the placenta from the uterine wall. The placenta and remaining fetal membranes (e.g., the amnion) collectively are referred to as the **afterbirth.** The expulsion of the afterbirth typically is completed within 30 minutes. The obstetrician or other birth practitioner carefully examines the afterbirth to make sure all portions of the placenta have been expelled from the uterus, because fragments of placenta left in the uterus can lead to extensive bleeding or other postpartum complications.

 WHAT DID YOU LEARN?

30 What are the three stages of labor?

31 Why is the placental stage a critical part of labor?

29.7 Postnatal Changes for the Newborn

 LEARNING OBJECTIVES

1. Describe the respiratory events that occur as the newborn adjusts to life outside of the uterus.

2. Compare and contrast the fetal circulatory pattern with the newborn circulatory pattern.

Once the fetus is expelled from the uterus, it is now known as a newborn, or **neonate** (nē′ō-nāt; *neo* = new, *natalis* = relating to birth). A variety of respiratory and cardiovascular changes must occur quickly after birth in order for the neonate to adjust to life outside of the uterus.

Prior to birth, respiratory gases were exchanged between maternal and fetal circulation at the placenta. The fetal lungs are not fully inflated because they are not yet fully functional. However, within about 10 seconds after being born, the neonate typically takes its first breath. This first breath is thought to be caused by the central nervous system responding to the change in environment and temperature. This process may be facilitated by a general respiratory acidosis (caused by clamping of the umbilical vessels and constriction of the umbilical vessels prior to birth), but note that the first

breath typically occurs regardless of whether the umbilical vessels have been clamped or not.

Once this first breath is taken, the lungs become inflated and the surfactant that is present in the alveoli keeps the alveoli patent (open) (see section 23.3d). Thus, every breath after the first is easier now that the alveoli remain patent. Premature infants born earlier than 28 weeks are not producing sufficient levels of surfactant to keep their alveoli patent, so these infants may need to placed on a ventilator until their lungs mature.

 WHAT DO YOU THINK?

3 In the past, obstetricians would lift a neonate by its feet and slap it on the back, allegedly to start it breathing. The baby's crying was supposed to be an indication that the procedure had worked. Was this procedure necessary? Could it actually be harmful?

The fetal circulatory pattern was described in detail in section 20.12a. Recall that because the fetal lungs are not functional, other pathways (i.e., ductus arteriosus, foramen ovale) shunt blood away from the nonfunctional lungs and directly to the fetal circulation. As a result, the fetal cardiovascular system has some structures that are modified or that cease to function once the fetus is born.

At birth, the fetal circulation begins to change into the postnatal pattern. When the neonate takes its first breath, pulmonary resistance drops, and the pulmonary arteries dilate. As a result, pressure on the right side of the heart decreases and the pressure is then greater on the left side of the heart, which handles the systemic circulation. The specific changes are described in section 20.12b.

 WHAT DID YOU LEARN?

32 What event must occur when the neonate takes its first breath?

33 How does the fetal circulation change once the neonate begins breathing?

29.8 Changes in the Mother After Delivery

Postpartum refers to the time period after giving birth. It is a time when a woman's body must undergo further transformative changes to both feed the neonate and return to pre-pregnancy form and function. Here we describe most of the postpartum changes that typically occur over the first 6 weeks after giving birth.

29.8a Hormonal Changes

 LEARNING OBJECTIVE

1. Compare and contrast the hormonal levels of a woman prior to birth and after birth.

Within a few days after giving birth, estrogen and progesterone levels plummet, because the uterine lining no longer needs to be maintained for pregnancy. Many experts believe this precipitous drop in sex hormone levels may account for the *baby blues*, feelings of varying degrees of sadness and depression some women experience immediately after giving birth.

As estrogen and progesterone levels plummet, the integumentary system is affected. Recall that high levels of these hormones prevented the normal cyclical hair loss. After giving birth, the hair reverts back to its normal cyclical growth and hair loss cycle. And in fact, some of the hair that was prevented from falling out during pregnancy may fall out rather abruptly after giving birth for some women. A peak in hair loss may be experienced by some women about 3 to 4 months after delivery, and it may take up to 12 months for the normal cyclical hair growth and loss cycle to resume. Thus, many new mothers may experience temporary hair thinning during this time.

The decrease in progesterone also affects the respiratory system. Without the high levels of progesterone, the chemoreceptors are less sensitive to CO_2 levels. As a result, respiratory rate, tidal volume, and pulmonary ventilation return to pre-pregnancy levels.

Additionally, the levels of corticotropin-releasing hormone (CRH) drop dramatically, now that there is no longer a placenta producing copious amounts of this hormone. Recent research has suggested that high levels of CRH during pregnancy are associated with an increased risk of **postpartum depression,** a serious disorder in which the mother experiences severe depression and possibly suicidal thoughts; unlike baby blues, this condition should be treated as soon as possible.

Prolactin levels and oxytocin levels drop after birth as well. However, because both of these hormones are involved in lactation, periodic surges occur in these hormone levels each time a baby nurses. These surges are described in greater detail in section 29.8c.

 WHAT DID YOU LEARN?

 34 Which hormone levels drop in a postpartum woman, and what are the effects of this reduction in hormone levels?

29.8b Blood Volume and Fluid Changes

 **LEARNING OBJECTIVE**

2. List the various ways that the mother loses the excess fluids gained during pregnancy.

A pregnant woman retains a great deal of fluid throughout the 9 months of pregnancy. In addition to the fluid retained in the amniotic sac and some excess fluid found in the interstitial spaces, most additional fluid is due to the increased blood volume acquired during pregnancy. After a woman gives birth, she no longer has need for this additional fluid and must expel it in a relatively quick and efficient manner. The amniotic fluid is quickly expelled by the end of the first stage of labor. But what about the rest?

A portion of the blood volume, as well as mucus and hypertrophied endometrial tissue, is released from the uterus as **lochia** (lō′kē-ă; *lochos* = childbirth). Lochia is similar to a menstrual period, in that blood and some endometrial tissue is expelled from the uterus via the vagina. However, lochia results in much heavier bleeding than a typical menstrual period, because the uterine lining buildup occurred over a 9-month period, instead of a typical 28-day cycle. Thus, the first 5 days of lochia typically results in very heavy bleeding, after which it lightens but progresses for at least 2 to 3 weeks. For some women, it may take 4 to 6 weeks before the flow finally stops. As the blood volume decreases, the woman's cardiac output returns to pre-pregnancy levels.

A woman also may expel excess fluids via increased urination. The decline in CRH after birth results in a decline in aldosterone (see section 25.4c), which precipitates the overall drop in blood volume and interstitial fluid levels. The lymphatic system cycles some of the excess interstitial fluid into the blood circulation, where it may be filtered by the kidneys and secreted as urine. Within about 24 hours after giving birth, most women experience copious, frequent urination, the result of the kidneys "working overtime," so to speak, in filtering out this excess fluid. Urination levels typically return to normal by the end of the first week after birth.

Another common symptom new mothers experience is profuse sweating for the first 2 weeks after giving birth. This abundant sweating is yet another way the body eliminates the excess fluid gained during pregnancy.

 WHAT DID YOU LEARN?

 35 What are four ways by which excess fluid is eliminated from a postpartum woman's body?

29.8c Lactation

 LEARNING OBJECTIVE

3. Describe the process by which lactation occurs.

Lactation (lak-tā′shŭn; *lactatio* = suckle) refers to the production and release of breast milk from the mammary glands. Prolactin is produced by the anterior pituitary and is responsible for milk production (see section 17.7c). In nonpregnant women and in men, the secretion of significant amounts of prolactin is inhibited by dopamine secreted by cells in the hypothalamus.

High levels of estrogen positively influence the secretion of prolactin, so as estrogen levels rise during pregnancy, so do prolactin levels. Both estrogen and prolactin cause mammary gland acini proliferation and branching of the lactiferous ducts. Paradoxically, the high levels of estrogen and progesterone are also responsible for preventing breast milk secretion until after birth. It isn't until levels of estrogen and progesterone drop that prolactin works unopposed to stimulate breast milk production.

During late pregnancy and for the first few days after birth, the substance produced by the mammary glands isn't breast milk per se. It is a watery, yellowish milklike substance called **colostrum** (kō-los′trŭm; *foremilk*), and it has lower concentrations of fat than true breast milk but is rich in immunoglobulins, especially immunoglobulin A (IgA). By drinking colostrum, the infant acquires passive immunity from the mother. IgA resists breakdown in the infant's stomach and is believed to protect the infant against ingested pathogens. Colostrum also has a laxative effect, and facilitates the infant having its first bowel movement shortly after birth.

A few days postpartum, the true breast milk starts to be produced. It has a higher fat content than colostrum, and it contains several growth factors, essential fatty acids (needed for optimal brain growth and development), specific enzymes to aid in digestion of the milk, and an array of immunoglobulins. Breast milk typically is more easily digestible for an

INTEGRATE

CONCEPT CONNECTION

Many organ systems work in tandem to help bring the mother's body back to a pre-pregnancy homeostasis. Excess fluids are removed from the cardiovascular system and excreted by both the urinary system and the integumentary system. The endocrine system secretes oxytocin to promote uterine contractions and shrinkage to a near pre-pregnancy state. The immune system produces immunoglobulins that are found in abundance in the newly produced breast milk. Many of the pregnancy digestive system challenges, such as heartburn, constipation, and morning sickness, resolve once the baby is born. Finally, if the mother is breastfeeding, the skeletal system will be called upon to supply calcium for the breast milk.

INTEGRATE

CONCEPT CONNECTION

Recall from section 22.8c that there are five classes of immunoglobulins (or antibodies). The fetus produces only IgM antibodies, and during pregnancy the mother supplies the fetus with IgG, the only antibody that crosses the placenta. After birth, the nursing mother supplies the newborn with IgG, IgA, and IgM antibodies in her breast milk. It isn't until the baby is about 2 months old that he or she is able to produce enough IgA and IgM on his or her own.

infant than other types of breast milk substitutes (e.g., cow's milk, soy "milk") and it remains the optimal source of nutrition for an infant, if the mother is able to breastfeed. One vitamin that is *not* abundant in breast milk is vitamin D, and recently physicians have recommended that infants of breastfeeding mothers receive a vitamin D supplement.

WHAT DO YOU THINK?

4 The recommendation that breastfeeding infants receive vitamin D supplementation is a recent one. Can you think of any change in behaviors that may have made this recommendation necessary? (Hint: How does the body produce vitamin D?)

Lactation requires both prolactin for breast milk production and oxytocin for breast milk secretion (see section 28.3f). Although prolactin levels drop after birth by about 50%, surges in prolactin continue as long as the baby continues to breastfeed. Thus, the continual production and release of breast milk is a positive feedback mechanism that is maintained by regular breastfeeding (**figure 29.16a**).

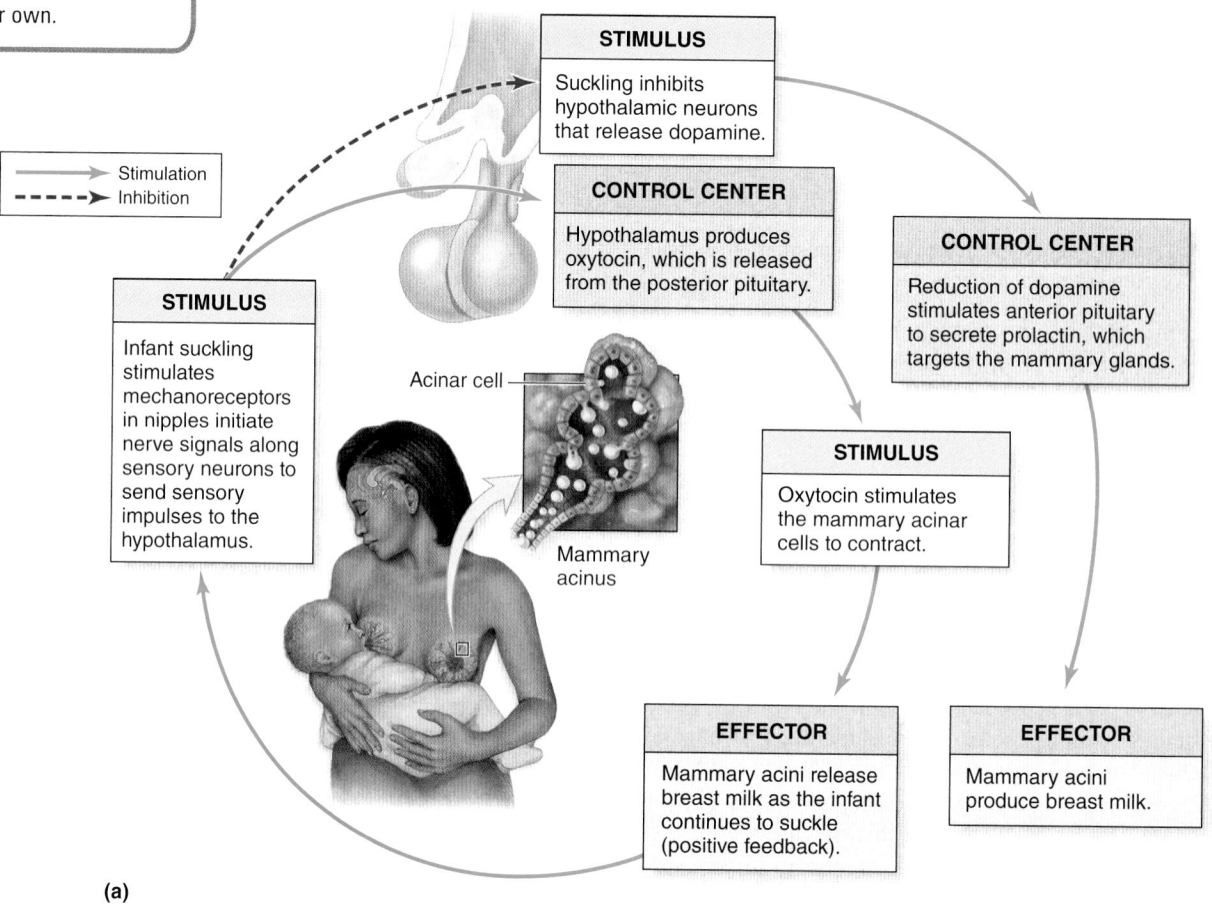

(a)

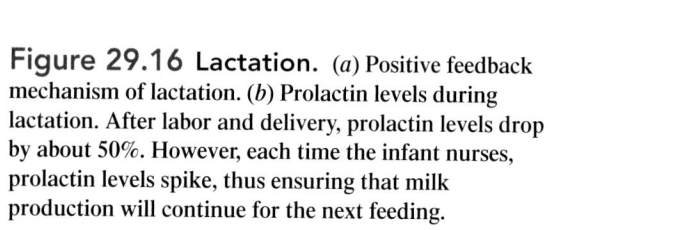

(b)

Figure 29.16 Lactation. (*a*) Positive feedback mechanism of lactation. (*b*) Prolactin levels during lactation. After labor and delivery, prolactin levels drop by about 50%. However, each time the infant nurses, prolactin levels spike, thus ensuring that milk production will continue for the next feeding.

The release of breast milk is referred to as **milk letdown,** or the *letdown reflex*, because it involves this positive feedback mechanism. When the infant suckles at the breast, mechanoreceptors in the nipple and areola are stimulated, and they send nerve signals along sensory neurons to the hypothalamus. The hypothalamus is stimulated to produce oxytocin, and the posterior pituitary releases this oxytocin into the blood (see section 17.7b). The oxytocin targets special cells in the mammary glands called myoepithelial cells, which surround the mammary acini. Specifically, oxytocin stimulates the myoepithelial cells to contract, thereby releasing breast milk from the mammary acini. As milk is released from the breast, the infant may continue to nurse. Milk will continue to be released from the breast as long as the infant continues to nurse. Once the infant stops suckling (and the nipple and areola are no longer being mechanically stimulated), oxytocin levels drop and milk letdown stops.

The milk letdown reflex may be initiated in some mothers simply in response to hearing a baby cry. Note that since a baby usually cries when it is hungry, the mother's milk letdown response may have adapted to prepare for the eventual feeding of this baby.

Note in figure 29.16*a* that as the infant breastfeeds, dopamine release is inhibited by the hypothalamus. This inhibition of dopamine secretion stimulates the anterior pituitary to secrete large amounts of prolactin. Figure 29.16*b* illustrates that spikes or peaks in prolactin production occur each time a baby breastfeeds. This prolactin promotes new breast milk production, so a new supply of breast milk will be available to the baby at the next feeding.

Most women who regularly breastfeed (i.e., more than four to five feedings a day) often do not ovulate at this time. Researchers believe that either the hormones or the sensory stimuli involved with breastfeeding inhibit release of gonadotropin-releasing hormone (GnRH) from the hypothalamus. If GnRH is not released, then FSH and LH won't be released by the anterior pituitary, and so ovulation is prevented. This mechanism may have evolved as a way of spacing births, so that a nursing mother's body won't be additionally taxed by a new pregnancy. However, breastfeeding is not a reliable form of birth control because ovulation can and does occur in breastfeeding mothers, especially those that may have reduced numbers of breastfeedings, and it is difficult to determine that ovulation has occurred. Thus, most lactating women are encouraged to use some form of birth control if they do not want to become pregnant again at that time.

 WHAT DID YOU LEARN?

36 How does the positive feedback mechanism in breastfeeding ensure that the infant receives breast milk?

29.8d Uterine Changes

 LEARNING OBJECTIVE

4. Explain the mechanisms by which the uterus returns close to its pre-pregnancy size.

The uterus spends the 9 months of pregnancy undergoing hyperplasia and hypertrophy. It takes 6 weeks following birth for the uterus to shrink to close to its pre-pregnancy size. Oxytocin facilitates this shrinkage by stimulating uterine contractions. These contractions tend to be most severe the first week after giving birth and are referred to as **afterpains.** Afterpains may be most severe and noticeable when a woman breastfeeds, since oxytocin is involved in milk ejection. After the first week, these contractions will continue but typically will be less noticeable.

The spike in oxytocin that occurs each time a woman breastfeeds not only expels the milk but stimulates uterine contractions. Thus, regular breastfeeding facilitates rapid, efficient shrinkage of the uterus. Although these contractions occur in women who don't breastfeed, they may not be as frequent or as efficient, and the uterus may not return to its pre-pregnancy size as quickly.

The changes that occur in the mother to both accommodate a pregnancy and return the body back to normal are quite remarkable. Refer to **figure 29.17** to review these processes.

 WHAT DID YOU LEARN?

37 How are prolactin and oxytocin related to both the uterine changes and lactation?

Figure 29.17 Anatomic and Physiologic Changes That Occur in the Mother. (*a*) During pregnancy and (*b*) postpartum (after birth).

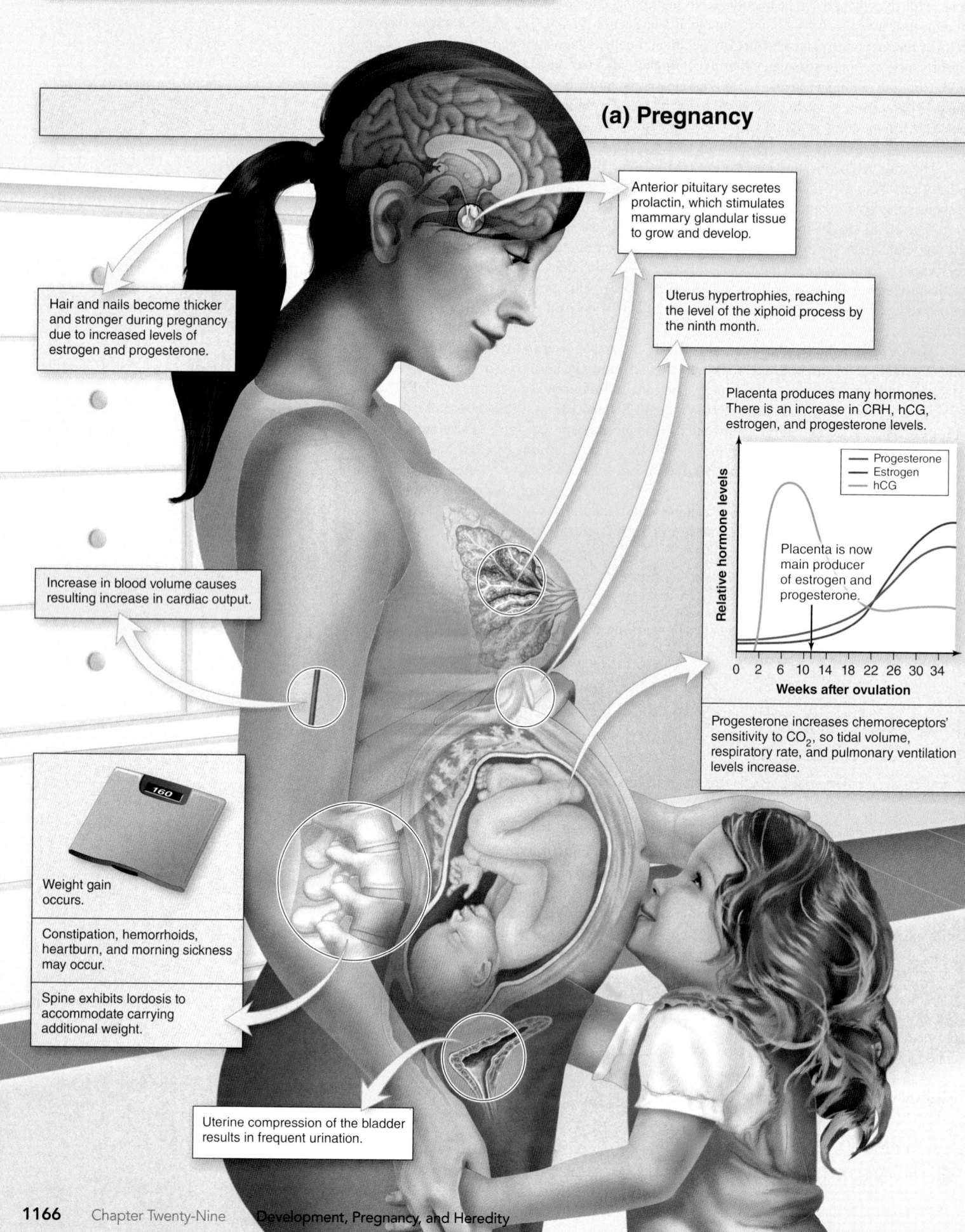

(a) Pregnancy

Anterior pituitary secretes prolactin, which stimulates mammary glandular tissue to grow and develop.

Uterus hypertrophies, reaching the level of the xiphoid process by the ninth month.

Hair and nails become thicker and stronger during pregnancy due to increased levels of estrogen and progesterone.

Placenta produces many hormones. There is an increase in CRH, hCG, estrogen, and progesterone levels.

Placenta is now main producer of estrogen and progesterone.

Increase in blood volume causes resulting increase in cardiac output.

Progesterone increases chemoreceptors' sensitivity to CO_2, so tidal volume, respiratory rate, and pulmonary ventilation levels increase.

Weight gain occurs.

Constipation, hemorrhoids, heartburn, and morning sickness may occur.

Spine exhibits lordosis to accommodate carrying additional weight.

Uterine compression of the bladder results in frequent urination.

Graph labels: Relative hormone levels — Progesterone, Estrogen, hCG; Weeks after ovulation: 0 2 6 10 14 18 22 26 30 34

(b) Postpartum

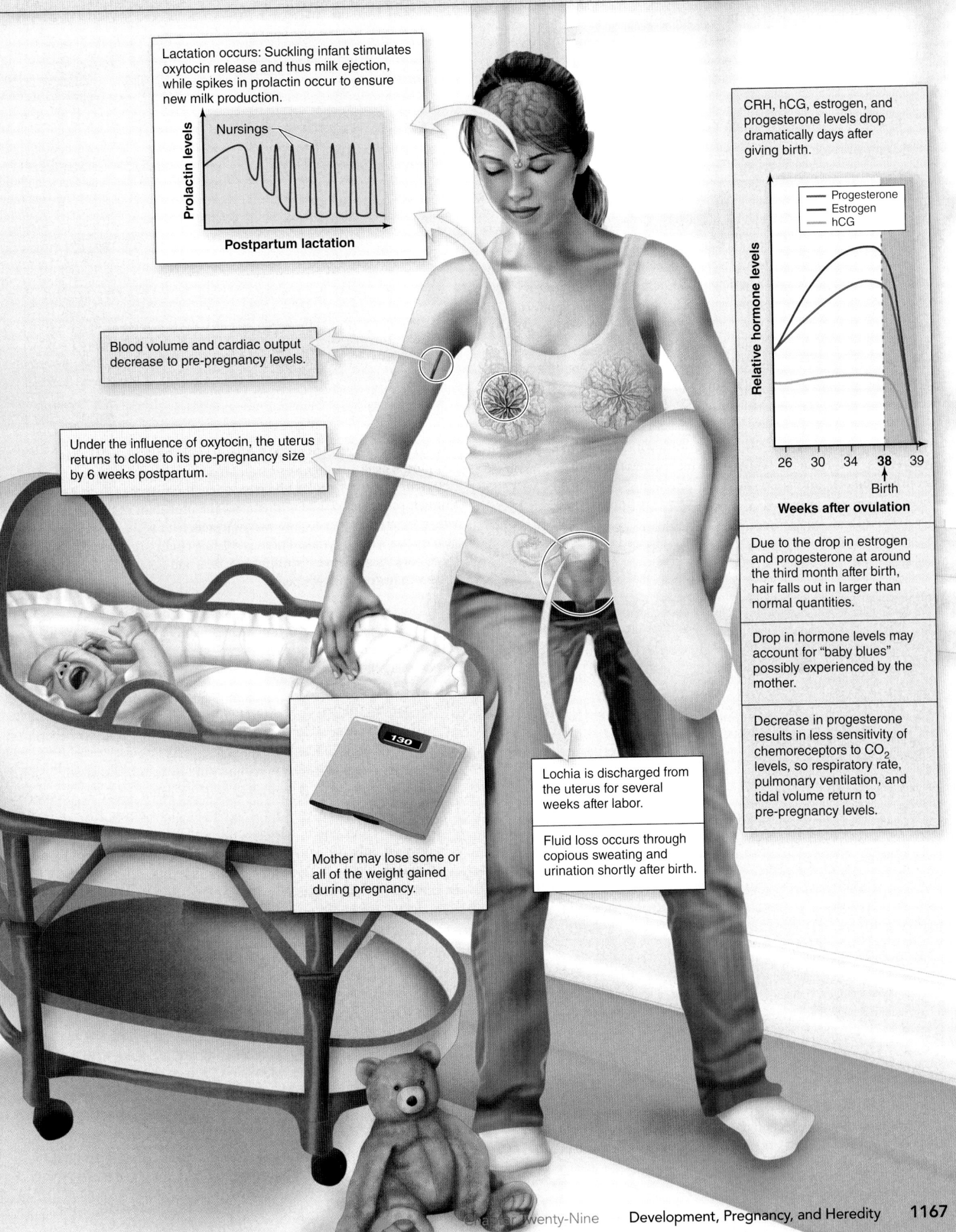

Lactation occurs: Suckling infant stimulates oxytocin release and thus milk ejection, while spikes in prolactin occur to ensure new milk production.

Postpartum lactation

Prolactin levels

Nursings

Blood volume and cardiac output decrease to pre-pregnancy levels.

Under the influence of oxytocin, the uterus returns to close to its pre-pregnancy size by 6 weeks postpartum.

Mother may lose some or all of the weight gained during pregnancy.

130

Lochia is discharged from the uterus for several weeks after labor.

Fluid loss occurs through copious sweating and urination shortly after birth.

CRH, hCG, estrogen, and progesterone levels drop dramatically days after giving birth.

Relative hormone levels

— Progesterone
— Estrogen
— hCG

26   30   34   **38**   39

Birth

Weeks after ovulation

Due to the drop in estrogen and progesterone at around the third month after birth, hair falls out in larger than normal quantities.

Drop in hormone levels may account for "baby blues" possibly experienced by the mother.

Decrease in progesterone results in less sensitivity of chemoreceptors to CO_2 levels, so respiratory rate, pulmonary ventilation, and tidal volume return to pre-pregnancy levels.

29.9 Heredity

Heredity is the transmission of genetic characteristics from parent to child. **Genetics** is the field of biology that studies heredity and its transmission patterns. Here we provide a basic overview of human genetics and introduce you to some of the concepts and vocabulary necessary to understand patterns of inheritance.

29.9a Overview of Human Genetics

LEARNING OBJECTIVE

1. Become familiar with common genetic terminology.

We have already discussed the mechanisms of somatic and sex cell division in previous chapters (see sections 4.9 and 28.2, respectively). Recall that human cells have 23 pairs of chromosomes. The display of the chromosome pairs, which are ordered and arranged by size and similar features (e.g., centromere location, banding patterns), is called the **karyotype** (kar′ē-ō-tīp; *karyon* = nucleus) (**figure 29.18**). Recall from section 28.2 that the paired chromosomes are called **homologous chromosomes** (or *homologues*), as they contain genes for equivalent biological characteristics, such as eye color. Twenty-two of these pairs of chromosomes are called **autosomes;** they have no genes that determine the sex of the individual. The last two chromosomes are called the **sex chromosomes,** which contain genes that specify the sex of an individual. (In mammals, two X chromosomes occur in females, and an X and a Y are found in males.)

Genes were described in section 4.7b as discrete units of DNA that provide the instructions for the production of specific proteins. Genes are located on chromosomes; the specific place where each gene is located on a chromosome is called its **locus** (lō′kūs; pl. = *loci*).

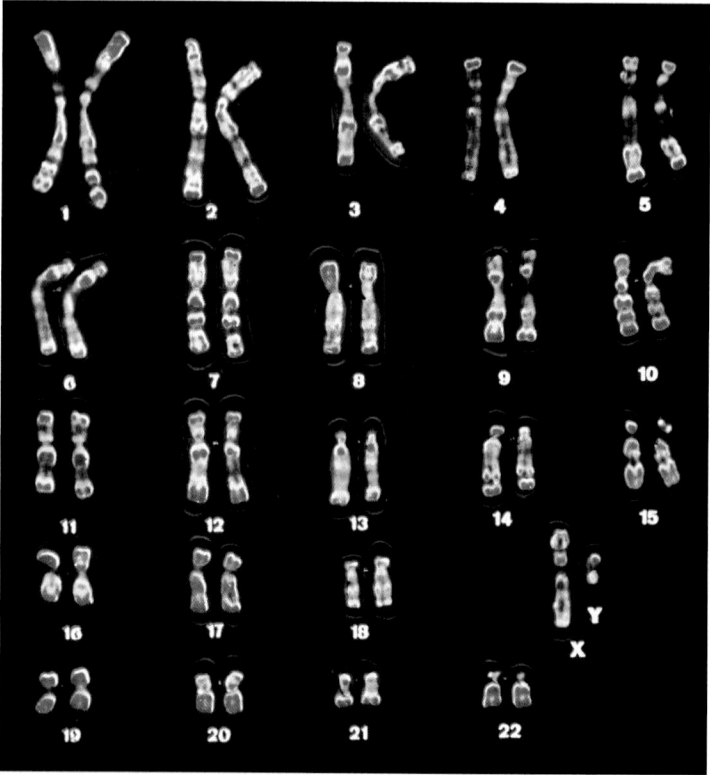

Figure 29.18 Karyotype. A karyotype is a conventional representation of pairs of homologous chromosomes as they appear in metaphase of the cell cycle and arranged by size and other structural features.

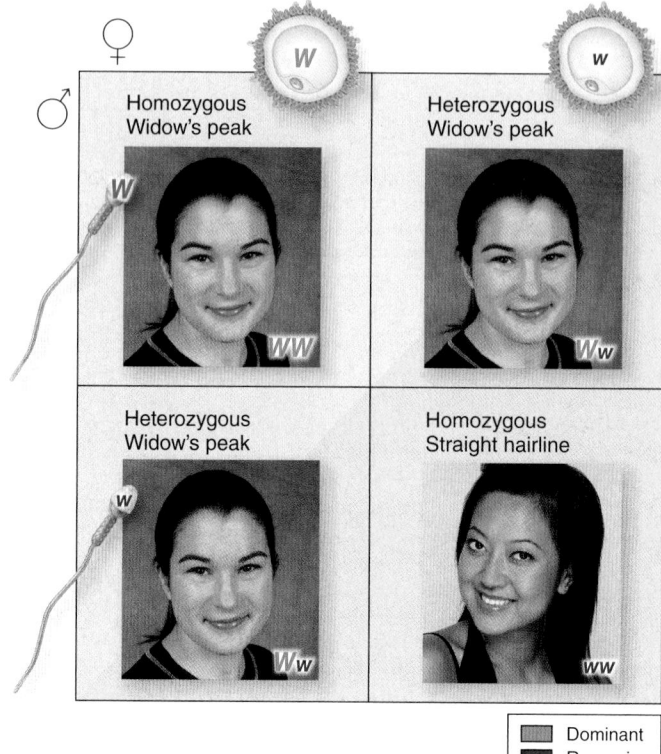

Dominant
Recessive

Figure 29.19 Dominant Versus Recessive Alleles.

A widow's peak is a dominant trait, and the allele for this trait is designated as *W*. A straight hairline allele is designated as *w*. The Punnett square shown here illustrates the probabilities that two parents who are heterozygous for the widow's peak trait will have offspring that also have a widow's peak. There is a 1 in 4 chance that a child born to these parents will have a straight hairline, and a 3 in 4 chance that a child will have a widow's peak.

Variants of one gene found at the same locus on homologous chromosomes are called **alleles** (ă-lēl′; *allelon* = reciprocally). For simple patterns of inheritance, you inherit from your parents a combination of two alleles that determines the expression of a particular trait. For example, the two alleles at a locus may determine whether you have the A or O blood type, or whether your hair is straight or wavy.

How do two alleles code for one trait? In the most straightforward situation, one allele is **dominant,** and the other allele is **recessive.** The dominant allele **expresses,** or physically shows, the trait (e.g., blood type), while the recessive trait is masked. The recessive allele is expressed only if it is present on both homologous chromosomes.

Conventionally, a dominant allele is expressed by a capital letter, and the recessive allele is represented by a lowercase letter. For example, having a widow's peak, in which the hairline on the forehead comes to a point or V in the center, is a dominant trait. So the allele for a widow's peak would be shown as *W*. The recessive allele for a strait forehead hairline would be represented by a *w*. **Figure 29.19** illustrates how these two traits would be expressed. The box used to sort out the inheritance pattern of alleles is called a **Punnett square,** and it shows the specific gene combinations resulting from two parents. The Punnett square can also give the probability that a particular gene combination can occur.

If identical alleles are present, the individual is said to be **homozygous** (hō-mō-zī′gŭs; *homos* = same, *zygotes* = yoke) for the trait. So in this example, *WW* (widow's peak) and *ww* (straight hairline) are both homozygous, because the alleles are identical. Individuals who are **heterozygous** for a trait have both a dominant allele and a recessive allele (e.g., *Ww*), but only the dominant allele is expressed. Expression of the recessive allele may appear to skip generations, because its phenotype is masked by the dominant allele.

In a mating between heterozygous individuals, or heterozygotes, there is a 1 in 4 chance (or 25%) the child would be homozygous for a widow's peak, a 2 in 4 chance (50%) of being heterozygous for a widow's peak, and a 1 in 4 chance (25%) the child would have a straight hairline. If we simply look at how the children may appear, then there is a 3 in 4 chance (75%) that a child would have a widow's peak (includes both *WW* and *Ww*), and a 1 in 4 chance (25%) that a child would have a straight hairline (*ww*).

The genetic makeup of an individual is called the **genotype** (jen′ō-tīp; *genos* = birth). So in this example, individuals may have the genotype *WW*, *Ww*, or *ww*. The physical expression of the genotype, however, is called the **phenotype** (fē′nō-tīp; *phaino* = to appear, show forth). The phenotype for a *WW* and a *Ww* genotype is to exhibit a widow's peak, whereas the phenotype for a *ww* genotype is a straight hairline. Thus, genotype and phenotype describe the difference between the genetic constitution and the actual observed traits of an individual.

 WHAT DID YOU LEARN?

38 What would be the results in terms of the genotype and phenotype if one parent is heterozygous for the widow's peak and the other parent is homozygous for a straight hairline?

29.9b Patterns of Inheritance

 **LEARNING OBJECTIVE**

2. Compare and contrast the types of inheritance patterns.

Not all types of inheritance follow the relatively simple pattern just described. In fact, inheritance of most traits involves the interaction of multiple genes. Here we describe some of the typical patterns of inheritance.

Strict dominant-recessive inheritance has been illustrated in figure 29.19. This type of inheritance is also referred to as *Mendelian inheritance* after Gregor Mendel, who first described this pattern. The dominant allele is always expressed in the phenotype, regardless of whether the individual is homozygous or heterozygous for that trait. Examples of traits that follow this strict dominant-recessive inheritance are found in **table 29.4**. Relatively few human traits follow this pattern of inheritance, but instead may involve the interaction of multiple genes (and thus multiple alleles). In addition, the expression of many traits may be affected by environmental factors.

For some traits, the phenotype of two heterozygous alleles is intermediate between the phenotypes of homozygous dominant or recessive alleles. This trait is said to exhibit **incomplete dominance.** Thus, three phenotypes are possible: (1) the phenotype produced by homozygous dominant alleles, (2) the phenotype produced by homozygous recessive alleles, and (3) the phenotype produced by heterozygous alleles.

One example of incomplete dominance is the sickle cell trait, which affects the shape of erythrocytes. Most individuals have two identical alleles (each designated *A*) that code for the normal hemoglobin A in erythrocytes. The sickling allele (designated *s*) produces an abnormal form of hemoglobin, hemoglobin S. This hemoglobin crystallizes when oxygen levels are low and results in the erythrocyte becoming brittle and forming a sickle-shape. The sickling allele is recessive, and individuals who are homozygous recessive exhibit **sickle cell disease,** also known as *sickle cell anemia* (see Clinical View: "Anemia," in section 18.3). Erythrocytes in these individuals are inefficient; they

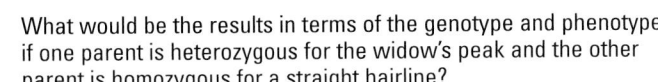 Table 29.4	Traits That Follow a Strict Dominant-Recessive Inheritance	
Dominant Trait	**Recessive Trait**
Dimples in cheeks	Absence of dimples
Cleft chin	Uncleft chin
Widow's peak	Straight hairline
Normal skin and hair pigmentation	Albinism (absence of skin and hair pigmentation)
Huntington disease (leads to progressive neurodegeneration and death)	Absence of Huntington disease
Astigmatism	Nondistorted vision
Achondroplasia (a type of dwarfism)	Normal endochondral ossification of bones
Rh⁺ blood type (e.g., O⁺)	Rh⁻ blood type (e.g., O⁻)
Free earlobes	Attached earlobes
Polydactyly (extra digits)	Normal number of digits
Extensive flexibility in thumb joints ("hitchhiker's thumb")	Inability to extend distal phalanx of thumb past 180 degrees

may fragment in blood vessels and obstruct them, and they don't have as long a life span as normal erythrocytes. Individuals with sickle cell anemia may experience greater morbidity and mortality than individuals who code for normal hemoglobin A.

Heterozygous individuals are said to carry the **sickle cell trait,** whereby under very low oxygen conditions, some (but not all) of their erythrocytes may develop this sickle shape. Researchers have found that individuals with the sickle cell trait experience less mortality from malaria than do individuals without the trait, so the sickle cell trait is somewhat adaptive for individuals who live in malaria-prone environments.

In some cases, two or more alleles appear to be equally dominant, and the trait is said to exhibit **codominant inheritance.** Both alleles are expressed in the phenotype. An example of codominance is the ABO blood group. Blood types A and B are codominant, so if an individual receives an *A* allele from one parent and a *B* allele from

 **INTEGRATE**

CLINICAL VIEW
Familial Hypercholesteremia

Another example of incomplete dominance in humans is the condition **familial hypercholesterolemia (FH)**. In this clinical disorder, a person who has two abnormal alleles lacks receptors for low-density lipoprotein (LDL) and cannot transport LDL from the blood; thus, the individual has very high serum cholesterol and often develops heart disease early in life. These individuals are at greater risk of heart attack at a young age. In contrast, the heterozygote for this condition has about half the number of receptors for LDL. This individual exhibits some LDL transport-related problems related to slightly elevated cholesterol blood levels, but these levels may be managed with diet, exercise, and statin medications.

the other parent, the individual will have an AB blood type. A third allele, *i*, is recessive; the combination *ii* results in the O blood type (see section 18.3b).

Polygenic inheritance, also known as *multiple gene inheritance,* occurs when multiple genes interact to produce a phenotypic trait. These multiple genes may be on the same chromosome or on different chromosomes. Most human traits are the result of polygenic inheritance. For example, eye color was originally believed to be a strict dominant-recessive inheritance involving a single gene, but we now know that eye color is the result of interactions of many genes. Other traits such as height, skin color, predisposition to certain diseases, and many others, are traits under polygenic influence.

 WHAT DID YOU LEARN?

 39 How does codominant inheritance differ from incomplete dominance?

29.9c Sex-Linked Inheritance

 LEARNING OBJECTIVE

3. Describe sex-linked inheritance and give a clinical example of this type of inheritance.,

Sex-linked traits are traits expressed by genes on the X or Y chromosomes. You can see from figure 29.18 that the X and Y chromosomes are physically different—and these differences are also represented in the number of genes each carries. The National Institutes of Health estimates there are between 900 and 1400 genes on the X chromosome, whereas the Y chromosome contains somewhere between 70 and 200 genes. Further, since only males have a Y chromosome, most of the Y chromosome genes code for male-specific development. In contrast, most of the genes on the X chromosome are not involved in sex determination. Thus, although either sex chromosome may be responsible for a sex-linked trait, it is far more common for the X chromosome to be involved.

X-linked traits may be recessive or dominant. An **X-linked recessive trait** is always expressed in a male, because he has only one X chromosome. In contrast, an X-linked recessive trait is expressed in a female *only* if she has two recessive alleles for that trait; typically, the probability of this occurring is very low. A woman is said to be a **carrier** if she has one X-linked recessive allele only and does not exhibit any phenotypic effects. This carrier may pass on the X-linked recessive allele to her offspring; if she passes it on to a female child, the female child also will be a carrier. If she passes the allele to a male child, he will express the X-linked trait.

One example of an X-linked recessive trait is *color blindness,* where in the most common form an individual has trouble distinguishing among different shades of red and green (see Clinical View: "Color Blindness" in section 16.4d). Women are rarely color-blind, because they would have to inherit a recessive allele each from their mother and their father. More often, a woman may be a carrier for the red-green color-blindness allele, but her other, normal allele (on her second X chromosome) allows her to see colors normally. In contrast, if a man inherits a red-green color-blindness allele, he will be color-blind because he has only one X chromosome (and thus, no additional normal allele to counteract the color-blindness allele).

Another example of an X-linked recessive trait is hemophilia A, which is a disorder of blood clotting (see Clinical View: "Bleeding and Blood Clotting Disorders" in section 18.4c). Individuals with this disorder bleed profusely after an injury and may die from this

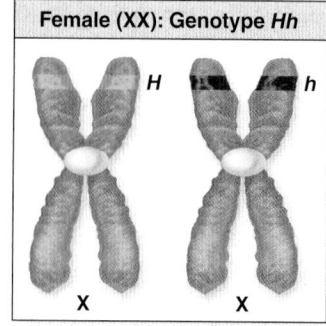
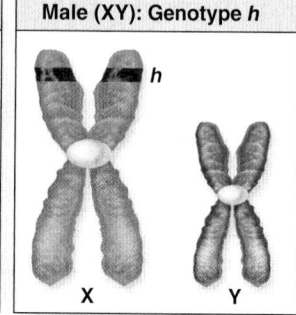

Female (XX): Genotype *Hh*	Male (XY): Genotype *h*
Normal clotting	Hemophiliac

(a)

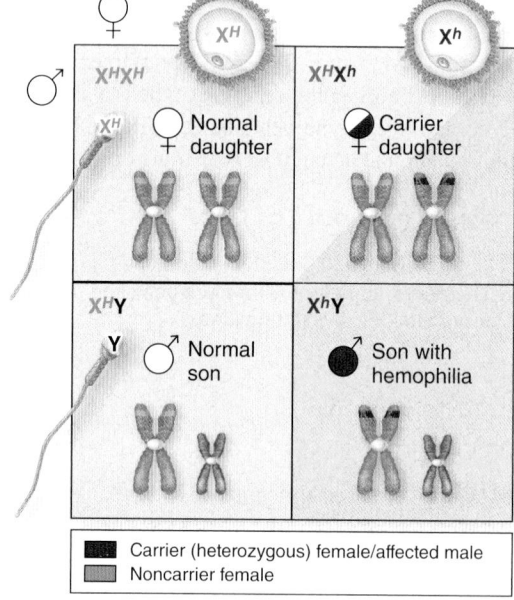

(b)

Figure 29.20 Hemophilia A. (*a*) Hemophilia A is an X-linked recessive trait that leads to abnormal blood clotting and severe bleeding. (*b*) A Punnett square illustrates the possible transmission paths from a heterozygous female and a normal male to their male and female offspring.

extensive bleeding. The normal allele is typically designated *H* and the hemophilic allele is designated *h,* as illustrated in **figure 29.20.** Females may be carriers, in that they will have one *h* allele and one normal *H* allele. Thus, although they do not have hemophilia, they could pass the hemophilic gene to their offspring. A male child who inherits the *h* allele will develop hemophilia, because he has only one X chromosome.

X-linked dominant traits are relatively rare. An X-linked dominant is expressed in both males and females who carry it; however, men typically are more severely affected because they have no normal recessive allele to counteract the effects of the dominant allele. In fact, many male zygotes or embryos with an X-linked dominant disorder are spontaneously aborted (miscarried) because a high rate of lethality is present with these disorders.

 WHAT DID YOU LEARN?

 40 If a woman homozygous for color blindness has children with a man having normal vision, what would be the genotypes and phenotypes of her offspring?

29.9d Penetrance and Environmental Influences on Heredity

LEARNING OBJECTIVE

4. Explain how the environment may influence genetic expression.

The expression of the genotype may be influenced by a variety of factors. Sometimes an individual does not display a particular phenotype, despite having the expected genotype for that phenotypical pattern. The term **penetrance** refers to the percentage of a population with a particular genotype that exhibits the expected phenotype. For example, the autosomal dominant disorder *hereditary pancreatitis* has a penetrance level of 80%. This means that 20% of individuals having this genotype will experience none of the symptoms seen in hereditary pancreatitis.

The environment has variable influence on many genetic traits, especially during embryonic and fetal development. If a pregnant woman is exposed to teratogens such as radiation, drugs, viruses, or alcohol, any of these teratogens have the potential to cause harm to the fetus and interfere with its normal phenotypic development. Mothers that ingest large amounts of alcohol while pregnant are more likely to give birth to children with **fetal alcohol syndrome,** which is characterized by mental retardation and distinctive facial features. The alcohol was a teratogen that affected the phenotypic pattern of the fetus.

However, these changes do not occur in the gametes of offspring, and thus they do not get passed on.

The environment continues to affect phenotype well after birth. Poor nutrition, an environmental effect, may negatively affect an individual's brain development and growth patterns. In the early twentieth century, researchers studied identical twins who were raised in separate households beginning at a young age. Twins with significant environmental differences (e.g., one child received adequate nutrition and an enriched environment, while the other had deficiencies in the environment) displayed phenotypic differences such as dramatic differences in height. Also, some individuals may have alleles associated with a risk of developing certain cancers (e.g., individuals with a mutation of the *BRCA1* gene are at greater risk of developing breast cancer; see Clinical View: "Breast Cancer" in section 28.3f). However, just because an individual has that mutant allele does not guarantee she (or he) will develop cancer. It is the combination of the genetics plus the individual's environment that determine whether an individual will develop cancer. Thus, while our genotype may be determined prior to birth, the phenotypic expression of these genes may be affected by a variety of external factors.

WHAT DID YOU LEARN?

 How would nutrition affect growth where hereditary factors are equal?

CHAPTER SUMMARY

29.1 Overview of the Prenatal Period 1138	• Embryology is the study of development between the fertilization of the secondary oocyte and birth. • The prenatal period is subdivided into the pre-embryonic period (the 2 weeks after fertilization), the embryonic period (between 3 and 8 weeks after fertilization), and the fetal period (the remaining 30 weeks prior to birth).
29.2 Pre-Embryonic Period 1139	• A pre-embryo develops during the first 2 weeks after fertilization. **29.2a Fertilization 1140** • Fertilization is the process whereby two sex cells fuse to form a zygote, which is the first cell of a new organism. • The phases of fertilization are corona radiata penetration, zona pellucida penetration, and fusion of the sperm and oocyte plasma membranes. • When a sperm penetrates the secondary oocyte, the secondary oocyte completes meiosis II and becomes an ovum. • The pronuclei of the ovum and sperm fuse and form a single diploid cell called a zygote. **29.2b Cleavage 1141** • The series of mitotic divisions of the zygote is called cleavage. Once the zygote starts dividing, it is referred to as a pre-embryo. • At the 16-cell stage, the ball of cells is called a morula. • Upon arrival in the uterine lumen, the morula has developed into a single central fluid-filled cavity and is called a blastocyst. • Cells within the blastocyst form the embryoblast (inner cell mass), which gives rise to the embryo proper. The outer ring of cells forms the trophoblast, which contributes to the placenta. **29.2c Implantation 1143** • Implantation consists of attachment and embedding of the blastocyst into the endometrial wall, and cell changes in the trophoblast and uterine epithelium. **29.2d Formation of the Bilaminar Germinal Disc and Extraembryonic Membranes 1144** • During the implantation of the blastocyst into the endometrium of the uterus, cells of the embryoblast differentiate into two layers: the hypoblast and the epiblast. Together, these two layers form a flat disc called a bilaminar germinal disc. • The extraembryonic membranes provide support to the embryo and fetus. **29.2e Development of the Placenta 1145** • The main functions of the placenta are the exchange of respiratory gases between fetal and maternal blood, transmission of maternal antibodies, and hormone production.

(continued on next page)

29.3 Embryonic Period 1146

- The embryonic period extends from week 3 to week 8.

29.3a Gastrulation and Formation of the Primary Germ Layers 1147

- Gastrulation produces three primary germ layers: ectoderm, mesoderm, and endoderm.
- Derivatives of the ectoderm include the nervous system, most exocrine glands, tooth enamel, epidermis, and sense organs.
- Mesoderm forms most muscle, connective tissue, much of the cardiovascular system, urinary system, and reproductive system.
- Endoderm gives rise to the inner lining of the digestive, respiratory, and urinary tracts as well as the thyroid, parathyroid, and thymus glands, portions of the palatine tonsils, and most of the liver, gallbladder, and pancreas.

29.3b Folding of the Embryonic Disc 1148

- The embryonic disc undergoes cephalocaudal and transverse folding, beginning late in the third week.

29.3c Organogenesis 1151

- Almost all of organogenesis (organ development) occurs during the embryonic period.
- During the embryonic period, the embryo is very sensitive to teratogens, substances that can cause birth defects or the death of the embryo.

29.4 Fetal Period 1151

- The time from the beginning of the third month to birth is known as the fetal period. It is characterized by maturation of tissues and organs and rapid growth.

29.5 Effects of Pregnancy on the Mother 1154

- Although certain anatomic and physiologic changes typically occur in a pregnant woman, a great deal of variation exists in how individual women experience pregnancy.

29.5a The Course of Pregnancy 1154

- During the first trimester, the zygote develops into an embryo and then into an early fetus.
- During the second trimester, the fetus grows and the mother's tissues undergo expansion.
- During the third trimester, the fetus undergoes rapid growth, and the mother's body prepares for eventual labor and delivery.

29.5b Hormonal Changes 1154

- Human chorionic gonadotropin peaks during the first trimester and then its levels drop thereafter.
- Estrogen, progesterone, and human placental lactogen (HPL) levels increase dramatically during the second and third trimesters.
- Prolactin and oxytocin levels rise in the later months of pregnancy.

29.5c Uterine and Mammary Gland Changes 1155

- The uterus enlarges throughout pregnancy, reaching the level of the xiphoid process by the ninth month. The uterine lining grows and proliferates throughout the pregnancy.
- The mammary glands enlarge, and the acini begin producing milk in the later months of pregnancy.

29.5d Digestive System, Nutrient, and Metabolic Changes 1156

- The hormone HPL is responsible for the mother metabolizing fatty acids versus glucose, so more glucose remains in the blood for the fetus.
- Some women may experience heartburn, constipation, and morning sickness during part or all of their pregnancy.

29.5e Cardiovascular and Respiratory System Changes 1157

- Plasma volume increases during pregnancy, along with cardiac output (due to an increase in both heart rate and stroke volume).
- Tidal volume increases during pregnancy, although respiratory rate increases only slightly.

29.5f Urinary System Changes 1157

- Glomerular filtration rate (GFR) and urine output increase during pregnancy, and during parts of pregnancy, urination may become more frequent.

29.6 Labor (Parturition) and Delivery 1158

- Labor is the expulsion of the fetus and placenta from the uterus.

29.6a Factors That Lead to Labor 1158

- Rising estrogen levels counteract the calming influence of progesterone on the uterine myometrium, and stimulate the production of oxytocin receptors on the myometrium. Both events facilitate uterine irritability and uterine contractions.

29.6b False Labor 1158

- False labor is irregular uterine contractions that do *not* result in the three stages of labor and the expulsion of the fetus.

29.6c Initiation of True Labor 1159

- True labor is defined as uterine contractions that increase in intensity and regularity and result in changes to the cervix.
- The fetus and the mother secrete oxytocin, which stimulates uterine contractions and also causes the placenta to release prostaglandins, which further stimulate uterine contractions.
- A positive feedback cycle is initiated, whereby the uterine contractions stimulate cervical stretching, which causes increased release of oxytocin and more intense and frequent uterine contractions.

29.6d Stages of True Labor   1160

- The dilation stage begins with the onset of rhythmic uterine contractions and is complete when the cervix dilates to 10 centimeters in diameter.

- The expulsion stage begins when the cervix is fully dilated and ends when the baby is expelled from the mother's body.

- The placental stage is where the placenta and fetal membranes are expelled from the uterus.

- The first breath is initiated by changes in the environment and temperature, which stimulate the neonate's central nervous system.

- The fetal cardiovascular system contains some structures that are modified or cease to function once the baby is born.

- After a woman gives birth, a variety of anatomic and physiologic changes occur to accommodate lactation and bring the mother's body back to its pre-pregnancy form.

29.8a Hormonal Changes   1162

- Estrogen, progesterone, and corticotropin-releasing hormone (CRH) levels all decrease after pregnancy.

- Prolactin levels decrease significantly after pregnancy, but prolactin levels will spike each time the baby nurses.

29.8b Blood Volume and Fluid Changes   1163

- Blood volume decreases after pregnancy, as the mother eliminates excess fluids.

- The mother loses excess fluids via the expulsion of amniotic fluid (during labor), discharge of lochia from the uterus, frequent urination, and copious sweating.

29.8c Lactation   1163

- Lactation occurs via a positive feedback mechanism: The infant suckling stimulates the mother's hypothalamus to release oxytocin, which facilitates milk ejection.

- Breastfeeding stimulates the hypothalamus, which stimulates the anterior pituitary to release prolactin, so new milk will be produced for the next feeding.

29.8d Uterine Changes   1165

- Within the first 6 weeks postpartum, the uterus shrinks to close to its pre-pregnancy size.

- Oxytocin facilitates uterine contractions that help the uterus shrink.

- Heredity is the transmission of genetic characteristics from parent to child.

29.9a Overview of Human Genetics   1168

- Humans have 23 pairs of chromosomes: 22 pairs are autosomes and one pair is the sex chromosomes.

- Alternate forms of a gene are called alleles.

- A dominant allele is one that is expressed if present; a recessive allele is expressed only if present in both alleles.

- The genetic makeup of an individual forms the genotype; the physical expression of the genotype is called the phenotype.

- A homozygous genotype has two identical alleles for the trait; a heterozygous genotype has two different alleles for the trait.

29.9b Patterns of Inheritance   1169

- Strict dominant-recessive inheritance is where the dominant phenotype is expressed, unless the individual is homozygous recessive.

- Incomplete dominance is where there are multiple phenotypes, depending upon whether the individual is homozygous dominant, heterozygous dominant, or homozygous recessive.

- Codominance is where two alleles are equally dominant, and both are expressed in the phenotype.

- Polygenic inheritance is where multiple genes interact to produce a phenotypic trait. Most human traits fall under polygenic inheritance.

29.9c Sex-Linked Inheritance   1170

- Sex-linked traits are traits expressed by genes on the X or Y chromosomes.

- X-linked recessive traits are always expressed in a male offspring, whereas a female offspring is typically a carrier and does not exhibit any phenotypic effects.

- X-linked dominant traits may be exhibited in both males and females, but typically males experience more severe effects.

29.9d Penetrance and Environmental influences on Heredity   1171

- Penetrance is the percentage of a population with a particular genotype that exhibits the expected phenotype.

- The environment has a variable influence on many genetic traits, especially during embryonic and fetal development.

Do You Know the Basics?

_____ 1. The outer layer of the blastocyst that attaches to the wall of the uterus at implantation is called the
- a. amnion.
- b. yolk sac.
- c. cembryoblast.
- d. trophoblast.

_____ 2. At about day 3 after fertilization, the cells of the pre-embryo adhere tightly to each other and increase their surface contact in a process called
- a. implantation.
- b. compaction.
- c. gastrulation.
- d. neurulation.

_____ 3. During gastrulation, cells from the _____ layer of the bilaminar germinal disc migrate and form the three primary germ layers.
- a. notochord
- b. hypoblast
- c. epiblast
- d. mesoblast

_____ 4. The cells of the embryoblast differentiate into the _____ and the _____ .
- a. epiblast; hypoblast
- b. cytotrophoblast; syncytiotrophoblast
- c. amnioblast; epiblast
- d. epiblast; cytotrophoblast

_____ 5. Which of the following is *not* an extraembryonic membrane?
- a. amnion
- b. endoderm
- c. chorion
- d. yolk sac

_____ 6. All of the following cardiovascular events occur during pregnancy *except*
- a. plasma volume increases.
- b. blood pressure increases during the second trimester.
- c. cardiac output increases by 30–50%.
- d. stroke volume increases.

_____ 7. After a woman gives birth, what happens to her levels of estrogen, progesterone, and prolactin?
- a. Progesterone, estrogen, and prolactin levels all decrease.
- b. Progesterone and prolactin levels decrease, and estrogen levels increase.
- c. Progesterone, estrogen, and prolactin levels all increase.
- d. Estrogen and prolactin levels increase, and progesterone levels decrease.

_____ 8. Freckles are considered to be a dominant trait. If a mother without freckles and a father without freckles have a child, what are the odds that the child will have freckles?
- a. 1 in 4 chance (25%)
- b. 2 in 4 chance (50%)
- c. 3 in 4 chance (75%)
- d. It is impossible for the child to develop freckles (0% chance).

_____ 9. Skin color is a trait that is determined by
- a. strict dominant-recessive inheritance.
- b. incomplete dominant inheritance.
- c. codominant inheritance.
- d. polygenic inheritance.

_____ 10. A woman is a carrier for the color-blindness gene, found on the X chromosome. She does not have color blindness, however. What are the chances that her female child will also inherit this gene?
- a. no chance (0%)
- b. 1 in 4 chance (25%)
- c. 2 in 4 chance (50%)
- d. 100% chance

11. Briefly describe the process of fertilization, mentioning a hallmark event that occurs in each phase.

12. List the five regions of the mesoderm, and identify some major body parts derived from each region.

13. Explain why teratogens are especially harmful to the developing organism during the embryonic period. What events occur during this period?

14. Describe the differences between the embryonic period and the fetal period.

15. List the major hormone levels that change during pregnancy and their major functions during pregnancy.

16. List the three stages of labor, and describe a major event that occurs in each stage.

17. Explain the positive feedback mechanism involved with lactation and milk letdown, and describe the function of the hormones oxytocin and prolactin during this process.

18. Describe the various ways by which the mother's body expels the excess fluids gained during pregnancy.

19. Compare and contrast strict dominant-recessive inheritance, incomplete dominance, codominance, and polygenic inheritance, and give an example of each.

20. Explain the difference between X-linked recessive traits and X-linked dominant traits. What are the possibilities that a female child will be affected and express each of the traits?

Can You Apply What You've Learned?

Use the following paragraph to answer questions 1–5.

Ashley is a 29-year-old pregnant woman who is in the second trimester of her pregnancy. At her regular OB/GYN checkup, the doctor records the following about Ashley:

Heart rate: 92

Urinalysis: Negative for bacteria and protein in urine

Weight gain since beginning of pregnancy: 15 pounds

Blood pressure: 115/79 mm Hg

Uterine fundus height: At a midpoint between the umbilicus and the xiphoid process of the sternum

Blood glucose level: 95

1. Which of the following medical assessments is abnormal and would be of concern for the physician?

   a. Blood pressure is above normal.

   b. Urinalysis results are abnormal.

   c. Ashley's blood glucose levels are above normal.

   d. All of the medical findings are within normal ranges for a pregnant woman.

2. The physician determines that Ashley is in her twentieth week of pregnancy. Which of the following best describes the development of the fetus at this time (refer to table 29.3)?

   a. The crown-rump length is 19 centimeters.

   b. The heart and lungs have fully developed, but the GI tract has yet to form.

   c. The testes have descended into the scrotum.

   d. The average weight of the fetus at this time is about 2.5 kilograms.

3. Months pass as Ashley has an uneventful pregnancy. Eventually, Ashley goes into labor and goes to the hospital to give birth. As Ashley progresses through the first stage of labor, all of the following occur *except*

   a. the fetus's head pushes against the cervix, promoting the cervical dilation.

   b. oxytocin levels decrease as labor progresses.

   c. the amniotic sac may rupture.

   d. contractions become more frequent and more intense.

4. Ashley gives birth to a 7.5-pound baby boy. In the first few weeks after giving birth, Ashley notices she develops cramping in the pelvic region whenever she breastfeeds her child. What is the likely cause of the cramping?

   a. The suckling stimulates oxytocin production, which both releases the milk and causes uterine contractions.

   b. Breastfeeding results in a rise in prolactin, which also causes cramping of the abdominal muscles.

   c. The drop in estrogen and progesterone levels causes the pelvic cramping.

   d. The tugging on the breast (caused by the suckling) pulls on the abdominal muscles, causing the cramping.

5. As Ashley cares for her son, she notices some problems in his development. During one of the baby's checkups, Ashley voices her concerns to the pediatrician. The pediatrician performs a series of assessments on the baby and orders a genetic workup. The genetic test results demonstrate that the baby has Duchenne muscular dystrophy, which is an X-linked recessive trait.

   Based on your knowledge of heredity, if the baby boy has Duchenne muscular dystrophy, then what else must be true?

   a. The baby's father is a carrier of the Duchenne muscular dystrophy allele and passed along this trait to the baby.

   b. Duchenne muscular dystrophy is an autosomal-dominant disorder.

   c. Ashley is a carrier of the Duchenne muscular dystrophy allele but does not display symptoms because she has two X chromosomes, one of which is normal.

   d. If Ashley had a baby girl (instead of a baby boy), the baby girl would be at equal risk for having Duchenne muscular dystrophy.

Can You Synthesize What You've Learned?

1. In the late 1960s, a number of pregnant women in Europe and Canada were prescribed a drug called thalidomide. Many of these women gave birth to children with amelia (no limbs) or meromelia (malformed upper and/or lower limbs). It was later discovered that thalidomide is a teratogen that can cause limb defects in an unborn baby. Based on this information, during what period of their pregnancy (pre-embryonic, embryonic, or fetal) do you think these women took thalidomide? During which of these periods would thalidomide cause the most harm to limb development?

2. A 22-year-old woman consumes large quantities of alcohol at a party and loses consciousness. Three weeks later, she misses her second consecutive period, and a pregnancy test is positive. Should she be concerned about the effects of her binge-drinking episode on her baby?

3. Simon and Carla are siblings who recently discovered that their father was diagnosed with Huntington disease. This disease is autosomal dominant and results in gradual neurologic degeneration and death, usually within 5 years of diagnosis. There is no cure for the disease. Their mother does not have the disease. Both Simon and Carla discuss the possibility of getting tested to see if they will develop Huntington disease as well. Based on your knowledge of heredity and genetics, what is the likelihood that Simon or Carla will develop Huntington disease?

INTEGRATE

ONLINE STUDY TOOLS

The following study aids may be accessed through Connect.

Clinical Case Study: Queen Victoria's Family Tree

Interactive Questions: This chapter's content is served up in a number of multimedia question formats for student study.

LearnSmart: Topics and terminology include overview of prenatal development; pre-embryonic period; embryonic period; fetal

period; effects of pregnancy on the mother; labor (parturition) and delivery; postnatal changes for the newborn; changes to the mother after delivery; heredity

Anatomy & Physiology Revealed: Ovulation through implantation

Animations: Fetal development and risk

Answers to "What Do You Think?"

CHAPTER 1

1. When you digest a meal, you are utilizing primarily catabolic chemical reactions, because the main goal is to break down larger molecules (such as starches in bread) into smaller molecules (such as simple sugars) that can be absorbed.

2. If you didn't have the lubricating serous fluid, there would be increased friction and it would be quite painful whenever your organs moved. For example, the illness *pleurisy* (inflammation of the pleura) makes it very painful to breathe, because the pleural membranes are inflamed and the serous fluid cannot appropriately lubricate the membranes.

CHAPTER 2

1. The chemical shorthand for oxygen is written as: $^{16}_{8}O$

2. Magnesium is in column IIA. The atomic structures of the elements in column IIA all contain two electrons in the outer shell. Consequently, a magnesium ion will donate two electrons and will develop a (+2) charge, written as Mg^{2+}.

3. A fatty acid is a nonpolar molecule because it contains predominantly C–C and C–H covalent nonpolar bonds. Nonpolar molecules are hydrophobic and do not dissolve in water.

4. Stomach acid with a pH of 2 is 100,000 times more acidic than pure water with a pH of 7 (a pH of 2 is 5 logarithmic units from a pH of 7; thus it is calculated as $10 \times 10 \times 10 \times 10 \times 10$). Without protection (a mucus coating and rapid

turnover of cells lining the stomach), the stomach wall would be destroyed.

5. Protein function is dependent upon the retention of their normal three-dimensional shape. An increase in temperature can weaken the intramolecular attractions between the amino acids in the primary structure of the protein strand, causing the protein to unfold or denature such that it can no longer function normally.

CHAPTER 3

1. NAD^+ is reduced to become NADH because NAD^+ has gained electrons (2 e^- and H^+) to become NADH.

2. Recall from section 2.3a that isomers are molecules that contain the same number and kind of atoms, but in a different arrangement. An isomerase converts a molecule to one of its isomers (different forms) by rearranging the atoms within the molecule.

3. During glycolysis, there is a net of 2 ATP molecules formed (2 ATP "invested" in steps 1 and 3; then 4 ATP generated as steps 7 and 10 [which occurs twice]); also, 2 NADH molecules are formed in step 6 (which occurs twice). So, the net energy transfer forms 2 ATP molecules and 2 NADH molecules.

4. The enzymatic pathway of the citric acid cycle is considered a "cycle" because it begins and ends with oxaloacetic acid (oxaloacetic acid is involved in the first step and then regenerated in the last step).

5. ATP total:
   Glycolysis   2 ATP   2 NADH
   Intermediate Stage   2 NADH (per original glucose)
   Citric Acid Cycle   2 ATP   6 NADH   2 $FADH_2$ (per original glucose)
   Conversion: NADH = 3 ATP; $FADH_2$ = 2 ATP
   Glycolysis   8 ATP
   Intermediate Stage   6 ATP
   Citric Acid Cycle   24 ATP
                        38 ATP

Thus, the total number of ATP is 38 ATP. However, 8 ATP molecules are used to produced the 38 ATP molecules. To determine the net ATP produced from one glucose molecule, it is necessary to subtract energy required (ATP molecules) needed for shipping (1) pyruvate from the cytosol into mitochondria, (2) phosphate and ADP into mitochondria for its use in ATP synthesis, and (3) NADH produced during glycolysis into mitochondria. The *net* ATP production is **30 ATP** from one glucose molecule.

CHAPTER 4

1. The difference in water concentration between a cell (with 0.9% solute is 99.1% water) that is immersed in pure water (with 0% solutes and 100% water) is: 100 − 99.1 = 0.9%.

    The difference in water concentration between a cell (with 0.9% solute is 99.1% water) that is immersed in 0.2% NaCl solution (with 0.2% solutes and 99.8% water) is: 99.8 − 99.1 = 0.7%.

    Thus, there is a greater difference (or steeper gradient) in the water concentration in the first set-up (cell in pure water), thus greater amounts of water will move generating a larger osmotic pressure.

2. The function of lysosomes is to digest unwanted or unneeded organic molecules within the cell. Without lysosomes these substances would accumulate and interfere with normal cell function. A cell that lacked lysosomes would be unable to undergo autophagy or autolysis. A cell cannot survive without lysosomes.

3. If a gene is mutated, the mRNA that is produced from it during transcription will have a different ribonucleotide sequence. The abnormal mRNA may result in the production of a protein during translation with an abnormal amino acid sequence. Because the

sequence and interactions of the amino acids influences the folding and ultimately the three-dimensional shape of the protein, the protein may not function normally because its function is dependent upon its shape.

4. DNA replication forms a new molecule of DNA using deoxyribonucleotides. It involves all DNA being copied. During transcription a new RNA molecule is formed using ribonucleotides. Only a small segment of the DNA is opened and "read."

CHAPTER 5

1. If epithelium contained blood vessels, the "gatekeeper" function of selective permeability would be compromised. Materials would be able to enter the body by entering the blood without passing through the epithelial cells.

2. All types of stratified epithelium (stratified squamous, stratified columnar, stratified cuboidal) and transitional epithelium are suited for protection, because they have multiple layers of cells.

3. You've damaged dense regular connective tissue when you sprain your ankle.

4. The striations in skeletal and cardiac muscle indicate the organized pattern of protein filaments in sarcomeres. Smooth muscle has these same protein filaments, but they are not arranged in the same ordered pattern. Thus, smooth muscle lacks striations.

CHAPTER 6

1. Thick skin is found on the palms of the hands and soles of the feet. Secretions from sebaceous glands would make these areas slippery, which would interfere with grasping items and walking. The presence of hair in these areas would similarly interfere with these functions.

2. These children already were not getting enough vitamin D in their diet, and they were spending all of their daylight hours indoors. Since the children were not exposed to much sunlight, their skin could not synthesize vitamin D from the UV rays of the sun. Without adequate amounts of vitamin D, the children developed rickets. Rickets is on the rise in the United States again among poor urban children who spend little time outdoors and who drink soda instead of milk.

3. When we are frightened or nervous, the sympathetic division of the autonomic nervous system is stimulated, which in turn stimulates sweat production and secretion. This is why our palms and other regions of the epidermis may become moist in nervous or frightening situations.

CHAPTER 7

1. The numerous complex steps in endochondral bone formation ensure that a working bone may be formed for a newborn and later develop into a working adult bone. Having a periosteal bone collar, epiphyseal plates, and constant bone remodeling ensures that the bone can reshape itself, grow in both width and length, and develop a medullary cavity so that it will not weigh too much.

2. A physician will look for active epiphyseal plates (indicating the bones are still growing in length). The epiphyseal plate becomes ossified once an individual reaches maturity. Presence of an epiphyseal line indicates that lengthwise bone growth is complete.

3. Testosterone initially accelerates bone growth. However, bone deposition by osteoblasts occurs at a greater rate than cartilage growth within the epiphyseal plate. This ultimately results in the premature closure of the growth plate. Because the young boy was taking anabolic steroids prior to puberty his growth plates fused at a younger age than would have occurred if he had not taken them. Consequently, he is shorter than he would have been.

CHAPTER 8

1. The shape of the skull is very complex, and it would be unlikely for a single bone to be formed in this shape. Multiple bones connected by sutures allow the size of the skull to increase as the child grows. If the skull were a single bone, it would be very difficult for it to change shape and increase in size during growth.

2. Male sex hormones and increased growth beginning at puberty cause the skull to become more robust, have more prominent bony features and a more squared-off jaw.

3. A typical cervical vertebra (C_3–C_6) has transverse foramina and a bifid spinous process. Both the lumbar and thoracic vertebra lack the bifid spinous process and transverse foramina.

4. The acetabulum of the os coxae is deeper than the glenoid cavity of the scapula. Consequently, the pelvic girdle maintains a stronger, more tightly fitting bony connection with the femur of the lower limbs than the pectoral girdle (scapula and clavicle) does with the humerus of the upper limbs.

5. The medial and lateral malleoli of the leg are similar to the styloid processes of the radius and ulna.

CHAPTER 9

1. A synchondrosis is a synarthrosis because as the bone ends are growing, they must not be allowed to move in relation to one another. For example, if the epiphysis and diaphysis move along the epiphyseal plate, bone growth is compromised and the bone may become misshapen.

2. The plane joint, also called a gliding joint, is the least mobile of the two joints. The uniaxial gliding joint occurs where two bones have limited side-to-side movements in a single plane. This is more stable than the multiaxial ball-and-socket joint that consists of a bone with a spherical head articulating in a shallow cuplike socket.

3. When sitting upright in a chair, both the hip joints and knee joints are flexed. Flexion is defined as a decrease in the angle between the bones in an anterior-posterior plane.

CHAPTER 10

1. The structures of skeletal muscle from largest to smallest are: muscle, fascicle, muscle fiber.

2. If the muscle is contracted: (a) the width of the A band remains the same as it represents the length of the myosin filaments, which do not change in length; (b) the width of the H zone shortens as the actin and myosin myofilaments slide past one another; (c) the Z discs in one sarcomere move closer together as the sarcomere shortens, and (d) the width of the I band decreases.

3. As long as excess calcium ions are present, the cell would stay contracted. ATP is needed to release the myosin head from the actin binding sites that would allow the muscle to return to a relaxed state.

4. Creatine kinase levels in the blood would be predicted to be directly correlated with the amount of muscle tissue damage. The greater the amount of skeletal muscle tissue damage, the higher the blood levels of creatine kinase.

5. The duration of a muscle twitch of an extrinsic eye muscle is shorter than it is in the gastrocnemius muscle because the extrinsic eye muscle is composed predominantly of fast glycolytic fibers and the gastrocnemius is composed of a mixture of fast glycolytic fibers and slow oxidative fibers.

CHAPTER 11

1. Since the omohyoid attaches to the scapula (which is part of the shoulder), the prefix *omo-* means "shoulder."

2. When you breathe deeply, the diaphragm contracts and pushes inferiorly on the GI tract (abdominal viscera). If these viscera are bulging with food, the diaphragm has less room to contracting fully, making it hard to take deep breaths.

3. The brachialis is an anterior arm muscle. Because anterior arm muscles tend to flex the elbow joint, we can surmise that the brachialis flexes the elbow joint.

4. Remember that there are leg muscles that also move the toes. In this case, the extensor digitorum longus (a leg muscle) attaches to toes 2–5 and helps move them all.

CHAPTER 12

1. The generation of IPSPs, which are produced in receptive segments either by the loss of K^+ or the gain of Cl^-, make it less likely that a nerve signal will be sent because the resting membrane potential is made relatively more negative inside the neuron, which changes the membrane potential so that it is further away from the threshold (the value that must be reached for a nerve signal to be initiated).

2. A drug that blocks acetylcholinesterase (an enzyme that breaks down acetylcholine molecules in the synaptic cleft) increases the amount of time acetylcholine molecules remain in the synaptic cleft. This allows an increase in the duration of stimulation of neurons of the brain. Given that acetylcholine increases attention (or mental alertness), a drug that increases the half-life of acetylcholine in the synaptic cleft would be predicted to increase attention.

CHAPTER 13

1. Although all the meningeal layers give some support and protection to the brain, the dura mater provides the most support. This layer is the thickest and most durable, and it also forms the cranial dural septa that support the brain components.

2. Cutting the corpus callosum dramatically reduces communication between the right and left cerebral hemispheres, but some communication is maintained between the hemispheres via the much smaller anterior and posterior commissures.

3. If the primary somatosensory cortex was damaged, general somatic sensory information would not be perceived. An individual likely would experience anesthesia (numbness) in the region that transmits sensations to the damaged cortex. In contrast, if the somatosensory association area was damaged, a person could still detect sensations, but would be unable to tell the *difference* in those sensations. A person may perceive that an object is in one's hand, but would be unable to tell if that object was smooth or rough, cold or hot, round or square.

4. If there were no thalamus, the cerebrum would still receive sensory stimuli, but the information would not be decoded first. So, the cerebrum would not be able to distinguish taste information from touch or vision information. In addition, there would be no filtering of sensory information so the person would experience sensory overload.

5. Severe injury to the medulla oblongata would most likely cause death, because the medulla oblongata is responsible for basic reflex and life functions, including breathing and heartbeat.

CHAPTER 14

1. The two layers of the cranial dura mater split to form the dural venous sinuses, which are large veins that drain blood away from the brain. The spinal cord lacks dural venous sinuses, so it does not need two layers of dura.

2. An anterior ramus is larger than a posterior ramus because the posterior rami only innervate deep back muscles and the skin of the back, whereas the anterior rami innervate almost all other body structures (e.g., the limbs and the anterior and lateral trunk).

3. A nerve plexus houses axons from several different spinal nerves. Thus, damage to a single segment of the spinal cord or damage to a single spinal nerve generally does not result in complete loss of innervation to a particular muscle or region of skin.

CHAPTER 15

1. The pterygopalatine ganglion is nicknamed the "hay fever ganglion" because when it is overstimulated, it causes some of the classic allergic reactions, including watery eyes, runny and itchy nose, sneezing, and scratchy throat.

2. Sympathetic innervation causes vasoconstriction of most blood vessels. When blood vessels are constricted, it takes more force and pressure to pump blood through the vessels, so blood pressure rises.

3. Nicotine binds to nicotinic receptors. Many neurons in the brain and peripheral nervous system are affected, but the overall effect usually is an increase in blood pressure and heart rate, and vasoconstriction (in other words, a general stimulation of the sympathetic division). Nicotine also works on the reward centers of the brain, stimulating a release of the neurotransmitter dopamine, which is responsible for the pleasurable, relaxed sensations most smokers get while lighting up. These effects are short lived, so a smoker must smoke fairly continuously to continue to get the increased dopamine in his or her system.

CHAPTER 16

1. Olfaction plays a major role in detecting tastes. Most taste is due to our perception of the odor, rather than the taste, of the food. If your nose is stuffed up and you can't smell the food, your taste perception is impaired as well.

2. A deer sacrifices depth perception for a wider field of vision, thus it has a greater visual range.

3. Air pressure increases when the airplane descends, which is why you may feel greater pressure outside your ears. The "popping" noise results from your auditory tubes opening and equalizing the pressure on either side of the tympanic membrane.

CHAPTER 17

1. Malfunction of the liver and kidneys can result in an imbalance of hormones within the blood. One would predict an increase in blood concentration of a hormone given the role of the liver and kidneys in eliminating hormones, wastes, and other metabolites.

2. If a drug binds to a specific hormone receptor of a cell to initiate its affect, the cell will down regulate receptors. Over time, more of the drug will be required for the same response.

3. Increased lipolysis is caused by growth hormone. This results in loss of stored adipose connective tissue and the "thinning out" experienced by many during puberty.

4. Iodine is absorbed into the blood and ultimately used for thyroid hormone synthesis.

5. The affected individual with hyperthyroidism would have a high temperature, elevated pulse, elevated breathing rate, and a thin body.

6. Elevated insulin levels typically results in decreased levels of all nutrients in the blood and an increase in the storage form of these molecules within the cells of the body. Thus, elevated serum insulin leads to protein anabolism in the muscle cells, resulting in increased muscle bulk. An insulin overdose may lead to dangerously low serum glucose levels and ATP levels. If ATP cannot be generated at the rate needed, death could result. (It also can result in a potassium imbalance, as described in chapter 25.)

CHAPTER 18

1. A woman who has 5 L of blood would donate approximately 10% of her blood. Low-weight individuals are not allowed to donate blood because these individuals have relatively less blood volume, and by donating 1 pint of blood they would be donating much more than 10% of their total blood volume.

2. By losing its nucleus and other organelles the erythrocyte has more space for hemoglobin, enabling the erythrocyte to bind to and transport greater amounts of O_2 and CO_2 more efficiently. Because the erythrocyte has lost its nucleus it will no longer be able to synthesize proteins or divide via mitosis. The oxygen being transported in an erythrocyte will not be used up by aerobic cellular respiration (since erythrocytes don't have mitochondria).

3. A person with type O⁻ blood is considered a "universal donor" because his or her erythrocytes have no surface antigens. Without surface antigens, the type O⁻ erythrocytes will not be destroyed through agglutination and hemolysis by antibodies in the recipient's plasma. Likewise, a person with type AB⁺ blood is considered a "universal recipient" because his or her blood plasma has no antibodies to the ABO or Rh blood types. Thus, the AB⁺ recipient may receive any type of blood and not worry about the donor's erythrocytes being destroyed.

4. The formation of the platelet plug is an example of positive feedback. The platelets release chemicals that stimulate more platelets to arrive at the site of injury and release their chemicals.

5. An individual who is confined to a wheelchair may have impaired blood flow, as the skeletal muscle pump is less effective. If blood is not moving through the blood vessels properly and begins to pool in the vessels, the clotting cascade may be initiated.

CHAPTER 19

1. The aorta and pulmonary veins are vessels attached to the heart that contain oxygenated blood. The aorta is an artery and the pulmonary veins are veins.

2. Coronary arteries are compressed more often as heart rate increases.

3. Slow Na⁺ voltage-gated channels are unique to nodal cells. These channels open and Na⁺ enters the nodal cells. When threshold is reached, Ca²⁺ voltage-gated channels open and Ca²⁺ enters the cells and causes depolarization.

4. Ca²⁺ channel blockers decrease the intracellular calcium in a cardiac muscle cell, thus decreasing the contractility of the heart. Ca²⁺ channel blockers may also block calcium movement into conduction cells, decreasing heart rate. (Blood pressure is lowered as a result.)

CHAPTER 20

1. A decrease in blood pressure (blood hydrostatic pressure) is dependent on the pumping action of the heart, resistance that occurs as blood moves through the blood vessels, and the total blood volume. A significant decrease in any of these (e.g., congestive heart failure, severe hemorrhage) can result in lowering blood pressure to such an extent that blood hydrostatic pressure is not sufficient to overcome blood colloid osmotic pressure. When this occurs, filtration (and capillary exchange) ceases.

2. A smoker likely would have elevated blood pressure, since nicotine increases firing rate of the SA node and increased contractility of myocardial cells, which increases cardiac output. Nicotine also causes vasoconstriction of arterioles, which increases resistance. Both an increase in cardiac output and an increase in resistance result in increasing blood pressure.

3. ACE inhibitors are a medication given to treat high blood pressure (hypertension). These medications interfere with the production of angiotensin II (a potent vasoconstrictor), thus helping to prevent an increase in resistance. Angiotensin II also regulates blood volume by influencing urine production. A decrease in angiotensin II would increase fluid loss from the blood and blood volume decreases. By preventing an increase in resistance and decreasing blood volume, ACE inhibitors help to lower blood pressure.

4. If the left ulnar artery were cut, the left hand and fingers could still receive blood via the left radial artery, since both vessels contribute to the superficial and deep palmar arches.

CHAPTER 21

1. The removal of lymph nodes interferes with the drainage of fluid from specific regions of the body and will result in lymphedema.

2. The spleen, located in the left upper quadrant, is a soft organ that is not protected by bone. It could easily be ruptured by the pressure of the seatbelt pushing against the abdomen in an automobile accident.

CHAPTER 22

1. Inflammation can be observed if the area is red and swollen. It also may feel warm to the touch.

2. If an antigen mutates, it is likely that it will not be recognized by the same lymphocytes. Each lymphocyte recognizes a specific molecular shape of an antigen. (An example of a virus that mutates often is HIV, the virus that causes AIDS. It is for this reason that vaccines have not been produced against HIV.)

3. Helper T cells play an essential role in both the innate and adaptive immune responses. These cells function to activate B cells, cytotoxic T cells, as well as macrophages and NK cells. Without the proper functioning of all of these cells, an individual is highly susceptible to infection.

CHAPTER 23

1. A deviated nasal septum occurs when the nasal septum—the bone and cartilage that divide the nasal cavity of the nose in half—is off-center so one side of the nasal cavity is larger than the other. This alters the normal flow of air through the nose, and if the narrower side becomes blocked, nasal congestion and sinus problems may result.

2. The epithelium changes because a stratified squamous epithelium is more sturdy and protective against smoke than a pseudostratified, ciliated columnar epithelium. Unfortunately, since a stratified squamous epithelium lacks both goblet cells and cilia, there is less mucus produced, and no cilia are present to propel particles away from the bronchi towards the pharynx. Thus, the main way to eliminate the particles is by coughing, leading to the chronic "smoker's cough."

3. The phrenic nerves extend from each cervical plexus to the diaphragm. Since the cervical plexus includes axons of C3–C5, spinal cord injuries at or above C2 prevent nerve stimulation of the diaphragm, and breathing stops. Spinal cord injuries between C6 and T12 result in nerve signals being sent along the phrenic nerves but not along all intercostal nerves. The degree of loss will depend on where the level of injury has occurred. The further down the spinal cord, the closer the breathing will be to normal (since additional intercostal nerves will be functioning). Spinal cord injuries below T12 should not normally interfere with normal quiet breathing, because these represent injuries that do not interfere with nerve signals being relayed along the phrenic or intercostal nerves.

4. Epinephrine causes bronchodilation; consequently resistance is decreased, with a subsequent increase in airflow.

5. No, not all of the air that we breathe in is available for gas exchange. Only the air that reaches the alveoli is available for gas exchange. There is no gas exchange with air in the anatomic dead space (air that remains in the conducting zone).

6. Blood PO_2 is lower in systemic capillaries than when it left pulmonary capillaries, because bronchial veins dump small amounts of deoxygenated blood into the pulmonary veins prior to the blood being returned to the heart and subsequently pumped by the left ventricle through the systemic circuit.

7. Since hemoglobin is 98% saturated with oxygen if PO_2 = 104 mm Hg (normal value at sea level), additional administered oxygen can only increase the % saturation of hemoglobin 2%. This can occur only with an increase from 1 atm to 3 atm, pressures that are only generally reached in hyperbaric oxygen chambers. (In addition, the limiting factor is usually not the ability to load oxygen in the lungs but the ability to deliver the oxygen to the body's tissues.) Therefore, it would not be expected that the athletic performance would be improved.

8. During exercise, the PO_2 decreases as cells engage in elevated levels of cellular respiration. The decrease in PO_2 results in more oxygen delivered to the cells. As a result the oxygen reserve (the amount remaining attached to hemoglobin) is lower.

9. Exercise increases venous return of blood and lymph by the action of both the muscular pump and respiratory pump. Subsequently, more blood enters the atria triggering the atrial (Bainbridge) reflex that causes increased sympathetic output from the cardiac center to the SA node, thus increasing heart rate.

CHAPTER 24

1. A loss of kidney function could: (a) impair the ability of the kidneys to filter and eliminate wastes; (b) induce anemia as erythropoietin production may be altered; (c) decrease urine production with an accompanying increase in blood volume and blood pressure; (d) lead to a disruption of the normal pH balance in the blood as the kidneys play a role in maintaining acid-base (pH) balance.

2. A substance that is not filtered remains in the blood and exits the glomerulus by the efferent arteriole.

3. Cirrhosis of the liver may lead to a drop of blood colloid osmotic pressure (OP_g). OP_g opposes the filtration pressure and draws fluid back into the glomerulus. A decrease in OP_g will induce less reabsorption of fluid and therefore more filtrate will be formed.

4. Glycosuria is the excretion of glucose in the urine and it is a classical sign of diabetes mellitus. Glucose molecules are not reabsorbed and they act as an osmotic diuretic, pulling water into the tubular fluid and causing loss of fluid in the urine.

5. An individual with renal disease that results in damage to the filtration membrane will have a lower concentration of plasma proteins because tubule cellular structures involved in protein uptake will be saturated and unable to reclaim all proteins that have been filtered. If renal filtration is lower than normal, there is a decrease in filtration of small plasma proteins resulting in an increase in plasma proteins.

6. If a substance in the blood is filtered and not all of it is completely reabsorbed back into the blood, the level of the substance in the blood decreases. One of the critical functions of the kidney is to regulate blood levels of specific substances.

7. When we are lying down, gravity is unable to passively transport urine to the urinary bladder. Thus, peristalsis is needed so that urine can be moved actively from the ureters to the urinary bladder, no matter the position of the body.

CHAPTER 25

1. This practice should *not* be encouraged. It is possible to "fool" the thirst center that you are hydrated simply by taking fluid into the mouth without swallowing the fluid, because even the brief presence of water in the mouth may be misinterpreted as increased salivary secretions because mucous membranes in the mouth become moist. Thus, the 'apparent' hydrated state of the body is not the correct state.

2. If individuals have high blood pressure, their physicians would want to recommend the following to reduce their blood pressure. Reduction of sodium intake would decrease Na^+ concentration in blood plasma in order to reestablish normal Na^+ concentration in the blood. The normal Na^+ concentration is reestablished and plasma blood volume would be reduced, thus decreasing blood pressure.

3. Since insulin increases K^+ uptake by cells, insulin injection could result in low blood K^+ (hypokalemia). Severe hypokalemia could be life threatening (fatal), as potassium is required for cells, especially nerve and muscle cells, to function normally.

CHAPTER 26

1. Saliva cleanses the mouth in a variety of ways. As saliva washes over the tongue and teeth, it helps remove foreign materials and buildup. In addition, saliva contains antibacterial substances including lysozyme and antibodies (IgA), which inhibit bacterial growth in the oral cavity. Consequently, a person who has a dry mouth is not able to cleanse the mouth well and is more likely to develop dental problems as a result of build-up of bacterial and foreign material.

2. The stomach is lined with specialized surface mucous cells that continuously secrete an alkaline product containing mucin onto the gastric surface. The mucin hydrates to form a 1–3 millimeter mucus layer that prevents acid from coming in contact with the stomach wall. In addition, the stomach epithelial lining is constantly regenerating to replace any cells that are damaged.

3. The duodenum has many circular folds so that movement of digested materials can be slowed down and may be adequately mixed with pancreatic juice and bile. By the time chyme reaches the ileum, most nutrient absorption has occurred and the circular folds are not needed to slow digested materials' movement.

4. When the gallbladder is removed, bile is still produced by the liver. Thus, if the gallbladder has been removed, the more concentrated bile is no longer available for facilitating the mechanical digestion of a fatty meal. Therefore, patients that have their gallbladder removed are instructed to limit their fat intake.

CHAPTER 27

1. Water soluble vitamins include vitamin A, vitamin D, vitamin E, and vitamin K. Fat soluble vitamins are stored in body fat. Excessive consumption of fat-soluble vitamins may induce toxicity (hypervitaminosis).

2. A lipid profile that includes both HDL and LDL levels is more informative than a total cholesterol count. HDLs are considered the "good cholesterol" because they function to transport lipids from the peripheral tissue to the liver. LDLs function to transport cholesterol to the cells for storage or use within the cell. LDL is considered "bad" cholesterol as some of this lipid tends to be deposited on the inner walls of arteries, increasing risk for heart disease and high blood pressure. The ratio provides the information regarding the relative amounts of HDLs and LDLs and is a better measure of cardiovascular risk.

3. A man that is 6 feet tall and 200 pounds would have a higher BMR than a man that is 5 feet 9 inches tall weighing 160 pounds. BMR is the amount of energy that an individual uses when their body is at rest. One of the most important variables in determining BMR is body surface area. A larger individual has a greater surface area of skin and thus loses more energy as heat. When more heat is lost, the body must increase its metabolic rate to maintain body temperature.

CHAPTER 28

1. A woman can become pregnant as long as she has one remaining functioning ovary. However, she may not ovulate every month, since the ovaries typically "take turns" ovulating each month.

2. Stress, age, medications, and body weight all can affect a woman's monthly uterine (menstrual) cycle. Stress and excessively lean body mass can lead to amenorrhea (absence of periods).

3. If a man's testes were removed, the adrenal glands could still produce a small amount of androgens. However, because the testes produce the overwhelming majority of androgens, the small amount produced by the adrenal glands would have little effect.

4. If a man has a vasectomy, sperm still form in the seminiferous tubule and then mature in the epididymis. However, since the sperm are not ejaculated, they die, and their components are broken down and resorbed in the epididymis. An individual who has had a vasectomy ejaculates seminal fluid only, not semen (which contains sperm).

CHAPTER 29

1. If two sperm penetrate the secondary oocyte, the condition called polyspermy occurs. In this case, the fertilized cell contains 23 triplets of chromosomes, instead of the normal 23 pairs of chromosomes. In humans, a cell with 69 chromosomes (3 sets of 23 chromosomes) will not survive.

2. In the last trimester, the uterus continues to expand and compress abdominal organs. The uterus also prevents the diaphragm from moving too far inferiorly, such as is needed when one takes a deep breath. The uterus and fetus are physical barriers that prevent full lung expansion, and thus deep breathing, for some women.

3. It is unnecessary to slap newborns to "make" them breathe, because the breathing reflex generally occurs in about 10 seconds in response to the change in environment and temperature. It can be harmful because crying may actually make the baby's breathing more difficult, to say nothing of the shock and fear response that could occur.

4. As parents have become more diligent about protecting infants' skin from sun exposure, one potential negative effect is that these infants cannot synthesize vitamin D from the UV rays. Because breast milk is low in vitamin D, and these infants aren't synthesizing enough vitamin D from sunlight, supplementation of this vitamin is recommended.

glossary

Pronunciation Key

Pronouncing a word correctly is as important as knowing its spelling and its contextual meaning. The mastery of all three allows a student to take ownership of the word. The system employed in this text is basic and consists of the following conventions:

1. Vowels marked with a line above the letter are pronounced as follows:
   - ā day, base
   - ē be, feet
   - ī pie, ivy
   - ō so, pole
   - ū unit, cute
2. Vowels marked with the breve (˘) are pronounced as follows:
   - ă above, about
   - ĕ genesis, bet
   - ĭ it, sip
   - ŏ collide
   - ŭ cut, bud
3. Vowels not marked are pronounced as follows:
   - a mat
   - o not, ought
   - e term
4. Other phonetic symbols used include the following:
   - ah father
   - aw fall
   - oo food
   - ow cow
   - oy void
5. For consonants, the following key was employed:
   - b bad
   - ch child
   - d dog
   - dh this
   - f fit
   - g got
   - h hit
   - j jive
   - k keep
   - ks tax
   - kw quit
   - l learn
   - m mice
   - n no
   - ng ring
   - p put
   - r right
   - s so
   - sh shoe
   - t tight
   - th thin
   - v very
   - w wet
   - y yes
   - z zero
   - zh measure

6. The principal stressed syllables are followed by a prime ('); single-syllable words do not have a stress mark. Nonstressed syllables are separated by a hyphen.
7. Acceptable alternate pronunciations are given as needed.

A

abduction (ab-dŭk′shŭn) Movement of a body part away from the median plane of the body.

absolute refractory (rē-frak′tōr-ē) **period** Time period when an excitable cell cannot be restimulated to respond.

absorption (ab-sōrp-shŭn) Process of moving substances, such as products of digestion, into the blood or lymph.

accommodation (ā-kom′ŏ-dā′shŭn) Changing the lens shape in order to focus on a nearby object.

acetylcholine (a-sē′til-kō′lēn) (**ACh**) Neurotransmitter produced by the central and peripheral nervous systems.

acid Substance that releases a hydrogen ion when added to a solution.

acid-base balance Maintaining hydrogen ion concentration of body fluids within normal limits; also called pH balance.

acidosis (as-i-dō′sis) Condition in which the pH of arterial blood is below 7.35.

acrosome (ak′rō-sōm) A cap-like structure on the anterior two-thirds of the sperm nucleus that contains digestive enzymes for penetrating an oocyte.

actin (ak′-tin) Contractile protein forming the major part of the thin filaments in a sarcomere.

action potential Self-propagating change in membrane potential occurring in excitable cells (e.g., neurons, muscle cells).

active immunity (i-myū′ni-tē) Activation of the immune system by a vaccine or by exposure to the naturally occurring infectious agent. Offers long-term protection because memory cells are formed. *Compare to passive immunity.*

active site Region of an enzyme where substrate binds.

active transport Method of transporting a substance across the membrane, against its concentration gradient.

acute (ă-kyūt′) Takes place over a short period of time. *Compare to chronic.*

adaptation (ad-ap-tā′shŭn) Advantageous change of an organ or tissue to meet new conditions.

adduction (ăd-dŭk′shŭn) Medial movement of a body part toward the midline.

adenosine triphosphate (ă-den′ō-sēn trī-fos′fāt) (**ATP**) Stores and releases chemical energy in a cell; composed of adenine, ribose, and three phosphate groups.

adipocyte (ad′i-pō-sīt) Fat storage cell.

adrenergic (ad-rĕ-ner′jik) Relating to nerve cells that release norepinephrine neurotransmitter.

adventitia (ad-ven-tish′ă) Outermost covering of an organ.

aerobic (ār-o′-bik) **respiration** Breakdown of glucose or other nutrients to produce ATP, water, and carbon dioxide. The process requires oxygen.

afferent (af′er-ent) Inflowing or going toward a center.

agglutination (ă-glū-ti-nā′shŭn) Process by which cells clump due to cross-linking by antibodies.

agonist (ag′on-ist) Muscle that contracts to produce a particular movement; also called prime mover.

albumin (al-bū′min) Plasma protein important in regulating fluid balance.

alimentary (al-i-men′ter-ē) Relating to food or nutrition.

alkalosis (al-kă-lō′-sis) Condition in which the pH of arterial blood is above 7.45.

allele (ă-lēl′) Variation of a gene found on the same locus of homologous chromosomes.

allergen (al′er-gen) Noninfectious substance that elicits an excessive response by the immune system (allergic reaction).

alveolus (al-vē′ō-lŭs; pl., alveoli, -ō-lī) Small cavity. Air sac in the lungs; also, a milk-secreting portion of a mammary gland.

amino (ă-mē′nō) **acid** Organic molecule used to build proteins; contains both an amine group and carboxyl group.

amnion (am′nē-on) Extra-embryonic membrane that envelops the embryo.

amphiarthrosis (am′fē-ar-thrō′sis) A slightly movable joint.

amphipathic (am-fē-path′ik) Molecule that contains a hydrophobic region and a hydrophilic region.

ampulla (am-pul′lă; pl., ampullae, -lē) Saccular dilation of a canal or duct, such as the ductus deferens in the male reproductive system.

anabolism (ă-nab′ō-lizm) Formation of large, complex molecules from simple molecules.

anaerobic (an-ār-ō′bik) Process that does not require oxygen.

anastomosis (ă-nas′tō-mō′sis; pl., anastomoses, -sēz) Union of two structures, such as blood vessels, to supply the same region.

anatomic (ă-na-tom′ik) **position** Standing upright, arms at the sides with palms forward and feet flat on the floor; reference position for naming body regions.

anatomy (ă-nat′ŏ-mē) Study of structures in the human body.

androgen (an′drō-jen) Generic term for a hormone that stimulates the activity of accessory male sex organs or the development of male sex characteristics.

anemia (ă-nē′mē-ă) Any condition in which the number of erythrocytes is below normal.

aneurysm (an′ū-rizm) Ballooning of an artery due to a weakened vessel wall; it is susceptible to rupture, leading to severe bleeding.

angiogenesis (an′jē-ō-jen′ě-sis) Process of forming new blood vessels.

angioplasty (an′jē-ō-plas′tē) The reopening of a blood vessel through a variety of means.

angiotensin II (an′jē-ō-ten′sin) A peptide hormone, derived from angiotensin I, that increases blood pressure.

anion (an′ī-on) Negatively charged ion; e.g., Cl⁻.

antagonist (an-tag′ō-nist) Muscle (or hormone) that opposes or resists the action of another.

antebrachium (an-te-brā′kē-ŭm) Forearm.

anterior Toward the front of the body.

antibody (an'tē-bod-ē) Immunoglobulin that binds to a specific antigen; released by plasma cells (activated B lymphocytes).

anticodon (antē-kō'don) Group of three nucleotide bases on a transfer RNA molecule; base pairs with a complementary codon on messenger RNA.

antidiuretic (an'tē-dī-ū-ret'-ik) **hormone (ADH)** Hormone released by the posterior pituitary gland; increases water reabsorption in the kidney and reduces urine production.

antigen (an'ti-jen) Substance that causes a state of sensitivity or responsiveness and reacts with antibodies or immune cells of the affected subject.

antigen-presenting cell (APC) Immune cell that presents (displays) an antigen to T-lymphocytes; e.g., a macrophage.

aorta (ā-ōr'tă) Main trunk of the systemic arterial system, beginning at the left ventricle and ending when it forks at its inferior end to form the common iliac arteries.

aortic (ā-ōr'tik) **body** Structure composed of neurons sensitive to changing levels of blood pH, O_2, and CO_2; located in the aortic arch.

aperture (ap'er-chūr) Open gap or hole.

apex (ā'peks) Extremity of a conical or pyramidal structure; e.g., the inferior, conical end of the heart.

apical (ap'i-kăl) Related to the tip or extremity of a conical or pyramidal structure; opposite of basal.

apnea (ap'nē ă) Cessation of breathing while sleeping.

apocrine (ap'ō-krin) **gland** Gland that releases a substance by pinching off the apical membrane of a cell and a portion of its cytoplasm; e.g., apocrine sweat glands.

aponeurosis (ap'ō-nū-rō'sis; pl., aponeuroses, -sēz) Fibrous sheet or flat, expanded tendon.

apoptosis (ap'op-tō'sis) Programmed cell death.

appendicular (ap'en-dik'-ū-lăr) Relating to an appendage or limb; e.g., the appendicular skeleton.

appositional (ap-ō-zish'ŭn-ăl) Being placed or fitted together; e.g., appositional growth of bone.

aquaporins (ak'wă-pōr-in) Protein channels in the plasma membrane allowing the passage of water.

arachnoid mater (ă-rak'noyd mah'ter) Spider weblike meningeal layer; located between the dura mater and the pia mater.

arcuate (ar'kū-āt) Having a shape that is arched or bowed.

arteriole (ar-tēr'ē-ōl) The smallest type of artery.

artery (ar'ter-ē) Blood vessel conveying blood away from the heart.

articular (ar-tik'ū-lăr) Relating to a joint.

articulation (ar-tik-ū-lā'shŭn) Joint or connection between bones.

astrocyte (as'trō-sīt) Largest and most abundant glial cell of the nervous system.

atherosclerosis (ath'er-ō-skler-ō'sis) Disease in which an artery wall thickens, leaving a smaller lumen for blood flow.

atlas (at'las) The first cervical vertebra.

atom The smallest particle that displays properties of an element; composed of electrons, protons, and neutrons (except in hydrogen).

atomic mass unit (AMU) Mass of a specific atom.

atomic number Indicates the number of protons in one atom of a specific element. Value appears above each atomic symbol in the periodic table.

ATP (adenosine triphosphate) Chemical that transfers energy within a cell.

atrioventricular node (ā'trē-ō-ven-trik'ū-lar nōd) **(AV)** Group of specialized heart cells, located in the inferior right atrium, that sends action potentials to the AV bundle.

atrium (ā'trē-ŭm; pl., atria, ā'trē-ă) Chamber or cavity to which are connected other chambers or passageways; e.g., the heart has both a right atrium and a left atrium that are thin-walled, superior chambers that receive blood returning to the heart.

atrophy (at'rō-fē) Wasting of tissues, organs, or the entire body.

auricle (aw'ri-kl) The external ear; also called the pinna. Also, flaplike extensions on the anterior part of each atrium.

autoimmune (aw-tō-i-mūn') **disease** Disease in which the immune system attacks self-antigens as if they were foreign.

autolysis (aw-tol'i-sis) Digestion of cells by enzymes present within the cell itself.

autonomic (aw-tō-nom'ik) **nervous system** The part of the nervous system that regulates processes that occur below the conscious level; e.g., the activity of cardiac muscle, smooth muscle, and glands.

autophagy (aw-tof'ā-jē) Segregation and disposal of damaged organelles within a cell.

autoregulation Intrinsic ability of an organ to regulate its activity.

axial (ak'sē-ăl) Relating to or situated in the central part of the body—the head, neck, and trunk; e.g., the axial skeleton.

axis (ak'sis) The second cervical vertebra.

axolemma (ak'sō-lem'ă) Plasma membrane of an axon.

axon (ak'son) Process of a nerve cell that conducts nerve signals away from the cell body.

axoplasm (ak'sō-plazm) Cytoplasm within the axon.

B

B-lymphocyte (lim'fō-sīt) Cell involved in humoral immunity; matures into a plasma cell.

baroreceptor (bar'ō-rē-sep'ter, -tōr) Any sensor of pressure changes.

basal nuclei (nū'klē'ī) *See cerebral nuclei.*

base Substance that accepts a hydrogen ion.

basement membrane Thin layer at the basal surface of epithelial tissue that helps to attach it to underlying connective tissue.

basophil (bā'sō-fil) The least common of the white blood cells. Basophils release proinflammatory agents (e.g., histamine, heparin).

benign (bē-nīn') Term that denotes the mild character of an illness or the nonmalignant character of a neoplasm.

bile (bīl) Fluid secreted by the liver, stored and discharged from the gallbladder into the duodenum.

bilirubin (bil-i-rū'bin) Waste product derived from the breakdown of hemoglobin and excreted in bile.

blood-brain barrier Structure formed by capillary endothelial cells and astrocytes that regulates what can enter the interstitial fluid in the brain: helps prevent transport of harmful substances from the blood into the brain.

blood pressure Measure of the force of blood pushing against the blood vessel wall; commonly measured in the brachial artery.

bolus (bō'lŭs) A single quantity of something, such as a mass of food swallowed.

brachial (brā'kē-ăl) Relating to the region between the shoulder and the elbow.

bradycardia (brad-ē-kar'dē-ă) A resting heart rate below 60 beats per minute.

brainstem Brain region composed of the midbrain, pons, and medulla oblongata.

bronchiole (brong'kē-ōl) Small tubules that branch from bronchi in the lungs.

bronchus (brong'kŭs) Airways that deliver air from the trachea to the bronchioles.

buffer Substance that minimizes a change in pH even after an acid or base is added.

bursa (ber'să; pl., bursae, ber'sē) Closed, fluid-filled sac lined with a synovial membrane; usually found in areas subject to friction.

C

calcification (kal'si-fi-kā'shŭn) Process in which structures in the body become hardened as a result of deposited calcium salts; normally occurs only in the formation of bone and teeth.

callus (kal'ŭs) Composite mass of cells and extracellular matrix that forms at a fracture site to establish continuity between the bone ends.

calorie The amount of energy required to raise the temperature of 1 g of water by 1°C.

calyx (kā'liks; pl., calyces or calices, kal'i-sēz) Cup-shaped structure.

canaliculus (kan-ă-lik'ū-lŭs; pl., canaliculi, -lī) Small canal or channel.

cancer Disease involving a malignant neoplasm that can metastasize (spread to other organs).

capacitation (kă-pas'i-tā'shŭn) A period of conditioning whereby the sperm cell membrane is modified while in the female reproductive tract prior to being able to fertilize the secondary oocyte.

capillary (kap'i-lār-ē) The smallest blood vessel, its thin walls allow movement of substances between blood and interstitial fluid.

carbohydrate An organic molecule composed of carbon, hydrogen, and oxygen.

cardiac cycle Events taking place in the heart during one heart beat.

cardiac output (CO) Volume of blood ejected by the ventricle in 1 minute; calculated by multiplying heart rate times stroke volume.

carotene (kar'ō-tēn) Class of yellow-red pigments widely distributed in plants and animals.

carotid (ka-ro'tid) **body** Contains neurons sensitive to changing levels of blood pH, CO_2, and O_2; located in the common carotid artery.

carotid sinus (sī′nŭs) Structure composed of neurons sensitive to changes in blood pressure; located in the internal carotid arteries.

carpal (kar′păl) Relating to the wrist.

carrier Protein that bonds to a molecule then alters its shape in order to transport the molecule across the plasma membrane; also, a person with one recessive mutant allele (when in a homozygous form causes a phenotypical change) who does not express the trait but can pass the mutation on to their offspring.

cartilaginous (kar-ti′laj′i-nŭs) Relating to or consisting of cartilage.

catabolism (kă-tab′ō-lizm) Breakdown of complex molecules into simple molecules.

catalyst (kat′ă-list) Substance that speeds up a chemical reaction.

cataract (kat′ă-rakt) Complete or partial opacity of the lens.

catecholamines (kat-ĕ-kōl′ă-mēnz) Class of neurotransmitters (includes epinephrine, norepinephrine, and dopamine).

cation (kat′-ī-on) Ion with a positive charge (e.g., Na^+).

cauda equina (kaw′dă ē-kwī′nă) Spinal nerve roots within the vertebral canal inferior to the tapered inferior end of the spinal cord proper.

cecum (sē′kŭm) Blind pouch forming the first part of the large intestine.

cell Basic structural and functional unit of a living organism.

cell-mediated immunity Immune response involving T-lymphocytes.

cellular respiration Multistep metabolic pathway in which organic molecules are disassembled in a controlled manner by a series of enzymes to eventually form ATP.

central nervous system (CNS) Composed of the brain and spinal cord.

centriole (sen′trē-ōl) Organelle that participates in the separation of chromosome pairs during cell division.

centromere (sen′trō-mēr) The nonstaining constriction of a chromosome that is the point of attachment of the spindle fiber.

cephalic (se-fal′ik) Relating to the head area.

cerebellum (ser-e-bel′ŭm) The second largest part of the brain;

develops posteriorly to the pons in the metencephalon.

cerebral cortex (se-rē′bral kor′teks) Superficial layer of gray matter in the cerebrum.

cerebral nuclei Paired irregular masses of gray matter buried deep within central white matter in the basal region of the cerebral hemispheres inferior to the floor of the lateral ventricle.

cerebrospinal (sĕ-rē′brō-spī-năl) **fluid (CSF)** A clear, colorless fluid that circulates in the ventricles and subarachnoid space to protect and support the brain and spinal cord; fluid produced by ependymal cells.

cerebrovascular (sĕ-rē′brō-vas′kū-lăr) **accident (CVA)** Altered brain activity caused by reduced blood supply to the brain. It may result in permanent damage.

cerebrum (ser′ē-brŭm, ser′ĕ-brŭm) The largest, most superior part of the brain; composed of the left and right cerebral hemispheres; location of conscious thought processes and origin of all complex intellectual functions.

cerumen (sĕ-rū′men) Soft, waxy secretion of the ceruminous gland; found in the external auditory meatus.

cervical (ser′vĭ-kal) Relating to the neck.

chemical energy Energy stored in the chemical bonds of a molecule.

chemical equilibrium (ē-kwi-lib′rē-ŭm) State of a chemical reaction in which there is no net change in formation of products or reactants.

chemical reaction Process during which chemical bonds of a molecule are broken or new ones are formed.

chemical synapse (sin′aps) Junction between two communicating cells (neurons, and neurons and effectors): neurotransmitter molecule travels across this type of synapse.

chemoreceptor (kē′mō-rē-sep′tor) Cell that detects specific chemicals in a fluid; e.g., taste receptors.

chemotaxis (kē-mo-tak′sis) Movement in response to chemicals.

cholecalciferol (kō′lē-kal-sif′er-ol) Vitamin D.

cholecystokinin (kō′lē-sis-to-kī′nin) **(CCK)** Hormone released from the duodenum in response to lipid-rich chyme; stimulates secretion from the pancreas and gallbladder.

cholesterol Type of steroid found in the plasma membrane.

cholinergic (kol-in-er′jik) Relating to neurons that use acetylcholine as their neurotransmitter.

chondroblast (kon′drō-blast) Actively mitotic form of a matrix-forming cell found in developing and growing cartilage.

chondrocyte (kon′drō-sīt) Mature, nondividing cartilage cell.

chorion (kō′rē-on) Multilayered, outermost extraembryonic membrane; together with the functional layer of the endometrium it forms the placenta, the site through which nourishment and waste are exchanged between mother and developing fetus; attachment to the uterus.

chromatid (krō′mă-tid) One of the two strands of a chromosome joined by a centromere.

chromatin (krō′ma-tin) Genetic material of the nucleus in a nondividing cell.

chromosome (krō′mō-sōm) The most organized level of genetic material; a single long molecule of DNA and associated proteins; becomes visible only when the cell is dividing.

chronic (krŏn′ik) Takes place over a long period of time. *Compare to acute.*

chronic obstructive pulmonary disease (COPD) Group of respiratory diseases involving obstruction of the airways.

chyle (kīl) Chylomicron-containing lymph drained from gastrointestinal tract.

chyme (kīm) Mixture of partially digested food and gastric juice that is pastelike.

cilium (sil′ē-ŭm; pl., cilia, -ă) Motile extension of a cell surface containing cytoplasm and microtubules.

circumduction (ser-kŭm-dŭk′shŭn) Movement of a body part in a circular direction.

cisterna (sis-ter′nă; pl., cisternae, -ter′nē) Enclosed microscopic space (e.g., between the parallel membranes of the Golgi apparatus).

citric acid cycle A cyclic metabolic pathway that occurs in the matrix of mitochondria during which energy in the bonds of acetyl CoA is transferred to form ATP, NADH, and FADH.

cleavage (klēv′ij) Series of mitotic cell divisions occurring in the zygote immediately following its fertilization.

climacteric (klī-mak′ter-ik, klī-mak-ter′ik) Changes occurring in the male and female reproductive systems, beginning at around 50 years, resulting in

altered levels of reproductive hormones.

coagulation (kō-ag-yū-lā′-shun) Formation of a blood clot.

codon (kō′don) Group of three nucleotide bases in a messenger RNA molecule.

coenzyme (kō-en′zīm) Organic chemical assisting in enzyme function.

cofactor Chemical that aids in enzyme function.

cognition (kog-ni′shŭn) Mental activities associated with thinking, learning, and memory.

coitus (kō-i′tŭs; koy′tus) Sexual union between a male and a female.

collagen (kol′lă-jen) **fibers** Strong, flexible protein found in many connective tissues.

colloid (kol′oyd) Opaque mixture composed of water and solute (usually protein); substance within thyroid follicles.

columnar (kol-ŭm′năr) Relating to epithelial cells that are taller than they are wide.

commissure (kom′i-shūr) Bundle of axons passing from one side to the other in the brain or spinal cord.

complement (kom′plē-ment) Group of plasma proteins working together during an innate immune response; may lead to increased inflammation and cytolysis of the invading cell.

concentration gradient (grā′dē-ent) Difference in the concentration of a substance between two areas.

conducting zone Respiratory system passageways transporting air from the nose to the terminal bronchioles. *Compare to respiratory zone.*

conductivity (kon-dŭk-tiv′i-tē) Property of sending an electrical change along the cell membrane.

cone Cone-shaped cell in the retina that provides color vision; works best in bright light.

congenital (kon-jen′i-tăl) Present at birth.

conjunctiva (kon-jŭnk-tī′vă) Mucous membrane covering the anterior surface of the eyeball and the posterior surface of the eyelids.

contralateral (kon-tră-lat′er-ăl) Relating to the opposite side.

control center Structure in a feedback cycle interpreting information from a receptor and sending information to an effector to produce a response.

cornea (kōr′nē-ă) Transparent structure that forms the anterior surface of the eye.

coronal (kōr′ō-năl) A vertical plane that divides the body into anterior and posterior parts; also called frontal plane.

coronary (kōr′o-nār-ē) Denoting the blood vessels or other structures and activities related to the heart.

cortex (kor′teks) Outer region of an organ (e.g., cerebral cortex or adrenal cortex).

corticosteroid (kōr-ti-kō-stēr′-oyd) Hormones secreted by the adrenal cortex (e.g., cortisol).

costal (kos′tăl) Relating to a rib.

covalent (kō-vāl′ent) **bond** Chemical bond formed when nearby atoms share electrons.

cranial (krā′nē-ăl) Relating to the skull.

cranium (krā′nē-ŭm) Region of the skull composed of the frontal, parietal, temporal, occipital, ethmoid, and sphenoid bone.

creatine (krē′-ă-tēn; -tin) Chemical providing muscle cells with energy; present as creatine phosphate.

creatinine (krē-at′i-nēn) Nitrogenous waste product resulting from the breakdown of creatine and excreted in urine. Used to estimate glomerular filtration rate since it is not reabsorbed by the kidney.

cross-section Transverse section.

cubital (kū′bi-tal) Relating to the elbow.

cuboidal (kū-boy′dăl) Relating to cells that are cube-shaped.

current Movement of charged particles, such as ions.

cutaneous (kū-tā′nē-ŭs) Relating to the skin.

cyclic adenosine monophosphate (sī′klik ă-den′ō-sēn mon-ō-fos′-făt) **(cAMP)** A second messenger used when some hormones act on a target cell. Formed when adenylate cyclase reacts with ATP.

cytokine (sī′tō-kīn) Protein that regulates the intensity and duration of the immune system.

cytokinesis (sī′tō-ki-nē′sis) Division of the cytoplasm during cell division.

cytology (sī-tol′ō-jē) Study of cells.

cytoplasm (sī′tō-plazm) All cellular contents contained between the plasma membrane and the nucleus; includes cytosol, organelles, and inclusions.

cytoskeleton (sī-tō-skel′ĕ-ton) Organized network of protein filaments and hollow tubules that provide organization, support, and movement throughout the cell.

cytosol (sī′tō-sol) The viscous, syruplike fluid medium with dissolved solutes in the cytoplasm.

cytotoxic T (T_C) lymphocyte (sī′tō-tok′-sik līm′fō′sīt) A class of T-lymphocytes that releases chemicals toxic to cells; also called CD8 cells.

D

deamination (dē-am-i-nā′shŭn) Process of removing the amine group of an amino acid.

deciduous (dē-sid′ū-ŭs) Not permanent; e.g., deciduous teeth.

decussation (dē-kŭ-sā′shŭn) Any crossing over or intersection of parts.

defecation (def-ĕ-kā′shŭn) Discharge of feces from the rectum.

deglutition (dē-glū-tish′ŭn) Swallowing.

dehydration synthesis Chemical reaction in which water is formed during formation of a complex molecule.

denaturation (dē-nā-tū-rā′shŭn) A change in a protein's complex three-dimensional shape into a simpler shape; may occur with changes in pH or increased temperature.

dendrite (den′drīt) Process of a neuron that receives and conducts graded potentials toward the cell body.

dendritic (den-dri′tik) **cell** Phagocytic cells of the skin and mucous membranes.

dentition (den-tish′ŭn) Natural teeth in the dental arch, considered collectively.

deoxyribonucleic (dē-oks′ē-rībō-nū-klēic′) **acid (DNA)** A double-stranded nucleic acid, composed of deoxyribonucleotide monomers; directs protein synthesis.

depolarization (dē-pō′lăr-i-zā-shŭn) Change in membrane potential or voltage to a more positive value.

depression (dē-presh′ŭn) Downward or inward displacement of a body part.

dermatome (der′mă-tōm) Specific segment of skin supplied by a single spinal nerve. Also, during embryonic development, the cells that form the connective tissue of the skin.

dermis (der′mis) Connective tissue layer of skin internal to the epidermis; contains blood and lymph vessels, nerves and nerve endings, glands, and usually hair follicles.

desmosome (dez′mō-sōm) One type of adhesion between two epithelial cells; a type of

intercellular junction that holds cells together at a single point (like a button).

detrusor (dē-trū′ser, -sōr) Muscle that acts to expel urine from the bladder.

diabetes (dī-ă-bē′tēz) **insipidus** Disease involving reduced ADH release or reduced kidney response to ADH; leads to excessive urine production.

diabetes mellitus (me-lī′tŭs) Disease involving reduced insulin release or reduced tissue response to insulin; leads to elevated blood and urine glucose levels.

diapedesis (dī′ă-pē-dē′sis) Passage of blood or its formed elements through the intact blood vessel wall.

diaphragm Muscle that separates the thoracic cavity and abdomino-pelvic cavity; aids in breathing.

diaphysis (dī-af′i-sis; pl., diaphyses, -sēz) Elongated, usually cylindrical part of a long bone between its two ends; the shaft of a long bone.

diarthrosis (dī-ar-thrō′sis; pl., diarthroses, -sēz) A freely movable (synovial) joint.

diastole (dī-as′tō-lē) The relaxation phase of a heart chamber.

diastolic pressure (dī-ă-stol′ik) Blood pressure measured in an artery during diastole.

diencephalon (dī-en-sef′ă-lon) Brain region deep to the cerebrum; contains the thalamus, hypothalamus, and epithalamus.

diffusion (di-fū′zhŭn) Random movement of molecules or particles down their concentration gradient.

diploë (dip′lō-ē) Central layer of spongy bone between the two layers of compact bone (outer and inner plates) of the flat cranial bones.

diploid (dip′loyd) State of a cell containing pairs of homologous chromosomes. In humans, the diploid number of chromosomes is 46 (23 pairs).

disaccharide (dī-sak′ă-rīd) Carbohydrate composed of two monosaccharides; e.g., sucrose.

dislocation (dis-lō-kā′shŭn) Complete displacement of an organ (e.g., a bone) from its normal position.

distal (dis′tăl) Far from the point of attachment at the trunk.

diuretic (dī-ū-ret′ik) Agent that increases the excretion of urine.

diverticulum (dī′ver-tik′yū-lŭm; pl., diverticula) A pouch or sac opening from a tubular or saccular organ, such as the gut or bladder; small bulge in the intestinal wall.

dorsal (dōr′săl) Toward the back.

dorsiflexion (dōr-si-flek′shŭn) Upward movement of the foot or toes, or of the hand or fingers.

duodenum (dū-ō-dē′nŭm, dū-od′ē-nŭm) First section of small intestine.

dura mater (dū′ră mah′ter) Tough, fibrous membrane forming the outer covering of the central nervous system.

E

ectoderm (ek′tō-derm) Outermost of the three primary germ layers of the embryo.

ectopic (ek-top′ik) Out of place; e.g., in an ectopic pregnancy, the pre-embryo implants in the uterine tube rather than in the uterus.

edema (e-dē′mă) Localized swelling of a tissue.

effector (ē-fek′tŏr, -tōr) Peripheral tissue or organ that responds to nervous or hormonal stimulation.

efferent (ef′er-ent) Outgoing or moving away from a center.

eicosanoids (ī′kō′-să-noydz) Local hormones derived from fatty acids.

ejaculation (ē-jak-ū-lā′shŭn) Expulsion of semen from the penis.

elastic fiber Protein that allows tissues to stretch and return to their original shape; found in connective tissues.

electrical energy Movement of charged particles (such as ions).

electrical synapse (sin′aps) Junction between two communicating cells. Signal travels across this type of synapse by moving through gap junctions.

electrochemical gradient Electrical charge difference across a membrane.

electrolyte (ē-lek′trō-līt) Chemical that dissociates when added to water and can conduct an electrical current; includes salts, bases, and acids.

electrolyte balance Maintenance of the appropriate levels of electrolytes (such as sodium ions) in body fluids.

electron Subatomic particle with a negative charge; found orbiting the nucleus of an atom.

element Substance composed of only one type of atom.

elevation (el-ĕ-vā′shŭn) Superior movement of a body part.

embolus (em-bō′lŭs) Dislodged blood clot or air bubble traveling through the blood.

embryo (em′brē-ō) Organism in the early stages of development; in humans, the embryonic stage extends from the third to the eighth week of development.

embryology (em-brē-ol′ō-jē) Study of the origin and development of the organism, from fertilization of the secondary oocyte until birth.

end-diastolic (dī-ă-stol′ik) **volume (EDV)** Volume of blood in the ventricle at the end of diastole (relaxation).

end-systolic volume (ESV) Volume of blood in the ventricle at the end of systole (contraction).

endergonic (en′der-gon′ik) **reaction** Chemical reaction that requires the input of energy.

endocardium (en-dō-kar′dē-ŭm) Covering of the internal surface of the heart wall and external surface of heart valves.

endochondral ossification (en-dō-kon′drăl os′i-fi-kā′shŭn) Bone formation that takes place within hyaline cartilage; used to form most bones of the body.

endocrine (en′dō-krin) Hormonal secretions that are transported by the blood. *Compare to exocrine.*

endocytosis (en′dō-sī-tō′sis) Movement of substances from the extracellular environment into the cell through the formation of a vesicle.

endoderm (en′dō-derm) Innermost of the three primary germ layers of the embryo.

endolymph (en′dō-limf) Fluid within the membranous labyrinth of the inner ear.

endometrium (en′dō-mē′trē-ŭm) Mucous membrane forming the inner layer of the uterine wall.

endomysium (en′dō-miz′ē-ŭm, -mis′ē-ŭm) Areolar connective tissue layer surrounding a muscle fiber.

endoneurium (en-dō-nū′rē-ŭm) Areolar connective tissue of a peripheral nerve that surrounds the axons.

endoplasmic reticulum (en′-dō-plas′mik re-tik′ū-lum) **(ER)** Organelle composed of an extensive network of connected membranes; involved in synthesis, transport, and storage of macro-molecules, and detoxification of drugs: present as smooth ER or rough ER.

endosteum (en-dos′tē-ŭm) Layer of cells lining the inner surface of bone in the medullary cavity.

endothelium (en-dō-thē′lē-ŭm) The simple squamous epithelium that lines the lumen of blood vessels, lymphatic vessels, and heart chambers and valves.

energy Capacity to do work.

enzyme (en′zīm) Protein that catalyzes a chemical reaction by lowering the activation energy.

eosinophil (ē-ō-sin′ō-fil) White blood cell that destroys parasitic worms and phagocytizes antibody-antigen complexes.

ependymal (ep-en′di-mal) Relating to the cellular lining of the brain ventricles and central canal of the spinal cord; assists in production and circulation of CSF.

epicardium (ep-i-kar′dē-ŭm) The visceral (outermost) layer of the heart; also called the serous pericardium.

epidermis (ep-i-derm′is) The epithelium (keratinized, stratified squamous) of the integument.

epimysium (ep-i-mis′ē-ŭm) A layer of dense, irregular connective tissue surrounding a skeletal muscle.

epinephrine (ep′i-nef′rin) Hormone released by the adrenal medulla during activation of the sympathetic nervous system.

epineurium (ep-i-nū′rē-ŭm) Outermost supporting connective tissue layer of peripheral nerves.

epiphyseal (ep-i-fiz′ē-ăl) **line** The remnant of the epiphyseal plate that remains when long bone growth ceases; thin, defined area of compact bone.

epiphyseal plate Layer of hyaline cartilage located between a long bone diaphysis and epiphysis; allows longitudinal growth of the bone.

epiphysis (e-pif′i-sis; pl., epiphyses, -sēz) Expanded, knobby region at the end of a long bone.

erection (ē-rek′shŭn) Erectile tissues in the penis fill with blood and cause the penis to enlarge and become firm.

erythrocyte (ě-rith′rō-sīt) Mature red blood cell.

erythropoiesis (ě-rith′rō-poy-ē′sis) Formation of erythrocytes.

erythropoietin (ě-rith-rō-poy′ě-tin) Protein that stimulates erythropoiesis.

esophagus (ē-sof′ă-gŭs) Portion of the gastrointestinal tract between the pharynx and the stomach.

excitability The ability of a cell to respond to a stimulus with an action potential.

excretion (eks-krē′-shŭn) Process of removing waste products from the body.

exergonic (ek′ser-gon′ik) **reaction** Chemical reaction in which chemical energy is released.

exocrine (ek′sō-krin) Glandular secretions delivered to an apical or luminal surface through a duct.

exocytosis (ek′sō-sī-to′sis) Process whereby secreting granules or droplets are released from a cell.

exon (ek′-son) A portion of a DNA molecule; that codes for a section of the future messenger RNA molecule; these "coding" regions in pre-messenger RNA are joined together to form mature messenger RNA.

expiration (eks-pi-rā′shŭn) To breathe out; exhalation.

extension (eks-ten′shŭn) Movement that increases the articulating angle.

exteroceptor (eks′ter-ō-sep′ter, -tōr) Peripheral end organ of the afferent nerves in the skin or mucous membrane that responds to external stimulation.

extracellular (eks-tră-sel′ū-lăr) **fluid (ECF)** Fluid located outside of cells.

extracellular matrix (mā′triks) Protein fibers and ground substance in the extracellular space of connective tissue.

extrinsic (eks-trin′sik) Originates outside of an organ (e.g., extrinsic eye muscles).

F

facet (fas′et, fă-set′) Small, flat, shallow articulating surface; smooth area on a bone.

facilitated diffusion Passive transport process using carrier proteins or channel proteins to move a chemical across the plasma membrane.

falciform (fal′si-fōrm) Having a crescent or sickle shape (e.g., falciform ligament).

falx cerebri (falks se-rē′brē) Portion of the dura mater septa that projects into the longitudinal fissure between the right and left hemispheres of the brain.

fascia (fash′ē-ă; pl., fasciae, -ē-ē) Sheath of fibrous connective tissue that envelops the body internal to the skin; encloses muscles, and separates their various layers or groups.

fascicle (fas′i-kl) Band or bundle of muscle or nerve fibers.

fauces (faw′sēz) Space between the oral cavity and the pharynx.

feces (fē′sēz) Material discharged from the GI tract during defecation.

fertilization (fer′til-i-zā′shŭn) Process of sperm penetration of the secondary oocyte.

fetal (fē′tăl) Relating to the fetus; in humans, the fetal period extends from the eighth week after conception until birth.

fibrin (fī′brin) Fibrous protein that creates a web during blood clot formation.

fibroblast (fī′brō-blast) Large, flat, connective tissue cells with tapered ends that produce the fibers and ground substance components of the extracellular matrix.

fibrosis (fī-brō′sis) Formation of fibrous connective tissue as a repair or reactive process.

filtrate (fil′trāt) The materials that pass through a filter.

fimbria (fim′brē-ă; pl., fimbriae, -brē-ē) Any fringelike structure (e.g., the fimbriae of the infundibulum enclose the ovary at the time of ovulation).

fissure (fish′ur) Deep furrow, cleft, or slit.

flagellum (flă-jel′-ŭm; pl., flagella, -ă) Whiplike locomotory organelle that arises from within the cell and extends outside it; permits a sperm cell to move.

flexion (flek′shŭn) Movement that decreases the angle between the articulating bones.

flexure (flek′sher) Bend in an organ or structure.

follicle Roughly spherical group of cells enclosing a cavity.

fontanelle (fon′tă-nel′) One of several membranous intervals at the margins of the cranial bones in an infant.

foramen (fō-rā′men) Hole in a bone (e.g., foramen magnum, obturator foramen).

formed elements Erythrocytes, leukocytes, and platelets found in blood.

fornix (fōr′niks; pl., fornices, -ni-sēz) Arch-shaped structure.

fossa (fos′ă; pl., fossae, -fos′ē) Depression, often more or less longitudinal in shape, below the level of the surface of a part.

G

G protein Specific protein that acquires its energy from guanosine triphosphate; when activated by a membrane receptor, it relays the signal to another membrane protein and alters the activity of that protein.

gamete (gam′ēt) A sex cell with the haploid number of chromosomes.

gametogenesis (gam′ě-tō-jen′ě-sis) Formation and development of gametes (sex cells).

ganglion (gang′glē-on; pl., ganglia, -glē-ă) Group of nerve cell bodies in the peripheral nervous system.

gastric (gas'trik) Relating to the stomach.

gastrulation (gas-trū-lā'shŭn) Formation of the three primary germ layers.

gene Segment of DNA containing information to direct synthesis of a specific protein; functional unit of DNA.

genotype (jen'ō-tīp) The genetic constitution of an individual that, along with environmental influences, contributes to the phenotype.

gestation (jes-tā'shŭn) Pregnancy.

gland Organ or individual cells that secrete a substance.

glomerulus (glō-mer'yū-lŭs) A capillary network within the renal corpuscle in a kidney.

glucocorticoid (glū-kō-kōr'ti-kōyd) Group of hormones released from the adrenal cortex; help regulate glucose levels (e.g., cortisol, corticosteroid).

gluconeogenesis (glū-kō-nē'ō-jen'ĕsis) Formation of glucose from a noncarbohydrate source.

glucose (glū'kōs) Primary nutrient source for producing cellular energy; one of the monosaccharides.

glycerol (glis'er-ol) Chemical component of a triglyceride molecule.

glycocalyx (glī-kō-kā'liks) Filamentous coating on the apical surface of certain types of cells.

glycogen (glī'kō-jen) Polysaccharide formed from glucose monomers.

glycogenesis (glī'kō-jen'ĕ-sis) Formation of glycogen from glucose.

glycogenolysis (glī'kō-je-nol'i-sis) Breakdown of glycogen into glucose.

glycolipid (glī-kō-lip'id) Lipid with an attached carbohydrate group found on the extracellular side of the plasma membrane.

glycolysis (glī-kol'i-sis) Stage of cellular respiration in which glucose is partially catabolized to form pyruvic acid and to form ATP molecules (also called anaerobic cellular respiration).

glycosuria (glī-kō-sū'rē-ă) Presence of glucose in urine; symptom of diabetes mellitus (also called glucosuria).

goblet cells Unicellular glands that secrete mucin.

Golgi (gol'jē) **apparatus** Series of saclike membranes that act as a center to package, sort, and modify molecules arriving from the endoplasmic reticulum in a transport vesicle.

gonad (gō'nad) Organ that produces sex cells and sex hormones; the ovaries in a female and the testes in a male.

gonadocorticoids (gō-na'dō-kōr'ti-kōyd) Group of sex hormones including estrogen and androgen.

graded potential Small deflection in the resting membrane potential that decreases in intensity as it travels; may be a depolarization or hyperpolarization.

gradient A difference from one area to another; e.g., concentration gradient, pressure gradient.

gray matter Brain or spinal cord tissue composed of neuron cell bodies, dendrites, and unmyelinated axons.

gustation (gŭs-tā'shŭn) Sense of taste.

gyrus (jī'rŭs; pl., gyri, -rī) Prominent rounded surface elevation that helps form the cerebral hemispheres.

H

hair cell Specialized sensory cell in the inner ear.

half-life Time required to reduce substance by one-half of its original quantity.

haploid (hap'loyd) The number of chromosomes in a sperm or secondary oocyte. In humans, the haploid number is 23.

hapten (hap'ten) Small substance that attaches to another molecule to initiate an immune reaction.

helper T (T_H) lymphocyte (lim'fō-sīt) A T-lymphocyte that releases cytokines to regulate humoral immunity and cellular immunity as well as cells of innate immunity; also called CD4 cells.

hematocrit (hē'mă-tō-krit, hem'ă-) Percentage of whole blood attributed to erythrocytes.

hematoma (hē-mă-tō'mă) Mass of blood outside of the blood vessels (e.g., subdural hematoma).

heme (hēm) Portion of a hemoglobin molecule that transports oxygen.

hemocytoblast (hē'mō-sī'tō-blast) Immature cells that produce all types of formed elements in blood.

hemoglobin (hē-mō-glō'bin) A red-pigmented protein that transports oxygen and carbon dioxide; responsible for characteristic bright red color of arterial blood.

hemolysis (hē-mol'i-sis) The process of rupture and destruction of erythrocytes.

hemopoiesis (hē'mō-poy-ē'sis) Formation and development of blood cells.

hemorrhage (hem'ŏ-rij) Abnormal escape of blood from a blood vessel.

hemostasis (hē'mō-stā'sĭs) Process of stopping bleeding; steps include vascular spasm, platelet plug formation, and coagulation.

heparin (hep'a-rĭn) Chemical that prevents blood clot formation; released by basophils and mast cells.

hepatic (he-pat'ik) Relating to the liver.

hernia (her'nē-ă) Protrusion of a part through the structures normally surrounding or containing it.

heterozygous (het'er-ō-zī'gŭs) Having both dominant and recessives alleles for a trait.

hiatus (hī-ā'tŭs) Opening.

hillock (hil'lok) Any small elevation or prominence.

hilum (hī'lŭm) The part of an organ where structures such as blood vessels, nerves, and lymph vessels enter or leave.

histamine (his'tă-mēn) Chemical that increases capillary permeability and causes vasodilation; released by mast cells and basophils.

histology (his-tol'ō-jē) Study of tissues formed by cells and cell products.

homeostasis (hō'mē-ō-stā'sis, -os'tă-sis) State of equilibrium in the body with respect to various functions and the chemical composition of fluids and tissues.

homologous (hō-mol'ō-gŭs) Alike in certain critical attributes.

homozygous (hō-mō-zī'gŭs) Having two identical alleles for a trait.

hormone (hōr'mōn) Chemical formed in one part of the body and carried by the blood to another part of the body where it can affect cellular activity.

humoral (hyū'mōr-ăl) **immunity** Immune response involving B-lymphocytes; involves release of antibodies from plasma cells.

hyaline (hī'a-lin, -lēn) Clear, homogeneous substance; a type of cartilage.

hydrogen bond A weak attraction formed when a hydrogen atom of one molecule is attracted to a slightly negative atom within either the same molecule or a different molecule.

hydrolysis (hī-drol'i-sis) Chemical reaction in which water is used during the breakdown of a complex molecule.

hydrophilic (hī-drō-fil'ik) **molecule** Substance that does dissolve in water.

hydrophobic (hī-drō-fōb'ik) **molecule** Substance that does not dissolve in water.

hydroxyapatite (hī-drok'sē-ap-ă-tīt) Natural mineral structure that the crystal lattice of bone and teeth closely resembles.

hyperextension (hī'per-eks-ten'shŭn) Extension of a body part beyond 180 degrees.

hyperplasia (hī-per-plā'zhē-ă) Increase in the number of cells in a tissue.

hyperpolarization (hī'pĕr-pō'lăr-i-zā-shŭn) Change in the membrane potential to a value more negative than the resting potential.

hypertension (hī'pĕr-ten'shŭn). Persistently elevated blood pressure.

hypertonic (hī'pĕr-ton'ik) **solution** Solution with a lower water concentration and higher solute concentration than that of the cytosol.

hypertrophy (hī-pĕr'trō-fē) Increase in the size of cells in a tissue.

hyperventilation (hī-pĕr-ven'ti-lā'shŭn) Increase in the breathing rate or depth.

hypodermis (hī-pō-der'mis) Subcutaneous layer of tissue internal to and not part of the integument.

hypogastric (hī-pō-gas'trik) Relating to the hypogastrium, or pubic region.

hypothalamus (hī'pō-thal'ă-mŭs) Region of the brain in the diencephalon; regulates body temperature, autonomic nervous system, and endocrine system.

hypotonic (hī'pō-ton'ik) **solution** Solution with a lower solute concentration and a higher water concentration than that of the cytosol.

hypoventilation Decrease in the breathing rate or depth.

hypoxia (hī-pok'sē-ă) Decreased oxygen level.

I

idiopathic (id'ē-ō-path'ik) Describing a disease of unknown cause.

immunity (i-myū'ni-tē) Ability of the body to protect itself against foreign cells and chemicals; includes humoral immunity and innate immunity.

immunocompetence (im'ū-nō-kom'pē-tens) Ability of immune cells to recognize and bind to an antigen.

immunoglobulin (im'-ū-nō-glob'-ū-lin) **(Ig)** *See antibody.*

implantation Embedding of the pre-embryo in the uterine wall.

inclusion A temporary store of molecules in the cytosol.

infarction (in-fark'-shŭn) Tissue death resulting from reduced or absent blood supply; e.g., myocardial infarction.

inferior Toward the feet.

inflammation A nonspecific, localized immune response to tissue injury.

infundibulum (in-fŭn-dib'-ū-lŭm) Funnel-shaped structure or passage.

ingestion (in-jes'-chŭn) Introduction of food and drink into the GI tract.

inguinal (ing'-gwi-năl) Relating to the groin area.

innervation (in-er-vā'-shŭn) Supply of axons functionally connected with a structure of the body.

inorganic Chemical substance that is not organic (e.g., water, acids, bases).

insertion (in-ser'-shŭn) The usually distal and more movable attachment of a muscle.

inspiration (in-spi-rā'-shŭn) To breathe in; inhalation.

integument (in-teg'-ū-ment) Cover enveloping the body; includes the epidermis, dermis, and all derivatives of the epidermis; also called skin or the cutaneous membrane.

intercalated disc (in-ter'-kă-lā-ted disk) Complex junction that interconnects cardiac muscle cells in the heart wall to mechanically and electrically link these cells.

interneuron (in'-ter-nū'-ron) Type of neuron that resides within the CNS and coordinates activity between sensory and motor neurons.

interoceptor (in'-ter-ō-sep'-ter) Small sensory receptors within the walls of the viscera.

interphase First phase of the cell cycle during which the cell carries out normal activities and prepares for cell division.

interstitial (in-ter-stish'-ăl) Relating to spaces within a tissue or organ, but not a body cavity; e.g., interstitial fluid occupies the extracellular environment of the cells.

intervertebral (in-ter-ver'-te-brăl) Between vertebrae.

intracellular fluid (ICF) *See cytosol.*

intramembranous (in'-tră-mem'-brā-nŭs) **ossification** Bone

formation that takes place within a membrane; the formation of flat bones.

intraperitoneal (in'-tră-per'-i-tō-nē'-ăl) Location of a structure or organ that is completely covered with visceral peritoneum.

intrinsic (in-trin'-sik) Originates inside of an organ; e.g., intrinsic muscles of the hand.

intrinsic factor Chemical, produced by stomach cells, required for absorption of vitamin B_{12}.

intron (in'-tron) A portion of DNA that lies between two exons; noncoding regions in pre-messenger RNA. These are removed during pre-messenger RNA processing.

invaginate (in-vaj'-i-nāt) To infold or insert a structure within itself.

ion (ī'-on) An atom with a net positive or negative charge.

ionic bond Chemical bond formed when a cation is electrostatically attracted to an anion.

ipsilateral (ip-si-lat'-er-ăl) Relating to the same side.

ischemia (is-kē'-mē-ă) Reduced blood flow; may lead to hypoxia.

isomer (ī'-sō-mer) Molecules composed of the same number and types of atoms but with a different arrangement; e.g., glucose and galactose.

isometric (ī-sō-met'-rik) **contraction** Muscle contraction during which its length does not change because tension does not exceed resistance (load).

isotonic (ī-sō-ton'-ik) **contraction** Muscle contraction during which tension exceeds the resistance and the muscle fibers shorten, resulting in movement.

isotonic solution Contains the same relative solute and water concentration as the cytosol.

isotope (ī'-sō-tōp) Atoms of an element that have a different number of neutrons.

K

keratin (ker'-ă-tin) Protein that strengthens the epidermis, hair, and nails; produced by keratinocytes.

keratinization (ker'-ă-tin-i-zā'-shŭn) Development of a horny layer due to cells filled with the protein keratin.

keratinocyte (ke-rat'-i-nō-sīt) The most abundant cell type in the epidermis and found throughout the epidermis; cells that produce keratin.

ketone (kē'-tōn) **bodies** Waste product of fatty acid breakdown; also called ketoacids.

ketoacidosis (kē'-tō-as-i-dō'-sis) Accumulation of ketone bodies in the blood; a symptom of diabetes mellitus.

kinase (kī'-nās) Type of enzyme that transfers a phosphate group from one molecule to another.

kinetic (ki-net'-ik) **energy** Energy of movement. *Compare to potential energy.*

L

labia (lā'-bē-ă; sing. labium) Tissue fold having the appearance of a lip.

lacrimal (lak'-ri-măl) Relating to tears.

lactate Chemical produced during glycolysis.

lactation (lak-tā'-shŭn) Production of milk.

lacteal (lak'-tē-al) Tiny vessel that transports lymph with lipids and lipid-soluble vitamins from the small intestine.

lacuna (lă-kū'-nă; pl., lacunae, -kū'-nē) Small space, cavity, or depression.

lamella (lă-mel'-ă; pl., lamellae, -mel'-ē) Layer of bone connective tissue; forms concentric rings.

lamina (lam'-i-nă) Thin layer (i.e., epithelial tissue basal lamina).

lanugo (lă-nū'-gō) Fine, soft, unpigmented fetal hair.

larynx (lar'-ingks) Organ of voice production that lies between the pharynx and the trachea.

lateral (lat'-er-ăl) Away from the midline.

lateralization (lat'-er-al-ī-zā'-shŭn) Process whereby certain asymmetries of structure and function occur.

lemniscus (lem-nis'-kŭs; pl., lemnisci, -nis'-ī) Bundle of axons ascending from sensory relay nuclei to the thalamus.

lesion (lē'-zhŭn) Pathologic change in a tissue.

lethargy (leth'-ar-jē) Mild impairment of consciousness characterized by reduced awareness and alertness.

leukocyte (lū'-kō-sīt) Any one of several types of white blood cells.

leukopoiesis (lū'-kō-poy-ē'-sis) Formation and development of leukocytes.

ligament (lig'-ă-ment) Band or sheet of dense regular fibrous tissue that connects structures; usually bones.

ligand (lig'-and, lī'-gand) Chemical released from one cell that binds to receptors on another cell.

lingual (lin'-gwăl) Relating to the tongue.

lipase (lip'-ās) Enzyme that digests triglycerides.

lipid Group of organic molecules including triglycerides, phospholipids, steroids, and ecosanoids.

lipogenesis (līp'-o-jĕn'-e-sis) Formation of triglycerides from glycerol and fatty acids.

lipolysis (li-pol'-i-sis) Breakdown of triglycerides to glycerol and fatty acids.

lobe (lōb) Subdivision of an organ, bounded by some structural demarcation.

lumbar (lŭm'-bar) Relating to the lower back.

lumen (lū'-men) The space inside a structure, such as where blood is transported within a blood vessel.

lymph (limf) Usually transparent fluid found in lymph vessels; derived from interstitial fluid.

lymphocyte (lim'-fō-sīt) White blood cell involved in immune responses (i.e. T-lymphocyte, B-lymphocyte).

lysosome (lī'-sō-sōm) Organelle containing digestive enzymes.

lysozyme (lī'-sō-zīm) Enzyme that attacks the bacterial cell wall. Present in saliva, tears, sweat, and nasal secretions.

M

macromolecules Large complex molecules (e.g., carbohydrates, nucleic acids, proteins).

macrophage (mak'-rō-fāj) Phagocytic cell derived from monocytes and can activate the immune system.

major histocompatibility (his'-tō-kom-pat'-i-bil'-i-tē) **complex (MHC)** Plasma membrane proteins used to mark cells as self and to present antigens.

malignant (mă-lig'-nant) In reference to a neoplasm, having the property of invasiveness and spread.

mass number Total number of neutrons and protons in the nucleus of an atom.

mast cells Resident cells that secrete heparin and histamine during inflammation; often found near blood vessels.

mastication (mas-ti-kā'-shŭn) Chewing ingested food in preparation for swallowing.

matrix (mā'-triks; pl., matrices, mā'-tri-sēz) Surrounding substance within which cells or structures are contained or embedded.

meatus (mē-ā′tŭs) Passageway traveling through a bone (i.e., external acoustic meatus).

mechanoreceptor (mek′ă-nō-rē-sep′tŏr) Sensory receptor that responds to touch, vibration, pressure, or distortion.

medial (mē′dē-ăl) Toward the midline.

mediastinum (me′dē-as-tī′nŭm) Median space of the thoracic cavity between the lungs.

medulla (me-dūl′ă) Inner region of an organ; e.g., adrenal medulla.

medulla oblongata (ob-long-gah′tă) Brain region that transmits information between the spinal cord and higher brain centers; controls heart rate, blood pressure, breathing.

meiosis (mī-ō′sis) Process of sex cell division that results in four gametes, each with the haploid number of chromosomes.

melanin (mel′ă-nin) Any of the dark brown-black pigments or yellow-red pigments that occur in the skin, hair, and retina.

melanocyte (mel′ă-nō-sīt) Pigment-producing cell in the basal layer of the epidermis.

membrane potential Difference in charges across the plasma membrane.

memory cells Lymphocytes that produce a strong, rapid response during the second exposure to an invader.

menarche (me-nar′kē) Time of the first menstrual period.

meninx (mē′ninks; pl., meninges, -jēz) Any one of the membranes covering the brain and spinal cord.

meniscus (mě-nis′kŭs; pl., menisci, mě-nis′sī) Crescent-shaped fibrocartilage found in certain joints.

menopause (men′ō-pawz) Permanent cessation of menses.

menses (men′sēz) Periodic physiologic hemorrhage from the uterine mucous membrane; commonly referred to as the menstrual period.

menstruation (men-strū-ā′shŭn) Cyclic endometrial shedding and discharge of bloody fluid.

mesenchyme (mez′en-kīm) An embryonic connective tissue.

mesentery (mes′en-ter-ē) Fan-shaped fold of peritoneum attaching the small intestine to the posterior abdominal wall; may refer to other membranes associated with organs of the abdominal cavity.

mesoderm (mez′ō-derm) The middle of the three primary germ layers of the embryo.

mesothelium (mez-ō-thē′lē-ŭm) The simple squamous epithelium that lines serous cavities.

metabolic acid Acid produced as a byproduct of metabolism (e.g., ketoacid).

metabolic rate Amount of energy used in a given time period.

metabolic water Water formed during dehydration reactions and during aerobic respiration; contributes approximately 8% of total fluid intake.

metabolism (mě-tab′ō-li-zem) All chemical reactions taking place in the body, including both anabolic and catabolic reactions.

metaphysis (mě-taf′i-sis; pl., metaphyses, -sēz) Section of a long bone between the epiphysis and the diaphysis.

metaplasia (met-ă-plā′zē-ă) Abnormal change in a tissue.

metastasis (mě-tas′tă-sis) The movement or spread of malignant cells from one part of the body to another.

microfilament (mī-krō-fil′ă-ment) Smallest structural protein of the cytoskeleton.

microglia (mī-krog′lē-ă) Category of small glial cells in the central nervous system; wander and exhibit phagocytic activity.

microscopy (mī-kros′kŏ-pē) Investigation of very small objects by means of a microscope.

microtubule (mī-krō-too′bŭl) Hollow cylinders of tubulin protein that are part of the cytoskeleton; able to lengthen and shorten.

microvilli (mī-krō-vil′ī; sing., microvillus -vil′ŭs) Microscopic extensions of the plasma membrane that increase the surface area for secretion or absorption.

micturition (mik-chū-rish′ŭn) Urination.

midsagittal (mid-saj′-i-tăl) Vertical plane cutting the body or body part into an even left and right half.

mineralocorticoid (min′er-al-ō-kōr′ti-kŏyd) Group of hormones released from the adrenal cortex that help regulate electrolyte levels in the body.

mitochondrion (mī-tō-kon′drē-on; pl., mitochondria, -ă) Organelle associated with the production of ATP.

mitosis (mī-tō′sis) Process of somatic cell division.

mixed nerve Nerve composed of both sensory and motor neurons.

molarity (mō-lar′i-tē) Number of moles in 1 liter of solution.

mole Unit composed of 6.023 × 10^{23} particles; its mass in grams is equal to the element's atomic mass or the compound's molecular mass.

molecule Composed of atoms or ions held together by an attraction or chemical bond.

monocyte (mon′ō-sīt) White blood cell that develops into a macrophage and phagocytizes bacteria and viruses.

monosaccharide (mon-ō-sak′ă-rīd) The simplest carbohydrate molecules; e.g., glucose, ribose.

mons (monz) Anatomic prominence or slight elevation above the general level of the surface; e.g., mons pubis.

motor (efferent) nerves Composed of neurons that take information away from the central nervous system.

motor unit One motor neuron and all the muscle cells it innervates.

mucin (mū′sin) An often protective secretion containing carbohydrate-rich glycoproteins that forms mucus when hydrated.

mucosa (mū-kō′să) Mucous membrane that lines various body structures.

multipotent (mŭl-tip′ŏ-tent) **stem cells** Adult stem cell that can differentiate into a few specific cell types.

muscle fiber Muscle cell.

muscularis (mūs-kū-lā′ris) Muscular layer in the wall of a hollow organ or tubular structure.

mutation An alteration in a segment of DNA; may lead to production of an abnormal protein.

myelin (mī′ĕ-lin) Lipoprotein-aceous material of the myelin sheath.

myoblast (mī′ō-blast) Undifferentiated muscle cell with the potential of becoming a muscle fiber.

myocardium (mī-ō-kar′dē-ŭm) Middle layer of the heart wall, consisting of cardiac muscle.

myofibril (mī-ō-fī′bril) Bundles of myofilaments within skeletal and cardiac muscle cells.

myofilament (mī-ō-fil′ă-ment) A protein filament that makes up the myofibrils in skeletal muscle.

myoglobin (mī-ō-glō′bin) Oxygen-carrying and -storing molecule in muscle.

myometrium (mī-ō-mē′trē-ŭm) Middle tunic (muscular wall) of the uterus.

N

naris (nā′ris; pl., nares, -res) Anterior opening to the nasal cavity.

necrosis (nē-krō′sis) Pathologic death of cells or a portion of a tissue or organ.

negative feedback Control mechanism that keeps a variable within normal levels (limits amount of product formed). *Compare to positive feedback*

neoplasia (nē-ō-plā′zē-ă) Process that results in the formation of a neoplasm or abnormal growth.

neoplasm (nē′ō-plazm) An abnormal tissue that grows by cell proliferation more rapidly than normal.

nephron (nef′ron) Functional filtration unit in the kidney; composed of a renal corpuscle, a proximal and distal convoluted tubule, and nephron loop.

nerve A bundle of peripheral nervous system axons.

nerve plexus (plek′sŭs) Group of interconnecting spinal nerves (e.g., brachial plexus, lumbar plexus).

nerve signal Propagation or conduction of an action potential along a neuron.

net filtration pressure The aggregate pressure that causes fluid to move out across a blood capillary wall; calculated by subtracting net osmotic pressure from net hydrostatic pressure.

neurilemma (nūr-i-lem′a) The delicate outer membrane sheath around an axon.

neurofibril (nūr-o-fi′bril) Filamentous protein structure in a neuron composed of neurofilaments and microtubules.

neurofilament (nūr-o-fil′a- ment) Group of intermediate-sized filaments in a neuron.

neuroglia (nū-rog′lē-a) Nervous system cells that support neurons.

neuromuscular (nūr-ō-mus′ku-lar) Relationship between a nerve and a muscle; e.g., a neuromuscular junction where a neuron and a muscle cell interface.

neuron (nūr′on) Functional unit of the nervous system.

neurotoxin (nūr-ō-tok′sin) Chemical that alters normal activity of a neuron.

neurotransmitter (nū′rō-trans-mit′er) Chemical released from a neuron that delivers information to another cell.

neurulation (nū′rū-lā′shun) Formation of the neural plate and its closure to form the neural tube.

neutron (nū′tron) Subatomic particle with a neutral charge; found in the nucleus of an atom.

neutrophil (nū′trō-fil) The most common of the white blood cells. Cells that phagocytize bacteria.

nociceptor (nō-si-sep′ter, -tōr) Peripheral sensory receptor for the detection of painful stimuli.

nonpolar molecules Molecules containing nonpolar covalent chemical bonds.

norepinephrine (nōr-ep-i-nef′rin) Neurotransmitter released by the sympathetic nervous system; hormones released from adrenal medulla.

nucleic acid (nū-klē′ik) An organic molecule composed of nucleotide monomers; in DNA and RNA. These store genetic information in the cell.

nucleolus (nū′klē′ō-lŭs; pl., nucleoli, -lī) Spherical, dark body within the nucleus where subunits of ribosomes are made.

nucleotide (nū′klē-ō-tīd) Building blocks of DNA and RNA; composed of a nitrogenous base, a phosphate group, and a sugar.

nucleus (nū′klē-ŭs) Cellular organelle housing DNA or a group of cell bodies in the central nervous system.

nutrient Chemicals obtained from food that are necessary for normal function in cells.

O

oblique (ob-lēk′) Slanted, at an angle.

occult (ŏ-kŭlt′, ok′ŭlt) Hidden or concealed.

octet rule Tendency of atoms to maintain an outer shell with eight electrons.

olfaction (ol-fak′shŭn) Sense of smell.

oligodendrocyte (ol′i-gō-den′drō-sīt) Category of large glial cells in the central nervous system that wrap around and insulate axons.

omentum (ō-men′tŭm) Fold of peritoneum from the stomach to either cover the region between the liver and stomach, or extends to cover most abdominal organs on their anterior surface.

oocyte (ō′-ō-sīt) Female gamete released from the ovary during ovulation.

oogenesis (ō-ō-jen′ē-sis) Formation and development of oocytes.

oogonium (ō-ō-gō′nē-ŭm; pl., oogonia) Primitive germ cell of the oocyte.

opposition (op′pō-si′shŭn) Movement of the thumb across the palm to touch the palmar side of the fingertips.

optic Relating to the eye.

orbit Bony socket housing the eye.

organ Composed of two or more tissue types that perform a specific function for the body.

organ system A group of organs working together to coordinate and perform specific function(s).

organelle (or′gă-nel) Complex, organized structures in the cytoplasm of a cell with unique characteristic shapes; called "little organs."

organic molecule Molecule containing carbon atoms; e.g., carbohydrates, proteins, lipids.

organism A living being.

organogenesis (ōr′gă-nō-jen′ē-sis) Formation of organs during development.

orifice (or′i-fis) Aperture or opening.

origin (ōr′i-jin) The usually proximal and less movable of the two points of attachment of a muscle.

os (os; pl., ossa, os′ă) An opening into a hollow organ or canal.

osmosis (os-mō′sis) Process by which water moves through a semipermeable membrane from a hypotonic solution to a hypertonic one.

osmotic pressure Pressure exerted by water movement across a membrane as it moves toward an area of lower water concentration.

osseous (os′ē-ŭs) Bony.

ossicle (os′i-kl) Small bone within the ear.

ossification (os′i-fi-kā′shŭn) Bone formation; also called osteogenesis.

osteoblast (os′tē-ō-blast) Bone-forming cell.

osteoclast (os′tē-ō-klast) Large cell type that functions in the absorption and removal of bone connective tissue.

osteocyte (os′tē-ō-sīt) Bone cell in a lacuna within bone matrix.

osteogenesis (os′tē-ō-jen′ē-sis) Production of new bone.

osteon (os′tē-on) Functional unit of compact bone tissue; also called a Haversian system.

osteopenia (os′tē-ō-pē′nē-ă) Decreased calcification or density of bone.

osteoporosis (os′tē-ō-pō-rō′sis) Medical condition characterized by decreased bone mass and increased susceptibility to fracture.

osteoprogenitor (os′tē-ō-prō-jen′i-ter) Precursor to bone cells.

ovulation (ov′ū-lā′shŭn) Release of a secondary oocyte from the ovarian follicle.

ovum (ō′vŭm; pl., ova, -vă) Female sex cell that has been fertilized and has completed meiosis II.

oxidation (ok-si-dā′shŭn) Occurs when a chemical loses an electron.

oxidation-reduction reaction Exchange reaction involving transfer of electrons from one chemical to another; also called a redox reaction.

oxidative phosphorylation (ok-si-dā-tiv fos′fōr-i-lā′shŭn) Process of forming ATP by using coenzymes to transfer energy.

P

palate (pal′ăt) Partition between the oral and nasal cavities; the roof of the mouth.

palpation (pal-pā′shŭn) Using the sense of touch to identify or examine internal body structures.

papilla (pă-pil′ă; pl., papillae, -pil′ē) Small, nipplelike process.

paracrine (par′a-krĭn) Hormonal secretions that stimulate neighboring cells; also called a local hormone (does not enter the blood). *Compare to endocrine.*

parietal (pă-rī′ē-tăl) Relating to the wall of any ventral body cavity.

partial pressure Pressure exerted by a gas in a mixture of gases.

passive immunity Activation of the immune system by administration of antibodies produced by another individual; offers short-term protection since memory cells are not formed. *Compare to active immunity.*

patent (pa′tent, pā′tent) Open or unblocked.

pathogen (path′ō-jen) Disease-causing substance or organism.

pathologic (path-ō-loj′-ik) Pertaining to disease.

pectoral (pek′tŏ-răl) Relating to the chest area.

pedal (ped′ăl) Relating to the foot.

peptide (pep′tīd) **bond** Chemical bond joining amino acid monomers.

perfusion (per-fyū′zhŭn) Blood flow through a tissue; measured in milliliters per minute per gram (mL/min/g).

pericardium (per-i-kar′dē-ŭm) Fibroserous membrane covering the heart.

perichondrium (per′i-kon′drē-ŭm) Layer of dense irregular connective tissue around the surface of cartilage.

perilymph (per′i-limf) Fluid within the osseous labyrinth that surrounds and protects the membranous labyrinth.

perimetrium (per-i-mē′trē-ŭm) Serous coat of the uterus.

perimysium (per-i-mis′ē-ŭm, -miz′ē-ŭm) Fibrous sheath enveloping each of the fascicles of skeletal muscle fibers.

perineurium (per-i-nū′rē-ŭm) Fibrous sheath of peripheral nerves that surrounds the nerve fascicles.

periosteum (per-ē-os′tē-ŭm) Thick, fibrous membrane covering the entire external surface of a bone, except for the articular cartilage on the epiphyses.

peripheral (pĕ-rif′-ē-răl) **nervous system (PNS)** Composed of neurons and neuroglial cells found outside of the central nervous system.

peristalsis (per-i-stal′sis) Rhythmic contractions that propel material through tubes in the gastrointestinal tract.

peritoneum (per′i-tō-nē′ŭm) Serous sac that lines the abdomino-pelvic cavity and covers most of the viscera within.

permeability (per′mē-ă-bil′i-tē) Capacity of a membrane to allow passage of a substance.

peroxisome (per-ok′si-sōm) Membrane-bound organelle containing oxidative enzymes.

pH Value indicating the relative hydrogen ion (H^+) and hydroxide ion (OH^-) concentrations of a solution.

phagocytosis (fāg′ō-sī-tō′sis) A form of endocytosis by which cells ingest and digest solid substances.

phalanx (fā′langks; pl., phalanges, -jēz) Long bone of a digit.

pharynx (far′ingks) Funnel-shaped muscular tube extending from the posterior nasal cavity to the esophagus and larynx; composed of the nasopharynx, oropharynx, and laryngopharynx.

phenotype (fē′nō-tīp) Observable characteristics of an individual.

phospholipid (fos-fō-lip′id) Lipid that forms bilayers of the plasma membrane.

phosphorylation (fos′fōr-i-lā′shŭn) Process of adding a phosphate group to another chemical.

photoreceptor Specialized receptor cells in the eye that detect light. *See rod; cone.*

physiology (fiz-ē-ol′ō-jē) Study of how body parts function together.

pinocytosis (pin′ō-sī-tō′sis) A form of endocytosis by which cells ingest liquid.

placenta (plă-sen′tă) Organ of exchange between the embryo or fetus and the mother.

plasma (plaz′mă) The liquid portion of blood.

plasma cell Cell that produces and releases specific antibodies; derived from B-lymphocytes.

plasma membrane Barrier separating the intracellular fluid from the interstitial fluid.

platelet (plăt′let) Irregularly shaped cell fragment that participates in blood clotting.

pleura (plŭr′ă; pl., pleurae, ploor′ē) Serous membranes enveloping the lungs and lining the internal walls of the thoracic cavity.

pleuripotent (plŭr′i′pō-tent) **cells** Embryonic stem cells that can differentiate into any cell type (except cells of the placenta).

plexus (plek′sŭs) Network of nerves, blood vessels, or lymph vessels.

polar (po′lăr) **molecule** Refers to partial electrical charges on different portions of a molecule.

polycythemia (pol′ē-sī-thē′mē-ă) More than the normal number of erythrocytes.

polymer (pol′-i-mer) Molecule composed of many repeating subunits.

polypeptide (pol-ē-pep′tīd) Chain of 21–199 amino acids linked by peptide bonds.

polysaccharide (pol-ē-sak′ă-rīd) Class of carbohydrates composed of three or more monosaccharide monomers; e.g., glycogen.

portal (pōr′tăl) **system** Unique organization of blood vessels sending blood from a capillary bed through a vein to another capillary bed.

positive feedback Control mechanism that increases the original change in a variable. *Compare to negative feedback.*

posterior Toward the back of the body.

postsynaptic (pōst-si-nap′tik) On the distal side of a synaptic cleft.

potential energy Stored energy.

pressure gradient Difference in pressure between two adjacent areas.

presynaptic (prē-si-nap′tik) On the proximal side of a synaptic cleft.

prion (prī′on) Small disease-causing protein.

pronation (prō-nā′shŭn) Rotational movement of the forearm such that the hand faces posteriorly or inferiorly.

proprioceptor (prō′prē-ō-sep′ter) Sensory receptor that detects position change or state of contraction.

prostaglandin (pros-tă-glan′din) Local hormone derived from a fatty acid.

protein Organic molecule composed of one or more chains of amino acid monomers.

proton (prō′ton) Subatomic particle with a positive charge; found in the nucleus of an atom.

protraction (prō-trak′shun) Movement of a body part anteriorly in a horizontal plane (i.e., thrusting out of the mandible [chin]).

protuberance (prō-tū′ber-ans) Swelling or knoblike outgrowth.

proximal (prok′si-măl) Near the point of attachment to the trunk.

puberty Time period when reproductive organs become fully functional and secondary sex characteristics become more prominent.

pulmonary (pŭl′mō-năr-ē) Relating to the lung.

pulse (pŭls) Rhythmic dilation of an artery resulting from passage of increased blood volume during heart contraction.

pupil (pū′pl) Circular hole in the center of the iris of the eye.

pus (pŭs) Substance composed of dead pathogens, white blood cells, and cellular debris.

pyrogen (pī′rō-jen) Fever-producing substance.

R

radioactivity Emission of high energy radiation from a radioisotope.

radioisotope (rā′dē-ō-ī′sō-tōp) Unstable isotope that emits high-energy radiation (e.g., gamma rays) until it decays into a stable isotope.

ramus (rā′mŭs; pl., rami, rā′mī) One of the primary divisions of a nerve or blood vessel. Also, part of an irregularly shaped bone that forms an angle.

raphe (rā′fē) Line of union between two contiguous, bilaterally symmetrical structures; e.g., the raphe of the scrotum.

receptor Structure that detects a stimulus.

reduction Occurs when a chemical gains an electron.

referred pain Pain sensed, not from the organ, but from an unrelated region of the body.

reflex Fast, predetermined response to a stimulus.

refraction (rē-frak′shŭn) Bending of light waves as they pass through a material.

refractory (rē-frak′tōr-ē) **period** Time period when a cell is either unresponsive or less responsive to a stimulus.

renal (rē′năl) Relating to the kidney.

repolarization (rē′pō-lăr-i-zā′shŭn) Change in membrane potential from a depolarized value back to the resting value.

respiration Exchange of O_2 and CO_2 between air and systemic body cells.

respiratory zone Respiratory system passageways transporting air from respiratory bronchioles to the alveoli; airways that participate in gas exchange. *Compare to conducting zone.*

resting membrane potential (RMP) Voltage measured across the plasma membrane of an excitable cell at rest.

reticular (re-tik′ū-lăr) **fibers** Proteins that form a flexible network in tissues; found in some connective tissues.

reticulocyte (re-tik′ū-lō-sīt) Immature erythrocyte.

retroperitoneal (re′trō-per′i-tō nē′ăl) External or posterior to the peritoneum.

ribonucleic (rī′bō-nū-klē′ik) **acid (RNA)** Nucleic acid composed of ribonucleotide monomers; used to direct protein synthesis based on instructions in DNA. *See messenger RNA, ribosomal RNA, and transfer RNA.*

ribosome (rī′bō-sōm) Ribonucleoprotein structure that is the site of protein synthesis; organelle in the cytoplasm.

rod Rod-shaped cell in retina that detects light and can work even in dim light.

rotation (rō-tā′shŭn) Movement of a part around its axis.

rouleau (rū-lō′; pl., rouleaux) Aggregation of erythrocytes in single file.

ruga (rū′gă; pl., rugae, rū′gē) Fold, ridge, or crease.

S

sacroiliac (sā-krō-il′ē-ak) Pertaining to the sacrum and the ilium (e.g., sacroiliac joint).

sacrum (sā′krŭm) Next-to-last group of bones in the vertebral column; formed of five fused vertebrae.

sagittal (saj′i-tăl) Relating to a plane cutting the body into a left and a right part.

salt Substance formed by an ionic bond.

saltatory (sal′tă-tōr-ē) **conduction** Transmission of an action potential along the plasma membrane of a myelinated axon.

sarcolemma (sar′kō-lem′ă) Plasma membrane of a muscle cell.

sarcomere (sar′kō-mēr) Functional unit of skeletal muscle.

sarcoplasm (sar′kō-plazm) Cytoplasm of a muscle cell.

sarcoplasmic reticulum (sar-ko′-plaz′mik re-tik′ū-lŭm) **(SR)** Endoplasmic reticulum found in muscle cells; stores calcium ions required for contraction.

sclera (skler′ă) Portion of the fibrous layer forming the outer covering of the eyeball, except for the cornea; "white" of the eye.

scoliosis (skō-lē-ō′sis) Abnormal lateral and rotational curvature of the vertebral column.

sebum (sē′bŭm) Secretion of a sebaceous gland.

second messenger Intracellular chemical modifying activity within a cell after a first messenger binds to the plasma membrane.

selective permeability Ability to regulate what can cross a membrane.

semen (sē′men) Secretion composed of sperm and seminal fluid.

semipermeable (sem′ē-per′mē-ă-bl) Freely permeable to some molecules, but relatively non-permeable to other molecules.

sensation (sen-sā′shun) Conscious perception of a stimulus.

sensory (afferent) nerve Nerves sending information to the central nervous system.

septum (sep′tŭm; pl., septa) Wall of tissue separating nearby chambers; e.g., nasal septum.

serosa (se-rō′să) Outermost coat of a visceral structure that lies in a closed body cavity.

serous (sēr′ŭs) Producing a substance that has a watery consistency.

serum (sēr′ŭm) Clear, watery fluid that remains after clotting proteins are removed from plasma.

set point Normal value of a variable.

simple diffusion Passive transport process used when a chemical slips between plasma membrane phospholipids to enter or leave a cell.

sinoatrial (sī′nō-ā′trē-ăl) **(SA) node** Group of cells in right atrium that set the heart rate.

sinus (sī′nŭs) Cavity or hollow space; also, a channel for the passage of blood or lymph.

sinusoid (sī′nŭ-soyd) Resembling a sinus; extremely permeable capillary in certain organs, such as the liver or the spleen.

skeleton (skel′ē-tŏn) Bony framework of the body; collectively, all the bones of the body.

solute (sol′ūt) Substance dissolving in a solvent.

solution A homogeneous mixture composed of a solvent and a solute.

solvent (sol′vent) Substance (e.g., water) holding a solute in solution.

somite (sō′mīt) One of the paired segments consisting of cell masses formed in the early embryonic mesoderm on the sides of the neural tube.

sonography (sŏ-nog′ră-fē) Radiographic technique using ultrasound waves.

sperm (spermatozoon) Male gamete produced in the testes.

spermatogenesis (sper′mă-tō-jen′ĕ-sis) Process by which stem cells called spermatogonia become sperm.

sphincter (sfingk′ter) Muscle that encircles a duct, tube, or orifice such that its contraction constricts the lumen or orifice and restricts movement through the tube.

sphygmomanometer (sfig′mō-mă-nom′ĕ-ter) Instrument for measuring arterial blood pressure.

splanchnic (splangk′nik) Relating to the viscera.

sprain Tearing or overstretching ligaments without fracturing the nearby bone; results in localized pain and swelling.

squamous (skwā′mŭs) Referring to cell or area that is flat.

stem cells Unspecialized, immature cells.

stenosis (ste-nō′sis) Inability of a tube or valve to open completely; e.g., valvular stenosis.

steroid (stēr′oyd, ster′oyd) Group of lipids including bile salts, cholesterol, and some hormones; molecule is composed of four attached hydrocarbon rings.

stimulus Change in a regulated variable; event that provokes a cellular response.

stria (strī′ă; pl., striae, strī′ē) Stripes, bands, streaks, or lines distinguished by a difference in color, texture, or elevation from surrounding tissue.

stupor (stū′per) State of impaired consciousness from which the individual can be aroused only by continual stimulation.

subcutaneous (sŭb-kū-tā′nē-ŭs) Immediately internal to the integument.

subluxation (sŭb-lŭk-sā′shŭn) Incomplete dislocation of a body part from its normal position.

substrate (sŭb′strāt) Substance binding to the active site of an enzyme; reactant in a chemical reaction.

sulcus (sŭl′kŭs; pl., sulci, sŭl′sī) Groove or furrow (e.g., on the surface of the brain).

summation (sŭm-ā′shŭn) Accumulation of stimuli occurring in neurons or muscle cells; e.g., wave summation.

superficial (sū-per-fish′ăl) Toward the surface.

supination (sū′pi-nā′shŭn) Rotation of the forearm such that the hand faces anteriorly or superiorly.

suspension Mixture of a solvent with large materials that do not dissolve (e.g., blood).

suture (sū′chūr) Synarthrosis in which cranial bones are united by a dense regular connective tissue membrane.

symphysis (sim′fi-sis; pl., symphyses, -sēz) Cartilaginous joint in which the two bones are separated by a pad of fibrocartilage.

synapse (sin′aps) Functional contact of a nerve cell with another nerve cell, effector, or receptor.

synaptic vesicle (si-nap′tik ves′i-kl) Package of plasma membrane enclosing neurotransmitter molecules in the synaptic knob.

syndesmosis (sin′dez-mō′sis; pl., syndesmoses, -sēz) Fibrous joint in which the opposing surfaces of articulating bones are united by ligaments.

synergist (sin′er-jist) Structure, muscle, agent, or process that aids the action of another.

synostosis (sin-os-tō′sis; pl., synostoses, -sēz) Osseous union between two bones that were initially separate.

system Group of interacting structures working as a whole entity.

systemic (sis-tem′ik) Relating to the entire organism as opposed to any of its individual parts.

systole (sis′tō-lē) Contraction of the heart.

systolic (sis-tol′ik) **pressure** Blood pressure measured in an artery during systole.

T

T-lymphocyte (lim′fō-sīt) Immune cell that matures in the thymus gland.

tachycardia (tak′i-kar′dē-ă) Resting heart rate above 100 beats per minute.

tactile (tak′til) Relating to touch.

tendon (ten′dŏn) Cord of dense regular connective tissue that connects muscle to bone.

teratogen (ter′ă-tō-jen, te-ră′tō-jēn) Substance that may cause death or malformation of an embryo.

testis (tes′tis; pl., testes, -tēz) Male organ for producing gametes and androgens.

tetanus (tet′ă-nŭs) Sustained contraction that may lead to spastic paralysis.

thalamus (thal′ă-mŭs) Region of the brain in the diencephalon involved in motor control and relaying sensory information to higher brain centers.

thermoreceptor (ther′mō-rē-sep′ter, -tōr) Receptor that is sensitive to heat.

thoracic (thō-ras′ik) Relating to the thorax, the area between the neck and the abdomen.

thrombopoiesis (throm′bō-poy-ē′sis) Formation of blood platelets.

thrombus (throm′bŭs) Blood clot formed within a blood vessel during hypercoagulation; may become an embolus if dislodged.

tissue Groups of similar types of cells performing a common function for the body.

tomography (tō-mog′ră-fē) A radiographic image of a plane constructed by means of reciprocal linear or curved motion of the x-ray tube and film cassette; used in producing a CT scan.

totipotent (tō-tip′ŏ-tent) **cells** Embryonic stem cells that can differentiate into any cell of the body or placenta.

trabecula (tră-bek′ū-lă; pl., trabeculae, -lē) Meshwork (e.g., trabeculae within spongy bone).

tract (trakt) Bundle of central nervous system axons traveling along the same path.

transcription (tran-skrip′shŭn) Copying information from DNA to form an RNA molecule.

transducer (tranz-dū′ser) Device or organ designed to convert energy from one form to another.

translation Process involving RNA and ribosomes to produce a new protein.

transverse (trans-vers′) Relating to a plane cutting a structure into a superior and inferior portion.

transverse (T) tubule (tū′būl) Invaginations of the sarcolemma that allow action potentials to approach and stimulate the sarcoplasmic reticulum.

triglyceride (trī-glis′er-īd) Lipid providing long-term energy storage in adipose connective tissue; composed of glycerol and three fatty acids.

trophoblast (trof′ō-blast, trō′fō- blast) Cell layer covering the blastocyst that will allow the embryo to receive nourishment from the mother.

tubercle (tū′ber-kl) Nodule or slight elevation from the surface of a bone that allows attachment of a tendon or ligament.

tuberosity (tū′ber-os′i-tē) Large tubercle or rounded elevation, as on the surface of a bone.

tumor Mass of new tissue cells resulting from abnormal cell division.

tunic (tū′nik) One of the covering layers of a body part, such as a blood vessel or muscle.

U

ultrastructure (ŭl-tră-strŭk′chūr) Cell structure viewable via the electron microscope.

umbilical (ŭm-bil′i-kăl) Relating to the umbilicus (navel).

unipotent (ū-ni′pŏ-tent) **stem cells** Adult stem cell that can differentiate into only one specific cell type.

unmyelinated (ŭn-mī′ĕ-li-nā-ted) Not covered by a myelin sheath.

urea (ū-rē′ă) A nitrogenous waste product produced by amino acid metabolism and eliminated in urine.

ureter (ū-rē′ter, ū′rē-ter) Tubes that connect the kidney to the urinary bladder.

urethra (ū-rē′thră) Tube that extends from the bladder and conducts urine to the exterior of the body.

uric (ūr′ik) **acid** Waste product of nucleic acid metabolism eliminated in urine.

urinalysis (ū-ri-nal′i-sis) Analysis of the urine to help assess the state of a person's health.

urine (ūr′in) Fluid and dissolved substances excreted by the kidney.

V

vaccine A weakened pathogen, or part of a pathogen, used to stimulate the immune system and produce memory cells; provides artificially acquired active immunity.

varicose (var′i-kōs) **vein** Vein that is extremely dilated and follows a twisting path due to poorly functioning venous valves.

vascular (vas′kū-lăr) Relating to blood vessels.

vasoconstriction (vă′sō-kon-strik′shŭn) Narrowing of blood vessel lumen.

vasodilation (vā′sō-dī-lā′shŭn) Increased blood vessel lumen.

vein (vān) Blood vessel carrying blood toward the heart.

vellus (vel′ŭs) Fine, nonpigmented hair covering most of the fetal body.

ventral (ven′trăl) Toward the belly.

ventricle (ven′tri-kl) Cavity within an organ such as the heart or the brain.

venules (ven′ūl) The smallest veins.

vertebra (ver′tĕ-bră; pl., vertebrae, -bră) Segment of the vertebral column.

vertebral (ver′tĕ-brăl) Relating to a vertebra or the vertebral column.

vesicle (ves′i-kl) Closed cellular structure within the cytoplasm surrounded by a single membrane.

vestibule (ves′ti-byūl) Small space near the entryway of a canal (e.g., nasal vestibule, inner ear vestibule)

viscera (vis′er-ă) Internal organs of the body; especially those in the thoracic and abdominopelvic cavities.

viscosity (vis-kos′i-tē) Thickness of a solution; provides resistance to fluid flow.

volatile (vol′ă-til) **acid** In the body, an acid produced from carbon dioxide (e.g., carbonic acid).

voltage Measure of potential energy in separated charges; expressed in volts.

vulva (vŭl′vă) External genitalia of the female.

W

white matter Brain or spinal cord tissue that derives its color from myelin in myelinated axons.

Y

yolk sac Extraembryonic membrane; site of early blood cell formation in the embryo.

Z

zygote (zī′gōt) Diploid cell resulting from the union of a sperm and a secondary oocyte.

credits

PHOTO CREDITS

About the Authors

Page iv: © Jan McKinley

Chapter 1

Opener: © Marmaduke St. John/Alamy RF; 1.1 (top left organelle): © Dennis Kunkel Microscopy, Inc./Phototake; 1.1 (top right organelle): © Keith R. Porter/Science Source; 1.1 (bottom left organelle): © Don W. Fawcett/Science Source; 1.1 (bottom right organelle): © EM Research Services, Newcastle University; 1.4a: © McGraw-Hill Education/Eric Wise, photographer; 1.4b: © James Cavallini/Science Source; 1.4c: © Trevor Lush/Getty Images; 1.4d: © ISM/Phototake; p. 24 (X-ray): © Medical Body Scans/Science Source; p. 24 (Sonogram): © ATL/Science Source; p. 24 (CT scan): © SIU BioMed Com/CustomMedical; p. 25 (DSA): © Athenais/Phototake; p. 25 (MRI): © Alfred Pasieka/Science Source; p. 25 (PET): © Hank Morgan/Science Source.

Chapter 2

Opener: Brenda J. Jones/CDC; p. 33: © Frédéric Astier/Science Source.

Chapter 3

Opener: James Gathany/CDC; p. 86: © Creatas/PunchStock RF.

Chapter 4

Opener: © Science Photo Library/agefotostock RF; 4.1a: © McGraw-Hill Education/Al Telser, photographer; 4.1b: © VVG/SPL/Science Source; 4.1c: © Eye of Science/Science Source; 4.5b: © Don W. Fawcett/Science Source; 4.13a-c, 4.22: © Dennis Kunkel Microscopy, Inc./Phototake; 4.23a: © Biophoto Associates/Science Source; 4.24: © Science Source; 4.25, 4.26: © Don W. Fawcett/Science Source; 4.27b: © Dennis Kunkel Microscopy, Inc./Phototake; 4.29: © Don W. Fawcett/Science Source; 4.30 (greyscale 3D proteasome model): © Edward P. Morris; 4.31, 4.34a: © Don W. Fawcett/Science Source; 4.41a: © Michael Abbey/Science Source; 4.41b-d: © Carolina Biological Supply Company/Phototake; 4.41e: © Michael Abbey/Science Source.

Chapter 5

Opener: © Antenna/Getty Images RF; Table 5.2a: © Dr. Thomas Caceci, Virginia-Maryland Regional College of Veterinary Medicine; Table 5.2b-c, e: © McGraw-Hill Education/Al Telser, photographer; Table 5.2d: © Victor P. Eroschenko; Table 5.2f: © Alvin Telser, Ph.D.; Table 5.3a-d, & e (1): © McGraw-Hill Education/Al Telser, photographer; Table 5.3e (2): © Victor P. Eroschenko; p. 164: © Comstock/Jupiterimages RF; Table 5.4a: © McGraw-Hill Education/Al Telser, photographer; Table 5.4b: © Ed Reschke; Table 5.5a-c: © McGraw-Hill Education/Al Telser, photographer; p. 167 (Marfan syndrome): © The Marfan Foundation; Table 5.6a: © Ed Reschke; Table 5.6b: © McGraw-Hill Education/Dennis Strete, photographer; Table 5.6c: © McGraw-Hill Education/Al Telser, photographer; Table 5.7a-b: © Ed Reschke; Table 5.7c: © McGraw-Hill Education/Al Telser, photographer; Table 5.8: © McGraw-Hill Education/Dennis Strete, photographer; Table 5.9: © McGraw-Hill Education/Al Telser, photographer; Table 5.10a: © Ed Reschke/Getty Images; Table 5.10b-c: © Victor P. Eroschenko; Table 5.11: © Carolina Biological Supply Company/Phototake; p. 179 (Red bone marrow cells): © Leonard Lessin/Science Source; p. 180 (Dry gangrene): © Dr. P. Marazzi/Science Source; p. 180 (Gas gangrene): © Stevie Grand/Science Source.

Chapter 6

Opener: © agefotostock/SuperStock; 6.2a: © Ed Reschke/Getty Images; p. 188 (microscope): © Steve Cole/Getty Images RF; 6.3a-b: © Carolina Biological Supply Company/Phototake; 6.4b: © John Burbidge/Science Source; p. 190 (sunscreen): © McGraw-Hill Education/Jill Braaten, photographer; p. 192 (tattoo): © Ingram Publishing RF; p. 198 (Onychomycosis): © Dr. P. Marazzi/Science Source; p. 198 (Yellow nail syndrome): © Logical Images, Inc.; p. 198 (Spoon nails): © John Radcliffe Hospital/Science Source; p. 198 (Beau's lines): © Dr. P. Marazzi/Science Source; 6.10b (1-2): © McGraw-Hill Education/Al Telser, photographer; 6.10c: © Science Source; 6.11b: © ISM/Phototake; 6.11c: © Alvin Telser, Ph.D.; 6.11d: © McGraw-Hill Education/Al Telser, photographer; p. 203: © Dr. Zara/Science Source; p. 205 (First-degree): © Sheila Terry/Science Source; p. 205 (Second-degree): © Dr. P. Marazzi/Science Source; p. 205 (Third-degree): © John Radcliffe Hospital/Science Source; p. 206: © Science Photo Library RF/Getty Images; Table 6.2 (Basal, Squamous): © Dr. P. Marazzi/Science Source; Table 6.2 (Malignant): © James Stevenson/Science Source.

Chapter 7

Opener: © AJPhoto/Hôpital Américain/Science Source; 7.4: © Susumu Nishinaga/Science Source; 7.5b: © Dr. M. Laurent, University Hospitals Leuven, Belgium. Image is available under a creative commons attribution license; 7.6c: © Alvin Telser, Ph.D.; p. 218 (X-ray): © SPL/Science Source; p. 218 (full bone, bent bone, shattered): © Trent Stephens; p. 220 (archery target): © C Squared Studios/Getty Images RF; 7.8a: © Carolina Biological Supply Company/Phototake; 7.8b: © Andrew Syred/Science Source; 7.8c: © Biophoto Associates/Science Source; 7.9 (top): © McGraw-Hill Education/Al Telser, photographer; p. 225: © David Hunt/Smithsonian Institution; 7.11 (Ten-week fetus): © Biophoto Associates/Science Source; 7.11 (Sixteen-week fetus): © Tissuepix/Science Source; 7.11 (Skeleton): © MShieldsPhotos/Alamy; 7.11 (Humerus): © Bone Clones; 7.11 (X-ray): © ZEPHYR/SPL/Getty Images RF; 7.12a: © McGraw-Hill Education/Al Telser, photographer; 7.12b: © Yoav Levy/Phototake; p. 232 (X-ray): © Dr. LR/Science Source; p. 233 (Normal bone): © SPL/Science Source; p. 233 (Osteoporotic bone): © Professor Pietro M. Motta/Science Source; 7.16 (Comminuted): © Medical-on-Line/Alamy; 7.16 (Compression): © ISM/Phototake; 7.16 (Transverse): © Wellcome Photo Library, Wellcome Images; p. 235 (bone scan): John Pape/MedPix™. Website: rad.usuhs.edu/medpix.

Chapter 8

Opener: © Dennis MacDonald/PhotoEdit; 8.4, 8.6: © McGraw-Hill Education/Christine Eckel, photographer; p. 255 (Cleft lip): © Dr. M.A. Ansary/SPL/Science Source; p. 255 (Cleft palate): © Wellcome Image Library/CustomMedical; p. 257 (left, middle): Dr. John A. Jane, Sr., David D. Weaver Professor of Neurosurgery, Department of Neurological Surgery, University of Virginia Health System, Charlottesville, Virginia; p. 257 (right): Used with permission and copyright of Cranial Technologies, Inc.; 8.12b: © McGraw-Hill Education/Photo and Dissection by Christine Eckel; Table 8.4, 8.16b, Table 8.5, 8.21c, 8.23c, 8.25c, 8.26c-d: © McGraw-Hill Education/Christine Eckel, photographer; 8.28b: © Image Source/Getty Images RF; Table 8.6 (top): © David Hunt/Smithsonian Institution; Table 8.6 (bottom): © VideoSurgery/Science Source; 8.33a-b: © McGraw-Hill Education/Christine Eckel, photographer; p. 288 (Bunion): © Dr. P. Marazzi/Science Source; p. 288 (Pes cavus): © SPL/Science Source; p. 288 (congenital clubfoot): © Medical-on-Line/Alamy; p. 288 (Pes planus): © Bart's Medical Library/Phototake; p. 288 (Metatarsal stress fractures): © ISM/Phototake; p. 292: © Science Photo Library/Science Source.

Chapter 9

Opener: © Russell Illig/Getty Images RF; 9.8a, d: © McGraw-Hill Education/Eric Wise, photographer; 9.8b-c, e, 9.9a-c: © McGraw-Hill Education/JW Ramsey, photographer; 9.9d: © McGraw-Hill Education/Eric Wise, photographer; 9.10, 9.11: © McGraw-Hill Education/JW Ramsey, photographer; 9.12a, d-e: © McGraw-Hill Education/Eric Wise, photographer; 9.12b-c: © McGraw-Hill Education/JW Ramsey, photographer; 9.15a: © McGraw-Hill Education/Photo and Dissection by Christine Eckel; p. 316 (left): © & Courtesy of Dr. Mike Langran/www.ski-injury.com; p. 316 (right): © Collection CNRI/Phototake; p. 317: © sot/Photodisc/Getty Images RF; 9.17d: © McGraw-Hill Education/Photo and Dissection by Christine Eckel; p. 323 (Arthroscopic view): © CNRI/Science Source; p. 326 (left): © John Watney/Science Source; p. 326 (right): © CNRI/Science Source.

Chapter 10

Opener: © Tony McConnell/Science Source; p. 338: © McGraw-Hill Education/Mark A.S. Dierker, photographer; 10.6b: © Dr. Thomas Caceci, Virginia-Maryland Regional College of Veterinary Medicine; 10.14: © John Heuser, Washington University School of Medicine, St. Louis, MO; 10.15a-b: © Dr. H.E. Huxley; 10.20: © Dr. Gladden Willis/Visuals Unlimited/Corbis; p. 357 (top): © Ed Scott/agefotostock RF; p. 359: © George Doyle/Getty Images RF; p. 360: © Ingram Publishing RF; 10.27 (blood vessel): © McGraw-Hill Education/Christine Eckel, photographer; 10.27 (bronchiole): © Dr. Gladden Willis/Visuals Unlimited/Corbis; 10.27 (large intestine): © Victor P. Eroschenko; 10.27 (urinary bladder): © McGraw-Hill Education/Al Telser, photographer; 10.27 (uterus): © Custom Medical Stock Photo/Newscom; Table 10.2 (Skeletal Muscle): © Ed Reschke/Getty Images; Table 10.2 (Cardiac Muscle, Smooth Muscle): © Victor P. Eroschenko.

Chapter 11

Opener: © Fotosearch RM/agefotostock; p. 376: © Dr. P. Marazzi/SPL/Science Source; 11.6: © McGraw-Hill Education/JW Ramsey, photographer; p. 389: © SC Photo; 11.15b, 11.16b: © McGraw-Hill Education/Photos and Dissections by Christine Eckel; 11.22a: © Elsa/Getty Images; 11.22b: © Brita Meng Outzen/AP Photo; 11.22c: © Mike Fiala/Getty Images; p. 409: © Karl Weatherly/Getty Images RF; 11.27a: © McGraw-Hill Education/Photo and Dissection by Christine Eckel; 11.28: © McGraw-Hill Education/JW Ramsey, photographer; 11.29a: © McGraw-Hill Education/Photo and Dissection by Christine Eckel; p. 424: © Itstock/Jupiterimages RF; 11.32a, 11.33a: © McGraw-Hill Education/Photos and Dissections by Christine Eckel; p. 427: © Digital Vision RF; 11.35a: © McGraw-Hill Education/Photo and Dissection by Christine Eckel.

Chapter 12

Opener: © BSIP/Science Source; 12.2b: © Ed Reschke/Getty Images; 12.4b: © Dr. Richard Kessel & Dr. Randy Kardon/Corbis; p. 447: © Simon Fraser/Science Source; 12.9b: © Dr. Donald Fawcett/Visuals Unlimited/Corbis.

12.16: © Science VU/Lewis-Everhart-Zeevi/Visuals Unlimited/Corbis; 12.22b: © Don W. Fawcett/T. Reese/ Science Source.

Chapter 13

Opener: Courtesy Kristine Queck; 13.1a-c: © McGraw-Hill Education/Photos and Dissections by Christine Eckel; p. 489 (anencephaly): © O.J. Staats M.D./Custom Medical Stock Photo/Newscom; p. 489 (spina bifida): © Wellcome Image Library/CustomMedical; 13.6: © McGraw-Hill Education/Photo and Dissection by Christine Eckel; p. 496 (hydrocephalus): © Ansary/Custom Medical Stock Photo/Newscom; 13.8a: © McGraw-Hill Education/Photo and Dissection by Christine Eckel; p. 508: © Kenneth Lambert/AP Photo; 13.28: © Daniel Mihailescu/AFP/Getty Images; p. 522 (normal): © Stevie Grand/Science Source; p. 522 (Alzheimer): © James Cavallini/Custom Medical Stock Photo/Newscom; 13.31b (1-2): © WDCN/Univ. College London/Science Source; 13.31b (3): © National Cancer Institute/Science Source; 13.32: © McGraw-Hill Education/Rebecca Gray, photographer.

Chapter 14

Opener: © Creatas Images/PunchStock RF; 14.1b: © McGraw-Hill Education/Photo and Dissection by Christine Eckel; 14.1c: From: Anatomy & Physiology Revealed, © McGraw-Hill Education/The University of Toledo, photography and dissection; 14.4b: © Dr. David Phillips/Science Source; p. 552: © Dr. Valerie Dean O'Loughlin; 14.17b-14.19b: © McGraw-Hill Education/ Photos and Dissections by Christine Eckel.

Chapter 15

Opener: © Tek Image/Science Source; 15.7: © McGraw-Hill Education/Photo and Dissection by Christine Eckel; p. 587: © Medical-on-Line/Alamy.

Chapter 16

Opener: © Corbis RF; p. 607: © Getty Images/Stockbyte RF; 16.6d: © McGraw-Hill Education/Al Telser, photographer; 16.8a: © McGraw-Hill Education/JW Ramsey, photographer; 16.12c: © McGraw-Hill Education/ Al Telser, photographer; 16.13a: © Paul Whitten/Science Source; p. 621: © Steve Mason/Getty Images RF; p. 622 (eye without cataract): © James P. Gilman, CRA/ Phototake; p. 622 (normal vision & image seen through cataract): © David Buffington/Getty Images RF; p. 622 (eye with cataract): © Dr. P. Marazzi/Science Source; 16.16: © Charles D. Winters/Science Source; p. 626: © Steve Allen/Getty Images RF; p. 635: © ISM/Phototake; 16.26d: © Dr. John D. Cunningham/Visuals Unlimited/ Corbis; p. 646: © McGraw-Hill Education/Jacques Cornell, photographer.

Chapter 17

Opener: © Lea Paterson/Science Source; p. 674: © Incredible Features/Barcroft Media/Getty Images; 17.15a: © McGraw-Hill Education/Photo and Dissection by Christine Eckel; 17.15b: © McGraw-Hill Education/ Al Telser, photographer; p. 678 (left): © Chris Barry/ Phototake; p. 678 (right): © Scott Camazine/Science Source; 17.18a: © McGraw-Hill Education/Photo and Dissection by Christine Eckel; 17.18c: © McGraw-Hill Education/Al Telser, photographer; p. 682: Courtesy of the Cushing's Support and Research Foundation, www.CSRF. net and Kathy Carbone; p. 683: © fStop/Getty Images RF; 17.21b: © Ed Reschke/Getty Images; p. 686: © Ian Hooton/SPL/Getty Images RF.

Chapter 18

Opener: © Arno Massee/Science Source; 18.2, 18.4a: © McGraw-Hill Education/Al Telser, photographer; 18.5b: © Ed Reschke/Getty Images; p. 714: © Eye of Science/ Science Source; p. 716: © liquidlibrary/PictureQuest RF;

18.10b: © Jean Claude Revy/ISM/Phototake; Table 18.7 (Lymphocyte, Neutrophil, Eosinophil, Basophil): © McGraw-Hill Education/Al Telser, photographer; Table 18.7 (Monocyte): © Herve Conge/Phototake; 18.11: Courtesy of John Weisel.

Chapter 19

Opener: © Robin Nelson/PhotoEdit; 19.7a-b: © McGraw-Hill Education/Photos and Dissections by Christine Eckel; p. 744: © ISM/Phototake; 19.11d: © Victor P. Eroschenko; 19.12b: © Dr. Carlos Baptista and Roy Schneider, University of Toledo HSC; 19.13c: © Ralph Hutchings/ Visuals Unlimited/Corbis; p. 749: © McGraw-Hill Education/Ken Karp, photographer.

Chapter 20

Opener: © Mark Harmel/Getty Images; 20.2: © Dr. Thomas Caceci, Virginia-Maryland Regional College of Veterinary Medicine; 20.4a-c & p. 780 (Normal artery): © McGraw-Hill Education/Al Telser, photographer; p. 780 (Athero-sclerotic artery): © CNRI/Science Source; p. 796: © Bart's Medical Library/Phototake; 20.25: © Victor P. Eroschenko.

Chapter 21

Opener: © Marty Heitner/The Image Works; p. 838: © Andy Crumo, TDR, WHO/Science Source; 21.5b: © McGraw-Hill Education/Al Telser, photographer; 21.5c: © Dr. Thomas Caceci, Virginia-Maryland Regional College of Veterinary Medicine; 21.6c: © Ed Reschke/ Getty Images; p. 842: © NYU Franklin Research Fund/ Phototake; 21.7a-b: © McGraw-Hill Education/Photos and Dissections by Christine Eckel; 21.7d: © McGraw-Hill Education/Al Telser, photographer; p. 844 (Acute tonsillitis): © Dr. P. Marazzi/Science Source; 21.8b: © Dr. Thomas Caceci, Virginia-Maryland Regional College of Veterinary Medicine; 21.8c: © McGraw-Hill Education/ Al Telser, photographer.

Chapter 22

Opener: © PHANIE/Science Source; Table 22.1 (1): Dr. William A. Clark/CDC; Table 22.1 (2, 5): CDC; Table 22.1 (3): Dr. Godon Roberstad/CDC; Table 22.1 (4): Janice Haney Carr/CDC; p. 857 (bottom left): © Adam Gault/ Getty Images RF; p. 861: © Image Source/Getty Images RF; p. 882: © Blend Images/Jupiterimages RF.

Chapter 23

Opener: © Science Photo Library/Alamy RF; 23.3d: © McGraw-Hill Education/Photo and Dissection by Christine Eckel; 23.7c: © ISM/Phototake; p. 899: © C. Richard Stasney, M.D.; 23.8b: © Lester V. Bergman/ Corbis; 23.8c: © SPL/Science Source; 23.11b: © McGraw-Hill Education/Al Telser, photographer; 23.11c: © Dr. David Phillips/Visuals Unlimited/Corbis; p. 909 (top left): © McGraw-Hill Education/Al Telser, photographer; p. 909 (top right, bottom right): © Ralph Hutchings/Visuals Unlimited/Corbis; p. 909 (bottom left): © Mike Peres RBP SPAS/CustomMedical; p. 928a: © CNRI/Science Source; p. 928b: © McGraw-Hill Education/Al Telser, photographer.

Chapter 24

Opener: © Javier Larrea/agefotostock; 24.1a: © McGraw-Hill Education/Photo and Dissection by Christine Eckel; 24.3: © McGraw-Hill Education/Rebecca Gray, photographer/Don Kincaid, dissections; 24.6, 24.25b: © McGraw-Hill Education/Al Telser, photographer; p. 979: © SPL/Science Source; 24.26b: © Dr. Frederick Skvara/ Visuals Unlimited/Corbis.

Chapter 25

Opener: © CC Studio/Science Source; 25.5a: © Bart's Medical Library/Phototake; p. 994 (bottom): © Cade Martin/agefotostock RF.

Chapter 26

Opener: © By Ian Miles-Flashpoint Pictures/Alamy; 26.3a-b: © McGraw-Hill Education/Photos and Dissections by Christine Eckel; 26.5c: © McGraw-Hill Education/Al Telser, photographer; 26.7b: © Alfred Pasieka/Science Source; p. 1031 (normal esophagus): © Gastrolab/Science Source; p. 1031 (Barrett esophagus): © Caliendo/Custom Medical Stock Photo/Newscom; 26.9b: © Victor P. Eroschenko; 26.9c: © McGraw-Hill Education/ Photo and Dissection by Christine Eckel; 26.10b: © McGraw-Hill Education/Al Telser, photographer; p. 1038: © Javier Domingo/Phototake; 26.17a: © McGraw-Hill Education/Al Telser, photographer; 26.17b: © Dr. Lee Peachey; p. 1046 (left): © Dr. Joseph William/Phototake; p. 1046 (right): © Ida Wyman/ Phototake; 26.20c (top): © McGraw-Hill Education/ Al Telser, photographer; 26.20c (bottom): © Victor P. Eroschenko; 26.20c (bottom): © Dr. Gladden Willis/Visuals Unlimited/Corbis; 26.21b: © McGraw-Hill Education/ Photo and Dissection by Christine Eckel; 26.22 (top): © Carolina Biological Supply Company/Phototake; 26.22 (bottom): © Carolina Biological/Visuals Unlimited/Corbis; p. 1050: © Medicimage/Phototake; 26.24b: © McGraw-Hill Education/Al Telser, photographer; p. 1053 (bottom): © Gastrolab/SPL/Science Source; p. 1057: © Weeping Willow/Getty Images RF.

Chapter 27

Opener: © BURGER/PHANIE/Science Source; p. 1070: © Fuse/Getty Images RF; 27.1: USDA; p. 1088: © Annie Griffiths Belt/Getty Images.

Chapter 28

Opener: © Jason Edwards/National Geographic Creative; p. 1098: © Hattie Young/Science Source; 28.3b: © McGraw-Hill Education/Photo and Dissection by Christine Eckel; 28.6 (Primordial follicle, Primary follicle, Corpus luteum, Corpus albicans): © McGraw-Hill Education/Al Telser, photographer; 28.6 (Secondary follicle, Mature follicle): © Ed Reschke/Getty Images; 28.6 (Ovulated secondary oocyte): © Dr. Francisco Gaytan; 28.9b-c: © McGraw-Hill Education/Al Telser, photographer; p. 1110: From: Operational Obstetrics & Gynecology - 2nd Edition. The Health Care of Women in Military Settings. CAPT Michael John Hughey, MC, USNR NAVMEDPUB 6300-2C. January 1, 2000; p. 1110 (Normal Pap smear): © Carolina Biological/Visuals Unlimited/ Corbis; p. 1110 (Abnormal Pap smear): © Parviz M. Pour/Science Source; 28.10: © McGraw-Hill Education/ Al Telser, photographer; p. 1115: © Visuals Unlimited/ Corbis; p. 1116: © McGraw-Hill Education/Jill Braaten, photographer; 28.15: © McGraw-Hill Education/Photo and Dissection by Christine Eckel; 28.16b-28.18a: © Dr. Thomas Caceci, Virginia-Maryland Regional College of Veterinary Medicine; 28.19b-c: © McGraw-Hill Education/ Al Telser, photographer; 28.19d: © Ed Reschke/Getty Images.

Chapter 29

Opener: © BSIP/Science Source; 29.2a: © David M. Phillips/Science Source; p. 1153 (top): © Splash News/ Newscom; p. 1153 (bottom): © AJP/Hop Américain/ Science Source; p. 1157: © PunchStock/BananaStock RF; 29.15c: © D. Van Rossum/Science Source; © INSADCO Photography/Alamy; p. 1163: © Blend Images/Getty Images RF; 29.18: © CNRI/Science Source; 29.19 (Widow's peak): © Aaron Haupt/Science Source; 29.19 (Straight hairline): © Brownstock/Alamy RF.

Note: Page numbers followed by f and t indicate figures and tables, respectively.

I-27

Word part	Meaning	Example
osmo-	push	osmosis (movement of water through a semipermeable membrane into a region of higher solute concentration)
-osis	condition, state	osteoporosis (atrophy of bone characterized by decreased bone mass)
-ose	full of	adipose (fatty; consisting of, resembling, or relating to fat)
-oma	tumor	carcinoma (cancerous tumor derived from epithelial cells)
oligo-	few, little	oliguria (reduced urine output)
-ole	little	arteriole (minute artery-like vessel)
neuro-	nerve	neurilemma (delicate outer membrane sheath around an axon)
nephro-	kidney	nephrology (branch of medicine concerned with kidney diseases)
neo-	new	neonatal (newborn)
natri-	sodium	hyponatremia (abnormally low concentration of sodium ions in blood)
morph-	form, shape	morphology (the study of structure)
mono-	one, single	monosaccharide (a simple sugar)
micro-	small	microbe (any minute organism)
metabol-	change	metabolism (sum of chemical and physical changes occurring in tissue)
meta-	after, beyond	metastasis (spread of a disease from one part of the body to another)
meso-	in the middle	mesoderm (the middle primary germ layer in the early embryo)
-mers, -meres	parts	polymers (large molecules made of smaller units called monomers)
medi-	middle	medial (toward the midline)
mamm-, mast-	breast	mammary glands (milk to feed young is produced by mammary glands)
macro-	large	macrophage (large, phagocytic leukocyte)
-lysis	loosening, dissolution	hemolysis (dissolution or destruction of erythrocytes)
lute-	yellow	corpus luteum (yellow body in the ovary)
lucid-	clear, not obscured	stratum lucidum (thin, clear layer of dead skin cells in the epidermis)
-logy	study	urology (study of the urinary system)
lip-	fat, fatty	lipid (an operational term denoting solubility characteristics; "fat soluble")
leuko-	white	leukocyte (white blood cell)
-let	small	platelets (small, irregularly-shaped cell fragments of a megakaryocyte)
lacto-	milk	lactose (milk sugar)
kine-	motion, movement	kinetic (pertaining to motion or movement)
kali-	potassium	hypokalemia (condition of low concentration of potassium ions in blood)
juxta-	near, adjacent	juxtaglomerular (near the glomerulus)
jug-	to join	conjugated (formed by the union of two compounds)
-itis	inflammation	neuritis (inflammation of a nerve)
-ite	resembling	dendrite (similar branched projections of neuron)
iso-	equal, like	isotonic (same osmotic pressure)
-issimus	the most	longissimus (the longest)
intra-	within	intracellular (within the cell)
inter-	among, between	interosseous (between two bones)
infra-	below	infraspinous (below a spinous process)
-in	'suffix' for names	globulin (a serum protein produced in the liver)
idio-	distinctive	idiopathic (disease of unknown cause)
-icle, -icul	small	ossicles (a small bone)
-ic	pertaining to	isotonic (solutions with equal osmotic pressure)
-ia	condition	hypocalcemia (state of low serum calcium levels in the blood)
hypo-	deficient, below	hypoglycemia (low blood glucose levels)
hyper-	excessive, above	hypertrophy (increase in bulk of a part or organ)
hydro-	water	hydroadipsia (absence of thirst for water)
homo-, homeo-	same	homeostasis (state of equilibrium in the body)
holo-	whole, entire	hologynic (related to characters manifest only in females)
histo-	tissue	histology (study of tissues)
hetero-	other, different	heterozygous (organism has different alleles for a trait on the chromosome pair)
hepato-	liver	hepatitis (inflammation of the liver)
hem-, hemato-	blood	hematology (study of blood and blood-forming tissues)
gyn-	female, woman	gynecology (medical specialty concerned with female genital tract)
gradi-	walk, step	retrograde (having a backward motion or direction)
glyco-	sugar, sweet	glycolysis (breakdown of glucose)
gloss-, glosso-	tongue	hypoglossal (below the tongue)
glob	small drop	globin (the protein of hemoglobin)
germ-	sprout, bud	germ (a primordium; the earliest structure within an embryo)
-genesis, -genic	origin, produce	gluconeogenesis (glucose formed from non-carbohydrates)
-form	shape	fusiform (spindle-shaped; tapering at both ends)
-ferent	carry	afferent (directed toward the center or internal region)
fer-	to carry	efferent (directed away from an organ or region)
ex-, exo-	out of, from	exhale (to breathe out); exocrine (gland that secretes onto body surface)
eu-	good	aneuploidy (abnormal number of chromosomes in cells)
epi-	upon, following	epicardium (visceral layer of the serous pericardium covering heart)

roots, combining forms, prefixes, and suffixes

Many terms used in the biological sciences are compound words: that is, words made up of one or more word roots and appropriate prefixes and/or suffixes. Less than 400 roots, prefixes, and suffixes make up more than 90% of the medical vocabulary. These combining forms are most often derived from ancient Latin or Greek. Prefixes are placed before the root term and suffixes are added after. The following list includes the most common forms used in anatomy and physiology and medicine, and an example for each. This list, and the word origin information found throughout the text, is intended to facilitate learning an often unnecessarily complex-sounding vocabulary. Exclusively a learning tool, the entries are by intention brief. If you learn them, you will find your progress in your anatomy and physiology course swift, steady, and strong (the three "s'es" of success).

Form	Meaning	Example
a-	not, without	asymptomatic (absence of symptoms)
ab-	away from	abstinence (to hold back from)
-ac, -al	pertaining to	cardiac (the heart), myocardial (heart muscle)
aden-, adeno-	gland	adenoma (tumor of a gland)
aero-	air, oxygen	aerobic (with oxygen)
af-	toward	afferent (moving toward)
amphi-	both, around	amphipathic (molecule contains both hydrophilic and hydrophobic properties)
an-	not, without	anesthesia (absence of sensation)
andro-	male	androgens (male hormones)
angi-, angio-	vessel	angiopathy (disease of blood vessels)
ante-	before	antepartum (before birth)
anti-	against	anticoagulant (prevents blood clotting)
-ary	associated with	urinary (associated with urine)
-asia, -asis	condition, state of	homeostasis (state of metabolic balance)
auto-	self	autolysis (self breakdown)
baro-	weight, pressure	baroreceptor (receptor for pressure changes)
bene-	good, well	benign (not malignant)
bi-	twice, double	bicuspid (two cusps)
-blast	precursor, germ	osteoblast (bone-producing cell)
brachi-	arm	brachial (of the arm)
brady-	slow	bradycardia (slow heart rate)
carcin-	cancer	carcinogenic (causing cancer)
cardio-	heart	cardiogram (register of heart activity)
cata-	down, break down	catabolism (metabolic pathways to break down molecules)
caud-	tail	caudal (by the tail)
-cel	little, process	pedicel (process of a podocyte)
cephal-, cephalo-	head	cephalic (cranial)
cerebro-	brain	cerebrospinal (relating to the brain and spinal cord)
chondro-	cartilage, gristle	chondrocyte (cartilage cell)
-cide	kill	spermicide (agent that kills sperm)
circum-	around	circumduction (movement of a part in a circular direction)
-clast	break	osteoclast (cell that breaks down bone)
-cle	little	corpuscle (a small mass or body)
co-, com- (con)	with, together	gray commissure (connects right/left horns)
contra-	against, opposed	contralateral (opposite side)
corpus-	body or mass	corpus luteum (yellow endocrine body)
costo-	rib	intercostals (muscles between the ribs)
crani-	skull	cranial cavity (the brain occupies the cranial cavity)
crino-	separate, secrete	endocrinology (the science of hormone secretions and their physiology)
cysto-, cysti-	sac, bladder	cystoscope (instrument for examining the interior of the bladder)
cyto-, cyte-	cell	erythrocyte (red blood cell), cytology (study of cells)
-de	from, away	dehydration (excess loss of body fluid)
demi-	half	costal demifacet (half-moon facet on vertebra for rib articulation)
derm-	skin	dermatology (study of skin)
di-, diplo-	two	diploid (two sets of chromosomes)
dia-	across, through, separate	diaphragm (skeletal muscle extending across the inferior side of the rib cage)
-dis	apart	dissect (to separate into pieces)
-durus	hard	dura mater (tough, fibrous outer membrane of CNS)
dys-	painful, difficult, bad	dysuria (painful urination)
e-, ec-, ef-, ex-	out, from	efferent (carries away from), excretion (eliminate from)
ecto-	outside, outer	ectocardia (congenital displacement of heart)
-ectomy	to cut out	appendectomy (surgical removal of appendix)
ede-, -edem	swelling	edema (accumulation of excessive amount of fluid in cells or tissues)
-el, -elle	small	organelle (tiny cellular structure that performs specific function)
electro-	electric, electricity	electrocardiograph (instrument for recording electrical currents in the heart)
-emia	blood	anemia (reduced number of red blood cells in the blood)
endo-	within, inner	endocardium (innermost tunic of the heart)

osse-, osteo-	bone	osteoblast (bone-forming cell)
-ous	much, full of	nitrogenous (containing nitrogen)
oxy-	oxygen	hypoxia (condition of decreased normal levels of oxygen in inspired gases)
para-	near, beside	paranasal (adjacent to the nose)
-pathy	disease	neuropathy (nerve disease)
pelv-	basin	renal pelvis (funnel-shaped expansion of the upper end of the ureter)
penia	deficiency	leukopenia (deficiency in the number of leukocytes in the blood)
peri-	around	periosteum (fibrous membrane covering surface of the bone)
phag-	eat	phagocytosis (ingestion and digestion of solid substances by cells)
phil-	have an affinity for	lipophilic (associates with fat)
phobo-	fearing, repelled by	hydrophobic (tending not to dissolve in water)
physio-	nature, natural cause	physiology (the science of the function of living systems)
physis	growth	diaphysis (the shaft of a long bone)
plasm-	formed	cytoplasm (gel-like substance between cell membrane and the nucleus)
plegia	paralyze, stroke	paraplegia (paralysis of lower extremities)
pneumo-	air, gas, lungs	pneumothorax (air in the pleural cavity)
poie, -poiesis	produce, formation of	erythropoietin (hormone that stimulates erythrocyte production)
poly-	many	polycythemia (relative increase in erythrocytes)
pro-	before in time, place	proacrosomal (early stage in the development of the acrosome)
pseudo-	false	pseudostratified (appears to be layered but it is not)
quad-	fourfold	quadriceps femoris (four-headed muscle of anterior side of the thigh)
rami-	branch	ramus (primary division of a nerve or blood vessel)
rect-	straight	rectus abdominis (straight muscle of abdomen)
reno-	kidney	renal (of the kidney)
reti-	network	reticular (fibers that crosslink to form a fine meshwork in soft tissues)
retro-	backward, behind	retroperitoneal (external or posterior to the peritoneum)
sclero-	hard	arteriosclerosis (hardening of the arteries)
semi-	half	semilunar (half-moon shaped)
serrate	saw-edged, toothed	serratus anterior (muscle of thorax)
somato-	body	somatotropin (protein hormone from anterior pituitary: growth hormone)
splanchno-	viscera	splanchnic (a term used to describe visceral organs)
spleno-	spleen	splenius capitis (a broad, straplike muscle in the back of the neck)
stasi, stati-	put, remain	homeostasis (regulate and maintain stable, constant environment)
steno-	narrow	stenosis (narrowing of opening or a canal)
stria-	stripe, band	striated (marked with stripes)
sub-	under	subcutaneous (beneath the skin)
super-, supra-	above, upper	suprarenal (superior to the kidney)
sym-, syn-	together, with	symphysis (growing together)
tachy-	rapid, fast	tachycardia (rapid beating of the heart)
therm-	heat	thermometer (tool to measure temperature)
thorac-	chest	thoracic cavity (body cavity housing the heart and lungs)
-tomy	cut, incise	appendectomy (removal of appendix)
tono-	tension, pressure	isotonic (solutions possessing the same osmotic pressure)
topo-	place, topical	ectopic (an organ that is not in the proper position or place)
trans-	across	transdermal (entering through the skin)
tri-	three	triceps brachii (three-headed muscle)
troph-	food, nourishment	trophoblast (cells providing nutrients to the embryo)
-tropic	having an affinity for	gonadotropic (effecting the gonads)
tunica-	layer, coat	tunica intima (inner part of blood vessel wall)
-ul	small	tubule (very small tube or tubular structure)
uni-	one	unicellular (single cell)
vas-	vessel	vasodilation (widening of lumen of blood vessel)
villo-	hair	microvilli (small hair-like projections of some epithelial cell membranes)
viscer-	internal organ	visceral (of the internal organs)
vivi-	life, alive	in vivo (means "within the living body")
zyg-	yoked, paired, union	azygos (unpaired anatomical structure that occurs singly)